# Tools for Your Success!

## MyMathLab®

MyMathLab® is a series of text-specific, easily customizable online courses for Prentice Hall textbooks in mathematics and statistics. Powered by CourseCompass™ (Pearson Education's online teaching and learning environment) and MathXL® (our online homework, tutorial, and assessment system), MyMathLab gives you the tools you need to deliver all or a portion of your course online, whether your students are in a lab setting or working from home. MyMathLab provides a rich and flexible set of course materials, featuring free-response exercises that are algorithmically generated for unlimited practice and mastery. Students can also use online tools, such as video lectures, animations, and a multimedia textbook, to independently improve their understanding and performance. Instructors can use MyMathLab's homework and test managers to select and assign online exercises correlated directly to the textbook, and they can also create and assign their own online exercises and import TestGen tests for added flexibility. MyMathLab's online gradebook—designed specifically for mathematics and statistics—automatically tracks students' homework and test results and gives the instructor control over how to calculate final grades. Instructors can also add offline (paper-and-pencil) grades to the gradebook. MyMathLab is available to qualified adopters. For more information, visit our website at www.mymathlab.com or contact your Prentice Hall sales representative. (MyMathLab must be set up and assigned by your instructor.)

## MathXL®

MathXL® is a powerful online homework, tutorial, and assessment system that accompanies Prentice Hall textbooks in mathematics or statistics. With MathXL, instructors can create, edit, and assign online homework and tests using algorithmically generated exercises correlated at the objective level to the textbook. They can also create and assign their own online exercises and import TestGen tests for added flexibility. All student work is tracked in MathXL's online gradebook. Students can take chapter tests in MathXL and receive personalized study plans based on their test results. The study plan diagnoses weaknesses and links students directly to tutorial exercises for the objectives they need to study and retest. Students can also access supplemental animations and video clips directly from selected exercises. MathXL is available to qualified adopters. For more information, visit our website at www.mathxl.com, or contact your Prentice Hall sales representative. (MathXL must be set up and assigned by your instructor.)

# More Tools for Your Success!

## STUDENT STUDY PACK

Includes:

- the *Student Solutions Manual*
- access to the *Prentice Hall Tutor Center*
- the *CD Lecture Series Videos*

## Student Solutions Manual

Provides full, worked-out solutions to:

- the odd-numbered section exercises
- all (even and odd) exercises in the Cumulative Review Exercises (within the section exercises), Mid-Chapter Tests, Chapter Review Exercises, Chapter Practice Tests, and Cumulative Review Tests

## Chapter Test Prep Video CD

Provides step-by-step video solutions to each problem in each *Chapter Practice Test* in the textbook.

## CD Lecture Series Videos

Each section includes:

- a 20-minute mini-lecture
- several worked-out section exercises (identified by a CD icon)

## Prentice Hall Tutor Center

Staffed by developmental math faculty, the Tutor Center provides live tutorial support via phone, fax, or email.

# ALGEBRA

## For College Students

# ALGEBRA
## For College Students
Third Edition

**Allen R. Angel**
Monroe Community College

with assistance from
**Richard Semmler**
Northern Virginia Community College
**Donna R. Petrie**
Monroe Community College

PEARSON
Prentice Hall
Upper Saddle River, New Jersey 07458

**Library of Congress Cataloging-in-Publication Data**
Angel, Allen R.
Algebra for college students—3rd ed.
Allen R. Angel, with assistance from Richard Semmler
p. cm,
Includes index.
ISBN 0-13-612908-0 (SE)—ISBN 0-13-614277-X (AIE)
1. Algebra—Textbooks. I. Semmler, Richard. II. Title.
QA152.3.A528 2008
512.9—dc22 2006053258

Executive Editor: *Paul Murphy*
Project Manager: *Dawn Nuttall*
Vice President and Editorial Director, Mathematics: *Christine Hoag*
Production Editor: *Lynn Savino Wendel*
Executive Managing Editor: *Kathleen Schiaparelli*
Senior Managing Editor: *Linda Mihatov Behrens*
Media Project Manager, Developmental Math: *Audra J. Walsh*
Media Production Editor: *Jessica Barna*
Assistant Managing Editor, Science and Math Supplements: *Karen Bosch*
Manufacturing Buyer: *Maura Zaldivar*
Manufacturing Manager: *Alexis Heydt-Long*
Director of Marketing: *Patrice Jones*
Senior Marketing Manager: *Kate Valentine*
Marketing Assistant: *Jennifer de Leeuwerk*
Editorial Assistant/Print Supplements Editor: *Abigail Rethore*
Editor in Chief, Development: *Carol Trueheart*
Art Director: *John Christiana*
Interior Designer: *Studio Indigo*
Cover Designer: *Michael J. Fruhbeis*
Art Editor: *Thomas Benfatti*
Creative Director: *Juan R. López*
Director of Creative Services: *Paul Belfanti*
Director, Image Resource Center: *Melinda Patelli*
Manager, Rights and Permissions: *Zina Arabia*
Manager, Visual Research: *Beth Brenzel*
Image Permission Coordinator: *Craig A. Jones*
Photo Researcher: *Teri Stratford*
Compositor: *Prepare, Inc.*
Art Studios: *Precision Graphics and Laserwords*


Pearson Prentice Hall
Pearson Education, Inc.
Upper Saddle River, NJ 07458



Printed in the United States of America

10 9 8 7 6 5 4 3 2 1

ISBN 0-13-614277-X (AIE)
ISBN 0-13-612908-0 (Student Edition)

Pearson Education, LTD., *London*
Pearson Education Australia PTY, Limited, *Sydney*
Pearson Education Singapore, Pte. Ltd
Pearson Education North Asia Ltd, *Hong Kong*
Pearson Education Canada, Ltd., *Toronto*
Pearson Educación de Mexico, S.A. de C.V.
Pearson Education –Japan, *Tokyo*
Pearson Education Malaysia, Pte. Ltd

To my mother,
Sylvia Angel-Baumgarten

And to the memory of my step-father,
Lenny Baumgarten

And to the memory of my mother-in-law,
Ruth Pollinger

# Contents

Preface xi

To the Student xvii

## 1 Basic Concepts 1

1.1 Study Skills for Success in Mathematics, and Using a Calculator 2
1.2 Sets and Other Basic Concepts 6
1.3 Properties of and Operations with Real Numbers 17
1.4 Order of Operations 28
**Mid-Chapter Test: Sections 1.1–1.4 40**
1.5 Exponents 40
1.6 Scientific Notation 50
Chapter 1 Summary 57
Chapter 1 Review Exercises 61
Chapter 1 Practice Test 64

## 2 Equations and Inequalities 65

2.1 Solving Linear Equations 66
2.2 Problem Solving and Using Formulas 77
2.3 Applications of Algebra 87
**Mid-Chapter Test: Sections 2.1–2.3 100**
2.4 Additional Application Problems 100
2.5 Solving Linear Inequalities 110
2.6 Solving Equations and Inequalities Containing Absolute Values 125
Chapter 2 Summary 135
Chapter 2 Review Exercises 138
Chapter 2 Practice Test 140
Cumulative Review Test 141

## 3 Graphs and Functions 143

3.1 Graphs 144
3.2 Functions 158
3.3 Linear Functions: Graphs and Applications 173
3.4 The Slope-Intercept Form of a Linear Equation 184
**Mid-Chapter Test: Sections 3.1–3.4 198**
3.5 The Point-Slope Form of a Linear Equation 199
3.6 The Algebra of Functions 208
3.7 Graphing Linear Inequalities 218
Chapter 3 Summary 222
Chapter 3 Review Exercises 225
Chapter 3 Practice Test 229
Cumulative Review Test 231

## 4 Systems of Equations and Inequalities 232


**4.1** Solving Systems of Linear Equations in Two Variables 233
**4.2** Solving Systems of Linear Equations in Three Variables 245
**4.3** Systems of Linear Equations: Applications and Problem Solving 252
**Mid-Chapter Test: Sections 4.1–4.3 265**
**4.4** Solving Systems of Equations Using Matrices 266
**4.5** Solving Systems of Equations Using Determinants and Cramer's Rule 275
**4.6** Solving Systems of Linear Inequalities 282
Chapter 4 Summary 288
Chapter 4 Review Exercises 293
Chapter 4 Practice Test 295
Cumulative Review Test 296


## 5 Polynomials and Polynomial Functions 297


**5.1** Addition and Subtraction of Polynomials 298
**5.2** Multiplication of Polynomials 308
**5.3** Division of Polynomials and Synthetic Division 317
**5.4** Factoring a Monomial from a Polynomial and Factoring by Grouping 327
**Mid-Chapter Test: Sections 5.1–5.4 334**
**5.5** Factoring Trinomials 335
**5.6** Special Factoring Formulas 346
**5.7** A General Review of Factoring 354
**5.8** Polynomial Equations 358
Chapter 5 Summary 370
Chapter 5 Review Exercises 374
Chapter 5 Practice Test 379
Cumulative Review Test 380


## 6 Rational Expressions and Equations 381


**6.1** The Domains of Rational Functions and Multiplication and Division of Rational Expressions 382
**6.2** Addition and Subtraction of Rational Expressions 392
**6.3** Complex Fractions 403
**6.4** Solving Rational Equations 409
**Mid-Chapter Test: Sections 6.1–6.4 421**
**6.5** Rational Equations: Applications and Problem Solving 421
**6.6** Variation 432
Chapter 6 Summary 440
Chapter 6 Review Exercises 443
Chapter 6 Practice Test 446
Cumulative Review Test 447

## 7 Roots, Radicals, and Complex Numbers 448

**7.1** Roots and Radicals 449
**7.2** Rational Exponents 457
**7.3** Simplifying Radicals 465
**7.4** Adding, Subtracting, and Multiplying Radicals 472
**Mid-Chapter Test: Sections 7.1–7.4 479**
**7.5** Dividing Radicals 480
**7.6** Solving Radical Equations 489
**7.7** Complex Numbers 500
Chapter 7 Summary 508
Chapter 7 Review Exercises 512
Chapter 7 Practice Test 515
Cumulative Review Test 516

## 8 Quadratic Functions 517

**8.1** Solving Quadratic Equations By Completing the Square 518
**8.2** Solving Quadratic Equations by the Quadratic Formula 527
**8.3** Quadratic Equations: Applications and Problem Solving 539
**Mid-Chapter Test: Sections 8.1–8.3 548**
**8.4** Writing Equations in Quadratic Form 549
**8.5** Graphing Quadratic Functions 555
**8.6** Quadratic and Other Inequalities in One Variable 572
Chapter 8 Summary 582
Chapter 8 Review Exercises 585
Chapter 8 Practice Test 588
Cumulative Review Test 589

## 9 Exponential and Logarithmic Functions 591

**9.1** Composite and Inverse Functions 592
**9.2** Exponential Functions 603
**9.3** Logarithmic Functions 611
**9.4** Properties of Logarithms 618
**Mid-Chapter Test: Sections 9.1–9.4 623**
**9.5** Common Logarithms 624
**9.6** Exponential and Logarithmic Equations 630
**9.7** Natural Exponential and Natural Logarithmic Functions 637
Chapter 9 Summary 648
Chapter 9 Review Exercises 652
Chapter 9 Practice Test 655
Cumulative Review Test 655

## 10 Characteristics of Functions and Their Graphs 657

**10.1** Continuous and Discontinuous Functions 658
**10.2** Characteristics of Graphs of Polynomial Functions 665
**10.3** Finding the Rational Zeros of a Polynomial Function 675
**Mid-Chapter Test: Sections 10.1–10.3 688**
**10.4** Finding the Real Zeros of a Polynomial Function 688
**10.5** Characteristics of Graphs of Rational Functions 696
Chapter 10 Summary 708
Chapter 10 Review Exercises 712
Chapter 10 Practice Test 715
Cumulative Review Test 716

## 11 Conic Sections 717

**11.1** The Parabola and the Circle 718
**11.2** The Ellipse 729
**Mid-Chapter Test: Sections 11.1–11.2 735**
**11.3** The Hyperbola 735
**11.4** Nonlinear Systems of Equations and Their Applications 742
**11.5** Nonlinear Systems of Inequalities 751
Chapter 11 Summary 755
Chapter 11 Review Exercises 759
Chapter 11 Practice Test 761
Cumulative Review Test 762

## 12 Sequences, Series, and Probability 763

**12.1** Sequences and Series 764
**12.2** Arithmetic Sequences and Series 771
**12.3** Geometric Sequences and Series 778
**Mid-Chapter Test: Sections 12.1–12.3 788**
**12.4** Mathematical Induction 789
**12.5** The Binomial Theorem 796
**12.6** The Counting Principle, Permutations, and Combinations 801
**12.7** Probability 812
Chapter 12 Summary 821
Chapter 12 Review Exercises 825
Chapter 12 Practice Test 828
Cumulative Review Test 829

Appendix 830
Answers A1
Applications Index I1
Subject Index I5
Photo Credits P1

# Preface

This book was written for college students who have successfully completed a first course in elementary algebra. My primary goal was to write a book that students can read, understand, and enjoy. To achieve this goal I have used short sentences, clear explanations, and many detailed, worked-out examples. I have tried to make the book relevant to college students by using practical applications of algebra throughout the text.

## Features of the Text

**Full-Color Format** Color is used pedagogically in the following ways:

- Important definitions and procedures are color screened.
- Color screening or color type is used to make other important items stand out.
- Artwork is enhanced and clarified with use of multiple colors.
- The full-color format allows for easy identification of important features by students.
- The full-color format makes the text more appealing and interesting to students.

**Readability** One of the most important features of the text is its readability. The book is very readable, even for those with weak reading skills. Short, clear sentences are used and more easily recognized, and easy-to-understand language is used whenever possible.

**Accuracy** Accuracy in a mathematics text is essential. To ensure accuracy in this book, mathematicians from around the country have read the pages carefully for typographical errors and have checked all the answers.

**Connections** Many of our students do not thoroughly grasp new concepts the first time they are presented. In this text we encourage students to make connections. That is, we introduce a concept, then later in the text briefly reintroduce it and build upon it. Often an important concept is used in many sections of the text. Students are reminded where the material was seen before, or where it will be used again. This also serves to emphasize the importance of the concept. Important concepts are also reinforced throughout the text in the Cumulative Review Exercises and Cumulative Review Tests.

**Chapter Opening Application** Each chapter begins with a real-life application related to the material covered in the chapter. By the time students complete the chapter, they should have the knowledge to work the problem.

**Goals of This Chapter** This feature on the chapter opener page gives students a preview of the chapter and also indicates where this material will be used again in other chapters of the book. This material helps students see the connections between various topics in the book and the connection to real-world situations.

**The Use of Icons** At the begining of each exercise set the icon for MathXL®, MathXL, and for MyMathLab®, MyMathLab, are illustrated. Both of these icons will be explained shortly.

**Keyed Section Objectives** Each section opens with a list of skills that the student should learn in that section. The objectives are then keyed to the appropriate portions of the sections with blue numbers such as 1.

**Problem Solving** Polya's five-step problem-solving procedure is discussed in Section 2.2. Throughout the book, problem solving and Polya's problem-solving procedure are emphasized.

**Practical Applications** Practical applications of algebra are stressed throughout the text. Students need to learn how to translate application problems into algebraic symbols. The problem-solving approach used throughout this text gives students ample practice in setting up and solving application problems. The use of practical applications motivates students.

**Detailed, Worked-Out Examples** A wealth of examples have been worked out in a step-by-step, detailed manner. Important steps are highlighted in color, and no steps are omitted until after the student has seen a sufficient number of similar examples.

**Now Try Exercises** In each section, after each example, students are asked to work an exercise that parallels the example given in the text. These Now Try Exercises make the students *active*, rather than passive, learners and they reinforce the concepts as students work the exercises. Through these exercises, students have the opportunity to immediately apply what they have learned. After each example, Now Try Exercises are indicated in green type such as ▶ **Now Try Exercise 27.** They are also indicated in green type in the exercise sets, such as 27.

**Study Skills Section** Many students taking this course have poor study skills in mathematics. Section 1.1, the first section of this text, discusses the study skills needed to be successful in mathematics. This section should be very beneficial for your students and should help them to achieve success in mathematics.

**Helpful Hints** The Helpful Hint boxes offer useful suggestions for problem solving and other varied topics. They are set off in a special manner so that students will be sure to read them.

**Helpful Hints—Study Tip** The Helpful Hint—Study Tip boxes offer valuable information on items related to studying and learning the material.

**Avoiding Common Errors** Errors that students often make are illustrated. The reasons why certain procedures are wrong are explained, and the correct procedure for working the problem is illustrated. These Avoiding Common Errors boxes will help prevent your students from making those errors we see so often.

**Using Your Calculator** The Using Your Calculator boxes, placed at appropriate locations in the text, are written to reinforce the algebraic topics presented in the section and to give the student pertinent information on using a scientific calculator to solve algebraic problems.

**Using Your Graphing Calculator** Using Your Graphing Calculator boxes are placed at appropriate locations throughout the text. They reinforce the algebraic topics taught and sometimes offer alternate methods of working problems. This book is designed to give the instructor the option of using or not using a graphing calculator in his or her course. Some of the Using Your Graphing Calculator boxes contain graphing calculator exercises, whose answers appear in the answer section of the book. The illustrations shown in the Using Your Graphing Calculator boxes are from a Texas Instruments TI-84 Plus calculator. The Using Your Graphing Calculator boxes are written assuming that the student has no prior graphing calculator experience.

## Exercise Sets

The exercise sets are broken into three main categories: Concept/Writing Exercises, Practice the Skills, and Problem Solving. Many exercise sets also contain Challenge Problems and/or Group Activities. Each exercise set is graded in difficulty. The early problems help develop the student's confidence, and then students are eased gradually into the more difficult problems. A sufficient number and variety of examples are given in each section for the student to successfully complete even the more difficult exercises. The number of exercises in each section is more than ample for student assignments and practice.

**Concept/Writing Exercises** Most exercise sets include exercises that require students to write out the answers in words. These exercises improve students' understanding and comprehension of the material. Many of these exercises involve problem solving and conceptualization and help develop better reasoning and critical thinking skills. Writing exercises are indicated by the symbol ✎.

**Problem Solving Exercises** These exercises have been added to help students become better thinkers and problem solvers. Many of these exercises involve real-life applications of algebra. It is important for students to be able to apply what they learn to real-life situations. Many problem solving exercises help with this.

**Challenge Problems** These exercises, which are part of many exercise sets, provide a variety of problems. Many were written to stimulate student thinking. Others provide additional applications of algebra or present material from future sections of the book so that students can see and learn the material on their own before it is covered in class. Others are more challenging than those in the regular exercise set.

**CD Lecture Exercises** The exercises that are worked out in detail on the CD Lecture Videos are marked with the CD icon, 💿. This will prove helpful for your students.

**Cumulative Review Exercises** All exercise sets (after the first two) contain questions from previous sections in the chapter and from previous chapters. These Cumulative Review Exercises will reinforce topics that were previously covered and help students retain the earlier material, while they are learning the new material. For the students' benefit, Cumulative Review Exercises are keyed to the section where the material is covered, using brackets, such as [3.4].

**Group Activities** Many exercise sets have group activity exercises that lead to interesting group discussions. Many students learn well in a cooperative learning atmosphere, and these exercises will get students talking mathematics to one another.

**Mid-Chapter Tests** In the middle of each chapter is a new Mid-Chapter Test. Students should take each Mid-Chapter Test make sure they understand the material presented in the chapter up to that point. In the student answers, brackets such as [2.3] are used to indicate the section where the material was first presented.

**Chapter Summary** At the end of each chapter is a newly formated, comprehensive chapter summary that includes important chapter facts and examples illustrating these important facts.

**Chapter Review Exercises** At the end of each chapter are review exercises that cover all types of exercises presented in the chapter. The review exercises are keyed, using color numbers and brackets, such as [1.5], to the sections where the material was first introduced.

**Chapter Practice Tests** The comprehensive end-of-chapter practice test will enable the students to see how well they are prepared for the actual class test. The section where the material was first introduced is indicated in brackets in the student answers.

**Cumulative Review Tests** These tests, which appear at the end of each chapter after the first, test the students' knowledge of material from the beginning of the book to the end of that chapter. Students can use these tests for review, as well as for preparation for the final exam. These exams, like the Cumulative Review Exercises, will serve to reinforce topics taught earlier. In the answer section, after each answers, the section where that material was covered is given using brackets.

**Answers** The *odd answers* are provided for the exercise sets. *All answers* are provided for the Using Your Graphing Calculator Exercises, Cumulative Review Exercises, Mid-Chapter Tests, Chapter Review Exercises, Chapter Practice Tests, and Cumulative Review Tests. Answers are not provided for the Group Activity exercises since we want students to reach agreement by themselves on the answers to these exercises.

## National Standards

Recommendations of the *Curriculum and Evaluation Standards for School Mathematics*, prepared by the National Council of Teachers of Mathematics (NCTM), and *Beyond Crossroads: Implementing Mathematics Standards in the First Two Years of College*, prepared by the American Mathematical Association of Two Year Colleges (AMATYC), are incorporated into this edition.

## Prerequisite

The prerequisite for this course is a working knowledge of elementary algebra. Although some elementary algebra topics are briefly reviewed in the text, students should have a basic understanding of elementary algebra before taking this course.

## Modes of Instruction

The format and readability of this book lends itself to many different modes of instruction. The constant reinforcement of concepts will result in greater understanding and retention of the material by your students.

The features of the text and the large variety of supplements available make this text suitable for many types of instructional modes including:

- lecture
- distance learning
- self-paced instruction
- modified lecture
- cooperative or group study
- learning laboratory

## Changes in the Third Edition

When I wrote the third edition, I considered many letters and reviews I got from students and faculty alike. I would like to thank all of you who made suggestions for improving the third edition. I would also like to thank the many instructors and students who wrote to inform me of how much they enjoyed, appreciated, and learned from the text. Some of the changes made in the third edition of the text include:

- A *Chapter Test Prep Video CD* now comes with the book. This video CD shows the worked-out solution to each exercise in the Chapter Practice Test for each chapter. This is yet another aid to improve student learning and understanding.
- *Every* example in the book now has a corresponding Now Try Exercise associated with it. Students are encouraged to work the exercise immediately after they finish studying the respective example. This gives students an opportunity to reinforce the concepts or topics covered in the example.
- A new feature called *Mid-Chapter Test* has been added to the middle of each chapter. These tests are designed to see how well students understand the topics covered in the first part of the chapter. If a student misses a question, the student should review the appropriate material. The section where the material was introduced is given in brackets next to the answer in the back of the book.
- More *Helpful Hints* and *Avoiding Common Errors* boxes have been added where appropriate.
- The Chapter Summary has been rewritten to include examples of important facts and concepts covered in the chapter. The left-hand column gives the fact or concept, and the right-hand column gives an example of the fact or concept. The new chapter summary should be an aid to students in reviewing the chapter and preparing for a test.
- New and exciting examples and exercises have been added throughout the book.
    - New exercises were added to many exercise sets to ensure that every example in the book now has exercises that correspond to that given example.
    - In some sections, more difficult exercises have been added at the end, or easier exercises have been added at the beginning of the exercise set so that there is a continuous increase in the level of difficulty of the exercises.
    - Every effort has been made to include applications that are of interest to students.

- Variables other than $x$ and $y$ are used more often in examples and exercises.

- *Goals of This Chapter* have replaced *A Look Ahead*. The information provided gives students an overview of what they will see and are expected to learn in the chapter.
- *Using Your Graphing Calculator* boxes now show keystrokes for the TI-84 Plus calculator. Note that the same keystrokes are appropriate for the TI-83 Plus calculator.
- The *Mathematics in Action* feature has been removed to conserve space.
- More art and photos have been added to the text to make the material either more understandable or more interesting for students.
- The basic colors used in the text have been softened to make the text more attractive and easier for students to read.

## Supplements for the Third Edition

### FOR INSTRUCTORS

#### Printed Supplements

Annotated Instructor's Edition (0-13-614277-X)

Contains all the content found in the student edition, plus the following.

- Answers to exercises are printed on the same text page with graphing answers in a special Graphing Answer Section in the back of the text.
- *Teaching Tips* throughout the text are placed at key points in the margin.
- **NEW!** An extra *Instructor Example* is now provided in the margin next to each student example. These extra examples are meant to be used as additional examples in the classroom.

Instructor's Solutions Manual (0-13-238360-8)

This manual contains complete solutions to every exercise in the text.

Instructor's Resource Manual with Tests (0-13-238362-4)

- **NEW!** Now includes a Mini-Lecture for every section of the text.
- Provides several test forms, both free response and multiple choice, for every chapter, as well as cumulative tests and final exams.
- Answers to all items also included.

#### Media Supplements

TestGen (0-13-614278-8)

TestGen enables instructors to build, edit, print, and administer tests using a computerized bank of questions developed to cover all the objectives of the text. TestGen is algorithmically based, allowing instructors to create multiple but equivalent versions of the same question or test with the click of a button. Instructors can also modify test bank questions or add new questions. Tests can be printed or administered online. The software is available on a dual-platform Windows/Macintosh CD-ROM.

CD Lecture Series Videos—Lab Pack (0-13-238399-3)

For each section, there is about 20 minutes of lecture along with several of the section exercises worked out. The exercises that are worked out are identified in the student edition by the icon .

NEW! MyMathLab® Instructor Version (0-13-147898-2)

MyMathLab® is a series of text-specific, easily customizable online courses for Prentice Hall textbooks in mathematics and statistics. Powered by CourseCompass™ (Pearson Education's online teaching and learning environment) and MathXL® (our online homework, tutorial, and assessment system), MyMathLab gives you the tools you need to deliver all or a portion of your course online, whether your students are in a lab setting or working from home.

NEW! MathXL® Instructor Version (0-13-147895-8)

MathXL® is a powerful online homework, tutorial, and assessment system that accompanies Prentice Hall textbooks in mathematics or statistics. With MathXL, instructors can create, edit, and assign online homework and tests using algorithmically generated exercises correlated at the objective level to the textbook.

### FOR STUDENTS

NEW! Student Study Pack (0-13-233334-1)

Includes

- The *Student Solutions Manual*
- Access to the *Prentice Hall Tutor Center*
- The *CD Lecture Series Videos*

#### Printed Supplements

Student Solutions Manual (0-13-238405-1)

Provides full, worked-out solutions to:

- The odd-numbered section exercises
- All (even and odd) exercises in the Cumulative Review Exercises (within the section exercises), Mid-Chapter Tests, Chapter Review Exercises, Chapter Practice Tests, and Cumulative Review Tests

#### Media Supplements

NEW! Chapter Test Prep Video CD (0-13-614279-6)

Provides step-by-step video solutions to each problem in each *Chapter Practice Test* in the textbook. Packaged with a new text, inside the back cover.

### NEW! MathXL® Tutorials on CD (0-13-159212-2)

This interactive tutorial CD-ROM provides algorithmically generated practice exercises that are correlated at the objective level to the exercises in the textbook. Every practice exercise is accompanied by an example and a guided solution designed to involve students in the solution process. Selected exercises may also include a video clip to help students visualize concepts. The software provides helpful feedback for incorrect answers and can generate printed summaries of students' progress.

### Prentice Hall Tutor Center: *www.prenhall.com/tutorcenter* (0-13-064604-0)

Staffed by developmental math faculty, the Tutor Center provides live tutorial support via phone, fax, or e-mail. Tutors are available Sunday through Thursday 5 pm EST to midnight, 5 days a week, 7 hours a day. The Tutor Center may be accessed through a registration number that may be bundled with a new text or purchased separately with a used book. Comes automatically within *MyMathLab*.

### InterAct Math Tutorial Web site: *www.interactmath.com*

Get practice and tutorial help online! This interactive tutorial Web site provides algorithmically generated practice exercises that correlate directly to the exercises in the textbook. Students can retry an exercise as many times as they like with new values each time for unlimited practice and mastery. Every exercise is accompanied by an interactive guided solution that provides helpful feedback for incorrect answers, and students can also view a worked-out sample problem that steps them through an exercise similar to the one they're working on.

## Acknowledgments

Writing a textbook is a long and time-consuming project. Many people deserve thanks for encouraging and assisting me with this project. Most importantly, my special thanks goes to my wife Kathy and sons, Robert and Steven. Without their constant encouragement and understanding, this project would not have become a reality. I would also like to thank my daughter-in-law, Kathy, for her support.

I would like to thank Richard Semmler of Northern Virginia Community College. Richard has worked with me throughout this project, and with many of my other projects throughout the years. He has helped me in too many ways to list, and he has always been there to help when I needed assistance with my books. Richard, I truly thank you.

I would also like to thank Donna Petrie of Monroe Community College for assisting me with this book. Donna was extremely conscientious and helpful in numerous ways.

I want to thank Rafiq Ladhani and his team at Edutorial for accuracy reviewing the pages and checking all answers.

I would like to thank several people at Prentice Hall, including Paul Murphy, Executive Editor; Dawn Nuttall, Project Manager; Thomas Benfatti, Art Editor; John Christiana, Art Director; and Lynn Savino Wendel, Production Editor, for their many valuable suggestions and conscientiousness with this project.

I want to thank those who worked with me on the print supplements for this book.

- Student and Instructor's Solutions Manuals:
  Randy Gallaher and Kevin Bodden, Lewis and Clark Community College, IL
- Instructor's Resource Manual:
  Randy Gallaher and Kevin Bodden, Lewis and Clark Community College, IL

I would like to thank the following reviewers of the last two editions for their thoughtful comments and suggestions:

Laura Adkins, *Missouri Southern State College, MO*
Arthur Altshiller, *Los Angeles Valley College, CA*
Jacob Amidon, *Cayuga Community College, NY*
Sheila Anderson, *Housatonic Community College, CT*
Peter Arvanites, *State University of New York–Rockland Community College, NY*
Jannette Avery, *Monroe Community College, NY*
Mary Lou Baker, *Columbia State Community College, TN*
Jon Becker, *Indiana University, IN*
Paul Boisvert, *Oakton Community College, IL*
Beverly Broomell, *Suffolk County Community College, NY*
Lavon Burton, *Abilene Christian University, TX*
Marc Campbell, *Daytona Beach Community College, FL*
Mitzi Chaffer, *Central Michigan University, MI*
Terry Cheng, *Irvine Valley College, CA*
Ted Corley, *Arizona State University and Glendale Community College, AZ*
Charles Curtis, *Missouri Southern State College, MO*
Joseph de Guzman, *Riverside City College (Norco), CA*
Marla Dresch Butler, *Gavilan Community College, CA*
Gary Egan, *Monroe Community College, NY*
Mark W. Ernsthausen, *Monroe Community College, NY*
Elizabeth Farber, *Bucks County Community College, PA*
Warrene Ferry, *Jones County Junior College, MS*
Christine Fogal, *Monroe Community College, NY*
Gary Glaze, *Spokane Falls Community College, WA*
James Griffiths, *San Jacinto College, TX*
Kathy Gross, *Cayuga Community College, NY*
Abdollah Hajikandi, *State University of New York–Buffalo, NY*
Cynthia Harrison, *Baton Rouge Community College, LA*
Mary Beth Headlee, *Manatee Community College, FL*
Kelly Jahns, *Spokane Community College, WA*
Judy Kasabian, *El Camino College, CA*
Maryanne Kirkpatrick, *Laramie County Community College, WY*
Marcia Kleinz, *Atlantic Cape Community College, NJ*
Shannon Lavey, *Cayuga Community College, NY*

Kimberley A. Martello, *Monroe Community College, NY*
Shywanda Moore, *Meridian Community College, MS*
Catherine Moushon, *Elgin Community College, IL*
Kathy Nickell, *College of DuPage, IL*
Shelle Patterson, *Moberly Area Community College, MO*
Patricia Pifko, *Housatonic Community College, CT*
Dennis Reissig, *Suffolk County Community College, NY*
Linda Retterath, *Mission College, CA*
Dale Rohm, *University of Wisconsin–Stevens Point, WI*
Troy Rux, *Spokane Falls Community College, WA*
Hassan Saffari, *Prestonburg Community College, KY*
Rick Silvey, *St. Mary College, KS*
Julia Simms, *Southern Illinois University–Edwardsville, IL*
Linda Smoke, *Central Michigan University, MI*
Jed Soifer, *Atlantic Cape Community College, NJ*
Richard C. Stewart, *Monroe Community College, NY*
Elizabeth Suco, *Miami–Dade Community College, FL*
Harold Tanner, *Orangeburg–Calhoun Technological College, SC*
Dale Thielker, *Ranken Technological College, MO*
Ken Wagman, *Gavilan Community College, CA*
Patrick Ward, *Illinois Central College, IL*
Robert E. White, *Allan Hancock College, CA*
Cindy Wilson, *Henderson State University, AZ*

# To the Student

Algebra is a course that cannot be learned by observation. To learn algebra you must become an active participant. You must read the text, pay attention in class, and, most importantly, you must work the exercises. The more exercises you work, the better.

The text was written with you in mind. Short, clear sentences are used, and many examples are given to illustrate specific points. The text stresses useful applications of algebra. Hopefully, as you progress through the course, you will come to realize that algebra is not just another math course that you are required to take, but a course that offers a wealth of useful information and applications.

This text makes full use of color. The different colors are used to highlight important information. Important procedures, definitions, and formulas are placed within colored boxes.

The boxes marked **Helpful Hints** should be studied carefully, for they stress important information. The boxes marked **Avoiding Common Errors** should also be studied carefully. These boxes point out errors that students commonly make, and provide the correct procedures for doing these problems.

After each example you will see a Now Try Exercise reference, such as ▸ **Now Try Exercise 27**. The exercise indicated is very similar to the example given in the book. You may wish to try the indicated exercise after you read the example to make sure you truly understand the example. In the exercise set, the Now Try exercises are written in green, such as 27.

In the exercise sets, the exercises marked with a pencil, ✎, indicate writing exercises—that is, exercises that require a written answer. The exercises marked with a CD, 💿, indicate that these exercises are worked out on the CD Lecture Videos.

Ask your professor early in the course to explain the policy on when the calculator may be used. Pay particular attention to the **Using Your Calculator** boxes. You should also read the **Using Your Graphing Calculator** boxes even if you are not using a graphing calculator in class. You may find the information presented here helps you better understand the algebraic concepts.

Other questions you should ask your professor early in the course include: What supplements are available for use? Where can help be obtained when the professor is not available? Supplements that may be available include: the Student Solutions Manual; the CD Lecture Series Videos; the Chapter Test Prep Video CD; MathXL® MathXL; MyMathLab MyMathLab; and the Prentice Hall Mathematics Tutor Center. All these items are discussed under the heading of Supplements in Section 1.1, as well as in the Preface.

You may wish to form a study group with other students in your class. Many students find that working in small groups provides an excellent way to learn the material. By discussing and explaining the concepts and exercises to one another you reinforce your own understanding. Once guidelines and procedures are determined by your group, make sure to follow them.

One of the first things you should do is to read Section 1.1, Study Skills for Success in Mathematics. Read this section slowly and carefully, and pay particular attention to the advice and information given. Occasionally, refer back to this section. This could be the most important section of the book. Carefully read the material on doing your homework and on attending class.

At the end of all Exercise Sets (after the first two) are **Cumulative Review Exercises**. You should work these problems on a regular basis, even if they are not assigned. These problems are from earlier sections and chapters of the text, and they will refresh your memory and reinforce those topics. If you have a problem when working these exercises, read the appropriate section of the text or study your notes that correspond to that material. The section of the text where the Cumulative Review Exercise was introduced is indicated in brackets, [ ], to the left of the exercise. After reviewing the material, if you still have a problem, make an appointment to see your professor. Working the Cumulative Review Exercises throughout the semester will also help prepare you to take your final exam.

Near the middle of each chapter is a **Mid-Chapter Test**. You should take each Mid-Chapter Test to make sure you understand the material up to that point. The section where the material was first introduced is given in brackets after the answer in the answer section of the book.

At the end of each chapter are a **Chapter Summary**, **Chapter Review Exercises**, a **Chapter Practice Test**, and a **Cumulative Review Test**. Before each examination you should review this material carefully, take the Chapter Practice Test (you may want to review the *Chapter Test Prep Video CD* also). If you do well on the Chapter Practice Test, you should do well on the class test. The questions in the Review Exercises are marked to indicate the section in which that material was first introduced. If you have a problem with a Review Exercise question, reread the section indicated. You may also wish to take the Cumulative Review Test that appears at the end of every chapter.

In the back of the text there is an **answer section** that contains the answers to the *odd-numbered* exercises, including the Challenge Problems. Answers to *all* Using Your Graphing Calculator Exercises, Cumulative Review Exercises, Mid-Chapter Tests, Chapter Review Exercises, Chapter Practice Tests, and Cumulative Review Tests are provided. Answers to the Group Activity exercises are not provided, for we wish students to reach agreement by themselves on answers to these exercises. The answers should be used only to check your work. For the Mid-Chapter Tests, Chapter Practice Tests, and Cumulative Review Tests, after each answer the section number where that type of exercise was covered is provided.

I have tried to make this text as clear and error free as possible. No text is perfect, however. If you find an error in the text, or an example or section that you believe can be improved, I would greatly appreciate hearing from you. If you enjoy the text, I would also appreciate hearing from you. You can submit comments at *http://247.prenhall.com*.

*Allen R. Angel*

# 1 Basic Concepts

## GOALS OF THIS CHAPTER

In this chapter, we review algebra concepts that are central to your success in this course. Throughout this chapter, and in the entire book, we use real-life examples to show how mathematics is relevant in your daily life. In Section 1.1, we present some advice to help you establish effective study skills and habits. Other topics discussed in this chapter are sets, real numbers, and exponents.

1.1 Study Skills for Success in Mathematics, and Using a Calculator

1.2 Sets and Other Basic Concepts

1.3 Properties of and Operations with Real Numbers

1.4 Order of Operations

Mid-Chapter Test: Sections 1.1–1.4

1.5 Exponents

1.6 Scientific Notation

Chapter 1 Summary

Chapter 1 Review Exercises

Chapter 1 Practice Test

**HAVE YOU EVER** asked yourself, "When am I going to use algebra?" In this chapter and throughout this book, we use algebra to study real-life applications. These applications range from the NASCAR Nextel Cup series in Exercise 101 to natural disasters in Exercise 102, both on page 14. We will find that mathematics can be used in virtually every aspect of our lives.

# 1.1 Study Skills for Success in Mathematics, and Using a Calculator

1 Have a positive attitude.

2 Prepare for and attend class.

3 Prepare for and take examinations.

4 Find help.

5 Learn to use a calculator.

You need to acquire certain study skills that will help you to complete this course successfully. These study skills will also help you succeed in any other mathematics courses you may take.

It is important for you to realize that this course is the foundation for more advanced mathematics courses. If you have a thorough understanding of algebra, you will find it easier to be successful in later mathematics courses.

## 1 Have a Positive Attitude

You may be thinking to yourself, "I hate math" or "I wish I did not have to take this class." You may have heard the term *math anxiety* and feel that you fall into this category. The first thing you need to do to be successful in this course is to change your attitude to a more positive one. You must be willing to give this course and yourself a fair chance.

Based on past experiences in mathematics, you may feel this will be difficult. However, mathematics is something you need to work at. Many of you taking this course are more mature now than when you took previous mathematics courses. Your maturity and your desire to learn are extremely important and can make a tremendous difference in your ability to succeed in mathematics. I believe you can be successful in this course, but you also need to believe it.

## 2 Prepare for and Attend Class

### Preview the Material

Before class, you should spend a few minutes previewing any new material in the textbook. You do not have to understand everything you read yet. Just get a feeling for the definitions and concepts that will be discussed. This quick preview will help you to understand what your instructor is explaining during class. After the material is explained in class, read the corresponding sections of the text slowly and carefully, word by word.

### Read the Text

A mathematics text is not a novel. Mathematics textbooks should be read slowly and carefully. If you do not understand what you are reading, reread the material. When you come across a new concept or definition, you may wish to underline or highlight it so that it stands out. This way, when you look for it later, it will be easier to find. When you come across a worked-out example, read and follow the example carefully. Do not just skim it. Try working out the example yourself on another sheet of paper. Also, work the **Now Try Exercises** that appear after each example. The Now Try Exercises are designed so that you have the opportunity to immediately apply new ideas. Make notes of anything that you do not understand to ask your instructor.

### Do the Homework

*Two very important commitments that you must make to be successful in this course are to attend class and do your homework regularly*. Your assignments must be worked conscientiously and completely. Mathematics cannot be learned by observation. You need to practice what you have heard in class. By doing homework you truly learn the material.

Don't forget to check the answers to your homework assignments. Answers to the odd-numbered exercises are in the back of this book. In addition, the answers to all the Cumulative Review Exercises, Mid-Chapter Tests, Chapter Review Exercises, Chapter Practice Tests, and Cumulative Review Tests are provided. For the Mid-Chapter Tests, Chapter Practice Tests, and Cumulative Review Tests, the section where the material was first introduced is provided in brackets after each answer. Answers to the Group Activity Exercises are not provided because we want you to arrive at the answers as a group.

If you have difficulty with some of the exercises, mark them and do not hesitate to ask questions about them in class. You should not feel comfortable until you understand all the concepts needed to work every assigned problem.

When you do your homework, make sure that you write it neatly and carefully. Pay particular attention to copying signs and exponents correctly. Do your homework in a step-by-step manner. This way you can refer back to it later and still understand what was written.

### Attend and Participate in Class

You should attend every class. Generally, the more absences you have, the lower your grade will be. Every time you miss a class, you miss important information. If you must miss a class, contact your instructor ahead of time and get the reading assignment and homework.

While in class, pay attention to what your instructor is saying. If you do not understand something, ask your instructor to repeat or explain the material. If you do not ask questions, your instructor will not know that you have a problem understanding the material.

In class, take careful notes. Write numbers and letters clearly so that you can read them later. It is not necessary to write down every word your instructor says. Copy down the major points and the examples that do not appear in the text. You should not be taking notes so frantically that you lose track of what your instructor is saying.

### Study

Study in the proper atmosphere. Study in an area where you are not constantly disturbed so that your attention can be devoted to what you are reading. The area where you study should be well ventilated and well lit. You should have sufficient desk space to spread out all your materials. Your chair should be comfortable. You should try to minimize distractions while you are studying. You should not study for hours on end. Short study breaks are a good idea.

When studying, you should not only understand how to work a problem, you should also know why you follow the specific steps you do to work the problem. If you do not have an understanding of why you follow the specific process, you will not be able to solve similar problems.

### Time Management

It is recommended that students spend at least 2 hours studying and doing homework for every hour of class time. Some students require more time than others. Finding the necessary time to study is not always easy. The following are some suggestions that you may find helpful.

1. Plan ahead. Determine when you will have time to study and do your homework. Do not schedule other activities for these time periods. Try to space these periods evenly over the week.
2. Be organized so that you will not have to waste time looking for your books, pen, calculator, or notes.
3. Use a calculator to perform tedious calculations.
4. When you stop studying, clearly mark where you stopped in the text.
5. Try not to take on added responsibilities. You must set your priorities. If your education is a top priority, as it should be, you may have to cut the time spent on other activities.
6. If time is a problem, do not overburden yourself with too many courses. Consider taking fewer credits. If you do not have sufficient time to study, your understanding and your grades in all of your courses may suffer.

## 3 Prepare for and Take Examinations

### Study for an Exam

If you do some studying each day, you should not need to cram the night before an exam. If you wait until the last minute, you will not have time to seek the help you may need. To review for an exam,

1. Read your class notes.
2. Review your homework assignments.
3. Study the formulas, definitions, and procedures you will need for the exam.
4. Read the Avoiding Common Errors boxes and Helpful Hint boxes carefully.
5. Read the summary at the end of each chapter.
6. Work the review exercises at the end of each chapter. If you have difficulties, restudy those sections. If you still have trouble, seek help.
7. Work the Mid-Chapter Tests and the Chapter Practice Tests.
8. Rework quizzes previously given if the material covered in the quizzes will be included on the test.
9. Work the Cumulative Review Test if material from earlier chapters will be included on the test.

### Take an Exam

Make sure that you get a good night's sleep the day before the test. If you studied properly, you should not have to stay up late the night before to prepare for the test. Arrive at the exam site early so that you have a few minutes to relax before the exam. If you need to rush to get to the exam, you will start out nervous and anxious. After you receive the exam, do the following:

1. Carefully write down any formulas or ideas that you want to remember.
2. Look over the entire exam quickly to get an idea of its length and to make sure that no pages are missing. You will need to pace yourself to make sure that you complete the entire exam. Be prepared to spend more time on problems worth more points.
3. Read the test directions carefully.
4. Read each problem carefully. Answer each question completely and make sure that you have answered the specific question asked.
5. Starting with number 1, work each question in order. If you come across a question that you are not sure of, do not spend too much time on it. Continue working the questions that you understand. After completing all other questions, go back and finish those questions you were not sure of. Do not spend too much time on any one question.
6. Attempt each problem. You may be able to earn at least partial credit.
7. Work carefully and write clearly so that your instructor can read your work. Also, it is easy to make mistakes when your writing is unclear.
8. Check your work and your answers if you have time.
9. Do not be concerned if others finish the test before you. Do not be disturbed if you are the last to finish. Use all your extra time to check your work.

## 4 Find Help

### Use the Supplements

This text comes with many supplements. Find out from your instructor early in the semester which of these supplements are available and which supplements might be beneficial for you to use. Reading supplements should not replace reading the text. Instead supplements should enhance your understanding of the material. If you miss a class, you may want to review the video on the topic you missed before attending the next class.

There are many supplements available. The supplements that may be available to you are: the Student Solutions Manual which works out the odd section exercises as well as all the end-of-chapter exercises; the CD Lecture Series Videos which show about 20 minutes of lecture per section and include the worked out solutions to the exercises marked with this icon; the Chapter Test Prep Video CD, which works out every problem in every Chapter Practice Test; MathXL MathXL®, a powerful online tutorial and homework system, which is also available on CD; MyMathLab MyMathlab, the online course which houses MathXL plus a variety of other supplements; and the Prentice Hall Mathematics Tutor Center, which provides live tutorial support via phone, fax, or e-mail.

### Seek Help

One thing I stress with my own students is to *get help as soon as you need it!* Do not wait! In mathematics, one day's material is usually based on the previous day's material. So if you don't understand the material today, you may not be able to understand the material tomorrow.

Where should you seek help? There are often a number of places to obtain help on campus. You should try to make a friend in the class with whom you can study. Often you can help one another. You may wish to form a study group with other students in your class. Discussing the concepts and homework with your peers will reinforce your own understanding of the material.

You should not hesitate to visit your instructor when you are having problems with the material. Be sure you read the assigned material and attempt the homework before meeting with your instructor. Come prepared with specific questions to ask.

Often other sources of help are available. Many colleges have a mathematics laboratory or a mathematics learning center where tutors are available to help students. Ask your instructor early in the semester if any tutors are available, and find out where the tutors are located. Then use these tutors as needed.

## 5 Learn to Use a Calculator

Many instructors require their students to purchase and to use a calculator in class. You should find out as soon as possible which calculator, if any, your instructor expects you to use. If you plan on taking additional mathematics courses, you should determine which calculator will be required in those courses and consider purchasing that calculator for use in this course if its use is permitted by your instructor. Many instructors require a scientific calculator and many others require a graphing calculator.

In this book we provide information about both types of calculators. Always read and save the user's manual for whatever calculator you purchase.

## EXERCISE SET 1.1

MathXL® MyMathLab

*Do you know all of the following information? If not, ask your instructor as soon as possible.*

1. What is your instructor's name?
2. What are your instructor's office hours?
3. Where is your instructor's office located?
4. How can you best reach your instructor?
5. Where can you obtain help if your instructor is not available?
6. What supplements are available to assist you in learning?
7. Does your instructor recommend or require a specific calculator? If so, which one?
8. When can you use a calculator? Can it be used in class, on homework, on tests?
9. What is your instructor's attendance policy?
10. Why is it important that you attend every class possible?
11. Do you know the name and phone number of a friend in class?
12. For each hour of class time, how many hours outside class are recommended for homework and studying?
13. List what you should do to be properly prepared for each class.
14. Explain how a mathematics textbook should be read.
15. Write a summary of the steps you should follow when taking an exam.

indicates an exercise worked out on the CD Lecture Series Videos.

indicates a writing exercise. That is, an exercise that requires a written answer.

**16.** Having a positive attitude is very important for success in this course. Are you beginning this course with a positive attitude? It is important that you do!

**17.** You need to make a commitment to spend the time necessary to learn the material, to do the homework, and to attend class regularly. Explain why you believe this commitment is necessary to be successful in this course.

**18.** What are your reasons for taking this course?

**19.** What are your goals for this course?

**20.** Have you given any thought to studying with a friend or a group of friends? Can you see any advantages in doing so? Can you see any disadvantages in doing so?

## 1.2 Sets and Other Basic Concepts

1 Identify sets.

2 Identify and use inequalities.

3 Use set builder notation.

4 Find the union and intersection of sets.

5 Identify important sets of numbers.

Let's start with some important definitions. When a letter is used to represent various numbers it is called a **variable**. For instance, if $t =$ the time, in hours, that a car is traveling, then $t$ is a variable since the time is constantly changing as the car is traveling. We often use the letters $x$, $y$, $z$, and $t$ to represent variables. However, other letters may be used. When presenting properties or rules, the letters $a$, $b$, and $c$ are often used as variables.

If a letter represents one particular value it is called a **constant**. For example, if $s =$ the number of seconds in a minute, then $s$ represents a constant because there are always 60 seconds in a minute. The number of seconds in a minute does not vary. In this book, letters representing both variables and constants are italicized.

The term **algebraic expression**, or simply **expression**, will be used often in the text. An expression is any combination of numbers, variables, exponents, mathematical symbols (other than equal signs), and mathematical operations.

### 1 Identify Sets

Sets are used in many areas of mathematics, so an understanding of sets and set notation is important. A **set** is a collection of objects. The objects in a set are called **elements** of the set. Sets are indicated by means of braces, { }, and are often named with capital letters. When the elements of a set are listed within the braces, as illustrated below, the set is said to be in **roster form**.

$$A = \{a, b, c\}$$
$$B = \{\text{yellow, green, blue, red}\}$$
$$C = \{1, 2, 3, 4, 5\}$$

Set $A$ has three elements, set $B$ has four elements, and set $C$ has five elements. The symbol $\in$ is used to indicate that an item is an element of a set. Since 2 is an element of set $C$, we may write $2 \in C$; this is read "2 is an element of set $C$."

A set may be finite or infinite. Sets $A$, $B$, and $C$ each have a finite number of elements and are therefore *finite sets*. In some sets it is impossible to list all the elements. These are *infinite sets*. The following set, called the set of **natural numbers** or **counting numbers**, is an example of an infinite set.

$$N = \{1, 2, 3, 4, 5, \ldots\}$$

The three dots after the last comma, called an *ellipsis*, indicate that the set continues in the same manner.

Another important infinite set is the integers. The set of **integers** follows.

$$I = \{\ldots, -4, -3, -2, -1, 0, 1, 2, 3, 4, \ldots\}$$

Notice that the set of integers includes both positive and negative integers and the number 0.

If we write

$$D = \{1, 2, 3, 4, 5, \ldots, 163\}$$

we mean that the set continues in the same manner until the number 163. Set $D$ is the set of the first 163 natural numbers. $D$ is therefore a finite set.

A special set that contains no elements is called the **null set**, or **empty set**, written { } or Ø. For example, the set of students in your class under 8 years of age is the null or empty set.

## 2 Identify and Use Inequalities

Before we introduce a second method of writing a set, called *set builder notation*, we will introduce the inequality symbols.

**Inequality Symbols**

$>$ is read "is greater than."

$\geq$ is read "is greater than or equal to."

$<$ is read "is less than."

$\leq$ is read "is less than or equal to."

$\neq$ is read "is not equal to."

**Inequalities can be explained using the real number line (Fig. 1.1).**

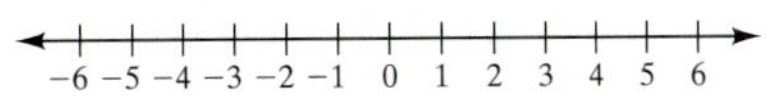


FIGURE 1.1

The number $a$ is greater than the number $b$, $a > b$, when $a$ is to the right of $b$ on the number line (**Fig. 1.2**). We can also state that the number $b$ is less than $a$, $b < a$, when $b$ is to the left of $a$ on the number line. The inequality $a \neq b$ means either $a < b$ or $a > b$.

FIGURE 1.2

**EXAMPLE 1** ▸ Insert either $>$ or $<$ in the shaded area between the numbers to make each statement true.

**a)** $6 \quad 2$ **b)** $-7 \quad 1$ **c)** $-4 \quad -5$

*Solution* Draw a number line and indicate the location of the numbers in parts **a)**, **b)**, and **c)** as illustrated in **Figure 1.3**.

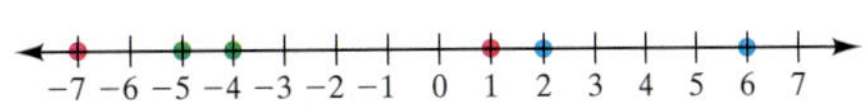


FIGURE 1.3

**a)** $6 > 2$ Note that 6 is to the right of 2 on the number line.

**b)** $-7 < 1$ Note that $-7$ is to the left of 1 on the number line.

**c)** $-4 > -5$ Note that $-4$ is to the right of $-5$ on the number line.

▸ **Now Try Exercise 29**

*Remember that the symbol used in an inequality, if it is true, always points to the smaller of the two numbers.*

We use the notation $x > 2$, read "$x$ is greater than 2," to represent *all* real numbers greater than 2. We use the notation $x \leq -3$, read "$x$ is less than or equal to $-3$," to represent all real numbers that are less than or equal to $-3$. The notation $-4 \leq x < 3$ means all real numbers that are greater than or equal to $-4$ and also less than 3. In the inequalities $x > 2$ and $x \leq -3$, the 2 and the $-3$ are called **endpoints**. In the inequality $-4 \leq x < 3$, the $-4$ and 3 are the endpoints. The solutions to inequalities that use either $<$ or $>$ do not include the endpoints, but the solutions to inequalities that use either $\leq$ or $\geq$ do include the endpoints. When inequalities are illustrated on the number line, a solid circle is used to show that the endpoint is included in the answer, and an open circle is used to show that the endpoint is not included. Following are some illustrations of how certain inequalities are indicated on the number line.

| Inequality | Inequality Indicated on the Number Line |
|---|---|
| $x > 2$ | 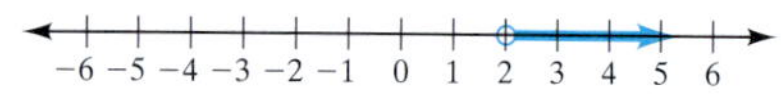 |
| $x \leq -1$ | 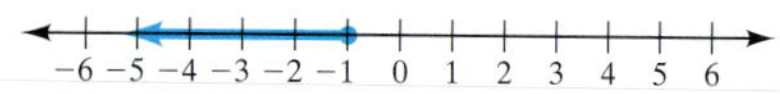 |
| $-4 \leq x < 3$ | 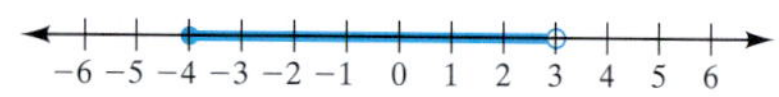 |

Some students misunderstand the word *between*. The word *between* indicates that the endpoints are not included in the answer. For example, the set of natural numbers between 2 and 6 is {3, 4, 5}. If we wish to include the endpoints, we can use the word *inclusive*. For example, the set of natural numbers between 2 and 6 inclusive is {2, 3, 4, 5, 6}.

## 3 Use Set Builder Notation

Now that we have introduced the inequality symbols we will discuss another method of indicating a set, called **set builder notation**. An example of set builder notation is

$$E = \{x | x \text{ is a natural number greater than } 7\}$$

This is read "Set $E$ is the set of all elements $x$, such that $x$ is a natural number greater than 7." In roster form, this set is written

$$E = \{8, 9, 10, 11, 12, \ldots\}$$

The general form of set builder notation is

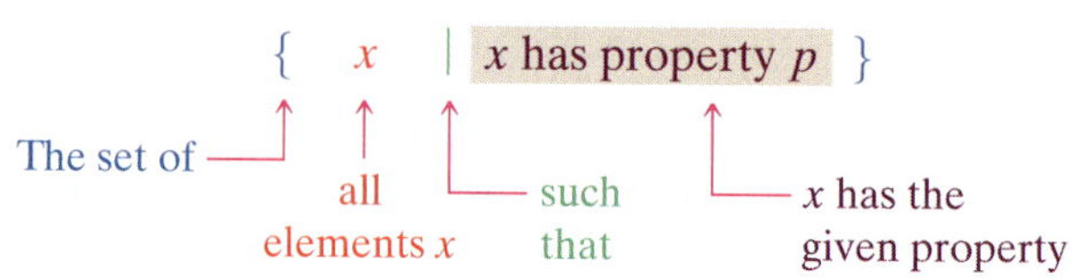


We often will use the variable $x$ when using set builder notation, although any variable can be used.

Two condensed ways of writing set $E = \{x | x \text{ is a natural number greater than } 7\}$ in set builder notation follow.

$$E = \{x | x > 7 \text{ and } x \in N\} \quad \text{or} \quad E = \{x | x \geq 8 \text{ and } x \in N\}$$

The set $A = \{x | -3 < x \leq 4 \text{ and } x \in I\}$ is the set of integers greater than $-3$ and less than or equal to 4. The set written in roster form is $\{-2, -1, 0, 1, 2, 3, 4\}$. Notice that the endpoint $-3$ is not included in the set but the endpoint 4 is included.

How do the sets $B = \{x | x > 2 \text{ and } x \in N\}$ and $C = \{x | x > 2\}$ differ? Can you write each set in roster form? Can you illustrate both sets on the number line? Set $B$ contains only the natural numbers greater than 2, that is, $\{3, 4, 5, 6, \ldots\}$. Set $C$ contains not only the natural numbers greater than 2 but also fractions and decimal numbers greater than 2. If you attempted to write set $C$ in roster form, where would you begin? What is the smallest number greater than 2? Is it 2.1 or 2.01 or 2.001? Since there is no smallest number greater than 2, this set cannot be written in roster form. On the top of the next page we illustrate these two sets on the number line. We have also illustrated two other sets.

| Set | Set Indicated on the Number Line |
|---|---|
| $\{x\|x > 2 \text{ and } x \in N\}$ | 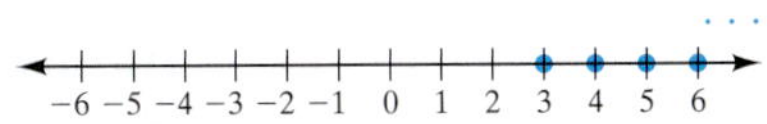 |
| $\{x\|x > 2\}$ | 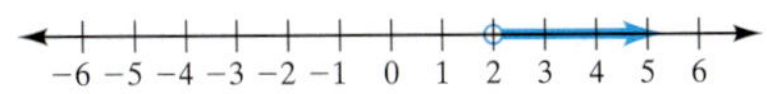 |
| $\{x\|-1 \le x < 4 \text{ and } x \in I\}$ |  |
| $\{x\|-1 \le x < 4\}$ | 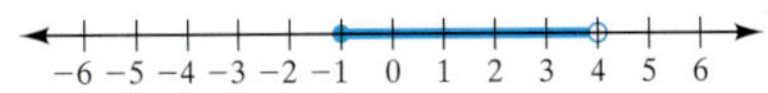 |

Another method of indicating inequalities, called *interval notation*, will be discussed in Section 2.5.

## 4 Find the Union and Intersection of Sets

Just as *operations* such as addition and multiplication are performed on numbers, operations can be performed on sets. Two set operations are *union* and *intersection*.

**Union**

The **union** of set $A$ and set $B$, written $A \cup B$, is the set of elements that belong to either set $A$ *or* set $B$.

Because the word *or*, as used in this context, means belonging to set $A$ or set $B$ or both sets, the union is formed by combining, or joining together, the elements in set $A$ with those in set $B$. If an item is an element in either set $A$, or set $B$, or in both sets, then it is an element in the union of the sets. If an element appears in both sets, we only list it once when we write the union of two sets.

**Examples of Union of Sets**

$A = \{1, 2, 3, 4, 5\}$, $B = \{3, 4, 5, 6, 7\}$, $A \cup B = \{1, 2, 3, 4, 5, 6, 7\}$

$A = \{a, b, c, d, e\}$, $B = \{x, y, z\}$, $A \cup B = \{a, b, c, d, e, x, y, z\}$

In set builder notation we can express $A \cup B$ as

$$A \cup B = \{x|x \in A \text{ or } x \in B\}$$

**Intersection**

The **intersection** of set $A$ and set $B$, written $A \cap B$, is the set of all elements that are common to both set $A$ *and* set $B$.

Because the word *and*, as used in this context, means belonging to *both* set $A$ and set $B$, the intersection is formed by using only those elements that are in both set $A$ and set $B$. If an item is an element in only one of the two sets, then it is not an element in the intersection of the sets.

**Examples of Intersection of Sets**

$A = \{1, 2, 3, 4, 5\}$, $B = \{3, 4, 5, 6, 7\}$, $A \cap B = \{3, 4, 5\}$

$A - \{a, b, c, d, e\}$, $B = \{x, y, z\}$, $A \cap B = \{\ \}$

Note that in the last example, sets $A$ and $B$ have no elements in common. Therefore, their intersection is the empty set. In set builder notation we can express $A \cap B$ as

$$A \cap B = \{x|x \in A \text{ and } x \in B\}$$

## 5 Identify Important Sets of Numbers

At this point we have all the necessary information to discuss important sets of real numbers. In the box below, we describe these sets and provide letters that are often used to represent these sets of numbers.

**Important Sets of Real Numbers**

| | |
|---|---|
| **Real numbers** | $\mathbb{R} = \{x \mid x \text{ is a point on the number line}\}$ |
| **Natural or counting numbers** | $N = \{1, 2, 3, 4, 5, \ldots\}$ |
| **Whole numbers** | $W = \{0, 1, 2, 3, 4, 5, \ldots\}$ |
| **Integers** | $I = \{\ldots, -3, -2, -1, 0, 1, 2, 3, \ldots\}$ |
| **Rational numbers** | $Q = \left\{\frac{p}{q} \middle\| p \text{ and } q \text{ are integers, } q \neq 0\right\}$ |
| **Irrational numbers** | $H = \{x \mid x \text{ is a real number that is not rational}\}$ |

Let us briefly look at the rational, irrational, and real numbers. A **rational number** is any number that can be represented as a quotient of two integers, with the denominator not 0.

### Examples of Rational Numbers

$$\frac{3}{5}, \quad -\frac{2}{3}, \quad 0, \quad 1.63, \quad 7, \quad -17, \quad \sqrt{4}$$

Notice that 0, or any other integer, is also a rational number since it can be written as a fraction with a denominator of 1. For example, $0 = \frac{0}{1}$ and $7 = \frac{7}{1}$.

The number 1.63 can be written $\frac{163}{100}$ and is thus a quotient of two integers. Since $\sqrt{4} = 2$ and 2 is an integer, $\sqrt{4}$ is a rational number. *Every rational number when written as a decimal number will be either a repeating or a terminating decimal number.*

| Examples of Repeating Decimals | Examples of Terminating Decimals |
|---|---|
| $\frac{2}{3} = 0.6666\ldots$ *The 6 repeats.* | $\frac{1}{2} = 0.5$ |
| $\frac{1}{7} = 0.142857142857\ldots$ *The block 142857 repeats.* | $\frac{9}{4} = 2.25$ |

To show that a digit or group of digits repeat, we can place a bar above the digit or group of digits that repeat. For example, we may write

$$\frac{2}{3} = 0.\overline{6} \quad \text{and} \quad \frac{1}{7} = 0.\overline{142857}$$

Although $\sqrt{4}$ is a rational number, the square roots of most integers are not. Most square roots will be neither terminating nor repeating decimals when expressed as decimal numbers and are **irrational numbers**. Some irrational numbers are $\sqrt{2}$, $\sqrt{3}$, $\sqrt{5}$, and $\sqrt{6}$. Another irrational number is pi, $\pi$. When we give a decimal value for an irrational number, we are giving only an *approximation* of the value of the irrational number. The symbol $\approx$ means "is approximately equal to."

$$\pi \approx 3.14 \qquad \sqrt{2} \approx 1.41 \qquad \sqrt{3} \approx 1.73 \qquad \sqrt{10} \approx 3.16$$

The **real numbers** are formed by taking the *union* of the rational numbers and the irrational numbers. Therefore, any real number must be either a rational number or an irrational number. The symbol $\mathbb{R}$ is often used to represent the set of real numbers. **Figure 1.4** illustrates various real numbers on the number line.

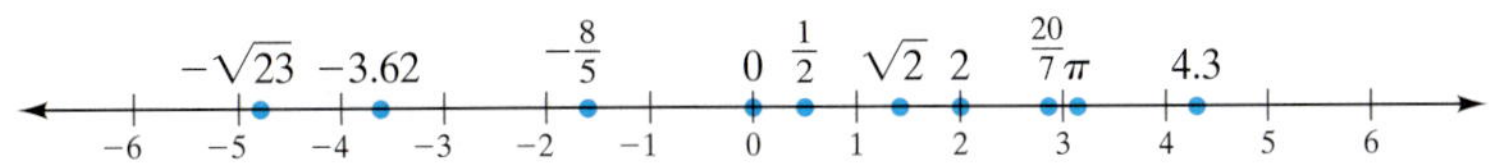


FIGURE 1.4

A first set is a **subset** of a second set when every element of the first set is also an element of the second set. For example, the set of natural numbers, $\{1, 2, 3, 4, \dots\}$, is a subset of the set of whole numbers, $\{0, 1, 2, 3, 4, \dots\}$, because every element in the set of natural numbers is also an element in the set of whole numbers. **Figure 1.5** illustrates the relationships between the various subsets of the real numbers. In **Figure 1.5a**, you see that the set of natural numbers is a subset of the set of whole numbers, of the set of integers, and of the set of rational numbers. Therefore, every natural number must also be a whole number, an integer, and a rational number. Using the same reasoning, we can see that the set of whole numbers is a subset of the set of integers and of the set of rational numbers, and that the set of integers is a subset of the set of rational numbers.

Looking at **Figure 1.5b**, we see that the positive integers, 0, and the negative integers form the integers, that the integers and noninteger rational numbers form the rational numbers, and so on.

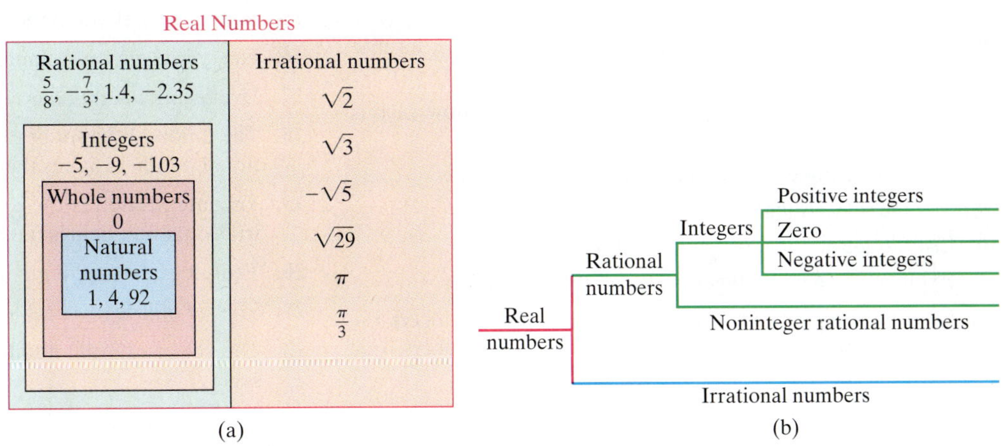


FIGURE 1.5

**EXAMPLE 2** ▸ Consider the following set:

$$\left\{-3, 0, \frac{5}{7}, 12.25, \sqrt{7}, -\sqrt{11}, \frac{22}{7}, 5, 7.1, -54, \pi\right\}$$

List the elements of the set that are

**a)** natural numbers. **b)** whole numbers. **c)** integers.
**d)** rational numbers. **e)** irrational numbers. **f)** real numbers.

### Solution

**a)** Natural numbers: 5 **b)** Whole numbers: 0, 5 **c)** Integers: −3, 0, 5, −54

**d)** Rational numbers can be written in the form $p/q$, $q \neq 0$. Each of the following can be written in this form and is a rational number.

$$-3, 0, \frac{5}{7}, 12.25, \frac{22}{7}, 5, 7.1, -54$$

**e)** Irrational numbers are real numbers that are not rational. The following numbers are irrational.

$$\sqrt{7}, -\sqrt{11}, \pi$$

**f)** All of the numbers in the set are real numbers. The union of the rational numbers and the irrational numbers forms the real numbers.

$$-3, 0, \frac{5}{7}, 12.25, \sqrt{7}, -\sqrt{11}, \frac{22}{7}, 5, 7.1, -54, \pi$$

▸ **Now Try Exercise 49**

Not all numbers are real numbers. Some numbers that we discuss later in the text that are not real numbers are complex numbers and imaginary numbers.

# EXERCISE SET 1.2

MathXL® MyMathLab

## Concept/Writing Exercises

1. What is a variable?
2. What is an algebraic expression?
3. What is a set?
4. What do we call the objects in a set?
5. What is the null or empty set?
6. Is the set of natural or counting numbers a finite or infinite set? Explain.
7. List the five inequality symbols and write down how each is read.
8. Give an example of a set that is empty.
9. List the set of integers *between* 3 and 7.
10. List the set of integers *between* −1 and 3 *inclusive*.
11. Explain why every integer is also a rational number.
12. Describe the counting numbers, whole numbers, integers, rational numbers, irrational numbers, and real numbers. Explain the relationships among the sets of numbers.

*In Exercises 13–22, indicate whether each statement is true or false.*

13. Every natural number is a whole number.
14. Every whole number is a natural number.
15. Some rational numbers are integers.
16. Every integer is a rational number.
17. Every rational number is an integer.
18. The union of the set of rational numbers with the set of irrational numbers forms the set of real numbers.
19. The intersection of the set of rational numbers and the set of irrational numbers is the empty set.
20. The set of natural numbers is a finite set.
21. The set of integers between $\pi$ and 4 is the null set.
22. The set of rational numbers between 3 and $\pi$ is an infinite set.

## Practice the Skills

*Insert either* < *or* > *in the shaded area to make each statement true.*

23. 5 ▒ 3
24. −1 ▒ 8
25. 0 ▒ −2
26. −3 ▒ 3
27. −1 ▒ −1.01
28. 2 ▒ −3
29. −5 ▒ −3
30. −8 ▒ −1
31. −14.98 ▒ −14.99
32. −3.4 ▒ −3.2
33. 1.7 ▒ 1.9
34. −1.1 ▒ −1.9
35. $-\pi$ ▒ −4
36. −723 ▒ −655
37. $-\frac{7}{8}$ ▒ $-\frac{10}{11}$
38. $-\frac{4}{7}$ ▒ $-\frac{5}{9}$

*In Exercises 39–48, list each set in roster form.*

39. $A = \{x \mid -1 < x < 1 \text{ and } x \in I\}$
40. $B = \{y \mid y \text{ is an odd natural number less than } 6\}$
41. $C = \{z \mid z$ is an even integer greater than 16 and less than or equal to $20\}$
42. $D = \{x \mid x \geq -3 \text{ and } x \in I\}$
43. $E = \{x \mid x < 3 \text{ and } x \in W\}$
44. $F = \left\{x \,\middle|\, -\frac{6}{5} \leq x < \frac{15}{4} \text{ and } x \in N\right\}$
45. $H = \{x \mid x \text{ is a whole number multiple of } 7\}$
46. $L = \{x \mid x \text{ is an integer greater than } -5\}$
47. $J = \{x \mid x > 0 \text{ and } x \in I\}$
48. $K = \{x \mid x \text{ is a whole number between 9 and 10}\}$

A green numbered exercise, such as 29. indicates a Now Try Exercise.

**49.** Consider the set $\left\{-2, 4, \frac{1}{2}, \frac{5}{9}, 0, \sqrt{2}, \sqrt{8}, -1.23, \frac{78}{79}\right\}$.
List the elements that are
**a)** natural numbers.
**b)** whole numbers.
**c)** integers.
**d)** rational numbers.
**e)** irrational numbers.
**f)** real numbers.

**50.** Consider the set $\left\{2, 4, -5.33, \frac{9}{2}, \sqrt{5}, \sqrt{2}, -100, -7, 4.7\right\}$.
List the elements that are:
**a)** whole numbers.
**b)** natural numbers.
**c)** rational numbers.
**d)** integers.
**e)** irrational numbers.
**f)** real numbers.

*Find $A \cup B$ and $A \cap B$ for each set A and B.*

**51.** $A = \{1, 2, 3\}, B = \{4, 5, 6\}$

**52.** $A = \{1, 2, 3, 4, 5\}, B = \{2, 4, 6\}$

**53.** $A = \{-3, -1, 1, 3\}, B = \{-4, -3, -2, -1, 0\}$

**54.** $A = \{-3, -2, -1, 0\}, B = \{-1, 0, 1, 2\}$

**55.** $A = \{\ \}, B = \{2, 4, 6, 8, 10\}$

**56.** $A = \{2, 4, 6\}, B = \{2, 4, 6, 8, \ldots\}$

**57.** $A = \{0, 10, 20, 30\}, B = \{5, 15, 25\}$

**58.** $A = \{1, 3, 5\}, B = \{1, 3, 5, 7, \ldots\}$

**59.** $A = \{-1, 0, 1, e, i, \pi\}, B = \{-1, 0, 1\}$

**60.** $A = \left\{1, \frac{1}{2}, \frac{1}{4}, \frac{1}{6}, \ldots\right\}, B = \left\{\frac{1}{4}, \frac{1}{6}, \frac{1}{8}\right\}$

*Describe each set.*

**61.** $A = \{1, 2, 3, 4, \ldots\}$

**62.** $B = \{2, 4, 6, 8, \ldots\}$

**63.** $C = \{0, 3, 6, 9, \ldots\}$

**64.** $A = \{a, b, c, d, \ldots, z\}$

**65.** $B = \{\ldots, -5, -3, -1, 1, 3, 5, \ldots\}$

**66.** $C = \{$Alabama, Alaska, . . . , Wyoming$\}$

*In Exercises 67 and 68, **a)** write out how you would read each set; **b)** write the set in roster form.*

**67.** $A = \{x | x < 7 \text{ and } x \in N\}$

**68.** $B = \{x | x$ is one of the last five capital letters in the English alphabet$\}$

*Illustrate each set on a number line.*

**69.** $\{x | x \geq 0\}$

**70.** $\{w | w > -5\}$

**71.** $\{z | z \leq 2\}$

**72.** $\{y | y < 4\}$

**73.** $\{p | -6 \leq p < 3\}$

**74.** $\{x | -1.67 \leq x < 5.02\}$

**75.** $\{q | q > -3 \text{ and } q \in N\}$

**76.** $\{x | -1.93 \leq x \leq 2 \text{ and } x \in I\}$

**77.** $\{r | r \leq \pi \text{ and } r \in W\}$

**78.** $\left\{x \middle| \frac{5}{12} < x \leq \frac{7}{12} \text{ and } x \in N\right\}$

*Express in set builder notation each set of numbers that are indicated on the number line.*

**79.** 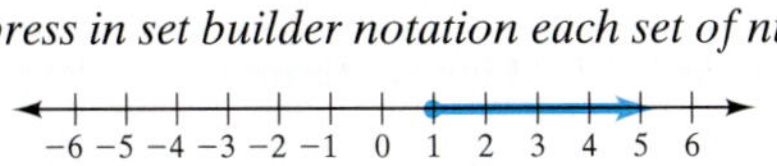

**80.** 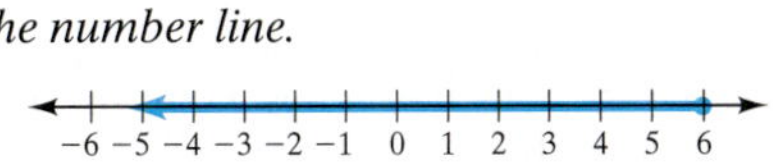

**81.** 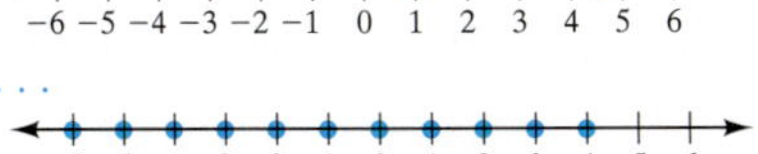

**82.** 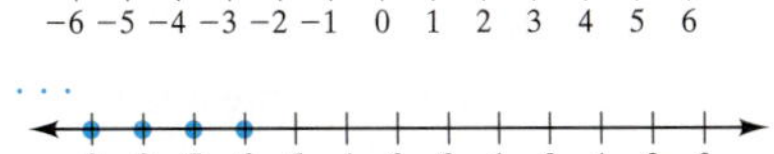

**83.** 

**84.** 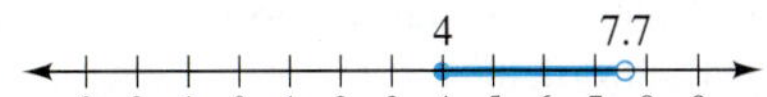

**85.** 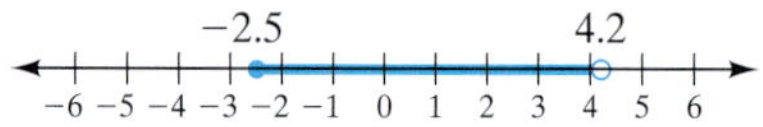


**86.** 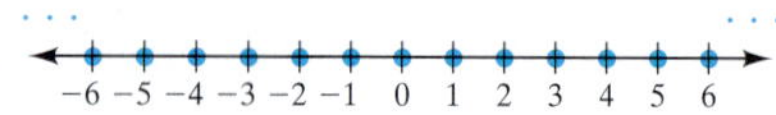


**87.** 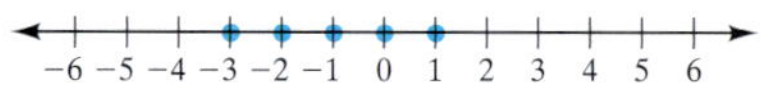


**88.** 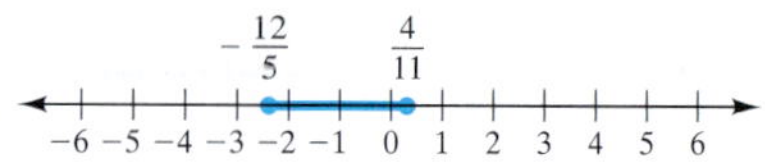


*Refer to the box on page 10 for the meanings of* $\mathbb{R}$, *N, W, I, Q, and H. Then determine whether the first set is a subset of the second set for each pair of sets.*

**89.** $N, W$ **90.** $W, Q$ **91.** $W, N$ **92.** $I, Q$

**93.** $Q, \mathbb{R}$ **94.** $Q, H$ **95.** $Q, I$ **96.** $H, \mathbb{R}$

## Problem Solving

**97.** Construct a set that contains five rational numbers between 1 and 2.

**98.** Construct a set that contains five rational numbers between 0 and 1.

**99.** Determine two sets $A$ and $B$ such that $A \cup B = \{2, 4, 5, 6, 8, 9\}$ and $A \cap B = \{4, 5, 9\}$.

**100.** Determine two sets $A$ and $B$ such that $A \cup B = \{3, 5, 7, 8, 9\}$ and $A \cap B = \{5, 7\}$.

**101. NASCAR Nextel Cup** The 2004 NASCAR Nextel Cup series consisted of 36 races held between February and November. Two such races were the Pocono 500 held on June 14 and the Ford 400 held on November 20. The tables below show the top six finishers in both races.

Pocono 500

| Position | Driver |
|---|---|
| 1 | Jimmie Johnson |
| 2 | Jerry Mayfield |
| 3 | Bobby Labonte |
| 4 | Jeff Gordon |
| 5 | Kurt Busch |
| 6 | Dale Earnhardt, Jr. |

Ford 400

| Position | Driver |
|---|---|
| 1 | Greg Biffle |
| 2 | Jimmie Johnson |
| 3 | Jeff Gordon |
| 4 | Tony Stewart |
| 5 | Kurt Busch |
| 6 | Brendan Gaughan |

*Source:* www.NASCAR.com

**a)** Find the set of drivers who had a top 6 finish in the Pocono 500 *or* the Ford 400.

**b)** Does part **a)** represent the union or intersection of the drivers?

**c)** Find the set of drivers who had a top 6 finish in the Pocono 500 *and* the Ford 400.

**d)** Does part **c)** represent the union or intersection of the drivers?

**102. Disasters** The tables below give estimates of the six deadliest earthquakes and the six deadliest natural disasters.

Six Deadliest Earthquakes

| Deaths | Magnitude | Location | Year |
|---|---|---|---|
| 255,000 | 7.8–8.2 | Tangshan, China | 1976 |
| 200,000 | 8.3 | Xining, China | 1927 |
| 200,000 | 8.6 | Gansu, China | 1920 |
| 175,000 | 9.0 | Asia/Africa | 2004 |
| 143,000 | 8.3 | Kwanto, Japan | 1923 |
| 110,000 | 7.3 | Turkmenistan | 1948 |

Six Deadliest Natural Disasters

| Deaths | Event | Location | Year |
|---|---|---|---|
| 3.7 million | Flood | Huang He River, China | 1931 |
| 300,000 | Cyclone | Bangladesh | 1970 |
| 255,000 | Earthquake | Tangshan, China | 1976 |
| 200,000 | Earthquake | Xining, China | 1927 |
| 200,000 | Earthquake | Gansu, China | 1920 |
| 175,000 | Earthquake/ Tsunami | Asia/Africa | 2004 |

*Source:* www.msnbc.com/modules/tables/worstquakesofcentury, Associated Press, Reuters, U.S. Geological Survey, *The World Almanac*, *The Washington Post* (12/29/2004)

**a)** Find the set of the location of the six deadliest earthquakes *or* the location of the six deadliest natural disasters.

**b)** Does part **a)** represent the union or intersection of the categories?

**c)** Find the set of the location of the six deadliest earthquakes *and* the location of the set of the six deadliest natural disasters.

**d)** Does part **c)** represent the union or intersection of the categories?

**103. Algebra Tests** The table below shows the students who had a grade of A on the first two tests in an intermediate algebra class. (Assume every student had a different first name.)

| First Test | Second Test |
|---|---|
| Albert | Linda |
| Carmen | Jason |
| Frank | David |
| Linda | Frank |
| Barbara | Earl |
| | Kate |
| | Ingrid |

**a)** Find the set of students who had a grade of A on the first *or* second tests.

**b)** Does part **a)** represent the union or intersection of the students?

**c)** Find the set of students who had a grade of A on the first and second tests.

**d)** Does part **c)** represent the union or intersection of the students?

**104. Running Races** The table below shows the runners who participated in a 3-kilometer (km) race and a 5-kilometer race. (Assume every runner had a different first name.)

| 3 Kilometers | 5 Kilometers |
|---|---|
| Adam | Luan |
| Kim | Betty |
| Luan | Darnell |
| Ngo | Ngo |
| Carmen | Frances |
| Earl | George |
| Martha | Adam |

**a)** Find the set of runners who participated in a 3-km *or* a 5-km race.

**b)** Does part **a)** represent the union or intersection of the runners?

**c)** Find the set of runners who participated in a 3-km *and* a 5-km race.

**d)** Does part **c)** represent the union or intersection of runners?

**105. Populous Countries** The table below shows the five most populous countries in 1950 and in 2005 and the five countries expected to be most populous in 2050. This information was taken from the U.S. Census Bureau Web site.

| 1950 | 2005 | 2050 |
|---|---|---|
| China | China | India |
| India | India | China |
| United States | United States | United States |
| Russia | Indonesia | Indonesia |
| Japan | Brazil | Nigeria |

**a)** Find the set of the five most populous countries in 2005 *or* 2050.

**b)** Find the set of the five most populous countries in 1950 *or* 2050.

**c)** Find the set of the five most populous countries in 1950 *and* 2005.

**d)** Find the set of the five most populous countries in 2005 *and* 2050.

**e)** Find the set of the five most populous countries in 1950 *and* 2005 *and* 2050.

**106. Writing Contest** The table below shows the students from an English class who participated in three writing contests in a local high school. (Assume every student had a different first name.)

| First Contest | Second Contest | Third Contest |
|---|---|---|
| Jill | Tom | Pat |
| Sam | Shirley | Richard |
| Tom | Bob | Arnold |
| Pat | Donna | Donna |
| Shirley | Sam | Kate |
| Richard | Jill | |
| | Kate | |

**a)** Find the set of students who participated in the first contest *or* the second contest.

**b)** Find the set of students who participated in the second contest *or* the third contest.

**c)** Find the set of students who participated in the first contest *and* the second contest.

**d)** Find the set of students who participated in the first contest *and* the third contest.

**e)** Find the set of students who participated in the first contest *and* the second contest *and* the third contest.

**107. Cub Scouts** The Cub Scouts in Pack 108 must complete four achievements to earn their Wolf Badge. Doug Wedding, their den leader, has the following table in his record book. A *Yes* indicates that the Cub Scout has completed that achievement.

Let $A$ = the set of scouts who have completed Achievement 1: *Feats of Skill.*

Let $B$ = the set of scouts who have completed Achievement 2: *Your Flag.*

Let $C$ = the set of scouts who have completed Achievement 3: *Cooking and Eating.*

Let $D$ = the set of scouts who have completed Achievement 4: *Making Choices.*

**a)** Give each of the sets $A$, $B$, $C$, and $D$ using the roster method.

**b)** Determine the set $A \cap B \cap C \cap D$, that is, find the set of elements that are common to all four sets.

**c)** Which scouts have met all the requirements to receive their Wolf Badge?

| | Achievement | | | |
|---|---|---|---|---|
| **Scout** | **1** | **2** | **3** | **4** |
| Alex | Yes | Yes | Yes | Yes |
| James | Yes | Yes | No | No |
| George | No | Yes | No | Yes |
| Connor | No | Yes | No | Yes |
| Stephen | No | No | Yes | No |

**108. Goods and Services** The following graph shows the percentage weight given to different goods and services in the consumer price index for December 2005.

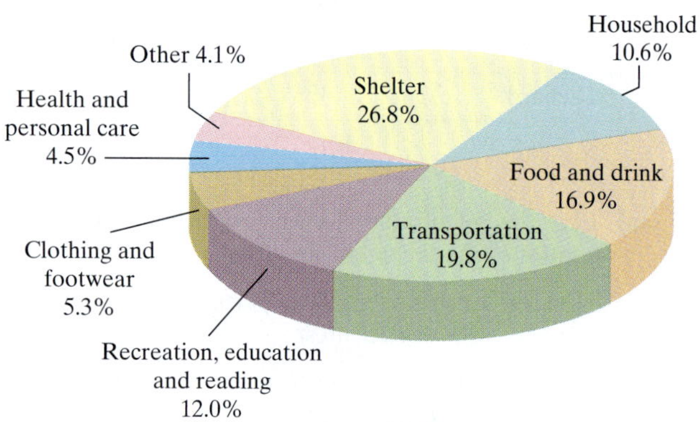


*Source:* U.S. Bureau of Labor Statistics

**a)** List the set of goods and services that have a weight of 21% or greater.

**b)** List the set of goods and services that have a weight of less than 6%.

**109.** The following diagram is called a *Venn diagram*. From the diagram determine the following sets:

**a)** $A$

**b)** $B$

**c)** $A \cup B$

**d)** $A \cap B$

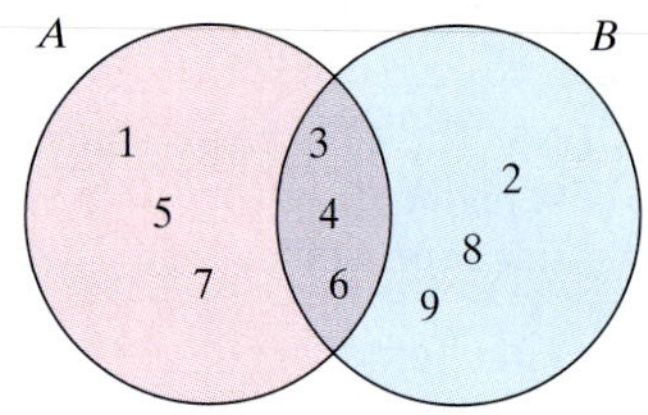


**110.** Use the following Venn diagram to determine the following sets:

**a)** $A$

**b)** $B$

**c)** $A \cup B$

**d)** $A \cap B$

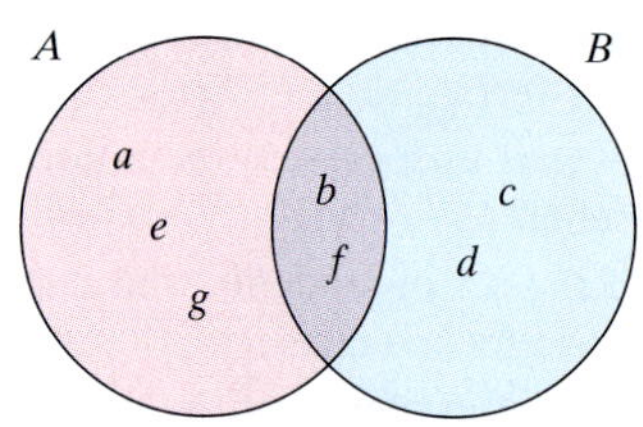


**111. a)** Explain the difference between the following sets of numbers: $\{x|x > 1 \text{ and } x \in N\}$ and $\{x|x > 1\}$.

**b)** Write the first set given in roster form.

**c)** Can you write the second set in roster form? Explain your answer.

**112.** Repeat Exercise 111 for the sets $\{x|2 < x < 6 \text{ and } x \in N\}$ and $\{x|2 < x < 6\}$.

**113. NASCAR Nextel Cup** Draw a Venn diagram for the data given in Exercise 101.

## Challenge Problems

**114. a)** Write the decimal numbers equivalent to $\frac{1}{9}, \frac{2}{9}$, and $\frac{3}{9}$.

**b)** Write the fractions equivalent to $0.\overline{4}$, $0.\overline{5}$, and $0.\overline{6}$.

**c)** What is $0.\overline{9}$ equal to? Explain how you determined your answer.

## Group Activity

**115. Newspaper Preferences** The Venn diagram that follows shows the results of a survey given to 45 people. The diagram shows the number of people in the survey who read the *New York Post*, the *New York Daily News*, and *The Wall Street Journal*.

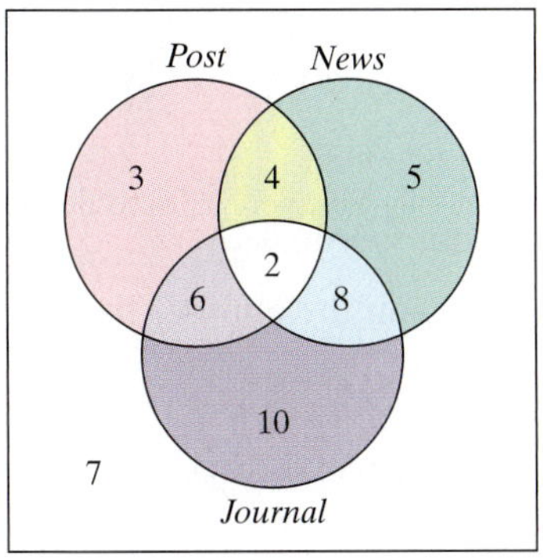


**a)** Group member 1: Determine the number surveyed who read *both* the *News* and the *Post*, that is, $News \cap Post$.

**b)** Group member 2: Determine the number who read both the *Post* and the *Journal*, that is, $Post \cap Journal$.

**c)** Group member 3: Determine the number who read both the *News* and the *Journal*, that is, $News \cap Journal$.

**d)** Share your answer with the other members of the group and see if the group agrees with your answer.

**e)** As a group, determine the number of people who read all three papers.

**f)** As a group, determine the number of people who do not read any of the three papers.

# 1.3 Properties of and Operations with Real Numbers

1. Evaluate absolute values.
2. Add real numbers.
3. Subtract real numbers.
4. Multiply real numbers.
5. Divide real numbers.
6. Use the properties of real numbers.

To succeed in algebra, you must understand how to add, subtract, multiply, and divide real numbers. Before we can explain addition and subtraction of real numbers, we need to discuss absolute value.

Two numbers that are the same distance from 0 on the number line but in opposite directions are called **additive inverses**, or **opposites**, of each other. For example, 3 is the additive inverse of $-3$, and $-3$ is the additive inverse of 3. The number 0 is its own additive inverse. The sum of a number and its additive inverse is 0. What are the additive inverses of $-56.3$ and $\frac{76}{5}$? Their additive inverses are 56.3 and $-\frac{76}{5}$, respectively. Notice that the additive inverse of a positive number is a negative number and the additive inverse of a negative number is a positive number.

**Additive Inverse**

For any real number $a$, its additive inverse is $-a$.

Consider the number $-5$. Its additive inverse is $-(-5)$. Since we know this number must be positive, this implies that $-(-5) = 5$. This is an example of the double negative property.

**Double Negative Property**

For any real number $a$, $-(-a) = a$.

By the double negative property, $-(-7.4) = 7.4$ and $-\left(-\frac{12}{5}\right) = \frac{12}{5}$.

## 1 Evaluate Absolute Values

The **absolute value** of a number is its distance from the number 0 on a number line. The symbol $|\ |$ is used to indicate absolute value.

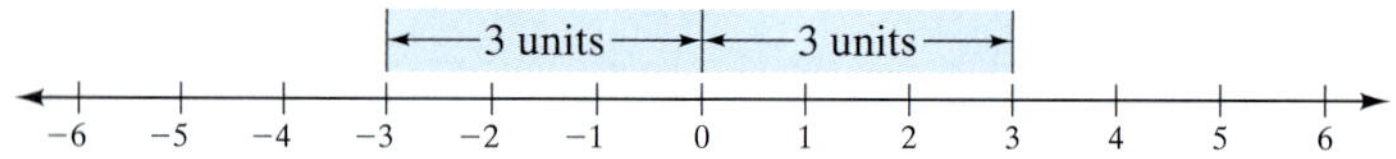


FIGURE 1.6

Consider the numbers 3 and −3 (**Fig. 1.6**). Both numbers are 3 units from 0 on the number line. Thus

$$|3| = 3 \quad \text{and} \quad |-3| = 3$$

**EXAMPLE 1** ▸ Evaluate. **a)** $|7|$ **b)** $|-8.2|$ **c)** $|0|$

*Solution*

**a)** $|7| = 7$, since 7 is 7 units from 0 on the number line.

**b)** $|-8.2| = 8.2$, since −8.2 is 8.2 units from 0 on the number line.

**c)** $|0| = 0$.

**The absolute value of any nonzero number will always be a positive number, and the absolute value of 0 is 0.**

To find the absolute value of a real number without using a number line, use the following definition.

▸ **Now Try Exercise 23**

**Absolute Value**

If $a$ represents any real number, then

$$|a| = \begin{cases} a & \text{if } a \geq 0 \\ -a & \text{if } a < 0 \end{cases}$$

The definition of absolute value indicates that the absolute value of any nonnegative number is the number itself, and the absolute value of any negative number is the additive inverse (or opposite) of the number. The absolute value of a number can be found by using the definition, as illustrated below.

| | |
|---|---|
| $\lvert 6.3\rvert = 6.3$ | Since 6.3 is greater than or equal to 0, its absolute value is 6.3. |
| $\lvert 0\rvert = 0$ | Since 0 is greater than or equal to 0, its absolute value is 0. |
| $\lvert -12\rvert = -(-12) = 12$ | Since −12 is less than 0, its absolute value is −(−12) or 12. |

**EXAMPLE 2** ▸ Evaluate using the definition of absolute value.

**a)** $-|5|$ **b)** $-|-6.43|$

*Solution*

**a)** We are finding the opposite of the absolute value of 5. Since the absolute value of 5 is positive, its opposite must be negative.

$$-|5| = -(5) = -5$$

**b)** We are finding the opposite of the absolute value of −6.43. Since the absolute value of −6.43 is positive, its opposite must be negative.

$$-|-6.43| = -(6.43) = -6.43$$

▸ **Now Try Exercise 31**

**EXAMPLE 3** ▸ Insert <, >, or = in the shaded area between the two values to make each statement true.

**a)** $|8| \;\square\; |-8|$ **b)** $|-1| \;\square\; -|-3|$

*Solution*

**a)** Since both $|8|$ and $|-8|$ equal 8, we have $|8| = |-8|$.

**b)** Since $|-1| = 1$ and $-|-3| = -3$, we have $|-1| > -|-3|$.

▸ **Now Try Exercise 39**

## 2 Add Real Numbers

We first discuss how to add two numbers with the same sign, either both positive or both negative, and then we will discuss how to add two numbers with different signs, one positive and the other negative.

**To Add Two Numbers with the Same Sign (Both Positive or Both Negative)**

Add their absolute values and place the common sign before the sum.

**The sum of two positive numbers will be a positive number, and the sum of two negative numbers will be a negative number.**

**EXAMPLE 4** ▸ Evaluate $-4 + (-7)$.

*Solution* Since both numbers being added are negative, the sum will be negative. To find the sum, add the absolute values of these numbers and then place a negative sign before the value.

$$|-4| = 4 \qquad |-7| = 7$$

Now add the absolute values.

$$|-4| + |-7| = 4 + 7 = 11$$

Since both numbers are negative, the sum must be negative. Thus,

$$-4 + (-7) = -11$$

▸ **Now Try Exercise 55**

**To Add Two Numbers with Different Signs (One Positive and the Other Negative)**

Subtract the smaller absolute from the larger absolute value. The answer has the sign of the number with the larger absolute value.

**The sum of a positive number and a negative number may be either positive, negative, or zero.** The sign of the answer will be the same as the sign of the number with the larger absolute value.

**EXAMPLE 5** ▸ Evaluate $5 + (-9)$.

*Solution* Since the numbers being added are of opposite signs, we subtract the smaller absolute value from the larger absolute value. First we take each absolute value.

$$|5| = 5 \qquad |-9| = 9$$

Now we find the difference, $9 - 5 = 4$. The number $-9$ has a larger absolute value than the number 5, so their sum is negative.

$$5 + (-9) = -4$$

▸ **Now Try Exercise 53**

**EXAMPLE 6** ▸ Evaluate. **a)** $1.3 + (-2.7)$ **b)** $-\frac{7}{8} + \frac{5}{6}$

*Solution*

**a)** $1.3 + (-2.7) = -1.4$

**b)** Begin by writing both fractions with the least common denominator, 24.

$$-\frac{7}{8} + \frac{5}{6} = -\frac{21}{24} + \frac{20}{24} = \frac{(-21) + 20}{24} = \frac{-1}{24} = -\frac{1}{24}$$

▸ Now Try Exercise 59

**EXAMPLE 7** ▸ **Depth of Ocean Trenches** The Palau Trench in the Pacific Ocean lies 26,424 feet below sea level. The deepest ocean trench, the Mariana Trench, is 9416 feet deeper than the Palau Trench (see **Fig. 1.7**). Find the depth of the Mariana Trench.

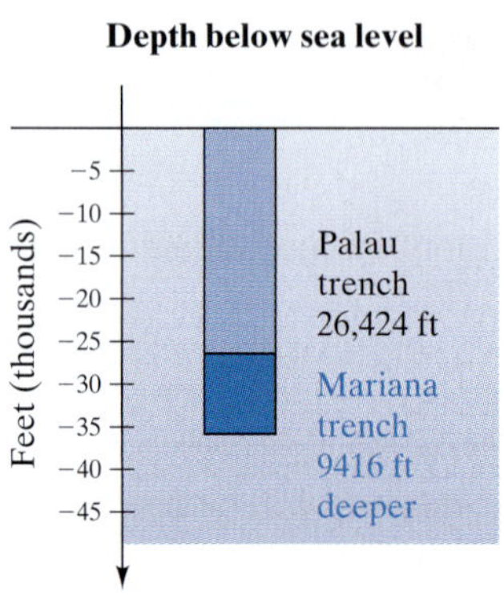


FIGURE 1.7

*Solution* Consider distance below sea level to be negative. Therefore, the total depth is

$$-26{,}424 + (-9416) = -35{,}840 \text{ feet}$$

or 35,840 feet below sea level.

▸ Now Try Exercise 137

## 3 Subtract Real Numbers

Every subtraction problem can be expressed as an addition problem using the following rule.

**Subtraction of Real Numbers**

$$a - b = a + (-b)$$

**To subtract $b$ from $a$, add the opposite (or additive inverse) of $b$ to $a$.**

For example, $5 - 7$ means $5 - (+7)$. To subtract $5 - 7$, add the opposite of $+7$, which is $-7$, to 5.

$$5 - 7 = 5 + (-7)$$

subtract positive 7 add negative 7

Since $5 + (-7) = -2$, then $5 - 7 = -2$.

**EXAMPLE 8** ▸ Evaluate. **a)** $3 - 8$ **b)** $-6 - 4$

*Solution* **a)** $3 - 8 = 3 + (-8) = -5$ **b)** $-6 - 4 = -6 + (-4) = -10$

▸ Now Try Exercise 89

**EXAMPLE 9** ▸ Evaluate $8 - (-15)$.

*Solution* In this problem, we are subtracting a negative number. The procedure to subtract remains the same.

$$8 - (-15) = 8 + 15 = 23$$

subtract negative 15 add positive 15

Thus, $8 - (-15) = 23$.

▸ Now Try Exercise 91

By studying Example 9 and similar problems, we can see that for any real numbers $a$ and $b$,

$$a - (-b) = a + b$$

We can use this principle to evaluate problems such as $8 - (-15)$ and other problems where we *subtract a negative quantity*.

**Double Negative Property**

$$-(-a) = a$$

**EXAMPLE 10** ▸ Evaluate $-4 - (-11)$.

*Solution* $-4 - (-11) = -4 + 11 = 7$

▸ Now Try Exercise 57

**EXAMPLE 11** ▸ **a)** Subtract 35 from $-42$. **b)** Subtract $-\frac{3}{5}$ from $-\frac{5}{9}$.

*Solution*

**a)** $-42 - 35 = -77$

**b)** $-\frac{5}{9} - \left(-\frac{3}{5}\right) = -\frac{5}{9} + \frac{3}{5} = -\frac{25}{45} + \frac{27}{45} = \frac{2}{45}$

▸ Now Try Exercise 109

**EXAMPLE 12** ▸ **Extreme Temperatures** The hottest temperature ever recorded in the United States was 134°F, which occurred at Greenland Ranch, California, in Death Valley on July 10, 1913. The coldest temperature ever recorded in the United States was −79.8°F, which occurred at Prospect Creek Camp, Alaska, in the Endicott Mountains on January 23, 1971 (see **Fig 1.8**). Determine the difference between these two temperatures. (*Source:* Learning Network Internet Web site

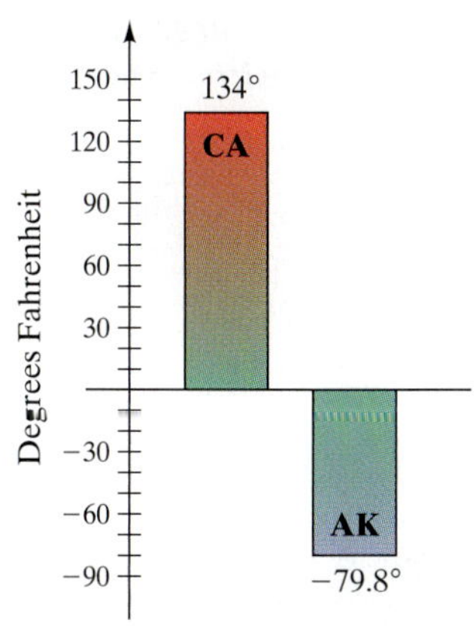

FIGURE 1.8

*Solution* To find the difference, we subtract.

$$134° - (-79.8°) = 134° + 79.8° = 213.8°$$

▸ Now Try Exercise 135

Addition and subtraction are often combined in the same problem, as in the following examples. Unless parentheses are present, if the expression involves only addition and subtraction, we add and subtract from left to right. When parentheses are used, we add and subtract within the parentheses first. Then we add and subtract from left to right.

**EXAMPLE 13** ▸ Evaluate $-15 + (-37) - (5 - 9)$.

*Solution*

$$\begin{aligned} -15 + (-37) - (5 - 9) &= -15 + (-37) - (-4) \\ &= -15 - 37 + 4 \\ &= -52 + 4 = -48 \end{aligned}$$

▸ Now Try Exercise 95

**EXAMPLE 14** ▸ Evaluate $2 - |-3| + 4 - (6 - |-7|)$.

*Solution* Begin by replacing the numbers in absolute value signs with their numerical equivalents; then evaluate.

$$\begin{aligned} 2 - |-3| + 4 - (6 - |-7|) &= 2 - 3 + 4 - (6 - 7) \\ &= 2 - 3 + 4 - (-1) \\ &= 2 - 3 + 4 + 1 \\ &= -1 + 4 + 1 \\ &= 3 + 1 = 4 \end{aligned}$$

▸ Now Try Exercise 69

## 4 Multiply Real Numbers

The following rules are used in determining the product when two numbers are multiplied.

### Multiply Two Real Numbers

1. To multiply two numbers with **like signs**, either both positive or both negative, multiply their absolute values. The answer is **positive**.
2. To multiply two numbers with **unlike signs**, one positive and the other negative, multiply their absolute values. The answer is **negative**.

**EXAMPLE 15** ▸ Evaluate. **a)** $(4.2)(-1.6)$ **b)** $(-18)\left(-\frac{1}{2}\right)$.

*Solution*

**a)** $(4.2)(-1.6) = -6.72$ *The numbers have unlike signs.*

**b)** $(-18)\left(-\frac{1}{2}\right) = 9$ *The numbers have like signs, both negative.*

▸ **Now Try Exercise 75**

**EXAMPLE 16** ▸ Evaluate $4(-2)(-3)(1)$.

*Solution* $4(-2)(-3)(1) = (-8)(-3)(1) = 24(1) = 24$

▸ **Now Try Exercise 77**

**When multiplying more than two numbers, the product will be *negative* when there is an *odd* number of negative numbers. The product will be *positive* when there is an *even* number of negative numbers.**

The multiplicative property of zero indicates that the product of 0 and any number is 0.

### Multiplicative Property of Zero

For any number $a$,

$$a \cdot 0 = 0 \cdot a = 0$$

By the multiplicative property of zero, $5(0) = 0$ and $(-7.3)(0) = 0$.

**EXAMPLE 17** ▸ Evaluate $9(5)(-2.63)(0)(4)$.

*Solution* If one or more of the factors is 0, the product is 0. Thus, $9(5)(-2.63)(0)(4) = 0$. Can you explain why the product of any number of factors will be 0 if any factor is 0?

▸ **Now Try Exercise 111**

## 5 Divide Real Numbers

The rules for the division of real numbers are similar to those for multiplication of real numbers.

### Divide Two Real Numbers

1. To divide two numbers with **like signs**, either both positive or both negative, divide their absolute values. The answer is **positive**.
2. To divide two numbers with **unlike signs**, one positive and the other negative, divide their absolute values. The answer is **negative**.

**EXAMPLE 18** Evaluate. **a)** $-24 \div 4$ **b)** $-6.45 \div (-0.4)$

*Solution*

**a)** $\dfrac{-24}{4} = -6$ *The numbers have unlike signs.*

**b)** $\dfrac{-6.45}{-0.4} = 16.125$ *The numbers have like signs.*

▸ **Now Try Exercise 81**

**EXAMPLE 19** Evaluate $\dfrac{-3}{8} \div \left|\dfrac{-2}{5}\right|$.

*Solution* Since $\left|\dfrac{-2}{5}\right|$ is equal to $\dfrac{2}{5}$, we write

$$\frac{-3}{8} \div \left|\frac{-2}{5}\right| = \frac{-3}{8} \div \frac{2}{5}$$

Now invert the divisor and proceed as in multiplication.

$$\frac{-3}{8} \div \frac{2}{5} = \frac{-3}{8} \cdot \frac{5}{2} = \frac{-3 \cdot 5}{8 \cdot 2} = \frac{-15}{16} \text{ or } -\frac{15}{16}$$

▸ **Now Try Exercise 85**

When the denominator of a fraction is a negative number, we usually rewrite the fraction with a positive denominator. To do this, we use the following fact.

**Sign of a Fraction**

For any number $a$ and any nonzero number $b$,

$$\frac{a}{-b} = \frac{-a}{b} = -\frac{a}{b}$$

Thus, when we have a quotient of $\dfrac{1}{-2}$, we rewrite it as either $\dfrac{-1}{2}$ or $-\dfrac{1}{2}$.

## 6 Use the Properties of Real Numbers

We have already discussed the double negative property and the multiplicative property of zero. **Table 1.1** lists other basic properties for the operations of addition and multiplication on the real numbers.

**TABLE 1.1**

| For real numbers *a*, *b*, and *c* | Addition | Multiplication |
|---|---|---|
| Commutative property | $a + b = b + a$ | $ab = ba$ |
| Associative property | $(a + b) + c = a + (b + c)$ | $(ab)c = a(bc)$ |
| Identity property | $a + 0 = 0 + a = a$ (0 is called the **additive identity element.**) | $a \cdot 1 = 1 \cdot a = a$ (1 is called the **multiplicative identity element.**) |
| Inverse property | $a + (-a) = (-a) + a = 0$ ($-a$ is called the **additive inverse** or **opposite** of $a$.) | $a \cdot \frac{1}{a} = \frac{1}{a} \cdot a = 1$ ($1/a$ is called the **multiplicative inverse** or **reciprocal** of $a$, $a \neq 0$.) |
| Distributive property (of multiplication over addition) | $a(b + c) = ab + ac$ | |

Note that the commutative property involves a change in *order*, and the associative property involves a change in *grouping*.

The distributive property also applies when there are more than two numbers within the parentheses.

$$a(b + c + d + \cdots + n) = ab + ac + ad + \cdots + an$$

This expanded form of the distributive property is often called the *extended distributive property*. However, when using the extended distributive property, we will just refer to it as the distributive property.

**EXAMPLE 20** ▸ Name each property illustrated.

**a)** $7 \cdot m = m \cdot 7$ **b)** $(a + 8) + 2b = a + (8 + 2b)$

**c)** $4s + 5t = 5t + 4s$ **d)** $2v(w + 3) = 2v \cdot w + 2v \cdot 3$

*Solution*

**a)** Commutative property of multiplication: change of order, $7 \cdot m = m \cdot 7$

**b)** Associative property of addition: change of grouping, $(a + 8) + 2b = a + (8 + 2b)$

**c)** Commutative property of addition: change of order, $4s + 5t = 5t + 4s$

**d)** Distributive property: $2v(w + 3) = 2v \cdot w + 2v \cdot 3$

▸ Now Try Exercise 123

In Example 20 **d)** the expression $2v \cdot w + 2v \cdot 3$ can be simplified to $2vw + 6v$ using the properties of the real numbers. Can you explain why?

**EXAMPLE 21** ▸ Name each property illustrated.

**a)** $9 \cdot 1 = 9$ **b)** $x + 0 = x$

**c)** $4 + (-4) = 0$ **d)** $1(x + y) = x + y$

*Solution*

**a)** Identity property of multiplication

**b)** Identity property of addition

**c)** Inverse property of addition

**d)** Identity property of multiplication

▸ Now Try Exercise 125

**EXAMPLE 22** ▸ Write the additive inverse (or opposite) and multiplicative inverse (or reciprocal) of each of the following.

**a)** $-3$ **b)** $\frac{2}{3}$

*Solution*

**a)** The additive inverse is 3. The multiplicative inverse is $\frac{1}{-3} = -\frac{1}{3}$.

**b)** The additive inverse is $-\frac{2}{3}$. The multiplicative inverse is $\frac{1}{\frac{2}{3}} = \frac{3}{2}$.

▸ Now Try Exercise 131

## EXERCISE SET 1.3

MathXL® MyMathLab

### Concept/Writing Exercises

1. What are additive inverses or opposites?
2. Give an example of the double negative property.
3. Will the absolute value of every real number be a positive number? Explain.
4. Give the definition of absolute value.

*In Exercises 5–10, find the unknown number(s). Explain how you determined your answer.*

5. All numbers $a$ such that $|a| = |-a|$
6. All numbers $a$ such that $|a| = a$
7. All numbers $a$ such that $|a| = 6$
8. All numbers $a$ such that $|a| = -a$
9. All numbers $a$ such that $|a| = -9$
10. All numbers $x$ such that $|x - 3| = |3 - x|$
11. Explain how to add two numbers with the same sign.
12. Explain how to add two numbers with different signs.
13. Explain how to subtract real numbers.
14. Explain how the rules for multiplication and division of real numbers are similar.
15. List two other ways that the fraction $\frac{a}{-b}$ may be written.
16. a) Write the associative property of multiplication.
    b) Explain the property.
17. a) Write the commutative property of addition.
    b) Explain the property.
18. a) Write the distributive property of multiplication over addition.
    b) Explain the property.
19. Using an example, explain why addition is not distributive over multiplication. That is, explain why $a + (b \cdot c) \neq (a + b) \cdot (a + c)$.
20. Give an example of the extended distributive property.

### Practice the Skills

*Evaluate each absolute value expression.*

21. $|5|$
22. $|-8|$
23. $|-7|$
24. $|1.9|$
25. $\left|-\frac{7}{8}\right|$
26. $|-8.61|$
27. $|0|$
28. $-|1|$
29. $-|-7|$
30. $-|-\pi|$
31. $-\left|\frac{5}{9}\right|$
32. $-\left|-\frac{7}{15}\right|$

*Insert <, >, or = in the shaded area to make each statement true.*

33. $|-9| \;\square\; |9|$
34. $|-4| \;\square\; |6|$
35. $|-8| \;\square\; -8$
36. $|-10| \;\square\; -5$
37. $|-\pi| \;\square\; -3$
38. $-|-1| \;\square\; -1$
39. $|-7| \;\square\; -|2|$
40. $-|9| \;\square\; -|13|$
41. $-(-3) \;\square\; -|-3|$
42. $|-(-4)| \;\square\; -4$
43. $|19| \;\square\; |-25|$
44. $-|-1| \;\square\; |-2|$

*List the values from smallest to largest.*

45. $-1, -2, |-3|, 4, -|5|$
46. $-8, -12, -|9|, -|20|, -|-18|$
47. $-32, |-7|, 15, -|4|, 4$
48. $\pi, -\pi, |-3|, -|-3|, -2, |-2|$
49. $-6.1, |-6.3|, -|-6.5|, 6.8, |6.4|$
50. $-2.1, -2, -2.4, |-2.8|, -|2.9|$
51. $\frac{1}{3}, \left|-\frac{1}{2}\right|, -2, \left|\frac{3}{5}\right|, \left|-\frac{3}{4}\right|$
52. $\left|-\frac{5}{2}\right|, \frac{3}{5}, |-3|, \left|-\frac{5}{3}\right|, \left|-\frac{2}{3}\right|$

*Evaluate each addition and subtraction problem.*

53. $7 + (-4)$
54. $-2 + 5$
55. $-12 + (-10)$
56. $-2.18 - 3.14$
57. $-9 - (-5)$
58. $-12 - (-4)$
59. $\frac{4}{5} - \frac{6}{7}$
60. $-\frac{5}{12} - \left(-\frac{7}{8}\right)$
61. $-14.21 - (-13.22)$
62. $-1 - \frac{7}{16}$
63. $10 - (-2.31) + (-4.39)$
64. $-|7.31| - (-3.28) + 5.76$
65. $9.9 - |8.5| - |17.6|$
66. $|11 - 4| - 8$
67. $|17 - 12| - |3|$
68. $|12 - 5| - |5 - 12|$
69. $-|-3| - |7| + (6 + |-2|)$
70. $|-4| - |-4| - |-4 - 4|$
71. $\left(\frac{3}{5} + \frac{3}{4}\right) - \frac{1}{2}$
72. $\frac{4}{5} - \left(\frac{3}{4} - \frac{2}{3}\right)$

*Evaluate each multiplication and division problem.*

**73.** $-5 \cdot 8$

**74.** $(-9)(-3)$

**75.** $-4\left(-\frac{5}{16}\right)$

**76.** $-4\left(-\frac{3}{4}\right)\left(-\frac{1}{2}\right)$

**77.** $(-1)(-2)(-1)(2)(-3)$

**78.** $(-2.1)(-7.8)(-9.1)$

**79.** $(-1.1)(3.4)(8.3)(-7.6)$

**80.** $-16 \div 8$

**81.** $-55 \div (-5)$

**82.** $-4 \div \left(-\frac{1}{4}\right)$

**83.** $-\frac{5}{9} \div \frac{-5}{9}$

**84.** $\left|-\frac{1}{2}\right| \cdot \left|\frac{-3}{4}\right|$

**85.** $\left(-\frac{3}{4}\right) \div |-16|$

**86.** $\left|\frac{3}{8}\right| \div (-4)$

**87.** $\left|\frac{-7}{6}\right| \div \left|\frac{-1}{2}\right|$

**88.** $\frac{-5}{9} \div |-5|$

*Evaluate.*

**89.** $10 - 14$

**90.** $-12 - 15$

**91.** $7 - (-13)$

**92.** $-\frac{1}{8} + \left(-\frac{1}{16}\right)$

**93.** $3\left(-\frac{2}{3}\right)\left(-\frac{5}{2}\right)$

**94.** $(-3.2)(4.9)(-2.73)$

**95.** $-14.4 - (-9.6) - 15.8$

**96.** $(1.32 - 2.76) - (-3.85 + 4.28)$

**97.** $9 - (6 - 5) - (-2 - 1)$

**98.** $(4.2)(-1)(-9.6)(3.8)$

**99.** $-|12| \cdot \left|\frac{-1}{2}\right|$

**100.** $-\left|\frac{-24}{5}\right| \cdot \left|\frac{3}{8}\right|$

**101.** $\left|\frac{-9}{4}\right| \div \left|\frac{-4}{9}\right|$

**102.** $(-|3| + |5|) - (1 - |-9|)$

**103.** $5 - |-7| + 3 - |-2|$

**104.** $\left(\frac{3}{8} - \frac{4}{7}\right) - \left(-\frac{1}{2}\right)$

**105.** $\left(-\frac{3}{5} - \frac{4}{9}\right) - \left(-\frac{2}{3}\right)$

**106.** $(|-4| - 3) - (3 \cdot |-5|)$

**107.** $(25 - |32|)(-7 - 4)$

**108.** $\left[(-2)\left|-\frac{1}{2}\right|\right] \div \left|-\frac{1}{4}\right|$

**109.** Subtract 29 from $-10$.

**110.** Subtract $-\frac{1}{2}$ from $-\frac{2}{3}$.

**111.** $7(3)(0)(-15.2)$

**112.** $16(-5)(-10)(0)$

*Name each property illustrated.*

**113.** $r + s = s + r$

**114.** $5(v + w) = 5v + 5w$

**115.** $b \cdot 0 = 0$

**116.** $c \cdot d = d \cdot c$

**117.** $(x + 3) + 6 = x + (3 + 6)$

**118.** $x + 0 = x$

**119.** $x = 1 \cdot x$

**120.** $x(y + z) = xy + xz$

**121.** $2(xy) = (2x)y$

**122.** $(2x \cdot 3y) \cdot 4y = 2x \cdot (3y \cdot 4y)$

**123.** $4(x + y + 2) = 4x + 4y + 8$

**124.** $-(-1) = 1$

**125.** $5 + 0 = 5$

**126.** $4 \cdot \frac{1}{4} = 1$

**127.** $3 + (-3) = 0$

**128.** $(x + y) = 1(x + y)$

**129.** $-(-x) = x$

**130.** $x + (-x) = 0$

*List both the additive inverse and the multiplicative inverse for each problem.*

**131.** 6

**132.** $-13$

**133.** $-\frac{22}{7}$

**134.** $-\frac{3}{5}$

## Problem Solving

**135. Temperature Change** The most unusual temperature change according to the *Guinness Book of World Records* occurred from 7:30 A.M. to 7:32 A.M. on January 22, 1943, in Spearfish, South Dakota. During these 2 minutes the temperature changed from $-4$°F to 45°F. Determine the increase in temperature in these 2 minutes.

**136. The Film Gold** During the production of the documentary film *Gold*, the film crew experienced severe changes in temperature. In a South African gold mine 3 miles below the surface of the earth, the temperature was 140°F. On a mountain near Cuzco, Peru, the temperature was 40°F. Determine the difference in temperature between these two filming sites. *Source:* History Channel Web site

**137. Submarine Dive** A submarine dives 358.9 feet. A short time later the submarine comes up 210.7 feet. Find the submarine's final depth from its starting point. (Consider distance in a downward direction as negative.)

**138. Checking Account** Sharon Koch had a balance of −\$32.64 in her checking account when she deposited a check for \$99.38. What is her new balance?

**139. Extreme Temperatures** The lowest temperature ever recorded in the United States was −79.8°F on January 23, 1971, in Prospect Creek, Alaska. The lowest temperature in the contiguous states (all states except Alaska and Hawaii) was −69.7°F on January 20, 1954, in Rogers Pass, Montana. Find the difference in these temperatures.

**140. Estimated Taxes** In 2006, Joanne Beebe made four quarterly estimated income tax payments of \$3000 each. When she completed her year 2006 income tax forms, she found her total tax was \$10,125.

**a)** Will Joanne be entitled to a refund or will she owe more taxes? Explain.

**b)** How much of a refund will she receive or how much more in taxes will she owe?

**141. Stock Prices** Ron Blackwood purchased 100 shares of Home Depot stock at \$30.30 per share. Six months later Ron sold all 100 shares for a price of \$42.37 per share. What was Ron's total gain or loss for this transaction?

**142. Book Contract** Samuel Pritchard signed a contract with a publishing company that called for an advance payment of \$60,000 on the sale of his book *Moon Spray*. When the book is published and sales begin, the publishers will automatically deduct this advance from the author's royalties.

**a)** Six months after the release of the book, the author's royalties totaled \$47,600 before the advance was deducted. Determine how much money he will receive from or owe to the publisher.

**b)** After 1 year, the author's royalties are \$87,500. Determine how much money he will receive from or owe to the publishing company.

**143.** Write your own realistic word problem that involves subtracting a positive number from a negative number. Indicate the answer to your word problem.

**144.** Write your own realistic word problem that involves subtracting a negative number from a negative number. Indicate the answer to your problem.

**145. Small Businesses** The average first-year expenditures and the average first-year incomes of small start-up businesses is shown on the bar graph below. Estimate the average first-year profit by subtracting the average first-year expenditures from the average first-year income.

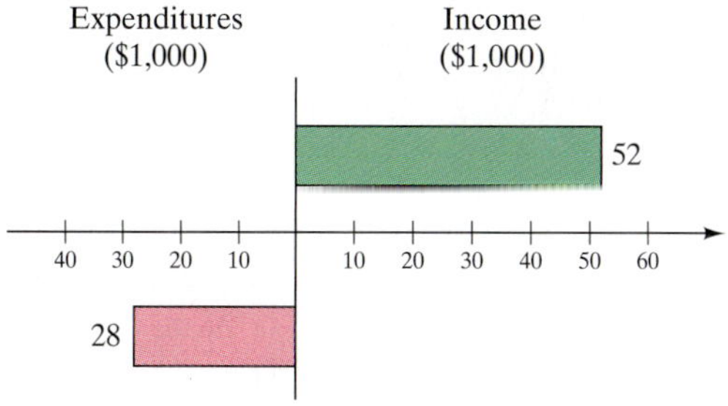


## Challenge Problems

**146.** Evaluate $1 - 2 + 3 - 4 + \cdots + 99 - 100$. (*Hint:* Group in pairs of two numbers.)

**147.** Evaluate $1 + 2 - 3 + 4 + 5 - 6 + 7 + 8 - 9 + 10 + 11 - 12 + \cdots + 22 + 23 - 24$. (*Hint:* Examine in groups of three numbers.)

**148.** Evaluate $\dfrac{(1)\cdot|-2|\cdot(-3)\cdot|4|\cdot(-5)}{|-1|\cdot(-2)\cdot|-3|\cdot(4)\cdot|-5|}$.

**149.** Evaluate $\dfrac{(1)(-2)(3)(-4)(5)\cdots(97)(-98)}{(-1)(2)(-3)(4)(-5)\cdots(-97)(98)}$.

## Cumulative Review Exercises

[1.2] **150.** Answer true or false: Every irrational number is a real number.

**151.** List the set of natural numbers.

**152.** Consider the set $\left\{3, 4, -2, \frac{5}{6}, \sqrt{11}, 0\right\}$. List the elements that are

**a)** integers,

**b)** rational numbers,

**c)** irrational numbers,

**d)** real numbers.

**153.** $A = \{4, 7, 9, 12\}$; $B = \{1, 4, 7, 15\}$. Find

**a)** $A \cup B$

**b)** $A \cap B$

**154.** Illustrate $\{x | -4 < x \leq 5\}$ on a number line.

# 1.4 Order of Operations

1. Evaluate exponential expressions.
2. Evaluate square and higher roots.
3. Evaluate expressions using the order of operations.
4. Evaluate expressions containing variables.
5. Evaluate expressions on a graphing calculator.

Before we discuss the order of operations we need to speak briefly about exponents and roots. We will discuss exponents in greater depth in Sections 1.5 and 7.2.

## 1 Evaluate Exponential Expressions

In a multiplication problem, the numbers or expressions that are multiplied are called **factors**. If $a \cdot b = c$, then $a$ and $b$ are factors of $c$. For example, since $2 \cdot 3 = 6$, both 2 and 3 are factors of 6. The number 1 is a factor of every number and expression. Can you explain why?

The quantity $3^2$ is called an **exponential expression**. In the expression, the 3 is called the **base** and the 2 is called the **exponent**. The expression $3^2$ is read "three squared" or "three to the second power." Note that

$$3^2 = \underbrace{3 \cdot 3}_{\text{2 factors of 3}}$$

The expression $5^3$ is read "five cubed" or "five to the third power." Note that

$$5^3 = \underbrace{5 \cdot 5 \cdot 5}_{\text{3 factors of 5}}$$

In general, the base $b$ to the $n$th power is written $b^n$. For any natural number $n$,

$$b^n = \underbrace{b \cdot b \cdot b \cdot b \cdot \cdots \cdot b}_{n \text{ factors of } b}$$

Note that $0^0$ is *undefined*.

**EXAMPLE 1** ▸ Evaluate. **a)** $(0.5)^3$ **b)** $(-3)^5$ **c)** $1^{27}$ **d)** $\left(-\frac{4}{7}\right)^3$

*Solution*

**a)** $(0.5)^3 = (0.5)(0.5)(0.5) = 0.125$

**b)** $(-3)^5 = (-3)(-3)(-3)(-3)(-3) = -243$

**c)** $1^{27} = 1$; 1 raised to any power will equal 1. Why?

**d)** $\left(-\frac{4}{7}\right)^3 = \left(-\frac{4}{7}\right)\left(-\frac{4}{7}\right)\left(-\frac{4}{7}\right) = -\frac{64}{343}$

▸ Now Try Exercise 19

### Helpful Hint *Study Tip*

Be very careful when writing or copying exponents. Since exponents are small it is very easy to write or copy an exponent and then later not recognize what you have written. Some exponents that may be easily confused if not clearly written are 1 and 7, 2 and 3, 3 and 5, 4 and 9, 5 and 6, and 5 and 8.

It is not necessary to write exponents of 1. Whenever we encounter a numerical value or a variable without an exponent, we assume that it has an exponent of 1. Thus, 3 means $3^1$, $x$ means $x^1$, $x^3y$ means $x^3y^1$, and $-xy$ means $-x^1y^1$.

Students often evaluate expressions containing $-x^2$ incorrectly. The expression $-x^2$ means $-(x^2)$, not $(-x)^2$. Note that $-5^2$ means $-(5^2) = -(5 \cdot 5) = -25$ while $(-5)^2$ means $(-5)(-5) = 25$. *In general, $-x^m$ means $-(x^m)$, not $(-x)^m$.* The expression $-x^2$ is read *negative x squared* or *the opposite of $x^2$*. The expression $(-x)^2$ is read *negative x quantity squared*.

**EXAMPLE 2** ▸ Evaluate $-x^2$ for each value of $x$. **a)** 6 **b)** $-6$

*Solution*

**a)** $-x^2 = -(6)^2 = -36$

**b)** $-x^2 = -(-6)^2 = -(36) = -36$

▸ **Now Try Exercise 41**

**EXAMPLE 3** ▸ Evaluate $-5^2 + (-5)^2 - 4^3 + (-4)^3$.

*Solution* First, we evaluate each exponential expression. Then we add or subtract, working from left to right.

$$\begin{aligned} -5^2 + (-5)^2 - 4^3 + (-4)^3 &= -(5^2) + (-5)^2 - (4^3) + (-4)^3 \\ &= -25 + 25 - 64 + (-64) \\ &= -25 + 25 - 64 - 64 \\ &= -128 \end{aligned}$$

▸ **Now Try Exercise 59**

**USING YOUR CALCULATOR** **Evaluating Exponential Expressions on a Scientific and a Graphing Calculator**

On both scientific and graphing calculators the $\boxed{x^2}$ key can be used to square a number. Below we show the sequence of keys to press to evaluate $5^2$.

Scientific calculator: $5 \ \boxed{x^2} \ 25$ ← answer displayed

Graphing calculator: $5 \ \boxed{x^2} \ \boxed{\text{ENTER}} \ 25$ ← answer displayed

To evaluate exponential expressions with other exponents, you can use the $\boxed{y^x}$ or $\boxed{\wedge}$ key. Most scientific calculators have a $\boxed{y^x}$ key,* whereas graphing calculators use the $\boxed{\wedge}$ key. To evaluate exponential expressions using these keys, first enter the base, then press either the $\boxed{y^x}$ or $\boxed{\wedge}$ key, and then enter the exponent. For example, to evaluate $6^4$ we do the following:

Scientific calculator: $6 \ \boxed{y^x} \ 4 \ \boxed{=} \ 1296$ ← answer displayed

Graphing calculator: $6 \ \boxed{\wedge} \ 4 \ \boxed{\text{ENTER}} \ 1296$ ← answer displayed

*Some calculators have $\boxed{x^y}$ or $\boxed{a^b}$ keys instead of an $\boxed{y^x}$ key.

## 2 Evaluate Square and Higher Roots

The symbol used to indicate a root, $\sqrt{\phantom{x}}$, is called a **radical sign**. The number or expression inside the radical sign is called the **radicand**. In $\sqrt{25}$, the radicand is 25. The **principal or positive square root** of a positive number $a$, written $\sqrt{a}$, is the positive number that when multiplied by itself gives $a$. For example, the principal square root of 4 is 2, written $\sqrt{4} = 2$, because $2 \cdot 2 = 4$. In general, $\sqrt{a} = b$ if $b \cdot b = a$. Whenever we use the words *square root*, we are referring to the "principal square root."

**EXAMPLE 4** ▸ Evaluate. **a)** $\sqrt{25}$ **b)** $\sqrt{\frac{81}{4}}$ **c)** $\sqrt{0.64}$ **d)** $-\sqrt{49}$

### Solution

**a)** $\sqrt{25} = 5$, since $5 \cdot 5 = 25$.

**b)** $\sqrt{\frac{81}{4}} = \frac{9}{2}$, since $\frac{9}{2} \cdot \frac{9}{2} = \frac{81}{4}$.

**c)** $\sqrt{0.64} = 0.8$, since $(0.8)(0.8) = 0.64$.

**d)** $-\sqrt{49}$ means $-(\sqrt{49})$. We determine that $\sqrt{49} = 7$, since $7 \cdot 7 = 49$. Therefore, $-\sqrt{49} = -7$.

▸ Now Try Exercise 21

The square root of 4, $\sqrt{4}$, is a rational number since it is equal to 2. The square roots of other numbers, such as $\sqrt{2}$, $\sqrt{3}$, and $\sqrt{5}$, are irrational numbers. The decimal values of such numbers can never be given exactly since irrational numbers are nonterminating, nonrepeating decimal numbers. The approximate value of $\sqrt{2}$ and other irrational numbers can be found with a calculator.

$$\sqrt{2} \approx 1.414213562 \quad \text{From a calculator}$$

In this section we introduce square roots; cube roots, symbolized by $\sqrt[3]{\phantom{x}}$; and higher roots. The number used to indicate the root is called the **index**.

index ↘ ↙ radical sign

$\sqrt[n]{a}$ ← radicand

The index of a square root is 2. However, we generally do not show the index 2. Therefore, $\sqrt{a} = \sqrt[2]{a}$.

The concept used to explain square roots can be expanded to explain cube roots and higher roots. The cube root of a number $a$ is written $\sqrt[3]{a}$.

$$\sqrt[3]{a} = b \quad \text{if} \quad \underbrace{b \cdot b \cdot b}_{\text{3 factors of } b} = a$$

For example, $\sqrt[3]{8} = 2$, because $2 \cdot 2 \cdot 2 = 8$. The expression $\sqrt[n]{a}$ is read "the $n$th root of $a$."

$$\sqrt[n]{a} = b \quad \text{if} \quad \underbrace{b \cdot b \cdot b \cdot \cdots \cdot b}_{n \text{ factors of } b} = a$$

**EXAMPLE 5** ▸ Evaluate. **a)** $\sqrt[3]{125}$ **b)** $\sqrt[4]{81}$ **c)** $\sqrt[5]{32}$

### Solution

**a)** $\sqrt[3]{125} = 5$, since $5 \cdot 5 \cdot 5 = 125$

**b)** $\sqrt[4]{81} = 3$, since $3 \cdot 3 \cdot 3 \cdot 3 = 81$

**c)** $\sqrt[5]{32} = 2$, since $2 \cdot 2 \cdot 2 \cdot 2 \cdot 2 = 32$

▸ Now Try Exercise 25

**EXAMPLE 6** ▸ Evaluate. **a)** $\sqrt[4]{256}$ **b)** $\sqrt[3]{\frac{1}{27}}$ **c)** $\sqrt[3]{-8}$ **d)** $-\sqrt[3]{8}$

*Solution*

**a)** $\sqrt[4]{256} = 4$, since $4 \cdot 4 \cdot 4 \cdot 4 = 256$.

**b)** $\sqrt[3]{\frac{1}{27}} = \frac{1}{3}$, since $\left(\frac{1}{3}\right)\left(\frac{1}{3}\right)\left(\frac{1}{3}\right) = \frac{1}{27}$.

**c)** $\sqrt[3]{-8} = -2$, since $(-2)(-2)(-2) = -8$.

**d)** $-\sqrt[3]{8}$ means $-(\sqrt[3]{8})$. We determine that $\sqrt[3]{8} = 2$, since $2 \cdot 2 \cdot 2 = 8$. Therefore, $-\sqrt[3]{8} = -2$.

▸ **Now Try Exercise 27**

Note that in Example 6 **c)** the cube root of a negative number is negative. Why is this so? We will discuss radicals in more detail in Chapter 7.

## USING YOUR CALCULATOR Evaluating Roots on a Scientific Calculator

The square roots of numbers can be found on calculators with a square-root key, $\boxed{\sqrt{x}}$. To evaluate $\sqrt{25}$ on most calculators that have this key, press

$$25 \;\boxed{\sqrt{x}}\; 5 \quad \leftarrow \text{answer displayed}$$

Higher roots can be found on calculators that contain either the $\boxed{\sqrt[x]{y}}$ key or the $\boxed{y^x}$ key.* To evaluate $\sqrt[4]{625}$ on a calculator with a $\boxed{\sqrt[x]{y}}$ key, do the following:

$$625 \;\boxed{\sqrt[x]{y}}\; 4 \;\boxed{=}\; 5 \quad \leftarrow \text{answer displayed}$$

Note that the number within the radical sign (the radicand), 625, is entered, then the $\boxed{\sqrt[x]{y}}$ key is pressed, and then the root (or index) 4 is entered. When the $\boxed{=}$ key is pressed, the answer 5 is displayed.

To evaluate $\sqrt[4]{625}$ on a calculator with a $\boxed{y^x}$ key, use the inverse key as follows:

$$625 \;\boxed{\text{INV}}\; \boxed{y^x}\; 4 \;\boxed{=}\; 5 \quad \leftarrow \text{answer displayed}$$

* Calculator keys vary. Some calculators have $\boxed{x^y}$ or $\boxed{a^b}$ keys instead of the $\boxed{y^x}$ key, and some calculators have a $\boxed{2^{\text{nd}}}$ or $\boxed{\text{shift}}$ key instead of the $\boxed{\text{INV}}$ key.

## USING YOUR GRAPHING CALCULATOR Evaluating Roots on a Graphing Calculator

To find the square root on a graphing calculator, use $\sqrt{\ }$. The $\sqrt{\ }$ appears above the $\boxed{x^2}$ key, so you will need to press the $\boxed{2^{\text{nd}}}$ key to evaluate square roots. For example, to evaluate $\sqrt{25}$ press

$$\boxed{2^{\text{nd}}}\; \boxed{x^2}\; 25 \;\boxed{\text{ENTER}}\; 5 \leftarrow \text{answer displayed}$$

When you press $\boxed{2^{\text{nd}}}\;\boxed{x^2}$, the Texas Instruments TI-84 Plus generates $\sqrt{\ }($. Then you insert the radicand, then the right parentheses, and press $\boxed{\text{ENTER}}$. To learn how to find cube and higher roots, refer to your graphing calculator manual. With the TI-84 Plus, you can use the $\boxed{\text{MATH}}$ key. When you press this key you get a number of options including 4 and 5, which are shown below.

$$4{:}\sqrt[3]{\ }( \qquad 5{:}\sqrt[x]{y}$$

Option 4 can be used to find cube roots and option 5 can be used to find higher roots, as shown in the following examples.

**EXAMPLE** Evaluate $\sqrt[3]{120}$.

*Solution*

$$\boxed{\text{MATH}}\; 4 \; 120 \;\boxed{)}\; \boxed{\text{ENTER}}\; 4.932424149 \quad \leftarrow \text{answer displayed}$$

(4: select option 4; 120: enter radicand)

*(continued on the next page)*

To find the root with an index greater than 3, first enter the index, then press the MATH key, and then press option 5.

**EXAMPLE** Evaluate $\sqrt[4]{625}$.

*Solution*

4 MATH 5 625 ) ENTER 5

(index: 4; select option: 5; enter radicand: 625; answer displayed: 5)

We will show another way to find roots on a graphing calculator in Section 7.2 when we discuss rational exponents.

## 3 Evaluate Expressions Using the Order of Operations

You will often have to evaluate expressions containing multiple operations. To do so, follow the **order of operations** indicated below.

### Order of Operations

To evaluate mathematical expressions, use the following order:

1. First, evaluate the expressions within grouping symbols, including parentheses, ( ), brackets, [ ], braces, { }, and absolute value, | |. If the expression contains nested grouping symbols (one pair of grouping symbols within another pair), evaluate the expression in the innermost grouping symbols first.
2. Next, evaluate all terms containing exponents and radicals.
3. Next, evaluate all multiplications or divisions in the order in which they occur, working from left to right.
4. Finally, evaluate all additions or subtractions in the order in which they occur, working from left to right.

*It should be noted that a fraction bar acts as a grouping symbol.* Thus, when evaluating expressions containing a fraction bar, we work separately above and below the fraction bar.

Brackets are often used in place of parentheses to help avoid confusion. For example, the expression $7((5 \cdot 3) + 6)$ is easier to follow when written $7[(5 \cdot 3) + 6]$. Remember to evaluate the innermost group first.

**EXAMPLE 7** ▸ Evaluate $6 + 3 \cdot 5^2 - 10$.

*Solution* We will use shading to indicate the order in which the operations are to be evaluated. Since there are no parentheses, we first evaluate $5^2$.

$$6 + 3 \cdot 5^2 - 10 = 6 + 3 \cdot 25 - 10$$

Next, we perform multiplications or divisions from left to right.

$$= 6 + 75 - 10$$

Finally, we perform additions or subtractions from left to right.

$$= 81 - 10$$
$$= 71$$

▸ Now Try Exercise 67

**EXAMPLE 8** ▸ Evaluate $10 + \{6 - [4(5 - 2)]\}^2$.

*Solution* First, evaluate the expression within the innermost parentheses. Then continue according to the order of operations.

$$\begin{aligned} 10 + \{6 - [4(5 - 2)]\}^2 &= 10 + \{6 - [4(3)]\}^2 \\ &= 10 + [6 - (12)]^2 \\ &= 10 + (-6)^2 \\ &= 10 + 36 \\ &= 46 \end{aligned}$$

▸ Now Try Exercise 77

**EXAMPLE 9** ▸ Evaluate $\dfrac{6 \div \frac{1}{2} + 5|7 - 3|}{1 + (3 - 5) \div 2}$.

*Solution* Remember that the fraction bar acts as a grouping symbol. Work separately above the fraction bar and below the fraction bar.

$$\begin{aligned} \frac{6 \div \frac{1}{2} + 5|7 - 3|}{1 + (3 - 5) \div 2} &= \frac{6 \div \frac{1}{2} + 5|4|}{1 + (-2) \div 2} \\ &= \frac{12 + 20}{1 + (-1)} \\ &= \frac{32}{0} \end{aligned}$$

Since division by 0 is not possible, the original expression is **undefined**.

▸ Now Try Exercise 83

## 4 Evaluate Expressions Containing Variables

To evaluate mathematical expressions, we use the order of operations just given. Example 10 is an application problem in which we use the order of operations.

**EXAMPLE 10** ▸ **Alternate Remedies** Frustration with traditional medicine has led many Americans to try alternate remedies such as vitamins, herbs, and other supplements available without a doctor's prescription. The approximate sales of such supplements between 1997 and 2004, in billions of U.S. dollars, can be estimated by the equation

$$\text{sales} = -0.063x^2 + 1.62x + 9.5$$

where $x$ represents years since 1997. In the expression on the right side of the equal sign, substitute 1 for $x$ to estimate the sales of supplements in 1998, 2 for $x$ to estimate the sales of supplements in 1999, and so on.

Estimate the sales of supplements during the years **a)** 1998 and **b)** 2002.

*Solution*

**a)** We will substitute 1 for $x$ to estimate the sales of supplements in 1998.

$$\begin{aligned} \text{sales} &= -0.063x^2 + 1.62x + 9.5 \\ &= -0.063(1)^2 + 1.62(1) + 9.5 \\ &= -0.063 + 1.62 + 9.5 \\ &= 11.057 \end{aligned}$$

Therefore, in 1998 approximately \$11.057 billion worth of supplements was sold in the United States.

**b)** The year 2002 corresponds to the number 5. We can obtain the 5 by subtracting 1997 from 2002. Therefore, to estimate the sales of supplements in 2002, we substitute 5 for $x$ in the equation.

$$\begin{aligned} \text{sales} &= -0.063x^2 + 1.62x + 9.5 \\ &= -0.063(5)^2 + 1.62(5) + 9.5 \\ &= -0.063(25) + 8.1 + 9.5 \\ &= 16.025 \end{aligned}$$

The answer is reasonable: From the information given we expected to see an increase. In 2002 approximately \$16.025 billion worth of supplements was sold in the United States.

▸ **Now Try Exercise 121**

---

**EXAMPLE 11** ▸ Evaluate $-x^3 - xy - y^2$ when $x = -2$ and $y = 5$.

*Solution* Substitute $-2$ for each $x$ and 5 for each $y$ in the expression. Then evaluate.

$$\begin{aligned} -x^3 - xy - y^2 &= -(-2)^3 - (-2)(5) - (5)^2 \\ &= -(-8) - (-10) - 25 \\ &= 8 + 10 - 25 \\ &= -7 \end{aligned}$$

▸ **Now Try Exercise 101**

---

## 5 Evaluate Expressions on a Graphing Calculator

Throughout this book, the material presented on **graphing calculators** (or graphers) will often reinforce the concepts presented. Therefore, even if you do not have or use a graphing calculator, you should read the material related to graphing calculators whenever it appears. You may find that it truly helps your understanding of the concepts. Some graphing calculator information will be given as regular text, and other information about graphing calculators will be given in Using Your Graphing Calculator boxes such as the one on page 31.

The information presented in this book is not meant to replace the manual that comes with your graphing calculator. Because of space limitations in this book, your graphing calculator manual may provide more detailed information on some tasks we discuss. Your manual will also illustrate many other uses for your grapher beyond what we discuss in this course. Keystrokes to use vary from calculator to calculator. When we illustrate keystrokes and screens, they will be for Texas Instruments TI-83 Plus and TI-84 Plus calculators. Although the screens and keystrokes are the same for both a TI-83 Plus and a TI-84 Plus, during further discussions we will refer to the TI-84 Plus. *We suggest that you carefully read the manual that came with your graphing calculator to determine the sequence of keystrokes to use to accomplish specific tasks.*

Many graphing calculators can store an expression (or equation) and then evaluate the expression for various values of the variable or variables without your having to reenter the expression each time. This is very valuable in both science and mathematics courses. For example, in Chapter 3 when we graph, we will need to evaluate an expression for various values of the variable.

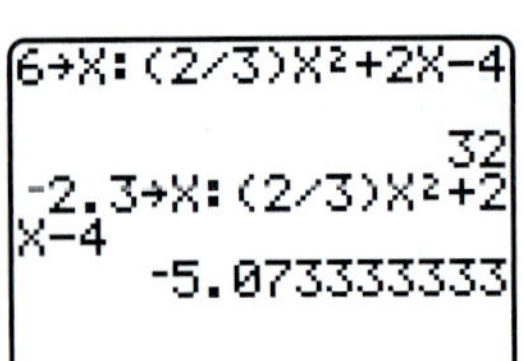

FIGURE 1.9

**Figure 1.9** displays the screen of a TI-84 Plus graphing calculator showing the expression $\frac{2}{3}x^2 + 2x - 4$ being evaluated for $x = 6$ and $x = -2.3$.

On this calculator screen, $6 \rightarrow X$ shows that we assigned a value of 6 to X. The expression being evaluated, $(2/3)X^2 + 2X - 4$, is shown after the colon. The 32 shown on the right side of the screen (or window) is the value of the expression when $X = 6$. On the next line, on the left side of the screen, we see $-2.3 \rightarrow X$, which shows that a value of $-2.3$ has been assigned to X. We see that the value of the expression is $-5.073333333$ when $X = -2.3$. After you have entered the expression to be evaluated it is not necessary to reenter the expression to evaluate it for a different value of the variable. Read your graphing calculator manual to learn how to evaluate an expression for various values of the variable without having to reenter the expression each time.

On the TI-84 Plus, after evaluating an expression for one value of the variable, you can press [2nd] [ENTER] to display the previously assigned value and the expression to be evaluated. Then you can replace the value that was previously assigned to X with the new value to be assigned to X. After doing this and pressing [ENTER] the new answer will be displayed.

The calculator screen displayed in **Figure 1.9** illustrates two important points about graphing calculators.

1. Notice the parentheses around the 2/3. Some graphing calculators interpret $2/3x^2$ as $2/(3x^2)$. To evaluate $\frac{2}{3}x^2$ on such calculators, you must use parentheses around the 2/3. You should learn how your calculator evaluates expressions like $2/3x^2$. *Whenever you are in doubt, use parentheses to prevent possible errors.*
2. In the display, you will notice that the negative sign preceding the 2.3 is slightly smaller and higher than the subtraction sign preceding the 4 in the expression. Graphing calculators generally have both a negative sign key, [(−)], and a subtraction sign key, [−]. You must be sure to use the correct key or you will get an error. The negative sign key is used to enter a negative number. The subtraction key is used to subtract one quantity from another. To enter the expression $-x - 4$ on a graphing calculator, you might press

   [(−)] [X, T, Θ, n] [−] [4]

   ↑ negative sign (under [(−)]); ↑ subtraction (under [−])

Remember that $-x - 4$ means $-1x - 4$. By beginning with [(−)] you are entering the coefficient $-1$. Different calculators use different keys to enter the variable $x$. The key shown after the negative sign is the key used on the TI-84 Plus calculator.

**EXAMPLE 12 ▸ Average Sale Price of Homes** Low interest rates, easy credit, and strong demand from the middle class drove up the average sales price of homes around the country from 1992 to 2006. The average sales price of a home, in thousands of U.S. dollars, during this time can be estimated by

$$\text{average sales price} = 0.71x^2 + 2.16x + 145.39$$

where $x$ represents years since 1992. In the expression on the right side of the equal sign, substitute 1 for $x$ to estimate the average sales price of a home in 1993, 2 for $x$ to estimate the average sales price of a home in 1994, and so on. Use a graphing calculator, if available, to estimate the average sales price of a home in **a)** 1995 and **b)** 2006.

*Source:* National Association of Realtors

## *Solution*

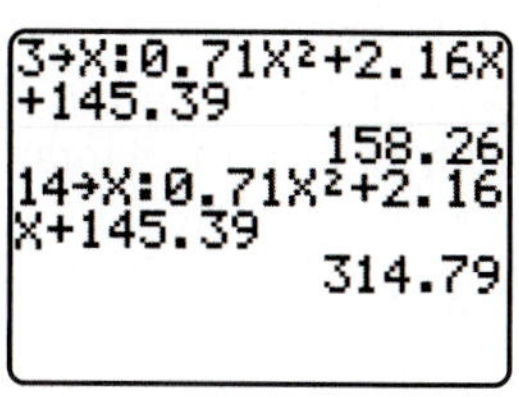


FIGURE 1.10

**a)** The year 1995 corresponds to $x = 3$, so we first assign $x$ a value of 3, then enter the expression, and press [ENTER]. **Figure 1.10** shows the screen for a TI-84 Plus calculator with the expression evaluated for $x = 3$. From the screen we see that the average sales price of a home in 1995 was approximately \$158.26 thousand or \$158,260.

**b)** Since $2006 - 1992 = 14$ the year 2006 corresponds to $x = 14$. We first assign $x$ a value of 14 and then reenter the expression and press [ENTER]. From **Figure 1.10** we see that the average sales price of a home in 2006 was approximately \$314.79 thousand or \$314,790.

▸ **Now Try Exercise 122**

## Helpful Hint

Always review your calculator screen to make sure that no keys were pressed incorrectly and no keys were omitted. Note that it is not necessary to enter the 0 before the decimal point in terms such as $-0.71x^2$.

## EXERCISE SET 1.4

MathXL® MyMathLab

### Concept/Writing Exercises

1. Consider the expression $a^n$.
   a) What is the $a$ called?
   b) What is the $n$ called?
2. What is the meaning of $a^n$?
3. Consider the radical expression $\sqrt[n]{a}$.
   a) What is the $n$ called?
   b) What is the $a$ called?
4. What does it mean if $\sqrt[n]{a} = b$?
5. What is the principal square root of a positive number?
6. Explain why $\sqrt{-4}$ cannot be a real number.
7. Explain why an odd root of a negative number will be negative.
8. Explain why an odd root of a positive number will be positive.
9. Explain the order of operations to follow when evaluating a mathematical expression. See page 32.
10. a) Explain step-by-step how you would evaluate
$$\frac{5 - 18 \div 3^2}{4 - 3 \cdot 2}$$
    b) Evaluate the expression.
11. a) Explain step-by-step how you would evaluate $16 \div 2^2 + 6 \cdot 4 - 24 \div 6$.
    b) Evaluate the expression.
12. a) Explain step-by-step how you would evaluate $\{5 - [4 - (3 - 8)]\}^2$.
    b) Evaluate the expression.

### Practice the Skills

*Evaluate each expression without using a calculator.*

13. $3^2$
14. $(-4)^3$
15. $-3^2$
16. $-4^3$
17. $(-3)^2$
18. $\left(\frac{1}{2}\right)^3$
19. $-\left(\frac{3}{5}\right)^4$
20. $(0.3)^2$
21. $\sqrt{49}$
22. $\sqrt{144}$
23. $-\sqrt{36}$
24. $-\sqrt{0.64}$
25. $\sqrt[3]{-27}$
26. $\sqrt[3]{\frac{-216}{343}}$
27. $\sqrt[3]{0.001}$
28. $\sqrt[4]{\frac{1}{16}}$

*Use a calculator to evaluate each expression. Round answers to the nearest thousandth.*

29. $(0.35)^4$
30. $-(1.7)^{3.9}$
31. $\left(-\frac{13}{12}\right)^8$
32. $\left(\frac{5}{7}\right)^7$
33. $(6.721)^{5.9}$
34. $\sqrt{78}$
35. $\sqrt[3]{26}$
36. $-\sqrt[4]{72.8}$
37. $\sqrt[5]{362.65}$
38. $-\sqrt{\frac{8}{9}}$
39. $-\sqrt[3]{\frac{20}{53}}$
40. $\sqrt[3]{-\frac{15}{19}}$

*Evaluate* **a)** $x^2$ *and* **b)** $-x^2$ *for each given value of x.*

41. 3
42. 4
43. 10
44. −2
45. −1
46. −6
47. $\frac{1}{3}$
48. $-\frac{4}{5}$

*Evaluate* **a)** $x^3$ *and* **b)** $-x^3$ *for each given value of x.*

49. 3
50. −3
51. −5
52. −1
53. −2
54. 4
55. $\frac{2}{5}$
56. $-\frac{3}{4}$

*Evaluate each expression.*

57. $4^2 + 2^3 - 2^2 - 3^3$
58. $(-1)^2 + (-1)^3 - 1^4 + 1^5$
59. $-2^2 - 2^3 + 1^{10} + (-2)^3$
60. $(-3)^3 - 2^2 - (-2)^2 + (6 - 6)^2$
61. $(1.5)^2 - (3.9)^2 + (-2.1)^3$
62. $(3.7)^2 - (0.8)^2 + (2.4)^3$
63. $\left(-\frac{1}{2}\right)^4 - \left(\frac{1}{2}\right)^2 + \left(-\frac{1}{2}\right)^3$
64. $\left(\frac{3}{4}\right)^2 - \frac{1}{4} - \left(-\frac{3}{8}\right)^2 + \left(\frac{1}{4}\right)^3$

*Evaluate each expression.*

65. $3 + 5 \cdot 8$
66. $(2 - 7) \div 5 + 3$
67. $18 - 6 \div 6 + 8$

**68.** $4 \cdot 3 \div 6 - 2^2$

**69.** $\frac{3}{4} \div \frac{1}{2} - 2 + 5 \div 10$

**70.** $3 \cdot 6 \div 18 + \frac{4}{5}$

**71.** $\frac{1}{2} \cdot \frac{2}{3} \div \frac{3}{4} - \frac{1}{6} \cdot \left(-\frac{1}{3}\right)$

**72.** $3[4 + (-2)(8)] + 3^3$

**73.** $10 \div [(3 + 2^2) - (2^4 - 8)]$

**74.** $[3 - (4 - 2^3)^2]^2$

**75.** $5\left(\sqrt[3]{27} + \sqrt[5]{32}\right) \div \frac{\sqrt{100}}{2}$

**76.** $\{5 + [4^2 - 3(2 - 7)] - 5\}^2$

**77.** $\{[(12 - 15) - 3] - 2\}^2$

**78.** $3\{6 - [(25 \div 5) - 2]\}^3$

**79.** $4[5(16 - 6) \div (25 \div 5)^2]^2$

**80.** $\dfrac{15 \div 3 + 7 \cdot 2}{\sqrt{25} \div 5 + 8 \div 2}$

**81.** $\dfrac{4 - (2 + 3)^2 - 6}{4(3 - 2) - 3^2}$

**82.** $-2\left|-3 - \frac{2}{3}\right| + 5$

**83.** $\dfrac{8 + 4 \div 2 \cdot 3 + 4}{5^2 - 3^2 \cdot 2 - 7}$

**84.** $\dfrac{5(-3) + 4 \cdot 7 - 3^2}{-6 + \sqrt{4(2^2 - 1)}}$

**85.** $\dfrac{8 - [4 - (3 - 1)^2]}{5 - (-3)^2 + 4 \div 2}$

**86.** $12 - 15 \div |5| - (|4| - 2)^2$

**87.** $-2|-3| - \sqrt{36} \div |2| + 3^2$

**88.** $\dfrac{4 - |-12| \div |3|}{2(4 - |5|) + 9}$

**89.** $\dfrac{6 - |-4| - 4|8 - 5|}{5 - 6 \cdot 2 \div |-6|}$

**90.** $-\frac{1}{4}[8 - |-6| \div 3 - 4]^2$

**91.** $\frac{2}{5}\left[\sqrt[3]{27} - |-9| + 4 - 3^2\right]^2$

**92.** $\dfrac{3(12 - 9)^2}{-3^2} - \dfrac{2(3^2 - 4^2)}{4 - (-2)}$

**93.** $\dfrac{24 - 5 - 4^2}{|-8| + 4 - 2(3)} + \dfrac{4 - (-3)^2 + |4|}{3^2 - 4 \cdot 3 + |-7|}$

**94.** $\dfrac{-2 - 8 \div 4^2 \cdot |8|}{|8| - \sqrt{64}} + \dfrac{[(8 - 3)^2 - 7]^2}{2^2 + 16}$

*Evaluate each expression for the given value or values.*

**95.** $5x^2 + 4x$ when $x = 2$

**96.** $5x^2 - 2x + 7$ when $x = 3$

**97.** $-9x^2 + 3x - 29$ when $x = -1$

**98.** $3(x - 2)^2$ when $x = \frac{1}{4}$

**99.** $16(x + 5)^3 - 25(x + 5)$ when $x = -4$

**100.** $-6x + 3y^2$ when $x = 2, y = 4$

**101.** $6x^2 + 3y^3 - 15$ when $x = 1, y = -3$

**102.** $4x^2 - 3y - 10$ when $x = 4, y = -2$

**103.** $3(a + b)^2 + 4(a + b) - 6$ when $a = 4, b = -1$

**104.** $-9 - \{2x - [5x - (2x + 1)]\}$ when $x = 3$

**105.** $-8 - \{x - [2x - (x - 3)]\}$ when $x = 4$

**106.** $\dfrac{(x - 3)^2}{9} + \dfrac{(y + 5)^2}{16}$ when $x = 4, y = 3$

**107.** $\dfrac{-b + \sqrt{b^2 - 4ac}}{2a}$ when $a = 6, b = -11, c = 3$

**108.** $\dfrac{-b - \sqrt{b^2 - 4ac}}{2a}$ when $a = 2, b = 1, c = -10$

## Problem Solving

*In Exercises 109–114, write an algebraic expression for each problem. Then evaluate the expression for the given value of the variable or variables.*

**109.** Multiply the variable $y$ by 7. From this product subtract 14. Now divide this difference by 2. Find the value of this expression when $y = 6$.

**110.** Subtract 4 from $z$. Multiply this difference by 5. Now square this product. Find the value of this expression when $z = 10$.

**111.** Six is added to the product of 3 and $x$. This expression is then multiplied by 6. Nine is then subtracted from this product. Find the value of the expression when $x = 3$.

**112.** The sum of $x$ and $y$ is multiplied by 2. Then 5 is subtracted from this product. This expression is then squared. Find the value of the expression when $x = 2$ and $y = -3$.

**113.** Three is added to $x$. This sum is divided by twice $y$. This quotient is then squared. Finally, 3 is subtracted from this expression. Find the value of the expression when $x = 5$ and $y = 2$.

**114.** Four is subtracted from $x$. This sum is divided by $10y$. The quotient is then cubed. Finally, 19 is added to this expression. Find the value of the expression when $x = 64$ and $y = 3$.

*Use a calculator to answer Exercises 115–128.*

**115. Riding a Bike** Frank Kelso can ride his bike at a rate of 8.2 miles per hour on the *C & O* Tow Path in Maryland. The distance, in miles, traveled after riding his bike $x$ hours is determined by

$$\text{distance} = 8.2x$$

How far has Frank traveled in

**a)** 3 hours?

**b)** 7 hours?

**116. Salary** On January 2, 2006, Mary Ferguson started a new job with an annual salary of \$32,550. Her boss agreed to give her a raise of \$1,200 per year for the next 20 years. Her salary, in dollars, is determined by

$$\text{salary} = 32{,}550 + 1{,}200x$$

where $x$ is the number of years since 2006. Substitute 1 for $x$ to determine her salary in 2007, 2 for $x$ to determine her salary in 2008, and so on. Find Mary's salary in

**a)** 2010.

**b)** 2020.

**117. Throwing a Ball** Cuong Chapman threw a baseball upward from a dormitory window. The height of the ball above the ground, in feet, is determined by

$$\text{height} = -16x^2 + 72x + 22$$

where $x$ is the number of seconds after the baseball is thrown from the window. Determine the height of the ball

**a)** 2 seconds

**b)** 4 seconds

after it is thrown out the window.

**118. Velocity** See Exercise 117. After the ball is thrown out the window, its velocity (or speed), in feet per second, is determined by

$$\text{velocity} = -32x + 72$$

Find the velocity of the ball

**a)** 2 seconds

**b)** 4 seconds

after it is thrown out the window.

**119. Spending Money** The amount of money consumers spend on holiday gifts during the holiday season has been on the rise in recent years. The amount, in dollars, spent on holiday gifts by an individual can be estimated by

$$\text{spending} = 26.865x + 488.725$$

where $x$ is the number of years since 2002. Substitute 1 for $x$ to find the amount spent in 2003, 2 for $x$ to find the amount spent in 2004, and so on. Assuming this trend continues, determine the amount each consumer will spend on holiday gifts in

**a)** 2007.

**b)** 2015.

*Source:* BIG research for the National Retail Federation, *USA Today* (12/22/2004).

**120. Centenarians** People who live to be 100 years old or older are known as centenarians. According to the U.S. Census Bureau, centenarians are the world's fastest-growing age group. The approximate number of centenarians living in the United States between the years of 1995 and 2050, in thousands, can be estimated by

$$\text{number of centenarians} = 0.30x^2 - 3.69x + 92.04$$

where $x$ represents years since 1995. Substitute 1 for $x$ to find the number of centenarians in 1996, 2 for $x$ to find the number of centenarians in 1997, and so on.

**a)** Estimate the number of centenarians that were living in the United States in 2005.

**b)** Estimate the number of centenarians that will be living in the United States in 2050.

*Source:* U.S. Census Bureau

**121. Public Transportation** Increased gas prices and growing congestion in America's major cities have caused a boom in public transportation usage. Between 1992 and 2004, the approximate number of public transportation trips per year in the United States, in billions, can be estimated using

$$\text{number of trips} = 0.065x^2 - 0.39x + 8.47$$

where $x$ represents years since 1992. Substitute 1 for $x$ to estimate the number of trips made in 1993, 2 for $x$ to estimate the number of trips made in 1994, and so on.

**a)** Estimate the number of trips made using public transportation in 2000.

**b)** Assume this trend continues. Estimate the number of trips that will be made in 2010.

*Source:* American Public Transportation Association

The trolley is one form of public transportation in San Francisco.

**122. Inflation** Inflation was on the decline for the years from 2002 to 2004. In 2005, it was on the rise. The inflation rate, as a percent, for the years 2002–2005, can be estimated by

$$\text{inflation} = 0.35x^2 - 1.37x + 2.93$$

where $x$ is the number of years since 2002. Substitute 1 for $x$ to find the inflation rate in 2003, 2 for $x$ to find the inflation rate in 2004, and so on. Assuming this trend continues, find the inflation rate in

**a)** 2005.

**b)** 2007.

*Source:* Treasury Department, Commerce Department, *The Wall Street Journal* (1/18/2005)

**123. Auctions** In recent years, sales from auctions have been on the rise. Sales, in billions of dollars, can be estimated by

$$\text{sales} = 13.5x + 189.83$$

where $x$ is the number of years since 2002. Substitute 1 for $x$ to find the sales from auctions in 2002, 2 for $x$ to find the sales from auctions in 2004, and so on. Assuming this trend continues, find the sales from auctions in

**a)** 2010.

**b)** 2018.

*Source:* National Auctioneers Association, *USA Today* (2/23/2005)

**124. Carbon Dioxide** Since 1905 the amount of carbon dioxide ($CO_2$) has been increasing. The total production of $CO_2$ from all countries except the United States, Canada, and western Europe (measured in millions of metric tons) can be approximated by

$$CO_2 = 0.073x^2 - 0.39x + 0.55$$

where $x$ represents each 10-year period since 1905. Substitute 1 for $x$ to calculate the $CO_2$ production in 1915, 2 for $x$ to calculate the $CO_2$ production in 1925, 3 for $x$ in 1935, and so on.

**a)** Find the approximate amount of $CO_2$ produced by all countries except the United States, Canada, and western Europe in 1945.

**b)** Assume this trend continues, find the approximate amount of $CO_2$ produced by all countries except the United States, Canada, and western Europe in 2005.

**125. Latchkey Kids** The number of *latchkey kids*, children who care for themselves while their parents are working, increases with age. The percent of children of different ages, from 5 to 14 years old, who are latchkey kids can be approximated by

$$\text{percent of children} = 0.23x^2 - 1.98x + 4.42$$

The $x$ value represents the age of the children. For example, substitute 5 for $x$ to get the percent of all 5-year-olds who are latchkey kids, substitute 6 for $x$ to get the percent of all 6-year-olds who are latchkey kids, and so on.

**a)** Find the percent of all 10-year-olds who are latchkey kids.

**b)** Find the percent of all 14-year-olds who are latchkey kids.

**126. Newspaper Readers** The number of Americans who read a daily newspaper is steadily going down. The percent reading a daily newspaper can be approximated by

$$\text{percent} = -6.2x + 82.2$$

where $x$ represents each 10-year period since 1960. Substitute 1 for $x$ to get the percent for 1970, 2 for $x$ to get the percent for 1980, 3 for $x$ to get the percent for 1990, and so forth.

**a)** Find the percent of U.S. adults who read a daily newspaper in 1970.

**b)** Assuming that this trend continues, find the percent of U.S. adults who will read a daily newspaper in 2010.

**127. Organically Grown Food** Increasing fears of pesticides and genetically altered crops have led many people to purchase organically grown food. From 1990 to 2007, sales, in billions of U.S. dollars, of organically grown food can be estimated by

$$\text{sales} = 0.062x^2 + 0.020x + 1.18$$

where $x$ represents years since 1990. Substitute 1 for $x$ to estimate the sales of organically grown food in 1991, 2 for $x$ to estimate sales in 1992, and so on.

**a)** Estimate the sales of organically grown food in 1991.

**b)** Estimate the sales of organically grown food in 2007.

**128. Cellular Phones** Use of cellular phones is currently on the rise. The number of cellular subscribers, in millions, can be approximated by

$$\text{number of subscribers} = 0.42x^2 - 3.44x + 5.80$$

where $x$ represents years since 1982. Substitute 1 for $x$ to get the number of subscribers in 1983, 2 for $x$ to get the number of subscribers in 1984, and so on.

**a)** Find the number of people who used cell phones in 1989.

**b)** Find the number of people who will use cell phones in 2009.

## Cumulative Review Exercises

[1.2] **129.** $A = \{a, b, c, d, f\}$, $B = \{b, c, f, g, h\}$. Find

**a)** $A \cap B$,

**b)** $A \cup B$.

[1.3] *In Exercises 130–132, the letter a represents a real number. For what values of a will each statement be true?*

**130.** $|a| = |-a|$

**131.** $|a| = a$

**132.** $|a| = 6$

**133.** List from smallest to largest: $-|6|, -4, |-5|, -|-2|, 0$.

**134.** Name the following property: $(7 + 3) + 9 = 7 + (3 + 9)$.

## Mid-Chapter Test: 1.1–1.4

*To find out how well you understand the chapter material covered to this point, take this brief test. The answers, and the section where the material was initially discussed, are given in the back of the book. Review any questions that you answered incorrectly.*

**1.** Where is your instructor's office? What are your instructor's office hours?

**2.** Given $A = \{-3, -2, -1, 0, 1, 2\}$ and $B = \{-1, 1, 3, 5\}$, find $A \cup B$ and $A \cap B$.

**3.** Describe the set $D = \{0, 5, 10, 15, \ldots\}$.

**4.** Illustrate the set $\{x | x \geq 3\}$ on a number line.

**5.** Insert $<$ or $>$ in the shaded area of $\frac{3}{5}$ ▢ $\frac{4}{9}$ to make a true statement.

**6.** Express 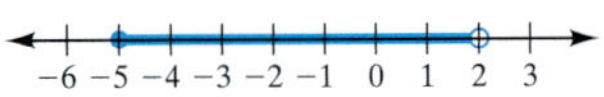 in set builder notation.

**7.** Is $W$ a subset of $N$? Explain.

**8.** List the values from smallest to largest: $-15, |-17|, |-6|, 7$.

*Evaluate each expression.*

**9.** $7 - 2.3 - (-4.5)$

**10.** $\left(\frac{2}{5} + \frac{1}{3}\right) - \frac{1}{2}$

**11.** $(5)(-2)(3.2)(-8)$

**12.** $\left|-\frac{8}{13}\right| \div (-2)$

**13.** Evaluate $(7 - |-2|) - (-8 + |16|)$.

**14.** Name the property illustrated by $5(x + y) = 5x + 5y$.

**15.** Simplify $\sqrt{0.81}$.

**16.** Evaluate

**a)** $x^2$ and

**b)** $-x^2$ for $x = -6$.

**17. a)** List the order of operations.

**b)** Evaluate $4 - 2 \cdot 3^2$ and explain how you determined your answer.

*Evaluate each expression.*

**18.** $5 \cdot 4 \div 10 + 2^5 - 8$.

**19.** $\frac{1}{4}\{[(12 \div 4)^2 - 7]^3 \div 2\}^2$

**20.** $\dfrac{\sqrt{16} + \left(\sqrt{49} - 6\right)^4}{\sqrt[3]{-27} - (4 - 3^2)}$

## 1.5 Exponents

1. Use the product rule for exponents.
2. Use the quotient rule for exponents.
3. Use the negative exponent rule.
4. Use the zero exponent rule.
5. Use the rule for raising a power to a power.
6. Use the rule for raising a product to a power.
7. Use the rule for raising a quotient to a power.

In the previous section we introduced exponents. In this section we discuss the rules of exponents. We begin with the product rule for exponents.

### 1 Use the Product Rule for Exponents

Consider the multiplication $x^3 \cdot x^5$. We can simplify this expression as follows:

$$x^3 \cdot x^5 = (x \cdot x \cdot x) \cdot (x \cdot x \cdot x \cdot x \cdot x) = x^8$$

This problem could also be simplified using the **product rule for exponents.***

**Product Rule for Exponents**

If $m$ and $n$ are natural numbers and $a$ is any real number, then

$$a^m \cdot a^n = a^{m+n}$$

*The rules given in this section also apply for rational or fractional exponents. Rational exponents will be discussed in Section 7.2. We will review these rules again at that time.

*To multiply exponential expressions, maintain the common base and add the exponents.*

$$x^3 \cdot x^5 = x^{3+5} = x^8$$

**EXAMPLE 1** ▸ Simplify. **a)** $2^3 \cdot 2^4$ **b)** $d^2 \cdot d^5$ **c)** $h \cdot h^9$

*Solution*

**a)** $2^3 \cdot 2^4 = 2^{3+4} = 2^7 = 128$ **b)** $d^2 \cdot d^5 = d^{2+5} = d^7$

**c)** $h \cdot h^9 = h^1 \cdot h^9 = h^{1+9} = h^{10}$

▸ Now Try Exercise 13

## 2 Use the Quotient Rule for Exponents

Consider the division $x^7 \div x^4$. We can simplify this expression as follows:

$$\frac{x^7}{x^4} = \frac{\overset{1}{\cancel{x}} \cdot \overset{1}{\cancel{x}} \cdot \overset{1}{\cancel{x}} \cdot \overset{1}{\cancel{x}} \cdot x \cdot x \cdot x}{\underset{1}{\cancel{x}} \cdot \underset{1}{\cancel{x}} \cdot \underset{1}{\cancel{x}} \cdot \underset{1}{\cancel{x}}} = x \cdot x \cdot x = x^3$$

This problem could also be simplified using the **quotient rule for exponents.**

**Quotient Rule for Exponents**

If $a$ is any nonzero real number and $m$ and $n$ are nonzero integers, then

$$\frac{a^m}{a^n} = a^{m-n}$$

*To divide expressions in exponential form, maintain the common base and subtract the exponents.*

$$\frac{x^7}{x^4} = x^{7-4} = x^3$$

**EXAMPLE 2** ▸ Simplify. **a)** $\frac{6^4}{6^2}$ **b)** $\frac{x^7}{x^3}$ **c)** $\frac{y^2}{y^5}$

*Solution* **a)** $\frac{6^4}{6^2} = 6^{4-2} = 6^2 = 36$ **b)** $\frac{x^7}{x^3} = x^{7-3} = x^4$ **c)** $\frac{y^2}{y^5} = y^{2-5} = y^{-3}$

▸ Now Try Exercise 15

## 3 Use the Negative Exponent Rule

Notice in Example 2 **c)** that the answer contains a negative exponent. Let's do part **c)** again by dividing out common factors.

$$\frac{y^2}{y^5} = \frac{\overset{1}{\cancel{y}} \cdot \overset{1}{\cancel{y}}}{\underset{1}{\cancel{y}} \cdot \underset{1}{\cancel{y}} \cdot y \cdot y \cdot y} = \frac{1}{y^3}$$

By dividing out common factors and using the result from Example 2 **c)**, we can reason that $y^{-3} = \frac{1}{y^3}$. This is an example of the negative exponent rule.

**Negative Exponent Rule**

For any nonzero real number $a$ and any whole number $m$,

$$a^{-m} = \frac{1}{a^m}$$

*An expression raised to a negative exponent is equal to 1 divided by the expression with the sign of the exponent changed.*

**EXAMPLE 3** ▸ Write each expression without negative exponents.

**a)** $7^{-2}$ **b)** $8a^{-4}$ **c)** $\dfrac{1}{c^{-5}}$

*Solution*

**a)** $7^{-2} = \dfrac{1}{7^2} = \dfrac{1}{49}$ **b)** $8a^{-4} = 8 \cdot \dfrac{1}{a^4} = \dfrac{8}{a^4}$

**c)** $\dfrac{1}{c^{-5}} = 1 \div c^{-5} = 1 \div \dfrac{1}{c^5} = \dfrac{1}{1} \cdot \dfrac{c^5}{1} = c^5$

▸ Now Try Exercise 37

## Helpful Hint

In Example 3 **c)** we showed that $\dfrac{1}{c^{-5}} = c^5$. In general, for any nonzero real number $a$ and any whole number $m$, $\dfrac{1}{a^{-m}} = a^m$. When a factor of the numerator or the denominator is raised to any power, the factor can be moved to the other side of the fraction bar provided the sign of the exponent is changed. Thus, for example

$$\frac{2a^{-3}}{b^2} = \frac{2}{a^3b^2} \qquad \frac{a^{-2}b^4}{c^{-3}} = \frac{b^4c^3}{a^2}$$

NOTE: When using this procedure, the sign of the base does not change, only the sign of the exponent. For example,

$$-c^{-3} = \frac{1}{-c^3} = -\frac{1}{c^3}$$

Generally, we do not leave exponential expressions with negative exponents. *When we indicate that an exponential expression is to be simplified, we mean that the answer should be written without negative exponents.*

**EXAMPLE 4** ▸ Simplify. **a)** $\dfrac{5xz^2}{y^{-4}}$ **b)** $4^{-2}x^{-1}y^2$ **c)** $-3^3x^2y^{-6}$

*Solution*

**a)** $\dfrac{5xz^2}{y^{-4}} = 5xy^4z^2$

**b)** $4^{-2}x^{-1}y^2 = \dfrac{1}{4^2} \cdot \dfrac{1}{x^1} \cdot y^2 = \dfrac{y^2}{16x}$

**c)** $-3^3x^2y^{-6} = -(3^3)x^2 \cdot \dfrac{1}{y^6} = -\dfrac{27x^2}{y^6}$

▸ Now Try Exercise 41

Notice that the expressions in Example 4 do not involve addition or subtraction. The presence of a plus sign or a minus sign makes for a very different problem as we will see in our next example.

**EXAMPLE 5** ▸ Simplify. **a)** $4^{-1} + 6^{-1}$ **b)** $2 \cdot 3^{-2} + 7 \cdot 6^{-2}$

*Solution*

**a)**
$$\begin{aligned} 4^{-1} + 6^{-1} &= \frac{1}{4} + \frac{1}{6} && \text{Negative exponent rule} \\ &= \frac{3}{12} + \frac{2}{12} && \text{Rewrite with the LCD, 12.} \\ &= \frac{3+2}{12} = \frac{5}{12} \end{aligned}$$

**b)** $2 \cdot 3^{-2} + 7 \cdot 6^{-2} = 2 \cdot \frac{1}{3^2} + 7 \cdot \frac{1}{6^2}$ *Negative exponent rule*

$= \frac{2}{1} \cdot \frac{1}{9} + \frac{7}{1} \cdot \frac{1}{36}$

$= \frac{2}{9} + \frac{7}{36}$

$= \frac{8}{36} + \frac{7}{36}$ *Rewrite with the LCD, 36.*

$= \frac{8+7}{36} = \frac{15}{36} = \frac{5}{12}$

▶ **Now Try Exercise 75**

## 4 Use the Zero Exponent Rule

The next rule we will study is the **zero exponent rule**. Any nonzero number divided by itself is 1. Therefore,

$$\frac{x^5}{x^5} = 1.$$

By the quotient rule for exponents,

$$\frac{x^5}{x^5} = x^{5-5} = x^0.$$

Since $x^0 = \frac{x^5}{x^5}$ and $\frac{x^5}{x^5} = 1$, then

$$x^0 = 1.$$

**Zero Exponent Rule**

If $a$ is any nonzero real number, then

$$a^0 = 1$$

The zero exponent rule illustrates that *any nonzero real number with an exponent of 0 equals 1*. We must specify that $a \neq 0$ because $0^0$ is not a real number.

**EXAMPLE 6** ▶ Simplify (assume that the base is not 0).

**a)** $162^0$ **b)** $7p^0$ **c)** $-y^0$ **d)** $-(8x + 9y)^0$

*Solution*

**a)** $162^0 = 1$

**b)** $7p^0 = 7 \cdot p^0 = 7 \cdot 1 = 7$

**c)** $-y^0 = -1 \cdot y^0 = -1 \cdot 1 = -1$

**d)** $-(8x + 9y)^0 = -1 \cdot (8x + 9y)^0 = -1 \cdot 1 = -1$

▶ **Now Try Exercise 33**

## 5 Use the Rule for Raising a Power to a Power

Consider the expression $(x^3)^2$. We can simplify this expression as follows:

$$(x^3)^2 = x^3 \cdot x^3 = x^{3+3} = x^6$$

This problem could also be simplified using the rule for **raising a power to a power** (also called the **power rule**).

**Raising a Power to a Power (the Power Rule)**

If $a$ is a real number and $m$ and $n$ are integers, then

$$(a^m)^n = a^{m \cdot n}$$

*To raise an exponential expression to a power, maintain the base and multiply the exponents.*

$$(x^3)^2 = x^{3 \cdot 2} = x^6$$

**EXAMPLE 7** ▸ Simplify (assume that the base is not 0).

**a)** $(2^2)^4$ **b)** $(z^{-5})^4$ **c)** $(2^{-3})^2$

*Solution*

**a)** $(2^2)^4 = 2^{2 \cdot 4} = 2^8 = 256$

**b)** $(z^{-5})^4 = z^{-5 \cdot 4} = z^{-20} = \dfrac{1}{z^{20}}$

**c)** $(2^{-3})^2 = 2^{-3 \cdot 2} = 2^{-6} = \dfrac{1}{2^6} = \dfrac{1}{64}$

▸ **Now Try Exercise 81**

**Helpful Hint**

Students often confuse the *product rule*

$$a^m \cdot a^n = a^{m+n}$$

with the *power rule*

$$(a^m)^n = a^{m \cdot n}$$

For example, $(x^3)^2 = x^6$, not $x^5$.

## 6 Use the Rule for Raising a Product to a Power

Consider the expression $(xy)^2$. We can simplify this expression as follows:

$$(xy)^2 = (xy)(xy) = x \cdot x \cdot y \cdot y = x^2y^2$$

This expression could also be simplified using the rule for **raising a product to a power**.

**Raising a Product to a Power**

If $a$ and $b$ are real numbers and $m$ is an integer, then

$$(ab)^m = a^m b^m$$

*To raise a product to a power, raise all factors within the parentheses to the power outside the parentheses.*

**EXAMPLE 8** ▸ Simplify. **a)** $(-9x^3)^2$ **b)** $(3x^{-5}y^4)^{-3}$

*Solution*

**a)** $(-9x^3)^2 = (-9)^2(x^3)^2 = 81x^6$

**b)** $(3x^{-5}y^{4})^{-3} = 3^{-3}(x^{-5})^{-3}(y^{4})^{-3}$ *Raise a product to a power.*

$= \dfrac{1}{3^3} \cdot x^{15} \cdot y^{-12}$ *Negative exponent rule, Power rule*

$= \dfrac{1}{27} \cdot x^{15} \cdot \dfrac{1}{y^{12}}$ *Negative exponent rule*

$= \dfrac{x^{15}}{27y^{12}}$

▸ **Now Try Exercise 93**

## 7 Use the Rule for Raising a Quotient to a Power

Consider the expression $\left(\dfrac{x}{y}\right)^2$. We can simplify this expression as follows:

$$\left(\frac{x}{y}\right)^2 = \frac{x}{y} \cdot \frac{x}{y} = \frac{x \cdot x}{y \cdot y} = \frac{x^2}{y^2}$$

This expression could also be simplified using the rule for **raising a quotient to a power**.

**Raising a Quotient to a Power**

If $a$ and $b$ are real numbers and $m$ is an integer, then

$$\left(\frac{a}{b}\right)^m = \frac{a^m}{b^m}, \qquad b \neq 0$$

*To raise a quotient to a power, raise all factors in the parentheses to the exponent outside the parentheses.*

**EXAMPLE 9** ▸ Simplify. **a)** $\left(\dfrac{5}{x^2}\right)^3$ **b)** $\left(\dfrac{2x^{-2}}{y^3}\right)^{-4}$

*Solution*

**a)** $\left(\dfrac{5}{x^2}\right)^3 = \dfrac{5^3}{(x^2)^3} = \dfrac{125}{x^6}$

**b)** $\left(\dfrac{2x^{-2}}{y^3}\right)^{-4} = \dfrac{2^{-4}(x^{-2})^{-4}}{(y^3)^{-4}}$ *Raise a quotient to a power.*

$= \dfrac{2^{-4}x^8}{y^{-12}}$ *Power rule*

$= \dfrac{x^8y^{12}}{2^4}$ *Negative exponent rule*

$= \dfrac{x^8y^{12}}{16}$

▸ **Now Try Exercise 99**

Consider $\left(\dfrac{a}{b}\right)^{-n}$. Using the rule for raising a quotient to a power, we get

$$\left(\frac{a}{b}\right)^{-n} = \frac{a^{-n}}{b^{-n}} = \frac{b^n}{a^n} = \left(\frac{b}{a}\right)^n$$

Using this result, we see that when we have a rational number raised to a negative exponent, we can take the reciprocal of the base and change the sign of the exponent as follows

$$\left(\frac{8}{9}\right)^{-3} = \left(\frac{9}{8}\right)^3 \qquad \left(\frac{x^2}{y^3}\right)^{-4} = \left(\frac{y^3}{x^2}\right)^4$$

Now we will work some examples that combine a number of properties. Whenever the same variable appears above and below the fraction bar, we generally move the variable with the *lesser exponent* to the opposite side of the fraction bar. This will result in the exponent on the variable being positive when the product rule is applied. Examples 10 and 11 illustrate this procedure.

**EXAMPLE 10** ▸ Simplify. **a)** $\left(\dfrac{15x^2y^4}{5x^2y}\right)^2$ **b)** $\left(\dfrac{5x^4y^{-2}}{10xy^3z^{-1}}\right)^{-3}$

*Solution* Exponential expressions can often be simplified in more than one way. In general, it will be easier to first simplify the expression within the parentheses.

**a)** $\left(\dfrac{15x^2y^4}{5x^2y}\right)^2 = (3y^3)^2 = 9y^6$

**b)**
$$\left(\frac{5x^4y^{-2}}{10xy^3z^{-1}}\right)^{-3} = \left(\frac{x^4 \cdot x^{-1}z}{2y^3 \cdot y^2}\right)^{-3}$$
Move $x$, $y^{-2}$, and $z^{-1}$ to the other side of the fraction bar and change the signs of their exponents.

$$= \left(\frac{x^3z}{2y^5}\right)^{-3}$$
Product rule

$$= \left(\frac{2y^5}{x^3z}\right)^{3}$$
Take the reciprocal of the expression inside the parentheses and change the sign of the exponent.

$$= \frac{2^3y^{5\cdot 3}}{x^{3\cdot 3}z^3}$$
Raise a quotient to a power.

$$= \frac{8y^{15}}{x^9z^3}$$

▸ **Now Try Exercise 109**

**EXAMPLE 11** ▸ Simplify $\dfrac{(2p^{-3}q^5)^{-2}}{(p^{-5}q^4)^{-3}}$.

*Solution* First, use the power rule. Then simplify further.

$$\frac{(2p^{-3}q^5)^{-2}}{(p^{-5}q^4)^{-3}} = \frac{2^{-2}p^6q^{-10}}{p^{15}q^{-12}}$$
Power rule

$$= \frac{q^{-10} \cdot q^{12}}{2^2p^{15} \cdot p^{-6}}$$
Move $2^{-2}$, $p^6$, and $q^{-12}$ to the other side of the fraction bar and change the signs of their exponents.

$$= \frac{q^{-10+12}}{4p^{15-6}}$$
Product rule

$$= \frac{q^2}{4p^9}$$

▸ **Now Try Exercise 115**

**Summary of Rules of Exponents**

For all real numbers $a$ and $b$ and all integers $m$ and $n$:

| Rule | Formula | Condition |
|---|---|---|
| Product rule | $a^m \cdot a^n = a^{m+n}$ | |
| Quotient rule | $\dfrac{a^m}{a^n} = a^{m-n}$, | $a \neq 0$ |
| Negative exponent rule | $a^{-m} = \dfrac{1}{a^m}$, | $a \neq 0$ |
| Zero exponent rule | $a^0 = 1$, | $a \neq 0$ |
| Raising a power to a power | $(a^m)^n = a^{m \cdot n}$ | |
| Raising a product to a power | $(ab)^m = a^mb^m$ | |
| Raising a quotient to a power | $\left(\dfrac{a}{b}\right)^m = \dfrac{a^m}{b^m}$, | $b \neq 0$ |

## EXERCISE SET 1.5

MathXL® MyMathLab

### Concept/Writing Exercises

1. a) Give the product rule for exponents.
   b) Explain the product rule.
2. a) Give the quotient rule for exponents.
   b) Explain the quotient rule.
3. a) Give the zero exponent rule.
   b) Explain the zero exponent rule.
4. a) Give the negative exponent rule.
   b) Explain the negative exponent rule.
5. a) Give the rule for raising a product to a power.
   b) Explain the rule for raising a product to a power.
6. a) Give the rule for raising a power to a power.
   b) Explain the rule for raising a power to a power.
7. a) Give the rule for raising a quotient to a power.
   b) Explain the rule for raising a quotient to a power.
8. What is the exponent on a variable or coefficient if none is shown?
9. If $x^{-1} = 5$, what is the value of $x$? Explain.
10. If $x^{-1} = y^2$, what is $x$ equal to? Explain.
11. a) Explain the difference between the opposite of $x$ and the reciprocal of $x$.
    For parts **b)** and **c)** consider

$$x^{-1}, \quad -x, \quad \frac{1}{x}, \quad \frac{1}{x^{-1}},$$

    b) Which represent (or are equal to) the *reciprocal* of $x$?
    c) Which represent the *opposite* (or *additive inverse*) of $x$?
12. Explain why $-2^{-2} \neq \frac{1}{(-2)^2}$.

### Practice the Skills

*Evaluate each expression.*

13. $2^3 \cdot 2^2$
14. $3^2 \cdot 3^3$
15. $\frac{3^7}{3^5}$
16. $\frac{8^4}{8^3}$
17. $9^{-2}$
18. $5^{-2}$
19. $\frac{1}{5^{-3}}$
20. $\frac{1}{3^{-2}}$
21. $15^0$
22. $19^0$
23. $(2^3)^2$
24. $(3^2)^2$
25. $(2 \cdot 4)^2$
26. $(6 \cdot 5)^2$
27. $\left(\frac{4}{7}\right)^2$
28. $\left(\frac{2}{5}\right)^4$

*Evaluate each expression.*

29. a) $3^{-2}$ b) $(-3)^{-2}$ c) $-3^{-2}$ d) $-(-3)^{-2}$
30. a) $4^{-3}$ b) $(-4)^{-3}$ c) $-4^{-3}$ d) $-(-4)^{-3}$
31. a) $\left(\frac{1}{2}\right)^{-1}$ b) $\left(-\frac{1}{2}\right)^{-1}$ c) $-\left(\frac{1}{2}\right)^{-1}$ d) $-\left(-\frac{1}{2}\right)^{-1}$
32. a) $\left(\frac{3}{4}\right)^{-2}$ b) $\left(-\frac{3}{4}\right)^{-2}$ c) $-\left(\frac{3}{4}\right)^{-2}$ d) $-\left(-\frac{3}{4}\right)^{-2}$

*Simplify each expression and write the answer without negative exponents. Assume that all bases represented by variables are nonzero.*

33. a) $5x^0$ b) $-5x^0$ c) $(-5x)^0$ d) $-(-5x)^0$
34. a) $4y^0$ b) $(4y)^0$ c) $-4y^0$ d) $(-4y)^0$
35. a) $3xyz^0$ b) $(3xyz)^0$ c) $3x(yz)^0$ d) $3(xyz)^0$
36. a) $x^0 + y^0$ b) $(x + y)^0$ c) $x + y^0$ d) $x^0 + y$

*Simplify each expression and write the answer without negative exponents.*

37. $7y^{-3}$
38. $\frac{1}{x^{-1}}$
39. $\frac{9}{x^{-4}}$
40. $\frac{8}{5y^{-2}}$
41. $\frac{2a}{b^{-3}}$
42. $\frac{10x^4}{y^{-1}}$
43. $\frac{13m^{-2}n^{-3}}{2}$
44. $\frac{10x^{-3}}{z^4}$
45. $\frac{5x^{-2}y^{-3}}{z^{-4}}$
46. $\frac{15ab^5}{3c^{-3}}$
47. $\frac{9^{-1}x^{-1}}{y}$
48. $\frac{8^{-1}z}{x^{-1}y^{-1}}$

*Simplify each expression and write the answer without negative exponents.*

**49.** $2^5 \cdot 2^{-7}$ **50.** $a^3 \cdot a^5$ **51.** $x^6 \cdot x^{-4}$ **52.** $x^{-4} \cdot x^3$

**53.** $\frac{8^5}{8^3}$ **54.** $\frac{4^2}{4^{-2}}$ **55.** $\frac{7^{-5}}{7^{-3}}$ **56.** $\frac{x^{-9}}{x^2}$

**57.** $\frac{m^{-6}}{m^5}$ **58.** $\frac{p^0}{p^{-8}}$ **59.** $\frac{5w^{-2}}{w^{-7}}$ **60.** $\frac{x^{-4}}{x^{-6}}$

**61.** $3a^{-2} \cdot 4a^{-6}$ **62.** $(-7v^4)(-3v^{-5})$ **63.** $(-3p^{-2})(-p^3)$ **64.** $(2x^{-3}y^{-4})(6x^{-4}y^7)$

**65.** $(5r^2s^{-2})(-2r^5s^2)$ **66.** $(-6p^{-4}q^6)(2p^3q)$ **67.** $(2x^4y^7)(4x^3y^{-5})$ **68.** $\frac{24x^3y^2}{8xy}$

**69.** $\frac{33x^5y^{-4}}{11x^3y^2}$ **70.** $\frac{6x^{-2}y^3z^{-2}}{-2x^4y}$ **71.** $\frac{9xy^{-4}z^3}{-3x^{-2}yz}$ **72.** $\frac{(x^{-2})(4x^2)}{x^3}$

*Evaluate each expression.*

**73. a)** $4(a + b)^0$ **b)** $4a^0 + 4b^0$ **c)** $(4a + 4b)^0$ **d)** $-4a^0 + 4b^0$

**74. a)** $-2^0 + (-2)^0$ **b)** $-2^0 - (-2)^0$ **c)** $-2^0 + 2^0$ **d)** $-2^0 - 2^0$

**75. a)** $4^{-1} - 3^{-1}$ **b)** $4^{-1} + 3^{-1}$ **c)** $2 \cdot 4^{-1} + 3 \cdot 5^{-1}$ **d)** $(2 \cdot 4)^{-1} + (3 \cdot 5)^{-1}$

**76. a)** $5^{-2} + 4^{-1}$ **b)** $5^{-2} - 4^{-1}$ **c)** $3 \cdot 5^{-2} + 2 \cdot 4^{-1}$ **d)** $(3 \cdot 5)^{-2} - (2 \cdot 4)^{-1}$

*Simplify each expression and write the answer without negative exponents.*

**77.** $(3^2)^2$ **78.** $(5^2)^{-1}$ **79.** $(3^2)^{-2}$ **80.** $(x^2)^{-3}$

**81.** $(b^{-3})^{-2}$ **82.** $(-c)^4$ **83.** $(-c)^3$ **84.** $(-x)^{-2}$

**85.** $(-4x^{-3})^2$ **86.** $-10(x^{-3})^2$ **87.** $5^{-1} + 2^{-1}$ **88.** $4^{-2} + 8^{-1}$

**89.** $3 \cdot 4^{-2} + 9 \cdot 8^{-1}$ **90.** $5 \cdot 2^{-3} + 7 \cdot 4^{-2}$ **91.** $\left(\frac{4b}{3}\right)^{-2}$ **92.** $(-10m^3n^2)^3$

**93.** $(4x^2y^{-2})^2$ **94.** $(4x^2y^3)^{-3}$ **95.** $(5p^2q^{-4})^{-3}$ **96.** $(8s^{-3}t^{-4})^2$

**97.** $(-3g^{-4}h^3)^{-3}$ **98.** $9(x^2y^{-1})^{-4}$ **99.** $\left(\frac{3j}{4k^2}\right)^2$ **100.** $\left(\frac{3x^2y^4}{z}\right)^3$

**101.** $\left(\frac{2r^4s^5}{r^2}\right)^3$ **102.** $\left(\frac{5m^5n^6}{10m^4n^7}\right)^3$ **103.** $\left(\frac{4xy}{y^3}\right)^{-3}$ **104.** $\left(\frac{7x^{-2}}{xy}\right)^{-2}$

**105.** $\left(\frac{5x^{-2}y}{x^{-5}}\right)^3$ **106.** $\left(\frac{4x^2y}{x^{-5}}\right)^{-3}$ **107.** $\left(\frac{10x^2y}{5xz}\right)^{-3}$ **108.** $\left(\frac{4xy}{z^{-2}}\right)^3$

**109.** $\left(\frac{x^8y^{-2}}{x^{-2}y^3}\right)^2$ **110.** $\left(\frac{x^2y^{-3}z^5}{x^{-1}y^2z^3}\right)^{-1}$ **111.** $\left(\frac{4x^{-1}y^{-2}z^3}{2xy^2z^{-3}}\right)^{-2}$ **112.** $\left(\frac{6x^4y^{-6}z^4}{2xy^{-6}z^{-2}}\right)^{-2}$

**113.** $\left(\frac{-a^3b^{-1}c^{-3}}{4ab^3c^{-4}}\right)^{-3}$ **114.** $\frac{(2x^{-1}y^{-2})^{-3}}{(5x^{-1}y^3)^2}$ **115.** $\frac{(3x^{-4}y^2)^3}{(2x^3y^5)^3}$ **116.** $\frac{(2xy^2z^{-3})^2}{(9x^{-1}yz^2)^{-1}}$

## Problem Solving

*Simplify each expression. Assume that all variables represent nonzero integers.*

**117.** $x^{2a} \cdot x^{5a+3}$ **118.** $y^{2m+3} \cdot y^{5m-7}$ **119.** $w^{2a-5} \cdot w^{3a-2}$ **120.** $d^{-4x+7} \cdot d^{5x-6}$

**121.** $\frac{x^{2w+3}}{x^{w-4}}$ **122.** $\frac{y^{5m-1}}{y^{7m-1}}$ **123.** $(x^{3p+5})(x^{2p-3})$ **124.** $(s^{2t-3})(s^{-t+5})$

**125.** $x^{-m}(x^{3m+2})$ **126.** $y^{3b+2} \cdot y^{2b+4}$ **127.** $\frac{30m^{a+b}n^{b-a}}{6m^{a-b}n^{a+b}}$ **128.** $\frac{24x^{c+3}y^{d+4}}{8x^{c-4}y^{d+6}}$

**129.** **a)** For what values of $x$ is $x^4 > x^3$?
**b)** For what values of $x$ is $x^4 < x^3$?
**c)** For what values of $x$ is $x^4 = x^3$?
**d)** Why can you not say that $x^4 > x^3$?

**130.** Is $3^{-8}$ greater than or less than $2^{-8}$? Explain.

**131.** **a)** Explain why $(-1)^n = 1$ for any even number $n$.
**b)** Explain why $(-1)^n = -1$ for any odd number $n$.

**132.** **a)** Explain why $(-12)^{-8}$ is positive.
**b)** Explain why $(-12)^{-7}$ is negative.

**133.** **a)** Is $\left(-\frac{2}{3}\right)^{-2}$ equal to $\left(\frac{2}{3}\right)^{-2}$?
**b)** Will $(x)^{-2}$ equal $(-x)^{-2}$ for all real numbers $x$ except 0? Explain your answer.

**134.** **a)** Is $\left(-\frac{2}{3}\right)^{-3}$ equal to $\left(\frac{2}{3}\right)^{-3}$?
**b)** Will $(x)^{-3}$ equal $(-x)^{-3}$ for any nonzero real number $x$? Explain.
**c)** What is the relationship between $(-x)^{-3}$ and $(x)^{-3}$ for any nonzero real number $x$?

*Determine what exponents must be placed in the shaded area to make each expression true. Each shaded area may represent a different exponent. Explain how you determined your answer.*

**135.** $\left(\dfrac{x^2y^{-2}}{x^{-3}y^{\blacksquare}}\right)^2 = x^{10}y^2$

**136.** $\left(\dfrac{x^{-2}y^3z}{x^4y^{\blacksquare}z^{-3}}\right)^3 = \dfrac{z^{12}}{x^{18}y^6}$

**137.** $\left(\dfrac{x^{\blacksquare}y^5z^{-2}}{x^4y^{\blacksquare}z}\right)^{-1} = \dfrac{x^5z^3}{y^2}$

## Challenge Problems

*We will learn in Section 7.2 that the rules of exponents given in this section also apply when the exponents are rational numbers. Using this information and the rules of exponents, evaluate each expression.*

**138.** $\left(\dfrac{x^{1/2}}{x^{-1}}\right)^{3/2}$

**139.** $\left(\dfrac{x^{5/8}}{x^{1/4}}\right)^3$

**140.** $\left(\dfrac{x^4}{x^{-1/2}}\right)^{-1}$

**141.** $\dfrac{x^{1/2}y^{-3/2}}{x^5y^{5/3}}$

**142.** $\left(\dfrac{x^{1/2}y^4}{x^{-3}y^{5/2}}\right)^2$

## Group Activity

*Discuss and answer Exercise 143 as a group.*

**143. Doubling a Penny** On day 1 you are given a penny. On each following day, you are given double the amount you were given on the previous day.

**a)** Write down the amounts you would be given on each of the first 6 days.

**b)** Express each of these numbers as an exponential expression with a base of 2.

**c)** By looking at the pattern, determine an exponential expression for the number of cents you will receive on day 10.

**d)** Write a general exponential expression for the number of cents you will receive on day $n$.

**e)** Write an exponential expression for the number of cents you will receive on day 30.

**f)** Calculate the value of the expression in part **e)**. Use a calculator if one is available.

**g)** Determine the amount found in part **f)** in dollars.

**h)** Write a general exponential expression for the number of dollars you will receive on day $n$.

## Cumulative Review Exercises

[1.2] **144.** If $A = \{3, 4, 6\}$ and $B = \{1, 2, 5, 9\}$, find
**a)** $A \cup B$ and
**b)** $A \cap B$.

**145.** Illustrate the following set on the number line: $\{x \mid -3 \le x < 2\}$.

[1.4] **146.** Evaluate $8 + |12| \div |-3| - 4 \cdot 2^2$.

**147.** Evaluate $\sqrt[3]{-125}$.

# 1.6 Scientific Notation

1 Write numbers in scientific notation.

2 Change numbers in scientific notation to decimal form.

3 Use scientific notation in problem solving.

## 1 Write Numbers in Scientific Notation

Scientists and engineers often deal with very large and very small numbers. For example, the frequency of an FM radio signal may be 14,200,000,000 hertz (or cycles per second) and the diameter of a hydrogen atom is about 0.0000000001 meter. Because it is difficult to work with many zeros, scientists often express such numbers with exponents. For example, the number 14,200,000,000 might be written as $1.42 \times 10^{10}$, and 0.0000000001 as $1 \times 10^{-10}$. Numbers such as $1.42 \times 10^{10}$ and $1 \times 10^{-10}$ are in a form called **scientific notation**. In scientific notation, numbers are expressed as $a \times 10^n$, where $1 \le a < 10$ and $n$ is an integer. When a power of 10 has no numerical coefficient showing, as in $10^5$, we assume that the numerical coefficient is 1. Thus, $10^5$ means $1 \times 10^5$ and $10^{-4}$ means $1 \times 10^{-4}$.

The diameter of this galaxy is about $1 \times 10^{21}$ meters.

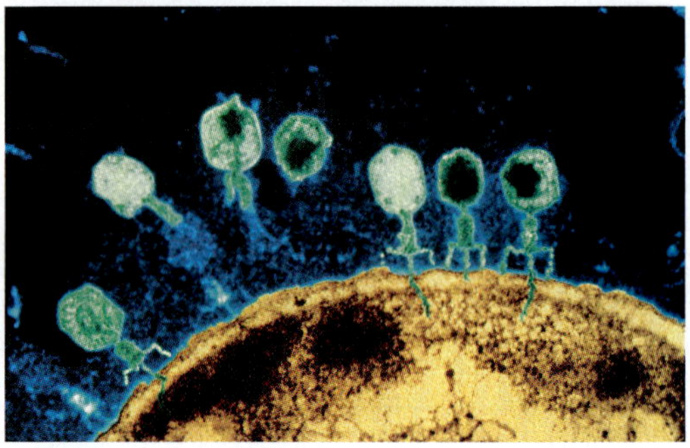

The diameter of these viruses (shown in green) is about $1 \times 10^{-7}$ meters.

### Examples of Numbers in Scientific Notation

$3.2 \times 10^6$ $\quad$ $4.176 \times 10^3$ $\quad$ $2.64 \times 10^{-2}$

The following shows the number 32,400 changed to scientific notation.

$$32.400 = 3.24 \times 10{,}000$$
$$= 3.24 \times 10^4 \qquad (10{,}000 = 10^4)$$

The are four zeros in 10,000, the same number as the exponent in $10^4$. The procedure for writing a number in scientific notation follows.

**To Write a Number in Scientific Notation**

1. Move the decimal point in the number to the right of the first nonzero digit. This gives a number greater than or equal to 1 and less than 10.
2. Count the number of places you moved the decimal point in step 1. If the original number is 10 or greater, the count is to be considered positive. If the original number is less than 1, the count is to be considered negative.
3. Multiply the number obtained in step 1 by 10 raised to the count (power) found in step 2.

**EXAMPLE 1** ▸ Write the following numbers in scientific notation.

**a)** 68,900 **b)** 0.000572 **c)** 0.0074

*Solution*

**a)** The decimal point in 68,900 is to the right of the last zero.

$$68{,}900. = 6.89 \times 10^4$$

The decimal point is moved four places. Since the original number is greater than 10, the exponent is positive.

**b)** $0.000572 = 5.72 \times 10^{-4}$

The decimal point is moved four places. Since the original number is less than 1, the exponent is negative.

**c)** $0.0074 = 7.4 \times 10^{-3}$

▸ **Now Try Exercise 11**

## 2 Change Numbers in Scientific Notation to Decimal Form

Occasionally, you may need to convert a number written in scientific notation to its decimal form. The procedure to do so follows.

**To Convert a Number in Scientific Notation to Decimal Form**

1. Observe the exponent on the base 10.
2. **a)** If the exponent is positive, move the decimal point in the number to the right the same number of places as the exponent. It may be necessary to add zeros to the number. This will result in a number greater than or equal to 10.
   **b)** If the exponent is 0, the decimal point in the number does not move from its present position. Drop the factor $10^0$. This will result in a number greater than or equal to 1 but less than 10.
   **c)** If the exponent is negative, move the decimal point in the number to the left the same number of places as the exponent. It may be necessary to add zeros. This will result in a number less than 1.

**EXAMPLE 2** ▸ Write the following numbers without exponents.

**a)** $2.1 \times 10^4$ **b)** $8.73 \times 10^{-3}$ **c)** $1.45 \times 10^8$

*Solution*

**a)** Move the decimal point four places to the right.

$$2.1 \times 10^4 = 2.1 \times 10{,}000 = 21{,}000$$

**b)** Move the decimal point three places to the left.

$$8.73 \times 10^{-3} = 0.00873$$

**c)** Move the decimal point eight places to the right.

$$1.45 \times 10^8 = 145{,}000{,}000$$

▸ **Now Try Exercise 25**

## 3 Use Scientific Notation in Problem Solving

We can use the rules of exponents when working with numbers written in scientific notation, as illustrated in the following applications.

**EXAMPLE 3** ▸ **Public Debt per Person** The public debt is the total amount owed by the U.S. federal government to lenders in the form of government bonds. On July 1, 2005, the U.S. public debt was approximately \$7,858,000,000,000 (7 trillion, 858 billion dollars). The U.S. population on that date was approximately 296,000,000.

**a)** Find the average U.S. debt per person in the United States (the per capita debt).

**b)** On July 1, 1982, the U.S. debt was approximately \$1,142,000,000,000. How much larger was the debt in 2005 than in 1982?

**c)** How many times greater was the debt in 2005 than in 1982?

## Solution

**a)** To find the per capita debt, we divide the public debt by the population.

$$\frac{7{,}858{,}000{,}000{,}000}{296{,}000{,}000} = \frac{7.858 \times 10^{12}}{2.96 \times 10^{8}} \approx 2.65 \times 10^{12-8} \approx 2.65 \times 10^{4} \approx 26{,}500$$

Thus, the per capita debt was about $26,500. This means that if the citizens of the United States wished to "chip in" and pay off the federal debt, it would take about $26,500 for every man, woman, and child in the United States.

**b)** We need to find the difference in the debt between 2005 and 1982.

$$\begin{aligned} 7{,}858{,}000{,}000{,}000 - 1{,}142{,}000{,}000{,}000 &= 7.858 \times 10^{12} - 1.142 \times 10^{12} \\ &(7.858 - 1.142) \times 10^{12} \\ &6.716 \times 10^{12} \\ &6{,}716{,}000{,}000{,}000 \end{aligned}$$

The U.S. public debt was $6,716,000,000,000 greater in 2005 than in 1982.

**c)** To find out how many times greater the 2005 public debt was, we divide the 2005 debt by the 1982 debt as follows:

$$\frac{7{,}858{,}000{,}000{,}000}{1{,}142{,}000{,}000{,}000} = \frac{7.858 \times 10^{12}}{1.142 \times 10^{12}} \approx 6.88$$

Thus, the 2005 public debt was about 6.88 times greater than the 1982 public debt.

▶ **Now Try Exercise 87**

**EXAMPLE 4** ▶ **Tax Collections** The data for the graph in **Figure 1.11** is taken from the U.S. Census Bureau Web site. The graphs shows the cumulative state government tax collections in 2004. We have given the amounts collected in scientific notation.

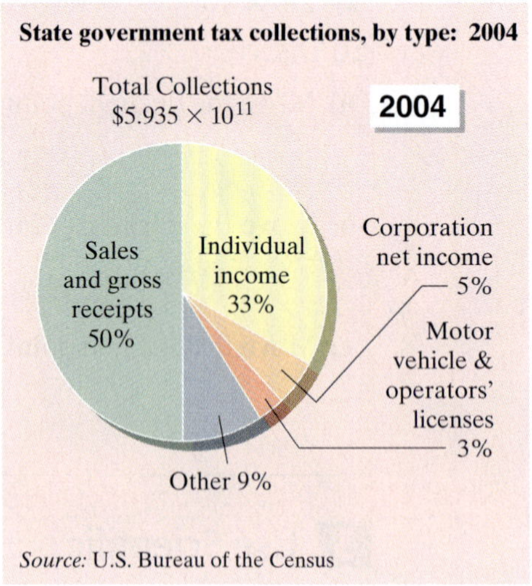


FIGURE 1.11

**a)** Determine, using scientific notation, how much money was collected from individual income taxes in 2004.

**b)** Determine, using scientific notation, how much more money was collected from sales and gross receipts than from corporation net income taxes.

## Solution

**a)** In 2004, 33% of the $\$5.935 \times 10^{11}$ was collected from individual income taxes. In decimal form, 33% is 0.33, and in scientific notation, 33% is $3.3 \times 10^{-1}$. To determine 33% of $\$5.935 \times 10^{11}$, we multiply using scientific notation.

$$\begin{aligned}\text{individual income tax collected} &= (3.3 \times 10^{-1})(5.935 \times 10^{11})\\ &= (3.3 \times 5.935)(10^{-1} \times 10^{11})\\ &= 19.5855 \times 10^{-1+11}\\ &= 19.5855 \times 10^{10}\\ &= 1.95855 \times 10^{11}\end{aligned}$$

Thus, about $\$1.95855 \times 10^{11}$ or \$195,855,000,000 was collected from individual income taxes in 2004.

**b)** In 2004, 50% was collected from sales and gross receipts and 5% was collected from corporation net income taxes. To determine how much more money was collected from sales and gross receipts than from corporation net income taxes, we first determine the difference in the two percents.

$$\text{difference} = 50\% - 5\% = 45\%$$

To find 45% of $\$5.935 \times 10^{11}$, we change 45% to scientific notation and then we multiply.

$$\begin{aligned}45\% &= 0.45 = 4.5 \times 10^{-1}\\ \text{difference in tax collections} &= (4.5 \times 10^{-1})(5.935 \times 10^{11})\\ &= (4.5 \times 5.935)(10^{-1} \times 10^{11})\\ &= 26.7075 \times 10^{10}\\ &= 2.67075 \times 10^{11}\end{aligned}$$

Thus, about $\$2.67075 \times 10^{11}$ or \$267,075,000,000 more money was collected from sales and gross receipts than from corporation net income taxes.

▶ Now Try Exercise 95

## USING YOUR CALCULATOR

On a scientific or graphing calculator the product of (8,000,000)(400,000) might be displayed as $3.2^{12}$ or 3.2E12. Both of these represent $3.2 \times 10^{12}$, which is 3,200,000,000,000.

To enter numbers in scientific notation on either a scientific or graphing calculator, you generally use the [EE] or [EXP] keys. To enter $4.6 \times 10^{8}$, you would press either 4.6 [EE] 8 or 4.6 [EXP] 8. Your calculator screen might then show $4.6^{08}$ or 4.6E8.

On the TI-84 Plus the EE appears above the [,] key. So to enter (8,000,000)(400,000) in scientific notation you would press

answer displayed

8 [2nd] [,] 6 [×] 4 [2nd] [,] 5 [ENTER] 3.2E12

(8 [2nd] [,]: to get EE; 4 [2nd] [,]: to get EE)

# EXERCISE SET 1.6

MathXL® MyMathLab

## Concept/Writing Exercises

1. What is the form of a number in scientific notation?

2. Can $1 \times 10^{n}$ ever be a negative number for any positive integer $n$? Explain.

3. Which is greater, $1 \times 10^{-2}$ or $1 \times 10^{-3}$? Explain.

4. Can $1 \times 10^{-n}$ ever be a negative number for any positive integer $n$? Explain.

## Practice the Skills

*Express each number in scientific notation.*

**5.** 3700 **6.** 860 **7.** 0.041 **8.** 0.000000718

**9.** 760,000 **10.** 9,260,000,000 **11.** 0.00000186 **12.** 0.00000914

**13.** 5,780,000 **14.** 0.0000723 **15.** 0.000106 **16.** 452,000,000

*Express each number without exponents.*

**17.** $3.1 \times 10^4$ **18.** $5 \times 10^8$ **19.** $2.13 \times 10^{-5}$ **20.** $5.78 \times 10^{-5}$

**21.** $9.17 \times 10^{-1}$ **22.** $5.4 \times 10^1$ **23.** $8 \times 10^6$ **24.** $7.6 \times 10^4$

**25.** $2.03 \times 10^5$ **26.** $9.25 \times 10^{-6}$ **27.** $1 \times 10^6$ **28.** $1 \times 10^{-8}$

*Express each value without exponents.*

**29.** $(4 \times 10^5)(6 \times 10^2)$ **30.** $(7.6 \times 10^{-3})(1.2 \times 10^{-1})$ **31.** $\dfrac{8.4 \times 10^{-6}}{4 \times 10^{-4}}$

**32.** $\dfrac{8.5 \times 10^3}{1.7 \times 10^{-2}}$ **33.** $\dfrac{9.45 \times 10^{-3}}{3.5 \times 10^2}$ **34.** $(5.2 \times 10^{-3})(4.1 \times 10^5)$

**35.** $(8.2 \times 10^5)(1.4 \times 10^{-2})$ **36.** $(6.3 \times 10^4)(3.7 \times 10^{-8})$ **37.** $\dfrac{1.68 \times 10^4}{5.6 \times 10^7}$

**38.** $\dfrac{7.2 \times 10^{-2}}{3.6 \times 10^{-6}}$ **39.** $(9.1 \times 10^{-4})(7.4 \times 10^{-4})$ **40.** $\dfrac{8.6 \times 10^{-8}}{4.3 \times 10^{-6}}$

*Express each value in scientific notation.*

**41.** $(0.03)(0.0005)$ **42.** $(2500)(7000)$ **43.** $\dfrac{35{,}000{,}000}{7000}$

**44.** $\dfrac{560{,}000}{0.0008}$ **45.** $\dfrac{0.00069}{23{,}000}$ **46.** $\dfrac{0.000012}{0.000006}$

**47.** $(47{,}000)(35{,}000{,}000)$ **48.** $\dfrac{0.0000286}{0.00143}$ **49.** $\dfrac{1008}{0.0021}$

**50.** $\dfrac{0.018}{160}$ **51.** $\dfrac{0.00153}{0.00051}$ **52.** $(0.0015)(0.00038)$

*Express each value in scientific notation. Round decimal numbers to the nearest thousandth.*

**53.** $(4.78 \times 10^9)(1.96 \times 10^5)$ **54.** $\dfrac{4.44 \times 10^3}{1.11 \times 10^1}$

**55.** $(7.23 \times 10^{-3})(1.46 \times 10^5)$ **56.** $(5.71 \times 10^5)(4.7 \times 10^{-3})$

**57.** $\dfrac{4.36 \times 10^{-4}}{8.17 \times 10^{-7}}$ **58.** $\dfrac{6.45 \times 10^{25}}{3.225 \times 10^{15}}$

**59.** $(4.89 \times 10^{15})(6.37 \times 10^{-41})$ **60.** $(4.36 \times 10^{-6})(1.07 \times 10^{-6})$

**61.** $(8.32 \times 10^3)(9.14 \times 10^{-31})$ **62.** $\dfrac{3.71 \times 10^{11}}{4.72 \times 10^{-9}}$

**63.** $\dfrac{1.5 \times 10^{35}}{4.5 \times 10^{-26}}$ **64.** $(4.9 \times 10^5)(1.347 \times 10^{31})$

**Scientific Notation** *In Exercises 65–78, write each italicized number in scientific notation.*

**65.** It cost NASA more than *$850 million* to send the rovers *Spirit* and *Opportunity* to mars.

**66.** The distance between the sun and earth is about *93 million* miles.

**67.** The average cost for a 30-second ad in Super Bowl XXXIX was *$2.4 million*.

**68.** According to the U.S. Census Bureau, the world population in 2050 will be about *9.2 billion* people.

**69.** According to the 2005 *World Almanac and Fact Book*, the richest man in the world is Bill Gates of Microsoft Corporation, who is worth about *$ 52.8 billion*.

**70.** The 2006 U.S. federal budget was about *$2.56 trillion*.

**71.** In 2006, the U.S. debt was about *$9.1 trillion*.

**72.** The speed of light is about *186,000* miles per second.

73. One centimeter = *0.00001* hectometer.

74. One millimeter = *0.000001* kiloliter.

75. One inch ≈ *0.0000158* mile.

76. One ounce ≈ *0.00003125* ton.

77. One milligram = *0.000000001* metric ton.

78. A certain computer can compute one computation in *0.0000001* second.

## Problem Solving

79. Explain how you can quickly divide a number given in scientific notation by

**a)** 10,

**b)** 100,

**c)** 1 million.

**d)** Divide $6.58 \times 10^{-4}$ by 1 million. Leave your answer in scientific notation.

80. Explain how you can quickly multiply a number given in scientific notation by

**a)** 10,

**b)** 100,

**c)** 1 million.

**d)** Multiply $7.59 \times 10^{7}$ by 1 million. Leave your answer in scientific notation.

81. **Science Experiment** During a science experiment you find that the correct answer is $5.25 \times 10^{4}$.

**a)** If you mistakenly write the answer as $4.25 \times 10^{4}$, by how much is your answer off?

**b)** If you mistakenly write the answer as $5.25 \times 10^{5}$, by how much is your answer off?

**c)** Which of the two errors is the more serious? Explain.

82. **Earth Orbit**

**a)** Earth completes its $5.85 \times 10^{8}$-mile orbit around the sun in 365 days. Find the distance traveled per day.

**b)** Earth's speed is about eight times faster than a bullet. Estimate the speed of a bullet in miles per hour.

83. **Distance to the Sun** The distance to the sun is 93,000,000 miles. If a spacecraft travels at a speed of 3100 miles per hour, how long will it take for it to reach the sun?

84. **Universe** We have proof that there are at least 1 sextillion, $10^{21}$, stars in the Universe.

**a)** Write that number without exponents.

**b)** How many million stars is this? Explain how you determined your answer to part **b)**.

85. **U.S. and World Population** The population of the United States on September 1, 2006, was estimated to be $2.995 \times 10^{8}$. On this day, the population of the world was about $6.536 \times 10^{9}$.
*Source:* U.S. Census Bureau

**a)** How many people lived outside of the United States in 2005?

**b)** What percentage of the world's population lived in the United States in 2005?

86. **New River George Bridge** The New River George bridge, shown below, is 3030.5 feet long. It was completed in 1977 near Fayetteville, West Virginia, and is the world's longest single arch steel span. Its total weight is $8.80 \times 10^{7}$ pounds and its heaviest single piece weighs $1.84 \times 10^{5}$ pounds.

**a)** How many times greater is the total weight of the bridge than the weight of the heaviest single piece?

**b)** What is the difference in weight between the total weight of the bridge and the weight of the heaviest single piece?

87. **Gross Domestic Product** Gross Domestic Product (GDP) is a measure of economic activity. GDP is the total amount of goods and services produced in a country in a year. In 2005, the GDP for the United States was about \$11.728 trillion and the population of the United States was about 296.5 million.
*Source:* U.S. Treasury Web site

**a)** Write each of these two numbers in scientific notation.

**b)** Determine the GDP *per capita* by dividing the GDP by the population of the United States.

88. **Gross Domestic Product** In 2003, the GDP (see Exercise 87) of the world was about \$36.356 trillion and the population of the world was about 6.3 billion people.
*Source:* U.S. Treasury Web site and www.en.wikipedia.org/wiki

**a)** Write each of these numbers in scientific notation.

**b)** Determine the GDP *per capita* by dividing the GDP by the population of the world.

89. **Population Density** The population density (people per square kilometer) is determined by dividing the population of a country by its land area. Find the population density of China if its population in 2005 was $1.29 \times 10^{9}$ people and its

land area was $9.8 \times 10^6$ square kilometers. (Round your answer to the nearest unit.)

**90. Population Density** Find the population density (see Exercise 89) of India if its population in 2005 was $1.095 \times 10^9$ people and its land area is $3.2 \times 10^6$ square kilometers. (Round your answer to the nearest unit.)

**91. Recycling Plastic** In the United States only about 5% of the $4.2 \times 10^9$ pounds of used plastic is recycled annually.

**a)** How many pounds are recycled annually?

**b)** How many pounds are not recycled annually?

**92. Distance to Proxima Centauri** The distance from earth to the sun is approximately 150 million kilometers. The next closest star to earth is Proxima Centauri. It is about 268,000 times farther away from earth than is the sun. Approximate the distance of Proxima Centauri from earth. Write your answer in scientific notation.

*Source:* NASA Web site

Proxima Centauri

**93. Most Populous Countries** In 2005, the six most populous countries accounted for 3,242,000,000 people out of the world's total population of 6,446,000,000. The six most populous countries in 2005 are shown in the following graph, along with each country's population.

**Six most populous countries** (population in millions)

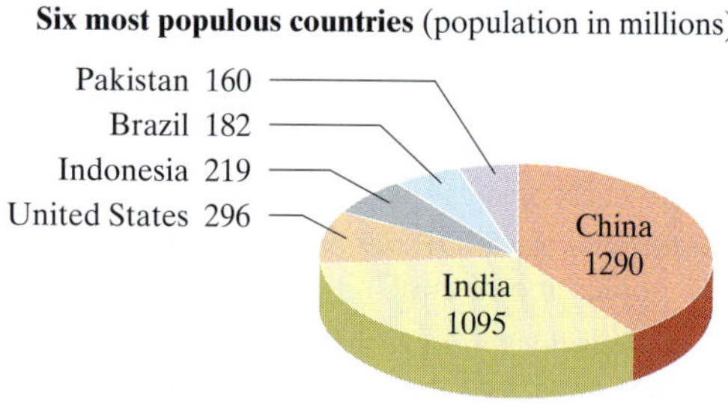


*Source:* U.S. Census Bureau
*Note:* China includes mainland China and Taiwan.

**a)** How many more people lived in China than in the United States?

**b)** What percent of the world's population lived in China?

**c)** If the area of China is $3.70 \times 10^6$ square miles, determine China's population density (people per square mile).

**d)** If the area of the United States is $3.62 \times 10^6$ square miles, determine the population density in the United States.*

**94. World Population** The entire span of human history was required for the world's population to reach $6.52 \times 10^9$ people in the year 2006. At current rates, the world's population will double in about 62 years.

**a)** Estimate the world's population in 2068.

**b)** Assuming 365 days in a year, estimate the average number of additional people added to the world's population each day between 2006 and 2068.

**95. Federal Outlays** The following graph appeared on page 81 of the 2005 Internal Revenue Service Form 1040 tax booklet. The graph shows the distribution of outlays (expenditures) of the federal government in Fiscal Year (FY) 2004. The total for the outlay of the federal government in FY 2004 was $\$2.3 \times 10^{12}$.

**Outlays**

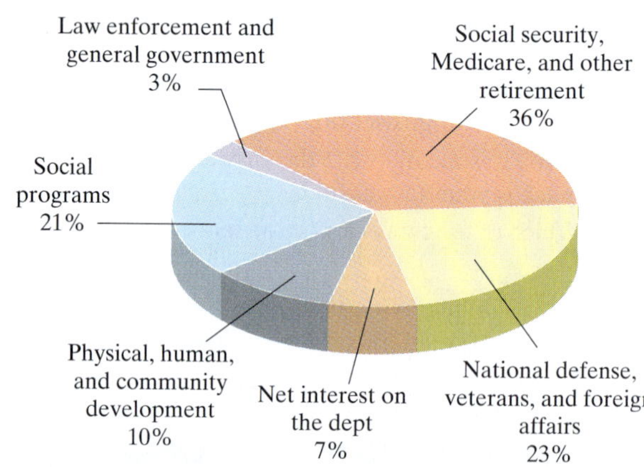


Use this circle graph to answer the following questions. Write all answers in scientific notation.

**a)** What was the outlay in FY 2004 for law enforcement and general government?

**b)** What was the outlay in FY 2004 for Social Security, Medicans, and other retirement programs?

**c)** What was the total outlay in FY 2004 for all programs other than the net interest on the national debt?

**96. Football Revenue in the NFL** In 2004, the 32 professional football teams in the NFL generated more than \$5 billion in revenue. The four teams generating the most revenue were the Washington Redskins, Dallas Cowboys, Philadelphia Eagles, and Houston Texans. The total revenue from these four teams was $\$8.49 \times 10^8$. The graph on the next page shows the percent distribution of the $\$8.49 \times 10^8$ among these four teams.

* As of July 1, 2005, the region with the greatest population density is Macau, with a population density of 45,978 people per square mile. The country with the greatest population density is Monaco, with a population density of 42,172 people per square mile.

Use this graph to answer the following questions.

**Four NFL Teams Generating The Most Revenue, total of $\$8.49 \times 10^8$**

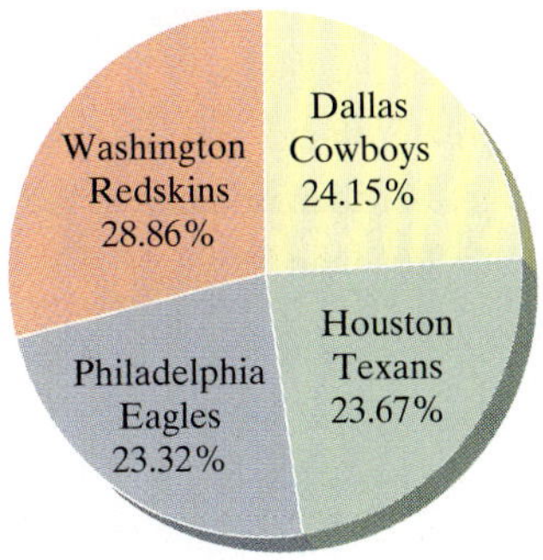


*Source:* NFL, *Forbes Magazine*, *The Washington Post* (1/8/2005)

**a)** Determine the revenue for the Dallas Cowboys and for the Houston Texans. Express answers in scientic notation.

**b)** What is the difference in revenue between the Dallas Cowboys and the Houston Texans?

**c)** If the total revenue from the 32 teams was \$5 billion in 2004, what percent of the total revenue did these 4 teams have? Express your answer to the nearest percent.

**97. Land Area** The land area, in square kilometers, for the five largest countries on our planet is given in the graph below.

**Land Area (in millions of square kilometers) of the Five Largest Countries**

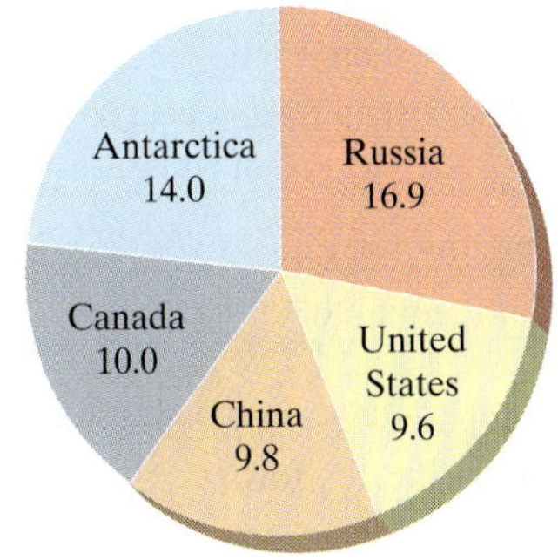


*Source:* www.world-gazetteer.com

**a)** What is the total land area of the five largest countries? Write the answer in scientific notation.

**b)** How much more land area is there in Antartica than in the United States? Write the answer in scientific notation.

## Challenge Problems

**98. Light-Year** A *light-year* is the distance that light travels in 1 year.

**a)** Find the number of miles in a light-year if light travels at $1.86 \times 10^5$ miles *per second*.

**b)** If earth is 93,000,000 miles from the sun, how long does it take for light from the sun to reach earth?

**c)** Our galaxy, the Milky Way, is about $6.25 \times 10^{16}$ miles across. If a spacecraft could travel at half the speed of light, how long would it take for the craft to travel from one end of the galaxy to the other?

# Chapter 1 Summary

| IMPORTANT FACTS AND CONCEPTS | EXAMPLES |
|---|---|
| **Section 1.2** | |
| A **variable** is a letter used to represent various numbers.<br>A **constant** is a letter use to represent a particular value.<br>An **algebraic expression** (or **expression**) is any combination of numbers, variables, exponents, mathematical symbols, and operations. | $x$ and $y$ are commonly used for variables.<br>If $h$ is the number of hours in a day, then $h = 24$, a constant.<br>$3x^2(x - 2) + 2x$ is an algebraic expression. |
| A **set** is a collection of objects. The **objects** are called elements.<br>**Roster form** is a set having elements listed inside a pair of braces.<br>A first set is a **subset** of a second set when every element of the first set is also an element of the second set.<br>**Null set**, or **empty set**, symbolized { } or ∅, has no elements. | If $A = \{\text{blue, green, red}\}$, then blue, green, and red are elements of $A$.<br>$\{1, 3, 5\}$ is a subset of $\{1, 2, 3, 4, 5\}$<br>The set of living people over 200 years of age is an empty set. |

| IMPORTANT FACTS AND CONCEPTS | EXAMPLES |
|---|---|

**Section 1.2 (continued)**

### Inequality Symbols

$>$ is read "is greater than."

$\geq$ is read "is greater than or equal to."

$<$ is read "is less than."

$\leq$ is read "is less than or equal to."

$\neq$ is read "is not equal to."

$6 > 2$ is read 6 is greater than 2

$5 \geq 5$ is read 5 is greater than or equal to 5

$-4 < 3$ is read $-4$ is less than 3

$-10 \leq -1$ is read $-10$ is less than or equal to $-1$

$-5 \neq 17$ is read $-5$ is not equal to 17

Inequalities can be graphed on a real number line.

$x > 3$

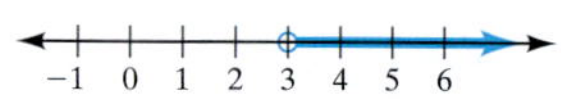

**Set builder notation** has the form

$\{\ x \mid x \text{ has property } p\ \}$

The set of — all elements $x$ — such that — $x$ has the given property

$\{x|-1 \leq x < 2\}$

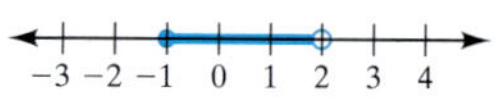

### Important Sets of Real Numbers

Real numbers: $\mathbb{R} = \{x|x \text{ is a point on a number line}\}$

Natural or counting numbers: $N = \{1, 2, 3, 4, 5, \ldots\}$

Whole numbers: $W = \{0, 1, 2, 3, 4, 5, \ldots\}$

Integers: $I = \{\ldots, -3, -2, -1, 0, 1, 2, 3, \ldots\}$

Rational numbers: $Q = \left\{\dfrac{p}{q}\middle| p \text{ and } q \text{ are integers, } q \neq 0\right\}$

Irrational numbers: $H = \{x|x \text{ is a real number that is not rational}\}$

The **union** of set $A$ and set $B$, written $A \cup B$, is the set of elements that belong to either set $A$ *or* set $B$.

The **intersection** of set $A$ and set $B$, written $A \cap B$, is the set of all elements that are common to both set $A$ *and* set $B$.

Given $A = \{1, 2, 3, 5, 7\}$ and $B = \{3, 4, 5, 6, 7\}$

then $A \cup B = \{1, 2, 3, 4, 5, 6, 7\}$

$A \cap B = \{3, 5, 7\}$.

**Section 1.3**

### Additive Inverse

For any real number $a$, its additive inverse is $-a$.

$-8$ is the additive inverse of 8.

### Double Negative Property

For any real number $a$, $-(-a) = a$.

$-(-5) = 5$

### Absolute Value

If $a$ represents any real number, then

$$|a| = \begin{cases} a & \text{if } a \geq 0 \\ -a & \text{if } a < 0 \end{cases}$$

$|9| = 9, \quad |-9| = 9$

To add two numbers with the **same sign** (both positive or both negative), add their absolute values and place the common sign before the sum.

Add $-6 + (-8)$.

$|-6| = 6$ and $|-8| = 8$

$|-6| + |-8| = 6 + 8 = 14$

Thus, $-6 + (-8) = -14$.

| IMPORTANT FACTS AND CONCEPTS | EXAMPLES |
|---|---|
| **Section 1.3 (continued)** | |
| To add two numbers with **different signs** (one positive and the other negative) subtract the smaller absolute value from the larger absolute value. The answer has the sign of the number with the larger absolute value. | Add $8 + (-2)$. $8 + (-2) = \|8\| - \|-2\| = 8 - 2 = 6$ Thus, $8 + (-2) = 6$. |
| **Subtraction of Real Numbers** $a - b = a + (-b)$ | $-14 - 10 = -14 + (-10) = -24$ |
| To multiply two numbers with **like signs**, either both positive or both negative, multiply their absolute values. The answer is **positive**. | $(-1.6)(-8.9) = 14.24$ |
| To multiply two numbers with **unlike signs**, one positive and the other negative, multiply their absolute values. The answer is **negative**. | $21\left(-\frac{1}{7}\right) = -3$ |
| **Multiplicative Property of Zero** For any number $a$, $a \cdot 0 = 0 \cdot a = 0$ | $0 \cdot 5 = 0$ |
| **Division By Zero** For any real number $a \neq 0$, then $\frac{a}{0}$ is not defined. | $\frac{7}{0}$ is not defined. |
| **Divide Two Real Numbers** **1.** To divide two numbers with **like signs**, either both positive or both negative, divide their absolute values. The answer is **positive**. | $\frac{-8}{-2} = 4$ |
| **2.** To divide two numbers with **unlike signs**, one positive and the other negative, divide their absolute values. The answer is **negative**. | $\frac{-21}{7} = -3$ |
| **Properties of Real Numbers.** For real numbers $a$, $b$, and $c$, | |
| **Commutative Property** $a + b = b + a$ $a \cdot b = b \cdot a$ | $6 + 7 = 7 + 6$ $3 \cdot 16 = 16 \cdot 3$ |
| **Associative Property** $(a + b) + c = a + (b + c)$ $(ab)c = a(bc)$ | $(5 + 4) + 11 = 5 + (4 + 11)$ $(8 \cdot 2) \cdot 15 = (8 \cdot 2) \cdot 15$ |
| **Identity Property** $a + 0 = 0 + a = a$ $a \cdot 1 = 1 \cdot a = a$ | $31 + 0 = 0 + 31 = 31$ $6 \cdot 1 = 1 \cdot 6 = 6$ |
| **Inverse Property** $a + (-a) = (-a) + a = 0$ $a \cdot \frac{1}{a} = \frac{1}{a} \cdot a = 1$ | $18 + (-18) = -18 + 18 = 0$ $14 \cdot \frac{1}{14} = \frac{1}{14} \cdot 14 = 1$ |
| **Distributive Property** $a(b + c) = ab + ac$ | $9(x + 10) = 9 \cdot x + 9 \cdot 10$ |

| IMPORTANT FACTS AND CONCEPTS | EXAMPLES |
|---|---|
| **Section 1.4** | |
| **Factors** are numbers or expressions that are multiplied. | In $3 \cdot 5 = 15$, the 3 and 5 are factors of 15. |
| For any natural number $n$, $b^n$ is an exponential expression such that $$b^n = \underbrace{b \cdot b \cdot b \cdot \cdots \cdot b}_{n\text{ factors}}$$ | $(-2)^5 = (-2)(-2)(-2)(-2)(-2) = -32$ |
| **Square root** of a number $$\sqrt{a} = b \text{ if } b^2 = a$$ | $\sqrt{36} = 6$ since $6^2 = 36$ |
| **Cube root** of a number $$\sqrt[3]{a} = b \quad \text{if} \quad b^3 = a$$ | $\sqrt[3]{64} = 4$ since $4^3 = 64$ |
| ***n*th root** of a number $$\sqrt[n]{a} = b \quad \text{if} \quad b^n = a$$ | $\sqrt[4]{625} = 5$ since $5^4 = 625$ |
| **Order of Operations**<br>To evaluate mathematical expressions, use the following order:<br>**1.** First, evaluate the expressions within grouping symbols, including parentheses, ( ), brackets, [ ], braces, { }, and absolute value, \| \|. If the expression contains nested grouping symbols (one pair of grouping symbols within another pair), evaluate the expression in the innermost grouping symbols first.<br>**2.** Next, evaluate all terms containing exponents and radicals.<br>**3.** Next, evaluate all multiplications or divisions in the order in which they occur, working from left to right.<br>**4.** Finally, evaluate all additions or subtractions in the order in which they occur, working from left to right. | Evaluate $4 + 3 \cdot 9^2 - \sqrt{121}$.<br>$4 + 3 \cdot 9^2 - \sqrt{121} = 4 + 3 \cdot 81 - 11$<br>$= 4 + 243 - 11$<br>$= 247 - 11$<br>$= 236$ |
| **Section 1.5** | |
| **Product Rule for Exponents**<br>If $m$ and $n$ are natural numbers and $a$ is any real number, then $$a^m \cdot a^n = a^{m+n}$$ | $x^8 \cdot x^{15} = x^{8+15} = x^{23}$ |
| **Quotient Rule for Exponents**<br>If $a$ is any nonzero real number and $m$ and $n$ are nonzero integers, then $$\frac{a^m}{a^n} = a^{m-n}$$ | $\frac{z^{21}}{z^{14}} = z^{21-14} = z^7$ |
| **Negative Exponent Rule**<br>For any nonzero real number $a$ and any whole number $m$, $$a^{-m} = \frac{1}{a^m}$$ | $y^{-13} = \frac{1}{y^{13}}$ |
| **Raising a Power to a Power (the Power Rule)**<br>If $a$ is a real number and $m$ and $n$ are integers, then $$(a^m)^n = a^{m \cdot n}$$ | $(c^{-8})^{-5} = c^{(-8)(-5)} = c^{40}$ |
| **Zero Exponent Rule**<br>If $a$ is any nonzero real number, then $$a^0 = 1$$ | $7x^0 = 7 \cdot 1 = 7$ |

| IMPORTANT FACTS AND CONCEPTS | EXAMPLES |
|---|---|
| **Section 1.5 (continued)** | |
| **Raising a Product to a Power**<br>If $a$ and $b$ are real numbers and $m$ is an integer, then<br>$(ab)^m = a^m b^m$ | $(8x^6)^2 = 8^2(x^6)^2 = 64x^{12}$ |
| **Raising a Quotient to a Power**<br>If $a$ and $b$ are real numbers and $m$ is an integer, then<br>$\left(\frac{a}{b}\right)^m = \frac{a^m}{b^m}, \quad b \neq 0$<br>and<br>$\left(\frac{a}{b}\right)^{-m} = \left(\frac{b}{a}\right)^m, \quad a \neq 0, b \neq 0$ | $\left(\frac{2}{r}\right)^3 = \frac{2^3}{r^3} = \frac{8}{r^3}$<br>$\left(\frac{6}{x^3}\right)^{-5} = \left(\frac{x^3}{6}\right)^5 = \frac{(x^3)^5}{6^5} = \frac{x^{15}}{6^5}$ |
| **Section 1.6** | |
| A number written in **scientific notation** has the form $a \times 10^n$ where $1 \leq a < 10$ and $n$ is an integer. | $5.2 \times 10^7, \quad 1.036 \times 10^{-8}$ |
| **To Write a Number in Scientific Notation**<br>**1.** Move the decimal point in the number to the right of the first nonzero digit.<br>**2.** Count the number of places you moved the decimal point in step 1. If the original number is 10 or greater, the count is positive. If the original number is less than 1, the count is negative.<br>**3.** Multiply the number obtained in step 1 by 10 raised to the count (power) found in step 2. | $12{,}900 = 1.29 \times 10^4$<br>$0.035 = 3.5 \times 10^{-2}$ |
| **To Convert a Number in Scientific Notation to Decimal Form**<br>**1.** Observe the exponent on the base 10.<br>**2. a)** If the exponent is positive, move the decimal point in the number to the right the same number of places as the exponent.<br>**b)** If the exponent is negative, move the decimal point in the number to the left the same number of places as the exponent. | $3.08 \times 10^3 = 3080$<br>$8.76 \times 10^{-4} = 0.000876$ |

## Chapter 1 Review Exercises

[1.2] *List each set in roster form.*

**1.** $A = \{x|x$ is a natural number between 3 and 9$\}$

**2.** $B = \{x|x$ is a whole number multiple of 3$\}$

*Let $N$ = set of natural numbers, $W$ = set of whole numbers, $I$ = set of integers, $Q$ = set of rational numbers, $H$ = set of irrational numbers, and $\mathbb{R}$ = set of real numbers. Determine whether the first set is a subset of the second set for each pair of sets.*

**3.** $N, W$

**4.** $Q, \mathbb{R}$

**5.** $Q, H$

**6.** $H, \mathbb{R}$

*Consider the set of numbers* $\left\{-2, 4, 6, \frac{1}{2}, \sqrt{7}, \sqrt{3}, 0, \frac{15}{27}, -\frac{1}{5}, 1.47\right\}$. *List the elements of the set that are*

**7.** natural numbers.

**8.** whole numbers.

**9.** integers.

**10.** rational numbers.

**11.** irrational numbers.

**12.** real numbers.

*Indicate whether each statement is true or false.*

**13.** $\frac{0}{1}$ is not a real number.

**14.** $0, \frac{3}{5}, -2$, and 4 are all rational numbers.

**15.** A real number cannot be divided by 0.

**16.** Every rational number and every irrational number is a real number.

*Find $A \cup B$ and $A \cap B$ for each set A and B.*

**17.** $A = \{1, 2, 3, 4, 5, 6\}, B = \{2, 4, 6, 8, 10\}$

**18.** $A = \{3, 5, 7, 9\}, B = \{2, 4, 6, 8\}$

**19.** $A = \{1, 3, 5, 7, \ldots\}, B = \{2, 4, 6, 8, \ldots\}$

**20.** $A = \{4, 6, 9, 10, 11\}, B = \{3, 5, 9, 10, 12\}$

*Illustrate each set on the number line.*

**21.** $\{x|x > 5\}$

**22.** $\{x|x \leq -2\}$

**23.** $\{x|-1.3 < x \leq 2.4\}$

**24.** $\left\{x \middle| \frac{2}{3} \leq x < 4 \text{ and } x \in N\right\}$

**[1.3]** *Insert either <, >, or = in the shaded area between the two numbers to make each statement true.*

**25.** $-3 \;\square\; 0$

**26.** $-4 \;\square\; -3.9$

**27.** $1.06 \;\square\; 1.6$

**28.** $|-8| \;\square\; 8$

**29.** $|-4| \;\square\; |-10|$

**30.** $13 \;\square\; |-9|$

**31.** $\left|-\frac{2}{3}\right| \;\square\; \frac{3}{5}$

**32.** $-|-2| \;\square\; -6$

*Write the numbers in each list from smallest to largest.*

**33.** $\pi, -\pi, -3, 3$

**34.** $0, \frac{3}{5}, 2.7, |-3|$

**35.** $|-10|, |-5|, 3, -2$

**36.** $|-3|, -7, |-7|, -3$

**37.** $-4, 6, -|-3|, 5$

**38.** $|1.6|, |-2.3|, -2, 0$

*Name each property illustrated.*

**39.** $-7(x + 4) = -7x - 28$

**40.** $rs = sr$

**41.** $(x + 5) + 2 = x + (5 + 2)$

**42.** $q + 0 = 0$

**43.** $5(rs) = (5r)s$

**44.** $-(-6) = 6$

**45.** $9(0) = 0$

**46.** $a + (-a) = 0$

**47.** $x \cdot \frac{1}{x} = 1$

**48.** $k + l = 1 \cdot (k + l)$

**[1.3, 1.4]** *Evaluate.*

**49.** $8 + 3^2 - \sqrt{36} \div 2$

**50.** $-4 \div (-2) + 16 - \sqrt{81}$

**51.** $(7 - 9) - (-3 + 5) + 15$

**52.** $2|-7| - 4|-6| + 5$

**53.** $(6 - 9) \div (9 - 6) + 2$

**54.** $|6 - 3| \div 3 + 4 \cdot 8 - 12$

**55.** $\sqrt{9} + \sqrt[3]{64} + \sqrt[5]{32}$

**56.** $3^2 - 6 \cdot 9 + 4 \div 2^2 - 5$

**57.** $4 - (2 - 9)^0 + 3^2 \div 1 + 3$

**58.** $5^2 + (-2 + 2^2)^3 + 1$

**59.** $-3^2 + 14 \div 2 \cdot 3 - 6$

**60.** $\{[(12 \div 4)^2 - 1]^2 \div 16\}^3$

**61.** $\dfrac{9 + 7 \div (3^2 - 2) + 6 \cdot 8}{\sqrt{81} + \sqrt{1} - 10}$

**62.** $\dfrac{-(5 - 7)^2 - 3(-2) + |-6|}{18 - 9 \div 3 \cdot 5}$

*Evaluate.*

**63.** Evaluate $2x^2 + 3x + 8$ when $x = 2$

**64.** Evaluate $5a^2 - 7b^2$ when $a = -3$ and $b = -4$.

**65. Political Campaigning** The cost of political campaigning has changed dramatically since 1952. The amount spent, in millions of dollars, on all U.S. elections—including local, state, and national offices; political parties; political action committees; and ballot issues—is approximated by

$$\text{dollars spent} = 50.86x^2 - 316.75x + 541.48$$

where $x$ represents each 4-year period since 1948. Substitute 1 for $x$ to get the amount spent in 1952, 2 for $x$ to get the amount spent in 1956, 3 for $x$ to get the amount spent in 1960, and so on.

**a)** Find the amount spent for elections in 1976.

**b)** Find the amount projected to be spent for elections in 2008.

**66. Railroad Traffic** Railroad traffic has been increasing steadily since 1965. Most of this is due to the increase in trains used to transport goods by container. We can approximate the amount of freight carried in ton-miles (1 ton-mile equals 1 ton of freight hauled 1 mile) by

$$\text{freight hauled} = 14.04x^2 + 1.96x + 712.05$$

where $x$ represents each 5-year period since 1960. Substitute 1 for $x$ to get the amount hauled in 1965, 2 for $x$ to get the amount hauled in 1970, 3 for $x$ for 1975, and so forth.

**a)** Find the amount of freight hauled by trains in 1980.

**b)** Find the amount of freight projected to be hauled by trains in 2010.

[1.5] *Simplify each expression and write the answer without negative exponents.*

**67.** $2^3 \cdot 2^2$ **68.** $x^2 \cdot x^3$ **69.** $\frac{a^{12}}{a^4}$ **70.** $\frac{y^{12}}{y^5}$

**71.** $\frac{b^7}{b^{-2}}$ **72.** $c^3 \cdot c^{-6}$ **73.** $5^{-2} \cdot 5^{-1}$ **74.** $8x^0$

**75.** $(-9m^3)^2$ **76.** $\left(\frac{4}{7}\right)^{-1}$ **77.** $\left(\frac{2}{3}\right)^{-3}$ **78.** $\left(\frac{x}{y^2}\right)^{-1}$

**79.** $(5xy^3)(-3x^2y)$ **80.** $(2v^3w^{-4})(7v^{-6}w)$ **81.** $\frac{6x^{-3}y^5}{2x^2y^{-2}}$ **82.** $\frac{12x^{-3}y^{-4}}{4x^{-2}y^5}$

**83.** $\frac{g^3h^{-6}j^{-9}}{g^{-2}h^{-1}j^5}$ **84.** $\frac{21m^{-3}n^{-2}}{7m^{-4}n^2}$ **85.** $\left(\frac{4a^2b}{a}\right)^3$ **86.** $\left(\frac{x^5y}{-3y^2}\right)^2$

**87.** $\left(\frac{p^3q^{-1}}{p^{-4}q^5}\right)^2$ **88.** $\left(\frac{-2ab^{-3}}{c^2}\right)^3$ **89.** $\left(\frac{5xy^3}{z^2}\right)^{-2}$ **90.** $\left(\frac{9m^{-2}n}{3mn}\right)^{-3}$

**91.** $(-2m^2n^{-3})^2$ **92.** $\left(\frac{15x^5y^{-3}z^{-2}}{-3x^4y^{-4}z^3}\right)^4$ **93.** $\left(\frac{2x^{-1}y^5z^4}{3x^4y^{-2}z^{-2}}\right)^{-2}$ **94.** $\left(\frac{8x^{-2}y^{-2}z}{-x^4y^{-4}z^3}\right)^{-1}$

[1.6] *Express each number in scientific notation.*

**95.** 0.0000742 **96.** 460,000 **97.** 183,000 **98.** 0.000001

*Simplify each expression and express the answer without exponents.*

**99.** $(25 \times 10^{-3})(1.2 \times 10^6)$ **100.** $\frac{27 \times 10^3}{9 \times 10^5}$ **101.** $\frac{4{,}000{,}000}{0.02}$ **102.** $(0.004)(500{,}000)$

**103. Online Adverstising** The three companies with the greatest amount spent on online advertising in 2004 are listed below.

| Company | Amount Spent |
|---|---|
| SBC Communications | $\$2.86 \times 10^7$ |
| Netflix | $\$2.69 \times 10^7$ |
| Dell Computers | $\$2.23 \times 10^7$ |

**a)** How much more did SBC Communications spend than Netflix?

**b)** How much more did Netflix spend than Dell Computers?

**c)** How many times larger is the amount SBC Communications spent than Dell Computer?

**104. Voyager** On February 17, 1998, the *Voyager 1* spacecraft became the most distant explorer in the solar system, breaking the record of the *Pioneer 10*. The 28-year-old *Voyager 1* has traveled more than $1.4 \times 10^{10}$ kilometers from earth (about 150 times the distance from earth to the sun).

**a)** Represent $1.4 \times 10^{10}$ as an integer.

**b)** How many billion kilometers has *Voyager 1* traveled?

**c)** Assuming that *Voyager 1* traveled about the same number of kilometers in each of the 28 years, how many kilometers did it average in a year?

**d)** If 1 kilometer $\approx$ 0.6 miles, how far, in miles, did *Voyager 1* travel?

## Chapter 1 Practice Test

*To find out how well you understand the chapter material, take this practice test. The answers, and the section where the material was initially discussed, are given in the back of the book. Each problem is also fully worked out on the* **Chapter Test Prep Video CD.** *Review any questions that you answered incorrectly.*

1. List $A = \{x|x$ is a natural number greater than or equal to $6\}$ in roster form.

*Indicate whether each statement is true or false.*

2. Every real number is a rational number.
3. The union of the set of rational numbers and the set of irrational numbers is the set of real numbers.

*Consider the set of numbers* $\left\{-\frac{3}{5}, 2, -4, 0, \frac{19}{12}, 2.57, \sqrt{8}, \sqrt{2}, -1.92\right\}$. *List the elements of the set that are*

4. rational numbers.
5. real numbers.

*Find* $A \cup B$ *and* $A \cap B$ *for sets A and B.*

6. $A = \{8, 10, 11, 14\}, B = \{5, 7, 8, 9, 10\}$
7. $A = \{1, 3, 5, 7, \ldots\}, B = \{3, 5, 7, 9, 11\}$

*In Exercises 8 and 9, illustrate each set on the number line.*

8. $\{x|-2.3 \le x < 5.2\}$
9. $\left\{x \middle| -\frac{5}{2} < x < \frac{6}{5} \text{ and } x \in I\right\}$
10. List from smallest to largest: $|3|, -|4|, -2, 9$.

*Name each property illustrated.*

11. $(x + y) + 8 = x + (y + 8)$
12. $3x + 4y = 4y + 3x$

*Evaluate each expression.*

13. $\{6 - [7 - 3^2 \div (3^2 - 2 \cdot 3)]\}$
14. $2^4 + 4^2 \div 2^3 \cdot \sqrt{25} + 7$
15. $\dfrac{-3|4 - 8| \div 2 + 6}{-\sqrt{36} + 18 \div 3^2 + 4}$
16. $\dfrac{-6^2 + 3(4 - |6|) \div 6}{4 - (-3) + 12 \div 4 \cdot 5}$
17. Evaluate $-x^2 + 2xy + y^2$ when $x = 2$ and $y = 3$.
18. **Cannonball** To celebrate July 4, a cannon is fired upwards from a fort overlooking the ocean below. The heigh, $h$, in feet, of the cannonball above sea level at any time $t$, in seconds, can be determined by the formula $h = -16t^2 + 120t + 200$. Find the height of the cannonball above sea level **a)** 1 second after the cannon is fired, **b)** 5 seconds after the cannon is fired.

*Simplify each expression and write the answer without negative exponents.*

19. $3^{-2}$
20. $\left(\dfrac{4m^{-3}}{n^2}\right)^2$
21. $\dfrac{24a^2b^{-3}c^0}{30a^3b^2c^{-2}}$
22. $\left(\dfrac{-3x^3y^{-2}}{x^{-1}y^5}\right)^{-3}$
23. Convert 389,000,000 to scientific notation.
24. Simplify $\dfrac{3.12 \times 10^6}{1.2 \times 10^{-2}}$ and write the number without exponents.
25. **World Population**
   - **a)** In 2050, the world population is expected to be about 9.2 billion people. Write this number in scientific notation.
   - **b)** The graph below shows the expected distribution of the 2050 world population for the three age groups 0–14, 15–64, and 65 and older. Use scientific notation to determine the number of people in each of these age groups in 2050.

**Expected Distribution of World Population by Age**

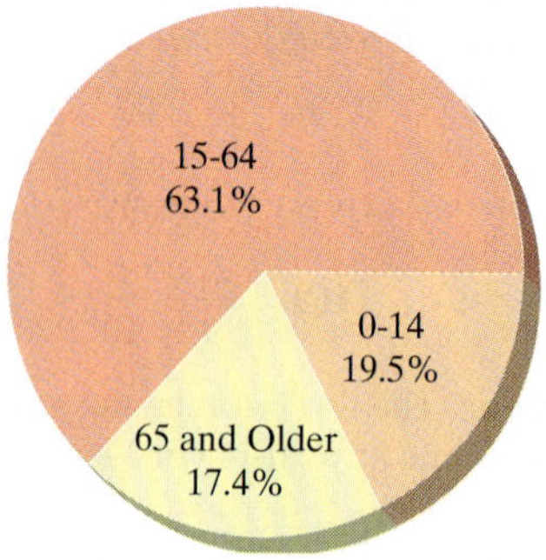

# 2 Equations and Inequalities

## GOALS FOR THIS CHAPTER

In this chapter, we focus on solving linear equations and inequalities, and on using linear equations, formulas, and inequalities to solve real-life problems. We will also introduce a powerful problem-solving technique that we will use through this textbook. We will witness the power of algebra as a problem-solving tool in a multitude of areas, including real estate, chemistry, business, banking, physics, and personal finance.


2.1 Solving Linear Equations

2.2 Problem Solving and Using Formulas

2.3 Applications of Algebra

Mid-Chapter Test: Sections 2.1–2.3

2.4 Additional Application Problems

2.5 Solving Linear Inequalities

2.6 Solving Equations and Inequalities Containing Absolute Values

Chapter 2 Summary

Chapter 2 Review Exercises

Chapter 2 Practice Test

Cumulative Review Test

FOR MANY PEOPLE, SELECTING a long-distance telephone plan is an important decision to make. Some telephone companies have plans with a monthly fee plus a low rate for every minute a long-distance call is made. Other plans do not have a monthly fee but may charge a higher rate for every minute a long-distance call is made. Which plan do you choose?

In Example 4 on page 90 we will investigate two such plans offered by the BellSouth Telephone Company.

# 2.1 Solving Linear Equations

1. Identify the reflexive, symmetric, and transitive properties.
2. Combine like terms.
3. Solve linear equations.
4. Solve equations containing fractions.
5. Identify conditional equations, contradictions, and identities.
6. Understand the concepts to solve equations.

## 1 Identify the Reflexive, Symmetric, and Transitive Properties

In elementary algebra you learned how to solve linear equations. We review these procedures briefly in this section. Before we do so, we need to introduce three useful properties of equality: the *reflexive property*, the *symmetric property*, and the *transitive property*.

**Properties of Equality**

For all real numbers $a$, $b$, and $c$:

1. $a = a$. — *reflexive property*
2. If $a = b$, then $b = a$. — *symmetric property*
3. If $a = b$ and $b = c$, then $a = c$. — *transitive property*

**Examples of the Reflexive Property**

$$7 = 7$$
$$x + 5 = x + 5$$

**Examples of the Symmetric Property**

If $x = 3$, then $3 = x$.

If $y = x + 9$, then $x + 9 = y$.

**Examples of the Transitive Property**

If $x = a$ and $a = 4y$, then $x = 4y$.

If $a + b = c$ and $c = 4d$, then $a + b = 4d$.

In this book, we will often use these properties without referring to them by name.

## 2 Combine Like Terms

When an algebraic expression consists of several parts, the parts that are added are called the **terms** of the expression. The expression $3x^2 - 6x - 2$, which can be written $3x^2 + (-6x) + (-2)$, has three terms; $3x^2$, $-6x$, and $-2$. The expression

$$6x^2 - 3(x + y) - 4 + \frac{x + 2}{8}$$

has four terms: $6x^2$, $-3(x + y)$, $-4$, and $\frac{x + 2}{8}$.

| Expression | Terms |
|---|---|
| $\frac{1}{2}x^2 - 3x - 7$ | $\frac{1}{2}x^2,\quad -3x,\quad -7$ |
| $-5x^3 + 3x^2y - 2$ | $-5x^3,\quad 3x^2y,\quad -2$ |
| $4(x + 3) + 2x + \frac{1}{5}(x - 2) + 1$ | $4(x + 3),\quad 2x,\quad \frac{1}{5}(x - 2),\quad 1$ |

The numerical part of a term that precedes the variable is called its **numerical coefficient** or simply its **coefficient**. In the term $6x^2$, the 6 is the numerical coefficient.

When the numerical coefficient is 1 or $-1$, we generally do not write the numeral 1. For example, $x$ means $1x$, $-x^2y$ means $-1x^2y$, and $(x + y)$ means $1(x + y)$.

| Term | Numerical Coefficient |
|---|---|
| $\frac{5k}{9}$ | $\frac{5}{9}$ |
| $-4(x + 2)$ | $-4$ |
| $\frac{x - 2}{7}$ | $\frac{1}{7}$ |
| $-(x + y)$ | $-1$ |

*Note that* $\frac{x - 2}{7}$ *means* $\frac{1}{7}(x - 2)$ *and* $-(x + y)$ *means* $-1(x + y)$.

When a term consists of only a number, that number is called a **constant**. For example, in the expression $x^2 - 4$, the $-4$ is a constant.

The **degree of a term** with whole number exponents is the sum of the exponents on the variables in the term. For example, $3x^2$ is a second-degree term, and $-4x$ is a first-degree term ($-4x$ means $-4x^1$). The number 3 can be written as $3x^0$, so the number 3 (and every other nonzero constant) has degree 0. The term 0 is said to have no degree. The term $4xy^5$ is a sixth-degree term since the sum of the exponents is $1 + 5$ or 6. The term $6x^3y^5$ is an eighth-degree term since $3 + 5 = 8$.

**Like terms** are terms that have the same variables with the same exponents. For example, $3x$ and $5x$ are like terms, $2x^2$ and $-3x^2$ are like terms, and $3x^2y$ and $-2x^2y$ are like terms. Terms that are not like terms are said to be **unlike terms**. All constants are considered like terms.

To **simplify an expression** means to combine all like terms in the expression. To combine like terms, we can use the distributive property.

### Examples of Combining Like Terms

$$8x - 2x = (8 - 2)x = 6x$$

$$3x^2 - 5x^2 = (3 - 5)x^2 = -2x^2$$

$$-7x^2y + 3x^2y = (-7 + 3)x^2y = -4x^2y$$

$$4(x - y) - (x - y) = 4(x - y) - 1(x - y) = (4 - 1)(x - y) = 3(x - y)$$

When simplifying expressions, we can rearrange the terms by using the commutative and associative properties discussed in Chapter 1.

**EXAMPLE 1** ▸ Simplify. If an expression cannot be simplified, so state.

**a)** $-2x + 5 + 3x - 7$ **b)** $7x^2 - 2x^2 + 3x + 4$ **c)** $2x - 3y + 5x - 6y + 3$

### *Solution*

**a)** $-2x + 5 + 3x - 7 = \underbrace{-2x + 3x}_{x} + \underbrace{5 - 7}_{-2}$ *Place like terms together.*

This expression simplifies to $x - 2$.

**b)** $7x^2 - 2x^2 + 3x + 4 = 5x^2 + 3x + 4$

**c)** $2x - 3y + 5x - 6y + 3 = 2x + 5x - 3y - 6y + 3$ *Place like terms together.*

$= 7x - 9y + 3$

▸ **Now Try Exercise 39**

**EXAMPLE 2** ▸ Simplify $-2(a + 7) - [-3(a - 1) + 8]$.

*Solution*

$$\begin{aligned} -2(a + 7) - [-3(a - 1) + 8] &= -2(a + 7) - 1[-3(a - 1) + 8] \\ &= -2a - 14 - 1[-3a + 3 + 8] && \textit{Distributive property} \\ &= -2a - 14 - 1[-3a + 11] && \textit{Combine like terms.} \\ &= -2a - 14 + 3a - 11 && \textit{Distributive property} \\ &= a - 25 && \textit{Combine like terms.} \end{aligned}$$

▸ **Now Try Exercise 55**

## 3 Solve Linear Equations

An **equation** is a mathematical statement of equality. *An equation must contain an equal sign* and a mathematical expression on each side of the equal sign.

**Examples of Equations**

$$x + 8 = -7$$
$$2x^2 - 4 = -3x + 13$$

The numbers that make an equation a true statement are called the **solutions** of the equation. The **solution set** of an equation is the set of real numbers that make the equation true.

| Equation | Solution | Solution Set |
|---|---|---|
| $2x + 3 = 9$ | 3 | $\{3\}$ |

Two or more equations with the same solution set are called **equivalent equations**. Equations are generally solved by starting with the given equation and producing a series of simpler equivalent equations.

**Example of Equivalent Equations**

| Equations | Solution Set |
|---|---|
| $2x + 3 = 9$ | $\{3\}$ |
| $2x = 6$ | $\{3\}$ |
| $x = 3$ | $\{3\}$ |

In this section, we will discuss how to solve **linear equations in one variable**. A linear equation is an equation that can be written in the form $ax + b = c, a \neq 0$.

*To solve equations, we use the addition and multiplication properties of equality to isolate the variable on one side of the equal sign.*

**Addition Property of Equality**

If $a = b$, then $a + c = b + c$ for any $a$, $b$, and $c$.

The addition property of equality states that the same number can be added to both sides of an equation without changing the solution to the original equation. Since subtraction is defined in terms of addition, *the addition property of equality also allows us to subtract the same number from both sides of an equation.*

**Multiplication Property of Equality**

If $a = b$, then $a \cdot c = b \cdot c$ for any $a$, $b$, and $c$.

The multiplication property of equality states that both sides of an equation can be multiplied by the same number without changing the solution. Since division is defined in terms of multiplication, *the multiplication property of equality also allows us to divide both sides of an equation by the same nonzero number.*

To solve an equation, we will often have to use a combination of properties to isolate the variable. Our goal is to get the variable all by itself on one side of the equation (to isolate the variable). A general procedure to solve linear equations follows.

**To Solve Linear Equations**

1. **Clear fractions.** If the equation contains fractions, eliminate the fractions by multiplying both sides of the equation by the least common denominator.
2. **Simplify each side separately.** Simplify each side of the equation as much as possible. Use the distributive property to clear parentheses and combine like terms as needed.
3. **Isolate the variable term on one side.** Use the addition property to get all terms with the variable on one side of the equation and all constant terms on the other side. It may be necessary to use the addition property a number of times to accomplish this.
4. **Solve for the variable.** Use the multiplication property to get an equation having just the variable (with a coefficient of 1) on one side.
5. **Check.** Check by substituting the value obtained in step 4 back into the original equation.

**EXAMPLE 3** ▸ Solve the equation $2x + 9 = 14$.

*Solution*

$$2x + 9 = 14$$

$$2x + 9 - 9 = 14 - 9 \qquad \text{Subtract 9 from both sides.}$$

$$2x = 5$$

$$\frac{\overset{1}{\cancel{2}}x}{\underset{1}{\cancel{2}}} = \frac{5}{2} \qquad \text{Divide both sides by 2.}$$

$$x = \frac{5}{2}$$

**Check**

$$2x + 9 = 14$$

$$\cancel{2}\left(\frac{5}{\cancel{2}}\right) + 9 \stackrel{?}{=} 14$$

$$5 + 9 \stackrel{?}{=} 14$$

$$14 = 14 \qquad \text{True}$$

Since the value checks, the solution is $\frac{5}{2}$.

▸ **Now Try Exercise 61**

Whenever an equation contains like terms on the same side of the equal sign, combine the like terms before using the addition or multiplication properties.

**EXAMPLE 4** ▸ Solve the equation $-2b + 8 = 3b - 7$.

*Solution*

$$-2b + 8 = 3b - 7$$

$$-2b + 2b + 8 = 3b + 2b - 7 \qquad \text{Add } 2b \text{ to both sides.}$$

$$8 = 5b - 7$$

$$8 + 7 = 5b - 7 + 7 \qquad \text{Add 7 to both sides.}$$

$$15 = 5b$$

$$\frac{15}{5} = \frac{5b}{5} \qquad \text{Divide both sides by 5.}$$

$$3 = b$$

▸ **Now Try Exercise 63**

Example 5 contains decimal numbers. We work this problem following the procedure given earlier.

**EXAMPLE 5** ▸ Solve the equation $4(x - 3.1) = 2.1(x - 4) + 3.5x$.

*Solution*

$$\begin{aligned}
4(x - 3.1) &= 2.1(x - 4) + 3.5x \\
4(x) - 4(3.1) &= 2.1(x) - 2.1(4) + 3.5x && \text{Distributive property} \\
4x - 12.4 &= 2.1x - 8.4 + 3.5x \\
4x - 12.4 &= 5.6x - 8.4 && \text{Combine like terms.} \\
4x - 12.4 + 8.4 &= 5.6x - 8.4 + 8.4 && \text{Add 8.4 to both sides.} \\
4x - 4.0 &= 5.6x \\
4x - 4x - 4.0 &= 5.6x - 4x && \text{Subtract } 4x \text{ from both sides.} \\
-4.0 &= 1.6x \\
\frac{-4.0}{1.6} &= \frac{1.6x}{1.6} && \text{Divide both sides by 1.6.} \\
-2.5 &= x
\end{aligned}$$

The solution is $-2.5$.

▸ **Now Try Exercise 111**

To save space, we will not always show the check of our answers. You should, however, check all your answers. When the equation contains decimal numbers, using a calculator to solve and check the equation may save you some time.

**USING YOUR CALCULATOR** **Checking Solutions by Substitution**

Solutions to equations may be checked using a calculator. To check, substitute your solution into both sides of the equation to see whether you get the same value (there may sometimes be a slight difference in the last digits). The graphing calculator screen in **Figure 2.1** shows that both sides of the equation given in Example 5 equal $-22.4$ when $-2.5$ is substituted for $x$. Thus the solution $-2.5$ checks.

$$4(x - 3.1) = 2.1(x - 4) + 3.5x$$
$$4(-2.5 - 3.1) = 2.1(-2.5 - 4) + 3.5(-2.5)$$

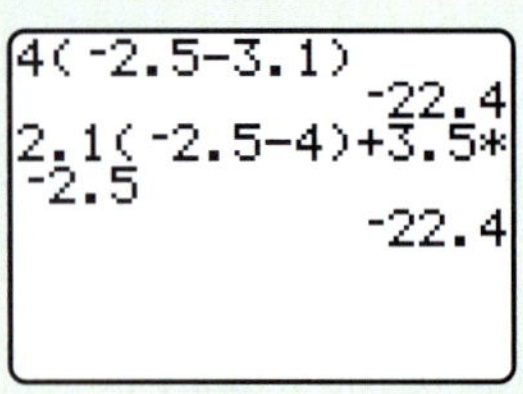


← Value of the left side of the equation

← Value of the right side of the equation

FIGURE 2.1

**EXERCISES**

Use your calculator to determine whether the given number is a solution to the equation.

**1.** $5.2(x - 3.1) = 2.3(x - 5.2)$; 1.4

**2.** $-2.3(4 - x) = 3.5(x - 6.1)$; 10.125

Now we will work an example that contains nested parentheses.

**EXAMPLE 6** ▸ Solve the equation $7c - 15 = -2[6(c - 3) - 4(2 - c)]$.

*Solution*

$$\begin{aligned}
7c - 15 &= -2[6(c - 3) - 4(2 - c)] \\
7c - 15 &= -2[6c - 18 - 8 + 4c] && \text{Distributive property} \\
7c - 15 &= -2[10c - 26] && \text{Combine like terms.} \\
7c - 15 &= -20c + 52 && \text{Distributive property} \\
7c + 20c - 15 &= -20c + 20c + 52 && \text{Add } 20c \text{ to both sides.} \\
27c - 15 &= 52
\end{aligned}$$

$$27c - 15 + 15 = 52 + 15 \quad \text{Add 15 to both sides.}$$
$$27c = 67$$
$$\frac{27c}{27} = \frac{67}{27} \quad \text{Divide both sides by 27.}$$
$$c = \frac{67}{27}$$

▸ Now Try Exercise 91

Notice that the solutions to Examples 5 and 6 are not integers. You should not expect the solutions to equations always to be integer values.

In solving equations, some of the intermediate steps can often be omitted. Now we will illustrate how this may be done.

| Solution | | Shortened Solution |
|---|---|---|
| **a)** $x + 4 = 6$ | | **a)** $x + 4 = 6$ |
| $x + 4 - 4 = 6 - 4$ | ⟵ *Do this step mentally.* | $x = 2$ |
| $x = 2$ | | |
| **b)** $3x = 6$ | | **b)** $3x = 6$ |
| $\frac{3x}{3} = \frac{6}{3}$ | ⟵ *Do this step mentally.* | $x = 2$ |
| $x = 2$ | | |

## 4 Solve Equations Containing Fractions

When an equation contains fractions, we begin by multiplying *both* sides of the equation by the least common denominator. The **least common denominator** (LCD) of a set of denominators (also called the **least common multiple**, LCM) is the smallest number that each of the denominators divides into without remainder. For example, if the denominators of two fractions are 5 and 6, then 30 is the least common denominator since 30 is the smallest number that both 5 and 6 divide into without remainder.

When you multiply both sides of the equation by the LCD, *each term* in the equation will be multiplied by the least common denominator. *After this step is performed, the equation should not contain any fractions.*

**EXAMPLE 7** ▸ Solve the equation $5 - \frac{2a}{3} = -9$.

*Solution* The least common denominator is 3. Multiply both sides of the equation by 3 and then use the distributive property on the left side of the equation. *This process will eliminate all fractions from the equation.*

$$5 - \frac{2a}{3} = -9$$
$$3\left(5 - \frac{2a}{3}\right) = 3(-9) \quad \text{Multiply both sides by 3.}$$
$$3(5) - \overset{1}{\cancel{3}}\left(\frac{2a}{\underset{1}{\cancel{3}}}\right) = -27 \quad \text{Distributive property}$$
$$15 - 2a = -27$$
$$15 - 15 - 2a = -27 - 15 \quad \text{Subtract 15 from both sides.}$$
$$-2a = -42$$
$$\frac{-2a}{-2} = \frac{-42}{-2} \quad \text{Divide both sides by } -2.$$
$$a = 21$$

▸ Now Try Exercise 97

**EXAMPLE 8** ▸ Solve the equation $\frac{1}{2}(x + 4) = \frac{1}{3}x$.

*Solution* Begin by multiplying both sides of the equation by 6, the LCD of 2 and 3.

$$6\left[\frac{1}{2}(x+4)\right] = 6\left(\frac{1}{3}x\right) \qquad \textit{Multiply both sides by 6.}$$
$$3(x+4) = 2x \qquad \textit{Simplify.}$$
$$3x + 12 = 2x \qquad \textit{Distributive property}$$
$$3x - 2x + 12 = 2x - 2x \qquad \textit{Subtract 2x from both sides.}$$
$$x + 12 = 0$$
$$x + 12 - 12 = 0 - 12 \qquad \textit{Subtract 12 from both sides.}$$
$$x = -12$$

▸ **Now Try Exercise 99**

We will be discussing equations containing fractions further in Section 6.4.

### Helpful Hint

The equation in Example 8 may also be written as $\frac{x+4}{2} = \frac{x}{3}$. Can you explain why?

### USING YOUR GRAPHING CALCULATOR

Equations in one variable may be solved graphically using a graphing calculator. In Section 3.3 we discuss how this is done. You may wish to review that material now.

## 5 Identify Conditional Equations, Contradictions, and Identities

All equations discussed so far have been true for only one value of the variable. Such equations are called **conditional equations**. Some equations are never true and have no solution; these are called **contradictions**. Other equations, called **identities** have an infinite number of solutions. **Table 2.1** summarizes these types of linear equations and their corresponding number of solutions.

**TABLE 2.1**

| Type of linear equation | Solution |
|---|---|
| Conditional equation | One |
| Contradiction | None (solution set: ∅) |
| Identity | Infinite number (solution set: ℝ) |

The solution set of a conditional equation contains the solution given in set braces. For example, the solution set to Example 8 is $\{-12\}$. The solution set of a contradiction is the empty set or null set, { } or ∅. The solution set of an identity is the set of real numbers, ℝ.

**EXAMPLE 9** ▸ Determine whether the equation $5(d - 7) + 4d + 3 = 3(3d - 10) - 2$ is a conditional equation, a contradiction, or an identity. Give the solution set for the equation.

*Solution*

$$5(d-7) + 4d + 3 = 3(3d-10) - 2$$
$$5d - 35 + 4d + 3 = 9d - 30 - 2 \qquad \textit{Distributive property}$$
$$9d - 32 = 9d - 32 \qquad \textit{Combine like terms.}$$

Since we obtain the same expression on both sides of the equation, it is an identity. This equation is true for all real numbers. Its solution set is $\mathbb{R}$.

▶ Now Try Exercise 125

**EXAMPLE 10** ▶ Determine whether $2(3m + 1) = 6m + 3$ is a conditional equation, a contradiction, or an identity. Give the solution set for the equation.

*Solution*

$$
\begin{aligned}
2(3m + 1) &= 6m + 3 \\
6m + 2 &= 6m + 3 && \text{Distributive property} \\
6m - 6m + 2 &= 6m - 6m + 3 && \text{Subtract } 6m \text{ from both sides.} \\
2 &= 3
\end{aligned}
$$

Since $2 = 3$ is never a true statement, this equation is a contradiction. Its solution set is $\varnothing$.

▶ Now Try Exercise 119

## 6 Understand the Concepts to Solve Equations

The numbers or variables that appear in equations do not affect the procedures used to solve the equations. In the following example, which does not contain any numbers or letters, we will solve the equation using the concepts and procedures that have been presented.

**EXAMPLE 11** ▶ Assume in the following equation that $\odot$ represents the variable that we are solving for and that all the other symbols represent nonzero real numbers. Solve the equation for $\odot$.

$$\square\odot + \triangle = \#$$

*Solution* To solve for $\odot$ we need to isolate the $\odot$. We use the addition and multiplication properties to solve for $\odot$.

$$
\begin{aligned}
\square\odot + \triangle &= \# \\
\square\odot + \triangle - \triangle &= \# - \triangle && \text{Subtract } \triangle \text{ from both sides.} \\
\square\odot &= \# - \triangle \\
\frac{\square\odot}{\square} &= \frac{\# - \triangle}{\square} && \text{Divide both sides by } \square. \\
\odot &= \frac{\# - \triangle}{\square}
\end{aligned}
$$

Thus the solution is $\odot = \dfrac{\# - \triangle}{\square}$.

▶ Now Try Exercise 143

Consider the equation $5x + 7 = 12$. If we let $5 = \square$, $x = \odot$, $7 = \triangle$, and $12 = \#$, the equation has the same form as the equation in Example 11. Therefore, the solution will be of the same form.

| Equation | Solution |
|---|---|
| $\square\odot + \triangle = \#$ | $\odot = \dfrac{\# - \triangle}{\square}$ |
| $5x + 7 = 12$ | $x = \dfrac{12 - 7}{5} = \dfrac{5}{5} = 1$ |

If you solve the equation $5x + 7 = 12$, you will see that its solution is 1. Thus the procedure used to solve an equation is not dependent on the numbers or variables given in the equation.

# EXERCISE SET 2.1

MathXL® MyMathLab

## Concept/Writing Exercises

1. What are the terms of an expression?
2. Determine the coefficient of each term.

   a) $x^2y^5$ b) $-a^3b^7$ c) $-\frac{m-7n}{5}$.
3. Determine the coefficient of each term.

   a) $\frac{x+y}{4}$ b) $-(p+3)$ c) $-\frac{3(x+2)}{5}$
4. How do you find the degree of a term?
5. a) What are like terms?

   b) Are $3x$ and $3x^2$ like terms? Explain.
6. What is an equation?
7. Is 4 the solution to the equation $2x + 3 = x + 5$? Explain.
8. Is 8 the solution to the equation $x + 1 = 2x - 7$? Explain.
9. State the addition property of equality.
10. State the multiplication property of equality.
11. a) How many solutions does an identity have?

    b) If a linear equation is an identity, what is its solution set?
12. a) What is a contradiction?

    b) What is the solution set of a contradiction?
13. a) Explain step-by-step how you would solve the equation
    $$5x - 2(x - 4) = 2(x - 2)$$
    b) Solve this equation.
14. a) Explain step-by-step how you would solve the equation
    $$\frac{1}{6} = \frac{2}{3}n - \frac{1}{8}$$
    b) Solve this equation.

## Practice the Skills

*Name each indicated property.*

15. If $x = 13$, then $13 = x$.
16. If $m + 2 = 3$, then $3 = m + 2$.
17. If $b = c$ and $c = 9$, then $b = 9$.
18. If $x + 1 = a$ and $a = 2y$, then $x + 1 = 2y$.
19. $a + c = a + c$
20. If $r = 4$, then $r + 3 = 4 + 3$.
21. If $x = 8$, then $x - 8 = 8 - 8$.
22. If $2x = 4$, then $3(2x) = 3(4)$.
23. If $5x = 4$, then $\frac{1}{5}(5x) = \frac{1}{5}(4)$.
24. If $a + 2 = 4$, then $a + 2 - 2 = 4 - 2$.
25. If $\frac{t}{4} + \frac{1}{3} = \frac{5}{6}$, then $12\left(\frac{t}{4} + \frac{1}{3}\right) = 12\left(\frac{5}{6}\right)$.
26. If $x - 3 = x + y$ and $x + y = z$, then $x - 3 = z$.

*Give the degree of each term.*

27. $5c^3$
28. $-6y^2$
29. $3ab$
30. $\frac{1}{2}x^4y$
31. $6$
32. $-3$
33. $-5r$
34. $18p^2q^3$
35. $5a^2b^4c$
36. $m^4n^6$
37. $3x^5y^6z$
38. $-2x^4y^7z^8$

*Simplify each expression. If an expression cannot be simplified, so state.*

39. $7r + 3b - 11x + 12y$
40. $3x^2 + 4x + 5$
41. $5x^2 - 11x + 10x - 5$
42. $11a - 12b - 4c + 5a$
43. $10.6c^2 - 2.3c + 5.9c - 1.9c^2$
44. $7y + 3x - 7 + 5x - 2y$
45. $w^3 + w^2 - w + 1$
46. $b + b^2 - 4b + b^2 + 3b$
47. $8pq - 9pq + p + q$
48. $7x^3y^2 + 11y^3x^2$
49. $12\left(\frac{1}{6} + \frac{d}{4}\right) + 5d$
50. $4.3 - 3.2x - 2(x - 2)$
51. $3\left(x + \frac{1}{2}\right) - \frac{1}{3}x + 5$
52. $6n + 0.6(n - 3) - 5(n + 0.7)$
53. $4 - [6(3x + 2) - x] + 4$
54. $3(a + c) - 4(a + c) - 3$
55. $9x - [3x - (5x - 4y)] - 2y$
56. $-2[3x - (2y - 1) - 5x] + y$
57. $5b - \{7[2(3b - 2) - (4b + 9)] - 2\}$
58. $2\{[3a - (2b - 5a)] - 3(2a - b)\}$
59. $-\{[2rs - 3(r + 2s)] - 2(2r^2 - s)\}$
60. $p^2q + 4pq - [-(pq + 4p^2q) + pq]$

*Solve each equation.*

61. $5a - 1 = 14$
62. $5x + 3 - 2x = 9$
63. $5x - 9 = 3(x - 2)$
64. $5s - 3 = 2s + 6$
65. $4x - 8 = -4(2x - 3) + 4$
66. $8w + 7 = -3w - 15$

**67.** $-6(z-1)=-5(z+2)$
**68.** $7(x-1)=3(x+2)$
**69.** $-3(t-5)=2(t-5)$
**70.** $4(2x-4)=-2(x+3)$
**71.** $3x+4(2-x)=4x+5$
**72.** $6(3-q)=-4(q+1)$
**73.** $2-(x+5)=4x-8$
**74.** $4x-2(3x-7)=2x-6$
**75.** $p-(p+4)=4(p-1)+2p$
**76.** $8x+2(x-4)=8x+12$
**77.** $-3(y-1)+2y=4(y-3)$
**78.** $5r-13-6r=3(r+5)-16$
**79.** $6-(n+3)=3n+5-2n$
**80.** $8-3(2a-4)=5+3a-4a$
**81.** $4(2x-2)-3(x+7)=-4$
**82.** $-2(3w+6)-(4w-3)=21$
**83.** $-4(3-4x)-2(x-1)=12x$
**84.** $-4(2z-6)=-3(z-4)+z$
**85.** $5(a+3)-a=-(4a-6)+1$
**86.** $3(2x-4)+3(x+1)=9$
**87.** $5(x-2)-14x=x-5$
**88.** $3[6-(h+2)]-6=4(-h+7)$
**89.** $2[3x-(4x-6)]=5(x-6)$
**90.** $-z-6z+3=4-[6-z-(3-2z)]$
**91.** $4\{2-[3(c+1)-2(c+1)]\}=-2c$
**92.** $3\{[(x-2)+4x]-(x-3)\}=4-(x-12)$
**93.** $-\{4(d+3)-5[3d-2(2d+7)]-8\}=-10d-6$
**94.** $-3(6-4x)=4-\{5x-[6x-(4x-(3x+2))]\}$

*Solve each equation. Leave your answer as a fraction if it is not an integer value.*

**95.** $\frac{s}{4}=-16$
**96.** $\frac{15c+3}{9}=2$
**97.** $\frac{4x-2}{3}=-6$
**98.** $\frac{1}{2}(6r-10)=7$
**99.** $\frac{3}{4}t+\frac{7}{8}t=39$
**100.** $\frac{1}{4}(x-2)=\frac{1}{3}(2x+6)$
**101.** $\frac{1}{2}(x-2)=\frac{1}{3}(x+2)$
**102.** $\frac{1}{2}x+2=\frac{1}{8}x-1$
**103.** $4-\frac{3}{4}a=7$
**104.** $x-2=\frac{3}{4}(x+4)$
**105.** $\frac{1}{2}=\frac{4}{5}x-\frac{1}{4}$
**106.** $\frac{1}{3}x+\frac{5}{6}=2x$
**107.** $\frac{1}{4}(x+3)=\frac{1}{3}(x-2)+1$
**108.** $\frac{5}{6}m-\frac{5}{12}=\frac{7}{8}m+\frac{2}{3}$

*Solve each equation. Round answers to the nearest hundredth.*

**109.** $0.4n+4.7=5.1n$
**110.** $0.2(x-30)=1.6x$
**111.** $4.7x-3.6(x-1)=4.9$
**112.** $6.1p-4.5(3-2p)=15.7$
**113.** $5(z+3.41)=-7.89(2z-4)-5.67$
**114.** $0.05(2000+2x)=0.04(2500-6x)$
**115.** $0.6(500-2.4x)=3.6(2x-4000)$
**116.** $0.42x-x=5.1(x+3)$
**117.** $1000(7.34q+14.78)=100(3.91-4.21q)$
**118.** $0.6(14x-8000)=-0.4(20x+12{,}000)+20.6x$

*Find the solution set for each exercise. Then indicate whether the equation is conditional, an identity, or a contradiction.*

**119.** $3(y+3)-4(2y-7)=-5y+2$
**120.** $9x+12-8x=-6(x-2)+7x$
**121.** $4(2x-3)+15=-6(x-4)+12x-21$
**122.** $-5(c+3)+4(c-2)=2(c+2)$
**123.** $4-\left(\frac{2}{3}x+2\right)=2\left(-\frac{1}{3}x+1\right)$
**124.** $7-\left(\frac{1}{2}x+4\right)=3\left(-\frac{1}{6}x+2\right)$
**125.** $6(x-1)=-3(2-x)+3x$
**126.** $0.6(z+5)-0.5(z+2)=0.1(z-23)$
**127.** $0.8z-0.3(z+10)=0.5(z+1)$
**128.** $4(2-3x)=-[6x-(8-6x)]$

## Problem Solving

**129. Population Density** The population density of the United States has been steadily increasing since 2000. The population density of the United States can be estimated using the equation

$$P=0.82t+78.5$$

where $P$ is the population density, measured in people per square mile, and $t$ is the number of years since 2000. Use $t=1$ for 2001, $t=2$ for 2002, and so on. If the population density continues to increase at its current rate,

**a)** determine the population density of the United States in 2008.

**b)** during what year will the population density of the United States reach 100 people per square mile?

**130. Sleeping Babies** Dr. Richard Ferber, a pediatric sleep expert, has developed a method* to help children, 6 months of age or older, sleep through the night. Often called "Ferberizing," it calls for parents to wait for increasing lengths of time before entering the child's room at night to comfort the crying child. The suggested waiting time depends on how many nights the parents have been using the method and may be found using the equation

$$W = 5n + 5$$

where $W$ is the waiting time in minutes and $n$ is the number of the night. For example, on the first night, $n = 1$, on the second night, $n = 2$, and so on.

**a)** How long should parents wait on the first night?

**b)** How long should parents wait on the fourth night?

**c)** On what night should parents wait 30 minutes?

**d)** On what night should parents wait 40 minutes?

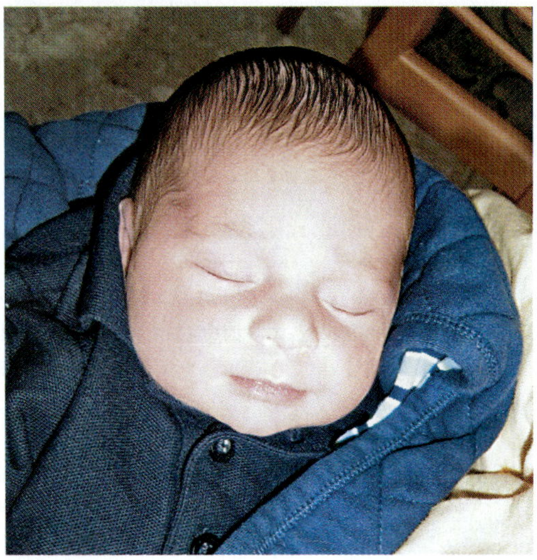

**131. American Automakers' Market Share** In recent years American automakers have been losing market share to Asian and European automakers. The percent of the total cars sold in the United States made by American automakers for the years 2004 through 2006 can be estimated using the equation

$$M = -1.26x + 61.48$$

where $M$ is the percent of the total cars sold in the United States made by American automakers and $x$ is the number of years since 2004. Use $x = 1$ for 2005, $x = 2$ for 2006, and so on.

**a)** What is the percent of the total cars sold in the United States made by American automakers in 2006?

**b)** If this trend continues, during what year will the percent of total cars sold in the United States made by American automakers be 53.92%?

**132. Annuities** Annuities are life insurance contracts that guarantee future payments. One type of annuity, called a variable annuity, is a retirement account that lets someone invest in mutual funds and defer payment of taxes until withdrawals are taken at a later time. The number of variable annuities sold has been growing steadily. Variable annuity sales can be approximated by the equation

$$S = 10x + 150$$

where $S$ represents total sales of variable annuities (in billions of dollars) and $x$ is the number of years since 2004. Use $x = 1$ for 2005, $x = 2$ for 2006, and so on.

**a)** Find the total sales of variable annuities in 2005.

**b)** In what year will annuity sales reach the 200-billion-dollar mark?

**133. Boston Marathon** Since 1940, the male winners of the Boston Marathon have generally been decreasing their time for completing the race. The time, in hours, to finish the race can be approximated using the equation

$$t = 2.405 - 0.005x$$

where $t$ is the finishing time and $x$ is the number of years since 1940. Use $x = 1$ for 1941, $x = 2$ for 1942, and so on.

**a)** Estimate the winning time of the Boston Marathon in 1941.

**b)** Estimate the winning time of the Boston Marathon in 2005.

**134.** Consider the equation $x = 4$. Give three equivalent equations. Explain why the equations are equivalent.

**135.** Consider the equation $2x = 5$. Give three equivalent equations. Explain why the equations are equivalent.

**136.** Make up an equation that is an identity. Explain how you created the equation.

**137.** Make up an equation that is a contradiction. Explain how you created the contradiction.

**138.** Create an equation with three terms to the left of the equal sign and two terms to the right of the equal sign that is equivalent to the equation $3m + 1 = m + 5$.

**139.** Create an equation with two terms to the left of the equal sign and three terms to the right of the equal sign that is equivalent to the equation $\frac{1}{2}p + 3 = 6$.

*Before trying this method, parents should first consult with their pediatrician.

140. Consider the equation $-3(x + 2) + 5x + 12 = n$. What real number must $n$ be for the solution of the equation to be 6? Explain how you determined your answer.

141. Consider the equation $2(a + 5) + n = 4a - 8$. What real number must $n$ be for the solution of the equation to be $-2$? Explain how you determined your answer.

142. Consider the equation $\frac{n}{6} + \frac{x}{4} = 2$. What real number must $n$ be for the solution of the equation to be $x = 2$? Explain how you determined your answer.

*Solve each equation for the given symbol. Assume that the symbol you are solving for represents the variable and that all other symbols represent nonzero real numbers. See Example 11.*

143. Solve $*\triangle - \square = \odot$ for $\triangle$.

144. Solve $\triangle(\odot + \square) = \otimes$ for $\triangle$.

145. Solve $\odot\square + \triangle = \otimes$ for $\odot$.

146. Solve $\triangle(\odot + \square) = \otimes$ for $\square$.

## Cumulative Review Exercises

[1.3] 147. a) Explain how to find the absolute value of a number.

b) Write the definition of absolute value.

[1.4] *Evaluate.*

148. a) $-3^2$ b) $(-3)^2$

149. $\sqrt[3]{-125}$

150. $\left(-\frac{2}{7}\right)^2$

# 2.2 Problem Solving and Using Formulas

1 Use the problem-solving procedure.

2 Solve for a variable in an equation or formula.

George Polya

## 1 Use the Problem-Solving Procedure

One of the main reasons for studying mathematics is that we can use it to solve everyday problems. To solve most real-life application problems mathematically, we need to be able to express the problem in mathematical symbols using expressions or equations, and when we do this, we are creating a **mathematical model** of the situation.

In this section, we present a problem-solving procedure and discuss formulas. A **formula** is an equation that is a mathematical model of a real-life situation. Throughout the book we will be problem solving. When we do so, we will determine an equation or formula that represents or models the real-life situation.

We will now give the general five-step problem-solving procedure developed by George Polya and presented in his book *How to Solve It*. You can approach any problem by following this general procedure.

### Guidelines for Problem Solving

1. **Understand the problem.**
   - Read the problem **carefully** at least twice. In the first reading, get a general overview of the problem. In the second reading, determine a) exactly what you are being asked to find and b) what information the problem provides.
   - If possible, make a sketch to illustrate the problem. Label the information given.
   - List the information in a table if it will help in solving the problem.
2. **Translate the problem into mathematical language.**
   - This will generally involve expressing the problem algebraically.
   - Sometimes this involves selecting a particular formula to use, whereas other times it is a matter of generating your own equation. It may be necessary to check other sources for the appropriate formula to use.
3. **Carry out the mathematical calculations necessary to solve the problem.**
4. **Check the answer obtained in step 3.**
   - Ask yourself: "Does the answer make sense?" "Is the answer reasonable?" If the answer is not reasonable, recheck your method for solving the problem and your calculations.
   - Check the solution in the original problem if possible.
5. **Answer the question.** Make sure you have answered the question asked. State the answer clearly.

The following examples show how to apply the guidelines for problem solving. We will sometimes provide the steps in the examples to illustrate the five-step process. However, in some problems it may not be possible or necessary to list every step.

As was stated in step 2 of the problem-solving process—*translate the problem into mathematical language*—we will sometimes need to find and use a *formula*. We will show how to do that in this section. In Section 2.3 we will explain how to develop *equations* to solve real-life application problems.

**EXAMPLE 1 ▸ A Personal Loan** Diane Basile makes a \$5000, 4% simple interest personal loan to her brother, Bob Basile, for a period of 5 years.

**a)** At the end of 5 years, what interest will Bob pay Diane?

**b)** When Bob settles his loan at the end of 5 years, how much money, in total, must he pay Diane?

*Solution* **a)** Understand When a person borrows money using a simple interest loan, the person must repay the simple interest and principal (the original amount borrowed) at the maturity date of the loan. For example, if a simple interest loan is for 5 years, then 5 years after the loan is made, the principal plus the interest must be repaid. We are told in the problem that the simple interest rate is 4% and the loan is for 5 years.

Translate Many business mathematics and investment books include the **simple interest formula**:

$$\text{interest} = \text{principal} \cdot \text{rate} \cdot \text{time} \quad \text{or} \quad i = prt$$

which can be used to find the simple interest, $i$. In the formula, $p$ is the principal, $r$ is the simple interest rate (always changed to decimal form when used in the formula), and $t$ is time. The time and rate must be in the same units. For example, if the rate is 4% per *year*, then the time must be in *years*. In this problem, $p = \$5000$, $r = 0.04$, and $t = 5$. We obtain the simple interest, $i$, by substituting these values in the simple interest formula.

Carry Out

$$\begin{aligned} i &= prt \\ &= 5000(0.04)(5) \\ &= 1000 \end{aligned}$$

Check The answer appears reasonable in that Bob will pay \$1000 for the use of \$5000 of Diane's money for 5 years.

Answer **a)** The simple interest owed is \$1000.
**b)** At the end of 5 years, Bob must pay the principal he borrowed, \$5000, plus the interest determined in part **a)**, \$1000. Thus, when Bob settles his loan, he must pay Diane \$6000.

▸ Now Try Exercise 67

**EXAMPLE 2 ▸ A Money Market Account** Christine Fogel receives a tax refund of \$1425 and invests this money to help pay for her son's first semester of college. She invests this money in a certificate of deposit at a 3% annual interest rate compounded monthly for 18 months.

**a)** How much money will the certificate of deposit be worth in 18 months?

**b)** How much interest will she earn during the 18 months?

*Solution* **a)** Understand Before you can understand the problem, you must understand what compound interest is. Compound interest means that you get interest on your investment for one period. Then in the next period you get interest paid on your investment, plus interest paid on the interest that was paid in the first period. This process then continues for each period. In many real-life situations, and in the workplace, you may need to do some research to answer questions asked.

We are given that \$1425 is invested for 18 months and the interest rate is 3% compounded monthly.

**Translate** If you look in a business mathematics book or speak to a person involved with finance, you will learn of the compound interest formula:

$$A = p\left(1 + \frac{r}{n}\right)^{nt}$$

The compound interest formula is used by financial institutions to compute the accumulated amount (or the balance), $A$, in a savings account or other investment that earns compound interest. In the formula, $p$ represents the principal (or the initial investment), $r$ represents the interest rate written in decimal form, $n$ represents the number of times per year the interest is compounded, and $t$ represents time measured in years. In this problem, $p = \$1425$, $r = 0.03$, $t = 1.5$ (18 months is 1.5 years), and since the interest is compounded monthly, $n = 12$. Substitute these values into the formula and evaluate.

$$A = p\left(1 + \frac{r}{n}\right)^{nt}$$

$$= 1425\left(1 + \frac{0.03}{12}\right)^{12(1.5)}$$

**Carry Out**

$$= 1425(1 + 0.0025)^{18}$$

$$= 1425(1.0025)^{18}$$

$$\approx 1425(1.04596912) \quad \text{From a calculator}$$

$$\approx 1490.51 \quad \text{Rounded to the nearest cent}$$

**Check** The answer \$1490.51 is reasonable, since it is more than Christine originally invested.

**Answer** Christine's certificate of deposit will be worth \$1490.51 at the end of 18 months.

**b) Understand** The interest will be the difference between the original amount invested and the value of the certificate of deposit at the end of 18 months.

**Translate** $\text{interest} = \left(\begin{array}{c}\text{value of the certificate}\\ \text{of deposit after 18 months}\end{array}\right) - \left(\begin{array}{c}\text{amount originally}\\ \text{invested}\end{array}\right)$

**Carry Out** $= 1490.51 - 1425 = 65.51$

**Check** The amount of interest is reasonable and the arithmetic is easily checked.

**Answer** The interest gained in the 18-month period will be \$65.51.

▸ **Now Try Exercise 77**

Often a formula contains subscripts. **Subscripts** are numbers (or other variables) placed below and to the right of variables. They are used to help clarify a formula. For example, if a formula contains two velocities, the original velocity and the final velocity, these velocities might be symbolized as $V_0$ and $V_f$, respectively. Subscripts are read using the word "sub." For example, $V_f$ is read "$V$ sub $f$" and $x_2$ is read "$x$ sub 2." The formula in Example 3 contains subscripts.

**EXAMPLE 3** ▸ **Comparing Investments** Sharon Griggs is in the 25% federal income tax bracket. She is trying to decide whether to invest in tax-free municipal bonds with a rate of 2.24% or in a taxable certificate of deposit with a rate of 3.70%.

**a)** Determine the taxable rate equivalent to a 2.24% tax-free rate for Sharon.

**b)** If both investments were for the same period of time, which investment would provide Sharon with the greater return on her investment?

*Solution* **a)** Understand Some interest we receive, such as from municipal bonds, is tax free. This means that we do not have to pay federal taxes on the interest received. Other interest we receive, such as from savings accounts or certificates of deposit, is taxable on our federal income tax returns. Paying taxes on the interest has the effect of reducing the amount of money we actually get to keep from the interest. We need to find the taxable interest rate that is equivalent to a 2.24% tax-free rate for Sharon (or for anyone in a 25% federal income tax bracket).

Translate A formula found in many investment books and some government publications that may be used to compare taxable and tax-free interest rates is

$$T_f = T_a(1 - F)$$

where $T_f$ is the tax-free rate, $T_a$ is the taxable rate, and $F$ is the federal income tax bracket. To determine the equivalent taxable rate, $T_a$, we substitute the appropriate values in the formula and solve for $T_a$.

$$T_f = T_a(1 - F)$$

$$0.0224 = T_a(1 - 0.25)$$

Carry Out

$$0.0224 = T_a(0.75)$$

$$\frac{0.0224}{0.75} = T_a$$

$$0.0299 \approx T_a \quad \text{Rounded to four places}$$

Check The answer, 0.0299 or 2.99%, appears reasonable because it is larger than 2.24%, which is what we expected.

Answer A taxable investment yielding about 2.99% would give Sharon about the same interest as a 2.24% tax-free investment.

We are asked to determine which investment will provide Sharon with the greater return on her investment. We can compare the equivalent taxable rate of the municipal bonds to the taxable interest rate of the certificate of deposit. The rate that is higher will provide Sharon with the greater return on her investment.

As we saw in part **a)**, the equivalent taxable rate of the municipal bonds is 2.99%. The taxable rate of the certificate of deposit is 3.70%. Therefore the certificate of deposit paying 3.70% will give Sharon a greater return on her investment than the tax-free municipal bond paying 2.24%.

**▸ Now Try Exercise 83**

## 2 Solve for a Variable in an Equation or Formula

There are many occasions when you might be given an equation or formula that is solved for one variable but you want to solve it for a different variable. Suppose in Example 3 we wanted to find the equivalent taxable rate, $T_a$, for many different tax-free rates and many different income tax brackets. We could solve each individual problem as was done in Example 3. However, it would be much quicker to solve the formula $T_f = T_a(1 - F)$ for $T_a$ and then substitute the appropriate values in the formula solved for $T_a$. We will do this in Example 8.

We will begin by solving equations for the variable $y$. We will need to do this in Chapter 3 when we discuss graphing. Since formulas are equations, the same procedure we use to solve for a variable in an equation will be used to solve for a variable in a formula.

When you are given an equation (or formula) that is solved for one variable and you want to solve it for a different variable, treat each variable in the equation, except the one you are solving for, as if it were a constant. Then *isolate the variable* you are solving for using the procedures similar to those used to solve equations.

**EXAMPLE 4** ▸ Solve the equation $5x - 8y = 32$ for $y$.

*Solution* We will solve for the variable $y$ by isolating the term containing the $y$ on the left side of the equation.

$$5x - 8y = 32$$

$$5x - 5x - 8y = -5x + 32 \qquad \text{Subtract } 5x \text{ from both sides.}$$

$$-8y = -5x + 32$$

$$\frac{-8y}{-8} = \frac{-5x + 32}{-8} \qquad \text{Divide both sides by } -8.$$

$$y = \frac{-5x + 32}{-8}$$

$$y = \frac{-1(5x - 32)}{-1(-8)} \qquad \text{Multiply the numerator and the denominator by } -1.$$

$$y = \frac{5x - 32}{8} \quad \text{or} \quad y = \frac{5}{8}x - 4$$

▸ **Now Try Exercise 29**

**EXAMPLE 5** ▸ Solve the equation $2y - 3 = \frac{1}{2}(x + 3y)$ for $y$.

*Solution* Since this equation contains a fraction, we begin by multiplying both sides of the equation by the least common denominator, 2. We then isolate the variable $y$ by collecting all terms containing the variable $y$ on one side of the equation and all other terms on the other side of the equation.

$$2y - 3 = \frac{1}{2}(x + 3y)$$

$$2(2y - 3) = 2\left[\frac{1}{2}(x + 3y)\right] \qquad \text{Multiply both sides by the LCD, 2.}$$

$$4y - 6 = x + 3y \qquad \text{Distributive property}$$

$$4y - 3y - 6 = x + 3y - 3y \qquad \text{Subtract } 3y \text{ from both sides.}$$

$$y - 6 = x$$

$$y - 6 + 6 = x + 6 \qquad \text{Add 6 to both sides.}$$

$$y = x + 6$$

▸ **Now Try Exercise 35**

Now let's solve for a variable in a formula. Remember: Our goal is to isolate the variable for which we are solving. We use the same general procedure we used in Examples 4 and 5.

**EXAMPLE 6** ▸ The formula for the perimeter of a rectangle is $P = 2l + 2w$, where $l$ is the length and $w$ is the width of the rectangle (see **Fig. 2.2**). Solve this formula for the width, $w$.

*Solution* Since we are solving for $w$, we must isolate the $w$ on one side of the equation.

$$P = 2l + 2w$$

$$P - 2l = 2l - 2l + 2w \qquad \text{Subtract } 2l \text{ from both sides.}$$

$$P - 2l = 2w$$

$$\frac{P - 2l}{2} = \frac{2w}{2} \qquad \text{Divide both sides by 2.}$$

$$\frac{P - 2l}{2} = w$$

Thus, $w = \dfrac{P - 2l}{2}$ or $w = \dfrac{P}{2} - \dfrac{2l}{2} = \dfrac{P}{2} - l$.

▸ **Now Try Exercise 49**

Rectangle

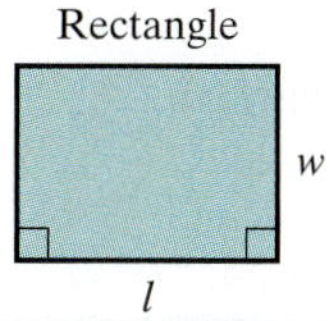


FIGURE 2.2

**EXAMPLE 7** ▸ A formula used to find the area of a trapezoid is $A = \frac{1}{2}h(b_1 + b_2)$, where $h$ is the height and $b_1$ and $b_2$ are the lengths of the bases of the trapezoid (see **Fig. 2.3**). Solve this formula for $b_2$.

*Solution* We begin by multiplying both sides of the equation by the LCD, 2, to clear fractions.

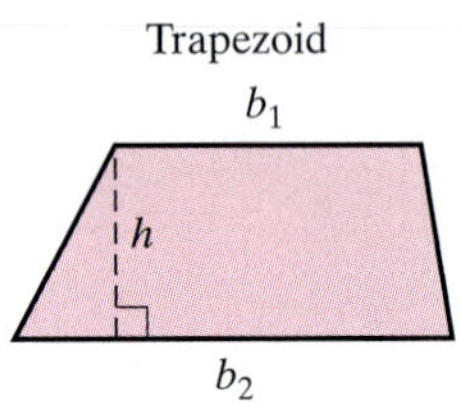


FIGURE 2.3

$$A = \frac{1}{2}h(b_1 + b_2)$$

$$2 \cdot A = 2\left[\frac{1}{2}h(b_1 + b_2)\right] \qquad \text{Multiply both sides by 2.}$$

$$2A = h(b_1 + b_2)$$

$$\frac{2A}{h} = \frac{h(b_1 + b_2)}{h} \qquad \text{Divide both sides by } h.$$

$$\frac{2A}{h} = b_1 + b_2$$

$$\frac{2A}{h} - b_1 = b_1 - b_1 + b_2 \qquad \text{Subtract } b_1 \text{ from both sides.}$$

$$\frac{2A}{h} - b_1 = b_2$$

▸ **Now Try Exercise 57**

**EXAMPLE 8** ▸ In Example 3 we introduced the formula $T_f = T_a(1 - F)$.

**a)** Solve this formula for $T_a$.

**b)** John and Dorothy Cutter are in the 33% income tax bracket. What is the equivalent taxable yield of a 2.6% tax-free yield?

*Solution*

**a)** We wish to solve this formula for $T_a$. Therefore we treat all other variables in the equation as if they were constants. Since $T_a$ is multiplied by $(1 - F)$, to isolate $T_a$ we divide both sides of the equation by $1 - F$.

$$T_f = T_a(1 - F)$$

$$\frac{T_f}{1 - F} = \frac{T_a(1 - F)}{1 - F} \qquad \text{Divide both sides by } 1 - F.$$

$$\frac{T_f}{1 - F} = T_a \quad \text{or} \quad T_a = \frac{T_f}{1 - F}$$

**b)** Substitute the appropriate values into the formula found in part **a)**.

$$T_a = \frac{T_f}{1 - F}$$

$$T_a = \frac{0.026}{1 - 0.33} = \frac{0.026}{0.67} \approx 0.039$$

Thus, the equivalent taxable yield would be about 3.9%.

▸ **Now Try Exercise 63**

# EXERCISE SET 2.2

MathXL® MyMathLab

## Concept/Writing Exercises

1. What is a formula?
2. What is a mathematical model?
3. Outline the five-step problem-solving process that we will use to work out problems.
4. When we are solving a formula for a variable, we need to isolate the variable. Explain what this means.
5. Consider the equation $16 = 2l + 2(3)$ and the formula $P = 2l + 2w$.
   **a)** Solve the equation for $l$.
   **b)** Solve the formula for $l$.
   **c)** Was the procedure used to solve the formula for $l$ any different from the procedure used to solve the equation for $l$?
   **d)** In the formula solved for $l$ in part **b)** substitute 16 for $P$ and 3 for $w$, and then find the value of $l$. How does it compare with your answer in part **a)**? Explain why this is so.
6. **a)** What are subscripts?
   **b)** How is $x_0$ read?
   **c)** How is $v_f$ read?

## Practice the Skills

*Evaluate the following formulas for the values given. Use the* $\pi$ *key on your calculator for* $\pi$ *when needed. Round answers to the nearest hundredth.*

7. $E = IR$ when $I = 63, R = 100$ (a formula known as *Ohm's Law* used when studying electricity)
8. $C = 2\pi r$ when $r = 12$ (a formula for finding the circumference of a circle)
9. $R = R_1 + R_2$ when $R_1 = 100, R_2 = 200$ (a formula used when studying electricity)
10. $A = \frac{1}{2}bh$ when $b = 7, h = 6$ (a formula for finding the area of a triangle)
11. $A = \pi r^2$ when $r = 8$ (a formula for finding the area of a circle)
12. $P_1 = \frac{T_1P_2}{T_2}$ when $T_1 = 150, T_2 = 300, P_2 = 200$ (a chemistry formula that relates temperature and pressure of gases)
13. $\bar{x} = \frac{x_1 + x_2 + x_3}{3}$ when $x_1 = 40, x_2 = 90, x_3 = 80$ (a formula for finding the average of three numbers)
14. $A = \frac{1}{2}h(b_1 + b_2)$ when $h = 15, b_1 = 20, b_2 = 28$ (a formula for finding the area of a trapezoid)
15. $A = P + Prt$ when $P = 160, r = 0.05, t = 2$ (a banking formula that yields the total amount in an account after the interest is added)
16. $E = a_1p_1 + a_2p_2$ when $a_1 = 10, p_1 = 0.2, a_2 = 100, p_2 = 0.3$ (a statistics formula for finding the expected value of an event)
17. $m = \frac{y_2 - y_1}{x_2 - x_1}$ when $y_2 = 4, y_1 = -3, x_2 = -2, x_1 = -6$ (a formula for finding the slope of a straight line; we will discuss this formula in Chapter 3)
18. $F = G\frac{m_1m_2}{r^2}$ when $G = 0.5, m_1 = 100, m_2 = 200, r = 4$ (a physics formula that gives the force of attraction between two masses that are separated by a distance, $r$)
19. $R_T = \frac{R_1R_2}{R_1 + R_2}$ when $R_1 = 100, R_2 = 200$ (a formula in electronics for finding the total resistance in a parallel circuit containing two resistors)
20. $d = \sqrt{(x_2 - x_1)^2 + (y_2 - y_1)^2}$ when $x_2 = 5, x_1 = -3, y_2 = -6, y_1 = 3$ (a formula for finding the distance between two points on a straight line; we will discuss this formula in Chapter 10)
21. $x = \frac{-b + \sqrt{b^2 - 4ac}}{2a}$ when $a = 2, b = -5, c = -12$ (from the quadratic formula, we will discuss the quadratic formula in Chapter 8)
22. $x = \frac{-b - \sqrt{b^2 - 4ac}}{2a}$ when $a = 2, b = -5, c = -12$ (from the quadratic formula)
23. $A = p\left(1 + \frac{r}{n}\right)^{nt}$ when $p = 100, r = 0.06, n = 1, t = 3$ (the compound interest formula; see Example 2)
24. $z = \frac{\bar{x} - \mu}{\frac{\sigma}{\sqrt{n}}}$ when $\bar{x} = 78, \mu = 66, \sigma = 15, n = 25$ (a statistics formula for finding the standard, or $z$ score, of a sample mean, $\bar{x}$)

*Solve each equation for y (see Examples 4 and 5).*

**25.** $3x + y = 5$

**26.** $8x + 3y = 9$

**27.** $x - 7y = 13$

**28.** $-6x + 5y = 25$

**29.** $6x - 2y = 16$

**30.** $9x = 7y + 23$

**31.** $\frac{3}{4}x - y = 5$

**32.** $\frac{x}{4} - \frac{y}{6} = 2$

**33.** $3(x - 2) + 3y = 6x$

**34.** $y - 4 = \frac{2}{3}(x + 6)$

**35.** $y + 1 = -\frac{4}{3}(x - 9)$

**36.** $\frac{1}{5}(x + 3y) = \frac{4}{7}(2x - 1)$

*Solve each equation for the indicated variable (see Examples 6–8).*

**37.** $d = rt$, for $t$

**38.** $i = prt$, for $t$

**39.** $C = \pi d$, for $d$

**40.** $A = lw$, for $l$

**41.** $P = 2l + 2w$, for $l$

**42.** $P = 2l + 2w$, for $w$

**43.** $V = lwh$, for $h$

**44.** $V = \pi r^2 h$, for $h$

**45.** $A = P + Prt$, for $r$

**46.** $Ax + By = C$, for $y$

**47.** $V = \frac{1}{3}lwh$, for $l$

**48.** $A = \frac{1}{2}bh$, for $b$

**49.** $y = mx + b$, for $m$

**50.** $IR + Ir = E$, for $R$

**51.** $y - y_1 = m(x - x_1)$, for $m$

**52.** $z = \frac{x - \mu}{\sigma}$, for $\sigma$

**53.** $z = \frac{x - \mu}{\sigma}$, for $\mu$

**54.** $y = \frac{kx}{z}$, for $z$

**55.** $P_1 = \frac{T_1 P_2}{T_2}$, for $T_2$

**56.** $F = \frac{mv^2}{r}$, for $m$

**57.** $A = \frac{1}{2}h(b_1 + b_2)$, for $h$

**58.** $D = \frac{x_1 + x_2 + x_3}{n}$, for $n$

**59.** $S = \frac{n}{2}(f + l)$ for $n$

**60.** $S = \frac{n}{2}(f + l)$, for $l$

**61.** $C = \frac{5}{9}(F - 32)$, for $F$

**62.** $F = \frac{9}{5}C + 32$, for $C$

**63.** $F = \frac{km_1 m_2}{d^2}$, for $m_1$

**64.** $F = \frac{km_1 m_2}{d^2}$ for $m_2$

## Problem Solving

*In Exercises 65–88, round your answer to two decimal places, when appropriate.*

**65. Currency Exchange**

**a)** According to the Universal Converter Internet Web site, on February 5, 2005, \$1 of U.S. currency could be exchanged for 9.11 Mexican pesos. Write a formula using $d$ for U.S. dollars and $p$ for Mexican pesos that can be used to convert from dollars to pesos.

**b)** Write a formula that can be used to convert from Mexican pesos to U.S. dollars.

**c)** Explain how you determined your answers to parts **a)** and **b)**.

**66. Speed of the Titanic** Ships at sea measure their speed in knots. For example, when the *Titanic* struck the iceberg, its speed was about 20.5 knots. One knot is 1 nautical mile per hour. A nautical mile is about 6076 feet. When measuring speed in miles per hour, a mile is 5280 ft.

**a)** Determine a formula for converting a speed in knots ($k$) to a speed in miles per hour ($m$).

**b)** Explain how you determined this formula.

**c)** Determine the speed, in miles per hour, at which the *Titanic* struck the iceberg.

*In Exercises 67–70, use the simple interest formula $i = prt$. See Example 1.*

**67. A Personal Loan** Edison Tan loaned his colleague, Ken Pothoven, \$1100 at a simple interest rate of 7% per year for 4 years. Determine the simple interest Ken must pay Edison when he repays the loan at the end of 4 years.

**68. Determine the Rate** Steve Marino borrowed \$500 from his credit union for 2 years. The simple interest that he paid was \$52.90. What simple interest rate was Jeff charged?

**69. Determine the Length of the Loan** Mary Haran loaned her daughter, Dawn, \$20,000 at a simple interest rate of 3.75% per year. At the end of the loan period, Dawn repaid Mary the original \$20,000 plus \$4875 interest. Determine the length of the loan.

**70. A Certificate of Deposit** Erin Grabish received \$2000 for speaking at a financial planning seminar. Erin invested the money in a certificate of deposit for 2 years. When she redeemed the certificate of deposit, she received \$2166. What simple interest rate did Erin receive on this certificate of deposit?

*In Exercises 71–76, if you are not sure of the formula to use, refer to Appendix A.*

**71. Dartboard Area** Marc Mazzoni, dart-throwing champion in the state of Michigan, practices on a dartboard with concentric circles as shown in the figure.

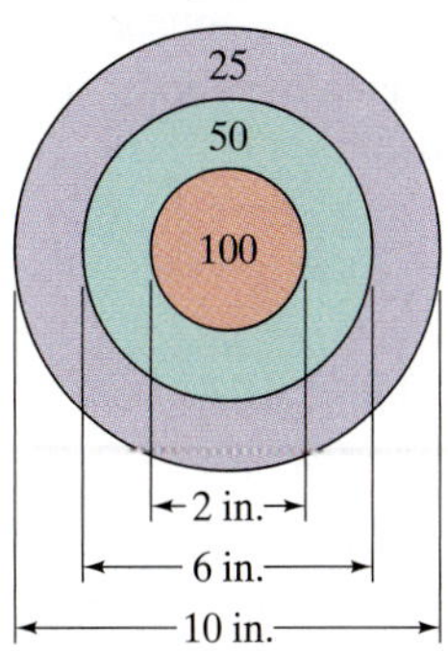


**a)** Find the area of the circle marked 100.

**b)** Find the area of the entire dartboard.

**72. Planning a Sandbox** Betsy Nixon is planning to build a rectangular sandbox for her daughter. She has 38 feet of wood to use for the sides. If the length is to be 11 feet, what is the width to be?

**73. Concrete Driveway Volume** Anthony Palmiotto, is laying concrete for a driveway. The driveway is to be 15 feet long by 10 feet wide by 6 inches deep.

**a)** Find the volume of concrete needed in cubic feet.

**b)** If 1 cubic yard = 27 cubic feet, how many cubic yards of concrete are needed?

**c)** If the concrete costs \$35 per cubic yard, what is the cost of the concrete? Concrete must be purchased in whole cubic yards.

**74. Helipad Area** A helipad in Raleigh, North Carolina, has two concentric circles as shown in the figure.

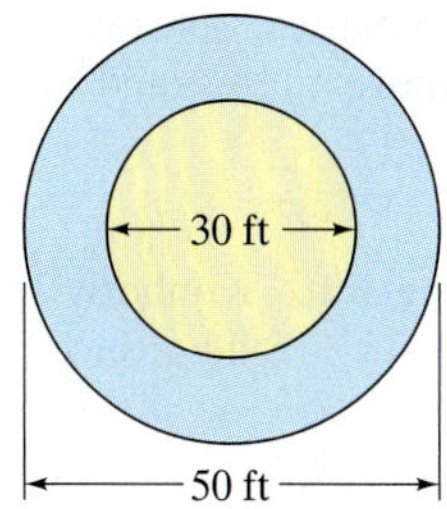


Find the area of the blue region in the figure.

**75. Ice Cream Containers** Gil and Lori's Delicious Ice Cream Company sells ice cream in two containers, a cylindrical tub and a rectangular solid box as shown in the figure. Which container holds more ice cream, and what is the difference in their volumes?

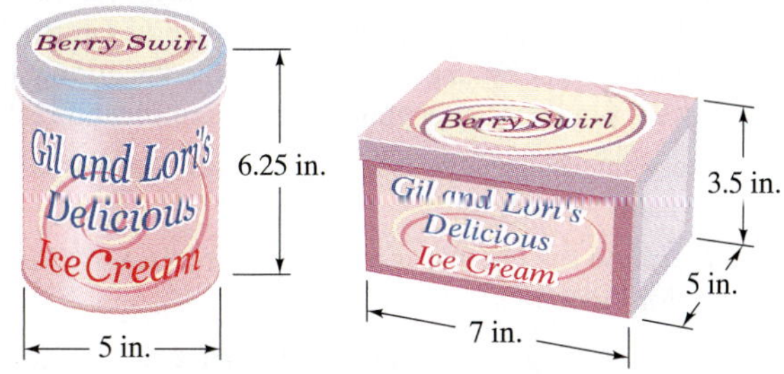


**76. Bucket Capacity** Sandra Hakanson has a bucket in which she wishes to mix some detergent. The dimensions of the bucket are shown in the figure.

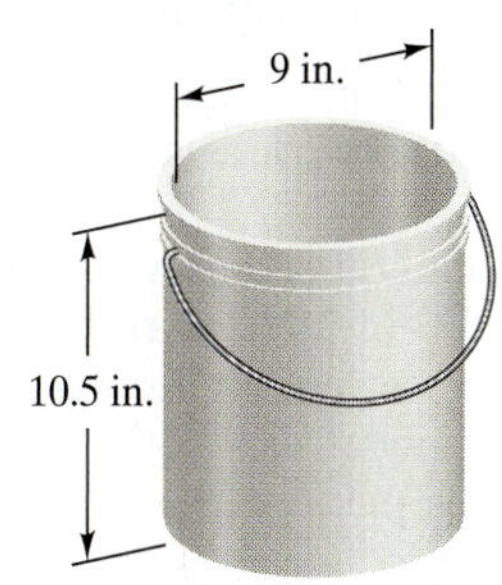


**a)** Find the capacity of the bucket in cubic inches.

**b)** If 231 cubic inches = 1 gallon, what is the capacity of the bucket in gallons?

**c)** If the instructions on the bottle of detergent say to add 1 ounce per gallon of water, how much detergent should Sandra add to a full bucket of water?

*For Exercises 77–80, refer to Example 2.*

77. **A Savings Account** Beth Rechsteiner invested \$10,000 in a savings account paying 6% interest compounded quarterly. How much money will she have in her account at the end of 2 years?

78. **Monthly Compounding** Vigay Patel invested \$8500 in a savings account paying 3.2% interest compounded monthly. How much money will he have in his account at the end of 4 years?

79. **A Certificate of Deposit** Heather Kazakoff invests \$4390 in a certificate of deposit paying 4.1% interest compounded semiannually. How much will the certificate of deposit be worth after 36 months?

80. **Comparing Accounts** James Misenti has \$1500 to invest for 1 year. He has a choice of a credit union account that pays 4.5% annual simple interest and a bank account that pays 4% interest compounded quarterly. Determine which account would pay more interest and by how much.

*For Exercises 81–84, refer to Example 3.*

81. **Equivalent Taxable Rate** Kimberly Morse-Austin is a student and is in the 15% federal income tax bracket. She is considering investing \$825 in a tax-free bond mutual fund paying 3.5% simple interest. Determine the taxable rate equivalent to a 3.5% tax-free rate.

82. **Comparing Investments** Dave Ostrow is in the 35% federal income tax bracket and is considering two investments: a tax-free municipal bond paying 3% simple interest or a taxable certificate of deposit paying 4.5% simple interest. Which investment yields the greatest return?

83. **Father and Son Investing** Anthony Rodriquez is in the 35% federal income tax bracket and his son Angelo is in the 28% federal income tax bracket. They are each considering a tax-free mutual fund currently yielding 4.6% simple interest.
   a) Determine the taxable rate equivalent to a 4.6% tax-free rate for Anthony.
   b) Determine the taxable rate equivalent to a 4.6% tax-free rate for Angelo.

84. **Investment Comparison** Marissa Felberty is considering investing \$9200 in a 6.75% taxable account or in a 5.5% tax-free account. If she is in the 25% tax bracket, which investment will yield the greater return?

*Exercises 85–88 contain a variety of situations. Solve each exercise.*

85. **Weight Loss** A nutritionist explains to Robin Thomas that a person loses weight by burning more calories than he or she eats. For example, Robin, a 5′6″ woman who weighs 132 pounds, will stay about the same weight with normal exercise if she has a daily diet of 2400 calories. If she burns more than 2400 calories daily, her weight loss can be approximated by the mathematical model $w = 0.02c$, where $w$ is the *weekly* weight loss and $c$ is the number of calories per *day* burned *above* 2400 calories.
   a) Find Robin's weekly weight loss if she exercises and burns off 2600 calories per day.
   b) How many calories would Robin have to burn off in a day to lose 2 pounds in a week?

86. **Stress Test** A person given a stress test is generally told that should the heart rate reach a certain point, the test will be stopped. The maximum allowable heart rate depends on the person's age. The maximum allowable heart rate, $m$, in beats per minute, can be approximated by the equation $m = -0.875x + 190$, where $x$ represents the patient's age from 1 through 99. Using this mathematical model, find
   a) the maximum heart rate for a 50-year-old.
   b) the age of a person whose maximum heart rate is 160 beats per minute.

87. **Balancing a Portfolio** Some financial planners recommend the following rule of thumb for investors. The percent of stocks in your total portfolio should be equal to 100 minus your age. The remainder is to be put into bonds or cash.
   a) Construct mathematical models for the percent to be kept in stocks (use $S$ for percent in stock and $a$ for a person's age).
   b) Using this rule of thumb, find the percent to be kept in stocks for a 60-year-old.

88. **Body Mass Index** The body mass index is a standard way of evaluating a person's body weight in relation to their height. To determine your body mass index (BMI) using metric measurements, divide your weight, in kilograms, by your height, in meters squared. An accurate shortcut for calculating BMI using pounds and inches, is to multiply your weight in pounds by 705, then divide by the square of your height, in inches.
   a) Create a formula for finding a person's BMI using kilograms and meters.
   b) Create a formula for finding a person's BMI when the weight is given in pounds and the height is given in inches.
   c) Determine your BMI.

## Challenge Problem

89. Solve the formula $r = \dfrac{s/t}{t/u}$ for **a)** $s$, **b)** $u$.

## Cumulative Review Exercises

[1.4] **90.** Evaluate $-\sqrt{3^2 + 4^2} + |3 - 4| - 6^2$.

**91.** Evaluate $\dfrac{7 + 9 \div (2^3 + 4 \div 4)}{|3 - 7| + \sqrt{5^2 - 3^2}}$.

**92.** Evaluate $a^3 - 3a^2b + 3ab^2 - b^3$ when $a = -2$, $b = 3$.

[2.1] **93.** Solve the equation $\frac{1}{4}t + \frac{1}{2} = 1 - \frac{1}{8}t$.

# 2.3 Applications of Algebra

1 Translate a verbal statement into an algebraic expression or equation.

2 Use the problem-solving procedure.

## 1 Translate a Verbal Statement into an Algebraic Expression or Equation

The next few sections will present some of the many uses of algebra in real-life situations. Whenever possible, we include other relevant applications throughout the text.

Perhaps the most difficult part of solving a word problem is translating the problem into an equation. This is step 2 in the problem-solving procedure presented in Section 2.2. Before we represent problems as equations, we give some examples of phrases represented as algebraic expressions.

| Phrase | Algebraic Expression |
|---|---|
| a number increased by 8 | $x + 8$ |
| twice a number | $2x$ |
| 7 less than a number | $x - 7$ |
| one-ninth of a number | $\frac{1}{9}x$ or $\frac{x}{9}$ |
| 2 more than 3 times a number | $3x + 2$ |
| 4 less than 6 times a number | $6x - 4$ |
| 12 times the sum of a number and 5 | $12(x + 5)$ |

The variable $x$ was used in these algebraic expressions, but any variable could have been used to represent the unknown quantity.

**EXAMPLE 1** Express each phrase as an algebraic expression.

**a)** the radius, $r$, decreased by 9 centimeters

**b)** 5 less than twice the distance, $d$

**c)** 7 times a number, $n$, increased by 8

*Solution*

**a)** $r - 9$ **b)** $2d - 5$ **c)** $7n + 8$

**Now Try Exercise 3**

### Helpful Hint *Study Tip*

It is important that you prepare for the remainder of the chapter carefully. Make sure to read the book and examples carefully. *Attend class every day, and, most of all, work all the exercises assigned to you.*

As you read through the examples in the rest of the chapter, think about how they can be expanded to other, similar problems. For example, in Example 1 **a)** we stated that the radius, $r$, decreased by 9 centimeters, can be represented by $r - 9$. You can generalize this to other, similar problems. For example, a weight, $w$, decreased by 15 pounds, can be represented as $w - 15$.

**EXAMPLE 2** ▸ Write each phrase as an algebraic expression.

**a)** the cost of purchasing $x$ shirts at \$4 each
**b)** the distance traveled in $t$ hours at 65 miles per hour
**c)** the number of cents in $n$ nickels
**d)** an 8% commission on sales of $x$ dollars

*Solution*

**a)** We can reason like this: one shirt would cost 1(4) dollars; two shirts, 2(4) dollars; three shirts, 3(4) dollars; four shirts, 4(4) dollars, and so on. Continuing this reasoning process, we can see that $x$ shirts would cost $x(4)$ or $4x$ dollars. We can use the same reasoning process to complete each of the other parts.
**b)** $65t$
**c)** $5n$
**d)** $0.08x$ (8% is written as 0.08 in decimal form.)

▸ Now Try Exercise 7

## Helpful Hint

When we are asked to find a percent, we are always finding the percent of some quantity. Therefore, when a percent is listed, it is **always** multiplied by a number or a variable. In the following examples we use the variable $c$, but any letter could be used to represent the variable.

| Phrase | How Written |
|---|---|
| 6% of a number | $0.06c$ |
| the cost of an item increased by a 7% tax | $c + 0.07c$ |
| the cost of an item reduced by 35% | $c - 0.35c$ |

Sometimes in a problem two numbers are related to each other. We often represent one of the numbers as a variable and the other as an expression containing that variable. We generally let the less complicated description be represented by the variable and write the second (more complex expression) in terms of the variable. In the following examples, we use $x$ for the variable.

| Phrase | One Number | Second Number |
|---|---|---|
| Dawn's age now and Dawn's age in 3 years | $x$ | $x + 3$ |
| one number is 9 times the other | $x$ | $9x$ |
| one number is 4 less than the other | $x$ | $x - 4$ |
| a number and the number increased by 16% | $x$ | $x + 0.16x$ |
| a number and the number decreased by 10% | $x$ | $x - 0.10x$ |
| the sum of two numbers is 19 | $x$ | $19 - x$ |
| a 13-foot board cut into two lengths | $x$ | $13 - x$ |
| \$10,000 shared by two people | $x$ | $10{,}000 - x$ |

The last three examples may not be obvious. Consider "The sum of two numbers is 10." When we add $x$ and $10 - x$, we get $x + (10 - x) = 10$. When a 6-foot board is cut into two lengths, the two lengths will be $x$ and $6 - x$. For example, if one length is 2 feet, the other must be $6 - 2$ or 4 feet.

## Helpful Hint

Suppose you read the following sentence in an application problem: "A 12-foot rope is cut into two pieces." You probably know you should let $x$ (or some other variable) represent the length of the first piece. What you may not be sure of is whether you should let $x - 12$ or $12 - x$ represent the length of the second piece. To help you decide, it can be helpful to use specific numbers to establish a pattern. In this example, you might use a pattern similar to the one below to help you.

| If the First Piece Is ... | Then the Second Piece Is ... |
|---|---|
| 2 feet | 10 feet = 12 feet − 2 feet |
| 5 feet | 7 feet = 12 feet − 5 feet |

From this pattern you can see that if the first piece is $x$ feet, then the second piece is $12 - x$ feet.

**EXAMPLE 3** ▸ For each relation, select a variable to represent one quantity and express the second quantity in terms of the first.

**a)** The speed of the second train is 1.8 times the speed of the first.

**b)** \$90 is shared by David and his brother.

**c)** It takes Tom 3 hours longer than Roberta to complete the task.

**d)** Hilda has \$5 more than twice the amount of money Hector has.

**e)** The length of a rectangle is 7 units less than 3 times its width.

*Solution*

**a)** speed of first train, $s$; speed of second train, $1.8s$

**b)** amount David has, $x$; amount brother has, $90 - x$

**c)** Roberta, $t$; Tom, $t + 3$

**d)** Hector, $x$; Hilda, $2x + 5$

**e)** width, $w$; length, $3w - 7$

▸ **Now Try Exercise 11**

The word *is* in a word problem often means **is equal to** and is represented by an equal sign, =.

| Verbal Statement | Algebraic Equation |
|---|---|
| 4 less than 6 times a number *is* 17 | $6x - 4 = 17$ |
| a number decreased by 4 *is* 5 more than twice the number | $x - 4 = 2x + 5$ |
| the product of two consecutive integers *is* 72 | $x(x + 1) = 72$ |
| a number increased by 15% *is* 90 | $x + 0.15x = 90$ |
| a number decreased by 12% *is* 52 | $x - 0.12x = 52$ |
| the sum of a number and the number increased by 4% *is* 324 | $x + (x + 0.04x) = 324$ |
| the cost of renting a VCR for $x$ days at \$18 per day *is* \$120 | $18x = 120$ |

## 2 Use the Problem-Solving Procedure

There are many types of word problems, and the general problem-solving procedure given in Section 2.2 can be used to solve all types. We now present the five-step problem-solving procedure again so you can easily refer to it. We have included some additional information under step 2, since in this section we are going to emphasize translating word problems into equations.

**Problem-Solving Procedure for Solving Application Problems**

1. **Understand the problem.** Identify the quantity or quantities you are being asked to find.
2. **Translate the problem into mathematical language** (express the problem as an equation).
   **a)** Choose a variable to represent one quantity, and **write down exactly what it represents**. Represent any other quantity to be found in terms of this variable.
   **b)** Using the information from step a), write an equation that represents the word problem.
3. **Carry out the mathematical calculations** (solve the equation).
4. **Check the answer** (using the original wording of the problem).
5. **Answer the question asked.**

Sometimes we will combine steps or not show some steps in the problem-solving procedure due to space limitations. Even if we do not show a check to a problem, you should always check the problem to make sure that your answer is reasonable and makes sense.

**EXAMPLE 4 ▸ Long-Distance Plans** BellSouth's Preferred Rate Telephone Plan requires the customer to pay a \$3.95 monthly fee and 6.9¢ per minute for any long-distance call made. BellSouth's Basic Service Telephone Plan does not have a monthly fee, but customers pay 18¢ per minute for any long-distance call made. Determine the number of minutes a customer would need to spend on long-distance calls for the two plans to cost the same.

*Solution* **Understand** We are given two plans in which one has a monthly fee and the other does not. We are asked to find the *number of minutes* of long-distance calls that would result in both plans having the same total cost. To solve the problem, we will set the cost of the two plans equal to one another and solve for the number of minutes.

**Translate** Let $n$ = number of minutes of long-distance calling.

Then $0.069n$ = cost for $n$ minutes at 6.9¢ per minute

and $0.18n$ = cost for $n$ minutes at 18¢ per minute.

Cost of the Preferred Rate Plan = Cost of the Basic Service Plan

monthly fee + calling cost = calling cost

$$3.95 + 0.069n = 0.18n$$

**Carry Out**

$$3.95 = 0.111n$$

$$\frac{3.95}{0.111} = \frac{0.111n}{0.111}$$

$$35.59 \approx n$$

**Check** The number of minutes is reasonable and the arithmetic is easily checked.

**Answer** If about 36 minutes were used per month, both plans would have about the same total cost.

▸ Now Try Exercise 33

CDC Buildings

**EXAMPLE 5 ▸ CDC Spending** In 2004, the Centers for Disease Control and Prevention (CDC) had a budget of \$4.440 billion. The 2004 budget was a 2.3% increase from the 2003 CDC budget. Determine the 2003 CDC Budget.

*Solution* **Understand** We need to determine the 2003 CDC budget. We will use the facts that the budget increased by 2.3% from 2003 to 2004 and that the 2004 budget was \$4.440 billion to solve the problem.

**Translate** Let $x$ = the 2003 CDC budget.

Then $0.023x$ = the increase in the CDC budget from 2003 to 2004.

$$\begin{pmatrix}\text{2003 CDC}\\\text{budget}\end{pmatrix} + \begin{pmatrix}\text{increase in the budget}\\\text{from 2003 to 2004}\end{pmatrix} = (\text{2004 CDC budget})$$

$$x + 0.023x = 4.440$$

**Carry Out**

$$\begin{aligned} x + 0.023x &= 4.440 \\ 1.023x &= 4.440 \\ x &\approx 4.340 \end{aligned}$$

**Check and Answer** The number obtained is less than the CDC budget for 2004, which is what we expected. The 2003 CDC budget was about \$4.340 billion.

*Source:* www.cdc.gov/fm.

▸ Now Try Exercise 41

**EXAMPLE 6 ▸ Land Area** The total land area of the four countries Gibraltar, Nauru, Bermuda, and Norfolk Island is 116 km$^2$ (square kilometers). The land area of Gibraltar is $\frac{1}{3}$ the land area of Nauru. The land area of Norfolk Island is $\frac{5}{3}$ the land area of Nauru. The land area of Bermuda is 10 km$^2$ less than 3 times the land area of Nauru. Determine the land area of each of these four countries.

*Solution* **Understand** We need to determine the land area (in km$^2$) of Gibraltar, Nauru, Bermuda, and Norfolk Island. Observe that the land area of the countries can be determined from the land area of Nauru. Therefore, we will let the unknown variable be the land area of Nauru. Then we can represent the land area of the other three countries using this variable. Also, notice that the total land area of the four countries is 116 km$^2$.

**Translate**

$$\begin{aligned} \text{Let } a &= \text{the land area of Nauru} \\ \text{and } \frac{1}{3}a &= \text{the land area of Gibraltar} \\ \text{and } \frac{5}{3}a &= \text{the land area of Norfolk Island} \\ \text{and } 3a - 10 &= \text{the land area of Bermuda.} \end{aligned}$$

$$\begin{pmatrix}\text{land area of}\\\text{Nauru}\end{pmatrix} + \begin{pmatrix}\text{land area of}\\\text{Gibraltar}\end{pmatrix} + \begin{pmatrix}\text{land area of}\\\text{Norfolk Island}\end{pmatrix} + \begin{pmatrix}\text{land area of}\\\text{Bermuda}\end{pmatrix} = \begin{pmatrix}\text{total}\\\text{land area}\end{pmatrix}$$

$$a + \frac{1}{3}a + \frac{5}{3}a + (3a - 10) = 116$$

**Carry Out**

$$\begin{aligned} a + \frac{1}{3}a + \frac{5}{3}a + (3a - 10) &= 116 \\ a + 2a + 3a - 10 &= 116 \\ 6a - 10 &= 116 \\ 6a &= 126 \\ a &= 21 \end{aligned}$$

**Check and Answer** The land area of Nauru is 21 km$^2$. The land of Gibraltar is $\frac{1}{3}(21) = 7$ km$^2$. The land area of Norfolk Island is $\frac{5}{3}(21) = 35$ km$^2$. The land area of Bermuda is $(3 \cdot 21) - 10 = 63 - 10 = 53$ km$^2$. The total land area of these four countries is $(21 + 7 + 35 + 53) = 116$ km$^2$, so the answer checks.
*Source:* www.worldgazateer.com

▸ Now Try Exercise 53

**EXAMPLE 7 ▸ Daytona Beach** Erin Grabish took her family to visit Daytona Beach, Florida. They stayed for one night at a Holiday Inn. When they made their hotel reservation, they were quoted a rate of \$95 per night before tax. When they checked out, their total bill was \$110.85, which included the room tax and a \$3.50 charge for a candy bar (from the in-room bar). Determine the tax rate for the room.

*Solution* **Understand** Their total bill consists of their room rate, the room tax, and the \$3.50 cost for the candy bar. The room tax is determined by multiplying the cost of the room by the tax rate for the room. We are asked to find the tax rate for the room.

The Daytona 500 race

**Translate**

let $t$ = tax rate for the room

then $0.01t$ = room tax rate as a decimal

$$\text{room cost} + \text{room tax} + \text{candy bar} = \text{total}$$

$$95 + 95(0.01t) + 3.50 = 110.85$$

**Carry Out**

$$95 + 0.95t + 3.50 = 110.85$$

$$0.95t + 98.50 = 110.85$$

$$0.95t = 12.35$$

$$t = 13$$

**Check and Answer** If you substitute 13 for $t$ in the equation, you will see that the answer checks. The room tax rate is 13%.

▸ Now Try Exercise 47

**EXAMPLE 8 ▸ Home Mortage** Mary Shapiro is buying her first home and she is considering two banks for a \$60,000 mortgage. Citicorp is charging a 6.50% interest rate with no points for a 30-year loan. (A point is a one-time charge of 1% of the amount of the mortgage.) The monthly mortgage payments for the Citicorp mortgage would be \$379.24. Citicorp is also charging a \$200 application fee. Bank of America Corporation is charging a 6.00% interest rate with 2 points for a 30-year loan. The monthly mortgage payments for Bank of America would be \$359.73 and the cost of the points that Mary would need to pay at the time of closing is 0.02(\$60,000) = \$1200. Bank of America has waived its application fee.

**a)** How long would it take for the total payments of the Citicorp mortgage to equal the total payments of the Bank of America mortgage?

**b)** If Mary plans to keep her house for 20 years, which mortgage would result in the lower total cost?

*Solution* **a) Understand** Citicorp is charging a higher interest rate and a small application fee but no points. Bank of America is charging a lower rate and no application fee but 2 points. We need to determine the number of months when the total payments of the two loans would be equal.

**Translate** Let $x$ = number of months.

Then $379.24x$ = cost of mortgage payments for $x$ months with the Citicorp mortgage

and $359.73x$ = cost of mortgage payments for $x$ months with the Bank of America Corporation.

total cost with Citicorp = total cost with Bank of America

| mortgage payments | + | application fee | = | mortgage payments | + | points |
|---|---|---|---|---|---|---|
| $379.24x$ | + | 200 | = | $359.73x$ | + | 1200 |

**Carry Out**

$$379.24x + 200 = 359.73x + 1200$$
$$379.24x = 359.73x + 1000$$
$$19.51x = 1000$$
$$x \approx 51.26$$

**Answer** The cost would be the same in about 51.26 months or about 4.3 years.

**b)** The total cost would be the same at about 4.3 years. Prior to the 4.3 years, the cost of the loan with Bank of America would be more because of the initial \$1200 charge for points. However, after the 4.3 years the cost with Bank of America would be less because of the lower monthly payment. If we evaluate the total cost with Citicorp over 20 years (240 monthly payments), we obtain \$91,217.60. If we evaluate the total cost with Bank of America over 20 years, we obtain \$87,535.20. Therefore, Mary will save \$3682.40 over the 20-year period with Bank of America.

▸ **Now Try Exercise 49**

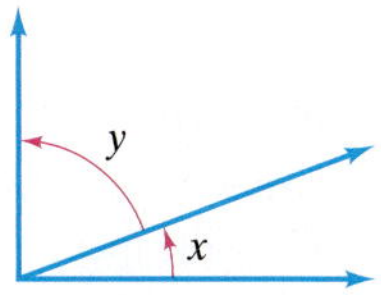


FIGURE 2.4

Let's now look at two examples that involve angles. In Example 9 we use complementary angles. **Complementary angles** are two angles whose sum measures 90° (see **Fig. 2.4**).

In **Figure 2.4**, angle $x$ (symbolized $\measuredangle x$) and angle $y$ ($\measuredangle y$) are complementary angles since their sum measures 90°.

**EXAMPLE 9** ▸ **Complementary Angles** If angles $A$ and $B$ are complementary angles and angle $B$ is 42° greater than angle $A$, determine the measures of angle $A$ and angle $B$.

*Solution* **Understand** The sum of the measures of the two angles must be 90° because they are complementary angles. We will use this fact to set up an equation. Since angle $B$ is described in terms of angle $A$, we will let $x$ represent the measure of angle $A$.

**Translate** Let $x$ = the measure of angle $A$.

Then $x + 42$ = the measure of angle $B$.

| the measure of angle $A$ | + | the measure of angle $B$ | = 90° |
|---|---|---|---|
| $x$ | + | $x + 42$ | = 90 |

**Carry Out**

$$2x + 42 = 90$$
$$2x = 48$$
$$x = 24$$

**Check and Answer** Since $x = 24$, the measure of angle $A$ is 24°. The measure of angle $B = x + 42 = 24 + 42 = 66$. Thus, angle $B$ has a measure of 66°. Note that angle $B$ is 42° greater than angle $A$, and the sum of the measures of both angles is $24° + 66° = 90°$.

▸ **Now Try Exercise 21**

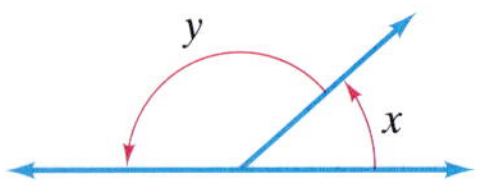


FIGURE 2.5

In Example 10 we use supplementary angles. **Supplementary angles** are two angles whose sum measures 180° (see **Fig. 2.5**).

In **Figure 2.5**, angle $x$ and angle $y$ are supplementary angles since their sum measures 180°.

**EXAMPLE 10 ▸ Supplementary Angles** If angles $C$ and $D$ are supplementary angles and the measure of angle $C$ is 6° greater than twice the measure of angle $D$, determine the measures of angle $C$ and angle $D$.

*Solution* **Understand** The sum of the measures of the two angles must be 180° because they are supplementary angles. Since angle $C$ is described in terms of angle $D$, we will let $x$ represent the measure of angle $D$.

**Translate**

$$\text{Let } x = \text{the measure of angle } D.$$
$$\text{Then } 2x + 6 = \text{the measure of angle } C.$$
$$\text{the measure of angle } C + \text{the measure of angle } D = 180°$$
$$2x + 6 \quad + \quad x \quad = 180$$

**Carry Out**

$$3x + 6 = 180$$
$$3x = 174$$
$$x = 58$$

**Check and Answer** Since $x = 58$, the measure of angle $D$ is 58°. The measure of angle $C = 2x + 6 = 2(58) + 6 = 122$. Thus the measure of angle $C = 122°$. Note that the measure of angle $C$ is 6° more than twice the measure of angle $D$, and the sum of the measures of the angles is $122° + 58° = 180°$.

▸ **Now Try Exercise 23**

## Helpful Hint *Study Tip*

Here are some suggestions if you are having some difficulty with application problems.

1. Instructor—Make an appointment to see your instructor. Make sure you have read the material in the book and attempted all the homework problems. Go with specific questions for your instructor.
2. Video CDs—Find out if the video CDs that accompany this book are available at your college. If so, view the videos that go with this chapter. Using the pause control, you can watch the videos at your own pace.
3. Tutoring—If your college learning center offers free tutoring, as many colleges do, you may wish to take advantage of tutoring.
4. Study Group—Form a study group with classmates. Exchange phone numbers and e-mail addresses. You may be able to help one another.
5. Student's Solutions Manual—If you get stuck on an exercise, you may want to use the Student's Solutions Manual to help you understand a problem. Do not use the Solutions Manual in place of working the exercises. In general, the Solutions Manual should be used only to check your work.
6. MyMathLab—MyMathLab provides free-response exercises correlated to the text that are algorithmically generated for unlimited practice and mastery. In addition, online tools such as video lectures, animations, and a multimedia textbook are available to help you understand the material. Check with your instructor to determine if MyMathLab is available.
7. MathXL®—MathXL is a powerful online homework, tutorial, and assessment system correlated specifically to this text. You can take chapter tests in MathXL and receive a personalized study plan based on your test results. The study plan links directly to tutorial exercises for the objectives you need to study or retest. Check with your instructor to determine if MathXL is available.
8. Prentice Hall Mathematics Tutor—Once the program has been initiated by your instructor, you can get individual tutoring by phone, fax, or email.

It is important that you keep trying! Remember, the more you practice, the better you will become at solving application problems.

## EXERCISE SET 2.3

MathXL® MyMathLab

### Practice the Skills

*In Exercises 1–10, express each phrase as an algebraic expression.*

1. 3 less than a number, x
2. 17 more than 4 times a number, $m$
3. the volume, $v$, increased by 6 meters
4. 11 times a number $n$, decreased by 7.5
5. the distance, $d$, increased by 2 miles
6. 7 times a number, $p$, increased by 8
7. the cost of buying $y$ books at $19.95 each
8. the number of cents in $q$ quarters
9. 9.6% commission on selling houses for $x$ dollars
10. the amount of interest earned in one year at a rate of 3.5% on $d$ dollars

*In Exercises 11–20, select one variable to represent one quantity and express the second quantity in terms of the first.*

11. A 12-foot piece of wood is cut into 2 pieces.
12. One angle of a triangle is 7° more than another angle.
13. The length of a rectangle is 29 meters longer than the width.
14. A 17-hour task is shared between Robin and Tom.
15. $165 is shared between Max and Lora.
16. George can paint a house twice as fast as Jason.
17. Nora can jog 1.3 miles per hour faster than Betty.
18. The speed limit on an express way is 30 miles per hour faster than the speed limit on a local road.
19. The cost for electricity has increased by 22%.
20. The price for a refrigerator has increased by 6%.

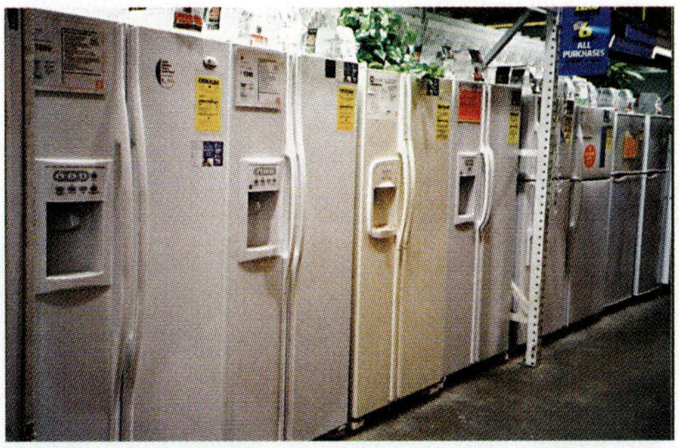

### Problem Solving

*In Exercises 21–72, write an equation that can be used to solve the problem. Find the solution to the problem.*

21. **Complementary Angles** Angles $A$ and $B$ are complementary angles. Determine the measures of angles $A$ and $B$ if angle $A$ is four times the size of angle $B$. See Example 9.

22. **Complementary Angles** Angles $C$ and $D$ are complementary angles. Determine the measures of angles $C$ and $D$ if angle $D$ is 15° less than twice angle $C$.

23. **Supplementary Angles** Angles $A$ and $B$ are supplementary angles. Find the measures of angles $A$ and $B$ if angle $B$ is 4 times the size of angle $A$. See Example 10.

24. **Supplementary Angles** Angle $A$ and angle $B$ are supplementary angles. Find the measure of each angle if angle $A$ is 30° greater than angle $B$.

25. **Angles in a Triangle** The sum of the measures of the angles of a triangle is 180°. Find the three angles of a triangle if one angle is 20° greater than the smallest angle and the third angle is twice the smallest angle.

26. **Angles in a Triangle** Find the measures of the three angles of a triangle if one angle is twice the smallest angle and the third angle is 60° greater than the smallest angle.

27. **History Honor Society** One benefit of membership in a national honor society is a 25% discount on all history magazine subscriptions. Thomas used this discount to order an annual subscription to *American Heritage* magazine and paid $24. What was the cost of a regular subscription?

28. **A New Suit** Matthew Stringer is shopping for a new suit. At K & G Menswear, he finds that the sale price of a suit that was reduced by 25% is $187.50. Find the regular price of the suit.

29. **Bus Pass** Kate Spence buys a monthly bus pass, which entitles her to unlimited bus travel, for $45 per month. Without the pass each bus ride costs $1.80. How many rides per month would Kate have to take so that the cost of the rides without the bus pass is equal to the total cost of the rides with the bus pass?

30. **Laundry Costs** It costs Bill Winschief $12.50 a week to wash and dry his clothes at the corner laundry. If a washer and dryer cost a total of $940, how many weeks will it take for the laundry cost to equal the cost of a washer and dryer? (Disregard energy cost.)

31. **Truck Rental** The cost of renting a truck is $35 a day plus 20¢ per mile. How far can Tanya Richardson drive in 1 day if she has only $80?

32. **Waitress Pay** Candice Colton is a banquet waitress. She is paid \$2.63 per hour plus 15% of the total cost of the food and beverages she serves during the banquet. If, during a 5-hour shift, Candice earns \$400, what was the total cost of the food and beverages she served?

33. **Playing Golf** Albert Sanchez has two options for membership in a golf club. A social membership costs \$1775 in annual dues. In addition, he would pay a \$50 green fee and a \$25 golf cart fee every time he played. A golf membership costs \$2425 in annual dues. With this membership, Albert would only pay a \$25 golf cart fee when he played. How many times per year would Albert need to golf for the two options to cost the same?

34. **George Washington Bridge Tolls** Going into New York using the George Washington Bridge, customers must pay a toll (there is no toll to return to New Jersey). They can pay \$6 cash or they can pay \$5 (during peak hours) using the E Z Pass system. The E Z Pass system is a prepaid plan that also requires a one-time \$12 account activation fee. How many trips to New York would a person need to make (during peak hours) so that the total amount spent with the E Z Pass is equal to the amount spent for tolls without the E Z Pass?

35. **Toll Bridge** Mr. and Mrs. Morgan live in a resort island community attached to the mainland by a toll bridge. The toll is \$2.50 per car going to the island, but there is no toll coming from the island. Island residents can purchase a monthly pass for \$20, which permits them to cross the toll bridge from the mainland for only 50¢ each time. How many times a month would the Morgans have to go to the island from the mainland for the cost of the monthly pass to equal the regular toll cost?

36. **Sales Tax** The sales tax rate in North Carolina is 4.5%. What is the maximum price that Don and Betty Lichtenberg can spend on a new computer desk if the total cost of the desk, including sales tax, is \$650?

37. **Renting an Apartment** The DuVall family is renting an apartment in Southern California. For 2007, the rent is \$1720 per month. The monthly rent in 2007 is 7.5% higher than the monthly rent in 2006. Determine the monthly rent in 2006.

38. **Retirement Funds** Eva Chang makes regular contributions of \$5000 annually to a retirement plan. She has some of her contribution going to the growth stock fund and some going to the global equities fund. Her contribution to the growth stock fund is \$250 less than twice what she contributes to the global equities fund. How much does she contribute to each fund?

39. **Girl Scouts** To make money for the organization, the Girl Scouts have their annual cookie drive. This year, the total sales from two districts, the southeast district and the northwest district, totaled \$4.6 million. If the sales from the southeast district were \$0.31 million more than the sales from the northwest district, find the sales from each district.

40. **Franchise Values** At the end of the 2004 National Football League season, the Washington Redskins and the Dallas Cowboys had the highest franchise values. The total value of the two franchises was \$2.023 billion. The value of the Washington Redskins was 19.2% more than the value of the Dallas Cowboys. Determine the franchise value of each team.

41. **Personal Income** Personal income has been on the increase since 1980. The average personal income in 2004 was \$29,367. This represents about a 232% increase in average personal income from 1980. Determine the average personal income in 1980.

*Source:* U.S. Bureau of Economic Analysis

42. **Amtrak Budget** Amtrak has approved budgets for the Fiscal Years 2006 and 2007. The 2007 budget for Amtrak is \$3.242 billion. This is 0.527% more than the 2006 budget. Determine the Amtrak budget for 2006.

43. **Minimum Wage Increase** From 1980 to 2005, the minimum hourly wage increased about 66.13% to \$5.15 per hour. What was the minimum hourly wage in 1980?

44. **Bones and Steel** According to *Health* magazine, the stress a bone can withstand in pounds per square inch is 6000 pounds more than 3 times the amount that steel can withstand. If the difference between the amount of stress a bone and steel can withstand is 18,000 pounds per square inch, find the stress that both steel and a bone can withstand.
bone:

45. **Pollen** There are 57 major sources of pollen in the United States. These pollen sources are categorized as grasses, weeds, and trees. If the number of weeds is 5 less than twice the number of grasses, and the number of trees is 2 more than twice the number of grasses, find the number of grasses, weeds, and trees that are major pollen sources.

46. **Car Antitheft System** By purchasing and installing a LoJack antitheft system, Pola Sommers can save 15% of the price of her auto insurance. The LoJack system costs \$743.65 to purchase and install. If Pola's annual insurance before installing the LoJack system is \$849.44, in how many years would the LoJack system pay for itself?

47. **Ordering Lunch** After Valerie Fandl is seated in a restaurant, she realizes that she only has \$20.00. If she must pay 7% sales tax and wishes to leave a 15% tip on the total bill (meal plus tax), what is the maximum price of the lunch she can order?

48. **Hotel Tax Rate** While on vacation in Milwaukee, the Ahmeds were quoted a hotel room price of \$85 per night plus tax. They stayed one night and watched a movie that cost \$9.25. Their total bill came to \$106.66. What was the tax rate?

49. **Comparing Mortgages** The Chos are purchasing a new home and are considering a 30-year \$70,000 mortgage with two different banks. Madison Savings is charging 9.0% with 0 points and First National is charging 8.5% with 2 points. First National is also charging a \$200 application fee, whereas Madison is charging none. The monthly mortgage payments with Madison would be \$563.50 and the monthly mortgage payments with First National would be \$538.30.

   **a)** After how many months would the total payments for the two banks be the same?

   **b)** If the Chos plan to keep their house for 30 years, which mortgage would have the lower total cost? (See Example 8.)

50. **Payment Plan** The Midtown Tennis Club offers two payment plans for its members. Plan 1 is a monthly fee of \$25 plus \$10 per hour of court rental time. Plan 2 has no monthly fee, but court time costs \$18.50 per hour. How many hours would Mrs. Levin have to play per month so that plan 1 becomes advantageous?

51. **Refinancing a Mortgage** Dung Nguyen is considering refinancing his house at a lower interest rate. He has an 11.875% mortgage, is presently making monthly principal and interest payments of \$510, and has 20 years left on his mortgage. Because interest rates have dropped, Countrywide Mortgage Corporation is offering him a rate of 9.5%, which would result in principal and interest payments of \$420.50 for 20 years. However, to get this mortgage, his closing cost would be \$2500.

   **a)** How many months after refinancing would he spend the same amount on his new mortgage plus closing cost as he would have spent on the original mortgage?

   **b)** If he plans to spend the next 20 years in the house, would he save money by refinancing?

52. **Dinner Seminars** Heather Jockson, a financial planner, is sponsoring dinner seminars. She must pay for the dinners of those attending out of her own pocket. She chooses a restaurant that seats 40 people and charges her \$9.50 per person. If she earns 12% commission of sales made, determine how much in sales she must make from these 40 people

   **a)** to break even.

   **b)** to make a profit of \$500.

53. **2004 Olympic Medals** In the 2004 Summer Olympics held in Athens, Greece, the United States, China, Russia, Australia, and Germany won a total 355 medals (gold, silver, bronze). Australia won 1 more medal than Germany. Russia won 4 fewer than twice the number of medals Germany won. China won 15 more medals than Germany. Finally, the United States won 7 more than twice the number of medals

Germany won. Determine the number of medals won by the United States, China, Russia, Australia, and Germany in the 2004 Summer Olympics.

*Source:* athens2004.com

**54. Test Grades** On a recent test in an intermediate algebra class, 34 students received grades of A, B, C, or D. There were twice as many C's as D's. There were 2 more B's than D's and there were 2 more than twice as many A's than D's. Determine the number of A's, B's, C's, and D's from this test.

**55. Plants and Animals** Approximately 1,500,000 species worldwide have been categorized as either plants, animals, or insects. Insects are often subdivided into beetles and insects that are not beetles. There are about 100,000 more plant than animal species. There are 290,000 more nonbeetle insects than animals. The number of beetles is 140,000 less than twice the number of animals. Find the number of animal, plant, nonbeetle insect, and beetle species.

**56. Price of Gas** From June 2005, through November 2005, the average cost of a gallon of gasoline increased by 36%. If the cost of a gallon of gasoline was \$2.69 on November 1, 2005, determine the cost on June 1, 2005.

**57. Perimeter of a Triangle** John is developing a game that contains a triangular game board. The perimeter of the triangular board is 36 inches. Find the length of the three sides if one side is 3 inches greater than the smallest side, and the third side is 3 inches less than twice the length of the smallest side.

**58. Angles of a Triangle** A rectangular piece of paper is cut from opposite corners to form a triangle. One angle of the triangle measures 12° greater than the smallest angle. The third angle measures 27° less than three times the smallest angle. If the sum of the interior angles of a triangle measure 180°, determine the measures of the three angles.

**59. Triangular Garden** The perimeter of a triangular garden is 60 feet. Find the length of the three sides if one side is 4 feet greater than twice the length of the smallest side, and the third side is 4 feet less than 3 times the length of the smallest side.

**60. Stairway Railing** A stairway railing has a design that contains triangles. In one of the triangles one angle measures 20° less than twice the smallest angle. The third angle measures 25° greater than twice the smallest angle. Determine the measures of the three angles.

**61. Fence Dimensions** Greg Middleton, a landscape architect, wishes to fence in two equal areas as illustrated in the figure. If both areas are squares and the total length of fencing used is 91 meters, find the dimensions of each square.

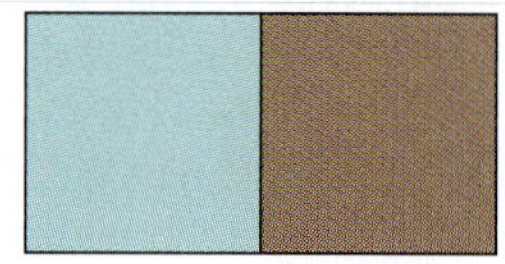

**62. Sandbox Construction** Edie Hall is planning to build a rectangular sandbox for her children. She wants its length to be 3 feet more than its width. Find the length and width of the sandbox if only 22 feet of lumber are available to form the frame. Use $P = 2l + 2w$.

**63. Bookcase Dimensions** Eric Krassow wishes to build a bookcase with four shelves (including the top) as shown in the figure. The width of the bookcase is to be 3 feet more than the height. If only 30 feet of wood are available to build the bookcase, what will be the dimensions of the bookcase?

**64. Fence Dimensions** Collette Siever wishes to fence in three rectangular areas along a river bank as illustrated in the figure. Each rectangle is to have the same dimensions, and the length of each rectangle is to be 1 meter greater than its width (along the river). Find the length and width of each rectangle if the total amount of fencing used is 114 meters.

**65. Price Reductions** During the first week of a going-out-of-business sale, Sam's General Store reduces all prices by 10%. The second week of the sale, Sam's reduces all items by an additional \$5. If Sivle Yelserp bought a calculator for \$49 during the second week of the sale, find the original price of the calculator.

**66. Farm Divisions** Deborah Schmidt's farm is divided into three regions. The area of one region is twice as large as the area of the smallest region, and the area of the third region is 4 acres less than three times the area of the smallest region. If the total acreage of the farm is 512 acres, find the area of each of the three regions.

**67. Selling Paintings** J. P. Richardson sells each of his paintings for \$500. The gallery where he displays his work charges him \$1350 a month plus a 10% commission on sales. How many paintings must J. P. sell in a month to break even?

**68. Comparing Toy Sales** Kristen Hodge is shopping for a certain bicycle for her niece and knows that Toys "R" Us and Wal-Mart sell the bike for the same price. On December 26, Toys "R" Us has the bike on sale for 37% off the original price and Wal-Mart has the bike on sale for \$50 off the original price. After visiting both stores, Kristen discovers that the sale prices for the bike at the two stores are also the same.

**a)** Determine the original price of the bike.

**b)** Determine the sale price of the bike.

**69. Incandescent Bulbs** The cost of purchasing incandescent bulbs for use over a 9750-hour period is \$9.75. The energy cost of incandescent bulbs over this period is \$73. The cost of one equivalent fluorescent bulb that lasts about 9750 hours is \$20. By using a fluorescent bulb instead of incandescent bulbs for 9750 hours, the total savings of purchase price plus energy cost is \$46.75. What is the energy cost of using the fluorescent bulb over this period?

**70. Dinner Bill** The five members of the Newton family are going out to dinner with the three members of the Lee family. Before dinner, they decide that the Newtons will pay $\frac{5}{8}$ of the bill (before tip) and the Lees will pay $\frac{3}{8}$ plus the entire 15% tip. If the total bill including the 15% tip comes to \$184.60, how much will be paid by each family?

**71. Earning an A** To find the average of a set of test scores, we divide the sum of the test scores by the number of test scores. On her first four algebra tests, Paula West's scores were 88, 92, 97, and 96.

**a)** Write an equation that can be used to determine the grade Paula needs to obtain on her fifth test to have a 90 average.

**b)** Explain how you determined your equation.

**c)** Solve the equation and determine the score.

**72. Physics Exam Average** Francis Timoney's grades on five physics exams were 70, 83, 97, 84, and 74.

**a)** If the final exam will count twice as much as each exam, what grade does Francis need on the final exam to have an 80 average?

**b)** If the highest possible grade on the final exam is 100 points, is it possible for Francis to obtain a 90 average? Explain.

**73. a)** Make up your own realistic word problem involving percents. Represent this word problem as an equation.

**b)** Solve the equation and answer the word problem.

**74. a)** Make up your own realistic word problem involving money. Represent this word problem as an equation.

**b)** Solve the equation and answer the word problem.

## Challenge Problems

**75. Truck Rental** The Elmers Truck Rental Agency charges \$28 per day plus 15 ¢ a mile. If Martina Estaban rented a small truck for 3 days and the total bill was \$121.68, including a 4% sales tax, how many miles did she drive?

**76. Money Market** On Monday Sophia Murkovic purchased shares in a money market fund. On Tuesday the value of the shares went up 5%, and on Wednesday the value fell 5%. How much did Sophia pay for the shares on Monday if she sold them on Thursday for \$59.85?

## Group Activity

*Discuss and answer Exercise 77 as a group.*

**77. a)** Have each member of the group pick a number. Then multiply the number by 2, add 33, subtract 13, divide by 2, and subtract the number each started with. Record each answer.

**b)** Now compare answers. If you did not all get the same answer, check each other's work.

**c)** As a group, explain why this procedure will result in an answer of 10 for any real number $n$ selected.

## Cumulative Review Exercises

[1.3] *Evaluate.*

**78.** $2 + \left|-\frac{3}{5}\right|$

**79.** $-6.4 - (-3.7)$

**80.** $\left|-\frac{5}{8}\right| \div |-4|$

**81.** $5 - |-3| - |12|$

[1.5] **82.** Simplify $(2x^4y^{-6})^{-3}$.

## Mid-Chapter Test: 2.1–2.3

*To find out how well you understand the chapter material to this point, take this brief test. The answers, and the section where the material was initially discussed, are given in the back of the book. Review any questions that you answered incorrectly.*

**1.** Give the degree of $6x^5y^7$.

*Simplify each expression.*

**2.** $3x^2 + 7x - 9x + 2x^2 - 11$

**3.** $2(a - 1.3) + 4(1.1a - 6) + 17$

*Solve each equation.*

**4.** $7x - 9 = 5x - 21$

**5.** $\frac{3}{4}y + \frac{1}{2} = \frac{7}{8}y - \frac{5}{4}$

**6.** $3p - 2(p + 6) = 4(p + 1) - 5$

**7.** $0.6(a - 3) - 3(0.4a + 2) = -0.2(5a + 9) - 4$

*Find the solution set for each equation. Then indicate whether the equation is conditional, an identify, or 0 contradiction.*

**8.** $4x + 15 - 9x = -7(x - 2) + 2x + 1$

**9.** $-3(3x + 1) = -[4x + (6x - 5)] + x + 7$

**10.** Evaluate $A = \frac{1}{2}hb$, where $h = 10$ and $b = 16$.

**11.** Evaluate $R_T = \frac{R_1R_2}{R_1 + R_2}$, where $R_1 = 100$ and $R_2 = 50$.

**12.** Solve $y = 7x + 13$ for $x$.

**13.** Solve $A = \frac{2x_1 + x_2 + x_3}{n}$ for $x_3$.

**14.** Robert invested $700 into a certificate of deposit earning 6% interest compounded quarterly. How much is the certificate of deposit worth 5 years later?

*Solve each exercise.*

**15.** Angles $A$ and $B$ are complementary angles. Determine the measures of angles $A$ and $B$ if angle $A$ is 6° more than twice angle $B$.

**16.** The cost of renting a ladder is $15 plus $1.75 per day. How many days did Tom Lang rent the ladder if the total cost was $32.50?

**17.** The perimeter of a triangle is 100 feet. The longest side is four times the length of the shortest side and the other side is 10 feet longer than the shortest side. Find the lengths of the three sides of the triangle.

**18.** Tien bought a pair of shoes for $36.00. With tax, the cost was $37.62. Find the tax rate.

**19.** The population of a small town is increasing by 52 people per month. If the current population is 5693 people, how many months ago was the population 3613 people?

**20.** When asked to solve the equation $\frac{1}{2}x + \frac{1}{3} = \frac{1}{4}x - \frac{1}{2}$, Mary Dunwell claimed that to elimate fractions, the left side should be multiplied by 6 and the right should be multiplied by 8. This is incorrect. Why is this incorrect? Explain your answer. What *single* number should be used to eliminate fractions from the *entire* equation? Solve the equation correctly.

# 2.4 Additional Application Problems

**1** Solve motion problems.

**2** Solve mixture problems.

In this section we discuss two additional types of application problems, motion and mixture problems. We have placed them in the same section because they are solved using similar procedures.

## 1 Solve Motion Problems

A formula with many useful applications is

**Motion Formula**

$$\text{amount} = \text{rate} \cdot \text{time}$$

The "amount" in this formula can be a measure of many different quantities, depending on the rate. For example, if the rate is measuring *distance* per unit time, the amount will be distance. If the rate is measuring *volume* per unit time, the amount will be volume, and so on.

When applying this formula, we must make sure that the units are consistent. For example, when speaking about a copier, if the rate is given in copies per *minute*, the time must be given in *minutes*. Problems that can be solved using this formula are called **motion problems** because they involve movement, at a constant rate, for a certain period of time.

A nurse giving a patient an intravenous injection may use this formula to determine the drip rate of the fluid being injected. A company drilling for oil or water may use this formula to determine the amount of time needed to reach its goal.

When the motion formula is used to calculate distance, the word *amount* is replaced with the word *distance* and the formula is called the **distance formula**.

**Distance Formula**

The distance formula is

$$\text{distance} = \text{rate} \cdot \text{time}$$
$$\text{or } d = rt$$

When a motion problem has two different rates, it is often helpful to put the information into a table to help analyze the problem.

**EXAMPLE 1** ▸ **Ships at Sea** The aircraft carrier USS *John F. Kennedy* and the nuclear-powered submarine USS *Memphis* leave from the Puget Sound Naval Yard at the same time heading for the same destination in the Indian Ocean. The aircraft carrier travels at its maximum speed of 34.5 miles per hour and the submarine travels submerged at its maximum speed of 20.2 miles per hour. The aircraft carrier and submarine are to travel at these speeds until they are 100 miles apart. How long will it take for the aircraft carrier and submarine to be 100 miles apart? (See **Fig. 2.6.**)

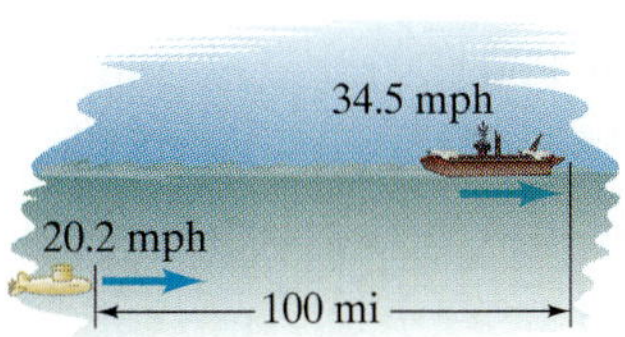


FIGURE 2.6

*Solution* **Understand** We wish to find out how long it will take for the difference between their distances to be 100 miles. To solve this problem, we will use the distance formula, $d = rt$. When we first introduced the problem-solving procedure, we indicated that to help understand a problem it might be useful to put the given information into a table, and that is what we will do now.

Let $t$ = time.

| | Rate | Time | Distance |
|---|---|---|---|
| Aircraft carrier | 34.5 | $t$ | $34.5t$ |
| Submarine | 20.2 | $t$ | $20.2t$ |

**Translate** The difference between their distances is 100 miles. Thus,

$$\text{aircraft carrier's distance} - \text{submarine's distance} = 100$$
$$34.5t - 20.2t = 100$$

**Carry Out**

$$14.3t = 100$$
$$t \approx 6.99$$

**Answer** The aircraft carrier and submarine will be 100 miles apart in about 7 hours.

▸ Now Try Exercise 3

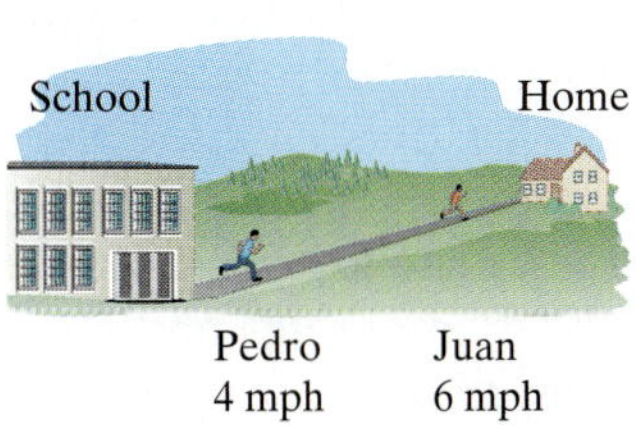


Juan arrives at home $\frac{1}{2}$ hour before Pedro

FIGURE 2.7

**EXAMPLE 2** ▸ **Running Home** To get in shape for the upcoming track season, Juan and Pedro Santiago begin running home after school. Juan runs at a rate of 6 mph and Pedro runs at a rate of 4 mph. When they leave the same school at the same time, Juan arrives home $\frac{1}{2}$ hour before Pedro (see **Fig. 2.7**).

**a)** How long does it take for Pedro to reach home?

**b)** How far do Juan and Pedro live from school?

*Solution* **a) Understand** Both boys will run the same distance. However, since Juan runs at a faster rate than Pedro does, Juan's time will be less than Pedro's time by $\frac{1}{2}$ hour.

$$\text{Let } t = \text{Pedro's time to run home.}$$

$$\text{Then } t - \frac{1}{2} = \text{Juan's time to run home.}$$

| Runner | Rate | Time | Distance |
|---|---|---|---|
| Pedro | 4 | $t$ | $4t$ |
| Juan | 6 | $t - \frac{1}{2}$ | $6\left(t - \frac{1}{2}\right)$ |

**Translate** When the boys are home they will both have run the same distance from school. So

$$\text{Pedro's distance} = \text{Juan's distance}$$

$$4t = 6\left(t - \frac{1}{2}\right)$$

**Carry Out**

$$4t = 6t - 3$$

$$-2t = -3$$

$$t = \frac{3}{2}$$

**Answer** Pedro will take $1\frac{1}{2}$ hours to reach home.

**b)** The distance can be determined using either Juan's or Pedro's rate and time. We will multiply Pedro's rate by Pedro's time to find the distance.

$$d = rt = 4\left(\frac{3}{2}\right) = \frac{12}{2} = 6 \text{ miles}$$

Therefore, Juan and Pedro live 6 miles from their school.

▸ **Now Try Exercise 9**

In Example 2, would the answer have changed if we had let $t$ represent the time Juan is running rather than the time Pedro is running? Try it and see.

**EXAMPLE 3** ▸ **Soda Pop Production** A Coca-Cola bottling machine fills and caps bottles. The machine can be run at two different rates. At the faster rate the machine fills and caps 600 more bottles per hour than it does at the slower rate. The machine is turned on for 4.8 hours on the slower rate, then it is changed to the faster rate for another 3.2 hours. During these 8 hours a total of 25,920 bottles were filled and capped. Find the slower rate and the faster rate.

*Solution* **Understand** This problem uses a number of bottles, an amount, in place of a distance. However, the problem is worked in a similar manner. We will use the formula, amount = rate · time. We are given that there are two different rates, and we are asked to find these rates. We will use the fact that the amount filled at the slower rate plus the amount filled at the faster rate equals the total amount filled.

$$\text{Let } r = \text{the slower rate.}$$

$$\text{Then } r + 600 = \text{the faster rate.}$$

| | Rate | Time | Amount |
|---|---|---|---|
| Slower rate | $r$ | 4.8 | $4.8r$ |
| Faster rate | $r + 600$ | 3.2 | $3.2(r + 600)$ |

**Translate** amount filled at slower rate + amount filled at faster rate = 25,920

$$4.8r + 3.2(r + 600) = 25,920$$

**Carry Out**

$$4.8r + 3.2r + 1920 = 25,920$$
$$8r + 1920 = 25,920$$
$$8r = 24,000$$
$$r = 3000$$

**Answer** The slower rate is 3000 bottles per hour. The faster rate is $r + 600$ or $3000 + 600 = 3600$ bottles per hour.

▸ Now Try Exercise 11

## 2 Solve Mixture Problems

Any problem where two or more quantities are combined to produce a different quantity or where a single quantity is separated into two or more different quantities may be considered a **mixture problem**. As we did when working with motion problems, we will use tables to help organize the information. Examples 4 and 5 are mixture problems that involve money.

**EXAMPLE 4** ▸ **Two Investments** Bettie Truitt sold her boat for \$15,000. Bettie loaned some of this money to her friend Kathy Testone. The loan was for 1 year with a simple interest rate of 4.5%. Bettie put the balance into a money market account at her credit union that yielded 3.75% simple interest. A year later, while working on her taxes, Bettie found that she had earned a total of \$637.50 from the two investments, but could not remember how much money she had loaned to Kathy. Determine the amount Bettie loaned to Kathy.

*Solution* **Understand and Translate** To work this problem, we will use the simple interest formula, interest = principal · rate · time. We know that part of the investment is made at 4.5% and the rest is made at 3.75% simple interest. We are asked to determine the amount that Bettie loaned to Kathy.

Let $p$ = amount loaned to Kathy at 4.5%.
Then $15,000 - p$ = amount invested at 3.75%.

Notice that the sum of the two amounts equals the total amount invested, \$15,000. We will find the amount loaned to Kathy with the aid of a table.

| Investment | Principal | Rate | Time | Interest |
|---|---|---|---|---|
| Loan to Kathy | $p$ | 0.045 | 1 | $0.045p$ |
| Money market | $15,000 - p$ | 0.0375 | 1 | $0.0375(15,000 - p)$ |

Since the total interest collected is \$637.50, we write

interest from 4.5% loan + interest from 3.75% account = total interest

$$0.045p + 0.0375(15,000 - p) = 637.50$$

**Carry Out**

$$0.045p + 0.0375(15,000 - p) = 637.50$$
$$0.045p + 562.50 - 0.0375p = 637.50$$
$$0.0075p + 562.50 = 637.50$$
$$0.0075p = 75$$
$$p = 10,000$$

**Answer** Therefore, the loan to Kathy was for \$10,000, and \$15,000 − $p$ or \$15,000 − \$10,000 = \$5000 was invested in the money market account.

▸ **Now Try Exercise 15**

**EXAMPLE 5 ▸ Hot Dog Stand Sales** Matt's Hot Dog Stand in Chicago sells hot dogs for \$2.00 each and beef tacos for \$2.25 each. If the sales for the day total \$585.50 and 278 items were sold, how many of each item were sold?

*Solution* **Understand and Translate** We are asked to find the number of hot dogs and beef tacos sold.

$$\text{Let } x = \text{number of hot dogs sold.}$$
$$\text{Then } 278 - x = \text{number of beef tacos sold.}$$

| Item | Cost of Item | Number of Items | Total Sales |
|---|---|---|---|
| Hot dogs | 2.00 | $x$ | $2.00x$ |
| Beef tacos | 2.25 | $278 - x$ | $2.25(278 - x)$ |

$$\text{total sales of hot dogs} + \text{total sales of beef tacos} = \text{total sales}$$
$$2.00x + 2.25(278 - x) = 585.50$$

**Carry Out**

$$2.00x + 625.50 - 2.25x = 585.50$$
$$-0.25x + 625.50 = 585.50$$
$$-0.25x = -40$$
$$x = \frac{-40}{-0.25} = 160$$

**Answer** Therefore, 160 hot dogs and $278 - 160 = 118$ beef tacos were sold.

▸ **Now Try Exercise 17**

In Example 5 we could have multiplied both sides of the equation by 100 to eliminate the decimal numbers, and then solved the equation.

Example 6 is a mixture problem that involves the mixing of two solutions.

**EXAMPLE 6 ▸ Medicine Mixture** George Devenney, a chemist, has both 6% and 15% lithium citrate solutions. He wishes to make 0.5 liter of an 8% lithium citrate solution. How much of each solution must he mix?

*Solution* **Understand and Translate** We are asked to find the amount of each solution to be mixed.

$$\text{Let } x = \text{number of liters of 6\% solution.}$$
$$\text{Then } 0.5 - x = \text{number of liters of 15\% solution.}$$

The amount of lithium citrate in a solution is found by multiplying the percent strength of lithium citrate in the solution by the volume of the solution. We will draw a sketch of the problem (see **Fig. 2.8**) and then construct a table.

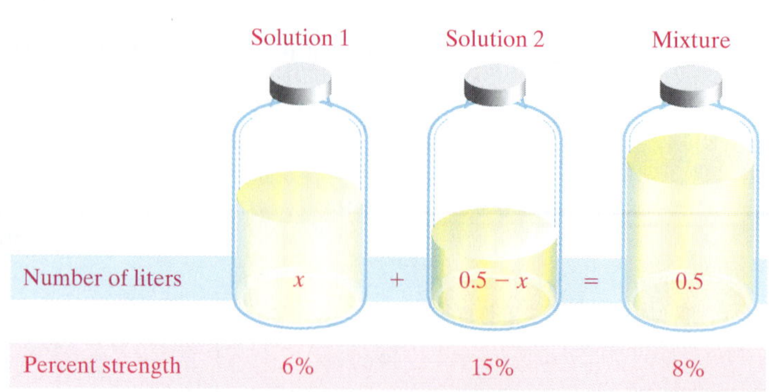


FIGURE 2.8

| Solution | Strength of Solution | Number of Liters | Amount of Lithium Citrate |
|---|---|---|---|
| 1 | 0.06 | $x$ | $0.06x$ |
| 2 | 0.15 | $0.5 - x$ | $0.15(0.5 - x)$ |
| Mixture | 0.08 | 0.5 | $0.08(0.5)$ |

$$\begin{pmatrix}\text{amount of}\\ \text{lithium citrate}\\ \text{in 6\% solution}\end{pmatrix} + \begin{pmatrix}\text{amount of}\\ \text{lithium citrate}\\ \text{in 15\% solution}\end{pmatrix} = \begin{pmatrix}\text{amount of lithium citrate}\\ \text{in mixture}\end{pmatrix}$$

$$0.06x \quad + \quad 0.15(0.5 - x) \quad = \quad 0.08(0.5)$$

**Carry Out**

$$\begin{aligned} 0.06x + 0.15(0.5 - x) &= 0.08(0.5)\\ 0.06x + 0.075 - 0.15x &= 0.04\\ 0.075 - 0.09x &= 0.04\\ -0.09x &= -0.035\\ x &= \frac{-0.035}{-0.09} \approx 0.39 \end{aligned}$$

(to nearest hundredth)

George must mix 0.39 liter of the 6% solution and $0.5 - x$ or $0.5 - 0.39 = 0.11$ liter of the 15% solution to make 0.5 liter of an 8% solution.

▸ Now Try Exercise 21

## EXERCISE SET 2.4

MathXL® MyMathLab

### Practice the Skills and Problem Solving

*In Exercises 1–14, write an equation that can be used to solve the motion problem. Solve the equation and answer the question asked.*

1. **A Hike in the Rockies** Two friends, Don O'Neal and Judy McElroy, go hiking in the Rocky Mountains. While hiking they come across Bear Lake. They wonder what the distance around the lake is and decide to find out. Don knows he walks at 5 mph and Judy knows she walks at 4.5 mph. If they start walking at the same time in opposite directions around the lake, and meet in 1.2 hours, what is the distance around the lake?

2. **Earthquake Shock Waves** An earthquake occurs in a desert of California. The shock waves travel outward in a circular path, similar to when a pebble is dropped into a pond. If the $p$-wave (one kind of shock wave) travels at 2.4 miles per second, how long would it take for the wave to have a diameter of 60 miles? (See the figure.)

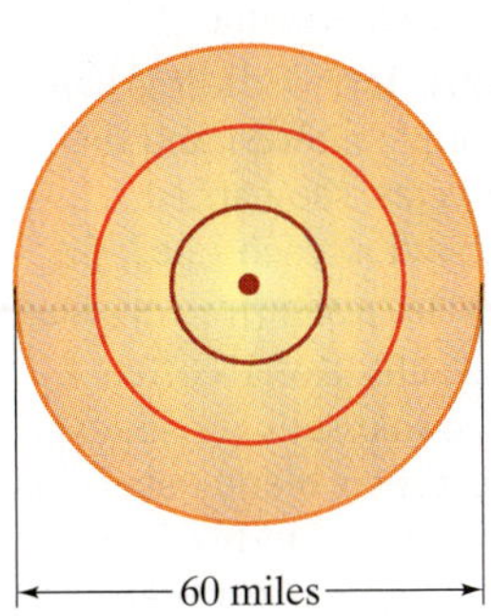


3. **Balloon Flights** Each year Albuquerque, New Mexico, has a hot air balloon festival during which people can obtain rides in hot air balloons. Suppose part of the Diaz family goes in one balloon and other members of the family go in another balloon. Because they fly at different altitudes and carry different weights, one balloon travels at 14 miles per hour while the other travels in the same direction at 11 miles per hour. In how many hours will they be 12 miles apart?

4. **Bikes** Paul and Frank are on the same bike path 39.15 miles apart. They are going to ride their bikes toward each other until they meet. Frank starts pedaling $1\frac{1}{2}$ hours before Paul. Paul rides 1.8 miles per hour faster than Frank. If they meet 3 hours after Paul starts riding, find the speed of each biker.

5. **Cornfield** Rodney Joseph and Dennis Clarence are gleaning (or collecting) corn from a cornfield that is 1.5 miles long. Rodney starts on one end and is gleaning corn at a rate of 0.15 miles per hour. Dennis starts on the other end and is gleaning corn at a rate of 0.10 miles per hour. If they start at the same time and continue to work at the same rate, how long will it be before Rodney and Dennis meet?

6. **Photocopying** To make a large number of copies, Ruth Cardiff uses two photocopiers. One copier can produce copies at a rate of 42 copies per minute. The other copier can produce copies at a rate of 52 copies per minute. If Ruth starts both machines at the same time, how long will it take the two machines to produce a total of 1316 copies?

7. **Charity Race** The Alpha Delta Pi Sorority raises money for the Ronald McDonald House by holding an annual "Roll for Ronald" race in College Station, Texas. Mary Lou Baker rides a bicycle and travels at twice the speed of Wayne Siegert, who is on rollerblades. Mary and Wayne begin the race at the same time and after 3 hours, Mary is 18 miles ahead of Wayne.

   **a)** What is Wayne's speed?

   **b)** What is Mary's speed?

8. **Canyon Hiking** Jennifer Moyers hikes down to the bottom of Bryce Canyon, camps overnight, and returns the next day. Her hiking speed down averages 3.5 miles per hour and her return trip averages 2.1 miles per hour. If she spent a total of 16 hours hiking, find

   **a)** how long it took her to reach the bottom of the canyon.

   **b)** the total distance traveled.

9. **Catching Up** Luis Nunez begins a long-distance walk at a rate of 4 mph. Forty-five minutes after he leaves, his wife, Kristin, realizes that Luis has forgotten his wallet. Kristin gets on her bicycle and begins riding at a rate of 24 mph along the same path that Luis took.

   **a)** How long will it take for Kristin to catch Luis?

   **b)** How far from their house will Kristin catch Luis?

10. **Snooty the Manatee** The South Florida Museum in Bradenton is home to Snooty the manatee, who lives in a 60,000-gallon tank. Once a year, his keepers drain and refill Snooty's tank. The tank has two inlet valves that have the same flow rate. To fill the tank, the first valve is opened for a period of 17 hours. During this 17-hour period the second valve is opened for 7 hours. Determine the flow rate of the two inlet valves in gallons per hour.

11. **Packing Spaghetti** Two machines are packing spaghetti into boxes. The smaller machine can package 400 boxes per hour and the larger machine can package 600 boxes per hour. If the larger machine is on for 2 hours before the smaller machine is turned on, how long after the smaller machine is turned on will a total of 15,000 boxes of spaghetti be boxed?

12. **Snail Races** As part of their preschool science project, Mrs. Joy Pribble's class is holding snail races. The first snail, Zippy, is known to move at a rate of 5 inches per hour. The second snail, Lightning, is known to move at 4.5 inches per hour. If the snails race along a straight path, and if Zippy finishes the race 0.25 hour before Lightning,

   **a)** determine the time it takes Lightning to run the race.

   **b)** determine the time it takes Zippy to run the race.

   **c)** what is the distance covered by both snails?

13. **Trip** Linda Smoke began driving to Pizza Hut at a speed of 35 miles per *hour*. Fifteen *minutes* later, her husband discovered that she had forgotten her wallet with the money to pay for the pizzas. He tried to catch up to her. If he traveled at 50 miles per hour, how long did it take for the husband to catch up to Linda?

14. **Walkie-Talkie Range** A set of Maxon RS446 walkie-talkies has a range of about two miles. Alice Burstein and Mary Kalscheur begin walking along a nature trail heading in opposite directions carrying their walkie-talkies. If Alice walks at a rate of 3.8 mph and Mary walks at a rate of 4.2 mph, how long will it take for them to be out of the range of the walkie-talkies?

*In Exercises 15–28, write an equation that can be used to solve the mixture problem. Solve each equation and answer the question asked.*

15. **Two Investments** Bill Palow invested $30,000 in two separate accounts paying 3% and 4.1% annual simple interest. If Bill earned a total of $1091.73 from the two investments, how much was invested in each account.

16. **Two Investments** Terry Edwards invested $3000 for two years, part at 3.5% simple interest and the rest at 2.5% simple interest. After two years, she earned a total interest of $190. How much was invested at each rate?

17. **Mixing Coffee** Joan Smith is the owner of a Starbucks Coffee Shop. She sells Kona coffee for $6.20 per pound and amaretto coffee that sells for $5.80 per pound. She finds that by mixing these two blends she creates a coffee that sells well. If she uses 18 pounds of amaretto coffee in the blend and wishes to sell the mixture at $6.10 per pound, how many

pounds of the Kona coffee should she mix with the amaretto coffee?

18. **Mixing Nuts** J. B. Davis owns a nut shop. He sells almonds for \$6 per pound and walnuts for \$5.20 per pound. He receives a special request from a customer who wants to purchase 30 pounds of a mixture of almonds and walnuts for \$165. Determine how many pounds of almonds and walnuts should be mixed.

19. **Investing an Inheritance** Don Beville has inherited \$250,000 and wishes to invest his inheritance in shares of Johnson & Johnson stock and AOL Time Warner stock. He wishes to buy twice as many shares of AOL as shares of Johnson & Johnson. Recently, the price of Johnson & Johnson stock was \$56.88 per share and the price of AOL stock was \$27.36.
    **a)** If Don wishes to buy shares in blocks of 100, how many shares of each company can he buy?
    **b)** How much money will he have left after making the purchases?

20. **Sulfuric Acid Solutions** Read Wickham, a chemistry teacher, needs a 5% sulfuric acid solution for an upcoming chemistry laboratory. When checking the storeroom, he realizes he has only 8 ounces of a 25% sulfuric acid solution. There is not enough time for him to order more, so he decides to make a 5% sulfuric acid solution by very carefully adding water to the 25% sulfuric acid solution. Determine how much water Read must add to the 25% solution to reduce it to a 5% solution.

21. **Vinegar Solutions** Distilled white vinegar purchased in supermarkets generally has a 5% acidity level. To make her sauerbraten, Chef Judy Ackermary marinates veal overnight in a special 8% distilled vinegar that she creates herself. To create the 8% solution she mixes a regular 5% vinegar solution with a 12% vinegar solution that she purchases by mail. How many ounces of the 12% vinegar solution should she add to 40 ounces of the 5% vinegar solution to get the 8% vinegar solution?

22. **Hydrogen Peroxide Solution** David Robertson works as a chemical engineer for US Peroxide Corporation. David has 2500 gallons of a commercial-grade hydrogen peroxide solution that is 60% pure hydrogen peroxide. How much distilled water (which is 0% hydrogen peroxide) will David need to add to this solution to create a new solution that is 25% pure hydrogen peroxide?

23. **Horseradish Sauces** Sally Finkelstein has a recipe that calls for horseradish sauce that is 45% pure horseradish. At the grocery store she finds one horseradish sauce that is 30% pure horseradish and another that is 80% pure horseradish. How many teaspoons of each of these horseradish sauces should Sally mix together to get 4 teaspoons of horseradish sauce that is 45% pure horseradish?

24. **Grass Seed Mixture** The Pearlman Nursery sells two types of grass seeds in bulk. The lower quality seeds have a germination rate of 76%, but the germination rate of the higher quality seeds is not known. Seven pounds of the higher quality seeds are mixed with 14 pounds of the lower quality seeds. If a later analysis of the mixture reveals that the mixture's germination rate was 80%, what is the germination rate of the higher quality seed?

25. **Acid Solutions** Two acid solutions are available to a chemist. One is a 20% sulfuric acid solution, but the label that indicates the strength of the other sulfuric acid solution is missing. Two hundred milliliters of the 20% acid solution and 100 milliliters of the solution with the unknown strength are mixed together. Upon analysis, the mixture was found to have a 25% sulfuric acid concentration. Find the strength of the solution with the missing label.

26. **A Tax Strategy** Some states allow a husband and wife to file individual state tax returns even though they file a joint federal return. It is usually to the taxpayers' advantage to do this when both the husband and wife work. The smallest amount of tax owed (or the largest refund) will occur when the husband's and wife's taxable incomes are the same.

    Mr. Juenger's 2005 taxable income was \$28,200 and Mrs. Juengers' taxable income for that year was \$32,450. The Juengers' total tax deductions for the year were \$6400. This deduction can be divided between Mr. and Mrs. Juenger in any way they wish. How should the \$6400 be divided between them to result in each person's having the same taxable income?

27. **Candy Mixture** A supermarket is selling two types of candies, orange slices and strawberry leaves. The orange slices cost \$1.29 per pound and the strawberry leaves cost \$1.79 per pound. How many pounds of each should be mixed to get a 12-pound mixture that sells for \$17.48?

28. **Octane Ratings** The octane rating of gasoline indicates the percent of pure octane in the gasoline. For example, most regular gasoline has an 87-octane rating, which means that this gasoline is 87% octane (and 13% of some other non-octane fuel such as pentane). Blake De Young owns a gas station and has 850 gallons of 87-octane gasoline. How many gallons of 93-octane gasoline must he mix with this 87-octane gasoline to obtain 89-octane gasoline?

*In Exercises 29–46, write an equation that can be used to solve the motion or mixture problem. Solve each equation and answer the questions asked.*

**29. Route 66** The famous highway U.S. Route 66 connects Chicago and Los Angeles and extends 2448 miles. Julie Turley begins in Chicago and drives at an average rate of 45 mph on Route 66 toward Los Angeles. At the same time Kamilia Nemri begins in Los Angeles and drives on Route 66 at an average rate of 50 mph toward Chicago. If Julie and Kamilia maintain these average speeds, how long after they start will they meet?

**30. Meeting at a Restaurant** Mike Mears and Scott Greenhalgh live 110 miles from each other. They frequently meet for lunch at a restaurant that is between Mike's house and Scott's house. Leaving at the same time from their respective houses, Mike takes 1 hour and 30 minutes and Scott takes 1 hour and 15 minutes to get to the restaurant. If they each drive at the same speed,

**a)** determine their speed.

**b)** how far from Scott's house is the restaurant?

**31. Sump Pump Rates** Gary Egan needs to drain his 15,000-gallon inground swimming pool to have it resurfaced. He uses two pumps to drain the pool. One drains 10 gallons of water a minute while the other drains 20 gallons of water a minute. If the pumps are turned on at the same time and remain on until the pool is drained, how long will it take for the pool to be drained?

**32. Two Investments** Chuy Carreon invested \$8000 for 1 year, part at 3% and part at 5% simple interest. How much was invested in each account if the same amount of interest was received from each account?

**33. Antifreeze Solution** How many quarts of pure antifreeze should Doreen Kelly add to 10 quarts of a 20% antifreeze solution to make a 50% antifreeze solution?

**34. Trip to Hawaii** A jetliner flew from Chicago to Los Angeles at an average speed of 500 miles per hour. Then it continued on over the Pacific Ocean to Hawaii at an average speed of 550 miles per hour. If the entire trip covered 5200 miles and the part over the ocean took twice as long as the part over land, how long did the entire trip take?

**35. Refueling a Jet** An Air Force jet is going on a long-distance flight and will need to be refueled in midair over the Pacific Ocean. A refueling plane that carries fuel can travel much farther than the jet but flies at a slower speed. The refueling plane and jet will leave from the same base, but the refueling plane will leave 2 hours before the jet. The jet will fly at 800 mph and the refueling plane will fly at 520 miles per hour.

**a)** How long after the jet takes off will the two planes meet?

**b)** How far from the base will the refueling take place?

**36. Working Two Jobs** Hal Turziz works two part-time jobs. One job pays \$7.50 an hour and the other pays \$8.25 an hour. Last week Hal earned a total of \$190.50 and worked for a total of 24 hours. How many hours did Hal work at each job?

**37. Painting Sales** Joseph DeGuizman, an artist, sells both large paintings and small paintings. He sells his small paintings for \$60 and his large paintings for \$180. At the end of the week he determined that the total amount he made by selling 12 paintings was \$1200. Determine the number of small and the number of large paintings that he sold.

**38. Trip to Work** Vince Jansen lives 35 miles from work. Due to construction, he must drive the first 15 minutes at a speed 10 mph slower than he does on the rest of the trip. If the entire trip takes him 45 minutes, determine Vince's speed on each part of the trip to work.

**39. Alcohol Solution** Herb Garrett has an 80% methyl alcohol solution. He wishes to make a gallon of windshield washer solution by mixing his methyl alcohol solution with water. If 128 ounces, or a gallon, of windshield washer fluid should contain 6% methyl alcohol, how much of the 80% methyl alcohol solution and how much water must be mixed?

**40. Mowing the Lawn** Richard Stewart mows part of his lawn in second gear and part in third gear. It took him 2 hours to mow the entire lawn and the odometer on his tractor shows that he covered 13.8 miles while cutting the grass. If he averages 4.2 miles per hour in second gear and 7.8 miles per hour in third gear, how long did he cut in each gear?

**41. Meatloaf** Lori Sullivan is making a meatloaf by combining chopped sirloin with veal. The sirloin contains 1.2 grams of fat per ounce, and the veal contains 0.3 gram of fat per ounce. If she wants her 64-ounce mixture to have only 0.8 gram of fat per ounce, find how much sirloin and how much veal she must use.

**42. Milk Mixture** Sundance Dairy has 200 quarts of whole milk containing 6% butterfat. How many quarts of low-fat milk containing 1.5% butterfat should be added to produce milk containing 2.4% butterfat?

**43. Comparing Transportation** George Young can ride his bike to work in $\frac{3}{4}$ hour. If he takes his car to work, the trip takes $\frac{1}{6}$ hour. If George drives his car an average of 14 miles per hour faster than he rides his bike, determine the distance he travels to work.

**44. Milk Carton Machine** An old machine that folds and seals milk cartons can produce 50 milk cartons per minute. A new machine can produce 70 milk cartons per minute. The old machine has made 1000 milk cartons when the new machine is turned on. If both machines then continue working, how long after the new machine is turned on will the new machine produce the same total number of milk cartons as the old machine?

**45. Ocean Salinity** The salinity (salt content) of the Atlantic Ocean averages 37 parts per thousand. If 64 ounces of water is collected and placed in the sun, how many ounces of pure water would need to evaporate to raise the salinity to 45 parts per thousand? (Only the pure water is evaporated; the salt is left behind.)

**46. Two Rockets** Two rockets are launched from the Kennedy Space Center. The first rocket, launched at noon, will travel at 8000 miles per hour. The second rocket will be launched some time later and travel at 9500 miles per hour. When should the second rocket be launched if the rockets are to meet at a distance of 38,000 miles from earth?

a) Explain how to find the solution to this problem.

b) Find the solution to the problem.

**47.** a) Make up your own realistic motion problem that can be represented as an equation.

b) Write the equation that represents your problem.

c) Solve the equation, and then find the answer to your problem.

**48.** a) Make up your own realistic mixture problem that can be represented as an equation.

b) Write the equation that represents your problem.

c) Solve the equation, and then find the answer to your problem.

## Challenge Problems

**49. Distance to Calais** The Chunnel (the underwater tunnel from Folkestone, England, to Calais, France) is 31 miles long. A person can board France's TGV bullet train in Paris and travel nonstop through the Chunnel and arrive in London 3 hours later. The TGV averages about 130 miles per hour from Paris to Calais. It then reduces its speed to an average of 90 miles per hour through the 31-mile Chunnel. When leaving the Chunnel in Folkestone it travels only at an average of about 45 miles per hour for the 68-mile trip from Folkestone to London because of outdated tracks. Using this information, determine the distance from Paris to Calais, France.

**50. Race Cars** Two cars labeled $A$ and $B$ are in a 500-lap race. Each lap is 1 mile. The lead car, $A$, is averaging 125 miles per hour when it reaches the halfway point. Car $B$ is exactly 6.2 laps behind.

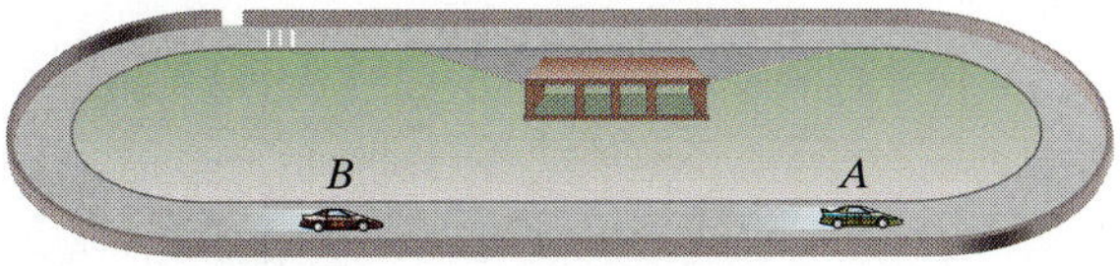


a) Find the average speed of car $B$.

b) When car $A$ reaches the halfway point, how far behind, in seconds, is car $B$ from car $A$?

**51. Antifreeze Solution** The radiator of an automobile has a capacity of 16 quarts. It is presently filled with a 20% antifreeze solution. How many quarts must be drained and replaced with pure antifreeze to make the radiator contain a 50% antifreeze solution?

## Cumulative Review Exercises

[1.6] **52.** Express the quotient in scientific notation. $\dfrac{2.52 \times 10^{17}}{3.6 \times 10^{4}}$

*Solve.*

[2.1] **53.** $0.6x + 0.22 = 0.4(x - 2.3)$

**54.** $\dfrac{2}{3}x + 8 = x + \dfrac{25}{4}$

[2.2] **55.** Solve the equation $\dfrac{3}{5}(x - 2) = \dfrac{2}{7}(2x + 3y)$ for $y$.

[2.3] **56. Truck Rental** Hertz/Penske Truck Rental Agency charges \$30 per day plus 14¢ a mile. Budget Truck Rental Agency charges \$16 per day plus 24¢ a mile for the same truck. What distance would you have to drive in 1 day to make the cost of renting from Hertz/Penske equal to the cost of renting from Budget?

# 2.5 Solving Linear Inequalities

1. Solve inequalities.
2. Graph solutions on a number line, interval notation, and solution sets.
3. Solve compound inequalities involving *and*.
4. Solve compound inequalities involving *or*.

## 1 Solve Inequalities

Inequalities and set builder notation were introduced in Section 1.2. You may wish to review that section now. The inequality symbols follow.*

**Inequality Symbols**

| | |
|---|---|
| $>$ | is greater than |
| $\geq$ | is greater than or equal to |
| $<$ | is less than |
| $\leq$ | is less than or equal to |

A mathematical expression containing one or more of these symbols is called an **inequality**. The direction of the inequality symbol is sometimes called the **order** or **sense** of the inequality.

### Examples of Inequalities in One Variable

$$2x + 3 \leq 5 \qquad 4x > 3x - 5 \qquad 1.5 \leq -2.3x + 4.5 \qquad \frac{1}{2}x + 3 \geq 0$$

*To solve an inequality, we must isolate the variable on one side of the inequality symbol.* To isolate the variable, we use the same basic techniques used in solving equations.

**Properties Used to Solve Inequalities**

1. If $a > b$, then $a + c > b + c$.
2. If $a > b$, then $a - c > b - c$.
3. If $a > b$, and $c > 0$, then $ac > bc$.
4. If $a > b$, and $c > 0$, then $\dfrac{a}{c} > \dfrac{b}{c}$.
5. If $a > b$, and $c < 0$, then $ac < bc$.
6. If $a > b$, and $c < 0$, then $\dfrac{a}{c} < \dfrac{b}{c}$.

The first two properties state that the same number can be added to or subtracted from both sides of an inequality. The third and fourth properties state that both sides of

*$\neq$, is not equal to, is also an inequality. $\neq$ means $<$ or $>$. Thus, $2 \neq 3$ means $2 < 3$ or $2 > 3$.

an inequality can be multiplied or divided by any positive real number. The last two properties indicate that **when both sides of an inequality are multiplied or divided by a negative number, the direction of the inequality symbol reverses.**

| Example of Multiplication by a Negative Number | Example of Division by a Negative Number |
|---|---|
| $4 > -2$ | $10 \geq -4$ |
| $-1(4) < -1(-2)$ | $\frac{10}{-2} \leq \frac{-4}{-2}$ |
| $-4 < 2$ | $-5 \leq 2$ |

*Multiply both sides of the inequality by −1 and reverse the direction of the inequality symbol.*

*Divide both sides of the inequality by −2 and reverse the direction of the inequality symbol.*

## Helpful Hint

Do not forget to reverse the direction of the inequality symbol when multiplying or dividing both sides of the inequality by a negative number.

| Inequality | Direction of Inequality Symbol |
|---|---|
| $-3x < 6$ | $\frac{-3x}{-3} > \frac{6}{-3}$ |
| $-\frac{x}{2} > 5$ | $(-2)\left(-\frac{x}{2}\right) < (-2)(5)$ |

**EXAMPLE 1** ▶ Solve the inequalities. **a)** $5x - 7 \geq -17$ **b)** $-6x + 4 < -14$

*Solution*

**a)**

$$5x - 7 \geq -17$$

$$5x - 7 + 7 \geq -17 + 7 \quad \textit{Add 7 to both sides.}$$

$$5x \geq -10$$

$$\frac{5x}{5} \geq \frac{-10}{5} \quad \textit{Divide both sides by 5.}$$

$$x \geq -2$$

The solution set is $\{x|x \geq -2\}$. Any real number greater than or equal to −2 will satisfy the inequality.

**b)**

$$-6x + 4 < -14$$

$$-6x + 4 - 4 < -14 - 4 \quad \textit{Subtract 4 from both sides.}$$

$$-6x < -18$$

$$\frac{-6x}{-6} > \frac{-18}{-6} \quad \textit{Divide both sides by −6 and reverse the direction of the inequality.}$$

$$x > 3$$

The solution set is $\{x|x > 3\}$. Any number greater than 3 will satisfy the inequality.

▶ **Now Try Exercise 17**

## 2 Graph Solutions on a Number Line, Interval Notation, and Solution Sets

The solution to an inequality can be indicated on a number line or written as a solution set, as was explained in Section 1.2. The solution can also be written in interval notation, as illustrated on the next page. Most instructors have a preferred way to indicate the solution to an inequality.

Recall that *a solid circle on the number line indicates that the endpoint is part of the solution, and an open circle indicates that the endpoint is not part of the solution. In interval notation, brackets,* [ ], *are used to indicate that the endpoints are part of the solution and parentheses,* ( ), *indicate that the endpoints are not part of the solution.* The symbol $\infty$ is read "infinity"; it indicates that the solution set continues indefinitely. Whenever $\infty$ is used in interval notation, a *parenthesis* must be used on the corresponding side of the interval notation.

| Solution of Inequality | Solution Set Indicated on Number Line | Solution Set Represented in Interval Notation |
|---|---|---|
| $x \geq 5$ | 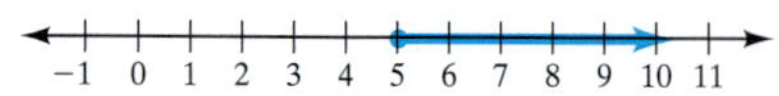 | $[5, \infty)$ |
| $x < 3$ | 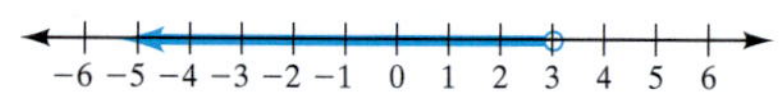 | $(-\infty, 3)$ |
| $2 < x \leq 6$ | 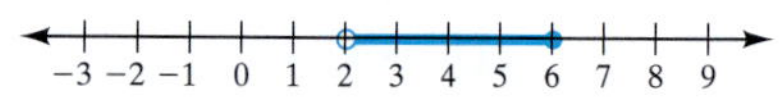 | $(2, 6]$ |
| $-6 \leq x \leq -1$ | 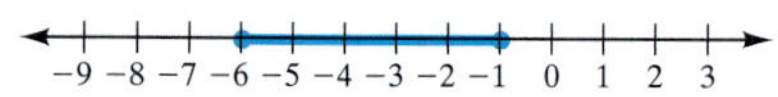 | $[-6, -1]$ |
| $x > a$ |  | $(a, \infty)$ |
| $x \geq a$ |  | $[a, \infty)$ |
| $x < a$ | 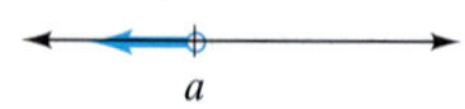 | $(-\infty, a)$ |
| $x \leq a$ |  | $(-\infty, a]$ |
| $a < x < b$ |  | $(a, b)$ |
| $a \leq x \leq b$ |  | $[a, b]$ |
| $a < x \leq b$ |  | $(a, b]$ |
| $a \leq x < b$ |  | $[a, b)$ |

In the next example, we will solve an inequality which has fractions.

**EXAMPLE 2 ▸** Solve the following inequality and give the solution both on a number line and in interval notation.

$$\frac{1}{4}z - \frac{1}{2} < \frac{2z}{3} + 2$$

*Solution* We can eliminate fractions from an inequality by multiplying both sides of the inequality by the least common denominator, LCD, of the fractions. In this case we multiply both sides of the inequality by 12. We then solve the resulting inequality as we did in the previous example.

$$\frac{1}{4}z - \frac{1}{2} < \frac{2z}{3} + 2$$

$$12\left(\frac{1}{4}z - \frac{1}{2}\right) < 12\left(\frac{2z}{3} + 2\right)$$ Multiply both sides by the LCD, 12.

$$3z - 6 < 8z + 24$$ Distributive property

$$3z - 8z - 6 < 8z - 8z + 24$$ Subtract $8z$ from both sides.

$$-5z - 6 < 24$$

$$-5z - 6 + 6 < 24 + 6$$ Add 6 to both sides.

$$-5z < 30$$

$$\frac{-5z}{-5} > \frac{30}{-5}$$ Divide both sides by $-5$ and reverse the direction of inequality symbol.

$$z > -6$$

[1.3] *Name each illustrated property.*

**101.** $(3x + 8) + 4y = 3x + (8 + 4y)$

**102.** $5x + y = y + 5x$

[2.2] **103.** Solve the formula $R = L + (V - D)r$ for $V$.

# 2.6 Solving Equations and Inequalities Containing Absolute Values

1 Understand the geometric interpretation of absolute value.

2 Solve equations of the form $|x| = a, a > 0$.

3 Solve inequalities of the form $|x| < a, a > 0$.

4 Solve inequalities of the form $|x| > a, a > 0$.

5 Solve inequalities of the form $|x| > a$ or $|x| < a, a < 0$.

6 Solve inequalities of the form $|x| > 0$ or $|x| < 0$.

7 Solve equations of the form $|x| = |y|$.

## 1 Understand the Geometric Interpretation of Absolute Value

In Section 1.3 we introduced the concept of absolute value. We stated that the absolute value of a number may be considered the distance (without sign) from the number 0 on the number line. The absolute value of 3, written $|3|$, is 3 since it is 3 units from 0 on the number line. Similarly, the absolute value of $-3$, written $|-3|$, is also 3 since it is 3 units from 0 on the number line.

Consider the equation $|x| = 3$; what values of $x$ make this equation true? We know that $|3| = 3$ and $|-3| = 3$. The solutions to $|x| = 3$ are 3 and $-3$. When solving the equation $|x| = 3$, we are finding the values whose distances are exactly 3 units from 0 on the number line (see **Fig. 2.14a**).

Now consider the inequality $|x| < 3$. To solve this inequality, we need to find the set of values whose distances are less than 3 units from 0 on the number line. These are the values of $x$ between $-3$ and 3 (see **Fig. 2.14b**).

To solve the inequality $|x| > 3$, we need to find the set of values whose distances are greater than 3 units from 0 on the number line. These are the values that are either less than $-3$ or greater than 3 (see **Fig. 2.14c**).

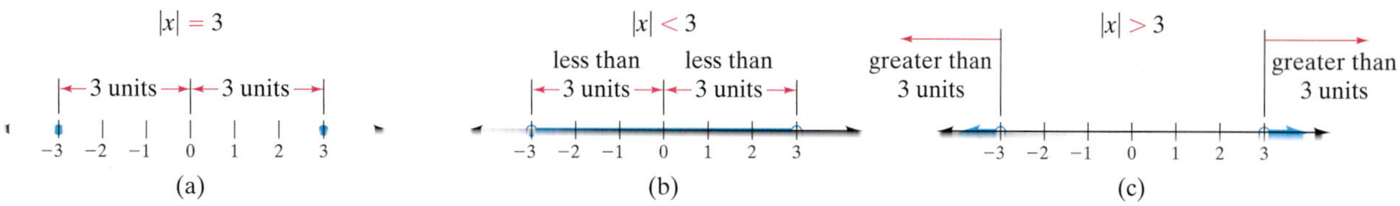


FIGURE 2.14

In this section we will solve equations and inequalities such as the following:

$$|2x - 1| = 5 \qquad |2x - 1| \le 5 \qquad |2x - 1| > 5$$

The geometric interpretation of $|2x - 1| = 5$ is similar to $|x| = 3$. When solving $|2x - 1| = 5$, we are determining the set of values that result in $2x - 1$ being exactly 5 units away from 0 on the number line.

The geometric interpretation of $|2x - 1| \le 5$ is similar to the geometric interpretation of $|x| \le 3$. When solving $|2x - 1| \le 5$, we are determining the set of values that result in $2x - 1$ being less than or equal to 5 units from 0 on the number line.

The geometric interpretation of $|2x - 1| > 5$ is similar to that of $|x| > 3$. When solving $|2x - 1| > 5$, we are determining the set of values that result in $2x - 1$ being greater than 5 units from 0 on the number line.

We will be solving absolute value equations and inequalities algebraically in the remainder of this section. We will first solve absolute value equations, then we will solve absolute value inequalities. We will end the section by solving absolute value equations where both sides of the equation contain an absolute value, for example, $|x + 3| = |2x - 5|$.

## 2 Solve Equations of the Form $|x| = a, a > 0$

When solving an equation of the form $|x| = a$, $a > 0$, we are finding the values that are exactly $a$ units from 0 on the number line. The following procedure may be used to solve such problems.

**To Solve Equations of the Form $|x| = a$**

If $|x| = a$ and $a > 0$, then $x = a$ or $x = -a$.

**EXAMPLE 1** ▶ Solve each equation.

**a)** $|x| = 7$ **b)** $|x| = 0$ **c)** $|x| = -7$

*Solution*

**a)** Using the procedure, we get $x = 7$ or $x = -7$. The solution set is $\{-7, 7\}$.

**b)** The only real number whose absolute value equals 0 is 0. Thus, the solution set for $|x| = 0$ is $\{0\}$.

**c)** The absolute value of a number is never negative, so there are no solutions to this equation. The solution set is $\varnothing$.

▶ **Now Try Exercise 15**

**EXAMPLE 2** ▶ Solve the equation $|2w - 1| = 5$.

*Solution* At first this might not appear to be of the form $|x| = a$. However, if we let $2w - 1$ be $x$ and 5 be $a$, you will see that the equation is of this form. We are looking for the values of $w$ such that $2w - 1$ is exactly 5 units from 0 on a number line. Thus, the quantity $2w - 1$ must be equal to 5 or $-5$.

$$2w - 1 = 5 \quad \text{or} \quad 2w - 1 = -5$$
$$2w = 6 \qquad\qquad 2w = -4$$
$$w = 3 \qquad\qquad w = -2$$

**Check** $w = 3$

$$|2w - 1| = 5$$
$$|2(3) - 1| \stackrel{?}{=} 5$$
$$|6 - 1| \stackrel{?}{=} 5$$
$$|5| \stackrel{?}{=} 5$$
$$5 = 5 \quad \textit{True}$$

$w = -2$

$$|2w - 1| = 5$$
$$|2(-2) - 1| \stackrel{?}{=} 5$$
$$|-4 - 1| \stackrel{?}{=} 5$$
$$|-5| \stackrel{?}{=} 5$$
$$5 = 5 \quad \textit{True}$$

The solutions 3 and $-2$ each result in $2w - 1$ being 5 units from 0 on the number line. The solution set is $\{-2, 3\}$.

▶ **Now Try Exercise 21**

Consider the equation $|2w - 1| - 3 = 2$. The first step in solving this equation is to isolate the absolute value term. We do this by adding 3 to both sides of the equation. This results in the equation we solved in Example 2.

## 3 Solve Inequalities of the Form $|x| < a, a > 0$

Now let's look at inequalities of the form $|x| < a$. Consider $|x| < 3$. This inequality represents the set of values that are less than 3 units from 0 on a number line (see **Fig. 2.14b** on page 125). The solution set is $\{x|-3 < x < 3\}$. The solution set to an inequality of the form $|x| < a$ is the set of values that are *less than a units from 0 on a number line.*

We can use the same reasoning process to solve more complicated problems, as shown in Example 3.

**EXAMPLE 3** ▸ Solve the inequality $|2x - 3| < 5$.

*Solution* The solution to this inequality will be the set of values such that the distance between $2x - 3$ and 0 on a number line will be less than 5 units (see **Fig. 2.15**). Using **Figure 2.15**, we can see that $-5 < 2x - 3 < 5$.

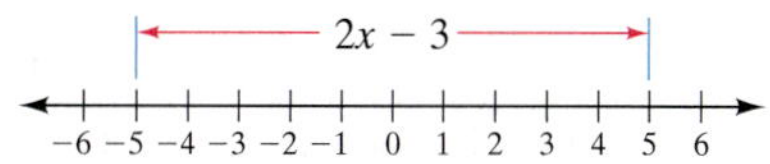


FIGURE 2.15

Solving, we get

$$-5 < 2x - 3 < 5$$
$$-2 < 2x < 8$$
$$-1 < x < 4$$

The solution set is $\{x|-1 < x < 4\}$. When $x$ is any number between $-1$ and 4, the expression $2x - 3$ will represent a number that is less than 5 units from 0 on a number line (or a number between $-5$ and 5).

▸ **Now Try Exercise 33**

To solve inequalities of the form $|x| < a$, we can use the following procedure.

**To Solve Inequalities of the Form $|x| < a$**

If $|x| < a$ and $a > 0$, then $-a < x < a$.

**EXAMPLE 4** ▸ Solve the inequality $|2x + 1| \le 9$ and graph the solution on a number line.

*Solution* Since this inequality is of the form $|x| \le a$, we write

$$-9 \le 2x + 1 \le 9$$
$$-10 \le 2x \le 8$$
$$-5 \le x \le 4$$

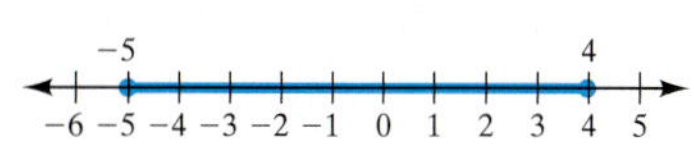


Any value of $x$ greater than or equal to $-5$ and less than or equal to 4 would result in $2x + 1$ being less than or equal to 9 units from 0 on a number line.

▸ **Now Try Exercise 75**

**EXAMPLE 5** ▸ Solve the inequality $|7.8 - 4x| - 5.3 < 14.1$ and graph the solution on a number line.

*Solution* First isolate the absolute value by adding 5.3 to both sides of the inequality. Then solve as in the previous examples.

$$|7.8 - 4x| - 5.3 < 14.1$$
$$|7.8 - 4x| < 19.4$$
$$-19.4 < 7.8 - 4x < 19.4$$
$$-27.2 < -4x < 11.6$$
$$\frac{-27.2}{-4} > \frac{-4x}{-4} > \frac{11.6}{-4}$$
$$6.8 > x > -2.9 \quad \text{or} \quad -2.9 < x < 6.8$$

The solution set is $\{x|-2.9 < x < 6.8\}$. The solution set in interval notation is $(-2.9, 6.8)$.

▸ **Now Try Exercise 43**

## 4 Solve Inequalities of the Form $|x| > a, a > 0$

Now we look at inequalities of the form $|x| > a$. Consider $|x| > 3$. This inequality represents the set of values that are greater than 3 units from 0 on a number line (see **Fig. 2.14c** on page 125). The solution set is $\{x|x < -3 \text{ or } x > 3\}$. The solution set to $|x| > a$ is the set of values that are *greater than a units from* 0 on a number line.

**EXAMPLE 6** ▸ Solve the inequality $|2x - 3| > 5$ and graph the solution on a number line.

*Solution* The solution to $|2x - 3| > 5$ is the set of values such that the distance between $2x - 3$ and 0 on a number line will be greater than 5. The quantity $2x - 3$ must be either less than $-5$ or greater than 5 (see **Fig. 2.16**).

FIGURE 2.16

Since $2x - 3$ must be either less than $-5$ or greater than 5, we set up and solve the following compound inequality:

$$\begin{aligned} 2x - 3 &< -5 \quad \text{or} \quad & 2x - 3 &> 5 \\ 2x &< -2 & 2x &> 8 \\ x &< -1 & x &> 4 \end{aligned}$$

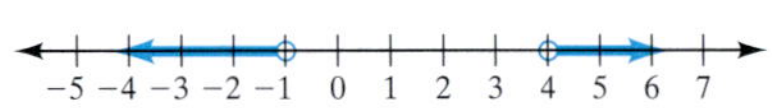


The solution set to $|2x - 3| > 5$ is $\{x|x < -1 \text{ or } x > 4\}$. When $x$ is any number less than $-1$ or greater than 4, the expression $2x - 3$ will represent a number that is greater than 5 units from 0 on a number line (or a number less than $-5$ or greater than 5).

▸ **Now Try Exercise 51**

To solve inequalities of the form $|x| > a$, we can use the following procedure.

**To Solve Inequalities of the Form $|x| > a$**

If $|x| > a$ and $a > 0$, then $x < -a$ or $x > a$.

**EXAMPLE 7** ▸ Solve the inequality $|2x - 1| \geq 7$ and graph the solution on a number line.

*Solution* Since this inequality is of the form $|x| \geq a$, we use the procedure given above.

$$\begin{aligned} 2x - 1 &\leq -7 \quad \text{or} \quad & 2x - 1 &\geq 7 \\ 2x &\leq -6 & 2x &\geq 8 \\ x &\leq -3 & x &\geq 4 \end{aligned}$$

Any value of $x$ less than or equal to $-3$, or greater than or equal to 4, would result in $2x - 1$ representing a number that is greater than or equal to 7 units from 0 on a number line. The solution set is $\{x|x \leq -3 \text{ or } x \geq 4\}$. In interval notation, the solution is $(-\infty, -3] \cup [4, \infty)$.

▸ **Now Try Exercise 53**

**EXAMPLE 8** ▸ Solve the inequality $\left|\frac{3x - 4}{2}\right| \geq 9$ and graph the solution on a number line.

*Solution* Since this inequality is of the form $|x| \geq a$, we write

$$\frac{3x - 4}{2} \leq -9 \quad \text{or} \quad \frac{3x - 4}{2} \geq 9$$

Now multiply both sides of each inequality by the least common denominator, 2. Then solve each inequality.

$$2\left(\frac{3x - 4}{2}\right) \leq -9 \cdot 2 \qquad \text{or} \qquad 2\left(\frac{3x - 4}{2}\right) \geq 9 \cdot 2$$

$$3x - 4 \leq -18 \qquad\qquad 3x - 4 \geq 18$$

$$3x \leq -14 \qquad\qquad 3x \geq 22$$

$$x \leq -\frac{14}{3} \qquad\qquad x = \ \geq \frac{22}{3}$$

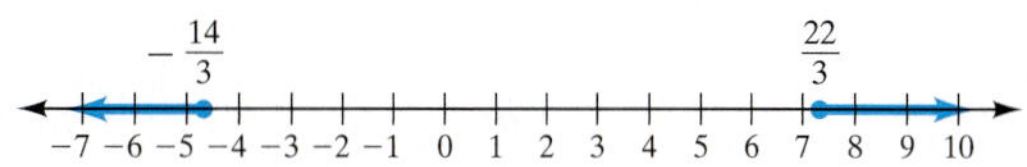


▸ **Now Try Exercise 57**

## Helpful Hint

Some general information about equations and inequalities containing absolute value follows. For real numbers $a$, $b$, and $c$, where $a \neq 0$ and $c > 0$:

| Form of Equation or Inequality | The Solution Will Be: | Solution on a Number Line: |
|---|---|---|
| $\lvert ax + b\rvert = c$ | Two distinct numbers, $p$ and $q$ | 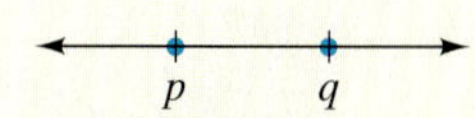  |
| $\lvert ax + b\rvert < c$ | The set of numbers between two numbers, $p < x < q$ | 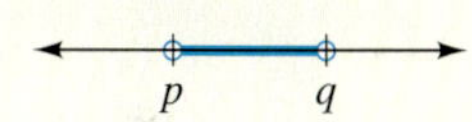  |
| $\lvert ax + b\rvert > c$ | The set of numbers less than one number or greater than a second number, $x < p$ or $x > q$ | 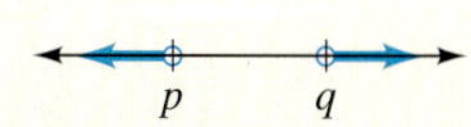  |

## 5 Solve Inequalities of the Form $|x| > a$ or $|x| < a, a < 0$

We have solved inequalities of the form $|x| < a$ where $a > 0$. Now let us consider what happens in an absolute value inequality when $a < 0$. Consider the inequality $|x| < -3$. Since $|x|$ will always have a value greater than or equal to 0 for any real number $x$, this inequality can never be true, and the solution is the empty set, $\varnothing$. Whenever we have an absolute value inequality of this type, the solution will be the empty set.

**EXAMPLE 9** ▸ Solve the inequality $|6x - 8| + 5 < 3$.

*Solution* Begin by subtracting 5 from both sides of the inequality.

$$|6x - 8| + 5 < 3$$
$$|6x - 8| < -2$$

Since $|6x - 8|$ will always be greater than or equal to 0 for any real number $x$, this inequality can never be true. Therefore, the solution is the empty set, $\varnothing$.

▸ **Now Try Exercise 41**

Now consider the inequality $|x| > -3$. Since $|x|$ will always have a value greater than or equal to 0 for any real number $x$, this inequality will always be true. Since every value of $x$ will make this inequality a true statement, the solution is the set of all real numbers, $\mathbb{R}$. Whenever we have an absolute value inequality of this type, the solution will be the set of all real numbers, $\mathbb{R}$.

**EXAMPLE 10 ▸** Solve the inequality $|5x + 3| + 4 \geq -9$.

*Solution* Begin by subtracting 4 from both sides of the inequality.

$$|5x + 3| + 4 \geq -9$$
$$|5x + 3| \geq -13$$

Since $|5x + 3|$ will always be greater than or equal to 0 for any real number $x$, this inequality is true for all real numbers. Thus, the solution is the set of all real numbers, $\mathbb{R}$.

**▸ Now Try Exercise 59**

## 6 Solve Inequalities of the Form $|x| > 0$ or $|x| < 0$

Now let us discuss inequalities where one side of the inequality is 0. The only value that satisfies the equation $|x - 5| = 0$ is 5, since 5 makes the expression inside the absolute value sign 0. Now consider $|x - 5| \leq 0$. Since the absolute value can never be negative, this inequality is true only when $x = 5$. The inequality $|x - 5| < 0$ has no solution. Can you explain why? What is the solution to $|x - 5| \geq 0$? Since any value of $x$ will result in the absolute value being greater than or equal to 0, the solution is the set of all real numbers, $\mathbb{R}$. What is the solution to $|x - 5| > 0$? The solution is every real number except 5. Can you explain why 5 is excluded from the solution?

**EXAMPLE 11 ▸** Solve each inequality. **a)** $|x + 2| > 0$ **b)** $|3x - 8| \leq 0$

*Solution*

**a)** The inequality will be true for every value of $x$ except $-2$. The solution set is $\{x|x < -2 \text{ or } x > -2\}$.

**b)** Determine the number that makes the absolute value equal to 0 by setting the expression within the absolute value sign equal to 0 and solving for $x$.

$$3x - 8 = 0$$
$$3x = 8$$
$$x = \frac{8}{3}$$

The inequality will be true only when $x = \frac{8}{3}$. The solution set is $\left\{\frac{8}{3}\right\}$.

**▸ Now Try Exercise 61**

## 7 Solve Equations of the Form $|x| = |y|$

Now we will discuss absolute value equations where an absolute value appears on both sides of the equation. To solve equations of the form $|x| = |y|$, use the procedure that follows.

**To Solve Equations of the Form $|x| = |y|$**

If $|x| = |y|$, then $x = y$ or $x = -y$.

When solving an absolute value equation with an absolute value expression on each side of the equal sign, the two expressions must have the same absolute value. Therefore, *the expressions must be equal to each other or be opposites of each other.*

**EXAMPLE 12** ▸ Solve the equation $|z + 3| = |2z - 7|$.

*Solution* If we let $z + 3$ be $x$ and $2z - 7$ be $y$, this equation is of the form $|x| = |y|$. Using the procedure given on the previous page, we obtain the two equations

$$z + 3 = 2z - 7 \quad \text{or} \quad z + 3 = -(2z - 7)$$

Now solve each equation.

$$\begin{aligned} z + 3 &= 2z - 7 \\ 3 &= z - 7 \\ 10 &= z \end{aligned} \qquad \text{or} \qquad \begin{aligned} z + 3 &= -(2z - 7) \\ z + 3 &= -2z + 7 \\ 3z + 3 &= 7 \\ 3z &= 4 \\ z &= \frac{4}{3} \end{aligned}$$

**Check** $z = 10$

$$\begin{aligned} |z + 3| &= |2z - 7| \\ |10 + 3| &\stackrel{?}{=} |2(10) - 7| \\ |13| &\stackrel{?}{=} |20 - 7| \\ |13| &\stackrel{?}{=} |13| \\ 13 &= 13 \quad \textit{True} \end{aligned}$$

$z = \frac{4}{3}$

$$\begin{aligned} |z + 3| &= |2z - 7| \\ \left|\frac{4}{3} + 3\right| &\stackrel{?}{=} \left|2\left(\frac{4}{3}\right) - 7\right| \\ \left|\frac{13}{3}\right| &\stackrel{?}{=} \left|\frac{8}{3} - \frac{21}{3}\right| \\ \left|\frac{13}{3}\right| &\stackrel{?}{=} \left|-\frac{13}{3}\right| \\ \frac{13}{3} &= \frac{13}{3} \quad \textit{True} \end{aligned}$$

The solution set is $\left\{10, \frac{4}{3}\right\}$.

▸ **Now Try Exercise 63**

**EXAMPLE 13** ▸ Solve the equation $|4x - 7| = |6 - 4x|$.

*Solution*

$$\begin{aligned} 4x - 7 &= 6 - 4x \\ 8x - 7 &= 6 \\ 8x &= 13 \\ x &= \frac{13}{8} \end{aligned} \qquad \text{or} \qquad \begin{aligned} 4x - 7 &= -(6 - 4x) \\ 4x - 7 &= -6 + 4x \\ -7 &= -6 \quad \textit{False} \end{aligned}$$

Since the equation $4x - 7 = -(6 - 4x)$ results in a false statement, the absolute value equation has only one solution. A check will show that the solution set is $\left\{\frac{13}{8}\right\}$.

▸ **Now Try Exercise 69**

**Summary of Procedures for Solving Equations and Inequalities Containing Absolute Value**

For $a > 0$,

If $|x| = a$, then $x = a$ or $x = -a$.

If $|x| < a$, then $-a < x < a$.

If $|x| > a$, then $x < -a$ or $x > a$.

If $|x| = |y|$, then $x = y$ or $x = -y$.

## EXERCISE SET 2.6

MathXL® MyMathLab

### Concept/Writing Exercises

1. How do we solve equations of the form $|x| = a, a > 0$?
2. For each of the following equations, find the solution set and explain how you determined your answer.
   **a)** $|x| = -2$
   **b)** $|x| = 0$
   **c)** $|x| = 2$
3. How do we solve inequalities of the form $|x| < a, a > 0$?
4. How do you check to see whether $-7$ is a solution to $|2x + 3| = 11$? Is $-7$ a solution?
5. How do we solve inequalities of the form $|x| > a, a > 0$?
6. What is the solution to $|x| < 0$? Explain your answer.
7. What is the solution to $|x| > 0$? Explain your answer.
8. Suppose $m$ and $n$ $(m < n)$ are two distinct solutions to the equation $|ax + b| = c$. Indicate the solutions, using both inequality symbols and the number line, to each inequality. (See the Helpful Hint on page 129.)
   **a)** $|ax + b| < c$
   **b)** $|ax + b| > c$
9. Explain how to solve an equation of the form $|x| = |y|$.
10. How many solutions will $|ax + b| = k, a \neq 0$ have if
   **a)** $k < 0$,
   **b)** $k = 0$,
   **c)** $k > 0$?
11. How many solutions are there to the following equations or inequalities if $a \neq 0$ and $k > 0$?
   **a)** $|ax + b| = k$
   **b)** $|ax + b| < k$
   **c)** $|ax + b| > k$
12. Match each absolute value equation or inequality labeled **a)** through **e)** with the graph of its solution set, labeled A.–E.
   **a)** $|x| = 4$ A. 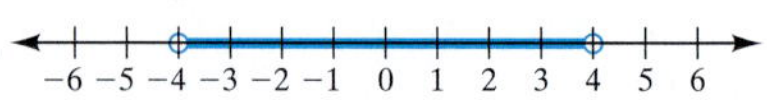
   **b)** $|x| < 4$ B. 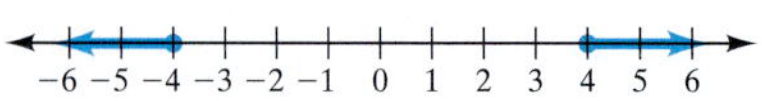
   **c)** $|x| > 4$ C. 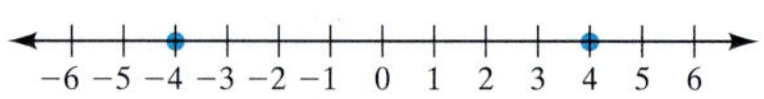
   **d)** $|x| \geq 4$ D. 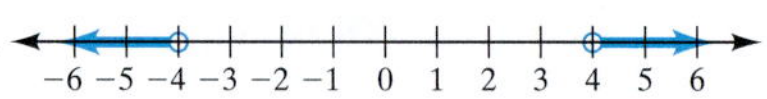
   **e)** $|x| \leq 4$ E. 
13. Match each absolute value equation or inequality, labeled **a)** through **e)**, with its solution set labeled A.–E.
   **a)** $|x| = 5$ A. $\{x|x \leq -5 \text{ or } x \geq 5\}$
   **b)** $|x| < 5$ B. $\{x|-5 < x < 5\}$
   **c)** $|x| > 5$ C. $\{x|-5 \leq x \leq 5\}$
   **d)** $|x| \leq 5$ D. $\{-5, 5\}$
   **e)** $|x| \geq 5$ E. $\{x|x < -5 \text{ or } x > 5\}$
14. Suppose $|x| < |y|$ and $x < 0$ and $y < 0$.
   **a)** Which of the following must be true: $x < y, x > y$, or $x = y$?
   **b)** Give an example to support your answer to part **a)**.

### Practice the Skills

*Find the solution set for each equation.*

15. $|a| = 2$
16. $|b| = 17$
17. $|c| = \frac{1}{2}$
18. $|x| = 0$
19. $|d| = -\frac{5}{6}$
20. $|l + 4| = 6$
21. $|x + 5| = 8$
22. $|3 + y| = \frac{3}{5}$
23. $|4.5q + 31.5| = 0$
24. $|4.7 - 1.6z| = 14.3$
25. $|5 - 3x| = \frac{1}{2}$
26. $|6(y + 4)| = 24$
27. $\left|\frac{x - 3}{4}\right| = 5$
28. $\left|\frac{3z + 5}{6}\right| - 2 = 7$
29. $\left|\frac{x - 3}{4}\right| + 8 = 8$
30. $\left|\frac{5x - 3}{2}\right| + 5 = 9$

*Find the solution set for each inequality.*

31. $|w| < 11$
32. $|p| \leq 9$
33. $|q + 5| \leq 8$
34. $|7 - x| < 6$
35. $|5b - 15| < 10$
36. $|x - 3| - 7 < -2$
37. $|2x + 3| - 5 \leq 10$
38. $|4 - 3x| - 4 < 11$
39. $|3x - 7| + 8 < 14$
40. $\left|\frac{2x - 1}{9}\right| \leq \frac{5}{9}$
41. $|2x - 6| + 5 \leq 1$
42. $|2x - 3| < -10$
43. $\left|\frac{1}{2}j + 4\right| < 7$
44. $\left|\frac{k}{4} - \frac{3}{8}\right| < \frac{7}{16}$
45. $\left|\frac{x - 3}{2}\right| - 4 \leq -2$
46. $\left|7x - \frac{1}{2}\right| < 0$

# Chapter 2 Summary

| IMPORTANT FACTS AND CONCEPTS | EXAMPLES |
| --- | --- |
| **Section 2.1** | |
| **Properties of Equality**<br>For all real numbers $a$, $b$, and $c$: | |
| **1.** $a = a$. *reflexive property* | $9 = 9$ |
| **2.** If $a = b$, then $b = a$. *symmetric property* | If $x = 10$, then $10 = x$. |
| **3.** If $a = b$ and $b = c$, then $a = c$. *transitive property* | If $y = a + b$ and $a + b = 4t$, then $y = 4t$. |
| **Terms** are the parts being added in an algebraic expression. | In the expression $9x^2 - 2x + \frac{1}{5}$, the terms are $9x^2$, $-2x$, and $\frac{1}{5}$. |
| The **coefficient** is the numerical part of a term that precedes the variable. | Term: $15x^4y$ — Coefficient: 15 |
| The **degree of a term** with whole number exponents is the sum of the exponents on the variables. | Term: $17xy^5$ — Degree: $1 + 5 = 6$ |
| **Like terms** are terms that have the same variables with the same exponents. **Unlike terms** are terms that are not like terms. | Like Terms: $2x, 7x$; $9x^2, -5x^2$<br>Unlike Terms: $3x, 4y$; $10x^2, 2x^{10}$ |
| **To simplify an expression** means to combine all the like terms. | $3x^2 + 12x - 5 + 7x^2 - 12x + 1 = 10x^2 - 4$ |
| An **equation** is a mathematical statement of equality. | $x + 15 = 36$ |
| The **solution** of an equation is the number(s) that make the equation a true statement. | The solution to $\frac{1}{2}x + 1 = 7$ is 12. |
| A **linear equation in one variable** is an equation that has the form $ax + b = c, a \neq 0$ | $8x - 3 = 17$ |
| **Addition Property of Equality**<br>If $a = b$, then $a + c = b + c$ for any $a$, $b$, and $c$. | If $5x - 7 = 19$, then $5x - 7 + 7 = 19 + 7$. |
| **Multiplication Property of Equality**<br>If $a = b$, then $a \cdot c = b \cdot c$ for any $a$, $b$, and $c$. | If $\frac{1}{3}x = 2$, then $3 \cdot \frac{1}{3}x = 3 \cdot 2$. |
| **To Solve Linear Equations**<br>**1.** Clear fractions.<br>**2.** Simplify each side separately.<br>**3.** Isolate the variable term on one side.<br>**4.** Solve for the variable.<br>**5.** Check.<br>See page 69 for more detail. | Solve the equation $\frac{1}{2}x + 7 = \frac{4}{3}x - 3$.<br>$\frac{1}{2}x + 7 = \frac{4}{3}x - 3$<br>$6\left(\frac{1}{2}x + 7\right) = 6\left(\frac{4}{3}x - 3\right)$<br>$3x + 42 = 8x - 18$<br>$42 = 5x - 18$<br>$60 = 5x$<br>$12 = x$<br>A check shows that 12 is the solution. |
| **Section 2.2** | |
| A **mathematical model** is a real-life application expressed mathematically. | The speed of a car, $s$, increased by 20 mph is 60 mph.<br>Model: $s + 20 = 60$. |

| IMPORTANT FACTS AND CONCEPTS | EXAMPLES |
|---|---|
| **Section 2.2 (continued)** | |
| A **formula** is an equation that is a mathematical model for a real-life situation. | $A = l \cdot w$ |
| A **conditional equation** is an equation that has only one real value solution. | $2x + 4 = 5$ |
| A **contradiction** is an equation that has no solution (solution set is $\varnothing$). | $2x + 6 = 2x + 8$ |
| An **identity** is an equation that has an infinite number of solutions (solution set is $\mathbb{R}$). | $3x + 6 = 3(x + 2)$ |
| **Guidelines for Problem Solving**<br>**1.** Understand the problem.<br>**2.** Translate the problem into mathematical language.<br>**3.** Carry out the mathematical calculations necessary to solve the problem.<br>**4.** Check the answer obtained in step 3.<br>**5.** Answer the question.<br>See page 77 for more detail. | Max Johnson made a \$2000, 3% simple interest personal loan to Jill Johnson for 6 years. At the end of 6 years, what interest will Jill pay to Max?<br>Understand: This is a simple interest problem.<br>Translate: $i = prt$<br>Carry Out: $= 2000(0.03)(6)$<br>$= 360$<br>Check: The answer appears reasonable.<br>Answer: The simple interest owed is \$360. |
| **Simple interest formula** is $i = prt$. | Find the simple interest on a 2-year, \$1000, 6% simple interest loan.<br>$i = (1000)(0.06)(2) = 120$<br>The simple interest is \$120. |
| **Compound interest formula** is $A = p\left(1 + \frac{r}{n}\right)^{nt}$. | Find the amount in a savings account for a deposit of \$6500 paying 4.8% interest compounded semiannually for 10 years.<br>$A = 6500\left(1 + \frac{0.048}{2}\right)^{2 \cdot 10}$<br>$\approx 10{,}445.10$<br>The amount in the savings account is \$10,445.10. |
| To solve an equation (or formula) for a variable means to isolate that variable. | Solve the equation $3x + 7y = 2$ for $y$.<br>$7y = -3x + 2$<br>$y = -\frac{3}{7}x + \frac{2}{7}$ |
| **Section 2.3** | |
| Phrases can be translated into algebraic expressions. | Phrase: 4 more than 7 times a number — Algebraic Expression: $7x + 4$ |
| **Complementary angles** are two angles whose sum measures 90°. | If angle $A = 62°$ and angle $B = 28°$, then angles $A$ and $B$ are complementary angles. |
| **Supplementary angles** are two angles whose sum measures 180°. | If angle $A = 103°$ and angle $B = 77°$, then angles $A$ and $B$ are supplementary angles. |
| **Section 2.4** | |
| A **general motion problem formula** is amount = rate · time. | Find the amount of gas pumped when gas is pumped for 3 minutes at 6 gallons per minute.<br>$A = 6 \cdot 3 = 18$ gallons |

| IMPORTANT FACTS AND CONCEPTS | EXAMPLES |
|---|---|
| **Section 2.4 (continued)** | |
| **Distance formula** is distance = rate · time. | Find the distance traveled when a car travels at 60 miles per hour for 5 hours. <br> $D = 60 \cdot 5 = 300$ miles |
| A **mixture problem** is any problem where two or more qualities are combined to produce a different quantity or where a single quantity is separated into two or more different quantities. | If 4 liters of a 10% solution is mixed with 8 liters of a 16% solution, find the strength of the mixture. <br> $4(0.10) + 8(0.16) = 12(x)$ <br> or $x = 14\%$ |
| **Section 2.5** | |
| **Properties Used to Solve Inequalities** <br> **1.** If $a > b$, then $a + c > b + c$. <br> **2.** If $a > b$, then $a - c > b - c$. <br> **3.** If $a > b$, and $c > 0$, then $ac > bc$. <br> **4.** If $a > b$, and $c > 0$, then $\frac{a}{c} > \frac{b}{c}$. <br> **5.** If $a > b$, and $c < 0$, then $ac < bc$. <br> **6.** If $a > b$, and $c < 0$, then $\frac{a}{c} < \frac{b}{c}$. | **1.** If $6 > 5$, then $6 + 3 > 5 + 3$. <br> **2.** If $6 > 5$, then $6 - 3 > 5 - 3$. <br> **3.** If $7 > 3$, then $7 \cdot 4 > 3 \cdot 4$. <br> **4.** If $7 > 3$, then $\frac{7}{4} > \frac{3}{4}$. <br> **5.** If $9 > 2$, then $9(-3) < 2(-3)$. <br> **6.** If $9 > 2$, then $\frac{9}{-3} < \frac{2}{-3}$. |
| A **compound inequality** is formed by joining two inequalities with the word *and* or *or*. | $x \le 7$ and $x > 5$ <br> $x < -1$ or $x \ge 4$ |
| To find the solution set of an inequality containing the word *and* take the **intersection** of the solution sets of the two inequalities. | Solve $x \le 7$ and $x > 5$. <br> The intersection of $\{x\|x \le 7\}$ and $\{x\|x > 5\}$ is $\{x\|5 < x \le 7\}$ or $(5, 7]$. |
| To find the solution set of an inequality containing the word *or*, take the **union** of the solution sets of the two inequalities. | Solve $x < -1$ or $x \ge 4$. <br> The union of $\{x\|x < -1\}$ or $\{x\|x \ge 4\}$ is $\{x\|x < -1 \text{ or } x \ge 4\}$ or $(-\infty, -1) \cup [4, \infty)$. |
| **Section 2.6** | |
| **To Solve Equations of the Form $\|x\| = a$** <br> If $\|x\| = a$ and $a > 0$, then $x = a$ or $x = -a$. | Solve $\|x\| = 6$. <br> $\|x\| = 6$ gives $x = 6$ or $x = -6$. |
| **To Solve Inequalities of the Form $\|x\| < a$** <br> If $\|x\| < a$ and $a > 0$, then $-a < x < a$. | Solve $\|3x + 1\| < 13$. <br> $-13 < 3x + 1 < 13$ <br> $-\frac{14}{3} < x < 4$ <br> $\left\{x\middle\|-\frac{14}{3} < x < 4\right\}$ or $\left(-\frac{14}{3}, 4\right)$ |
| **To Solve Inequalities of the Form $\|x\| > a$** <br> If $\|x\| > a$ and $a > 0$, then $x < -a$ or $x > a$. | Solve $\|2x - 3\| \ge 5$. <br> $2x - 3 \le -5$ or $2x - 3 \ge 5$ <br> $2x \le -2$ $\quad$ $2x \ge 8$ <br> $x \le -1$ $\quad$ $x \ge 4$ <br> $\{x\|x \le -1 \text{ or } x \ge 4\}$ or $(-\infty, -1] \cup [4, \infty)$ |
| If $\|x\| > a$ and $a < 0$, the solution set is $\mathbb{R}$. <br> If $\|x\| < a$ and $a < 0$, the solution set is $\varnothing$. | $\|x\| > -7$, the solution set is $\mathbb{R}$. <br> $\|x\| < -7$, the solution set is $\varnothing$. |
| **To Solve Equations of the Form $\|x\| = \|y\|$.** <br> If $\|x\| = \|y\|$, then $x = y$ or $x = -y$. | Solve $\|x\| = \|3\|$. <br> $x = 3$ or $x = -3$. |

## Chapter 2 Review Exercises

[2.1] *State the degree of each term.*

**1.** $15a^3b^5$

**2.** $-5x$

**3.** $-21xyz^5$

*Simplify each expression. If an expression cannot be simplified, so state.*

**4.** $a(a + 3) - 4(a - 1)$

**5.** $x^2 + 2xy + 6x^2 - 13$

**6.** $b^2 + b - 9$

**7.** $2[-(x - y) + 3x] - 5y + 10$

*Solve each equation. If an equation has no solution, so state.*

**8.** $5(c + 4) - 2c = -(c - 4)$

**9.** $3(x + 1) - 3 = 4(x - 5)$

**10.** $3 + \frac{x}{2} = \frac{5}{6}$

**11.** $\frac{1}{2}(3t + 4) = \frac{1}{3}(4t + 1)$

**12.** $2\left(\frac{x}{2} - 4\right) = 3\left(x + \frac{1}{3}\right)$

**13.** $3x - 7 = 9x + 8 - 6x$

**14.** $2(x - 6) = 5 - \{2x - [4(x - 2) - 9]\}$

[2.2] *Evaluate each formula for the given values.*

**15.** $m = \frac{y_2 - y_1}{x_2 - x_1}$ when $y_2 = 4, y_1 = -3, x_2 = -8, x_1 = 6$

**16.** $x = \frac{-b + \sqrt{b^2 - 4ac}}{2a}$ when $a = 8, b = 10, c = -3$

**17.** $h = \frac{1}{2}at^2 + v_0t + h_0$ when $a = -32, v_0 = 0, h_0 = 85, t = 1$

**18.** $z = \frac{\bar{x} - \mu}{\frac{\sigma}{\sqrt{n}}}$ when $\bar{x} = 50, \mu = 54, \sigma = 5, n = 25$

*Solve each equation for the indicated variable.*

**19.** $E = IR$, for $R$

**20.** $P = 2l + 2w$, for $w$

**21.** $A = \pi r^2h$, for $h$

**22.** $A = \frac{1}{2}bh$, for $h$

**23.** $y = mx + b$, for $m$

**24.** $2x - 3y = 5$, for $y$

**25.** $R_T = R_1 + R_2 + R_3$, for $R_2$

**26.** $S = \frac{3a + b}{2}$, for $a$

**27.** $K = 2(d + l)$, for $l$

[2.3] *In Exercises 28–32, write an equation that can be used to solve each problem. Solve the problem and check your answer.*

**28. Calendar Sale** On February 1, all Hallmark calendars go on sale for 75% off the original price. If Caroline Collins purchases a calendar on sale for $7.50, what was the original price of the calendar?

**29. Population Increase** A small town's population is increasing by 350 people per year. If the present population is 4750, how long will it take for the population to reach 7200?

**30. Commission Salary** Celeste Nossiter's salary is $300 per week plus 6% commission of sales. How much in sales must Celeste make to earn $708 in a week?

**31. Car Rental Comparison** At the Kansas City airport, the cost to rent a Ford Focus from Hertz is $24.99 per day with unlimited mileage. The cost to rent the same car from Avis is $19.99 per day plus $0.10 per mile that the car is driven. If Cathy Panik needs to rent a car for 3 days, determine the number of miles she would need to drive in order for the cost of the car rental to be the same from both companies.

**32. Sale** At a going-out-of-business sale, furniture is selling at 40% off the regular price. In addition, green-tagged items are reduced by an additional $20. If Alice Barr purchased a green-tagged item and paid $136, find the item's regular price.

[2.4] *In Exercises 33–37, solve the following motion and mixture problems.*

**33. Investing a Bonus** After Ty Olden received a $5000 bonus at work, he invested some of the money in a money market account yielding 3.5% simple interest and the rest in a certificate of deposit yielding 4.0% simple interest. If the total amount of interest that Mr. Olden earned for the year was $187.15, determine the amount invested in each investment.

**34. Fertilizer Solutions** Dale Klitzke has liquid fertilizer solutions that are 20% and 60% nitrogen. How many gallons of each of these solutions should Dale mix to obtain 250 gallons of a solution that is 30% nitrogen?

**35. Two Trains** Two trains leave Portland, Oregon, at the same time traveling in opposite directions. One train travels at 60 miles per hour and the other at 80 miles per hour. In how many hours will they be 910 miles apart?

**36. Space Shuttles** Space Shuttle 2 takes off 0.5 hour after Shuttle 1 takes off. If Shuttle 2 travels 300 miles per hour faster than Shuttle 1 and overtakes Shuttle 1 exactly 5 hours after Shuttle 2 takes off, find

**a)** the speed of Shuttle 1 and

**b)** the distance from the launch pad when Shuttle 2 overtakes Shuttle 1.

**37. Mixing Coffee** Tom Tomlins, the owner of a gourmet coffee shop, has two coffees, one selling for \$6.00 per pound and the other for \$6.80 per pound. How many pounds of each type of coffee should he mix to make 40 pounds of coffee to sell for \$6.50 per pound?

[2.3, 2.4] *Solve.*

**38. Electronics Sale** At Circuit City, the price of a cordless telephone has been reduced by 20%. If the sale price is \$28.80 determine the original price.

**39. Jogging** Nicolle Ryba jogged for a distance and then turned around and walked back to her starting point. While jogging she averaged 7.2 miles per hour, and while walking she averaged 2.4 miles per hour. If the total time spent jogging and walking was 4 hours, find

**a)** how long she jogged, and

**b)** the total distance she traveled.

**40. Angle Measures** Find the measures of three angles of a triangle if one angle measures 25° greater than the smallest angle and the other angle measures 5° less than twice the smallest angle.

**41. Swimming Pool** Two hoses are filling a swimming pool. The hose with the larger diameter supplies 1.5 times as much water as the hose with the smaller diameter. The larger hose is on for 2 hours before the smaller hose is turned on. If 5 hours after the larger hose is turned on there are 3150 gallons of water in the pool, find the rate of flow from each hose.

**42. Complementary Angles** One complementary angle has a measure that is 30° less than twice the measure of the other angle. Determine the measures of the two angles.

**43. Blue Dye** A clothier has two blue dye solutions, both made from the same concentrate. One solution is 6% blue dye and the other is 20% blue dye. How many ounces of the 20% solution must be mixed with 10 ounces of the 6% solution to result in the mixture being a 12% blue dye solution?

**44. Two Investments** David Alevy invests \$12,000 in two savings accounts. One account is paying 10% simple interest and the other account is paying 6% simple interest. If in one year, the same interest is earned on each account, how much was invested at each rate?

**45. Fitness Center** The West Ridge Fitness Center has two membership plans. The first plan is a flat \$40-per-month fee plus \$1.00 per visit. The second plan is \$25 per month plus a \$4.00 per visit charge. How many visits would Jeff Feazell have to make per month to make it advantageous for him to select the first plan?

**46. Trains in Alaska** Two trains leave Anchorage at the same time along parallel tracks heading in opposite directions. The faster train travels 10 miles per hour faster than the slower train. Find the speed of each train if the trains are 270 miles apart after 3 hours.

[2.5] *Solve the inequality. Graph the solution on a real number line.*

**47.** $3z + 9 \leq 15$

**48.** $8 - 2w > -4$

**49.** $2x + 1 > 6$

**50.** $26 \leq 4x + 5$

**51.** $\dfrac{4x + 3}{3} > -5$

**52.** $2(x - 1) > 3x + 8$

**53.** $-4(x - 2) \geq 6x + 8 - 10x$

**54.** $\dfrac{x}{2} + \dfrac{3}{4} > x - \dfrac{x}{2} + 1$

*Write an inequality that can be used to solve each problem. Solve the inequality and answer the question.*

**55. Weight Limit** A canoe can safely carry a total weight of 560 pounds. If Bob and Kathy together weigh a total of 300 pounds, what is the maximum number of 40-pound boxes they can carry in their canoe?

**56. Phone Booth Call** Michael Lamb, a telephone operator, informs a customer in a phone booth that the charge for calling Omaha, Nebraska, is \$4.50 for the first 3 minutes and 95¢ for each additional minute or any part thereof. How long can the customer talk if he has \$8.65?

**57. Fitness Center** A fitness center guarantees that customers will lose a minimum of 5 pounds the first week and $1\frac{1}{2}$ pounds each additional week. Find the maximum amount of time needed to lose 27 pounds.

**58. Exam Scores** Patrice Lee's first four exam scores are 94, 73, 72, and 80. If a final average greater than or equal to 80 and less than 90 is needed to receive a final grade of B in the course, what range of scores on the fifth and last exam will result in Patrice's receiving a B in the course? Assume a maximum score of 100.

*Solve each inequality. Write the solution in interval notation.*

**59.** $1 < x - 4 < 7$

**60.** $8 < p + 11 \leq 16$

**61.** $3 < 2x - 4 < 12$

**62.** $-12 < 6 - 3x < -2$

**63.** $-1 < \frac{5}{9}x + \frac{2}{3} \leq \frac{11}{9}$

**64.** $-8 < \frac{4 - 2x}{3} < 0$

*Find the solution set to each compound inequality.*

**65.** $h \leq 1$ and $7h - 4 > -25$

**66.** $2x - 1 > 5$ or $3x - 2 \leq 10$

**67.** $4x - 5 < 11$ and $-3x - 4 \geq 8$

**68.** $\frac{7 - 2g}{3} \leq -5$ or $\frac{3 - g}{9} > 1$

[2.5, 2.6] *Find the solution set to each equation or inequality.*

**69.** $|a| = 2$

**70.** $|x| < 8$

**71.** $|x| \geq 9$

**72.** $|l + 5| = 13$

**73.** $|x - 2| \geq 5$

**74.** $|4 - 2x| = 5$

**75.** $|-2q + 9| < 7$

**76.** $\left|\frac{2x - 3}{5}\right| = 1$

**77.** $\left|\frac{x - 4}{3}\right| < 6$

**78.** $|4d - 1| = |6d + 9|$

**79.** $|2x - 3| + 4 \geq -17$

*Solve each inequality. Give the solution in interval notation.*

**80.** $|3c + 8| - 6 \leq 1$

**81.** $3 < 2x - 5 \leq 11$

**82.** $-6 \leq \frac{3 - 2x}{4} < 5$

**83.** $2p - 5 < 7$ and $9 - 3p \leq 15$

**84.** $x - 3 \leq 4$ or $2x - 5 > 7$

**85.** $-10 < 3(x - 4) \leq 18$

## Chapter 2 Practice Test

*To find out how well you understand the chapter material, take this practice test. The answers, and the section where the material was initially discussed, are given in the back of the book. Each problem is also fully worked out on the* **Chapter Test Prep Video CD.** *Review any questions that you answered incorrectly.*

**1.** State the degree of the term $-3a^2bc^4$.

*Simplify.*

**2.** $2p - 3q + 2pq - 6p(q - 3) - 4p$

**3.** $7q - \{2[3 - 4(q + 7)] + 5q\} - 8$

*In Exercises 4–8, solve the equation.*

**4.** $7(d + 2) = 3(2d - 4)$

**5.** $\frac{r}{12} + \frac{1}{3} = \frac{4}{9}$

**6.** $-2(x + 3) = 4\{3[x - (3x + 7)] + 2\}$

**7.** $7x - 6(2x - 4) = 3 - (5x - 6)$

**8.** $-\frac{1}{2}(4x - 6) = \frac{1}{3}(3 - 6x) + 2$

**9.** Find the value of $S_n$ for the given values.

$$S_n = \frac{a_1(1 - r^n)}{1 - r}, a_1 = 3, r = \frac{1}{3}, n = 3$$

**10.** Solve $c = \frac{a - 5b}{2}$ for $b$.

**11.** Solve $A = \frac{1}{2}h(b_1 + b_2)$ for $b_2$.

*In Exercises 12–16, write an equation that can be used to solve each problem. Solve the equation and answer the question asked.*

**12. Golf Club Discount** Find the cost of a set of golf clubs, before tax, if the cost of the clubs plus 7% tax is \$668.75.

**13. Health Club Costs** The cost of joining a health club is \$240 per year, plus \$2 per visit (for towel cleaning and toiletry expenses). If Bill Rush wishes to spend a total of \$400 per year for the health club, how many visits can he make?

**14. Bicycle Travel** Jeffrey Chang and Roberto Fernandez start at the same point and bicycle in opposite directions. Jeffrey's speed is 15 miles per hour and Roberto's speed is 20 miles per hour. In how many hours will the two men be 147 miles apart?

**15. Salt Solution** How many liters of a 12% salt solution must be added to 10 liters of a 25% salt solution to get a 20% salt solution?

**16. Two Investments** June White has \$12,000 to invest. She places part of her money in a savings account paying 8% simple interest and the balance in a savings account paying 7% simple interest. If the total interest from the two accounts at the end of 1 year is \$910, find the amount placed in each account.

*Solve each inequality and graph the solution on a number line.*

**17.** $3(2q + 4) < 5(q - 1) + 7$

**18.** $\dfrac{6 - 2x}{5} \geq -12$

*Solve each inequality and write the solution in interval notation.*

**19.** $x - 3 \leq 4$ and $2x + 1 > 10$

**20.** $7 \leq \dfrac{2u - 5}{3} < 9$

*Find the solution set to the following equations.*

**21.** $|2b + 5| = 9$

**22.** $|2x - 3| = \left|\dfrac{1}{2}x - 10\right|$

*Find the solution set to the following inequalities.*

**23.** $|4z + 12| = 0$

**24.** $|2x - 3| + 6 > 11$

**25.** $\left|\dfrac{2x - 3}{8}\right| \leq \dfrac{1}{4}$

## Cumulative Review Test

*Take the following test and check your answers with those given in the back of the book. Review any questions that you answered incorrectly. The section where the material was covered is indicated after the answer.*

**1.** If $A = \{1, 3, 5, 7, 9, 11, 13, 15\}$ and $B = \{2, 3, 5, 7, 11, 13\}$, find

**a)** $A \cup B$

**b)** $A \cap B$

**2.** Name each indicated property.

**a)** $9x + y = y + 9x$

**b)** $(2x)y = 2(xy)$

**c)** $4(x + 3) = 4x + 12$

*Evaluate.*

**3.** $-4^3 + (-6)^2 \div (2^3 - 2)^2$

**4.** $a^2b^3 + ab^2 - 3b$ when $a = -1$ and $b = -2$

**5.** $\dfrac{8 - \sqrt[3]{27} \cdot 3 \div 9}{|-5| - [5 - (12 \div 4)]^2}$

*In Exercises 6 and 7, simplify.*

**6.** $(5x^4y^3)^{-2}$

**7.** $\left(\dfrac{4m^2n^{-4}}{m^{-3}n^2}\right)^2$

**8. Comparing State Sizes** Rhode Island has a land area of about $1.045 \times 10^3$ square miles. Alaska has a land area of about $5.704 \times 10^5$ square miles. How many times larger is the land area of Alaska than that of Rhode Island?

*In Exercises 9–11, solve the equation.*

**9.** $-3(y + 7) = 2(-2y - 8)$

**10.** $1.2(x - 3) = 2.4x - 4.98$

**11.** $\dfrac{2m}{3} - \dfrac{1}{6} = \dfrac{4}{9}m$

**12.** Explain the difference between a conditional linear equation, an identity, and a contradiction. Give an example of each.

**13.** Evaluate the formula $x = \dfrac{-b + \sqrt{b^2 - 4ac}}{2a}$ for $a = 3, b = -8$, and $c = -3$.

**14.** Solve the formula $y - y_1 = m(x - x_1)$ for $x$.

**15.** Solve the inequality $-4 < \frac{5x - 2}{3} < 2$ and give the answer

**a)** on a number line,

**b)** as a solution set, and

**c)** in interval notation.

*In Exercises 16 and 17, find the solution set.*

**16.** $|3h - 1| = 8$

**17.** $|2x - 4| - 6 \geq 18$

**18. Baseball Sale** One week after the World Series, Target marks the price of all baseball equipment down by 40%. If Maxwell Allen purchases a Louisville Slugger baseball bat for $21 on sale, what was the original price of the bat?

**19. Two Cars** Two cars leave Newark, New Jersey, at the same time traveling in opposite directions. The car traveling north is moving 20 miles per hour faster than the car traveling south. If the two cars are 300 miles apart after 3 hours, find the speed of each car.

**20. Mixed Nuts** Molly Fitzgerald, owner of Molly's Nut House, has cashews that cost $6.50 per pound and peanuts that cost $2.50 per pound. If she wishes to make 40 pounds of a mixture of cashews and peanuts that sells for $4.00 per pound, how many pounds of cashews and how many pounds of peanuts should Molly mix together?

# 3 Graphs and Functions

## GOALS OF THIS CHAPTER

The two primary goals of this chapter are to provide you with a good understanding of graphing and functions. Graphing is a key element to this and many mathematics courses. Functions are closely related to graphing and are a unifying concept throughout mathematics. We will use both graphing and functions throughout the rest of this book.


3.1 Graphs

3.2 Functions

3.3 Linear Functions: Graphs and Applications

3.4 The Slope-Intercept Form of a Linear Equation

Mid-Chapter Test: Sections 3.1–3.4

3.5 The Point-Slope Form of a Linear Equation

3.6 The Algebra of Functions

3.7 Graphing Linear Inequalities

Chapter 3 Summary

Chapter 3 Review Exercises

Chapter 3 Practice Test

Cumulative Review Test

WE SEE GRAPHS DAILY in newspapers and magazines. You will see many such graphs in this chapter. For example, in Exercise 74 on page 172, a graph is used to show the growth in the shipment of LCD monitors.

# 3.1 Graphs

1. Plot points in the Cartesian coordinate system.
2. Draw graphs by plotting points.
3. Graph nonlinear equations.
4. Use a graphing calculator.
5. Interpret graphs.

René Descartes

## 1 Plot Points in the Cartesian Coordinate System

Many algebraic relationships are easier to understand if we can see a visual picture of them. A graph is a picture that shows the relationship between two or more variables in an equation. Before learning how to construct a graph, you must know the Cartesian coordinate system.

The **Cartesian** (or **rectangular**) **coordinate system**, named after the French mathematician and philosopher René Descartes (1596–1650), consists of two axes (or number lines) in a plane drawn perpendicular to each other (**Fig. 3.1**). Note how the two axes yield four **quadrants**, labeled with capital Roman numerals I, II, III, and IV.

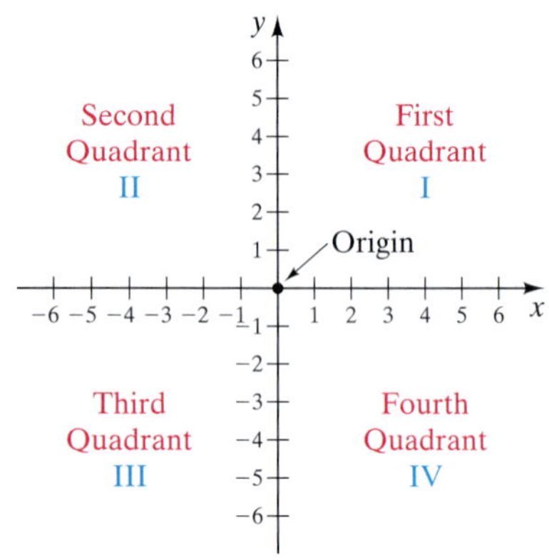


FIGURE 3.1

The horizontal axis is called the ***x*-axis**. The vertical axis is called the ***y*-axis**. The point of intersection of the two axes is called the **origin**. Starting from the origin and moving to the right, the numbers increase; moving to the left, the numbers decrease. Starting from the origin and moving up, the numbers increase; moving down, the numbers decrease. Note that the $x$-axis and $y$-axis are simply number lines, one horizontal and the other vertical.

An **ordered pair** $(x, y)$ is used to give the two **coordinates** of a point. If, for example, the $x$-coordinate of a point is 2 and the $y$-coordinate is 3, the ordered pair representing that point is $(2, 3)$. The $x$-coordinate is always the first coordinate listed in the ordered pair. To plot a point, find the $x$-coordinate on the $x$-axis and the $y$-coordinate on the $y$-axis. Then suppose there was an imaginary vertical line from the $x$-coordinate and an imaginary horizontal line from the $y$-coordinate. The point is placed where the two imaginary lines intersect.

For example, the point corresponding to the ordered pair $(2, 3)$ is plotted in **Figure 3.2**. The phrase "the point corresponding to the ordered pair $(2, 3)$" is often abbreviated "the point $(2, 3)$." For example, if we write "the point $(-1, 5)$," it means the point corresponding to the ordered pair $(-1, 5)$. The ordered pairs $A$ at $(-2, 3)$, $B$ at $(0, 2)$, $C$ at $(4, -1)$, and $D$ at $(-4, 0)$ are plotted in **Figure 3.3**.

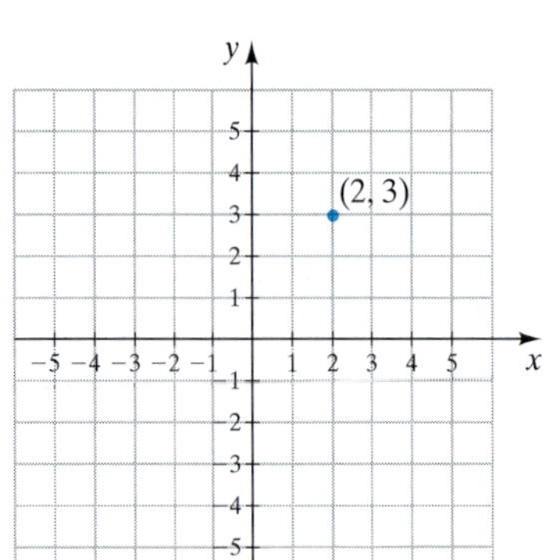


FIGURE 3.2

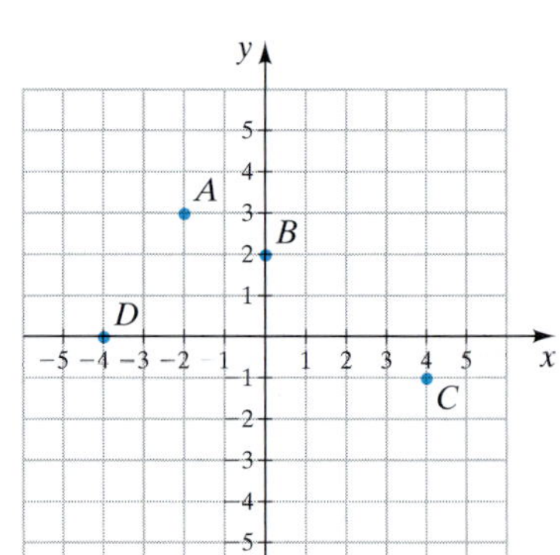


FIGURE 3.3

**EXAMPLE 1** ▸ Plot the following points on the same set of axes.

**a)** $A(1, 4)$ **b)** $B(5, 5)$ **c)** $C(0, 2)$
**d)** $D(-3, 0)$ **e)** $E(-3, -1)$ **f)** $F(2, -4)$

*Solution* See **Figure 3.4**. Notice that point $(1, 4)$ is a different point than $(4, 1)$. Also notice that when the $x$-coordinate is 0, as in part **c)**, the point is on the $y$-axis. When the $y$-coordinate is 0, as in part **d)**, the point is on the $x$-axis.

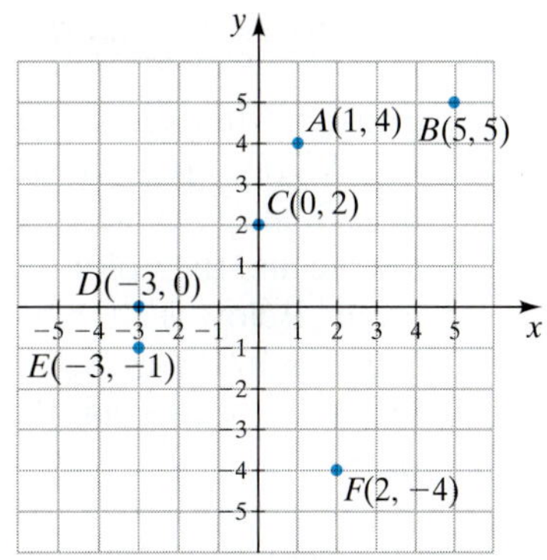


FIGURE 3.4

▸ Now Try Exercise 7

## 2 Draw Graphs by Plotting Points

In Chapter 2, we solved equations that contained one variable. Now we will discuss equations that contain two variables. If an equation contains two variables, its solutions are pairs of numbers.

**EXAMPLE 2** ▸ Determine whether the following ordered pairs are solutions to the equation $y = 2x - 3$.

**a)** $(1, -1)$ **b)** $\left(\frac{1}{2}, -2\right)$
**c)** $(4, 6)$ **d)** $(-1, -5)$

*Solution* We substitute the first number in the ordered pair for $x$ and the second number for $y$. If the substitutions result in a true statement, the ordered pair is a solution to the equation. If the substitutions result in a false statement, the ordered pair is not a solution to the equation.

**a)**
$$y = 2x - 3$$
$$-1 \stackrel{?}{=} 2(1) - 3$$
$$-1 \stackrel{?}{=} 2 - 3$$
$$-1 = -1 \quad \textit{True}$$

**b)**
$$y = 2x - 3$$
$$-2 \stackrel{?}{=} 2\left(\frac{1}{2}\right) - 3$$
$$-2 \stackrel{?}{=} 1 - 3$$
$$-2 = -2 \quad \textit{True}$$

**c)**
$$y = 2x - 3$$
$$6 \stackrel{?}{=} 2(4) - 3$$
$$6 \stackrel{?}{=} 8 - 3$$
$$6 = 5 \quad \textit{False}$$

**d)**
$$y = 2x - 3$$
$$-5 \stackrel{?}{=} 2(-1) - 3$$
$$-5 \stackrel{?}{=} -2 - 3$$
$$-5 \stackrel{?}{=} 5 \quad \textit{True}$$

Thus the ordered pairs $(1, -1)$, $\left(\frac{1}{2}, -2\right)$, and $(-1, -5)$ are solutions to the equation $y = 2x - 3$. The ordered pair $(4, 6)$ is not a solution.

▸ Now Try Exercise 17

There are many other solutions to the equation in Example 2. In fact, there are an infinite number of solutions. One method that may be used to find solutions to an equation like $y = 2x - 3$ is to substitute values for $x$ and find the corresponding values of $y$. For example, to find the solution to the equation $y = 2x - 3$ when $x = 0$ we substitute 0 for $x$ and solve for $y$.

$$y = 2x - 3$$
$$y = 2(0) - 3$$
$$y = 0 - 3$$
$$y = -3$$

Thus, another solution to the equation is $(0, -3)$.

A **graph** is an illustration of the set of points whose coordinates satisfy the equation. Sometimes when drawing a graph, we list some points that satisfy the equation in a table and then plot those points. We then draw a line through the points to obtain the graph. Below is a table of some points that satisfy the equation $y = 2x - 3$. The graph is drawn in **Figure 3.5**. Note that the equation $y = 2x - 3$ contains an infinite number of solutions and that the line continues indefinitely in both directions (as indicated by the arrows).

In **Figure 3.5**, the four points are in a straight line. Points that are in a straight line are said to be **collinear**. The graph is said to be **linear** because it is a straight line. Any equation whose graph is a straight line is called a **linear equation**. The equation $y = 2x - 3$ is an example of a linear equation. Linear equations are also called **first-degree equations** since the greatest exponent that appears on any variable is 1. In Examples 3 and 4, we graph linear equations.

| $x$ | $y$ | $(x, y)$ |
|---|---|---|
| $-1$ | $-5$ | $(-1, -5)$ |
| $0$ | $-3$ | $(0, -3)$ |
| $\frac{1}{2}$ | $-2$ | $\left(\frac{1}{2}, -2\right)$ |
| $1$ | $-1$ | $(1, -1)$ |

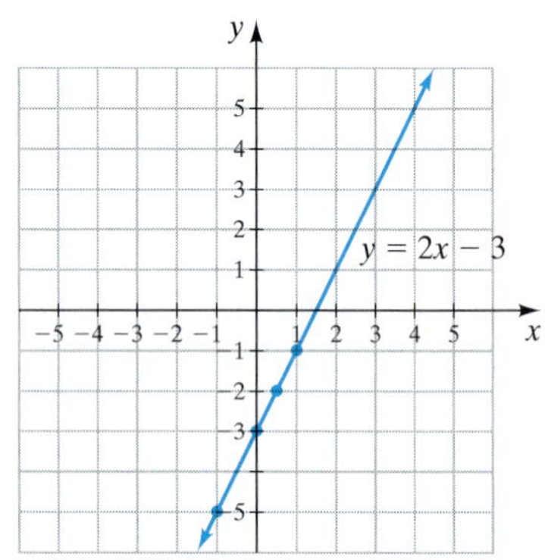


FIGURE 3.5

## Helpful Hint *Study Tip*

In this chapter, and in several upcoming chapters, you will be plotting points and drawing graphs using the Cartesian coordinate system. Sometimes students have trouble drawing accurate graphs. The following are some suggestions to improve the quality of the graphs you produce.

1. For your homework, use graph paper to draw your graphs. This will help you maintain a consistent scale throughout your graph. Ask your teacher if you may use graph paper on your tests and quizzes.
2. Use a ruler or straightedge to draw axes and lines. Your axes and lines will look much better and will be much more accurate if drawn with a ruler or straightedge.
3. If you are not using graph paper, use a ruler to make a consistent scale on your axes. It is impossible to get an accurate graph when axes are marked with an uneven scale.
4. Use a pencil instead of a pen. You may make a mistake as you draw your graph. A mistake can often be fixed quickly with a pencil and eraser, and you will not have to start over from the beginning. Make sure your pencil is sharp.
5. You will need to practice to improve your graphing skills. Work all of the homework problems that you are assigned. To check your graphs on even-numbered exercises, you may want to use a graphing calculator.

**EXAMPLE 3** ▸ Graph $y = x$.

*Solution* We first find some ordered pairs that are solutions by selecting values of $x$ and finding the corresponding values of $y$. We will select 0, some positive values, and some negative values for $x$. In general, we will choose numbers close to 0, so that the ordered pairs will fit on the axes. The graph is illustrated in **Figure 3.6**.

| $x$ | $y$ | $(x, y)$ |
|---|---|---|
| −2 | −2 | (−2, −2) |
| −1 | −1 | (−1, −1) |
| 0 | 0 | (0, 0) |
| 1 | 1 | (1, 1) |
| 2 | 2 | (2, 2) |

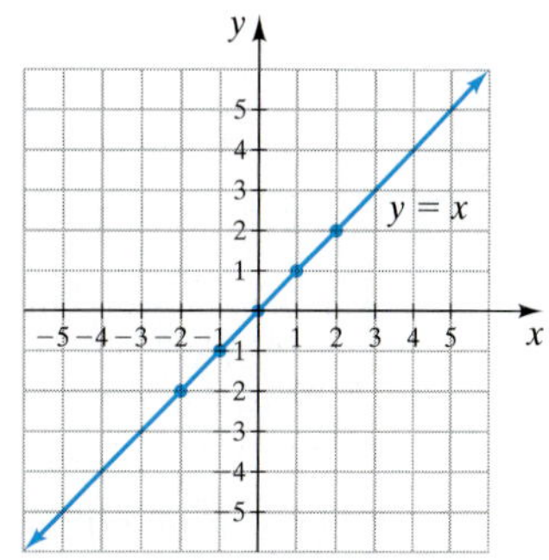


FIGURE 3.6

1. Select values for $x$.
2. Compute $y$.
3. Ordered pairs.
4. Plot the points and draw the graph.

▸ **Now Try Exercise 27**

When graphing linear equations that contain fractions, we often choose values for $x$ that are multiples of the denominator of the $x$-term. This selection often results in the values of $y$ being integer values. This is illustrated in Example 4.

**EXAMPLE 4** ▸ Graph $y = -\frac{1}{3}x + 1$.

*Solution* We will select some values for $x$, find the corresponding values of $y$, and then draw the graph. When we select values for $x$, we will select some positive values, some negative values, and 0. The graph is illustrated in **Figure 3.7**. (To conserve space, we will not always list a column in the table for ordered pairs.)

| $x$ | $y$ |
|---|---|
| −6 | 3 |
| −3 | 2 |
| 0 | 1 |
| 3 | 0 |
| 6 | −1 |

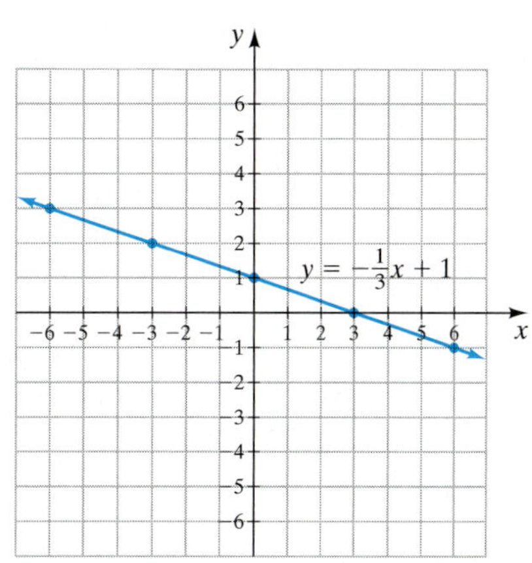


FIGURE 3.7

1. Select values for $x$.
2. Compute $y$.
3. Plot the points and draw the graph.

▸ **Now Try Exercise 35**

In Example 4, notice that we selected values of $x$ that were multiples of 3 so we would not have to work with fractions.

If we are asked to graph an equation not solved for $y$, such as $x + 3y = 3$, our first step will be to solve the equation for $y$. For example, if we solve $x + 3y = 3$ for $y$ using the procedure discussed in Section 2.2, we obtain

$$x + 3y = 3$$

$$3y = -x + 3 \qquad \text{Subtract } x \text{ from both sides.}$$

$$y = \frac{-x + 3}{3} \qquad \text{Divide both sides by 3.}$$

$$y = \frac{-x}{3} + \frac{3}{3} = -\frac{1}{3}x + 1$$

The resulting equation, $y = -\frac{1}{3}x + 1$, is the same equation we graphed in Example 4. Therefore, the graph of $x + 3y = 3$ is also illustrated in **Figure 3.7**.

## 3 Graph Nonlinear Equations

There are many equations whose graphs are not straight lines. Such equations are called **nonlinear equations**. To graph nonlinear equations by plotting points, we follow the same procedure used to graph linear equations. However, since the graphs are not straight lines, we may need to plot more points to draw the graphs.

**EXAMPLE 5** ▸ Graph $y = x^2 - 4$.

*Solution* We select some values for $x$ and find the corresponding values of $y$. Then we plot the points and connect them with a smooth curve. When we substitute values for $x$ and evaluate the right side of the equation, we follow the order of operations discussed in Section 1.4. For example, if $x = -3$, then $y = (-3)^2 - 4 = 9 - 4 = 5$. The graph is shown in **Figure 3.8**.

| $x$ | $y$ |
|---|---|
| −3 | 5 |
| −2 | 0 |
| −1 | −3 |
| 0 | −4 |
| 1 | −3 |
| 2 | 0 |
| 3 | 5 |

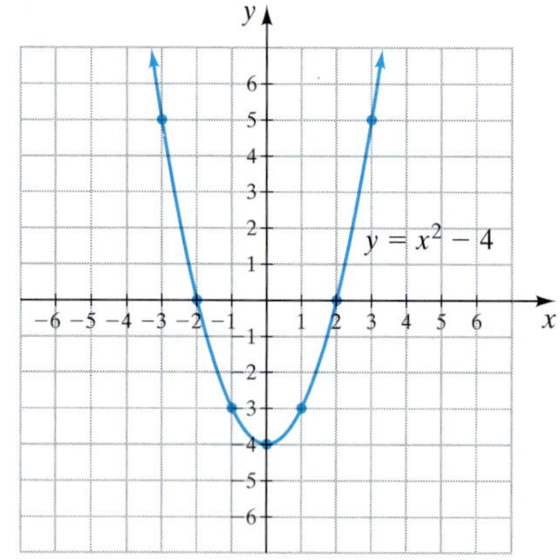


FIGURE 3.8

If we substituted 4 for $x$, $y$ would equal 12. When $x = 5$, $y = 21$. Notice that this graph rises steeply as $x$ moves away from the origin.

▸ Now Try Exercise 41

**EXAMPLE 6** ▸ Graph $y = \frac{1}{x}$.

*Solution* We begin by selecting values for $x$ and finding the corresponding values of $y$. We then plot the points and draw the graph. Notice that if we substitute 0 for $x$, we obtain $y = \frac{1}{0}$. Since $\frac{1}{0}$ is undefined, we cannot use 0 as a first coordinate. There will be no part of the graph at $x = 0$. We will plot points to the left of $x = 0$, and points to the right of $x = 0$ separately. Select points close to 0 to see what happens to the graph as $x$ gets close to $x = 0$. Note, for example, that when $x = -\frac{1}{2}$, $y = \frac{1}{-\frac{1}{2}} = -2$. This graph has two branches, one to the left of the $y$-axis and one to the right of the $y$-axis, as shown in **Figure 3.9**.

| $x$ | $y$ |
|---|---|
| $-3$ | $-\frac{1}{3}$ |
| $-2$ | $-\frac{1}{2}$ |
| $-1$ | $-1$ |
| $-\frac{1}{2}$ | $-2$ |
| $\frac{1}{2}$ | $2$ |
| $1$ | $1$ |
| $2$ | $\frac{1}{2}$ |
| $3$ | $\frac{1}{3}$ |

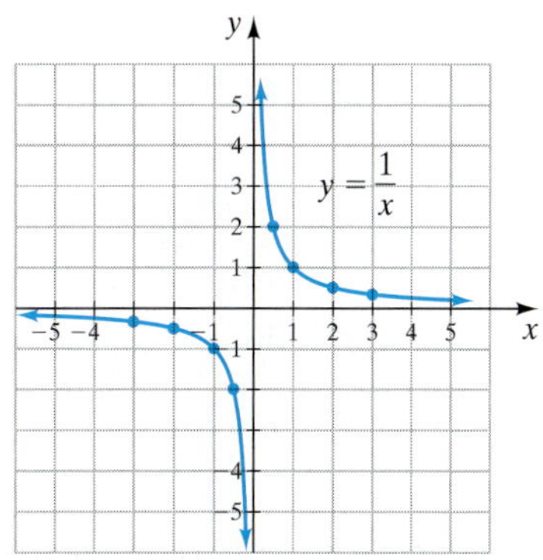


FIGURE 3.9

▸ **Now Try Exercise 51**

In the graph for Example 6, notice that for values of $x$ far to the right of 0, or far to the left of 0, the graph approaches the $x$-axis but does not touch it. For example when $x = 1000$, $y = 0.001$ and when $x = -1000$, $y = -0.001$. Can you explain why $y$ can never have a value of 0?

**EXAMPLE 7** ▸ Graph $y = |x|$.

*Solution* Recall that $|x|$ is read "the absolute value of $x$." Absolute values were discussed in Section 1.3. To graph this absolute value equation, we select some values for $x$ and find the corresponding values of $y$. For example, if $x = -4$, then $y = |-4| = 4$. Then we plot the points and draw the graph.

Notice that this graph is V-shaped, as shown in **Figure 3.10**.

| $x$ | $y$ |
|---|---|
| $-4$ | 4 |
| $-3$ | 3 |
| $-2$ | 2 |
| $-1$ | 1 |
| 0 | 0 |
| 1 | 1 |
| 2 | 2 |
| 3 | 3 |
| 4 | 4 |

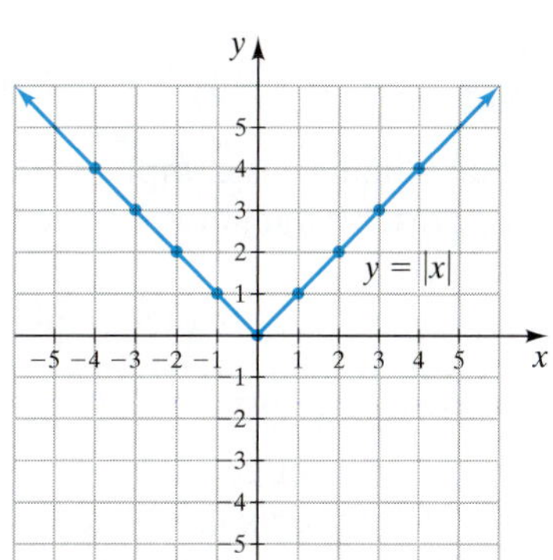


FIGURE 3.10

▸ **Now Try Exercise 45**

## Avoiding Common Errors

When graphing nonlinear equations, many students do not plot enough points to get a true picture of the graph. For example, when graphing $y = \frac{1}{x}$ many students consider only integer values of $x$. Following is a table of values for the equation and two graphs that contain the points indicated in the table.

| $x$ | $-3$ | $-2$ | $-1$ | $1$ | $2$ | $3$ |
|---|---|---|---|---|---|---|
| $y$ | $-\frac{1}{3}$ | $-\frac{1}{2}$ | $-1$ | $1$ | $\frac{1}{2}$ | $\frac{1}{3}$ |

CORRECT

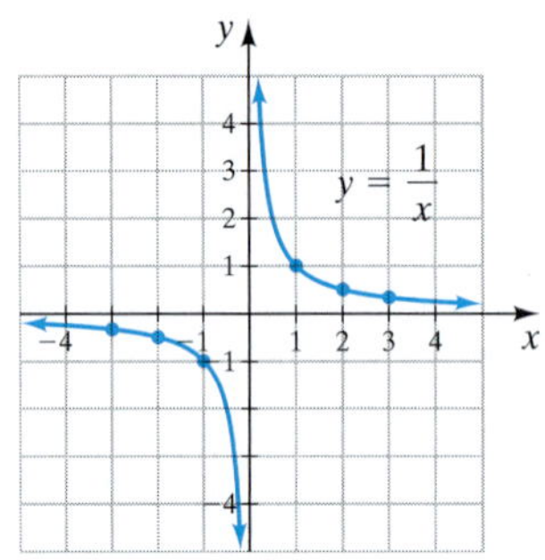


FIGURE 3.11

INCORRECT

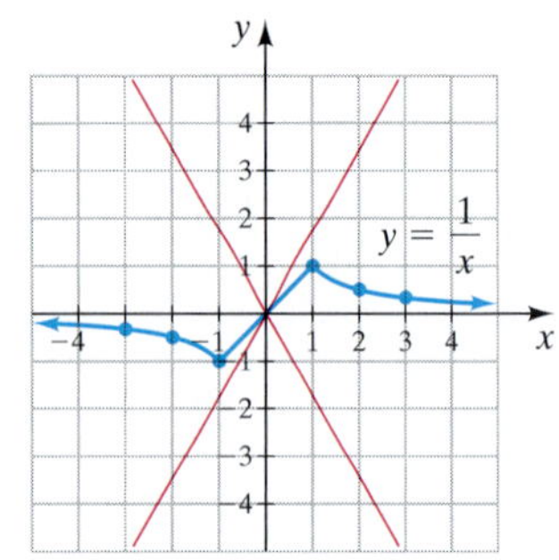


FIGURE 3.12

If you select and plot fractional values of $x$ near 0, as was done in Example 6, you get the graph in **Figure 3.11**. The graph in **Figure 3.12** cannot be correct because the equation is not defined when $x$ is 0 and therefore the graph cannot cross the $y$-axis. Whenever you plot a graph that contains a variable in the denominator, select values for the variable that are very close to the value that makes the denominator 0 and observe what happens. For example, when graphing $y = \frac{1}{x - 3}$ you should use values of $x$ close to 3, such as 2.9 and 3.1 or 2.99 and 3.01, and see what values you obtain for $y$.

Also, when graphing nonlinear equations, it is a good idea to consider both positive and negative values. For example, if you used only positive values of $x$ when graphing $y = |x|$, the graph would appear to be a straight line going through the origin, instead of the V-shaped graph shown in **Figure 3.10** on page 149.

## 4 Use a Graphing Calculator

If an equation is complex, finding ordered pairs can be time consuming. In this section we present a general procedure that can be used to graph equations using a **graphing calculator**.

A primary use of a graphing calculator is to graph equations. A **graphing calculator window** is the rectangular screen in which a graph is displayed.

In this book all graphing calculator windows will be from a TI-83 Plus or TI-84 Plus graphing calculators. Both calculators display the same window. In the *Using Your Graphing Calculator* boxes used throughout this book, we will sometimes indicate that the keystrokes and windows shown are for a TI-84 Plus graphing calculator. The same keystrokes and windows shown also apply to the TI-83 Plus graphing calculator even though we may not indicate this in the boxes.

**Figure 3.13** shows a TI-84 Plus calculator window with some information illustrated; **Figure 3.14** shows the meaning of the information given in **Figure 3.13**.

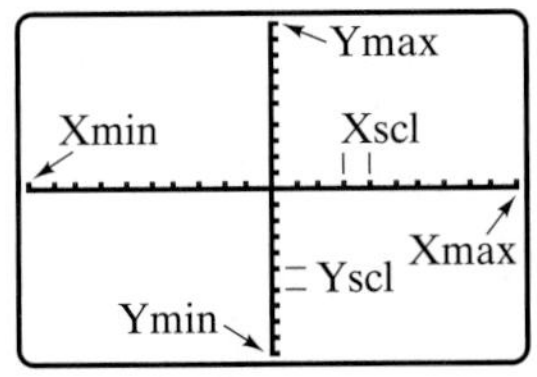


FIGURE 3.13

FIGURE 3.14

The $x$-axis on the *standard calculator screen* goes from $-10$ (the minimum value of $x$, Xmin) to 10 (the maximum value of $x$, Xmax) with a scale of 1. Therefore each tick mark represents 1 unit (Xscl $= 1$). The $y$-axis goes from $-10$ (the minimum value of $y$, Ymin) to 10 (the maximum value of $y$, Ymax) with a scale of 1 (Yscl $= 1$).

Since the window is rectangular, the distances between tick marks on the standard window are greater on the horizontal axis than on the vertical axis.

When graphing you will often need to change these window values. Read your graphing calculator manual to learn how to change the window setting. On the TI-84 Plus, you press the WINDOW key and then change the settings.

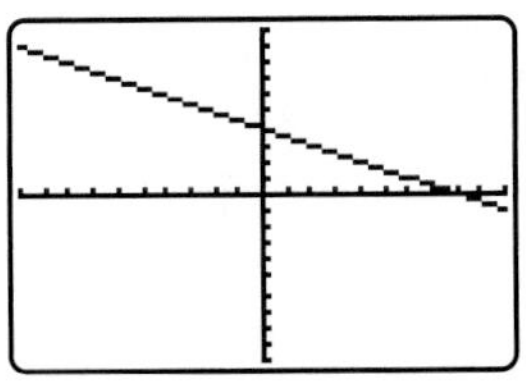

$-10, 10, 1, -10, 10, 1$

FIGURE 3.15

Since the graphing calculator (or grapher) does not display the $x$- and $y$-values in the window, we will occasionally list a set of values below the screen. **Figure 3.15** shows a TI-84 Plus calculator window with the equation $y = -\frac{1}{2}x + 4$ graphed. Below the window we show six numbers, which represent in order: Xmin, Xmax, Xscl, Ymin, Ymax, and Yscl. The Xscl and Yscl represent the scale on the $x$- and $y$-axes, respectively. When we are showing the standard calculator window, we will generally not show these values below the window.

To graph the equation $y = -\frac{1}{2}x + 4$ on a TI-84 Plus, you would press

Y = (−) ( 1 ÷ 2 ) X, T, Θ, $n$ + 4

Then when you press GRAPH, the equation is graphed. The X, T, Θ, $n$ key can be used to enter any of the symbols shown on the key. In this book this key will always be used to enter the variable $x$.

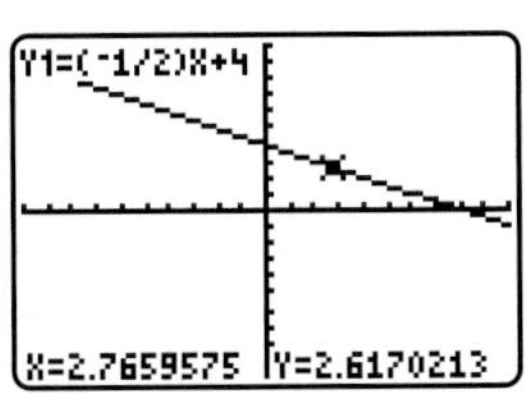


FIGURE 3.16

Most graphing calculators offer a **TRACE feature** that allows you to investigate individual points after a graph is displayed. Often the TRACE key is pressed to access this feature. After pressing the TRACE key, you can move the flashing cursor along the line by pressing the arrow keys. As the flashing cursor moves along the line, the values of $x$ and $y$ change to correspond with the position of the cursor. **Figure 3.16** shows the graph in **Figure 3.15** after the TRACE key has been pressed and the right arrow key has been pressed a few times.

Many graphing calculators also provide a **TABLE feature** that will illustrate a table of ordered pairs for any equation entered. On the TI-84 Plus, since TABLE appears above the GRAPH key, to obtain a table you press 2nd GRAPH.

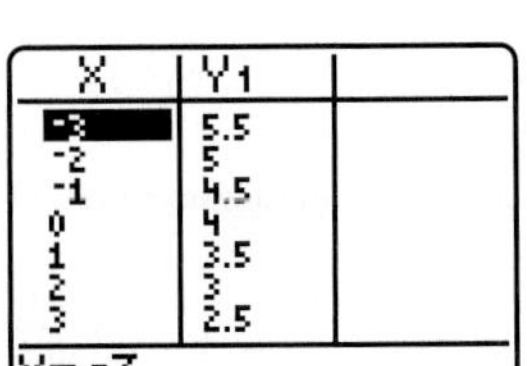


FIGURE 3.17

**Figure 3.17** shows a table of values for the equation $y = -\frac{1}{2}x + 4$. You can scroll up and down the table by using the arrow keys.

Using TBLSET (for Table setup), you can control the $x$-values that appear in the table. For example, if you want the table to show values of $x$ in tenths, you could do this using TBLSET.

This section is just a brief introduction to graphing equations, the TRACE feature, and the TABLE feature on a graphing calculator. You should read your graphing calculator manual to learn how to fully utilize these features.

## 5 Interpret Graphs

We see many different types of graphs daily in newspapers, in magazines, on television, and so on. Throughout this book, we present a variety of graphs. Since being able to draw and interpret graphs is very important, we will study this further in Section 3.2. In Example 8 you must understand and interpret graphs to answer the question.

**EXAMPLE 8** ▸ When Jim Herring went to see his mother in Cincinnati, he boarded a Southwest Airlines plane. The plane sat on the runway for 20 minutes and then took off. The plane flew at about 600 miles per hour for about 2 hours. It then reduced its speed to about 300 miles per hour and circled the Cincinnati Airport for about 15 minutes before it came in for a landing. After landing, the plane taxied to the gate and stopped. Which graph in **Figures 3.18a–3.18d** best illustrates this situation?

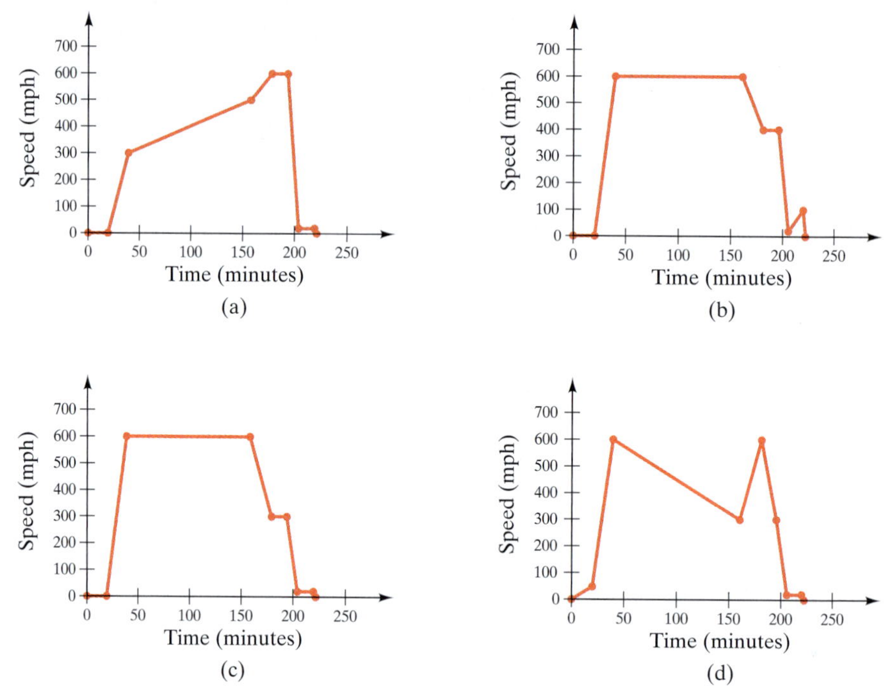


FIGURE 3.18

*Solution* The graph that depicts the situation described is (c), reproduced with annotations in **Figure 3.19**. The graph shows speed versus time, with time on the horizontal axis. While the plane sat on the runway for 20 minutes its speed was 0 miles per hour (the horizontal line at 0 from 0 to 20 minutes). After 20 minutes the plane took off, and its speed increased to 600 miles per hour (the near-vertical line going from 0 to 600

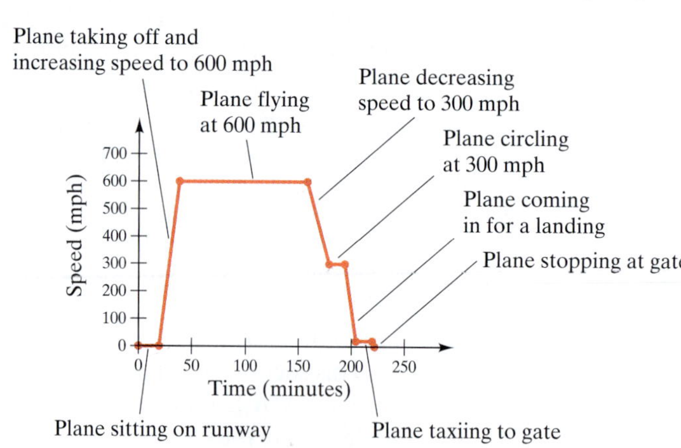


FIGURE 3.19

mph). The plane then flew at about 600 miles per hour for 2 hours (the horizontal line at about 600 mph). It then slowed down to about 300 miles per hour (the near-vertical line from 600 mph to 300 mph). Next the plane circled at about 300 miles per hour for about 15 minutes (the horizontal line at about 300 mph). The plane then came in for a landing (the near-vertical line from about 300 mph to about 20 mph). It then taxied to the gate (the horizontal line at about 20 mph). Finally, it stopped at the gate (the near-vertical line when the speed dropped to 0 mph).

▸ Now Try Exercise 81

## EXERCISE SET 3.1

MathXL® MyMathLab

### Concept/Writing Exercises

1. a) What does the graph of any linear equation look like?
   b) How many points are needed to graph a linear equation? Explain.
2. How many solutions does a linear equation in two variables have?
3. What does it mean when a set of points is collinear?
4. When graphing the equation $y = \frac{1}{x}$, what value cannot be substituted for $x$? Explain.

### Practice the Skills

*List the ordered pairs corresponding to the indicated points.*

5.

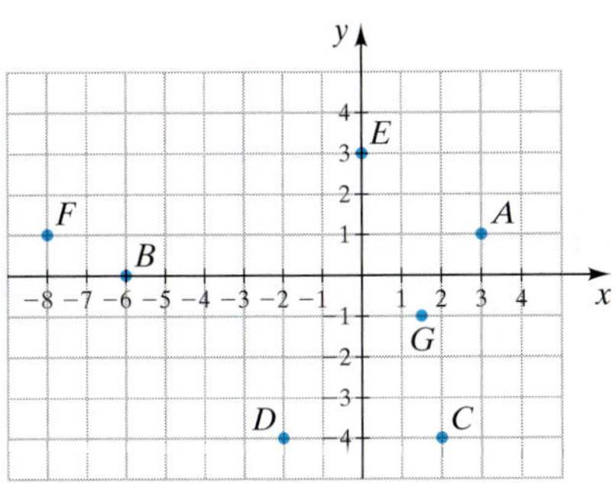


6.

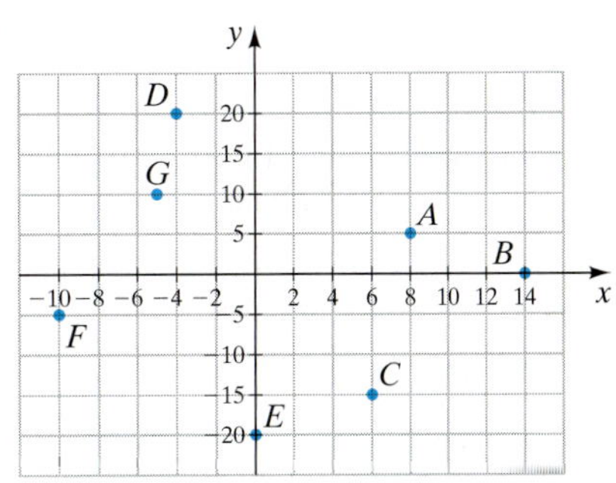


7. Graph the following points on the same axes.
   $A(4, 2)$ $B(-6, 2)$ $C(0, -1)$ $D(-2, 0)$
8. Graph the following points on the same axes.
   $A(-4, -2)$ $B(3, 2)$ $C(2, -3)$ $D(-3, 3)$

*Determine the quadrant in which each point is located.*

9. $(3, 5)$
10. $(-9, 1)$
11. $(4, -3)$
12. $(36, 43)$
13. $(-12, 18)$
14. $(-31, -8)$
15. $(-11, -19)$
16. $(8, -52)$

*Determine whether the given ordered pair is a solution to the given equation.*

17. $(2, 21)$; $y = 2x - 5$
18. $(1, 1)$; $2x + 3y = 6$
19. $(-4, -2)$; $y = |x| + 3$
20. $(1, -5)$; $y = x^2 + x - 7$
21. $(-2, 5)$; $s = 2r^2 - r - 5$
22. $\left(\frac{1}{4}, \frac{11}{4}\right)$; $y = |x - 3|$
23. $(2, 1)$; $-a^2 + 2b^2 = -2$
24. $(-10, -2)$; $|p| - 3|q| = 4$
25. $\left(\frac{1}{2}, \frac{5}{2}\right)$; $2x^2 + 6x - y = 0$
26. $\left(-3, \frac{7}{2}\right)$; $2m^2 + 3n = 2$

*Graph each equation.*

27. $y = x + 1$
28. $y = 3x$
29. $y = -3x - 5$
30. $y = -2x + 2$
31. $y = 2x + 4$
32. $y = x + 2$
33. $y = \frac{1}{2}x$
34. $y = -\frac{1}{3}x$
35. $y = \frac{1}{2}x - 1$
36. $y = -\frac{1}{2}x - 3$
37. $y = -\frac{1}{3}x + 2$
38. $y = -\frac{1}{3}x + 4$

**39.** $y = x^2$

**40.** $y = x^2 - 2$

**41.** $y = -x^2$

**42.** $y = -x^2 + 4$

**43.** $y = |x| + 1$

**44.** $y = |x| + 2$

**45.** $y = -|x|$

**46.** $y = -|x| - 3$

**47.** $y = x^3$

**48.** $y = -x^3$

**49.** $y = x^3 + 1$

**50.** $y = \frac{1}{x}$

**51.** $y = -\frac{1}{x}$

**52.** $x^2 = 1 + y$

**53.** $x = |y|$

**54.** $x = y^2$

*In Exercises 55–62, use a calculator to obtain at least eight points that are solutions to the equation. Then graph the equation by plotting the points.*

**55.** $y = x^3 - x^2 - x + 1$

**56.** $y = -x^3 + x^2 + x - 1$

**57.** $y = \frac{1}{x + 1}$

**58.** $y = \frac{1}{x} + 1$

**59.** $y = \sqrt{x}$

**60.** $y = \sqrt{x + 4}$

**61.** $y = \frac{1}{x^2}$

**62.** $y = \frac{|x^2|}{2}$

**63.** Is the point represented by the ordered pair $\left(\frac{1}{3}, \frac{1}{12}\right)$ on the graph of the equation $y = \frac{x^2}{x + 1}$? Explain.

**64.** Is the point represented by the ordered pair $\left(-\frac{1}{2}, -\frac{3}{5}\right)$ on the graph of the equation $y = \frac{x^2 + 1}{x^2 - 1}$? Explain.

**65. a)** Plot the points $A(2, 7)$, $B(2, 3)$, and $C(6, 3)$, and then draw $\overline{AB}$, $\overline{AC}$, and $\overline{BC}$. ($\overline{AB}$ represents the line segment from $A$ to $B$.)

**b)** Find the area of the figure.

**66. a)** Plot the points $A(-4, 5)$, $B(2, 5)$, $C(2, -3)$, and $D(-4, -3)$, and then draw $\overline{AB}$, $\overline{BC}$, $\overline{CD}$, and $\overline{DA}$.

**b)** Find the area of the figure.

**67. Golf Courses** The following graph shows that the average golf course length at the majors has been on the increase in recent years.

**Average Golf Course Length at Majors**

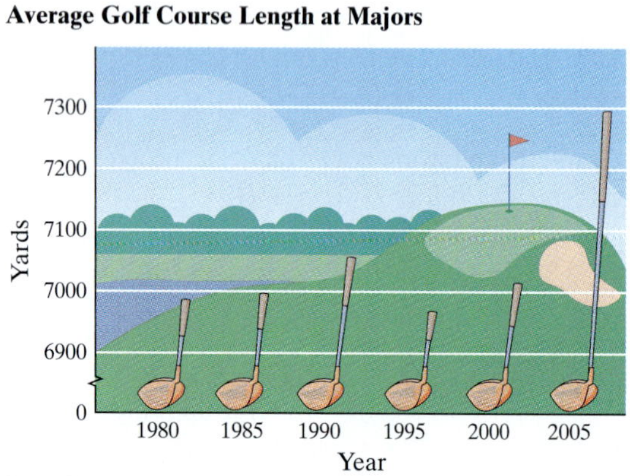


*Source*: Rees Jones Inc., PGA Tour, *USA TODAY* research

**a)** Estimate the average length of the golf course at the majors in 1980.

**b)** Estimate the average length of the golf course at the majors in 2005.

**c)** In which years was the average length of the golf courses at the majors greater than 7000 yards?

**d)** Does the increase in the average length of golf courses in the majors from 1995 to 2005 appear to be linear? Explain.

**68. E-Commerce** The following graph shows that E-commerce (sales on the Internet) has been constantly rising. The graph shows the sales, in the first quarter of each year, for the years from 2000 to 2005.

**E-Commerce Soars**

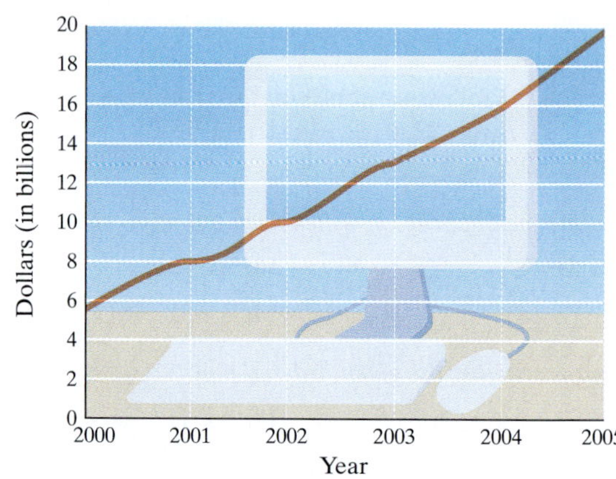


*Source*: Census Bureau, *USA TODAY* (8/9/05)

**a)** Estimate Internet sales in the first quarter of 2000.

**b)** Estimate Internet sales in the first quarter of 2005.

**c)** In what years was Internet sales in the first quarter greater than \$12 billion?

**d)** Does the increase in Internet sales in the first quarter from 2000 to 2005 appear to be approximately linear? Explain.

*We will discuss many of the concepts introduced in Exercises 69–76 in Section 3.4.*

**69.** Graph $y = x + 1$, $y = x + 3$, and $y = x - 1$ on the same axes.

**a)** What do you notice about the graphs of the equations and the values where the graphs intersect the $y$-axis?

**b)** Do all the graphs seem to have the same slant (or slope)?

**70.** Graph $y = \frac{1}{2}x$, $y = \frac{1}{2}x + 3$, and $y = \frac{1}{2}x - 4$ on the same axes.

**a)** What do you notice about the graphs of equations and the values where the graphs intersect the $y$-axis?

**b)** Do all of these graphs seem to have the same slant (or slope)?

**71.** Graph $y = 2x$. Determine the *rate of change* of $y$ with respect to $x$. That is, by how many units does $y$ change compared to each unit change in $x$?

**72.** Graph $y = 4x$. Determine the rate of change of $y$ with respect to $x$.

**73.** Graph $y = 3x + 2$. Determine the rate of change of $y$ with respect to $x$.

**74.** Graph $y = \frac{1}{2}x$. Determine the rate of change of $y$ with respect to $x$.

**75.** The ordered pair $(3, -7)$ represents one point on the graph of a linear equation. If $y$ increases 4 units for each unit increase in $x$ on the graph, find two other solutions to the equation.

**76.** The ordered pair $(1, -4)$ represents one point on the graph of a linear equation. If $y$ increases 3 units for each unit increase in $x$ on the graph, find two other solutions to the equation.

*Match Exercises 77–80 with the corresponding graph of elevation above sea level versus time, labeled a–d, below.*

**77.** Mary Leeseberg walked for 5 minutes on level ground. Then for 5 minutes she climbed a slight hill. Then she walked on level ground for 5 minutes. Then for the next 5 minutes she climbed a steep hill. During the next 10 minutes she descended uniformly until she reached the height at which she had started.

**78.** Don Gordon walked on level ground for 5 minutes. Then he walked down a steep hill for 10 minutes. For the next 5 minutes he walked on level ground. For the next 5 minutes he walked back up to his starting height. For the next 5 minutes he walked on level ground.

**79.** Nancy Johnson started out by walking up a steep hill for 5 minutes. For the next 5 minutes she walked down a steep hill to an elevation lower than her starting point. For the next 10 minutes she walked on level ground. For the next 10 minutes she walked up a slight hill, at which time she reached her starting elevation.

**80.** James Condor started out by walking up a hill for 5 minutes. For the next 10 minutes he walked down a hill to an elevation equal to his starting elevation. For the next 10 minutes he walked on level ground. For the next 5 minutes he walked downhill.

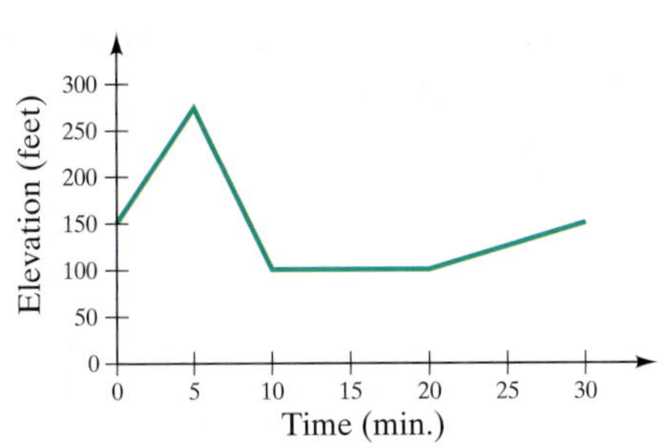

(a)

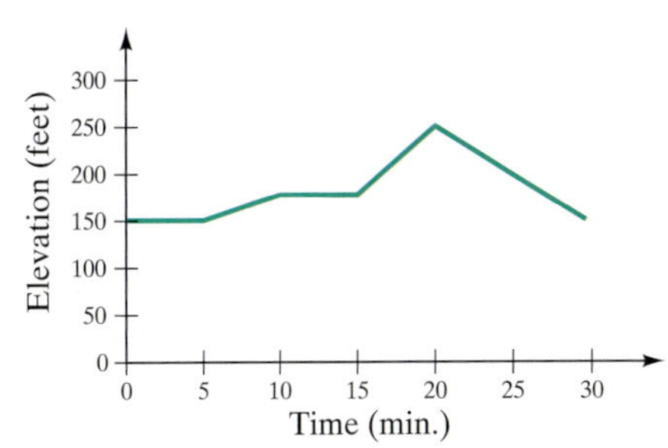

(c)

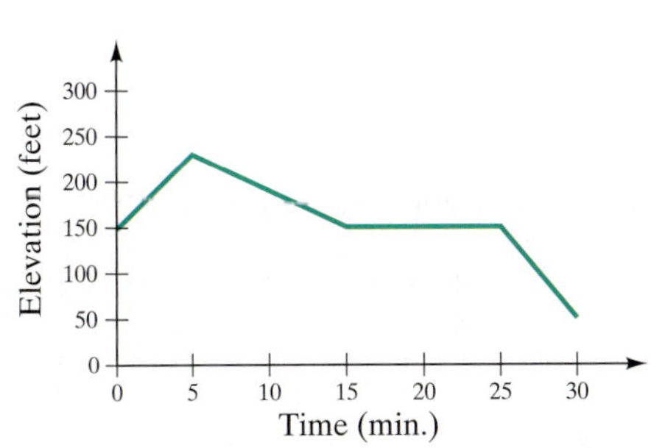

(b)

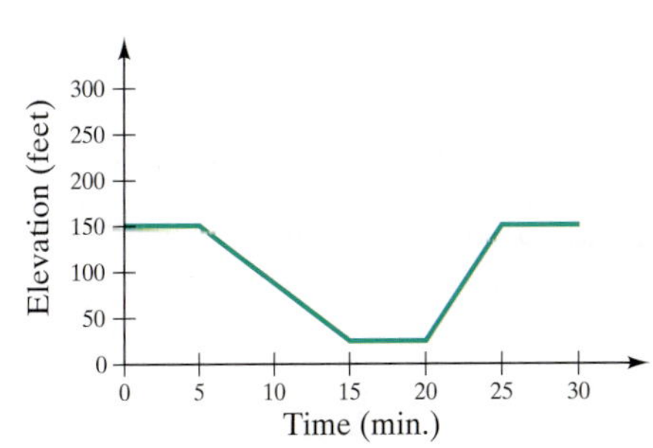

(d)

*Match Exercises 81–84 with the corresponding graph of speed versus time, labeled a–d, below.*

**81.** To go to work, Cletidus Hunt walked for 3 minutes, waited for the train for 5 minutes, rode the train for 15 minutes, then walked for 7 minutes.

**82.** To go to work, Tyrone Williams drove in stop-and-go traffic for 5 minutes, then drove on the expressway for 20 minutes, then drove in stop-and-go traffic for 5 minutes.

**83.** To go to work, Sheila Washington drove on a country road for 10 minutes, then drove on a highway for 12 minutes, then drove in stop-and-go traffic for 8 minutes.

**84.** To go to work, Brenda Pinkney rode her bike uphill for 10 minutes, then rode downhill for 15 minutes, then rode on a level street for 5 minutes.

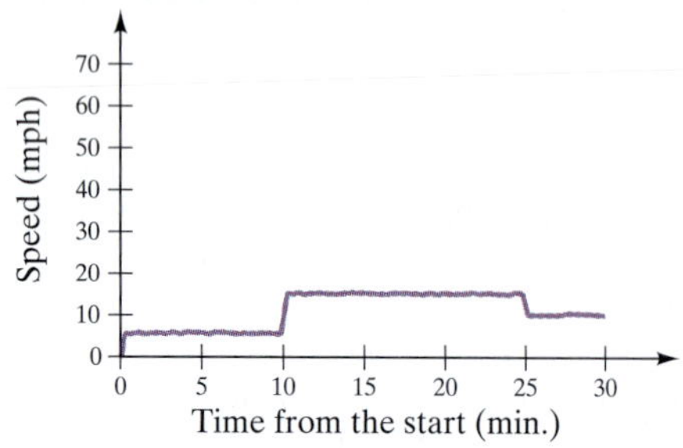


(a)

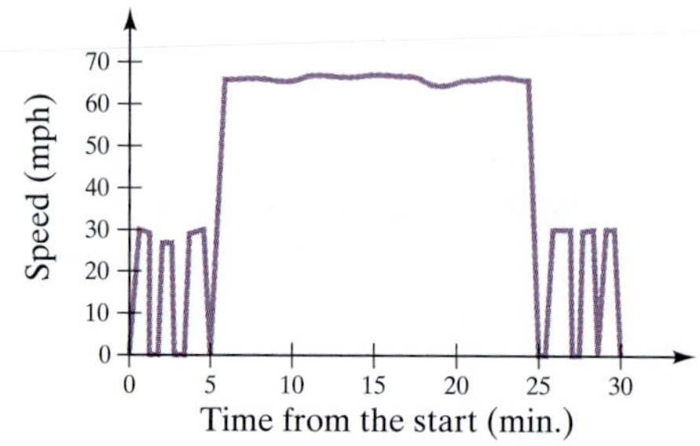


(c)

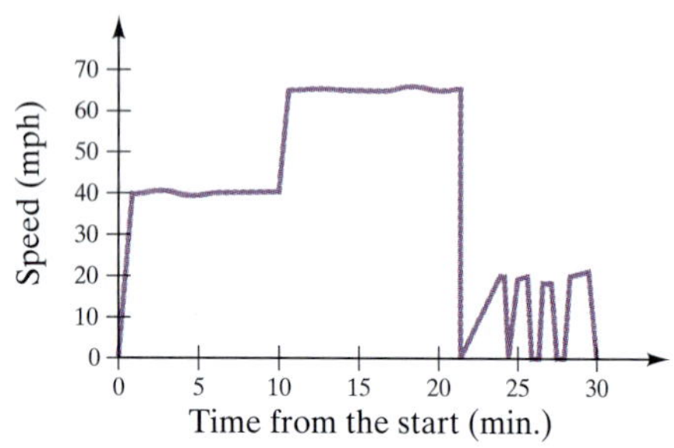


(b)

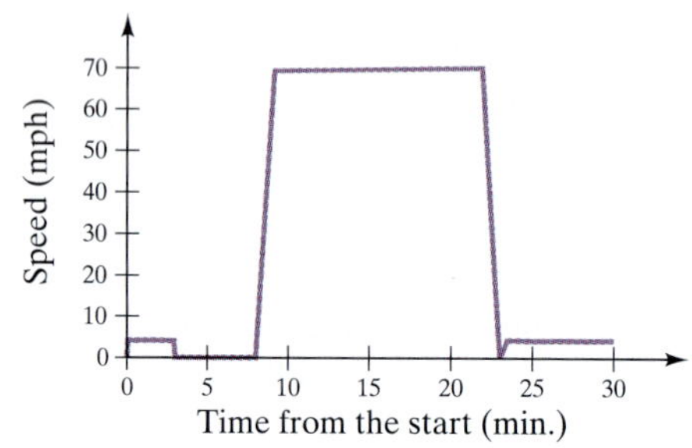


(d)

*Match Exercises 85–88 with the corresponding graph of speed versus time, labeled a–d below.*

**85.** Christina Dwyer walked for 5 minutes to warm up, jogged for 20 minutes, and then walked for 5 minutes to cool down.

**86.** Annie Droullard went for a leisurely bike ride at a constant speed for 30 minutes.

**87.** Michael Odu took a 30-minute walk through his neighborhood. He stopped very briefly on 7 occasions to pick up trash.

**88.** Richard Dai walked through his neighborhood and stopped 3 times to chat with his neighbors. He was gone from his house a total of 30 minutes.

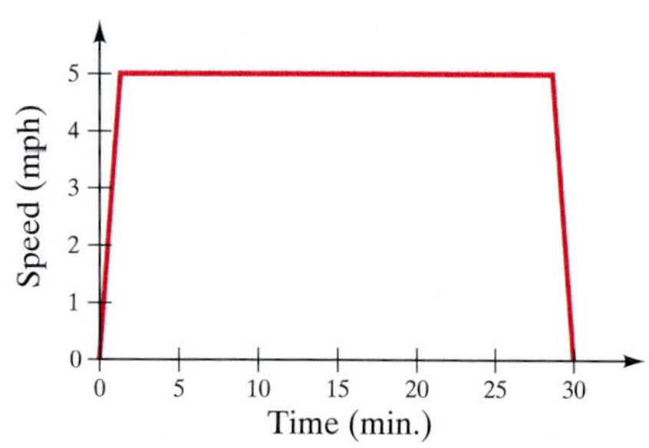


(a)

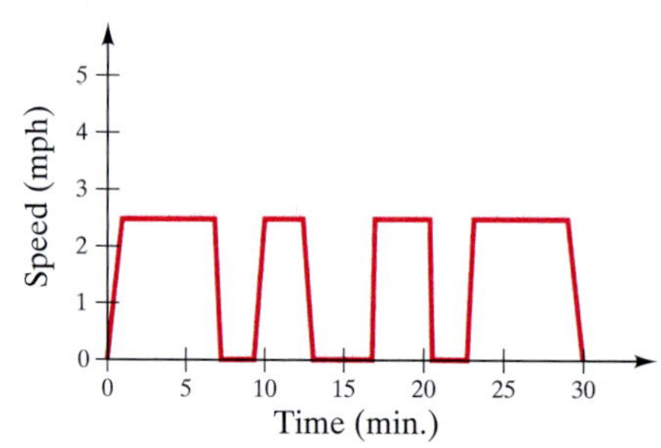


(c)

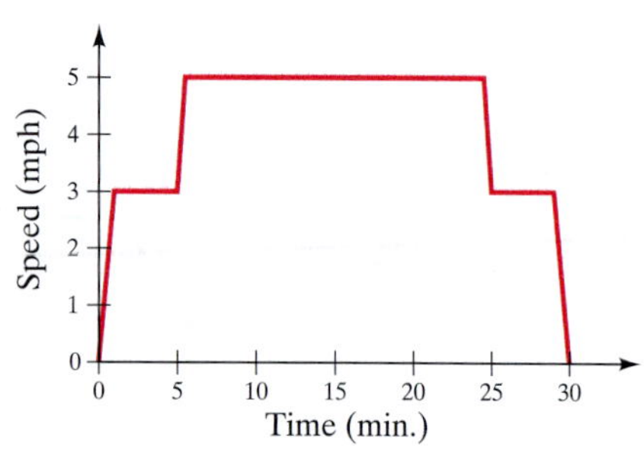


(b)

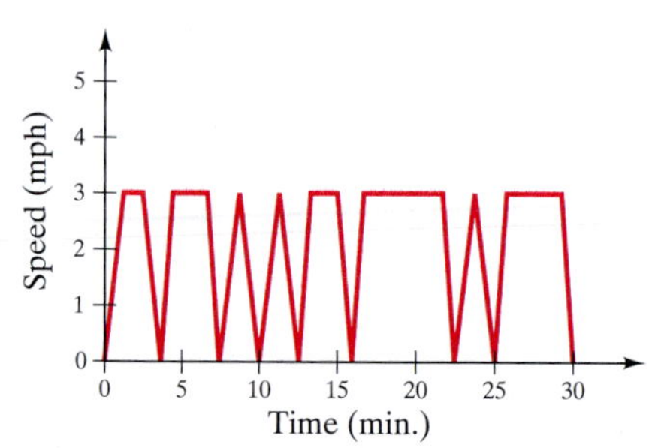


(d)

*Match Exercises 89–92 with the corresponding graph of distance traveled versus time, labeled a–d. Recall from Chapter 2 that distance = rate × time. Selected distances are indicated on the graphs.*

**89.** Train A traveled at a speed of 40 mph for 1 hour, then 80 mph for 2 hours, and then 60 mph for 3 hours.

**90.** Train C traveled at a speed of 80 mph for 2 hours, then stayed in a station for 1 hour, and then traveled 40 mph for 3 hours.

**91.** Train B traveled at a speed of 20 mph for 2 hours, then 60 mph for 3 hours, and then 80 mph for 1 hour.

**92.** Train D traveled at 30 mph for 1 hour, then 65 mph for 2 hours, and then 30 mph for 3 hours.

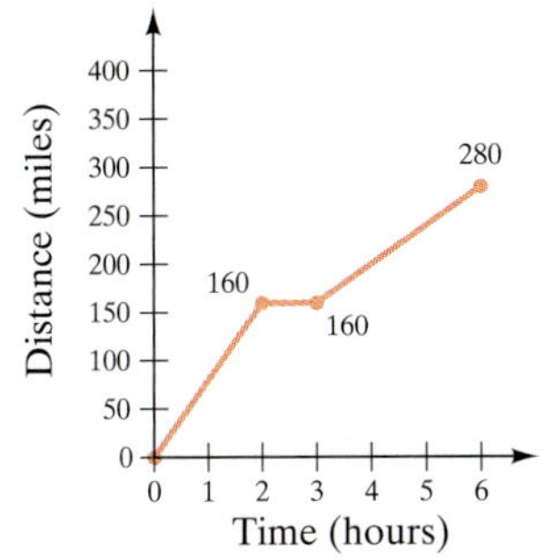

(a)

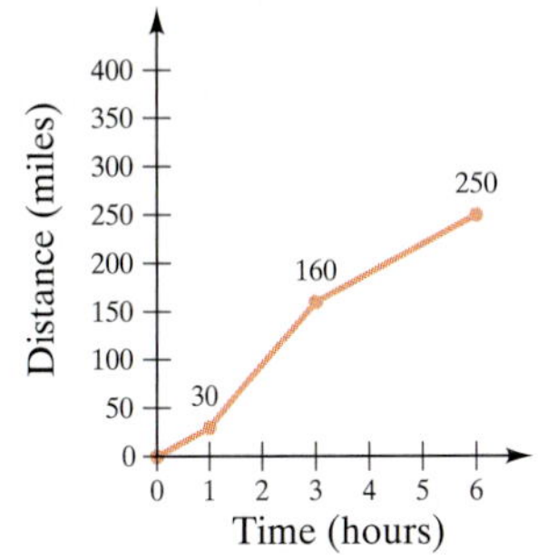

(c)

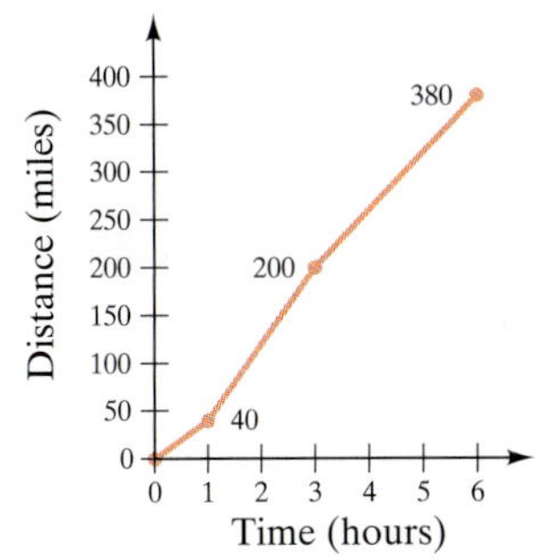

(b)

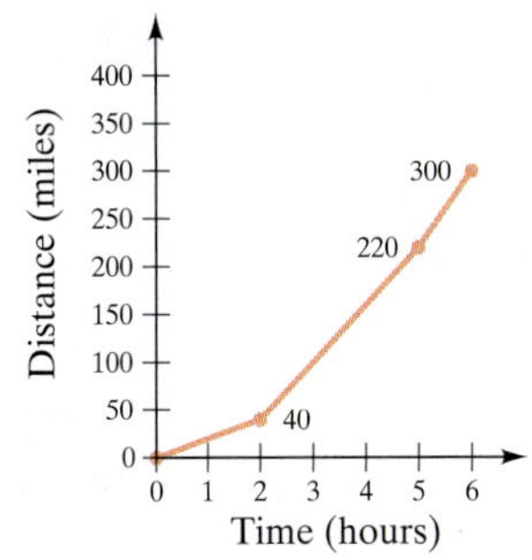

(d)

*Use a graphing calculator to graph each function. Make sure you select values for the window that will show the curvature of the graph. Then, if your calculator can display tables, display a table of values in which the x-values extend by units, from 0 to 6.*

**93.** $y = 2x - 3$

**94.** $y = \frac{1}{3}x + 2$

**95.** $y = x^2 - 2x - 8$

**96.** $y = -x^2 + 16$

**97.** $y = x^3 - 2x + 4$

**98.** $y = 2x^3 - 6x^2 - 1$

## Challenge Problems

*Graph each equation.*

**99.** $y = |x - 2|$

**100.** $x = y^2 + 2$

## Group Activity

*Discuss and work Exercises 101–102 as a group.*

**101. a)** Group member 1: Plot the points $(-2, 4)$ and $(6, 8)$. Determine the *midpoint* of the line segment connecting these points.

Group member 2: Follow the above instructions for the points $(-3, -2)$ and $(5, 6)$.

Group member 3: Follow the above instructions for the points $(4, 1)$ and $(-2, 4)$.

**b)** As a group, determine a formula for the midpoint of the line segment connecting the points $(x_1, y_1)$ and $(x_2, y_2)$. (*Note*: We will discuss the midpoint formula further in Chapter 10.)

**102.** Three points on a parallelogram are $A(3, 5)$, $B(8, 5)$, and $C(-1, -3)$.

**a)** Individually determine a fourth point $D$ that completes the parallelogram.

**b)** Individually compute the area of your parallelogram.

**c)** Compare your answers. Did you all get the same answers? If not, why not?

**d)** Is there more than one point that can be used to complete the parallelogram? If so, give the points and find the corresponding areas of each parallelogram.

## Cumulative Review Exercises

[2.2] **103.** Evaluate $\dfrac{-b + \sqrt{b^2 - 4ac}}{2a}$ for $a = 2, b = 7$, and $c = -15$.

[2.3] **104. Truck Rental** Hertz Truck Rental charges a daily fee of \$60 plus 10¢ a mile. National Automobile Rental Agency charges a daily fee of \$50 plus 24¢ a mile for the same truck. What distance would you have to drive in 1 day to make the cost of renting from Hertz equal to the cost of renting from National?

[2.5] **105.** Solve the inequality $-1 \leq \dfrac{4 - 3x}{2} < 5$. Write the solution in set builder notation.

[2.6] **106.** Find the solution set for the inequality $|3x + 2| > 7$.

# 3.2 Functions

1. Understand relations.
2. Recognize functions.
3. Use the vertical line test.
4. Understand function notation.
5. Applications of functions in daily life.

## 1 Understand Relations

In real life we often find that one quantity is related to a second quantity. For example, the amount you spend for oranges is related to the number of oranges you purchase. The speed of a sailboat is related to the speed of the wind. And the income tax you pay is related to the income you earn.

Suppose oranges cost 30 cents apiece. Then one orange costs 30 cents, two oranges cost 60 cents, three oranges cost 90 cents, and so on. We can list this information, or relationship, as a set of ordered pairs by listing the number of oranges first and the cost, in cents, second. The ordered pairs that represent this situation are (1, 30), (2, 60), (3, 90), and so on. An equation that represents this situation is $c = 30n$, where $c$ is the cost, in cents, and $n$ is the number of oranges. Since the cost depends on the number of oranges, we say that the cost is the *dependent variable* and the number of oranges is the *independent variable*.

Now consider the equation $y = 2x + 3$. Some ordered pairs that satisfy this equation are $(-2, -1), (-1, 1), (0, 3), (1, 5), (2, 7)$, and so on. In this equation, the value obtained for $y$ depends on the value selected for $x$. Therefore, $x$ is the *independent variable* and $y$ is the *dependent variable*. Note that in this example, unlike with the oranges, there is no physical connection between $x$ and $y$. The variable $x$ is the independent variable and $y$ is the dependent variable simply because of their placement in the equation.

For an equation in variables $x$ and $y$, if the value of $y$ depends on the value of $x$, then $y$ is the **dependent variable** and $x$ is the **independent variable**. Since related quantities can be represented as ordered pairs, the concept of a **relation** can be defined as follows.

**Relation**

A **relation** is any set of ordered pairs.

Since the equation $y = 2x + 3$ can be represented as a set of ordered pairs, it is a relation.

## 2 Recognize Functions

We now develop the idea of a **function**—one of the most important concepts in mathematics. A function is a special type of relation in which each element in one set (called the domain) corresponds to *exactly one* element in a second set (called the range).

Consider the oranges that cost 30 cents apiece that we just discussed. We can illustrate the number of oranges and the cost of the oranges using **Figure 3.20**.

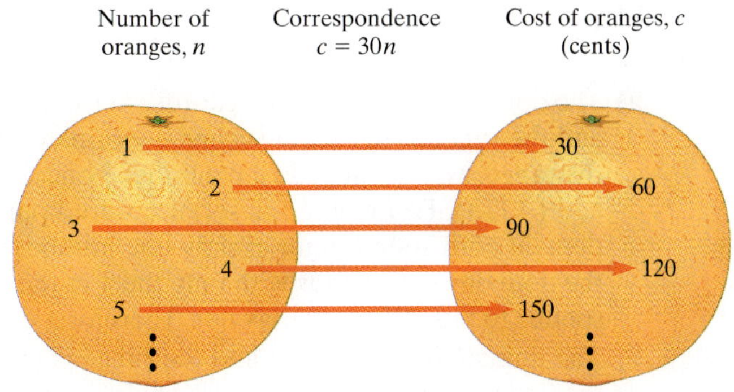


FIGURE 3.20

Notice that each number in the set of numbers of oranges, $n$, corresponds to (or is mapped to) exactly one number in the set of cost of oranges, $c$. Therefore, this correspondence is a function. The set consisting of the number of oranges, $\{1, 2, 3, 4, 5, \ldots\}$, is called the **domain**. The set consisting of the costs in cents, $\{30, 60, 90, 120, 150, \ldots\}$, is called the **range**. In general, the set of values for the independent variable is called the **domain**. The set of values for the dependent variable is called the **range**; see **Figure 3.21**.

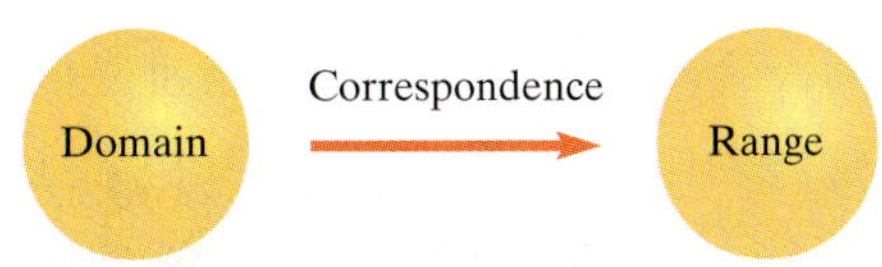


FIGURE 3.21

**EXAMPLE 1** ▸ Determine whether each correspondence is a function.

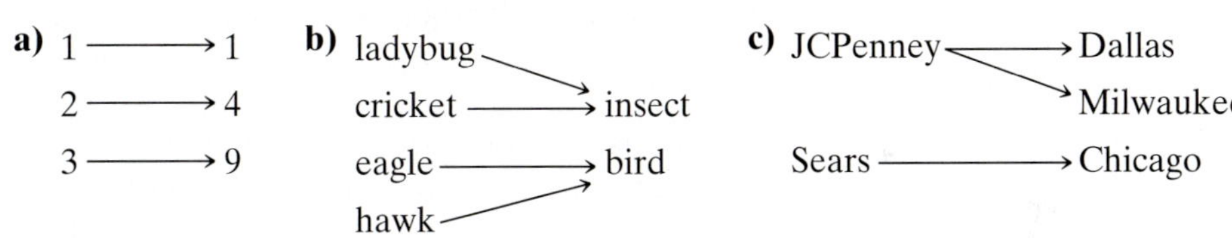


*Solution*

**a)** For a correspondence to be a function, each element in the domain must correspond with exactly one element in the range. Here the domain is $\{1, 2, 3\}$ and the range is $\{1, 4, 9\}$. Since each element in the domain corresponds to exactly one element in the range, this correspondence is a function.

**b)** Here the domain is {ladybug, cricket, eagle, hawk} and the range is {insect, bird}. Even though the domain has four elements and the range has two elements, each element in the domain corresponds with exactly one element in the range. Thus, this correspondence is a function.

**c)** Here the domain is {JCPenney, Sears} and the range is {Dallas, Milwaukee, Chicago}. Notice that JCPenney corresponds to both Dallas and Milwaukee. Therefore each element in the domain *does not* correspond to exactly one element in the range. Thus, this correspondence is a relation but *not* a function.

▸ **Now Try Exercise 17**

Now we will formally define function.

**Function**

A **function** is a correspondence between a first set of elements, the domain, and a second set of elements, the range, such that each element of the domain corresponds to *exactly one* element in the range.

**EXAMPLE 2** ▶ Which of the following relations are functions?

**a)** $\{(1, 4), (2, 3), (3, 5), (-1, 3), (0, 6)\}$

**b)** $\{(-1, 3), (4, 2), (3, 1), (2, 6), (3, 5)\}$

*Solution*

**a)** The domain is the set of first coordinates in the set of ordered pairs, $\{1, 2, 3, -1, 0\}$, and the range is the set of second coordinates, $\{4, 3, 5, 6\}$. Notice that when listing the range, we only include the number 3 once, even though it appears in both $(2, 3)$ and $(-1, 3)$. Examining the set of ordered pairs, we see that each number in the domain corresponds with exactly one number in the range. For example, the 1 in the domain corresponds with only the 4 in the range, and so on. No $x$-value corresponds to more than one $y$-value. Therefore, this relation *is a function.*

**b)** The domain is $\{-1, 4, 3, 2\}$ and the range is $\{3, 2, 1, 6, 5\}$. Notice that 3 appears as the first coordinate in two ordered pairs even though it is listed only once in the set of elements that represent the domain. Since the ordered pairs $(3, 1)$ and $(3, 5)$ have *the same first coordinate* and a different second coordinate, each value in the domain does not correspond to exactly one value in the range. Therefore, this relation is *not a function.*

▶ **Now Try Exercise 23**

Example 2 leads to an alternate definition of function.

**Function**

A **function** is a set of ordered pairs in which no *first* coordinate is repeated.

If the second coordinate in a set of ordered pairs repeats, the set of ordered pairs may still be a function, as in Example 2 **a)**. However, if two or more ordered pairs contain the same first coordinate, as in Example 2 **b)**, the set of ordered pairs is not a function.

Let us now consider the equation $y = 2x + 3$, which was introduced on page 158. As mentioned earlier, some ordered pairs that satisfy this equation are $(-2, -1)$, $(-1, 1)$, $(0, 3)$, $(1, 5)$, and $(2, 7)$. Notice that each value of $x$ that we substitute into the equation gives a unique value of $y$. Therefore, the equation $y = 2x + 3$ is not only a relation, it is also a function. In general, if we are given a linear function with variables $x$ and $y$, where $x$ is the independent variable, as in $y = 2x + 3$, *then for each value of $x$ there is exactly one value of $y$.* We will discuss this idea further in this section.

## 3 Use the Vertical Line Test

The **graph of a function or relation** is the graph of its set of ordered pairs. The two sets of ordered pairs in Example 2 **a)** and **b)** are graphed in **Figures 3.22a** and **3.22b** respectively. Notice that in the function in **Figure 3.22a** it is not possible to draw a vertical line that intersects two points. We should expect this because, in a function, each $x$-value must correspond to exactly one $y$-value. In **Figure 3.22b** we *can* draw a vertical line through the points $(3, 1)$ and $(3, 5)$. This shows that each $x$-value does not correspond to exactly one $y$-value, and the graph does not represent a function.

Function

Not a function

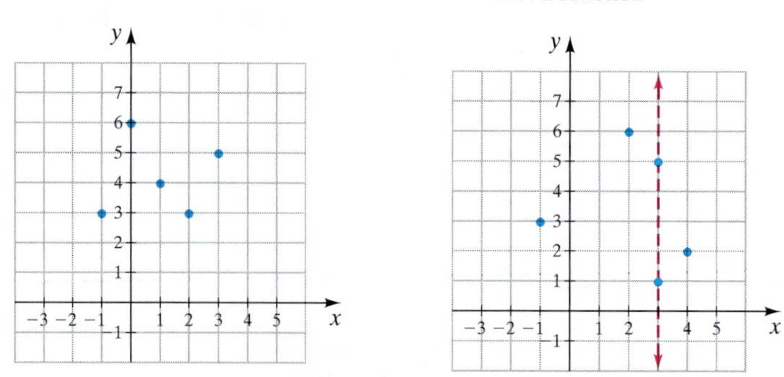


FIGURE 3.22 (a) First set of ordered pairs (b) Second set of ordered pairs

This method of determining whether a graph represents a function is called the **vertical line test.**

**Vertical Line Test**

If a vertical line can be drawn through any part of the graph and the line intersects another part of the graph, the graph does not represent a function. If a vertical line cannot be drawn to intersect the graph at more than one point, the graph represents a function.

We use the vertical line test to show that **Figure 3.23b** represents a function and **Figures 3.23a** and **3.23c** do not represent functions.

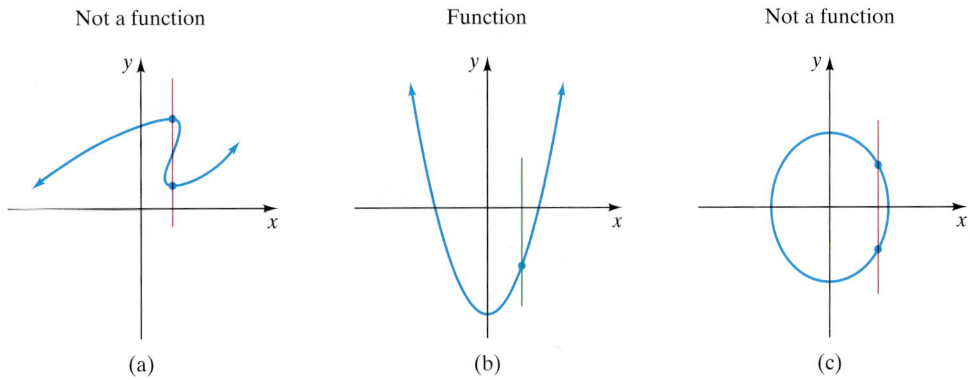


FIGURE 3.23

**EXAMPLE 3** ▸ Use the vertical line test to determine whether the following graphs represent functions. Also determine the domain and range of each function or relation.

**a)**

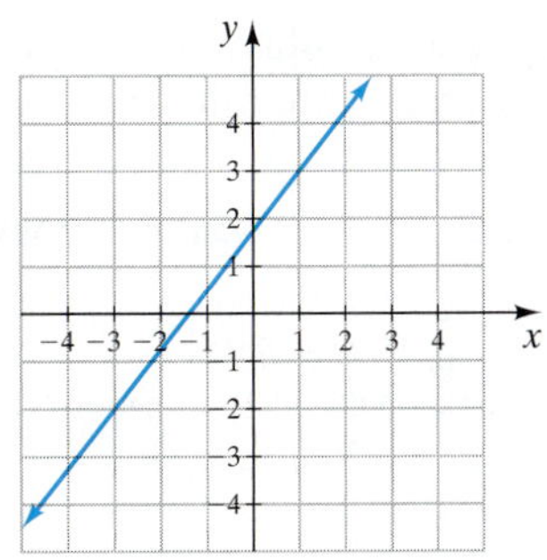


FIGURE 3.24

**b)**

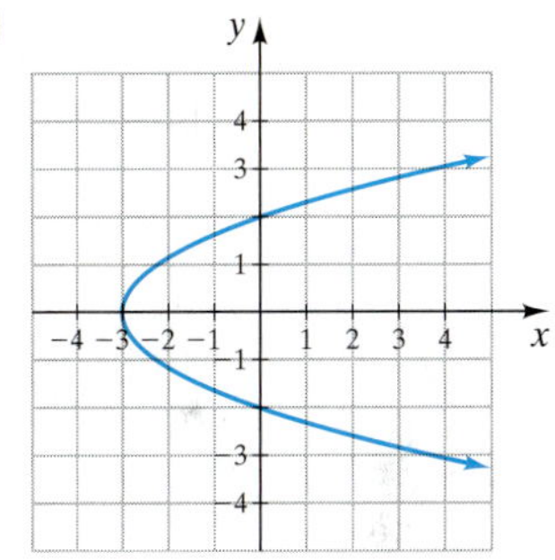


FIGURE 3.25

## *Solution*

**a)** A vertical line cannot be drawn to intersect the graph in **Figure 3.24** at more than one point. Thus this is the graph of a function. Since the line extends indefinitely in both directions, every value of $x$ will be included in the domain. The domain is the set of real numbers.

$$\text{Domain: } \mathbb{R} \quad \text{or} \quad (-\infty, \infty)$$

The range is also the set of real numbers since all values of $y$ are included on the graph.

$$\text{Range: } \mathbb{R} \quad \text{or} \quad (-\infty, \infty)$$

**b)** Since a vertical line can be drawn to intersect the graph in **Figure 3.25** at more than one point, this is *not* the graph of a function. The domain of this relation is the set of values greater than or equal to $-3$.

$$\text{Domain: } \{x|x \geq -3\} \quad \text{or} \quad [-3, \infty)$$

The range is the set of $y$-values, which can be any real number.

$$\text{Range: } \mathbb{R} \quad \text{or} \quad (-\infty, \infty)$$

▸ **Now Try Exercise 33**

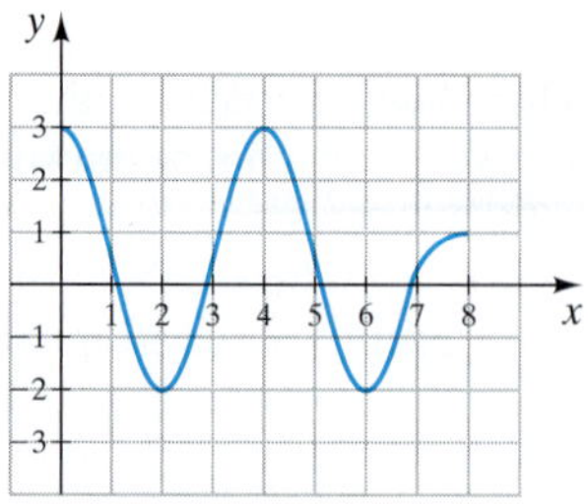


FIGURE 3.26

**EXAMPLE 4** ▶ Consider the graph shown in **Figure 3.26**.

**a)** What member of the range is paired with 4 in the domain?

**b)** What members of the domain are paired with $-2$ in the range?

**c)** What is the domain of the function?

**d)** What is the range of the function?

*Solution*

**a)** The range is the set of $y$-values. The $y$-value paired with the $x$-value of 4 is 3.

**b)** The domain is the set of $x$-values. The $x$-values paired with the $y$-value of $-2$ are 2 and 6.

**c)** The domain is the set of $x$-values, 0 through 8. Thus the domain is

$$\{x|0 \le x \le 8\} \quad \text{or} \quad [0, 8]$$

**d)** The range is the set of $y$-values, $-2$ through 3. Thus, the range is

$$\{y|-2 \le y \le 3\} \quad \text{or} \quad [-2, 3]$$

▶ **Now Try Exercise 39**

**EXAMPLE 5** ▶ **Figure 3.27** illustrates a graph of speed versus time of a man out for a walk and run. Write a story about the man's outing that corresponds to this function.

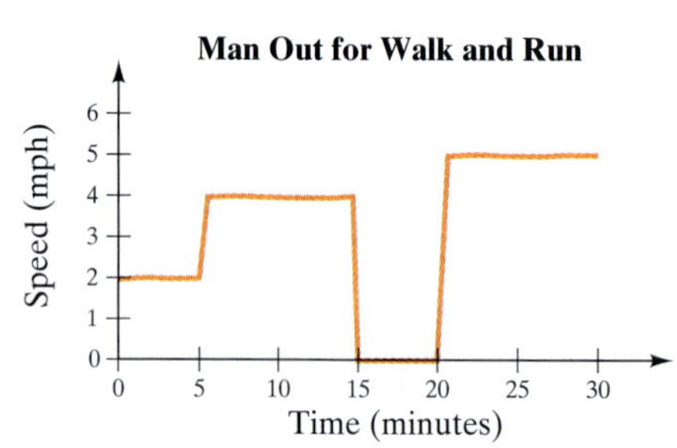


FIGURE 3.27

*Solution* **Understand** The horizontal axis is time and the vertical axis is speed. When the graph is horizontal it means the person is traveling at the constant speed indicated on the vertical axis. The near-vertical lines that increase with time (or have a positive slope, as will be discussed later) indicate an increase in speed, whereas the near-vertical lines that decrease with time (or have a negative slope) indicate a decrease in speed.

**Answer** Here is one possible interpretation of the graph. The man walks for about 5 minutes at a speed of about 2 miles per hour. Then the man speeds up to about 4 miles per hour and walks fast or runs at about this speed for about 10 minutes. Then the man slows down and stops, and then rests for about 5 minutes. Finally, the man speeds up to about 5 miles per hour and runs at this speed for about 10 minutes.

▶ **Now Try Exercise 65**

## 4 Understand Function Notation

In Section 3.1 we graphed a number of equations, as summarized in **Table 3.1**. If you examine each equation in the table, you will see that they are all functions, since their graphs pass the vertical line test.

**TABLE 3.1** Example

| Section 3.1 example | Equation graphed | Graph | Does the graph represent a function? | Domain | Range |
|---|---|---|---|---|---|
| 3 | $y = x$ | 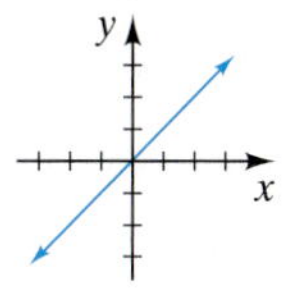 | Yes | $(-\infty, \infty)$ | $(-\infty, \infty)$ |
| 4 | $y = -\frac{1}{3}x + 1$ | 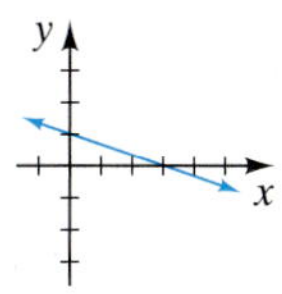 | Yes | $(-\infty, \infty)$ | $(-\infty, \infty)$ |
| 5 | $y = x^2 - 4$ | 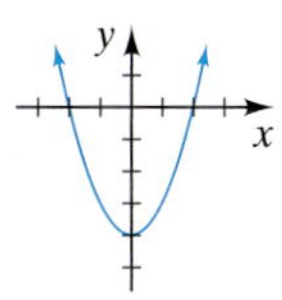 | Yes | $(-\infty, \infty)$ | $[-4, \infty)$ |
| 6 | $y = \frac{1}{x}$ | 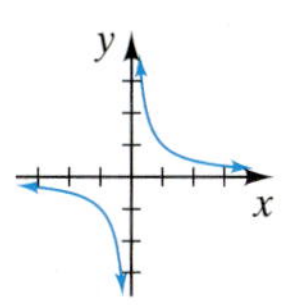 | Yes | $(-\infty, 0) \cup (0, \infty)$ | $(-\infty, 0) \cup (0, \infty)$ |
| 7 | $y = \lvert x \rvert$ | 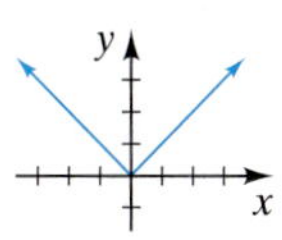 | Yes | $(-\infty, \infty)$ | $[0, \infty)$ |

Since the graph of each equation shown represents a function, we may refer to each equation in the table as a function. When we refer to an equation in variables $x$ and $y$ as a function, it means that the graph of the equation satisfies the criteria for a function. That is, each $x$-value corresponds to exactly one $y$-value, and the graph of the equation passes the vertical line test.

Not all equations are functions, as you will see in a later chapter in this book, Conic Sections, where we discuss equations of circles and ellipses. However, until we get to this chapter, all equations that we discuss will be functions.

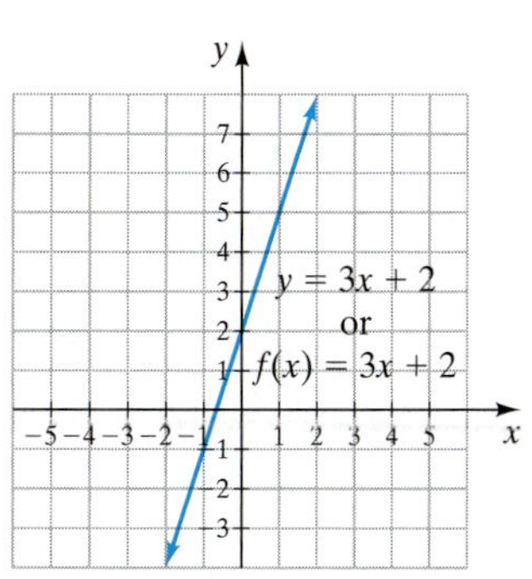


FIGURE 3.28

Consider the equation $y = 3x + 2$. By applying the vertical line test to its graph (**Fig. 3.28**), we can see that the graph represents a function. When an equation in variables $x$ and $y$ is a function, we often write the equation using **function notation**, $f(x)$, read "$f$ of $x$." Since the equation $y = 3x + 2$ is a function, and the value of $y$ depends on the value of $x$, we say that ***y* is a function of *x***. When we are given a linear equation in variables $x$ and $y$, *that is solved for* $y$, we can write the equation in function notation by substituting $f(x)$ for $y$. In this case, we can write the equation in function notation as $f(x) = 3x + 2$. The notation $f(x)$ represents the dependent variable *and does not mean $f$ times $x$*. Other letters may be used to indicate functions. For example, $g(x)$ and $h(x)$ also represent functions of $x$, and in Section 5.1 we will use $P(x)$ to represent polynomial functions.

Functions written in function notation are also equations since they contain an equal sign. We may refer to $y = 3x + 2$ as either an equation or a function. Similarly, we may refer to $f(x) = 3x + 2$ as either a function or an equation.

If $y$ is a function of $x$, the notation $f(5)$, read "$f$ of 5," means the value of $y$ when $x$ is 5. To evaluate a function for a specific value of $x$, substitute that value for $x$ in the function. For example, if $f(x) = 3x + 2$, then $f(5)$ is found as follows:

$$f(x) = 3x + 2$$
$$f(5) = 3(5) + 2 = 17$$

Therefore, when $x$ is 5, $y$ is 17. The ordered pair $(5, 17)$ would appear on the graph of $y = 3x + 2$.

### Helpful Hint

Linear equations that are not solved for $y$ can be written using function notation by solving the equation for $y$, then replacing $y$ with $f(x)$. For example, the equation $-9x + 3y = 6$ becomes $y = 3x + 2$ when solved for $y$. We can therefore write $f(x) = 3x + 2$.

**EXAMPLE 6** ▸ If $f(x) = -4x^2 + 3x - 2$, find

**a)** $f(2)$ **b)** $f(-1)$ **c)** $f(a)$

*Solution*

**a)** $f(x) = -4x^2 + 3x - 2$

$f(2) = -4(2)^2 + 3(2) - 2 = -4(4) + 6 - 2 = -16 + 6 - 2 = -12$

**b)** $f(-1) = -4(-1)^2 + 3(-1) - 2 = -4(1) - 3 - 2 = -4 - 3 - 2 = -9$

**c)** To evaluate the function at $a$, we replace each $x$ in the function with an $a$.

$$f(x) = -4x^2 + 3x - 2$$
$$f(a) = -4a^2 + 3a - 2$$

▸ **Now Try Exercise 45**

**EXAMPLE 7** ▸ Determine each indicated function value.

**a)** $g(-2)$ for $g(t) = \dfrac{1}{t + 8}$

**b)** $h(5)$ for $h(s) = 2|s - 6|$

**c)** $j(-3)$ for $j(r) = \sqrt{22 - r}$

*Solution* In each part, substitute the indicated value into the function and evaluate.

**a)** $g(-2) = \dfrac{1}{-2 + 8} = \dfrac{1}{6}$

**b)** $h(5) = 2|5 - 6| = 2|-1| = 2(1) = 2$

**c)** $j(-3) = \sqrt{22 - (-3)} = \sqrt{22 + 3} = \sqrt{25} = 5$

▸ **Now Try Exercise 49**

## 5 Applications of Functions in Daily Life

Many of the applications that we discussed in Chapter 2 were functions. However, we had not defined a function at that time. Now we examine additional applications of functions.

**EXAMPLE 8 ▸ Business Jets** The graph in **Figure 3.29** is taken from the November 11, 2004, issue of *USA Today*. The graph shows the number of business jets manufactured for the years from 1994 through 2004, projected through 2013.

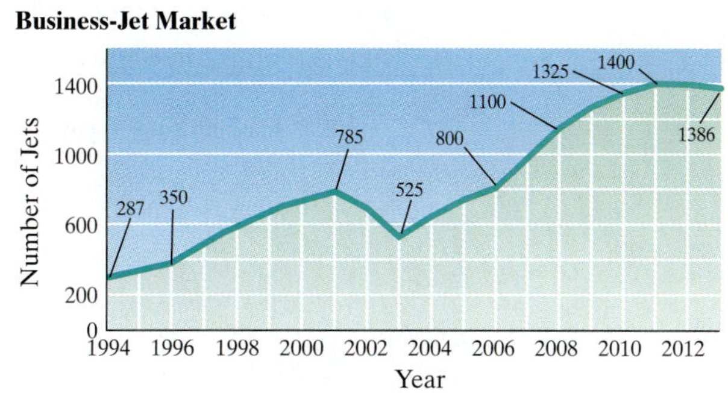


FIGURE 3.29

*Source*: Forecast International, USA TODAY (11/11/04)

**a)** Explain why the graph in **Figure 3.29** represents a function.

**b)** Determine the number of business jets projected to be manufactured in 2010.

**c)** Determine the projected percent increase in the number of business jets to be manufactured from 2003 to 2011.

**d)** Determine the percent decrease in the number of business jets manufactured from 2001 to 2003.

*Solution*

**a)** The graph represents a function since each year corresponds to a specific number of business jets manufactured. Notice that the graph passes the vertical line test.

**b)** In 2010, the graph shows that 1325 business jets should be manufactured. If we let the function be represented by $J$, then $J(2010) = 1325$.

**c)** We will follow the problem-solving procedure to solve this problem.

**Understand and Translate** We need to determine the percent increase in the number of business jets to be manufactured from 2003 to 2011. To do this, use the formula

$$\text{percent change (increase or decrease)} = \frac{\begin{pmatrix}\text{value in}\\ \text{latest period}\end{pmatrix} - \begin{pmatrix}\text{value in}\\ \text{previous period}\end{pmatrix}}{\text{value in previous period}}$$

The latest period is 2011 and the previous period is 2003. Substituting the values, we get

$$\text{percent change} = \frac{1400 - 525}{525}$$

**Carry Out**

$$= \frac{875}{525} \approx 1.667 = 166.7\%$$

**Check and Answer** Our calculations appear correct. There is projected to be about a 166.7% increase in the number of business jets manufactured from 2003 to 2011.

**d)** To find the percent decrease from 2001 to 2003, we follow the same procedure as in part **c)**. The latest period is 2003 and the previous period is 2001.

$$\text{percent change (increase or decrease)} = \frac{\begin{pmatrix}\text{value in}\\ \text{latest period}\end{pmatrix} - \begin{pmatrix}\text{value in}\\ \text{previous period}\end{pmatrix}}{\text{value in previous period}}$$

$$= \frac{525 - 785}{785} = \frac{-260}{785} \approx -0.331 = -33.1\%$$

The negative sign preceding the 33.1% indicates a percent decrease. Thus, there was about a 33.1% decrease in the manufacture of business jets from 2001 to 2003.

▸ **Now Try Exercise 71**

**EXAMPLE 9 ▸ Immigration** The size of the U.S. foreign-born population is at an all-time high. The graph in **Figure 3.30** shows the U.S. foreign-born population, in millions, from 1890 to 2004 and projected to 2010.

**a)** Using the graph in **Figure 3.30**, explain why this set of points represents a function.

**b)** Using the graph in **Figure 3.31**, estimate the foreign-born population in 2008.

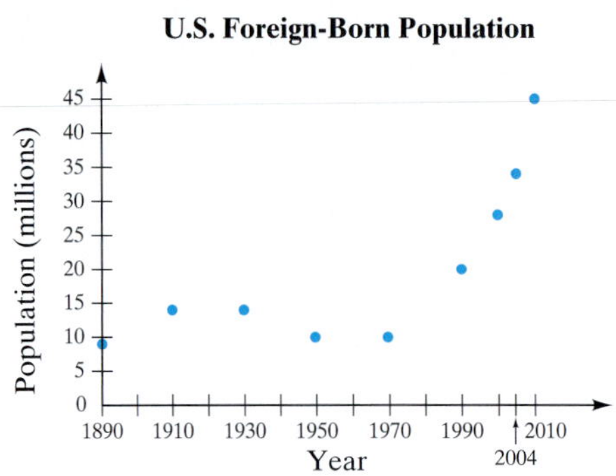


*Source:* U.S. Census Bureau, USA Today (3/8/05)

FIGURE 3.30

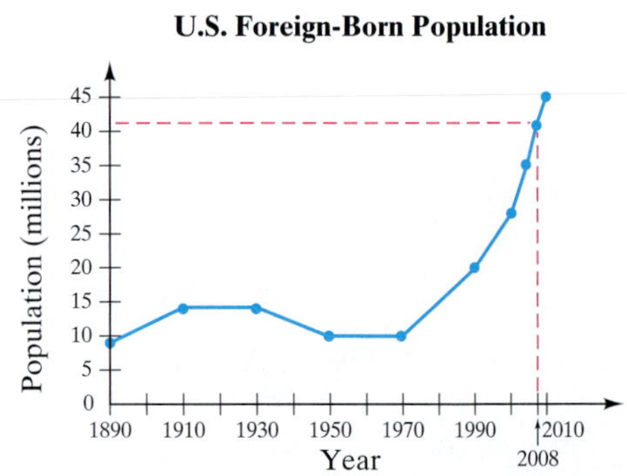


FIGURE 3.31

## Solution

**a)** Since each year corresponds with exactly one population, this set of points represents a function. Notice that this graph passes the vertical line test.

**b)** We can connect the points with straight line segments as in **Figure 3.31**. Then we can estimate from the graph that there were about 41 million foreign-born Americans in 2008. If we call the function $f$, then $f(2008) = 41$.

▸ **Now Try Exercise 75**

In Section 2.2 we learned to use formulas. Consider the formula for the area of a circle, $A = \pi r^2$. In the formula, $\pi$ is a constant that is approximately 3.14. For each specific value of the radius, $r$, there corresponds exactly one area, $A$. Thus the area of a circle is a function of its radius. We may therefore write

$$A(r) = \pi r^2$$

Often formulas are written using function notation like this.

**EXAMPLE 10 ▸** The Celsius temperature, $C$, is a function of the Fahrenheit temperature, $F$.

$$C(F) = \frac{5}{9}(F - 32)$$

Determine the Celsius temperature that corresponds to 50°F.

*Solution* We need to find $C(50)$. We do so by substitution.

$$C(F) = \frac{5}{9}(F - 32)$$

$$C(50) = \frac{5}{9}(50 - 32)$$

$$= \frac{5}{9}(18) = 10$$

Therefore, 50°F = 10°C.

▸ **Now Try Exercise 55**

In Example 10, $F$ is the independent variable and $C$ is the dependent variable. If we solved the function for $F$, we would obtain $F(C) = \frac{9}{5}C + 32$. In this formula, $C$ is the independent variable and $F$ is the dependent variable.

# EXERCISE SET 3.2

MathXL® MyMathLab

## Concept/Writing Exercises

1. What is a function?
2. What is a relation?
3. Are all functions also relations? Explain.
4. Are all relations also functions? Explain.
5. Explain how to use the vertical line test to determine if a relation is a function.
6. What is the domain of a function?
7. What is the range of a function?
8. What are the domain and range of the function $f(x) = 2x + 1$? Explain your answer.
9. Consider the function $y = \frac{1}{x}$. What is its domain and range? Write the answer using set notation. Explain.
10. What are the domain and range of a function of the form $f(x) = ax + b, a \neq 0$? Explain your answer.
11. Consider the absolute value function $y = |x|$. What is its domain and range? Write the answer using set notation. Explain.
12. What is a dependent variable?
13. What is an independent variable?
14. How is "$f(x)$" read?

## Practice the Skills

*In Exercises 15–20,* **a)** *determine if the relation illustrated is a function.* **b)** *Give the domain and range of each function or relation.*

**15.** twice a number

3 → 6
5 → 10
11 → 22

**16.** nicknames

Robert → Bobby
Robert → Rob
Margaret → Peggy
Margaret → Maggie

**17.** number of siblings

Cameron → 3
Tyrone → 6
Vishnu → 6

**18.** a number squared

4 → 16
5 → 25
7 → 49

**19.** cost of a stamp

1990 → 20
2001 → 34
2002 → 34
2002 → 37

**20.** absolute value

$|-8|$ → 8
$|8|$ → 8
$|0|$ → 0

*In Exercises 21–28,* **a)** *determine which of the following relations are also functions.* **b)** *Give the domain and range of each relation or function.*

**21.** $\{(1, 4), (2, 2), (3, 5), (4, 3), (5, 1)\}$

**22.** $\{(1, 0), (4, 2), (9, 3), (1, -1), (4, -2), (9, -3)\}$

**23.** $\{(3, -1), (5, 0), (1, 2), (4, 4), (2, 2), (7, 9)\}$

**24.** $\{(-1, 1), (0, -3), (3, 4), (4, 5), (-2, -2)\}$

**25.** $\{(1, 4), (2, 5), (3, 6), (2, 2), (1, 1)\}$

**26.** $\{(6, 3), (-3, 4), (0, 3), (5, 2), (3, 5), (2, 8)\}$

**27.** $\{(0, 3), (1, 3), (2, 2), (1, -1), (2, -7)\}$

**28.** $\{(3, 5), (2, 5), (1, 5), (0, 5), (-1, 5)\}$

*In Exercises 29–40,* **a)** *determine whether the graph illustrated represents a function.* **b)** *Give the domain and range of each function or relation.* **c)** *Approximate the value or values of x where $y = 2$.*

**29.**

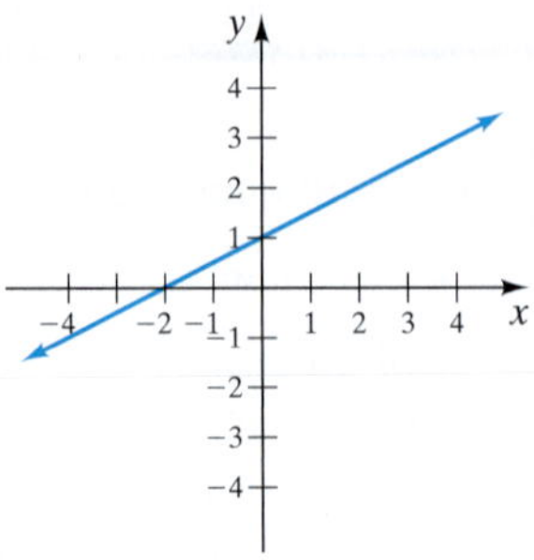

**30.**

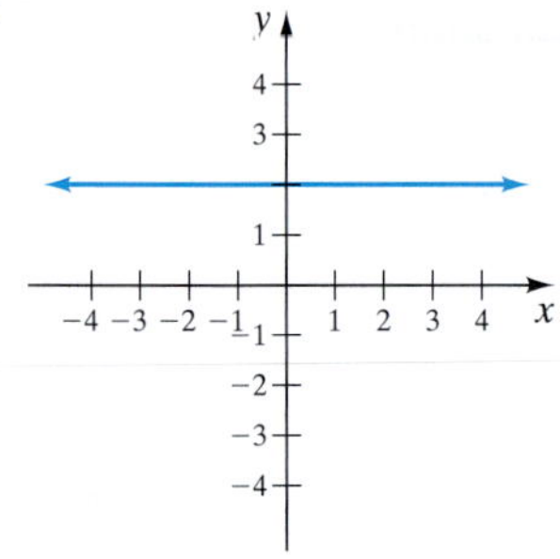

**31.**

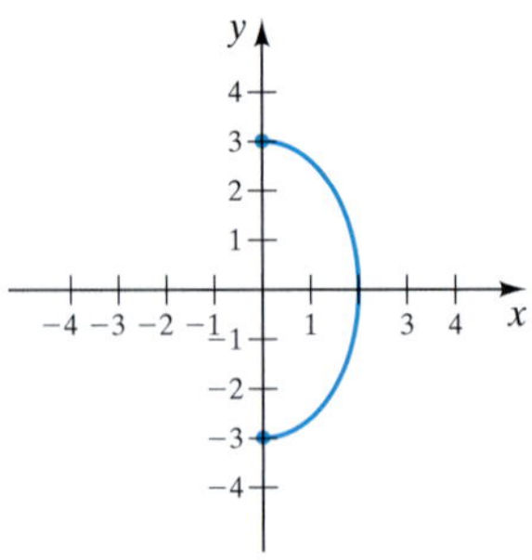

**32.**

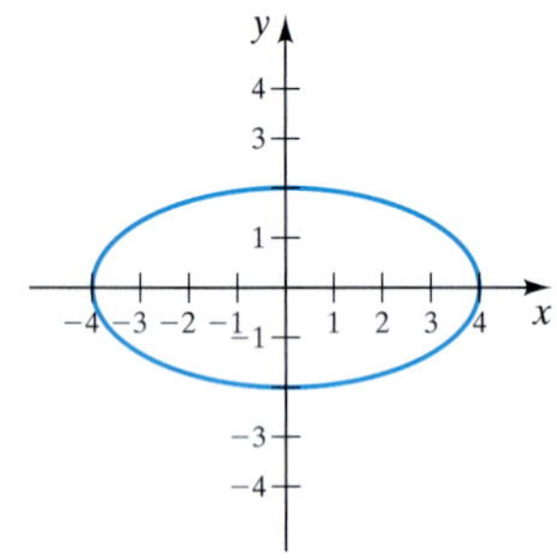

**33.**

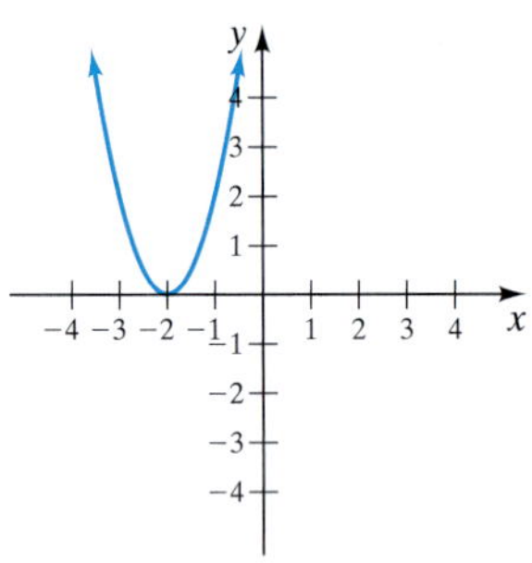

**34.**

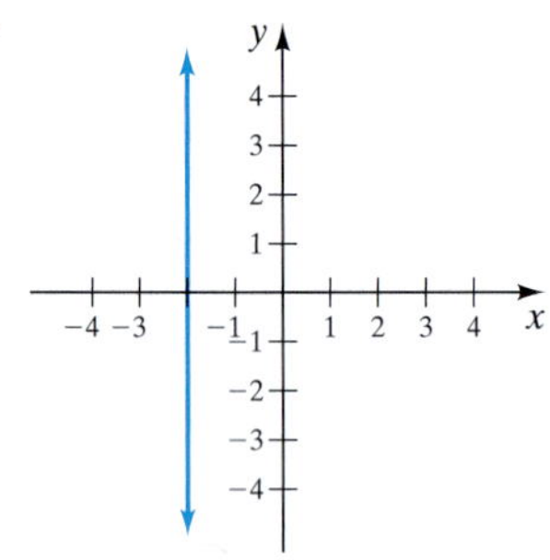

**35.**

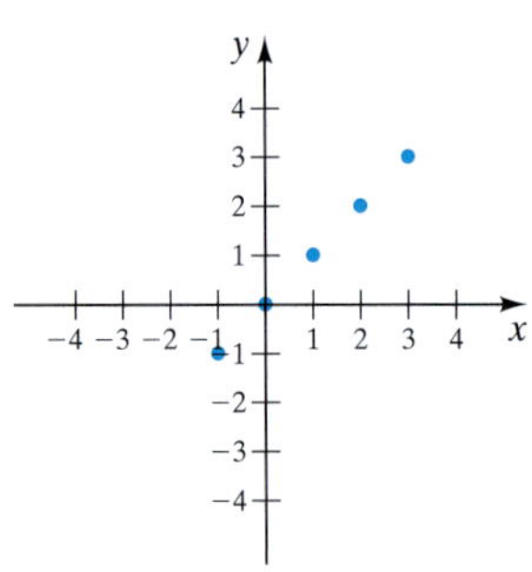

**36.**

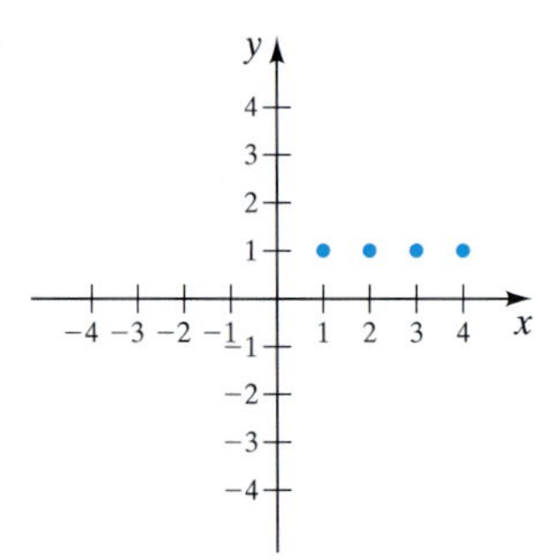

**37.**

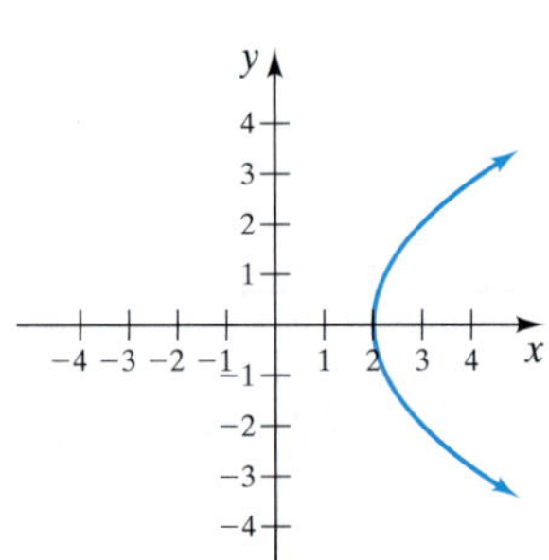

**38.**

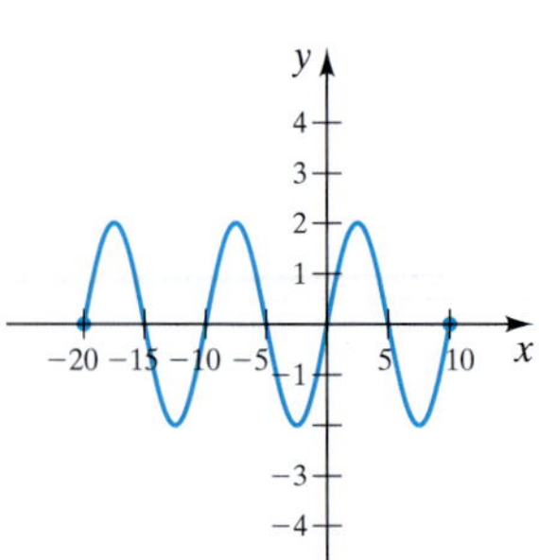

39.

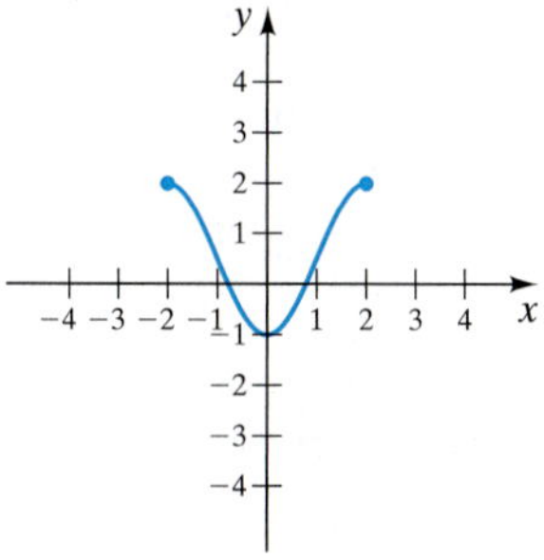

40.

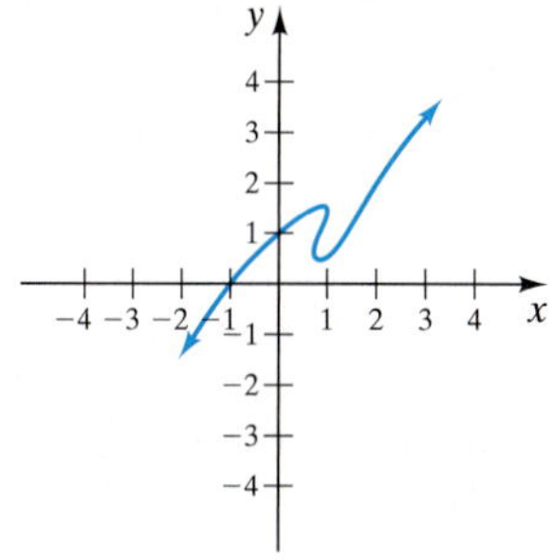

*Evaluate each function at the indicated values.*

41. $f(x) = -2x + 7$; find
**a)** $f(2)$.
**b)** $f(-3)$.

42. $f(a) = \frac{1}{3}a + 4$; find
**a)** $f(0)$.
**b)** $f(-12)$.

43. $h(x) = x^2 - x - 6$; find
**a)** $h(0)$.
**b)** $h(-1)$.

44. $g(x) = -2x^2 + 7x - 11$; find
**a)** $g(2)$.
**b)** $g\left(\frac{1}{2}\right)$.

45. $r(t) = -t^3 - 2t^2 + t + 4$; find
**a)** $r(1)$.
**b)** $r(-2)$.

46. $g(t) = 4 - 3t + 16t^2 - 2t^3$; find
**a)** $g(0)$.
**b)** $g(3)$.

47. $h(z) = |5 - 2z|$; find
**a)** $h(6)$.
**b)** $h\left(\frac{5}{2}\right)$.

48. $q(x) = -2|x + 8| + 13$; find
**a)** $q(0)$.
**b)** $q(-4)$.

49. $s(t) = \sqrt{t + 3}$; find
**a)** $s(-3)$.
**b)** $s(6)$.

50. $f(t) = \sqrt{5 - 2t}$; find
**a)** $f(-2)$.
**b)** $f(2)$.

51. $g(x) = \frac{x^3 - 2}{x - 2}$; find
**a)** $g(0)$.
**b)** $g(2)$.

52. $h(x) = \frac{x^2 + 4x}{x + 6}$; find
**a)** $h(-3)$.
**b)** $h\left(\frac{2}{5}\right)$.

## Problem Solving

53. **Area of a Rectangle** The formula for the area of a rectangle is $A = lw$. If the length of a rectangle is 6 inches, then the area is a function of its width, $A(w) = 6w$. Find the area when the width is
**a)** 4 feet.
**b)** 6.5 feet.

54. **Simple Interest** The formula for the simple interest earned for a period of 1 year is $i = pr$, where $p$ is the principal invested and $r$ is the simple interest rate. If \$1000 is invested, the simple interest earned in 1 year is a function of the simple interest rate, $i(r) = 1000r$. Determine the simple interest earned in 1 year if the interest rate is
**a)** 2.5%.
**b)** 4.25%.

55. **Area of a Circle** The formula for the area of a circle is $A = \pi r^2$. The area is a function of the radius.
**a)** Write this function using function notation.
**b)** Determine the area when the radius is 12 yards.

56. **Perimeter of a Square** The formula for the perimeter of a square is $P = 4s$ where $s$ represents the length of any one of the sides of the square.
**a)** Write this function using function notation.
**b)** Determine the perimeter of a square with sides of length 7 meters.

57. **Temperature** The formula for changing Fahrenheit temperature into Celsius temperature is $C = \frac{5}{9}(F - 32)$. The Celsius temperature is a function of Fahrenheit temperature.
**a)** Write this function using function notation.
**b)** Find the Celsius temperature that corresponds to $-31°$F.

**58. Volume of a Cylinder** The formula for the volume of a right circular cylinder is $V = \pi r^2 h$. If the height, $h$, is 3 feet, then the volume is a function of the radius, $r$.

**a)** Write this formula in function notation, where the height is 3 feet.

**b)** Find the volume if the radius is 2 feet.

**59. Sauna Temperature** The temperature, $T$, in degrees Celsius, in a sauna $n$ minutes after being turned on is given by the function $T(n) = -0.03n^2 + 1.5n + 14$. Find the sauna's temperature after

**a)** 3 minutes. **b)** 12 minutes.

**60. Stopping Distance** The stopping distance, $d$, in meters for a car traveling $v$ kilometers per hour is given by the function $d(v) = 0.18v + 0.01v^2$. Find the stopping distance for the following speeds:

**a)** 60 km/hr **b)** 25 km/hr

**61. Air Conditioning** When an air conditioner is turned on maximum in a bedroom at 80°, the temperature, $T$, in the room after $A$ minutes can be approximated by the function $T(A) = -0.02A^2 - 0.34A + 80, 0 \leq A \leq 15$.

**a)** Estimate the room temperature 4 minutes after the air conditioner is turned on.

**b)** Estimate the room temperature 12 minutes after the air conditioner is turned on.

**62. Accidents** The number of accidents, $n$, in 1 month involving drivers $x$ years of age can be approximated by the function $n(x) = 2x^2 - 150x + 4000$. Find the approximate number of accidents in 1 month that involved

**a)** 18-year-olds.

**b)** 25-year-olds.

**63. Oranges** The total number of oranges, $T$, in a square pyramid whose base is $n$ by $n$ oranges is given by the function

$$T(n) = \frac{1}{3}n^3 + \frac{1}{2}n^2 + \frac{1}{6}n$$

Find the number of oranges if the base is

**a)** 6 by 6 oranges.

**b)** 8 by 8 oranges.

**64. Rock Concert** If the cost of a ticket to a rock concert is increased by $x$ dollars, the estimated increase in revenue, $R$, in thousands of dollars is given by the function $R(x) = 24 + 5x - x^2, x < 8$. Find the increase in revenue if the cost of the ticket is increased by

**a)** \$1.

**b)** \$4.

*Review Example 5 before working Exercises 65–70.*

**65. Heart Rate** The following graph shows a person's heart rate while doing exercise. Write a story that this graph may represent.

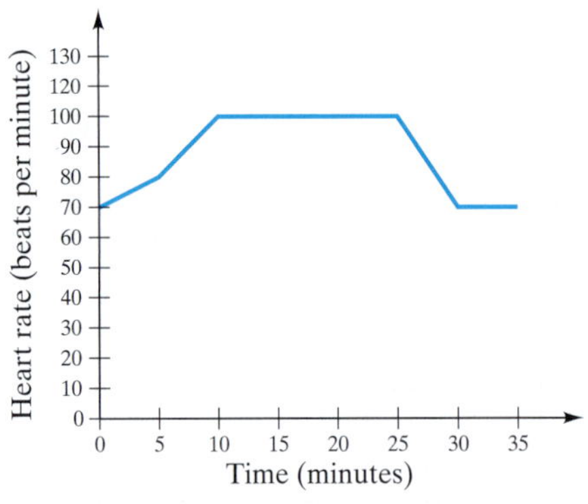


**66. Water Level** The following graph shows the water level at a certain point during a flood. Write a story that this graph may represent.

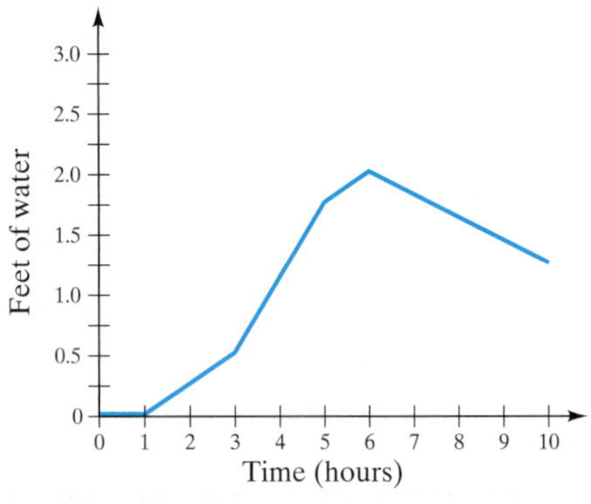

## Group Activity

*In many real-life situations, more than one function may be needed to represent a problem. This often occurs where two or more different rates are involved. For example, when discussing federal income taxes, there are different tax rates. When two or more functions are used to represent a problem, the function is called a* ***piecewise function****. Following are two examples of piecewise functions and their graphs.*

$$f(x) = \begin{cases} -x + 2, & 0 \le x < 4 \\ 2x - 10, & 4 \le x < 8 \end{cases}$$

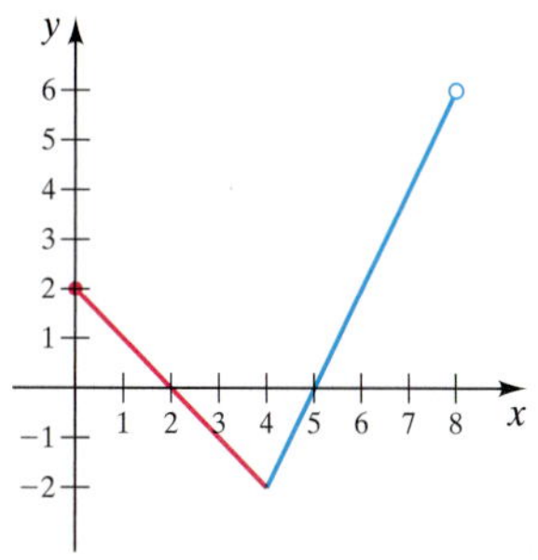


$$f(x) = \begin{cases} 2x - 1, & -2 \le x < 2 \\ x - 2, & 2 \le x < 4 \end{cases}$$

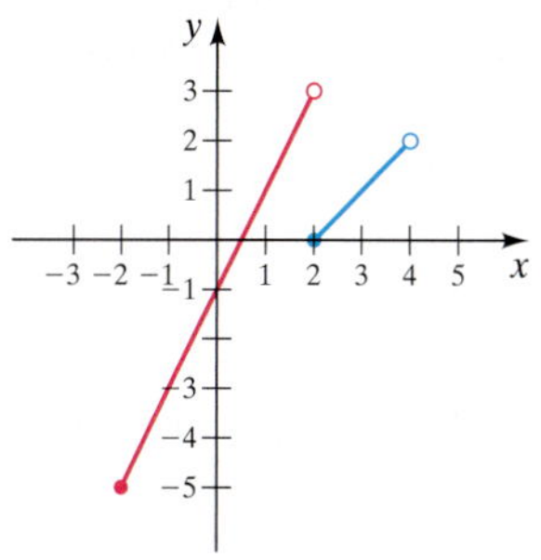


*As a group, graph the following piecewise functions.*

**78.** $f(x) = \begin{cases} x + 3, & -1 \le x < 2 \\ 7 - x, & 2 \le x < 4 \end{cases}$

**79.** $g(x) = \begin{cases} 2x + 3, & -3 < x < 0 \\ -3x + 1, & 0 \le x < 2 \end{cases}$

## Cumulative Review Exercises

[2.1] **80.** Solve $3x - 2 = \frac{1}{3}(3x - 3)$.

[2.2] **81.** Solve the following formula for $p_2$.

$$E = a_1p_1 + a_2p_2 + a_3p_3$$

[2.5] **82.** Solve the inequality $\frac{3}{5}(x - 3) > \frac{1}{4}(3 - x)$ and indicate the solution

**a)** on the number line;

**b)** in interval notation; and

**c)** in set builder notation.

[2.6] **83.** Solve $\left|\frac{x - 4}{3}\right| + 9 = 11$.

# 3.3 Linear Functions: Graphs and Applications

**1** Graph linear functions.

**2** Graph linear functions using intercepts.

**3** Graph equations of the form $x = a$ and $y = b$.

**4** Study applications of functions.

**5** Solve linear equations in one variable graphically.

## 1 Graph Linear Functions

In Section 3.1 we graphed linear equations. To graph the linear equation $y = 2x + 4$, we can make a table of values, plot the points, and draw the graph, as shown in **Figure 3.32**. Notice that this graph represents a function since it passes the vertical line test.

| $x$ | $y$ |
|---|---|
| −2 | 0 |
| 0 | 4 |
| 1 | 6 |

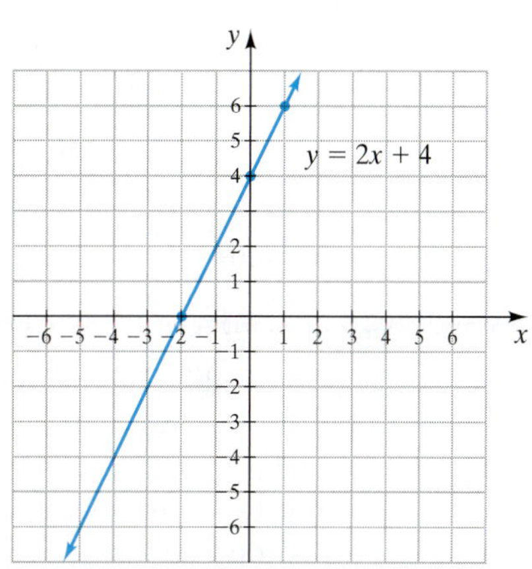


FIGURE 3.32

We may write the equation graphed in **Figure 3.32** using function notation as $f(x) = 2x + 4$. This is an example of a linear function. A **linear function** is a function of the form $f(x) = ax + b$. The graph of any linear function is a straight line. The domain of any function is the set of real numbers for which the function is a real number. The domain of any linear function is the set of all real numbers, $\mathbb{R}$: Any real number, $x$, substituted into a linear function will result in $f(x)$ being a real number. We will discuss domains of functions further in Section 3.6.

To graph a linear function, we treat $f(x)$ as $y$ and follow the same procedure used to graph linear equations.

### Helpful Hint

When graphing a linear function, remember $y = f(x)$.

**EXAMPLE 1** ▸ Graph $f(x) = \frac{1}{2}x - 1$.

*Solution* We construct a table of values by substituting values for $x$ and finding corresponding values of $f(x)$ or $y$. Then we plot the points and draw the graph, as illustrated in **Figure 3.33**.

| $x$ | $f(x)$ |
|---|---|
| −2 | −2 |
| 0 | −1 |
| 2 | 0 |

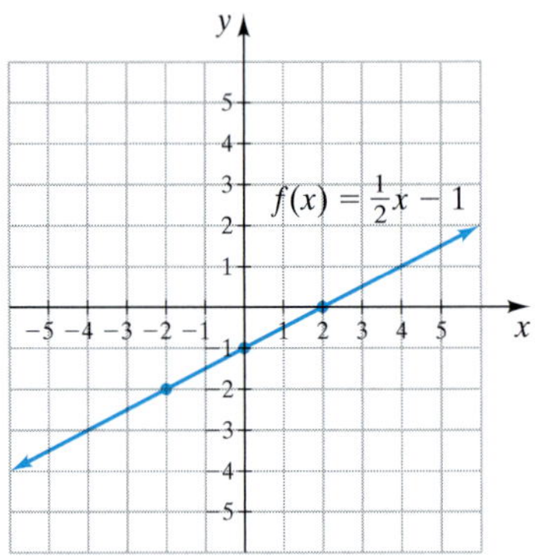


FIGURE 3.33

▸ Now Try Exercise 33

Note that the vertical axis in **Figure 3.33** may also be labeled as $f(x)$ instead of $y$. In this book we will continue to label it $y$.

## 2 Graph Linear Functions Using Intercepts

Linear equations are not always given in the form $y = ax + b$. The equation $2x + 3y = 6$ is an example of a linear equation given in *standard form*.

**Standard Form of a Linear Equation**

The **standard form of a linear equation** is

$$ax + by = c$$

where $a$, $b$, and $c$ are real numbers, and $a$ and $b$ are not both 0.

**Examples of Linear Equations in Standard Form**

$$2x + 3y = 4 \qquad -x + 5y = -2$$

Sometimes when an equation is given in standard form it may be easier to draw the graph using the $x$- and $y$-intercepts. Let's examine two points on the graph shown in **Figure 3.32**. Note that the graph crosses the $x$-axis at the point $(-2, 0)$. Therefore, $(-2, 0)$ is called the ***x*-intercept**. Sometimes we say that the $x$-intercept is *at* −2 (on the $x$-axis), the $x$-coordinate of the ordered pair.

The graph crosses the $y$-axis at the point $(0, 4)$. Therefore, $(0, 4)$ is called the ***y*-intercept**. Sometimes we say that the $y$-intercept is *at* 4 (on the $y$-axis), the $y$-coordinate of the ordered pair.

Below we explain how the $x$- and $y$-intercepts may be determined algebraically.

**To Find the $x$- and $y$-Intercepts**

To find the $y$-intercept, set $x = 0$ and solve for $y$.

To find the $x$-intercept, set $y = 0$ and solve for $x$.

To graph a linear equation or linear function using the $x$- and $y$-intercepts, find the intercepts and plot the points. Then draw a straight line through the points. When graphing linear equations or linear functions using the intercepts, you must be very careful. If either of your points is plotted wrong, your graph will be wrong.

**EXAMPLE 2** ▸ Graph $5x = 10y - 20$ using the $x$- and $y$-intercepts.

*Solution* To find the $y$-intercept (the point where the graph crosses the $y$-axis), set $x = 0$ and solve for $y$.

$$\begin{aligned} 5x &= 10y - 20 \\ 5(0) &= 10y - 20 \\ 0 &= 10y - 20 \\ 20 &= 10y \\ 2 &= y \end{aligned}$$

The graph crosses the $y$-axis at $y = 2$. The ordered pair representing the $y$-intercept is $(0, 2)$.

To find the $x$-intercept (the point where the graph crosses the $x$-axis), set $y = 0$ and solve for $x$.

$$\begin{aligned} 5x &= 10y - 20 \\ 5x &= 10(0) - 20 \\ 5x &= -20 \\ x &= -4 \end{aligned}$$

The graph crosses the $x$-axis at $x = -4$. The ordered pair representing the $x$-intercept is $(-4, 0)$. Now plot the intercepts and draw the graph (**Fig. 3.34**).

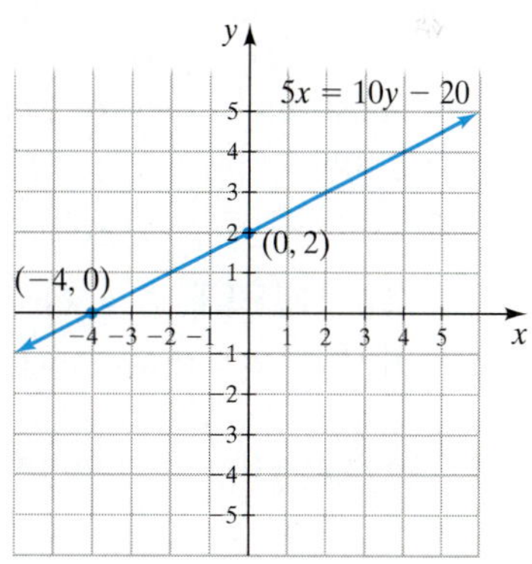


FIGURE 3.34

▸ **Now Try Exercise 23**

**EXAMPLE 3** ▸ Graph $f(x) = -\frac{1}{3}x - 1$ using the $x$- and $y$-intercepts.

*Solution* Treat $f(x)$ the same as $y$. To find the $y$-intercept, set $x = 0$ and solve for $f(x)$.

$$f(x) = -\frac{1}{3}x - 1$$

$$f(x) = -\frac{1}{3}(0) - 1 = -1$$

The $y$-intercept is $(0, -1)$.

To find the $x$-intercept, set $f(x) = 0$ and solve for $x$.

$$f(x) = -\frac{1}{3}x - 1$$

$$0 = -\frac{1}{3}x - 1$$

$$3(0) = 3\left(-\frac{1}{3}x - 1\right) \quad \text{Multiply both sides by 3.}$$

$$0 = -x - 3 \quad \text{Distributive property}$$

$$x = -3 \quad \text{Add x to both sides.}$$

The $x$-intercept is $(-3, 0)$. The graph is shown in **Figure 3.35**.

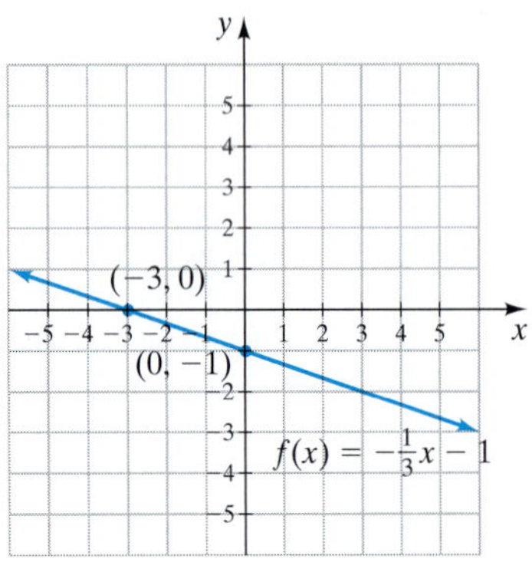


FIGURE 3.35

▶ **Now Try Exercise 17**

Graphs of the form $ax + by = 0$ go through the origin and have the same $x$- and $y$-intercept, (0, 0). To graph such equations we can use the intercept as one point and substitute values for $x$ and find the corresponding values of $y$ to get other points on the graph.

**EXAMPLE 4** ▶ Graph $-6x + 4y = 0$.

*Solution* If we substitute $x = 0$ we find that $y = 0$. Thus the graph goes through the origin. We will select $x = -2$ and $x = 2$ and substitute these values into the equation, one at a time, to find two other points on the graph.

| Let $x = -2$. | Let $x = 2$. |
|---|---|
| $-6x + 4y = 0$ | $-6x + 4y = 0$ |
| $-6(-2) + 4y = 0$ | $-6(2) + 4y = 0$ |
| $12 + 4y = 0$ | $-12 + 4y = 0$ |
| $4y = -12$ | $4y = 12$ |
| $y = -3$ | $y = 3$ |
| ordered pairs: $(-2, -3)$ | $(2, 3)$ |

Two other points on the graph are at $(-2, -3)$ and $(2, 3)$. The graph of $-6x + 4y = 0$ is shown in **Figure 3.36** on page 177.

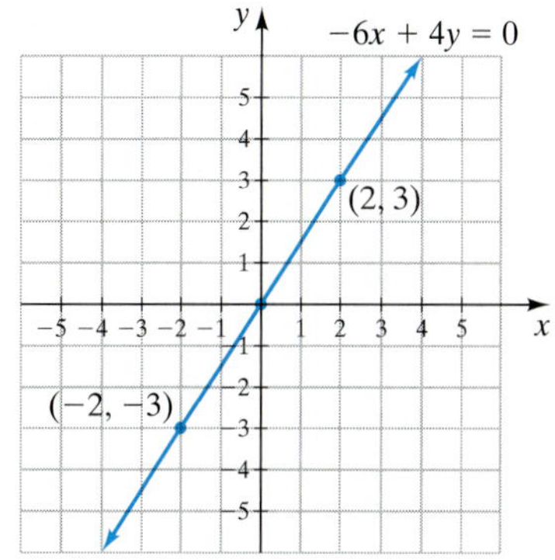


FIGURE 3.36

▸ Now Try Exercise 35

## USING YOUR GRAPHING CALCULATOR

Sometimes it may be difficult to estimate the intercepts of a graph accurately. When this occurs, you might want to use a graphing calculator. We demonstrate how in the following example.

**EXAMPLE** Determine the $x$- and $y$-intercepts of the graph of $y = 1.3(x - 3.2)$.

*Solution* Press the [Y=] key, and then assign $1.3(x - 3.2)$ to $y$. Then press the [GRAPH] key to graph the function $y = 1.3(x - 3.2)$, as shown in **Figure 3.37a**.

From the graph it may be difficult to determine the intercepts. One way to find the $y$-intercept is to use the TRACE feature, which was discussed in Section 3.1. **Figure 3.37b** shows a TI-84 Plus screen after the [TRACE] key is pressed. Notice the $y$-intercept is at $-4.16$.

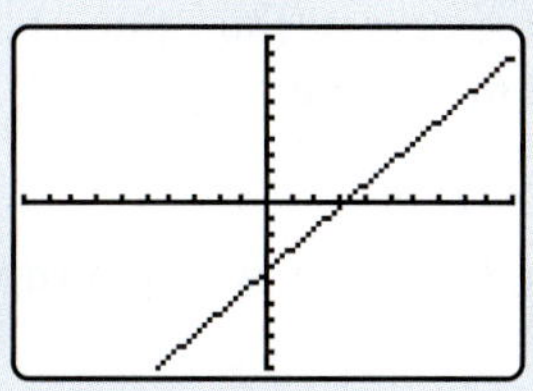

FIGURE 3.37a

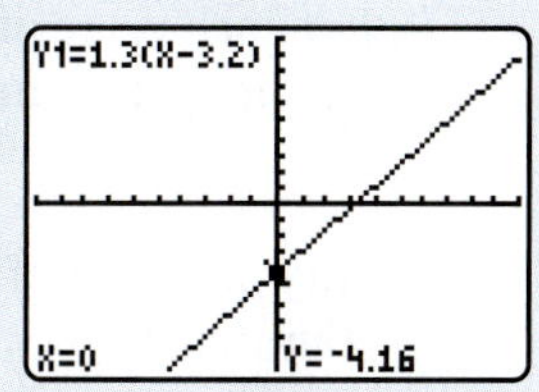


FIGURE 3.37b

Some graphing calculators have the ability to find the $x$-intercepts of a function by pressing just a few keys. A **zero** (or **root**) of a function is a value of $x$ such that $f(x) = 0$. A zero (or root) of a function is the $x$-coordinate of the $x$-intercept of the graph of the function. Read your calculator manual to learn how to find the zeros or roots of a function. On a TI-84 Plus you press the keys [2nd] [TRACE] to get to the CALC menu (which stands for calculate). Then you choose option 2, *zero*. Once the zero feature has been selected, the calculator will display

Left bound?

At this time, move the cursor along the curve until it is to the *left* of the zero. Then press [ENTER]. The calculator now displays

Right bound?

Move the cursor along the curve until it is to the *right* of the zero. Then press [ENTER]. The calculator now displays

Guess?

Now press [ENTER] for the third time and the zero is displayed at the bottom of the screen, as in **Figure 3.38**. Thus the $x$-intercept of the function is at 3.2. For practice at finding the intercepts on your calculator, work Exercises 69–72.

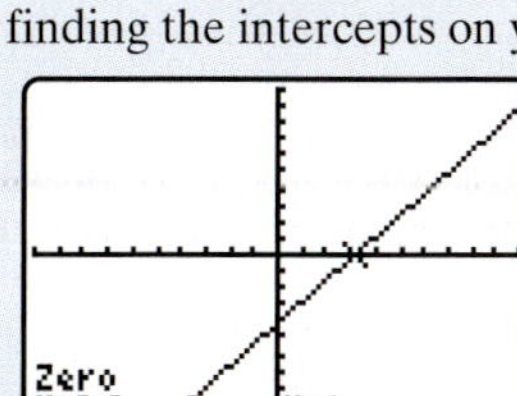


FIGURE 3.38

## 3 Graph Equations of the Form $x = a$ and $y = b$

Examples 5 and 6 illustrate how equations of the form $x = a$ and $y = b$, where $a$ and $b$ are constants, are graphed.

**EXAMPLE 5** ▸ Graph the equation $y = -3$.

*Solution* This equation can be written as $y = -3 + 0x$. Thus, for any value of $x$ selected, $y$ is $-3$. The graph of $y = -3$ is illustrated in **Figure 3.39**.

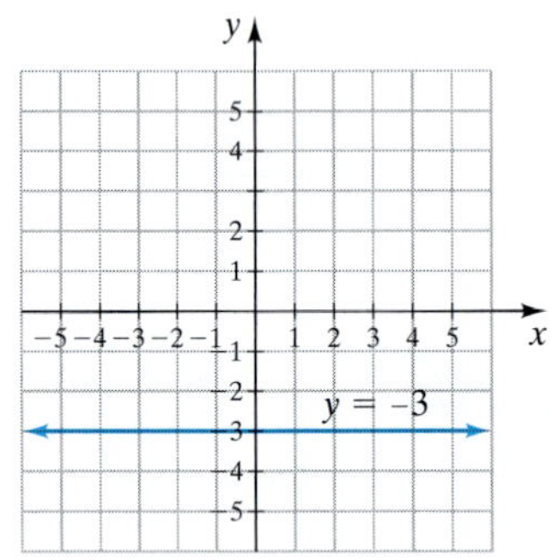


FIGURE 3.39

▸ Now Try Exercise 43

**Equation of a Horizontal Line**

The graph of any equation of the form $y = b$ will always be a horizontal line for any real number $b$.

Notice that the graph of $y = -3$ is a function since it passes the vertical line test. For each value of $x$ selected, the value of $y$, or the value of the function, is $-3$. This is an example of a **constant function**. We may write

$$f(x) = -3$$

Any equation of the form $y = b$ or $f(x) = b$, where $b$ represents a constant, is a constant function.

**EXAMPLE 6** ▸ Graph the equation $x = 2$.

*Solution* This equation can be written as $x = 2 + 0y$. Thus, for every value of $y$ selected, $x$ will have a value of 2 (**Fig. 3.40**).

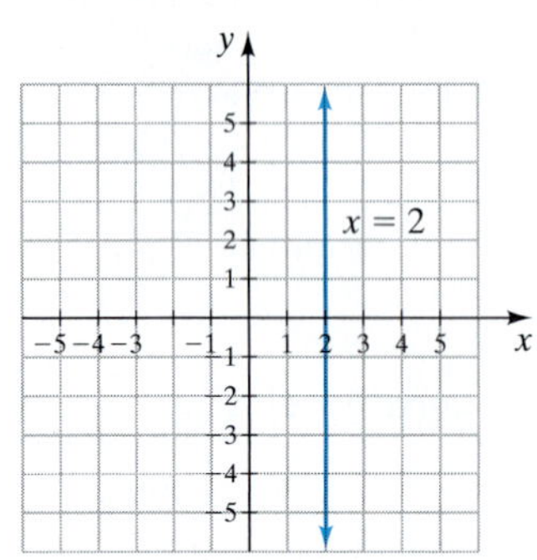


FIGURE 3.40

Now Try Exercise 41

**Equation of a Vertical Line**

The graph of any equation of the form $x = a$ will always be a vertical line for any real number $a$.

Notice that the graph of $x = 2$ does not represent a function since it does not pass the vertical line test. For $x = 2$ there is more than one value of $y$. In fact, when $x = 2$ there are an infinite number of values for $y$.

## 4 Study Applications of Functions

Graphs are often used to show the relationship between variables. The axes of a graph do not have to be labeled $x$ and $y$. They can be any designated variables. Consider the following example.

**EXAMPLE 7 ▸ Tire Store Profit** The yearly profit, $p$, of a tire store can be estimated by the function $p(n) = 20n - 30{,}000$, where $n$ is the number of tires sold per year.

**a)** Draw a graph of profit versus tires sold for up to and including 6000 tires.

**b)** Estimate the number of tires that must be sold for the company to break even.

**c)** Estimate the number of tires sold if the company has a \$70,000 profit.

*Solution* **a) Understand** The profit, $p$, is a function of the number of tires sold, $n$. The horizontal axis will therefore be labeled Number of tires sold (the independent variable) and the vertical axis will be labeled Profit (the dependent variable). Since the minimum number of tires that can be sold is 0, negative values do not have to be listed on the horizontal axis. The horizontal axis will therefore go from 0 to 6000 tires.

We will graph this equation by determining and plotting the intercepts.

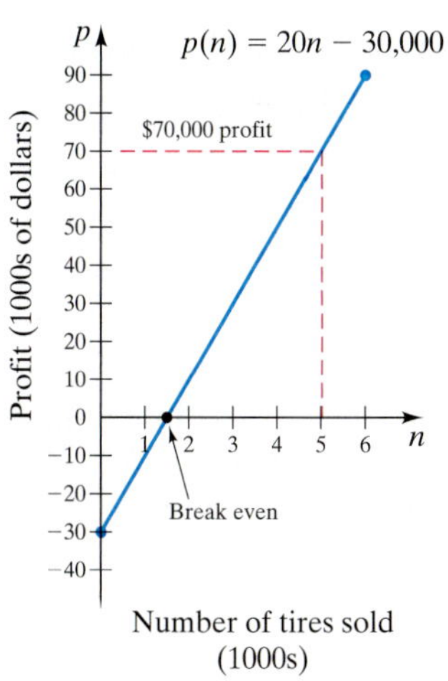


FIGURE 3.41

**Translate and Carry Out** To find the $p$-intercept, we set $n = 0$ and solve for $p(n)$.

$$p(n) = 20n - 30{,}000$$
$$p(n) = 20(0) - 30{,}000 = -30{,}000$$

Thus, the $p$-intercept is $(0, -30{,}000)$.

To find the $n$-intercept, we set $p(n) = 0$ and solve for $n$.

$$p(n) = 20n - 30{,}000$$
$$0 = 20n - 30{,}000$$
$$30{,}000 = 20n$$
$$1500 = n$$

Thus the $n$-intercept is $(1500, 0)$.

**Answer** Now we use the $p$- and $n$-intercepts to draw the graph (see **Fig. 3.41**).

**b)** The break-even point is the number of tires that must be sold for the company to have neither a profit nor a loss. The break-even point is where the graph intersects the $n$-axis, for this is where the profit, $p$, is 0. To break even, approximately 1500 tires must be sold.

**c)** To make \$70,000, approximately 5000 tires must be sold (shown by the dashed red line in **Fig. 3.41**).

▸ **Now Try Exercise 51**

Sometimes it is difficult to read an exact answer from a graph. To determine the exact number of tires needed to break even in Example 7, substitute 0 for $p(n)$ in the function $p(n) = 20n - 30{,}000$ and solve for $n$. To determine the exact number of tires needed to obtain a \$70,000 profit, substitute 70,000 for $p(n)$ and solve the equation for $n$.

**EXAMPLE 8 ▸ Toy Store Sales** Andrew Gestrich is the owner of a toy store. His monthly salary consists of \$200 plus 10% of the store's sales for that month.

**a)** Write a function expressing his monthly salary, $m$, in terms of the store's sales, $s$.

**b)** Draw a graph of his monthly salary for sales up to and including \$20,000.

**c)** If the store's sales for the month of April are \$15,000, what will Andrew's salary be for April?

| $s$ | $m$ |
|---|---|
| 0 | 200 |
| 10,000 | 1200 |
| 20,000 | 2200 |

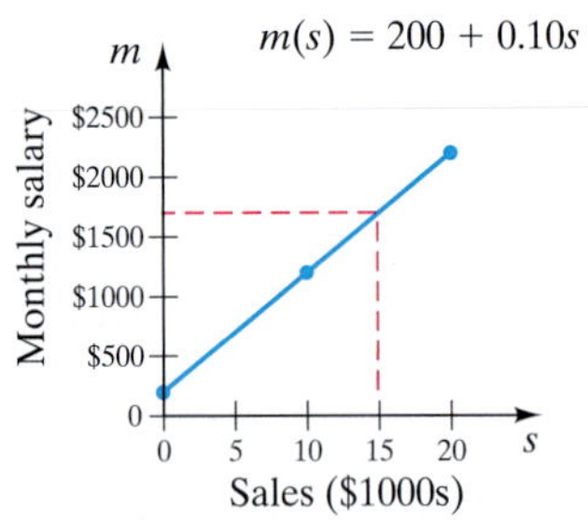


FIGURE 3.42

### Solution

**a)** Andrew's monthly salary is a function of sales. His monthly salary, $m$, consists of \$200 plus 10% of the sales, $s$. Ten percent of $s$ is $0.10s$. Thus the function for finding his salary is

$$m(s) = 200 + 0.10s$$

**b)** Since monthly salary is a function of sales, Sales will be represented on the horizontal axis and Monthly salary will be represented on the vertical axis. Since sales can never be negative, the monthly salary can never be negative. Thus both axes will be drawn with only positive numbers. We will draw this graph by plotting points. We select values for $s$, find the corresponding values of $m$, and then draw the graph. We can select values of $s$ that are between \$0 and \$20,000 (**Fig. 3.42**).

**c)** By reading our graph carefully, we can estimate that when the store's sales are \$15,000, Andrew's monthly salary is about \$1700.

▶ **Now Try Exercise 53**

## 5 Solve Linear Equations in One Variable Graphically

Earlier we discussed the graph of $f(x) = 2x + 4$. In **Figure 3.43** below we illustrate the graph of $f(x)$ along with the graph of $g(x) = 0$. Notice that the two graphs intersect at $(-2, 0)$. We can obtain the $x$-coordinate of the ordered pair by solving the equation $f(x) = g(x)$. Remember $f(x)$ and $g(x)$ both represent $y$, and by solving this equation for $x$ we are obtaining the value of $x$ where the $y$'s are equal.

$$\begin{aligned} f(x) &= g(x) \\ \overbrace{2x + 4} &= \overbrace{0} \\ 2x &= -4 \\ x &= -2 \end{aligned}$$

Note that we obtain $-2$, the $x$-coordinate in the ordered pair at the point of intersection.

Now let's find the $x$-coordinate of the point at which the graphs of $f(x) = 2x + 4$ and $g(x) = 2$ intersect. We solve the equation $f(x) = g(x)$.

$$\begin{aligned} f(x) &= g(x) \\ \overbrace{2x + 4} &= \overbrace{2} \\ 2x &= -2 \\ x &= -1 \end{aligned}$$

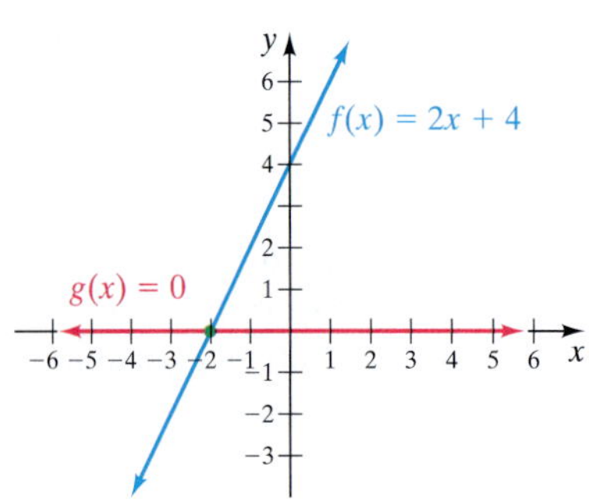


FIGURE 3.43

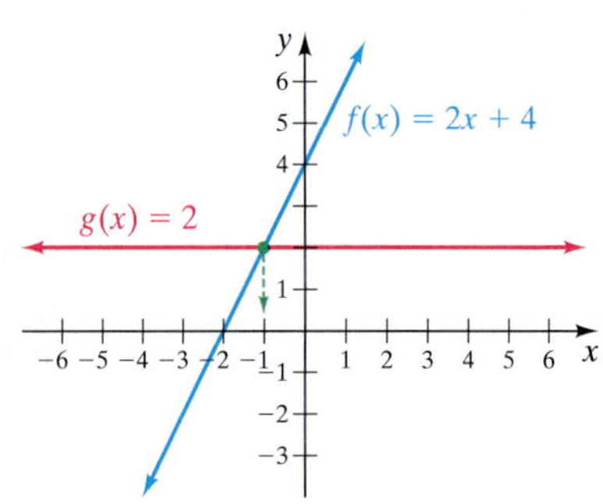


FIGURE 3.44

The $x$-coordinate of the point of intersection of the two graphs is $-1$, as shown in **Figure 3.44**. Notice that $f(-1) = 2(-1) + 4 = 2$.

In general, if we are given an equation in one variable, we can regard each side of the equation as a separate function. To obtain the solution to the equation, we can graph the two functions. The $x$-coordinate of the point of intersection will be the solution to the equation.

What are the $y$-intercepts of $y = 2x + 3$, $y = 2x$ (or $y = 2x + 0$), and $y = 2x - 3$? The $y$-intercepts are at $(0, 3)$, $(0, 0)$, and $(0, -3)$, respectively. Notice that the graph of $y = 2x + 3$ is the graph of $y = 2x$ shifted, or **translated**, 3 units up and $y = 2x - 3$ is the graph of $y = 2x$ translated 3 units down. All three lines are **parallel**; that is, they do not intersect no matter how far they are extended.

Using this information, can you guess what the $y$-intercept of $y = 2x + 4$ will be? How about the $y$-intercept of $y = 2x - \frac{5}{3}$? If you answered $(0, 4)$ and $\left(0, -\frac{5}{3}\right)$, respectively, you answered correctly. In fact, the graph of an equation of the form $y = 2x + b$ will have a $y$-intercept of $(0, b)$.

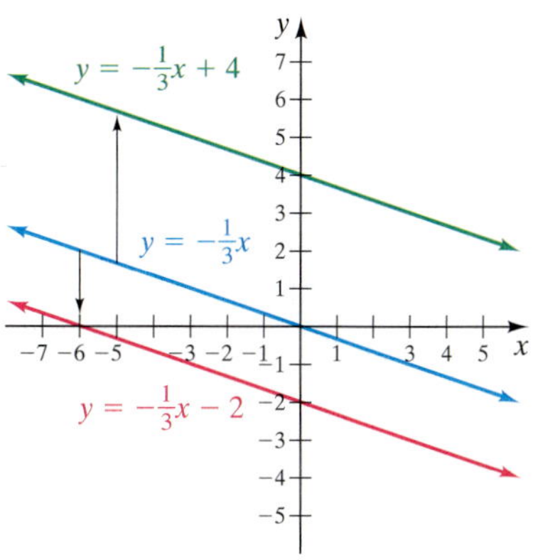


FIGURE 3.51

Now consider the graphs of the equations $y = -\frac{1}{3}x + 4$, $y = -\frac{1}{3}x$, and $y = -\frac{1}{3}x - 2$, shown in **Figure 3.51**. The $y$-intercepts of the three lines are $(0, 4)$, $(0, 0)$, and $(0, -2)$, respectively. The graph of $y = -\frac{1}{3}x + b$ will have a $y$-intercept of $(0, b)$.

By looking at the preceding equations, their graphs, and $y$-intercepts, can you determine the $y$-intercept of the graph of $y = mx + b$, where $m$ and $b$ are real numbers? If you answered $(0, b)$, you answered correctly. In general, the graph of $y = mx + \mathbf{b}$, where $m$ and $b$ are real numbers, has a $y$-intercept $(0, \mathbf{b})$.

If we look at the graphs in **Figure 3.50**, we see that the slopes (or slants) of the three lines appear to be the same. If we look at the graphs in **Figure 3.51**, we see that the slopes of those three lines appear to be the same, but their slope is different from the slope of the three lines in **Figure 3.50**.

If we consider the equation $y = mx + b$, where the $b$ determines the $y$-intercept of the line, we can reason that the $m$ is responsible for the slope (or the slant) of the line.

## 2 Find the Slope of a Line

Now let's discuss slope. The **slope of a line** is the ratio of the vertical change (or rise) to the horizontal change (or run) between any two points on a line. Consider the graph of $y = 2x$ (the blue line in **Figure 3.50**, repeated in **Figure 3.52a**). Two points on this line are $(1, 2)$ and $(3, 6)$. Let's find the slope of the line through these points. If we draw a line parallel to the $x$-axis through the point $(1, 2)$ and a line parallel to the $y$-axis through the point $(3, 6)$, the two lines intersect at $(3, 2)$. (See **Fig. 3.52b**.)

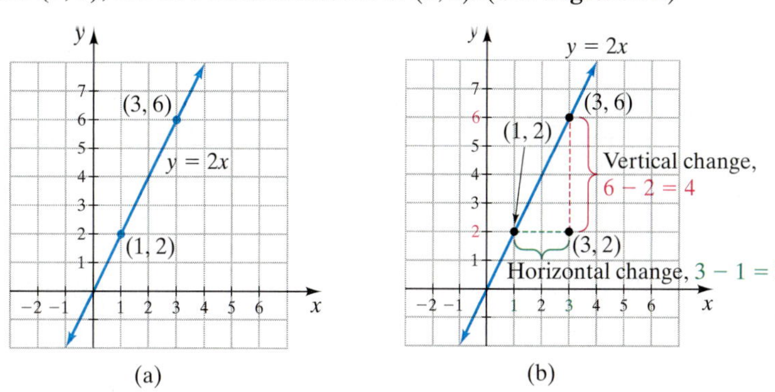


FIGURE 3.52

From **Figure 3.52b** we can determine the slope of the line. The vertical change (along the $y$-axis) is $6 - 2$, or 4 units. The horizontal change (along the $x$-axis) is $3 - 1$, or 2 units.

$$\text{slope} = \frac{\text{vertical change}}{\text{horizontal change}} = \frac{4}{2} = 2$$

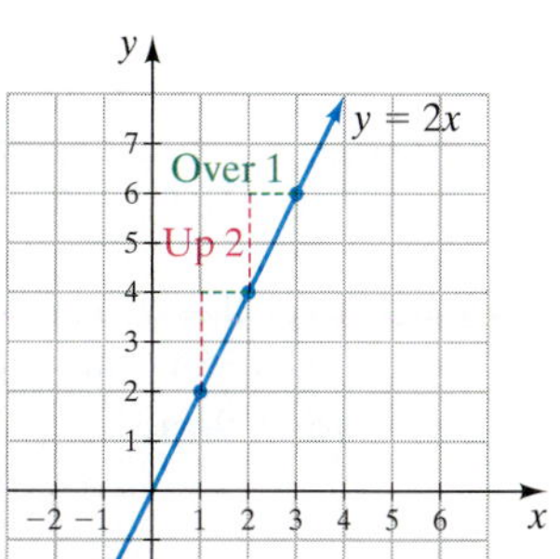


FIGURE 3.53

Thus, the slope of the line through the points $(3, 6)$ and $(1, 2)$ is 2. By examining the line connecting these two points, we can see that for each 2 units the graph moves up the $y$-axis, it moves 1 unit to the right on the $x$-axis (**Fig. 3.53**).

We have determined that the slope of the graph of $y = 2x$ is 2. If you were to compute the slope of the other two lines in **Figure 3.50**, you would find that the graphs of $y = 2x + 3$ and $y = 2x - 3$ also have a slope of 2.

Can you guess what the slope of the graphs of the equations $y = -3x + 2$, $y = -3x$, and $y = -3x - 2$ is? The slope of all three lines is $-3$. In general, the slope of an equation of the form $y = mx + b$ is $m$.*

Now let's determine the procedure to find the slope of a line passing through the two points $(x_1, y_1)$ and $(x_2, y_2)$. Consider **Figure 3.54**. The vertical change can be found by subtracting $y_1$ from $y_2$. The horizontal change can be found by subtracting $x_1$ from $x_2$.

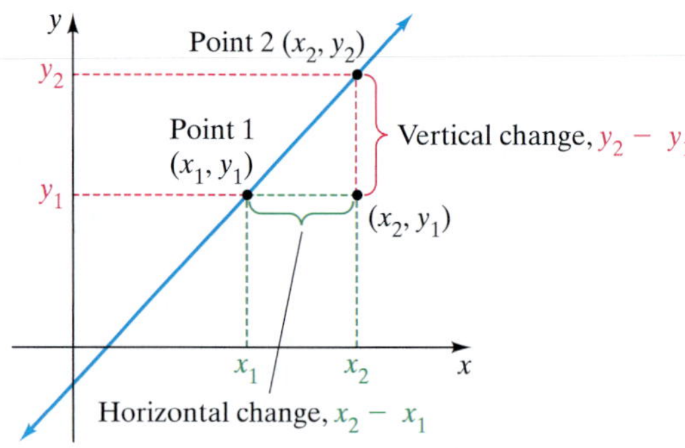


FIGURE 3.54

**Slope**

The **slope** of the line through the distinct points $(x_1, y_1)$ and $(x_2, y_2)$ is

$$\text{slope} = \frac{\text{change in } y \text{ (vertical change)}}{\text{change in } x \text{ (horizontal change)}} = \frac{y_2 - y_1}{x_2 - x_1}$$

provided that $x_1 \neq x_2$.

It makes no difference which two points on the line are selected when finding the slope of a line. It also makes no difference which point you label $(x_1, y_1)$ or $(x_2, y_2)$. As mentioned before, the letter $m$ is used to represent the slope of a line. The Greek capital letter delta, $\Delta$, is used to represent the words *the change in*. Thus, the slope is sometimes indicated as

$$m = \frac{\Delta y}{\Delta x} = \frac{y_2 - y_1}{x_2 - x_1}$$

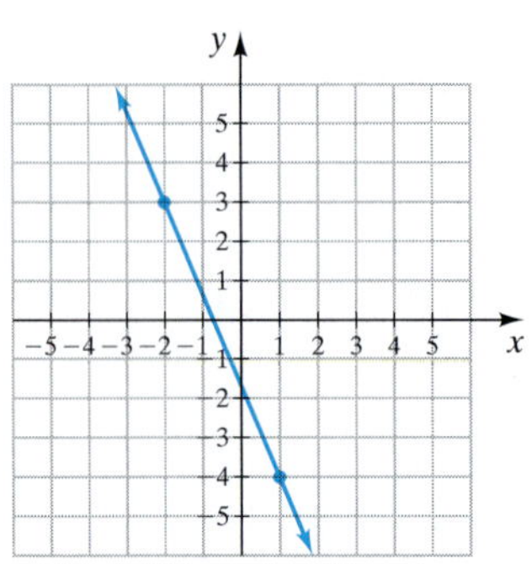


FIGURE 3.55

**EXAMPLE 1** ▸ Find the slope of the line in **Figure 3.55**.

*Solution* Two points on the line are $(-2, 3)$ and $(1, -4)$. Let $(x_2, y_2) = (-2, 3)$ and $(x_1, y_1) = (1, -4)$. Then

$$m = \frac{y_2 - y_1}{x_2 - x_1} = \frac{3 - (-4)}{-2 - 1} = \frac{3 + 4}{-3} = -\frac{7}{3}$$

The slope of the line is $-\frac{7}{3}$. Note that if we had let $(x_1, y_1) = (-2, 3)$ and $(x_2, y_2) = (1, -4)$, the slope would still be $-\frac{7}{3}$. Try it and see.

▸ **Now Try Exercise 35**

A line that rises going from left to right (**Fig. 3.56a** on page 187) has a **positive slope**. A line that neither rises nor falls going from left to right (**Fig. 3.56b**) has **zero slope**. A line that falls going from left to right (**Fig. 3.56c**) has a **negative slope**.

*The letter $m$ is traditionally used for slope. It is believed $m$ comes from the French word *monter*, which means "to climb."

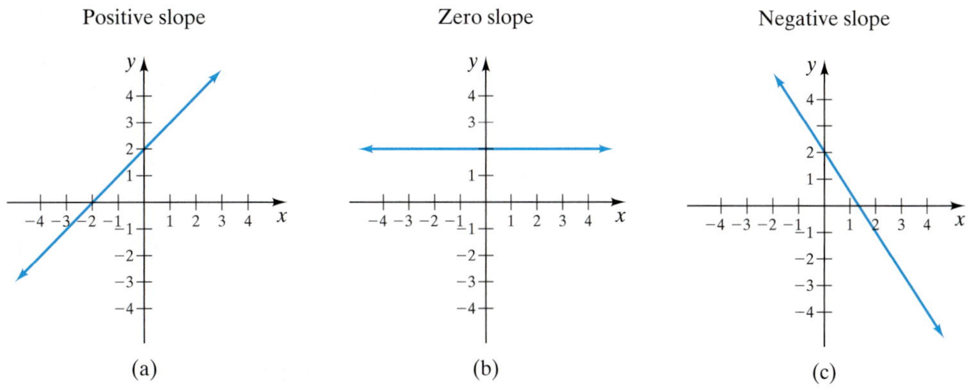


FIGURE 3.56

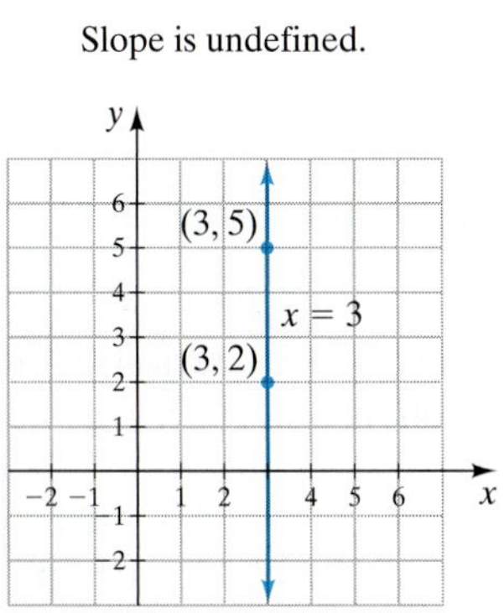


FIGURE 3.57

Consider the graph of $x = 3$ (**Fig. 3.57**). What is its slope? The graph is a vertical line and goes through the points (3, 2) and (3, 5). Let the point (3, 5) represent $(x_2, y_2)$ and let (3, 2) represent $(x_1, y_1)$. Then the slope of the line is

$$m = \frac{y_2 - y_1}{x_2 - x_1} = \frac{5 - 2}{3 - 3} = \frac{3}{0}$$

Since it is meaningless to divide by 0, we say that the slope of this line is undefined. **The slope of any vertical line is undefined.**

### Helpful Hint

When students are asked to give the slope of a horizontal or a vertical line, they often answer incorrectly. When asked for the slope of a horizontal line, your response should be "the slope is 0." If you give your answer as "no slope," your instructor may well mark it wrong since these words may have various interpretations. When asked for the slope of a vertical line, your answer should be "the slope is undefined." Again, if you use the words "no slope," this may be interpreted differently by your instructor and marked wrong.

## 3 Recognize Slope as a Rate of Change

Sometimes it is helpful to describe slope as a *rate of change*. Consider a slope of $\frac{5}{3}$. This means that the $y$-value increases 5 units for each 3-unit increase in $x$. Equivalently, we can say that the $y$-value increases $\frac{5}{3}$ units, or $1.\overline{6}$ units, for each 1-unit increase in $x$. When we give the change in $y$ per unit change in $x$ we are giving the slope as a **rate of change**. When discussing real-life situations or when creating mathematical models, it is often useful to discuss slope as a rate of change.

**EXAMPLE 2 ▸ Public Debt** The following table of values and the corresponding graph (**Fig. 3.58**) illustrate the U.S. public debt in billions of dollars from 1910 through 2005.

| Year | U.S. Public Debt (billions of dollars) |
|---|---|
| 1910 | 1.1 |
| 1930 | 16.1 |
| 1950 | 256.1 |
| 1970 | 370.1 |
| 1990 | 3323.3 |
| 2002 | 5957.2 |
| 2005 | 7832.6 |

*Source:* U.S. Dept. of the Treasury, Bureau of Public Debt.

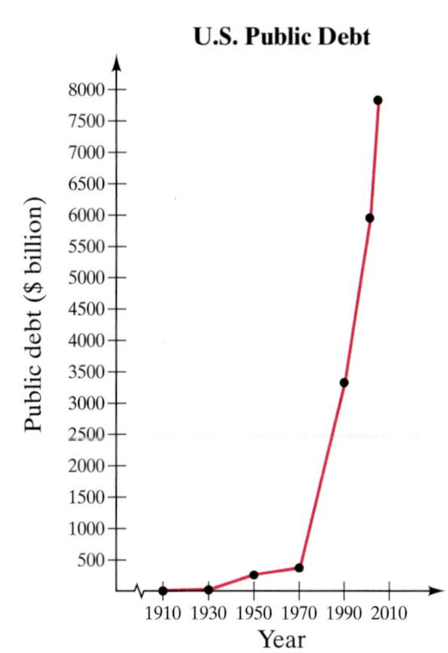


FIGURE 3.58

**a)** Determine the slope of the line segments between 1910 and 1930 and between 2002 and 2005.

**b)** Compare the two slopes found in part **a)** and explain what this means in terms of the U.S. public debt.

*Solution* **Understand** **a)** To find the slope between any 2 years, find the ratio of the change in debt to the change in years.

**Slope from 1910 to 1930**

$$m = \frac{16.1 - 1.1}{1930 - 1910} = \frac{15}{20} = 0.75$$

The U.S. public debt from 1910 to 1930 increased at a rate of \$0.75 billion per year.

**Slope from 2002 to 2005**

$$m = \frac{7832.6 - 5957.2}{2005 - 2002} = \frac{1875.4}{3} \approx 625.13$$

The U.S. public debt from 2002 to 2005 increased at a rate of about \$625.13 billion per year.

**b)** Slope measures a rate of change. Comparing the slopes for the two periods shows that there was a much greater increase in the average rate of change in the public debt from 2002 to 2005 than from 1910 to 1930. The slope of the line segment from 2002 to 2005 is greater than the slope of any other line segment on the graph. This indicates that the public debt from 2002 to 2005 grew at a faster rate than at any other time period illustrated.

▸ Now Try Exercise 69

## 4 Write Linear Equations in Slope-Intercept Form

We have already shown that for an equation of the form $y = mx + b$, $m$ represents the slope and $b$ represents the $y$-intercept. For this reason a linear equation written in the form $y = mx + b$ is said to be in **slope-intercept form**.

**Slope-Intercept Form**

The **slope-intercept form of a linear equation** is

$$y = mx + b$$

where ***m* is the slope** of the line and **(0, *b*) is the *y*-intercept** of the line.

**Examples of Equations in Slope-Intercept Form**

$$y = 3x - 6 \qquad y = \frac{1}{2}x + \frac{3}{2}$$

Slope → $m$; $y$-intercept is $(0, b)$ → $b$

$$y = mx + b$$

| Equation | Slope | $y$-Intercept |
|---|---|---|
| $y = 3x - 6$ | 3 | $(0, -6)$ |
| $y = \frac{1}{2}x + \frac{3}{2}$ | $\frac{1}{2}$ | $\left(0, \frac{3}{2}\right)$ |

**Writing an Equation in Slope-Intercept Form**

**To write an equation in slope-intercept form**, solve the equation for $y$.

**EXAMPLE 3** ▸ Determine the slope and $y$-intercept of the graph of the equation $-5x + 2y = 8$.

*Solution* Write the equation in slope-intercept form by solving the equation for $y$.

$$\begin{aligned} -5x + 2y &= 8 \\ 2y &= 5x + 8 \\ y &= \frac{5x + 8}{2} \\ y &= \frac{5x}{2} + \frac{8}{2} \\ y &= \frac{5}{2}x + 4 \end{aligned}$$

The slope is $\frac{5}{2}$; the $y$-intercept is $(0, 4)$.

▸ Now Try Exercise 43

## 5 Graph Linear Equations Using the Slope and the $y$-Intercept

One reason for studying the slope-intercept form of a line is that it can be useful in drawing the graph of a linear equation, as illustrated in Example 4.

**EXAMPLE 4** ▸ Graph $2y + 4x = 6$ using the $y$-intercept and slope.

*Solution* Begin by solving for $y$ to get the equation in slope-intercept form.

$$\begin{aligned} 2y + 4x &= 6 \\ 2y &= -4x + 6 \\ y &= -2x + 3 \end{aligned}$$

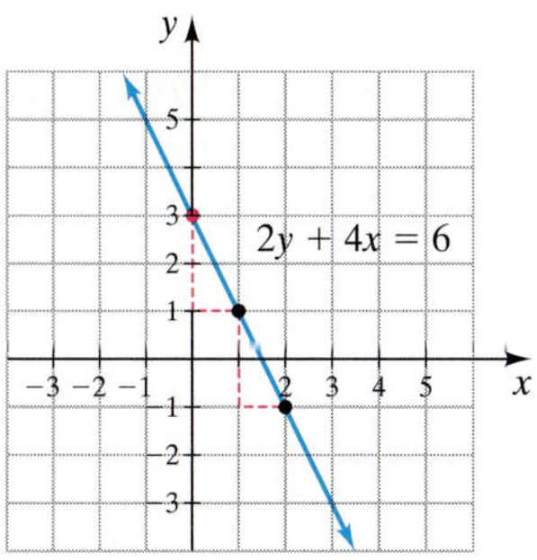

FIGURE 3.59

The slope is $-2$ and the $y$-intercept is $(0, 3)$. Place a point at 3 on the $y$-axis (**Fig. 3.59**). Then use the slope to obtain a second point. The slope is negative; therefore, the graph must fall as it goes from left to right. Since the slope is $-2$, the ratio of the vertical change to the horizontal change must be 2 to 1 (remember, 2 means $\frac{2}{1}$). Thus, if you start at $y = 3$ and move down 2 units and to the right 1 unit, you will obtain a second point on the graph.

Continue this process of moving 2 units down and 1 unit to the right to get a third point. Now draw a line through the three points to get the graph.

▸ Now Try Exercise 45

In Example 4, we chose to move down and to the right to get the second and third points. We could have also chosen to move up and to the left to get the second and third points.

**EXAMPLE 5** ▸ Graph $f(x) = \frac{4}{3}x - 3$ using the $y$-intercept and slope.

*Solution* Since $f(x)$ is the same as $y$, this function is in slope-intercept form. The $y$-intercept is $(0, -3)$ and the slope is $\frac{4}{3}$. Place a point at $-3$ on the $y$-axis. Then, since the slope is positive, obtain the second and third points by moving up 4 units and to the right 3 units. The graph is shown in **Figure 3.60**.

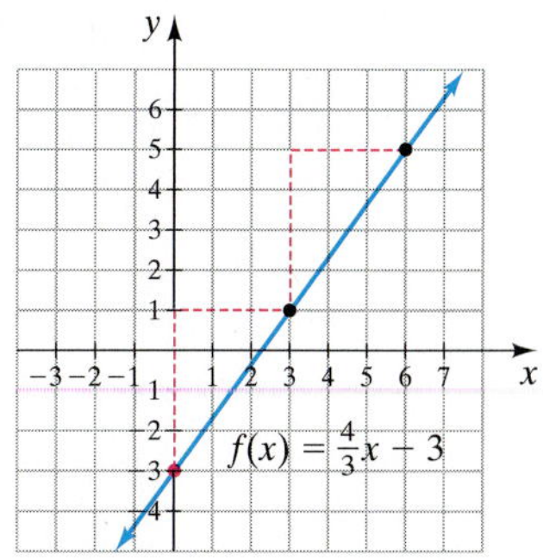

FIGURE 3.60

▸ Now Try Exercise 51

## 6 Use the Slope-Intercept Form to Construct Models from Graphs

Often we can use the slope-intercept form of a linear equation to determine a function that models a real-life situation. Example 6 shows how this may be done.

**EXAMPLE 6** ▸ **Newspapers** Consider the purple graph in **Figure 3.61**, which shows the declining number of adults who read the daily newspaper. Notice that the graph is somewhat linear. The dashed red line is a linear function which was drawn to approximate the purple graph.

**a)** Write a linear function to represent the dashed red line.

**b)** Assuming this trend continues, use the function determined in part **a)** to estimate the percent of adults who will read a newspaper in 2012.

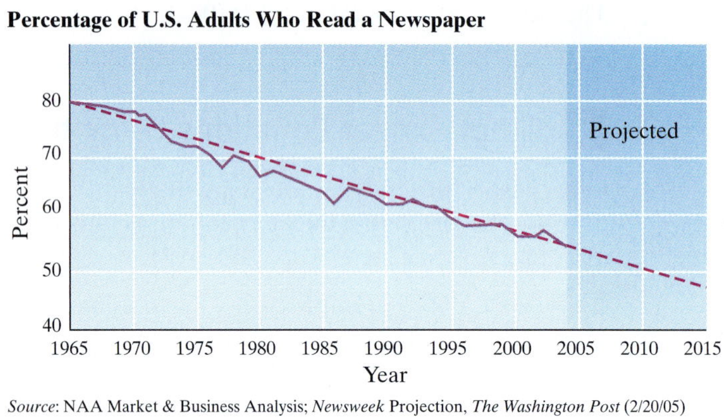


FIGURE 3.61

### Solution

**a)** To make the numbers easier to work with, we will select 1965 as a *reference year*. Then we can replace 1965 with 0, 1966 with 1, 1967 with 2, and so on. Then 2004 would be 39 and 2005 would be 40 (see **Fig. 3.62**).

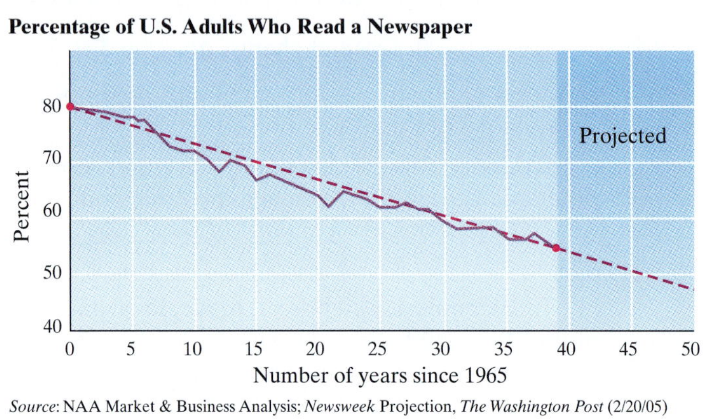


FIGURE 3.62

We will select two points an the graph that will allow us to find the slope of the graph. If we call the vertical axis $y$ and the horizontal axis $x$, then the $y$-intercept is 80. Thus, one point on the graph is (0,80). In 2004, or year 39 in **Figure 3.62**, it appears that about 55% of the adult population read a daily newspaper. Let's select (39, 55) as a second point on the graph of the straight line **Figure 3.62**. We designate (39, 55) as $(x_2, y_2)$ and (0, 80) as $(x_1, y_1)$.

$$\text{slope} = \frac{\text{change in percent}}{\text{change in year}} = \frac{y_2 - y_1}{x_2 - x_1} = \frac{55 - 80}{39 - 0} = \frac{-25}{39} \approx -0.641$$

Since the slope is approximately $-0.641$ and the $y$-intercept is $(0, 80)$, the equation of the straight line is $y = -0.641x + 80$. This equation in function notation is

$f(x) = -0.641x + 80$. To use this function remember that $x = 0$ represents 1965, $x = 1$ represents 1966, and so on. Note that $f(x)$, the percent, is a function of $x$, the number of years since 1965.

**b)** To determine the approximate percent of readers in 2012, and since $2012 - 1965 = 47$, we substitute 47 for $x$ in the function.

$$\begin{aligned} f(x) &= -0.641x + 80 \\ f(47) &= -0.641(47) + 80 \\ &= -30.127 + 80 \\ &= 49.873 \end{aligned}$$

Thus, if the current trend continues, about 49.9% of adults will read a daily newspaper in 2012.

▸ Now Try Exercise 73

## EXERCISE SET 3.4



### Concept/Writing Exercises

1. Explain how to find the slope of a line from its graph.
2. Explain what it means when the slope of a line is negative.
3. Explain what it means when the slope of a line is positive.
4. What is the slope of a horizontal line? Explain.
5. Why is the slope of a vertical line undefined?
6. **a)** Using the slope formula, $m = \frac{y_2 - y_1}{x_2 - x_1}$, determine the slope of the line that contains the points (3, 4) and (6, 10). Use (3, 4) as $(x_1, y_1)$.

   **b)** Calculate the slope again, but this time use (6, 10) as $(x_1, y_1)$.

   **c)** When finding the slope using the formula, will your answer be the same regardless of which of the two points you designate as $(x_1, y_1)$? Explain.
7. Explain how to write an equation given in standard form in slope-intercept form.
8. In the equation $y = mx + b$, what does the $m$ represent? What does the $b$ represent?
9. **a)** What does it mean when a graph is translated down 4 units?

   **b)** If the $y$-intercept of a graph is (0, −3) and the graph is translated down 5 units, what will be its new $y$-intercept?
10. **a)** What does it mean when a graph is translated up 6 units?

    **b)** If the $y$-intercept of a graph is (0, 2) and the graph is translated up 6 units, what will be its new $y$-intercept?
11. What does it mean when slope is given as a rate of change?
12. Explain how to graph a linear equation using its slope and $y$-intercept.

### Practice the Skills

*Find the slope of the line through the given points. If the slope of the line is undefined, so state.*

**13.** (3, 5) and (0, 11) **14.** (3, 4) and (6, 5) **15.** (5, 2) and (1, 4)

**16.** (−3, 7) and (7, −3) **17.** (−3, 5) and (1, 1) **18.** (2, 6) and (2, −3)

**19.** (4, 2) and (4, −6) **20.** (8, −4) and (−1, −2) **21.** (−3, 4) and (−1, 4)

**22.** (2, 8) and (−5, 8) **23.** (0, 3) and (9, −3) **24.** (0, −6) and (−5, −3)

*Solve for the given variable if the line through the two given points is to have the given slope.*

**25.** (3, 2) and (4, $b$), $m = 1$ **26.** (−4, 3) and (−2, $b$), $m = -3$ **27.** (5, 0) and (1, $k$), $m = \frac{1}{2}$

**28.** (5, $d$) and (9, 2), $m = -\frac{3}{4}$ **29.** ($x$, 2) and (3, −4), $m = 2$ **30.** (−2, −3) and ($x$, 5), $m = \frac{1}{2}$

**31.** (12, −4) and ($r$, 2), $m = -\frac{1}{2}$ **32.** (−4, −4) and ($x$, −1), $m = -\frac{3}{5}$

*Find the slope of the line in each of the figures. If the slope of the line is undefined, so state. Then write an equation of the given line.*

**33.**

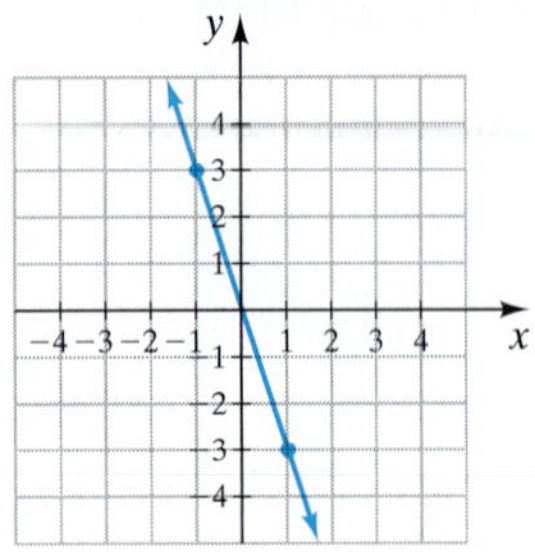

**34.**

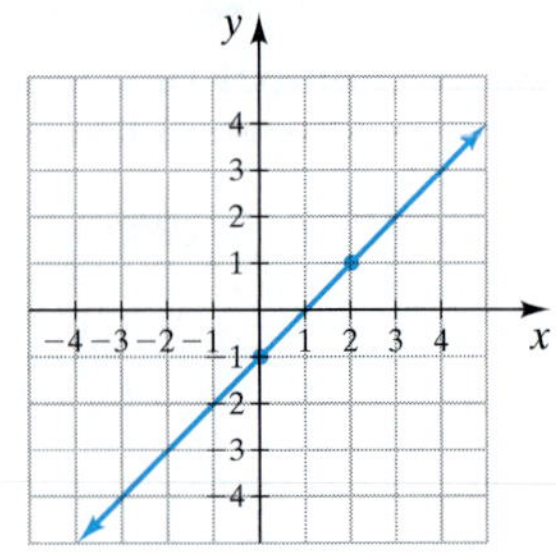

**35.**

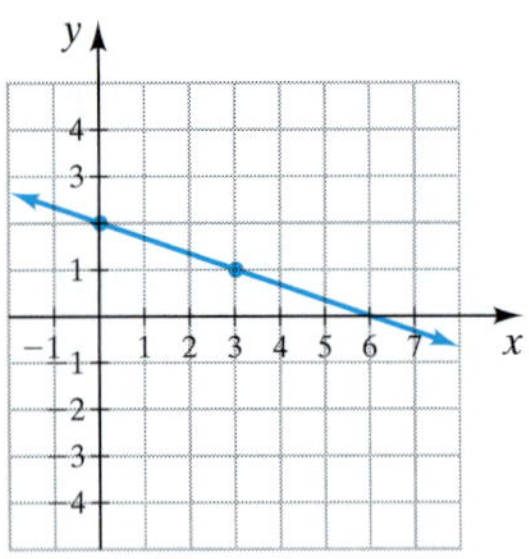

**36.**

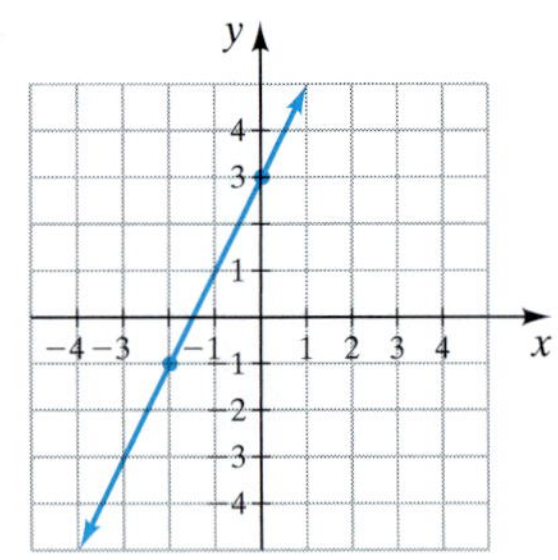

**37.**

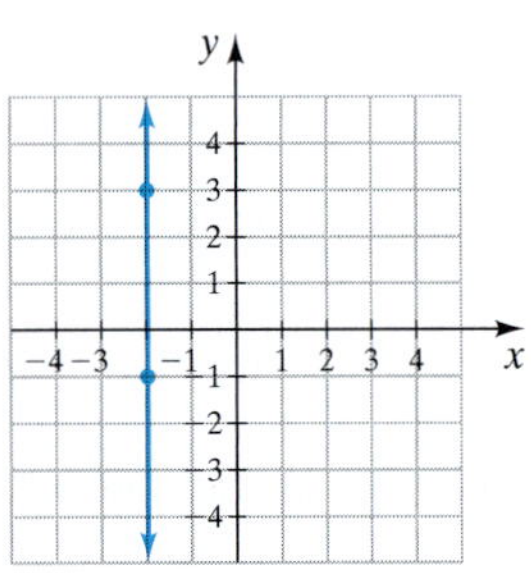

**38.**

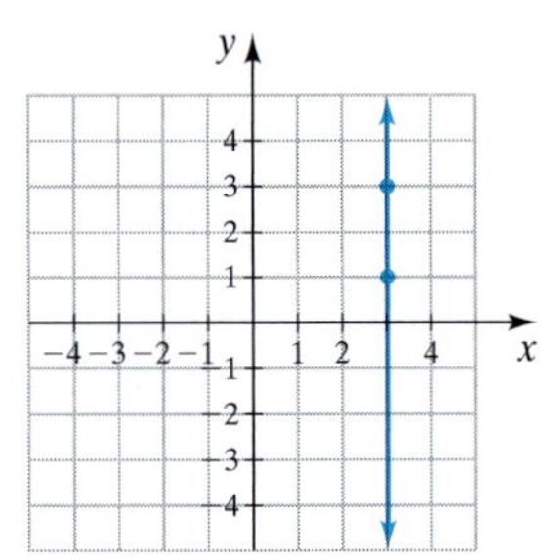

**39.**

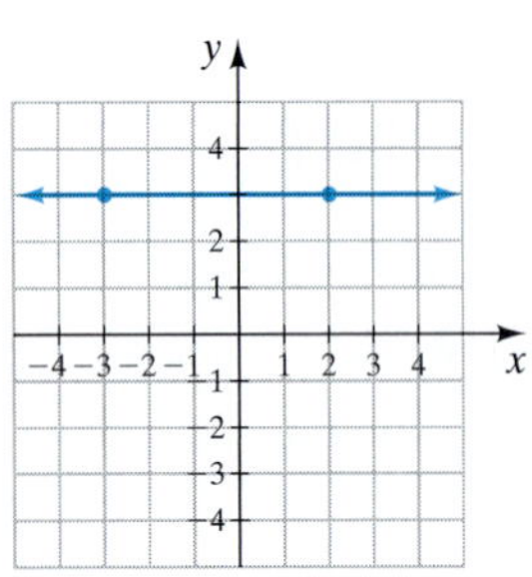

**40.**

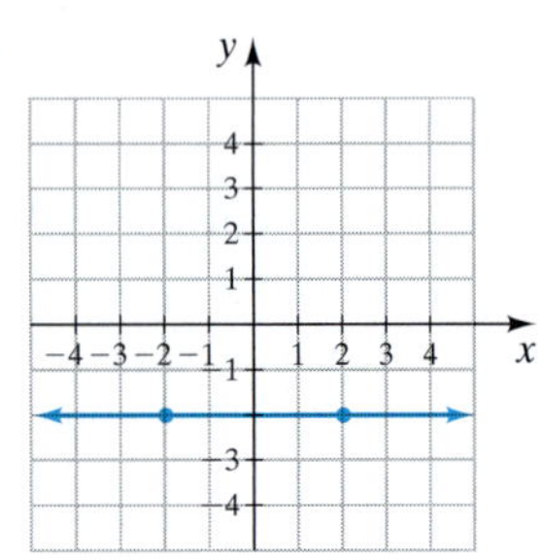

**41.**

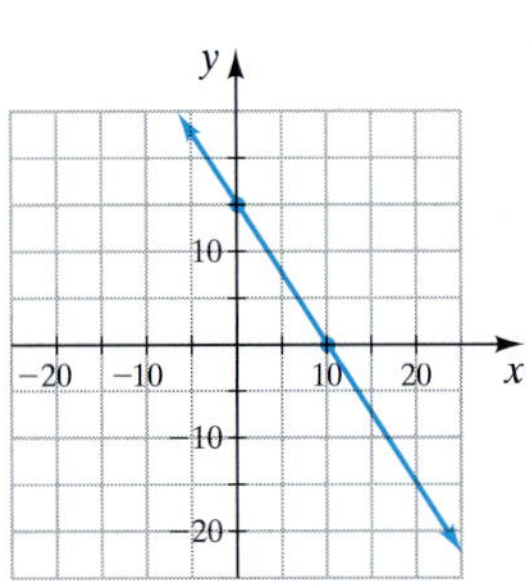

**42.**

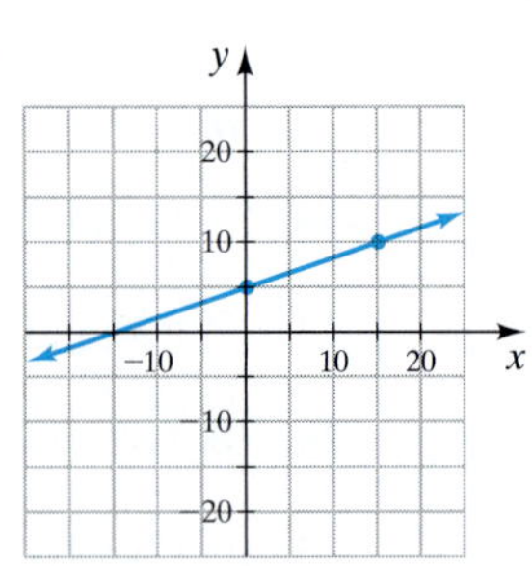

*Write each equation in slope-intercept form (if not given in that form). Determine the slope and the y-intercept and use them to draw the graph of the linear equation.*

**43.** $y = -x + 2$

**44.** $-2x + y = 6$

**45.** $5x + 15y = 30$

**46.** $-2x = 3y + 6$

**47.** $-50x + 20y = 40$

**48.** $60x = -30y + 60$

*Use the slope and y-intercept to graph each function.*

**49.** $f(x) = -2x + 1$ **50.** $g(x) = \frac{2}{3}x - 4$ **51.** $h(x) = -\frac{3}{4}x + 2$ **52.** $h(x) = -\frac{2}{5}x + 4$

## Problem Solving

**53.** Given the equation $y = mx + b$, for the values of $m$ and $b$ given, match parts **a)–d)** with the appropriate graphs labeled 1–4.

**a)** $m > 0, b < 0$ **b)** $m < 0, b < 0$ **c)** $m < 0, b > 0$ **d)** $m > 0, b > 0$

**1.**
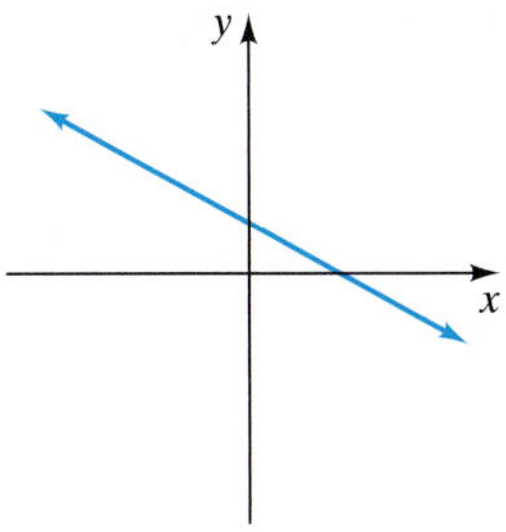


**2.**
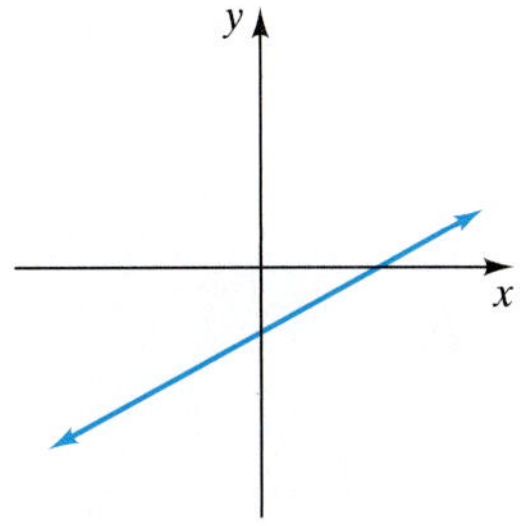


**3.**
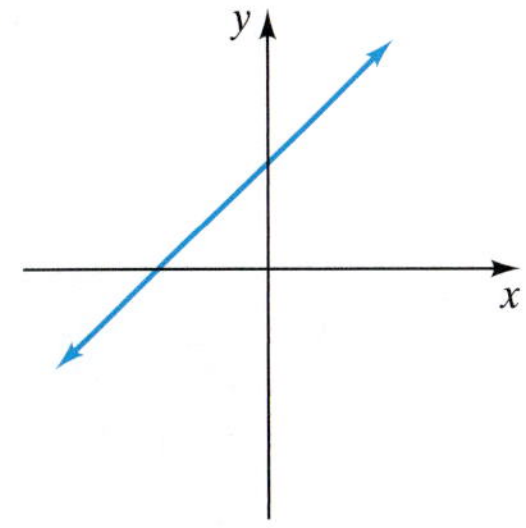


**4.**
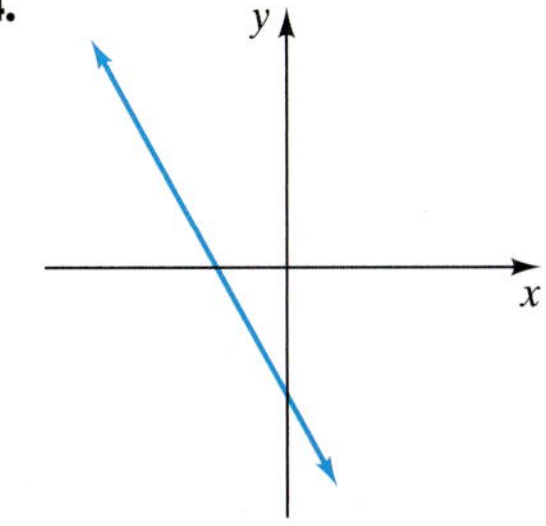


**54.** Given the equation $y = mx + b$, for the values of $m$ and $b$ given, match parts **a)–d)** with the appropriate graphs labeled 1–4.

**a)** $m = 0, b > 0$ **b)** $m = 0, b < 0$ **c)** $m$ is undefined, $x$-intercept $< 0$ **d)** $m$ is undefined, $x$-intercept $> 0$

**1.**
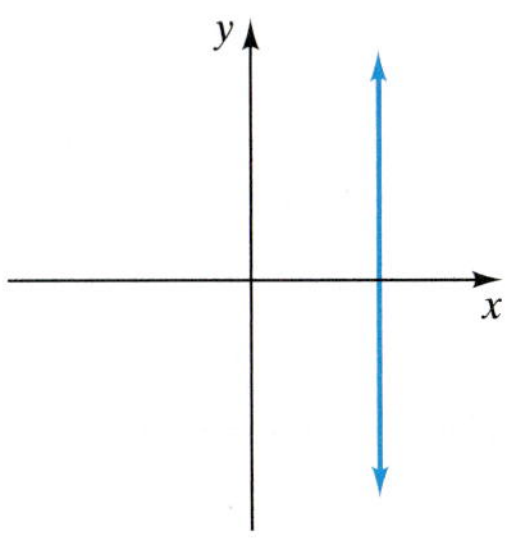


**2.**
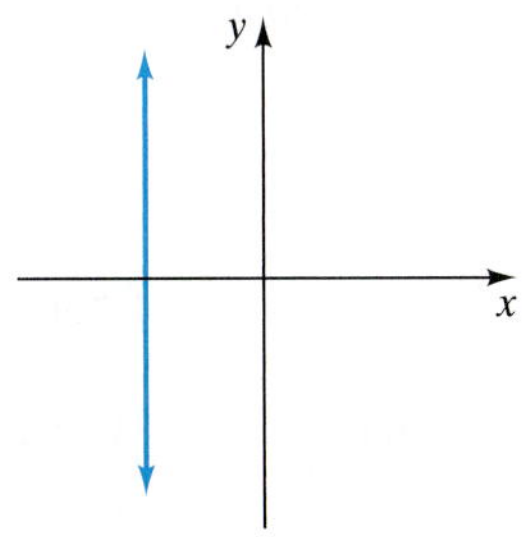


**3.**
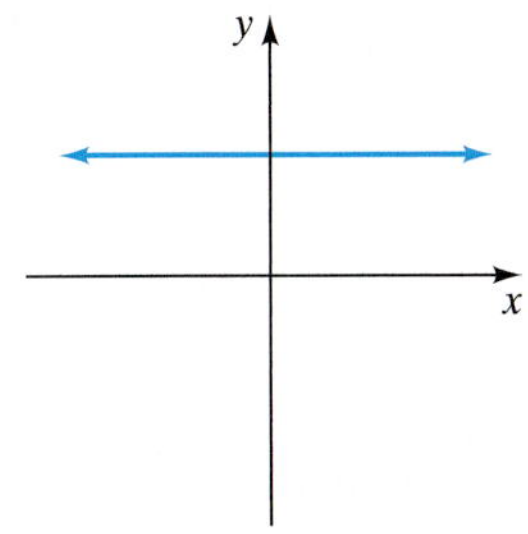


**4.**
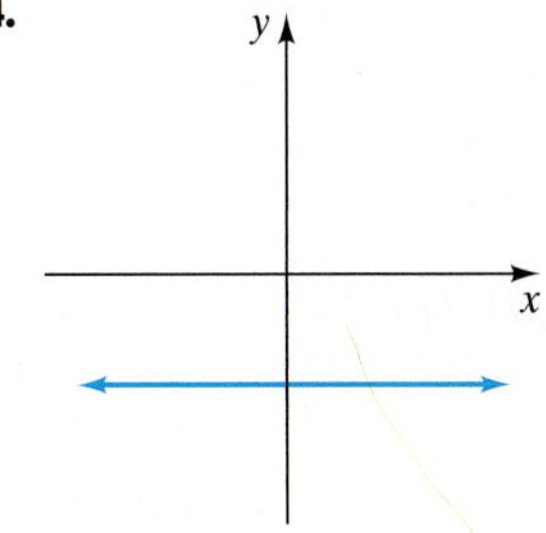


**55.** We will be discussing parallel lines in the next section. Based on what you have read in this section, explain how you could determine (without graphing) that the graphs of two equations are parallel.

**56.** How can you determine whether two straight lines are parallel?

**57.** If one point on a graph is $(6, 3)$ and the slope of the line is $\frac{4}{3}$, find the $y$-intercept of the graph.

**58.** If one point on a graph is $(9, 2)$ and the slope of the line is $m = \frac{2}{3}$, find the $y$-intercept of the graph.

**59.** In the following graph, the green line is a translation of the blue line.

**a)** Determine the equation of the blue line.

**b)** Use the equation of the blue line to determine the equation of the green line.

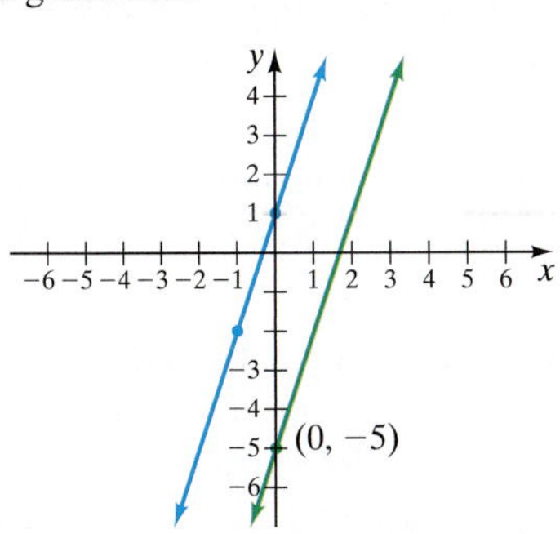


**60.** In the following graph, the red line is a translation of the blue line.

**a)** Determine the equation of the blue line.

**b)** Use the equation of the blue line to determine the equation of the red line.

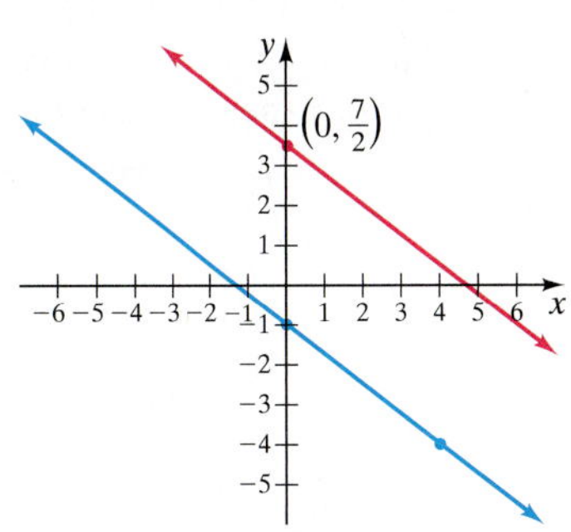


**61.** The graph of $y = x - 1$ is translated up 5 units. Determine

**a)** the slope of the translated graph.

**b)** the $y$-intercept of the translated graph.

**c)** the equation of the translated graph.

**62.** The graph of $y = -\frac{3}{2}x + 3$ is translated down 6 units. Determine

**a)** the slope of the translated graph.

**b)** the $y$-intercept of the translated graph.

**c)** the equation of the translated graph.

63. The graph of $3x - 2y = 6$ is translated down 4 units. Find the equation of the translated graph.

64. The graph of $-3x - 5y = 15$ is translated up 3 units. Find the equation of the translated graph.

65. If a line passes through the points $(6, 4)$ and $(-4, 2)$, find the change of $y$ with respect to a 1-unit change in $x$.

66. If a line passes through the points $(-3, -4)$ and $(5, 2)$, find the change of $y$ with respect to a 1-unit change in $x$.

**TV Sales** *For Exercises 67 and 68, use the graphs below. The graph on the left shows the projected digital TV sales (in millions) and the graph on the right shows the projected analog TV sales (in millions) for the years from 2004 to 2008.*

**TV Sales**

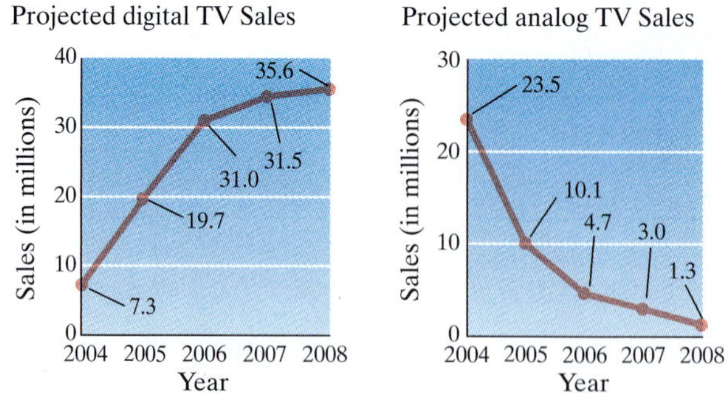


*Source*: Consumer Electronics Association, *USA Today* (1/5/05)

67. **a)** For the graph of digital TV sales, determine the slope of the line segment from 2005 to 2006.
**b)** Is the slope of the line segment positive or negative?
**c)** Find the average rate of change from 2004 to 2008.

68. **a)** For the graph of analog TV sales, determine the slope of the line segment from 2005 to 2006.
**b)** Is the slope of the line segment positive or negative?
**c)** Find the average rate of change from 2004 to 2008.

69. **Amtrak Expenses** The National Railroad and Passenger Corporation, better known as Amtrak, continues to face economic struggles. The following table gives the expenses, in millions of dollars, of Amtrak for selected years.

| Year | Amtrak Expenses (in millions of dollars) |
|---|---|
| 1995 | \$ 2257 |
| 2000 | \$ 2876 |
| 2004 | \$ 3133 |
| *2008 | \$ 3260 |

*Source:* Amtrak Fiscal Year 2004 Annual Report
*Projected

**a)** Plot these points on a graph.
**b)** Connect these points using line segments.
**c)** Determine the slopes of each of the three line segments.
**d)** During which period was there the greatest average rate of change? Explain.

70. **Demand for Steel** The world demand for steel has been on the rise in recent years. The following table gives the world demand for steel, in millions of metric tons, for the years from 2001 to 2004.

### World Demand for Steel

| Year | Demand (in millions of metric tons) |
|---|---|
| 2001 | 740 |
| 2002 | 810 |
| 2003 | 880 |
| 2004 | 950 |

*Source:* "World Steel Dynamics," *Wall Street Journal (12/8/04)*

**a)** Plot these points on a graph.
**b)** Determine the slope of each line segment.
**c)** Is this graph an example of a linear function? Explain.
**d)** Determine a linear function that can be used to estimate the world demand for steel, $d$, from 2001 to 2004. Let $t$ represent the number of years since 2001. (That is, 2001 corresponds to $t = 0$, 2002 corresponds to $t = 1$, and so on.)
**e)** Assuming this trend continues for the next 20 years, find the world demand for steel in 2016.
**f)** Assuming this trend continues, in which year will the demand reach 1230 metric tons?

71. **Heart Rate** The following bar graph shows the maximum recommended heart rate, in beats per minute, under stress for men of different ages. The bars are connected by a straight line.

**a)** Use the straight line to determine a function that can be used to estimate the maximum recommended heart rate, $h$, for $0 \le x \le 50$, where $x$ is the number of years after age 20.

**b)** Using the function from part **a)**, determine the maximum recommended heart rate for a 34-year-old man.

**Heart Rate vs. Age**

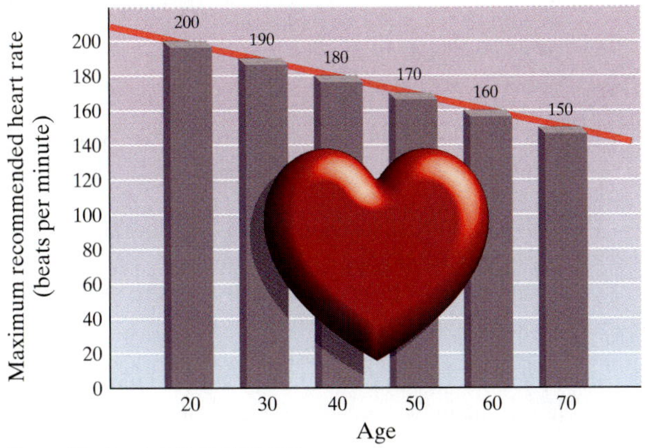


*Source:* The American Geriatric Society

**72. Poverty Threshold** The federal government defines the poverty threshold as an estimate of the annual family income necessary to have what society defines as a minimally acceptable standard of living. The following bar graph shows the poverty threshold for a family of four for the years 2000 through 2004.

**U.S. Poverty Threshold for a Family of Four**

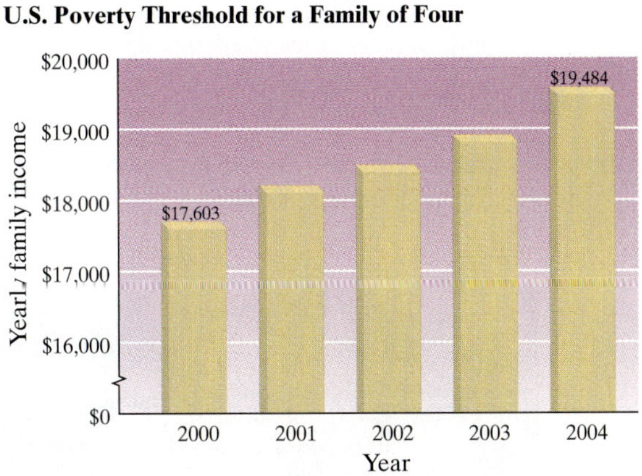


*Source*: U.S. Bureau of the Census: www.census.gov/hhes/poverty

**a)** Determine a linear function that can be used to estimate the poverty threshold for a family of four, $P$, from 2000 through 2004. Let $t$ represent the number of years since 2000. (In other words, 2000 corresponds to $t = 0$, 2001 corresponds to $t = 1$, and so on.)

**b)** Using the function from part **a)**, determine the poverty threshold in 2003. Compare your answer with the graph to see whether the graph supports your answer.

**c)** Assuming this trend continues, determine the poverty threshold for a family of four in the year 2010.

**d)** Assuming this trend continues, in which year will the poverty threshold for a family of four reach \$20,424.50?

**73. Medicaid Spending** The following graph shows the amount of money spent on Medicaid for the years from 1997 to 2004.

**Medicaid Spending**

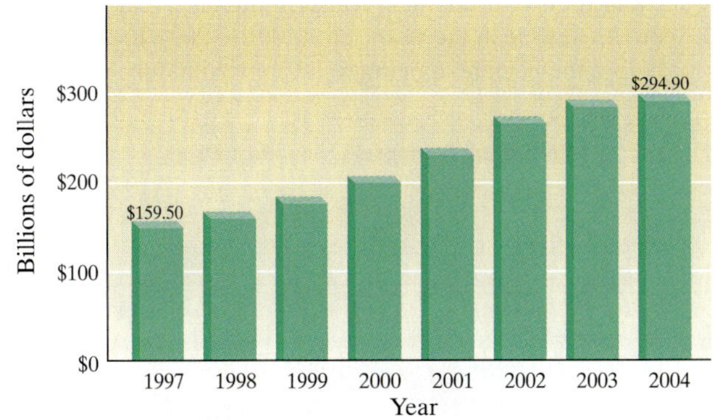


*Source*: Centers for Medicaid and Medicare Services, USA Today (8/2/05)

**a)** Using 1997 as a reference year, determine a linear function that can be used to estimate Medicaid spending (in billions of dollars), $M$, for the years from 1997 to 2004. In this function, let $t$ represent the number of years since 1997.

**b)** Using the function from part **a)**, estimate Medicaid spending for the year 2003. Compare your answer with the graph to see whether the graph supports your answer.

**c)** Assuming this trend continues, what will be the Medicaid spending in 2010?

**d)** Assuming this trend continues, during what year will Medicaid spending reach \$340 billion?

**74. Purchasing Power of the Dollar** The purchasing power of the dollar is measured by comparing the current price of items to the price of those same items in 1982. From the chart below you will see that the purchasing power of the dollar has steadily declined for the years 1990 through 2003. This means that \$1 buys less each year.

**Purchasing Power of the Dollar**

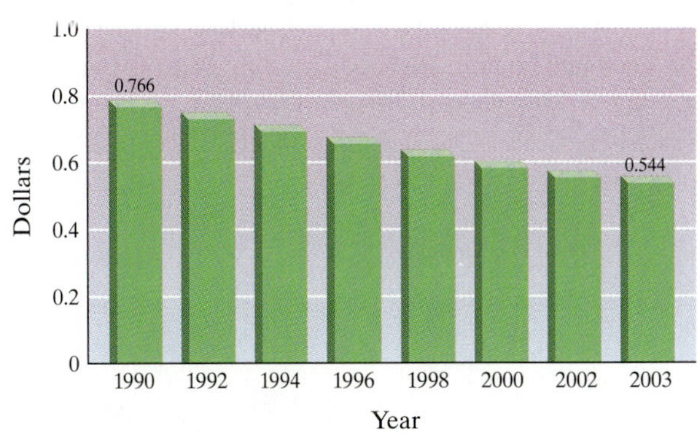


*Source:* U.S. Bureau of Economic Analysis

**a)** Using 1990 as the reference year, determine a linear function that can be used to estimate the purchasing power, $P$, for the years 1990 through 2003. In the function, let $t$ represent the number of years since 1990.

**b)** Using the function from part **a)**, estimate the purchasing power of the dollar in 1994. Compare your answer with the graph to see whether the graph supports your answer.

**c)** Assuming this trend continues, what would be the purchasing power of the dollar in 2006?

**d)** Assuming this trend continues, when would the purchasing power of the dollar reach \$0.426?

**75. Teenagers Using Illicit Drugs** The percent of teenagers who claim to have used illicit drugs (in the last 30 days) has been on the decline in the years from 2001 to 2004. From the graph below, the decline appears to be approximately linear.

**Percent of Teenagers Using Illicit Drugs**

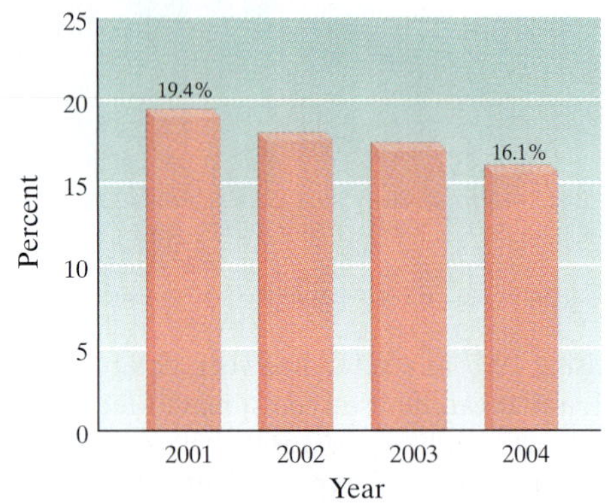


*Source*: University of Michigan, 2004 Monitoring the Future Study, *The Washington Post* (12/22/04)

a) Using 2001 as a reference year, determine a linear function that can be used to estimate the percent of teenagers using illicit drugs, $P$, for the years 2001 through 2004. In this function, let $t$ represent the number of years since 2001.

b) Is the slope of the linear function positive or negative? Explain.

c) Using the function from part **a)**, estimate the percent of teenagers using illicit drugs in 2003. Compare your answer with the graph to see whether the graph supports your answer.

d) Assuming this trend continues, what would be the percent of teenagers using illicit drugs in 2010?

**76. Personal Income** Personal income was on the rise every month from June 2003 to November 2004. From the graph below, the increase appears to be approximately linear.

**Personal Income**

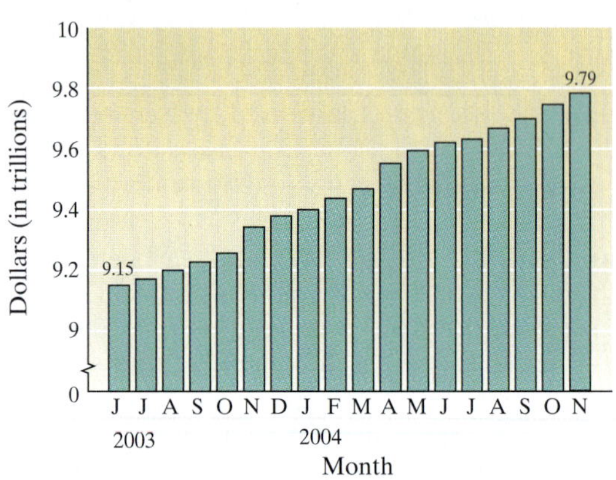


*Source*: Commerce Department, *The New York Times* (12/24/04)

a) Using June 2003 as a reference point, determine a linear function that can be used to estimate the personal income (in trillions of dollars), I, for the months June 2003 through November 2004. In this function, let $t$ represent the number of months since June 2003 (that is, $t = 0$ corresponds to June 2003, $t = 1$ corresponds to July 2003, $t = 6$ corresponds to December 2003, $t = 17$ corresponds to November 2004, and so on).

b) Is the slope of this linear function positive or negative? Explain.

c) Using the function from part **a)**, estimate the personal income (in trillions of dollars) for February 2004 ($t = 8$). Compare your answer with the graph to see whether the graph supports your answer.

d) Assuming this trend continues, what would be the personal income in December 2005 ($t \approx 30$)?

**77. Median Home Sale Price** The median home sale price in the United States has been rising approximately linearly since 1995. The median home sale price in 1995 was \$110,500. The median home sale price in 2004 was \$185,200. Let P be the median home sale price and let $t$ be the number of years since 1995. *Source:* National Association of Realtors

a) Determine a function $P(t)$ that fits this data.

b) Use the function from part **a)** to estimate the median home sale price in 2000.

c) If this trend continues, estimate the median home sale price in 2010.

d) If this trend continues, in which year will the median home sale price reach \$200,000?

**78. Social Security** The number of workers per social security beneficiary has been declining approximately linearly since 1970. In 1970 there were 3.7 workers per beneficiary. In 2050 it is projected there will be 2.0 workers per beneficiary. Let $W$ be the workers per social security beneficiary and $t$ be the number of years since 1970.

a) Find a function $W(t)$ that fits the data.

b) Estimate the number of workers per beneficiary in 2020.

*Suppose you are attempting to graph the equations shown and you get the screens shown. Explain how you know that you have made a mistake in entering each equation. The standard window setting is used on each graph.*

**79.** $y = 3x + 6$

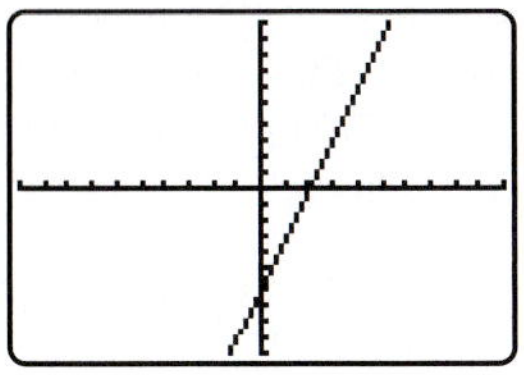

**80.** $y = -2x - 4$

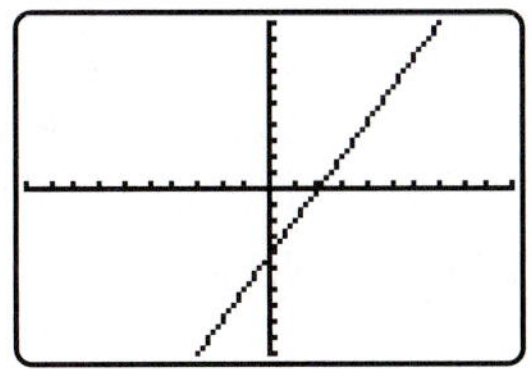

**81.** $y = \frac{1}{2}x + 4$

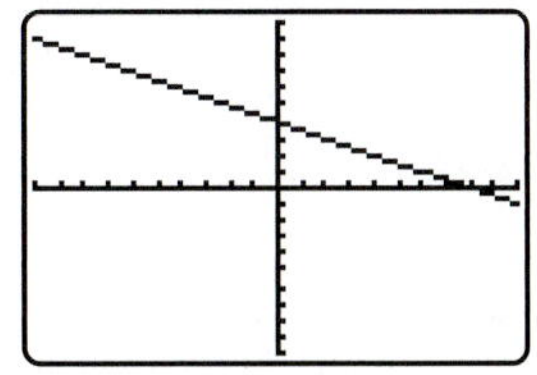

**82.** $y = -4x - 1$

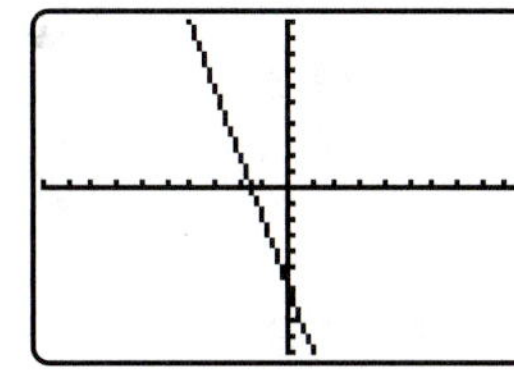

## Challenge Problems

**83. Castle** The photo below is the Castle at Chichén Itzá, Mexico. Each side of the castle has a stairway consisting of 91 steps. The steps of the castle are quite narrow and steep, which makes them hard to climb. The total vertical distance of the 91 steps is 1292.2 inches. If a straight line were to be drawn connecting the tips of the steps, the absolute value of the slope of this line would be 2.21875. Find the average height and width of a step.

**84.** A **tangent line** is a straight line that touches a curve at a single point (the tangent line may cross the curve at a different point if extended). **Figure 3.63** shows three tangent lines to the curve at points $a$, $b$, and $c$. Note that the tangent line at point $a$ has a positive slope, the tangent line at point $b$ has a slope of 0, and the tangent line at point $c$ has a negative slope. Now consider the curve in **Figure 3.64**. Imagine that tangent lines are drawn at all points on the curve except at endpoints $a$ and $e$. Where on the curve in **Figure 3.64** would the tangent lines have a positive slope, a slope of 0, and a negative slope?

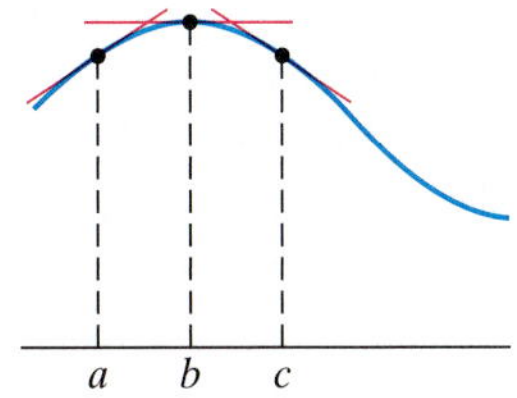


FIGURE 3.63

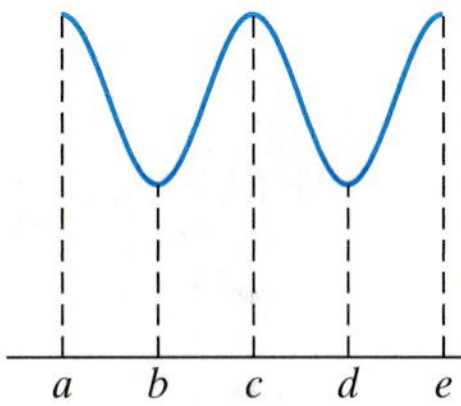


FIGURE 3.64

## Group Activity

**85.** The following graph from *Consumer Reports* shows the depreciation on a typical car. The initial purchase price is represented as 100%.

**a)** Group member 1: Determine the 1-year period in which a car depreciates most. Estimate from the graph the percent a car depreciates during this period.

**b)** Group member 2: Determine between which years the depreciation appears linear or nearly linear.

**c)** Group member 3: Determine between which 2 years the depreciation is the lowest.

**d)** As a group, estimate the slope of the line segment from year 0 to year 1. Explain what this means in terms of rate of change.

**Typical Depreciation Curve**

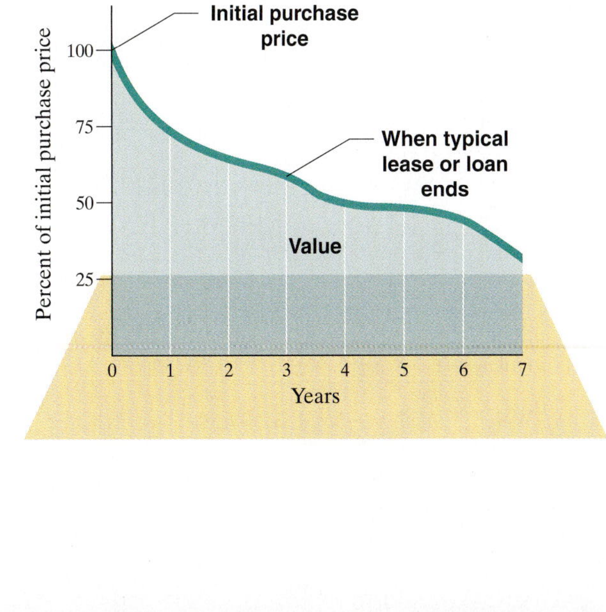

## Cumulative Review Exercises

[1.4] **86.** Evaluate $\dfrac{-6^2 - 32 \div 2 \div |-8|}{5 - 3 \cdot 2 - 4 \div 2^2}$.

*Solve each equation.*

[2.1] **87.** $\frac{1}{4}(x + 3) + \frac{1}{5}x = \frac{2}{3}(x - 2) + 1$

**88.** $2.6x - (-1.4x + 3.4) = 6.2$

[2.4] **89. Trains** Two trains leave Chicago, Illinois, traveling in the same direction along parallel tracks. The first train leaves 3 hours before the second, and its speed is 15 miles per hour faster than the second. Find the speed of each train if they are 270 miles apart 3 hours after the second train leaves Chicago.

[2.6] **90.** Solve

**a)** $|2x + 1| > 5$. **b)** $|2x + 1| < 5$.

## Mid-Chapter Test: 3.1–3.4

*To find out how well you understand the chapter material to this point, take this brief test. The answers, and the section where the material was initially discussed, are given in the back of the book. Review any questions you answered incorrectly.*

**1.** In which quadrant is the point $(-3.5, -4.2)$ located?

*Graph each equation.*

**2.** $y = 3x + 2$

**3.** $y = -x^2 + 3$

**4.** $y = |x| - 4$

**5.** $y = \sqrt{x - 4}$

**6. a)** What is a relation?

**b)** What is a function?

**c)** Is every relation a function? Explain.

**d)** Is every function a relation? Explain.

*In Exercises 7–9, determine which of the following relations are also functions. Give the domain and range of each relation or function.*

**7.** $\{(1, 5), (2, -3), (7, -1), (-5, 6)\}$

**8.**

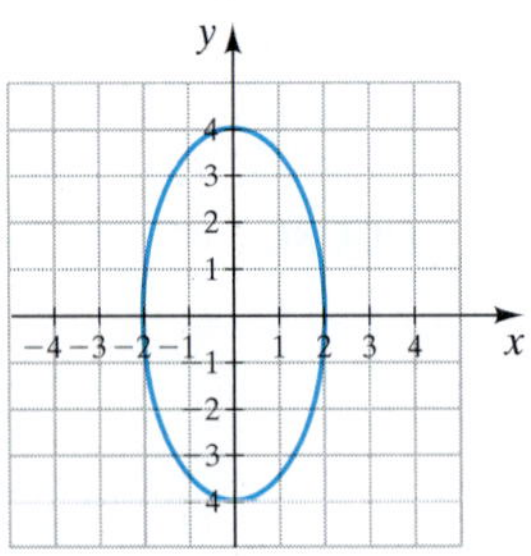

**9.**

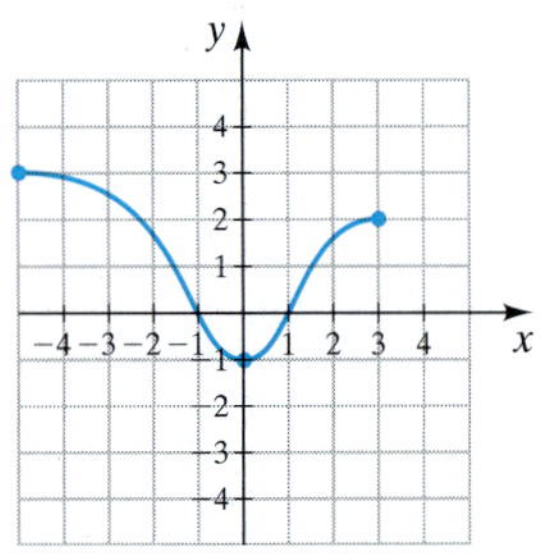

**10.** If $g(x) = 2x^2 + 8x - 13$, find $g(-2)$.

**11.** The height, $h$, in feet of an apple thrown from the top of a building is

$$h(t) = -6t^2 + 3t + 150$$

where $t$ is time in seconds. Find the height of the apple 3 seconds after it is thrown.

**12.** Write the equation $7(x + 3) + 2y = 3(y - 1) + 18$ in standard form.

*Graph each equation.*

**13.** $x + 3y = -3$

**14.** $x = -4$

**15.** $y = 5$

**16. Profit** The daily profit, in dollars, for a shoe company is $p(x) = 30x - 660$, where $x$ is the number of pairs of shoes manufactured and sold.

**a)** Draw a graph of profit versus the number of pairs of shoes sold (for up to 40 pairs).

**b)** Determine the number of pairs of shoes that must be sold for the company to break even.

**c)** Determine the number of pairs of shoes that must be sold for the company to make a daily profit of $360.

**17.** Find the slope of the line passing through $(9, -2)$ and $(-7, 8)$.

**18.** Write the equation of the line given in the graph shown.

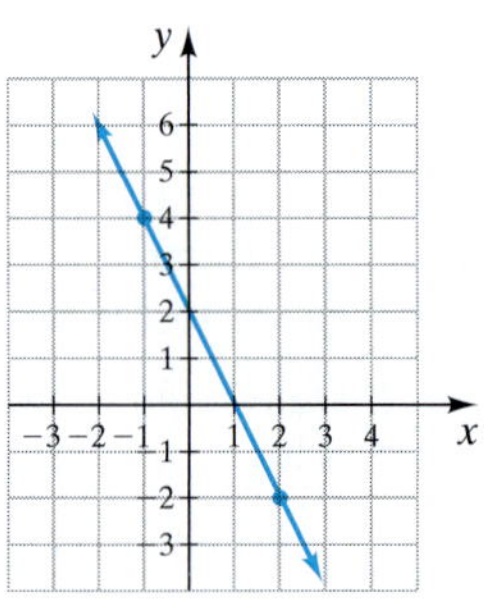

**19.** Write the equation $-3x + 2y = 18$ in slope-intercept form. Determine the slope and $y$-intercept.

**20.** If the graph of $y = 5x - 3$ is translated up 4 units, determine

**a)** the slope of the translated graph.

**b)** the $y$-intercept of the translated graph.

**c)** the equation of the translated graph.

# 3.5 The Point-Slope Form of a Linear Equation

1. Understand the point-slope form of a linear equation.
2. Use the point-slope form to construct models from graphs.
3. Recognize parallel and perpendicular lines.

## 1 Understand the Point-Slope Form of a Linear Equation

In the previous section we learned how to use the *slope-intercept form* of a line to determine the equation of a line when the slope and $y$-intercept of the line are known. In this section we learn how to use the **point-slope form** of a line to determine the equation of a line when the slope and a point on the line are known. The point-slope form can be developed from the expression for the slope between any two points $(x, y)$ and $(x_1, y_1)$ on a line, as shown in **Figure 3.65**.

$$m = \frac{y - y_1}{x - x_1}$$

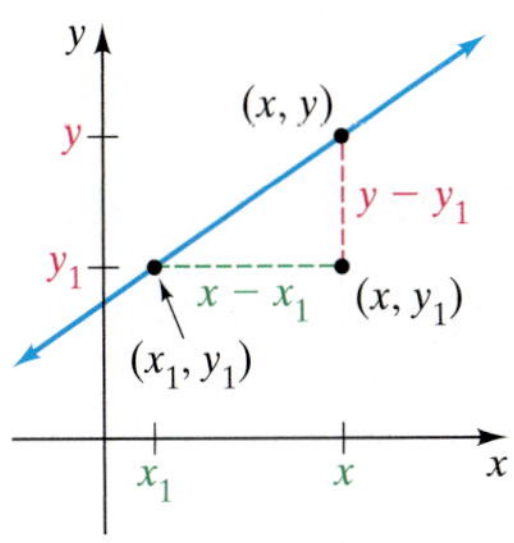

FIGURE 3.65

Multiplying both sides of the equation by $x - x_1$, we obtain

$$y - y_1 = m(x - x_1)$$

**Point-Slope Form**

The **point-slope form of a linear equation** is

$$y - y_1 = m(x - x_1)$$

where $m$ **is the slope** of the line and $(x_1, y_1)$ is a point on the line.

**EXAMPLE 1** ▸ Write, in slope-intercept form, the equation of the line that passes through the point $(1, 4)$ and has slope $-3$.

*Solution* Since we are given the slope of the line and a point on the line, we can write the equation in point-slope form. We can then solve the equation for $y$ to write the equation in slope-intercept form. The slope, $m$, is $-3$. The point on the line, $(x_1, y_1)$, is $(1, 4)$. Substitute $-3$ for $m$, 1 for $x_1$, and 4 for $y_1$ in the point-slope form.

$$y - y_1 = m(x - x_1)$$

$$y - 4 = -3(x - 1) \qquad \text{Point-slope form}$$

$$y - 4 = -3x + 3$$

$$y = -3x + 7 \qquad \text{Slope-intercept form}$$

The graph of $y = -3x + 7$ has a slope of $-3$ and passes through the point $(1, 4)$.

▸ Now Try Exercise 5

In Example 1 we used the point-slope form to get the equation of a line when we were given a point on the line and the slope of the line. The point-slope form can also be used to find the equation of a line when we are given two points on the line. We show how to do this in Example 2.

**EXAMPLE 2** ▸ Write, in slope-intercept form, the equation of the line that passes through the points $(2, 3)$ and $(1, 4)$.

*Solution* Although we are not given the slope of the line, we can use the two given points to determine the slope. We can then proceed as we did in Example 1. We will let $(2, 3)$ be $(x_1, y_1)$ and $(1, 4)$ be $(x_2, y_2)$.

$$m = \frac{y_2 - y_1}{x_2 - x_1} = \frac{4 - 3}{1 - 2} = \frac{1}{-1} = -1$$

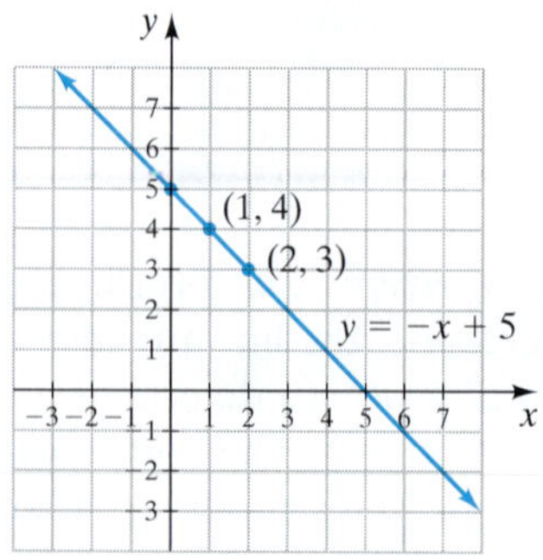


FIGURE 3.66

The slope, $m$, is $-1$. Now we must choose one of the two given points to use as $(x_1, y_1)$ in the point-slope form of the equation of a line. We will choose $(2, 3)$. Substitute $-1$ for $m$, 2 for $x_1$, and 3 for $y_1$ in the point-slope form.

$$y - y_1 = m(x - x_1)$$
$$y - 3 = -1(x - 2)$$
$$y - 3 = -x + 2$$
$$y = -x + 5$$

The graph of $y = -x + 5$ is shown in **Figure 3.66**. Notice that the $y$-intercept of this line is at 5, the slope is $-1$, and the line goes through the points $(2, 3)$ and $(1, 4)$.

Note that we could have also chosen the point $(1, 4)$ to substitute into the point-slope form. Had we done this we would still have obtained the equation $y = -x + 5$. You should verify this now.

▸ Now Try Exercise 11

## 2 Use the Point-Slope Form to Construct Models from Graphs

Now let's look at an application where we use the point-slope form to determine a function that models a given situation.

**EXAMPLE 3** ▸ **Burning Calories** The number of calories burned in 1 hour riding a bicycle is a linear function of the speed of the bicycle. The average person riding at 12 mph will burn about 564 calories in 1 hour and while riding at 18 mph will burn about 846 calories in 1 hour. This information is shown in **Figure 3.67**.

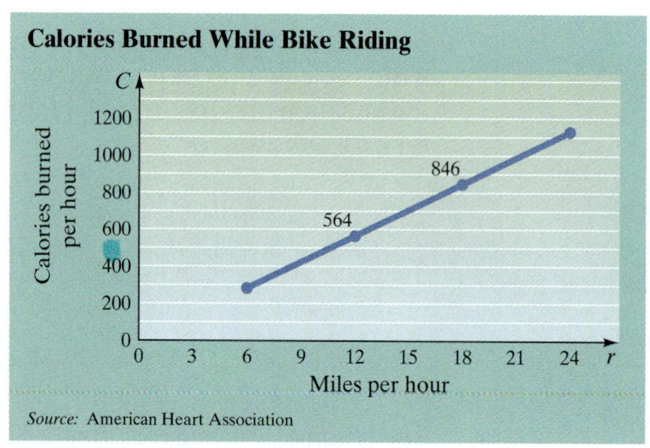


FIGURE 3.67

**a)** Determine a linear function that can be used to estimate the number of calories, $C$, burned in 1 hour when a bicycle is ridden at $r$ mph, for $6 \leq r \leq 24$.

**b)** Use the function determined in part **a)** to estimate the number of calories burned in 1 hour when a bicycle is ridden at 20 mph.

**c)** Use the function determined in part **a)** to estimate the speed at which a bicycle should be ridden to burn 800 calories in 1 hour.

*Solution* **a) Understand and Translate** In this example, instead of using the variables $x$ and $y$ as we used in Examples 1 and 2, we use the variables $r$ (for rate or speed) and $C$ (for calories). Regardless of the variables used, the procedure used to determine the equation of the line remains the same. To find the necessary function, we will use the points $(12, 564)$ and $(18, 846)$ and proceed as we did in Example 2. We will first calculate the slope and then use the point–slope form to determine the equation of the line.

**Carry Out**

$$m = \frac{C_2 - C_1}{r_2 - r_1}$$
$$= \frac{846 - 564}{18 - 12} = \frac{282}{6} = 47$$

Now we write the equation using the point-slope form. We will choose the point (12, 564) for $(r_1, C_1)$.

$$C - C_1 = m(r - r_1)$$
$$C - 564 = 47(r - 12) \quad \text{Point-slope form}$$
$$C - 564 = 47r - 564$$
$$C = 47r \quad \text{Slope-intercept form}$$

**Answer** Since the number of calories burned, $C$, is a function of the rate, $r$, the function we are seeking is

$$C(r) = 47r$$

**b)** To estimate the number of calories burned in 1 hour while riding at 20 mph, we substitute 20 for $r$ in the function.

$$C(r) = 47r$$
$$C(20) = 47(20) = 940$$

Therefore, 940 calories are burned while riding at 20 mph for 1 hour.

**c)** To estimate the speed at which a bicycle should be ridden to burn 800 calories in 1 hour, we substitute 800 for $C(r)$ in the function.

$$C(r) = 47r$$
$$800 = 47r$$
$$\frac{800}{47} = r$$
$$r \approx 17.02$$

Thus the bicycle would need to be ridden at about 17.02 mph to burn 800 calories in 1 hour.

▸ Now Try Exercise 53

In Example 3, the function we determined was $C(r) = 47r$. The graph of this function has a slope of 47 and a $y$-intercept at (0, 0). If the graph in **Figure 3.67** on page 200 was extended to the left, it would intersect the origin. This makes sense since a rate of 0 miles per hour would result in 0 calories being burned by riding in 1 hour.

Parallel lines

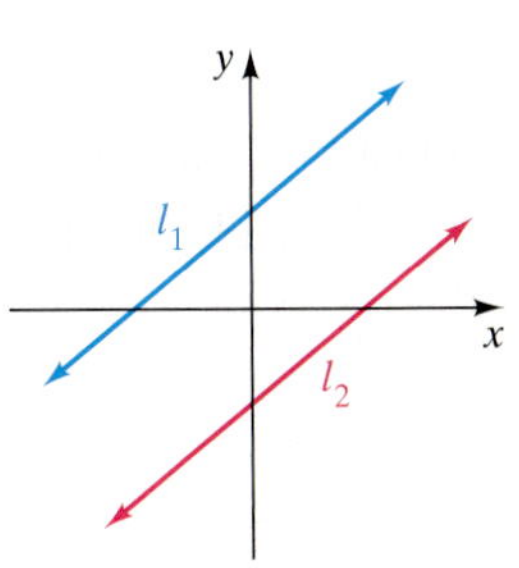


FIGURE 3.68

## 3 Recognize Parallel and Perpendicular Lines

**Figure 3.68** illustrates two *parallel* lines.

**Parallel Lines**

Two lines are **parallel** when they have the same slope.

All vertical lines are parallel even though their slope is undefined.

**Figure 3.69** illustrates perpendicular lines. Two lines are *perpendicular* when they intersect at right (or 90°) angles.

Perpendicular lines

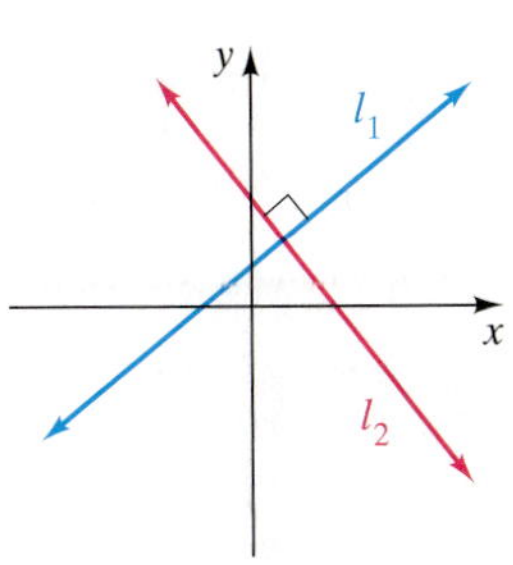


FIGURE 3.69

**Perpendicular Lines**

Two lines are **perpendicular** when their slopes are *negative reciprocals*.

For any nonzero number $a$, its **negative reciprocal** is $\frac{-1}{a}$ or $-\frac{1}{a}$. For example, the negative reciprocal of 2 is $\frac{-1}{2}$ or $-\frac{1}{2}$. The product of any nonzero number and its negative reciprocal is $-1$.

$$a\left(-\frac{1}{a}\right) = -1$$

Note that any vertical line is perpendicular to any horizontal line even though the negative reciprocal cannot be applied. (Why not?)

**EXAMPLE 4** ▸ Two points on $l_1$ are $(8, 5)$ and $(4, -1)$. Two points on $l_2$ are $(0, 2)$ and $(6, -2)$. Determine whether $l_1$ and $l_2$ are parallel lines, perpendicular lines, or neither.

*Solution* Determine the slopes of $l_1$ and $l_2$.

$$m_1 = \frac{5 - (-1)}{8 - 4} = \frac{6}{4} = \frac{3}{2} \qquad m_2 = \frac{2 - (-2)}{0 - 6} = \frac{4}{-6} = -\frac{2}{3}$$

Since their slopes are different, $l_1$ and $l_2$ are not parallel. To see whether the lines are perpendicular, we need to determine whether the slopes are negative reciprocals. If $m_1 m_2 = -1$, the slopes are negative reciprocals and the lines are perpendicular.

$$m_1 m_2 = \frac{3}{2}\left(-\frac{2}{3}\right) = -1$$

Since the product of the slopes equals $-1$, the lines are perpendicular.

▸ Now Try Exercise 15

**EXAMPLE 5** ▸ Consider the equation $2x + 4y = 8$. Determine the equation of the line that has a $y$-intercept of 5 and is **a)** parallel to the given line and **b)** perpendicular to the given line.

*Solution*

**a)** If we know the slope of a line and its $y$-intercept, we can use the slope-intercept form, $y = mx + b$, to write the equation. We begin by solving the given equation for $y$.

$$2x + 4y = 8$$
$$4y = -2x + 8$$
$$y = \frac{-2x + 8}{4}$$
$$y = -\frac{1}{2}x + 2$$

Two lines are parallel when they have the same slope. Therefore, the slope of the line parallel to the given line must be $-\frac{1}{2}$. Since its slope is $-\frac{1}{2}$ and its $y$-intercept is 5, its equation must be

$$y = -\frac{1}{2}x + 5$$

The graphs of $2x + 4y = 8$ and $y = -\frac{1}{2}x + 5$ are shown in **Figure 3.70** on page 203.

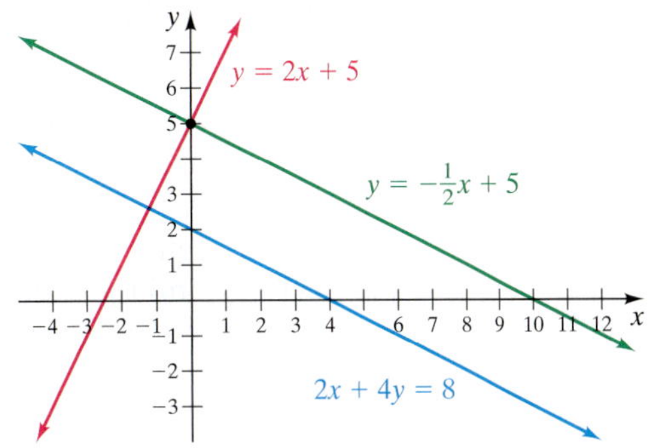


FIGURE 3.70

**b)** Two lines are perpendicular when their slopes are negative reciprocals. We know that the slope of the given line is $-\frac{1}{2}$. Therefore, the slope of the perpendicular line must be $-1\Big/\left(-\frac{1}{2}\right)$ or 2. The line perpendicular to the given line has a $y$-intercept of 5. Thus the equation is

$$y = 2x + 5$$

**Figure 3.70** also shows the graph of $y = 2x + 5$.

▸ **Now Try Exercise 35**

---

**EXAMPLE 6** ▸ Consider the equation $5y = -10x + 7$.

**a)** Determine the equation of a line that passes through $\left(4, \frac{1}{3}\right)$ that is perpendicular to the graph of the given equation. Write the equation in standard form.

**b)** Write the equation determined in part **a)** using function notation.

## *Solution*

**a)** Determine the slope of the given line by solving the equation for $y$.

$$5y = -10x + 7$$
$$y = \frac{-10x + 7}{5}$$
$$y = -2x + \frac{7}{5}$$

Since the slope of the given line is −2, the slope of a line perpendicular to it must be the negative reciprocal of −2, which is $\frac{1}{2}$. The line we are seeking must pass through the point $\left(4, \frac{1}{3}\right)$. Using the point-slope form, we obtain

$$y - y_1 = m(x - x_1)$$
$$y - \frac{1}{3} = \frac{1}{2}(x - 4) \qquad \textit{Point-slope form}$$

Now multiply both sides of the equation by the least common denominator, 6, to eliminate fractions.

$$6\left(y - \frac{1}{3}\right) = 6\left[\frac{1}{2}(x - 4)\right]$$
$$6y - 2 = 3(x - 4)$$
$$6y - 2 = 3x - 12$$

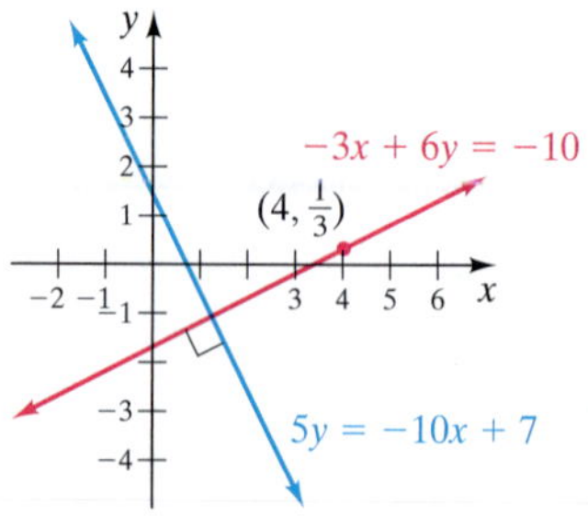


FIGURE 3.71

Now write the equation in standard form.

$$-3x + 6y - 2 = -12$$
$$-3x + 6y = -10 \quad \text{Standard form}$$

Note that $3x - 6y = 10$ is also an acceptable answer (see **Fig. 3.71**).

**b)** To write the equation using function notation, we solve the equation determined in part **a)** for $y$, and then replace $y$ with $f(x)$.

We will leave it to you to show that the function is $f(x) = \frac{1}{2}x - \frac{5}{3}$.

▸ **Now Try Exercise 39**

## Helpful Hint

The following chart summarizes the three forms of a linear equation we have studied and mentions when each may be useful.

| | |
|---|---|
| Standard form: | Useful when finding the intercepts of a graph |
| $ax + by = c$ | Will be used in Chapter 4, Systems of Equations and Inequalities |
| Slope-intercept form: | Used to find the slope and $y$-intercept of a line |
| $y = mx + b$ | Used to find the equation of a line given its slope and $y$-intercept |
| | Used to determine if two lines are parallel or perpendicular |
| | Used to graph a linear equation |
| Point-slope form: | Used to find the equation of a line when given the slope of a line and a point on the line |
| $y - y_1 = m(x - x_1)$ | Used to find the equation of a line when given two points on a line |

## EXERCISE SET 3.5

MathXL® MyMathLab

### Concept/Writing Exercises

1. Give the point-slope form of a linear equation.
2. How can we determine whether two lines are parallel?
3. How can we determine whether two lines are perpendicular?
4. Why can't the negative reciprocal test be used to determine whether a vertical line is perpendicular to a horizontal line?

### Practice the Skills

*Use the point-slope form to find the equation of a line with the properties given. Then write the equation in slope-intercept form.*

5. Slope $= 2$, through $(3, 1)$
6. Slope $= -3$, through $(1, -2)$
7. Slope $= -\frac{1}{2}$, through $(4, -1)$
8. Slope $= -\frac{7}{8}$, through $(-8, -2)$
9. Slope $= \frac{1}{2}$, through $(-1, -5)$
10. Slope $= -\frac{3}{2}$, through $(7, -4)$
11. Through $(2, -3)$ and $(-6, 9)$
12. Through $(4, -2)$ and $(1, 9)$
13. Through $(4, -3)$ and $(6, -2)$
14. Through $(1, 0)$ and $(-4, -1)$

*Two points on $l_1$ and two points on $l_2$ are given. Determine whether $l_1$ is parallel to $l_2$, $l_1$ is perpendicular to $l_2$, or neither.*

15. $l_1$: $(2, 0)$ and $(0, 2)$; $l_2$: $(3, 0)$ and $(0, 3)$
16. $l_1$: $(7, 6)$ and $(3, 9)$; $l_2$: $(5, -1)$ and $(9, -4)$
17. $l_1$: $(4, 6)$ and $(5, 7)$; $l_2$: $(-1, -1)$ and $(1, 4)$
18. $l_1$: $(-3, 4)$ and $(4, -3)$; $l_2$: $(-5, -6)$ and $(6, -5)$
19. $l_1$: $(3, 2)$ and $(-1, -2)$; $l_2$: $(2, 0)$ and $(3, -1)$
20. $l_1$: $(3, 5)$ and $(9, 1)$; $l_2$: $(4, 0)$ and $(6, 3)$

*Determine whether the two equations represent lines that are parallel, perpendicular, or neither.*

**21.** $y = \frac{1}{5}x + 9$
$y = -5x + 2$

**22.** $2x + 3y = 11$
$y = -\frac{2}{3}x + 4$

**23.** $4x + 2y = 8$
$8x = 4 - 4y$

**24.** $2x - y = 4$
$3x + 6y = 18$

**25.** $2x - y = 4$
$-x + 4y = 4$

**26.** $6x + 2y = 8$
$4x - 5 = -y$

**27.** $y = \frac{1}{2}x - 6$
$-4y = 8x + 15$

**28.** $2y - 8 = -5x$
$y = -\frac{5}{2}x - 2$

**29.** $y = \frac{1}{2}x + 6$
$-2x + 4y = 8$

**30.** $-4x + 6y = 11$
$2x - 3y = 5$

**31.** $x - 2y = -9$
$y = x + 6$

**32.** $\frac{1}{2}x - \frac{3}{4}y = 1$
$\frac{3}{5}x + \frac{2}{5}y = -1$

*Find the equation of a line with the properties given. Write the equation in the form indicated.*

**33.** Through (2, 5) and parallel to the graph of $y = 2x + 4$ (slope-intercept form)

**34.** Through $(-1, 6)$ and parallel to the graph of $4x - 2y = 6$ (slope-intercept form)

**35.** Through $(-3, -5)$ and parallel to the graph of $2x - 5y = 7$ (standard form)

**36.** Through $(-1, 4)$ and perpendicular to the graph of $y = -2x - 1$ (standard form)

**37.** With $x$-intercept $(3, 0)$ and $y$-intercept $(0, 5)$ (slope-intercept form)

**38.** Through $(-2, -1)$ and perpendicular to the graph of $f(x) = -\frac{1}{5}x + 1$ (function notation)

**39.** Through $(5, -2)$ and perpendicular to the graph of $y = \frac{1}{3}x + 1$ (function notation)

**40.** Through $(-3, 5)$ and perpendicular to the line with $x$-intercept $(2, 0)$ and $y$-intercept $(0, 2)$ (standard form)

**41.** Through $(6, 2)$ and perpendicular to the line with $x$-intercept $(2, 0)$ and $y$-intercept $(0, -3)$ (slope-intercept form)

**42.** Through the point $(1, 2)$ and parallel to the line through the points $(3, 5)$ and $(-2, 3)$ (function notation)

## Problem Solving

**43. Treadmill** The number of calories burned in 1 hour on a treadmill is a function of the speed of the treadmill. The average person walking on a treadmill (at 0° incline) at a speed of 2.5 miles per hour will burn about 210 calories. At 6 miles per hour the average person will burn about 370 calories. Let $C$ be the calories burned in 1 hour and $s$ be the speed of the treadmill.

**a)** Determine a linear function $C(s)$ that fits the data.

**b)** Estimate the calories burned by the average person on a treadmill in 1 hour at a speed of 5 miles per hour.

**44. Inclined Treadmill** The number of calories burned for 1 hour on a treadmill going at a constant speed is a function of the incline of the treadmill. At 4 miles per hour an average person on a 5° incline will burn 525 calories. At 4 mph on a 15° incline the average person will burn 880 calories. Let $C$ be the calories burned and $d$ be the degrees of incline of the treadmill.

**a)** Determine a linear function $C(d)$ that fits the data.

**b)** Determine the number of calories burned by the average person in 1 hour on a treadmill going 4 miles per hour and at a 9° incline.

**45. Demand for DVD Players** The *demand* for a product is the number of items the public is willing to buy at a given price. Suppose the demand, $d$, for DVD players sold in 1 month is a linear function of the price, $p$, for $\$150 \le p \le \$400$. If the price is \$200, then 50 DVD players will be sold each month. If the price is \$300, only 30 DVD players will be sold.

**a)** Using ordered pairs of the form $(p, d)$, write an equation for the demand, $d$, as a function of price, $p$.

**b)** Using the function from part **a)**, determine the demand when the price of the DVD players is \$260.

**c)** Using the function from part **a)**, determine the price charged if the demand for DVD players is 45.

**46. Demand for New Sandwiches** The marketing manager of Arby's restaurants determines that the demand, $d$, for a new chicken sandwich is a linear function of the price, $p$, for $\$0.80 \le p \le \$4.00$. If the price is \$1.00, then 530 chicken sandwiches will be sold each month. If the price is \$2.00, only 400 chicken sandwiches will be sold each month.

**a)** Using ordered pairs of the form $(p, d)$, write an equation for the demand, $d$, as a function of price, $p$.

**b)** Using the function from part **a)**, determine the demand when the price of the chicken sandwich is \$2.60.

**c)** Using the function from part **a)**, determine the price charged if the demand for chicken sandwiches is 244 chicken sandwiches.

**47. Supply of Kites** The *supply* of a product is the number of items a seller is willing to sell at a given price. The maker of a new kite for children determines that the number of kites she is willing to supply, $s$, is a linear function of the selling price $p$ for $\$2.00 \le p \le \$4.00$. If a kite sells for \$2.00, then 130 per month will be supplied. If a kite sells for \$4.00, then 320 per month will be supplied.

**a)** Using ordered pairs of the form $(p, s)$, write an equation for the supply, $s$, as a function of price, $p$.

**b)** Using the function from part **a)**, determine the supply when the price of a kite is \$2.80.

**c)** Using the function from part **a)**, determine the price paid if the supply is 225 kites.

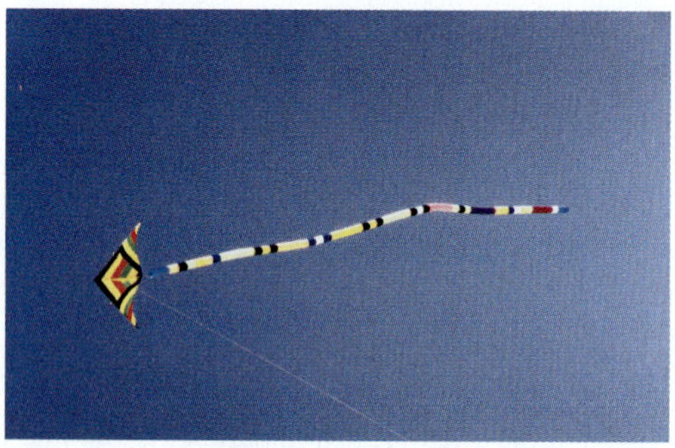

**48. Supply of Baby Strollers** The manufacturer of baby strollers determines that the supply, $s$, is a linear function of the selling price, $p$, for $\$200 \le p \le \$300$. If a stroller sells for \$210.00, then 20 strollers will be supplied per month. If a stroller sells for \$230.00, then 30 strollers will be supplied per month.

**a)** Using ordered pairs of the form $(p, s)$, write an equation for the supply, $s$, as a function of price, $p$.

**b)** Using the function from part **a)**, determine the supply when the price of a stroller is \$220.00.

**c)** Using the function from part **a)**, determine the selling price if the supply is 35 strollers.

**49. High School Play** The income, $i$, from a high school play is a linear function of the number of tickets sold, $t$. When 80 tickets are sold, the income is \$1000. When 200 tickets are sold, the income is \$2500.

**a)** Use these data to write the income, $i$, as a function of the number of tickets sold, $t$.

**b)** Using the function from part **a)**, determine the income if 120 tickets are sold.

**c)** If the income is \$2200, how many tickets were sold?

**50. Gas Mileage of a Car** The gas mileage, $m$, of a specific car is a linear function of the speed, $s$, at which the car is driven, for $30 \le s \le 60$. If the car is driven at a rate of 30 mph, the car's gas mileage is 35 miles per gallon. If the car is driven at 60 mph, the car's gas mileage is 20 miles per gallon.

**a)** Use this data to write the gas mileage, $m$, as a function of speed, $s$.

**b)** Using the function from part **a)**, determine the gas mileage if the car is driven at a speed of 48 mph.

**c)** Using the function from part **a)**, determine the speed at which the car must be driven to get gas mileage of 40 miles per gallon.

**51. Auto Registration** The registration fee, $r$, for a vehicle in a certain region is a linear function of the weight of the vehicle, $w$, for $1000 \le w \le 6000$ pounds. When the weight is 2000 pounds, the registration fee is \$30. When the weight is 4000 pounds, the registration fee is \$50.

**a)** Use these data to write the registration fee, $r$, as a function of the weight of the vehicle, $w$.

**b)** Using the function from part **a)**, determine the registration fee for a 2006 Ford Mustang if the weight of the vehicle is 3613 pounds.

**c)** If the cost of registering a vehicle is \$60, determine the weight of the vehicle.

**52. Lecturer Salary** Suppose the annual salary of a lecturer at Chaumont University is a linear function of the number of years of teaching experience. A lecturer with 9 years of teaching experience is paid \$41,350. A lecturer with 15 years of teaching experience is paid \$46,687.

**a)** Use this data to write the annual salary of a lecturer, $s$, as a function of the number of years of teaching experience, $n$.

**b)** Using the function from part **a)**, determine the annual salary of a lecturer with 10 years of teaching experience.

**c)** Using the function from part **a)**, estimate the number of years of teaching experience a lecturer must have to obtain an annual salary of \$44,908.

**53. Life Expectancy** As seen in the following graph, the expected number of remaining years of life of a person, $y$, *approximates* a linear function. The expected number of remaining years is a function of the person's current age, $a$, for $30 \le a \le 80$. For example, from the graph we see that a person who is currently 50 years old has a life expectancy of 36.0 *more* years.

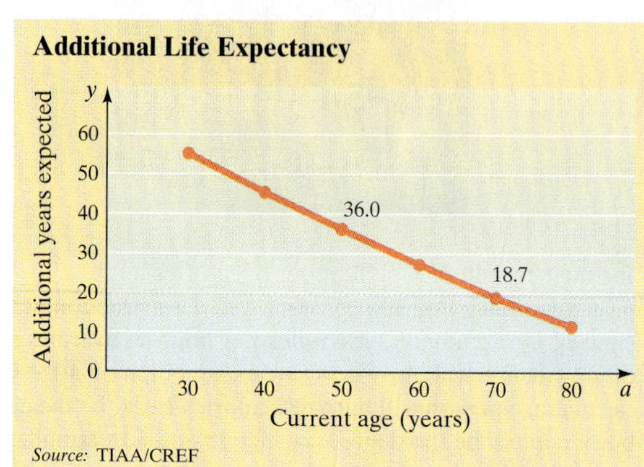

a) Using the two points on the graph, determine the function $y(a)$ that can be used to approximate the graph.

b) Using the function from part **a)**, estimate the life expectancy of a person who is currently 37 years old.

c) Using the function from part **a)**, estimate the current age of a person who has a life expectancy of 25 years.

**54. Guarneri del Gesù Violin** Handcrafted around 1735, Guarneri del Gesù violins are extremely rare and extremely valuable. The graph below shows that the projected value, $v$, of a Guarneri del Gesù violin is a linear function of the age, $a$, in years, of the violin, for $261 \le a \le 290$.

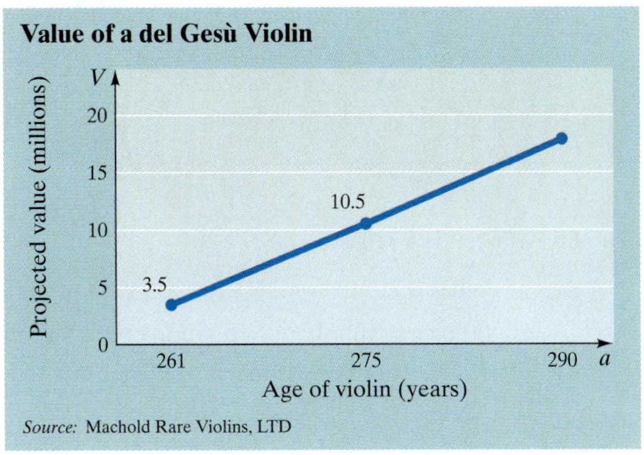


a) Determine the function $v(a)$ represented by this line.

b) Using the function from part **a)**, determine the projected value of a 265-year-old Guarneri del Gesù violin.

c) Using the function from part **a)**, determine the age of a Guarneri del Gesù violin with a projected value of \$15 million.

Guarneri del Gesù, "Sainton," 1741

**55. Boys' Weights** Parents may recognize the following diagram from visits to the pediatrician's office. The diagram shows percentiles for boys' heights and weights from birth to age 36 months. Overall, the graphs shown are not linear functions. However, certain portions of the graphs can be approximated with a linear function. For example, the graph representing the 95th percentile of boys' weights (the top red line) from age 18 months to age 36 months is approximately linear.

**Boys: Birth to 36 months**
Length-for-Age and Weight-for-Age Percentiles

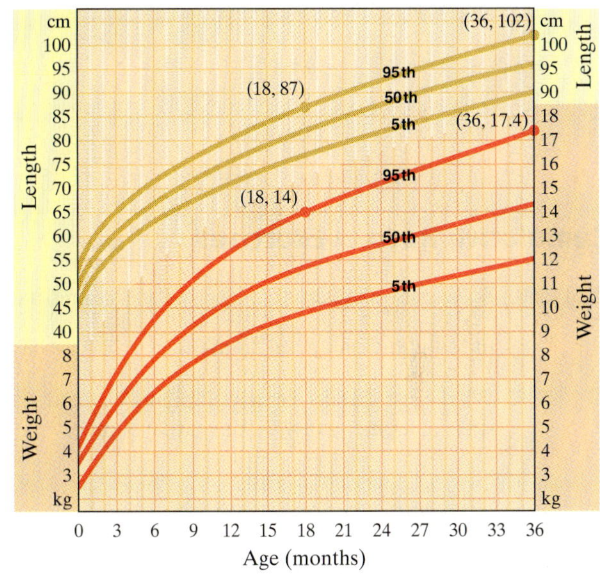


*Source:* National Center for Health Statistics

a) Use the points shown on the graph of the 95th percentile to write weight, $w$, as a linear function of age, $a$, for boys between 18 and 36 months old.

b) Using the function from part **a)**, estimate the weight of a 22-month-old boy who is in the 95th percentile for weight. Compare your answer with the graph to see whether the graph supports your answer.

**56. Boys' Lengths** The diagram in Exercise 55 shows that the graph representing the 95th percentile of boys' lengths (the top yellow line) from age 18 months to age 36 months is approximately linear.

a) Use the points shown on the graph of the 95th percentile to write length, $l$, as a linear function of age, $a$, for boys between age 18 and 36 months.

b) Using the function from part **a)**, estimate the length of a 21-month-old boy who is in the 95th percentile. Compare your answer with the graph to see whether the graph supports your answer.

## Group Activity

**57.** The graph on page 208 shows the growth of the circumference of a girl's head. The red line is the average head circumference of all girls for the given age while the green lines represent the upper and lower limits of the normal range. Discuss and answer the following questions as a group.

a) Explain why the graph of the average head circumference represents a function.

b) What is the independent variable? What is the dependent variable?

c) What is the domain of the graph of the average head circumference? What is the range of the average head circumference graph?

d) What interval is considered normal for girls of age 18?

**e)** For this graph, is head circumference a function of age or is age a function of head circumference? Explain your answer.

**f)** Estimate the average girl's head circumference at age 10 and at age 14.

**g)** This graph appears to be nearly linear. Determine an equation or function that can be used to estimate the orange line between (2, 48) and (18, 55).

**Head Circumference**

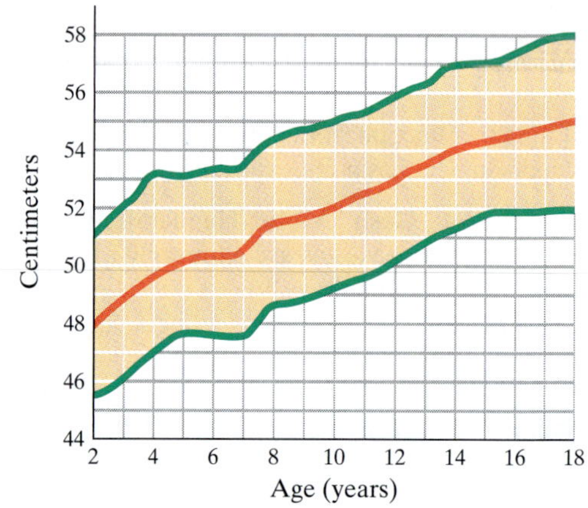


*Source:* National Center for Health Statistics

## Cumulative Review Exercises

[2.5] **58.** Solve the inequality $6 - \frac{1}{2}x > 2x + 5$ and indicate the solution in interval notation.

**59.** What must you do when multiplying or dividing both sides of an inequality by a negative number?

[3.2] **60.** **a)** What is a relation?

**b)** What is a function?

**c)** Draw a graph that is a relation but not a function.

**61.** Find the domain and range of the function $\{(4, 7), (5, -4), (3, 2), (6, -1)\}$.

# 3.6 The Algebra of Functions

**1** Find the sum, difference, product, and quotient of functions.

**2** Graph the sum of functions.

## 1 Find the Sum, Difference, Product, and Quotient of Functions

Let's discuss some ways that functions can be combined. If we let $f(x) = x - 3$ and $g(x) = x^2 + 2x$, we can find $f(5)$ and $g(5)$ as follows.

$$f(x) = x - 3 \qquad g(x) = x^2 + 2x$$
$$f(5) = 5 - 3 = 2 \qquad g(5) = 5^2 + 2(5) = 35$$

If we add $f(x) + g(x)$, we get

$$\begin{aligned} f(x) + g(x) &= (x - 3) + (x^2 + 2x) \\ &= x^2 + 3x - 3 \end{aligned}$$

This new function formed by the sum of $f(x)$ and $g(x)$ is designated as $(f + g)(x)$. Therefore, we may write

$$(f + g)(x) = x^2 + 3x - 3$$

We find $(f + g)(5)$ as follows.

$$\begin{aligned} (f + g)(5) &= 5^2 + 3(5) - 3 \\ &= 25 + 15 - 3 = 37 \end{aligned}$$

Notice that

$$f(5) + g(5) = (f + g)(5)$$
$$2 + 35 = 37 \qquad \textit{True}$$

In fact, for any real number substituted for $x$ you will find that

$$f(x) + g(x) = (f + g)(x)$$

Similar notation exists for subtraction, multiplication, and division of functions.

**Operations on Functions**

If $f(x)$ represents one function, $g(x)$ represents a second function, and $x$ is in the domain of both functions, then the following operations on functions may be performed:

**Sum of functions:** $(f + g)(x) = f(x) + g(x)$

**Difference of functions:** $(f - g)(x) = f(x) - g(x)$

**Product of functions:** $(f \cdot g)(x) = f(x) \cdot g(x)$

**Quotient of functions:** $(f/g)(x) = \dfrac{f(x)}{g(x)}$, provided that $g(x) \neq 0$

**EXAMPLE 1** ▸ If $f(x) = x^2 + x - 6$ and $g(x) = x - 3$, find

**a)** $(f + g)(x)$ **b)** $(f - g)(x)$

**c)** $(g - f)(x)$ **d)** Does $(f - g)(x) = (g - f)(x)$?

*Solution* To answer parts **a)–c)**, we perform the indicated operation.

**a)**
$$\begin{aligned}(f + g)(x) &= f(x) + g(x)\\ &= (x^2 + x - 6) + (x - 3)\\ &= x^2 + x - 6 + x - 3\\ &= x^2 + 2x - 9\end{aligned}$$

**b)**
$$\begin{aligned}(f - g)(x) &= f(x) - g(x)\\ &= (x^2 + x - 6) - (x - 3)\\ &= x^2 + x - 6 - x + 3\\ &= x^2 - 3\end{aligned}$$

**c)**
$$\begin{aligned}(g - f)(x) &= g(x) - f(x)\\ &= (x - 3) - (x^2 + x - 6)\\ &= x - 3 - x^2 - x + 6\\ &= -x^2 + 3\end{aligned}$$

**d)** By comparing the answers to parts **b)** and **c)**, we see that

$$(f - g)(x) \neq (g - f)(x)$$

▸ Now Try Exercise 11

**EXAMPLE 2** ▸ If $f(x) = x^2 - 4$ and $g(x) = x - 2$, find

**a)** $(f - g)(6)$ **b)** $(f \cdot g)(5)$ **c)** $(f/g)(8)$

*Solution*

**a)**
$$\begin{aligned}(f - g)(x) &= f(x) - g(x)\\ &= (x^2 - 4) - (x - 2)\\ &= x^2 - x - 2\\ (f - g)(6) &= 6^2 - 6 - 2\\ &= 36 - 6 - 2\\ &= 28\end{aligned}$$

We could have also found the solution as follows:

$$f(x) = x^2 - 4 \qquad g(x) = x - 2$$
$$f(6) = 6^2 - 4 = 32 \qquad g(6) = 6 - 2 = 4$$
$$\begin{aligned}(f - g)(6) &= f(6) - g(6)\\ &= 32 - 4 = 28\end{aligned}$$

**b)** We will find $(f \cdot g)(5)$ using the fact that

$$(f \cdot g)(5) = f(5) \cdot g(5)$$

$$f(x) = x^2 - 4 \qquad g(x) = x - 2$$

$$f(5) = 5^2 - 4 = 21 \qquad g(5) = 5 - 2 = 3$$

Thus $f(5) \cdot g(5) = 21 \cdot 3 = 63$. Therefore, $(f \cdot g)(5) = 63$. We could have also found $(f \cdot g)(5)$ by multiplying $f(x) \cdot g(x)$ and then substituting 5 into the product. We will discuss how to do this in Section 5.2.

**c)** We will find $(f/g)(8)$ by using the fact that

$$(f/g)(8) = f(8)/g(8)$$

$$f(x) = x^2 - 4 \qquad g(x) = x - 2$$

$$f(8) = 8^2 - 4 = 60 \qquad g(8) = 8 - 2 = 6$$

Then $f(8)/g(8) = 60/6 = 10$. Therefore, $(f/g)(8) = 10$. We could have also found $(f/g)(8)$ by dividing $f(x)/g(x)$ and then substituting 8 into the quotient. We will discuss how to do this in Chapter 5.

**▸ Now Try Exercise 31**

Notice that we included the phrase "and $x$ is in the domain of both functions" in the Operations on Functions box on page 209 . As we stated earlier, the domain of a function is the set of values that can be used for the independent variable. For example, the domain of the function $f(x) = 2x^2 - 6x + 5$ is all real numbers, because when $x$ is any real number $f(x)$ will also be a real number. The domain of $g(x) = \frac{1}{x - 8}$ is all real numbers except 8, because when $x$ is any real number except 8, the function $g(x)$ is a real number. When $x$ is 8, the function is not a real number because $\frac{1}{0}$ is undefined. We will discuss the domain of functions further in Section 6.1.

## 2 Graph the Sum of Functions

Now we will explain how we can graph the sum, difference, product, or quotient of two functions. **Figure 3.72** shows two functions, $f(x)$ and $g(x)$.

To graph the sum of $f(x)$ and $g(x)$, or $(f + g)(x)$, we use $(f + g)(x) = f(x) + g(x)$. The table on the next page gives the integer values of $x$ from $-2$ to 4, the values of $f(-2)$ through $f(4)$, and the values of $g(-2)$ through $g(4)$. These values are taken directly from **Figure 3.72**. The values of $(f + g)(-2)$ through $(f + g)(4)$ are determined by adding the values of $f(x)$ and $g(x)$. The graph of $(f + g)(x) = f(x) + g(x)$ is illustrated in green in **Figure 3.73**.

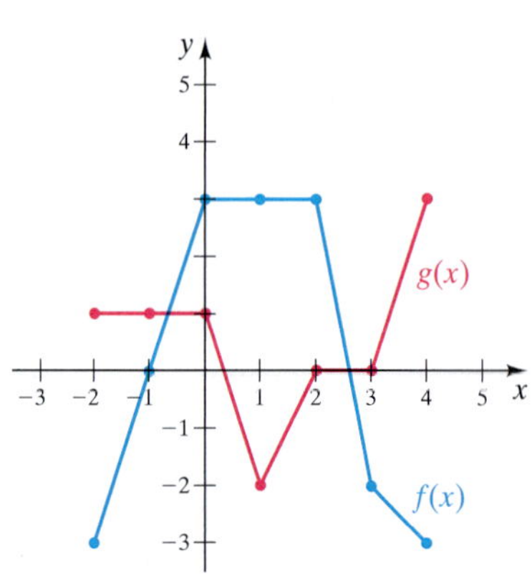


FIGURE 3.72

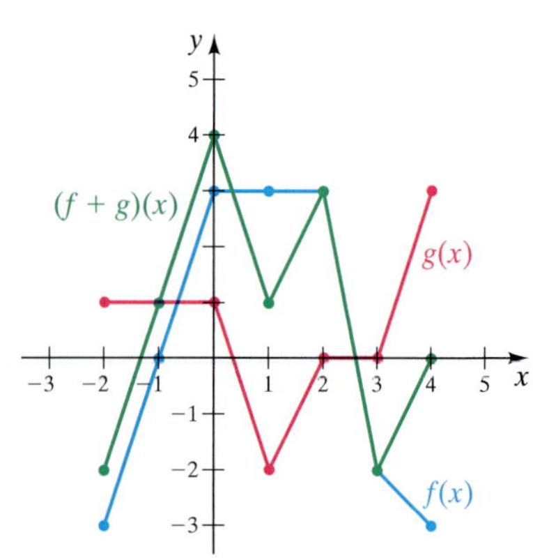


FIGURE 3.73

| $x$ | $f(x)$ | $g(x)$ | $(f + g)(x)$ |
|---|---|---|---|
| $-2$ | $-3$ | $1$ | $-3 + 1 = -2$ |
| $-1$ | $0$ | $1$ | $0 + 1 = 1$ |
| $0$ | $3$ | $1$ | $3 + 1 = 4$ |
| $1$ | $3$ | $-2$ | $3 + (-2) = 1$ |
| $2$ | $3$ | $0$ | $3 + 0 = 3$ |
| $3$ | $-2$ | $0$ | $-2 + 0 = -2$ |
| $4$ | $-3$ | $3$ | $-3 + 3 = 0$ |

We could graph the difference, product, or quotient of the two functions using a similar technique. For example, to graph the product function $(f \cdot g)(x)$, we would evaluate $(f \cdot g)(-2)$ as follows:

$$\begin{aligned}(f \cdot g)(-2) &= f(-2) \cdot g(-2) \\ &= (-3)(1) = -3\end{aligned}$$

Thus, the graph of $(f \cdot g)(x)$ would have an ordered pair at $(-2, -3)$. Other ordered pairs would be determined by the same procedure.

In newspapers, magazines, and on the Internet we often find graphs that show the sum of two or more functions. Graphs that show the sum of functions are generally indicated in one of three ways: line graphs, bar graphs, or stacked (or cumulative) line graphs. Examples 3 through 5 show the three general methods. Each of these examples will use the same data pertaining to cholesterol.

**EXAMPLE 3** ▸ **Line Graph** Jim Silverstone has kept a record of his bad cholesterol (low-density lipoprotein, or LDL) and his good cholesterol (high-density lipoprotein or HDL) from 2002 through 2006. **Table 3.2** shows his LDL and his HDL for these years.

**TABLE 3.2** Cholesterol

| | 2002 | 2003 | 2004 | 2005 | 2006 |
|---|---|---|---|---|---|
| **LDL** | 220 | 240 | 140 | 235 | 130 |
| **HDL** | 30 | 40 | 70 | 35 | 40 |

**a)** Explain why the data consisting of the years and the LDL values are a function, and the data consisting of the years and the HDL values are also a function.

**b)** Draw a line graph that shows the LDL, the HDL, and the total cholesterol from 2002 through 2006. The total cholesterol is the sum of the LDL and the HDL.

**c)** If $L$ represents the amount of LDL and $H$ represents the amount of HDL, show that $(L + H)(2006) = 170$.

**d)** By looking at the graph drawn in part **b)**, determine the years in which the LDL was less than 180.

### Solution

**a)** The data consisting of the years and the LDL values are a function because for each year there is exactly one LDL value. Note that the year is the independent variable, and the LDL value is the dependent variable. The data consisting of the years and the HDL values are a function for the same reason.

**b)** For any given year, the total cholesterol is the sum of the LDL and HDL for that year. For example, for 2005, to find the cholesterol, we add $235 + 35 = 270$. The graph in **Figure 3.74** shows LDL, HDL, and total cholesterol for the years 2002 through 2006.

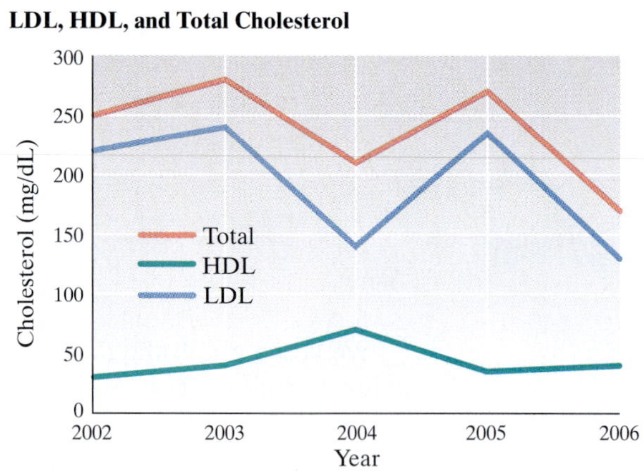


FIGURE 3.74

**c)** To find LDL + HDL, or the total cholesterol, we add the two values for 2006.

$$(L + H)(2006) = L(2006) + H(2006)$$
$$= 130 + 40 = 170$$

**d)** By looking at the graph drawn in part **b)**, we see that the years in which the LDL was less than 180 are 2004 and 2006.

▸ Now Try Exercise 63a

**EXAMPLE 4 ▸ Bar Graph**

**a)** Using the data given in **Table 3.2** on page 211, draw a bar graph that shows the LDL, HDL, and total cholesterol for the years 2002 through 2006.

**b)** If $L$ represents the amount of LDL and $H$ represents the amount of HDL, use the graph drawn in part **a)** to determine $(L + H)(2003)$.

**c)** By observing the graph drawn in part **a)**, determine in which years the total cholesterol was less than 220.

**d)** By observing the graph drawn in part **a)**, estimate the HDL in 2004.

*Solution*

**a)** To obtain a bar graph showing the total cholesterol, we add the HDL to the LDL for each given year. For example, for 2002, we start by drawing a bar up to 220 to represent the LDL. Directly on top of that bar we add a second bar of 30 units to represent the HDL. This brings the total bar to $220 + 30$ or 250 units. We use the same procedure for each year from 2002 to 2006. The bar graph is shown in **Figure 3.75**.

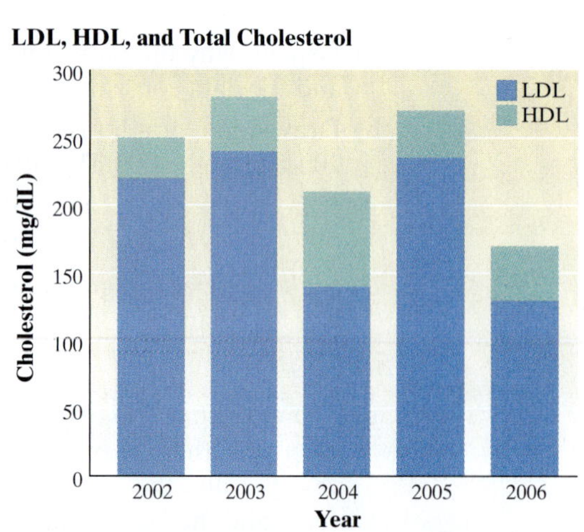


FIGURE 3.75

**b)** By observing the graph in **Figure 3.75**, we see that the $(L + H)(2003)$, or the total cholesterol for 2003, is about 280.

**c)** By observing the graph, we see that the total cholesterol was less than 220 in 2004 and 2006.

**d)** For 2004, the HDL bar begins at about 140 and ends at about 210. The difference in these amounts, $210 - 140 = 70$, represents the amount of HDL in 2004. Therefore, the HDL in 2004 was about 70.

▸ Now Try Exercise 63b

## EXAMPLE 5 ▸ Stacked Line Graph

**a)** Using the data from **Table 3.2** on page 211, draw a stacked (or cumulative) line graph that shows the LDL, HDL, and total cholesterol for the years 2002–2006.

**b)** Using the graph drawn in part **a)**, determine which years the total cholesterol was greater than or equal to 200.

**c)** Using the graph drawn in part **a)**, estimate the amount of HDL in 2006.

**d)** Using the graph from part **a)**, determine the years in which the LDL was greater than or equal to 180 and the total cholesterol was less than or equal to 250.

### Solution

**a)** To obtain a stacked line graph, draw the line to represent the LDL. This will be the same line that was drawn to represent the LDL in **Figure 3.74** on page 212. On top of this line draw a line to represent the HDL. One way to obtain the line for the HDL is to work year by year, and then connect the points for each year with straight-line segments. For example, in 2002, the HDL would start at 220, the LDL amount, and be increased by 30, the HDL amount, to get a total of 250. Use this procedure for each year. The graph is shown in **Figure 3.76**.

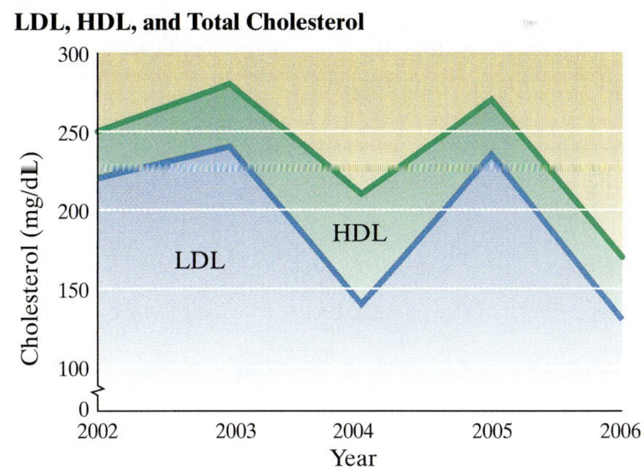


FIGURE 3.76

**b)** By looking at the graph, we see that the total cholesterol indicated by the dark green line, was greater than or equal to 200 in 2002, 2003, 2004, and 2005.

**c)** By looking at the HDL area of the graph, we can see that in 2006 the HDL starts at around 130 and ends at about 170. If we subtract, we obtain $170 - 130 = 40$. Therefore, the HDL in 2006 is about 40. If we let $H$ represent the amount of HDL, then $H(2006) \approx 40$.

**d)** By looking at the graph, we can determine that the only year in which the LDL was greater than or equal to 180 and the total cholesterol was less than or equal to 250 was 2002.

▸ Now Try Exercise 63c

### USING YOUR GRAPHING CALCULATOR

Graphing calculators can graph the sums, differences, products, and quotients of functions. One way to do this is to enter the individual functions. Then, following the instructions that come with your calculator, you can add, subtract, multiply, or divide the functions. For example, the screen in **Figure 3.77** shows a TI-84 Plus ready to graph $Y_1 = x - 3$, $Y_2 = 2x + 4$, and the sum of the functions, $Y_3 = Y_1 + Y_2$. On the TI-84 Plus, to get $Y_3 = Y_1 + Y_2$, you press the [VARS] key. Then you move the cursor to Y-VARS, and the you select 1: Function. Next you press [1] to enter $Y_1$. Next you press [+]. Then press [VARS] and go to Y-VARS, and choose 1: Function. Finally, press [2] to enter $Y_2$. **Figure 3.78** shows the graphs of the two functions, and the graph of the sum of the functions.

FIGURE 3.77

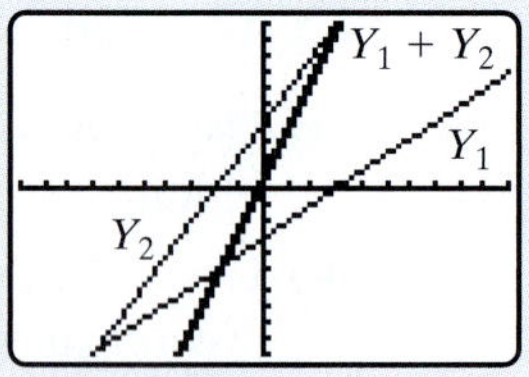


FIGURE 3.78

## EXERCISE SET 3.6

MathXL® MyMathLab

### Concept/Writing Exercises

1. Does $f(x) + g(x) = (f + g)(x)$ for all values of $x$?

2. Does $f(x) - g(x) = (f - g)(x)$ for all values of $x$?

3. What restriction is placed on the property $f(x)/g(x) = (f/g)(x)$? Explain.

4. Does $(f + g)(x) = (g + f)(x)$ for all values of $x$? Explain and give an example to support your answer.

5. Does $(f - g)(x) = (g - f)(x)$ for all values of $x$? Explain and give an example to support your answer.

6. If $f(2) = 9$ and $g(2) = -3$, determine
   **a)** $(f + g)(2)$ **b)** $(f - g)(2)$
   **c)** $(f \cdot g)(2)$ **d)** $(f/g)(2)$

7. If $f(-2) = -3$ and $g(-2) = 5$, find
   **a)** $(f + g)(-2)$ **b)** $(f - g)(-2)$
   **c)** $(f \cdot g)(-2)$ **d)** $(f/g)(-2)$

8. If $f(7) = 10$ and $g(7) = 0$, determine
   **a)** $(f + g)(7)$ **b)** $(f - g)(7)$
   **c)** $(f \cdot g)(7)$ **d)** $(f/g)(7)$

### Practice the Skills

*For each pair of functions, find* **a)** $(f + g)(x)$, **b)** $(f + g)(a)$, *and* **c)** $(f + g)(2)$.

9. $f(x) = x + 5, g(x) = x^2 + x$

10. $f(x) = x^2 - x - 8, g(x) = x^2 + 1$

11. $f(x) = -3x^2 + x - 4, g(x) = x^3 + 3x^2$

12. $f(x) = 4x^3 + 2x^2 - x - 1, g(x) = x^3 - x^2 + 2x + 6$

13. $f(x) = 4x^3 - 3x^2 - x, g(x) = 3x^2 + 4$

14. $f(x) = 3x^2 - x + 2, g(x) = 6 - 4x^2$

*Let* $f(x) = x^2 - 4$ *and* $g(x) = -5x + 3$. *Find the following.*

15. $f(2) + g(2)$

16. $f(5) + g(5)$

17. $f(4) - g(4)$

18. $f\left(\frac{1}{4}\right) - g\left(\frac{1}{4}\right)$

19. $f(3) \cdot g(3)$

20. $f(-1) \cdot g(-1)$

21. $\dfrac{f\left(\frac{3}{5}\right)}{g\left(\frac{3}{5}\right)}$

22. $f(-1)/g(-1)$

23. $g(-3) - f(-3)$

24. $g(6) \cdot f(6)$

25. $g(0)/f(0)$

26. $f(2)/g(2)$

*Let $f(x) = 2x^2 - x$ and $g(x) = x - 6$. Find the following.*

**27.** $(f + g)(x)$ **28.** $(f + g)(a)$ **29.** $(f + g)(2)$

**30.** $(f + g)(-3)$ **31.** $(f - g)(-2)$ **32.** $(f - g)(1)$

**33.** $(f \cdot g)(0)$ **34.** $(f \cdot g)(3)$ **35.** $(f/g)(-1)$

**36.** $(f/g)(6)$ **37.** $(g/f)(5)$ **38.** $(g - f)(4)$

**39.** $(g - f)(x)$ **40.** $(g - f)(r)$

## Problem Solving

*Using the graph on the right, find the value of the following.*

**41.** $(f + g)(0)$ **42.** $(f - g)(0)$

**43.** $(f \cdot g)(2)$ **44.** $(f/g)(1)$

**45.** $(g - f)(-1)$ **46.** $(g + f)(-3)$

**47.** $(g/f)(4)$ **48.** $(g \cdot f)(-1)$

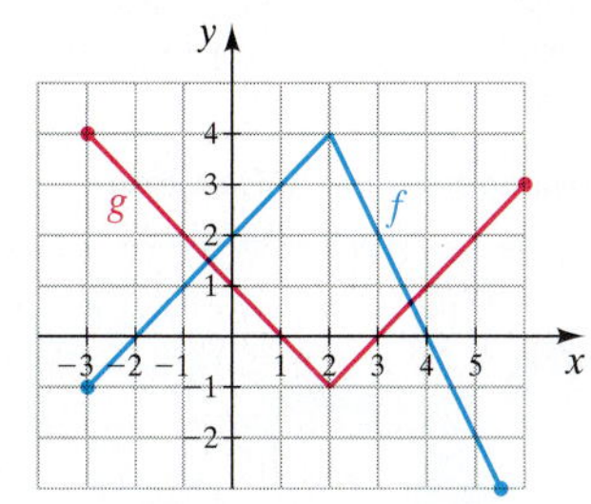


*Using the graph on the right, find the value of the following.*

**49.** $(f + g)(-2)$ **50.** $(f - g)(-1)$

**51.** $(f \cdot g)(1)$ **52.** $(g - f)(3)$

**53.** $(f/g)(4)$ **54.** $(g/f)(5)$

**55.** $(g/f)(2)$ **56.** $(g \cdot f)(0)$

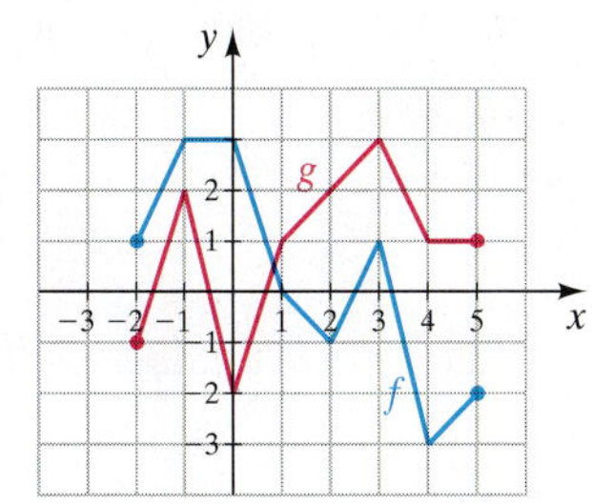


**57. Retirement Account** The following graph shows the amount of money Sharon and Frank Dangman have contributed to a joint retirement account for the years 2002 to 2006.

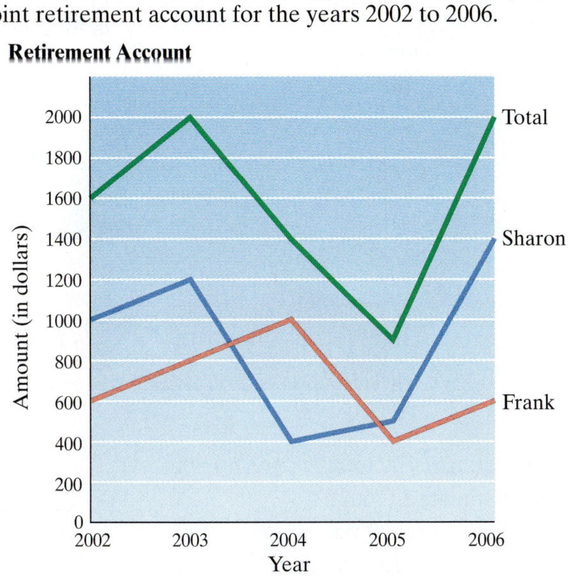


**a)** In which year did Frank contribute $1000?

**b)** In 2006, estimate how much more Sharon contributed to the retirement account than Frank contributed.

**c)** For this five-year period, estimate the total amount Sharon and Frank contributed to the joint retirement account.

**d)** Estimate $(F + S)(2005)$.

**58. Genetically Modified Crops** Worldwide production of genetically modified (or transgenic) crops—in both developing nations and industrial nations—is rapidly increasing. The following graph shows the land area devoted to genetically modified crops for developing nations, industrial nations, and total worldwide from 1995 to 2003. The total is determined by adding the amounts of both the developing and industrial nations. Land area is given in millions of hectares. A hectare is a metric system unit that is approximately equal to 2.471 acres.

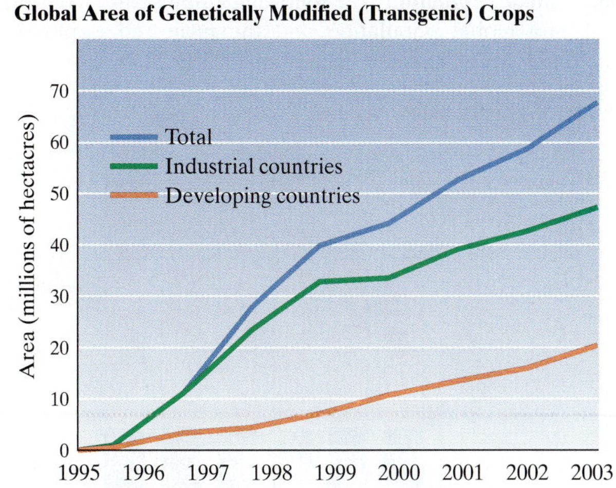


*Source*: Crop Biotech Net and www.isaaa.org/ko/global

**a)** Estimate the area in developing nations devoted to genetically modified crops in 2002.

**b)** Estimate the area in industrial nations devoted to genetically modified crops in 2002.

**c)** In what years from 1995 to 2003 was the total area devoted to genetically modified crops less than 23 million hectares?

**d)** In what years from 1995 to 2003 was the total area devoted to genetically modified crops greater than 50 million hectares?

**59. Oil Consumption in China** China's thirst for crude oil has been on the rise in recent years. The following bar graph shows China's total consumption, $C$, in millions of barrels, of crude oil per day. The red bars at the bottom represent China's import, $I$, of crude oil per day. The pink bars represent the crude oil produced in China per day for the years from 1995 to 2003.

**Oil Consumption Per Day in China**

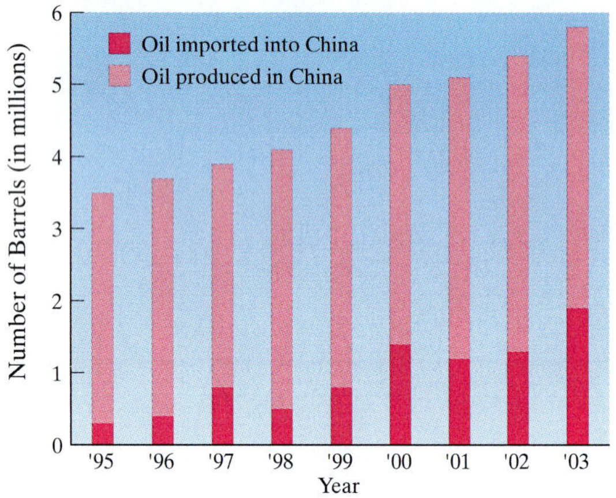


*Sources:* Bloomberg Financial Markets: BP Statistical Review: Bloomberg News: Customs General Administration of China, *New York Times* (12/23/04)

**a)** In what year was the import of crude oil to China the greatest? What was the amount imported each day?

**b)** In what years did the import of crude oil to China decrease from the year before?

**c)** Estimate $I(2002)$.

**d)** Estimate the amount of crude oil produced in China per day in 2003.

**60. Global Population** The following graph shows the projected total global population and the projected population of children 0–14 years of age from 2002 to 2050.

**Global Population**

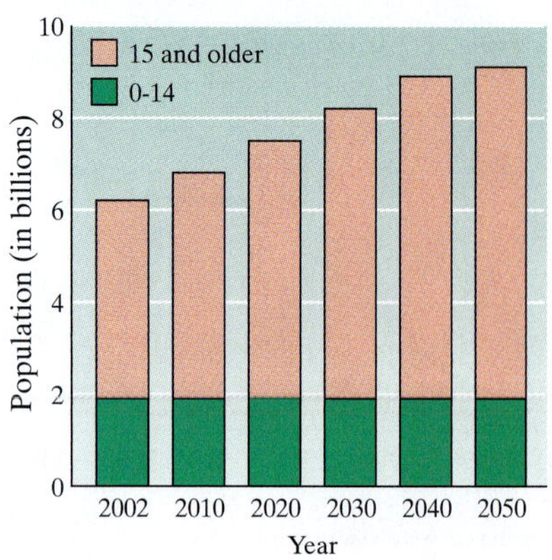


*Source:* U.S. Census Bureau, International Programs Center, International Data Base

**a)** Estimate the projected global population in 2050.

**b)** Estimate the projected number of children 0–14 years of age in 2050.

**c)** Estimate the projected number of people 15 years of age and older in 2050.

**d)** Estimate the projected difference in the total global population between 2002 and 2050.

**61. House Sales** In many regions of the country, houses sell better in the summer than at other times of the year. The graph below shows the total sales of houses in the town of Fuller from 2002 to 2006. The graph also shows the sale of houses in the summer, $S$, and in other times of the year, $Y$.

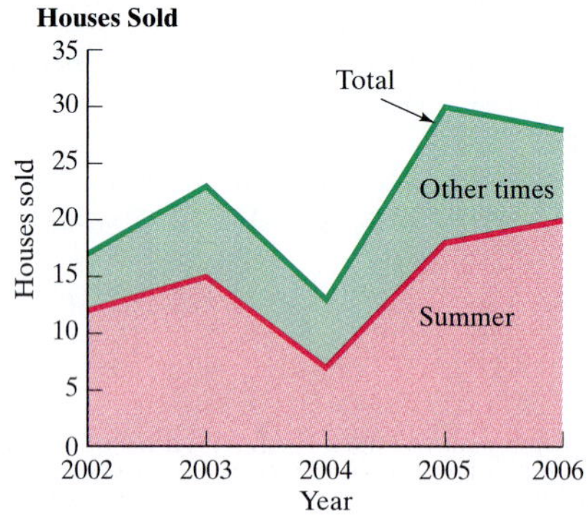


**a)** Estimate the number of houses sold in the summer of 2006.

**b)** Estimate the number of houses sold at other times in 2006.

**c)** Estimate $Y(2005)$.

**d)** Estimate $(S + Y)(2003)$.

**62. Income** Mark Whitaker owns a business where he does landscaping in the summer and snow removal in the winter. The graph below shows the total income, $T$, for the years 2002–2006 broken down into his landscaping income, $L$, and his snow removal income, $S$.

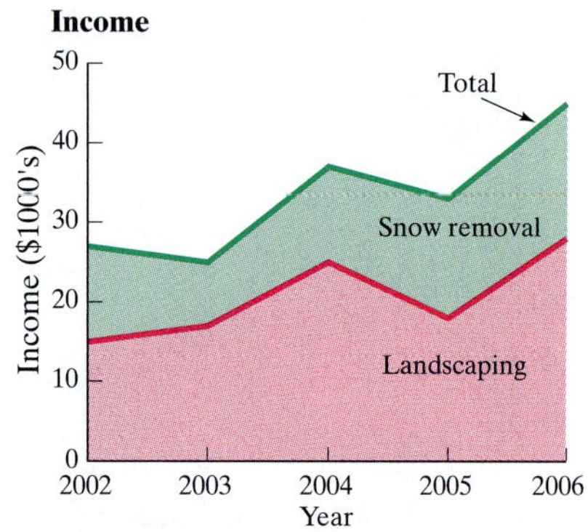


**a)** Estimate the total income for 2006.

**b)** Estimate $L(2002)$.

**c)** Estimate $S(2005)$.

**d)** Estimate $(L + S)(2003)$.

**63. Income** The chart below shows Mr. and Mrs. Abrams's income for the years from 2002 to 2006.

| | **2002** | **2003** | **2004** | **2005** | **2006** |
|---|---|---|---|---|---|
| **Mr. Abrams** | \$15,500 | \$17,000 | \$8,000 | \$25,000 | \$20,000 |
| **Mrs. Abrams** | \$4,500 | \$18,000 | \$28,000 | \$7,000 | \$22,500 |

a) Draw a line graph illustrating Mr. Abram's income, Mrs. Abram's income, and their total income for the years 2002–2006. See Example 3.

b) Draw a bar graph illustrating the given information. See Example 4.

c) Draw a stacked line graph illustrating the given information. See Example 5.

**64. Telephone Bills** The chart below shows Kelly Lopez's home telephone bills and cellular telephone bills (rounded to the nearest $10) for the years from 2002 to 2006.

| | 2002 | 2003 | 2004 | 2005 | 2006 |
|---|---|---|---|---|---|
| **Home** | $40 | $50 | $60 | $50 | $0 |
| **Cellular** | $80 | $50 | $20 | $50 | $60 |

a) Draw a line graph illustrating the home telephone bills, the cellular phone bills, and the total phone bills for the years 2002–2006.

b) Draw a bar graph illustrating the given information.

c) Draw a stacked line graph illustrating the given information.

**65. Taxes** Maria Cisneros pays both federal and state income taxes. The chart shows the amount of income taxes she paid to the federal government and to her state government (rounded to the nearest $100) for the years from 2002 to 2006.

| | 2002 | 2003 | 2004 | 2005 | 2006 |
|---|---|---|---|---|---|
| **Federal** | $4000 | $5000 | $3000 | $6000 | $6500 |
| **State** | $1600 | $2000 | $0 | $1700 | $1200 |

a) Draw a line graph illustrating the amount spent on federal taxes, the amount spent on state taxes, and the total amount spent on these two taxes for the years 2002–2006.

b) Draw a bar graph illustrating the given information.

c) Draw a stacked line graph illustrating the given information.

**66. College Tuition** The Olmert family has twin children, Justin and Kelly, who are attending different colleges. The tuition for Justin's and Kelly's colleges are given (to the nearest $1000) in the chart below for the years from 2004 to 2007.

| | 2004 | 2005 | 2006 | 2007 |
|---|---|---|---|---|
| **Justin** | $12,000 | $6000 | $8000 | $9000 |
| **Kelly** | $2000 | $8000 | $8000 | $5000 |

a) Draw a line graph illustrating the given information, including the total tuition spent on college for both Justin and Kelly for the years 2004–2007.

b) Draw a bar graph illustrating the given information.

c) Draw a stacked line graph illustrating the given information.

*For Exercises 67–72, let f and g represent two functions that are graphed on the same axes.*

**67.** If, at $a$, $(f + g)(a) = 0$, what must be true about $f(a)$ and $g(a)$?

**68.** If, at $a$, $(f \cdot g)(a) = 0$, what must be true about $f(a)$ and $g(a)$?

**69.** If, at $a$, $(f - g)(a) = 0$, what must be true about $f(a)$ and $g(a)$?

**70.** If, at $a$, $(f - g)(a) < 0$, what must be true about $f(a)$ and $g(a)$?

**71.** If, at $a$, $(f/g)(a) < 0$, what must be true about $f(a)$ and $g(a)$?

**72.** If, at $a$, $(f \cdot g)(a) < 0$, what must be true about $f(a)$ and $g(a)$?

*Graph the following functions on your graphing calculator.*

**73.** $y_1 = 2x + 3$
$y_2 = -x + 4$
$y_3 = y_1 + y_2$

**74.** $y_1 = x - 3$
$y_2 = 2x$
$y_3 = y_1 - y_2$

**75.** $y_1 = x$
$y_2 = x + 5$
$y_3 = y_1 \cdot y_2$

**76.** $y_1 = 2x^2 - 4$
$y_2 = x$
$y_3 = y_1/y_2$

## Group Activity

**77. SAT Scores** The following graph shows the average math and verbal scores of entering college classes on the SAT college entrance exam for the years 1992 through 2004. Let $f$ represent the math scores, and $g$ represent the verbal scores, and let $t$ represent the year. As a group, draw a graph that represents $(f + g)(t)$.

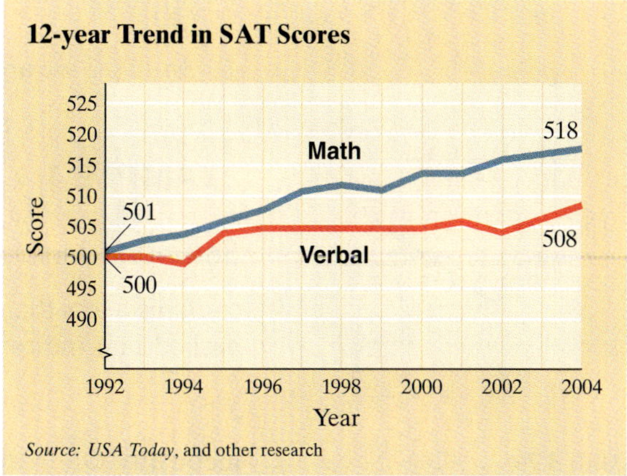

## Cumulative Review Exercises

[1.5] **78.** Evaluate $(-4)^{-3}$.

[1.6] **79.** Express 2,960,000 in scientific notation.

[2.2] **80.** Solve the formula $A = \frac{1}{2}bh$ for $h$.

[2.3] **81. Washing Machine** The cost of a washing machine, including a 6% sales tax, is \$477. Determine the pre-tax cost of the washing machine.

[3.1] **82.** Graph $y = |x| - 2$.

[3.3] **83.** Graph $3x - 4y = 12$.

# 3.7 Graphing Linear Inequalities

**1** Graph linear inequalities in two variables.

## 1 Graph Linear Inequalities in Two Variables

A **linear inequality** results when the equal sign in a linear equation is replaced with an inequality sign.

### Examples of Linear Inequalities in Two Variables

$$2x + 3y > 2 \qquad 3y < 4x - 9$$
$$-x - 2y \le 3 \qquad 5x \ge 2y - 7$$

A line divides a plane into three regions: the line itself and the two **half-planes** on either side of the line. The line is called the **boundary**. Consider the linear equation $2x + 3y = 6$. The graph of this line, the boundary line, divides the plane into the set of points that satisfy the inequality $2x + 3y < 6$ from the set of points that satisfy the inequality $2x + 3y > 6$. An inequality may or may not include the boundary line. Since the inequality $2x + 3y \le 6$ means $2x + 3y < 6$ or $2x + 3y = 6$, the inequality $2x + 3y \le 6$ contains the boundary line. Similarly, the inequality $2x + 3y \ge 6$ contains the boundary line. The graph of the inequalities $2x + 3y < 6$ and $2x + 3y > 6$ do not contain the boundary line. Now let's discuss how to graph linear inequalities.

**To Graph a Linear Inequality in Two Variables**

1. Replace the inequality symbol with an equal sign.
2. Draw the graph of the equation in step 1. If the original inequality contains a $\ge$ or $\le$ symbol, draw the graph using a solid line. If the original inequality contains a $>$ or $<$ symbol, draw the graph using a dashed line.
3. Select any point not on the line and determine if this point is a solution to the original inequality. If the point selected is a solution, shade the region on the side of the line containing this point. If the selected point does not satisfy the inequality, shade the region on the side of the line not containing this point.

In step 3 we are deciding which set of points satisfies the given inequality.

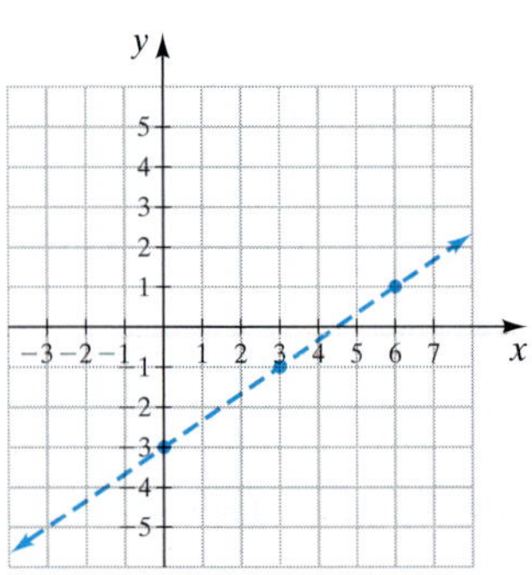

FIGURE 3.79

**EXAMPLE 1** ▸ Graph the inequality $y < \frac{2}{3}x - 3$.

*Solution* First graph the equation $y = \frac{2}{3}x - 3$. Since the original inequality contains a less than sign, $<$, use a dashed line when drawing the graph (**Fig. 3.79**). The dashed line indicates that the points on this line are not solutions to the inequality $y < \frac{2}{3}x - 3$. Select a point not on the line and determine if this point satisfies the inequality. Often the easiest point to use is the origin, (0, 0).

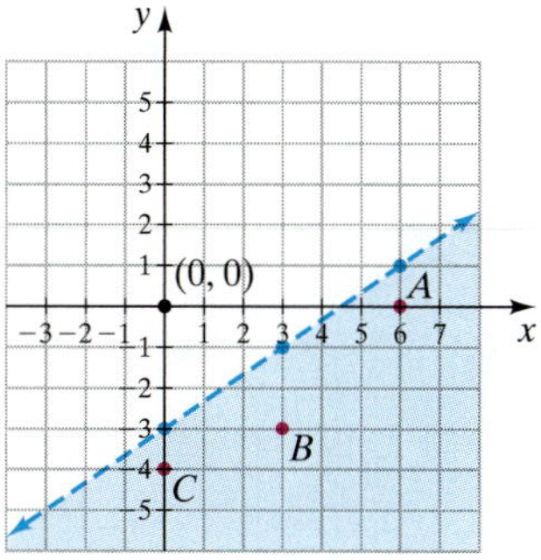


FIGURE 3.80

**Check Point (0, 0)**

$$y < \frac{2}{3}x - 3$$

$$0 \overset{?}{<} \frac{2}{3}(0) - 3$$

$$0 \overset{?}{<} 0 - 3$$

$$0 < -3 \qquad \textit{False}$$

Since 0 is not less than $-3$, the point $(0, 0)$ does not satisfy the inequality. The solution will be all points on the other side of the line from the point $(0, 0)$. Shade in this region (**Fig. 3.80**). Every point in the shaded area satisfies the given inequality. Let's check a few selected points $A$, $B$, and $C$.

| Point $A$ | Point $B$ | Point $C$ |
|---|---|---|
| $(6, 0)$ | $(3, -3)$ | $(0, -4)$ |
| $y < \frac{2}{3}x - 3$ | $y < \frac{2}{3}x - 3$ | $y < \frac{2}{3}x - 3$ |
| $0 \overset{?}{<} \frac{2}{3}(6) - 3$ | $-3 \overset{?}{<} \frac{2}{3}(3) - 3$ | $-4 \overset{?}{<} \frac{2}{3}(0) - 3$ |
| $0 \overset{?}{<} 4 - 3$ | $-3 \overset{?}{<} 2 - 3$ | $-4 \overset{?}{<} 0 - 3$ |
| $0 < 1$ *True* | $-3 < -1$ *True* | $-4 < -3$ *True* |

▸ **Now Try Exercise 15**

**EXAMPLE 2** ▸ Graph the inequality $y \geq -\frac{1}{2}x$.

*Solution* First, we graph the equation $y = -\frac{1}{2}x$. Since the inequality is $\geq$, we use a solid line to indicate that the points on the line are solutions to the inequality (**Fig. 3.81**). Since the point $(0, 0)$ is on the line, we cannot select that point to find the solution. Let's arbitrarily select the point $(3, 1)$.

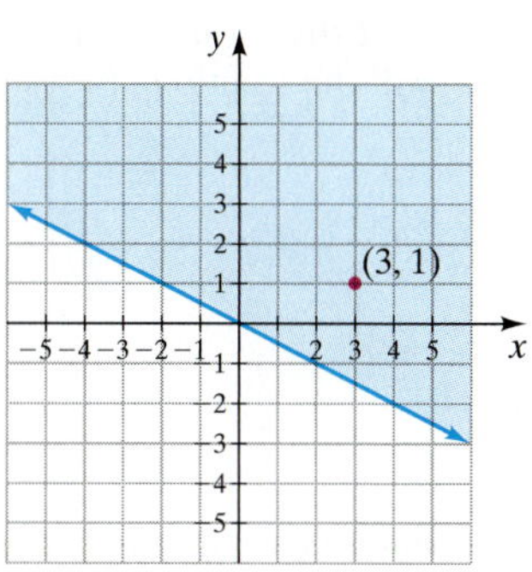


FIGURE 3.81

**Checkpoint (3, 1)**

$$y \geq -\frac{1}{2}x$$

$$1 \overset{?}{\geq} -\frac{1}{2}(3)$$

$$1 \geq -\frac{3}{2} \qquad \textit{True}$$

Since the point $(3, 1)$ satisfies the inequality, every point on the same side of the line as $(3, 1)$ will also satisfy the inequality $y \geq -\frac{1}{2}x$. Shade this region as indicated. Every point in the shaded region, as well as every point on the line, satisfies the inequality.

▸ **Now Try Exercise 9**

**EXAMPLE 3** ▸ Graph the inequality $3x - 2y < -6$.

*Solution* First, we graph the equation $3x - 2y = -6$. Since the inequality is $<$, we use a dashed line when drawing the graph **Fig. 3.82**. Substituting the checkpoint $(0, 0)$ into the inequality results in a false statement.

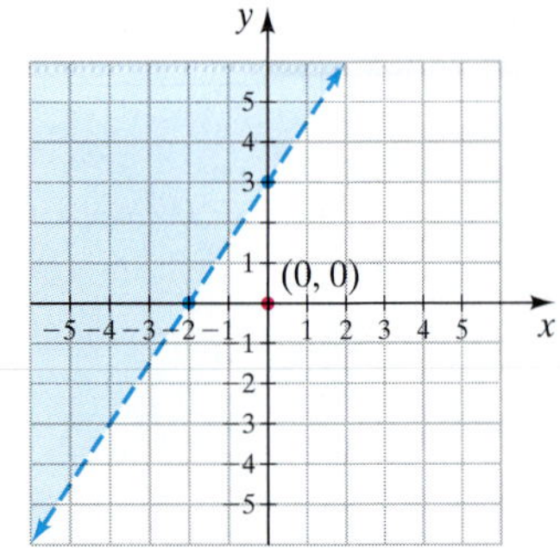

FIGURE 3.82

Checkpoint (0, 0)

$$3x - 2y < -6$$

$$3(0) - 2(0) \overset{?}{<} -6$$

$$0 < -6 \quad \text{False}$$

The solution is, therefore, that part of the plane that does not contain the origin.

▸ Now Try Exercise 23

## USING YOUR GRAPHING CALCULATOR

A graphing calculator can also display graphs of inequalities. The procedure to display the graphs varies from calculator to calculator. In **Figure 3.83** we show the graph of $y > 2x + 3$. Read your graphing calculator manual and learn how to display graphs of inequalities.

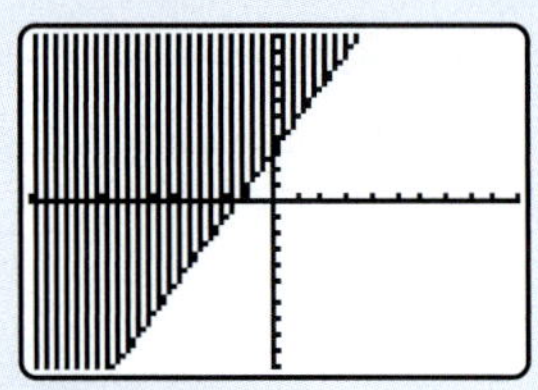

FIGURE 3.83

# EXERCISE SET 3.7

MathXL® MyMathLab

## Concept/Writing Exercises

1. When graphing an inequality containing $>$ or $<$, why are points on the line not solutions to the inequality?
2. When graphing an inequality containing $\geq$ or $\leq$, why are points on the line solutions to the inequality?
3. When graphing a linear inequality, when can $(0, 0)$ not be used as a check point?
4. When graphing a linear inequality of the form $y > ax + b$ where $a$ and $b$ are real numbers, will the solution always be above the line? Explain.

## Practice the Skills

*Graph each inequality.*

5. $x > 1$
6. $x \geq 4$
7. $y < -2$
8. $y < x$
9. $y \geq -\frac{1}{2}x$
10. $y < \frac{1}{2}x$
11. $y < 2x + 1$
12. $y \geq 3x - 1$
13. $y > 2x - 1$
14. $y \leq -x + 4$
15. $y \geq \frac{1}{2}x - 3$
16. $y < 3x + 2$
17. $2x + 3y > 6$
18. $2x - 3y \geq 12$
19. $y \leq -3x + 5$
20. $y \leq \frac{2}{3}x + 3$
21. $2x + y < 4$
22. $3x - 4y \leq 12$
23. $10 \geq 5x - 2y$
24. $-x - 2y > 4$

## Problem Solving

**25. Life Insurance** The monthly rates for \$100,000 of life insurance for women from the American General Financial Group increases approximately linearly from age 35 through age 50. The rate for a 35-year-old woman is \$10.15 per month and the rate for a 50-year-old woman is \$16.45 per month.

**a)** Draw a graph that fits this data.

**b)** On the graph, darken the part of the graph where the rate is less than or equal to \$15 per month.

**c)** Estimate the age at which the rate first exceeds \$15 per month.

**26. Consumer Price Index** The consumer price index (CPI) is a measure of inflation. Since 1990, the CPI has been increasing approximately linearly. The CPI in 1990 was 130.7 and in 2005 the CPI was 190.7.

*Source:* U.S. Bureau of the Census

**a)** Draw a graph that fits this data.

**b)** On the graph, darken the part of the graph where the CPI is greater than or equal to 171.

**c)** Estimate the first year in which the CPI was greater than or equal to 171.

**27. Fewer Smokers** The percentage of Americans 18 and over who smoke has been decreasing approximately linearly since 1997. In 1997, approximately 29.2% of Americans 18 and older smoked. In 2004, approximately 20.9% of those 18 and older smoked.

*Source:* Centers for Disease Control and Prevention

**a)** Draw a graph that fits these data.

**b)** On the graph, darken the part of the graph where the percentage of Americans 18 and older is less than or equal to 25%.

**c)** Estimate the first year that the percentage of Americans 18 and older was less than 23%.

**28. China Travel** The number of Chinese travelers has increased approximately linearly from 1993 through 2005. In 1993, there were about 3.7 million Chinese travelers. In 2005, there were about 17.9 million Chinese travelers.

*Source:* Travel Industry Association of America.

**a)** Draw a graph that fits these data.

**b)** On the graph, darken the part of the graph where the number of Chinese travelers is greater than or equal to 10 million.

**c)** Estimate the first year in which the number of Chinese travelers is greater than or equal to 12 million.

**29. a)** Graph $f(x) = 2x - 4$.

**b)** On the graph, shade the region bounded by $f(x)$, $x = 2$, $x = 4$, and the $x$-axis.

**30. a)** Graph $g(x) = -x + 4$.

**b)** On the graph, shade the region bounded by $g(x)$, $x = 1$, and the $x$- and $y$-axes.

## Challenge Problems

*Graph each inequality.*

**31.** $y < |x|$

**32.** $y \geq x^2$

**33.** $y < x^2 - 4$

## Cumulative Review Exercises

[2.1] **34.** Solve the equation $9 - \frac{5x}{3} = -6$.

[2.2] **35.** If $C = \bar{x} + Z\frac{\sigma}{\sqrt{n}}$, find $C$ when $\bar{x} = 80$, $Z = 1.96$, $\sigma = 3$, and $n = 25$.

[2.3] **36. Store Sale** Olie's Records and Stuff is going out of business. The first week all items are being reduced by 10%. The second week all items are being reduced by an additional \$2. If during the second week Bob Frieble purchases a CD for \$12.15, find the original cost of the CD.

[3.2] **37.** $f(x) = -x^2 + 5$; find $f(-3)$.

[3.3] **38.** Write an equation of the line that passes through the point $(8, -2)$ and is perpendicular to the line whose equation is $2x - y = 4$.

[3.4] **39.** Determine the slope of the line through $(-2, 7)$ and $(2, -1)$.

# Chapter 3 Summary

| IMPORTANT FACTS AND CONCEPTS | EXAMPLES |
|---|---|

## Section 3.1

The **Cartesian (or rectangular) coordinate** system consists of two axes drawn perpendicular to each other. The **y-axis** is the horizontal axis. The **y-axis** is the vertical axis. The **origin** is the point of intersection of the two axes. The two axes yield four **quadrants** (I, II, III, IV). An **ordered pair (*x*, *y*)** is used to give the two coordinates of a point.

Plot the following points on the same set of axes.

$$A(2,3),\ B(-2,3),\ C(-4,-1),\ D(3,-3),\ E(5,0)$$

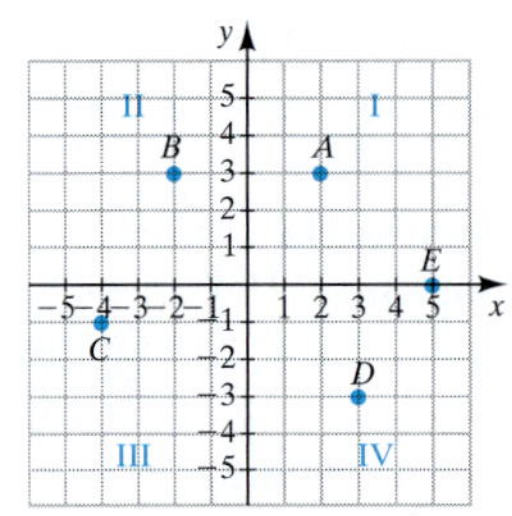


A **graph** is an illustration of the set of points whose coordinates satisfy the equation. Points in a straight line are said to be **collinear**.

A **linear equation** is an equation whose graph is a straight line. A linear equation is also called a **first-degree equation**.

$y = 2x + 1$ is a linear equation whose graph is illustrated below.

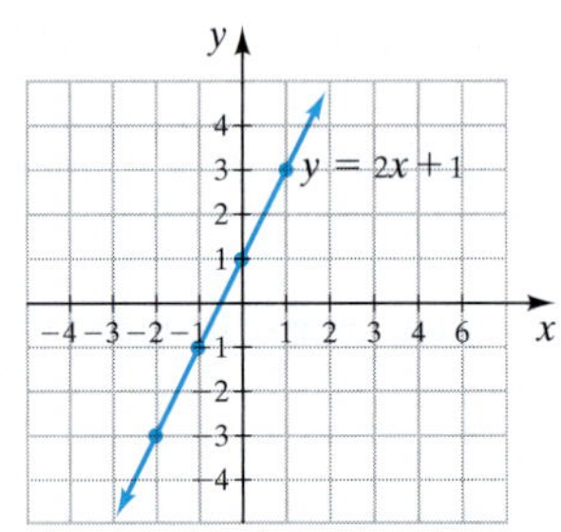


The points $(1, 3)$, $(0, 1)$, $(-1, -1)$, and $(-2, -3)$ are collinear.

A **nonlinear equation** is an equation whose graph is not a straight line.

$y = x^2 + 2$ is a nonlinear equation whose graph is illustrated below.

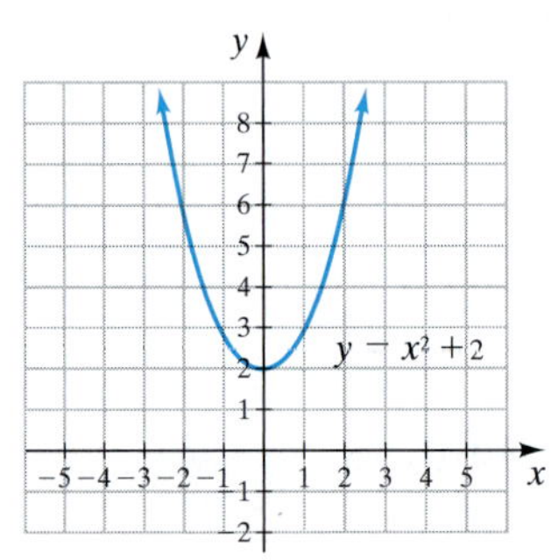


## Section 3.2

For an equation in variables $x$ and $y$, if the value of $y$ depends on the value of $x$, then $y$ is the **dependent variable** and $x$ is the **independent variable**.

In the equation $y = 2x^2 + 3x - 4$, $x$ is the independent variable and $y$ is the dependent variable.

A **relation** is any set of ordered pairs.
A **function** is a correspondence between a first set of elements, the **domain**, and a second set of elements, the **range**, such that each element of the domain corresponds to exactly one element of the range.

### Alternate Definition:

A **function** is a set of ordered pairs in which no first coordinate is repeated.

$\{(1,2), (2,3), (1,4)\}$ is a relation, but not a function.
$\{(1,6), (2,7), (3,10)\}$ is a relation. It is also a function since each element in the domain corresponds to exactly one element in the range.

$$\text{domain: } \{1, 2, 3\},\ \text{range} = \{6, 7, 10\}$$

| IMPORTANT FACTS AND CONCEPTS | EXAMPLES |
|---|---|

## Section 3.2 (continued)

The **vertical line test** can be used to determine if a graph represents a function.

If a vertical line can be drawn through any part of the graph and the line intersects another part of the graph, the graph does not represent a function. If a vertical line cannot be drawn to intersect the graph at more than one point, the graph represents a function.

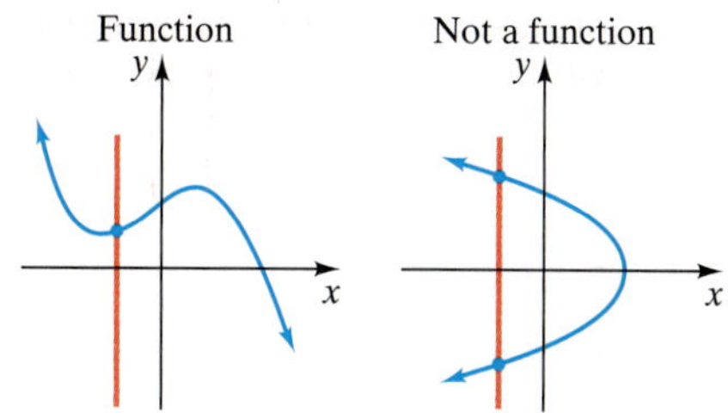


**Function notation** can be used to write an equation when $y$ is a function of $x$. For function notation, replace $y$ with $f(x)$, $g(x)$, $h(x)$, and so on.

$y = 7x - 9$ can be written as $f(x) = 7x - 9$

Given $y = f(x)$, to find $f(a)$, replace each $x$ with $a$.

Let $f(x) = x^2 + 2x - 8.$

Then $f(1) = 1^2 + 2(1) - 8 = -5$

$f(a) = a^2 + 2a - 8.$

## Section 3.3

A **linear function** is a function of the form $f(x) = ax + b$. The graph of a linear function is a straight line.

Graph $f(x) = \frac{1}{3}x - 2.$

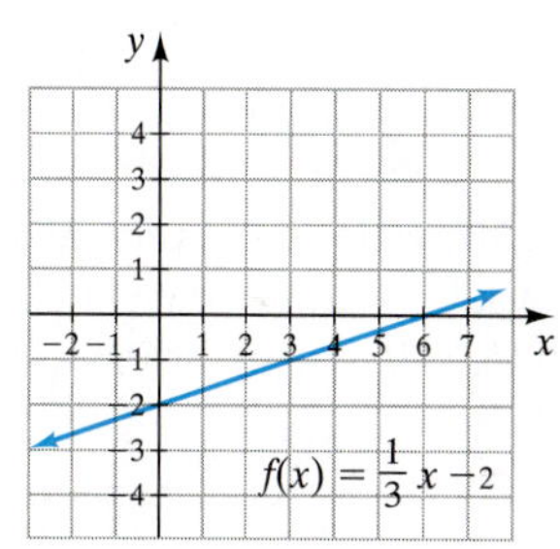


The **standard form of a linear equation** is $ax + by = c$, where $a$, $b$, and $c$ are real numbers, and $a$ and $b$ are not both zero.

$$3x + 5y = 7, \qquad -2x + \frac{1}{3}y = \frac{1}{8}$$

The **$x$-intercept** is the point where the graph crosses the $x$-axis. To find the $x$-intercept, set $y = 0$ and solve for $x$.

The **$y$-intercept** is the point where the graph crosses the $y$-axis. To find the $y$-intercept, set $x = 0$ and solve for $y$.

Graph $2x - 3y = 12$ using the $x$- and $y$- intercepts.

For $x$-intercept, let $y = 0$.

$$2x - 3y = 12$$
$$2x - 3(0) = 12$$
$$2x = 12$$
$$x = 6$$

Therefore, the $x$-intercept is $(6, 0)$.

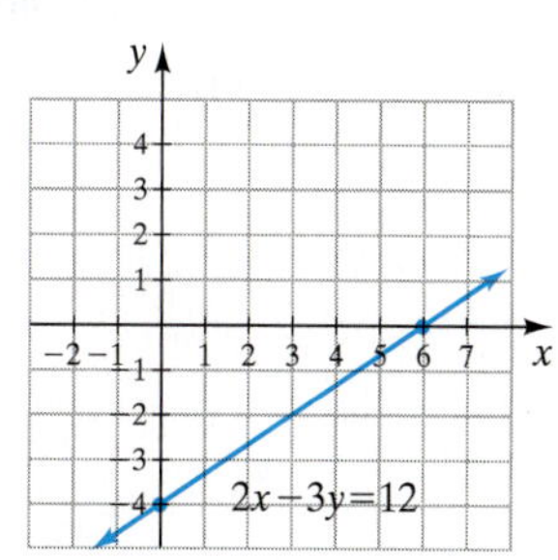


For $y$-intercept, let $x = 0$.

$$2x - 3y = 12$$
$$2(0) - 3y = 12$$
$$-3y = 12$$
$$y = -4$$

Therefore, the $y$-intercept is $(0, -4)$.

**IMPORTANT FACTS AND CONCEPTS** | **EXAMPLES**

## Section 3.3 (continued)

The graph of any equation of the form $y = b$ (or function of the form $f(x) = b$) will always be a horizontal line for any real number $b$. The function $f(x) = b$ is called the **constant function**.

Graph $y = 5$ (or $f(x) = 5$).

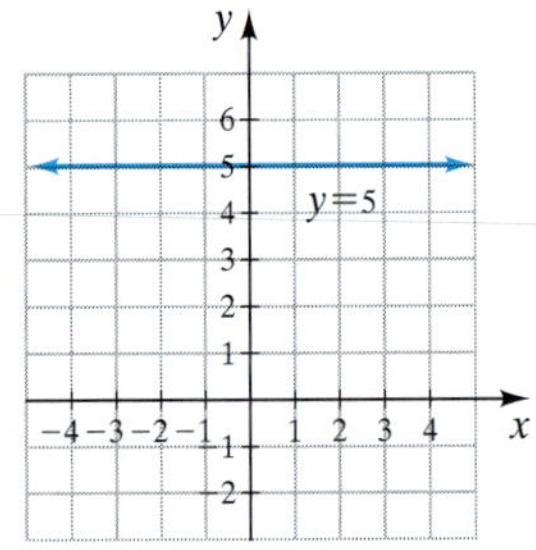


The graph of any equation of the form $x = a$ will always be a vertical line for any real number $a$.

Graph $x = -3.5$.

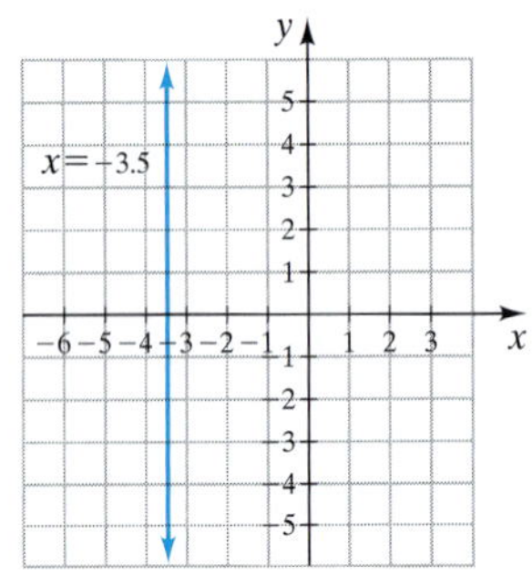


## Section 3.4

The **slope of a line** is the ratio of the vertical change (or rise) to the horizontal change (or run) between any two points.

The slope of the line through the distinct points $(x_1, y_1)$ and $(x_2, y_2)$ is

$$\text{slope} = \frac{\text{change in } y \text{ (vertical change)}}{\text{change in } x \text{ (horizontal change)}} = \frac{y_2 - y_1}{x_2 - x_1}$$

provided that $x_1 \neq x_2$.

The slope of the line through $(-1, 3)$ and $(7, 5)$ is

$$m = \frac{5 - 3}{7 - (-1)} = \frac{2}{8} = \frac{1}{4}.$$

A line that rises from left to right has a **positive slope**.

A line that falls from left to right has a **negative slope**.

A horizontal line has **zero slope**.

The slople of a vertical line is **undefined**.

Positive slope

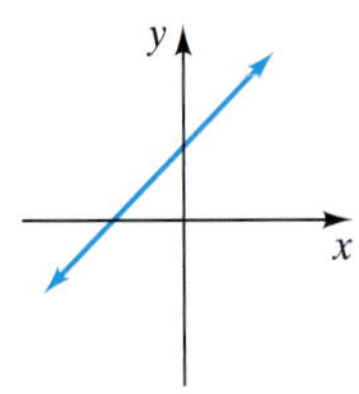

Negative slope

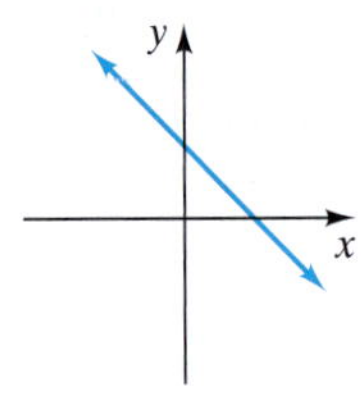

Zero slope

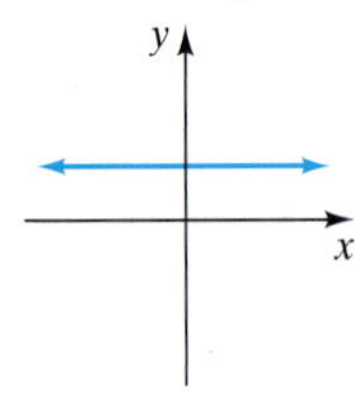

Slope is undefined

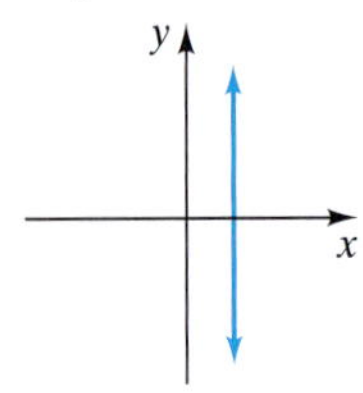

The **slope-intercept form of a linear equation** is

$$y = mx + b$$

where $m$ is the slope of the line and $(0, b)$ is the $y$-intercept of the line.

$y = 7x - 1, \quad y = -3x + 10$

| IMPORTANT FACTS AND CONCEPTS | EXAMPLES |
|---|---|
| **Section 3.5** | |
| The **point-slope form of a linear equation** is $$y - y_1 = m(x - x_1)$$ where $m$ is the slope of the line and $(x_1, y_1)$ is a point on the line. | If $m = 9$ and $(x_1, y_1)$ is $(5, 2)$, then $$y - 2 = 9(x - 5)$$ |
| Two lines are **parallel** if they have the same slope. | The graphs of $y = 2x + 4$ and $y = 2x + 7$ are parallel since both graphs have the same slope of 2, but different $y$-intercepts. |
| Two lines are **perpendicular** if their slopes are negative reciprocals. For any real number $a \neq 0$, its negative reciprocal is $-\frac{1}{a}$. | The graphs of $y = 3x - 5$ and $y = -\frac{1}{3}x + 8$ are perpendicular since one graph has a slope of 3 and the other graph has a slope of $-\frac{1}{3}$. The number $-\frac{1}{3}$ is the negative reciprocal of 3. |
| **Section 3.6** | |
| **Operations on Functions** | If $f(x) = x^2 + 2x - 5$ and $g(x) = x - 3$, then |
| **Sum of functions:** $(f + g)(x) = f(x) + g(x)$ | $(f + g)(x) = f(x) + g(x) = (x^2 + 2x - 5) + (x - 3)$ $= x^2 + 3x - 8$ |
| **Difference of functions:** $(f - g)(x) = f(x) - g(x)$ | $(f - g)(x) = f(x) - g(x) = (x^2 + 2x - 5) - (x - 3)$ $= x^2 + x - 2$ |
| **Product of functions:** $(f \cdot g)(x) = f(x) \cdot g(x)$ | $(f \cdot g)(x) = f(x) \cdot g(x)$ $= (x^2 + 2x - 5)(x - 3)$ $= x^3 - x^2 - 11x + 15$ |
| **Quotient of functions:** $(f/g)(x) = \frac{f(x)}{g(x)},\ g(x) \neq 0$ | $(f/g)(x) = \frac{f(x)}{g(x)} = \frac{x^2 + 2x - 5}{x - 3},\quad x \neq 3$ |
| **Section 3.7** | |
| A **linear inequality** results when the equal sign of a linear equation is replaced with an inequality sign. | $3x - 4y > 1,\quad 2x + 5y \leq -4$ |
| **To Graph a Linear Inequality in Two Variables**<br>**1.** Replace the inequality symbol with an equal sign.<br>**2.** Draw the graph of the equation in step 1. If the original inequality is $\geq$ or $\leq$ draw a solid line. If the original inequality is a $>$ or $<$ draw a dashed line.<br>**3.** Select any point not on the line. If the point selected is a solution, shade the region on the side of the line containing this point. If the selected point does not satisfy the inequality, shade the region on the side of the line not containing this point. | Graph $y > -x + 1$.<br>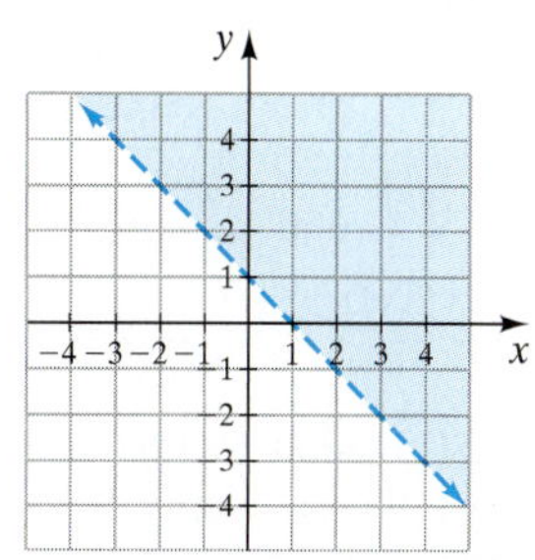 |



## Chapter 3 Review Exercises

[3.1] **1.** Plot the ordered pairs on the same axes.

**a)** $A(5, 3)$ **b)** $B(0, -3)$ **c)** $C\left(5, \frac{1}{2}\right)$ **d)** $D(-4, 2)$ **e)** $E(-6, -1)$ **f)** $F(-2, 0)$

*Graph each equation.*

**2.** $y = \frac{1}{2}x$ **3.** $y = -2x - 1$ **4.** $y = \frac{1}{2}x + 3$ **5.** $y = -\frac{3}{2}x + 1$ **6.** $y = x^2$

**7.** $y = x^2 - 1$ **8.** $y = |x|$ **9.** $y = |x| - 1$ **10.** $y = x^3$ **11.** $y = x^3 + 4$

[3.2]

**12.** Define function.

**13.** Is every relation a function? Is every function a relation? Explain.

*Determine whether the following relations are functions. Explain your answers.*

**14.**

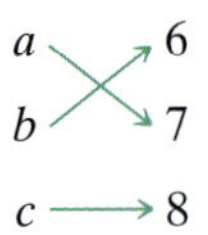


**15.** $\{(2, 5), (3, -4), (5, -9), (6, -1), (2, -2)\}$

*For Exercises 16–19,* **a)** *determine whether the following graphs represent functions;* **b)** *determine the domain and range of each relation or function.*

**16.**

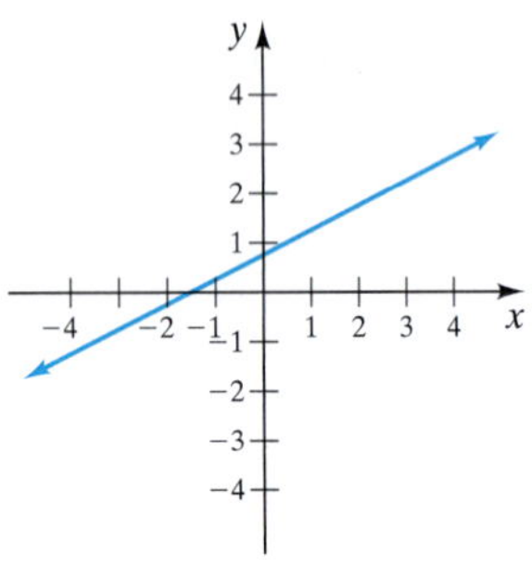


**17.**

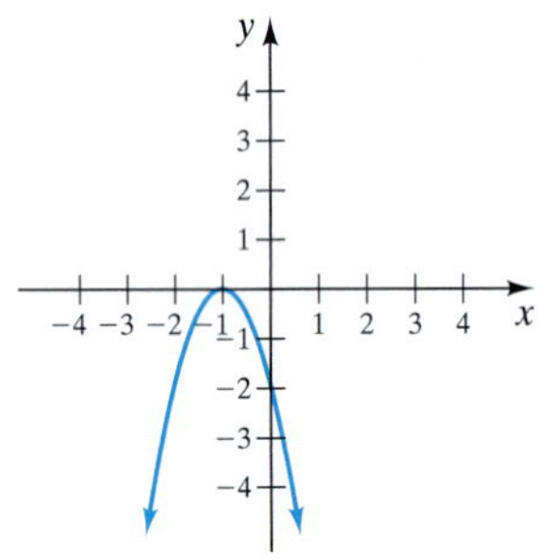


**18.**

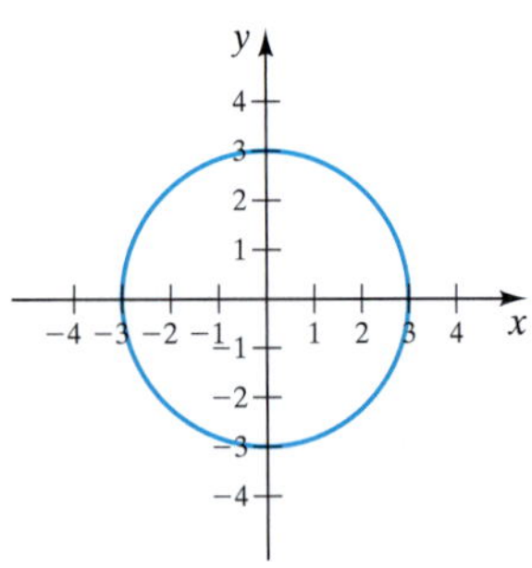


**19.**

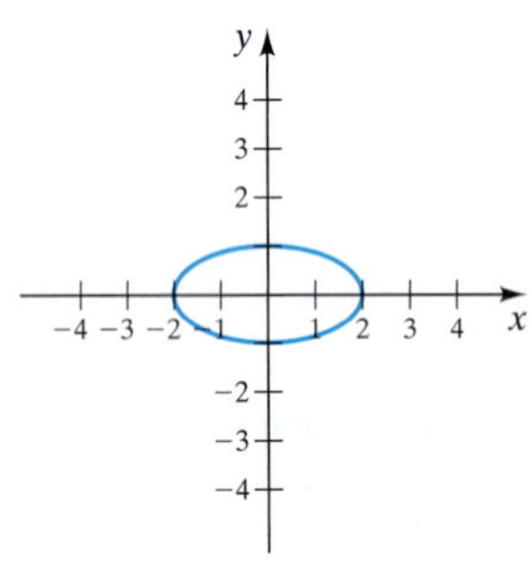


**20.** If $f(x) = -x^2 + 3x - 4$, find

**a)** $f(2)$ and

**b)** $f(h)$.

**21.** If $g(t) = 2t^3 - 3t^2 + 6$, find

**a)** $g(-1)$ and

**b)** $g(a)$.

**22. Speed of Car** Jane Covillion goes for a ride in a car. The following graph shows the car's speed as a function of time. Make up a story that corresponds to this graph.

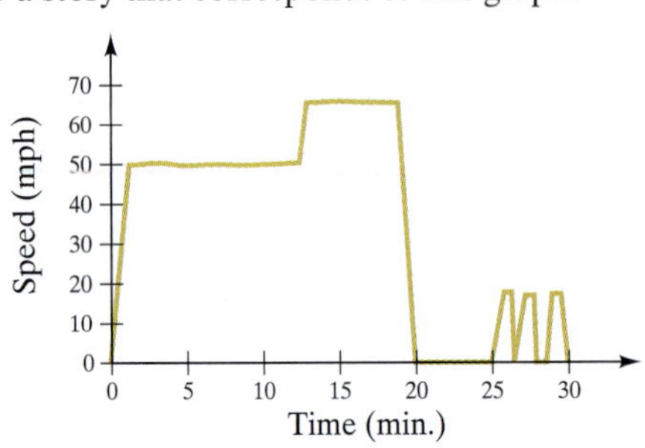


**23. Apple Orchard** The number of baskets of apples, $N$, that are produced by $x$ trees in a small orchard ($x \le 100$) is given by the function $N(x) = 40x - 0.2x^2$. How many baskets of apples are produced by

**a)** 30 trees?

**b)** 50 trees?

**24. Falling Ball** If a ball is dropped from the top of a 196-foot building, its height above the ground, $h$, at any time, $t$, can be found by the function $h(t) = -16t^2 + 196, 0 \le t \le 3.5$. Find the height of the ball at

**a)** 1 second.

**b)** 3 seconds.

[3.3] *Graph each equation using intercepts.*

**25.** $3x - 4y = 6$

**26.** $\frac{1}{3}x = \frac{1}{8}y + 10$

*Graph each equation or function.*

**27.** $f(x) = 4$

**28.** $x = -2$

**29. Bagel Company** The yearly profit, $p$, of a bagel company can be estimated by the function $p(x) = 0.1x - 5000$, where $x$ is the number of bagels sold per year.

**a)** Draw a graph of profits versus bagels sold for up to and including 250,000 bagels.

**b)** Estimate the number of bagels that must be sold for the company to break even.

**c)** Estimate the number of bagels sold if the company has $22,000 profit.

**30. Interest** Draw a graph illustrating the interest on a $12,000 loan for a 1-year period for various interest rates up to and including 20%. Use interest = principal · rate · time.

[3.4] *Determine the slope and y-intercept of the graph represented by the given equation.*

**31.** $y = \frac{1}{2}x - 5$

**32.** $f(x) = -2x + 3$

**33.** $3x + 5y = 13$

**34.** $3x + 4y = 10$

**35.** $x = -7$

**36.** $f(x) = 8$

*Determine the slope of the line through the two given points.*

**37.** $(2, -5), (6, 7)$

**38.** $(-2, 3)(4, 1)$

*Find the slope of each line. If the slope is undefined, so state. Then write the equation of the line.*

**39.**

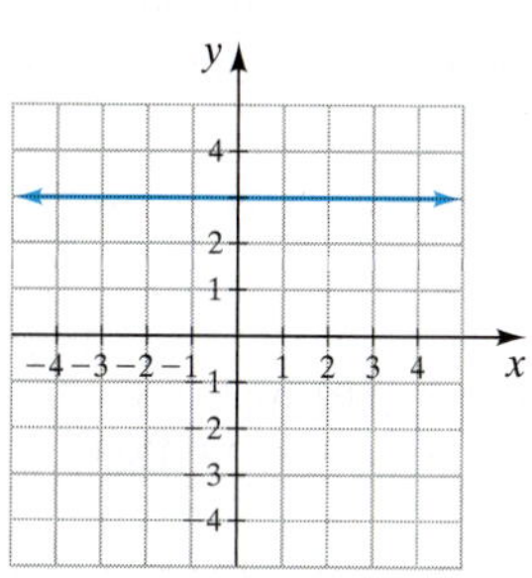

**40.**

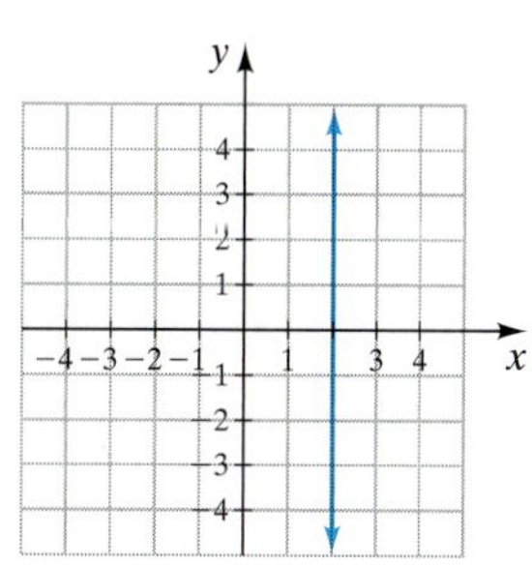

**41.**

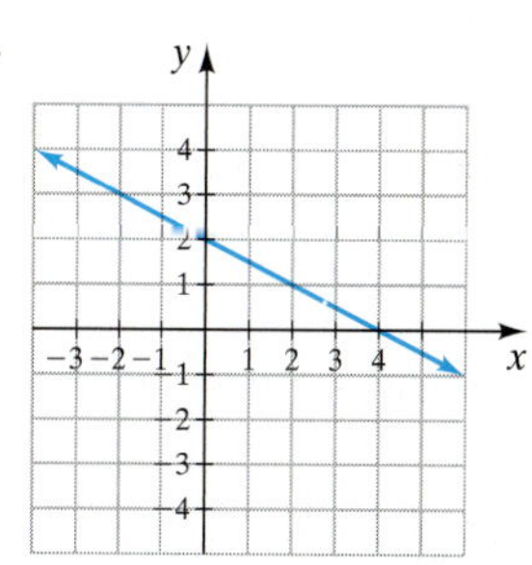

**42.** If the graph of $y = -2x + 5$ is translated down 4 units, determine

**a)** the slope of the translated graph.

**b)** the $y$-intercept of the translated graph.

**c)** the equation of the translated graph.

**43.** If one point on a graph is $(-6, -4)$ and the slope is $\frac{2}{3}$, find the $y$-intercept of the graph.

**44. Typhoid Fever** The following chart shows the number of reported cases of typhoid fever in the United States for select years from 1970 through 2000.

**a)** Plot each point and draw line segments from point to point.

**b)** Compute the slope of the line segments.

**c)** During which 10-year period did the number of reported cases of typhoid fever increase the most?

| Year | Number of reported typhoid fever cases |
|---|---|
| 1970 | 346 |
| 1980 | 510 |
| 1990 | 552 |
| 2000 | 317 |

**Source:** U.S. Dept. of Health and Human Services

**45. Social Security** The following graph shows the number of social security beneficiaries from 1980 projected through 2070. Use the slope-intercept form to find the function $n(t)$ (represented by the straight line) that can be used to represent this data.

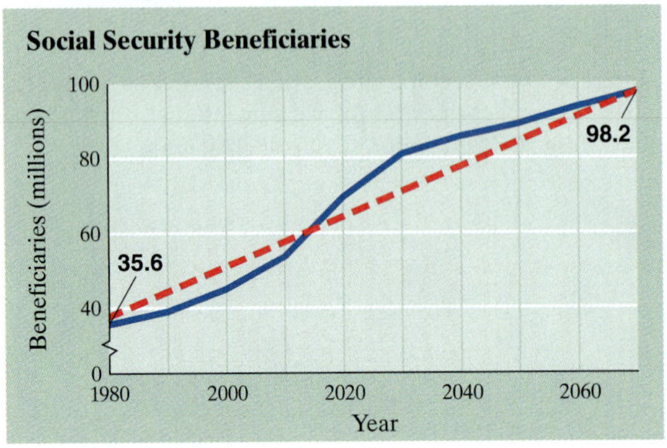


[3.5] *Determine whether the two given lines are parallel, perpendicular, or neither.*

**46.** $2x - 3y = 10$

$y = \frac{2}{3}x - 5$

**47.** $2x - 3y = 7$

$-3x - 2y = 8$

**48.** $4x - 2y = 13$

$-2x + 4y = -9$

*Find the equation of the line with the properties given. Write each answer in slope–intercept form.*

**49.** Slope $= \frac{1}{2}$, through $(4, 9)$

**50.** Through $(-3, 1)$ and $(4, -6)$

**51.** Through $(0, 6)$ and parallel to the graph of $y = -\frac{2}{3}x + 1$

**52.** Through $(2, 8)$ and parallel to the graph whose equation is $5x - 2y = 7$

**53.** Through $(-3, 1)$ and perpendicular to the graph whose equation is $y = \frac{3}{5}x + 5$

**54.** Through $(4, 5)$ and perpendicular to the graph whose equation is $4x - 2y = 8$

*Two points on $l_1$ and two points on $l_2$ are given. Determine whether $l_1$ is parallel to $l_2$, $l_1$ is perpendicular to $l_2$, or neither.*

**55.** $l_1$: $(5, 3)$ and $(0, -3)$; $l_2$: $(1, -1)$ and $(2, -2)$

**56.** $l_1$: $(3, 2)$ and $(2, 3)$; $l_2$: $(4, 1)$ and $(1, 4)$

**57.** $l_1$: $(7, 3)$ and $(4, 6)$; $l_2$: $(5, 2)$ and $(6, 3)$

**58.** $l_1$: $(-3, 5)$ and $(2, 3)$; $l_2$: $(-4, -2)$ and $(-1, 2)$

**59. Insurance Rates** The monthly rates for \$100,000 of life insurance from the General Financial Group for men increases approximately linearly from age 35 through age 50. The rate for a 35-year-old man is \$10.76 per month and the rate for a 50-year-old man is \$19.91 per month. Let $r$ be the rate and let $a$ be the age of a man between 35 and 50 years of age.

**a)** Determine a linear function $r(a)$ that fits these data.

**b)** Using the function in part **a)**, estimate the monthly rate for a 40 year-old man.

**60. Burning Calories** The number of calories burned in 1 hour of swimming, when swimming between 20 and 50 yards per minute, is a linear function of the speed of the swimmer. A person swimming at 30 yards per minute will burn about 489 calories in 1 hour. While swimming at 50 yards per minute a person will burn about 525 calories in 1 hour. This information is shown in the following graph.

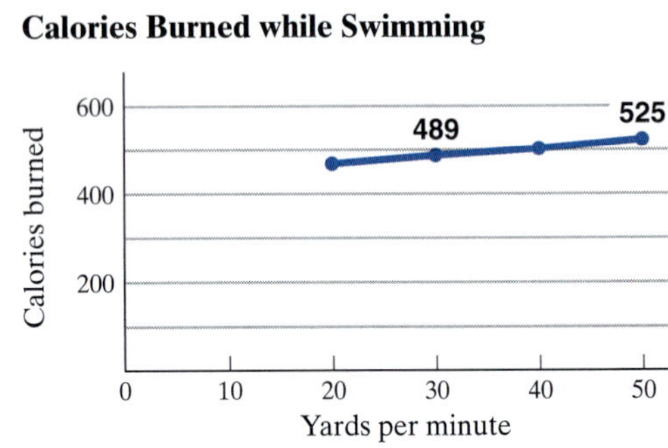


Source: *Health Magazine* Web Site, www.health.com

**a)** Determine a linear function that can be used to estimate the number of calories, $C$, burned in 1 hour when a person swims at $r$ yards per minute.

**b)** Use the function determined in part **a)** to determine the number of calories burned in 1 hour when a person swims at 40 yards per minute.

**c)** Use the function determined in part **a)** to estimate the speed at which a person needs to swim to burn 600 calories in 1 hour.

[3.6] *Given $f(x) = x^2 - 3x + 4$ and $g(x) = 2x - 5$, find the following.*

**61.** $(f + g)(x)$

**62.** $(f + g)(4)$

**63.** $(g - f)(x)$

**64.** $(g - f)(-1)$

**65.** $(f \cdot g)(-1)$

**66.** $(f \cdot g)(3)$

**67.** $(f/g)(1)$

**68.** $(f/g)(2)$

**69. Female Population** According to the U.S. Census, the female population is expected to grow worldwide. The following graph shows the female population worldwide for selected years from 2002 to 2050.

**Global Female Population**

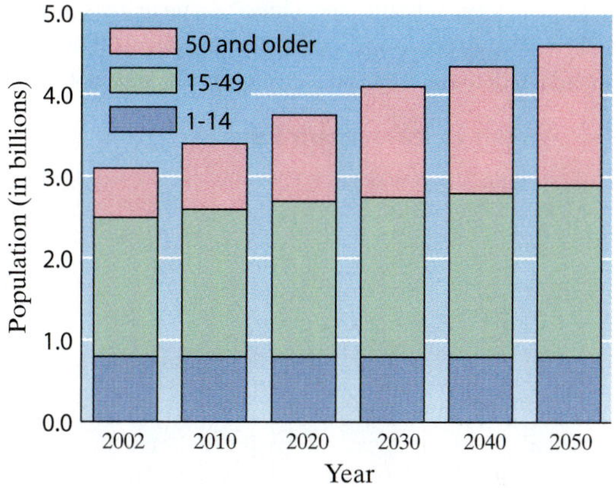


*Source*: U.S. Census Bureau, International Programs Center, International Data Base.

**a)** Estimate the projected female population worldwide in 2050.

**b)** Estimate the projected number of women 15–49 years of age in 2050.

**c)** Estimate the number of women who are projected to be in the 50 years and older age group in 2010.

**d)** Estimate the projected percent increase in the number of women 50 years and older from 2002 to 2010.

**70. Retirement Income** Ginny Jennings recently retired from her full-time job. The following graph shows her retirement income for the years 2003–2006.

**Ginny's Retirement Income**

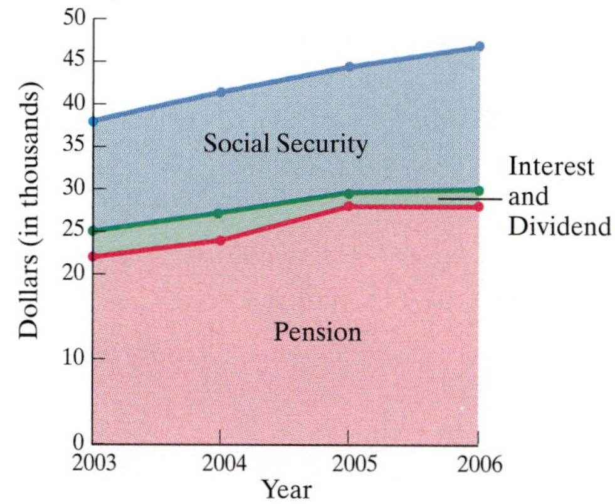


**a)** Estimate Ginny's total retirement income in 2006.

**b)** Estimate Ginny's pension income in 2005.

**c)** Estimate Ginny's interest and dividend income in 2003.

[3.7] *Graph each inequality.*

**71.** $y \geq -5$

**72.** $x < 4$

**73.** $y \leq 4x - 3$

**74.** $y < \frac{1}{3}x - 2$

## Chapter 3 Practice Test

*To find out how well you understand the chapter material, take this practice test. The answers, and the section where the material was initially discussed, are given in the back of the book. Each problem is also fully worked out on the* **Chapter Test Prep Video CD**. *Review any questions that you answered incorrectly.*

**1.** Graph $y = -2x + 1$.

**2.** Graph $y = \sqrt{x}$.

**3.** Graph $y = x^2 - 4$.

**4.** Graph $y = |x|$.

**5.** Define *function*.

**6.** Is the following set of ordered pairs a function? Explain your answer.

$$\{(3, 1), (-2, 6), (4, 6), (5, 2), (7, 3)\}$$

*In Exercises 7 and 8, determine whether the graphs represent functions. Give the domain and range of the relation or function.*

**7.**

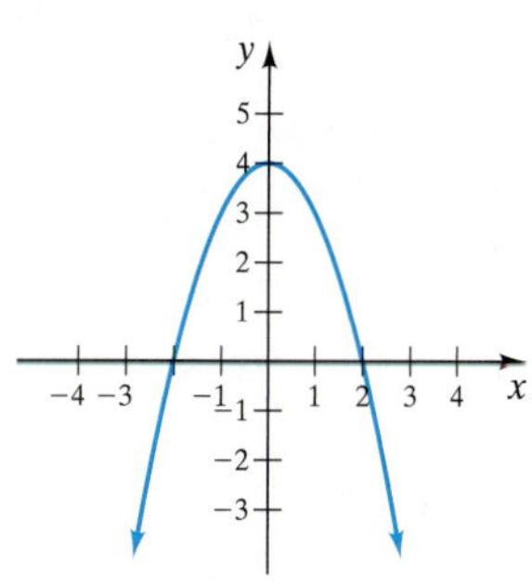

**8.**

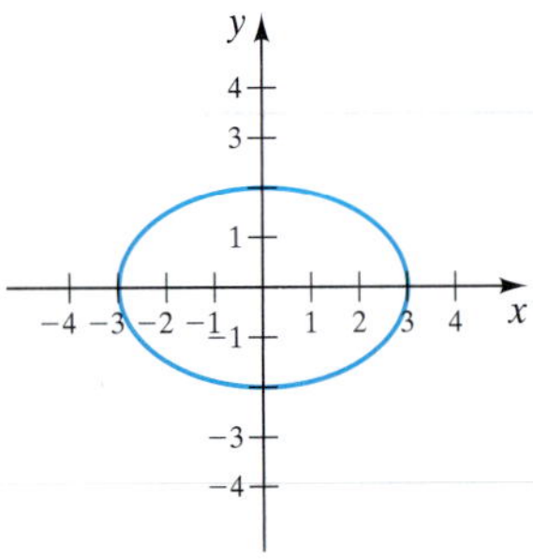


**9.** If $f(x) = 3x^2 - 6x + 5$, find $f(-2)$.

*In Exercises 10 and 11, graph the equation using the x- and y-intercepts.*

**10.** $-20x + 10y = 40$

**11.** $\frac{x}{5} - \frac{y}{4} = 1$

**12.** Graph $f(x) = -3$

**13.** Graph $x = 4$.

**14. Profit Graph** The yearly profit, $p$, for Zico Publishing Company on the sales of a particular book can be estimated by the function $p(x) = 10.2x - 50{,}000$, where $x$ is the number of books produced and sold.

**a)** Draw a graph of profit versus books sold for up to and including 30,000 books.

**b)** Use function $p(x)$ to estimate the number of books that must be sold for the company to break even.

**c)** Use function $p(x)$ to estimate the number of books that the company must sell to make a \$100,000 profit.

**15.** Determine the slope and $y$-intercept of the graph of the equation $4x - 3y = 15$

**16.** Write the equation, in slope-intercept form, of the line that goes through the points $(3, 2)$ and $(4, 5)$.

**17.** Determine the equation, in slope-intercept form, of the line that goes through the point $(6, -5)$ and is perpendicular to the graph of $y = \frac{1}{2}x + 1$.

**18. U.S. Population** Determine the function represented by the red line on the graph that can be used to estimate the projected U.S. population, $p$, from 2000 to 2050. Let 2000 be the reference year so that 2000 corresponds to $t = 0$.

**U.S. Population Projections 2000–2050**

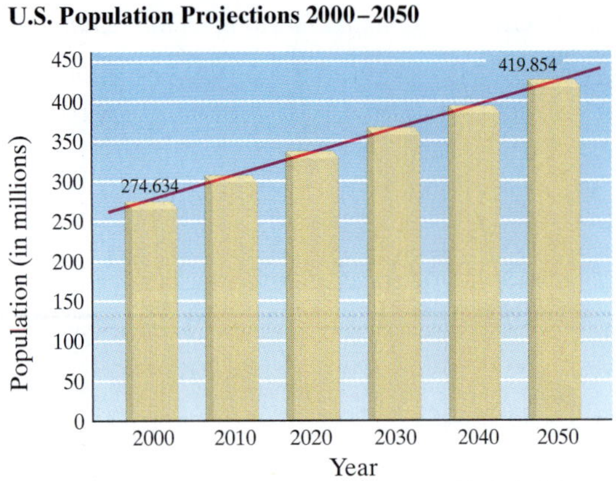


*Source:* U.S. Bureau of the Census, Statistical Abstract of the United States: 2004-2005

**19.** Determine whether the graphs of the two equations are parallel, perpendicular, or neither. Explain your answer.

$$2x - 3y = 12$$
$$4x + 10 = 6y$$

**20. Heart Disease** Deaths due to heart disease has been declining approximately linearly since the year 2000. The bar graph below shows the number of deaths, per 100,000 deaths, due to heart disease in selected years, projected for 2006–2010.

**Heart Disease Death Rate**

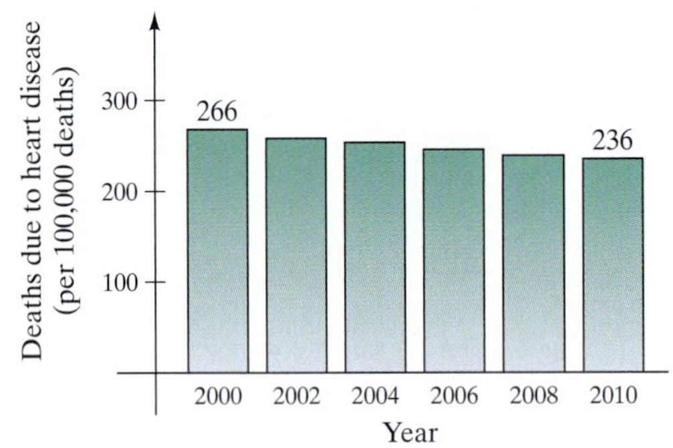


*Source:* U.S. Dept. of Health and Human Services

**a)** Let $r$ be the number of deaths due to heart disease per 100,000 deaths, and let $t$ represent the years since 2000. Write the linear function $r(t)$ that can be used to approximate the data.

**b)** Use the function from part **a)** to estimate the death rate due to heart disease in 2006.

**c)** Assuming this trend continues until the year 2020, estimate the death rate due to heart disease in 2020.

*In Exercises 21–23, if $f(x) = 2x^2 - x$ and $g(x) = x - 6$, find*

**21.** $(f + g)(3)$

**22.** $(f/g)(-1)$

**23.** $f(a)$

**24. Paper Use** The following graph shows paper use in 1995 and projected paper use from 1995 through 2015.

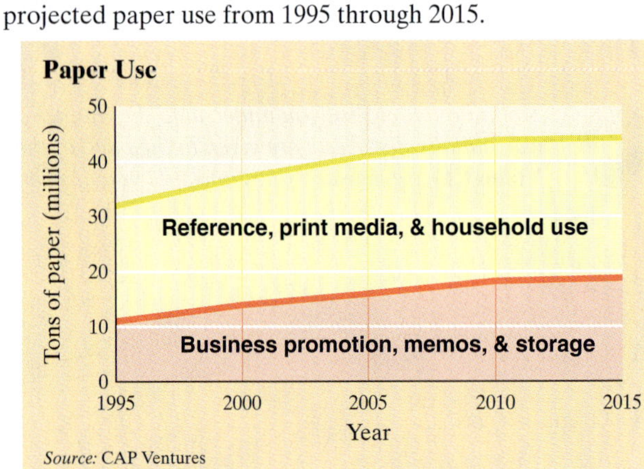


**a)** Estimate the total number of tons of paper to be used in 2010.

**b)** Estimate the number of tons of paper to be used by businesses in 2010.

**c)** Estimate the number of tons of paper to be used for reference, print media, and household use in 2010.

**25.** Graph $y < 3x - 2$.

## Cumulative Review Test

*Take the following test and check your answers with those given in the back of the book. Review any questions that you answered incorrectly. The section where the material was covered is indicated after the answer.*

**1.** For $A = \{1, 3, 5, 7, 9\}$ and $B = \{2, 3, 5, 7, 11, 14\}$, determine

**a)** $A \cap B$.

**b)** $A \cup B$.

**2.** Consider the set $\{-6, -4, \frac{1}{3}, 0, \sqrt{3}, 4.67, \frac{37}{2}, -\sqrt{5}\}$ List the elements of the set that are

**a)** natural numbers.

**b)** real numbers.

**3.** Evaluate $10 - \{3[6 - 4(6^2 \div 4)]\}$.

*Simplify.*

**4.** $\left(\frac{5x^2}{y^{-3}}\right)^2$

**5.** $\left(\frac{3x^4y^{-2}}{6xy^3}\right)^3$

**6.** **Natural Gas Consumption** The total consumption of natural gas in 2003 was 21.8 trillion cubic feet ($2.18 \times 10^{13}$). The following pie chart shows the breakdown of consumption by sector.

**Natural Gas Consumption by Sector**
($2.18 \times 10^{13}$ cubic feet)

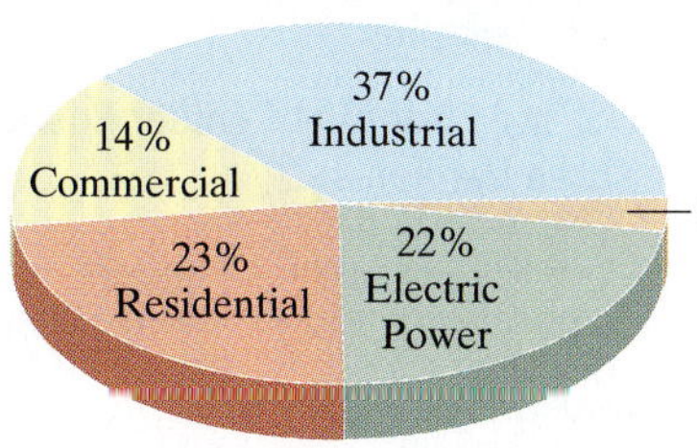


(Total is 99% due to rounding.)

*Source:* Energy Information Administration

Answer the following questions using scientific notation.

**a)** What was the amount of natural gas consumption by the commercial sector in 2003?

**b)** How much more natural gas was consumed by the industrial sector than by the transportation sector in 2003?

**c)** If the consumption of natural gas is expected to increase by a total of 10% from 2003 to 2006, what will be the consumption of natural gas in 2006?

*In Exercises 7 and 8, solve the equations.*

**7.** $2(x + 4) - 5 = -3[x - (2x + 1)]$

**8.** $\frac{4}{5} - \frac{x}{3} = 10$

**9.** Simplify $7x - \{4 - [2(x - 4)] - 5\}$.

**10.** Solve $A = \frac{1}{2}h(b_1 + b_2)$ for $b_1$.

**11.** **Hydrogen Peroxide Solutions** How many gallons of 15% hydrogen peroxide solution must be mixed with 10 gallons of 4% hydrogen peroxide solution to get a 10% hydrogen peroxide solution?

**12.** Solve the inequality $4(x - 4) < 8(2x + 3)$.

**13.** Solve the inequality $-1 < 3x - 7 < 11$

**14.** Determine the solution set of $|3x + 5| = |2x - 10|$.

**15.** Determine the solution set of $|2x - 1| \leq 3$.

**16.** Graph $y = -\frac{3}{2}x - 4$.

**17. a)** Determine whether the following graph represents a function.

**b)** Find the domain and range of the graph.

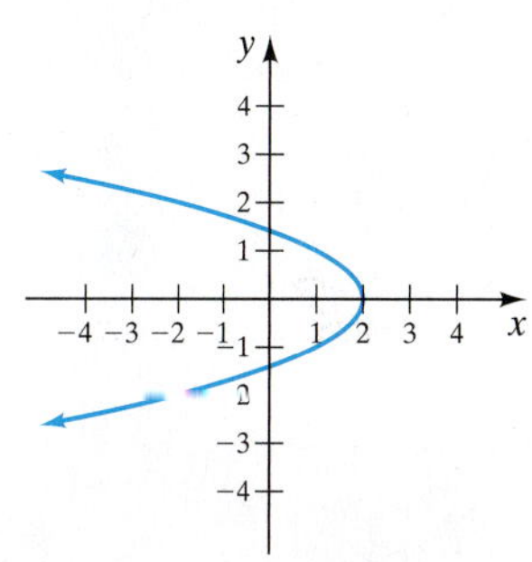


**18.** Determine the slope of the line through the points $(-5, 3)$ and $(4, -1)$.

**19.** Determine whether the graphs of the two given equations are parallel, perpendicular, or neither.

$$2x - 5y = 8$$
$$5x - 2y = 12$$

**20.** If $f(x) = x^2 + 3x - 2$ and $g(x) = 4x - 9$, find $(f + g)(x)$.

# Systems of Equations and Inequalities

# 4

## GOALS OF THIS CHAPTER

In this chapter we solve systems of linear equations using the following methods: by graphing, by substitution, by the addition method, by using matrices, and by using determinants and Cramer's rule. We also solve systems of linear *inequalities*. Throughout the chapter, especially in Section 4.3, there are many real-life applications. The chapter covers essential topics used by businesses to consider the relationships among variables involved in the day-to-day operations of a business.


4.1 Solving Systems of Linear Equations in Two Variables

4.2 Solving Systems of Linear Equations in Three Variables

4.3 Systems of Linear Equations: Applications and Problem Solving

Mid-Chapter Test: Sections 4.1–4.3

4.4 Solving Systems of Equations Using Matrices

4.5 Solving Systems of Equations Using Determinants and Cramer's Rule

4.6 Solving Systems of Linear Inequalities

Chapter 4 Summary

Chapter 4 Review Exercises

Chapter 4 Practice Test

Cumulative Review Test


SYSTEMS OF EQUATIONS ARE frequently used to solve real-life problems. For example, in Example 6 on page 255 we use a system of equations to determine how much of two given solutions a chemist must mix to get a third solution with the desired chemical composition.

# 4.1 Solving Systems of Linear Equations in Two Variables

1 Solve systems of linear equations graphically.

2 Solve systems of linear equations by substitution.

3 Solve systems of linear equations using the addition method.

It is often necessary to find a common solution to two or more linear equations. We refer to these equations as a **system of linear equations**. For example,

$$\left.\begin{aligned}(1)\ y &= x + 5\\(2)\ y &= 2x + 4\end{aligned}\right\}\quad \text{System of linear equations}$$

A **solution to a system of equations** is an ordered pair or pairs that satisfy all equations in the system. The only solution to the system above is (1, 6).

| Check in Equation (1) | Check in Equation (2) |
|---|---|
| $(1, 6)$ | $(1, 6)$ |
| $y = x + 5$ | $y = 2x + 4$ |
| $6 \stackrel{?}{=} 1 + 5$ | $6 \stackrel{?}{=} 2(1) + 4$ |
| $6 = 6$ *True* | $6 = 6$ *True* |

The ordered pair (1, 6) satisfies *both* equations and is the solution to the system of equations.

A system of equations may consist of more than two equations. If a system consists of three equations in three variables, such as $x$, $y$, and $z$, the solution will be an **ordered triple** of the form $(x, y, z)$. If the ordered triple $(x, y, z)$ is a solution to the system, it must satisfy all three equations in the system. Systems with three variables are discussed in Section 4.2. Systems of equations may have more than three variables, but we will not discuss them in this book.

## 1 Solve Systems of Linear Equations Graphically

To solve a system of linear equations in two variables graphically, graph both equations in the system on the same axes. The solution to the system will be the ordered pair (or pairs) common to both lines, or the point of intersection of both lines in the system.

When two lines are graphed, three situations are possible, as illustrated in **Figure 4.1** below. In **Figure 4.1a**, lines 1 and 2 intersect at exactly one point. This system of equations has *exactly one solution*. This is an example of a *consistent* system of equations. A **consistent system of equations** is a system of equations that has a solution.

Lines 1 and 2 of **Figure 4.1b** are different but parallel lines. The lines do not intersect, and this system of equations has *no solution*. This is an example of an *inconsistent* system of equations. An **inconsistent system of equations** is a system of equations that has no solution.

In **Figure 4.1c**, lines 1 and 2 are actually the same line. In this case, every point on the line satisfies both equations and is a solution to the system of equations. This system has *an infinite number of solutions*. This is an example of a *dependent* system of equations. In a dependent system of linear equations both equations represent the same line. A **dependent system of equations** is a system of equations that has an infinite number of solutions. *Note that a dependent system is also a consistent system since it has solutions.*

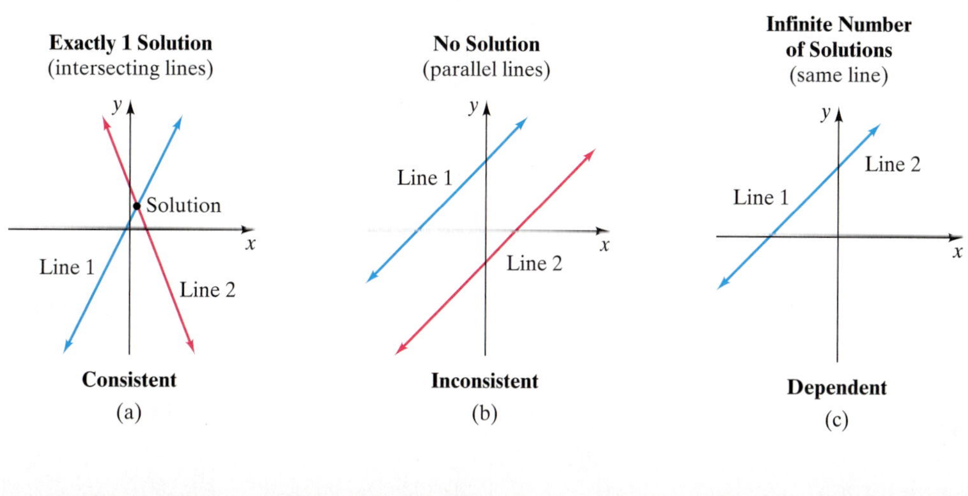


FIGURE 4.1

We can determine if a system of linear equations is consistent, inconsistent, or dependent by writing each equation in slope-intercept form and comparing the slopes and $y$-intercepts. If the slopes of the lines are different (**Fig. 4.1a**), the system is consistent. If the slopes are the same but the $y$-intercepts are different (**Fig. 4.1b**), the system is inconsistent, and if both the slopes and the $y$-intercepts are the same (**Fig. 4.1c**), the system is dependent.

**EXAMPLE 1 ▸** Without graphing the equations, determine whether the following system of equations is consistent, inconsistent, or dependent.

$$3x - 4y = 8$$
$$-9x + 12y = -24$$

*Solution* Write each equation in slope-intercept form.

$$3x - 4y = 8 \qquad\qquad -9x + 12y = -24$$
$$-4y = -3x + 8 \qquad\qquad 12y = 9x - 24$$
$$y = \frac{3}{4}x - 2 \qquad\qquad y = \frac{3}{4}x - 2$$

Since the equations have the same slope, $\frac{3}{4}$, and the same $y$-intercept, $(0, -2)$, the equations represent the same line. Therefore, the system is dependent, and there are an infinite number of solutions.

▸ **Now Try Exercise 19**

**EXAMPLE 2 ▸** Solve the following system of equations graphically.

$$y = x + 2$$
$$y = -x + 4$$

*Solution* Graph both equations on the same axes (**Fig. 4.2**). The solution is the point of intersection of the two lines, $(1, 3)$.

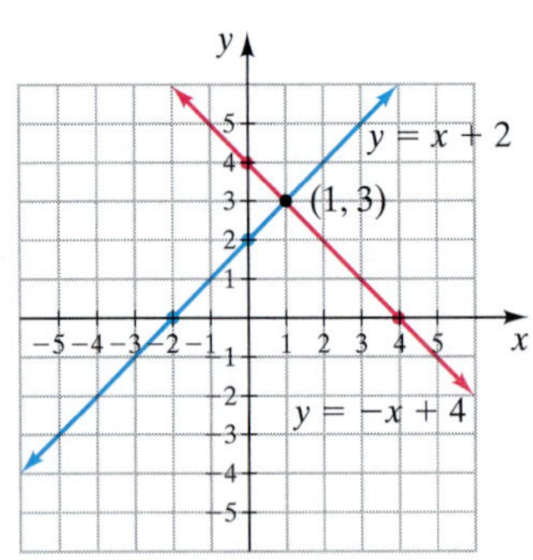


FIGURE 4.2

▸ **Now Try Exercise 25**

The system of equations in Example 2 could be represented in function notation as

$$f(x) = x + 2$$
$$g(x) = -x + 4$$

### USING YOUR GRAPHING CALCULATOR

In the Using Your Graphing Calculator box on page 181, Section 3.3, we discussed using a graphing calculator to find the intersection of two graphs. Let's use the information provided in that Graphing Calculator box to solve a system of equations.

**EXAMPLE** Use your graphing calculator to solve the system of equations. Round the solution to the nearest hundredth.

$$-2.6x - 5.2y = -15.3$$
$$-8.6x + 3.7y = -12.5$$

*Solution* First solve each equation for $y$.

$$-2.6x - 5.2y = -15.3$$
$$-2.6x = 5.2y - 15.3$$
$$-2.6x + 15.3 = 5.2y$$
$$\frac{-2.6x + 15.3}{5.2} = y$$

$$-8.6x + 3.7y = -12.5$$
$$3.7y = 8.6x - 12.5$$
$$y = \frac{8.6x - 12.5}{3.7}$$

Now let $Y_1 = \frac{-2.6x + 15.3}{5.2}$ and $Y_2 = \frac{8.6x - 12.5}{3.7}$. The graphs $Y_1$ and $Y_2$ are illustrated in **Figure 4.3**.

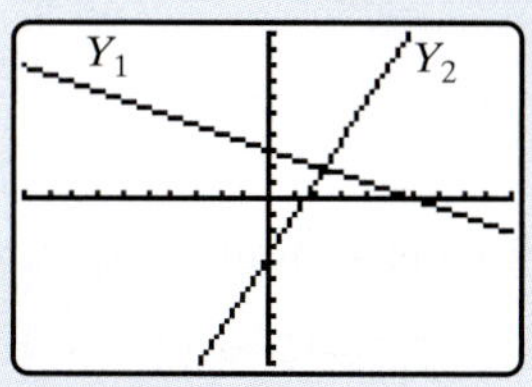


FIGURE 4.3

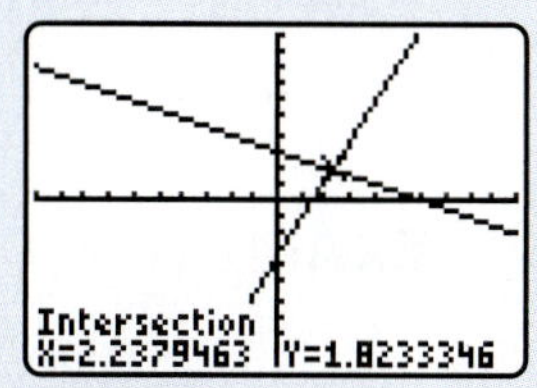


FIGURE 4.4

**Figure 4.4** shows that the intersection of the two graphs occurs at (2.24, 1.82), rounded to the nearest hundredth.

**EXERCISES**

*Use your graphing calculator to determine the solution to each system. Round your answers to the nearest hundredth.*

**1.** $2x + 3y = 8$
$-3x + 4y = -5$

**2.** $5x - 6y = 9$
$-3x + 5y = 8$

**3.** $3.4x - 5.6y = 10.2$
$5.8x + 1.4y = -33.6$

**4.** $-2.3x + 7.9y = 88.3$
$-5.3x - 2.7y = -16.5$

## 2 Solve Systems of Linear Equations by Substitution

Often, an exact solution to a system of equations may be difficult to find on a graph. A graphing calculator may not give an exact answer either. When an exact answer is necessary, the system should be solved algebraically, by either the substitution method or the addition (elimination) method. We discuss **substitution** first.

### To Solve a Linear System of Equations by Substitution

1. Solve for a variable in either equation. (If possible, solve for a variable with a numerical coefficient of 1 to avoid working with fractions.)
2. Substitute the expression found for the variable in step 1 *into the other equation*. This will result in an equation containing only one variable.
3. Solve the equation obtained in step 2 to find the value of this variable.
4. Substitute the value found in step 3 into the equation from step 1. Solve the equation to find the remaining variable.
5. Check your solution in *all* equations in the system.

**EXAMPLE 3** ▶ Solve the following system of equations by substitution.

$$y = 3x - 5$$
$$y = -4x + 9$$

*Solution* Since both equations are already solved for $y$, we can substitute $3x - 5$ for $y$ in the second equation and then solve for the remaining variable, $x$.

$$3x - 5 = -4x + 9$$
$$7x - 5 = 9$$
$$7x = 14$$
$$x = 2$$

Now find $y$ by substituting $x = 2$ into either of the original equations. We will use the first equation.

$$y = 3x - 5$$
$$y = 3(2) - 5$$
$$y = 6 - 5 = 1$$

A check will show that the solution to the system of equations is (2, 1).

▶ **Now Try Exercise 39**

**EXAMPLE 4** ▶ Solve the following system of equations by substitution.

$$2x + y = 11$$
$$x + 3y = 18$$

*Solution* Begin by solving for one of the variables in either of the equations. You may solve for either of the variables; however, if you solve for a variable with a numerical coefficient of 1, you may avoid working with fractions. In this system the $y$-term in $2x + y = 11$ and the $x$-term in $x + 3y = 18$ both have numerical coefficient 1.

Let's solve for $y$ in $2x + y = 11$.

$$2x + y = 11$$
$$y = -2x + 11$$

Next, substitute $-2x + 11$ for $y$ in the *other equation*, $x + 3y = 18$, and solve for the remaining variable, $x$.

$$x + 3y = 18$$
$$x + 3(-2x + 11) = 18 \quad \text{Substitute } -2x + 11 \text{ for } y.$$
$$x - 6x + 33 = 18$$
$$-5x + 33 = 18$$
$$-5x = -15$$
$$x = 3$$

Finally, substitute $x = 3$ into the equation $y = -2x + 11$ and solve for $y$.

$$y = -2x + 11$$
$$y = -2(3) + 11 = 5$$

The solution is the ordered pair (3, 5). Check this solution.

▶ **Now Try Exercise 41**

If, when solving a system of equations by either substitution or the addition method, you arrive at an equation that is false, such as $5 = 6$ or $0 = 3$, the system is inconsistent and has no solution. If you obtain an equation that is always true, such as $6 = 6$ or $0 = 0$, the system is dependent and has an infinite number of solutions.

### Helpful Hint

Students sometimes successfully solve for one of the variables and forget to solve for the other. Remember that a solution must contain a numerical value for each variable in the system.

## 3 Solve Systems of Linear Equations Using the Addition Method

A third and often the easiest method of solving a system of equations is the **addition** (or elimination) **method**. The object of this process is to obtain two equations whose sum will be an equation containing only one variable. Keep in mind that your immediate goal is to obtain one equation containing only one unknown.

**EXAMPLE 5** ▸ Solve the following system of equations using the addition method.

$$2x + 5y = 3$$
$$3x - 5y = 17$$

*Solution* Note that one equation contains $+5y$ and the other contains $-5y$. By adding the equations, we can eliminate the variable $y$ and obtain one equation containing only one variable, $x$.

$$\begin{aligned} 2x + 5y &= 3 \\ 3x - 5y &= 17 \\ \hline 5x \quad\quad &= 20 \end{aligned}$$

Now solve for the remaining variable, $x$.

$$\frac{5x}{5} = \frac{20}{5}$$
$$x = 4$$

Finally, solve for $y$ by substituting 4 for $x$ in either of the original equations.

$$\begin{aligned} 2x + 5y &= 3 \\ 2(4) + 5y &= 3 \\ 8 + 5y &= 3 \\ 5y &= -5 \\ y &= -1 \end{aligned}$$

A check will show that the solution is $(4, -1)$.

▸ Now Try Exercise 53

### To Solve a Linear System of Equations Using the Addition (or Elimination) Method

1. If necessary, rewrite each equation in standard form, that is, with the terms containing variables on the left side of the equal sign and the constant on the right side of the equal sign.
2. If necessary, multiply one or both equations by a constant(s) so that when the equations are added, the sum will contain only one variable.
3. Add the respective sides of the equations. This will result in a single equation containing only one variable.
4. Solve for the variable in the equation obtained in step 3.
5. Substitute the value found in step 4 into either of the original equations. Solve that equation to find the value of the remaining variable.
6. Check your solution in all equations in the system.

In step 2 of the procedure, we indicate that it may be necessary to multiply both sides of an equation by a constant. To help avoid confusion, we will number our equations using parentheses, such as (*eq.* 1) or (*eq.* 2).

In Example 6 we will solve the same system we solved in Example 4, but this time we will use the addition method.

**EXAMPLE 6** ▸ Solve the following system of equations using the addition method.

$$2x + y = 11 \quad (eq.\ 1)$$
$$x + 3y = 18 \quad (eq.\ 2)$$

*Solution* The object of the addition process is to obtain two equations whose sum will be an equation containing only one variable. To eliminate the variable $x$, we multiply (eq. 2) by $-2$ and then add the two equations.

$$2x + y = 11 \quad (eq.\ 1)$$
$$-2x - 6y = -36 \quad (eq.\ 2) \quad \text{Multiplied by } -2$$

Now add.

$$\begin{aligned} 2x + y &= 11 \\ -2x - 6y &= -36 \\ \hline -5y &= -25 \\ y &= 5 \end{aligned}$$

Now solve for $x$ by substituting 5 for $y$ in either of the original equations.

$$\begin{aligned} 2x + y &= 11 \\ 2x + 5 &= 11 \quad \text{Substitute 5 for } y. \\ 2x &= 6 \\ x &= 3 \end{aligned}$$

The solution is (3, 5). Note that we could have first eliminated the variable $y$ by multiplying (eq. 1) by $-3$ and then adding.

▸ **Now Try Exercise 61**

Sometimes both equations must be multiplied by different numbers in order for one of the variables to be eliminated. This procedure is illustrated in Example 7.

**EXAMPLE 7** ▸ Solve the following system of equations using the addition method.

$$4x + 3y = 7 \quad (eq.\ 1)$$
$$3x - 7y = -3 \quad (eq.\ 2)$$

*Solution* We can eliminate the variable $x$ by multiplying (eq. 1) by $-3$ and (eq. 2) by 4.

$$\begin{aligned} -12x - 9y &= -21 \quad (eq.\ 1) \quad \text{Multiplied by } -3 \\ 12x - 28y &= -12 \quad (eq.\ 2) \quad \text{Multiplied by } 4 \\ \hline -37y &= -33 \quad \text{Sum of equations} \\ y &= \frac{33}{37} \end{aligned}$$

We can now find $x$ by substituting $\frac{33}{37}$ for $y$ in one of the original equations and solving for $x$. If you try this, you will see that, although it can be done, it gets messy. An easier method to solve for $x$ is to go back to the original equations and eliminate the variable $y$.

$$\begin{aligned} 28x + 21y &= 49 \quad (eq.\ 1) \quad \text{Multiplied by } 7 \\ 9x - 21y &= -9 \quad (eq.\ 2) \quad \text{Multiplied by } 3 \\ \hline 37x &= 40 \quad \text{Sum of equations} \\ x &= \frac{40}{37} \end{aligned}$$

The solution is $\left(\frac{40}{37}, \frac{33}{37}\right)$.

▸ Now Try Exercise 67

In Example 7, the same solution could be obtained by multiplying (*eq.* 1) by 3 and (*eq.* 2) by −4 and then adding. Try it now and see.

**EXAMPLE 8** ▸ Solve the following system of equations using the addition method.

$$2x + y = 11 \quad (eq.\ 1)$$
$$\frac{1}{18}x + \frac{1}{6}y = 1 \quad (eq.\ 2)$$

*Solution* When a system of equations contains fractions, it is generally best to *clear*, or remove, fractions. In (*eq.* 2), if we multiply both sides of the equation by the least common denominator, 18, we obtain

$$18\left(\frac{1}{18}x + \frac{1}{6}y\right) = 18(1)$$
$$18\left(\frac{1}{18}x\right) + 18\left(\frac{1}{6}y\right) = 18(1)$$
$$x + 3y = 18 \quad (eq.\ 3)$$

This system now simplifies to

$$2x + y = 11 \quad (eq.\ 1)$$
$$x + 3y = 18 \quad (eq.\ 3)$$

This is the same system of equations we solved in Example 6. Thus the solution to this system is (3, 5), the same as obtained in Example 6.

▸ Now Try Exercise 51

**EXAMPLE 9** ▸ Solve the following system of equations using the addition method.

$$0.2x + 0.1y = 1.1 \quad (eq.\ 1)$$
$$x + 3y = 18 \quad (eq.\ 2)$$

*Solution* When a system of equations contains decimal numbers, it is generally best to *clear*, or remove, the decimal numbers. In (*eq.* 1), if we multiply both sides of the equation by 10 we obtain

$$10(0.2x) + 10(0.1y) = 10(1.1)$$
$$2x + y = 11 \quad (eq.\ 3)$$

The system of equations is now simplified to

$$2x + y = 11 \quad (eq.\ 3)$$
$$x + 3y = 18 \quad (eq.\ 2)$$

This is the same system of equations we solved in Example 6. Thus the solution to this system is (3, 5), the same as obtained in Example 6.

▸ Now Try Exercise 69

**EXAMPLE 10** ▸ Solve the following system of equations using the addition method.

$$x - 3y = 4 \quad (eq.\ 1)$$
$$-2x + 6y = 1 \quad (eq.\ 2)$$

*Solution* We begin by multiplying (*eq.* 1) by 2.

$$\begin{array}{rl} 2x - 6y = 8 & (eq.\ 1) \quad \text{Multiplied by 2} \\ -2x + 6y = 1 & (eq.\ 2) \\ \hline 0 = 9 & \text{False} \end{array}$$

*Since 0 = 9 is a false statement, this system has no solution. The system is inconsistent and the graphs of these equations are parallel lines.*

▸ **Now Try Exercise 59**

**EXAMPLE 11** ▸ Solve the following system of equations using the addition method.

$$x - \frac{1}{2}y = 2$$
$$y = 2x - 4$$

*Solution* First, align the $x$- and $y$-terms in the second equation on the left side of the equation.

$$\begin{array}{rl} x - \frac{1}{2}y = 2 & (eq.\ 1) \\ -2x + y = -4 & (eq.\ 2) \end{array}$$

Now proceed as in previous examples.

$$\begin{array}{rl} 2x - y = 4 & (eq.\ 1) \quad \text{Multiplied by 2} \\ -2x + y = -4 & (eq.\ 2) \\ \hline 0 = 0 & \text{True} \end{array}$$

*Since 0 = 0 is a true statement, the system is dependent and has an infinite number of solutions. Both equations represent the same line.* Notice that if you multiply both sides of (*eq.* 1) by 2 you obtain (*eq.* 2).

▸ **Now Try Exercise 63**

We have illustrated three methods that can be used to solve a system of linear equations: graphing, substitution, and the addition method. When you are given a system of equations, which method should you use to solve the system? When you need an exact solution, graphing may not be the best to use. Of the two algebraic methods, the addition method may be the easiest to use if there are no numerical coefficients of 1 in the system. If one or more of the variables have a coefficient of 1, you may wish to use either method. We will present a fourth method, using matrices, in Section 4.4, and a fifth method, using determinants, in Section 4.5.

## EXERCISE SET 4.1

MathXL® MyMathLab

### Concept/Writing Exercises

1. What is a solution to a system of linear equations?
2. What is the solution to a system of linear equations in three variables called?
3. What is a dependent system of equations?
4. What is an inconsistent system of equations?
5. What is a consistent system of equations?
6. Explain how to determine the solution of a linear system graphically.
7. Explain how you can determine, without graphing or solving, whether a system of two linear equations is consistent, inconsistent or dependent.
8. When solving a system of linear equations using the addition (or elimination) method, what is the object of the process?
9. When solving a linear system by addition, how can you tell if the system is dependent?
10. When solving a linear system by addition, how can you tell if the system is inconsistent?

## Practice the Skills

*Determine which, if any, of the given ordered pairs or ordered triples satisfy the system of linear equations.*

**11.** $y = 2x + 4$
$y = 2x - 1$
**a)** $(0, 4)$
**b)** $(3, 10)$

**12.** $3x - 5y = 12$
$y = \frac{3}{4}x - 3$
**a)** $(4, 0)$ **b)** $(7, 2)$

**13.** $x + y = 25$
$0.25x + 0.45y = 7.50$
**a)** $(5, 20)$
**b)** $(18.75, 6.25)$

**14.** $y = \frac{x}{3} - \frac{7}{3}$
$5x - 35 = 15y$
**a)** $(1, -2)$
**b)** $(7, 0)$

**15.** $x + 2y - z = -5$
$2x - y + 2z = 8$
$3x + 3y + 4z = 5$
**a)** $(3, 1, -2)$
**b)** $(1, -2, 2)$

**16.** $4x + y - 3z = 1$
$2x - 2y + 6z = 11$
$-6x + 3y + 12z = -4$
**a)** $(2, -1, -2)$
**b)** $\left(\frac{1}{2}, 2, 1\right)$

*Write each equation in slope-intercept form. Without graphing the equations, state whether the system of equations is consistent, inconsistent, or dependent. Also indicate whether the system has exactly one solution, no solution, or an infinite number of solutions.*

**17.** $-7x + 3y = 1$
$3y + 12 = -6x$

**18.** $x - \frac{1}{2}y = 4$
$2x - y = 7$

**19.** $\frac{x}{3} + \frac{y}{4} = 1$
$4x + 3y = 12$

**20.** $\frac{x}{3} + \frac{y}{4} = 1$
$2x - 3y = 12$

**21.** $3x - 3y = 9$
$2x - 2y = -4$

**22.** $2x = 3y + 4$
$6x - 9y = 12$

**23.** $y = \frac{3}{2}x + \frac{1}{2}$
$3x - 2y = -\frac{5}{2}$

**24.** $x - y = 3$
$\frac{1}{4}x - 2y = -6$

*Determine the solution to each system of equations graphically. If the system is inconsistent or dependent, so state.*

**25.** $y = x + 5$
$y = -x + 3$

**26.** $y = 2x + 8$
$y = -3x - 12$

**27.** $y = 4x - 1$
$3y = 12x + 9$

**28.** $x + y = 1$
$3x - y = -5$

**29.** $2x + 3y = 6$
$4x = -6y + 12$

**30.** $y = -2x - 1$
$x + 2y = 4$

**31.** $5x + 3y = 13$
$x = 2$

**32.** $2x - 5y = 10$
$y = \frac{2}{5}x - 2$

**33.** $y = -5x + 5$
$y = 2x - 2$

**34.** $4x - y = 9$
$x - 3y = 16$

**35.** $x - \frac{1}{2}y = -2$
$2y = 4x - 6$

**36.** $y = -\frac{1}{3}x - 1$
$3y = 4x - 18$

*Find the solution to each system of equations by substitution.*

**37.** $x + 3y = -1$
$y = x + 1$

**38.** $3x - 2y = -7$
$y = 2x + 6$

**39.** $x = 2y + 3$
$y = x$

**40.** $y = 3x - 16$
$x = y$

**41.** $a + 3b = 5$
$2a - b = 3$

**42.** $m + 2n = 4$
$m + \frac{1}{2}n = 4$

**43.** $5x + 6y = 6.7$
$3x - 2y = 0.1$

**44.** $x = 0.5y + 1.7$
$10x - y = 1$

**45.** $a - \frac{1}{2}b = 2$
$b = 2a - 4$

**46.** $x + 3y = -2$
$y = -\frac{1}{3}x - \frac{2}{3}$

**47.** $5x - 2y = -7$
$y = \frac{5}{2}x + 1$

**48.** $y = \frac{2}{3}x - 1$
$2x - 3y = 5$

**49.** $5x - 4y = -7$
$x - \frac{3}{5}y = -2$

**50.** $6s + 3t = 4$
$s = \frac{1}{2}t$

**51.** $\frac{1}{2}x - \frac{1}{3}y = 2$
$\frac{1}{4}x + \frac{2}{3}y = 6$

**52.** $\frac{1}{2}x + \frac{1}{3}y = 3$
$\frac{1}{5}x + \frac{1}{8}y = 1$

*Solve each system of equations using the addition method.*

53. $x + y = 9$
$x - y = -3$

54. $-x + y = 4$
$x - 2y = 6$

55. $4x - 3y = 1$
$5x + 3y = -10$

56. $2x - 5y = 6$
$-4x + 10y = -1$

57. $10m - 2n = 6$
$-5m + n = -8$

58. $4r - 3s = 2$
$2r + s = 6$

59. $2c - 5d = 1$
$-4c + 10d = 6$

60. $2v - 3w = 8$
$3v - 6w = 1$

61. $7p - 3q = 4$
$2p + 5q = 7$

62. $5s - 3t = 7$
$t = s + 1$

63. $5a - 10b = 15$
$a = 2b + 3$

64. $2x - 7y = 3$
$-5x + 3y = 7$

65. $2x - y = 8$
$3x + y = 6$

66. $5x + 4y = 6$
$2x = -5y - 1$

67. $3x - 4y = 5$
$2x = 5y - 3$

68. $4x + 5y = 3$
$2x - 3y = 4$

69. $0.2x - 0.5y = -0.4$
$-0.3x + 0.4y = -0.1$

70. $0.15x - 0.40y = 0.65$
$0.60x + 0.25y = -1.1$

71. $2.1m - 0.6n = 8.4$
$-1.5m - 0.3n = -6.0$

72. $-0.25x + 0.10y = 1.05$
$-0.40x - 0.625y = -0.675$

73. $\frac{1}{2}x - \frac{1}{3}y = 1$
$\frac{1}{4}x - \frac{1}{9}y = \frac{2}{3}$

74. $\frac{1}{5}x + \frac{1}{2}y = 4$
$\frac{2}{3}x - y = \frac{8}{3}$

75. $\frac{1}{3}x = 4 - \frac{1}{4}y$
$3x = 4y$

76. $\frac{2}{3}x - 4 = \frac{1}{2}y$
$x - 3y = \frac{1}{3}$

## Problem Solving

77. **a)** Write a system of equations that would be most easily solved by substitution.
**b)** Explain why substitution would be the easiest method to use.
**c)** Solve the system by substitution.

78. **a)** Write a system of equations that would be most easily solved using the addition method.
**b)** Explain why the addition method would be the easiest method to use.
**c)** Solve the system using the addition method.

79. **Salaries** In January 2006, Mary Jones started a new job with an annual salary of \$38,000. Her boss agreed to increase her salary by \$1000 each January in the years to come. Her salary is determined by the equation $y = 1000t + 38{,}000$, where $t$ is the number of years since 2006. (See red line in graph.) Also, in January 2006, Wynn Nguyen started a new job with an annual salary of \$45,500. Her boss agreed to increase her salary by \$500 each January in the years to come. Her salary is determined by the equation $y = 45{,}500 + 500t$, where $t$ is the number of years since 2006. (See blue line in graph.) Solve the system of equations to determine the year both salaries will be the same. What will be the salary in that year?

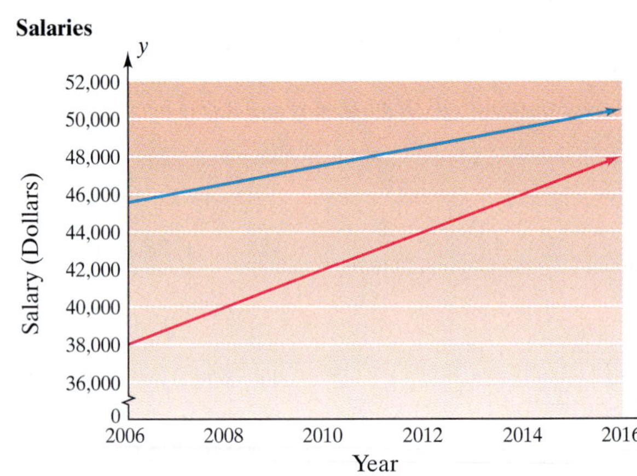


80. **Martha Stewart Revenue** The graph on the top of the next page shows the revenue, in millions of dollars, from publishing sales and Internet sales in the Martha Stewart Living Omnimedia Company for the years from 2002 to 2004.

**Martha Stewart Sales**

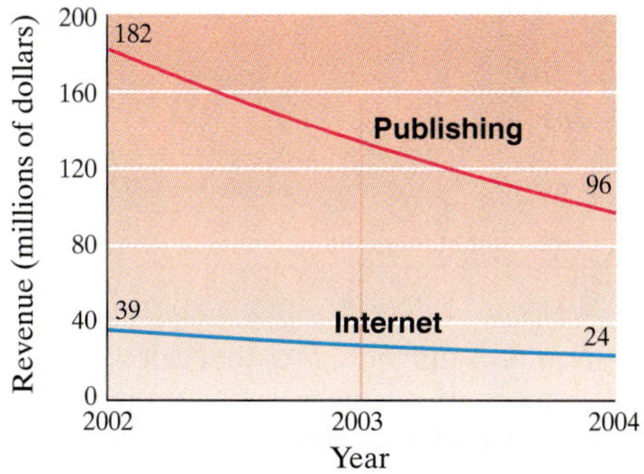


*Source:* CSI, Martha Stewart Living Omnimedia Company, *USA Today (3/4/2005)*

The graph shows that publishing sales and internet sales are decreasing approximately linearly. The revenue, in millions of dollars, for publishing sales (red line) can be approximated by the function $p(t) = -43t + 182$ and for Internet sales (blue line) can be approximated by the function $I(t) = -7.5t + 39$, where $t$ is the number of years since 2002. Assuming this trend continues, solve the system of equations to determine the year when the revenue will be the same for publishing and Internet sales. What will be the revenue in that year?

**81.** Explain how you can tell by observation that the following system is dependent.

$$2x + 3y = 1$$
$$4x + 6y = 2$$

**82.** Explain how you can tell by observation that the following system is inconsistent.

$$-x + 3y = 5$$
$$2x - 6y = -13$$

**83.** The solutions of a system of linear equations include $(-4, 3)$ and $(-6, 11)$.

**a)** How many other solutions does the system have? Explain.

**b)** Determine the slope of the line containing $(-4, 3)$ and $(-6, 11)$. Determine the equation of the line containing these points. Then determine the $y$-intercept.

**c)** Does this line represent a function?

**84.** The solutions of a system of linear equations include $(-5, 1)$ and $(-5, -4)$.

**a)** How many other solutions does the system have? Explain.

**b)** Determine the slope of the line containing $(-5, 1)$ and $(-5, -4)$. Determine the equation of the line containing these points. Does this graph have a $y$-intercept? Explain.

**c)** Does this line represent a function?

**85.** Construct a system of equations that is dependent. Explain how you created your system.

**86.** Construct a system of equations that is inconsistent. Explain how you created your system.

*In Exercises 87 and 88,* **a)** *create a system of linear equations that has the solution indicated and* **b)** *explain how you determined your solution.*

**87.** $(2, 5)$

**88.** $(-3, 4)$

**89.** The solution to the following system of equations is $(2, -3)$. Find $A$ and $B$.

$$Ax + 4y = -8$$
$$3x - By = 21$$

**90.** The solution to the following system of equations is $(-5, 3)$. Find $A$ and $B$.

$$3x + Ay = -3$$
$$Bx - 2y = -16$$

**91.** If $(2, 6)$ and $(-1, -6)$ are two solutions of $f(x) = mx + b$, find $m$ and $b$.

**92.** If $(3, -5)$ and $(-2, 10)$ are two solutions of $f(x) = mx + b$, find $m$ and $b$.

**93.** Suppose you graph a system of two linear equations on your graphing calculator, but only one line shows in the window. What are two possible explanations for this?

**94.** Suppose you graph a system of linear equations on your graphing calculator and get the following.

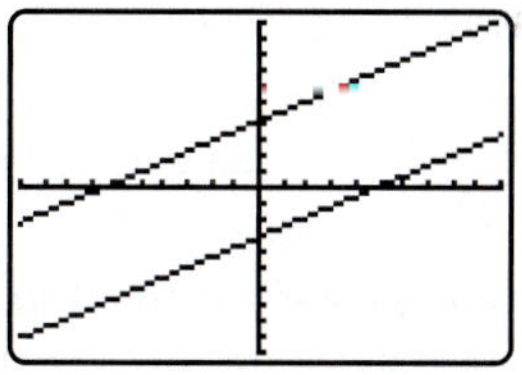

**a)** By observing the window, can you be sure that this system is inconsistent? Explain.

**b)** What can you do on your graphing calculator to determine whether the system is inconsistent?

## Challenge Problems

*Solve each system of equations.*

**95.** $\dfrac{x+2}{2} - \dfrac{y+4}{3} = 4$

$\dfrac{x+y}{2} = \dfrac{1}{2} + \dfrac{x-y}{3}$

**96.** $\dfrac{5x}{2} + 3y = \dfrac{9}{2} + y$

$\dfrac{1}{4}x - \dfrac{1}{2}y = 6x + 12$

*Solve each system of equations. (Hint:* $\dfrac{3}{a} = 3 \cdot \dfrac{1}{a} = 3x$ *if* $x = \dfrac{1}{a}$*.)*

**97.** $\dfrac{3}{a} + \dfrac{4}{b} = -1$

$\dfrac{1}{a} + \dfrac{6}{b} = 2$

**98.** $\dfrac{6}{x} + \dfrac{1}{y} = -1$

$\dfrac{3}{x} - \dfrac{2}{y} = -3$

*By solving for x and y, determine the solution to each system of equations. In all equations, $a \neq 0$ and $b \neq 0$. The solution will contain either a, b, or both letters.*

**99.** $4ax + 3y = 19$
$-ax + y = 4$

**100.** $ax = 2 - by$
$-ax + 2by - 1 = 0$

## Group Activity

*Discuss and answer Exercises 101 and 102 as a group.*

**101. Trends** The graph below appeared in both the *Journal of the American Medical Association* and *Scientific American*. The red line indicates the long-term trend of firearm deaths, and the purple line indicates the long-term trend in motor vehicle deaths. The black lines indicate the short-term trends in deaths from firearms and motor vehicles.

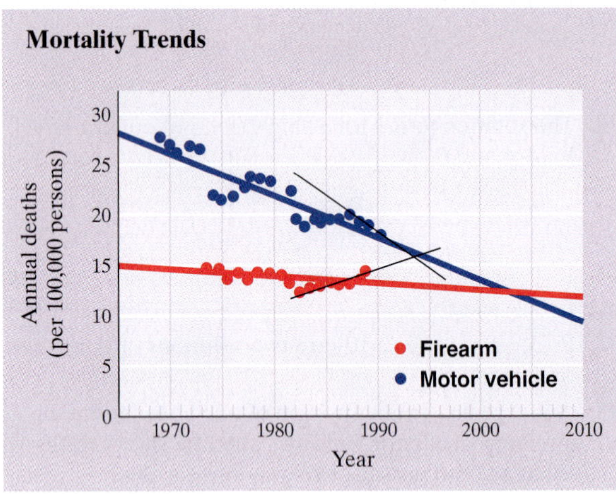


**a)** Discuss the long-term trend in motor vehicle deaths.

**b)** Discuss the long-term trend in firearm deaths.

**c)** Discuss the short-term trend in motor vehicle deaths compared with the long-term trend in motor vehicle deaths.

**d)** Discuss the short-term trend in firearm deaths compared with the long-term trend in firearm deaths.

**e)** Using the long-term trends, estimate when the number of deaths from firearms will equal the number of deaths from motor vehicles.

**f)** Repeat part **e)** using the short-term trends.

**g)** Determine a function, $M(t)$, that can be used to estimate the number of deaths per 100,000 people (long term) from motor vehicles from 1965 through 2010.

**h)** Determine a function, $F(t)$, that can be used to estimate the number of deaths per 100,000 people (long term) from firearms from 1965 through 2010.

**i)** Solve the system of equations formed in parts **g)** and **h)**. Does the solution agree with the solution in part **e)**? If not, explain why.

**102. Sales of SUVs** As the graph shows, since 2002 the sales of truck-based SUVs have been declining approximately linearly, while the sales of crossover SUVs have been increasing approximately linearly.

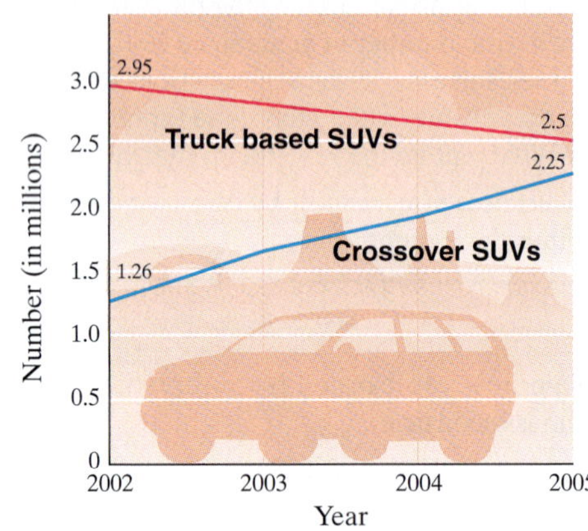


*Source*: Ford Motor Company

We will use the information from Chapter 3 to determine the linear equations that approximate both curves in the graph.

**a)** Using the values for 2002 and 2005, determine the equation of a straight line that can be used to approximate the sales of truck-based SUVs (the red curve). Use the format of the equation $s(t) = mt + b$, where $s$ is the sales, in millions, and $t$ is the year since 2002, and $b$ is the sales in 2002.

**b)** Using the values for 2002 and 2005, determine the equation of the straight line that can be used to approximate the sales of crossover SUVs (the blue curve).

**c)** Solve the system of equations using the equations you obtained in parts **a)** and **b)**. Round values to the nearest hundredth.

## Cumulative Review Exercises

[1.2] **103.** Explain the difference between a rational number and an irrational number.

[1.2] **104. a)** Are all rational numbers real numbers?

**b)** Are all irrational numbers real numbers?

[2.1] **105.** Solve the equation $\frac{1}{2}(x - 7) = \frac{3}{4}(2x + 1)$.

[2.2] **106.** Find all numbers such that $|x - 6| = |6 - x|$.

[2.2] **107.** Evaluate $A = p\left(1 + \frac{r}{n}\right)^{nt}$, when $p = 500$, $r = 0.04$, $n = 2$, and $t = 1$.

[3.5] **108.** Is the following relation a function? Explain your answer. $\{(-3, 4), (7, 2), (-4, 5), (5, 0), (-3, -1)\}$

[3.6] **109.** Let $f(x) = x + 3$ and $g(x) = x^2 - 9$. Find $(f/g)(3)$.

# 4.2 Solving Systems of Linear Equations in Three Variables

1. Solve systems of linear equations in three variables.
2. Learn the geometric interpretation of a system of equations in three variables.
3. Recognize inconsistent and dependent systems.

## 1 Solve Systems of Linear Equations in Three Variables

The equation $2x - 3y + 4z = 8$ is an example of a linear equation in three variables. The solution to a linear equation in three variables is an *ordered triple* of the form $(x, y, z)$. One solution to the equation given is $(1, 2, 3)$. Check now to verify that $(1, 2, 3)$ is a solution to the equation.

To solve systems of linear equations with three variables, we can use either substitution or the addition method, both of which were discussed in Section 4.1.

**EXAMPLE 1** ▸ Solve the following system by substitution.

$$\begin{aligned} x &= -3 \\ 3x + 4y &= 7 \\ -2x - 3y + 5z &= 19 \end{aligned}$$

*Solution* Since we know that $x = -3$, we substitute $-3$ for $x$ in the equation $3x + 4y = 7$ and solve for $y$.

$$\begin{aligned} 3x + 4y &= 7 \\ 3(-3) + 4y &= 7 \\ -9 + 4y &= 7 \\ 4y &= 16 \\ y &= 4 \end{aligned}$$

Now we substitute $x = -3$ and $y = 4$ into the last equation and solve for $z$.

$$\begin{aligned} -2x - 3y + 5z &= 19 \\ -2(-3) - 3(4) + 5z &= 19 \\ 6 - 12 + 5z &= 19 \\ -6 + 5z &= 19 \\ 5z &= 25 \\ z &= 5 \end{aligned}$$

**Check** $x = -3, y = 4, z = 5$. The solution must be checked in *all three* original equations.

$$\begin{array}{lll} x = -3 & 3x + 4y = 7 & -2x - 3y + 5z = 19 \\ -3 = -3 \quad \textit{True} & 3(-3) + 4(4) \stackrel{?}{=} 7 & -2(-3) - 3(4) + 5(5) \stackrel{?}{=} 19 \\ & 7 = 7 \quad \textit{True} & 19 = 19 \quad \textit{True} \end{array}$$

The solution is the ordered triple $(-3, 4, 5)$. Remember that the ordered triple lists the $x$-value first, the $y$-value second, and the $z$-value third.

▸ **Now Try Exercise 3**

Not every system of linear equations in three variables can be solved by substitution. When such a system cannot be solved using substitution, we can find the solution by the addition method, as illustrated in Example 2.

**EXAMPLE 2** ▸ Solve the following system of equations using the addition method.

$$\begin{aligned} 3x + 2y + z &= 4 \quad (\text{eq. } 1) \\ 2x - 3y + 2z &= -7 \quad (\text{eq. } 2) \\ x + 4y - z &= 10 \quad (\text{eq. } 3) \end{aligned}$$

**Solution** To solve this system of equations, we must first obtain two equations containing the same two variables. We do so by selecting two equations and using the addition method to eliminate one of the variables. For example, by adding (*eq.* 1) and (*eq.* 3), the variable $z$ will be eliminated. Then we use a different pair of equations [either (*eq.* 1) and (*eq.* 2) or (*eq.* 2) and (*eq.* 3)] and use the addition method to eliminate the *same* variable that was eliminated previously. If we multiply (*eq.* 1) by $-2$ and add it to (*eq.* 2), the variable $z$ will again be eliminated. We will then have two equations containing only two unknowns. Let us begin by adding (*eq.* 1) and (*eq.* 3).

$$\begin{aligned} 3x + 2y + z &= 4 && (eq.\ 1) \\ x + 4y - z &= 10 && (eq.\ 3) \\ \hline 4x + 6y \quad &= 14 && \text{Sum of equations, } (eq.\ 4) \end{aligned}$$

Now let's use a different pair of equations and again eliminate the variable $z$.

$$\begin{aligned} -6x - 4y - 2z &= -8 && (eq.\ 1) \quad \text{Multiplied by } -2 \\ 2x - 3y + 2z &= -7 && (eq.\ 2) \\ \hline -4x - 7y \quad &= -15 && \text{Sum of equations, } (eq.\ 5) \end{aligned}$$

We now have a system consisting of two equations with two unknowns, (*eq.* 4) and (*eq.* 5). If we add these two equations, the variable $x$ will be eliminated.

$$\begin{aligned} 4x + 6y &= 14 && (eq.\ 4) \\ -4x - 7y &= -15 && (eq.\ 5) \\ \hline -y &= -1 && \text{Sum of equations} \\ y &= 1 \end{aligned}$$

Next we substitute $y = 1$ into either one of the two equations containing only two variables [(*eq.* 4) or (*eq.* 5)] and solve for $x$.

$$\begin{aligned} 4x + 6y &= 14 && (eq.\ 4) \\ 4x + 6(1) &= 14 && \text{Substitute 1 for } y \text{ in } (eq.\ 4). \\ 4x + 6 &= 14 \\ 4x &= 8 \\ x &= 2 \end{aligned}$$

Finally, we substitute $x = 2$ and $y = 1$ into any of the original equations and solve for $z$.

$$\begin{aligned} 3x + 2y + z &= 4 && (eq.\ 1) \\ 3(2) + 2(1) + z &= 4 && \text{Substitute 2 for } x \text{ and 1 for } y \text{ in } (eq.\ 1). \\ 6 + 2 + z &= 4 \\ 8 + z &= 4 \\ z &= -4 \end{aligned}$$

The solution is the ordered triple $(2, 1, -4)$. Check this solution in *all three* original equations.

**▸ Now Try Exercise 15**

In Example 2 we chose first to eliminate the variable $z$ by using (*eq.* 1) and (*eq.* 3) and then (*eq.* 1) and (*eq.* 2). We could have elected to eliminate either the variable $x$ or the variable $y$ first. For example, we could have eliminated variable $x$ by multiplying (*eq.* 3) by $-2$ and then adding it to (*eq.* 2). We could also eliminate the variable $x$ by multiplying (*eq.* 3) by $-3$ and then adding it to (*eq.* 1). Try solving the system in Example 2 by first eliminating the variable $x$.

**EXAMPLE 3** ▸ Solve the following system of equations.

$$\begin{aligned} 2x - 3y + 2z &= -1 \quad (eq.\ 1) \\ x + 2y \quad\quad &= 14 \quad (eq.\ 2) \\ x \quad\quad - 5z &= -11 \quad (eq.\ 3) \end{aligned}$$

*Solution* The third equation does not contain $y$. We will therefore work to obtain another equation that does not contain $y$. We will use (*eq.* 1) and (*eq.* 2) to do this.

$$\begin{aligned} 4x - 6y + 4z &= -2 \quad (eq.\ 1) \text{ Multiplied by 2} \\ 3x + 6y \quad\quad &= 42 \quad (eq.\ 2) \text{ Multiplied by 3} \\ \hline 7x \quad\quad + 4z &= 40 \quad \text{Sum of equations, } (eq.\ 4) \end{aligned}$$

We now have two equations containing only the variables $x$ and $z$.

$$\begin{aligned} 7x + 4z &= 40 \quad (eq.\ 4) \\ x - 5z &= -11 \quad (eq.\ 3) \end{aligned}$$

Let's now eliminate the variable $x$.

$$\begin{aligned} 7x + 4z &= 40 \quad (eq.\ 4) \\ -7x + 35z &= 77 \quad (eq.\ 3) \text{ Multiplied by } -7 \\ \hline 39z &= 117 \quad \text{Sum of equations} \\ z &= 3 \end{aligned}$$

Now we solve for $x$ by using one of the equations containing only the variables $x$ and $z$. We substitute 3 for $z$ in (*eq.* 3).

$$\begin{aligned} x - 5z &= -11 \quad (eq.\ 3) \\ x - 5(3) &= -11 \quad \text{Substitute 3 for } z \text{ in } (eq.\ 3). \\ x - 15 &= -11 \\ x &= 4 \end{aligned}$$

Finally, we solve for $y$ using any of the original equations that contains $y$.

$$\begin{aligned} x + 2y &= 14 \quad (eq.\ 2) \\ 4 + 2y &= 14 \quad \text{Substitute 4 for } x \text{ in } (eq.\ 2). \\ 2y &= 10 \\ y &= 5 \end{aligned}$$

The solution is the ordered triple (4, 5, 3).

**Check** (eq. 1)

$$\begin{aligned} 2x - 3y + 2z &= -1 \\ 2(4) - 3(5) + 2(3) &\stackrel{?}{=} -1 \\ 8 - 15 + 6 &\stackrel{?}{=} -1 \\ -1 &= -1 \quad \text{True} \end{aligned}$$

(eq. 2)

$$\begin{aligned} x + 2y &= 14 \\ 4 + 2(5) &\stackrel{?}{=} 14 \\ 4 + 10 &\stackrel{?}{=} 14 \\ 14 &= 14 \quad \text{True} \end{aligned}$$

(eq. 3)

$$\begin{aligned} x - 5z &= -11 \\ 4 - 5(3) &\stackrel{?}{=} -11 \\ 4 - 15 &\stackrel{?}{=} -11 \\ -11 &= -11 \quad \text{True} \end{aligned}$$

▸ **Now Try Exercise 11**

## Helpful Hint

If an equation in a system contains fractions, eliminate the fractions by multiplying each term in the equation by the least common denominator. Then continue to solve the system. If, for example, one equation in the system is $\frac{3}{4}x - \frac{5}{8}y + z = \frac{1}{2}$, multiply both sides of the equation by 8 to obtain the equivalent equation $6x - 5y + 8z = 4$.

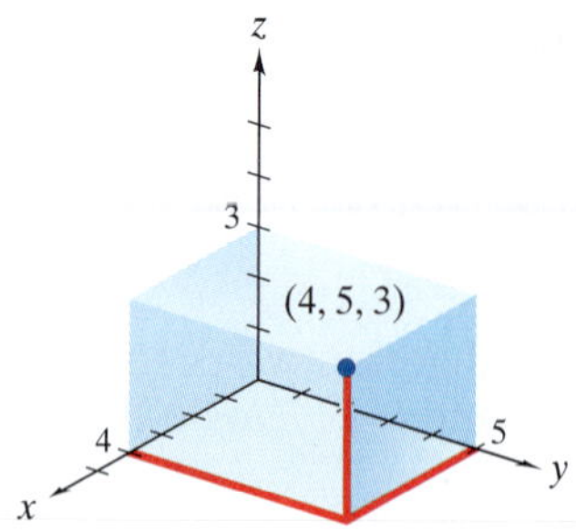

FIGURE 4.5

## 2 Learn the Geometric Interpretation of a System of Equations in Three Variables

When we have a system of linear equations in two variables, we can find its solution graphically using the Cartesian coordinate system. A linear equation in three variables, $x$, $y$, and $z$, can be graphed on a coordinate system with three axes drawn perpendicular to each other (see **Fig. 4.5**).

A point plotted in this three-dimensional system would appear to be a point in space. If we were to graph an equation such as $x + 2y + 3z = 4$, we would find that its graph would be a plane, not a line. In Example 3 we indicated the solution to be the ordered triple (4, 5, 3). This means that the three planes, one from each of the three given equations, all intersect at the point (4, 5, 3). In general, the ordered triple that is the solution to a system of equations in three variables is the point at which the three planes intersect. **Figure 4.5** shows the location of this point of intersection of the three planes. The drawing in Exercise 39 illustrates three planes intersecting at a point.

## 3 Recognize Inconsistent and Dependent Systems

We discussed inconsistent and dependent systems of equations in Section 4.1. Systems of linear equations in three variables may also be inconsistent or dependent. When solving a system of linear equations in three variables, if you obtain a false statement like $3 = 0$, the system is inconsistent and has no solution. This means that at least two of the planes are parallel, so the three planes cannot intersect. (See Exercises 37 and 38.)

When solving a system of linear equations in three variables, if you obtain the true statement $0 = 0$, it indicates that the system is dependent and has an infinite number of solutions. This may happen when all three equations represent the same plane or when the intersection of the planes is a line, as in the drawing in Exercise 40. Examples 4 and 5 illustrate an inconsistent system and a dependent system, respectively.

**EXAMPLE 4** ▸ Solve the following system of equations.

$$\begin{aligned} -3x + 5y + z &= -3 \quad (eq.\ 1) \\ 6x - 10y - 2z &= 1 \quad (eq.\ 2) \\ 7x - 4y + 11z &= -6 \quad (eq.\ 3) \end{aligned}$$

*Solution* We will begin by eliminating the variable $x$ from $(eq.\ 1)$ and $(eq.\ 2)$.

$$\begin{aligned} -6x + 10y + 2z &= -6 \quad (eq.\ 1) \quad \text{Multiplied by 2} \\ 6x - 10y - 2z &= 1 \quad (eq.\ 2) \\ \hline 0 &= -5 \quad \text{False} \end{aligned}$$

Since we obtained the false statement $0 = -5$, this system is inconsistent and has no solution.

▸ **Now Try Exercise 31**

**EXAMPLE 5** ▸ Solve the following system of equations.

$$\begin{aligned} x - y + z &= 1 \quad (eq.\ 1) \\ x + 2y - z &= 1 \quad (eq.\ 2) \\ x - 4y + 3z &= 1 \quad (eq.\ 3) \end{aligned}$$

*Solution* We will begin by eliminating the variable $x$ from $(eq.\ 1)$ and $(eq.\ 2)$ and then from $(eq.\ 1)$ and $(eq.\ 3)$.

$$\begin{aligned} -x + y - z &= -1 \quad (eq.\ 1) \quad \text{Multiplied by } -1 \\ x + 2y - z &= 1 \quad (eq.\ 2) \\ \hline 3y - 2z &= 0 \quad \text{Sum of equations, } (eq.\ 4) \end{aligned}$$

$$\begin{array}{rll} x - y + z = 1 & (eq.\ 1) & \\ -x + 4y - 3z = -1 & (eq.\ 3) & \text{Multiplied by } -1 \\ \hline 3y - 2z = 0 & & \text{Sum of equations, } (eq.\ 5) \end{array}$$

Now we eliminate the variable $y$ using $(eq.\ 4)$ and $(eq.\ 5)$.

$$\begin{array}{rll} -3y + 2z = 0 & (eq.\ 4) & \text{Multiplied by } -1 \\ 3y - 2z = 0 & (eq.\ 5) & \\ \hline 0 = 0 & & \text{True} \end{array}$$

Since we obtained the true statement $0 = 0$, this system is dependent and has an infinite number of solutions.

Recall from Section 4.1 that systems of equations that are dependent are also consistent since they have a solution.

▶ Now Try Exercise 33

## EXERCISE SET 4.2

MathXL® MyMathLab

### Concept/Writing Exercises

1. What will be the graph of an equation such as $3x - 4y + 2z = 1$?

2. Assume that the solution to a system of linear equations in three variables is $(1, 3, 5)$. What does this mean geometrically?

### Practice the Skills

*Solve by substitution.*

3. $x = 1$
$2x - y = 4$
$-3x + 2y - 2z = 1$

4. $-x + 3y - 5z = -7$
$2y - z = -1$
$z = 3$

5. $5x - 6z = -17$
$3x - 4y + 5z = -1$
$2z = -6$

6. $2x - 5y = 12$
$-3y = -9$
$2x - 3y + 4z = 8$

7. $x + 2y = 6$
$3y = 9$
$x + 2z = 12$

8. $x - y + 5z = -4$
$3x - 2z = 6$
$4z = 2$

*Solve using the addition method.*

9. $x - 2y = -3$
$3x + 2y = 7$
$2x - 4y + z = -6$

10. $x - y + 2z = 1$
$y - 4z = 2$
$-2x + 2y - 5z = 2$

11. $2y + 4z = 2$
$x + y + 2z = -2$
$2x + y + z = 2$

12. $2x + y - 8 = 0$
$3x - 4z = -3$
$2x - 3z = 1$

13. $3p + 2q = 11$
$4q - r = 6$
$6p + 7r = 4$

14. $3s + 5t = -12$
$2t - 2u = 2$
$-s + 6u = -2$

15. $p + q + r = 4$
$p - 2q - r = 1$
$2p - q - 2r = -1$

16. $x - 2y + 3z = -7$
$2x - y - z = 7$
$-4x + 3y + 2z = -14$

17. $2x - 2y + 3z = 5$
$2x + y - 2z = -1$
$4x - y - 3z = 0$

18. $2x - y - 2z = 3$
$x - 3y - 4z = 2$
$x + y + 2z = -1$

19. $r - 2s + t = 2$
$2r + 3s - t = -3$
$2r - s - 2t = 1$

20. $3a - 3b + 4c = -1$
$a - 2b + 2c = 2$
$2a - 2b - c = 3$

**21.** $2a + 2b - c = 2$
$3a + 4b + c = -4$
$5a - 2b - 3c = 5$

**22.** $x - 2y + 2z = 3$
$2x - 3y + 2z = 5$
$x + y + 6z = -2$

**23.** $-x + 3y + z = 0$
$-2x + 4y - z = 0$
$3x - y + 2z = 0$

**24.** $x + y + z = 0$
$-x - y + z = 0$
$-x + y + z = 0$

**25.** $-\frac{1}{4}x + \frac{1}{2}y - \frac{1}{2}z = -2$
$\frac{1}{2}x + \frac{1}{3}y - \frac{1}{4}z = 2$
$\frac{1}{2}x - \frac{1}{2}y + \frac{1}{4}z = 1$

**26.** $\frac{2}{3}x + y - \frac{1}{3}z = \frac{1}{3}$
$\frac{1}{2}x + y + z = \frac{5}{2}$
$\frac{1}{4}x - \frac{1}{4}y + \frac{1}{4}z = \frac{3}{2}$

**27.** $x - \frac{2}{3}y - \frac{2}{3}z = -2$
$\frac{2}{3}x + y - \frac{2}{3}z = \frac{1}{3}$
$\frac{1}{4}x + y + \frac{1}{4}z = \frac{3}{4}$

**28.** $\frac{1}{8}x + \frac{1}{4}y + z = 2$
$\frac{1}{3}x + \frac{1}{4}y + z = \frac{17}{6}$
$-\frac{1}{4}x + \frac{1}{3}y - \frac{1}{2}z = -\frac{5}{6}$

**29.** $0.2x + 0.3y + 0.3z = 1.1$
$0.4x - 0.2y + 0.1z = 0.4$
$-0.1x - 0.1y + 0.3z = 0.4$

**30.** $0.6x - 0.4y + 0.2z = 2.2$
$-0.1x - 0.2y + 0.3z = 0.9$
$-0.2x - 0.1y - 0.3z = -1.2$

*Determine whether the following systems are inconsistent, dependent, or neither.*

**31.** $2x + y + 2z = 1$
$x - 2y - z = 0$
$3x - y + z = 2$

**32.** $2p - 4q + 6r = 8$
$-p + 2q - 3r = 6$
$3p + 4q + 5r = 8$

**33.** $x - 4y - 3z = -1$
$-3x + 12y + 9z = 3$
$2x - 10y - 7z = 5$

**34.** $5a - 4b + 2c = 5$
$-10a + 8b - 4c = -10$
$-7a - 4b + c = 7$

**35.** $x + 3y + 2z = 6$
$x - 2y - z = 8$
$-3x - 9y - 6z = -7$

**36.** $2x - 2y + 4z = 2$
$-3x + y = -9$
$2x - y + z = 5$

## Problem Solving

*An equation in three variables represents a plane. Consider a system of equations consisting of three equations in three variables. Answer the following questions.*

**37.** If the three planes are parallel to one another as illustrated in the figure, how many points will be common to all three planes? Is the system consistent or inconsistent? Explain your answer.

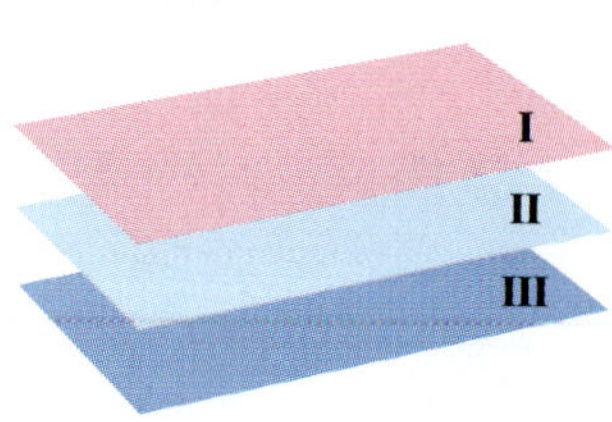


**38.** If two of the planes are parallel to each other and the third plane intersects each of the other two planes, how many points will be common to all three planes? Is the system consistent or inconsistent? Explain your answer.

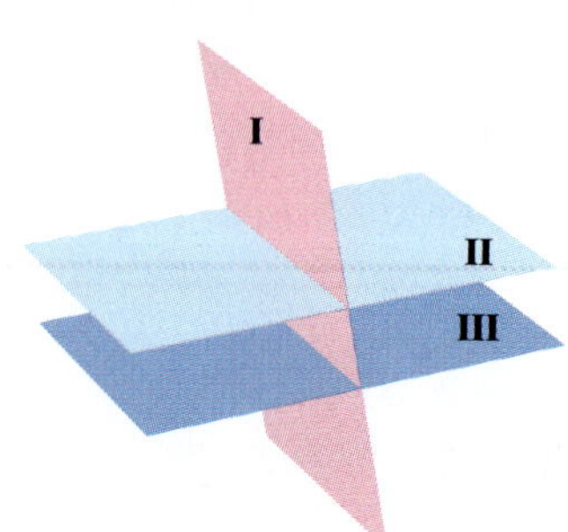

**39.** If the three planes are as illustrated in the figure, how many points will be common to all three planes? Is the system consistent or inconsistent? Explain your answer.

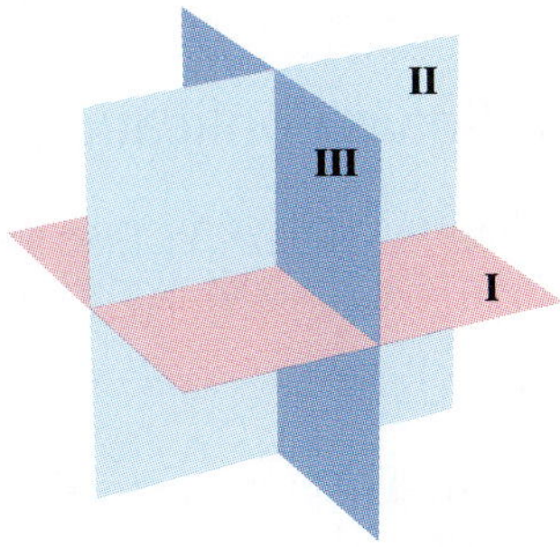


**40.** If the three planes are as illustrated in the figure, how many points will be common to all three planes? Is the system dependent? Explain your answer.

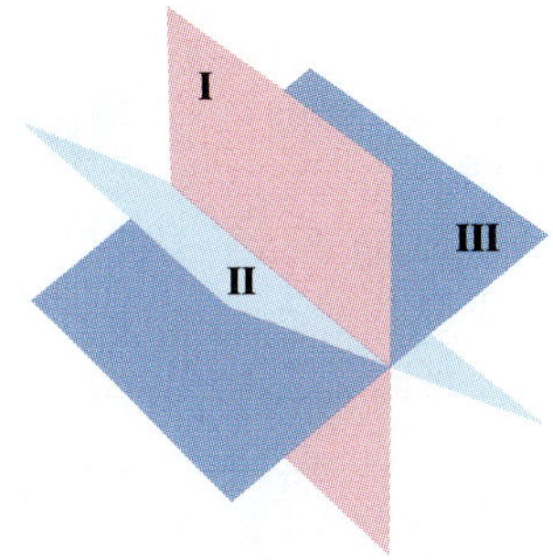


**41.** Is it possible for a system of linear equations in three variables to have exactly

**a)** no solution,

**b)** one solution,

**c)** two solutions? Explain your answer.

**42.** In a system of linear equations in three variables, if the graphs of two equations are parallel planes, is it possible for the system to be

**a)** consistent,

**b)** dependent,

**c)** inconsistent? Explain your answer.

**43.** Three solutions to the equation $Ax + By + Cz = 1$ are $(-1, 2, -1)$, $(-1, 1, 2)$, and $(1, -2, 2)$. Determine the values of $A$, $B$, and $C$ and write the equation using the numerical values found.

**44.** Three solutions to the equation $Ax + By + Cz = 14$ are $(3, -1, 2)$, $(2, -2, 1)$, and $(-5, 3, -24)$. Find the values of $A$, $B$, and $C$ and write the equation using the numerical values found.

*In Exercises 45 and 46, write a system of linear equations in three variables that has the given solution. Explain how you determined your answer.*

**45.** $(3, 1, 6)$

**46.** $(-2, 5, 3)$

**47. a)** Find the values of $a$, $b$, and $c$ such that the points $(1, -1)$, $(-1, -5)$, and $(3, 11)$ lie on the graph of $y = ax^2 + bx + c$.

**b)** Find the quadratic equation whose graph passes through the three points indicated. Explain how you determined your answer.

**48. a)** Find the values of $a$, $b$, and $c$ such that the points $(1, 7)$, $(-2, -5)$, and $(3, 5)$ lie on the graph of $y = ax^2 + bx + c$.

**b)** Find the quadratic equation whose graph passes through the three points indicated. Explain how you determined your answer.

## Challenge Problems

*Find the solution to the following systems of equations.*

**49.**
$$\begin{aligned} 3p + 4q &= 11 \\ 2p + r + s &= 9 \\ q - s &= -2 \\ p + 2q - r &= 2 \end{aligned}$$

**50.**
$$\begin{aligned} 3a + 2b - c &= 0 \\ 2a + 2c + d &= 5 \\ a + 2b - d &= -2 \\ 2a - b + c + d &= 2 \end{aligned}$$

## Cumulative Review Exercises

[2.2] **51. Cross-Country Skiing** Margie Steiner begins skiing along a trail at 3 miles per hour. Ten minutes $\left(\frac{1}{6}\text{hour}\right)$ later, her husband, David, begins skiing along the same trail at 5 miles per hour.

**a)** How long after David leaves will he catch up to Margie?

**b)** How far from the starting point will they be when they meet?

[2.6] *Determine each solution set.*

**52.** $\left|4 - \frac{2x}{3}\right| > 5$

**53.** $\left|\frac{3x - 4}{2}\right| + 1 < 7$

**54.** $\left|3x + \frac{1}{5}\right| = -5$

# 4.3 Systems of Linear Equations: Applications and Problem Solving

1 Use systems of equations to solve applications.

2 Use linear systems in three variables to solve applications.

## 1 Use Systems of Equations to Solve Applications

Many of the applications solved in earlier chapters using only one variable can now be solved using two variables. Following are some examples showing how applications can be represented by systems of equations.

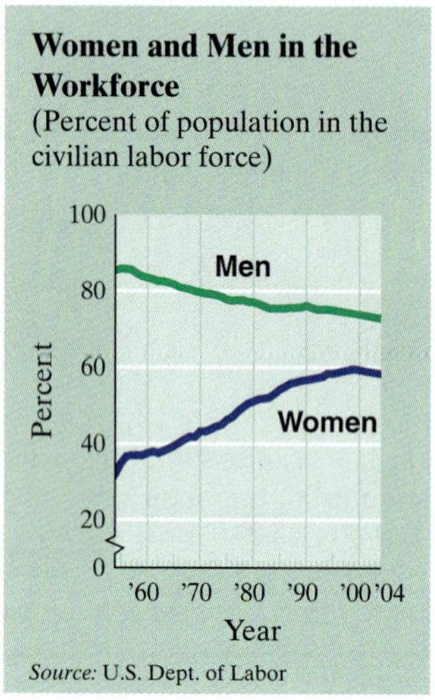


FIGURE 4.6

**EXAMPLE 1** ▸ **A Changing Workforce** The graph in **Figure 4.6** indicates that the percent of males in the workforce is decreasing while the percent of females is increasing. The function $m(t) = -0.25t + 85.4$, where $t$ = years since 1955, can be used to estimate the percent of males in the workforce, and the function $w(t) = 0.52t + 35.7$ can be used to estimate the percent of women in the workforce. If this trend continues, determine when the percent of women in the workforce will equal the percent of men.

*Solution* **Understand and Translate** Consider the two functions given above as the system of equations. To determine when the percent of women will equal the percent of men, we can set the two functions equal to one another and solve for time, $t$.

**Carry Out**

$$\text{percent of women} = \text{percent of men}$$

$$0.52t + 35.7 = -0.25t + 85.4$$

$$0.77t = 49.7$$

$$t \approx 64.5$$

**Answer** If this trend continues the percent of women in the workforce will equal the percent of men about 64.5 years after 1955. Since 1955 + 64.5 = 2019.5, the percents will be equal in the year 2019.

▸ **Now Try Exercise 39**

**EXAMPLE 2** ▸ **Land Area** The combined land area of Grenada and Guam is 890 square kilometers. The land area of Guam is 200 square kilometers more than the land area of Grenada. Find the land area of Guam and Grenada.

*Solution* **Understand** We need to determine the land area of Guam and Grenada.

**Translate** Let $a$ = land area of Guam

$b$ = land area of Grenada.

Uruano Beach, Guam

Since the total land area of Grenada and Guam is 890 square kilometers, the first equation is

$$a + b = 890$$

Since the land area of Guam is 200 square kilometers greater than the land area of Grenada, the second equation is

$$a = b + 200$$

The system of equations is

$$a + b = 890 \quad (eq.\ 1)$$

$$a = b + 200 \quad (eq.\ 2)$$

**Carry Out** We will use the substitution method, discussed in Section 4.1, to solve this system of equations.

Using (*eq.* 2), substitute $b + 200$ for $a$ into the first equation to obtain

$$\begin{aligned} a + b &= 890 && \text{First equation} \\ (b + 200) + b &= 890 && \text{Substitute } b + 200 \text{ for } a. \\ 2b + 200 &= 890 && \text{Simplify.} \\ 2b &= 690 && \text{Subtract 200 from both sides.} \\ b &= 345 && \text{Divide both sides by 2.} \end{aligned}$$

Thus, $b = 345$. To determine the value for $a$, substitute 345 for $b$ into (*eq.* 2).

$$\begin{aligned} a &= b + 200 \\ a &= 345 + 200 \\ &= 545 \end{aligned}$$

**Answer** The land area of Guam is 545 square kilometers and the land area of Grenada is 345 square kilometers.

▸ **Now Try Exercise 1**

**EXAMPLE 3** ▸ **Canoe Speed** The Burnhams are canoeing on the Suwannee River. They travel at an average speed of 4.75 miles per hour when paddling with the current and 2.25 miles per hour when paddling against the current. Determine the speed of the canoe in still water and the speed of the current.

***Solution*** **Understand** When they are traveling with the current, the canoe's speed is the canoe's speed in still water *plus* the current's speed. When traveling against the current, the canoe's speed is the canoe's speed in still water *minus* the current's speed.

**Translate** Let $s$ = speed of the canoe in still water

$c$ = speed of the current.

The system of equations is:

speed of the canoe traveling with the current: $s + c = 4.75$

speed of the canoe traveling against the current: $s - c = 2.25$

**Carry Out** We will use the addition method, as we discussed in Section 4.1, to solve this system of equations.

$$\begin{aligned} s + c &= 4.75 \\ s - c &= 2.25 \\ \hline 2s \quad &= 7.00 \\ s &= 3.5 \end{aligned}$$

The speed of the canoe in still water is 3.5 miles per hour. We now determine the speed of the current.

$$\begin{aligned} s + c &= 4.75 \\ 3.5 + c &= 4.75 \\ c &= 1.25 \end{aligned}$$

**Answer** The speed of the current is 1.25 miles per hour, and the speed of the canoe in still water is 3.5 miles per hour.

▸ **Now Try Exercise 13**

**EXAMPLE 4** ▸ **Salary** Yamil Bermudez, a salesman at Hancock Appliances, receives a weekly salary plus a commission, which is a percentage of his sales. One week, with sales of \$3000, his total take-home pay was \$850. The next week, with sales of \$4000, his total take-home pay was \$1000. Find his weekly salary and his commission rate.

***Solution*** **Understand** Yamil's take-home pay consists of his weekly salary plus commission. We are given information about two specific weeks that we can use to find his weekly salary and his commission rate.

**Translate** Let $s$ = his weekly salary
$r$ = his commission rate.

In week 1, his commission on \$3000 is $3000r$, and in week 2 his commission on \$4000 is $4000r$. Thus the system of equations is

$$\text{salary} + \text{commission} = \text{take-home salary}$$

$$\begin{aligned} &\text{1st week} \quad s + 3000r = 850 \\ &\text{2nd week} \quad s + 4000r = 1000 \end{aligned} \quad \text{System of equations}$$

**Carry Out**

$$\begin{aligned} -s - 3000r &= -850 \quad \text{1st week multiplied by } -1 \\ s + 4000r &= 1000 \quad \text{2nd week} \\ \hline 1000r &= 150 \quad \text{Sum of equations} \\ r &= \frac{150}{1000} \\ r &= 0.15 \end{aligned}$$

Yamil's commission rate is 15%. Now we find his weekly salary by substituting 0.15 for $r$ in either equation.

$$\begin{aligned} s + 3000r &= 850 \\ s + 3000(0.15) &= 850 \quad \text{Substitute 0.15 for } r \text{ in the 1st-week equation.} \\ s + 450 &= 850 \\ s &= 400 \end{aligned}$$

**Answer** Yamil's weekly salary is \$400 and his commission rate is 15%.

▶ **Now Try Exercise 15**

**EXAMPLE 5 ▶ Riding Horses** Ben Campbell leaves his ranch riding on his horse at 5 miles per hour. One-half hour later, Joe Campbell leaves the same ranch and heads along the same route on his horse at 8 miles per hour.

**a)** How long after Joe leaves the ranch will he catch up to Ben?

**b)** When Joe catches up to Ben, how far from the ranch will they be?

*Solution* **a) Understand** When Joe catches up to Ben, they both will have traveled the same distance. Joe will have traveled the distance in $\frac{1}{2}$ hour less time since he left $\frac{1}{2}$ hour after Ben. We will use the formula distance = rate · time to solve this problem.

**Translate** Let $b$ = time traveled by Ben
$j$ = time traveled by Joe.

We will set up a table to organize the given information.

| | Rate | Time | Distance |
|---|---|---|---|
| Ben | 5 | $b$ | $5b$ |
| Joe | 8 | $j$ | $8j$ |

Since both Ben and Joe cover the same distance, we write

$$\text{Ben's distance} = \text{Joe's distance}$$

$$5b = 8j$$

Our second equation comes from the fact that Joe is traveling for $\frac{1}{2}$ hour less time than Ben. Therefore, $j = b - \frac{1}{2}$. Thus our system of equations is:

$$5b = 8j$$
$$j = b - \frac{1}{2}$$

**Carry Out** We will solve this system of equations using substitution. Since $j = b - \frac{1}{2}$, substitute $b - \frac{1}{2}$ for $j$ in the first equation and solve for $b$.

$$5b = 8j$$
$$5b = 8\left(b - \frac{1}{2}\right)$$
$$5b = 8b - 4$$
$$-3b = -4$$
$$b = \frac{-4}{-3} = 1\frac{1}{3}$$

Therefore, the time Ben has been traveling is $1\frac{1}{3}$ hours. To get the time Joe has been traveling, we will subtract $\frac{1}{2}$ hour from Ben's time.

$$j = b - \frac{1}{2}$$
$$j = 1\frac{1}{3} - \frac{1}{2}$$
$$j = \frac{4}{3} - \frac{1}{2} = \frac{8}{6} - \frac{3}{6} = \frac{5}{6}$$

**Answer** Joe will catch up to Ben $\frac{5}{6}$ of an hour (or 50 minutes) after Joe leaves the ranch.

**b)** We can use either Ben's or Joe's distance to determine the distance traveled from the ranch. We will use Joe's distance.

$$d = 8j = 8\left(\frac{5}{6}\right) = \frac{\overset{4}{\cancel{8}}}{1} \cdot \frac{5}{\underset{3}{\cancel{6}}} = \frac{20}{3} = 6\frac{2}{3}$$

Thus, Joe will catch up to Ben when they are $6\frac{2}{3}$ miles from the ranch.

▸ Now Try Exercise 33

**EXAMPLE 6 ▸ Mixing Solutions** Chung Song, a chemist with Johnson and Johnson, wishes to create a new household cleaner containing 30% trisodium phosphate (TSP). Chung needs to mix a 16% TSP solution with a 72% TSP solution to get 6 liters of a 30% TSP solution. How many liters of the 16% solution and of the 72% solution will he need to mix?

*Solution* **Understand** To solve this problem we use the fact that the amount of TSP in a solution is found by multiplying the percent strength of the solution by the number of liters (the volume) of the solution. Chung needs to mix a 16% solution and a 72% solution to obtain 6 liters of a solution whose strength, 30%, is between the strengths of the two solutions being mixed.

**Translate** Let $x$ = number of liters of the 16% solution

$y$ = number of liters of the 72% solution.

We will draw a sketch (**Fig. 4.7** on page 256) and then use a table to help analyze the problem.

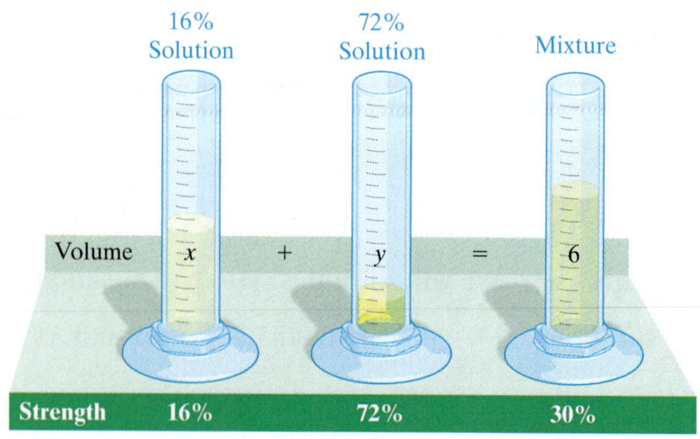


FIGURE 4.7

| Solution | Strength of Solution | Number of Liters | Amount of TSP |
|---|---|---|---|
| 16% solution | 0.16 | $x$ | $0.16x$ |
| 72% solution | 0.72 | $y$ | $0.72y$ |
| Mixture | 0.30 | 6 | 0.30(6) |

Since the sum of the volumes of the 16% solution and the 72% solution is 6 liters, our first equation is

$$x + y = 6$$

The second equation comes from the fact that the solutions are mixed.

$$\begin{pmatrix}\text{amount of TSP}\\ \text{in 16\% solution}\end{pmatrix} + \begin{pmatrix}\text{amount of TSP}\\ \text{in 72\% solution}\end{pmatrix} = \begin{pmatrix}\text{amount of TSP}\\ \text{in mixture}\end{pmatrix}$$

$$0.16x \quad + \quad 0.72y \quad = \quad 0.30(6)$$

Therefore, the system of equations is

$$\begin{aligned} x + y &= 6 \\ 0.16x + 0.72y &= 0.30(6) \end{aligned}$$

**Carry Out** Solving $x + y = 6$ for $y$, we get $y = -x + 6$. Substituting $-x + 6$ for $y$ in the second equation gives us

$$\begin{aligned} 0.16x + 0.72y &= 0.30(6) \\ 0.16x + 0.72(-x + 6) &= 0.30(6) \\ 0.16x - 0.72x + 4.32 &= 1.8 \\ -0.56x + 4.32 &= 1.8 \\ -0.56x &= -2.52 \\ x &= \frac{-2.52}{-0.56} = 4.5 \end{aligned}$$

Therefore, Chung must use 4.5 liters of the 16% solution. Since the two solutions must total 6 liters, he must use $6 - 4.5$ or 1.5 liters of the 72% solution.

▸ **Now Try Exercise 17**

In Example 6, the equation $0.16x + 0.72y = 0.30(6)$ could have been simplified by multiplying both sides of the equation by 100. This would give the equation $16x + 72y = 30(6)$ or $16x + 72y = 180$. Then the system of equations would be $x + y = 6$ and $16x + 72y = 180$. If you solve this system, you should obtain the same solution. Try it and see.

## 2 Use Linear Systems in Three Variables to Solve Applications

Now let us look at some applications that involve three equations and three variables.

**EXAMPLE 7** ▸ **Bank Loans** Tiny Tots Toys must borrow \$25,000 to pay for an expansion. It is not able to obtain a loan for the total amount from a single bank, so it takes out loans from three different banks. It borrows some of the money at a bank that charges it 8% interest. At the second bank, it borrows \$2000 more than one-half the amount borrowed from the first bank. The interest rate at the second bank is 10%. The balance of the \$25,000 is borrowed from a third bank, where Tiny Tots pays 9% interest. The total annual interest Tiny Tots Toys pays for the three loans is \$2220. How much does it borrow at each rate?

*Solution* **Understand** We are asked to determine how much is borrowed at each of the three different rates. Therefore, this problem will contain three variables, one for each amount borrowed. Since the problem will contain three variables, we will need to determine three equations to use in our system of equations.

**Translate** Let $x$ = amount borrowed at first bank
$y$ = amount borrowed at second bank
$z$ = amount borrowed at third bank.

Since the total amount borrowed is \$25,000 we know that

$$x + y + z = 25{,}000 \qquad \text{Total amount borrowed is \$25,000.}$$

At the second bank, Tiny Tots Toys borrows \$2000 more than one-half the amount borrowed from the first bank. Therefore, our second equation is

$$y = \frac{1}{2}x + 2000 \qquad \text{Second, } y\text{, is \$2000 more than } \frac{1}{2} \text{ of first, } x.$$

Our last equation comes from the fact that the total annual interest charged by the three banks is \$2220. The interest at each bank is found by multiplying the interest rate by the amount borrowed.

$$0.08x + 0.10y + 0.09z = 2220 \qquad \text{Total interest is \$2220.}$$

Thus, our system of equations is

$$x + y + z = 25{,}000 \qquad (1)$$

$$y = \frac{1}{2}x + 2000 \qquad (2)$$

$$0.08x + 0.10y + 0.09z = 2220 \qquad (3)$$

Both sides of equation (2) can be multiplied by 2 to remove fractions.

$$2(y) = 2\left(\frac{1}{2}x + 2000\right)$$

$$2y = x + 4000 \qquad \text{Distributive property}$$

$$-x + 2y = 4000 \qquad \text{Subtract } x \text{ from both sides.}$$

The decimals in equation (3) can be removed by multiplying both sides of the equation by 100. This gives

$$8x + 10y + 9z = 222{,}000$$

Our simplified system of equations is therefore

$$x + y + z = 25{,}000 \qquad (eq. 1)$$

$$-x + 2y = 4000 \qquad (eq. 2)$$

$$8x + 10y + 9z = 222{,}000 \qquad (eq. 3)$$

**Carry Out** There are various ways of solving this system. Let's use (*eq.* 1) and (*eq.* 3) to eliminate the variable $z$.

$$\begin{aligned} -9x - 9y - 9z &- -225{,}000 & &(eq.\ 1) \quad \text{Multiplied by } -9 \\ 8x + 10y + 9z &= 222{,}000 & &(eq.\ 3) \\ \hline -x + y &= -3{,}000 & &\text{Sum of equations, } (eq.\ 4) \end{aligned}$$

Now we use (*eq.* 2) and (*eq.* 4) to eliminate the variable $x$ and solve for $y$.

$$\begin{aligned} x - 2y &= -4000 & &(eq.\ 2) \quad \text{Multiplied by } -1 \\ -x + y &= -3000 & &(eq.\ 4) \\ \hline -y &= -7000 & &\text{Sum of equations} \\ y &= 7000 \end{aligned}$$

Now that we know the value of $y$ we can solve for $x$.

$$\begin{aligned} -x + 2y &= 4000 & &(eq.\ 2) \\ -x + 2(7000) &= 4000 & &\text{Substitute 7000 for } y \text{ in } (eq.\ 2). \\ -x + 14{,}000 &= 4000 \\ -x &= -10{,}000 \\ x &= 10{,}000 \end{aligned}$$

Finally, we solve for $z$.

$$\begin{aligned} x + y + z &= 25{,}000 & &(eq.\ 1) \\ 10{,}000 + 7000 + z &= 25{,}000 \\ 17{,}000 + z &= 25{,}000 \\ z &= 8000 \end{aligned}$$

**Answer** Tiny Tots Toys borrows \$10,000 at 8%, \$7000 at 10%, and \$8000 at 9% interest.

**▸ Now Try Exercise 55**

**EXAMPLE 8 ▸ Inflatable Boats** Hobson, Inc., has a small manufacturing plant that makes three types of inflatable boats: one-person, two-person, and four-person models. Each boat requires the service of three departments: cutting, assembly, and packaging. The cutting, assembly, and packaging departments are allowed to use a total of 380, 330, and 120 person-hours per week, respectively. The time requirements for each boat and department are specified in the following table. Determine how many of each type of boat Hobson must produce each week for its plant to operate at full capacity.

| | Time (person-hr) per Boat | | |
|---|---|---|---|
| **Department** | **One-Person Boat** | **Two-Person Boat** | **Four-Person Boat** |
| Cutting | 0.6 | 1.0 | 1.5 |
| Assembly | 0.6 | 0.9 | 1.2 |
| Packaging | 0.2 | 0.3 | 0.5 |

*Solution* **Understand** We are told that three different types of boats are produced and we are asked to determine the number of each type produced. Since this problem involves three amounts to be found, the system will contain three equations in three variables.

**Translate** We will use the information given in the table.

Let $x$ = number of one-person boats
$y$ = number of two-person boats
$z$ = number of four-person boats.

The total number of cutting hours for the three types of boats must equal 380 person-hours.

$$0.6x + 1.0y + 1.5z = 380$$

The total number of assembly hours must equal 330 person-hours.

$$0.6x + 0.9y + 1.2z = 330$$

The total number of packaging hours must equal 120 person-hours.

$$0.2x + 0.3y + 0.5z = 120$$

Therefore, the system of equations is

$$\begin{aligned} 0.6x + 1.0y + 1.5z &= 380 \\ 0.6x + 0.9y + 1.2z &= 330 \\ 0.2x + 0.3y + 0.5z &= 120 \end{aligned}$$

Multiplying each equation in the system by 10 will eliminate the decimal numbers and give a simplified system of equations.

$$\begin{aligned} 6x + 10y + 15z &= 3800 \quad (eq.\ 1) \\ 6x + 9y + 12z &= 3300 \quad (eq.\ 2) \\ 2x + 3y + 5z &= 1200 \quad (eq.\ 3) \end{aligned}$$

**Carry Out** Let's first eliminate the variable $x$ using $(eq.\ 1)$ and $(eq.\ 2)$, and then $(eq.\ 1)$ and $(eq.\ 3)$.

$$\begin{aligned} 6x + 10y + 15z &= 3800 \quad (eq.\ 1) \\ -6x - 9y - 12z &= -3300 \quad (eq.\ 2) \quad \text{Multiplied by } -1 \\ \hline y + 3z &= 500 \quad \text{Sum of equations, } (eq.\ 4) \end{aligned}$$

$$\begin{aligned} 6x + 10y + 15z &= 3800 \quad (eq.\ 1) \\ -6x - 9y - 15z &= -3600 \quad (eq.\ 3) \quad \text{Multiplied by } -3 \\ \hline y &= 200 \quad \text{Sum of equations, } (eq.\ 5) \end{aligned}$$

Note that when we added the last two equations, both variables $x$ and $z$ were eliminated at the same time. Now we know the value of $y$ and can solve for $z$.

$$\begin{aligned} y + 3z &= 500 \quad (eq.\ 4) \\ 200 + 3z &= 500 \quad \text{Substitute 200 for } y. \\ 3z &= 300 \\ z &= 100 \end{aligned}$$

Finally, we find $x$.

$$\begin{aligned} 6x + 10y + 15z &= 3800 \quad (eq.\ 1) \\ 6x + 10(200) + 15(100) &= 3800 \\ 6x + 2000 + 1500 &= 3800 \\ 6x + 3500 &= 3800 \\ 6x &= 300 \\ x &= 50 \end{aligned}$$

**Answer** Hobson should produce 50 one-person boats, 200 two-person boats, and 100 four-person boats per week.

▸ **Now Try Exercise 59**

## EXERCISE SET 4.3

MathXL® MyMathLab

### Practice the Skills/Problem Solving

1. **Land Area** The combined land area of the countries of Georgia and Ireland is 139,973 square kilometers. Ireland is larger by 573 square kilometers. Determine the land area of each country.

Cliffs of Moher, Ireland

2. **Daytona 500 Wins** As of this writing, Richard Petty has won the Daytona 500 race the greatest number of times and Dale Yarborough has won the second greatest number of Daytona 500 races. Petty's number of wins is one less than twice Yarborough's number of wins. The total number of wins by the two drivers is 11. Determine the number of wins by Petty and by Yarborough.

3. **Fat Content** A nutritionist finds that a large order of fries at McDonald's has more fat than a McDonald's quarter-pound hamburger. The fries have 4 grams more than three times the amount of fat that the hamburger has. The difference in the fat content between the fries and the hamburger is 46 grams. Find the fat content of the hamburger and of the fries.

4. **Theme Parks** The two most visited theme parks in the United States in 2004 were Walt Disney's Magic Kingdom in Florida and Disneyland in California. The total number of visitors to these parks was 28.4 million people. The number of people who visited the Magic Kingdom was 1.8 million more than the number of people who visited Disneyland. How many people visited each of these parks in 2004? *Source:* www.coastergrotto.com

5. **Hot Dog Stand** At Big Al's hot dog stand, 2 hot dogs and 3 sodas cost \$7. The cost of 4 hot dogs and 2 sodas is \$10. Determine the cost of a hot dog and the cost of a soda.

6. **Water and Pretzel** At a professional football game, the cost of 2 bottles of water and 3 pretzels is \$16.50. The cost of 4 bottles of water and 1 pretzel is \$15.50. Determine the cost of a bottle of water and the cost of a pretzel.

7. **Digital Cameras** Ashley Dawn just bought a new digital camera, a 128-megabyte memory card, and a 512-megabyte memory card. The 512-MB memory card can store four times as many photos as the 128-MB memory card. Together the two memory cards can store 360 photos (of fine quality). Determine how many photos each memory card can store.

8. **Photo Printers** The July 2005 *Consumer Reports* magazine featured an article on photo printers that compared the costs to print photos with each printer. The most expensive of the $4 \times 6$ snapshot printers was the Olympus P-5100. The least expensive was the Epson Picture Mate. To print one photo on both printers would cost \$0.80. The cost to print a photo on the Olympus is \$0.20 more than twice the cost to print a photo on the Epson. Determine the cost to print a photo on each printer.

9. **Complementary Angles** Two angles are **complementary angles** if the sum of their measures is 90°. (See Section 2.3.) If the measure of the larger of two complementary angles is 15° more than two times the measure of the smaller angle, find the measures of the two angles.

10. **Complementary Angles** The difference between the measures of two complementary angles is 46°. Determine the measures of the two angles.

11. **Supplementary Angles** Two angles are **supplementary angles** if the sum of their measures is 180°. (See Section 2.3.) Find the measures of two supplementary angles if the measure of one angle is 28° less than three times the measure of the other.

12. **Supplementary Angles** Determine the measures of two supplementary angles if the measure of one angle is three and one half times larger than the measure of the other angle.

13. **Rowing Speed** The Heart O'Texas Rowing Team, while practicing in Austin, Texas rowed an average of 15.6 miles per hour with the current and 8.8 miles per hour against the current. Determine the team's rowing speed in still water and the speed of the current.

14. **Flying Speed** Jung Lee, in his Piper Cub airplane, flew an average of 121 miles per hour with the wind and 87 miles per hour against the wind. Determine the speed of the airplane in still air and the speed of the wind.

15. **Salary Plus Commission** Don Lavigne, an office equipment sales representative, earns a weekly salary plus a commission on his sales. One week his total compensation on sales of \$4000 was \$660. The next week his total compensation on sales of \$6000 was \$740. Find Don's weekly salary and his commission rate.

16. **Truck Rental** A truck rental agency charges a daily fee plus a mileage fee. Hugo was charged \$85 for 2 days and 100 miles and Christina was charged \$165 for 3 days and 400 miles. What is the agency's daily fee, and what is the mileage fee?

17. **Lavender Oil** Pola Sommers, a massage therapist, needs 3 ounces of a 20% lavender oil solution. She has only 5% and 30% lavender oil solutions available. How many ounces of each should Pola mix to obtain the desired solution?

18. **Fertilizer Solutions** Frank Ditlman needs to apply a 10% liquid nitrogen solution to his rose garden, but he only has a 4% liquid nitrogen solution and a 20% liquid nitrogen solution available. How much of the 4% solution and how much of the 20% solution should Frank mix together to get 10 gallons of the 10% solution?

19. **Weed Killer** Round-Up Concentrate Grass and Weed Killer consists of an 18% active ingredient glyphosate (and 82% inactive ingredients). The concentrate is to be mixed with water and the mixture applied to weeds. If the final mixture is to contain 0.9% active ingredient, how much concentrate and how much water should be mixed to make 200 gallons of the final mixture?

20. **Lawn Fertilizer** Scott's Winterizer Lawn Fertilizer is 22% nitrogen. Schultz's Lime with Lawn Fertilizer is 4% nitrogen. William Weaver, owner of Weaver's Nursery, wishes to mix these two fertilizers to make 400 pounds of a special 10% nitrogen mixture for midseason lawn feeding. How much of each fertilizer should he mix?

21. **Birdseed** Birdseed costs \$0.59 a pound and sunflower seeds cost \$0.89 a pound. Angela Leinenbachs' pet store wishes to make a 40-pound mixture of birdseed and sunflower seeds that sells for \$0.76 per pound. How many pounds of each type of seed should she use?

22. **Coffee** Franco Manue runs a grocery store. He wishes to mix 30 pounds of coffee to sell for a total cost of \$170. To obtain the mixture, he will mix coffee that sells for \$5.20 per pound with coffee that sells for \$6.30 per pound. How many pounds of each coffee should he use?

23. **Amtrak** Ann Marie Whittle has been pricing Amtrak fares for a group to visit New York. Three adults and four children would cost a total of \$159. Two adults and three children would cost a total of \$112. Determine the price of an adult ticket and a child's ticket.

24. **Buffalo Wings** The Wing House sells both regular size and jumbo size orders of Buffalo chicken wings. Three regular orders and five jumbo orders of wings cost \$67. Four regular and four jumbo orders of wings cost \$64. Determine the cost of a regular order of wings and a jumbo order of wings.

25. **Savings Accounts** Mr. and Mrs. Gamton invest a total of \$10,000 in two savings accounts. One account pays 5% interest and the other 6%. Find the amount placed in each account if the accounts receive a total of \$540 in interest after 1 year. Use interest = principal · rate · time.

26. **Investments** Louis Okonkwo invested \$30,000, part at 9% and part at 5%. If he had invested the entire amount at 6.5%, his total annual interest would be the same as the sum of the annual interest received from the two other accounts. How much was invested at each interest rate?

27. **Milk** Becky Slaats is a plant engineer at Velda Farms Dairy Cooperative. She wishes to mix whole milk, which is 3.25% fat, and skim milk, which has no fat, to obtain 260 gallons of a mixture of milk that contains 2% fat. How many gallons of whole milk and how many gallons of skim milk should Becky mix to obtain the desired mixture?

**28. Quiche Lorraine** Lambert Marx's recipe for quiche lorraine calls for 2 cups (16 ounces) of light cream that is 20% butterfat. It is often difficult to find light cream with 20% butterfat at the supermarket. What is commonly found is heavy cream, which is 36% butterfat, and half-and-half, which is 10.5% butterfat. How much of the heavy cream and how much of the half-and-half should Lambert mix to obtain the mixture necessary for the recipe?

**29. Birdseed** By ordering directly through *www.birdseed.com*, the Carters can purchase Season's Choice birdseed for \$1.79 per pound and Garden Mix birdseed for \$1.19 per pound. If they wish to purchase 20 pounds and spend \$28 on birdseed, how many pounds of each type should they buy?

**30. Juice** The Healthy Favorites Juice Company sells apple juice for 8.3¢ an ounce and raspberry juice for 9.3¢ an ounce. The company wishes to market and sell 8-ounce cans of apple-raspberry juice for 8.7¢ an ounce. How many ounces of each should be mixed?

**31. Car Travel** Two cars start at the same point in Alexandria, Virginia, and travel in opposite directions. One car travels 5 miles per hour faster than the other car. After 4 hours, the two cars are 420 miles apart. Find the speed of each car.

**32. Road Construction** Kip Ortiz drives from Atlanta to Louisville, a distance of 430 miles. Due to road construction and heavy traffic, during the first part of his trip, Kip drives at an average rate of 50 miles per hour. During the rest of his trip he drives at an average rate of 70 miles per hour. If his total trip takes 7 hours, how many hours does he drive at each speed?

**33. Avon Conference** Cabrina Wilson and Dabney Jefferson are Avon representatives who are attending a conference in Seattle. After the conference, Cabrina drives home to Boise at an average speed of 65 miles per hour and Dabney drives home to Portland at an average speed of 50 miles per hour. If the sum of their driving times is 11.4 hours and if the sum of the distances driven is 690 miles, determine the time each representative spent driving home.

**34. Exercise** For her exercise routine, Cynthia Harrison rides a bicycle for half an hour and then rollerblades for half an hour. Cynthia rides the bicycle at a speed that is twice the speed at which she rollerblades. If the total distance covered is 12 miles, determine the speed at which she bikes and rollerblades.

**35. Animal Diet** Animals in an experiment are on a strict diet. Each animal is to receive, among other nutrients, 20 grams of protein and 6 grams of carbohydrates. The scientist has only two food mixes available of the following compositions. How many grams of each mix should be used to obtain the right diet for a single animal?

| Mix | Protein (%) | Carbohydrate (%) |
|---|---|---|
| Mix *A* | 10 | 6 |
| Mix *B* | 20 | 2 |

**36. Chair Manufacturing** A company makes two models of chairs. Information about the construction of the chairs is given in the table shown. On a particular day the company allocated 46.4 person-hours for assembling and 8.8 person-hours for painting. How many of each chair can be made?

| Model | Time to Assemble | Time to Paint |
|---|---|---|
| Model *A* | 1 hr | 0.5 hr |
| Model *B* | 3.2 hr | 0.4 hr |

**37. Brass Alloy** By weight, one alloy of brass is 70% copper and 30% zinc. Another alloy of brass is 40% copper and 60% zinc. How many grams of each of these alloys need to be melted and combined to obtain 300 grams of a brass alloy that is 60% copper and 40% zinc?

**38. Silver Alloy** Sterling silver is 92.5% pure silver. How many grams of pure (100%) silver and how many grams of sterling silver must be mixed to obtain 250 g of a 94% silver alloy?

**39. Internal Revenue Service** The following graph shows the number of paper Form 1040 tax returns and the number of online Forms 1040, 1040A, 1040EZ tax returns filed with the IRS in the years from 2002 to 2005, and projected to 2010. If $t$ represents the number of years since 2002, the number of paper form 1040 tax returns, in millions, filed with the IRS can be estimated by the function $P(t) = -2.73t + 58.37$ and the number of online Forms 1040, 1040A, 1040EZ tax returns, in millions, filed with the IRS can be estimated by the function $o(t) = 1.95t + 10.58$. Assuming this trend continues, solve this system of equations to determine the year that the number of paper Form 1040 tax returns will be the same as the number of online Forms 1040, 1040A, 1040EZ tax returns.

**Federal Tax Return Method**

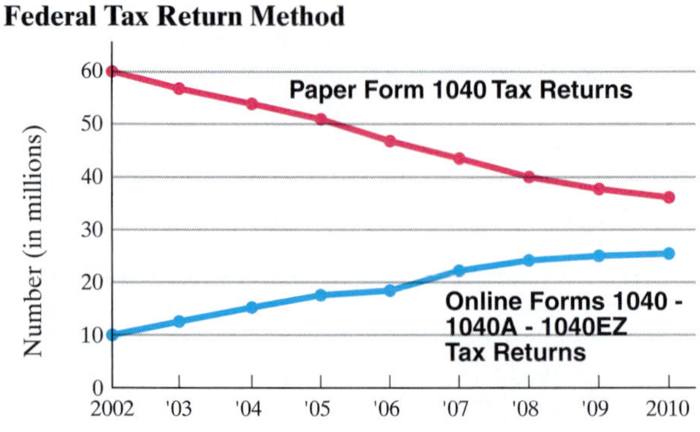


*Source:* www.irs.gov/pubs

**40. Walking and Jogging** Cuong Tham tries to exercise every day. He walks at 3 miles per hour and then jogs at 5 miles per hour. If it takes him 0.9 hours to travel a total of 3.5 miles, how long does he jog?

**41. Texas Driving** Tom Johnson and Melissa Acino started driving at the same time in different cars from Oklahoma City. They both traveled south on Route 35. When Melissa reached the Dallas/Ft. Worth area, a distance of 150 miles, Tom had only reached Denton, Texas, a distance of 120 miles. If Melissa averaged 15 miles per hour faster than Tom, find the average speed of each car.

**42. Photocopy Costs** At a local copy center two plans are available.

Plan 1: 10¢ per copy

Plan 2: an annual charge of $120 plus 4¢ per copy

**a)** Represent this information as a system of equations.

**b)** Graph the system of equations for up to 4000 copies made.

**c)** From the graph, estimate the number of copies a person would have to make in a year for the two plans to have the same total cost.

**d)** Solve the system algebraically. If your answer does not agree with your answer in part **c)**, explain why.

*In Exercises 43–62 Solve each problem using a system of three equations in three unknowns.*

**43. Mail Volume** The average American household receives 24 pieces of mail each week. The number of bills and statements is two less than twice the number of pieces of personal mail. The number of advertisements is two more than five times the number of pieces of personal mail. How many pieces of personal mail, bills and statements, and advertisements does the average family get each week? *Source:* Arthur D. Little, Inc.

**44. Submarine Personnel** A 141-person crew is standard on a Los Angeles class submarine. The number of chief petty officers (enlisted) is four more than the number of commissioned officers. The number of other enlisted men is three less than eight times the number of commissioned officers. Determine the number of commissioned officers, chief petty officers, and other enlisted people on the submarine.

**45. College Football Bowl Games** Through 2004, the Universities of Alabama, Tennessee, and Texas have had the most appearances in college football bowl games. These three schools have had a total of 141 bowl appearances. Alabama has had 8 more appearances than Texas. Together, the number of appearances by Tennessee and Texas is 37 more than the number of appearances by Alabama. Determine the number of bowl appearances for each school and complete the following diagram.

**Schools with the Most Appearances in a College Football Bowl Game** (Includes 2004 Season):

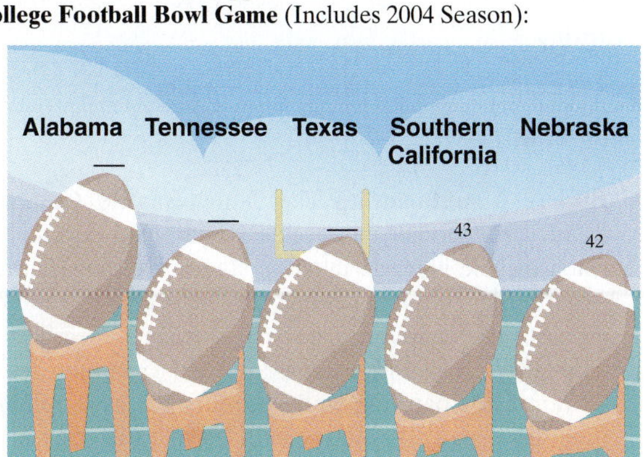


*Source*: NCAA, USA Today (12/22/04)

**46. 2004 Summer Olympics** In the 2004 Summer Olympics in Greece, the countries earning the most gold medals were the United States, China, and Russia. Together, these three countries earned a total of 94 gold medals. The United States earned 3 more gold medals than China. Together the number of gold medals earned by the United States and Russia is 2 less than twice the number of gold medals earned by China. Determine the number of gold medals each country earned and complete the following diagram. *Source:* www.athens2004.com

2004 Summer Olympics

| Rank by Gold | Country | Number of Gold Medals |
|---|---|---|
| 1 | USA | |
| 2 | China | |
| 3 | Russia | |
| 4 | Australia | 17 |
| 5 | Japan | 16 |
| 6 | Germany | 14 |

**47. Top-10 Finishes on PGA Tour** In the five years from 2000 through 2004, the three golfers with the most top-10 finishes in the Professional Golfers Association, or PGA, tour were Vijay Singh, Tiger Woods, and Phil Mickelson. Together, during these years, these three golfers had a total of 191 top-10 finishes. Tiger Woods had 8 more top-10 finishes than Phil Mickelson. Vijay Singh had 12 more top-10 finishes than Phil Mickelson. Determine the number of top-10 finishes for each golfer and complete the following diagram.

**Swinging into the Top 10**

*Source*: PGA Tour, USA Today (1/12/05)

**48. Nascar Racing** The Nascar Nextel Cup Series consist of 36 races starting with the Daytona 500 in February and ending with the Ford 400 in Miami in November. In 2005, the top three drivers with the most points were, in order, Tony Stewart, Greg Biffle, and Carl Edwards. These three drivers had a total of 15 wins in the Nextel Cup series. Stewart had one more win than Edwards, and Biffle had two more wins than Edwards. Determine the number of wins each driver had.

**49. New England Snowstorm** During the last week of January 2005, New England had a record-breaking snowstorm that lasted several days. The coastal regions were hardest hit, with some cities receiving more than 3 feet of snow. In Massachusetts, the cities of Haverhill, Plymouth, and Salem had the most snow. The total snowfall for these three cities was 112.5 inches. Salem and Plymouth had the same amount of snow. Both had 1.5 inches more than Haverhill. Determine the snowfall amounts for each of these cities.

**50. Football** In the 2004 regular season of the National Football League (NFL), 19 players scored 100 or more points. The three players scoring the most points were Adam Vinatieri (NE), Jason Elam (DEN), and Jeff Reed (PIT). These three players scored a total of 394 points. Vinatieri scored 17 more points than Reed. Together, Vinatieri and Reed scored 7 more points than double the number of points scored by Elam. Determine the number of points Vinatieri Elam, and Reed scored. *Source:* www.nfl.com/stats/leaders

Jeff Reid of the Pittsburgh Steelers

**51. Super Bowls** Super Bowl XXXIX was held on February 6, 2005, in Jacksonville, Florida. Over the years, the states of Florida, California, and Louisiana, in this order, have hosted the most Super Bowls. These three states hosted a total of 32 Super Bowls. Florida hosted 3 more Super Bowls than Louisiana. Together, Florida and Louisiana hosted one less than twice the number California hosted. Determine the number of Super Bowls hosted by each of these three states. *Source:* NFL, *USA Today* (2/1/05)

**52. Concert Tickets** Three kinds of tickets for a Soggy Bottom Boys concert are available: up-front, main-floor, and balcony. The most expensive tickets, the up-front tickets, are twice as expensive as balcony tickets. Balcony tickets are \$10 less than main-floor tickets and \$30 less than up-front tickets. Determine the price of each kind of ticket.

**53. Triangle** The sum of the measures of the angles of a triangle is 180°. The smallest angle of the triangle has a measure $\frac{2}{3}$ the measure of the second smallest angle. The largest angle has a measure that is 30° less than three times the measure of the second smallest angle. Determine the measure of each angle.

**54. Triangle** The largest angle of a triangle has a measure that is 10° less than three times the measure of the second smallest angle. The measure of the smallest angle is equal to the difference between the measure of the largest angle and twice the measure of the second smallest angle. Determine the measures of the three angles of the triangle.

**55. Investments** Tam Phan received a check for \$10,000. She decided to divide the money (not equally) into three different investments. She placed part of her money in a savings account paying 3% interest. The second amount, which was twice the first amount, was placed in a certificate of deposit paying 5% interest. She placed the balance in a money market fund paying 6% interest. If Tam's total interest over the period of 1 year was \$525.00, how much was placed in each account?

**56. Bonus** Nick Pfaff, an attorney, divided his \$15,000 holiday bonus check among three different investments. With some of the money, he purchased a municipal bond paying 5.5% simple interest. He invested twice the amount of money that he paid for the municipal bond in a certificate of deposit paying 4.5% simple interest. Nick placed the balance of the money in a money market account paying 3.75% simple interest. If Nick's total interest for 1 year was \$692.50, how much was placed in each account?

**57. Hydrogen Peroxide** A 10% solution, a 12% solution, and a 20% solution of hydrogen peroxide are to be mixed to get 8 liters of a 13% solution. How many liters of each must be mixed if the volume of the 20% solution must be 2 liters less than the volume of the 10% solution?

**58. Sulfuric Acid** An 8% solution, a 10% solution, and a 20% solution of sulfuric acid are to be mixed to get 100 milliliters of a 12% solution. If the *volume of acid* from the 8% solution is to equal half the *volume of acid* from the other two solutions, how much of each solution is needed?

**59. Furniture Manufacturing** Donaldson Furniture Company produces three types of rocking chairs: the children's model, the standard model, and the executive model. Each chair is made in three stages: cutting, construction, and finishing. The time needed for each stage of each chair is given in the following chart. During a specific week the company has available a maximum of 154 hours for cutting, 94 hours for construction, and 76 hours for finishing. Determine how many of each type of chair the company should make to be operating at full capacity.

| Stage | Children's | Standard | Executive |
|---|---|---|---|
| Cutting | 5 hr | 4 hr | 7 hr |
| Construction | 3 hr | 2 hr | 5 hr |
| Finishing | 2 hr | 2 hr | 4 hr |

**60. Bicycle Manufacturing** The Jamis Bicycle Company produces three models of bicycles: Dakar, Komodo, and Aragon. Each bicycle is made in three stages: welding, painting, and assembling. The time needed for each stage of each bicycle is given in the chart on page 265. During a specific week, the company has available a maximum of 133 hours for welding, 78 hours for painting, and 96 hours for assembling. Determine how many of each type of bicycle the company should make to be operating at full capacity.

| Stage | Dakar | Komodo | Aragon |
|---|---|---|---|
| Welding | 2 | 3 | 4 |
| Painting | 1 | 2 | 2.5 |
| Assembling | 1.5 | 2 | 3 |

**61. Current Flow** In electronics it is necessary to analyze current flow through paths of a circuit. In three paths ($A$, $B$, and $C$) of a circuit, the relationships are the following:

$$I_A + I_B + I_C = 0$$
$$-8I_B + 10I_C = 0$$
$$4I_A - 8I_B = 6$$

where $I_A$, $I_B$, and $I_C$ represent the current in paths $A$, $B$, and $C$, respectively. Determine the current in each path of the circuit.

**62. Forces on a Beam** In physics we often study the forces acting on an object. For three forces, $F_1$, $F_2$, and $F_3$, acting on a beam, the following equations were obtained.

$$3F_1 + F_2 - F_3 = 2$$
$$F_1 - 2F_2 + F_3 = 0$$
$$4F_1 - F_2 + F_3 = 3$$

Find the three forces.

## Group Activity

*Discuss and answer Exercise 63 as a group.*

**63. Two Cars** A *nonlinear system of equations* is a system of equations containing at least one equation that is not linear. (Nonlinear systems of equations will be discussed in Chapter 10.) The graph shows a nonlinear system of equations. The curves represent speed versus time for two cars.

**a)** Are the two curves functions? Explain.

**b)** Discuss the meaning of this graph.

**c)** At time $t = 0.5$ hr, which car is traveling at a greater speed? Explain your answer.

**d)** Assume the two cars start at the same position and are traveling in the same direction. Which car, $A$ or $B$, traveled farther in 1 hour? Explain your answer.

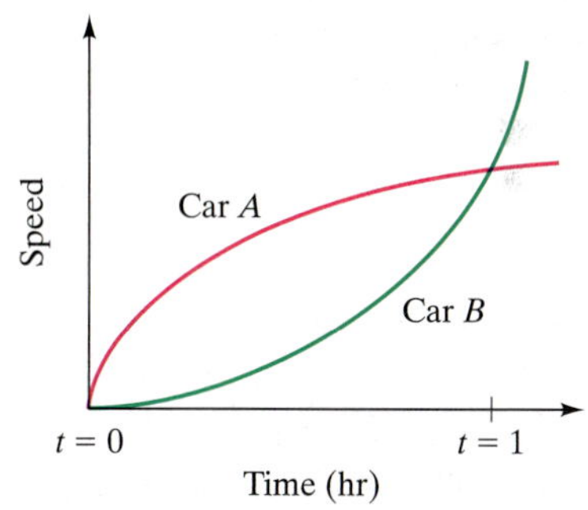


## Cumulative Review Exercises

[1.4] **64.** Evaluate $\frac{1}{2}x + \frac{2}{5}xy + \frac{1}{8}y$ when $x = -2$, $y = 5$.

[2.1] **65.** Solve $4 - 2[(x - 5) + 2x] = -(x + 6)$.

[3.2] **66.** Explain how to determine whether a graph represents a function.

[3.5] **67.** Write an equation of the line that passes through points $(6, -4)$ and $(2, -8)$.

## Mid-Chapter Test: 4.1–4.3

*To find out how well you understand the chapter material to this point, take this brief test. The answers, and the section where the material was initially discussed, are given in the back of the book. Review any questions that you answered incorrectly.*

**1.** For the following system of equations:

**a)** Write each equation in slope-intercept form.

**b)** Without graphing the equations, state whether the system is consistent, inconsistent, or dependent.

**c)** Indicate whether the system has exactly one solution, no solution, or an infinite number of solutions.

$$7x - y = 13$$
$$2x + 3y = 9$$

*Solve each system of equations using the graphing method.*

**2.** $y = 2x$
$y = -x + 3$

**3.** $x + y = -4$
$3x - 2y = 3$

*Solve each system of equations using the substitution method.*

**4.** $2x + 5y = -3$
$x - 2y = -6$

**5.** $4x - 3y = 8$
$2x + y = -1$

*Solve each system of equations using the addition method.*

**6.** $x = 4y - 19$
$7x + 5y = -1$

**7.** $3x + 4y = 3$
$9x + 5y = \frac{11}{2}$

*Solve each system of equations by any method. If the system is inconsistent or dependent, so state.*

**8.** $\frac{1}{3}a - \frac{1}{4}b = -1$
$\frac{1}{2}a + \frac{1}{6}b = 5$

**9.** $3m - 2n = 1$
$n = \frac{3}{2}m - 7$

**10.** $8x - 16y = 24$
$x = 2y + 3$

*Solve each system of equations.*

**11.** $x + y + z = 2$
$2x - y + 2z = -2$
$3x + 2y + 6z = 1$

**12.** $2x - y - z = 1$
$3x + 5y + 2z = 12$
$-6x - 4y + 5z = 3$

**13.** When asked to solve the system of equations

$$x + y + z = 4$$
$$-x + 2y + 2z = 5$$
$$7x + 5y - z = -2$$

Frank Dumont claimed that the solution was only $x = 1$.

This is incorrect. Why is it incorrect? Explain your answer.

Then solve the system completely.

**14. Cashews and Pecans** A local nut shop sells cashews for $12 per pound and pecans for $6 per pound. How many pounds of each type should William Pritchard buy to have a 15-pound mixture that sells for $10 per pound?

**15. Sum of Numbers** The sum of three numbers is 32. The largest number is four times the smallest number. The sum of the two smaller numbers is 8 less than the largest number. Find the three numbers.

# 4.4 Solving Systems of Equations Using Matrices

1 Write an augmented matrix.
2 Solve systems of linear equations.
3 Solve systems of linear equations in three variables.
4 Recognize inconsistent and dependent systems.

## 1 Write an Augmented Matrix

A **matrix** is a rectangular array of numbers within brackets. The plural of *matrix* is **matrices**. Examples of matrices are

$$\begin{bmatrix} 4 & 6 \\ 9 & -2 \end{bmatrix} \quad \begin{bmatrix} 5 & 7 & 2 \\ -1 & 3 & 4 \end{bmatrix}$$

The numbers inside the brackets are referred to as **elements** of the matrix.

The matrix on the left contains 2 rows and 2 columns and is called a 2 by 2 ($2 \times 2$) matrix. The matrix on the right contains 2 rows and 3 columns and is a 2 by 3 ($2 \times 3$) matrix. The number of rows is the first dimension given, and the number of columns is the second dimension given when describing the dimensions of a matrix. A **square matrix** has the same number of rows as columns. Thus, the matrix on the left is a square matrix.

In this section, we will use matrices to solve systems of linear equations. The first step in solving a system of two linear equations using matrices is to write each equation in the form $ax + by = c$. The next step is to write the **augmented matrix**, which is made up of two smaller matrices separated by a vertical line. The numbers on the left of the vertical line are the coefficients of the variables in the system of equations, and the numbers on the right are the constants. For the system of equations

$$a_1x + b_1y = c_1$$
$$a_2x + b_2y = c_2$$

the augmented matrix is written

$$\left[\begin{array}{cc|c} a_1 & b_1 & c_1 \\ a_2 & b_2 & c_2 \end{array}\right]$$

Following is a system of equations and its augmented matrix.

| System of Equations | Augmented Matrix |
|---|---|
| $-x + \frac{1}{6}y = 4$<br>$-3x - 5y = -\frac{1}{2}$ | $\left[\begin{array}{cc\|c} -1 & \frac{1}{6} & 4 \\ -3 & -5 & -\frac{1}{2} \end{array}\right]$ |

Notice that the bar in the augmented matrix separates the numerical coefficients from the constants. Since the matrix is just a shortened way of writing the system of equations, we can solve a linear system using matrices in a manner very similar to solving a system of equations using the addition method.

## 2 Solve Systems of Linear Equations

To solve a system of two linear equations using matrices, we rewrite the augmented matrix in **triangular form**,

$$\left[\begin{array}{cc|c} 1 & a & p \\ 0 & 1 & q \end{array}\right]$$

where the $a$, $p$, and $q$ are constants. From this type of augmented matrix we can write an equivalent system of equations. This matrix represents the linear system

$$\begin{aligned} 1x + ay &= p \\ 0x + 1y &= q \end{aligned} \quad \text{or} \quad \begin{aligned} x + ay &= p \\ y &= q \end{aligned}$$

For example,

$$\left[\begin{array}{cc|c} 1 & 2 & 4 \\ 0 & 1 & 5 \end{array}\right] \quad \text{represents} \quad \begin{aligned} x + 2y &= 4 \\ y &= 5 \end{aligned}$$

Note that the system above on the right can be easily solved by substitution. Its solution is $(-6, 5)$.

We use **row transformations** to rewrite the augmented matrix in triangular form. We will use three row transformation procedures.

**Procedures for Row Transformations**

1. All the numbers in a row may be multiplied (or divided) by any nonzero real number. (This is the same as multiplying both sides of an equation by a nonzero real number.)
2. All the numbers in a row may be multiplied by any nonzero real number. These products may then be added to the corresponding numbers in any other row. (This is equivalent to eliminating a variable from a system of equations using the addition method.)
3. The order of the rows may be switched. (This is equivalent to switching the order of the equations in a system of equations.)

Generally, when changing an element in the augmented matrix to 1 we use row transformation procedure 1, and when changing an element to 0 we use row transformation procedure 2. *Work by columns starting from the left.* Start with the first column, first row.

**EXAMPLE 1** ▸ Solve the following system of equations using matrices.

$$\begin{aligned} 2x - 3y &= 10 \\ 4x + 5y &= 9 \end{aligned}$$

*Solution* First we write the augmented matrix.

$$\left[\begin{array}{cc|c} 2 & -3 & 10 \\ 4 & 5 & 9 \end{array}\right]$$

Our goal is to obtain a matrix of the form $\left[\begin{array}{cc|c} 1 & a & p \\ 0 & 1 & q \end{array}\right]$ We begin by using row transformation procedure 1 to change the 2 in the first column, first row, to 1. To do so, we multiply the first row of numbers by $\frac{1}{2}$. (We abbreviate this multiplication as $\frac{1}{2}R_1$ and place it to the right of the matrix in the same row where the operation was performed. This may help you follow the process more clearly.)

$$\left[\begin{array}{cc|c} 2\left(\frac{1}{2}\right) & -3\left(\frac{1}{2}\right) & 10\left(\frac{1}{2}\right) \\ 4 & 5 & 9 \end{array}\right] \frac{1}{2}R_1$$

This gives

$$\left[\begin{array}{cc|c} 1 & -\frac{3}{2} & 5 \\ 4 & 5 & 9 \end{array}\right]$$

The next step is to obtain 0 in the first column, second row. At present, 4 is in this position. We do this by multiplying the numbers in row 1 by $-4$, and adding the products to the numbers in row 2. (This is abbreviated $-4R_1 + R_2$.)

The numbers in the first row multiplied by $-4$ are

$$1(-4) \quad -\frac{3}{2}(-4) \quad 5(-4)$$

Now we add these products to their respective numbers in the second row. This gives

$$\left[\begin{array}{cc|c} 1 & -\frac{3}{2} & 5 \\ 4 + 1(-4) & 5 + \left(-\frac{3}{2}\right)(-4) & 9 + 5(-4) \end{array}\right] -4R_1 + R_2$$

Now we have

$$\left[\begin{array}{cc|c} 1 & -\frac{3}{2} & 5 \\ 0 & 11 & -11 \end{array}\right]$$

To obtain 1 in the second column, second row, we multiply the second row of numbers by $\frac{1}{11}$.

$$\left[\begin{array}{cc|c} 1 & -\frac{3}{2} & 5 \\ 0\left(\frac{1}{11}\right) & 11\left(\frac{1}{11}\right) & -11\left(\frac{1}{11}\right) \end{array}\right] \frac{1}{11}R_2$$

$$\left[\begin{array}{cc|c} 1 & -\frac{3}{2} & 5 \\ 0 & 1 & -1 \end{array}\right]$$

The matrix is now in the form we are seeking. The equivalent triangular system of equations is

$$x - \frac{3}{2}y = 5$$
$$y = -1$$

Now we can solve for $x$ using substitution.

$$x - \frac{3}{2}y = 5$$
$$x - \frac{3}{2}(-1) = 5$$
$$x + \frac{3}{2} = 5$$
$$x = \frac{7}{2}$$

A check will show that the solution to the system is $\left(\frac{7}{2}, -1\right)$.

▶ Now Try Exercise 19

## 3 Solve Systems of Linear Equations in Three Variables

Now we will use matrices to solve a system of three linear equations in three variables. We use the same row transformation procedures used when solving a system of two linear equations. Our goal is to obtain an augmented matrix in the triangular form

$$\left[\begin{array}{ccc|c} 1 & a & b & p \\ 0 & 1 & c & q \\ 0 & 0 & 1 & r \end{array}\right]$$

where $a, b, c$, and $p, q$ and $r$ are constants. This matrix represents the following system of equations.

$$\begin{aligned} 1x + ay + bz &= p \\ 0x + 1y + cz &= q \\ 0x + 0y + 1z &= r \end{aligned} \qquad \text{or} \qquad \begin{aligned} x + ay + bz &= p \\ y + cz &= q \\ z &= r \end{aligned}$$

When constructing the augmented matrix, *work by columns, from the left-hand column to the right-hand column. Always complete one column before moving to the next column. In each column, first obtain the 1 in the indicated position, and then obtain the zeros.* Example 2 illustrates this procedure.

### Helpful Hint *Study Tip*

When using matrices, be careful to keep all the numbers lined up neatly in rows and columns. One slight mistake in copying numbers from one matrix to another will lead to an incorrect, and often frustrating, attempt at solving a system of equations.

For example, the system of equations, $\begin{aligned} x - 3y + z &= 3 \\ 4x + 2y - 5z &= 20 \\ -5x - y - 4z &= 13 \end{aligned}$, when correctly represented with the augmented matrix, $\left[\begin{array}{ccc|c} 1 & -3 & 1 & 3 \\ 4 & 2 & -5 & 20 \\ -5 & -1 & -4 & 13 \end{array}\right]$, leads to the solution $(1, -2, -4)$.

However, a matrix that looks quite similar, $\left[\begin{array}{ccc|c} 1 & -3 & 1 & 3 \\ 4 & -1 & -5 & 20 \\ -5 & 2 & -4 & 13 \end{array}\right]$, leads to the incorrect ordered triple of $\left(-\frac{25}{53}, -\frac{130}{53}, -\frac{206}{53}\right)$.

**EXAMPLE 2** ▸ Solve the following system of equations using matrices.

$$\begin{aligned} x - 2y + 3z &= -7 \\ 2x - y - z &= 7 \\ -x + 3y + 2z &= -8 \end{aligned}$$

*Solution* First write the augmented matrix.

$$\left[\begin{array}{rrr|r} 1 & -2 & 3 & -7 \\ 2 & -1 & -1 & 7 \\ -1 & 3 & 2 & -8 \end{array}\right]$$

Our next step is to use row transformations to change the first column to $\begin{matrix}1\\0\\0\end{matrix}$. Since the number in the first column, first row, is already a 1, we will work with the 2 in the first column, second row. Multiplying the numbers in the first row by $-2$ and adding those products to the respective numbers in the second row will result in the 2 changing to 0. The matrix is now

$$\left[\begin{array}{rrr|r} 1 & -2 & 3 & -7 \\ 0 & 3 & -7 & 21 \\ -1 & 3 & 2 & -8 \end{array}\right] -2R_1 + R_2$$

Continuing down the first column, we now change the $-1$ in the third row to 0. By multiplying the numbers in the first row by 1, and then adding the products to the third row, we get

$$\left[\begin{array}{rrr|r} 1 & -2 & 3 & -7 \\ 0 & 3 & -7 & 21 \\ 0 & 1 & 5 & -15 \end{array}\right] 1R_1 + R_3$$

Now we work with the second column. We wish to change the numbers in the second column to the form $\begin{matrix}a\\1\\0\end{matrix}$ where $a$ represents a number. Since there is presently a 1 in the third row, second column, and we want a 1 in the second row, second column, we switch the second and third rows of the matrix. This gives

$$\left[\begin{array}{rrr|r} 1 & -2 & 3 & -7 \\ 0 & 1 & 5 & -15 \\ 0 & 3 & -7 & 21 \end{array}\right] \text{Switch } R_2 \text{ and } R_3.$$

Continuing down the second column, we now change the 3 in the third row to 0 by multiplying the numbers in the second row by $-3$ and adding those products to the third row. This gives

$$\left[\begin{array}{rrr|r} 1 & -2 & 3 & -7 \\ 0 & 1 & 5 & -15 \\ 0 & 0 & -22 & 66 \end{array}\right] -3R_2 + R_3$$

Now we work with the third column. We wish to change the numbers in the third column to the form $\begin{matrix}b\\c\\1\end{matrix}$ where $b$ and $c$ represent numbers. We must change the $-22$ in the third row to 1. We can do this by multiplying the numbers in the third row by $-\frac{1}{22}$. This results in the following.

$$\left[\begin{array}{ccc|c} 1 & -2 & 3 & -7 \\ 0 & 1 & 5 & -15 \\ 0 & 0 & 1 & -3 \end{array}\right] \quad -\tfrac{1}{22}R_3$$

This matrix is now in the desired form. From this matrix we obtain the system of equations

$$\begin{aligned} x - 2y + 3z &= -7 \\ y + 5z &= -15 \\ z &= -3 \end{aligned}$$

The third equation gives us the value of $z$ in the solution. Now we can solve for $y$ by substituting $-3$ for $z$ in the second equation.

$$\begin{aligned} y + 5z &= -15 \\ y + 5(-3) &= -15 \\ y - 15 &= -15 \\ y &= 0 \end{aligned}$$

Now we solve for $x$ by substituting 0 for $y$ and $-3$ for $z$ in the first equation.

$$\begin{aligned} x - 2y + 3z &= -7 \\ x - 2(0) + 3(-3) &= -7 \\ x - 0 - 9 &= -7 \\ x - 9 &= -7 \\ x &= 2 \end{aligned}$$

The solution is $(2, 0, -3)$. Check this now by substituting the appropriate values into each of the original equations.

▸ **Now Try Exercise 33**

## 4 Recognize Inconsistent and Dependent Systems

When solving a system of two equations, if you obtain an augmented matrix in which one row of numbers on the left side of the vertical line is all zeros but a zero does not appear in the same row on the right side of the vertical line, the system is inconsistent and has no solution. For example, a system of equations that yields the following augmented matrix is an inconsistent system.

$$\left[\begin{array}{cc|c} 1 & 2 & 5 \\ 0 & 0 & 3 \end{array}\right] \quad \longleftarrow \text{Inconsistent system}$$

The second row of the matrix represents the equation

$$0x + 0y = 3$$

which is never true.

If you obtain a matrix in which a 0 appears across an entire row, the system of equations is dependent. For example, a system of equations that yields the following augmented matrix is a dependent system.

$$\left[\begin{array}{cc|c} 1 & -3 & -4 \\ 0 & 0 & 0 \end{array}\right] \quad \longleftarrow \text{Dependent system}$$

The second row of the matrix represents the equation

$$0x + 0y = 0$$

which is always true.

Similar rules hold for systems with three equations.

$$\left[\begin{array}{ccc|c} 1 & 3 & 7 & 5 \\ 0 & 0 & 0 & -1 \\ 0 & 1 & -2 & 3 \end{array}\right] \longleftarrow \text{Inconsistent system}$$

$$\left[\begin{array}{ccc|c} 1 & 3 & -1 & 2 \\ 0 & 0 & 0 & 0 \\ 0 & 5 & 6 & -4 \end{array}\right] \longleftarrow \text{Dependent system}$$

### USING YOUR GRAPHING CALCULATOR

Many graphing calculators have the ability to work with matrices. Such calculators have the ability to perform row operations on matrices. These graphing calculators can therefore be used to solve systems of equations using matrices.

Read the instruction manual that came with your graphing calculator to see if it can handle matrices. If so, learn how to use your graphing calculator to solve systems of equations using matrices.

## EXERCISE SET 4.4

MathXL® MyMathLab

### Concept/Writing Exercises

1. What is a square matrix?
2. Explain how to construct an augmented matrix.
3. If you obtain the following augmented matrix when solving a system of equations, what would be your next step in completing the process? Explain.

$$\left[\begin{array}{cc|c} 1 & 3 & 6 \\ 0 & -2 & 14 \end{array}\right]$$

4. If you obtained the following augmented matrix when solving a system of equations, what would be your next step in completing the process? Explain your answer.

$$\left[\begin{array}{ccc|c} 1 & -3 & 7 & -1 \\ 0 & -1 & 5 & 3 \\ 2 & 6 & 4 & -8 \end{array}\right]$$

5. If you obtained the following augmented matrix when solving a system of linear equations, what would be your next step in completing the process? Explain your answer.

$$\left[\begin{array}{ccc|c} 1 & 4 & -7 & 7 \\ 0 & 5 & 2 & -1 \\ 0 & 1 & 6 & -2 \end{array}\right]$$

6. If you obtained the following augmented matrix when solving a system of linear equations, what would be your next step in completing the process? Explain your answer.

$$\left[\begin{array}{ccc|c} 1 & 3 & -2 & 1 \\ 0 & 1 & 2 & -3 \\ 0 & 0 & -4 & -12 \end{array}\right]$$

7. When solving a system of linear equations by matrices, if two rows are identical, will the system be consistent, dependent, or inconsistent?
8. When solving a system of equations using matrices, how will you know if the system is
   **a)** dependent,
   **b)** inconsistent?

### Practice the Skills

*Perform each row transformation indicated and write the new matrix.*

**9.** $\left[\begin{array}{cc|c} 5 & -10 & -25 \\ 3 & -7 & -4 \end{array}\right]$ Multiply numbers in the first row by $\frac{1}{5}$.

**10.** $\left[\begin{array}{cc|c} 1 & 8 & 3 \\ 0 & 4 & -3 \end{array}\right]$ Multiply numbers in the second row by $\frac{1}{4}$.

**11.** $\left[\begin{array}{ccc|c} 4 & 7 & 2 & -1 \\ 3 & 2 & 1 & -5 \\ 1 & 1 & 3 & -8 \end{array}\right]$ Switch row 1 and row 3.

**12.** $\left[\begin{array}{rrr|r} 1 & 5 & 7 & 2 \\ 0 & 8 & -1 & -6 \\ 0 & 1 & 3 & -4 \end{array}\right]$ Switch row 2 and row 3.

**13.** $\left[\begin{array}{rr|r} 1 & 3 & 12 \\ -4 & 11 & -6 \end{array}\right]$ Multiply numbers in the first row by 4 and add the products to the second row.

**14.** $\left[\begin{array}{rr|r} 1 & 5 & 6 \\ \frac{1}{2} & 10 & -4 \end{array}\right]$ Multiply numbers in the first row by $-\frac{1}{2}$ and add the products to the second row.

**15.** $\left[\begin{array}{rrr|r} 1 & 0 & 8 & \frac{1}{4} \\ 5 & 2 & 2 & -2 \\ 6 & -3 & 1 & 0 \end{array}\right]$ Multiply numbers in the first row by $-5$ and add the products to the second row.

**16.** $\left[\begin{array}{rrr|r} 1 & 2 & -1 & 6 \\ 0 & 1 & 5 & 0 \\ 0 & 0 & 3 & 12 \end{array}\right]$ Multiply numbers in the third row by $\frac{1}{3}$.

*Solve each system using matrices.*

**17.** $x + 3y = 3$
$-x + y = -3$

**18.** $x + 2y = 10$
$3x - y = 9$

**19.** $x + 3y = -2$
$-2x - 7y = 3$

**20.** $3x + 6y = 0$
$2x - y = 10$

**21.** $5a - 10b = -10$
$2a + b = 1$

**22.** $3s - 2t = 1$
$-2s + 4t = -6$

**23.** $2x - 5y = -6$
$-4x + 10y = 12$

**24.** $-2m - 4n = 7$
$3m + 6n = -8$

**25.** $12x + 2y = 2$
$6x - 3y = -11$

**26.** $4r + 2s = -10$
$-2r + s = -7$

**27.** $-3x + 6y = 5$
$2x - 4y = 7$

**28.** $8x = 4y + 12$
$-2x + y = -3$

**29.** $12x - 8y = 6$
$-3x + 4y = -1$

**30.** $2x - 3y = 3$
$-5x + 9y = -7$

**31.** $10m = 8n + 15$
$16n = -15m - 2$

**32.** $8x = 9y + 4$
$16x - 27y = 11$

*Solve each system using matrices.*

**33.** $x - 3y + 2z = 5$
$2x + 3y - 4z = -3$
$-3x + y - 2z = -11$

**34.** $a - 3b + 4c = 7$
$4a + b + c = -2$
$-2a - 3b + 5c = 12$

**35.** $x + 2y = 5$
$y - z = -1$
$2x - 3z = 0$

**36.** $3a - 5c = 3$
$a + 2b = -6$
$7b - 4c = 5$

**37.** $x - 2y + 4z = 5$
$-3x + 4y - 2z = -8$
$4x + 5y - 4z = -3$

**38.** $3x + 5y + 2z = 3$
$-x - y - z = -2$
$2x - 2y + 5z = 11$

**39.** $2x - 5y + z = 1$
$3x - 5y + z = 3$
$-4x + 10y - 2z = -2$

**40.** $x + 2y + 3z = 1$
$4x + 5y + 6z = -3$
$7x + 8y + 9z = 0$

**41.** $4p - q + r = 4$
$-6p + 3q - 2r = -5$
$2p + 5q - r = 7$

**42.** $-4r + 3s - 6t = 14$
$4r + 2s - 2t = -3$
$2r - 5s - 8t = -23$

**43.** $2x - 4y + 3z = -12$
$3x - y + 2z = -3$
$-4x + 8y - 6z = 10$

**44.** $3x - 2y + 4z = -1$
$5x + 2y - 4z = 9$
$-6x + 4y - 8z = 2$

**45.** $5x - 3y + 4z = 22$
$-x - 15y + 10z = -15$
$-3x + 9y - 12z = -6$

**46.** $9x - 4y + 5z = -2$
$-9x + 5y - 10z = -1$
$9x + 3y + 10z = 1$

## Problem Solving

**47.** When solving a system of linear equations using matrices, if two rows of matrices are switched, will the solution to the system change? Explain.

**48.** You can tell whether a system of two equations in two variables is consistent, dependent, or inconsistent by comparing the slopes and $y$-intercepts of the graphs of the equations. Can you tell, without solving, if a system of three equations in three variables is consistent, dependent, or inconsistent? Explain.

*Solve using matrices.*

**49. Angles of a Roof** In a triangular cross section of a roof, the largest angle is 55° greater than the smallest angle. The largest angle is 20° greater than the remaining angle. Find the measure of each angle.

**50. Right Angle** A right angle is divided into three smaller angles. The largest of the three angles is twice the smallest. The remaining angle is 10° greater than the smallest angle. Find the measure of each angle.

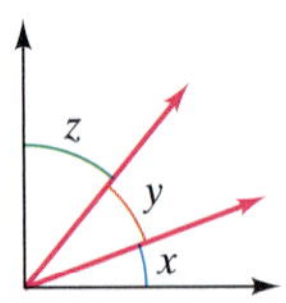


**51. Bananas** Sixty-five percent of the world's bananas are controlled by Chiquita, Dole, or Del Monte (all American companies). Chiquita, the largest, controls 12% more bananas than Del Monte. Dole, the second largest, controls 3% less than twice the percent that Del Monte controls. Determine the percents to be placed in each sector of the circle graph shown.

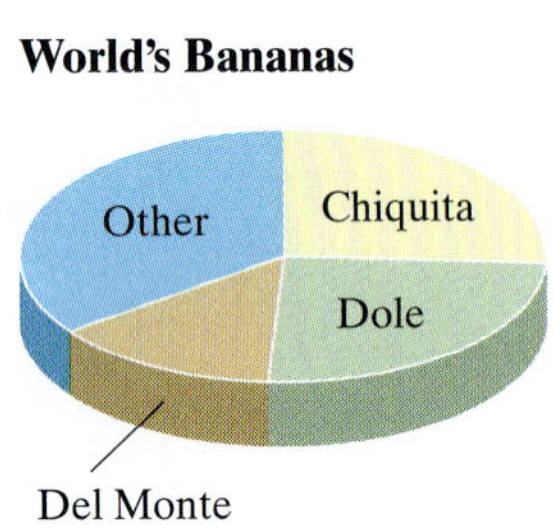


**52. Impact Upon Businesses** A sample of CEOs in a TEC International survey were asked to list the most important changes that could be made from 2004 to 2006 to strengthen their companies. The top three responses were, in order: reduce taxes, reform health care insurance, and strengthen the U.S. dollar. Seventy-seven percent of all the CEOs selected one of these three items as their top choice. Four percent more CEOs selected reducing taxes than reforming health care insurance. Reducing taxes was also two percent higher than three times the percent who selected strengthening the U.S. dollar. Determine the percent of CEOs that selected reducing taxes, reforming health care insurance, and strengthening the U.S. dollar. Then complete the graph below.

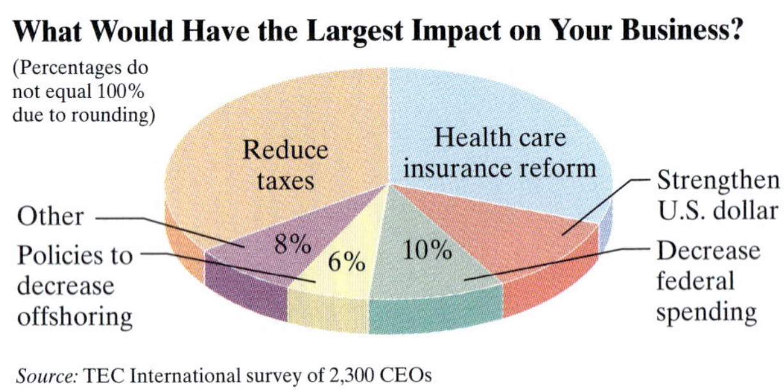


## Cumulative Review Exercises

[1.2] **53.** $A = \{1, 2, 4, 6, 9\}$; $B = \{3, 4, 5, 6, 10\}$. Find

**a)** $A \cup B$;

**b)** $A \cap B$.

[2.5] **54.** Indicate the inequality $-1 < x \leq 4$

**a)** on a number line,

**b)** as a solution set, and

**c)** in interval notation.

[3.2] **55.** What does a graph represent?

[3.4] **56.** If $f(x) = -2x^2 + 3x - 6$, find $f(-5)$.

# 4.5 Solving Systems of Equations Using Determinants and Cramer's Rule

1 Evaluate a determinant of a $2 \times 2$ matrix.

2 Use Cramer's rule.

3 Evaluate a determinant of a $3 \times 3$ matrix.

4 Use Cramer's rule with systems in three variables.

## 1 Evaluate a Determinant of a 2 × 2 Matrix

We have discussed various ways of solving a system of linear equations, including: graphing, substitution, the addition (or elimination) method, and matrices. A system of linear equations may also be solved using determinants.

Associated with every square matrix is a number called its **determinant**. For a $2 \times 2$ matrix, its determinant is defined as follows.

**Determinant**

The **determinant** of a $2 \times 2$ matrix $\begin{bmatrix} a_1 & b_1 \\ a_2 & b_2 \end{bmatrix}$ is denoted $\begin{vmatrix} a_1 & b_1 \\ a_2 & b_2 \end{vmatrix}$ and is evaluated as

$$\begin{vmatrix} a_1 & b_1 \\ a_2 & b_2 \end{vmatrix} = a_1b_2 - a_2b_1$$

**EXAMPLE 1** ▸ Evaluate each determinant.

**a)** $\begin{vmatrix} 2 & -1 \\ 3 & -5 \end{vmatrix}$ **b)** $\begin{vmatrix} 2 & 3 \\ -1 & 4 \end{vmatrix}$

*Solution*

**a)** $a_1 = 2, a_2 = 3, b_1 = -1, b_2 = -5$

$$\begin{vmatrix} 2 & -1 \\ 3 & -5 \end{vmatrix} = 2(-5) - (3)(-1) = -10 + 3 = -7$$

**b)** $$\begin{vmatrix} 2 & 3 \\ -1 & 4 \end{vmatrix} = (2)(4) - (-1)(3) = 8 + 3 = 11$$

▸ Now Try Exercise 7

## 2 Use Cramer's Rule

If we begin with the equations

$$a_1x + b_1y = c_1$$
$$a_2x + b_2y = c_2$$

we can use the addition method to show that

$$x = \frac{c_1b_2 - c_2b_1}{a_1b_2 - a_2b_1} \quad \text{and} \quad y = \frac{a_1c_2 - a_2c_1}{a_1b_2 - a_2b_1}$$

(see Challenge Problem 65 on page 281). Notice that the *denominators* of $x$ and $y$ are both $a_1b_2 - a_2b_1$. Following is the determinant that yields this denominator. We have labeled this denominator $D$.

$$D = \begin{vmatrix} a_1 & b_1 \\ a_2 & b_2 \end{vmatrix} = a_1b_2 - a_2b_1$$

The *numerators* of $x$ and $y$ are different. Following are two determinants, labeled $D_x$ and $D_y$, that yield the numerators of $x$ and $y$.

$$D_x = \begin{vmatrix} c_1 & b_1 \\ c_2 & b_2 \end{vmatrix} = c_1b_2 - c_2b_1 \qquad D_y = \begin{vmatrix} a_1 & c_1 \\ a_2 & c_2 \end{vmatrix} = a_1c_2 - a_2c_1$$

We use determinants $D$, $D_x$, and $D_y$ in Cramer's rule. **Cramer's rule** can be used to solve systems of linear equations.

**Cramer's Rule for Systems of Linear Equations**

For a system of linear equations of the form

$$a_1x + b_1y = c_1$$
$$a_2x + b_2y = c_2$$

$$x = \frac{\begin{vmatrix} c_1 & b_1 \\ c_2 & b_2 \end{vmatrix}}{\begin{vmatrix} a_1 & b_1 \\ a_2 & b_2 \end{vmatrix}} = \frac{D_x}{D} \quad \text{and} \quad y = \frac{\begin{vmatrix} a_1 & c_1 \\ a_2 & c_2 \end{vmatrix}}{\begin{vmatrix} a_1 & b_1 \\ a_2 & b_2 \end{vmatrix}} = \frac{D_y}{D}, \quad D \neq 0$$

## Helpful Hint

The elements in determinant $D$ are the numerical coefficients of the $x$ and $y$ terms in the two given equations, listed in the same order they are listed in the equations. To obtain the determinant $D_x$ from determinant $D$, replace the coefficients of the $x$-terms (the values in the first column) with the constants of the two given equations. To obtain the determinant $D_y$ from determinant $D$, replace the coefficients of the $y$-terms (the values in the second column) with the constants of the two given equations.

**EXAMPLE 2** ▸ Solve the following system of equations using Cramer's rule.

$$3x + 5y = 7$$
$$4x - y = -6$$

*Solution* Both equations are given in the desired form, $ax + by = c$. When labeling $a$, $b$, and $c$, we will refer to $3x + 5y = 7$ as equation 1 and $4x - y = -6$ as equation 2 (in the subscripts).

$$\begin{array}{ccc} a_1 & b_1 & c_1 \\ \downarrow & \downarrow & \downarrow \\ 3x & + 5y & = 7 \\ 4x & - 1y & = -6 \\ \uparrow & \uparrow & \uparrow \\ a_2 & b_2 & c_2 \end{array}$$

We now determine $D$, $D_x$, $D_y$.

$$D = \begin{vmatrix} a_1 & b_1 \\ a_2 & b_2 \end{vmatrix} = \begin{vmatrix} 3 & 5 \\ 4 & -1 \end{vmatrix} = 3(-1) - 4(5) = -3 - 20 = -23$$

$$D_x = \begin{vmatrix} c_1 & b_1 \\ c_2 & b_2 \end{vmatrix} = \begin{vmatrix} 7 & 5 \\ -6 & -1 \end{vmatrix} = 7(-1) - (-6)(5) = -7 + 30 = 23$$

$$D_y = \begin{vmatrix} a_1 & c_1 \\ a_2 & c_2 \end{vmatrix} = \begin{vmatrix} 3 & 7 \\ 4 & -6 \end{vmatrix} = 3(-6) - 4(7) = -18 - 28 = -46$$

Now we find the values of $x$ and $y$.

$$x = \frac{D_x}{D} = \frac{23}{-23} = -1$$

$$y = \frac{D_y}{D} = \frac{-46}{-23} = 2$$

Thus the solution is $x = -1$, $y = 2$ or the ordered pair $(-1, 2)$. A check will show that this ordered pair satisfies both equations.

▸ **Now Try Exercise 15**

When the determinant $D = 0$, Cramer's rule cannot be used since division by 0 is undefined. You must then use a different method to solve the system. Or you may evaluate $D_x$ and $D_y$ to determine whether the system is dependent or inconsistent.

**When $D = 0$**

If $D = 0$, $D_x = 0$, $D_y = 0$, then the system is dependent.

If $D = 0$ and either $D_x \neq 0$ or $D_y \neq 0$, then the system is inconsistent.

## 3 Evaluate a Determinant of a 3 × 3 Matrix

For the determinant

$$\begin{vmatrix} a_1 & b_1 & c_1 \\ a_2 & b_2 & c_2 \\ a_3 & b_3 & c_3 \end{vmatrix}$$

the **minor determinant** of $a_1$ is found by crossing out the elements in the same row and column in which the element $a_1$ appears. The remaining elements form the minor determinant of $a_1$. The minor determinants of other elements are found similarly.

$$\begin{vmatrix} a_1 & b_1 & c_1 \\ a_2 & b_2 & c_2 \\ a_3 & b_3 & c_3 \end{vmatrix} \qquad \begin{vmatrix} b_2 & c_2 \\ b_3 & c_3 \end{vmatrix} \qquad \text{Minor determinant of } a_1$$

$$\begin{vmatrix} a_1 & b_1 & c_1 \\ a_2 & b_2 & c_2 \\ a_3 & b_3 & c_3 \end{vmatrix} \qquad \begin{vmatrix} b_1 & c_1 \\ b_3 & c_3 \end{vmatrix} \qquad \text{Minor determinant of } a_2$$

$$\begin{vmatrix} a_1 & b_1 & c_1 \\ a_2 & b_2 & c_2 \\ a_3 & b_3 & c_3 \end{vmatrix} \qquad \begin{vmatrix} b_1 & c_1 \\ b_2 & c_2 \end{vmatrix} \qquad \text{Minor determinant of } a_3$$

To evaluate determinants of a 3 × 3 matrix, we use minor determinants. The following box shows how such a determinant may be evaluated by **expansion by the minors of the first column**.

**Expansion of the Determinant by the Minors of the First Column**

Minor determinant of $a_1$ ↓    Minor determinant of $a_2$ ↓    Minor determinant of $a_3$ ↓

$$\begin{vmatrix} a_1 & b_1 & c_1 \\ a_2 & b_2 & c_2 \\ a_3 & b_3 & c_3 \end{vmatrix} = a_1\begin{vmatrix} b_2 & c_2 \\ b_3 & c_3 \end{vmatrix} - a_2\begin{vmatrix} b_1 & c_1 \\ b_3 & c_3 \end{vmatrix} + a_3\begin{vmatrix} b_1 & c_1 \\ b_2 & c_2 \end{vmatrix}$$

**EXAMPLE 3** ▸ Evaluate $\begin{vmatrix} 4 & -2 & 6 \\ 3 & 5 & 0 \\ 1 & -3 & -1 \end{vmatrix}$ using expansion by the minors of the first column.

*Solution* We will follow the procedure given in the box.

$$\begin{vmatrix} 4 & -2 & 6 \\ 3 & 5 & 0 \\ 1 & -3 & -1 \end{vmatrix} = 4\begin{vmatrix} 5 & 0 \\ -3 & -1 \end{vmatrix} - 3\begin{vmatrix} -2 & 6 \\ -3 & -1 \end{vmatrix} + 1\begin{vmatrix} -2 & 6 \\ 5 & 0 \end{vmatrix}$$

$$= 4[5(-1) - (-3)0] - 3[(-2)(-1) - (-3)6] + 1[(-2)0 - 5(6)]$$

$$= 4(-5 + 0) - 3(2 + 18) + 1(0 - 30)$$

$$= 4(-5) - 3(20) + 1(-30)$$

$$= -20 - 60 - 30$$

$$= -110$$

The determinant has a value of $-110$.

▸ Now Try Exercise 13

## 4 Use Cramer's Rule with Systems in Three Variables

Cramer's rule can be extended to systems of equations in three variables as follows.

**Cramer's Rule for a System of Equations in Three Variables**

To solve the system

$$\begin{aligned} a_1x + b_1y + c_1z &= d_1 \\ a_2x + b_2y + c_2z &= d_2 \\ a_3x + b_3y + c_3z &= d_3 \end{aligned}$$

with

$$D = \begin{vmatrix} a_1 & b_1 & c_1 \\ a_2 & b_2 & c_2 \\ a_3 & b_3 & c_3 \end{vmatrix} \qquad D_x = \begin{vmatrix} d_1 & b_1 & c_1 \\ d_2 & b_2 & c_2 \\ d_3 & b_3 & c_3 \end{vmatrix}$$

$$D_y = \begin{vmatrix} a_1 & d_1 & c_1 \\ a_2 & d_2 & c_2 \\ a_3 & d_3 & c_3 \end{vmatrix} \qquad D_z = \begin{vmatrix} a_1 & b_1 & d_1 \\ a_2 & b_2 & d_2 \\ a_3 & b_3 & d_3 \end{vmatrix}$$

then

$$x = \frac{D_x}{D} \quad y = \frac{D_y}{D} \quad z = \frac{D_z}{D}, \quad D \neq 0$$

Note that the denominators of the expressions for $x$, $y$, and $z$ are all the same determinant, $D$. Note that the $d$'s replace the $a$'s, the numerical coefficients of the $x$-terms, in $D_x$. The $d$'s replace the $b$'s, the numerical coefficients of the $y$-terms, in $D_y$. And the $d$'s replace the $c$'s, the numerical coefficients of the $z$-terms, in $D_z$.

**EXAMPLE 4** ▸ Solve the following system of equations using determinants.

$$\begin{aligned} 3x - 2y - \phantom{2}z &= -6 \\ 2x + 3y - 2z &= \phantom{-}1 \\ x - 4y + \phantom{2}z &= -3 \end{aligned}$$

*Solution*

$$\begin{array}{llll} a_1 = 3 & b_1 = -2 & c_1 = -1 & d_1 = -6 \\ a_2 = 2 & b_2 = \phantom{-}3 & c_2 = -2 & d_2 = \phantom{-}1 \\ a_3 = 1 & b_3 = -4 & c_3 = \phantom{-}1 & d_3 = -3 \end{array}$$

We will use expansion by the minor determinants of the first column to evaluate $D$, $D_x$, $D_y$, and $D_z$.

$$D = \begin{vmatrix} 3 & -2 & -1 \\ 2 & 3 & -2 \\ 1 & -4 & 1 \end{vmatrix} = 3\begin{vmatrix} 3 & -2 \\ -4 & 1 \end{vmatrix} - 2\begin{vmatrix} -2 & -1 \\ -4 & 1 \end{vmatrix} + 1\begin{vmatrix} -2 & -1 \\ 3 & -2 \end{vmatrix}$$

$$= 3(-5) - 2(-6) + 1(7)$$

$$= -15 + 12 + 7 = 4$$

$$D_x = \begin{vmatrix} -6 & -2 & -1 \\ 1 & 3 & -2 \\ -3 & -4 & 1 \end{vmatrix} = -6\begin{vmatrix} 3 & -2 \\ -4 & 1 \end{vmatrix} - 1\begin{vmatrix} -2 & -1 \\ -4 & 1 \end{vmatrix} + (-3)\begin{vmatrix} -2 & -1 \\ 3 & -2 \end{vmatrix}$$

$$= -6(-5) - 1(-6) - 3(7)$$

$$= 30 + 6 - 21 = 15$$

$$D_y = \begin{vmatrix} 3 & -6 & -1 \\ 2 & 1 & -2 \\ 1 & -3 & 1 \end{vmatrix} = 3\begin{vmatrix} 1 & -2 \\ -3 & 1 \end{vmatrix} - 2\begin{vmatrix} -6 & -1 \\ -3 & 1 \end{vmatrix} + 1\begin{vmatrix} -6 & -1 \\ 1 & -2 \end{vmatrix}$$

$$= 3(-5) - 2(-9) + 1(13)$$

$$= -15 + 18 + 13 = 16$$

$$D_z = \begin{vmatrix} 3 & -2 & -6 \\ 2 & 3 & 1 \\ 1 & -4 & -3 \end{vmatrix} = 3\begin{vmatrix} 3 & 1 \\ -4 & -3 \end{vmatrix} - 2\begin{vmatrix} -2 & -6 \\ -4 & -3 \end{vmatrix} + 1\begin{vmatrix} -2 & -6 \\ 3 & 1 \end{vmatrix}$$

$$= 3(-5) - 2(-18) + 1(16)$$

$$= -15 + 36 + 16 = 37$$

We found that $D = 4$, $D_x = 15$, $D_y = 16$, and $D_z = 37$. Therefore,

$$x = \frac{D_x}{D} = \frac{15}{4} \qquad y = \frac{D_y}{D} = \frac{16}{4} = 4 \qquad z = \frac{D_z}{D} = \frac{37}{4}$$

The solution to the system is $\left(\frac{15}{4}, 4, \frac{37}{4}\right)$. Note the ordered triple lists $x$, $y$, and $z$ in this order.

**▸ Now Try Exercise 33**

When we have a system of equations in three variables in which one or more equations are missing a variable, we insert the variable with a coefficient of 0. Thus,

$$\begin{aligned} 2x - 3y + 2z &= -1 \\ x + 2y \quad\quad &= 14 \\ x \quad\quad - 3z &= -5 \end{aligned} \quad \text{is written} \quad \begin{aligned} 2x - 3y + 2z &= -1 \\ x + 2y + 0z &= 14 \\ x + 0y - 3z &= -5 \end{aligned}$$

## Helpful Hint

When evaluating determinants, if any two rows (or columns) are identical, or identical except for opposite signs, the determinant has a value of 0. For example,

$$\begin{vmatrix} 5 & -2 \\ 5 & -2 \end{vmatrix} = 0 \quad \text{and} \quad \begin{vmatrix} 5 & -2 \\ -5 & 2 \end{vmatrix} = 0$$

$$\begin{vmatrix} 5 & -3 & 4 \\ 2 & 6 & 5 \\ 5 & -3 & 4 \end{vmatrix} = 0 \quad \text{and} \quad \begin{vmatrix} 5 & -3 & 4 \\ -5 & 3 & -4 \\ 6 & 8 & 2 \end{vmatrix} = 0$$

As with determinants of a $2 \times 2$ matrix, when the determinant $D = 0$, Cramer's rule cannot be used since division by 0 is undefined. You must then use a different method to solve the system. Or you may evaluate $D_x$, $D_y$, and $D_z$ to determine whether the system is dependent or inconsistent.

**When $D = 0$**

If $D = 0$, $D_x = 0$, $D_y = 0$, and $D_z = 0$, then the system is dependent.

If $D = 0$ and $D_x \neq 0$, $D_y \neq 0$, or $D_z \neq 0$, then the system is inconsistent.

## USING YOUR GRAPHING CALCULATOR

In Section 4.4 we mentioned that some graphing calculators can handle matrices. Graphing calculators with matrix capabilities can also find determinants of square matrices. Read your graphing calculator manual to learn if your calculator can find determinants. If so, learn how to do so on your calculator.

## EXERCISE SET 4.5

MathXL® MyMathLab

### Concept/Writing Exercises

1. Explain how to evaluate a $2 \times 2$ determinant.
2. Explain how to evaluate a $3 \times 3$ determinant by expansion by the minors of the first column.
3. Explain how you can determine whether a system of three linear equations is inconsistent using determinants.
4. Explain how you can determine whether a system of three linear equations is dependent using determinants.
5. While solving a system of two linear equations using Cramer's rule, you determine that $D = 4$, $D_x = 12$, and $D_y = -2$. What is the solution to this system?
6. While solving a system of three linear equations using Cramer's rule, you determine that $D = -2$, $D_x = 8$, $D_y = 14$, and $D_z = -2$. What is the solution to this system?

### Practice the Skills

*Evaluate each determinant.*

7. $\begin{vmatrix} 2 & 4 \\ 1 & 5 \end{vmatrix}$

8. $\begin{vmatrix} 3 & 5 \\ -1 & -2 \end{vmatrix}$

9. $\begin{vmatrix} \frac{1}{2} & 3 \\ 2 & -4 \end{vmatrix}$

10. $\begin{vmatrix} 13 & -\frac{2}{3} \\ -1 & 0 \end{vmatrix}$

11. $\begin{vmatrix} 3 & 2 & 0 \\ 0 & 5 & 3 \\ -1 & 4 & 2 \end{vmatrix}$

12. $\begin{vmatrix} 4 & 1 & 1 \\ 0 & 0 & 3 \\ 2 & 2 & 9 \end{vmatrix}$

13. $\begin{vmatrix} 2 & 3 & 1 \\ 1 & -3 & -6 \\ -4 & 5 & 9 \end{vmatrix}$

14. $\begin{vmatrix} 5 & -8 & 6 \\ 3 & 0 & 4 \\ -5 & -2 & 1 \end{vmatrix}$

*Solve each system of equations using determinants.*

15. $x + 3y = 1$
$-2x - 3y = 4$

16. $2x + 4y = -2$
$-5x - 2y = 13$

17. $-x - 2y = 2$
$x + 3y = -6$

18. $2r + 3s = -9$
$3r + 5s = -16$

19. $6x = 4y + 7$
$8x - 1 = -3y$

20. $6x + 3y = -4$
$9x + 5y = -6$

21. $5p - 7q = -21$
$-4p + 3q = 22$

22. $4x = -5y - 2$
$-2x = y + 4$

23. $x + 5y = 3$
$2x - 6 = -10y$

24. $9x + 6y = -3$
$6x + 4y = -2$

25. $3r = -4s - 6$
$3s = -5r + 1$

26. $x = y - 1$
$3y = 2x + 9$

27. $5x - 5y = 3$
$-x + y = -4$

28. $2x - 5y = -3$
$-4x + 10y = 7$

29. $6.3x - 4.5y = -9.9$
$-9.1x + 3.2y = -2.2$

30. $-1.1x + 8.3y = 36.5$
$3.5x + 1.6y = -4.1$

*Solve each system using determinants.*

31. $x + y + z = 3$
$-3y + 4z = 15$
$-3x + 4y - 2z = -13$

32. $2x + 3y = 4$
$3x + 7y - 4z = -3$
$x - y + 2z = 9$

33. $3x - 5y - 4z = -4$
$4x + 2y = 1$
$6y - 4z = -11$

34. $2x + 5y + 3z = 2$
$6x - 9y = 5$
$3y + 2z = 1$

35. $x + 4y - 3z = -6$
$2x - 8y + 5z = 12$
$3x + 4y - 2z = -3$

36. $2x + y - 2z = 4$
$2x + 2y - 4z = 1$
$-6x + 8y - 4z = 1$

37. $a - b + 2c = 3$
$a - b + c = 1$
$2a + b + 2c = 2$

38. $-2x + y + 8 = -2$
$3x + 2y + z = 3$
$x - 3y - 5z = 5$

39. $a + 2b + c = 1$
$a - b + 3c = 2$
$2a + b + 4c = 3$

40. $4x - 2y + 6z = 2$
$-6x + 3y - 9z = -3$
$2x - 7y + 11z = -5$

41. $1.1x + 2.3y - 4.0z = -9.2$
$-2.3x + 4.6z = 6.9$
$-8.2y - 7.5z = -6.8$

42. $4.6y - 2.1z = 24.3$
$-5.6x + 1.8y = -5.8$
$2.8x - 4.7y - 3.1z = 7.0$

**43.** $-6x + 3y - 12z = -13$
$5x + 2y - 3z = 1$
$2x - y + 4z = -5$

**44.** $x - 2y + z = 2$
$4x - 6y + 2z = 3$
$2x - 3y + z = 0$

**45.** $2x + \frac{1}{2}y - 3z = 5$
$-3x + 2y + 2z = 1$
$4x - \frac{1}{4}y - 7z = 4$

**46.** $\frac{1}{4}x - \frac{1}{2}y + 3z = -3$
$2x - 3y + 2z = -1$
$\frac{1}{6}x + \frac{1}{3}y - \frac{1}{3}z = 1$

**47.** $0.3x - 0.1y - 0.3z = -0.2$
$0.2x - 0.1y + 0.1z = -0.9$
$0.1x + 0.2y - 0.4z = 1.7$

**48.** $0.6u - 0.4v + 0.5w = 3.1$
$0.5u + 0.2v + 0.2w = 1.3$
$0.1u + 0.1v + 0.1w = 0.2$

## Problem Solving

**49.** Given a determinant of the form $\begin{vmatrix} a_1 & b_1 \\ a_2 & b_2 \end{vmatrix}$, how will the value of the determinant change if the $a$'s are switched with each other and the $b$'s are switched with each other, $\begin{vmatrix} a_2 & b_2 \\ a_1 & b_1 \end{vmatrix}$? Explain your answer.

**50.** Given a determinant of the form $\begin{vmatrix} a_1 & b_1 \\ a_2 & b_2 \end{vmatrix}$, how will the value of the determinant change if the $a$'s are switched with the $b$'s, $\begin{vmatrix} b_1 & a_1 \\ b_2 & a_2 \end{vmatrix}$? Explain your answer.

**51.** In a $2 \times 2$ determinant, if the rows are the same, what is the value of the determinant?

**52.** If all the numbers in one row or one column of a $2 \times 2$ determinant are 0, what is the value of the determinant?

**53.** If all the numbers in one row or one column of a $3 \times 3$ determinant are 0, what is the value of the determinant?

**54.** Given a $3 \times 3$ determinant, if all the numbers in one row are multiplied by $-1$, will the value of the new determinant change? Explain.

**55.** Given a $3 \times 3$ determinant, if the new first and second rows are switched, will the value of the determinant change? Explain.

**56.** In a $3 \times 3$ determinant, if any two rows are the same, can you make a generalization about the value of the determinant?

**57.** In a $3 \times 3$ determinant, if the numbers in the first row are multiplied by $-1$ and the numbers in the second row are multiplied by $-1$, will the value of the new determinant change? Explain.

**58.** In a $3 \times 3$ determinant, if the numbers in the second row are multiplied by $-1$ and the numbers in the third row are multiplied by $-1$, will the value of the new determinant change? Explain.

**59.** In a $3 \times 3$ determinant, if the numbers in the second row are multiplied by 2, will the value of the new determinant change? Explain.

**60.** In a $3 \times 3$ determinant, if the numbers in the first row are multiplied by 3 and the numbers in the third row are multiplied by 4, will the value of the new determinant change? Explain.

*Solve for the given letter.*

**61.** $\begin{vmatrix} 4 & 6 \\ -2 & y \end{vmatrix} = 32$

**62.** $\begin{vmatrix} b-3 & -4 \\ b+2 & -6 \end{vmatrix} = 14$

**63.** $\begin{vmatrix} 4 & 7 & y \\ 3 & -1 & 2 \\ 4 & 1 & 5 \end{vmatrix} = -35$

**64.** $\begin{vmatrix} 3 & 2 & -2 \\ 0 & 5 & -6 \\ -1 & x & -7 \end{vmatrix} = -31$

## Challenge Problems

**65.** Use the addition method to solve the following system for **a)** $x$ and **b)** $y$.

$$a_1x + b_1y = c_1$$
$$a_2x + b_2y = c_2$$

## Cumulative Review Exercises

[2.5] **66.** Solve the inequality $3(x - 2) < \frac{4}{5}(x - 4)$ and indicate the solution in interval notation.

*Graph* $3x + 4y = 8$ *using the indicated method.*

[3.2] **67.** By plotting points

**68.** Using the $x$- and $y$-intercepts

[3.3] **69.** Using the slope and $y$-intercept

# 4.6 Solving Systems of Linear Inequalities

1. Solve systems of linear inequalities.
2. Solve linear programming problems.
3. Solve systems of linear inequalities containing absolute value.

## 1 Solve Systems of Linear Inequalities

In Section 3.7 we showed how to graph linear inequalities in two variables. In Section 4.1 we learned how to solve systems of equations graphically. In this section we show how to solve **systems of linear inequalities** graphically.

**To Solve a System of Linear Inequalities**

Graph each inequality on the same axes. The solution is the set of points whose coordinates satisfy all the inequalities in the system.

**EXAMPLE 1** Determine the solution to the following system of inequalities.

$$y < -\frac{1}{2}x + 2$$
$$x - y \le 4$$

*Solution* First graph the inequality $y < -\frac{1}{2}x + 2$ (**Fig. 4.8**). Now on the same axes graph the inequality $x - y \le 4$ (**Fig. 4.9**). The solution is the set of points common to the graphs of both inequalities. It is the part of the graph that contains both shadings. The dashed line is not part of the solution, but the part of the solid line that satisfies both inequalities is part of the solution.

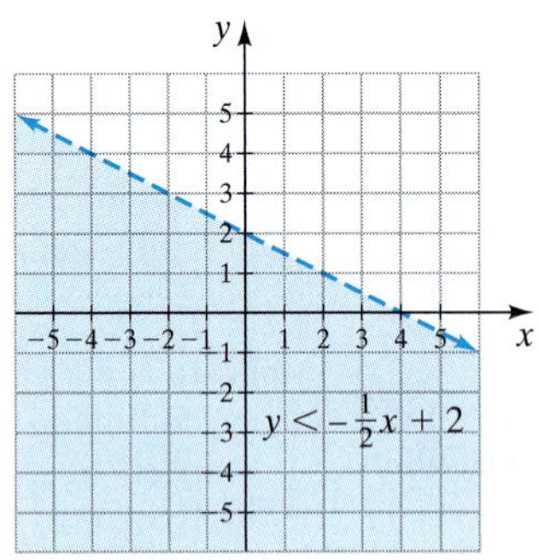


FIGURE 4.8

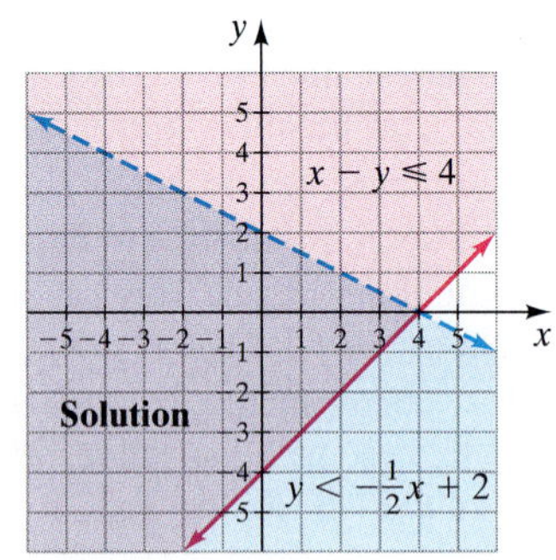


FIGURE 4.9

Now Try Exercise 5

**EXAMPLE 2** Determine the solution to the following system of inequalities.

$$3x - y < 6$$
$$2x + 2y \ge 5$$

*Solution* Graph $3x - y < 6$ (see **Fig. 4.10**). Graph $2x + 2y \ge 5$ on the same axes (**Fig. 4.11**). The solution is the part of the graph that contains both shadings and the part of the solid line that satisfies both inequalities.

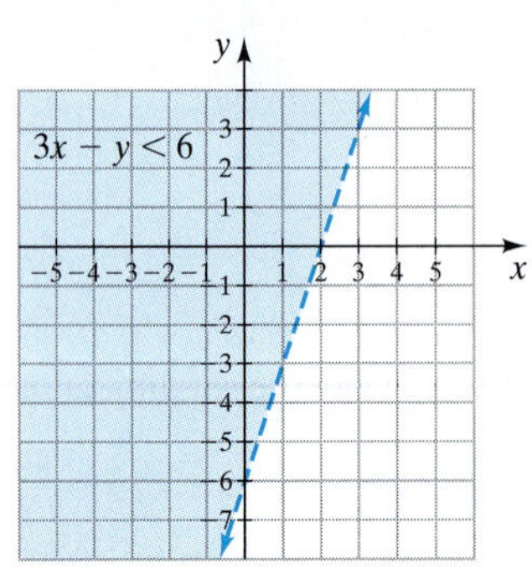


FIGURE 4.10

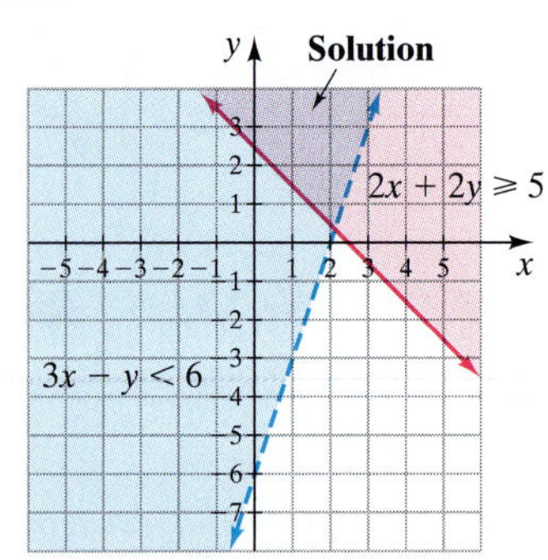


FIGURE 4.11

Now Try Exercise 7

**EXAMPLE 3** ▶ Determine the solution to the following system of inequalities.

$$y > -1$$
$$x \leq 4$$

*Solution* The solution is illustrated in **Figure 4.12**.

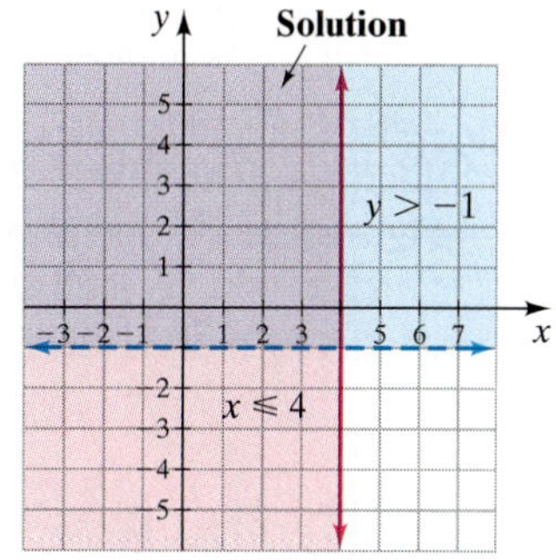


FIGURE 4.12

▶ **Now Try Exercise 15**

## 2 Solve Linear Programming Problems

There is a mathematical process called **linear programming** for which you often have to graph more than two linear inequalities on the same axes. These inequalities are called **constraints**. The following two examples illustrate how to determine the solution to a system of more than two inequalities.

**EXAMPLE 4** ▶ Determine the solution to the following system of inequalities.

$$x \geq 0$$
$$y \geq 0$$
$$2x + 3y \leq 12$$
$$2x + y \leq 8$$

*Solution* The first two inequalities, $x \geq 0$ and $y \geq 0$, indicate that the solution must be in the first quadrant because that is the only quadrant where both $x$ and $y$ are positive. **Figure 4.13** illustrates the graphs of the four inequalities.

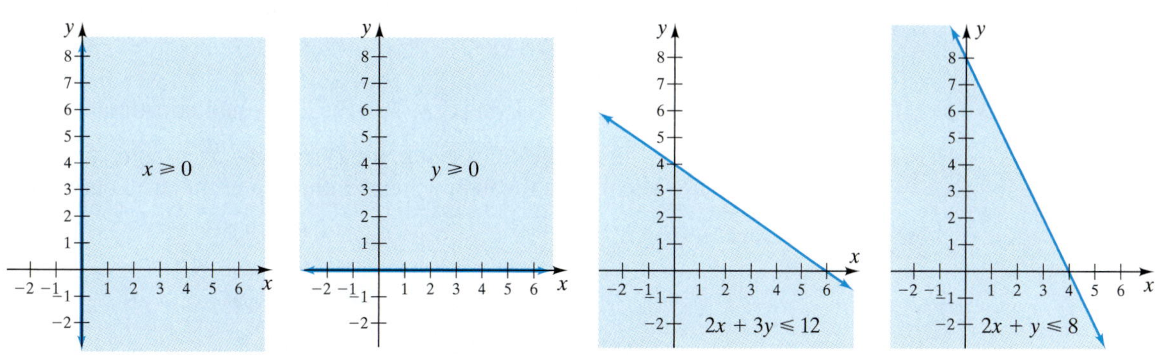


FIGURE 4.13

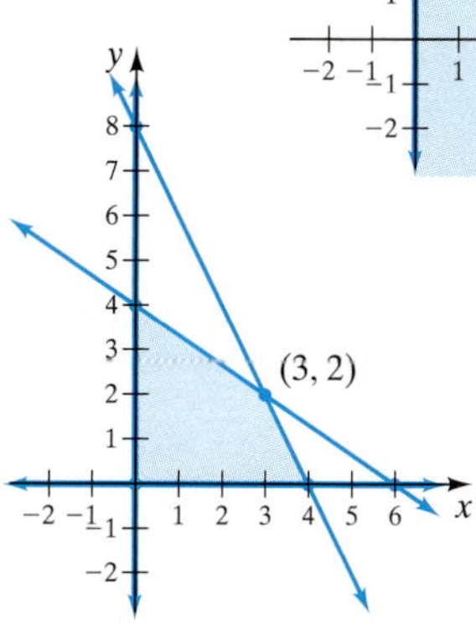


FIGURE 4.14

**Figure 4.14** illustrates the graphs on the same axes and the solution to the system of inequalities. Note that every point in the shaded area and every point on the lines that form the polygonal region is part of the answer.

▶ **Now Try Exercise 23**

**EXAMPLE 5** ▸ Determine the solution to the following system of inequalities.

$$\begin{aligned} x &\geq 0 \\ y &\geq 0 \\ x &\leq 15 \\ 8x + 8y &\leq 160 \\ 4x + 12y &\leq 180 \end{aligned}$$

*Solution* The first two inequalities indicate that the solution must be in the first quadrant. The third inequality indicates that $x$ must be a value less than or equal to 15. **Figure 4.15a** indicates the graphs of corresponding equations and shows the region that satisfies all the inequalities in the system. **Figure 4.15b** indicates the solution to the system of inequalities.

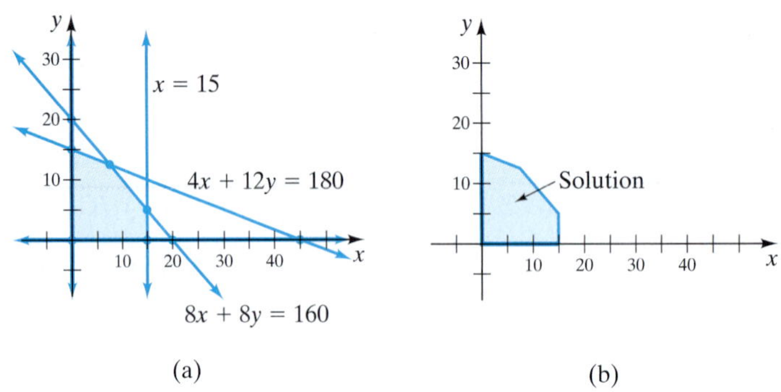


FIGURE 4.15

▸ Now Try Exercise 29

## 3 Solve Systems of Linear Inequalities Containing Absolute Value

Now we will graph *systems of linear inequalities containing absolute value* in the Cartesian coordinate system. Before we do some examples, let us recall the rules for absolute value inequalities that we learned in Section 2.6. Recall the following:

**Solving Absoute Value Inequalities**

If $|x| < a$ and $a > 0$, then $-a < x < a$.

If $|x| > a$ and $a > 0$, then $x < -a$ or $x > a$.

**EXAMPLE 6** ▸ Graph $|x| < 3$ in the Cartesian coordinate system.

*Solution* From the rules given, we know that $|x| < 3$ means $-3 < x < 3$. We draw dashed vertical lines through $-3$ and 3 and shade the area between the two (**Fig. 4.16**).

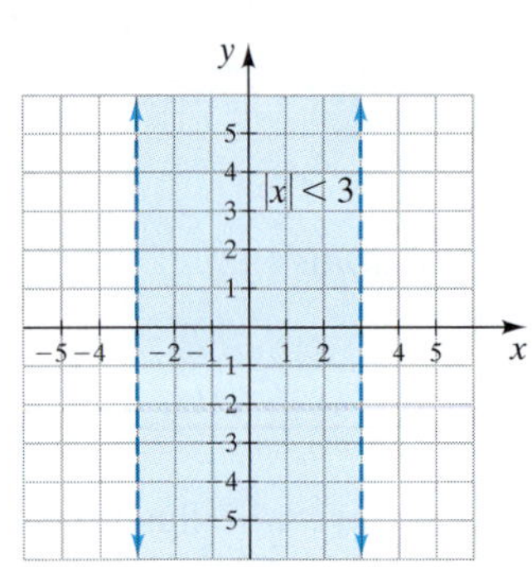


FIGURE 4.16

▸ Now Try Exercise 33

**EXAMPLE 7** ▸ Graph $|y + 1| > 3$ in the Cartesian coordinate system.

*Solution* From the rules given, we know that $|y + 1| > 3$ means $y + 1 < -3$ or $y + 1 > 3$. First we solve each inequality.

$$y + 1 < -3 \quad \text{or} \quad y + 1 > 3$$
$$y < -4 \qquad\qquad y > 2$$

Now we graph both inequalities and take the *union* of the two graphs. The solution is the shaded area in **Figure 4.17**.

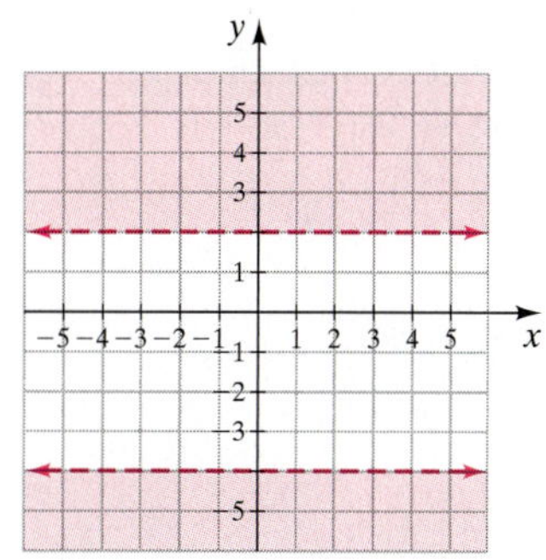


FIGURE 4.17

▸ Now Try Exercise 35

**EXAMPLE 8** ▸ Graph the following system of inequalities.

$$|x| < 3$$
$$|y + 1| > 3$$

*Solution* We draw both inequalities on the same axes. Therefore, we combine the graph drawn in Example 6 with the graph drawn in Example 7 (see **Fig. 4.18**). The points common to both inequalities form the solution to the system.

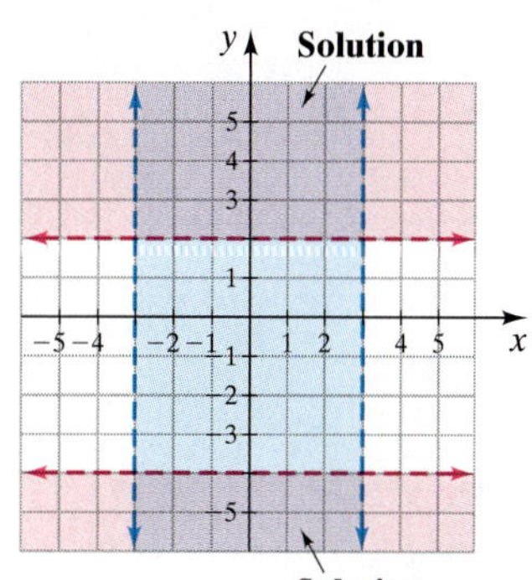


FIGURE 4.18

▸ Now Try Exercise 41

## EXERCISE SET 4.6

MathXL® MyMathLab

### Concept/Writing Exercises

1. Explain how to find the solution to a system of linear inequalities graphically.
2. If in a system of two inequalities, one inequality contains $<$ and the other inequality contains $\geq$, is the point of intersection of the two boundary lines of the inequalities in the solution set? Explain.
3. If in a system of two inequalities, one inequality contains $\leq$ and the other inequality contains $\geq$, is the point of intersection of the two boundary lines of the inequalities in the solution set? Explain.
4. If in a system of two inequalities, one inequality contains $<$ and the other inequality contains $>$, is the point of intersection of the two boundary lines of the inequalities in the solution set? Explain.

## Practice the Skills

*Determine the solution to each system of inequalities.*

5. $2x - y < 4$
   $y \geq -x + 2$

6. $y \leq -2x + 1$
   $y > -3x$

7. $y < 3x - 2$
   $y \leq -2x + 3$

8. $y \geq 2x - 5$
   $y > -3x + 5$

9. $y < x$
   $y \geq 3x + 2$

10. $-3x + 2y \geq -5$
    $y \leq -4x + 7$

11. $-2x + 3y < -5$
    $3x - 8y > 4$

12. $-4x + 3y \geq -4$
    $y > -3x + 3$

13. $-4x + 5y < 20$
    $x \geq -3$

14. $y \geq -\frac{2}{3}x + 1$
    $y > -4$

15. $x \leq 4$
    $y \geq -2$

16. $x \geq 0$
    $x - 3y < 6$

17. $5x + 2y > 10$
    $3x - y > 3$

18. $3x + 2y > 8$
    $x - 5y < 5$

19. $-2x > y + 4$
    $-x < \frac{1}{2}y - 1$

20. $y \leq 3x - 2$
    $\frac{1}{3}y < x + 1$

21. $y < 3x - 4$
    $6x \geq 2y + 8$

22. $\frac{1}{2}x + \frac{1}{2}y \geq 2$
    $2x - 3y \leq -6$

*Determine the solution to each system of inequalities. Use the method discussed in Examples 4 and 5.*

23. $x \geq 0$
    $y \geq 0$
    $2x + 3y \leq 6$
    $4x + y \leq 4$

24. $x \geq 0$
    $y \geq 0$
    $x + y \leq 6$
    $7x + 4y \leq 28$

25. $x \geq 0$
    $y \geq 0$
    $2x + 3y \leq 8$
    $4x + 2y \leq 8$

26. $x \geq 0$
    $y \geq 0$
    $3x + 2y \leq 18$
    $2x + 4y \leq 20$

27. $x \geq 0$
    $y \geq 0$
    $3x + y \leq 9$
    $2x + 5y \leq 10$

28. $x \geq 0$
    $y \geq 0$
    $5x + 4y \leq 16$
    $x + 6y \leq 18$

29. $x \geq 0$
    $y \geq 0$
    $x \leq 4$
    $x + y \leq 6$
    $x + 2y \leq 8$

30. $x \geq 0$
    $y \geq 0$
    $x \leq 4$
    $2x + 3y \leq 18$
    $4x + 2y \leq 20$

31. $x \geq 0$
    $y \geq 0$
    $x \leq 15$
    $30x + 25y \leq 750$
    $10x + 40y \leq 800$

32. $x \geq 0$
    $y \geq 0$
    $x \leq 15$
    $40x + 25y \leq 1000$
    $5x + 30y \leq 900$

*Determine the solution to each inequality.*

33. $|x| < 2$

34. $|x| > 1$

35. $|y - 2| \leq 4$

36. $|y| \geq 2$

*Determine the solution to each system of inequalities.*

37. $|y| > 2$
    $y \leq x + 3$

38. $|x| > 1$
    $y \leq 3x + 2$

39. $|y| < 4$
    $y \geq -2x + 2$

40. $|x - 2| \leq 3$
    $x - y > 2$

41. $|x + 2| < 3$
    $|y| > 4$

42. $|x - 2| > 1$
    $y > -2$

43. $|x - 3| \leq 4$
    $|y + 2| \leq 1$

44. $|x + 1| \leq 2$
    $|y - 3| \leq 1$

## Problem Solving

**45. Tax Returns** The following graph shows the percent of all U.S. federal tax returns filed electronically and by paper for the years 2001–2005 and projected to the year 2009. The information for the graph was obtained from the IRS Web site.

**Federal Tax Return Method**

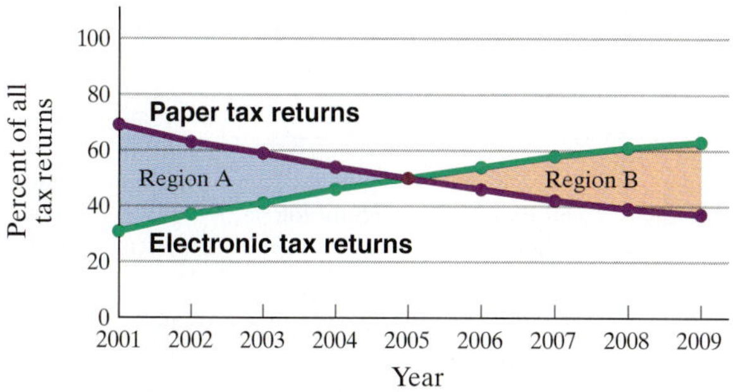


*Source:* Internal Revenue Service: www.irs.gov/pubs

Let $P(r)$ represent the paper tax returns (purple line) and $E(t)$ represent the electronic tax returns (green line). The regions between the two curves are identified as either Region A, shaded in blue, or Region B shaded in orange.

**a)** Which region, A or B, is a solution to the system of inequalities?

$$y \le P(t)$$
$$y \ge E(t)$$

**b)** Which region, A or B, is a solution to the system of inequalities?

$$y \ge P(t)$$
$$y \le E(t)$$

**46. Income from Department Stores** The following graph shows the annual net income, in millions of dollars, for the Federated Department Stores (Macy's, Bloomingdale's, Burdines) and the May Department Stores (Hecht's, Lord & Taylor, Filene's) for the years 1999–2003.

**Annual Net Income for Federated Department Stores and May Department Stores**

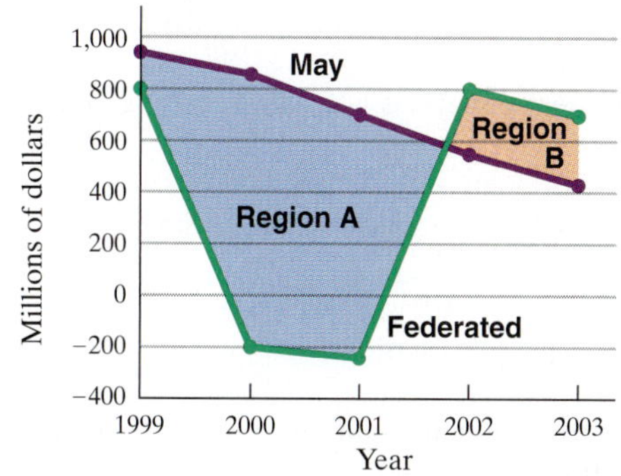


*Source:* Thomson Financial, the companies, *The Washington Post* (1/21/05)

Let $P(t)$ represent the income for the May Department Stores (purple line) and $Q(t)$ represent the income for the Federated Department stores (green line). The regions between the two curves are identified as either Region A, shaded in blue, or Region B, shaded in orange.

**a)** Which region, A or B, is a solution to the system of inequalities?

$$y \ge P(t)$$
$$y \le Q(t)$$

**b)** Which region, A or B, is a solution to the system of inequalities?

$$y \le P(t)$$
$$y \ge Q(t)$$

**47.** Is it possible for a system of linear inequalities to have no solution? Explain. Make up an example to support your answer.

**48.** Is it possible for a system of two linear inequalities to have exactly one solution? Explain. If you answer yes, make up an example to support your answer.

*Without graphing, determine the number of solutions in each indicated system of inequalities. Explain your answers.*

**49.** $3x - y \le 4$
$3x - y > 4$

**50.** $2x + y < 6$
$2x + y > 6$

**51.** $5x - 2y \le 3$
$5x - 2y \ge 3$

**52.** $5x - 3y > 5$
$5x - 3y > -1$

**53.** $2x - y < 7$
$3x - y < -2$

**54.** $x + y \le 0$
$x - y \ge 0$

## Challenge Problems

*Determine the solution to each system of inequalities.*

**55.** $y \ge x^2$
$y \le 4$

**56.** $y < 4 - x^2$
$y > -5$

**57.** $y < |x|$
$y < 4$

**58.** $y \ge |x - 2|$
$y \le -|x - 2|$

## Cumulative Review Exercises

[2.2] **59.** A formula for levers in physics is $f_1d_1 + f_2d_2 = f_3d_3$. Solve this formula for $f_2$.

[3.2] *State the domain and range of each function.*

**60.** $\{(4, 3), (5, -2), (-1, 2), (0, -5)\}$

**61.** $f(x) = \frac{2}{3}x - 4$

**62.**

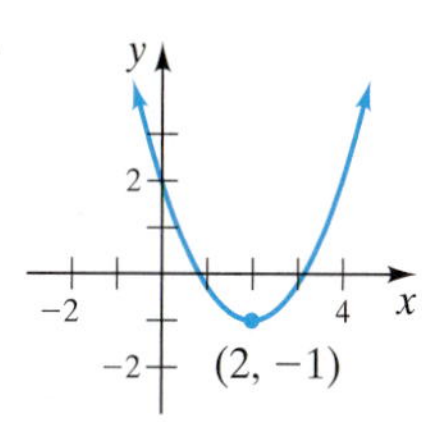

# Chapter 4 Summary

**IMPORTANT FACTS AND CONCEPTS** | **EXAMPLES**

## Section 4.1

A **system of linear equations** is a system having two or more linear equations.

A **solution** to a system of linear equations is the ordered pair or pairs that satisfy all equations in the system.

System of equations $\begin{cases} y = 3x + 1 \\ y = \frac{1}{2}x + 6 \end{cases}$

The solution to the above system of equations is (2, 7).

---

A **consistent system of equations** is a system of equations that has one solution.

An **inconsistent system of equations** is a system of equations that has no solution.

A **dependent system of equations** is a system of equations that has an infinite number of solutions.

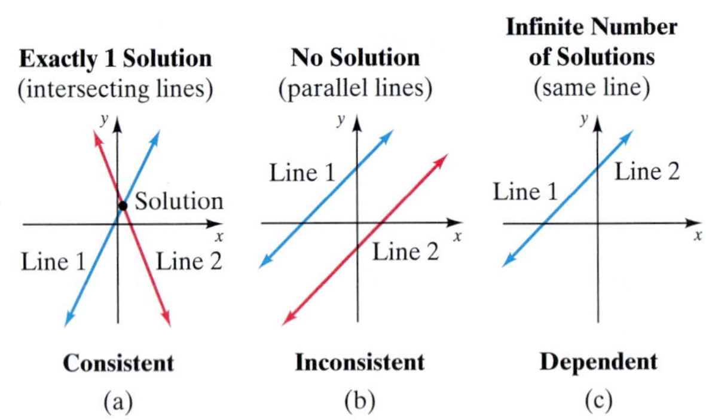


---

### To Solve a System of Linear Equations by Graphing

1. Graph both lines.
2. Determine the point(s) of intersection, if it exists.
3. Check your solution in all equations of the system.

Solve the system of equations graphically.

$$y = x - 4$$
$$y = -x + 6$$

Graph both lines on the same set of axes.

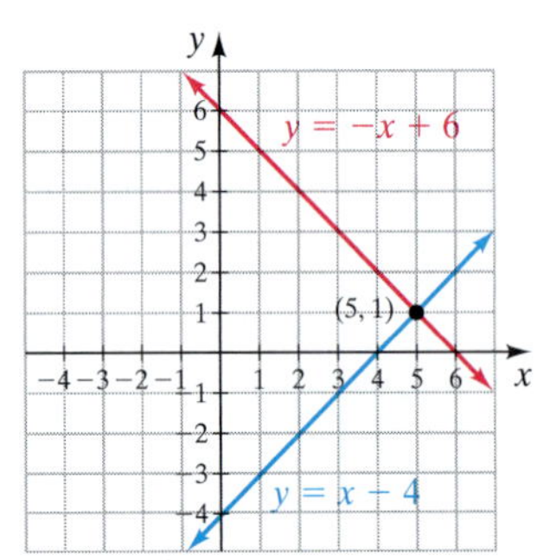


A check shows that (5, 1) is a solution to the system of equations.

---

### To Solve a Linear System of Equations by Substitution

1. Solve for a variable in either equation.
2. Substitute the expression found for the variable in step 1 into the other equation.
3. Solve the equation obtained in step 2 to find the value of this variable.
4. Substitute the value found in step 3 into the equation from step 1. Solve the equation to find the remaining variable.
5. Check your solution in all equations in the system.

Solve the system of equations by substitution.

$$y = -2x - 1$$
$$5x + 6y = 8$$

Substitute $y = -2x - 1$ in the second equation:

$$\begin{aligned} 5x + 6y &= 8 \\ 5x + 6(-2x - 1) &= 8 \\ 5x - 12x - 6 &= 8 \\ -7x - 6 &= 8 \\ -7x &= 14 \\ x &= -2. \end{aligned}$$

Substitute $x = -2$ into $y = -2x - 1$ to obtain

$$\begin{aligned} y &= -2x - 1 \\ y &= -2(-2) - 1 = 4 - 1 = 3 \end{aligned}$$

A check shows that $(-2, 3)$ is a solution to the system of equations.

| IMPORTANT FACTS AND CONCEPTS | EXAMPLES |
|---|---|

**Section 4.1 (continued)**

**To Solve a Linear System of Equations Using the Addition (or Elimination) Method**

1. If necessary, rewrite each equation in standard form.
2. If necessary, multiply one or both equations by a constant(s) so that when the equations are added, the sum will contain only one variable.
3. Add the respective sides of the equations.
4. Solve for the variable in the equation obtained in step 3.
5. Substitute the value found in step 4 into either of the original equations. Solve that equation to find the value of the remaining variable.
6. Check your solution in all equations in the system.

Solve the system of equations using the addition method.

$$2x + y = 4 \quad (eq.\ 1)$$
$$x - 2y = 2 \quad (eq.\ 2)$$

$$\begin{aligned} 4x + 2y &= 8 \quad (eq.\ 1)\ \text{Multiplied by 2} \\ x - 2y &= 2 \\ \hline 5x &= 10 \quad \text{Sum of equations.} \\ x &= 2 \end{aligned}$$

Now solve for $y$ using (*eq.* 1).

$$2(2) + y = 4$$
$$y = 0$$

The solution is $(2, 0)$.

**Section 4.2**

To solve a system of three linear equations, use the substitution method or addition method.

Solve the system of equations.

$$x - y + 3z = -1 \quad (eq.\ 1)$$
$$4y - 7z = 2 \quad (eq.\ 2)$$
$$z = 2 \quad (eq.\ 3)$$

Substitute 2 for $z$ in (eq. 2) to obtain the value for y.

$$4y - 7z = 2$$
$$4y - 7(2) = 2$$
$$4y = 16$$
$$y = 4.$$

Substitute 4 for $y$ and 2 for $z$ in (eq. 1) to obtain the value for $x$.

$$x - y + 3z = -1$$
$$x - 4 + 3(2) = -1$$
$$x = -3.$$

A check shows that $(-3, 4, 2)$ is a solution to the system of equations.

**Section 4.3**

**Applications:**
Systems of two linear equations in two unknowns.

The sum of the areas of two circles is 180 square meters. The difference of their areas is 20 square meters. Determine the area of each circle.

**Solution**

Let $x$ be the area of the larger circle and $y$ be the area of the smaller circle.

The two equations for this systems are

$$\begin{aligned} x + y &= 180 \leftarrow \text{Sum of areas} \\ x - y &= 20 \leftarrow \text{Difference of areas.} \\ \hline 2x &= 200 \\ x &= 100 \end{aligned}$$

Substitute 100 for $x$ in the first equation to get

$$x + y = 180$$
$$100 + y = 180$$
$$y = 80$$

The area of the larger circle is 100 square meters and the area of the smaller circle is 80 square meters.

| IMPORTANT FACTS AND CONCEPTS | EXAMPLES |
|---|---|

**Section 4.4**

A **matrix** is a rectangular array of numbers within brackets. The numbers inside the brackets are called **elements**.

$$\begin{bmatrix} 8 & 1 & 4 \\ -3 & 0 & 2 \end{bmatrix}, \quad \begin{bmatrix} 5 & 0 \\ -2 & 8 \\ 6 & -11 \end{bmatrix}$$

A **square matrix** has the same number of rows and columns.

$$\begin{bmatrix} 5 & -1 \\ 8 & 2 \end{bmatrix}, \quad \begin{bmatrix} 3 & 0 & -6 \\ -1 & 5 & 2 \\ 9 & 10 & -7 \end{bmatrix}$$

An **augmented matrix** is a matrix separated by a vertical line. For a system of equations, the coefficients of the variables are placed on the left side of the vertical line and the constants are placed on the right side in the augmented matrix.

The **triangular form** of an augmented matrix is

$$\left[\begin{array}{cc|c} 1 & a & p \\ 0 & 1 & q \end{array}\right]$$

SYSTEM / AUGMENTED MATRIX

$$\begin{aligned} 2x - 3y &= 8 \\ 5x + 7y &= -4 \end{aligned} \qquad \left[\begin{array}{cc|c} 2 & -3 & 8 \\ 5 & 7 & -4 \end{array}\right]$$

triangular form $\left[\begin{array}{cc|c} 1 & -6 & 2 \\ 0 & 1 & 9 \end{array}\right]$

**Row transformations** can be used to rewrite a matrix into triangular form.

### Procedures for Row Transformations

1. All the numbers in a row may be multiplied (or divided) by any nonzero real number.
2. All the numbers in a row may be multiplied by any nonzero real number. These products may then be added to the corresponding numbers in any other row.
3. The order of the rows may be switched.

Solve the system of equations

$$\begin{aligned} x + 4y &= -7 \\ 6x - 5y &= 16 \end{aligned}$$

The augmented matrix is

$$\left[\begin{array}{cc|c} 1 & 4 & -7 \\ 6 & -5 & 16 \end{array}\right] = \left[\begin{array}{cc|c} 1 & 4 & -7 \\ 0 & -29 & 58 \end{array}\right] -6R_1 + R_2$$

$$= \left[\begin{array}{cc|c} 1 & 4 & -7 \\ 0 & 1 & -2 \end{array}\right] -\frac{1}{29}R_2$$

The equivalent system of equations is

$$\begin{aligned} x + 4y &= -7 \\ y &= -2 \end{aligned}$$

Substitute $-2$ for $y$ into the first equation.

$$\begin{aligned} x + 4(-2) &= -7 \\ x - 8 &= -7 \\ x &= 1. \end{aligned}$$

The solution is $(1, -2)$.

A system of equations is **inconsistent** and has **no solution** if you obtain an augmented matrix in which one row of numbers has zeros on the left side of the vertical line and a nonzero number on the right side of the vertical line.

$$\left[\begin{array}{ccc|c} 1 & 2 & -3 & 23 \\ 0 & 0 & 0 & 8 \\ -1 & 7 & 6 & 9 \end{array}\right]$$

The second row shows this system is inconsistent and has no solution.

A system of equations is **dependent** and has an **infinite number of solutions** if you obtain an augmented matrix in which a 0 appears across an entire row.

$$\left[\begin{array}{ccc|c} 1 & 6 & -1 & 15 \\ 0 & 0 & 0 & 0 \\ 3 & 5 & 8 & -12 \end{array}\right]$$

The second row shows this system is dependent and has an infinite number of solutions.

| IMPORTANT FACTS AND CONCEPTS | EXAMPLES |
|---|---|
| **Section 4.5** | |
| The **determinant** of a 2 × 2 matrix $\begin{bmatrix} a_1 & b_1 \\ a_2 & b_2 \end{bmatrix}$ is denoted $\begin{vmatrix} a_1 & b_1 \\ a_2 & b_2 \end{vmatrix}$ and is evaluated as $$\begin{vmatrix} a_1 & b_1 \\ a_2 & b_2 \end{vmatrix} = a_1b_2 - a_2b_1$$ | $$\begin{vmatrix} 3 & -2 \\ 5 & 1 \end{vmatrix} = (3)(1) - (5)(-2) = 3 + 10 = 13$$ |
| **Cramer's Rule for Systems of Linear Equations** <br> For a system of linear equations of the form $$a_1x + b_1y = c_1$$ $$a_2x + b_2y = c_2$$ $$x = \frac{\begin{vmatrix} c_1 & b_1 \\ c_2 & b_2 \end{vmatrix}}{\begin{vmatrix} a_1 & b_1 \\ a_2 & b_2 \end{vmatrix}} = \frac{D_x}{D} \quad \text{and} \quad y = \frac{\begin{vmatrix} a_1 & c_1 \\ a_2 & c_2 \end{vmatrix}}{\begin{vmatrix} a_1 & b_1 \\ a_2 & b_2 \end{vmatrix}} = \frac{D_y}{D}, \quad D \neq 0$$ | Solve the system of equations. $$2x + y = 6$$ $$4x - 3y = -13$$ $$D = \begin{vmatrix} 2 & 1 \\ 4 & -3 \end{vmatrix} = -10$$ $$D_x = \begin{vmatrix} 6 & 1 \\ -13 & -3 \end{vmatrix} = -5 \qquad D_y = \begin{vmatrix} 2 & 6 \\ 4 & -13 \end{vmatrix} = -50$$ Then $$x = \frac{D_x}{D} = \frac{-5}{-10} = \frac{1}{2}, \quad y = \frac{D_y}{D} = \frac{-50}{-10} = 5$$ The solution is $\left(\frac{1}{2}, 5\right)$. |
| For the determinant $$\begin{vmatrix} a_1 & b_1 & c_1 \\ a_2 & b_2 & c_2 \\ a_3 & b_3 & c_3 \end{vmatrix}$$ The **minor determinant of $a_1$** is found by crossing out the elements in the same row and column containing the element $a_1$. <br><br> **Expansion of the Determinant by the Minors of the First Column** <br> Minor determinant of $a_1$ ↓   Minor determinant of $a_2$ ↓   Minor determinant of $a_3$ ↓ $$\begin{vmatrix} a_1 & b_1 & c_1 \\ a_2 & b_2 & c_2 \\ a_3 & b_3 & c_3 \end{vmatrix} = a_1\begin{vmatrix} b_2 & c_2 \\ b_3 & c_3 \end{vmatrix} - a_2\begin{vmatrix} b_1 & c_1 \\ b_3 & c_3 \end{vmatrix} + a_3\begin{vmatrix} b_1 & c_1 \\ b_2 & c_2 \end{vmatrix}$$ | For $\begin{vmatrix} 6 & 2 & -1 \\ 0 & 3 & 5 \\ 7 & 1 & 9 \end{vmatrix}$, the minor determinant of $a_1$ is $\begin{vmatrix} 3 & 5 \\ 1 & 9 \end{vmatrix}$. <br><br> Evaluate $\begin{vmatrix} 2 & 0 & 3 \\ -1 & -5 & 2 \\ 1 & 6 & -4 \end{vmatrix}$ using expansion by minors of the first column. $$\begin{vmatrix} 2 & 0 & 3 \\ -1 & -5 & 2 \\ 3 & 6 & -4 \end{vmatrix} = 2\begin{vmatrix} -5 & 2 \\ 6 & -4 \end{vmatrix} - (-1)\begin{vmatrix} 0 & 3 \\ 6 & -4 \end{vmatrix} + 3\begin{vmatrix} 0 & 3 \\ -5 & 2 \end{vmatrix}$$ $$= 2(8) + 1(-18) + 3(15)$$ $$= 16 - 18 + 45$$ $$= 43$$ |

| IMPORTANT FACTS AND CONCEPTS | EXAMPLES |
|---|---|

## Section 4.5 (continued)

**Cramer's Rule for a System of Equations in Three Variables**

To solve the system

$$a_1x + b_1y + c_1z = d_1$$
$$a_2x + b_2y + c_2z = d_2$$
$$a_3x + b_3y + c_3z = d_3$$

with

$$D = \begin{vmatrix} a_1 & b_1 & c_1 \\ a_2 & b_2 & c_2 \\ a_3 & b_3 & c_3 \end{vmatrix} \quad D_x = \begin{vmatrix} d_1 & b_1 & c_1 \\ d_2 & b_2 & c_2 \\ d_3 & b_3 & c_3 \end{vmatrix}$$

$$D_y = \begin{vmatrix} a_1 & d_1 & c_1 \\ a_2 & d_2 & c_2 \\ a_3 & d_3 & c_3 \end{vmatrix} \quad D_z = \begin{vmatrix} a_1 & b_1 & d_1 \\ a_2 & b_2 & d_2 \\ a_3 & b_3 & d_3 \end{vmatrix}$$

then

$$x = \frac{D_x}{D} \quad y = \frac{D_y}{D} \quad z = \frac{D_z}{D}, \quad D \neq 0$$

Solve the system of equations.

$$2x + y + z = 0$$
$$4x - y + 3z = -9$$
$$6x + 2y + 5z = -8$$

$$D = \begin{vmatrix} 2 & 1 & 1 \\ 4 & -1 & 3 \\ 6 & 2 & 5 \end{vmatrix} = -10 \quad D_x = \begin{vmatrix} 0 & 1 & 1 \\ -9 & -1 & 3 \\ -8 & 2 & 5 \end{vmatrix} = -5$$

$$D_y = \begin{vmatrix} 2 & 0 & 1 \\ 4 & -9 & 3 \\ 6 & -8 & 5 \end{vmatrix} = -20 \quad D_z = \begin{vmatrix} 2 & 1 & 0 \\ 4 & -1 & -9 \\ 6 & 2 & -8 \end{vmatrix} = 30$$

Then

$$x = \frac{D_x}{D} = \frac{-5}{-10} = \frac{1}{2}, \quad y = \frac{D_y}{D} = \frac{-20}{-10} = 2 \quad z = \frac{D_z}{D} = \frac{30}{-10} = -3$$

The solution is $\left(\frac{1}{2}, 2, -3\right)$.

## Section 4.6

To solve a system of linear inequalities, graph each inequality on the same axes. The solution is the set of points whose coordinates satisfy all the inequalities in the system.

Determine the solution to the system of inequalities.

$$y < -\frac{1}{3}x + 1$$
$$x - y \leq 2$$

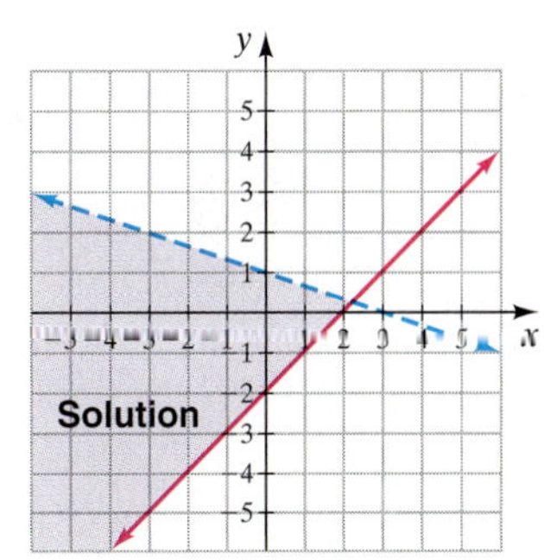


**Linear programming** is a process where more than two linear inequalities are graphed on the same axes.

Determine the solution to the system of inequalities.

$$x \geq 0$$
$$y \geq 0$$
$$x \leq 8$$
$$x + y \leq 10$$
$$x + 2y \leq 16$$

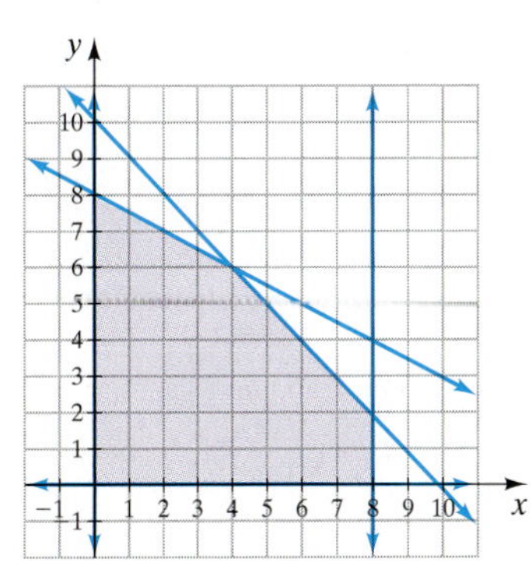

| IMPORTANT FACTS AND CONCEPTS | EXAMPLES |
|---|---|
| **Section 4.6 (continued)** | |
| For systems of linear inequalities with absolute values:<br>If $\lvert x \rvert < a$ and $a > 0$, then $-a < x < a$.<br>If $\lvert x \rvert > a$ and $a > 0$, then $x < -a$ or $x > a$. | Determine the solution to the system of inequalities.<br>$\lvert x \rvert < 2$<br>$\lvert y - 1 \rvert > 3$<br>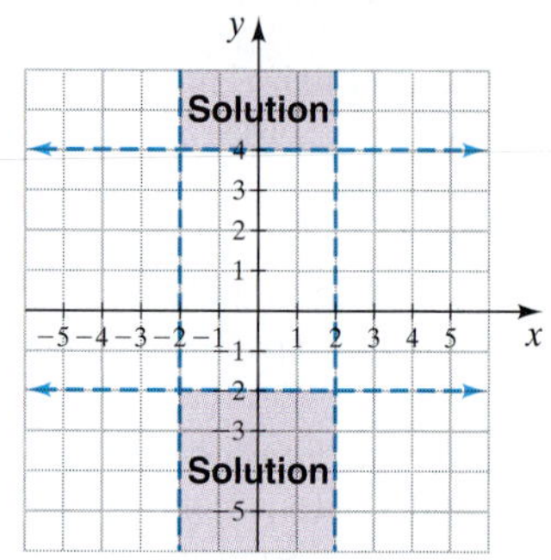 |

## Chapter 4 Review Exercises

[4.1] *Write each equation in slope-intercept form. Without graphing or solving the system of equations, state whether the system of linear equations is consistent, inconsistent, or dependent. Also indicate whether the system has exactly one solution, no solution, or an infinite number of solutions.*

**1.** $2x - 3y = -1$, $-4x + 6y = 1$

**2.** $4x - 5y = 8$, $3x + 4y = 9$

**3.** $y = \frac{1}{3}x + 4$, $x + 2y = 8$

**4.** $6x = 5y - 8$, $4x = 6y + 10$

*Determine the solution to each system of equations graphically. If the system is inconsistent or dependent, so state.*

**5.** $y = x + 3$, $y = 2x + 5$

**6.** $x = -5$, $y = 3$

**7.** $3x + 3y = 12$, $2x - y = -4$

**8.** $3y - 3x = -9$, $\frac{1}{2}x - \frac{1}{2}y = \frac{3}{2}$

*Find the solution to each system of equations by subsitution.*

**9.** $y = -4x + 2$, $y = 3x - 12$

**10.** $4x - 3y = -1$, $y = 2x + 1$

**11.** $a = 2b - 8$, $2b - 5a = 0$

**12.** $2x + y = 12$, $\frac{1}{2}x - \frac{3}{4}y = 1$

*Find the solution to each system of equations using the addition method.*

**13.** $x - 2y = 5$, $2x + 2y = 4$

**14.** $-2x - y = 5$, $2x + 2y = 6$

**15.** $2a + 3b = 7$, $a - 2b = -7$

**16.** $0.4x - 0.3y = 1.8$, $-0.7x + 0.5y = -3.1$

**17.** $4r - 3s = 8$, $2r + 5s = 8$

**18.** $-2m + 3n = 15$, $3m + 3n = 10$

**19.** $x + \frac{3}{5}y = \frac{11}{5}$, $x - \frac{3}{2}y = -2$

**20.** $4x + 4y = 16$, $y = 4x - 3$

**21.** $y = -\frac{3}{4}x + \frac{5}{2}$, $x + \frac{5}{4}y = \frac{7}{2}$

**22.** $2x - 5y = 12$, $x - \frac{4}{3}y = -2$

**23.** $2x + y = 4$, $3x + \frac{3}{2}y = 6$

**24.** $2x = 4y + 5$, $2y = x - 7$

[4.2] *Determine the solution to each system of equations using substitution or the addition method.*

**25.** $x - 2y - 4z = 13$, $3y + 2z = -2$, $5z = -20$

**26.** $2a + b - 2c = 5$, $3b + 4c = 1$, $3c = -6$

**27.** $x + 2y + 3z = 3$, $-2x - 3y - z = 5$, $3x + 3y + 7z = 2$

**28.** $-x - 4y + 2z = 1$, $2x + 2y + z = 0$, $-3x - 2y - 5z = 5$

**29.** $3y - 2z = -4$
$3x - 5z = -7$
$2x + y = 6$

**30.** $a + 2b - 5c = 19$
$2a - 3b + 3c = -15$
$5a - 4b - 2c = -2$

**31.** $x - y + 3z = 1$
$-x + 2y - 2z = 1$
$x - 3y + z = 2$

**32.** $-2x + 2y - 3z = 6$
$4x - y + 2z = -2$
$2x + y - z = 4$

[4.3] *Express each problem as a system of linear equations and use the method of your choice to find the solution to the problem.*

**33. Ages** Luan Baker is 10 years older than his niece, Jennifer Miesen. If the sum of their ages is 66, find Luan's age and Jennifer's age.

**34. Air Speed** An airplane can travel 560 miles per hour with the wind and 480 miles per hour against the wind. Determine the speed of the plane in still air and the speed of the wind.

**35. Mixing Solutions** Sally Dove has two acid solutions, as illustrated. How much of each must she mix to get 6 liters of a 40% acid solution?

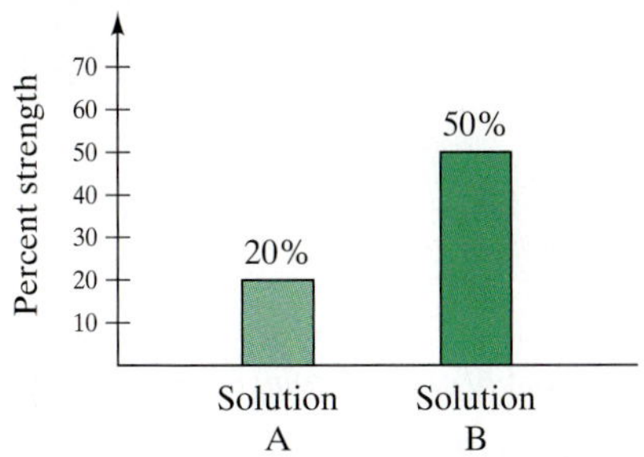


**36. Ice Hockey** The admission at an ice hockey game is \$15 for adults and \$11 for children. A total of 650 tickets were sold. Determine how many children's tickets and how many adult tickets were sold if a total of \$8790 was collected.

**37. Return to Space** John Glenn was the first American astronaut to go into orbit around the Earth. Many years later he returned to space. The second time he returned to space he was 5 years younger than twice his age when he went into space for the first time. The sum of his ages for both times he was in space is 118. Find his age each time he was in space.

**38. Savings Accounts** Jorge Minez has a total of \$40,000 invested in three different savings accounts. He has some money invested in one account that gives 7% interest. The second account has \$5000 less than the first account and gives 5% interest. The third account gives 3% interest. If the total annual interest that Jorge receives in a year is \$2300, find the amount in each account.

[4.4] *Solve each system of equations using matrices.*

**39.** $x + 5y = 1$
$-2x - 8y = -6$

**40.** $2x - 5y = 1$
$2x + 4y = 10$

**41.** $3y = 6x - 12$
$4x = 2y + 8$

**42.** $2x - y - z = 5$
$x + 2y + 3z = -2$
$3x - 2y + z = 2$

**43.** $3a - b + c = 2$
$2a - 3b + 4c = 4$
$a + 2b - 3c = -6$

**44.** $x + y + z = 3$
$3x + 4y = -1$
$y - 3z = -10$

[4.5] *Solve each system of equations using determinants.*

**45.** $7x - 8y = -10$
$-5x + 4y = 2$

**46.** $x + 4y = 5$
$5x + 3y = -9$

**47.** $9m + 4n = -1$
$7m - 2n = -11$

**48.** $p + q + r = 5$
$2p + q - r = -5$
$3p + 2q - 3r = -12$

**49.** $-2a + 3b - 4c = -7$
$2a + b + c = 5$
$-2a - 3b + 4c = 3$

**50.** $y + 3z = 4$
$-x - y + 2z = 0$
$x + 2y + z = 1$

[4.6] *Graph the solution to each system of inequalities.*

**51.** $-x + 3y > 6$
$2x - y \le 2$

**52.** $5x - 2y \le 10$
$3x + 2y > 6$

**53.** $y > 2x + 3$
$y < -x + 4$

**54.** $x > -2y + 4$
$y < -\frac{1}{2}x - \frac{3}{2}$

*Determine the solution to the system of inequalities.*

**55.** $x \ge 0$
$y \ge 0$
$x + y \le 6$
$4x + y \le 8$

**56.** $x \ge 0$
$y \ge 0$
$2x + y \le 6$
$4x + 5y \le 20$

**57.** $|x| \le 3$
$|y| > 2$

**58.** $|x| > 4$
$|y - 2| \le 3$

## Chapter 4 Practice Test

*To find out how well you understand the chapter material, take this practice test. The answers, and the section where the material was initially discussed, are given in the back of the book. Each problem is also fully worked out on the* **Chapter Test Prep Video CD.** *Review any questions that you answered incorrectly.*

1. Define **a)** a consistent system of equations, **b)** a dependent system of equations, and **c)** an inconsistent system of equations.

*Determine, without solving the system, whether the system of equations is consistent, inconsistent, or dependent. State whether the system has exactly one solution, no solution, or an infinite number of solutions.*

2. $5x + 2y = 4$
   $6x = 3y - 7$

3. $5x + 3y = 9$
   $2y = -\frac{10}{3}x + 6$

4. $5x - 4y = 6$
   $-10x + 8y = -10$

*Solve each system of equations by the method indicated.*

5. $y = 3x - 2$
   $y = -2x + 8$
   graphically

6. $y = -x + 6$
   $y = 2x + 3$
   graphically

7. $y = 4x - 3$
   $y = 5x - 4$
   substitution

8. $4a + 7b = 2$
   $5a + b = -13$
   substitution

9. $8x + 3y = 8$
   $6x + y = 1$
   addition

10. $0.3x = 0.2y + 0.4$
    $-1.2x + 0.8y = -1.6$
    addition

11. $\frac{3}{2}a + b = 6$
    $a - \frac{5}{2}b = -4$
    addition

12. $x + y + z = 2$
    $-2x - y + z = 1$
    $x - 2y - z = 1$
    addition

13. Write the augmented matrix for the following system of equations.
    $-2x + 3y + 7z = 5$
    $3x - 2y + z = -2$
    $x - 6y + 9z = -13$

14. Consider the following augmented matrix.
$$\left[\begin{array}{rrr|r} 6 & -2 & 4 & 4 \\ 4 & 3 & 5 & 6 \\ 2 & -1 & 4 & -3 \end{array}\right]$$
    Show the results obtained by multiplying the elements in the third row by $-2$ and adding the products to their corresponding elements in the second row.

*Solve each system of equations using matrices.*

15. $2x + 7y = 1$
    $3x + 5y = 7$

16. $x - 2y + z = 7$
    $-2x - y - z = -7$
    $4x + 5y - 2z = 3$

*Evaluate each determinant.*

17. $\begin{vmatrix} 3 & -1 \\ 5 & -2 \end{vmatrix}$

18. $\begin{vmatrix} 8 & 2 & -1 \\ 3 & 0 & 5 \\ 6 & -3 & 4 \end{vmatrix}$

*Solve each system of equations using determinants and Cramer's rule.*

19. $4x + 3y = -6$
    $-2x + 5y = 16$

20. $2r - 4s + 3t = -1$
    $-3r + 5s - 4t = 0$
    $-2r + s - 3t = -2$

*Use the method of your choice to find the solution to each problem.*

21. **Sunflower Seed Mixture** Agway Gardens has sunflower seeds, in a barrel, that sell for \$0.49 per pound and gourmet bird seed mix that sells for \$0.89 per pound. How much of each must be mixed to get a 20-pound mixture that sells for \$0.73 per pound?

22. **Mixing Solutions** Tyesha Blackwell, a chemist, has 6% and 15% solutions of sulfuric acid. How much of each solution should she mix to get 10 liters of a 9% solution?

23. **Sum of Numbers** The sum of three numbers is 29. The greatest number is four times the smallest number. The remaining number is 1 more than twice the smallest number. Find the three numbers.

*Determine the solution to each system of inequalities.*

24. $3x + 2y < 9$
    $-2x + 5y \le 10$

25. $|x| > 3$
    $|y| \le 1$

## Cumulative Review Test

*Take the following test and check your answers with those given in the back of the book. Review any questions that you answered incorrectly. The section where the material was covered is indicated after the answer.*

**1.** Evaluate $48 \div \left\{4\left[3 + \left(\frac{5 + 10}{5}\right)^2\right] - 32\right\}$.

**2.** Consider the following set of numbers.

$$\left\{\frac{1}{2}, -4, 9, 0, \sqrt{3}, -4.63, 1\right\}$$

List the elements of the set that are

**a)** natural numbers;

**b)** rational numbers;

**c)** real numbers.

**3.** Write the following numbers in order from smallest to largest.

$$-1, |-4|, \frac{3}{4}, \frac{5}{8}, -|-8|, |-12|$$

*Solve.*

**4.** $-[3 - 2(x - 4)] = 3(x - 6)$

**5.** $\frac{2}{3}x - \frac{5}{6} = 2$

**6.** $|2x - 3| - 5 = 4$

**7.** Solve the formula $M = \frac{1}{2}(a + x)$ for $x$.

**8.** Find the solution set of the inequality.

$$0 < \frac{3x - 2}{4} \leq 8$$

**9.** Simplify $\left(\frac{3x^2y^{-2}}{y^3}\right)^{-2}$.

**10.** Graph $2y = 3x - 8$.

**11.** Write in slope-intercept form the equation of the line that is parallel to the graph of $2x - 3y = 8$ and passes through the point $(2, 3)$

**12.** Graph the inequality $6x - 3y < 12$.

**13.** Determine which of the following graphs represent functions. Explain.

**a)**

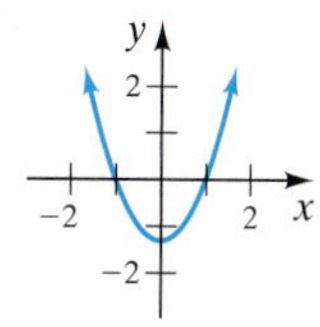


**b)**

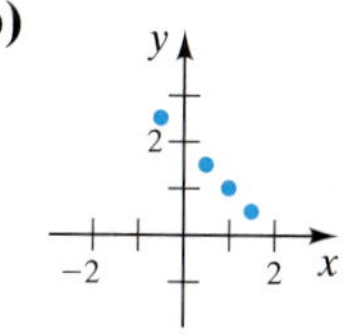


**c)**

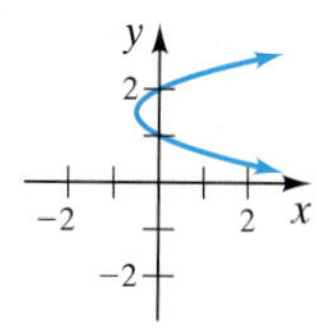


**14.** If $f(x) = \frac{x + 3}{x^2 - 9}$, find

**a)** $f(-4)$ **b)** $f(h)$ **c)** $f(3)$.

*Solve each system of equations.*

**15.** $3x + y = 6$
$y = 4x - 1$

**16.** $2p + 3q = 11$
$-3p - 5q = -16$

**17.** $x - 2y = 0$
$2x + z = 7$
$y - 2z = -5$

**18. Angles of a Triangle** If the largest angle of a triangle is nine times the measure of the smallest angle, and the middle-sized angle is 70° greater than the measure of the smallest angle, find the measure of the three angles.

**19. Walking and Jogging** Mark Simmons walks at 4 miles per hour and Judy Bolin jogs at 6 miles per hour. Mark begins walking $\frac{1}{2}$ hour before Judy starts jogging. If Judy jogs on the same path that Mark walks, how long after Judy begins jogging will she catch up to Mark?

**20. Rock Concert** There are two different prices of seats at a rock concert. The higher priced seats sell for $20 and the less expensive seats sell for $16. If a total of 1000 tickets are sold and the total ticket sales are $18,400, how many of each type of ticket are sold?

# 5 Polynomials and Polynomial Functions

## GOALS OF THIS CHAPTER

In the first part of this chapter, we discuss polynomials and polynomial functions. We then turn our attention to factoring. *You must have a thorough understanding of factoring to work the problems in many of the remaining chapters*. Pay particular attention to how to use factoring to find the $x$-intercepts of a quadratic function. We will refer back to this topic later in the course.


5.1 Addition and Subtraction of Polynomials

5.2 Multiplication of Polynomials

5.3 Division of Polynomials and Synthetic Division

5.4 Factoring a Monomial from a Polynomial and Factoring by Grouping

Mid-Chapter Test: Sections 5.1–5.4

5.5 Factoring Trinomials

5.6 Special Factoring Formulas

5.7 A General Review of Factoring

5.8 Polynomial Equations

Chapter 5 Summary

Chapter 5 Review Exercises

Chapter 5 Practice Test

Cumulative Review Test

WHEN AN OBJECT IS projected upward or dropped, its height above the ground at any time can be represented by a polynomial function. In Exercise 91 on page 305, we determine the height of an object from the ground six seconds after it is dropped from the top of the Empire State Building.

# 5.1 Addition and Subtraction of Polynomials

1 Find the degree of a polynomial.

2 Evaluate polynomial functions.

3 Understand graphs of polynomial functions.

4 Add and subtract polynomials.

## 1 Find the Degree of a Polynomial

Recall from Chapter 2 that the parts that are added or subtracted in a mathematical expression are called **terms**. The **degree of a term** with whole number exponents is the sum of the exponents of the variables, if there are variables. Nonzero constants have degree 0, and the term 0 has no degree.

A **polynomial** is a finite sum of terms in which all variables have whole number exponents and no variable appears in a denominator. The expression $3x^2 + 2x + 6$ is an example of a *polynomial in one variable, x*. The expression $x^2y - 2x + 3$ is an example of a *polynomial in two variables, x* and *y*. The expressions $x^{1/2}$ and $\frac{1}{x}$ (or $x^{-1}$) are *not* polynomials because the variables do not have whole number exponents. The expression $\frac{1}{x - 1}$ is *not* a polynomial because a variable appears in the denominator.

The **leading term** of a polynomial is the term of highest degree. The **leading coefficient** is the coefficient of the leading term.

**EXAMPLE 1** ▸ For each polynomial give the number of terms, the degree of the polynomial, the leading term, and the leading coefficient.

**a)** $2x^5 - 3x^2 + 6x - 9$ **b)** $8x^2y^4 - 6xy^3 + 3xy^2z^4$

*Solution* We will organize the answers in tabular form.

| Polynomial | Number of terms | Degree of polynomial | Leading term | Leading coefficient |
|---|---|---|---|---|
| **a)** $2x^5 - 3x^2 + 6x - 9$ | 4 | 5 (from $2x^5$) | $2x^5$ | 2 |
| **b)** $8x^2y^4 - 6xy^3 + 3xy^2z^4$ | 3 | 7 (from $3xy^2z^4$) | $3xy^2z^4$ | 3 |

▸ **Now Try Exercise 29**

Polynomials are classified according to the number of terms they have, as indicated in the following chart.

| Polynomial type | Description | Examples |
|---|---|---|
| **Monomial** | A polynomial of one term | $4x^2,\ 6x^2y,\ 3,\ -2xyz^5,\ 7$ |
| **Binomial** | A polynomial of two terms | $x^2 + 1,\ 2x^2 - y,\ 6x^3 - 5y^2$ |
| **Trinomial** | A polynomial of three terms | $x^3 + 6x - 8,\ x^2y - 9x + y^2$ |

Polynomials containing more than three terms are not given specific names. *Poly* is a prefix meaning *many*. A polynomial is called **linear** if it is of degree 0 or 1. A polynomial in one variable is called **quadratic** if it is of degree 2, and **cubic** if it is of degree 3.

| Type of Polynomial | Examples |
|---|---|
| Linear | $2x - 4,\ \ 5$ |
| Quadratic | $3x^2 + x - 6,\ \ 4x^2 - 8$ |
| Cubic | $-4x^3 + 3x^2 + 5,\ \ 2x^3 + 7x$ |

The polynomials $2x^3 + 4x^2 - 6x + 3$ and $4x^2 - 3xy + 5y^2$ are examples of polynomials in **descending order** of the variable $x$ because the exponents on the variable $x$ descend (or get lower) as the terms go from left to right. Polynomials are often written in descending order of a given variable.

**EXAMPLE 2** ▸ Write each polynomial in descending order of the variable $x$.

**a)** $5x + 4x^2 - 6$ **b)** $xy - 6x^2 + 8y^2$

*Solution*

**a)** $5x + 4x^2 - 6 = 4x^2 + 5x - 6$

**b)** $xy - 6x^2 + 8y^2 = -6x^2 + xy + 8y^2$

▸ **Now Try Exercise 25**

## 2 Evaluate Polynomial Functions

The expression $2x^3 + 6x^2 + 3$ is a polynomial. If we write $P(x) = 2x^3 + 6x^2 + 3$, then we have a polynomial function. In a **polynomial function**, the expression used to describe the function is a polynomial. To evaluate a polynomial function, we use substitution, just as we did to evaluate other functions in Chapter 3.

**EXAMPLE 3** ▸ For the polynomial function $P(x) = 4x^3 - 6x^2 - 2x + 7$, find

**a)** $P(0)$ **b)** $P(3)$ **c)** $P(-2)$

*Solution*

**a)**
$$\begin{aligned} P(x) &= 4x^3 - 6x^2 - 2x + 7 \\ P(0) &= 4(0)^3 - 6(0)^2 - 2(0) + 7 \\ &= 0 - 0 - 0 + 7 = 7 \end{aligned}$$

**b)**
$$\begin{aligned} P(3) &= 4(3)^3 - 6(3)^2 - 2(3) + 7 \\ &= 4(27) - 6(9) - 6 + 7 = 55 \end{aligned}$$

**c)**
$$\begin{aligned} P(-2) &= 4(-2)^3 - 6(-2)^2 - 2(-2) + 7 \\ &= 4(-8) - 6(4) + 4 + 7 = -45 \end{aligned}$$

▸ **Now Try Exercise 35**

Businesses, governments, and other organizations often need to track and make projections about things such as sales, profits, changes in the population, effectiveness of new drugs, and so on. To do so, they often use graphs and functions. Example 4 gives one such example.

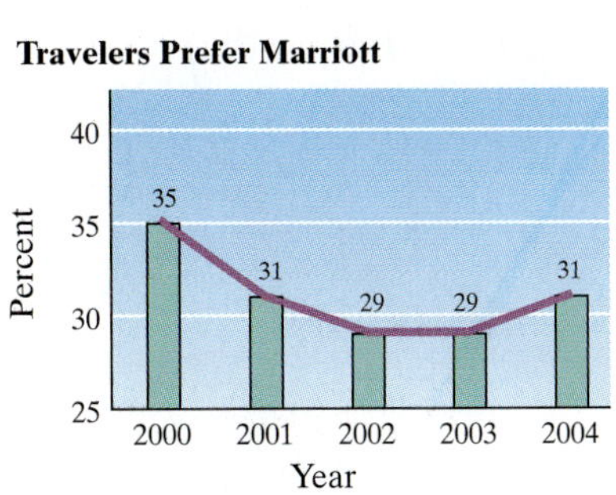


Source: The Washington Post (3/14/2005)

FIGURE 5.1

**EXAMPLE 4** ▸ **Travelers Staying at Marriott** The bar graph in **Figure 5.1** shows the percent of travelers who considered Marriott their first choice among major hotels for the years from 2000 to 2004. The polynomial function that can be used to approximate the percent of travelers who considered Marriott their first choice is

$$M(t) = t^2 - 5t + 35$$

where $t$ is the number of years since 2000 and $0 \le t \le 4$.

**a)** Use this function to estimate the percent of travelers who considered Marriott their first choice in 2004.

**b)** Compare your answer from part **a)** to the bar graph. Does the bar graph support your answer?

**c)** If this trend continues beyond 2004, estimate the percent of travelers who will consider Marriott their first choice in 2008.

*Solution* **a) Understand** We need to determine the value of $t$ to substitute into this function. Since $t$ is the number of years since 2000, the year 2004 corresponds to $t = 4$. Thus, to estimate the percent of travelers who considered Marriott their first choice, we evaluate $M(4)$.

**Translate and Carry Out**

$$\begin{aligned} M(t) &= t^2 - 5t + 35 \\ M(4) &= 4^2 - 5 \cdot 4 + 35 \\ &= 16 - 20 + 35 \\ &= 31 \end{aligned}$$

**Check and Answer:** The percent of travelers who considered Marriott their first choice in 2004 was about 31%.

**b)** From part **a)** we see that about 31% of the travelers considered Marriott as their first choice in 2004. In the bar graph above, the bar for year 2004 has a height of 31 which means about 31% of the travelers considered Marriott as their first choice. Since both values are the same, we conclude that the bar graph supports the outcome from part **a)**.

**c) Understand** To estimate the percent of travelers who will consider Marriott their first choice in 2008, observe that 2008 is 8 years since 2000. Thus, $t = 8$ and we substitute 8 for $t$ in the polynomial function.

**Translate and Carry Out**

$$\begin{aligned} M(t) &= t^2 - 5t + 35 \\ M(8) &= 8^2 - 5 \cdot 8 + 35 = 64 - 40 + 35 = 59 \end{aligned}$$

**Check and Answer** If this trend were to continue, in 2008 about 59% of the travelers would consider Marriott as their first choice among hotel brands.

▶ **Now try Exercise 103**

## 3 Understand Graphs of Polynomial Functions

The graphs of all polynomial functions are smooth, continuous curves. **Figure 5.2** shows a graph of a quadratic polynomial function. The graphs of all quadratic polynomial functions with a *positive leading coefficient* will have the shape of the graph in **Figure 5.2**.

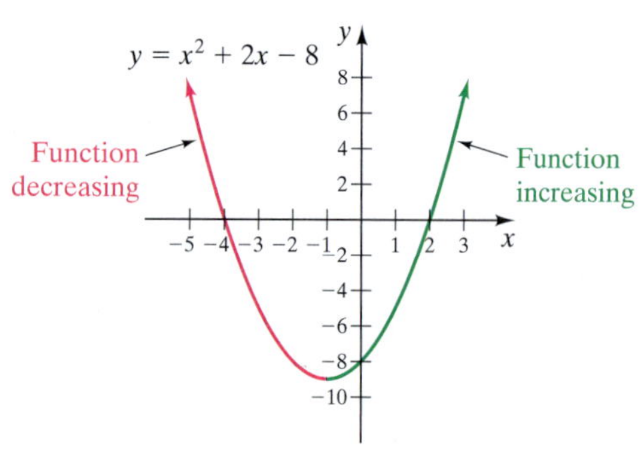


FIGURE 5.2

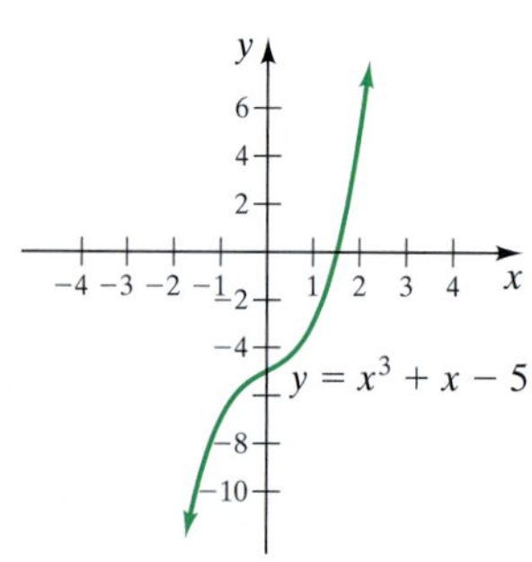


FIGURE 5.3

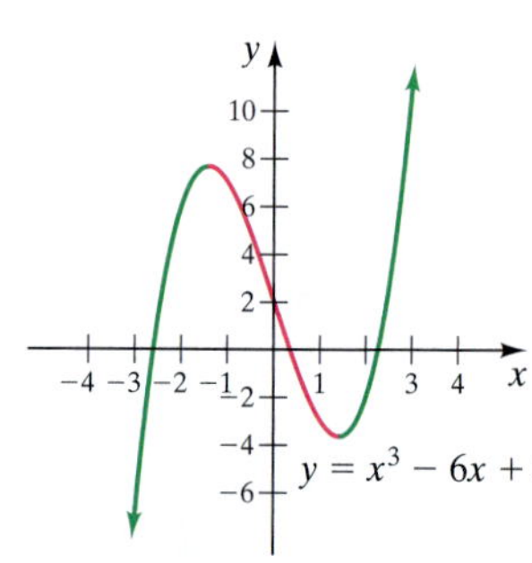


FIGURE 5.4

The graph of a cubic polynomial function with a *positive leading coefficient* may have the shape of the graph in either **Figure 5.3** or **Figure 5.4**. Notice that *whenever the leading coefficient in a polynomial function is positive, the polynomial function will increase (or move upward as x increases—the green part of the curve) to the right of some value of x*. For example, in **Figure 5.2**, the graph continues increasing to the right of $x = -1$. In **Figure 5.3**, the graph is continuously increasing, and in **Figure 5.4**, the graph is increasing to the right of about $x = 1.4$.

A quadratic polynomial function with a negative leading coefficient is shown in **Figure 5.5**, and cubic polynomial functions with negative leading coefficients are shown in **Figure 5.6** and **Figure 5.7**. In **Figure 5.5**, the quadratic function is decreasing to the right of $x = 2$. In **Figure 5.6**, the cubic function is continuously decreasing, and in **Figure 5.7** the cubic function is decreasing to the right of about $x = 1.2$.

*Polynomial functions with a negative leading coefficient will decrease (or move downward as x increases—the red part of the curve) to the right of some value of x.*

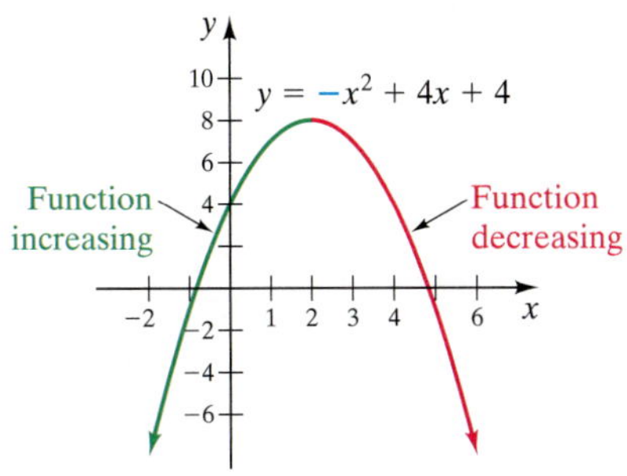


FIGURE 5.5

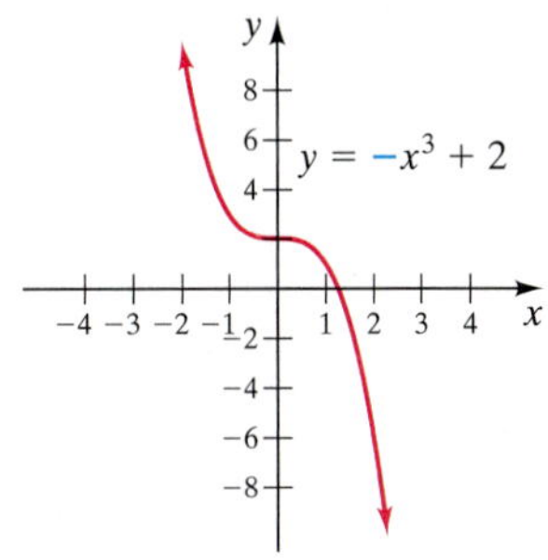


FIGURE 5.6

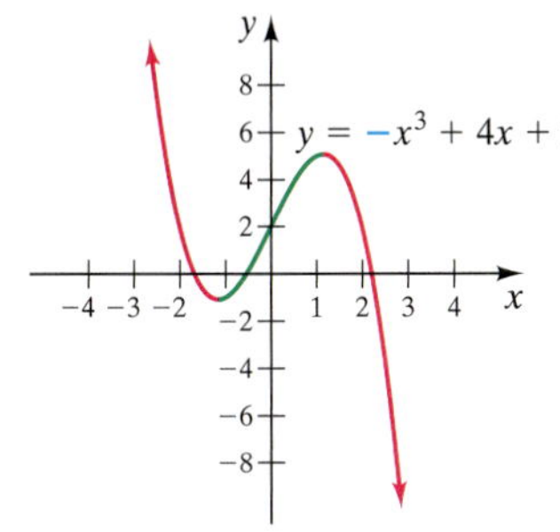


FIGURE 5.7

Why does the leading coefficient determine whether a function will increase or decrease to the right of some value of $x$? The leading coefficient is the coefficient of the term with the greatest exponent on the variable. As $x$ increases, this term will eventually dominate all the other terms in the function. So if the coefficient of this term is positive, the function will *eventually* increase as $x$ increases. If the leading coefficient is negative, the function will *eventually* decrease as $x$ increases. This information, along with checking the $y$-intercept of the graph, can be useful in determining whether a graph is correct or complete. Read the Using Your Graphing Calculator box that follows, even if you are not using a graphing calculator. Also work Exercises 99 through 102 on pages 305 and 306.

## USING YOUR GRAPHING CALCULATOR

Whenever you graph a polynomial function on your grapher, make sure your screen shows every change in direction of your graph. For example, suppose you graph $y = 0.1x^3 - 2x^2 + 5x - 8$ on your grapher. Using the standard window, you get the graph shown in **Figure 5.8**.

However, from our preceding discussion you should realize that since the leading coefficient, 0.1, is positive, the graph must increase to the right of some value of $x$. The graph in **Figure 5.8** does not show this. If you change your window as shown in **Figure 5.9**, you will get the graph shown. Now you can see how the graph increases to the right of about $x = 12$. When graphing, the $y$-intercept is often helpful in determining the values to use for the range. Recall that to find the $y$-intercept, we set $x = 0$ and solve for $y$. For example, if graphing $y = 4x^3 + 6x^2 + x - 180$, the $y$-intercept will be at $-180$.

$y = 0.1x^3 - 2x^2 + 5x - 8$

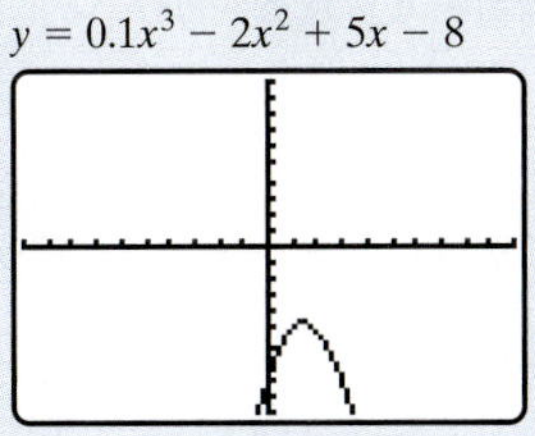

FIGURE 5.8

$y = 0.1x^3 - 2x^2 + 5x - 8$

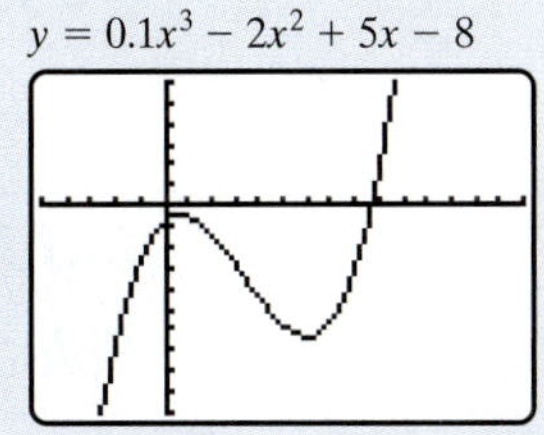

FIGURE 5.9 [−10, 30, 2, −100, 60, 10]

### EXERCISES

*Use your grapher to graph each polynomial. Make sure your window shows all changes in direction of the graph.*

**1.** $y = 0.2x^3 + 5.1x^2 - 6.2x + 9.3$

**2.** $y = 4.1x^3 - 19.6x^2 + 5.4x - 60.2$

## 4 Add and Subtract Polynomials

When we found sums and differences of functions in Section 3.6, we added and subtracted polynomials, although they were not called polynomials at that time. To *add or subtract polynomials*, remove parentheses if any are present, then combine like terms.

**EXAMPLE 5** Simplify $(4x^2 - 6x + 3) + (2x^2 + 5x - 1)$.

*Solution*

$$\begin{aligned} &(4x^2 - 6x + 3) + (2x^2 + 5x - 1) \\ &= 4x^2 - 6x + 3 + 2x^2 + 5x - 1 && \text{Remove parentheses.} \\ &= \underbrace{4x^2 + 2x^2}\ \underbrace{- 6x + 5x}\ \underbrace{+ 3 - 1} && \text{Rearrange terms.} \\ &= \quad 6x^2 \qquad -x \qquad +2 && \text{Combine like terms.} \end{aligned}$$

**Now Try Exercise 45**

**EXAMPLE 6** Simplify $(3x^2y - 4xy + y) + (x^2y + 2xy + 8y - 2)$.

*Solution*

$$\begin{aligned} &(3x^2y - 4xy + y) + (x^2y + 2xy + 8y - 2) \\ &= 3x^2y - 4xy + y + x^2y + 2xy + 8y - 2 && \text{Remove parentheses.} \\ &= \underbrace{3x^2y + x^2y}\ \underbrace{- 4xy + 2xy}\ \underbrace{+ y + 8y} - 2 && \text{Rearrange terms.} \\ &= \quad 4x^2y \qquad -2xy \qquad +9y \quad -2 && \text{Combine like terms.} \end{aligned}$$

**Now Try Exercise 51**

### Helpful Hint

Recall that $-x$ means $-1 \cdot x$. Thus $-(2x^2 - 4x + 6)$ means $-1(2x^2 - 4x + 6)$, and the distributive property applies. When you subtract one polynomial from another, the *signs of every term* of the polynomial being subtracted must change. For example,

$$\begin{aligned} x^2 - 6x + 3 - (2x^2 - 4x + 6) &= x^2 - 6x + 3 - 1(2x^2 - 4x + 6) \\ &= x^2 - 6x + 3 - 2x^2 + 4x - 6 \\ &= -x^2 - 2x - 3 \end{aligned}$$

**EXAMPLE 7** Subtract $(-x^2 - 2x + 11)$ from $(x^3 + 4x + 6)$.

*Solution*

$$\begin{aligned} &(x^3 + 4x + 6) - (-x^2 - 2x + 11) \\ &= (x^3 + 4x + 6) - 1(-x^2 - 2x + 11) && \text{Insert 1.} \\ &= x^3 + 4x + 6 + x^2 + 2x - 11 && \text{Distributive property} \\ &= x^3 + x^2 + 4x + 2x + 6 - 11 && \text{Rearrange terms.} \\ &= x^3 + x^2 + 6x - 5 && \text{Combine like terms.} \end{aligned}$$

**Now Try Exercise 67**

**EXAMPLE 8** Simplify $x^2y - 4xy^2 + 5 - (2x^2y - 3y^2 + 11)$.

*Solution*

$$\begin{aligned} &x^2y - 4xy^2 + 5 - 1(2x^2y - 3y^2 + 11) && \text{Insert 1.} \\ &= x^2y - 4xy^2 + 5 - 2x^2y + 3y^2 - 11 && \text{Distributive property} \\ &= x^2y - 2x^2y - 4xy^2 + 3y^2 + 5 - 11 && \text{Rearrange terms.} \\ &= -x^2y - 4xy^2 + 3y^2 - 6 && \text{Combine like terms.} \end{aligned}$$

Note that $-x^2y$ and $-4xy^2$ are not like terms since the variables have different exponents. Also, $-4xy^2$ and $3y^2$ are not like terms since $3y^2$ does not contain the variable $x$.

**Now Try Exercise 55**

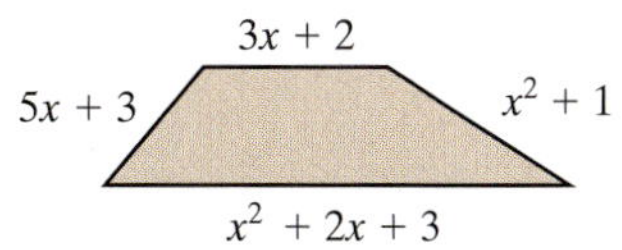


FIGURE 5.10

**EXAMPLE 9 ▸ Perimeter** Find an expression for the perimeter of the quadrilateral in **Figure 5.10**.

*Solution* The perimeter is the sum of the lengths of the sides of the figure. For a quadrilateral, the perimeter is the sum of the lengths of the four sides.

$$\begin{aligned} \text{Perimeter} &= (x^2 + 2x + 3) + (x^2 + 1) + (5x + 3) + (3x + 2) && \textit{Sum of the sides} \\ &= x^2 + 2x + 3 + x^2 + 1 + 5x + 3 + 3x + 2 && \textit{Remove parentheses.} \\ &= x^2 + x^2 + 2x + 5x + 3x + 3 + 1 + 3 + 2 && \textit{Rearrange terms.} \\ &= 2x^2 + 10x + 9 && \textit{Combine like terms.} \end{aligned}$$

The perimeter of the quadrilateral is $2x^2 + 10x + 9$.

▸ **Now Try Exercise 79**

# EXERCISE SET 5.1

MathXL® MyMathLab

## Concept/Writing Exercises

1. What are terms of a mathematical expression?
2. What is the degree of a nonzero constant?
3. What is a polynomial?
4. What is the leading term of a polynomial?
5. What is the leading coefficient of a polynomial?
6. a) How do you determine the degree of a term?
   b) What is the degree of $6x^4y^3z$?
7. a) How do you determine the degree of a polynomial?
   b) What is the degree of $-4x^4 + 6x^3y^4 + z^5$?
8. What does it mean when a polynomial is in descending order of the variable $x$?
9. a) When is a polynomial linear?
   b) Give an example of a linear polynomial.
10. a) When is a polynomial quadratic?
    b) Give an example of a quadratic polynomial.
11. a) When is a polynomial cubic?
    b) Give an example of a cubic polynomial.
12. When one polynomial is being subtracted from another, what happens to the signs of all the terms of the polynomial being subtracted?
13. Write a fifth-degree trinomial in $x$ in descending order that lacks fourth-, third-, and second-degree terms.
14. Write a seventh-degree polynomial in $y$ in descending order that lacks fifth-, third-, and second-degree terms.

## Practice the Skills

*Determine whether each expression is a polynomial. If the polynomial has a specific name, for example, "monomial" or "binomial," give the name. If the expression is not a polynomial, explain why it is not.*

15. $-6$
16. $4x^{-1}$
17. $7z$
18. $5x^2 - 6x + 9$
19. $5z^{-3}$
20. $8x^2 - 2x + 9y^2$
21. $3x^{1/2} + 2xy$
22. $2xy + 5y^2$

*Write each polynomial in descending order of the variable x. If the polynomial is already in descending order, so state. Give the degree of each polynomial.*

23. $-5 + 2x - x^2$
24. $-3x - 9 + 8x^2$
25. $9y^2 + 3xy + 10x^2$
26. $-2 + x - 8x^2 + 4x^3$
27. $-2x^4 + 5x^2 - 4$
28. $5xy^2 + 3x^2y - 9 - 2x^3$

*Give* **a)** *the degree of each polynomial and* **b)** *its leading coefficient.*

29. $x^4 + 3x^6 - 2x - 13$
30. $17x^4 + 13x^5 - x^7 + 4x^3$
31. $4x^2y^3 + 6xy^4 + 9xy^5$
32. $-a^4b^3c^2 + 9a^8b^9c^4 - 5a^7c^{20}$
33. $-\frac{1}{3}m^4n^5p^8 + \frac{3}{5}m^3p^6 - \frac{5}{9}n^4p^6q$
34. $-0.6x^2y^3z^2 + 2.9xyz^9 - 1.3x^8y^4$

*Evaluate each polynomial function at the given value.*

35. Find $P(2)$ if $P(x) = x^2 - 6x + 5$.

36. Find $P(-1)$ if $P(x) = 4x^2 + 6x - 11$.

37. Find $P\left(\frac{1}{2}\right)$ if $P(x) = 2x^2 - 3x - 6$.

38. Find $P\left(\frac{1}{3}\right)$ if $P(x) = \frac{1}{2}x^3 - x^2 + 6$.

39. Find $P(0.4)$ if $P(x) = 0.2x^3 + 1.6x^2 - 2.3$.

40. Find $P(-1.2)$ if $P(x) = -1.6x^3 - 4.6x^2 - 0.1x$.

*In Exercises 41–62, simplify.*

41. $(x^2 + 3x - 1) + (6x - 5)$

42. $(5b^2 - 8b + 7) - (2b^2 - 3b - 5)$

43. $(x^2 - 8x + 11) - (5x + 9)$

44. $(2x - 13) - (3x^2 - 4x + 16)$

45. $(4y^2 + 9y - 1) - (2y^2 + 10)$

46. $(5n^2 - 7) + (9n^2 + 3n + 12)$

47. $\left(-\frac{5}{9}a + 6\right) + \left(-\frac{2}{3}a^2 - \frac{1}{4}a - 1\right)$

48. $(6y^2 - 9y + 4) - (-2y^2 - y - 8)$

49. $(1.4x^2 + 1.6x - 8.3) - (4.9x^2 + 3.7x + 11.3)$

50. $(-12.4x^2y - 6.2xy + 9.3y^2) - (-5.3x^2y + 1.6xy - 10.4y^2)$

51. $\left(-\frac{1}{3}x^3 + \frac{1}{4}x^2y + 8xy^2\right) + \left(-x^3 - \frac{1}{2}x^2y + xy^2\right)$

52. $\left(-\frac{3}{5}xy^2 + \frac{5}{8}\right) - \left(-\frac{1}{2}xy^2 + \frac{3}{5}\right)$

53. $(3a - 6b + 5c) - (-2a + 4b - 8c)$

54. $(9r + 7s - t) + (-2r - 2s - 3t)$

55. $(3a^2b - 6ab + 5b^2) - (4ab - 6b^2 - 5a^2b)$

56. $(3x^2 - 5y^2 - 2xy) - (4x^2 + 8y^2 - 9xy)$

57. $(8r^2 - 5t^2 + 2rt) + (-6rt + 2t^2 - r^2)$

58. $(a^2 - b^2 + 5ab) + (-3b^2 - 2ab + a^2)$

59. $6x^2 - 5x - [3x - (4x^2 - 9)]$

60. $3xy^2 - 2x - [-(4xy^2 + 3x) - 6xy]$

61. $5w - 6w^2 - [(3w - 2w^2) - (4w + w^2)]$

62. $-[-(5r^2 - 3r) - (2r - 3r^2) - 2r^2]$

63. Subtract $(4x - 11)$ from $(7x + 8)$.

64. Subtract $(-x^2 + 3x + 5)$ from $(4x^2 - 6x + 2)$.

65. Add $-2x^2 + 4x - 12$ and $-x^2 - 2x$.

66. Subtract $(5x^2 - 6)$ from $(2x^2 - 9x + 8)$.

67. Subtract $0.2a^2 - 3.9a + 26.4$ from $-5.2a^2 - 9.6a$.

68. Add $6x^2 + 12xy$ and $-2x^2 + 4xy + 3y$.

69. Subtract $\left(5x^2y + \frac{5}{9}\right)$ from $\left(-\frac{1}{2}x^2y + xy^2 + \frac{3}{5}\right)$.

70. Subtract $(6x^2y + 7xy)$ from $(2x^2y + 12xy)$.

*Simplify. Assume that all exponents represent natural numbers.*

71. $(3x^{2r} - 7x^r + 1) + (2x^{2r} - 3x^r + 2)$

72. $(8x^{2r} - 5x^r + 4) + (6x^{2r} + x^r + 3)$

73. $(x^{2s} - 8x^s + 6) - (2x^{2s} - 4x^s - 13)$

74. $(5a^{2m} - 6a^m + 4) - (2a^{2m} + 7)$

75. $(7b^{4n} - 5b^{2n} + 1) - (3b^{3n} - b^{2n})$

76. $(-3r^{3a} + r^a - 6) - (-2r^{3a} - 8r^{2a} + 6)$

## Problem Solving

**Perimeter** *In Exercises 77–82, find an expression for the perimeter of each figure. See Example 9.*

77.

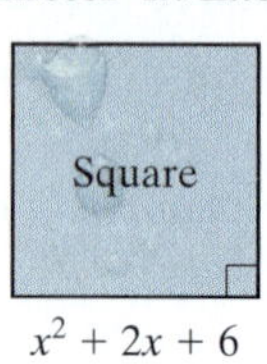


78.

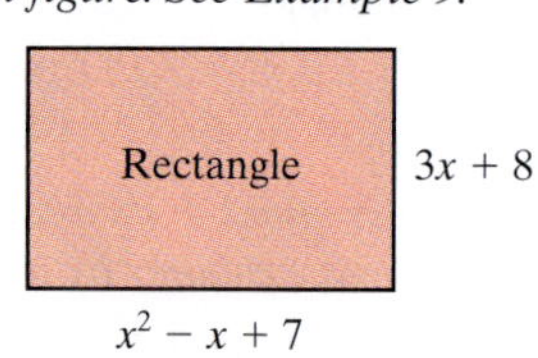


79.

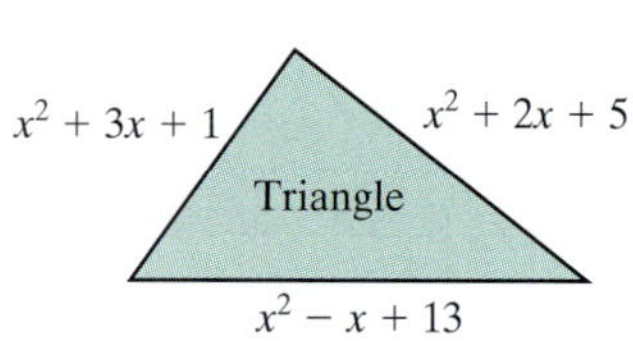


80.

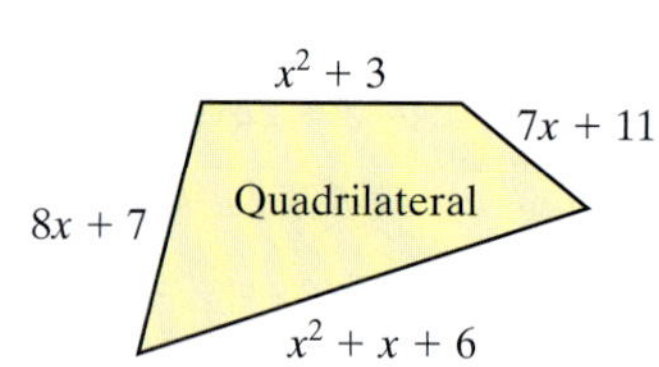


81.

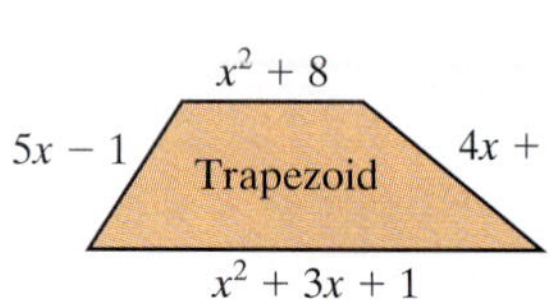


82.

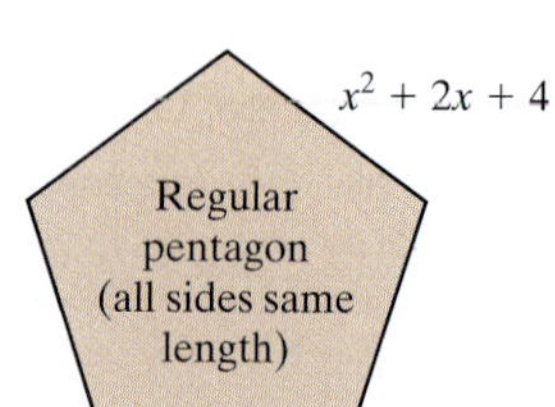

**83.** Is the sum of two trinomials always a trinomial? Explain, and give an example to support your answer.

**84.** Is the sum of two binomials always a binomial? Explain, and give an example to support your answer.

**85.** Is the sum of two quadratic polynomials always a quadratic polynomial? Explain, and give an example to support your answer.

**86.** Is the difference of two cubic polynomials always a cubic polynomial? Explain, and give an example to support your answer.

**87. Area** The area of a square is a function of its side, $s$, where $A(s) = s^2$. Find the area of a square if its side is 12 meters.

**88. Volume** The volume of a cube is a function of its side, $s$, where $V(s) = s^3$. Find the volume of a cube if its side is 7 centimeters.

**89. Area** The area of a circle is a function of its radius, where $A(r) = \pi r^2$. Find the area of a circle if its radius is 6 inches. Use the $\boxed{\pi}$ key on your calculator.

**90. Volume** The volume of a sphere is a function of its radius where $V(r) = \frac{4}{3}\pi r^3$. A circular balloon is being blown up. Find its volume when its radius is 4 inches.

**91. Height** When an object is dropped from the Empire State Building (height 1250 feet), the object's height, $h$, in feet, from the ground at time, $t$, in seconds, after being dropped can be determined by

$$h = P(t) = -16t^2 + 1250$$

Find the distance an object is from the ground 6 seconds after being dropped.

**92. Spelling Bee** The number of ways that the first-, second-, and third-place winners in a spelling bee can be selected from $n$ participants is given by $P(n) = n^3 - 3n^2 + 2n$. If there are seven participants, how many ways can the winner and first and second runners-up be selected?

**93. Committees** The number of different committees of 2 students where the 2 students are selected from a class of $n$ students is given by $c(n) = \frac{1}{2}(n^2 - n)$. If a biology class has 15 students, how many different committees having 2 students can be selected?

**94. Committees** The number of different committees of 3 students where the 3 students are selected from a class of $n$ students is given by $c(n) = \frac{1}{6}n^3 - \frac{1}{2}n^2 + \frac{1}{3}n$. If an art class has 10 students, how many different committees having 3 students can be selected?

**95. Savings Account** On January 2, 2006, Jorge Sanchez deposited $650 into a savings account that pays simple interest at a rate of $24 each year. The amount in the account is a function of time given by $A(t) = 650 + 24t$, where $t$ is the number of years after 2006. Find the amount in the account in **a)** 2007. **b)** 2021.

**96. Financing** Frank Gunther just bought a new car. After making the down payment, the amount to be financed is $23,250. Using a 0% (or interest-free) loan, the monthly payment is $387.50. The amount still owed on the car is a function of time given by $A(t) = \$23{,}250 - \$387.50t$, where $t$ is the number of months after Frank bought the car. How much is still owed **a)** 2 months, **b)** 15 months after Frank bought the car?

**Profit** *The profit of a company is found by subtracting its cost from its revenue. In Exercises 97 and 98, R(x) represents the company's revenue when selling x items and C(x) represents the company's cost when producing x items.* **a)** *Determine a profit function P(x).* **b)** *Evaluate P(x) when x = 100.*

**97.** $R(x) = 2x^2 - 60x$,
$C(x) = 8050 - 420x$

**98.** $R(x) = 5.5x^2 - 80.3x$
$C(x) = 1.2x^2 + 16.3x + 12{,}040.6$

*In Exercises 99–102, determine which of the graphs—**a)**, **b)**, or **c)**—is the graph of the given equation. Explain how you determined your answer.*

**99.** $y = x^2 + 3x - 4$

**a)**

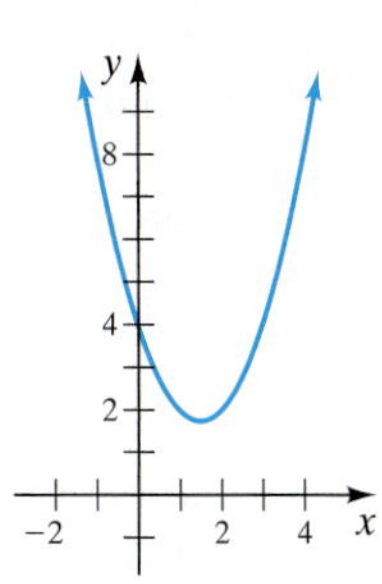

**b)**

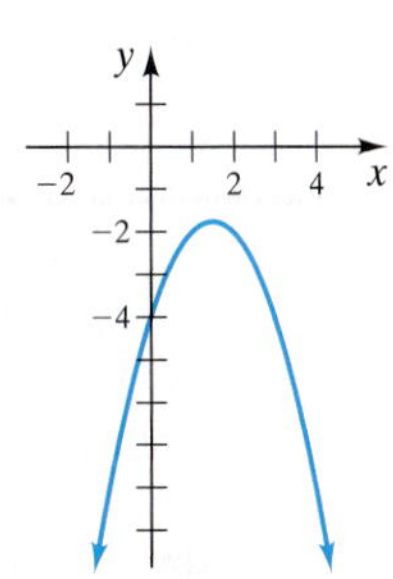

**c)**

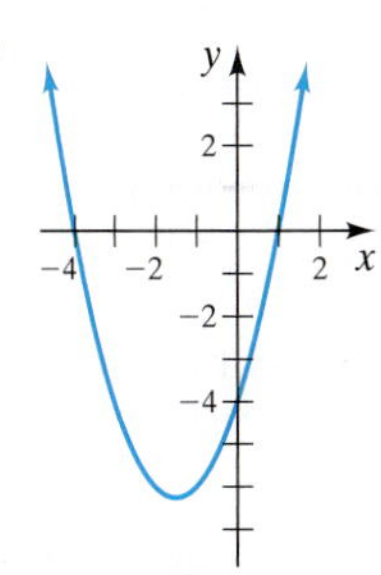

**100.** $y = x^3 + 2x^2 - 4$

a) 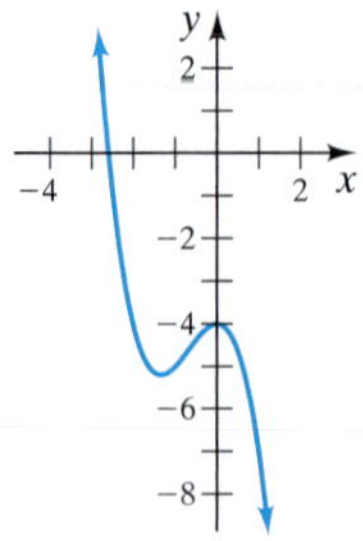


b) 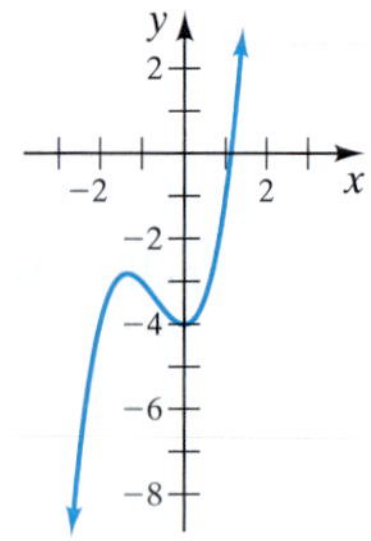


c) 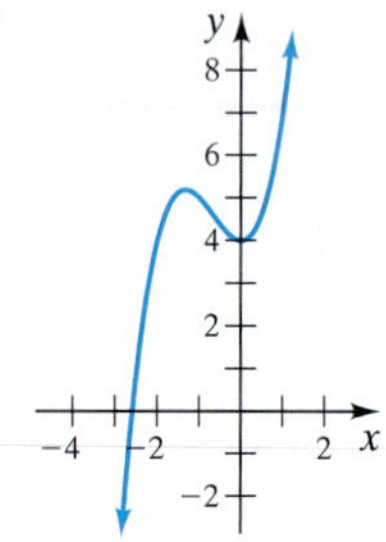


**101.** $y = -x^3 + 2x - 6$

a) 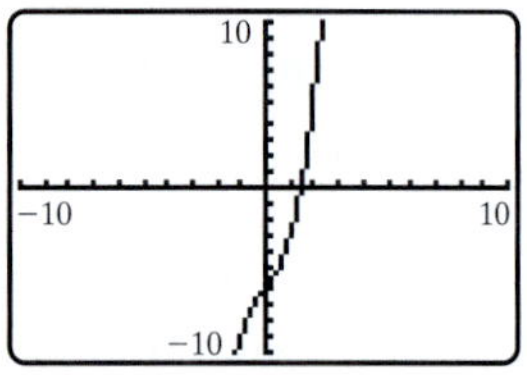


b) 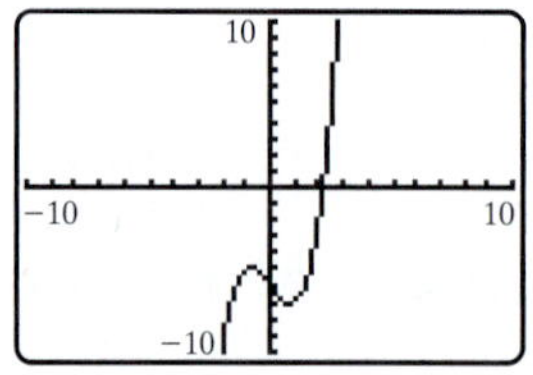


c) 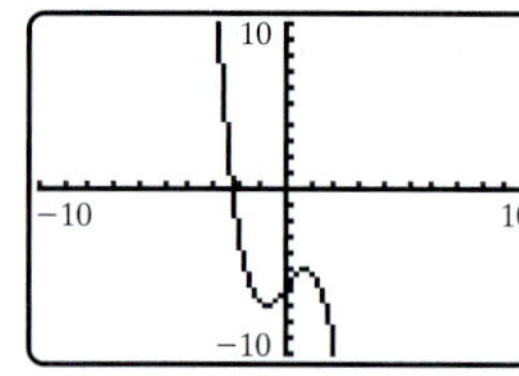


**102.** $y = x^3 + 4x^2 - 5$

a) 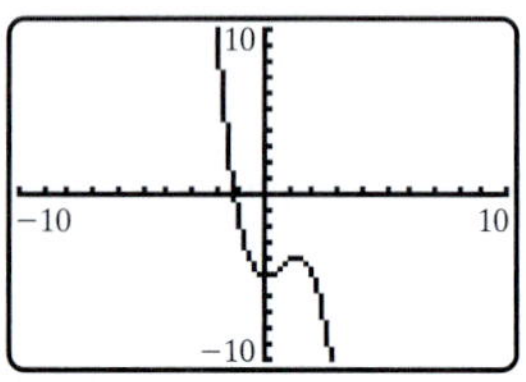


b) 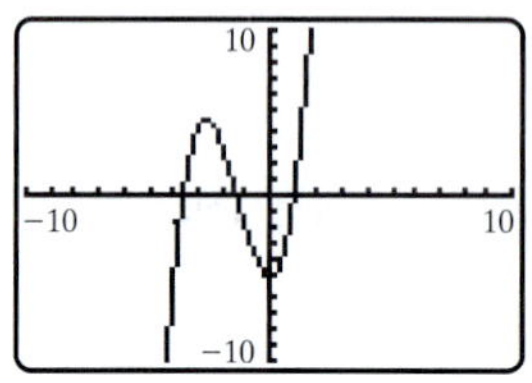


c) 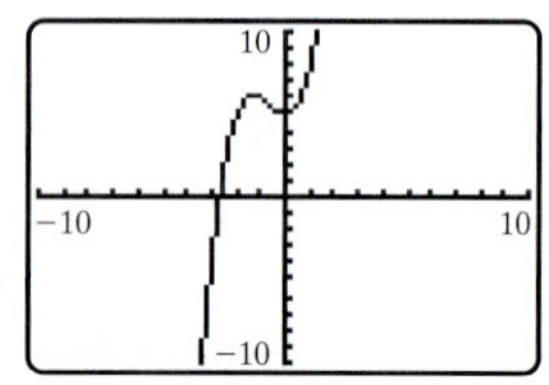


**103. Capital Spending** The graph shows the expenditure by oil companies for new oil and natural gas projects for the years 2001 to 2004. The expenditure, $E(t)$, in billions of dollars can be approximated by the function

$$E(t) = 7t^2 - 7.8t + 81.2$$

where $t$ is the number of years since 2001.

**a)** Use this function to estimate the expenditure by oil companies in 2004.

**b)** Compare your answer from part **a)** to the bar graph. Does the graph support your answer?

**c)** If this trend continues, estimate the expenditure of oil companies for new oil and natural gas projects in 2007.

**Expenditures of Oil Companies**

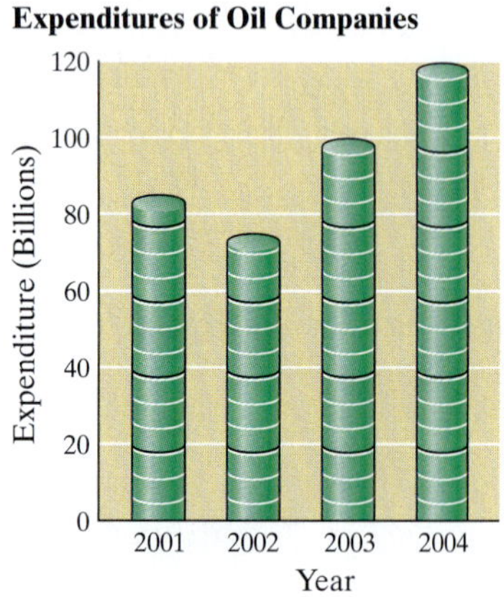


*Source*: John S. Herald, Inc., *The Washington Post (3/14/2005)*

**104. Inclined Plane** A ball rolls down an inclined plane. The distance, $d(t)$, in feet, the ball has traveled is given by the function

$$d(t) = 2.36t^2$$

where $t$ is time in seconds, $0 \le t \le 5$.

Find the distance the ball has traveled down the inclined plane in

**a)** 1 second.

**b)** 3 seconds.

**c)** 5 seconds.

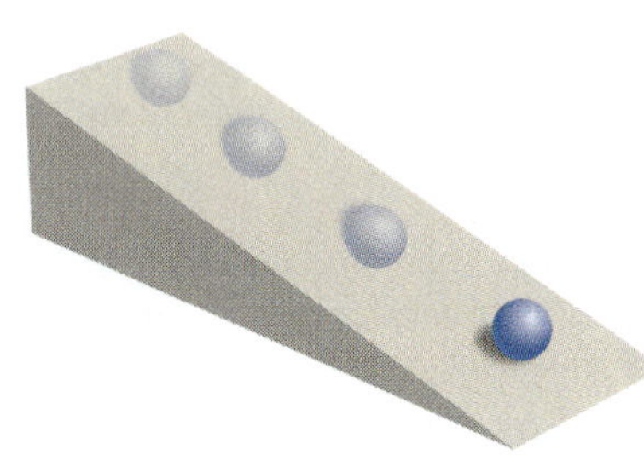

**105. Inflation** Inflation erodes our purchasing power. Because of inflation, we will pay more for the same goods in the future than we pay for them today. The function $C(t) = 0.31t^2 + 0.59t + 9.61$, where $t$ is years since 1997, approximates the cost, in thousands of dollars, for purchasing in the future what \$10,000 would purchase in 1997. This function is based on a 6% annual inflation rate and $0 \le t \le 25$. Estimate the cost in 2012 for goods that cost \$10,000 in 1997.

**106. Drug-Free Schools** The function $f(a) = -2.32a^2 + 76.85a - 559.87$ can be used to estimate the percent of students who say their school is not drug free. In this function, $a$ represents the student's age, where $12 \le a \le 17$. Use this function to estimate the percent of 13-year-olds who say their school is not drug free.

*Answer Exercises 107 and 108 using a graphing calculator if you have one. If you do not have a graphing calculator, draw the graphs in part* **a)** *by plotting points. Then answer parts* **b)** *through* **e)**.

**107.** **a)** Graph

$$y_1 = x^3$$
$$y_2 = x^3 - 3x^2 - 3$$

**b)** In both graphs, for values of $x > 3$, do the functions increase or decrease as $x$ increases?

**c)** When the leading term of a polynomial function is $x^3$, the polynomial must increase for $x > a$, where $a$ is some real number greater than 0. Explain why this must be so.

**d)** In both graphs, for values of $x < -3$, do the functions increase or decrease as $x$ decreases?

**e)** When the leading term of a polynomial function is $x^3$, the polynomial must decrease for $x < a$, where $a$ is some real number less than 0. Explain why this must be so.

**108.** **a)** Graph

$$y_1 = x^4$$
$$y_2 = x^4 - 6x^2$$

**b)** In each graph, for values of $x > 3$, are the functions increasing or decreasing as $x$ increases?

**c)** When the leading term of a polynomial function is $x^4$, the polynomial must increase for $x > a$, where $a$ is some real number greater than 0. Explain why this must be so.

**d)** In each graph, for values of $x < -3$, are the functions increasing or decreasing as $x$ decreases?

**e)** When the leading term of a polynomial function is $x^4$, the polynomial must increase for $x < a$, where $a$ is some real number less than 0. Explain why this must be so.

## Challenge Problems

*Determine which of the graphs—***a), b),** *or* **c)***—is the graph of the given equation. Explain how you determined your answer.*

**109.** $y = -x^4 + 3x^3 - 5$

**a)**

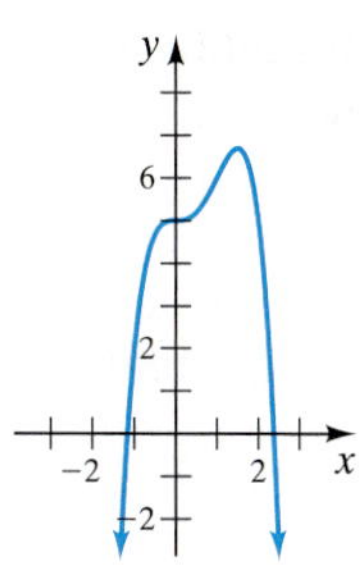

**b)**

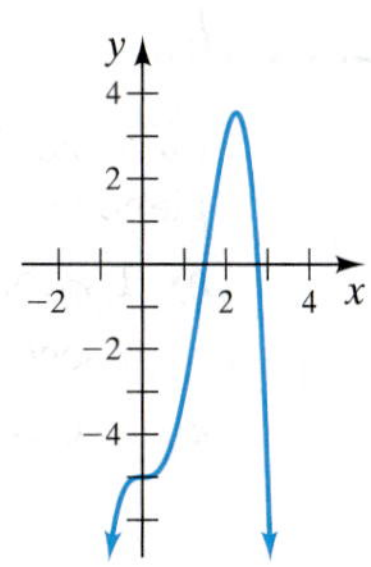

**c)**

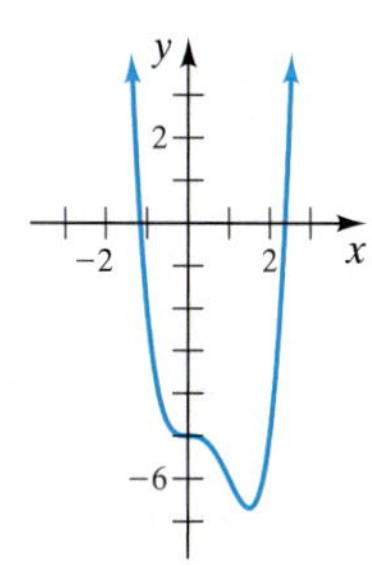

**110.** $y = 2x^4 + 9x^2 - 5$

**a)**

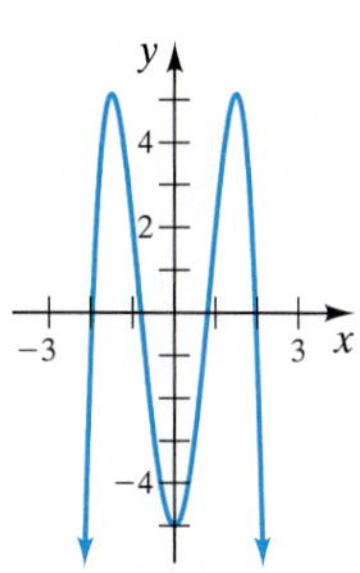

**b)**

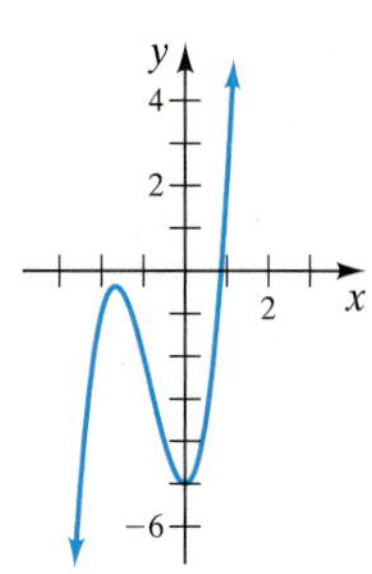

**c)**

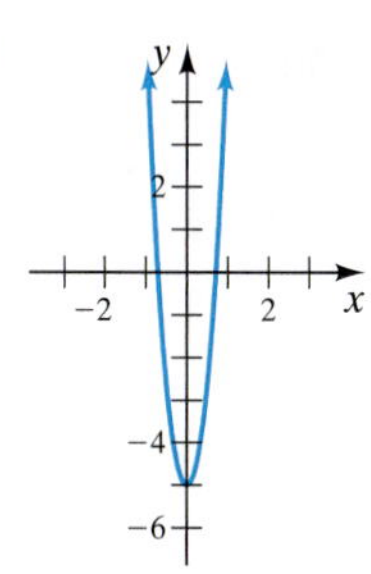

## Group Activity

*Discuss and answer Exercises 111 and 112 as a group.*

**111.** If the leading term of a polynomial function is $3x^3$, which of the following could possibly be the graph of the polynomial? Explain. Consider what happens for large positive values of $x$ and for large negative values of $x$.

**a)**

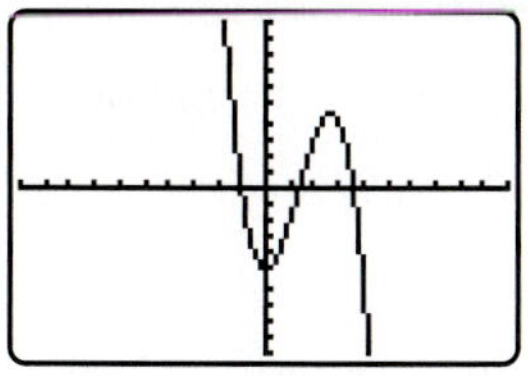

**b)**

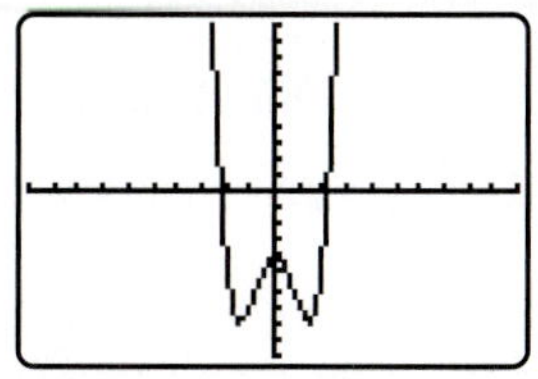

**c)**

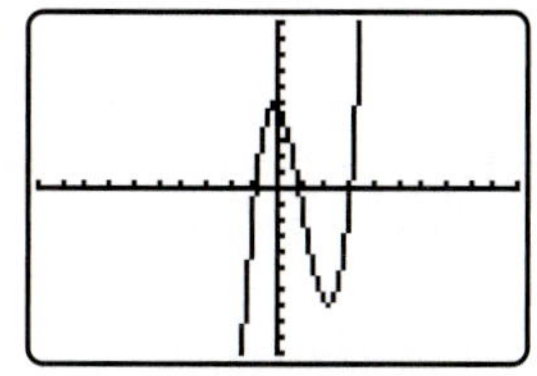

**112.** If the leading term of a polynomial is $-2x^4$, which of the following could possibly be the graph of the polynomial? Explain.

**a)** 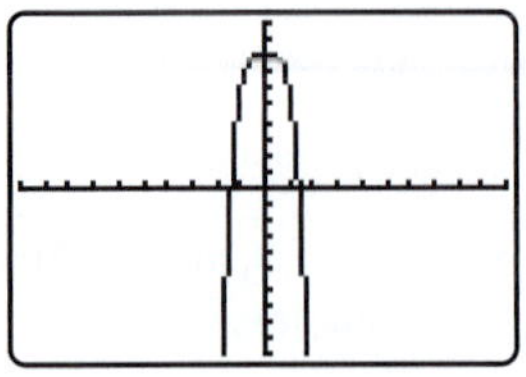

**b)** 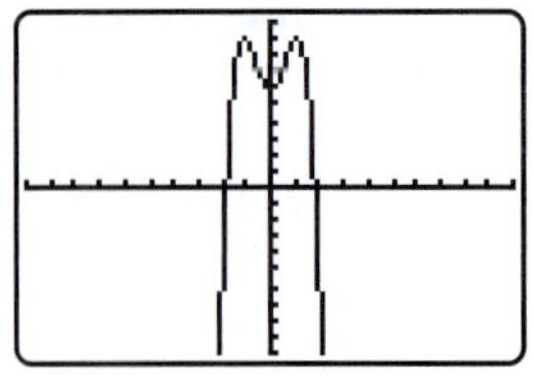

**c)** 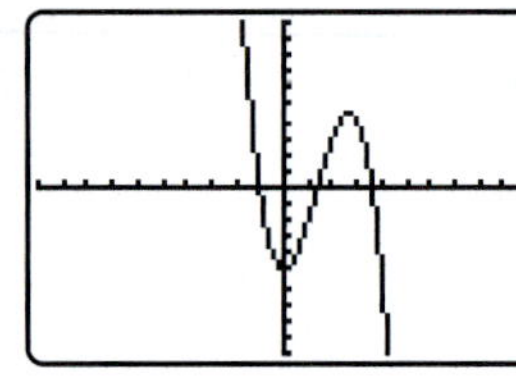

## Cumulative Review Exercises

[1.4] **113.** Evaluate $\sqrt[4]{81}$.

[2.1] **114.** Solve $1 = \frac{8}{5}x - \frac{1}{2}$.

[2.4] **115. Molding Machines** An older molding machine can produce 40 plastic buckets in 1 hour. A newer machine can produce 50 buckets in 1 hour. How long will it take the two machines to produce a total of 540 buckets?

[3.4] **116.** Find the slope of the line through the points $(10, -4)$ and $(-1, -2)$.

[4.2] **117.** Solve the system of equations.

$$\begin{aligned} -4s + 3t &= 16 \\ 4t - 2u &= 2 \\ -s + 6u &= -2 \end{aligned}$$

# 5.2 Multiplication of Polynomials

1 Multiply a monomial by a polynomial.

2 Multiply a binomial by a binomial.

3 Multiply a polynomial by a polynomial.

4 Find the square of a binomial.

5 Find the product of the sum and difference of the same two terms.

6 Find the product of polynomial functions.

## 1 Multiply a Monomial by a Polynomial

In Section 3.6 we added and subtracted functions, but we did not multiply polynomial functions. After this section you will be able to find the product of functions, that is, $(f \cdot g)(x)$.

To multiply polynomials, you must remember that *each term of one polynomial must multiply each term of the other polynomial.* This results in monomials multiplying monomials. To multiply monomials, we use the product rule of exponents, which was presented in Section 1.5.

### Helpful Hint *Study Tip*

In this chapter you will be working with exponents. The rules for exponents were covered in Section 1.5. The rules for exponents that you will need to work problems in this chapter are presented again for your convenience, along with examples, prior to where you need to use them. Below we present the product rule for exponents, and in Section 5.3 we present the quotient rule for exponents and the zero exponent rule. If, after reading the examples, you would like additional examples of the use of these rules, review Section 1.5.

**Product Rule for Exponents**

$$a^m \cdot a^n = a^{m+n}$$

In Example 1, we review how to multiply monomials using the Product Rule for Exponents. Also, in Example 1, we use the word *factors*. Recall that any expressions that are *multiplied* are called *factors*.

### Multiply a Monomial by a Monomial

**EXAMPLE 1** ▸ Multiply.

**a)** $(4x^2)(5x^3)$ **b)** $(3x^2y)(4x^5y^3)$ **c)** $(-2a^4b^7)(-3a^8bc^5)$

*Solution* We use the Product Rule for Exponents to multiply the factors.

**a)** $(4x^2)(5x^3) = 4 \cdot 5 \cdot x^2 \cdot x^3$ *Remove parentheses and rearrange terms.*

$= 20x^{2+3}$ *Product Rule, $x^2 \cdot x^3 = x^{2+3}$*

$= 20x^5$

**b)** $(3x^2y)(4x^5y^3) = 3 \cdot 4 \cdot x^2 \cdot x^5 \cdot y \cdot y^3$ *Remove parentheses, rearrange terms.*

$= 12x^{2+5}y^{1+3}$ *Product Rule*

$= 12x^7y^4$

**c)** $(-2a^4b^7)(-3a^8b^3c^5) = (-2)(-3)a^4 \cdot a^8 \cdot b^7 \cdot b^3 \cdot c^5$ *Remove parentheses, rearrange terms.*

$= 6a^{4+8}b^{7+3}c^5$ *Product rule*

$= 6a^{12}b^{10}c^5$

**▸ Now Try Exercise 9**

In Example 1**a)**, both $4x^2$ and $5x^3$ are *factors* of the product $20x^5$. In Example 1**b)**, both $3x^2y$ and $4x^5y^3$ are *factors* of the product $12x^7y^4$.

### Multiply a Monomial by a Polynomial

When multiplying a monomial by a binomial, we can use the distributive property. When multiplying a monomial by a polynomial that contains more than two terms we can use the **expanded form of the distributive property**.

**Distributive Property, Expanded Form**

$$a(b + c + d + \cdots + n) = ab + ac + ad + \cdots + an$$

In Example 2**a)** we multiply a monomial by a binomial and in Examples 2**b)** and 2**c)** we multiply a monomial by a trinomial.

**EXAMPLE 2 ▸** Multiply.

**a)** $3x^2\left(\frac{1}{6}x^3 - 5x^2\right)$ **b)** $2xy(3x^2y + 6xy^2 + 9)$ **c)** $0.4x(0.3x^3 + 0.7xy^2 - 0.2y^4)$

*Solution*

**a)** $3x^2\left(\frac{1}{6}x^3 - 5x^2\right) = 3x^2\left(\frac{1}{6}x^3\right) - 3x^2(5x^2) = \frac{1}{2}x^5 - 15x^4$

**b)** $2xy(3x^2y + 6xy^2 + 9) = (2xy)(3x^2y) + (2xy)(6xy^2) + (2xy)(9)$

$= 6x^3y^2 + 12x^2y^3 + 18xy$

**c)** $0.4x(0.3x^3 + 0.7xy^2 - 0.2y^4)$

$= (0.4x)(0.3x^3) + (0.4x)(0.7xy^2) - (0.4x)(0.2y^4)$

$= 0.12x^4 + 0.28x^2y^2 - 0.08xy^4$

**▸ Now Try Exercise 13**

## 2 Multiply a Binomial by a Binomial

Consider multiplying $(a + b)(c + d)$. Treating $(a + b)$ as a single term and using the distributive property, we get

$$(a + b)(c + d) = (a + b)c + (a + b)d$$
$$= ac + bc + ad + bd$$

When multiplying a binomial by a binomial, each term of the first binomial must be multiplied by each term of the second binomial, and then all like terms are combined.

Binomials can be multiplied vertically as well as horizontally.

**EXAMPLE 3** ▸ Multiply $(3x + 2)(x - 5)$.

*Solution* We will multiply vertically. List the binomials in descending order of the variable one beneath the other. It makes no difference which one is placed on top. Then multiply each term of the top binomial by each term of the bottom binomial, as shown. Remember to align like terms so that they can be added.

$$\begin{array}{rrrl} & 3x + 2 & & \\ & x - 5 & & \\ \hline -5(3x+2) \longrightarrow & -15x - 10 & & \text{Multiply the top binomial by } -5. \\ x(3x+2) \rightarrow 3x^2 + & 2x & & \text{Multiply the top binomial by } x. \\ \hline 3x^2 - & 13x - 10 & & \text{Add like terms in columns.} \end{array}$$

In Example 3, the binomials $3x + 2$ and $x - 5$ are *factors* of the trinomial $3x^2 - 13x - 10$.

▸ **Now Try Exercise 21**

### The FOIL Method

A convenient way to multiply two binomials is called the **FOIL method**. To multiply two binomials using the FOIL method, list the binomials side by side. The word FOIL indicates that you multiply the First terms, Outer terms, Inner terms, and Last terms of the two binomials. This procedure is illustrated in Example 4, where we multiply the same two binomials we multiplied in Example 3.

**EXAMPLE 4** ▸ Multiply $(3x + 2)(x - 5)$ using the FOIL method.

*Solution*

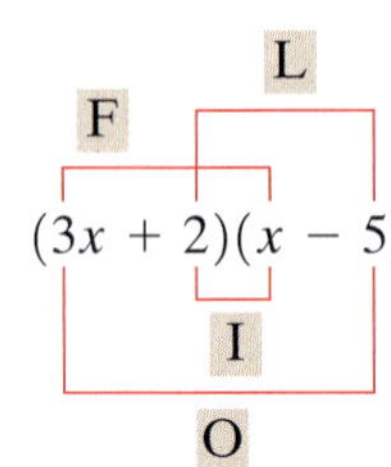


$$\begin{array}{l} \quad\ \ \text{F} \qquad\quad \text{O} \qquad\quad \text{I} \qquad\quad \text{L} \\ (3x)(x) + (3x)(-5) + (2)(x) + (2)(-5) \\ = 3x^2 - 15x + 2x - 10 = 3x^2 - 13x - 10 \end{array}$$

▸ **Now Try Exercise 25**

We performed the multiplications following the FOIL order. However, any order could be followed as long as each term of one binomial is multiplied by each term of the other binomial. We use FOIL rather than OILF or any other combination of letters because it is easier to remember.

## 3 Multiply a Polynomial by a Polynomial

When multiplying a trinomial by a binomial or a trinomial by a trinomial, every term of the first polynomial must be multiplied by every term of the second polynomial. It is helpful for aligning terms to place each polynomial in descending order, if not given that way.

**EXAMPLE 5** ▸ Multiply $x^2 + 1 - 4x$ by $2x^2 - 3$.

*Solution* Since the trinomial is not in descending order, rewrite it as $x^2 - 4x + 1$.

Place the longer polynomial on top, then multiply. Make sure you align like terms as you multiply so that the terms can be added more easily.

$$\begin{array}{lrrrrrl}
 & & & x^2 & -4x & +1 & \text{Trinomial written in descending order} \\
 & & & 2x^2 & & -3 & \\
\hline
-3(x^2 - 4x + 1) \longrightarrow & & & -3x^2 & +12x & -3 & \text{Multiply top expression by } -3. \\
2x^2(x^2 - 4x + 1) \longrightarrow & 2x^4 & -8x^3 & +2x^2 & & & \text{Multiply top expression by } 2x^2. \\
\hline
 & 2x^4 & -8x^3 & -x^2 & +12x & -3 & \text{Add like terms in columns.}
\end{array}$$

▸ **Now Try Exercise 35**

**EXAMPLE 6** ▸ Multiply $3x^2 + 6xy - 5y^2$ by $x + 4y$.

*Solution*

$$\begin{array}{lrrrrl}
 & & 3x^2 & +6xy & -5y^2 & \\
 & & & x & +4y & \\
\hline
4y(3x^2 + 6xy - 5y^2) \longrightarrow & & 12x^2y & +24xy^2 & -20y^3 & \text{Multiply top expression by } 4y. \\
x(3x^2 + 6xy - 5y^2) \longrightarrow & 3x^3 & +6x^2y & -5xy^2 & & \text{Multiply top expression by } x. \\
\hline
 & 3x^3 & +18x^2y & +19xy^2 & -20y^3 & \text{Add like terms in columns.}
\end{array}$$

▸ **Now Try Exercise 31**

## 4 Find the Square of a Binomial

Now we will study some special formulas. We must often *square a binomial*, so we have special formulas for doing so.

**Square of a Binomial**

$$(a + b)^2 = a^2 + 2ab + b^2$$
$$(a - b)^2 = a^2 - 2ab + b^2$$

If you forget the formulas, you can easily derive them by multiplying $(a + b)(a + b)$ and $(a - b)(a - b)$.

Examples 7 and 8 illustrate the use of the square of a binomial formula.

**EXAMPLE 7** ▸ Expand. **a)** $(3x + 7)^2$ **b)** $(4x^2 - 3y)^2$

*Solution*

**a)** $(3x + 7)^2 = (3x)^2 + 2(3x)(7) + (7)^2$
$= 9x^2 + 42x + 49$

**b)** $(4x^2 - 3y)^2 = (4x^2)^2 - 2(4x^2)(3y) + (3y)^2$
$= 16x^4 - 24x^2y + 9y^2$

▸ **Now Try Exercise 45**

Squaring binomials, as in Example 7, can also be done using the FOIL method.

**Avoiding Common Errors**

Remember the middle term when squaring a binomial.

| CORRECT | INCORRECT |
|---|---|
| $(x + 2)^2 = (x + 2)(x + 2)$ <br> $= x^2 + 4x + 4$ | ~~$(x + 2)^2 = x^2 + 4$~~ |
| $(x - 3)^2 = (x - 3)(x - 3)$ <br> $= x^2 - 6x + 9$ | ~~$(x - 3)^2 = x^2 + 9$~~ |

**EXAMPLE 8** ▸ Expand $[x + (y - 1)]^2$.

*Solution* This problem looks more complicated than the previous example, but it is worked the same way as any other square of a binomial. Treat $x$ as the first term and $(y - 1)$ as the second term. Use the formula twice.

$$\begin{aligned}[x + (y - 1)]^2 &= (x)^2 + 2(x)(y - 1) + (y - 1)^2 \\ &= x^2 + (2x)(y - 1) + y^2 - 2y + 1 \\ &= x^2 + 2xy - 2x + y^2 - 2y + 1\end{aligned}$$

None of the six terms are like terms, so no terms can be combined. Note that $(y - 1)^2$ is also the square of a binomial and was expanded as such.

▸ **Now Try Exercise 51**

## 5 Find the Product of the Sum and Difference of the Same Two Terms

Below we multiply $(x + 6)(x - 6)$ using the FOIL method.

$$(x + 6)(x - 6) = x^2 - 6x + 6x - (6)(6) = x^2 - 6^2$$

Note that the outer and inner products add to 0. By examining this example, we see that the product of the sum and difference of the same two terms is the difference of the squares of the two terms.

**Product of the Sum and Difference of the Same Two Terms**

$$(a + b)(a - b) = a^2 - b^2$$

In other words, to multiply two binomials that differ only in the sign between their two terms, subtract the square of the second term from the square of the first term. Note that $a^2 - b^2$ represents a **difference of two squares**.

**EXAMPLE 9** ▸ Multiply.

**a)** $\left(3x + \frac{4}{5}\right)\left(3x - \frac{4}{5}\right)$ **b)** $(0.2x + 0.3z^2)(0.2x - 0.3z^2)$

*Solution* Each is a product of the sum and difference of the same two terms. Therefore,

**a)** $\left(3x + \frac{4}{5}\right)\left(3x - \frac{4}{5}\right) = (3x)^2 - \left(\frac{4}{5}\right)^2 = 9x^2 - \frac{16}{25}$

**b)** $\begin{aligned}(0.2x + 0.3z^2)(0.2x - 0.3z^2) &= (0.2x)^2 - (0.3z^2)^2 \\ &= 0.04x^2 - 0.09z^4\end{aligned}$

▸ **Now Try Exercise 27**

**EXAMPLE 10** ▸ Multiply $(5x + y^4)(5x - y^4)$.

*Solution* $(5x + y^4)(5x - y^4) = (5x)^2 - (y^4)^2 = 25x^2 - y^8$

▸ **Now Try Exercise 49**

**EXAMPLE 11** ▸ Multiply $[4x + (3y + 2)][4x - (3y + 2)]$.

*Solution* We treat $4x$ as the first term and $3y + 2$ as the second term. Then we have the sum and difference of the same two terms.

$$\begin{aligned}[4x + (3y + 2)][4x - (3y + 2)] &= (4x)^2 - (3y + 2)^2 \\ &= 16x^2 - (9y^2 + 12y + 4) \\ &= 16x^2 - 9y^2 - 12y - 4\end{aligned}$$

▸ **Now Try Exercise 55**

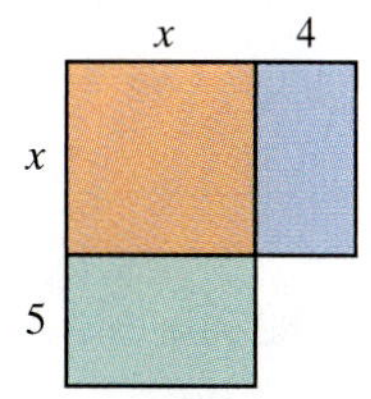


FIGURE 5.11

**EXAMPLE 12** ▸ **Area**

**Figure 5.11** consists of a square and two rectangles. Find a polynomial expression for the total area of the figure.

*Solution* To find the total area, we find the areas of the three regions and then add.

$$\text{Area of the orange square} = x \cdot x = x^2$$
$$\text{Area of the blue rectangle} = x \cdot 4 = 4x$$
$$\text{Area of the green rectangle} = x \cdot 5 = 5x$$

The total area is the sum of these three quantities.

$$\text{Total area} = x^2 + 4x + 5x = x^2 + 9x.$$

▸ Now Try Exercise 85

## 6 Find the Product of Polynomial Functions

Earlier we learned that for functions $f(x)$ and $g(x)$, $(f \cdot g)(x) = f(x) \cdot g(x)$. Let's work one example involving multiplication of polynomial functions now.

**EXAMPLE 13** ▸ Let $f(x) = x + 4$ and $g(x) = x - 2$. Find

**a)** $f(3) \cdot g(3)$ **b)** $(f \cdot g)(x)$ **c)** $(f \cdot g)(3)$

*Solution*

**a)** Both $f(x)$ and $g(x)$ are polynomial functions since the expressions to the right of the equal signs are polynomials.

$$f(x) = x + 4 \qquad g(x) = x - 2$$
$$f(3) = 3 + 4 = 7 \qquad g(3) = 3 - 2 = 1$$
$$f(3) \cdot g(3) = 7 \cdot 1 = 7$$

**b)** From Section 3.6, we know that

$$\begin{aligned}(f \cdot g)(x) &= f(x) \cdot g(x)\\ &= (x + 4)(x - 2)\\ &= x^2 - 2x + 4x - 8\\ &= x^2 + 2x - 8\end{aligned}$$

**c)** To evaluate $(f \cdot g)(3)$, substitute 3 for each $x$ in $(f \cdot g)(x)$.

$$\begin{aligned}(f \cdot g)(x) &= x^2 + 2x - 8\\ (f \cdot g)(3) &= 3^2 + 2(3) - 8\\ &= 9 + 6 - 8 = 7\end{aligned}$$

Note that in part **c)** we found $(f \cdot g)(3) = 7$ and in part **a)** we found $f(3) \cdot g(3) = 7$. Thus $(f \cdot g)(3) = f(3) \cdot g(3)$, which is what we expected from what we learned in Section 3.6.

▸ Now Try Exercise 79

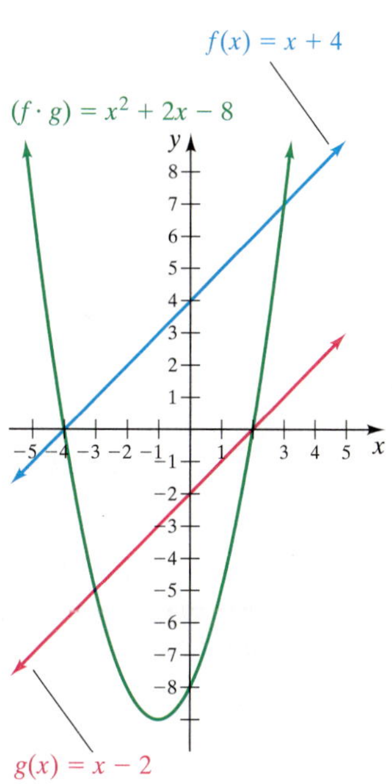


FIGURE 5.12

In Example 13, we found that if $f(x) = x + 4$ and $g(x) = x - 2$, then $(f \cdot g)(x) = x^2 + 2x - 8$. The graphs of $y = f(x) = x + 4$, $y = g(x) = x - 2$, and $y = (f \cdot g)(x) = x^2 + 2x - 8$ are shown in **Figure 5.12**. We see from the graphs that $f(3) = 7$, $g(3) = 1$, and $(f \cdot g)(3) = 7$, which is what we expected from Example 13. Every point on $y = x^2 + 2x - 8$ can be determined the same way. For example, $f(-4) = 0$ and $g(-4) = -6$. Since $0(-6) = 0$, $(f \cdot g)(-4) = 0$. Also $f(2) = 6$ and $g(2) = 0$; thus $(f \cdot g)(2) = 6 \cdot 0 = 0$. Notice in **Figure 5.12** that the product of two linear functions gives a quadratic function.

## EXERCISE SET 5.2

MathXL® MyMathLab

### Concept/Writing Exercises

1. a) Explain how to multiply two binomials using the FOIL method.
   b) Make up two binomials and multiply them using the FOIL method.
   c) Multiply the same two binomials using LIOF (last, inner, outer, first).
   d) Compare the results of parts **b)** and **c)**. If they are not the same, explain why.
2. a) Explain how to multiply a monomial by a polynomial.
   b) Multiply $3x(4x^2 - 6x - 7)$ using your procedure from part **a)**.
3. a) Explain how to multiply a polynomial by a polynomial.
   b) Multiply $4 + x$ by $x^2 - 6x + 3$ using your procedure from part **a)**.
4. a) Explain how to expand $(2x - 3)^2$ using the formula for the square of a binomial.
   b) Expand $(2x - 3)^2$ using your procedure from part **a)**.
5. a) What is meant by the product of the sum and difference of the same two terms?
   b) Give an example of a problem that is the product of the sum and difference of the same two terms.
   c) How do you multiply the product of the sum and difference of the same two terms?
   d) Multiply the example you gave in part **b)** using your procedure from part **c)**.
6. Will the product of two binomials always be a
   a) binomial?
   b) trinomial? Explain.
7. Will the product of two first-degree polynomials always be a second-degree polynomial? Explain.
8. a) Given $f(x)$ and $g(x)$, explain how you would find $(f \cdot g)(x)$.
   b) If $f(x) = x - 8$ and $g(x) = x + 8$, find $(f \cdot g)(x)$.

### Practice the Skills

*Multiply.*

9. $(4xy)(6xy^4)$
10. $(-2xy^4)(9x^4y^6)$
11. $\left(\frac{5}{9}x^2y^5\right)\left(\frac{1}{5}x^5y^3z^2\right)$
12. $2y^3(3y^2 + 2y - 8)$
13. $-3x^2y(-2x^4y^2 + 5xy^3 + 4)$
14. $3x^4(2xy^2 + 5x^7 - 9y)$
15. $\frac{2}{3}yz(3x + 4y - 12y^2)$
16. $\frac{1}{2}x^2y(4x^5y^2 + 3x - 7y^2)$
17. $0.3(2x^2 - 5x + 11y)$
18. $0.8(0.2a + 0.9b - 1.3c)$
19. $0.3a^5b^4(9.5a^6b - 4.6a^4b^3 + 1.2ab^5)$
20. $4.6m^2n(1.3m^4n^2 - 2.6m^3n^3 + 5.9n^4)$

*Multiply the following binomials.*

21. $(4x - 6)(3x - 5)$
22. $(2x - 1)(7x + 5)$
23. $(4 - x)(3 + 2x^2)$
24. $(5x + y)(6x - y)$
25. $\left(\frac{1}{2}x + 2y\right)\left(2x - \frac{1}{3}y\right)$
26. $\left(\frac{1}{3}a + \frac{1}{4}b\right)\left(\frac{1}{2}a - b\right)$
27. $(0.3a + 0.5b)(0.3a - 0.5b)$
28. $(4.6r - 5.8s)(0.2r - 2.3s)$

*Multiply the following polynomials.*

29. $(x^2 + 3x + 1)(x - 4)$
30. $(x + 3)(2x^2 - x - 8)$
31. $(a - 3b)(2a^2 - ab + 2b^2)$
32. $(7p - 3)(-2p^2 - 4p + 1)$
33. $(x^3 - x^2 + 3x + 7)(x + 1)$
34. $(2x - 1)(x^3 + 3x^2 - 5x + 6)$
35. $(5x^3 + 4x^2 - 6x + 2)(x + 5)$
36. $(a^3 - 2a^2 + 5a - 6)(2a^2 - 5a - 3)$
37. $(3m^2 - 2m + 4)(m^2 - 3m - 5)$
38. $(2a^2 - 6a + 3)(3a^2 - 5a - 2)$
39. $(2x - 1)^3$
40. $(3x + y)^3$
41. $(5r^2 - rs + 2s^2)(2r^2 - s^2)$
42. $(4x^2 - 5xy + y^2)(x^2 - 2y^2)$

*Multiply using either the formula for the square of a binomial or for the product of the sum and difference of the same two terms.*

43. $(x + 2)(x + 2)$
44. $(y - 5)(y - 5)$
45. $(2x - 7)(2x - 7)$
46. $(3z + 4)(3z + 4)$
47. $(4x - 3y)^2$
48. $(2a + 5b)^2$
49. $(5m^2 + 2n)(5m^2 - 2n)$
50. $(5p^2 + 6q^2)(5p^2 - 6q^2)$

51. $[y + (4 - 2x)]^2$

52. $[(a + b) + 9]^2$

53. $[5x + (2y + 1)]^2$

54. $[4 - (p - 3q)]^2$

55. $[a + (b + 4)][a - (b + 4)]$

56. $[2x + (y + 5)][2x - (y + 5)]$

*Multiply.*

57. $2xy(x^2 + xy + 12y^2)$

58. $3a^2b^2\left(\frac{1}{3}ab - \frac{1}{9}b^6\right)$

59. $\frac{1}{2}xy^2(4x^2 + 3xy - 7y^4)$

60. $-\frac{3}{5}x^2y\left(-\frac{2}{3}xy^4 + \frac{1}{9}xy^2 + 4\right)$

61. $-\frac{3}{5}xy^3z^2\left(-xy^2z^5 - 5xy + \frac{1}{9}xz^7\right)$

62. $\frac{2}{3}x^2y^4\left(\frac{3}{5}xy^3 - \frac{1}{8}x^4y + 2xy^3z^5\right)$

63. $(3a + 4)(7a - 6)$

64. $(5p - 9q)(4p - 11q)$

65. $\left(8x + \frac{1}{5}\right)\left(8x - \frac{1}{5}\right)$

66. $\left(5a - \frac{1}{7}\right)\left(5a + \frac{1}{7}\right)$

67. $\left(x - \frac{1}{2}y\right)^3$

68. $\left(\frac{1}{2}m - n\right)^3$

69. $(x + 3)(2x^2 + 4x - 3)$

70. $(5a + 4)(a^2 - a + 3)$

71. $(2p - 3q)(3p^2 + 4pq - 2q^2)$

72. $(2m + n)(3m^2 - mn + 2n^2)$

73. $[(3x + 2) + y][(3x + 2) - y]$

74. $[a + (3b + 5)][a - (3b + 5)]$

75. $(a + b)(a - b)(a^2 - b^2)$

76. $(2a + 3)(2a - 3)(4a^2 + 9)$

77. $(x - 4)(6 + x)(2x - 8)$

78. $(3x - 5)(5 - 2x)(3x + 8)$

*For the functions given, find* **a)** $(f \cdot g)(x)$ *and* **b)** $(f \cdot g)(4)$.

79. $f(x) = x - 5, g(x) = x + 6$

80. $f(x) = 2x - 3, g(x) = x - 1$

81. $f(x) = 2x^2 + 6x - 4, g(x) = 5x + 3$

82. $f(x) = 4x^2 + 7, g(x) = 2 - x$

83. $f(x) = -x^2 + 3x, g(x) = x^2 + 2$

84. $f(x) = -x^2 + 2x + 7, g(x) = x^2 - 1$

## Problem Solving

**Area** *In Exercises 85–88, find a polynomial expression for the total area of each figure.*

**85.**

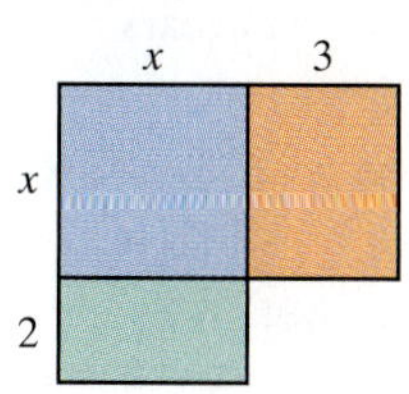


**86.**

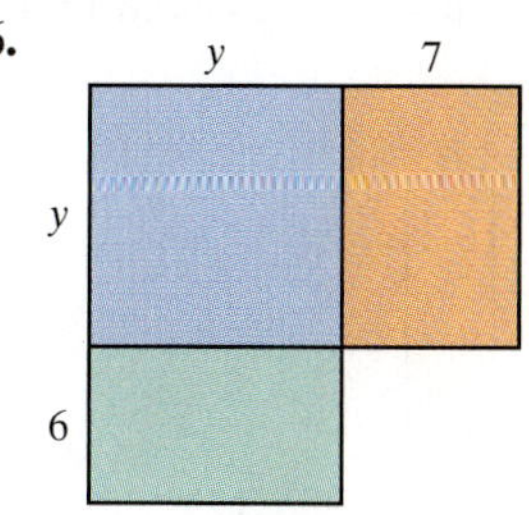


**87.**

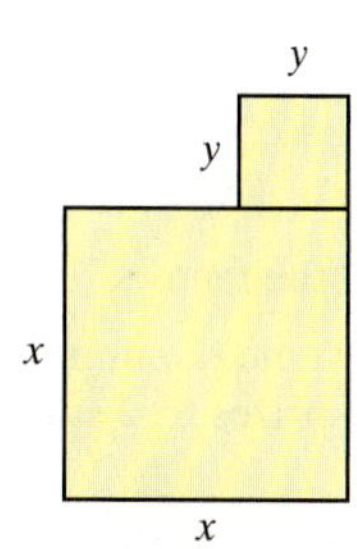


**88.**

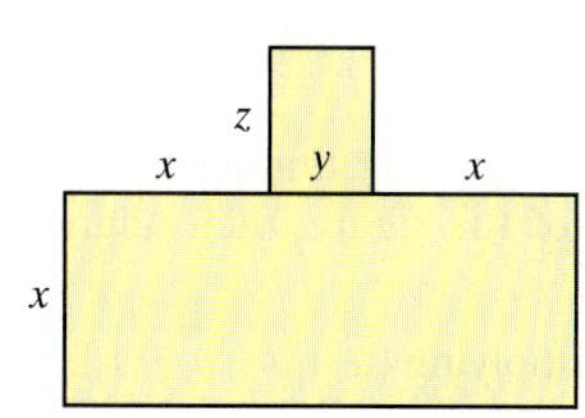


**Area** *In Exercises 89 and 90,* **a)** *find the area of the rectangle by finding the area of the four sections and adding them, and* **b)** *multiply the two sides and compare the product with your answer to part* **a)**.

**89.**

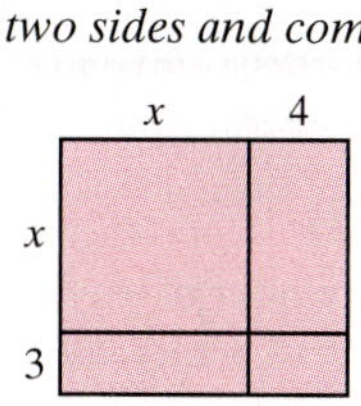


**90.**

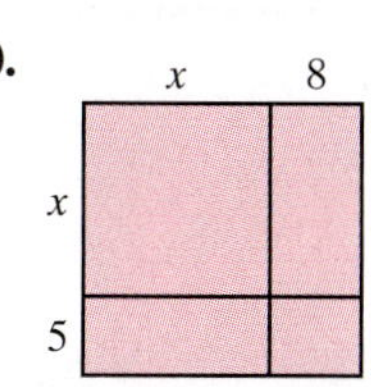

**Area** *Write a polynomial expression for the area of each figure. All angles in the figures are right angles.*

**91.**

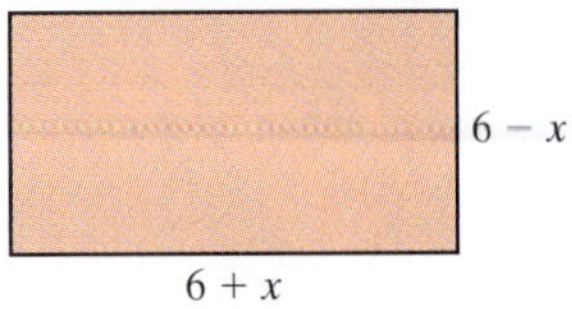


**92.**

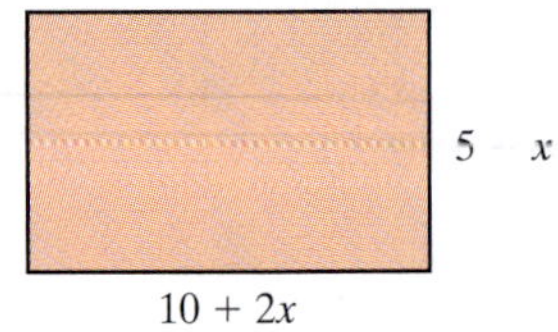


**Area** *In Exercises 93 and 94,* **a)** *write a polynomial expression for the area of the shaded portion of the figure.* **b)** *The area of the shaded portion is indicated above each figure. Find the area of the larger and smaller rectangles.*

**93.**

Area of shaded region = 67 sq in.

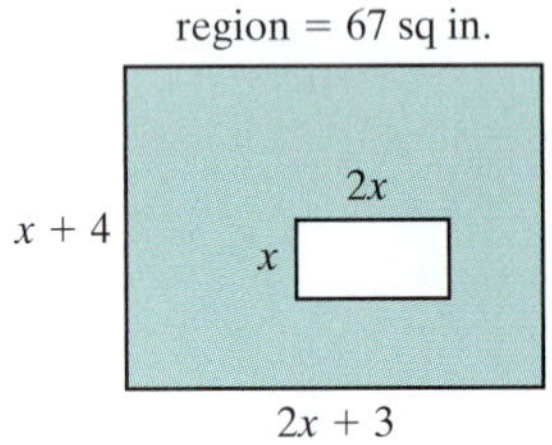


**94.**

Area of shaded region = 139 sq in.

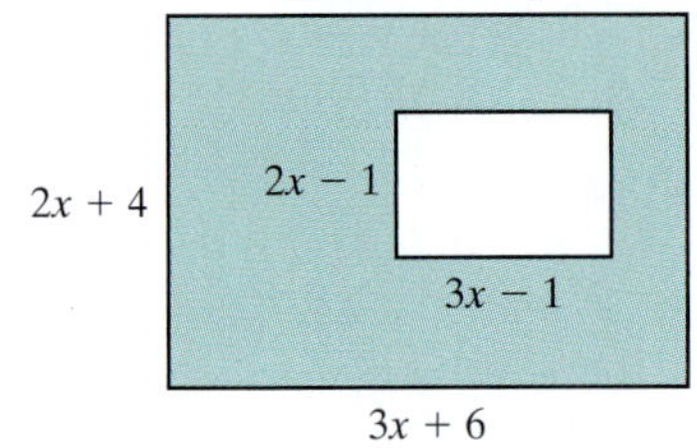


**95.** Write two binomials whose product is $x^2 - 49$. Explain how you determined your answer.

**96.** Write two binomials whose product is $4x^2 - 9$. Explain how you determined your answer.

**97.** Write two binomials whose product is $x^2 + 12x + 36$. Explain how you determined your answer.

**98.** Write two binomials whose product is $16y^2 - 8y + 1$. Explain how you determined your answer.

**99.** Consider the expression $a(x - n)^3$. Write this expression as a product of factors.

**100.** Consider the expression $P(1 - r)^4$. Write this expression as a product of factors.

**101. Area** The expression $(a + b)^2$ can be represented by the following figure.

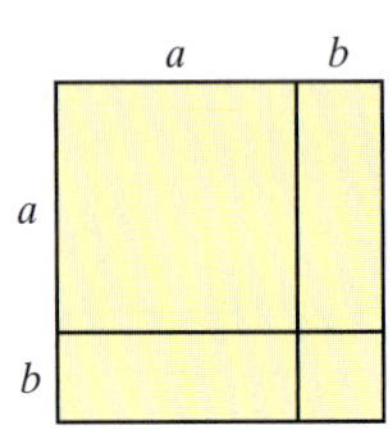


**a)** Explain why this figure represents $(a + b)^2$.

**b)** Find $(a + b)^2$ using the figure by finding the area of each of the four parts of the figure, then adding the areas together.

**c)** Simplify $(a + b)^2$ by multiplying $(a + b)(a + b)$.

**d)** How do the answers in parts **b)** and **c)** compare? If they are not the same, explain why.

**102. Volume** The expression $(a + b)^3$ can be represented by the following figure.

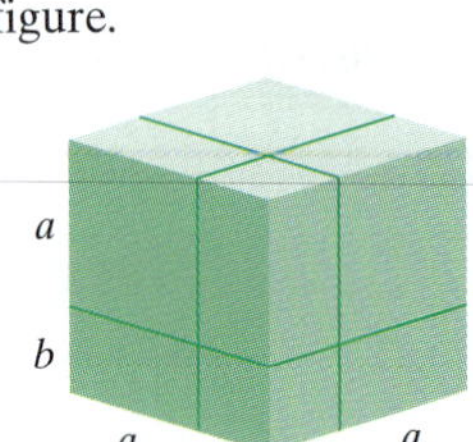


**a)** Explain why this figure represents $(a + b)^3$.

**b)** Find $(a + b)^3$ by adding the volume of the eight parts of the figure.

**c)** Simplify $(a + b)^3$ by multiplying.

**d)** How do the answers in parts **b)** and **c)** compare? If they are not the same, explain why.

**103. Compound Interest** The compound interest formula is

$$A = P\left(1 + \frac{r}{n}\right)^{nt}$$

where $A$ is the amount, $P$ is the principal invested, $r$ is the annual rate of interest, $n$ is the number of times the interest is compounded annually, and $t$ is the time in years.

**a)** Simplify this formula for $n = 1$.

**b)** Find the value of $A$ if $P = \$1000$, $n = 1$, $r = 6\%$, and $t = 2$ years.

**104. Compound Interest** Use the formula given in Exercise 103 to find $A$ if $P = \$4000$, $n = 2$, $r = 8\%$, and $t = 2$ years.

**105. Line-Up** The number of ways a teacher can award different prizes to 2 students in a class having $n$ students is given by the formula $P(n) = n(n - 1)$.

**a)** Use this formula to determine the number of ways a teacher can award different prizes to 2 students in a class having 11 students.

**b)** Rewrite the formula by multiplying the factors.

**c)** Use the result from part **b)** to determine the number of ways a teacher can award different prizes to 2 students in a class having 11 students.

**d)** Are the results from parts **a)** and **c)** the same? Explain.

**106. Horse Racing** The number of ways horses can finish first, second, and third, in a race having $n$ horses, is given by the formula

$$P(n) = n(n - 1)(n - 2)$$

**a)** Use this formula to determine the number of ways horses can finish first, second, and third in a race having 7 horses.

**b)** Rewrite the formula by multiplying the factors.

**c)** Use the result from part **b)** to determine the number of ways horses can finish first, second, and third in a race having 7 horses.

**d)** Are the results from parts **a)** and **c)** the same? Explain.

**107.** If $f(x) = x^2 - 3x + 5$, find $f(a + b)$ by substituting $(a + b)$ for each $x$ in the function.

**108.** If $f(x) = 2x^2 - x + 3$, find $f(a + b)$.

*In Exercises 109–114, simplify. Assume that all variables represent natural numbers.*

**109.** $3x^t(5x^{2t-1} + 6x^{3t})$

**110.** $5k^{r+2}(4k^{r+2} - 3k^r - k)$

**111.** $(6x^m - 5)(2x^{2m} - 3)$

**112.** $(x^{3n} - y^{2n})(x^{2n} + 2y^{4n})$

**113.** $(y^{a-b})^{a+b}$

**114.** $(a^{m+n})^{m+n}$

*In Exercises 115 and 116, perform the polynomial multiplication.*

**115.** $(x - 3y)^4$

**116.** $(2a - 4b)^4$

**117. a)** Explain how a multiplication in one variable, such as $(x^2 + 2x + 3)(x + 2) = x^3 + 4x^2 + 7x + 6$, may be checked using a graphing calculator.

**b)** Check the multiplication given in part **a)** using your grapher.

**118. a)** Show that the multiplication $(x^2 - 4x - 5)(x - 1) \neq x^3 + 6x^2 - 5x + 5$ using your grapher.

**b)** Multiply $(x^2 - 4x - 5)(x - 1)$.

**c)** Check your answer in part **b)** on your grapher.

## Challenge Problems

*Multiply.*

**119.** $[(y + 1) - (x + 2)]^2$

**120.** $[(a - 2) - (a + 1)]^2$

## Cumulative Review Exercises

[1.3] **121.** Evaluate $\frac{4}{5} - \left(\frac{3}{4} - \frac{2}{3}\right)$.

[1.5] **122.** Simplify $\left(\frac{2r^4s^5}{r^2}\right)^3$.

[2.5] **123.** Solve the inequality $-12 < 3x - 5 \leq -1$ and give the solution in interval notation.

[3.2] **124.** If $g(x) = x^2 + 2x + 3$, find $g\left(\frac{1}{2}\right)$.

# 5.3 Division of Polynomials and Synthetic Division

1. Divide a polynomial by a monomial.
2. Divide a polynomial by a binomial.
3. Divide polynomials using synthetic division.
4. Use the Remainder Theorem.

## 1 Divide a Polynomial by a Monomial

In division of polynomials, division by 0 is not permitted. When we are given a division problem containing a variable in the denominator, *we always assume that the denominator is nonzero.*

To divide a polynomial by a monomial, we use the fact that

$$\frac{A + B}{C} = \frac{A}{C} + \frac{B}{C}$$

If the polynomial has more than two terms, we expand this procedure.

**To Divide a Polynomial by a Monomial**

Divide each term of the polynomial by the monomial.

To divide a polynomial by a monomial, we will need to use two rules of exponents that were presented in Section 1.5, the quotient rule for exponents and the zero exponent rule. We state the two rules below and then give examples reviewing both rules.

Quotient rule for exponents: $\frac{a^m}{a^n} = a^{m-n}, \quad a \neq 0$

Zero exponent rule: $a^0 = 1, \quad a \neq 0$

**EXAMPLE 1** ▸ Divide. **a)** $\frac{x^{10}}{x^4}$ **b)** $\frac{5x^3y^5}{2xy^2}$

*Solution* We will use the quotient rule to divide.

**a)** $\frac{x^{10}}{x^4} = x^{10-4}$ *Quotient rule*

$= x^6$

**b)** $\frac{5x^3y^5}{2xy^2} = \frac{5}{2} \cdot \frac{x^3}{x} \cdot \frac{y^5}{y^2}$

$= \frac{5}{2}x^{3-1}y^{5-2}$ *Quotient rule*

$= \frac{5x^2y^3}{2}$

▸ **Now Try Exercise 11**

**EXAMPLE 2** ▸ Divide. **a)** $\frac{p^4}{p^4}$ **b)** $\frac{8r^6s^7}{3rs^7}$

*Solution* We will use both the quotient rule and the zero exponent rule to divide.

**a)** $\frac{p^4}{p^4} = p^{4-4}$ *Quotient rule*

$= p^0$

$= 1$ *Zero exponent rule*

**b)** $\frac{8r^6s^7}{3rs^7} = \frac{8}{3} \cdot \frac{r^6}{r} \cdot \frac{s^7}{s^7}$

$= \frac{8}{3}r^{6-1}s^{7-7}$ *Quotient rule*

$= \frac{8}{3}r^5s^0$

$= \frac{8}{3}r^5(1)$ *Zero exponent rule*

$= \frac{8}{3}r^5$ or $\frac{8r^5}{3}$

▸ **Now Try Exercise 17**

In Example 2, both $\frac{8}{3}r^5$ and $\frac{8r^5}{3}$ are acceptable answers. Now we are ready to divide a polynomial by a monomial.

**EXAMPLE 3** ▸ Divide $\frac{4x^2 - 8x - 7}{2x}$.

*Solution*

$$\frac{4x^2 - 8x - 7}{2x} = \frac{4x^2}{2x} - \frac{8x}{2x} - \frac{7}{2x}$$

$$= 2x - 4 - \frac{7}{2x}$$

▸ **Now Try Exercise 25**

**EXAMPLE 4** ▸ Divide $\dfrac{4y - 6x^4y^3 - 3x^5y^2 + 5x}{2xy^2}$.

*Solution*

$$\frac{4y - 6x^4y^3 - 3x^5y^2 + 5x}{2xy^2} = \frac{4y}{2xy^2} - \frac{6x^4y^3}{2xy^2} - \frac{3x^5y^2}{2xy^2} + \frac{5x}{2xy^2}$$

$$= \frac{2}{xy} - 3x^3y - \frac{3x^4}{2} + \frac{5}{2y^2}$$

▸ **Now Try Exercise 31**

## 2 Divide a Polynomial by a Binomial

We divide a polynomial by a binomial in much the same way as we perform long division. In a division problem, the expression we are dividing is called the *dividend* and the expression we are dividing by is called the *divisor*.

**EXAMPLE 5** ▸ Divide $\dfrac{x^2 + 7x + 10}{x + 2}$.

*Solution* Rewrite the division problem as

$$x + 2\overline{)x^2 + 7x + 10}$$

Divide $x^2$ (the first term in the dividend $x^2 + 7x + 10$) by $x$ (the first term in the divisor $x + 2$).

$$\frac{x^2}{x} = x$$

Place the quotient, $x$, above the term containing $x$ in the dividend.

$$\begin{array}{r} x \phantom{+ 7x + 10} \\ x + 2\overline{)x^2 + 7x + 10} \end{array}$$

Next, multiply the $x$ by $x + 2$ as you would do in long division and place the product under the dividend, aligning like terms.

$$\begin{array}{rl} \text{Times} \quad x \phantom{+ 7x + 10} & \\ x + 2\overline{)x^2 + 7x + 10} & \\ \text{Equals} \quad x^2 + 2x \phantom{+ 10} & \longleftarrow x(x + 2) \end{array}$$

Now subtract $x^2 + 2x$ from $x^2 + 7x$.

$$\begin{array}{r} x \phantom{+ 7x + 10} \\ x + 2\overline{)x^2 + 7x + 10} \\ \underline{-(x^2 + 2x)} \phantom{+ 10} \\ 5x \phantom{+ 10} \end{array}$$

Now bring down the next term, +10.

$$\begin{array}{r} x \phantom{+ 7x + 10} \\ x + 2\overline{)x^2 + 7x + 10} \\ \underline{x^2 + 2x} \phantom{+ 10} \\ 5x + 10 \end{array}$$

Divide $5x$ by $x$.

$$\frac{5x}{x} = +5$$

Place $+5$ above the constant in the dividend and multiply 5 by $x + 2$. Finish the problem by subtracting.

Times

$$\begin{array}{rl} x + 5 & \\ x + 2\overline{)\,x^2 + 7x + 10} & \\ \underline{x^2 + 2x\phantom{+10}} & \\ 5x + 10 & \\ \underline{5x + 10} & \leftarrow 5(x+2) \\ 0 & \leftarrow \text{Remainder} \end{array}$$

Equals

Thus, $\dfrac{x^2 + 7x + 10}{x + 2} = x + 5$. There is no remainder.

▸ **Now Try Exercise 35**

In Example 5, there was no remainder. Therefore, $x^2 + 7x + 10 = (x + 2)(x + 5)$. Note that $x + 2$ and $x + 5$ are both *factors* of $x^2 + 7x + 10$. In a division problem, if there is no remainder, then both the divisor and quotient are factors of the dividend.

When writing an answer in a division problem when there is a remainder, write the remainder over the divisor and add this expression to the quotient. For example, suppose that the remainder in Example 5 was 4. Then the answer would be written $x + 5 + \dfrac{4}{x + 2}$. If the remainder in Example 5 was $-7$, the answer would be written $x + 5 + \dfrac{-7}{x + 2}$, which we would rewrite as $x + 5 - \dfrac{7}{x + 2}$.

**EXAMPLE 6** ▸ Divide $\dfrac{6x^2 - 7x + 3}{2x + 1}$.

*Solution* In this example we will subtract mentally and not show the change of sign in the subtractions.

$$\begin{array}{rl} 3x - 5 & \\ 2x + 1\overline{)\,6x^2 - 7x + 3} & \\ \underline{6x^2 + 3x\phantom{+3}} & \leftarrow 3x(2x+1) \\ -10x + 3 & \\ \underline{-10x - 5} & \leftarrow -5(2x+1) \\ 8 & \leftarrow \text{Remainder} \end{array}$$

Thus, $\dfrac{6x^2 - 7x + 3}{2x + 1} = 3x - 5 + \dfrac{8}{2x + 1}$.

▸ **Now Try Exercise 45**

When dividing a polynomial by a binomial, the answer may be *checked* by multiplying the divisor by the quotient and then adding the remainder. You should obtain the polynomial you began with. To check Example 6, we do the following:

$$(2x + 1)(3x - 5) + 8 = 6x^2 - 10x + 3x - 5 + 8 = 6x^2 - 7x + 3$$

Since we got the polynomial we began with, our division is correct.

**When you are dividing a polynomial by a binomial, you should list both the polynomial and binomial in descending order. If a term of any degree is missing, it is often helpful to include that term with a numerical coefficient of 0.** For example, when dividing $(6x^2 + x^3 - 4) \div (x - 2)$, we rewrite the problem as $(x^3 + 6x^2 + 0x - 4) \div (x - 2)$ before beginning the division.

**EXAMPLE 7** ▸ Divide $(4x^2 - 12x + 3x^5 - 17)$ by $(-2 + x^2)$.

*Solution* Write both the dividend and divisor in descending powers of the variable $x$. This gives $(3x^5 + 4x^2 - 12x - 17) \div (x^2 - 2)$. Where a power of $x$ is missing, add that power of $x$ with a coefficient of 0, then divide.

$$
\begin{array}{r l}
 & 3x^3 \qquad + 6x + 4 \\
x^2 + 0x - 2 \overline{)} & 3x^5 + 0x^4 + 0x^3 + 4x^2 - 12x - 17 \\
 & \underline{3x^5 + 0x^4 - 6x^3} \longleftarrow 3x^3(x^2 + 0x - 2) \\
 & \qquad 6x^3 + 4x^2 - 12x \\
 & \qquad \underline{6x^3 + 0x^2 - 12x} \longleftarrow 6x(x^2 + 0x - 2) \\
 & \qquad\qquad 4x^2 + 0x - 17 \\
 & \qquad\qquad \underline{4x^2 + 0x - 8} \longleftarrow 4(x^2 + 0x - 2) \\
 & \qquad\qquad\qquad -9 \longleftarrow \text{Remainder}
\end{array}
$$

In obtaining the answer, we performed the divisions

$$\frac{3x^5}{x^2} = 3x^3 \qquad \frac{6x^3}{x^2} = 6x \qquad \frac{4x^2}{x^2} = 4$$

The quotients $3x^3$, $6x$, and 4 were placed above their like terms in the dividend. The answer is $3x^3 + 6x + 4 - \frac{9}{x^2 - 2}$. You should check this answer for yourself by multiplying the divisor by the quotient and adding the remainder.

▸ **Now Try Exercise 55**

## 3 Divide Polynomials Using Synthetic Division

When a polynomial is divided by a binomial of the form $x - a$, the division process can be greatly shortened by a process called **synthetic division**. Consider the following examples. In the example on the right, we use only the numerical coefficients.

$$
\begin{array}{r l}
 & 2x^2 + 5x - 4 \\
x - 3 \overline{)} & 2x^3 - x^2 - 19x + 18 \\
 & \underline{2x^3 - 6x^2} \\
 & \quad 5x^2 - 19x \\
 & \quad \underline{5x^2 - 15x} \\
 & \qquad -4x + 18 \\
 & \qquad \underline{-4x + 12} \\
 & \qquad\qquad 6
\end{array}
\qquad\qquad
\begin{array}{r l}
 & 2 + 5 - 4 \\
1 - 3 \overline{)} & 2 - 1 - 19 + 18 \\
 & \underline{2 - 6} \\
 & \quad 5 - 19 \\
 & \quad \underline{5 - 15} \\
 & \qquad -4 + 18 \\
 & \qquad \underline{-4 + 12} \\
 & \qquad\qquad 6
\end{array}
$$

Note that the variables do not play a role in determining the numerical coefficients of the quotient. This division problem can be done more quickly and easily using synthetic division.

Following is an explanation of how we use synthetic division. Consider again the division

$$\frac{2x^3 - x^2 - 19x + 18}{x - 3}$$

1. Write the dividend in descending powers of $x$. Then list the numerical coefficients of each term in the dividend. If a term of any degree is missing, place 0 in the appropriate position to serve as a placeholder. In the problem above, the numerical coefficients of the dividend are

$$2 \quad -1 \quad -19 \quad 18$$

2. When dividing by a binomial of the form $x - a$, place $a$ to the left of the line of numbers from step 1. In this problem, we are dividing by $x - 3$; thus, $a = 3$. We write

$$\underline{3|}\quad 2 \quad -1 \quad -19 \quad 18$$

3. Leave some space under the row of coefficients, then draw a horizontal line. Bring down the first coefficient on the left as follows:

$$\begin{array}{r|rrrr} 3 & 2 & -1 & -19 & 18 \\ & & & & \\ \hline & 2 & & & \end{array}$$

4. Multiply the 3 by the number brought down, the 2, to get 6. Place the 6 under the next coefficient, the $-1$. Then add $-1 + 6$ to get 5.

$$\begin{array}{r|rrrr} 3 & 2 & -1 & -19 & 18 \\ & & 6 & & \\ \hline & 2 & 5 & & \end{array}$$

5. Multiply the 3 by the sum 5 to get 15. Place 15 under $-19$. Then add to get $-4$. Repeat this procedure as illustrated.

$$\begin{array}{r|rrrr} 3 & 2 & -1 & -19 & 18 \\ & & 6 & 15 & -12 \\ \hline & 2 & 5 & -4 & 6 \end{array}$$

In the last row, the first three numbers are the numerical coefficients of the quotient, as shown in the long division. The last number, 6, is the remainder obtained by long division. The quotient must be one less than the degree of the dividend since we are dividing by $x - 3$. The original dividend was a third-degree polynomial. Therefore, the quotient must be a second-degree polynomial. Use the first three numbers from the last row as the coefficients of a second-degree polynomial in $x$. This gives $2x^2 + 5x - 4$, which is the quotient. The last number, 6, is the remainder. Therefore,

$$\frac{2x^3 - x^2 - 19x + 18}{x - 3} = 2x^2 + 5x - 4 + \frac{6}{x - 3}$$

**EXAMPLE 8** ▸ Use synthetic division to divide.

$$(5 - x^2 + x^3) \div (x + 2)$$

*Solution* First, list the terms of the dividend in descending order of $x$.

$$(x^3 - x^2 + 5) \div (x + 2)$$

Since there is no first-degree term, insert 0 as a placeholder when listing the numerical coefficients. Since $x + 2 = x - (-2)$, $a = -2$.

$$\begin{array}{r|rrrr} -2 & 1 & -1 & 0 & 5 \\ & & -2 & 6 & -12 \\ \hline & 1 & -3 & 6 & -7 \end{array} \leftarrow \text{Remainder}$$

Since the dividend is a third-degree polynomial, the quotient must be a second-degree polynomial. The answer is $x^2 - 3x + 6 - \dfrac{7}{x + 2}$.

▸ **Now Try Exercise 61**

**EXAMPLE 9** ▸ Use synthetic division to divide.

$$(3x^4 + 11x^3 - 20x^2 + 7x + 35) \div (x + 5)$$

*Solution*

| | | | | | | |
|---|---|---|---|---|---|---|
| −5 | 3 | 11 | −20 | 7 | 35 | |
| | | −15 | 20 | 0 | −35 | |
| | 3 | −4 | 0 | 7 | 0 | ← Remainder |

Since the dividend is of the fourth degree, the quotient must be of the third degree. The quotient is $3x^3 - 4x^2 + 0x + 7$ with no remainder. This can be simplified to $3x^3 - 4x^2 + 7$.

▸ **Now Try Exercise 71**

In Example 9, since there was no remainder, both $x + 5$ and $3x^3 - 4x + 7$ are *factors* of $3x^4 + 11x^3 - 20x^2 + 7x + 35$. Furthermore, since both are factors,

$$(x + 5)(3x^3 - 4x^2 + 7) = 3x^4 + 11x^3 - 20x^2 + 7x + 35$$

## 4 Use the Remainder Theorem

In Example 8, when we divided $x^3 - x^2 + 5$ by $x + 2$, we found that the remainder was $-7$. If we write $x + 2$ as $x - (-2)$ and evaluate the polynomial function $P(x) = x^3 - x^2 + 5$ at $-2$, we obtain $-7$.

$$P(x) = x^3 - x^2 + 5$$
$$P(-2) = (-2)^3 - (-2)^2 + 5 = -8 - 4 + 5 = -7$$

Is it just a coincidence that $P(-2)$, the value of the function at $-2$, is the same as the remainder when the function $P(x)$ is divided by $x - (-2)$? The answer is no. It can be shown that for any polynomial function $P(x)$, the value of the function at $a$, $P(a)$, has the same value as the remainder when $P(x)$ is divided by $x - a$.

To obtain the remainder when a polynomial $P(x)$ is divided by a binomial of the form $x - a$, we can use the **Remainder Theorem**.

**Remainder Theorem**

If the polynomial $P(x)$ is divided by $x - a$, the remainder is equal to $P(a)$.

**EXAMPLE 10** ▸ Use the Remainder Theorem to find the remainder when $3x^4 + 6x^3 - 2x + 4$ is divided by $x + 4$.

*Solution* First we write the divisor $x + 4$ in the form $x - a$. Since $x + 4 = x - (-4)$, we evaluate $P(-4)$.

$$\begin{aligned} P(x) &= 3x^4 + 6x^3 - 2x + 4 \\ P(-4) &= 3(-4)^4 + 6(-4)^3 - 2(-4) + 4 \\ &= 3(256) + 6(-64) + 8 + 4 \\ &= 768 - 384 + 8 + 4 = 396 \end{aligned}$$

Thus, when $3x^4 + 6x^3 - 2x + 4$ is divided by $x + 4$, the remainder is 396.

▸ **Now Try Exercise 87**

Using synthetic division, we will show that the remainder in Example 10 is indeed 396.

| | | | | | | |
|---|---|---|---|---|---|---|
| −4 | 3 | 6 | 0 | −2 | 4 | |
| | | −12 | 24 | −96 | 392 | ← Remainder |
| | 3 | −6 | 24 | −98 | 396 | |

If we were to graph the polynomial $P(x) = 3x^4 + 6x^3 - 2x + 4$, the value of $P(x)$, or $y$, at $x = -4$ would be 396.

**EXAMPLE 11** ▸ Use the Remainder Theorem to determine whether $x - 5$ is a factor of $6x^2 - 25x - 25$.

*Solution* Let $P(x) = 6x^2 - 25x - 25$. If $P(5) = 0$, then the remainder of $(6x^2 - 25x - 25)/(x - 5)$ is 0 and $x - 5$ is a factor of the polynomial. If $P(5) \neq 0$, then there is a remainder and $x - 5$ is not a factor.

$$\begin{aligned} P(x) &= 6x^2 - 25x - 25 \\ P(5) &= 6(5)^2 - 25(5) - 25 \\ &= 6(25) - 25(5) - 25 \\ &= 150 - 125 - 25 = 0 \end{aligned}$$

Since $P(5) = 0$, $x - 5$ is a factor of $6x^2 - 25x - 25$. Note that $6x^2 - 25x - 25 = (x - 5)(6x + 5)$.

▸ Now Try Exercise 89

## EXERCISE SET 5.3

MathXL® MyMathLab

### Concept/Writing Exercises

1. a) Explain how to divide a polynomial by a monomial.
   b) Divide $\dfrac{5x^4 - 6x^3 - 4x^2 - 12x + 1}{3x}$ using the procedure you gave in part **a)**.
2. a) Explain how to divide a trinomial in $x$ by a binomial in $x$.
   b) Divide $2x^2 - 12 + 5x$ by $x + 4$ using the procedure you gave in part **a)**.
3. A trinomial divided by a binomial has a remainder of 0. Is the quotient a factor of the trinomial? Explain.
4. a) Explain how the answer may be checked when dividing a polynomial by a binomial.
   b) Use your explanation in part **a)** to check whether the following division is correct.
   $$\frac{8x^2 + 2x - 15}{4x - 5} = 2x + 3$$
   c) Check to see whether the following division is correct.
   $$\frac{6x^2 - 23x + 13}{3x - 4} = 2x - 5 - \frac{8}{3x - 4}$$
5. When dividing a polynomial by a polynomial, before you begin the division, what should you do to the polynomials?
6. Explain why $\dfrac{x - 1}{x}$ is not a polynomial.
7. a) Describe how to divide a polynomial by $(x - a)$ using synthetic division.
   b) Divide $x^2 + 3x - 4$ by $x - 5$ using the procedure you gave in part **a)**.
8. a) State the Remainder Theorem.
   b) Find the remainder when $x^2 - 6x - 4$ is divided by $x - 1$, using the procedure you stated in part **a)**.
9. In the division problem $\dfrac{x^2 + 11x + 21}{x + 2} = x + 9 + \dfrac{3}{x + 2}$, is $x + 9$ a factor of $x^2 + 11x + 21$? Explain.
10. In the division problem $\dfrac{x^2 - 3x - 28}{x + 4} = x - 7$, is $x - 7$ a factor of $x^2 - 3x - 28$? Explain.

### Practice the Skills

*Divide.*

11. $\dfrac{x^9}{x^7}$ 12. $\dfrac{m^{13}}{m^3}$ 13. $\dfrac{a^{11}}{a^7}$ 14. $\dfrac{b^8}{b^3}$ 15. $\dfrac{z^{16}}{z^8}$

16. $\dfrac{q^6}{q^6}$ 17. $\dfrac{12r^7s^{10}}{3rs^8}$ 18. $\dfrac{7y^{14}z^7}{4y^{11}z}$ 19. $\dfrac{15x^{18}y^{19}}{3x^{10}y^8}$ 20. $\dfrac{21a^8b^{17}}{9a^7b^{10}}$

*Divide.*

21. $\dfrac{4x + 18}{2}$ 22. $\dfrac{9x + 8}{3}$ 23. $\dfrac{4x^2 + 2x}{2x}$

24. $\dfrac{12x^2 - 8x - 24}{4}$ 25. $\dfrac{5y^3 + 6y^2 - 12y}{3y}$ 26. $\dfrac{21y^5 + 14y^2}{7y^4}$

27. $\dfrac{4x^5 - 6x^4 + 12x^3 - 8x^2}{4x^2}$ 28. $\dfrac{15x^3y - 25xy^3}{5xy}$ 29. $\dfrac{8x^2y^2 - 10xy^3 - 5y}{2y^2}$

30. $\dfrac{4x^{13} + 12x^9 - 11x^7}{4x^6}$

31. $\dfrac{9x^2y - 12x^3y^2 + 15y^3}{2xy^2}$

32. $\dfrac{a^2b^2c - 6abc^2 + 5a^3b^5}{2abc^2}$

33. $\dfrac{3xyz + 6xyz^2 - 9x^3y^5z^7}{6xy}$

34. $\dfrac{6abc^3 - 5a^2b^3c^4 + 13ab^5c}{3ab^2c^3}$

*Divide using long division.*

35. $\dfrac{x^2 + 3x + 2}{x + 1}$

36. $\dfrac{x^2 + x - 20}{x + 5}$

37. $\dfrac{6x^2 + 16x + 8}{3x + 2}$

38. $\dfrac{2x^2 + 13x + 20}{x + 4}$

39. $\dfrac{6x^2 + x - 2}{2x - 1}$

40. $\dfrac{12x^2 - 17x - 7}{3x + 1}$

41. $\dfrac{x^2 + 6x + 3}{x + 1}$

42. $\dfrac{a^2 - a - 17}{a + 3}$

43. $\dfrac{2b^2 + b - 8}{b - 2}$

44. $\dfrac{2c^2 + c + 1}{2c + 5}$

45. $\dfrac{8x^2 + 6x - 25}{2x - 3}$

46. $\dfrac{8z^2 - 18z - 7}{4z + 1}$

47. $\dfrac{4x^2 - 36}{2x - 6}$

48. $\dfrac{16p^2 - 9}{4p + 3}$

49. $\dfrac{x^3 + 3x^2 + 5x + 4}{x + 1}$

50. $\dfrac{-a^3 - 6a^2 + 2a - 4}{a - 1}$

51. $\dfrac{4y^3 + 12y^2 + 7y - 9}{2y + 3}$

52. $\dfrac{9b^3 - 3b^2 - 3b + 4}{3b + 2}$

53. $(4a^3 - 5a) \div (2a - 1)$

54. $(2x^3 + 6x + 33) \div (x + 4)$

55. $\dfrac{3x^5 + 2x^2 - 12x - 4}{x^2 - 2}$

56. $\dfrac{4b^5 - 18b^3 + 14b^2 + 18b - 21}{2b^2 - 3}$

57. $\dfrac{3x^4 + 4x^3 - 32x^2 - 5x - 20}{3x^3 - 8x^2 - 5}$

58. $\dfrac{3a^4 - 9a^3 + 13a^2 - 11a + 4}{a^2 - 2a + 1}$

59. $\dfrac{2c^4 - 8c^3 + 19c^2 - 33c + 15}{c^2 - c + 5}$

60. $\dfrac{2y^5 + 2y^4 - 3y^3 - 15y^2 + 18}{2y^2 - 3}$

*Use synthetic division to divide.*

61. $(x^2 + 7x + 6) \div (x + 1)$

62. $(x^2 - 7x + 6) \div (x - 1)$

63. $(x^2 + 5x + 6) \div (x + 2)$

64. $(x^2 - 5x + 6) \div (x - 2)$

65. $(x^2 - 11x + 28) \div (x - 4)$

66. $(x^2 + 17x + 72) \div (x + 9)$

67. $(x^2 + 5x - 14) \div (x - 3)$

68. $(x^2 - 2x - 39) \div (x + 5)$

69. $(3x^2 \quad 7x \quad 10) \div (x \quad 4)$

70. $(2b^2 - 9b + 1) \div (b - 6)$

71. $(4x^3 - 3x^2 + 2x) \div (x - 1)$

72. $(z^3 - 7z^2 - 13z + 25) \div (z - 2)$

73. $(3c^3 + 7c^2 - 4c + 16) \div (c + 3)$

74. $(3y^4 - 25y^2 - 29) \div (y - 3)$

75. $(y^4 - 1) \div (y - 1)$

76. $(a^4 - 16) \div (a - 2)$

77. $\dfrac{x^4 + 16}{x + 4}$

78. $\dfrac{z^4 + 81}{z + 3}$

79. $\dfrac{x^5 + x^4 - 9}{x + 1}$

80. $\dfrac{a^7 - 2a^6 + 13}{a - 2}$

81. $\dfrac{b^5 + 4b^4 - 14}{b + 1}$

82. $\dfrac{z^5 - 3z^3 - 7z}{z - 2}$

83. $(3x^3 + 2x^2 - 4x + 1) \div \left(x - \dfrac{1}{3}\right)$

84. $(8x^3 - 6x^2 - 5x + 3) \div \left(x + \dfrac{3}{4}\right)$

85. $(2x^4 - x^3 + 2x^2 - 3x + 7) \div \left(x - \dfrac{1}{2}\right)$

86. $(9y^3 + 9y^2 - y + 2) \div \left(y + \dfrac{2}{3}\right)$

*Determine the remainder for the following divisions using the Remainder Theorem. If the divisor is a factor of the dividend, so state.*

87. $(4x^2 - 5x + 6) \div (x - 2)$

88. $(-2x^2 + 3x - 2) \div (x + 3)$

89. $(x^3 - 2x^2 + 4x - 8) \div (x - 2)$

90. $(x^4 + 3x^3 + x^2 + 22x + 8) \div (x + 4)$

91. $(-2x^3 - 6x^2 + 2x - 4) \div \left(x - \dfrac{1}{2}\right)$

92. $(-5x^3 - 6) \div \left(x - \dfrac{1}{5}\right)$

## Problem Solving

**93. Area** The area of a rectangle is $6x^2 - 8x - 8$. If the length is $2x - 4$, find the width.

**94. Area** The area of a rectangle is $15x^2 - 29x - 14$. If the width is $5x + 2$, find the length.

**Area and Volume** *In Exercises 95 and 96, how many times greater is the area or volume of the figure on the right than the figure on the left? Explain how you determined your answer.*

**95.**

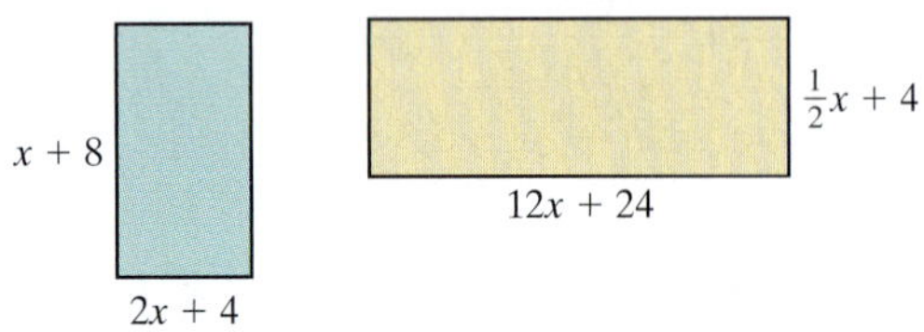


**96.**

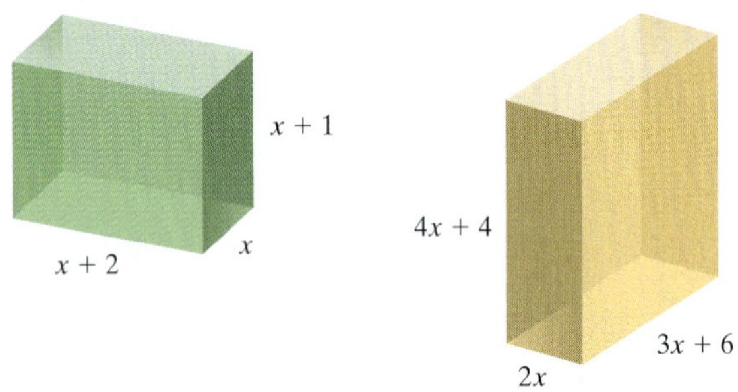


**97.** Is it possible to divide a binomial by a monomial and obtain a monomial as a quotient? Explain.

**98. a)** Is the sum, difference, and product of two polynomials always a polynomial?

**b)** Is the quotient of two polynomials always a polynomial? Explain.

**99.** Explain how you can determine using synthetic division if an expression of the form $x - a$ is a factor of a polynomial in $x$.

**100.** Given $P(x) = ax^2 + bx + c$ and a value $d$ such that $P(d) = 0$, explain why $d$ is a solution to the equation $ax^2 + bx + c = 0$.

**101.** If $\dfrac{P(x)}{x - 4} = x + 2$, find $P(x)$.

**102.** If $\dfrac{P(x)}{2x + 4} = 2x - 3$, find $P(x)$.

**103.** If $\dfrac{P(x)}{x + 4} = x + 5 + \dfrac{6}{x + 4}$, find $P(x)$.

**104.** If $\dfrac{P(x)}{2x - 3} = 2x - 1 - \dfrac{8}{2x - 3}$, find $P(x)$.

*In Exercises 105 and 106, divide.*

**105.** $\dfrac{2x^3 - x^2y - 7xy^2 + 2y^3}{x - 2y}$

**106.** $\dfrac{x^3 + y^3}{x + y}$

*In Exercises 107 and 108, divide. The answers contain fractions.*

**107.** $\dfrac{2x^2 + 2x - 2}{2x - 3}$

**108.** $\dfrac{3x^3 - 7}{3x - 2}$

**109. Volume** The volume of the box that follows is $2r^3 + 4r^2 + 2r$. Find $w$ in terms of $r$.

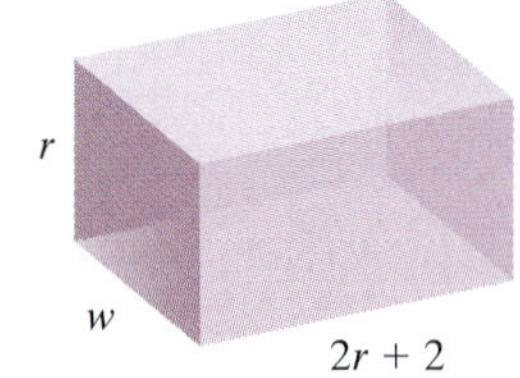


**110. Volume** The volume of the box that follows is $6a^3 + a^2 - 2a$. Find $b$ in terms of $a$.

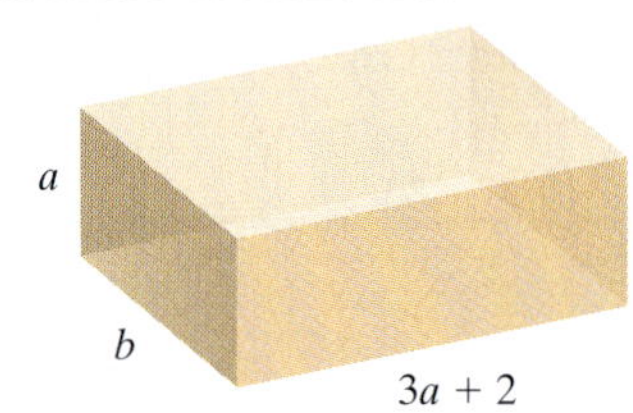


**111.** When a polynomial is divided by $x - 3$, the quotient is $x^2 - 3x + 4 + \dfrac{5}{x - 3}$. What is the polynomial? Explain how you determined your answer.

**112.** When a polynomial is divided by $2x - 3$, the quotient is $2x^2 + 6x - 5 + \dfrac{5}{2x - 3}$. What is the polynomial? Explain how you determined your answer.

*In Exercises 113 and 114, divide. Assume that all variables in the exponents are natural numbers.*

**113.** $\dfrac{4x^{n+1} + 2x^n - 3x^{n-1} - x^{n-2}}{2x^n}$

**114.** $\dfrac{3x^n + 6x^{n-1} - 2x^{n-2}}{2x^{n-1}}$

**115.** Is $x - 1$ a factor of $x^{100} + x^{99} + \cdots + x^1 + 1$? Explain.

**116.** Is $x + 1$ a factor of $x^{100} + x^{99} + \cdots + x^1 + 1$? Explain.

**117.** Is $x + 1$ a factor of $x^{99} + x^{98} + \cdots + x^1 + 1$? Explain.

**118.** Divide $0.2x^3 - 4x^2 + 0.32x - 0.64$ by $x - 0.4$.

## Cumulative Review Exercises

[1.6] **119.** Divide $\dfrac{8.45 \times 10^{23}}{4.225 \times 10^{13}}$ and express the answer in scientific notation.

[2.3] **120. Triangle** Find the three angles of a triangle if one angle is twice the smallest angle and the third angle is 60° greater than the smallest angle.

[2.6] **121.** Find the solution set for $\left|\dfrac{5x - 3}{2}\right| + 4 = 8$.

[3.6] **122.** Let $f(x) = x^2 - 4$ and $g(x) = -5x + 3$. Find $f(6) \cdot g(6)$.

[5.1] **123.** Add $(6r + 5s - t) + (-3r - 2s - 7t)$.

# 5.4 Factoring a Monomial from a Polynomial and Factoring by Grouping

1. Find the greatest common factor.
2. Factor a monomial from a polynomial.
3. Factor a common binomial factor.
4. Factor by grouping.

*Factoring is the opposite of multiplying.* To factor an expression means to write it as a product of other expressions. For example, in Section 5.2 we learned to perform the following multiplications:

$$3x^2(6x + 3y + 4x^3) = 18x^3 + 9x^2y + 12x^5$$

and

$$(6x + 3y)(2x - 5y) = 12x^2 - 24xy - 15y^2$$

In this section, we learn how to determine the *factors* of a given expression. For example, we will learn how to perform each factoring illustrated below.

$$18x^3 + 9x^2y + 12x^5 = 3x^2(6x + 3y + 4x^3)$$

and

$$12x^2 - 24xy - 15y^2 = (6x + 3y)(2x - 5y)$$

## 1 Find the Greatest Common Factor

To factor a monomial from a polynomial, we factor the **greatest common factor** (GCF) from each term in the polynomial. The GCF is the product of the factors common to all terms in the polynomial. For example, the GCF for $6x + 21$ is 3 since 3 is the largest number that is a factor of both $6x$ and 21. To factor, we use the distributive property.

$$6x + 21 = 3(2x + 7)$$

The 3 and the $2x + 7$ are *factors* of the polynomial $6x + 21$.

Consider the terms $x^3$, $x^4$, $x^5$, and $x^6$. The GCF of these terms is $x^3$, since $x^3$ is the highest power of $x$ that divides all four terms.

**EXAMPLE 1** ▸ Find the GCF of the following terms.

**a)** $y^{12}, y^4, y^9, y^7$ **b)** $x^3y^2, xy^4, x^5y^6$ **c)** $6x^2y^3z, 9x^3y^4, 24x^2z^5$

*Solution*

**a)** Note that $y^4$ is the highest power of $y$ common to all four terms. The GCF is, therefore, $y^4$.

**b)** The highest power of $x$ that is common to all three terms is $x$ (or $x^1$). The highest power of $y$ that is common to all three terms is $y^2$. Thus, the GCF of the three terms is $xy^2$.

**c)** The GCF is $3x^2$. Since $y$ does not appear in $24x^2z^5$, it is not part of the GCF. Since $z$ does not appear in $9x^3y^4$, it is not part of the GCF.

▸ Now Try Exercise 3

**EXAMPLE 2** ▸ Find the GCF of the following terms.

$$6(x - 2)^2, 5(x - 2), 18(x - 2)^7$$

*Solution* The three numbers 6, 5, and 18 have no common factor other than 1. The highest power of $(x - 2)$ common to all three terms is $(x - 2)$. Thus, the GCF of the three terms is $(x - 2)$.

▸ Now Try Exercise 5

## 2 Factor a Monomial from a Polynomial

When we factor a monomial from a polynomial, we are factoring out the greatest common factor. *The first step in any factoring problem is to determine and then factor out the GCF.*

**To Factor a Monomial from a Polynomial**

1. Determine the greatest common factor of all terms in the polynomial.
2. Write each term as the product of the GCF and another factor.
3. Use the distributive property to *factor out* the GCF.

**EXAMPLE 3** ▸ Factor $15x^4 - 5x^3 + 25x^2$.

*Solution* The GCF is $5x^2$. Write each term as the product of the GCF and another product. Then factor out the GCF.

$$\begin{aligned}15x^4 - 5x^3 + 25x^2 &= 5x^2 \cdot 3x^2 - 5x^2 \cdot x + 5x^2 \cdot 5\\ &= 5x^2(3x^2 - x + 5)\end{aligned}$$

▸ Now Try Exercise 15

**To check the factoring process, multiply the factors using the distributive property. The product should be the expression with which you began.** For instance, in Example 3,

Check
$$\begin{aligned}5x^2(3x^2 - x + 5) &= 5x^2(3x^2) + 5x^2(-x) + 5x^2(5)\\ &= 15x^4 - 5x^3 + 25x^2\end{aligned}$$

**EXAMPLE 4** ▸ Factor $20x^3y^3 + 6x^2y^4 - 12xy^5$.

*Solution* The GCF is $2xy^3$. Write each term as the product of the GCF and another product. Then factor out the GCF.

$$\begin{aligned}20x^3y^3 + 6x^2y^4 - 12xy^5 &= 2xy^3 \cdot 10x^2 + 2xy^3 \cdot 3xy - 2xy^3 \cdot 6y^2\\ &= 2xy^3(10x^2 + 3xy - 6y^2)\end{aligned}$$

Check
$$2xy^3(10x^2 + 3xy - 6y^2) = 20x^3y^3 + 6x^2y^4 - 12xy^5$$

▸ Now Try Exercise 19

When the leading coefficient of a polynomial is negative, we generally factor out a common factor with a negative coefficient. This results in the leading coefficient of the remaining polynomial being positive.

**EXAMPLE 5** ▸ Factor. **a)** $-12a - 18$ **b)** $-2b^3 + 6b^2 - 16b$

*Solution* Since the leading coefficients in parts **a)** and **b)** are negative, we factor out common factors with a negative coefficient.

**a)** $-12a - 18 = -6(2a + 3)$ *Factor out $-6$.*

**b)** $-2b^3 + 6b^2 - 16b = -2b(b^2 - 3b + 8)$ *Factor out $-2b$.*

▸ Now Try Exercise 27

**EXAMPLE 6** ▸ **Throwing a Ball** When a ball is thrown upward with a velocity of 32 feet per second from the top of a 160-foot-tall building, its distance, $d$, from the ground at any time, $t$, can be determined by the function $d(t) = -16t^2 + 32t + 160$.

**a)** Determine the ball's distance from the ground after 3 seconds—that is, find $d(3)$.

**b)** Factor out the GCF from the right side of the function.

**c)** Evaluate $d(3)$ in factored form.

**d)** Compare your answers to parts **a)** and **c)**.

*Solution*

**a)**
$$\begin{aligned} d(t) &= -16t^2 + 32t + 160 \\ d(3) &= -16(3)^2 + 32(3) + 160 && \text{Substitute 3 for } t. \\ &= -16(9) + 96 + 160 \\ &= 112 \end{aligned}$$

The distance is 112 feet.

**b)** Factor $-16$ from the three terms on the right side of the equal sign.

$$d(t) = -16(t^2 - 2t - 10)$$

**c)**
$$\begin{aligned} d(t) &= -16(t^2 - 2t - 10) \\ d(3) &= -16[3^2 - 2(3) - 10] && \text{Substitute 3 for } t. \\ &= -16(9 - 6 - 10) \\ &= -16(-7) \\ &= 112 \end{aligned}$$

**d)** The answers are the same. You may find the calculations in part **c)** easier than the calculations in part **a)**.

▶ **Now Try Exercise 65**

## 3 Factor a Common Binomial Factor

Sometimes factoring involves factoring a binomial as the greatest common factor, as illustrated in Examples 7 through 9.

**EXAMPLE 7** ▶ Factor $3x(5x - 6) + 4(5x - 6)$.

*Solution* The GCF is $(5x - 6)$. Factoring out the GCF gives

$$3x(5x - 6) + 4(5x - 6) = (5x - 6)(3x + 4)$$

▶ **Now Try Exercise 37**

In Example 7, we could have also placed the common factor on the right to obtain

$$3x(5x - 6) + 4(5x - 6) = (3x + 4)(5x - 6)$$

The factored forms $(5x - 6)(3x + 4)$ and $(3x + 4)(5x - 6)$ are equivalent because of the commutative property of multiplication and both are correct. Generally, when we list the answer to an example or exercise, we will place the common term that has been factored out on the left.

**EXAMPLE 8** ▶ Factor $9(2x - 5) + 6(2x - 5)^2$.

*Solution* The GCF is $3(2x - 5)$. Rewrite each term as the product of the GCF and another factor.

$$\begin{aligned} 9(2x - 5) + 6(2x - 5)^2 &= 3(2x - 5)\cdot 3 + 3(2x - 5)\cdot 2(2x - 5) \\ &= 3(2x - 5)[3 + 2(2x - 5)] && \text{Factor out } 3(2x - 5). \\ &= 3(2x - 5)[3 + 4x - 10] && \text{Distributive property} \\ &= 3(2x - 5)(4x - 7) && \text{Simplify.} \end{aligned}$$

▶ **Now Try Exercise 39**

**EXAMPLE 9** ▶ Factor $(3x - 4)(a + b) - (x - 1)(a + b)$.

*Solution* The binomial $a + b$ is the GCF of the two terms. We therefore factor it out.

$$\begin{aligned}(3x - 4)(a + b) - (x - 1)(a + b) &= (a + b)[(3x - 4) - (x - 1)] && \text{Factor out } (a + b).\\ &= (a + b)(3x - 4 - x + 1) && \text{Simplify.}\\ &= (a + b)(2x - 3) && \text{Factors}\end{aligned}$$

▶ **Now Try Exercise 43**

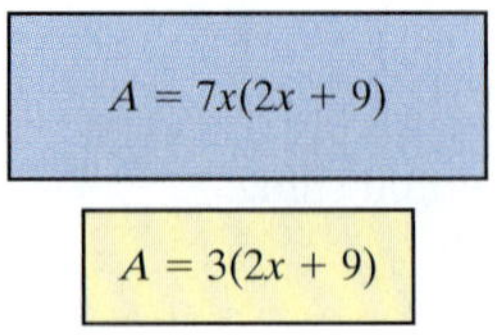

FIGURE 5.13

**EXAMPLE 10** ▶ **Area** In **Figure 5.13**, the area of the large rectangle is $7x(2x + 9)$ and the area of the small rectangle is $3(2x + 9)$. Find an expression, in factored form, for the difference of the areas for these two rectangles.

*Solution* To find the difference of the areas, subtract the area of the small rectangle from the area of the large rectangle.

$$\begin{aligned}&7x(2x + 9) - 3(2x + 9) && \text{Subtract areas.}\\ &= (2x + 9)(7x - 3) && \text{Factor out } (2x + 9).\end{aligned}$$

The difference of the areas for the two rectangles is $(2x + 9)(7x - 3)$.

▶ **Now Try Exercise 59**

## 4 Factor by Grouping

When a polynomial contains *four terms*, it may be possible to factor the polynomial by grouping. To **factor by grouping**, remove common factors from groups of terms. This procedure is illustrated in the following example.

**EXAMPLE 11** ▶ Factor $ax + ay + bx + by$.

*Solution* There is no factor (other than 1) common to all four terms. However, $a$ is common to the first two terms and $b$ is common to the last two terms. Factor $a$ from the first two terms and $b$ from the last two terms.

$$ax + ay + bx + by = a(x + y) + b(x + y)$$

Now $(x + y)$ is common to both terms. Factor out $(x + y)$.

$$a(x + y) + b(x + y) = (x + y)(a + b)$$

Thus, $ax + ay + bx + by = (x + y)(a + b)$ or $(a + b)(x + y)$.

▶ **Now Try Exercise 49**

**To Factor Four Terms by Grouping**

1. Determine if all four terms have a common factor. If so, factor out the greatest common factor from each term.
2. Arrange the four terms into two groups of two terms each. Each group of two terms must have a GCF.
3. Factor the GCF from each group of two terms.
4. If the two terms formed in step 3 have a GCF, factor it out.

**EXAMPLE 12** ▶ Factor $x^3 - 5x^2 + 2x - 10$ by grouping.

*Solution* There are no factors common to all four terms. However, $x^2$ is common to the first two terms and 2 is common to the last two terms. Factor $x^2$ from the first two terms and factor 2 from the last two terms.

$$\begin{aligned}x^3 - 5x^2 + 2x - 10 &= x^2(x - 5) + 2(x - 5)\\ &= (x - 5)(x^2 + 2)\end{aligned}$$

▶ **Now Try Exercise 55**

In Example 12, $(x^2 + 2)(x - 5)$ is also an acceptable answer. Would the answer to Example 12 change if we switch the order of $2x$ and $-5x^2$? Let's try it in Example 13.

**EXAMPLE 13** ▶ Factor $x^3 + 2x - 5x^2 - 10$.

*Solution* There is no factor common to all four terms. Factor $x$ from the first two terms and $-5$ from the last two terms.

$$x^3 + 2x - 5x^2 - 10 = x(x^2 + 2) - 5(x^2 + 2)$$
$$= (x^2 + 2)(x - 5)$$

Notice that we got equivalent results in Examples 12 and 13.

▶ Now Try Exercise 51

## Helpful Hint

When factoring four terms by grouping, if the *first* and *third* terms are positive, you must factor a positive expression from both the first two terms and the last two terms to obtain a factor common to the remaining two terms (see Example 12). If the *first* term is positive and the *third* term is negative, you must factor a positive expression from the first two terms and a negative expression from the last two terms to obtain a factor common to the remaining two terms (see Example 13).

The first step in any factoring problem is to determine whether all the terms have a common factor. If so, begin by factoring out the common factor. For instance, to factor $x^4 - 5x^3 + 2x^2 - 10x$, we first factor out $x$ from each term. Then we factor the remaining four terms by grouping, as was done in Example 12.

$$x^4 - 5x^3 + 2x^2 - 10x = x(x^3 - 5x^2 + 2x - 10)$$ *Factor out the GCF, x, from all four terms.*

$$= x(x - 5)(x^2 + 2)$$ *Factors from Example 12*

# EXERCISE SET 5.4

MathXL® MyMathLab

## Concept/Writing Exercises

1. What is the first step in *any* factoring problem?
2. What is the greatest common factor of the terms of an expression?
3. a) Explain how to find the greatest common factor of the terms of a polynomial.
   b) Using your procedure from part **a)**, find the greatest common factor of the polynomial
   $$6x^2y^5 - 2x^3y + 12x^9y^3$$
   c) Factor the polynomial in part **b)**.
4. Determine the GCF of the following terms:
   $$x^4y^6, x^3y^5, xy^6, x^2y^4$$
   Explain how you determined your answer.
5. Determine the GCF of the following terms:
   $$12(x - 4)^3, 6(x - 4)^6, 3(x - 4)^9$$
   Explain how you determined your answer.
6. When a term of a polynomial is itself the GCF, what is written in place of that term when the GCF is factored out? Explain.
7. a) Explain how to factor a polynomial of four terms by grouping.
   b) Factor $6x^3 - 2xy^3 + 3x^2y^2 - y^5$ by your procedure from part **a)**.
8. What is the first step when factoring $-x^2 + 8x - 15$? Explain your answer.

## Practice the Skills

*Factor out the greatest common factor.*

9. $7n + 14$
10. $15p + 25$
11. $2x^2 - 4x + 10$
12. $6x^2 - 12x + 27$
13. $12y^2 - 16y + 28$
14. $12x^3 - 8x^2 - 6x$
15. $9x^4 - 3x^3 + 11x^2$
16. $45y^{12} + 60y^{10}$
17. $-24a^7 + 9a^6 - 3a^2$
18. $-16c^5 - 12c^4 + 6c^3$
19. $3x^2y + 6x^2y^2 + 3xy$
20. $24a^2b^2 + 16ab^4 + 72ab^3$
21. $80a^5b^4c - 16a^4b^2c^2 + 24a^2c$
22. $36xy^2z^3 + 36x^3y^2z + 9x^2yz$
23. $9p^4q^5r - 3p^2q^2r^2 + 12pq^5r^3$
24. $24m^6 + 8m^4 - 4m^3n$
25. $-22p^2q^2 - 16pq^3 + 26r$
26. $-15y^3z^5 - 28y^3z^6 + 9xy^2z^2$

*Factor out a factor with a negative coefficient.*

27. $-8x + 4$
28. $-20a - 30$
29. $-x^2 - 4x + 22$
30. $-y^5 - 6y^2 - 4$
31. $-3r^2 - 6r + 9$
32. $-12t^2 + 48t - 60$
33. $-6r^4s^3 + 4r^2s^4 + 2rs^5$
34. $-5p^6q^3 - 10p^4q^4 + 25pq^7$
35. $-a^4b^2c + 5a^3bc^2 + a^2b$
36. $-20x^5y^3z - 4x^4yz^2 - 8x^2y^5$

*Factor.*

37. $x(a + 3) + 1(a + 3)$
38. $y(b - 2) - 5(b - 2)$
39. $7x(x - 4) + 2(x - 4)^2$
40. $4y(y + 1) - 7(y + 1)^2$
41. $(x - 2)(3x + 5) - (x - 2)(5x - 4)$
42. $(z + 4)(z + 3) + (z - 1)(z + 3)$
43. $(2a + 4)(a - 3) - (2a + 4)(2a - 1)$
44. $(6b - 1)(b + 4) + (6b - 1)(2b + 5)$
45. $x^2 + 4x - 5x - 20$
46. $a^2 + 3a - 6a - 18$
47. $8y^2 - 4y - 20y + 10$
48. $18m^2 + 30m + 9m + 15$
49. $am + an + bm + bn$
50. $cx - cy - dx + dy$
51. $x^3 - 3x^2 + 4x - 12$
52. $2z^3 + 4z^2 - 5z - 10$
53. $10m^2 - 12mn - 25mn + 30n^2$
54. $12x^2 + 9xy - 4xy - 3y^2$
55. $5a^3 + 15a^2 - 10a - 30$
56. $2r^4 - 2r^3 - 7r^2 + 7r$
57. $c^5 - c^4 + c^3 - c^2$
58. $b^4 - b^3 - b + b^2$

## Problem Solving

**Area** *In Exercises 59–62, A represents an expression for the area of the figure. Find an expression, in factored form, for the difference of the areas of the geometric figures. See Example 10.*

59.
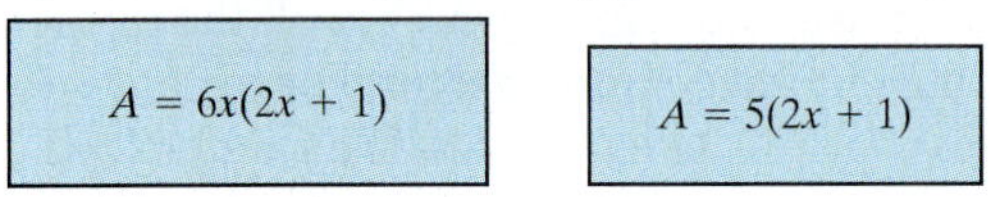


60.
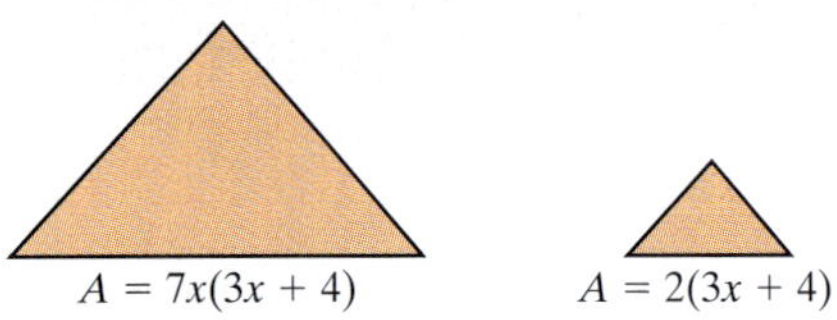


61.
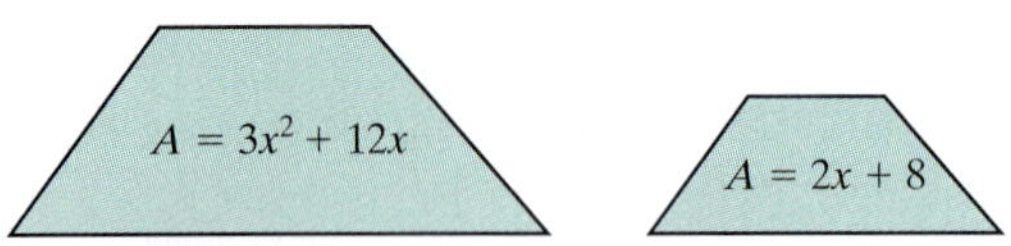


62.
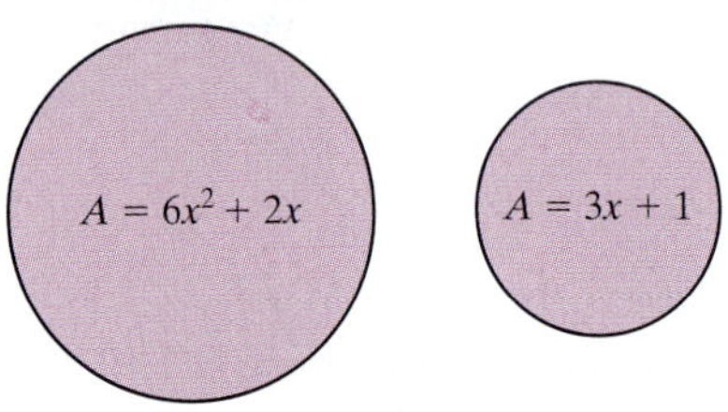


**Volume** *In Exercises 63 and 64, V represents an expression for the volume of the figure. Find an expression, in factored form, for the difference of the volumes of the geometric solids.*

63.
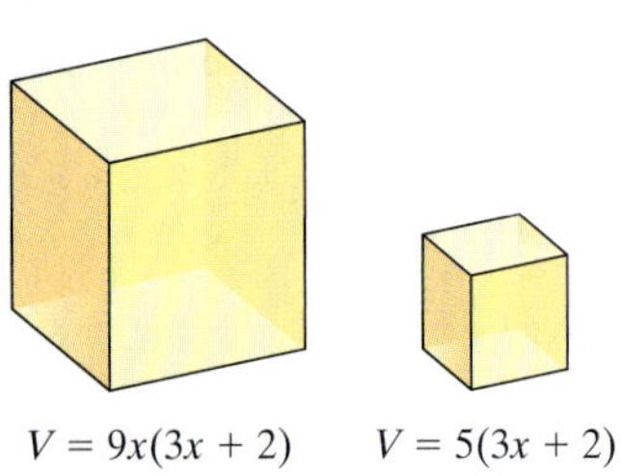


64.
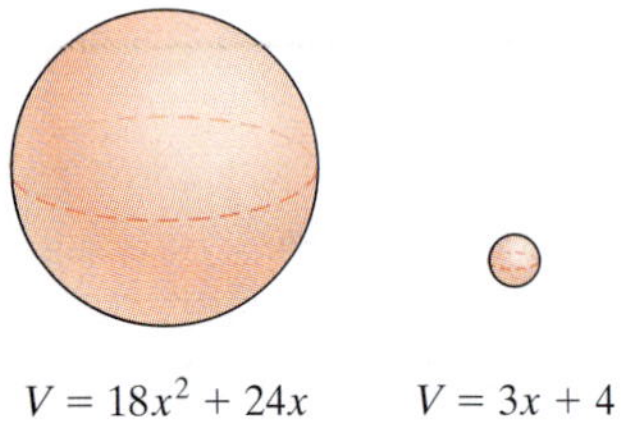


65. **Flare** When a flare is shot upward with a velocity of 80 feet per second, its height, $h$, in feet, above the ground at $t$ seconds can be found by the function $h(t) = -16t^2 + 80t$.

**a)** Find the height of the flare 3 seconds after it was shot.

**b)** Express the function with the right side in factored form.

**c)** Evaluate $h(3)$ using the factored form from part **b)**.

66. **Jump Shot** When a basketball player shoots a jump shot, the height, $h$, in feet, of the ball above the ground at any time $t$, under certain circumstances, may be found by the function $h(t) = -16t^2 + 20t + 8$.

**a)** Find the height of the ball at 1 second.

**b)** Express the function with the right side in factored form.

**c)** Evaluate $h(1)$ using the factored form in part **b)**.

67. **Skating Rink** The area of the skating rink with semicircular ends shown is $A = \pi r^2 + 2rl$.

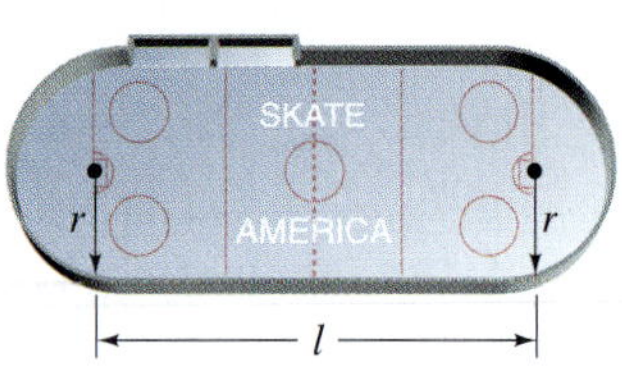


**a)** Find $A$ when $r = 20$ feet and $l = 40$ feet

**b)** Write the area $A$ in factored form.

**c)** Find $A$ when $r = 20$ feet and $l = 40$ feet using the factored form in part **b)**.

**68. Area** The formula for finding the area of a trapezoid may be given as $A = \frac{1}{2}hb_1 + \frac{1}{2}hb_2$. Express this formula in factored form.

**69. Purchase a Computer** Fred Yang just bought a computer for \$975. With a 0% (interest-free) loan, Fred must pay \$75 each month until the computer is paid off. The amount still owed on the computer is a function of time, where.

$$A(t) = 975 - 75t$$

and $t$ is the number of months after Fred bought the computer.

**a)** Determine the amount still owed 6 months after Fred bought the computer.

**b)** Write the function in factored form.

**c)** Use the result from part **b)** to determine the amount still owed 6 months after Fred bought the computer.

**70. Salary** On January 2, 2006, Jill Ferguson started a new job with an annual salary of \$33,000. Her salary will increase by \$1500 each year. Thus, her salary is a function of the number of years she works, where

$$S(n) = 33{,}000 + 1500n$$

and $n$ is the number of years after 2006.

**a)** Determine Jill's salary in 2010.

**b)** Write the function in factored form.

**c)** Use the result from part **b)** to determine Jill's salary in 2010.

**71. Price of Cars** When the 2006 cars came out, the list price of one model increased by 6% over the list price of the 2005 model. Then in a special sale, the prices of all 2006 cars were reduced by 6%. The sale price can be represented by $(x + 0.06x) - 0.06(x + 0.06x)$, where $x$ is the list price of the 2005 model.

**a)** Factor out $(x + 0.06x)$ from each term.

**b)** Is the sale price more or less than the price of the 2005 model?

*Read Exercise 71 before working Exercises 72–74.*

**72. Price of Dress** A dress is reduced by 10%, and then the sale price is reduced by another 10%.

**a)** Write an expression for the final price of the dress.

**b)** How does the final price compare with the regular price of the dress? Use factoring in obtaining your answer.

**73. Price of Mower** The price of a Toro lawn mower is increased by 15%. Then at a Fourth of July sale the price is reduced by 20%.

**a)** Write an expression for the final price of the lawn mower.

**b)** How does the sale price compare with the regular price? Use factoring in obtaining your answer.

**74. Find Price** In which of the following, **a)** or **b)**, will the final price be lower, and by how much?

**a)** Decreasing the price of an item by 6%, then increasing that price by 8%

**b)** Increasing the price of an item by 6%, then decreasing that price by 8%

*Factor.*

**75.** $5a(3x + 2)^5 + 4(3x + 2)^4$

**76.** $4p(2r - 3)^7 - 3(2r - 3)^6$

**77.** $4x^2(x - 3)^3 - 6x(x - 3)^2 + 4(x - 3)$

**78.** $12(p + 2q)^4 - 40(p + 2q)^3 + 12(p + 2q)^2$

**79.** $ax^2 + 2ax - 3a + bx^2 + 2bx - 3b$

**80.** $6a^2 - a^2c + 18a - 3ac + 6ab - abc$

*Factor. Assume that all variables in the exponents represent natural numbers.*

**81.** $x^{6m} - 2x^{4m}$

**82.** $x^{2mn} + x^{6mn}$

**83.** $3x^{4m} - 2x^{3m} + x^{2m}$

**84.** $r^{y+4} + r^{y+3} + r^{y+2}$

**85.** $a^rb^r + c^rb^r - a^rd^r - c^rd^r$

**86.** $6a^kb^k - 2a^kc^k - 9b^k + 3c^k$

**87. a)** Does $6x^3 - 3x^2 + 9x = 3x(2x^2 - x + 3)$?

**b)** If the above factoring is correct, what should be the value of $6x^3 - 3x^2 + 9x - [3x(2x^2 - x + 3)]$ for any value of $x$? Explain.

**c)** Select a value for $x$ and evaluate the expression in part **b)**. Did you get what you expected? If not, explain why.

**88. a)** Determine whether the following factoring is correct.

$$\begin{aligned} 3(x - 2)^2 - 6(x - 2) &= 3(x - 2)[(x - 2) - 2] \\ &= 3(x - 2)(x - 4) \end{aligned}$$

**b)** If the above factoring is correct, what should be the value of $3(x - 2)^2 - 6(x - 2) - [3(x - 2)(x - 4)]$ for any value of $x$? Explain.

**c)** Select a value for $x$ and evaluate the expression in part **b)**. Did you get what you expected? If not, explain why.

**89.** Consider the factoring $8x^3 - 16x^2 - 4x = 4x(2x^2 - 4x - 1)$.

**a)** If we let

$$y_1 = 8x^3 - 16x^2 - 4x$$
$$y_2 = 4x(2x^2 - 4x - 1)$$

and graph each function, what should happen? Explain.

**b)** On your graphing calculator, graph $y_1$ and $y_2$ as given in part **a)**.

**c)** Did you get the results you expected?

**d)** When checking a factoring process by this technique, what does it mean if the graphs do not overlap? Explain.

**90.** Consider the factoring $2x^4 - 6x^3 - 8x^2 = 2x^2(x^2 - 3x - 4)$.

**a)** Enter

$$y_1 = 2x^4 - 6x^3 - 8x^2$$
$$y_2 = 2x^2(x^2 - 3x - 4)$$

into your calculator.

**b)** If you use the TABLE feature of your calculator, how would you expect the table of values for $y_1$ to compare with the table of values for $y_2$? Explain.

**c)** Use the TABLE feature to show the values for $y_1$ and $y_2$ for values of $x$ from 0 to 6.

**d)** Did you get the results you expected?

**e)** When checking a factoring process using the TABLE feature, what does it mean if the values of $y_1$ and $y_2$ are different?

## Cumulative Review Exercises

[1.4] **91.** Evaluate $\dfrac{\left(\left|\frac{1}{2}\right| - \left|-\frac{1}{3}\right|\right)^2}{-\left|\frac{1}{3}\right| \cdot \left|-\frac{2}{5}\right|}$.

[2.1] **92.** Solve $3(2x - 4) + 3(x + 1) = 9$.

[3.1] **93.** Graph $y = x^2 - 1$.

[4.3] **94. Exercise** Jason Richter tries to exercise every day. He walks at 3 mph and then jogs at 5 mph. If it takes him 0.9 hr to travel a total of 3.5 mi, how long does he jog?

[5.2] **95.** Multiply $(7a - 3)(-2a^2 - 4a + 1)$.

## Mid-Chapter Test: 5.1–5.4

*To find out how well you understand the chapter material to this point, take this brief test. The answers, and the section where the material was initially discussed, are given in the back of the book. Review any questions that you answered incorrectly.*

**1.** Write the polynomial $-7 + 2x + 5x^4 - 1.5x^3$ in descending order. Give the degree of this polynomial

**2.** Evaluate $P\left(\frac{1}{2}\right)$ given $P(x) = 8x^2 - 7x + 3$.

**3.** Simplify $(2n^2 - n - 12) + (-3n^2 - 6n + 8)$.

**4.** Subtract $(7x^2y - 10xy)$ from $(-9x^2y + 4xy)$.

**5.** Find a polynomial expression, in simplified form, for the perimeter of the triangle.

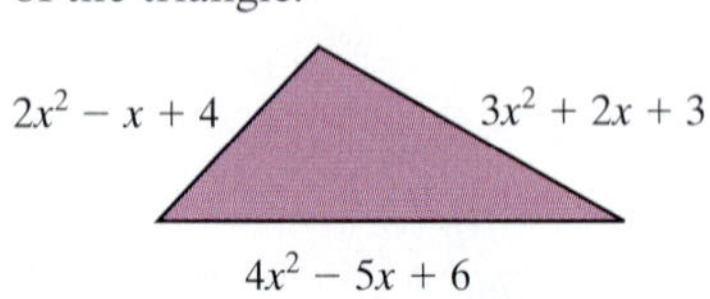


*Multiply.*

**6.** $2x^5(3xy^4 + 5x^2 - 7x^3y)$

**7.** $(7x - 6y)(3x + 2y)$

**8.** $(3x + 1)(2x^3 - x^2 + 5x + 9)$

**9.** $\left(8p - \frac{1}{5}\right)\left(8p + \frac{1}{5}\right)$

**10.** $(4m - 3n)(3m^2 + 2mn - 6n^2)$

**11.** Write $x^2 - 14x + 49$ as the square of a binomial. Explain how you determined your answer.

*Divide.*

**12.** $\dfrac{4x^4y^3 + 6x^2y^2 - 11x}{2x^2y^2}$

**13.** $\dfrac{12x^2 + 23x + 7}{4x + 1}$

**14.** $\dfrac{2y^3 - y^2 + 7y - 10}{2y - 3}$

*Use synthetic division to divide.*

**15.** $\dfrac{x^2 - x - 72}{x + 8}$

**16.** $\dfrac{3a^4 - 2a^3 - 14a^2 + 11a + 2}{a - 2}$

**17.** Factor out the greatest common factor in $32b^3c^3 + 16b^2c + 24b^5c^4$.

*Factor completely.*

**18.** $7b(2x + 9) - 3c(2x + 9)$

**19.** $2b^4 - b^3c + 4b^3c - 2b^2c^2$

**20.** $5a(3x - 2)^5 - 4(3x - 2)^6$

# 5.5 Factoring Trinomials

1 Factor trinomials of the form $x^2 + bx + c$.

2 Factor out a common factor.

3 Factor trinomials of the form $ax^2 + bx + c, a \neq 1$, using trial and error.

4 Factor trinomials of the form $ax^2 + bx + c, a \neq 1$, using grouping.

5 Factor trinomials using substitution.

## 1 Factor Trinomials of the Form $x^2 + bx + c$

In this section we learn how to **factor trinomials** of the form $ax^2 + bx + c, a \neq 0$. Notice that $a$ represents the coefficient of the $x$-squared term, $b$ represents the coefficient of the $x$-term, and $c$ represents the constant term.

| Trinomials | Coefficients |
|---|---|
| $3x^2 + 2x - 5$ | $a = 3, \quad b = 2, \quad c = -5$ |
| $-\frac{1}{2}x^2 - 4x + 3$ | $a = -\frac{1}{2}, \quad b = -4, \quad c = 3$ |

**To Factor Trinomials of the Form $x^2 + bx + c$ (note: $a = 1$)**

1. Find two numbers (or factors) whose product is $c$ and whose sum is $b$.
2. The factors of the trinomial will be of the form

$$(x + \square)(x + \square)$$

One factor determined in step 1 — Other factor determined in step 1

If the numbers determined in step 1 are, for example, 3 and $-5$, the factors would be written $(x + 3)(x - 5)$. This procedure is illustrated in the following examples.

**EXAMPLE 1** Factor $x^2 - x - 12$.

*Solution* $a = 1, b = -1, c = -12$. We must find two numbers whose product is $c$, which is $-12$, and whose sum is $b$, which is $-1$. We begin by listing the factors of $-12$, trying to find a pair whose sum is $-1$.

| Factors of $-12$ | Sum of Factors |
|---|---|
| $(1)(-12)$ | $1 + (-12) = -11$ |
| $(2)(-6)$ | $2 + (-6) = -4$ |
| $(3)(-4)$ | $3 + (-4) = -1$ |
| $(4)(-3)$ | $4 + (-3) = 1$ |
| $(6)(-2)$ | $6 + (-2) = 4$ |
| $(12)(-1)$ | $12 + (-1) = 11$ |

The numbers we are seeking are 3 and $-4$ because their product is $-12$ and their sum is $-1$. Now we factor the trinomial using the 3 and $-4$.

$$x^2 - x - 12 = (x + 3)(x - 4)$$

One factor of $-12$ — Other factor of $-12$

Now Try Exercise 13

Notice in Example 1 that we listed all the factors of $-12$. However, after the two factors whose product is $c$ and whose sum is $b$ are found, there is no need to go further in listing the factors. The factors were listed here to show, for example, that $(2)(-6)$ is a different set of factors than $(-2)(6)$. Note that as the positive factor increases the sum of the factors increases.

### Helpful Hint

Consider the factors $(2)(-6)$ and $(-2)(6)$ and the sums of these factors.

| Factors | Sum of Factors |
|---|---|
| $2(-6)$ | $2 + (-6) = -4$ |
| $-2(6)$ | $-2 + 6 = 4$ |

Notice that if the signs of each number in the product are changed, the sign of the sum of factors is changed. We can use this fact to more quickly find the factors we are seeking. If, when seeking a specific sum, you get the opposite of that sum, change the sign of each factor to get the sum you are seeking.

**EXAMPLE 2** ▸ Factor $p^2 - 7p + 6$.

*Solution* We must find two numbers whose product is 6 and whose sum is $-7$. Since the sum of two negative numbers is a negative number, and the product of two negative numbers is a positive number, both numbers must be negative numbers. The negative factors of 6 are $(-1)(-6)$ and $(-2)(-3)$. As shown below, the numbers we are looking for are $-1$ and $-6$.

| Factors of 6 | Sum of Factors |
|---|---|
| $(-1)(-6)$ | $-1 + (-6) = -7$ |
| $(-2)(-3)$ | $-2 + (-3) = -5$ |

Therefore,

$$p^2 - 7p + 6 = (p - 1)(p - 6)$$

Since the factors may be placed in any order, $(p - 6)(p - 1)$ is also an acceptable answer.

▸ **Now Try Exercise 23**

### Helpful Hint

***Checking Factoring***

Factoring problems can be checked by multiplying the factors obtained. If the factoring is correct, you will obtain the polynomial you started with. To check Example 2, we will multiply the factors using the FOIL method.

$$(p - 1)(p - 6) = p^2 - 6p - p + 6 = p^2 - 7p + 6$$

Since the product of the factors is the trinomial we began with, our factoring is correct. You should always check your factoring.

The procedure used to factor trinomials of the form $x^2 + bx + c$ can be used on other trinomials, as in the following example.

**EXAMPLE 3** ▸ Factor $x^2 + 2xy - 15y^2$.

*Solution* We must find two numbers whose product is $-15$ and whose sum is 2. The two numbers are 5 and $-3$.

| Factors of $-15$ | Sum of Factors |
|---|---|
| $5(-3)$ | $5 + (-3) = 2$ |

Since the last term of the trinomial contains $y^2$, the second term of each factor must contain $y$.

$$x^2 + 2xy - 15y^2 = (x + 5y)(x - 3y)$$

Check
$$(x + 5y)(x - 3y) = x^2 - 3xy + 5xy - 15y^2$$
$$= x^2 + 2xy - 15y^2$$

▸ **Now Try Exercise 75**

## 2 Factor Out a Common Factor

*The first step when factoring any trinomial is to determine whether all three terms have a common factor.* If so, factor out that common factor. Then factor the remaining polynomial.

**EXAMPLE 4** ▶ Factor $3x^4 - 6x^3 - 72x^2$.

*Solution* The factor $3x^2$ is common to all three terms of the trinomial. Factor it out first.

$$3x^4 - 6x^3 - 72x^2 = 3x^2(x^2 - 2x - 24) \qquad \text{Factor out } 3x^2.$$

The $3x^2$ that was factored out is a part of the answer but plays no further part in the factoring process. Now continue to factor $x^2 - 2x - 24$. Find two numbers whose product is $-24$ and whose sum is $-2$. The numbers are $-6$ and 4.

$$3x^2(x^2 - 2x - 24) = 3x^2(x - 6)(x + 4)$$

Therefore, $3x^4 - 6x^3 - 72x^2 = 3x^2(x - 6)(x + 4)$.

▶ **Now Try Exercise 33**

## 3 Factor Trinomials of the Form $ax^2 + bx + c, a \neq 1$, Using Trial and Error

Now we will look at some examples of factoring trinomials of the form

$$ax^2 + bx + c, \quad a \neq 1$$

Two methods of factoring this type of trinomial will be illustrated. The first method, trial and error, involves trying various combinations until the correct combination is found. The second method makes use of factoring by grouping, a procedure that was presented in Section 5.4.

Let's first discuss the trial-and-error method of factoring trinomials. This procedure is sometimes called the FOIL (or Reverse FOIL) method. As an aid in our explanation, we will multiply $(2x + 3)(x + 1)$ using the FOIL method.

Product of first terms

Product of last terms

$$(2x + 3)(x + 1) = \overset{F}{2x(x)} + \overset{O}{2x(1)} + \overset{I}{3(x)} + \overset{L}{3(1)} = 2x^2 + 5x + 3$$

Sum of products of outer and inner terms

Therefore, if you are factoring the trinomial $2x^2 + 5x + 3$, you should realize that the product of the first terms of the factors must be $2x^2$, the product of the last terms must be 3, and the sum of the products of the outer and inner terms must be $5x$.

To factor $2x^2 + 5x + 3$, we begin as shown here.

$$2x^2 + 5x + 3 = (2x \quad\ )(x \quad\ ) \qquad \text{The product of the first terms is } 2x^2.$$

Now we fill in the second terms using positive integers whose product is 3. Only positive integers will be considered since the product of the last terms is positive, and the sum of the products of the outer and inner terms is also positive. The two possibilities are

$$\left.\begin{array}{l}(2x + 1)(x + 3)\\(2x + 3)(x + 1)\end{array}\right\} \quad \text{The product of the last terms is 3.}$$

To determine which factoring is correct, we find the sum of the products of the outer and inner terms. If either has a sum of $5x$, the middle term of the trinomial, that factoring is correct.

$$(2x + 1)(x + 3) = 2x^2 + 6x + x + 3 = 2x^2 + 7x + 3 \quad \text{Wrong middle term}$$

$$(2x + 3)(x + 1) = 2x^2 + 2x + 3x + 3 = 2x^2 + 5x + 3 \quad \text{Correct middle term}$$

Therefore, the factors of $2x^2 + 5x + 3$ are $2x + 3$ and $x + 1$. Thus,

$$2x^2 + 5x + 3 = (2x + 3)(x + 1).$$

Note that if we had begun factoring by writing

$$2x^2 + 5x + 3 = (x \quad\quad)(2x \quad\quad)$$

we could have also obtained the correct factors.

Following are guidelines for the **trial and error** method of factoring a trinomial where $a \neq 1$ and the three terms have no common factors.

**To Factor Trinomials of the Form $ax^2 + bx + c, a \neq 1$, Using Trial and Error**

1. Write all pairs of factors of the coefficient of the squared term, $a$.
2. Write all pairs of factors of the constant, $c$.
3. Try various combinations of these factors until the correct middle term, $bx$, is found.

**EXAMPLE 5** ▸ Factor $3t^2 - 13t + 10$.

*Solution* First we determine that the three terms have no common factor. Next we determine that $a$ is 3 and the only factors of 3 are 1 and 3. Therefore, we write

$$3t^2 - 13t + 10 = (3t \quad\quad)(t \quad\quad)$$

The number 10 has both positive and negative factors. However, since the product of the last terms must be positive $(+10)$, and the sum of the products of the outer and inner terms must be negative $(-13)$, the two factors of 10 must both be negative. (Why?) The negative factors of 10 are $(-1)(-10)$ and $(-2)(-5)$. Below is a list of the possible factors. We look for the factors that give us the correct middle term, $-13t$.

| Possible Factors | Sum of Products of Outer and Inner Terms | |
|---|---|---|
| $(3t - 1)(t - 10)$ | $-31t$ | |
| $(3t - 10)(t - 1)$ | $-13t$ | ← Correct middle term |
| $(3t - 2)(t - 5)$ | $-17t$ | |
| $(3t - 5)(t - 2)$ | $-11t$ | |

Thus, $3t^2 - 13t + 10 = (3t - 10)(t - 1)$.

▸ Now Try Exercise 35

The following Helpful Hint is very important. Study it carefully.

## Helpful Hint

### *Factoring by Trial and Error*

When factoring a trinomial of the form $ax^2 + bx + c$, the sign of the constant term, $c$, is very helpful in finding the solution. If $a > 0$, then:

1. When the constant term, $c$, is positive, and the numerical coefficient of the $x$ term, $b$, is positive, both numerical factors will be positive.

   Example $x^2 + 7x + 12 = (x + 3)(x + 4)$

   (Positive, Positive; Positive, Positive)

2. When $c$ is positive and $b$ is negative, both numerical factors will be negative.

   Example $x^2 - 5x + 6 = (x - 2)(x - 3)$

   (Negative, Positive; Negative, Negative)

   Whenever the constant, $c$, is positive (as in the two examples above) the sign in both factors will be the same as the sign in the $x$-term of the trinomial.

3. When $c$ is negative, one of the numerical factors will be positive and the other will be negative.

   Example $x^2 + x - 6 = (x + 3)(x - 2)$

   (Negative; Positive, Negative)

**EXAMPLE 6** ▸ Factor $8x^2 + 8x - 30$.

*Solution* First we check to see whether the three terms have a common factor. We notice that 2 can be factored out.

$$8x^2 + 8x - 30 = 2(4x^2 + 4x - 15)$$

The factors of 4, the leading coefficient, are $4 \cdot 1$ and $2 \cdot 2$. Therefore, the factoring will be of the form $(4x \quad)(x \quad)$ or $(2x \quad)(2x \quad)$. It makes no difference whether you start with the first set of factors or the last set of factors. We will generally start with the medium-sized factors first, so we will start with $(2x \quad)(2x \quad)$. If using these factors does not give our answer, we will work with the other set of factors. The factors of $-15$ are $(1)(-15)$, $(3)(-5)$, $(5)(-3)$, and $(15)(-1)$. We want our middle term to be $4x$.

| Possible Factors | Sum of Products of Outer and Inner Terms |
|---|---|
| $(2x + 1)(2x - 15)$ | $-28x$ |
| $(2x + 3)(2x - 5)$ | $-4x$ |
| $(2x + 5)(2x - 3)$ | $4x$ |

Since we found the set of factors that gives the correct $x$-term, we can stop. Thus,

$$8x^2 + 8x - 30 = 2(2x + 5)(2x - 3)$$

▸ Now Try Exercise 37

In Example 6, if we compare the second and third set of factors we see they are the same except for the signs of the second terms. Notice that when the signs of the second term in each factor are switched the sum of the products of the outer and inner terms also changes sign.

## USING YOUR GRAPHING CALCULATOR

The graphing calculator can be used to check factoring problems. To check the factoring in Example 6,

$$8x^2 + 8x - 30 = 2(2x + 5)(2x - 3)$$

we let $Y_1 = 8x^2 + 8x - 30$ and $Y_2 = 2(2x + 5)(2x - 3)$. Then we use the TABLE feature to compare results, as in **Figure 5.14**.

| X | Y1 | Y2 |
|---|---|---|
| -3 | 18 | 18 |
| -2 | -14 | -14 |
| -1 | -30 | -30 |
| 0 | -30 | -30 |
| 1 | -14 | -14 |
| 2 | 18 | 18 |
| 3 | 66 | 66 |

X=0

FIGURE 5.14

Since $Y_1$ and $Y_2$ have the same values for each value of $X$, a mistake has not been made. This procedure can only tell you if a mistake has been made; it cannot tell you if you have factored completely. For example, $8x^2 + 8x - 30$ and $(4x + 10)(2x - 3)$ will give the same set of values.

### EXERCISES

*Use your graphing calculator to determine whether each trinomial is factored correctly.*

**1.** $30x^2 + 37x - 84 \stackrel{?}{=} (6x - 7)(5x + 12)$

**2.** $72x^2 + 20x - 35 \stackrel{?}{=} (9x - 5)(8x + 7)$

**EXAMPLE 7 ▸** Factor $6x^2 - 11xy - 10y^2$.

*Solution* The factors of 6 are either $6 \cdot 1$ or $2 \cdot 3$. Therefore, the factors of the trinomial may be of the form $(6x \quad)(x \quad)$ or $(2x \quad)(3x \quad)$. We will begin with the middle-sized factors. We write

$$6x^2 - 11xy - 10y^2 = (2x \quad)(3x \quad)$$

The factors of $-10$ are $(-1)(10)$, $(1)(-10)$, $(-2)(5)$, and $(2)(-5)$. Since there are eight factors of $-10$, there will be eight pairs of possible factors to try. Can you list them? The correct factorization is

$$6x^2 - 11xy - 10y^2 = (2x - 5y)(3x + 2y)$$

▸ **Now Try Exercise 51**

In Example 7, we were fortunate to find the correct factors by using the form $(2x \quad)(3x \quad)$. If we had not found the correct factors using these, we would have tried $(6x \quad)(x \quad)$.

When factoring a trinomial whose leading coefficient is negative, we start by factoring out a negative number. For example,

$$\begin{aligned} -24x^3 - 60x^2 + 36x &= -12x(2x^2 + 5x - 3) && \text{Factor out } -12x. \\ &= -12x(2x - 1)(x + 3) \end{aligned}$$

and

$$\begin{aligned} -3x^2 + 8x + 16 &= -1(3x^2 - 8x - 16) && \text{Factor out } -1. \\ &= -(3x + 4)(x - 4) \end{aligned}$$

**EXAMPLE 8 ▸ Area of Shaded Region** In **Figure 5.15**, find an expression, in factored form, for the area of the shaded region.

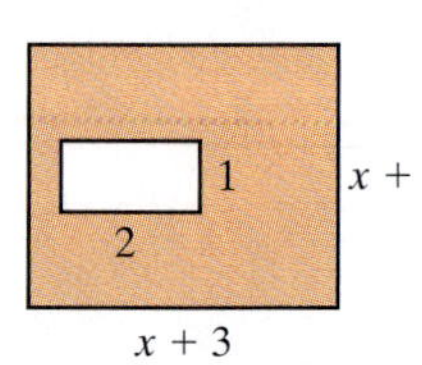


FIGURE 5.15

*Solution* To find the area of the shaded region, we need to subtract the area of the small rectangle from the area of the large rectangle. Recall that the area of a rectangle is length · width.

$$\begin{aligned}\text{Area of large rectangle} &= (x + 3)(x + 2)\\ &= x^2 + 2x + 3x + 6\\ &= x^2 + 5x + 6\\ \text{Area of small rectangle} &= (2)(1) = 2\\ \text{Area of shaded region} &= \text{large area} - \text{small area}\\ &= x^2 + 5x + 6 - 2\\ &= x^2 + 5x + 4 && \textit{Simplify.}\\ &= (x + 4)(x + 1) && \textit{Factor.}\end{aligned}$$

The area of the shaded region is $(x + 4)(x + 1)$.

▸ **Now Try Exercise 89**

## 4 Factor Trinomials of the Form $ax^2 + bx + c, a \neq 1$, Using Grouping

Now we will discuss the **grouping** method of factoring trinomials of the form $ax^2 + bx + c, a \neq 1$.

**To Factor Trinomials of the Form $ax^2 + bx + c, a \neq 1$, Using Grouping**

1. Find two numbers whose product is $a \cdot c$ and whose sum is $b$.
2. Rewrite the middle term, $bx$, using the numbers found in step 1.
3. Factor by grouping.

**EXAMPLE 9** ▸ Factor $2x^2 - 5x - 12$.

*Solution* We see that $a = 2, b = -5$, and $c = -12$. We must find two numbers whose product is $a \cdot c$ or $2(-12) = -24$, and whose sum is $b$, $-5$. The two numbers are $-8$ and 3 because $(-8)(3) = -24$ and $-8 + 3 = -5$. Now rewrite the middle term, $5x$, using $8x$ and $3x$.

$$2x^2 - 5x - 12 = 2x^2 \overbrace{- 8x + 3x}^{-5x} - 12$$

Now factor by grouping as explained in Section 5.4. Factor out $2x$ from the first two terms and 3 from the last two terms.

$$\begin{aligned}2x^2 - 5x + 12 &= 2x^2 - 8x + 3x - 12\\ &= 2x(x - 4) + 3(x - 4)\\ &= (x - 4)(2x + 3) && \textit{Factor out } (x - 4).\end{aligned}$$

Thus $2x^2 - 5x - 12 = (x - 4)(2x + 3)$.

▸ **Now Try Exercise 61**

Note in Example 9 that we wrote $-5x$ as $-8x + 3x$. As we show below, the same factors would be obtained if we wrote $-5x$ as $3x - 8x$. Therefore, it makes no difference which factor is listed first when factoring by grouping. Below we factor $x$ from the first two terms and $-4$ from the last two terms.

$$\begin{aligned}2x^2 - 5x - 12 &= 2x^2 \overbrace{+ 3x - 8x}^{-5x} - 12\\ &= x(2x + 3) - 4(2x + 3)\\ &= (2x + 3)(x - 4) && \textit{Factor out } (2x + 3).\end{aligned}$$

**EXAMPLE 10** ▸ Factor $12a^2 - 19ab + 5b^2$.

*Solution* We must find two numbers whose product is $(12)(5) = 60$ and whose sum is $-19$. Since the product of the numbers is positive and their sum is negative, the two numbers must both be negative. (Why?)

The two numbers are $-15$ and $-4$ because $(-15)(-4) = 60$ and $-15 + (-4) = -19$. Now rewrite the middle term, $-19ab$, using $-15ab$ and $-4ab$. Then factor by grouping.

$$\begin{aligned} 12a^2 - 19ab + 5b^2 &= 12a^2 \overbrace{- 15ab - 4ab}^{-19ab} + 5b^2 \\ &= 3a(4a - 5b) - b(4a - 5b) \\ &= (4a - 5b)(3a - b) \end{aligned}$$

▸ **Now Try Exercise 45**

Try Example 10 again, this time writing $-19ab$ as $-4ab - 15ab$. If you do it correctly, you should get the same factors.

It is important for you to realize that not every trinomial can be factored by the methods presented in this section. In Sections 8.1 and 8.2, we will give some procedures that can be used to factor polynomials that cannot be factored using only the integers (or over the set of integers). A polynomial that cannot be factored (over a specific set of numbers) is called a **prime polynomial**.

**EXAMPLE 11** ▸ Factor $2x^2 + 6x + 5$.

*Solution* When you try to factor this polynomial, you will see that it cannot be factored using either trial and error or grouping. This polynomial is prime over the set of integers.

▸ **Now Try Exercise 47**

## 5 Factor Trinomials Using Substitution

Sometimes a more complicated trinomial can be factored by substituting one variable for another. The next three examples illustrate **factoring using substitution**.

**EXAMPLE 12** ▸ Factor $y^4 - y^2 - 6$.

*Solution* If we can rewrite this expression in the form $ax^2 + bx + c$, it will be easier to factor. Since $(y^2)^2 = y^4$, if we substitute $x$ for $y^2$, the trinomial becomes

$$\begin{aligned} y^4 - y^2 - 6 &= (y^2)^2 - y^2 - 6 \\ &= x^2 - x - 6 \qquad \text{Substitute } x \text{ for } y^2. \end{aligned}$$

Now factor $x^2 - x - 6$.

$$= (x + 2)(x - 3)$$

Finally, substitute $y^2$ in place of $x$ to obtain

$$= (y^2 + 2)(y^2 - 3) \qquad \text{Substitute } y^2 \text{ for } x.$$

Thus, $y^4 - y^2 - 6 = (y^2 + 2)(y^2 - 3)$. Note that $x$ was substituted for $y^2$, and then $y^2$ was substituted back for $x$.

▸ **Now Try Exercise 65**

**EXAMPLE 13** ▸ Factor $3z^4 - 17z^2 - 28$.

*Solution* Let $x = z^2$. Then the trinomial can be written

$$\begin{aligned} 3z^4 - 17z^2 - 28 &= 3(z^2)^2 - 17z^2 - 28 \\ &= 3x^2 - 17x - 28 && \text{Substitute } x \text{ for } z^2. \\ &= (3x + 4)(x - 7) && \text{Factor.} \end{aligned}$$

Now substitute $z^2$ for $x$.

$$= (3z^2 + 4)(z^2 - 7) \qquad \text{Substitute } z^2 \text{ for } x.$$

Thus, $3z^4 - 17z^2 - 28 = (3z^2 + 4)(z^2 - 7)$.

▸ **Now Try Exercise 69**

**EXAMPLE 14** ▸ Factor $2(x + 5)^2 - 5(x + 5) - 12$.

*Solution* We will again use a substitution, as in Examples 12 and 13. By substituting $a = x + 5$ in the equation, we obtain

$$\begin{aligned} &2(x + 5)^2 - 5(x + 5) - 12 \\ &= 2a^2 - 5a - 12 && \text{Substitute } a \text{ for } (x + 5). \end{aligned}$$

Now factor $2a^2 - 5a - 12$.

$$= (2a + 3)(a - 4)$$

Finally, replace $a$ with $x + 5$ to obtain

$$\begin{aligned} &= [2(x + 5) + 3][(x + 5) - 4] && \text{Substitute } (x + 5) \text{ for } a. \\ &= [2x + 10 + 3][x + 1] \\ &= (2x + 13)(x + 1) \end{aligned}$$

Thus, $2(x + 5)^2 - 5(x + 5) - 12 = (2x + 13)(x + 1)$. Note that $a$ was substituted for $x + 5$, and then $x + 5$ was substituted back for $a$.

▸ **Now Try Exercise 73**

In Examples 12 and 13 we used $x$ in our substitution, whereas in Example 14 we used $a$. The letter selected does not affect the final answer.

## EXERCISE SET 5.5

MathXL® MyMathLab

### Concept/Writing Exercises

1. When factoring any trinomial, what should the first step always be?
2. On a test, Tom Phu wrote the following factoring and did not receive full credit. Explain why Tom's factoring is not complete.
$$15x^2 - 21x - 18 = (5x + 3)(3x - 6)$$
3. **a)** Explain the step-by-step procedure to factor $6x^2 + x - 12$.
   **b)** Factor $6x^2 + x - 12$ using the procedure you explained in part **a)**.
4. **a)** Explain the step-by-step procedure to factor $8x^2 - 20x - 12$.
   **b)** Factor $8x^2 - 20x - 12$ using the procedure you explained in part **a)**.
5. Has $2x^2 + 8x + 6 = (x + 3)(2x + 2)$ been factored completely? If not, give the complete factorization. Explain.
6. Has $x^3 - 3x^2 - 10x = (x^2 + 2x)(x - 5)$ been factored completely? If not, give the complete factorization. Explain.
7. Has $3x^3 + 6x^2 - 24x = x(x + 4)(3x - 6)$ been factored completely? If not, give the complete factorization. Explain.
8. Has $x^4 + 11x^3 + 30x^2 = x^2(x + 5)(x + 6)$ been factored completely? If not, give the complete factorization. Explain.

*When factoring a trinomial of the form $ax^2 + bx + c$, what will be the signs between the terms in the binomial factors if:*

**9.** $a > 0, b > 0$, and $c > 0$

**10.** $a > 0, b > 0$, and $c < 0$

**11.** $a > 0, b < 0$, and $c < 0$

**12.** $a > 0, b < 0$, and $c > 0$

## Practice the Skills

*Factor each trinomial completely. If the polynomial is prime, so state.*

**13.** $x^2 + 7x + 12$

**14.** $a^2 - 2a - 15$

**15.** $b^2 + 8b - 9$

**16.** $y^2 - 9y + 20$

**17.** $z^2 + 4z + 4$

**18.** $c^2 - 12c + 36$

**19.** $r^2 + 24r + 144$

**20.** $y^2 - 18y + 81$

**21.** $x^2 + 30x - 64$

**22.** $x^2 + 11x - 210$

**23.** $x^2 - 13x - 30$

**24.** $p^2 - 6p - 19$

**25.** $-a^2 + 18a - 45$

**26.** $-x^2 - 15x - 56$

**27.** $x^2 + xy + 7y^2$

**28.** $a^2 + 7ab + 12b^2$

**29.** $-2m^2 - 14m - 20$

**30.** $-3x^2 - 12x - 9$

**31.** $4r^2 + 12r - 16$

**32.** $b^2 - 12bc - 45c^2$

**33.** $x^3 + 3x^2 - 18x$

**34.** $x^4 + 14x^3 + 33x^2$

**35.** $5a^2 - 8a + 3$

**36.** $4w^2 + 9w + 2$

**37.** $3x^2 - 3x - 6$

**38.** $-3b^2 - 14b + 5$

**39.** $6c^2 - 13c - 63$

**40.** $30z^2 - 71z + 35$

**41.** $8b^2 - 2b - 3$

**42.** $4a^2 + 43a + 30$

**43.** $6c^2 + 11c - 10$

**44.** $5z^2 - 11z + 6$

**45.** $16p^2 - 16pq - 12q^2$

**46.** $6r^4 + 5r^3 - 4r^2$

**47.** $4x^2 + 4xy + 9y^2$

**48.** $6r^2 + 7rs + 8s^2$

**49.** $18a^2 + 18ab - 8b^2$

**50.** $9y^2 - 104y - 48$

**51.** $8x^2 + 30xy - 27y^2$

**52.** $32x^2 - 22xy + 3y^2$

**53.** $100b^2 - 90b + 20$

**54.** $x^5y - 3x^4y - 18x^3y$

**55.** $a^3b^5 - a^2b^5 - 12ab^5$

**56.** $a^3b + 2a^2b - 35ab$

**57.** $3b^4c - 18b^3c^2 + 27b^2c^3$

**58.** $6p^3q^2 - 24p^2q^3 - 30pq^4$

**59.** $8m^8n^3 + 4m^7n^4 - 24m^6n^5$

**60.** $18x^2 + 9x - 20$

**61.** $30x^2 - x - 20$

**62.** $36x^2 - 23x - 8$

**63.** $8x^4y^5 + 24x^3y^5 - 32x^2y^5$

**64.** $8b^3c^2 + 28b^2c^3 + 12bc^4$

*Factor each trinomial completely.*

**65.** $x^4 + x^2 - 6$

**66.** $y^4 + y^2 - 12$

**67.** $b^4 + 9b^2 + 20$

**68.** $c^4 + 8c^2 + 12$

**69.** $6a^4 + 5a^2 - 25$

**70.** $(2x + 1)^2 + 2(2x + 1) - 15$

**71.** $4(x + 1)^2 + 8(x + 1) + 3$

**72.** $(2y + 3)^2 - (2y + 3) - 6$

**73.** $6(a + 2)^2 - 7(a + 2) - 5$

**74.** $6(p - 5)^2 + 11(p - 5) + 3$

**75.** $x^2y^2 + 9xy + 14$

**76.** $a^2b^2 + 6ab - 27$

**77.** $2x^2y^2 - 9xy - 11$

**78.** $3b^2c^2 - bc - 2$

**79.** $2y^2(2 - y) - 7y(2 - y) + 5(2 - y)$

**80.** $2y^2(y + 3) + 13y(y + 3) + 15(y + 3)$

**81.** $2p^2(p - 4) + 7p(p - 4) + 6(p - 4)$

**82.** $3x^2(x - 1) + 5x(x - 1) - 2(x - 1)$

**83.** $a^6 - 7a^3 - 30$

**84.** $2y^6 - 9y^3 - 5$

**85.** $x^2(x + 5) + 3x(x + 5) + 2(x + 5)$

**86.** $x^2(x + 6) - x(x + 6) - 30(x + 6)$

**87.** $5a^5b^2 - 8a^4b^3 + 3a^3b^4$

**88.** $2x^4y^6 + 3x^3y^5 - 9x^2y^4$

## Problem Solving

**Area** *In Exercises 89–92, find an expression, in factored form, for the area of the shaded region. See Example 8.*

**89.**

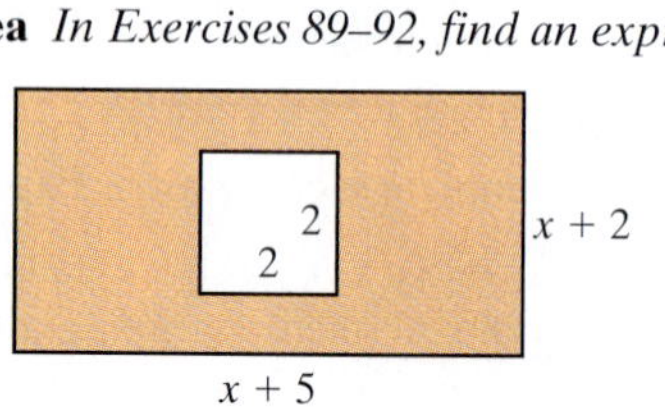


**90.**

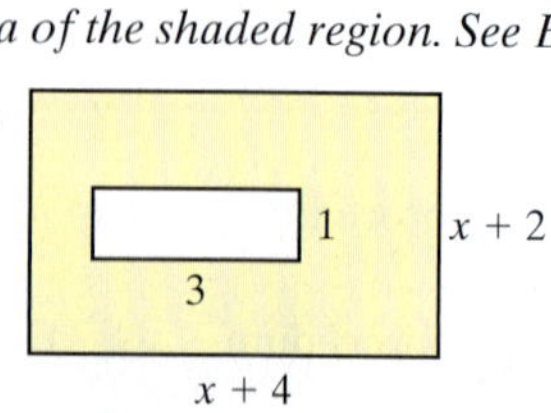

**91.**

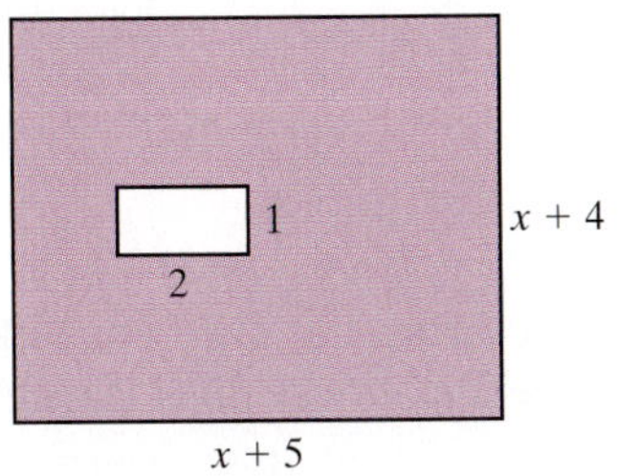


**92.**

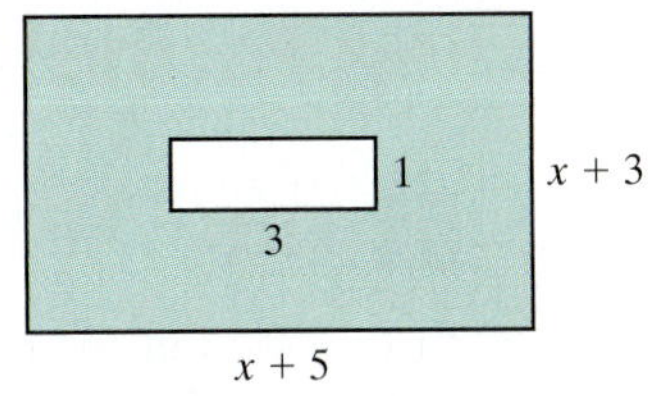


**93.** If the factors of a polynomial are $(2x + 3y)$ and $(x - 4y)$, find the polynomial. Explain how you determined your answer.

**94.** If the factors of a polynomial are 3, $(4x + 5)$, and $(2x + 3)$, find the polynomial. Explain how you determined your answer.

**95.** If we know that one factor of the polynomial $x^2 + 4x - 21$ is $x - 3$, how can we find the other factor? Find the other factor.

**96.** If we know that one factor of the polynomial $x^2 - xy - 6y^2$ is $x - 3y$, how can we find the other factor? Find the other factor.

**97. a)** Which of the following do you think would be more difficult to factor by trial and error? Explain your answer.

$$30x^2 + 23x - 40 \quad \text{or} \quad 49x^2 - 98x + 13$$

**b)** Factor both trinomials.

**98. a)** Which of the following do you think would be more difficult to factor by trial and error? Explain your answer.

$$48x^2 + 26x - 35 \quad \text{or} \quad 35x^2 - 808x + 69$$

**b)** Factor both trinomials.

**99.** Find all integer values of $b$ for which $2x + bx - 5$ is factorable.

**100.** Find all integer values of $b$ for which $3x^2 + bx - 7$ is factorable.

**101.** If $x^2 + bx + 5$ is factorable, what are the only two possible values of $b$? Explain.

**102.** If $x^2 + bx + c$ is factorable and $c$ is a prime number, what are the only two possible factors of $b$? Explain.

*Consider the trinomial $ax^2 + bx + c$. We will learn later in the course that if the expression $b^2 - 4ac$, called the* **discriminant**, *is not a perfect square, the trinomial cannot be factored over the set of integers.* **Perfect squares** *are 1, 4, 9, 16, 25, 49, and so on. The square root of a perfect square is a whole number. For Exercises 103–106,* **a)** *find the value of $b^2 - 4ac$.* **b)** *If $b^2 - 4ac$ is a perfect square, factor the polynomial; if $b^2 - 4ac$ is not a perfect square, indicate that the polynomial cannot be factored.*

**103.** $x^2 - 8x + 15$

**104.** $6y^2 - 5y - 6$

**105.** $x^2 - 4x + 6$

**106.** $3t^2 - 6t + 2$

**107.** Construct a trinomial of the form $x^2 + (c + 1)x + c$, where $c$ is a real number, that is factorable.

**108.** Construct a trinomial of the form $x^2 - (c + 1)x + c$, where $c$ is a real number, that is factorable.

*In Exercises 109–114, factor completely. Assume that the variables in the exponents represent positive integers.*

**109.** $4a^{2n} - 4a^n - 15$

**110.** $a^2(a + b) - 2ab(a + b) - 3b^2(a + b)$

**111.** $x^2(x + y)^2 - 7xy(x + y)^2 + 12y^2(x + y)^2$

**112.** $3m^2(m - 2n) - 4mn(m - 2n) - 4n^2(m - 2n)$

**113.** $x^{2n} + 3x^n - 10$

**114.** $9r^{4y} + 3r^{2y} - 2$

**115.** Consider $x^2 + 2x - 8 = (x + 4)(x - 2)$.

**a)** Explain how you can check this factoring using graphs on your graphing calculator.

**b)** Check the factoring as you explained in part **a)** to see whether it is correct.

**116.** Consider $6x^3 - 11x^2 - 10x = x(2x - 5)(3x + 2)$.

**a)** Explain how you can check this factoring using the TABLE feature of a graphing calculator.

**b)** Check the factoring as you explained in part **a)** to see whether it is correct.

## Cumulative Review Exercises

[2.2] **117.** Solve $F = \frac{9}{5}C + 32$ for $C$.

[3.3] **118.** Graph $y = -3x + 4$.

[4.5] **119.** Evaluate the determinant $\begin{vmatrix} 3 & -2 & -1 \\ 2 & 3 & -2 \\ 1 & -4 & 1 \end{vmatrix}$.

[5.2] **120.** Multiply $[(x + y) + 6]^2$.

[5.3] **121.** Factor $2x^3 + 4x^2 - 5x - 10$.

# 5.6 Special Factoring Formulas

1 Factor the difference of two squares.

2 Factor perfect square trinomials.

3 Factor the sum and difference of two cubes.

## 1 Factor the Difference of Two Squares

In this section we present some special formulas for factoring the difference of two squares, perfect square trinomials, and the sum and difference of two cubes. It will be to your advantage to memorize these formulas.

The expression $x^2 - 9$ is an example of the difference of two squares.

$$x^2 - 9 = (x)^2 - (3)^2$$

To factor the difference of two squares, it is convenient to use the **difference of two squares formula**. This formula was first presented in Section 5.2.

**Difference of Two Squares**

$$a^2 - b^2 = (a + b)(a - b)$$

**EXAMPLE 1** ▸ Factor the following difference of squares.

**a)** $x^2 - 16$ **b)** $25x^2 - 36y^2$

*Solution* Rewrite each expression as a difference of two squares. Then use the formula.

**a)** $x^2 - 16 = (x)^2 - (4)^2$
$= (x + 4)(x - 4)$

**b)** $25x^2 - 36y^2 = (5x)^2 - (6y)^2$
$= (5x + 6y)(5x - 6y)$

▸ Now Try Exercise 13

**EXAMPLE 2** ▸ Factor the following differences of squares.

**a)** $x^6 - y^4$ **b)** $2z^4 - 162x^6$

*Solution* Rewrite each expression as a difference of two squares. Then use the formula.

**a)** $x^6 - y^4 = (x^3)^2 - (y^2)^2$
$= (x^3 + y^2)(x^3 - y^2)$

**b)** $2z^4 - 162x^6 = 2(z^4 - 81x^6)$
$= 2[(z^2)^2 - (9x^3)^2]$
$= 2(z^2 + 9x^3)(z^2 - 9x^3)$

▸ Now Try Exercise 21

**EXAMPLE 3** ▸ Factor $x^4 - 81y^4$.

*Solution* $x^4 - 81y^4 = (x^2)^2 - (9y^2)^2$
$= (x^2 + 9y^2)(x^2 - 9y^2)$

Note that $(x^2 - 9y^2)$ is also a difference of two squares. We use the difference of two squares formula a second time to obtain

$= (x^2 + 9y^2)[(x)^2 - (3y)^2]$
$= (x^2 + 9y^2)(x + 3y)(x - 3y)$

▸ Now Try Exercise 69

**EXAMPLE 4** ▸ Factor $(x - 5)^2 - 4$ using the formula for the difference of two squares.

*Solution* First we express $(x - 5)^2 - 4$ as a difference of two squares.

$$\begin{aligned}(x - 5)^2 - 4 &= (x - 5)^2 - 2^2 \\ &= [(x - 5) + 2][(x - 5) - 2] \\ &= (x - 3)(x - 7)\end{aligned}$$

▸ Now Try Exercise 25

Note: **It is not possible to factor the sum of two squares of the form $a^2 + b^2$ over the set of real numbers.**

For example, it is not possible to factor $x^2 + 4$ since $x^2 + 4 = x^2 + 2^2$, which is a sum of two squares.

## 2 Factor Perfect Square Trinomials

In Section 5.2 we saw that

$$(a + b)^2 = a^2 + 2ab + b^2$$
$$(a - b)^2 = a^2 - 2ab + b^2$$

If we reverse the left and right sides of these two formulas, we obtain two **special factoring formulas**.

**Perfect Square Trinomials**

$$a^2 + 2ab + b^2 = (a + b)^2$$
$$a^2 - 2ab + b^2 = (a - b)^2$$

These two trinomials are called **perfect square trinomials** since each is the square of a binomial. *To be a perfect square trinomial, the first and last terms must be the squares of some expression and the middle term must be twice the product of the first and last terms.* When you are given a trinomial to factor, determine whether it is a perfect square trinomial before you attempt to factor it by the procedures explained in Section 5.5. If it is a perfect square trinomial, you can factor it using the formulas given above.

### Examples of Perfect Square Trinomials

$$y^2 + 6y + 9 \quad \text{or} \quad y^2 + 2(y)(3) + 3^2$$
$$9a^2b^2 - 24ab + 16 \quad \text{or} \quad (3ab)^2 - 2(3ab)(4) + 4^2$$
$$(r + s)^2 + 10(r + s) + 25 \quad \text{or} \quad (r + s)^2 + 2(r + s)(5) + 5^2$$

Now let's factor some perfect square trinomials.

**EXAMPLE 5** ▸ Factor $x^2 - 8x + 16$.

*Solution* Since the first term, $x^2$, and the last term, 16, or $4^2$, are squares, this trinomial might be a perfect square trinomial. To determine whether it is, take twice the product of $x$ and 4 to see if you obtain $8x$.

$$2(x)(4) = 8x$$

Since $8x$ is the middle term and since the sign of the middle term is negative, factor as follows:

$$x^2 - 8x + 16 = (x - 4)^2$$

▸ Now Try Exercise 29

**EXAMPLE 6** ▸ Factor $9x^4 - 12x^2 + 4$.

*Solution* The first term is a square, $(3x^2)^2$, as is the last term, $2^2$. Since $2(3x^2)(2) = 12x^2$, we factor as follows:

$$9x^4 - 12x^2 + 4 = (3x^2 - 2)^2$$

▸ **Now Try Exercise 37**

**EXAMPLE 7** ▸ Factor $(a + b)^2 + 12(a + b) + 36$.

*Solution* The first term, $(a + b)^2$, is a square. The last term, 36 or $6^2$, is a square. The middle term is $2(a + b)(6) = 12(a + b)$. Therefore, this is a perfect square trinomial. Thus,

$$(a + b)^2 + 12(a + b) + 36 = [(a + b) + 6]^2 = (a + b + 6)^2$$

▸ **Now Try Exercise 39**

**EXAMPLE 8** ▸ Factor $x^2 - 6x + 9 - y^2$.

*Solution* Since $x^2 - 6x + 9$ is a perfect square trinomial, which can be expressed as $(x - 3)^2$, we write

$$(x - 3)^2 - y^2$$

Now $(x - 3)^2 - y^2$ is a difference of two squares; therefore

$$\begin{aligned}(x - 3)^2 - y^2 &= [(x - 3) + y][(x - 3) - y] \\ &= (x - 3 + y)(x - 3 - y)\end{aligned}$$

Thus, $x^2 - 6x + 9 - y^2 = (x - 3 + y)(x - 3 - y)$.

▸ **Now Try Exercise 45**

The polynomial in Example 8 has four terms. In Section 5.4 we learned to factor polynomials with four terms by grouping. If you study Example 8, you will see that no matter how you arrange the four terms they cannot be arranged so that the first two terms have a common factor and the last two terms have a common factor. Whenever a polynomial of four terms cannot be factored by grouping, try to rewrite three of the terms as the square of a binomial and then factor using the difference of two squares formula.

**EXAMPLE 9** ▸ Factor $4a^2 + 12ab + 9b^2 - 25$.

*Solution* We first notice that this polynomial of four terms cannot be factored by grouping. We next look to see if three terms of the polynomial can be expressed as the square of a binomial. Since this can be done, we write the three terms as the square of a binomial. We complete our factoring using the difference of two squares formula.

$$\begin{aligned}4a^2 + 12ab + 9b^2 - 25 &= (2a + 3b)^2 - 5^2 \\ &= [(2a + 3b) + 5][(2a + 3b) - 5] \\ &= (2a + 3b + 5)(2a + 3b - 5)\end{aligned}$$

▸ **Now Try Exercise 47**

## 3 Factor the Sum and Difference of Two Cubes

Earlier in this section we factored the difference of two squares. Now we will factor the sum and difference of two cubes. Consider the product of $(a + b)(a^2 - ab + b^2)$.

$$\begin{array}{rrrr} & a^2 & - ab & + b^2 \\ & & a & + b \\ \hline & a^2b & - ab^2 & + b^3 \\ a^3 & - a^2b & + ab^2 & \\ \hline a^3 & & & + b^3 \end{array}$$

Thus, $a^3 + b^3 = (a + b)(a^2 - ab + b^2)$. Using multiplication, we can also show that $a^3 - b^3 = (a - b)(a^2 + ab + b^2)$. Formulas for factoring **the sum and the difference of two cubes** appear in the following boxes.

**Sum of Two Cubes**

$$a^3 + b^3 = (a + b)(a^2 - ab + b^2)$$

**Difference of Two Cubes**

$$a^3 - b^3 = (a - b)(a^2 + ab + b^2)$$

**EXAMPLE 10** ▸ Factor the following sum of cubes $x^3 + 64$.

*Solution* Rewrite $x^3 + 64$ as a sum of two cubes, $x^3 + 4^3$. Let $x$ correspond to $a$ and 4 to $b$. Then factor using the sum of two cubes formula.

$$\begin{aligned} a^3 + b^3 &= (a + b)(a^2 - a\ b + b^2) \\ &\downarrow \\ x^3 + 4^3 &= (x + 4)[x^2 - x(4) + 4^2] \\ &= (x + 4)(x^2 - 4x + 16) \end{aligned}$$

Thus, $x^3 + 64 = (x + 4)(x^2 - 4x + 16)$.

▸ Now Try Exercise 51

**EXAMPLE 11** ▸ Factor the following difference of cubes $27x^3 - 8y^6$.

*Solution* We first observe that $27x^3$ and $8y^6$ have no common factors other than 1. Since we can express both $27x^3$ and $8y^6$ as cubes, we can factor using the difference of two cubes formula.

$$\begin{aligned} 27x^3 - 8y^6 &= (3x)^3 - (2y^2)^3 \\ &= (3x - 2y^2)[(3x)^2 + (3x)(2y^2) + (2y^2)^2] \\ &= (3x - 2y^2)(9x^2 + 6xy^2 + 4y^4) \end{aligned}$$

Thus, $27x^3 - 8y^6 = (3x - 2y^2)(9x^2 + 6xy^2 + 4y^4)$.

▸ Now Try Exercise 57

**EXAMPLE 12** ▸ Factor $8y^3 - 64x^6$.

*Solution* First factor out 8, which is common to both terms.

$$8y^3 - 64x^6 = 8(y^3 - 8x^6)$$

Next factor $y^3 - 8x^6$ by writing it as a difference of two cubes.

$$\begin{aligned} 8(y^3 - 8x^6) &= 8[(y)^3 - (2x^2)^3] \\ &= 8(y - 2x^2)[y^2 + y(2x^2) + (2x^2)^2] \\ &= 8(y - 2x^2)(y^2 + 2x^2y + 4x^4) \end{aligned}$$

Thus, $8y^3 - 64x^6 = 8(y - 2x^2)(y^2 + 2x^2y + 4x^4)$.

▸ Now Try Exercise 59

**EXAMPLE 13** ▸ Factor $(x-2)^3+125$.

*Solution* Write $(x-2)^3+125$ as a sum of two cubes, then use the sum of two cubes formula to factor.

$$\begin{aligned}(x-2)^3+(5)^3 &= [(x-2)+5][(x-2)^2-(x-2)(5)+(5)^2]\\ &= (x-2+5)(x^2-4x+4-5x+10+25)\\ &= (x+3)(x^2-9x+39)\end{aligned}$$

▸ **Now Try Exercise 65**

### Helpful Hint

The square of a binomial has a 2 as part of the middle term of the trinomial.

$$(a+b)^2=a^2+2ab+b^2$$
$$(a-b)^2=a^2-2ab+b^2$$

The sum or the difference of two cubes has a factor similar to the trinomial in the square of the binomial. However, the middle term does not contain a 2.

$$a^3+b^3=(a+b)(a^2-ab+b^2)$$
$$a^3-b^3=(a-b)(a^2+\underbrace{ab}_{\text{not }2ab}+b^2)$$

**EXAMPLE 14** ▸ **Volume** Using the cubes in **Figure 5.16**, find an expression, in factored form, for the difference of the volumes.

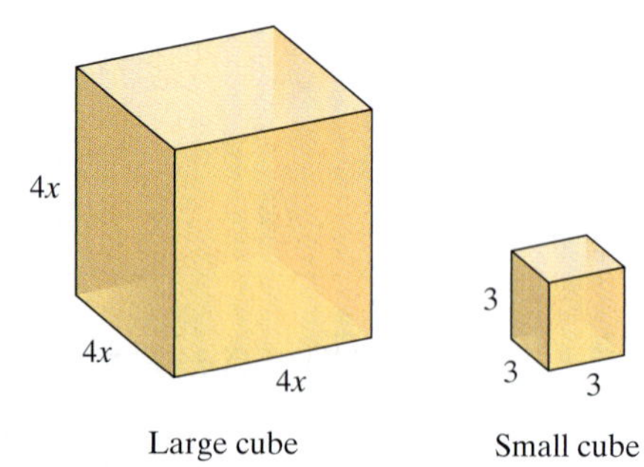


FIGURE 5.16

*Solution* To find the difference of the volumes, subtract the volume of the small cube from the volume of the large cube.

$$\begin{aligned}\text{Volume of large cube} &= (4x)^3\\ \text{Volume of small cube} &= 3^3\\ \text{Difference of the volumes} &= (4x)^3-3^3 && \textit{Subtract volumes.}\\ &= (4x-3)[(4x)^2+(4x)3+3^2] && \textit{Factor.}\\ &= (4x-3)(16x^2+12x+9) && \textit{Simplify.}\end{aligned}$$

The difference of the volumes for the two cubes is $(4x-3)(16x^2+12x+9)$.

▸ **Now Try Exercise 87**

## EXERCISE SET 5.6

MathXL® MyMathLab

### Concept/Writing Exercises

1. a) Explain how to factor the difference of two squares.

   b) Factor $x^2-16$ using the procedure you explained in part **a)**.

2. Explain why a sum of two squares, $a^2+b^2$, cannot be factored over the set of real numbers.

3. Explain how to determine whether a trinomial is a perfect square trinomial.

4. a) Explain how to factor a perfect square trinomial.
   b) Factor $x^2 + 12x + 36$ using the procedure you explained in part **a)**.
5. Give the formula for factoring the sum of two cubes.
6. Give the formula for factoring the difference of two cubes.
7. Is $x^2 + 14x - 49 = (x + 7)(x - 7)$ factored correctly? Explain.
8. Is $x^2 + 14x + 49 = (x + 7)^2$ factored correctly? Explain.
9. Is $x^2 - 81 = (x - 9)^2$ factored correctly? Explain.
10. Is $x^2 - 64 = (x + 8)(x - 8)$ factored correctly? Explain.

## Practice the Skills

*Use the difference of two squares formula or the perfect square trinomial formula to factor each polynomial.*

11. $x^2 - 81$
12. $x^2 - 25$
13. $a^2 - 100$
14. $1 - 9x^2$
15. $1 - 49b^2$
16. $x^2 - 81z^2$
17. $25 - 16y^4$
18. $49 - 144b^4$
19. $\frac{1}{100} - y^2$
20. $\frac{1}{25} - z^2$
21. $x^2y^2 - 121c^2$
22. $5a^2c^2 - 20x^2y^2$
23. $0.04x^2 - 0.09$
24. $0.16p^2 - 0.81q^2$
25. $36 - (x - 6)^2$
26. $144 - (a + b)^2$
27. $a^2 - (3b + 2)^2$
28. $(2c + 3)^2 - 9$
29. $x^2 + 10x + 25$
30. $b^2 - 18b + 81$
31. $49 - 14t + t^2$
32. $4 + 4a + a^2$
33. $36p^2q^2 + 12pq + 1$
34. $4x^2 - 20xy + 25y^2$
35. $0.81x^2 - 0.36x + 0.04$
36. $0.25x^2 - 0.40x + 0.16$
37. $y^4 + 4y^2 + 4$
38. $b^4 - 16b^2 + 64$
39. $(a + b)^2 + 6(a + b) + 9$
40. $(x + y)^2 + 2(x + y) + 1$
41. $(y - 3)^2 + 8(y - 3) + 16$
42. $a^4 - 2a^2b^2 + b^4$
43. $x^2 + 6x + 9 - y^2$
44. $p^2 + 2pq + q^2 - 16r^2$
45. $25 - (x^2 + 4x + 4)$
46. $49 - (c^2 - 8c + 16)$
47. $9a^2 - 12ab + 4b^2 - 9$
48. $(4a - 3b)^2 - (2a + 5b)^2$
49. $y^4 - 6y^2 + 9$
50. $z^6 + 14z^3 + 49$

*Factor using the sum or difference of two cubes formula.*

51. $a^3 + 125$
52. $x^3 - 27$
53. $64 - a^3$
54. $8 - b^3$
55. $p^3 - 27a^3$
56. $w^3 - 216$
57. $27y^3 - 8x^3$
58. $6x^3 + 48y^3$
59. $16a^3 - 54b^3$
60. $2b^3 - 250c^3$
61. $x^6 + y^9$
62. $16x^6 - 250y^3$
63. $(x + 1)^3 + 1$
64. $(a - 3)^3 + 8$
65. $(a - b)^3 - 27$
66. $(2x + y)^3 - 64$
67. $b^3 - (b + 3)^3$
68. $(m - n)^3 - (m + n)^3$

*Factor using a special factoring formula.*

69. $a^4 - 4b^4$
70. $121y^4 - 49x^2$
71. $49 - 64x^2y^2$
72. $25y^2 - 81x^2$
73. $(x + y)^2 - 16$
74. $25x^4 - 81y^6$
75. $x^3 - 64$
76. $3a^2 - 36a + 108$
77. $9x^2y^2 + 24xy + 16$
78. $a^4 + 12a^2 + 36$
79. $a^4 + 2a^2b^2 + b^4$
80. $8y^3 - 125x^6$
81. $x^2 - 2x + 1 - y^2$
82. $16x^2 - 8xy + y^2 - 4$
83. $(x + y)^3 + 1$
84. $4r^2 + 4rs + s^2 - 9$
85. $(m + n)^2 - (2m - n)^2$
86. $(r + p)^3 + (r - p)^3$

## Problem Solving

**Volume** *In Exercises 87–90, find an expression, in factored form, for the difference of the volumes of the two cubes. See Example 14.*

87.

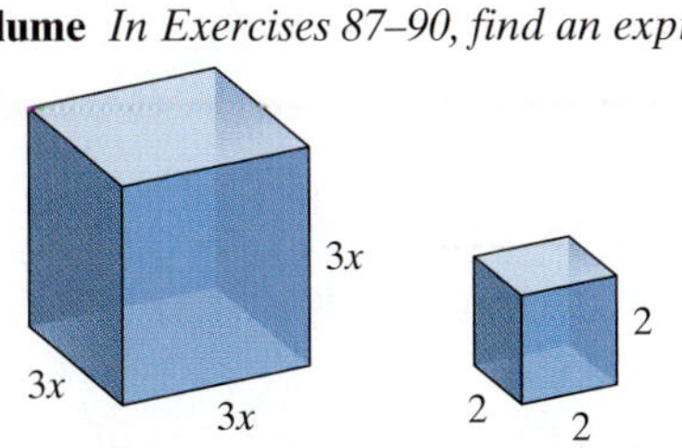


88.

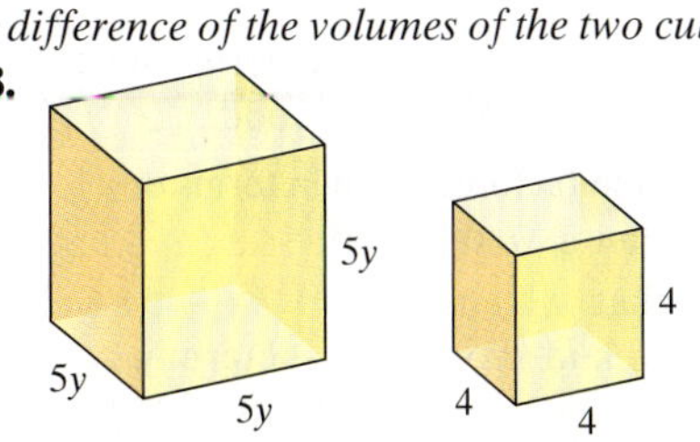

**89.**

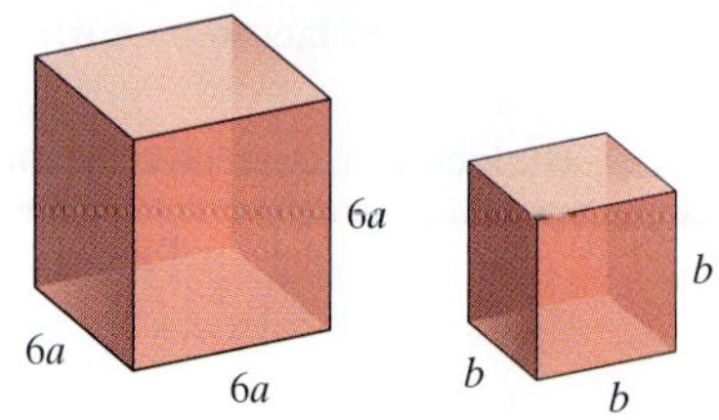


**90.**

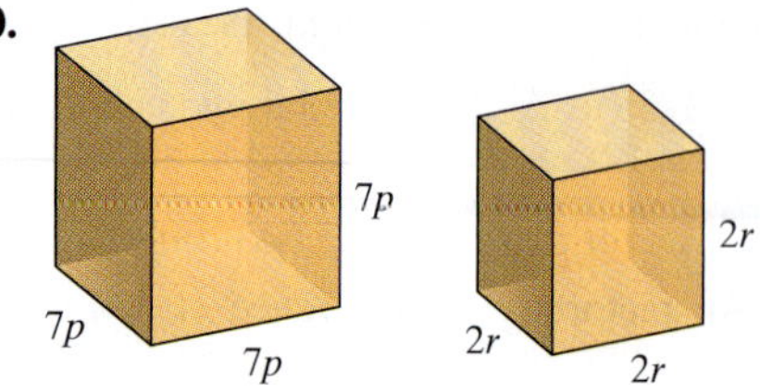


**Volume** *In Exercises 91 and 92, find an expression, in factored form, for the sum of the volumes of the two cubes.*

**91.**

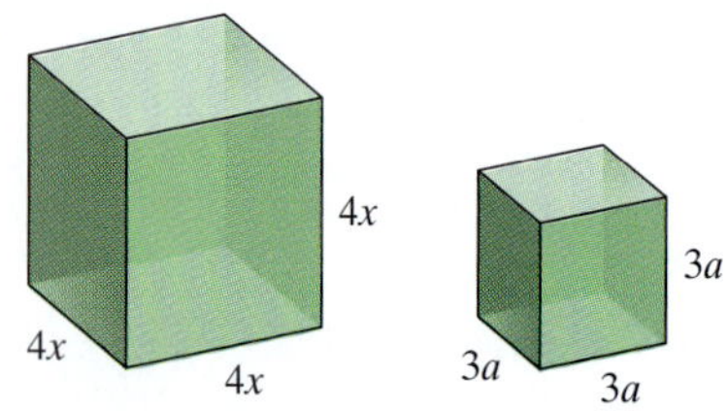


**92.**

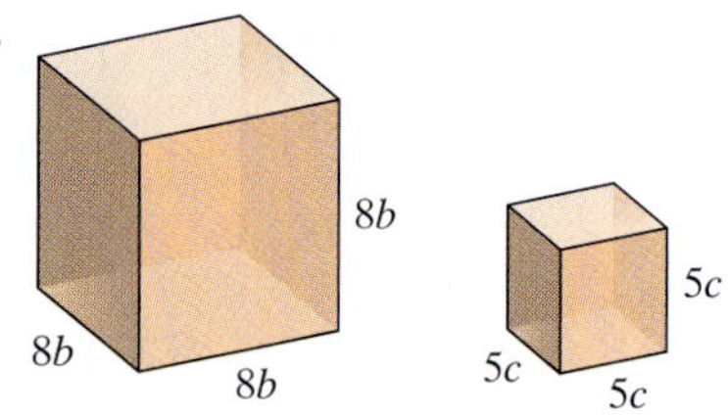


**Area or Volume** *In Exercises 93–97,* **a)** *find the area or volume of the shaded figure by subtracting the smaller area or volume from the larger. The formula to find the area or volume is given under the figure.* **b)** *Write the expression obtained in part* **a)** *in factored form. Part of the GCF in Exercises 94, 96, and 97 is* $\pi$.

**93.** Squares

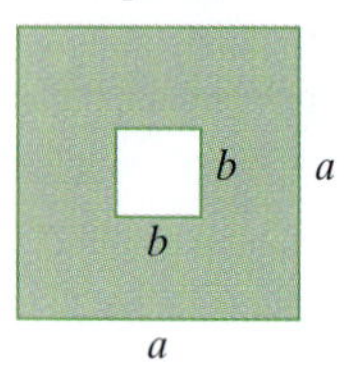


$A = s^2$

**94.** Circles

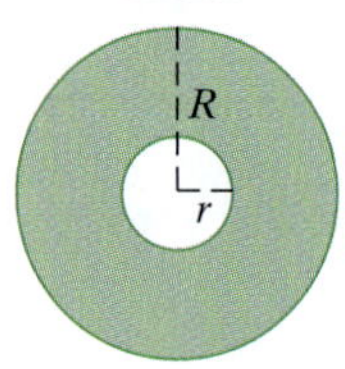


$A = \pi r^2$

**95.** Rectangular solid

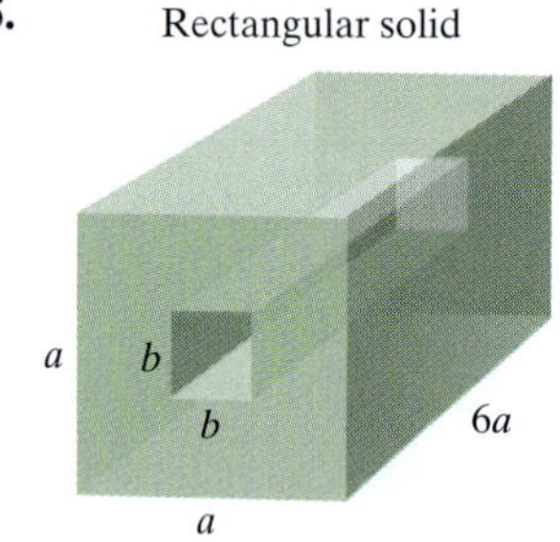


$V = lwh$

**96.** Cylinder

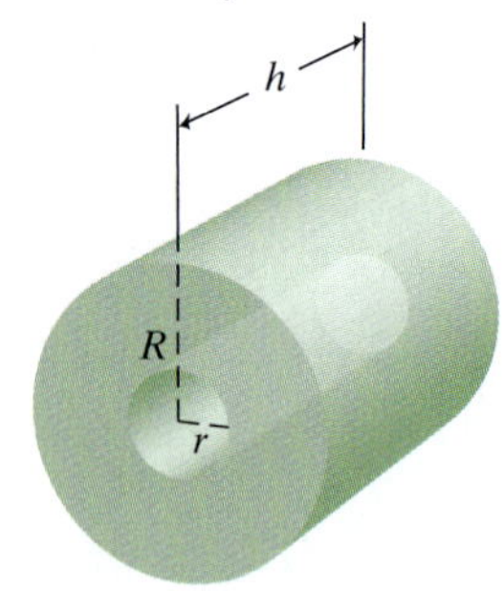


$V = \pi r^2 h$

**97.** Sphere

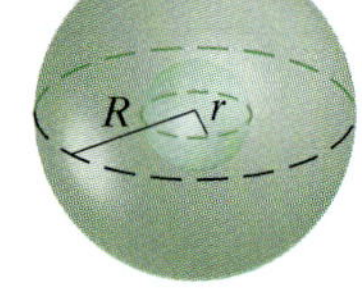


$V = \frac{4}{3}\pi r^3$

**98. Area and Volume** A circular hole is cut from a cube of wood, as shown in the figure.

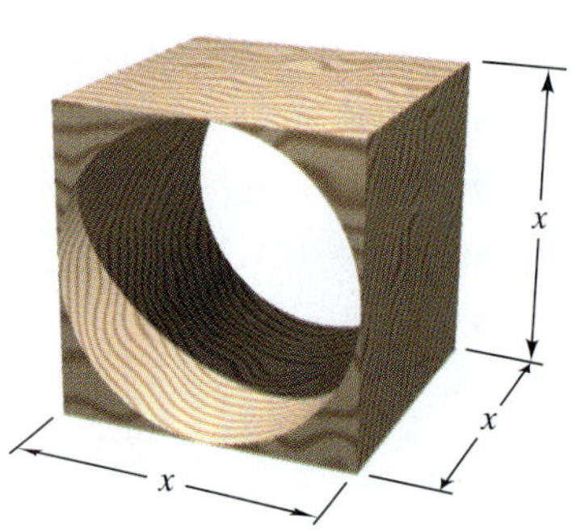


**a)** Write an expression in factored form in terms of $x$ for the cross-sectional area of the remaining wood.

**b)** Write an expression in factored form in terms of $x$ for the volume of the remaining wood.

**99.** Find two values of $b$ that will make $4x^2 + bx + 9$ a perfect square trinomial. Explain how you determined your answer.

**100.** Find two values of $c$ that will make $16x^2 + cx + 4$ a perfect square trinomial. Explain how you determined your answer.

**101.** Find the value of $c$ that will make $25x^2 + 20x + c$ a perfect square trinomial. Explain how you determined your answer.

**102.** Find the value of $d$ that will make $49x^2 - 42x + d$ a perfect square trinomial. Explain how you determined your answer.

**103. Area** A formula for the area of a square is $A = s^2$, where $s$ is a side. Suppose the area of a square is as given below,

$A(x) = 25x^2 - 30x + 9$

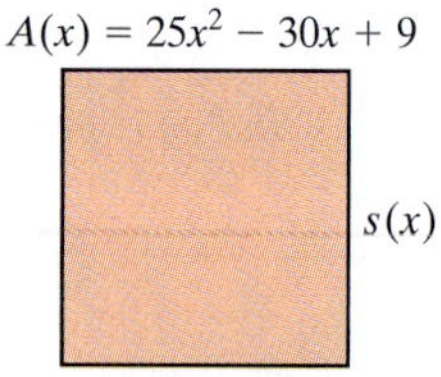


**a)** explain how to find the length of side $x$, $s(x)$,

**b)** find $s(x)$,

**c)** find $s(2)$.

**104. Area** The formula for the area of a circle is $A = \pi r^2$, where $r$ is the radius. Suppose the area of a circle is as given below,

$$A(x) = 9\pi x^2 + 12\pi x + 4\pi$$

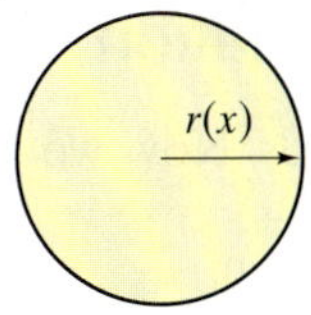


**a)** explain how to find the radius, $r(x)$,
**b)** find $r(x)$,
**c)** find $r(4)$.

**105.** Factor $x^4 + 64$ by writing the expression as $(x^4 + 16x^2 + 64) - 16x^2$, which is a difference of two squares.

**106.** Factor $x^4 + 4$ by adding and subtracting $4x^2$. (See Exercise 105.)

**107.** If $P(x) = x^2$, use the difference of two squares to simplify $P(a + h) - P(a)$.

**108.** If $P(x) = x^2$, use the difference of two squares to simplify $P(a + 1) - P(a)$.

**109. Sum of Areas** The figure shows how we *complete the square*. The sum of the areas of the three parts of the square that are shaded in blue and purple is

$$x^2 + 4x + 4x \quad \text{or} \quad x^2 + 8x$$

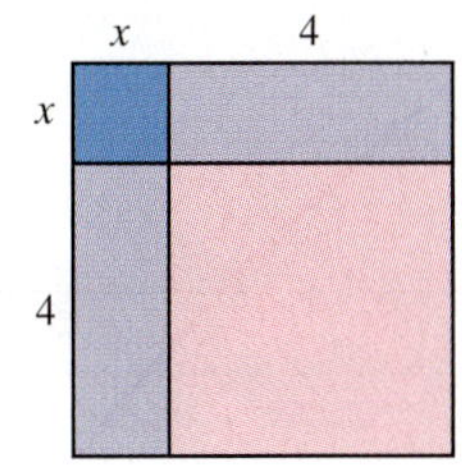


**a)** Find the area of the fourth part (in pink) to complete the square.
**b)** Find the sum of the areas of the four parts of the square.
**c)** This process has resulted in a perfect square trinomial in part **b)**. Write this perfect square trinomial as the square of a binomial.

**110.** Factor $(m - n)^3 - (9 - n)^3$.

*Factor completely.*

**111.** $64x^{4a} - 9y^{6a}$

**112.** $16p^{8w} - 49p^{6w}$

**113.** $a^{2n} - 16a^n + 64$

**114.** $144r^{8k} + 48r^{4k} + 4$

**115.** $x^{3n} - 8$

**116.** $27x^{3m} + 64x^{6m}$

*In Exercises 117 and 118, use your graphing calculator to check the factoring. Indicate if the factoring is correct or not correct. Explain your answers.*

**117.** $2x^2 - 18 \stackrel{?}{=} 2(x + 3)(x - 3)$

**118.** $8x^3 + 27 \stackrel{?}{=} 2x(4x^2 + 5x + 9)$

## Challenge Problem

**119.** The expression $x^6 - 1$ can be factored using either the difference of two squares or the difference of two cubes. At first the factors do not appear the same. But with a little algebraic manipulation they can be shown to be equal. Factor $x^6 - 1$ using **a)** the difference of two squares and **b)** the difference of two cubes. **c)** Show that these two answers are equal by factoring the answer obtained in part **a)** completely. Then multiply the two binomials together and the two trinomials together.

## Group Activity

*Discuss and answer Exercise 120 as a group.*

**120.** Later in the book we will need to construct perfect square trinomials. Examine some perfect square trinomials with a leading coefficient of 1.

**a)** Explain how $b$ and $c$ are related if the trinomial $x^2 + bx + c$ is a perfect square trinomial.
**b)** Construct a perfect square trinomial if the first two terms are $x^2 + 6x$.
**c)** Construct a perfect square trinomial if the first two terms are $x^2 - 10x$.
**d)** Construct a perfect square trinomial if the first two terms are $x^2 - 14x$.

## Cumulative Review Exercises

[2.1] **121.** Simplify $-2[3x - (2y - 1) - 5x] + 3y$.

[3.6] **122.** If $f(x) = x^2 - 3x + 6$ and $g(x) = 5x - 2$, find $(g - f)(-1)$.

[4.4] **123. Angles** A right angle is divided into three smaller angles. The largest of the three angles is twice the smallest. The remaining angle is 10° greater than the smallest angle. Find the measure of each angle.

[5.4] **124.** Factor out the greatest common factor of $45y^{12} + 60y^{10}$.

**125.** Factor $12x^2 - 9xy + 4xy - 3y^2$.

# 5.7 A General Review of Factoring

**1** Factor polynomials using a combination of techniques.

## 1 Factor Polynomials Using a Combination of Techniques

We have presented a number of factoring methods. Now we will combine problems and techniques from the previous sections.

A general procedure to factor any polynomial follows.

**To Factor a Polynomial**

1. Determine whether all the terms in the polynomial have a greatest common factor other than 1. If so, factor out the GCF.
2. If the polynomial has two terms, determine whether it is a difference of two squares or a sum or difference of two cubes. If so, factor using the appropriate formula.
3. If the polynomial has three terms, determine whether it is a perfect square trinomial. If so, factor accordingly. If it is not, factor the trinomial using trial and error, grouping, or substitution as explained in Section 5.5.
4. If the polynomial has more than three terms, try factoring by grouping. If that does not work, see if three of the terms are the square of a binomial.
5. As a final step, examine your factored polynomial to see if any factors listed have a common factor and can be factored further. If you find a common factor, factor it out at this point.

The following examples illustrate how to use the procedure.

**EXAMPLE 1** ▸ Factor $2x^4 - 50x^2$.

*Solution* First, check for a greatest common factor other than 1. Since $2x^2$ is common to both terms, factor it out.

$$2x^4 - 50x^2 = 2x^2(x^2 - 25) = 2x^2(x + 5)(x - 5)$$

Note that $x^2 - 25$ is factored as a difference of two squares.

▸ **Now Try Exercise 3**

**EXAMPLE 2** ▸ Factor $3x^2y^2 - 24xy^2 + 48y^2$.

*Solution* Begin by factoring the GCF, $3y^2$, from each term.

$$3x^2y^2 - 24xy^2 + 48y^2 = 3y^2(x^2 - 8x + 16) = 3y^2(x - 4)^2$$

Note that $x^2 - 8x + 16$ is a perfect square trinomial. If you did not recognize this, you would still obtain the correct answer by factoring the trinomial into $(x - 4)(x - 4)$.

▸ **Now Try Exercise 27**

**EXAMPLE 3** ▸ Factor $24x^2 - 6xy + 40xy - 10y^2$.

*Solution* As always, begin by determining if all the terms in the polynomial have a common factor. In this example, 2 is common to all terms. Factor out the 2; then factor the remaining four-term polynomial by grouping.

$$\begin{aligned} 24x^2 - 6xy + 40xy - 10y^2 &= 2(12x^2 - 3xy + 20xy - 5y^2) \\ &= 2[3x(4x - y) + 5y(4x - y)] \\ &= 2(4x - y)(3x + 5y) \end{aligned}$$

▸ **Now Try Exercise 31**

**EXAMPLE 4** ▸ Factor $12a^2b - 18ab + 24b$.

*Solution* $$12a^2b - 18ab + 24b = 6b(2a^2 - 3a + 4)$$

Since $2a^2 - 3a + 4$ cannot be factored, we stop here.

▸ Now Try Exercise 7

**EXAMPLE 5** ▸ Factor $2x^4y + 54xy$.

*Solution*
$$\begin{aligned} 2x^4y + 54xy &= 2xy(x^3 + 27) \\ &= 2xy(x + 3)(x^2 - 3x + 9) \end{aligned}$$

Note that $x^3 + 27$ was factored as a sum of two cubes.

▸ Now Try Exercise 19

**EXAMPLE 6** ▸ Factor $3x^2 - 18x + 27 - 3y^2$.

*Solution* Factor out 3 from all four terms.

$$3x^2 - 18x + 27 - 3y^2 = 3(x^2 - 6x + 9 - y^2)$$

Now try factoring by grouping. Since the four terms within parentheses cannot be factored by grouping, check to see whether any three of the terms can be written as the square of a binomial. Since this can be done, express $x^2 - 6x + 9$ as $(x - 3)^2$ and then use the difference of two squares formula. Thus,

$$\begin{aligned} 3x^2 - 18x + 27 - 3y^2 &= 3[(x - 3)^2 - y^2] \\ &= 3[(x - 3 + y)(x - 3 - y)] \\ &= 3(x - 3 + y)(x - 3 - y) \end{aligned}$$

▸ Now Try Exercise 43

### Helpful Hint *Study Tip*

In this section, we have reviewed all the techniques for factoring expressions. If you are still having difficulty with factoring, you should study the material in Sections 5.4–5.6 again.

## EXERCISE SET 5.7

MathXL® MyMathLab

### Concept/Writing Exercises

1. Explain the possible procedures that may be used to factor a polynomial of **a)** two terms **b)** three terms **c)** four terms.
2. What is the first step in the factoring process?

### Practice the Skills

*Factor each polynomial completely.*

3. $3x^2 - 75$
4. $4x^2 - 24x + 36$
5. $10s^2 + 19s - 15$
6. $-8r^2 + 26r - 15$
7. $6x^3y^2 + 10x^2y^3 + 14x^2y^2$
8. $24m^3n - 12m^2n^2 + 16mn^3$
9. $0.8x^2 - 0.072$
10. $0.5x^2 - 0.08$
11. $6x^5 - 54x$
12. $8x^2y^2z^2 - 32x^2y^2$
13. $3x^6 - 3x^5 + 12x^5 - 12x^4$
14. $2x^2y^2 + 6xy^2 - 10xy^2 - 30y^2$
15. $5x^4y^2 + 20x^3y^2 + 15x^3y^2 + 60x^2y^2$
16. $6x^2 - 15x - 9$
17. $x^4 - x^2y^2$
18. $5x^3 + 135$
19. $x^7y^2 - x^4y^2$
20. $x^4 - 81$

21. $x^5 - 16x$

22. $12x^2y^2 + 33xy^2 - 9y^2$

23. $4x^6 + 32y^3$

24. $8x^4 - 4x^3 - 4x^3 + 2x^2$

25. $5(a + b)^2 - 20$

26. $12x^3y^2 + 4x^2y^2 - 40xy^2$

27. $6x^2 + 36xy + 54y^2$

28. $3x^2 - 30x + 75$

29. $(x + 2)^2 - 4$

30. $5y^4 - 45x^6$

31. $6x^2 + 24xy - 3xy - 12y^2$

32. $pq - 8q + pr - 8r$

33. $(y + 5)^2 + 4(y + 5) + 4$

34. $(x + 1)^2 - (x + 1) - 6$

35. $b^4 + 2b^2 + 1$

36. $45a^4 - 30a^3 + 5a^2$

37. $x^3 + \frac{1}{64}$

38. $8y^3 - \frac{1}{27}$

39. $6y^3 + 14y^2 + 4y$

40. $3x^3 + 2x^2 - 27x - 18$

41. $a^3b - 81ab^3$

42. $x^6 + y^6$

43. $49 - (x^2 + 2xy + y^2)$

44. $x^2 - 2xy + y^2 - 25$

45. $24x^2 - 34x + 12$

46. $40x^2 + 52x - 12$

47. $18x^2 + 39x - 15$

48. $7(a - b)^2 + 4(a - b) - 3$

49. $x^4 - 16$

50. $(x + 5)^2 - 12(x + 5) + 36$

51. $5bc - 10cx - 7by + 14xy$

52. $16y^4 - 9y^2$

53. $3x^4 - x^2 - 4$

54. $x^2 + 16x + 64 - 100y^2$

55. $z^2 - (x^2 - 12x + 36)$

56. $4a^3 + 32$

57. $2(y + 4)^2 + 5(y + 4) - 12$

58. $x^6 + 15x^3 + 54$

59. $a^2 + 12ab + 36b^2 - 16c^2$

60. $y^2 - y^4$

61. $10x^4y + 25x^3y - 15x^2y$

62. $4x^2y^2 + 12xy + 9$

63. $x^4 - 2x^2y^2 + y^4$

64. $12r^2s^2 + rs - 1$

## Problem Solving

*Match Exercises 65–72 with the items labeled* **a)** *through* **h)** *on the right.*

65. $a^2 + b^2$

66. $a^2 - b^2$

67. $a^2 + 2ab + b^2$

68. $a^3 + b^3$

69. $a^3 - b^3$

70. $a^2 - 2ab + b^2$

71. a factor of $a^3 + b^3$

72. a factor of $a^3 - b^3$

**a)** $(a + b)(a^2 - ab + b^2)$

**b)** $(a - b)^2$

**c)** $a^2 - ab + b^2$

**d)** $(a + b)^2$

**e)** not factorable

**f)** $(a - b)(a^2 + ab + b^2)$

**g)** $(a + b)(a - b)$

**h)** $a^2 + ab + b^2$

**Perimeter** *In Exercises 73 and 74, find an expression, in factored form, for the perimeter of each figure.*

73.

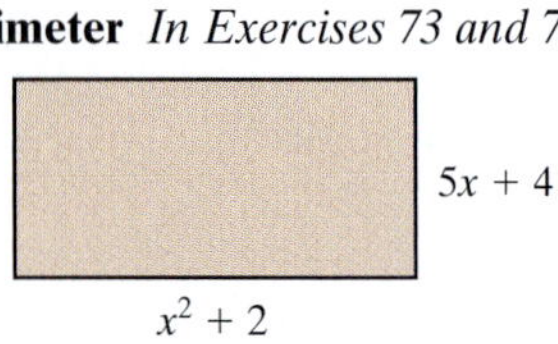


74.

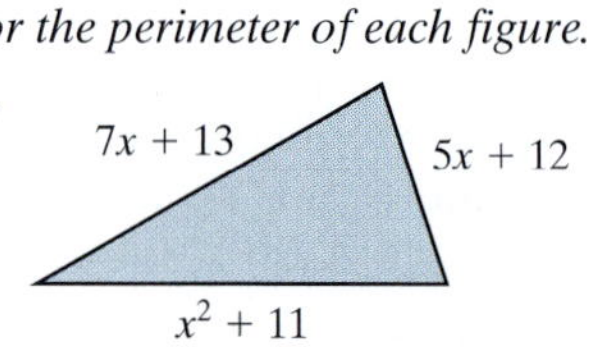


**Area** *In Exercises 75–78, find an expression, in factored form, for the area of the shaded region for each figure.*

75.

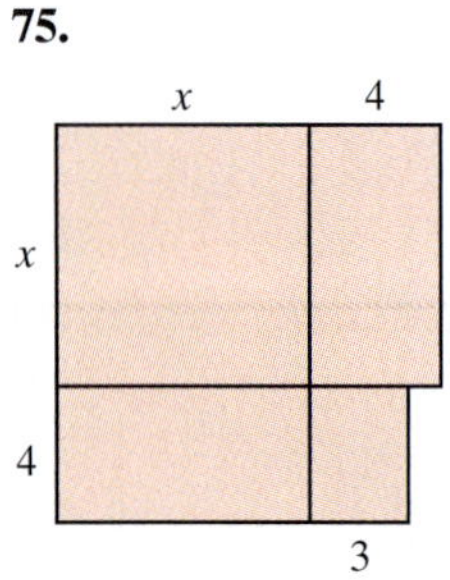


76.

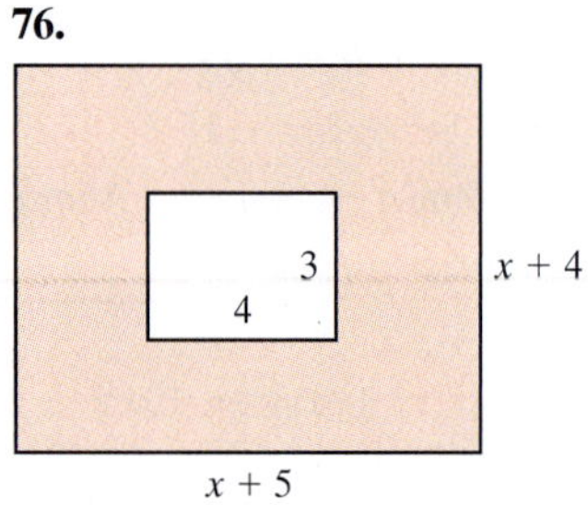


77.

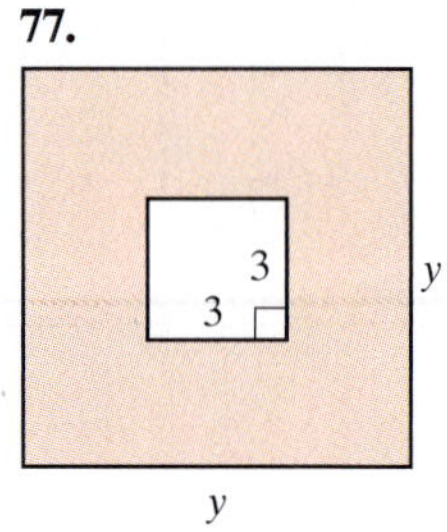


78.

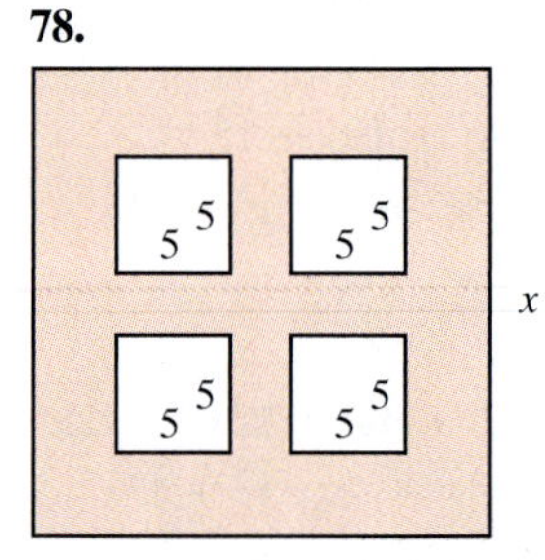

**Volume** *In Exercises 79 and 80, find an expression, in factored form, for the difference in the volumes of the cubes.*

**79.**

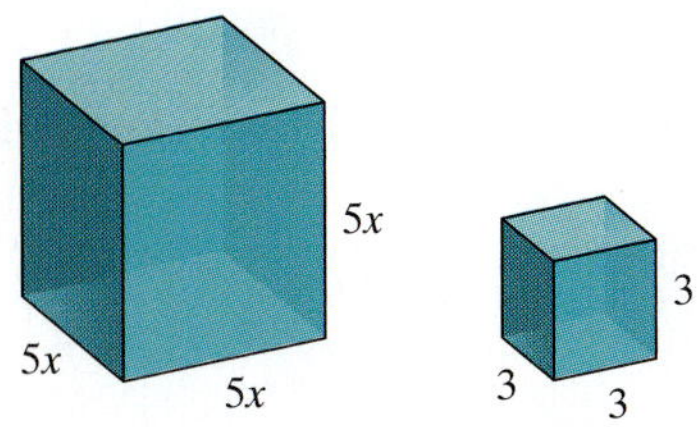


**80.**

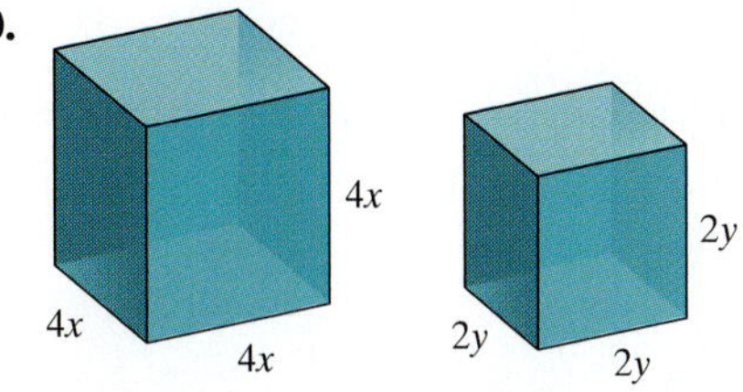


**Area** *In Exercises 81–84,* **a)** *write an expression for the shaded area of the figure, and* **b)** *write the expression in factored form.*

**81.**

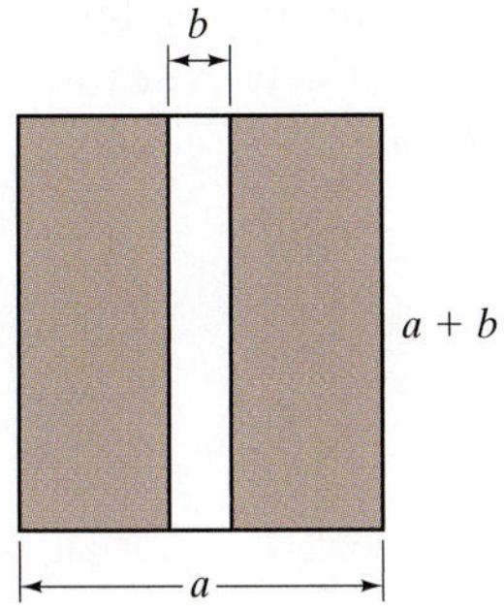


**82.**

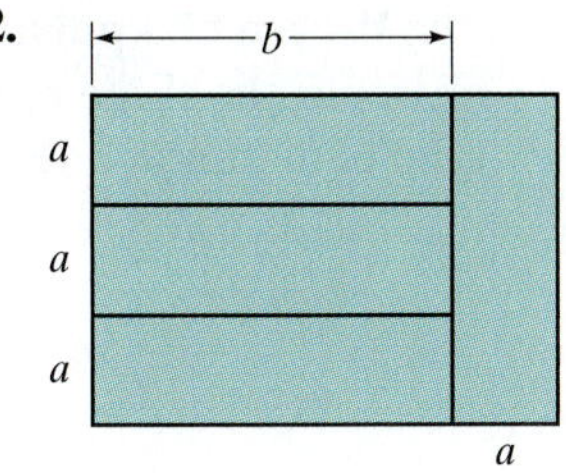


**83.**

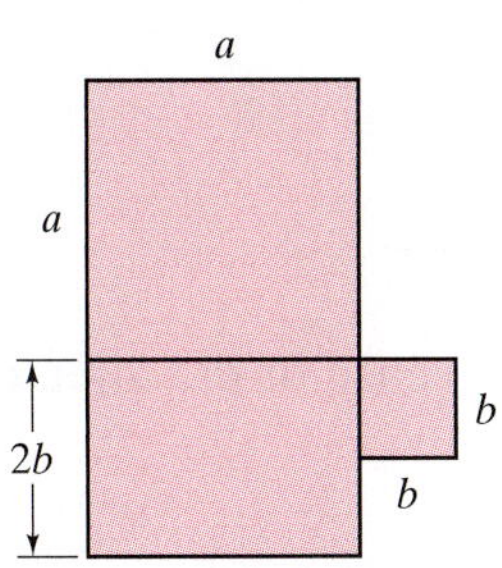


**84.**

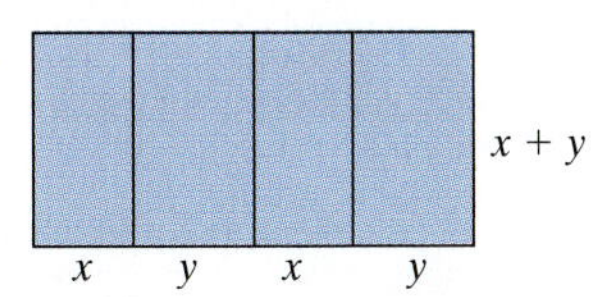


**85. Surface Area**

**a)** Write an expression for the surface area of the four sides of the box shown (omit top and bottom).

**b)** Write the expression in factored form.

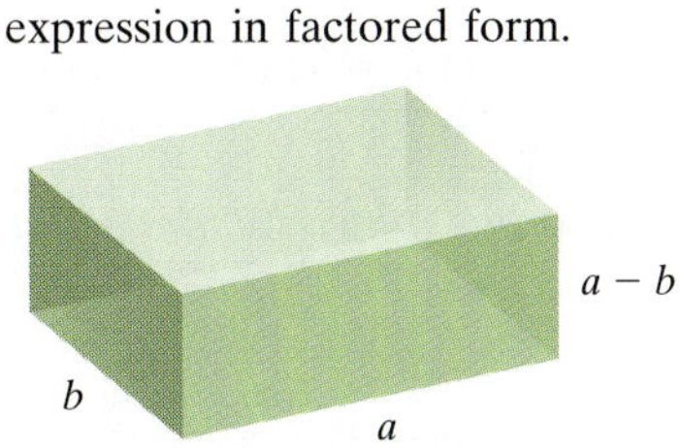


**86.** Explain how the formula for factoring the *difference* of two cubes can be used to factor $x^3 + 27$.

**87. a)** Explain how to construct a perfect square trinomial.

**b)** Construct a perfect square trinomial and then show its factors.

## Challenge Problems

*We have worked only with positive integer exponents in this chapter. However, fractional exponents and negative integer exponents may also be factored out of an expression. The expressions below are not polynomials.* **a)** *Factor out the variable with the lowest (or most negative) exponent from each expression. (Fractional exponents are discussed in Section 7.2.)* **b)** *Factor completely.*

**88.** $x^{-2} - 5x^{-3} + 6x^{-4}$, factor out $x^{-4}$

**89.** $x^{-3} - 2x^{-4} - 3x^{-5}$, factor out $x^{-5}$

**90.** $x^{5/2} + 3x^{3/2} - 4x^{1/2}$, factor out $x^{1/2}$

**91.** $5x^{1/2} + 2x^{-1/2} - 3x^{-3/2}$, factor out $x^{-3/2}$

## Cumulative Review Exercises

[2.1] **92.** Solve $6(x + 4) - 4(3x + 3) = 6$.

[2.6] **93.** Find the solution set for $\left|\frac{6 + 2z}{3}\right| > 2$.

[4.3] **94. Mixing Coffees** Dennis Reissig runs a grocery store. He wishes to mix 30 pounds of coffee to sell for a total cost of \$170. To obtain the mixture, he will mix coffee that sells for \$5.20 per pound with coffee that sells for \$6.30 per pound. How many pounds of each coffee should he use?

[5.2] **95.** Multiply $(5x + 4)(x^2 - x + 4)$.

[5.4] **96.** Factor $2x^3 + 6x^2 - 5x - 15$.

# 5.8 Polynomial Equations

1. Use the zero-factor property to solve equations.
2. Use factoring to solve equations.
3. Use factoring to solve applications.
4. Use factoring to find the $x$-intercepts of a quadratic function.

Whenever two polynomials are set equal to each other, we have a **polynomial equation**.

Examples of Polynomial Equations

$$x^2 + 2x = x - 5$$
$$y^3 + 3y - 2 = 0$$
$$4x^4 + 2x^2 = -3x + 2$$

The **degree of a polynomial equation** is the same as the degree of its highest term. For example, the three equations above have degree 2, 3, and 4, respectively. A second-degree equation in one variable is often called a **quadratic equation**.

Examples of Quadratic Equations

$$3x^2 + 6x - 4 = 0$$
$$5x = 2x^2 - 4$$
$$(x + 4)(x - 3) = 0$$

Any quadratic equation can be written in **standard form**.

**Standard Form of a Quadratic Equation**

$$ax^2 + bx + c = 0, \quad a \neq 0$$

where $a$, $b$, and $c$ are real numbers.

Before going any further, make sure that you can rewrite each of the three quadratic equations given above in standard form, with $a > 0$.

## 1 Use the Zero-Factor Property to Solve Equations

To solve equations using factoring, we use the **zero-factor property**.

**Zero-Factor Property**

For all real numbers $a$ and $b$, if $a \cdot b = 0$, then either $a = 0$ or $b = 0$, or both $a$ and $b = 0$.

The zero-factor property states that *if the product of two factors equals 0, one (or both) of the factors must be 0.*

**EXAMPLE 1** ▸ Solve the equation $(x + 5)(x - 3) = 0$.

*Solution* Since the product of the factors equals 0, according to the zero-factor property, one or both factors must equal 0. Set each factor equal to 0 and solve each equation separately.

$$x + 5 = 0 \quad \text{or} \quad x - 3 = 0$$
$$x = -5 \qquad\qquad x = 3$$

Thus, if $x$ is either $-5$ or 3, the product of the factors is 0.

Check

$x = -5$

$$(x + 5)(x - 3) = 0$$
$$(-5 + 5)(-5 - 3) \stackrel{?}{=} 0$$
$$0(-8) \stackrel{?}{=} 0$$
$$0 = 0 \quad \text{True}$$

$x = 3$

$$(x + 5)(x - 3) = 0$$
$$(3 + 5)(3 - 3) \stackrel{?}{=} 0$$
$$8(0) \stackrel{?}{=} 0$$
$$0 = 0 \quad \text{True}$$

▸ Now Try Exercise 21

## 2 Use Factoring to Solve Equations

Following is a procedure that can be used to obtain the solution to an equation by factoring.

**To Solve an Equation by Factoring**

1. Use the addition property to remove all terms from one side of the equation. This will result in one side of the equation being equal to 0.
2. Combine like terms in the equation and then factor.
3. Set each factor *containing a variable* equal to 0, solve the equations, and find the solutions.
4. Check the solutions in the *original* equation.

**EXAMPLE 2** ▸ Solve the equation $4x^2 = 24x$.

*Solution* First, make the right side of the equation equal to 0 by subtracting $24x$ from both sides of the equation. Then factor the left side of the equation.

$$4x^2 - 24x = 0$$
$$4x(x - 6) = 0$$

Now set each factor equal to 0 and solve for $x$.

$$4x = 0 \quad \text{or} \quad x - 6 = 0$$
$$x = 0 \qquad\qquad x = 6$$

A check will show that the numbers 0 and 6 both satisfy the equation $4x^2 = 24x$.

▸ Now Try Exercise 27

**Avoiding Common Errors**

The zero-factor property can be used only when one side of the equation is equal to 0.

| CORRECT | INCORRECT |
|---|---|
| $(x - 4)(x + 3) = 0$ | $(x - 4)(x + 3) = 2$ |
| $x - 4 = 0$ or $x + 3 = 0$ | $x - 4 = 2$ or $x + 3 = 2$ |
| $x = 4$ $\quad$ $x = -3$ | $x = 6$ $\quad$ $x = -1$ |

In the incorrect process illustrated on the right, the zero-factor property cannot be used since the right side of the equation is not equal to 0. Example 3 shows how to solve such problems correctly.

**EXAMPLE 3** ▸ Solve the equation $(x - 1)(3x + 2) = 4x$.

*Solution* Since the right side of the equation is not 0, we cannot use the zero-factor property yet. Instead, we begin by multiplying the factors on the left side of the equation. Then we subtract $4x$ from both sides of the equation to obtain 0 on the right side. Then we can factor and solve the equation.

$$(x - 1)(3x + 2) = 4x$$
$$3x^2 - x - 2 = 4x \qquad \text{Multiply the factors.}$$
$$3x^2 - 5x - 2 = 0 \qquad \text{Make one side 0.}$$
$$(3x + 1)(x - 2) = 0 \qquad \text{Factor the trinomial.}$$
$$3x + 1 = 0 \quad \text{or} \quad x - 2 = 0 \qquad \text{Zero-factor property}$$
$$3x = -1 \qquad x = 2 \qquad \text{Solve the equations.}$$
$$x = -\frac{1}{3}$$

The solutions are $-\frac{1}{3}$ and 2. Check these values in the original equation.

▸ Now Try Exercise 41

**EXAMPLE 4** ▸ Solve the equation $3x^2 + 2x + 12 = -13x$

*Solution*

$$3x^2 + 2x + 12 = -13x$$

$$3x^2 + 15x + 12 = 0 \quad \text{Make one side 0.}$$

$$3(x^2 + 5x + 4) = 0 \quad \text{Factor out 3.}$$

$$3(x + 4)(x + 1) = 0 \quad \text{Factor the trinomial.}$$

$$x + 4 = 0 \quad \text{or} \quad x + 1 = 0 \quad \text{Zero-factor property}$$

$$x = -4 \qquad\qquad x = -1 \quad \text{Solve for } x.$$

Since the factor 3 does not contain a variable, we do not have to set it equal to 0. Only the numbers $-4$ and $-1$ satisfy the equation $3x^2 + 2x + 12 = -13x$.

▸ **Now Try Exercise 45**

## Helpful Hint

When solving an equation whose leading term has a negative coefficient, we generally make the coefficient positive by multiplying both sides of the equation by $-1$. This makes the factoring process easier, as in the following example.

$$-x^2 + 5x + 6 = 0$$

$$-1(-x^2 + 5x + 6) = -1 \cdot 0$$

$$x^2 - 5x - 6 = 0$$

Now we can solve the equation $x^2 - 5x - 6 = 0$ by factoring.

$$(x - 6)(x + 1) = 0$$

$$x - 6 = 0 \quad \text{or} \quad x + 1 = 0$$

$$x = 6 \qquad\qquad x = -1$$

The numbers 6 and $-1$ both satisfy the original equation, $-x^2 + 5x + 6 = 0$.

The equations in Examples 1 through 4 were all quadratic equations that were rewritten in the form $ax^2 + bx + c = 0$ and solved by factoring. Other methods that can be used to solve quadratic equations include completing the square and the quadratic formula; we discuss these methods in Chapter 8.

The zero-factor property can be extended to three or more factors, as illustrated in Example 5.

**EXAMPLE 5** ▸ Solve the equation $2p^3 + 5p^2 - 3p = 0$.

*Solution* First factor, then set each factor containing a $p$ equal to 0.

$$2p^3 + 5p^2 - 3p = 0$$

$$p(2p^2 + 5p - 3) = 0 \quad \text{Factor out } p.$$

$$p(2p - 1)(p + 3) = 0 \quad \text{Factor the trinomial.}$$

$$p = 0 \quad \text{or} \quad 2p - 1 = 0 \quad \text{or} \quad p + 3 = 0 \quad \text{Zero-factor property}$$

$$2p = 1 \qquad\qquad p = -3 \quad \text{Solve for } p.$$

$$p = \frac{1}{2}$$

The numbers $0, \frac{1}{2}$, and $-3$ are all solutions to the equation.

▸ **Now Try Exercise 49**

Note that the equation in Example 5 is not a quadratic equation because the exponent in the leading term is 3, not 2. This is a **cubic**, or **third-degree, equation**.

**EXAMPLE 6** ▸ For the function $f(x) = 2x^2 - 13x - 16$, find all values of $a$ for which $f(a) = 8$.

*Solution* First we rewrite the function as $f(a) = 2a^2 - 13a - 16$. Since $f(a) = 8$, we write

$$2a^2 - 13a - 16 = 8 \qquad \text{Set } f(a) \text{ equal to 8.}$$

$$2a^2 - 13a - 24 = 0 \qquad \text{Make one side 0.}$$

$$(2a + 3)(a - 8) = 0 \qquad \text{Factor the trinomial.}$$

$$2a + 3 = 0 \quad \text{or} \quad a - 8 = 0 \qquad \text{Zero-factor property}$$

$$2a = -3 \qquad a = 8 \qquad \text{Solve for } a.$$

$$a = -\frac{3}{2}$$

If you check these answers, you will find that $f\left(-\frac{3}{2}\right) = 8$ and $f(8) = 8$.

▸ **Now Try Exercise 69**

## 3 Use Factoring to Solve Applications

Now let us look at some applications that use factoring in their solution.

**EXAMPLE 7** ▸ **Triangle** At an exhibition, a large canvas tent is to have an entrance in the shape of a triangle (see **Fig. 5.17**).

Find the base and height of the entrance if the height is to be 3 feet less than twice the base and the total area of the entrance is 27 square feet.

*Solution* **Understand** Let's draw a picture of the entrance and label it with the given information (**Fig. 5.18**).

FIGURE 5.17

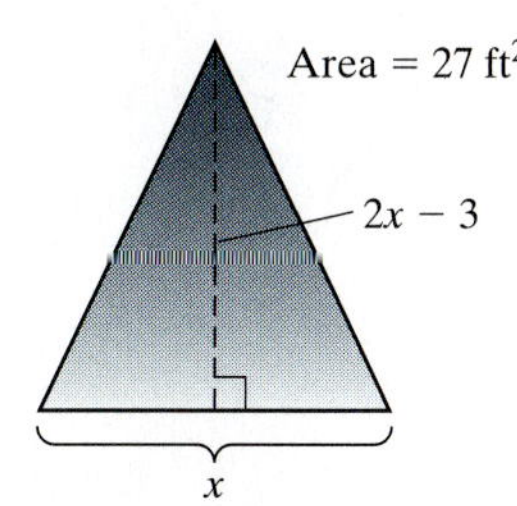


FIGURE 5.18

**Translate** We use the formula for the area of a triangle to solve the problem.

$$A = \frac{1}{2}(\text{base})(\text{height})$$

$$27 = \frac{1}{2}(x)(2x - 3) \qquad \text{Substitute expressions for base, height, and area.}$$

**Carry Out**

$$2(27) = 2\left[\frac{1}{2}(x)(2x - 3)\right] \qquad \text{Multiply both sides by 2 to remove fractions.}$$

$$54 = x(2x - 3)$$

$$54 = 2x^2 - 3x$$

$$\text{or} \quad 2x^2 - 3x - 54 = 0 \qquad \text{Make one side 0.}$$

$$(2x + 9)(x - 6) = 0 \qquad \text{Factor the trinomial.}$$

$$2x + 9 = 0 \quad \text{or} \quad x - 6 = 0 \qquad \text{Zero-factor property}$$

$$2x = -9 \qquad x = 6 \qquad \text{Solve for } x.$$

$$x = -\frac{9}{2}$$

**Answer** Since the dimensions of a geometric figure cannot be negative, we can eliminate $x = -\frac{9}{2}$ as an answer to our problem. Therefore,

$$\text{base} = x = 6 \text{ feet}$$
$$\text{height} = 2x - 3 = 2(6) - 3 = 9 \text{ feet}$$

▶ **Now Try Exercise 99**

**EXAMPLE 8 ▶ Height of a Cannonball** A cannon sits on top of a 288-foot cliff overlooking a lake. A cannonball is fired upward at a speed of 112 feet per second. The height, $h$, in feet, of the cannonball above the lake at any time, $t$, is determined by the function

$$h(t) = -16t^2 + 112t + 288$$

Find the time it takes for the cannonball to hit the water after the cannon is fired.

*Solution* **Understand** We will draw a picture to help analyze the problem (see **Fig. 5.19**). When the cannonball hits the water, its height above the water is 0 feet.

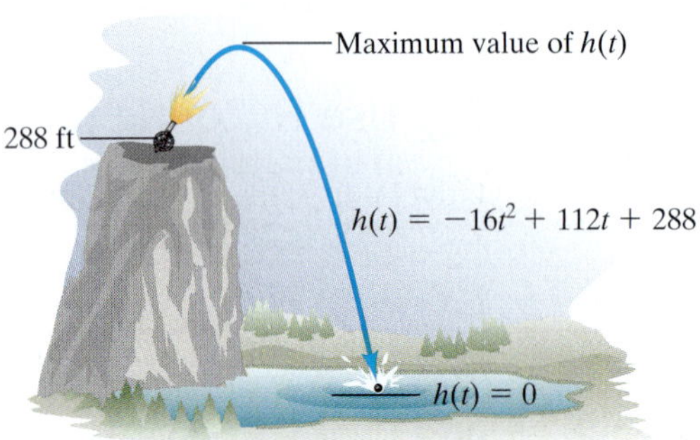


FIGURE 5.19

**Translate** To solve the problem we need to find the time, $t$, when $h(t) = 0$. To do so we set the given function equal to 0 and solve for $t$.

$$-16t^2 + 112t + 288 = 0 \quad \text{Set } h(t) = 0.$$
$$-16(t^2 - 7t - 18) = 0 \quad \text{Factor out } -16.$$
$$-16(t + 2)(t - 9) = 0 \quad \text{Factor the trinomial.}$$
$$t + 2 = 0 \quad \text{or} \quad t - 9 = 0 \quad \text{Zero-factor property}$$
$$t = -2 \qquad\qquad t = 9 \quad \text{Solve for } t.$$

**Answer** Since $t$ is the number of seconds, $-2$ is not a possible answer. The cannonball will hit the water in 9 seconds.

▶ **Now Try Exercise 105**

## Pythagorean Theorem

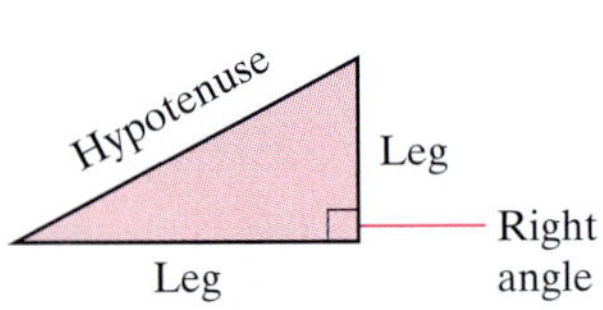


FIGURE 5.20

Our next application uses the Pythagorean Theorem. Consider a right triangle (see **Fig. 5.20**). The two shorter sides of a right triangle are called the **legs** and the side opposite the right angle is called the **hypotenuse**. The **Pythagorean Theorem** expresses the relationship between the legs of the triangle and its hypotenuse.

**Pythagorean Theorem**

The square of the length of the hypotenuse of a right triangle is equal to the sum of the squares of the lengths of the two legs; that is,

$$\text{leg}^2 + \text{leg}^2 = \text{hyp}^2$$

If $a$ and $b$ represent the lengths of the legs and $c$ represents the length of the hypotenuse, then

$$a^2 + b^2 = c^2$$

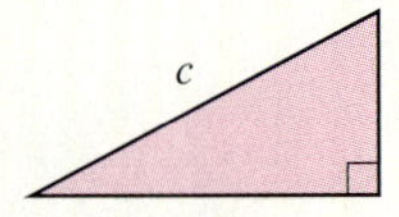

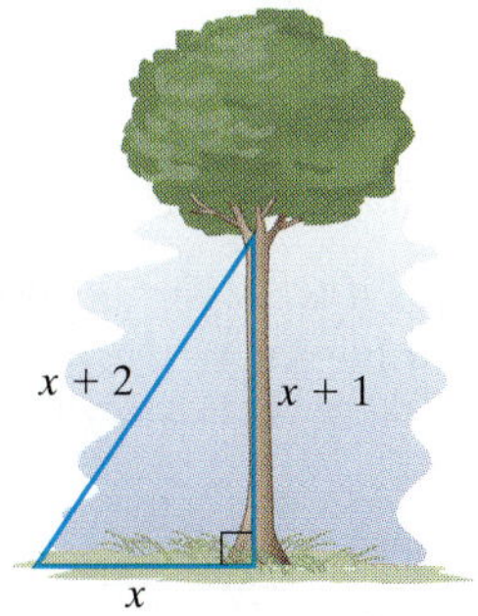


FIGURE 5.21

**EXAMPLE 9** ▸ **Tree Wires** Jack Keating places a guy wire on a tree to help it grow straight. The location of the stake and where the wire attaches to the tree are given in **Figure 5.21**. Find the length of the wire. Notice that the length of the wire is the hypotenuse of a right triangle formed by the tree and the ground.

**Solution** **Understand** To solve this problem we use the Pythagorean Theorem. From the figure, we see that the legs are $x$ and $x + 1$ and the hypotenuse is $x + 2$.

**Translate**

$$\text{leg}^2 + \text{leg}^2 = \text{hyp}^2 \qquad \textit{Pythagorean Theorem}$$

$$x^2 + (x+1)^2 = (x+2)^2 \qquad \textit{Substitute expressions for legs and hypotenuse.}$$

**Carry Out**

$$x^2 + x^2 + 2x + 1 = x^2 + 4x + 4 \qquad \textit{Square terms.}$$

$$2x^2 + 2x + 1 = x^2 + 4x + 4 \qquad \textit{Simplify.}$$

$$x^2 - 2x - 3 = 0 \qquad \textit{Make one side 0.}$$

$$(x-3)(x+1) = 0 \qquad \textit{Factor.}$$

$$x - 3 = 0 \quad \text{or} \quad x + 1 = 0 \qquad \textit{Solve.}$$

$$x = 3 \qquad\qquad x = -1$$

**Answer** From the diagram we know that $x$ cannot be a negative value. Therefore the only possible answer is 3. The stake is placed 3 feet from the tree. The wire attaches to the tree $x + 1$, or 4, feet from the ground. The length of the wire is $x + 2$, or 5, feet.

▸ **Now Try Exercise 109**

## USING YOUR GRAPHING CALCULATOR

The applications given in this section and the exercise set have been written so that the quadratic equations are factorable. In real life, quadratic equations generally are not factorable (over the set of integers) and need to be solved in other ways. We will discuss methods to solve quadratic equations that are not factorable in Sections 8.1 and 8.2.

You can find approximate solutions to quadratic equations that are not factorable using your graphing calculator. Consider the following real-life example.

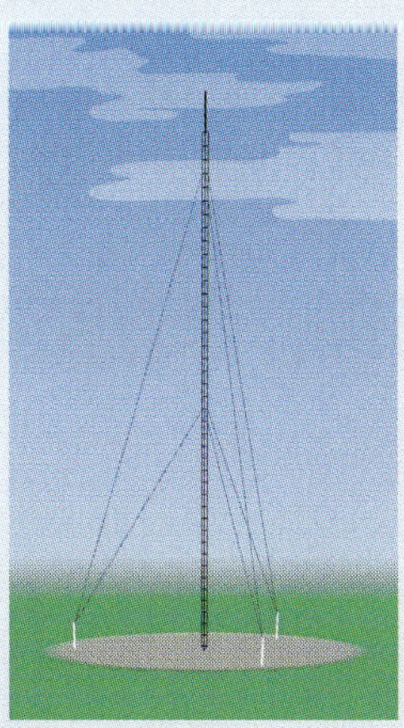

**EXAMPLE** **Cellular Antennas** The number of relay antennas for cellular phones in the United States has been increasing. From 1996 through 2002, the number of cellular relay antennas, $N$, in thousands, in the United States can be closely approximated by the function

$$N(t) = -1.45t^2 + 21.88t + 25.44$$

where $t$ is the number of years since 1996. Determine the year in which the number of cellular antennas reached 80,000.

**Solution** **Understand and Translate** To answer this question we need to set the function $N(t)$ equal to 80 and solve for $t$.

$$-1.45t^2 + 21.88t + 25.44 = 80 \qquad \textit{Set } N(t) = 80.$$

We cannot solve this equation by factoring, but we can solve it using a graphing calculator. To do so, let's call one side of the equation $Y_1$ and the other side $Y_2$.

$$Y_1 = -1.45x^2 + 21.88x + 25.44$$

$$Y_2 = 80$$

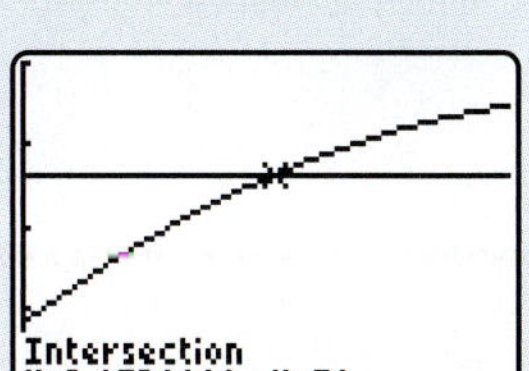


[0, 6, 1, 0, 120, 30]

FIGURE 5.22

**Carry Out** Now graph the two functions on your graphing calculator and use the TRACE and ZOOM keys, or other keys (for example the CALC key with option 5, *intersect*, on the TI-84 Plus) to obtain your answer. **Figure 5.22** illustrates the screen from a TI-84 Plus showing that the $x$-coordinate of the intersection of the equations is about $x = 3.1520$.

**Answer** Therefore, there were about 80,000 relay antennas about 3 years after 1996, or in 1999.

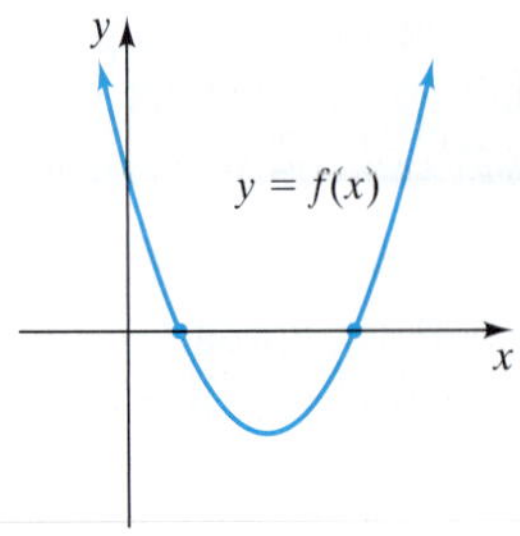


FIGURE 5.23

## 4 Use Factoring to Find the $x$-Intercepts of a Quadratic Function

Consider the graph in **Figure 5.23**.

At the $x$-intercepts, the value of the function, or $y$, is 0. Thus if we wish to find the ***x*-intercepts of a graph**, we can set the function equal to 0 and solve for $x$.

**EXAMPLE 10** ▸ Find the $x$-intercepts of the graph of $y = x^2 - 2x - 8$.

*Solution* At the $x$-intercepts $y$ has a value of 0. Thus to find the $x$-intercepts we write

$$x^2 - 2x - 8 = 0$$
$$(x - 4)(x + 2) = 0$$
$$x - 4 = 0 \quad \text{or} \quad x + 2 = 0$$
$$x = 4 \qquad\qquad x = -2$$

The solutions of $x^2 - 2x - 8 = 0$ are 4 and $-2$. The $x$-intercepts of the graph of $y = x^2 - 2x - 8$ are $(4, 0)$ and $(-2, 0)$, as illustrated in **Figure 5.24**.

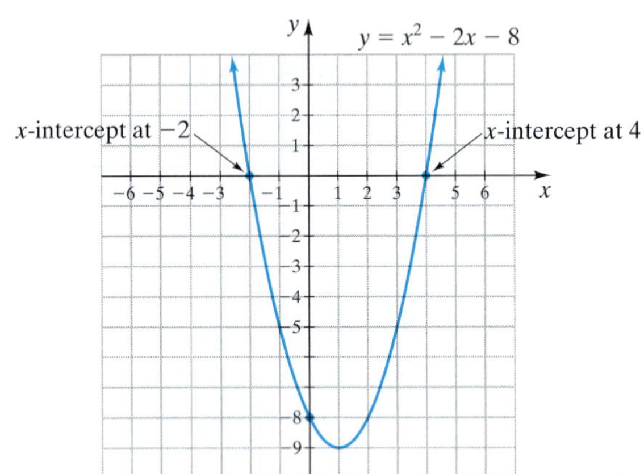


FIGURE 5.24

▸ **Now Try Exercise 75**

If we know the $x$-intercepts of a graph, we can work backward to find the equation of the graph. Read the Using Your Graphing Calculator box that follows to learn how this is done.

### USING YOUR GRAPHING CALCULATOR

Determine the equation of the graph in **Figure 5.25**.

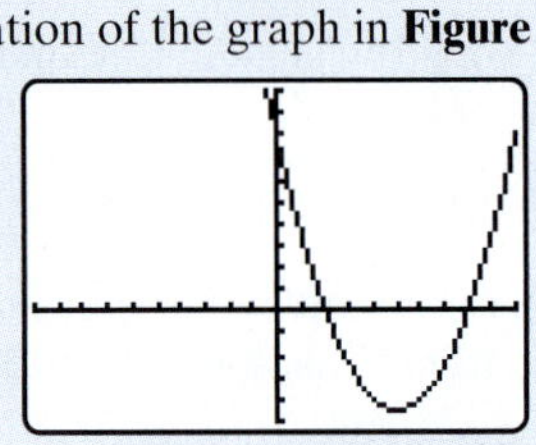

$[-10, 10, 1, -10, 20, 2]$ FIGURE 5.25

If we assume the intercepts are integer values, then the $x$-intercepts are at 2 and 8. Therefore,

| ***x*-Intercepts At** | **Factors** | **Possible Equation of Graph** |
|---|---|---|
| 2 and 8 | $(x - 2)(x - 8)$ | $y = (x - 2)(x - 8)$ <br> or $y = x^2 - 10x + 16$ |

Since the $y$-intercept of the graph in **Figure 5.25** is at 16, $y = x^2 - 10x + 16$ is the equation of the graph. Example 11 explains why we used the words *possible equation of the graph*. (*continued on the next page*)

### EXERCISES

*Write the equation of each graph illustrated. Assume that all x-intercepts are integer values and that the standard window is shown.*

**1.** 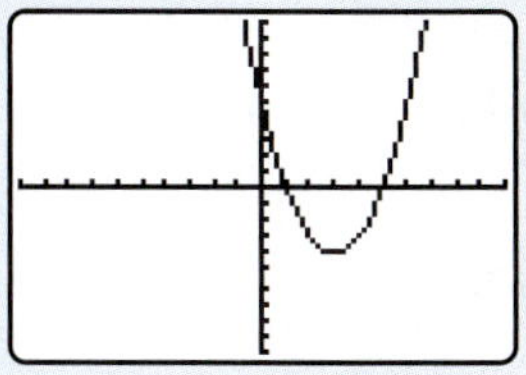

**2.** 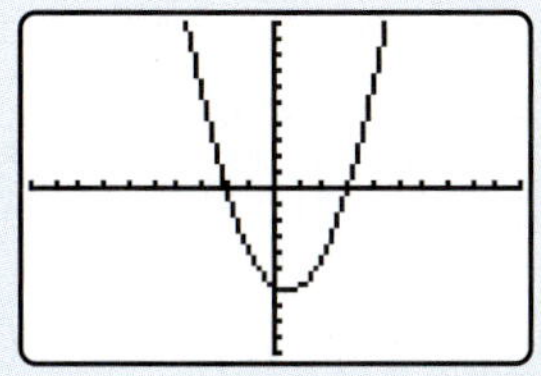

**3.** 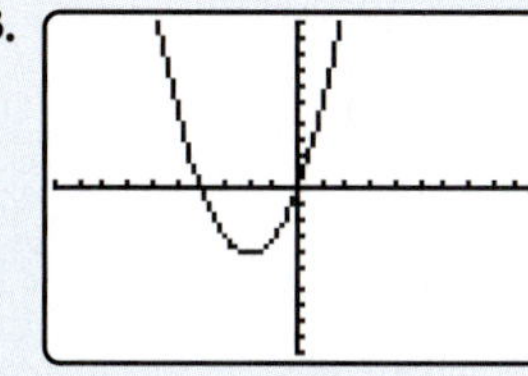

**EXAMPLE 11** ▸ Write an equation whose graph will have $x$-intercepts at $-2$ and 4.

*Solution:* If the $x$-intercepts are at $-2$ and 4, then one set of factors that yields these $x$-intercepts is $(x + 2)$ and $(x - 4)$, respectively. Therefore, one equation that will have $x$-intercepts at $-2$ and 4 is

$$y = (x + 2)(x - 4) \text{ or } y = x^2 - 2x - 8.$$

Note that other equations may have graphs with the same $x$-intercepts. For example, the graph of $y = 2(x^2 - 2x - 8)$ or $y = 2x^2 - 4x - 16$ also has $x$-intercepts at $-2$ and 4. In fact, the graph of $y = a(x^2 - 2x - 8)$, for any nonzero real number $a$, will have $x$-intercepts at $-2$ and 4.

▸ **Now Try Exercise 93**

In Example 11, although the $x$-intercepts of the graph of $y = a(x^2 - 2x - 8)$ will always be at $-2$ and 4, the $y$-intercept of the graph will depend on the value of $a$. For example, if $a = 1$, the $y$-intercept will be at 1(8) or 8. If $a = 2$, the $y$-intercept will be at 2(8) or 16 and so on.

## EXERCISE SET 5.8

MathXL® MyMathLab

### Concept/Writing Exercises

1. How do you determine the degree of a polynomial function?
2. What is a quadratic equation?
3. What is the standard form of a quadratic equation?
4. **a)** Explain the zero-factor property.
   **b)** Solve the equation $(3x - 7)(2x + 3) = 0$ using the zero-factor property.
5. **a)** Explain why the equation $(x + 3)(x + 4) = 2$ *cannot* be solved by writing $x + 3 = 2$ or $x + 4 = 2$.
   **b)** Solve the equation $(x + 3)(x + 4) = 2$.
6. When a constant is factored out of an equation, why is it not necessary to set that constant equal to 0 when solving the equation?
7. **a)** Explain how to solve a polynomial equation using factoring.
   **b)** Solve the equation $-x - 20 = -12x^2$ using the procedure in part **a)**.
8. **a)** What is the first step in solving the equation $-x^2 + 2x + 35 = 0$?
   **b)** Solve the equation in part **a)**.
9. **a)** What are the two shorter sides of a right triangle called?
   **b)** What is the longest side of a right triangle called?
10. Give the Pythagorean Theorem and explain its meaning.
11. If the graph of $y = x^2 + 10x + 16$ has $x$-intercepts at $-8$ and $-2$, what is the solution to the equation $x^2 + 10x + 16 = 0$? Explain.
12. If the solutions to the equation $2x^2 - 15x + 18 = 0$ are $\frac{3}{2}$ and 6, what are the $x$-intercepts of the graph of $y = 2x^2 - 15x + 18$? Explain.
13. Is it possible for a quadratic function to have no $x$-intercepts? Explain.
14. Is it possible for a quadratic function to have only one $x$-intercept? Explain.
15. Is it possible for a quadratic function to have two $x$-intercepts? Explain.
16. Is it possible for a quadratic function to have three $x$-intercepts? Explain.

## Practice the Skills

*Solve.*

17. $x(x + 3) = 0$
18. $x(x - 4) = 0$
19. $4x(x - 1) = 0$
20. $8x(x + 6) = 0$
21. $2(x + 1)(x - 7) = 0$
22. $3(a - 5)(a + 2) = 0$
23. $x(x - 9)(x - 4) = 0$
24. $2a(a + 3)(a + 8) = 0$
25. $(3x - 2)(7x - 1) = 0$
26. $(2x + 3)(4x + 5) = 0$
27. $4x^2 = 12x$
28. $3y^2 = -24y$
29. $x^2 + 5x = 0$
30. $2a^2 - 8a = 0$
31. $-x^2 + 6x = 0$
32. $-3x^2 - 21x = 0$
33. $3x^2 = 27x$
34. $18a^2 = -36a$
35. $a^2 + 6a + 5 = 0$
36. $x^2 - 6x + 5 = 0$
37. $x^2 + x - 12 = 0$
38. $b^2 + b - 72 = 0$
39. $x^2 + 8x + 16 = 0$
40. $c^2 - 10c = -25$
41. $(2x + 5)(x - 1) = 12x$
42. $a(a + 2) = 48$
43. $2y^2 = -y + 6$
44. $3a^2 = -a + 2$
45. $3x^2 - 6x - 72 = 0$
46. $2a^2 + 18a + 40 = 0$
47. $x^3 - 3x^2 = 18x$
48. $x^3 = -19x^2 + 42x$
49. $4c^3 + 4c^2 - 48c = 0$
50. $3b^3 - 8b^2 - 3b = 0$
51. $18z^3 = 15z^2 + 12z$
52. $12a^3 = 16a^2 + 3a$
53. $x^2 - 25 = 0$
54. $2y^2 = 98$
55. $4x^2 = 9$
56. $49c^2 = 81$
57. $4y^3 - 36y = 0$
58. $3x^4 - 48x^2 = 0$
59. $-x^2 = 2x - 99$
60. $-x^2 + 16x = 63$
61. $(x + 7)^2 - 16 = 0$
62. $(x - 6)^2 - 4 = 0$
63. $(2x + 5)^2 - 9 = 0$
64. $(x + 1)^2 - 3x = 7$
65. $6a^2 - 12 - 4a = 19a - 32$
66. $4(a^2 - 3) = 6a + 4(a + 3)$
67. $2b^3 + 16b^2 = -30b$
68. $(a - 1)(3a + 2) = 4a$

69. For $f(x) = 3x^2 + 7x + 9$, find all values of $a$ for which $f(a) = 7$.

70. For $f(x) = 4x^2 - 11x + 2$, find all values of $a$ for which $f(a) = -4$.

71. For $g(x) = 10x^2 - 31x + 16$ find all values of $a$ for which $g(a) = 1$.

72. For $g(x) = 6x^2 + x - 3$, find all values of $a$ for which $g(a) = -2$.

73. For $r(x) = x^2 - x$, find all values of $a$ for which $r(a) = 30$.

74. For $r(x) = 10x^2 - 19x - 5$ find all values of $a$ for which $r(a) = -11$.

*Use factoring to find the x-intercepts of the graphs of each equation (see Example 10).*

75. $y = x^2 - 10x + 24$
76. $y = x^2 - 13x + 42$
77. $y = x^2 + 16x + 64$
78. $y = 15x^2 - 14x - 8$
79. $y = 12x^3 - 46x^2 + 40x$
80. $y = 12x^3 - 39x^2 + 30x$

**Right Triangle** *In Exercises 81–86, use the Pythagorean Theorem to find x.*

81.

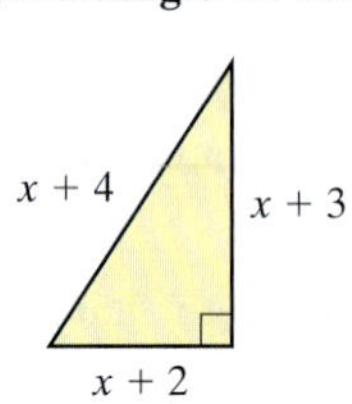


82.

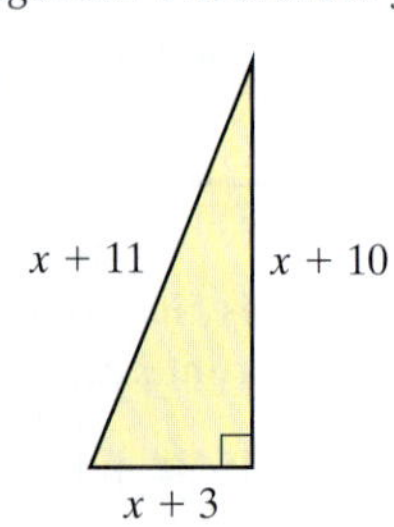


83.

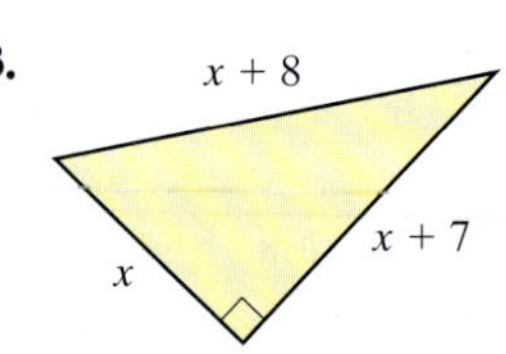

**84.**

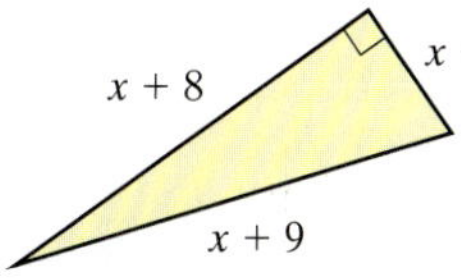


**85.**

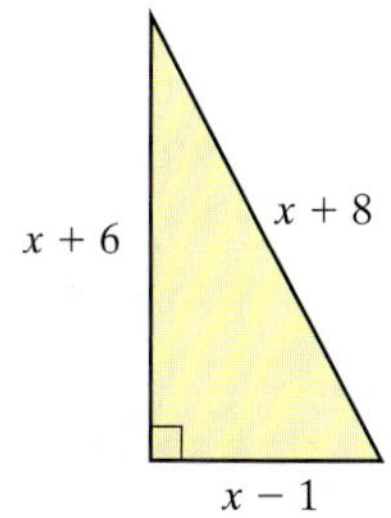


**86.**

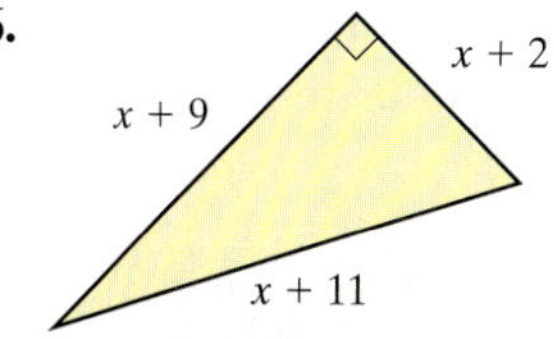


## Problem Solving

*In Exercises 87–90, determine the x-intercepts of each graph; then match the equation with the appropriate graph labeled* **a)–d)**.

**87.** $y = x^2 - 5x + 6$ **88.** $y = x^2 - x - 6$ **89.** $y = x^2 + 5x + 6$ **90.** $y = x^2 + x - 6$

**a)** 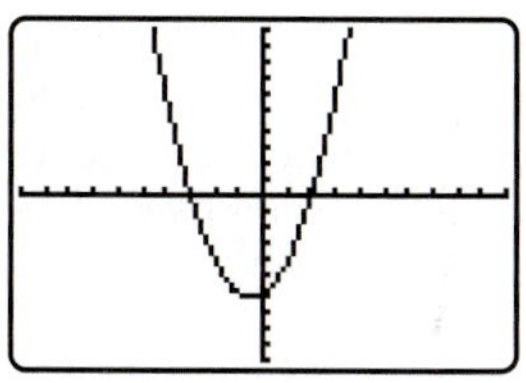

**b)** 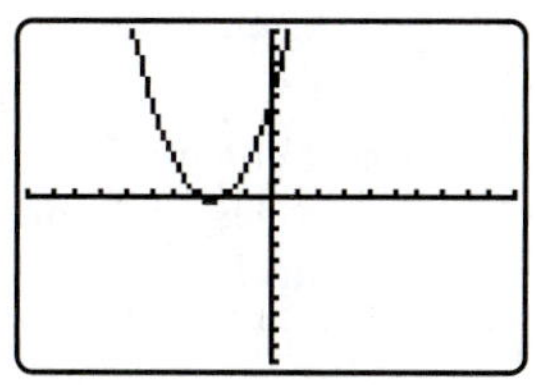

**c)** 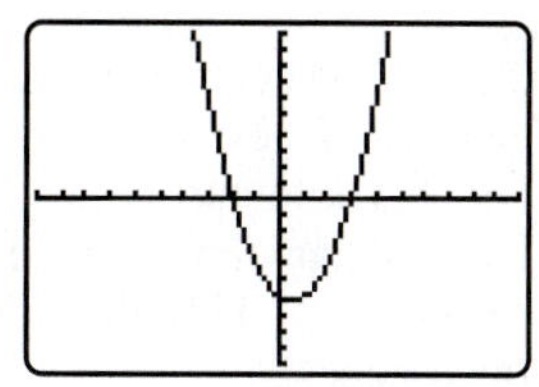

**d)** 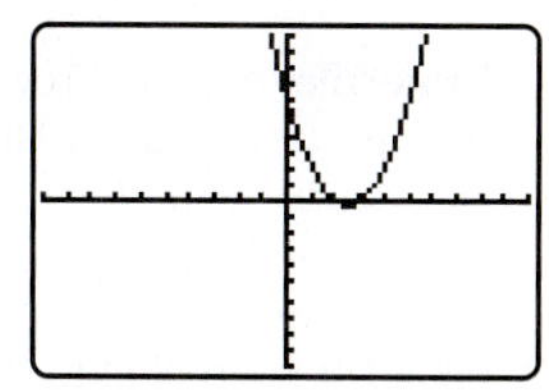

*Write an equation whose graph will have x-intercepts at the given values.*

**91.** 1 and 5

**92.** 3 and −7

**93.** 4 and −2

**94.** $\frac{3}{2}$ and 6

**95.** $-\frac{5}{6}$ and 2

**96.** −0.4 and 2.6

**97. Rectangular Coffee Table** A coffee table is rectangular. If the length of its surface area is 1 foot greater than twice its width and the surface area of the tabletop is 10 square feet, find its length and width.

**98. Rectangular Shed** The floor of a shed has an area of 60 square feet. Find the length and width if the length is 2 feet less than twice its width.

**99. Triangular Sail** A sailboat sail is triangular with a height 6 feet greater than its base. If the sail's area is 80 square feet, find its base and height.

**100. Triangular Tent** A triangular tent has a height that is 4 feet less than its base. If the area of a side is 70 square feet, find the base and height of the tent.

**101. Rectangle** Frank Bullock's garden is surrounded by a uniform-width walkway. The garden and the walkway together cover an area of 320 square feet. If the dimensions of the garden are 12 feet by 16 feet, find the width of the walkway.

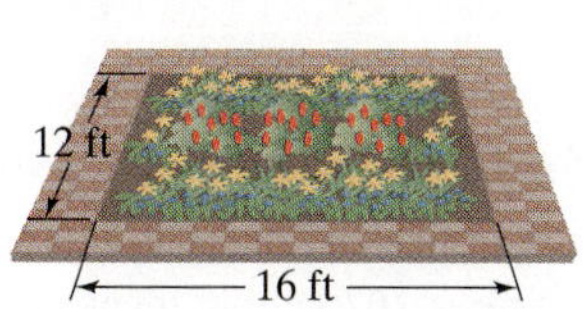


**102. Picture Frame** The outside dimensions of a picture frame are 28 cm and 23 cm. The area of the picture itself is 414 square centimeters. Find the width of the frame itself.

**103. Vegetable Garden** Sally Yang's rectangular vegetable garden is 20 feet by 30 feet. In addition to mulching her garden, she wants to put mulch around the outside of her garden in a uniform width. If she has enough mulch to cover an area of 936 square feet, how wide should the mulch border be?

**104. Square Garden** Ronnie Tucker has a square garden. He adds a 2-foot-wide walkway around his garden. If the total area of the walkway and garden is 196 square feet, find the dimensions of the garden.

**105. Water Sculpture** In a building at Navy Pier in Chicago, a water fountain jet shoots short spurts of water over a walkway. The water spurts reach a maximum height, then come down into a pond of water on the other side of the walkway. The height above the jet, $h$, of a spurt of water $t$ seconds after leaving the jet can be found by the function $h(t) = -16t^2 + 32t$. Find the time it takes for the spurt of water to return to the jet's height; that is, when $h(t) = 0$.

**106. Projectile** A model rocket will be launched from a hill 80 feet above sea level. The launch site is next to the ocean (sea level), and the rocket will fall into the ocean. The rocket's distance, $s$, above sea level at any time, $t$, is found by the equation $s(t) = -16t^2 + 64t + 80$. Find the time it takes for the rocket to strike the ocean.

**107. Bicycle Riding** Two cyclists, Bob and Tim, start at the same point. Bob rides west and Tim rides north. At some time, they are 13 miles apart. If Bob traveled 7 miles farther than Tim, determine how far each person traveled.

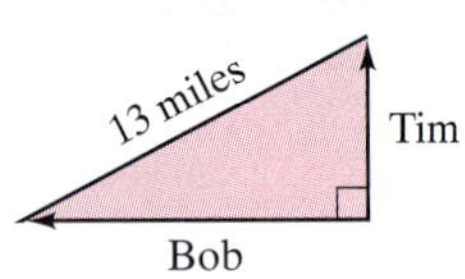


**108. Picture Frame** April is making a rectangular picture frame for her mother. The diagonal of the frame is 20 inches. Find the dimensions of the frame if its length is 4 inches greater than its width.

**109. Tent Wires** A tent has wires attached to it to help stabilize it. A wire is attached to the ground 12 feet from the tent. The length of wire used is 8 feet greater than the height from the ground where the wire is attached. How long is the wire?

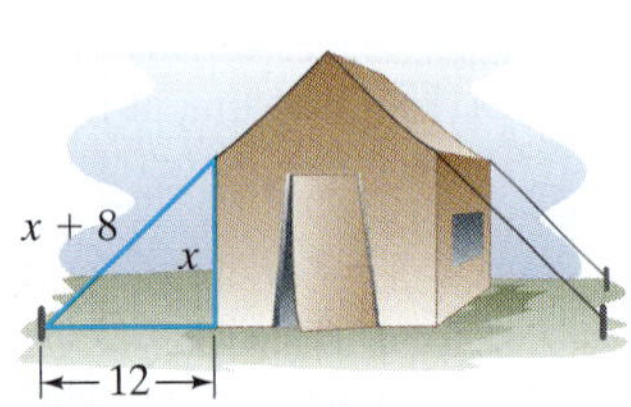


**110. Car in Mud** Suppose two cars, indicated by points $A$ and $B$ in the figure, are pulling a third car, $C$, out of the mud. Determine the distance from car $A$ to car $B$.

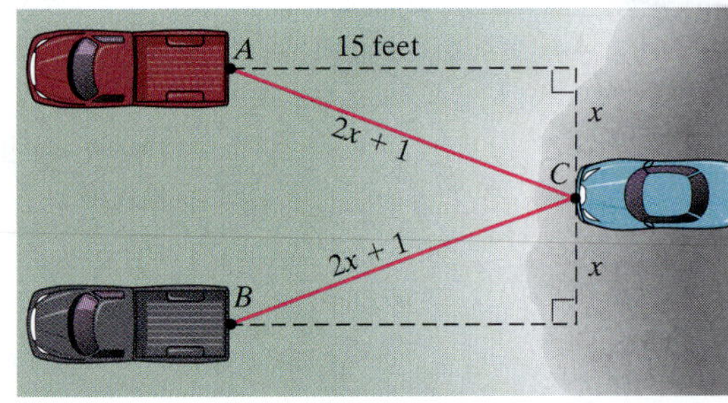


**111. Bicycle Shop** The Energy Conservatory Bicycle Shop has a monthly revenue equation $R(x) = 70x - x^2$ and a monthly cost equation $C(x) = 17x + 150$, where $x$ is the number of bicycles sold and $x \geq 10$. Find the number of bicycles that must be sold for the company to break even; that is, where revenue equals costs.

**112. Silk Plant** Edith Hall makes silk plants and sells them to various outlets. Her company has a revenue equation $R(x) = 40x - x^2$ and a cost equation $C(x) = 14x + 25$, where $x$ is the number of plants sold and $x \geq 5$. Find the number of plants that must be sold for the company to break even.

**113. Making a Box** Monique Siddiq is making a box by cutting out 2-in.-by-2-in. squares from a square piece of cardboard and folding up the edges to make a 2-inch-high box. What size piece of cardboard does Monique need to make a 2-inch-high box with a volume of 162 cubic inches?

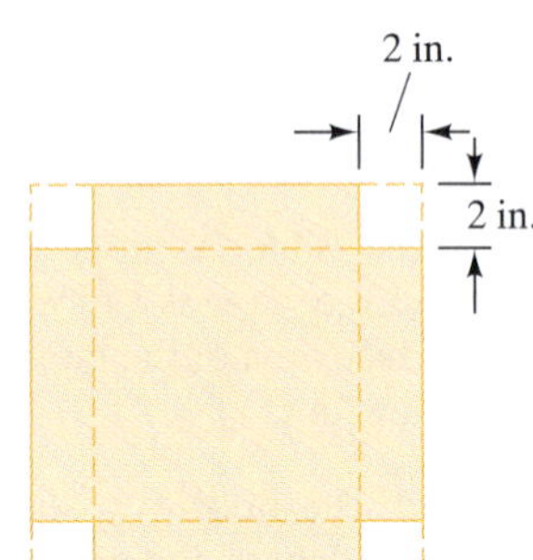


**114. Making a Box** A rectangular box is to be formed by cutting squares from each corner of a rectangular piece of tin and folding up the sides. The box is to be 3 inches high, the length is to be twice the width, and the volume of the box is to be 96 cubic inches. Find the length and width of the box.

**115. Cube** A solid cube with dimensions $a^3$ has a rectangular solid with dimensions $ab^2$ removed.

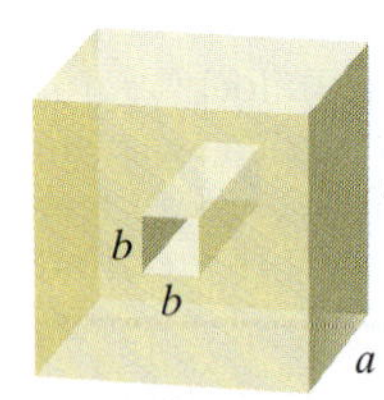


**a)** Write a formula for the remaining volume, $V$.

**b)** Factor the right side of the formula in part **a)**.

**c)** If the volume is 1620 cubic inches and $a$ is 12 in., find $b$.

**116. Circular Steel Blade** A circular steel blade has a hole cut out of its center as shown in the figure.

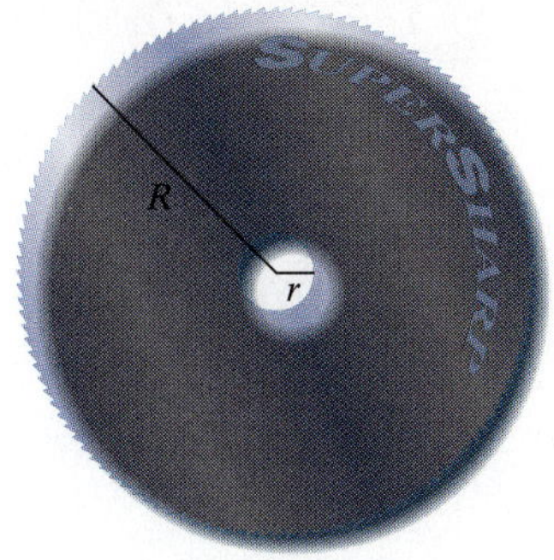


a) Write a formula for the remaining area of the blade.

b) Factor the right side of the formula in part **a)**.

c) Find $A$ if $R = 10$ cm and $r = 3$ cm.

**117.** Consider the following graph of a quadratic function.

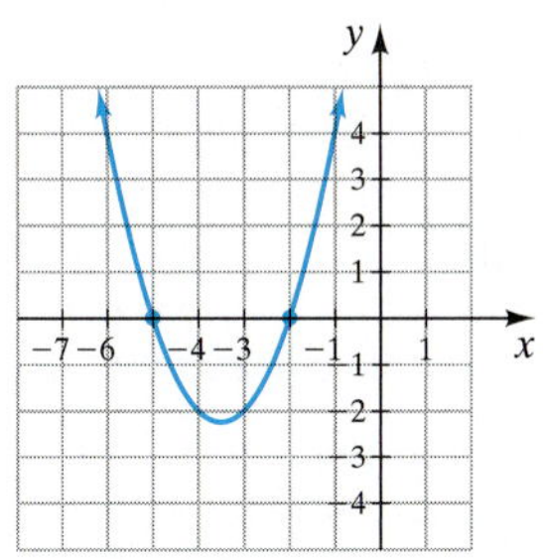

a) Write a quadratic function that has the $x$-intercepts indicated.

b) Write a quadratic equation in one variable that has solutions of $-2$ and $-5$.

c) How many different quadratic functions can have $x$-intercepts of $-2$ and $-5$? Explain.

d) How many different quadratic equations in one variable can have solutions of $-2$ and $-5$? Explain.

**118.** The graph of the equation $y = x^2 + 4$ is illustrated below.

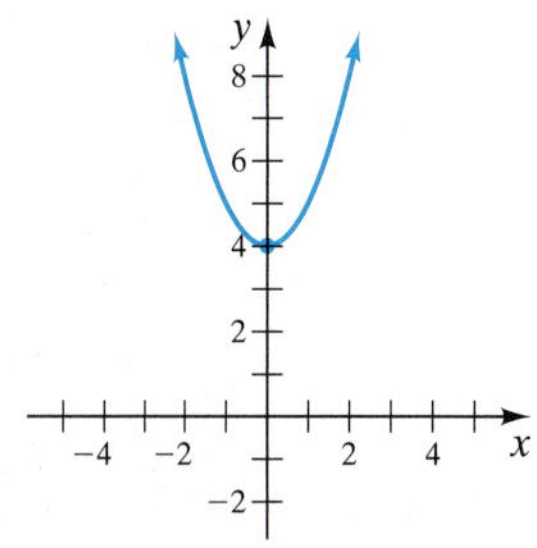

a) How many $x$-intercepts does the graph have?

b) How many real solutions does the equation $x^2 + 4 = 0$ have? Explain your answer.

**119.** Consider the quadratic function

$$P(x) = ax^2 + bx + c, a > 0.$$

a) The graph of this type of function may have no $x$-intercepts, one $x$-intercept, or two $x$-intercepts. Sketch each of these possibilities.

b) How many possible real solutions may the equation $ax^2 + bx + c = 0, a > 0$ have? Explain your answer to part **b)** by using the sketches in part **a)**.

**120. Stopping Distance** A typical car's stopping distance on dry pavement, $d$, in feet, can be approximated by the function $d(s) = 0.034s^2 + 0.56s - 17.11$, where $s$ is the speed of the car before braking and $60 \le s \le 80$ miles per hour. How fast is the car going if it requires 190 ft to stop after the brakes are applied?

**121. Stopping Distance** A typical car's stopping distance on wet pavement, $d$, in feet, can be approximated by the function $d(s) = -0.31s^2 + 59.82s - 2180.22$, where $s$ is the speed of the car before braking and $60 \le s \le 80$ miles per hour. How fast is the car going if it requires 545 feet for the car to stop after the brakes are applied?

## Challenge Problems

*Solve.*

**122.** $x^4 - 17x^2 + 16 = 0$

**123.** $x^4 - 13x^2 = -36$

**124.** $x^6 - 9x^3 + 8 = 0$

## Group Activity

*In more advanced mathematics courses you may need to solve an equation for $y'$ (read "y prime"). When doing so, treat the $y'$ as a different variable from y. Individually solve each equation for $y'$. Compare your answers and as a group obtain the correct answers.*

**125.** $xy' + yy' = 1$

**126.** $xy - xy' = 3y' + 2$

**127.** $2xyy' - xy = x - 3y'$

## Cumulative Review Exercises

[1.5] **128.** Simplify $(4x^{-2}y^3)^{-2}$.

[2.5] **129.** Solve the inequality and graph the solution on the number line.

$$-1 < \frac{4(3x - 2)}{3} \le 5$$

[4.1] **130.** Solve the system of equations.

$$3x + 4y = 2$$
$$2x = -5y - 1$$

[5.2] **131.** If $f(x) = -x^2 + 3x$ and $g(x) = x^2 + 5$, find $(f \cdot g)(4)$.

[5.7] **132.** Factor $(x + 1)^2 - (x + 1) - 6$.

# Chapter 5 Summary

| IMPORTANT FACTS AND CONCEPTS | EXAMPLES |
|---|---|
| **Section 5.1** | |
| **Terms** are parts that are added or subtracted in a mathematical expression. | The terms of $-3x^2 + 1.6x + 15$ are $-3x^2$, $1.6x$, and $15$. |
| A **polynomial** is a finite sum of terms in which all variables have whole number exponents and no variable appears in the denominator. | $9x^7 - 3x^5 + 4x - \frac{1}{2}$ is a polynomial. |
| The **degree of a term** is the sum of the exponents of the variables. | The term $3x^2y^9$ is of degree 11. |
| The **leading term** of a polynomial is the term of highest degree. The **leading coefficient** is the coefficient of the leading term. | In the polynomial $9x^7 - 3x^5 + 4x - \frac{1}{2}$, the leading term is $9x^7$ and the leading coefficient is 9. |
| A **monomial** is a polynomial of one term.<br>A **binomial** is a polynomial of two terms.<br>A **trinomial** is a polynomial of three terms. | $-13mn^2p^3$<br>$x^4 - 1$<br>$1.9x^3 - 28.3x^2 - 101.5x$ |
| A polynomial is **linear** if it is of degree 0 or 1.<br>A polynomial in one variable is **quadratic** if it is of degree 2.<br>A polynomial in one variable is **cubic** if it is of degree 3. | $19, 8y + 17$<br>$x^2 - 5x + 16$<br>$-4x^3 + 11x^2 - 9x + 6$ |
| A **polynomial function** has the form $y = P(x)$. To evaluate $P(a)$, replace $x$ by $a$. | $P(x) = 2x^2 - x + 3$ is a polynomial function. To evaluate $P(x)$ at $x = 10$,<br>$P(10) = 2(10)^2 - 10 + 3$<br>$= 200 - 10 + 3 = 193$ |
| To **add or subtract polynomials**, combine like terms. | $(5x^2 - 9x + 10) + (2x^2 + 17x - 8)$<br>$= 5x^2 - 9x + 10 + 2x^2 + 17x - 8 = 7x^2 + 8x + 2$ |
| **Section 5.2** | |
| To **multiply polynomials**, multiply each term of one polynomial by each term of the other polynomial. | $3a(a - 2) = 3a \cdot a - 3a \cdot 2$<br>$= 3a^2 - 6a$ |
| **Distributive Property, Expanded Form**<br>$a(b + c + d + \cdots + n) = ab + ac + ad + \cdots + an$ | $x(2x^2 + 8x - 5) = 2x^3 + 8x^2 - 5x$ |
| To **multiply two binomials**, use the **FOIL method:** multiply the **F**irst terms, **O**uter terms, **I**nner terms, **L**ast terms. | $(3x - 1)(4x + 9) = 12x^2 + 27x - 4x - 9$<br>$= 12x^2 + 23x - 9$ |
| To multiply a polynomial by a polynomial, you can use the vertical format. | Multiply $(2x^2 - x + 8)(5x + 1)$.<br>$\begin{array}{r} 2x^2 - \phantom{5}x + 8 \\ 5x + 1 \\ \hline 2x^2 - \phantom{5}x + 8 \\ 10x^3 - 5x^2 + 40x \phantom{{}+8} \\ \hline 10x^3 - 3x^2 + 39x + 8 \end{array}$ |

| IMPORTANT FACTS AND CONCEPTS | EXAMPLES |
|---|---|
| **Section 5.2 (continued)** | |
| **Square of a Binomial**<br>$(a+b)^2 = a^2 + 2ab + b^2$<br>$(a-b)^2 = a^2 - 2ab + b^2$ | $(7x+4)^2 = (7x)^2 + 2(7x)(4) + 4^2$<br>$= 49x^2 + 56x + 16$<br>$\left(\frac{1}{2}m - 3\right)^2 = \left(\frac{1}{2}m\right)^2 - 2\left(\frac{1}{2}m\right)(3) + 3^2 = \frac{1}{4}m^2 - 3m + 9$ |
| **Product of the Sum and Difference of the Same Two Terms**<br>$(a+b)(a-b) = a^2 - b^2$ | $(5c+6)(5c-6) = (5c)^2 - 6^2 = 25c^2 - 36$ |
| **Section 5.3** | |
| To divide a polynomial by a monomial, divide each term of the polynomial by the monomial. | $\frac{6y + 10x^2y^5 - 17x^9y^8}{2xy^2} = \frac{6y}{2xy^2} + \frac{10x^2y^5}{2xy^2} - \frac{17x^9y^8}{2xy^2}$<br>$= \frac{3}{xy} + 5xy^3 - \frac{17x^8y^6}{2}$ |
| To divide two polynomials, use long division. | Divide $(8x^2 + 6x - 9) \div (2x + 1)$.<br>$\begin{array}{r} 4x + 1 \\ 2x+1\overline{)8x^2 + 6x - 9} \\ \underline{8x^2 + 4x} \\ 2x - 9 \\ \underline{2x + 1} \\ -10 \end{array}$<br>Thus, $\frac{8x^2 + 6x - 9}{2x + 1} = 4x + 1 - \frac{10}{2x + 1}$ |
| To divide a polynomial by a binomial of the form $x - a$, use **synthetic division**. | Use synthetic division to divide<br>$(x^3 + 2x^2 - 11x + 5) \div (x + 4)$<br>$\begin{array}{r\|rrrr} -4 & 1 & 2 & -11 & 5 \\ & & -4 & 8 & 12 \\ \hline & 1 & -2 & -3 & 17 \end{array}$<br>Thus, $\frac{x^3 + 2x^2 - 11x + 5}{x + 4} = x^2 - 2x - 3 + \frac{17}{x + 4}$ |
| **Remainder Theorem**<br>If the polynomial $P(x)$ is divided by $x - a$, the remainder is $P(a)$. | Find the remainder when<br>$2x^3 - 6x^2 - 11x + 29$ is divided by $x + 2$.<br>Let $P(x) = 2x^3 - 6x^2 - 11x + 29$; then<br>$P(-2) = 2(-2)^3 - 6(-2)^2 - 11(-2) + 29$<br>$= -16 - 24 + 22 + 29$<br>$= 11$.<br>The remainder is 11. |
| **Section 5.4** | |
| The **greatest common factor** (GCF) is the product of the factors common to all terms in the polynomial. | The GCF of $z^5, z^4, z^9, z^2$ is $z^2$.<br>The GCF of $9(x-4)^3, 6(x-4)^{10}$ is $3(x-4)^3$. |

| IMPORTANT FACTS AND CONCEPTS | EXAMPLES |
|---|---|

**Section 5.4 (continued)**

**To Factor a Monomial from a Polynomial**

1. Determine the greatest common factor of all terms in the polynomial.
2. Write each term as the product of the GCF and another factor.
3. Use the distributive property to *factor out* the GCF.

$$35x^6 + 15x^4 + 5x^3 = 5x^3(7x^3) + 5x^3(3x) + 5x^3(1)$$
$$= 5x^3(7x^3 + 3x + 1)$$

$$4n(7n + 10) - 13(7n + 10) = (7n + 10)(4n - 13)$$

**To Factor Four Terms by Grouping**

1. Determine if all four terms have a common factor. If so, factor out the GCF from each term.
2. Arrange the four terms into two groups of two terms each. Each group of two terms must have a GCF.
3. Factor the GCF from each group of two terms.
4. If the two terms formed in step 3 have a GCF, factor it out.

$$cx + cy + dx + dy = c(x + y) + d(x + y)$$
$$= (x + y)(c + d)$$

$$x^3 + 6x^2 - 5x - 30 = x^2(x + 6) - 5(x + 6)$$
$$= (x + 6)(x^2 - 5)$$

**Section 5.5**

**To Factor Trinomials of the Form $x^2 + bx + c$**

1. Find two numbers (or factors) whose product is $c$ and whose sum is $b$.
2. The factors of the trinomial will be of the form

$$(x + \square)(x + \square)$$

One factor determined in step 1 — Other factor determined in step 1

Factor $m^2 - m - 42$.

The factors of $-42$ whose sum is $-1$ are $-7$ and 6. Note that $(-7)(6) = -42$ and $-7 + 6 = -1$. Therefore,

$$m^2 - m - 42 = (m - 7)(m + 6)$$

**To Factor Trinomials of the Form $ax^2 + bx + c, a \neq 1$, Using Trial and Error**

1. Write all pairs of factors of the coefficient of the squared term, $a$.
2. Write all pairs of factors of the constant, $c$.
3. Try various combinations of these factors until the correct middle term, $bx$, is found.

$$4t^2 + 9t + 5 = (4t + 5)(t + 1)$$

Notice that $4t + 5t = 9t$.

$$2a^2 - 15ab + 28b^2 = (2a - 7b)(a - 4b)$$

Notice that $-8ab - 7ab = -15ab$.

**To Factor Trinomials of the Form $ax^2 + bx + c, a \neq 1$, Using Grouping**

1. Find two numbers whose product is $a \cdot c$ and whose sum is $b$.
2. Rewrite the middle term, $bx$, using the numbers found in step 1.
3. Factor by grouping.

Factor $2y^2 + 9y - 18$ by grouping.

Two numbers whose product is $-36$ and whose sum is 9 are 12 and $-3$. Therefore;

$$2y^2 + 9y - 18 = 2y^2 + 12y - 3y - 18$$
$$= 2y(y + 6) - 3(y + 6)$$
$$= (y + 6)(2y - 3)$$

A **prime polynomial** is a polynomial that cannot be factored.

$x^2 + 5x + 9$ is a prime polynomial.

**Factoring by substitution** occurs when one variable is substituted for another variable or expression.

Factor $a^6 - 2a^3 - 3$.

$$a^6 - 2a^3 - 3 = (a^3)^2 - 2a^3 - 3$$
$$= x^2 - 2x - 3 \quad \text{Substitute } x \text{ for } a^3.$$
$$= (x - 3)(x + 1)$$
$$= (a^3 - 3)(a^3 + 1) \quad \text{Substitute } a^3 \text{ for } x.$$

| IMPORTANT FACTS AND CONCEPTS | EXAMPLES |
|---|---|
| **Section 5.5 (continued)** | |
| **Difference of Two Squares**<br>$a^2 - b^2 = (a + b)(a - b)$ | $x^2 - 49 = x^2 - 7^2 = (x + 7)(x - 7)$ |
| **Perfect Square Trinomials**<br>$a^2 + 2ab + b^2 = (a + b)^2$<br>$a^2 - 2ab + b^2 = (a - b)^2$ | $d^2 + 8d + 16 = d^2 + 2(d)(4) + 4^2 = (d + 4)^2$<br>$4m^2 - 12m + 9 = (2m)^2 - 2(2m)(3) + 3^2 = (2m - 3)^2$ |
| **Sum of Two Cubes**<br>$a^3 + b^3 = (a + b)(a^2 - ab + b^2)$ | $y^3 + 8 = y^3 + 2^3 = (y + 2)(y^2 - 2y + 4)$ |
| **Difference of Two Cubes**<br>$a^3 - b^3 = (a - b)(a^2 + ab + b^2)$ | $27z^3 - 64x^3 = (3z)^3 - (4x)^3 = (3z - 4x)(9z^2 + 12xz + 16x^2)$ |
| **Section 5.7** | |
| **To Factor a Polynomial**<br>**1.** Determine whether all the terms in the polynomial have a greatest common factor other than 1. If so, factor out the GCF.<br>**2.** If the polynomial has two terms, determine whether it is a difference of two squares or a sum or difference of two cubes. If so, factor using the appropriate formula.<br>**3.** If the polynomial has three terms, determine whether it is a perfect square trinomial. If so, factor accordingly. If it is not, factor the trinomial using trial and error, grouping, or substitution as explained in Section 5.5.<br>**4.** If the polynomial has more than three terms, try factoring by grouping. If that does not work, see if three of the terms are the square of a binomial.<br>**5.** As a final step, examine your factored polynomial to see if any factors listed have a common factor and can be factored further. If you find a common factor, factor it out at this point. | $2x^7 + 16x^6 + 24x^5 = 2x^5(x^2 + 8x + 12)$<br>$= 2x^5(x + 6)(x + 2)$<br><br>$36a^6 - 100a^4b^2 = 4a^4(9a^2 - 25b^2)$<br>$= 4a^4[(3a)^2 - (5b)^2]$<br>$= 4a^4(3a + 5b)(3a - 5b)$<br><br>$125m^3 - 64 = (5m)^3 - 4^3$<br>$= (5m - 4)(25m^2 + 20m + 16)$ |
| **Section 5.8** | |
| A **polynomial equation** is formed when two polynomials are set equal to each other. | $x^2 - 5x = 2x + 7$ |
| A **quadratic equation** is a second-degree polynomial equation (in one variable).<br><br>**Standard Form of a Quadratic Equation**<br>$ax^2 + bx + c = 0, \quad a \neq 0$<br>where $a$, $b$, and $c$ are real numbers. | $2x^2 - 6x + 11 = 0$<br>$x^2 - 4 = x + 2$<br><br>$x^2 - 3x + 5 = 0$ is a quadratic equation in standard form. |

**IMPORTANT FACTS AND CONCEPTS** | **EXAMPLES**

**Section 5.8 (continued)**

### Zero-Factor Property

For all real numbers $a$ and $b$, if $a \cdot b = 0$, then either $a = 0$ or $b = 0$, or both $a$ and $b = 0$.

Solve $(x + 6)(x - 1) = 0$.

$$x + 6 = 0 \quad \text{or} \quad x - 1 = 0$$
$$x = -6 \qquad\qquad x = 1$$

The solutions are $-6$ and 1.

### To Solve an Equation by Factoring

1. Use the addition property to remove all terms from one side of the equation. This will result in one side of the equation being equal to 0.
2. Combine like terms in the equation and then factor.
3. Set each factor *containing a variable* equal to 0, solve the equations, and find the solutions.
4. Check the solutions in the *original* equation.

Solve $3x^2 + 13x - 4 = 2x$.

$$3x^2 + 11x - 4 = 0$$
$$(3x - 1)(x + 4) = 0$$
$$3x - 1 = 0 \quad \text{or} \quad x + 4 = 0$$
$$x = \frac{1}{3} \quad \text{or} \quad x = -4$$

A check shows $\frac{1}{3}$ and $-4$ are the solutions.

### Pythagorean Theorem

In a right triangle, if $a$ and $b$ represent the lengths of the legs and $c$ represents the length of the hypotenuse, then

$$\text{leg}^2 + \text{leg}^2 = \text{hyp}^2$$
$$a^2 + b^2 = c^2$$

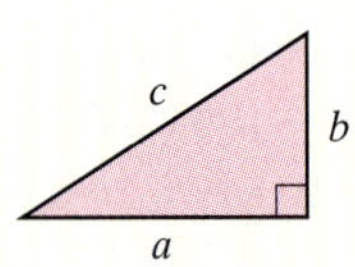


Find the length of the hypotenuse in the following right triangle.

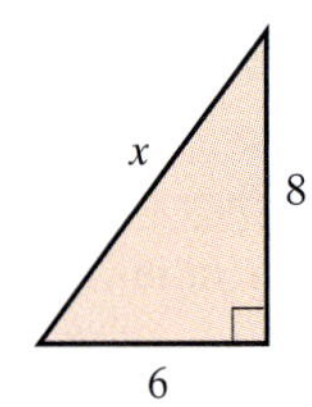


$$\text{leg}^2 + \text{leg}^2 = \text{hyp}^2$$
$$6^2 + 8^2 = x^2$$
$$36 + 64 = x^2$$
$$100 = x^2$$
$$10 = x$$

*Note*: $-10$ is not a possible answer.

## Chapter 5 Review Exercises

[5.1] *Determine whether each expression is a polynomial. If the expression is a polynomial,* **a)** *give the special name of the polynomial if it has one,* **b)** *write the polynomial in descending order of the variable x, and* **c)** *give the degree of the polynomial.*

**1.** $3x^2 + 9$

**2.** $5x + 4x^3 - 7$

**3.** $8x - x^{-1} + 6$

**4.** $-3 - 10x^2y + 6xy^3 + 2x^4$

*Perform each indicated operation.*

**5.** $(x^2 - 5x + 8) + (2x + 6)$

**6.** $(7x^2 + 2x - 5) - (2x^2 - 9x - 1)$

**7.** $(2a - 3b - 2) - (-a + 5b - 9)$

**8.** $(4x^3 - 4x^2 - 2x) + (2x^3 + 4x^2 - 7x + 13)$

**9.** $(3x^2y + 6xy - 5y^2) - (4y^2 + 3xy)$

**10.** $(-8ab + 2b^2 - 3a) + (-b^2 + 5ab + a)$

**11.** Add $x^2 - 3x + 12$ and $4x^2 + 10x - 9$.

**12.** Subtract $3a^2b - 2ab$ from $-7a^2b - ab$.

**13.** Find $P(2)$ if $P(x) = 2x^2 - 3x + 19$

**14.** Find $P(-3)$ if $P(x) = x^3 - 3x^2 + 4x - 10$

**Perimeter** *In Exercises 15 and 16, find a polynomial expression for the perimeter of each figure.*

**15.**

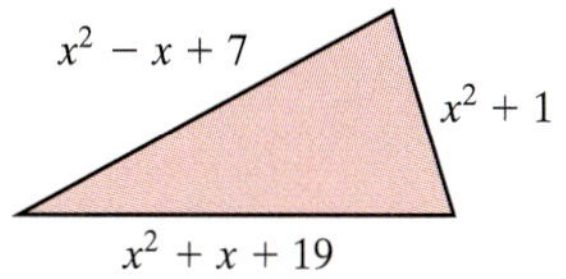


**16.**

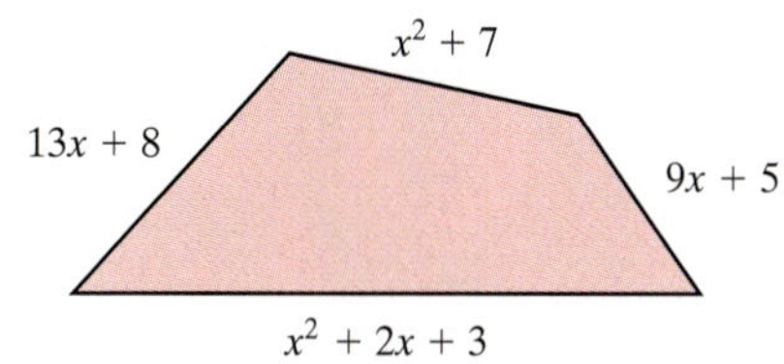


*Use the following graph in Exercises 17 and 18. The graph shows social security receipts and outlays from 1997 through 2025.*

**Social Security Receipts and Outlays**

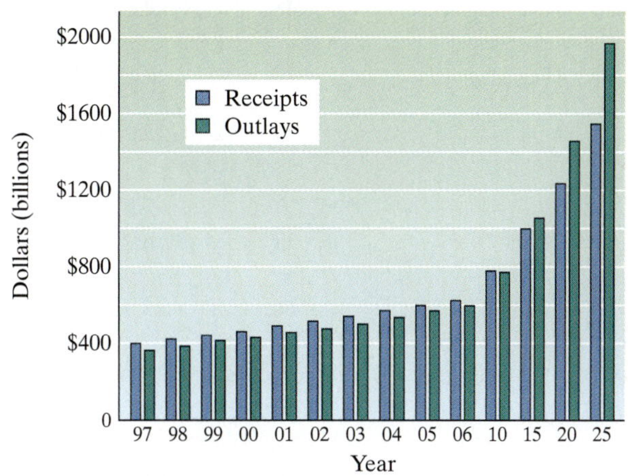


*Source:* Social Security Administration

**17. Social Security Receipts**
The function $R(t) = 0.78t^2 + 20.28t + 385.0$, where $t$ is years since 1997 and $0 \le t \le 28$, gives an approximation of social security receipts, $R(t)$, in billions of dollars.

**a)** Using the function provided, estimate the receipts in 2010.

**b)** Compare your answer in part **a)** with the graph. Does the graph support your answer?

**18. Social Security Outlays**
The function $G(t) = 1.74t^2 + 7.32t + 383.91$, where $t$ is years since 1997 and $0 \le t \le 28$, gives an approximation of social security outlays, $G(t)$, in billions of dollars.

**a)** Using the function provided, estimate the outlays in 2010.

**b)** Compare your answer in part **a)** with the graph. Does the graph support your answer?

[5.2] *Multiply.*

**19.** $2x(3x^2 - 7x + 5)$

**20.** $-3xy^2(x^3 + xy^4 - 4y^5)$

**21.** $(3x - 5)(2x + 9)$

**22.** $(5a + 1)(10a - 3)$

**23.** $(x + 8y)^2$

**24.** $(a - 11b)^2$

**25.** $(2xy - 1)(5x + 4y)$

**26.** $(2pq - r)(3pq + 7r)$

**27.** $(2a + 9b)^2$

**28.** $(4x - 3y)^2$

**29.** $(7x + 5y)(7x - 5y)$

**30.** $(2a - 5b^2)(2a + 5b^2)$

**31.** $(4xy + 6)(4xy - 6)$

**32.** $(9a^2 - 2b^2)(9a^2 + 2b^2)$

**33.** $[(x + 3y) + 2]^2$

**34.** $[(2p - q) - 5]^2$

**35.** $(3x^2 + 4x - 6)(2x - 3)$

**36.** $(4x^3 + 6x - 2)(x + 3)$

**Area** *In Exercises 37 and 38, find an expression for the total area of each figure.*

**37.**

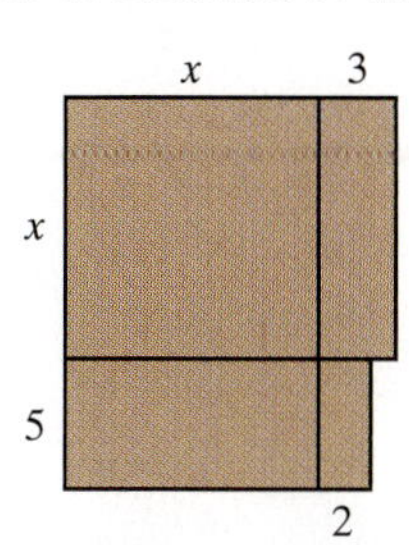


**38.**

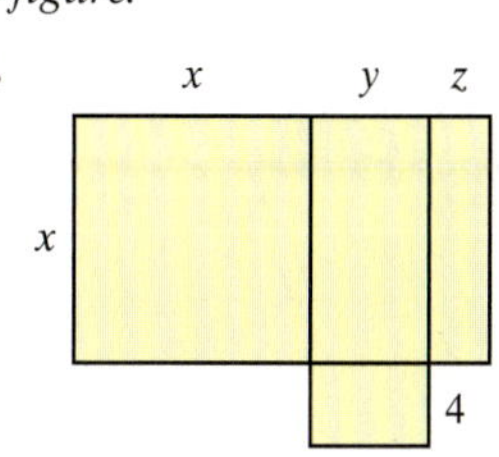

*For each pair of functions, find* **a)** $(f \cdot g)(x)$ *and* **b)** $(f \cdot g)(3)$.

**39.** $f(x) = x + 1, g(x) = x - 3$

**40.** $f(x) = 2x - 4, g(x) = x^2 - 3$

**41.** $f(x) = x^2 + x - 3, g(x) = x - 2$

**42.** $f(x) = x^2 - 2, g(x) = x^2 + 2$

[5.3] *Divide.*

**43.** $\dfrac{4x^7y^5}{20xy^3}$

**44.** $\dfrac{3s^5t^8}{12s^5t^3}$

**45.** $\dfrac{45pq - 25q^2 - 15q}{5q}$

**46.** $\dfrac{7a^2 - 16a + 32}{4}$

**47.** $\dfrac{2x^3y^2 + 8x^2y^3 + 12xy^4}{8xy^3}$

**48.** $(8x^2 + 14x - 15) \div (2x + 5)$

**49.** $(2x^4 - 3x^3 + 4x^2 + 17x + 7) \div (2x + 1)$

**50.** $(4a^4 - 7a^2 - 5a + 4) \div (2a - 1)$

**51.** $(x^2 + x - 22) \div (x - 3)$

**52.** $(4x^3 + 12x^2 + x - 9) \div (2x + 3)$

*Use synthetic division to obtain each quotient.*

**53.** $(3x^3 - 2x^2 + 10) \div (x - 3)$

**54.** $(2y^5 - 10y^3 + y - 2) \div (y + 1)$

**55.** $(x^5 - 18) \div (x - 2)$

**56.** $(2x^3 + x^2 + 5x - 3) \div \left(x - \dfrac{1}{2}\right)$

*Determine the remainder of each division problem using the Remainder Theorem. If the divisor is a factor of the dividend, so state.*

**57.** $(x^2 - 4x + 13) \div (x - 3)$

**58.** $(2x^3 - 6x^2 + 3x) \div (x + 4)$

**59.** $(3x^3 - 6) \div \left(x - \dfrac{1}{3}\right)$

**60.** $(2x^4 - 6x^2 - 8) \div (x + 2)$

[5.4] *Factor out the greatest common factor in each expression.*

**61.** $4x^2 + 8x + 32$

**62.** $15x^5 + 6x^4 - 12x^5y^3$

**63.** $10a^3b^3 - 14a^2b^6$

**64.** $24xy^4z^3 + 12x^2y^3z^2 - 30x^3y^2z^3$

*Factor by grouping.*

**65.** $5x^2 - xy + 30xy - 6y^2$

**66.** $12a^2 + 8ab + 15ab + 10b^2$

**67.** $(2x - 5)(2x + 1) - (2x - 5)(x - 8)$

**68.** $7x(3x - 7) + 3(3x - 7)^2$

**Area** *In Exercises 69 and 70, A represents the area of the figure. Find an expression, in factored form, for the difference of the areas of the geometric figures.*

**69.**

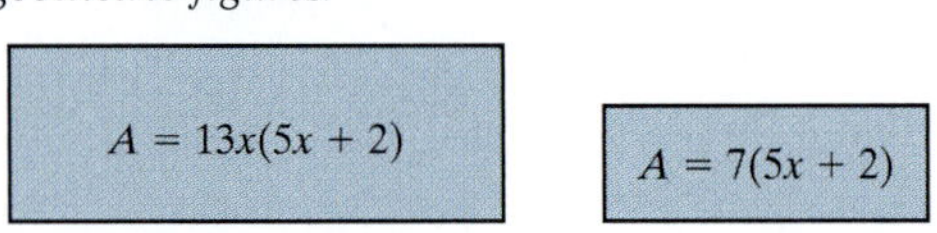


**70.**

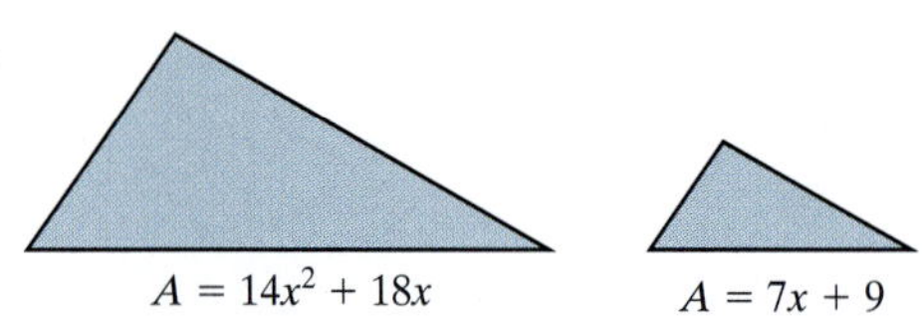


**Volume** *In Exercises 71 and 72, V represents the volume of the figure. Find an expression, in factored form, for the difference of the volumes of the geometric figures.*

**71.**

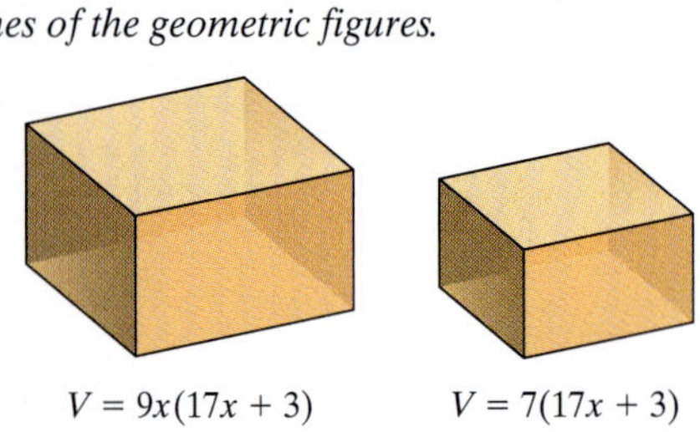


**72.**

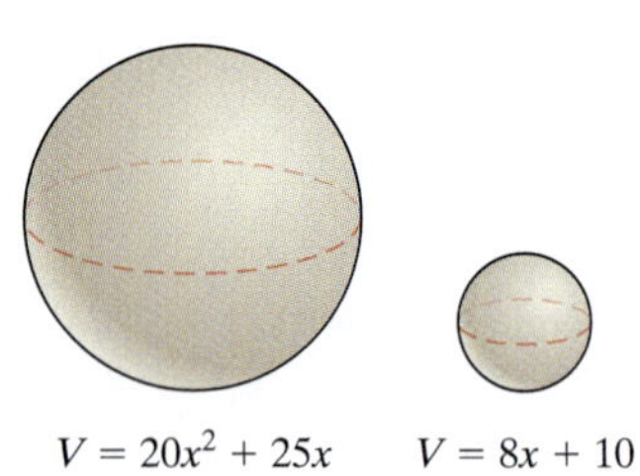


[5.5] *Factor each trinomial.*

**73.** $x^2 + 9x + 18$

**74.** $x^2 + 3x - 10$

**75.** $x^2 - 3x - 28$

**76.** $x^2 - 10x + 16$

**77.** $-x^2 + 12x + 45$

**78.** $-x^2 + 13x - 12$

**79.** $2x^3 + 13x^2 + 6x$

**80.** $8x^4 + 10x^3 - 25x^2$

**81.** $4a^5 - 9a^4 + 5a^3$

**82.** $12y^5 + 61y^4 + 5y^3$

**83.** $x^2 - 15xy - 54y^2$

**84.** $6p^2 - 19pq + 10q^2$

**85.** $x^4 + 10x^2 + 21$

**86.** $x^4 + 2x^2 - 63$

**87.** $(x + 3)^2 + 10(x + 3) + 24$

**88.** $(x - 4)^2 - (x - 4) - 20$

**Area** *In Exercises 89 and 90, find an expression, in factored form, for the area of the shaded region in each figure.*

**89.**

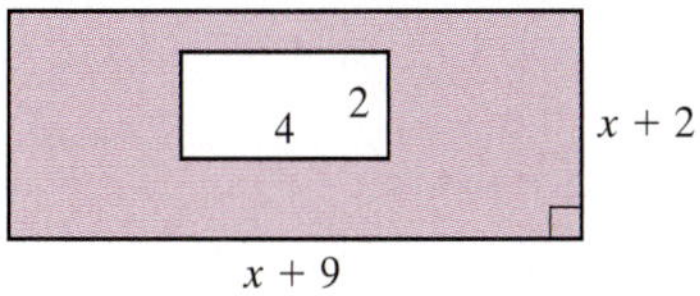


**90.**

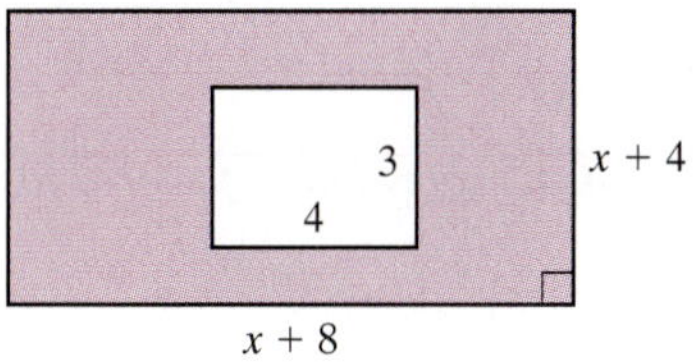


[5.6] *Use a special factoring formula to factor the following.*

**91.** $x^2 - 36$

**92.** $x^2 - 121$

**93.** $x^4 - 81$

**94.** $x^4 - 16$

**95.** $4a^2 + 4a + 1$

**96.** $16y^2 - 24y + 9$

**97.** $(x + 2)^2 - 16$

**98.** $(3y - 1)^2 - 36$

**99.** $p^4 + 18p^2 + 81$

**100.** $m^4 - 20m^2 + 100$

**101.** $x^2 + 8x + 16 - y^2$

**102.** $a^2 + 6ab + 9b^2 - 36c^2$

**103.** $16x^2 + 8xy + y^2$

**104.** $36b^2 - 60bc + 25c^2$

**105.** $x^3 - 27$

**106.** $y^3 + 64z^3$

**107.** $125x^3 - 1$

**108.** $8a^3 + 27b^3$

**109.** $y^3 - 64z^3$

**110.** $(x - 2)^3 - 27$

**111.** $(x + 1)^3 - 8$

**112.** $(a + 4)^3 + 1$

**Area** *In Exercises 113 and 114, find an expression, in factored form, for the area of the shaded region in each figure.*

**113.**

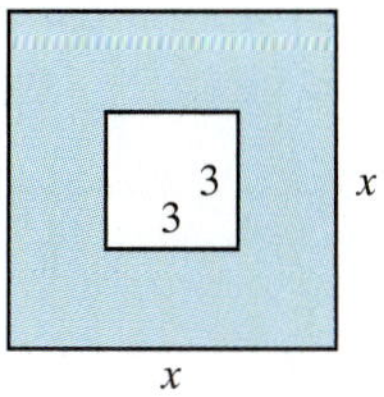


**114.**

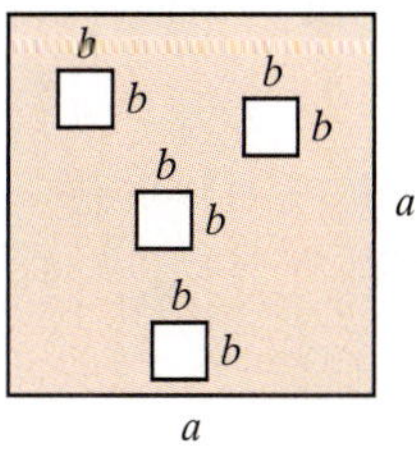


**115. Volume** Find an expression, in factored form, for the difference in volumes of the two cubes below.

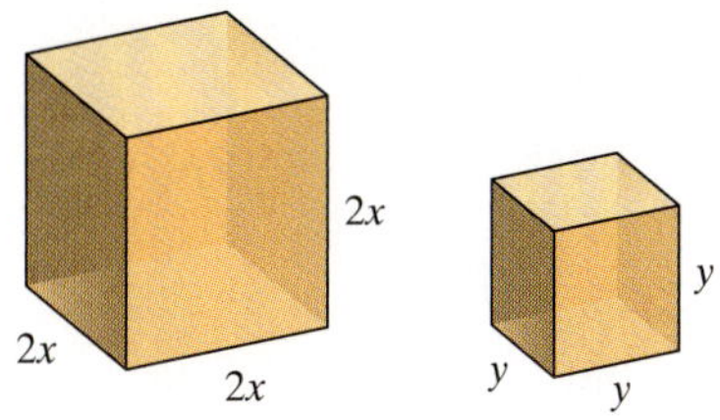


**116. Volume** Find an expression, in factored form, for the volume of the shaded region in the figure below.

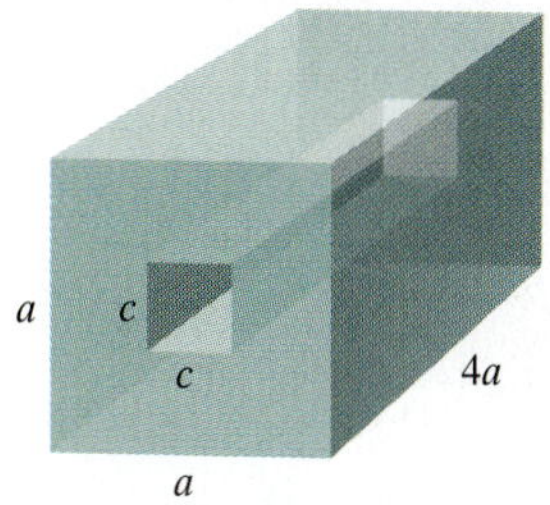


[5.4–5.7] *Factor completely.*

**117.** $x^2y^4 - 2xy^4 - 15y^4$

**118.** $5x^3 - 30x^2 + 40x$

**119.** $3x^3y^4 + 18x^2y^4 - 6x^2y^4 - 36xy^4$

**120.** $3y^5 - 75y$

**121.** $4x^3y + 32y$

**122.** $5x^4y + 20x^3y + 20x^2y$

**123.** $6x^3 - 21x^2 - 12x$

**124.** $x^2 + 10x + 25 - z^2$

**125.** $5x^3 + 40y^3$

**126.** $x^2(x + 6) + 3x(x + 6) - 4(x + 6)$

**127.** $4(2x + 3)^2 - 12(2x + 3) + 5$

**128.** $4x^4 + 4x^2 - 3$

**129.** $(x + 1)x^2 - (x + 1)x - 2(x + 1)$

**130.** $9ax - 3bx + 21ay - 7by$

**131.** $6p^2q^2 - 5pq - 6$

**132.** $9x^4 - 12x^2 + 4$

**133.** $16y^2 - (x^2 + 4x + 4)$

**134.** $6(2a + 3)^2 - 7(2a + 3) - 3$

**135.** $6x^4y^5 + 9x^3y^5 - 27x^2y^5$

**136.** $x^3 - \frac{8}{27}y^6$

**Area** *In Exercises 137–142, find an expression, in factored form, for the area of the shaded region in each figure.*

**137.**

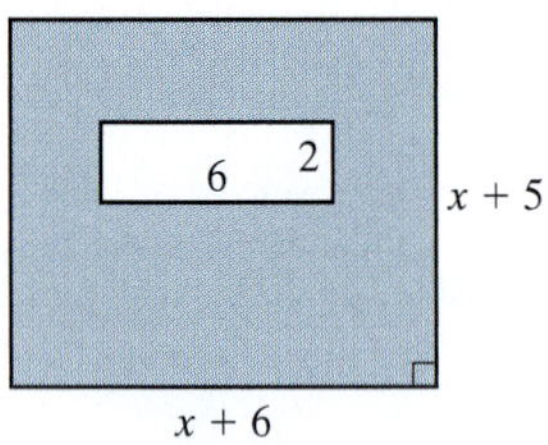


**138.**

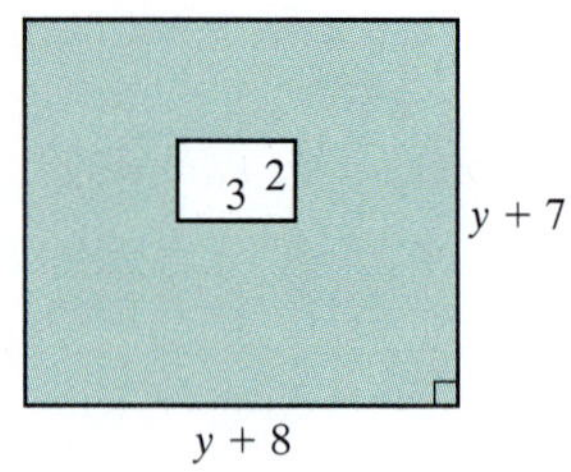


**139.**

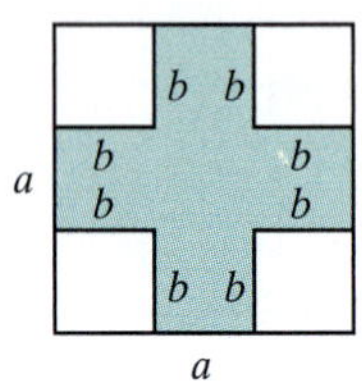


**140.**

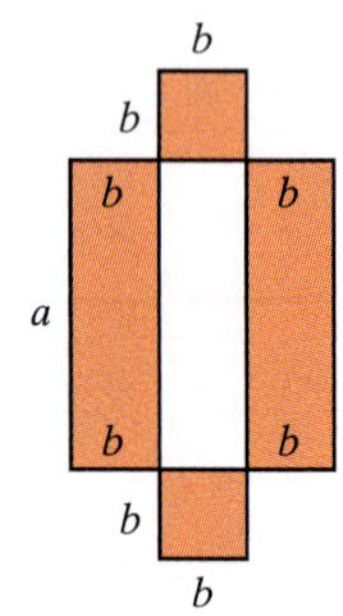


**141.**

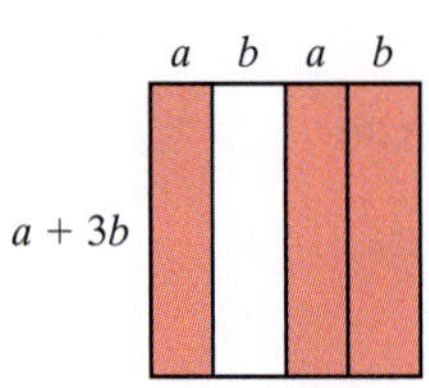


**142.**

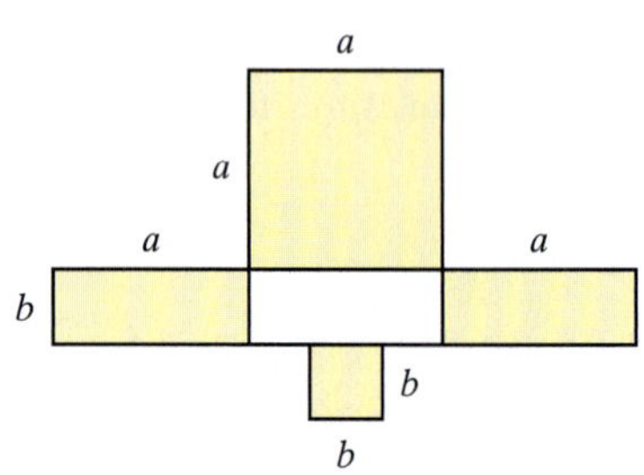


[5.8] *Solve.*

**143.** $(x - 2)(4x + 1) = 0$

**144.** $(2x + 5)(3x + 10) = 0$

**145.** $4x^2 = 8x$

**146.** $12x^2 + 16x = 0$

**147.** $x^2 + 7x + 12 = 0$

**148.** $a^2 + a - 30 = 0$

**149.** $x^2 = 8x - 7$

**150.** $c^3 - 6c^2 + 8c = 0$

**151.** $5x^2 = 80$

**152.** $x(x + 3) = 2(x + 4) - 2$

**153.** $12d^2 = 13d + 4$

**154.** $20p^2 - 6 = 7p$

*Use factoring to find the x-intercepts of the graph of each equation.*

**155.** $y = 2x^2 - 6x - 36$

**156.** $y = 20x^2 - 49x + 30$

*Write an equation whose graph will have x-intercepts at the given values.*

**157.** $-4$ and $6$

**158.** $-\frac{5}{2}$ and $-\frac{1}{6}$

*In Exercises 159–163, answer the question.*

**159. Carpeting** The area of Fred Bank's rectangular carpet is 108 square feet. Find the length and width of the carpet if the length is 3 feet greater than the width.

**160. Triangular Sign** The base of a large triangular sign is 5 feet more than twice the height. Find the base and height if the area of the triangle is 26 square feet.

**161. Square** One square has a side 4 inches longer than the side of a second square. If the area of the larger square is 49 square inches, find the length of a side of each square.

**162. Velocity** A rocket is projected upward from the top of a 144-foot-tall building with a velocity of 128 feet per second. The rocket's distance from the ground, $s$, at any time, $t$, in seconds, is given by the formula $s(t) = -16t^2 + 128t + 144$. Find the time it takes for the rocket to strike the ground.

**163. Telephone Pole** Two guy wires are attached to a telephone pole to help stabilize it. One wire is attached to the ground $x$ feet from the base of the pole. The height of the pole is $x + 31$ and the length of the wire is $x + 32$. Find $x$.

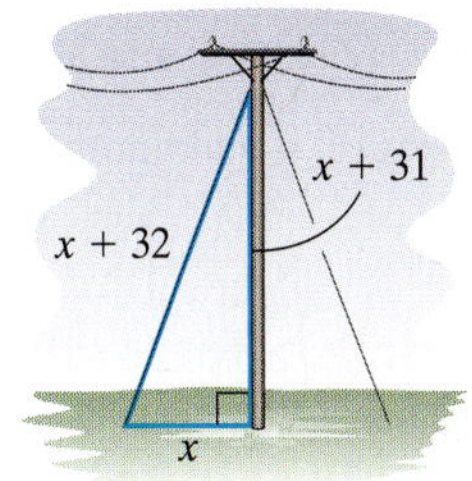


## Chapter 5 Practice Test

*To find out how well you understand the chapter material, take this practice test. The answers, and the section where the material was initially discussed, are given in the back of the book. Each problem is also fully worked out on the* **Chapter Test Prep Video CD**. *Review any questions that you answered incorrectly.*

1. **a)** Give the specific name of the following polynomial.
$$-4x^2 + 3x - 6x^4$$
**b)** Write the polynomial in descending powers of the variable $x$.
**c)** State the degree of the polynomial.
**d)** What is the leading coefficient of the polynomial?

*Perform each operation.*

2. $(7x^2y - 5y^2 + 4x) - (3x^2y + 9y^2 - 6y)$
3. $2x^3y^2(-4x^5y + 12x^3y^2 - 6x)$
4. $(2a - 3b)(5a + b)$
5. $(2x^2 + 3xy - 6y^2)(2x + y)$
6. $(12x^6 - 15x^2y + 21) \div 3x^2$
7. $(2x^2 - 7x + 9) \div (2x + 3)$
8. Use synthetic division to obtain the quotient. $(3x^4 - 12x^3 - 60x + 1) \div (x - 5)$
9. Use the Remainder Theorem to find the remainder when $2x^3 - 6x^2 - 5x + 8$ is divided by $x + 3$.

*Factor completely.*

10. $12x^3y + 10x^2y^4 - 14xy^3$
11. $x^3 - 2x^2 - 3x$
12. $2a^2 + 4ab + 3ab + 6b^2$
13. $2b^4 + 5b^2 - 18$
14. $4(x - 5)^2 + 20(x - 5)$
15. $(x + 4)^2 + 2(x + 4) - 3$
16. $27p^3q^6 - 8q^6$
17. If $f(x) = 3x - 4$ and $g(x) = x - 5$, find **a)** $(f \cdot g)(x)$ and **b)** $(f \cdot g)(2)$

**Area** *In Exercises 18 and 19, find an expression, in factored form, for the area of the shaded region.*

18.

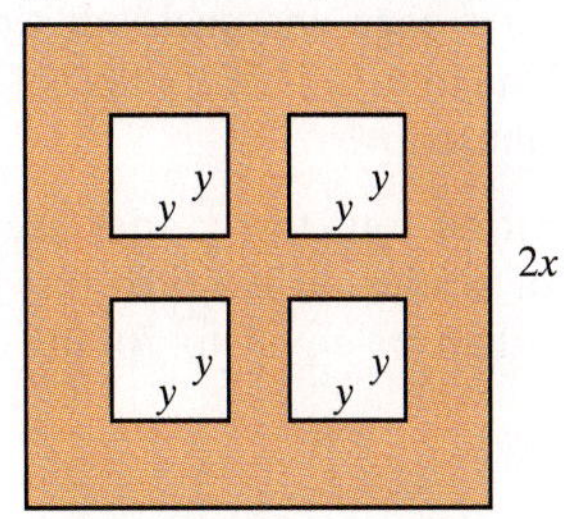


19.

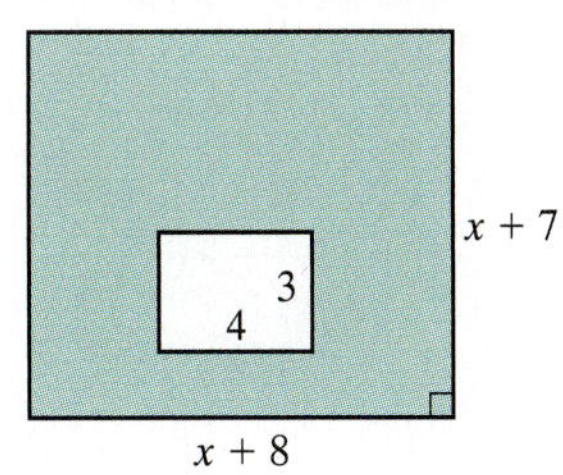


*Solve.*

20. $7x^2 + 25x - 12 = 0$
21. $x^3 + 3x^2 - 10x = 0$
22. Use factoring to find the $x$-intercepts of the graph of the equation $y = 8x^2 + 10x - 3$.
23. Find an equation whose graph has $x$-intercepts at 2 and 7.
24. **Area** The area of a triangle is 22 square meters. If the base of the triangle is 3 meters greater than 2 times the height, find the base and height of the triangle.
25. **Baseball** A baseball is projected upward from the top of a 448-foot-tall building with an initial velocity of 48 feet per second. The distance, $s$, of the baseball from the ground at any time, $t$, in seconds, is given by the equation $s(t) = -16t^2 + 48t + 448$. Find the time it takes for the baseball to strike the ground.

## Cumulative Review Test

*Take the following test and check your answers with those given in the back of the book. Review any questions that you answered incorrectly. The section where the material was covered is indicated after the answer.*

**1.** Find $A \cup B$ for $A = \{2, 4, 6, 8\}$ and $B = \{3, 5, 6, 8\}$.

**2.** Illustrate $\{x | x \le -5\}$ on a real number line.

**3.** Divide $\left|\frac{3}{8}\right| \div (-4)$.

**4.** Evaluate $(-3)^3 - 2^2 - (-2)^2 + (9 - 8)^2$.

**5.** Simplify $\left(\frac{2r^4s^5}{r^2}\right)^3$.

**6.** Solve $4(2x - 2) - 3(x + 7) = -4$.

**7.** Solve $k = 2(d + e)$ for $e$.

**8. Landscaping** Craig Campanella, a landscape architect, wishes to fence in two equal areas as illustrated in the figure. If both areas are squares and the total length of fencing used is 91 meters, find the dimensions of each square.

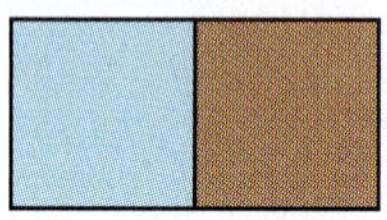

**9. Making Copies** Cecil Winthrop has a manuscript. He needs to make 6 copies before sending it to his editor in Boston. The first copy costs 15 cents per page and each additional copy costs 5 cents per page. If the total bill before tax is \$248, how many pages are in the manuscript?

**10. Test Average** Todd Garner's first four test grades are 68, 72, 90, and 86. What range of grades on his fifth test will result in an average greater than or equal to 70 and less than 80?

**11.** Is $(4, 1)$ a solution to the equation $3x + 2y = 13$?

**12.** Write the equation $2 = 6x - 3y$ in standard form.

**13.** Find the slope of the line passing through the points $(8, -4)$ and $(-1, -2)$.

**14.** If $f(x) = 2x^3 - 4x^2 + x + 16$, find $f(-4)$.

**15.** Graph the inequality $2x - y \le 6$.

**16.** Solve the system of equations.

$$\frac{1}{5}x + \frac{1}{2}y = 4$$
$$\frac{2}{3}x - y = \frac{8}{3}$$

**17.** Solve the system of equations.

$$x - 2y = 2$$
$$2x + 3y = 11$$
$$-y + 4z = 7$$

**18.** Evaluate the determinant.

$$\begin{vmatrix} 8 & 5 \\ -2 & 1 \end{vmatrix}$$

**19.** Divide $(2x^3 - 9x + 15) \div (x - 6)$.

**20.** Factor $64x^3 - 27y^3$.

# 6 Rational Expressions and Equations

## GOALS OF THIS CHAPTER

*Rational expressions* are expressions that contain fractions and *rational equations* are equations that contain rational expressions. In this chapter you will learn how to work with rational expressions, and solve rational equations. To be successful in this chapter, you must have an understanding of the factoring techniques discussed in Chapter 5.


6.1 The Domains of Rational Functions and Multiplication and Division of Rational Expressions

6.2 Addition and Subtraction of Rational Expressions

6.3 Complex Fractions

6.4 Solving Rational Equations

Mid-Chapter Test: Sections 6.1–6.4

6.5 Rational Equations: Applications and Problem Solving

6.6 Variation

Chapter 6 Summary

Chapter 6 Review Exercises

Chapter 6 Practice Test

Cumulative Review Test


WHEN TWO OR MORE people perform a task, it takes less time than for each person to do the task separately. For example, on pages 427 and 428, we will determine how long it takes two people working together to pick apples (Exercise 11), clean gutters (Exercise 13), or pull weeds (Exercise 14) when we know how long it takes for each person to complete the task separately.

# 6.1 The Domains of Rational Functions and Multiplication and Division of Rational Expressions

1 Find the domains of rational functions.

2 Simplify rational expressions.

3 Multiply rational expressions.

4 Divide rational expressions.

## 1 Find the Domains of Rational Functions

*To understand rational expressions, you must have a thorough understanding of the factoring techniques discussed in Chapter 5.* A **rational expression** is an expression of the form $\frac{p}{q}$, where $p$ and $q$ are polynomials and $q \neq 0$.

### Examples of Rational Expressions

$$\frac{2}{x}, \quad \frac{x+3}{x}, \quad \frac{x^2+4x}{x-6}, \quad \frac{a}{a^2-4}, \quad \frac{t^2-5t+7}{t^3+t^2-9t}$$

Note that the denominator of a rational expression cannot equal 0 because division by 0 is undefined. In the expression $\frac{x+3}{x}$, $x$ cannot equal 0, since the denominator would then equal 0. In $\frac{x^2+4x}{x-6}$ $x$ cannot equal 6 because that would result in the denominator having a value of 0. What values of $a$ cannot be used in the expression $\frac{a}{a^2-4}$? If you answered 2 and $-2$, you answered correctly.

**Whenever we write a rational expression containing a variable in the denominator, we always assume that the value or values of the variable that make the denominator 0 are excluded. For example, if we write $\frac{5}{x-3}$, we assume $x \neq 3$, even though we do not specifically write it.**

In Section 5.1, we discussed polynomial functions. Now we introduce rational functions. A **rational function** is a function of the form $f(x) = \frac{p}{q}$ or $y = \frac{p}{q}$, where $p$ and $q$ are polynomials and $q \neq 0$.

### Examples of Rational Functions

$$f(x) = \frac{4}{x} \qquad y = \frac{x^2+2}{x+3} \qquad T(a) = \frac{a+9}{a^2-4} \qquad h(x) = \frac{7x-8}{2x+1}$$

The **domain** of a rational function will be the set of values that can be used to replace the variable. For example, in the rational function $f(x) = \frac{x+2}{x-3}$, the domain will be all real numbers except 3, written $\{x|x \neq 3\}$. If $x$ were 3, the denominator would be 0, and division by 0 is undefined.

**EXAMPLE 1** ▸ For the functions $f(x)$ and $g(x)$, find the domain of $\left(\frac{f}{g}\right)(x)$.

**a)** $f(x) = x^2, g(x) = x^2 - 4$

**b)** $f(x) = x - 3, g(x) = x^2 + 2x - 15$

**c)** $f(x) = x, g(x) = x^2 + 8$

*Solution*

**a)** Since $f(x)$ and $g(x)$ are polynomial functions, each of their domains is the set of all real numbers. The domain of the quotient of functions $\left(\frac{f}{g}\right)(x)$ will therefore be all real numbers for which the denominator of the quotient is not equal to 0. From Section 3.6 we know that

$$\left(\frac{f}{g}\right)(x) = \frac{f(x)}{g(x)}.$$

Therefore,

$$\left(\frac{f}{g}\right)(x) = \frac{x^2}{x^2-4} \qquad \text{Substitute expressions for } f(x) \text{ and } g(x).$$

$$= \frac{x^2}{(x+2)(x-2)} \qquad \text{Factor the denominator.}$$

From this factored form we see that $x$ cannot be 2 or $-2$. Therefore the domain is all real numbers except 2 or $-2$. The domain may be expressed as $\{x|x \neq 2 \text{ and } x \neq -2\}$.

**b)**

$$\left(\frac{f}{g}\right)(x) = \frac{f(x)}{g(x)}$$

$$= \frac{x - 3}{x^2 + 2x - 15} \quad \text{Substitute expressions for } f(x) \text{ and } g(x).$$

$$= \frac{x - 3}{(x - 3)(x + 5)} \quad \text{Factor the denominator.}$$

Notice that the $x - 3$ in the numerator will divide out with the $x - 3$ in the denominator. However, when determining the domain of the quotient of functions, we do so *before* we simplify the expression. Since the denominator cannot be 0, $x$ cannot have values of 3 or $-5$. The domain is $\{x|x \neq 3 \text{ and } x \neq -5\}$.

**c)**

$$\left(\frac{f}{g}\right)(x) = \frac{f(x)}{g(x)}$$

$$= \frac{x}{x^2 + 8}$$

Since there is no value of $x$ that makes the denominator 0, the domain is all real numbers. The domain can be written as $\{x|x \text{ is a real number}\}$.

▸ Now Try Exercise 21

## USING YOUR GRAPHING CALCULATOR

If you have a graphing calculator, you may wish to experiment by graphing some rational functions. This will give you some idea of the wide variety of graphs of rational functions.

If you graph $y = \dfrac{x^2}{x^2 - 4}$ from Example 1**a)** on a graphing calculator, the display might look like that in **Figure 6.1**.

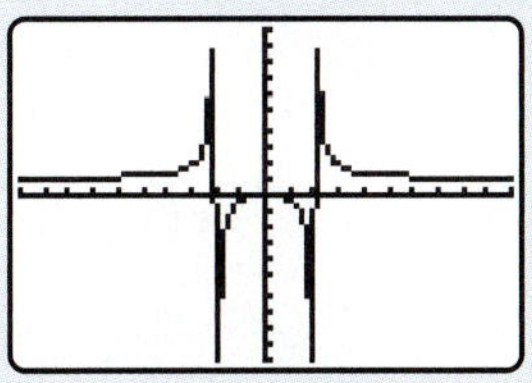

FIGURE 6.1

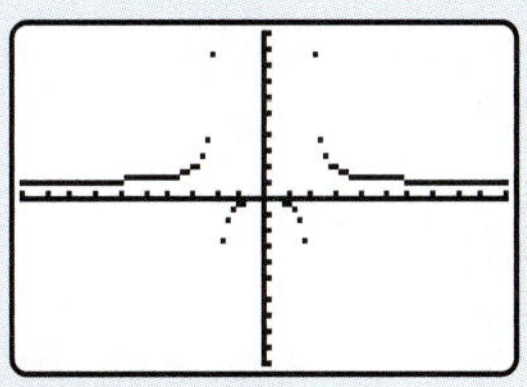

FIGURE 6.2

The domain of this function is all real numbers except 2 and $-2$.

Notice what appear to be vertical lines at $x = -2$ and $x = 2$, the values of $x$ where the function is undefined. This calculator is in a mode called *connected mode*. When a calculator is in connected mode, it connects all points it plots, going from the point with the smallest $x$-coordinate to the next larger one. Just to the left of $-2$, the value of $y$ is a large positive number, and just to the right of $-2$, the value of $y$ is a large negative number. The vertical line is the calculator's attempt to connect the point with this large positive $y$-value to the point with this large negative $y$-value. A similar situation occurs at $x = 2$.

You may sometimes wish to have your calculator in *dot mode*. When the calculator is in dot mode it displays unconnected points that have been calculated. Read the manual that comes with your calculator to learn how to change from connected mode to dot mode, and vice versa. The graph in **Figure 6.2** shows the same graph as in **Figure 6.1** except that this time the calculator is in dot mode.

## 2 Simplify Rational Expressions

When we work problems containing rational expressions, we must make sure that we write the answer in lowest terms. A rational expression is **simplified** when the numerator and denominator have no common factors other than 1. The fraction $\frac{6}{9}$ is not simplified because the 6 and 9 both contain the common factor of 3. When the 3 is factored out, the simplified fraction is $\frac{2}{3}$.

$$\frac{6}{9} = \frac{\overset{1}{\cancel{3}} \cdot 2}{\underset{1}{\cancel{3}} \cdot 3} = \frac{2}{3}$$

The rational expression $\frac{ab - b^2}{2b}$ is not simplified because both the numerator and denominator have a common factor, $b$. To simplify this expression, factor $b$ from each term in the numerator; then divide it out.

$$\frac{ab - b^2}{2b} = \frac{\cancel{b}(a - b)}{2\cancel{b}} = \frac{a - b}{2}$$

Thus $\frac{ab - b^2}{2b}$ becomes $\frac{a - b}{2}$ when simplified.

**To Simplify Rational Expressions**

1. Factor both numerator and denominator as completely as possible.
2. Divide both the numerator and the denominator by any common factors.

**EXAMPLE 2** Simplify. **a)** $\frac{x^2 + 5x + 4}{x + 4}$ **b)** $\frac{3x^3 - 3x^2}{x^3 - x}$

*Solution*

**a)** Factor the numerator; then divide out the common factor.

$$\frac{x^2 + 5x + 4}{x + 4} = \frac{\cancel{(x + 4)}(x + 1)}{\cancel{x + 4}} = x + 1$$

The rational expression simplifies to $x + 1$.

**b)** Factor the numerator and denominator. Then divide out common factors.

$$\frac{3x^3 - 3x^2}{x^3 - x} = \frac{3x^2(x - 1)}{x(x^2 - 1)}$$

$$= \frac{3\overset{x}{\cancel{x^2}}\cancel{(x - 1)}}{\cancel{x}(x + 1)\cancel{(x - 1)}} \quad \text{Factor } x^2 - 1.$$

$$= \frac{3x}{x + 1}$$

The rational expression simplifies to $\frac{3x}{x + 1}$.

**Now Try Exercise 33**

When the terms in a numerator differ only in sign from the terms in a denominator, we can factor out $-1$ from either the numerator or denominator. *When $-1$ is factored from a polynomial, the sign of each term in the polynomial changes.* For example,

$$-2x + 3 = -1(2x - 3) = -(2x - 3)$$

$$6 - 5x = -1(-6 + 5x) = -(5x - 6)$$

$$-3x^2 + 8x - 6 = -1(3x^2 - 8x + 6) = -(3x^2 - 8x + 6)$$

**EXAMPLE 3** ▸ Simplify $\dfrac{27x^3 - 8}{2 - 3x}$.

*Solution*

$$\frac{27x^3 - 8}{2 - 3x} = \frac{(3x)^3 - (2)^3}{2 - 3x}$$ *Write the numerator as a difference of two cubes.*

$$= \frac{(3x - 2)(9x^2 + 6x + 4)}{2 - 3x}$$ *Factor; recall that $a^3 - b^3 = (a - b)(a^2 + ab + b^2)$.*

$$= \frac{\cancel{(3x - 2)}(9x^2 + 6x + 4)}{-1\cancel{(3x - 2)}}$$ *Factor −1 from the denominator and divide out common factors.*

$$= \frac{9x^2 + 6x + 4}{-1}$$

$$= -(9x^2 + 6x + 4) \quad \text{or} \quad -9x^2 - 6x - 4$$

▸ **Now Try Exercise 41**

## Avoiding Common Errors

INCORRECT: $\dfrac{\overset{x}{\cancel{x^2}} + 6}{\underset{1}{\cancel{x}}}$ INCORRECT: $\dfrac{x + \overset{2}{\cancel{8}}}{\underset{1}{\cancel{4}}}$

Remember that you can divide out only common **factors**. Therefore, the expressions $\dfrac{x^2 + 6}{x}$ and $\dfrac{x + 8}{4}$ cannot be simplified. Only when expressions are *multiplied* can they be factors. Neither of the expressions above can be simplified from their original form.

CORRECT:

$$\frac{x^2 - 4}{x - 2} = \frac{(x + 2)\cancel{(x - 2)}}{\cancel{x - 2}} = x + 2$$

INCORRECT:

$$\frac{\overset{x}{\cancel{x^2}} - \overset{2}{\cancel{4}}}{\underset{1}{\cancel{x}} - \underset{1}{\cancel{2}}}$$

## 3 Multiply Rational Expressions

Now that we know how to simplify a rational expression, we can discuss multiplication of rational expressions.

### To Multiply Rational Expressions

To multiply rational expressions, use the following rule:

$$\frac{a}{b} \cdot \frac{c}{d} = \frac{a \cdot c}{b \cdot d}, \quad b \neq 0, d \neq 0$$

To multiply rational expressions, follow these steps.

1. Factor all numerators and denominators as far as possible.
2. Divide out any common factors.
3. Multiply using the above rule.
4. Simplify the answer when possible. (This step is not necessary if step 2 is done completely!)

If all common factors were factored out in step 2, your answer in step 4 should be in simplified form. However, if you missed a common factor in step 2, you can factor it out in step 4 to obtain an answer that is simplified.

**EXAMPLE 4** ▸ Multiply. **a)** $\dfrac{x-5}{6x}\cdot\dfrac{x^2-2x}{x^2-7x+10}$ **b)** $\dfrac{2x-3}{x-4}\cdot\dfrac{x^2-8x+16}{3-2x}$

*Solution*

**a)** $$\frac{x-5}{6x}\cdot\frac{x^2-2x}{x^2-7x+10} = \frac{\cancel{x-5}}{6\cancel{x}}\cdot\frac{\cancel{x}\cancel{(x-2)}}{\cancel{(x-2)}\cancel{(x-5)}}$$ *Factor; divide out common factors.*

$$= \frac{1}{6}$$

**b)** $$\frac{2x-3}{x-4}\cdot\frac{x^2-8x+16}{3-2x} = \frac{2x-3}{x-4}\cdot\frac{(x-4)(x-4)}{3-2x}$$ *Factor.*

$$= \frac{\cancel{2x-3}}{\cancel{x-4}}\cdot\frac{\cancel{(x-4)}(x-4)}{-1\cancel{(2x-3)}}$$ *Factor −1 from denominator; divide out common factors.*

$$= \frac{x-4}{-1}$$

$$= -(x-4) \quad \text{or} \quad -x+4 \quad \text{or} \quad 4-x$$

▸ **Now Try Exercise 61**

**EXAMPLE 5** ▸ Multiply $\dfrac{x^2-y^2}{x+y}\cdot\dfrac{x+4y}{2x^2-xy-y^2}$.

*Solution*

$$\frac{x^2-y^2}{x+y}\cdot\frac{x+4y}{2x^2-xy-y^2} = \frac{\cancel{(x+y)}\cancel{(x-y)}}{\cancel{x+y}}\cdot\frac{x+4y}{(2x+y)\cancel{(x-y)}}$$ *Factor; divide out common factors.*

$$= \frac{x+4y}{2x+y}$$

▸ **Now Try Exercise 55**

**EXAMPLE 6** ▸ Multiply $\dfrac{ab-ac+bd-cd}{ab+ac+bd+cd}\cdot\dfrac{b^2+bc+bd+cd}{b^2+bd-bc-cd}$.

*Solution* Factor both numerators and denominators by grouping. Then divide out common factors.

$$\frac{ab-ac+bd-cd}{ab+ac+bd+cd}\cdot\frac{b^2+bc+bd+cd}{b^2+bd-bc-cd}$$

$$= \frac{a(b-c)+d(b-c)}{a(b+c)+d(b+c)}\cdot\frac{b(b+c)+d(b+c)}{b(b+d)-c(b+d)}$$ *Distributive property*

$$= \frac{(b-c)(a+d)}{(b+c)(a+d)}\cdot\frac{(b+c)(b+d)}{(b+d)(b-c)}$$ *Factor out common factors.*

$$= \frac{\cancel{(b-c)}\cancel{(a+d)}}{\cancel{(b+c)}\cancel{(a+d)}}\cdot\frac{\cancel{(b+c)}\cancel{(b+d)}}{\cancel{(b+d)}\cancel{(b-c)}} = 1$$ *Divide out common factors.*

▸ **Now Try Exercise 75**

## 4 Divide Rational Expressions

Now let's discuss division of rational expressions.

**To Divide Rational Expressions**

To divide rational expressions, use the following rule:

$$\frac{a}{b} \div \frac{c}{d} = \frac{a}{b} \cdot \frac{d}{c} = \frac{a \cdot d}{b \cdot c}, \quad b \neq 0, c \neq 0, d \neq 0$$

To divide rational expressions, invert the divisor (the second or bottom fraction) and proceed as when multiplying rational expressions.

**EXAMPLE 7** ▸ Divide $\dfrac{18x^4}{5y^3} \div \dfrac{3x^5}{25y}$.

*Solution*

$$\frac{18x^4}{5y^3} \div \frac{3x^5}{25y} = \frac{\overset{6}{\cancel{18}}\,\cancel{x^4}}{\cancel{5}\,\underset{y^2}{\cancel{y^3}}} \cdot \frac{\overset{5}{\cancel{25}}\,\cancel{y}}{\cancel{3}\,\underset{x}{\cancel{x^5}}}$$ Invert divisor; divide out common factors.

$$= \frac{6 \cdot 5}{y^2 x} = \frac{30}{xy^2}$$

▸ Now Try Exercise 51

In Example 7, all the numerators and denominators were monomials. When the numerators or denominators are binomials or trinomials, we factor them, if possible, in order to divide out common factors. This procedure is illustrated in Example 8.

**EXAMPLE 8** ▸ Divide. **a)** $\dfrac{x^2 - 25}{x + 7} \div \dfrac{x - 5}{x + 7}$ **b)** $\dfrac{12a^2 - 22a + 8}{3a} \div \dfrac{3a^2 + 2a - 8}{8a^2 + 16a}$

*Solution*

**a)** $$\frac{x^2 - 25}{x + 7} \div \frac{x - 5}{x + 7} = \frac{x^2 - 25}{x + 7} \cdot \frac{x + 7}{x - 5}$$ Invert divisor.

$$= \frac{(x + 5)\cancel{(x - 5)}}{\cancel{x + 7}} \cdot \frac{\cancel{x + 7}}{\cancel{x - 5}}$$ Factor; divide out common factors.

$$= x + 5$$

**b)** $$\frac{12a^2 - 22a + 8}{3a} \div \frac{3a^2 + 2a - 8}{8a^2 + 16a}$$

$$= \frac{12a^2 - 22a + 8}{3a} \cdot \frac{8a^2 + 16a}{3a^2 + 2a - 8}$$ Invert divisor.

$$= \frac{2(6a^2 - 11a + 4)}{3a} \cdot \frac{8a(a + 2)}{(3a - 4)(a + 2)}$$ Factor.

$$= \frac{2\cancel{(3a - 4)}(2a - 1)}{3\cancel{a}} \cdot \frac{8\cancel{a}\cancel{(a + 2)}}{\cancel{(3a - 4)}\cancel{(a + 2)}}$$ Factor further; divide out common factors.

$$= \frac{16(2a - 1)}{3}$$

▸ Now Try Exercise 59

**EXAMPLE 9** ▸ Divide $\dfrac{x^4 - y^4}{x - y} \div \dfrac{x^2 + xy}{x^2 - 2xy + y^2}$.

*Solution*

$$\frac{x^4 - y^4}{x - y} \div \frac{x^2 + xy}{x^2 - 2xy + y^2}$$

$$= \frac{x^4 - y^4}{x - y} \cdot \frac{x^2 - 2xy + y^2}{x^2 + xy} \quad \text{Invert divisor.}$$

$$= \frac{(x^2 + y^2)(x^2 - y^2)}{x - y} \cdot \frac{(x - y)(x - y)}{x(x + y)} \quad \text{Factor.}$$

$$= \frac{(x^2 + y^2)\cancel{(x + y)}\cancel{(x - y)}}{\cancel{x - y}} \cdot \frac{(x - y)(x - y)}{x\cancel{(x + y)}} \quad \text{Factor further; divide out common factors.}$$

$$= \frac{(x^2 + y^2)(x - y)^2}{x}$$

▸ Now Try Exercise 69

**Helpful Hint** *Study Tip*

Throughout this chapter, you will need to factor polynomials. It is important that you understand the factoring techniques covered in Chapter 5. If you are having difficulty factoring, review factoring now.

## EXERCISE SET 6.1

MathXL® MyMathLab

### Concept/Writing Exercises

1. a) What is a rational expression?
   b) Give your own example of a rational expression.
2. Explain why $\dfrac{\sqrt{x}}{x + 3}$ is not a rational expression.
3. a) What is a rational function?
   b) Give your own example of a rational function.
4. Explain why $f(x) = \dfrac{2}{\sqrt{x + 4}}$ is not a rational function.
5. a) What is the domain of a rational function?
   b) What is the domain of $f(x) = \dfrac{3}{x^2 - 25}$?
6. a) Explain how to simplify a rational expression.
   b) Using the procedure stated in part **a)**, simplify
   $$\frac{6x^2 + 19x + 10}{4x^2 - 25}$$
7. a) Explain how to simplify a rational expression where the numerator and denominator differ only in sign.
   b) Using the procedure explained in part **a)**, simplify
   $$\frac{3x^2 - 2x - 7}{-3x^2 + 2x + 7}$$
8. a) Explain how to multiply rational expressions.
   b) Using the procedure given in part **a)**, multiply
   $$\frac{6a^2 + a - 1}{3a^2 + 2a - 1} \cdot \frac{3a^2 + 4a + 1}{6a^2 + 5a + 1}$$
9. a) Explain how to divide rational expressions.
   b) Using the procedure given in part **a)**, divide
   $$\frac{r + 2}{r^2 + 9r + 18} \div \frac{(r + 2)^2}{r^2 + 5r + 6}$$
10. Consider $f(x) = \dfrac{x}{x}$. Will $f(x) = 1$ for all values of $x$? Explain.

### Practice the Skills

*Determine the values that are excluded in the following expressions.*

11. $\dfrac{4x}{5x - 20}$
12. $\dfrac{x + 2}{x^2 - 64}$
13. $\dfrac{4}{2x^2 - 15x + 25}$
14. $\dfrac{2}{(x - 6)^2}$
15. $\dfrac{x - 3}{x^2 + 12}$
16. $\dfrac{-2}{49 - r^2}$
17. $\dfrac{x^2 + 81}{x^2 - 81}$
18. $\dfrac{x^2 - 36}{x^2 + 36}$

*Determine the domain of each function.*

**19.** $f(p) = \dfrac{p + 1}{p - 2}$

**20.** $f(z) = \dfrac{3}{-18z + 9}$

**21.** $y = \dfrac{5}{x^2 + x - 6}$

**22.** $y = \dfrac{9}{x^2 + 4x - 21}$

**23.** $f(a) = \dfrac{3a^2 - 6a + 4}{2a^2 + 3a - 2}$

**24.** $f(x) = \dfrac{10 - 3x}{x^3 + 8x}$

**25.** $g(x) = \dfrac{x^2 - x + 8}{x^2 + 4}$

**26.** $h(x) = \dfrac{x^3 - 64x}{x^2 + 81}$

**27.** $m(a) = \dfrac{a^2 + 36}{a^2 - 36}$

**28.** $k(b) = \dfrac{b^2 - 36}{b^2 + 36}$

*Simplify each rational expression.*

**29.** $\dfrac{x - xy}{x}$

**30.** $\dfrac{x^2 - 5x}{x}$

**31.** $\dfrac{5x^2 - 20xy}{15x}$

**32.** $\dfrac{x^2 + 7x}{x^2 - 2x}$

**33.** $\dfrac{x^3 - x}{x^2 - 1}$

**34.** $\dfrac{4x^2y + 12xy + 18x^3y^3}{10xy^2}$

**35.** $\dfrac{5r - 8}{8 - 5r}$

**36.** $\dfrac{4x^2 - 16x^4 + 6x^5y}{14x^3y^2}$

**37.** $\dfrac{p^2 - 2p - 24}{6 - p}$

**38.** $\dfrac{4x^2 - 9}{2x^2 - x - 3}$

**39.** $\dfrac{a^2 - 3a - 10}{a^2 + 5a + 6}$

**40.** $\dfrac{y^2 - 10yz + 24z^2}{y^2 - 5yz + 4z^2}$

**41.** $\dfrac{8x^3 - 125y^3}{2x - 5y}$

**42.** $\dfrac{64x^3 - 27z^3}{3z - 4x}$

**43.** $\dfrac{(x + 6)(x - 3) + (x + 6)(x - 2)}{2(x + 6)}$

**44.** $\dfrac{(2x - 1)(x + 4) + (2x - 1)(x + 1)}{3(2x - 1)}$

**45.** $\dfrac{a^2 + 7a - ab - 7b}{a^2 - ab + 5a - 5b}$

**46.** $\dfrac{xy - yw + xz - zw}{xy + yw + xz + zw}$

**47.** $\dfrac{x^2 - x - 12}{x^3 + 27}$

**48.** $\dfrac{a^3 - b^3}{a^2 - b^2}$

*Multiply or divide as indicated. Simplify all answers.*

**49.** $\dfrac{2x}{5y} \cdot \dfrac{y^3}{6}$

**50.** $\dfrac{32x^2}{y^4} \cdot \dfrac{5x^3}{8y^2}$

**51.** $\dfrac{9x^3}{1} \div \dfrac{3}{16y^2}$

**52.** $\dfrac{10m^4}{49x^5y^7} \div \dfrac{25m^5}{21x^{12}y^5}$

**53.** $\dfrac{3 - r}{r - 3} \cdot \dfrac{r - 9}{9 - r}$

**54.** $\dfrac{7a + 7b}{5} \div \dfrac{a^2 - b^2}{a - b}$

**55.** $\dfrac{x^2 + 3x - 10}{4x} \cdot \dfrac{x^2 - 3x}{x^2 - 5x + 6}$

**56.** $\dfrac{p^2 + 7p + 10}{p + 5} \cdot \dfrac{1}{p + 2}$

**57.** $\dfrac{r^2 + 10r + 21}{r + 7} \div \dfrac{(r^2 - 5r - 24)}{r^3}$

**58.** $(x - 3) \div \dfrac{x^2 + 3x - 18}{x^3}$

**59.** $\dfrac{x^2 + 12x + 35}{x^2 + 4x - 5} \div \dfrac{x^2 + 3x - 28}{7x - 7}$

**60.** $\dfrac{x + 1}{x^2 - 17x + 30} \div \dfrac{8x + 8}{x^2 + 7x - 18}$

**61.** $\dfrac{a - b}{9a + 9b} \div \dfrac{a^2 - b^2}{a^2 + 2a + 1}$

**62.** $\dfrac{2x^2 + 8xy + 8y^2}{x^2 + 4xy + 4y^2} \cdot \dfrac{2x^2 + 7xy + 6y^2}{4x^2 + 14xy + 12y^2}$

**63.** $\dfrac{3x^2 - x - 4}{4x^2 + 5x + 1} \cdot \dfrac{2x^2 - 5x - 12}{6x^2 + x - 12}$

**64.** $\dfrac{6x^3 - x^2 - x}{2x^2 + x - 1} \cdot \dfrac{x^2 - 1}{x^3 - 2x^2 + x}$

**65.** $\dfrac{x + 2}{x^3 - 8} \cdot \dfrac{(x - 2)^2}{x^2 + 4}$

**66.** $\dfrac{x^4 - y^8}{x^2 + y^4} \div \dfrac{x^2 - y^4}{x^2}$

**67.** $\dfrac{x^2 - y^2}{x^2 - 2xy + y^2} \div \dfrac{(x + y)^2}{(x - y)^2}$

**68.** $\dfrac{(x^2 - y^2)^2}{(x^2 - y^2)^3} \div \dfrac{x^2 + y^2}{x^4 - y^4}$

**69.** $\dfrac{2x^4 + 4x^2}{6x^2 + 14x + 4} \div \dfrac{x^2 + 2}{3x^2 + x}$

**70.** $\dfrac{8a^3 - 1}{4a^2 + 2a + 1} \div \dfrac{a^2 - 2a + 1}{(a - 1)^2}$

**71.** $\dfrac{(a - b)^3}{a^3 - b^3} \cdot \dfrac{a^2 - b^2}{(a - b)^2}$

**72.** $\dfrac{r^2 - 16}{r^3 - 64} \div \dfrac{r^2 + 8r + 16}{r^2 + 4r + 16}$

73. $\dfrac{4x + y}{5x + 2y} \cdot \dfrac{10x^2 - xy - 2y^2}{8x^2 - 2xy - y^2}$

74. $\dfrac{2x^3 - 7x^2 + 3x}{x^2 + 2x - 3} \cdot \dfrac{x^2 + 3x}{(x - 3)^2}$

75. $\dfrac{ac - ad + bc - bd}{ac + ad + bc + bd} \cdot \dfrac{pc + pd - qc - qd}{pc - pd + qc - qd}$

76. $\dfrac{2p^2 + 2pq - pq^2 - q^3}{p^3 + p^2 + pq^2 + q^2} \div \dfrac{p^3 + p + p^2q + q}{p^3 + p + p^2 + 1}$

77. $\dfrac{3r^2 + 17rs + 10s^2}{6r^2 + 13rs - 5s^2} \div \dfrac{6r^2 + rs - 2s^2}{6r^2 - 5rs + s^2}$

78. $\dfrac{x^3 - 4x^2 + x - 4}{x^5 - x^4 + x^3 - x^2} \cdot \dfrac{2x^3 + 2x^2 + x + 1}{2x^3 - 8x^2 + x - 4}$

## Problem Solving

79. Make up a rational expression that is undefined at $x = 2$ and $x = -3$. Explain how you determined your answer.

80. Make up a rational expression that is undefined at $x = 4$ and $x = -5$. Explain how you determined your answer.

81. Consider the rational function $f(x) = \dfrac{1}{x}$. Explain why this function can never equal 0.

82. Consider the rational function $g(x) = \dfrac{2}{x + 3}$. Explain why this function can never equal 0.

83. Consider the rational function $f(x) = \dfrac{x - 4}{x^2 - 36}$. For what value of $x$, if any, will this function **a)** equal 0? **b)** be undefined? Explain.

84. Consider the function $f(x) = \dfrac{x - 2}{x^2 - 81}$. For what value of $x$, if any, will this function **a)** equal 0? **b)** be undefined? Explain.

85. Give a function that is undefined at $x = 3$ and $x = -1$ and has a value of 0 at $x = 2$. Explain how you determined your answer.

86. Give a function that is undefined at $x = -4$ and $x = -2$ and has a value of 0 at $x = 5$. Explain how you determined your answer.

*Determine the polynomial to be placed in the shaded area to give a true statement. Explain how you determined your answer.*

87. $\dfrac{\boxed{\quad}}{x^2 + 2x - 15} = \dfrac{1}{x - 3}$

88. $\dfrac{\boxed{\quad}}{3x + 2} = x - 3$

89. $\dfrac{y^2 - y - 20}{\boxed{\quad}} = \dfrac{y + 4}{y + 1}$

90. $\dfrac{\boxed{\quad}}{6p^2 + p - 15} = \dfrac{2p - 1}{2p - 3}$

*Determine the polynomial to be placed in the shaded area to give a true statement. Explain how you determined your answer.*

91. $\dfrac{x^2 - x - 12}{x^2 + 2x - 3} \cdot \dfrac{\boxed{\quad}}{x^2 - 2x - 8} = 1$

92. $\dfrac{x^2 - 4}{(x + 2)^2} \cdot \dfrac{2x^2 + x - 6}{\boxed{\quad}} = \dfrac{x - 2}{2x + 5}$

93. $\dfrac{x^2 - 9}{2x^2 + 3x - 2} \div \dfrac{2x^2 - 9x + 9}{\boxed{\quad}} = \dfrac{x + 3}{2x - 1}$

94. $\dfrac{4r^2 - r - 18}{\boxed{\quad}} \div \dfrac{4r^3 - 9r^2}{6r^2 - 9r + 3} = \dfrac{3(r - 1)}{r^2}$

95. **Area** Consider the rectangle below. Its area is $3a^2 + 7ab + 2b^2$ and its length is $2a + 4b$. Find its width, $w$, in terms of $a$ and $b$, by dividing its area by its length.

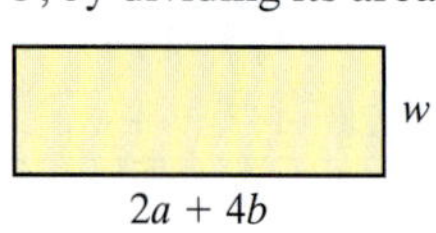


96. **Area** Consider the rectangle below. Its area is $a^2 + 2ab + b^2$ and its length is $3a + 3b$. Find its width, $w$, in terms of $a$ and $b$, by dividing its area by its length.

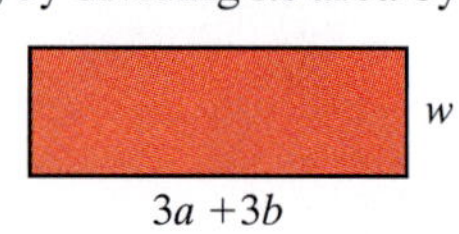


97. **Area** Consider the triangle below. If its area is $a^2 + 4ab + 3b^2$ and its base is $a + 3b$, find its height $h$. Use area $= \dfrac{1}{2}(\text{base})(\text{height})$.

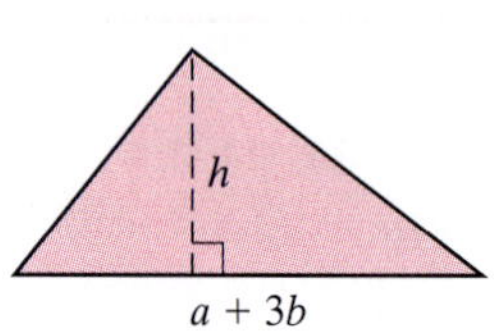


98. **Area** Consider the trapezoid below. If its area is $a^2 + 2ab + b^2$, find its height, $h$. Use area $= \dfrac{1}{2}h(a + b)$.

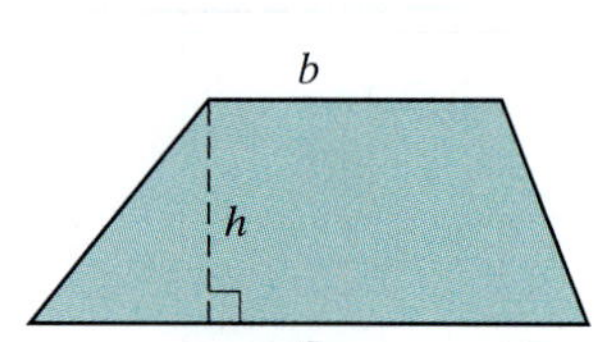

*Perform each indicated operation.*

**99.** $\left(\dfrac{2x^2 - 3x - 14}{2x^2 - 9x + 7} \div \dfrac{6x^2 + x - 15}{3x^2 + 2x - 5}\right) \cdot \dfrac{6x^2 - 7x - 3}{2x^2 - x - 3}$

**100.** $\left(\dfrac{a^2 - b^2}{2a^2 - 3ab + b^2} \cdot \dfrac{2a^2 - 7ab + 3b^2}{a^2 + ab}\right) \div \dfrac{ab - 3b^2}{a^2 + 2ab + b^2}$

**101.** $\dfrac{5x^2(x - 1) - 3x(x - 1) - 2(x - 1)}{10x^2(x - 1) + 9x(x - 1) + 2(x - 1)} \cdot \dfrac{2x + 1}{x + 3}$

**102.** $\dfrac{x^2(3x - y) - 5x(3x - y) - 24(3x - y)}{x^2(3x - y) - 9x(3x - y) + 8(3x - y)} \cdot \dfrac{x - 1}{x + 3}$

**103.** $\dfrac{(x - p)^n}{x^{-2}} \div \dfrac{(x - p)^{2n}}{x^{-6}}$

**104.** $\dfrac{x^{-3}}{(a - b)^r} \div \dfrac{x^{-5}}{(a - b)^{r+2}}$

*Simplify.*

**105.** $\dfrac{x^{5y} + 3x^{4y}}{3x^{3y} + x^{4y}}$

**106.** $\dfrac{m^{2x} - m^x - 2}{m^{2x} - 4}$

*For Exercises 107–110,*

**a)** *Determine the domain of the function.*

**b)** *Graph the function in connected mode.*

**c)** *Is the function increasing, decreasing, or remaining the same as x gets closer and closer to 2, approaching 2 from the left side?*

**d)** *Is the function increasing, decreasing, or remaining the same as x gets closer and closer to 2, approaching 2 from the right side?*

**107.** $f(x) = \dfrac{1}{x - 2}$

**108.** $f(x) = \dfrac{x}{x - 2}$

**109.** $f(x) = \dfrac{x^2}{x - 2}$

**110.** $f(x) = \dfrac{x - 2}{x - 2}$

**111.** Consider the rational function $f(x) = \dfrac{1}{x}$.

**a)** Determine the domain of the function.

**b)** Complete the following table for the function.

| $x$ | −10 | −1 | −0.5 | −0.1 | −0.01 | 0.01 | 0.1 | 0.5 | 1 | 10 |
|---|---|---|---|---|---|---|---|---|---|---|
| $y$ | | | | | | | | | | |

**c)** Draw the graph of $f(x) = \dfrac{1}{x}$. Consider what happens to the function as $x$ gets closer and closer to 0, approaching 0 from both the left and right sides.

**d)** Can this fraction ever have a value of 0? Explain your answer.

## Group Activity

**112.** Consider the rational function $f(x) = \dfrac{x^2 - 4}{x - 2}$.

**a)** As a group, determine the domain of this function.

**b)** Have each member of the group individually complete the following table for the function.

| $x$ | −2 | −1 | 0 | 1 | 1.9 | 1.99 | 2.01 | 2.1 | 3 | 4 | 5 | 6 |
|---|---|---|---|---|---|---|---|---|---|---|---|---|
| $y$ | | | | | | | | | | | | |

**c)** Compare your answers to part **b)** and agree on the correct table values.

**d)** As a group draw the graph of $f(x) = \dfrac{x^2 - 4}{x - 2}$. Is the function defined when $x = 2$?

**e)** Can this fraction ever have a value of 0? If so, for what value(s) of $a$ is $f(a) = 0$?

## Cumulative Review Exercises

[2.2] **113.** Solve $6(x - 2) + 6y = 12x$ for $y$.

[2.5] **114.** Solve $4 + \dfrac{4x}{3} < 6$ and give the answer in interval notation.

[2.6] **115.** Solve $\left|\dfrac{2x - 4}{12}\right| = 5$.

[3.2] **116.** Let $f(x) = |6 - 3x| - 2$. Find $f(1.3)$.

[4.1] **117.** Solve the system of equations.

$$3x + 4y = 2$$
$$2x + 5y = -1$$

[5.6] **118.** Factor $9x^2 + 6xy + y^2 - 4$.

# 6.2 Addition and Subtraction of Rational Expressions

1 Add and subtract expressions with a common denominator.

2 Find the least common denominator (LCD).

3 Add and subtract expressions with unlike denominators.

4 Study an application of rational expressions.

## 1 Add and Subtract Expressions with a Common Denominator

When adding (or subtracting) two rational expressions with a common denominator, we add (or subtract) the numerators while keeping the common denominator.

### To Add or Subtract Rational Expressions

To add or subtract rational expressions, use the following rules.

ADDITION

$$\frac{a}{c} + \frac{b}{c} = \frac{a+b}{c}, \quad c \neq 0$$

SUBTRACTION

$$\frac{a}{c} - \frac{b}{c} = \frac{a-b}{c}, \quad c \neq 0$$

To add or subtract rational expressions with a common denominator,

1. Add or subtract the expressions using the rules given above.
2. Simplify the expression if possible.

**EXAMPLE 1** Add.

**a)** $\dfrac{3}{x+6} + \dfrac{x-4}{x+6}$ **b)** $\dfrac{x^2+3x-2}{(x+5)(x-3)} + \dfrac{4x+12}{(x+5)(x-3)}$

*Solution*

**a)** Since the denominators are the same, we add the numerators and keep the common denominator.

$$\frac{3}{x+6} + \frac{x-4}{x+6} = \frac{3+(x-4)}{x+6} \qquad \textit{Add numerators.}$$

$$= \frac{x-1}{x+6}$$

**b)** $$\frac{x^2+3x-2}{(x+5)(x-3)} + \frac{4x+12}{(x+5)(x-3)} = \frac{x^2+3x-2+(4x+12)}{(x+5)(x-3)} \qquad \textit{Add numerators.}$$

$$= \frac{x^2+7x+10}{(x+5)(x-3)} \qquad \textit{Combine like terms.}$$

$$= \frac{\cancel{(x+5)}(x+2)}{\cancel{(x+5)}(x-3)} \qquad \textit{Factor; divide out common factors.}$$

$$= \frac{x+2}{x-3}$$

Now Try Exercise 11

*When subtracting rational expressions, be sure to subtract the entire numerator of the fraction being subtracted.* Read the Avoiding Common Errors box that follows very carefully.

## Avoiding Common Errors

The error presented here is sometimes made by students. Study the information presented so that you will not make this error.

How do you simplify this problem?

$$\frac{4x}{x-2} - \frac{2x+1}{x-2}$$

CORRECT

$$\frac{4x}{x-2} - \frac{2x+1}{x-2} = \frac{4x-(2x+1)}{x-2} = \frac{4x-2x-1}{x-2} = \frac{2x-1}{x-2}$$

INCORRECT

$$\frac{4x}{x-2} - \frac{2x+1}{x-2} = \frac{4x-2x+1}{x-2} = \frac{2x+1}{x-2}$$

The procedure on the right side is incorrect because the *entire numerator*, $2x + 1$, must be subtracted from $4x$. Instead, only $2x$ was subtracted. Note that **the sign of each term** (not just the first term) **in the numerator of the fraction being subtracted must change**. Note that $-(2x + 1) = -2x - 1$, by the distributive property.

**EXAMPLE 2** Subtract $\frac{a}{a-6} - \frac{a^2-4a-6}{a-6}$.

*Solution*

$$\frac{a}{a-6} - \frac{a^2-4a-6}{a-6} = \frac{a-(a^2-4a-6)}{a-6} \quad \text{Subtract numerators.}$$

$$= \frac{a-a^2+4a+6}{a-6}$$

$$= \frac{-a^2+5a+6}{a-6} \quad \text{Combine like terms.}$$

$$= \frac{-(a^2-5a-6)}{a-6} \quad \text{Factor out } -1.$$

$$= \frac{-(a-6)(a+1)}{a-6} \quad \text{Factor; divide out common factors.}$$

$$= -(a+1) \text{ or } -a-1$$

Now Try Exercise 13

## 2 Find the Least Common Denominator (LCD)

To add or subtract two numerical fractions with *unlike denominators*, we must first obtain a common denominator. Obtaining a common denominator may involve writing numerical values as products of prime numbers. A **prime number** is a number greater than 1 that has only two divisors, itself and 1. Some prime numbers are 2, 3, 5, 7, 11, 13, and 17. Following we show how the numbers 36 and 48 are written as a product of prime numbers:

$$36 = 2 \cdot 2 \cdot 3 \cdot 3 = 2^2 \cdot 3^2$$
$$48 = 2 \cdot 2 \cdot 2 \cdot 2 \cdot 3 = 2^4 \cdot 3$$

We may need to write numerical coefficients as products of prime numbers to find the **least common denominator** of a rational expression.

**To Find the Least Common Denominator (LCD) of Rational Expressions**

1. Write each nonprime coefficient (other than 1) of monomials that appear in denominators as a product of prime numbers.
2. Factor each denominator completely. Any factors that occur more than once should be expressed as powers. For example, $(x + 5)(x + 5)$ should be expressed as $(x + 5)^2$.
3. List all different factors (other than 1) that appear in any of the denominators. When the same factor appears in more than one denominator, write the factor with the highest power that appears.
4. The least common denominator is the product of all the factors found in step 2.

**EXAMPLE 3** ▸ Find the LCD of each expression.

**a)** $\dfrac{3}{5x} - \dfrac{2}{x^2}$ **b)** $\dfrac{1}{18x^3y} + \dfrac{5}{27x^2y^3}$ **c)** $\dfrac{3}{x} - \dfrac{2y}{x + 5}$ **d)** $\dfrac{7}{x^2(x + 1)} + \dfrac{3z}{x(x + 1)^3}$

*Solution*

**a)** The factors that appear in the denominators are 5 and $x$. List each factor with its highest power. The LCD is the product of these factors.

Highest power of $x$

$$\text{LCD} = 5 \cdot x^2 = 5x^2$$

**b)** The numerical coefficients written as products of prime numbers are $18 = 2 \cdot 3^2$ and $27 = 3^3$. The variable factors are $x$ and $y$. Using the highest powers of the factors, we obtain the LCD.

$$\text{LCD} = 2 \cdot 3^3 \cdot x^3y^3 = 54x^3y^3$$

**c)** The factors are $x$ and $x + 5$. Note that the $x$ in the second denominator, $x + 5$, is not a factor of that denominator since the operation is addition rather than multiplication.

$$\text{LCD} = x(x + 5)$$

**d)** The factors are $x$ and $x + 1$. The highest power of $x$ is 2 and the highest power of $x + 1$ is 3.

$$\text{LCD} = x^2(x + 1)^3$$

▸ **Now Try Exercise 31**

Sometimes it is necessary to factor all denominators to obtain the LCD. This is illustrated in the next example.

**EXAMPLE 4** ▸ Find the LCD of each expression.

**a)** $\dfrac{3}{2x^2 - 4x} + \dfrac{8x}{x^2 - 4x + 4}$ **b)** $\dfrac{4x}{x^2 - x - 12} - \dfrac{6x^2}{x^2 - 7x + 12}$

*Solution*

**a)** Factor both denominators.

$$\frac{3}{2x^2 - 4x} + \frac{8x}{x^2 - 4x + 4} = \frac{3}{2x(x - 2)} + \frac{8x}{(x - 2)^2}$$

The factors are 2, $x$, and $x - 2$. Multiply the factors raised to the highest power that appears for each factor.

$$\text{LCD} = 2 \cdot x \cdot (x - 2)^2 = 2x(x - 2)^2$$

**b)** Factor both denominators.

$$\frac{4x}{x^2 - x - 12} - \frac{6x^2}{x^2 - 7x + 12} = \frac{4x}{(x + 3)(x - 4)} - \frac{6x^2}{(x - 3)(x - 4)}$$

$$\text{LCD} = (x + 3)(x - 4)(x - 3)$$

Note that although $x - 4$ is a common factor of each denominator, the highest power of the factor that appears in either denominator is 1.

▶ Now Try Exercise 29

## 3 Add and Subtract Expressions with Unlike Denominators

The procedure used to add or subtract rational expressions with unlike denominators is given below.

**To Add or Subtract Rational Expressions with Unlike Denominators**

1. Determine the LCD.
2. Rewrite each fraction as an equivalent fraction with the LCD. This is done by multiplying both the numerator and denominator of each fraction by any factors needed to obtain the LCD.
3. Leave the denominator in factored form, but multiply out the numerator.
4. Add or subtract the numerators while maintaining the LCD.
5. When it is possible to reduce the fraction by factoring the numerator, do so.

**EXAMPLE 5** ▶ Add. **a)** $\frac{2}{x} + \frac{9}{y}$ **b)** $\frac{5}{4a^2} + \frac{3}{14ab^3}$

### *Solution*

**a)** First we determine the LCD.

$$\text{LCD} = xy$$

Now we write each fraction with the LCD. We do this by multiplying *both* numerator and denominator of each fraction by any factors needed to obtain the LCD.

In this problem, the first fraction must be multiplied by $\frac{y}{y}$ and the second fraction must be multiplied by $\frac{x}{x}$.

$$\frac{2}{x} + \frac{9}{y} = \frac{y}{y} \cdot \frac{2}{x} + \frac{9}{y} \cdot \frac{x}{x} = \frac{2y}{xy} + \frac{9x}{xy}$$

By multiplying both the numerator and denominator by the same factor, we are in effect multiplying by 1, which does not change the value of the fraction, only its appearance. Thus, the new fraction is equivalent to the original fraction.

Now we add the numerators while leaving the LCD alone.

$$\frac{2y}{xy} + \frac{9x}{xy} = \frac{2y + 9x}{xy} \quad \text{or} \quad \frac{9x + 2y}{xy}$$

Therefore, $\frac{2}{x} + \frac{9}{y} = \frac{9x + 2y}{xy}$.

**b)** The LCD of 4 and 14 is 28. The LCD of the two fractions is $28a^2b^3$. We must write each fraction with the denominator $28a^2b^3$. To do this, we multiply the first fraction by $\frac{7b^3}{7b^3}$ and the second fraction by $\frac{2a}{2a}$.

$$\frac{5}{4a^2} + \frac{3}{14ab^3} = \frac{7b^3}{7b^3} \cdot \frac{5}{4a^2} + \frac{3}{14ab^3} \cdot \frac{2a}{2a}$$ *Multiply to obtain LCD.*

$$= \frac{35b^3}{28a^2b^3} + \frac{6a}{28a^2b^3}$$

$$= \frac{35b^3 + 6a}{28a^2b^3}$$ *Add numerators.*

▸ **Now Try Exercise 39**

**EXAMPLE 6** ▸ Subtract $\frac{x+2}{x-4} - \frac{x+5}{x+4}$.

*Solution* The LCD is $(x - 4)(x + 4)$. Write each fraction with the denominator $(x - 4)(x + 4)$.

$$\frac{x+2}{x-4} - \frac{x+5}{x+4} = \frac{x+4}{x+4} \cdot \frac{x+2}{x-4} - \frac{x+5}{x+4} \cdot \frac{x-4}{x-4}$$ *Multiply to obtain LCD.*

$$= \frac{(x+4)(x+2)}{(x+4)(x-4)} - \frac{(x+5)(x-4)}{(x+4)(x-4)}$$

$$= \frac{x^2+6x+8}{(x+4)(x-4)} - \frac{x^2+x-20}{(x+4)(x-4)}$$ *Multiply binomials in numerators.*

$$= \frac{x^2+6x+8-(x^2+x-20)}{(x+4)(x-4)}$$ *Subtract numerators.*

$$= \frac{x^2+6x+8-x^2-x+20}{(x+4)(x-4)}$$

$$= \frac{5x+28}{(x+4)(x-4)}$$ *Combine like terms.*

▸ **Now Try Exercise 45**

**EXAMPLE 7** ▸ Add $\frac{2}{x-3} + \frac{x+5}{3-x}$.

*Solution* Note that each denominator is the opposite, or additive inverse, of the other. (The terms of one denominator differ only in sign from the terms of the other denominator.) When this special situation arises, we can multiply the numerator and denominator of either one of the fractions by $-1$ to obtain the LCD.

$$\frac{2}{x-3} + \frac{x+5}{3-x} = \frac{2}{x-3} + \frac{-1}{-1} \cdot \frac{(x+5)}{(3-x)}$$ *Multiply to obtain LCD.*

$$= \frac{2}{x-3} + \frac{-x-5}{x-3}$$

$$= \frac{2-x-5}{x-3}$$ *Add numerators.*

$$= \frac{-x-3}{x-3}$$ *Combine like terms.*

Because there are no common factors in both the numerator and denominator, $\frac{-x-3}{x-3}$ cannot be simplified further.

▸ **Now Try Exercise 43**

**EXAMPLE 8** ▸ Subtract $\dfrac{3x+4}{2x^2-5x-12} - \dfrac{2x-3}{5x^2-18x-8}$.

*Solution* Factor the denominator of each expression.

$$\frac{3x+4}{2x^2-5x-12} - \frac{2x-3}{5x^2-18x-8} = \frac{3x+4}{(2x+3)(x-4)} - \frac{2x-3}{(5x+2)(x-4)}$$

The LCD is $(2x+3)(x-4)(5x+2)$.

$$\frac{3x+4}{(2x+3)(x-4)} - \frac{2x-3}{(5x+2)(x-4)}$$

$$= \frac{5x+2}{5x+2}\cdot\frac{3x+4}{(2x+3)(x-4)} - \frac{2x-3}{(5x+2)(x-4)}\cdot\frac{2x+3}{2x+3} \quad \textit{Multiply to obtain LCD.}$$

$$= \frac{15x^2+26x+8}{(5x+2)(2x+3)(x-4)} - \frac{4x^2-9}{(5x+2)(2x+3)(x-4)} \quad \textit{Multiply numerators.}$$

$$= \frac{15x^2+26x+8-(4x^2-9)}{(5x+2)(2x+3)(x-4)} \quad \textit{Subtract numerators.}$$

$$= \frac{15x^2+26x+8-4x^2+9}{(5x+2)(2x+3)(x-4)}$$

$$= \frac{11x^2+26x+17}{(5x+2)(2x+3)(x-4)} \quad \textit{Combine like terms.}$$

▸ **Now Try Exercise 49**

**EXAMPLE 9** ▸ Perform the indicated operations.

$$\frac{x-1}{x-2} - \frac{x+1}{x+2} + \frac{x-6}{x^2-4}$$

*Solution* First, factor $x^2-4$. The LCD of the three fractions is $(x+2)(x-2)$.

$$\frac{x-1}{x-2} - \frac{x+1}{x+2} + \frac{x-6}{x^2-4}$$

$$= \frac{x-1}{x-2} - \frac{x+1}{x+2} + \frac{x-6}{(x+2)(x-2)}$$

$$= \frac{x+2}{x+2}\cdot\frac{x-1}{x-2} - \frac{x+1}{x+2}\cdot\frac{x-2}{x-2} + \frac{x-6}{(x+2)(x-2)} \quad \textit{Multiply to obtain LCD.}$$

$$= \frac{x^2+x-2}{(x+2)(x-2)} - \frac{x^2-x-2}{(x+2)(x-2)} + \frac{x-6}{(x+2)(x-2)} \quad \textit{Multiply numerators.}$$

$$= \frac{x^2+x-2-(x^2-x-2)+(x-6)}{(x+2)(x-2)} \quad \textit{Subtract and add numerators.}$$

$$= \frac{x^2+x-2-x^2+x+2+x-6}{(x+2)(x-2)}$$

$$= \frac{3x-6}{(x+2)(x-2)} \quad \textit{Combine like terms.}$$

$$= \frac{3\cancel{(x-2)}}{(x+2)\cancel{(x-2)}} \quad \textit{Factor; divide out common factors.}$$

$$= \frac{3}{x+2}$$

▸ **Now Try Exercise 67**

## USING YOUR GRAPHING CALCULATOR

In Example 9, we found that

$$\frac{x-1}{x-2} - \frac{x+1}{x+2} + \frac{x-6}{x^2-4} = \frac{3}{x+2}$$

Suppose on a graphing calculator we let

$$Y_1 = \frac{x-1}{x-2} - \frac{x+1}{x+2} + \frac{x-6}{x^2-4}$$

$$Y_2 = \frac{3}{x+2}$$

If we use the TABLE feature on a graphing calculator, how will the values of $Y_1$ and $Y_2$ compare? The function $Y_1$ is not defined at $x = -2$ and $x = 2$. The function $Y_2$ is not defined at $x = -2$. For all values of $x$ other than $-2$ and 2, the values of $Y_1$ and $Y_2$ should be the same if we have not made a mistake. Following is a table of values for $Y_1$ and $Y_2$, for values of $x$ from $-3$ to 3.

| X | Y1 | Y2 |
|---|---|---|
| -3 | -3 | -3 |
| -2 | ERROR | ERROR |
| -1 | 3 | 3 |
| 0 | 1.5 | 1.5 |
| 1 | 1 | 1 |
| 2 | ERROR | .75 |
| 3 | .6 | .6 |

X=-3

The graphs of $Y_1$ and $Y_2$ are shown in **Figures 6.3** and **6.4** respectively. We illustrated the graphs in this format (rather than on a graphing calculator screen) to show more detail. The open circle on the graph in **Figure 6.3** is not shown on a graphing calculator. Notice that the graph of $Y_1$ has an open circle at 2 because $Y_1$ is not defined at $x = 2$. Since $Y_2$ is defined at $x = 2$, **Figure 6.4** does not include this open circle. Neither function is defined at $x = -2$.

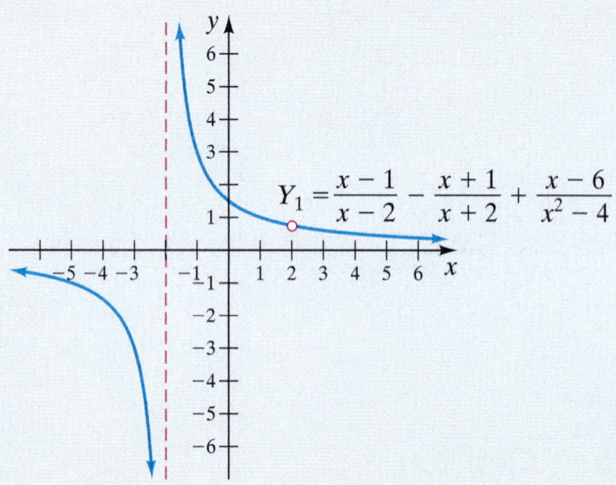


FIGURE 6.3

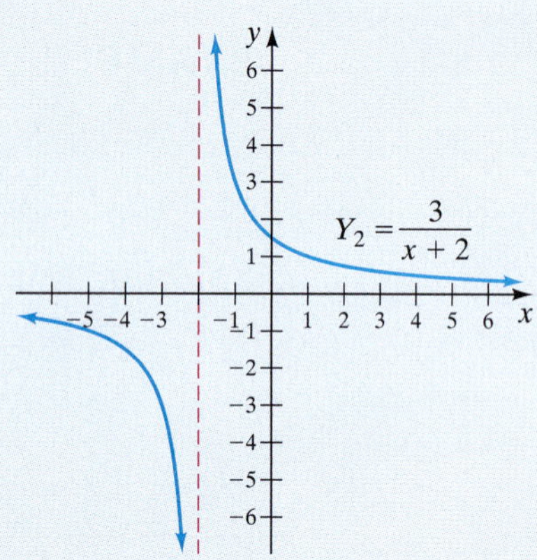


FIGURE 6.4

### Helpful Hint *Study Tip*

Now that we have discussed the operations of addition, subtraction, multiplication, and division of rational expressions, let's quickly summarize the procedures.

To add or subtract rational expressions, obtain the LCD. Express each fraction with the LCD. Then add or subtract the numerators and write this result over the LCD.

To multiply rational expressions, factor each expression completely, divide out common factors, multiply numerators, and multiply denominators.

To divide rational expressions, multiply the first (or top) fraction by the reciprocal of the second (or bottom) fraction. Then, factor each expression completely, divide out common factors, multiply the numerators, and multiply the denominators.

## 4 Study an Application of Rational Expressions

Section 6.5 deals with applications of rational expressions. At this time, we will introduce an application that involves adding and subtracting rational expressions or functions.

In economics, we study revenue, cost, and profit. If $R(x)$ is a revenue function and $C(x)$ is a cost function, then the profit function, $P(x)$, is

$$P(x) = R(x) - C(x)$$

where $x$ is the number of items manufactured and sold by a company. We will use this information in Example 10.

**EXAMPLE 10** ▸ **Sailboats** The Don Perrione Sailboat Company builds and sells at least six sailboats each week.

Suppose $$R(x) = \frac{6x - 7}{x + 2} \quad \text{and} \quad C(x) = \frac{4x - 13}{x + 3}$$

where $x$ is the number of sailboats sold. Determine the profit function.

*Solution* **Understand and Translate** To determine the profit function, we subtract the cost function from the revenue function.

$$P(x) = R(x) - C(x)$$

$$P(x) = \frac{6x - 7}{x + 2} - \frac{4x - 13}{x + 3}$$

The LCD is $(x + 2)(x + 3)$.

**Carry Out**

$$= \frac{x + 3}{x + 3} \cdot \frac{6x - 7}{x + 2} - \frac{4x - 13}{x + 3} \cdot \frac{x + 2}{x + 2}$$ *Multiply to obtain LCD.*

$$= \frac{6x^2 + 11x - 21}{(x + 3)(x + 2)} - \frac{4x^2 - 5x - 26}{(x + 3)(x + 2)}$$ *Multiply numerators.*

$$= \frac{(6x^2 + 11x - 21) - (4x^2 - 5x - 26)}{(x + 3)(x + 2)}$$ *Subtract numerators.*

$$= \frac{6x^2 + 11x - 21 - 4x^2 + 5x + 26}{(x + 3)(x + 2)}$$

$$= \frac{2x^2 + 16x + 5}{(x + 3)(x + 2)}$$ *Combine like terms.*

**Answer** The profit function is $P(x) = \dfrac{2x^2 + 16x + 5}{(x + 3)(x + 2)}$.

▸ **Now Try Exercise 81**

## EXERCISE SET 6.2

MathXL® MyMathLab

### Concept/Writing Exercises

1. a) What is the least common denominator of two or more rational expressions?
   b) Explain how to find the LCD.
   c) Using the procedure you gave in part **b)**, find the LCD of

$$\frac{5}{64x^2 - 121} \quad \text{and} \quad \frac{1}{8x^2 - 27x + 22}$$

2. a) Explain how to add or subtract two rational expressions.
   b) Add $\dfrac{4}{x + 2} + \dfrac{x}{3x^2 - 4x - 20}$ following the procedure you gave in part **a)**.

*In Exercises 3 and 4,* **a)** *explain why the subtraction is not correct, and* **b)** *show the correct subtraction.*

3. $$\frac{x^2 - 4x}{(x + 3)(x - 2)} - \frac{x^2 + x - 2}{(x + 3)(x - 2)} \neq \frac{x^2 - 4x - x^2 + x - 2}{(x + 3)(x - 2)}$$

4. $$\frac{x - 5}{(x + 4)(x - 3)} - \frac{x^2 - 6x + 5}{(x + 4)(x - 3)} \neq \frac{x - 5 - x^2 - 6x + 5}{(x + 4)(x - 3)}$$

## Practice the Skills

*Add or subtract.*

5. $\frac{3x}{x+2}+\frac{5}{x+2}$

6. $\frac{3x}{x+4}+\frac{12}{x+4}$

7. $\frac{7x}{x-5}-\frac{2}{x-5}$

8. $\frac{10x}{x-6}-\frac{60}{x-6}$

9. $\frac{x}{x+3}+\frac{9}{x+3}-\frac{2}{x+3}$

10. $\frac{2x}{x+7}+\frac{17}{x+7}-\frac{3}{x+7}$

11. $\frac{5x-6}{x-8}+\frac{2x-5}{x-8}$

12. $\frac{-4x+6}{x^2+x-6}+\frac{5x-3}{x^2+x-6}$

13. $\frac{x^2-2}{x^2+6x-7}-\frac{-4x+19}{x^2+6x-7}$

14. $\frac{-x^2}{x^2+5xy-14y^2}+\frac{x^2+xy-2y^2}{x^2+5xy-14y^2}$

15. $\frac{x^3-12x^2+45x}{x(x-8)}-\frac{x^2+5x}{x(x-8)}$

16. $\frac{3r^2+15r}{r^3+2r^2-8r}+\frac{2r^2+5r}{r^3+2r^2-8r}$

17. $\frac{3x^2-x}{2x^2-x-21}+\frac{2x-8}{2x^2-x-21}-\frac{x^2-2x+27}{2x^2-x-21}$

18. $\frac{2x^2+9x-15}{2x^2-13x+20}-\frac{3x+10}{2x^2-13x+20}-\frac{3x-5}{2x^2-13x+20}$

*Find the least common denominator.*

19. $\frac{5}{2a^2}+\frac{9}{3a^3}$

20. $\frac{1}{9x^2}-\frac{8}{6x^5}$

21. $\frac{-4}{8x^2y^2}+\frac{7}{5x^4y^6}$

22. $\frac{x+12}{16x^2y}-\frac{x^2}{3x^3}$

23. $\frac{2}{3a^4b^2}+\frac{7}{2a^3b^5}$

24. $\frac{1}{x-1}-\frac{x}{x-3}$

25. $\frac{4x}{x+3}+\frac{6}{x+9}$

26. $\frac{4}{(r-7)(r+2)}-\frac{r+8}{r-7}$

27. $5z^2+\frac{9z}{z-6}$

28. $\frac{b^2+3}{18b}-\frac{b-7}{12(b+8)}$

29. $\frac{x}{x^4(x-2)}-\frac{x+9}{x^2(x-2)^3}$

30. $\frac{x+2}{(x-3)^3(x+4)^2}+\frac{x-7}{(x+4)^4(x-9)}$

31. $\frac{a-2}{a^2-5a-24}+\frac{3}{a^2+11a+24}$

32. $\frac{3x-5}{6x^2+13xy+6y^2}+\frac{3}{3x^2+5xy+2y^2}$

33. $\frac{x}{2x^2-7x+3}+\frac{x-3}{4x^2+4x-3}-\frac{x^2+1}{2x^2-3x-9}$

34. $\frac{3}{x^2+3x-4}-\frac{4}{4x^2+5x-9}+\frac{x+2}{4x^2+25x+36}$

*Add or subtract.*

35. $\frac{2}{3r}+\frac{8}{r}$

36. $\frac{9}{x^2}+\frac{3}{2x}$

37. $\frac{5}{12x}-\frac{1}{4x^2}$

38. $\frac{5x}{4y}+\frac{7}{6xy}$

39. $\frac{3}{8x^4y}+\frac{1}{5x^2y^3}$

40. $\frac{7}{4xy^3}+\frac{1}{6x^2y}$

41. $\frac{b}{a-b}-\frac{a+b}{b}$

42. $\frac{4x}{3xy}+11$

43. $\frac{a}{a-b}-\frac{a}{b-a}$

44. $\frac{9}{b-2}+\frac{3b}{2-b}$

45. $\frac{4x}{x-4}+\frac{x+3}{x+1}$

46. $\frac{x}{x^2-9}-\frac{4(x-3)}{x+3}$

47. $\frac{3}{a+2}+\frac{3a+1}{a^2+4a+4}$

48. $\frac{2m+9}{m-5}-\frac{4}{m^2-3m-10}$

49. $\frac{x}{x^2+2x-8}+\frac{x+1}{x^2-3x+2}$

50. $\frac{-x^2+5x}{(x-5)^2}+\frac{x+8}{x-5}$

51. $\frac{5x}{x^2-9x+8}-\frac{3(x+2)}{x^2-6x-16}$

52. $\frac{2}{(2p-3)(p+4)}-\frac{3}{(p+4)(p-4)}$

53. $4-\frac{x-1}{x^2+3x-10}$

54. $\frac{3x}{2x-3}+\frac{3x+6}{2x^2+x-6}$

55. $\frac{3a+2}{4a+1}-\frac{3a+6}{4a^2+9a+2}$

56. $\frac{7}{3q^2+q-4}+\frac{9q+2}{3q^2-2q-8}$

57. $\frac{x-y}{x^2-4xy+4y^2}+\frac{x-3y}{x^2-4y^2}$

58. $\frac{x+2y}{x^2-xy-2y^2}-\frac{y}{x^2-3xy+2y^2}$

59. $\frac{2r}{r-4}-\frac{2r}{r+4}+\frac{64}{r^2-16}$

60. $\frac{4}{p+1}+\frac{3}{p-1}+\frac{p+4}{p^2-1}$

**61.** $\dfrac{-4}{x^2+2x-3} - \dfrac{1}{x+3} + \dfrac{1}{x-1}$

**62.** $\dfrac{2}{x^2-16} + \dfrac{x+1}{x^2+8x+16} + \dfrac{3}{x-4}$

**63.** $\dfrac{3}{3x-2} - \dfrac{1}{x-4} + 5$

**64.** $\dfrac{x}{3x+4} + \dfrac{3x+2}{x-5} - \dfrac{7x^2+24x+28}{3x^2-11x-20}$

**65.** $2 - \dfrac{1}{8r^2+2r-15} + \dfrac{r+2}{4r-5}$

**66.** $\dfrac{x}{x^2-10x+24} - \dfrac{3}{x-6} + 1$

**67.** $\dfrac{3}{5x+6} + \dfrac{x^2-x}{5x^2-4x-12} - \dfrac{4}{x-2}$

**68.** $\dfrac{3}{x^2-13x+36} + \dfrac{4}{2x^2-7x-4} + \dfrac{1}{2x^2-17x-9}$

**69.** $\dfrac{3m}{6m^2+13mn+6n^2} + \dfrac{2m}{4m^2+8mn+3n^2}$

**70.** $\dfrac{(x-y)^2}{x^3-y^3} + \dfrac{2}{x^2+xy+y^2}$

**71.** $\dfrac{5r-2s}{25r^2-4s^2} - \dfrac{2r-s}{10r^2-rs-2s^2}$

**72.** $\dfrac{6}{(2r-1)^2} + \dfrac{2}{2r-1} - 3$

**73.** $\dfrac{2}{2x+3y} - \dfrac{4x^2-6xy+9y^2}{8x^3+27y^3}$

**74.** $\dfrac{4}{4x-5y} - \dfrac{3x^2+2y^2}{64x^3-125y^3}$

## Problem Solving

**75.** When two rational expressions are being added or subtracted, should the numerators of the expressions being added or subtracted be factored? Explain.

**76.** Are the fractions $\dfrac{x-3}{4-x}$ and $-\dfrac{x-3}{x-4}$ equivalent? Explain.

**77.** Are the fractions $\dfrac{8-x}{3-x}$ and $\dfrac{x-8}{x-3}$ equivalent? Explain.

**78.** If $f(x)$ and $g(x)$ are both rational functions, will $(f+g)(x)$ always be a rational function?

**79.** If $f(x) = \dfrac{x+2}{x-3}$ and $g(x) = \dfrac{x}{x+4}$, find

**a)** the domain of $f(x)$.
**b)** the domain of $g(x)$.
**c)** $(f+g)(x)$.
**d)** the domain of $(f+g)(x)$.

**80.** If $f(x) = \dfrac{x+1}{x^2-9}$ and $g(x) = \dfrac{x}{x-3}$, find

**a)** the domain of $f(x)$.
**b)** the domain of $g(x)$.
**c)** $(f+g)(x)$.
**d)** the domain of $(f+g)(x)$.

**Profit** *In Exercises 81–84, find the profit function, P(x). (See Example 10.)*

**81.** $R(x) = \dfrac{4x-5}{x+1}$ and $C(x) = \dfrac{2x-7}{x+2}$

**82.** $R(x) = \dfrac{5x-2}{x+2}$ and $C(x) = \dfrac{3x-4}{x+1}$

**83.** $R(x) = \dfrac{8x-3}{x+2}$ and $C(x) = \dfrac{5x-8}{x+3}$

**84.** $R(x) = \dfrac{7x-10}{x+3}$ and $C(x) = \dfrac{5x-8}{x+4}$

*The dashed red lines in the figures below are called* **asymptotes**. *The asymptotes are not a part of the graph but are used to show values that the graph approaches, but does not touch. In Exercises 85 and 86, determine the domain and range of the rational function shown.*

**85.**

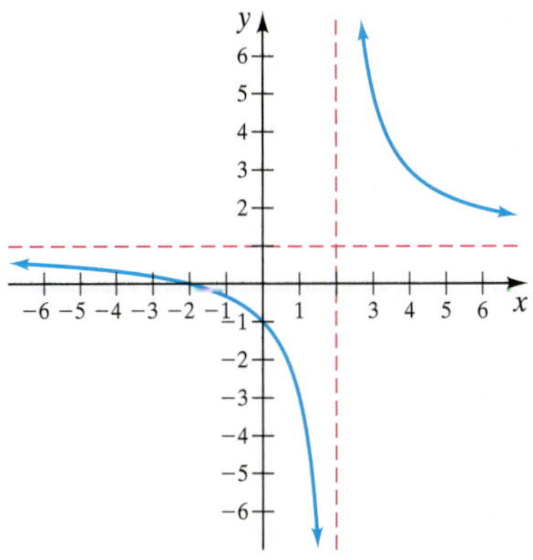


**86.**

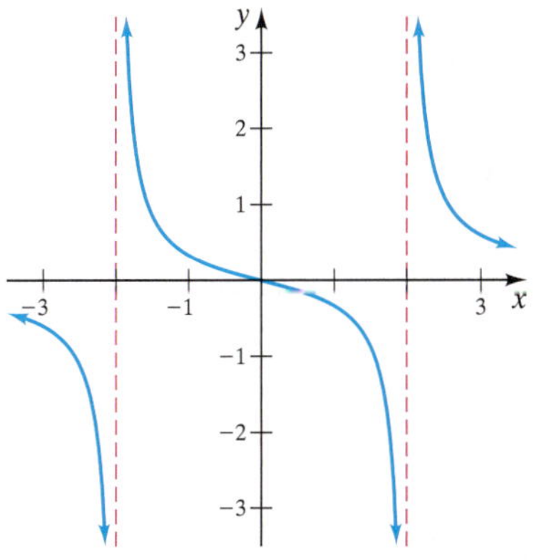

*In Exercises 87–90, use* $f(x) = \dfrac{x}{x^2 - 4}$ *and* $g(x) = \dfrac{2}{x^2 + x - 6}$. *Find the following.*

**87.** $(f + g)(x)$

**88.** $(f - g)(x)$

**89.** $(f \cdot g)(x)$

**90.** $(f/g)(x)$

**91.** Show that $\dfrac{a}{b} + \dfrac{c}{d} = \dfrac{ad + bc}{bd}$.

**92.** Show that $x^{-1} + y^{-1} = \dfrac{x + y}{xy}$.

**Area and Perimeter** *Consider the rectangles below. Find* **a)** *the perimeter, and* **b)** *the area.*

**93.**

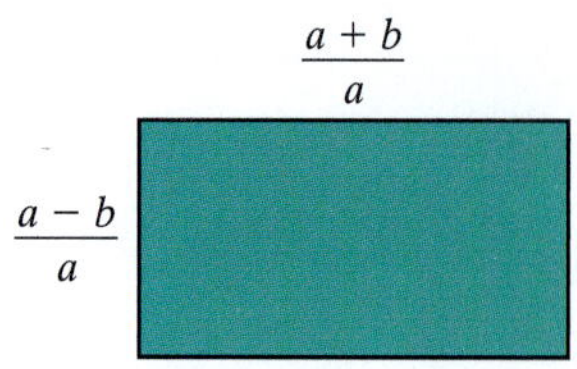


**94.**

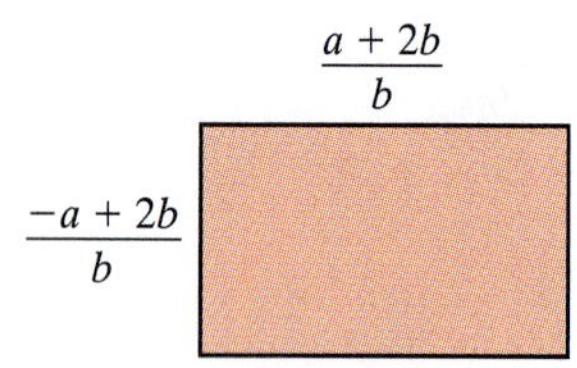


*Determine the polynomial to be placed in the shaded area to give a true statement. Explain how you determined your answer.*

**95.** $\dfrac{5x^2 - 6}{x^2 - x - 1} - \dfrac{\qquad\qquad}{x^2 - x - 1} = \dfrac{-2x^2 + 6x - 12}{x^2 - x - 1}$

**96.** $\dfrac{r^2 - 6}{r^2 - 5r + 6} - \dfrac{\qquad\qquad}{r^2 - 5r + 6} = \dfrac{1}{r - 2}$

*Perform the indicated operations.*

**97.** $\left(3 + \dfrac{1}{x + 3}\right)\left(\dfrac{x + 3}{x - 2}\right)$

**98.** $\left(\dfrac{3}{r + 1} - \dfrac{4}{r - 2}\right)\left(\dfrac{r - 2}{r + 10}\right)$

**99.** $\left(\dfrac{5}{a - 5} - \dfrac{2}{a + 3}\right) \div (3a + 25)$

**100.** $\left(\dfrac{x^2 + 4x - 5}{2x^2 + x - 3} \cdot \dfrac{2x + 3}{x + 1}\right) - \dfrac{2}{x + 2}$

**101.** $\left(\dfrac{x + 5}{x - 3} - x\right) \div \dfrac{1}{x - 3}$

**102.** $\left(\dfrac{x + 5}{x^2 - 25} + \dfrac{1}{x + 5}\right)\left(\dfrac{2x^2 - 13x + 15}{4x^2 - 6x}\right)$

**103.** The weighted average of two values $a$ and $b$ is given by $a\left(\dfrac{x}{n}\right) + b\left(\dfrac{n - x}{n}\right)$, where $\dfrac{x}{n}$ is the weight given to $a$ and $\dfrac{n - x}{n}$ is the weight given to $b$.

**a)** Express this sum as a single fraction.

**b)** On exam $a$ you received a grade of 60 and on exam $b$ you received a grade of 92. If exam $a$ counts $\dfrac{2}{5}$ of your final grade and exam $b$ counts $\dfrac{3}{5}$, determine your weighted average.

**104.** Show that $\left(\dfrac{x}{y}\right)^{-1} + \left(\dfrac{y}{x}\right)^{-1} + (xy)^{-1} = \dfrac{x^2 + y^2 + 1}{xy}$.

*In Exercises 105 and 106, perform the indicated operation.*

**105.** $(a - b)^{-1} + (a - b)^{-2}$

**106.** $\left(\dfrac{a - b}{a}\right)^{-1} - \left(\dfrac{a + b}{a}\right)^{-1}$

*Use your graphing calculator to determine whether the following additions are correct.*

**107.** $\dfrac{x - 3}{x + 4} + \dfrac{x}{x^2 - 2x - 24} \stackrel{?}{=} \dfrac{x^2 - 10x + 18}{(x + 4)(x - 6)}$

**108.** $\dfrac{x - 2}{x^2 - 25} + \dfrac{x - 2}{2x^2 + 17x + 35} \stackrel{?}{=} \dfrac{3x^2 - 4x - 4}{(x + 5)(x - 5)(2x + 7)}$

## Challenge Problems

**109.** Express each sum as a single fraction.

**a)** $1 + \dfrac{1}{x}$

**b)** $1 + \dfrac{1}{x} + \dfrac{1}{x^2}$

**c)** $1 + \dfrac{1}{x} + \dfrac{1}{x^2} + \dfrac{1}{x^3} + \dfrac{1}{x^4}$

**d)** $1 + \dfrac{1}{x} + \dfrac{1}{x^2} + \cdots + \dfrac{1}{x^n}$

**110.** Let $f(x) = \dfrac{1}{x}$. Find $f(a + h) - f(a)$.

**111.** Let $g(x) = \dfrac{1}{x + 1}$. Find $g(a + h) - g(a)$.

## Cumulative Review Exercises

[2.4] **112. Filling Boxes** A cereal box machine fills cereal boxes at a rate of 80 per minute. Then the machine is slowed down and fills cereal boxes at a rate of 60 per minute. If the sum of the two time periods is 14 minutes and the number of cereal boxes filled at the higher rate is the same as the number filled at the lower rate, determine **a)** how long the machine is used at the faster rate, and **b)** the total number of cereal boxes filled over the 14-minute period.

[2.6] **113.** Solve for $x$ and give the solution in set notation. $|x - 3| - 6 < -1$

[3.4] **114.** Find the slope of the line passing through the points $(-2, 3)$ and $(7, -3)$.

[4.5] **115.** Evaluate the determinant $\begin{vmatrix} -1 & 3 \\ 5 & -4 \end{vmatrix}$.

[5.3] **116.** Divide $\dfrac{6x^2 - 5x + 6}{2x + 3}$.

[5.8] **117.** Solve $3p^2 = 22p - 7$.

## 6.3 Complex Fractions

1 Recognize complex fractions.

2 Simplify complex fractions by multiplying by a common denominator.

3 Simplify complex fractions by simplifying the numerator and denominator.

### 1 Recognize Complex Fractions

A **complex fraction** is one that has a rational (or fractional) expression in its numerator or its denominator or both its numerator and denominator.

**Examples of Complex Fractions**

$$\frac{\frac{2}{3}}{5}, \quad \frac{\frac{x+1}{x}}{4x}, \quad \frac{x}{\frac{y}{x+1}}, \quad \frac{\frac{a+b}{a}}{\frac{a-b}{b}}, \quad \frac{9 + \frac{1}{x}}{\frac{1}{x^2} + \frac{8}{x}}$$

The expression above the **main fraction line** is the numerator, and the expression below the main fraction line is the denominator of the complex fraction.

$$\frac{a+b}{a} \longleftarrow \text{numerator of complex fraction}$$

$$\text{———} \longleftarrow \text{main fraction line}$$

$$\frac{a-b}{b} \longleftarrow \text{denominator of complex fraction}$$

We will explain two methods that can be used to simplify complex fractions. To simplify a complex fraction means to write the expression without a fraction in its numerator and its denominator.

## 2 Simplify Complex Fractions by Multiplying by a Common Denominator

The first method involves multiplying both the numerator and denominator of the complex fraction by a common denominator.

**To Simplify a Complex Fraction by Multiplying by a Common Denominator**

1. Find the least common denominator of all fractions appearing within the complex fraction. This is the LCD of the complex fraction.
2. Multiply both the numerator and denominator of the complex fraction by the LCD of the complex fraction found in step 1.
3. Simplify when possible.

In step 2, you are actually multiplying the complex fraction by $\frac{\text{LCD}}{\text{LCD}}$, which is equivalent to multiplying the fraction by 1.

**EXAMPLE 1** ▸ Simplify $\dfrac{\dfrac{4}{x^2} - \dfrac{3}{x}}{\dfrac{x^2}{5}}$.

*Solution* The denominators in the complex fraction are $x^2$, $x$, and 5. Therefore, the LCD of the complex fraction is $5x^2$. Multiply the numerator and denominator by $5x^2$.

$$\frac{\dfrac{4}{x^2} - \dfrac{3}{x}}{\dfrac{x^2}{5}} = \frac{5x^2\left(\dfrac{4}{x^2} - \dfrac{3}{x}\right)}{5x^2\left(\dfrac{x^2}{5}\right)}$$ *Multiply the numerator and denominator by $5x^2$.*

$$= \frac{5x^2\left(\dfrac{4}{x^2}\right) - 5x^2\left(\dfrac{3}{x}\right)}{5x^2\left(\dfrac{x^2}{5}\right)}$$ *Distributive property*

$$= \frac{20 - 15x}{x^4}$$ *Simplify.*

▸ **Now Try Exercise 13**

**EXAMPLE 2** ▸ Simplify $\dfrac{a + \dfrac{3}{b}}{b + \dfrac{3}{a}}$.

*Solution* Multiply the numerator and denominator of the complex fraction by its LCD, $ab$.

$$\frac{a + \frac{3}{b}}{b + \frac{3}{a}} = \frac{ab\left(a + \frac{3}{b}\right)}{ab\left(b + \frac{3}{a}\right)}$$ Multiply the numerator and denominator by ab.

$$= \frac{a^2b + 3a}{ab^2 + 3b}$$ Distributive property

$$= \frac{a(ab + 3)}{b(ab + 3)} = \frac{a}{b}$$ Factor and simplify.

▸ Now Try Exercise 17

**EXAMPLE 3** ▸ Simplify $\dfrac{a^{-1} + ab^{-2}}{ab^{-2} - a^{-2}b^{-1}}$.

*Solution* First rewrite each expression without negative exponents.

$$\frac{a^{-1} + ab^{-2}}{ab^{-2} - a^{-2}b^{-1}} = \frac{\frac{1}{a} + \frac{a}{b^2}}{\frac{a}{b^2} - \frac{1}{a^2b}}$$

$$= \frac{a^2b^2\left(\frac{1}{a} + \frac{a}{b^2}\right)}{a^2b^2\left(\frac{a}{b^2} - \frac{1}{a^2b}\right)}$$ Multiply the numerator and denominator by $a^2b^2$, the LCD of the complex fraction.

$$= \frac{a^2b^2\left(\frac{1}{a}\right) + a^2b^2\left(\frac{a}{b^2}\right)}{a^2b^2\left(\frac{a}{b^2}\right) - a^2b^2\left(\frac{1}{a^2b}\right)}$$ Distributive property

$$= \frac{ab^2 + a^3}{a^3 - b}$$

▸ Now Try Exercise 43

In Example 3, although we could factor an $a$ from both terms in the numerator of the answer, we could not simplify the answer further by dividing out common factors. So, we leave the answer in the form given.

## 3 Simplify Complex Fractions by Simplifying the Numerator and Denominator

Complex fractions can also be simplified as follows:

**To Simplify a Complex Fraction by Simplifying the Numerator and the Denominator**

1. Add or subtract as necessary to get one rational expression in the numerator.
2. Add or subtract as necessary to get one rational expression in the denominator.
3. Invert the denominator of the complex fraction and multiply by the numerator of the complex fraction.
4. Simplify when possible.

Example 4 will show how Example 1 can be simplified by this second method.

**EXAMPLE 4** ▸ Simplify $\dfrac{\dfrac{4}{x^2} - \dfrac{3}{x}}{\dfrac{x^2}{5}}$.

*Solution* Subtract the fractions in the numerator to get one rational expression in the numerator. The common denominator of the fractions in the numerator is $x^2$.

$$\frac{\frac{4}{x^2} - \frac{3}{x}}{\frac{x^2}{5}} = \frac{\frac{4}{x^2} - \frac{3}{x} \cdot \frac{x}{x}}{\frac{x^2}{5}}$$ *Obtain common denominator in numerator.*

$$= \frac{\frac{4}{x^2} - \frac{3x}{x^2}}{\frac{x^2}{5}}$$

$$= \frac{\frac{4 - 3x}{x^2}}{\frac{x^2}{5}}$$

$$= \frac{4 - 3x}{x^2} \cdot \frac{5}{x^2}$$ *Invert denominator and multiply.*

$$= \frac{5(4 - 3x)}{x^4}$$

$$\text{or} \quad \frac{20 - 15x}{x^4}$$

This is the same answer obtained in Example 1.

▸ Now Try Exercise 13

### Helpful Hint

Some students prefer to use the second method when the complex fraction consists of a single fraction over a single fraction, such as

$$\frac{\frac{x + 3}{18}}{\frac{x - 8}{6}}$$

For more complex fractions, many students prefer the first method because you do not have to add fractions.

## EXERCISE SET 6.3

MathXL® MyMathLab

### Concept/Writing Exercises

1. What is a complex fraction?

2. We have indicated two procedures for evaluating complex fractions. Which procedure do you prefer? Why?

## Practice the Skills

*Simplify.*

3. $\dfrac{\frac{15a}{b^2}}{\frac{b^3}{5}}$

4. $\dfrac{\frac{10x^2y^4}{3z^3}}{\frac{5xy}{9z^5}}$

5. $\dfrac{\frac{36x^4}{5y^4z^5}}{\frac{9xy^2}{15z^5}}$

6. $\dfrac{\frac{40x^3}{7y^5z^5}}{\frac{8x^2y^2}{28x^4z^5}}$

7. $\dfrac{\frac{10x^3y^2}{9yz^4}}{\frac{40x^4y^7}{27y^2z^8}}$

8. $\dfrac{\frac{3a^4b^3}{7b^4c}}{\frac{15a^2b^6}{14ac^7}}$

9. $\dfrac{1 - \frac{x}{y}}{3x}$

10. $\dfrac{2 + \frac{a}{b}}{5b}$

11. $\dfrac{x - \frac{x}{y}}{\frac{8 + x}{y}}$

12. $\dfrac{a + \frac{2a}{b}}{\frac{7 + a}{b}}$

13. $\dfrac{x + \frac{5}{y}}{1 + \frac{x}{y}}$

14. $\dfrac{\frac{4}{x} + \frac{2}{x^2}}{2 + \frac{1}{x}}$

15. $\dfrac{\frac{2}{a} + \frac{1}{2a}}{a + \frac{a}{2}}$

16. $\dfrac{3 - \frac{1}{y}}{2 - \frac{1}{y}}$

17. $\dfrac{\frac{a^2}{b} - b}{\frac{b^2}{a} - a}$

18. $\dfrac{x - \frac{4}{y}}{y - \frac{4}{x}}$

19. $\dfrac{\frac{x}{y} - \frac{y}{x}}{\frac{x + y}{x}}$

20. $\dfrac{\frac{1}{m} + \frac{9}{m^2}}{2 + \frac{1}{m^2}}$

21. $\dfrac{\frac{a}{b} - 6}{\frac{-a}{b} + 6}$

22. $\dfrac{7 - \frac{x}{y}}{\frac{x}{y} - 7}$

23. $\dfrac{\frac{4x + 8}{3x^2}}{\frac{4x^3}{9}}$

24. $\dfrac{\frac{x^2 - y^2}{x}}{\frac{x + y}{x^4}}$

25. $\dfrac{\frac{a}{a + 1} - 1}{\frac{2a + 1}{a - 1}}$

26. $\dfrac{\frac{x}{4} - \frac{1}{x}}{1 + \frac{x + 4}{x}}$

27. $\dfrac{1 + \frac{x}{x + 1}}{\frac{2x + 1}{x - 1}}$

28. $\dfrac{\frac{2}{x - 1} + 2}{\frac{2}{x + 1} - 2}$

29. $\dfrac{\frac{a + 1}{a - 1} + \frac{a - 1}{a + 1}}{\frac{a + 1}{a - 1} - \frac{a - 1}{a + 1}}$

30. $\dfrac{\frac{a - 2}{a + 2} - \frac{a + 2}{a - 2}}{\frac{a - 2}{a + 2} + \frac{a + 2}{a - 2}}$

31. $\dfrac{\frac{5}{5 - x} + \frac{6}{x - 5}}{\frac{3}{x} + \frac{2}{x - 5}}$

32. $\dfrac{\frac{2}{m} + \frac{1}{m^2} + \frac{3}{m - 1}}{\frac{6}{m - 1}}$

33. $\dfrac{\frac{3}{x^2} - \frac{1}{x} + \frac{2}{x - 2}}{\frac{1}{x}}$

34. $\dfrac{\frac{2}{x^2 + x - 20} + \frac{3}{x^2 - 6x + 8}}{\frac{2}{x^2 + 3x - 10} + \frac{3}{x^2 + 2x - 24}}$

35. $\dfrac{\frac{2}{a^2 - 3a + 2} + \frac{2}{a^2 - a - 2}}{\frac{2}{a^2 - 1} + \frac{2}{a^2 + 4a + 3}}$

36. $\dfrac{\frac{1}{x^2 + 5x + 4} + \frac{2}{x^2 + 2x - 8}}{\frac{2}{x^2 - x - 2} + \frac{1}{x^2 - 5x + 6}}$

*Simplify.*

37. $2a^{-2} + b$

38. $6a^{-2} + b^{-1}$

39. $(a^{-1} + b^{-1})^{-1}$

40. $\dfrac{a^{-1} + b^{-1}}{\frac{5}{ab}}$

41. $\dfrac{a^{-1} + 1}{b^{-1} - 1}$

42. $\dfrac{x^{-1} - y^{-1}}{x^{-1} + y^{-1}}$

43. $\dfrac{a^{-2} - ab^{-1}}{ab^{-2} + a^{-1}b^{-1}}$

44. $\dfrac{xy^{-1} + x^{-1}y^{-2}}{x^{-1} - x^{-2}y^{-1}}$

45. $\dfrac{\frac{9a}{b} + a^{-1}}{\frac{b}{a} + a^{-1}}$

46. $\dfrac{x^{-2} + \frac{3}{x}}{3x^{-1} + x^{-2}}$

47. $\dfrac{a^{-1} + b^{-1}}{(a + b)^{-1}}$

48. $\dfrac{4a^{-1} - b^{-1}}{(a - b)^{-1}}$

49. $5x^{-1} - (3y)^{-1}$

50. $\dfrac{\frac{7}{x} + \frac{1}{y}}{(x - y)^{-1}}$

51. $\dfrac{\frac{2}{xy} - \frac{8}{y} + \frac{5}{x}}{3x^{-1} - 4y^{-2}}$

52. $\dfrac{4m^{-1} + 3n^{-1} + (2mn)^{-1}}{\frac{5}{m} + \frac{7}{n}}$

## Problem Solving

**Area** *For Exercises 53–56, the area and width of each rectangle are given. Find the length, l, by dividing the area, A, by the width, w.*

**53.** $A = \dfrac{x^2 + 12x + 35}{x + 3}$ $w = \dfrac{x^2 + 6x + 5}{x^2 + 5x + 6}$ $l$

**54.** $A = \dfrac{x^2 + 10x + 16}{x + 4}$ $(x > 1)$ $w = \dfrac{x^2 + 11x + 24}{x^2 + 3x - 4}$ $l$

**55.** $A = \dfrac{x^2 + 11x + 28}{x + 5}$ $(x > 1)$ $w = \dfrac{x^2 + 8x + 7}{x^2 + 4x - 5}$ $l$

**56.** $A = \dfrac{x^2 + 17x + 72}{x + 3}$ $(x > 2)$ $w = \dfrac{x^2 + 11x + 18}{x^2 + x - 6}$ $l$

**57. Automobile Jack** The efficiency of a jack, $E$, is given by the formula

$$E = \frac{\frac{1}{2}h}{h + \frac{1}{2}}$$

where $h$ is determined by the pitch of the jack's thread.

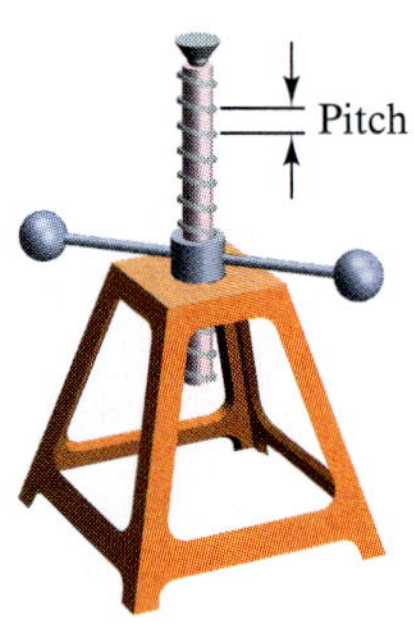


Determine the efficiency of a jack whose values of $h$ are:

**a)** $\frac{2}{5}$ **b)** $\frac{1}{3}$

**58. Resistors** If two resistors with resistances $R_1$ and $R_2$ are connected in parallel, their combined resistance, $R_T$, can be found from the formula

$$R_T = \frac{1}{\frac{1}{R_1} + \frac{1}{R_2}}$$

Simplify the right side of the formula.

**59. Resistors** If three resistors with resistances $R_1$, $R_2$, and $R_3$ are connected in parallel, their combined resistance, $R_T$, can be found from the formula

$$R_T = \frac{1}{\frac{1}{R_1} + \frac{1}{R_2} + \frac{1}{R_3}}$$

Simplify the right side of this formula.

**60. Optics** A formula used in the study of optics is

$$f = (p^{-1} + q^{-1})^{-1}$$

where $p$ is the object's distance from a lens, $q$ is the image distance from the lens, and $f$ is the focal length of the lens. Express the right side of the formula without any negative exponents.

**61.** If $f(x) = \frac{1}{x}$, find $f(f(a))$.

**62.** If $f(x) = \frac{2}{x + 2}$, find $f(f(a))$.

## Challenge Problems

*For each function, find* $\dfrac{f(a + h) - f(a)}{h}$.

**63.** $f(x) = \dfrac{1}{x}$

**64.** $f(x) = \dfrac{5}{x}$

**65.** $f(x) = \dfrac{1}{x + 1}$

**66.** $f(x) = \dfrac{6}{x - 1}$

**67.** $f(x) = \dfrac{1}{x^2}$

**68.** $f(x) = \dfrac{3}{x^2}$

*Simplify.*

**69.** $\dfrac{1}{2a + \dfrac{1}{2a + \dfrac{1}{2a}}}$

**70.** $\dfrac{1}{x + \dfrac{1}{x + \dfrac{1}{x + 1}}}$

**71.** $\dfrac{1}{2 + \dfrac{1}{2 + \dfrac{1}{2}}}$

## Cumulative Review Exercises

[1.4] **72.** Evaluate $\dfrac{\left|-\frac{3}{9}\right| - \left(-\frac{5}{9}\right) \cdot \left|-\frac{3}{8}\right|}{|-5 - (-3)|}$.

[2.5] **73.** Solve $\frac{3}{5} < \frac{-x - 5}{3} < 6$ and give the solution in interval notation.

[2.6] **74.** Solve $|x - 1| = |2x - 4|$.

[3.5] **75.** Determine if the two lines represented by the following equations are parallel, perpendicular, or neither.

$$6x + 2y = 5$$
$$4x - 9 = -2y$$

# 6.4 Solving Rational Equations

1. Solve rational equations.
2. Check solutions.
3. Solve proportions.
4. Solve problems involving rational functions.
5. Solve applications using rational expressions.
6. Solve for a variable in a formula containing rational expressions.

## 1 Solve Rational Equations

In Sections 6.1 through 6.3 we presented techniques to add, subtract, multiply, and divide rational expressions. In this section, we present a method for solving rational equations. A **rational equation** is an equation that contains at least one rational expression.

**To Solve Rational Equations**

1. Determine the LCD of all rational expressions in the equation.
2. Multiply *both* sides of the equation by the LCD. This will result in every term in the equation being multiplied by the LCD.
3. Remove any parentheses and combine like terms on each side of the equation.
4. Solve the equation using the properties discussed in earlier sections.
5. Check the solution in the *original* equation.

In step 2, we multiply both sides of the equation by the LCD to eliminate fractions. In some examples we will not show the check to save space.

**EXAMPLE 1** Solve $\frac{3x}{4} + \frac{1}{2} = \frac{2x - 3}{4}$.

*Solution* Multiply both sides of the equation by the LCD, 4. Then use the distributive property, which results in each term in the equation being multiplied by the LCD.

$$4\left(\frac{3x}{4} + \frac{1}{2}\right) = \frac{2x - 3}{4} \cdot 4 \qquad \text{Multiply both sides by 4.}$$
$$4\left(\frac{3x}{4}\right) + 4\left(\frac{1}{2}\right) = 2x - 3 \qquad \text{Distributive property}$$
$$3x + 2 = 2x - 3$$
$$x + 2 = -3$$
$$x = -5$$

A check will show that $-5$ is the solution.

Now Try Exercise 15

## 2 Check Solutions

**Whenever a variable appears in any denominator, you must check your apparent solution in the original equation. When checking, if an apparent solution makes any denominator equal to 0, that value is not a solution to the equation.** Such values are called **extraneous roots** or **extraneous solutions**. An extraneous root is a number obtained when solving an equation that is not a solution to the original equation.

**EXAMPLE 2** Solve $2 - \frac{4}{x} = \frac{1}{3}$.

*Solution* Multiply both sides of the equation by the LCD, $3x$.

$$3x\left(2 - \frac{4}{x}\right) = \left(\frac{1}{3}\right) \cdot 3x \quad \text{Multiply both sides by } 3x.$$

$$3x(2) - 3x\left(\frac{4}{x}\right) = \left(\frac{1}{3}\right)3x \quad \text{Distributive property}$$

$$6x - 12 = x$$

$$5x - 12 = 0$$

$$5x = 12$$

$$x = \frac{12}{5}$$

**Check**

$$2 - \frac{4}{x} = \frac{1}{3}$$

$$2 - \frac{4}{\frac{12}{5}} \stackrel{?}{=} \frac{1}{3} \quad \text{Substitute } \frac{12}{5} \text{ for } x.$$

$$2 - \frac{20}{12} \stackrel{?}{=} \frac{1}{3}$$

$$\frac{1}{3} = \frac{1}{3} \quad \text{True}$$

**Now Try Exercise 21**

**EXAMPLE 3** Solve $x - \frac{6}{x} = -5$.

*Solution*

$$x \cdot \left(x - \frac{6}{x}\right) = -5 \cdot x \quad \text{Multiply both sides by the LCD, } x.$$

$$x(x) - x\left(\frac{6}{x}\right) = -5x \quad \text{Distributive property}$$

$$x^2 - 6 = -5x$$

$$x^2 + 5x - 6 = 0$$

$$(x - 1)(x + 6) = 0$$

$$x - 1 = 0 \quad \text{or} \quad x + 6 = 0$$

$$x = 1 \qquad\qquad x = -6$$

Checks of 1 and $-6$ will show that they are both solutions to the equation.

**Now Try Exercise 35**

**EXAMPLE 4** Solve $\frac{3x}{x^2 - 4} + \frac{1}{x - 2} = \frac{2}{x + 2}$.

*Solution* First factor the denominator $x^2 - 4$, then find the LCD.

$$\frac{3x}{(x + 2)(x - 2)} + \frac{1}{x - 2} = \frac{2}{x + 2}$$

The LCD is $(x + 2)(x - 2)$. Multiply both sides of the equation by the LCD, and then use the distributive property. This process will eliminate the fractions from the equation.

$$(x+2)(x-2)\cdot\left[\frac{3x}{(x+2)(x-2)}+\frac{1}{x-2}\right]=\frac{2}{x+2}\cdot(x+2)(x-2)$$

$$\cancel{(x+2)}\cancel{(x-2)}\cdot\frac{3x}{\cancel{(x+2)}\cancel{(x-2)}}+(x+2)\cancel{(x-2)}\cdot\frac{1}{\cancel{x-2}}=\frac{2}{\cancel{x+2}}\cdot\cancel{(x+2)}(x-2)$$

$$3x+(x+2)=2(x-2)$$
$$4x+2=2x-4$$
$$2x+2=-4$$
$$2x=-6$$
$$x=-3$$

A check will show that $-3$ is the solution.

▸ **Now Try Exercise 39**

---

**EXAMPLE 5** ▸ Solve $\dfrac{22}{2p^2-9p-5}-\dfrac{3}{2p+1}=\dfrac{2}{p-5}$.

*Solution* Factor the denominator, then determine the LCD.

$$\frac{22}{(2p+1)(p-5)}-\frac{3}{2p+1}=\frac{2}{p-5}$$

Multiply both sides of the equation by the LCD, $(2p+1)(p-5)$.

$$\cancel{(2p+1)}\cancel{(p-5)}\cdot\frac{22}{\cancel{(2p+1)}\cancel{(p-5)}}-\cancel{(2p+1)}(p-5)\cdot\frac{3}{\cancel{2p+1}}=\frac{2}{\cancel{p-5}}\cdot(2p+1)\cancel{(p-5)}$$

$$22-3(p-5)=2(2p+1)$$
$$22-3p+15=4p+2$$
$$37-3p=4p+2$$
$$35=7p$$
$$5=p$$

The solution appears to be 5. However, since a variable appears in a denominator, this solution must be checked.

**Check**

$$\frac{22}{2p^2-9p-5}-\frac{3}{2p+1}=\frac{2}{p-5}$$

$$\frac{22}{2(5)^2-9(5)-5}-\frac{3}{2(5)+1}\stackrel{?}{=}\frac{2}{5-5} \quad \text{Substitute 5 for } p.$$

$$\text{Undefined}\longrightarrow\frac{22}{0}-\frac{3}{11}=\frac{2}{0}\longleftarrow\text{Undefined}$$

Since 5 makes a denominator 0 and division by 0 is undefined, 5 is an extraneous solution. Therefore, you should write "**no solution**" as your answer.

▸ **Now Try Exercise 43**

---

In Example 5, the only possible solution is 5. However, when $p = 5$, the denominator of $\dfrac{2}{p-5}$ is 0. Therefore, 5 cannot be a solution. We did not actually have to show the complete check, but we did so in this example for clarity.

## Helpful Hint

Remember, whenever you solve an equation where a variable appears in any denominator, you must check the apparent solution to make sure it is not an extraneous solution. If the apparent solution makes any denominator 0, then it is an extraneous solution and not a true solution to the equation.

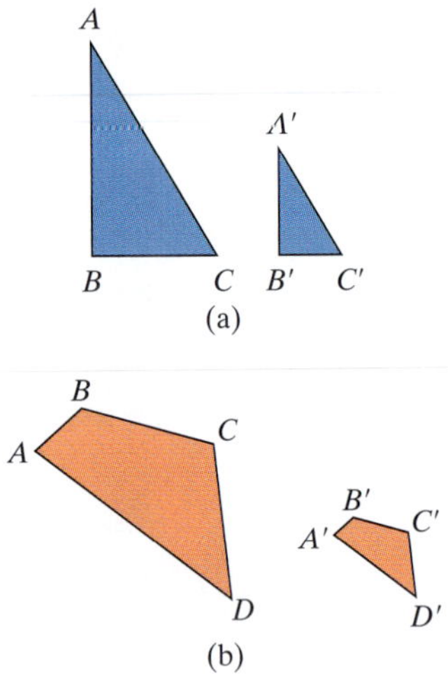


FIGURE 6.5

## 3 Solve Proportions

**Proportions** are equations of the form $\frac{a}{b} = \frac{c}{d}$. Proportions are one type of rational equation. Proportions may be solved by *cross multiplication* as follows. If $\frac{a}{b} = \frac{c}{d}$, then $ad = bc$, $b \neq 0$, $d \neq 0$. Proportions may also be solved by multiplying both sides of the proportion by the least common denominator. In Examples 6 and 7, we solve proportions by multiplying both sides by the LCD. We then ask you to determine the solutions, if possible, using cross multiplication. *When solving a proportion where the denominator of one or more of the ratios contains a variable, you must check to make sure that your solution is not extraneous.*

Proportions are often used when working with similar figures. **Similar figures** are figures whose corresponding angles are equal and whose corresponding sides are in proportion. **Figure 6.5** illustrates two sets of similar figures.

In **Figure 6.5a**, the ratio of the length of side $AB$ to the length of side $BC$ is the same as the ratio of the length of side $A'B'$ to the length of side $B'C'$. That is,

$$\frac{AB}{BC} = \frac{A'B'}{B'C'}$$

In a pair of similar figures, if the length of a side is unknown, it can often be found by using proportions, as illustrated in Example 6.

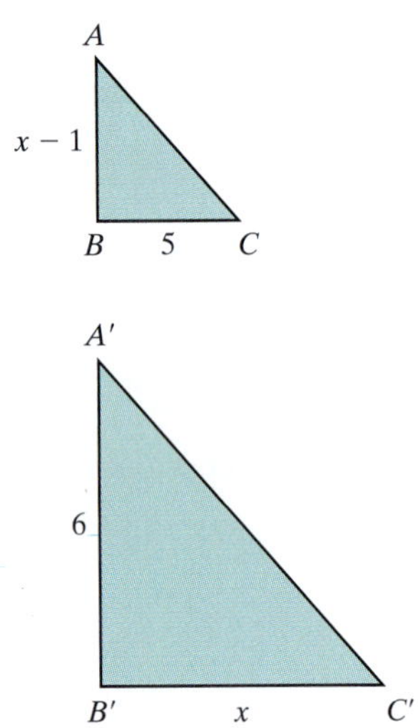


FIGURE 6.6

**EXAMPLE 6 ▸ Similar Triangles** Triangles $ABC$ and $A'B'C'$ in **Figure 6.6** are similar figures. Find the length of sides $AB$ and $B'C'$.

*Solution* We can set up a proportion and then solve for $x$. Then we can find the lengths.

$$\frac{AB}{BC} = \frac{A'B'}{B'C'}$$

$$\frac{x-1}{5} = \frac{6}{x}$$

$$5x \cdot \frac{x-1}{5} = \frac{6}{x} \cdot 5x \qquad \textit{Multiply both sides by the LCD, 5x.}$$

$$x(x-1) = 6 \cdot 5$$

$$x^2 - x = 30$$

$$x^2 - x - 30 = 0$$

$$(x-6)(x+5) = 0 \qquad \textit{Factor the trinomial.}$$

$$x - 6 = 0 \quad \text{or} \quad x + 5 = 0$$

$$x = 6 \qquad\qquad x = -5$$

Since the length of the side of a triangle cannot be a negative number, $-5$ is not a possible answer. Substituting 6 for $x$, we see that the length of side $B'C'$ is 6 and the length of side $AB$ is $6 - 1$ or 5.

**Check**

$$\frac{AB}{BC} = \frac{A'B'}{B'C'}$$

$$\frac{5}{5} \stackrel{?}{=} \frac{6}{6}$$

$$1 = 1 \qquad \textit{True}$$

**▸ Now Try Exercise 49**

The answer to Example 6 could also be obtained using cross multiplication. Try solving Example 6 using cross multiplication now.

**EXAMPLE 7** ▸ Solve $\dfrac{x^2}{x-3} = \dfrac{9}{x-3}$.

*Solution* This equation is a proportion. We will solve this equation by multiplying both sides of the equation by the LCD, $x - 3$.

$$\cancel{(x-3)} \cdot \frac{x^2}{\cancel{x-3}} = \frac{9}{\cancel{x-3}} \cdot \cancel{(x-3)}$$

$$x^2 = 9$$

$$x^2 - 9 = 0$$

$$(x+3)(x-3) = 0 \qquad \text{Factor the difference of two squares.}$$

$$x + 3 = 0 \quad \text{or} \quad x - 3 = 0$$

$$x = -3 \qquad\qquad x = 3$$

**Check**

$x = -3$

$$\frac{x^2}{x-3} = \frac{9}{x-3}$$

$$\frac{(-3)^2}{-3-3} \stackrel{?}{=} \frac{9}{-3-3}$$

$$\frac{9}{-6} \stackrel{?}{=} \frac{9}{-6}$$

$$-\frac{3}{2} = -\frac{3}{2} \qquad \text{True}$$

$x = 3$

$$\frac{x^2}{x-3} = \frac{9}{x-3}$$

$$\frac{3^2}{3-3} \stackrel{?}{=} \frac{9}{3-3}$$

$$\frac{9}{0} \stackrel{?}{=} \frac{9}{0} \qquad \longleftarrow \text{Undefined}$$

Since $x = 3$ results in a denominator of 0, 3 is *not* a solution to the equation. It is an extraneous root. The only solution to the equation is $-3$.

▸ **Now Try Exercise 45**

In Example 7, what would you obtain if you began by cross multiplying? Try it and see.

## 4 Solve Problems Involving Rational Functions

Now we will work a problem that involves a rational function.

**EXAMPLE 8** ▸ Consider the function $f(x) = x - \dfrac{2}{x}$. Find all $a$ for which $f(a) = 1$.

*Solution* Since $f(a) = a - \dfrac{2}{a}$, we need to find all values for which $a - \dfrac{2}{a} = 1$, $a \neq 0$. We begin by multiplying both sides of the equation by $a$, the LCD.

$$a \cdot \left(a - \frac{2}{a}\right) = a \cdot 1$$

$$a^2 - 2 = a$$

$$a^2 - a - 2 = 0$$

$$(a-2)(a+1) = 0$$

$$a - 2 = 0 \quad \text{or} \quad a + 1 = 0$$

$$a = 2 \qquad\qquad a = -1$$

**Check**

$$f(x) = x - \frac{2}{x}$$

$$f(2) = 2 - \frac{2}{2} = 2 - 1 = 1$$

$$f(-1) = -1 - \frac{2}{(-1)} = -1 + 2 = 1$$

For $a = 2$ or $a = -1$, $f(a) = 1$.

▸ **Now Try Exercise 53**

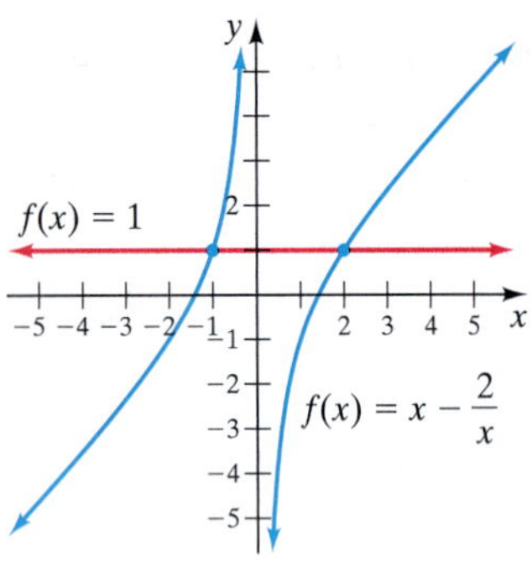


FIGURE 6.7

We used $f(x) = x - \frac{2}{x}$ in Example 8. **Figure 6.7** shows the graph of $f(x) = x - \frac{2}{x}$. In this course you will not have to graph functions like this. We illustrate this graph to reinforce the answer obtained in Example 8.

Notice the function is undefined at $x = 0$. Also notice that when $x = -1$ or $x = 2$, it appears that $f(x) = 1$. This was what we should have expected from the results obtained in Example 8.

Example 8 also could have been solved using your graphing calculator by setting $y_1 = x - \frac{2}{x}$ and $y_2 = 1$ and finding the $x$-coordinate of the intersections of the two graphs.

## 5 Solve Applications Using Rational Expressions

Now let's look at an application of rational equations.

**EXAMPLE 9** ▸ **Total Resistance** In electronics, the total resistance, $R_T$, of resistors connected in a parallel circuit is determined by the formula

$$\frac{1}{R_T} = \frac{1}{R_1} + \frac{1}{R_2} + \frac{1}{R_3} + \cdots + \frac{1}{R_n}$$

where $R_1, R_2, R_3, \ldots, R_n$ are the resistances of the individual resistors (measured in ohms, symbolized $\Omega$) in the circuit. Find the total resistance if two resistors, one of 100 ohms and the other of 300 ohms, are connected in a parallel circuit. See **Figure 6.8**.

*Solution* Since there are only two resistances, use the formula

$$\frac{1}{R_T} = \frac{1}{R_1} + \frac{1}{R_2}$$

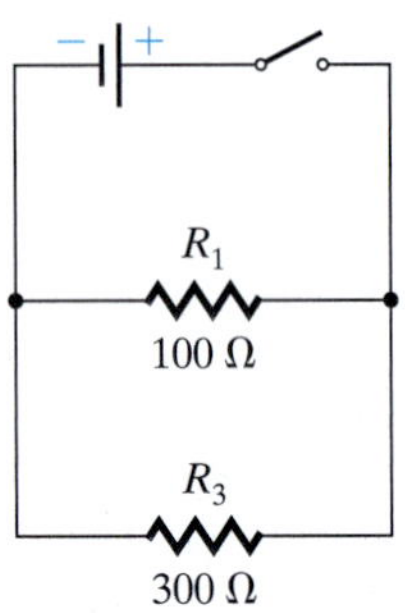


FIGURE 6.8

Let $R_1 = 100$ ohms and $R_2 = 300$ ohms; then

$$\frac{1}{R_T} = \frac{1}{100} + \frac{1}{300}$$

Multiply both sides of the equation by the LCD, $300R_T$.

$$300R_T \cdot \frac{1}{R_T} = 300R_T\left(\frac{1}{100} + \frac{1}{300}\right)$$

$$300\cancel{R_T} \cdot \frac{1}{\cancel{R_T}} = \overset{3}{\cancel{300}}R_T\left(\frac{1}{\cancel{100}}\right) + \cancel{300}R_T\left(\frac{1}{\cancel{300}}\right)$$

$$300 = 3R_T + R_T$$

$$300 = 4R_T$$

$$R_T = \frac{300}{4} = 75$$

Thus, the total resistance of the parallel circuit is 75 ohms. Notice that the resistors actually have less resistance when connected in a parallel circuit than separately.

▸ **Now Try Exercise 95**

## 6 Solve for a Variable in a Formula Containing Rational Expressions

Sometimes you may need to solve for a variable in a formula where the variable you are solving for occurs in more than one term. When this happens, it may be possible to solve for the variable by using factoring. To do so, collect all the terms containing the variable you are solving for on one side of the equation and all other terms on the other side of the equation. Then factor out the variable you are solving for. This process is illustrated in Examples 10 through 12.

**EXAMPLE 10 ▸ Optics** A formula used in the study of optics is $\frac{1}{p} + \frac{1}{q} = \frac{1}{f}$. In the formula, $p$ is the distance of an object from a lens or a mirror, $q$ is the distance of the image from the lens or mirror, and $f$ is the focal length of the lens or mirror. For people wearing glasses, the image distance is the distance from the lens to their retina. See **Figure 6.9**. Solve this formula for $f$.

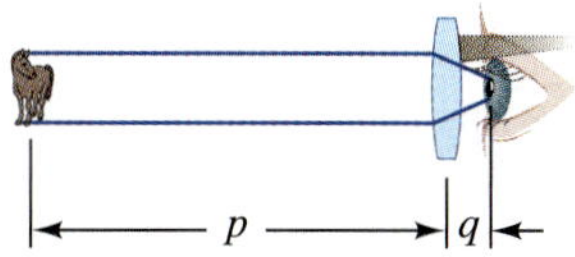

FIGURE 6.9

*Solution* Our goal is to isolate the variable $f$. We begin by multiplying both sides of the equation by the least common denominator, $pqf$, to eliminate fractions.

$$\frac{1}{p} + \frac{1}{q} = \frac{1}{f}$$

$$pqf\left(\frac{1}{p} + \frac{1}{q}\right) = pqf\left(\frac{1}{f}\right) \qquad \text{Multiply both sides by the LCD, } pqf.$$

$$pqf\left(\frac{1}{p}\right) + pqf\left(\frac{1}{q}\right) = pqf\left(\frac{1}{f}\right) \qquad \text{Distributive property}$$

$$qf + pf = pq \qquad \text{Simplify.}$$

$$f(q + p) = pq \qquad \text{Factor out } f.$$

$$\frac{f(q + p)}{q + p} = \frac{pq}{q + p} \qquad \text{Divide both sides by } q + p.$$

$$f = \frac{pq}{q + p} \quad \text{or} \quad f = \frac{pq}{p + q}$$

▸ Now Try Exercise 69

**EXAMPLE 11 ▸ Banking** A formula used in banking is $A = P + Prt$, where $A$ represents the amount that must be repaid to the bank when $P$ dollars are borrowed at simple interest rate, $r$, for time, $t$, in years. Solve this equation for $P$.

*Solution* Since both terms containing $P$ are by themselves on the right side of the equation, we factor $P$ from both terms.

$$A = P + Prt \qquad P \text{ is in both terms.}$$

$$A = P(1 + rt) \qquad \text{Factor out } P.$$

$$\frac{A}{1 + rt} = \frac{P(1 + rt)}{1 + rt} \qquad \text{Divide both sides by } 1 + rt \text{ to isolate } P.$$

$$\frac{A}{1 + rt} = P$$

Thus, $P = \frac{A}{1 + rt}$.

▸ Now Try Exercise 73

**EXAMPLE 12** ▸ **Physics** A formula used for levers in physics is $d = \frac{fl}{f + w}$. Solve this formula for $f$.

*Solution* We begin by multiplying both sides of the formula by $f + w$ to clear fractions. Then we will rewrite the expression with all terms containing $f$ on one side of the equal sign and all terms not containing $f$ on the other side of the equal sign.

$$d = \frac{fl}{f + w}$$

$$d(f + w) = \frac{fl}{\cancel{(f + w)}} \cancel{(f + w)} \quad \text{Multiply by } f + w \text{ to clear fractions.}$$

$$d(f + w) = fl$$

$$df + dw = fl \quad \text{Distributive property}$$

$$df - df + dw = fl - df \quad \text{Isolate terms containing } f \text{ on the right side of the equation.}$$

$$dw = fl - df$$

$$dw = f(l - d) \quad \text{Factor out } f.$$

$$\frac{dw}{l - d} = \frac{f\cancel{(l - d)}}{\cancel{l - d}} \quad \text{Isolate } f \text{ by dividing both sides by } l - d.$$

$$\frac{dw}{l - d} = f$$

Thus, $f = \frac{dw}{l - d}$.

▸ Now Try Exercise 79

## Avoiding Common Errors

After solving equations, as was done in this section, some students forget to keep their common denominator when adding and subtracting rational expressions. Remember, we multiply both sides of an *equation* by the LCD to remove the common denominator. If we are *adding or subtracting rational expressions*, we write the fractions with the LCD then add or subtract the numerators while *keeping the common denominator*.

For example, consider the addition problem

$$\frac{x}{x + 7} + \frac{3}{x + 7}$$

CORRECT

$$\frac{x}{x + 7} + \frac{3}{x + 7} = \frac{x + 3}{x + 7}$$

INCORRECT (crossed out)

$$\frac{x}{x + 7} + \frac{3}{x + 7} = (x + 7)\left(\frac{x}{x + 7} + \frac{3}{x + 7}\right) = x + 3$$

## EXERCISE SET 6.4

MathXL® MyMathLab

### Concept/Writing Exercises

1. What is an extraneous root?

2. Under what circumstances is it necessary to check your answers for extraneous roots?

3. Consider the equation $\frac{x}{4} - \frac{x}{3} = 2$ and the expression $\frac{x}{4} - \frac{x}{3} + 2$.

   a) What is the first step in solving the equation? Explain what effect the first step will have on the equation.

   b) Solve the equation.

   c) What is the first step in simplifying the expression? Explain what effect the first step has when simplifying the expression.

   d) Simplify the expression.

4. Consider the equation $\frac{x}{2} - \frac{x}{3} = 3$ and the expression $\frac{x}{2} - \frac{x}{3} + 3$.
   a) What is the first step in solving the equation?
   b) Solve the equation.
   c) What is the first step in simplifying the expression? Explain what effect the first step has when simplifying the expression.
   d) Simplify the expression.

5. What are similar figures?

6. **a)** Explain how to solve a rational equation.
   **b)** Solve $\frac{3}{x-4} + \frac{1}{x+4} = \frac{4}{x^2-16}$ following your procedure in part **a)**.

7. Tom Kelly solved an equation containing the term $\frac{7}{x-3}$ and obtained the answer $x = 3$. Can this be correct? Explain.

8. Geurfino Muldo solved an equation containing the term $\frac{21x}{x^2-16}$ and obtained the answer $x = 4$. Can this be correct? Explain.

## Practice the Skills

*Solve each equation and check your solution.*

9. $\frac{5}{x} = 1$

10. $\frac{12}{x} = 3$

11. $\frac{11}{b} = 2$

12. $\frac{1}{4} = \frac{z+2}{12}$

13. $\frac{6x+7}{5} = \frac{2x+9}{3}$

14. $\frac{a+2}{7} = \frac{a-3}{2}$

15. $\frac{3x}{8} + \frac{1}{4} = \frac{2x-3}{8}$

16. $\frac{3x}{10} + \frac{2}{5} = \frac{4x-3}{5}$

17. $\frac{z}{3} - \frac{3z}{4} = -\frac{5z}{12}$

18. $\frac{w}{2} + \frac{2w}{3} = \frac{7w}{6}$

19. $\frac{3}{4} - x = 2x$

20. $\frac{2}{y} + \frac{1}{2} = \frac{5}{2y}$

21. $\frac{2}{r} + \frac{5}{3r} = 1$

22. $3 + \frac{2}{x} = \frac{1}{4}$

23. $\frac{x-2}{x-5} = \frac{3}{x-5}$

24. $\frac{c+3}{c+1} = \frac{5}{2}$

25. $\frac{5y-2}{7} = \frac{15y-2}{28}$

26. $\frac{3}{x+1} = \frac{2}{x-3}$

27. $\frac{5.6}{-p-6.2} = \frac{2}{p}$

28. $\frac{4.5}{y-3} = \frac{6.9}{y+3}$

29. $\frac{m+1}{m+10} = \frac{m-2}{m+4}$

30. $\frac{x-3}{x+1} = \frac{x-6}{x+5}$

31. $x - \frac{4}{3x} = -\frac{1}{3}$

32. $x + \frac{2}{x} = \frac{27}{x}$

33. $\frac{2x-1}{3} - \frac{x}{4} = \frac{7.4}{6}$

34. $\frac{15}{x} + \frac{9x-7}{x+2} = 9$

35. $x + \frac{6}{x} = -7$

36. $b - \frac{8}{b} = -7$

37. $2 - \frac{5}{2b} = \frac{2b}{b+1}$

38. $\frac{3z-2}{z+1} = 4 - \frac{z+2}{z-1}$

39. $\frac{1}{w-3} + \frac{1}{w+3} = \frac{-5}{w^2-9}$

40. $\frac{6}{x+3} + \frac{5}{x+4} = \frac{12x+31}{x^2+7x+12}$

41. $\frac{8}{x^2-9} = \frac{2}{x-3} - \frac{4}{x+3}$

42. $a - \frac{a}{4} + \frac{a}{5} = 19$

43. $\frac{y}{2y+2} + \frac{2y-16}{4y+4} = \frac{2y-3}{y+1}$

44. $\frac{2}{w-5} = \frac{22}{2w^2-9w-5} - \frac{3}{2w+1}$

45. $\frac{x^2}{x-5} = \frac{25}{x-5}$

46. $\frac{x^2}{x-9} = \frac{81}{x-9}$

47. $\frac{5}{x^2+4x+3} + \frac{2}{x^2+x-6} = \frac{3}{x^2-x-2}$

48. $\frac{2}{x^2+2x-8} - \frac{1}{x^2+9x+20} = \frac{4}{x^2+3x-10}$

**Similar Figures** *For each pair of similar figures, find the length of the two unknown sides (that is, those two sides whose lengths involve the variable x).*

49.

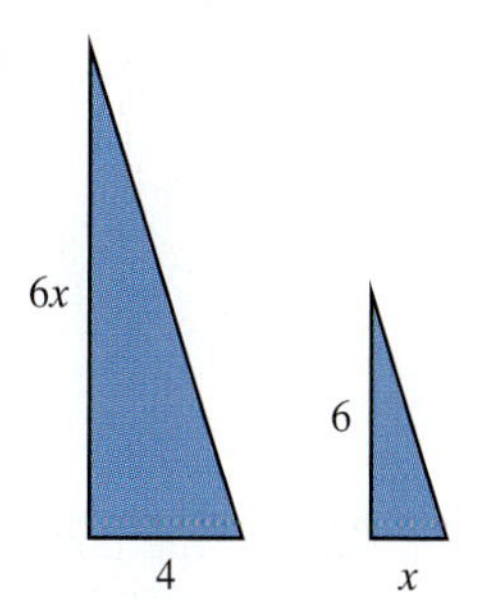


50.

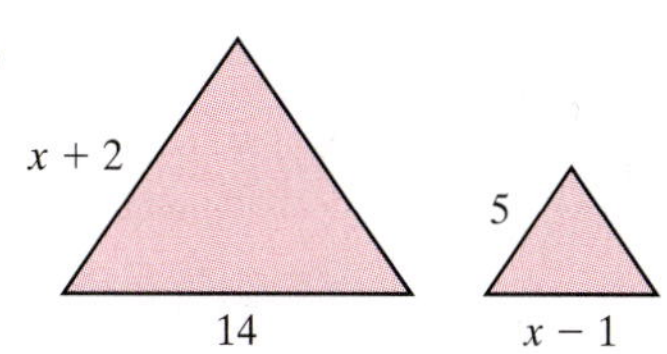


51.

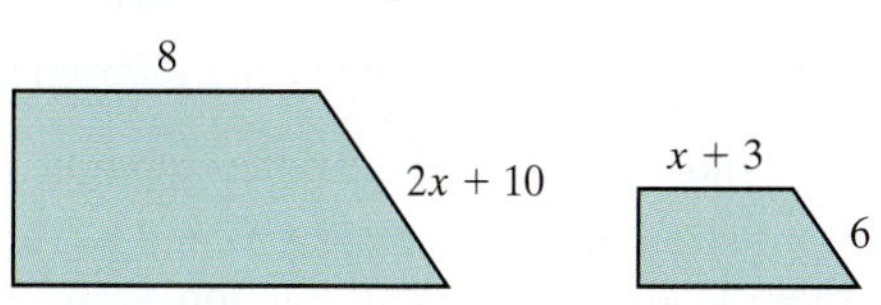


52.

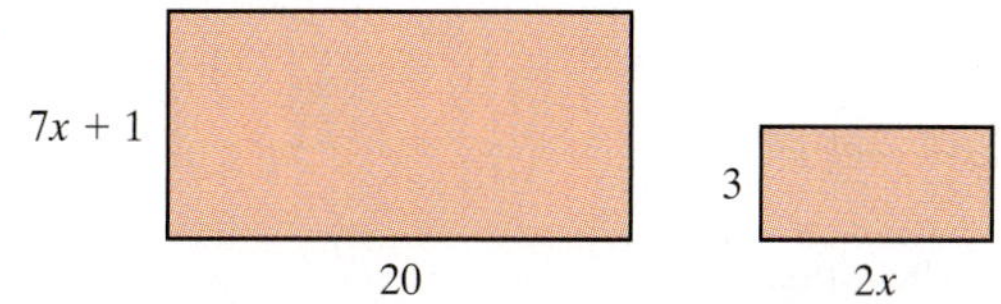

*For each rational function given, find all values a for which $f(a)$ has the indicated value.*

**53.** $f(x) = 2x - \dfrac{4}{x}, f(a) = -2$

**54.** $f(x) = 3x - \dfrac{5}{x}, f(a) = -14$

**55.** $f(x) = \dfrac{x-2}{x+5}, f(a) = \dfrac{3}{5}$

**56.** $f(x) = \dfrac{x+3}{x+5}, f(a) = \dfrac{4}{7}$

**57.** $f(x) = \dfrac{6}{x} + \dfrac{6}{2x}, f(a) = 6$

**58.** $f(x) = \dfrac{4}{x} - \dfrac{3}{2x}, f(a) = 4$

*Solve each formula for the indicated variable.*

**59.** $\dfrac{V_1}{V_2} = \dfrac{P_2}{P_1}$, for $P_2$ (chemistry)

**60.** $T_a = \dfrac{T_f}{1-f}$, for $f$ (investment formula)

**61.** $\dfrac{V_1}{V_2} = \dfrac{P_2}{P_1}$, for $V_2$ (chemistry)

**62.** $S = \dfrac{a}{1-r}$, for $r$ (mathematics)

**63.** $m = \dfrac{y - y_1}{x - x_1}$, for $y$ (slope)

**64.** $m = \dfrac{y - y_1}{x - x_1}$, for $x_1$ (slope)

**65.** $z = \dfrac{x - \bar{x}}{s}$, for $x$ (statistics)

**66.** $z = \dfrac{x - \bar{x}}{s}$, for $s$ (statistics)

**67.** $d = \dfrac{fl}{f + w}$, for $w$ (physics)

**68.** $\dfrac{1}{p} + \dfrac{1}{q} = \dfrac{1}{f}$, for $p$ (optics)

**69.** $\dfrac{1}{p} + \dfrac{1}{q} = \dfrac{1}{f}$, for $q$ (optics)

**70.** $\dfrac{1}{R_T} = \dfrac{1}{R_1} + \dfrac{1}{R_2}$, for $R_T$ (electronics)

**71.** $at_2 - at_1 + v_1 = v_2$, for $a$ (physics)

**72.** $2P_1 - 2P_2 - P_1P_c = P_2P_c$, for $P_c$ (economics)

**73.** $a_n = a_1 + nd - d$, for $d$ (mathematics)

**74.** $S_n - S_nr = a_1 - a_1r^n$, for $S_n$ (mathematics)

**75.** $F = \dfrac{Gm_1m_2}{d^2}$, for $G$ (physics)

**76.** $\dfrac{P_1V_1}{T_1} = \dfrac{P_2V_2}{T_2}$, for $T_2$ (physics)

**77.** $\dfrac{P_1V_1}{T_1} = \dfrac{P_2V_2}{T_2}$, for $T_1$ (physics)

**78.** $A = \dfrac{1}{2}h(a + b)$, for $h$ (mathematics)

**79.** $\dfrac{S - S_0}{V_0 + gt} = t$, for $V_0$ (physics)

**80.** $\dfrac{E}{e} = \dfrac{R + r}{r}$, for $e$ (engineering)

*Simplify each expression in **a)** and solve each equation in **b)**.*

**81. a)** $\dfrac{2}{x-2} + \dfrac{5}{x^2-4}$

**b)** $\dfrac{2}{x-2} + \dfrac{5}{x^2-4} = 0$

**82. a)** $\dfrac{4}{x+3} + \dfrac{5}{2x+6} + \dfrac{1}{2}$

**b)** $\dfrac{4}{x+3} + \dfrac{5}{2x+6} = \dfrac{1}{2}$

**83. a)** $\dfrac{b+3}{b} - \dfrac{b+4}{b+5} - \dfrac{15}{b^2+5b}$

**b)** $\dfrac{b+3}{b} - \dfrac{b+4}{b+5} = \dfrac{15}{b^2+5b}$

**84. a)** $\dfrac{4x+3}{x^2+11x+30} - \dfrac{3}{x+6} + \dfrac{2}{x+5}$

**b)** $\dfrac{4x+3}{x^2+11x+30} - \dfrac{3}{x+6} = \dfrac{2}{x+5}$

## Problem Solving

**85.** What restriction must be added to the statement "If $ac = bc$, then $a = b$."? Explain.

**86.** Consider $\dfrac{x-2}{x-5} = \dfrac{3}{x-5}$.

**a)** Solve the equation.

**b)** If you subtract $\dfrac{3}{x-5}$ from both sides of the equation, you get $\dfrac{x-2}{x-5} - \dfrac{3}{x-5} = 0$. Simplify the difference on the left side of the equation and solve the equation.

**c)** Use the information obtained in parts **a)** and **b)** to construct another equation that has no solution.

**EXAMPLE 5 ▸ Flying an Airplane** Sally Sestani owns a single-engine Cessna airplane. When making her preflight plan, she finds that there is a 20-mile-per-hour wind moving from east to west at the altitude she plans to fly. If she travels west (with the wind), she will be able to travel 400 miles in the same amount of time that she would be able to travel 300 miles flying east (against the wind). See **Figure 6.10**. Assuming that if it were not for the wind, the plane would fly at the same speed going east or west, find the speed of the plane in still air.

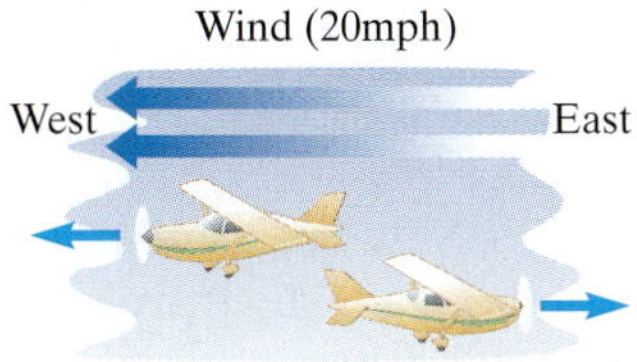


FIGURE 6.10

*Solution* **Understand** Let $x$ = speed of the plane in still air. We will set up a table to help answer the question.

| Plane | Distance | Rate | Time |
|---|---|---|---|
| Against wind | 300 | $x - 20$ | $\frac{300}{x - 20}$ |
| With wind | 400 | $x + 20$ | $\frac{400}{x + 20}$ |

**Translate** Since the times are the same, we set up and solve the following equation:

$$\frac{300}{x - 20} = \frac{400}{x + 20}$$

**Carry Out**

$$\begin{aligned} 300(x + 20) &= 400(x - 20) && \text{Cross multiply.} \\ 300x + 6000 &= 400x - 8000 \\ 6000 &= 100x - 8000 \\ 14{,}000 &= 100x \\ 140 &= x \end{aligned}$$

**Answer** The speed of the plane in still air is 140 miles per hour.

▸ **Now Try Exercise 41**

**EXAMPLE 6 ▸ Paddling a Water Bike** Becky and Al Ryckman go out on a water bike. When paddling against the current (going out from shore), they average 2 miles per hour. Coming back (going toward shore), paddling with the current, they average 3 miles per hour. If it takes $\frac{1}{4}$ hour longer to paddle out from shore than to paddle back, how far out did they paddle?

*Solution* **Understand** In this problem, the times going out and coming back are not the same. It takes $\frac{1}{4}$ hour longer to paddle out than to paddle back. Therefore, to make the times equal, we can add $\frac{1}{4}$ hour to the time to paddle back (or subtract $\frac{1}{4}$ hour from the time to paddle out). Let $x$ = the distance they paddle out from shore.

| Bike | Distance | Rate | Time |
|---|---|---|---|
| Going out | $x$ | 2 | $\frac{x}{2}$ |
| Coming back | $x$ | 3 | $\frac{x}{3}$ |

**Translate** time coming back + $\frac{1}{4}$ hour = time going out

$$\frac{x}{3} + \frac{1}{4} = \frac{x}{2}$$

**Carry Out**

$$12\left(\frac{x}{3}+\frac{1}{4}\right) = 12\cdot\frac{x}{2} \qquad \text{Multiply by the LCD, 12.}$$

$$12\left(\frac{x}{3}\right)+12\left(\frac{1}{4}\right) = 12\left(\frac{x}{2}\right) \qquad \text{Distributive property}$$

$$4x + 3 = 6x$$

$$3 = 2x$$

$$1.5 = x$$

**Answer** Therefore, they paddle out 1.5 miles from shore.

▶ **Now Try Exercise 53**

---

**EXAMPLE 7** ▶ **Taking a Trip** Dawn Puppel lives in Buffalo, New York, and travels to college in South Bend, Indiana. The speed limit on some of the roads she travels is 55 miles per hour, while on others it is 65 miles per hour. The total distance traveled by Dawn is 490 miles. If Dawn follows the speed limits and the total trip takes 8 hours, how long does she drive at 55 miles per hour and how long does she drive at 65 miles per hour?

*Solution* **Understand and Translate** Let $x$ = number of miles driven at 55 mph.
Then $490 - x$ = number of miles driven at 65 mph.

| Speed Limit | Distance | Rate | Time |
|---|---|---|---|
| 55 mph | $x$ | 55 | $\frac{x}{55}$ |
| 65 mph | $490 - x$ | 65 | $\frac{490 - x}{65}$ |

Since the total time is 8 hours, we write

$$\frac{x}{55}+\frac{490-x}{65} = 8$$

**Carry Out** The LCD of 55 and 65 is 715.

$$715\left(\frac{x}{55}+\frac{490-x}{65}\right) = 715\cdot 8$$

$$715\left(\frac{x}{55}\right)+715\left(\frac{490-x}{65}\right) = 5720$$

$$13x + 11(490 - x) = 5720$$

$$13x + 5390 - 11x = 5720$$

$$2x + 5390 = 5720$$

$$2x = 330$$

$$x = 165$$

**Answer** The number of miles driven at 55 mph is 165 miles. Then the time driven at 55 mph is $\frac{165}{55} = 3$ hours, and the time driven at 65 mph is $\frac{490-165}{65} = \frac{325}{65} = 5$ hours.

▶ **Now Try Exercise 59**

---

Notice that in Example 7 the answer to the question was not the value obtained for $x$. The value obtained was a distance, and the question asked us to find the time. *When working word problems, you must read and work the problems very carefully and make sure you answer the questions that were asked.*

## EXERCISE SET 6.5

MathXL® MyMathLab

### Practice the Skills and Problem Solving

1. **Painting a Wall** Two brothers take exactly the same time to paint a wall. If they paint the wall together, will the total time needed to paint the wall be less than $\frac{1}{2}$ the time, equal to $\frac{1}{2}$ the time, or greater than $\frac{1}{2}$ the time it takes each brother separately to paint the wall? Explain.

2. **Tractors in a Field** Two tractors, a larger one and a smaller one, work together to level a field. In the same amount of time, the larger tractor levels more land than the smaller one. Will the smaller tractor take more or less time working alone than the two take working together? Explain.

3. a) **Working Together** Bill and Bob are planning to do a task together. Bill can do the task in 7 hours and Bob can do it in 9 hours. Let $x$ = the time for Bill and Bob to do the task together. Complete the table below.

| Worker | Rate of Work | Time Worked | Part of Task Completed |
|---|---|---|---|
| Bill | | $x$ | |
| Bob | | $x$ | |

b) Study the examples in this section. Then, write the equation that can be used to solve for $x$. (Do not solve.)

c) Working together, will it take them more or less than 7 hours to complete the task? Explain.

4. a) **Working Together** Juanita and Sally are planning to do a task together. Sally can do the task in 3.6 hours and Juanita can do the task in 5.2 hours. Let $x$ = the time for Sally and Juanita to do the task together. Complete the table below.

| Worker | Rate of Work | Time Worked | Part of Task Completed |
|---|---|---|---|
| Sally | | $x$ | |
| Juanita | | $x$ | |

b) Study the examples in this section. Then, write the equation that can be used to solve for $x$. (Do not solve.)

c) Working together, will it take them more or less than 3.6 hours to complete the task? Explain.

*Exercises 5–36 involve work problems. Answer the question asked. When necessary, round answers to the nearest hundredth.*

5. **Totem Pole** It takes Marilyn Mays 2 months to carve a totem pole. It takes Larry Gilligan 6 months to carve the same totem pole. How long will it take them working together to carve the totem pole?

6. **Window Washers** Fran Thompson can wash windows in the lobby at the Days Inn in 3 hours. Jill Franks can wash the same windows in 4.5 hours. How long will it take them working together to wash the windows?

7. **Shampooing a Carpet** Jason La Rue can shampoo the carpet on the main floor of the Sheraton Hotel in 3 hours. Tom Lockheart can shampoo the same carpet in 6 hours. If they work together, how long will it take them to shampoo the carpet?

8. **Printing Checks** At the Merck Corporation it takes one computer 3 hours to print checks for its employees and a second computer 7 hours to complete the same job. How long will it take the two computers working together to complete the job?

9. **Dairy Farm** On a small dairy farm, Jin Chenge can milk 10 cows in 30 minutes. His son, Ming, can milk the same cows in 50 minutes. How long will it take them working together to milk the 10 cows?

10. **Mowing Lawns** Julio and Marcella Lopez mow lawns during the summer months. Using a self-propelled hand mower, Julio can mow a large area in 9 hours. Using a riding mower, Marcella can mow the same lawn in 4 hours. How long will it take them working together to mow the lawn?

11. **Picking Apples** In an apple orchard in Sodus, New York, Kevin Bamard can pick 25 bushels of apples in 6 hours. His young son takes twice as long to pick 25 bushels of apples. How long will it take them working together to pick 25 bushels of apples?

12. **Picking Strawberries** In a strawberry patch in Jacksonville, North Carolina, Amanda Heinl can pick 80 quarts of strawberries in 10 hours. Her sister, Emily, can pick 80 quarts of strawberries in 15 hours. How long will it take them working together to pick 80 quarts of strawberries?

**13. Cleaning Gutters** At a housing complex in Altoona, Pennsylvania, Olga Palmieri can clean the gutters on 28 houses in 4.5 days. Her co-worker, Jien-Ping, can clean the same gutters in 5.5 days. How long will it take them working together to clean the gutters on the 28 houses?

**14. Pulling Weeds** At a farm near Portland, Maine, Val Short can pull weeds from a row of potatoes in 70 minutes. His friend, Jason, can pull the same weeds in 80 minutes. How long will it take them working together to pull the weeds from this row of potatoes?

**15. Plowing a Field** Wanda Garner can plow a field used for corn in 4 hours. Shawn Robinson can plow the same field in 6 hours. How long will it take them working together to plow the field?

**16. Painting** Karen Sharp and Hephner Bennett are painters. Karen can paint a living room in a Habitat for Humanity house in 6 hours. Hephner can paint the same room in 4.5 hours. How long will it take them working together to paint the living room?

**17. Filling a Pool** A $\frac{1}{2}$-inch-diameter hose can fill a swimming pool in 8 hours. A $\frac{4}{5}$-inch-diameter hose can fill the same pool in 5 hours. How long will it take to fill the pool when both hoses are used?

**18. Milk Tank** At a dairy plant, a milk tank can be filled in 6 hours (using the in-valve). Using the out-valve, the tank can be emptied in 8 hours. If both valves are open and milk is being pumped into the tank, how long will it take to fill the tank?

**19. Oil Refinery** An oil refinery has large tanks to hold oil. Each tank has an inlet valve and an outlet valve. The tank can be filled with oil in 20 hours when the inlet valve is wide open and the outlet valve is closed. The tank can be emptied in 25 hours when the outlet valve is wide open and the inlet valve is closed. If a new tank is placed in operation and both the inlet valve and outlet valve are wide open, how long will it take to fill the tank?

**20. Cabinet Makers** Laura Burton and Marcia Kleinz are cabinet makers. Laura can make a set of kitchen cabinets by herself in 10 hours. If Laura and Marcia work together, they can make the same set of cabinets in 8 hours. How long will it take Marcia working by herself to make the cabinets?

**21. Archeology** Dr. Indiana Jones and his father, Dr. Henry Jones, are working on a dig near the Forum in Rome. Indiana and his father working together can unearth a specific plot of land in 2.6 months. Indiana can unearth the entire area by himself in 3.9 months. How long would it take Henry to unearth the entire area by himself?

**22. Digging a Trench** Arthur Altshiller and Sally Choi work for General Telephone. Together it takes them 2.4 hours to dig a trench where a wire is to be laid. If Arthur can dig the trench by himself in 3.2 hours, how long would it take Sally by herself to dig the trench?

**23. Jellyfish Tanks** Wade Martin and Shane Wheeler work together at the Monterey Aquarium. It takes Wade 50 minutes to clean the jellyfish tanks. Since Shane is new to the job, it takes him longer to perform the same task. When working together, they can perform the task in 30 minutes. How long does it take Shane to do the task by himself?

**24. Planting Flowers** Maria Vasquez and LaToya Johnson plant petunias in a botanical garden. It takes Maria twice as long as LaToya to plant the flowers. Working together, they can plant the flowers in 10 hours. How long does it take LaToya to plant the flowers by herself?

**25. Filling a Washtub** When only the cold water valve is opened, a washtub will fill in 8 minutes. When only the hot water valve is opened, the washtub will fill in 12 minutes. When the drain of the washtub is open, it will drain completely in 7 minutes. If both the hot and cold water valves are open and the drain is open, how long will it take for the washtub to fill?

**26. Irrigating Crops** A large tank is being used on Jed Saifer's farm to irrigate the crops. The tank has two inlet pipes and one outlet pipe. The two inlet pipes can fill the tank in 8 and 12 hours, respectively. The outlet pipe can empty the tank in 15 hours. If the tank is empty, how long would it take to fill the tank when all three valves are open?

27. **Pumping Water** The Rushville fire department uses three pumps to remove water from flooded basements. The three pumps can remove all the water from a flooded basement in 6 hours, 5 hours, and 4 hours, respectively. If all three pumps work together how long will it take to empty the basement?

28. **Installing Windows** Adam, Frank, and Willy are experts at installing windows in houses. Adam can install five living room windows in a house in 10 hours. Frank can do the same job in 8 hours, and Willy can do it in 6 hours. If all three men work together, how long will it take them to install the windows?

29. **Roofing a House** Gary Glaze requires 15 hours to put a new roof on a house. His apprentice, Anna Gandy, can reroof the house by herself in 20 hours. After working alone on a roof for 6 hours, Gary leaves for another job. Anna takes over and completes the job. How long does it take Anna to complete the job?

30. **Filling a Tank** Two pipes are used to fill an oil tanker. When the larger pipe is used alone, it takes 60 hours to fill the tanker. When the smaller pipe is used alone, it takes 80 hours to fill the tanker. The large pipe begins filling the tanker. After 20 hours, the large pipe is closed and the smaller pipe is opened. How much longer will it take to finish filling the tanker using only the smaller pipe?

*Exercises 31–40 involve number problems. Answer the question asked.*

31. What number multiplied by the numerator and added to the denominator of the fraction $\frac{4}{3}$ makes the resulting fraction $\frac{5}{2}$?

32. What number added to the numerator and multiplied by the denominator of the fraction $\frac{4}{5}$ makes the resulting fraction $\frac{1}{15}$?

33. One number is twice another. The sum of their reciprocals is $\frac{3}{4}$. Find the numbers.

34. The sum of the reciprocals of two consecutive integers is $\frac{11}{30}$. Find the two integers.

35. The sum of the reciprocals of two consecutive even integers is $\frac{5}{12}$. Find the two integers.

36. When a number is added to both the numerator and denominator of the fraction $\frac{7}{9}$, the resulting fraction is $\frac{5}{6}$. Find the number added.

37. When 3 is added to twice the reciprocal of a number, the sum is $\frac{31}{10}$. Find the number.

38. The reciprocal of 3 less than a certain number is twice the reciprocal of 6 less than twice the number. Find the number(s).

39. If three times a number is added to twice the reciprocal of the number, the answer is 5. Find the number(s).

40. If three times the reciprocal of a number is subtracted from twice the reciprocal of the square of the number, the difference is $-1$. Find the number(s).

*Exercises 41–61 involve motion problems. Answer the question asked. When necessary, round answers to the nearest hundredth.*

41. **Gondola** When Angelo Burnini rows his gondola in still water (with no current) in Venice, Italy, he travels at 3 mph. When he rows at the same rate in the Grand Canal, it takes him the same amount of time to travel 2.4 miles downstream (with the current) as it does to travel 2.3 miles upstream (against the current). Find the current of the river.

42. **Auto Train** The Amtrak Auto Train travels nonstop from Lorton, Virginia to Sanford, Florida. June White wishes to bring two cars down to Florida for the winter, so she decides to send one by auto train and drive the other down. The train travels 600 kilometers in the time it takes for her to drive 400 kilometers. If the average speed of the train is 40 kilometers greater than the speed of June's car, find the speeds of the train and car.

43. **Moving Sidewalk** The moving sidewalk at Chicago's O'Hare International Airport moves at a speed of 2.0 feet per second. Walking on the moving sidewalk, Nancy Killian walks 120 feet in the same time that it takes her to walk 52 feet without the moving sidewalk. How fast does Nancy walk?

44. **Moving Sidewalk** The moving sidewalk at the Philadelphia International Airport moves at a speed of 1.8 feet per second. Nathan Trotter walks 100 feet on the moving sidewalk, then turns around on the moving sidewalk and walks at the same speed 40 feet in the opposite direction. If the time walking in each direction was the same, find Nathan's walking speed.

**45. Skiing** Bonnie Hellier and Clide Vincent go cross-country skiing in the Adirondack Mountains. Clide is an expert skier who averages 10 miles per hour. Bonnie averages 6 miles per hour. If it takes Bonnie 1/2 hour longer to ski the trail, how long is the trail?

**46. Outing in a Park** Ruth and Jerry Mackin go for an outing in Memorial Park in Houston, Texas. Ruth jogs while Jerry rollerblades. Jerry rollerblades 2.9 miles per hour faster than Ruth jogs. When Jerry has rollerbladed 5.7 miles, Ruth has jogged 3.4 miles. Find Ruth's jogging speed.

**47. Visiting a Resort** Phil Mahler drove 60 miles to Yosemite National Park. He spent twice as much time visiting the park as it took him to drive to the park. The total time for driving and visiting the park was 5 hours. Find the average speed driving to Yosemite National Park.

**48. Boating Trip** Ray Packerd starts out on a boating trip at 8 A.M. Ray's boat can go 20 miles per hour in still water. How far downstream can Ray go if the current is 5 miles per hour and he wishes to go down and back in 4 hours?

**49. Football Game** At a football game the Carolina Panthers have the ball on their own 20-yard line. Jake Delhomme passes the ball to Steve Smith, who catches the ball and runs into the end zone for a touchdown. Assume that the ball, when passed, traveled at 14.7 yards per second and that after Steve caught the ball, he ran at 5.8 yards per second into the end zone. If the play, from the time Jake released the ball to the time Steve reached the end zone, took 10.6 seconds, how far from where Jake threw the ball did Steve catch it? Assume the entire play went down the center of the field.

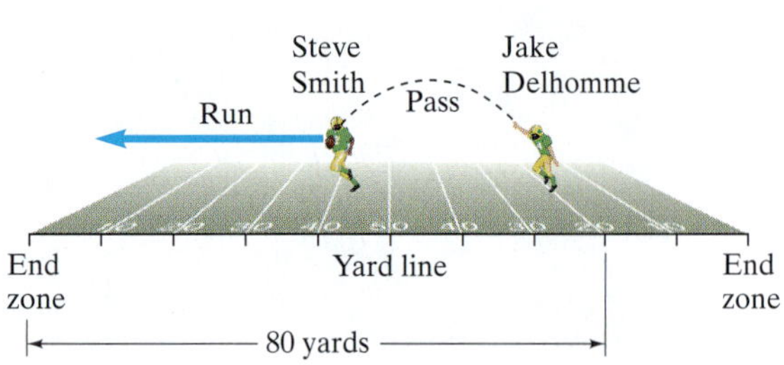


**50. Skyline Drive** In one day, Pauline Shannon drove from Front Royal, Virginia, to Asheville, North Carolina, a distance of 492 miles. For part of the trip she drove at a steady rate of 50 miles per hour, but in some areas she drove at a steady rate of 35 miles per hour. If the total time of the trip was 11.13 hours, how far did she travel at each speed?

**51. Subway Trains** The number 4 train in the New York City subway system goes from Woodlawn/Jerome Avenue in the Bronx to Flatbush Avenue in Brooklyn. The one-way distance between these two stops is 24.2 miles. On this route, two tracks run parallel to each other, one for the local train and the other for the express train. The local and express trains leave Woodlawn/Jerome Avenue at the same time. When the express reaches the end of the line at Flatbush Avenue, the local is at Wall Street, 7.8 miles from Flatbush. If the express averages 5.2 miles per hour faster than the local, find the speeds of the two trains.

**52. Riding a Horse** Each morning, Ron Lucky takes his horse, Beauty, for a ride on Pfeiffer Beach in Big Sur, California. He typically rides Beauty for 5.4 miles, then walks Beauty 2.3 miles. His speed when riding is 4.2 times his speed when walking. If his total outing takes 1.5 hours, find the rate at which he walks Beauty.

**53. Traveling** A car and a train leave Union Station in Washington, D.C., at the same time for Rochester, New York, 390 miles away. If the speed of the car averages twice the speed of the train and the car arrives 6.5 hours before the train, find the speed of the car and the speed of the train.

**54. Traveling** A train and car leave from the Pasadena, California, railroad station at the same time headed for the state fair in Sacramento. The car averages 50 miles per hour and the train averages 70 miles per hour. If the train arrives at the fair 2 hours ahead of the car, find the distance from the railroad station to the state fair.

**55. Traveling** Two friends drive from Dallas going to El Paso, a distance of 600 miles. Mary Ann Zilke travels by highway and arrives at the same destination 2 hours ahead of Carla Canola, who took a different route. If the average speed of Mary Ann's car was 10 miles per hour faster than Carla's car, find the average speed of Mary Ann's car.

**56. Sailboat Race** In a 30-mile sailboat race, the winning boat, the Buccaneer, finishes 10 minutes ahead of the second-place boat, the Raven. If the Buccaneer's average speed was 2 miles per hour faster than the Raven's, find the average speed of the Buccaneer.

**57. Helicopter Ride** Kathy Angel took a helicopter ride to the top of the Mt. Cook Glacier in New Zealand. The flight covered a distance of 60 kilometers. Kathy stayed on top of the glacier for $\frac{1}{2}$ hour. She then flew to the town of Te Anu, 140 kilometers away. The helicopter averaged 20 kilometers per hour faster going to Te Anu than on the flight up to the top of the glacier. The total time involved in the outing was 2 hours. Find the average speed of the helicopter going to the glacier.

58. **Sailboats** Two sailboats, the *Serendipity* and the *Zerwilliker*, start at the same point and at the same time on Lake Michigan, heading to the same restaurant on the lake. The *Serendipity* sails an average of 5.2 miles per hour and the *Zerwilliker* sails an average of 4.6 miles per hour. If the *Serendipity* arrives at the restaurant 0.4 hour ahead of the *Zerwilliker*, find the distance from their starting point to the restaurant.

59. **Riding a Bike** Robert Wiggins rides his bike from DuPont Circle in Washington, D.C., to Mount Vernon in Virginia. It takes $2\frac{1}{2}$ hours to complete the 17-mile trip from DuPont circle to Mount Vernon. On the slow part of the trip, Robert pedals his bike at a rate of 6 miles per hour. On the fast part of the trip, he pedals at 10 miles per hour. How long does Robert pedal at 6 miles per hour and how long does he pedal at 10 miles per hour?

60. **Roller Blading and Jogging** Sharon McGhee roller blades and jogs on a trail that is 38 miles long. On the part that is paved, she roller blades at a rate of 11 miles per hour. On the part that is not paved, she jogs at a rate of 7 miles per hour. The entire trip takes 4 hours to complete. How long does Sharon rollerblade and how long does she jog?

61. **Suspension Bridge** The Capilano suspension bridge in Vancouver, Canada, is 450 feet long. Phil and Heim started walking across the bridge at the same time. Heim's speed was 2 feet per minute faster than Phil's. If Heim finished walking across the bridge $2\frac{1}{2}$ minutes before Phil, find Phil's average speed in feet per minute.

62. **Inclined Railroad** A trip up Mount Pilatus, near Lucerne, Switzerland, involves riding an inclined railroad up to the top of the mountain, spending time at the top, then coming down the opposite side of the mountain in an aerial tram. The distance traveled up the mountain is 7.5 kilometers and the distance traveled down the mountain is 8.7 kilometers. The speed coming down the mountain is 1.2 times the speed going up. If the Lieblichs stayed at the top of the mountain for 3 hours and the total time of their outing was 9 hours, find the speed of the inclined railroad.

63. **Launching Rockets** Two rockets are to be launched at the same time from NASA headquarters in Houston, Texas, and are to meet at a space station many miles from Earth. The first rocket is to travel at 20,000 miles per hour and the second rocket will travel at 18,000 miles per hour. If the first rocket is scheduled to reach the space station 0.6 hour before the second rocket, how far is the space station from NASA headquarters?

64. Make up your own word problem and find the solution.

65. Make up your own motion problem and find the solution.

66. Make up your own number problem and find the solution.

## Challenge Problem

67. An officer flying a California Highway Patrol aircraft determines that a car 10 miles ahead of her is speeding at 90 miles per hour.
    **a)** If the aircraft is traveling 150 miles per hour, how long, in minutes, will it take for the aircraft to reach the car?
    **b)** How far will the car have traveled in the time it takes the aircraft to reach it?
    **c)** If the pilot wishes to reach the car in exactly 8 minutes, how fast must the airplane fly?

## Cumulative Review Exercises

[1.5] 68. Simplify $\dfrac{(2x^{-2}y^{-2})^{-3}}{(3x^{-1}y^{3})^{2}}$.

[1.6] 69. Express 9,260,000,000 in scientific notation.

[2.3] 70. **Weekly Salary** Sandy Ivey receives a flat weekly salary of \$240 plus 12% commission on the total dollar volume of all sales she makes. What must her dollar sales volume be in a week for her to earn \$540?

[3.1] 71. Graph $y = |x| - 2$.

[5.4] 72. Factor $2a^4 - 2a^3 - 5a^2 + 5a$.

# 6.6 Variation

1 Solve direct variation problems.

2 Solve inverse variation problems.

3 Solve joint variation problems.

4 Solve combined variation problems.

In Sections 6.4 and 6.5, we saw many applications of equations containing rational expressions. In this section, we see still more.

## 1 Solve Direct Variation Problems

Many scientific formulas are expressed as variations. A **variation** is an equation that relates one variable to one or more other variables using the operations of multiplication or division (or both operations). There are essentially three types of variation problems: direct, inverse, and joint variation.

In **direct variation**, the two related variables will both increase together or both decrease together; that is, as one increases so does the other, and as one decreases so does the other.

Consider a car traveling at 30 miles an hour. The car travels 30 miles in 1 hour, 60 miles in 2 hours, and 90 miles in 3 hours. Notice that as the time increases, the distance traveled increases.

The formula used to calculate distance traveled is

$$\text{distance} = \text{rate} \cdot \text{time}$$

Since the rate is a constant, 30 miles per hour, the formula can be written

$$d = 30t$$

We say that distance varies directly as time or that distance is directly proportional to time. This is an example of a direct variation.

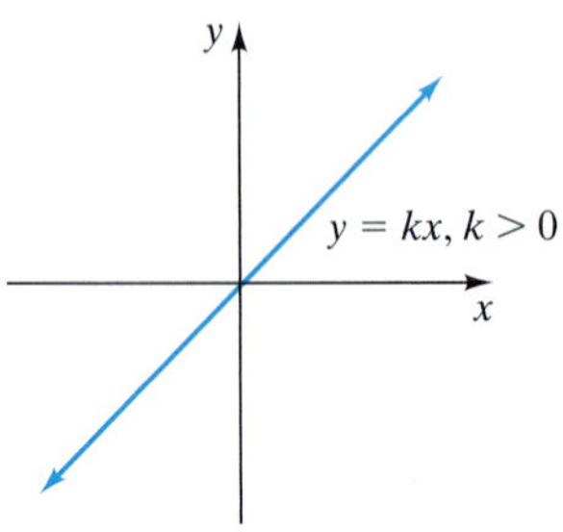

FIGURE 6.11

### Direct Variation

If a variable $y$ varies directly as a variable $x$, then

$$y = kx$$

where $k$ is the **constant of proportionality** (or the variation constant).

The graph of $y = kx$, $k > 0$, is always a straight line that goes through the origin (see **Fig. 6.11**). The slope of the line depends on the value of $k$. The greater the value of $k$, the greater the slope.

**EXAMPLE 1 ▸ Circle** The circumference of a circle, $C$, is directly proportional to (or varies directly as) its radius, $r$. Write the equation for the circumference of a circle if the constant of proportionality, $k$, is $2\pi$.

*Solution*

$C = kr$ ($C$ varies directly as $r$)

$C = 2\pi r$ (constant of proportionality is $2\pi$)

▸ Now Try Exercise 11

**EXAMPLE 2 ▸ Administering a Drug** The amount, $a$, of the drug theophylline given to patients is directly proportional to the patient's mass, $m$, in kilograms.

**a)** Write this variation as an equation.

**b)** If 150 mg is given to a boy whose mass is 30 kg, find the constant of proportionality.

**c)** How much of the drug should be given to a patient whose mass is 62 kg?

*Solution* **a)** We are told this is a direct variation. That is, the greater a person's mass the more of the drug that will need to be given. We therefore set up a direct variation.

$$a = km$$

**b)** Understand and Translate To determine the value of the constant of proportionality, we substitute the given values for the amount and mass. We then solve for $k$.

$$a = km$$

$$150 = k(30) \quad \text{Substitute the given values.}$$

Carry Out

$$5 = k$$

Answer Thus, $k = 5$ mg. Five milligrams of the drug should be given for each kilogram of a person's mass.

**c)** Understand and Translate Now that we know the constant of proportionality we can use it to determine the amount of the drug to use for a person's mass. We set up the variation and substitute the values of $k$ and $m$.

$$a = km$$

$$a = 5(62) \quad \text{Substitute the given values.}$$

Carry Out

$$a = 310$$

Answer Thus, 310 mg of theophylline should be given to a person whose mass is 62 kg.

▶ Now Try Exercise 57

---

**EXAMPLE 3** ▶ $y$ varies directly as the square of $z$. If $y$ is 80 when $z$ is 20, find $y$ when $z$ is 45.

*Solution* Since $y$ varies directly as the *square of* $z$, we begin with the formula $y = kz^2$. Since the constant of proportionality is not given, we must first find $k$ using the given information.

$$y = kz^2$$

$$80 = k(20)^2 \quad \text{Substitute the given values.}$$

$$80 = 400k \quad \text{Solve for } k.$$

$$\frac{80}{400} = \frac{400k}{400}$$

$$0.2 = k$$

We now use $k = 0.2$ to find $y$ when $z$ is 45.

$$y = kz^2$$

$$y = 0.2(45)^2 \quad \text{Substitute the given values.}$$

$$y = 405$$

Thus, when $z$ equals 45, $y$ equals 405.

▶ Now Try Exercise 35

---

## 2 Solve Inverse Variation Problems

A second type of variation is **inverse variation**. When two quantities vary inversely, it means that as one quantity increases, the other quantity decreases, and vice versa.

To explain inverse variation, we again use the formula, distance = rate · time. If we solve for time, we get time $= \dfrac{\text{distance}}{\text{rate}}$. Assume that the distance is fixed at 120 miles; then

$$\text{time} = \frac{120}{\text{rate}}$$

At 120 miles per hour, it would take 1 hour to cover this distance. At 60 miles an hour, it would take 2 hours. At 30 miles an hour, it would take 4 hours. Note that as the rate (or speed) decreases the time increases, and vice versa.

The equation above can be written

$$t = \frac{120}{r}$$

This equation is an example of an inverse variation. The time and rate are inversely proportional. The constant of proportionality is 120.

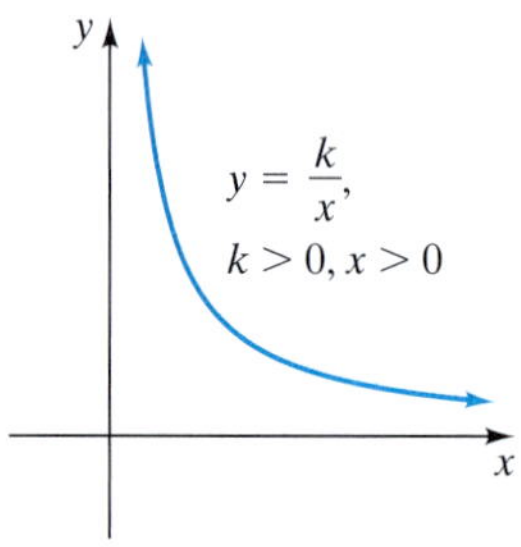


FIGURE 6.12

**Inverse Variation**

If a variable $y$ varies inversely as a variable $x$, then

$$y = \frac{k}{x} \qquad (\text{or } xy = k)$$

where $k$ is the constant of proportionality.

Two quantities vary inversely, or are inversely proportional, when as one quantity increases the other quantity decreases. The graph of $y = \frac{k}{x}$, for $k > 0$ and $x > 0$, will have the shape illustrated in **Figure 6.12**. The graph of an inverse variation is not defined at $x = 0$ because 0 is not in the domain of the function $y = \frac{k}{x}$.

**EXAMPLE 4** ▸ **Melting Ice** The amount of time, $t$, it takes a block of ice to melt in water is inversely proportional to the water's temperature, $T$.

**a)** Write this variation as an equation.

**b)** If a block of ice takes 15 minutes to melt in 60°F water, determine the constant of proportionality.

**c)** Determine how long it will take a block of ice of the same size to melt in 50°F water.

*Solution* **a)** The hotter the water temperature, the shorter the time for the block of ice to melt. The inverse variation is

$$t = \frac{k}{T}$$

**b)** Understand and Translate To determine the constant of proportionality, we substitute the values for the temperature and time and solve for $k$.

$$t = \frac{k}{T}$$

$$15 = \frac{k}{60} \qquad \textit{Substitute the given values.}$$

Carry Out

$$900 = k$$

Answer The constant of proportionality is 900.

**c)** Understand and Translate Now that we know the constant of proportionality, we can use it to determine how long it will take for the same size block of ice to melt in 50°F water. We set up the proportion, substitute the values for $k$ and $T$, and solve for $t$.

$$t = \frac{k}{T}$$

$$t = \frac{900}{50} \qquad \textit{Substitute the given values.}$$

Carry Out

$$t = 18$$

Answer It will take 18 minutes for the block of ice to melt in the 50°F water.

▸ **Now Try Exercise 61**

**EXAMPLE 5 ▸ Lighting** The illuminance, $I$, of a light source varies inversely as the square of the distance, $d$, from the source. Assuming that the illuminance is 75 units at a distance of 4 meters, find the formula that expresses the relationship between the illuminance and the distance.

*Solution* **Understand and Translate** Since the illuminance varies inversely as the *square* of the distance, the general form of the equation is

$$I = \frac{k}{d^2} \qquad (\text{or } Id^2 = k)$$

To find $k$, we substitute the given values for $I$ and $d$.

$$75 = \frac{k}{4^2} \qquad \textit{Substitute the given values.}$$

**Carry Out**

$$75 = \frac{k}{16} \qquad \textit{Solve for } k.$$

$$(75)(16) = k$$

$$1200 = k$$

**Answer** The formula is $I = \dfrac{1200}{d^2}$.

▸ **Now Try Exercise 65**

## 3 Solve Joint Variation Problems

One quantity may vary as a product of two or more other quantities. This type of variation is called **joint variation.**

**Joint Variation**

If $y$ varies jointly as $x$ and $z$, then

$$y = kxz$$

where $k$ is the constant of proportionality.

**EXAMPLE 6 ▸ Area of a Triangle** The area, $A$, of a triangle varies jointly as its base, $b$, and height, $h$. If the area of a triangle is 48 square inches when its base is 12 inches and its height is 8 inches, find the area of a triangle whose base is 15 inches and height is 40 inches.

*Solution* **Understand and Translate** First write the joint variation; then substitute the known values and solve for $k$.

$$A = kbh$$

$$48 = k(12)(8) \qquad \textit{Substitute the given values.}$$

**Carry Out**

$$48 = k(96) \qquad \textit{Solve for } k.$$

$$\frac{48}{96} = k$$

$$k = \frac{1}{2}$$

Now solve for the area of the given triangle.

$$A = kbh$$

$$= \frac{1}{2}(15)(40) \qquad \textit{Substitute the given values.}$$

$$= 300$$

**Answer** The area of the triangle is 300 square inches.

▸ **Now Try Exercise 69**

**Summary of Variations**

| DIRECT | INVERSE | JOINT |
|---|---|---|
| $y = kx$ | $y = \dfrac{k}{x}$ | $y = kxz$ |

## 4 Solve Combined Variation Problems

Often in real-life situations one variable varies as a combination of variables. The following examples illustrate the use of **combined variations.**

**EXAMPLE 7** ▸ **Pretzel Shop** The owners of a Auntie Anne's Pretzel Shop find that their weekly sales of pretzels, $S$, vary directly as their advertising budget, $A$, and inversely as their pretzel price, $P$. When their advertising budget is \$400 and the price is \$1, they sell 6200 pretzels.

**a)** Write an equation of variation expressing $S$ in terms of $A$ and $P$. Include the value of the constant.

**b)** Find the expected sales if the advertising budget is \$600 and the price is \$1.20.

*Solution* **a)** Understand and Translate We begin with the equation

$$S = \frac{kA}{P}$$

$$6200 = \frac{k(400)}{1} \qquad \textit{Substitute the given values.}$$

Carry Out

$$6200 = 400k \qquad \textit{Solve for k.}$$

$$15.5 = k$$

Answer Therefore, the equation for the sales of pretzels is $S = \dfrac{15.5A}{P}$.

**b)** Understand and Translate Now that we know the combined variation equation, we can use it to determine the expected sales for the given values.

$$S = \frac{15.5A}{P}$$

$$= \frac{15.5(600)}{1.20} \qquad \textit{Substitute the given values.}$$

Carry Out

$$= 7750$$

Answer They can expect to sell 7750 pretzels.

▸ Now Try Exercise 71

**EXAMPLE 8** ▸ **Electrostatic Force** The electrostatic force, $F$, of repulsion between two positive electrical charges is jointly proportional to the two charges, $q_1$ and $q_2$, and inversely proportional to the square of the distance, $d$, between the two charges. Express $F$ in terms of $q_1$, $q_2$, and $d$.

*Solution*

$$F = \frac{kq_1q_2}{d^2}$$

▸ Now Try Exercise 75

## EXERCISE SET 6.6

MathXL® MyMathLab

### Concept/Writing Exercises

1. **a)** Explain what it means when two items vary directly.
   **b)** Give your own example of two quantities that vary directly.
   **c)** Write the direct variation for your example in part **b)**.
2. **a)** Explain what it means when two items vary inversely.
   **b)** Give your own example of two quantities that vary inversely.
   **c)** Write the inverse variation for your example in part **b)**.
3. What is meant by joint variation?
4. What is meant by combined variation?
5. **a)** In the equation $y = \frac{17}{x}$, as $x$ increases, does the value for $y$ increase or decrease?
   **b)** Is this an example of direct or inverse variation? Explain.
6. **a)** In the equation $z = 0.8x^3$, as $x$ increases, does the value for $z$ increase or decrease?
   **b)** Is this an example of direct or inverse variation? Explain.

**Variation** *Use your intuition to determine whether the variation between the indicated quantities is direct or inverse.*

7. The speed and the distance covered by a person riding a bike on the Mount Vernon bike path in Alexandria, Virginia
8. The number of pages Tom can read in a 2-hour period and his reading speed
9. The speed of an athlete and the time it takes him to run a 10-kilometer race
10. Barbara's weekly salary and the amount of money withheld for state income taxes
11. The radius of a circle and its area
12. The side of a cube and its volume
13. The radius of a balloon and its volume
14. The diameter of a circle and its circumference
15. The diameter of a hose and the volume of water coming out of the hose
16. The weight of a rocket (due to Earth's gravity) and its distance from Earth
17. The time it takes an ice cube to melt in water and the temperature of the water
18. The distance between two cities on a map and the actual distance between the two cities
19. The shutter opening of a camera and the amount of sunlight that reaches the film
20. The cubic-inch displacement in liters and the horsepower of an engine

21. The length of a board and the force needed to break the board at the center
22. The number of calories eaten and the amount of exercise required to burn off those calories
23. The light illuminating an object and the distance the light is from the object
24. The number of calories in a cheeseburger and the size of the cheeseburger

### Practice the Skills

*For Exercises 25–32,* **a)** *write the variation and* **b)** *find the quantity indicated.*

25. $x$ varies directly as $y$. Find $x$ when $y = 12$ and $k = 6$.
26. $C$ varies directly as the square of $Z$. Find $C$ when $Z = 9$ and $k = \frac{3}{4}$.
27. $y$ varies directly as $R$. Find $y$ when $R = 180$ and $k = 1.7$.
28. $x$ varies inversely as $y$. Find $x$ when $y = 25$ and $k = 5$.
29. $R$ varies inversely as $W$. Find $R$ when $W = 160$ and $k = 8$.
30. $L$ varies inversely as the square of $P$. Find $L$ when $P = 4$ and $k = 100$.
31. $A$ varies directly as $B$ and inversely as $C$. Find $A$ when $B = 12$, $C = 4$, and $k = 3$.
32. $A$ varies jointly as $R_1$ and $R_2$ and inversely as the square of $L$. Find $A$ when $R_1 = 120$, $R_2 = 8$, $L = 5$, and $k = \frac{3}{2}$.

*For Exercises 33–42,* **a)** *write the variation and* **b)** *find the quantity indicated.*

33. $x$ varies directly as $y$. If $x$ is 12 when $y$ is 3, find $x$ when $y$ is 5.
34. $Z$ varies directly as $W$. If $Z$ is 7 when $W$ is 28, find $Z$ when $W$ is 140.
35. $y$ varies directly as the square of $R$. If $y$ is 5 when $R$ is 5, find $y$ when $R$ is 10.
36. $P$ varies directly as the square of $Q$. If $P$ is 32 when $Q$ is 4, find $P$ when $Q$ is 7.
37. $S$ varies inversely as $G$. If $S$ is 12 when $G$ is 0.4, find $S$ when $G$ is 5.
38. $C$ varies inversely as $J$. If $C$ is 7 when $J$ is 0.7, find $C$ when $J$ is 12.

**39.** $x$ varies inversely as the square of $P$. If $x$ is 4 when $P$ is 5, find $x$ when $P$ is 2.

**40.** $R$ varies inversely as the square of $T$. If $R$ is 3 when $T$ is 6, find $R$ when $T$ is 2.

**41.** $F$ varies jointly as $M_1$ and $M_2$ and inversely as $d$. If $F$ is 20 when $M_1 = 5, M_2 = 10,$ and $d = 0.2$, find $F$ when $M_1 = 10, M_2 = 20,$ and $d = 0.4$.

**42.** $F$ varies jointly as $q_1$ and $q_2$ and inversely as the square of $d$. If $F$ is 8 when $q_1 = 2, q_2 = 8,$ and $d = 4$, find $F$ when $q_1 = 28, q_2 = 12,$ and $d = 2$.

## Problem Solving

**43.** Assume $a$ varies directly as $b$. If $b$ is doubled, how will it affect $a$? Explain.

**44.** Assume $a$ varies directly as $b^2$. If $b$ is doubled, how will it affect $a$? Explain.

**45.** Assume $y$ varies inversely as $x$. If $x$ is doubled, how will it affect $y$? Explain.

**46.** Assume $y$ varies inversely as $a^2$. If $a$ is doubled, how will it affect $y$? Explain.

*In Exercises 47–52, use the formula* $F = \dfrac{km_1m_2}{d^2}$.

**47.** If $m_1$ is doubled, how will it affect $F$?

**48.** If $m_1$ is quadrupled and $d$ is doubled, how will it affect $F$?

**49.** If $m_1$ is doubled and $m_2$ is halved, how will it affect $F$?

**50.** If $d$ is halved, how will it affect $F$?

**51.** If $m_1$ is halved and $m_2$ is quadrupled, how will it affect $F$?

**52.** If $m_1$ is doubled, $m_2$ is quadrupled, and $d$ is quadrupled, how will it affect $F$?

*In Exercises 53 and 54, determine if the variation is of the form* $y = kx$ *or* $y = \dfrac{k}{x}$, *and find k.*

**53.**

| $x$ | $y$ |
|---|---|
| 2 | $\frac{5}{2}$ |
| 5 | 1 |
| 10 | $\frac{1}{2}$ |
| 20 | $\frac{1}{4}$ |

**54.**

| $x$ | $y$ |
|---|---|
| 6 | 2 |
| 9 | 3 |
| 15 | 5 |
| 27 | 9 |

**55. Profit** The profit from selling lamps is directly proportional to the number of lamps sold. When 150 lamps are sold, the profit is $2542.50. Find the profit when 520 lamps are sold.

**56. Profit** The profit from selling stereos is directly proportional to the number of stereos sold. When 65 stereos are sold, the profit is $4056. Find the profit when 80 stereos are sold.

**57. Antibiotic** The recommended dosage, $d$, of the antibiotic drug vancomycin is directly proportional to a person's weight. If Phuong Kim, who is 132 pounds, is given 2376 milligrams, find the recommended dosage for Nathan Brown, who weighs 172 pounds.

**58. Dollars and Pesos** Converting American dollars to Mexican pesos is a direct variation. The more dollars you convert the more pesos you receive. Last week, Carlos Manuel converted $275 into 2433.75 pesos. Today, he received $400 from his aunt. If the conversion rate is unchanged when he converts the $400 into pesos, how many pesos will be receive?

**59. Hooke's Law** Hooke's law states that the length a spring will stretch, $S$, varies directly with the force (or weight), $F$, attached to the spring. If a spring stretches 1.4 inches when 20 pounds is attached, how far will it stretch when 15 pounds is attached?

**60. Distance** When a car travels at a constant speed, the distance traveled, $d$, is directly proportional to the time, $t$. If a car travels 150 miles in 2.5 hours, how far will the same car travel in 4 hours?

**61. Pressure and Volume** The volume of a gas, $V$, varies inversely as its pressure, $P$. If the volume, $V$, is 800 cubic centimeters when the pressure is 200 millimeters (mm) of mercury, find the volume when the pressure is 25 mm of mercury.

**62. Building a Brick Wall** The time, $t$, required to build a brick wall varies inversely as the number of people, $n$, working on it. If it takes 8 hours for five bricklayers to build a wall, how long will it take four bricklayers to build a wall?

**63. Running a Race** The time, $t$, it takes a runner to cover a specified distance is inversely proportional to the runner's speed. If Jann Avery runs at an average of 6 miles per hour, she will finish a race in 2.6 hours. How long will it take Jackie Donofrio who runs at 5 miles per hour, to finish the same race?

64. **Pitching a Ball** When a ball is pitched in a professional baseball game, the time, $t$, it takes for the ball to reach home plate varies inversely with the speed, $s$, of the pitch.* A ball pitched at 90 miles per hour takes 0.459 second to reach the plate. How long will it take a ball pitched at 75 miles per hour to reach the plate?

65. **Intensity of Light** The intensity, $I$, of light received at a source varies inversely as the square of the distance, $d$, from the source. If the light intensity is 20 foot-candles at 15 feet, find the light intensity at 10 feet.

66. **Tennis Ball** When a tennis player serves the ball, the time it takes for the ball to hit the ground in the service box is inversely proportional to the speed the ball is traveling. If Andy Roddick serves at 122 miles per hour, it takes 0.21 second for the ball to hit the ground after striking his racquet. How long will it take the ball to hit the ground if he serves at 80 miles per hour?

Andy Roddick

67. **Stopping Distance** Assume that the stopping distance of a van varies directly with the square of the speed. A van traveling 40 miles per hour can stop in 60 feet. If the van is traveling 56 miles per hour, what is its stopping distance?

68. **Falling Rock** A rock is dropped from the top of a cliff. The distance it falls in feet is directly proportional to the square of the time in seconds. If the rock falls 4 feet in $\frac{1}{2}$ second, how far will it fall in 3 seconds?

69. **Volume of a Pyramid** The volume, $V$, of a pyramid varies jointly as the area of its base, $B$, and its height, $h$ (see the figure). If the volume of the pyramid is 160 cubic meters when the area of its base is 48 square meters and its height is 10 meters, find the volume of a pyramid when the area of its base is 42 square meters and its height is 9 meters.

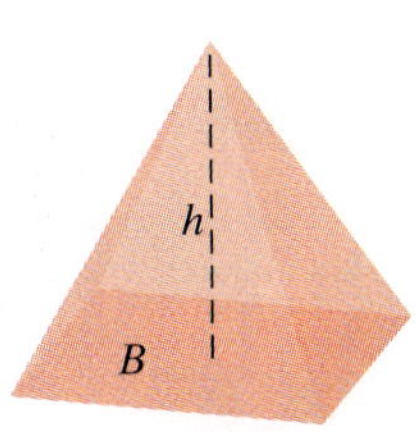


70. **Mortgage Payment** The monthly mortgage payment, $P$, you pay on a mortgage varies jointly as the interest rate, $r$, and the amount of the mortgage, $m$. If the monthly mortgage payment on a \$50,000 mortgage at a 7% interest rate is \$332.50, find the monthly payment on a \$66,000 mortgage at 7%.

71. **DVD Rental** The weekly DVD rentals, $R$, at Busterblock Video vary directly with their advertising budget, $A$, and inversely with the daily rental price, $P$. When their advertising budget is \$400 and the rental price is \$2 per day, they rent 4600 DVDs per week. How many DVDs would they rent per week if they increased their advertising budget to \$500 and raised their rental price to \$2.50?

72. **Electrical Resistance** The electrical resistance of a wire, $R$, varies directly as its length, $L$, and inversely as its cross-sectional area, $A$. If the resistance of a wire is 0.2 ohm when the length is 200 feet and its cross-sectional area is 0.05 square inch, find the resistance of a wire whose length is 5000 feet with a cross-sectional area of 0.01 square inch.

73. **Weight of an Object** The weight, $w$, of an object in Earth's atmosphere varies inversely with the square of the distance, $d$, between the object and the center of Earth. A 140-pound person standing on Earth is approximately 4000 miles from Earth's center. Find the weight (or gravitational force of attraction) of this person at a distance 100 miles from Earth's surface.

74. **Wattage Rating** The wattage rating of an appliance, $W$, varies jointly as the square of the current, $I$, and the resistance, $R$. If the wattage is 3 watts when the current is 0.1 ampere and the resistance is 100 ohms, find the wattage when the current is 0.4 ampere and the resistance is 250 ohms.

75. **Phone Calls** The number of phone calls between two cities during a given time period, $N$, varies directly as the populations $p_1$ and $p_2$ of the two cities and inversely as the distance, $d$, between them. If 100,000 calls are made between two cities 300 miles apart and the populations of the cities are 60,000 and 200,000, how many calls are made between two cities with populations of 125,000 and 175,000 that are 450 miles apart?

76. **Water Bill** In a specific region of the country, the amount of a customer's water bill, $W$, is directly proportional to the average daily temperature for the month, $T$, the lawn area, $A$, and the square root of $F$, where $F$ is the family size, and inversely proportional to the number of inches of rain, $R$.

In one month, the average daily temperature is 78°F and the number of inches of rain is 5.6. If the average family of four, who has 1000 square feet of lawn, pays \$68 for water, estimate the water bill in the same month for the average family of six, who has 1500 square feet of lawn.

*A ball slows down on its way to the plate due to wind resistance. For a 95-mph pitch, the ball is about 8 mph faster when it leaves the pitcher's hand than when it crosses the plate.

**77. Intensity of Illumination** An article in the magazine *Outdoor and Travel Photography* states, "If a surface is illuminated by a point-source of light (a flash), the intensity of illumination produced is inversely proportional to the square of the distance separating them."

If the subject you are photographing is 4 feet from the flash, and the illumination on this subject is $\frac{1}{16}$ of the light of the flash, what is the intensity of illumination on an object that is 7 feet from the flash?

**78. Force of Attraction** One of Newton's laws states that the force of attraction, $F$, between two masses is directly proportional to the masses of the two objects, $m_1$ and $m_2$, and inversely proportional to the square of the distance, $d$, between the two masses.

a) Write the formula that represents Newton's law.

b) What happens to the force of attraction if one mass is doubled, the other mass is tripled, and the distance between the objects is halved?

**79. Pressure on an Object** The pressure, $P$, in pounds per square inch (psi) on an object $x$ feet below the sea is 14.70 psi plus the product of a constant of proportionality, $k$, and the number of feet, $x$, the object is below sea level (see the figure). The 14.70 represents the weight, in pounds, of the column of air (from sea level to the top of the atmosphere) standing over a 1-inch-by-1-inch square of ocean. The $kx$ represents the weight, in pounds, of a column of water 1 inch by 1 inch by $x$ feet.

a) Write a formula for the pressure on an object $x$ feet below sea level.

b) If the pressure gauge in a submarine 60 feet deep registers 40.5 psi, find the constant $k$.

c) A submarine is built to withstand a pressure of 160 psi. How deep can the submarine go?

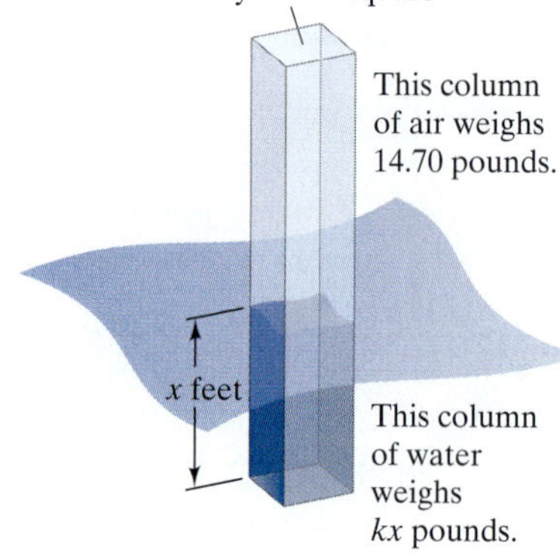


## Cumulative Review Exercises

[2.2] **80.** Solve the formula $V = \frac{4}{3}\pi r^2 h$ for $h$.

[3.6] **81.** Let $f(x) = x^2 - 4$ and $g(x) = -5x + 1$. Find $f(-4) \cdot g(-2)$.

[5.2] **82.** Multiply $(7x - 3)(-2x^2 - 4x + 5)$.

[5.7] **83.** Factor $(x + 1)^2 - (x + 1) - 6$.

# Chapter 6 Summary

| IMPORTANT FACTS AND CONCEPTS | EXAMPLES |
|---|---|
| **Section 6.1** | |
| A **rational expression** is an expression of the form $\frac{p}{q}$, where $p$ and $q$ are polynomials and $q \neq 0$. | $\frac{7}{x}$, $\frac{x^2 - 5}{x + 1}$, $\frac{t^2 - t + 1}{3t^2 + 5t - 7}$ |
| A **rational function** is a function of the form $f(x) = \frac{p}{q}$ or $y = \frac{p}{q}$, where $p$ and $q$ are polynomials and $q \neq 0$. | $y = \frac{x - 8}{x + 9}$, $f(x) = \frac{2x^2 + 4x + 1}{9x^2 - x + 3}$ |
| The **domain** of a rational function is the set of values that can be used to replace the variable. | The domain of $f(x) = \frac{x + 9}{x - 2}$ is $\{x \mid x \neq 2\}$. |

| IMPORTANT FACTS AND CONCEPTS | EXAMPLES |
|---|---|

**Section 6.1 (continued)**

**To Simplify Rational Expressions**

1. Factor both numerator and denominator completely.
2. Divide both the numerator and the denominator by any common factors.

$$\frac{x^3-8}{x^2-4}=\frac{\cancel{(x-2)}(x^2+2x+4)}{(x+2)\cancel{(x-2)}}=\frac{x^2+2x+4}{x+2}$$

**To Multiply Rational Expressions**

To multiply rational expressions, factor all numerators and denominators, and then use the following rule:

$$\frac{a}{b}\cdot\frac{c}{d}=\frac{a\cdot c}{b\cdot d},\quad b\neq 0, d\neq 0$$

$$\frac{2x^2+x-1}{x^2-1}\cdot\frac{x+1}{2x-1}=\frac{\cancel{(x+1)}\cancel{(2x-1)}}{\cancel{(x+1)}(x-1)}\cdot\frac{x+1}{\cancel{2x-1}}=\frac{x+1}{x-1}$$

**To Divide Rational Expressions**

$$\frac{a}{b}\div\frac{c}{d}=\frac{a}{b}\cdot\frac{d}{c}=\frac{a\cdot d}{b\cdot c},\quad b\neq 0, c\neq 0, d\neq 0$$

$$\frac{x^2+4x+3}{x+1}\div\frac{x+3}{x}=\frac{\cancel{(x+3)}\cancel{(x+1)}}{\cancel{x+1}}\cdot\frac{x}{\cancel{x+3}}=x$$

**Section 6.2**

**To Add or Subtract Rational Expressions**

Addition

$$\frac{a}{c}+\frac{b}{c}=\frac{a+b}{c},\quad c\neq 0$$

Subtraction

$$\frac{a}{c}-\frac{b}{c}=\frac{a-b}{c},\quad c\neq 0$$

$$\frac{x}{x^2-49}-\frac{7}{x^2-49}=\frac{x-7}{x^2-49}=\frac{\cancel{x-7}}{(x+7)\cancel{(x-7)}}=\frac{1}{x+7}$$

**To Find the Least Common Denominator (LCD) of Rational Expressions**

1. Write each nonprime coefficient (other than 1) of monomials that appear in denominators as a product of prime numbers.
2. Factor each denominator completely.
3. List all different factors of each denominator. When the same factor appears in more than one denominator, write the factor with the highest power that appears.
4. The least common denominator is the product of all the factors found in step 2.

The LCD of $\frac{7}{9x^2y}+\frac{17}{3xy^3}$ is $3\cdot 3\cdot x^2\cdot y^3=9x^2y^3$.

The LCD of $\frac{1}{x^2-36}-\frac{4x+3}{x^2+13x+42}$

is $(x+6)(x-6)(x+7)$. Note that $x^2-36=(x+6)(x-6)$ and $x^2+13x+42=(x+6)(x+7)$.

**To Add or Subtract Rational Expressions with Unlike Denominators**

1. Determine the LCD.
2. Rewrite each fraction as an equivalent fraction with the LCD.
3. Leave the denominator in factored form, but multiply out the numerator.
4. Add or subtract the numerators while maintaining the LCD.
5. When it is possible to reduce the fraction by factoring the numerator, do so.

Add $\frac{2a}{x^2y}+\frac{b}{xy^3}$.

The LCD is $x^2y^3$.

$$\frac{2a}{x^2y}+\frac{b}{xy^3}=\frac{y^2}{y^2}\cdot\frac{2a}{x^2y}+\frac{b}{xy^3}\cdot\frac{x}{x}$$

$$=\frac{2ay^2}{x^2y^3}+\frac{bx}{x^2y^3}$$

$$=\frac{2ay^2+bx}{x^2y^3}$$

**IMPORTANT FACTS AND CONCEPTS** | **EXAMPLES**

## Section 6.3

A **complex fraction** is one that has a rational expression in its numerator or its denominator or both its numerator and denominator.

$$\frac{\frac{2x}{x-1}}{\frac{x^2}{x+1}}, \quad \frac{7-\frac{6}{y}}{\frac{1}{y^2}+\frac{8}{y^3}}$$

### To Simplify a Complex Fraction by Multiplying by a Common Denominator

1. Find the LCD of all fractions appearing within the complex fraction.
2. Multiply both the numerator and denominator of the complex fraction by the LCD of the complex fraction found in step 1.
3. Simplify when possible.

Simplify $\dfrac{1+\frac{1}{x}}{x}$.

The LCD is $x$.

$$\frac{1+\frac{1}{x}}{x} = \frac{x}{x}\cdot\frac{1+\frac{1}{x}}{x} = \frac{x(1)+x\left(\frac{1}{x}\right)}{x(x)} = \frac{x+1}{x^2}$$

### To Simplify a Complex Fraction by Simplifying the Numerator and the Denominator

1. Add or subtract as necessary to get one rational expression in the numerator.
2. Add or subtract as necessary to get one rational expression in the denominator.
3. Invert the denominator of the complex fraction and multiply by the numerator of the complex fraction.
4. Simplify when possible.

Simplify $\dfrac{1+\frac{1}{x}}{x}$.

$$\frac{1+\frac{1}{x}}{x} = \frac{\frac{x+1}{x}}{x} = \frac{x+1}{x}\cdot\frac{1}{x} = \frac{x+1}{x^2}$$

## Section 6.4

### To Solve Rational Equations

1. Determine the LCD of all rational expressions in the equation.
2. Multiply *both* sides of the equation by the LCD.
3. Remove any parentheses and combine like terms on each side of the equation.
4. Solve the equation using the properties discussed in earlier sections.
5. Check the solution in the *original* equation.

Solve $\dfrac{5}{x}+1=\dfrac{11}{x}$.

Multiply both sides by the LCD of $x$.

$$x\left(\frac{5}{x}+1\right) = x\left(\frac{11}{x}\right)$$
$$x\cdot\frac{5}{x}+x\cdot 1 = x\cdot\frac{11}{x}$$
$$5+x=11$$
$$x=6$$

The answer checks.

**Proportions** are equations of the form $\dfrac{a}{b}=\dfrac{c}{d}$.

$\dfrac{2}{7}=\dfrac{9}{x}$ is a proportion.

**Similar figures** are figures whose corresponding angles are equal and whose corresponding sides are in proportion.

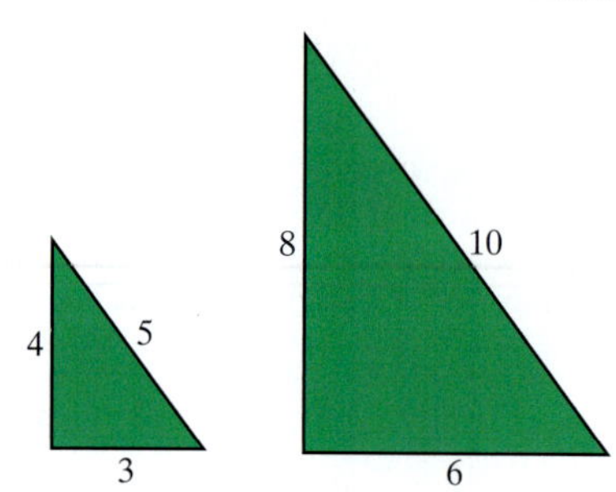


are similar figures.

| IMPORTANT FACTS AND CONCEPTS | EXAMPLES |
|---|---|
| **Section 6.5** | |
| **Applications**<br>**Work Problems:**<br>A work problem is a problem where two or more machines or people work together to complete a task. | Carlos and Ali plant flowers in a garden. Carlos can plant a tray of flowers in 30 minutes. Ali can plant the same tray in 20 minutes. How long will it take them working together to plant the tray of flowers? |
| **Number Problems:**<br>A number problem is a problem where one number is related to another number. | When the reciprocal of a number is subtracted from 5, the result is the reciprocal of twice the number. Find the number. |
| **Motion problems:**<br>A motion problem is a problem that involves time, rate, and distance. | Tom starts out on a canoe trip at noon. He can row at 5 miles per hour in still water. How far downstream can he go if the current is 2 miles per hour and he goes down and back in 4 hours? |
| **Section 6.6** | |
| **Direct Variation**<br>If a variable $y$ varies directly as a variable $x$, then $y = kx$, where $k$ is the constant of proportionality. | $y = 3x$ |
| **Inverse Variation**<br>If a variable $y$ varies inversely as a variable $x$, then<br>$y = \dfrac{k}{x}$ (or $xy = k$)<br>where $k$ is the constant of proportionality. | $y = \dfrac{3}{x}$ |
| **Joint Variation**<br>If $y$ varies jointly as $x$ and $z$, then<br>$y = kxz$<br>where $k$ is the constant of proportionality. | $y = 3xz$ |

## Chapter 6 Review Exercises

[6.1] *Determine the value or values of the variable that must be excluded in each rational expression.*

**1.** $\dfrac{3}{x-5}$ **2.** $\dfrac{x}{x+1}$ **3.** $\dfrac{-2x}{x^2+9}$

*Determine the domain of each rational function.*

**4.** $y = \dfrac{5}{(x+3)^2}$ **5.** $f(x) = \dfrac{x+6}{x^2}$ **6.** $f(x) = \dfrac{x^2-2}{x^2+4x-12}$

*Simplify each expression.*

**7.** $\dfrac{x^2+xy}{x+y}$ **8.** $\dfrac{x^2-36}{x+6}$ **9.** $\dfrac{7-5x}{5x-7}$ **10.** $\dfrac{x^2+5x-6}{x^2+4x-12}$

**11.** $\dfrac{2x^2-6x+5x-15}{2x^2+7x+5}$ **12.** $\dfrac{a^3-8b^3}{a^2-4b^2}$ **13.** $\dfrac{27x^3+y^3}{9x^2-y^2}$ **14.** $\dfrac{2x^2+x-6}{x^3+8}$

[6.2] *Find the least common denominator of each expression.*

**15.** $\frac{6x}{x+4} - \frac{3}{x}$

**16.** $\frac{3x+1}{x+2y} + \frac{7x-2y}{x^2-4y^2}$

**17.** $\frac{19x-5}{x^2+2x-35} + \frac{3x-2}{x^2-3x-10}$

**18.** $\frac{3}{(x+2)^2} - \frac{6(x+3)}{x^2-4} - \frac{4x}{x+3}$

[6.1, 6.2] *Perform each indicated operation.*

**19.** $\frac{30x^2y^3}{3z} \cdot \frac{6z^3}{5xy^3}$

**20.** $\frac{x}{x-9} \cdot \frac{9-x}{6}$

**21.** $\frac{18x^2y^4}{xz^5} \div \frac{2x^2y^4}{x^4z^{10}}$

**22.** $\frac{11}{3x} + \frac{2}{x^2}$

**23.** $\frac{4x-4y}{x^2y} \cdot \frac{y^3}{16x}$

**24.** $\frac{4x^2-11x+4}{x-3} - \frac{x^2-4x+10}{x-3}$

**25.** $\frac{6}{xy} + \frac{3y}{5x^2}$

**26.** $\frac{x+2}{x-1} \cdot \frac{x^2+3x-4}{x^2+6x+8}$

**27.** $\frac{3x^2-7x+4}{3x^2-14x-5} - \frac{x^2+2x+9}{3x^2-14x-5}$

**28.** $5 + \frac{a+2}{a+1}$

**29.** $7 - \frac{b+1}{b-1}$

**30.** $\frac{a^2-b^2}{a+b} \cdot \frac{a^2+2ab+b^2}{a^3+a^2b}$

**31.** $\frac{1}{a^2+8a+15} \div \frac{3}{a+5}$

**32.** $\frac{a+c}{c} - \frac{a-c}{a}$

**33.** $\frac{4x^2+8x-5}{2x+5} \cdot \frac{x+1}{4x^2-4x+1}$

**34.** $(a+b) \div \frac{a^2-2ab-3b^2}{a-3b}$

**35.** $\frac{x^2-3xy-10y^2}{6x} \div \frac{x+2y}{24x^2}$

**36.** $\frac{a+1}{2a} + \frac{3}{4a+8}$

**37.** $\frac{x-2}{x-5} - \frac{3}{x+5}$

**38.** $\frac{x+4}{x^2-4} - \frac{3}{x-2}$

**39.** $\frac{x+1}{x-3} \cdot \frac{x^2+2x-15}{x^2+7x+6}$

**40.** $\frac{2}{x^2-x-6} - \frac{3}{x^2-4}$

**41.** $\frac{4x^2-16y^2}{9} \div \frac{(x+2y)^2}{12}$

**42.** $\frac{a^2+5a+6}{a^2+4a+4} \cdot \frac{3a+6}{a^4+3a^3}$

**43.** $\frac{x+5}{x^2-15x+50} - \frac{x-2}{x^2-25}$

**44.** $\frac{x+2}{x^2-x-6} + \frac{x-3}{x^2-8x+15}$

**45.** $\frac{1}{x+3} - \frac{2}{x-3} + \frac{6}{x^2-9}$

**46.** $\frac{a-4}{a-5} - \frac{3}{a+5} - \frac{10}{a^2-25}$

**47.** $\frac{x^3+64}{2x^2-32} \div \frac{x^2-4x+16}{2x+12}$

**48.** $\frac{a^2-b^4}{a^2+2ab^2+b^4} \div \frac{3a-3b^2}{a^2+3ab^2+2b^4}$

**49.** $\left(\frac{x^2-x-56}{x^2+14x+49} \cdot \frac{x^2+4x-21}{x^2-9x+8}\right) + \frac{3}{x^2+8x-9}$

**50.** $\left(\frac{x^2-8x+16}{2x^2-x-6} \cdot \frac{2x^2-7x-15}{x^2-2x-24}\right) \div \frac{x^2-9x+20}{x^2+2x-8}$

**51.** If $f(x) = \frac{x+1}{x+2}$ and $g(x) = \frac{x}{x+4}$, find

**a)** the domain of $f(x)$.

**b)** the domain of $g(x)$.

**c)** $(f+g)(x)$.

**d)** the domain of $(f+g)(x)$.

**52.** If $f(x) = \frac{x}{x^2-9}$ and $g(x) = \frac{x+4}{x-3}$, find

**a)** the domain of $f(x)$.

**b)** the domain of $g(x)$.

**c)** $(f+g)(x)$.

**d)** the domain of $(f+g)(x)$.

[6.3] *Simplify each complex fraction.*

**53.** $\dfrac{\frac{9a^2b}{2c}}{\frac{6ab^4}{4c^3}}$

**54.** $\dfrac{\frac{2}{x}+\frac{4}{y}}{\frac{x}{y}+y^2}$

**55.** $\dfrac{\frac{3}{y}-\frac{1}{y^2}}{7+\frac{1}{y^2}}$

**56.** $\dfrac{a^{-1}+5}{a^{-1}+\frac{1}{a}}$

**57.** $\dfrac{x^{-2}+\frac{3}{x}}{\frac{1}{x^2}-\frac{1}{x}}$

**58.** $\dfrac{\frac{1}{x^2-3x-18}+\frac{2}{x^2-2x-15}}{\frac{3}{x^2-11x+30}+\frac{1}{x^2-9x+20}}$

**Area** *In Exercises 59 and 60, the area and width of each rectangle are given. Find the length, l, by dividing the area, A, by the width, w.*

**59.**

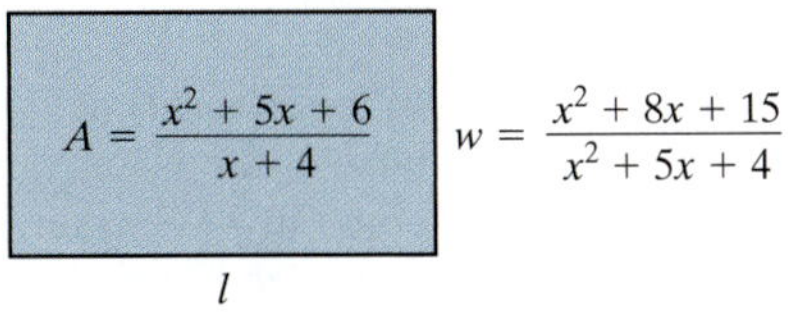


**60.**

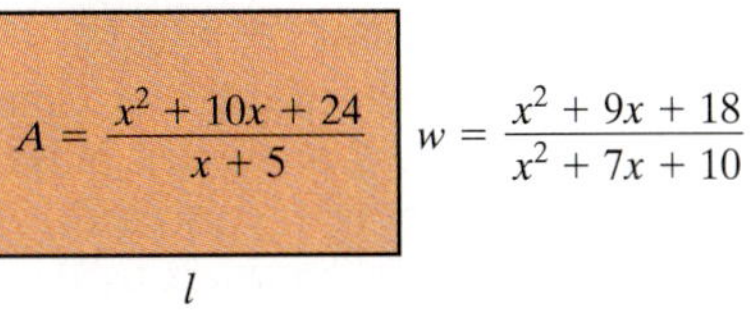


[6.4] *In Exercises 61–70, solve each equation.*

**61.** $\frac{2}{x} = \frac{5}{9}$

**62.** $\frac{x}{1.5} = \frac{x - 4}{4.5}$

**63.** $\frac{3x + 4}{5} = \frac{2x - 8}{3}$

**64.** $\frac{x}{4.8} + \frac{x}{2} = 1.7$

**65.** $\frac{2}{y} + \frac{1}{5} = \frac{3}{y}$

**66.** $\frac{2}{x + 4} - \frac{3}{x - 4} = \frac{-11}{x^2 - 16}$

**67.** $\frac{x}{x^2 - 9} + \frac{2}{x + 3} = \frac{4}{x - 3}$

**68.** $\frac{7}{x^2 - 5} + \frac{3}{x + 5} = \frac{4}{x - 5}$

**69.** $\frac{x - 3}{x - 2} + \frac{x + 1}{x + 3} = \frac{2x^2 + x + 1}{x^2 + x - 6}$

**70.** $\frac{x + 1}{x + 3} + \frac{x + 2}{x - 4} = \frac{2x^2 - 18}{x^2 - x - 12}$

**71.** Solve $\frac{1}{a} + \frac{1}{b} = \frac{1}{c}$ for $b$.

**72.** Solve $z = \frac{x - \overline{x}}{s}$ for $\overline{x}$.

**73. Resistors** Three resistors of 100, 200, and 600 ohms are connected in parallel. Find the total resistance of the circuit. Use the formula $\frac{1}{R_T} = \frac{1}{R_1} + \frac{1}{R_2} + \frac{1}{R_3}$.

**74. Focal Length** What is the focal length, $f$, of a curved mirror if the object distance, $p$, is 6 centimeters and the image distance, $q$, is 3 centimeters? Use the formula $\frac{1}{p} + \frac{1}{q} = \frac{1}{f}$.

**Triangles** *In Exercises 75 and 76, each pair of triangles is similar. Find the lengths of the unknown sides.*

**75.**

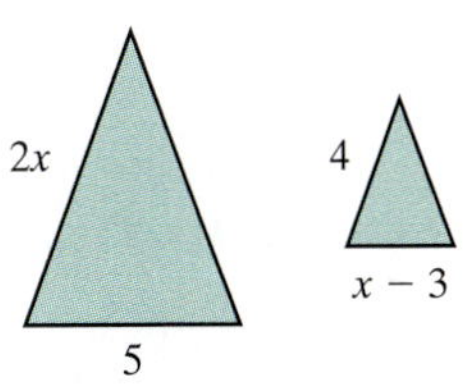


**76.**

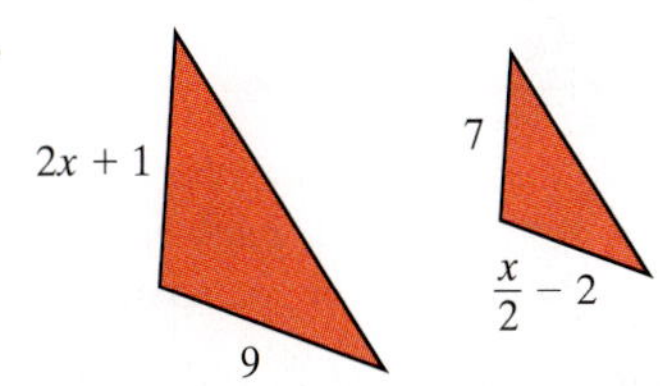


[6.5] *In Exercises 77–82, answer the question asked. When necessary, round answers to the nearest hundredth.*

**77. Picking String Beans** Sanford and Jerome work on a farm near Oklahoma City, Oklahoma. Sanford can pick a basket of string beans in 40 minutes while Jerome can do the same task in 30 minutes. How long will it take them to pick a basket of string beans if they work together?

**78. Planting a Garden** Sam and Fran plan to plant a flower garden. Together, they can plant a garden in 4.2 hours. If Sam by himself can plant the same garden in 6 hours, how long will it take Fran to plant the garden by herself?

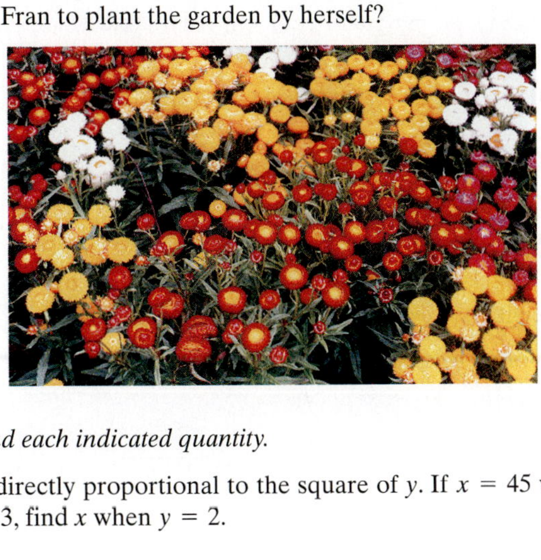

**79. Fractions** What number added to the numerator and subtracted from the denominator of the fraction $\frac{1}{11}$ makes the result equal to $\frac{1}{2}$?

**80. Fractions** When the reciprocal of twice a number is subtracted from 1, the result is the reciprocal of three times the number. Find the number.

**81. Traveling by Boat** Paul Webster's motorboat can travel 15 miles per hour in still water. Traveling with the current of a river, the boat can travel 20 miles in the same time it takes to go 10 miles against the current. Find the current.

**82. Flying a Plane** A small plane and a car start from the same location, at the same time, heading toward the same town 450 miles away. The speed of the plane is three times the speed of the car. The plane arrives at the town 6 hours ahead of the car. Find the speeds of the car and the plane.

[6.6] *Find each indicated quantity.*

**83.** $x$ is directly proportional to the square of $y$. If $x = 45$ when $y = 3$, find $x$ when $y = 2$.

**84.** $W$ is directly proportional to the the square of $L$ and inversely proportional to $A$. If $W = 4$ when $L = 2$ and $A = 10$, find $W$ when $L = 5$ and $A = 20$.

**85.** $z$ is jointly proportional to $x$ and $y$ and inversely proportional to the square of $r$. If $z = 12$ when $x = 20$, $y = 8$, and $r = 8$, find $z$ when $x = 10$, $y = 80$, and $r = 3$.

**86. Electric Surcharge** The Potomac Electric Power Company places a space occupancy surcharge, $s$, on electric bills that is directly proportional to the amount of energy used, $e$. If the surcharge is \$7.20 when 3600 kilowatt-hours are used, what is the surcharge when 4200 kilowatt-hours are used?

**87. Free Fall** The distance, $d$, an object falls in free fall is directly proportional to the square of the time, $t$. If a person falls 16 feet in 1 second, how far will the person fall in 10 seconds? Disregard wind resistance.

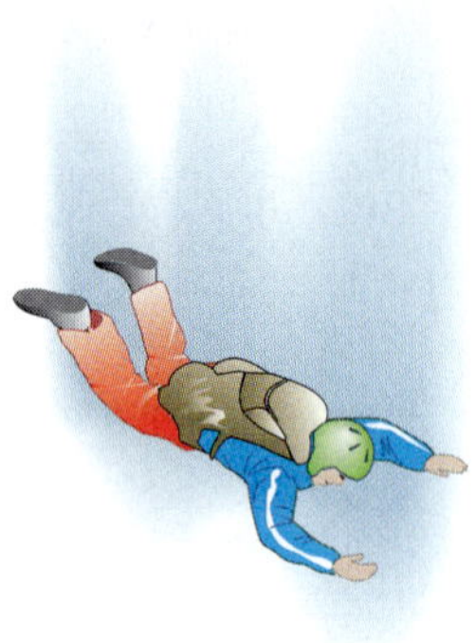

**88. Area** The area, $A$, of a circle varies directly with the square of its radius, $r$. If the area is 78.5 when the radius is 5, find the area when the radius is 8.

**89. Melting Ice Cube** The time, $t$, for an ice cube to melt is inversely proportional to the temperature of the water it is in. If it takes an ice cube 1.7 minutes to melt in 70°F water, how long will it take an ice cube of the same size to melt in 50°F water?

## Chapter 6 Practice Test

*To find out how well you understand the chapter material, take this practice test. The answers, and the section where the material was initially discussed, are given in the back of the book. Each problem is also fully worked out on the* **Chapter Test Prep Video CD**. *Review any questions that you answered incorrectly.*

**1.** Determine the values that are excluded in the expression $\dfrac{x+4}{x^2+3x-28}$.

**2.** Determine the domain of the function $f(x) = \dfrac{x^2+7}{2x^2+7x-4}$.

*Simplify each expression.*

**3.** $\dfrac{10x^7y^2 + 16x^2y + 22x^3y^3}{2x^2y}$

**4.** $\dfrac{x^2 - 4xy - 12y^2}{x^2 + 3xy + 2y^2}$

*In Exercise 5–14, perform the indicated operation.*

**5.** $\dfrac{3xy^4}{6x^2y^3} \cdot \dfrac{2x^2y^4}{x^5y^7}$

**6.** $\dfrac{x+1}{x^2-7x-8} \cdot \dfrac{x^2-x-56}{x^2+9x+14}$

**7.** $\dfrac{7a+14b}{a^2-4b^2} \div \dfrac{a^3+a^2b}{a^2-2ab}$

**8.** $\dfrac{x^3+y^3}{x+y} \div \dfrac{x^2-xy+y^2}{x^2+y^2}$

**9.** $\dfrac{5}{x+1} + \dfrac{2}{x^2}$

**10.** $\dfrac{x-1}{x^2-9} - \dfrac{x}{x^2-2x-3}$

**11.** $\dfrac{m}{12m^2+4mn-5n^2} + \dfrac{2m}{12m^2+28mn+15n^2}$

**12.** $\dfrac{x+1}{4x^2-4x+1} + \dfrac{3}{2x^2+5x-3}$

**13.** $\dfrac{x^3-8}{x^2+5x-14} \div \dfrac{x^2+2x+4}{x^2+10x+21}$

**14.** If $f(x) = \dfrac{x-3}{x+5}$ and $g(x) = \dfrac{x}{2x+3}$, find

**a)** $(f+g)(x)$.

**b)** the domain of $(f+g)(x)$.

**15. Area** If the area of a rectangle is $\dfrac{x^2+11x+30}{x+2}$ and the length is $\dfrac{x^2+9x+18}{x+3}$, find the width.

*For Exercises 16–18, simplify.*

**16.** $\dfrac{\frac{1}{x}+\frac{2}{y}}{\frac{1}{x}-\frac{3}{y}}$

**17.** $\dfrac{\frac{a^2-b^2}{ab}}{\frac{a+b}{b^2}}$

18. $\dfrac{\dfrac{7}{x} - \dfrac{6}{x^2}}{4 - \dfrac{1}{x}}$

*Solve each equation.*

19. $\dfrac{x}{5} - \dfrac{x}{4} = -1$

20. $\dfrac{x}{x-8} + \dfrac{6}{x-2} = \dfrac{x^2}{x^2 - 10x + 16}$

21. Solve $A = \dfrac{2b}{C - d}$ for $C$.

22. **Wattage Rating** The wattage rating of an appliance, $W$, varies jointly as the square of the current, $I$, and the resistance, $R$. If the wattage is 10 when the current is 1 ampere and the resistance is 1000 ohms, find the wattage when the current is 0.5 ampere and the resistance is 300 ohms.

23. $R$ varies directly as $P$ and inversely as the square of $T$. If $R = 30$ when $P = 40$ and $T = 2$, find $R$ when $P = 50$ and $T = 5$.

24. **Washing Windows** Paul Weston can wash the windows of a house in 10 hours. His friend, Nancy Delaney, can wash the same windows in 8 hours. How long will it take them together to wash the windows of this house?

25. **Rollerblading** Cameron Barnette and Ashley Elliot start rollerblading at the same time at the beginning of a trail. Cameron averages 8 miles per hour, while Ashley averages 5 miles per hour. If it takes Ashley $\frac{1}{2}$ hour longer than Cameron to reach the end of the trail, how long is the trail?

## Cumulative Review Test

*Take the following test and check your answers with those given in the back of the book. Review any questions that you answered incorrectly. The section where the material was covered is indicated after the answer.*

1. Illustrate the set $\left\{x \middle| -\frac{5}{3} < x \le \frac{19}{4}\right\}$ on the number line.

2. Evaluate $-3x^3 - 2x^2y + \frac{1}{2}xy^2$ when $x = 2$ and $y = \frac{1}{2}$.

3. Solve the equation $2(x + 1) = \frac{1}{2}(x - 5)$

4. **Distance Learning** The Internet has made it possible for schools to offer degrees through online distance learning. The following diagram shows the degrees most often granted through online programs in 2003.

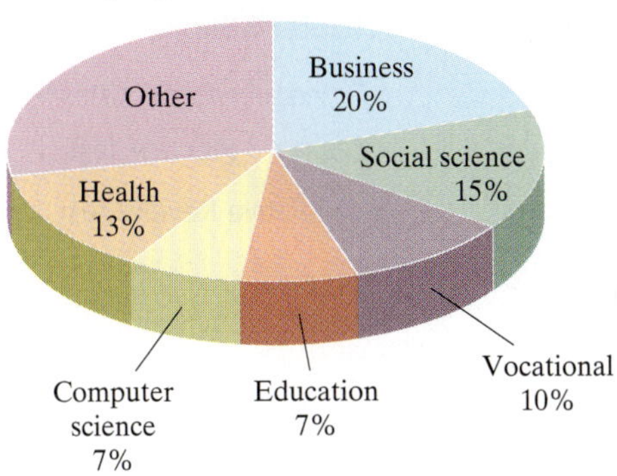


*Source:* The CEO Forum and Market Data Research

a) What percent makes up the "Other" category?

b) If approximately 220,000 degrees were granted through online programs, approximately how many were awarded in business?

5. Evaluate $4x^2 - 3y - 8$ when $x = 4$ and $y = -2$.

6. Simplify $\left(\dfrac{6x^5y^6}{12x^4y^7}\right)^3$.

7. Solve $F = \dfrac{mv^2}{r}$ for $m$.

8. **Simple Interest** Carmalla Banjanie invested \$3000 in a certificate of deposit for 1 year. When she redeemed the certificate, she received \$3180. What was the simple interest rate?

9. **Meeting for a Picnic** Dawn and Paula leave their homes at 8 A.M. planning to meet for a picnic at a point between them. If Dawn travels at 60 miles per hour and Paula travels at 50 miles per hour, and they live 330 miles apart, how long will it take for them to meet for the picnic?

10. Solve $\left|\dfrac{3x + 5}{3}\right| - 3 = 6$.

11. Graph $y = x^2 - 2$.

12. Let $f(x) = \sqrt{2x + 7}$. Evaluate $f(9)$.

13. Find the slope of the line passing through $(2, -4)$ and $(-5, -3)$.

14. Find an equation of a line through $\left(\frac{1}{2}, 1\right)$ that is parallel to the graph of $2x + 3y - 9 = 0$. Write the equation in standard form.

15. Solve the system of equations:

$$10x - y = 2$$
$$4x + 3y = 11$$

16. Multiply $(3x^2 - 5y)(3x^2 + 5y)$.

17. Factor $3x^2 - 30x + 75$.

18. Graph $y = |x| + 2$.

19. Add $\dfrac{7}{3x^2 + x - 4} + \dfrac{9x + 2}{3x^2 - 2x - 8}$.

20. Solve $\dfrac{3y - 2}{y + 1} = 4 - \dfrac{y + 2}{y - 1}$.

# Roots, Radicals, and Complex Numbers

# 7

## GOALS OF THIS CHAPTER

In this chapter, we explain how to add, subtract, multiply, and divide radical expressions. We also graph radical functions, solve radical equations, and introduce imaginary numbers and complex numbers.

Make sure that you understand the three requirements for simplifying radical expressions as discussed in Section 7.5.


7.1 Roots and Radicals

7.2 Rational Exponents

7.3 Simplifying Radicals

7.4 Adding, Subtracting, and Multiplying Radicals

Mid-Chapter Test: Sections 7.1–7.4

7.5 Dividing Radicals

7.6 Solving Radical Equations

7.7 Complex Numbers

Chapter 7 Summary

Chapter 7 Review Exercises

Chapter 7 Practice Test

Cumulative Review Test


MANY SCIENTIFIC FORMULAS, INCLUDING many that pertain to real-life situations, contain radical expressions. In Exercise 135 on page 487, we will see how a radical is used to determine the relationship between the illumination on an object and the distance the object is from a light source.

# 7.1 Roots and Radicals

1. Find square roots.
2. Find cube roots.
3. Understand odd and even roots.
4. Evaluate radicals using absolute value.

In this chapter we expand on the concept of radicals introduced in Chapter 1. In the expression $\sqrt{x}$, the $\sqrt{\phantom{x}}$ is called the **radical sign**. The expression within the radical sign is called the **radicand**.

Radical sign

$$\sqrt{x}$$

Radicand

The entire expression, including the radical sign and radicand, is called the **radical expression**. Another part of the radical expression is its index. The **index** (plural *indices* or *indexes*) gives the "root" of the expression. Square roots have an index of 2. The index of a square root is generally not written. Thus,

$$\sqrt{x} \quad \text{means} \quad \sqrt[2]{x}$$

## 1 Find Square Roots

Every positive number has two square roots, a principal or positive square root and a negative square root. For any positive number $x$, the positive square root is written $\sqrt{x}$, and the negative square root is written $-\sqrt{x}$.

| Number | Principal or Positive Square Root | Negative Square Root |
|---|---|---|
| 25 | $\sqrt{25}$ | $-\sqrt{25}$ |
| 19 | $\sqrt{19}$ | $-\sqrt{19}$ |

**Principal Square Root**

The **principal square root** of a positive number $a$, written $\sqrt{a}$, is the *positive* number $b$ such that $b^2 = a$.

**Examples**

$\sqrt{25} = 5$ since $5^2 = 5 \cdot 5 = 25$

$\sqrt{0.49} = 0.7$ since $(0.7)^2 = (0.7)(0.7) = 0.49$

$\sqrt{\frac{4}{9}} = \frac{2}{3}$ since $\left(\frac{2}{3}\right)^2 = \left(\frac{2}{3}\right)\left(\frac{2}{3}\right) = \frac{4}{9}$

Remember that $-\sqrt{25}$ means the opposite of $\sqrt{25}$. Because $\sqrt{25} = 5$, $-\sqrt{25} = -5$.

**In this book whenever we use the words *square root* we will be referring to the principal or positive square root**. If you are asked to find the value of $\sqrt{25}$, your answer will be 5.

In Chapter 1, we indicated that a rational number is one that can be represented as either a terminating or repeating decimal number. If you use the square root key on your calculator, $\boxed{\sqrt{\phantom{x}}}$, to evaluate the three examples above, you will find they are all terminating or repeating decimal numbers. Thus, they are all *rational numbers*. Many radicals, such as $\sqrt{2}$ and $\sqrt{19}$, are not rational numbers. When evaluating $\sqrt{2}$ and $\sqrt{19}$ on a calculator, the results are nonterminating, nonrepeating decimal numbers. Thus, $\sqrt{2}$ and $\sqrt{19}$ are *irrational numbers*.

| Radical | Calculator Results | |
|---|---|---|
| $\sqrt{2}$ | 1.414213562 | Nonterminating, nonrepeating decimal |
| $\sqrt{19}$ | 4.35889894 | Nonterminating, nonrepeating decimal |

Now let's consider $\sqrt{-25}$. Since the square of any real number will always be greater than or equal to 0, there is no real number that when squared equals $-25$. For this reason, $\sqrt{-25}$ is *not a real number*. Since the square of any real number cannot be negative, *the square root of a negative number is not a real number*. If you evaluate $\sqrt{-25}$ on a calculator, you will get an error message. We will discuss numbers like $\sqrt{-25}$ later in this chapter.

| Radical | Calculator Results | |
|---|---|---|
| $\sqrt{-25}$ | Error | $\sqrt{-25}$ *is not a real number.* |
| $\sqrt{-3}$ | Error | $\sqrt{-3}$ *is not a real number.* |

## Helpful Hint

Do not confuse $-\sqrt{36}$ with $\sqrt{-36}$. Because $\sqrt{36} = 6$, $-\sqrt{36} = -6$. However, $\sqrt{-36}$ is not a real number. The square root of a negative number is not a real number.

$$\sqrt{36} = 6$$
$$-\sqrt{36} = -6$$
$\sqrt{-36}$ is not a real number.

### The Square Root Function

When graphing square root functions, functions of the form $f(x) = \sqrt{x}$, we must always remember that the radicand, $x$, cannot be negative. Thus, the domain of $f(x) = \sqrt{x}$ is $\{x|x \geq 0\}$, or $[0, \infty)$ in interval notation. To graph $f(x) = \sqrt{x}$, we can select some convenient values of $x$ and find the corresponding values of $f(x)$ or $y$ and then plot the ordered pairs, as shown in **Figure 7.1**.

| $x$ | $y$ |
|---|---|
| 0 | 0 |
| 1 | 1 |
| 4 | 2 |
| 6 | ≈2.4 |
| 9 | 3 |

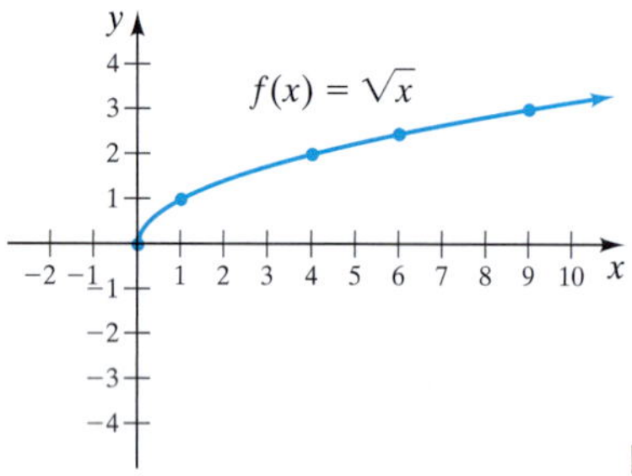


FIGURE 7.1

Since the value of $f(x)$ can never be negative, the range of $f(x) = \sqrt{x}$ is $\{y|y \geq 0\}$, or $[0, \infty)$ in interval notation.

By studying **Figure 7.1**, do you think you can graph $g(x) = -\sqrt{x}$? The graph of $g(x) = -\sqrt{x}$ would be similar to the graph in **Figure 7.1**, but the graph would be drawn below the $x$-axis. Can you explain why? What about the graph of $h(x) = \sqrt{x - 4}$? For the graph of $h(x) = \sqrt{x - 4}$, you would only select values of $x \geq 4$ since the radicand cannot be negative. The domain of $h(x) = \sqrt{x - 4}$ is $\{x|x \geq 4\}$ or $[4, \infty)$.

To evaluate radical functions, it may be necessary to use a calculator.

**EXAMPLE 1** ▸ For each function, find the indicated value(s).

**a)** $f(x) = \sqrt{11x - 2}$, $f(6)$ **b)** $g(r) = -\sqrt{-3r + 1}$, $g(-5)$ and $g(7)$

*Solution*

**a)** $f(6) = \sqrt{11(6) - 2}$ *Substitute 6 for x.*
$= \sqrt{64}$
$= 8$

**b)** $g(-5) = -\sqrt{-3(-5) + 1}$ *Substitute −5 for r.*
$= -\sqrt{16}$
$= -4$

$g(7) = -\sqrt{-3(7) + 1}$ *Substitute 7 for r.*
$= -\sqrt{-20}$ *Not a real number.*

Thus, $g(7)$ is not a real number.

▸ Now Try Exercise 77

## 2 Find Cube Roots

The index of a cube root is 3. We introduced cube roots in Section 1.4, where we explained how to find cube roots on a calculator. You may wish to review that material now.

**Cube Root**

The **cube root** of a number $a$, written $\sqrt[3]{a}$, is the number $b$ such that $b^3 = a$.

**Examples**

$\sqrt[3]{8} = 2$ since $2^3 = 8$

$\sqrt[3]{-27} = -3$ since $(-3)^3 = -27$

For each real number, there is only one cube root. The cube root of a positive number is positive and the cube root of a negative number is negative. The cube root function, $f(x) = \sqrt[3]{x}$, has all real numbers as its domain.

**EXAMPLE 2** ▸ For each function, find the indicated value(s).

**a)** $f(x) = \sqrt[3]{10x + 34}, f(3)$ **b)** $g(r) = \sqrt[3]{12r - 20}, g(-4)$ and $g(1)$

*Solution*

**a)** $f(3) = \sqrt[3]{10(3) + 34}$ *Substitute 3 for x.*

$= \sqrt[3]{64} = 4$

**b)** $g(-4) = \sqrt[3]{12(-4) - 20}$ *Substitute −4 for r.*

$= \sqrt[3]{-68}$

$\approx -4.081655102$ *From a calculator*

$g(1) = \sqrt[3]{12(1) - 20}$ *Substitute 1 for r.*

$= \sqrt[3]{-8}$

$= -2$

▸ **Now Try Exercise 83**

### The Cube Root Function

**Figure 7.2** shows the graph of $y = \sqrt[3]{x}$. To obtain this graph, we substituted values for $x$ and found the corresponding values of $f(x)$ or $y$.

| $x$ | $y$ |
|---|---|
| −8 | −2 |
| −1 | −1 |
| 0 | 0 |
| 1 | 1 |
| 8 | 2 |

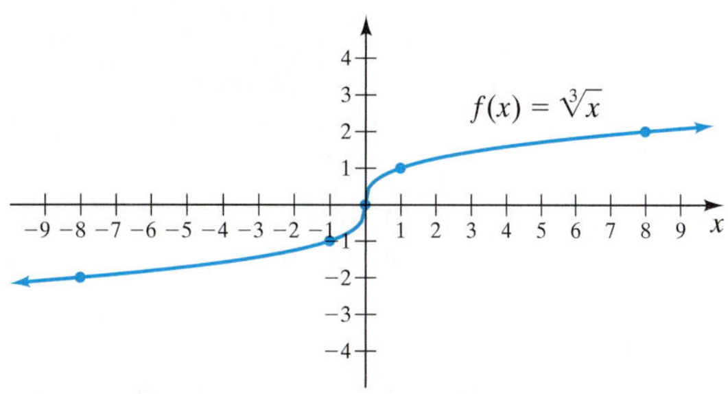


FIGURE 7.2

Notice that both the domain and range are all real numbers, ℝ. We will ask you to graph cube root functions on your graphing calculator in the exercise set.

## 3 Understand Odd and Even Roots

Up to this point we have discussed square and cube roots. Other radical expressions have different indices. For example, in the expression $\sqrt[5]{xy}$, (read "the fifth root of $xy$") the index is 5 and the radicand is $xy$.

Radical expressions that have indices of 2, 4, 6, . . . , or any even integer are **even roots**. Square roots are even roots since their index is 2. Radical expressions that have indices of 3, 5, 7, . . . , or any odd integer are **odd roots**.

**Even Indices**

The $n$th root of $a$, $\sqrt[n]{a}$, where $n$ is an *even index* and $a$ is a nonnegative real number, is the nonnegative real number $b$ such that $b^n = a$.

### Examples of Even Roots

$\sqrt{9} = 3$ since $3^2 = 3 \cdot 3 = 9$

$\sqrt[4]{16} = 2$ since $2^4 = 2 \cdot 2 \cdot 2 \cdot 2 = 16$

$\sqrt[6]{729} = 3$ since $3^6 = 3 \cdot 3 \cdot 3 \cdot 3 \cdot 3 \cdot 3 = 729$

$\sqrt[4]{\frac{1}{256}} = \frac{1}{4}$ since $\left(\frac{1}{4}\right)^4 = \left(\frac{1}{4}\right)\left(\frac{1}{4}\right)\left(\frac{1}{4}\right)\left(\frac{1}{4}\right) = \frac{1}{256}$

Any real number when raised to an even power results in a positive real number. Thus, *when the index of a radical is even, the radicand must be nonnegative for the radical to be a real number.*

**Helpful Hint**

There is an important difference between $-\sqrt[4]{16}$ and $\sqrt[4]{-16}$. The number $-\sqrt[4]{16}$ is the opposite of $\sqrt[4]{16}$. Because $\sqrt[4]{16} = 2$, $-\sqrt[4]{16} = -2$. However, $\sqrt[4]{-16}$ is not a real number since no real number when raised to the fourth power equals $-16$.

$$-\sqrt[4]{16} = -(\sqrt[4]{16}) = -2$$

$\sqrt[4]{-16}$ is not a real number.

**Odd Indices**

The $n$th root of $a$, $\sqrt[n]{a}$, where $n$ is an *odd index* and $a$ is *any real number*, is the real number $b$ such that $b^n = a$.

### Examples of Odd Roots

$\sqrt[3]{8} = 2$ since $2^3 = 2 \cdot 2 \cdot 2 = 8$

$\sqrt[3]{-8} = -2$ since $(-2)^3 = (-2)(-2)(-2) = -8$

$\sqrt[5]{243} = 3$ since $3^5 = 3 \cdot 3 \cdot 3 \cdot 3 \cdot 3 = 243$

$\sqrt[5]{-243} = -3$ since $(-3)^5 = (-3)(-3)(-3)(-3)(-3) = -243$

*An odd root of a positive number is a positive number, and an odd root of a negative number is a negative number.*

It is important to realize that a radical with an even index must have a nonnegative radicand if it is to be a real number. A radical with an odd index will be a real number with any real number as its radicand. Note that $\sqrt[n]{0} = 0$, regardless of whether $n$ is an odd or even index.

**EXAMPLE 3** ▸ Indicate whether or not each radical expression is a real number. If the expression is a real number, find its value.

**a)** $\sqrt[4]{-81}$ **b)** $-\sqrt[4]{81}$ **c)** $\sqrt[5]{-32}$ **d)** $-\sqrt[5]{-32}$

*Solution*

**a)** Not a real number. Even roots of negative numbers are not real numbers.

**b)** Real number, $-\sqrt[4]{81} = -(\sqrt[4]{81}) = -(3) = -3$

**c)** Real number, $\sqrt[5]{-32} = -2$ since $(-2)^5 = -32$

**d)** Real number, $-\sqrt[5]{-32} = -(-2) = 2$

▸ Now Try Exercise 21

**Table 7.1** summarizes the information about even and odd roots.

**Table 7.1**

| | $n$ **Is Even** | $n$ **Is Odd** |
|---|---|---|
| $a > 0$ | $\sqrt[n]{a}$ is a positive real number. | $\sqrt[n]{a}$ is a positive real number. |
| $a < 0$ | $\sqrt[n]{a}$ is not a real number. | $\sqrt[n]{a}$ is a negative real number. |
| $a = 0$ | $\sqrt[n]{0} = 0$ | $\sqrt[n]{0} = 0$ |

## 4 Evaluate Radicals Using Absolute Value

You may think that $\sqrt{a^2} = a$, but this is not necessarily true. Below we evaluate $\sqrt{a^2}$ for $a = 2$ and $a = -2$. You will see that when $a = -2$, $\sqrt{a^2} \neq a$.

$a = 2$: $\quad \sqrt{a^2} = \sqrt{2^2} = \sqrt{4} = 2 \quad$ *Note that* $\sqrt{2^2} = 2$.

$a = -2$: $\quad \sqrt{a^2} = \sqrt{(-2)^2} = \sqrt{4} = 2 \quad$ *Note that* $\sqrt{(-2)^2} \neq -2$.

By examining these examples and other examples we could make up, we can reason that $\sqrt{a^2}$ *will always be a positive real number* for any nonzero real number $a$. Recall from Section 1.3 that the *absolute value* of any real number $a$, or $|a|$, is also a positive number for any nonzero number. We use these facts to reason that

**Radicals and Absolute Value**

For any real number $a$,

$$\sqrt{a^2} = |a|$$

This indicates that the principal square root of $a^2$ is the absolute value of $a$.

**EXAMPLE 4** ▸ Use absolute value to evaluate.

**a)** $\sqrt{9^2}$ **b)** $\sqrt{0^2}$ **c)** $\sqrt{(15.7)^2}$

### Solution

**a)** $\sqrt{9^2} = |9| = 9$ **b)** $\sqrt{0^2} = |0| = 0$ **c)** $\sqrt{(15.7)^2} = |15.7| = 15.7$

▸ **Now Try Exercise 41**

When simplifying a square root, if the radicand contains a variable and we are not sure that the radicand is positive, we need to use absolute value signs when simplifying.

**EXAMPLE 5** ▸ Simplify.

**a)** $\sqrt{(x+8)^2}$ **b)** $\sqrt{16x^2}$ **c)** $\sqrt{25y^6}$ **d)** $\sqrt{a^2 - 6a + 9}$

### Solution

Each square root has a radicand that contains a variable. Since we do not know the value of the variable, we do not know whether it is positive or negative. Therefore, we must use absolute value signs when simplifying.

**a)** $\sqrt{(x+8)^2} = |x+8|$

**b)** Write $16x^2$ as $(4x)^2$, then simplify.

$$\sqrt{16x^2} = \sqrt{(4x)^2} = |4x|$$

**c)** Write $25y^6$ as $(5y^3)^2$, then simplify.

$$\sqrt{25y^6} = \sqrt{(5y^3)^2} = |5y^3|$$

**d)** Notice that $a^2 - 6a + 9$ is a perfect square trinomial. Write the trinomial as the square of a binomial, then simplify.

$$\sqrt{a^2 - 6a + 9} = \sqrt{(a-3)^2} = |a-3|$$

▸ **Now Try Exercise 63**

If you have a square root whose radicand contains a variable and are given instructions like "Assume all variables represent positive values and the radicand is nonnegative," then it is not necessary to use the absolute value sign when simplifying.

**EXAMPLE 6** ▸ Simplify. Assume all variables represent positive values and the radicand is nonnegative.

**a)** $\sqrt{64x^2}$ **b)** $\sqrt{81p^4}$ **c)** $\sqrt{49x^6}$ **d)** $\sqrt{4x^2 - 12xy + 9y^2}$

*Solution*

**a)** $\sqrt{64x^2} = \sqrt{(8x)^2} = 8x$ — Write $64x^2$ as $(8x)^2$.

**b)** $\sqrt{81p^4} = \sqrt{(9p^2)^2} = 9p^2$ — Write $81p^4$ as $(9p^2)^2$.

**c)** $\sqrt{49x^6} = \sqrt{(7x^3)^2} = 7x^3$ — Write $49x^6$ as $(7x^3)^2$.

**d)** $\sqrt{4x^2 - 12xy + 9y^2} = \sqrt{(2x - 3y)^2}$ — Write $4x^2 - 12xy + 9y^2$ as $(2x - 3y)^2$.

$= 2x - 3y$

▸ **Now Try Exercise 67**

We only need to be concerned about adding absolute value signs when discussing square (and other even) roots. We do not need to use absolute value signs when the index is odd.

## EXERCISE SET 7.1

MathXL® MyMathLab

### Concept/Writing Exercises

1. **a)** How many square roots does every positive real number have? Name them.
   **b)** Find all square roots of the number 49.
   **c)** In this text, when we refer to "the square root," which square root are we referring to?
   **d)** Find the square root of 49.
2. **a)** What are even roots? Give an example of an even root.
   **b)** What are odd roots? Give an example of an odd root.
3. Explain why $\sqrt{-81}$ is not a real number.
4. Will a radical expression with an odd index and a real number as the radicand always be a real number? Explain your answer.
5. Will a radical expression with an even index and a real number as the radicand always be a real number? Explain your answer.
6. **a)** To what is $\sqrt{a^2}$ equal?
   **b)** To what is $\sqrt{a^2}$ equal if we know $a \geq 0$?
7. **a)** Evaluate $\sqrt{a^2}$ for $a = 1.3$.
   **b)** Evaluate $\sqrt{a^2}$ for $a = -1.3$.
8. **a)** Evaluate $\sqrt{a^2}$ for $a = 5.72$.
   **b)** Evaluate $\sqrt{a^2}$ for $a = -5.72$.
9. **a)** Evaluate $\sqrt[3]{27}$.
   **b)** Evaluate $-\sqrt[3]{27}$.
   **c)** Evaluate $\sqrt[3]{-27}$.
10. **a)** Evaluate $\sqrt[4]{16}$
    **b)** Evaluate $-\sqrt[4]{16}$
    **c)** Evaluate $\sqrt[4]{-16}$.

### Practice the Skills

*Evaluate each radical expression if it is a real number. Use a calculator to approximate irrational numbers to the nearest hundredth. If the expression is not a real number, so state.*

**11.** $\sqrt{36}$ **12.** $-\sqrt{36}$ **13.** $\sqrt[3]{-64}$ **14.** $\sqrt[3]{125}$

**15.** $\sqrt[3]{-125}$ **16.** $-\sqrt[3]{-125}$ **17.** $\sqrt[5]{-1}$ **18.** $-\sqrt[5]{-1}$

**19.** $\sqrt[5]{1}$ **20.** $\sqrt[6]{64}$ **21.** $\sqrt[6]{-64}$ **22.** $\sqrt[4]{-81}$

**23.** $\sqrt[3]{-343}$ **24.** $\sqrt{121}$ **25.** $\sqrt{-36}$ **26.** $\sqrt{45.3}$

**27.** $\sqrt{-45.3}$ **28.** $\sqrt{53.9}$ **29.** $\sqrt{\frac{1}{25}}$ **30.** $\sqrt{-\frac{1}{25}}$

**31.** $\sqrt[3]{\frac{1}{8}}$ **32.** $\sqrt[3]{-\frac{1}{8}}$ **33.** $\sqrt{\frac{4}{49}}$ **34.** $\sqrt[3]{\frac{8}{27}}$

**35.** $\sqrt[3]{-\frac{8}{27}}$ **36.** $\sqrt[4]{-8.9}$ **37.** $-\sqrt[4]{18.2}$ **38.** $\sqrt[5]{93}$

*Use absolute value to evaluate.*

**39.** $\sqrt{7^2}$ **40.** $\sqrt{(-7)^2}$ **41.** $\sqrt{19^2}$ **42.** $\sqrt{(-19)^2}$

**43.** $\sqrt{119^2}$ **44.** $\sqrt{(-119)^2}$ **45.** $\sqrt{(235.23)^2}$ **46.** $\sqrt{(-201.5)^2}$

**47.** $\sqrt{(0.06)^2}$ **48.** $\sqrt{(-0.19)^2}$ **49.** $\sqrt{\left(\frac{12}{13}\right)^2}$ **50.** $\sqrt{\left(-\frac{101}{319}\right)^2}$

*Write as an absolute value.*

**51.** $\sqrt{(x-4)^2}$ **52.** $\sqrt{(a+10)^2}$ **53.** $\sqrt{(x-3)^2}$ **54.** $\sqrt{(7a-11b)^2}$

**55.** $\sqrt{(3x^2-1)^2}$ **56.** $\sqrt{(7y^2-3y)^2}$ **57.** $\sqrt{(6a^3-5b^4)^2}$ **58.** $\sqrt{(9y^4-2z^3)^2}$

*Use absolute value to simplify. You may need to factor first.*

**59.** $\sqrt{a^{14}}$ **60.** $\sqrt{y^{22}}$ **61.** $\sqrt{z^{32}}$ **62.** $\sqrt{x^{200}}$

**63.** $\sqrt{a^2-8a+16}$ **64.** $\sqrt{x^2-12x+36}$ **65.** $\sqrt{9a^2+12ab+4b^2}$ **66.** $\sqrt{4x^2+20xy+25y^2}$

*Simplify. Assume that all variables represent positive values and that the radicand is nonnegative.*

**67.** $\sqrt{49x^2}$ **68.** $\sqrt{100a^4}$ **69.** $\sqrt{16c^6}$ **70.** $\sqrt{121z^8}$

**71.** $\sqrt{x^2+4x+4}$ **72.** $\sqrt{9a^2-6a+1}$ **73.** $\sqrt{4x^2+4xy+y^2}$ **74.** $\sqrt{16b^2-40bc+25c^2}$

*Find the indicated value of each function. Use your calculator to approximate irrational numbers. Round irrational numbers to the nearest thousandth.*

**75.** $f(x)=\sqrt{5x-6},\ f(2)$ **76.** $f(c)=\sqrt{7c+1},\ f(5)$ **77.** $q(x)=\sqrt{76-3x},\ q(4)$

**78.** $q(b)=\sqrt{9b+34},\ q(-1)$ **79.** $t(a)=\sqrt{-15a-9},\ t(-6)$ **80.** $f(a)=\sqrt{14a-36},\ f(4)$

**81.** $g(x)=\sqrt{64-8x},\ g(-3)$ **82.** $p(x)=\sqrt[3]{8x+9},\ p(2)$ **83.** $h(x)=\sqrt[3]{9x^2+4},\ h(4)$

**84.** $k(c)=\sqrt[4]{16c-5},\ k(6)$ **85.** $f(x)=\sqrt[3]{-2x^2+x-6},\ f(-3)$ **86.** $t(x)=\sqrt[4]{2x^3-3x^2+6x},\ t(2)$

## Problem Solving

**87.** Find $f(81)$ if $f(x)=x+\sqrt{x}+7$.

**88.** Find $g(25)$ if $g(x)=x^2+\sqrt{x}-13$.

**89.** Find $t(18)$ if $t(x)=\frac{x}{2}+\sqrt{2x}-4$.

**90.** Find $m(36)$ if $m(x)=\frac{x}{3}+\sqrt{4x}+10$.

**91.** Find $k(8)$ if $k(x)=x^2+\sqrt{\frac{x}{2}}-21$.

**92.** Find $r(45)$ if $r(x)=\frac{x}{9}+\sqrt{\frac{x}{5}}+13$.

**93.** Select a value for $x$ for which $\sqrt{(2x+1)^2}\neq 2x+1$.

**94.** Select a value for $x$ for which $\sqrt{(5x-3)^2}\neq 5x-3$.

**95.** For what values of $x$ will $\sqrt{(x-1)^2}=x-1$? Explain how you determined your answer.

**96.** For what values of $x$ will $\sqrt{(x+3)^2}=x+3$? Explain how you determined your answer.

**97.** For what values of $x$ will $\sqrt{(2x-6)^2}=2x-6$? Explain how you determined your answer.

**98.** For what values of $x$ will $\sqrt{(3x-8)^2}=3x-8$? Explain how you determined your answer.

**99. a)** For what values of $a$ is $\sqrt{a^2}=|a|$?

**b)** For what values of $a$ is $\sqrt{a^2}=a$?

**c)** For what values of $a$ is $\sqrt[3]{a^3}=a$?

**100.** Under what circumstances is the expression $\sqrt[n]{x}$ not a real number?

**101.** Explain why the expression $\sqrt[n]{x^n}$ is a real number for any real number $x$.

**102.** Under what circumstances is the expression $\sqrt[n]{x^m}$ not a real number?

**103.** Find the domain of $\frac{\sqrt{x+5}}{\sqrt[3]{x+5}}$. Explain how you determined your answer.

**104.** Find the domain of $\frac{\sqrt[3]{x-2}}{\sqrt[6]{x+1}}$. Explain how you determined your answer.

*By considering the domains of the functions in Exercises 105 through 108, match each function with its graph, labeled* **a)** *through* **d)**.

**105.** $f(x)=\sqrt{x}$ **106.** $f(x)=\sqrt{x^2}$ **107.** $f(x)=\sqrt{x-5}$ **108.** $f(x)=\sqrt{x+5}$

**a)**

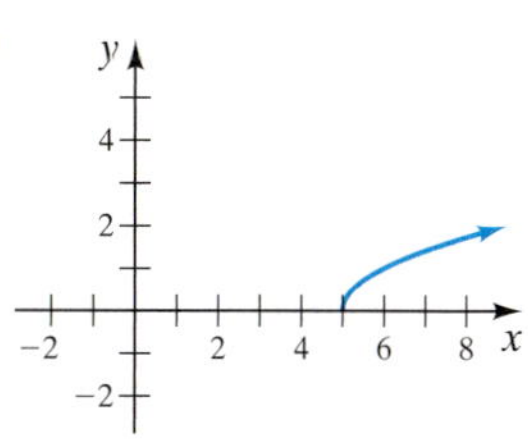


**b)**

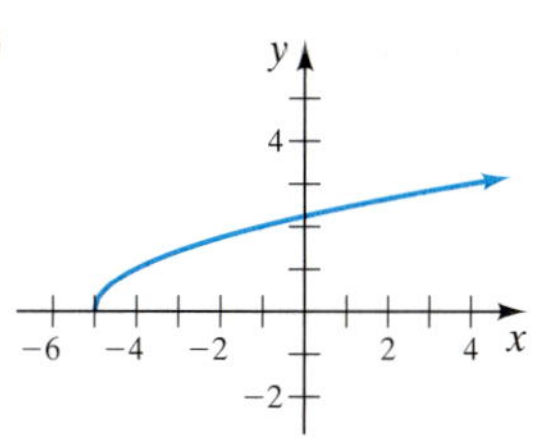


**c)**

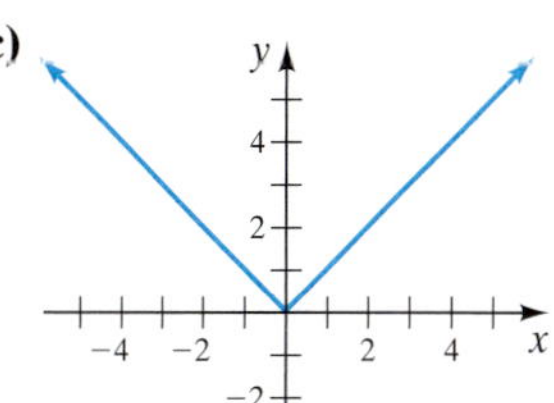


**d)**

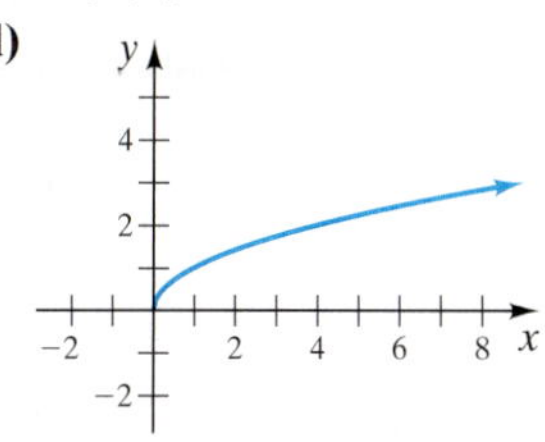

**109.** Give a radical function whose domain is $\{x|x \geq 8\}$.

**110.** Give a radical function whose domain is $\{x|x \leq 5\}$.

**111.** If $f(x) = -\sqrt{x}$, can $f(x)$ ever be

**a)** greater than 0,

**b)** equal to 0,

**c)** less than 0?

Explain your answers.

**112.** If $f(x) = \sqrt{x+5}$, can $f(x)$ ever be

**a)** less than 0,

**b)** equal to 0,

**c)** greater than 0?

Explain your answers.

**113. Velocity of an Object** The velocity, $V$, of an object, in feet per second, after it has fallen a distance, $h$, in feet, can be found by the formula $V = \sqrt{64.4h}$. A pile driver is a large mass that is used as a hammer to drive pilings into soft earth to support a building or other structure.

With what velocity will the hammer hit the piling if it falls from

**a)** 20 feet above the top of the piling?

**b)** 40 feet above the top of the piling?

**114. Wave Action** Scripps Institute of Oceanography in La Jolla, California, developed the formula for relating wind speed, $u$, in knots, with the height, $H$, in feet, of the waves the wind produces in certain areas of the ocean. This formula is

$$u = \sqrt{\frac{H}{0.026}}$$

If waves produced by a storm have a height of 15 feet, what is the wind speed producing the waves?

**115.** Graph $f(x) = \sqrt{x+1}$.

**116.** Graph $g(x) = -\sqrt{x}$.

**117.** Graph $g(x) = \sqrt{x} + 1$.

**118.** Graph $f(x) = \sqrt{x-2}$.

*For Exercises 119–124, use your graphing calculator.*

**119.** Check the graph drawn in Exercise 115.

**120.** Check the graph drawn in Exercise 117.

**121.** Determine whether the domain you gave in Exercise 103 is correct.

**122.** Determine whether the domain you gave in Exercise 104 is correct.

**123.** Graph $y = \sqrt[3]{x} + 4$.

**124.** Graph $f(x) = \sqrt[3]{2x-3}$.

## Group Activity

*In this activity, you will determine the conditions under which certain properties of radicals are true. We will discuss these properties later in this chapter. Discuss and answer these exercises as a group.*

**125.** The property $\sqrt[n]{a} \cdot \sqrt[n]{b} = \sqrt[n]{ab}$, called the *multiplication property for radicals*, is true for certain real numbers $a$ and $b$. By substituting values for $a$ and $b$, determine under what conditions this property is true.

**126.** The property $\frac{\sqrt[n]{a}}{\sqrt[n]{b}} = \sqrt[n]{\frac{a}{b}}$, called the *division property for radicals*, is true for certain real numbers $a$ and $b$. By substituting values for $a$ and $b$, determine under what conditions this property is true.

## Cumulative Review Exercises

*Factor.*

[5.4] **127.** $9ax - 3bx + 12ay - 4by$

[5.5] **128.** $3x^3 - 18x^2 + 24x$

**129.** $8x^4 + 10x^2 - 3$

[5.6] **130.** $x^3 - \frac{8}{27}y^3$

# 7.2 Rational Exponents

1. Change a radical expression to an exponential expression.
2. Simplify radical expressions.
3. Apply the rules of exponents to rational and negative exponents.
4. Factor expressions with rational exponents.

## 1 Change a Radical Expression to an Exponential Expression

In this section we discuss changing radical expressions to exponential expressions and vice versa. When you see a rational exponent, you should realize that the expression can be written as a radical expression by using the following procedure.

**Exponential Form of $\sqrt[n]{a}$**

$$\sqrt[n]{a} = a^{1/n}$$

When $a$ is nonnegative, $n$ can be any index.
When $a$ is negative, $n$ must be odd.

**For the remainder of this chapter, unless you are instructed otherwise, assume that all variables in radicands represent nonnegative real numbers and that the radicand is nonnegative.** With this assumption, we will not need to state that the variable is nonnegative whenever we have a radical with an even index. This will allow us to write many answers without absolute value signs.

**EXAMPLE 1** ▸ Write each expression in exponential form (with rational exponents).

**a)** $\sqrt{7}$ **b)** $\sqrt[3]{15ab}$ **c)** $\sqrt[7]{-4x^2y^5}$ **d)** $\sqrt[8]{\dfrac{5x^7}{2z^{11}}}$

### Solution

**a)** $\sqrt{7} = 7^{1/2}$ Recall that the index of a square root is 2.

**b)** $\sqrt[3]{15ab} = (15ab)^{1/3}$ **c)** $\sqrt[7]{-4x^2y^5} = (-4x^2y^5)^{1/7}$ **d)** $\sqrt[8]{\dfrac{5x^7}{2z^{11}}} = \left(\dfrac{5x^7}{2z^{11}}\right)^{1/8}$

▸ Now Try Exercise 19

Exponential expressions can be converted to radical expressions by reversing the procedure.

**EXAMPLE 2** ▸ Write each expression in radical form (without rational exponents).

**a)** $9^{1/2}$ **b)** $(-8)^{1/3}$ **c)** $y^{1/4}$ **d)** $(10x^2y)^{1/7}$ **e)** $5rs^{1/2}$

### Solution

**a)** $9^{1/2} = \sqrt{9} = 3$ **b)** $(-8)^{1/3} = \sqrt[3]{-8} = -2$

**c)** $y^{1/4} = \sqrt[4]{y}$ **d)** $(10x^2y)^{1/7} = \sqrt[7]{10x^2y}$ **e)** $5rs^{1/2} = 5r\sqrt{s}$

▸ Now Try Exercise 33

## 2 Simplify Radical Expressions

We can expand the preceding rule so that radicals of the form $\sqrt[n]{a^m}$ can be written as exponential expressions. Consider $a^{2/3}$. We can write $a^{2/3}$ as $(a^{1/3})^2$ or $(a^2)^{1/3}$. This suggests $a^{2/3} = (\sqrt[3]{a})^2 = \sqrt[3]{a^2}$.

**Exponential Form of $\sqrt[n]{a^m}$**

For any nonnegative number $a$, and integers $m$ and $n$,

$$\sqrt[n]{a^m} = \left(\sqrt[n]{a}\right)^m = a^{m/n}$$

Power ($m$) ← Index ($n$)

This rule can be used to change an expression from radical form to exponential form and vice versa. When changing a radical expression to exponential form, the *power* is placed in the *numerator*, and the *index or root* is placed in the *denominator* of the rational exponent. Thus, for example, $\sqrt[3]{x^4}$ can be written $x^{4/3}$. Also $(\sqrt[5]{y})^2$ can be written $y^{2/5}$. Additional examples follow.

**Examples**

$$\sqrt{y^3} = y^{3/2} \qquad \sqrt[3]{z^2} = z^{2/3} \qquad \sqrt[5]{2^8} = 2^{8/5}$$
$$(\sqrt{p})^3 = p^{3/2} \qquad (\sqrt[4]{x})^3 = x^{3/4} \qquad (\sqrt[4]{7})^3 = 7^{3/4}$$

By this rule, for nonnegative values of the variable we can write

$$\sqrt{x^5} = (\sqrt{x})^5 \qquad (\sqrt[4]{p})^3 = \sqrt[4]{p^3}$$

**EXAMPLE 3** ▸ Write each expression in exponential form (with rational exponents) and then simplify.

**a)** $\sqrt[4]{x^{12}}$ **b)** $(\sqrt[3]{y})^{15}$ **c)** $(\sqrt[6]{x})^{12}$

*Solution*

**a)** $\sqrt[4]{x^{12}} = x^{12/4} = x^3$ **b)** $(\sqrt[3]{y})^{15} = y^{15/3} = y^5$

**c)** $(\sqrt[6]{x})^{12} = x^{12/6} = x^2$

▸ Now Try Exercise 45

Exponential expressions with rational exponents can be converted to radical expressions by reversing the procedure. The *numerator* of the rational exponent is the *power*, and the *denominator* of the rational exponent is the *index or root* of the radical expression. Here are some examples.

**Examples**

$$x^{1/2} = \sqrt{x} \qquad 5^{1/3} = \sqrt[3]{5}$$
$$7^{2/3} = \sqrt[3]{7^2} \text{ or } (\sqrt[3]{7})^2 \qquad y^{3/10} = \sqrt[10]{y^3} \text{ or } (\sqrt[10]{y})^3$$
$$x^{9/5} = \sqrt[5]{x^9} \text{ or } (\sqrt[5]{x})^9 \qquad z^{10/3} = \sqrt[3]{z^{10}} \text{ or } (\sqrt[3]{z})^{10}$$

Notice that you may choose, for example, to write $6^{2/3}$ as either $\sqrt[3]{6^2}$ or $(\sqrt[3]{6})^2$.

**EXAMPLE 4** ▸ Write each expression in radical form (without rational exponents).

**a)** $x^{2/5}$ **b)** $(3ab)^{5/4}$

*Solution*

**a)** $x^{2/5} = \sqrt[5]{x^2}$ or $(\sqrt[5]{x})^2$ **b)** $(3ab)^{5/4} = \sqrt[4]{(3ab)^5}$ or $(\sqrt[4]{3ab})^5$

▸ Now Try Exercise 35

**EXAMPLE 5** ▸ Simplify.

**a)** $4^{3/2}$ **b)** $\sqrt[6]{(49)^3}$ **c)** $\sqrt[4]{(xy)^{20}}$ **d)** $(\sqrt[15]{z})^5$

*Solution*

**a)** Sometimes an expression with a rational exponent can be simplified more easily by writing the expression as a radical, as illustrated.

$$\begin{aligned} 4^{3/2} &= (\sqrt{4})^3 && \text{Write as a radical.} \\ &= (2)^3 \\ &= 8 \end{aligned}$$

**b)** Sometimes a radical expression can be simplified more easily by writing the expression with rational exponents, as illustrated in parts **b)** through **d)**.

$$\begin{aligned} \sqrt[6]{(49)^3} &= 49^{3/6} && \text{Write with a rational exponent.} \\ &= 49^{1/2} && \text{Reduce exponent.} \\ &= \sqrt{49} && \text{Write as a radical.} \\ &= 7 && \text{Simplify.} \end{aligned}$$

**c)** $\sqrt[4]{(xy)^{20}} = (xy)^{20/4} = (xy)^5$

**d)** $(\sqrt[15]{z})^5 = z^{5/15} = z^{1/3}$ or $\sqrt[3]{z}$

▸ **Now Try Exercise 51**

Now let's consider $\sqrt[5]{x^5}$. When written in exponential form, this is $x^{5/5} = x^1 = x$. This leads to the following rule.

**Exponential form of $\sqrt[n]{a^n}$**

For any nonnegative real number $a$,

$$\sqrt[n]{a^n} = (\sqrt[n]{a})^n = a^{n/n} = a$$

In the preceding box, we specified that $a$ was nonnegative. If $n$ is an even index and $a$ is a negative real number, $\sqrt[n]{a^n} = |a|$ and not $a$. For example, $\sqrt[6]{(-5)^6} = |-5| = 5$. *Since we are assuming, except where noted otherwise, that variables in radicands represent nonnegative real numbers*, we may write $\sqrt[6]{x^6} = x$ and not $|x|$. This assumption also lets us write $\sqrt{x^2} = x$ and $(\sqrt[4]{z})^4 = z$.

**Examples**

$$\sqrt{3^2} = 3 \qquad \sqrt[4]{y^4} = y$$

$$\sqrt[6]{(xy)^6} = xy \qquad (\sqrt[5]{z})^5 = z$$

## 3 Apply the Rules of Exponents to Rational and Negative Exponents

In Section 1.5, we introduced and discussed the rules of exponents. In that section we only used exponents that were whole numbers. The rules still apply when the exponents are rational numbers. Let's review those rules now.

**Rules of Exponents**

For all real numbers $a$ and $b$ and all rational numbers $m$ and $n$,

| Rule | |
|---|---|
| Product rule | $a^m \cdot a^n = a^{m+n}$ |
| Quotient rule | $\dfrac{a^m}{a^n} = a^{m-n}, \quad a \neq 0$ |
| Negative exponent rule | $a^{-m} = \dfrac{1}{a^m}, \quad a \neq 0$ |
| Zero exponent rule | $a^0 = 1, \quad a \neq 0$ |
| Raising a power to a power | $(a^m)^n = a^{m \cdot n}$ |
| Raising a product to a power | $(ab)^m = a^m b^m$ |
| Raising a quotient to a power | $\left(\dfrac{a}{b}\right)^m = \dfrac{a^m}{b^m}, \quad b \neq 0$ |

Using these rules, we will now work some problems in which the exponents are rational numbers.

**EXAMPLE 6** ▸ Evaluate. **a)** $8^{-2/3}$ **b)** $(-27)^{-5/3}$ **c)** $(-32)^{-6/5}$

*Solution*

**a)** Begin by using the negative exponent rule.

$$8^{-2/3} = \frac{1}{8^{2/3}} \quad \text{Negative exponent rule}$$

$$= \frac{1}{(\sqrt[3]{8})^2} \quad \text{Write the denominator as a radical.}$$

$$= \frac{1}{2^2} \quad \text{Simplify the denominator.}$$

$$= \frac{1}{4}$$

**b)** $(-27)^{-5/3} = \dfrac{1}{(-27)^{5/3}} = \dfrac{1}{(\sqrt[3]{-27})^5} = \dfrac{1}{(-3)^5} = -\dfrac{1}{243}$

**c)** $(-32)^{-6/5} = \dfrac{1}{(-32)^{6/5}} = \dfrac{1}{(\sqrt[5]{-32})^6} = \dfrac{1}{(-2)^6} = \dfrac{1}{64}$

▸ **Now Try Exercise 81**

Note that Example 6 **a)** could have been evaluated as follows:

$$8^{-2/3} = \frac{1}{8^{2/3}} = \frac{1}{\sqrt[3]{8^2}} = \frac{1}{\sqrt[3]{64}} = \frac{1}{4}$$

However, it is generally easier to evaluate the root before applying the power.

Consider the expression $(-16)^{3/4}$. This can be rewritten as $(\sqrt[4]{-16})^3$. Since $(\sqrt[4]{-16})^3$ is not a real number, the expression $(-16)^{3/4}$ is not a real number.

In Chapter 1, we indicated that

$$\left(\frac{a}{b}\right)^{-n} = \left(\frac{b}{a}\right)^{n}$$

We use this fact in the following example.

**EXAMPLE 7** ▸ Evaluate. **a)** $\left(\dfrac{9}{25}\right)^{-1/2}$ **b)** $\left(\dfrac{27}{8}\right)^{-1/3}$

*Solution*

**a)** $\left(\dfrac{9}{25}\right)^{-1/2} = \left(\dfrac{25}{9}\right)^{1/2} = \sqrt{\dfrac{25}{9}} = \dfrac{5}{3}$

**b)** $\left(\dfrac{27}{8}\right)^{-1/3} = \left(\dfrac{8}{27}\right)^{1/3} = \sqrt[3]{\dfrac{8}{27}} = \dfrac{2}{3}$

▸ **Now Try Exercise 83**

**Helpful Hint**

How do the expressions $-25^{1/2}$ and $(-25)^{1/2}$ differ?
Recall that $-x^2$ means $-(x^2)$. The same principle applies here.

$$-25^{1/2} = -(25)^{1/2} = -\sqrt{25} = -5$$

$(-25)^{1/2} = \sqrt{-25}$, which is not a real number.

**EXAMPLE 8** ▸ Simplify each expression and write the answer without negative exponents.

**a)** $a^{1/2} \cdot a^{-2/3}$ **b)** $(6x^2y^{-4})^{-1/2}$ **c)** $3.2x^{1/3}(2.4x^{1/2} + x^{-1/4})$ **d)** $\left(\dfrac{9x^{-4}z^{2/5}}{z^{-3/5}}\right)^{1/8}$

## Solution

**a)**
$$a^{1/2} \cdot a^{-2/3} = a^{(1/2)-(2/3)} \quad \text{Product rule}$$
$$= a^{-1/6} \quad \text{Find the LCD and subtract the exponents.}$$
$$= \frac{1}{a^{1/6}} \quad \text{Negative exponent rule}$$

**b)**
$$(6x^2y^{-4})^{-1/2} = 6^{-1/2}x^{2(-1/2)}y^{-4(-1/2)} \quad \text{Raise the product to a power.}$$
$$= 6^{-1/2}x^{-1}y^2 \quad \text{Multiply the exponents.}$$
$$= \frac{y^2}{6^{1/2}x}\left(\text{or } \frac{y^2}{x\sqrt{6}}\right) \quad \text{Negative exponent rule}$$

**c)** Begin by using the distributive property.
$$3.2x^{1/3}(2.4x^{1/2} + x^{-1/4}) = (3.2x^{1/3})(2.4x^{1/2}) + (3.2x^{1/3})(x^{-1/4}) \quad \text{Distributive property}$$
$$= (3.2)(2.4)(x^{(1/3)+(1/2)}) + 3.2x^{(1/3)-(1/4)} \quad \text{Product rule}$$
$$= 7.68x^{5/6} + 3.2x^{1/12}$$

**d)**
$$\left(\frac{9x^{-4}z^{2/5}}{z^{-3/5}}\right)^{1/8} = (9x^{-4}z^{(2/5)-(-3/5)})^{1/8} \quad \text{Quotient rule}$$
$$= (9x^{-4}z)^{1/8} \quad \text{Subtract the exponents.}$$
$$= 9^{1/8}x^{-4(1/8)}z^{1/8} \quad \text{Raise the product to a power.}$$
$$= 9^{1/8}x^{-4/8}z^{1/8} \quad \text{Multiply the exponents.}$$
$$= 9^{1/8}x^{-1/2}z^{1/8} \quad \text{Simplify exponent.}$$
$$= \frac{9^{1/8}z^{1/8}}{x^{1/2}} \quad \text{Negative exponent rule}$$

▸ **Now Try Exercise 105**

**EXAMPLE 9** ▸ Simplify. **a)** $\sqrt[15]{(7y)^5}$ **b)** $\left(\sqrt[4]{a^2b^3c}\right)^{20}$ **c)** $\sqrt[4]{\sqrt[3]{x}}$

## Solution

**a)**
$$\sqrt[15]{(7y)^5} = (7y)^{5/15} \quad \text{Write with a rational exponent.}$$
$$= (7y)^{1/3} \quad \text{Simplify the exponent.}$$
$$= \sqrt[3]{7y} \quad \text{Write as a radical.}$$

**b)**
$$\left(\sqrt[4]{a^2b^3c}\right)^{20} = (a^2b^3c)^{20/4} \quad \text{Write with a rational exponent.}$$
$$= (a^2b^3c)^5$$
$$= a^{10}b^{15}c^5 \quad \text{Raise the product to a power.}$$

**c)**
$$\sqrt[4]{\sqrt[3]{x}} = \sqrt[4]{x^{1/3}} \quad \text{Write } \sqrt[3]{x} \text{ as } x^{1/3}.$$
$$= (x^{1/3})^{1/4} \quad \text{Write with a rational exponent.}$$
$$= x^{1/12} \quad \text{Raise the power to a power.}$$
$$= \sqrt[12]{x} \quad \text{Write as a radical.}$$

▸ **Now Try Exercise 53**

**USING YOUR CALCULATOR Finding Roots or Expressions with Rational Exponents on a Scientific or Graphing Calculator**

There are often many ways to evaluate an expression like $(\sqrt[5]{845})^3$ or $845^{3/5}$ on a calculator. The procedure to use varies from calculator to calculator. One general method is to write the expression with a rational exponent and use the $\boxed{y^x}$ or $\boxed{a^x}$ or $\boxed{\wedge}$ key with parentheses keys, as shown below.*

**Scientific Calculator**

To evaluate $845^{3/5}$, press

Answer displayed

$845\ \boxed{y^x}\ \boxed{(}\ 3\ \boxed{\div}\ 5\ \boxed{)}\ \boxed{=}\ 57.03139903$

To evaluate $845^{-3/5}$, press

Answer displayed

$845\ \boxed{y^x}\ \boxed{(}\ 3\ \boxed{+/-}\ \boxed{\div}\ 5\ \boxed{)}\ \boxed{=}\ 0.017534201$

**Graphing Calculator**

To evaluate $845^{3/5}$, press the following keys.

Answer displayed

$845\ \boxed{\wedge}\ \boxed{(}\ 3\ \boxed{\div}\ 5\ \boxed{)}\ \boxed{\text{ENTER}}\ 57.03139903$

To evaluate $845^{-3/5}$, press the following keys.

Answer displayed

$845\ \boxed{\wedge}\ \boxed{(}\ \boxed{(-)}\ 3\ \boxed{\div}\ 5\ \boxed{)}\ \boxed{\text{ENTER}}\ .0175342008$

*Keystrokes to use vary from calculator to calculator. Read your calculator manual to learn how to evaluate exponential expressions.

## 4 Factor Expressions with Rational Exponents

In higher-level math courses, you may have to factor out a variable with a rational exponent. To factor a rational expression, factor out the term with the smallest (or most negative) exponent.

**EXAMPLE 10** ▸ Factor $x^{2/5} + x^{-3/5}$.

*Solution* The smallest of the two exponents is $-3/5$. Therefore, we will factor $x^{-3/5}$ from both terms. To find the new exponent on the variable that originally had the greater exponent, we subtract the exponent that was factored out from the original exponent.

Original exponent — Exponent factored out

$$x^{2/5} + x^{-3/5} = x^{-3/5}\left(x^{2/5-(-3/5)} + 1\right)$$
$$= x^{-3/5}(x^1 + 1)$$
$$= x^{-3/5}(x + 1)$$
$$= \frac{x+1}{x^{3/5}}$$

We can check our factoring by multiplying.

$$x^{-3/5}(x + 1) = x^{-3/5} \cdot x + x^{-3/5} \cdot 1$$
$$= x^{(-3/5)+1} + x^{-3/5}$$
$$= x^{2/5} + x^{-3/5}$$

Since we obtained the original expression, the factoring is correct.

▸ **Now Try Exercise 135**

# EXERCISE SET 7.2

MathXL® MyMathLab

## Concept/Writing Exercises

1. a) Under what conditions is $\sqrt[n]{a}$ a real number?
   b) When $\sqrt[n]{a}$ is a real number, how can it be expressed with rational exponents?
2. a) Under what conditions is $\sqrt[n]{a^m}$ a real number?
   b) Under what conditions is $(\sqrt[n]{a})^m$ a real number?
   c) When $\sqrt[n]{a^m}$ is a real number, how can it be expressed with rational exponents?
3. a) Under what conditions is $\sqrt[n]{a^n}$ a real number?
   b) When $n$ is even and $a \geq 0$, what is $\sqrt[n]{a^n}$ equal to?
   c) When $n$ is odd, what is $\sqrt[n]{a^n}$ equal to?
   d) When $n$ is even and $a$ may be any real number, what is $\sqrt[n]{a^n}$ equal to?
4. a) Explain the difference between $-16^{1/2}$ and $(-16)^{1/2}$.
   b) Evaluate each expression in part **a)** if possible.
5. a) Is $(xy)^{1/2} = xy^{1/2}$? Explain.
   b) Is $(xy)^{-1/2} = \dfrac{x^{1/2}}{y^{-1/2}}$? Explain.
6. a) Is $\sqrt[6]{(3y)^3} = (3y)^{6/3}$? Explain.
   b) Is $\sqrt{(ab)^4} = (ab)^2$? Explain.

## Practice the Skills

*In this exercise set, assume that all variables represent positive real numbers. Write each expression in exponential form.*

7. $\sqrt{a^3}$
8. $\sqrt{y^7}$
9. $\sqrt{9^5}$
10. $\sqrt[3]{y}$
11. $\sqrt[3]{z^5}$
12. $\sqrt[3]{x^{11}}$
13. $\sqrt[3]{7^{10}}$
14. $\sqrt[5]{9^{11}}$
15. $\sqrt[4]{9^7}$
16. $(\sqrt{x})^9$
17. $(\sqrt[3]{y})^{14}$
18. $\sqrt{ab^5}$
19. $\sqrt[4]{a^3b}$
20. $\sqrt[3]{x^4y}$
21. $\sqrt[4]{x^9z^5}$
22. $\sqrt[6]{y^{11}z}$
23. $\sqrt[6]{3a + 8b}$
24. $\sqrt[9]{3x + 5z^4}$
25. $\sqrt[5]{\dfrac{2x^6}{11y^7}}$
26. $\sqrt[4]{\dfrac{3a^8}{11b^5}}$

*Write each expression in radical form.*

27. $a^{1/2}$
28. $b^{2/3}$
29. $c^{5/2}$
30. $19^{1/2}$
31. $18^{5/3}$
32. $y^{17/6}$
33. $(24x^3)^{1/2}$
34. $(85a^3)^{5/2}$
35. $(11b^2c)^{3/5}$
36. $(8x^3y^2)^{7/4}$
37. $(6a + 5b)^{1/5}$
38. $(8x^2 + 9y)^{7/3}$
39. $(b^3 - d)^{-1/3}$
40. $(7x^2 - 2y^3)^{-1/6}$

*Simplify each radical expression by changing the expression to exponential form. Write the answer in radical form when appropriate.*

41. $\sqrt{a^6}$
42. $\sqrt[4]{a^8}$
43. $\sqrt[3]{x^9}$
44. $\sqrt[4]{x^{12}}$
45. $\sqrt[6]{y^2}$
46. $\sqrt[8]{b^4}$
47. $\sqrt[6]{y^3}$
48. $\sqrt[12]{z^4}$
49. $(\sqrt{19.3})^2$
50. $\sqrt[4]{(6.83)^4}$
51. $(\sqrt[3]{xy^2})^{15}$
52. $(\sqrt[4]{a^4bc^3})^{40}$
53. $(\sqrt[8]{xyz})^4$
54. $(\sqrt[9]{a^2bc^4})^3$
55. $\sqrt{\sqrt{x}}$
56. $\sqrt{\sqrt[3]{a}}$
57. $\sqrt{\sqrt[4]{y}}$
58. $\sqrt[3]{\sqrt[4]{b}}$
59. $\sqrt[3]{\sqrt[3]{x^2y}}$
60. $\sqrt[4]{\sqrt[3]{7y}}$
61. $\sqrt{\sqrt[5]{a^9}}$
62. $\sqrt[5]{\sqrt[4]{ab}}$

*Evaluate if possible. If the expression is not a real number, so state.*

63. $25^{1/2}$
64. $121^{1/2}$
65. $64^{1/3}$
66. $81^{1/4}$
67. $64^{2/3}$
68. $27^{2/3}$
69. $(-49)^{1/2}$
70. $(-64)^{1/4}$
71. $\left(\dfrac{25}{9}\right)^{1/2}$
72. $\left(\dfrac{100}{49}\right)^{1/2}$
73. $\left(\dfrac{1}{8}\right)^{1/3}$
74. $\left(\dfrac{1}{32}\right)^{1/5}$
75. $-81^{1/2}$
76. $(-81)^{1/2}$
77. $-64^{1/3}$
78. $(-64)^{1/3}$
79. $64^{-1/3}$
80. $49^{-1/2}$
81. $16^{-3/2}$
82. $64^{-2/3}$
83. $\left(\dfrac{64}{27}\right)^{-1/3}$
84. $(-81)^{3/4}$
85. $(-100)^{3/2}$
86. $-\left(\dfrac{25}{49}\right)^{-1/2}$
87. $121^{1/2} + 169^{1/2}$
88. $49^{-1/2} + 36^{-1/2}$
89. $343^{-1/3} + 16^{-1/2}$
90. $16^{-1/2} - 256^{-3/4}$

*Simplify. Write the answer in exponential form without negative exponents.*

**91.** $x^4 \cdot x^{1/2}$ **92.** $x^6 \cdot x^{1/2}$ **93.** $\dfrac{x^{1/2}}{x^{1/3}}$ **94.** $x^{-6/5}$

**95.** $(x^{1/2})^{-2}$ **96.** $(a^{-1/3})^{-1/2}$ **97.** $(9^{-1/3})^0$ **98.** $\dfrac{x^4}{x^{-1/2}}$

**99.** $\dfrac{5y^{-1/3}}{60y^{-2}}$ **100.** $x^{-1/2}x^{-2/5}$ **101.** $4x^{5/3}3x^{-7/2}$ **102.** $(x^{-4/5})^{1/3}$

**103.** $\left(\dfrac{3}{24x}\right)^{1/3}$ **104.** $\left(\dfrac{52}{2x^4}\right)^{1/3}$ **105.** $\left(\dfrac{22x^{3/7}}{2x^{1/2}}\right)^2$ **106.** $\left(\dfrac{x^{-1/3}}{x^{-2}}\right)^2$

**107.** $\left(\dfrac{a^4}{4a^{-2/5}}\right)^{-3}$ **108.** $\left(\dfrac{27z^{1/4}y^3}{3z^{1/4}}\right)^{1/2}$ **109.** $\left(\dfrac{x^{3/4}y^{-3}}{x^{1/2}y^2}\right)^4$ **110.** $\left(\dfrac{250a^{-3/4}b^5}{2a^{-2}b^2}\right)^{2/3}$

*Multiply.*

**111.** $4z^{-1/2}(2z^4 - z^{1/2})$ **112.** $-3a^{-4/9}(5a^{1/9} - a^2)$ **113.** $5x^{-1}(x^{-4} + 4x^{-1/2})$

**114.** $-9z^{3/2}(z^{3/2} - z^{-3/2})$ **115.** $-6x^{5/3}(-2x^{1/2} + 3x^{1/3})$ **116.** $\dfrac{1}{2}x^{-2}(10x^{4/3} - 38x^{-1/2})$

*Use a calculator to evaluate each expression. Give the answer to the nearest hundredth.*

**117.** $\sqrt{180}$ **118.** $\sqrt[3]{168}$ **119.** $\sqrt[5]{402.83}$ **120.** $\sqrt[4]{1096}$

**121.** $93^{2/3}$ **122.** $38.2^{3/2}$ **123.** $1000^{-1/2}$ **124.** $8060^{-3/2}$

## Problem Solving

**125.** Under what conditions will $\sqrt[n]{a^n} = (\sqrt[n]{a})^n = a$?

**126.** By selecting values for $a$ and $b$, show that $(a^2 + b^2)^{1/2}$ *is not equal to* $a + b$.

**127.** By selecting values for $a$ and $b$, show that $(a^{1/2} + b^{1/2})^2$ *is not equal to* $a + b$.

**128.** By selecting values for $a$ and $b$, show that $(a^3 + b^3)^{1/3}$ is not equal to $a + b$.

**129.** By selecting values for $a$ and $b$, show that $(a^{1/3} + b^{1/3})^3$ is not equal to $a + b$.

**130.** Determine whether $\sqrt[3]{\sqrt{x}} = \sqrt{\sqrt[3]{x}}$, $x \geq 0$.

*Factor. Write the answer without negative exponents.*

**131.** $x^{3/2} + x^{1/2}$ **132.** $x^{1/4} - x^{5/4}$ **133.** $y^{1/3} - y^{7/3}$

**134.** $x^{-1/2} + x^{1/2}$ **135.** $y^{-2/5} + y^{8/5}$ **136.** $a^{6/5} + a^{-4/5}$

*In Exercises 137 through 142, use a calculator where appropriate.*

**137. Growing Bacteria** The function, $B(t) = 2^{10} \cdot 2^t$, approximates the number of bacteria in a certain culture after $t$ hours.

**a)** The initial number of bacteria is determined when $t = 0$. What is the initial number of bacteria?

**b)** How many bacteria are there after $\frac{1}{2}$ hour?

**138. Carbon Dating** Carbon dating is used by scientists to find the age of fossils, bones, and other items. The formula used in carbon dating is $P = P_0 2^{-t/5600}$, where $P_0$ represents the original amount of carbon 14 ($C_{14}$) present and $P$ represents the amount of $C_{14}$ present after $t$ years. If 10 milligrams (mg) of $C_{14}$ is present in an animal bone recently excavated, how many milligrams will be present in 5000 years?

**139. Retirement Plans** Each year more and more people contribute to their companies' 401(k) retirement plans. The total assets, $A(t)$, in U.S. 401(k) plans, in billions of dollars, can be approximated by the function $A(t) = 2.69t^{3/2}$, where $t$ is years since 1993 and $1 \leq t \leq 16$. (Therefore, this function holds from 1994 through 2009.) Estimate the total assets in U.S. 401(k) plans in **a)** 2000 and **b)** 2009.

**140. Internet Sales** Retail Internet sales are increasing annually. The total amount, $I(t)$, in billions of dollars, of Internet sales can be approximated by the function $I(t) = 0.25t^{5/3}$, where $t$ is years since 1999 and $1 \leq t \leq 9$. Find the total Internet sales in **a)** 2000 and **b)** 2008.

**141.** Evaluate $(3^{\sqrt{2}})^{\sqrt{2}}$. Explain how you determined your answer.

**142.** **a)** On your calculator, evaluate $3^{\pi}$.

**b)** Explain why your value from part **a)** does or does not make sense.

**143.** Find the domain of $f(x) = (x - 7)^{1/2}(x + 3)^{-1/2}$.

**144.** Find the domain of $f(x) = (x + 4)^{1/2}(x - 3)^{-1/2}$.

**145.** Assume that $x$ can be any real number. Simplify $\sqrt[n]{(x - 6)^{2n}}$ if

**a)** $n$ is even.

**b)** $n$ is odd.

*Determine the index to be placed in the shaded area to make the statement true. Explain how you determined your answer.*

**146.** $\sqrt[4]{\sqrt[\square]{\sqrt{x}}} = x^{1/24}$

**147.** $\sqrt[4]{\sqrt[5]{\sqrt[\square]{\sqrt[3]{z}}}} = z^{1/120}$

**148.** **a)** Write $f(x) = \sqrt{2x + 3}$ in exponential form.

**b)** On your grapher, check that the answer you gave in part **a)** is correct by graphing both $f(x)$ as given and the function you gave in exponential form.

## Cumulative Review Exercises

[3.2] **149.** Determine which of the following relations are also functions.

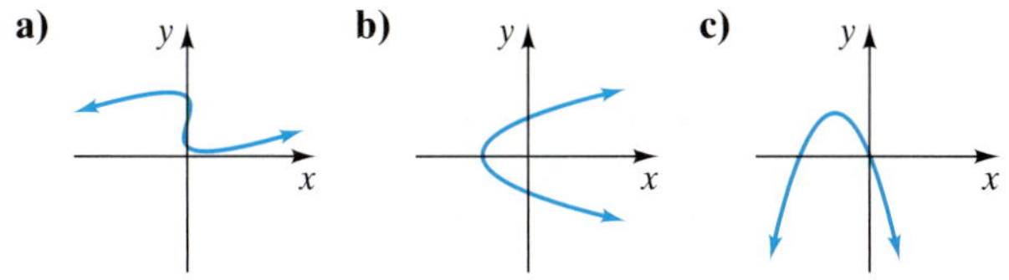


[6.3] **150.** Simplify $\dfrac{a^{-2} + ab^{-1}}{ab^{-2} - a^{-2}b^{-1}}$.

[6.4] **151.** Solve $\dfrac{3x - 2}{x + 4} = \dfrac{2x + 1}{3x - 2}$.

[6.5] **152.** **Flying a Plane** Amy Mayfield can fly her plane 500 miles against the wind in the same time it takes her to fly 560 miles with the wind. If the wind blows at 25 miles per hour, find the speed of the plane in still air.

# 7.3 Simplifying Radicals

1 Understand perfect powers.

2 Simplify radicals using the product rule for radicals.

3 Simplify radicals using the quotient rule for radicals.

## 1 Understand Perfect Powers

In this section, we will simplify radicals using the **product rule for radicals** and the **quotient rule for radicals**. However, before we introduce these rules let's look at **perfect powers**, which will help us with our discussion.

A number or expression is a **perfect square** if it is the square of an expression. Examples of perfect squares are illustrated below.

| | | | | | | |
|---|---|---|---|---|---|---|
| Perfect squares | 1, | 4, | 9, | 16, | 25, | 36, ... |
| | ↓ | ↓ | ↓ | ↓ | ↓ | ↓ |
| Square of a number | $1^2$, | $2^2$, | $3^2$, | $4^2$, | $5^2$, | $6^2$, ... |

Variables with exponents may also be perfect squares, as illustrated below.

| | | | | | |
|---|---|---|---|---|---|
| Perfect squares | $x^2$, | $x^4$, | $x^6$, | $x^8$, | $x^{10}$, ... |
| | ↓ | ↓ | ↓ | ↓ | ↓ |
| Square of an expression | $(x)^2$, | $(x^2)^2$, | $(x^3)^2$, | $(x^4)^2$, | $(x^5)^2$, ... |

Notice that the exponents on the variables in the perfect squares are all multiples of 2.

Just as there are perfect squares, there are perfect cubes. A number or expression is a **perfect cube** if it can be written as the cube of an expression. Examples of perfect cubes are illustrated below.

| | | | | | | |
|---|---|---|---|---|---|---|
| Perfect cubes | 1, | 8, | 27, | 64, | 125, | 216, ... |
| | ↓ | ↓ | ↓ | ↓ | ↓ | ↓ |
| Cube of a number | $1^3$, | $2^3$, | $3^3$, | $4^3$, | $5^3$, | $6^3$, ... |

| | | | | | |
|---|---|---|---|---|---|
| Perfect cubes | $x^3$, | $x^6$, | $x^9$, | $x^{12}$, | $x^{15}$, ... |
| | ↓ | ↓ | ↓ | ↓ | ↓ |
| Cube of an expression | $(x)^3$, | $(x^2)^3$, | $(x^3)^3$, | $(x^4)^3$, | $(x^5)^3$, ... |

Notice that the exponents on the variables in the perfect cubes are all multiples of 3.

We can expand our discussion to perfect powers of a variable for any radicand. In general, the radicand $x^n$ is a perfect power *when n is a multiple of the index* of the radicand (or where $n$ is divisible by the index).

**Example**

Perfect powers of $x^n$ for index $n$ $\quad x^n, \quad x^{2n}, \quad x^{3n}, \quad x^{4n}, \quad x^{5n}, \ldots$

For example, if the index of a radical expression is 5, then $x^5$, $x^{10}$, $x^{15}$, $x^{20}$, and so on, are perfect powers of the index.

### Helpful Hint

A quick way to determine if a radicand $x^n$ is a perfect power for an index is to determine if the exponent $n$ is divisible by the index of the radical. For example, consider $\sqrt[5]{x^{20}}$. Since the exponent, 20, is divisible by the index, 5, $x^{20}$ is a perfect fifth power. Now consider $\sqrt[6]{x^{20}}$. Since the exponent, 20, is not divisible by the index, 6, $x^{20}$ is not a perfect sixth power. However, $x^{18}$ and $x^{24}$ are both perfect sixth powers since 6 divides both 18 and 24.

Notice that the square root of a perfect square simplifies to an expression without a radical sign, the cube root of a perfect cube simplifies to an expression without a radical sign, and so on.

**Examples**

$$\sqrt{36} = \sqrt{6^2} = 6^{2/2} = 6$$
$$\sqrt[3]{27} = \sqrt[3]{3^3} = 3^{3/3} = 3$$
$$\sqrt{x^6} = x^{6/2} = x^3$$
$$\sqrt[3]{z^{12}} = z^{12/3} = z^4$$
$$\sqrt[5]{n^{35}} = n^{35/5} = n^7$$

Now we are ready to discuss the product rule for radicals.

## 2 Simplify Radicals Using the Product Rule for Radicals

To introduce the **product rule for radicals**, observe that $\sqrt{4}\cdot\sqrt{9} = 2\cdot 3 = 6$. Also, $\sqrt{4\cdot 9} = \sqrt{36} = 6$. We see that $\sqrt{4}\cdot\sqrt{9} = \sqrt{4\cdot 9}$. This is one example of the product rule for radicals.

### Product Rule for Radicals

For nonnegative real numbers $a$ and $b$,

$$\sqrt[n]{a}\cdot\sqrt[n]{b} = \sqrt[n]{ab}$$

**Examples of the Product Rule for Radicals**

$$\sqrt{20} = \begin{cases} \sqrt{1}\cdot\sqrt{20} \\ \sqrt{2}\cdot\sqrt{10} \\ \sqrt{4}\cdot\sqrt{5} \end{cases} \qquad \sqrt[3]{20} = \begin{cases} \sqrt[3]{1}\cdot\sqrt[3]{20} \\ \sqrt[3]{2}\cdot\sqrt[3]{10} \\ \sqrt[3]{4}\cdot\sqrt[3]{5} \end{cases}$$

$\sqrt{20}$ can be factored into any of these forms. $\qquad$ $\sqrt[3]{20}$ can be factored into any of these forms.

$$\sqrt{x^7} = \begin{cases} \sqrt{x}\cdot\sqrt{x^6} \\ \sqrt{x^2}\cdot\sqrt{x^5} \\ \sqrt{x^3}\cdot\sqrt{x^4} \end{cases} \qquad \sqrt[3]{x^7} = \begin{cases} \sqrt[3]{x}\cdot\sqrt[3]{x^6} \\ \sqrt[3]{x^2}\cdot\sqrt[3]{x^5} \\ \sqrt[3]{x^3}\cdot\sqrt[3]{x^4} \end{cases}$$

$\sqrt{x^7}$ can be factored into any of these forms. $\qquad$ $\sqrt[3]{x^7}$ can be factored into any of these forms.

Now that we have introduced the product rule for radicals, we will use this rule to simplify radicals. Here is a general procedure that can be used to simplify radicals using the product rule.

### To Simplify Radicals Using the Product Rule

1. If the radicand contains a coefficient other than 1, write it as a product of two numbers, one of which is the largest perfect power for the index.
2. Write each variable factor as a product of two factors, one of which is the largest perfect power of the variable for the index.
3. Use the product rule to write the radical expression as a product of radicals. Place all the perfect powers (numbers and variables) under the same radical.
4. Simplify the radical containing the perfect powers.

If we are simplifying a *square* root, we will write the radicand as the product of the largest *perfect square* and another number. If we are simplifying a *cube* root, we will write the radicand as the product of the largest *perfect cube* and another number, and so on.

**EXAMPLE 1** ▸ Simplify. **a)** $\sqrt{32}$ **b)** $\sqrt{60}$ **c)** $\sqrt[3]{54}$ **d)** $\sqrt[4]{96}$

*Solution* The radicands in this example contain no variables. We will follow step 1 of the procedure.

**a)** Since we are evaluating a square root, we look for the largest perfect square that divides 32. The largest perfect square that divides, or is a factor of, 32 is 16.

$$\sqrt{32} = \sqrt{16 \cdot 2} = \sqrt{16}\sqrt{2} = 4\sqrt{2}$$

**b)** The largest perfect square that is a factor of 60 is 4.

$$\sqrt{60} = \sqrt{4 \cdot 15} = \sqrt{4}\sqrt{15} = 2\sqrt{15}$$

**c)** The largest perfect cube that is a factor of 54 is 27.

$$\sqrt[3]{54} = \sqrt[3]{27 \cdot 2} = \sqrt[3]{27}\sqrt[3]{2} = 3\sqrt[3]{2}$$

**d)** The largest perfect fourth power that is a factor of 96 is 16.

$$\sqrt[4]{96} = \sqrt[4]{16 \cdot 6} = \sqrt[4]{16}\sqrt[4]{6} = 2\sqrt[4]{6}$$

▸ Now Try Exercise 19

### Helpful Hint

In Example 1 **a)**, if you first thought that 4 was the largest perfect square that divided 32, you could proceed as follows:

$$\sqrt{32} = \sqrt{4 \cdot 8} = \sqrt{4}\sqrt{8} = 2\sqrt{8}$$
$$= 2\sqrt{4 \cdot 2} = 2\sqrt{4}\sqrt{2} = 2 \cdot 2\sqrt{2} = 4\sqrt{2}$$

Note that the final result is the same, but you must perform more steps. The lists of perfect squares and perfect cubes on page 465 can help you determine the largest perfect square or perfect cube that is a factor of a radicand.

In Example 1 **b)**, $\sqrt{15}$ can be factored as $\sqrt{5 \cdot 3}$; however, since neither 5 nor 3 is a perfect square, $\sqrt{15}$ cannot be simplified.

When the radicand is a perfect power for the index, the radical can be simplified by writing it in exponential form, as in Example 2.

**EXAMPLE 2** ▸ Simplify. **a)** $\sqrt{x^4}$ **b)** $\sqrt[3]{x^{12}}$ **c)** $\sqrt[5]{z^{40}}$

*Solution*

**a)** $\sqrt{x^4} = x^{4/2} = x^2$ **b)** $\sqrt[3]{x^{12}} = x^{12/3} = x^4$ **c)** $\sqrt[5]{z^{40}} = z^{40/5} = z^8$

▸ Now Try Exercise 33

**EXAMPLE 3** ▸ Simplify. **a)** $\sqrt{x^9}$ **b)** $\sqrt[5]{x^{23}}$ **c)** $\sqrt[4]{y^{33}}$

*Solution* Because the radicands have coefficients of 1, we start with step 2 of the procedure.

**a)** The largest perfect square less than or equal to $x^9$ is $x^8$.

$$\sqrt{x^9} = \sqrt{x^8 \cdot x} = \sqrt{x^8} \cdot \sqrt{x} = x^{8/2}\sqrt{x} = x^4\sqrt{x}$$

**b)** The largest perfect fifth power less than or equal to $x^{23}$ is $x^{20}$.

$$\sqrt[5]{x^{23}} = \sqrt[5]{x^{20} \cdot x^3} = \sqrt[5]{x^{20}}\sqrt[5]{x^3} = x^{20/5}\sqrt[5]{x^3} = x^4\sqrt[5]{x^3}$$

**c)** The largest perfect fourth power less than or equal to $y^{33}$ is $y^{32}$.

$$\sqrt[4]{y^{33}} = \sqrt[4]{y^{32} \cdot y} = \sqrt[4]{y^{32}}\sqrt[4]{y} = y^{32/4}\sqrt[4]{y} = y^8\sqrt[4]{y}$$

▸ **Now Try Exercise 39**

If you observe the answers to Example 3, you will see that the exponent on the variable in the radicand is always less than the index. **When a radical is simplified, the radicand does not have a variable with an exponent greater than or equal to the index.**

In Example 3 **b**), we simplified $\sqrt[5]{x^{23}}$. If we divide 23, the exponent in the radicand, by 5, the index, we obtain

$$\begin{array}{r} 4 \\ 5\overline{)23} \\ \underline{20} \\ 3 \end{array} \quad \begin{array}{l} \longleftarrow \text{Quotient} \\ \\ \\ \longleftarrow \text{Remainder} \end{array}$$

Notice that $\sqrt[5]{x^{23}}$ simplifies to $x^4\sqrt[5]{x^3}$ and

$$\text{Quotient} \longrightarrow x^4\sqrt[5]{x^3} \longleftarrow \text{Remainder}$$

When simplifying a radical, if you divide the exponent within the radical by the index, the quotient will be the exponent on the variable outside the radical sign and the remainder will be the exponent on the variable within the radical sign. Simplify Example 3 **c**) using this technique now.

**EXAMPLE 4** ▸ Simplify. **a)** $\sqrt{x^{12}y^{17}}$ **b)** $\sqrt[4]{x^6y^{23}}$

*Solution*

**a)** $x^{12}$ is a perfect square. The largest perfect square that is a factor of $y^{17}$ is $y^{16}$. Write $y^{17}$ as $y^{16} \cdot y$.

$$\begin{aligned} \sqrt{x^{12}y^{17}} &= \sqrt{x^{12} \cdot y^{16} \cdot y} = \sqrt{x^{12}y^{16}}\sqrt{y} \\ &= \sqrt{x^{12}}\sqrt{y^{16}}\sqrt{y} \\ &= x^{12/2}y^{16/2}\sqrt{y} \\ &= x^6y^8\sqrt{y} \end{aligned}$$

**b)** We begin by finding the largest perfect fourth power factors of $x^6$ and $y^{23}$. For an index of 4, the largest perfect power that is a factor of $x^6$ is $x^4$. The largest perfect power that is a factor of $y^{23}$ is $y^{20}$.

$$\begin{aligned} \sqrt[4]{x^6y^{23}} &= \sqrt[4]{x^4 \cdot x^2 \cdot y^{20} \cdot y^3} \\ &= \sqrt[4]{x^4y^{20} \cdot x^2y^3} \\ &= \sqrt[4]{x^4y^{20}}\sqrt[4]{x^2y^3} \\ &= xy^5\sqrt[4]{x^2y^3} \end{aligned}$$

▸ **Now Try Exercise 51**

Often the steps where we change the radical expression to exponential form are done mentally, and those steps are not illustrated. For instance, in Example 4 **b**), we changed $\sqrt[4]{x^4y^{20}}$ to $xy^5$ mentally and did not show the intermediate steps.

**EXAMPLE 5** ▸ Simplify. **a)** $\sqrt{80x^5y^{12}z^3}$ **b)** $\sqrt[3]{54x^{17}y^{25}}$

### Solution

**a)** The largest perfect square that is a factor of 80 is 16. The largest perfect square that is a factor of $x^5$ is $x^4$. The expression $y^{12}$ is a perfect square. The largest perfect square that is a factor of $z^3$ is $z^2$. Place all the perfect squares under the same radical, and then simplify.

$$\begin{aligned}\sqrt{80x^5y^{12}z^3} &= \sqrt{16\cdot 5\cdot x^4\cdot x\cdot y^{12}\cdot z^2\cdot z}\\ &= \sqrt{16x^4y^{12}z^2\cdot 5xz}\\ &= \sqrt{16x^4y^{12}z^2}\cdot\sqrt{5xz}\\ &= 4x^2y^6z\sqrt{5xz}\end{aligned}$$

**b)** The largest perfect cube that is a factor of 54 is 27. The largest perfect cube that is a factor of $x^{17}$ is $x^{15}$. The largest perfect cube that is a factor of $y^{25}$ is $y^{24}$.

$$\begin{aligned}\sqrt[3]{54x^{17}y^{25}} &= \sqrt[3]{27\cdot 2\cdot x^{15}\cdot x^2\cdot y^{24}\cdot y}\\ &= \sqrt[3]{27x^{15}y^{24}\cdot 2x^2y}\\ &= \sqrt[3]{27x^{15}y^{24}}\cdot\sqrt[3]{2x^2y}\\ &= 3x^5y^8\sqrt[3]{2x^2y}\end{aligned}$$

▸ **Now Try Exercise 57**

## Helpful Hint

In Example 4 **b)**, we showed that

$$\sqrt[4]{x^6y^{23}} = xy^5\sqrt[4]{x^2y^3}$$

As mentioned on page 468, this radical can also be simplified by dividing the exponents on the variables in the radicand, 6 and 23, by the index, 4, and observing the quotients and remainders.

| Quotient | Quotient | Remainder | Remainder |
|---|---|---|---|
| $6 \div 4$ | $23 \div 4$ | $6 \div 4$ | $23 \div 4$ |

$$\sqrt[4]{x^6y^{23}} = x^1y^5\sqrt[4]{x^2y^3}$$

Can you explain why this procedure works? You may wish to use this procedure to work or check certain problems.

Now let's introduce the quotient rule for radicals.

## 3 Simplify Radicals Using the Quotient Rule for Radicals

In mathematics we sometimes need to simplify a quotient of two radicals. To do so we use the **quotient rule for radicals**.

**Quotient Rule for Radicals**

For nonnegative real numbers $a$ and $b$,

$$\frac{\sqrt[n]{a}}{\sqrt[n]{b}} = \sqrt[n]{\frac{a}{b}}, \quad b \neq 0$$

**Examples of the Quotient Rule for Radicals**

$$\frac{\sqrt{18}}{\sqrt{3}} = \sqrt{\frac{18}{3}} \qquad \sqrt{\frac{9}{25}} = \frac{\sqrt{9}}{\sqrt{25}}$$

$$\frac{\sqrt{x^3}}{\sqrt{x}} = \sqrt{\frac{x^3}{x}} \qquad \sqrt{\frac{x^4}{y^2}} = \frac{\sqrt{x^4}}{\sqrt{y^2}}$$

$$\frac{\sqrt[3]{y^5}}{\sqrt[3]{y^2}} = \sqrt[3]{\frac{y^5}{y^2}} \qquad \sqrt[3]{\frac{z^9}{27}} = \frac{\sqrt[3]{z^9}}{\sqrt[3]{27}}$$

Examples 6 and 7 illustrate how to use the quotient rule to simplify radical expressions.

**EXAMPLE 6** ▸ Simplify. **a)** $\frac{\sqrt{75}}{\sqrt{3}}$ **b)** $\frac{\sqrt[3]{24x}}{\sqrt[3]{3x}}$ **c)** $\frac{\sqrt[3]{x^4y^7}}{\sqrt[3]{xy^{-5}}}$

*Solution* In each part we use the quotient rule to write the quotient of radicals as a single radical. Then we simplify.

**a)** $\frac{\sqrt{75}}{\sqrt{3}} = \sqrt{\frac{75}{3}} = \sqrt{25} = 5$

**b)** $\frac{\sqrt[3]{24x}}{\sqrt[3]{3x}} = \sqrt[3]{\frac{24x}{3x}} = \sqrt[3]{8} = 2$

**c)** $\frac{\sqrt[3]{x^4y^7}}{\sqrt[3]{xy^{-5}}} = \sqrt[3]{\frac{x^4y^7}{xy^{-5}}}$ *Quotient rule for radicals*

$= \sqrt[3]{x^3y^{12}}$ *Simplify the radicand.*

$= xy^4$

▸ **Now Try Exercise 93**

In Section 7.1, when we introduced radicals we indicated that $\sqrt{\frac{4}{9}} = \frac{2}{3}$ since $\frac{2}{3} \cdot \frac{2}{3} = \frac{4}{9}$. The quotient rule may be helpful in evaluating square roots containing fractions as illustrated in Example 7 **a)**.

**EXAMPLE 7** ▸ Simplify. **a)** $\sqrt{\frac{121}{25}}$ **b)** $\sqrt[3]{\frac{8x^4y}{27xy^{10}}}$ **c)** $\sqrt[4]{\frac{18xy^5}{3x^9y}}$

*Solution* In each part we first simplify the radicand, if possible. Then we use the quotient rule to write the given radical as a quotient of radicals.

**a)** $\sqrt{\frac{121}{25}} = \frac{\sqrt{121}}{\sqrt{25}} = \frac{11}{5}$

**b)** $\sqrt[3]{\frac{8x^4y}{27xy^{10}}} = \sqrt[3]{\frac{8x^3}{27y^9}} = \frac{\sqrt[3]{8x^3}}{\sqrt[3]{27y^9}} = \frac{2x}{3y^3}$

**c)** $\sqrt[4]{\frac{18xy^5}{3x^9y}} = \sqrt[4]{\frac{6y^4}{x^8}} = \frac{\sqrt[4]{6y^4}}{\sqrt[4]{x^8}} = \frac{\sqrt[4]{y^4}\sqrt[4]{6}}{x^2} = \frac{y\sqrt[4]{6}}{x^2}$

▸ **Now Try Exercise 97**

### Avoiding Common Errors

The following simplifications are correct because the numbers and variables divided out are not within square roots.

CORRECT

$$\frac{\overset{2}{\cancel{6}}\sqrt{2}}{\underset{1}{\cancel{3}}} = 2\sqrt{2}$$

CORRECT

$$\frac{\cancel{x}\sqrt{2}}{\cancel{x}} = \sqrt{2}$$

An expression within a square root cannot be divided by an expression not within the square root.

CORRECT

$$\frac{\sqrt{2}}{2} \quad \textit{Cannot be simplified further}$$

$$\frac{\sqrt{x^3}}{x} = \frac{\sqrt{x^2}\sqrt{x}}{x} = \frac{\cancel{x}\sqrt{x}}{\cancel{x}} = \sqrt{x}$$

INCORRECT

$$\frac{\sqrt{\cancel{2}^{1}}}{\underset{1}{\cancel{2}}} = \sqrt{1} = 1$$

$$\frac{\sqrt{x^{\cancel{3}^{2}}}}{\cancel{x}} = \sqrt{x^2} = x$$

## EXERCISE SET 7.3

MathXL® MyMathLab

### Concept/Writing Exercises

1. a) How do you obtain the numbers that are perfect squares?
   b) List the first six perfect squares.
2. a) How do you obtain the numbers that are perfect cubes?
   b) List the first six perfect cube numbers.
3. a) How do you obtain numbers that are perfect fifth powers?
   b) List the first five perfect fifth-power numbers.
4. State the product rule for radicals.
5. When we gave the product rule, we stated that for nonnegative real numbers $a$ and $b$, $\sqrt[n]{a}\cdot\sqrt[n]{b} = \sqrt[n]{ab}$. Why is it necessary to specify that both $a$ and $b$ are nonnegative real numbers?
6. State the quotient rule for radicals.
7. When we gave the quotient rule, we stated that for nonnegative real numbers $a$ and $b$, $\frac{\sqrt[n]{a}}{\sqrt[n]{b}} = \sqrt[n]{\frac{a}{b}}$, $b \neq 0$. Why is it necessary to state that both $a$ and $b$ are nonnegative real numbers?
8. In the quotient rule discussed in Exercise 7, why can the denominator never equal 0?

### Practice the Skills

*In this exercise set, assume that all variables represent positive real numbers.*

*Simplify.*

9. $\sqrt{8}$
10. $\sqrt{28}$
11. $\sqrt{24}$
12. $\sqrt{18}$
13. $\sqrt{32}$
14. $\sqrt{12}$
15. $\sqrt{50}$
16. $\sqrt{72}$
17. $\sqrt{75}$
18. $\sqrt{300}$
19. $\sqrt{40}$
20. $\sqrt{600}$
21. $\sqrt[3]{16}$
22. $\sqrt[3]{24}$
23. $\sqrt[3]{54}$
24. $\sqrt[3]{81}$
25. $\sqrt[3]{32}$
26. $\sqrt[3]{108}$
27. $\sqrt[3]{40}$
28. $\sqrt[4]{80}$
29. $\sqrt[4]{48}$
30. $\sqrt[4]{162}$
31. $-\sqrt[5]{64}$
32. $-\sqrt[5]{243}$
33. $\sqrt[3]{b^9}$
34. $6\sqrt{y^{12}}$
35. $\sqrt[3]{x^6}$
36. $\sqrt[5]{y^{20}}$
37. $\sqrt{x^3}$
38. $-\sqrt{x^5}$
39. $\sqrt{a^{11}}$
40. $\sqrt[3]{b^{13}}$
41. $8\sqrt[3]{z^{32}}$
42. $\sqrt[3]{a^7}$
43. $\sqrt[4]{b^{23}}$
44. $\sqrt[5]{z^7}$
45. $\sqrt[6]{x^9}$
46. $\sqrt[7]{y^{15}}$
47. $3\sqrt[5]{y^{23}}$
48. $\sqrt{24x^3}$
49. $2\sqrt{50y^9}$
50. $\sqrt{75a^7b^{11}}$
51. $\sqrt[3]{x^3y^7}$
52. $\sqrt{x^5y^9}$
53. $\sqrt[5]{a^6b^{23}}$
54. $-\sqrt{20x^6y^7z^{12}}$
55. $\sqrt{24x^{15}y^{20}z^{27}}$
56. $\sqrt[3]{16x^3y^6}$
57. $\sqrt[3]{81a^6b^8}$
58. $\sqrt[3]{128a^{10}b^{11}c^{12}}$
59. $\sqrt[4]{32x^8y^9z^{19}}$
60. $\sqrt[4]{48x^{11}y^{21}}$
61. $\sqrt[4]{81a^8b^9}$
62. $-\sqrt[4]{32x^{18}y^{31}}$
63. $\sqrt[5]{32a^{10}b^{12}}$
64. $\sqrt[6]{64x^{12}y^{23}z^{50}}$

*Simplify.*

65. $\sqrt{\frac{75}{3}}$
66. $\sqrt{\frac{36}{4}}$
67. $\sqrt{\frac{81}{100}}$
68. $\sqrt{\frac{8}{50}}$
69. $\frac{\sqrt{27}}{\sqrt{3}}$
70. $\frac{\sqrt{72}}{\sqrt{2}}$
71. $\frac{\sqrt{3}}{\sqrt{48}}$
72. $\frac{\sqrt{15}}{\sqrt{60}}$
73. $\sqrt[3]{\frac{3}{24}}$
74. $\sqrt[3]{\frac{2}{54}}$
75. $\frac{\sqrt[3]{3}}{\sqrt[3]{81}}$
76. $\frac{\sqrt[3]{32}}{\sqrt[3]{4}}$
77. $\sqrt[4]{\frac{3}{48}}$
78. $\frac{\sqrt[4]{243}}{\sqrt[4]{3}}$
79. $\sqrt[5]{\frac{96}{3}}$
80. $\frac{\sqrt[5]{2}}{\sqrt[5]{64}}$
81. $\sqrt{\frac{r^4}{4}}$
82. $\sqrt{\frac{100a^8}{49b^6}}$
83. $\sqrt{\frac{16x^4}{25y^{10}}}$
84. $\sqrt{\frac{49a^8b^{10}}{121c^{14}}}$
85. $\sqrt[3]{\frac{c^6}{64}}$
86. $\sqrt[3]{\frac{27x^6}{y^{12}}}$
87. $\sqrt[3]{\frac{a^8b^{12}}{b^{-8}}}$
88. $\sqrt[4]{\frac{16x^{16}y^{32}}{81x^{-4}}}$
89. $\frac{\sqrt{24}}{\sqrt{3}}$
90. $\frac{\sqrt{64x^5}}{\sqrt{2x^3}}$
91. $\frac{\sqrt{27x^6}}{\sqrt{3x^2}}$
92. $\frac{\sqrt{72x^3y^5}}{\sqrt{8x^3y^7}}$
93. $\frac{\sqrt{48x^6y^9}}{\sqrt{6x^2y^6}}$
94. $\frac{\sqrt{300a^{10}b^{11}}}{\sqrt{2ab^4}}$
95. $\sqrt[3]{\frac{5xy}{8x^{13}}}$
96. $\sqrt[3]{\frac{64a^5b^{12}}{27a^{14}b^5}}$
97. $\sqrt[3]{\frac{25x^2y^9}{5x^8y^2}}$
98. $\sqrt[3]{\frac{54xy^4z^{17}}{18x^{13}z^4}}$
99. $\sqrt[4]{\frac{10x^4y}{81x^{-8}}}$
100. $\sqrt[4]{\frac{3a^6b^5}{16a^{-6}b^{13}}}$

## Problem Solving

101. Prove $\sqrt{a \cdot b} = \sqrt{a}\sqrt{b}$ by converting $\sqrt{a \cdot b}$ to exponential form.

102. Will the product of two radicals always be a radical? Give an example to support your answer.

103. Will the quotient of two radicals always be a radical? Give an example to support your answer.

104. Prove $\sqrt[n]{\frac{a}{b}} = \frac{\sqrt[n]{a}}{\sqrt[n]{b}}$ by converting $\sqrt[n]{\frac{a}{b}}$ to exponential form.

105. a) Will $\frac{\sqrt[n]{x}}{\sqrt[n]{x}}$ always equal 1?

b) If your answer to part **a)** was no, under what conditions does $\frac{\sqrt[n]{x}}{\sqrt[n]{x}}$ equal 1?

## Cumulative Review Exercises

[2.2] 106. Solve the formula $F = \frac{9}{5}C + 32$ for $C$.

[2.6] 107. Solve for $x$: $\left|\frac{2x-4}{5}\right| = 12$

[5.3] 108. Divide $\frac{15x^{12} - 5x^9 + 20x^6}{5x^6}$.

[5.6] 109. Factor $(x-3)^3 + 8$.

# 7.4 Adding, Subtracting, and Multiplying Radicals

1 Add and subtract radicals.

2 Multiply radicals.

## 1 Add and Subtract Radicals

**Like radicals** are radicals having the same radicand and index. **Unlike radicals** are radicals differing in either the radicand or the index.

| Examples of Like Radicals | Examples of Unlike Radicals | |
|---|---|---|
| $\sqrt{5}, 3\sqrt{5}$ | $\sqrt{5}, \sqrt[3]{5}$ | *Indices differ.* |
| $6\sqrt{7}, -2\sqrt{7}$ | $\sqrt{6}, \sqrt{7}$ | *Radicands differ.* |
| $\sqrt{x}, 5\sqrt{x}$ | $\sqrt{x}, \sqrt{2x}$ | *Radicands differ.* |
| $\sqrt[3]{2x}, -4\sqrt[3]{2x}$ | $\sqrt{x}, \sqrt[3]{x}$ | *Indices differ.* |
| $\sqrt[4]{x^2y^5}, -\sqrt[4]{x^2y^5}$ | $\sqrt[3]{xy}, \sqrt[3]{x^2y}$ | *Radicands differ.* |

Like radicals are added and subtracted in exactly the same way that like terms are added or subtracted. To add or subtract like radicals, add or subtract their numerical coefficients and multiply this sum or difference by the like radical.

**Examples of Adding and Subtracting Like Radicals**

$$3\sqrt{6} + 2\sqrt{6} = (3 + 2)\sqrt{6} = 5\sqrt{6}$$
$$5\sqrt{x} - 7\sqrt{x} = (5 - 7)\sqrt{x} = -2\sqrt{x}$$
$$\sqrt[3]{4x^2} + 5\sqrt[3]{4x^2} = (1 + 5)\sqrt[3]{4x^2} = 6\sqrt[3]{4x^2}$$
$$4\sqrt{5x} - y\sqrt{5x} = (4 - y)\sqrt{5x}$$

**EXAMPLE 1** ▸ Simplify.

**a)** $6 + 4\sqrt{2} - \sqrt{2} + 7$ **b)** $2\sqrt[3]{x} + 8x + 4\sqrt[3]{x} - 3$

*Solution*

**a)**
$$\begin{aligned} 6 + 4\sqrt{2} - \sqrt{2} + 7 &= 6 + 7 + 4\sqrt{2} - \sqrt{2} \\ &= 13 + (4 - 1)\sqrt{2} \\ &= 13 + 3\sqrt{2} \quad (\text{or } 3\sqrt{2} + 13) \end{aligned}$$

*Place like terms together.*

**b)** $2\sqrt[3]{x} + 8x + 4\sqrt[3]{x} - 3 = 6\sqrt[3]{x} + 8x - 3$

▸ Now Try Exercise 15

It is sometimes possible to convert unlike radicals into like radicals by simplifying one or more of the radicals, as was discussed in Section 7.3.

**EXAMPLE 2** ▸ Simplify $\sqrt{3} + \sqrt{27}$.

*Solution* Since $\sqrt{3}$ and $\sqrt{27}$ are unlike radicals, they cannot be added in their present form. We can simplify $\sqrt{27}$ to obtain like radicals.

$$\begin{aligned} \sqrt{3} + \sqrt{27} &= \sqrt{3} + \sqrt{9}\sqrt{3} \\ &= \sqrt{3} + 3\sqrt{3} = 4\sqrt{3} \end{aligned}$$

▸ Now Try Exercise 19

**To Add or Subtract Radicals**

1. Simplify each radical expression.
2. Combine like radicals (if there are any).

**EXAMPLE 3** ▸ Simplify.

**a)** $5\sqrt{24} + \sqrt{54}$ **b)** $2\sqrt{45} - \sqrt{80} + \sqrt{20}$ **c)** $\sqrt[3]{27} + \sqrt[3]{81} - 7\sqrt[3]{3}$

*Solution*

**a)**
$$\begin{aligned} 5\sqrt{24} + \sqrt{54} &= 5 \cdot \sqrt{4} \cdot \sqrt{6} + \sqrt{9} \cdot \sqrt{6} \\ &= 5 \cdot 2\sqrt{6} + 3\sqrt{6} \\ &= 10\sqrt{6} + 3\sqrt{6} = 13\sqrt{6} \end{aligned}$$

**b)**
$$\begin{aligned} 2\sqrt{45} - \sqrt{80} + \sqrt{20} &= 2 \cdot \sqrt{9} \cdot \sqrt{5} - \sqrt{16} \cdot \sqrt{5} + \sqrt{4} \cdot \sqrt{5} \\ &= 2 \cdot 3\sqrt{5} - 4\sqrt{5} + 2\sqrt{5} \\ &= 6\sqrt{5} - 4\sqrt{5} + 2\sqrt{5} = 4\sqrt{5} \end{aligned}$$

**c)**
$$\begin{aligned} \sqrt[3]{27} + \sqrt[3]{81} - 7\sqrt[3]{3} &= 3 + \sqrt[3]{27} \cdot \sqrt[3]{3} - 7\sqrt[3]{3} \\ &= 3 + 3\sqrt[3]{3} - 7\sqrt[3]{3} = 3 - 4\sqrt[3]{3} \end{aligned}$$

▸ Now Try Exercise 23

**EXAMPLE 4** ▸ Simplify. **a)** $\sqrt{x^2} - \sqrt{x^2y} + x\sqrt{y}$ **b)** $\sqrt[3]{x^{13}y^2} - \sqrt[3]{x^4y^8}$

### Solution

**a)** $\sqrt{x^2} - \sqrt{x^2y} + x\sqrt{y} = x - \sqrt{x^2}\cdot\sqrt{y} + x\sqrt{y}$

$$= x - x\sqrt{y} + x\sqrt{y}$$
$$= x$$

**b)** $\sqrt[3]{x^{13}y^2} - \sqrt[3]{x^4y^8} = \sqrt[3]{x^{12}}\cdot\sqrt[3]{xy^2} - \sqrt[3]{x^3y^6}\cdot\sqrt[3]{xy^2}$

$$= x^4\sqrt[3]{xy^2} - xy^2\sqrt[3]{xy^2}$$

Now factor out the common factor, $\sqrt[3]{xy^2}$.

$$= (x^4 - xy^2)\sqrt[3]{xy^2}$$

▸ **Now Try Exercise 35**

### Helpful Hint

The product rule and quotient rule for radicals presented in Section 7.3 are

$$\sqrt[n]{a}\cdot\sqrt[n]{b} = \sqrt[n]{ab} \qquad \frac{\sqrt[n]{a}}{\sqrt[n]{b}} = \sqrt[n]{\frac{a}{b}}$$

Students often incorrectly assume similar properties exist for addition and subtraction. They do not. To illustrate this, let $n$ be a square root (index 2), $a = 9$, and $b = 16$.

$$\sqrt[n]{a} + \sqrt[n]{b} \neq \sqrt[n]{a+b}$$
$$\sqrt{9} + \sqrt{16} \neq \sqrt{9+16}$$
$$3 + 4 \neq \sqrt{25}$$
$$7 \neq 5$$

We will now discuss multiplication of radicals.

## 2 Multiply Radicals

To multiply radicals, we use the product rule given earlier. After multiplying, we can often simplify the new radical (see Examples 5 and 6).

**EXAMPLE 5** ▸ Multiply and simplify.

**a)** $\sqrt{6x^3}\sqrt{8x^6}$ **b)** $\sqrt[3]{2x}\sqrt[3]{4x^2}$ **c)** $\sqrt[4]{4x^{11}y}\sqrt[4]{16x^6y^{22}}$

### Solution

**a)** $\sqrt{6x^3}\sqrt{8x^6} = \sqrt{6x^3\cdot 8x^6}$ *Product rule for radicals*

$= \sqrt{48x^9}$

$= \sqrt{16x^8}\sqrt{3x}$ *$16x^8$ is a perfect square.*

$= 4x^4\sqrt{3x}$

**b)** $\sqrt[3]{2x}\sqrt[3]{4x^2} = \sqrt[3]{2x\cdot 4x^2}$ *Product rule for radicals*

$= \sqrt[3]{8x^3}$ *$8x^3$ is a perfect cube.*

$= 2x$

**c)** $\sqrt[4]{4x^{11}y}\sqrt[4]{16x^6y^{22}} = \sqrt[4]{4x^{11}y\cdot 16x^6y^{22}}$ *Product rule for radicals*

$= \sqrt[4]{64x^{17}y^{23}}$

$= \sqrt[4]{16x^{16}y^{20}}\sqrt[4]{4xy^3}$ *The largest perfect fourth root factors are 16, $x^{16}$, and $y^{20}$.*

$= 2x^4y^5\sqrt[4]{4xy^3}$

▸ **Now Try Exercise 47**

*Remember, as stated earlier, when a radical is simplified, the radicand does not have any variable with an exponent greater than or equal to the index.*

**EXAMPLE 6** ▸ Multiply and simplify $\sqrt{2x}(\sqrt{8x} - \sqrt{50})$.

*Solution* Begin by using the distributive property.

$$\begin{aligned}\sqrt{2x}(\sqrt{8x} - \sqrt{50}) &= (\sqrt{2x})(\sqrt{8x}) + (\sqrt{2x})(-\sqrt{50})\\ &= \sqrt{16x^2} - \sqrt{100x}\\ &= 4x - \sqrt{100}\sqrt{x}\\ &= 4x - 10\sqrt{x}\end{aligned}$$

▸ **Now Try Exercise 53**

Note in Example 6 that the same result could be obtained by first simplifying $\sqrt{8x}$ and $\sqrt{50}$ and then multiplying. You may wish to try this now.

Now we will multiply two binomial factors. To multiply binomial factors, each term in one factor must be multiplied by each term in the other factor. This can be accomplished using the FOIL method that was discussed earlier.

**EXAMPLE 7** ▸ Multiply $(\sqrt{x} - \sqrt{y})(\sqrt{x} - y)$.

*Solution* We will multiply using the FOIL method.

$$\begin{aligned}&\overset{\text{F}}{(\sqrt{x})(\sqrt{x})} + \overset{\text{O}}{(\sqrt{x})(-y)} + \overset{\text{I}}{(-\sqrt{y})(\sqrt{x})} + \overset{\text{L}}{(-\sqrt{y})(-y)}\\ &= \sqrt{x^2} - y\sqrt{x} - \sqrt{xy} + y\sqrt{y}\\ &= x - y\sqrt{x} - \sqrt{xy} + y\sqrt{y}\end{aligned}$$

▸ **Now Try Exercise 63**

**EXAMPLE 8** ▸ Simplify. **a)** $(2\sqrt{6} - \sqrt{3})^2$ **b)** $(\sqrt[3]{x} - \sqrt[3]{2y^2})(\sqrt[3]{x^2} - \sqrt[3]{8y})$

*Solution*

**a)** $(2\sqrt{6} - \sqrt{3})^2 = (2\sqrt{6} - \sqrt{3})(2\sqrt{6} - \sqrt{3})$

Now multiply the factors using the FOIL method.

$$\begin{aligned}&\overset{\text{F}}{(2\sqrt{6})(2\sqrt{6})} + \overset{\text{O}}{(2\sqrt{6})(-\sqrt{3})} + \overset{\text{I}}{(-\sqrt{3})(2\sqrt{6})} + \overset{\text{L}}{(-\sqrt{3})(-\sqrt{3})}\\ &= 4(6) - 2\sqrt{18} - 2\sqrt{18} + 3\\ &= 24 - 2\sqrt{18} - 2\sqrt{18} + 3\\ &= 27 - 4\sqrt{18}\\ &= 27 - 4\sqrt{9}\sqrt{2}\\ &= 27 - 12\sqrt{2}\end{aligned}$$

**b)** Multiply the factors using the FOIL method.

$$\begin{aligned}(\sqrt[3]{x} - \sqrt[3]{2y^2})(\sqrt[3]{x^2} - \sqrt[3]{8y}) &= \overset{\text{F}}{(\sqrt[3]{x})(\sqrt[3]{x^2})} + \overset{\text{O}}{(\sqrt[3]{x})(-\sqrt[3]{8y})} + \overset{\text{I}}{(-\sqrt[3]{2y^2})(\sqrt[3]{x^2})} + \overset{\text{L}}{(-\sqrt[3]{2y^2})(-\sqrt[3]{8y})}\\ &= \sqrt[3]{x^3} - \sqrt[3]{8xy} - \sqrt[3]{2x^2y^2} + \sqrt[3]{16y^3}\\ &= \sqrt[3]{x^3} - \sqrt[3]{8}\sqrt[3]{xy} - \sqrt[3]{2x^2y^2} + \sqrt[3]{8y^3}\sqrt[3]{2}\\ &= x - 2\sqrt[3]{xy} - \sqrt[3]{2x^2y^2} + 2y\sqrt[3]{2}\end{aligned}$$

▸ **Now Try Exercise 99**

**EXAMPLE 9** ▸ Multiply $(3 + \sqrt{6})(3 - \sqrt{6})$.

*Solution* We can multiply using the FOIL method.

$$\begin{aligned}(3 + \sqrt{6})(3 - \sqrt{6}) &= \overset{F}{3(3)} + \overset{O}{3(-\sqrt{6})} + \overset{I}{(\sqrt{6})(3)} + \overset{L}{(\sqrt{6})(-\sqrt{6})} \\ &= 9 - 3\sqrt{6} + 3\sqrt{6} - \sqrt{36} \\ &= 9 - \sqrt{36} \\ &= 9 - 6 = 3\end{aligned}$$

▸ **Now Try Exercise 59**

Note that in Example 9, we multiplied *the sum and difference of the same two terms.* Recall from Section 5.6 that $(a + b)(a - b) = a^2 - b^2$. If we let $a = 3$ and $b = \sqrt{6}$, then we can multiply as follows.

$$\begin{aligned}(a + b)(a - b) &= a^2 - b^2 \\ (3 + \sqrt{6})(3 - \sqrt{6}) &= 3^2 - (\sqrt{6})^2 \\ &= 9 - 6 \\ &= 3\end{aligned}$$

When multiplying the sum and difference of the same two terms, you may obtain the answer using the difference of the squares of the two terms. We will see additional multiplications of this type in Section 7.5.

**EXAMPLE 10** ▸ If $f(x) = \sqrt[3]{x^2}$ and $g(x) = \sqrt[3]{x^4} + \sqrt[3]{x^2}$, find **a)** $(f \cdot g)(x)$ and **b)** $(f \cdot g)(6)$.

*Solution*

**a)** From Section 3.6, we know that $(f \cdot g)(x) = f(x) \cdot g(x)$.

$$\begin{aligned}(f \cdot g)(x) &= f(x) \cdot g(x) \\ &= \sqrt[3]{x^2}\left(\sqrt[3]{x^4} + \sqrt[3]{x^2}\right) && \text{Substitute given values.} \\ &= \sqrt[3]{x^2}\,\sqrt[3]{x^4} + \sqrt[3]{x^2}\,\sqrt[3]{x^2} && \text{Distributive property} \\ &= \sqrt[3]{x^6} + \sqrt[3]{x^4} && \text{Product rule for radicals} \\ &= x^2 + x\sqrt[3]{x} && \text{Simplify radicals.}\end{aligned}$$

**b)** To compute $(f \cdot g)(6)$, substitute 6 for $x$ in the answer obtained in part **a)**.

$$\begin{aligned}(f \cdot g)(x) &= x^2 + x\sqrt[3]{x} \\ (f \cdot g)(6) &= 6^2 + 6\sqrt[3]{6} && \text{Substitute 6 for } x. \\ &= 36 + 6\sqrt[3]{6}\end{aligned}$$

▸ **Now Try Exercise 77**

**EXAMPLE 11** ▸ Simplify $f(x)$ if **a)** $f(x) = \sqrt{x + 3}\,\sqrt{x + 3}$, $x \ge -3$ and **b)** $f(x) = \sqrt{3x^2 - 30x + 75}$; assume that the variable may be any real number.

*Solution*

$$\begin{aligned}\textbf{a)}\ f(x) &= \sqrt{x + 3}\,\sqrt{x + 3} \\ &= \sqrt{(x + 3)(x + 3)} && \text{Product rule for radicals} \\ &= \sqrt{(x + 3)^2} \\ &= x + 3\end{aligned}$$

Since we are told that $x \ge -3$, we can use the product rule. Note that the radicand will be a nonnegative number for any $x \ge -3$ and we can write the answer as $x + 3$ rather than $|x + 3|$.

**b)** $f(x) = \sqrt{3x^2 - 30x + 75}$

$= \sqrt{3(x^2 - 10x + 25)}$ *Factor out 3.*

$= \sqrt{3(x-5)^2}$ *Write as the square of a binomial.*

$= \sqrt{3}\sqrt{(x-5)^2}$ *Product rule for radicals*

$= \sqrt{3}|x - 5|$

Since the variable could be any real number, we write our answer with absolute value signs. If we had been told that $x - 5$ was nonnegative, then we could have written our answer as $\sqrt{3}(x-5)$.

**▸ Now Try Exercise 105**

## EXERCISE SET 7.4

MathXL® MyMathLab

### Concept/Writing Exercises

1. What are like radicals?
2. a) Explain how to add like radicals.
   b) Using the procedure in part **a)**, add $\frac{3}{5}\sqrt{5} + \frac{5}{4}\sqrt{5}$.
3. Use a calculator to estimate $\sqrt{3} + 3\sqrt{2}$.
4. Use a calculator to estimate $2\sqrt{3} + \sqrt{5}$.
5. Does $\sqrt{a} + \sqrt{b} = \sqrt{a+b}$? Explain your answer and give an example supporting your answer.
6. Since $64 + 36 = 100$, does $\sqrt{64} + \sqrt{36} = \sqrt{100}$? Explain your answer.

### Practice the Skills

*In this exercise set, assume that all variables represent positive real numbers.*

*Simplify.*

7. $\sqrt{3} - \sqrt{3}$
8. $2\sqrt{6} - \sqrt{6}$
9. $6\sqrt{5} - 2\sqrt{5}$
10. $3\sqrt{2} + 7\sqrt{2} - 11$
11. $2\sqrt{3} - 2\sqrt{3} - 4\sqrt{3} + 5$
12. $6\sqrt[3]{7} - 8\sqrt[3]{7}$
13. $2\sqrt[4]{y} - 9\sqrt[4]{y}$
14. $3\sqrt[5]{a} + 7 + 5\sqrt[5]{a} - 2$
15. $3\sqrt{5} - \sqrt[3]{x} + 6\sqrt{5} + 3\sqrt[3]{x}$
16. $9 + 4\sqrt[4]{a} - 7\sqrt[4]{a} + 5$
17. $5\sqrt{x} - 8\sqrt{y} + 3\sqrt{x} + 2\sqrt{y} - \sqrt{x}$
18. $8\sqrt{a} + 4\sqrt[3]{b} + 7\sqrt{a} - 12\sqrt[3]{b}$

*Simplify.*

19. $\sqrt{5} + \sqrt{20}$
20. $\sqrt{75} + \sqrt{108}$
21. $-6\sqrt{75} + 5\sqrt{125}$
22. $3\sqrt{250} + 4\sqrt{160}$
23. $-4\sqrt{90} + 3\sqrt{40} + 2\sqrt{10}$
24. $3\sqrt{40x^2y} + 2x\sqrt{490y}$
25. $\sqrt{500xy^2} + y\sqrt{320x}$
26. $5\sqrt{8} + 2\sqrt{50} - 3\sqrt{72}$
27. $2\sqrt{5x} - 3\sqrt{20x} - 4\sqrt{45x}$
28. $3\sqrt{27c^2} - 2\sqrt{108c^2} - \sqrt{48c^2}$
29. $3\sqrt{50a^2} - 3\sqrt{72a^2} - 8a\sqrt{18}$
30. $4\sqrt[3]{5} - 5\sqrt[3]{40}$
31. $\sqrt[3]{108} + \sqrt[3]{32}$
32. $3\sqrt[3]{16} + \sqrt[3]{54}$
33. $\sqrt[3]{27} - 5\sqrt[3]{8}$
34. $3\sqrt{45x^3} + \sqrt{5x}$
35. $2\sqrt[3]{a^4b^2} + 4a\sqrt[3]{ab^2}$
36. $5y\sqrt[4]{48x^5} - x\sqrt[4]{3x^5y^4}$
37. $\sqrt{4r^7s^5} + 3r^2\sqrt{r^3s^5} - 2rs\sqrt{r^5s^3}$
38. $x\sqrt[3]{27x^5y^2} - x^2\sqrt[3]{x^2y^2} + 4\sqrt[3]{x^8y^2}$
39. $\sqrt[3]{128x^8y^{10}} - 2x^2y\sqrt[3]{16x^2y^7}$
40. $5\sqrt[3]{320x^5y^8} + 3x\sqrt[3]{135x^2y^8}$

*Simplify.*

41. $\sqrt{3}\sqrt{27}$
42. $\sqrt[3]{2}\sqrt[3]{4}$
43. $\sqrt[3]{4}\sqrt[3]{14}$
44. $\sqrt[3]{3}\sqrt[3]{54}$
45. $\sqrt{9m^3n^7}\sqrt{3mn^4}$
46. $\sqrt[3]{5ab^2}\sqrt[3]{25a^4b^{12}}$
47. $\sqrt[3]{9x^7y^{10}}\sqrt[3]{6x^4y^3}$
48. $\sqrt[4]{3x^9y^{12}}\sqrt[4]{54x^4y^7}$
49. $\sqrt[5]{x^{24}y^{30}z^9}\sqrt[5]{x^{13}y^8z^7}$
50. $\sqrt[4]{8x^4yz^3}\sqrt[4]{2x^2y^3z^7}$
51. $\left(\sqrt[3]{2x^3y^4}\right)^2$
52. $\sqrt{2}(\sqrt{6} + \sqrt{18})$
53. $\sqrt{5}(\sqrt{5} - \sqrt{3})$
54. $\sqrt{3}(\sqrt{12} + \sqrt{8})$
55. $\sqrt[3]{y}\left(2\sqrt[3]{y} - \sqrt[3]{y^8}\right)$
56. $\sqrt{3y}\left(\sqrt{27y^2} - \sqrt{y}\right)$
57. $2\sqrt[3]{x^4y^5}\left(\sqrt[3]{8x^{12}y^4} + \sqrt[3]{16xy^9}\right)$
58. $\sqrt[5]{16x^7y^6}\left(\sqrt[5]{2x^6y^9} - \sqrt[5]{10x^3y^7}\right)$
59. $(8 + \sqrt{5})(8 - \sqrt{5})$
60. $(9 - \sqrt{5})(9 + \sqrt{5})$
61. $(\sqrt{6} + x)(\sqrt{6} - x)$
62. $(\sqrt{x} + y)(\sqrt{x} - y)$
63. $(\sqrt{7} - \sqrt{z})(\sqrt{7} + \sqrt{z})$
64. $(3\sqrt{a} - 5\sqrt{b})(3\sqrt{a} + 5\sqrt{b})$

**65.** $(\sqrt{3}+4)(\sqrt{3}+5)$
**66.** $(1+\sqrt{5})(8+\sqrt{5})$
**67.** $(3-\sqrt{2})(4-\sqrt{8})$
**68.** $(5\sqrt{6}+3)(4\sqrt{6}-1)$
**69.** $(4\sqrt{3}+\sqrt{2})(\sqrt{3}-\sqrt{2})$
**70.** $(\sqrt{3}+7)^2$
**71.** $(2\sqrt{5}-3)^2$
**72.** $(\sqrt{y}+\sqrt{6z})(\sqrt{2z}-\sqrt{8y})$
**73.** $(2\sqrt{3x}-\sqrt{y})(3\sqrt{3x}+\sqrt{y})$
**74.** $(\sqrt[3]{9}+\sqrt[3]{2})(\sqrt[3]{3}+\sqrt[3]{4})$
**75.** $(\sqrt[3]{4}-\sqrt[3]{6})(\sqrt[3]{2}-\sqrt[3]{36})$
**76.** $(\sqrt[3]{4x}-\sqrt[3]{2y})(\sqrt[3]{4x}+\sqrt[3]{10})$

*In Exercises 77–82, $f(x)$ and $g(x)$ are given. Find $(f \cdot g)(x)$.*

**77.** $f(x)=\sqrt{2x}, g(x)=\sqrt{8x}-\sqrt{32}$
**78.** $f(x)=\sqrt{6x}, g(x)=\sqrt{6x}-\sqrt{10x}$
**79.** $f(x)=\sqrt[3]{x}, g(x)=\sqrt[3]{x^5}+\sqrt[3]{x^4}$
**80.** $f(x)=\sqrt[3]{2x^2}, g(x)=\sqrt[3]{4x}+\sqrt[3]{32x^2}$
**81.** $f(x)=\sqrt[4]{3x^2}, g(x)=\sqrt[4]{9x^4}-\sqrt[4]{x^7}$
**82.** $f(x)=\sqrt[4]{2x^3}, g(x)=\sqrt[4]{8x^5}-\sqrt[4]{5x^6}$

*Simplify. These exercises are a combination of the types of exercises presented earlier in this exercise set.*

**83.** $\sqrt{24}$
**84.** $\sqrt{300}$
**85.** $\sqrt{125}-\sqrt{20}$
**86.** $4\sqrt{7}+2\sqrt{63}-2\sqrt{28}$
**87.** $(3\sqrt{2}-4)(\sqrt{2}+5)$
**88.** $(\sqrt{5}+\sqrt{2})(\sqrt{2}+\sqrt{20})$
**89.** $\sqrt{6}(5-\sqrt{2})$
**90.** $3\sqrt[3]{81}+4\sqrt[3]{24}$
**91.** $\sqrt{150}\sqrt{3}$
**92.** $\sqrt[4]{2}\sqrt[4]{40}$
**93.** $\sqrt[3]{80x^{11}}$
**94.** $\sqrt[3]{x^9y^{11}z}$
**95.** $\sqrt[6]{128ab^{17}c^9}$
**96.** $\sqrt[5]{14x^4y^2}\sqrt[5]{3x^4y^3}$
**97.** $2b\sqrt[4]{a^4b}+ab\sqrt[4]{16b}$
**98.** $2\sqrt[3]{24a^3y^4}+4a\sqrt[3]{81y^4}$
**99.** $\left(\sqrt[3]{x^2}-\sqrt[3]{y}\right)\left(\sqrt[3]{x}-2\sqrt[3]{y^2}\right)$
**100.** $(\sqrt[3]{a}+5)\left(\sqrt[3]{a^2}-6\right)$
**101.** $\sqrt[3]{3ab^2}\left(\sqrt[3]{4a^4b^3}-\sqrt[3]{8a^5b^4}\right)$
**102.** $\sqrt[4]{4st^2}\left(\sqrt[4]{2s^5t^6}+\sqrt[4]{5s^9t^2}\right)$

*Simplify the following. In Exercises 105 and 106, assume the variable can be any real number. See Example 11.*

**103.** $f(x)=\sqrt{2x-5}\,\sqrt{2x-5}, x \ge -\dfrac{5}{2}$
**104.** $g(a)=\sqrt{3a+7}\,\sqrt{3a+7}, a \ge -\dfrac{7}{3}$
**105.** $h(r)=\sqrt{4r^2-32r+64}$
**106.** $f(b)=\sqrt{20b^2+60b+45}$

## Problem Solving

*Find the perimeter and area of the following figures. Write the perimeter and area in radical form with the radicals simplified.*

**107.**

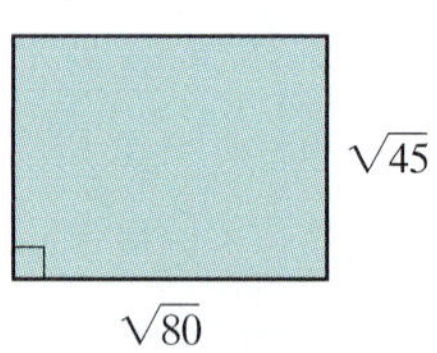


**108.**

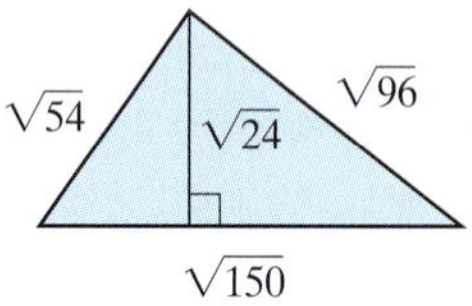


**109.**

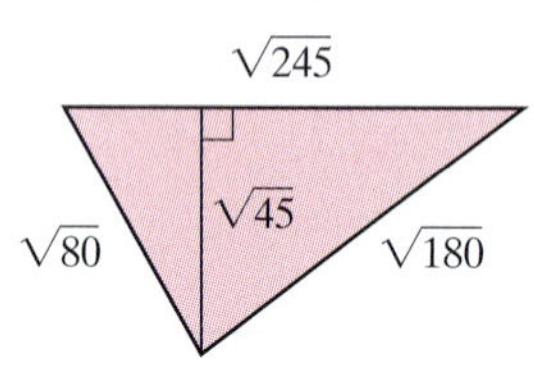


**110.**

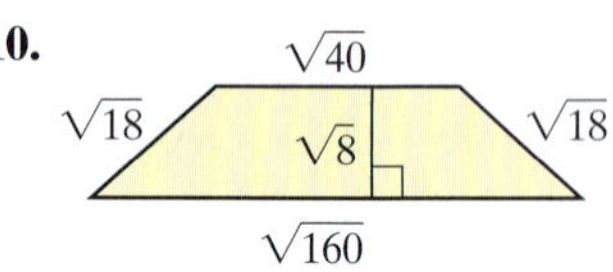


**111.** Will the sum of two radicals always be a radical? Give an example to support your answer.

**112.** Will the difference of two radicals always be a radical? Give an example to support your answer.

**113. Skid Marks** Law enforcement officials sometimes use the formula $s=\sqrt{30FB}$ to determine a car's speed, $s$, in miles per hour, from a car's skid marks. The $F$ in the formula represents the "road factor," which is determined by the road's surface, and the $B$ represents the braking distance, in feet. Officer Jenkins is investigating an accident. Find the car's speed if the skid marks are 80 feet long and **a)** the road was dry asphalt, whose road factor is 0.85, **b)** the road was wet gravel, whose road factor is 0.52.

**114. Water through a Fire Hose** The rate at which water flows through a particular fire hose, $R$, in gallons per minute, can be approximated by the formula $R=28d^2\sqrt{P}$, where $d$ is the diameter of the nozzle, in inches, and $P$ is the nozzle pressure, in pounds per square inch. If a nozzle has a diameter of 2.5 inches and the nozzle pressure is 80 pounds per square inch, find the flow rate.

**115. Height of Girls** The function $f(t) = 3\sqrt{t} + 19$ can be used to approximate the median height, $f(t)$, in inches, for U.S. girls of age $t$, in months, where $1 \le t \le 60$. Estimate the median height of girls at age **a)** 36 months and **b)** 40 months.

**116. Standard Deviation** In statistics, the standard deviation of the population, $\sigma$, read "sigma," is a measure of the spread of a set of data about the mean of the data. The greater the spread, the greater the standard deviation. One formula used to determine sigma is $\sigma = \sqrt{npq}$, where $n$ represents the sample size, $p$ represents the percent chance (or probability) that something specific happens, and $q$ represents the percent chance (or probability) that the specific thing does not happen. In a sample of 600 people who purchase airline tickets, the percent that showed up for their flight, $p$, was 0.93, and the percent that did not show up for their flight, $q$, was 0.07. Use this information to find $\sigma$.

**117.** The graph of $f(x) = \sqrt{x}$ is shown.

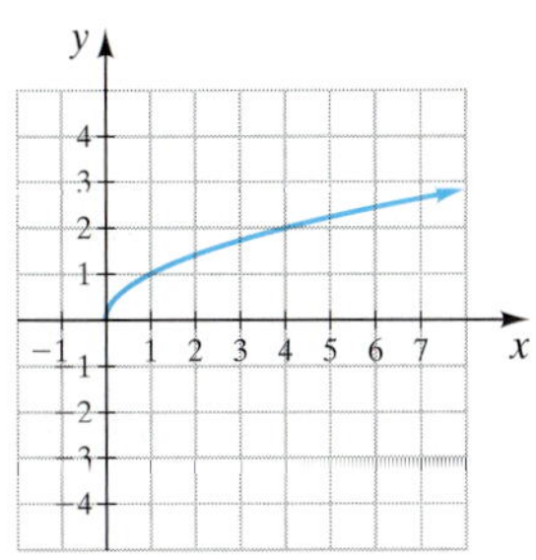


a) If $g(x) = 2$, sketch the graph of $(f + g)(x)$.

b) What effect does adding 2 have to the graph of $f(x)$?

**118.** The graph of $f(x) = -\sqrt{x}$ is shown.

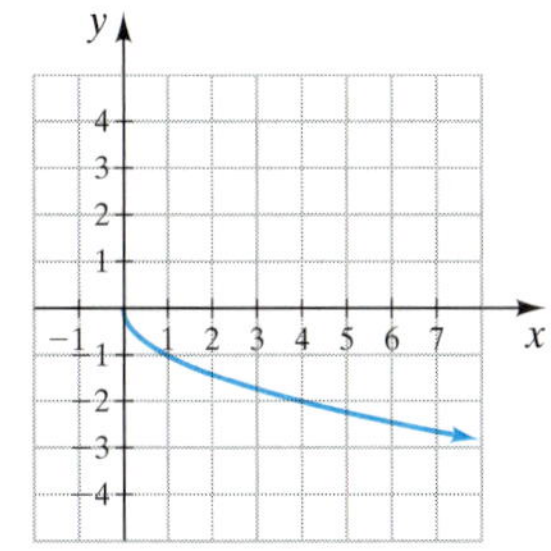


a) If $g(x) = 3$, sketch the graph of $(f + g)(x)$.

b) What effect does adding 3 have to the graph of $f(x)$?

**119.** You are given that $f(x) = \sqrt{x}$ and $g(x) = \sqrt{x} - 2$.

a) Sketch the graph of $(f - g)(x)$. Explain how you determined your answer.

b) What is the domain of $(f - g)(x)$?

**120.** You are given that $f(x) = \sqrt{x}$ and $g(x) = -\sqrt{x} - 3$.

a) Sketch the graph of $(f + g)(x)$. Explain how you determined your answer.

b) What is the domain of $(f + g)(x)$?

**121.** Graph the function $f(x) = \sqrt{x^2}$.

**122.** Graph the function $f(x) = \sqrt{x^2} - 4$.

## Cumulative Review Exercises

[1.2] **123.** What is a rational number?

[1.3] **124.** What is a real number?

**125.** What is an irrational number?

**126.** What is the definition of $|a|$?

[2.2] **127.** Solve the formula $E = \frac{1}{2}mv^2$ for $m$.

[2.5] **128.** Solve the inequality $-4 < 2x - 3 \le 7$ and indicate the solution **a)** on the number line; **b)** in internal notation; **c)** in set builder notation.

## Mid-Chapter Test: 7.1–7.4

*To find out how well you understand the chapter material to this point, take this brief test. The answers, and the section where the material was initially discussed, are given in the back of the book. Review any questions that you answered incorrectly.*

*Find the indicated root.*

**1.** $\sqrt{121}$

**2.** $\sqrt[3]{-\frac{27}{64}}$

*Use absolute value to evaluate.*

**3.** $\sqrt{(-16.3)^2}$

**4.** $\sqrt{(3a^2 - 4b^3)^2}$

**5.** Find $g(16)$ if $g(x) = \frac{x}{8} + \sqrt{4x} - 7$.

**6.** Write $\sqrt[5]{7a^4b^3}$ in exponential form.

**7.** Evaluate $-49^{1/2} + 81^{3/4}$.

*Simplify each expression.*

**8.** $\left(\sqrt[4]{a^2b^3c}\right)^{20}$

**9.** $7x^{-5/2} \cdot 2x^{3/2}$

**10.** Multiply $8x^{-2}(x^3 + 2x^{-1/2})$.

*Simplify each radical.*

**11.** $\sqrt{32x^4y^9}$

**12.** $\sqrt[6]{64a^{13}b^{23}c^{15}}$

**13.** $\dfrac{\sqrt[3]{3}}{\sqrt[3]{81}}$

**14.** $\dfrac{\sqrt{20x^5y^{12}}}{\sqrt{180x^{15}y^7}}$

*Simplify.*

**15.** $2\sqrt{x} - 3\sqrt{y} + 9\sqrt{x} + 15\sqrt{y}$

**16.** $2\sqrt{90x^2y} + 3x\sqrt{490y}$

**17.** $(x + \sqrt{5})(2x - 3\sqrt{5})$

**18.** $2\sqrt{3a}\left(\sqrt{27a^2} - 5\sqrt{4a}\right)$

**19.** $3b\sqrt[4]{a^5b} + 2ab\sqrt[4]{16ab}$

**20.** When simplifying the following square roots, in which parts will the answer contain an absolute value? Explain your answer and simplify parts **a)** and **b)**.

**a)** $\sqrt{(x-3)^2}$

**b)** $\sqrt{64x^2}, x \geq 0$

# 7.5 Dividing Radicals

1 Rationalize denominators.

2 Rationalize a denominator using the conjugate.

3 Understand when a radical is simplified.

4 Use rationalizing the denominator in an addition problem.

5 Divide radical expressions with different indices.

## 1 Rationalize Denominators

We introduced the quotient rule for radicals in Section 7.3. Now we will use the quotient rule to work other division problems and to rationalize denominators.

When the denominator of a fraction contains a radical, we generally simplify the expression by **rationalizing the denominator**. To rationalize a denominator is to remove all radicals from the denominator. When adding radicals, it may be necessary to rationalize denominators, as will be illustrated in Example 6.

**To Rationalize a Denominator**

Multiply both the numerator and the denominator of the fraction by a radical that will result in the radicand in the denominator becoming a perfect power.

When both the numerator and denominator are multiplied by the same radical expression, you are in effect multiplying the fraction by 1, which does not change its value.

**EXAMPLE 1** ▸ Simplify. **a)** $\dfrac{1}{\sqrt{5}}$ **b)** $\dfrac{x}{4\sqrt{3}}$ **c)** $\dfrac{11}{\sqrt{2x}}$ **d)** $\dfrac{\sqrt[3]{16a^4}}{\sqrt[3]{b}}$

*Solution* To simplify each expression, we must rationalize the denominators. We do so by multiplying both the numerator and denominator by a radical that will result in the denominator becoming a perfect power for the given index.

**a)** $$\frac{1}{\sqrt{5}} = \frac{1}{\sqrt{5}} \cdot \frac{\sqrt{5}}{\sqrt{5}} = \frac{\sqrt{5}}{\sqrt{25}} = \frac{\sqrt{5}}{5}$$

**b)** $$\frac{x}{4\sqrt{3}} = \frac{x}{4\sqrt{3}} \cdot \frac{\sqrt{3}}{\sqrt{3}} = \frac{x\sqrt{3}}{4 \cdot 3} = \frac{x\sqrt{3}}{12}$$

**c)** There are two factors in the radicand, 2 and $x$. We must make each factor a perfect square. Since $2^2$ or 4 is a perfect square and $x^2$ is a perfect square, we multiply both numerator and denominator by $\sqrt{2x}$.

$$\frac{11}{\sqrt{2x}} = \frac{11}{\sqrt{2x}} \cdot \frac{\sqrt{2x}}{\sqrt{2x}} = \frac{11\sqrt{2x}}{\sqrt{4x^2}} = \frac{11\sqrt{2x}}{2x}$$

**d)** There are no common factors in the numerator and denominator. Before we rationalize the denominator, let's simplify the numerator.

$$\frac{\sqrt[3]{16a^4}}{\sqrt[3]{b}} = \frac{\sqrt[3]{8a^3}\,\sqrt[3]{2a}}{\sqrt[3]{b}} \qquad \textit{Product rule for radicals}$$

$$= \frac{2a\sqrt[3]{2a}}{\sqrt[3]{b}} \qquad \textit{Simplify the numerator.}$$

Now we rationalize the denominator. Since the denominator is a cube root, we need to make the radicand a perfect cube. Since the denominator contains $b$ and we want $b^3$, we need two more factors of $b$, or $b^2$. We therefore multiply both numerator and denominator by $\sqrt[3]{b^2}$.

$$= \frac{2a\sqrt[3]{2a}}{\sqrt[3]{b}} \cdot \frac{\sqrt[3]{b^2}}{\sqrt[3]{b^2}}$$

$$= \frac{2a\sqrt[3]{2ab^2}}{\sqrt[3]{b^3}}$$

$$= \frac{2a\sqrt[3]{2ab^2}}{b}$$

▸ **Now Try Exercise 15**

---

**EXAMPLE 2** ▸ Simplify. **a)** $\sqrt{\frac{5}{7}}$ **b)** $\sqrt[3]{\frac{x}{2y^2}}$ **c)** $\sqrt[4]{\frac{32x^9y^6}{3z^2}}$

*Solution* In each part, we will use the quotient rule to write the radical as a quotient of two radicals.

**a)** $\sqrt{\frac{5}{7}} = \frac{\sqrt{5}}{\sqrt{7}} \cdot \frac{\sqrt{7}}{\sqrt{7}} = \frac{\sqrt{35}}{\sqrt{49}} = \frac{\sqrt{35}}{7}$

**b)** $\sqrt[3]{\frac{x}{2y^2}} = \frac{\sqrt[3]{x}}{\sqrt[3]{2y^2}}$

The denominator is $\sqrt[3]{2y^2}$ and we want to change it to $\sqrt[3]{2^3y^3}$. We now multiply both the numerator and denominator by the cube root of an expression that will make the radicand in the denominator $\sqrt[3]{2^3y^3}$. Since $2 \cdot 2^2 = 2^3$ and $y^2 \cdot y = y^3$, we multiply both numerator and denominator by $\sqrt[3]{2^2y}$.

$$\frac{\sqrt[3]{x}}{\sqrt[3]{2y^2}} = \frac{\sqrt[3]{x}}{\sqrt[3]{2y^2}} \cdot \frac{\sqrt[3]{2^2y}}{\sqrt[3]{2^2y}}$$

$$= \frac{\sqrt[3]{x}\,\sqrt[3]{4y}}{\sqrt[3]{2^3y^3}}$$

$$= \frac{\sqrt[3]{4xy}}{2y}$$

**c)** After using the quotient rule, we simplify the numerator.

$$\sqrt[4]{\frac{32x^9y^6}{3z^2}} = \frac{\sqrt[4]{32x^9y^6}}{\sqrt[4]{3z^2}} \qquad \textit{Quotient rule for radicals}$$

$$= \frac{\sqrt[4]{16x^8y^4}\,\sqrt[4]{2xy^2}}{\sqrt[4]{3z^2}} \qquad \textit{Product rule for radicals}$$

$$= \frac{2x^2y\sqrt[4]{2xy^2}}{\sqrt[4]{3z^2}} \qquad \textit{Simplify the numerator.}$$

Now we rationalize the denominator. To make the radicand in the denominator a perfect fourth power, we need to get each factor to a power of 4. Since the denominator contains one factor of 3, we need 3 more factors of 3, or $3^3$. Since there are two factors of $z$, we need 2 more factors of $z$, or $z^2$. Thus we will multiply both numerator and denominator by $\sqrt[4]{3^3z^2}$.

$$= \frac{2x^2y\sqrt[4]{2xy^2}}{\sqrt[4]{3z^2}} \cdot \frac{\sqrt[4]{3^3z^2}}{\sqrt[4]{3^3z^2}}$$

$$= \frac{2x^2y\sqrt[4]{2xy^2}\sqrt[4]{27z^2}}{\sqrt[4]{3z^2}\sqrt[4]{3^3z^2}}$$

$$= \frac{2x^2y\sqrt[4]{54xy^2z^2}}{\sqrt[4]{3^4z^4}} \qquad \text{Product rule for radicals}$$

$$= \frac{2x^2y\sqrt[4]{54xy^2z^2}}{3z}$$

*Note:* There are no perfect fourth power factors of 54, and each exponent in the radicand is less than the index.

▶ **Now Try Exercise 53**

## 2 Rationalize a Denominator Using the Conjugate

When the denominator of a rational expression is a binomial that contains a radical, we rationalize the denominator. We do this by multiplying both the numerator and the denominator of the fraction by the **conjugate** of the denominator. The conjugate of a binomial is a binomial having the same two terms with the sign of the second term changed.

| Expression | Conjugate |
|---|---|
| $9 + \sqrt{2}$ | $9 - \sqrt{2}$ |
| $8\sqrt{3} - \sqrt{5}$ | $8\sqrt{3} + \sqrt{5}$ |
| $\sqrt{x} + \sqrt{y}$ | $\sqrt{x} - \sqrt{y}$ |
| $6a - \sqrt{b}$ | $6a + \sqrt{b}$ |

When a binomial is multiplied by its conjugate, the outer and inner products will sum to 0. We multiplied radicals containing binomial factors in Section 7.4. We will work one more example of multiplication of radical expressions in Example 3.

**EXAMPLE 3** ▶ Multiply $(6 + \sqrt{3})(6 - \sqrt{3})$.

*Solution* Multiply using the FOIL method.

$$\begin{aligned}
(6 + \sqrt{3})(6 - \sqrt{3}) &= \overset{\text{F}}{6(6)} + \overset{\text{O}}{6(-\sqrt{3})} + \overset{\text{I}}{6(\sqrt{3})} + \overset{\text{L}}{\sqrt{3}(-\sqrt{3})} \\
&= 36 - 6\sqrt{3} + 6\sqrt{3} - \sqrt{9} \\
&= 36 - \sqrt{9} \\
&= 36 - 3 \\
&= 33
\end{aligned}$$

▶ **Now Try Exercise 57**

In Example 3, we would get the same result using the formula for the product of the sum and difference of the same two terms. The product results in the difference of two squares, $(a + b)(a - b) = a^2 - b^2$. In Example 3, if we let $a = 6$ and $b = \sqrt{3}$, then using the formula we get the following.

$$\begin{aligned}(a + b)(a - b) &= a^2 - b^2 \\ \downarrow \quad \downarrow \quad \downarrow \quad \downarrow \quad &\quad \downarrow \quad \downarrow \\ (6 + \sqrt{3})(6 - \sqrt{3}) &= 6^2 - (\sqrt{3})^2 \\ &= 36 - 3 \\ &= 33\end{aligned}$$

Now let's work an example where we rationalize a denominator containing two terms.

**EXAMPLE 4** ▸ Simplify. **a)** $\dfrac{13}{4 + \sqrt{3}}$ **b)** $\dfrac{6}{\sqrt{5} - \sqrt{2}}$ **c)** $\dfrac{a - \sqrt{b}}{a + \sqrt{b}}$

*Solution* In each part, we rationalize the denominator by multiplying the numerator and the denominator by the conjugate of the denominator.

**a)**
$$\begin{aligned}\frac{13}{4 + \sqrt{3}} &= \frac{13}{4 + \sqrt{3}} \cdot \frac{4 - \sqrt{3}}{4 - \sqrt{3}} \\ &= \frac{13(4 - \sqrt{3})}{(4 + \sqrt{3})(4 - \sqrt{3})} \\ &= \frac{13(4 - \sqrt{3})}{16 - 3} \\ &= \frac{\overset{1}{\cancel{13}}(4 - \sqrt{3})}{\underset{1}{\cancel{13}}} \text{ or } 4 - \sqrt{3}\end{aligned}$$

**b)**
$$\begin{aligned}\frac{6}{\sqrt{5} - \sqrt{2}} &= \frac{6}{\sqrt{5} - \sqrt{2}} \cdot \frac{\sqrt{5} + \sqrt{2}}{\sqrt{5} + \sqrt{2}} \\ &= \frac{6(\sqrt{5} + \sqrt{2})}{5 - 2} \\ &= \frac{\overset{2}{\cancel{6}}(\sqrt{5} + \sqrt{2})}{\underset{1}{\cancel{3}}} \\ &= 2(\sqrt{5} + \sqrt{2}) \quad \text{or} \quad 2\sqrt{5} + 2\sqrt{2}\end{aligned}$$

**c)**
$$\begin{aligned}\frac{a - \sqrt{b}}{a + \sqrt{b}} &= \frac{a - \sqrt{b}}{a + \sqrt{b}} \cdot \frac{a - \sqrt{b}}{a - \sqrt{b}} \\ &= \frac{a^2 - a\sqrt{b} - a\sqrt{b} + \sqrt{b^2}}{a^2 - b} \\ &= \frac{a^2 - 2a\sqrt{b} + b}{a^2 - b}\end{aligned}$$

Remember that you cannot divide out $a^2$ or $b$ because they are terms, not factors.

▸ **Now Try Exercise 75**

Now that we have illustrated how to rationalize denominators, let's discuss the criteria a radical must meet to be considered simplified.

## 3 Understand when a Radical Is Simplified

After you have simplified a radical expression, you should check it to make sure that it is simplified as far as possible.

**A Radical Expression Is Simplified When the Following Are All True**

1. No perfect powers are factors of the radicand and all exponents in the radicand are less than the index.
2. No radicand contains a fraction.
3. No denominator contains a radical.

**EXAMPLE 5** ▸ Determine whether the following expressions are simplified. If not, explain why. Simplify the expressions if not simplified.

**a)** $\sqrt{48x^5}$ **b)** $\sqrt{\frac{1}{2}}$ **c)** $\frac{1}{\sqrt{6}}$

*Solution*

**a)** This expression is not simplified because 16 is a perfect square factor of 48 and because $x^4$ is a perfect square factor of $x^5$. Notice that the exponent on the variable in the radicand, 5, is greater than the index, 2. Whenever the exponent on the variable in the radicand is greater than the index, the radicand has a perfect power factor of the variable, and the radical needs to be simplified further. Let's simplify the radical.

$$\sqrt{48x^5} = \sqrt{16x^4 \cdot 3x} = \sqrt{16x^4} \cdot \sqrt{3x} = 4x^2\sqrt{3x}$$

**b)** This expression is not simplified since the radicand contains the fraction $\frac{1}{2}$. This violates item 2. We simplify by first using the quotient rule, and then we rationalize the denominator, as follows.

$$\sqrt{\frac{1}{2}} = \frac{\sqrt{1}}{\sqrt{2}} \cdot \frac{\sqrt{2}}{\sqrt{2}} = \frac{\sqrt{2}}{2}$$

**c)** This expression is not simplified since the denominator, $\sqrt{6}$, contains a radical. This violates item 3. We simplify by rationalizing the denominator, as follows.

$$\frac{1}{\sqrt{6}} = \frac{1}{\sqrt{6}} \cdot \frac{\sqrt{6}}{\sqrt{6}} = \frac{\sqrt{6}}{6}$$

▸ **Now Try Exercise 7**

## 4 Use Rationalizing the Denominator in an Addition Problem

Now let's work an addition problem that requires rationalizing the denominator. This example makes use of the methods discussed in Sections 7.3 and 7.4 to add and subtract radicals.

**EXAMPLE 6** ▸ Simplify $4\sqrt{2} - \frac{3}{\sqrt{8}} + \sqrt{32}$.

*Solution* Begin by rationalizing the denominator and by simplifying $\sqrt{32}$.

$$4\sqrt{2} - \frac{3}{\sqrt{8}} + \sqrt{32} = 4\sqrt{2} - \frac{3}{\sqrt{8}} \cdot \frac{\sqrt{2}}{\sqrt{2}} + \sqrt{16}\sqrt{2}$$ *Rationalize denominator.*

$$= 4\sqrt{2} - \frac{3\sqrt{2}}{\sqrt{16}} + 4\sqrt{2}$$ *Product rule*

$$= 4\sqrt{2} - \frac{3}{4}\sqrt{2} + 4\sqrt{2}$$ *Write $\frac{3\sqrt{2}}{\sqrt{16}}$ as $\frac{3}{4}\sqrt{2}$.*

$$= \left(4 - \frac{3}{4} + 4\right)\sqrt{2}$$ *Simplify.*

$$= \frac{29\sqrt{2}}{4}$$

▸ **Now Try Exercise 115**

## 5 Divide Radical Expressions with Different Indices

Now we will divide radical expressions where the radicals have different indices. To divide such problems, write each radical in exponential form. Then use the rules of exponents, with the rational exponents, as was discussed in Section 7.2, to simplify the expression. Example 7 illustrates this procedure.

**EXAMPLE 7** ▸ Simplify. **a)** $\dfrac{\sqrt[5]{(m+n)^7}}{\sqrt[3]{(m+n)^4}}$ **b)** $\dfrac{\sqrt[3]{a^5b^4}}{\sqrt{a^2b}}$

*Solution* Begin by writing the numerator and denominator with rational exponents.

**a)**
$$\frac{\sqrt[5]{(m+n)^7}}{\sqrt[3]{(m+n)^4}} = \frac{(m+n)^{7/5}}{(m+n)^{4/3}} \quad \textit{Write with rational exponents.}$$
$$= (m+n)^{(7/5)-(4/3)} \quad \textit{Quotient rule for exponents}$$
$$= (m+n)^{1/15}$$
$$= \sqrt[15]{m+n} \quad \textit{Write as a radical.}$$

**b)**
$$\frac{\sqrt[3]{a^5b^4}}{\sqrt{a^2b}} = \frac{(a^5b^4)^{1/3}}{(a^2b)^{1/2}} \quad \textit{Write with rational exponents.}$$
$$= \frac{a^{5/3}b^{4/3}}{ab^{1/2}} \quad \textit{Raise the product to a power.}$$
$$= a^{(5/3)-1}b^{(4/3)-(1/2)} \quad \textit{Quotient rule for exponents}$$
$$= a^{2/3}b^{5/6}$$
$$= a^{4/6}b^{5/6} \quad \textit{Write the fractions with denominator 6.}$$
$$= (a^4b^5)^{1/6} \quad \textit{Rewrite using the laws of exponents.}$$
$$= \sqrt[6]{a^4b^5} \quad \textit{Write as a radical.}$$

▸ Now Try Exercise 133

## EXERCISE SET 7.5

MathXL® MyMathLab

### Concept/Writing Exercises

1. **a)** What is the conjugate of a binomial?
   **b)** What is the conjugate of $x - \sqrt{3}$?
2. What does it mean to rationalize a denominator?
3. **a)** Explain how to rationalize a denominator that contains a radical expression of one term.
   **b)** Rationalize $\dfrac{4}{\sqrt{3y}}$ using the procedure you specified in part **a)**.
4. **a)** Explain how to rationalize a denominator that contains a binomial in which one or both terms is a radical expression.
   **b)** Rationalize $\dfrac{\sqrt{2}+\sqrt{5}}{\sqrt{2}-\sqrt{5}}$ using the procedure you specified in part **a)**.
5. What are the three conditions that must be met for a radical expression to be simplified?
6. Explain why each of the following is not simplified.
   **a)** $\sqrt{x^5}$ **b)** $\sqrt{\dfrac{1}{2}}$ **c)** $\dfrac{1}{\sqrt{3}}$

*Simplify. In this exercise set, assume all variables represent positive real numbers.*

7. $\dfrac{1}{\sqrt{3}}$
8. $\dfrac{1}{\sqrt{6}}$
9. $\dfrac{4}{\sqrt{5}}$
10. $\dfrac{3}{\sqrt{7}}$
11. $\dfrac{6}{\sqrt{6}}$
12. $\dfrac{17}{\sqrt{17}}$
13. $\dfrac{1}{\sqrt{z}}$
14. $\dfrac{y}{\sqrt{y}}$

15. $\dfrac{p}{\sqrt{2}}$ 16. $\dfrac{m}{\sqrt{13}}$ 17. $\dfrac{\sqrt{y}}{\sqrt{7}}$ 18. $\dfrac{\sqrt{19}}{\sqrt{q}}$

19. $\dfrac{6\sqrt{3}}{\sqrt{6}}$ 20. $\dfrac{15x}{\sqrt{x}}$ 21. $\dfrac{\sqrt{x}}{\sqrt{y}}$ 22. $\dfrac{2\sqrt{3}}{\sqrt{a}}$

23. $\sqrt{\dfrac{5m}{8}}$ 24. $\dfrac{9\sqrt{3}}{\sqrt{y^3}}$ 25. $\dfrac{2n}{\sqrt{18n}}$ 26. $\sqrt{\dfrac{120x}{4y^3}}$

27. $\sqrt{\dfrac{18x^4y^3}{2z^3}}$ 28. $\sqrt{\dfrac{7pq^4}{2r}}$ 29. $\sqrt{\dfrac{20y^4z^3}{3xy^{-4}}}$ 30. $\sqrt{\dfrac{5xy^6}{3z}}$

31. $\sqrt{\dfrac{48x^6y^5}{3z^3}}$ 32. $\sqrt{\dfrac{45y^{12}z^{10}}{2x}}$

*Simplify.*

33. $\dfrac{1}{\sqrt[3]{2}}$ 34. $\dfrac{1}{\sqrt[3]{4}}$ 35. $\dfrac{8}{\sqrt[3]{y}}$ 36. $\dfrac{2}{\sqrt[3]{a^2}}$

37. $\dfrac{1}{\sqrt[4]{3}}$ 38. $\dfrac{z}{\sqrt[4]{4}}$ 39. $\dfrac{a}{\sqrt[4]{8}}$ 40. $\dfrac{8}{\sqrt[4]{z}}$

41. $\dfrac{5}{\sqrt[4]{z^2}}$ 42. $\dfrac{13}{\sqrt[4]{z^3}}$ 43. $\dfrac{10}{\sqrt[5]{y^3}}$ 44. $\dfrac{x}{\sqrt[5]{y^4}}$

45. $\dfrac{2}{\sqrt[7]{a^4}}$ 46. $\sqrt[3]{\dfrac{4x}{y}}$ 47. $\sqrt[3]{\dfrac{1}{2x}}$ 48. $\sqrt[3]{\dfrac{7c}{9y^2}}$

49. $\dfrac{5m}{\sqrt[4]{2}}$ 50. $\dfrac{3}{\sqrt[4]{a}}$ 51. $\sqrt[4]{\dfrac{5}{3x^3}}$ 52. $\sqrt[4]{\dfrac{2x^3}{4y^2}}$

53. $\sqrt[3]{\dfrac{3x^2}{2y^2}}$ 54. $\sqrt[3]{\dfrac{15x^6y^7}{2z^2}}$ 55. $\sqrt[3]{\dfrac{14xy^2}{2z^2}}$ 56. $\sqrt[6]{\dfrac{r^4s^9}{2r^5}}$

*Multiply.*

57. $(5 - \sqrt{6})(5 + \sqrt{6})$ 58. $(7 + \sqrt{3})(7 - \sqrt{3})$ 59. $(8 + \sqrt{2})(8 - \sqrt{2})$ 60. $(6 - \sqrt{7})(6 + \sqrt{7})$

61. $(2 - \sqrt{10})(2 + \sqrt{10})$ 62. $(3 + \sqrt{17})(3 - \sqrt{17})$ 63. $(\sqrt{a} - \sqrt{b})(\sqrt{a} + \sqrt{b})$ 64. $(\sqrt{x} - \sqrt{y})(\sqrt{x} + \sqrt{y})$

65. $(2\sqrt{x} - 3\sqrt{y})(2\sqrt{x} + 3\sqrt{y})$ 66. $(5\sqrt{c} - 4\sqrt{d})(5\sqrt{c} + 4\sqrt{d})$

*Simplify by rationalizing the denominator.*

67. $\dfrac{2}{\sqrt{3} + 1}$ 68. $\dfrac{4}{\sqrt{2} + 1}$ 69. $\dfrac{1}{2 + \sqrt{3}}$ 70. $\dfrac{3}{5 - \sqrt{7}}$

71. $\dfrac{5}{\sqrt{2} - 7}$ 72. $\dfrac{6}{\sqrt{2} + \sqrt{3}}$ 73. $\dfrac{\sqrt{5}}{2\sqrt{5} - \sqrt{6}}$ 74. $\dfrac{1}{\sqrt{17} - \sqrt{8}}$

75. $\dfrac{3}{6 + \sqrt{x}}$ 76. $\dfrac{4\sqrt{5}}{\sqrt{a} - 3}$ 77. $\dfrac{4\sqrt{x}}{\sqrt{x} - y}$ 78. $\dfrac{\sqrt{8x}}{x + \sqrt{y}}$

79. $\dfrac{\sqrt{2} - 2\sqrt{3}}{\sqrt{2} + 4\sqrt{3}}$ 80. $\dfrac{\sqrt{c} - \sqrt{2d}}{\sqrt{c} - \sqrt{d}}$ 81. $\dfrac{\sqrt{a^3} + \sqrt{a^7}}{\sqrt{a}}$ 82. $\dfrac{2\sqrt{xy} - \sqrt{xy}}{\sqrt{x} + \sqrt{y}}$

83. $\dfrac{4}{\sqrt{x + 2} - 3}$ 84. $\dfrac{8}{\sqrt{y - 3} + 6}$

*Simplify. These exercises are a combination of the types of exercises presented earlier in this exercise set.*

85. $\sqrt{\dfrac{x}{16}}$ 86. $\sqrt[4]{\dfrac{x^4}{16}}$ 87. $\sqrt{\dfrac{2}{9}}$ 88. $\sqrt{\dfrac{a}{b}}$

89. $(\sqrt{7} + \sqrt{6})(\sqrt{7} - \sqrt{6})$ 90. $\sqrt[3]{\dfrac{1}{16}}$ 91. $\sqrt{\dfrac{24x^3y^6}{5z}}$ 92. $\dfrac{5}{4 - \sqrt{y}}$

93. $\sqrt{\dfrac{28xy^4}{2x^3y^4}}$ 94. $\dfrac{8x}{\sqrt[3]{5y}}$ 95. $\dfrac{1}{\sqrt{a} + 7}$ 96. $\dfrac{\sqrt{x}}{\sqrt{x} + 6\sqrt{y}}$

97. $-\dfrac{7\sqrt{x}}{\sqrt{98}}$ 98. $\sqrt{\dfrac{2xy^4}{50xy^2}}$ 99. $\sqrt[4]{\dfrac{3y^2}{2x}}$ 100. $\sqrt{\dfrac{49x^2y^5}{3z}}$

**101.** $\sqrt[3]{\dfrac{32y^{12}z^{10}}{2x}}$ **102.** $\dfrac{\sqrt{3}+2}{\sqrt{2}+\sqrt{3}}$ **103.** $\dfrac{\sqrt{ar}}{\sqrt{a}-2\sqrt{r}}$ **104.** $\sqrt[4]{\dfrac{2}{9x}}$

**105.** $\dfrac{\sqrt[3]{6x}}{\sqrt[3]{5xy}}$ **106.** $\dfrac{\sqrt[3]{16m^2n}}{\sqrt[3]{2mn^2}}$ **107.** $\sqrt[4]{\dfrac{2x^7y^{12}z^4}{3x^9}}$ **108.** $\dfrac{9}{\sqrt{y+9}-\sqrt{y}}$

*Simplify.*

**109.** $\dfrac{1}{\sqrt{2}}+\dfrac{\sqrt{2}}{2}$ **110.** $\dfrac{1}{\sqrt{3}}+\dfrac{\sqrt{3}}{3}$ **111.** $\sqrt{5}-\dfrac{2}{\sqrt{5}}$

**112.** $\dfrac{\sqrt{6}}{2}-\dfrac{2}{\sqrt{6}}$ **113.** $4\sqrt{\dfrac{1}{6}}+\sqrt{24}$ **114.** $5\sqrt{3}-\dfrac{3}{\sqrt{3}}+2\sqrt{18}$

**115.** $5\sqrt{2}-\dfrac{2}{\sqrt{8}}+\sqrt{50}$ **116.** $\dfrac{2}{3}+\dfrac{1}{\sqrt{3}}+\sqrt{75}$ **117.** $\sqrt{\dfrac{1}{2}}+7\sqrt{2}+\sqrt{18}$

**118.** $\dfrac{1}{2}\sqrt{18}-\dfrac{3}{\sqrt{2}}-9\sqrt{50}$ **119.** $\dfrac{2}{\sqrt{50}}-3\sqrt{50}-\dfrac{1}{\sqrt{8}}$ **120.** $\dfrac{\sqrt{3}}{3}+\dfrac{5}{\sqrt{3}}+\sqrt{12}$

**121.** $\sqrt{\dfrac{3}{8}}+\sqrt{\dfrac{3}{2}}$ **122.** $2\sqrt{\dfrac{8}{3}}-4\sqrt{\dfrac{100}{6}}$ **123.** $-2\sqrt{\dfrac{x}{y}}+3\sqrt{\dfrac{y}{x}}$

**124.** $-5x\sqrt{\dfrac{y}{y^2}}+9x\sqrt{\dfrac{1}{y}}$ **125.** $\dfrac{3}{\sqrt{a}}-\sqrt{\dfrac{9}{a}}+2\sqrt{a}$ **126.** $6\sqrt{x}+\dfrac{1}{\sqrt{x}}+\sqrt{\dfrac{1}{x}}$

*Simplify.*

**127.** $\dfrac{\sqrt{(a+b)^4}}{\sqrt[3]{a+b}}$ **128.** $\dfrac{\sqrt[3]{c+2}}{\sqrt[4]{(c+2)^3}}$ **129.** $\dfrac{\sqrt[5]{(a+2b)^4}}{\sqrt[3]{(a+2b)^2}}$ **130.** $\dfrac{\sqrt[6]{(r+3)^5}}{\sqrt[3]{(r+3)^5}}$

**131.** $\dfrac{\sqrt[3]{r^2s^4}}{\sqrt{rs}}$ **132.** $\dfrac{\sqrt{a^2b^4}}{\sqrt[3]{ab^2}}$ **133.** $\dfrac{\sqrt[5]{x^4y^6}}{\sqrt[3]{(xy)^2}}$ **134.** $\dfrac{\sqrt[6]{4m^8n^4}}{\sqrt[4]{m^4n^2}}$

## Problem Solving

**135. Illumination of a Light** Under certain conditions the formula

$$d=\sqrt{\frac{72}{I}}$$

is used to show the relationship between the illumination on an object, $I$, in lumens per meter, and the distance, $d$, in meters, the object is from the light source. If the illumination on a person standing near a light source is 5.3 lumens per meter, how far is the person from the light source?

**136. Strength of a Board** When sufficient pressure is applied to a particular particle board, the particle board will break (or rupture). The thicker the particle board the greater will be the pressure that will need to be applied before the board breaks. The formula

$$T=\sqrt{\frac{0.05\,LB}{M}}$$

relates the thickness of a specific particle board, $T$, in inches, the board's length, $L$, in inches, the board's load that will cause the board to rupture, $B$, in pounds, and the modulus of rupture, $M$, in pounds per square inch. The modulus of rupture is a constant determined by sample tests on the specific type of particle board.

Find the thickness of a 36-inch-long particle board if the modulus of rupture is 2560 pounds per square inch and the board ruptures when 800 pounds are applied.

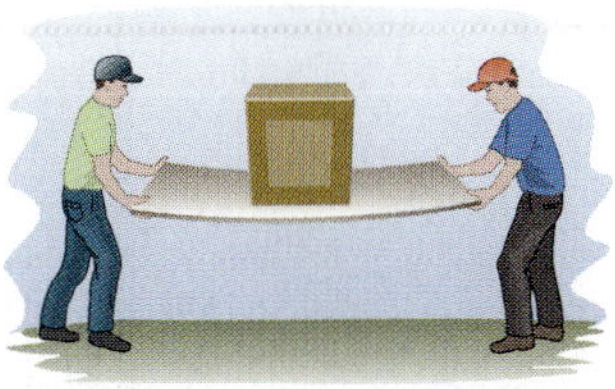

**137. Volume of a Fish Tank** A new restaurant wants to have a spherical fish tank in its lobby. The radius, $r$, in inches, of a spherical tank can be found by the formula

$$r=\sqrt[3]{\frac{3V}{4\pi}}$$

where $V$ is the volume of the tank in cubic inches. Find the radius of a spherical tank whose volume is 7238.23 cubic inches.

**138. Consecutive Numbers** If we consider the set of consecutive natural numbers $1, 2, 3, 4, \ldots, n$ to be the population, the standard deviation, $\sigma$, which is a measure of the spread of the data from the mean, can be calculated by the formula

$$\sigma=\sqrt{\frac{n^2-1}{12}}$$

where $n$ represents the number of natural numbers in the population. Find the standard deviation for the first 100 consecutive natural numbers.

**139. U.S. Farms** The number of farms in the United States is declining annually (however, the size of the remaining farms is increasing). A function that can be used to estimate the number of farms, $N(t)$, in millions, is

$$N(t) = \frac{6.21}{\sqrt[4]{t}}$$

where $t$ is years since 1959 and $1 \le t \le 50$. Estimate the number of farms in the United States in **a)** 1960 and **b)** 2008.

**140. Infant Mortality Rate** The U.S. infant mortality rate has been declining steadily. The infant mortality rate, $N(t)$, defined as deaths per 1000 live births, can be estimated by the function

$$N(t) = \frac{28.46}{\sqrt[3]{t^2}}$$

where $t$ is years since 1969 and $1 \le t \le 37$. Estimate the infant mortality rate in **a)** 1970 and **b)** 2006.

**141.** Which is greater, $\frac{2}{\sqrt{2}}$ or $\frac{3}{\sqrt{3}}$? Explain.

**142.** Which is greater, $\frac{\sqrt{3}}{2}$ or $\frac{2}{\sqrt{3}}$? Explain.

**143.** Which is greater, $\frac{1}{\sqrt{3}+2}$ or $2+\sqrt{3}$? (Do not use a calculator.) Explain how you determined your answer.

**144.** Which is greater, $\frac{1}{\sqrt{3}} + \sqrt{75}$ or $\frac{2}{\sqrt{12}} + \sqrt{48} + 2\sqrt{3}$? (Do not use a calculator.) Explain how you determined your answer.

**145.** Consider the functions $f(x) = x^{a/2}$ and $g(x) = x^{b/3}$.

**a)** List three values for $a$ that will result in $x^{a/2}$ being a perfect square.

**b)** List three values for $b$ that will result in $x^{b/3}$ being a perfect cube.

**c)** If $x \ge 0$, find $(f \cdot g)(x)$.

**d)** If $x \ge 0$, find $(f/g)(x)$.

*Rationalize each denominator.*

**146.** $\frac{1}{\sqrt{a+b}}$

**147.** $\frac{3}{\sqrt{2a-3b}}$

*In higher math courses, it may be necessary to rationalize the numerators of radical expressions. Rationalize the numerators of the following expressions. (Your answers will contain radicals in the denominators.)*

**148.** $\frac{\sqrt{7}}{3}$ **149.** $\frac{5-\sqrt{5}}{6}$ **150.** $\frac{6\sqrt{x}-\sqrt{3}}{x}$ **151.** $\frac{\sqrt{x+h}-\sqrt{x}}{h}$

## Group Activity

**Similar Figures** *The following two exercises will reinforce many of the concepts presented in this chapter. Work each problem as a group. Make sure each member of the group understands each step in obtaining the solution. The figures in each exercise are similar. For each exercise, use a proportion to find the length of side x. Write the answer in radical form with a rationalized denominator.*

**152.**

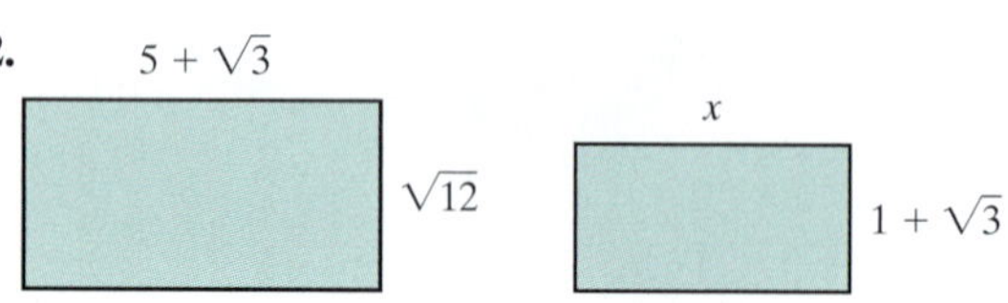


**153.**

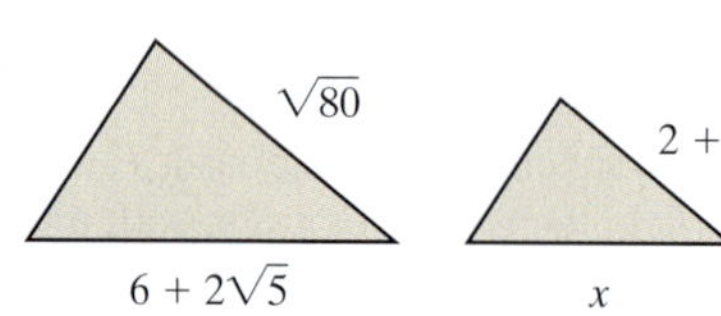


## Cumulative Review Exercises

[2.2] **154.** Solve the equation $A = \frac{1}{2}h(b_1 + b_2)$ for $b_2$.

[2.4] **155. Moving Vehicles** Two cars leave from West Point at the same time traveling in opposite directions. One travels 10 miles per hour faster than the other. If the two cars are 270 miles apart after 3 hours, find the speed of each car.

[5.2] **156.** Multiply $(x-2)(4x^2+9x-2)$.

[6.4] **157.** Solve $\frac{x}{2} - \frac{4}{x} = -\frac{7}{2}$.

# 7.6 Solving Radical Equations

1. Solve equations containing one radical.
2. Solve equations containing two radicals.
3. Solve equations containing two radical terms and a nonradical term.
4. Solve applications using radical equations.
5. Solve for a variable in a radicand.

## 1 Solve Equations Containing One Radical

A **radical equation** is an equation that contains a variable in a radicand.

Examples of Radical Equations

$$\sqrt{x} = 5, \qquad \sqrt[3]{y+4} = 9, \qquad \sqrt{x-2} = 7 + \sqrt{x+8}$$

**To Solve Radical Equations**

1. Rewrite the equation so that one radical containing a variable is by itself (isolated) on one side of the equation.
2. Raise each side of the equation to a power equal to the index of the radical.
3. Combine like terms.
4. If the equation still contains a term with a variable in a radicand, repeat steps 1 through 3.
5. Solve the resulting equation for the variable.
6. Check all solutions in the original equation for extraneous solutions.

Recall from Section 6.4 that an extraneous solution is a number obtained when solving an equation that is not a solution to the original equation.

The following examples illustrate the procedure for solving radical equations.

**EXAMPLE 1** ▸ Solve the equation $\sqrt{x} = 5$.

*Solution* The square root containing the variable is already by itself on one side of the equation. Square both sides of the equation.

$$\sqrt{x} = 5$$
$$(\sqrt{x})^2 = (5)^2$$
$$x = 25$$

Check
$$\sqrt{x} = 5$$
$$\sqrt{25} \stackrel{?}{=} 5$$
$$5 = 5 \quad \text{True}$$

▸ Now Try Exercise 11

**EXAMPLE 2** ▸ Solve.

**a)** $\sqrt{x-4} - 6 = 0$ **b)** $\sqrt[3]{x} + 10 = 8$ **c)** $\sqrt{x} + 3 = 0$

*Solution* The first step in each part will be to isolate the term containing the radical.

**a)**
$$\sqrt{x-4} - 6 = 0$$
$$\sqrt{x-4} = 6 \quad \text{Isolate the radical containing the variable.}$$
$$(\sqrt{x-4})^2 = 6^2 \quad \text{Square both sides.}$$
$$x - 4 = 36 \quad \text{Solve for the variable.}$$
$$x = 40$$

A check will show that 40 is the solution.

**b)**
$$\sqrt[3]{x} + 10 = 8$$
$$\sqrt[3]{x} = -2 \quad \text{Isolate the radical containing the variable.}$$
$$(\sqrt[3]{x})^3 = (-2)^3 \quad \text{Cube both sides.}$$
$$x = -8$$

A check will show that −8 is the solution.

**c)**

$$\sqrt{x} + 3 = 0$$

$$\sqrt{x} = -3 \qquad \text{Isolate the radical containing the variable.}$$

$$(\sqrt{x})^2 = (-3)^2 \qquad \text{Square both sides.}$$

$$x = 9$$

**Check**

$$\sqrt{x} + 3 = 0$$

$$\sqrt{9} + 3 \stackrel{?}{=} 0$$

$$3 + 3 \stackrel{?}{=} 0$$

$$6 = 0 \qquad \text{False}$$

A check shows that 9 is not a solution. The answer to part **c)** is "no real solution." You may have realized there was no real solution to the problem when you obtained the equation $\sqrt{x} = -3$, because $\sqrt{x}$ cannot equal a negative real number.

▸ **Now Try Exercise 17**

## Helpful Hint

Don't forget to check your solutions in the original equation. When you raise both sides of an equation to a power you may introduce extraneous solutions.

Consider the equation $x = 2$. Note what happens when you square both sides of the equation.

$$x = 2$$

$$x^2 = 2^2$$

$$x^2 = 4$$

Note that the equation $x^2 = 4$ has two solutions, +2 and −2. Since the original equation $x = 2$ has only one solution, 2, we introduced the extraneous solution, −2.

**EXAMPLE 3** ▸ Solve $\sqrt{2x - 3} = x - 3$.

*Solution* Since the radical is already isolated, we square both sides of the equation. Then we solve the resulting quadratic equation.

$$(\sqrt{2x - 3})^2 = (x - 3)^2$$

$$2x - 3 = (x - 3)(x - 3)$$

$$2x - 3 = x^2 - 6x + 9$$

$$0 = x^2 - 8x + 12$$

Now we factor and use the zero-factor property.

$$x^2 - 8x + 12 = 0$$

$$(x - 6)(x - 2) = 0$$

$$x - 6 = 0 \quad \text{or} \quad x - 2 = 0$$

$$x = 6 \qquad\qquad x = 2$$

**Check**

| $x = 6$ | $x = 2$ |
|---|---|
| $\sqrt{2x - 3} = x - 3$ | $\sqrt{2x - 3} = x - 3$ |
| $\sqrt{2(6) - 3} \stackrel{?}{=} 6 - 3$ | $\sqrt{2(2) - 3} \stackrel{?}{=} 2 - 3$ |
| $\sqrt{9} \stackrel{?}{=} 3$ | $\sqrt{1} \stackrel{?}{=} -1$ |
| $3 = 3$ True | $1 = -1$ False |

Thus, 6 is a solution, but 2 is not a solution to the equation. The 2 is an extraneous solution because 2 satisfies the equation $(\sqrt{2x - 3})^2 = (x - 3)^2$, but not the original equation, $\sqrt{2x - 3} = x - 3$.

▸ **Now Try Exercise 43**

## USING YOUR GRAPHING CALCULATOR

In Example 3, we found the solution to $\sqrt{2x - 3} = x - 3$ to be 6. If we let $Y_1 = \sqrt{2x - 3}$ and $Y_2 = x - 3$ and graph $Y_1$ and $Y_2$ on a graphing calculator, we get **Figure 7.3**. Notice the graphs appear to intersect at $x = 6$, which is what we expect.

The table of values in **Figure 7.4** shows that the $y$-coordinate at the point of intersection is 3. In the table, ERROR appears under $Y_1$ for the values of $x$ of 0 and 1. For any values less than $\frac{3}{2}$, the value of $2x - 3$ is negative and therefore $\sqrt{2x - 3}$ is not a real number. The domain of function $Y_1$ is $\left\{x \middle| x \geq \frac{3}{2}\right\}$, which may be found by solving the inequality $2x - 3 \geq 0$.

You can use your graphing calculator to either solve or check radical equations.

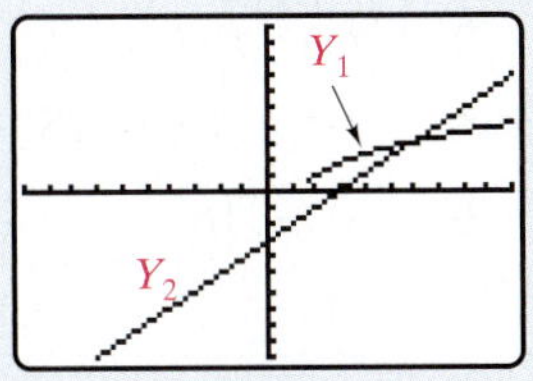


FIGURE 7.3

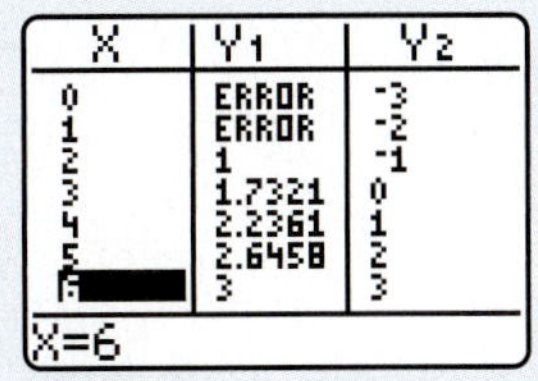

| X | Y1 | Y2 |
|---|---|---|
| 0 | ERROR | -3 |
| 1 | ERROR | -2 |
| 2 | 1 | -1 |
| 3 | 1.7321 | 0 |
| 4 | 2.2361 | 1 |
| 5 | 2.6458 | 2 |
| 6 | 3 | 3 |

X=6

FIGURE 7.4

### EXERCISES

*Use your graphing calculator to determine whether the indicated value is the solution to the radical equation. If it is not the solution, use your grapher to determine the solution.*

**1.** $\sqrt{2x + 9} = 5(x - 7), 8$

**2.** $\sqrt{3x + 4} = \sqrt{x + 12}, 6$

**EXAMPLE 4** ▸ Solve $x - 2\sqrt{x} - 3 = 0$.

*Solution* First, isolate the radical term by writing the radical term by itself on one side of the equation.

$$x - 2\sqrt{x} - 3 = 0$$
$$-2\sqrt{x} = -x + 3$$
$$2\sqrt{x} = x - 3$$

Now square both sides of the equation.

$$(2\sqrt{x})^2 = (x - 3)^2$$
$$4x = x^2 - 6x + 9$$
$$0 = x^2 - 10x + 9$$
$$0 = (x - 1)(x - 9)$$
$$x - 1 = 0 \quad \text{or} \quad x - 9 = 0$$
$$x = 1 \qquad\qquad x = 9$$

**Check**

| $x = 1$ | | $x = 9$ | |
|---|---|---|---|
| $x - 2\sqrt{x} - 3 = 0$ | | $x - 2\sqrt{x} - 3 = 0$ | |
| $1 - 2\sqrt{1} - 3 \stackrel{?}{=} 0$ | | $9 - 2\sqrt{9} - 3 \stackrel{?}{=} 0$ | |
| $1 - 2(1) - 3 \stackrel{?}{=} 0$ | | $9 - 2(3) - 3 \stackrel{?}{=} 0$ | |
| $1 - 2 - 3 \stackrel{?}{=} 0$ | | $9 - 6 - 3 \stackrel{?}{=} 0$ | |
| $-4 = 0$ | False | $3 - 3 \stackrel{?}{=} 0$ | |
| | | $0 = 0$ | True |

The solution is 9. The value 1 is an extraneous solution.

▸ Now Try Exercise 41

## 2 Solve Equations Containing Two Radicals

Now we will look at some equations that contain two radicals.

**EXAMPLE 5** ▸ Solve $\sqrt{9x^2 + 6} = 3\sqrt{x^2 + x - 2}$.

*Solution* Since the two radicals appear on different sides of the equation, we square both sides of the equation.

$$\left(\sqrt{9x^2 + 6}\right)^2 = \left(3\sqrt{x^2 + x - 2}\right)^2 \quad \text{Square both sides.}$$
$$9x^2 + 6 = 9(x^2 + x - 2)$$
$$9x^2 + 6 = 9x^2 + 9x - 18 \quad \text{Distributive property.}$$
$$6 = 9x - 18 \quad 9x^2 \text{ was subtracted from both sides.}$$
$$24 = 9x$$
$$\frac{8}{3} = x$$

A check will show that $\frac{8}{3}$ is the solution.

▸ **Now Try Exercise 27**

In higher mathematics courses, equations are sometimes given using exponents rather than radicals. Example 6 illustrates such an equation.

**EXAMPLE 6** ▸ For $f(x) = 3(x - 2)^{1/3}$ and $g(x) = (17x - 14)^{1/3}$, find all values of $x$ for which $f(x) = g(x)$.

*Solution* You should realize that alternate ways of writing $f(x)$ and $g(x)$ are $f(x) = 3\sqrt[3]{x - 2}$ and $g(x) = \sqrt[3]{17x - 14}$. We could therefore work this example using radicals, but we will work instead with rational exponents. We set the two functions equal to each other and solve for $x$.

$$f(x) = g(x)$$
$$3(x - 2)^{1/3} = (17x - 14)^{1/3}$$
$$[3(x - 2)^{1/3}]^3 = [(17x - 14)^{1/3}]^3 \quad \text{Cube both sides.}$$
$$3^3(x - 2) = 17x - 14$$
$$27(x - 2) = 17x - 14$$
$$27x - 54 = 17x - 14$$
$$10x - 54 = -14$$
$$10x = 40$$
$$x = 4$$

A check will show that the solution is 4. If you substitute 4 into both $f(x)$ and $g(x)$, you will find they both simplify to $3\sqrt[3]{2}$. Check this now.

▸ **Now Try Exercise 69**

In Example 6, if you solve the equation $3\sqrt[3]{x - 2} = \sqrt[3]{17x - 14}$ you will obtain the solution 4. For additional practice, do this now.

## 3 Solve Equations Containing Two Radical Terms and a Nonradical Term

When a radical equation contains two radical terms and a third nonradical term, you will sometimes need to raise both sides of the equation to a given power twice to obtain the solution. First, isolate one radical term. Then raise both sides of the equation to the given power. This will eliminate one of the radicals. Next, isolate the remaining radical on one side of the equation. Then raise both sides of the equation to the given power a second time. This procedure is illustrated in Example 7.

**EXAMPLE 7** ▸ Solve $\sqrt{5x-1} - \sqrt{3x-2} = 1$.

*Solution* We must isolate one radical term on one side of the equation. We will begin by adding $\sqrt{3x-2}$ to both sides of the equation to isolate $\sqrt{5x-1}$. Then we will square both sides of the equation and combine like terms.

$$\sqrt{5x-1} = 1 + \sqrt{3x-2} \qquad \text{Isolate } \sqrt{5x-1}.$$
$$(\sqrt{5x-1})^2 = (1 + \sqrt{3x-2})^2 \qquad \text{Square both sides.}$$
$$5x - 1 = (1 + \sqrt{3x-2})(1 + \sqrt{3x-2}) \qquad \text{Write as a product.}$$
$$5x - 1 = 1 + \sqrt{3x-2} + \sqrt{3x-2} + (\sqrt{3x-2})^2 \qquad \text{Multiply.}$$
$$5x - 1 = 1 + 2\sqrt{3x-2} + 3x - 2 \qquad \text{Combine like terms; simplify.}$$
$$5x - 1 = 3x - 1 + 2\sqrt{3x-2} \qquad \text{Combine like terms.}$$
$$2x = 2\sqrt{3x-2} \qquad \text{Isolate the radical term.}$$
$$x = \sqrt{3x-2} \qquad \text{Both sides were divided by 2.}$$

We have isolated the remaining radical term. We now square both sides of the equation again and solve for $x$.

$$x = \sqrt{3x-2}$$
$$x^2 = (\sqrt{3x-2})^2 \qquad \text{Square both sides.}$$
$$x^2 = 3x - 2$$
$$x^2 - 3x + 2 = 0$$
$$(x-2)(x-1) = 0$$
$$x - 2 = 0 \quad \text{or} \quad x - 1 = 0$$
$$x = 2 \qquad\qquad x = 1$$

A check will show that both 2 and 1 are solutions of the equation.

▸ Now Try Exercise 61

**EXAMPLE 8** ▸ For $f(x) = \sqrt{5x-1} - \sqrt{3x-2}$, find all values of $x$ for which $f(x) = 1$.

*Solution* Substitute 1 for $f(x)$. This gives

$$1 = \sqrt{5x-1} - \sqrt{3x-2}$$

Since this is the same equation we solved in Example 7, the answers are $x = 2$ and $x = 1$. Verify for yourself that $f(2) = 1$ and $f(1) = 1$.

▸ Now Try Exercise 121

### Avoiding Common Errors

In Chapter 5, we stated that $(a+b)^2 \neq a^2 + b^2$. Be careful when squaring a binomial like $1 + \sqrt{x}$. Look at the following computations carefully so that you do not make the mistake shown on the right.

CORRECT

$$(1+\sqrt{x})^2 = (1+\sqrt{x})(1+\sqrt{x})$$
F O I L
$$= 1 + \sqrt{x} + \sqrt{x} + \sqrt{x}\sqrt{x}$$
$$= 1 + 2\sqrt{x} + x$$

INCORRECT

$$(1+\sqrt{x})^2 = 1^2 + (\sqrt{x})^2$$
$$= 1 + x$$

## 4 Solve Applications Using Radical Equations

Now we will look at a few of the many applications of radicals.

**EXAMPLE 9 ▸ The Green Monster** In Fenway Park, where the Boston Red Sox play baseball, the distance from home plate down the third base line to the bottom of the wall in left field is 310 feet. In left field at the end of the baseline there is a green wall perpendicular to the ground that is 37 feet tall. This green wall is commonly known as *the Green Monster* (see photo). Determine the distance from home plate to the top of the Green Monster along the third base line.

*Solution* **Understand** **Figure 7.5** illustrates the problem. We need to find the distance from home plate to the top of the wall in left field.

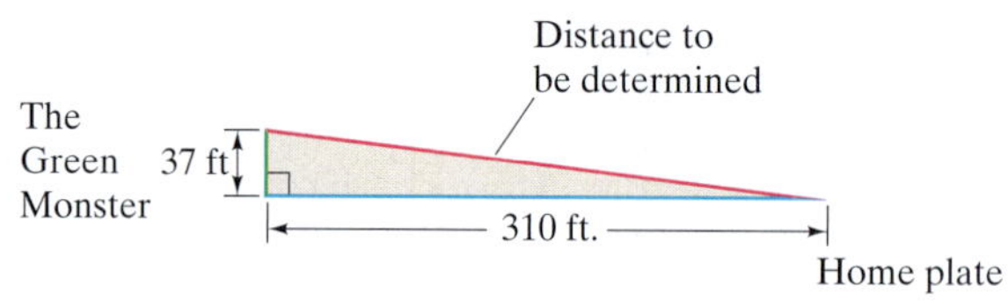


FIGURE 7.5

**Translate** To solve the problem we use the Pythagorean Theorem, which was discussed earlier, $\text{leg}^2 + \text{leg}^2 = \text{hyp}^2$, or $a^2 + b^2 = c^2$.

$$310^2 + 37^2 = c^2 \quad \textit{Substitute known values.}$$

**Carry Out**

$$96{,}100 + 1369 = c^2$$

$$97{,}469 = c^2$$

$$\sqrt{97{,}469} = \sqrt{c^2} \quad \textit{Take the square root of both sides.}$$

$$\sqrt{97{,}469} = c \quad \textit{* See footnote.}$$

$$312.20 \approx c$$

**Answer** The distance from home plate to the top of the wall is about 312.20 feet.

▸ **Now Try Exercise 99**

**EXAMPLE 10 ▸ Period of a Pendulum** The length of time it takes for a pendulum to make one complete swing back and forth is called the *period* of the pendulum. See **Figure 7.6**. The period of a pendulum, $T$, in seconds, can be found by the formula $T = 2\pi\sqrt{\frac{L}{32}}$, where $L$ is the length of the pendulum in feet. Find the period of a pendulum if its length is 5 feet.

FIGURE 7.6

*Solution* Substitute 5 for $L$ and 3.14 for $\pi$ in the formula. If you have a calculator that has a $\pi$ key, use it to enter $\pi$.

$$T = 2\pi\sqrt{\frac{L}{32}}$$

$$\approx 2(3.14)\sqrt{\frac{5}{32}}$$

$$\approx 2(3.14)\sqrt{0.15625} \approx 2.48$$

Thus, the period is about 2.48 seconds. If you have a grandfather clock with a 5-foot pendulum, it will take about 2.48 seconds for it to swing back and forth.

▸ **Now Try Exercise 103**

*$c^2 = 97{,}469$ has two solutions: $c = \sqrt{97{,}469}$ and $c = -\sqrt{97{,}469}$. Since we are solving for a length, which must be a positive quantity, we use the positive root.

## 5 Solve for a Variable in a Radicand

You may be given a formula and be asked to solve for a variable in a radicand. To do so, follow the same general procedure used to solve a radical equation. Begin by isolating the radical expression. Then raise both sides of the equation to the same power as the index of the radical. This procedure is illustrated in Example 11 **b)**.

**EXAMPLE 11** ▸ **Error of Estimation** A formula in statistics for finding the maximum error of estimation is $E = Z\dfrac{\sigma}{\sqrt{n}}$.

**a)** Find $E$ if $Z = 1.28$, $\sigma = 10$, and $n = 36$.

**b)** Solve this equation for $n$.

### Solution

**a)** $E = Z\dfrac{\sigma}{\sqrt{n}} = 1.28\left(\dfrac{10}{\sqrt{36}}\right) = 1.28\left(\dfrac{10}{6}\right) \approx 2.13$

**b)** First multiply both sides of the equation by $\sqrt{n}$ to eliminate fractions. Then isolate $\sqrt{n}$. Finally, solve for $n$ by squaring both sides of the equation.

$$E = Z\frac{\sigma}{\sqrt{n}}$$

$$\sqrt{n}(E) = \left(Z\frac{\sigma}{\sqrt{n}}\right)\sqrt{n} \qquad \text{Eliminate fractions.}$$

$$\sqrt{n}(E) = Z\sigma$$

$$\sqrt{n} = \frac{Z\sigma}{E} \qquad \text{Isolate the radical term.}$$

$$(\sqrt{n})^2 = \left(\frac{Z\sigma}{E}\right)^2 \qquad \text{Square both sides.}$$

$$n = \left(\frac{Z\sigma}{E}\right)^2 \quad \text{or} \quad n = \frac{Z^2\sigma^2}{E^2}$$

▸ **Now Try Exercise 75**

## EXERCISE SET 7.6

MathXL® MyMathLab

### Concept/Writing Exercises

1. **a)** Explain how to solve a radical equation.
   **b)** Solve $\sqrt{2x + 26} - 2 = 4$ using the procedure you gave in part **a)**.
2. Consider the equation $\sqrt{x + 3} = -\sqrt{2x - 1}$. Explain why this equation can have no real solution.
3. Consider the equation $-\sqrt{x^2} = \sqrt{(-x)^2}$. By studying the equation, can you determine its solution? Explain.
4. Consider the equation $\sqrt[3]{x^2} = -\sqrt[3]{x^2}$. By studying the equation, can you determine its solution? Explain.
5. Explain without solving the equation how you can tell that $\sqrt{x - 3} + 4 = 0$ has no solution.
6. Why is it necessary to check solutions to radical equations?
7. Does the equation $\sqrt{x} = 5$ have one or two solutions? Explain.
8. Does the equation $x^2 = 9$ have one or two solutions? Explain.

### Practice the Skills

*Solve and check your solution(s). If the equation has no real solution, so state.*

9. $\sqrt{x} = 4$
10. $\sqrt{x} = 13$
11. $\sqrt{x} = -9$
12. $\sqrt[3]{x} = 4$
13. $\sqrt[3]{x} = -4$
14. $\sqrt{a} + 5 = 0$
15. $\sqrt{2x + 3} = 5$
16. $\sqrt[3]{7x - 6} = 4$
17. $\sqrt[3]{3x} + 4 = 7$
18. $2\sqrt{4x + 5} = 14$
19. $\sqrt[3]{2x + 29} = 3$
20. $\sqrt[3]{6x + 2} = -4$
21. $\sqrt[4]{x} = 3$
22. $\sqrt[4]{x} = -3$
23. $\sqrt[4]{x + 10} = 3$

**24.** $\sqrt[4]{3x-2}=2$
**25.** $\sqrt[4]{2x+1}+6=2$
**26.** $\sqrt{2x}+7=13$
**27.** $\sqrt{x+8}=\sqrt{x-8}$
**28.** $\sqrt{r+5}+7=10$
**29.** $2\sqrt[3]{x-1}=\sqrt[3]{x^2+2x}$
**30.** $\sqrt[3]{6t-1}=\sqrt[3]{2t+3}$
**31.** $\sqrt[4]{x+8}=\sqrt[4]{2x}$
**32.** $\sqrt[4]{3x-1}+4=0$
**33.** $\sqrt{5x+1}-6=0$
**34.** $\sqrt{x^2+12x+3}=-x$
**35.** $\sqrt{m^2+6m-4}=m$
**36.** $\sqrt{x^2+3x+12}=x$
**37.** $\sqrt{5c+1}-9=0$
**38.** $\sqrt{b^2-2}=b+4$
**39.** $\sqrt{z^2+5}=z+1$
**40.** $\sqrt{x}+6x=1$
**41.** $\sqrt{2y+5}+5-y=0$
**42.** $\sqrt{4x+1}=\frac{1}{2}x+2$
**43.** $\sqrt{5x+6}=2x-6$
**44.** $\sqrt{4b+5}+b=10$
**45.** $(2a+9)^{1/2}-a+3=0$
**46.** $(3x+4)^{1/2}-x=-2$
**47.** $(2x^2+4x+9)^{1/2}=(2x^2+9)^{1/2}$
**48.** $(2x+1)^{1/2}+7=x$
**49.** $(r+4)^{1/3}=(3r+10)^{1/3}$
**50.** $(7x+6)^{1/3}+4=0$
**51.** $(5x+7)^{1/4}=(9x+1)^{1/4}$
**52.** $(5b+3)^{1/4}=(2b+17)^{1/4}$
**53.** $\sqrt[4]{x+5}=-2$
**54.** $\sqrt{x^2+x-1}=-\sqrt{x+3}$

*Solve. You will have to square both sides of the equation twice to eliminate all radicals.*

**55.** $\sqrt{4x+1}=\sqrt{2x}+1$
**56.** $3\sqrt{b}-1=\sqrt{b+21}$
**57.** $\sqrt{3a+1}=\sqrt{a-4}+3$
**58.** $\sqrt{x+1}=2-\sqrt{x}$
**59.** $\sqrt{x+3}=\sqrt{x}-3$
**60.** $\sqrt{y+1}=2+\sqrt{y-7}$
**61.** $\sqrt{x+7}=6-\sqrt{x-5}$
**62.** $\sqrt{b-3}=4-\sqrt{b+5}$
**63.** $\sqrt{4x-3}=2+\sqrt{2x-5}$
**64.** $\sqrt{r+10}+2+\sqrt{r-5}=0$
**65.** $\sqrt{y+1}=\sqrt{y+10}-3$
**66.** $3+\sqrt{x+1}=\sqrt{3x+12}$

*For each pair of functions, find all real values of x where $f(x)=g(x)$.*

**67.** $f(x)=\sqrt{x+8},\ g(x)=\sqrt{2x+1}$
**68.** $f(x)=\sqrt{x^2-6x+10},\ g(x)=\sqrt{x-2}$
**69.** $f(x)=\sqrt[3]{5x-19},\ g(x)=\sqrt[3]{6x-23}$
**70.** $f(x)=(14x-8)^{1/2},\ g(x)=2(3x+2)^{1/2}$
**71.** $f(x)=2(8x+24)^{1/3},\ g(x)=4(2x-2)^{1/3}$
**72.** $f(x)=2\sqrt{x+2},\ g(x)=8-\sqrt{x+14}$

*Solve each formula for the indicated variable.*

**73.** $p=\sqrt{2v}$, for $v$
**74.** $l=\sqrt{4r}$, for $r$
**75.** $v=\sqrt{2gh}$, for $g$
**76.** $v=\sqrt{\frac{2E}{m}}$, for $E$
**77.** $v=\sqrt{\frac{FR}{M}}$, for $F$
**78.** $\omega=\sqrt{\frac{a_0}{b_0}}$, for $b_0$
**79.** $x=\sqrt{\frac{m}{k}}V_0$, for $m$
**80.** $T=2\pi\sqrt{\frac{L}{32}}$, for $L$
**81.** $r=\sqrt{\frac{A}{\pi}}$, for $A$
**82.** $r=\sqrt[3]{\frac{3V}{4\pi}}$, for $V$

## Problem Solving

*Use the Pythagorean Theorem to find the length of the unknown side of each triangle. Write the answer as a radical in simplified form.*

**83.**

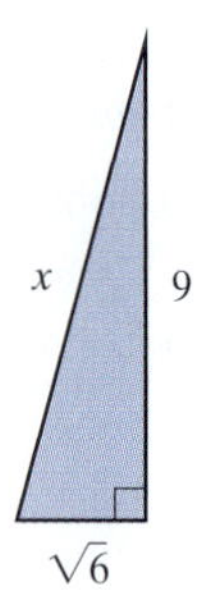


**84.**

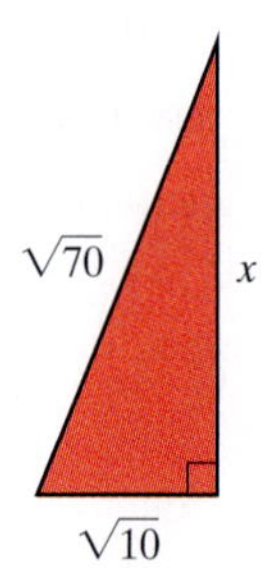


**85.**

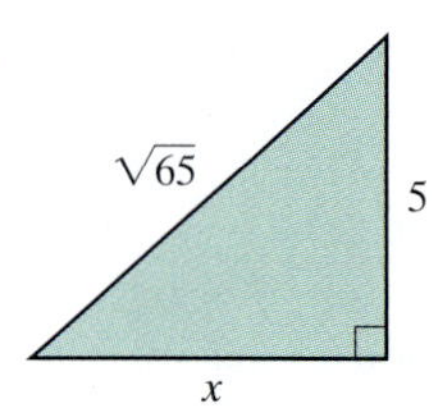


**86.**

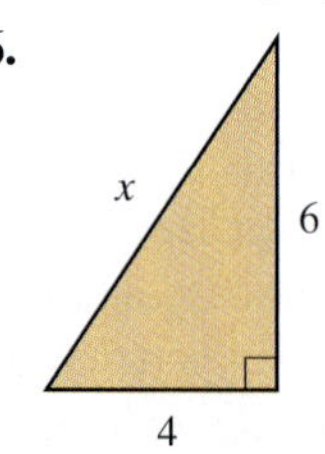


*Solve. You will need to square both sides of the equation twice.*

**87.** $\sqrt{x+5}-\sqrt{x}=\sqrt{x-3}$
**88.** $\sqrt{2x}-\sqrt{x-4}=\sqrt{12-x}$
**89.** $\sqrt{4y+6}+\sqrt{y+5}=\sqrt{y+1}$
**90.** $\sqrt{2b-2}+\sqrt{b-5}=\sqrt{4b}$
**91.** $\sqrt{c+1}+\sqrt{c-2}=\sqrt{3c}$
**92.** $\sqrt{2t-1}+\sqrt{t-4}=\sqrt{3t+1}$
**93.** $\sqrt{a+2}-\sqrt{a-3}=\sqrt{a-6}$
**94.** $\sqrt{r-1}-\sqrt{r+6}=\sqrt{r-9}$

*Solve. You will need to square both sides of the equation twice.*

**95.** $\sqrt{2-\sqrt{x}}=\sqrt{x}$
**96.** $\sqrt{6+\sqrt{x+4}}=\sqrt{2x-1}$
**97.** $\sqrt{2+\sqrt{x+1}}=\sqrt{7-x}$
**98.** $\sqrt{1+\sqrt{x-1}}=\sqrt{x-6}$

**99. Baseball Diamond** A regulation baseball diamond is a square with 90 feet between bases. How far is second base from home plate?

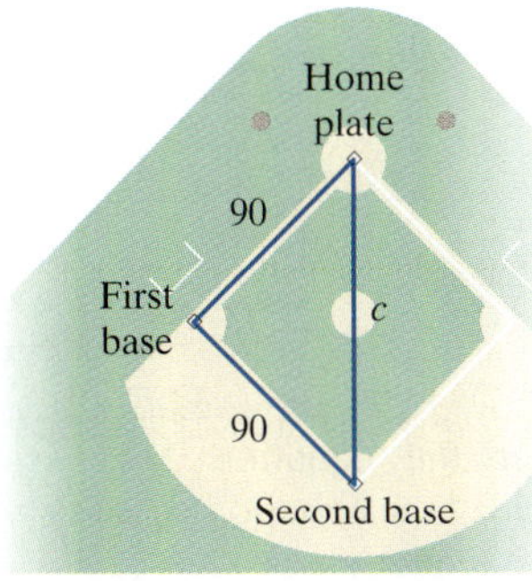


**100. Wire from Telephone Pole** A telephone pole is at a right, or 90°, angle with the ground as shown in the figure. Find the length of the wire that connects to the pole 40 feet above the ground and is anchored to the ground 20 feet from the base of the pole.

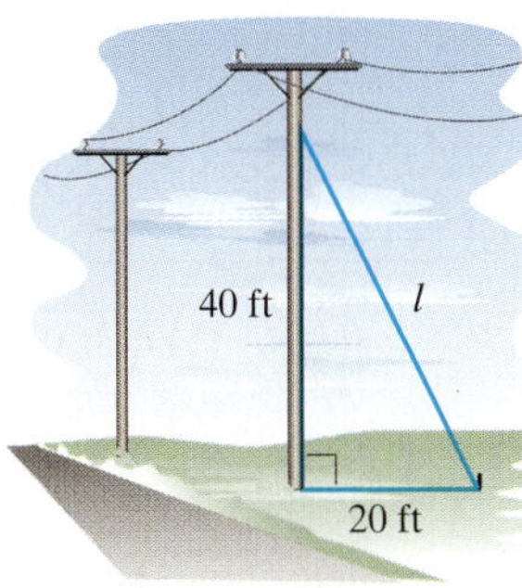


**101. Side of a Garden** When you are given the area of a square, the length of a side can be found by the formula $s = \sqrt{A}$. Find the side of Tom Kim's square garden if it has an area of 169 square feet.

**102. Radius of Basketball Hoop** When you are given the area of a circle, its radius can be found by the formula $r = \sqrt{A/\pi}$.

a) Find the radius of a basketball hoop if the area enclosed by the hoop is 254.47 square inches.

b) If the diameter of a basketball is 9 inches, what is the minimum distance possible between the hoop and the ball when the center of the ball is in the center of the hoop?

**103. Period of a Pendulum** The formula for the period of a pendulum is

$$T = 2\pi\sqrt{\frac{l}{g}}$$

where $T$ is the period in seconds, $l$ is its length in feet, and $g$ is the acceleration of gravity. On Earth, gravity is 32 ft/sec$^2$. The formula when used on Earth becomes

$$T = 2\pi\sqrt{\frac{l}{32}}$$

a) Find the period of a pendulum whose length is 8 feet.

b) If the length of a pendulum is doubled, what effect will this have on the period? Explain.

c) The gravity on the Moon is 1/6 that on Earth. If a pendulum has a period of 2 seconds on Earth, what will be the period of the same pendulum on the Moon?

**104. Diagonal of a Suitcase** A formula for the length of a diagonal from the upper corner of a box to the opposite lower corner is $d = \sqrt{L^2 + W^2 + H^2}$, where $L$, $W$, and $H$ are the length, width, and height, respectively.

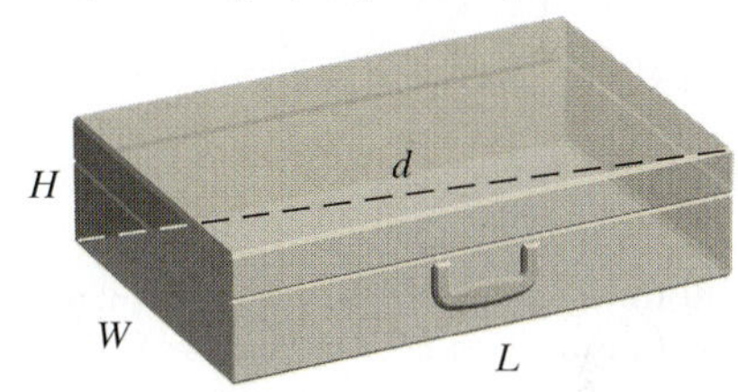


a) Find the length of the diagonal of a suitcase of length 22 inches, width 15 inches, and height 12 inches.

b) If the length, width, and height are all doubled, how will the diagonal change?

c) Solve the formula for $W$.

**105. Blood Flowing in an Artery** The formula

$$r = \sqrt[4]{\frac{8\mu l}{\pi R}}$$

is used in determining movement of blood through arteries. In the formula, $R$ represents the resistance to blood flow, $\mu$ is the viscosity of blood, $l$ is the length of the artery, and $r$ is the radius of the artery. Solve this equation for $R$.

**106. Falling Object** The formula

$$t = \frac{\sqrt{19.6s}}{9.8}$$

can be used to tell the time, $t$, in seconds, that an object has been falling if it has fallen $s$ meters. Suppose an object has been dropped from a helicopter and has fallen 100 meters. How long has it been in free fall?

**107. Earth Days** For any planet in our solar system, its "year" is the time it takes for the planet to revolve once around the Sun. The number of Earth days in a given planet's year, $N$, is approximated by the formula $N = 0.2(\sqrt{R})^3$, where $R$ is the mean distance of the planet to the Sun in millions of kilometers. Find the number of Earth days in the year of the planet Earth, whose mean distance to the Sun is 149.4 million kilometers.

**108. Earth Days** Find the number of Earth days in the year of the planet Mercury, whose mean distance to the Sun is 58 million kilometers. See Exercise 107.

**109. Forces on a Car** When two forces, $F_1$ and $F_2$, pull at right angles to each other as illustrated below, the resultant, or the effective force, $R$, can be found by the formula $R = \sqrt{F_1^2 + F_2^2}$. Two cars are trying to pull a third out of the mud, as illustrated. If car $A$ is exerting a force of 60 pounds and car $B$ is exerting a force of 80 pounds, find the resulting force on the car stuck in the mud.

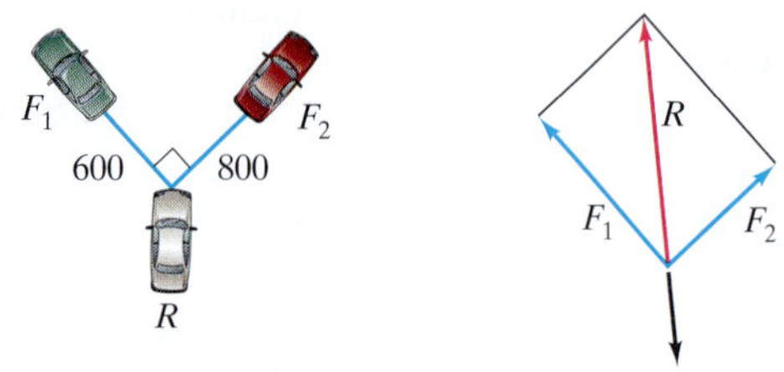


**110. Escape Velocity** The escape velocity, or the velocity needed for a spacecraft to escape a planet's gravitational field, is found by the formula $v_e = \sqrt{2gR}$, where $g$ is the force of gravity of the planet and $R$ is the radius of the planet. Find the escape velocity for Earth, in meters per second, where $g = 9.75$ meters per second squared and $R = 6{,}370{,}000$ meters.

**111. Motion of a Wave** A formula used in the study of shallow water wave motion is $c = \sqrt{gH}$, in which $c$ is wave velocity, $H$ is water depth, and $g$ is the acceleration due to gravity. Find the wave velocity if the water's depth is 10 feet. (Use $g = 32$ ft/sec$^2$.)

**112. Diagonal of a Box** The top of a rectangular box measures 20 inches by 32 inches. Find the length of the diagonal for the top of the box.

**113. Flower Garden** A rectangular flower garden measures 25 meters by 32 meters. Find the length of the diagonal for the garden.

**114. Speed of Sound** When sound travels through air (or any gas), the velocity of the sound wave is dependent on the air (or gas) temperature. The velocity, $v$, in meters per second, at air temperature, $t$, in degrees Celsius, can be found by the formula

$$v = 331.3\sqrt{1 + \frac{t}{273}}$$

Find the speed of sound in air whose temperature is 20°C (equivalent to 68°F).

*A formula that we have already mentioned and that we will be discussing in more detail shortly is the quadratic formula*

$$x = \frac{-b \pm \sqrt{b^2 - 4ac}}{2a}$$

**115.** Find $x$ when $a = 1, b = 0, c = -4$.

**116.** Find $x$ when $a = 1, b = 1, c = -12$.

**117.** Find $x$ when $a = -1, b = 4, c = 5$.

**118.** Find $x$ when $a = 2, b = 5, c = -12$.

*Given $f(x)$, find all values of $x$ for which $f(x)$ is the indicated value.*

**119.** $f(x) = \sqrt{x - 5}, f(x) = 5$

**120.** $f(x) = \sqrt[3]{2x + 3}, f(x) = 3$

**121.** $f(x) = \sqrt{3x^2 - 11} + 7, f(x) = 15$

**122.** $f(x) = 8 + \sqrt[3]{x^2 + 152}, f(x) = 14$

**123. a)** Consider the equation $\sqrt{4x - 12} = x - 3$. Setting each side of the equation equal to $y$ yields the following system of equations.

$$y = \sqrt{4x - 12}$$
$$y = x - 3$$

The graphs of the equations in the system are illustrated in the figure.

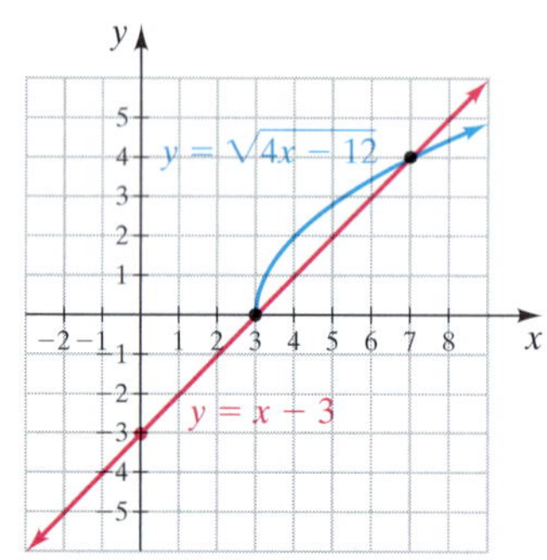


From the graphs, determine the values that appear to be solutions to the equation $\sqrt{4x - 12} = x - 3$. Explain how you determined your answer.

**b)** Substitute the values found in part **a)** into the original equation and determine whether they are the solutions to the equation.

**c)** Solve the equation $\sqrt{4x - 12} = x - 3$ algebraically and see if your solution agrees with the values obtained in part **a)**.

**124.** If the graph of a radical function, $f(x)$, does not intersect the $x$-axis, then the equation $f(x) = 0$ has no real solutions. Explain why.

**125.** Suppose we are given a rational function $g(x)$. If $g(4) = 0$, then the graph of $g(x)$ must intersect the $x$-axis at 4. Explain why.

**126.** The graph of the equation $y = \sqrt{x - 3} + 2$ is illustrated in the figure.

**a)** What is the domain of the function?

**b)** How many real solutions does the equation $\sqrt{x - 3} + 2 = 0$ have? List all the real solutions. Explain how you determined your answer.

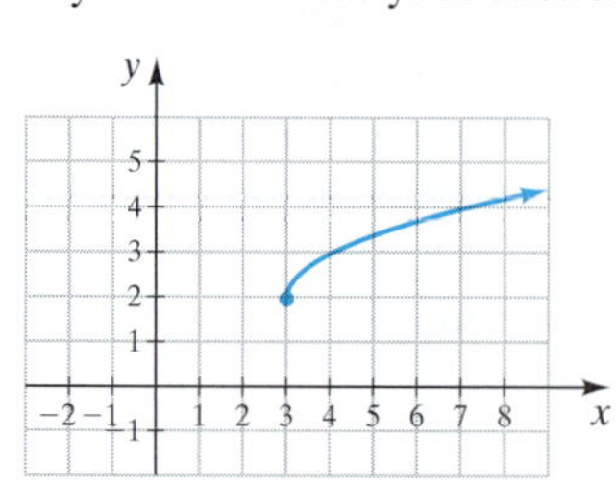

**127. Confidence Interval** In statistics, a "confidence interval" is a range of values that is likely to contain the true value of the population. For a "95% confidence interval," the lower limit of the range, $L_1$, and the upper limit of the range, $L_2$, can be found by the formulas

$$L_1 = p - 1.96\sqrt{\frac{p(1-p)}{n}}$$

$$L_2 = p + 1.96\sqrt{\frac{p(1-p)}{n}}$$

where $p$ represents the percent obtained from a sample and $n$ is the size of the sample. Francesco, a statistician, takes a sample of 36 families and finds that 60% of those surveyed use an answering machine in their home. He can be 95% certain that the true percent of families that use an answering machine in their home is between $L_1$ and $L_2$. Find the values of $L_1$ and $L_2$. Use $p = 0.60$ and $n = 36$ in the formulas.

**128. Quadratic Mean** The *quadratic mean* (or *root mean square, RMS*) is often used in physical applications. In power distribution systems, for example, voltages and currents are usually referred to in terms of their RMS values. The quadratic mean of a set of scores is obtained by squaring each score and adding the results (signified by $\Sigma x^2$), then dividing the value obtained by the number of scores, $n$, and then taking the square root of this value. We may express the formula as

$$\text{quadratic mean} = \sqrt{\frac{\Sigma x^2}{n}}$$

Find the quadratic mean of the numbers 2, 4, and 10.

*In Exercises 129 and 130, solve the equation.*

**129.** $\sqrt{x^2 + 49} = (x^2 + 49)^{1/2}$

**130.** $\sqrt{x^2 - 16} = (x^2 - 16)^{1/2}$

*In Exercises 131–134, use your graphing calculator to solve the equations. Round your solutions to the nearest tenth.*

**131.** $\sqrt{x + 8} = \sqrt{3x + 5}$

**132.** $\sqrt{10x - 16} - 15 = 0$

**133.** $\sqrt[3]{5x^2 - 6} - 4 = 0$

**134.** $\sqrt[3]{5x^2 - 22} = \sqrt[3]{4x + 83}$

## Challenge Problems

*Solve.*

**135.** $\sqrt{\sqrt{x + 25} - \sqrt{x}} = 5$

**136.** $\sqrt{\sqrt{x + 9} + \sqrt{x}} = 3$

*Solve each equation for n.*

**137.** $z = \dfrac{\overline{x} - \mu}{\dfrac{\sigma}{\sqrt{n}}}$

**138.** $z = \dfrac{p' - p}{\sqrt{\dfrac{pq}{n}}}$

## Group Activity

*Discuss and answer Exercise 139 as a group.*

**139. Heron's Formula** The area of a triangle is $A = \frac{1}{2}bh$. If the height is not known but we know the lengths of the three sides, we can use Heron's formula to find the area, $A$. Heron's formula is

$$A = \sqrt{S(S-a)(S-b)(S-c)}$$

where $a$, $b$, and $c$ are the lengths of the three sides and

$$S = \frac{a + b + c}{2}$$

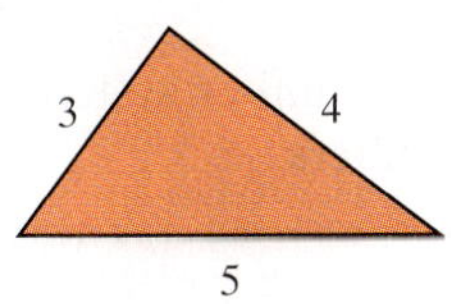


**a)** Have each group member use Heron's formula to find the area of a triangle whose sides are 3 inches, 4 inches, and 5 inches.

**b)** Compare your answers for part **a)**. If any member of the group did not get the correct answer, make sure they understand their error.

**c)** Have each member of the group do the following.

**1.** Draw a triangle on the following grid. Place each vertex of the triangle at the intersection of two grid lines.

**2.** Measure with a ruler the length of each side of your triangle.

**3.** Use Heron's formula to find the area of your triangle.

**4.** Compare and discuss your work from part **c)**.

## Cumulative Review Exercises

[2.2] **140.** Solve the formula $P_1P_2 - P_1P_3 = P_2P_3$ for $P_2$.

[6.1] **141.** Simplify $\dfrac{x(x-5)+x(x-2)}{2x-7}$.

*Perform each indicated operation.*

[6.1] **142.** $\dfrac{4a^2-9b^2}{4a^2+12ab+9b^2} \cdot \dfrac{6a^2b}{8a^2b^2-12ab^3}$

**143.** $(t^2-2t-15) \div \dfrac{t^2-9}{t^2-3t}$

[6.2] **144.** $\dfrac{2}{x+3} - \dfrac{1}{x-3} + \dfrac{2x}{x^2-9}$

[6.4] **145.** Solve $2 + \dfrac{3x}{x-1} = \dfrac{8}{x-1}$.

# 7.7 Complex Numbers

1. Recognize a complex number.
2. Add and subtract complex numbers.
3. Multiply complex numbers.
4. Divide complex numbers.
5. Find powers of $i$.

## 1 Recognize a Complex Number

In Section 7.1 we mentioned that the square roots of negative numbers, such as $\sqrt{-4}$, are not real numbers. Numbers like $\sqrt{-4}$ are called **imaginary numbers**. Such numbers are called imaginary because when they were introduced many mathematicians refused to believe that they existed. Although they do not belong to the set of real numbers, the imaginary numbers, by definition, do exist and are very useful in mathematics and science.

Every imaginary number has $\sqrt{-1}$ as a factor. The $\sqrt{-1}$, called the **imaginary unit**, is often denoted by the letter $i$.

**Imaginary Unit**

$$i = \sqrt{-1}$$

To write the square root of a negative number in terms of $i$, use the following property.

**Square Root of a Negative Number**

For any positive real number $n$,

$$\sqrt{-n} = \sqrt{-1}\,\sqrt{n} = i\sqrt{n}$$

Therefore, we can write

$$\sqrt{-4} = \sqrt{-1}\,\sqrt{4} = i2 \quad \text{or} \quad 2i$$
$$\sqrt{-9} = \sqrt{-1}\,\sqrt{9} = i3 \quad \text{or} \quad 3i$$
$$\sqrt{-7} = \sqrt{-1}\,\sqrt{7} = i\sqrt{7}$$

In this book we will generally write $i\sqrt{7}$ rather than $\sqrt{7}i$ to avoid confusion with $\sqrt{7i}$. Also, $3\sqrt{5}i$ is written as $3i\sqrt{5}$.

**Examples**

$$\sqrt{-81} = 9i \qquad \sqrt{-6} = i\sqrt{6}$$
$$\sqrt{-49} = 7i \qquad \sqrt{-10} = i\sqrt{10}$$

The real number system is a part of a larger number system, called the *complex number system*. Now we will discuss **complex numbers**.

**Complex Number**

Every number of the form

$$a + bi$$

where $a$ and $b$ are real numbers, is a **complex number**.

Every real number and every imaginary number are also complex numbers. A complex number has two parts: a real part, $a$, and an imaginary part, $b$.

Real part → $a + bi$ ← Imaginary part

If $b = 0$, the complex number is a real number. If $a = 0$, the complex number is a *pure imaginary number*.

### Examples of Complex Numbers

| | | |
|---|---|---|
| $3 + 2i$ | $a = 3, b = 2$ | |
| $5 - i\sqrt{6}$ | $a = 5, b = -\sqrt{6}$ | |
| $4$ | $a = 4, b = 0$ | (real number, $b = 0$) |
| $8i$ | $a = 0, b = 8$ | (imaginary number, $a = 0$) |
| $-i\sqrt{7}$ | $a = 0, b = -\sqrt{7}$ | (imaginary number, $a = 0$) |

We stated that all real numbers and imaginary numbers are also complex numbers. The relationship between the various sets of numbers is illustrated in **Figure 7.7**.

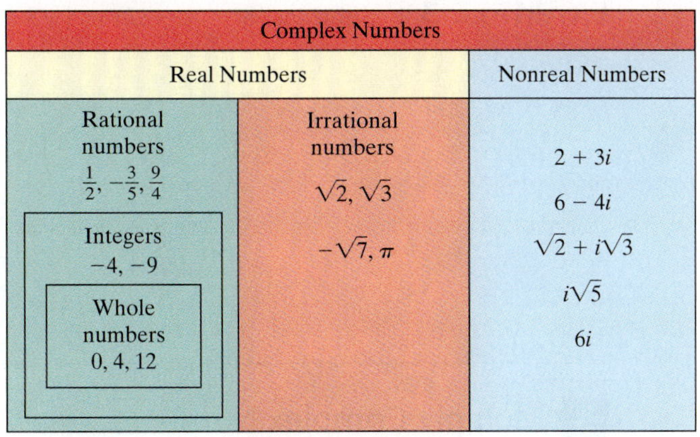


FIGURE 7.7

**EXAMPLE 1** ▸ Write each complex number in the form $a + bi$.

**a)** $7 + \sqrt{-36}$ **b)** $4 - \sqrt{-12}$ **c)** $19$ **d)** $\sqrt{-50}$ **e)** $6 + \sqrt{10}$

### Solution

**a)** $7 + \sqrt{-36} = 7 + \sqrt{-1}\sqrt{36}$
$= 7 + i6 \quad \text{or} \quad 7 + 6i$

**b)** $4 - \sqrt{-12} = 4 - \sqrt{-1}\sqrt{12}$
$= 4 - \sqrt{-1}\sqrt{4}\sqrt{3}$
$= 4 - i(2)\sqrt{3} \quad \text{or} \quad 4 - 2i\sqrt{3}$

**c)** $19 = 19 + 0i$

**d)** $\sqrt{-50} = 0 + \sqrt{-50}$
$= 0 + \sqrt{-1}\sqrt{25}\sqrt{2}$
$= 0 + i(5)\sqrt{2} \quad \text{or} \quad 0 + 5i\sqrt{2}$

**e)** Both 6 and $\sqrt{10}$ are real numbers. Written as a complex number, the answer is $(6 + \sqrt{10}) + 0i$.

▸ Now Try Exercise 23

Complex numbers can be added, subtracted, multiplied, and divided. To perform these operations, we use the definitions that $i = \sqrt{-1}$ and

**Definition of $i^2$**

$$i^2 = -1$$

## 2 Add and Subtract Complex Numbers

We now explain how to add or subtract complex numbers.

**To Add or Subtract Complex Numbers**

1. Change all imaginary numbers to *bi* form.
2. Add (or subtract) the real parts of the complex numbers.
3. Add (or subtract) the imaginary parts of the complex numbers.
4. Write the answer in the form $a + bi$.

**EXAMPLE 2** ▸ Add $(9 + 15i) + (-6 - 2i) + 18$.

*Solution*

$$\begin{aligned}(9 + 15i) + (-6 - 2i) + 18 &= 9 + 15i - 6 - 2i + 18\\ &= 9 - 6 + 18 + 15i - 2i \quad \text{Rearrange terms.}\\ &= 21 + 13i \quad \text{Combine like terms.}\end{aligned}$$

▸ Now Try Exercise 27

**EXAMPLE 3** ▸ Subtract $(8 - \sqrt{-27}) - (-3 + \sqrt{-48})$.

*Solution*

$$\begin{aligned}(8 - \sqrt{-27}) - (-3 + \sqrt{-48}) &= (8 - \sqrt{-1}\sqrt{27}) - (-3 + \sqrt{-1}\sqrt{48})\\ &= (8 - \sqrt{-1}\sqrt{9}\sqrt{3}) - (-3 + \sqrt{-1}\sqrt{16}\sqrt{3})\\ &= (8 - 3i\sqrt{3}) - (-3 + 4i\sqrt{3})\\ &= 8 - 3i\sqrt{3} + 3 - 4i\sqrt{3}\\ &= 8 + 3 - 3i\sqrt{3} - 4i\sqrt{3}\\ &= 11 - 7i\sqrt{3}\end{aligned}$$

▸ Now Try Exercise 35

## 3 Multiply Complex Numbers

Now let's discuss how to multiply complex numbers.

**To Multiply Complex Numbers**

1. Change all imaginary numbers to *bi* form.
2. Multiply the complex numbers as you would multiply polynomials.
3. Substitute $-1$ for each $i^2$.
4. Combine the real parts and the imaginary parts. Write the answer in $a + bi$ form.

**EXAMPLE 4** ▸ Multiply.

**a)** $5i(6 - 2i)$ **b)** $\sqrt{-9}(\sqrt{-3} + 8)$ **c)** $(2 - \sqrt{-18})(\sqrt{-2} + 5)$

*Solution*

**a)**
$$\begin{aligned}5i(6 - 2i) &= 5i(6) + 5i(-2i) \quad \text{Distributive property}\\ &= 30i - 10i^2\\ &= 30i - 10(-1) \quad \text{Replace } i^2 \text{ with } -1.\\ &= 30i + 10 \quad \text{or} \quad 10 + 30i\end{aligned}$$

**b)**
$$\begin{aligned}\sqrt{-9}(\sqrt{-3} + 8) &= 3i(i\sqrt{3} + 8) \quad \text{Change imaginary numbers to } bi \text{ form.}\\ &= 3i(i\sqrt{3}) + 3i(8) \quad \text{Distributive property}\\ &= 3i^2\sqrt{3} + 24i\\ &= 3(-1)\sqrt{3} + 24i \quad \text{Replace } i^2 \text{ with } -1.\\ &= -3\sqrt{3} + 24i\end{aligned}$$

**c)** $(2 - \sqrt{-18})(\sqrt{-2} + 5) = (2 - \sqrt{-1}\,\sqrt{18})(\sqrt{-1}\,\sqrt{2} + 5)$
$$= (2 - \sqrt{-1}\,\sqrt{9}\,\sqrt{2})(\sqrt{-1}\,\sqrt{2} + 5)$$
$$= (2 - 3i\sqrt{2})(i\sqrt{2} + 5)$$

Now use the FOIL method to multiply.

$$\begin{aligned}(2 - 3i\sqrt{2})(i\sqrt{2} + 5) &= (2)(i\sqrt{2}) + (2)(5) + (-3i\sqrt{2})(i\sqrt{2}) + (-3i\sqrt{2})(5)\\ &= 2i\sqrt{2} + 10 - 3i^2(2) - 15i\sqrt{2}\\ &= 2i\sqrt{2} + 10 - 3(-1)(2) - 15i\sqrt{2}\\ &= 2i\sqrt{2} + 10 + 6 - 15i\sqrt{2}\\ &= 16 - 13i\sqrt{2}\end{aligned}$$

▶ Now Try Exercise 45

## Avoiding Common Errors

What is $\sqrt{-4} \cdot \sqrt{-2}$?

**CORRECT**

$$\begin{aligned}\sqrt{-4} \cdot \sqrt{-2} &= 2i \cdot i\sqrt{2}\\ &= 2i^2\sqrt{2}\\ &= 2(-1)\sqrt{2}\\ &= -2\sqrt{2}\end{aligned}$$

**INCORRECT**

$$\begin{aligned}\sqrt{-4} \cdot \sqrt{-2} &= \sqrt{8}\\ &= \sqrt{4} \cdot \sqrt{2}\\ &= 2\sqrt{2}\end{aligned}$$

Recall that $\sqrt{a} \cdot \sqrt{b} = \sqrt{ab}$ only for *nonnegative* real numbers $a$ and $b$.

## 4 Divide Complex Numbers

The **conjugate of a complex number** $a + bi$ is $a - bi$. For example,

| Complex Number | Conjugate |
|---|---|
| $3 + 7i$ | $3 - 7i$ |
| $1 - i\sqrt{3}$ | $1 + i\sqrt{3}$ |
| $2i$ (or $0 + 2i$) | $-2i$ (or $0 - 2i$) |

When a complex number is multiplied by its conjugate using the FOIL method, the inner and outer products will sum to 0, and the result is a real number. For example,

$$\begin{aligned}(5 + 3i)(5 - 3i) &= 25 - 15i + 15i - 9i^2\\ &= 25 - 9i^2\\ &= 25 - 9(-1)\\ &= 25 + 9 = 34\end{aligned}$$

Now we explain how to divide complex numbers.

### To Divide Complex Numbers

1. Change all imaginary numbers to $bi$ form.
2. Rationalize the denominator by multiplying both the numerator and denominator by the conjugate of the denominator.
3. Write the answer in $a + bi$ form.

**EXAMPLE 5** ▶ Divide $\dfrac{9+i}{i}$.

*Solution* Begin by multiplying both numerator and denominator by $-i$, the conjugate of $i$.

$$\frac{9+i}{i}\cdot\frac{-i}{-i}=\frac{(9+i)(-i)}{-i^2}$$

$$=\frac{-9i-i^2}{-i^2} \qquad \text{Distributive property}$$

$$=\frac{-9i-(-1)}{-(-1)} \qquad \text{Replace } i^2 \text{ with } -1.$$

$$=\frac{-9i+1}{1}$$

$$=1-9i$$

▶ **Now Try Exercise 59**

**EXAMPLE 6** ▶ Divide $\dfrac{3+2i}{4-i}$.

*Solution* Multiply both numerator and denominator by $4+i$, the conjugate of $4-i$.

$$\frac{3+2i}{4-i}\cdot\frac{4+i}{4+i}=\frac{12+3i+8i+2i^2}{16-i^2}$$

$$=\frac{12+11i+2(-1)}{16-(-1)}$$

$$=\frac{10+11i}{17} \quad \text{or} \quad \frac{10}{17}+\frac{11}{17}i$$

▶ **Now Try Exercise 65**

**EXAMPLE 7** ▶ **Impedance** A concept needed for the study of electronics is *impedance*. Impedance affects the current in a circuit. The impedance, $Z$, in a circuit is found by the formula $Z=\dfrac{V}{I}$, where $V$ is voltage and $I$ is current. Find $Z$ when $V=1.6-0.3i$ and $I=-0.2i$, where $i=\sqrt{-1}$.

*Solution* $Z=\dfrac{V}{I}=\dfrac{1.6-0.3i}{-0.2i}$. Now multiply both numerator and denominator by the conjugate of the denominator, $0.2i$.

$$Z=\frac{1.6-0.3i}{-0.2i}\cdot\frac{0.2i}{0.2i}=\frac{0.32i-0.06i^2}{-0.04i^2}$$

$$=\frac{0.32i+0.06}{0.04}$$

$$=\frac{0.32i}{0.04}+\frac{0.06}{0.04}$$

$$=8i+1.5 \quad \text{or} \quad 1.5+8i$$

▶ **Now Try Exercise 127**

Most algebra books use $i$ as the imaginary unit. However, most electronics books use $j$ as the imaginary unit because $i$ is often used to represent current.

## 5 Find Powers of $i$

Using $i = \sqrt{-1}$ and $i^2 = -1$, we can find other **powers of $i$**. For example,

$$i^3 = i^2 \cdot i = -1 \cdot i = -i \qquad i^6 = i^4 \cdot i^2 = 1(-1) = -1$$
$$i^4 = i^2 \cdot i^2 = (-1)(-1) = 1 \qquad i^7 = i^4 \cdot i^3 = 1(-i) = -i$$
$$i^5 = i^4 \cdot i^1 = 1 \cdot i = i \qquad i^8 = i^4 \cdot i^4 = (1)(1) = 1$$

Note that successive powers of $i$ rotate through the four values $i$, $-1$, $-i$, and 1 (see **Fig. 7.8**).

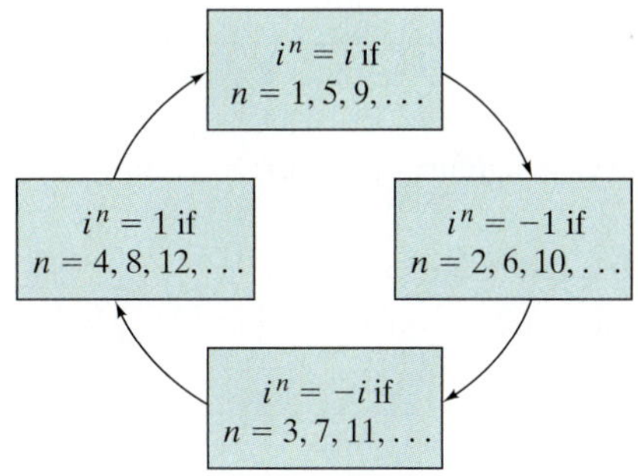


FIGURE 7.8

**EXAMPLE 8** ▸ Evaluate. **a)** $i^{35}$ **b)** $i^{101}$

*Solution* Write each expression as a product of factors such that the exponent of one factor is the largest multiple of 4 less than or equal to the given exponent. Then write this factor as $i^4$ raised to some power. Since $i^4$ has a value of 1, the expression $i^4$ raised to a power will also have a value of 1.

**a)** $i^{35} = i^{32} \cdot i^3 = (i^4)^8 \cdot i^3 = 1 \cdot i^3 = 1(-i) = -i$

**b)** $i^{101} = i^{100} \cdot i^1 = (i^4)^{25} \cdot i = 1 \cdot i = i$

▸ Now Try Exercise 101

### Helpful Hint

A quick way of evaluating $i^n$ is to divide the exponent by 4 and observe the remainder.

If the remainder is 0, the value is 1. If the remainder is 2, the value is $-1$.
If the remainder is 1, the value is $i$. If the remainder is 3, the value is $-i$.

For Example 8 **a)**

$$\begin{array}{r} 8 \\ 4\overline{)35} \\ \underline{32} \\ 3 \end{array}$$

Answer is $-i$.

$i^{35} = (i^4)^8 \cdot i^3 = (1)^8 \cdot i^3 = 1 \cdot i^3 = i^3 = -i$

For 8 **b)**

$$\begin{array}{r} 25 \\ 4\overline{)101} \\ \underline{8} \\ 21 \\ \underline{20} \\ 1 \end{array}$$

← Answer is $i$.

**EXAMPLE 9** ▸ Let $f(x) = x^2$. Find **a)** $f(6i)$ **b)** $f(3 + 7i)$.

*Solution*

**a)** $f(x) = x^2$

$f(6i) = (6i)^2 = 36i^2 = 36(-1) = -36$

**b)** $f(x) = x^2$

$$\begin{aligned} f(3 + 7i) = (3 + 7i)^2 &= (3)^2 + 2(3)(7i) + (7i)^2 \\ &= 9 + 42i + 49i^2 \\ &= 9 + 42i + 49(-1) \\ &= 9 + 42i - 49 \\ &= -40 + 42i \end{aligned}$$

▸ Now Try Exercise 117

## EXERCISE SET 7.7

MathXL® MyMathLab

1. a) What does $i$ equal?
   b) What does $i^2$ equal?
2. Write $\sqrt{-n}$ using $i$.
3. Are all of the following complex numbers? If any are not complex numbers, explain why.
   a) $9$ b) $-\frac{1}{2}$ c) $4 - \sqrt{-2}$
   d) $7 - 3i$ e) $4.2i$ f) $11 + \sqrt{3}$
4. What does $i^4$ equal?
5. Is every real and every imaginary number a complex number?
6. Is every complex number a real number?
7. What is the conjugate of $a + bi$?
8. a) Is $i \cdot i$ a real number? Explain.
   b) Is $i \cdot i \cdot i$ a real number? Explain.
9. List, if possible, a number that is *not*
   a) a rational number.
   b) an irrational number.
   c) a real number.
   d) an imaginary number.
   e) a complex number.
10. Write a paragraph or two explaining the relationship between the real numbers, imaginary numbers, and complex numbers. Include in your discussion how the various sets of numbers relate to each other.

### Practice the Skills

*Write each expression as a complex number in the form $a + bi$.*

11. $7$
12. $3i$
13. $\sqrt{25}$
14. $\sqrt{-100}$
15. $21 - \sqrt{-36}$
16. $\sqrt{3} + \sqrt{-3}$
17. $\sqrt{-24}$
18. $\sqrt{49} - \sqrt{-49}$
19. $8 - \sqrt{-12}$
20. $\sqrt{-9} + \sqrt{-81}$
21. $3 + \sqrt{-98}$
22. $\sqrt{-9} + 7i$
23. $12 - \sqrt{-25}$
24. $10 + \sqrt{-32}$
25. $7i - \sqrt{-45}$
26. $\sqrt{144} + \sqrt{-96}$

*Add or subtract.*

27. $(19 - i) + (2 + 9i)$
28. $(22 + i) - 5(11 - 3i) + 4$
29. $(8 - 3i) + (-8 + 3i)$
30. $(7 - \sqrt{-4}) - (-1 - \sqrt{-16})$
31. $(1 + \sqrt{-1}) + (-18 - \sqrt{-169})$
32. $(16 - i\sqrt{3}) + (17 - \sqrt{-3})$
33. $(\sqrt{3} + \sqrt{2}) + (3\sqrt{2} - \sqrt{-8})$
34. $(8 - \sqrt{2}) - (5 + \sqrt{-15})$
35. $(5 - \sqrt{-72}) + (6 + \sqrt{-8})$
36. $(29 + \sqrt{-75}) + (\sqrt{-147})$
37. $(\sqrt{4} - \sqrt{-45}) + (-\sqrt{25} + \sqrt{-5})$
38. $(\sqrt{20} - \sqrt{-12}) + (2\sqrt{5} + \sqrt{-75})$

*Multiply.*

39. $2(3 - i)$
40. $-7(5 + 3i\sqrt{5})$
41. $i(4 + 9i)$
42. $3i(6 - i)$
43. $\sqrt{-9}(6 + 11i)$
44. $\frac{1}{2}i\left(\frac{1}{3} - 18i\right)$
45. $\sqrt{-16}(\sqrt{3} - 7i)$
46. $-\sqrt{-24}(\sqrt{6} - \sqrt{-3})$
47. $\sqrt{-27}(\sqrt{3} - \sqrt{-3})$
48. $\sqrt{-32}(\sqrt{2} + \sqrt{-8})$
49. $(3 + 2i)(1 + i)$
50. $(6 - 2i)(3 + i)$
51. $(10 - 3i)(10 + 3i)$
52. $(-4 + 3i)(2 - 5i)$
53. $(7 + \sqrt{-2})(5 - \sqrt{-8})$
54. $(\sqrt{4} - 3i)(4 + \sqrt{-4})$
55. $\left(\frac{1}{2} - \frac{1}{3}i\right)\left(\frac{1}{4} + \frac{2}{3}i\right)$
56. $\left(\frac{3}{5} - \frac{1}{4}i\right)\left(\frac{2}{3} + \frac{2}{5}i\right)$

*Divide.*

57. $\frac{8}{3i}$
58. $\frac{5}{4i}$
59. $\frac{2 + 3i}{2i}$
60. $\frac{7 - 3i}{2i}$
61. $\frac{6}{2 - i}$
62. $\frac{9}{5 + i}$
63. $\frac{3}{1 - 2i}$
64. $\frac{13}{-3 - 4i}$
65. $\frac{6 - 3i}{4 + 2i}$
66. $\frac{4 - 3i}{4 + 3i}$
67. $\frac{4}{6 - \sqrt{-4}}$
68. $\frac{2}{3 + \sqrt{-5}}$
69. $\frac{\sqrt{2}}{5 + \sqrt{-12}}$
70. $\frac{\sqrt{6}}{\sqrt{3} - \sqrt{-9}}$
71. $\frac{\sqrt{10} + \sqrt{-3}}{5 - \sqrt{-20}}$
72. $\frac{12 - \sqrt{-12}}{\sqrt{3} + \sqrt{-5}}$
73. $\frac{\sqrt{-75}}{\sqrt{-3}}$
74. $\frac{\sqrt{-30}}{\sqrt{-2}}$
75. $\frac{\sqrt{-32}}{\sqrt{-18}\sqrt{8}}$
76. $\frac{\sqrt{-40}\sqrt{-20}}{\sqrt{-4}}$

*Perform each indicated operation. These exercises are a combination of the types of exercises presented earlier in this exercise set.*

**77.** $(9 - 2i) + (3 - 5i)$

**78.** $\left(\frac{1}{2} + 2i\right) - \left(\frac{3}{5} - \frac{2}{3}i\right)$

**79.** $(\sqrt{50} - \sqrt{2}) - (\sqrt{-12} - \sqrt{-48})$

**80.** $(8 - \sqrt{-6}) - (2 - \sqrt{-24})$

**81.** $5.2(4 - 3.2i)$

**82.** $\sqrt{-6}(\sqrt{3} - \sqrt{-10})$

**83.** $(9 + 2i)(3 - 5i)$

**84.** $(\sqrt{3} + 2i)(\sqrt{6} - \sqrt{-8})$

**85.** $\dfrac{11 + 4i}{2i}$

**86.** $\dfrac{1}{4 + 3i}$

**87.** $\dfrac{6}{\sqrt{3} - \sqrt{-4}}$

**88.** $\dfrac{5 - 2i}{3 + 2i}$

**89.** $\left(11 - \frac{5}{9}i\right) - \left(4 - \frac{3}{5}i\right)$

**90.** $\frac{8}{7}\left(4 - \frac{2}{5}i\right)$

**91.** $\left(\frac{2}{3} - \frac{1}{5}i\right)\left(\frac{3}{5} - \frac{3}{4}i\right)$

**92.** $\sqrt{\frac{4}{9}}\left(\sqrt{\frac{25}{36}} - \sqrt{-\frac{4}{25}}\right)$

**93.** $\dfrac{\sqrt{-48}}{\sqrt{-12}}$

**94.** $\dfrac{-6 - 2i}{2 + \sqrt{-5}}$

**95.** $(5.23 - 6.41i) - (9.56 + 4.5i)$

**96.** $(\sqrt{-6} + 3)(\sqrt{-15} + 5)$

*For each imaginary number, indicate whether its value is i, −1, −i, or 1.*

**97.** $i^6$ **98.** $i^{63}$ **99.** $i^{160}$ **100.** $i^{231}$

**101.** $i^{93}$ **102.** $i^{103}$ **103.** $i^{811}$ **104.** $i^{1213}$

## Problem Solving

**105.** Consider the complex number $2 + 3i$.

**a)** Find the additive inverse.

**b)** Find the multiplicative inverse. Write the answer in simplified form.

**106.** Consider the complex number $4 - 5i$.

**a)** Find the additive inverse.

**b)** Find the multiplicative inverse. Write the answer in simplified form.

*In Exercises 107–110, answer true or false. Support your answer with an example.*

**107.** The product of two pure imaginary numbers is always a real number.

**108.** The sum of two pure imaginary numbers is always an imaginary number.

**109.** The product of two complex numbers is always a real number.

**110.** The sum of two complex numbers is always a complex number.

**111.** What values of $n$ will result in $i^n$ being a real number? Explain.

**112.** What values of $n$ will result in $i^{2n}$ being a real number? Explain.

**113.** If $f(x) = x^2$, find $f(2i)$.

**114.** If $f(x) = x^2$, find $f(4i)$.

**115.** If $f(x) = x^4 - 2x$, find $f(2i)$.

**116.** If $f(x) = x^3 - 4x^2$, find $f(5i.)$

**117.** If $f(x) = x^2 + 2x$, find $f(3 + i)$.

**118.** If $f(x) = \dfrac{x^2}{x - 2}$, find $f(4 - i)$.

*Evaluate each expression for the given value of x.*

**119.** $x^2 - 2x + 5, x = 1 + 2i$

**120.** $x^2 - 2x + 5, x = 1 - 2i$

**121.** $x^2 + 2x + 7, x = -1 + i\sqrt{5}$

**122.** $x^2 + 2x + 9, x = -1 - i\sqrt{5}$

*In Exercises 123–126, determine whether the given value of x is a solution to the equation.*

**123.** $x^2 - 4x + 5 = 0, x = 2 - i$

**124.** $x^2 - 4x + 5 = 0, x = 2 + i$

**125.** $x^2 - 6x + 11 = 0, x = -3 + i\sqrt{3}$

**126.** $x^2 - 6x + 15 = 0, x = 3 - i\sqrt{3}$

**127. Impedance** Find the impedance, $Z$, using the formula $Z = \dfrac{V}{I}$ when $V = 1.8 + 0.5i$ and $I = 0.6i$. See Example 7.

**128. Impedance** Refer to Exercise 127. Find the impedance when $V = 2.4 - 0.6i$ and $I = -0.4i$.

**129. Impedance** Under certain conditions, the total impedance, $Z_T$, of a circuit is given by the formula

$$Z_T = \frac{Z_1 Z_2}{Z_1 + Z_2}$$

Find $Z_T$ when $Z_1 = 2 - i$ and $Z_2 = 4 + i$.

**130. Impedance** Refer to Exercise 129. Find $Z_T$ when $Z_1 = 3 - i$ and $Z_2 = 5 + i$.

**131.** Determine whether $i^{-1}$ is equal to $i, -1, -i$, or 1. Show your work.

**132.** Determine whether $i^{-5}$ is equal to $i, -1, -i$, or 1. Show your work.

*In Chapter 8, we will use the quadratic formula* $x = \dfrac{-b \pm \sqrt{b^2 - 4ac}}{2a}$ *to solve equations of the form* $ax^2 + bx + c = 0$. ***(a)*** *Use the quadratic formula to solve the following quadratic equations.* ***(b)*** *Check each of the two solutions by substituting the values found for x (one at a time) back into the original equation. In these exercises, the ± (read "plus or minus") results in two distinct complex answers.*

**133.** $x^2 - 2x + 6 = 0$

**134.** $x^2 - 4x + 6 = 0$

*Given the complex numbers* $a = 5 + 2i\sqrt{3}, b = 1 + i\sqrt{3}$, *evaluate each expression.*

**135.** $a + b$

**136.** $a - b$

**137.** $ab$

**138.** $\dfrac{a}{b}$

## Cumulative Review Exercises

[4.3] **139. Mixture** Berreda Coughlin, a grocer in Dallas, has two coffees, one selling for $5.50 per pound and the other for $6.30 per pound. How many pounds of each type of coffee should he mix to make 40 pounds of coffee to sell for $6.00 per pound?

[5.3] **140.** Divide $\dfrac{8c^2 + 6c - 35}{4c + 9}$.

[6.2] **141.** Add $\dfrac{b}{a - b} + \dfrac{a + b}{b}$.

[6.4] **142.** Solve $\dfrac{x}{4} + \dfrac{1}{2} = \dfrac{x - 1}{2}$.

# Chapter 7 Summary

| IMPORTANT FACTS AND CONCEPTS | EXAMPLES |
|---|---|
| **Section 7.1** | |
| A **radical expression** has the form $\sqrt[n]{x}$, where $n$ is the index and $x$ is the radicand. | In the radical expression $\sqrt[3]{x}$, 3 is the index and $x$ is the radicand. |
| The **principal square root** of a positive number $a$, written $\sqrt{a}$, is the positive number $b$ such that $b^2 = a$. | $\sqrt{81} = 9$, since $9^2 = 81$<br>$\sqrt{0.36} = 0.6$ since $(0.6)^2 = 0.36$ |
| The **square root function** is $f(x) = \sqrt{x}$. Its domain is $[0, \infty)$ and its range is $[0, \infty)$. | 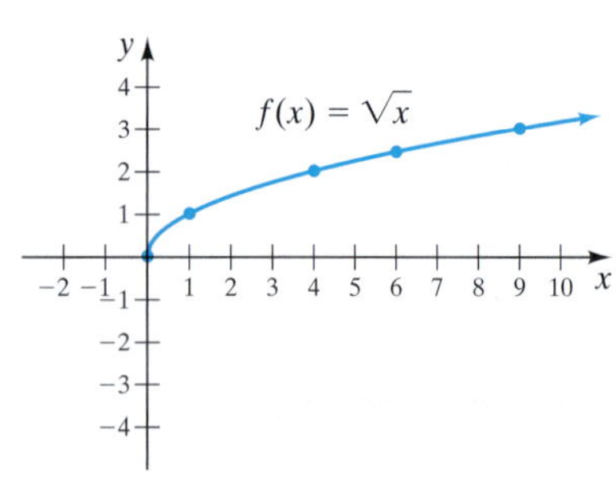 |
| The **cube root** of a number $a$, written $\sqrt[3]{a}$, is the number $b$ such that $b^3 = a$. | $\sqrt[3]{27} = 3$ since $3^3 = 27$<br>$\sqrt[3]{-125} = -5$ since $(-5)^3 = -125$ |

| IMPORTANT FACTS AND CONCEPTS | EXAMPLES |
|---|---|
| **Section 7.1 (continued)** | |
| The **cube root function** is $f(x) = \sqrt[3]{x}$. Its domain is $(-\infty, \infty)$ or $\mathbb{R}$ and its range is $(-\infty, \infty)$ or $\mathbb{R}$. | 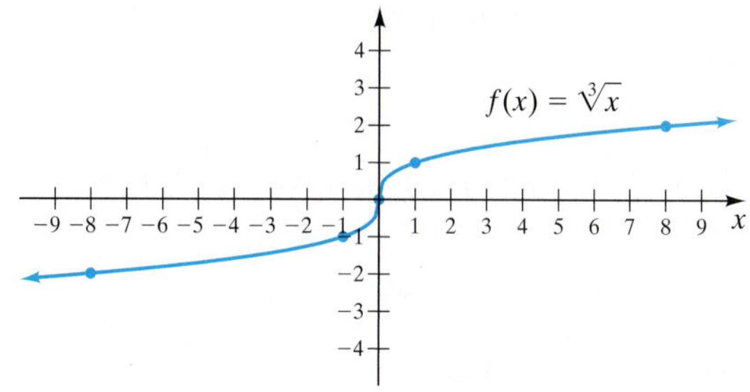  |
| The *n*th root of *a*, $\sqrt[n]{a}$, where *a* is an **even index** and *a* is a nonnegative real number, is the nonnegative number *b* such that $b^n = a$. | $\sqrt{4} = 2$ since $2^2 = 2 \cdot 2 = 4$<br>$\sqrt[4]{81} = 3$ since $3^4 = 3 \cdot 3 \cdot 3 \cdot 3 = 81$ |
| The *n*th root of *a*, $\sqrt[n]{a}$, where *n* is an **odd index** and *a* is any real number, is the real number *b* such that $b^n = a$. | $\sqrt[3]{27} = 3$ since $3^3 = 3 \cdot 3 \cdot 3 = 27$<br>$\sqrt[5]{-32} = -2$ since $(-2)^5 = (-2)(-2)(-2)(-2)(-2) = -32$ |
| For any real number *a*, $\sqrt{a^2} = \lvert a \rvert$. | $\sqrt{(-6)^2} = \lvert -6 \rvert = 6$<br>$\sqrt{(y + 8)^2} = \lvert y + 8 \rvert$ |
| **Section 7.2** | |
| **Rational Exponent**<br>$$\sqrt[n]{a} = a^{1/n}$$<br>When *a* is nonnegative, *n* can be any index.<br>When *a* is negative, *n* must be odd. | $\sqrt{17} = 17^{1/2}$<br>$\sqrt[4]{21x^3y^2} = (21x^3y^2)^{1/4}$ |
| For any nonnegative number *a*, and integers *m* and *n*,<br>$$\sqrt[n]{a^m} = \left(\sqrt[n]{a}\right)^m = a^{m/n}$$ (Power: *m*; Index: *n*) | $\sqrt[4]{z^9} = (\sqrt[4]{z})^9 = z^{9/4}$ |
| For any nonnegative real number *a*,<br>$$\sqrt[n]{a^n} = (\sqrt[n]{a})^n = a^{n/n} = a$$ | $\sqrt[4]{y^4} = y, \quad \sqrt[8]{14^8} = 14$ |
| **Rules of Exponents**<br>For all real numbers *a* and *b* and all rational numbers *m* and *n*, | |
| Product rule $\quad a^m \cdot a^n = a^{m+n}$ | $x^{1/3} \cdot x^{4/3} = x^{(1/3)+(4/3)} = x^{5/3}$ |
| Quotient rule $\quad \dfrac{a^m}{a^n} = a^{m-n}, \quad a \neq 0$ | $\dfrac{x^{4/5}}{x^{1/2}} = x^{(4/5)-(1/2)} = x^{(8/10)-(5/10)} = x^{3/10}$ |
| Negative exponent rule $\quad a^{-m} = \dfrac{1}{a^m}, \quad a \neq 0$ | $x^{-1/7} = \dfrac{1}{x^{1/7}}$ |
| Zero exponent rule $\quad a^0 = 1, \quad a \neq 0$ | $m^0 = 1$ |
| Raising a power to a power $\quad (a^m)^n = a^{m \cdot n}$ | $(c^{1/8})^{16} = c^{(1/8) \cdot 16} = c^2$ |
| Raising a product to a power $\quad (ab)^m = a^m b^m$ | $(p^3q^4)^{1/8} = p^{3/8}q^{1/2}$ |
| Raising a quotient to a power $\quad \left(\dfrac{a}{b}\right)^m = \dfrac{a^m}{b^m}, \quad b \neq 0$<br>$\left(\dfrac{a}{b}\right)^{-n} = \left(\dfrac{b}{a}\right)^n = \dfrac{b^n}{a^n}, \quad a \neq 0, b \neq 0$ | $\left(\dfrac{81}{49}\right)^{-1/2} = \left(\dfrac{49}{81}\right)^{1/2} = \dfrac{49^{1/2}}{81^{1/2}} = \dfrac{7}{9}$ |

| IMPORTANT FACTS AND CONCEPTS | EXAMPLES |
|---|---|
| **Section 7.3** | |
| A number or expression is a **perfect square** if it is the square of an expression.<br><br>A number or expression is a **perfect cube** if it is the cube of an expression. | Perfect squares: $49 \quad 81 \quad x^{12} \quad y^{50}$<br>$\downarrow \quad \downarrow \quad \downarrow \quad \downarrow$<br>Square of a number or expression: $7^2 \quad 9^2 \quad (x^6)^2 \quad (y^{25})^2$<br>Perfect cubes: $27 \quad -27 \quad y^{18} \quad z^{30}$<br>$\downarrow \quad \downarrow \quad \downarrow \quad \downarrow$<br>Cube of a number or expression: $3^3 \quad (-3)^3 \quad (y^6)^3 \quad (z^{10})^3$ |
| **Product Rule for Radicals**<br>For nonnegative real numbers $a$ and $b$,<br>$\sqrt[n]{a}\cdot\sqrt[n]{b} = \sqrt[n]{ab}$ | $\sqrt{2}\cdot\sqrt{8} = \sqrt{16} = 4, \quad \sqrt[3]{2x^3} = \sqrt[3]{x^3}\cdot\sqrt[3]{2} = x\sqrt[3]{2}$ |
| **To Simplify Radicals Using the Product Rule**<br>**1.** If the radicand contains a coefficient other than 1, write it as a product of two numbers, one of which is the largest perfect power for the index.<br>**2.** Write each variable factor as a product of two factors, one of which is the largest perfect power of the variable for the index.<br>**3.** Use the product rule to write the radical expression as a product of radicals. Place all the perfect powers (numbers and variables) under the same radical.<br>**4.** Simplify the radical containing the perfect powers. | $\sqrt{24} = \sqrt{4\cdot 6} = \sqrt{4}\,\sqrt{6} = 2\sqrt{6}$<br><br>$\sqrt[3]{16x^5y^9} = \sqrt[3]{8x^3y^9\cdot 2x^2}$<br>$= \sqrt[3]{8x^3y^9}\,\sqrt[3]{2x^2}$<br>$= 2xy^3\sqrt[3]{2x^2}$ |
| **Quotient Rule for Radicals**<br>For nonnegative real numbers $a$ and $b$,<br>$\dfrac{\sqrt[n]{a}}{\sqrt[n]{b}} = \sqrt[n]{\dfrac{a}{b}}, \quad b \neq 0$ | $\dfrac{\sqrt{32}}{\sqrt{2}} = \sqrt{\dfrac{32}{2}} = \sqrt{16} = 4, \quad \sqrt[3]{\dfrac{x^6}{y^{12}}} = \dfrac{\sqrt[3]{x^6}}{\sqrt[3]{y^{12}}} = \dfrac{x^2}{y^4}$ |
| **Section 7.4** | |
| **Like radicals** are radicals with the same radicand and index.<br>**Unlike radicals** are radicals with a different radicand or index. | Like Radicals: $\sqrt{3},\ 12\sqrt{3}$; $2\sqrt[4]{xy^3},\ -3\sqrt[4]{xy^3}$<br>Unlike Radicals: $\sqrt{3},\ 7\sqrt[4]{3}$; $\sqrt[5]{xy^3},\ x\sqrt[5]{y^3}$ |
| **To Add or Subtract Radicals**<br>**1.** Simplify each radical expression.<br>**2.** Combine like radicals (if there are any). | $\sqrt{27} + \sqrt{48} - 2\sqrt{75} = \sqrt{9}\cdot\sqrt{3} + \sqrt{16}\cdot\sqrt{3} - 2\cdot\sqrt{25}\cdot\sqrt{3}$<br>$= 3\sqrt{3} + 4\sqrt{3} - 10\sqrt{3}$<br>$= -3\sqrt{3}$ |
| **To Multiply Radicals**<br>Use the product rule.<br>$\sqrt[n]{a}\cdot\sqrt[n]{b} = \sqrt[n]{ab}$ | $\sqrt[4]{8c^2}\,\sqrt[4]{4c^3} = \sqrt[4]{32c^5} = \sqrt[4]{16c^4}\,\sqrt[4]{2c}$<br>$= 2c\sqrt[4]{2c}$ |
| **Section 7.5** | |
| To **rationalize a denominator** multiply both the numerator and the denominator of the fraction by a radical that will result in the radicand in the denominator becoming a perfect power. | $\dfrac{6}{\sqrt{3x}}\cdot\dfrac{\sqrt{3x}}{\sqrt{3x}} = \dfrac{6\sqrt{3x}}{\sqrt{9x^2}} = \dfrac{6\sqrt{3x}}{3x} = \dfrac{2\sqrt{3x}}{x}$ |

| IMPORTANT FACTS AND CONCEPTS | EXAMPLES |
|---|---|
| **Section 7.5 (continued)** | |
| **A Radical Expression Is Simplified When the Following Are All True**<br>**1.** No perfect powers are factors of the radicand and all exponents in the radicand are less than the index.<br>**2.** No radicand contains a fraction.<br>**3.** No denominator contains a radical. | Not Simplified / Simplified<br>**1.** $\sqrt{x^3}$ → $x\sqrt{x}$<br>**2.** $\sqrt{\frac{1}{2}}$ → $\frac{\sqrt{2}}{2}$<br>**3.** $\frac{1}{\sqrt{2}}$ → $\frac{\sqrt{2}}{2}$ |
| **Section 7.6** | |
| **To Solve Radical Equations**<br>**1.** Rewrite the equation so that one radical containing a variable is by itself (isolated) on one side of the equation.<br>**2.** Raise each side of the equation to a power equal to the index of the radical.<br>**3.** Combine like terms.<br>**4.** If the equation still contains a term with a variable in a radicand, repeat steps 1 through 3.<br>**5.** Solve the resulting equation for the variable.<br>**6.** Check all solutions in the original equation for extraneous solutions. | Solve $\sqrt{x} - 8 = 0$.<br>$\sqrt{x} - 8 = 0$<br>$\sqrt{x} = 8$<br>$(\sqrt{x})^2 = 8^2$<br>$x = 64$<br>A check shows that 64 is the solution. |
| **Section 7.7** | |
| The **imaginary unit**, $i$ is defined as $i = \sqrt{-1}$. (Also, $i^2 = -1$.) | $\sqrt{-25} = \sqrt{25}\sqrt{-1} = 5i$ |
| **Imaginary Number**<br>For any positive number $n$,<br>$\sqrt{-n} = i\sqrt{n}$. | $\sqrt{-19} = i\sqrt{19}$ |
| A **complex number** is a number of the form $a + bi$, where $a$ and $b$ are real numbers. | $3 + 2i$ and $26 - 15i$ are complex numbers. |
| **To Add or Subtract Complex Numbers**<br>**1.** Change all imaginary numbers to $bi$ form.<br>**2.** Add (or subtract) the real parts of the complex numbers.<br>**3.** Add (or subtract) the imaginary parts of the complex numbers.<br>**4.** Write the answer in the form $a + bi$. | Add $(8 - 3i) + (12 + 5i)$.<br>$(8 - 3i) + (12 + 5i)$<br>$= 8 + 12 - 3i + 5i$<br>$= 20 + 2i$ |
| **To Multiply Complex Numbers**<br>**1.** Change all imaginary numbers to $bi$ form.<br>**2.** Multiply the complex numbers as you would multiply polynomials.<br>**3.** Substitute $-1$ for each $i^2$.<br>**4.** Combine the real parts and the imaginary parts. Write the answer in $a + bi$ form. | Multiply $(7 + 2i\sqrt{3})(5 - 4i\sqrt{3})$.<br>$(7 + 2i\sqrt{3})(5 - 4i\sqrt{3})$<br>$= 35 - 28i\sqrt{3} + 10i\sqrt{3} - 8(i^2)(3)$<br>$= 35 - 28i\sqrt{3} + 10i\sqrt{3} + 24$<br>$= 59 - 18i\sqrt{3}$ |
| The **conjugate of a complex number** $a + bi$ is $a - bi$. | Complex Number / Conjugate<br>$14 + 2i$ → $14 - 2i$<br>$17 - 8i$ → $17 + 8i$ |

| IMPORTANT FACTS AND CONCEPTS | EXAMPLES |
|---|---|
| **Section 7.7 (continued)** | |
| **To Divide Complex Numbers**<br>1. Change all imaginary numbers to *bi* form.<br>2. Rationalize the denominator by multiplying both the numerator and denominator by the conjugate of the denominator.<br>3. Write the answer in $a + bi$ form. | Divide $\frac{2-i}{5+3i}$.<br>$\frac{2-i}{5+3i} \cdot \frac{5-3i}{5-3i} = \frac{10-6i-5i+3i^2}{25-9i^2} = \frac{7-11i}{34}$ |
| **Powers of i**<br>$i^2 = -1, i^3 = -i, i^4 = 1, i^5 = i$ | $i^{38} = i^{36} \cdot i^2 = (i^4)^9 \cdot i^2 = 1^9 \cdot (-1) = -1$<br>$i^{63} = i^{60} \cdot i^3 = (i^4)^{15} \cdot i^3 = 1^{15}(-i) = -i$ |

## Chapter 7 Review Exercises

[7.1] *Evaluate.*

1. $\sqrt{100}$
2. $\sqrt[3]{-27}$
3. $\sqrt[3]{-125}$
4. $\sqrt[4]{256}$

*Use absolute value to evaluate.*

5. $\sqrt{(-8)^2}$
6. $\sqrt{(38.2)^2}$

*Write as an absolute value.*

7. $\sqrt{x^2}$
8. $\sqrt{(x-3)^2}$
9. $\sqrt{(x-y)^2}$
10. $\sqrt{(x^2-4x+12)^2}$
11. Let $f(x) = \sqrt{10x+9}$. Find $f(4)$.
12. Let $k(x) = 2x + \sqrt{\frac{x}{3}}$. Find $k(27)$.
13. Let $g(x) = \sqrt[3]{2x+3}$. Find $g(4)$ and round the answer to the nearest tenth.
14. **Area** The area of a square is 144 square meters. Find the length of its side.

*For the remainder of these review exercises, assume that all variables represent positive real numbers.*

[7.2] *Write in exponential form.*

15. $\sqrt{x^7}$
16. $\sqrt[3]{x^5}$
17. $(\sqrt[4]{y})^{13}$
18. $\sqrt[7]{6^{-2}}$

*Write in radical form.*

19. $x^{1/2}$
20. $a^{4/5}$
21. $(8m^2n)^{7/4}$
22. $(x+y)^{-5/3}$

*Simplify each radical expression by changing the expression to exponential form. Write the answer in radical form when appropriate.*

23. $\sqrt[3]{4^6}$
24. $\sqrt{x^{12}}$
25. $(\sqrt[4]{9})^8$
26. $\sqrt[20]{a^5}$

*Evaluate if possible. If the expression is not a real number, so state.*

27. $-36^{1/2}$
28. $(-36)^{1/2}$
29. $\left(\frac{64}{27}\right)^{-1/3}$
30. $64^{-1/2} + 8^{-2/3}$

*Simplify. Write the answer without negative exponents.*

31. $x^{3/5} \cdot x^{-1/3}$
32. $\left(\frac{64}{y^9}\right)^{1/3}$
33. $\left(\frac{a^{-6/5}}{a^{2/5}}\right)^{2/3}$
34. $\left(\frac{20x^5y^{-3}}{4y^{1/2}}\right)^2$

*Multiply.*

35. $a^{1/2}(5a^{3/2} - 3a^2)$
36. $4x^{-2/3}\left(x^{-1/2} + \frac{11}{4}x^{2/3}\right)$

*Factor each expression. Write the answer without negative exponents.*

37. $x^{2/5} + x^{7/5}$
38. $a^{-1/2} + a^{3/2}$

*For each function, find the indicated value of the function. Use your calculator to evaluate irrational numbers. Round irrational numbers to the nearest thousandth.*

39. If $f(x) = \sqrt{6x-11}$, find $f(6)$.
40. If $g(x) = \sqrt[3]{9x-17}$, find $g(4)$.

*Graph the following functions.*

**41.** $f(x) = \sqrt{x}$

**42.** $f(x) = \sqrt{x} - 4$

[7.2–7.5] *Simplify.*

**43.** $\sqrt{48}$

**44.** $\sqrt[3]{128}$

**45.** $\sqrt{\dfrac{49}{9}}$

**46.** $\sqrt[3]{\dfrac{8}{125}}$

**47.** $-\sqrt{\dfrac{81}{49}}$

**48.** $\sqrt[3]{-\dfrac{27}{125}}$

**49.** $\sqrt{32}\,\sqrt{2}$

**50.** $\sqrt[3]{32}\,\sqrt[3]{2}$

**51.** $\sqrt{18x^2y^3z^4}$

**52.** $\sqrt{75x^3y^7}$

**53.** $\sqrt[3]{54a^7b^{10}}$

**54.** $\sqrt[3]{125x^8y^9z^{16}}$

**55.** $\left(\sqrt[6]{x^2y^3z^5}\right)^{42}$

**56.** $\left(\sqrt[5]{2ab^4c^6}\right)^{15}$

**57.** $\sqrt{5x}\,\sqrt{8x^5}$

**58.** $\sqrt[3]{2x^2y}\,\sqrt[3]{4x^9y^4}$

**59.** $\sqrt[3]{2x^4y^5}\,\sqrt[3]{16x^4y^4}$

**60.** $\sqrt[4]{4x^4y^7}\,\sqrt[4]{4x^5y^9}$

**61.** $\sqrt{3x}\left(\sqrt{12x} - \sqrt{20}\right)$

**62.** $\sqrt[3]{2x^2y}\left(\sqrt[3]{4x^4y^7} + \sqrt[3]{9x}\right)$

**63.** $\sqrt{\sqrt{a^3b^2}}$

**64.** $\sqrt{\sqrt[3]{x^5y^2}}$

**65.** $\left(\dfrac{4r^2p^{1/3}}{r^{1/2}p^{4/3}}\right)^3$

**66.** $\left(\dfrac{6y^{2/5}z^{1/3}}{x^{-1}y^{3/5}}\right)^{-1}$

**67.** $\sqrt{\dfrac{3}{5}}$

**68.** $\sqrt[3]{\dfrac{7}{9}}$

**69.** $\sqrt[4]{\dfrac{5}{4}}$

**70.** $\dfrac{x}{\sqrt{10}}$

**71.** $\dfrac{8}{\sqrt{x}}$

**72.** $\dfrac{m}{\sqrt[3]{25}}$

**73.** $\dfrac{10}{\sqrt[3]{y^2}}$

**74.** $\dfrac{9}{\sqrt[4]{z}}$

**75.** $\sqrt[3]{\dfrac{x^3}{27}}$

**76.** $\dfrac{\sqrt[3]{2x^{10}}}{\sqrt[3]{16x^7}}$

**77.** $\sqrt{\dfrac{32x^2y^5}{2x^8y}}$

**78.** $\sqrt[4]{\dfrac{48x^9y^{15}}{3xy^3}}$

**79.** $\sqrt{\dfrac{6x^4}{y}}$

**80.** $\sqrt{\dfrac{12a}{7b}}$

**81.** $\sqrt{\dfrac{18x^4y^5}{3z}}$

**82.** $\sqrt{\dfrac{125x^2y^5}{3z}}$

**83.** $\sqrt[3]{\dfrac{108x^3y^7}{2y^3}}$

**84.** $\sqrt[3]{\dfrac{3x}{5y}}$

**85.** $\sqrt[3]{\dfrac{9x^5y^3}{x^6}}$

**86.** $\sqrt[3]{\dfrac{y^6}{5x^2}}$

**87.** $\sqrt[4]{\dfrac{2a^2b^{11}}{a^5b}}$

**88.** $\sqrt[4]{\dfrac{3x^2y^6}{8x^3}}$

**89.** $(3 - \sqrt{2})(3 + \sqrt{2})$

**90.** $(\sqrt{x} + y)(\sqrt{x} - y)$

**91.** $(x - \sqrt{y})(x + \sqrt{y})$

**92.** $(\sqrt{3} + 2)^2$

**93.** $(\sqrt{x} - \sqrt{3y})(\sqrt{x} + \sqrt{5y})$

**94.** $(\sqrt[3]{2x} - \sqrt[3]{3y})(\sqrt[3]{3x} - \sqrt[3]{2y})$

**95.** $\dfrac{6}{2 + \sqrt{5}}$

**96.** $\dfrac{x}{4 + \sqrt{x}}$

**97.** $\dfrac{a}{4 - \sqrt{b}}$

**98.** $\dfrac{x}{\sqrt{y} - 7}$

**99.** $\dfrac{\sqrt{x}}{\sqrt{x} + \sqrt{y}}$

**100.** $\dfrac{\sqrt{x} - 3\sqrt{y}}{\sqrt{x} - \sqrt{y}}$

**101.** $\dfrac{2}{\sqrt{a - 1} - 2}$

**102.** $\dfrac{5}{\sqrt{y + 2} - 3}$

**103.** $\sqrt[3]{x} + 10\sqrt[3]{x} - 2\sqrt[3]{x}$

**104.** $\sqrt{3} + \sqrt{27} - \sqrt{192}$

**105.** $\sqrt[3]{16} - 5\sqrt[3]{54} + 3\sqrt[3]{64}$

**106.** $\sqrt{2} - \dfrac{3}{\sqrt{32}} + \sqrt{50}$

**107.** $9\sqrt{x^5y^6} - \sqrt{16x^7y^8}$

**108.** $8\sqrt[3]{x^7y^8} - \sqrt[3]{x^4y^2} + 3\sqrt[3]{x^{10}y^2}$

*In Exercises 109 and 110, $f(x)$ and $g(x)$ are given. Find $(f \cdot g)(x)$.*

**109.** $f(x) = \sqrt{3x},\ g(x) = \sqrt{6x} - \sqrt{15}$

**110.** $f(x) = \sqrt[3]{2x^2},\ g(x) = \sqrt[3]{4x^4} + \sqrt[3]{16x^5}$

*Simplify. In Exercise 112, assume that the variable can be any real number.*

**111.** $f(x) = \sqrt{2x + 7}\,\sqrt{2x + 7},\ x \geq -\dfrac{7}{2}$

**112.** $g(a) = \sqrt{20a^2 + 100a + 125}$

*Simplify.*

**113.** $\dfrac{\sqrt[3]{(x + 5)^5}}{\sqrt{(x + 5)^3}}$

**114.** $\dfrac{\sqrt[3]{a^3b^2}}{\sqrt[4]{a^4b}}$

**Perimeter and Area** *For each figure, find* **a)** *the perimeter and* **b)** *the area. Write the perimeter and area in radical form with the radicals simplified.*

**115.**

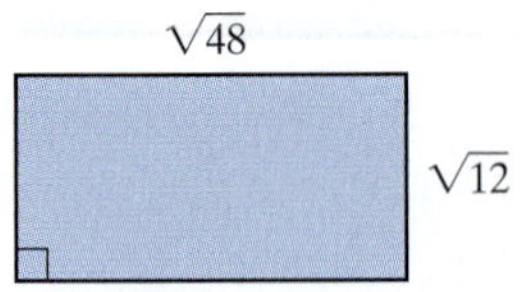


Rectangle

**116.**

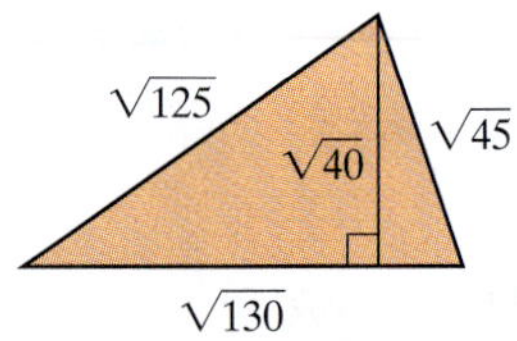


**117.** The graph of $f(x) = \sqrt{x} + 2$ is given.

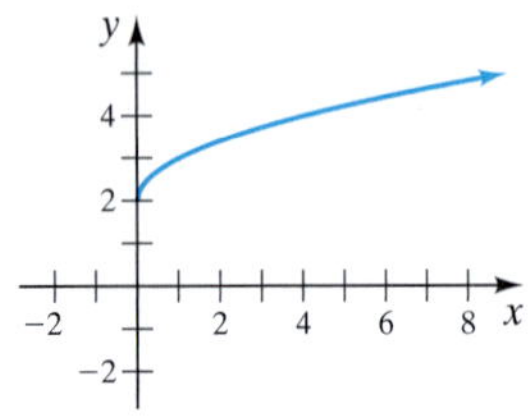


**a)** For $g(x) = -3$, sketch the graph of $(f + g)(x)$.

**b)** What is the domain of $(f + g)(x)$?

**118.** The graph of $f(x) = -\sqrt{x}$ is given.

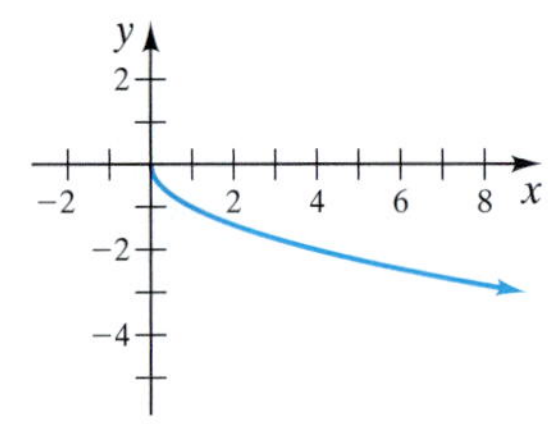


**a)** For $g(x) = \sqrt{x} + 2$, sketch the graph of $(f + g)(x)$.

**b)** What is the domain of $(f + g)(x)$?

[7.6] *Solve each equation and check your solutions.*

**119.** $\sqrt{x} = 9$

**120.** $\sqrt{x} = -4$

**121.** $\sqrt[3]{x} = 4$

**122.** $\sqrt[3]{x} = -5$

**123.** $7 + \sqrt{x} = 10$

**124.** $7 + \sqrt[3]{x} = 12$

**125.** $\sqrt{3x + 4} = \sqrt{5x + 14}$

**126.** $\sqrt{x^2 + 2x - 8} = x$

**127.** $\sqrt[3]{x - 9} = \sqrt[3]{5x + 3}$

**128.** $(x^2 + 7)^{1/2} = x + 1$

**129.** $\sqrt{x} + 3 = \sqrt{3x + 9}$

**130.** $\sqrt{6x - 5} - \sqrt{2x + 6} - 1 = 0$

*For each pair of functions, find all values of x for which* $f(x) = g(x)$.

**131.** $f(x) = \sqrt{3x + 4}, g(x) = 2\sqrt{2x - 4}$

**132.** $f(x) = (4x + 5)^{1/3}, g(x) = (6x - 7)^{1/3}$

*Solve the following for the indicated variable.*

**133.** $V = \sqrt{\dfrac{2L}{w}}$, for $L$

**134.** $r = \sqrt{\dfrac{A}{\pi}}$, for $A$

*Find the length of the unknown side of each right triangle. Write the answer as a radical in simplified form.*

**135.**

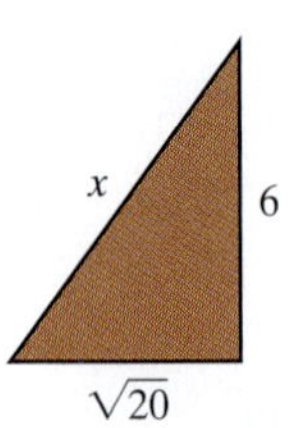


**136.**

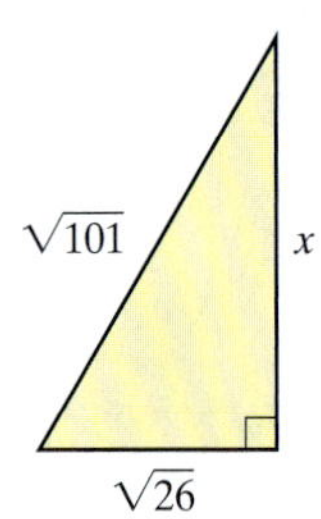


*Solve.*

**137. Telephone Pole** How long a wire does a phone company need to reach the top of a 5-meter telephone pole from a point on the ground 2 meters from the base of the pole?

**138. Velocity** Use the formula $v = \sqrt{2gh}$ to find the velocity of an object after it has fallen 20 feet ($g = 32$ ft/s$^2$).

**139. Pendulum** Use the formula

$$T = 2\pi\sqrt{\frac{L}{32}}$$

to find the period of a pendulum, $T$, if its length, $L$, is 64 feet.

**140. Kinetic and Potential Energy** There are two types of energy: kinetic and potential. Potential energy is due to position and kinetic energy is due to motion. For example, if you hold a billiard ball above the ground it has potential energy. If you let go of the ball the potential energy is changed to kinetic energy as the ball drops. The formula

$$V = \sqrt{\frac{2K}{m}}$$

can be used to determine the velocity, $V$, in meters per second, when a mass, $m$, in kilograms, has a kinetic energy, $K$, in joules. A 0.145-kg baseball is thrown. If the kinetic energy of the moving ball is 45 joules, at what speed is the ball moving?

**141. Speed of Light** Albert Einstein found that if an object at rest, with mass $m_0$, is made to travel close to the speed of light, its mass increases to $m$, where

$$m = \frac{m_0}{\sqrt{1 - \frac{v^2}{c^2}}}$$

In the formula, $v$ is the velocity of the moving object and $c$ is the speed of light.* In an accelerator used for cancer therapy, particles travel at speeds of $0.98c$, that is, at 98% the speed of light. At a speed of $0.98c$, determine a particle's mass, $m$, in terms of its rest mass, $m_0$. Use $v = 0.98c$ in the above formula.

[7.7] *Write each expression as a complex number in the form $a + bi$.*

**142.** $5$ **143.** $-8$ **144.** $7 - \sqrt{-256}$ **145.** $9 + \sqrt{-16}$

*Perform each indicated operation.*

**146.** $(3 + 2i) + (10 - i)$ **147.** $(9 - 6i) - (3 - 4i)$ **148.** $(\sqrt{3} + \sqrt{-5}) + (11\sqrt{3} - \sqrt{-7})$

**149.** $\sqrt{-6}(\sqrt{6} + \sqrt{-6})$ **150.** $(4 + 3i)(2 - 3i)$ **151.** $(6 + \sqrt{-3})(4 - \sqrt{-15})$

**152.** $\frac{8}{3i}$ **153.** $\frac{2 + \sqrt{3}}{2i}$ **154.** $\frac{4}{3 + 2i}$ **155.** $\frac{\sqrt{3}}{5 - \sqrt{-6}}$

*Evaluate each expression for the given value of x.*

**156.** $x^2 - 2x + 9, x = 1 + 2i\sqrt{2}$ **157.** $x^2 - 2x + 12, x = 1 - 2i$

*For each imaginary number, indicate whether its value is $i$, $-1$, $-i$, or 1.*

**158.** $i^{33}$ **159.** $i^{59}$ **160.** $i^{404}$ **161.** $i^{802}$

*The speed of light is $3.00 \times 10^8$ meters per second. However, you do not need this information to solve this problem.

## Chapter 7 Practice Test

*To find out how well you understand the chapter material, take this practice test. The answers, and the section where the material was initially discussed, are given in the back of the book. Each problem is also fully worked out on the* **Chapter Test Prep Video CD.** *Review any questions that you answered incorrectly.*

**1.** Write $\sqrt{(5x - 3)^2}$ as an absolute value.

**2.** Simplify $\left(\frac{x^{2/5} \cdot x^{-1}}{x^{3/5}}\right)^2$.

**3.** Factor $x^{-2/3} + x^{4/3}$.

**4.** Graph $g(x) = \sqrt{x} + 1$.

*In Exercises 5–14, simplify. Assume that all variables represent positive real numbers.*

**5.** $\sqrt{54x^7y^{10}}$

**6.** $\sqrt[3]{25x^5y^2}\,\sqrt[3]{10x^6y^8}$

**7.** $\sqrt{\frac{7x^6y^3}{8z}}$

**8.** $\frac{9}{\sqrt[3]{x}}$

**9.** $\frac{\sqrt{3}}{3 + \sqrt{27}}$

**10.** $2\sqrt{24} - 6\sqrt{6} + 3\sqrt{54}$

**11.** $\sqrt[3]{8x^3y^5} + 4\sqrt[3]{x^6y^8}$

**12.** $(\sqrt{3} - 2)(6 - \sqrt{8})$

**13.** $\sqrt[4]{\sqrt{x^5y^3}}$

**14.** $\frac{\sqrt[4]{(7x + 2)^5}}{\sqrt[3]{(7x + 2)^2}}$

*In Exercises 15–17, solve the equation.*

**15.** $\sqrt{2x + 19} = 3$

**16.** $\sqrt{x^2 - x - 12} = x + 3$

**17.** $\sqrt{a - 8} = \sqrt{a} - 2$

**18.** For $f(x) = (9x + 37)^{1/3}$ and $g(x) = 2(2x + 2)^{1/3}$, find all values of $x$ such that $f(x) = g(x)$.

**19.** Solve the formula $w = \frac{\sqrt{2gh}}{4}$ for $g$.

**20. Falling Object** The velocity, $V$, in feet per second, after an object has fallen a distance, $h$, in feet, can be found by the formula $V = \sqrt{64.4h}$. Find the velocity of a pen after it has fallen 200 feet.

**21. Ladder** A ladder is placed against a house. If the base of the ladder is 5 feet from the house and the ladder rests on the house 12 feet above the ground, find the length of the ladder.

**22. Springs** A formula used in the study of springs is

$$T = 2\pi\sqrt{\frac{m}{k}}$$

where $T$ is the period of a spring (the time for the spring to stretch and return to its rest point), $m$ is the mass on the spring, in kilograms, and $k$ is the spring's constant, in newtons/meter. A mass of 1400 kilograms rests on a spring. Find the period of the spring if the spring's constant is 65,000 newtons/meter.

**23.** Multiply $(6 - \sqrt{-4})(2 + \sqrt{-16})$.

**24.** Divide $\frac{5 - i}{7 + 2i}$.

**25.** Evaluate $x^2 + 6x + 12$ for $x = -3 + i$.

## Cumulative Review Test

*Take the following test and check your answers with those given in the back of the book. Review any questions that you answered incorrectly. The section where the material was covered is indicated after the answer.*

**1.** Solve $\frac{1}{5}(x - 3) = \frac{3}{4}(x + 3) - x$.

**2.** Solve $3(x - 4) = 6x - (4 - 5x)$.

**3. Sweater** When the price of a sweater is decreased by 60%, it costs \$16. Find the original price of the sweater.

**4.** Find the solution set of $|3 - 2x| < 5$.

**5.** Graph $y = \frac{3}{2}x - 3$.

**6.** Determine whether the graphs of the given equations are parallel lines, perpendicular lines, or neither.

$$y = 3x - 8$$
$$6y = 18x + 12$$

**7.** Given $f(x) = x^2 - 3x + 4$ and $g(x) = 2x - 9$, find $(g - f)(x)$.

**8.** Find the equation of the line through $(1, -4)$ that is perpendicular to the graph of $3x - 2y = 6$.

**9.** Solve the system of equations.

$$x + 2y = 12$$
$$4x = 8$$
$$3x - 4y + 5z = 20$$

**10.** Evaluate the determinant.

$$\begin{vmatrix} 3 & -6 & -1 \\ 2 & 1 & -2 \\ 1 & 3 & 1 \end{vmatrix}$$

**11. Volume** The volume of the box that follows is $6r^3 + 5r^2 + r$. Find $w$ in terms of $r$.

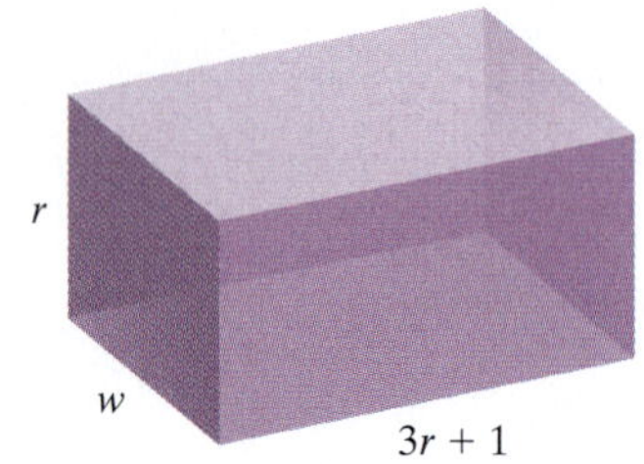

**12.** Multiply $(5xy - 3)(5xy + 3)$.

**13.** Solve $\sqrt{2x^2 + 7} + 3 = 8$.

**14.** Factor $4x^3 - 9x^2 + 5x$.

**15.** Factor $(x + 1)^3 - 27$.

**16.** Solve $8x^2 - 3 = -10x$.

**17.** Multiply $\frac{4x + 4y}{x^2y} \cdot \frac{y^3}{12x}$.

**18.** Add $\frac{x - 4}{x - 5} - \frac{3}{x + 5} - \frac{10}{x^2 - 25}$.

**19.** Solve $\frac{4}{x} - \frac{1}{6} = \frac{1}{x}$.

**20. Falling Object** The distance, $d$, an object drops in free fall is directly proportional to the square of the time, $t$. If an object falls 16 feet in 1 second, how far will the object fall in 5 seconds?

# 8 Quadratic Functions

## GOALS OF THIS CHAPTER

We introduced quadratic functions in Section 5.1. Now we expand on the concepts. We introduce completing the square and the quadratic formula. After studying these sections, you will know three techniques for solving quadratic equations: factoring (when possible), completing the square, and the quadratic formula. We will also discuss techniques for graphing quadratic functions and study nonlinear inequalities in one variable.

8.1 Solving Quadratic Equations by Completing the Square

8.2 Solving Quadratic Equations by the Quadratic Formula

8.3 Quadratic Equations: Applications and Problem Solving

Mid-Chapter Test: Sections 8.1–8.3

8.4 Writing Equations in Quadratic Form

8.5 Graphing Quadratic Functions

8.6 Quadratic and Other Inequalities in One Variable

Chapter 8 Summary

Chapter 8 Review Exercises

Chapter 8 Practice Test

Cumulative Review Test

THERE ARE MANY REAL-LIFE situations that can be represented or approximated with quadratic equations. As you read through this chapter, you will see several real-life applications of quadratic equations and quadratic functions. For instance, in Exercises 101 and 102 on page 538, we will use quadratic equations and the quadratic formula to determine the time it takes for a drop of water on top of a waterfall to reach the bottom of the waterfall.

# 8.1 Solving Quadratic Equations by Completing the Square

1. Use the square root property to solve equations.
2. Understand perfect square trinomials.
3. Solve quadratic equations by completing the square.

In this section, we introduce two concepts, the square root property and completing the square. The square root property will be used in several sections of this book.

In Section 5.8, we solved quadratic, or second-degree, equations by factoring. Quadratic equations that cannot be solved by factoring can be solved by completing the square, or by the quadratic formula, which is presented in Section 8.2.

## 1 Use the Square Root Property to Solve Equations

In Section 7.1, we stated that every positive number has two square roots. Thus far, we have been using only the positive square root. In this section, we use both the positive and negative square roots of a number.

| Positive Square Root of 25 | Negative Square Root of 25 |
|---|---|
| $\sqrt{25} = 5$ | $-\sqrt{25} = -5$ |

A convenient way to indicate the two square roots of a number is to use the plus or minus symbol, $\pm$. For example, the square roots of 25 can be indicated by $\pm 5$, read "plus or minus 5." The equation $x^2 = 25$ has two solutions, the two square roots of 25, which are $\pm 5$. If you check each root, you will see that each value satisfies the equation. The **square root property** can be used to find the solutions to equations of the form $x^2 = a$.

**Square Root Property**

If $x^2 = a$, where $a$ is a real number, then $x = \pm\sqrt{a}$.

**EXAMPLE 1** Solve the following equations.

**a)** $x^2 - 9 = 0$ **b)** $x^2 + 10 = 85$

*Solution*

**a)** Add 9 to both sides of the equation to isolate the variable.

$$\begin{aligned} x^2 - 9 &= 0 \\ x^2 &= 9 && \text{Isolate the variable.} \\ x &= \pm\sqrt{9} && \text{Square root property} \\ &= \pm 3 \end{aligned}$$

Check the solutions in the original equation.

| $x = 3$ | $x = -3$ |
|---|---|
| $x^2 - 9 = 0$ | $x^2 - 9 = 0$ |
| $3^2 - 9 \stackrel{?}{=} 0$ | $(-3)^2 - 9 \stackrel{?}{=} 0$ |
| $0 = 0$ True | $0 = 0$ True |

In both cases the check is true, which means that both 3 and $-3$ are solutions to the equation.

**b)**

$$\begin{aligned} x^2 + 10 &= 85 \\ x^2 &= 75 && \text{Isolate the variable.} \\ x &= \pm\sqrt{75} && \text{Square root property} \\ &= \pm\sqrt{25}\sqrt{3} && \text{Simplify.} \\ &= \pm 5\sqrt{3} \end{aligned}$$

The solutions are $5\sqrt{3}$ and $-5\sqrt{3}$.

▸ Now Try Exercise 13

Not all quadratic equations have real solutions, as is illustrated in Example 2.

**EXAMPLE 2** ▸ Solve the equation $x^2 + 7 = 0$.

*Solution*

$$x^2 + 7 = 0$$
$$x^2 = -7 \quad \text{Isolate the variable.}$$
$$x = \pm\sqrt{-7} \quad \text{Square root property}$$
$$= \pm i\sqrt{7}$$

The solutions are $i\sqrt{7}$ and $-i\sqrt{7}$, both of which are imaginary numbers.

▸ Now Try Exercise 15

**EXAMPLE 3** ▸ Solve **a)** $(a - 5)^2 = 32$ **b)** $(z + 3)^2 + 28 = 0$.

*Solution*

**a)** Since the term containing the variable is already isolated, begin by using the square root property.

$$(a - 5)^2 = 32$$
$$a - 5 = \pm\sqrt{32} \quad \text{Square root property}$$
$$a = 5 \pm \sqrt{32} \quad \text{Add 5 to both sides.}$$
$$= 5 \pm \sqrt{16}\sqrt{2} \quad \text{Simplify.}$$
$$= 5 \pm 4\sqrt{2}$$

The solutions are $5 + 4\sqrt{2}$ and $5 - 4\sqrt{2}$.

**b)** Begin by subtracting 28 from both sides of the equation to isolate the term containing the variable.

$$(z + 3)^2 + 28 = 0$$
$$(z + 3)^2 = -28$$

Now use the square root property.

$$z + 3 = \pm\sqrt{-28} \quad \text{Square root property.}$$
$$z = -3 \pm \sqrt{-28} \quad \text{Subtract 3 from both sides.}$$
$$= -3 \pm \sqrt{28}\sqrt{-1}$$
$$= -3 \pm i\sqrt{4}\sqrt{7} \quad \text{Simplify } \sqrt{28} \text{ and replace } \sqrt{-1} \text{ with } i.$$
$$= -3 \pm 2i\sqrt{7}$$

The solutions are $-3 + 2i\sqrt{7}$ and $-3 - 2i\sqrt{7}$. Note that the solutions to the equation $(z + 3)^2 + 28 = 0$ are not real numbers. The solutions are complex numbers.

▸ Now Try Exercise 23

## 2 Understand Perfect Square Trinomials

Now that we know the square root property we can focus our attention on completing the square. To understand this procedure, you need to know how to form perfect square trinomials. Perfect square trinomials were introduced in Section 5.6. Recall that a **perfect square trinomial** is a trinomial that can be expressed as the square of a binomial. Some examples follow.

| Perfect Square Trinomials | | Factors | | Square of a Binomial |
|---|---|---|---|---|
| $x^2 + 8x + 16$ | = | $(x + 4)(x + 4)$ | = | $(x + 4)^2$ |
| $x^2 - 8x + 16$ | = | $(x - 4)(x - 4)$ | = | $(x - 4)^2$ |
| $x^2 + 10x + 25$ | = | $(x + 5)(x + 5)$ | = | $(x + 5)^2$ |
| $x^2 - 10x + 25$ | = | $(x - 5)(x - 5)$ | = | $(x - 5)^2$ |

In a perfect square trinomial with a leading coefficient of 1, there is a relationship between the coefficient of the first-degree term and the constant term. In such trinomials the constant term is the square of one-half the coefficient of the first degree term.

Let's examine some perfect square trinomials for which the leading coefficient is 1.

$$x^2 + 8x + 16 = (x + 4)^2$$
$$\left[\tfrac{1}{2}(8)\right]^2 = (4)^2$$

$$x^2 - 10x + 25 = (x - 5)^2$$
$$\left[\tfrac{1}{2}(-10)\right]^2 = (-5)^2$$

When a perfect square trinomial with a leading coefficient of 1 is written as the square of a binomial, the constant in the binomial is one-half the coefficient of the first-degree term in the trinomial. For example,

$$x^2 + 8x + 16 = (x + 4)^2$$
$$\tfrac{1}{2}(8)$$

$$x^2 - 10x + 25 = (x - 5)^2$$
$$\tfrac{1}{2}(-10)$$

## 3 Solve Quadratic Equations by Completing the Square

Now we introduce completing the square. To solve a quadratic equation by **completing the square**, we add a constant to both sides of the equation so that the remaining trinomial is a perfect square trinomial. Then we use the square root property to solve the resulting equation. We will now summarize the procedure.

**To Solve a Quadratic Equation by Completing the Square**

1. Use the multiplication (or division) property of equality if necessary to make the leading coefficient 1.
2. Rewrite the equation with the constant by itself on the right side of the equation.
3. Take one-half the numerical coefficient of the first-degree term, square it, and add this quantity to both sides of the equation.
4. Replace the trinomial with the square of a binomial.
5. Use the square root property to take the square root of both sides of the equation.
6. Solve for the variable.
7. Check your solutions in the *original* equation.

**EXAMPLE 4** ▸ Solve the equation $x^2 + 6x + 5 = 0$ by completing the square.

*Solution* Since the leading coefficient is 1, step 1 is not necessary.

Step 2: Move the constant, 5, to the right side of the equation by subtracting 5 from both sides of the equation.

$$x^2 + 6x + 5 = 0$$
$$x^2 + 6x = -5$$

**Step 3:** Determine the square of one-half the numerical coefficient of the first-degree term, 6.

$$\frac{1}{2}(6) = 3, \qquad 3^2 = 9$$

Add this value to both sides of the equation.

$$x^2 + 6x + 9 = -5 + 9$$
$$x^2 + 6x + 9 = 4$$

**Step 4:** By following this procedure, we produce a perfect square trinomial on the left side of the equation. The expression $x^2 + 6x + 9$ is a perfect square trinomial that can be expressed as $(x + 3)^2$.

$\frac{1}{2}$ *the numerical coefficient of the first-degree term is* $\frac{1}{2}(6) = +3$.

$$(x + 3)^2 = 4$$

**Step 5:** Use the square root property.

$$x + 3 = \pm\sqrt{4}$$
$$x + 3 = \pm 2$$

**Step 6:** Finally, solve for $x$ by subtracting 3 from both sides of the equation.

$$x + 3 - 3 = -3 \pm 2$$
$$x = -3 \pm 2$$
$$x = -3 + 2 \quad \text{or} \quad x = -3 - 2$$
$$x = -1 \qquad\qquad x = -5$$

**Step 7:** Check both solutions in the original equation.

| $x = -1$ | $x = -5$ |
|---|---|
| $x^2 + 6x + 5 = 0$ | $x^2 + 6x + 5 = 0$ |
| $(-1)^2 + 6(-1) + 5 \stackrel{?}{=} 0$ | $(-5)^2 + 6(-5) + 5 \stackrel{?}{=} 0$ |
| $1 - 6 + 5 \stackrel{?}{=} 0$ | $25 - 30 + 5 \stackrel{?}{=} 0$ |
| $0 = 0$ *True* | $0 = 0$ *True* |

Since each number checks, both $-1$ and $-5$ are solutions to the original equation.

**▸ Now Try Exercise 49**

## Helpful Hint

When solving the equation $x^2 + bx + c = 0$ by completing the square, we obtain $x^2 + bx + \left(\frac{b}{2}\right)^2$ on the left side of the equation and a constant on the right side of the equation. We then replace $x^2 + bx + \left(\frac{b}{2}\right)^2$ with $\left(x + \frac{b}{2}\right)^2$. In the figure on the next page we show why

$$x^2 + bx + \left(\frac{b}{2}\right)^2 = \left(x + \frac{b}{2}\right)^2$$

The figure is a square with sides of length $x + \frac{b}{2}$. The area is therefore $\left(x + \frac{b}{2}\right)^2$. The area of the square can also be determined by adding the areas of the four sections as follows:

$$x^2 + \frac{b}{2}x + \frac{b}{2}x + \left(\frac{b}{2}\right)^2 = x^2 + bx + \left(\frac{b}{2}\right)^2$$

*(continued on the next page)*

Comparing the areas, we see that $x^2 + bx + \left(\frac{b}{2}\right)^2 = \left(x + \frac{b}{2}\right)^2$.

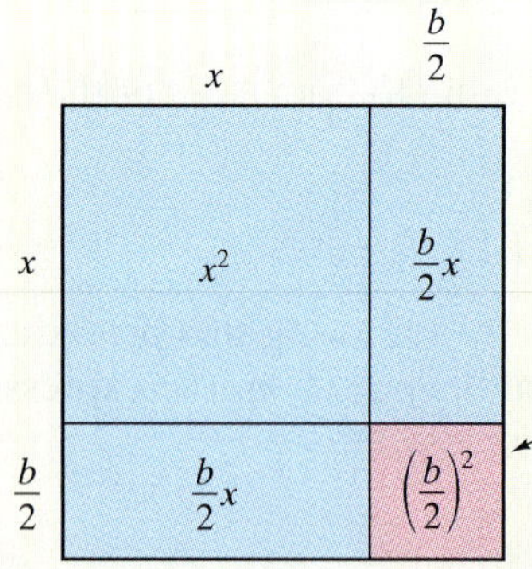


The area of this piece represents the term we add to each side of the equation when we complete the square.

**EXAMPLE 5** ▸ Solve the equation $-x^2 = -3x - 18$ by completing the square.

*Solution* The numerical coefficient of the squared term must be 1, not $-1$. Therefore, begin by multiplying both sides of the equation by $-1$ to make the coefficient of the squared term equal to 1.

$$-x^2 = -3x - 18$$
$$-1(-x^2) = -1(-3x - 18)$$
$$x^2 = 3x + 18$$

Now move all terms except the constant to the left side of the equation.

$$x^2 - 3x = 18$$

Take half the numerical coefficient of the $x$-term, square it, and add this product to both sides of the equation. Then rewrite the left side of the equation as the square of a binomial.

$$\frac{1}{2}(-3) = -\frac{3}{2} \qquad \left(-\frac{3}{2}\right)^2 = \frac{9}{4}$$

$$x^2 - 3x + \frac{9}{4} = 18 + \frac{9}{4} \qquad \textit{Complete the square.}$$

$$\left(x - \frac{3}{2}\right)^2 = 18 + \frac{9}{4} \qquad \textit{Rewrite the trinomial as the square of a binomial.}$$

$$\left(x - \frac{3}{2}\right)^2 = \frac{72}{4} + \frac{9}{4}$$

$$\left(x - \frac{3}{2}\right)^2 = \frac{81}{4}$$

$$x - \frac{3}{2} = \pm\sqrt{\frac{81}{4}} \qquad \textit{Square root property}$$

$$x - \frac{3}{2} = \pm\frac{9}{2} \qquad \textit{Simplify.}$$

$$x = \frac{3}{2} \pm \frac{9}{2} \qquad \textit{Add } \frac{3}{2} \textit{ to both sides.}$$

$$x = \frac{3}{2} + \frac{9}{2} \quad \text{or} \quad x = \frac{3}{2} - \frac{9}{2}$$

$$x = \frac{12}{2} = 6 \qquad x = -\frac{6}{2} = -3$$

The solutions are 6 and $-3$.

▸ **Now Try Exercise 53**

In the following examples we will not show some of the intermediate steps.

**EXAMPLE 6** ▸ Solve $x^2 - 8x + 34 = 0$ by completing the square.

*Solution*

$$x^2 - 8x + 34 = 0$$

$$x^2 - 8x = -34 \qquad \textit{Move the constant term to the right side.}$$

$$x^2 - 8x + 16 = -34 + 16 \qquad \textit{Complete the square.}$$

$$(x - 4)^2 = -18 \qquad \textit{Write the trinomial as the square of a binomial.}$$

$$x - 4 = \pm\sqrt{-18} \qquad \textit{Square root property}$$

$$x - 4 = \pm 3i\sqrt{2} \qquad \textit{Simplify.}$$

$$x = 4 \pm 3i\sqrt{2} \qquad \textit{Solve for x.}$$

The solutions are $4 + 3i\sqrt{2}$ and $4 - 3i\sqrt{2}$.

▸ **Now Try Exercise 61**

---

**EXAMPLE 7** ▸ Solve the equation $-4m^2 + 8m + 32 = 0$ by completing the square.

*Solution*

$$-4m^2 + 8m + 32 = 0$$

$$-\frac{1}{4}(-4m^2 + 8m + 32) = -\frac{1}{4}(0) \qquad \textit{Multiply by } -\frac{1}{4} \textit{ to obtain a leading coefficient of 1.}$$

$$m^2 - 2m - 8 = 0$$

Now proceed as before.

$$m^2 - 2m = 8 \qquad \textit{Move the constant term to the right side.}$$

$$m^2 - 2m + 1 = 8 + 1 \qquad \textit{Complete the square.}$$

$$(m - 1)^2 = 9 \qquad \textit{Write the trinomial as the square of a binomial.}$$

$$m - 1 = \pm 3 \qquad \textit{Square root property}$$

$$m = 1 \pm 3 \qquad \textit{Solve for m.}$$

$$m = 1 + 3 \quad \text{or} \quad m = 1 - 3$$

$$m = 4 \qquad\qquad m = -2$$

▸ **Now Try Exercise 75**

---

If you were asked to solve the equation $-\frac{1}{4}x^2 + 2x - 8 = 0$ by completing the square, what would you do first? If you answered, "Multiply both sides of the equation by $-4$ to make the leading coefficient 1," you answered correctly. To solve the equation $\frac{2}{3}x^2 + 3x - 5 = 0$, you would first multiply both sides of the equation by $\frac{3}{2}$ to obtain a leading coefficient of 1.

Generally, quadratic equations that cannot be easily solved by factoring will be solved by the *quadratic formula*, which will be presented in the next section. We introduced completing the square because we use it to derive the quadratic formula in Section 8.2. We will use completing the square later in this chapter and in a later chapter.

**EXAMPLE 8 ▸ Compound Interest** The compound interest formula $A = p\left(1 + \frac{r}{n}\right)^{nt}$ can be used to find the amount, $A$, when an initial principal, $p$, is invested at an annual interest rate, $r$, compounded $n$ times a year for $t$ years.

**a)** Josh Adams initially invested \$1000 in a savings account where interest is compounded annually (once a year). If after 2 years the amount, or balance, in the account is \$1102.50, find the annual interest rate, $r$.

**b)** Trisha McDowell initially invested \$1000 in a savings account where interest is compounded quarterly. If after 3 years the amount in the account is \$1195.62, find the annual interest rate, $r$.

*Solution* **a) Understand** We are given the following information:

$$p = \$1000, \quad A = \$1102.50, \quad n = 1, \quad t = 2$$

We are asked to find the annual rate, $r$. To do so, we substitute the appropriate values into the formula and solve for $r$.

**Translate**

$$A = p\left(1 + \frac{r}{n}\right)^{nt}$$

$$1102.50 = 1000\left(1 + \frac{r}{1}\right)^{1(2)}$$

**Carry Out**

$$1102.50 = 1000(1 + r)^2$$

$$1.10250 = (1 + r)^2 \quad \text{Divide both sides by 1000.}$$

$$\sqrt{1.10250} = 1 + r \quad \text{Square root property; use principal root since } r \text{ must be positive.}$$

$$1.05 = 1 + r$$

$$0.05 = r \quad \text{Subtract 1 from both sides.}$$

**Answer** The annual interest rate is 0.05 or 5%.

**b) Understand** We are given

$$p = 1000, \quad A = \$1195.62, \quad n = 4, \quad t = 3$$

To find $r$, we substitute the appropriate values into the formula and solve for $r$.

**Translate**

$$A = p\left(1 + \frac{r}{n}\right)^{nt}$$

$$1195.62 = 1000\left(1 + \frac{r}{4}\right)^{4(3)}$$

$$1.19562 = \left(1 + \frac{r}{4}\right)^{12} \quad \text{Divide both sides by 1000.}$$

**Carry Out**

$$\sqrt[12]{1.19562} = 1 + \frac{r}{4} \quad \text{Take the 12th root of both sides (or raise both sides to the 1/12 power).}$$

$$1.015 \approx 1 + \frac{r}{4} \quad \text{Approximate } \sqrt[12]{1.19562} \text{ on a calculator.}$$

$$0.015 \approx \frac{r}{4} \quad \text{Subtract 1 from both sides.}$$

$$0.06 \approx r \quad \text{Multiply both sides by 4.}$$

**Answer** The annual interest rate is approximately 0.06 or 6%.

▸ **Now Try Exercise 103**

**Helpful Hint** *Study Tip*

In this chapter, you will be working with roots and radicals. This material was discussed in Chapter 7. If you do not remember how to evaluate or simplify radicals, review Chapter 7 now.

## EXERCISE SET 8.1

MathXL® MyMathLab

### Concept/Writing Exercises

1. Write the two square roots of 36.
2. Write the two square roots of 17.
3. Write the square root property.
4. What is the first step in completing the square?
5. Explain how to determine whether a trinomial is a perfect square trinomial.
6. Write a paragraph explaining how to construct a perfect square trinomial.
7. a) Is $x = 4$ the solution to $x - 4 = 0$? If not, what is the correct solution? Explain.
   b) Is $x = 2$ the solution to $x^2 - 4 = 0$? If not, what is the correct solution? Explain.
8. a) Is $x = -7$ the solution to $x + 7 = 0$? If not, what is the correct solution? Explain.
   b) Is $x = \pm\sqrt{7}$ the solution to $x^2 + 7 = 0$? If not, what is the correct solution? Explain.
9. What is the first step in solving the equation $2x^2 + 3x = 9$ by completing the square? Explain.
10. What is the first step in solving the equation $\frac{1}{7}x^2 + 12x = -4$ by completing the square? Explain.
11. When solving the equation $x^2 - 6x = 17$ by completing the square, what number do we add to both sides of the equation? Explain.
12. When solving the equation $x^2 + 10x = 39$ by completing the square, what number do we add to both sides of the equation? Explain.

### Practice the Skills

*Use the square root property to solve each equation.*

13. $x^2 - 25 = 0$
14. $x^2 - 49 = 0$
15. $x^2 + 49 = 0$
16. $x^2 - 24 = 0$
17. $x^2 + 24 = 0$
18. $y^2 - 10 = 51$
19. $y^2 + 10 = -51$
20. $(x - 3)^2 = 49$
21. $(p - 4)^2 = 16$
22. $(x + 3)^2 = 49$
23. $(x + 3)^2 + 25 = 0$
24. $(a - 3)^2 = 45$
25. $(a - 2)^2 + 45 = 0$
26. $(a + 2)^2 + 45 = 0$
27. $\left(b + \frac{1}{3}\right)^2 = \frac{4}{9}$
28. $\left(b - \frac{1}{3}\right)^2 = \frac{4}{9}$
29. $\left(b - \frac{2}{3}\right)^2 + \frac{4}{9} = 0$
30. $(x - 0.2)^2 = 0.64$
31. $(x + 0.8)^2 = 0.81$
32. $\left(x + \frac{1}{2}\right)^2 = \frac{16}{9}$
33. $(2a - 5)^2 = 18$
34. $(4y + 1)^2 = 12$
35. $\left(2y + \frac{1}{2}\right)^2 = \frac{4}{25}$
36. $\left(3x - \frac{1}{4}\right)^2 = \frac{9}{25}$

*Solve each equation by completing the square.*

37. $x^2 + 3x - 4 = 0$
38. $x^2 - 3x - 4 = 0$
39. $x^2 + 8x + 15 = 0$
40. $x^2 - 8x + 15 = 0$
41. $x^2 + 6x + 8 = 0$
42. $x^2 - 6x + 8 = 0$
43. $x^2 - 7x + 6 = 0$
44. $x^2 + 9x + 18 = 0$
45. $2x^2 + x - 1 = 0$
46. $3c^2 - 4c - 4 = 0$
47. $2z^2 - 7z - 4 = 0$
48. $4a^2 + 9a = 9$
49. $x^2 - 13x + 40 = 0$
50. $x^2 + x - 12 = 0$
51. $-x^2 + 6x + 7 = 0$
52. $-a^2 - 5a + 14 = 0$
53. $-z^2 + 9z - 20 = 0$
54. $-z^2 - 4z + 12 = 0$
55. $b^2 = 3b + 28$
56. $-x^2 = 6x - 27$
57. $x^2 + 10x = 11$
58. $-x^2 + 40 = -3x$
59. $x^2 - 4x - 10 = 0$
60. $x^2 - 6x + 2 = 0$
61. $r^2 + 8r + 5 = 0$
62. $a^2 + 4a - 8 = 0$
63. $c^2 - c - 3 = 0$
64. $p^2 - 5p = 4$
65. $x^2 + 3x + 6 = 0$
66. $z^2 - 5z + 7 = 0$
67. $9x^2 - 9x = 0$
68. $4y^2 + 12y = 0$
69. $-\frac{3}{4}b^2 - \frac{1}{2}b = 0$
70. $\frac{1}{3}a^2 - \frac{5}{3}a = 0$
71. $36z^2 - 6z = 0$
72. $x^2 = \frac{9}{2}x$

**73.** $-\frac{1}{2}p^2 - p + \frac{3}{2} = 0$

**74.** $2x^2 + 6x = 20$

**75.** $2x^2 = 8x + 64$

**76.** $3x^2 + 33x + 72 = 0$

**77.** $2x^2 + 18x + 4 = 0$

**78.** $\frac{2}{3}x^2 + \frac{4}{3}x + 1 = 0$

**79.** $\frac{3}{4}w^2 + \frac{1}{2}w - \frac{1}{2} = 0$

**80.** $\frac{3}{4}c^2 - 2c + 1 = 0$

**81.** $2x^2 - x = -5$

**82.** $\frac{5}{2}x^2 + \frac{3}{2}x - \frac{5}{4} = 0$

**83.** $-3x^2 + 6x = 6$

**84.** $x^2 + 2x = -5$

## Problem Solving

**Area** *In Exercises 85–88, the area, A, of each rectangle is given.* **a)** *Write an equation for the area.* **b)** *Solve the equation for x.*

**85.**
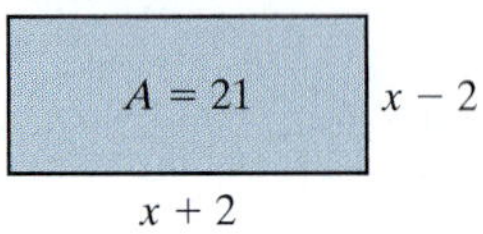


**86.**
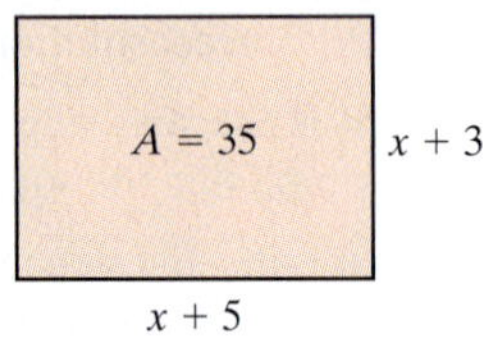


**87.**
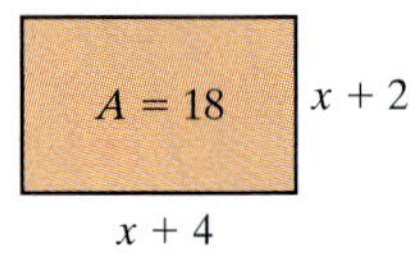


**88.**
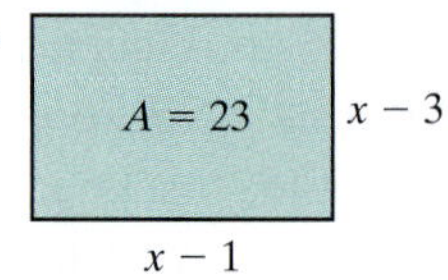


**89. Stopping Distance on Snow** The formula for approximating the stopping distance, $d$, in feet, for a specific car on snow is $d = \frac{1}{6}x^2$, where $x$ is the speed of the car, in miles per hour, before the brakes are applied. If the car's stopping distance was 150 feet, what was the car's speed before the brakes were applied?

**90. Stopping Distance on Dry Pavement** The formula for approximating the stopping distance, $d$, in feet, for a specific car on dry pavement is $d = \frac{1}{10}x^2$, where $x$ is the speed of the car, in miles per hour, before the brakes are applied. If the car's stopping distance was 40 feet, what was the car's speed before the brakes were applied?

**91. Integers** The product of two consecutive positive odd integers is 35. Find the two odd integers.

**92. Integers** The larger of two integers is 2 more than twice the smaller. Find the two numbers if their product is 12.

**93. Rectangular Garden** Donna Simm has marked off an area in her yard where she will plant tomato plants. Find the dimensions of the rectangular area if the length is 2 feet more than twice the width and the area is 60 square feet.

**94. Driveway** Manuel Cortez is planning to blacktop his driveway. Find the dimensions of the rectangular driveway if its area is 381.25 square feet and its length is 18 feet greater than its width.

**95. Patio** Bill Justice is designing a square patio whose diagonal is 6 feet longer than the length of a side. Find the dimensions of the patio.

**96. Wading Pool** The Lakeside Hotel is planning to build a shallow wading pool for children. If the pool is to be square and the diagonal of the square is 7 feet longer than a side, find the dimensions of the pool.

**97. Inscribed Triangle** When a triangle is inscribed in a semicircle where a diameter of the circle is a side of the triangle, the triangle formed is always a right triangle. If an isosceles triangle (two equal sides) is inscribed in a semicircle of radius 10 inches, find the length of the other two sides of the triangle.

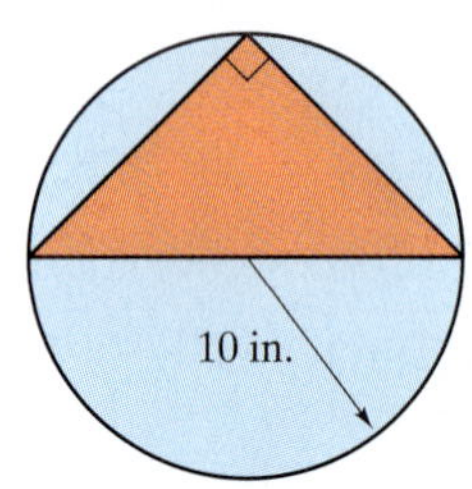


**98. Inscribed Triangle** Refer to Exercise 97. Suppose a triangle is inscribed in a semicircle whose diameter is 12 meters. If one side of the inscribed triangle is 6 meters, find the third side.

**99. Area of Circle** The area of a circle is $24\pi$ square feet. Use the formula $A = \pi r^2$ to find the radius of the circle.

**100. Area of Circle** The area of a circle is $16.4\pi$ square meters. Find the radius of the circle.

*Use the formula* $A = p\left(1 + \frac{r}{n}\right)^{nt}$ *to answer Exercises 101–104.*

**101. Savings Account** Frank Dipalo initially invested \$500 in a savings account where interest is compounded annually. If after 2 years the amount in the account is \$540.80, find the annual interest rate.

**102. Savings Account** Margret Chang initially invested \$1000 in a savings account where interest is compounded annually. If after 2 years the amount in the account is \$1102.50, find the annual interest rate.

**103. Savings Account** Steve Rodi initially invested \$1200 in a savings account where interest is compounded semiannually. If after 3 years the amount in the account is \$1432.86, find the annual interest rate.

**104. Savings Account** Angela Reyes initially invested \$1500 in a savings account where interest is compounded semiannually. If after 4 years the amount in the account is \$2052.85, find the annual interest rate.

**105. Surface Area and Volume** The surface area, $S$, and volume, $V$, of a right circular cylinder of radius, $r$, and height, $h$, are given by the formulas

$$S = 2\pi r^2 + 2\pi rh, \quad V = \pi r^2 h$$

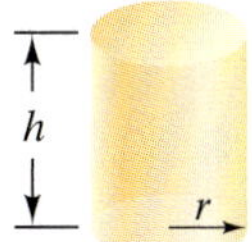


a) Find the surface area of the cylinder if its height is 10 inches and its volume is 160 cubic inches.

b) Find the radius if the height is 10 inches and the volume is 160 cubic inches.

c) Find the radius if the height is 10 inches and the surface area is 160 square inches.

## Group Activity

*Discuss and answer Exercise 106 as a group.*

**106.** On the following grid, the points $(x_1, y_1)$, $(x_2, y_2)$, and $(x_1, y_2)$ are plotted.

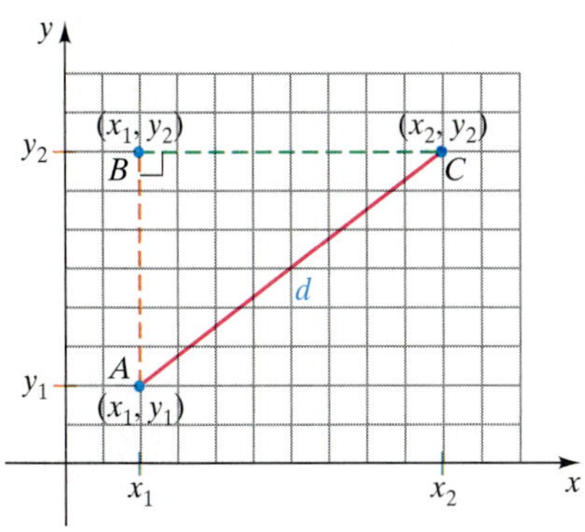


a) Explain why $(x_1, y_2)$ is placed where it is and not somewhere else on the graph.

b) Express the length of the orange dashed line in terms of $y_2$ and $y_1$. Explain how you determined your answer.

c) Express the length of the green dashed line in terms of $x_2$ and $x_1$.

d) Using the Pythagorean Theorem and the right triangle $ABC$, derive a formula for the distance, $d$, between points $(x_1, y_1)$ and $(x_2, y_2)$.* Explain how you determined the formula.

e) Use the formula you determined in part **d)** to find the distance of the line segment between the points (1, 4) and (3, 7).

## Cumulative Review Exercises

[2.1] **107.** Solve $-4(2z - 6) = -3(z - 4) + z$.

[2.4] **108. Investment** Thea Prettyman invested \$10,000 for 1 year, part at 7% and part at $6\frac{1}{4}$%. If she earned a total interest of \$656.50, how much was invested at each rate?

[2.6] **109.** Solve $|x + 3| = |2x - 7|$.

[3.4] **110.** Find the slope of the line through $(-2, 5)$ and $(0, 5)$.

[5.2] **111.** Multiply $(x - 2)(4x^2 + 9x - 3)$.

# 8.2 Solving Quadratic Equations by the Quadratic Formula

1 Derive the quadratic formula.

2 Use the quadratic formula to solve equations.

3 Determine a quadratic equation given its solutions.

4 Use the discriminant to determine the number of real solutions to a quadratic equation.

5 Study applications that use quadratic equations.

## 1 Derive the Quadratic Formula

The quadratic formula can be used to solve any quadratic equation. *It is the most useful and most versatile method of solving quadratic equations.* It is generally used in place of completing the square because of its efficiency.

The standard form of a quadratic equation is $ax^2 + bx + c = 0$, where $a$ is the coefficient of the squared term, $b$ is the coefficient of the first-degree term, and $c$ is the constant.

| Quadratic Equation in Standard Form | Values of Coefficients | | |
|---|---|---|---|
| $x^2 - 3x + 4 = 0$ | $a = 1,$ | $b = -3,$ | $c = 4$ |
| $1.3x^2 - 7.9 = 0$ | $a = 1.3,$ | $b = 0,$ | $c = -7.9$ |
| $-\frac{5}{6}x^2 + \frac{3}{8}x = 0$ | $a = -\frac{5}{6},$ | $b = \frac{3}{8},$ | $c = 0$ |

*The distance formula will be discussed in a later chapter.

We can derive the quadratic formula by starting with a quadratic equation in standard form and completing the square, as discussed in the preceding section.

$$ax^2 + bx + c = 0$$

$$\frac{ax^2}{a} + \frac{b}{a}x + \frac{c}{a} = 0 \qquad \textit{Divide both sides by a.}$$

$$x^2 + \frac{b}{a}x = -\frac{c}{a} \qquad \textit{Subtract c/a from both sides.}$$

$$x^2 + \frac{b}{a}x + \frac{b^2}{4a^2} = -\frac{c}{a} + \frac{b^2}{4a^2}$$

*Take 1/2 of b/a (that is, b/2a), and square it to get $b^2/4a^2$. Then add this expression to both sides.*

$$\left(x + \frac{b}{2a}\right)^2 = \frac{b^2}{4a^2} - \frac{c}{a}$$

*Rewrite the left side of the equation as the square of a binomial.*

$$\left(x + \frac{b}{2a}\right)^2 = \frac{b^2 - 4ac}{4a^2}$$

*Write the right side with a common denominator.*

$$x + \frac{b}{2a} = \pm\sqrt{\frac{b^2 - 4ac}{4a^2}} \qquad \textit{Square root property}$$

$$x + \frac{b}{2a} = \pm\frac{\sqrt{b^2 - 4ac}}{2a} \qquad \textit{Quotient rule for radicals}$$

$$x = -\frac{b}{2a} \pm \frac{\sqrt{b^2 - 4ac}}{2a} \qquad \textit{Subtract b/2a from both sides.}$$

$$x = \frac{-b \pm \sqrt{b^2 - 4ac}}{2a}$$

*Write with a common denominator to get the quadratic formula.*

## 2 Use the Quadratic Formula to Solve Equations

Now that we have derived the quadratic formula, we will use it to solve quadratic equations.

**To Solve a Quadratic Equation by the Quadratic Formula**

1. Write the quadratic equation in standard form, $ax^2 + bx + c = 0$, and determine the numerical values for $a$, $b$, and $c$.
2. Substitute the values for $a$, $b$, and $c$ into the quadratic formula and then evaluate the formula to obtain the solution.

**The Quadratic Formula**

$$x = \frac{-b \pm \sqrt{b^2 - 4ac}}{2a}$$

**EXAMPLE 1** ▸ Solve $x^2 + 2x - 8 = 0$ by the quadratic formula.

*Solution* In this equation, $a = 1$, $b = 2$, and $c = -8$.

$$x = \frac{-b \pm \sqrt{b^2 - 4ac}}{2a}$$

$$x = \frac{-2 \pm \sqrt{2^2 - 4(1)(-8)}}{2(1)}$$

$$= \frac{-2 \pm \sqrt{4 + 32}}{2}$$

$$= \frac{-2 \pm \sqrt{36}}{2}$$

$$= \frac{-2 \pm 6}{2}$$

$$x = \frac{-2 + 6}{2} \quad \text{or} \quad x = \frac{-2 - 6}{2}$$
$$x = \frac{4}{2} = 2 \qquad\qquad x = \frac{-8}{2} = -4$$

A check will show that both 2 and $-4$ are solutions to the equation. Note that the solutions to the equation $x^2 + 2x - 8 = 0$ are two real numbers.

▸ **Now Try Exercise 23**

The solution to Example 1 could also be obtained by factoring, as follows:

$$x^2 + 2x - 8 = 0$$
$$(x + 4)(x - 2) = 0$$
$$x + 4 = 0 \quad \text{or} \quad x - 2 = 0$$
$$x = -4 \qquad\qquad x = 2$$

When you are given a quadratic equation to solve and the method to solve it has not been specified, you may try solving by factoring first (as we discussed in Section 5.8). If the equation cannot be easily factored, use the quadratic formula.

When solving a quadratic equation using the quadratic formula, the calculations may be easier if the leading coefficient, $a$, is positive. Thus, if solving the quadratic equation $-x^2 + 3x = 2$, you may wish to rewrite the equation as $x^2 - 3x + 2 = 0$.

**EXAMPLE 2** ▸ Solve $-9x^2 = -6x + 1$ by the quadratic formula.

*Solution* Begin by adding $9x^2$ to both sides of the equation to obtain

$$0 = 9x^2 - 6x + 1$$
$$\text{or} \quad 9x^2 - 6x + 1 = 0$$
$$a = 9, \quad b = -6, \quad c = 1$$
$$x = \frac{-b \pm \sqrt{b^2 - 4ac}}{2a}$$
$$= \frac{-(-6) \pm \sqrt{(-6)^2 - 4(9)(1)}}{2(9)}$$
$$= \frac{6 \pm \sqrt{36 - 36}}{18} = \frac{6 \pm \sqrt{0}}{18} = \frac{6}{18} = \frac{1}{3}$$

Note that the solution to the equation $-9x^2 = -6x + 1$ is $\frac{1}{3}$, a single value. Some quadratic equations have just one value as the solution. This occurs when $b^2 - 4ac = 0$.

▸ **Now Try Exercise 39**

## Avoiding Common Errors

The entire numerator of the quadratic formula must be divided by $2a$.

| CORRECT | INCORRECT |
|---|---|
| $x = \frac{-b \pm \sqrt{b^2 - 4ac}}{2a}$ | ~~$x = -b \pm \frac{\sqrt{b^2 - 4ac}}{2a}$~~ <br> ~~$x = \frac{-b}{2a} \pm \sqrt{b^2 - 4ac}$~~ |

**EXAMPLE 3** ▸ Solve $p^2 + \frac{1}{3}p + \frac{5}{6} = 0$ by the quadratic formula.

*Solution* Do not let the change in variable worry you. The quadratic formula is used exactly the same way as when $x$ is the variable.

We could solve this equation using the quadratic formula with $a = 1$, $b = \frac{1}{3}$, and $c = \frac{5}{6}$. However, when a quadratic equation contains fractions, it is generally easier to begin by multiplying both sides of the equation by the least common denominator. In this example, the least common denominator is 6.

$$6\left(p^2 + \frac{1}{3}p + \frac{5}{6}\right) = 6(0)$$
$$6p^2 + 2p + 5 = 0$$

Now we can use the quadratic formula with $a = 6$, $b = 2$, and $c = 5$.

$$\begin{aligned}
p &= \frac{-b \pm \sqrt{b^2 - 4ac}}{2a} \\
&= \frac{-2 \pm \sqrt{2^2 - 4(6)(5)}}{2(6)} \\
&= \frac{-2 \pm \sqrt{-116}}{12} \\
&= \frac{-2 \pm \sqrt{-4}\sqrt{29}}{12} \\
&= \frac{-2 \pm 2i\sqrt{29}}{12} \\
&= \frac{\overset{1}{\cancel{2}}(-1 \pm i\sqrt{29})}{\underset{6}{\cancel{12}}} \\
&= \frac{-1 \pm i\sqrt{29}}{6}
\end{aligned}$$

The solutions are $\frac{-1 + i\sqrt{29}}{6}$ and $\frac{-1 - i\sqrt{29}}{6}$. Note that neither solution is a real number. Both solutions are complex numbers.

▸ **Now Try Exercise 53**

## Avoiding Common Errors

Some students use the quadratic formula correctly until the last step, where they make an error. Below are illustrated both the correct and incorrect procedures for simplifying an answer.

When *both* terms in the numerator *and* the denominator have a common factor, that common factor may be divided out, as follows:

CORRECT

$$\frac{2 + 4\sqrt{3}}{2} = \frac{\overset{1}{\cancel{2}}(1 + 2\sqrt{3})}{\underset{1}{\cancel{2}}} = 1 + 2\sqrt{3}$$

$$\frac{6 + 3\sqrt{3}}{6} = \frac{\overset{1}{\cancel{3}}(2 + \sqrt{3})}{\underset{2}{\cancel{6}}} = \frac{2 + \sqrt{3}}{2}$$

Below are some common errors. Study them carefully so you will not make them. Can you explain why each of the following procedures is incorrect?

INCORRECT

$$\frac{2 + 3}{2} = \frac{\overset{1}{\cancel{2}} + 3}{\underset{1}{\cancel{2}}} \qquad \frac{3 + 2\sqrt{5}}{2} = \frac{3 + \overset{1}{\cancel{2}}\sqrt{5}}{\underset{1}{\cancel{2}}}$$

$$\frac{3 + \sqrt{6}}{2} = \frac{3 + \sqrt{\cancel{6}^{3}}}{\underset{1}{\cancel{2}}} \qquad \frac{4 + 3\sqrt{5}}{2} = \frac{\overset{2}{\cancel{4}} + 3\sqrt{5}}{\underset{1}{\cancel{2}}}$$

Note that $(2 + 3)/2$ simplifies to $5/2$. However, $(3 + 2\sqrt{5})/2$, $(3 + \sqrt{6})/2$, and $(4 + 3\sqrt{5})/2$ cannot be simplified any further.

**EXAMPLE 4** ▸ Given $f(x) = 2x^2 + 4x$, find all real values of $x$ for which $f(x) = 5$.

*Solution* We wish to determine all real values of $x$ for which

$$2x^2 + 4x = 5$$

We can solve this equation with the quadratic formula. First, write the equation in standard form.

$$2x^2 + 4x - 5 = 0$$

Now, use the quadratic formula with $a = 2$, $b = 4$, and $c = -5$.

$$x = \frac{-b \pm \sqrt{b^2 - 4ac}}{2a}$$

$$= \frac{-4 \pm \sqrt{4^2 - 4(2)(-5)}}{2(2)} = \frac{-4 \pm \sqrt{56}}{4} = \frac{-4 \pm 2\sqrt{14}}{4}$$

Next, factor out 2 from both terms in the numerator, and then divide out the common factor.

$$x = \frac{\overset{1}{\cancel{2}}(-2 \pm \sqrt{14})}{\underset{2}{\cancel{4}}} = \frac{-2 \pm \sqrt{14}}{2}\,^{*}$$

Thus, the solutions are $\frac{-2 + \sqrt{14}}{2}$ and $\frac{-2 - \sqrt{14}}{2}$.

Note that the expression in Example 4, $2x^2 + 4x - 5$, is not factorable. Therefore, Example 4 could not be solved by factoring.

▸ **Now Try Exercise 69**

If all the numerical coefficients in a quadratic equation have a common factor, you should factor it out before using the quadratic formula. For example, consider the equation $3x^2 + 12x + 3 = 0$. Here $a = 3$, $b = 12$, and $c = 3$. If we use the quadratic formula, we would eventually obtain $x = -2 \pm \sqrt{3}$ as solutions. By factoring the equation before using the formula, we get

$$3x^2 + 12x + 3 = 0$$
$$3(x^2 + 4x + 1) = 0$$

If we consider $x^2 + 4x + 1 = 0$, then $a = 1$, $b = 4$, and $c = 1$. If we use these new values of $a$, $b$, and $c$ in the quadratic formula, we will obtain the identical solutions, $x = -2 \pm \sqrt{3}$. However, the calculations with these smaller values of $a$, $b$, and $c$ are simplified. Solve both equations now using the quadratic formula to convince yourself.

## 3 Determine a Quadratic Equation Given Its Solutions

If we are given the solutions of an equation, we can find the equation by working backward. This procedure is illustrated in Example 5.

**EXAMPLE 5** ▸ Determine an equation that has the following solutions:

**a)** $-5$ and 1 **b)** $3 + 2i$ and $3 - 2i$

*Solution*

**a)** If the solutions are $-5$ and 1 we write

$$x = -5 \quad \text{or} \quad x = 1$$

$$x + 5 = 0 \qquad x - 1 = 0 \qquad \text{Set equations equal to 0.}$$

$$(x + 5)(x - 1) = 0 \qquad \text{Zero-factor property}$$

$$x^2 - x + 5x - 5 = 0 \qquad \text{Multiply factors.}$$

$$x^2 + 4x - 5 = 0 \qquad \text{Combine like terms.}$$

*Solutions will be given in this form in the Answer Section.

Thus, the equation is $x^2 + 4x - 5 = 0$. Many other equations have solutions $-5$ and 1. In fact, any equation of the form $k(x^2 + 4x - 5) = 0$, where $k$ is a constant, has those solutions. Can you explain why?

**b)**

$$x = 3 + 2i \quad \text{or} \quad x = 3 - 2i$$

$$x - (3 + 2i) = 0 \qquad x - (3 - 2i) = 0 \qquad \textit{Set equations equal to 0.}$$

$$[x - (3 + 2i)][x - (3 - 2i)] = 0 \qquad \textit{Zero-factor property}$$

$$x \cdot x - x(3 - 2i) - x(3 + 2i) + (3 + 2i)(3 - 2i) = 0 \qquad \textit{Multiply.}$$

$$x^2 - 3x + 2xi - 3x - 2xi + (9 - 4i^2) = 0 \qquad \textit{Distributive property; multiply}$$

$$x^2 - 6x + 9 - 4i^2 = 0 \qquad \textit{Combine like terms.}$$

$$x^2 - 6x + 9 - 4(-1) = 0 \qquad \textit{Substitute } i^2 = -1.$$

$$x^2 - 6x + 13 = 0 \qquad \textit{Simplify.}$$

The equation $x^2 - 6x + 13 = 0$ has the complex solutions $3 + 2i$ and $3 - 2i$.

▸ **Now Try Exercise 75**

In Example 5 **a)**, we determined that the equation $x^2 + 4x - 5 = 0$ has solutions $-5$ and 1. Consider the graph of $f(x) = x^2 + 4x - 5$. The $x$-intercepts of the graph of $f(x)$ occur when $f(x) = 0$, or when $x^2 + 4x - 5 = 0$. Therefore, the $x$-intercepts of the graph of $f(x) = x^2 + 4x - 5$ are $(-5, 0)$ and $(1, 0)$, as shown in **Figure 8.1**. In Example 5 **b)**, we determined that the equation $x^2 - 6x + 13 = 0$ has no real solutions. Thus, the graph of $f(x) = x^2 - 6x + 13$ has no $x$-intercepts. The graph of $f(x) = x^2 - 6x + 13$ is shown in **Figure 8.2**.

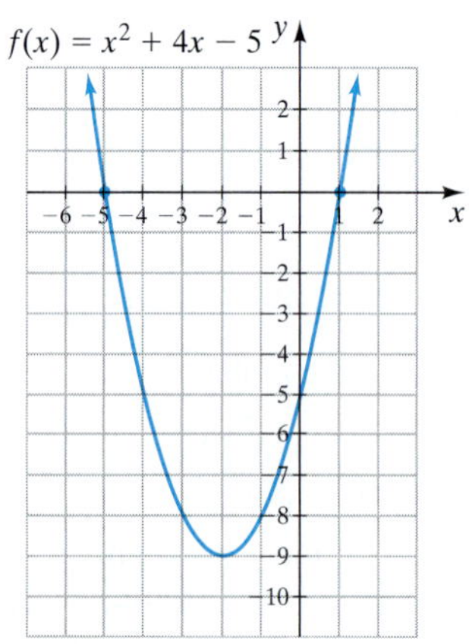


FIGURE 8.1

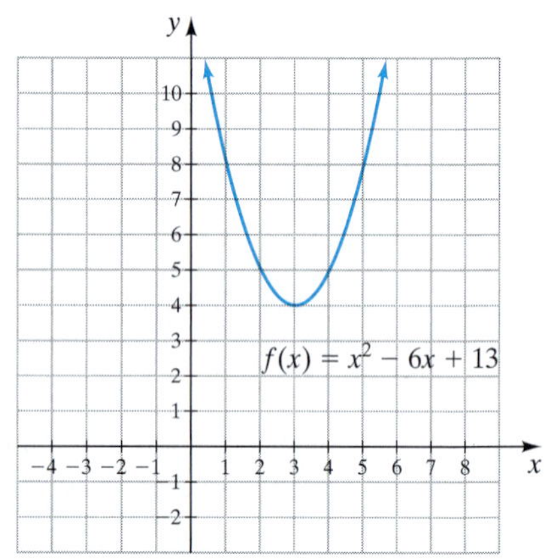


FIGURE 8.2

## 4 Use the Discriminant to Determine the Number of Real Solutions to a Quadratic Equation

The expression under the radical sign in the quadratic formula is called the **discriminant**.

$$\underbrace{b^2 - 4ac}_{\text{discriminant}}$$

The discriminant provides information to determine the number and kinds of solutions of a quadratic equation.

**Solutions of a Quadratic Equation**

For a quadratic equation of the form $ax^2 + bx + c = 0, a \neq 0$:

If $b^2 - 4ac > 0$, the quadratic equation has two distinct real number solutions.
If $b^2 - 4ac = 0$, the quadratic equation has a single real number solution.
If $b^2 - 4ac < 0$, the quadratic equation has no real number solution.

**EXAMPLE 6 ▸**

**a)** Find the discriminant of the equation $x^2 - 8x + 16 = 0$.

**b)** How many real number solutions does the given equation have?

**c)** Use the quadratic formula to find the solution(s).

*Solution*

**a)** $a = 1, \qquad b = -8, \qquad c = 16$

$$\begin{aligned} b^2 - 4ac &= (-8)^2 - 4(1)(16) \\ &= 64 - 64 = 0 \end{aligned}$$

**b)** Since the discriminant equals 0, there is a single real number solution.

**c)**

$$\begin{aligned} x &= \frac{-b \pm \sqrt{b^2 - 4ac}}{2a} \\ &= \frac{-(-8) \pm \sqrt{0}}{2(1)} = \frac{8 \pm 0}{2} = \frac{8}{2} = 4 \end{aligned}$$

The only solution is 4.

▸ **Now Try Exercise 9**

---

**EXAMPLE 7 ▸** Without actually finding the solutions, determine whether the following equations have two distinct real number solutions, a single real number solution, or no real number solution.

**a)** $2x^2 - 4x + 6 = 0$ **b)** $x^2 - 5x - 3 = 0$ **c)** $4x^2 - 12x = -9$

*Solution* We use the discriminant of the quadratic formula to answer these questions.

**a)** $b^2 - 4ac = (-4)^2 - 4(2)(6) = 16 - 48 = -32$

Since the discriminant is negative, this equation has no real number solution.

**b)** $b^2 - 4ac = (-5)^2 - 4(1)(-3) = 25 + 12 = 37$

Since the discriminant is positive, this equation has two distinct real number solutions.

**c)** First, rewrite $4x^2 - 12x = -9$ as $4x^2 - 12x + 9 = 0$.

$$b^2 - 4ac = (-12)^2 - 4(4)(9) = 144 - 144 = 0$$

Since the discriminant is 0, this equation has a single real number solution.

▸ **Now Try Exercise 15**

---

The discriminant can be used to find the number of real solutions to an equation of the form $ax^2 + bx + c = 0$. Since the $x$-intercepts of a quadratic function, $f(x) = ax^2 + bx + c$, occur where $f(x) = 0$, the discriminant can also be used to find the number of $x$-intercepts of a quadratic function. **Figure 8.3** shows the relationship between the discriminant and the number of $x$-intercepts for a function of the form $f(x) = ax^2 + bx + c$.

**Graphs of $f(x) = ax^2 + bx + c$**

If $b^2 - 4ac > 0$, $f(x)$ has two distinct $x$-intercepts. If $b^2 - 4ac = 0$, $f(x)$ has a single $x$-intercept. If $b^2 - 4ac < 0$, $f(x)$ has no $x$-intercepts.

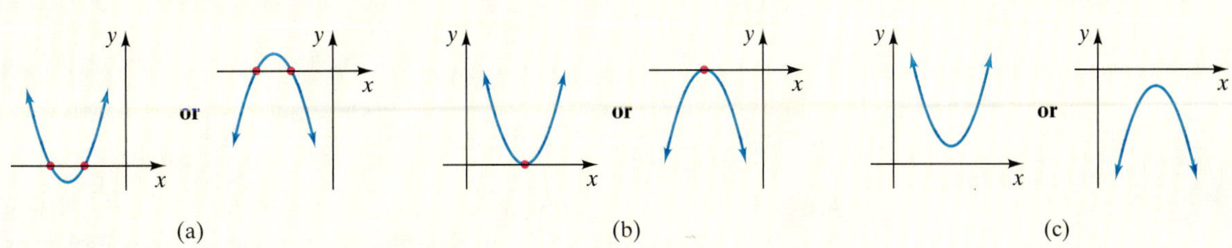


FIGURE 8.3

We will discuss graphing quadratic functions in detail in Section 8.5.

## 5 Study Applications That Use Quadratic Equations

We will now look at some applications of quadratic equations.

**EXAMPLE 8** ▸ **Cell Phones** Mary Olson owns a business that manufactures and sells cell phones. The revenue, $R(n)$, from selling the cell phones is determined by multiplying the number of cell phones by the cost per phone. Suppose the revenue from selling $n$ cell phones, $n \le 50$, is

$$R(n) = n(50 - 0.2n)$$

where $(50 - 0.2n)$ is the price per cell phone, in dollars.

**a)** Find the revenue when 30 cell phones are sold.

**b)** How many cell phones must be sold to have a revenue of \$480?

*Solution* **a)** To find the revenue when 30 cell phones are sold, we evaluate the revenue function for $n = 30$.

$$\begin{aligned} R(n) &= n(50 - 0.2n) \\ R(30) &= 30[50 - 0.2(30)] \\ &= 30(50 - 6) \\ &= 30(44) \\ &= 1320 \end{aligned}$$

The revenue from selling 30 cell phones is \$1320.

**b)** Understand We want to find the number of cell phones that need to be sold to have \$480 in revenue. Thus, we need to let $R(n) = 480$ and solve for $n$.

$$\begin{aligned} R(n) &= n(50 - 0.2n) \\ 480 &= n(50 - 0.2n) \\ 480 &= 50n - 0.2n^2 \\ 0.2n^2 - 50n + 480 &= 0 \end{aligned}$$

Now we can use the quadratic formula to solve the equation.

Translate $a = 0.2, \quad b = -50, \quad c = 480$

$$\begin{aligned} n &= \frac{-b \pm \sqrt{b^2 - 4ac}}{2a} \\ &= \frac{-(-50) \pm \sqrt{(-50)^2 - 4(0.2)(480)}}{2(0.2)} \end{aligned}$$

Carry Out

$$\begin{aligned} &= \frac{50 \pm \sqrt{2500 - 384}}{0.4} \\ &= \frac{50 \pm \sqrt{2116}}{0.4} \\ &= \frac{50 \pm 46}{0.4} \end{aligned}$$

$$n = \frac{50 + 46}{0.4} = 240 \quad \text{or} \quad n = \frac{50 - 46}{0.4} = 10$$

Answer Since the problem specified that $n \le 50$, the only acceptable solution is $n = 10$. Thus, to obtain \$480 in revenue, Mary must sell 10 cell phones.

▸ **Now Try Exercise 87**

An important formula in physics is $h = \frac{1}{2}gt^2 + v_0t + h_0$. When an object is projected upward from an initial height, $h_0$, with initial velocity of $v_0$, this formula can be used to find the height, $h$, of the object above the ground at any time, $t$. The $g$ in the formula is the acceleration of gravity. Since the acceleration of Earth's gravity is $-32$ ft/sec$^2$, we use $-32$ for $g$ in the formula when discussing Earth. This formula can also be used in describing projectiles on the Moon and other planets, but the value of $g$ in the formula will need to change for each planetary body. We will use this formula in Example 9.

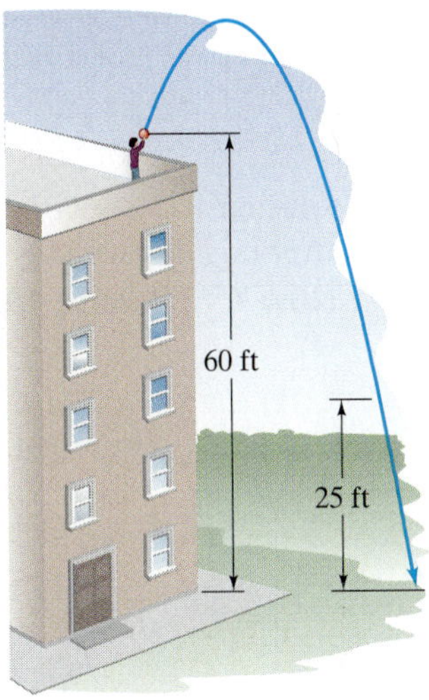

FIGURE 8.4

**EXAMPLE 9 ▸ Throwing a Ball** Betsy Farber is standing on top of a building and throws a ball upward from a height of 60 feet with an initial velocity of 30 feet per second. Use the formula $h = \frac{1}{2}gt^2 + v_0t + h_0$ to answer the following questions.

**a)** How long after the ball is thrown, to the nearest tenth of a second, will the ball be 25 feet above the ground?

**b)** How long after the ball is thrown will the ball strike the ground?

*Solution* **a)** Understand We will illustrate this problem with a diagram (see **Fig. 8.4**). Here $g = -32$, $v_0 = 30$, and $h_0 = 60$. We are asked to find the time, $t$, it takes for the ball to reach a height, $h$, of 25 feet above the ground. We substitute these values into the formula and then solve for $t$.

Translate
$$h = \frac{1}{2}gt^2 + v_0t + h_0$$
$$25 = \frac{1}{2}(-32)t^2 + 30t + 60$$

Carry Out Now we write the quadratic equation in standard form and solve for $t$ by using the quadratic formula.

$$0 = -16t^2 + 30t + 35$$
$$\text{or} \quad -16t^2 + 30t + 35 = 0$$
$$a = -16, \quad b = 30, \quad c = 35$$
$$t = \frac{-b \pm \sqrt{b^2 - 4ac}}{2a}$$
$$= \frac{-30 \pm \sqrt{(30)^2 - 4(-16)(35)}}{2(-16)}$$
$$= \frac{-30 \pm \sqrt{3140}}{-32}$$
$$t = \frac{-30 + \sqrt{3140}}{-32} \quad \text{or} \quad t = \frac{-30 - \sqrt{3140}}{-32}$$
$$\approx -0.8 \qquad\qquad \approx 2.7$$

Answer Since time cannot be negative, the only reasonable solution is 2.7 seconds. Thus, about 2.7 seconds after the ball is thrown upward, it will be 25 feet above the ground.

**b)** Understand We wish to find the time at which the ball strikes the ground. When the ball strikes the ground, its distance above the ground is 0. We substitute $h = 0$ into the formula and solve for $t$.

Translate
$$h = \frac{1}{2}gt^2 + v_0t + h_0$$
$$0 = \frac{1}{2}(-32)t^2 + 30t + 60$$

**Carry Out**

$$0 = -16t^2 + 30t + 60$$

$$a = -16, \quad b = 30, \quad c = 60$$

$$t = \frac{-b \pm \sqrt{b^2 - 4ac}}{2a}$$

$$= \frac{-30 \pm \sqrt{(30)^2 - 4(-16)(60)}}{2(-16)}$$

$$= \frac{-30 \pm \sqrt{4740}}{-32}$$

$$t = \frac{-30 + \sqrt{4740}}{-32} \quad \text{or} \quad t = \frac{-30 - \sqrt{4740}}{-32}$$

$$\approx -1.2 \qquad\qquad \approx 3.1$$

**Answer** Since time cannot be negative, the only reasonable solution is 3.1 seconds. Thus, the ball will strike the ground approximately 3.1 seconds after it is thrown.

▶ **Now Try Exercise 103**

## EXERCISE SET 8.2

MathXL® MyMathLab

### Concept/Writing Exercises

1. Give the quadratic formula. (You should memorize this formula.)
2. To solve the equation $3x + 2x^2 - 9 = 0$ using the quadratic formula, what are the values for $a$, $b$, and $c$?
3. To solve the equation $6x - 3x^2 + 8 = 0$ using the quadratic formula, what are the values for $a$, $b$, and $c$?
4. To solve the equation $4x^2 - 5x = 7$ using the quadratic formula, what are the values for $a$, $b$, and $c$?
5. Consider the two equations $-6x^2 + \frac{1}{2}x - 5 = 0$ and $6x^2 - \frac{1}{2}x + 5 = 0$. Must the solutions to these two equations be the same? Explain your answer.
6. Consider $12x^2 - 15x - 6 = 0$ and $3(4x^2 - 5x - 2) = 0$.
   **a)** Will the solution to the two equations be the same? Explain.
   **b)** Solve $12x^2 - 15x - 6 = 0$.
   **c)** Solve $3(4x^2 - 5x - 2) = 0$.
7. **a)** Explain how to find the discriminant.
   **b)** What is the discriminant for the equation $3x^2 - 6x + 10 = 0$?
   **c)** Write a paragraph or two explaining the relationship between the value of the discriminant and the number of real solutions to a quadratic equation. In your paragraph, explain *why* the value of the discriminant determines the number of real solutions.
8. Write a paragraph or two explaining the relationship between the value of the discriminant and the number of $x$-intercepts of $f(x) = ax^2 + bx + c$. In your paragraph, explain when the function will have no, one, and two $x$-intercepts.

### Practice the Skills

*Use the discriminant to determine whether each equation has two distinct real solutions, a single real solution, or no real solution.*

9. $x^2 + 3x + 1 = 0$
10. $2x^2 + x + 3 = 0$
11. $4z^2 + 6z + 5 = 0$
12. $-a^2 + 3a - 6 = 0$
13. $5p^2 + 3p - 7 = 0$
14. $2x^2 = 16x - 32$
15. $-5x^2 + 5x - 8 = 0$
16. $4.1x^2 - 3.1x - 2.8 = 0$
17. $x^2 + 10.2x + 26.01 = 0$
18. $\frac{1}{2}x^2 + \frac{2}{3}x + 10 = 0$
19. $b^2 = -3b - \frac{9}{4}$
20. $\frac{x^2}{3} = \frac{2x}{7}$

*Solve each equation by the quadratic formula.*

21. $x^2 - 9x + 18 = 0$
22. $x^2 + 9x + 18 = 0$
23. $a^2 - 6a + 8 = 0$
24. $a^2 + 6a + 8 = 0$
25. $x^2 = -6x + 7$
26. $-a^2 - 9a + 10 = 0$
27. $-b^2 = 4b - 20$
28. $a^2 - 16 = 0$
29. $b^2 - 64 = 0$
30. $2x^2 = 4x + 1$
31. $3w^2 - 4w + 5 = 0$
32. $x^2 - 6x = 0$

**33.** $c^2 - 5c = 0$

**34.** $-t^2 - t - 1 = 0$

**35.** $4s^2 - 8s + 6 = 0$

**36.** $-3r^2 = 9r + 6$

**37.** $a^2 + 2a + 1 = 0$

**38.** $y^2 + 16y + 64 = 0$

39. $16x^2 - 8x + 1 = 0$

**40.** $100m^2 + 20m + 1 = 0$

**41.** $x^2 - 2x - 1 = 0$

**42.** $2 - 3r^2 = -4r$

**43.** $-n^2 = 3n + 6$

**44.** $-9d - 3d^2 = 5$

**45.** $2x^2 + 5x - 3 = 0$

**46.** $(r - 3)(3r + 4) = -10$

**47.** $(2a + 3)(3a - 1) = 2$

**48.** $6x^2 = 21x + 27$

**49.** $\frac{1}{2}t^2 + t - 12 = 0$

**50.** $\frac{2}{3}x^2 = 8x - 18$

**51.** $9r^2 + 3r - 2 = 0$

**52.** $2x^2 - 4x - 2 = 0$

53. $\frac{1}{2}x^2 + 2x + \frac{2}{3} = 0$

**54.** $x^2 - \frac{11}{3}x = \frac{10}{3}$

**55.** $a^2 - \frac{a}{5} - \frac{1}{3} = 0$

**56.** $b^2 = -\frac{b}{2} + \frac{2}{3}$

**57.** $c = \frac{c - 6}{4 - c}$

**58.** $3y = \frac{5y + 6}{2y + 3}$

**59.** $2x^2 - 4x + 5 = 0$

**60.** $3a^2 - 4a = -5$

**61.** $y^2 + \frac{y}{2} = -\frac{3}{2}$

**62.** $2b^2 - \frac{7}{3}b + \frac{4}{3} = 0$

**63.** $0.1x^2 + 0.6x - 1.2 = 0$

**64.** $2.3x^2 - 5.6x - 0.4 = 0$

*For each function, determine all real values of the variable for which the function has the value indicated.*

**65.** $f(x) = x^2 - 2x + 5, f(x) = 5$

**66.** $g(x) = x^2 + 3x + 8, g(x) = 8$

**67.** $k(x) = x^2 - x - 15, k(x) = 15$

**68.** $p(r) = r^2 + 17r + 81, p(r) = 9$

69. $h(t) = 2t^2 - 7t + 6, h(t) = 2$

**70.** $t(x) = x^2 + 5x - 4, t(x) = 3$

**71.** $g(a) = 2a^2 - 3a + 16, g(a) = 14$

**72.** $h(x) = 6x^2 + 3x + 1, h(x) = -7$

*Determine an equation that has the given solutions.*

**73.** 2, 5

**74.** −3, 4

75. 1, −9

**76.** −2, −6

**77.** $-\frac{3}{5}, \frac{2}{3}$

**78.** $-\frac{1}{3}, -\frac{3}{4}$

**79.** $\sqrt{2}, -\sqrt{2}$

**80.** $\sqrt{5}, -\sqrt{5}$

**81.** $3i, -3i$

**82.** $8i, -8i$

**83.** $3 + \sqrt{2}, 3 - \sqrt{2}$

**84.** $5 - \sqrt{3}, 5 + \sqrt{3}$

**85.** $2 + 3i, 2 - 3i$

**86.** $5 - 4i, 5 + 4i$

## Problem Solving

*In Exercises 87–90,* **a)** *set up a revenue function, $R(n)$, that can be used to solve the problem, and then,* **b)** *solve the problem. See Example 8.*

87. **Selling Lamps** A business sells $n$ lamps, $n \leq 65$, at a price of $(10 - 0.02n)$ dollars per lamp. How many lamps must be sold to have a revenue of \$450?

**88. Selling Batteries** A business sells $n$ batteries, $n \leq 26$, at a price of $(25 - 0.1n)$ dollars per battery. How many batteries must be sold to have a revenue of \$460?

**89. Selling Chairs** A business sells $n$ chairs, $n \leq 50$, at a price of $(50 - 0.4n)$ dollars per chair. How many chairs must be sold to have a revenue of \$660?

**90. Selling Watches** A business sells $n$ watches, $n \leq 75$, at a price of $(30 - 0.15n)$ dollars per watch. How many watches must be sold to have a revenue of \$1260?

**91.** Give your own example of a quadratic equation that can be solved by the quadratic formula but not by factoring over the set of integers.

**92.** Are there any quadratic equations that **a)** can be solved by the quadratic formula that cannot be solved by completing the square? **b)** can be solved by completing the square that cannot be solved by factoring over the set of integers?

**93.** When solving a quadratic equation by the quadratic formula, if the discriminant is a perfect square, must the equation be factorable over the set of integers?

**94.** When solving a quadratic equation by the quadratic formula, if the discriminant is a natural number, must the equation be factorable over the set of integers?

*In Exercises 95–102, use a calculator as needed to give the solution in decimal form. Round irrational numbers to the nearest hundredth.*

**95. Numbers** Twice the square of a positive number increased by three times the number is 27. Find the number.

**96. Numbers** Three times the square of a positive number decreased by twice the number is 21. Find the number.

**97. Rectangular Garden** The length of a rectangular garden is 1 foot less than 3 times its width. Find the length and width if the area of the garden is 24 square feet.

**98. Rectangular Region** Lora Wallman wishes to fence in a rectangular region along a riverbank by constructing fencing as illustrated in the diagram. If she has only 400 feet of fencing and wishes to enclose an area of 15,000 square feet, find the dimensions of the rectangular region.

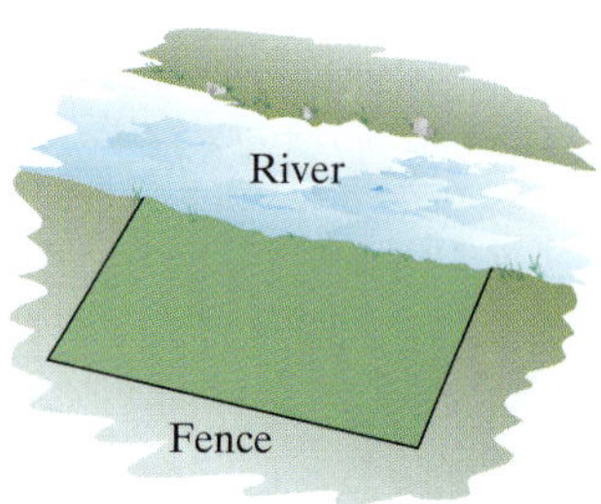


**99. Photo** John Williams, a professional photographer, has a 6-inch-by-8-inch photo. He wishes to reduce the photo by the same amount on each side so that the resulting photo will have half the area of the original photo. By how much will he have to reduce the length of each side?

**100. Rectangular Garden** Bart Simmons has a 12-meter-by-9-meter flower garden. He wants to build a gravel path of uniform width along the inside of the garden on each side so that the resulting garden will have half the area of the original garden. What will be the width of the gravel path?

**101. Waterfalls** When a drop of water (or other object) at the top of the Lower Falls in Yellowstone National Park goes over the top of the falls, the height, $h$, in feet, of the drop of water above the pool of water at the bottom of the falls can be determined by the equation $h = -16t^2 + 308$. In the equation, $t$ is the time, in seconds, after the drop goes over the falls. Determine the time it takes for the drop of water to reach the bottom of the falls (when $h = 0$).

**102. Waterfalls** When a drop of water (or other object) at the top of Niagara Falls goes over the top of the falls, the height, $h$, in feet, of the drop of water above the pool of water at the bottom of the falls can be determined by the equation $h = -16t^2 + 176$. In the equation, $t$ is the time, in seconds, after the drop goes over the falls. Determine the time it takes for the drop of water to reach the bottom of the falls (when $h = 0$).

*In Exercises 103 and 104, use the equation* $h = \frac{1}{2}gt^2 + v_0t + h_0$ *(refer to Example 9).*

**103. Throwing a Horseshoe** A horseshoe is thrown upward from an initial height of 80 feet with an initial velocity of 60 feet per second. How long after the horseshoe is projected upward

**a)** will it be 20 feet from the ground?

**b)** will it strike the ground?

**104. Gravity on the Moon** Gravity on the Moon is about one-sixth of that on Earth. Suppose Neil Armstrong is standing on a hill on the Moon 60 feet high. If he jumps upward with a velocity of 40 feet per second, how long will it take for him to land on the ground below the hill?

*Solve by the quadratic formula.*

**105.** $x^2 - \sqrt{5}x - 10 = 0$

**106.** $x^2 + 5\sqrt{6}x + 36 = 0$

## Challenge Problems

**107. Heating a Metal Cube** A metal cube expands when heated. If each edge increases 0.20 millimeter after being heated and the total volume increases by 6 cubic millimeters, find the original length of a side of the cube.

**108. Six Solutions** The equation $x^n = 1$ has $n$ solutions (including the complex solutions). Find the six solutions to $x^6 = 1$. (*Hint:* Rewrite the equation as $x^6 - 1 = 0$, then factor using the formula for the difference of two squares.)

**109. Throwing a Rock** Travis Hawley is on the fourth floor of an eight-story building and Courtney Prenzlow is on the roof. Travis is 60 feet above the ground while Courtney is 120 feet above the ground.

**a)** If Travis drops his rock out of a window, determine the time it takes for the rock to strike the ground.

**b)** If Courtney drops her rock off the roof, determine the time it takes for the rock to strike the ground.

**c)** If Travis throws a rock upward with an initial velocity of 100 feet per second at the same time that Courtney throws a rock upward at 60 feet per second, whose rock will strike the ground first? Explain.

**d)** Will the rocks ever be at the same distance above the ground? If so, at what time?

## Cumulative Review Exercises

[1.6] **110.** Evaluate $\dfrac{5.55 \times 10^3}{1.11 \times 10^1}$.

[3.2] **111.** If $f(x) = x^2 + 2x - 8$, find $f(3)$.

[4.1] **112.** Solve the system of equations.

$$3x + 4y = 2$$
$$2x = -5y - 1$$

[6.3] **113.** Simplify $2x^{-1} - (3y)^{-1}$.

[7.6] **114.** Solve $\sqrt{x^2 - 6x - 4} = x$.

# 8.3 Quadratic Equations: Applications and Problem Solving

1 Solve additional applications.

2 Solve for a variable in a formula.

## 1 Solve Additional Applications

We have already discussed a few applications of quadratic equations. In this section, we will explore several more applications. We will also discuss solving for a variable in a formula. We start by investigating a profit for a new company.

**EXAMPLE 1** ▸ **Company Profits** Laserox, a start-up company, projects that its annual profits, $p(n)$, in thousands of dollars, over the first 6 years of operation can be approximated by the function $p(n) = 1.2n^2 + 4n - 8$, where $n$ is the number of years completed.

**a)** Estimate the profit (or loss) of the company after the first year.

**b)** Estimate the profit (or loss) of the company after 6 years.

**c)** Estimate the time needed for the company to break even.

*Solution* **a)** To estimate the profit after 1 year, we evaluate the function at 1.

$$p(n) = 1.2n^2 + 4n - 8$$
$$p(1) = 1.2(1)^2 + 4(1) - 8 = -2.8$$

Thus, at the end of the first year the company projects a loss of \$2.8 thousand or a loss of \$2800.

**b)** $p(6) = 1.2(6)^2 + 4(6) - 8 = 59.2$
Thus, at the end of the sixth year the company's projected profit is \$59.2 thousand, or a profit of \$59,200.

**c)** Understand The company will break even when the profit is 0. Thus, to find the break-even point (no profit or loss) we solve the equation

$$1.2n^2 + 4n - 8 = 0$$

We can use the quadratic formula to solve this equation.

Translate $a = 1.2, \quad b = 4, \quad c = -8$

$$n = \frac{-b \pm \sqrt{b^2 - 4ac}}{2a}$$
$$= \frac{-4 \pm \sqrt{4^2 - 4(1.2)(-8)}}{2(1.2)}$$

**Carry Out**

$$= \frac{-4 \pm \sqrt{16 + 38.4}}{2.4}$$

$$= \frac{-4 \pm \sqrt{54.4}}{2.4}$$

$$\approx \frac{-4 \pm 7.376}{2.4}$$

$$n \approx \frac{-4 + 7.376}{2.4} \approx 1.4 \quad \text{or} \quad n \approx \frac{-4 - 7.376}{2.4} \approx -4.74$$

**Answer** Since time cannot be negative, the break-even time is about 1.4 years.

▸ **Now Try Exercise 29**

Now, let's consider another example that uses the quadratic formula to solve a quadratic equation.

**EXAMPLE 2** ▸ **Life Expectancy** The function $N(t) = 0.0054t^2 - 1.46t + 95.11$ can be used to estimate the average number of years of life expectancy remaining for a person of age $t$ years where $30 \le t \le 100$.

**a)** Estimate the remaining life expectancy of a person of age 40.

**b)** If a person has a remaining life expectancy of 14.3 years, estimate the age of the person.

*Solution* **a) Understand** We would expect that the older a person gets the shorter the remaining life expectancy. To determine the remaining life expectancy for a 40-year-old, we substitute 40 for $t$ in the function and evaluate.

**Translate**

$$N(t) = 0.0054t^2 - 1.46t + 95.11$$

$$N(40) = 0.0054(40)^2 - 1.46(40) + 95.11$$

**Carry Out**

$$= 0.0054(1600) - 58.4 + 95.11$$

$$= 8.64 - 58.4 + 95.11$$

$$= 45.35$$

**Answer and Check** The answer appears reasonable. Thus, on the average, a 40-year-old can expect to live another 45.35 years to an age of 85.35 years.

**b) Understand** Here we are given the remaining life expectancy, $N(t)$, and asked to find the age of the person, $t$. To solve this problem, we substitute 14.3 for $N(t)$ and solve for $t$. To solve for $t$, we will use the quadratic formula.

**Translate**

$$N(t) = 0.0054t^2 - 1.46t + 95.11$$

$$14.3 = 0.0054t^2 - 1.46t + 95.11$$

**Carry Out**

$$0 = 0.0054t^2 - 1.46t + 80.81$$

$$a = 0.0054, \quad b = -1.46, \quad c = 80.81$$

$$t = \frac{-b \pm \sqrt{b^2 - 4ac}}{2a}$$

$$= \frac{-(-1.46) \pm \sqrt{(-1.46)^2 - 4(0.0054)(80.81)}}{2(0.0054)}$$

$$= \frac{1.46 \pm \sqrt{2.1316 - 1.745496}}{0.0108}$$

$$= \frac{1.46 \pm \sqrt{0.386104}}{0.0108}$$

$$\approx \frac{1.46 \pm 0.6214}{0.0108}$$

$$t \approx \frac{1.46 + 0.6214}{0.0108} \quad \text{or} \quad t \approx \frac{1.46 - 0.6214}{0.0108}$$

$$\approx 192.72 \qquad\qquad \approx 77.65$$

**Answer** Since 192.72 is not a reasonable age, we can exclude that as a possibility. Thus, the average person who has a life expectancy of 14.3 years is about 77.65 years old.

▶ Now Try Exercise 31

### Motion Problems

We first discussed motion problems in Section 2.4. The motion problem we give here is solved using the quadratic formula.

**EXAMPLE 3 ▶ Motorboat Ride** Charles Curtis decides to go for a relaxing ride in his motorboat on the Potomac River. His trip starts in Bethesda, Maryland. He travels downstream for 12 miles with the current. He then turns around and heads back to the starting point going upstream against the current. The total time of his trip is 5 hours and the river current is 2 miles per hour. If during the entire trip he did not touch the throttle to change the speed, find the speed the boat would have traveled in still water.

*Solution* **Understand** We are asked to find the rate of the boat in still water. Let $r$ = the rate of the boat in still water. We know that the total time of the trip is 5 hours. Thus, the time downriver plus the time upriver must sum to 5 hours. Since distance = rate · time, we can find the time by dividing the distance by the rate.

| Direction | Distance | Rate | Time |
|---|---|---|---|
| Downriver (with current) | 12 | $r + 2$ | $\frac{12}{r+2}$ |
| Upriver (against current) | 12 | $r - 2$ | $\frac{12}{r-2}$ |

**Translate**

$$\text{time downriver} + \text{time upriver} = \text{total time}$$

$$\frac{12}{r+2} + \frac{12}{r-2} = 5$$

**Carry Out**

$$(r+2)(r-2)\left(\frac{12}{r+2} + \frac{12}{r-2}\right) = (r+2)(r-2)(5) \quad \text{Multiply by the LCD.}$$

$$(r+2)(r-2)\left(\frac{12}{r+2}\right) + (r+2)(r-2)\left(\frac{12}{r-2}\right) = (r+2)(r-2)(5) \quad \text{Distributive property}$$

$$12(r-2) + 12(r+2) = 5(r^2 - 4)$$

$$12r - 24 + 12r + 24 = 5r^2 - 20 \quad \text{Distributive property}$$

$$24r = 5r^2 - 20 \quad \text{Simplify.}$$

$$\text{or} \quad 5r^2 - 24r - 20 = 0$$

Using the quadratic formula with $a = 5$, $b = -24$, and $c = -20$, we obtain

$$r = \frac{24 \pm \sqrt{976}}{10}$$

$$r \approx 5.5 \quad \text{or} \quad r \approx -0.7$$

**Answer** Since the rate cannot be negative, the rate or speed of the boat in still water is about 5.5 miles per hour.

▶ Now Try Exercise 43

Notice that in real-life situations most answers are not integral values.

### Work Problems

Let's do an example involving a work problem. Work problems were discussed in Section 6.5. You may wish to review that section before studying the next example.

**EXAMPLE 4 ▸ Pumping Water** After a hurricane, the Durals needed to pump water from their flooded basement. They had one sump pump (used to pump out water) and borrowed a second from their local fire department. With both pumps working together, their basement would empty in about 6 hours. The fire department's pump had a higher horsepower, and it would empty their basement by itself in 2 hours less time than the Dural's pump would if it were working alone. How long would it take each pump to empty the basement if each were working alone?

*Solution* **Understand** Recall from Section 6.5 that the rate of work multiplied by the time worked gives the part of the task completed.

Let $t$ = number of hours for the Durals' (slower) pump to complete the job by itself,

then $t - 2$ = number of hours for the fire department's pump to complete the job by itself.

| Pump | Rate of Work | Time Worked | Part of Task Completed |
|---|---|---|---|
| Durals' pump | $\frac{1}{t}$ | 6 | $\frac{6}{t}$ |
| Fire department's pump | $\frac{1}{t-2}$ | 6 | $\frac{6}{t-2}$ |

**Translate**

$$\left(\begin{array}{c}\text{part of task}\\ \text{by Durals' pump}\end{array}\right) + \left(\begin{array}{c}\text{part of task}\\ \text{by fire department's pump}\end{array}\right) = 1$$

$$\frac{6}{t} + \frac{6}{t-2} = 1$$

**Carry Out**

$$t(t-2)\left(\frac{6}{t} + \frac{6}{t-2}\right) = t(t-2)(1) \quad \text{Multiply both sides by the LCD, } t(t-2).$$

$$\cancel{t}(t-2)\left(\frac{6}{\cancel{t}}\right) + t(\cancel{t-2})\left(\frac{6}{\cancel{t-2}}\right) = t^2 - 2t \quad \text{Distributive property}$$

$$6(t-2) + 6t = t^2 - 2t$$

$$6t - 12 + 6t = t^2 - 2t$$

$$t^2 - 14t + 12 = 0$$

Using the quadratic formula, we obtain

$$t = \frac{14 \pm \sqrt{148}}{2}$$

$$t \approx 13.1 \quad \text{or} \quad t \approx 0.9$$

**Answer** Both 13.1 and 0.9 satisfy the equation $\frac{6}{t} + \frac{6}{t-2} = 1$ (with some round-off involved). However, if we accept 0.9 as a solution, then the fire department's pump could complete the task in a negative time ($t - 2 = 0.9 - 2 = -1.1$ hours), which is not possible. Therefore, 0.9 hour is not an acceptable solution. The only solution is 13.1 hours. The Durals' pump takes approximately 13.1 hours by itself, and the fire department's pump takes approximately 13.1 − 2 or 11.1 hours by itself to empty the basement.

▸ **Now Try Exercise 45**

## 2 Solve for a Variable in a Formula

When the square of a variable appears in a *formula*, you may need to use the square root property to solve for the variable. However, *when you use the square root property in most formulas, you will use only the principal or positive root*, because you are generally solving for a quantity that cannot be negative.

**EXAMPLE 5**

**a)** The formula for the area of a circle is $A = \pi r^2$. Solve this equation for the radius, $r$.

**b)** *Newton's law of universal gravity* states that every particle in the universe attracts every other particle with a force proportional to the product of their masses and inversely proportional to the square of the distance between them. We may represent Newton's law as

$$F = G\frac{m_1 m_2}{r^2}$$

Solve the equation for $r$.

**Solution**

**a)**

$$A = \pi r^2$$

$$\frac{A}{\pi} = r^2 \quad \text{Isolate } r^2 \text{ by dividing both sides by } \pi.$$

$$\sqrt{\frac{A}{\pi}} = r \quad \text{Square root property}$$

**b)**

$$F = G\frac{m_1 m_2}{r^2}$$

$$r^2 F = G m_1 m_2 \quad \text{Multiply both sides of formula by } r^2.$$

$$r^2 = \frac{G m_1 m_2}{F} \quad \text{Isolate } r^2 \text{ by dividing both sides by } F.$$

$$r = \sqrt{\frac{G m_1 m_2}{F}} \quad \text{Square root property}$$

**Now Try Exercise 23**

In Example 5, since $r$ must be greater than 0, when we used the square root property, we listed only the principal or positive square root.

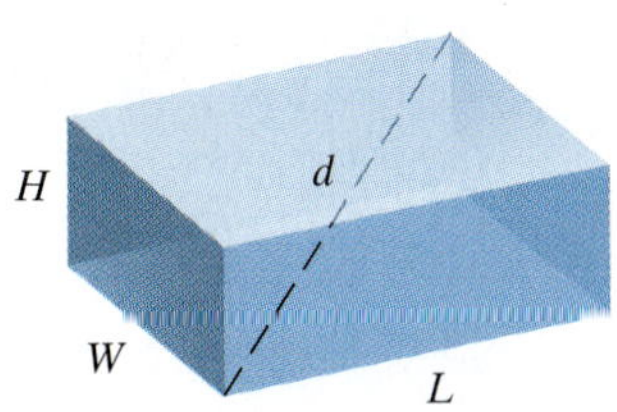


FIGURE 8.5

**EXAMPLE 6** **Diagonal of a Suitcase** The diagonal of a box can be calculated by the formula

$$d = \sqrt{L^2 + W^2 + H^2}$$

where $L$ is the length, $W$ is the width, and $H$ is the height of the box. See **Figure 8.5**.

**a)** Find the diagonal of a suitcase of length 30 inches, width 15 inches, and height 10 inches.

**b)** Solve the equation for the width, $W$.

**Solution** **a) Understand** To find the diagonal, we need to substitute the appropriate values into the formula and solve for the diagonal, $d$.

**Translate**

$$d = \sqrt{L^2 + W^2 + H^2}$$

$$d = \sqrt{(30)^2 + (15)^2 + (10)^2}$$

**Carry Out**

$$= \sqrt{900 + 225 + 100}$$

$$= \sqrt{1225}$$

$$= 35$$

**Answer** Thus, the diagonal of the suitcase is 35 inches.

**b)** Our first step in solving for $W$ is to square both sides of the formula.

$$d = \sqrt{L^2 + W^2 + H^2}$$

$$d^2 = \left(\sqrt{L^2 + W^2 + H^2}\right)^2 \quad \text{Square both sides.}$$

$$d^2 = L^2 + W^2 + H^2 \quad \text{Use } (\sqrt{a})^2 = a,\ a \ge 0.$$

$$d^2 - L^2 - H^2 = W^2 \quad \text{Isolate } W^2.$$

$$\sqrt{d^2 - L^2 - H^2} = W \quad \text{Square root property}$$

**Now Try Exercise 15**

**EXAMPLE 7** ▸ **Traffic Cones** The surface area of a right circular cone is

$$s = \pi r\sqrt{r^2 + h^2}$$

**a)** An orange traffic cone used on roads is 18 inches high with a radius of 12 inches. Find the surface area of the cone.

**b)** Solve the formula for $h$.

*Solution* **a)** Understand and Translate To find the surface area, we substitute the appropriate values into the formula.

$$s = \pi r\sqrt{r^2 + h^2}$$
$$= \pi(12)\sqrt{(12)^2 + (18)^2}$$

Carry Out

$$= 12\pi\sqrt{144 + 324}$$
$$= 12\pi\sqrt{468}$$
$$\approx 815.56$$

Answer The surface area is about 815.56 square inches.

**b)** To solve for $h$ we need to isolate $h$ on one side of the equation. There are various ways to solve the equation for $h$.

$$s = \pi r\sqrt{r^2 + h^2}$$

$$\frac{s}{\pi r} = \sqrt{r^2 + h^2} \quad \text{Divide both sides by } \pi r.$$

$$\left(\frac{s}{\pi r}\right)^2 = \left(\sqrt{r^2 + h^2}\right)^2 \quad \text{Square both sides.}$$

$$\frac{s^2}{\pi^2 r^2} = r^2 + h^2 \quad \text{Use } (\sqrt{a})^2 = a,\ a \geq 0.$$

$$\frac{s^2}{\pi^2 r^2} - r^2 = h^2 \quad \text{Subtract } r^2 \text{ from both sides.}$$

$$\sqrt{\frac{s^2}{\pi^2 r^2} - r^2} = h \quad \text{Square root property}$$

Other acceptable answers are $h = \sqrt{\dfrac{s^2 - \pi^2 r^4}{\pi^2 r^2}}$ and $h = \dfrac{\sqrt{s^2 - \pi^2 r^4}}{\pi r}$. Can you explain why?

▸ Now Try Exercise 27

## EXERCISE SET 8.3

MathXL® MyMathLab

### Concept/Writing Exercises

1. In general, when solving for a variable in a formula, whether you use the square root property or the quadratic formula, you use only the positive square root. Explain why.

2. Suppose $P = ☺^2 + \square^2$ is a real formula. Solving for ☺ gives $☺ = \sqrt{P - \square^2}$. If ☺ is to be a real number, what relationship must exist between $P$ and $\square$?

## Practice the Skills

*Solve for the indicated variable. Assume the variable you are solving for must be greater than 0.*

3. $A = s^2$, for $s$ (area of a square)

4. $A = (s + 1)^2$, for $s$ (area of a square)

5. $d = 4.9t^2$, for $t$ (distance an object has fallen)

6. $A = S^2 - s^2$, for $S$ (area between two squares)

7. $E = i^2r$, for $i$ (current in electronics)

8. $A = 4\pi r^2$, for $r$ (surface area of a sphere)

9. $d = 16t^2$, for $t$ (distance of a falling object)

10. $d = \frac{1}{9}x^2$, for $x$ (stopping distance on pavement)

11. $E = mc^2$, for $c$ (Einstein's famous energy formula)

12. $V = \pi r^2 h$, for $r$ (volume of a right circular cylinder)

13. $V = \frac{1}{3}\pi r^2 h$, for $r$ (volume of a right circular cone)

14. $d = \sqrt{L^2 + W^2}$, for $L$ (diagonal of a rectangle)

15. $d = \sqrt{L^2 + W^2}$, for $W$ (diagonal of a rectangle)

16. $a^2 + b^2 = c^2$, for $a$ (Pythagorean Theorem)

17. $a^2 + b^2 = c^2$, for $b$ (Pythagorean Theorem)

18. $d = \sqrt{L^2 + W^2 + H^2}$, for $L$ (diagonal of a box)

19. $d = \sqrt{L^2 + W^2 + H^2}$, for $H$ (diagonal of a box)

20. $A = P(1 + r)^2$, for $r$ (compound interest formula)

21. $h = -16t^2 + s_0$, for $t$ (height of an object)

22. $h = -4.9t^2 + s_0$, for $t$ (height of an object)

23. $E = \frac{1}{2}mv^2$, for $v$ (kinetic energy)

24. $f_x^2 + f_y^2 = f^2$, for $f_x$ (forces acting on an object)

25. $a = \dfrac{v_2^2 - v_1^2}{2d}$, for $v_1$ (acceleration of a vehicle)

26. $A = 4\pi(R^2 - r^2)$, for $R$ (surface area of two spheres)

27. $v' = \sqrt{c^2 - v^2}$, for $c$ (relativity; $v'$ is read "$v$ prime")

28. $L = L_0\sqrt{1 - \dfrac{v^2}{c^2}}$, for $v$ (art, a painting's contraction)

## Problem Solving

29. **Profit** The profit for the Hillside Tractor Company, which sells tractors, is $P(n) = 2.7n^2 + 9n - 3$, where $P(n)$ is in hundreds of dollars.
    a) Find the profit when 5 tractors are sold.
    b) How many tractors should be sold to have a profit of \$20,000?

30. **Profit** The profit for the Jacksons Appliances, which sells refrigerators, is $P(n) = 6.2n^2 + 6n - 3$ where $P(n)$ is in dollars.
    a) Find the profit when 7 refrigerators are sold.
    b) How many refrigerators should be sold to have a profit of \$675?

31. **Temperature** The temperature, $T$, in degrees Fahrenheit, in a car's radiator during the first 4 minutes of driving is a function of time, $t$. The temperature can be found by the formula $T = 6.2t^2 + 12t + 32, 0 \le t \le 4$.
    a) What is the car radiator's temperature at the instant the car is turned on?
    b) What is the car radiator's temperature after the car has been driven for 2 minutes?
    c) How long after the car has begun operating will the car radiator's temperature reach 120°F?

32. **School Enrollment** The function $N(t) = -0.043t^2 + 1.22t + 46.0$ can be used to estimate total U.S. elementary and secondary school enrollment, in millions, between the years 1990 and 2008. In the equation $t$ is years since 1989 and $1 \le t \le 19$.

    a) Estimate total enrollment in 1995.
    b) In what years is the total enrollment 54 million students?

33. **Downloaded Songs** The number of downloaded songs, in billions, from 2002 to 2006 and projected to 2008, can be estimated by the function $D = 0.04t^2 - 0.03t + 0.01$. In this function, $t$ is the number of years since 2002 and $0 \le t \le 6$. *Source:* Price Waterhouse Coopers, LLP, RIAA, *Newsweek* (7/11/05)
    a) Estimate the number of downloaded songs in 2006.
    b) In which year is 1 billion songs projected to be downloaded?

**34. Grade Point Average** At a college, records show that the average person's grade point average, $G$, is a function of the number of hours he or she studies and does homework per week, $h$. The grade point average can be estimated by the equation $G = 0.01h^2 + 0.2h + 1.2, 0 \leq h \leq 8$.

**a)** What is the GPA of the average student who studies for 0 hours a week?

**b)** What is the GPA of the average student who studies 3 hours per week?

**c)** To obtain a 3.2 GPA, how many hours per week would the average student need to study?

**35. Apple Production** The following graph shows the annual average yield per acre, of apple trees, for the years 2000 to 2004.

**Yield per Acre of Apple Trees**

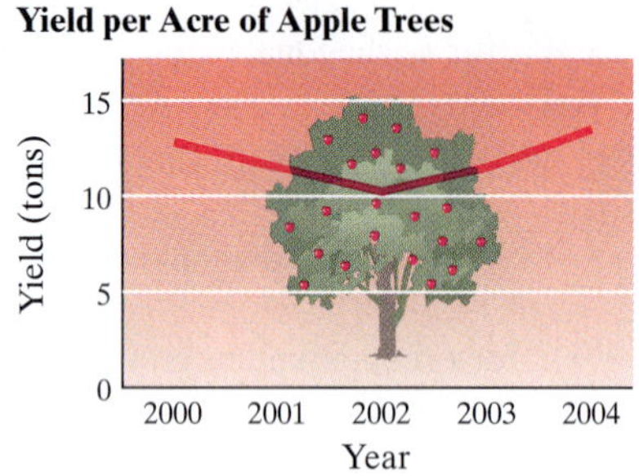


*Source:* National Agricultural Statistics, *USA Today* (9/15/05)

The average annual yield per acre of apple trees, in tons, can be estimated by the function $Y = 0.66t^2 - 2.49t + 12.93$. In this function, $t$ is the number of years since 2000 and $0 \leq t \leq 4$.

**a)** Estimate the yield per acre in 2003.

**b)** In which year was the yield per acre 13 billion tons?

**36. Drug-Free Schools** The following graph summarizes data on the percent of students at various ages who say their school is not drug free.

**Students Who Say Their School Is Not Drug Free**

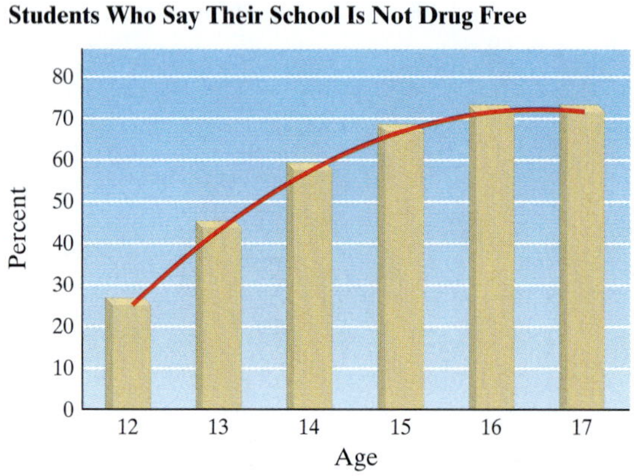


*Source:* National Center on Addiction and Substance Abuse

The function $f(a) = -2.32a^2 + 76.58a - 559.87$ can be used to estimate the percent of students who say their school is not drug free. In the function, $a$ represents the student's age, where $12 \leq a \leq 17$. Use the function to answer the following questions.

**a)** Estimate the percent of 14-year-olds who say their school is not drug free.

**b)** At what age do 70% of the students say their school is not drug free?

**37. Motorcycle Sales** The following graph shows the number of new motorcycles, in millions, sold in the United States for the years 1997 to 2004.

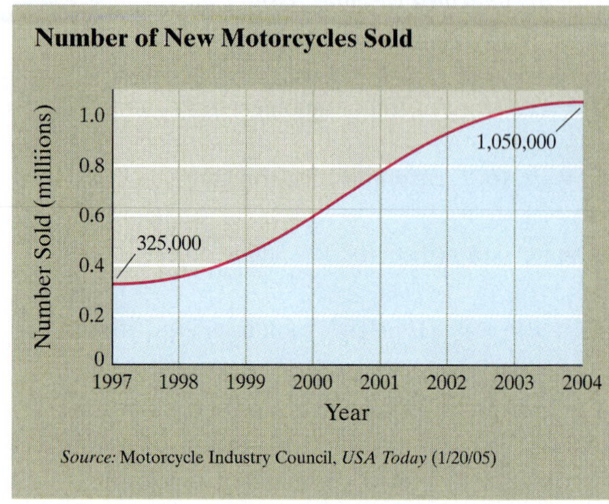


The number of new motorcycles, $m(t)$, in millions, sold in the United States can be estimated by the function $M = -0.00434t^2 + 0.142t + 0.315$. In this function, $t$ is the number of years since 1997.

**a)** If this trend continues, use this function to estimate the number of motorcycles that will be sold in the United States in 2007.

**b)** In what year will the number of new motorcycles sold in the United States be 1.4 million?

**38. Profit** A video store's weekly profit, $P$, in thousands of dollars, is a function of the rental price of the tapes, $t$. The profit equation is $P = 0.2t^2 + 1.5t - 1.2, 0 \leq t \leq 5$.

**a)** What is the store's weekly profit or loss if they charge \$3 per tape?

**b)** What is the weekly profit if they charge \$5 per tape?

**c)** At what tape rental price will their weekly profit be 1.4 thousand?

**39. Playground** The area of a children's rectangular playground is 600 square meters. The length is 10 meters longer than the width. Find the length and width of the playground.

**40. Travel** Hana Juarez drove for 80 miles in heavy traffic. She then reached the highway where she drove 260 miles at an average speed that was 25 miles per hour greater than the average speed she drove in heavy traffic. If the total trip took 6 hours, determine her average speed in heavy traffic and on the highway.

**41. Drilling a Well** Paul and Rima Jones, who live in Cedar Rapids, Iowa, want a well on their property. They hired the Ruth Cardiff Drilling Company to drill the well. The company had to drill 64 feet to hit water. The company informed the Jones that they had just ordered new drilling equipment that drills at an average of 1 foot per hour faster, and that with their new equipment, they would have hit water in 3.2 hours less time. Find the rate at which their present equipment drills.

**42. Car Carrier** Frank Sims, a truck driver, was transporting a heavy load of new cars on a car carrier from Detroit, Michigan, to Indianapolis, Indiana. On his return trip to Detroit, since his truck was lighter, he averaged 10 miles per hour faster than on his trip to Indianapolis. If the total distance traveled each way was 300 miles and the total time he spent driving was 11 hours, find his average speed going and returning.

**43. Runner** Latoya Williams, a long-distance runner, starts jogging at her house. She jogs 6 miles and then turns around and jogs back to her house. The first part of her jog is mostly uphill, so her speed averages 2 miles per hour less than her returning speed. If the total time she spends jogging is $1\frac{3}{4}$ hours, find her speed going and her speed returning.

**44. Red Rock Canyon** Kathy Nickell traveled from the Red Rock Canyon Conservation Area, just outside Las Vegas, to Phoenix, Arizona. The total distance she traveled was 300 miles. After she got to Phoenix, she figured out that had she averaged 10 miles per hour faster, she would have arrived 1 hour earlier. Find the average speed that Kathy drove.

Red Rock Canyon

**45. Build an Engine** Two mechanics, Bonita Rich and Pamela Pearson, take 6 hours to rebuild an engine when they work together. If each worked alone, Bonita, the more experienced mechanic, could complete the job 1 hour faster than Pamela. How long would it take each of them to rebuild the engine working alone?

**46. Riding a Bike** Ricky Bullock enjoys riding his bike from Washington, D.C., to Bethesda, Maryland, and back, a total of 30 miles on the Capital Crescent path. The trip to Bethesda is uphill most of the distance. The bike's average speed going to Bethesda is 5 miles per hour slower than the average speed returning to D.C. If the round trip takes 4.5 hours, find the average speed in each direction.

**47. Flying a Plane** Dole Rohm flew his single-engine Cessna airplane 80 miles with the wind from Jackson Hole, Wyoming, to above Blackfoot, Idaho. He then turned around and flew back to Jackson Hole against the wind. If the wind was a constant 30 miles per hour, and the total time going and returning was 1.3 hours, find the speed of the plane in still air.

**48. Ships** After a small oil spill, two cleanup ships are sent to siphon off the oil floating in Baffin Bay. The newer ship can clean up the entire spill by itself in 3 hours less time than the older ship takes by itself. Working together the two ships can clean up the oil spill in 8 hours. How long will it take the newer ship by itself to clean up the spill?

**49. Janitorial Service** The O'Connors own a small janitorial service. John requires $\frac{1}{2}$ hour more time to clean the Moose Club by himself than Chris does working by herself. If together they can clean the club in 6 hours, find the time required by each to clean the club.

**50. Electric Heater** A small electric heater requires 6 minutes longer to raise the temperature in an unheated garage to a comfortable level than does a larger electric heater. Together the two heaters can raise the garage temperature to a comfortable level in 42 minutes. How long would it take each heater by itself to raise the temperature in the garage to a comfortable level?

**51. Travel** Shywanda Moore drove from San Antonio, Texas, to Austin, Texas, a distance of 75 miles. She then stopped for 2 hours to see a friend in Austin before continuing her journey from Austin to Dallas, Texas, a distance of 195 miles. If she drove 10 miles per hour faster from San Antonio to Austin and the total time of the trip was 6 hours, find her average speed from San Antonio to Austin.

River Walk, San Antonio, Texas

**52. Travel** Lewis and his friend George are traveling from Nashville, Tennessee, to Baltimore, Maryland. Lewis travels by car while George travels by train. The train and car leave Nashville at the same time from the same point. During the trip Lewis and George speak by cellular phone, and Lewis informs George that he has just stopped for the evening after traveling for 500 miles. One and two-thirds hours later, George calls Lewis and informs him that the train had just reached Baltimore, a distance of 800 miles from Nashville. Assuming the train averaged 20 miles per hour faster than the car, find the average speed of the car and the train.

**53. Widescreen TVs** A widescreen television (see figure) has an aspect ratio of 16 : 9. This means that the ratio of the length to the height of the screen is 16 to 9. The figure drawn on the photo illustrates how the length and the height of the screen of a 40-inch widescreen television can be found. Determine the length and height of a 40-inch widescreen television.

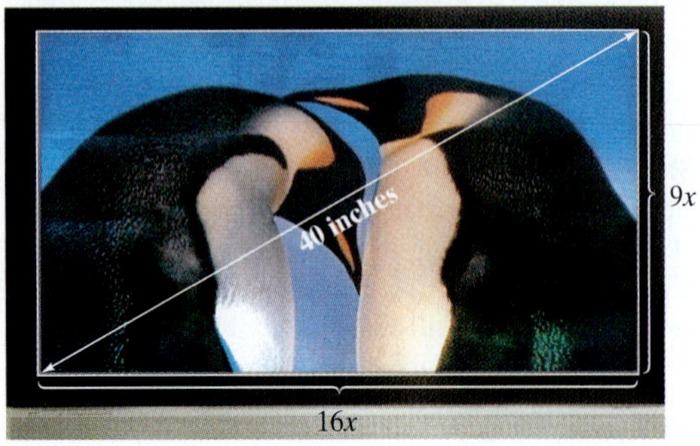


**54. Standard TVs** Many picture tube televisions have a screen aspect ratio of 4:3. Determine the length and height of a television screen that has an aspect ratio of 4:3 and whose diagonal is 36 inches. See Exercise 53.

**55.** Write your own motion problem and solve it.

**56.** Write your own work problem and solve it.

## Challenge Problems

**57. Area** The area of a rectangle is 18 square meters. When the length is increased by 2 meters and the width by 3 meters, the area becomes 48 square meters. Find the dimensions of the smaller rectangle.

**58. Area** The area of a rectangle is 35 square inches. When the length is decreased by 1 inch and the width is increased by 1 inch, the area of the new rectangle is 36 square inches. Find the dimensions of the original rectangle.

## Cumulative Review Exercises

[1.4] **59.** Evaluate $-[4(5-3)^3] + 2^4$.

[2.2] **60.** Solve $IR + Ir = E$ for $R$.

[6.2] **61.** Add $\frac{r}{r-4} - \frac{r}{r+4} + \frac{32}{r^2-16}$.

[7.2] **62.** Simplify $\left(\frac{x^{3/4}y^{-2}}{x^{1/2}y^2}\right)^8$.

[7.6] **63.** Solve $\sqrt{x^2 + 3x + 12} = x$.

## Mid-Chapter Test: 8.1–8.3

*To find out how well you understand the chapter material to this point, take this brief test. The answers, and the section where the material was initially discussed, are given in the back of the book. Review any questions that you answered incorrectly.*

*Use the square root property to solve each equation.*

**1.** $x^2 - 12 = 86$

**2.** $(a-3)^2 + 20 = 0$

**3.** $(2m+7)^2 = 36$

*Solve each equation by completing the square.*

**4.** $y^2 + 4y - 12 = 0$

**5.** $3a^2 - 12a - 30 = 0$

**6.** $4c^2 + c = -9$

**7. Patio** The patio of a house is a square where the diagonal is 6 meters longer than a side. Find the length of one side of the patio.

**8. a)** Give the formula for the discriminant of a quadratic equation.

**b)** Explain how to determine if a quadratic equation has two distinct real solutions, a single real solution, or no real solution.

**9.** Use the discriminant to determine if the equation $2b^2 - 6b - 11 = 0$ has two distinct real solutions, a single real solution, or no real solution.

*Solve each equation by the quadratic formula.*

**10.** $6n^2 + n = 15$

**11.** $p^2 = -4p + 8$

**12.** $3d^2 - 2d + 5 = 0$

*In Exercises 13 and 14, determine an equation that has the given solutions.*

**13.** $7, -2$

**14.** $2 + \sqrt{5}, 2 - \sqrt{5}$

**15. Lamps** A business sells $n$ lamps, $n \le 20$, at a price of $(60 - 0.5n)$ dollars per lamp. How many lamps must be sold to have revenue of $550?

*In Exercises 16–18, solve for the indicated variable. Assume all variables are positive.*

**16.** $y = x^2 - r^2$, for $r$

**17.** $A = \frac{1}{3}kx^2$, for $x$

**18.** $D = \sqrt{x^2 + y^2}$, for $y$

**19. Area** The length of a rectangle is two feet more than twice the width. Find the dimensions if its area is 60 square feet.

**20. Clocks** The profit from a company selling $n$ clocks is $p(n) = 2n^2 + n - 35$, where $p(n)$ is hundreds of dollars. How many clocks must be sold to have a profit of \$2000?

# 8.4 Writing Equations in Quadratic Form

1 Solve equations that are quadratic in form.

2 Solve equations with rational exponents.

## 1 Solve Equations That Are Quadratic in Form

Sometimes we need to solve an equation that is not a quadratic equation but can be rewritten in the form of a quadratic equation. We can then solve the equation in quadratic form by factoring, completing the square, or the quadratic formula.

**Equations Quadratic in Form**

An equation that can be written in the form $au^2 + bu + c = 0$ for $a \neq 0$, where $u$ is an algebraic expression, is called **quadratic in form**.

When you are given an equation quadratic in form, make a substitution to get the equation in the form $au^2 + bu + c = 0$. In general, if the exponents are positive, the substitution to make is to let $u$ be the middle term, without the numerical coefficient, when the expression is listed in descending order of the variable. For example,

| Equation Quadratic in Form | Substitution | Equation with Substitution |
|---|---|---|
| $y^4 - y^2 - 6 = 0$ | $u = y^2$ | $u^2 - u - 6 = 0$ |
| $2(x+5)^2 - 5(x+5) - 12 = 0$ | $u = x + 5$ | $2u^2 - 5u - 12 = 0$ |
| $x^{2/3} + 4x^{1/3} - 3 = 0$ | $u = x^{1/3}$ | $u^2 + 4u - 3 = 0$ |

To solve equations quadratic in form, we use the following procedure. We will illustrate this procedure in Example 1.

**To Solve Equations Quadratic in Form**

1. Make a substitution that will result in an equation of the form $au^2 + bu + c = 0$, $a \neq 0$, where $u$ is a function of the original variable.
2. Solve the equation $au^2 + bu + c = 0$ for $u$.
3. Replace $u$ with the function of the original variable from step 1 and solve the resulting equation for the original variable.
4. Check for extraneous solutions by substituting the apparent solutions into the original equation.

**EXAMPLE 1**

**a)** Solve $x^4 - 5x^2 + 4 = 0$.

**b)** Find the $x$-intercepts of the graph of the function $f(x) = x^4 - 5x^2 + 4$.

*Solution*

**a)** To obtain an equation that is quadratic in form, write $x^4$ as $(x^2)^2$.

$$x^4 - 5x^2 + 4 = 0$$

$$(x^2)^2 - 5x^2 + 4 = 0 \quad \text{Replace } x^4 \text{ with } (x^2)^2 \text{ to obtain an equation in desired form.}$$

Now let $u = x^2$. This gives an equation that is quadratic in form.

$$u^2 - 5u + 4 = 0 \qquad \text{Substitute } u \text{ for } x^2.$$
$$(u - 4)(u - 1) = 0 \qquad \text{Solve for } u.$$
$$u - 4 = 0 \quad \text{or} \quad u - 1 = 0$$
$$u = 4 \qquad u = 1$$
$$x^2 = 4 \qquad x^2 = 1 \qquad \text{Replace } u \text{ with } x^2.$$
$$x = \pm\sqrt{4} \qquad x = \pm\sqrt{1} \qquad \text{Solve for } x.$$
$$x = \pm 2 \qquad x = \pm 1$$

Check the four possible solutions in the original equation.

| $x = 2$ | $x = -2$ | $x = 1$ | $x = -1$ |
|---|---|---|---|
| $x^4 - 5x^2 + 4 = 0$ | $x^4 - 5x^2 + 4 = 0$ | $x^4 - 5x^2 + 4 = 0$ | $x^4 - 5x^2 + 4 = 0$ |
| $2^4 - 5(2)^2 + 4 \stackrel{?}{=} 0$ | $(-2)^4 - 5(-2)^2 + 4 \stackrel{?}{=} 0$ | $1^4 - 5(1)^2 + 4 \stackrel{?}{=} 0$ | $(-1)^4 - 5(-1)^2 + 4 \stackrel{?}{=} 0$ |
| $16 - 20 + 4 \stackrel{?}{=} 0$ | $16 - 20 + 4 \stackrel{?}{=} 0$ | $1 - 5 + 4 \stackrel{?}{=} 0$ | $1 - 5 + 4 \stackrel{?}{=} 0$ |
| $0 = 0$ | $0 = 0$ | $0 = 0$ | $0 = 0$ |
| True | True | True | True |

Thus, the solutions are 2, −2, 1, and −1.

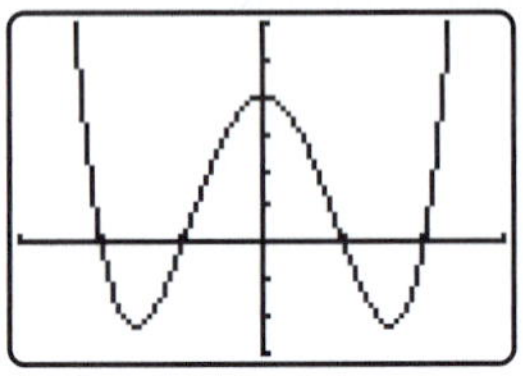

−3, 3, 1, −3, 6, 1

FIGURE 8.6

**b)** The $x$-intercepts occur where $f(x) = 0$. Therefore, the graph will cross the $x$-axis at the solutions to the equation $x^4 - 5x^2 + 4 = 0$.

From part **a)**, we know the solutions are 2, −2, 1, and −1. Thus, the $x$-intercepts are $(2, 0), (-2, 0), (1, 0)$, and $(-1, 0)$. **Figure 8.6** is the graph of $f(x) = x^4 - 5x^2 + 4 = 0$ as illustrated on a graphing calculator. Notice that the graph crosses the $x$-axis at $x = 2$, $x = -2$, $x = 1$, and $x = -1$.

▶ **Now Try Exercise 7**

**EXAMPLE 2** ▶ Solve $p^4 + 2p^2 = 8$.

### Solution

$$p^4 + 2p^2 - 8 = 0 \qquad \text{Set equation equal to 0.}$$
$$(p^2)^2 + 2p^2 - 8 = 0 \qquad \text{Write } p^4 \text{ as } (p^2)^2 \text{ to obtain equation in desired form.}$$

Now let $u = p^2$. This gives an equation that is quadratic in form.

$$u^2 + 2u - 8 = 0 \qquad \text{Substitute } u \text{ for } p^2.$$
$$(u + 4)(u - 2) = 0 \qquad \text{Solve the equation for } u.$$
$$u + 4 = 0 \quad \text{or} \quad u - 2 = 0$$
$$u = -4 \qquad u = 2$$

We are not finished. Since the variable in the original equation is $p$, we must solve for $p$, not $u$. Therefore, we substitute back $p^2$ for $u$ and solve for $p$.

$$p^2 = -4 \qquad p^2 = 2 \qquad \text{Replace } u \text{ with } p^2.$$
$$p = \pm\sqrt{-4} \qquad p = \pm\sqrt{2} \qquad \text{Solve for } p.$$
$$p = \pm 2i$$

Check the four possible solutions in the *original* equation.

| $p = 2i$ | $p = -2i$ | $p = \sqrt{2}$ | $p = -\sqrt{2}$ |
|---|---|---|---|
| $p^4 + 2p^2 = 8$ | $p^4 + 2p^2 = 8$ | $p^4 + 2p^2 = 8$ | $p^4 + 2p^2 = 8$ |
| $(2i)^4 + 2(2i)^2 \stackrel{?}{=} 8$ | $(-2i)^4 + 2(-2i)^2 \stackrel{?}{=} 8$ | $(\sqrt{2})^4 + 2(\sqrt{2})^2 \stackrel{?}{=} 8$ | $(-\sqrt{2})^4 + 2(-\sqrt{2})^2 \stackrel{?}{=} 8$ |
| $2^4i^4 + 2(2^2)(i^2) \stackrel{?}{=} 8$ | $(-2)^4i^4 + 2(-2)^2i^2 \stackrel{?}{=} 8$ | $4 + 2(2) \stackrel{?}{=} 8$ | $4 + 2(2) \stackrel{?}{=} 8$ |
| $16(1) + 8(-1) \stackrel{?}{=} 8$ | $16(1) + 8(-1) \stackrel{?}{=} 8$ | $8 = 8$ | $8 = 8$ |
| $16 - 8 = 8$ | $16 - 8 = 8$ | True | True |
| True | True | | |

Thus, the solutions are $2i$, $-2i$, $\sqrt{2}$, and $-\sqrt{2}$.

▶ **Now Try Exercise 17**

The solutions to equations like $p^4 + 2p^2 = 8$ will always check unless a mistake has been made. In equations like this, extraneous solutions will not be introduced. However, extraneous solutions *may* be introduced when working with rational exponents, as will be shown in Example 6.

### Helpful Hint

Students sometimes solve the equation for $u$ but then forget to complete the problem by solving for the original variable. Remember that if the original equation is in $x$ you must obtain values for $x$. If the original equation is in $p$ (as in Example 2) you must obtain values for $p$, and so on.

**EXAMPLE 3** ▶ Solve $4(2w + 1)^2 - 16(2w + 1) + 15 = 0$.

**Solution** If we let $u = 2w + 1$, the equation becomes

$$4(2w + 1)^2 - 16(2w + 1) + 15 = 0$$

$$4u^2 - 16u + 15 = 0 \quad \text{Substitute } u \text{ for } 2w + 1.$$

Now we can factor and solve.

$$(2u - 3)(2u - 5) = 0$$

$$2u - 3 = 0 \quad \text{or} \quad 2u - 5 = 0$$

$$2u = 3 \qquad\qquad 2u = 5$$

$$u = \frac{3}{2} \qquad\qquad u = \frac{5}{2}$$

We are not finished. Since the variable in the original equation is $w$, we must solve for $w$, not $u$. Therefore, we substitute back $2w + 1$ for $u$ and solve for $w$.

$$u = \frac{3}{2} \qquad\qquad u = \frac{5}{2}$$

$$2w + 1 = \frac{3}{2} \qquad\qquad 2w + 1 = \frac{5}{2} \quad \text{Substitute } 2w + 1 \text{ for } u.$$

$$2w = \frac{1}{2} \qquad\qquad 2w = \frac{3}{2}$$

$$w = \frac{1}{4} \qquad\qquad w = \frac{3}{4}$$

A check will show that both $\frac{1}{4}$ and $\frac{3}{4}$ are solutions to the original equation.

▶ **Now Try Exercise 29**

**EXAMPLE 4** ▶ Find the $x$-intercepts of the graph of the function $f(x) = 2x^{-2} + x^{-1} - 1$.

**Solution** The $x$-intercepts occur where $f(x) = 0$. Therefore, to find the $x$-intercepts we must solve the equation

$$2x^{-2} + x^{-1} - 1 = 0$$

This equation can be expressed as

$$2(x^{-1})^2 + x^{-1} - 1 = 0$$

When we let $u = x^{-1}$, the equation becomes

$$2u^2 + u - 1 = 0$$

$$(2u - 1)(u + 1) = 0$$

$$2u - 1 = 0 \quad \text{or} \quad u + 1 = 0$$

$$u = \frac{1}{2} \qquad\qquad u = -1$$

Now we substitute $x^{-1}$ for $u$.

$$x^{-1} = \frac{1}{2} \qquad \text{or} \qquad x^{-1} = -1$$
$$\frac{1}{x} = \frac{1}{2} \qquad\qquad \frac{1}{x} = -1$$
$$x = 2 \qquad\qquad x = -1$$

A check will show that both 2 and $-1$ are solutions to the original equation. Thus, the $x$-intercepts are $(2, 0)$ and $(-1, 0)$.

▸ **Now Try Exercise 61**

The equation in Example 4 could also be expressed as

$$\frac{2}{x^2} + \frac{1}{x} - 1 = 0$$

A second method to solve this equation is to multiply both sides of the equation by the least common denominator, $x^2$, then simplify.

$$x^2\left(\frac{2}{x^2} + \frac{1}{x} - 1\right) = x^2 \cdot 0$$
$$2 + x - x^2 = 0$$
$$x^2 - x - 2 = 0$$
$$(x - 2)(x + 1) = 0$$
$$x - 2 = 0 \qquad \text{or} \qquad x + 1 = 0$$
$$x = 2 \qquad\qquad x = -1$$

Many of the equations solved in this section may be solved by more than one method.

## 2 Solve Equations with Rational Exponents

When solving equations that are quadratic in form with rational exponents, we raise both sides of the equation to some power to eliminate the rational exponents. Recall that we did this in Section 7.6 when we solved radical equations. Whenever you raise both sides of an equation to a power, you may introduce extraneous solutions. **Therefore, whenever you raise both sides of an equation to a power, you must check all apparent solutions in the original equation to make sure that none are extraneous.** We will now work two examples showing how to solve equations that contain rational exponents. We use the same procedure as used earlier.

**EXAMPLE 5** ▸ Solve $x^{2/5} + x^{1/5} - 6 = 0$.

*Solution* This equation can be rewritten as

$$(x^{1/5})^2 + x^{1/5} - 6 = 0$$

Let $u = x^{1/5}$. Then the equation becomes

$$u^2 + u - 6 = 0$$
$$(u + 3)(u - 2) = 0$$
$$u + 3 = 0 \qquad \text{or} \qquad u - 2 = 0$$
$$u = -3 \qquad\qquad u = 2$$

Now substitute $x^{1/5}$ for $u$ and raise both sides of the equation to the fifth power to remove the rational exponents.

$$x^{1/5} = -3 \qquad \text{or} \qquad x^{1/5} = 2$$
$$(x^{1/5})^5 = (-3)^5 \qquad\qquad (x^{1/5})^5 = 2^5$$
$$x = -243 \qquad\qquad x = 32$$

The two *possible* solutions are −243 and 32. Remember that whenever you raise both sides of an equation to a power, as you did here, you need to check for extraneous solutions.

**Check** $x = -243$

$$\begin{aligned} x^{2/5} + x^{1/5} - 6 &= 0 \\ (-243)^{2/5} + (-243)^{1/5} - 6 &\stackrel{?}{=} 0 \\ (\sqrt[5]{-243})^2 + \sqrt[5]{-243} - 6 &\stackrel{?}{=} 0 \\ (-3)^2 - 3 - 6 &\stackrel{?}{=} 0 \\ 9 - 3 - 6 &\stackrel{?}{=} 0 \\ 0 &= 0 \quad \textit{True} \end{aligned}$$

$x = 32$

$$\begin{aligned} x^{2/5} + x^{1/5} - 6 &= 0 \\ (32)^{2/5} + (32)^{1/5} - 6 &\stackrel{?}{=} 0 \\ (\sqrt[5]{32})^2 + \sqrt[5]{32} - 6 &\stackrel{?}{=} 0 \\ 2^2 + 2 - 6 &\stackrel{?}{=} 0 \\ 4 + 2 - 6 &\stackrel{?}{=} 0 \\ 0 &= 0 \quad \textit{True} \end{aligned}$$

Since both values check, the solutions are −243 and 32.

▸ **Now Try Exercise 63**

**EXAMPLE 6** ▸ Solve $2p - \sqrt{p} - 10 = 0$.

*Solution* We can express this equation as

$$\begin{aligned} 2p - p^{1/2} - 10 &= 0 \\ 2(p^{1/2})^2 - p^{1/2} - 10 &= 0 \end{aligned}$$

If we let $u = p^{1/2}$, this equation is quadratic in form.

$$\begin{aligned} 2u^2 - u - 10 &= 0 \\ (2u - 5)(u + 2) &= 0 \end{aligned}$$

$$2u - 5 = 0 \quad \text{or} \quad u + 2 = 0$$

$$2u = 5 \qquad\qquad u = -2$$

$$u = \frac{5}{2}$$

However, since our original equation is in the variable $p$, we must solve for $p$. We substitute $p^{1/2}$ for $u$.

$$p^{1/2} = \frac{5}{2} \qquad\qquad p^{1/2} = -2$$

Now we square both sides of the equation.

$$(p^{1/2})^2 = \left(\frac{5}{2}\right)^2 \qquad\qquad (p^{1/2})^2 = (-2)^2$$

$$p = \frac{25}{4} \qquad\qquad p = 4$$

We must now check both apparent solutions in the original equation.

**Check** $p = \dfrac{25}{4}$

$$\begin{aligned} 2p - \sqrt{p} - 10 &= 0 \\ 2\left(\frac{25}{4}\right) - \sqrt{\frac{25}{4}} - 10 &\stackrel{?}{=} 0 \\ \frac{25}{2} - \frac{5}{2} - 10 &\stackrel{?}{=} 0 \\ 0 &= 0 \quad \textit{True} \end{aligned}$$

$p = 4$

$$\begin{aligned} 2p - \sqrt{p} - 10 &= 0 \\ 2(4) - \sqrt{4} - 10 &\stackrel{?}{=} 0 \\ 8 - 2 - 10 &\stackrel{?}{=} 0 \\ -4 &= 0 \quad \textit{False} \end{aligned}$$

Since 4 does not check, it is an extraneous solution. The only solution is $\frac{25}{4}$.

▸ **Now Try Exercise 25**

Example 6 could also be solved by writing the equation as $\sqrt{p} = 2p - 10$ and squaring both sides of the equation. Try this now. If you have forgotten how to do this, review Section 7.6.

## EXERCISE SET 8.4

MathXL® MyMathLab

### Concept/Writing Exercises

1. Explain how you can determine whether a given equation can be expressed as an equation that is quadratic in form.
2. When solving an equation that is quadratic in form, when is it essential to check your answer for extraneous solutions? Explain why.
3. To solve the equation $3x^4 - 5x^2 + 1 = 0$, what is the correct choice for $u$ to obtain an equation that is quadratic in form? Explain.
4. To solve the equation $2y^{4/3} + 9y^{2/3} - 7 = 0$, what is the correct choice for $u$ to obtain an equation that is quadratic in form? Explain.
5. To solve the equation $z^{-2} - z^{-1} = 56$, what is the correct choice for $u$ to obtain an equation that is quadratic in form? Explain.
6. To solve the equation $3\left(\frac{x+2}{x+3}\right)^2 + \left(\frac{x+2}{x+3}\right) - 9 = 0$, what is the correct choice for $u$ to obtain an equation that is quadratic in form? Explain.

### Practice the Skills

*Solve each equation.*

7. $x^4 - 10x^2 + 9 = 0$
8. $x^4 - 37x^2 + 36 = 0$
9. $x^4 + 17x^2 + 16 = 0$
10. $x^4 + 50x^2 + 49 = 0$
11. $x^4 - 13x^2 + 36 = 0$
12. $x^4 + 13x^2 + 36 = 0$
13. $a^4 - 7a^2 + 12 = 0$
14. $b^4 + 7b^2 + 12 = 0$
15. $4x^4 - 17x^2 + 4 = 0$
16. $9d^4 - 13d^2 + 4 = 0$
17. $r^4 - 8r^2 = -15$
18. $p^4 - 8p^2 = -12$
19. $z^4 - 7z^2 = 18$
20. $a^4 + a^2 = 42$
21. $-c^4 = 4c^2 - 5$
22. $9b^4 = 57b^2 - 18$
23. $\sqrt{x} = 2x - 6$
24. $x - 2\sqrt{x} = 8$
25. $x - \sqrt{x} = 6$
26. $x - 4 = -3\sqrt{x}$
27. $9x + 3\sqrt{x} = 2$
28. $8x + 2\sqrt{x} = 1$
29. $(x + 3)^2 + 2(x + 3) = 24$
30. $(x + 1)^2 + 4(x + 1) + 3 = 0$
31. $6(a - 2)^2 = -19(a - 2) - 10$
32. $10(z + 2)^2 = 3(z + 2) + 1$
33. $(x^2 - 3)^2 - (x^2 - 3) - 6 = 0$
34. $(a^2 - 1)^2 - 5(a^2 - 1) - 14 = 0$
35. $2(b + 3)^2 + 5(b + 3) - 3 = 0$
36. $(z^2 - 6)^2 + 2(z^2 - 6) - 24 = 0$
37. $18(x^2 - 5)^2 + 27(x^2 - 5) + 10 = 0$
38. $28(x^2 - 8)^2 - 23(x^2 - 8) - 15 = 0$
39. $a^{-2} + 4a^{-1} + 4 = 0$
40. $x^{-2} + 10x^{-1} + 25 = 0$
41. $12b^{-2} - 7b^{-1} + 1 = 0$
42. $5x^{-2} + 4x^{-1} - 1 = 0$
43. $2b^{-2} = 7b^{-1} - 3$
44. $10z^{-2} - 3z^{-1} - 1 = 0$
45. $x^{-2} + 9x^{-1} = 10$
46. $6a^{-2} = a^{-1} + 12$
47. $x^{-2} = 4x^{-1} + 12$
48. $x^{2/3} - 5x^{1/3} + 6 = 0$
49. $x^{2/3} - 4x^{1/3} = -3$
50. $x^{2/3} = 3x^{1/3} + 4$
51. $b^{2/3} - 9b^{1/3} + 18 = 0$
52. $c^{2/3} - 4 = 0$
53. $-2a - 5a^{1/2} + 3 = 0$
54. $r^{2/3} - 7r^{1/3} + 10 = 0$
55. $c^{2/5} + 3c^{1/5} + 2 = 0$
56. $x^{2/5} - 5x^{1/5} + 6 = 0$

*Find all x-intercepts of each function.*

57. $f(x) = x - 5\sqrt{x} + 4$
58. $g(x) = x - 15\sqrt{x} + 56$
59. $h(x) = x + 14\sqrt{x} + 45$
60. $k(x) = x + 7\sqrt{x} + 12$
61. $p(x) = 4x^{-2} - 19x^{-1} - 5$
62. $g(x) = 4x^{-2} + 12x^{-1} + 9$
63. $f(x) = x^{2/3} - x^{1/3} - 6$
64. $f(x) = x^{1/2} + 6x^{1/4} - 7$
65. $g(x) = (x^2 - 3x)^2 + 2(x^2 - 3x) - 24$
66. $g(x) = (x^2 - 6x)^2 - 5(x^2 - 6x) - 24$
67. $f(x) = x^4 - 29x^2 + 100$
68. $h(x) = x^4 - 4x^2 + 3$

## Problem Solving

69. Give a general procedure for solving an equation of the form $ax^4 + bx^2 + c = 0$.

70. Give a general procedure for solving an equation of the form $ax^{2n} + bx^n + c = 0$.

71. Give a general procedure for solving an equation of the form $ax^{-2} + bx^{-1} + c = 0$.

72. Give a general procedure for solving an equation of the form $a(x - r)^2 + b(x - r) - c = 0$.

73. Determine an equation of the form $ax^4 + bx^2 + c = 0$ that has solutions $\pm 2$ and $\pm 1$. Explain how you obtained your answer.

74. Determine an equation of the form $ax^4 + bx^2 + c = 0$ that has solutions $\pm 3$ and $\pm 2i$. Explain how you obtained your answer.

75. Determine an equation of the form $ax^4 + bx^2 + c = 0$ that has solutions $\pm\sqrt{2}$ and $\pm\sqrt{5}$. Explain how you obtained your answer.

76. Determine an equation of the form $ax^4 + bx^2 + c = 0$ that has solutions $\pm 2i$ and $\pm 5i$. Explain how you obtained your answer.

77. Is it possible for an equation of the form $ax^4 + bx^2 + c = 0$ to have exactly one imaginary solution? Explain.

78. Is it possible for an equation of the form $ax^4 + bx^2 + c = 0$ to have exactly one real solution? Explain.

79. Solve the equation $\frac{3}{x^2} - \frac{3}{x} = 60$ by
   a) multiplying both sides of the equation by the LCD.
   b) writing the equation with negative exponents.

80. Solve the equation $1 = \frac{2}{x} - \frac{2}{x^2}$ by
   a) multiplying both sides of the equation by the LCD.
   b) writing the equation with negative exponents.

*Find all real solutions to each equation.*

81. $15(r + 2) + 22 = -\frac{8}{r + 2}$

82. $2(p + 3) + 5 = \frac{3}{p + 3}$

83. $4 - (x - 1)^{-1} = 3(x - 1)^{-2}$

84. $3(x - 4)^{-2} = 16(x - 4)^{-1} + 12$

85. $x^6 - 9x^3 + 8 = 0$

86. $x^6 - 28x^3 + 27 = 0$

87. $(x^2 + 2x - 2)^2 - 7(x^2 + 2x - 2) + 6 = 0$

88. $(x^2 + 3x - 2)^2 - 10(x^2 + 3x - 2) + 16 = 0$

*Find all solutions to each equation.*

89. $2n^4 - 6n^2 - 3 = 0$

90. $3x^4 + 8x^2 - 1 = 0$

## Cumulative Review Exercises

[1.3] 91. Evaluate $\frac{4}{5} - \left(\frac{3}{4} - \frac{2}{3}\right)$.

[2.1] 92. Solve $3(x + 2) - 2(3x + 3) = -3$

[3.2] 93. State the domain and range for $y = (x - 3)^2$.

[7.3] 94. Simplify $\sqrt[3]{16x^3y^6}$.

[7.4] 95. Add $\sqrt{75} + \sqrt{48}$.

# 8.5 Graphing Quadratic Functions

1. Determine when a parabola opens upward or downward.
2. Find the axis of symmetry, vertex, and $x$-intercepts of a parabola.
3. Graph quadratic functions using the axis of symmetry, vertex, and intercepts.
4. Solve maximum and minimum problems.
5. Understand translations of parabolas.
6. Write functions in the form $f(x) = a(x - h)^2 + k$.

We graphed quadratic equations by plotting points in Section 3.2, and we had a brief discussion of the $x$-intercepts of the graphs of quadratic functions in Section 5.8. In this section we study the graphs of quadratic functions, called **parabolas**, in more depth. In objective 3, we will explain how to graph quadratic functions using the axis of symmetry, vertex, and intercepts. In objective 5, we will study patterns in the graphs of parabolas and use these patterns to determine translations, or shifts, that can be used to graph parabolas.

## 1 Determine When a Parabola Opens Upward or Downward

Parabolas have a shape that resembles, but is not the same as, the letter U. Parabolas may open upward or downward. For a quadratic function of the form $f(x) = ax^2 + bx + c$, the *sign* of the leading coefficient, $a$, determines whether a parabola opens upward or downward. *When $a > 0$, the parabola opens upward* (see **Fig. 8.7a**). *When $a < 0$, the parabola opens downward* (see **Fig. 8.7b**).

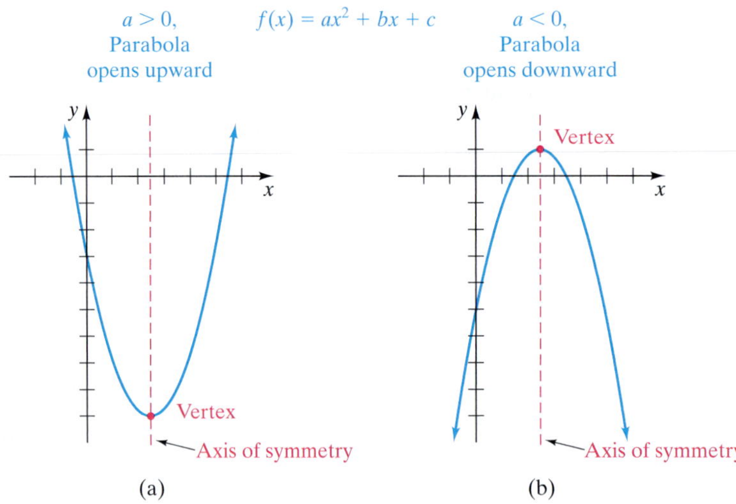


FIGURE 8.7

For a parabola that opens upward, the **vertex** is the lowest point on the curve. The minimum value of the function is the $y$-coordinate of the vertex. The minimum value is obtained when the $x$-coordinate of the vertex is substituted into the function. For a parabola that opens downward, the vertex is the highest point on the curve. The maximum value of the function is the $y$-coordinate of the vertex. The maximum value is obtained when the $x$-coordinate of the vertex is substituted into the function.

## 2 Find the Axis of Symmetry, Vertex, and $x$-Intercepts of a Parabola

Graphs of quadratic functions of the form $f(x) = ax^2 + bx + c$ will have **symmetry** about a vertical line through the vertex. This means that if we fold the paper along this imaginary line, called the **axis of symmetry**, the right and left sides of the graph will coincide (see **Fig. 8.7**). We will now give the equation for finding the axis of symmetry.

### To Find the Axis of Symmetry

For a function of the form $f(x) = ax^2 + bx + c$, the equation of the **axis of symmetry**, of the parabola is

$$x = -\frac{b}{2a}$$

Now we will derive the formula for the axis of symmetry, and find the coordinates of the vertex of a parabola, by beginning with a quadratic function of the form $f(x) = ax^2 + bx + c$ and completing the square on the first two terms.

$$\begin{aligned} f(x) &= ax^2 + bx + c \\ &= a\left(x^2 + \frac{b}{a}x\right) + c \qquad \text{Factor out } a. \end{aligned}$$

One half the coefficient of $x$ is $\frac{b}{2a}$. Its square is $\frac{b^2}{4a^2}$. Add and subtract this term inside the parentheses. The sum of these two terms is zero.

$$f(x) = a\left[x^2 + \frac{b}{a}x + \frac{b^2}{4a^2} - \frac{b^2}{4a^2}\right] + c$$

Now rewrite the function as follows.

$$f(x) = a\left[x^2 + \frac{b}{a}x + \left(\frac{b^2}{4a^2}\right)\right] - a\left(\frac{b^2}{4a^2}\right) + c$$

$$= a\left(x + \frac{b}{2a}\right)^2 - \frac{b^2}{4a} + c$$

Replace the trinomial with the square of a binomial.

$$= a\left(x + \frac{b}{2a}\right)^2 - \frac{b^2}{4a} + \frac{4ac}{4a}$$

Write fractions with a common denominator.

$$= a\left(x + \frac{b}{2a}\right)^2 + \frac{4ac - b^2}{4a}$$

Combine the last two terms; write with the variable a first.

$$= a\left[x - \left(-\frac{b}{2a}\right)\right]^2 + \frac{4ac - b^2}{4a}$$

The expression $\left[x - \left(-\frac{b}{2a}\right)\right]^2$ will always be greater than or equal to 0. (Why?) If $a > 0$, the parabola will open upward and have a minimum value. Since $\left[x - \left(-\frac{b}{2a}\right)\right]^2$ will have a minimum value when $x = -\frac{b}{2a}$, the minimum value of the graph will occur when $x = -\frac{b}{2a}$. If $a < 0$, the parabola will open downward and have a maximum value. The maximum value will occur when $x = -\frac{b}{2a}$. To determine the lowest, or highest, point on a parabola, substitute $-\frac{b}{2a}$ for $x$ in the function to find $y$. The resulting ordered pair will be the vertex of the parabola. Since the axis of symmetry is the vertical line through the vertex, its equation is found using the $x$-coordinate of the ordered pair. Thus, the equation of the axis of symmetry is $x = -\frac{b}{2a}$. Note that when $x = -\frac{b}{2a}$, the value of $f(x)$ is $\frac{4ac - b^2}{4a}$. Do you know why?

### To Find the Vertex of a Parabola

The parabola represented by the function $f(x) = ax^2 + bx + c$ will have axis of symmetry $x = -\frac{b}{2a}$ and vertex

$$\left(-\frac{b}{2a}, \frac{4ac - b^2}{4a}\right)$$

Since we often find the $y$-coordinate of the vertex by substituting the $x$-coordinate of the vertex into $f(x)$, the vertex may also be designated as

$$\left(-\frac{b}{2a}, f\left(-\frac{b}{2a}\right)\right)$$

The parabola given by the function $f(x) = ax^2 + bx + c$ will open upward when $a$ is greater than 0 and open downward when $a$ is less than 0.

Recall that to find the $x$-intercept of the graph of $f(x) = ax^2 + bx + c$, we set $f(x) = 0$ and solve the equation

$$ax^2 + bx + c = 0$$

This equation may be solved by factoring, the quadratic formula, or completing the square.

As we mentioned in Section 8.2, the discriminant, $b^2 - 4ac$, may be used to determine the *number of x-intercepts*. The following table summarizes information about the discriminant.

| Discriminant, $b^2 - 4ac$ | Number of $x$-Intercepts | Possible Graphs of $f(x) = ax^2 + bx + c$ |
|---|---|---|
| $> 0$ | Two | 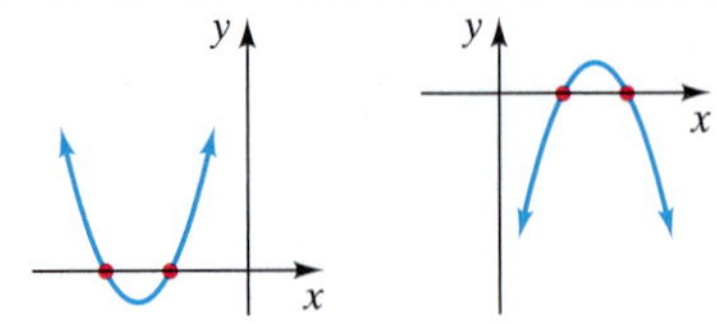 |
| $= 0$ | One | 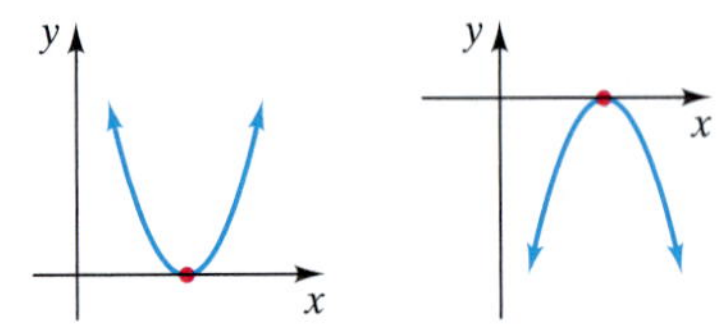 |
| $< 0$ | None | 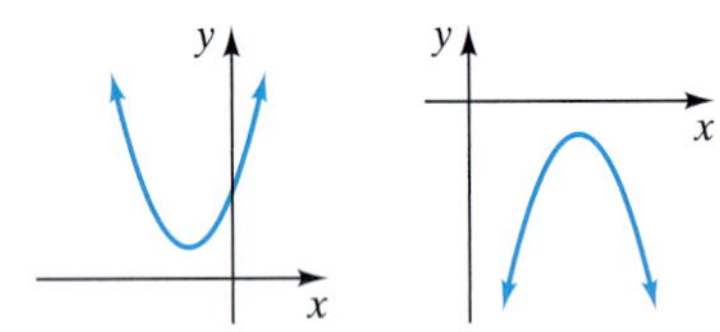 |

## 3 Graph Quadratic Functions Using the Axis of Symmetry, Vertex, and Intercepts

Now we will draw graphs of quadratic functions.

**EXAMPLE 1** ▸ Consider the equation $y = -x^2 + 8x - 12$.

**a)** Determine whether the parabola opens upward or downward.
**b)** Find the $y$-intercept.
**c)** Find the vertex.
**d)** Find the $x$-intercepts, if any.
**e)** Draw the graph.

### *Solution*

**a)** Since $a$ is $-1$, which is less than 0, the parabola opens downward.

**b)** To find the $y$-intercept, set $x = 0$ and solve for $y$.

$$y = -(0)^2 + 8(0) - 12 = -12$$

The $y$-intercept is $(0, -12)$.

**c)** First, find the $x$-coordinate, then find the $y$-coordinate of the vertex.

$$x = -\frac{b}{2a} = -\frac{8}{2(-1)} = 4$$

$$y = \frac{4ac - b^2}{4a} = \frac{4(-1)(-12) - 8^2}{4(-1)} = \frac{48 - 64}{-4} = 4$$

The vertex is at $(4, 4)$. The $y$-coordinate of the vertex could also be found by substituting 4 for $x$ in the function and finding the corresponding value of $y$, which is 4.

**d)** To find the $x$-intercepts, set $y = 0$.

$$0 = -x^2 + 8x - 12$$

$$\text{or}\quad x^2 - 8x + 12 = 0$$

$$(x - 6)(x - 2) = 0$$

$$x - 6 = 0 \quad \text{or} \quad x - 2 = 0$$

$$x = 6 \quad \text{or} \quad x = 2$$

Thus, the $x$-intercepts are $(2, 0)$ and $(6, 0)$. These values could also be found by the quadratic formula (or by completing the square).

**e)** Use all this information to draw the graph (**Fig. 8.8**).

▸ Now Try Exercise 15

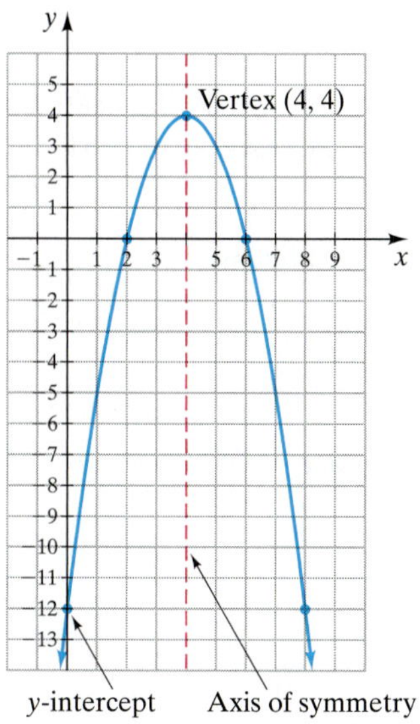


FIGURE 8.8

Notice that in Example 1, the equation is $y = -x^2 + 8x - 12$ and the $y$-intercept is $(0, -12)$. In general, for any equation of the form $y = ax^2 + bx + c$, the $y$-intercept will be $(0, c)$.

If you obtain irrational values when finding $x$-intercepts by the quadratic formula, use your calculator to estimate these values, and then plot these decimal values. For example, if you obtain $x = \frac{2 \pm \sqrt{10}}{2}$, you would evaluate $\frac{2 + \sqrt{10}}{2}$ and $\frac{2 - \sqrt{10}}{2}$ on your calculator and obtain 2.58 and $-0.58$, respectively, to the nearest hundredth. The $x$-intercepts would therefore be $(2.58, 0)$ and $(-0.58, 0)$.

**EXAMPLE 2** ▸ Consider the function $f(x) = 2x^2 + 6x + 5$.

**a)** Determine whether the parabola opens upward or downward.

**b)** Find the $y$-intercept.

**c)** Find the vertex.

**d)** Find the $x$-intercepts, if any.

**e)** Draw the graph.

*Solution*

**a)** Since $a$ is 2, which is greater than 0, the parabola opens upward.

**b)** Since $f(x)$ is the same as $y$, to find the $y$-intercept, set $x = 0$ and solve for $f(x)$, or $y$.

$$f(0) = 2(0)^2 + 6(0) + 5 = 5$$

The $y$-intercept is $(0, 5)$.

**c)**
$$x = -\frac{b}{2a} = -\frac{6}{2(2)} = -\frac{6}{4} = -\frac{3}{2}$$

$$y = \frac{4ac - b^2}{4a} = \frac{4(2)(5) - 6^2}{4(2)} = \frac{40 - 36}{8} = \frac{4}{8} = \frac{1}{2}$$

The vertex is $\left(-\frac{3}{2}, \frac{1}{2}\right)$. The $y$-coordinate of the vertex can also be found by evaluating $f\left(-\frac{3}{2}\right)$.

**d)** To find the $x$-intercepts, set $f(x) = 0$.

$$0 = 2x^2 + 6x + 5$$

This trinomial cannot be factored. To determine whether this equation has any real solutions, evaluate the discriminant.

$$b^2 - 4ac = 6^2 - 4(2)(5) = 36 - 40 = -4$$

Since the discriminant is less than 0, this equation has no real solutions. You should have expected this answer because the $y$-coordinate of the vertex is a positive number and therefore above the $x$-axis. Since the parabola opens upward, it cannot intersect the $x$-axis.

**e)** The graph is given in **Figure 8.9**.

▸ Now Try Exercise 39

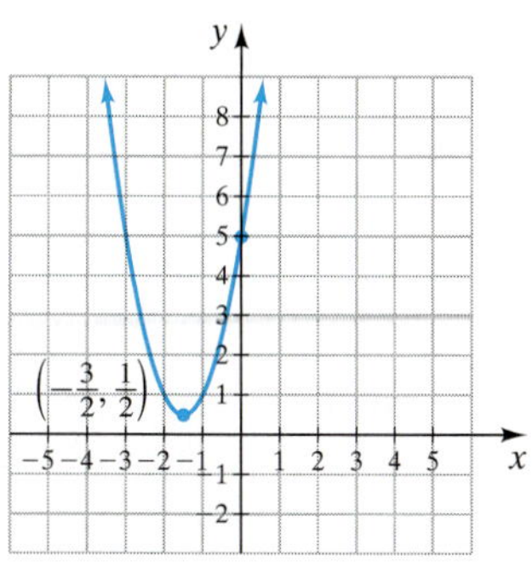


FIGURE 8.9

## 4 Solve Maximum and Minimum Problems

A parabola that opens upward has a **minimum value** at its vertex, as illustrated in **Figure 8.10a**. A parabola that opens downward has a **maximum value** at its vertex, as shown in **Figure 8.10b**. If you are given a function of the form $f(x) = ax^2 + bx + c$, the maximum or minimum value will occur at $-\frac{b}{2a}$, and the value will be $\frac{4ac - b^2}{4a}$. There are many real-life problems that require finding maximum and minimum values.

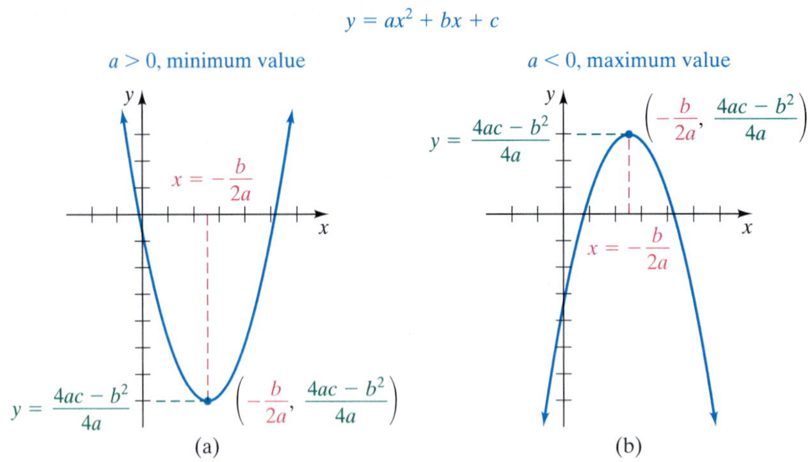


FIGURE 8.10

FIGURE 8.11

**EXAMPLE 3 ▸ Baseball** Tommy Magee plays baseball with the Yorktown Cardinals. In the seventh inning of a game against the Arlington Blue Jays, he hit the ball at a height of 3 feet above the ground (see **Fig. 8.11**). For this particular hit, the height of the ball above the ground, $f(t)$, in feet, at time, $t$, in seconds, can be estimated by the formula

$$f(t) = -16t^2 + 52t + 3$$

**a)** Find the maximum height attained by the baseball.

**b)** Find the time it takes for the baseball to reach its maximum height.

**c)** Find the time it takes for the baseball to strike the ground.

*Solution* **a) Understand** The baseball will follow the path of a parabola that opens downward ($a < 0$). The baseball will rise to a maximum height, then begin its fall back to the ground due to gravity. To find the maximum height, we use the formula $y = \frac{4ac - b^2}{4a}$.

**Translate** $\quad a = -16, \quad b = 52, \quad c = 3$

$$y = \frac{4ac - b^2}{4a}$$

**Carry Out**

$$= \frac{4(-16)(3) - (52)^2}{4(-16)}$$

$$= \frac{-192 - 2704}{-64}$$

$$= \frac{-2896}{-64}$$

$$= 45.25$$

**Answer** The maximum height attained by the baseball is 45.25 feet.

**b)** The baseball reaches its maximum height at

$$t = -\frac{b}{2a} = -\frac{52}{2(-16)} = -\frac{52}{-32} = \frac{13}{8} \text{ or } 1\frac{5}{8} \text{ or } 1.625 \text{ seconds}$$

**c)** Understand and Translate When the baseball strikes the ground, its height, $y$, above the ground is 0. Thus, to determine when the baseball strikes the ground, we solve the equation

$$-16t^2 + 52t + 3 = 0$$

We will use the quadratic formula to solve the equation.

$$t = \frac{-b \pm \sqrt{b^2 - 4ac}}{2a}$$

Carry Out

$$= \frac{-52 \pm \sqrt{(52)^2 - 4(-16)(3)}}{2(-16)}$$

$$= \frac{-52 \pm \sqrt{2704 + 192}}{-32}$$

$$= \frac{-52 \pm \sqrt{2896}}{-32}$$

$$\approx \frac{-52 \pm 53.81}{-32}$$

$$t \approx \frac{-52 + 53.81}{-32} \quad \text{or} \quad t \approx \frac{-52 - 53.81}{-32}$$

$$\approx -0.06 \text{ second} \qquad \approx 3.31 \text{ seconds}$$

Answer The only acceptable value is 3.31 seconds. The baseball strikes the ground in about 3.31 seconds. Notice in part **b)** that the time it takes the baseball to reach its maximum height, 1.625 seconds, is not quite half the total time the baseball was in flight, 3.31 seconds. The reason for this is that the baseball was hit from a height of 3 feet and not at ground level.

▸ Now Try Exercise 95

**EXAMPLE 4** ▸ **Area of a Rectangle** Consider the rectangle below where the length is $x + 3$ and the width is $10 - x$.

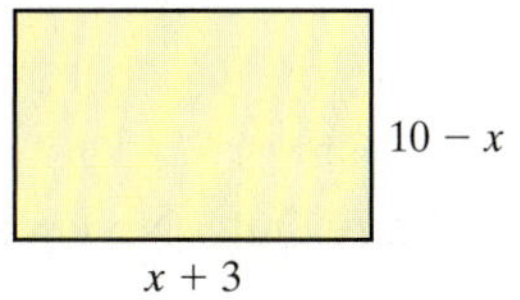


**a)** Find an equation for the area, $A(x)$.

**b)** Find the value for $x$ that gives the largest (maximum) area.

**c)** Find the maximum area.

*Solution* **a)** Area is length times width. The area function is

$$A(x) = (x + 3)(10 - x)$$
$$= -x^2 + 7x + 30$$

**b)** Understand and Translate The graph of the function is a parabola that opens downward. Thus, the maximum value occurs at the vertex. Therefore, the maximum area occurs at $x = -\frac{b}{2a}$.

Carry Out

$$x = -\frac{b}{2a} = -\frac{7}{2(-1)} = \frac{7}{2} = 3.5$$

Answer The maximum area occurs when $x$ is 3.5 units.

**c)** To find the maximum area, substitute 3.5 for each $x$ in the equation determined in part **a)**.

$$\begin{aligned} A(x) &= -x^2 + 7x + 30 \\ A(3.5) &= -(3.5)^2 + 7(3.5) + 30 \\ &= -12.25 + 24.5 + 30 \\ &= 42.25 \end{aligned}$$

Observe that for this rectangle, the length is $x + 3 = 3.5 + 3 = 6.5$ units and the width is $10 - x = 10 - 3.5 = 6.5$ units. The rectangle is actually a square, and its area is $(6.5)(6.5) = 42.25$ square units. Therefore, the maximum area is 42.25 square units.

▸ **Now Try Exercise 73**

In Example 4**c)**, the maximum area could have been determined by using the formula $y = \dfrac{4ac - b^2}{4a}$. Determine the maximum area now using this formula. You should obtain the same answer, 42.25 square units.

**EXAMPLE 5** ▸ **Rectangular Corral** John W. Brown is building a corral for newborn calves in the shape of a rectangle (see **Fig. 8.12**). If he plans to use 160 meters of fencing, find the dimensions of the corral that will give the greatest area.

FIGURE 8.12

*Solution* **Understand** We are given the perimeter of the corral, 160 meters. The formula for the perimeter of a rectangle is $P = 2l + 2w$. For this problem, $160 = 2l + 2w$. We are asked to maximize the area, $A$, where

$$A = lw$$

We need to express the area in terms of one variable, not two. To express the area in terms of $l$, we solve the perimeter formula, $160 = 2l + 2w$, for $w$, then make a substitution.

**Translate**

$$\begin{aligned} 160 &= 2l + 2w \\ 160 - 2l &= 2w \\ 80 - l &- w \end{aligned}$$

**Carry Out** Now we substitute $80 - l$ for $w$ into $A = lw$. This gives

$$\begin{aligned} A &= lw \\ A &= l(80 - l) \\ A &= -l^2 + 80l \end{aligned}$$

In this quadratic equation, $a = -1$, $b = 80$, and $c = 0$. The maximum area will occur at

$$l = -\frac{b}{2a} = -\frac{80}{2(-1)} = 40$$

**Answer** The length that will give the largest area is 40 meters. The width, $w = 80 - l$, will also be 40 meters. Thus, a square with dimensions 40 by 40 meters will give the largest area.

The largest area can be found by substituting $l = 40$ into the formula $A = l(80 - l)$ or by using $A = \dfrac{4ac - b^2}{4a}$. In either case, we obtain an area of 1600 square meters.

▸ **Now Try Exercise 93**

In Example 5, when we obtained the equation $A = -l^2 + 80l$, we could have completed the square as follows:

$$\begin{aligned} A &= -(l^2 - 80l) \\ &= -(l^2 - 80l + 1600 - 1600) \\ &= -(l^2 - 80l + 1600) + 1600 \\ &= -(l - 40)^2 + 1600 \end{aligned}$$

From this equation we can determine that the maximum area, 1600 square meters, occurs when the length is 40 meters.

## 5 Understand Translations of Parabolas

Now we will look at another method used to graph parabolas. With this method, you start with a graph of an equation of the form $f(x) = ax^2$ and **translate**, or shift, the position of the graph to obtain the graph of the function you are seeking. As a reference, **Figure 8.13a** shows the graphs of $f(x) = x^2$, $g(x) = 2x^2$, and $h(x) = \frac{1}{2}x^2$. **Figure 8.13b** shows the graphs of $f(x) = -x^2$, $g(x) = -2x^2$, and $h(x) = -\frac{1}{2}x^2$.

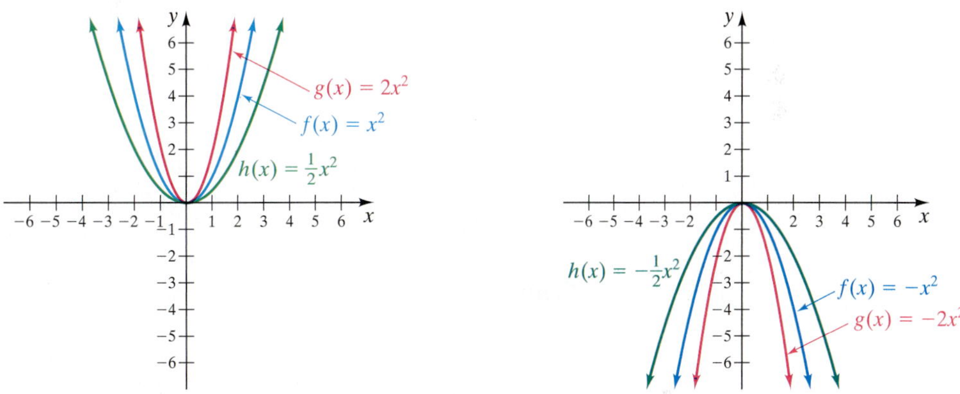


FIGURE 8.13 (a) (b)

You can verify that each of the graphs is correct by plotting points. Notice that in **Figures 8.13a** and **b** the *value of a* in $f(x) = ax^2$ *determines the width of the parabola*. As $|a|$ gets larger, the parabola gets narrower and as $|a|$ gets smaller, the parabola gets wider.

Now let's consider the three functions $f(x) = x^2$, $g(x) = (x - 2)^2$, and $h(x) = (x + 2)^2$. These functions are graphed in **Figure 8.14**. (You can verify that these are graphs of the three functions by plotting points.)

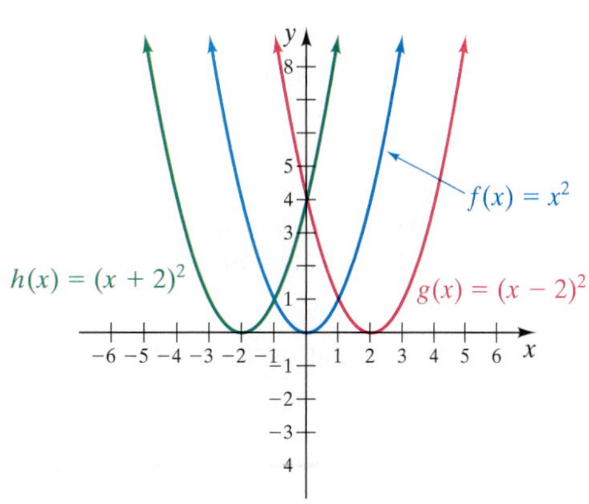


FIGURE 8.14

Notice that the graphs of $g(x)$ and $h(x)$ are identical in shape to the graph of $f(x)$ except that $g(x)$ has been translated, or shifted, 2 units to the right and $h(x)$ has been translated 2 units to the left. In general, the graph of $g(x) = a(x - h)^2$ will have the same shape as the graph of $f(x) = ax^2$. The graph of an equation of the form

$g(x) = a(x - h)^2$ will be shifted horizontally from the graph of $f(x) = ax^2$. *If h is a positive real number, the graph of* $g(x) = a(x - h)^2$ *will be shifted h units to the right of the graph of* $f(x) = ax^2$. *If h is a negative real number, the graph of* $g(x) = a(x - h)^2$ *will be shifted* $|h|$ *units to the left of the graph of* $f(x) = ax^2$.

Now consider the graphs of $f(x) = x^2$, $g(x) = x^2 + 3$ and $h(x) = x^2 - 3$ that are illustrated in **Figure 8.15**. You can verify that these are graphs of the three functions by plotting points.

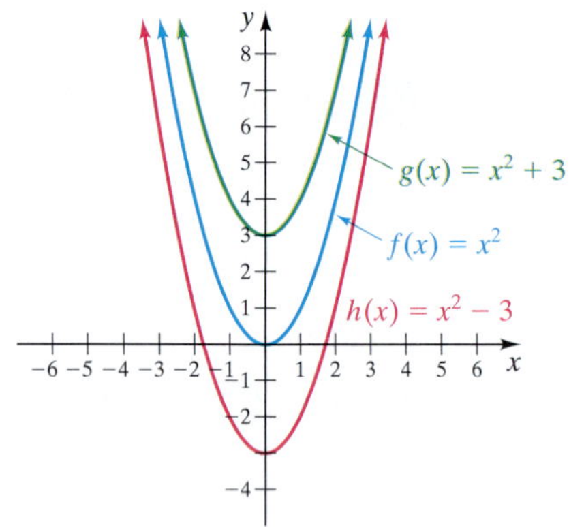


FIGURE 8.15

Notice that the graphs of $g(x)$ and $h(x)$ are identical to the graph of $f(x)$ except that $g(x)$ is translated 3 units up, and $h(x)$ is translated 3 units down. In general, *the graph of* $g(x) = ax^2 + k$ *is the graph of* $f(x) = ax^2$ *shifted k units up if k is a positive real number, and* $|k|$ *units down if k is a negative real number.*

Now consider the graphs of $f(x) = x^2$ and $g(x) = (x - 2)^2 + 3$, shown in **Figure 8.16**.

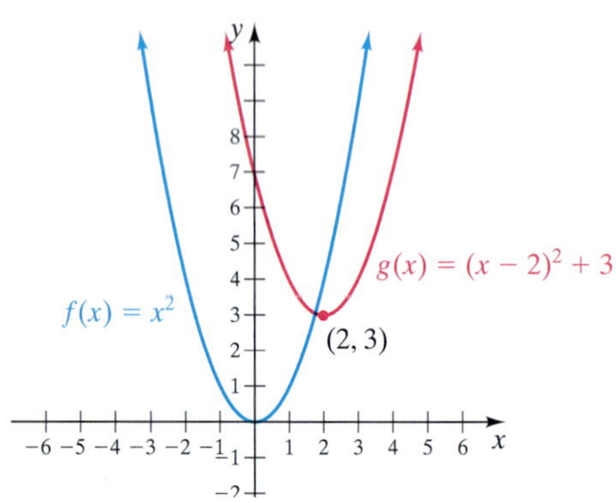


FIGURE 8.16

Notice that the graph of $g(x)$ has the same general shape as that of $f(x)$. The graph of $g(x)$ is the graph of $f(x)$ translated 2 units to the right and 3 units up. This graph and the discussion preceding it lead to the following important facts.

**Parabola Shifts**

For any function $f(x) = ax^2$, the graph of $g(x) = a(x - h)^2 + k$ will have the same shape as the graph of $f(x)$. The graph of $g(x)$ will be the graph of $f(x)$ shifted as follows:

- If $h$ is a positive real number, the graph will be shifted $h$ units to the right.
- If $h$ is a negative real number, the graph will be shifted $|h|$ units to the left.
- If $k$ is a positive real number, the graph will be shifted $k$ units up.
- If $k$ is a negative real number, the graph will be shifted $|k|$ units down.

Examine the graph of $g(x) = (x - 2)^2 + 3$ in **Figure 8.16**. Notice that its axis of symmetry is $x = 2$ and its vertex is $(2, 3)$.

**Axis of Symmetry and Vertex of a Parabola**

The graph of any function of the form

$$f(x) = a(x - h)^2 + k$$

will be a parabola with axis of symmetry $x = h$ and vertex at $(h, k)$.

| Example | Axis of Symmetry | Vertex | Parabola Opens |
|---|---|---|---|
| $f(x) = 2(x - 5)^2 + 7$ | $x = 5$ | $(5, 7)$ | upward, $a > 0$ |
| $f(x) = -\frac{1}{2}(x - 6)^2 - 3$ | $x = 6$ | $(6, -3)$ | downward, $a < 0$ |

Now consider $f(x) = 2(x + 5)^2 + 3$. We can rewrite this as $f(x) = 2[x - (-5)]^2 + 3$. Therefore, $h$ has a value of $-5$ and $k$ has a value of 3. The graph of this function has axis of symmetry $x = -5$ and vertex at $(-5, 3)$.

| Example | Axis of Symmetry | Vertex | Parabola Opens |
|---|---|---|---|
| $f(x) = 3(x + 4)^2 - 2$ | $x = -4$ | $(-4, -2)$ | upward, $a > 0$ |
| $f(x) = -\frac{1}{2}\left(x + \frac{1}{3}\right)^2 + \frac{1}{4}$ | $x = -\frac{1}{3}$ | $\left(-\frac{1}{3}, \frac{1}{4}\right)$ | downward, $a < 0$ |

Now we are ready to graph parabolas using translations.

**EXAMPLE 6** The graph of $f(x) = -2x^2$ is illustrated in **Figure 8.17**. Using this graph as a guide, graph $g(x) = -2(x + 3)^2 - 4$.

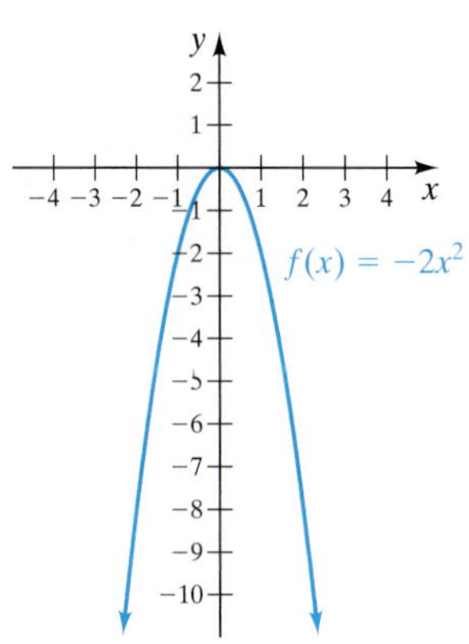


FIGURE 8.17

*Solution* The function $g(x)$ may be written $g(x) = -2[x - (-3)]^2 - 4$. Therefore, in the function, $h$ has a value of $-3$ and $k$ has a value of $-4$. The graph of $g(x)$ will therefore be the graph of $f(x)$ translated 3 units to the left (because $h = -3$) and 4 units down (because $k = -4$). The graphs of $f(x)$ and $g(x)$ are illustrated in **Figure 8.18**.

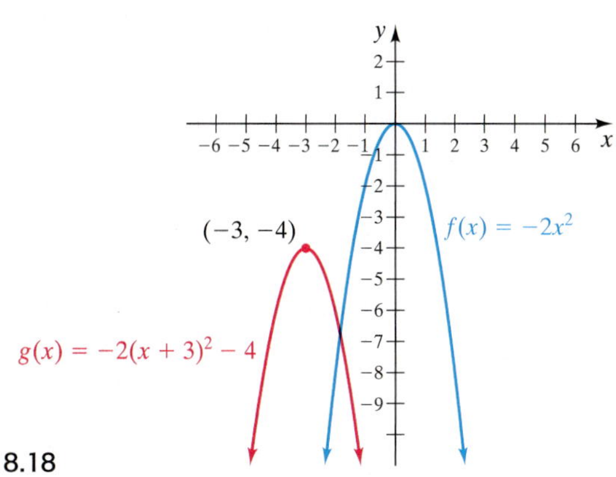


FIGURE 8.18

**Now Try Exercise 49**

In objective 2, we started with a function of the form $f(x) = ax^2 + bx + c$ and completed the square to obtain

$$f(x) = a\left[x - \left(-\frac{b}{2a}\right)\right]^2 + \frac{4ac - b^2}{4a}$$

We stated that the vertex of the parabola of this function is $\left(-\frac{b}{2a}, \frac{4ac - b^2}{4a}\right)$. Suppose we substitute $h$ for $-\frac{b}{2a}$ and $k$ for $\frac{4ac - b^2}{4a}$ in the function. We then get

$$f(x) = a(x - h)^2 + k$$

which we know is a parabola with vertex at $(h, k)$. Therefore, both functions $f(x) = ax^2 + bx + c$ and $f(x) = a(x - h)^2 + k$ yield the same vertex and axis of symmetry for any given function.

## 6 Write Functions in the Form $f(x) = a(x - h)^2 + k$

If we wish to graph parabolas using translations, we need to change the form of a function from $f(x) = ax^2 + bx + c$ to $f(x) = a(x - h)^2 + k$. To do this, we *complete the square* as was discussed in Section 8.1. By completing the square we obtain a perfect square trinomial, which we can represent as the square of a binomial. Examples 7 and 8 explain the procedure. We will use this procedure again in a later chapter, when we discuss conic sections.

**EXAMPLE 7** ▸ Given $f(x) = x^2 - 6x + 10$,

**a)** Write $f(x)$ in the form $f(x) = a(x - h)^2 + k$.

**b)** Graph $f(x)$.

### Solution

**a)** We use the $x^2$ and $-6x$ terms to obtain a perfect square trinomial.

$$f(x) = (x^2 - 6x) + 10$$

Now we take half the coefficient of the $x$-term and square it.

$$\left[\frac{1}{2}(-6)\right]^2 = 9$$

We then add this value, 9, within the parentheses. Since we are adding 9 within parentheses, we add $-9$ outside parentheses. Adding 9 and $-9$ to an expression is the same as adding 0, which does not change the value of the expression.

$$f(x) = (x^2 - 6x + 9) - 9 + 10$$

By doing this we have created a perfect square trinomial within the parentheses, plus a constant outside the parentheses. We express the perfect square trinomial as the square of a binomial.

$$f(x) = (x - 3)^2 + 1$$

The function is now in the form we are seeking.

**b)** Since $a = 1$, which is greater than 0, the parabola opens upward. The axis of symmetry of the parabola is $x = 3$ and the vertex is at $(3, 1)$. The $y$-intercept can be easily obtained by substituting $x = 0$ and finding $f(x)$. When $x = 0$, $f(x) = (-3)^2 + 1 = 10$. Thus, the $y$-intercept is at 10. By plotting the vertex, $y$-intercept, and a few other points, we obtain the graph in **Figure 8.19**. The figure also shows the graph of $y = x^2$ for comparison.

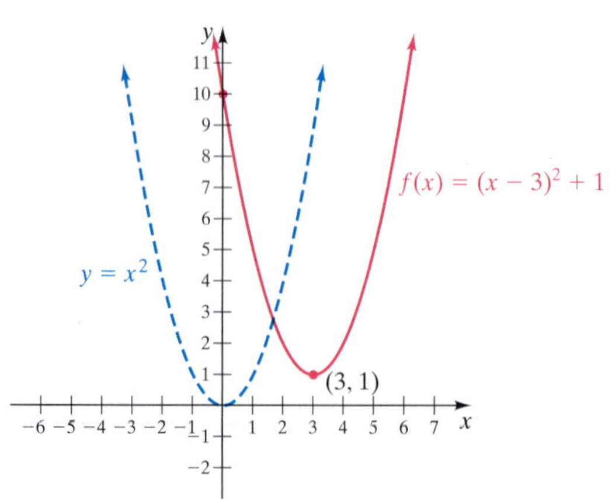


FIGURE 8.19

▸ **Now Try Exercise 59**

**EXAMPLE 8** ▸ Given $f(x) = -2x^2 - 10x - 13$,

**a)** Write $f(x)$ in the form $f(x) = a(x - h)^2 + k$.

**b)** Graph $f(x)$.

## Solution

**a)** When the leading coefficient is not 1, we factor out the leading coefficient from the terms containing the variable.

$$f(x) = -2(x^2 + 5x) - 13$$

Now we complete the square.

Half of coefficient of x-term squared

$$\left[\frac{1}{2}(5)\right]^2 = \frac{25}{4}$$

If we add $\frac{25}{4}$ within the parentheses, we are actually adding $-2\left(\frac{25}{4}\right)$ or $-\frac{25}{2}$, since each term in parentheses is multiplied by $-2$. Therefore, to compensate, we must add $\frac{25}{2}$ outside the parentheses.

$$f(x) = -2\left(x^2 + 5x + \frac{25}{4}\right) + \frac{25}{2} - 13$$

$$= -2\left(x + \frac{5}{2}\right)^2 - \frac{1}{2}$$

**b)** Since $a = -2$, the parabola opens downward. The axis of symmetry is $x = -\frac{5}{2}$ and the vertex is $\left(-\frac{5}{2}, -\frac{1}{2}\right)$. The $y$-intercept is at $f(0) = -13$. We plot a few points and draw the graph in **Figure 8.20**. In the figure, we also show the graph of $y = -2x^2$ for comparison.

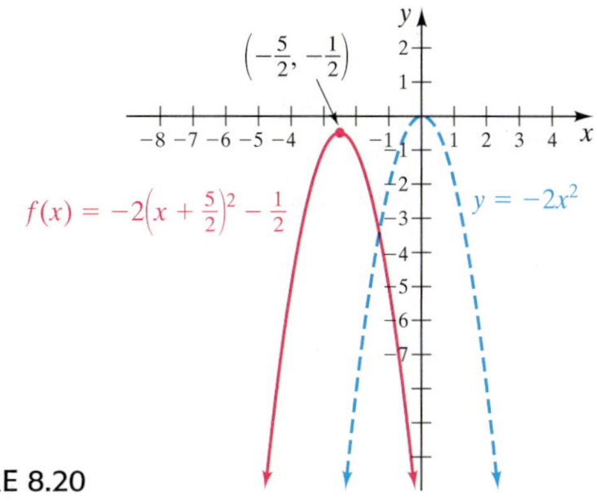


FIGURE 8.20

Notice that $f(x) = -2\left(x + \frac{5}{2}\right)^2 - \frac{1}{2}$ has no $x$-intercepts. Therefore, there are no real values of $x$ for which $f(x) = 0$.

▸ **Now Try Exercise 63**

A second way to change the equation from $f(x) = ax^2 + bx + c$ to $f(x) = a(x - h)^2 + k$ form is to let $h = -\frac{b}{2a}$ and $k = \frac{4ac - b^2}{4a}$. Find the values for $h$

and $k$ and then substitute the values obtained into $f(x) = a(x - h)^2 + k$. For example, for the function $f(x) = -2x^2 - 10x - 13$, in Example 8, $a = -2$, $b = -10$, and $c = -13$. Then

$$h = -\frac{b}{2a} = -\frac{-10}{2(-2)} = -\frac{5}{2}$$

$$k = \frac{4ac - b^2}{4a} = \frac{4(-2)(-13) - (-10)^2}{4(-2)} = -\frac{1}{2}$$

Therefore,

$$\begin{aligned} f(x) &= a(x - h)^2 + k \\ &= -2\left[x - \left(-\frac{5}{2}\right)\right]^2 - \frac{1}{2} \\ &= -2\left(x + \frac{5}{2}\right)^2 - \frac{1}{2} \end{aligned}$$

This answer checks with that obtained in Example 8.

## EXERCISE SET 8.5

MathXL® MyMathLab

### Concept/Writing Exercises

1. What is the graph of a quadratic equation called?
2. What is the vertex of a parabola?
3. What is the axis of symmetry of a parabola?
4. What is the equation of the axis of symmetry of the graph of $f(x) = ax^2 + bx + c$?
5. What is the vertex of the graph of $f(x) = ax^2 + bx + c$?
6. How many $x$-intercepts does a quadratic function have if the discriminant is **a)** $<0$, **b)** $=0$, **c)** $>0$?
7. For $f(x) = ax^2 + bx + c$, will $f(x)$ have a maximum or a minimum if **a)** $a > 0$, **b)** $a < 0$? Explain.
8. Explain how to find the $x$-intercepts of the graph of a quadratic function.
9. Explain how to find the $y$-intercept of the graph of a quadratic function.
10. Consider the graph of $f(x) = ax^2$. Explain how the shape of $f(x)$ changes as $|a|$ increases and as $|a|$ decreases.
11. Consider the graph of $f(x) = ax^2$. What is the general shape of $f(x)$ if **a)** $a > 0$, **b)** $a < 0$?
12. Will the graphs of $f(x) = ax^2$ and $g(x) = -ax^2$ have the same vertex for any nonzero real number $a$? Explain.
13. Does the function $f(x) = 3x^2 - 4x + 2$ have a maximum or a minimum value? Explain.
14. Does the function $g(x) = -\frac{1}{2}x^2 + 2x - 7$ have a maximum or a minimum value? Explain.

### Practice the Skills

**a)** *Determine whether the parabola opens upward or downward.* **b)** *Find the y-intercept.* **c)** *Find the vertex.* **d)** *Find the x-intercepts (if any).* **e)** *Draw the graph.*

15. $f(x) = x^2 + 8x + 15$
16. $g(x) = x^2 + 2x - 3$
17. $f(x) = x^2 - 4x + 3$
18. $h(x) = x^2 - 2x - 8$
19. $f(x) = -x^2 - 2x + 8$
20. $p(x) = -x^2 + 8x - 15$
21. $g(x) = -x^2 + 4x + 5$
22. $n(x) = -x^2 - 2x + 24$
23. $t(x) = -x^2 + 4x - 5$

**24.** $g(x) = x^2 + 6x + 13$

**25.** $f(x) = x^2 - 4x + 4$

**26.** $r(x) = -x^2 + 10x - 25$

**27.** $r(x) = x^2 + 2$

**28.** $f(x) = x^2 + 4x$

**29.** $l(x) = -x^2 + 5$

**30.** $g(x) = -x^2 + 6x$

**31.** $f(x) = -2x^2 + 4x - 8$

**32.** $g(x) = -2x^2 - 6x + 4$

**33.** $m(x) = 3x^2 + 4x + 3$

**34.** $p(x) = -2x^2 + 5x + 4$

**35.** $y = 3x^2 + 4x - 6$

**36.** $y = x^2 - 6x + 4$

**37.** $y = 2x^2 - x - 6$

**38.** $g(x) = -4x^2 + 6x - 9$

**39.** $f(x) = -x^2 + 3x - 5$

**40.** $h(x) = -2x^2 + 4x - 5$

*Using the graphs in* **Figures 8.13** *through* **8.16** *as a guide, graph each function and label the vertex.*

**41.** $f(x) = (x - 3)^2$

**42.** $f(x) = (x - 4)^2$

**43.** $f(x) = (x + 1)^2$

**44.** $f(x) = (x + 2)^2$

**45.** $f(x) = x^2 + 3$

**46.** $f(x) = x^2 + 5$

**47.** $f(x) = x^2 - 1$

**48.** $f(x) = x^2 - 4$

**49.** $f(x) = (x - 2)^2 + 3$

**50.** $f(x) = (x - 3)^2 - 4$

**51.** $f(x) = (x + 4)^2 + 4$

**52.** $h(x) = (x + 4)^2 - 1$

**53.** $g(x) = -(x + 3)^2 - 2$

**54.** $g(x) = (x - 1)^2 + 4$

**55.** $y = -2(x - 2)^2 + 2$

**56.** $y = -2(x - 3)^2 + 1$

**57.** $h(x) = -2(x + 1)^2 - 3$

**58.** $f(x) = -(x - 5)^2 + 2$

*In Exercises 59–68,* **a)** *express each function in the form* $f(x) = a(x - h)^2 + k$ *and* **b)** *draw the graph of each function and label the vertex.*

**59.** $f(x) = x^2 - 6x + 8$

**60.** $g(x) = x^2 + 6x + 2$

**61.** $g(x) = x^2 - x - 3$

**62.** $f(x) = x^2 - x + 1$

**63.** $f(x) = -x^2 - 4x - 6$

**64.** $h(x) = -x^2 + 6x + 1$

**65.** $g(x) = x^2 - 4x - 1$

**66.** $p(x) = x^2 - 2x - 6$

**67.** $f(x) = 2x^2 + 5x - 3$

**68.** $k(x) = 2x^2 + 7x - 4$

## Problem Solving

*Match the functions in Exercises 69–72 with the appropriate graphs labeled* **a)** *through* **d)**.

**a)**

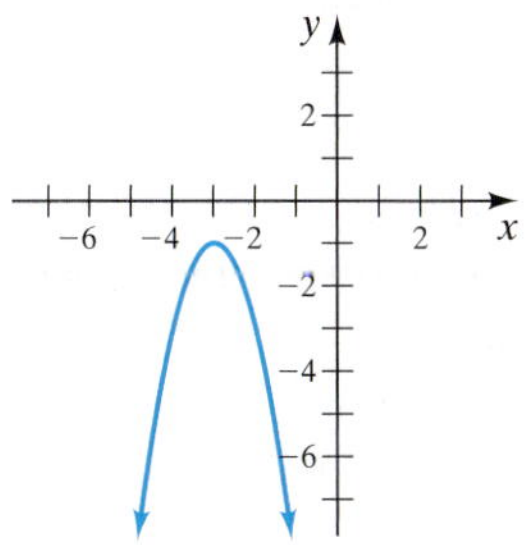


**b)**

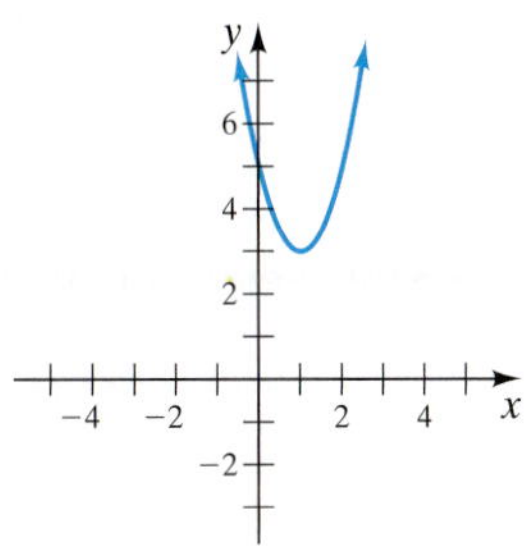


**c)**

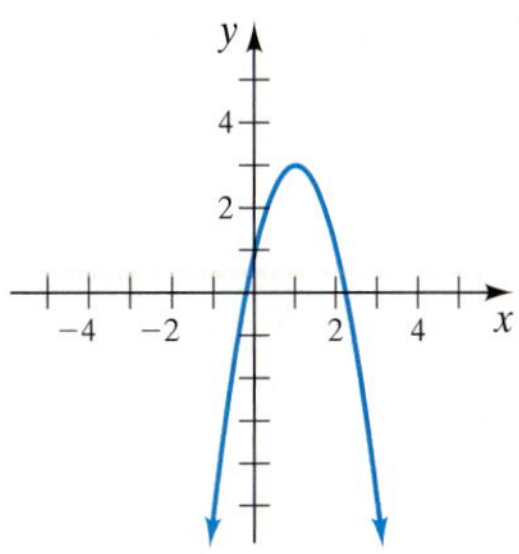


**d)**

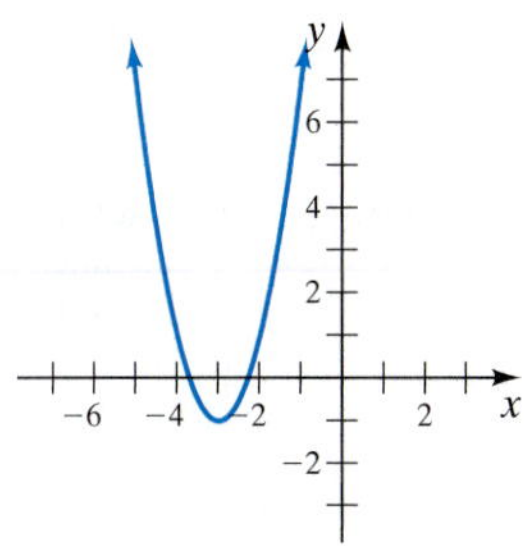


**69.** $f(x) = 2(x + 3)^2 - 1$

**70.** $f(x) = -2(x + 3)^2 - 1$

**71.** $f(x) = 2(x - 1)^2 + 3$

**72.** $f(x) = -2(x - 1)^2 + 3$

**Area** *For each rectangle,* **a)** *find the value for x that gives the maximum area and* **b)** *find the maximum area.*

**73.**

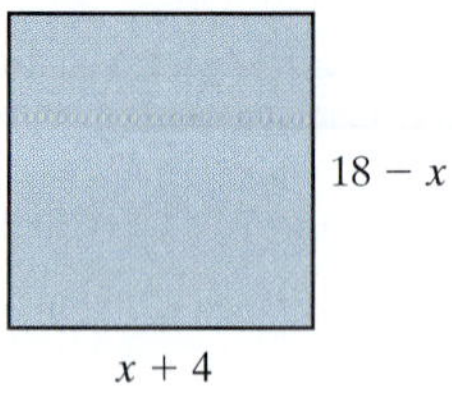


**74.**

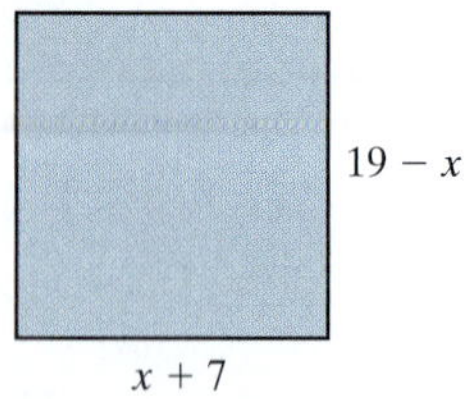


**75.**

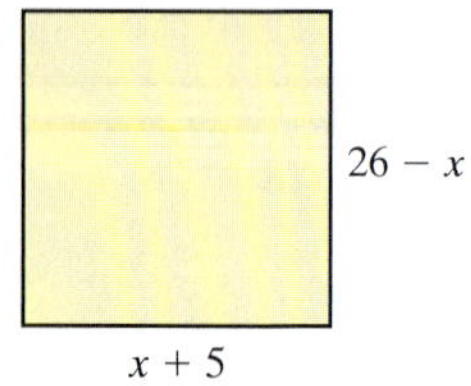


**76.**

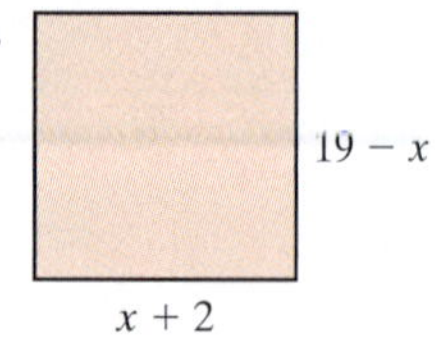


**77. Selling Batteries** The revenue function for selling $n$ batteries is $R(n) = n(8 - 0.02n) = -0.02n^2 + 8n$. Find **a)** the number of batteries that must be sold to obtain the maximum revenue and **b)** the maximum revenue.

**78. Selling Watches** The revenue function for selling $n$ watches is $R(n) = n(25 - 0.1n) = -0.1n^2 + 25n$. Find **a)** the number of watches that must be sold to obtain the maximum revenue and **b)** the maximum revenue.

**79. Enrollment** The enrollment in a high school in the Maplewood School District can be approximated by the function

$$N(t) = -0.043t^2 + 1.82t + 46.0$$

where $t$ is the number of years since 1989 and $1 \le t \le 22$. In what year will the maximum enrollment be obtained?

**80. Drug-Free Schools** The percent of students in schools in the United States who say their school is not drug free can be approximated by the function

$$f(a) = -2.32a^2 + 76.58a - 559.87$$

where $a$ is the student's age and $12 < a < 20$. What age has the highest percent of students who say their school is not drug free?

**81.** What is the distance between the vertices of the graphs of $f(x) = (x - 2)^2 + \frac{5}{2}$ and $g(x) = (x - 2)^2 - \frac{3}{2}$?

**82.** What is the distance between the vertices of the graphs of $f(x) = 2(x - 4)^2 - 3$ and $g(x) = -3(x - 4)^2 + 2$?

**83.** What is the distance between the vertices of the graphs of $f(x) = 2(x + 4)^2 - 3$ and $g(x) = -(x + 1)^2 - 3$?

**84.** What is the distance between the vertices of the graphs of $f(x) = -\frac{1}{3}(x - 3)^2 - 2$ and $g(x) = 2(x + 5)^2 - 2$?

**85.** Write the function whose graph has the shape of the graph of $f(x) = 2x^2$ and has a vertex at $(3, -2)$.

**86.** Write the function whose graph has the shape of the graph of $f(x) = -\frac{1}{2}x^2$ and has a vertex at $\left(\frac{2}{3}, -5\right)$.

**87.** Write the function whose graph has the shape of the graph of $f(x) = -4x^2$ and has a vertex at $\left(-\frac{3}{5}, -\sqrt{2}\right)$.

**88.** Write the function whose graph has the shape of the graph of $f(x) = \frac{3}{5}x^2$ and has a vertex at $(-\sqrt{3}, \sqrt{5})$.

**89.** Consider $f(x) = x^2 - 8x + 12$ and $g(x) = -x^2 + 8x - 12$.

**a)** Without graphing, can you explain how the graphs of the two functions compare?

**b)** Will the graphs have the same $x$-intercepts? Explain.

**c)** Will the graphs have the same vertex? Explain.

**d)** Graph both functions on the same axes.

**90.** By observing the leading coefficient in a quadratic function and by determining the coordinates of the vertex of its graph, explain how you can determine the number of $x$-intercepts the parabola has.

**91. Selling Tickets** The Johnson High School Theater Club is trying to set the price of tickets for a play. If the price is too low, they will not make enough money to cover expenses, and if the price is too high, not enough people will pay the price of a ticket. They estimate that their total income per concert, $I$, in hundreds of dollars, can be approximated by the formula

$$I = -x^2 + 24x - 44, 0 \le x \le 24$$

where $x$ is the cost of a ticket.

**a)** Draw a graph of income versus the cost of a ticket.

**b)** Determine the minimum cost of a ticket for the theater club to break even.

**c)** Determine the maximum cost of a ticket that the theater club can charge and break even.

**d)** How much should they charge to receive the maximum income?

**e)** Find the maximum income.

**92. Throwing an Object** An object is projected upward with an initial velocity of 192 feet per second. The object's distance above the ground, $d$, after $t$ seconds may be found by the formula $d = -16t^2 + 192t$.

**a)** Find the object's distance above the ground after 3 seconds.

**b)** Draw a graph of distance versus time.

**c)** What is the maximum height the object will reach?

**d)** At what time will it reach its maximum height?

**e)** At what time will the object strike the ground?

**93. Profit** The Fulton Bird House Company earns a weekly profit according to the function $f(x) = -0.4x^2 + 80x - 200$, where $x$ is the number of bird feeders built and sold.

**a)** Find the number of bird feeders that the company must sell in a week to obtain the maximum profit.

**b)** Find the maximum profit.

**94. Profit** The A. B. Bronson Company earns a weekly profit according to the function $f(x) = -1.2x^2 + 180x - 280$, where $x$ is the number of rocking chairs built and sold.

**a)** Find the number of rocking chairs that the company must sell in a week to obtain the maximum profit.

**b)** Find the maximum profit.

**95. Firing a Cannon** If a certain cannon is fired from a height of 9.8 meters above the ground, at a certain angle, the height of the cannonball above the ground, $h$, in meters, at time, $t$, in seconds, is found by the function

$$h(t) = -4.9t^2 + 24.5t + 9.8$$

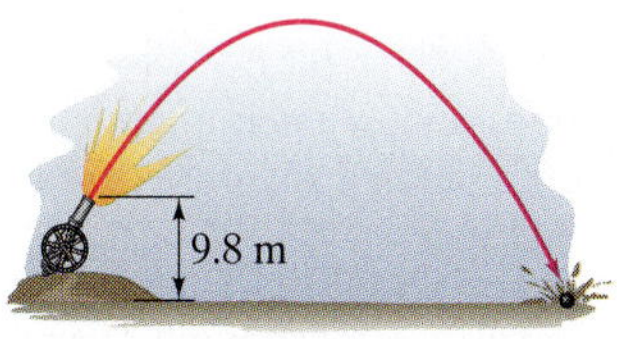


**a)** Find the maximum height attained by the cannonball.

**b)** Find the time it takes for the cannonball to reach its maximum height.

**c)** Find the time it takes for the cannonball to strike the ground.

**96. Throwing a Ball** Ramon Loomis throws a ball into the air with an initial velocity of 32 feet per second. The height of the ball at any time $t$ is given by the formula $h = 96t - 16t^2$. At what time does the ball reach its maximum height? What is the maximum height?

**97. Rent** The following graph shows the average monthly rent for an apartment in Maricopa County, Arizona (apartment complexes with 50 or more units), for the years 1994 to 2003.

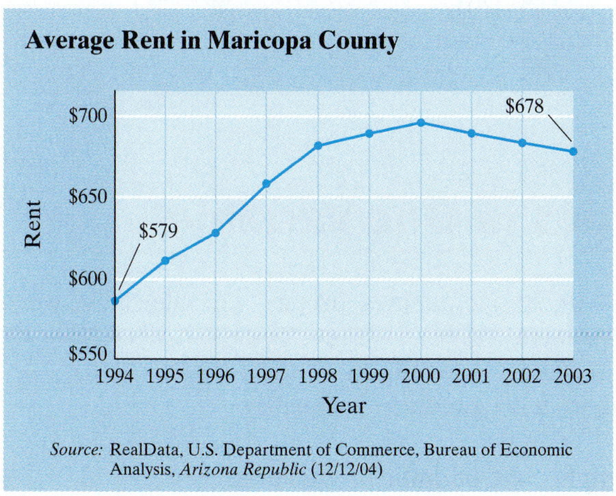


*Source:* RealData, U.S. Department of Commerce, Bureau of Economic Analysis, *Arizona Republic* (12/12/04)

The function $r(t) = -2.723t^2 + 35.273t + 579$ can be used to estimate the average monthly rent for an apartment in Maricopa County, where $t$ is the number of years since 1994.

**a)** If use assume the trend continues, estimate the average monthly rent for an apartment in Maricopa County in 2007.

**b)** In what year was the average monthly rent for an apartment a maximum?

**98. Canadian Dollar** The following graph shows the value of a Canadian dollar in U.S. dollars on February 22 of each year for the years from 2000 to 2005.

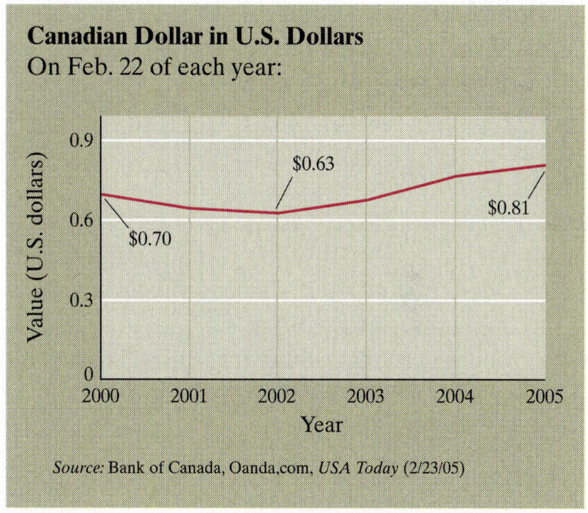


*Source:* Bank of Canada, Oanda,com, *USA Today* (2/23/05)

The function $C(t) = 0.019t^2 - 0.074t + 0.702$ can be used to estimate the value of a Canadian dollar in U.S. dollars on February 22 of each year, where $t$ is the number of years since 2000.

**a)** If we assume the trend continues, estimate the value of a Canadian dollar, in U.S. dollars, on February 22 in 2008.

**b)** In what year was the value of the Canadian dollar, in U.S. dollars, on February 22 at a minimum?

**99. Room in a House** Jake Kishner is designing plans for his house. What is the maximum possible area of a room if its perimeter is to be 80 feet?

**100. Greatest Area** What are the dimensions of a garden that will have the greatest area if its perimeter is to be 70 feet?

**101. Minimum Product** What is the minimum product of two numbers that differ by 8? What are the numbers?

**102. Minimum Product** What is the minimum product of two numbers that differ by 10? What are the numbers?

**103. Maximum Product** What is the maximum product of two numbers that add to 60? What are the numbers?

**104. Maximum Product** What is the maximum product of two numbers that add to 5? What are the numbers?

*The profit of a company, in dollars, is the difference between the company's revenue and cost. Exercises 105 and 106 give cost, $C(x)$, and revenue, $R(x)$, functions for a particular company. The x represents the number of items produced and sold to distributors. Determine* **a)** *the maximum profit of the company and* **b)** *the number of items that must be produced and sold to obtain the maximum profit.*

**105.** $C(x) = 2000 + 40x$
$R(x) = 800x - x^2$

**106.** $C(x) = 5000 + 12x$
$R(x) = 2000x - x^2$

## Challenge Problems

**107. Baseball** In Example 3 of this section, we used the function $f(t) = -16t^2 + 52t + 3$ to find that the maximum height, $f$, attained by a baseball hit by Tommy Magee was 45.25 feet. The ball reached this height at 1.625 seconds after the baseball was hit.
Review Example 3 now.

**a)** Write $f(t)$ in the form $f(t) = a(t - h)^2 + k$ by completing the square.

**b)** Using the function you obtained in part **a)**, determine the maximum height attained by the baseball, and the time after it was hit that the baseball attained its maximum value.

**c)** Are the answers you obtained in part **b)** the same answers obtained in Example 3? If not, explain why not.

## Group Activity

*Discuss and answer Exercise 108 as a group.*

**108. a)** Group member 1: Write two quadratic functions $f(x)$ and $g(x)$ so that the functions will not intersect.

**b)** Group member 2: Write two quadratic functions $f(x)$ and $g(x)$ so that neither function will have $x$-intercepts, and the vertices of the functions are on opposite sides of the $x$-axis.

**c)** Group member 3: Write two quadratic functions $f(x)$ and $g(x)$ so that both functions have the same vertex but one function opens upward and the other opens downward.

**d)** As a group, review each answer in parts **a)–c)** and decide whether each answer is correct. Correct any answer that is incorrect.

## Cumulative Review Exercises

[2.2] **109.** Find the area shaded blue in the figure.

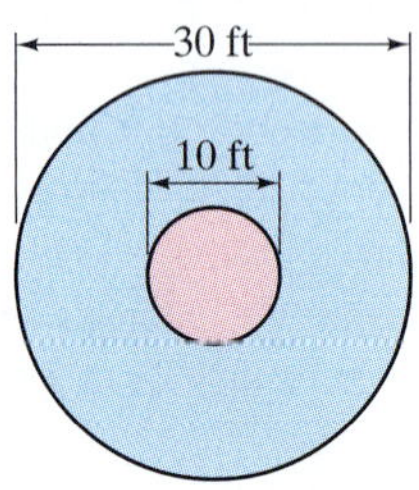


[3.7] **110.** Graph $y \leq \frac{2}{3}x + 3$.

[4.2] **111.** Solve the system of equations.

$$\begin{aligned} x - y &= -5 \\ 2x + 2y - z &= 0 \\ x + y + z &= 3 \end{aligned}$$

[4.5] **112.** Evaluate the determinant.

$$\begin{vmatrix} \frac{1}{2} & 3 \\ 2 & -4 \end{vmatrix}$$

[6.1] **113.** Divide $(x - 3) \div \dfrac{x^2 + 3x - 18}{x}$.

# 8.6 Quadratic and Other Inequalities in One Variable

**1** Solve quadratic inequalities.

**2** Solve other polynomial inequalities.

**3** Solve rational inequalities.

In Section 2.5, we discussed linear inequalities in one variable. Now we discuss quadratic inequalities in one variable.

When the equal sign in a quadratic equation of the form $ax^2 + bx + c = 0$ is replaced by an inequality sign, we get a **quadratic inequality**.

### Examples of Quadratic Inequalities

$$x^2 + x - 12 > 0, \qquad 2x^2 - 9x - 5 \leq 0$$

The **solution to a quadratic inequality** is the set of all values that make the inequality a true statement. For example, if we substitute 5 for $x$ in $x^2 + x - 12 > 0$, we obtain

$$x^2 + x - 12 > 0$$
$$5^2 + 5 - 12 \overset{?}{>} 0$$
$$18 > 0 \quad \textit{True}$$

The inequality is true when $x$ is 5, so 5 satisfies the inequality. However, 5 is not the only solution. There are other values that satisfy (or are solutions to) the inequality. Does 4 satisfy the inequality? Does 2 satisfy the inequality?

## 1 Solve Quadratic Inequalities

A number of methods can be used to find the solutions to quadratic inequalities. We will begin by introducing a **sign graph**. Consider the function $f(x) = x^2 + x - 12$. Its graph is shown in **Figure 8.21a**. **Figure 8.21b** shows, in red, that when $x < -4$ or $x > 3$, $f(x) > 0$ or $x^2 + x - 12 > 0$. It also shows, in green, that when $-4 < x < 3$, $f(x) < 0$ or $x^2 + x - 12 < 0$.

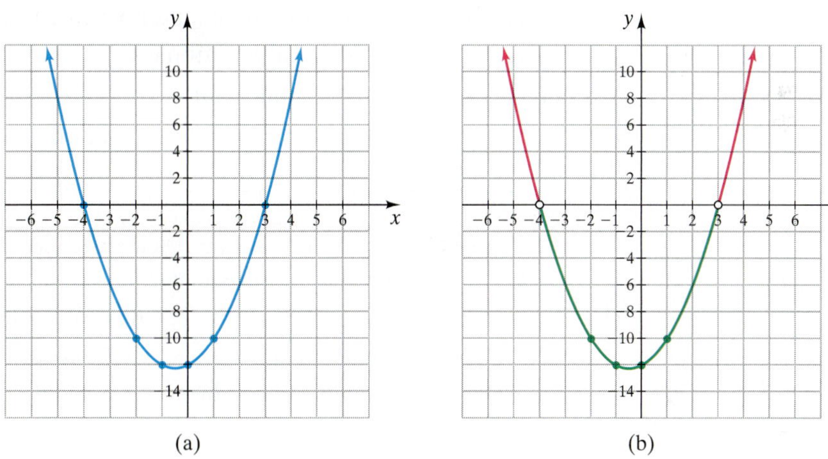

FIGURE 8.21

One way of finding the solutions to inequalities is to draw the graph and determine from the graph which values of the variable satisfy the inequality, as we just did. In many cases, it may be inconvenient or take too much time to draw the graph of a function, so we provide an alternate method to solve quadratic and other inequalities.

In Example 1, we will show how $x^2 + x - 12 > 0$ is solved using a number line. Then we will outline the procedure we used.

**EXAMPLE 1** ▸ Solve the inequality $x^2 + x - 12 > 0$. Give the solution **a)** on a number line, **b)** in interval notation, and **c)** in set builder notation.

*Solution* Set the inequality equal to 0 and solve the equation.

$$x^2 + x - 12 = 0$$
$$(x + 4)(x - 3) = 0$$
$$x + 4 = 0 \quad \text{or} \quad x - 3 = 0$$
$$x = -4 \qquad\qquad x = 3$$

The numbers obtained are called **boundary values**. The boundary values are used to break a number line up into intervals. If the original inequality is $<$ or $>$, the boundary values are not part of the intervals. If the original inequality is $\le$ or $\ge$, the boundary values are part of the intervals.

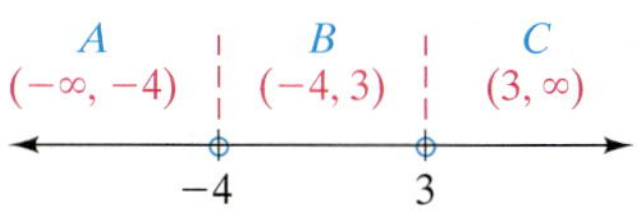


FIGURE 8.22

In **Figure 8.22**, we have labeled the intervals $A$, $B$, and $C$. Next, we select one test value in *each* interval. Then we substitute each of those numbers, one at a time, into either $x^2 + x - 12 > 0$ or $(x + 4)(x - 3) > 0$ and determine whether they result in a

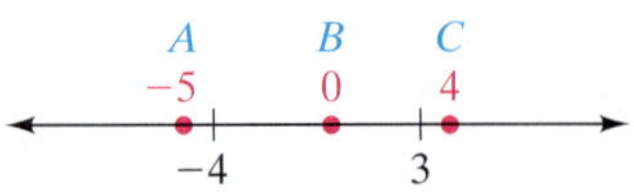


FIGURE 8.23

true statement. If the test value results in a true statement, all values in that interval will also satisfy the inequality. If the test value results in a false statement, no numbers in that interval will satisfy the inequality.

In this example, we will use the test values of −5 in interval $A$, 0 in interval $B$, and 4 in interval $C$ (see **Fig. 8.23**).

| Interval $A$ | Interval $B$ | Interval $C$ |
|---|---|---|
| $(-\infty, -4)$ | $(-4, 3)$ | $(3, \infty)$ |
| Test value, −5 | Test value, 0 | Test value, 4 |
| Is $x^2 + x - 12 > 0$? | Is $x^2 + x - 12 > 0$? | Is $x^2 + x - 12 > 0$? |
| $(-5)^2 - 5 - 12 \overset{?}{>} 0$ | $0^2 + 0 - 12 \overset{?}{>} 0$ | $4^2 + 4 - 12 \overset{?}{>} 0$ |
| $8 > 0$ | $-12 > 0$ | $8 > 0$ |
| *True* | *False* | *True* |

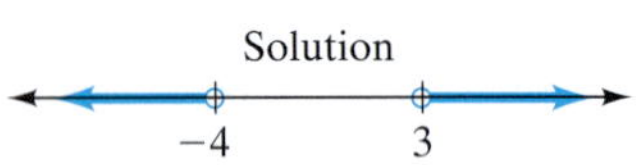


FIGURE 8.24

Since the test values in both intervals $A$ and $C$ satisfy the inequality, the solution is all real numbers in intervals $A$ or $C$. Since the inequality symbol is $>$, the values −4 and 3 are not included in the solution because they make the inequality equal to 0.

The answer to parts **a), b),** and **c)** follow.

**a)** The solution is illustrated on a number line in **Figure 8.24**.

**b)** The solution in interval notation is $(-\infty, -4) \cup (3, \infty)$.

**c)** The solution in set builder notation is $\{x | x < -4 \text{ or } x > 3\}$.

Note that the solution, in any form, is consistent with the graph in **Figure 8.21b**.

▸ **Now Try Exercise 15**

### To Solve Quadratic and Other Inequalities

1. Write the inequality as an equation and solve the equation.
2. If solving a rational inequality, determine the values that make any denominator 0.
3. Construct a number line. Mark each solution from step 1 and numbers obtained in step 2 on the number line. Mark the lowest value on the left, with values increasing from left to right.
4. Select a test value in each interval and determine whether it satisfies the inequality. Also test each boundary value.
5. Write the solution in the form requested by your instructor.

**EXAMPLE 2** ▸ Solve the inequality $x^2 - 4x \geq -4$. Give the solution **a)** on a number line, **b)** in interval notation, and **c)** in set builder notation.

*Solution* Write the inequality as an equation, then solve the equation.

$$x^2 - 4x = -4$$
$$x^2 - 4x + 4 = 0$$
$$(x - 2)(x - 2) = 0$$
$$x - 2 = 0 \quad \text{or} \quad x - 2 = 0$$
$$x = 2 \qquad\qquad x = 2$$

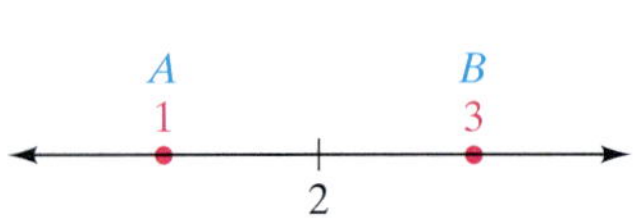


FIGURE 8.25

Since both factors are the same there is only one boundary value, 2 (see **Fig. 8.25**). Both test values, 1 and 3, result in true statements.

| Interval $A$ | Interval $B$ |
|---|---|
| $(-\infty, 2)$ | $(2, \infty)$ |
| Test value, 1 | Test value, 3 |
| $x^2 - 4x \geq -4$ | $x^2 - 4x \geq -4$ |
| $1^2 - 4(1) \overset{?}{\geq} -4$ | $3^2 - 4(3) \overset{?}{\geq} -4$ |
| $1 - 4 \overset{?}{\geq} -4$ | $9 - 12 \overset{?}{\geq} -4$ |
| $-3 \geq -4$ *True* | $-3 \geq -4$ *True* |

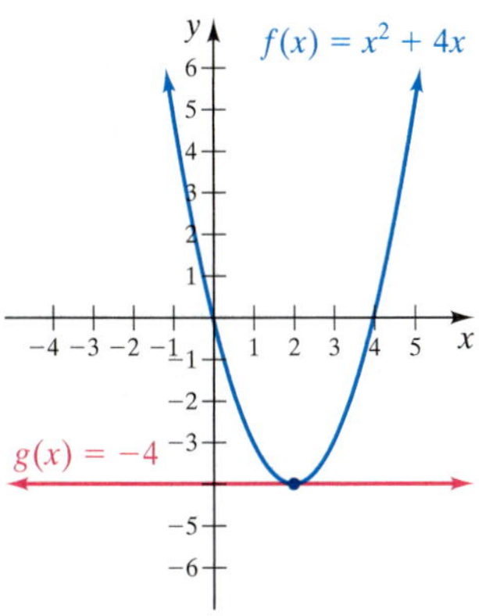

FIGURE 8.26

The solution set includes both intervals and the boundary value, 2. The solution set is the set of real numbers, $\mathbb{R}$. The answer to parts **a)**, **b)**, and **c)** follow.

**a)** 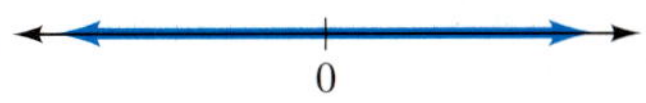

**b)** $(-\infty, \infty)$ **c)** $\{x|-\infty < x < \infty\}$

▸ **Now Try Exercise 11**

We can check the solution to Example 2 using graphing. Let $f(x) = x^2 - 4x$ and $g(x) = -4$. For $x^2 - 4x \geq -4$ to be true, we want $f(x) \geq g(x)$. The graphs of $f(x)$ and $g(x)$ are given in **Figure 8.26**.

Observe that $f(x) = g(x)$ at $x = 2$ and $f(x) > g(x)$ for all other values of $x$. Thus, $f(x) \geq g(x)$ for all values of $x$, and the solution set is the set of real numbers.

In Example 2, if we rewrite the inequality $x^2 - 4x \geq -4$ as $x^2 - 4x + 4 \geq 0$ and then as $(x - 2)^2 \geq 0$ we can see that the solution must be the set of real numbers, since $(x - 2)^2$ must be greater than or equal to 0 for any real number $x$. The solution to $x^2 - 4x < -4$ is the empty set, $\varnothing$. Can, you explain why?

**EXAMPLE 3** ▸ Solve the inequality $x^2 - 2x - 4 \leq 0$. Express the solution in interval notation.

*Solution* First we need to solve the equation $x^2 - 2x - 4 = 0$. Since this equation is not factorable, we use the quadratic formula to solve.

$$x = \frac{-b \pm \sqrt{b^2 - 4ac}}{2a}$$

$$= \frac{2 \pm \sqrt{4 - 4(1)(-4)}}{2(1)} = \frac{2 \pm \sqrt{20}}{2} = \frac{2 \pm 2\sqrt{5}}{2} = 1 \pm \sqrt{5}$$

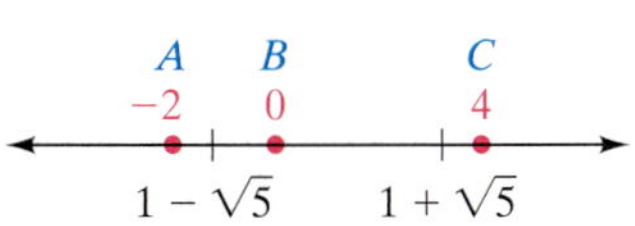

FIGURE 8.27

The boundary values are $1 - \sqrt{5}$ and $1 + \sqrt{5}$. The value of $1 - \sqrt{5}$ is about $-1.24$ and the value of $1 + \sqrt{5}$ is about 3.24. We will select test values of $-2$, 0, and 4 (see **Fig. 8.27**).

| Interval *A* | Interval *B* | Interval *C* |
|---|---|---|
| $(-\infty, 1 - \sqrt{5})$ | $(1 - \sqrt{5}, 1 + \sqrt{5})$ | $(1 + \sqrt{5}, \infty)$ |
| Test value, −2 | Test value, 0 | Test value, 4 |
| $x^2 \quad 2x \quad 4 \leq 0$ | $x^2 \quad 2x \quad 4 \leq 0$ | $x^2 \quad 2x \quad 4 \leq 0$ |
| $(-2)^2 - 2(-2) - 4 \overset{?}{\leq} 0$ | $0^2 - 2(0) - 4 \overset{?}{\leq} 0$ | $4^2 - 2(4) - 4 \overset{?}{\leq} 0$ |
| $4 + 4 - 4 \overset{?}{\leq} 0$ | $0 - 0 - 4 \overset{?}{\leq} 0$ | $16 - 8 - 4 \overset{?}{\leq} 0$ |
| $4 \leq 0$ | $-4 \leq 0$ | $4 \leq 0$ |
| *False* | *True* | *False* |

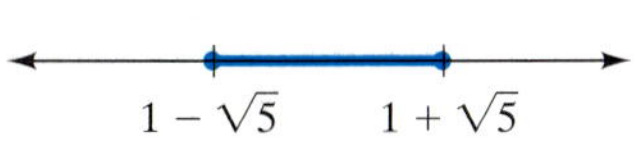

FIGURE 8.28

The boundary values are part of the solution because the inequality symbol is $\leq$ and the boundary values make the inequality equal to 0. Thus, the solution in interval notation is $[1 - \sqrt{5}, 1 + \sqrt{5}]$. The solution is illustrated on the number line in **Figure 8.28**.

▸ **Now Try Exercise 19**

## Helpful Hint

If $ax^2 + bx + c = 0$, with $a > 0$, has two distinct real solutions, then:

| Inequality of Form | Solution Is | Solution on Number Line |
|---|---|---|
| $ax^2 + bx + c \geq 0$ | End intervals |  |
| $ax^2 + bx + c \leq 0$ | Center interval | 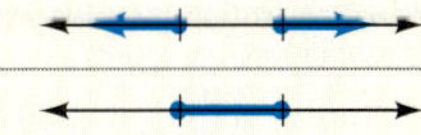 |

Example 1 is an inequality of the form $ax^2 + bx + c > 0$, and Example 3 is an inequality of the form $ax^2 + bx + c \leq 0$. Example 2 does not have two distinct real solutions, so this Helpful Hint does not apply.

## 2 Solve Other Polynomial Inequalities

A procedure similar to the one used earlier can be used to solve other **polynomial inequalities**, as illustrated in the following examples.

**EXAMPLE 4** ▸ Solve the inequality $(3x - 2)(x + 3)(x + 5) < 0$. Illustrate the solution on a number line and write the solution in both interval notation and set builder notation.

*Solution* We use the zero-factor property to solve the equation $(3x - 2)(x + 3)(x + 5) = 0$.

$$3x - 2 = 0 \quad \text{or} \quad x + 3 = 0 \quad \text{or} \quad x + 5 = 0$$
$$x = \frac{2}{3} \qquad\qquad x = -3 \qquad\qquad x = -5$$

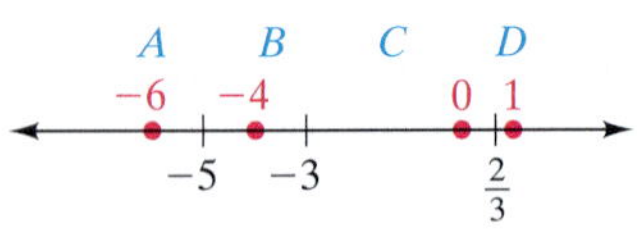

FIGURE 8.29

The solutions $-5$, $-3$, and $\frac{2}{3}$ break the number line into four intervals (see **Fig. 8.29**). The test values we will use are $-6$, $-4$, $0$, and $1$. We show the results in the following table.

| Interval | Test value | $(3x - 2)(x + 3)(x + 5)$ | $< 0$ |
|---|---|---|---|
| $A: (-\infty, -5)$ | $-6$ | $-60$ | True |
| $B: (-5, -3)$ | $-4$ | $14$ | False |
| $C: \left(-3, \frac{2}{3}\right)$ | $0$ | $-30$ | True |
| $D: \left(\frac{2}{3}, \infty\right)$ | $1$ | $24$ | False |

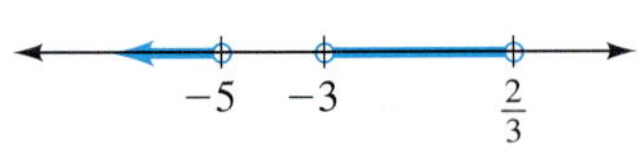

FIGURE 8.30

Since the original inequality symbol is $<$, the boundary values are not part of the solution. The solution, intervals $A$ and $C$, is illustrated on the number line in **Figure 8.30**. The solution in set builder notation is $\left\{x \,\middle|\, x < -5 \text{ or } -3 < x < \frac{2}{3}\right\}$.

The solution in interval notation is $(-\infty, -5) \cup \left(-3, \frac{2}{3}\right)$.

▸ **Now Try Exercise 27**

**EXAMPLE 5** ▸ Given $f(x) = 3x^3 - 3x^2 - 6x$, find all values of $x$ for which $f(x) \geq 0$. Illustrate the solution on a number line and give the solution in interval notation.

*Solution* We need to solve the inequality

$$3x^3 - 3x^2 - 6x \geq 0$$

We start by solving the equation $3x^3 - 3x^2 - 6x = 0$.

$$3x(x^2 - x - 2) = 0$$
$$3x(x - 2)(x + 1) = 0$$

$$3x = 0 \quad \text{or} \quad x - 2 = 0 \quad \text{or} \quad x + 1 = 0$$
$$x = 0 \qquad\qquad x = 2 \qquad\qquad x = -1$$

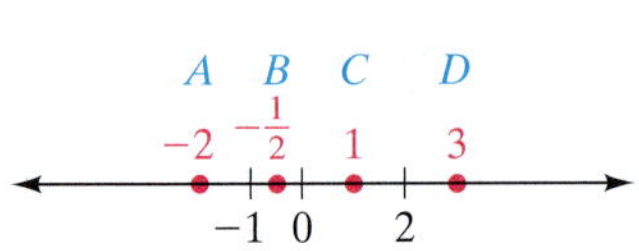

FIGURE 8.31

The solutions $-1$, $0$, and $2$ break the number line into four intervals (see **Fig. 8.31**). The test values that we will use are $-2$, $-\frac{1}{2}$, $1$, and $3$.

| Interval | Test Value | $3x^3 - 3x^2 - 6x$ | $\geq 0$ |
|---|---|---|---|
| $A: (-\infty, 1)$ | $-2$ | $-24$ | *False* |
| $B: (-1, 0)$ | $-\frac{1}{2}$ | $\frac{15}{8}$ | *True* |
| $C: (0, 2)$ | $1$ | $-6$ | *False* |
| $D: (2, \infty)$ | $3$ | $36$ | *True* |

Since the original inequality is $\geq$, the boundary values are part of the solution. The solution, intervals $B$ and $D$, is illustrated on the number line in **Figure 8.32a**. The solution in interval notation is $[-1, 0] \cup [2, \infty)$. **Figure 8.32b** shows the graph of $f(x) = 3x^3 - 3x^2 - 6x$. Notice $f(x) \geq 0$ for $-1 \leq x \leq 0$ and for $x \geq 2$, which agrees with our solution.

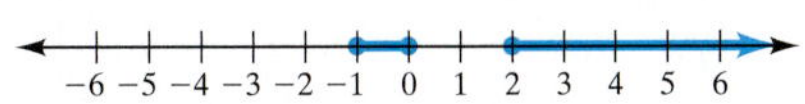


FIGURE 8.32a

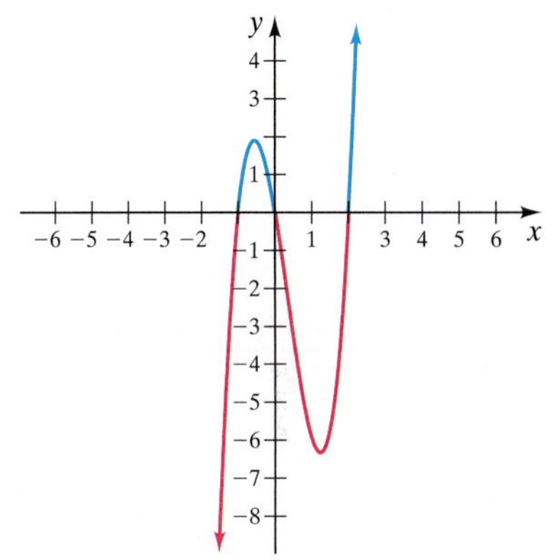


FIGURE 8.32b

▸ **Now Try Exercise 41**

In all the examples we have solved, the coefficient of the leading term has been a positive number.

Consider the inequality $-3x^3 + 3x^2 + 6x \leq 0$. Notice the coefficient of the leading term, $-3x^3$, is a negative number, $-3$. It is generally easier to solve an inequality where the coefficient of the leading term is a positive number. We can make the leading coefficient positive by multiplying both sides of the inequality by $-1$. When doing this, remember to reverse the inequality symbol.

$$-3x^3 + 3x^2 + 6x \leq 0$$
$$-1(-3x^3 + 3x^2 + 6x) \geq -1(0) \qquad \textit{Reverse inequality symbol.}$$
$$3x^3 - 3x^2 - 6x \geq 0$$

This inequality was solved in Example 5.

## 3 Solve Rational Inequalities

In Examples 6 and 7, we solve **rational inequalities**, which are inequalities that contain a rational expression.

**EXAMPLE 6** ▸ Solve the inequality $\frac{x-1}{x+3} \geq 2$ and graph the solution on a number line.

*Solution* Change the $\geq$ to $=$ and solve the resulting equation.

$$\frac{x-1}{x+3} = 2$$
$$(x+3) \cdot \frac{x-1}{x+3} = 2(x+3) \qquad \textit{Multiply both sides by } x + 3.$$
$$x - 1 = 2x + 6$$
$$-1 = x + 6$$
$$-7 = x$$

When solving rational inequalities, we also need to determine the value or values that make the denominator 0. We set the denominator equal to 0 and solve.

$$x + 3 = 0$$
$$x = -3$$

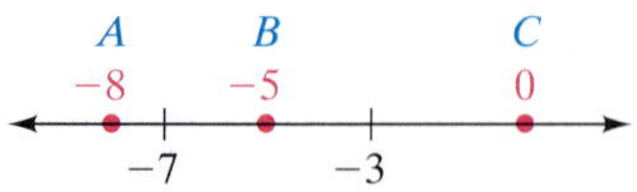


FIGURE 8.33

We use the solution to the equation, −7, and the value that makes the denominator 0, −3, to determine the intervals, shown in **Figure 8.33.** We will use −8, −5, and 0 as our test values.

| Interval $A$ | Interval $B$ | Interval $C$ |
|---|---|---|
| $(-\infty, -7)$ | $(-7, -3)$ | $(-3, \infty)$ |
| Test value, −8 | Test value, −5 | Test value, 0 |
| $\frac{x-1}{x+3} \geq 2$ | $\frac{x-1}{x+3} \geq 2$ | $\frac{x-1}{x+3} \geq 2$ |
| $\frac{-8-1}{-8+3} \overset{?}{\geq} 2$ | $\frac{-5-1}{-5+3} \overset{?}{\geq} 2$ | $\frac{0-1}{0+3} \overset{?}{\geq} 2$ |
| $\frac{9}{5} \geq 2$ *False* | $3 \geq 2$ *True* | $-\frac{1}{3} \geq 2$ *False* |

FIGURE 8.34

Only interval $B$ satisfies the inequality. Whenever we have a rational inequality we must be very careful to determine which boundary values are included in the solution. Remember we can never include in our solution any value that makes the denominator 0. Now check the boundary values −7 and −3. Since −7 results in the inequality $-2 \geq -2$, which is true, −7 is a solution. Since division by 0 is not permitted, −3 is not a solution. Thus, the solution is $[-7, -3)$. The solution is illustrated on the number line in **Figure 8.34**.

▸ **Now Try Exercise 81**

In Example 6, we solved $\frac{x-1}{x+3} \geq 2$. Suppose we graphed $f(x) = \frac{x-1}{x+3}$. For what values of $x$ would $f(x) \geq 2$? If you answered $-7 \leq x < -3$ you answered correctly. **Figure 8.35** shows the graph of $f(x) = \frac{x-1}{x+3}$ and the graph of $y = 2$. Notice $f(x) \geq 2$ when $-7 \leq x < -3$.

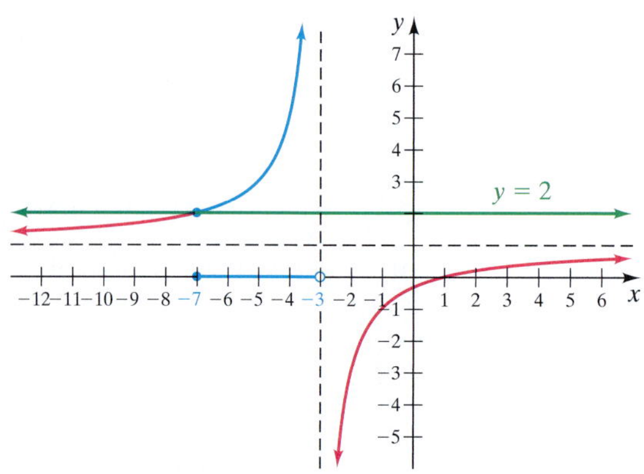


FIGURE 8.35

**EXAMPLE 7** ▸ Solve the inequality $\frac{(x-3)(x+4)}{x+1} \geq 0$. Graph the solution on a number line and give the solution in interval notation.

*Solution* The solutions to the equation $\frac{(x-3)(x+4)}{x+1} = 0$ are 3 and −4 since these are the values that make the numerator equal to 0. The equation is not defined at −1.

We therefore use the values 3, $-4$, and $-1$ to determine the intervals on the number line (see **Fig. 8.36**). Checking test values at $-5$, $-2$, 0, and 4, we find that the values in intervals $B$ and $D$, $-4 < x < -1$ and $x > 3$, satisfy the inequality. Check the test values yourself to verify this. The values 3 and $-4$ make the inequality equal to 0 and are part of the solution. The inequality is not defined at $-1$, so $-1$ is not part of the solution. The solution is $[-4, -1) \cup [3, \infty)$. The solution is illustrated on the number line in **Figure 8.37**.

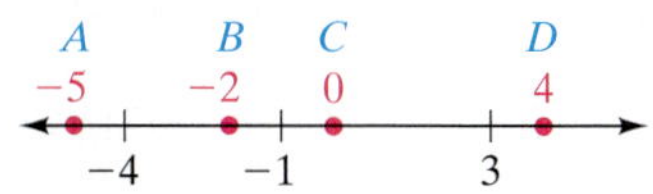

FIGURE 8.36

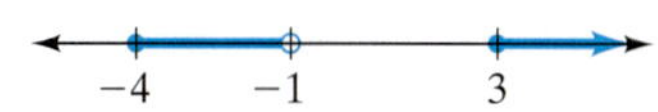

FIGURE 8.37

▸ **Now Try Exercise 71**

## EXERCISE SET 8.6

MathXL® MyMathLab

### Concept/Writing Exercises

**1.** The graph of $f(x) = x^2 - 7x + 10$ is given. Find the solution to **a)** $f(x) > 0$ and **b)** $f(x) < 0$.

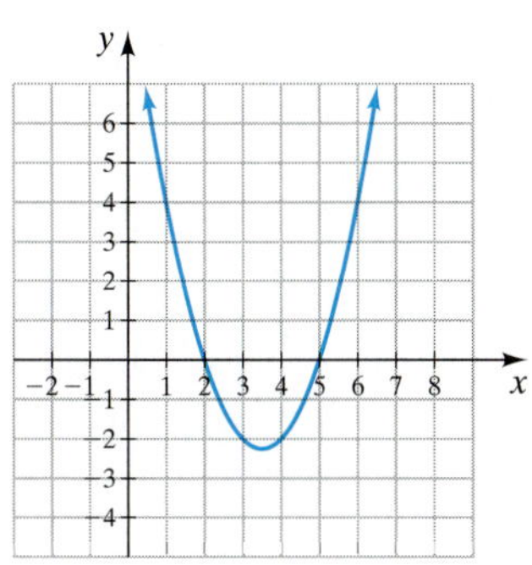

**2.** The graph of $f(x) = -x^2 - 4x + 5$ is given. Find the solution to **a)** $f(x) \geq 0$ and **b)** $f(x) \leq 0$.

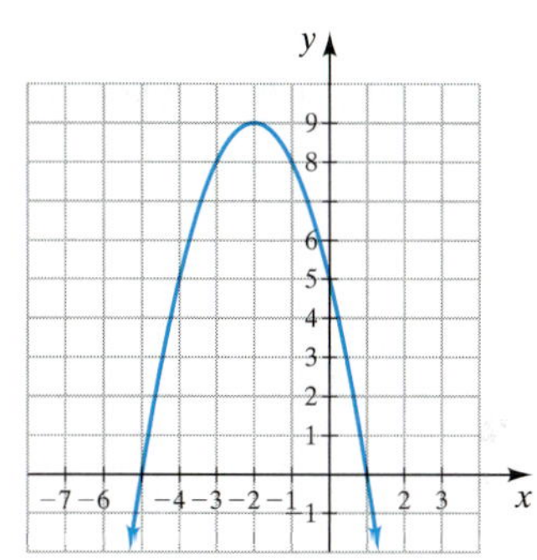

**3.** In solving the inequality $(x - 5)(x + 3) \geq 0$, are the boundary values 5 and $-3$ included in the solution set? Explain.

**4.** In solving the inequality $(x - 2)(x + 4) < 0$, are the boundary values 2 and $-4$ included in the solution set? Explain.

**5.** In solving the inequality $\frac{(x+2)(x-1)}{x+1} \leq 0$, are the boundary values $-2$ and 1 included in the solution set? Is the boundary value $-1$ included in the solution set? Explain.

**6.** In solving the inequality $\frac{(x+3)}{(x+4)(x-2)} \geq 0$, is the boundary value $-3$ included in the solution set? Are the boundary values $-4$ and 2 included in the solution set? Explain.

### Practice the Skills

*Solve each inequality and graph the solution on a number line.*

**7.** $x^2 - 2x - 8 \geq 0$

**8.** $x^2 - 2x - 8 < 0$

**9.** $x^2 + 7x + 6 > 0$

**10.** $x^2 + 8x + 7 < 0$

**11.** $n^2 - 6n + 9 \geq 0$

**12.** $x^2 - 8x \geq 0$

**13.** $x^2 - 16 < 0$

**14.** $r^2 - 5r < 0$

**15.** $2x^2 + 5x - 3 \geq 0$

**16.** $3n^2 - 7n \leq 6$

**17.** $5x^2 + 6x \leq 8$

**18.** $3x^2 + 5x - 3 \leq 0$

**19.** $2x^2 - 12x + 9 \leq 0$

**20.** $5x^2 \leq -20x - 4$

*Solve each inequality and give the solution in interval notation.*

**21.** $(x - 2)(x + 1)(x + 5) \geq 0$

**22.** $(x - 2)(x + 2)(x + 5) \leq 0$

**23.** $(a - 3)(a + 2)(a + 4) < 0$

**24.** $(r - 1)(r + 2)(r + 7) < 0$

**25.** $(2c + 5)(3c - 6)(c + 6) > 0$

**26.** $(a - 4)(a - 2)(a + 8) > 0$

**27.** $(3x + 5)(x - 3)(x + 1) > 0$

**28.** $(3c - 1)(c + 4)(3c + 6) \leq 0$

**29.** $(x + 2)(x + 2)(3x - 8) \geq 0$

**30.** $(x + 3)^2(4x - 7) \leq 0$

**31.** $x^3 - 6x^2 + 9x < 0$

**32.** $x^3 + 3x^2 - 40x > 0$

*For each function provided, find all values of x for which f(x) satisfies the indicated conditions. Graph the solution on a number line.*

**33.** $f(x) = x^2 - 6x, f(x) \geq 0$

**34.** $f(x) = x^2 - 7x, f(x) > 0$

**35.** $f(x) = x^2 + 4x, f(x) > 0$

**36.** $f(x) = x^2 + 8x, f(x) \leq 0$

**37.** $f(x) = x^2 - 14x + 48, f(x) < 0$

**38.** $f(x) = x^2 - 2x - 15, f(x) < 0$

**39.** $f(x) = 2x^2 + 9x - 1, f(x) \leq 5$

**40.** $f(x) = x^2 + 5x - 3, f(x) \leq 4$

**41.** $f(x) = 2x^3 + 9x^2 - 35x, f(x) \geq 0$

**42.** $f(x) = x^3 - 9x, f(x) \leq 0$

*Solve each inequality and give the solution in set builder notation.*

**43.** $\frac{x + 2}{x - 4} > 0$

**44.** $\frac{x + 2}{x - 4} \geq 0$

**45.** $\frac{x - 1}{x + 5} < 0$

**46.** $\frac{x - 1}{x + 5} \leq 0$

**47.** $\frac{x + 3}{x - 2} \geq 0$

**48.** $\frac{x - 4}{x + 6} > 0$

**49.** $\frac{a - 9}{a + 5} < 0$

**50.** $\frac{b + 7}{b + 1} \leq 0$

**51.** $\frac{c - 10}{c - 4} > 0$

**52.** $\frac{2d - 6}{d - 1} < 0$

**53.** $\frac{3y + 6}{y + 4} \leq 0$

**54.** $\frac{4z - 8}{z - 9} \geq 0$

**55.** $\frac{5a + 10}{3a - 1} \geq 0$

**56.** $\frac{x + 4}{x - 4} \leq 0$

**57.** $\frac{3x + 4}{2x - 1} < 0$

**58.** $\frac{k + 3}{k} \geq 0$

**59.** $\frac{3x + 8}{x - 2} \leq 0$

**60.** $\frac{4x - 2}{2x - 8} > 0$

*Solve each inequality and give the solution in interval notation.*

**61.** $\frac{(x + 1)(x - 6)}{x + 3} < 0$

**62.** $\frac{(x + 1)(x - 6)}{x + 3} \leq 0$

**63.** $\frac{(x - 2)(x + 3)}{x - 5} > 0$

**64.** $\frac{(x - 2)(x + 3)}{x - 5} \geq 0$

**65.** $\frac{(a - 1)(a - 7)}{a + 2} \geq 0$

**66.** $\frac{(b - 2)(b + 4)}{b} < 0$

**67.** $\frac{c}{(c - 3)(c + 8)} \leq 0$

**68.** $\frac{z - 5}{(z + 6)(z - 9)} \geq 0$

**69.** $\frac{x - 6}{(x + 4)(x - 1)} \leq 0$

**70.** $\frac{x + 9}{(x - 2)(x + 4)} > 0$

**71.** $\frac{(x - 3)(2x + 5)}{x - 4} \geq 0$

**72.** $\frac{r(r - 8)}{2r + 6} < 0$

*Solve each inequality and graph the solution on a number line.*

**73.** $\dfrac{2}{x-4} \ge 1$

**74.** $\dfrac{2}{x-4} > 1$

**75.** $\dfrac{3}{x-1} > -1$

**76.** $\dfrac{3}{x+1} \ge -1$

**77.** $\dfrac{5}{x+2} \le 1$

**78.** $\dfrac{5}{x+2} < 1$

**79.** $\dfrac{2p-5}{p-4} \le 1$

**80.** $\dfrac{2}{2a-1} > 2$

**81.** $\dfrac{4}{x+2} \ge 2$

**82.** $\dfrac{x+6}{x+2} > 1$

**83.** $\dfrac{w}{3w-2} > -2$

**84.** $\dfrac{x-1}{2x+6} \le -3$

**85.** The graph of $y = \dfrac{x^2-4x+4}{x-4}$ is illustrated. Determine the solution to the following inequalities.

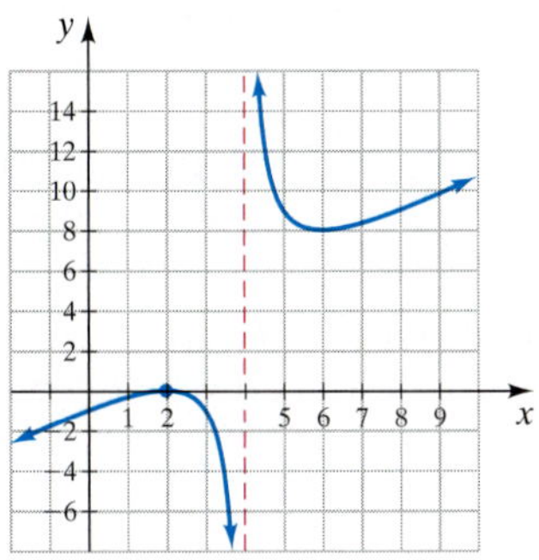


a) $\dfrac{x^2-4x+4}{x-4} > 0$

b) $\dfrac{x^2-4x+4}{x-4} < 0$

Explain how you determined your answer.

**86.** The graph of $y = \dfrac{x^2+x-6}{x-4}$ is illustrated. Determine the solution to the following inequalities.

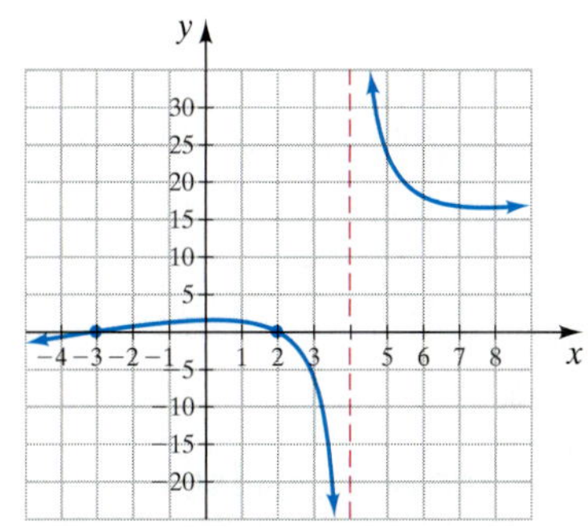


a) $\dfrac{x^2+x-6}{x-4} \ge 0$

b) $\dfrac{x^2+x-6}{x-4} < 0$

Explain how you determined your answer.

**87.** Write a quadratic inequality whose solution is

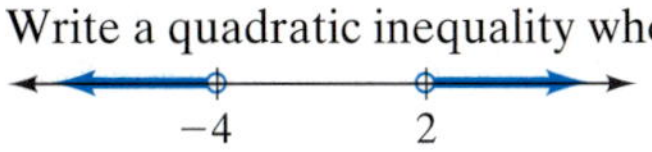


**88.** Write a quadratic inequality whose solution is

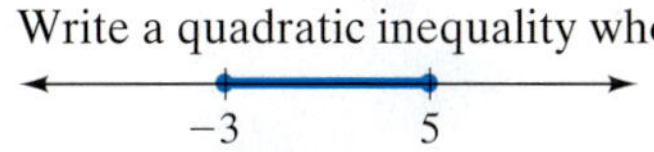


**89.** Write a rational inequality whose solution is

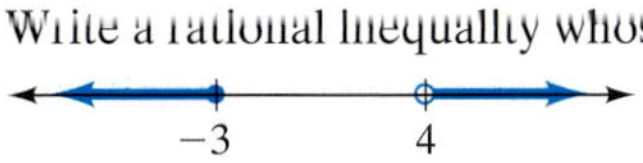


**90.** Write a rational inequality whose solution is

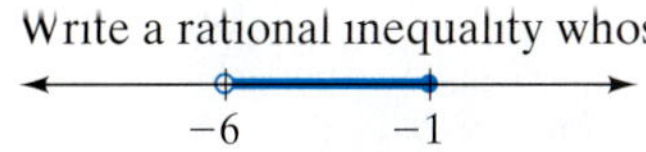


**91.** What is the solution to the inequality $(x+3)^2(x-1)^2 \ge 0$? Explain your answer.

**92.** What is the solution to the inequality $x^2(x-3)^2(x+4)^2 < 0$? Explain your answer.

**93.** What is the solution to the inequality $\dfrac{x^2}{(x+2)^2} \ge 0$? Explain your answer.

**94.** What is the solution to the inequality $\dfrac{x^2}{(x-3)^2} > 0$? Explain your answer.

**95.** If $f(x) = ax^2 + bx + c$ where $a > 0$ and the discriminant is negative, what is the solution to $f(x) < 0$? Explain.

**96.** If $f(x) = ax^2 + bx + c$ where $a < 0$ and the discriminant is negative, what is the solution to $f(x) > 2$? Explain.

## Challenge Problems

*Solve each inequality and graph the solution on the number line.*

**97.** $(x+1)(x-3)(x+5)(x+8) \ge 0$

**98.** $\dfrac{(x-4)(x+2)}{x(x+9)} \ge 0$

*Write a quadratic inequality with the following solutions. Many answers are possible. Explain how you determined your answers.*

**99.** $(-\infty, 0) \cup (3, \infty)$

**100.** $\{2\}$

**101.** $\varnothing$

**102.** $\mathbb{R}$

*In Exercises 103 and 104, solve each inequality and give the solution in interval notation. Use techniques from Section 8.5 to help you find the solution.*

**103.** $x^4 - 10x^2 + 9 > 0$

**104.** $x^4 - 26x^2 + 25 \leq 0$

*In Exercises 105 and 106, solve each inequality using factoring by grouping. Give the solution in interval notation.*

**105.** $x^3 + x^2 - 4x - 4 \geq 0$

**106.** $2x^3 + x^2 - 32x - 16 < 0$

## Group Activity

*Discuss and answer Exercises 107 and 108 as a group.*

**107.** Consider the number line below, where $a$, $b$, and $c$ are distinct real numbers.

**a)** In which intervals will the real numbers satisfy the inequality $(x - a)(x - b)(x - c) > 0$? Explain.

**b)** In which intervals will the real numbers satisfy the inequality $(x - a)(x - b)(x - c) < 0$? Explain.

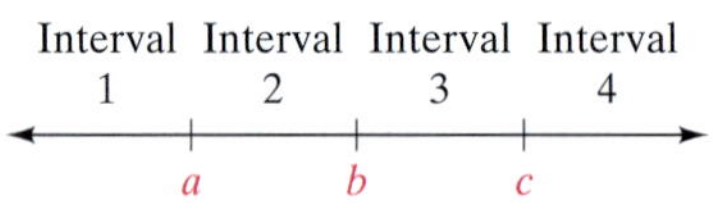


**108.** Consider the number line below where $a$, $b$, $c$, and $d$ are distinct real numbers.

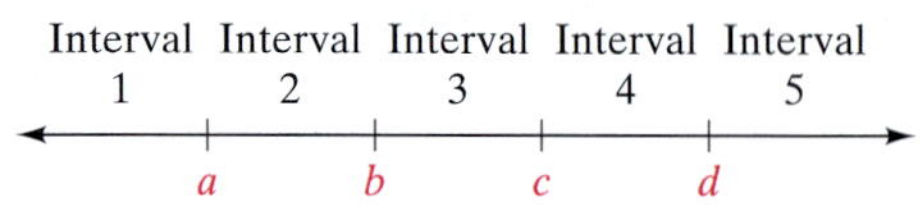


**a)** In which intervals do the real numbers satisfy the inequality $(x - a)(x - b)(x - c)(x - d) > 0$? Explain.

**b)** In which interval do the real numbers satisfy the inequality $(x - a)(x - b)(x - c)(x - d) < 0$? Explain.

## Cumulative Review Exercises

[2.4] **109. Antifreeze** How many quarts of a 100% antifreeze solution should Paul Simmons add to 10 quarts of a 20% antifreeze solution to make a 50% antifreeze solution?

[3.2] **110.** If $h(x) = \dfrac{x^2 + 4x}{x + 9}$, find $h(-3)$.

[5.1] **111.** Add $(6r + 5s - t) + (-3r - 2s - 8t)$.

[6.3] **112.** Simplify $\dfrac{1 + \dfrac{x}{x + 1}}{\dfrac{2x + 1}{x - 3}}$.

[7.7] **113.** Multiply $(3 - 4i)(6 + 5i)$.

# Chapter 8 Summary

| IMPORTANT FACTS AND CONCEPTS | EXAMPLES |
|---|---|
| **Section 8.1** | |
| **Square Root Property**<br>If $x^2 = a$, where $a$ is a real number, then $x = \pm\sqrt{a}$. | Solve $x^2 - 36 = 0$.<br>$x^2 - 36 = 0$<br>$x^2 = 36$<br>$x = \pm\sqrt{36} = \pm 6$<br>The solutions are $-6$ and 6. |
| A **perfect square trinomial** is a trinomial that can be expressed as a square of a binomial. | $x^2 - 10x + 25 = (x - 5)^2$ |

| IMPORTANT FACTS AND CONCEPTS | EXAMPLES |
|---|---|

**Section 8.1 (continued)**

### To Solve a Quadratic Equation by Completing the Square

1. Use the multiplication (or division) property of equality if necessary to make the leading coefficient 1.
2. Rewrite the equation with the constant by itself on the right side of the equation.
3. Take one-half the numerical coefficient of the first-degree term, square it, and add this quantity to both sides of the equation.
4. Replace the trinomial with the square of a binomial.
5. Use the square root property to take the square root of both sides of the equation.
6. Solve for the variable.
7. Check your solutions in the *original* equation.

Solve $x^2 + 4x - 12 = 0$ by completing the square.

$$x^2 + 4x - 12 = 0$$
$$x^2 + 4x = 12$$
$$x^2 + 4x + 4 = 12 + 4$$
$$(x + 2)^2 = 16$$
$$x + 2 = \pm\sqrt{16}$$
$$x + 2 = \pm 4$$
$$x = -2 \pm 4$$
$$x = -2 - 4 = -6 \quad \text{or} \quad x = -2 + 4 = 2$$

The solutions are $-6$ and 2.

**Section 8.2**

The **standard form of a quadratic equation** is $ax^2 + bx + c = 0, a \neq 0$.

$$x^2 - 5x + 17 = 0$$

### To Solve a Quadratic Equation by the Quadratic Formula

1. Write the quadratic equation in standard form, $ax^2 + bx + c = 0$, and determine the numerical values for $a$, $b$, and $c$.
2. Substitute the values for $a$, $b$, and $c$ into the quadratic formula and then evaluate the formula to obtain the solution.

The Quadratic Formula

$$x = \frac{-b \pm \sqrt{b^2 - 4ac}}{2a}$$

Solve $x^2 - 2x - 15 = 0$ by the quadratic formula.

$$a = 1, \quad b = -2, \quad c = -15$$
$$x = \frac{-b \pm \sqrt{b^2 - 4ac}}{2a}$$
$$= \frac{-(-2) \pm \sqrt{(-2)^2 - 4(1)(-15)}}{2(1)}$$
$$= \frac{2 \pm \sqrt{64}}{2}$$
$$= \frac{2 \pm 8}{2}$$
$$x = \frac{2+8}{2} = \frac{10}{2} = 5 \quad \text{or} \quad x = \frac{2-8}{2} = \frac{-6}{2} = -3$$

The solutions are 5 and $-3$.

### Solutions of a Quadratic Equation

For a quadratic equation of the form $ax^2 + bx + c = 0, a \neq 0$, the **discriminant** is $b^2 - 4ac$.

If $b^2 - 4ac > 0$, the quadratic equation has two distinct real number solutions.

If $b^2 - 4ac = 0$, the quadratic equation has a single real number solution.

If $b^2 - 4ac < 0$, the quadratic equation has no real number solution.

Determine the number of solutions of $3x^2 - x + 7 = 0$.

$$a = 3, b = -1, c = 7$$
$$b^2 - 4ac = (-1)^2 - 4(3)(7)$$
$$= 1 - 84$$
$$= -83$$

Since the discriminant is negative, the equation has no real number solution.

## IMPORTANT FACTS AND CONCEPTS — EXAMPLES

### Section 8.4

An equation that can be written in the form $au^2 + bu + c = 0$ for $a \neq 0$, where $u$ is an algebraic expression, is called **quadratic in form**.

**To Solve Equations Quadratic in Form**

1. Make a substitution that will result in an equation of the form $au^2 + bu + c = 0$, $a \neq 0$, where $u$ is a function of the original variable.
2. Solve the equation $au^2 + bu + c = 0$ for $u$.
3. Replace $u$ with the function of the original variable from step 1 and solve the resulting equation for the original variable.
4. Check for extraneous solutions by substituting the apparent solutions into the original equation.

Solve $x^4 - 17x^2 + 16 = 0$.

Let $u = x^2$.

Then, $u^2 - 17u + 16 = 0$

$$(u - 16)(u - 1) = 0$$

$$u - 16 = 0 \quad \text{or} \quad u - 1 = 0$$

$$u = 16 \qquad u = 1$$

$$x^2 = 16 \qquad x^2 = 1$$

$$x = \pm 4 \qquad x = \pm 1$$

A check will show the solutions are 4, −4, 1, and −1.

### Section 8.5

**Parabola**

Graphs of equations of the form $f(x) = ax^2 + bx + c$ are parabolas.

**a)** The parabola opens upward when $a > 0$ and downward when $a < 0$.

**b)** The axis of symmetry is the line $x = -\dfrac{b}{2a}$.

**c)** The vertex is the point $\left(-\dfrac{b}{2a}, \dfrac{4ac - b^2}{4a}\right)$ or $\left(-\dfrac{b}{2a}, f\left(-\dfrac{b}{2a}\right)\right)$.

**d)** The $y$-intercept is the point $(0, c)$.

**e)** To obtain the $x$-intercept(s), set $f(x) = 0$ and solve for $x$.

The graph of $f(x) = x^2 - 2x - 3$ is a parabola.

**a)** It opens upward since $a > 0$.

**b)** The axis of symmetry is $x = -\dfrac{-2}{2(1)} = 1$.

**c)** The vertex is $(1, -4)$.

**d)** The $y$-intercept is $(0, -3)$.

**e)**
$$x^2 - 2x - 3 = 0$$

$$(x - 3)(x + 1) = 0$$

$$x - 3 = 0 \quad \text{or} \quad x + 1 = 0$$

$$x = 3 \qquad x = -1$$

The $x$-intercepts are $(3, 0)$ and $(-1, 0)$.

The graph of $f(x) = x^2 - 2x - 3$ is shown below.

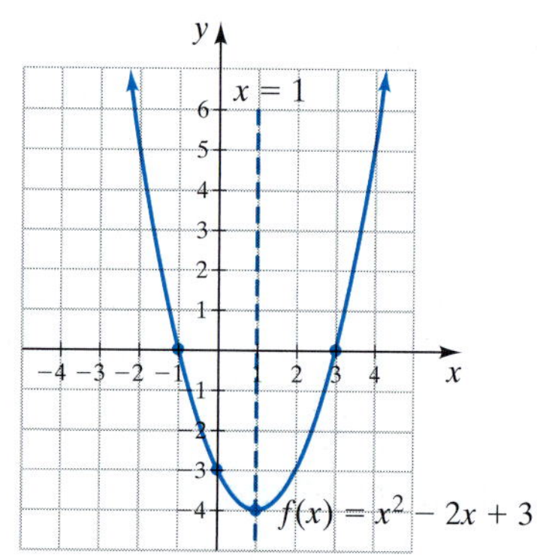


### Section 8.6

A **quadratic inequality** is obtained when the equal sign in the quadratic equation $ax^2 + bx + c = 0$ is replaced by an inequality sign.

The **solution to a quadratic inequality** is the set of all values that make the inequality a true statement.

$$x^2 - 5x + 7 > 0$$

| IMPORTANT FACTS AND CONCEPTS | EXAMPLES |
|---|---|

**Section 8.6 (continued)**

**To Solve Quadratic, Polynomial, and Rational Inequalities**

1. Write the inequality as an equation and solve the equation.
2. If solving a rational inequality, determine the values that make any denominator 0.
3. Construct a number line. Mark each solution from step 1 and numbers obtained in step 2 on the number line.
4. Select a test value in each interval and determine whether it satisfies the inequality. Also test each boundary value.
5. Write the solution in the form requested by your instructor.

Solve $(2x - 1)(x - 3)(x + 1) < 0$.

$$(2x - 1)(x - 3)(x + 1) < 0$$
$$(2x - 1)(x - 3)(x + 1) = 0$$
$$2x - 1 = 0 \quad \text{or} \quad x - 3 = 0 \quad \text{or} \quad x + 1 = 0$$
$$x = \frac{1}{2} \qquad x = 3 \qquad x = -1$$

The intervals and test values selected are shown below.

| Interval | Test value | $(2x - 1)(x - 3)(x + 1) < 0$ |
|---|---|---|
| $(-\infty, -1)$ | $-2$ | $-25$ |
| $\left(-1, \frac{1}{2}\right)$ | 0 | 3 |
| $\left(\frac{1}{2}, 3\right)$ | 1 | $-4$ |
| $(3, \infty)$ | 5 | 108 |

The solution is $x < -1$ or $\frac{1}{2} < x < 3$.

Solution on a number line:

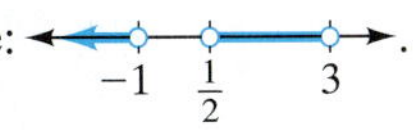


Solution in interval notation: $(-\infty, -1) \cup \left(\frac{1}{2}, 3\right)$

Solution in set builder notation:

$$\left\{x \middle| x < -1 \quad \text{or} \quad \frac{1}{2} < x < 3\right\}$$

## Chapter 8 Review Exercises

[8.1] *Use the square root property to solve each equation.*

1. $(x - 5)^2 = 24$
2. $(2x + 1)^2 = 60$
3. $\left(x - \frac{1}{3}\right)^2 = \frac{4}{9}$
4. $\left(2x - \frac{1}{2}\right)^2 = 4$

*Solve each equation by completing the square.*

5. $x^2 - 7x + 12 = 0$
6. $x^2 + 4x - 32 = 0$
7. $a^2 + 2a - 9 = 0$
8. $z^2 + 6z = 12$
9. $x^2 - 2x + 10 = 0$
10. $2r^2 - 8r = -64$

**Area** *In Exercises 11 and 12, the area, A, of each rectangle is given.* **a)** *Write an equation for the area.* **b)** *Solve the equation for x.*

11.

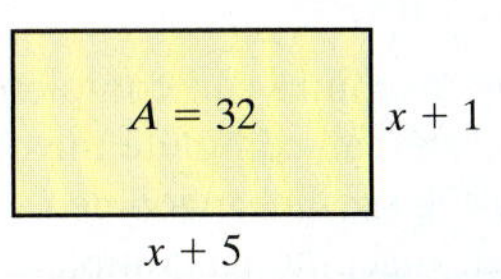


12.

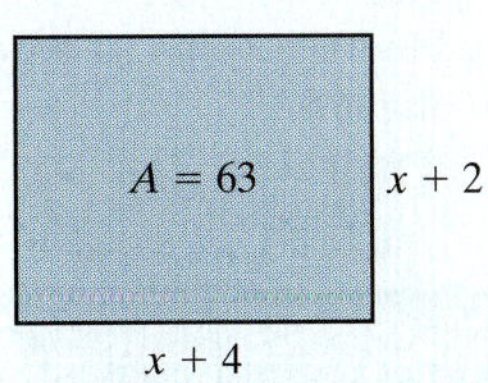


13. **Consecutive Integers** The product of two consecutive positive integers is 42. Find the two integers.

14. **Living Room** Ronnie Sampson just moved into a new house where the living room is a square whose diagonal is 7 feet longer than the length of a side. Find the dimensions of the living room.

[8.2] *Determine whether each equation has two distinct real solutions, a single real solution, or no real solution.*

**15.** $2x^2 - 5x - 1 = 0$

**16.** $3x^2 + 2x = -6$

**17.** $r^2 + 16r = -64$

**18.** $5x^2 - x + 2 = 0$

**19.** $u^2 - 14u = -49$

**20.** $\frac{1}{2}x^2 - 3x = 8$

*Solve each equation by the quadratic formula.*

**21.** $3x^2 + 4x = 0$

**22.** $x^2 - 11x = -18$

**23.** $r^2 = 3r + 40$

**24.** $7x^2 = 9x$

**25.** $6a^2 + a - 15 = 0$

**26.** $4x^2 + 11x = 3$

**27.** $x^2 + 8x + 5 = 0$

**28.** $b^2 + 4b = 8$

**29.** $2x^2 + 4x - 3 = 0$

**30.** $3y^2 - 6y = 8$

**31.** $x^2 - x + 13 = 0$

**32.** $x^2 - 2x + 11 = 0$

**33.** $2x^2 - \frac{5}{3}x = \frac{25}{3}$

**34.** $4x^2 + 5x - \frac{3}{2} = 0$

*For the given function, determine all real values of the variable for which the function has the value indicated.*

**35.** $f(x) = x^2 - 4x - 35, f(x) = 25$

**36.** $g(x) = 6x^2 + 5x, g(x) = 6$

**37.** $h(r) = 5r^2 - 7r - 10, h(r) = -8$

**38.** $f(x) = -2x^2 + 6x + 7, f(x) = -2$

*Determine an equation that has the given solutions.*

**39.** $3, -1$

**40.** $\frac{2}{3}, -2$

**41.** $-\sqrt{11}, \sqrt{11}$

**42.** $3 - 2i, 3 + 2i$

[8.1–8.3]

**43. Rectangular Garden** Sophia Yang is designing a rectangular flower garden. If the area is to be 96 square feet and the length is to be 4 feet greater than the width, find the dimensions of the garden.

**44. Triangle and Circle** Find the length of side $x$ in the figure.

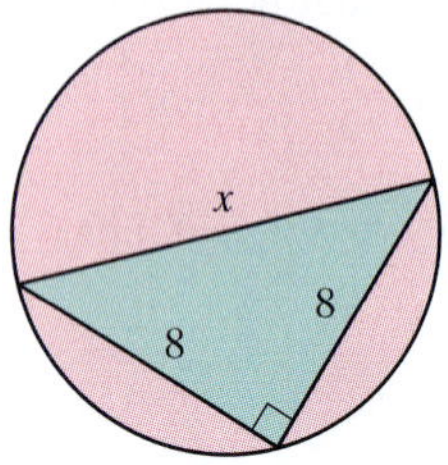


**45. Savings Account** Samuel Rivera invested \$1000 in a savings account where the interest is compounded annually. If after 2 years the amount in the account is \$1081.60, find the annual interest rate.

**46. Numbers** The larger of two positive numbers is 4 greater than the smaller. Find the two numbers if their product is 77.

**47. Rectangle** The length of a rectangle is 4 inches less than twice its width. Find the dimensions if its area is 96 square inches.

**48. Wheat Crop** The value, $V$, in dollars per acre of a wheat crop $d$ days after planting is given by the formula $V = 12d - 0.05d^2, 20 < d < 80$. Find the value of an acre of wheat 60 days after it has been planted.

**49. Expenditure by Oil Companies** The expenditure, $E(t)$, in billions of dollars, by oil companies for new oil and natural gas projects can be approximated by the equation $E(t) = 7t^2 - 7.8t + 82.2$, where $t$ is the number of years since 2001. **Source:** John S. Herald Inc., *Washington Post* (3/14/05).

**a)** Find the expenditure of oil companies for new oil and gas projects in 2004.

**b)** If this trend continues, in what year will the expenditure by oil companies be \$579 billion?

**50. Falling Object** The distance, $d$, in feet, that an object is from the ground $t$ seconds after being dropped from an airplane is given by the formula $d = -16t^2 + 784$.

**a)** Find the distance the object is from the ground 2 seconds after it has been dropped.

**b)** When will the object hit the ground?

**51. Oil Leak** A tractor has an oil leak. The amount of oil, $L(t)$, in milliliters per hour that leaks out is a function of the tractor's operating temperature, $t$, in degrees Celsius. The function is

$$L(t) = 0.0004t^2 + 0.16t + 20, 100°\text{C} \le t \le 160°\text{C}$$

**a)** How many milliliters of oil will leak out in 1 hour if the operating temperature of the tractor is 100°C?

**b)** If oil is leaking out at 53 milliliters per hour, what is the operating temperature of the tractor?

**52. Molding Machines** Two molding machines can complete an order in 12 hours. The larger machine can complete the order by itself in 1 hour less time than the smaller machine can by itself. How long will it take each machine to complete the order working by itself?

**53. Travel Time** Steve Forrester drove 25 miles at a constant speed, and then he increased his speed by 15 miles per hour for the next 65 miles. If the time required to travel 90 miles was 1.5 hours, find the speed he drove during the first 25 miles.

54. **Canoe Trip** Joan Banker canoed downstream going with the current for 3 miles, then turned around and canoed upstream against the current to her starting point. If the total time she spent canoeing was 4 hours and the current was 0.4 mile per hour, what is the speed she canoes in still water?

55. **Area** The area of a rectangle is 80 square units. If the length is $x$ units and the width is $x - 2$ units, find the length and the width. Round your answer to the nearest tenth of a unit.

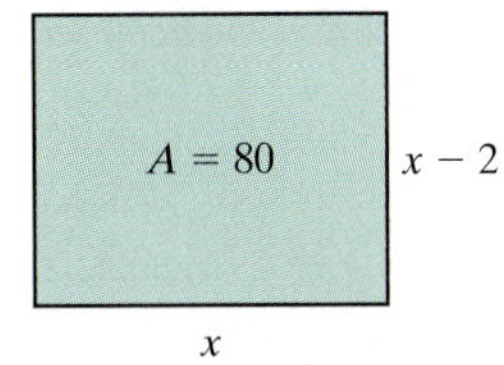


56. **Selling Tables** A business sells $n$ tables, $n \le 40$, at a price of $(60 - 0.3n)$ dollars per table. How many tables must be sold to have a revenue of \$1080?

*In Exercises 57–60, solve each equation for the variable indicated.*

57. $a^2 + b^2 = c^2$, for $a$ (Pythagorean Theorem)

58. $h = -4.9t^2 + c$, for $t$ (height of an object)

59. $v_x^2 + v_y^2 = v^2$, for $v_y$ (vectors)

60. $a = \dfrac{v_2^2 - v_1^2}{2d}$, for $v_2$

[8.4] *Solve each equation.*

61. $x^4 - 13x^2 + 36 = 0$

62. $x^4 - 21x^2 + 80 = 0$

63. $a^4 = 5a^2 + 24$

64. $3y^{-2} + 16y^{-1} = 12$

65. $3r + 11\sqrt{r} - 4 = 0$

66. $2p^{2/3} - 7p^{1/3} + 6 = 0$

67. $6(x - 2)^{-2} = -13(x - 2)^{-1} + 8$

68. $10(r + 1) = \dfrac{12}{r + 1} - 7$

*Find all x-intercepts of the given function.*

69. $f(x) = x^4 - 82x + 81$

70. $f(x) = 30x + 13\sqrt{x} - 10$

71. $f(x) = x - 6\sqrt{x} + 12.$

72. $f(x) = (x^2 - 6x)^2 - 5(x^2 - 6x) - 24$

[8.5] **a)** *Determine whether the parabola opens upward or downward.* **b)** *Find the y-intercept.* **c)** *Find the vertex.* **d)** *Find the x-intercepts if they exist.* **e)** *Draw the graph.*

73. $f(x) = x^2 + 5x$

74. $f(x) = x^2 - 2x - 8$

75. $g(x) = -x^2 - 2$

76. $g(x) = -2x^2 - x + 15$

77. **Selling Tickets** The Hamilton Outdoor Theater estimates that its total income, $I$, in hundreds of dollars, for its production of a play, can be approximated by the formula $I = -x^2 + 22x - 45, 2 \le x \le 20$, where $x$ is the cost of a ticket.

**a)** How much should the theater charge to maximize its income?

**b)** What is the maximum income?

78. **Tossing a Ball** Josh Vincent tosses a ball upward from the top of a 75-foot building. The height, $s(t)$, of the ball at any time $t$ can be determined by the function $s(t) = -16t^2 + 80t + 75$.

**a)** At what time will the ball attain its maximum height?

**b)** What is the maximum height?

*Graph each function.*

**79.** $f(x) = (x - 3)^2$ **80.** $f(x) = -(x + 2)^2 - 3$ **81.** $g(x) = -2(x + 4)^2 - 1$ **82.** $h(x) = \frac{1}{2}(x - 1)^2 + 3$

[8.6] *Graph the solution to each inequality on a number line.*

**83.** $x^2 + 4x + 3 \geq 0$

**84.** $x^2 + 3x - 10 \leq 0$

**85.** $x^2 \leq 11x - 20$

**86.** $3x^2 + 8x > 16$

**87.** $4x^2 - 9 \leq 0$

**88.** $6x^2 - 30 > 0$

*Solve each inequality and give the solution in set builder notation.*

**89.** $\frac{x + 1}{x - 5} > 0$

**90.** $\frac{x - 3}{x + 2} \leq 0$

**91.** $\frac{2x - 4}{x + 3} \geq 0$

**92.** $\frac{3x + 5}{x - 6} < 0$

**93.** $(x + 4)(x + 1)(x - 2) > 0$

**94.** $x(x - 3)(x - 6) \leq 0$

*Solve each inequality and give the solution in interval notation.*

**95.** $(3x + 4)(x - 1)(x - 3) \geq 0$

**96.** $2x(x + 2)(x + 4) < 0$

**97.** $\frac{x(x - 4)}{x + 2} > 0$

**98.** $\frac{(x - 2)(x - 8)}{x + 3} < 0$

**99.** $\frac{x - 3}{(x + 2)(x - 7)} \geq 0$

**100.** $\frac{x(x - 6)}{x + 3} \leq 0$

*Solve each inequality and graph the solution on a number line.*

**101.** $\frac{5}{x + 4} \geq -1$

**102.** $\frac{2x}{x - 2} \leq 1$

**103.** $\frac{2x + 3}{3x - 5} < 4$

## Chapter 8 Practice Test

*To find out how well you understand the chapter material, take this practice test. The answers, and the section where the material was initially discussed, are given in the back of the book. Each problem is also fully worked out on the* **Chapter Test Prep Video CD**. *Review any questions that you answered incorrectly.*

*Solve by completing the square.*

**1.** $x^2 + 2x - 15 = 0$ **2.** $a^2 + 7 = 6a$

*Solve by the quadratic formula.*

**3.** $x^2 - 6x - 16 = 0$ **4.** $x^2 - 4x = -11$

*Solve by the method of your choice.*

**5.** $3r^2 + r = 2$ **6.** $p^2 + 4 = -7p$

**7.** Write an equation that has $x$-intercepts $4, -\frac{2}{5}$.

**8.** Solve the formula $K = \frac{1}{2}mv^2$ for $v$.

**9. Cost** The cost, $c$, of a house in Duquoin, Illinois, is a function of the number of square feet, $s$, of the house. The cost of the house can be approximated by

$$c(s) = -0.01s^2 + 78s + 22{,}000, \quad 1300 \leq s \leq 3900$$

**a)** Estimate the cost of a 1600-square-foot house.

**b)** How large a house can Clarissa Skocy purchase if she wishes to spend \$160,000 on a house?

**10. Trip in a Park** Tom Ficks drove his 4-wheel-drive Jeep from Anchorage, Alaska, to the Chena River State Recreation Park, a distance of 520 miles. Had he averaged 15 miles per hour faster, the trip would have taken 2.4 hours less. Find the average speed that Tom drove.

Chena River State Recreation Park

*Solve.*

**11.** $2x^4 + 15x^2 - 50 = 0$

**12.** $3r^{2/3} + 11r^{1/3} - 42 = 0$

**13.** Find all $x$-intercepts of $f(x) = 16x - 24\sqrt{x} + 9$.

*Graph each function.*

**14.** $f(x) = (x - 3)^2 + 2$

**15.** $h(x) = -\frac{1}{2}(x - 2)^2 - 2$

**16.** Determine whether $6x^2 = 2x + 3$ has two distinct real solutions, a single real solution, or no real solution. Explain your answer.

**17.** Consider the quadratic equation $y = x^2 + 2x - 8$.

**a)** Determine whether the parabola opens upward or downward.

**b)** Find the $y$-intercept.

**c)** Find the vertex.

**d)** Find the $x$-intercepts (if they exist).

**e)** Draw the graph.

**18.** Write a quadratic equation whose $x$-intercepts are $(-7, 0), \left(\frac{1}{2}, 0\right)$.

*Solve each inequality and graph the solution on a number line.*

**19.** $x^2 - x \geq 42$

**20.** $\frac{(x + 5)(x - 4)}{x + 1} \geq 0$

*Solve the following inequality. Write the answer in* ***a)*** *interval notation and* ***b)*** *set builder notation.*

**21.** $\frac{x + 3}{x + 2} \leq -1$

**22. Carpet** The length of a rectangular Persian carpet is 3 feet greater than twice its width. Find the length and width of the carpet if its area is 65 square feet.

**23. Throwing a Ball** Jose Ramirez throws a ball upward from the top of a building. The distance, $d$, of the ball from the ground at any time, $t$, is $d = -16t^2 + 80t + 96$. How long will it take for the ball to strike the ground?

**24. Profit** The Leigh Ann Sims Company earns a weekly profit according to the function $f(x) = -1.4x^2 + 56x - 70$, where $x$ is the number of wood carvings made and sold each week.

**a)** Find the number of carvings the company must sell in a week to maximize its profit.

**b)** What is its maximum weekly profit?

**25. Selling Brooms** A business sells $n$ brooms, $n \leq 32$, at a price of $(10 - 0.1n)$ dollars per broom. How many brooms must be sold to have a revenue of \$210?

## Cumulative Review Test

*Take the following test and check your answers with those given in the back of the book. Review any questions that you answered incorrectly. The section where the material was covered is indicated after the answer.*

**1.** Evaluate $-4 \div (-2) + 18 - \sqrt{49}$.

**2.** Evaluate $2x^2 + 3x + 4$ when $x = 2$.

**3.** Express 2,540,000 in scientific notation.

**4.** Find the solution set for the equation $|4 - 2x| = 5$.

**5.** Simplify $6x - \{3 - [2(x - 2) - 5x]\}$.

**6.** Solve the equation $-\frac{1}{2}(4x - 6) = \frac{1}{3}(3 - 6x) + 2$.

**7.** Solve the inequality $-4 < \frac{x + 4}{2} < 6$. Write the solution in interval notation.

**8.** Find the slope and $y$-intercept of the graph of $9x + 7y = 15$.

**9. Small Orchard** The number of baskets of apples, $N$, that are produced by $x$ trees in a small orchard is given by the function $N(x) = -0.2x^2 + 40x$. How many baskets of apples are produced by 50 trees?

**10.** Write the equation in slope-intercept form of a line passing through the points $(6, 5)$ and $(4, 3)$.

**11. a)** Determine whether the following graph represents a function. Explain your answer.

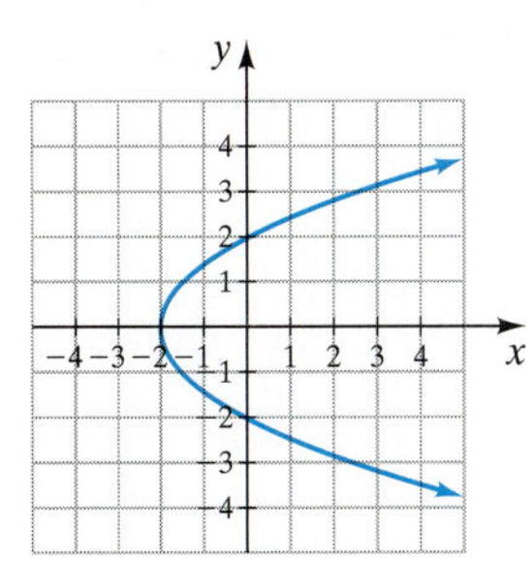


**b)** Determine the domain and range of the function or relation.

**12.** Graph each equation.

**a)** $x = -4$

**b)** $y = 2$

**13.** Evaluate the following determinant.

$$\begin{vmatrix} 4 & 0 & -2 \\ 3 & 5 & 1 \\ 1 & -1 & 7 \end{vmatrix}$$

**14.** Solve the system of equations.

$$4x - 3y = 10$$
$$2x + y = 5$$

**15.** Factor $(x + 3)^2 + 10(x + 3) + 24$.

**16.** **a)** Write an expression for the shaded area in the figure and **b)** write the expression in factored form.

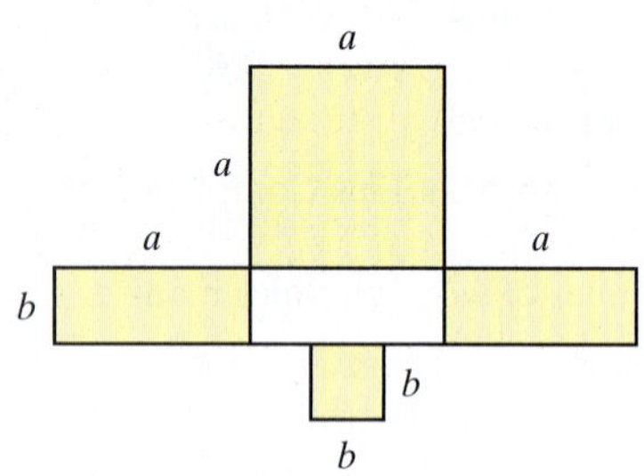


**17.** Add $\frac{x + 2}{x^2 - x - 6} + \frac{x - 3}{x^2 - 8x + 15}$.

**18.** Solve the equation

$$\frac{1}{a - 2} = \frac{4a - 1}{a^2 + 5a - 14} + \frac{2}{a + 7}.$$

**19.** **Wattage Rating** The wattage rating of an appliance, $w$, varies jointly as the square of the current, $I$, and the resistance, $R$. If the wattage is 12 when the current is 2 amperes and the resistance is 100 ohms, find the wattage when the current is 0.8 ampere and the resistance is 600 ohms.

**20.** Simplify $\frac{3 - 4i}{2 + 5i}$.

# 9 Exponential and Logarithmic Functions

## GOALS OF THIS CHAPTER

Exponential and logarithmic functions have a wide variety of uses, some of which you will see as you read through this chapter. You often read in newspaper and magazine articles that health care spending, use of the Internet, and the world population, to list just a few, are growing exponentially. By the time you finish this chapter you should have a clear understanding of just what this means.

We also introduce two special functions, the natural exponential function and the natural logarithmic function. Many natural phenomena, such as carbon dating, radioactive decay, and the growth of savings invested in an account compounding interest continuously, can be described by natural exponential functions.


9.1 Composite and Inverse Functions

9.2 Exponential Functions

9.3 Logarithmic Functions

9.4 Properties of Logarithms

Mid-Chapter Test: Sections 9.1–9.4

9.5 Common Logarithms

9.6 Exponential and Logarithmic Equations

9.7 Natural Exponential and Natural Logarithmic Functions

Chapter 9 Summary

Chapter 9 Review Exercises

Chapter 9 Practice Test

Cumulative Review Test

THE DEPARTMENT OF ECONOMIC and Social Affairs uses mathematical models to make estimates and projections of the world population. Currently, the world population is growing at about 1.13% per year. Since the population is growing by a percent rather than by a fixed amount, it is modeled by an exponential function rather than a linear one. In Exercise 79 on page 646, we will investigate the effect different rates have on the growth of the world population.

# 9.1 Composite and Inverse Functions

1. Find composite functions.
2. Understand one-to-one functions.
3. Find inverse functions.
4. Find the composition of a function and its inverse.

The focus of this chapter is logarithms. However, before we can study logarithms we discuss composite functions, one-to-one functions, and inverse functions. Let's start with composite functions.

## 1 Find Composite Functions

Often we come across situations in which one quantity is a function of one variable. That variable, in turn, is a function of some other variable. For example, the cost of advertising on a television show may be a function of the Nielsen rating of the show. The Nielsen rating, in turn, is a function of the number of people who watch the show. In the final outcome, the cost of advertising may be affected by the number of people who watch the show. Functions like this are called *composite functions*.

Let's consider another example. Suppose that 1 U.S. dollar can be converted into 1.20 Canadian dollars, and 1 Canadian dollar can be converted into 9.3 Mexican pesos. Using this information, we can convert 20 U.S. dollars into Mexican pesos. We have the following functions.

$$g(x) = 1.20x \text{ (U.S. dollars to Canadian dollars)}$$
$$f(x) = 9.3x \text{ (Canadian dollars to Mexican pesos)}$$

If we let $x = 20$, for \$20 U.S., then it can be converted into \$24 Canadian using function $g$:

$$g(x) = 1.20x$$
$$g(20) = 1.20(20) = \$24 \text{ Canadian}$$

The \$24 Canadian can, in turn, be converted into 223.20 Mexican pesos using function $f$:

$$f(x) = 9.3x$$
$$f(24) = 9.3(24) = 223.20 \text{ Mexican pesos}$$

Is there a way of finding this conversion without performing this string of calculations? The answer is yes. One U.S. dollar can be converted into Mexican pesos by substituting the $1.20x$ found in function $g(x)$ for the $x$ in $f(x)$. This gives a new function, $h$, which converts U.S. dollars directly into Mexican pesos.

$$g(x) = 1.20x \qquad f(x) = 9.3x$$
$$h(x) = f[g(x)]$$
$$= 9.3(1.20x) \quad \text{Substitute } g(x) \text{ for } x \text{ in } f(x).$$
$$= 11.16x$$

Thus, for each U.S. dollar, $x$, we get 11.16 Mexican pesos. If we substitute \$20 for $x$, we get 223.20 pesos, which is what we expected.

$$h(x) = 11.16x$$
$$h(20) = 11.16(20) = 223.20$$

Function $h$, called a **composition of $f$ with $g$**, is denoted $(f \circ g)$ and is read "$f$ composed with $g$" or "$f$ circle $g$." **Figure 9.1** shows how the composite function $h$ relates to functions $f$ and $g$.

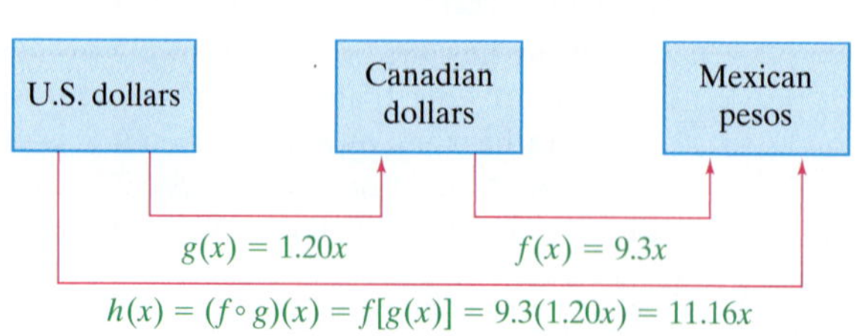


FIGURE 9.1

We now define the **composite function**.

**Composite Function**

The **composite function** $f \circ g$ is defined as

$$(f \circ g)(x) = f[g(x)]$$

When we are given $f(x)$ and $g(x)$, to find $(f \circ g)(x)$ we substitute $g(x)$ for $x$ in $f(x)$ to get $f[g(x)]$.

**EXAMPLE 1** ▸ Given $f(x) = x^2 - 2x + 3$ and $g(x) = x - 5$, find

**a)** $f(4)$ **b)** $f(a)$ **c)** $(f \circ g)(x)$ **d)** $(f \circ g)(3)$

*Solution*

**a)** To find $f(4)$, we substitute 4 for each $x$ in $f(x)$.

$$f(x) = x^2 - 2x + 3$$
$$f(4) = 4^2 - 2 \cdot 4 + 3 = 16 - 8 + 3 = 11$$

**b)** To find $f(a)$, we substitute $a$ for each $x$ in $f(x)$.

$$f(x) = x^2 - 2x + 3$$
$$f(a) = a^2 - 2a + 3$$

**c)** $(f \circ g)(x) = f[g(x)]$. To find $(f \circ g)(x)$, we substitute $g(x)$, which is $x - 5$, for each $x$ in $f(x)$.

$$f(x) = x^2 - 2x + 3$$
$$f[g(x)] = [g(x)]^2 - 2[g(x)] + 3$$

Since $g(x) = x - 5$, we substitute as follows

$$\begin{aligned} f[g(x)] &= (x - 5)^2 - 2(x - 5) + 3 \\ &= (x - 5)(x - 5) - 2x + 10 + 3 \\ &= x^2 - 10x + 25 - 2x + 13 \\ &= x^2 - 12x + 38 \end{aligned}$$

Therefore, the composite function of $f$ with $g$ is $x^2 - 12x + 38$.

$$(f \circ g)(x) = f[g(x)] = x^2 - 12x + 38$$

**d)** To find $(f \circ g)(3)$, we substitute 3 for $x$ in $(f \circ g)(x)$.

$$(f \circ g)(x) = x^2 - 12x + 38$$
$$(f \circ g)(3) = 3^2 - 12(3) + 38 = 11$$

▸ **Now Try Exercise 9**

How do you think we would determine $(g \circ f)(x)$ or $g[f(x)]$? If you answered, "Substitute $f(x)$ for each $x$ in $g(x)$," you answered correctly. Using $f(x)$ and $g(x)$ as given in Example 1, we find $(g \circ f)(x)$ as follows.

$$g(x) = x - 5, \qquad f(x) = x^2 - 2x + 3$$

$$\begin{aligned} g[f(x)] &= f(x) - 5 \\ g[f(x)] &= (x^2 - 2x + 3) - 5 \\ &= x^2 - 2x + 3 - 5 \\ &= x^2 - 2x - 2 \end{aligned}$$

Therefore, the composite function of $g$ with $f$ is $x^2 - 2x - 2$.

$$(g \circ f)(x) = g[f(x)] = x^2 - 2x - 2$$

By comparing the illustrations above, we see that in this example, $f[g(x)] \neq g[f(x)]$.

**EXAMPLE 2** ▸ Given $f(x) = x^2 + 4$ and $g(x) = \sqrt{x - 1}$, find

**a)** $(f \circ g)(x)$ **b)** $(g \circ f)(x)$

*Solution*

**a)** To find $(f \circ g)(x)$, we substitute $g(x)$, which is $\sqrt{x - 1}$ for each $x$ in $f(x)$. You should realize that $\sqrt{x - 1}$ is a real number only when $x \geq 1$.

$$f(x) = x^2 + 4$$
$$(f \circ g)(x) = f[g(x)] = (\sqrt{x - 1})^2 + 4 = x - 1 + 4 = x + 3, x \geq 1$$

Since values of $x < 1$ are not in the domain of $g(x)$, values of $x < 1$ are not in the domain of $(f \circ g)(x)$.

**b)** To find $(g \circ f)(x)$, we substitute $f(x)$, which is $x^2 + 4$, for each $x$ in $g(x)$.

$$g(x) = \sqrt{x - 1}$$
$$(g \circ f)(x) = g[f(x)] = \sqrt{(x^2 + 4) - 1} = \sqrt{x^2 + 3}$$

▸ **Now Try Exercise 19**

**EXAMPLE 3** ▸ Given $f(x) = x - 3$ and $g(x) = x + 7$, find

**a)** $(f \circ g)(x)$ **b)** $(f \circ g)(2)$ **c)** $(g \circ f)(x)$ **d)** $(g \circ f)(2)$

*Solution*

**a)**
$$f(x) = x - 3$$
$$(f \circ g)(x) = f[g(x)] = (x + 7) - 3 = x + 4$$

**b)** We find $(f \circ g)(2)$ by substituting 2 for each $x$ in $(f \circ g)(x)$.

$$(f \circ g)(x) = x + 4$$
$$(f \circ g)(2) = 2 + 4 = 6$$

**c)**
$$g(x) = x + 7$$
$$(g \circ f)(x) = g[f(x)] = (x - 3) + 7 = x + 4$$

**d)** Since $(g \circ f)(x) = x + 4$, $(g \circ f)(2) = 2 + 4 = 6$.

▸ **Now Try Exercise 11**

In general, $(f \circ g)(x) \neq (g \circ f)(x)$ as we saw at the end of Example 1. In Example 3, $(f \circ g)(x) = (g \circ f)(x)$, but this is only due to the specific functions used.

### Helpful Hint

Do not confuse finding the product of two functions with finding a composite function.

Product of functions $f$ and $g$: $(fg)(x)$ or $(f \cdot g)(x)$
Composite function of $f$ with $g$: $(f \circ g)(x)$

When multiplying functions $f$ and $g$, we can use a dot between the $f$ and $g$. When finding the composite function of $f$ with $g$, we use a small *open* circle.

## 2 Understand One-to-One Functions

Consider the following two sets of ordered pairs.

$$A = \{(1, 2), (3, 5), (4, 6), (-2, 1)\}$$
$$B = \{(1, 2), (3, 5), (4, 6), (-2, 5)\}$$

Both sets of ordered pairs, $A$ and $B$, are functions since each value of $x$ has a unique value of $y$. In set $A$, each value of $y$ also has a unique value of $x$, as shown in **Figure 9.2.** In set $B$, each value of $y$ does not have a unique value of $x$. In the ordered pairs (3, 5) and (−2, 5), the $y$-value 5 corresponds with two values of $x$, as shown in **Figure 9.3.**

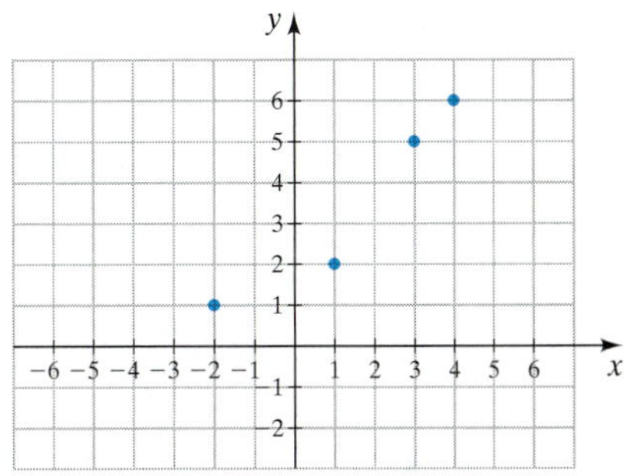

FIGURE 9.2

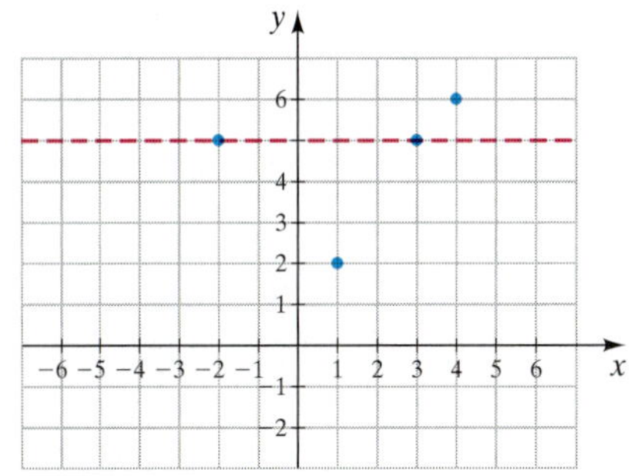

FIGURE 9.3

The set of ordered pairs in $A$ is an example of a *one-to-one function*. The set of ordered pairs in $B$ is not a one-to-one function. In a **one-to-one function**, each value in the range has a unique value in the domain. Thus, if $y$ is a one-to-one function of $x$, in addition to each $x$-value having a unique $y$-value (the definition of a function), each $y$-value must also have a unique $x$-value.

**One-to-One Function**

A function is a **one-to-one function** if each value in the range corresponds with exactly one value in the domain.

For a function to be a one-to-one function, its graph must pass not only a **vertical line test** (the test to ensure that it is a function) but also a **horizontal line test** (to test the one-to-one criteria).

Consider the function $f(x) = x^2$ (see **Fig. 9.4**). Note that it is a function since its graph passes the vertical line test. For each value of $x$, there is a unique value of $y$. Does each value of $y$ also have a unique value of $x$? The answer is no, as illustrated in **Figure 9.5**. Note that for the indicated value of $y$ there are two values of $x$, namely $x_1$ and $x_2$. If we limit the domain of $f(x) = x^2$ to values of $x$ greater than or equal to 0, then each $x$-value has a unique $y$-value and each $y$-value also has a unique $x$-value (see **Fig. 9.6**). The function $f(x) = x^2, x \geq 0$, **Figure 9.6**, is an example of a one-to-one function.

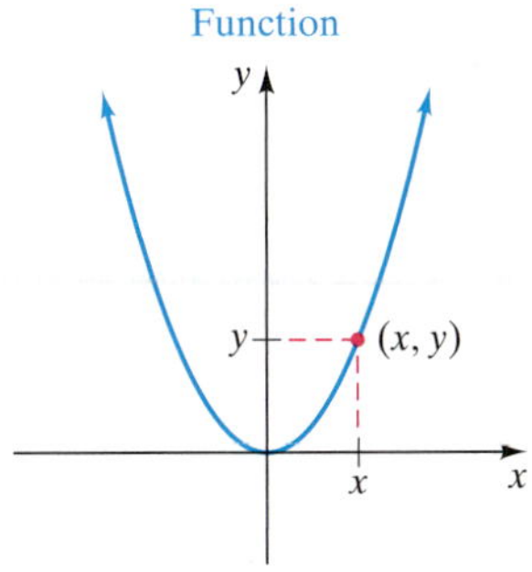


FIGURE 9.4

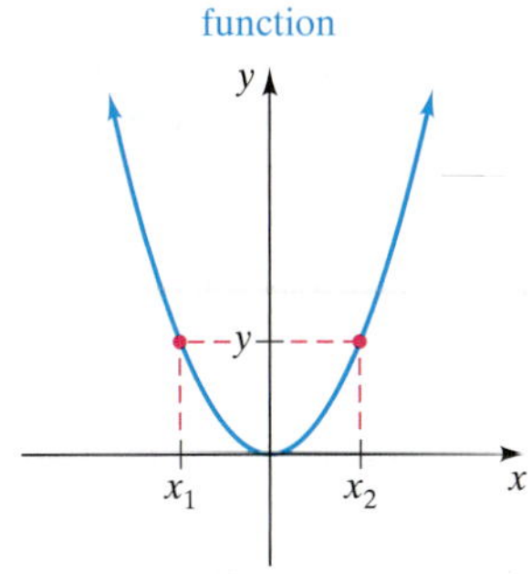


FIGURE 9.5

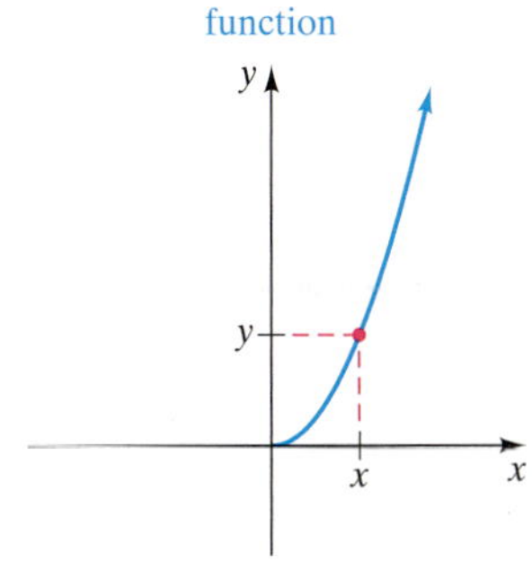


FIGURE 9.6

In **Figure 9.7**, the graphs in parts (a) through (e) are functions since they all pass the vertical line test. However, only the graphs in parts (a), (d), and (e) are one-to-one functions since they also pass the horizontal line test. The graph in part (f) is not a function; therefore, it is not a one-to-one function, even though it passes the horizontal line test.

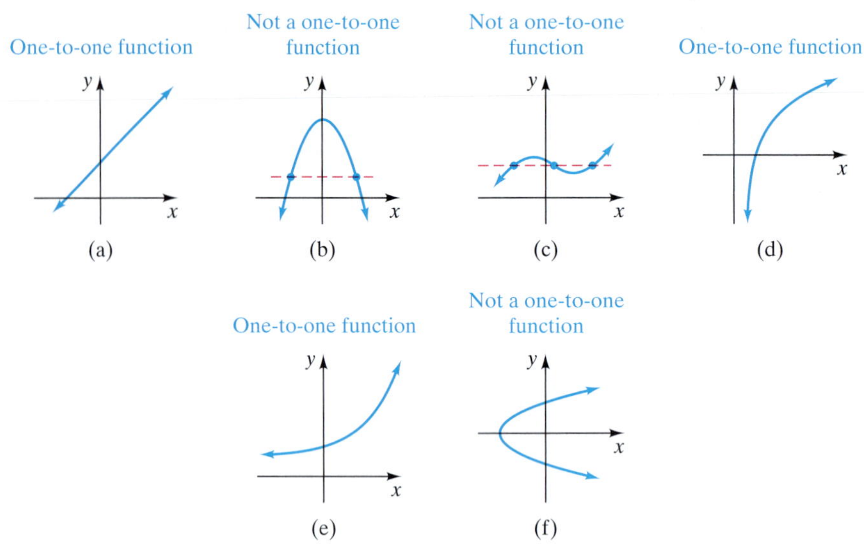


FIGURE 9.7

## 3 Find Inverse Functions

Now that we have discussed one-to-one functions, we can introduce inverse functions. **Only one-to-one functions have inverse functions**. If a function is one-to-one, its **inverse function** may be obtained by interchanging the first and second coordinates in each ordered pair of the function. Thus, for each ordered pair $(x, y)$ in the function, the ordered pair $(y, x)$ will be in the inverse function. For example,

| | |
|---|---|
| Function: | $\{(1, 4), (2, 0), (3, 7), (-2, 1), (-1, -5)\}$ |
| Inverse function: | $\{(4, 1), (0, 2), (7, 3), (1, -2), (-5, -1)\}$ |

**Note that the domain of the function becomes the range of the inverse function, and the range of the function becomes the domain of the inverse function.**

If we graph the points in the function and the points in the inverse function (**Fig. 9.8**), we see that the points are symmetric with respect to the line $y = x$.

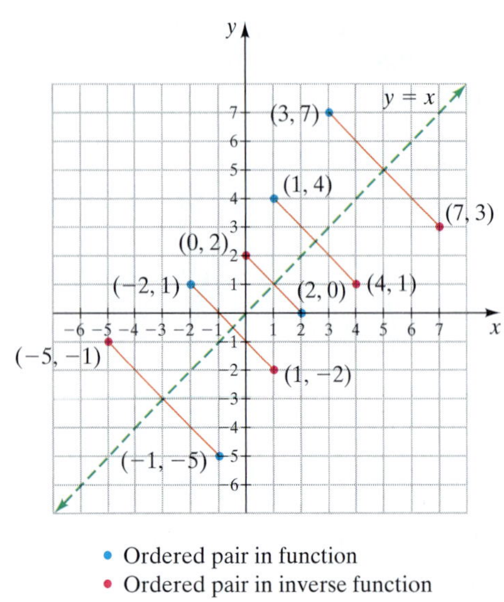


FIGURE 9.8

For a function $f(x)$, the notation $f^{-1}(x)$ represents its inverse function. Note that the $-1$ in the notation is *not* an exponent. Therefore, $f^{-1}(x) \neq \frac{1}{f(x)}$.

**Inverse Function**

If $f(x)$ is a one-to-one function with ordered pairs of the form $(x, y)$, its **inverse function**, $f^{-1}(x)$, is a one-to-one function with ordered pairs of the form $(y, x)$.

When a function $f(x)$ and its inverse function $f^{-1}(x)$ are graphed on the same axes, *$f(x)$ and $f^{-1}(x)$* are *symmetric about the line $y = x$* as seen in **Figure 9.8** on page 596.

When a one-to-one function is given as an equation, its inverse function can be found by the following procedure.

**To Find the Inverse Function of a One-to-One Function**

1. Replace $f(x)$ with $y$.
2. Interchange the two variables $x$ and $y$.
3. Solve the equation for $y$.
4. Replace $y$ with $f^{-1}(x)$ (this gives the inverse function using inverse function notation).

The following example will illustrate the procedure.

**EXAMPLE 4**

**a)** Find the inverse function of $f(x) = 4x + 2$.

**b)** On the same axes, graph both $f(x)$ and $f^{-1}(x)$.

*Solution* **a)** This function is one-to-one, therefore we will follow the four-step procedure.

| | | |
|---|---|---|
| | $f(x) = 4x + 2$ | Original function |
| Step 1 | $y = 4x + 2$ | Replace f(x) with y. |
| Step 2 | $x = 4y + 2$ | Interchange x and y. |
| Step 3 | $x - 2 = 4y$ | Solve for y. |
| | $\frac{x-2}{4} = y$ | |
| | or $y = \frac{x-2}{4}$ | |
| Step 4 | $f^{-1}(x) = \frac{x-2}{4}$ | Replace y with $f^{-1}(x)$. |

**b)** Below we show tables of values for $f(x)$ and $f^{-1}(x)$. The graphs of $f(x)$ and $f^{-1}(x)$ are shown in **Figure 9.9**.

| $x$ | $y = f(x)$ |
|---|---|
| 0 | 2 |
| 1 | 6 |

| $x$ | $y = f^{-1}(x)$ |
|---|---|
| 2 | 0 |
| 6 | 1 |

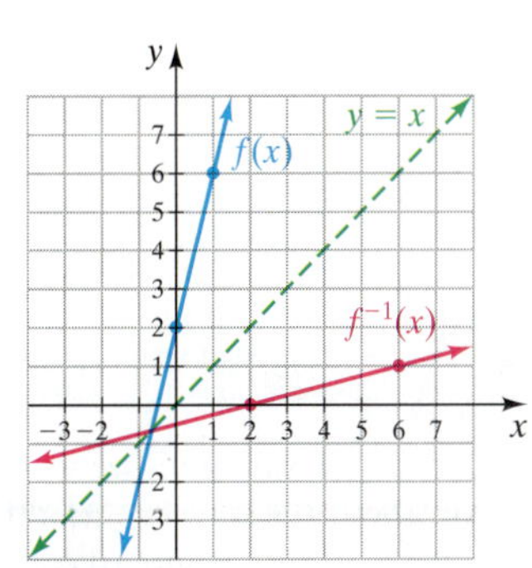


FIGURE 9.9

Note the symmetry of $f(x)$ and $f^{-1}(x)$ about the line $y = x$. Also note that both the domain and range of both $f(x)$ and $f^{-1}(x)$ are the set of real numbers, $\mathbb{R}$.

Now Try Exercise 67

In Chapter 7 when we solved equations containing cube roots, we cubed each side of the equation. To solve cubic equations, we raise each side of the equation to the one-third power, which is equivalent to taking the cube root of each side of the equation. Recall from Chapter 7 that $\sqrt[3]{a^3} = a$ for any real number $a$.

**EXAMPLE 5 ▸**

**a)** Find the inverse function of $f(x) = x^3 + 2$.

**b)** On the same axes, graph both $f(x)$ and $f^{-1}(x)$.

*Solution* **a)** This function is one-to-one; therefore we will follow the four-step procedure to find its inverse.

| | | |
|---|---|---|
| | $f(x) = x^3 + 2$ | *Original function* |
| **Step 1** | $y = x^3 + 2$ | *Replace f(x) with y.* |
| **Step 2** | $x = y^3 + 2$ | *Interchange x and y.* |
| **Step 3** | $x - 2 = y^3$ | *Solve for y.* |
| | $\sqrt[3]{x - 2} = \sqrt[3]{y^3}$ | *Take the cube root of both sides.* |
| | $\sqrt[3]{x - 2} = y$ | |
| | or $y = \sqrt[3]{x - 2}$ | |
| **Step 4** | $f^{-1}(x) = \sqrt[3]{x - 2}$ | *Replace y with $f^{-1}(x)$.* |

**b)** Below we show tables of values for $f(x)$ and $f^{-1}(x)$.

| $x$ | $y = f(x)$ |
|---|---|
| −2 | −6 |
| −1 | 1 |
| 0 | 2 |
| 1 | 3 |
| 2 | 10 |

| $x$ | $y = f^{-1}(x)$ |
|---|---|
| −6 | −2 |
| 1 | −1 |
| 2 | 0 |
| 3 | 1 |
| 10 | 2 |

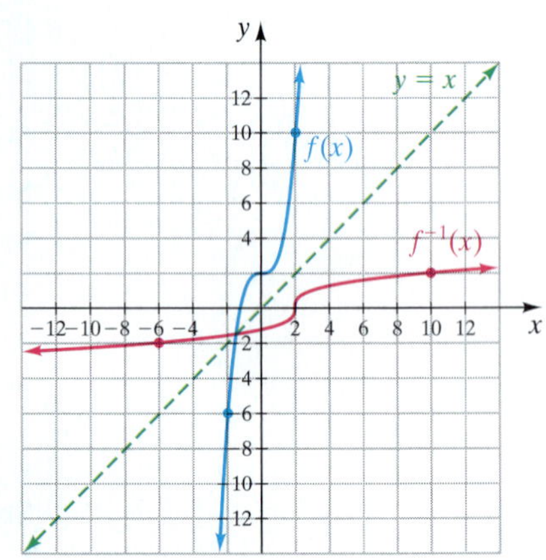


FIGURE 9.10

The graphs of $f(x)$ and $f^{-1}(x)$ are shown in **Figure 9.10**. Notice for each point $(a, b)$ on the graph of $f(x)$, the point $(b, a)$ appears on the graph of $f^{-1}(x)$. For example, the points (2, 10) and (−2, −6), indicated in blue, appear on the graph of $f(x)$, and the points (10, 2) and (−6, −2), indicated in red, appear on the graph of $f^{-1}(x)$.

**▸ Now Try Exercise 61**

**USING YOUR GRAPHING CALCULATOR**

In Example 5, we were given $f(x) = x^3 + 2$ and found that $f^{-1}(x) = \sqrt[3]{x - 2}$. The graphs of these functions are symmetric about the line $y = x$, although it may not appear that way on a graphing calculator. **Figure 9.11** on page 599 shows the two graphs using a standard calculator window.

Since the horizontal axis is longer than the vertical axis, and both axes have 10 tick marks, the graphs appear distorted. Many calculators have a feature to "square the axes." When this feature is used, the window is still rectangular, but the distance between the tick marks is equalized. To equalize the spacing between the tick marks on the vertical and horizontal axes on the

(*continued on the next page*)

TI-84 Plus calculator, press ZOOM then select option 5, ZSquare. **Figure 9.12** shows the graphs after this option is selected. A third illustration of the graphs can be obtained using ZOOM, option 4, ZDecimal. This option resets the $x$-axis to go from $-4.7$ to 4.7 and the $y$-axis to go from $-3.1$ to 3.1, as shown in **Figure 9.13**.

Standard window

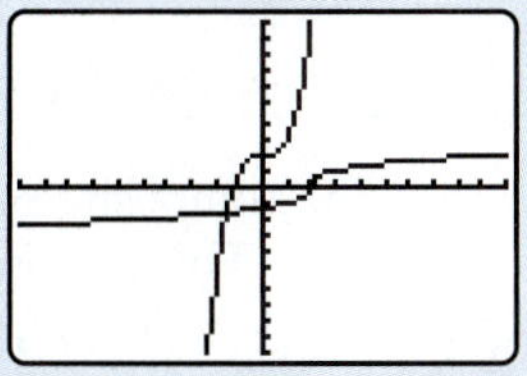

$-10, 10, 1, -10, 10, 1$

FIGURE 9.11

ZSquare window

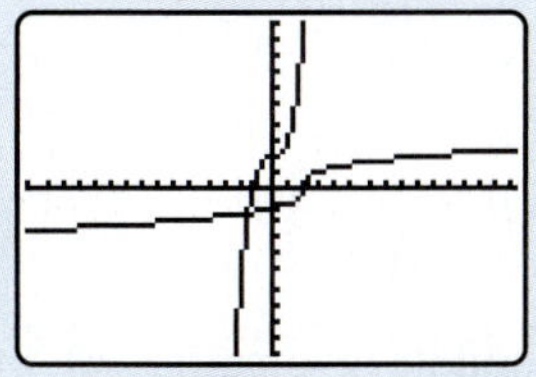

$\approx -15.2, \approx 15.2, 1, -10, 10, 1$

FIGURE 9.12

ZDecimal window

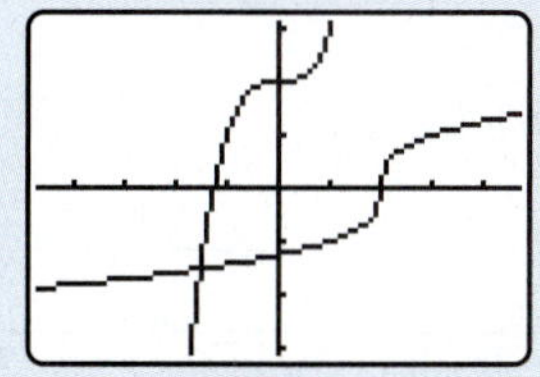

$-4.7, 4.7, 1, -3.1, 3.1, 1$

FIGURE 9.13

## 4 Find the Composition of a Function and Its Inverse

*If two functions $f(x)$ and $f^{-1}(x)$ are inverses of each other, $(f \circ f^{-1})(x) = x$ and $(f^{-1} \circ f)(x) = x$.*

**EXAMPLE 6** ▶ In Example 4, we determined that for $f(x) = 4x + 2$, $f^{-1}(x) = \dfrac{x - 2}{4}$. Show that

**a)** $(f \circ f^{-1})(x) = x$ **b)** $(f^{-1} \circ f)(x) = x$

*Solution*

**a)** To determine $(f \circ f^{-1})(x)$, substitute $f^{-1}(x)$ for each $x$ in $f(x)$.

$$f(x) = 4x + 2$$
$$(f \circ f^{-1})(x) = 4\left(\frac{x-2}{4}\right) + 2$$
$$= x - 2 + 2 = x$$

**b)** To determine $(f^{-1} \circ f)(x)$, substitute $f(x)$ for each $x$ in $f^{-1}(x)$.

$$f^{-1}(x) = \frac{x - 2}{4}$$
$$(f^{-1} \circ f)(x) = \frac{4x + 2 - 2}{4}$$
$$= \frac{4x}{4} = x$$

Thus, $(f \circ f^{-1})(x) = (f^{-1} \circ f)(x) = x$.

▶ **Now Try Exercise 77**

**EXAMPLE 7** ▶ In Example 5, we determined that $f(x) = x^3 + 2$ and $f^{-1}(x) = \sqrt[3]{x - 2}$ are inverse functions. Show that

**a)** $(f \circ f^{-1})(x) = x$ **b)** $(f^{-1} \circ f)(x) = x$

*Solution*

**a)** To determine $(f \circ f^{-1})(x)$, substitute $f^{-1}(x)$ for each $x$ in $f(x)$.

$$f(x) = x^3 + 2$$
$$(f \circ f^{-1})(x) = (\sqrt[3]{x - 2})^3 + 2$$
$$= x - 2 + 2 = x$$

**b)** To determine $(f^{-1} \circ f)(x)$, substitute $f(x)$ for each $x$ in $f^{-1}(x)$.

$$\begin{aligned} f^{-1}(x) &= \sqrt[3]{x - 2} \\ (f^{-1} \circ f)(x) &= \sqrt[3]{(x^3 + 2) - 2} \\ &= \sqrt[3]{x^3} = x \end{aligned}$$

Thus, $(f \circ f^{-1})(x) = (f^{-1} \circ f)(x) = x$.

▸ Now Try Exercise 79

Because a function and its inverse "undo" each other, the composite of a function with its inverse results in the given value from the domain. For example, for any function $f(x)$ and its inverse $f^{-1}(x)$, $(f^{-1} \circ f)(3) = 3$, and $(f \circ f^{-1})\left(-\frac{1}{2}\right) = -\frac{1}{2}$.

## EXERCISE SET 9.1

### Concept/Writing Exercises

1. Explain how to find $(f \circ g)(x)$ when you are given $f(x)$ and $g(x)$.
2. Explain how to find $(g \circ f)(x)$ when you are given $f(x)$ and $g(x)$.
3. **a)** What are one-to-one functions?
   **b)** Explain how you may determine whether a function is a one-to-one function.
4. Do all functions have inverse functions? If not, which functions do?
5. Consider the set of ordered pairs $\{(3, 5), (4, 2), (-1, 3), (0, -2)\}$.
   **a)** Is this set of ordered pairs a function? Explain.
   **b)** Does this function have an inverse? Explain.
   **c)** If this function has an inverse, give the inverse function. Explain how you determined your answer.
6. Suppose $f(x)$ and $g(x)$ are inverse functions.
   **a)** What is $(f \circ g)(x)$ equal to?
   **b)** What is $(g \circ f)(x)$ equal to?
7. What is the relationship between the domain and range of a function and the domain and range of its inverse function?
8. What is the value of $(f \circ f^{-1})(6)$? Explain.

### Practice the Skills

*For each pair of functions, find* **a)** $(f \circ g)(x)$, **b)** $(f \circ g)(4)$, **c)** $(g \circ f)(x)$, *and* **d)** $(g \circ f)(4)$.

9. $f(x) = x^2 + 1, g(x) = x + 2$
10. $f(x) = x^2 - 3, g(x) = x + 6$
11. $f(x) = x + 3, g(x) = x^2 + x - 4$
12. $f(x) = x + 2, g(x) = x^2 + 4x - 2$
13. $f(x) = \frac{1}{x}, g(x) = 2x + 3$
14. $f(x) = \frac{2}{x}, g(x) = x^2 + 1$
15. $f(x) = 3x + 1, g(x) = \frac{3}{x}$
16. $f(x) = x^2 - 5, g(x) = \frac{4}{x}$
17. $f(x) = x^2 + 1, g(x) = x^2 + 5$
18. $f(x) = x^2 - 4, g(x) = x^2 + 3$
19. $f(x) = x - 4, g(x) = \sqrt{x + 5}, x \geq -5$
20. $f(x) = \sqrt{x + 6}, x \geq -6, g(x) = x + 7$

*In Exercises 21–42, determine whether each function is a one-to-one function.*

**21.**

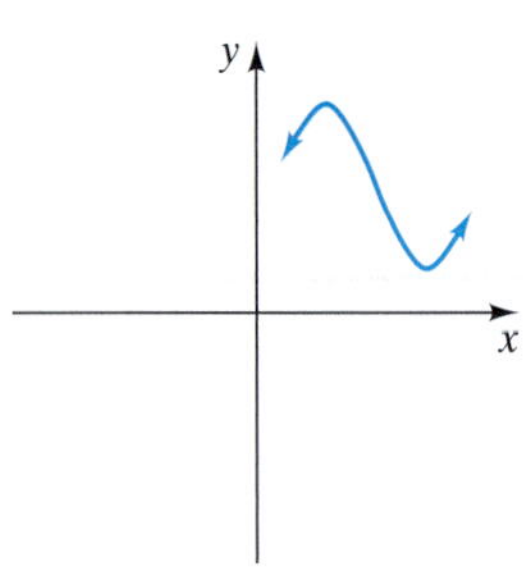


**22.**

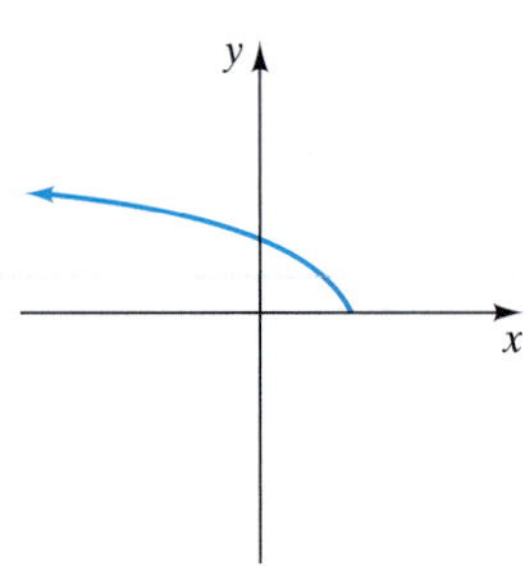


**23.**

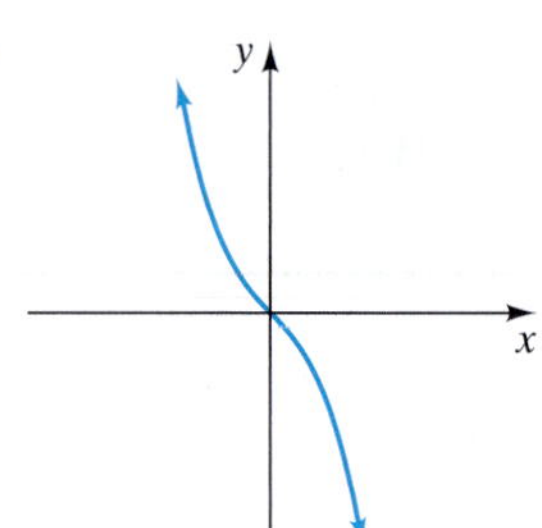


**24.**

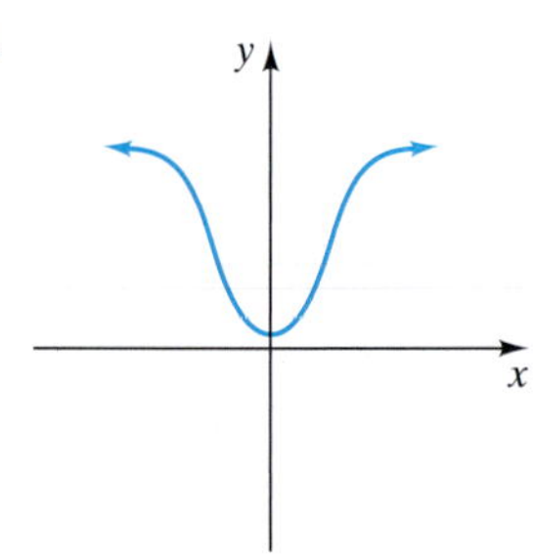

25. $\{(2, 4), (3, -7), (5, 3), (-6, 0)\}$
26. $\{(-4, 2), (2, 3), (4, 1), (0, 4)\}$
27. $\{(-4, 2), (5, 3), (0, 2), (4, 8)\}$
28. $\{(0, 5), (1, 4), (-3, 5), (4, 2)\}$
29. $y = 2x + 5$
30. $y = 3x - 8$
31. $y = x^2 - 1$
32. $y = -x^2 + 3$
33. $y = x^2 - 2x + 5$
34. $y = x^2 - 2x + 6, x \geq 1$
35. $y = x^2 - 9, x \geq 0$
36. $y = x^2 - 9, x \leq 0$
37. $y = \sqrt{x}$
38. $y = -\sqrt{x}$
39. $y = |x|$
40. $y = -|x|$
41. $y = \sqrt[3]{x}$
42. $y = x^3$

*In Exercises 43–48, for the given function, find the domain and range of both $f(x)$ and $f^{-1}(x)$.*

43. $\{(4, 0), (8, 9), (2, 7), (-1, 6), (-2, 4)\}$
44. $\left\{(-2, -3), (-4, 0), (5, 3), (6, 2), \left(2, \frac{1}{2}\right)\right\}$

45.
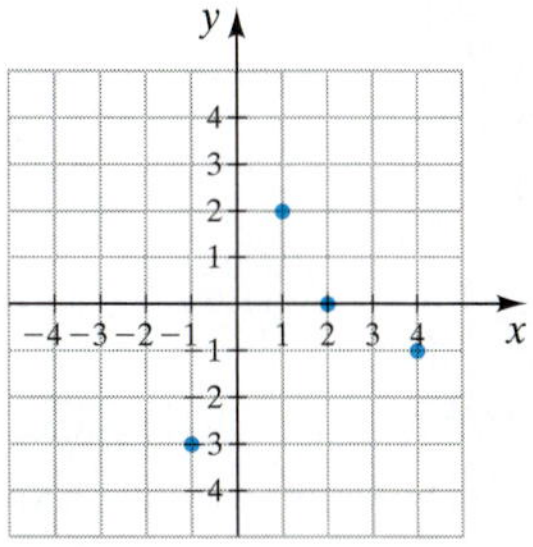

46.
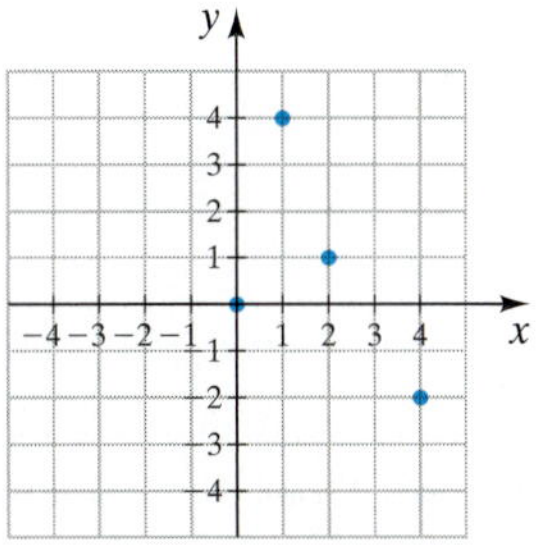

47.
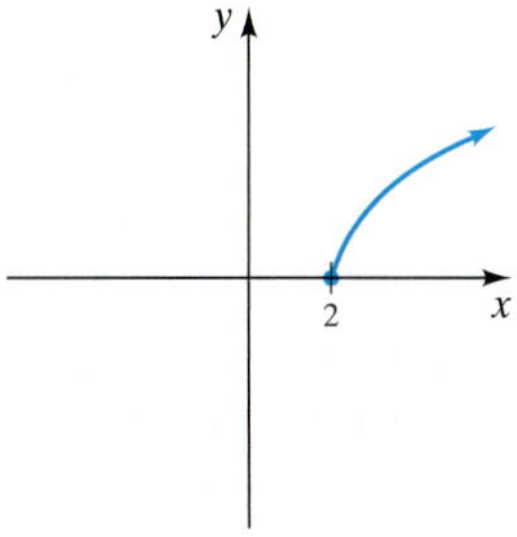

48.
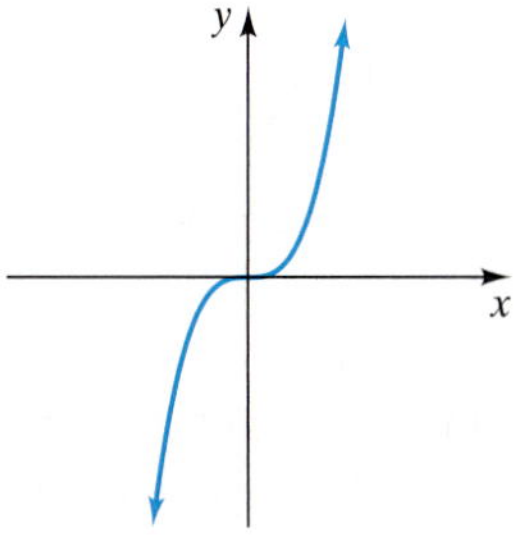

*For each function,* **a)** *determine whether it is one-to-one;* **b)** *if it is one-to-one, find its inverse function.*

49. $f(x) = x - 2$
50. $g(x) = x + 5$
51. $h(x) = 4x$
52. $k(x) = 2x - 7$
53. $p(x) = 3x^2$
54. $r(x) = |x|$
55. $t(x) = x^2 + 3$
56. $m(x) = -x^2 + x + 8$
57. $g(x) = \frac{1}{x}$
58. $h(x) = \frac{5}{x}$
59. $f(x) = x^2 + 10$
60. $g(x) = x^3 + 9$
61. $g(x) = x^3 - 6$
62. $f(x) = \sqrt{x}, x \geq 0$
63. $g(x) = \sqrt{x + 2}, x \geq -2$
64. $f(x) = x^2 - 3, x \geq 0$
65. $h(x) = x^2 - 4, x \geq 0$
66. $h(x) = |x|$

*For each one-to-one function,* **a)** *find $f^{-1}(x)$ and* **b)** *graph $f(x)$ and $f^{-1}(x)$ on the same axes.*

67. $f(x) = 2x + 8$
68. $f(x) = -3x + 6$
69. $f(x) = \sqrt{x}, x \geq 0$
70. $f(x) = -\sqrt{x}, x \geq 0$
71. $f(x) = \sqrt{x - 1}, x \geq 1$
72. $f(x) = \sqrt{x + 4}, x \geq -4$
73. $f(x) = \sqrt[3]{x}$
74. $f(x) = \sqrt[3]{x + 3}$
75. $f(x) = \frac{1}{x}, x > 0$
76. $f(x) = \frac{1}{x}$

*For each pair of inverse functions, show that $(f \circ f^{-1})(x) = x$ and $(f^{-1} \circ f)(x) = x$.*

77. $f(x) = x - 8, f^{-1}(x) = x + 8$
78. $f(x) = 7x + 3, f^{-1}(x) = \frac{x - 3}{7}$
79. $f(x) = \frac{1}{2}x + 3, f^{-1}(x) = 2x - 6$
80. $f(x) = -\frac{1}{3}x + 2, f^{-1}(x) = -3x + 6$
81. $f(x) = \sqrt[3]{x - 2}, f^{-1}(x) = x^3 + 2$
82. $f(x) = \sqrt[3]{x + 9}, f^{-1}(x) = x^3 - 9$
83. $f(x) = \frac{3}{x}, f^{-1}(x) = \frac{3}{x}$
84. $f(x) = \sqrt{x + 5}, f^{-1}(x) = x^2 - 5, x \geq 0$

## Problem Solving

**85.** Is $(f \circ g)(x) = (g \circ f)(x)$ for all values of $x$? Explain and give an example to support your answer.

**86.** Consider the functions $f(x) = \sqrt{x+5}, x \geq -5$, and $g(x) = x^2 - 5, x \geq 0$.

**a)** Show that $(f \circ g)(x) = (g \circ f)(x)$ for $x \geq 0$.

**b)** Explain why we need to stipulate that $x \geq 0$ for part **a)** to be true.

**87.** Consider the functions $f(x) = x^3 + 2$ and $g(x) = \sqrt[3]{x-2}$.

**a)** Show that $(f \circ g)(x) = (g \circ f)(x)$.

**b)** What are the domains of $f(x)$, $g(x)$, $(f \circ g)(x)$, and $(g \circ f)(x)$? Explain.

**88.** For the function $f(x) = x^3, f(2) = 2^3 = 8$. Explain why $f^{-1}(8) = 2$.

**89.** For the function $f(x) = x^4, x > 0, f(2) = 16$. Explain why $f^{-1}(16) = 2$.

**90.** The function $f(x) = 12x$ converts feet, $x$, into inches. Find the inverse function that converts inches into feet. In the inverse function, what do $x$ and $f^{-1}(x)$ represent?

**91.** The function $f(x) = 3x$ converts yards, $x$, into feet. Find the inverse function that converts feet into yards. In the inverse function, what do $x$ and $f^{-1}(x)$ represent?

**92.** The function $f(x) = \frac{22}{15}x$ converts miles per hour, $x$, into feet per second. Find the inverse function that converts feet per second into miles per hour.

**93.** The function $f(x) = \frac{5}{9}(x - 32)$ converts degrees Fahrenheit, $x$, to degrees Celsius. Find the inverse function that changes degrees Celsius into degrees Fahrenheit.

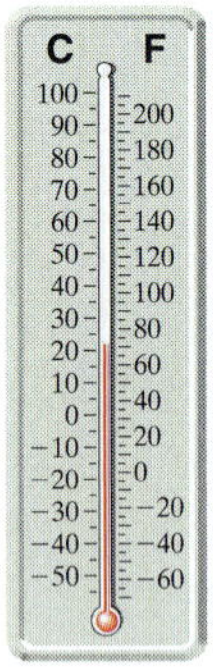


**94. a)** Does the function $f(x) = |x|$ have an inverse? Explain.

**b)** If the domain is limited to $x \geq 0$, does the function have an inverse? Explain.

**c)** Find the inverse function of $f(x) = |x|, x \geq 0$.

**Composition of Functions** *In Exercises 95–98, the functions $f(x)$ and $g(x)$ are given. Determine the composition $(f \circ g)(x)$. For the composition function, what does $x$ represent and what does $(f \circ g)(x)$ represent?*

**95.** $f(x) = 16x$ converts pounds, $x$, to ounces. $g(x) = 28.35x$ converts ounces, $x$ to grams.

**96.** $f(x) = 2000x$ converts tons, $x$, to pounds. $g(x) = 16x$ converts pounds, $x$, to ounces.

**97.** $f(x) = 3x$ converts yards, $x$, to feet. $g(x) = 0.305x$ converts feet, $x$, to meters.

**98.** $f(x) = 1760x$ converts miles, $x$, to yards. $g(x) = 0.915x$ converts yards, $x$, to meters.

*Use your graphing calculator to determine whether the following functions are inverses.*

**99.** $f(x) = 3x - 4, g(x) = \frac{x}{3} + \frac{4}{3}$

**100.** $f(x) = \sqrt{4 - x^2}, g(x) = \sqrt{4 - 2x}$

**101.** $f(x) = x^3 - 12, g(x) = \sqrt[3]{x + 12}$

**102.** $f(x) = x^5 + 5, g(x) = \sqrt[5]{x - 5}$

## Challenge Problems

**103. Area** When a pebble is thrown into a pond, the circle formed by the pebble hitting the water expands with time. The area of the expanding circle may be found by the formula $A = \pi r^2$. The radius, $r$, of the circle, in feet, is a function of time, $t$, in seconds. Suppose that the function is $r(t) = 2t$.

**a)** Find the radius of the circle at 3 seconds.

**b)** Find the area of the circle at 3 seconds.

**c)** Express the area as a function of time by finding $A \circ r$.

**d)** Using the function found in part **c)**, find the area of the circle at 3 seconds.

**e)** Do your answers in parts **b)** and **d)** agree? If not, explain.

**104. Surface Area** The surface area, $S$, of a spherical balloon of radius $r$, in inches, is found by $S(r) = 4\pi r^2$. If the balloon is being blown up at a constant rate by a machine, then the radius of the balloon is a function of time. Suppose that this function is $r(t) = 1.2t$, where $t$ is in seconds.

**a)** Find the radius of the balloon at 2 seconds.

**b)** Find the surface area at 2 seconds.

**c)** Express the surface area as a function of time by finding $S \circ r$.

**d)** Using the function found in part **c)**, find the surface area after 2 seconds.

**e)** Do your answers in parts **b)** and **d)** agree? If not, explain why not.

## Group Activity

*Discuss and answer Exercise 105 as a group.*

**105.** Consider the function $f(x) = 2^x$. This is an example of an *exponential function*, which we will discuss in the next section.

**a)** Graph this function by substituting values for $x$ and finding the corresponding values of $f(x)$.

**b)** Do you think this function has an inverse? Explain your answer.

**c)** Using the graph in part **a)**, draw the inverse function, $f^{-1}(x)$ on the same axes.

**d)** Explain how you obtained the graph of $f^{-1}(x)$.

## Cumulative Review Exercises

[1.3] **106.** Divide $\left|\frac{-9}{4}\right| \div \left|\frac{-4}{9}\right|$.

[3.5] **107.** Determine the equation of a line in standard form that passes through $\left(\frac{1}{2}, 3\right)$ and is parallel to the graph of $2x + 3y - 9 = 0$.

[6.3] **108.** Simplify $\dfrac{\frac{3}{x^2} - \frac{2}{x}}{\frac{x}{6}}$.

[6.4] **109.** Solve the formula $\frac{1}{f} = \frac{1}{p} + \frac{1}{q}$ for $p$.

[8.1] **110.** Solve $x^2 + 2x - 10 = 0$ by completing the square.

# 9.2 Exponential Functions

1 Graph exponential functions.

2 Solve applications of exponential functions.

## 1 Graph Exponential Functions

We often read about things that are growing exponentially. For example, you may read that the world population is growing exponentially or that the use of e-mail is growing exponentially. What does this indicate? The graph in **Figure 9.14** shows population growth worldwide. The graph in **Figure 9.15** also shows the shipment of smart handheld devices. Both graphs have the same general shape, as indicated by the curves. Both graphs are exponential functions and rise rapidly.

In the quadratic function $f(x) = x^2$, the variable is the base and the exponent is a constant. In the function $f(x) = 2^x$, the constant is the base and the variable is the exponent. The function $f(x) = 2^x$ is an example of an *exponential function*, which we define on page 604.

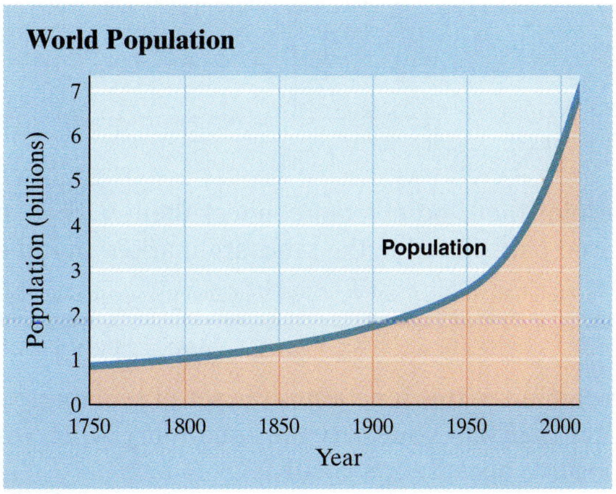


FIGURE 9.14

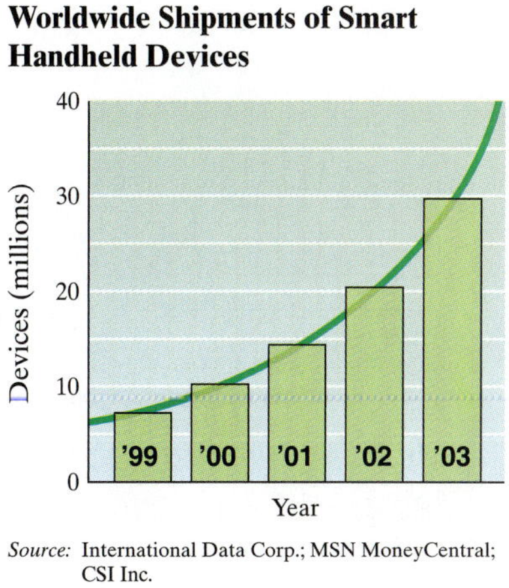


FIGURE 9.15

**Exponential Function**

For any real number $a > 0$ and $a \neq 1$,

$$f(x) = a^x$$

is an **exponential function**.

An exponential function is a function of the form $f(x) = a^x$, where $a$ is a positive real number not equal to 1. Notice the variable is in the exponent.

### Examples of Exponential Functions

$$f(x) = 2^x, \qquad g(x) = 5^x, \qquad h(x) = \left(\frac{1}{2}\right)^x$$

Since $y = f(x)$, functions of the form $y = a^x$ are also exponential functions. Exponential functions can be graphed by selecting values for $x$, finding the corresponding values of $y$ [or $f(x)$], and plotting the points.

Before we graph exponential functions, let's discuss some characteristics of the graphs of exponential functions.

**Graphs of Exponential Functions**

For all exponential functions of the form $y = a^x$ or $f(x) = a^x$, where $a > 0$ and $a \neq 1$,

1. The domain of the function is $(-\infty, \infty)$.
2. The range of the function is $(0, \infty)$.
3. The graph of the function passes through the points $\left(-1, \frac{1}{a}\right)$, $(0, 1)$, and $(1, a)$.

In most cases, a reasonably good exponential graph can be drawn from only the three points listed in item 3. When $a > 1$, the graph becomes almost horizontal to the left of $\left(-1, \frac{1}{a}\right)$ and somewhat vertical to the right of $(1, a)$; see Example 1. When $0 < a < 1$, the graph becomes almost horizontal to the right of $(1, a)$ and somewhat vertical to the left of $\left(-1, \frac{1}{a}\right)$; see Example 2.

**EXAMPLE 1** ▸ Graph the exponential function $y = 2^x$. State the domain and range of the function.

*Solution* The function is of the form $y = a^x$, where $a = 2$. First, construct a table of values. In the table, the three points listed in item 3 of the box are shown in red.

| $x$ | $-4$ | $-3$ | $-2$ | $-1$ | $0$ | $1$ | $2$ | $3$ | $4$ |
|---|---|---|---|---|---|---|---|---|---|
| $y$ | $\frac{1}{16}$ | $\frac{1}{8}$ | $\frac{1}{4}$ | $\frac{1}{2}$ | $1$ | $2$ | $4$ | $8$ | $16$ |

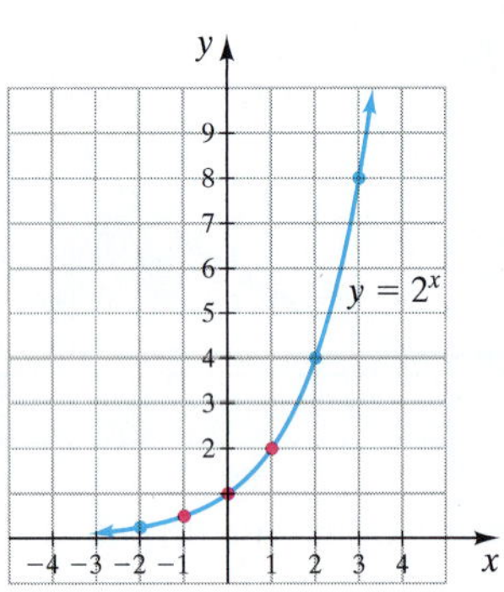

FIGURE 9.16

Now plot these points and connect them with a smooth curve (**Fig. 9.16**). The three ordered pairs in red in the table are marked in red on the graph.

Domain: $\mathbb{R}$

Range: $\{y \mid y > 0\}$

The domain of this function is the set of real numbers, $\mathbb{R}$. The range is the set of values greater than 0. If you study the equation $y = 2^x$, you should realize that $y$ must always be positive because 2 is positive.

▸ Now Try Exercise 7

**EXAMPLE 2** ▸ Graph $y = \left(\frac{1}{2}\right)^x$. State the domain and range of the function.

*Solution* This function is of the form $y = a^x$, where $a = \frac{1}{2}$. Construct a table of values and plot the curve (**Fig. 9.17**).

| $x$ | $-4$ | $-3$ | $-2$ | $-1$ | $0$ | $1$ | $2$ | $3$ | $4$ |
|---|---|---|---|---|---|---|---|---|---|
| $y$ | $16$ | $8$ | $4$ | $2$ | $1$ | $\frac{1}{2}$ | $\frac{1}{4}$ | $\frac{1}{8}$ | $\frac{1}{16}$ |

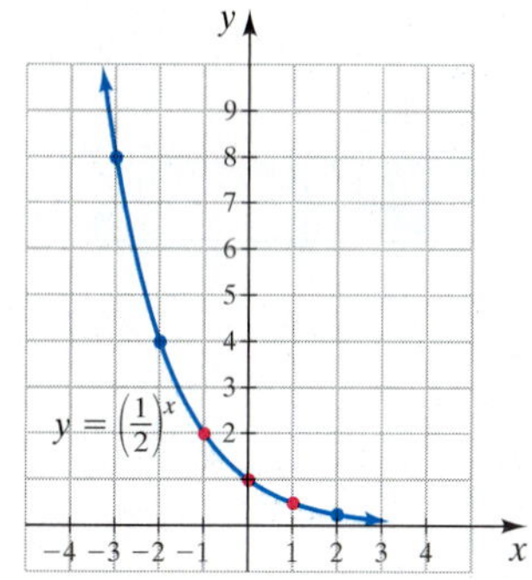


FIGURE 9.17

The domain is the set of real numbers, $\mathbb{R}$. The range is $\{y|y > 0\}$.

▸ **Now Try Exercise 13**

Note that the graphs in **Figures 9.16** and **9.17** are both graphs of one-to-one functions. *The graphs of exponential functions of the form* $y = a^x$ *are similar to* **Figure 9.16** *when* $a > 1$ *and similar to* **Figure 9.17** *when* $0 < a < 1$. Note that $y = 1^x$ is not a one-to-one function, so we exclude it from our discussion of exponential functions.

What will the graph of $y = 2^{-x}$ look like? Remember that $2^{-x}$ means $\frac{1}{2^x}$ *or* $\left(\frac{1}{2}\right)^x$. Thus, the graph of $y = 2^{-x}$ will be identical to the graph in **Figure 9.17.** Now consider the equation $y = \left(\frac{1}{2}\right)^{-x}$. This equation may be rewritten as $y = 2^x$ since $\left(\frac{1}{2}\right)^{-x} = \left(\frac{2}{1}\right)^x = 2^x$. Thus, the graph of $y = \left(\frac{1}{2}\right)^{-x}$ will be identical to the graph in **Figure 9.16**.

### USING YOUR GRAPHING CALCULATOR

In **Figure 9.18**, we show the graph of the function $y = 2^x$ on the standard window of a graphing calculator. In this chapter, we will sometimes use equations like $y = 2000(1.08)^x$. If you were to graph this function on a standard calculator window, you would not see any of the graph. Can you explain why? By observing the function, can you determine the $y$-intercept of the graph? To determine the $y$-intercept, substitute 0 for $x$. When you do so, you find that the $y$-intercept is at $2000(1.08)^0 = 2000(1) = 2000$. In **Figure 9.19**, we show the graph of $y = 2000(1.08)^x$.

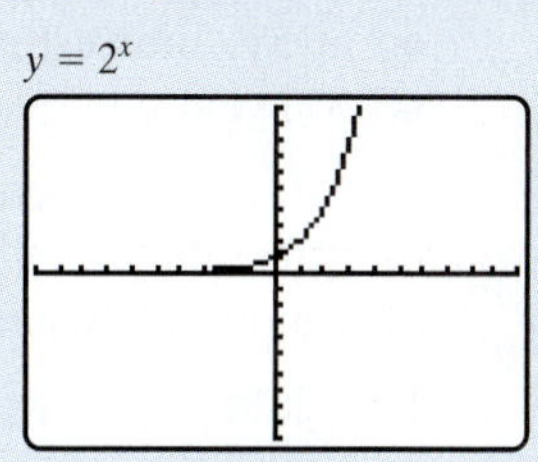


FIGURE 9.18

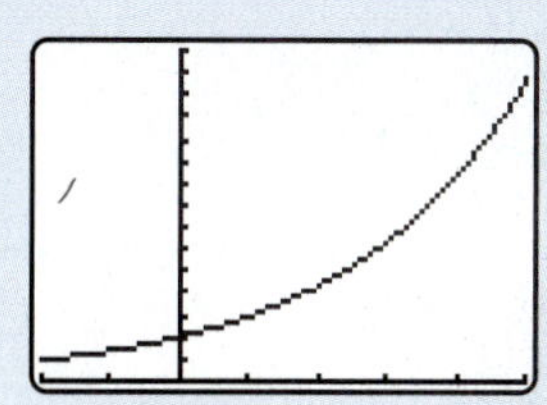


FIGURE 9.19

**EXAMPLE 3** ▸ **Pennies Add Up** Jennifer Hewlett told her young son that if he did his chores, she would give him 2 cents the first week and double the amount each week for the next 10 weeks. The number of cents her son would receive in any given week, $w$, can be determined by the function $n(w) = 2^w$. Determine the number of cents Jennifer would give her son in week 8.

*Solution* By evaluating $2^8$ on a calculator, we determine that in week 8 Jennifer would give her son 256 cents, or \$2.56.

▸ **Now Try Exercise 35**

## 2 Solve Applications of Exponential Functions

Exponential functions are often used to describe the growth and decay of certain quantities. The next four examples are illustrations of exponential functions.

**EXAMPLE 4 ▸ Value of a Jeep** Ronald Yates just bought a new Jeep for \$22,000. Assume the value of the Jeep depreciates at a rate of 20% per year. Therefore, the value of the Jeep is 80% of the previous year's value. One year from now, its value will be \$22,000(0.80). Two years from now, its value will be $\$22{,}000(0.80)(0.80) = \$22{,}000(0.80)^2$ and so on. Therefore, the formula for the value of the Jeep is

$$v(t) = 22{,}000(0.80)^t$$

where $t$ is time in years. Find the value of the Jeep **a)** 1 year from now and **b)** 5 years from now.

### Solution

**a)** To find the value 1 year from now, substitute 1 for $t$.

$$\begin{aligned} v(t) &= 22{,}000(0.80)^t \\ v(1) &= 22{,}000(0.80)^1 && \text{Substitute 1 for } t. \\ &= 17{,}600 \end{aligned}$$

One year from now, the value of the Jeep will be \$17,600.

**b)** To find the value 5 years from now, substitute 5 for $t$.

$$\begin{aligned} v(t) &= 22{,}000(0.80)^t \\ v(5) &= 22{,}000(0.80)^5 && \text{Substitute 5 for } t. \\ &= 22{,}000(0.32768) \\ &= 7208.96 \end{aligned}$$

Five years from now, the value of the Jeep will be \$7208.96.

▸ **Now Try Exercise 49**

**EXAMPLE 5 ▸ Compound Interest** We have seen the *compound interest formula* $A = p\left(1 + \frac{r}{n}\right)^{nt}$ in earlier chapters. When interest is compounded periodically (yearly, monthly, quarterly), this formula can be used to find the amount, $A$.

In the formula, $r$ is the interest rate, $p$ is the principal, $n$ is the number of compounding periods per year, and $t$ is the number of years. Suppose that \$10,000 is invested at 5% interest compounded quarterly for 6 years. Find the amount in the account after 6 years.

### Solution

**Understand** We are given that the principal, $p$, is \$10,000. We are also given that the interest rate, $r$, is 5%. Because the interest is compounded quarterly, the number of compounding periods, $n$, is 4. The money is invested for 6 years. Therefore, $t$ is 6.

**Translate** Now we substitute these values into the formula

$$\begin{aligned} A &= p\left(1 + \frac{r}{n}\right)^{nt} \\ &= 10{,}000\left(1 + \frac{0.05}{4}\right)^{4(6)} \end{aligned}$$

**Carry Out**

$$\begin{aligned} &= 10{,}000(1 + 0.0125)^{24} \\ &= 10{,}000(1.0125)^{24} \\ &\approx 10{,}000(1.347351) && \text{From a calculator} \\ &\approx 13{,}435.51 \end{aligned}$$

**Answer** The original \$10,000 has grown to about \$13,435.51 after 6 years.

▸ **Now Try Exercise 41**

**EXAMPLE 6 ▸ Carbon 14 Dating** Carbon 14 dating is used by scientists to find the age of fossils and artifacts. The formula used in carbon dating is

$$A = A_0 \cdot 2^{-t/5600}$$

where $A_0$ represents the amount of carbon 14 present when the fossil was formed and $A$ represents the amount of carbon 14 present after $t$ years. If 500 grams of carbon 14 were present when an organism died, how many grams will be found in the fossil 2000 years later?

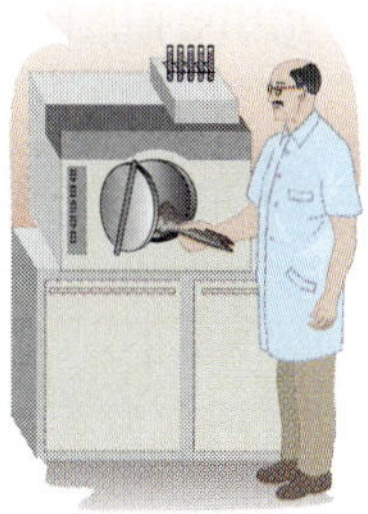

*Solution* **Understand** When the fossil died, 500 grams of carbon 14 were present. Therefore, $A_0 = 500$. To find out how many grams of carbon 14 will be present 2000 years later, we substitute 2000 for $t$ in the formula.

**Translate**

$$\begin{aligned} A &= A_0 \cdot 2^{-t/5600} \\ &= 500(2)^{-2000/5600} \end{aligned}$$

**Carry Out**

$$\begin{aligned} &\approx 500(0.7807092) \quad \text{From a calculator} \\ &\approx 390.35 \text{ grams} \end{aligned}$$

**Answer** After 2000 years, about 390.35 of the original 500 grams of carbon 14 are still present.

▸ **Now Try Exercise 43**

**EXAMPLE 7 ▸ Medicare Premiums** Monthly Medicare Part B premiums have been on the rise since 2000. The graph in **Figure 9.20** shows monthly Medicare Part B premiums for the years from 2000 to 2005.

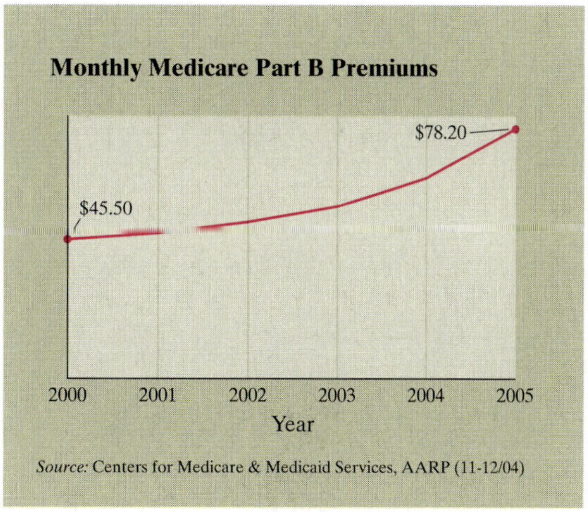


FIGURE 9.20

An exponential function that closely approximates this curve is $f(t) = 44.584(1.11)^t$. In this function, $f(t)$ is the monthly premium and $t$ is the number of years since 2000. Assume that the monthly Medicare Part B premiums continue to rise as in the past. Use this function to estimate the monthly Medicare Part B premiums in **a)** 2006 and **b)** 2010.

*Solution.* **a) Understand** In this function, $t$ is years since 2000. The year 2006 would represent $t = 6$ as 2006 is 6 years after 2000. To find the monthly premium in 2006, we need to evaluate the function for $t = 6$.

**Translate and Carry Out**

$$f(t) = 44.584(1.11)^t$$

$$f(6) = 44.584(1.11)^6 \approx \$83.39 \quad \text{From a calculator}$$

**Answer** Therefore, the monthly Medicare Part B premium in 2006 would be about \$83.39.

**b)** Since 2010 is 10 years after 2000, to find the monthly premium, we need to evaluate the function for $t = 10$.

$$f(t) = 44.584(1.11)^t$$
$$f(10) = 44.584(1.11)^{10} \approx \$126.59 \quad \text{From a calculator}$$

**Answer** Therefore, the monthly Medicare Part B premium in 2010 would be about \$126.59.

▸ Now Try Exercise 53

## EXERCISE SET 9.2



### Concept/Writing Exercises

1. What are exponential functions?
2. Consider the exponential function $y = 2^x$.
   a) As $x$ increases, what happens to $y$?
   b) Can $y$ ever be 0? Explain.
   c) Can $y$ ever be negative? Explain.
3. Consider the exponential function $y = \left(\frac{1}{2}\right)^x$.
   a) As $x$ increases, what happens to $y$?
   b) Can $y$ ever be 0? Explain.
   c) Can $y$ ever be negative? Explain.
4. Consider the exponential function $y = 2^{-x}$. Write an equivalent exponential function that does not contain a negative sign in the exponent. Explain how you obtained your answer.
5. Consider the equations $y = 2^x$ and $y = 3^x$.
   a) Will both graphs have the same or different $y$-intercepts? Determine their $y$-intercepts.
   b) How will the graphs of the two functions compare?
6. Consider the equations $y = \left(\frac{1}{2}\right)^x$ and $y = \left(\frac{1}{3}\right)^x$.
   a) Will both graphs have the same or different $y$-intercepts? Determine their $y$-intercepts.
   b) How will the graphs of the two functions compare?

### Practice the Skills

*Graph each exponential function.*

7. $y = 2^x$
8. $y = 3^x$
9. $y = \left(\frac{1}{2}\right)^x$
10. $y = \left(\frac{1}{3}\right)^x$
11. $y = 4^x$
12. $y = 5^x$
13. $y = \left(\frac{1}{4}\right)^x$
14. $y = \left(\frac{1}{5}\right)^x$
15. $y = 3^{-x}$
16. $y = 4^{-x}$
17. $y = \left(\frac{1}{3}\right)^{-x}$
18. $y = \left(\frac{1}{4}\right)^{-x}$
19. $y = 2^{x-1}$
20. $y = 2^{x+1}$
21. $y = \left(\frac{1}{3}\right)^{x+1}$
22. $y = \left(\frac{1}{3}\right)^{x-1}$
23. $y = 2^x + 1$
24. $y = 2^x - 1$
25. $y = 3^x - 1$
26. $y = 3^x + 2$

### Problem Solving

27. We stated earlier that, for exponential functions $f(x) = a^x$, the value of $a$ cannot equal 1.
   a) What does the graph of $f(x) = a^x$ look like when $a = 1$?
   b) Is $f(x) = a^x$ a function when $a = 1$?
   c) Does $f(x) = a^x$ have an inverse function when $a = 1$? Explain your answer.
28. How will the graphs of $y = a^x$ and $y = a^x + k, k > 0$, compare?
29. How will the graphs of $y = a^x$ and $y = a^x - k, k > 0$, compare?
30. For $a > 1$, how will the graphs of $y = a^x$ and $y = a^{x+1}$ compare?
31. For $a > 1$, how will the graphs of $y = a^x$ and $y = a^{x+2}$ compare?
32. a) Is $y = x^\pi$ an exponential function? Explain.
   b) Is $y = \pi^x$ an exponential function? Explain.

33. **U.S. Population** The following graph shows the growth in the U.S. population of people age 85 and older for the years from 1960 to 2000 and projected to 2050. The exponential function that closely approximates this graph is

$$f(t) = 0.592(1.042)^t$$

In this function, $f(t)$ is the population, in millions, of people age 85 and older and $t$ is the number of years since 1960. Assuming this trend continues, use this function to estimate the number of U.S. people age 85 and older in **a)** 2060. **b)** 2100.

**U.S. Population 85 and Older**

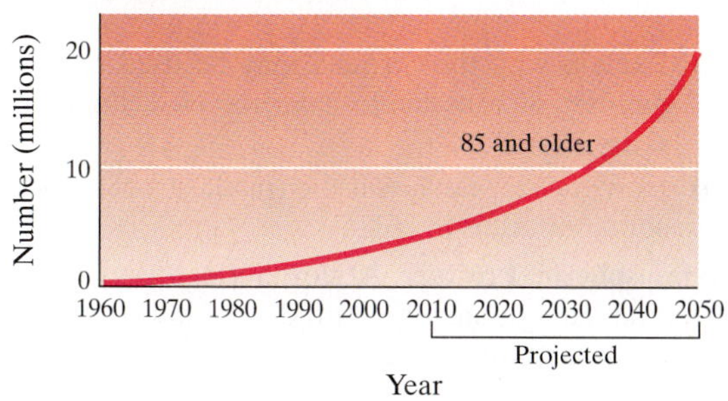


*Source:* U.S. Census Bureau, Decennial Census and Projections, Older Americans 2004, www.agingstats.gov

34. **World Population** Since about 1650 the world population has been growing exponentially. The exponential function that closely approximates the world population from 1650 and projected to 2015 is

$$f(t) = \frac{1}{2}(2.718)^{0.0072t}$$

In this function, $f(t)$ is the world population, in billions of people, and $t$ is the number of years since 1650. If this trend continues, estimate the world population in **a)** 2010. **b)** 2015.

35. **Doubling** If \$2 is doubled each day for 9 days, determine the amount on day 9.

36. **Doubling** If \$2 is doubled each day for 12 days, determine the amount on day 12.

37. **Simple and Compound Interest** The following graph indicates linear growth of \$100 invested at 7% simple interest and exponential growth at 7% interest compounded annually. In the formulas, $A$ represents the amount in dollars and $t$ represents the time in years.

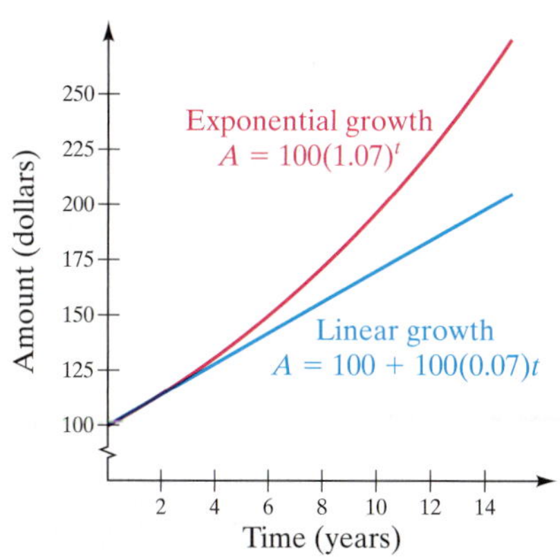


**a)** Use the graph to estimate the doubling time for \$100 invested at 7% simple interest.

**b)** Estimate the doubling time for \$100 invested at 7% interest compounded annually.

**c)** Estimate the difference in amounts after 10 years for \$100 invested by each method.

**d)** Most banks compound interest daily instead of annually. What effect does this have on the total amount? Explain.

38. **Outstanding Consumer Credit** The following graph shows the outstanding consumer credit, in trillions of dollars, for the years 2001 to 2004. The exponential function that closely approximates these data is

$$f(t) = 1.841(1.045)^t$$

In this function, $f(t)$ is the outstanding consumer credit, in trillions of dollars, and $t$ is the years since 2001.
Assuming that this trend continues, use this function to estimate the outstanding consumer credit in **a)** 2007. **b)** 2011.

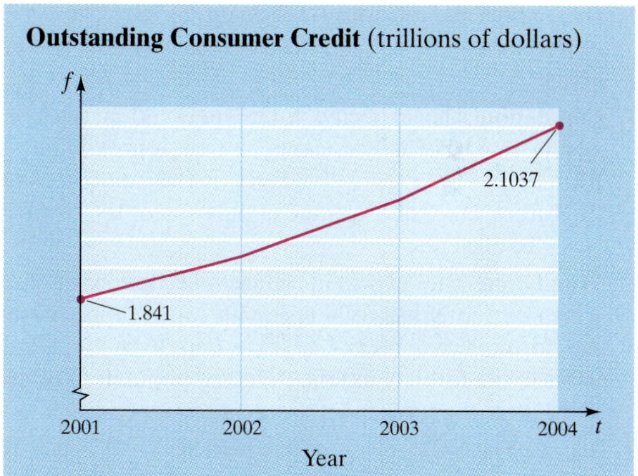


*Source:* www.federalreserve.gov/releases (based on 2004 dollars)

39. **Bacteria in a Petri Dish** Five bacteria are placed in a petri dish. The population will triple every day. The formula for the number of bacteria in the dish on day $t$ is

$$N(t) = 5(3)^t$$

where $t$ is the number of days after the five bacteria are placed in the dish. How many bacteria are in the dish 2 days after the four bacteria are placed in the dish?

40. **Bacteria in a Petri Dish** Refer to Exercise 39. How many bacteria are in the dish 6 days after the five bacteria are placed in the dish?

41. **Compound Interest** If Don Gecewicz invests \$5000 at 6% interest compounded quarterly, find the amount after 4 years (see Example 5).

42. **Compound Interest** If Don Treadwell invests \$8000 at 4% interest compounded quarterly, find the amount after 5 years.

43. **Carbon 14 Dating** If 12 grams of carbon 14 are originally present in a certain animal bone, how much will remain at the end of 1000 years? Use $A = A_0 \cdot 2^{-t/5600}$ (see Example 6).

44. **Carbon 14 Dating** If 60 grams of carbon 14 are originally present in the fossil Tim Jonas found at an archeological site, how much will remain after 10,000 years?

**45. Radioactive Substance** The amount of a radioactive substance present, in grams, at time $t$, in years, is given by the formula $y = 80(2)^{-0.4t}$. Find the number of grams present in **a)** 10 years. **b)** 100 years.

**46. Radioactive Substance** The amount of a radioactive substance present, in grams, at time $t$, in years, is given by the formula $y = 20(3)^{-0.6t}$. Find the number of grams present in 4 years.

**47. Population** The expected future population of Ackworth, which now has 2000 residents, can be approximated by the formula $y = 2000(1.2)^{0.1t}$, where $t$ is the number of years in the future. Find the expected population of the town in **a)** 10 years. **b)** 50 years.

**48. Population** The expected future population of Antwerp, which currently has 6800 residents, can be approximated by the formula $y = 6800(1.4)^{-0.2t}$, where $t$ is the number of years in the future. Find the expected population of the town 30 years in the future.

**49. Value of an SUV** The cost of a new SUV is \$24,000. If it depreciates at a rate of 18% per year, the value of the SUV in $t$ years can be approximated by the formula

$$V(t) = 24{,}000(0.82)^t$$

Find the value of the SUV in 4 years.

**50. Value of an ATV** The cost of a new all-terrain vehicle is \$6200. If it depreciates at a rate of 15% per year, the value of the ATV in $t$ years can be approximated by the formula

$$V(t) = 6200(0.85)^t$$

Find the value of the ATV in 10 years.

**51. Water Use** The average U.S. resident used about 580,000 gallons of water in 2005. Suppose that each year after 2005 the average resident is able to reduce the amount of water used by 5%. The amount of water used by the average resident $t$ years after 2005 could then be found by the formula $A = 580{,}000(0.95)^t$.

**a)** Explain why this formula may be used to find the amount of water used.

**b)** What would be the average amount of water used in the year 2009?

**52. Recycling Aluminum** Currently, about $\frac{2}{3}$ of all aluminum cans are recycled each year, while about $\frac{1}{3}$ are disposed of in landfills. The recycled aluminum is used to make new cans. Americans used about 190,000,000 aluminum cans in 2004. The number of new cans made each year from recycled 2004 aluminum cans $n$ years later can be estimated by the formula $A = 190{,}000{,}000\left(\frac{2}{3}\right)^n$.

**a)** Explain why the formula may be used to estimate the number of cans made from recycled aluminum cans $n$ years after 2004.

**b)** How many cans will be made from 2004 recycled aluminum cans in 2011?

**53. Atmospheric Pressure** Atmospheric pressure varies with altitude. The greater the altitude, the lower the pressure, as shown in the following graph.

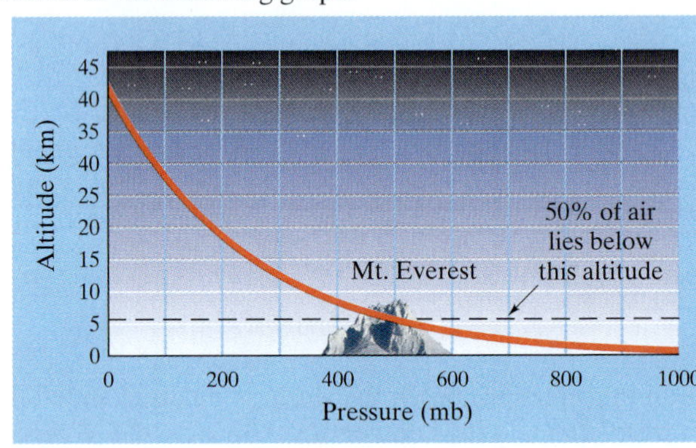


The equation $A = 41.97(0.996)^x$ can be used to estimate the altitude, $A$, in kilometers, for a given pressure, $x$, in millibars (mb). If the atmospheric pressure on top of Mt. Everest is about 389 mb, estimate the altitude of the top of Mt. Everest.

**54. Centenarians** Based on projections of the U.S. Census Bureau, the number of centenarians (people age 100 or older) will grow exponentially beyond 1995 (see the graph below). The function

$$f(t) = 71.24(1.045)^t$$

can be used to approximate the number of centenarians, in thousands, in the United States where $t$ is time in years since 1995. Use this function to estimate the number of centenarians in **a)** 2060. **b)** 2070.

**Number of Centenarians in the United States**

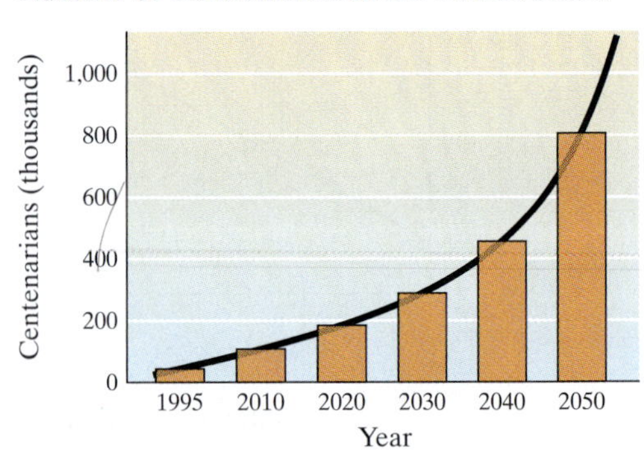


*Source:* U.S. Bureau of the Census; middle series projections

**55.** In Exercise 37, we graphed the amount for various years when \$100 is invested at 7% simple interest and at 7% interest compounded annually.

**a)** Use the compound interest formula given in Example 5 to determine the amount if \$100 is compounded daily at 7% for 10 years (assume 365 days per year).

**b)** Estimate the difference in the amount in 10 years for the \$100 invested at 7% simple interest versus the 7% interest compounded daily.

**56.** Graph $y = 2^x$ and $y = 3^x$ on the same window.

**57. a)** Graph $y = 3^{x-5}$.

**b)** Use your graphing calculator to solve the equation $4 = 3^{x-5}$. Round your answer to the nearest hundredth.

**58. a)** Graph $y = \left(\frac{1}{2}\right)^{2x+3}$.

**b)** Use your graphing calculator to solve the equation $-3 = \left(\frac{1}{2}\right)^{2x+3}$.

## Challenge Problem

**59.** Suppose Bob Jenkins gives Carol Dantuma \$1 on day 1, \$2 on day 2, \$4 on day 3, \$8 on day 4, and continues this doubling process for 30 days.

**a)** Determine how much Bob will give Carol on day 15.

**b)** Determine how much Bob will give Carol on day 20.

**c)** Express the amount, using exponential form, that Bob gives Carol on day $n$.

**d)** How much, in dollars, will Bob give Carol on day 30? Write the amount in exponential form. Then use a calculator to evaluate.

**e)** Express the total amount Bob gives Carol over the 30 days as a sum of exponential terms. (Do not find the actual value.)

## Group Activity

**60.** Functions that are exponential or are approximately exponential are commonly seen.

**a)** Have each member of the group individually determine a function not given in this section that may approximate an exponential function. You may use newspapers, books, or other sources.

**b)** As a group, discuss one another's functions. Determine whether each function presented is an exponential function.

**c)** As a group, write a paper that discusses each of the exponential functions and state why you believe each function is exponential.

## Cumulative Review Exercises

[5.1] **61.** Consider the polynomial

$$2.3x^4y - 6.2x^6y^2 + 9.2x^5y^2$$

**a)** Write the polynomial in descending order of the variable $x$.

**b)** What is the degree of the polynomial?

**c)** What is the leading coefficient?

[5.2] **62.** If $f(x) = x + 5$ and $g(x) = x^2 - 2x + 4$, find $(f \cdot g)(x)$.

[7.1] **63.** Write $\sqrt{a^2 - 8a + 16}$ as an absolute value.

[7.3] **64.** Simplify $\sqrt[4]{\dfrac{32x^5y^9}{2y^3z}}$.

# 9.3 Logarithmic Functions

1 Convert from exponential form to logarithmic form.

2 Graph logarithmic functions.

3 Compare the graphs of exponential and logarithmic functions.

4 Solve applications of logarithmic functions.

## 1 Convert from Exponential Form to Logarithmic Form

Now we are ready to introduce **logarithms**. Consider the exponential function $y = 2^x$. Recall from Section 9.1 that to find the inverse function, we interchange $x$ and $y$ and solve the resulting equation for $y$. Interchanging $x$ and $y$ gives the equation $x = 2^y$. But at this time we have no way of solving the equation $x = 2^y$ for $y$. To solve this equation for $y$, we introduce a new definition.

**Logarithms**

For all positive numbers $a$, where $a \neq 1$,

$$y = \log_a x \quad \text{means} \quad x = a^y$$

By the definition of logarithm, $x = 2^y$ means $y = \log_2 x$. We can therefore reason that $y = 2^x$ and $y = \log_2 x$ are inverse functions. In general, $y = a^x$ and $y = \log_a x$ are inverse functions.

In the equation $y = \log_a x$, the word *log* is an abbreviation for the word *logarithm*; $y = \log_a x$ is read "*y* is the logarithm of *x* to the base *a*." The letter *y* represents the logarithm, the letter *a* represents the base, and the letter *x* represents the number.

logarithm (exponent) → $y$; number → $x$; base → $a$

$$y = \log_a x \quad \text{means} \quad x = a^y$$

number → $x$; exponent → $y$; base → $a$

In words, the logarithm of the number *x* to the base *a* is the *exponent* to which the base *a* must be raised to equal the number *x*. In short, *a logarithm is an exponent.* For example,

$$2 = \log_{10} 100 \quad \text{means} \quad 100 = 10^2$$

In $\log_{10} 100 = 2$, the logarithm is 2, the base is 10, and the number is 100. The logarithm, 2, is the *exponent* to which the base, 10, must be raised to equal the number, 100. Note $10^2 = 100$.

Following are some examples of how an exponential expression can be converted to a logarithmic expression.

| Exponential Form | Logarithmic Form |
|---|---|
| $10^0 = 1$ | $\log_{10} 1 = 0$ |
| $4^2 = 16$ | $\log_4 16 = 2$ |
| $\left(\frac{1}{2}\right)^5 = \frac{1}{32}$ | $\log_{1/2} \frac{1}{32} = 5$ |
| $5^{-2} = \frac{1}{25}$ | $\log_5 \frac{1}{25} = -2$ |

Now let's do a few examples involving conversion from exponential form into logarithmic form, and vice versa.

**EXAMPLE 1** ▸ Write each equation in logarithmic form.

**a)** $3^4 = 81$ **b)** $\left(\frac{1}{5}\right)^3 = \frac{1}{125}$ **c)** $2^{-5} = \frac{1}{32}$

*Solution*

**a)** $\log_3 81 = 4$ **b)** $\log_{1/5} \frac{1}{125} = 3$ **c)** $\log_2 \frac{1}{32} = -5$

▸ **Now Try Exercise 31**

**EXAMPLE 2** ▸ Write each equation in exponential form.

**a)** $\log_7 49 = 2$ **b)** $\log_4 64 = 3$ **c)** $\log_{1/3} \frac{1}{81} = 4$

*Solution*

**a)** $7^2 = 49$ **b)** $4^3 = 64$ **c)** $\left(\frac{1}{3}\right)^4 = \frac{1}{81}$

▸ **Now Try Exercise 47**

**EXAMPLE 3** ▸ Write each equation in exponential form; then find the unknown value.

**a)** $y = \log_5 25$ **b)** $2 = \log_a 16$ **c)** $3 = \log_{1/2} x$

*Solution*

**a)** $5^y = 25$. Since $5^2 = 25$, $y = 2$.

**b)** $a^2 = 16$. Since $4^2 = 16$, $a = 4$. Note that $a$ must be greater than 0, so $-4$ is not a possible answer for $a$.

**c)** $\left(\frac{1}{2}\right)^3 = x$. Since $\left(\frac{1}{2}\right)^3 = \frac{1}{8}$, $x = \frac{1}{8}$.

▶ Now Try Exercise 65

## 2 Graph Logarithmic Functions

Now that we know how to convert from exponential form into logarithmic form and vice versa, we can graph logarithmic functions. Equations of the form $y = \log_a x$, $a > 0, a \neq 1$, and $x > 0$, are called **logarithmic functions**. The graphs of logarithmic functions pass the vertical line test. To graph a logarithmic function, change it to exponential form and then plot points. This procedure is illustrated in Examples 4 and 5.

Before we graph logarithmic functions, let's discuss some characteristics of the graphs of logarithmic functions.

**Graphs of Logarithmic Functions**

For all logarithmic functions of the form $y = \log_a x$ or $f(x) = \log_a x$, where $a > 0, a \neq 1$, and $x > 0$

1. The domain of the function is $(0, \infty)$.
2. The range of the function is $(-\infty, \infty)$.
3. The graph passes through the points $\left(\frac{1}{a}, -1\right)$, $(1, 0)$, and $(a, 1)$.

In most cases, a reasonably good logarithmic graph can be drawn from just the three points listed in item 3. When $a > 1$, the graph becomes almost vertical to the left of $\left(\frac{1}{a}, -1\right)$ and somewhat horizontal to the right of $(a, 1)$, see Example 4.
When $0 < a < 1$, the graph becomes almost vertical to the left of $(a, 1)$ and somewhat horizontal to the right of $\left(\frac{1}{a}, -1\right)$, see Example 5.

**EXAMPLE 4** ▶ Graph $y = \log_2 x$. State the domain and range of the function.

*Solution* This is an equation of the form $y = \log_a x$, where $a = 2$. $y = \log_2 x$ means $x = 2^y$. Using $x = 2^y$, construct a table of values. The table will be easier to develop by selecting values for $y$ and finding the corresponding values for $x$. In the table, the three points listed in item 3 in the box are shown in blue.

| $x$ | $\frac{1}{16}$ | $\frac{1}{8}$ | $\frac{1}{4}$ | $\frac{1}{2}$ | 1 | 2 | 4 | 8 | 16 |
|---|---|---|---|---|---|---|---|---|---|
| $y$ | $-4$ | $-3$ | $-2$ | $-1$ | 0 | 1 | 2 | 3 | 4 |

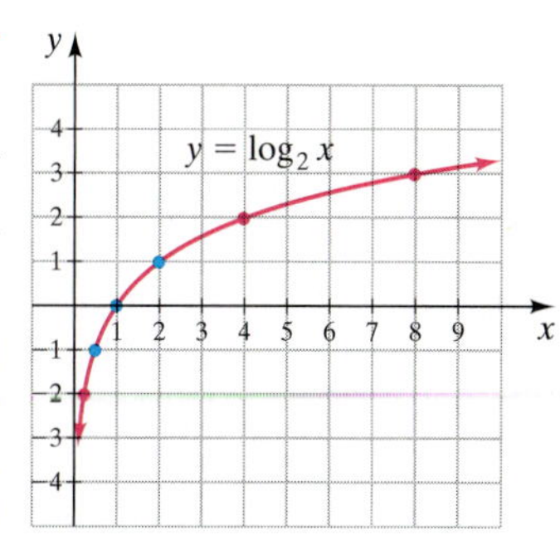

FIGURE 9.21

Now draw the graph (**Fig. 9.21**). The three ordered pairs in blue in the table are marked in blue on the graph. The domain, the set of $x$-values, is $\{x|x > 0\}$. The range, the set of $y$-values, is the set of all real numbers, $\mathbb{R}$.

▶ Now Try Exercise 11

**EXAMPLE 5** ▸ Graph $y = \log_{1/2} x$. State the domain and range of the function.

*Solution* This is an equation of the form $y = \log_a x$, where $a = \frac{1}{2}$. $y = \log_{1/2} x$ means $x = \left(\frac{1}{2}\right)^y$. Construct a table of values by selecting values for $y$ and finding the corresponding values of $x$.

| $x$ | 16 | 8 | 4 | 2 | 1 | $\frac{1}{2}$ | $\frac{1}{4}$ | $\frac{1}{8}$ | $\frac{1}{16}$ |
|---|---|---|---|---|---|---|---|---|---|
| $y$ | −4 | −3 | −2 | −1 | 0 | 1 | 2 | 3 | 4 |

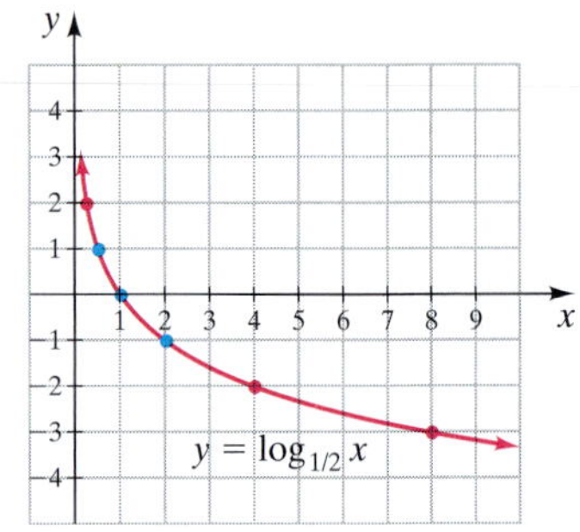

FIGURE 9.22

The graph is illustrated in **Figure 9.22**. The domain is $\{x|x > 0\}$. The range is the set of real numbers, $\mathbb{R}$.

▸ **Now Try Exercise 13**

If we study the domains in Examples 4 and 5, we see that the domains of both $y = \log_2 x$ and $y = \log_{1/2} x$ are $\{x|x > 0\}$. In fact, **for any logarithmic function $y = \log_a x$, the domain is $\{x|x > 0\}$.** Also note that the graphs in Examples 4 and 5 are both graphs of one-to-one functions.

## 3 Compare the Graphs of Exponential and Logarithmic Functions

Recall that to find inverse functions we switch $x$ and $y$ and solve the resulting equation for $y$. Consider $y = a^x$. If we switch $x$ and $y$, we get $x = a^y$. By our definition of logarithm, this function may be rewritten as $y = \log_a x$, which is an equation solved for $y$. Therefore, *$y = a^x$ and $y = \log_a x$ are inverse functions.* We may therefore write: if $f(x) = a^x$, then $f^{-1}(x) = \log_a x$.

In **Figure 9.23**, on page 615, we show general graphs of $y = a^x$ and $y = \log_a x$, $a > 1$, on the same axes. Notice they are symmetric about the line $y = x$. Also, notice the following boxed information.

**Graph Characteristics**

| | EXPONENTIAL FUNCTION | | LOGARITHMIC FUNCTION |
|---|---|---|---|
| | $y = a^x \ (a > 0, a \neq 1)$ | | $y = \log_a x \ (a > 0, a \neq 1)$ |
| Domain: | $(-\infty, \infty)$ | | $(0, \infty)$ |
| Range: | $(0, \infty)$ | | $(-\infty, \infty)$ |
| Points on graph: | $\left(-1, \frac{1}{a}\right)$ | x becomes y, | $\left(\frac{1}{a}, -1\right)$ |
| | $(0, 1)$ | y becomes x | $(1, 0)$ |
| | $(1, a)$ | | $(a, 1)$ |

From the information in the box, we can see that the range of the exponential function is the domain of the logarithmic function, and vice versa. We can also see that the $x$- and $y$-values in the ordered pairs are switched in the exponential and logarithmic functions.

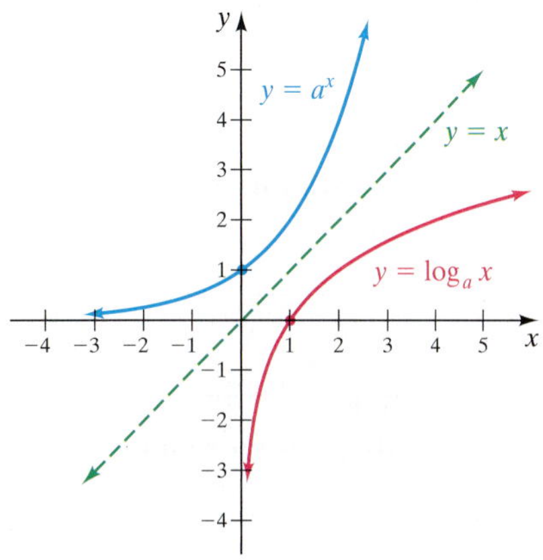


FIGURE 9.23

The graphs of $y = 2^x$ and $y = \log_2 x$ are illustrated in **Figure 9.24**. The graphs of $y = \left(\frac{1}{2}\right)^x$ and $y = \log_{1/2} x$ are illustrated in **Figure 9.25**. In each figure, the graphs are inverses of each other and are symmetric with respect to the line $y = x$.

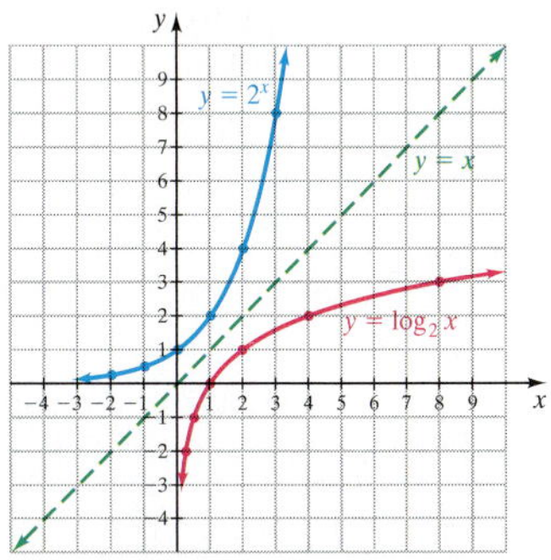


FIGURE 9.24

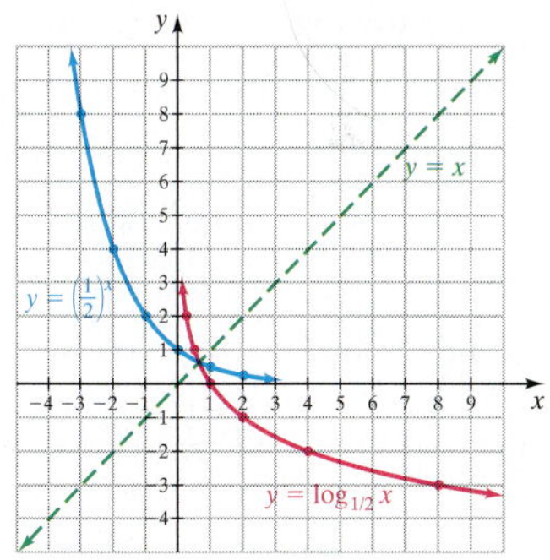


FIGURE 9.25

## 4 Solve Applications of Logarithmic Functions

We will see many applications of logarithms later, but let's look at one important application now.

**EXAMPLE 6** ▸ **Earthquakes** Logarithms are used to measure the magnitude of earthquakes. The Richter scale for measuring earthquakes was developed by Charles R. Richter. The magnitude, $R$, of an earthquake on the Richter scale is given by the formula

$$R = \log_{10} I$$

where $I$ represents the number of times greater (or more intense) the earthquake is than the smallest measurable activity that can be measured on a seismograph.

**a)** If an earthquake measures 4 on the Richter scale, how many times more intense is it than the smallest measurable activity?

**b)** How many times more intense is an earthquake that measures 5 on the Richter scale than an earthquake that measures 4?

*Solution* **a) Understand** The Richter number, $R$, is 4. To find how many times more intense the earthquake is than the smallest measurable activity, $I$, we substitute $R = 4$ into the formula and solve for $I$.

**Translate**

$$R = \log_{10} I$$
$$4 = \log_{10} I$$

**Carry Out**
$$10^4 = I \quad \text{Change to exponential form.}$$
$$10{,}000 = I$$

**Answer** Therefore, an earthquake that measures 4 on the Richter scale is 10,000 times more intense than the smallest measurable activity.

**b)**
$$5 = \log_{10} I$$
$$10^5 = I \quad \text{Change to exponential form.}$$
$$100{,}000 = I$$

Since $(10{,}000)(10) = 100{,}000$, an earthquake that measures 5 on the Richter scale is 10 times more intense than an earthquake that measures 4 on the Richter scale .

▸ **Now Try Exercise 113**

## EXERCISE SET 9.3

MathXL® MyMathLab

### Concept/Writing Exercises

1. Consider the logarithmic function $y = \log_a x$.
   a) What are the restrictions on $a$?
   b) What is the domain of the function?
   c) What is the range of the function?
2. Write $y = \log_a x$ in exponential form.
3. If some points on the graph of the exponential function $f(x) = a^x$ are $\left(-3, \frac{1}{27}\right), \left(-2, \frac{1}{9}\right), \left(-1, \frac{1}{3}\right)$, (0, 1), (1, 3), (2, 9), and (3, 27), list some points on the graph of the logarithmic function $g(x) = \log_a x$. Explain how you determined your answer.
4. For the logarithmic function $y = \log_a(x - 3)$, what must be true about $x$? Explain.
5. Discuss the relation between the graphs of $y = a^x$ and $y = \log_a x$ for $a > 0$ and $a \neq 1$.
6. What is the $x$-intercept of the graph of an equation of the form $y = \log_a x$?

### Practice the Skills

*Graph the logarithmic function.*

7. $y = \log_2 x$
8. $y = \log_3 x$
9. $y = \log_{1/2} x$
10. $y = \log_{1/3} x$
11. $y = \log_5 x$
12. $y = \log_4 x$
13. $y = \log_{1/5} x$
14. $y = \log_{1/4} x$

*Graph each pair of functions on the same axes.*

15. $y = 2^x, y = \log_{1/2} x$
16. $y = \left(\frac{1}{2}\right)^x, y = \log_2 x$
17. $y = 2^x, y = \log_2 x$
18. $y = \left(\frac{1}{2}\right)^x, y = \log_{1/2} x$

*Write each equation in logarithmic form.*

19. $2^3 = 8$
20. $3^5 = 243$
21. $3^2 = 9$
22. $2^6 = 64$
23. $16^{1/2} = 4$
24. $49^{1/2} = 7$
25. $8^{1/3} = 2$
26. $16^{1/4} = 2$
27. $\left(\frac{1}{2}\right)^5 = \frac{1}{32}$
28. $\left(\frac{1}{3}\right)^4 = \frac{1}{81}$
29. $2^{-3} = \frac{1}{8}$
30. $6^{-3} = \frac{1}{216}$
31. $4^{-3} = \frac{1}{64}$
32. $81^{1/2} = 9$
33. $64^{1/3} = 4$
34. $5^{-4} = \frac{1}{625}$
35. $8^{-1/3} = \frac{1}{2}$
36. $16^{-1/2} = \frac{1}{4}$
37. $81^{-1/4} = \frac{1}{3}$
38. $32^{-1/5} = \frac{1}{2}$
39. $10^{0.8451} = 7$
40. $10^{1.0792} = 12$
41. $e^2 = 7.3891$
42. $e^{-1/2} = 0.6065$
43. $a^n = b$
44. $c^b = w$

*Write each equation in exponential form.*

**45.** $\log_2 8 = 3$ **46.** $\log_5 125 = 3$ **47.** $\log_{1/3} \frac{1}{27} = 3$

**48.** $\log_{1/2} \frac{1}{64} = 6$ **49.** $\log_5 \frac{1}{25} = -2$ **50.** $\log_5 \frac{1}{625} = -4$

**51.** $\log_{49} 7 = \frac{1}{2}$ **52.** $\log_{64} 4 = \frac{1}{3}$ **53.** $\log_9 \frac{1}{81} = -2$

**54.** $\log_{10} \frac{1}{100} = -2$ **55.** $\log_{10} \frac{1}{1000} = -3$ **56.** $\log_{10} 1000 = 3$

**57.** $\log_6 216 = 3$ **58.** $\log_4 1024 = 5$ **59.** $\log_{10} 0.62 = -0.2076$

**60.** $\log_{10} 8 = 0.9031$ **61.** $\log_e 6.52 = 1.8749$ **62.** $\log_e 30 = 3.4012$

**63.** $\log_w s = -p$ **64.** $\log_r c = -a$

*Write each equation in exponential form; then find the unknown value.*

**65.** $\log_4 64 = y$ **66.** $\log_5 25 = y$ **67.** $\log_a 125 = 3$ **68.** $\log_a 81 = 4$

**69.** $\log_3 x = 3$ **70.** $\log_2 x = 5$ **71.** $\log_2 \frac{1}{16} = y$ **72.** $\log_8 \frac{1}{64} = y$

**73.** $\log_{1/2} x = 6$ **74.** $\log_{1/3} x = 4$ **75.** $\log_a \frac{1}{27} = -3$ **76.** $\log_9 \frac{1}{81} = y$

*Evaluate the following.*

**77.** $\log_{10} 1$ **78.** $\log_{10} 10$ **79.** $\log_{10} 100$ **80.** $\log_{10} 1000$

**81.** $\log_{10} \frac{1}{100}$ **82.** $\log_{10} \frac{1}{1000}$ **83.** $\log_{10} 10{,}000$ **84.** $\log_{10} 100{,}000$

**85.** $\log_4 256$ **86.** $\log_{13} 169$ **87.** $\log_3 \frac{1}{81}$ **88.** $\log_5 \frac{1}{125}$

**89.** $\log_8 \frac{1}{64}$ **90.** $\log_{14} \frac{1}{14}$ **91.** $\log_9 1$ **92.** $\log_{15} 1$

**93.** $\log_9 9$ **94.** $\log_{12} 12$ **95.** $\log_4 1024$ **96.** $\log_2 128$

## Problem Solving

**97.** If $f(x) = 5^x$, what is $f^{-1}(x)$?

**98.** If $f(x) = \log_6 x$, what is $f^{-1}(x)$?

**99.** Between which two integers must $\log_3 62$ lie? Explain.

**100.** Between which two integers must $\log_{10} 0.672$ lie? Explain.

**101.** Between which two integers must $\log_{10} 425$ lie? Explain.

**102.** Between which two integers must $\log_5 0.3256$ lie? Explain.

**103.** For $x > 1$, which will grow faster as $x$ increases, $2^x$ or $\log_{10} x$? Explain.

**104.** For $x > 1$, which will grow faster as $x$ increases, $x$ or $\log_{10} x$? Explain.

*Change to exponential form, then solve for x. We will discuss rules for solving problems like this in Section 9.4.*

**105.** $x = \log_{10} 10^6$ **106.** $x = \log_7 7^9$ **107.** $x = \log_b b^8$ **108.** $x = \log_e e^5$

*Change to logarithmic form, then solve for x. We will discuss rules for solving problems like this in Section 9.4.*

**109.** $x = 10^{\log_{10} 3}$ **110.** $x = 6^{\log_6 5}$ **111.** $x = b^{\log_b 9}$ **112.** $x = c^{\log_c 2}$

**113. Earthquake** If the magnitude of an earthquake is 7 on the Richter scale, how many times more intense is the earthquake than the smallest measurable activity? Use $R = \log_{10} I$ (see Example 6).

**114. Earthquake** If the magnitude of an earthquake is 5 on the Richter scale, how many times more intense is the earthquake than the smallest measurable activity? Use $R = \log_{10} I$.

**115. Earthquake** How many times more intense is an earthquake that measures 6 on the Richter scale than an earthquake that measures 2?

**116. Earthquake** How many times more intense is an earthquake that measures 4 on the Richter scale than an earthquake that measures 1?

**117.** Graph $y = \log_2 (x - 1)$.

**118.** Graph $y = \log_3 (x - 2)$.

### Cumulative Review Exercises

[5.4–5.7] *Factor.*

**119.** $2x^3 - 6x^2 - 36x$

**120.** $x^4 - 16$

**121.** $40x^2 + 52x - 12$

**122.** $6r^2s^2 + rs - 1$

## 9.4 Properties of Logarithms

1. Use the product rule for logarithms.
2. Use the quotient rule for logarithms.
3. Use the power rule for logarithms.
4. Use additional properties of logarithms.

### 1 Use the Product Rule for Logarithms

When finding the logarithm of an expression, the expression is called the **argument** of the logarithm. For example, in $\log_{10} 3$, the 3 is the argument, and in $\log_{10}(2x + 4)$, the $(2x + 4)$ is the argument. When the argument contains a variable, we assume that the argument represents a positive value. *Remember, only logarithms of positive numbers exist.*

To be able to do calculations using logarithms, you must understand their properties. The first property we discuss is the product rule for logarithms.

**Product Rule for Logarithms**

For positive real numbers $x$, $y$, and $a$, $a \neq 1$,

$$\log_a xy = \log_a x + \log_a y \qquad \textbf{Property 1}$$

This rule tells us that the logarithm of a product of two factors equals the sum of the logarithms of the factors.

To prove this property, we let $\log_a x = m$ and $\log_a y = n$. Remember, logarithms are exponents. Now we write each logarithm in exponential form.

$$\log_a x = m \quad \text{means} \quad a^m = x$$
$$\log_a y = n \quad \text{means} \quad a^n = y$$

By substitution and using the rules of exponents, we see that

$$xy = a^m \cdot a^n = a^{m+n}$$

We can now convert $xy = a^{m+n}$ into logarithmic form.

$$xy = a^{m+n} \quad \text{means} \quad \log_a xy = m + n$$

Finally, substituting $\log_a x$ for $m$ and $\log_a y$ for $n$, we obtain

$$\log_a xy = \log_a x + \log_a y$$

which is property 1.

**Examples of Property 1**

$$\log_3(6 \cdot 7) = \log_3 6 + \log_3 7$$
$$\log_4 3z = \log_4 3 + \log_4 z$$
$$\log_8 x^2 = \log_8(x \cdot x) = \log_8 x + \log_8 x \text{ or } 2\log_8 x$$

Property 1, the product rule, can be expanded to three or more factors, for example, $\log_a xyz = \log_a x + \log_a y + \log_a z$.

### 2 Use the Quotient Rule for Logarithms

Now we give the quotient rule for logarithms, which we refer to as property 2.

**Quotient Rule for Logarithms**

For positive real numbers $x$, $y$, and $a$, $a \neq 1$,

$$\log_a \frac{x}{y} = \log_a x - \log_a y \qquad \textbf{Property 2}$$

This rule tells us that the logarithm of a quotient equals the difference between the logarithm of the numerator and the logarithm of the denominator.

**Examples of Property 2**

$$\log_3 \frac{19}{4} = \log_3 19 - \log_3 4$$

$$\log_6 \frac{x}{3} = \log_6 x - \log_6 3$$

$$\log_5 \frac{z}{z+2} = \log_5 z - \log_5 (z+2)$$

## 3 Use the Power Rule for Logarithms

The next property we discuss is the power rule for logarithms.

**Power Rule for Logarithms**

If $x$ and $a$ are positive real numbers, $a \neq 1$, and $n$ is any real number, then

$$\log_a x^n = n \log_a x \qquad \textbf{Property 3}$$

This rule tells us that the logarithm of a number raised to a power equals the exponent times the logarithm of the number.

**Examples of Property 3**

$$\log_2 4^3 = 3 \log_2 4$$

$$\log_3 x^2 = 2 \log_3 x$$

$$\log_5 \sqrt{12} = \log_5 (12)^{1/2} = \frac{1}{2} \log_5 12$$

$$\log_8 \sqrt[5]{z+3} = \log_8 (z+3)^{1/5} = \frac{1}{5} \log_8 (z+3)$$

Properties 2 and 3 can be proved in a manner similar to that given for property 1 (see Exercises 79 and 80 on page 623).

**EXAMPLE 1** ▸ Use properties 1 through 3 to expand.

**a)** $\log_8 \frac{29}{43}$ **b)** $\log_4 (64 \cdot 180)$ **c)** $\log_{10} (22)^{1/5}$

*Solution*

**a)** $\log_8 \frac{29}{43} = \log_8 29 - \log_8 43$ *Quotient rule*

**b)** $\log_4 (64 \cdot 180) = \log_4 64 + \log_4 180$ *Product rule*

**c)** $\log_{10} (22)^{1/5} = \frac{1}{5} \log_{10} 22$ *Power rule*

▸ Now Try Exercise 11

Often we will have to use two or more of these properties in the same problem.

**EXAMPLE 2** ▸ Expand.

**a)** $\log_{10} 4(x+2)^3$ **b)** $\log_5 \frac{(4-a)^2}{3}$

**c)** $\log_5 \left(\frac{4-a}{3}\right)^2$ **d)** $\log_5 \frac{[x(x+4)]^3}{8}$

## Solution

**a)** $\log_{10} 4(x + 2)^3 = \log_{10} 4 + \log_{10}(x + 2)^3$ *Product rule*

$= \log_{10} 4 + 3\log_{10}(x + 2)$ *Power rule*

**b)** $\log_5 \dfrac{(4 - a)^2}{3} = \log_5 (4 - a)^2 - \log_5 3$ *Quotient rule*

$= 2\log_5(4 - a) - \log_5 3$ *Power rule*

**c)** $\log_5 \left(\dfrac{4 - a}{3}\right)^2 = 2\log_5\left(\dfrac{4 - a}{3}\right)$ *Power rule*

$= 2[\log_5(4 - a) - \log_5 3]$ *Quotient rule*

$= 2\log_5(4 - a) - 2\log_5 3$ *Distributive property*

**d)** $\log_5 \dfrac{[x(x + 4)]^3}{8} = \log_5 [x(x + 4)]^3 - \log_5 8$ *Quotient rule*

$= 3\log_5 x(x + 4) - \log_5 8$ *Power rule*

$= 3[\log_5 x + \log_5(x + 4)] - \log_5 8$ *Product rule*

$= 3\log_5 x + 3\log_5(x + 4) - \log_5 8$ *Distributive property*

**▸ Now Try Exercise 21**

### Helpful Hint

In Example 2**b**), when we expanded $\log_5 \dfrac{(4 - a)^2}{3}$, we first used the quotient rule. In Example 2**c**), when we expanded $\log_5\left(\dfrac{4 - a}{3}\right)^2$, we first used the power rule. Do you see the difference in the two problems? In $\log_5 \dfrac{(4 - a)^2}{3}$, just the numerator of the argument is squared; therefore, we use the quotient rule first. In $\log_5\left(\dfrac{4 - a}{3}\right)^2$, the entire argument is squared, so we use the power rule first.

**EXAMPLE 3 ▸** Write each of the following as the logarithm of a single expression.

**a)** $3\log_8(z + 2) - \log_8 z$

**b)** $\log_7(x + 1) + 2\log_7(x + 4) - 3\log_7(x - 5)$

## Solution

**a)** $3\log_8(z + 2) - \log_8 z = \log_8(z + 2)^3 - \log_8 z$ *Power rule*

$= \log_8 \dfrac{(z + 2)^3}{z}$ *Quotient rule*

**b)** $\log_7(x + 1) + 2\log_7(x + 4) - 3\log_7(x - 5)$

$= \log_7(x + 1) + \log_7(x + 4)^2 - \log_7(x - 5)^3$ *Power rule*

$= \log_7(x + 1)(x + 4)^2 - \log_7(x - 5)^3$ *Product rule*

$= \log_7 \dfrac{(x + 1)(x + 4)^2}{(x - 5)^3}$ *Quotient rule*

**▸ Now Try Exercise 39**

### Avoiding Common Errors

THE CORRECT RULES ARE

$$\log_a xy = \log_a x + \log_a y$$

$$\log_a \frac{x}{y} = \log_a x - \log_a y$$

Note that

$$\log_a (x + y) \neq \log_a x + \log_a y \qquad \log_a xy \neq (\log_a x)(\log_a y)$$

$$\log_a (x - y) \neq \log_a x - \log_a y \qquad \log_a \frac{x}{y} \neq \frac{\log_a x}{\log_a y}$$

## 4 Use Additional Properties of Logarithms

The last properties we discuss in this section will be used to solve equations in Section 9.6.

### Additional Properties of Logarithms

If $a > 0$, and $a \neq 1$, then

$\log_a a^x = x$ **Property 4**

and $a^{\log_a x} = x (x > 0)$ **Property 5**

| Examples of Property 4 | Examples of Property 5 |
|---|---|
| $\log_6 6^5 = 5$ | $3^{\log_3 7} = 7$ |
| $\log_9 9^x = x$ | $5^{\log_5 x} = x (x > 0)$ |

**EXAMPLE 4** ▸ Evaluate. **a)** $\log_5 25$ **b)** $\sqrt{16}^{\log_4 9}$

*Solution*

**a)** $\log_5 25$ may be written as $\log_5 5^2$. By property 4,

$$\log_5 25 = \log_5 5^2 = 2$$

**b)** $\sqrt{16}^{\log_4 9}$ may be written $4^{\log_4 9}$. By property 5,

$$\sqrt{16}^{\log_4 9} = 4^{\log_4 9} = 9$$

▸ **Now Try Exercise 55**

## EXERCISE SET 9.4

MathXL® MyMathLab

### Concept/Writing Exercises

1. Explain the product rule for logarithms.
2. Explain the quotient rule for logarithms.
3. Explain the power rule for logarithms.
4. Explain why we need to stipulate that $x$ and $y$ are positive real numbers when discussing the product and quotient rules.
5. Is $\log_a(xyz) = \log_a x + \log_a y + \log_a z$ a true statement? Explain.
6. Is $\log_b(x + y + z) = \log_b x + \log_b y + \log_b z$ a true statement? Explain.

### Practice the Skills

*Use properties 1–3 to expand.*

7. $\log_4 (3 \cdot 10)$
8. $\log_5 (4 \cdot 7)$
9. $\log_8 7(x + 3)$
10. $\log_9 x(x + 2)$
11. $\log_2 \frac{27}{11}$
12. $\log_5 (41 \cdot 9)$

13. $\log_{10} \frac{\sqrt{x}}{x - 9}$

14. $\log_5 3^8$

15. $\log_6 x^7$

16. $\log_9 12(4)^6$

17. $\log_4 (r + 7)^5$

18. $\log_8 b^3(b - 2)$

19. $\log_4 \sqrt{\frac{a^3}{a + 2}}$

20. $\log_9 (x - 6)^3 x^2$

21. $\log_3 \frac{d^6}{(a - 8)^4}$

22. $\log_7 x^2(x - 13)$

23. $\log_8 \frac{y(y + 4)}{y^3}$

24. $\log_{10} \left(\frac{z}{6}\right)^2$

25. $\log_{10} \frac{9m}{8n}$

26. $\log_5 \frac{\sqrt{a}\sqrt[3]{b}}{\sqrt[4]{c}}$

*Write as a logarithm of a single expression.*

27. $\log_5 2 + \log_5 8$

28. $\log_3 4 + \log_3 11$

29. $\log_2 9 - \log_2 5$

30. $\log_7 17 - \log_7 3$

31. $6 \log_4 2$

32. $\frac{1}{3} \log_8 7$

33. $\log_{10} x + \log_{10} (x + 3)$

34. $\log_5 (a + 1) - \log_5 (a + 10)$

35. $2 \log_9 z - \log_9 (z - 2)$

36. $3 \log_8 y + 2 \log_8 (y - 9)$

37. $4(\log_5 p - \log_5 3)$

38. $\frac{1}{2} [\log_6 (r - 1) - \log_6 r]$

39. $\log_2 n + \log_2 (n + 4) - \log_2 (n - 3)$

40. $2 \log_5 t + 5 \log_5 (t - 6) + \log_5 (3t + 7)$

41. $\frac{1}{2}[\log_5 (x - 8) - \log_5 x]$

42. $6 \log_7 (a + 3) + 2 \log_7 (a - 1) - \frac{1}{2} \log_7 a$

43. $2 \log_9 4 + \frac{1}{3} \log_9 (r - 6) - \frac{1}{2} \log_9 r$

44. $5 \log_6 (x + 3) - [2 \log_6 (7x + 1) + 3 \log_6 x]$

45. $4 \log_6 3 - [2 \log_6 (x + 3) + 4 \log_6 x]$

46. $2 \log_7 (m - 4) + 3 \log_7 (m + 3) - [5 \log_7 2 + 3 \log_7 (m - 2)]$

*Find the value by writing each argument using the numbers 2 and/or 5 and using the values* $\log_a 2 = 0.3010$ *and* $\log_a 5 = 0.6990$.

47. $\log_a 10$

48. $\log_a 2.5$

49. $\log_a 0.4$

50. $\log_a \frac{1}{8}$

51. $\log_a 25$

52. $\log_a \sqrt[3]{5}$

*Evaluate (see Example 4).*

53. $5^{\log_5 10}$

54. $\log_3 3$

55. $(2^3)^{\log_8 7}$

56. $\log_8 64$

57. $\log_3 27$

58. $2 \log_9 \sqrt{9}$

59. $5(\sqrt[3]{27})^{\log_3 5}$

60. $\frac{1}{2} \log_6 \sqrt[3]{6}$

## Problem Solving

61. For $x > 0$ and $y > 0$, is $\log_a \frac{x}{y} = \log_a xy^{-1} = \log_a x + \log_a y^{-1} = \log_a x + \log_a \frac{1}{y}$?

62. Read Exercise 61. By the quotient rule, $\log_a \frac{x}{y} = \log_a x - \log_a y$. Can we therefore conclude that $\log_a x - \log_a y = \log_a x + \log_a \frac{1}{y}$?

63. Use the product rule to show that

$$\log_a \frac{x}{y} = \log_a x + \log_a \frac{1}{y}$$

64. **a)** Explain why

$$\log_a \frac{3}{xy} \neq \log_a 3 - \log_a x + \log_a y$$

**b)** Expand $\log_a \frac{3}{xy}$ correctly.

65. Express $\log_a(x^2 - 4) - \log_a(x + 2)$ as a single logarithm and simplify.

66. Express $\log_a(x - 3) - \log_a(x^2 + 5x - 24)$ as a single logarithm and simplify.

67. Is $\log_a(x^2 + 8x + 16) = 2\log_a(x + 4)$? Explain.

68. Is $\log_a(4x^2 - 20x + 25) = 2\log_a(2x - 5)$? Explain.

*If* $\log_{10} x = 0.4320$, *find the following.*

69. $\log_{10} x^2$

70. $\log_{10} \sqrt[3]{x}$

71. $\log_{10} \sqrt[4]{x}$

72. $\log_{10} x^{11}$

*If* $\log_{10} x = 0.5000$ *and* $\log_{10} y = 0.2000$, *find the following.*

73. $\log_{10} xy$

74. $\log_{10}\left(\frac{x}{y}\right)$

75. Using the information given in the instructions for Exercises 73 and 74, is it possible to find $\log_{10}(x + y)$? Explain.

76. Are the graphs of $y = \log_b x^2$ and $y = 2\log_b x$ the same? Explain your answer by discussing the domain of each equation.

*Use properties 1–3 to expand.*

77. $\log_2 \frac{\sqrt[4]{xy}\sqrt[3]{a}}{\sqrt[5]{a - b}}$

78. $\log_3\left[\frac{(a^2 + b^2)(c^2)}{(a - b)(b + c)(c + d)}\right]^2$

79. Prove the quotient rule for logarithms.

80. Prove the power rule for logarithms.

## Group Activity

*Discuss and answer Exercise 81 as a group.*

81. Consider $\log_a \frac{\sqrt{x^4y}}{\sqrt{xy^3}}$, where $x > 0$ and $y > 0$.

**a)** Group member 1: Expand the expression using the quotient rule.

**b)** Group member 2: Expand the expression using the product rule.

**c)** Group member 3: First simplify $\frac{\sqrt{x^4y}}{\sqrt{xy^3}}$, then expand the resulting logarithm.

**d)** Check each other's work and make sure all answers are correct. Can this expression be simplified by all three methods?

## Cumulative Review Exercises

[2.5] 82. Solve the inequality $\frac{x - 4}{2} - \frac{2x - 5}{5} > 3$ and indicate the solution in

**a)** set builder notation.

**b)** interval notation.

[5.7] 83. **a)** Write an expression for the shaded area of the figure.

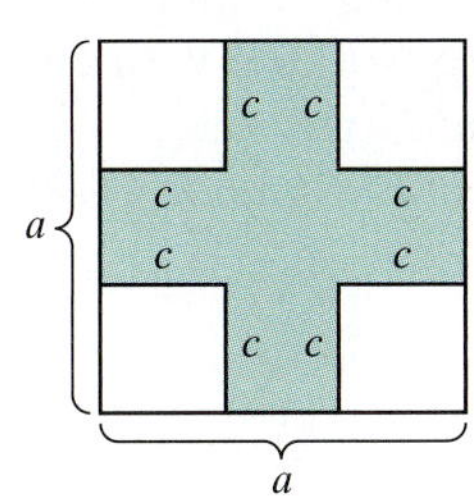


**b)** Write the expression in part **a)** in factored form.

[6.4] 84. Solve $\frac{15}{x} + \frac{9x - 7}{x + 2} = 9$ for $x$.

[7.7] 85. Multiply $(3i + 4)(2i - 5)$.

[8.4] 86. Solve $a - 6\sqrt{a} = 7$ for $a$.

## Mid-Chapter Test: 9.1–9.4

*To find out how well you understand the chapter material to this point, take this brief test. The answers, and the section where the material was initially discussed, are given in the back of the book. Review any questions you answered incorrectly.*

1. **a)** Explain how to find $(f \circ g)(x)$.

**b)** If $f(x) = 3x + 3$ and $g(x) = 2x + 5$, find $(f \circ g)(x)$.

2. Let $f(x) = x^2 + 5$ and $g(x) = \frac{6}{x}$; find

**a)** $(f \circ g)(x)$

**b)** $(f \circ g)(3)$

**c)** $(g \circ f)(x)$

**d)** $(g \circ f)(3)$

**3. a)** Explain what it means when a function is a one-to-one function.

**b)** Is the function represented by the following graph a one-to-one function? Explain.

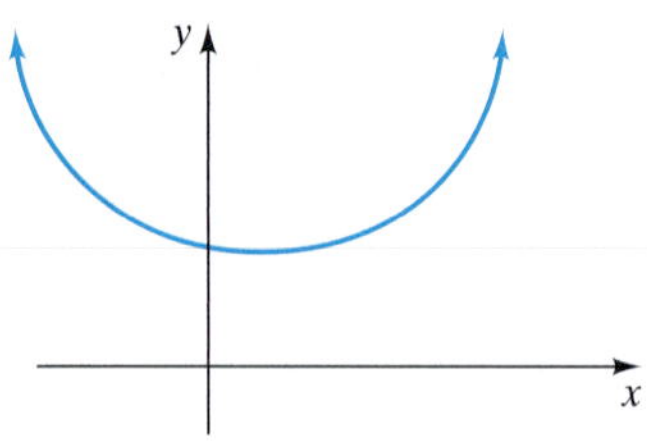


*In Exercises 4–6, for each function,* **a)** *determine whether it is a one-to-one function;* **b)** *if it is a one-to-one function, find its inverse function.*

**4.** $\{(-3, 2), (2, 3), (5, 1), (6, 8)\}$

**5.** $p(x) = \frac{1}{3}x - 5$

**6.** $k(x) = \sqrt{x - 4}, \quad x \geq 4$

**7.** Let $m(x) = -2x + 4$. Find $m^{-1}(x)$ and then graph $m(x)$ and $m^{-1}(x)$ on the same axes.

*Graph each exponential function.*

**8.** $y = 2^x$

**9.** $y = 3^{-x}$

**10.** Graph the logarithmic function $y = \log_2 x$.

**11. Bacteria** The number of bacteria in a petri-dish is $N(t) = 5(2)^t$, where $t$ is the number of hours after the 5 original bacteria are placed in the dish. How many bacteria are in the dish

**a)** 1 hour later?

**b)** 6 hours later?

**12.** Write $27^{2/3} = 9$ in logarithmic form.

**13.** Write $\log_2 \frac{1}{64} = -6$ in exponential form.

**14.** Evaluate $\log_5 125$.

**15.** Solve the equation $\log_{1/4} \frac{1}{16} = x$ for $x$.

**16.** Solve the equation $\log_x 64 = 3$ for $x$.

*Use properties 1–3 to write as a sum or difference of logarithms.*

**17.** $\log_9 x^2(x - 5)$

**18.** $\log_5 \frac{7m}{\sqrt{n}}$

*Write as a single logarithm.*

**19.** $3 \log_2 x + \log_2 (x + 7) - 4 \log_2 (x + 1)$

**20.** $\frac{1}{2}[\log_7 (x + 2) - \log_7 x]$

# 9.5 Common Logarithms

1 Find common logarithms of powers of 10.

2 Find common logarithms.

3 Find antilogarithms.

## 1 Find Common Logarithms of Powers of 10

The properties discussed in Section 9.4 can be used with any valid base (a real number greater than 0 and not equal to 1). However, since we are used to working in base 10, we will often use the base 10 when computing with logarithms. **Base 10 logarithms** are called **common logarithms**. When we are working with common logarithms, it is not necessary to list the base. Thus, $\log x$ means $\log_{10} x$.

The properties of logarithms written as common logarithms follow. For positive real numbers $x$ and $y$, and any real number $n$,

**1.** $\log xy = \log x + \log y$

**2.** $\log \frac{x}{y} = \log x - \log y$

**3.** $\log x^n = n \log x$

The logarithms of most numbers are irrational numbers. Even the values given by calculators are usually only approximations of the actual values. *Even though we are working with approximations when evaluating most logarithms, we generally write the logarithm with an equal sign.* Thus, rather than writing $\log 6 \approx 0.77815$, we will write $\log 6 = 0.77815$. The values given for logarithms are accurate to at least four decimal places.

In Chapter 1 we learned that 1 can be expressed as $10^0$ and 10 can be expressed as $10^1$. Since, for example, 5 is between 1 and 10, it must also be between $10^0$ and $10^1$.

$$1 < 5 < 10$$
$$10^0 < 5 < 10^1$$

The number 5 can be expressed as the base 10 raised to an exponent between 0 and 1. The number 5 is approximately equal to $10^{0.69897}$. As is done with logarithms, when writing exponential expressions, we often use the equal sign, even when the values are only approximations. Thus, for example, we will generally write $10^{0.69897} = 5$ rather than $10^{0.69897} \approx 5$.

If you look up log 5 on a calculator, as will be explained shortly, it will display a value about 0.69897. Notice that

$$\log 5 = 0.69897 \quad \text{and} \quad 5 = 10^{0.69897}$$

We can see that *the common logarithm*, 0.69897, *is the exponent* on the base 10. Now we are ready to define common logarithms.

**Common Logarithm**

The **common logarithm** of a positive real number is the *exponent* to which the base 10 is raised to obtain the number.

$$\text{If } \log N = L, \quad \text{then} \quad 10^L = N.$$

For example, if $\log 5 = 0.69897$, then $10^{0.69897} = 5$.

Now consider the number 50.

$$10 < 50 < 100$$
$$10^1 < 50 < 10^2$$

The number 50 can be expressed as the base 10 raised to an exponent between 1 and 2. The number $50 = 10^{1.69897}$; thus $\log 50 = 1.69897$.

## 2 Find Common Logarithms

To find common logarithms of numbers, we can use a calculator that has a logarithm key, [LOG].

**USING YOUR CALCULATOR** Finding Common Logarithms

Scientific Calculator

To find common logarithms, enter the number, then press the logarithm key. The answer will then be displayed.

| EXAMPLE | KEYS TO PRESS | ANSWER DISPLAYED |
|---|---|---|
| Find log 400. | 400 [LOG] | 2.60206 |
| Find log 0.0538. | 0.0538 [LOG] | −1.2692177 |

Graphing Calculator

On some graphing calculators, you first press the [LOG] key and then you enter the number. For example, on the TI-84 Plus, you would do the following:

| EXAMPLE | KEYS TO PRESS | ANSWER DISPLAYED |
|---|---|---|
| Find log 400. | [LOG] (400 [)] [ENTER] | 2.602059991 |

↑ Generated by calculator

**EXAMPLE 1** ▸ Find the exponent to which the base 10 must be raised to obtain the number 43,600.

*Solution* *We are asked to find the exponent, which is a logarithm.* We need to determine log 43,600. Using a calculator, we find

$$\log 43{,}600 = 4.6394865$$

Thus, the exponent is 4.6394865. Note that $10^{4.6394865} = 43{,}600$.

▸ **Now Try Exercise 7**

## 3 Find Antilogarithms

The question that should now be asked is, "If we know the common logarithm of a number, how do we find the number?" For example, if $\log N = 3.406$, what is $N$? To find $N$, the number, we need to determine the value of $10^{3.406}$. Since

$$10^{3.406} = 2546.830253$$

$N = 2546.830253$. The number 2546.830253 is the *antilogarithm* of 3.406.

When we find the value of the number from the logarithm, we say we are finding the **antilogarithm** or **inverse logarithm**. If the logarithm of $N$ is $L$, then $N$ is the antilogarithm or inverse logarithm of $L$.

**Antilogarithm**

If $\log N = L$, then $N = \text{antilog } L$.

When we are given the common logarithm, which is the exponent on the base 10, the *antilog is the number* obtained when the base 10 is raised to that exponent.

### Examples

| Number ↓ Exponent ↓ | Exponent ↓ Number ↓ |
|---|---|
| $\log 962 = 2.9831751$ | $\text{antilog } 2.9831751 = 962$ |
| $\log 0.00046 = -3.3372422$ | $\text{antilog } (-3.3372422) = 0.00046$ |

When finding an antilog, we start with the logarithm, or the exponent, and end with the number equal to 10 raised to that logarithm or exponent. If $\text{antilog } (-3.3372422) = 0.00046$, then $10^{-3.3372422} = 0.00046$.

**USING YOUR CALCULATOR** Finding Antilogarithms

**Scientific Calculator**

To find antilogarithms on a scientific calculator, enter the logarithm, and press the [2nd], [INV], or [Shift] key, depending upon which of these keys your calculator has. Then press the [LOG] key. After the [LOG] key is pressed, the antilog will be displayed.

| EXAMPLE | KEYS TO PRESS | ANSWER DISPLAYED |
|---|---|---|
| Find antilog 2.9831751. | 2.9831751 [INV] [LOG] | 962.00006* |
| Find antilog (−3.3372422). | 3.3372422 [+/−] [INV] [LOG] | 0.00046** |

When you are finding the antilog of a negative value, enter the value and then press the [+/−] key before pressing the inverse and logarithm keys.

*Some calculators give slightly different answers, depending on their electronics.
**Some calculators may display answers in scientific notation form.

**Graphing Calculator**

On most graphing calculators, you press the [2nd] then the [LOG] key before you enter the logarithm.

On the TI-84 Plus and on certain other calculators, $10^x$ is printed directly above the [LOG] key. The antilog is actually the value of $10^x$, where $x$ is the logarithm. When you press [2nd] [LOG], the TI-84 Plus displays 10^ followed by a left parentheses. You then enter the logarithm followed by the [ ) ] key. After you press [ENTER], the antilog is displayed.

| EXAMPLE | KEYS TO PRESS | ANSWER DISPLAYED |
|---|---|---|
| Find antilog 2.9831751. | [2nd] [LOG]† (2.9831751 [ ) ] [ENTER] | 962.0000619 |
| Find antilog (−3.3372422). | [2nd] [LOG] ([(−)] 3.3372422 [ ) ] [ENTER] | 4.599999664E⁻4†† |

†Left parenthesis is generated by the TI-84 Plus.
††Recall from scientific notation that this number is 0.0004599999664.

Since we generally do not need the accuracy given by most calculators, in the exercise set that follows we will round logarithms to four decimal places and antilogarithms to three **significant digits**. In a number written in decimal form, any zeros preceding the first nonzero digit are not significant digits. The first nonzero digit in a number, moving from left to right, is the first significant digit.

| Examples | |
|---|---|
| 0.0063402 | First significant digit is shaded. |
| 3.0424080 | First three significant digits are shaded. |
| 0.0000138483 | First three significant digits are shaded. |
| 206,435.05 | First four significant digits are shaded. |

Although most antilogarithms will be irrational numbers, when writing antilogarithms we will use an equal sign rather than an approximately equal to sign, just as we did when evaluating logarithms. All antilogarithms will be accurate to at least three significant digits.

**EXAMPLE 2 ▸** Find the value obtained when the base 10 is raised to the $-1.052$ power.

*Solution* We are asked to find the value of $10^{-1.052}$. Since we are given the exponent, or logarithm, we can find the value by taking the antilog of $-1.052$.

$$\text{antilog } (-1.052) = 0.0887156$$

Thus, $10^{-1.052} = 0.0887$ rounded to three significant digits.

▸ **Now Try Exercise 55**

**EXAMPLE 3 ▸** Find $N$ if $\log N = 4.192$.

*Solution* We are given the logarithm and asked to find the antilog, or the number $N$.

$$\text{antilog } 4.192 = 15{,}559.6563$$

Thus, $N = 15{,}559.6563$.

▸ **Now Try Exercise 33**

**EXAMPLE 4 ▸** Find the following antilogs and round to three significant digits.

**a)** antilog 6.827 **b)** antilog $(-2.35)$

*Solution*

**a)** Using a calculator, we find antilog $6.827 = 6{,}714{,}288.5$. Rounding to three significant digits, we get antilog $6.827 = 6{,}710{,}000$.

**b)** Using a calculator, we find antilog $(-2.35) = 0.0044668$. Rounding to three significant digits, we get antilog $(-2.35) = 0.00447$.

▸ **Now Try Exercise 25**

**EXAMPLE 5 ▸ Earthquake** The magnitude of an earthquake on the Richter scale is given by the formula $R = \log I$, where $I$ is the number of times more intense the quake is than the smallest measurable activity. How many times more intense is an earthquake measuring 6.2 on the Richter scale than the smallest measurable activity?

*Solution* We want to find the value for $I$. We are given that $R = 6.2$. Substitute 6.2 for $R$ in the formula $R = \log I$, and then solve for $I$.

$$R = \log I$$

$$6.2 = \log I \quad \text{Substitute 6.2 for } R.$$

To find $I$, we need to take the antilog of both sides of the equation.

$$\text{antilog } 6.2 = I$$
$$1{,}580{,}000 = I$$

Thus, this earthquake is about 1,580,000 times more intense than the smallest measurable activity.

▸ Now Try Exercise 85

## EXERCISE SET 9.5

MathXL® MyMathLab

### Concept/Writing Exercises

1. What are common logarithms?
2. Write $\log N = L$ in exponential form.
3. What are antilogarithms?
4. If $\log 793 = 2.8993$ what is antilog 2.8993?

### Practice the Skills

*Find the common logarithm of each number. Round the answer to four decimal places.*

5. 86
6. 352
7. 19,200
8. 1000
9. 0.0613
10. 941,000
11. 100
12. 0.000835
13. 3.75
14. 0.375
15. 0.0173
16. 0.00872

*Find the antilog of each logarithm. Round the answer to three significant digits.*

17. 0.2137
18. 1.3845
19. 4.6283
20. 3.5527
21. −1.7086
22. −3.7431
23. 0.0000
24. 5.5922
25. 2.7625
26. −0.1543
27. −4.1390
28. −2.8139

*Find each number N. Round N to three significant digits.*

29. $\log N = 2.0000$
30. $\log N = 1.4612$
31. $\log N = 3.3817$
32. $\log N = 1.9330$
33. $\log N = 4.1409$
34. $\log N = -2.103$
35. $\log N = -1.06$
36. $\log N = -3.1469$
37. $\log N = -0.6218$
38. $\log N = 1.5177$
39. $\log N = -0.1256$
40. $\log N = -1.3206$

*To what exponent must the base 10 be raised to obtain each value? Round your answer to four decimal places.*

41. 3560
42. 817,000
43. 0.0727
44. 0.00612
45. 243
46. 8.16
47. 0.00592
48. 73,700,000

*Find the value obtained when 10 is raised to the following exponents. Round your answer to three significant digits.*

49. 2.8316
50. 3.2473
51. −0.5186
52. −3.7081
53. −1.4802
54. 4.5619
55. 1.3503
56. −2.1918

*By changing the logarithm to exponential form, evaluate the common logarithm without the use of a calculator.*

57. $\log 1$
58. $\log 100$
59. $\log 0.1$
60. $\log 1000$
61. $\log 0.01$
62. $\log 10$
63. $\log 0.001$
64. 0.0001

*In Section 9.4, we stated that for $a > 0$, and $a \neq 1$, $\log_a a^x = x$ and $a^{\log_a x} = x\,(x > 0)$. Rewriting these properties using common logarithms $(a = 10)$, we obtain $\log 10^x = x$ and $10^{\log x} = x\,(x > 0)$, respectively. Use these properties to evaluate the following.*

65. $\log 10^7$
66. $\log 10^{3.4}$
67. $10^{\log 7}$
68. $10^{\log 3.4}$
69. $4 \log 10^{5.2}$
70. $8 \log 10^{1.2}$
71. $5(10^{\log 8.3})$
72. $2.3(10^{\log 5.2})$

### Problem Solving

73. On your calculator, you find log 462 and obtain the value 1.6646. Can this value be correct? Explain.
74. On your calculator, you find log 6250 and obtain the value 2.7589. Can this value be correct? Explain.

75. On your calculator, you find log 0.163 and obtain the value −2.7878. Can this value be correct? Explain.

76. On your calculator, you find $\log(-1.23)$ and obtain the value 0.08991. Can this value be correct? Explain.

77. Is $\log \frac{y}{4x} = \log y - \log 4 + \log x$? Explain.

78. Is $\log \frac{5x^2}{3} = 2(\log 5 + \log x) - \log 3$? Explain.

*If* $\log 25 = 1.3979$ *and* $\log 5 = 0.6990$, *find the answer if possible. If it is not possible to find the answer, indicate so. Do not find the logarithms on your calculator except to check answers.*

79. log 125

80. log 35

81. $\log \frac{1}{5}$

82. $\log \frac{1}{25}$

83. log 625

84. $\log \sqrt{5}$

*Solve Exercises 85–88 using* $R = \log I$ *(see Example 5). Round your answer to three significant digits.*

85. Find $I$ if $R = 3.4$

86. Find $I$ if $R = 4.9$

87. Find $I$ if $R = 5.7$

88. Find $I$ if $R = 8.1$

89. **Astronomy** In astronomy, a formula used to find the diameter, in kilometers, of minor planets (also called asteroids) is $\log d = 3.7 - 0.2g$, where $g$ is a quantity called the absolute magnitude of the minor planet. Find the diameter of a minor planet if its absolute magnitude is **a)** 11 and **b)** 20. **c)** Find the absolute magnitude of the minor planet whose diameter is 5.8 kilometers.

90. **Standardized Test** The average score on a standardized test is a function of the number of hours studied for the test. The average score, $f(x)$, in points, can be approximated by $f(x) = \log 0.3x + 1.8$, where $x$ is the number of hours studied for the test. The maximum possible score on the test is 4.0. Find the score received by the average person who studied for **a)** 15 hours. **b)** 55 hours.

91. **Learning Retention** Sammy Barcia just finished a course in physics. The percent of the course he will remember $t$ months later can be approximated by the function

$$R(t) = 94 - 46.8 \log(t + 1)$$

for $0 \le t \le 48$. Find the percent of the course Sammy will remember **a)** 2 months later. **b)** 48 months later.

92. **Learning Retention** Karen Frye just finished a course in psychology. The percent of the course she will remember $t$ months later can be approximated by the function

$$R(t) = 85 - 41.9 \log(t + 1)$$

for $0 \le t \le 48$. Find the percent of the course she will remember **a)** 10 months later. **b)** 25 months later.

93. **Earthquake** How many times more intense is an earthquake having a Richter scale number of 3.8 than the smallest measurable activity? See Example 5.

94. **Sport Utility Vehicles** Since 1992, sport utility vehicle (SUV) sales in the United States have been on the rise. The number of sales each year, $f(t)$, in millions, can be approximated by the function $f(t) = 0.98 + 1.97 \log(t + 1)$, where $t = 0$ represents 1992, $t = 1$ represents 1993, and so on. If this trend continues, estimate the number of SUVs sold in **a)** 2003. **b)** 2008.

95. **Energy of an Earthquake** A formula sometimes used to estimate the seismic energy released by an earthquake is $\log E = 11.8 + 1.5m_s$, where $E$ is the seismic energy and $m_s$ is the surface wave magnitude.

**a)** Find the energy released in an earthquake whose surface wave magnitude is 6.

**b)** If the energy released during an earthquake is $1.2 \times 10^{15}$, what is the magnitude of the surface wave?

96. **Sound Pressure** The sound pressure level, $s_p$, is given by the formula $s_p = 20 \log \frac{p_r}{0.0002}$, where $p_r$ is the sound pressure in dynes/cm$^2$.

**a)** Find the sound pressure level if the sound pressure is 0.0036 dynes/cm$^2$

**b)** If the sound pressure level is 10.0, find the sound pressure.

97. **Earthquake** The Richter scale, used to measure the strength of earthquakes, relates the magnitude, $M$, of the earthquake to the release of energy, $E$, in ergs, by the formula

$$M = \frac{\log E - 11.8}{1.5}$$

An earthquake releases $1.259 \times 10^{21}$ ergs of energy. What is the magnitude of such an earthquake on the Richter scale?

98. **pH of a Solution** The pH is a measure of the acidity or alkalinity of a solution. The pH of water, for example, is 7. In general, acids have pH numbers less than 7 and alkaline solutions have pH numbers greater than 7. The pH of a solution is defined as $pH = -\log[H_3O^+]$, where $H_3O^+$ represents the hydronium ion concentration of the solution. Find the pH of a solution whose hydronium ion concentration is $2.8 \times 10^{-3}$.

## Challenge Problems

**99.** Solve the formula $R = \log I$ for $I$.

**100.** Solve the formula $\log E = 11.8 + 1.5m$ for $E$.

**101.** Solve the formula $R = 26 - 41.9\log(t + 1)$ for $t$.

**102.** Solve the formula $f = 76 - \log x$ for $x$.

## Group Activity

**103.** In Section 9.7, we introduce the *change of base formula*, $\log_a x = \dfrac{\log_b x}{\log_b a}$, where $a$ and $b$ are bases and $x$ is a positive number.

**a)** Group member 1: Use the change of base formula to evaluate $\log_3 45$. (*Hint*: Let $b = 10$.)

**b)** Group member 2: Repeat part **a)** for $\log_5 30$.

**c)** Group member 3: Repeat part **a)** for $\log_6 40$.

**d)** As a group, use the fact that $\log_a x = \dfrac{\log_b x}{\log_b a}$, where $b = 10$, to graph the equation $y = \log_2 x$ for $x > 0$. Use a graphing calculator if available.

## Cumulative Review Exercises

[4.3] **104. Cars** Two cars start at the same point in Alexandria, Virginia, and travel in opposite directions. One car travels 5 miles per hour faster than the other car. After 4 hours, the two cars are 420 miles apart. Find the speed of each car.

[4.5] **105.** Solve the system of equations.

$$3r = -4s - 6$$
$$3s = -5r + 1$$

[5.8] **106.** Solve $3x^3 + 3x^2 - 36x = 0$ for $x$.

[7.1] **107.** Write $\sqrt{(3x^2 - y)^2}$ as an absolute value.

[8.6] **108.** Solve $(x - 5)(x + 4)(x - 2) \le 0$ and give the solution in interval notation.

# 9.6 Exponential and Logarithmic Equations

1 Solve exponential and logarithmic equations.

2 Solve applications.

## 1 Solve Exponential and Logarithmic Equations

In Sections 9.2 and 9.3 we introduced **exponential** and **logarithmic equations.** In this section we give more examples of their use and discuss further procedures for solving such equations.

To solve exponential and logarithmic equations, we often use the following properties.

**Properties for Solving Exponential and Logarithmic Equations**

**a.** If $x = y$, then $a^x = a^y$.

**b.** If $a^x = a^y$, then $x = y$.

**c.** If $x = y$, then $\log_b x = \log_b y$ $(x > 0, y > 0)$.

**d.** If $\log_b x = \log_b y$, then $x = y$ $(x > 0, y > 0)$. **Properties 6a–6d**

We will be referring to these properties when explaining the solutions to the examples in this section.

**EXAMPLE 1** Solve the equation $8^x = \dfrac{1}{2}$.

*Solution* To solve this equation, we will write both sides of the equation with the same base, 2, and then use property 6b.

$$8^x = \frac{1}{2}$$

$$(2^3)^x = \frac{1}{2} \qquad \text{Write 8 as } 2^3.$$

$$2^{3x} = 2^{-1} \qquad \text{Write } \frac{1}{2} \text{ as } 2^{-1}.$$

Using property 6b, we can write

$$3x = -1$$
$$x = -\frac{1}{3}$$

**Now Try Exercise 7**

When both sides of the exponential equation cannot be written as a power of the same base, we often begin by taking the logarithm of both sides of the equation, as in Example 2. In the following examples, we will round logarithms to the nearest ten-thousandth.

**EXAMPLE 2** ▸ Solve the equation $5^n = 28$.

*Solution* Take the logarithm of both sides of the equation and solve for $n$.

$$\log 5^n = \log 28$$

$$n \log 5 = \log 28 \quad \text{Power rule}$$

$$n = \frac{\log 28}{\log 5} \quad \text{Divide both sides by log 5.}$$

$$\approx \frac{1.4472}{0.6990} \approx 2.0704$$

▸ **Now Try Exercise 23**

Some logarithmic equations can be solved by expressing the equation in exponential form. **It is necessary to check logarithmic equations for extraneous solutions.** When checking a solution, if you obtain the logarithm of a nonpositive number, the solution is extraneous.

**EXAMPLE 3** ▸ Solve the equation $\log_2 (x + 3)^3 = 4$.

*Solution* Write the equation in exponential form.

$$(x + 3)^3 = 2^4 \quad \text{Write in exponential form.}$$

$$(x + 3)^3 = 16$$

$$x + 3 = \sqrt[3]{16} \quad \text{Take the cube root of both sides.}$$

$$x = -3 + \sqrt[3]{16} \quad \text{Solve for } x.$$

**Check**

$$\log_2 (x + 3)^3 = 4$$

$$\log_2 [(-3 + \sqrt[3]{16}) + 3]^3 \stackrel{?}{=} 4$$

$$\log_2 (\sqrt[3]{16})^3 \stackrel{?}{=} 4$$

$$\log_2 16 \stackrel{?}{=} 4 \quad (\sqrt[3]{16})^3 = 16$$

$$2^4 \stackrel{?}{=} 16 \quad \text{Write in exponential form.}$$

$$16 = 16 \quad \text{True}$$

▸ **Now Try Exercise 43**

Other logarithmic equations can be solved using the properties of logarithms given in earlier sections.

**EXAMPLE 4** ▸ Solve the equation $\log(3x+2) + \log 9 = \log(x+5)$.

*Solution*

$$\begin{aligned} \log(3x+2) + \log 9 &= \log(x+5) \\ \log[(3x+2)(9)] &= \log(x+5) && \text{Product rule} \\ (3x+2)(9) &= (x+5) && \text{Property 6d} \\ 27x + 18 &= x + 5 \\ 26x + 18 &= 5 \\ 26x &= -13 \\ x &= -\frac{1}{2} \end{aligned}$$

Check for yourself that the solution is $-\frac{1}{2}$.

▸ **Now Try Exercise 51**

**EXAMPLE 5** ▸ Solve the equation $\log x + \log(x+1) = \log 12$.

*Solution*

$$\begin{aligned} \log x + \log(x+1) &= \log 12 \\ \log x(x+1) &= \log 12 && \text{Product rule} \\ x(x+1) &= 12 && \text{Property 6d} \\ x^2 + x &= 12 \\ x^2 + x - 12 &= 0 \\ (x+4)(x-3) &= 0 \end{aligned}$$

$$x + 4 = 0 \quad \text{or} \quad x - 3 = 0$$
$$x = -4 \qquad\qquad x = 3$$

**Check** $x = -4$

$$\begin{aligned} \log x + \log(x+1) &= \log 12 \\ \log(-4) + \log(-4+1) &\stackrel{?}{=} \log 12 \\ \log(-4) + \log(-3) &\stackrel{?}{=} \log 12 \end{aligned}$$

*Stop.* ↑ ↑ *Logarithms of negative numbers are not real numbers.*

$x = 3$

$$\begin{aligned} \log x + \log(x+1) &= \log 12 \\ \log 3 + \log(3+1) &\stackrel{?}{=} \log 12 \\ \log 3 + \log 4 &\stackrel{?}{=} \log 12 \\ \log[(3)(4)] &\stackrel{?}{=} \log 12 \\ \log 12 &= \log 12 && \text{True} \end{aligned}$$

Thus, $-4$ is an extraneous solution. The only solution is 3.

▸ **Now Try Example 65**

**USING YOUR GRAPHING CALCULATOR**

We have shown how equations in one variable may be solved graphically. Logarithmic and exponential equations may also be solved graphically by graphing each side of the equation and finding the $x$-coordinate of the point of intersection of the two graphs. In Example 5, we found that the solution to the equation $\log x + \log(x+1) = \log 12$ was $x = 3$. **Figure 9.26** shows the graphical solution to this equation. The horizontal line is the graph of $y = \log 12$ since $\log 12$ is a constant. Notice that the $x$-coordinate of the point of intersection of the two graphs, 3, is the solution to the equation.

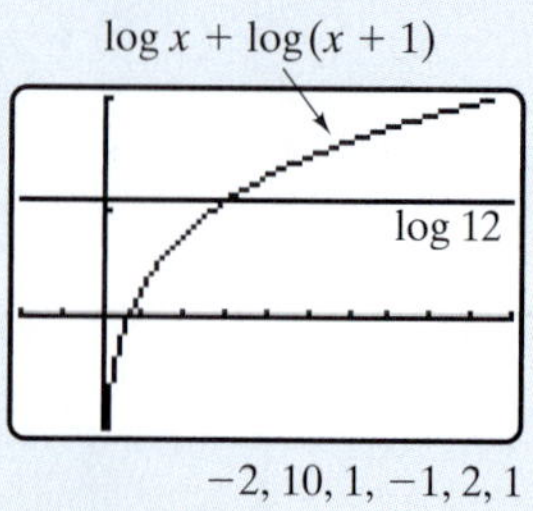


FIGURE 9.26

**EXAMPLE 6** ▸ Solve the equation $\log(3x - 5) - \log 5x = 1.23$.

*Solution*

$$\log(3x - 5) - \log 5x = 1.23$$

$$\log \frac{3x - 5}{5x} = 1.23 \quad \text{Quotient rule}$$

$$\frac{3x - 5}{5x} = \text{antilog } 1.23 \quad \text{Take the antilog of both sides.}$$

$$\frac{3x - 5}{5x} = 17.0 \quad \text{Rounded to three significant digits.}$$

$$3x - 5 = 5x(17.0) \quad \text{Multiply both sides by } 5x.$$

$$3x - 5 = 85x$$

$$-5 = 82x$$

$$x = -\frac{5}{82} \approx -0.061$$

**Check**

$$\log(3x - 5) - \log 5x = 1.23$$

$$\log[3(-0.061) - 5] - \log[(5)(-0.061)] \stackrel{?}{=} 1.23$$

$$\log(-5.183) - \log(-0.305) \stackrel{?}{=} 1.23 \quad \text{Stop.}$$

Since we have the logarithms of negative numbers, $-0.061$ is an extraneous solution. Thus, this equation has no solution. Its solution is the empty set, $\varnothing$.

▸ **Now Try Exercise 57**

## Helpful Hint

Below we show some of the steps used in the solutions of Example 3 and Example 6 from this section.

**Example 3**

$$\log_2 (x + 3)^3 = 4$$

$$(x + 3)^3 = 2^4 \quad \text{Write in exponential form.}$$

**Example 6**

$$\log \frac{3x - 5}{5x} = 1.23$$

$$\frac{3x - 5}{5x} = \text{antilog } 1.23 \quad \text{Take the antilog of both sides.}$$

Notice the steps we used were different in Examples 3 and 6. In Example 3, we write the equation in exponential form, while in Example 6, we take the antilog of both sides of the equation. In Example 6, we could have also written the second step (line) as $10^{1.23} = \frac{3x - 5}{5x}$, and then evaluated $10^{1.23}$ on a calculator to obtain 17.0 (rounded to three significant digits). Then we could finish the problem to find the solution. However, since Example 6 is given in base 10, we decided to just take the antilog of both sides. Antilogs of base 10 numbers are easily evaluated on a calculator. You may solve problems similar to Example 6 using either method.

## 2 Solve Applications

Now we will look at an application that involves an exponential equation.

**EXAMPLE 7** ▸ **Bacteria** If there are initially 1000 bacteria in a culture, and the number of bacteria doubles each hour, the number of bacteria after $t$ hours can be found by the formula

$$N = 1000(2)^t$$

How long will it take for the culture to grow to 30,000 bacteria?

### Solution

$$N = 1000(2)^t$$
$$30{,}000 = 1000(2)^t \quad \text{Substitute 30,000 for N.}$$
$$30 = (2)^t \quad \text{Divide both sides by 1000.}$$

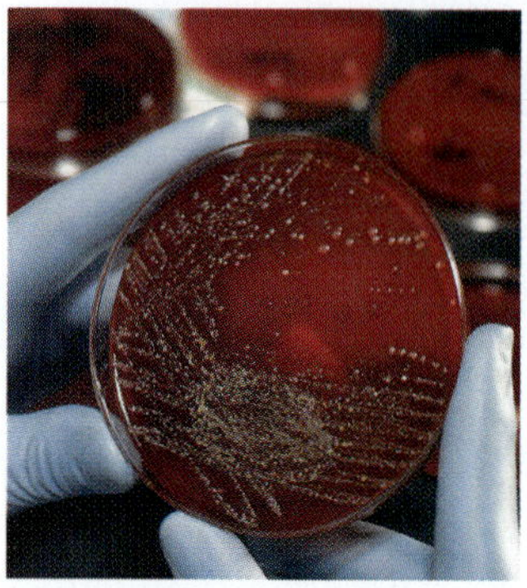

We want to find the value for $t$. To accomplish this we will use logarithms. Begin by taking the logarithm of both sides of the equation.

$$\log 30 = \log (2)^t$$
$$\log 30 = t \log 2 \quad \text{Power rule}$$
$$\frac{\log 30}{\log 2} = t \quad \text{Divide both sides by log 2.}$$
$$\frac{1.4771}{0.3010} \approx t$$
$$4.91 \approx t$$

It will take about 4.91 hours for the culture to grow to 30,000 bacteria.

▶ Now Try Exercise 69

## EXERCISE SET 9.6

MathXL® MyMathLab

### Concept/Writing Exercises

1. If $\log c = \log d$, then what is the relationship between $c$ and $d$?

2. If $c^r = c^s$, then what is the relationship between $r$ and $s$?

3. After solving a logarithmic equation, what must you do?

4. In properties 6c and 6d, we specify that both $x$ and $y$ must be positive. Explain why.

5. How can you tell quickly that $\log (x + 4) = \log (-2)$ has no real solution?

6. Can $x = -1$ be a solution of the equation $\log_3 x + \log_3(x - 8) = 2$? Explain.

### Practice the Skills

*Solve each exponential equation without using a calculator.*

7. $5^x = 125$
8. $2^x = 128$
9. $3^x = 81$
10. $4^x = 256$
11. $64^x = 8$
12. $81^x = 3$
13. $7^{-x} = \frac{1}{49}$
14. $6^{-x} = \frac{1}{216}$
15. $27^x = \frac{1}{3}$
16. $25^x = \frac{1}{5}$
17. $2^{x+2} = 64$
18. $3^{x-6} = 81$
19. $2^{3x-2} = 128$
20. $64^x = 4^{4x+1}$
21. $27^x = 3^{2x+3}$
22. $\left(\frac{1}{2}\right)^x = 16$

*Use a calculator to solve each equation. Round your answers to the nearest hundredth.*

23. $7^x = 50$
24. $1.05^x = 23$
25. $4^{x-1} = 35$
26. $2.3^{x-1} = 26.2$
27. $1.63^{x+1} = 25$
28. $4^x = 9^{x-2}$
29. $3^{x+4} = 6^x$
30. $5^x = 2^{x+5}$

*Solve each logarithmic equation. Use a calculator where appropriate. If the answer is irrational, round the answer to the nearest hundredth.*

31. $\log_{36} x = \frac{1}{2}$
32. $\log_{81} x = \frac{1}{2}$
33. $\log_{125} x = \frac{1}{3}$
34. $\log_{81} x = \frac{1}{4}$
35. $\log_2 x = -4$
36. $\log_7 x = -2$
37. $\log x = 2$
38. $\log x = 4$
39. $\log_2 (5 - 3x) = 3$
40. $\log_4 (3x + 7) = 3$
41. $\log_5 (x + 1)^2 = 2$
42. $\log_3 (a - 2)^2 = 2$
43. $\log_2 (r + 4)^2 = 4$
44. $\log_2 (p - 3)^2 = 6$
45. $\log (x + 8) = 2$
46. $\log (3x - 8) = 1$
47. $\log_2 x + \log_2 5 = 2$
48. $\log_3 2x + \log_3 x = 4$

49. $\log(r + 2) = \log(3r - 1)$

50. $\log 2a = \log(1 - a)$

51. $\log(2x + 1) + \log 4 = \log(7x + 8)$

52. $\log(x - 5) + \log 3 = \log(2x)$

53. $\log n + \log(3n - 5) = \log 2$

54. $\log(x + 4) - \log x = \log(x + 1)$

55. $\log 6 + \log y = 0.72$

56. $\log(x + 4) - \log x = 1.22$

57. $2\log x - \log 9 = 2$

58. $\log 6000 - \log(x + 2) = 3.15$

59. $\log x + \log(x - 3) = 1$

60. $2\log_2 x = 4$

61. $\log x = \frac{1}{3}\log 64$

62. $\log_7 x = \frac{3}{2}\log_7 9$

63. $\log_8 x = 4\log_8 2 - \log_8 8$

64. $\log_4 x + \log_4(6x - 7) = \log_4 5$

65. $\log_5(x + 3) + \log_5(x - 2) = \log_5 6$

66. $\log_7(x + 6) - \log_7(x - 3) = \log_7 4$

67. $\log_2(x + 3) - \log_2(x - 6) = \log_2 4$

68. $\log(x - 7) - \log(x + 3) = \log 6$

## Problem Solving

*Solve each problem. Round your answers to the nearest hundredth.*

69. **Bacteria** If the initial number of bacteria in the culture in Example 7 is 4500, when will the number of bacteria in the culture reach 50,000? Use $N = 4500(2)^t$.

70. **Bacteria** If after 4 hours the culture in Example 7 contains 2224 bacteria, how many bacteria were present initially?

71. **Radioactive Decay** The amount, $A$, of 200 grams of a certain radioactive material remaining after $t$ years can be found by the equation $A = 200(0.75)^t$. When will 80 grams remain?

72. **Radioactive Decay** The amount, $A$, of 70 grams of a certain radioactive material remaining after $t$ years can be found by the equation $A = 70(0.62)^t$. When will 10 grams remain?

73. **Savings Account** Paul Trapper invests \$2000 in a savings account earning interest at a rate of 5% compounded annually. How long will it take for the \$2000 to grow to \$4600? Use the compound interest formula, $A = p\left(1 + \frac{r}{n}\right)^{nt}$, which was discussed on page 606.

74. **Savings Account** If Tekar Werner invests \$600 in a savings account earning interest at a rate of 6% compounded semiannually, how long will it take for the \$600 to grow to \$1800?

75. **Infant Mortality Rate** The infant mortality rate (deaths per 1000 live births) in the United States has been decreasing since before 1959. Although it has fallen significantly, it is still higher than in many other nations. The U.S. infant mortality rate can be approximated by the function

$$f(t) = 26 - 12.1\log(t + 1)$$

where $t$ is the number of years since 1960 and $0 \le t \le 45$. Use this function to estimate the U.S. infant mortality rate in **a)** 1990. **b)** 2005.

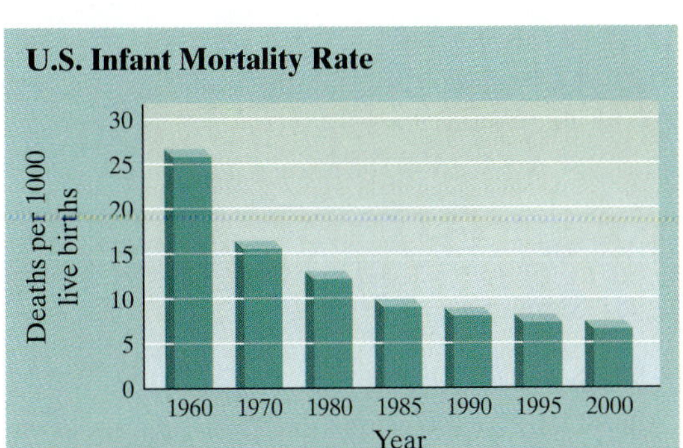


76. **Homicides** Since 1993, the number of homicides in New York City has been on the decline. The number of homicides can be approximated by the function

$$f(t) = 1997 - 1576\log(t + 1)$$

where $t$ is the number of years since 1993. If this trend continues, use this function to estimate the number of homicides in New York City in 2008.

77. **Depreciation** A machine purchased for business use can be depreciated to reduce income tax. The value of the machine at the end of its useful life is called its *scrap value*. When the machine depreciates by a constant percentage annually, its scrap value, $S$, is $S = c(1 - r)^n$, where $c$ is the original cost, $r$ is the annual rate of depreciation as a decimal, and $n$ is the useful life in years. Find the scrap value of a machine that costs \$50,000, has a useful life of 12 years, and has an annual depreciation rate of 15%.

78. **Depreciation** If the machine in Exercise 77 costs \$100,000, has a useful life of 15 years, and has an annual depreciation rate of 8%, find its scrap value.

79. **Power Gain of an Amplifier** The power gain, $P$, of an amplifier is defined as

$$P = 10\log\left(\frac{P_{out}}{P_{in}}\right)$$

where $P_{out}$ is the output power in watts and $P_{in}$ is the input power in watts. If an amplifier has an output power of 12.6 watts and an input power of 0.146 watts, find the power gain.

**80. Earthquake** Measured on the Richter scale, the magnitude, $R$, of an earthquake of intensity $I$ is defined by $R = \log I$, where $I$ is the number of times more intense the earthquake is than the minimum level for comparison.

**a)** How many times more intense was the 1906 San Francisco earthquake, which measured 8.25 on the Richter scale, than the minimum level for comparison?

**b)** How many times more intense is an earthquake that measures 8.3 on the Richter scale than one that measures 4.7?

**81. Magnitude of Sound** The decibel scale is used to measure the magnitude of sound. The magnitude $d$, in decibels, of a sound is defined to be $d = 10 \log I$, where $I$ is the number of times greater (or more intense) the sound is than the minimum intensity of audible sound.

**a)** An airplane engine (nearby) measures 120 decibels. How many times greater than the minimum level of audible sound is the airplane engine?

**b)** The intensity of the noise in a busy city street is 50 decibels. How many times greater is the intensity of the sound of the airplane engine than the sound of the city street?

**82.** In the following procedure, we begin with a true statement and end with a false statement. Can you find the error?

| | |
|---|---|
| $2 < 3$ | True |
| $2 \log (0.1) < 3 \log (0.1)$ | Multiply both sides by log (0.1). |
| $\log (0.1)^2 < \log (0.1)^3$ | Property 3 |
| $(0.1)^2 < (0.1)^3$ | Property 6d |
| $0.01 < 0.001$ | False |

**83.** Solve $8^x = 16^{x-2}$.

**84.** Solve $27^x = 81^{x-3}$.

**85.** Use equations that are quadratic in form to solve the equation $2^{2x} - 6(2^x) + 8 = 0$.

**86.** Use equations that are quadratic in form to solve the equation $2^{2x} - 18(2^x) + 32 = 0$.

*Change the exponential or logarithmic equation to the form $ax + by = c$, and then solve the system of equations.*

**87.** $2^x = 8^y$
$x + y = 4$

**88.** $3^{2x} = 9^{y+1}$
$x - 2y = -3$

**89.** $\log (x + y) = 2$
$x - y = 8$

**90.** $\log (x + y) = 3$
$2x - y = 5$

*Use your calculator to estimate the solutions to the nearest tenth. If a real solution does not exist, so state.*

**91.** $\log (x + 3) + \log x = \log 16$

**92.** $\log (3x + 5) = 2.3x - 6.4$

**93.** $5.6 \log (5x - 12) = 2.3 \log (x - 5.4)$

**94.** $5.6 \log (x + 12.2) - 1.6 \log (x - 4) = 20.3 \log (2x - 6)$

## Cumulative Review Exercises

[2.2] **95.** Consider the following two figures. Which has a greater volume, and by how much?

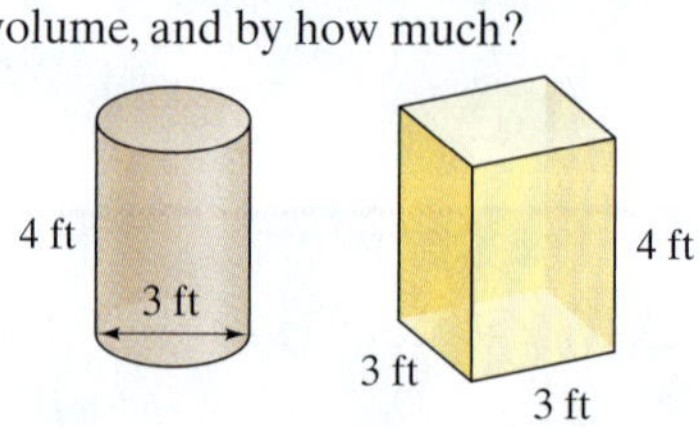


[3.6] **96.** Let $f(x) = x^2 - x$ and $g(x) = x - 1$. Find $(g - f)(3)$.

[4.6] **97.** Determine the solution set to the system of inequalities.

$$3x - 4y \le 6$$
$$y > -x + 4$$

[7.5] **98.** Simplify $\dfrac{2\sqrt{xy} - \sqrt{xy}}{\sqrt{x} + \sqrt{y}}$.

[8.3] **99.** Solve $E = mc^2$ for $c$.

[8.5] **100.** Determine the function for the parabola that has the shape of $f(x) = 2x^2$ and has its vertex at $(3, -5)$.

# 9.7 Natural Exponential and Natural Logarithmic Functions

1. Identify the natural exponential function.
2. Identify the natural logarithmic function.
3. Find values on a calculator.
4. Find natural logarithms using the change of base formula.
5. Solve natural logarithmic and natural exponential equations.
6. Solve applications.

The **natural exponential function** and *its inverse*, the **natural logarithmic** function, are exponential functions and logarithmic functions of the type presented in the previous sections. They share all the properties of exponential functions and logarithmic functions discussed earlier. The importance of these special functions lies in the many varied applications in real life of a unique irrational number designated by the letter $e$.

## 1 Identify the Natural Exponential Function

In Section 9.2 we indicated that exponential functions were of the form $f(x) = a^x$, $a > 0$ and $a \neq 1$. Now we introduce a very special exponential function. It is called the natural exponential function, and it uses the number $\boldsymbol{e}$. Like the irrational number $\pi$, the number $e$ is an irrational number whose value can only be approximated by a decimal number. The number $e$ plays a very important role in higher-level mathematics courses. The value of $e$ is approximately 2.7183. Now we define the natural exponential function.

**The Natural Exponential Function**

The natural exponential function is

$$f(x) = e^x$$

where $e \approx 2.7183$.

## 2 Identify the Natural Logarithmic Function

We discussed common logarithms in Section 9.5. Now we will discuss natural logarithms.

**Natural Logarithms**

**Natural logarithms** are logarithms to the base $e$. Natural logarithms are indicated by the letters ln.

$$\log_e x = \ln x$$

The notation $\ln x$ is read the "natural logarithm of $x$." The function $f(x) = \ln x$ is called the **natural logarithmic function**.

You must remember that the base of the natural logarithm is $e$. Thus, when you change a natural logarithm to exponential form, the base of the exponential expression will be $e$.

**Natural Logarithm in Exponential Form**

For $x > 0$ if $y = \ln x$, then $e^y = x$.

**EXAMPLE 1** ▸ Find the value of the expression by changing the natural logarithm to exponential form.

**a)** $\ln 1$ **b)** $\ln e$

*Solution*

**a)** Let $y = \ln 1$; then $e^y = 1$. Since any nonzero value to the 0th power equals 1, $y$ must equal 0. Thus, $\ln 1 = 0$.

**b)** Let $y = \ln e$; then $e^y = e$. For $e^y$ to equal $e$, $y$ must equal 1. Thus, $\ln e = 1$.

▸ **Now Try Exercise 1**

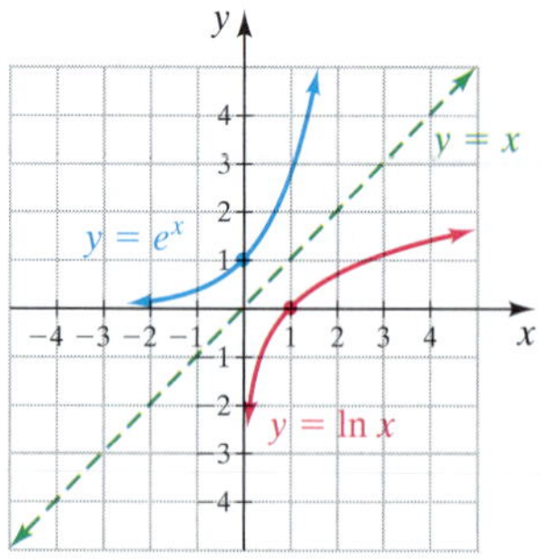


FIGURE 9.27

The functions $y = a^x$ and $y = \log_a x$ are inverse functions. Similarly, the functions $y = e^x$ and $y = \ln x$ are inverse functions. (Remember, $y = \ln x$ means $y = \log_e x$.) That is, if $f(x) = e^x$, then $f^{-1}(x) = \ln x$.

The graphs of $y = e^x$ and $y = \ln x$ are illustrated in **Figure 9.27**. Notice that the graphs are symmetric about the line $y = x$, which is what we expect of inverse functions.

Note that the graph of $y = e^x$ is similar to graphs of the form $y = a^x, a > 1$, and that the graph of $y = \ln x$ is similar to graphs of the form $y = \log_a x, a > 1$.

## 3 Find Values on a Calculator

Now we will learn how to find natural logarithms on a calculator.

### USING YOUR CALCULATOR Find Natural Logarithms

Natural logarithms can be found using a calculator that has a [LN] key. Natural logarithms are found in the same manner that we found common logarithms on a calculator, except we use the natural log key, [LN], instead of the common log key, [LOG].

**Scientific Calculator**

| EXAMPLE | KEYS TO PRESS | ANSWER DISPLAYED |
|---|---|---|
| Find ln 242. | 242 [LN] | 5.4889377 |
| Find ln 0.85. | .85 [LN] | −0.1625189 |

**Graphing Calculator***

On the TI-84 Plus, after the [LN] key is pressed, the calculator displays ln( on the screen.

| EXAMPLE | KEYS TO PRESS | ANSWER DISPLAYED |
|---|---|---|
| Find ln 242. | [LN] (242 [)] [ENTER] | 5.488937726 |
| Find ln 0.85. | [LN] (.85 [)] [ENTER] | −.1625189295 |

*The keys are for a TI-84 Plus. Read your manual for instructions on how to find natural logarithms on your calculator.

When finding the natural logarithm of a number, we are finding an exponent. The natural logarithm of a number is the exponent to which the base $e$ must be raised to obtain that number. For example,

$$\text{If } \ln 242 = 5.4889377, \text{ then } e^{5.4889377} = 242.$$

$$\text{If } \ln 0.85 = -0.1625189, \text{ then } e^{-0.1625189} = 0.85.$$

Since $y = \ln x$ and $y = e^x$ are inverse functions, we can use the inverse key [INV], in combination with the natural log key, [LN], to obtain values for $e^x$.

### USING YOUR CALCULATOR Finding Values of $e^x$

**Scientific Calculator**

To find values of $e^x$, first enter the exponent on $e$. Then press either [shift], [2nd], or [INV], depending on your calculator. Then press the natural log key, [LN]. After the [LN] key is pressed, the value of $e^x$ will be displayed.

| EXAMPLE | KEYS TO PRESS | ANSWER DISPLAYED |
|---|---|---|
| Find $e^{5.24}$. | 5.24 [INV] [LN] | 188.6701 |
| Find $e^{-1.639}$. | 1.639 [+/−] [INV] [LN] | 0.1941741 |

**Graphing Calculator***

On the TI-84 Plus, after [2nd] [LN] is pressed, the calculator displays e^( on the screen.

| EXAMPLE | KEYS TO PRESS | ANSWER DISPLAYED |
|---|---|---|
| Find $e^{5.24}$. | [2nd] [LN] (5.24 [)] [ENTER] | 188.6701024 |
| Find $e^{-1.639}$. | [2nd] [LN] ( [(−)] 1.639 [)] [ENTER] | .1941741194 |

*Keys are for a TI-84 Plus. Read your manual for instructions on how to evaluate natural exponential expressions on your calculator.

Remember $e$ is about 2.7183. When we evaluated $e^{5.24}$ or $(2.7183)^{5.24}$ in the previous calculator box we obtained a value close to 188.6701. If we found ln 188.6701 on a calculator, we would obtain a value close to 5.24. What do you think we would get if we evaluated ln 0.1941741 on a calculator? If you answered, "A value close to −1.639," you answered correctly.

**EXAMPLE 2** ▸ Find $N$ if **a)** $\ln N = 5.26$ and **b)** $\ln N = -0.0253$.

*Solution*

**a)** If we write $\ln N = 5.26$ in exponential form, we get $e^{5.26} = N$. Thus, we simply need to evaluate $e^{5.26}$ to determine $N$.

$$e^{5.26} = 192.48149 \qquad \text{From a calculator}$$

Thus, $N = 192.48149$

**b)** If we write $\ln N = -0.0253$ in exponential form, we get $e^{-0.0253} = N$.

$$e^{-0.0253} = 0.9750174 \qquad \text{From a calculator}$$

Thus, $N = 0.9750174$.

▸ **Now Try Exercise 19**

## 4 Find Natural Logarithms Using the Change of Base Formula

If you are given a logarithm in a base other than 10 or $e$, you will not be able to evaluate it on your calculator directly. When this occurs, you can use the **change of base formula**.

**Change of Base Formula**

For any logarithm bases $a$ and $b$, and positive number $x$,

$$\log_a x = \frac{\log_b x}{\log_b a}$$

We can prove the change of base formula by beginning with $\log_a x = m$.

$$\log_a x = m$$
$$a^m = x \qquad \text{Change to exponential form.}$$
$$\log_b a^m = \log_b x \qquad \text{By property 6c on page 630}$$
$$m \log_b a = \log_b x \qquad \text{Power rule}$$
$$(\log_a x)(\log_b a) = \log_b x \qquad \text{Substitution for } m$$
$$\log_a x = \frac{\log_b x}{\log_b a} \qquad \text{Divide both sides by } \log_b a.$$

In the change of base formula, 10 is often used in place of base $b$ because we can find common logarithms on a calculator. Replacing base $b$ with 10, we get

$$\log_a x = \frac{\log_{10} x}{\log_{10} a} \quad \text{or} \quad \log_a x = \frac{\log x}{\log a}$$

**EXAMPLE 3** ▸ Use the change of base formula to find $\log_3 24$.

*Solution* If we substitute 3 for $a$ and 24 for $x$ in $\log_a x = \frac{\log x}{\log a}$, we obtain

$$\log_3 24 = \frac{\log 24}{\log 3} \approx \frac{1.3802}{0.4771} \approx 2.8929$$

Note that $3^{2.8929} \approx 24$.

▸ **Now Try Exercise 23**

## Helpful Hint:

In Example 3, if we compute the actual values in the quotient $\frac{\log 24}{\log 3}$ using a calculator, the value is $\approx 2.8928$ instead of the value of $\approx 2.8929$ we obtained in Example 3. To obtain the answer of $\approx 2.8929$, we divided the *approximation* of the numerator, 1.3802, by the *approximation* of the denominator, 0.4771. When we give answers to the exercises involving the change of base formula in the answer section, we will *not* round the values in the numerator and denominator. We will only round the final answer.

We can use the same procedure as in Example 3 to find natural logarithms using the change of base formula. For example, to evaluate ln 20 (or $\log_e 20$), we can substitute $e$ for $a$ and 20 for $x$ in the formula $\log_a x = \frac{\log x}{\log a}$.

$$\log_e 20 = \frac{\log 20}{\log e} \approx \frac{1.3010}{0.4343} \approx 2.9956$$

Thus, $\ln 20 \approx 2.9956$. If you find ln 20 on a calculator, you will obtain a very close value.

Since $\log e \approx 0.4343$, to evaluate natural logarithms using common logarithms, we use the formula

$$\ln x = \frac{\log x}{\log e} \approx \frac{\log x}{0.4343}$$

**EXAMPLE 4** ▸ Use the change of base formula to find ln 95.

*Solution*

$$\ln 95 = \frac{\log 95}{\log e} \approx \frac{1.9777}{0.4343} \approx 4.5538$$

If you evaluate ln 95 on your calculator, you will obtain a value very close to 4.5538. Do so now and check your result.

▸ **Now Try Exercise 33**

## 5 Solve Natural Logarithmic and Natural Exponential Equations

The properties of logarithms discussed in Section 9.4 still hold true for natural logarithms. Following is a summary of these properties in the notation of natural logarithms.

**Properties for Natural Logarithms**

| | | |
|---|---|---|
| $\ln xy = \ln x + \ln y$ | $(x > 0 \text{ and } y > 0)$ | *Product rule* |
| $\ln \frac{x}{y} = \ln x - \ln y$ | $(x > 0 \text{ and } y > 0)$ | *Quotient rule* |
| $\ln x^n = n \ln x$ | $(x > 0)$ | *Power rule* |

Consider the expression $\ln e^x$, which means $\log_e e^x$. From property 4 on page 621, $\log_e e^x = x$. Thus, $\ln e^x = x$. Similarly, $e^{\ln x} = e^{\log_e x} = x$ by property 5. Although $\ln e^x = x$ and $e^{\ln x} = x$ are just special cases of properties 4 and 5, respectively, we will call these properties 7 and 8 so that we can make reference to them.

**Additional Properties for Natural Logarithms and Natural Exponential Expressions**

| | | |
|---|---|---|
| $\ln e^x = x$ | | **Property 7** |
| $e^{\ln x} = x,$ | $x > 0$ | **Property 8** |

Using property 7, $\ln e^x = x$, we can state, for example, that $\ln e^{kt} = kt$, and $\ln e^{-2.06t} = -2.06t$. Using property 8, $e^{\ln x} = x$, we can state, for example, that $e^{\ln (t+2)} = t + 2$ and $e^{\ln kt} = kt$.

**EXAMPLE 5** ▸ Solve the equation $\ln y - \ln (x + 9) = t$ for $y$.

*Solution*

$$\ln y - \ln (x + 9) = t$$

$$\ln \frac{y}{x + 9} = t \qquad \text{Quotient rule}$$

$$\frac{y}{x + 9} = e^t \qquad \text{Change to exponential form.}$$

$$y = e^t(x + 9) \qquad \text{Solve for } y.$$

▸ **Now Try Exercise 63**

**EXAMPLE 6** ▸ Solve the equation $225 = 450e^{-0.4t}$ for $t$.

*Solution* Begin by dividing both sides of the equation by 450 to isolate $e^{-0.4t}$.

$$\frac{225}{450} = \frac{450e^{-0.4t}}{450}$$

$$0.5 = e^{-0.4t}$$

Now take the natural logarithm of both sides of the equation to eliminate the exponential expression on the right side of the equation.

$$\ln 0.5 = \ln e^{-0.4t}$$

$$\ln 0.5 = -0.4t \qquad \text{Property 7}$$

$$-0.6931472 = -0.4t$$

$$\frac{-0.6931472}{-0.4} = t$$

$$1.732868 = t$$

▸ **Now Try Exercise 49**

**EXAMPLE 7** ▸ Solve the equation $P = P_0e^{kt}$ for $t$.

*Solution* We can follow the same procedure as used in Example 6.

$$P = P_0e^{kt}$$

$$\frac{P}{P_0} = \frac{P_0e^{kt}}{P_0} \qquad \text{Divide both sides by } P_0.$$

$$\frac{P}{P_0} = e^{kt}$$

$$\ln \frac{P}{P_0} = \ln e^{kt} \qquad \text{Take natural log of both sides.}$$

$$\ln P - \ln P_0 = \ln e^{kt} \qquad \text{Quotient rule}$$

$$\ln P - \ln P_0 = kt \qquad \text{Property 7}$$

$$\frac{\ln P - \ln P_0}{k} = t \qquad \text{Solve for } t.$$

▸ **Now Try Exercise 59**

## 6 Solve Applications

Now let's look at some applications that involve the natural exponential function and natural logarithms.

When a quantity $P$ increases or decreases at an *exponential rate*, a formula often used to find the value of $P$ after time $t$ is

$$P = P_0 e^{kt}$$

where $P_0$ is the initial or starting value and $k$ is the constant growth rate or decay rate compounded continuously. We will refer to this formula as the **exponential growth (or decay) formula**. In the formula, other letters may be used in place of $P$. When $k > 0$, $P$ increases as $t$ increases. When $k < 0$, $P$ decreases and gets closer to 0 as $t$ increases.

**EXAMPLE 8 ▸ Interest Compounded Continuously** Banks often credit compound interest continuously. When interest is compounded continuously, the balance, $P$, in the account at any time, $t$, can be calculated by the exponential growth formula $P = P_0 e^{kt}$, where $P_0$ is the principal initially invested and $k$ is the interest rate.

**a)** Suppose the interest rate is 6% compounded continuously and \$1000 is initially invested. Determine the balance in the account after 3 years.

**b)** How long will it take the account to double in value?

*Solution* **a) Understand and Translate** We are told that the principal initially invested, $P_0$, is \$1000. We are also given that the time, $t$, is 3 years and that the interest rate, $k$, is 6% or 0.06. We substitute these values into the given formula and solve for $P$.

$$P = P_0 e^{kt}$$

$$P = 1000e^{(0.06)(3)}$$

**Carry Out**

$$= 1000e^{0.18} = 1000(1.1972174) \quad \text{From a calculator}$$

$$\approx 1197.22$$

**Answer** After 3 years, the balance in the account is ≈ \$1197.22.

**b) Understand and Translate** For the value of the account to double, the balance in the account would have to reach \$2000. Therefore, we substitute 2000 for $P$ and solve for $t$.

$$P = P_0 e^{kt}$$

$$2000 = 1000e^{0.06t}$$

**Carry Out**

$$2 = e^{0.06t} \quad \text{Divide both sides by 1000.}$$

$$\ln 2 = \ln e^{0.06t} \quad \text{Take natural log of both sides.}$$

$$\ln 2 = 0.06t \quad \text{Property 7}$$

$$\frac{\ln 2}{0.06} = t$$

$$\frac{0.6931472}{0.06} = t$$

$$11.552453 \approx t$$

**Answer** Thus, with an interest rate of 6% compounded continuously, the account will double in about 11.6 years.

▸ **Now Try Exercise 69**

**EXAMPLE 9 ▸ Radioactive Decay** Strontium 90 is a radioactive isotope that decays exponentially at 2.8% per year. Suppose there are initially 1000 grams of strontium 90 in a substance.

**a)** Find the number of grams of strontium 90 left after 50 years.

**b)** Find the half-life of strontium 90.

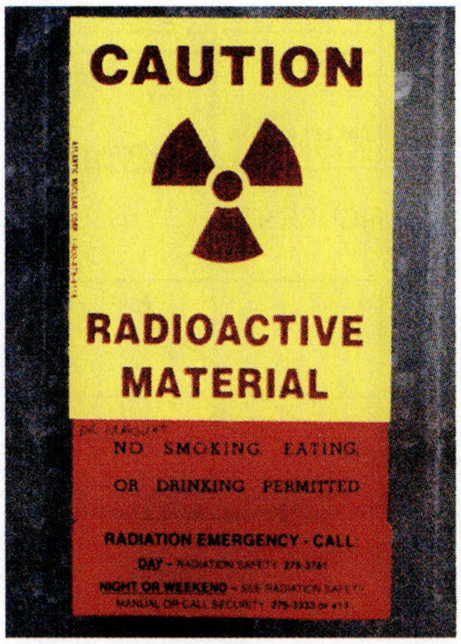


*Solution* **a) Understand** Since the strontium 90 is decaying over time, the value of $k$ in the formula $P = P_0e^{kt}$ is negative. Since the rate of decay is 2.8% per year, we use $k = -0.028$. Therefore, the formula we use is $P = P_0e^{-0.028t}$.

**Translate** $$P = P_0e^{-0.028t}$$
$$= 1000e^{-0.028(50)}$$

**Carry Out** $$= 1000e^{-1.4} = 1000(0.246597) = 246.597$$

**Answer** Thus, after 50 years, 246.597 grams of strontium 90 remain.

**b)** To find the half-life, we need to determine when 500 grams of strontium 90 are left.

$$P = P_0e^{-0.028t}$$
$$500 = 1000e^{-0.028t}$$
$$0.5 = e^{-0.028t} \quad \text{Divide both sides by 1000.}$$
$$\ln 0.5 = \ln e^{-0.028t} \quad \text{Take natural log of both sides.}$$
$$-0.6931472 = -0.028t \quad \text{Property 7}$$
$$\frac{-0.6931472}{-0.028} = t$$
$$24.755257 \approx t$$

Thus, the half-life of strontium 90 is about 24.8 years.

▸ Now Try Exercise 71

**EXAMPLE 10 ▸ Selling Toys** The formula for estimating the amount of money, $A$, spent on advertising a certain toy is $A = 350 + 650 \ln n$ where $n$ is the expected number of toys to be sold.

**a)** If the company wishes to sell 2200 toys, how much money should the company expect to spend on advertising?

**b)** How many toys can be expected to be sold if \$6000 is spent on advertising?

*Solution*

**a)** $$A = 350 + 650 \ln n$$
$$= 350 + 650 \ln 2200 \quad \text{Substitute 2200 for } n.$$
$$= 350 + 650(7.6962126)$$
$$= 5352.54$$

Thus, \$5352.54 should be expected to be spent on advertising.

**b) Understand and Translate** We are asked to find the number of toys expected to be sold, $n$, if \$6000 is spent on advertising. We substitute the given values into the equation and solve for $n$.

$$A = 350 + 650 \ln n$$

**Carry Out**
$$6000 = 350 + 650 \ln n \quad \text{Substitute 6000 for } A.$$
$$5650 = 650 \ln n \quad \text{Subtract 350 from both sides.}$$
$$\frac{5650}{650} = \ln n \quad \text{Divide both sides by 650.}$$
$$8.69231 \approx \ln n$$
$$e^{8.69231} \approx n \quad \text{Change to exponential form.}$$
$$5957 \approx n \quad \text{Obtain answer from a calculator.}$$

**Answer** Thus, about 5957 toys can be expected to be sold if \$6000 is spent on advertising.

▸ Now Try Exercise 75

## USING YOUR GRAPHING CALCULATOR

Equations containing natural logarithms and natural exponential functions can be solved on your graphing calculator. For example, to solve the equation $\ln x + \ln (x + 3) = \ln 8$, we set

$$Y_1 = \ln x + \ln (x + 3)$$

$$Y_2 = \ln 8$$

and find the intersection of the graphs, as shown in **Figure 9.28**.

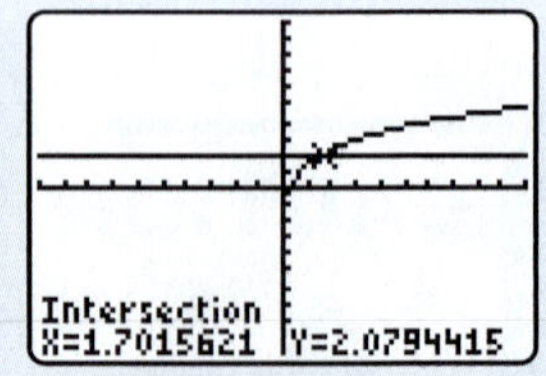


FIGURE 9.28

In **Figure 9.28**, we used the CALC, INTERSECT option to find the intersection of the graphs. The solution is the $x$-coordinate of the intersection. The solution to the equation is $x = 1.7016$, to the nearest ten-thousandth.

To solve the equation $4e^{0.3x} - 5 = x + 3$, we set

$$Y_1 = 4e^{0.3x} - 5$$

$$Y_2 = x + 3$$

and find the intersection of the graphs, as shown in **Figure 9.29**. This equation has two solutions since there are two intersections.

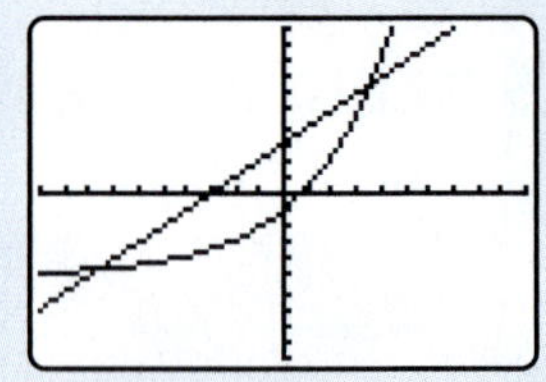

FIGURE 9.29

The solutions to the equation are approximately $x = -7.5896$ and $x = 3.5284$. In Exercises 93 through 97, we use the graphing calculator to check or solve equations.

# EXERCISE SET 9.7

MathXL® MyMathLab

## Concept/Writing Exercises

1. a) What is the base in the natural exponential function?
   b) What is the approximate value of $e$?
2. What is another way of writing $\log_e x$?
3. What is the domain of $\ln x$?
4. Under what conditions will $\ln x < 0$?
5. Give the change of base formula.
6. Is $n \log_e x = \ln x^n$? Explain.
7. To what is $\ln e^x$ equal?
8. To what is $e^{\ln x}$ equal?
9. What is the inverse of $\ln x$?
10. Under what circumstances will $P$ in the formula $P = P_0e^{kt}$ increase when $t$ increases?
11. Under what conditions will $P$ in the formula $P = P_0e^{kt}$ decrease when $t$ increases?
12. Is it possible to find the value of $\ln (-3.52)$? Explain.

## Practice the Skills

*Find the following values. Round values to four decimal places.*

**13.** $\ln 62$ **14.** $\ln 791$ **15.** $\ln 0.813$ **16.** $\ln 0.000568$

*Find the value of N. Round values to three significant digits.*

**17.** $\ln N = 1.6$ **18.** $\ln N = 5.2$ **19.** $\ln N = -2.85$

**20.** $\ln N = 0.543$ **21.** $\ln N = -0.0287$ **22.** $\ln N = -0.674$

*Use the change of base formula to find the value of the following logarithms. Do not round logarithms in the change of base formula. Write the answer rounded to the nearest ten-thousandth.*

**23.** $\log_3 56$ **24.** $\log_3 198$ **25.** $\log_2 21$ **26.** $\log_2 89$

**27.** $\log_4 11$ **28.** $\log_4 316$ **29.** $\log_5 82$ **30.** $\log_5 1893$

**31.** $\log_6 185$ **32.** $\log_6 806$ **33.** $\ln 51$ **34.** $\ln 3294$

**35.** $\log_5 0.463$ **36.** $\log_3 0.0365$

*Solve the following logarithmic equations.*

**37.** $\ln x + \ln (x - 1) = \ln 12$ **38.** $\ln (x + 4) + \ln (x - 2) = \ln 16$ **39.** $\ln x + \ln (x + 4) = \ln 5$

**40.** $\ln (x + 3) + \ln (x - 3) = \ln 40$ **41.** $\ln x = 5 \ln 2 - \ln 8$ **42.** $\ln x = \frac{3}{2}\ln 16$

**43.** $\ln (x^2 - 4) - \ln (x + 2) = \ln 4$ **44.** $\ln (x + 12) - \ln (x - 4) = \ln 5$

*Each of the following equations is in the form $P = P_0e^{kt}$. Solve each equation for the remaining variable. Remember, e is a constant. Write the answer rounded to the nearest ten-thousandth.*

**45.** $P = 120e^{2.3(1.6)}$

**46.** $900 = P_0e^{(0.4)(3)}$

**47.** $50 = P_0e^{-0.5(3)}$

**48.** $18 = 9e^{2t}$

**49.** $60 = 20e^{1.4t}$

**50.** $29 = 58e^{-0.5t}$

**51.** $86 = 43e^{k(3)}$

**52.** $15 = 75e^{k(4)}$

**53.** $20 = 40e^{k(2.4)}$

**54.** $100 = A_0e^{-0.02(3)}$

**55.** $A = 6000e^{-0.08(3)}$

**56.** $51 = 68e^{-0.04t}$

*Solve for the indicated variable.*

**57.** $V = V_0e^{kt}$, for $V_0$

**58.** $P = P_0e^{kt}$, for $P_0$

**59.** $P = 150e^{7t}$, for $t$

**60.** $361 = P_0e^{kt}$, for $t$

**61.** $A = A_0e^{kt}$, for $k$

**62.** $167 = R_0e^{kt}$, for $k$

**63.** $\ln y - \ln x = 2.3$, for $y$

**64.** $\ln y + 9 \ln x = \ln 2$, for $y$

**65.** $\ln y - \ln (x + 6) = 5$ for $y$

**66.** $\ln (x + 2) - \ln (y - 1) = \ln 5$, for $y$

## Problem Solving

*Use a calculator to solve.*

**67.** If $e^x = 12.183$, find the value of $x$. Explain how you obtained your answer.

**68.** To what exponent must the base $e$ be raised to obtain the value 184.93? Explain how you obtained your answer.

**69. Interest Compounded Continuously** If \$5000 is invested at 6% compounded continuously,

**a)** determine the balance in the account after 2 years.

**b)** How long would it take the value of the account to double? (See Example 8.)

**70. Interest Compounded Continuously** If \$3000 is invested at 3% compounded continuously,

**a)** determine the balance in the account after 30 years.

**b)** How long would it take the value of the account to double?

**71. Radioactive Decay** Refer to Example 9. Determine the amount of strontium 90 remaining after 20 years if there were originally 70 grams.

**72. Strontium 90** Refer to Example 9. Determine the amount of strontium 90 remaining after 40 years if there were originally 200 grams.

**73. Soft Drinks** For a certain soft drink, the percent of a target market, $f(t)$, that buys the soft drink is a function of the number of days, $t$, that the soft drink is advertised. The function that describes this relationship is $f(t) = 1 - e^{-0.04t}$.

**a)** What percent of the target market buys the soft drink after 50 days of advertising?

**b)** How many days of advertising are needed if 75% of the target market is to buy the soft drink?

**74. Trout in a Lake** In 2005, a lake had 300 trout. The growth in the number of trout is estimated by the function $g(t) = 300e^{0.07t}$ where $t$ is the number of years after 2005. How many trout will be in the lake in **a)** 2008? **b)** 2015?

**75. Walking Speed** It was found in a psychological study that the average walking speed, $f(P)$, of a person living in a city is a function of the population of the city. For a city of population $P$, the average walking speed in feet per second is given by $f(P) = 0.37 \ln P + 0.05$. The population of Nashville, Tennessee, is 972,000.

**a)** What is the average walking speed of a person living in Nashville?

**b)** What is the average walking speed of a person living in New York City, population 8,567,000?

**c)** If the average walking speed of the people in a certain city is 5.0 feet per second, what is the population of the city?

**76. Advertising** For a certain type of tie, the number of ties sold, $N(a)$, is a function of the dollar amount spent on advertising, $a$ (in thousands of dollars). The function that describes this relationship is $N(a) = 800 + 300 \ln a$.

**a)** How many ties were sold after \$1500 (or \$1.5 thousand) was spent on advertising?

**b)** How much money must be spent on advertising to sell 1000 ties?

**77.** Assume that the value of the island of Manhattan has grown at an exponential rate of 8% per year since 1626 when Peter Minuet of the Dutch West India Company purchased it for \$24. Then the value of Manhattan can be determined by the equation $V = 24e^{0.08t}$, where $t$ is the number of years since 1626. Determine the value of the island of Manhattan in 2008, that is when $t = 382$ years.

**78. Prescribing a Drug** The percent of doctors who accept and prescribe a new drug is given by the function $P(t) = 1 - e^{-0.22t}$, where $t$ is the time in months since the drug was placed on the market. What percent of doctors accept a new drug 2 months after it is placed on the market?

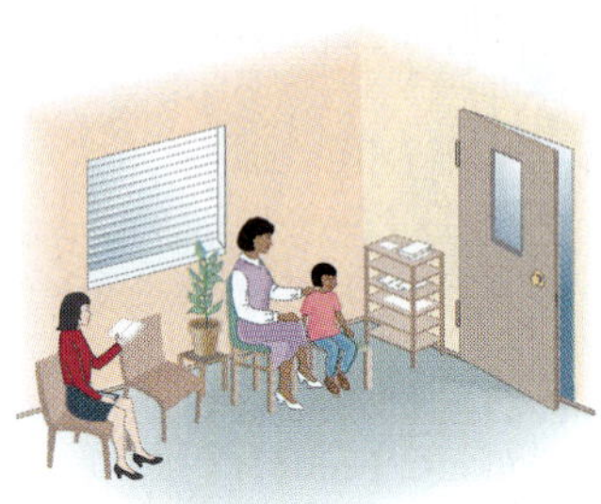

**79. World Population** The world population in January 2003 was estimated to be about 6.30 billion people. Let's assume that the world population will continue to grow exponentially at the current growth rate of about 1.3% per year. Then, the expected world population, in billions of people, in $t$ years, is given by the function

$$P(t) = 6.30e^{0.013t}$$

where $t$ is the number of years since 2003.

**a)** Estimate the world population in 2010.

**b)** In how many years will the world population double?

**80. Generic Drugs** Since 2001, the number of prescriptions for generic drugs used by members of the COVA Care Health Plan in the Commonwealth of Virginia has been growing exponentially (see graph). The number of prescriptions for generic drugs, $f(t)$, in 100,000s, can be approximated by the function $f(t) = 6.52e^{0.087t}$, where $t$ is the number of years since 2001. Assuming that this trend continues, use this function to estimate the number of prescriptions for generic drugs used by member of the COVA Health Care Plan in **a)** 2006. **b)** 2008.

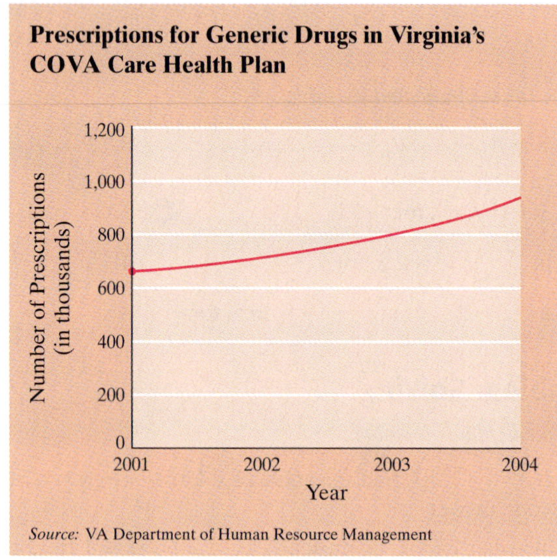


**81. Demand for Nurses** The demand for registered nurses is expected to grow exponentially from 2005 to 2020. (See graph). The demand for registered nurses, $d(t)$, in millions, can be approximated by the function $d(t) = 2.19e^{0.0164t}$, where $t$ is the number of years since 2005. Assuming that this trend continues, use this function to estimate the demand for registered nurses in **a)** 2025. **b)** 2040.

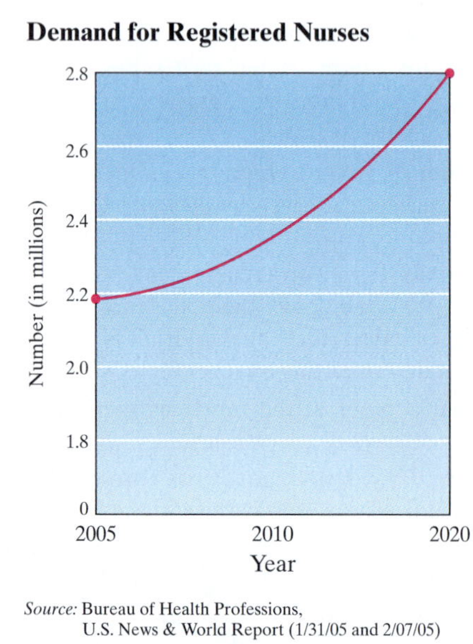


**82. Splenda® Products** Since 2000, the number of new products using Splenda each year has been growing exponentially (see graph on the next page). The number of new products using Splenda, $N(t)$, can be approximated by the function $N(t) = 163.21e^{0.481t}$, where $t$ is the number of years since

| IMPORTANT FACTS AND CONCEPTS | EXAMPLES |
|---|---|

## Section 9.4

**Product Rule for Logarithms**

For positive real numbers $x$, $y$, and $a$, $a \neq 1$,

$$\log_a xy = \log_a x + \log_a y \qquad \textit{Property 1}$$

Examples:

$$\log_5 (9 \cdot 13) = \log_5 9 + \log_5 13$$
$$\log_7 mn = \log_7 m + \log_7 n$$

**Quotient Rule for Logarithms**

For positive real numbers $x$, $y$, and $a$, $a \neq 1$,

$$\log_a \frac{x}{y} = \log_a x - \log_a y \qquad \textit{Property 2}$$

Examples:

$$\log_3 \frac{15}{4} = \log_3 15 - \log_3 4$$
$$\log_8 \frac{z+1}{z+3} = \log_8 (z+1) - \log_8 (z+3)$$

**Power Rule for Logarithms**

If $x$ and $a$ are positive real numbers, $a \neq 1$, and $n$ is any real number, then

$$\log_a x^n = n \log_a x \qquad \textit{Property 3}$$

Examples:

$$\log_9 23^5 = 5 \log_9 23$$
$$\log_6 \sqrt[3]{x+4} = \log_6 (x+4)^{1/3} = \frac{1}{3} \log_6 (x+4)$$

**Additional Properties of Logarithms**

If $a > 0$, and $a \neq 1$, then

$$\log_a a^x = x \qquad \textit{Property 4}$$

and $$a^{\log_a x} = x \; (x > 0) \qquad \textit{Property 5}$$

Examples:

$$\log_4 16 = \log_4 4^2 = 2$$
$$7^{\log_7 3} = 3$$

## Section 9.5

**Common Logarithm**

Base 10 logarithms are called common logarithms.

$\log x$ means $\log_{10} x$

Examples:

$\log 17$ means $\log_{10} 17$

$\log (b + c)$ means $\log_{10} (b + c)$

The **common logarithm** of a positive real number is the *exponent* to which the base 10 is raised to obtain the number.

If $\log N = L$, then $10^L = N$.

Examples:

If $\log 14 = 1.1461$, then $10^{1.1461} = 14$

If $\log 0.6 = -0.2218$, then $10^{-0.2218} = 0.6$.

To find a common logarithm, use a scientific or graphing calculator. We round the answer to four decimal places.

Examples:

$\log 183 = 2.2625$ (rounded to 4 places)

$\log 0.42 = -0.3768$ (rounded to 4 places)

**Antilogarithm**

If $\log N = L$, then $N = \text{antilog } L$.

To find antilogarithms, use a scientific or graphing calculator.

Examples:

If $\log 1890.1662 = 3.2765$, then antilog $3.2765 = 1890.1662$.

If $\log 0.0143 = -1.8447$, then antilog $(-1.8447) = 0.0143$.

## Section 9.6

**Properties for Solving Exponential and Logarithmic Equations**

**a)** If $x = y$, then $a^x = a^y$.

**b)** If $a^x = a^y$, then $x - y$.

**c)** If $x = y$, then $\log_b x = \log_b y \quad (x > 0, y > 0)$.

**d)** If $\log_b x = \log_b y$, then $x = y \quad (x > 0, y > 0)$.

*Properties 6a–6d*

Examples:

**a)** If $x = 5$, then $3^x = 3^5$.

**b)** If $3^x = 3^5$, then $x = 5$.

**c)** If $x = 2$, then $\log x = \log 2$.

**d)** If $\log x = \log 2$, then $x = 2$.

| IMPORTANT FACTS AND CONCEPTS | EXAMPLES |
| --- | --- |
| **Section 9.7** | |

The **natural exponential function** is

$$f(x) = e^x$$

where $e \approx 2.7183$.

**Natural logarithms** are logarithms to the base $e$. Natural logarithms are indicated by the letters ln.

$$\log_e x = \ln x$$

For $x > 0$, if $y = \ln x$, then $e^y = x$.

The **natural logarithmic function** is

$$g(x) = \ln x$$

where the base $e \approx 2.7183$.

To find natural exponential and natural logarithmic values, use a scientific or graphing calculator.

The natural exponential function, $f(x) = e^x$, and the natural logarithmic function, $g(x) = \ln x$, are inverses of each other.

*Examples:*

Graph $f(x) = e^x$ and $g(x) = \ln x$ on the same set of axes.

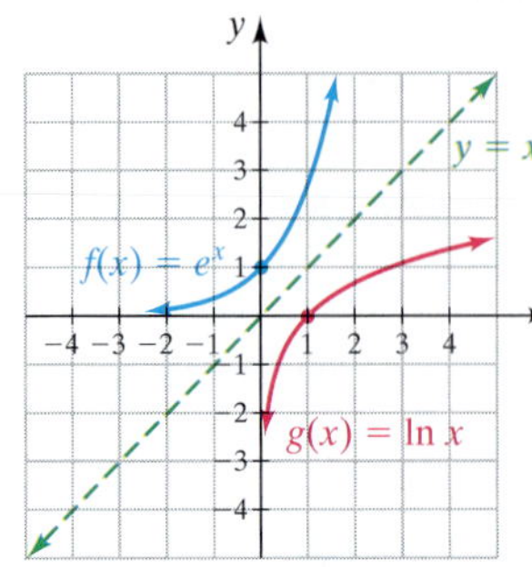


$\ln 5.83 = 1.7630$

If $\ln N = -2.09$, then $N = e^{-2.09} = 0.1237$.

### Change of Base Formula

For any logarithm bases $a$ and $b$, and positive number $x$,

$$\log_a x = \frac{\log_b x}{\log_b a}$$

*Example:*

$$\log_5 98 = \frac{\log 98}{\log 5} \approx \frac{1.9912}{0.6990} \approx 2.8486$$

### Properties for Natural Logarithms

| | | | Example |
| --- | --- | --- | --- |
| $\ln xy = \ln x + \ln y$ | $(x > 0 \text{ and } y > 0)$ | *Product rule* | $\ln 7 \cdot 30 = \ln 7 + \ln 30$ |
| $\ln \frac{x}{y} = \ln x - \ln y$ | $(x > 0 \text{ and } y > 0)$ | *Quotient rule* | $\ln \frac{x+1}{x+8} = \ln (x + 1) - \ln (x + 8)$ |
| $\ln x^n = n \ln x$ | $(x > 0)$ | *Power rule* | $\ln m^5 = 5 \ln m$ |

### Additional Properties for Natural Logarithms and Natural Exponential Expressions

| | | Example |
| --- | --- | --- |
| $\ln e^x = x$ | *Property 7* | $\ln e^{19} = 19$ |
| $e^{\ln x} = x, \quad x > 0$ | *Property 8* | $e^{\ln 2} = 2$ |

## Chapter 9 Review Exercises

[9.1] *Given $f(x) = x^2 - 3x + 4$ and $g(x) = 2x - 5$, find the following.*

**1.** $(f \circ g)(x)$ **2.** $(f \circ g)(3)$ **3.** $(g \circ f)(x)$ **4.** $(g \circ f)(-3)$

*Given $f(x) = 6x + 7$ and $g(x) = \sqrt{x - 3}$, $x \geq 3$, find the following.*

**5.** $(f \circ g)(x)$ **6.** $(g \circ f)(x)$

*Determine whether each function is a one-to-one function.*

**7.**

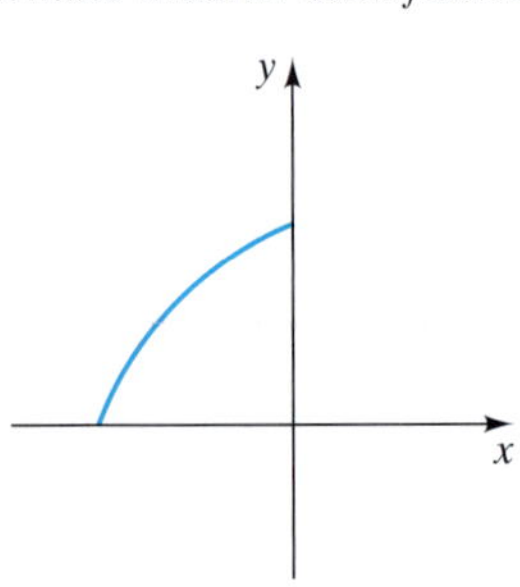


**8.**

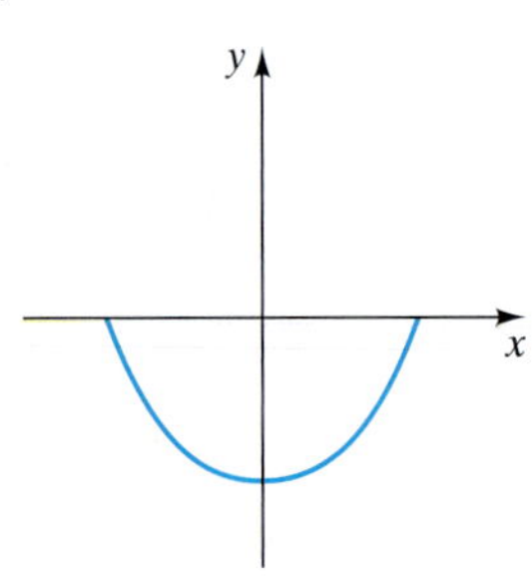


**9.** $\{(6, 2), (4, 0), (-5, 7), (3, 8)\}$

**10.** $\left\{(0, -2), (6, 1), (3, -2), \left(\frac{1}{2}, 4\right)\right\}$

**11.** $y = \sqrt{x + 8}, x \geq -8$

**12.** $y = x^2 - 9$

*In Exercises 13 and 14, for each function, find the domain and range of both $f(x)$ and $f^{-1}(x)$.*

**13.** $\{(5, 3), (6, 2), (-4, -3), (-1, 8)\}$

**14.**

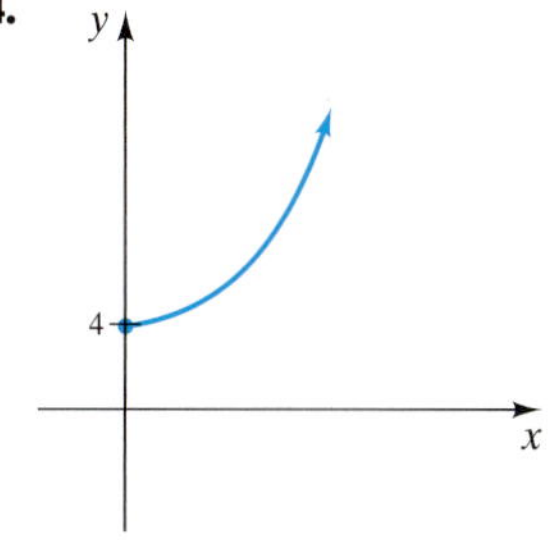


*In Exercises 15 and 16, find $f^{-1}(x)$ and graph $f(x)$ and $f^{-1}(x)$ on the same axes.*

**15.** $y = f(x) = 4x - 2$

**16.** $y = f(x) = \sqrt[3]{x - 1}$

**17. Yards to Feet** The function $f(x) = 36x$ converts yards, $x$, into inches. Find the inverse function that converts inches into yards. In the inverse function, what do $x$ and $f^{-1}(x)$ represent?

**18. Gallons to Quarts** The function $f(x) = 4x$ converts gallons, $x$, into quarts. Find the inverse function that converts quarts into gallons. In the inverse function, what do $x$ and $f^{-1}(x)$ represent?

[9.2] *Graph the following functions.*

**19.** $y = 2^x$

**20.** $y = \left(\frac{1}{2}\right)^x$

**21. Smart Handheld Devices** Since 1999, the number of worldwide shipments of smart handheld devices has been growing exponentially (see the graph on the right). The number of shipments, $f(t)$, in millions, can be approximated by the function $f(t) = 7.02e^{0.365t}$ where $t$ is the number of years since 1999. Use this function to estimate the number of worldwide shipments of these devices in **a)** 2003, **b)** 2005, **c)** 2008.

**Worldwide Shipments of Smart Handheld Devices**

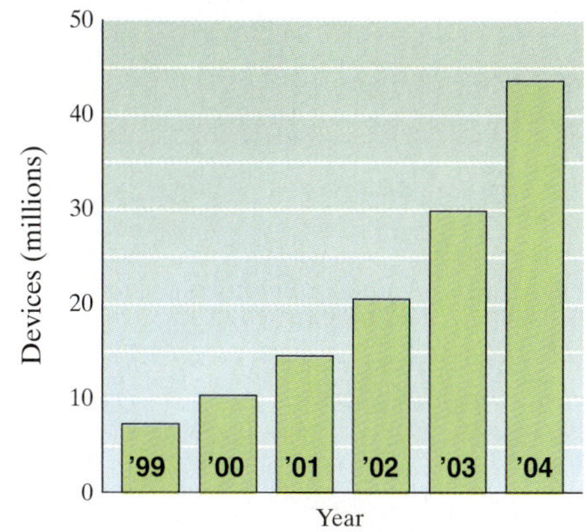


*Source:* International Data Corp.; MSN MoneyCentral; CSI Inc.; additional research

[9.3] *Write each equation in logarithmic form.*

**22.** $8^2 = 64$

**23.** $81^{1/4} = 3$

**24.** $5^{-3} = \frac{1}{125}$

*Write each equation in exponential form.*

**25.** $\log_2 32 = 5$

**26.** $\log_{1/4} \frac{1}{16} = 2$

**27.** $\log_6 \frac{1}{36} = -2$

*Write each equation in exponential form and find the missing value.*

**28.** $3 = \log_4 x$

**29.** $4 = \log_a 8$

**30.** $-3 = \log_{1/5} x$

*Graph the following functions.*

**31.** $y = \log_3 x$

**32.** $y = \log_{1/2} x$

[9.4] *Use the properties of logarithms to expand each expression.*

**33.** $\log_5 17^8$

**34.** $\log_3 \sqrt{x - 9}$

**35.** $\log \frac{6(a + 1)}{19}$

**36.** $\log \frac{x^4}{7(2x + 3)^5}$

*Write the following as the logarithm of a single expression.*

**37.** $5 \log x - 3 \log (x + 1)$

**38.** $4(\log 2 + \log x) - \log y$

**39.** $\frac{1}{3}[\ln x - \ln (x + 2)] - \ln 2$

**40.** $3 \ln x + \frac{1}{2}\ln (x + 1) - 6 \ln (x + 4)$

*Evaluate.*

**41.** $8^{\log_8 10}$

**42.** $\log_4 4^5$

**43.** $11^{\log_9 81}$

**44.** $9^{\log_8 \sqrt{8}}$

[9.5, 9.7] *Use a calculator to find each logarithm. Round your answers to the nearest ten-thousandth.*

**45.** $\log 819$

**46.** $\ln 0.0281$

*Use a calculator to find the antilog of each number. Give the antilog to three significant digits.*

**47.** 3.159

**48.** $-3.157$

*Use a calculator to find N. Round your answer to three significant digits.*

**49.** $\log N = 4.063$

**50.** $\log N = -1.2262$

*Evaluate.*

**51.** $\log 10^5$

**52.** $10^{\log 9}$

**53.** $7 \log 10^{3.2}$

**54.** $2(10^{\log 4.7})$

[9.6] *Solve without using a calculator.*

**55.** $625 = 5^x$

**56.** $49^x = \frac{1}{7}$

**57.** $2^{3x-1} = 32$

**58.** $27^x = 3^{2x+5}$

*Solve using a calculator. Round your answers to the nearest thousandth.*

**59.** $7^x = 152$

**60.** $3.1^x = 856$

**61.** $12.5^{x+1} = 381$

**62.** $3^{x+2} = 8^x$

*Solve the logarithmic equation.*

**63.** $\log_7 (2x - 3) = 2$

**64.** $\log x + \log (4x - 19) = \log 5$

**65.** $\log_3 x + \log_3 (2x + 1) = 1$

**66.** $\ln (x + 1) - \ln (x - 2) = \ln 4$

[9.7] *Solve each exponential equation for the remaining variable. Round your answer to the nearest thousandth.*

**67.** $50 = 25e^{0.6t}$

**68.** $100 = A_0 e^{-0.42(3)}$

*Solve for the indicated variable.*

**69.** $A = A_0 e^{kt}$, for $t$

**70.** $200 = 800e^{kt}$, for $k$

**71.** $\ln y - \ln x = 6$, for $y$

**72.** $\ln (y + 1) - \ln (x + 8) = \ln 3$, for $y$

*Use the change of base formula to evaluate. Write the answer rounded to the nearest ten-thousandth.*

**73.** $\log_2 196$

**74.** $\log_3 47$

[9.2–9.7]

**75. Compound Interest** Find the amount of money accumulated if Justine Elwood puts \$12,000 in a savings account yielding 6% interest per year for 8 years. Use $A = p(1 + r)^n$.

**76. Interest Compounded Continuously** If \$6000 is placed in a savings account paying 4% interest compounded continuously, find the time needed for the account to double in value.

**77. Bacteria** The bacteria *Escherichia coli* are commonly found in the bladders of humans. Suppose that 2000 bacteria are present at time 0. Then the number of bacteria present $t$ minutes later may be found by the function $N(t) = 2000(2)^{0.05t}$.

**a)** When will 50,000 bacteria be present?

**b)** Suppose that a human bladder infection is classified as a condition with 120,000 bacteria. When would a person develop a bladder infection if he or she started with 2000 bacteria?

**78. Atmospheric Pressure** The atmospheric pressure, $P$, in pounds per square inch at an elevation of $x$ feet above sea level can be found by the formula $P = 14.7e^{-0.00004x}$. Find the atmospheric pressure at the top of the half dome in Yosemite National Park, an elevation of 8842 feet.

**79. Remembering** A class of history students is given a final exam at the end of the course. As part of a research project, the students are also given equivalent forms of the exam each month for $n$ months. The average grade of the class after $n$ months may be found by the function $A(n) = 72 - 18 \log(n + 1), n \geq 0$.

**a)** What was the class average when the students took the original exam ($n = 0$)?

**b)** What was the class average for the exam given 3 months later?

**c)** After how many months was the class average 58.0?

## Chapter 9 Practice Test

*To find out how well you understand the chapter material, take this practice test. The answers, and the section where the material was initially discussed, are given in the back of the book. Each problem is also fully worked out on the* **Chapter Test Prep Video CD**. *Review any questions that you answered incorrectly.*

**1. a)** Determine whether the following function is a one-to-one function.

$$\{(4, 2), (-3, 8), (-1, 3), (6, -7)\}$$

**b)** List the set of ordered pairs in the inverse function.

**2.** Given $f(x) = x^2 - 3$ and $g(x) = x + 2$, find **a)** $(f \circ g)(x)$. **b)** $(f \circ g)(6)$.

**3.** Given $f(x) = x^2 + 8$ and $g(x) = \sqrt{x - 5}, x \geq 5$, find **a)** $(g \circ f)(x)$. **b)** $(g \circ f)(7)$.

*In Exercises 4 and 5,* **a)** *find $f^{-1}(x)$ and* **b)** *graph $f(x)$ and $f^{-1}(x)$ on the same axes.*

**4.** $y = f(x) = -3x - 5$

**5.** $y = f(x) = \sqrt{x - 1}, x \geq 1$

**6.** What is the domain of $y = \log_5 x$?

**7.** Evaluate $\log_4 \frac{1}{256}$.

**8.** Graph $y = 3^x$.

**9.** Graph $y = \log_2 x$.

**10.** Write $2^{-5} = \frac{1}{32}$ in logarithmic form.

**11.** Write $\log_5 125 = 3$ in exponential form.

*Write Exercises 12 and 13 in exponential form and find the missing value.*

**12.** $4 = \log_2 (x + 3)$

**13.** $y = \log_{64} 16$

**14.** Expand $\log_2 \frac{x^3(x - 4)}{x + 2}$.

**15.** Write as the logarithm of a single expression.

$$7 \log_6 (x - 4) + 2 \log_6 (x + 3) - \frac{1}{2} \log_6 x.$$

**16.** Evaluate $10 \log_9 \sqrt{9}$.

**17. a)** Find log 4620 rounded to 4 decimal places.

**b)** Find ln 0.0692 rounded to 4 decimal places.

**18.** Solve $3^x = 19$ for $x$.

**19.** Solve $\log 4x = \log (x + 3) + \log 2$ for $x$.

**20.** Solve $\log (x + 5) - \log (x - 2) = \log 6$ for $x$.

**21.** Find $N$, rounded to 4 decimal places, if $\ln N = 2.79$.

**22.** Evaluate $\log_6 40$, rounded to 4 decimal places, using the change of base formula.

**23.** Solve $100 = 250e^{-0.03t}$ for $t$, rounded to 4 decimal places.

**24. Savings Account** What amount of money accumulates if Kim Lee puts \$3500 in a savings account yielding 4% interest compounded quarterly for 10 years?

**25. Carbon 14** The amount of carbon 14 remaining after $t$ years is found by the formula $v = v_0e^{-0.0001205t}$, where $v_0$ is the original amount of carbon 14. If a fossil originally contained 60 grams of carbon 14, and now contains 40 grams of carbon 14, how old is the fossil?

## Cumulative Review Test

*Take the following test and check your answers with those given in the back of the book. Review any questions that you answered incorrectly. The section where the material is covered is indicated after the answer.*

**1.** Simplify $\frac{(2xy^2z^{-3})^2}{(3x^{-1}yz^2)^{-1}}$.

**2.** Evaluate $5^2 - (2 - 3^2)^2 + 4^3$.

**3. Dinner** Thomas Furgeson took his wife out to dinner at a nice restaurant. The cost of the meal before tax was \$92. If the total price, including tax, was \$98.90, find the tax rate.

**4.** Solve the inequality $-3 \leq 2x - 7 < 8$ and write the answer as a solution set and in interval notation.

**5.** Solve $2x - 3y = 8$ for $y$.

**6.** Let $h(x) = \frac{x^2 + 4x}{x + 6}$. Find $h(-4)$.

**7.** Find the slope of the line in the figure below. Then write the equation of the given line.

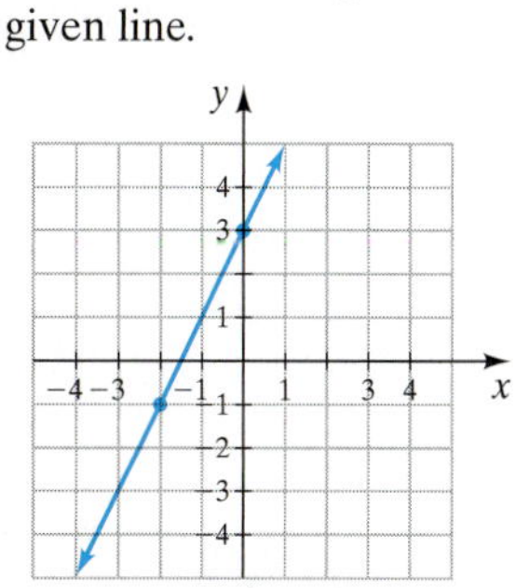

**8.** Graph $4x = 3y - 3$.

**9.** Graph $y \le \frac{1}{3}x + 6$.

**10.** Solve the system of equations

$$\frac{1}{2}x + \frac{1}{3}y = 13$$

$$\frac{1}{5}x + \frac{1}{8}y = 5$$

**11.** Divide $\dfrac{x^3 + 3x^2 + 5x + 9}{x + 1}$.

**12.** Factor $x^2 - 2xy + y^2 - 64$.

**13.** Solve $(2x + 1)^2 - 9 = 0$.

**14.** Solve $\dfrac{2x + 3}{x + 1} = \dfrac{3}{2}$.

**15.** Solve $a_n = a_1 + nd - d$, for $d$.

**16.** $L$ varies inversely as the square of $P$. Find $L$ when $P = 4$ and $k = 100$.

**17.** Simplify $4\sqrt{45x^3} + \sqrt{5x}$.

**18.** Solve $\sqrt{2a + 9} - a + 3 = 0$.

**19.** Solve $(x^2 - 5)^2 + 3(x^2 - 5) - 10 = 0$.

**20.** Let $g(x) = x^2 - 4x - 5$.

**a)** Express $g(x)$ in the form $g(x) = a(x - h)^2 + k$.

**b)** Draw the graph and label the vertex.

# 10 Characteristics of Functions and Their Graphs

## GOALS OF THIS CHAPTER

In this chapter we look at the general characteristics of functions. We begin with a discussion of continuous and discontinuous functions. Then we discuss the specific characteristics of polynomial functions and rational functions.

Earlier we saw that we could graph a function by plotting a sufficient number of points. In this chapter, we graph functions using a minimum number of points based on a function's key features. We also develop strategies for finding the number and location of real zeros of polynomial functions, and finding the asymptotes of rational functions, and using them to graph the functions.


10.1 Continuous and Discontinuous Functions

10.2 Characteristics of Graphs of Polynomial Functions

10.3 Finding the Rational Zeros of a Polynomial Function

Mid-Chapter Test: Sections 10.1–10.3

10.4 Finding the Real Zeros of a Polynomial Function

10.5 Characteristics of Graphs of Rational Functions

Chapter 10 Summary

Chapter 10 Review Exercises

Chapter 10 Practice Test

Cumulative Review Test


TO OPERATE A BUSINESS at a profit, its revenues must be greater than its costs. If a company produces $x$ items and $R(x)$ represents the company's revenue from these items and $C(x)$ represents the company's cost of producing these items, this profit can be represented by the function $P(x) = R(x) - C(x)$. In Example 7 on page 682, we write a polynomial function to represent the profit of a company selling computers.

# 10.1 Continuous and Discontinuous Functions

1 Learn new terminology.

2 Identify local maxima and minima.

3 Identify continuous and discontinuous functions.

4 Graph functions defined piecewise.

5 Use the greatest integer function.

## 1 Learn New Terminology

In this chapter we will discuss functions in more depth. We will expand our knowledge of some functions that were introduced earlier, such as polynomial functions and rational functions, and we will introduce some new functions such as piecewise functions and the greatest integer function. We will also discuss what it means for a graph to be continuous or discontinuous.

Let us start by examining the graph in **Figure 10.1**.

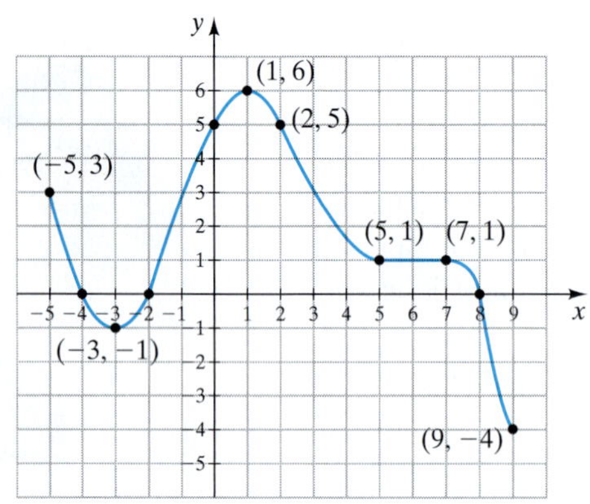


FIGURE 10.1

First notice that this graph represents a function since it passes the vertical line test. Starting at the left side of the $x$-axis moving to the right, the values of $x$ increase. If the function increases as the values of $x$ increase, we say the **function is increasing**. If the function decreases as the values of $x$ increase, we say the **function is decreasing**. On the graph we have indicated 11 points. We can use the $x$-values of the ordered pairs to designate intervals. For example, the interval from $x = -5$ to $x = -3$ may be indicated by $[-5, -3]$. Below we illustrate some intervals and indicate whether the function is increasing, decreasing, or staying constant in the interval.

| Interval | Function |
|---|---|
| $[-5, -3]$ | decreasing |
| $[-3, 1]$ | increasing |
| $[1, 5]$ | decreasing |
| $[5, 7]$ | constant |
| $[7, 9]$ | decreasing |

Do not confuse the terms increasing function and decreasing function with positive values of the function and negative values of the function. Consider the interval $[-5, -3]$. Within this interval the function is *positive* from $[-5, -4)$ since it is above the $x$-axis, and negative from $(-4, -3]$ since the function is below the $x$-axis. The function is 0 at $x = -4$.

## 2 Identify Local Maxima and Minima

In **Figure 10.1** the point at $(-3, -1)$ is the lowest point in a small interval centered around the point $x = -3$. The point $(-3, -1)$ is called a **local minimum**. The point $(1, 6)$ is the highest point in a small interval centered around the point $x = 1$. The point $(1, 6)$ is called a **local maximum**. Now look at the $x$-intercepts $(-4, 0)$, $(-2, 0)$, $(8, 0)$. The $x$-coordinates of the $x$-intercepts are called the **real zeros of the function**, for these are the real-number values of $x$ where the function has a value of zero. The real zeros of the function given in **Figure 10.1** are $-4$, $-2$, and 8. A function may have non-real zeros. At non-real zeros the values of $x$ are complex numbers.

The graph does not cross the $x$-axis at non-real zeros. We will discuss non-real zeros later in the chapter.

## 3 Identify Continuous and Discontinuous Functions

Now consider **Figures 10.2** and **10.3.**

The graph in **Figure 10.2** is an example of a **continuous curve** in the interval $[-5, 6]$. It is also a **smooth curve** since there are no sharp corners on the graph. Notice that you can trace the entire graph without removing the pencil from the paper. Now look at **Figure 10.3**. The graph is continuous in the interval $[-2, 7)$. At $x = 7$ there is a **gap** in the graph, which makes it discontinuous. We say that there is a **point of discontinuity** at $x = 7$, or that the graph is **discontinuous** at $x = 7$. Now look at the graph where $x = 9$. There is a ***hole*** in the graph at $x = 9$, which indicates that the function is not defined at $x = 9$. Therefore, there is also a point of discontinuity at $x = 9$. A continuous curve can have no gaps or holes. A smooth continuous curve can have no corners or cusps or sharp points.

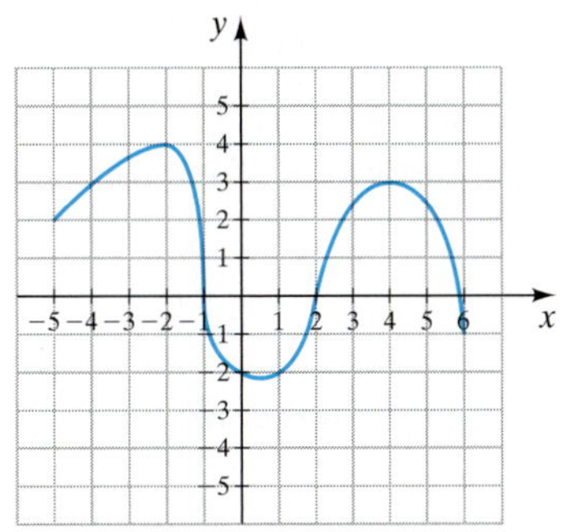

FIGURE 10.2

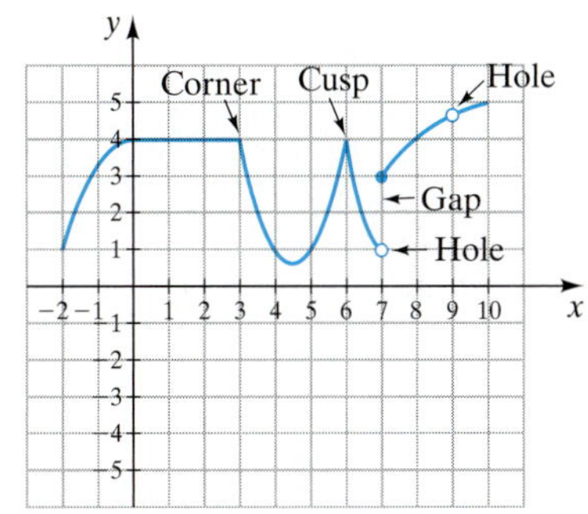


FIGURE 10.3

## 4 Graph Functions Defined Piecewise

Most of the functions we have discussed so far have been continuous functions. Linear, quadratic, cubic, logarithmic, and exponential functions are all continuous functions. It is important to recognize and graph discontinuous functions. We will now introduce and graph two special types of functions: piecewise functions and greatest integer functions. We begin with piecewise functions.

While most functions have a single equation that defines the behavior over its entire domain, **functions defined piecewise**, or **piecewise functions**, are defined using two or more equations. One familiar piecewise defined function is the function used to determine how much income tax we pay. The income tax owed is different for different taxable incomes. Each tax bracket has its own equation to determine the amount of tax owed as a function of taxable income. We present a function defined piecewise in Example 1.

**EXAMPLE 1** ▸ Graph the function

$$f(x) = \begin{cases} x \text{ if } x < 3 \\ -x \text{ if } x \geq 3 \end{cases}$$

and determine if the function is continuous or discontinuous.

*Solution* This function defined piecewise has two equations. To graph the function in the interval $(-\infty, 3)$ we use $f(x) = x$, and to graph the function in the interval $[3, \infty)$ we use $f(x) = -x$. The function $f(x) = x$ is defined for $x < 3$. If this function were defined at $x = 3$, the value of the function would be 3. Since this part of the function is not defined at 3, we place an open circle, or hole, at $(3, 3)$. We then continue to graph the equation for values of $x$ less than 3, see the red line in **Figure 10.4**. We now graph $f(x) = -x$ for $x \geq 3$. At $x = 3$, $f(x)$ has a value of $-3$. The point $(3, -3)$ is included on the graph of $f(x) = -x$, so we place a solid dot at $(3, -3)$. Now continue to graph $f(x) = -x$ for values of $x > 3$; see the green line in **Figure 10.4**.

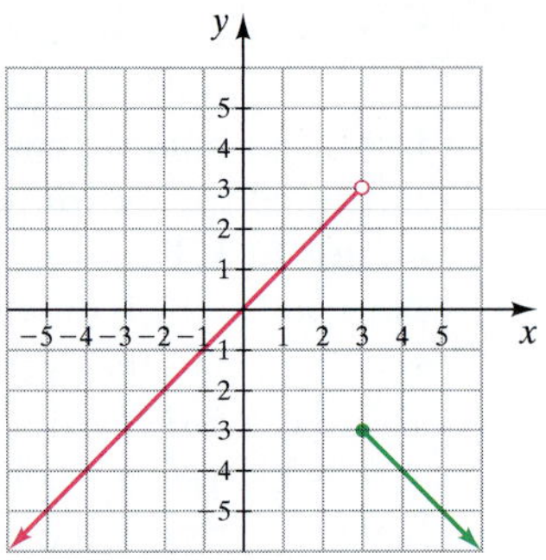

FIGURE 10.4

Notice the function has a point of discontinuity at $x = 3$.

▸ Now Try Exercise 31

**EXAMPLE 2** ▸ Graph the function

$$f(x) = \begin{cases} x - 3 \text{ if } x \leq 3 \\ -x + 3 \text{ if } x > 3 \end{cases}$$

and determine if the graph is continuous or discontinuous.

*Solution* First graph $f(x) = x - 3$ for values of $x$ less than or equal to 3. When $x$ is 3, $f(x) = 0$; see the red line in **Figure 10.5**.

Now graph $f(x) = -x + 3$ for values of $x$ greater than 3. If $x$ were allowed to be 3, $f(x)$ would be 0. The point $(3, 0)$ is already included on the graph. Now graph $f(x) = -x + 3$ for values greater than 3; see the green line in **Figure 10.5**.

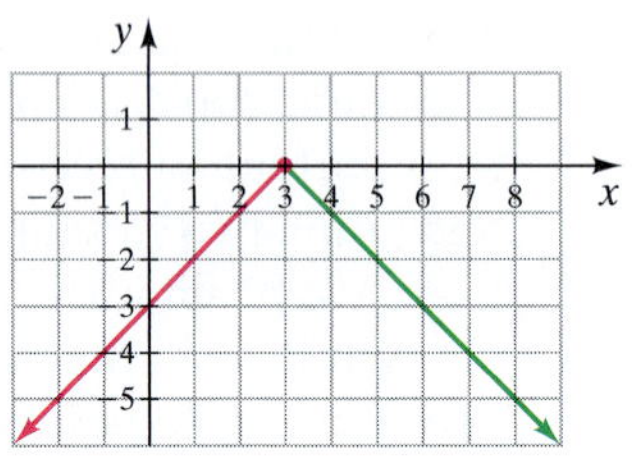

FIGURE 10.5

By observing the graph, we see it is continuous for all values of $x$. However, the graph is not smooth since there is a corner at $x = 3$.

▸ Now Try Exercise 35

Notice that the piecewise function in Example 1 is not continuous, but the piecewise function in Example 2 is continuous.

## 5 Use the Greatest Integer Function

A function that will always have a discontinuity is the greatest integer function. The **greatest integer function** is a function that truncates, or rounds down, a value to the greatest integer less than or equal to the value. A symbol used to indicate the greatest integer less than or equal to $x$ is $[\![x]\!]$. For example,

$$[\![8.7]\!] = 8, \quad \left[\!\!\left[\frac{19}{5}\right]\!\!\right] = 3, \quad [\![-6]\!] = -6 \quad \text{and} \quad [\![-4.03]\!] = -5.$$

**EXAMPLE 3** ▸ Graph $f(x) = [\![x]\!]$.

*Solution* For any value of $x$ in the interval $[0, 1)$, the value of $[\![x]\!] = 0$. For any value of $x$ in the interval $[1, 2)$, $[\![x]\!] = 1$, and so on. The values in each interval are constant, but the function jumps at integer values of $x$. The graph of the function is shown in **Figure 10.6**. Notice it consists of a series of line segments. The left endpoint of each line segment is included, and the right endpoint is excluded.

The domain of the function is $(-\infty, \infty)$. The range is the set of integers

$$\{\ldots, -3, -2, -1, 0, 1, 2, 3, \ldots\}$$

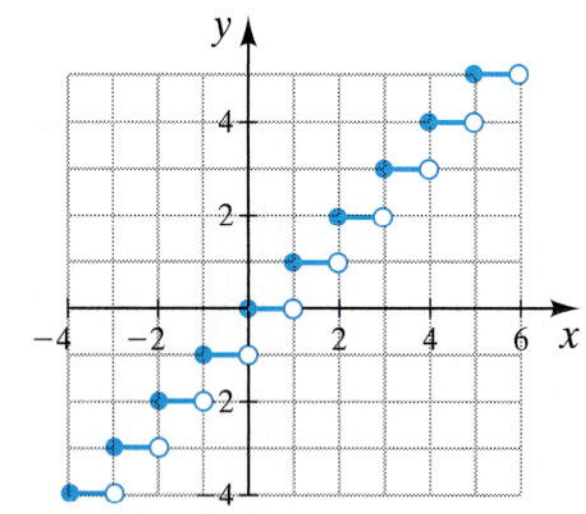

$f(x) = [\![x]\!]$ for $-4 \leq x < 6$

FIGURE 10.6

▸ **Now Try Exercise 39**

Notice the graph in **Figure 10.6** is made up of a series of horizontal lines that resemble steps. Such graphs are called **step functions**.

**EXAMPLE 4** ▸ Graph $f(x) = \left[\!\left[\frac{1}{2}x + 3\right]\!\right]$.

*Solution* We need to determine the intervals for the steps of this greatest integer function. To find the closed endpoint for $f(x) = 0$, set the expression $\frac{1}{2}x + 3$ equal to 0 and solve for $x$.

$$\frac{1}{2}x + 3 = 0$$

$$\frac{1}{2}x = -3$$

$$x = -6$$

To find the closed endpoint for $f(x) = 1$, set the expression equal to 1 and solve for $x$.

$$\frac{1}{2}x + 3 = 1$$

$$\frac{1}{2}x = -2$$

$$x = -4$$

Since the closed endpoint for $f(x) = 0$ is $-6$ and the closed endpoint for $f(x) = 1$ is $-4$, $f(x)$ must equal 0 in the interval $[-6, -4)$.

To find the closed endpoint for $f(x) = 2$, set the expression equal to 2 and solve for $x$.

$$\frac{1}{2}x + 3 = 2$$

$$\frac{1}{2}x - -1$$

$$x = -2$$

Since the closed endpoint for $f(x) = 1$ is $-4$ and the closed endpoint for $f(x) = 2$ is $-2$, $f(x)$ must equal 1 in the interval $[-4, -2)$.

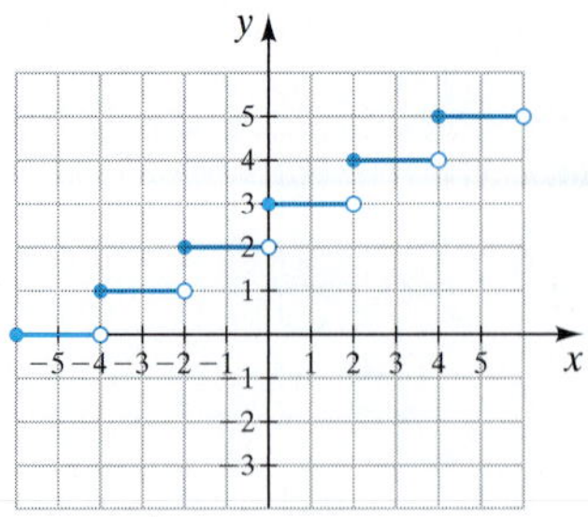


$f(x) = [\![\frac{1}{2}x + 3]\!]$ for $-6 \le x < 6$

FIGURE 10.7

We continue this process to obtain the following table of values.

| Interval | Value of $f(x)$ |
|---|---|
| $[-6, -4)$ | 0 |
| $[-4, -2)$ | 1 |
| $[-2, 0)$ | 2 |
| $[0, 2)$ | 3 |
| $[2, 4)$ | 4 |
| $[4, 6)$ | 5 |

Using this table, we get the graph in **Figure 10.7**.

▸ **Now Try Exercise 43**

The greatest integer function has many practical applications. Example 5 shows a practical application of the greatest integer function.

### EXAMPLE 5 ▸ Purchasing CDs

**a)** Write a greatest integer function that can be used to determine the maximum number of CD's that a person can purchase for $x$ dollars if each CD costs $14.

**b)** Graph the function.

*Solution*

**a)** A person could purchase 1 CD with $14, 2 CD's with $28, 3 CD's with $42, and so on. To determine the maximum number of CD's, divide the given dollar amount by 14, and round down. Thus, the greatest integer function is $f(x) = \left[\!\left[\frac{x}{14}\right]\!\right]$.

**b)** By selecting values for $x$, we can determine the following table.

| Interval | Value of $f(x)$ |
|---|---|
| $[0, 14)$ | 0 |
| $[14, 28)$ | 1 |
| $[28, 42)$ | 2 |
| $[42, 56)$ | 3 |
| $[56, 70)$ | 4 |

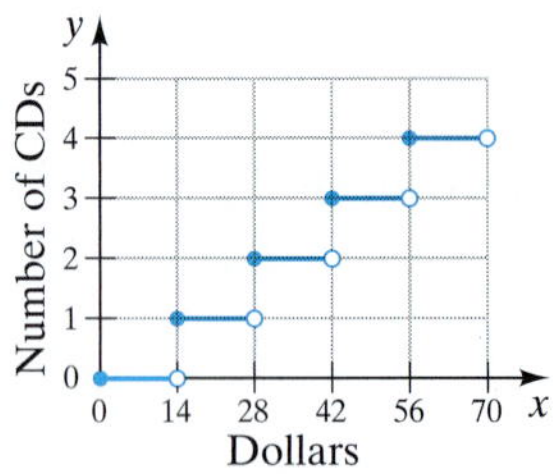


FIGURE 10.8

**Figure 10.8** shows the graph for values from $0 up to but not including $70.

▸ **Now Try Exercise 47**

In the Exercise Set you will be asked to determine whether the graphs of certain functions are continuous or discontinuous. On page 382 we discussed finding the domains of rational functions, and in the Using Your Graphing Calculator box we graphed a rational function. Every rational function will have a point of discontinuity at any value of $x$ for which the function is not defined. For example, in $f(x) = \frac{x}{x^2 - 4}$ the function is not defined at $x = 2$ and $x = -2$, and so the graph of $f(x)$ is discontinuous at those points. We will discuss the graphs of rational functions further in Section 10.5.

#### USING YOUR GRAPHING CALCULATOR

Graphing calculators can graph the greatest integer function. On the TI-84 Plus calculator, the greatest integer is given as [int (]. It is found under the CATALOG menu. Scroll down until you reach [int (]. It is also found using [NUM] of [MATH] menu. Other calculators may use a different representation. If you are using a different calculator, consult the manual to learn how to use the greatest integer function.

# EXERCISE SET 10.1

MathXL® MyMathLab

## Concept/Writing Exercises

1. Does every function have a local minimum or maximum? Explain.
2. If a function is defined for all real numbers, is it always continuous? Explain.
3. Is $f(x) = x^2 - 4$ a continuous function? Explain.
4. Is $f(x) = \begin{cases} 3x - 4 \text{ if } x < \dfrac{4}{3} \\ 4 - 3x \text{ if } x \geq \dfrac{4}{3} \end{cases}$ a continuous function? Explain.
5. Is $g(x) = \begin{cases} 2x - 1 \text{ if } x < -5 \\ 2x + 6 \text{ if } x \geq -5 \end{cases}$ a continuous function? Explain.
6. Describe the graph of a greatest integer function.

## Practice the Skills

*Classify each function as continuous or discontinuous.*

7.
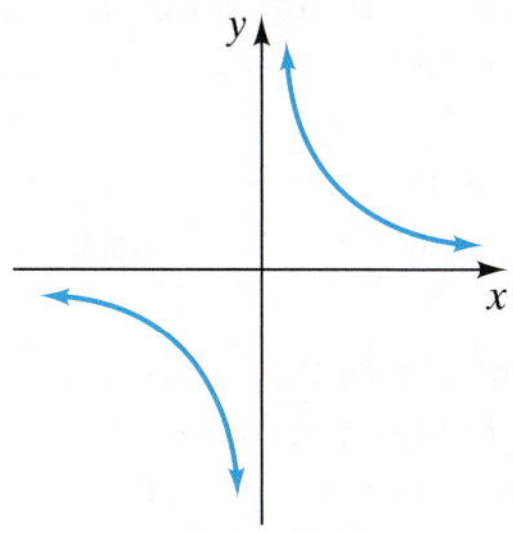

8.
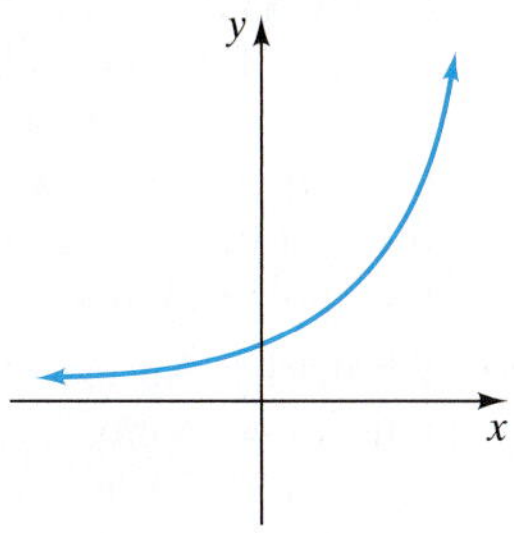

9.
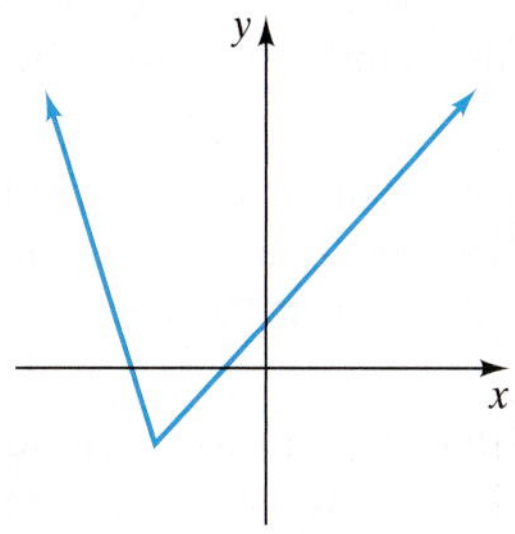

10.
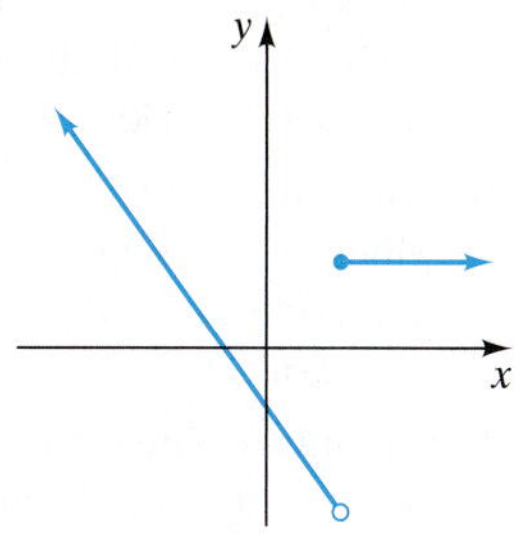

11.
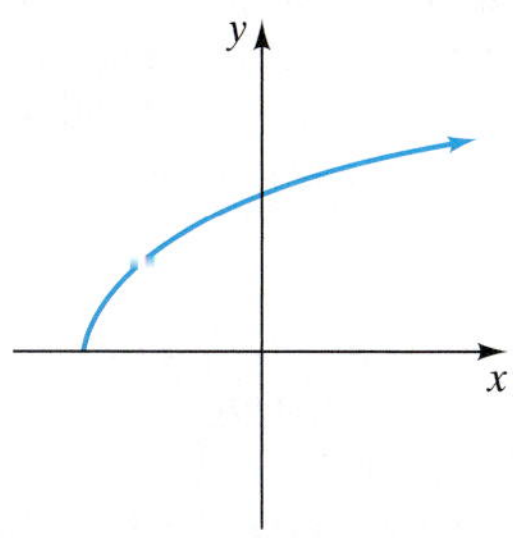

12.
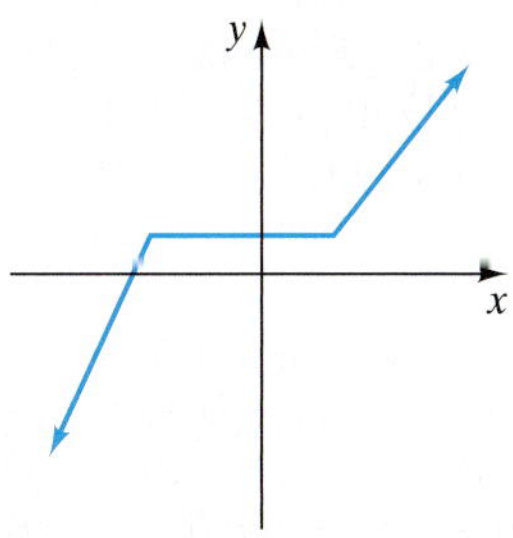

13.
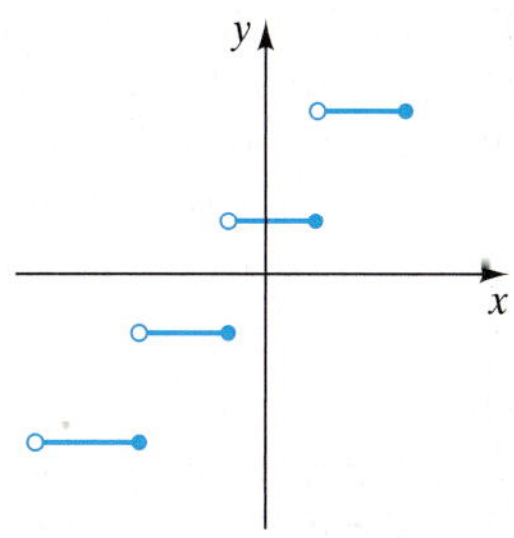

14.
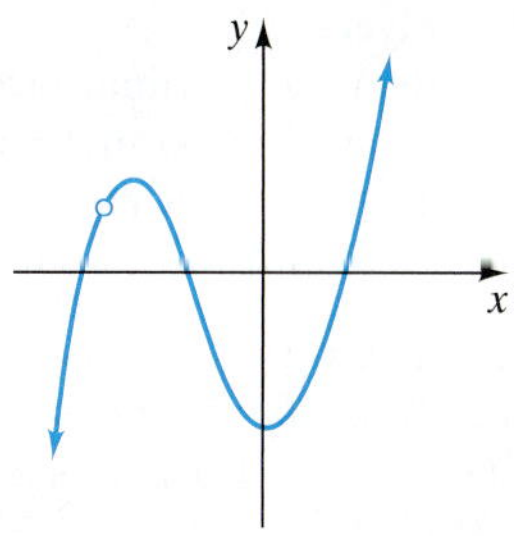

*Classify each function as continuous or discontinuous. If the function is not continuous, identify the point(s) of discontinuity.*

15. $f(x) = 5x^2 + 3x + 1$
16. $f(x) = x^3 + x - 4$
17. $g(x) = 7^x$
18. $g(x) = 2^x - 1$
19. $h(x) = \log_7 x$
20. $h(x) = \log_2(x + 3)$
21. $m(x) = \dfrac{1}{x + 5}$
22. $m(x) = \dfrac{1}{x^2 + 7}$
23. $j(x) = [\![2x]\!]$
24. $j(x) = \left[\!\!\left[\dfrac{x}{2}\right]\!\!\right]$
25. $f(x) = -x + 9$
26. $f(x) = 2(x - 11)^2$
27. $g(x) = \dfrac{3}{x - 3}$
28. $g(x) = \dfrac{5}{x + 6}$
29. $p(x) = \dfrac{1}{x^2 + 8}$
30. $p(x) = \dfrac{7}{x + 10}$

*Graph each piecewise function.*

31. $f(x) = \begin{cases} x + 5 \text{ if } x < -2 \\ -x + 5 \text{ if } x \geq -2 \end{cases}$
32. $h(x) = \begin{cases} \dfrac{1}{2}x \text{ if } x \leq 1 \\ 2x \text{ if } x > 1 \end{cases}$
33. $g(x) = \begin{cases} -x + 4 & \text{if } x \leq 0 \\ 4 & \text{if } x > 0 \end{cases}$

34. $r(x) = \begin{cases} 3 \text{ if } x < -1 \\ -3 \text{ if } x \geq -1 \end{cases}$

35. $h(x) = \begin{cases} -2x + 1 \text{ if } x < \frac{1}{2} \\ 2x - 1 \text{ if } x \geq \frac{1}{2} \end{cases}$

36. $f(x) = \begin{cases} -x + 4 \text{ if } x < 4 \\ -2x + 8 \text{ if } x > 4 \end{cases}$

37. $p(x) = \begin{cases} \frac{3}{2}x - 4 \text{ if } x \leq 2 \\ -\frac{2}{3}x + \frac{19}{3} \text{ if } x > 2 \end{cases}$

38. $g(x) = \begin{cases} 3x & \text{if } x < 5 \\ -\frac{1}{5}x + 16 & \text{if } x \geq 5 \end{cases}$

*Graph each greatest integer function.*

39. $f(x) = [\![3x]\!]$

40. $g(x) = [\![x - 5]\!]$

41. $h(x) = [\![-2x]\!]$

42. $f(x) = \left[\!\!\left[\frac{x + 2}{3}\right]\!\!\right]$

43. $g(x) = \left[\!\!\left[-\frac{1}{2}x + 1\right]\!\!\right]$

44. $h(x) = \left[\!\!\left[\frac{3}{4}x - 2\right]\!\!\right]$

45. $f(x) = \left[\!\!\left[\frac{2x + 7}{10}\right]\!\!\right]$

46. $g(x) = \left[\!\!\left[\frac{8 - x}{3}\right]\!\!\right]$

## Problem Solving

47. **Greeting Cards** Kelly Chen wants to stock up on greeting cards which are selling for \$1.25 a piece. Write a greatest integer function that represents the number of cards she can purchase for $x$ dollars.

48. **Cat Food** Don Jacobs wants to stock up on cans of cat food which are selling for \$0.89 per can. Write a greatest integer function that represents the number of cans he can purchase for $x$ dollars.

49. **Calendars** The office supply manager wants to purchase calendars which are selling for \$3.99 each. Write a greatest integer function that represents the number of calendars she can purchase for $x$ dollars.

50. **Flyers** When Teri Lovelace lost her cat Kundera, she made a flyer to distribute around the neighborhood. She went to Copy Cats to make copies of her flyer. Copies cost \$0.12 each. Write a greatest integer function that represents the number of copies she can make for $x$ dollars.

51. **Bikes** A bike rental company rents bikes by the hour. They charge \$5.00 per hour for the first 3 rental hours and \$3.00 per hour for each additional hour. Write a piecewise function to represent the cost of renting a bike as a function of hours it is rented. *Hint*: The cost, in dollars, for $x > 3$ hours will be the cost of the first 3 hours, \$15, plus \$$3(x - 3)$, where $x$ is the number of hours.

52. **Electricity** An electric company's monthly charge is \$9.27 plus an economy rate of 11.589 cents per kilowatt-hour for the first 345 kWhr and premium rate of 13.321 cents per kilowatt-hour for kilowatt-hours over 345 kWhr. Write a piecewise function to represent the monthly electric cost as a function of kilowatt-hours. *Hint*: The cost for 345 kWhr is $0.11589(345) + 9.27$, or \$49.25205. The cost for any kilowatt-hours over 345 ($x > 345$) will be the cost for the first 345 kWhr, \$49.25205, plus \$$0.13321(x - 345)$, where $x$ is the number of kilowatt-hours.

*A function $f(x)$ is an even function if $f(-x) = f(x)$ for all $x$ in the domain of $f$. A function $f(x)$ is an odd function if $f(-x) = -f(x)$ for all $x$ in the domain of $f$. For example, $f(x) = x^4 + 2$ is an even function because $f(-x) = (-x)^4 + 2 = x^4 + 2$. Thus $f(-x) = f(x)$. The function $f(x) = x^3 + x$ is an odd function because $f(-x) = (-x)^3 + (-x) = -x^3 - x$ and $-f(x) = -(x^3 + x) = -x^3 - x$. Therefore $f(-x) = -f(x)$. For Exercises 53 through 62 determine whether the function is an even function, an odd function, or neither.*

53. $f(x) = x^2 - 15$

54. $f(x) = 3x^3 - x$

55. $f(x) = 3x^3 + x^2 - 10$

56. $g(x) = 2x^4 - 6x^2 + \frac{1}{3}$

57. $g(x) = 3x^6 - 15x^2 + 4$

58. $g(x) = e^{2x} + 8$

59. $g(x) = e^{x^2} + 5$

60. $h(x) = 5x^5 + 3x^3 - 2x$

61. $h(x) = x^4 - x^3 + x^2$

62. $h(x) = 6x^5 - 3x^4 + \frac{5}{2}$

## Challenge Problems

*Graph each piecewise function.*

63. $f(x) = \begin{cases} 2x & \text{if } x < -1 \\ -x + 5 & \text{if } -1 \leq x < 2 \\ x + 1 & \text{if } x \geq 2 \end{cases}$

64. $h(x) = \begin{cases} -x - 4 & \text{if } x \leq -1 \\ 2x - 1 & \text{if } -1 < x < 2 \\ x + 1 & \text{if } x \geq 2 \end{cases}$

65. $p(x) = \begin{cases} 2x + 3 & \text{if } x < -1 \\ x + 1 & \text{if } -1 \leq x < 1 \\ 2x & \text{if } 1 \leq x < 3 \\ x - 1 & \text{if } x \geq 3 \end{cases}$

66. $g(x) = \begin{cases} -x + 3 & \text{if } x < 1 \\ 5 & \text{if } x = 1 \\ \sqrt{x} & \text{if } x > 1 \end{cases}$

## Group Activity

*We discussed even and odd functions in Exercises 53–62. Review that material now. Then solve Exercises 67 and 68.*

**67. a)** Explain why for every even function if the ordered pair $(x, y)$ is a point on the graph of the function, the ordered pair $(-x, y)$ will also be on the graph.

**b)** Explain why the graph of an even function will always be symmetric with respect to the $y$-axis.

**68. a)** Explain why for every odd function if the ordered pair $(x, y)$ is a point on the graph of the function, the ordered pair $(-x, -y)$ will also be on the graph.

**b)** Explain why the graph of an odd function will always be symmetric with respect to the origin.

## Cumulative Review Exercises

[3.1] **69.** Graph $y = -x^3 - 1$.

**70.** Graph $y = \dfrac{1}{x - 2}$.

[5.1] **71.** Consider the polynomial $4x^5 - 7x^4 - 6x^3 + x^2 - 9$.

**a)** Give the degree of the polynomial.

**b)** Give its leading coefficient.

[6.1] **72.** Determine the domain of the rational function $g(x) = \dfrac{x - 7}{x^2 - 4x - 21}$.

# 10.2 Characteristics of Graphs of Polynomial Functions

1 Define a polynomial function.

2 Define power functions.

3 Determine the $x$-intercepts of a graph and the zeros of a polynomial function.

4 Determine the maximum number of turning points of the graph of a polynomial function.

5 Graph polynomial functions.

## 1 Define a Polynomial Function

We introduced polynomial functions in Section 5.1. We will now give a definition of a polynomial function.

**Polynomial Function**

A **polynomial function** is a function of the form

$$P(x) = a_n x^n + a_{n-1}x^{n-1} + a_{n-2}x^{n-2} + \cdots + a_1 x + a_0$$

where the coefficients $a_n, a_{n-1}, a_{n-2}, \ldots, a_0$ are real numbers,* and all exponents on the variable are nonnegative integers. The domain of all polynomial functions is all real numbers.

Examples of polynomial functions are:

$$P(x) = x^5 + 5x^3 - 2x \quad \text{and} \quad f(x) = -4x^3 + 2x^2 + \frac{1}{2}x - 5$$

An important part of this chapter involves graphing polynomial functions. We will discuss a number of items to help us graph polynomial functions. In this section we first discuss the end behavior of polynomial functions. By **end behavior** we mean how the graph behaves for large positive values of $x$ and for large negative values of $x$. That is, whether the graph will be above or below the $x$-axis for large positive values of $x$ and for large negative values of $x$.

## 2 Define Power Functions

Functions of the form $f(x) = ax^n$ are called **power functions**. Power functions are also polynomial functions. A power function of degree $n$ can be used to determine the end behavior of a polynomial function of the form $f(x) = a_n x^n + a_{n-1}x^{n-1} + \cdots + a_0$. For example, the graph of $f(x) = x^5$ can be used to determine the end behavior of the

*A polynomial function can be defined using complex numbers for $a_n, a_{n-1}, a_{n-2}, \ldots, a_0$. We will not discuss such polynomials in this book.

graph of $f(x) = 6x^5 + 5x^4 + 2x^3 - 6$ because the leading term of the polynomial, $6x^5$, has degree 5. For your convenience, in **Figure 10.9** we give the graphs of power functions of the form $f(x) = x^n$ for $n = 1, 2, 3, 4, 5$, and 6.

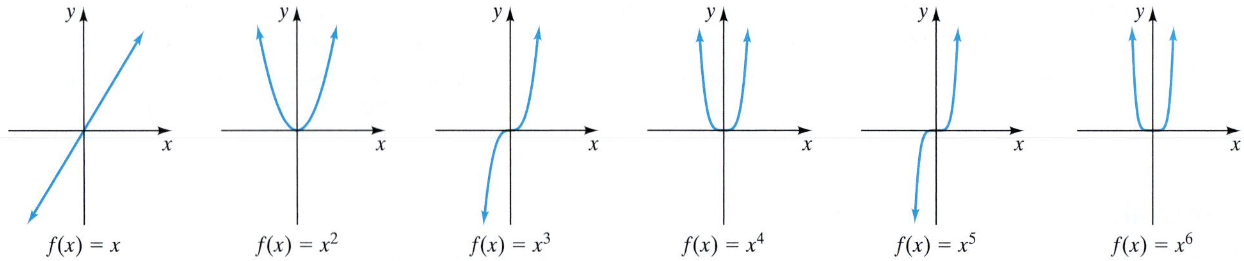


FIGURE 10.9

Consider the graph of $f(x) = x^4$. Suppose $x$ is a large positive value, then $f(x)$ will be a large positive value, and the graph will be above the $x$-axis. Suppose $x$ is a large negative value, then $f(x)$ will still be a large positive value, since a negative number raised to an even exponent is positive. For a large negative value of $x$ the function will be positive and the graph will be above the $x$-axis. Using this example, we can reason that the graph of a function of the form $f(x) = x^n$, where $x$ is an even number $(2, 4, 6, \ldots)$ will be above the $x$-axis for both large positive and large negative values of $x$.

Consider the graph of $f(x) = x^5$. Suppose that $x$ is a large positive value, then $f(x)$ will be a large positive value, and the graph will be above the $x$-axis. Suppose $x$ is a large negative value, then $f(x)$ will be a large negative value since a negative number raised to an odd exponent is negative. For a large negative value of $x$, the function will be negative, and its graph will be below the $x$-axis. Using this example, we can reason that the graph of a function of the form $f(x) = x^n$, where $x$ is an odd number $(1, 3, 5, \ldots)$, will be above the $x$-axis when $x$ is a large positive value and below the $x$-axis when $x$ is a large negative value. The graphs of $y = -x^n$ for $n = 1, 2, 3, 4, 5$, and 6 are shown in **Figure 10.10**.

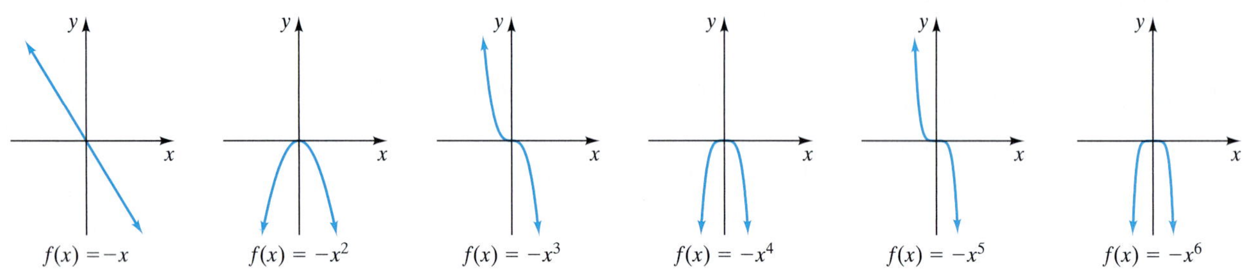


FIGURE 10.10

When the leading coefficient in a polynomial function is negative, the end behavior of the graph of the function can be determined from the graphs in **Figure 10.10**.

**EXAMPLE 1** ▸ By considering only the end behavior of the graphs of the given polynomial function, determine which, if any, of graphs 1, 2, or 3 could possibly be the graph of the polynomial function.

**a)** $P(x) = 3x^4 + 3x^3 - 6$

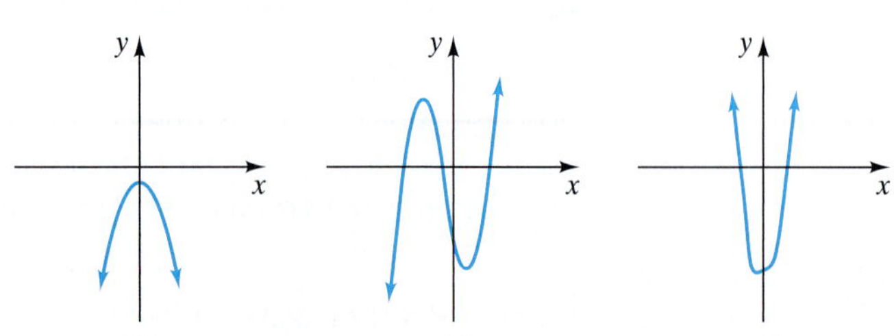

**b)** $P(x) = -2x^5 + 6x^4 - 3x + 2$

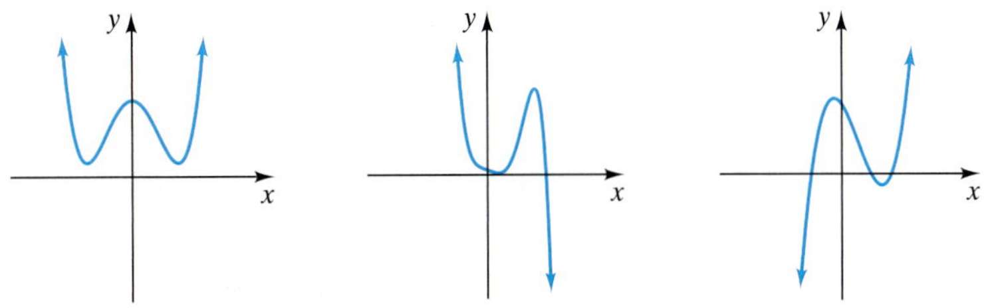

(1) (2) (3)

*Solution*

**a)** Since the leading term is $3x^4$, the graph has the same end behavior as the graph of $f(x) = x^4$. The graph must be above the $x$-axis for both large positive and large negative values of $x$. Therefore the only possible answer is 3.

**b)** Since the leading term is $-2x^5$, the graph has the same end behavior as the graph of $f(x) = -x^5$. The graph must be below the $x$-axis for large positive values of $x$ and above the $x$-axis for large negative values of $x$. Therefore the only possible answer is 2.

▸ **Now Try Exercise 7**

## 3 Determine the $x$-intercepts of a Graph and the Zeros of a Polynomial Function

Now we are ready to discuss how to find the $x$-intercepts of the graph of a polynomial function and the zeros of a polynomial function. A **zero of a polynomial function** is a value of $x$ where the function is equal to 0.

**Zero of a Polynomial Function**

If $P(x)$ is a polynomial function and $r$ is a real number for which $P(r) = 0$, then $r$ is called a **zero** of $P(x)$.

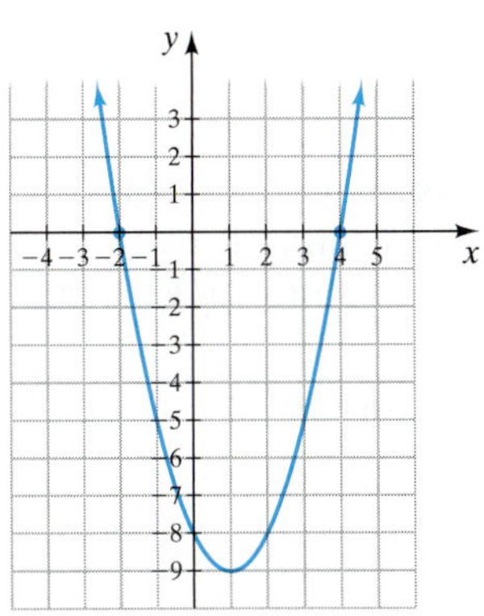

$P(x) = x^2 - 2x - 8$

FIGURE 10.11

Consider the quadratic function $P(x) = x^2 - 2x - 8$ which we graphed on page 364. Quadratic functions are one type of polynomial function. The graph is shown in **Figure 10.11** for your convenience.

The $x$-intercepts are $(-2, 0)$ and $(4, 0)$. The zeros of the function are therefore $-2$ and 4. We can obtain the zeros by setting the function equal to 0, and solving the resulting equation for $x$.

$$P(x) = x^2 - 2x - 8$$
$$0 = x^2 - 2x - 8$$
$$0 = (x + 2)(x - 4)$$
$$x + 2 = 0 \quad \text{or} \quad x - 4 = 0$$
$$x = -2 \qquad\qquad x = 4$$

As we just illustrated, if we can find the linear factors of a polynomial function, we can find the real zeros of the function by setting the product of the factors equal to zero and then using the zero-factor property. Also, we can work backward, if we know the real zeros of a function, then we can determine the $x$-intercepts of the graph of the function.

**$x$-Intercepts and Factors of a Polynomial Function**

If a real number $r$ is a zero of a polynomial function, that is $P(r) = 0$,

**a)** $(r, 0)$ is an $x$-intercept of the graph, and

**b)** $(x - r)$ is a factor of $P(x)$.

For example, consider $P(x) = x^3 + 6x^2 - x - 30$. Notice $P(2) = 0$.

$$P(x) = x^3 + 6x^2 - x - 30$$
$$P(2) = 8 + 24 - 2 - 30 = 0$$

Since $P(2) = 0$, then 2 is a zero of the polynomial function, (2, 0) is an $x$-intercept of the graph of $P(x)$, and $(x - 2)$ is a factor of $P(x)$.

**EXAMPLE 2** ▸ Consider the function $P(x) = x^3 - 2x^2 - 8x$.

**a)** Determine the zeros of the function.

**b)** Determine the $x$-intercepts of the graph of $P(x)$.

*Solution*

**a)** First factor the expression, as follows.

$$P(x) = x(x^2 - 2x - 8)$$
$$P(x) = x(x - 4)(x + 2)$$

Next set $P(x) = 0$ and solve for $x$.

$$0 = x(x - 4)(x + 2)$$
$$x = 0 \quad \text{or} \quad x - 4 = 0 \quad \text{or} \quad x + 2 = 0$$
$$x = 0 \qquad x = 4 \qquad x = -2$$

Thus, the zeros are 0, 4, −2.

**b)** The $x$-intercepts of the graph are (0, 0), (4, 0), and (−2, 0).

▸ **Now Try Exercise 21**

Consider the function $P(x) = (x + 5)(x - 4)^2$. This function can be written $P(x) = (x + 5)(x - 4)(x - 4)$. Since 4 makes two factors of $(x - 4)$ equal 0, we say that 4 is a zero of **multiplicity** 2.

**EXAMPLE 3** ▸ Determine the zeros of $P(x) = (x - 3)^2(x + 4)^3(x - 5)^6$ and give their multiplicities.

*Solution* The factor $(x - 3)^2$ is zero when $x = 3$. Since there are two factors of $(x - 3)$, its multiplicity is 2.

The factor $(x + 4)^3$ is zero when $x = -4$. Thus, −4 is a zero of multiplicity 3.

The factor $(x - 5)^6$ is zero when $x = 5$. Thus, 5 is a zero of multiplicity 6.

▸ **Now Try Exercise 25**

Let's now discuss $x$-intercepts a little further. Consider **Figure 10.12**.

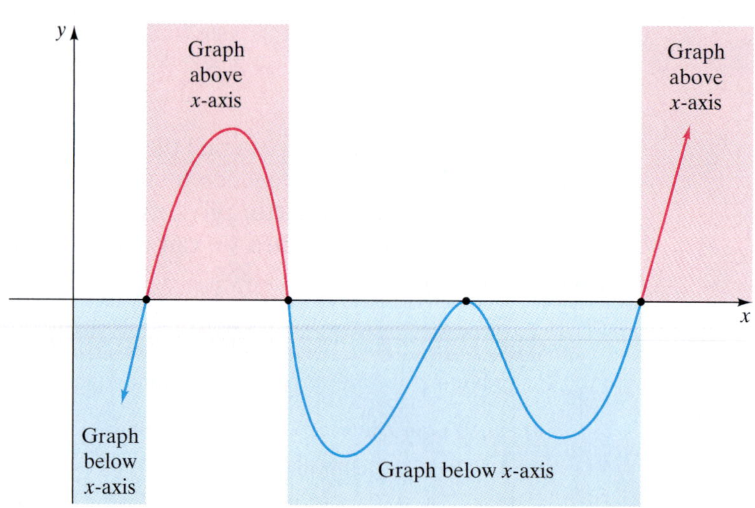


FIGURE 10.12

The graph shows four $x$-intercepts. That is, four values of $x$ where $P(x) = 0$. Notice at three of the four intercepts the graph crosses the $x$-axis, thus there is a change in the sign of $P(x)$. At one $x$-intercept, the graph touches, but does not cross, the $x$-axis. Notice that the function has the same sign, negative, on both sides of that $x$-intercept.

If we know the multiplicity of a zero we can determine whether the function crosses or just touches the $x$-axis at that zero.

**Zeros of Even or Odd Multiplicity**

For a real zero of an even multiplicity, the graph of $P(x)$ touches but does not cross the $x$-axis at that zero.
For a real zero of an odd multiplicity, the graph of $P(x)$ crosses the $x$-axis at that zero.

For a zero of an even multiplicity, the sign of $P(x)$ does not change from one side of the zero to the other side of the zero. For a zero of an odd multiplicity, the sign of $P(x)$ changes from one side of the zero to the other side of the zero.

## 4 Determine the Maximum Number of Turning Points of the Graph of a Polynomial Function

In **Figure 10.1** we introduced local maxima and local minima. Local maxima and local minima are points at which the graph changes direction (from increasing to decreasing or vice versa). Therefore such points are called **turning points**, see **Figure 10.13**.

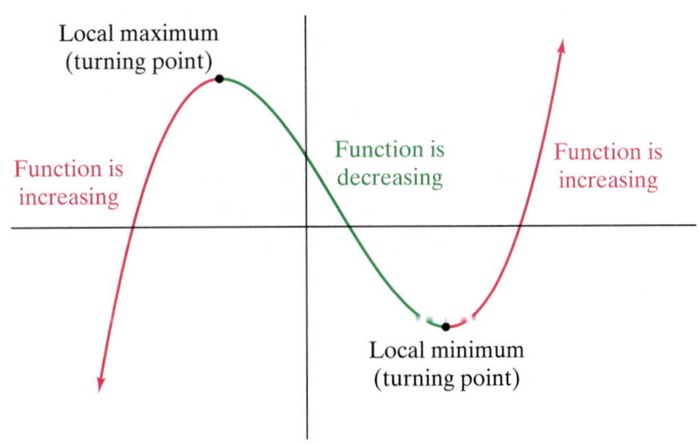


FIGURE 10.13

To find the turning points of a graph requires a knowledge of calculus, so we will not ask you to determine the location of turning points on the graphs. However, the maximum number of turning points on a graph can be found by the following theorem.

**Maximum Number of Turning Points**

The maximum number of turning points on the graph of a polynomial function of degree $n$ is $n - 1$.

For example, the graph of the polynomial function $P(x) = 3x^4 + 2x - 6$ can have at most 4 1 or 3 turning points.

## 5 Graph Polynomial Functions

In Examples 4 and 5 we will use our knowledge of end behavior, zeros of polynomial functions, and turning points to graph the functions.

**EXAMPLE 4 ▸** For the polynomial function $P(x) = (x + 5)(x + 2)(x - 2)$

**a)** Determine the $x$-intercepts and whether the graph crosses or touches the $x$-axis at each $x$-intercept.

**b)** Explain the end behavior of the graph of the function for large positive values of $x$ and for large negative values of $x$.

**c)** Determine the maximum number of turning points on the graph of the function.

**d)** Use the $x$-intercepts and test values to find the intervals on which the graph is above the $x$-axis, and the intervals on which the graph is below the $x$-axis.

**e)** Using all the information, connect the $x$-intercepts and the test points with a smooth continuous curve to sketch the graph of the function.

## Solution

**a)** To determine the $x$-intercepts, set the function equal to 0.

$$P(x) = (x + 5)(x + 2)(x - 2)$$
$$0 = (x + 5)(x + 2)(x - 2)$$
$$x + 5 = 0 \quad \text{or} \quad x + 2 = 0 \quad \text{or} \quad x - 2 = 0$$
$$x = -5 \qquad x = -2 \qquad x = 2$$

The $x$-intercepts are $(-5, 0)$, $(-2, 0)$ and $(2, 0)$. Since each of the zeros is of multiplicity one, the graph will cross the axis at each $x$-intercept.

**b)** If we were to multiply the factors, the leading term would be $x^3$. Therefore, the end behavior of the given function is the same as the end behavior of $f(x) = x^3$. The function will be positive and above the $x$-axis for large positive values of $x$ and negative and below the $x$-axis for large negative values of $x$.

**c)** Since the polynomial is of degree 3, the maximum number of turning points is $3 - 1$ or 2.

**d)** The $x$-intercepts break the graph into 4 regions as illustrated below. We will select a test value in each region. The test values we arbitrarily selected are $-6$, $-4$, 0 and 3, as shown in **Figure 10.14**.

FIGURE 10.14

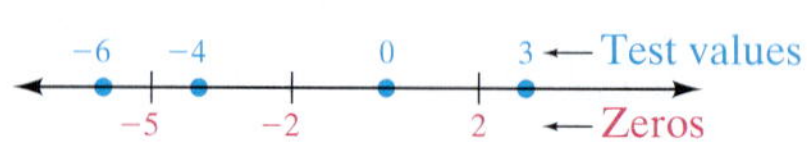


To determine the value of $P(x)$ at a test value, substitute the test value for each $x$ in the function. For example, for the test value $-6$, we get

$$P(x) = (x + 5)(x + 2)(x - 2)$$
$$P(-6) = (-6 + 5)(-6 + 2)(-6 - 2)$$
$$= (-1)(-4)(-8) = -32$$

Using the intervals and the test values, we complete the following table.

| Interval | Test Value | Value of $P(x)$ | Location of $P(x)$ with regard to the $x$-axis |
|---|---|---|---|
| $(-\infty, -5)$ | $-6$ | $-32$ | below |
| $(-5, -2)$ | $-4$ | 12 | above |
| $(-2, 2)$ | 0 | $-20$ | below |
| $(2, \infty)$ | 3 | 40 | above |

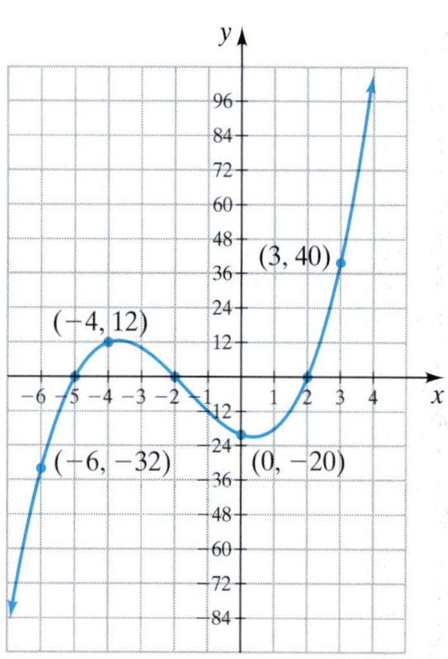


FIGURE 10.15

**e)** Plot the $x$-intercepts and the test points and draw a smooth continuous curve through the points. Notice that when $x = 0$, $P(x) = -20$, see **Figure 10.15**.

Notice that the end behavior is as we expected, and the function crosses the $x$-axis at each $x$-intercept. The graph has 2 turning points.

▸ **Now Try Exercise 49**

---

**EXAMPLE 5** ▸ Sketch the graph of $P(x) = 3x^4 - x^3 - 4x^2$.

*Solution* Begin by factoring the expression on the right of the equal sign.

$$\begin{aligned} P(x) &= 3x^4 - x^3 - 4x^2 \\ &= x^2(3x^2 - x - 4) \\ &= x^2(3x - 4)(x + 1) \end{aligned}$$

Now set $P(x) = 0$ and solve for $x$.

$$0 = x^2(3x - 4)(x + 1)$$

$$x^2 = 0 \quad \text{or} \quad 3x - 4 = 0 \quad \text{or} \quad x + 1 = 0$$

$$x = 0 \qquad\qquad x = \frac{4}{3} \qquad\qquad x = -1$$

Therefore, the $x$-intercepts are $(0, 0)$, $\left(\frac{4}{3}, 0\right)$ and $(-1, 0)$. Notice the zero at $x = 0$ is of multiplicity 2. Now use the $x$-values obtained above to set up intervals. When setting up the intervals, start with the smallest value of $x$ and work from the smallest toward the largest value of $x$. The test values we selected are shown in **Figure 10.16**.

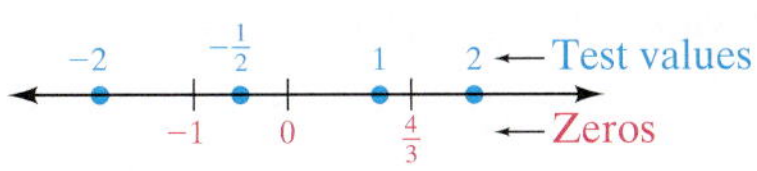


FIGURE 10.16

| Interval | Test Value | Value of $P(x)$ | Location of $P(x)$ with regard to the $x$-axis |
|---|---|---|---|
| $(-\infty, -1)$ | $-2$ | 40 | above |
| $(-1, 0)$ | $-\frac{1}{2}$ | $-\frac{11}{16}$ | below |
| $\left(0, \frac{4}{3}\right)$ | 1 | $-2$ | below |
| $\left(\frac{4}{3}, \infty\right)$ | 2 | 24 | above |

Now we plot the points and draw the graph, see **Figure 10.17**.

In **Figure 10.17** we did not illustrate the ordered pair $(-2, 40)$ since it is off the graph. Notice the end behavior of the function is the same as the end behavior of $f(x) = x^4$. Also notice that at $x = 0$ the graph touches, but does not cross the $x$-axis. We expect this behavior because this zero has multiplicity 2, an even number. The graph has 3 turning points which is the maximum number it can have.

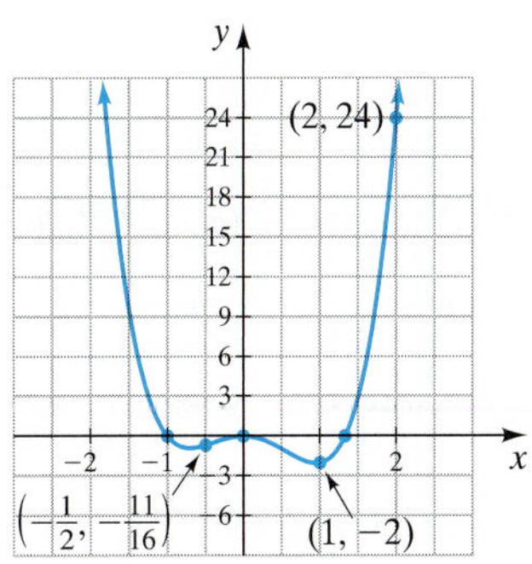


FIGURE 10.17

▸ **Now Try Exercise 65**

## EXERCISE SET 10.2

MathXL® MyMathLab

### Concept/Writing Exercises

1. Consider the polynomial function graphed below.

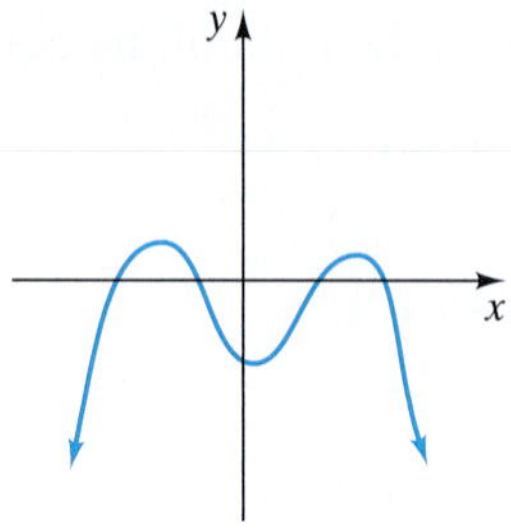

a) Is the constant term positive or negative? Explain.
b) Is the leading coefficient positive or negative? Explain.
c) Is the degree of the polynomial function even or odd? Explain.
d) Can the degree of the polynomial function be 6? Explain.
e) Can the degree of the polynomial function be 7? Explain.
f) How many real zeros does the polynomial function have? Explain.

2. Consider the polynomial function graphed below.

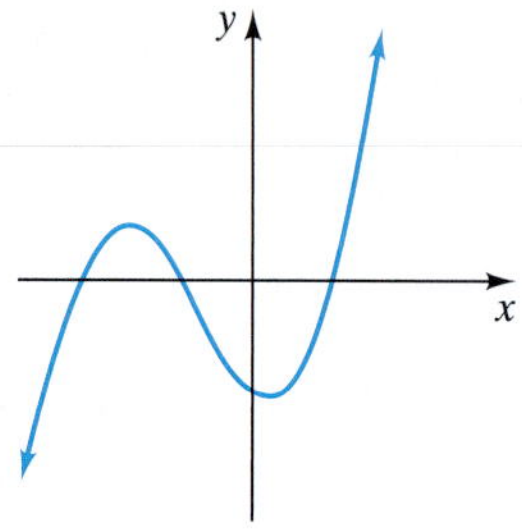

a) Is the constant term positive or negative? Explain.
b) Is the leading coefficient positive or negative? Explain.
c) Is the degree of the polynomial function even or odd? Explain.
d) Can the degree of the polynomial function be 1? Explain.
e) Can the degree of the polynomial function be 4? Explain.
f) How many real zeros does the polynomial function have? Explain.

3. a) Describe the end behavior of the graph of $f(x) = x^8$.
b) Describe the end behavior of the graph of $g(x) = x^9$.

4. a) If the zeros of a polynomial function are $x = 2$, $x = 1$, and $x = 5$, can the function be $P(x) = a(x + 2)(x + 1)(x + 5)$? Explain.
b) If the zeros of a polynomial function are $x = 3$, $x = -1$, and $x = -5$, can the function be $P(x) = a(x - 3)(x + 1)(x + 5)$? Explain.

### Practice the Skills

*For Exercises 5–8, determine which of the given graphs could be the graph of the given polynomial function. Explain your answer.*

5. $P(x) = -2x^3 + 5x - 8$

a)
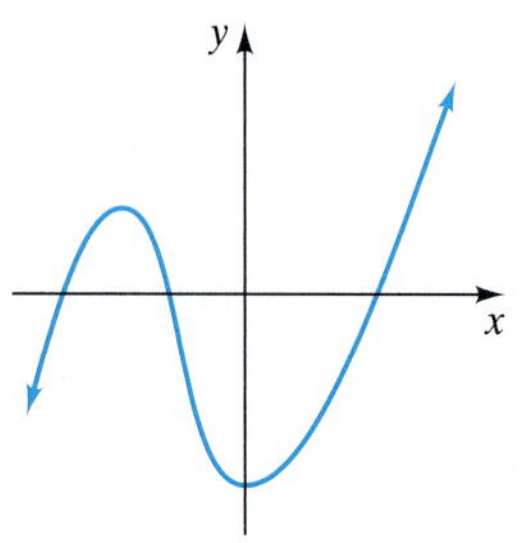

b)
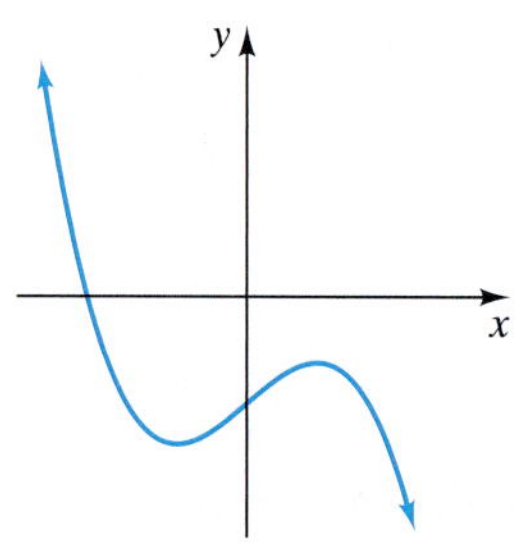

c)
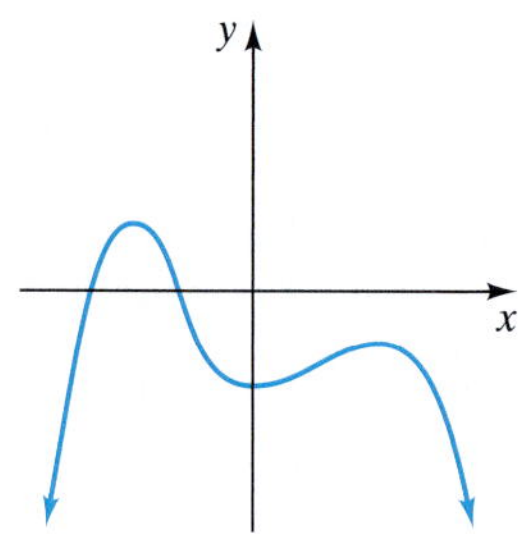

6. $P(x) = -\frac{1}{3}x^4 + 5x + 4$

a)
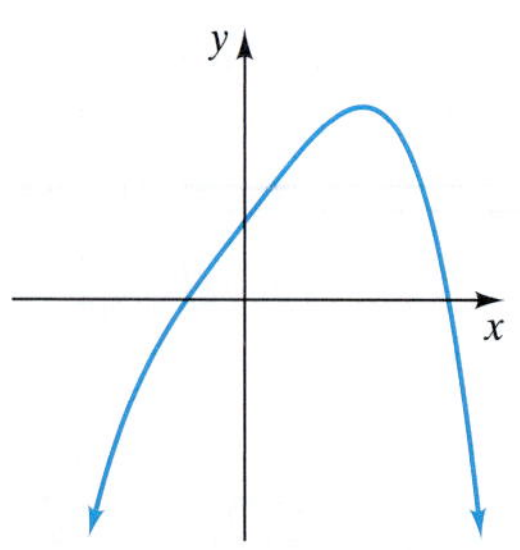

b)
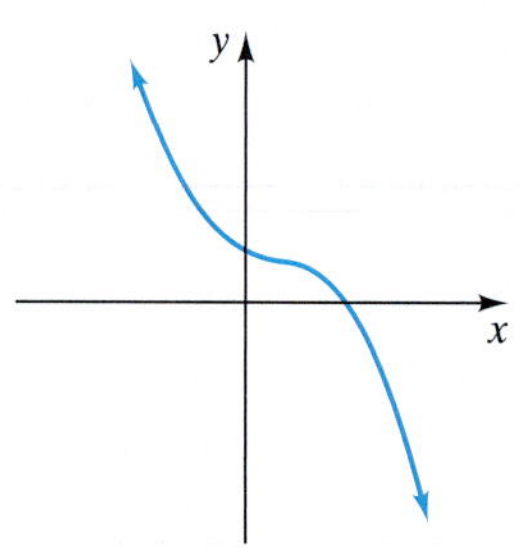

c)
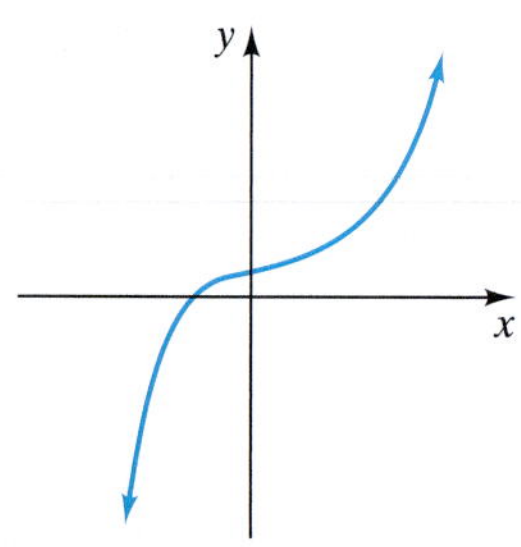

**7.** $P(x) = 3x^5 + 5x^3 - 25x^2 + 20$

**a)**

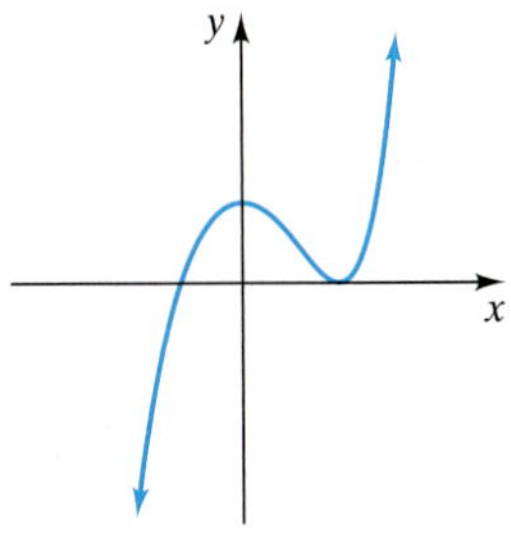


**b)**

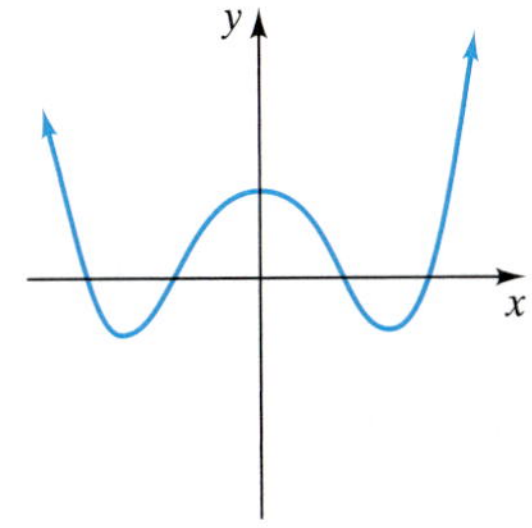


**c)**

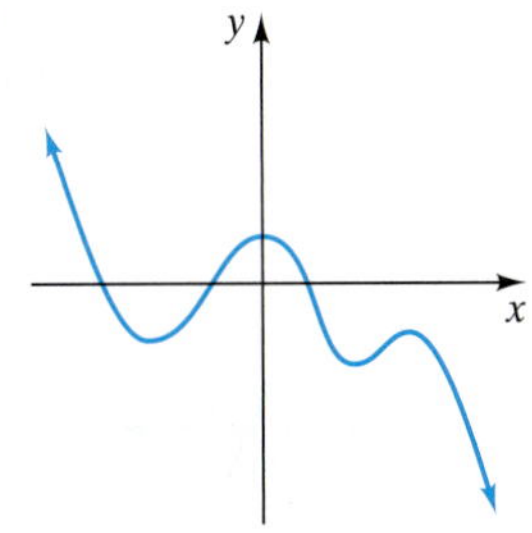


**8.** $P(x) = x^6 - 2x^5 + 4x^4 - 8x^3 - 16x^2 + 32x - 64$

**a)**

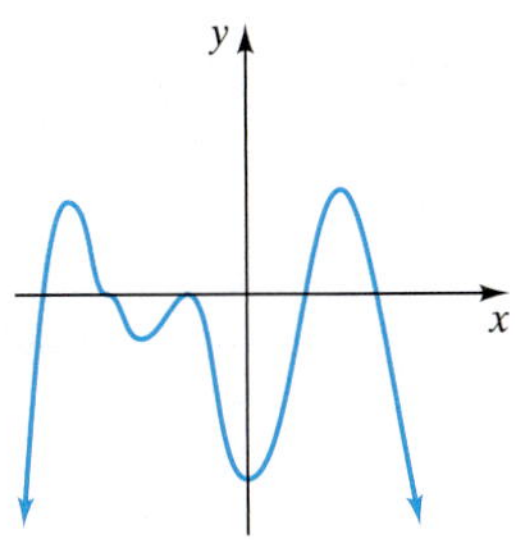


**b)**

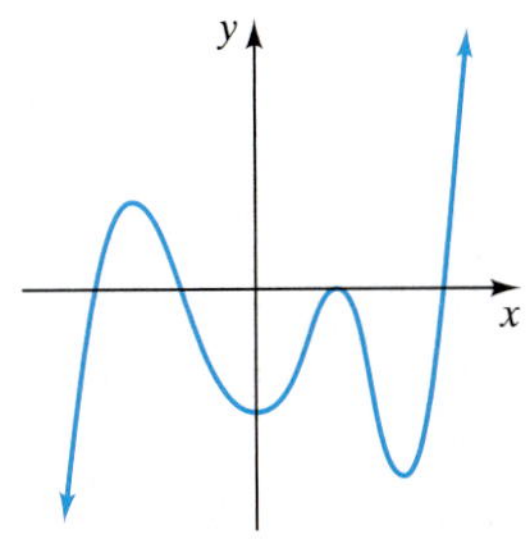


**c)**

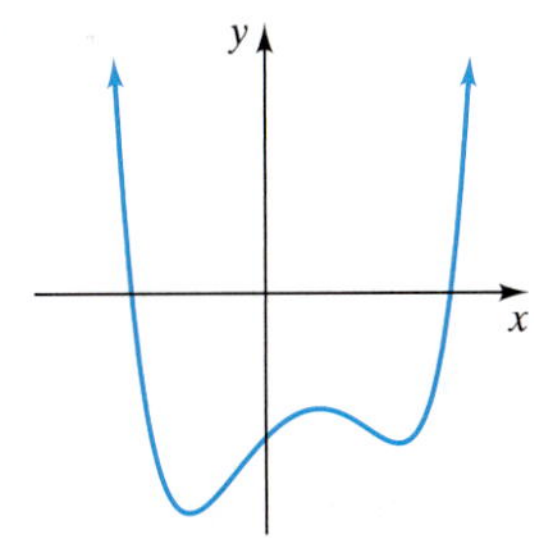


*Describe the location, either above the x-axis or below the x-axis, of the graph of each function for* **a)** *large positive values of x, and* **b)** *large negative values of x.*

**9.** $P(x) = x^7$

**10.** $f(x) = 2x^7$

**11.** $f(x) = -3x^7$

**12.** $P(x) = x^7 + x^6 + x^5 + x^4 + x + 10$

**13.** $f(x) = -\frac{1}{4}x^2 + 2x$

**14.** $f(x) = 4x^2 + 2x$

**15.** $f(x) = -3x^4 + x^3 - 2x + 17$

**16.** $P(x) = 7x^4 - 3x + 5$

**17.** $f(x) = -5x^7 + 2x^5 - x^4 + x^2 - x + 1$

**18.** $f(x) = -x^6 + 7x^4 - 2x^3 + x^2 - x + 3$

*Find the real zeros of each function and state their multiplicities if it is a number other than 1.*

**19.** $f(x) = x^2 - 2x - 15$

**20.** $g(x) = x^2 - 8x + 16$

**21.** $P(x) = x^3 + 6x^2 + 5x$

**22.** $h(x) = 2x^3 + 9x^2 - 5x$

**23.** $f(x) = x^2 - 25$

**24.** $g(x) = x^2 - 3$

**25.** $P(x) = (x + 7)(x - 2)^3$

**26.** $f(x) = (x - 9)^3$

**27.** $g(x) = (x + 3)(2x + 1)(3x - 1)$

**28.** $h(x) = (x + 7)(x + 1)(x - 3)^2$

**29.** $h(x) = (x + 2)^3(x - 4)^2$

**30.** $f(x) = (x + 6)(x - 5)(x^2 - 49)$

**31.** $P(x) = x(x + 1)^2(x - 1)^3(x - 6)^4$

**32.** $g(x) = x^4(x^2 - 25)(x^2 - 9)$

*Determine the maximum possible number of turning points for the graph of each function.*

**33.** $f(x) = 4x^5 + x^3 - 8x^2 + 64$

**34.** $g(x) = 3x^2 + 2x - 9$

**35.** $h(x) = \frac{1}{2}x - 7$

**36.** $P(x) = 5x^5 - x^7$

**37.** $g(x) = 3x^6 + x^5 - x^4 + 6x^2 - 5x + 12$

**38.** $h(x) = x^3 - 4x^2 + 9x - 24$

**39.** $f(x) = x(x - 5)(3x + 4)$

**40.** $P(x) = x^2(x^2 - 16)(x + 3)$

**41.** $h(x) = (x - 1)(x + 2)$

**42.** $f(x) = (x - 6)^2(4x - 1)$

**43.** $g(x) = (2x - 3)^3(x + 8)(x - 1)$

**44.** $P(x) = (x + 8)(x - 5)(x + 3)(x - 2)(x + 1)$

*Graph each function. See Example 4.*

**45.** $f(x) = x(x + 3)(x - 1)$

**46.** $f(x) = x^2(x + 3)(x + 1)$

**47.** $P(x) = (x + 3)^2(x - 1)$

**48.** $g(x) = (x + 2)(x + 1)(x - 3)$

**49.** $P(x) = (x + 3)(x - 3)(x - 5)$

**50.** $g(x) = x(x + 4)^2$

**51.** $g(x) = (x + 1)(x^2 - 36)$

**52.** $h(x) = (x + 3)(x + 1)(x - 2)(x - 4)$

## Problem Solving

*Determine the minimum degree of the polynomial function graphed.*

**53.**

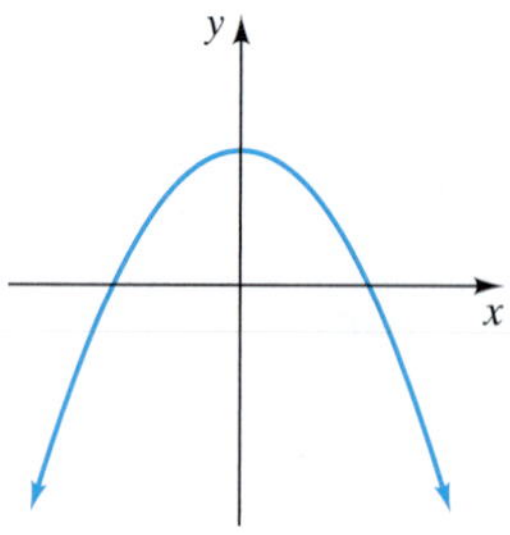


**54.**

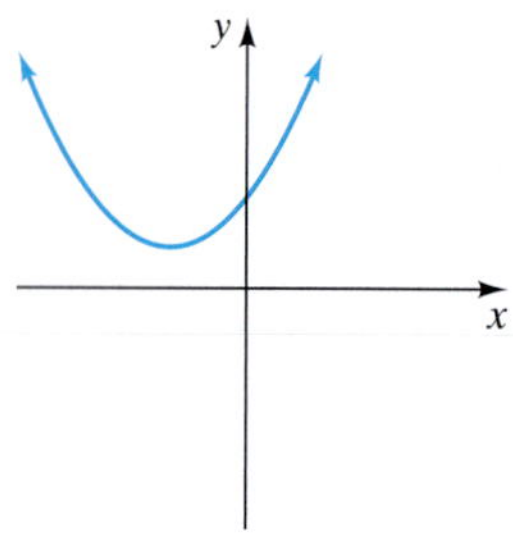


**55.**

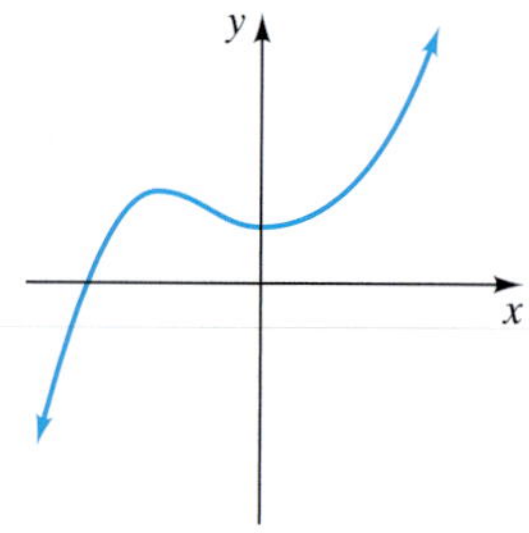


**56.**

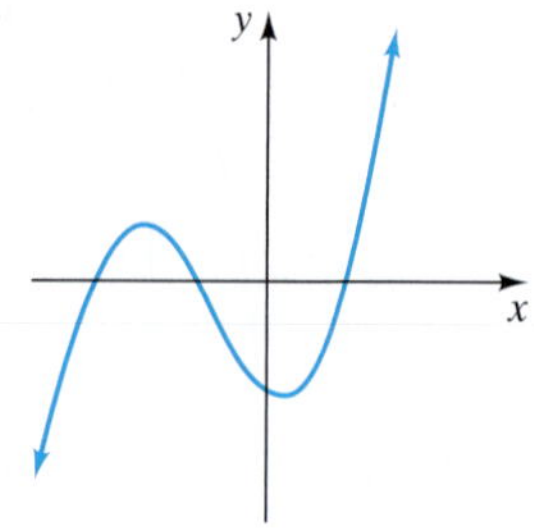


**57.**

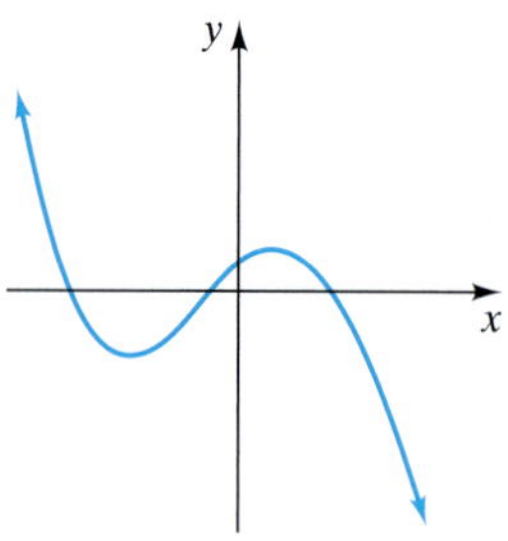


**58.**

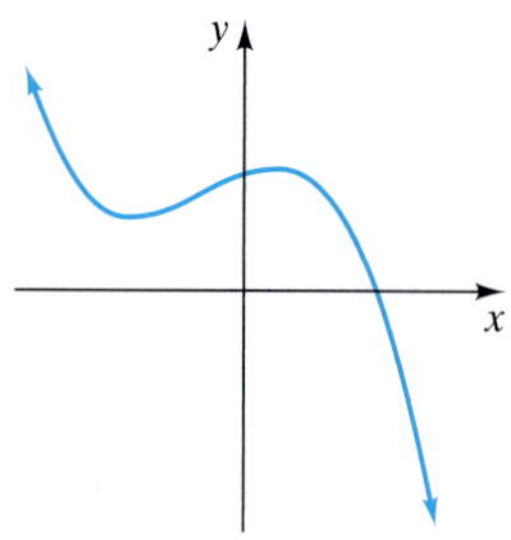


**59.**

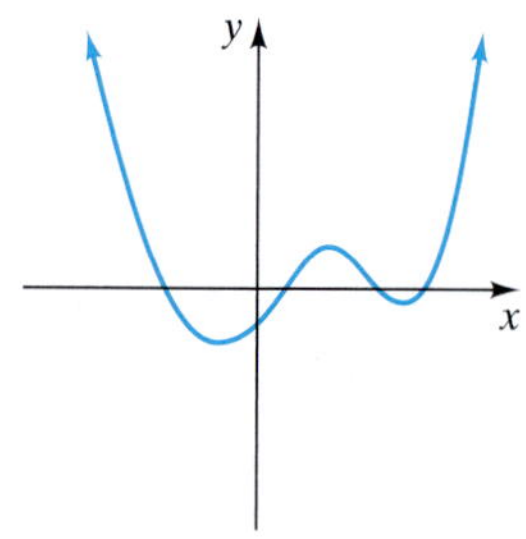


**60.**

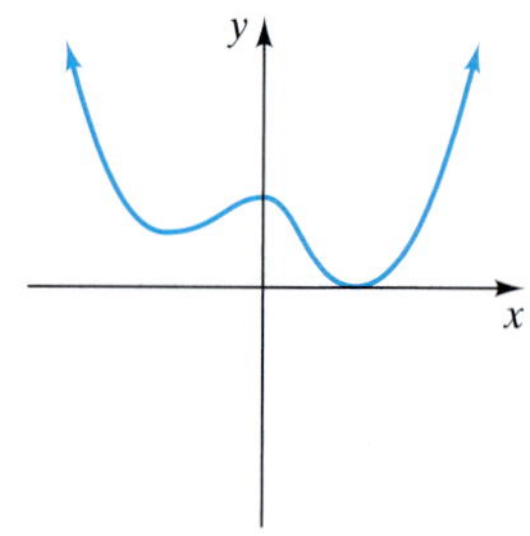


**61.**

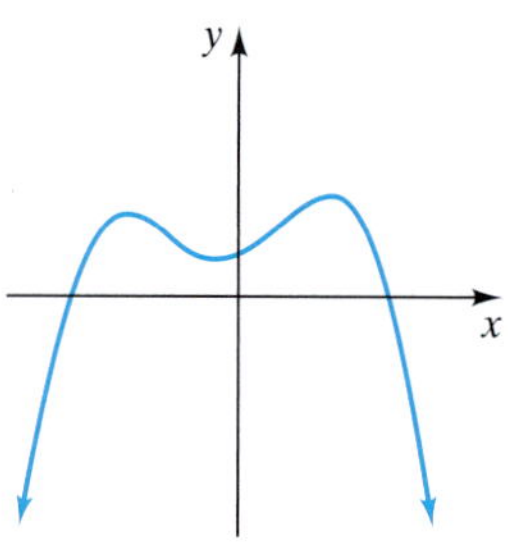


**62.**

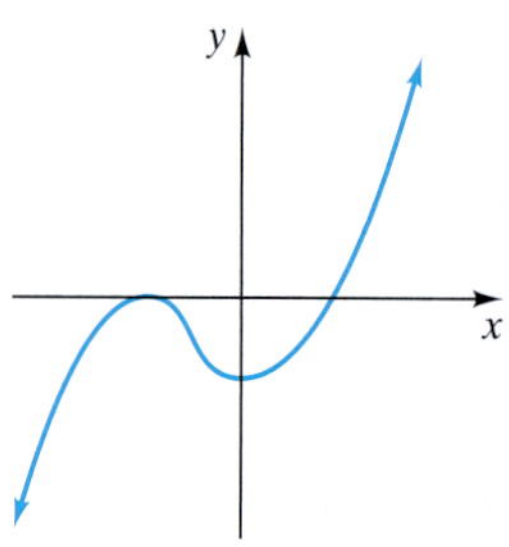


*Graph each function. See Example 5.*

**63.** $f(x) = x^3 - 2x^2 - 3x$

**64.** $g(x) = 2x^3 + 5x^2 - 3x$

**65.** $f(x) = 5x^3 + 4x^2 - x$

**66.** $g(x) = 6x^4 - 10x^3 - 4x^2$

**67.** $f(x) = x^3 - 4x^2 + 4x$

**68.** $g(x) = x^4 - 4x^3 + 4x^2$

*In the Using Your Graphing Calculator box on page 177 we explained how to find the zeros of a function using a graphing calculator. A similar procedure can be used to find the turning points (the local maxima or local minima) of a polynomial function. On the TI-84 Plus you use the CALC menu then select item 3:minimum or item 4:maximum. Read page 177 now. Then read your graphing calculator manual if further explanation is needed on how to find the local maxima and local minima of a polynomial function. Then for each of the following exercises:*

**a)** *On your graphing calculator, graph the polynomial function over the given interval.*

**b)** *Determine if the function has a local maximum or local minimum in the given interval.*

**c)** *Use your graphing calculator to find the turning point of the polynomial function in the given interval. Round the coordinates to the nearest hundredth when appropriate.*

**69.** $y = 3x^2 - 2x + 3, (-1, 1)$

**70.** $y = -\frac{1}{2}x^2 - 5x + 4, (-6, -4)$

**71.** $y = -x^3 + 2x - 3, (-2, 0)$

**72.** $y = x^3 + 3x^2 - 5, (-3, -1)$

## Group Activity

**73.** In Exercise 47, we graphed $P(x) = (x + 3)^2(x - 1)$ where the linear factor $(x + 3)$ is squared. Now we will look at graphs of functions having two linear factors, each of which is squared.

**a)** Each group member should graph the appropriate function given below.
Group Member 1: Graph $f(x) = (x + 1)^2(x - 1)^2$
Group Member 2: Graph $f(x) = (x + 2)^2(x - 1)^2$
Group Member 3: Graph $f(x) = (x + 2)^2(x + 1)^2$

**b)** Is $f(x) \geq 0$ for every value of $x$ for your graph?

**c)** Share your graph with the other members. As a group, what can you conclude from the three graphs?

**d)** As a group, graph $g(x) = (x + 2)^2(x + 1)^2(x - 1)^2$.

**e)** Discuss why the graph of $g(x)$ is similar to the three graphs of $f(x)$ from part **a)**.

### Cumulative Review Exercises

[5.3] **74.** Use synthetic division to divide

$(4x^4 - 3x^3 + 2x^2 - x - 250) \div (x - 3)$.

**75.** Use the Remainder Theorem to determine the remainder for the division

$(5x^3 - 7x^2 + 9x - 11) \div (x + 4)$.

[6.2] **76.** Add $\frac{3}{5x} + \frac{x}{4}$.

**77.** Subtract $\frac{3x - 1}{y - 4} - \frac{2x + 5}{4 - y}$.

# 10.3 Finding the Rational Zeros of a Polynomial Function

1 Determine the maximum number of real zeros of a polynomial function.

2 Use Descartes' rule of signs.

3 Determine the possible rational zeros of a polynomial function.

4 Use synthetic division to determine the rational zeros of a polynomial function.

5 Find complex zeros of a polynomial function.

## 1 Determine the Maximum Number of Real Zeros of a Polynomial Function

In the previous section we learned that the maximum number of turning points of the graph of a polynomial of degree $n$ is $n - 1$. Consider **Figure 10.18** which shows the graph of a cubic or third-degree polynomial function with 2 turning points, the maximum number possible. **Figure 10.19** shows the graph of a fourth-degree polynomial with 3 turning points, the maximum number possible.

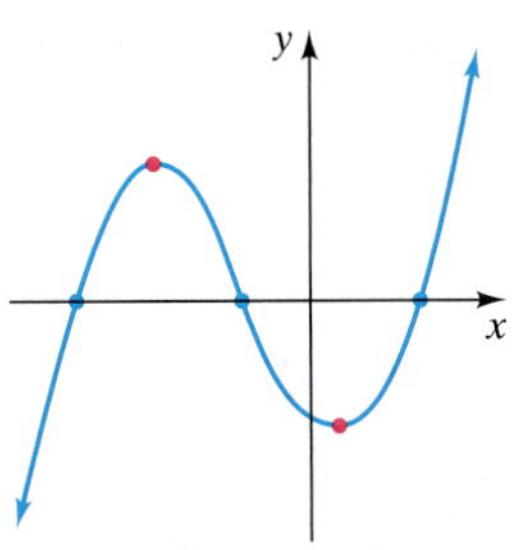


FIGURE 10.18

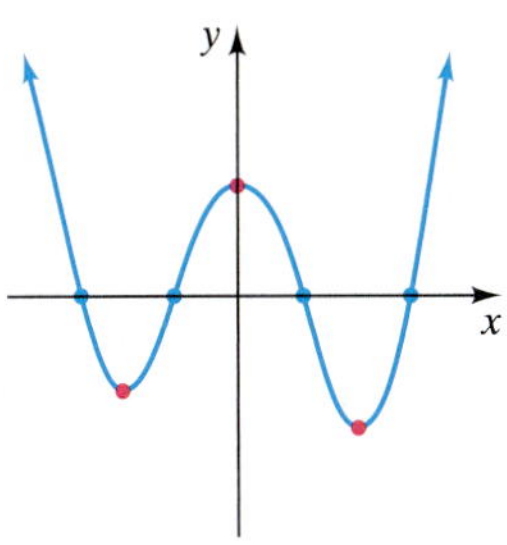


FIGURE 10.19

Notice that in **Figure 10.18**, the number of $x$-intercepts is 3, one more than the maximum number of turning points. In **Figure 10.19** the number of $x$-intercepts is 4, one more than the maximum number of turning points. By examining other graphs, we can determine that the maximum number of $x$-intercepts is one more than the maximum number of turning points. Since the maximum number of turning points for a polynomial of degree $n$ is $n - 1$, the maximum number of $x$-intercepts is $(n - 1) + 1$ or $n$.

**For a Polynomial Function $P(x)$ of Degree $n$**

**a)** the maximum number of $x$-intercepts of its graph is $n$, and

**b)** the maximum number of real zeros of $P(x)$ is $n$.

An **odd-degree polynomial function** will always have at least one real zero because one end of the graph will be above the $x$-axis and the other end will be below the $x$-axis, see **Figure 10.20**.

Since an **even-degree polynomial function** will have both ends of its graph either above the $x$-axis or below the $x$-axis, it is possible for the polynomial function to have no real zeros, see **Figure 10.21**.

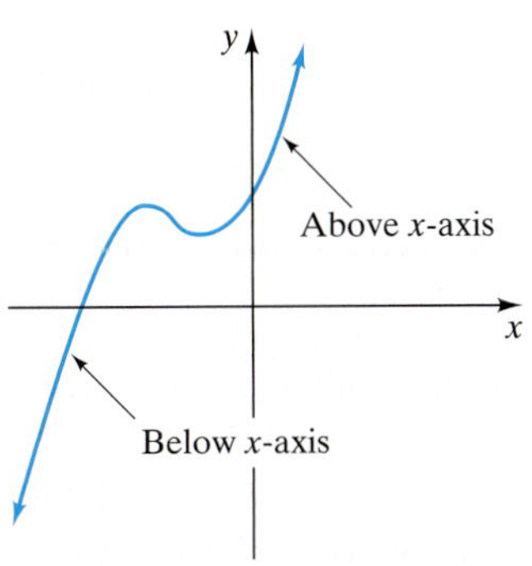


FIGURE 10.20

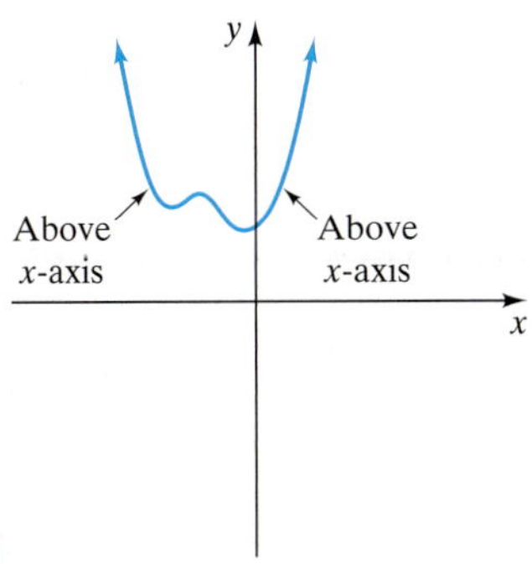


FIGURE 10.21

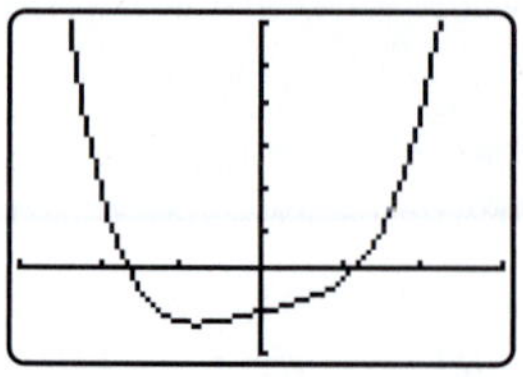

$-3, 3, 1, -8, 24, 4$

FIGURE 10.22

An even-degree polynomial function with a leading coefficient that has the opposite sign as its constant term has at least 2 real zeros. For example, $P(x) = x^4 + 2x - 4$ has a minimum of 2 real zeros since the leading coefficient, 1, is positive and the constant, $-4$, is negative. Note that $P(-2) = 8$, $P(0) = -4$ and $P(2) = 16$. Since there is a change of sign of $P(x)$ from $x = -2$ to $x = 0$, there must be a zero between these values. Similarly, since there is a change of sign of $P(x)$ between $x = 0$ and $x = 2$, there must be another zero between these values. See **Figure 10.22**.

**EXAMPLE 1 ▸** Determine the maximum and minimum number of $x$-intercepts of the graph of each polynomial function.

**a)** $P(x) = 4x^3 - 5x^2 + 9x + 7$

**b)** $P(x) = -x^4 + 2x^3 + 13x^2 - 5x + 20$

*Solution*

**a)** Because $P(x) = 4x^3 - 5x^2 + 9x + 7$ is a third-degree polynomial function, its graph has a maximum of three $x$-intercepts. Since its degree is an odd number, its graph has a minimum of one $x$-intercept.

**b)** Because $P(x) = -x^4 + 2x^3 + 13x^2 - 5x + 20$ is a fourth-degree polynomial function, its graph has a maximum of four $x$-intercepts. Since its degree is an even number, and its leading coefficient and constant term have opposite signs, its graph has at least two $x$-intercepts.

**▸ Now Try Exercise 11**

## 2 Use Descartes' Rule of Signs

We have recently shown that the maximum number of real zeros of an $n$th degree polynomial function is $n$. Therefore we know that the maximum number of positive zeros is $n$ and the maximum number of negative zeros is $n$.

Using **Descartes' Rule of Signs**, we can give an even better estimate of the maximum number of positive zeros and the maximum number of negative zeros.

**Descartes' Rule of Signs**

For any polynomial function $P(x)$ written in descending order,

1. the number of positive real zeros of $P(x)$ either equals the number of variations in sign of the nonzero coefficients of $P(x)$, or equals that number less an even integer.
2. the number of negative real zeros of $P(x)$ either equals the number of variations in sign of the nonzero coefficients of $P(-x)$ or equals that number less an even integer.

Although the Descartes' Rule of Signs is intended for higher degree polynomial functions, we will look at a linear function to help us develop our intuitive understanding of the rule.

The function $P(x) = 3x + 4$ has 1 zero. Notice that because both coefficients are positive there are no variations in sign in the coefficients of $P(x)$. Thus, the zero cannot be positive. Now let's look at the variations in $P(-x)$.

$$P(x) = 3x + 4$$
$$P(-x) = 3(-x) + 4$$
$$P(-x) = -3x + 4$$

Notice that there is 1 variation in sign in the coefficients, from minus to plus. By Descartes' Rule of Signs, the number of negative zeros is 1. Thus, the 1 real zero in $P(x)$ occurs at a negative value of $x$.

We will now look at a quadratic function. Quadratic functions have a maximum of 2 real zeros. The quadratic function $P(x) = (x - 2)(x - 4)$ has zeros at $x = 2$ and

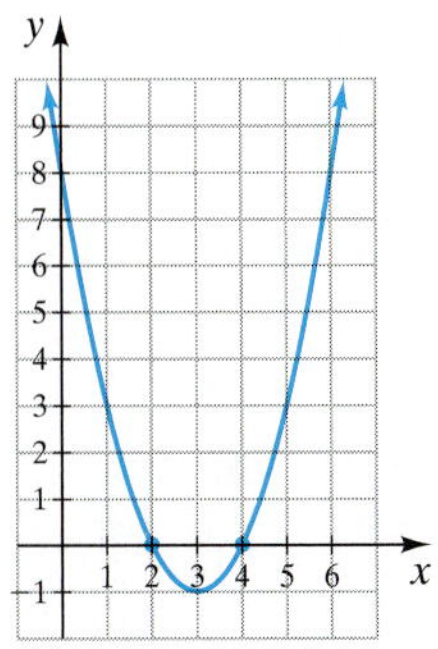

FIGURE 10.23

$x = 4$. From the graph of the function, shown in **Figure 10.23**, we see that the function has 2 positive real zeros and no negative real zeros.

Now let's see how Descartes' Rule of Signs could have been used to determine the possible number of real zeros of the function. When multiplied out we obtain $P(x) = x^2 - 6x + 8$. $P(x)$ has 2 variations in sign. By Descartes' Rule of Signs, we know that the function either has 2 or 0 positive zeros. Now let's look at the sign changes in the coefficients of $P(-x)$.

$$\begin{aligned} P(x) &= x^2 - 6x + 8 \\ P(-x) &= (-x)^2 - 6(-x) + 8 \\ P(-x) &= x^2 + 6x + 8 \end{aligned}$$

Because there are no variations in the sign of the coefficients of $P(-x)$, we know that the function does not have any negative zeros.

Now consider $P(x) = x^2 + x - 6$. Since this is an even-degree polynomial function whose leading coefficient, 1, is opposite in sign from the constant term, $-6$, we know the polynomial has two real zeros. Since there is one change in sign in $P(x)$ there is one positive real zero. $P(-x) = x^2 - x - 6$. Since there is one change of sign in $P(-x)$ there is one negative real zero. Thus, this polynomial has one positive real zero and one negative real zero. Use factoring now to verify the real zeros are $-3$ and 2.

Now we will use Descartes' Rule of Signs to determine the possible number of positive and negative real zeros in a sixth-degree polynomial function.

**EXAMPLE 2** ▸ Determine the possible number of positive real and negative real zeros of $P(x) = 3x^6 - 5x^5 + 3x^4 + x^2 - 10$.

*Solution* The function, $P(x) = 3x^6 - 5x^5 + 3x^4 + x^2 - 10$, is written in descending order. Since the nonzero coefficients of $P(x)$, which are $3, -5, 3, 1, -10$, have 3 sign changes, the graph of the function has either 3 or 1 positive real zeros. Now find $P(-x)$.

$$\begin{aligned} P(x) &= 3x^6 - 5x^5 + 3x^4 + x^2 - 10 \\ P(-x) &= 3(-x)^6 - 5(-x)^5 + 3(-x)^4 + (-x)^2 - 10 \\ &= 3x^6 + 5x^5 + 3x^4 + x^2 - 10 \end{aligned}$$

Since the nonzero coefficients of $P(-x)$, which are $3, 5, 3, 1, -10$, have 1 sign change, the function has 1 negative real zero.

▸ **Now Try Exercise 21**

## 3 Determine the Possible Rational Zeros of a Polynomial Function

Now that we know the possible number of positive and negative real zeros of a polynomial function, we will turn our attention to finding the zeros. A real zero of a polynomial function will either be a rational number or an irrational number. The set of possible rational zeros for a polynomial function can be determined from the leading coefficient and the constant term of the polynomial.

**Rational Zeros Theorem**

For any polynomial function $P(x) = a_n x^n + a_{n-1}x^{n-1} + \cdots + a_1 x + a_0$ where $a_n \neq 0$, $a_0 \neq 0$, and all coefficients are integers, the possible rational zeros of $P(x)$ are of the form $\frac{p}{q}$ where $p$ is a factor of $a_0$ and $q$ is a factor of $a_n$.

To develop our intuitive understanding of the **Rational Zeros Theorem**, we will use it to determine the possible rational zeros of the function $P(x) = 2x^2 + x - 3$. By the Rational Zeros Theorem, we know that the possible rational zeros are of the form

$\frac{p}{q}$ where $p$ is a factor of the constant term, $-3$, and $q$ is a factor of the leading coefficient, 2.

$$\text{Factors of } -3\text{: } \pm 1, \pm 3$$
$$\text{Factors of } 2\text{: } \pm 1, \pm 2$$

Now use the factors of $-3$ in the numerator and the factors of 2 in the denominator. For $p = \pm 1$ and $q = \pm 1$, we get $\pm 1$. For $p = \pm 3$ and $q = \pm 1$, we get $\pm 3$. For $p = \pm 1$ and $q = \pm 2$, we get $\pm\frac{1}{2}$. For $p = \pm 3$ and $q = \pm 2$, we get $\pm\frac{3}{2}$. Therefore the list of possible rational zeros is as follows:

$$\text{Possible rational zeros: } \pm 1, \pm 3, \pm\frac{1}{2}, \pm\frac{3}{2}$$

If we find the actual zeros of $P(x) = 2x^2 + x - 3$ using factoring and the zero factor property, we obtain the real zeros $-\frac{3}{2}$ and 1.

$$P(x) = 2x^2 + x - 3$$
$$P(x) = (2x + 3)(x - 1)$$
$$0 = (2x + 3)(x - 1)$$
$$2x + 3 = 0 \qquad \text{or} \qquad x - 1 = 0$$
$$2x = -3 \qquad\qquad x = 1$$
$$x = -\frac{3}{2}$$

Notice the real zeros are among the list of possible rational zeros.

The Rational Zeros Theorem is used to find the zeros of higher degree polynomial functions. In Example 3 we will use it to find the possible zeros of a fifth- degree polynomial function.

**EXAMPLE 3** ▸ Find the possible rational zeros of

$$P(x) = 3x^5 - 4x^4 - 6x^3 + 2x^2 - x + 8.$$

*Solution* The possible rational zeros of $P(x) = 3x^5 - 4x^4 - 6x^3 + 2x^2 - x + 8$ are of the form $\frac{p}{q}$ where $p$ is a factor of 8 and $q$ is a factor of 3.

$$\text{Factors of } 8\text{: } \pm 1, \pm 2, \pm 4, \pm 8$$
$$\text{Factors of } 3\text{: } \pm 1, \pm 3$$

$$\text{Possible rational zeros: } \pm 1, \pm 2, \pm 4, \pm 8, \pm\frac{1}{3}, \pm\frac{2}{3}, \pm\frac{4}{3}, \pm\frac{8}{3}.$$

▸ Now Try Exercise 35

## 4 Use Synthetic Division to Determine the Rational Zeros of a Polynomial Function

Now that we know how to find the possible rational zeros of a polynomial function, we will discuss procedures to find the actual zeros. We will discuss two methods that can be used to find zeros: the **remainder theorem** and **synthetic division**.

In Section 5.3 we gave a brief introduction to the remainder theorem. We will repeat it below.

**Remainder Theorem**

If the polynomial $P(x)$ is divided by $x - a$, the remainder is equal to $P(a)$.

We can use the remainder theorem with the factor theorem that follows to identify zeros of a polynomial function.

**Factor Theorem**

Let $P(x)$ be a polynomial function, then $x - a$ is a factor of $P(x)$ if and only if $P(a) = 0$.

If we use the remainder theorem for a specific value $a$, and obtain a remainder of 0, then that specific value, $a$, is a zero of the polynomial and $(x - a)$ is a factor of the polynomial.

**EXAMPLE 4** ▸ Determine which of the following rational numbers are zeros of the polynomial $P(x) = 3x^3 + 2x^2 - 19x + 6$.

**a)** 2 **b)** $-5$ **c)** $\frac{1}{2}$ **d)** $\frac{1}{3}$

*Solution* To determine if any of the values is a zero of the polynomial, we will evaluate the polynomial at the value. If the value of the polynomial is 0, then the given value is a zero of the polynomial. If it is not 0, then it is not a zero of the polynomial.

**a)**
$$\begin{aligned} P(x) &= 3x^3 + 2x^2 - 19x + 6 \\ P(2) &= 3(2)^3 + 2(2)^2 - 19(2) + 6 \\ &= 3(8) + 2(4) - 38 + 6 \\ &= 24 + 8 - 38 + 6 = 0 \end{aligned}$$

Since $P(2) = 0$, the number 2 is a zero of the polynomial function and $(x - 2)$ is a factor of the polynomial.

**b)**
$$\begin{aligned} P(x) &= 3x^3 + 2x^2 - 19x + 6 \\ P(-5) &= 3(-5)^3 + 2(-5)^2 - 19(-5) + 6 = -224 \end{aligned}$$

Since $P(-5) \neq 0$, $-5$ is not a zero of the polynomial function.

**c)**
$$\begin{aligned} P(x) &= 3x^3 + 2x^2 - 19x + 6 \\ P\left(\frac{1}{2}\right) &= 3\left(\frac{1}{2}\right)^3 + 2\left(\frac{1}{2}\right)^2 - 19\left(\frac{1}{2}\right) + 6 = -\frac{21}{8} \end{aligned}$$

Since $P\left(\frac{1}{2}\right) \neq 0$, $\frac{1}{2}$ is not a zero.

**d)**
$$\begin{aligned} P(x) &= 3x^3 + 2x^2 - 19x + 6 \\ P\left(\frac{1}{3}\right) &= 3\left(\frac{1}{3}\right)^3 + 2\left(\frac{1}{3}\right)^2 - 19\left(\frac{1}{3}\right) + 6 \\ &= 3\left(\frac{1}{27}\right) + 2\left(\frac{1}{9}\right) - \frac{19}{3} + 6 \\ &= \frac{1}{9} + \frac{2}{9} - \frac{57}{9} + \frac{54}{9} = 0 \end{aligned}$$

Since $P\left(\frac{1}{3}\right) = 0$, $\frac{1}{3}$ is a zero of the polynomial function and $\left(x - \frac{1}{3}\right)$ is a factor.

▸ Now Try Exercise 59

Another method to determine if a specific value is a zero of a polynomial uses synthetic division. Synthetic division was discussed in Section 5.3. Please review that material before continuing with this section.

We will do one synthetic division problem now.

**EXAMPLE 5** ▸ Divide $x^3 - 4x^2 - 11x + 30$ by $x + 3$ and determine the remainder.

*Solution* From Section 5.3 we know we write $x + 3$ as $x - (-3)$ and set up and work the synthetic division as follows.

$$\begin{array}{r|rrrr} -3 & 1 & -4 & -11 & 30 \\ & & -3 & 21 & -30 \\ \hline & 1 & -7 & 10 & 0 \end{array} \quad \leftarrow \text{remainder}$$

Since the number below the line to the far right is 0, the remainder when $x^3 - 4x^2 - 11x + 30$ is divided by $x + 3$ is 0, and $x + 3$ is a factor of $x^3 - 4x^2 - 11x + 30$.

▸ **Now Try Exercise 63**

In Example 5, suppose we were given the polynomial function $P(x) = x^3 - 4x^2 - 11x + 30$, rather than the polynomial $x^3 - 4x^2 - 11x + 30$. Because the polynomial divided by $x + 3$ has a remainder of 0, $x + 3$ must be a factor of the polynomial function, $P(x)$, and $-3$ must be a zero of the polynomial function. If we evaluate the polynomial function at $-3$ we obtain

$$\begin{aligned} P(x) &= x^3 - 4x^2 - 11x + 30 \\ P(-3) &= (-3)^3 - 4(-3)^2 - 11(-3) + 30 \\ &= -27 - 36 + 33 + 30 = 0 \end{aligned}$$

Thus, $-3$ is a zero and $(x + 3)$ is a factor of $P(x)$.

**EXAMPLE 6** ▸ Determine the rational zeros of $P(x) = x^4 + 4x^3 - 7x^2 - 34x - 24$.

*Solution* The possible rational zeros of $P(x) = x^4 + 4x^3 - 7x^2 - 34x - 24$ are of the form $\frac{p}{q}$ where $p$ is a factor of $-24$ and $q$ is a factor of 1.

Factors of $-24$: $\pm1, \pm2, \pm3, \pm4, \pm6, \pm8, \pm12, \pm24$

Factors of 1: $\pm1$

Possible rational zeros: $\pm1, \pm2, \pm3, \pm4, \pm6, \pm8, \pm12, \pm24$

We will use Descartes' Rule of Signs to determine the possible number of positive and negative zeros. Since $P(x)$ has only 1 variation in sign, it has 1 positive real zero.

$$\begin{aligned} P(x) &= x^4 + 4x^3 - 7x^2 - 34x - 24 \\ P(-x) &= x^4 - 4x^3 - 7x^2 + 34x - 24 \end{aligned}$$

Since $P(-x)$ has 3 variations in sign, it can have 3 or 1 negative real zeros.

Let us find the 1 positive zero first. We can use either synthetic division or the remainder theorem to find the zeros. In this example we will use synthetic division. We will try the possible rational zeros in increasing order, starting with 1, until we find the positive zero.

Using synthetic division

$$\begin{array}{r|rrrrr} 1 & 1 & 4 & -7 & -34 & -24 \\ & & 1 & 5 & -2 & -36 \\ \hline & 1 & 5 & -2 & -36 & -60 \end{array}$$

Since the remainder is $-60$, 1 is not a zero.

$$\begin{array}{r|rrrrr} 2 & 1 & 4 & -7 & -34 & -24 \\ & & 2 & 12 & 10 & -48 \\ \hline & 1 & 6 & 5 & -24 & -72 \end{array}$$

Since the remainder is $-72$, 2 is not a zero.

$$\begin{array}{r|rrrrr} 3 & 1 & 4 & -7 & -34 & -24 \\ & & 3 & 21 & 42 & 24 \\ \hline & 1 & 7 & 14 & 8 & 0 \end{array}$$

Since the remainder is 0, 3 is a zero of the polynomial function. From Section 5.3 we know that the polynomial can now be written in factored form as follows.

$$\begin{aligned} P(x) &= x^4 + 4x^3 - 7x^2 - 34x - 24 \\ &= (x - 3)(x^3 + 7x^2 + 14x + 8) \end{aligned}$$

Now we will find the negative zeros of $P(x) = x^4 + 4x^3 - 7x^2 - 34x - 24$. To do so we use the fact that the remaining zeros of $P(x)$ must also be zeros of the remaining reduced polynomial $Q(x) = x^3 + 7x^2 + 14x + 8$. Therefore, we work with $x^3 + 7x^2 + 14x + 8$ to find the remaining zeros of $P(x)$.

Now we find the possible rational zeros of $Q(x) = x^3 + 7x^2 + 14x + 8$.

Factors of 8: $\pm1, \pm2, \pm4, \pm8$

Factors of 1: $\pm1$

Possible rational zeros: $\pm1, \pm2, \pm4, \pm8$

We can eliminate 1 and 2 because they were not zeros of $P(x) = x^4 + 4x^3 - 7x^2 - 34x - 24$. In fact, we can eliminate all of the possible positive zeros because by Descartes' rule there is only 1 positive zero, and we have already found it. This leaves the following.

Possible rational zeros: $-1, -2, -4, -8$

Now use synthetic division with $x^3 + 7x^2 + 14x + 8$.

$$\begin{array}{r|rrrr} -1 & 1 & 7 & 14 & 8 \\ & & -1 & -6 & -8 \\ \hline & 1 & 6 & 8 & 0 \end{array}$$

Since the remainder is 0, $-1$ is a zero.

We can now factor the polynomial further as

$$\begin{aligned} P(x) &= (x - 3)(x^3 + 7x^2 + 14x + 8) \\ &= (x - 3)(x + 1)(x^2 + 6x + 8) \\ &= (x - 3)(x + 1)(x + 2)(x + 4) \end{aligned}$$

The function has 4 rational zeros $x = 3$, $x = -1$, $x = -2$, and $x = -4$. Notice there is 1 positive zero and 3 negative zeros, which agrees with what we expected by Descartes' Rule of Signs.

**▸ Now Try Exercise 47**

When analyzing a situation that can be modeled with a polynomial function, it can be useful to identify the zeros of the polynomial function. In Example 7, we use the zeros of a polynomial function to find the production requirement for a business to make a profit.

Businesses often deal with costs, $C(x)$, profits, $P(x)$, and revenue, $R(x)$. In each case, $x$ deals with the number of items manufactured and $C(x)$, $P(x)$, and $R(x)$ are represented in terms of money. Revenue is the money taken in from the selling of the items, cost is the money required to operate the business, and profit is the difference between these two. Thus, $P(x) = R(x) - C(x)$. The company makes a profit when more money is earned than is spent.

**EXAMPLE 7 ▸ Manufacturing Computers** The Big Screen Computer Company manufactures computers. The costs to operate the plant on a daily basis are given by $C(x) = x^2 + 5x + 9$ and the revenue is given by $R(x) = x^3 + 2x^2 - 4x$. Both $C$ and $R$ are expressed in thousands of dollars and $x$ is the number, in hundreds, of computers manufactured each day. Due to constraints at the plant, no more than 900 computers can be manufactured in one day.

**a)** Find $P(x)$.

**b)** Find $C(0)$, $R(0)$, and $P(0)$, and interpret each of these.

**c)** Find the rational zero(s) of $P(x)$.

**d)** How many computers must be manufactured each day to make a profit?

### Solution

**a)** Since profit is the difference between revenue and cost,

$$\begin{aligned} P(x) &= R(x) - C(x) \\ &= (x^3 + 2x^2 - 4x) - (x^2 + 5x + 9) \\ &= x^3 + x^2 - 9x - 9 \end{aligned}$$

**b)**
$$\begin{aligned} C(x) &= x^2 + 5x + 9 \\ C(0) &= 0^2 + 5 \cdot 0 + 9 \\ &= 9 \end{aligned}$$

It costs \$9000 to operate the plant daily even though no computers are being manufactured.

$$\begin{aligned} R(x) &= x^3 + 2x^2 - 4x \\ R(0) &= 0^3 + 2(0)^2 - 4 \cdot 0 \\ &= 0 \end{aligned}$$

There is no revenue when no computers are being manufactured.

$$\begin{aligned} P(x) &= x^3 + x^2 - 9x - 9 \\ P(0) &= 0^3 + 0^2 - 9 \cdot 0 - 9 \\ &= -9 \end{aligned}$$

The company shows a loss of \$9000 each day when no computers are manufactured. This amount is the money needed to operate the plant when no computers are being made.

**c)** The possible rational zeros of $P(x) = x^3 + x^2 - 9x - 9$ are of the form $\frac{p}{q}$ where $p$ is a factor of $-9$ and $q$ is a factor of 1.

Factors of $-9$: $\pm 1, \pm 3, \pm 9$

Factors of 1: $\pm 1$

Possible rational zeros: $\pm 1, \pm 3, \pm 9$

By Descartes' Rule of Signs, we know there is one positive real zero. We will try the possible positive rational zeros in increasing order, starting with 1.

$$\begin{array}{r|rrrr} 1 & 1 & 1 & -9 & -9 \\ & & 1 & 2 & -7 \\ \hline & 1 & 2 & -7 & -16 \end{array}$$

Since the remainder is $-16$, 1 is not a zero.

$$\begin{array}{r|rrrr} 3 & 1 & 1 & -9 & -9 \\ & & 3 & 12 & 9 \\ \hline & 1 & 4 & 3 & 0 \end{array}$$

Since the remainder is 0, 3 is a zero of the polynomial function. The polynomial can now be written in factored form as follows.

$$\begin{aligned} P(x) &= x^3 + x^2 - 9x - 9 \\ &= (x - 3)(x^2 + 4x + 3) \\ &= (x - 3)(x + 3)(x + 1) \end{aligned}$$

The rational zeros are $x = 3$, $x = -3$, and $x = -1$.

**d)** In this real-life problem, the number of computers, $x$, cannot be negative. Therefore we can discard $-3$ and $-1$ as $x$-intercepts of the graph of $P(x)$ for this problem. At $x = 3$, which represents 300 computers, the profit is \$0. This is called the *break-even point*. At the break-even point the revenue and cost are the same, and their difference is \$0.

To decide the number of computers to be made to make a profit let's look at the graph of $P(x)$ for $0 \le x \le 9$ in **Figure 10.24**. (Remember the company cannot manufacture more than 900 computers each day.)

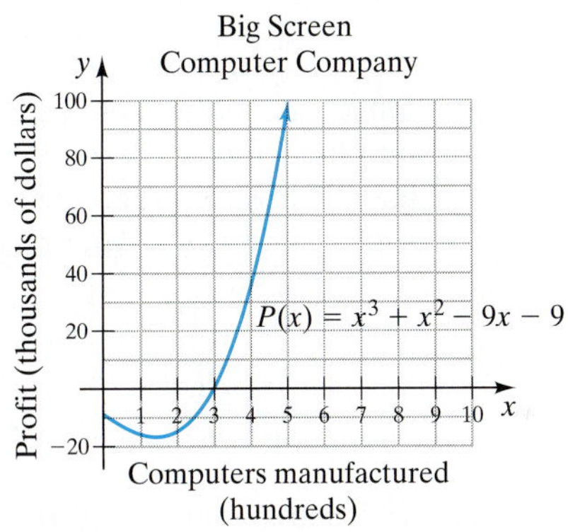


FIGURE 10.24

From the graph we see that profit is negative (representing a loss) when fewer than 300 computers are manufactured. On the other hand, a profit is shown when more than 300 computers are made. The manager should plan to manufacture more than 300 computers each day to make a profit.

▸ **Now Try Exercise 69**

## 5 Find Complex Zeros of a Polynomial Function

Thus far we have discussed only real zeros of polynomial functions. Polynomial functions can also have nonreal zeros, that is, complex numbers that are zeros of the function. There are no $x$-intercepts of the graph of the polynomial function at nonreal zeros. In Example 8 the polynomial function has nonreal zeros.

**EXAMPLE 8** ▸ Find all zeros of $P(x) = x^2 + 4$.

*Solution* We set $P(x) = 0$ and solve for $x$.

$$\begin{aligned} P(x) &= x^2 + 4 \\ 0 &= x^2 + 4 \\ \text{or } x^2 + 4 &= 0 \\ x^2 &= -4 \\ x &= \pm\sqrt{-4} \qquad \textit{Square root property} \\ x &= \pm 2i \end{aligned}$$

Thus the zeros are $2i$ and $-2i$. We can check the zeros in Example 8 using the remainder theorem as follows.

$$\begin{aligned} P(x) &- x^2 + 4 \\ P(2i) &= (2i)^2 + 4 \\ &= 4i^2 + 4 \\ &= 4(-1) + 4 \\ &= 0 \end{aligned} \qquad \begin{aligned} P(x) &= x^2 + 4 \\ P(-2i) &= (-2i)^2 + 4 \\ &= 4i^2 + 4 \\ &= 4(-1) + 4 \\ &= 0 \end{aligned}$$

Since $2i$ and $-2i$ are zeros of $P(x) = x^2 + 4$, $(x - 2i)$ and $(x + 2i)$ are factors of $x^2 + 4$. Thus we can factor $P(x)$ as follows. $P(x) = x^2 + 4 = (x - 2i)(x + 2i)$.

▶ **Now Try Exercise 67**

**EXAMPLE 9** ▶ Determine all zeros of $P(x) = x^5 + 5x^4 + 12x^3 + 24x^2 + 32x + 16$.

*Solution* The possible rational zeros of $P(x) = x^5 + 5x^4 + 12x^3 + 24x^2 + 32x + 16$ are of the form $\frac{p}{q}$ where $p$ is a factor of 16 and $q$ is a factor of 1.

Factors of 16: $\pm1, \pm2, \pm4, \pm8, \pm16$

Factors of 1: $\pm1$

Possible rational zeros: $\pm1, \pm2, \pm4, \pm8, \pm16$

Since the coefficients of $P(x) = x^5 + 5x^4 + 12x^3 + 24x^2 + 32x + 16$ are all positive numbers, the coefficients have no variation in sign. By Descartes' Rule of Signs, $P(x)$ has no positive real zeros. As a result the possible rational zeros can be reduced to $-1, -2, -4, -8, -16$. Let's now find $P(-x)$.

$$\begin{aligned} P(x) &= x^5 + 5x^4 + 12x^3 + 24x^2 + 32x + 16 \\ P(-x) &= (-x)^5 + 5(-x)^4 + 12(-x)^3 + 24(-x)^2 + 32(-x) + 16 \\ \text{or} \quad P(-x) &= -x^5 + 5x^4 - 12x^3 + 24x^2 - 32x + 16 \end{aligned}$$

There are 5 variations in sign in $P(-x)$. Thus there can be 5, 3, or 1 negative real zeros. Let's now try using synthetic division with $x = -1$.

$$\begin{array}{r|rrrrrr} -1 & 1 & 5 & 12 & 24 & 32 & 16 \\ & & -1 & -4 & -8 & -16 & -16 \\ \hline & 1 & 4 & 8 & 16 & 16 & 0 \end{array}$$

The remainder is 0, therefore $-1$ is a zero. Now factor the polynomial.

$$\begin{aligned} P(x) &= x^5 + 5x^4 + 12x^3 + 24x^2 + 32x + 16 \\ &= (x + 1)(x^4 + 4x^3 + 8x^2 + 16x + 16) \end{aligned}$$

We now use synthetic division to try to obtain a zero in the new nonlinear factor $x^4 + 4x^3 + 8x^2 + 16x + 16$. Notice that its leading coefficient, 1, and its constant term, 16, are the same as the original polynomial, so the possible rational zeros are the same. Let's now try $x = -2$.

$$\begin{array}{r|rrrrr} -2 & 1 & 4 & 8 & 16 & 16 \\ & & -2 & -4 & -8 & -16 \\ \hline & 1 & 2 & 4 & 8 & 0 \end{array}$$

Therefore, $-2$ is a zero. Now factor the polynomial further.

$$\begin{aligned} P(x) &= (x + 1)(x^4 + 4x^3 + 8x^2 + 16x + 16) \\ &= (x + 1)(x + 2)(x^3 + 2x^2 + 4x + 8) \end{aligned}$$

We continue to use synthetic division to try to obtain a zero in the new nonlinear factor $x^3 + 2x^2 + 4x + 8$. Notice that its leading coefficient is 1, and its constant term is 8, so $-16$ has been eliminated as a possible rational zero. Thus the possible negative real zeros are $-1, -2, -4$, and $-8$. We found $-1$ to be one real factor. However, it could have a multiplicity of 2, so we will try it again.

$$\begin{array}{r|rrrr} -1 & 1 & 2 & 4 & 8 \\ & & -1 & -1 & -3 \\ \hline & 1 & 1 & 3 & 5 \end{array}$$

Since the remainder is not 0, $-1$ is not a zero of multiplicity of 2. Let's try $-2$ again.

$$\begin{array}{r|rrrr} -2 & 1 & 2 & 4 & 8 \\ & & -2 & 0 & -8 \\ \hline & 1 & 0 & 4 & 0 \end{array}$$

Since the remainder is 0, $-2$ is a zero a second time. Thus, $-2$ is a zero of multiplicity 2. Let's now factor the polynomial further.

$$\begin{aligned} P(x) &= (x + 1)(x + 2)(x^3 + 2x^2 + 4x + 8) \\ &= (x + 1)(x + 2)(x + 2)(x^2 + 4) \end{aligned}$$

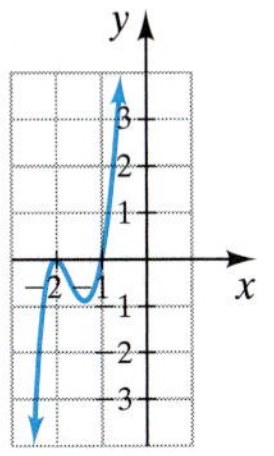

FIGURE 10.25

In Example 8 we found that $x^2 + 4 = (x + 2i)(x - 2i)$. Therefore we can factor the polynomial as follows.

$$P(x) = (x + 1)(x + 2)^2(x + 2i)(x - 2i)$$

The function has 3 real zeros: $-2$ of multiplicity 2 and $-1$, and two complex zeros: $2i$ and $-2i$. Since $-2$ is of multiplicity 2, the $x$-intercept of the graph touches the $x$-axis at $-2$ but does not pass through the $x$-axis. The graph of $P(x) = x^5 + 5x^4 + 12x^3 + 24x^2 + 32x + 16$ is shown in **Figure 10.25**. The $y$-intercept of the graph is at $(0, 16)$.

▶ **Now Try Exercise 79**

**USING YOUR GRAPHING CALCULATOR**

In this section we explained how to find zeros of polynomial functions. On page 177 we explained how the real zeros of a function can be found using a graphing calculator. You should read that material now. In this section, the emphasis is on finding the zeros algebraically. If you wish, you can use your graphing calculator to check your answers when finding zeros, or when graphing polynomial functions.

## EXERCISE SET 10.3

MathXL® MyMathLab

### Concept/Writing Exercises

1. What is the maximum number of $x$-intercepts of the graph of a fourth-degree polynomial function? Explain.
2. What is the maximum number of $x$-intercepts of the graph of a seventh-degree polynomial function? Explain.
3. What is the minimum number of $x$-intercepts of the graph of a sixth-degree polynomial function? Explain.
4. What is the minimum number of $x$-intercepts of the graph of a fifth-degree polynomial function? Explain.
5. If the coefficients of a polynomial function are all positive, how many positive $x$-intercepts does its graph have? Explain.
6. If the coefficients of a polynomial are positive for any term which is of even degree and negative for any term which is of odd degree, how many negative $x$-intercepts does the polynomial have? (Note that a constant term has degree 0.) Explain.

### Practice the Skills

*Determine the maximum and minimum number of x-intercepts for the graph of each polynomial function. Do not graph the function.*

7. $f(x) = x^2 + 8x + 5$
8. $f(x) = 3x^2 - 2x - 5$
9. $g(x) = 2x^3 + x^2 - 5x - 3$
10. $g(x) = -2x^3 - x^2 + 7x - 15$
11. $r(x) = 5x^4 - x^3 + 4x^2 + 2x + 17$
12. $r(x) = 5x^4 - x^3 - 7x^2 + 21x - 17$

13. $m(x) = 7x^5 - 2x^4 - x^3 + 9x - 10$

14. $m(x) = -3x^5 + x^4 - \frac{1}{2}x^3 + x - 11$

15. $r(x) = 8x^6 + 7x^5 - x^4 + 3x^3 - 2x^2 + 11x + 5$

16. $r(x) = -2x^6 + x^5 - 3x^4 + 13x^3 - 2x^2 - 7x + 8$

*Use Descartes' Rule of Signs to determine the possible number of positive and negative zeros of each polynomial function.*

17. $f(x) = x^3 + 2x^2 + 4x + 7$

18. $f(x) = x^3 - 3x^2 + 5x + 11$

19. $f(x) = -6x^3 + 2x^2 + 3x - 1$

20. $g(x) = 2x^4 + 7x^3 - 2x^2 - 6x + 5$

21. $g(x) = 5x^4 - 2x^3 + x^2 - 9x - 14$

22. $g(x) = 7x^4 - 3x^3 + x^2 - x + 17$

23. $g(x) = -3x^4 - x^3 - 6x^2 - 2x - 3$

24. $g(x) = 17x^4 + 7x^3 - 2x^2 - x - 12$

25. $r(x) = x^5 - x^4 + x^3 + 2x^2 + 7x - 3$

26. $r(x) = x^5 + x^4 - 6x^3 - 15x^2 + 12x + 16$

27. $r(x) = 7x^5 + 2x^4 + x^3 + 5x^2 + 9x + 20$

28. $r(x) = -7x^5 - 8x^4 - 4x^3 + 6x^2 + x + 19$

*List the possible rational zeros of each function.*

29. $f(x) = x^2 + 3x + 2$

30. $f(x) = x^2 - 5x + 6$

31. $f(x) = 3x^2 + 4x + 3$

32. $g(x) = x^3 - x^2 + 3x - 5$

33. $g(x) = x^3 - 6x^2 + 3x + 12$

34. $g(x) = 4x^3 - x^2 + 9x + 5$

35. $g(x) = 3x^3 - 7x^2 + 17x - 4$

36. $h(x) = 5x^4 - x^3 + 7x^2 - 19x + 6$

37. $h(x) = 6x^4 + 5x^3 - x^2 + 9x + 4$

38. $h(x) = 2x^4 - 4x^3 + x + 5$

39. $r(x) = 6x^5 + x^4 - x^2 + 11x + 3$

40. $r(x) = 8x^5 - 7x^4 + 3x^2 - x + 1$

41. $r(x) = 7x^5 - x^4 + 15x^3 - 18x^2 + 4$

42. $t(x) = 9x^5 + 23x^4 - 7x^2 + 3$

*Find all the rational zeros of each function.*

43. $f(x) = x^2 - 6x + 8$

44. $f(x) = x^2 + 3x - 18$

45. $f(x) = 3x^2 + 14x - 5$

46. $f(x) = 6x^2 + 5x + 1$

47. $g(x) = x^3 + 7x^2 - x - 7$

48. $g(x) = x^3 + 2x^2 - 5x - 6$

49. $g(x) = 2x^3 + x^2 - 8x - 4$

50. $g(x) = x^3 - 5x^2 + 9x - 45$

51. $r(x) = x^4 - 12x^2 - 64$

52. $r(x) = x^4 - 13x^2 + 36$

53. $r(x) = x^4 - 8x^2 + 16$

54. $r(x) = 6x^4 - x^3 + 5x^2 - x - 1$

55. $r(x) = x^4 + x^3 - 7x^2 - 13x - 6$

56. $r(x) = 4x^4 - 8x^3 - 7x^2 + 17x - 6$

57. $r(x) = x^4 - 8x^3 + 22x^2 - 24x + 9$

58. $t(x) = x^5 + 7x^4 - 5x^3 - 35x^2 + 4x + 28$

*Determine if the given values are zeros of the polynomial function.*

59. $f(x) = x^3 + x^2 - 10x + 8$
 **a)** 2 **b)** $-4$

60. $f(x) = x^3 + 8x^2 - 4x - 32$
 **a)** $-8$ **b)** 1

61. $g(x) = x^4 - 3x^3 - 3x^2 + 11x - 6$
 **a)** $-6$ **b)** $-1$

62. $g(x) = 6x^3 + 5x^2 - 2x - 1$
 **a)** $\frac{1}{2}$ **b)** $\frac{1}{3}$

63. Determine the remainder when $x^3 + 2x^2 - 5x + 11$ is divided by $x - 1$.

64. Determine the remainder when $x^4 + 6x^3 - 13x^2 - 7x + 17$ is divided by $x - 2$.

65. Determine the remainder when $2x^4 + 5x^3 - 4x^2 - 15x + 8$ is divided by $x + 3$.

66. Determine the remainder when $6x^5 + 7x^4 - 3x^3 - 4x^2 + 8x - 14$ is divided by $x + 1$.

## Problem Solving

*Find all the zeros (real and complex) of each function.*

67. $f(x) = x^2 + 1$

68. $x^2 + 49$

69. $f(x) = x^4 - 3x^2 - 4$

70. $g(x) = x^4 + 24x - 25$

71. $h(x) = x^4 + x^3 - x - 1$

72. $f(x) = x^4 - x^3 + x - 1$

73. $g(x) = x^4 + 3x^3 - x^2 - 12x - 12$

74. $g(x) = x^4 - 3x^3 - 5x^2 + 27x - 36$

75. $f(x) = x^5 - x^4 + 2x^3 - 2x^2 + x - 1$

76. $g(x) = x^5 + x^4 + x^3 + x^2 - 2x - 2$

*For Exercises 77–81 you are given R(x) and C(x) where x represents the number of items produced in hundreds. R(x) and C(x) represent revenue and cost in thousands of dollars. You may want to review Example 7 at this time.*

**a)** *Find P(x).*

**b)** *Find $P(0)$, $R(0)$, and $C(0)$ and interpret each of these.*

**c)** *Find the break-even point (value where $P(x) = 0$) in the given domain.*

**d)** *For what values of x in the given domain is $P(x) > 0$?*

**e)** *For what values of x in the given domain is $P(x) < 0$?*

**f)** *Interpret the results from parts* **d)** *and* **e)**.

77. $R(x) = x^3 + 5x^2 + x, C(x) = x^2 + 6,\ 0 \leq x \leq 3$

78. $R(x) = x^3 + 3x^2 - 5x, C(x) = x^2 + 4x + 18,\ 0 \leq x \leq 7$

79. $R(x) = x^3 + x^2 + x, C(x) = 3x^2 + 14x + 10,\ 0 \leq x \leq 8$

80. $R(x) = x^3 + 3x, C(x) = x^2 + 13x + 8,\ 0 \leq x \leq 5$

81. $R(x) = 2x^3 + x^2 + 2x, C(x) = 2x^2 + 7x + 2,\ 0 \leq x \leq 5$

## Group Activity

**82.** In this section we have been finding the rational and complex zeros of polynomial functions. Now you will work with your group to create polynomial functions with certain characteristics. For each description, do the following:

**a)** Write your polynomial function in factored form.

**b)** Multiply the factors of your polynomial function.

**c)** Determine if this is the only polynomial function having these characteristics. If it is not the only one with these characteristics, have every group member find another polynomial with the given characteristics.

**1.** Create a polynomial function with the rational zeros: $-4, -2, 0, 3$.

**2.** Create a polynomial function with the zeros: $-5$ of multiplicity 2, 6.

**3.** Create a polynomial function with the zeros: $-2$ of multiplicity 3, 1, 3.

## Cumulative Review Exercises

[7.2] **83.** Simplify $\left(\dfrac{y^{-2/3}}{y^{-6}}\right)^{1/2}$.

[7.3] **84.** Simplify $3\sqrt{72x^3y}$.

[7.5] **85.** Simplify $\dfrac{5}{\sqrt{5} - 10}$.

[7.6] **86.** Solve $\sqrt{5x + 6} = x$.

## Mid-Chapter Test: 10.1–10.3

*To find out how well you understand the chapter material to this point, take this brief test. The answers, and the section where the material was initially discussed, are given in the back of the book. Review any questions that you answered incorrectly.*

*Classify each function as continuous or discontinuous. If the function is not continuous, identify the point(s) of discontinuity.*

**1.** $f(x) = 3x^2 + 5x - 9$

**2.** $g(x) = \dfrac{1}{x^2 + 9}$

**3.** $h(x) = \dfrac{7}{x^2 - 9}$

*Graph each piecewise function.*

**4.** $m(x) = \begin{cases} 2x & \text{if } x \le -1 \\ 3x & \text{if } x > -1 \end{cases}$

**5.** $t(x) = \begin{cases} -x - 3 & \text{if } x < -3 \\ 2 & \text{if } x \ge -3 \end{cases}$

*Graph each greatest integer function.*

**6.** $n(x) = [\![2x]\!]$

**7.** $q(x) = \left[\!\!\left[\dfrac{x+1}{3}\right]\!\!\right]$

*Describe the location, either above the x-axis or below the x-axis, of the graph of each function for* **a)** *large positive values of x, and* **b)** *large negative values of x.*

**8.** $f(x) = x^4 - 2x^3 + 10x^2 + x - 9$

**9.** $f(x) = -3x^5 + x^4 + 2x^3 - 7x + 6$

*Find the real zeros of each function and state their multiplicities if it is a number other than 1.*

**10.** $f(x) = x^3 + 2x^2 - 24$

**11.** $h(x) = (x + 5)(x + 1)^2(x - 3)^4$

*Determine the maximum possible number of turning points for the graph of each function.*

**12.** $g(x) = 2x^3 - x^2 + 7x - 19$

**13.** $k(x) = (x - 5)(x - 3)(x^2 + 1)(x^2 - 10)$

*Graph each function.*

**14.** $f(x) = x(x + 3)(x - 1)$

**15.** $g(x) = x^2(x + 4)(x - 2)$

**16.** Determine the maximum and minimum number of $x$-intercepts for the graph of $f(x) = 5x^3 + 2x^2 - 6x - 7$.

**17.** Use Descartes' Rule of Signs to determine the possible number of positive and negative zeros of $t(x) = x^5 + 3x^4 - x^3 - 12x^2 + 5x + 11$.

**18.** List the possible rational zeros of $q(x) = 3x^4 - 8x^3 + 7x^2 - x + 4$.

*Find all the zeros (real and complex) of each function.*

**19.** $f(x) = x^4 - x^3 - 19x^2 - x - 20$

**20.** $g(x) = 10x^3 - 29x^2 - 5x + 6$

# 10.4 Finding the Real Zeros of a Polynomial Function

**1** Locate and determine the real zeros of a polynomial function.

**2** Find upper and lower bounds for the real zeros of a polynomial function.

**3** Graph a polynomial function using its key features.

**4** Estimate irrational zeros.

## 1 Locate and Determine the Real Zeros of a Polynomial Function

Since polynomial functions are continuous, they have no gaps or holes in their graphs. If we evaluate a function at two points $a$ and $b$, we know that the function has a value at every number between $f(a)$ and $f(b)$. For instance, if $a < b$, and $f(a) = 1$, and $f(b) = 2$, we know that there exists $c$ between $a$ and $b$ such that $f(c) = \sqrt{2}$ because $\sqrt{2}$ is between 1 and 2 (see **Figure 10.26**).

Using this concept, we know that a polynomial function $f$ has a real zero between any two numbers $a$ and $b$ for which the sign of $f(a)$ is opposite the sign of $f(b)$, see **Figure 10.27**.

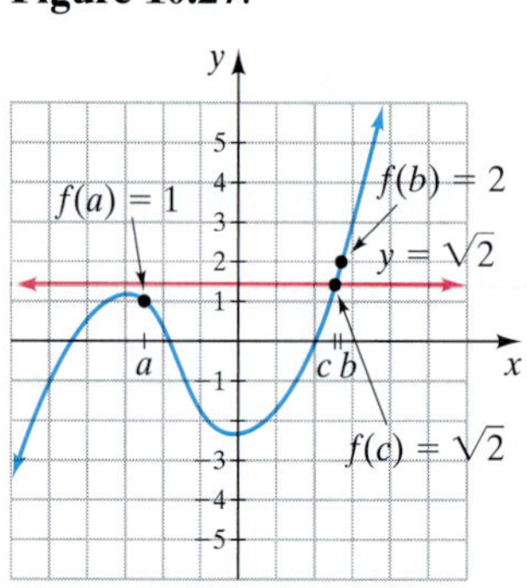


FIGURE 10.26

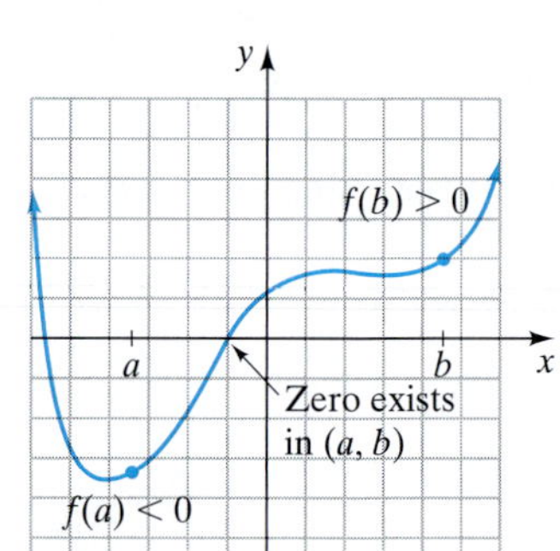


FIGURE 10.27

This fact is known as the Intermediate Value Theorem.

**Intermediate Value Theorem**

For a continuous function $f(x)$ defined over the interval $[a, b]$, if the sign of $f(a)$ is the opposite of the sign of $f(b)$, then a real zero exists in $(a, b)$.

The Intermediate Value Theorem is used to locate intervals where real zeros occur.

Suppose we check the possible rational zeros by evaluating the function at each possible rational zero, in order, from smallest possible rational zero to largest possible rational zero. If we find that the sign of the value of the function changes from one possible rational zero to the next, we know that an irrational zero exists between those 2 possible rational zeros.

**EXAMPLE 1** ▸ In which of the following intervals does the polynomial function, $f(x) = \frac{1}{2}x^4 - x^3 - 5$, have a real zero?

**a)** $(-2, -1)$ **b)** $(-1, 2)$

*Solution* A polynomial function will have a real zero in an interval if the sign of the function at one endpoint of the interval is opposite the sign of the function at the other endpoint of the interval.

**a)** We will substitute each endpoint in the function, and then evaluate the function to determine if there is a change of sign of $f(x)$.

$$\begin{aligned} f(x) &= \frac{1}{2}x^4 - x^3 - 5 \\ f(-2) &= \frac{1}{2}(-2)^4 - (-2)^3 - 5 \\ &= \frac{1}{2}(16) - (-8) - 5 \\ &= 8 + 8 - 5 \\ &= 11 \end{aligned}$$

$$\begin{aligned} f(x) &= \frac{1}{2}x^4 - x^3 - 5 \\ f(-1) &= \frac{1}{2}(-1)^4 - (-1)^3 - 5 \\ &= \frac{1}{2}(1) - (-1) - 5 \\ &= 0.5 + 1 - 5 \\ &= -3.5 \end{aligned}$$

Since the function changes sign in the interval $(-2, -1)$, it has at least one real zero in this interval.

**b)** From part **a)** we know that $f(-1) = -3.5$. We will now evaluate $f(2)$.

$$\begin{aligned} f(x) &= \frac{1}{2}x^4 - x^3 - 5 \\ f(2) &= \frac{1}{2}(2)^4 - (2)^3 - 5 \\ &= \frac{1}{2}(16) - 8 - 5 \\ &= 8 - 8 - 5 \\ &= -5 \end{aligned}$$

The sign of the function at $-1$ is the same as the sign of the function at 2, both are negative. From this information, we cannot determine whether a real zero exists in this interval.

▸ **Now Try Exercise 9**

It is important to recognize that the Intermediate Value Theorem does not imply that there is no zero in $(a, b)$, when $f(a)$ is the same sign as $f(b)$. As illustrated in **Figure 10.28a**, the function will have the same sign at both endpoints of the interval when the interval contains a zero of even multiplicity. The function may also have the same sign at both endpoints of the interval when the function has more than one zero in the interval (see **Figure 10.28b**).

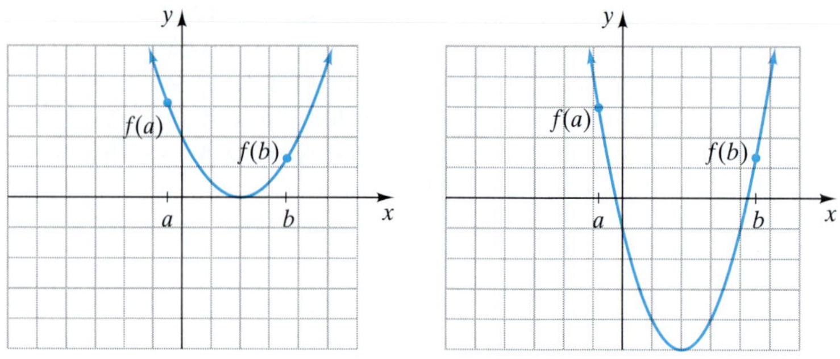


FIGURE 10.28 (a) (b)

## 2 Find Upper and Lower Bounds for the Real Zeros of a Polynomial Function

We have strategies for determining the possible number of positive and negative real zeros, for identifying all possible rational zeros, and for identifying intervals where a real zero occurs. We will now learn a strategy for recognizing a region over which all of the real zeros occur. We can identify such a region by finding an upper and lower bound for the zeros using the Boundedness Theorem.

### Boundedness Theorem

Let $P(x)$ be a polynomial function with real coefficients and a positive leading coefficient.

**a)** If $P(x)$ is synthetically divided by $x - b$, where $b > 0$ and all the numbers in the bottom row of the synthetic division are nonnegative, then $P(x)$ has no real zero greater than $b$. We call $b$ an **upper bound** for the real zeros of $P(x)$.

**b)** If $P(x)$ is synthetically divided by $x - c$ where $c < 0$ and all of the numbers in the bottom row of the synthetic division alternate in sign (with 0 considered positive or negative as explained below), then $P(x)$ has no real zero less than $c$. We call $c$ a **lower bound** for the real zeros of $P(x)$.

In part **b)** where a 0 occurs in the bottom row of numbers, treat it as a negative number when it follows a positive number, and as a positive number when it follows a negative number. For example, if the bottom row of numbers is $2\ -4\ 0\ 6\ -2$ we treat the 0 as a positive number since it follows $-4$. Thus we get $+ - + + -$. Since there are two pluses in a row we do not consider all the numbers alternating in sign. Now consider $4\ -3\ 0\ -2\ 1$. Since the 0 follows $-3$ we treat it as a positive number. Thus we get $+ - + - +$. Since these signs alternate, the numbers $4\ -3\ 0\ -2\ 1$ are considered alternating in sign.

If $a$ is a lower bound and $b$ is an upper bound for the zeros of $P(x)$, we say that the zeros of $P(x)$ are *bounded over* the interval $[a, b]$.

We use the Boundedness Theorem after we have completed a synthetic division to reduce the number of rational zeros which need to be checked.

**EXAMPLE 2** ▸ Determine if the real zeros of $f(x) = x^4 - 5x^3 + 7x^2 + 3x - 10$ are all included in the interval, or are bounded over the interval $[-2, 6]$.

*Solution* Use synthetic division with $x = -2$ for the possible lower bound and then use synthetic division with $x = 6$, the possible upper bound.

$$\begin{array}{r|rrrrr} -2 & 1 & -5 & 7 & 3 & -10 \\ & & -2 & 14 & -42 & 78 \\ \hline & 1 & -7 & 21 & -39 & 68 \end{array}$$

Since the bottom row of numbers in the synthetic division has alternating signs, $-2$ is a lower bound for the real zeros of the function. Therefore, there can be no real zero that is less than $-2$. Now use synthetic division with $x = 6$.

$$\begin{array}{r|rrrrr} 6 & 1 & -5 & 7 & 3 & -10 \\ & & 6 & 6 & 78 & 486 \\ \hline & 1 & 1 & 13 & 81 & 476 \end{array}$$

Since the bottom row of numbers in the synthetic division are all greater than or equal to zero, 6 is an upper bound for the real zeros of the function. Therefore, there can be no real zero that is greater than 6. Thus, all real zeros of this function are in the interval $[-2, 6]$.

▸ **Now Try Exercise 25**

The Boundedness Theorem applies when there is a positive leading coefficient. The real zeros of $f(x)$ and $-f(x)$ will be the same since the graphs of $f(x)$ and $-f(x)$ are symmetric with regard to the $x$-axis. Therefore, if the leading coefficient of a polynomial function is negative, multiply both sides of the equation by $-1$, and proceed as in Example 2.

## 3 Graph a Polynomial Function Using Its Key Features

With graphing calculators widely available, you may question the need to be skilled at graphing polynomial functions. The process of graphing polynomial functions using the theorems presented in this chapter gives you the opportunity to develop your reasoning skills to understand the basic characteristics of polynomial functions.

For the graph of a polynomial function to be complete, it needs an accurate $y$-intercept, accurate rational $x$-intercepts, reasonable estimates for irrational $x$-intercepts, an accurate data point in each interval defined by the $x$-intercepts, and a smooth curve connecting these points. Although the graph can be made more accurate by plotting more points in each region, the basic sketch of the graph will be complete.

**EXAMPLE 3** ▸ Consider $f(x) = x^3 + 3x^2 - x - 3$.

**a)** Determine the $y$-intercept of the graph.

**b)** Determine the possible number of positive zeros and the possible number of negative zeros of $f(x)$.

**c)** Determine the possible rational zeros of $f(x)$.

**d)** Determine all real zeros of $f(x)$.

**e)** Graph the function.

*Solution*

**a)** To find the $y$-intercept, substitute 0 for $x$.

$$f(x) = x^3 + 3x^2 - x - 3$$
$$f(0) = 0 + 0 - 0 - 3 = -3$$

The $y$-intercept is $(0, -3)$.

**b)** We find the possible number of positive and negative zeros using Descartes' Rule of Signs.

$$f(x) = x^3 + 3x^2 - x - 3 \quad \text{Has one variation in sign.}$$
$$f(-x) = -x^3 + 3x^2 + x - 3 \quad \text{Has two variations in sign.}$$

Since the number of variations in sign in $f(x)$ is 1, there is 1 positive real zero. Since the number of variations in sign in $f(-x)$ is 2, the number of real negative zeros is either 2 or zero.

**c)** The possible rational zeros are of the form $\frac{p}{q}$ where $p$ is a factor of $-3$ and $q$ is a factor of 1.

Factors of $-3$: $\pm1, \pm3$

Factors of 1: $\pm1$

Possible rational zeros: $\pm1, \pm3$

**d)** Since we definitely have 1 positive real zero we will look for that one first. We will use synthetic division with $x = 1$.

$$\begin{array}{r|rrrr} 1 & 1 & 3 & -1 & -3 \\ & & 1 & 4 & 3 \\ \hline & 1 & 4 & 3 & 0 \end{array}$$

Since the remainder is 0, 1 is a zero of $f(x)$ and $(x - 1)$ is a factor of $f(x)$. We have found the one positive zero. Notice the bottom row of the synthetic division contains all positive numbers.

This tells us that 1 is an upper bound. We can now factor $f(x)$ as follows.

$$f(x) = x^3 + 3x^2 - x - 3$$
$$f(x) = (x - 1)(x^2 + 4x + 3)$$
$$= (x - 1)(x + 3)(x + 1)$$

The other zeros, $-3$ and $-1$ can now be obtained using the zero factor property. The 3 zeros, from smallest to largest, are $-3$, $-1$, and 1.

**e)** We will set up a table to help in graphing the polynomial. From the 3 real zeros we get 3 $x$-intercepts on the graph, which break the graph into 4 intervals.

| Interval | Test value | Value of $f(x)$ |
|---|---|---|
| $(-\infty, -3)$ | $-4$ | $-15$ |
| $(-3, -1)$ | $-2$ | 3 |
| $(-1, 1)$ | 0 | $-3$ |
| $(1, \infty)$ | 2 | 15 |

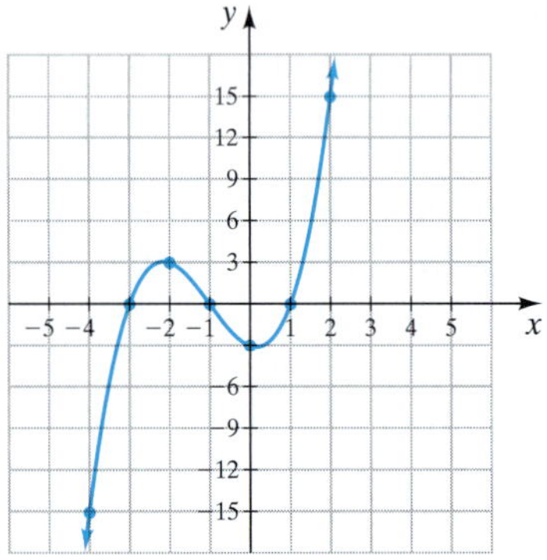

$f(x) = x^3 + 3x^2 - x - 3$

FIGURE 10.29

Now draw the graph, see **Figure 10.29**.

Notice the end behavior is the same end behavior as $f(x) = x^3$, which is what we expected.

▸ **Now Try Exercise 37**

**EXAMPLE 4** ▸ Graph $f(x) = x^4 - x^3 - 5x^2 - x + 4$.

*Solution* We will follow the same basic procedure we used in Example 3. By substituting 0 for $x$ we find that $(0, 4)$ is the $y$-intercept.

$$f(x) = x^4 - x^3 - 5x^2 - x + 4 \quad \text{Has 2 variations in sign.}$$
$$f(-x) = x^4 + x^3 - 5x^2 + x + 4 \quad \text{Has 2 variations in sign.}$$

Therefore the number of positive real zeros is either 2 or 0 and the number of negative real zeros is either 2 or 0.

Possible factors of 4: $\pm 1, \pm 2, \pm 4$

Possible factors of 1: $\pm 1$

Possible rational zeros: $\pm 1, \pm 2, \pm 4$

Let us find the positive zeros first. We will start with 1.

$$\begin{array}{r|rrrrr} 1 & 1 & -1 & -5 & -1 & 4 \\ & & 1 & 0 & -5 & -6 \\ \hline & 1 & 0 & -5 & -6 & -2 \end{array}$$

Therefore 1 is not a zero. Notice $f(1) = -2$. Also note that $f(0) = 4$. Therefore there must be a real positive irrational zero between $x = 0$ and $x = 1$. We now know that there is 1 real positive zero. By Descartes' Rule, we know that there must be a second real positive zero. Let's continue to find it. Let's try synthetic division with $x = 2$.

$$\begin{array}{r|rrrrr} 2 & 1 & -1 & -5 & -1 & 4 \\ & & 2 & 2 & -6 & -14 \\ \hline & 1 & 1 & -3 & -7 & -10 \end{array}$$

Therefore, 2 is not a zero. Let's try synthetic division with $x = 4$.

$$\begin{array}{r|rrrrr} 4 & 1 & -1 & -5 & -1 & 4 \\ & & 4 & 12 & 28 & 108 \\ \hline & 1 & 3 & 7 & 27 & 112 \end{array}$$

Therefore 4 is not a zero. Since $f(2) = -10$ and $f(4) = 112$, there must be a real irrational zero between $x = 2$ and $x = 4$. If we evaluate the function at 3, we find $f(3) = 10$. Therefore the positive irrational zero is in the interval (2, 3). We have determined there are 2 positive irrational zeros. Since we know we have only 2 positive real zeros, we will now look for the negative zeros. Let's use synthetic division with $x = -1$.

$$\begin{array}{r|rrrrr} -1 & 1 & -1 & -5 & -1 & 4 \\ & & -1 & 2 & 3 & -2 \\ \hline & 1 & -2 & -3 & 2 & 2 \end{array}$$

Therefore, $-1$ is not a zero. Let's try $x = -2$.

$$\begin{array}{r|rrrrr} -2 & 1 & -1 & -5 & -1 & 4 \\ & & -2 & 6 & -2 & 6 \\ \hline & 1 & -3 & 1 & -3 & 10 \end{array}$$

Therefore, $-2$ is not a zero. Since the signs in the bottom row of the synthetic division alternate in sign, there can be no real zero less than $-2$. Therefore we can eliminate $-4$ as a possibility. If the function has negative zeros, it has 2 which occur in the interval $(-2, 0)$ because $-2$ is a lower bound and 0 is an upper bound. From the $y$-intercept and the synthetic divisions we have already determined that the following points are on $f(x)$:

$$(0, 4), (1, -2), (2, -10), (4, 112), (-1, 2), (-2, 10).$$

Let us use these points and the point (3, 10) to sketch the graph. We know the end behavior of $f(x) = x^4$. Therefore the graph must be positive both for large positive values and large negative values of $x$. The graph is sketched in **Figure 10.30**. Note that there is an irrational zero between $x = 0$ and $x = 1$ and an irrational zero between $x = 2$ and $x = 3$, and that there are no negative real zeros. Next, we will learn how to estimate the irrational zeros of a function.

▶ **Now Try Exercise 49**

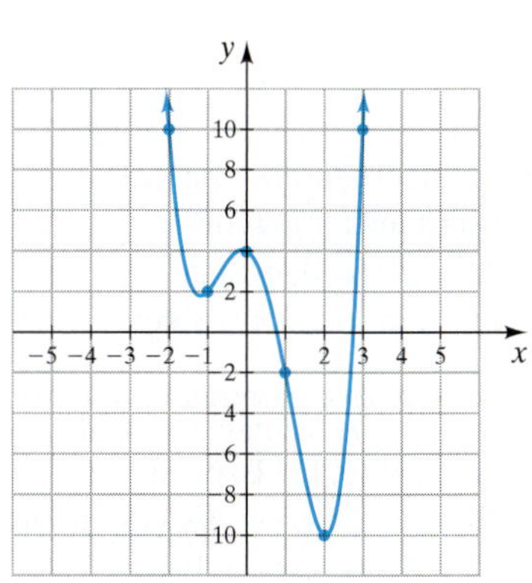


$f(x) = x^4 - x^3 - 5x^2 - x + 4$

FIGURE 10.30

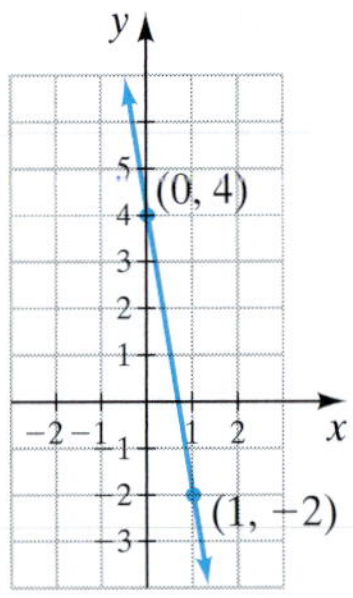


FIGURE 10.31

## 4 Estimate Irrational Zeros

In Example 4, we determined that one real zero of $f(x) = x^4 - x^3 - 5x^2 - x + 4$ was between $x = 0$ and $x = 1$. We can get close approximations to the actual irrational zero by evaluating the function for various values in the interval. If there is a change of sign in $f(x)$ between any 2 values of $x$, then there is a real zero between those values of $x$; we found in Example 4 that $f(0) = 4$ and $f(1) = -2$. Let's consider the graph between these 2 points, see **Figure 10.31**.

If the curve was straight, it appears that the $x$-intercept is about 0.7. Let's evaluate $f(x)$ for selected values of $x$ and determine the values of $f(x)$. The table below shows some values.

| $x$ | $f(x) = x^4 - x^3 - 5x^2 - x + 4$ |
|---|---|
| 0.6 | $f(0.6) = 0.6^4 - 0.6^3 - 5(0.6)^2 - 0.6 + 4 = 1.5136$ |
| 0.7 | $f(0.7) = 0.7^4 - 0.7^3 - 5(0.7)^2 - 0.7 + 4 = 0.7471$ |
| 0.8 | $f(0.8) = 0.8^4 - 0.8^3 - 5(0.8)^2 - 0.8 + 4 = -0.1024$ |

There is a change of sign in $f(x)$ between $x = 0.7$ and $x = 0.8$. Therefore the zero is between 0.7 and 0.8. If we desire more accuracy we could evaluate $f(x)$ for values between 0.7 and 0.8 and find where there is a change of sign in $f(x)$. If you evaluate additional values you will find that

$$f(0.77) \approx 0.1605$$
$$f(0.78) \approx 0.0736$$
$$f(0.79) \approx -0.0140.$$

Thus the irrational zero is between 0.78 and 0.79. Since $f(0.785) \approx 0.0299$ the irrational root is between 0.785 and 0.790. Thus the irrational root, estimated to the nearest hundredth, is 0.79. If you were to determine the irrational zero of $f(x) = x^4 - x^3 - 5x^2 - x + 4$ between $x = 2$ and $x = 3$, you would find that

$$f(2.75) \approx -0.1680$$
$$f(2.755) \approx -0.00716$$
$$f(2.76) \approx 0.1553.$$

Thus, the irrational zero between $x = 2$ and $x = 3$ is 2.76 to the nearest hundredth.

## EXERCISE SET 10.4

MathXL® MyMathLab

### Concept/Writing Exercises

1. Let $f(x) = x^3 - 2x^2 - 5x + 2$.
   **a)** Is $x = 5$ an upper bound for the real zeros?
   **b)** Is $x = 4$ an upper bound for the real zeros?
   **c)** Is $x = 3$ an upper bound for the real zeros?
   Explain your answers.

2. Let $f(x) = x^3 - 2x^2 - 5x + 2$.
   **a)** Is $x = -3$ a lower bound for the real zeros?
   **b)** Is $x = -2$ a lower bound for the real zeros?
   **c)** Is $x = -1$ a lower bound for the real zeros?
   Explain your answers.

3. Let $f(x) = x^3 - 2x^2 - 5x + 2$.
   **a)** Using the Intermediate Value Theorem, can you determine whether $f(x)$ has a zero in $(1, 2)$? Explain.
   **b)** Using the Intermediate Value Theorem, can you determine whether $f(x)$ has a zero in $(-2, -1)$? Explain.

4. Let $g(x) = x^4 - x^3 + 7x - 17$.
   **a)** Using the Intermediate Value Theorem, can you determine whether $g(x)$ has a zero in $(0, 1)$? Explain.
   **b)** Using the Intermediate Value Theorem, can you determine whether $g(x)$ has a zero in $(1, 2)$? Explain.
   **c)** Is $x = 3$ an upper bound for the real zeros of $g(x)$?

## Practice the Skills

*Use the Intermediate Value Theorem to determine if a real zero of $f(x)$ occurs in the given interval. If this cannot be determined, indicate so.*

**5.** $f(x) = x^2 - 5x + 1$ in $(2, 3)$

**6.** $f(x) = x^2 - 5x + 2$ in $(4, 5)$

**7.** $f(x) = 3x^2 - 2x - 4$ in $(1, 2)$

**8.** $f(x) = x^3 - 7x - 4$ in $(-4, -3)$

**9.** $f(x) = x^3 - 7x - 2$ in $(-3, -2)$

**10.** $f(x) = x^3 - 2x^2 - 5x + 2$ in $(-3, -2)$

**11.** $f(x) = x^3 - 2x^2 - 5x + 2$ in $(-2, -1)$

**12.** $f(x) = x^4 - x^3 + 7x - 17$ in $(2, 3)$

**13.** $f(x) = x^4 - x^3 + 7x - 15$ in $(1, 3)$

**14.** $f(x) = x^4 - x^3 + 7x - 17$ in $(-3, -2)$

**15.** $f(x) = x^4 - x^3 + 7x - 17$ in $(-2, -1)$

**16.** $f(x) = 2x^4 - 3x^3 + 8x^2 - 9x - 7$ in $(-2, -1)$

**17.** $f(x) = 2x^4 - 3x^3 + 8x^2 - 9x - 12$ in $(-1, 0)$

**18.** $f(x) = x^4 + 5x^3 - 12x^2 - 13x + 5$ in $(2, 3)$

**19.** $f(x) = x^5 - 2x^4 + 6x^3 - x^2 - 17$ in $(0, 1)$

**20.** $f(x) = x^5 - 2x^4 + 7x^3 - x^2 - 12$ in $(1, 2)$

*Use the Boundedness Theorem to determine if the zeros of $f(x)$ are bounded over the given interval. If this cannot be determined, indicate so.*

**21.** $f(x) = x^2 - 6x + 3$ in $[-1, 6]$

**22.** $f(x) = 3x^2 - x - 5$ in $[-1, 6]$

**23.** $f(x) = 5x^2 - 2x + 7$ in $[-2, 2]$

**24.** $g(x) = x^3 - 2x^2 - 5x + 11$ in $[-3, 2]$

**25.** $g(x) = x^3 + 2x^2 - 5x - 6$ in $[-2, 4]$

**26.** $g(x) = 2x^3 + x^2 - 8x + 7$ in $[-3, 5]$

**27.** $g(x) = x^3 - 5x^2 + 9x - 53$ in $[-5, 6]$

**28.** $g(x) = 2x^3 - 7x^2 + 6x + 8$ in $[-1, 1]$

**29.** $h(x) = x^4 + x^3 - 7x^2 - 13x - 6$ in $[-2, 2]$

**30.** $h(x) = x^4 + x^3 - 7x^2 - 13x - 6$ in $[-2, 5]$

**31.** $h(x) = 6x^4 - x^3 + 5x^2 - x + 10$ in $[-1, 2]$

**32.** $h(x) = 2x^4 - 3x^3 + 6x^2 - 7x - 15$ in $[-3, 1]$

**33.** $h(x) = x^4 + 5x^3 - 12x^2 - 13x + 2$ in $[-5, 3]$

**34.** $t(x) = x^5 - 2x^4 + 7x^3 - x^2 + 2x + 1$ in $[-2, 3]$

*Graph each polynomial function using the procedure in Examples 3 and 4.*

**35.** $f(x) = x^2 - x - 20$

**36.** $h(x) = x^2 + 2x - 8$

**37.** $g(x) = x^3 + 2x^2 - 5x - 6$

**38.** $g(x) = x^3 + 7x^2 - x - 7$

**39.** $g(x) = 2x^3 + x^2 - 8x - 4$

**40.** $g(x) = x^3 - 5x^2 + 9x - 45$

**41.** $g(x) = x^3 - x^2 - x - 2$

**42.** $g(x) = x^3 + x^2 + 5x + 5$

**43.** $g(x) = x^3 + x^2 - 8x - 12$

**44.** $g(x) = x^3 - 4x^2 - 3x + 18$

**45.** $g(x) = x^3 - 10x^2 + 17x + 28$

**46.** $g(x) = x^3 - 4x^2 - 11x + 30$

**47.** $h(x) = x^4 + 2x^3 - 16x^2 - 2x + 15$

**48.** $h(x) = x^4 - 20x^2 + 64$

**49.** $h(x) = x^4 + x^3 - 7x^2 - 13x - 6$

**50.** $h(x) = x^4 - 8x^3 + 22x^2 - 24x + 9$

**51.** $h(x) = 6x^4 - x^3 + 5x^2 - x - 1$

**52.** $h(x) = x^4 - 12x^2 - 64$

**53.** $h(x) = x^4 - x^3 - 5x^2 - x - 6$

**54.** $h(x) = x^5 + 7x^4 - 5x^3 - 35x^2 + 4x + 28$

## Problem Solving

*Find all real zeros (estimate the irrational zeros to the nearest hundredth).*

**55.** $f(x) = x^3 + x^2 + x - 4$

**56.** $g(x) = 2x^4 + x^2 - 1$

**57.** $h(x) = 2x^4 - 3x^3 + x^2 - 12$

**58.** $f(x) = 3x^3 - 2x^2 - 20$

**59.** $g(x) = 8x^4 - 2x^2 + 5x - 1$

**60.** $h(x) = x^4 + 8x^3 - x^2 + 2$

**61.** $f(x) = 2x^3 + 6x^2 - 8x + 2$

**62.** $g(x) = 3x^3 - 10x + 9$

## Cumulative Review Exercises

[6.1] **63.** Simplify $\dfrac{3x^2 + 9x^3 - 12x^4y}{6x^4y}$.

[6.2] **64.** Add $\dfrac{4}{x+2} + \dfrac{5}{x+3}$.

**65.** Subtract $\dfrac{3}{4x} - \dfrac{5}{6y}$.

[6.4] **66.** Solve $\dfrac{3}{y} + \dfrac{1}{3} = \dfrac{2}{3y}$.

# 10.5 Characteristics of Graphs of Rational Functions

1. Recognize the key features of graphs of rational functions.
2. Find vertical asymptotes.
3. Find holes in graphs of rational functions.
4. Find horizontal asymptotes.
5. Find oblique asymptotes.
6. Graph rational functions.

## 1 Recognize the Key Features of Graphs of Rational Functions

In Chapter 6 we briefly discussed rational functions. We will repeat the definition here for your convenience.

**Rational Function**

A **rational function** is a function of the form $f(x) = \dfrac{p}{q}$ where $p$ and $q$ are polynomials and $q \neq 0$.

A rational function is **proper** if the degree of its numerator is less than the degree of its denominator. If the degree of its numerator is greater than or equal to the degree of its denominator, the function is an **improper** rational function.

One example of a proper rational function is the reciprocal function, $f(x) = \dfrac{1}{x}$. In Example 6 of Section 3.1 we graphed this function by plotting points. Although we can always graph functions by plotting points, it is often easier to graph them using their key features.

We will now revisit the graph of the reciprocal function so that we can identify its key features, see **Figure 10.32**. Shortly we will discuss how to find the key features of rational functions and use them as tools for graphing.

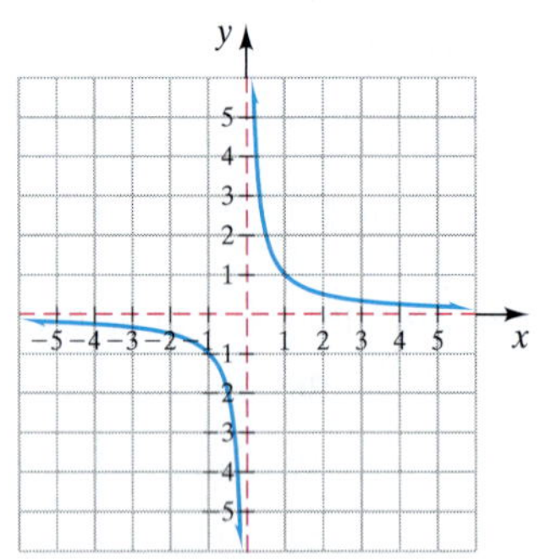


FIGURE 10.32

Notice that it appears that the graph has no $x$-intercept and no $y$-intercept. Instead, the graph appears to approach the $x$-axis and the $y$-axis. A line that a portion of a graph approaches but never reaches is called an **asymptote**. For instance the graph of $f(x) = \dfrac{1}{x}$ has a **horizontal asymptote** at $y = 0$ and a **vertical asymptote** at $x = 0$. The vertical and horizontal asymptotes of a rational function are two key features of its graph. The graph in **Figure 10.33** has a horizontal asymptote at $y = 2$ and a vertical asymptote at $x = 3$.

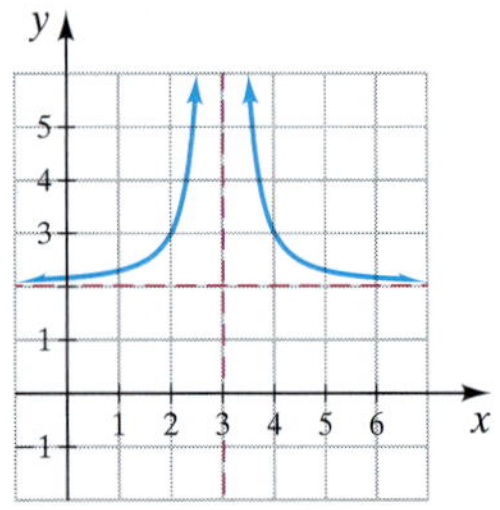


FIGURE 10.33

Before we formally define horizontal and vertical asymptotes, we need to introduce a new symbol. We often use an arrow pointing to the right to represent the word "approaches." For example, $x \to 4$ is read "as $x$ approaches 4," $p(x) \to \infty$ is read "as $p(x)$ approaches infinity," and $|f(x)| \to \infty$ is read "as the absolute value of $f(x)$ approaches infinity." Now let's define vertical and horizontal asymptotes.

**Asymptotes of Rational Functions**

For the rational function $f(x) = \dfrac{p(x)}{q(x)}$, written in the lowest terms,

**a)** if $f(x) \to a$ as $|x| \to \infty$, then the line $y = a$ is a **horizontal asymptote**; and

**b)** if $|f(x)| \to \infty$ as $x \to a$, then the line $x = a$ is a **vertical asymptote**.

A horizontal asymptote is the $y$-value that a function approaches as $|x| \to \infty$. A vertical asymptote is the $x$-value where $|f(x)|$ approaches $\infty$.

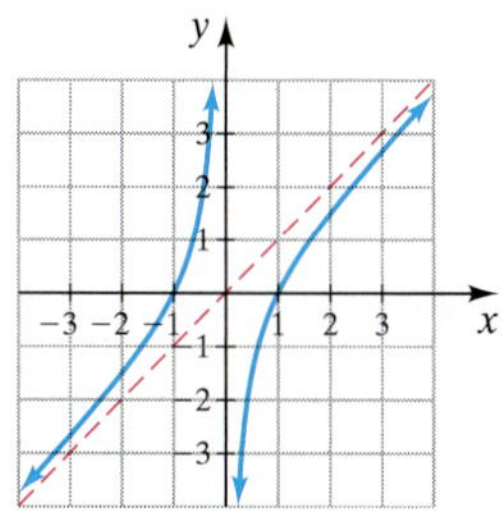

FIGURE 10.34

Some asymptotes are neither vertical nor horizontal. These are called **oblique asymptotes**. **Figure 10.34** illustrates an oblique asymptote. We will discuss oblique asymptotes in more depth shortly.

Consider the rational function $f(x) = \dfrac{x^2 - 4}{x + 2}$. Its graph is shown in **Figure 10.35**.

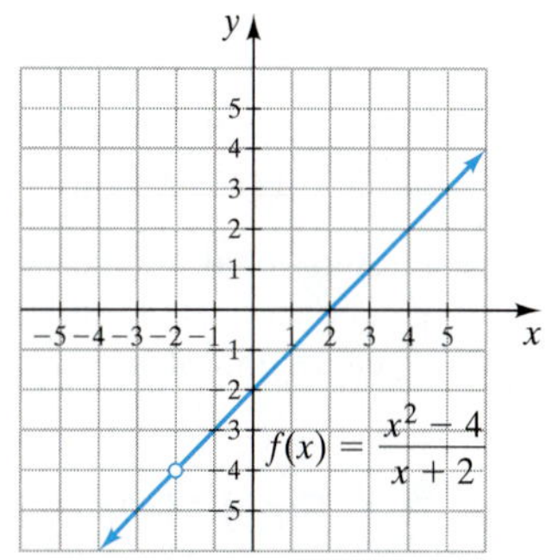


FIGURE 10.35

This graph appears linear with a **hole** at $x = -2$. Holes are another feature that may be present in the graph of a rational function. We will discuss holes in more detail shortly. Now let's learn how to find asymptotes.

## 2 Find Vertical Asymptotes

Rational functions are not defined for values of $x$ for which the denominator is equal to 0. Rational functions will have a point of discontinuity, where the function is not defined. At values of $x$ where the denominator is 0, the graph will either have a vertical asymptote or a hole. We will first discuss finding vertical asymptotes.

**Locating Vertical Asymptotes**

A rational function $f(x) = \dfrac{p(x)}{q(x)}$, in lowest terms, will have a vertical asymptote at $r$ if $r$ is a real zero of the denominator $q(x)$.

By the boxed information, if *after simplifying the rational function*, we find a value that results in the remaining denominator having a value of 0, then the graph of the rational function will have a vertical asymptote at that value.

**EXAMPLE 1** ▶ Find the vertical asymptotes, if any, of the graph of each rational function.

**a)** $f(x) = \dfrac{x}{x^2 - 4}$ **b)** $g(x) = \dfrac{x - 2}{x + 5}$ **c)** $R(x) = \dfrac{x^2}{x^2 + 3}$

**d)** $f(x) = \dfrac{x^2 - 4}{x + 2}$ **e)** $H(x) = \dfrac{x^2 - 9}{x^2 + 3x - 18}$

*Solution*

**a)** The function cannot be simplified further. Since both 2 and $-2$ make the denominator 0, the graph will have vertical asymptotes at $x = 2$ and $x = -2$.

**b)** The function cannot be simplified further. Since $-5$ makes the denominator 0, the graph has a vertical asymptote at $x = -5$.

**c)** The function cannot be simplified further. Since there is no value of $x$ that makes the denominator 0, the graph of this function contains no vertical asymptotes.

**d)** The function can be simplified by factoring as follows

$$f(x) = \frac{x^2 - 4}{x + 2} = \frac{\cancel{(x + 2)}(x - 2)}{\cancel{x + 2}} = x - 2$$

Since there are no values that make the denominator of the simplified expression 0, there are no vertical asymptotes.

**e)** The function can be simplified by factoring as follows

$$H(x) = \frac{x^2 - 9}{x^2 + 3x - 18} = \frac{(x + 3)\cancel{(x - 3)}}{(x + 6)\cancel{(x - 3)}} = \frac{x + 3}{x + 6}$$

Because $-6$ makes the denominator of the simplified expression 0, there is a vertical asymptote at $x = -6$.

▸ **Now Try Exercise 11**

## Helpful Hint

A rational function can have no vertical asymptotes, one vertical asymptote, or more than one vertical asymptote. The graph of a rational function will never intersect any of its vertical asymptotes.

## 3 Find Holes in Graphs of Rational Functions

If a value makes both the numerator and denominator of a rational function zero, then there will be either a hole or a vertical asymptote at that value. If, when simplified, the value makes the denominator of the simplified rational function zero, then there will be a vertical asymptote at that value. If the value does not make the denominator of the simplified rational function zero, then there will be a hole at that value.

Notice that in the function in Example 1 **d)**, $f(x) = \dfrac{x^2 - 4}{x + 2} = \dfrac{(x + 2)(x - 2)}{x + 2}$, the function is not defined at $x = -2$. When $x = -2$ both the numerator and denominator of the function are 0. When this occurs there may be a *hole* in the graph of the function. Therefore, the graph of the function may have a hole at $x = -2$. To determine if there is a hole, simplify the function.

$$f(x) = \frac{\cancel{(x + 2)}(x - 2)}{\cancel{x + 2}}$$

$$f(x) = x - 2,\ x \neq -2$$

Since the denominator of the simplified function, $x - 2$, is 1, the value $-2$ does not make the denominator of the simplified function 0. Therefore, there is a hole, and not a vertical asymptote, in the graph of the function at $x = -2$

Now look at the function in Example 1 **e)** $H(x) = \dfrac{x^2 - 9}{x^2 + 3x - 18} = \dfrac{(x + 3)(x - 3)}{(x + 6)(x - 3)}$. This function is not defined at $x = 3$ and $x = -6$. When $x = 3$ both the numerator and denominator are 0. Therefore, the graph of this function may have a hole at $x = 3$. To determine if there is a hole in the function, simplify the function

$$f(x) = \frac{(x + 3)\cancel{(x - 3)}}{(x + 6)\cancel{(x - 3)}}$$

$$f(x) = \frac{x + 3}{x + 6},\ x \neq 3,\ x \neq -6$$

Since the denominator of the simplified function, $x + 6$, is not 0 when $x = 3$, the graph of the function has a hole at $x = 3$.

Now consider the function $G(x) = \dfrac{x - 4}{(x - 4)^2}$. Since the value 4 makes both the numerator and denominator 0, there may be a hole in the function at $x = 4$. To determine if there is a hole, simplify the function

$$G(x) = \frac{x - 4}{(x - 4)^2} = \frac{\cancel{x - 4}}{(x - 4)\cancel{(x - 4)}}$$

$$G(x) = \frac{1}{x - 4},\ x \neq 4$$

Since 4 makes the denominator of the simplified function $\frac{1}{x-4}$ equal to 0, there is a vertical asymptote, and not a hole, in the graph of $G(x)$ at $x = 4$.

## 4 Find Horizontal Asymptotes

To determine horizontal asymptotes we will look at two categories of rational functions: proper rational functions, and improper rational functions where the numerator and denominator contain polynomials of the same degree. Consider the **proper rational function** $f(x) = \frac{x+1}{4x^2+9}$. We need to determine how the values of $f(x)$ behave as $x \to \infty$ or $x \to -\infty$. Suppose $x$ gets larger and larger. Then the numerator of $f(x)$, which is $x + 1$, can be approximated by the power function $y = x$. The denominator, $4x^2 + 2$, can be approximated by the power function $y = 4x^2$. Therefore, $f(x) = \frac{x+1}{4x^2+9} \approx \frac{x}{4x^2} = \frac{1}{4x}$.

The rational function $f(x) = \frac{1}{4x}$ will approach 0 as $x \to \infty$ or $x \to -\infty$. For example, if $x = 100$, $f(x) = \frac{1}{400} = 0.0025$ and if $x = -100$, $f(x) = \frac{1}{-400} = -0.0025$. Since the function approaches 0 as $x \to \infty$ or $x \to -\infty$ there is a horizontal asymptote at $y = 0$. Using the same reasoning we can determine the following theorem.

**Horizontal Asymptote at $y = 0$**

If a rational function is proper, the line $y = 0$ is a horizontal asymptote of its graph.

**EXAMPLE 2** ▸ Determine if the graph of the functions has a horizontal asymptote of $y = 0$.

**a)** $f(x) = \frac{x^2 - 6}{2x^3 + 9}$

**b)** $g(x) = \frac{x^3 + 5x - 2}{3x^6 - 6x}$

### Solution

**a)** Since the function is a proper rational function, its graph will have a horizontal asymptote at $y = 0$.

**b)** Since the function is a proper rational function, its graph will have a horizontal asymptote at $y = 0$.

▸ **Now Try Exercise 33**

Now we look at rational functions whose polynomials in the numerator and denominator have the same degree. Consider $f(x) = \frac{3x^2 + 4x}{5x^2 - 2x + 6}$. Both the numerator and denominator are of degree 2. Let us divide each term in the numerator and each term in the denominator by $x^2$, the variable and exponent in the leading terms.

$$f(x) = \frac{\frac{3x^2}{x^2} + \frac{4x}{x^2}}{\frac{5x^2}{x^2} - \frac{2x}{x^2} + \frac{6}{x^2}} = \frac{3 + \frac{4}{x}}{5 - \frac{2}{x} + \frac{6}{x^2}}$$

As $x \to \infty$ or $x \to -\infty$, $\frac{4}{x}$, $-\frac{2}{x}$, and $\frac{6}{x^2}$ will all approach 0, and therefore the value of the function will approach $\frac{3}{5}$. This graph will have horizontal asymptote at $y = \frac{3}{5}$. Using the same reasoning we can conclude the following.

**Horizonal Asymptote**

For a rational function where the numerator and denominator are of the same degree, the ratio of the leading coefficients of the simplified function will give the horizontal asymptote of the graph. That is, the graph of a rational function with simplified form $f(x) = \frac{a_n x^n + a_{n-1}x^{n-1} + \cdots + a_0}{b_n x^n + b_{n-1}x^{n-1} + \cdots + b_0}$ will have a horizontal asymptote at $y = \frac{a_n}{b_n}$.

**EXAMPLE 3** ▶ Determine the horizontal asymptote of the graphs of the following functions.

**a)** $T(x) = \dfrac{3x^6 + 5x^2 - x}{2x^6 - 9x^3 + 5}$

**b)** $R(x) = \dfrac{15x^4 - 6x^3 + 3x - 2}{4x^4 - 6x^2 + 3}$

*Solution*

**a)** Since the numerator and denominator are of the same degree, the graph of the function will have a horizontal asymptote at $y = \frac{3}{2}$.

**b)** The graph of $R(x)$ will have a horizontal asymptote at $y = \frac{15}{4}$.

▶ **Now Try Exercise 35**

## 5 Find Oblique Asymptotes

Now we consider a rational function where the polynomial in the numerator is exactly 1 degree higher than the polynomial in the denominator. For example, consider $R(x) = \dfrac{3x^2 + 2x - 6}{x + 2}$. The numerator is of degree 2 and the denominator is of degree 1. If we divide the denominator by the numerator we obtain

$$\begin{array}{r} 3x - 4 \\ x + 2\overline{)3x^2 + 2x - 6} \\ \underline{3x^2 + 6x} \\ -4x - 6 \\ \underline{-4x - 8} \\ 2 \end{array}$$

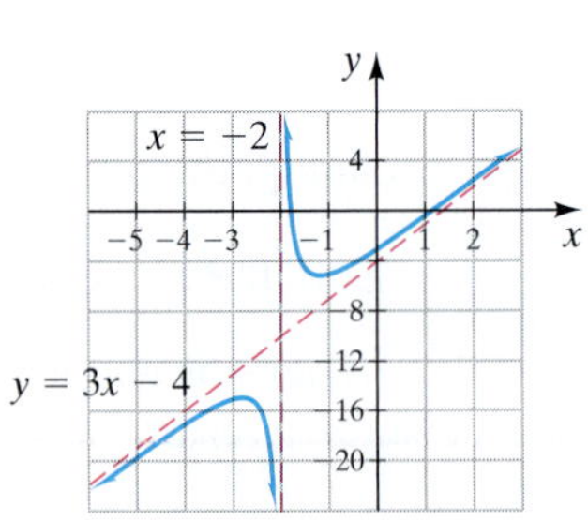


$R(x) = \dfrac{3x^2 + 2x - 6}{x + 2}$

FIGURE 10.36

Therefore, $R(x) = \dfrac{3x^2 + 2x - 6}{x + 2} = 3x - 4 + \dfrac{2}{x + 2}$.

For large values of $x$ the fraction $\dfrac{2}{x + 2}$ approaches 0 and $f(x)$ approaches $3x - 4$. Therefore there is an asymptote at $y = 3x - 4$. The asymptote at $y = 3x - 4$ is an oblique asymptote. The graph of $R(x)$ is shown in **Figure 10.36**.

Using the same procedure and reasoning we can conclude the following.

**Oblique Asymptote**

If the numerator of a rational function is of degree exactly 1 more than the degree of the denominator, to obtain the oblique asymptote (if it exists), divide the numerator by the denominator and disregard the remainder. The new quotient will be of the form $ax + b$. Set the quotient equal to $y$ to obtain the equation of an oblique asymptote of the form $y = ax + b$. The oblique asymptote will be a line.

**EXAMPLE 4** ▸ Determine if the function $g(x) = \dfrac{6x^3 + 4x^2 + 18x - 7}{2x^2 + 3}$ has an oblique asymptote. If so, give the equation of the oblique asymptote.

*Solution* Divide the numerator by the denominator

$$\begin{array}{r}
3x \quad + 2 \\
2x^2 + 0x + 3\overline{)6x^3 + 4x^2 + 18x \quad - 7} \\
\underline{6x^3 + 0x^2 + \phantom{1}9x \phantom{\quad - 7}} \\
4x^2 + \phantom{1}9x \quad - 7 \\
\underline{4x^2 + \phantom{1}0x \quad + 6} \\
9x - 13
\end{array}$$

Thus $g(x) = \dfrac{6x^3 + 4x^2 + 18x - 7}{2x^2 + 3} = 3x + 2 + \dfrac{9x - 13}{2x^2 + 3}$. If we drop the remainder and set the quotient equal to $y$, we get $y = 3x + 2$. The graph of $g(x)$ will have an oblique asymptote at $y = 3x + 2$. Note that there is no vertical asymptote since the denominator can never equal 0.

▸ Now Try Exercise 39

If the degree of the numerator of a rational function is 2 or more greater than the degree of the denominator, then the graph of the function has no horizontal or oblique asymptotes.

The graph of a rational function may have only 1 nonvertical asymptote (either horizontal or oblique). It is possible for the graph of a rational function to intersect the nonvertical asymptote. However, the graph of a rational function cannot intersect a vertical asymptote.

**Summary of What We Have Learned**

1. **Vertical Asymptotes** occur at values that make the denominator of a rational function, in its lowest terms, equal to 0.
2. **Holes** may occur at values that result in both the numerator and denominator having a value of 0.
3. **Horizontal Asymptotes**
   a) The graph of a proper rational function has a horizontal asymptote at $y = 0$.
   b) If the degree of the numerator of a rational function is the same as the degree of the denominator, the graph of the rational function has a horizontal asymptote equal to the ratio of the leading coefficients.
4. **Oblique Asymptotes** If the degree of the numerator of a rational function is exactly 1 more than the degree of the denominator, the graph of the rational function may have an oblique asymptote. The asymptote can be found by dividing the numerator by the denominator, disregarding the remainder, and setting the quotient equal to $y$.

## 6 Graph Rational Functions

Now we will use the information we have learned to graph rational functions. It may be helpful to plot some points from time to time to get a more accurate graph.

**EXAMPLE 5 ▸** Graph $f(x) = \dfrac{1}{x - 5}$.

*Solution* When $x = 0$, $f(x) = -\frac{1}{5}$. Thus the $y$-intercept is $\left(0, -\frac{1}{5}\right)$. Because 5 makes the denominator 0, there is a vertical asymptote at $x = 5$. Since this is a proper rational function, there is a horizontal asymptote at $y = 0$, see **Figure 10.37**. We use dashed lines to indicate the asymptotes because the asymptotes are not a part of the graph. They are only an aid in drawing the graph. To determine the behavior of the graph as it approaches the vertical asymptote, select values for $x$ approaching 5 *from the right*, that is, select values for $x$ getting closer and closer to 5, but always greater than 5. Selecting values for $x$ to the right of 5, we get the following table.

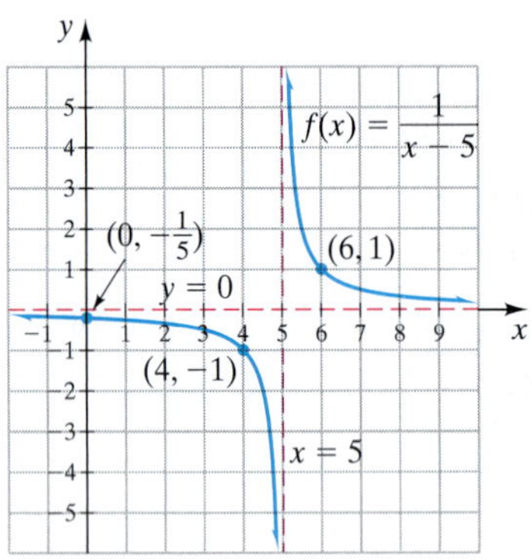

FIGURE 10.37

| $x$ | $f(x)$ |
|---|---|
| 6 | 1 |
| 5.5 | 2 |
| 5.1 | 10 |
| 5.01 | 100 |

We see that as we approach 5 from the right, $f(x)$ gets larger and larger. Therefore $f(x) \to \infty$ as $x \to 5$ from the right. Now select values for $x$ approaching $x = 5$ *from the left*. That is, select values for $x$ closer and closer to 5, but always less than 5. Selecting value for $x$ to the left of 5, we get the following table.

| $x$ | $f(x)$ |
|---|---|
| 4 | −1 |
| 4.5 | −2 |
| 4.9 | −10 |
| 4.99 | −100 |

We see that as we approach 5 from the left, $f(x)$ gets smaller and smaller. Therefore, $f(x) \to -\infty$ as $x \to 5$ from the left. We use the values in the table, the fact that a graph never crosses a vertical asymptote, and the fact that the horizontal asymptote is at $y = 0$ to draw the graph; see **Figure 10.37**.

▸ **Now Try Exercise 47**

### USING YOUR GRAPHING CALCULATOR

In some of the following examples the calculations may be time consuming. If you have a graphing calculator, you should use it to evaluate functions for various values of the variable.

On page 34 we explained how to evaluate an expression for various values of the variable. If you have forgotten how to do so, please read that material now.

Consider the function $f(x) = \dfrac{4x - 1}{2x + 6}$, which we will graph in Example 6. To evaluate the function at −4, −3.5, and −3.1, we enter the following.

```
−4→x: (4x − 1)/(2x + 6)
                    8.5
−3.5→x: (4x − 1)/(2x + 6)
                     15
−3.1→x: (4x − 1)/(2x + 6)
                     67
```

Notice that when $x = -4$, the function equals 8.5; when $x = -3.5$, the function equals 15; and when $x = -3.1$, the function equals 67.

**EXAMPLE 6** ▸ Graph $f(x) = \dfrac{4x - 1}{2x + 6}$.

*Solution* By substituting $x = 0$, we determine the $y$-intercept is $\left(0, -\dfrac{1}{6}\right)$. There is a vertical asymptote at $x = -3$, since $-3$ makes the denominator 0. Since the numerator and denominator are of the same degree, the horizontal asymptote is at $y = \dfrac{4}{2}$ or $y = 2$; see **Figure 10.38a**.

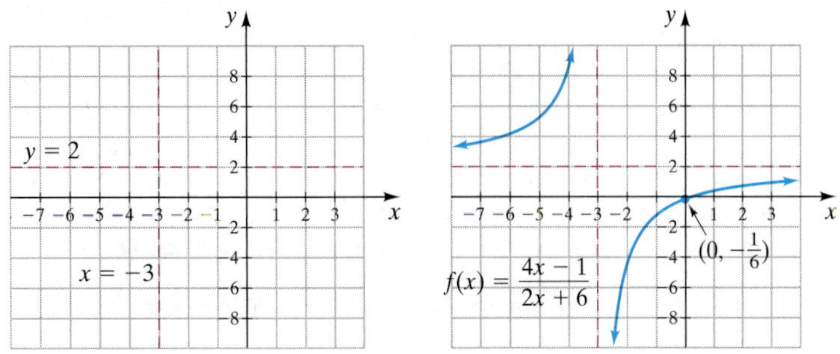


FIGURE 10.38 (a) (b)

Now plot a few points on each side of the vertical asymptote. Remember to work on only one side of the vertical asymptote at a time.

In $(-\infty, -3)$, approaching $-3$ from the left, yields the following table,

| $x$ | $f(x)$ |
|---|---|
| $-4$ | 8.5 |
| $-3.5$ | 15.0 |

In $(-3, \infty)$, approaching $-3$ from the right, yields the following table,

| $x$ | $f(x)$ |
|---|---|
| $-2$ | $-4.5$ |
| $-2.5$ | $-11$ |

We can see from the few points plotted and the vertical asymptote that as $x$ approaches $-3$ from the left, the graph approaches $\infty$, and as $x$ approaches $-3$ from the right, the graph approaches $-\infty$. Remember **a graph never crosses a vertical asymptote**. The graph is shown in **Figure 10.38b**.

▸ **Now Try Exercise 53**

**EXAMPLE 7** ▸ Graph $f(x) = \dfrac{x + 1}{(2x - 1)(x + 3)}$.

*Solution* When $x = 0$, $f(x) = -\dfrac{1}{3}$. Thus the $y$-intercept is $\left(0, -\dfrac{1}{3}\right)$. There are vertical asymptotes at $x = \dfrac{1}{2}$ and $x = -3$. Since the numerator is of degree 1 and the denominator is of degree 2, this is a proper rational function. Therefore, the horizontal asymptote is $y = 0$; see **Figure 10.39a**.

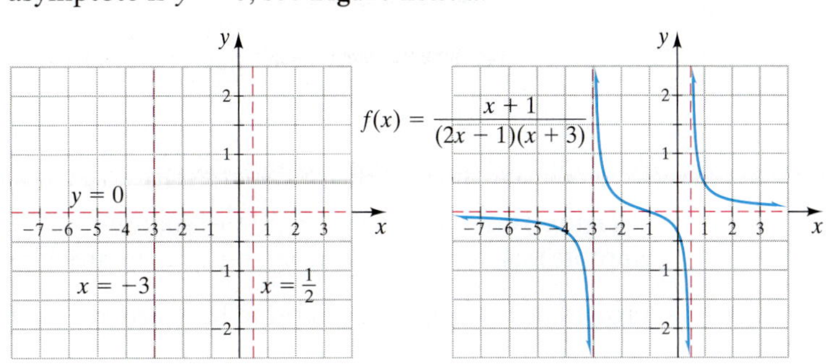


FIGURE 10.39 (a) (b)

Now let's select some values in each of the three intervals $(\infty, -3)$, $\left(-3, \frac{1}{2}\right)$, and $\left(\frac{1}{2}, \infty\right)$. We work with each interval individually.

In $(-\infty, -3)$ approaching $-3$ from the left, yields the following table,

| $x$ | $f(x)$ |
|---|---|
| $-4$ | $-0.333$ |
| $-3.5$ | $-0.625$ |
| $-3.1$ | $-2.9$ |
| $-3.01$ | $-28.6$ |

As $x$ approaches $-3$ from the left, the function approaches $-\infty$.

In the interval $\left(-3, \frac{1}{2}\right)$ approaching $-3$ from the right, yields the following table,

| $x$ | $f(x)$ |
|---|---|
| $-2$ | $0.2$ |
| $-2.5$ | $0.5$ |
| $-2.9$ | $2.79$ |
| $-2.99$ | $28.5$ |

As $x$ approaches $-3$ from the right, the function approaches $\infty$.

In the interval $\left(-3, \frac{1}{2}\right)$ approaching $\frac{1}{2}$ from the left, yields the following table,

| $x$ | $f(x)$ |
|---|---|
| $0$ | $-0.333$ |
| $0.4$ | $-2.06$ |
| $0.49$ | $-21.3$ |
| $0.499$ | $-214$ |

As $x$ approaches $\frac{1}{2}$ from the left, the function approaches $-\infty$.

In the interval $\left(\frac{1}{2}, \infty\right)$ approaching $\frac{1}{2}$ from the right, yields the following table,

| $x$ | $f(x)$ |
|---|---|
| $1.5$ | $0.28$ |
| $0.6$ | $2.22$ |
| $0.51$ | $21.5$ |
| $0.501$ | $214$ |

As $x$ approaches $\frac{1}{2}$ from the right, the function approaches $\infty$. The graph is illustrated in **Figure 10.39b** on page 703. Notice that the function approaches 0 (the horizontal asymptote) as $x$ approaches $\infty$ and as $x$ approaches $-\infty$. Also notice that the graph crosses the horizontal asymptote between $x = -3$ and $x = \frac{1}{2}$.

▶ **Now Try Exercise 49**

**EXAMPLE 8 ▸** Graph $f(x) = \dfrac{x^2 + 4}{x + 2}$.

*Solution* When $x = 0$, $f(x) = 2$. Thus the $y$-intercept of the graph is $(0, 2)$. This function is reduced to its lowest terms since the numerator cannot be factored over the set of real numbers. There is a vertical asymptote at $x = -2$ because $-2$ makes the denominator 0. Since the numerator is exactly 1 degree greater than the denominator, we can obtain the oblique asymptote by dividing.

$$\begin{array}{r} x - 2 \\ x + 2 \overline{\big)\, x^2 + 0x + 4} \\ \underline{x^2 + 2x} \quad\;\; \\ -2x + 4 \\ \underline{-2x - 4} \\ 8 \end{array}$$

The oblique asymptote is obtained from the quotient, $x - 2$. The oblique asymptote is $y = x - 2$, see **Figure 10.40a**.

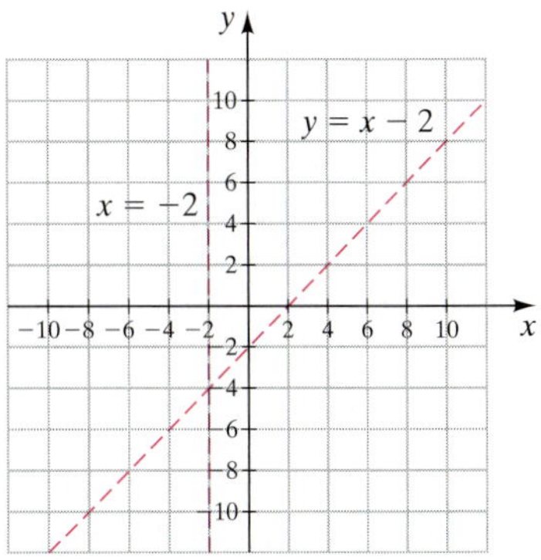

FIGURE 10.40a

Now select some values for $x$ in the intervals $(-\infty\ -2)$ and $(-2, \infty)$ and evaluate $f(x)$ at those values.

In the interval $(-\infty, -2)$ approaching $-2$ from the left, yields the following table,

| $x$ | $f(x)$ |
|---|---|
| −5 | −9.67 |
| −3 | −13 |
| −2.5 | −20.5 |
| −2.1 | −84.1 |
| −2.01 | −804 |

As $x$ approaches $-2$ from the left, the function approaches $-\infty$.

In the interval $(-2, \infty)$ approaching $-2$ from the right, yields the following table,

| $x$ | $f(x)$ |
|---|---|
| 5 | 4.14 |
| −1 | 5 |
| −1.5 | 12.5 |
| −1.9 | 76.1 |
| −1.99 | 796 |

As $x$ approaches $-2$ from the right, the function approaches $\infty$.

The graph is drawn in **Figure 10.40b**. Notice the graph approaches $x = -2$, but is not defined at $-2$. Also the graph approaches the oblique asymptote $y = x - 2$ as $x \to \infty$ and as $x \to -\infty$.

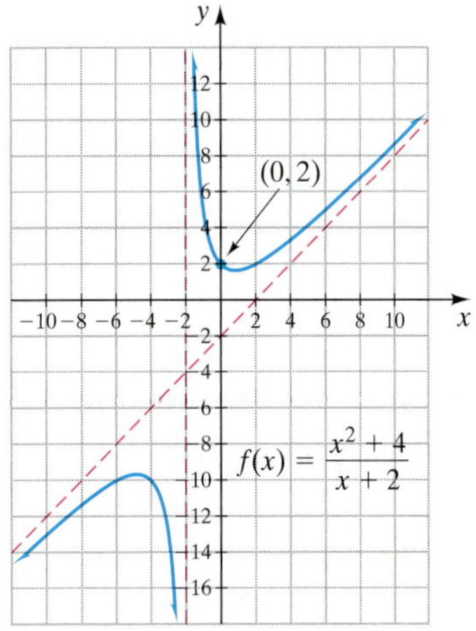

FIGURE 10.40b

▸ **Now Try Exercise 59**

### USING YOUR GRAPHING CALCULATOR

If you have a graphing calculator, you should graph each function we graph on your graphing calculator after you have graphed the function using the procedure discussed in this section. On page 383 we discussed using the dot mode to graph functions. You should reread that material now.

In **Figure 10.41** we illustrate the graph of $f(x) = \dfrac{x^2 + 4}{x + 2}$ which we graphed in **Figure 10.40b**.

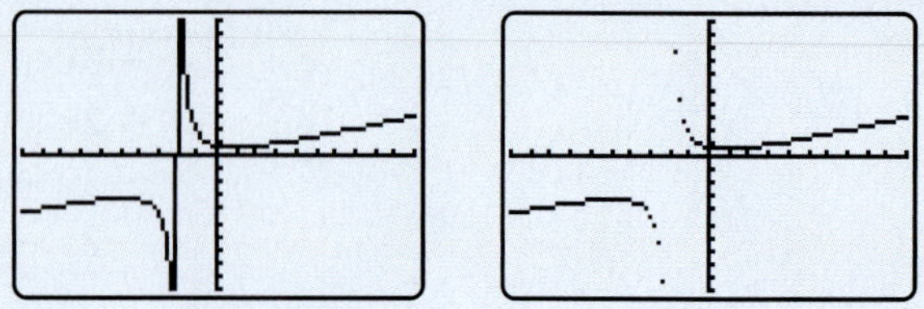

$-10, 10, 1, -30, 30, 3$
(a)

$-10, 10, 1, -30, 30, 3$
(b)

FIGURE 10.41

In **Figure 10.41a** we use the connected mode, and in **Figure 10.41b** we use the dot mode.

**EXAMPLE 9** ▸ Graph $f(x) = \dfrac{x^2 - 4}{x - 2}$.

*Solution* When $x = 0, f(x) = 2$. Thus, the $y$-intercept is $(0, 2)$. Notice that 2 makes both the numerator and denominator 0 and the function is not defined at 2. The function can be factored as follows.

$$f(x) = \frac{x^2 - 4}{x - 2} = \frac{(x + 2)\cancel{(x - 2)}}{\cancel{(x - 2)}} = x + 2, (x \neq 2)$$

Since $x + 2$ is defined for all values of $x$, there is no vertical asymptote. From the factored form of $f(x)$ we see that $f(x)$ is not defined at $x = 2$. However at $x = 2$, both the numerator and denominator have a value of 0. Therefore, there is a hole at 2 in the function $f(x) = x + 2, x \neq 2$. Following is a table of values. The graph is illustrated by **Figure 10.42**.

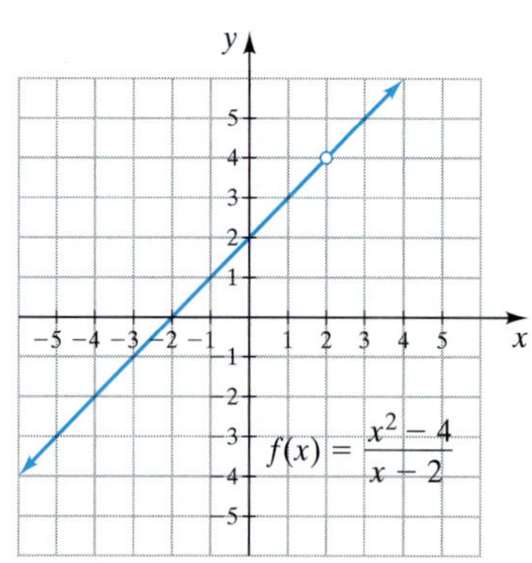


FIGURE 10.42

| $x$ | $f(x)$ |
|---|---|
| −2 | 0 |
| −1 | 1 |
| 0 | 2 |
| 1 | 3 |
| 2 | undefined |
| 3 | 5 |
| 4 | 6 |
| 5 | 7 |

▸ Now Try Exercise 57

## EXERCISE SET 10.5

MathXL® MyMathLab

### Concept/Writing Exercises

1. Do the graphs of all rational functions have at least 1 asymptote? Explain.

2. Can the graph of a rational function have more than 1 vertical asymptote? Explain.

3. Can the graph of a rational function have more than 1 horizontal asymptote? Explain.

4. Explain how to find the $y$-intercept of the graph of a rational function.

5. Explain how to determine the points of discontinuity of the graph of a rational function.

6. Explain why the graph of a proper rational function has a horizontal asymptote at $y = 0$.

7. Under what conditions could the graph of a rational function have an oblique asymptote?

8. Under what conditions will the graph of a rational function have neither a horizontal nor an oblique asymptote?

## Practice the Skills

*For each rational function, find all points of discontinuity and determine whether there is a hole or a vertical asymptote at each point.*

9. $f(x) = \dfrac{3 - x}{x}$

10. $g(x) = \dfrac{4 - 3x}{x - 9}$

11. $h(x) = \dfrac{x + 2}{x + 5}$

12. $f(x) = \dfrac{x - 6}{x - 7}$

13. $g(x) = \dfrac{x^2 - 25}{x + 5}$

14. $h(x) = \dfrac{x^2 - 7x + 10}{x - 5}$

15. $f(x) = \dfrac{x^2 + 8x - 9}{x - 1}$

16. $g(x) = \dfrac{6x^2 + x - 2}{3x + 2}$

17. $g(x) = \dfrac{4x - 7}{x^2 - 16}$

18. $h(x) = \dfrac{3x}{4x^2 - 2x}$

19. $f(x) = \dfrac{x - 1}{x^2 - 7x + 6}$

20. $g(x) = \dfrac{x + 3}{x^2 - 7x + 12}$

21. $h(x) = \dfrac{x + 7}{x^2 + 8x + 15}$

22. $f(x) = \dfrac{x - 4}{x^2 - x - 12}$

23. $g(x) = \dfrac{x + 5}{x^2 - 25}$

24. $h(x) = \dfrac{x^2 - 3x - 10}{x^2 - 8x + 15}$

25. $f(x) = \dfrac{x^2 + 4x + 3}{x^2 - 4x + 3}$

26. $g(x) = \dfrac{x^2 + 10x + 24}{x^2 - 36}$

27. $h(x) = \dfrac{x + 5}{x^2 + 5}$

28. $f(x) = \dfrac{5x - 10}{x^2 - x - 2}$

*Determine whether the graph of the rational function has a horizontal asymptote, an oblique asymptote, or neither. Give the equation for the asymptote if it exists.*

29. $f(x) = \dfrac{7}{x}$

30. $g(x) = \dfrac{5}{x - 6}$

31. $h(x) = \dfrac{6x - 1}{2x + 3}$

32. $f(x) = \dfrac{8x - 3}{4x + 11}$

33. $g(x) = \dfrac{12x + 5}{3x^2}$

34. $h(x) = \dfrac{16x - 5}{7x^2 + 3}$

35. $f(x) = \dfrac{4x^2 - 1}{x^2 - 9}$

36. $g(x) = \dfrac{3x^2 - 5x + 9}{2x^2 - x - 3}$

37. $f(x) = \dfrac{x^2 - 8x + 15}{x + 2}$

38. $f(x) = \dfrac{x^2 - 5x - 8}{x - 3}$

39. $f(x) = \dfrac{2x^2 + 8x + 12}{x - 2}$

40. $f(x) = \dfrac{3x^2 - 9x + 13}{x - 1}$

41. $f(x) = \dfrac{4x^2 - 6}{8x^3 - 1}$

42. $f(x) = \dfrac{8x^3 - 12x^2 + 3x - 1}{4x^2 - 1}$

43. $f(x) = \dfrac{x^3 + 2}{x^2 + 2x}$

44. $f(x) = \dfrac{x^2 + 2x}{x^3 + 5x}$

*Graph the rational function.*

45. $f(x) = \dfrac{1}{x + 2}$

46. $g(x) = \dfrac{1}{x - 3}$

47. $h(x) = \dfrac{5}{x - 1}$

48. $f(x) = -\dfrac{2}{x + 6}$

49. $h(x) = \dfrac{x + 1}{x(x + 4)}$

50. $f(x) = \dfrac{x}{(x + 1)(x - 2)}$

51. $g(x) = \dfrac{x + 1}{x - 1}$

52. $h(x) = \dfrac{6x - 5}{2x + 8}$

53. $f(x) = \dfrac{3x + 3}{2x + 4}$

54. $g(x) = \dfrac{2x + 4}{x - 1}$

55. $h(x) = \dfrac{15x^2 + 10}{3x^2}$

56. $f(x) = \dfrac{3 - 8x^2}{2x^2 + 1}$

57. $g(x) = \dfrac{x^2 - 3x - 4}{x - 2}$

58. $h(x) = \dfrac{x^2 - x - 12}{x - 2}$

59. $g(x) = \dfrac{x^2 + 2x - 8}{x + 4}$

**60.** $f(x) = \dfrac{x^2 + 9}{x - 5}$

**61.** $r(x) = \dfrac{x^2 - 5x + 6}{x - 2}$

**62.** $h(x) = \dfrac{2x^2 - x - 3}{x + 1}$

**63.** $g(x) = \dfrac{2x^3 - x^2 + x - 8}{x^2 - 2x + 1}$

**64.** $r(x) = \dfrac{x^3 - x^2 - x + 1}{x - 1}$

## Problem Solving

**65.** Write a rational function that has a horizontal asymptote at $y = 0$ and a vertical asymptote at $x = -4$. Explain how you determined your answer.

**66.** Write a rational function that has a horizontal asymptote at $y = 5$ and a vertical asymptote at $x = 3$. Explain how you determined your answer.

**67.** Write a rational function that is the graph of $f(x) = 2x - 5$ with a hole at the point $(4, 3)$. Explain how you determined your answer.

**68.** Write a rational function that is the graph of $f(x) = x^2 - 3$ with a hole at the point $(-3, 6)$. Explain how you determined your answer.

**69.** Write a rational function that has a horizontal asymptote at $y = 0$ and no vertical asymptote. Explain how you determined your answer.

**70.** Write a rational function that has an oblique asymptote at $y = 2x + 1$ and a vertical asymptote at $x = 3$. Explain how you determined your answer.

## Group Activity

**71.** In this section we have been graphing rational functions in which the degree of the numerator is at most 1 greater than the degree of the denominator. Now we will look at the graphs of rational functions in which the degree of the numerator is 2 greater than the degree of the denominator.

**a)** Graph the function $f(x) = \dfrac{x^4 + 1}{x^2}$.

**b)** Find the quotient of $x^4 + 1$ divided by $x^2$.

**c)** Graph the quotient (disregard the remainder) on the same coordinate axes.

**d)** What do you notice?

**e)** Explain why the graph of $f(x) = \dfrac{x^4 + 1}{x^2}$ is very similar to the graph of the quotient, $g(x) = x^2$, when $|x| \to \infty$.

## Cumulative Review Exercises

[9.1] **72.** Find the inverse of $f(x) = \dfrac{1}{2}x + 3$.

**73.** Find the inverse of $f(x) = \dfrac{5}{x}$.

[9.3] **74.** Solve $\log_a \dfrac{1}{144} = -2$ for $a$.

**75.** Solve $\log_2 x = -5$ for $x$.

# Chapter 10 Summary

| IMPORTANT FACTS AND CONCEPTS | EXAMPLES |
|---|---|
| **Section 10.1** | |

### Terminology

A **function is increasing** if it increases as the values of $x$ increases.

A **function is decreasing** if it decreases as the values of $x$ increases.

A **local minimum** is the lowest point in a small interval centered around the point.

A **local maximum** is the highest point in a small interval centered around the point.

The **real zeros of a function** are the real-number values where the function has a value of 0. These are the $x$-coordinates of the $x$-intercepts on the graph of the function.

Consider the following graph of a function.

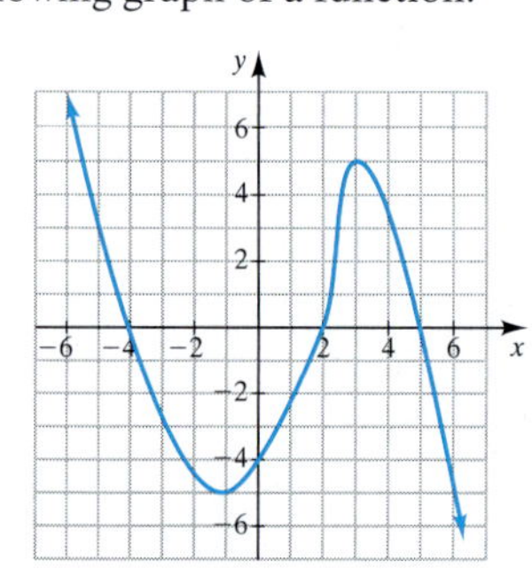


Function is increasing on $[-1, 3]$.
Function is decreasing on $(-\infty, -1] \cup [3, \infty)$.
Local minimum is at $(-1, -5)$.
Local maximum is at $(3, 5)$.
Real zeros are $-4, 2, 5$.
$x$-intercepts are at $(-4, 0), (2, 0), (5, 0)$.

| IMPORTANT FACTS AND CONCEPTS | EXAMPLES |
|---|---|

**Section 10.1 (continued)**

A function is **continuous** if its graph has no gaps or holes.
A function is **discontinuous** if its graphs has gaps or holes.

Continuous function

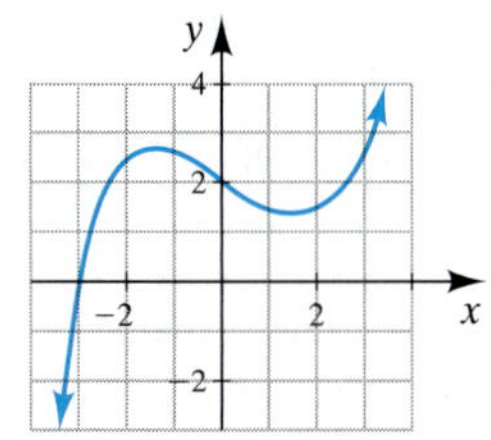


Discontinuous function
(hole at $x = -2$, gap at $x = 1$)

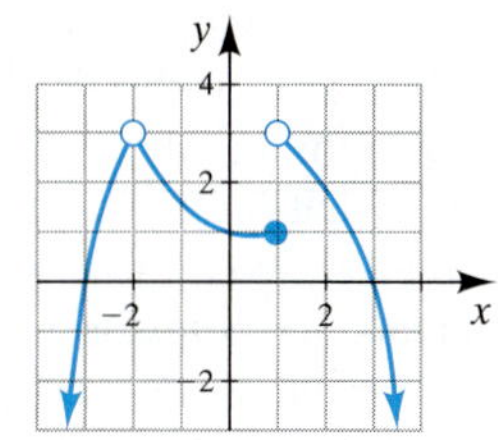


The **greatest integer function** (or *step function*), $f(x) = [\![x]\!]$, is a function that truncates, or rounds down, a value to the greatest integer less than or equal to the value.

Graph of $f(x) = [\![x]\!]$.

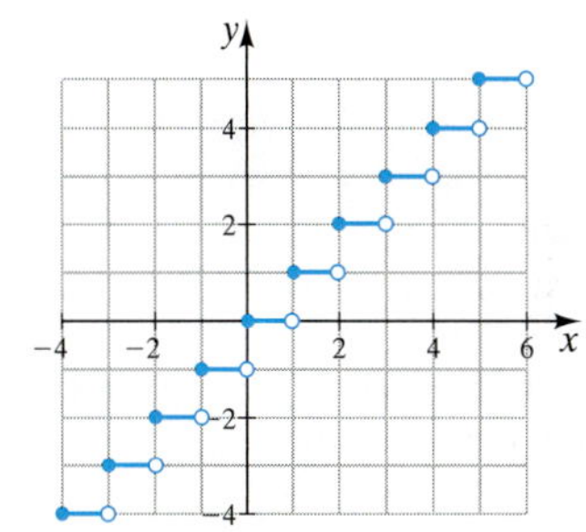


A **function defined piecewise** (or a piecewise function) is a function defined using two or more equations.

$g(x) = \begin{cases} x + 2 & \text{if } x < 1 \\ 2x + 3 & \text{if } x \geq 1 \end{cases}$ is a piecewise function.

**Section 10.2**

### Functions

A **polynomial function** is a function of the form

$$P(x) = a_n x^n + a_{n-1}x^{n-1} + a_{n-2}x^{n-2} + \cdots + a_1 x + a_0$$

where the coefficients $a_n, a_{n-1}, a_{n-2}, \ldots, a_0$ are real numbers, and all exponents on the variable are nonnegative integers. The domain of all polynomial functions is all real numbers.

$f(x) = 7x^3 - 2x^2 + 5x - 19$ and $g(x) = -x^4 + x^3 - 8x^2 + x + 3$ are examples of polynomial functions.

A **power function** is a function of the form $f(x) = ax^n$.

$f(x) = x^5$ is an example of a power function.

A **zero of a polynomial function** is a value of $x$ where the function is equal to zero.

If $P(x)$ is a polynomial function and $r$ is a real number for which $P(r) = 0$, then $r$ is called a **zero** of $P(x)$.

If a real number $r$ is a zero of a polynomial function, that is $P(r) = 0$, then

**a)** $(r, 0)$ is an $x$-intercept of the graph, and

**b)** $(x - r)$ is a factor of $P(x)$.

For the function $P(x) = x^2 - 2x - 35$, find its zeros and $x$-intercepts of its graph.

$$P(x) = x^2 - 2x - 35$$
$$0 = x^2 - 2x - 35 \quad \text{Set } P(x) = 0.$$
$$0 = (x + 5)(x - 7)$$
$$x + 5 = 0 \quad \text{or} \quad x - 7 = 0$$
$$x = -5 \qquad\qquad x = 7$$

Zeros are $-5$ and 7.
$x$-intercepts are $(-5, 0)$ and $(7, 0)$.

| IMPORTANT FACTS AND CONCEPTS | EXAMPLES |
|---|---|
| **Section 10.2 (continued)** | |
| A function of the form $P(x) = (x - a)^m$ has $a$ as a zero of **multiplicity** $m$. For a real zero of an even multiplicity, the graph of $P(x)$ touches but does not cross the $x$-axis at that zero. For a real zero of an odd multiplicity, the graph of $P(x)$ crosses the $x$-axis at that zero. | $P(x) = (x + 2)^3(x - 7)^4$ has $-2$ as a zero of multiplicity 3, and 7 as a zero of multiplicity 4. The graph of $P(x)$ touches the $x$-axis at $x = 7$ and crosses the $x$-axis at $x = -2$. |
| **Turning points** are points at which the graph changes direction (from increasing to decreasing or vice versa). These points occur at the local maxima and local minima. | 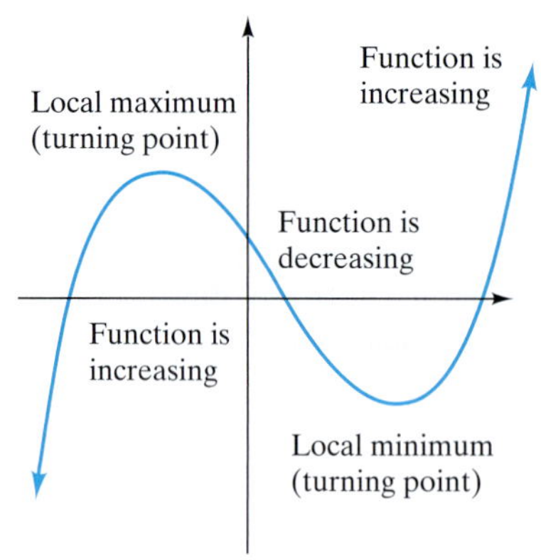  |
| The **maximum number** of turning points on the graph of a polynomial function of degree $n$ is $n - 1$. | The maximum number of turning points on the graph of $P(x) = x^4 - 2x^3 + 5x - 13$ is $4 - 1 = 3$. |
| **Section 10.3** | |
| **For a Polynomial Function $P(x)$ of Degree $n$** **a)** the maximum number of $x$-intercepts of its graph is $n$, and **b)** the maximum number of real zeros of $P(x)$ is $n$. | $P(x) = 5x^3 - x^2 - x + 7$ has a maximum number of 3 real zeros and its graph has a maximum number of 3 $x$-intercepts. |
| **Descartes' Rule of Signs** For any polynomial function $P(x)$ written in descending order, **1.** the number of **positive real zeros** of $P(x)$ either equals the number of variations in sign the nonzero coefficients of $P(x)$ or equals that number less an even integer. **2.** the number of **negative real zeros** of $P(x)$ either equals the number of variations in sign of the nonzero coefficients of $P(-x)$ or equals that number less an even integer. | Consider $P(x)$ and $P(-x)$, $P(x) = x^5 - 2x^4 + x^3 + 4x^2 + x - 7$ $P(-x) = -x^5 - 2x^4 - x^3 + 4x^2 - x - 7$ Since $P(x)$ has 3 changes in sign, $P(x)$ has 3 or 1 positive real zeros. Since $P(-x)$ has 2 changes in sign, $P(x)$ has 2 or 0 negative real zeros. |
| **Rational Zeros Theorem** For any polynomial function $P(x) = a_nx^n + a_{n-1}x^{n-1} + \cdots + a_1x + a_0$ where $a_n \neq 0$, $a_0 \neq 0$, and all coefficients are integers, the possible rational zeros of $P(x)$ are of the form $\frac{p}{q}$ where $p$ is a factor of $a_0$ and $q$ is a factor of $a_n$. | The possible rational zeros of $P(x) = 3x^3 + 2x^2 - x - 4$ are $\pm1, \pm2, \pm4, \pm\frac{1}{3}, \pm\frac{2}{3}, \pm\frac{4}{3}$. |
| **Remainder Theorem** If the polynomial $P(x)$ is divided by $x - a$, the remainder is equal to $P(a)$. | If $P(x) = x^4 - 3x^3 + 7x - 5$ is divided by $x - 1$, the remainder is 0. Thus, $P(1) = 0$. |
| **Factor Theorem** Let $P(x)$ be a polynomial function, then $x - a$ is a factor of $P(x)$ if and only if $P(a) = 0$. | Let $P(x) = x^4 - 3x^3 + 7x - 5$. Then, $P(1) = 1^4 - 3(1)^3 + 7(1) - 5 = 0$. Therefore, $(x - 1)$ is a factor of $P(x)$. |
| A nonreal zero of $P(x)$ is called a **complex zero**. | $3i$ and $-3i$ are complex zeros of $P(x) = x^2 + 9$. |

| IMPORTANT FACTS AND CONCEPTS | EXAMPLES |
|---|---|

## Section 10.4

### Intermediate Value Theorem

For a continuous function $f(x)$ defined over the interval $[a, b]$, if the sign of $f(a)$ is the opposite of the sign of $f(b)$, then a real zero exists in $(a, b)$.

$f(x) = x^2 - x - 9$ has a zero in the interval $(0, 4)$ since $f(0) = -9 < 0$ and $f(4) = 3 > 0$.

It is not possible to determine if $g(x) = x^2 + x + 7$ has a zero in the interval $(0, 4)$ since $g(0) = 7 > 0$ and $g(4) = 27 > 0$.

### Boundedness Theorem

Let $P(x)$ be a polynomial function with real coefficients and a positive leading coefficient.

**a)** If $P(x)$ is synthetically divided by $x - b$, where $b > 0$ and all the numbers in the bottom row of the synthetic division are nonnegative, then $P(x)$ has no real zero greater than $b$. We call $b$ an **upper bound** for the real zeros of $P(x)$.

**b)** If $P(x)$ is synthetically divided by $x - c$ where $c < 0$ and all of the numbers in the bottom row of the synthetic division alternate in sign, then $P(x)$ has no real zero less than $c$. We call $c$ a **lower bound** for the real zeros of $P(x)$.

$P(x) = x^3 - 3x^2 + 4x - 6$ has all the real zeros in the interval $[-2, 5]$.

| −2 | 1 | −3 | 4 | −6 | |
|---|---|---|---|---|---|
| | | −2 | 10 | −28 | |
| | 1 | −5 | 14 | −34 | ← *Alternating signs* |

| 5 | 1 | −3 | 4 | −6 | |
|---|---|---|---|---|---|
| | | 5 | 10 | 70 | |
| | 1 | 2 | 14 | 64 | ← *Positive numbers* |

## Section 10.5

A **rational function** is a function of the form $f(x) = \dfrac{p}{q}$ where $p$ and $q$ are polynomials and $q \neq 0$.

A rational function is **proper** if the degree of the numerator is less than the degree of the denominator.

A rational function is **improper** if the degree of the numerator is greater than or equal to the degree of the denominator.

$f(x) = \dfrac{3x^2 - x + 2}{4x^3 - x^2 + 7}$ is a rational function.

$g(x) = \dfrac{7x - 9}{3x^2 + x + 5}$ is a proper rational function.

$r(x) = \dfrac{4x^3 + 2x - 11}{5x + 1}$ is an improper rational function.

An **asymptote** is a line that a portion of a graph approaches but never reaches.

For the rational function $f(x) = \dfrac{p(x)}{q(x)}$, written in the lowest terms,

**a)** if $f(x) \to a$ as $|x| \to \infty$, then the line $y = a$ is a **horizontal asymptote**; and

**b)** if $|f(x)| \to \infty$ as $x \to a$, then the line $x = a$ is a **vertical asymptote**.

$f(x) = \dfrac{1}{x}$ has a horizontal asymptote at $y = 0$ and a vertical asymptote at $x = 0$.

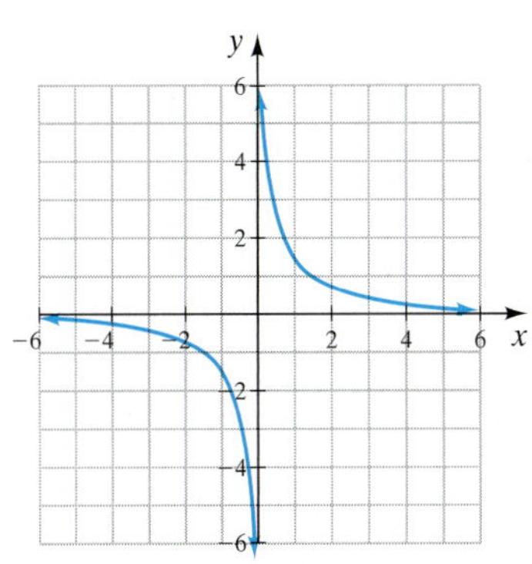


### Locating Vertical Asymptotes

A rational function $f(x) = \dfrac{p(x)}{q(x)}$, in lowest terms, will have a vertical asymptote at $r$ if $r$ is a real zero of the denominator $q(x)$.

$f(x) = \dfrac{3x + 5}{x - 4}$ has a vertical asymptote at $x = 4$.

$g(x) = \dfrac{x^2 - 5x + 7}{(x - 5)(x + 6)}$ has vertical asymptotes at $x = 5$ and $x = -6$.

If a rational function is proper, the line $y = 0$ is a horizontal asymptote of its graph.

$q(x) = \dfrac{7x - 1}{3x^2 - 8x + 9}$ has a horizontal asymptote at $y = 0$.

**IMPORTANT FACTS AND CONCEPTS** | **EXAMPLES**

**Section 10.5 (continued)**

For a rational function where the numerator and denominator are of the same degree, the ratio of the leading coefficients of the simplified function will give the horizontal asymptote of the graph. That is, the graph of a rational function with simplified form $f(x) = \dfrac{a_n x^n + a_{n-1}x^{n-1} + \cdots + a_0}{b_n x^n + b_{n-1}x^{n-1} + \cdots + b_0}$ will have a horizontal asymptote at $y = \dfrac{a_n}{b_n}$.

$r(x) = \dfrac{5x^5 - 3x^2 + 2x - 7}{11x^5 + x^4 - 6x^3 + 2}$ has a horizontal asymptote at $y = \dfrac{5}{11}$.

An **oblique asymptote** is an asymptote that is neither vertical nor horizontal.

If the numerator of a rational function is of degree exactly 1 more than the degree of the denominator, to obtain the oblique asymptote (if it exists), divide the numerator by the denominator and disregard the remainder. The new quotient will be of the form $ax + b$. Set the quotient equal to $y$ to obtain the equation of an oblique asymptote of the form $y = ax + b$. The oblique asymptote will be a line.

$g(x) = \dfrac{x^2 + 7x + 3}{x - 2}$ has an oblique asymptote of $y = x + 9$.

$$\begin{array}{r} x + 9 \quad \leftarrow \text{quotient} \\ x - 2 \overline{) x^2 + 7x + 3} \\ \underline{x^2 - 2x} \qquad \\ 9x + 3 \\ \underline{9x - 18} \\ 21 \end{array}$$

A **hole** may occur in the graph of a function when both the numerator and denominator have a value of 0.

The graph of $f(x) = \dfrac{x^2 - 9}{x - 3} = \dfrac{(x + 3)(x - 3)}{x - 3}$ has a hole at $x = 3$ since both the numerator and denominator are 0 when $x = 3$.

## Chapter 10 Review Exercises

[10.1] *Classify each function as continuous or discontinuous.*

1.

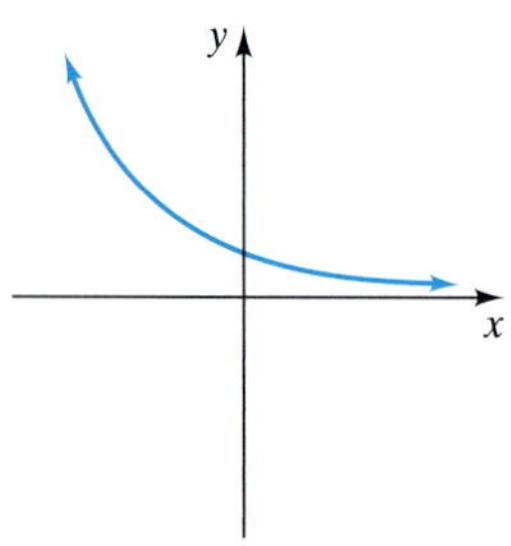


2.

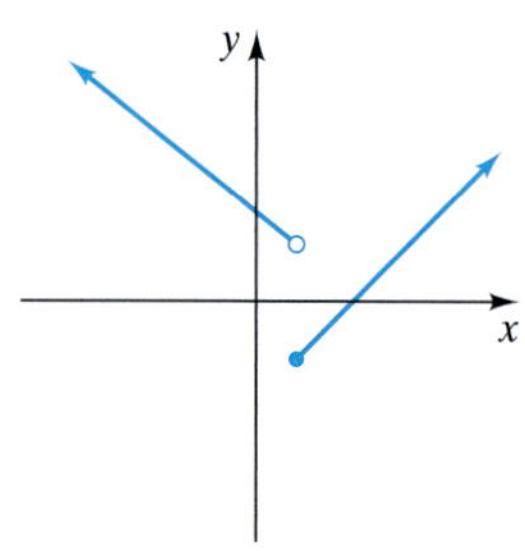


*Classify each function as continuous or discontinuous. If the function is not continuous, name the point(s) of discontinuity.*

**3.** $f(x) = -4x^3 + 8x^2 - 12x + 19$

**4.** $g(x) = \dfrac{x + 8}{x^2 + 2}$

**5.** $h(x) = \left[\!\left[\dfrac{x + 5}{4}\right]\!\right]$

**6.** $f(x) = \begin{cases} x - 2 & \text{if } x < 2 \\ 2 - x & \text{if } x \geq 2 \end{cases}$

*Graph each piecewise function.*

**7.** $f(x) = \begin{cases} 2x & \text{if } x < 1 \\ 3 & \text{if } x \geq 1 \end{cases}$

**8.** $g(x) = \begin{cases} 4 - x & \text{if } x \leq -2 \\ \frac{1}{4}x & \text{if } x > -2 \end{cases}$

**9.** $h(x) = \begin{cases} -4 & \text{if } x < 0 \\ 4 & \text{if } x \geq 0 \end{cases}$

**10.** $f(x) = \begin{cases} 3 & \text{if } x \leq 3 \\ x & \text{if } x > 3 \end{cases}$

*Graph each greatest integer function.*

**11.** $f(x) = [\![2x - 3]\!]$

**12.** $g(x) = \left[\!\!\left[\frac{1}{2}x + 1\right]\!\!\right]$

**13.** $h(x) = \left[\!\!\left[\frac{x}{12}\right]\!\!\right]$

**14.** $f(x) = [\![\sqrt{x}]\!]$

**15. Pencils** Sharon Bell wants to stock up on pencils, which cost \$0.37 a piece. Write a greatest integer function that represents the number of pencils she can purchase for $x$ dollars.

**16. Videos** Guillermo Lopez wants to rent some videos, which rent for \$2.25 per video. Write a greatest integer function that represents the number of videos he can rent for $x$ dollars.

[10.2] *Describe the location, either above the x-axis or below the x-axis, of the graph of $f(x)$ for* **a)** *large positive values of x, and* **b)** *large negative values of x.*

**17.** $f(x) = -2x^3 - 6x^2 + 7x - 3$

**18.** $f(x) = 2x^4 - x^3 + 7x^2 + 26x$

**19.** $f(x) = -x^6 + 3x^4 - 5x^2 + 11$

**20.** $f(x) = 7x^5 - 2x^4 - 5x^3 + 7x - 1$

*Find the real zeros of each function and state their multiplicities if it is a number other than 1.*

**21.** $f(x) = x^2 + 11x + 24$

**22.** $g(x) = x^3 + 5x^2 - 36x$

**23.** $h(x) = (x + 2)^3(2x - 1)$

**24.** $f(x) = x^2 - 7$

*Determine the maximum number of turning points for the graph of each function. Do not graph the function.*

**25.** $f(x) = 3x^2 + 6x - 16$

**26.** $g(x) = -2x^4 - 5x^3 + 6x + 11$

**27.** $h(x) = 2x^3 + 7x^2 - x + 9$

**28.** $f(x) = -5x^6 + 6x^5 + 13x - 2$

*Determine the minimum degree of the polynomial function graphed.*

**29.**

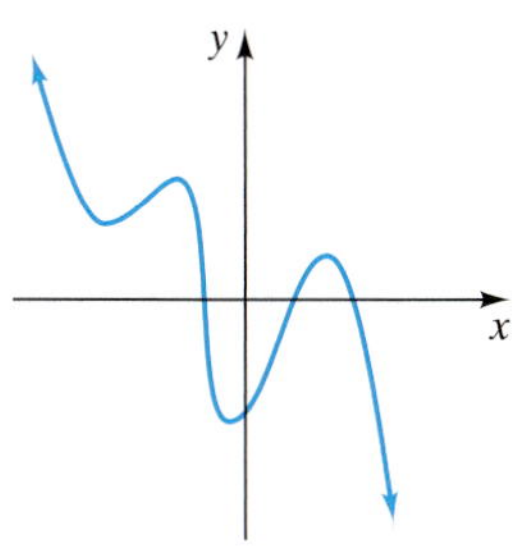


**30.**

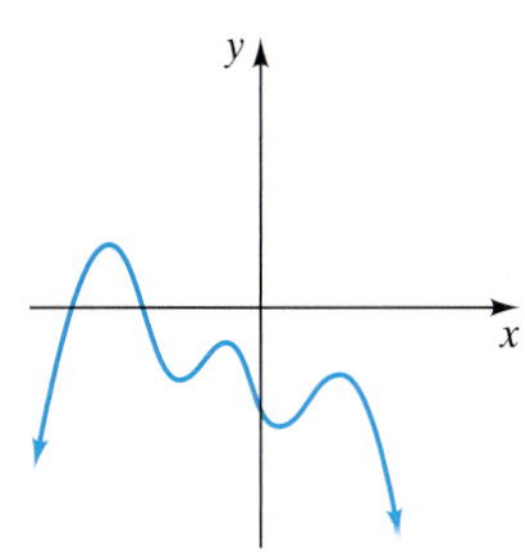


*Graph each function.*

**31.** $f(x) = (x + 4)(x + 1)(x - 5)$

**32.** $g(x) = (x + 4)(x + 1)^2(x - 5)$

**33.** $h(x) = x^4 - 7x^3 + 6x^2$

**34.** $f(x) = x^5 - 7x^4 + 6x^3$

[10.3] *Determine the maximum and minimum number of x-intercepts for the graph of each polynomial function.*

**35.** $f(x) = -x^3 + 5x^2 - 7x - 9$

**36.** $g(x) = x^4 - x^3 - x^2 + x - 9$

**37.** $h(x) = x^5 - x^4 - 2x^3 + 20$

**38.** $f(x) = -x^6 + 5x^4 + 3x^2 - 18$

*Use Descartes' Rule of Signs to determine*

**a)** *the possible number of positive real zeros of each polynomial function.*

**b)** *the possible number of negative real zeros of each polynomial function.*

**39.** $f(x) = x^2 - 12x + 36$

**40.** $g(x) = x^2 + 3x - 21$

**41.** $h(x) = x^3 + 5x^2 + 3x + 9$

**42.** $f(x) = x^4 - x^3 - 6x^2 + 2x + 3$

*List the possible rational zeros of each function.*

**43.** $f(x) = x^3 + 3x^2 - 9x + 12$

**44.** $g(x) = 2x^5 - 9x^2 - 16x - 5$

**45.** $h(x) = 3x^4 - 12x^3 - 8x^2 + 4x - 9$

**46.** $f(x) = x^6 - 5x^4 + 6x - 64$

*Determine the rational zeros of each function.*

**47.** $f(x) = x^3 + 10x^2 - x - 10$

**48.** $g(x) = x^4 - 4x^3 + 5x^2 - 4x + 4$

**49.** $h(x) = 3x^4 + 10x^3 - 5x^2 + 10x - 8$

**50.** $f(x) = x^4 + x^3 - 35x^2 - 57x + 90$

*Find all the zeros of each function.*

**51.** $f(x) = x^4 - 4x^3 + 11x^2 - 64x - 80$

**52.** $g(x) = x^4 + x^3 + 19x^2 + 25x - 150$

[10.4] *Use the Intermediate Value Theorem to determine if a real zero of* $g(x) = x^3 - 4x^2 - 5x + 13$ *occurs in the given interval. If this cannot be determined, indicate so.*

**53.** $(-3, -1)$

**54.** $(-1, 1)$

**55.** $(1, 5)$

**56.** $(4, 5)$

*Use the Boundedness Theorem to determine if the zeros of* $f(x)$ *are bounded over the given interval. If this cannot be determined, indicate so.*

**57.** $f(x) = x^3 - x^2 - 8$ in $[-1, 2]$

**58.** $g(x) = x^3 - x^2 - 11x - 6$ in $[-3, 3]$

**59.** $h(x) = x^4 - 5x^3 + 7x^2 - 12x - 96$ in $[-3, 6]$

**60.** $f(x) = x^5 - 3x^4 - 5x^2 + 18x + 3$ in $[-1, 4]$

*By determining the y-intercept, all x-intercepts, and using a test value in each interval determined by the x-intercepts, graph each polynomial function.*

**61.** $f(x) = 2x^3 + x^2 + 2x + 1$

**62.** $g(x) = x^3 + 2x^2 - 19x - 20$

**63.** $h(x) = x^4 - x^3 - 6x^2 + 4x + 8$

**64.** $h(x) = 4x^5 + 12x^4 - x - 3$

[10.5] *Find all points of discontinuity and determine whether there is a hole or a vertical asymptote at each point.*

**65.** $f(x) = \dfrac{4x}{x^2 - 25}$

**66.** $g(x) = \dfrac{x + 5}{x^2 - 25}$

**67.** $h(x) = \dfrac{x^2 - 16}{x + 4}$

**68.** $f(x) = \dfrac{x + 4}{x^2 + 16}$

*Determine if the graph of the rational function has a horizontal asymptote, an oblique asymptote, or neither. Give the equation of the horizontal or oblique asymptote if it exists.*

**69.** $f(x) = \dfrac{16x^5 + 5}{4x^7 - 12}$

**70.** $g(x) = \dfrac{24x^2 + x - 10}{8x^2 - 12x + 3}$

**71.** $h(x) = \dfrac{6x^2 - 5x + 7}{2x + 3}$

**72.** $f(x) = \dfrac{x^2 + 9x + 14}{x + 2}$

**73.** $g(x) = \dfrac{8x^2 + 6x - 25}{4x + 9}$

**74.** $h(x) = \dfrac{18x^3 - 12x^2 + 7x}{12x^3 - 9x^2 + 4x}$

*Graph the rational function.*

**75.** $f(x) = \dfrac{2}{x + 2}$

**76.** $g(x) = \dfrac{7x - 5}{2x + 1}$

**77.** $h(x) = \dfrac{x^2 + x - 20}{x - 4}$

**78.** $f(x) = \dfrac{x - 3}{x^2 + x - 2}$

**79.** $g(x) = \dfrac{x^2 + 5x - 12}{x - 3}$

**80.** $h(x) = \dfrac{2x^2 - 9x + 15}{x - 6}$

# 11 Conic Sections

## GOALS OF THIS CHAPTER

The focus of this chapter is on graphing conic sections. These include the circle, ellipse, parabola, and hyperbola. We have already discussed parabolas. In this chapter, we will learn more about them. We will also solve nonlinear systems of equations both algebraically and graphically.


11.1 The Parabola and the Circle

11.2 The Ellipse

Mid-Chapter Test: Sections 11.1–11.2

11.3 The Hyperbola

11.4 Nonlinear Systems of Equations and Their Applications

11.5 Nonlinear Systems of Inequalities

Chapter 11 Summary

Chapter 11 Review Exercises

Chapter 11 Practice Test

Cumulative Review Test

THE SHAPE OF AN ellipse gives it an unusual feature. Anything bounced off the wall of an elliptical shape from one focal point will ricochet to the other focal point. This feature has been used in architecture and medicine. One example is the National Statuary Hall in the Capitol Building, which has an elliptically shaped domed ceiling. If you whisper at one focal point, your whisper can be heard at the other focal point. Similarly, a ball hit from one focus of an elliptical billiard table will rebound to the other focal point. In Exercise 56 on page 734, you will determine the location of the foci of an elliptical billiard table.

# 11.1 The Parabola and the Circle

1 Identify and describe the conic sections.

2 Review parabolas.

3 Graph parabolas of the form $x = a(y - k)^2 + h$.

4 Learn the distance and midpoint formulas.

5 Graph circles with centers at the origin.

6 Graph circles with centers at $(h, k)$.

## 1 Identify and Describe the Conic Sections

In previous chapters, we discussed parabolas. A **parabola** is one type of conic section. Parabolas will be discussed further in this section. Other conic sections are circles, ellipses, and hyperbolas. Each of these shapes is called a conic section because each can be made by slicing a cone and observing the shape of the slice. The methods used to slice the cone to obtain each conic section are illustrated in **Figure 11.1**.

FIGURE 11.1

## 2 Review Parabolas

We discussed parabolas in Section 8.5. Example 1 will refresh your memory on how to graph parabolas in the forms $y = ax^2 + bx + c$ and $y = a(x - h)^2 + k$.

**EXAMPLE 1** ▸ Consider $y = 2x^2 + 4x - 6$.

**a)** Write the equation in $y = a(x - h)^2 + k$ form.

**b)** Determine whether the parabola opens upward or downward.

**c)** Determine the vertex of the parabola.

**d)** Determine the $y$-intercept of the parabola.

**e)** Determine the $x$-intercepts of the parabola.

**f)** Graph the parabola.

*Solution*

**a)** First, factor 2 from the two terms containing the variable to make the coefficient of the squared term 1. (Do not factor 2 from the constant, $-6$.) Then complete the square.

$$\begin{aligned} y &= 2x^2 + 4x - 6 \\ &= 2(x^2 + 2x) - 6 \\ &= 2(x^2 + 2x + 1) - 2 - 6 \quad \text{Complete the square.} \\ &= 2(x + 1)^2 - 8 \end{aligned}$$

**b)** The parabola opens upward because $a = 2$, which is greater than 0.

**c)** The vertex of the graph of an equation in the form $y = a(x - h)^2 + k$ is $(h, k)$. Therefore, the vertex of the graph of $y = 2(x + 1)^2 - 8$ is $(-1, -8)$. The vertex of a parabola can also be found using

$$\left(-\frac{b}{2a}, \frac{4ac - b^2}{4a}\right) \quad \text{or} \quad \left(-\frac{b}{2a}, f\left(-\frac{b}{2a}\right)\right)$$

Show that both of these procedures give $(-1, -8)$ as the vertex of the parabola now.

**d)** To determine the $y$-intercept, let $x = 0$ and solve for $y$.

$$\begin{aligned} y &= 2(x+1)^2 - 8 \\ &= 2(0+1)^2 - 8 \\ &= -6 \end{aligned}$$

The $y$-intercept is $(0, -6)$.

**e)** To determine the $x$-intercepts, let $y = 0$ and solve for $x$.

$$\begin{aligned} y &= 2(x+1)^2 - 8 \\ 0 &= 2(x+1)^2 - 8 && \text{Substitute 0 for } y. \\ 8 &= 2(x+1)^2 && \text{Add 8 to both sides.} \\ 4 &= (x+1)^2 && \text{Divide both sides by 2.} \\ \pm 2 &= x + 1 && \text{Square root property} \\ -1 \pm 2 &= x && \text{Subtract 1 from both sides.} \end{aligned}$$

$$x = -1 - 2 \quad \text{or} \quad x = -1 + 2$$
$$x = -3 \qquad\qquad x = 1$$

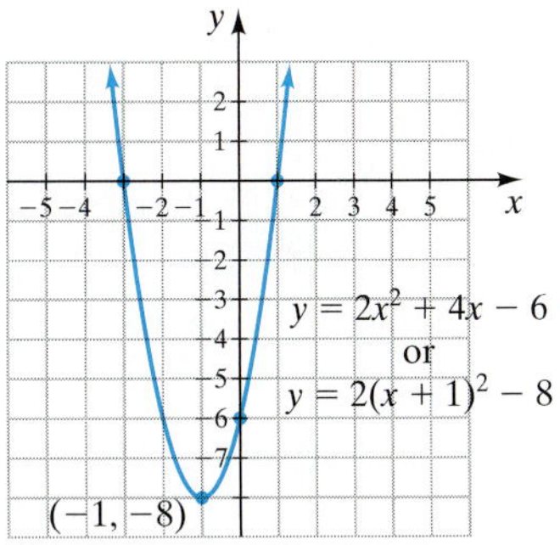

FIGURE 11.2

The $x$-intercepts are $(-3, 0)$ and $(1, 0)$. The $x$-intercepts could also be found by substituting 0 for $y$ in $y = 2x^2 + 4x - 6$ and solving for $x$ using factoring or the quadratic formula. Do this now and show that you get the same $x$-intercepts.

**f)** We use the vertex and the $x$- and $y$-intercepts to draw the graph, which is shown in **Figure 11.2**.

▸ **Now Try Exercise 19**

## 3 Graph Parabolas of the Form $x = a(y - k)^2 + h$

Parabolas can also open to the right or left. The graph of an equation of the form $x = a(y - k)^2 + h$ will be a parabola whose vertex is at the point $(h, k)$. If $a$ is a positive number, the parabola will open to the right, and if $a$ is a negative number, the parabola will open to the left. The four different forms of a parabola are shown in **Figure 11.3**.

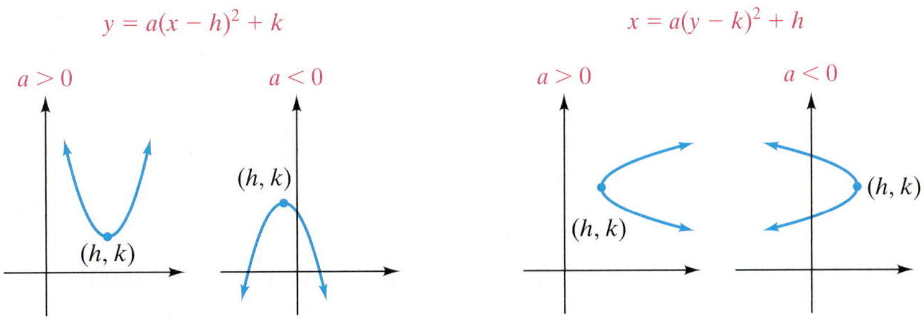

FIGURE 11.3

**Parabola with Vertex at $(h, k)$**

1. $y = a(x - h)^2 + k, a > 0$ (opens upward)
2. $y = a(x - h)^2 + k, a < 0$ (opens downward)
3. $x = a(y - k)^2 + h, a > 0$ (opens to the right)
4. $x = a(y - k)^2 + h, a < 0$ (opens to the left)

Note that equations of the form $y = a(x - h)^2 + k$ are functions since their graphs pass the vertical line test. However, equations of the form $x = a(y - k)^2 + h$ are not functions since their graphs do not pass the vertical line test.

**EXAMPLE 2** ▸ Sketch the graph of $x = -2(y + 4)^2 - 1$.

*Solution* The graph opens to the left since the equation is of the form $x = a(y - k)^2 + h$ and $a = -2$, which is less than 0. The equation can be expressed as $x = -2[y - (-4)]^2 - 1$. Thus, $h = -1$ and $k = -4$. The vertex of the graph is $(-1, -4)$. See **Figure 11.4.** If we set $y = 0$, we see that the $x$-intercept is at $-2(0 + 4)^2 - 1 = -2(16) - 1$ or $-33$. By substituting values for $y$ you can find the corresponding values of $x$. When $y = -2$, $x = -9$, and when $y = -6$, $x = -9$. These points are marked on the graph. Notice that this graph has no $y$-intercept.

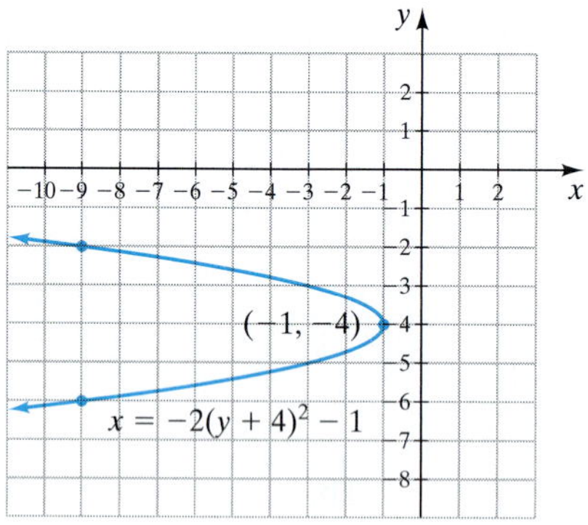


FIGURE 11.4

▸ **Now Try Exercise 31**

**EXAMPLE 3** ▸

**a)** Write the equation $x = 2y^2 + 12y + 13$ in the form $x = a(y - k)^2 + h$.

**b)** Graph $x = 2y^2 + 12y + 13$.

*Solution*

**a)** First factor 2 from the first two terms. Then, complete the square on the expression within the parentheses.

$$\begin{aligned} x &= 2y^2 + 12y + 13 \\ &= 2(y^2 + 6y) + 13 \\ &= 2(y^2 + 6y + 9) + (2)(-9) + 13 \\ &= 2(y^2 + 6y + 9) - 18 + 13 \\ &= 2(y + 3)^2 - 5 \end{aligned}$$

**b)** Since $a > 0$, the parabola opens to the right. Note that when $y = 0$, $x = 2(0)^2 + 12(0) + 13 = 13$. Thus, the $x$-intercept is $(13, 0)$. The vertex of the parabola is $(-5, -3)$. When $y = -6$, we find that $x = 13$. Thus, another point on the graph is $(13, -6)$. Using the quadratic formula, we can determine that the $y$-intercepts are about $(0, -4.6)$ and $(0, -1.4)$. The graph is shown in **Figure 11.5.**

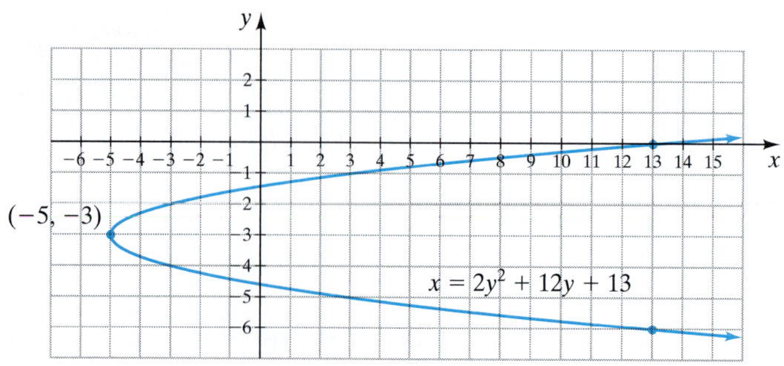


FIGURE 11.5

▸ **Now Try Exercise 45**

## 4 Learn the Distance and Midpoint Formulas

Now we will derive a formula to find the **distance** between two points on a line. We will use this formula shortly to develop the formula for a circle. Consider **Figure 11.6.**

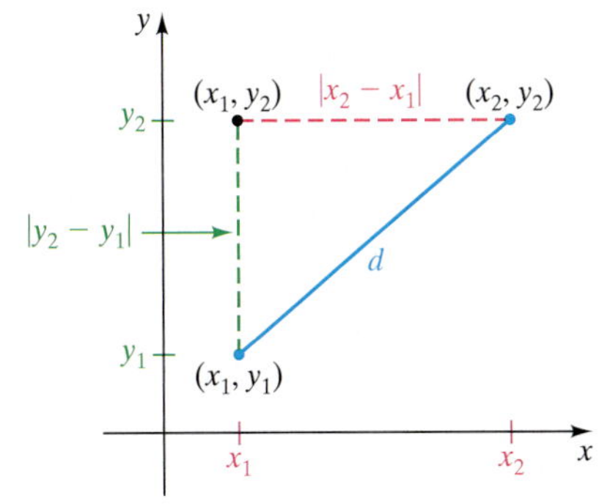


FIGURE 11.6

The horizontal distance between the two points $(x_1, y_2)$ and $(x_2, y_2)$, indicated by the red dashed line, is $|x_2 - x_1|$. We use the absolute value because we want the distance to be positive. If $x_1$ was larger than $x_2$, then $x_2 - x_1$ would be negative. The vertical distance between the points $(x_1, y_1)$ and $(x_1, y_2)$, indicated by the green dashed line, is $|y_2 - y_1|$. Using the Pythagorean Theorem where $d$ is the distance between the two points, we get

$$d^2 = |x_2 - x_1|^2 + |y_2 - y_1|^2$$

Since any nonzero number squared is positive, we do not need absolute value signs. We can therefore write

$$d^2 = (x_2 - x_1)^2 + (y_2 - y_1)^2$$

Using the square root property, with the principal square root, we get the distance between the points $(x_1, y_1)$ and $(x_2, y_2)$, which is $d = \sqrt{(x_2 - x_1)^2 + (y_2 - y_1)^2}$.

**Distance Formula**

The distance, $d$, between any two points $(x_1, y_1)$ and $(x_2, y_2)$ can be found by the distance formula:

$$d = \sqrt{(x_2 - x_1)^2 + (y_2 - y_1)^2}$$

The distance between any two points will always be a positive number. Can you explain why? When finding the distance, it makes no difference which point we designate as point 1, $(x_1, y_1)$, or point 2, $(x_2, y_2)$. Note that the square of any real number will always be greater than or equal to 0. For example, $(5 - 2)^2 = (2 - 5)^2 = 9$.

**EXAMPLE 4** ▸ Determine the distance between the points (4, 5) and (−2, 3).

*Solution* As an aid, we plot the points (**Fig. 11.7**). Label (4, 5) point 1 and (−2, 3) point 2. Thus, $(x_2, y_2)$ represents (−2, 3) and $(x_1, y_1)$ represents (4, 5). Now use the distance formula to find the distance, $d$.

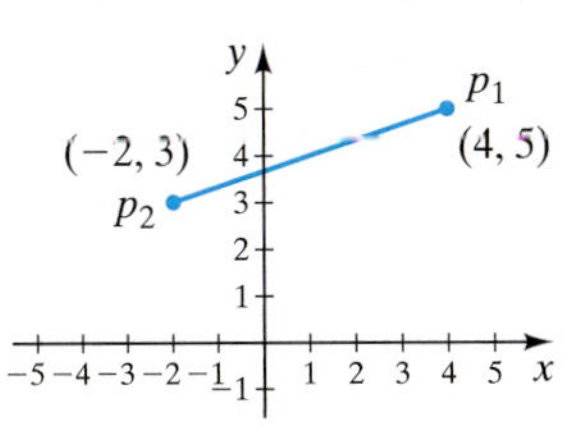


FIGURE 11.7

$$\begin{aligned} d &= \sqrt{(x_2 - x_1)^2 + (y_2 - y_1)^2} \\ &= \sqrt{(-2 - 4)^2 + (3 - 5)^2} \\ &= \sqrt{(-6)^2 + (-2)^2} \\ &= \sqrt{36 + 4} \\ &= \sqrt{40} \quad \text{or} \quad \approx 6.32 \end{aligned}$$

Thus, the distance between the points (4, 5) and (−2, 3) is $\sqrt{40}$ or about 6.32 units.

▸ Now Try Exercise 57

**Avoiding Common Errors**

Students will sometimes begin finding the distance correctly using the distance formula but will forget to take the square root of the sum $(x_2 - x_1)^2 + (y_2 - y_1)^2$ to obtain the correct answer. When taking the square root, remember that $\sqrt{a^2 + b^2} \neq a + b$.

It is often necessary to find the **midpoint** of a line segment between two given endpoints. To do this, we use the midpoint formula.

**Midpoint Formula**

Given any two points $(x_1, y_1)$ and $(x_2, y_2)$, the point halfway between the given points can be found by the midpoint formula:

$$\text{midpoint} = \left(\frac{x_1 + x_2}{2}, \frac{y_1 + y_2}{2}\right)$$

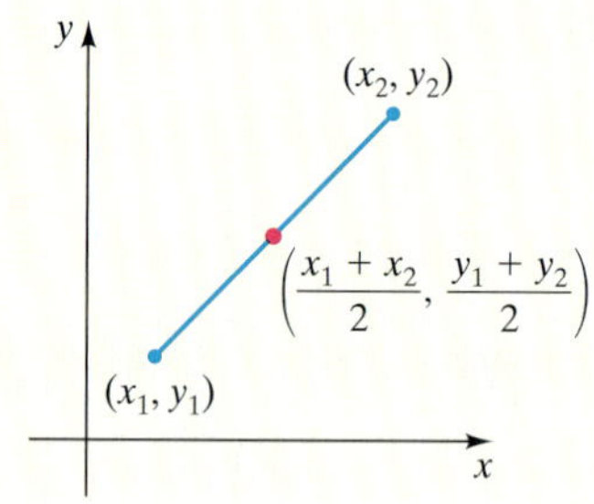


To find the midpoint, we take the average (the mean) of the $x$-coordinates and of the $y$-coordinates.

**EXAMPLE 5** ▸ A line segment through the center of a circle intersects the circle at the points $(-3, 6)$ and $(4, 1)$. Find the center of the circle.

*Solution* To find the center of the circle, we find the midpoint of the line segment between $(-3, 6)$ and $(4, 1)$. It makes no difference which points we label $(x_1, y_1)$ and $(x_2, y_2)$. We will let $(-3, 6)$ be $(x_1, y_1)$ and $(4, 1)$ be $(x_2, y_2)$. See **Figure 11.8.**

$$\text{midpoint} = \left(\frac{x_1 + x_2}{2}, \frac{y_1 + y_2}{2}\right)$$
$$= \left(\frac{-3 + 4}{2}, \frac{6 + 1}{2}\right) = \left(\frac{1}{2}, \frac{7}{2}\right)$$

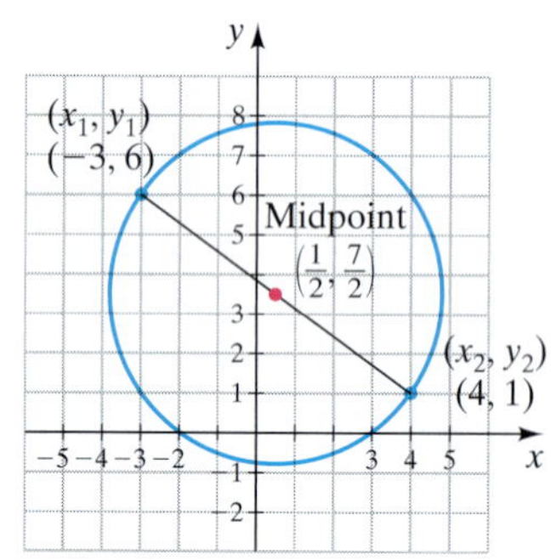


FIGURE 11.8

The point $\left(\frac{1}{2}, \frac{7}{2}\right)$ is halfway between the points $(-3, 6)$ and $(4, 1)$. It is also the center of the circle.

▸ Now Try Exercise 69

## 5 Graph Circles with Centers at the Origin

A **circle** may be defined as the set of points in a plane that are the same distance from a fixed point called its **center**.

The *standard form* of the equation of a circle whose center is at the origin may be derived using the distance formula. Let $(x, y)$ be a point on a circle of radius $r$ with center at $(0, 0)$. See **Figure 11.9**. Using the distance formula, we have

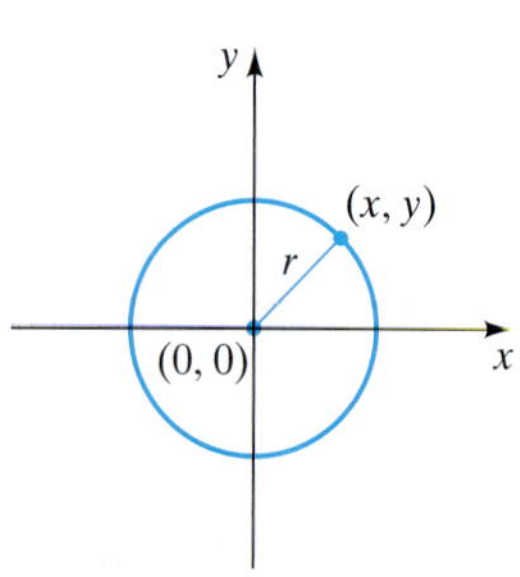


FIGURE 11.9

$$d = \sqrt{(x_2 - x_1)^2 + (y_2 - y_1)^2} \quad \text{Distance formula}$$

$$\text{or} \quad r = \sqrt{(x - 0)^2 + (y - 0)^2} \quad \text{Substitute } r \text{ for } d, (x, y) \text{ for } (x_2, y_2), \text{ and } (0, 0) \text{ for } (x_1, y_1).$$

$$r = \sqrt{x^2 + y^2} \quad \text{Simplify the radicand.}$$

$$r^2 = x^2 + y^2 \quad \text{Square both sides.}$$

**Circle with Its Center at the Origin and Radius $r$**

$$x^2 + y^2 = r^2$$

For example, $x^2 + y^2 = 16$ is a circle with its center at the origin and radius 4, and $x^2 + y^2 = 10$ is a circle with its center at the origin and radius $\sqrt{10}$. Note that $4^2 = 16$ and $(\sqrt{10})^2 = 10$.

**EXAMPLE 6** ▸ Graph the following equations.

**a)** $x^2 + y^2 = 64$ **b)** $y = \sqrt{64 - x^2}$ **c)** $y = -\sqrt{64 - x^2}$

### *Solution*

**a)** If we rewrite the equation as

$$x^2 + y^2 = 8^2$$

we see that the radius of the circle is 8. The graph is illustrated in **Figure 11.10**.

**b)** If we solve the equation $x^2 + y^2 = 64$ for $y$, we obtain

$$y^2 = 64 - x^2$$
$$y = \pm\sqrt{64 - x^2}$$

In the equation $y = \pm\sqrt{64 - x^2}$, the equation $y = +\sqrt{64 - x^2}$ or, simply, $y = \sqrt{64 - x^2}$, represents the top half of the circle, while the equation $y = -\sqrt{64 - x^2}$ represents the bottom half of the circle. Thus, the graph of $y = \sqrt{64 - x^2}$, where $y$ is the principal square root, lies above and on the $x$-axis. For any value of $x$ in the domain of the function, the value of $y$ must be greater than or equal to 0. Why? The graph is the semicircle shown in **Figure 11.11**.

**c)** The graph of $y = -\sqrt{64 - x^2}$ is also a semicircle. However, this graph lies below and on the $x$-axis. For any value of $x$ in the domain of the function, the value of $y$ must be less than or equal to 0. Why? The graph is shown in **Figure 11.12**.

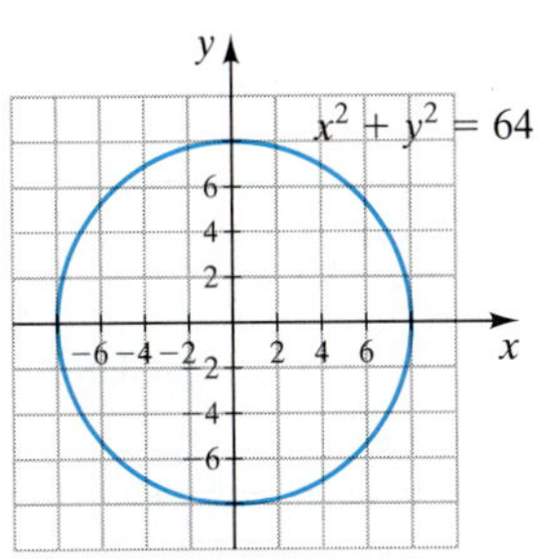


FIGURE 11.10

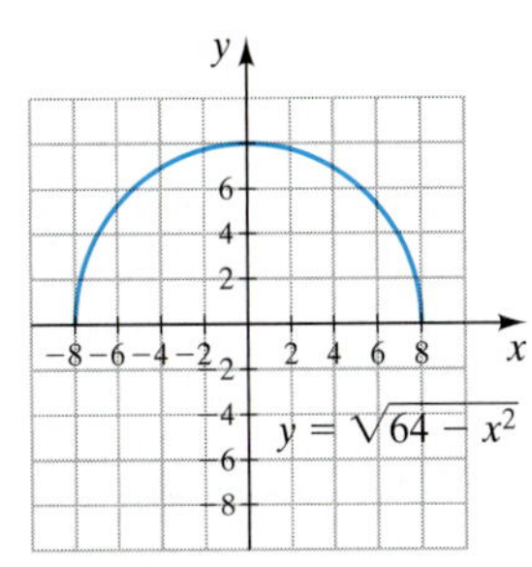


FIGURE 11.11

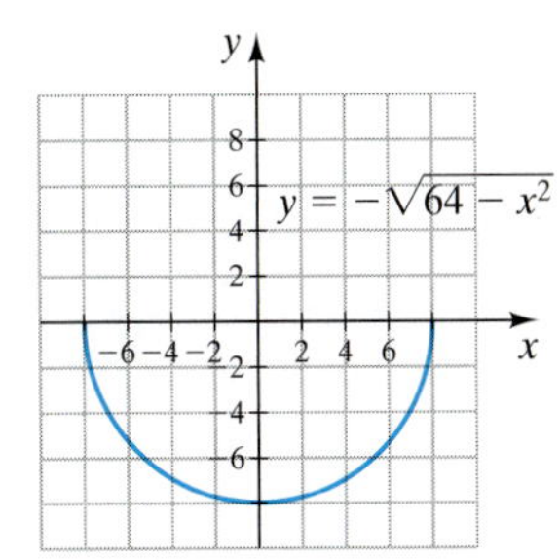


FIGURE 11.12

▸ **Now Try Exercise 101**

Consider the equations $y = \sqrt{64 - x^2}$ and $y = -\sqrt{64 - x^2}$ in Example **6 b)** and **6 c)**. If you square both sides of the equations and rearrange the terms, you will obtain $x^2 + y^2 = 64$. Try this now and see.

**USING YOUR GRAPHING CALCULATOR**

When using your calculator, you insert the function you wish to graph to the right of $y =$. Circles are not functions since they do not pass the vertical line test. To graph the equation $x^2 + y^2 = 64$, which is a circle of radius 8, we solve the equation for $y$ to obtain $y = \pm\sqrt{64 - x^2}$. We then graph the two functions $Y_1 = \sqrt{64 - x^2}$ and $Y_2 = -\sqrt{64 - x^2}$ on the same axes to obtain the circle. These graphs are illustrated in **Figure 11.13**. Because of the distortion (described in the Using Your Graphing Calculator box in Section 9.1), the graph does not appear to be a circle. When you use the SQUARE feature of the calculator, the figure appears as a circle (see **Fig. 11.14**).

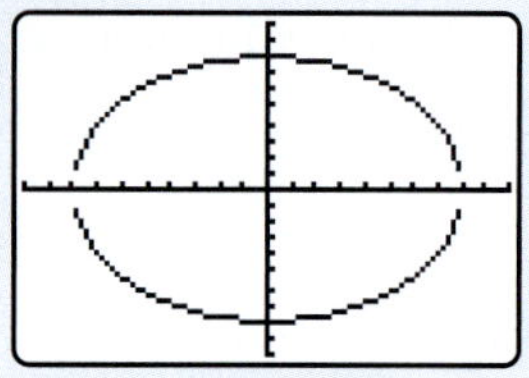

$-10, 10, 1, -10, 10, 1$

FIGURE 11.13

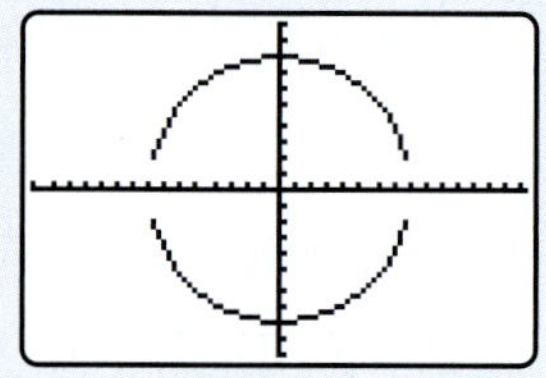

$\approx -15.2, \approx 15.2, 1, -10, 10, 1$

FIGURE 11.14

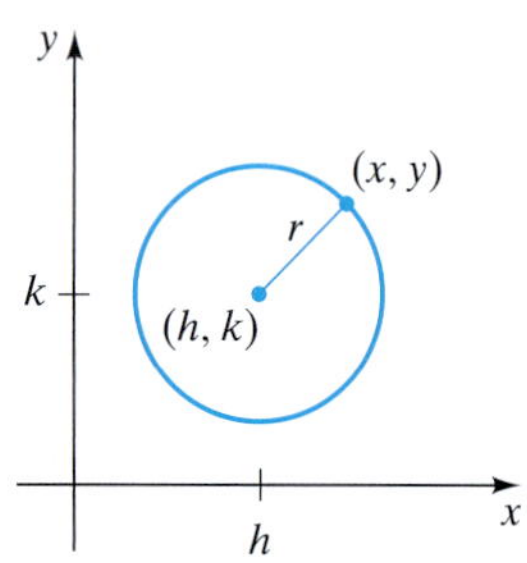


FIGURE 11.15

## 6 Graph Circles with Centers at $(h, k)$

The standard form of a circle with center at $(h, k)$ and radius $r$ can be derived using the distance formula. Let $(h, k)$ be the center of the circle and let $(x, y)$ be any point on the circle (see **Fig. 11.15**). If the radius, $r$, represents the distance between a point, $(x, y)$, on the circle and its center, $(h, k)$, then by the distance formula

$$r = \sqrt{(x - h)^2 + (y - k)^2}$$

We now square both sides of the equation to obtain the standard form of a circle with center at $(h, k)$ and radius $r$.

$$r^2 = (x - h)^2 + (y - k)^2$$

**Circle with Its Center at $(h, k)$ and Radius $r$**

$$(x - h)^2 + (y - k)^2 = r^2$$

**EXAMPLE 7** Determine the equation of the circle shown in **Figure 11.16**.

*Solution* The center is $(-3, 2)$ and the radius is 3.

$$(x - h)^2 + (y - k)^2 = r^2$$
$$[x - (-3)]^2 + (y - 2)^2 = 3^2$$
$$(x + 3)^2 + (y - 2)^2 = 9$$

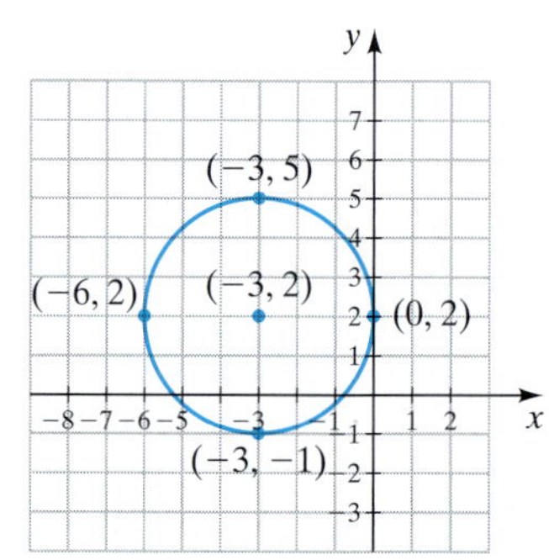


FIGURE 11.16

▸ **Now Try Exercise 87**

**EXAMPLE 8**

**a)** Show that the graph of the equation $x^2 + y^2 + 6x - 2y - 6 = 0$ is a circle.

**b)** Determine the center and radius of the circle and then draw the circle.

**c)** Find the area of the circle.

*Solution*

**a)** We will write this equation in standard form by completing the square. First we rewrite the equation, placing all the terms containing like variables together.

$$x^2 + 6x + y^2 - 2y - 6 = 0$$

Then we move the constant to the right side of the equation.

$$x^2 + 6x + y^2 - 2y = 6$$

Now we complete the square twice, once for each variable. We will first work with the variable $x$.

$$x^2 + 6x + 9 + y^2 - 2y = 6 + 9$$

Now we work with the variable $y$.

$$x^2 + 6x + 9 + y^2 - 2y + 1 = 6 + 9 + 1$$

or

$$\underbrace{x^2 + 6x + 9}_{(x+3)^2} + \underbrace{y^2 - 2y + 1}_{(y-1)^2} = 16$$

$$(x + 3)^2 + (y - 1)^2 = 16$$

$$(x + 3)^2 + (y - 1)^2 = 4^2$$

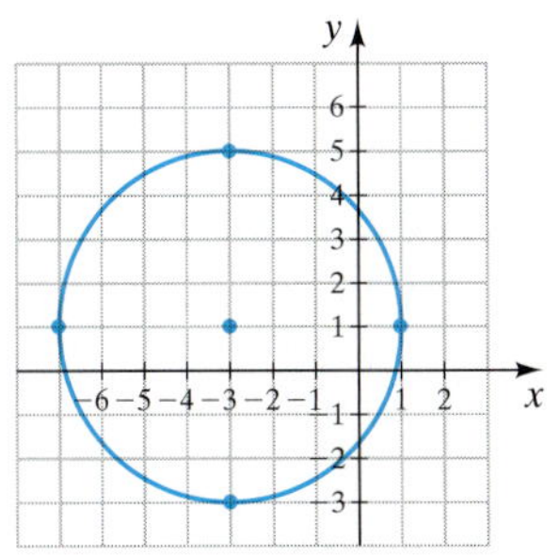

$x^2 + y^2 + 6x - 2y - 6 = 0$

FIGURE 11.17

**b)** The center of the circle is at $(-3, 1)$ and the radius is 4. The circle is sketched in **Figure 11.17**.

**c)** The area is

$$A = \pi r^2 = \pi(4)^2 = 16\pi \approx 50.3 \text{ square units}$$

**Now Try Exercise 111**

## EXERCISE SET 11.1

MathXL® MyMathLab

### Concept/Writing Exercises

1. Name the four conic sections. Draw a picture showing how each is formed.
2. Explain how to determine the direction a parabola will open by examining the equation.
3. Will all parabolas of the form $y = a(x - h)^2 + k, a > 0$ be functions? Explain. What will be the domain and range of $y = a(x - h)^2 + k, a > 0$?
4. Will all parabolas of the form $x = a(y - k)^2 + h, a > 0$ be functions? Explain. What will be the domain and range of $x = a(y - k)^2 + h, a > 0$?
5. How will the graphs of $y = 2(x - 3)^2 + 4$ and $y = -2(x - 3)^2 + 4$ compare?
6. Give the distance formula.
7. When the distance between two different points is found by using the distance formula, why must the distance always be a positive number?
8. Give the midpoint formula.
9. What is the definition of a circle?
10. What is the equation of a circle with center at $(h, k)$?
11. Is $x^2 - y^2 = 9$ an equation for a circle? Explain.
12. Is $-x^2 + y^2 = 25$ an equation for a circle? Explain.
13. Is $2x^2 + 3y^2 = 6$ an equation for a circle? Explain.
14. Is $x = y^2 - 6y + 3$ an equation for a parabola? Explain.
15. Is $x^2 = y^2 - 6y + 3$ an equation for a parabola? Explain.
16. Is $x = y + 2$ an equation for a parabola? Explain.

### Practice the Skills

*Graph each equation.*

17. $y = (x - 2)^2 + 3$
18. $y = (x - 2)^2 - 3$
19. $y = (x + 3)^2 + 2$
20. $y = (x + 3)^2 - 2$
21. $y = (x - 2)^2 - 1$
22. $y = (x + 2)^2 + 1$
23. $y = -(x - 1)^2 + 1$
24. $y = -(x + 4)^2 - 5$
25. $y = -(x + 3)^2 + 4$
26. $y = 2(x + 1)^2 - 3$
27. $y = -3(x - 5)^2 + 3$
28. $x = (y - 1)^2 + 1$
29. $x = (y - 4)^2 - 3$
30. $x = -(y - 2)^2 + 1$
31. $x = -(y - 5)^2 + 4$
32. $x = -2(y - 4)^2 + 4$
33. $x = -5(y + 3)^2 - 6$
34. $x = 3(y + 1)^2 + 5$
35. $y = -2\left(x + \frac{1}{2}\right)^2 + 6$
36. $y = -\left(x - \frac{5}{2}\right)^2 + \frac{1}{2}$

*In Exercises 37–50,* **a)** *Write the equation in the form* $y = a(x-h)^2 + k$ *or* $x = a(y-k)^2 + h$. **b)** *Graph the equation.*

**37.** $y = x^2 + 2x$

**38.** $y = x^2 - 2x$

**39.** $y = x^2 + 6x$

**40.** $y = x^2 - 4x$

**41.** $x = y^2 + 4y$

**42.** $x = y^2 - 6y$

**43.** $y = x^2 + 7x + 10$

**44.** $y = x^2 + 2x - 7$

**45.** $x = -y^2 + 6y - 9$

**46.** $x = -y^2 - 5y - 4$

**47.** $y = -x^2 + 4x - 4$

**48.** $y = 2x^2 - 4x - 4$

**49.** $x = -y^2 + 3y - 4$

**50.** $x = 3y^2 - 12y - 36$

*Determine the distance between each pair of points. Use a calculator where appropriate and round your answers to the nearest hundredth.*

**51.** $(5, -1)$ and $(5, -6)$

**52.** $(-7, 2)$ and $(-3, 2)$

**53.** $(-1, 6)$ and $(8, 6)$

**54.** $(1, 8)$ and $(4, 12)$

**55.** $(-1, -3)$ and $(4, 9)$

**56.** $(-4, -5)$ and $(2, 3)$

**57.** $(-4, -5)$ and $(5, -2)$

**58.** $(6, 7)$ and $(11, 0)$

**59.** $(3, -1)$ and $\left(\frac{1}{2}, 4\right)$

**60.** $\left(-\frac{1}{4}, 2\right)$ and $\left(-\frac{3}{2}, 6\right)$

**61.** $(-1.6, 3.5)$ and $(-4.3, -1.7)$

**62.** $(5.2, -3.6)$ and $(-1.6, 2.3)$

**63.** $(\sqrt{7}, \sqrt{3})$ and $(0, 0)$

**64.** $(-\sqrt{2}, -\sqrt{5})$ and $(0, 0)$

*Determine the midpoint of the line segment between each pair of points.*

**65.** $(1, 3)$ and $(5, 9)$

**66.** $(0, 8)$ and $(4, -6)$

**67.** $(-7, 2)$ and $(7, -2)$

**68.** $(4, 7)$ and $(1, -3)$

**69.** $(-1, 4)$ and $(4, 6)$

**70.** $(-2, -9)$ and $(-6, -3)$

**71.** $\left(3, \frac{1}{2}\right)$ and $(2, -4)$

**72.** $\left(\frac{5}{2}, 3\right)$ and $\left(2, \frac{9}{2}\right)$

**73.** $(\sqrt{3}, 2)$ and $(\sqrt{2}, 7)$

**74.** $(-\sqrt{7}, 8)$ and $(\sqrt{5}, \sqrt{3})$

*Write the equation of each circle with the given center and radius.*

**75.** Center $(0, 0)$, radius 4

**76.** Center $(0, 0)$, radius 7

**77.** Center $(2, 0)$, radius 5

**78.** Center $(-3, 0)$, radius 9

**79.** Center $(0, 5)$, radius 1

**80.** Center $(0, -6)$, radius 6

**81.** Center $(3, 4)$, radius 8

**82.** Center $(-5, 2)$, radius 2

**83.** Center $(7, -6)$, radius 10

**84.** Center $(-6, -1)$, radius 7

**85.** Center $(1, 2)$, radius $\sqrt{5}$

**86.** Center $(-7, -2)$, radius $\sqrt{13}$

*Write the equation of each circle. Assume the radius is a whole number.*

**87.**

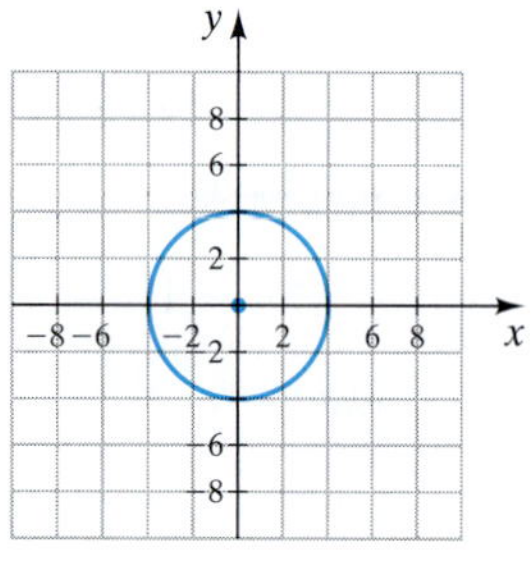


**88.**

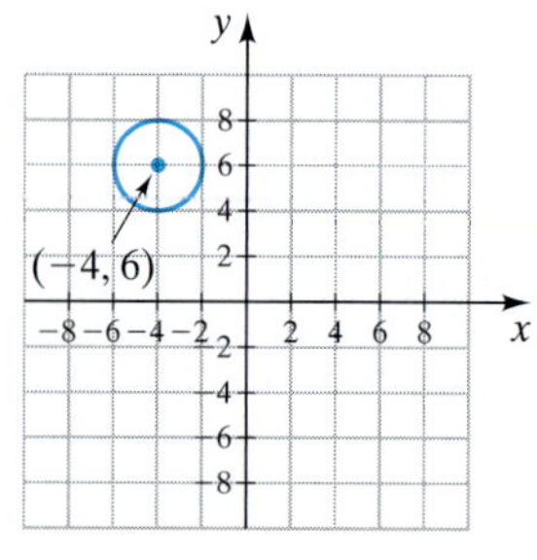

## Group Activity

*Discuss and answer Exercise 135 as a group.*

**135. Equation of a Parabola** The equation of a parabola can be found if three points on the parabola are known. To do so, start with $y = ax^2 + bx + c$. Then substitute the $x$- and $y$-coordinates of the first point into the equation. This will result in an equation in $a$, $b$, and $c$. Repeat the procedure for the other two points. This process yields a system of three equations in three variables. Next solve the system for $a$, $b$, and $c$. To find the equation of the parabola, substitute the values found for $a$, $b$, and $c$ into the equation $y = ax^2 + bx + c$.

Three points on a parabola are $(0, 12)$, $(3, -3)$, and $(-2, 32)$.

**a)** Individually, find a system of equations in three variables that can be used to find the equation of the parabola. Then compare your answers. If each member of the group does not have the same system, determine why.

**b)** Individually, solve the system and determine the values of $a$, $b$, and $c$. Then compare your answers.

**c)** Individually, write the equation of the parabola passing through $(0, 12)$, $(3, -3)$, and $(-2, 32)$. Then compare your answers.

**d)** Individually, write the equation in

$$y = a(x - h)^2 + k$$

form. Then compare your answers.

**e)** Individually, graph the equation in part **d)**. Then compare your answers.

## Cumulative Review Exercises

[1.5] **136.** Simplify $\dfrac{6x^{-3}y^4}{18x^{-2}y^3}$.

[2.5] **137.** Solve the inequality $-4 < 3x - 4 < 17$. Write the solution in interval notation.

[4.5] **138.** Evaluate the determinant.

$$\begin{vmatrix} 4 & 0 & 3 \\ 5 & 2 & -1 \\ 3 & 6 & 4 \end{vmatrix}$$

[5.2] **139. a)** Write expressions to represent each of the four areas shown in the figure.

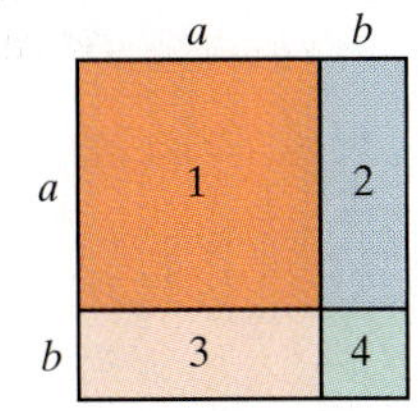


**b)** Express the total area shown as the square of a binomial.

[11.1] **140.** Graph $y = (x - 4)^2 + 1$.

# 11.2 The Ellipse

**1** Graph ellipses.

**2** Graph ellipses with centers at $(h, k)$.

## 1 Graph Ellipses

An **ellipse** may be defined as a set of points in a plane, the sum of whose distances from two fixed points is a constant. The two fixed points are called the **foci** (each is a focus) of the ellipse (see **Fig. 11.18**). In **Figure 11.18**, $F_1$ and $F_2$ represent the two foci.

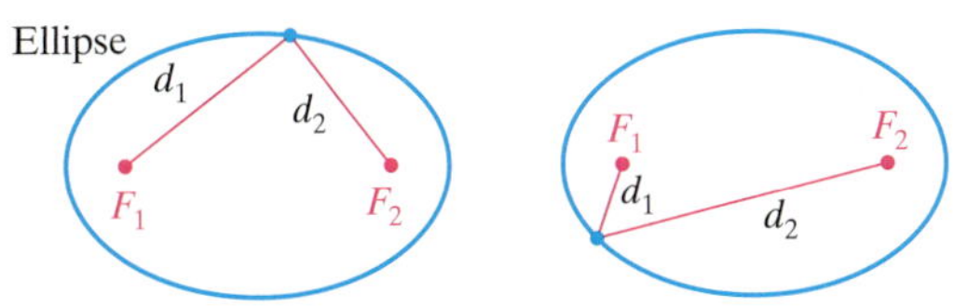


FIGURE 11.18

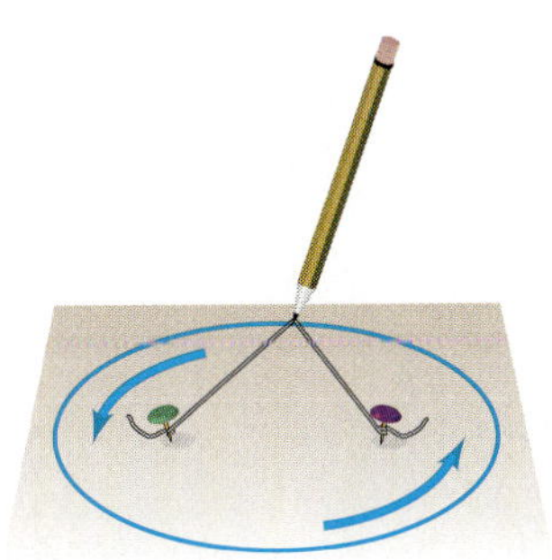

FIGURE 11.19

We can construct an ellipse using a length of string and two thumbtacks. Place the two thumbtacks fairly close together (**Fig. 11.19**). Then tie the ends of the string to the thumbtacks. With a pencil or pen pull the string taut, and, while keeping the string taut, draw the ellipse by moving the pencil around the thumbtacks.

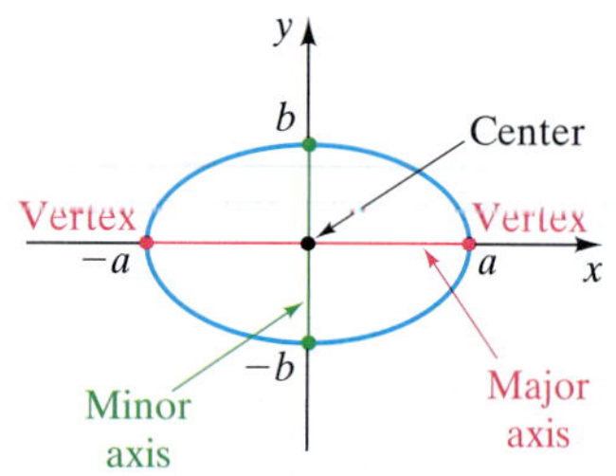


FIGURE 11.20

In **Figure 11.20**, the line segment from $-a$ to $a$ on the $x$-axis is the *longer* or **major axis** and the line segment from $-b$ to $b$ is the *shorter* or **minor axis** of the ellipse. The major axis of an ellipse may also be on the $y$-axis. **Figure 11.20** also shows the *center* of the ellipse and two vertices (the red dots). The vertices are the endpoints of the major axis.

The standard form of an ellipse with its center at the origin follows.

**Ellipse with Its Center at the Origin**

$$\frac{x^2}{a^2} + \frac{y^2}{b^2} = 1$$

where $(a, 0)$ and $(-a, 0)$ are the $x$-intercepts and $(0, b)$ and $(0, -b)$ are the $y$-intercepts.

Notice the $x$-intercepts are found using the constant in the denominator of the $x^2$-term, and the $y$-intercepts are found using the constant in the denominator of the $y^2$-term. If $a^2 > b^2$, the major axis of the ellipse will be along the $x$-axis. If $b^2 > a^2$, the major axis of the ellipse will be along the $y$-axis.

In Example 1, the major axis of the ellipse is along the $x$-axis.

**EXAMPLE 1** ▸ Graph $\dfrac{x^2}{9} + \dfrac{y^2}{4} = 1$.

*Solution* We can rewrite the equation as

$$\frac{x^2}{3^2} + \frac{y^2}{2^2} = 1$$

Thus, $a = 3$ and the $x$-intercepts are $(3, 0)$ and $(-3, 0)$. Since $b = 2$, the $y$-intercepts are $(0, 2)$ and $(0, -2)$. The ellipse is illustrated in **Figure 11.21**.

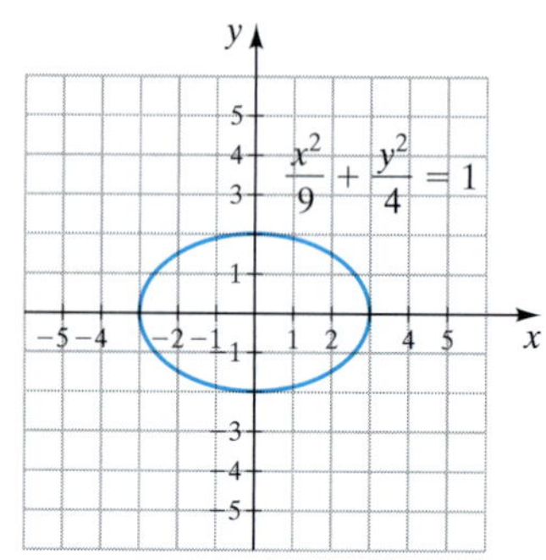


FIGURE 11.21

▸ **Now Try Exercise 15**

An equation may be written so that it may not be obvious that its graph is an ellipse. This is illustrated in Example 2.

**EXAMPLE 2** ▸ Graph $20x^2 + 9y^2 = 180$.

*Solution* To make the right side of the equation equal to 1, we divide both sides of the equation by 180. We then obtain an equation that we can recognize as an ellipse.

$$\frac{20x^2 + 9y^2}{180} = \frac{180}{180}$$

$$\frac{20x^2}{180} + \frac{9y^2}{180} = 1$$

$$\frac{x^2}{9} + \frac{y^2}{20} = 1$$

The equation can now be recognized as an ellipse in standard form.

$$\frac{x^2}{a^2} + \frac{y^2}{b^2} = 1$$

Since $a^2 = 9, a = 3$. We know that $b^2 = 20$; thus $b = \sqrt{20}$ (or approximately 4.47).

$$\frac{x^2}{3^2} + \frac{y^2}{(\sqrt{20})^2} = 1$$

The $x$-intercepts are $(3, 0)$ and $(-3, 0)$. The $y$-intercepts are $(0, -\sqrt{20})$ and $(0, \sqrt{20})$. The graph is illustrated in **Figure 11.22**. Note that the major axis lies along the $y$-axis.

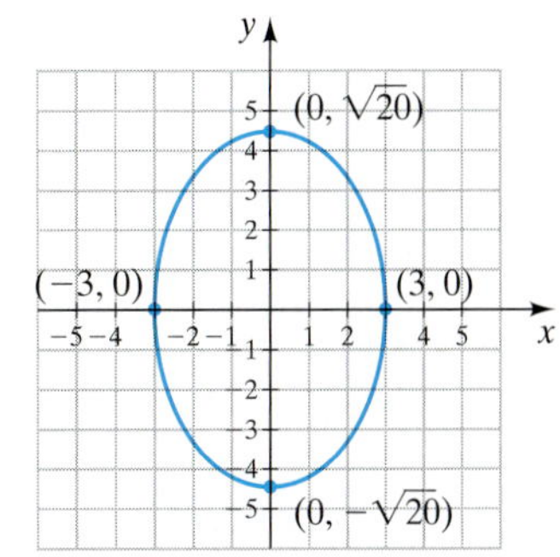


$20x^2 + 9y^2 = 180$

FIGURE 11.22

▸ **Now Try Exercise 19**

In Example 1, since $a^2 = 9$ and $b^2 = 4$ and $a^2 > b^2$, the major axis is along the $x$-axis. In Example 2, since $a^2 = 9$ and $b^2 = 20$ and $b^2 > a^2$, the major axis is along the $y$-axis. In the specific case where $a^2 = b^2$, the figure is a circle. Thus, the circle is a special case of an ellipse.

**EXAMPLE 3** ▸ Write the equation of the ellipse shown in **Figure 11.23**.

*Solution* The $x$-intercepts are $(-\sqrt{10}, 0)$ and $(\sqrt{10}, 0)$; thus, $a = \sqrt{10}$ and $a^2 = 10$. The $y$-intercepts are $(0, -12)$ and $(0, 12)$; thus, $b = 12$ and $b^2 = 144$.

$$\frac{x^2}{a^2} + \frac{y^2}{b^2} = 1$$

$$\frac{x^2}{10} + \frac{y^2}{144} = 1$$

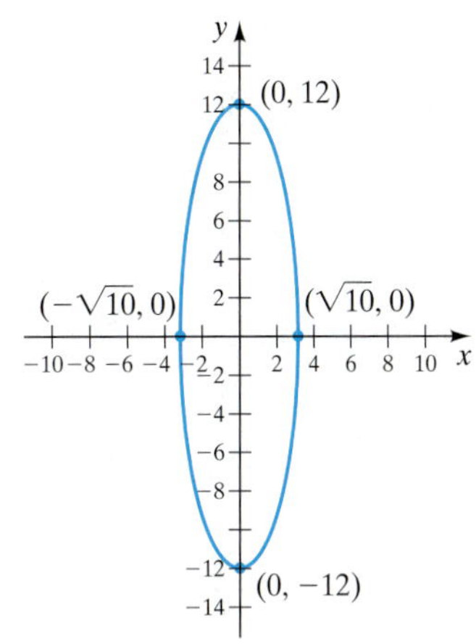


FIGURE 11.23

▸ **Now Try Exercise 45**

The formula for the **area of an ellipse** is $A = \pi ab$. In Example 1, where $a = 3$ and $b = 2$, the area is $A = \pi(3)(2) = 6\pi \approx 18.8$ square units.

In Example 2, where $a = 3$ and $b = \sqrt{20}$, the area is $A = \pi(3)(\sqrt{20}) = \pi(3)(2\sqrt{5}) = 6\pi\sqrt{5} \approx 42.1$ square units.

## 2 Graph Ellipses with Centers at $(h, k)$

Horizontal and vertical translations, similar to those used in Chapter 8, may be used to obtain the equation of an ellipse with center at $(h, k)$.

**Ellipse with Its Center at $(h, k)$**

$$\frac{(x-h)^2}{a^2} + \frac{(y-k)^2}{b^2} = 1$$

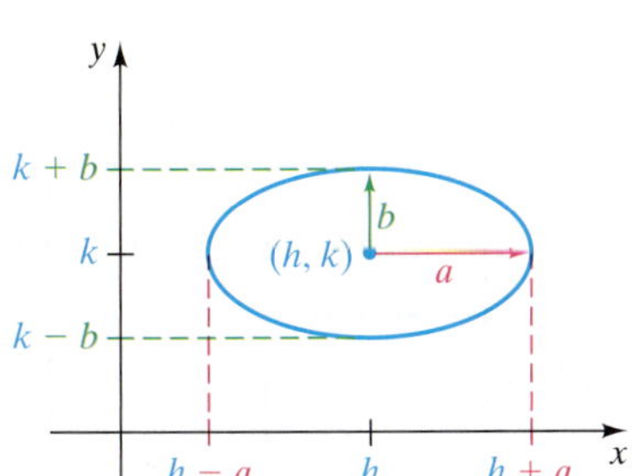


FIGURE 11.24

In the formula, the $h$ shifts the graph left or right from the origin and $k$ shifts the graph up or down from the origin, as shown in **Figure 11.24**.

**EXAMPLE 4** ▸ Graph $\frac{(x-2)^2}{25} + \frac{(y+3)^2}{16} = 1$.

*Solution* This is the graph of $\frac{x^2}{25} + \frac{y^2}{16} = 1$ or $\frac{x^2}{5^2} + \frac{y^2}{4^2} = 1$ translated so that its center is at $(2, -3)$. Note that $a = 5$ and $b = 4$. The graph is shown in **Figure 11.25.**

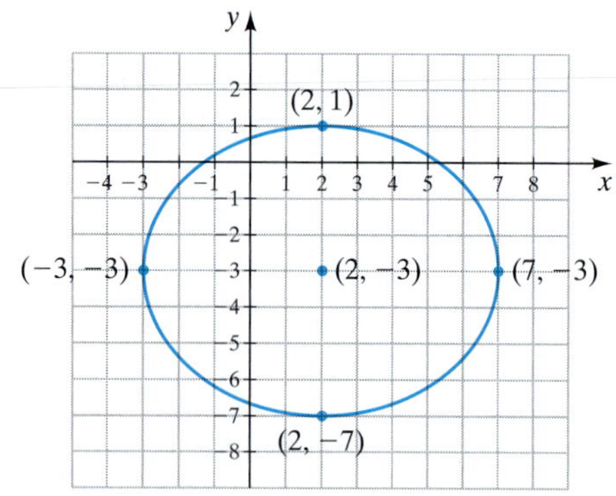


FIGURE 11.25

▸ **Now Try Exercise 33**

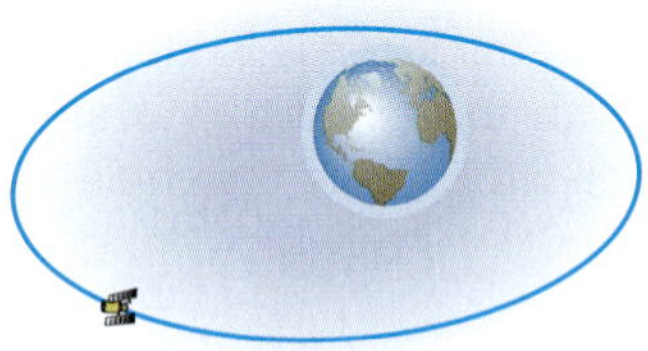

FIGURE 11.26

An understanding of ellipses is useful in many areas. Astronomers know that planets revolve in elliptical orbits around the Sun. Communications satellites move in elliptical orbits around Earth (see **Fig. 11.26**).

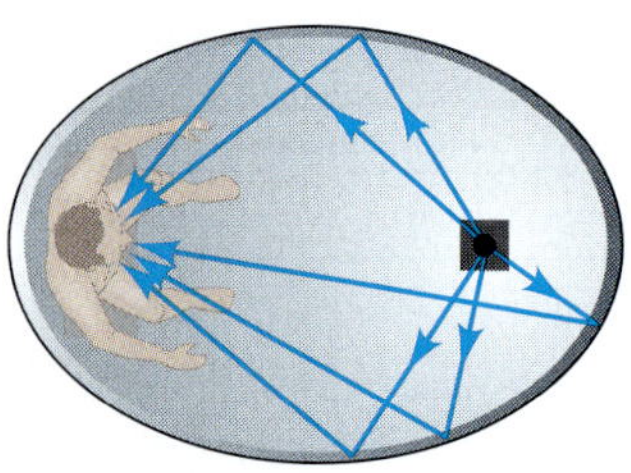

FIGURE 11.27

Ellipses are used in medicine to smash kidney stones. When a signal emerges from one focus of an ellipse, the signal is reflected to the other focus. In kidney stone machines, the person is situated so that the stone to be smashed is at one focus of an elliptically shaped chamber called a lithotripter (see **Fig. 11.27** and Exercises 57 and 58).

In certain buildings with ellipsoidal ceilings, a person standing at one focus can whisper something and a person standing at the other focus can clearly hear what the person whispered. There are many other uses for ellipses, including lamps that are made to concentrate light at a specific point.

## USING YOUR GRAPHING CALCULATOR

Ellipses are not functions. To graph ellipses on a graphing calculator, we solve the equation for $y$. This will give the two equations that we use to graph the ellipse.

In Example 1, we graphed $\frac{x^2}{9} + \frac{y^2}{4} = 1$. Solving this equation for $y$, we get

$$\frac{x^2}{9} + \frac{y^2}{4} = 1$$

$$36 \cdot \frac{x^2}{9} + 36 \cdot \frac{y^2}{4} = 1 \cdot 36 \qquad \textit{Multiply by the LCD.}$$

$$4x^2 + 9y^2 = 36$$

$$9y^2 = 36 - 4x^2$$

$$y^2 = \frac{36 - 4x^2}{9}$$

$$y^2 = \frac{4(9 - x^2)}{9} \qquad \textit{Factor 4 from the numerator.}$$

$$y = \pm\frac{2}{3}\sqrt{9 - x^2} \qquad \textit{Square root property}$$

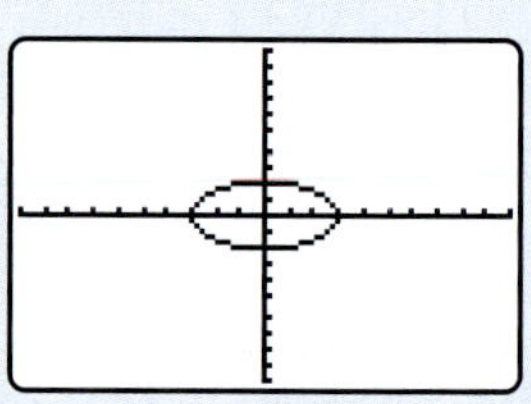

FIGURE 11.28

To graph the ellipse, we let $Y_1 = \frac{2}{3}\sqrt{9 - x^2}$ and $Y_2 = -\frac{2}{3}\sqrt{9 - x^2}$ and graph both equations. The graphs of $Y_1$ and $Y_2$ are illustrated in **Figure 11.28**.

# EXERCISE SET 11.2

MathXL® MyMathLab

## Concept/Writing Exercises

1. What is the definition of an ellipse?
2. What is the equation of an ellipse with its center at the origin?
3. What is the equation of an ellipse with its center at $(h, k)$?
4. Discuss the graphs of $\frac{x^2}{a^2} + \frac{y^2}{b^2} = 1$ when $a > b$, $a < b$, and $a = b$.
5. Explain why the circle is a special case of the ellipse.
6. In the formula $\frac{x^2}{a^2} + \frac{y^2}{b^2} = 1$, what do the $a$ and $b$ represent?
7. What is the first step in graphing the ellipse whose equation is $10x^2 + 36y^2 = 180$?
8. What is the first step in graphing the ellipse whose equation is $9x^2 + 25y^2 = 225$?
9. Is $\frac{x^2}{36} - \frac{y^2}{49} = 1$ an equation for an ellipse? Explain.
10. Is $-\frac{x^2}{49} + \frac{y^2}{81} = 1$ an equation for an ellipse? Explain.

## Practice the Skills

*Graph each equation.*

11. $\frac{x^2}{4} + \frac{y^2}{1} = 1$
12. $\frac{x^2}{1} + \frac{y^2}{4} = 1$
13. $\frac{x^2}{4} + \frac{y^2}{9} = 1$
14. $\frac{x^2}{9} + \frac{y^2}{4} = 1$
15. $\frac{x^2}{25} + \frac{y^2}{9} = 1$
16. $\frac{x^2}{100} + \frac{y^2}{16} = 1$
17. $\frac{x^2}{16} + \frac{y^2}{25} = 1$
18. $\frac{x^2}{81} + \frac{y^2}{49} = 1$
19. $x^2 + 16y^2 = 16$
20. $x^2 + 25y^2 = 25$
21. $49x^2 + y^2 = 49$
22. $9x^2 + 25y^2 = 225$
23. $9x^2 + 16y^2 = 144$
24. $25x^2 + 4y^2 = 100$
25. $25x^2 + 100y^2 = 400$
26. $100x^2 + 25y^2 = 400$
27. $x^2 + 2y^2 = 8$
28. $x^2 + 36y^2 = 36$
29. $\frac{x^2}{16} + \frac{(y-2)^2}{9} = 1$
30. $\frac{(x-1)^2}{16} + \frac{y^2}{1} = 1$
31. $\frac{(x-4)^2}{9} + \frac{(y+3)^2}{25} = 1$
32. $\frac{(x-3)^2}{25} + \frac{(y+2)^2}{49} = 1$
33. $\frac{(x+1)^2}{9} + \frac{(y-2)^2}{4} = 1$
34. $\frac{(x-3)^2}{16} + \frac{(y-4)^2}{25} = 1$
35. $(x+3)^2 + 9(y+1)^2 = 81$
36. $18(x-1)^2 + 2(y+3)^2 = 72$
37. $(x-5)^2 + 4(y+4)^2 = 4$
38. $4(x-2)^2 + 9(y+2)^2 = 36$
39. $12(x+4)^2 + 3(y-1)^2 = 48$
40. $16(x-2)^2 + 4(y+3)^2 = 16$

## Problem Solving

41. Find the area of the ellipse in Exercise 11.
42. Find the area of the ellipse in Exercise 15.
43. How many points are on the graph of $16x^2 + 25y^2 = 0$? Explain.
44. Consider the graph of the equation $\frac{x^2}{a^2} + \frac{y^2}{b^2} = 1$. Explain what will happen to the shape of the graph as the value of $b$ gets closer to the value of $a$. What is the shape of the graph when $a = b$?

*In Exercises 45–48, find the equation of the ellipse that has the four points as endpoints of the major and minor axes.*

45. $(3, 0), (-3, 0), (0, 4), (0, -4)$
46. $(6, 0), (-6, 0), (0, 5), (0, -5)$
47. $(2, 0), (-2, 0), (0, 3), (0, -3)$
48. $(1, 0), (-1, 0), (0, 7), (0, -7)$
49. How many points of intersection will the graphs of the equations $x^2 + y^2 = 49$ and $\frac{x^2}{16} + \frac{y^2}{25} = 1$ have? Explain.
50. How many points of intersection will the graphs of the equations $y = 2(x-2)^2 - 3$ and $\frac{(x-2)^2}{4} + \frac{(y+3)^2}{9} = 1$ have? Explain.

*In Exercises 51 and 52, write the following equation in standard form. Determine the center of each ellipse.*

51. $x^2 + 4y^2 + 6x + 16y - 11 = 0$
52. $x^2 + 4y^2 - 4x - 8y - 92 = 0$

**53. Art Gallery** An art gallery has an elliptical hall. The maximum distance from one focus to the wall is 90.2 feet and the minimum distance is 20.7 feet. Find the distance between the foci.

**54. Communications Satellite** A space shuttle transported a communications satellite to space. The satellite travels in an elliptical orbit around Earth. The maximum distance of the satellite from Earth is 23,200 miles and the minimum distance is 22,800 miles. Earth is at one focus of the ellipse. Find the distance from Earth to the other focus.

**55. Tunnel through a Mountain** The tunnel in the photo is the top half of an ellipse. The width of the tunnel is 20 feet and the height is 24 feet.

**a)** If you pictured a completed ellipse with the center of the ellipse being at the center of the road, determine the equation of the ellipse.

**b)** Find the area of the ellipse found in part **a)**.

**c)** Find the area of the opening of the tunnel.

**56. Billiard Table** An elliptical billiard table is 8 feet long by 5 feet wide. Determine the location of the foci. On such a table, if a ball is put at each focus and one ball is hit with enough force, it would hit the ball at the other focus no matter where it banks on the table.

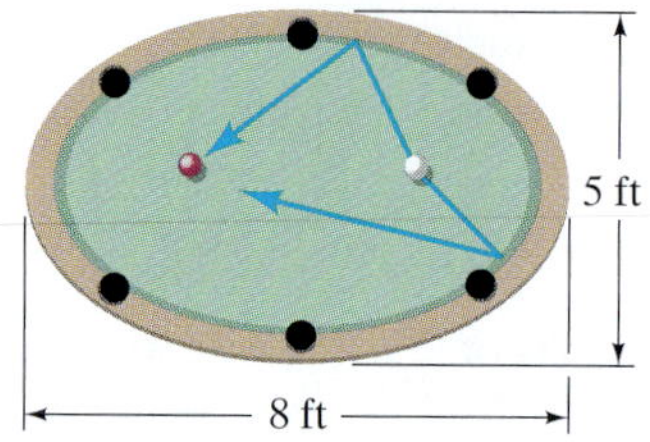


**57. Lithotripter Machine** Suppose the lithotripter machine described on page 732 is 6 feet long and 4 feet wide. Describe the location of the foci.

**58. Lithotripter** On page 732 we gave a brief introduction to the lithotripter, which uses ultrasound waves to shatter kidney stones. Do research and write a detailed report describing the procedure used to shatter kidney stones. Make sure that you explain how the waves are directed on the stone.

**59. Whispering Gallery** The National Statuary Hall in the Capitol Building in Washington, D.C., is a "whispering gallery." Do research and explain why one person standing at a certain point can whisper something and someone standing a considerable distance away can hear it.

**60.** Check your answer to Exercise 11 on your grapher.

**61.** Check your answer to Exercise 17 on your grapher.

## Challenge Problems

*Determine the equation of the ellipse that has the following four points as endpoints of the major and minor axes.*

**62.** $(-7, 3), (5, 3), (-1, 5), (-1, 1)$

**63.** $(-3, 2), (11, 2), (4, 5), (4, -1)$

## Group Activity

*Work Exercise 64 individually. Then compare your answers.*

**64. Tunnel** The photo shows an elliptical tunnel (with the bottom part of the ellipse not shown) near Rockefeller Center in New York City. The maximum width of the tunnel is 18 feet and the maximum height *from the ground to the top* is 10.5 feet.

**a)** If the *completed ellipse* would have a maximum height of 15 feet, how high from the ground is the center of the elliptical tunnel?

**b)** Consider the following graph, which could be used to represent the tunnel.

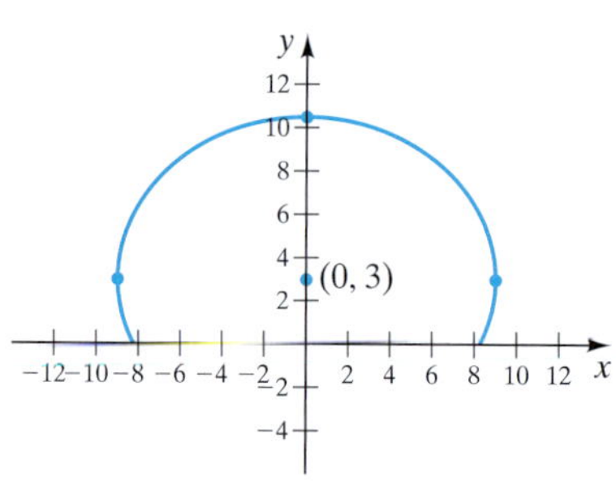


If the ellipse were continued, what would be the other $y$-intercept of the graph?

**c)** Write the equation of the ellipse, if completed, in part **b)**.

## Cumulative Review Exercises

[2.2] **65.** Solve the formula $S = \frac{n}{2}(f + l)$ for $l$.

[5.4] **66.** Divide $\frac{2x^2 + 2x - 7}{2x - 3}$.

[7.6] **67.** Solve $\sqrt{3b - 2} = 10 - b$.

[8.6] **68.** Solve $\frac{3x + 5}{x - 4} \le 0$, and give the solution in interval notation.

[9.7] **69.** Find $\log_8 321$.

## Mid-Chapter Test: 11.1–11.2

*To find out how well you understand the chapter material to this point, take this brief test. The answers, and the section where the material was initially discussed, are given in the back of the book. Review any questions you answered incorrectly.*

*Graph each equation.*

**1.** $y = (x - 2)^2 - 1$

**2.** $y = -(x + 1)^2 + 3$

**3.** $x = -(y - 4)^2 + 1$

**4.** $x = 2(y + 3)^2 - 2$

**5.** $y = x^2 + 6x + 10$

*Find the distance between each pair of points. Where appropriate, round your answers to the nearest hundredth.*

**6.** $(-7, 4)$ and $(-2, -8)$

**7.** $(5, -3)$ and $(2, 9)$

*Find the midpoint of the line segment between each pair of points.*

**8.** $(9, -1)$ and $(-11, 6)$

**9.** $\left(-\frac{5}{2}, 7\right)$ and $\left(8, \frac{1}{2}\right)$

**10.** Write the equation of the circle with center at $(-3, 2)$ and a radius of 5 units.

*Graph each equation.*

**11.** $x^2 + (y - 1)^2 = 16$

**12.** $y = \sqrt{36 - x^2}$

**13.** $x^2 + y^2 - 2x + 4y - 4 = 0$

**14.** What is the definition of a circle?

*Graph each equation.*

**15.** $\frac{x^2}{4} + \frac{y^2}{9} = 1$

**16.** $\frac{x^2}{81} + \frac{y^2}{25} = 1$

**17.** $\frac{(x - 1)^2}{49} + \frac{(y + 2)^2}{4} = 1$

**18.** $36(x + 3)^2 + (y - 4)^2 = 36$.

**19.** Find the area of the ellipse in Exercise 15.

**20.** Find the equation of the ellipse that has the four points $(8, 0)$, $(-8, 0)$, $(0, 5)$, and $(0, -5)$ as the endpoints of the major and minor axes.

# 11.3 The Hyperbola

1 Graph hyperbolas.

2 Review conic sections.

## 1 Graph Hyperbolas

A **hyperbola** is the set of points in a plane, the difference of whose distances from two fixed points (called foci) is a constant. A hyperbola is illustrated in **Figure 11.29a**. In the figure, for every point on the hyperbola, the difference $M - N$ is the same constant. A hyperbola may look like a pair of parabolas. However, the shapes are actually quite different. A hyperbola has two **vertices.** The point halfway between the two vertices is the **center** of the hyperbola. The line through the vertices is called the **transverse axis.** In **Figure 11.29b** the transverse axis lies along the $x$-axis, and in **Figure 11.29c**, the transverse axis lies along the $y$-axis.

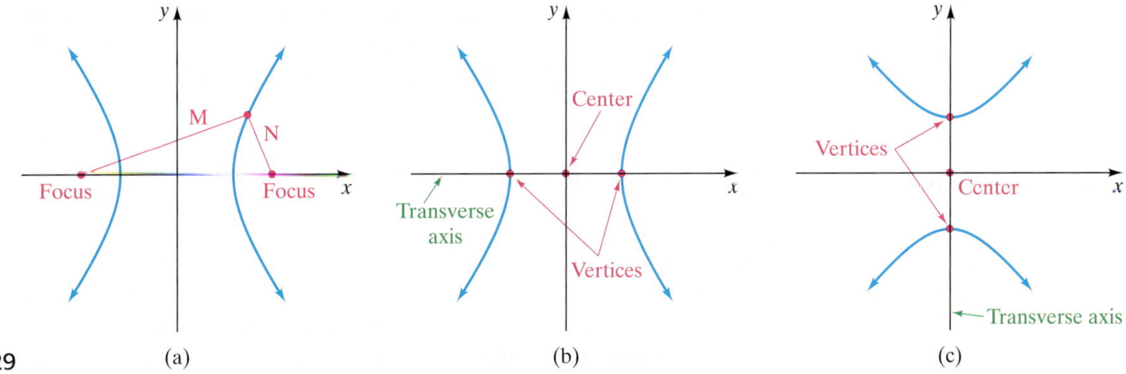


FIGURE 11.29

The dashed lines in **Figure 11.30** are called **asymptotes**. The asymptotes are not a part of the hyperbola but are used as an aid in graphing hyperbolas. (We will discuss asymptotes shortly.) Also given in **Figure 11.30** is the standard form of the equation for each hyperbola. In **Figure 11.30a**, both vertices are $a$ units from the origin. In **Figure 11.30b**, both vertices are $b$ units from the origin. Note that in the standard form of the equation, the denominator of the $x^2$ term is always $a^2$ and the denominator of the $y^2$ term is always $b^2$.

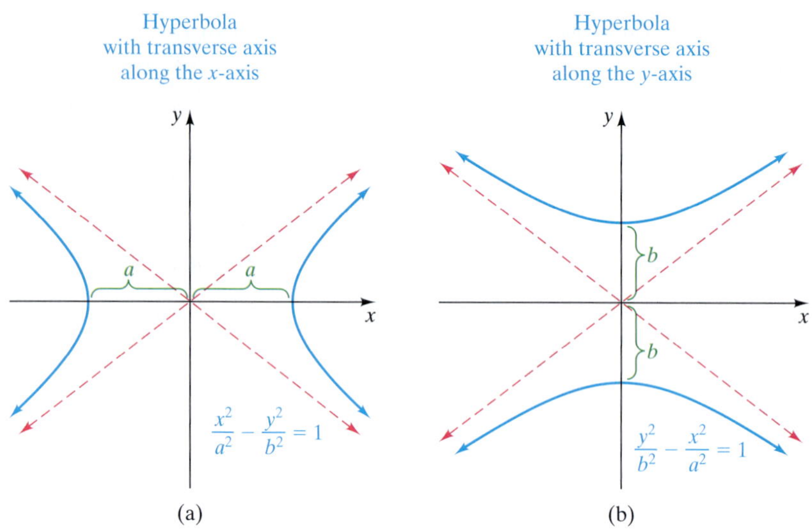


FIGURE 11.30

A hyperbola centered at the origin whose transverse axis is one of the coordinate axes has either $x$-intercepts (**Fig. 11.30a**) or $y$-intercepts (**Fig. 11.30b**), but not both. When a hyperbola is centered at the origin, the intercepts are the vertices of the hyperbola. When written in standard form, the intercepts will be on the axis indicated by the variable with the positive coefficient. The intercepts will be the positive and the negative square root of the denominator of the positive term.

| Examples | Intercepts on | Intercepts |
|---|---|---|
| $\frac{x^2}{49} - \frac{y^2}{16} = 1$ | $x$-axis | $(-7, 0)$ and $(7, 0)$ |
| $\frac{y^2}{16} - \frac{x^2}{49} = 1$ | $y$-axis | $(0, -4)$ and $(0, 4)$ |

Asymptotes can help you graph hyperbolas. The asymptotes are two straight lines that go through the center of the hyperbola (see **Fig. 11.30**). As the values of $x$ and $y$ get larger, the graph of the hyperbola approaches the asymptotes. The equations of the asymptotes of a hyperbola whose center is the origin are

$$y = \frac{b}{a}x \quad \text{and} \quad y = -\frac{b}{a}x$$

The asymptotes can be drawn quickly by plotting the four points $(a, b)$, $(-a, b)$, $(a, -b)$, and $(-a, -b)$, and then connecting these points with dashed lines to form a rectangle. Next, draw dashed lines through the opposite corners of the rectangle to obtain the asymptotes.

## Hyperbola with Its Center at the Origin

| TRANSVERSE AXIS ALONG *x*-AXIS (OPENS TO THE RIGHT AND LEFT) | TRANSVERSE AXIS ALONG *y*-AXIS (OPENS UPWARD AND DOWNWARD) |
| --- | --- |
| $\dfrac{x^2}{a^2} - \dfrac{y^2}{b^2} = 1$ | $\dfrac{y^2}{b^2} - \dfrac{x^2}{a^2} = 1$ |

**ASYMPTOTES**

$$y = \frac{b}{a}x \quad \text{and} \quad y = -\frac{b}{a}x$$

### EXAMPLE 1

**a)** Determine the equations of the asymptotes of the hyperbola with equation

$$\frac{x^2}{9} - \frac{y^2}{16} = 1$$

**b)** Draw the hyperbola using the asymptotes.

### Solution

**a)** The value of $a^2$ is 9; the positive square root of 9 is 3. The value of $b^2$ is 16; the positive square root of 16 is 4. The asymptotes are

$$y = \frac{b}{a}x \quad \text{and} \quad y = -\frac{b}{a}x$$

or

$$y = \frac{4}{3}x \quad \text{and} \quad y = -\frac{4}{3}x$$

**b)** To graph the hyperbola, we first graph the asymptotes. To graph the asymptotes, we can plot the points $(3, 4)$, $(-3, 4)$, $(3, -4)$, and $(-3, -4)$ and draw the rectangle as illustrated in **Figure 11.31**. The asymptotes are the dashed lines through the opposite corners of the rectangle.

Since the $x$-term in the original equation is positive, the graph intersects the $x$-axis. Since the denominator of the positive term is 9, the vertices are at $(3, 0)$ and $(-3, 0)$. Now draw the hyperbola by letting the hyperbola approach its asymptotes (**Fig. 11.32**). Note that the asymptotes are drawn using dashed lines since they are not part of the hyperbola. They are used merely to help draw the graph.

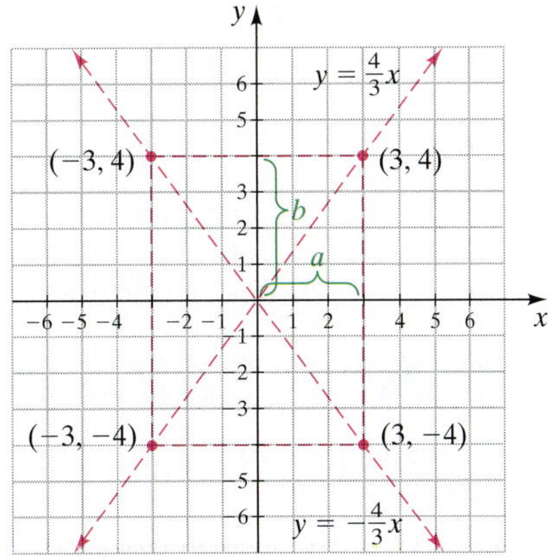


FIGURE 11.31

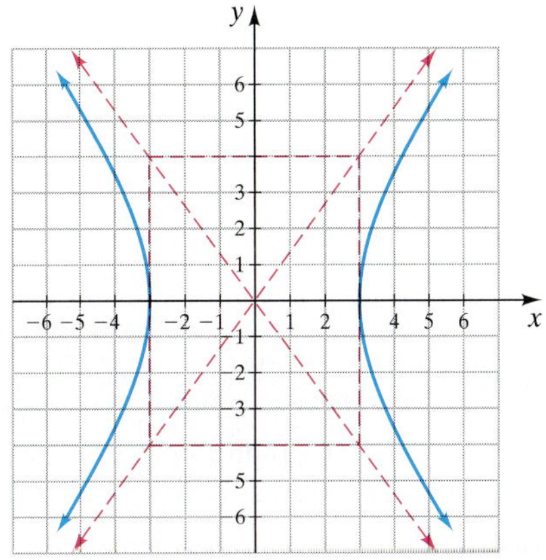


$\dfrac{x^2}{9} - \dfrac{y^2}{16} = 1$

FIGURE 11.32

Now Try Exercise 21

**EXAMPLE 2**

**a)** Show that the equation $-25x^2 + 4y^2 = 100$ is a hyperbola by expressing the equation in standard form.

**b)** Determine the equations of the asymptotes of the graph.

**c)** Draw the graph.

*Solution*

**a)** We divide both sides of the equation by 100 to obtain 1 on the right side of the equation.

$$\frac{-25x^2 + 4y^2}{100} = \frac{100}{100}$$

$$\frac{-25x^2}{100} + \frac{4y^2}{100} = 1$$

$$\frac{-x^2}{4} + \frac{y^2}{25} = 1$$

Rewriting the equation in standard form (positive term first), we get

$$\frac{y^2}{25} - \frac{x^2}{4} = 1$$

**b)** Since $a = 2$ and $b = 5$, the equations of the asymptotes are

$$y = \frac{5}{2}x \quad \text{and} \quad y = -\frac{5}{2}x$$

**c)** The graph intersects the $y$-axis at $(0, 5)$ and $(0, -5)$. **Figure 11.33a** illustrates the asymptotes, and **Figure 11.33b** illustrates the hyperbola.

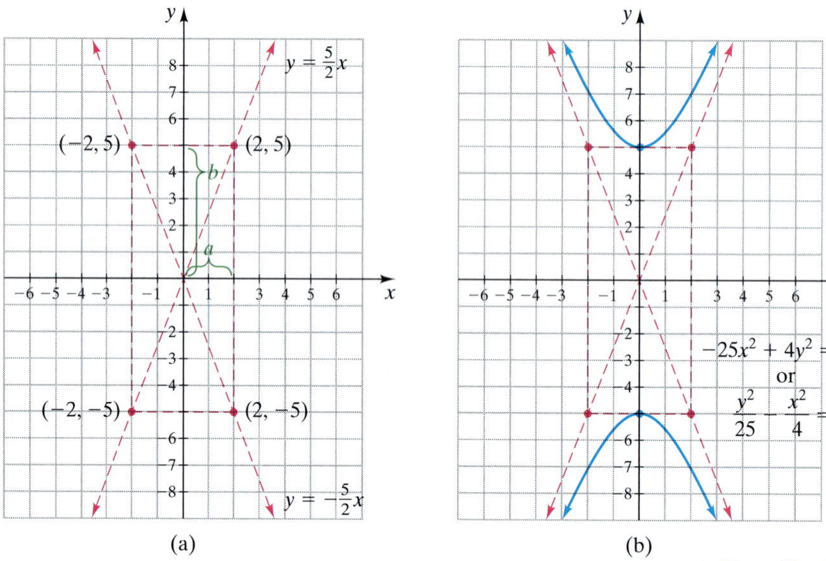


FIGURE 11.33

**Now Try Exercise 29**

We have discussed hyperbolas with their centers at the origin. Hyperbolas do not have to be centered at the origin. In this book, we will not discuss such hyperbolas.

**USING YOUR GRAPHING CALCULATOR**

We can graph hyperbolas just as we did circles and ellipses. To graph hyperbolas on the graphing calculator, solve the equation for $y$ and graph each part. Consider Example 1,

$$\frac{x^2}{9} - \frac{y^2}{16} = 1.$$

Show that if you solve this equation for $y$ you get $y = \pm\frac{4}{3}\sqrt{x^2 - 9}$. Let $Y_1 = \frac{4}{3}\sqrt{x^2 - 9}$ and $Y_2 = -\frac{4}{3}\sqrt{x^2 - 9}$. **Figures 11.34a, 11.34b, 11.34c,** and **11.34d**, on the next page, give the graphs of $Y_1$ and $Y_2$ for different window settings. The window settings used are indicated above each graph.

(*continued on the next page*)

| Standard setting | ZOOM: option 5 ZSquare setting | ZOOM: option 4, ZDecimal setting | Set window as shown below figure (called "friendly window settings") |
|---|---|---|---|

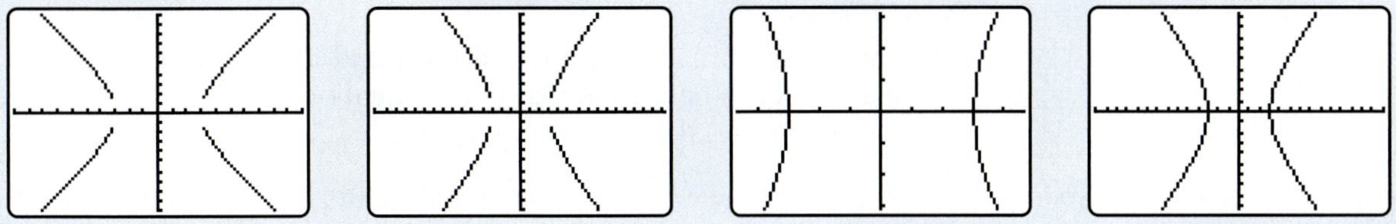

| −10, 10, 1, −10, 10, 1 | ≈−15.2, ≈15.2, 1, −10, 10, 1 | −4.7, 4.7, 1, −3.1, 3.1, 1 | −14.1, 14.1, 1, −9.3, 9.3, 1 |
|---|---|---|---|
| (a) | (b) | (c) | (d) |

FIGURE 11.34

In part (d), the "friendly window setting," the ratio of the length of the $x$-axis (28.2 units) to the length of the $y$-axis (18.6 units) is about 1.516. This is the same ratio as the length to the width of the display window of the calculator on the TI-84 Plus.

## 2 Review Conic Sections

The following chart summarizes conic sections.

| **Parabola** | **Circle** | **Ellipse** | **Hyperbola** |
|---|---|---|---|
| $y = a(x - h)^2 + k$ or $y = ax^2 + bx + c$ | $x^2 + y^2 = r^2$ | $\frac{x^2}{a^2} + \frac{y^2}{b^2} = 1$ | $\frac{x^2}{a^2} - \frac{y^2}{b^2} = 1$ |
| $a > 0$ 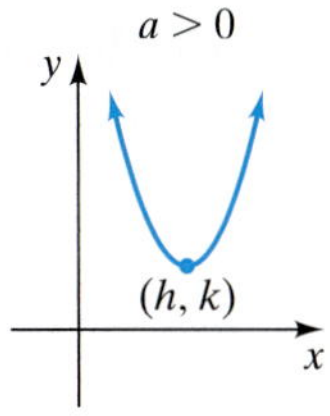 | 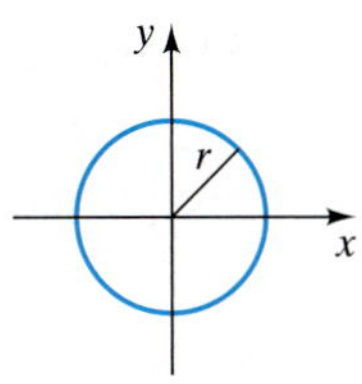 | 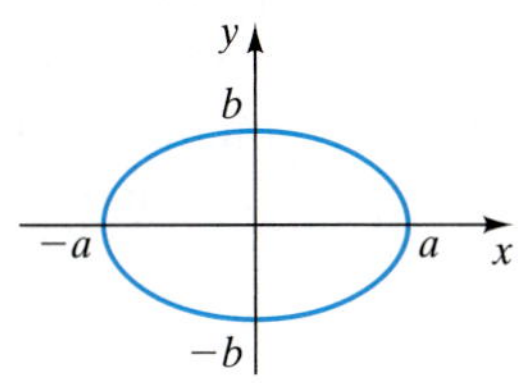 | 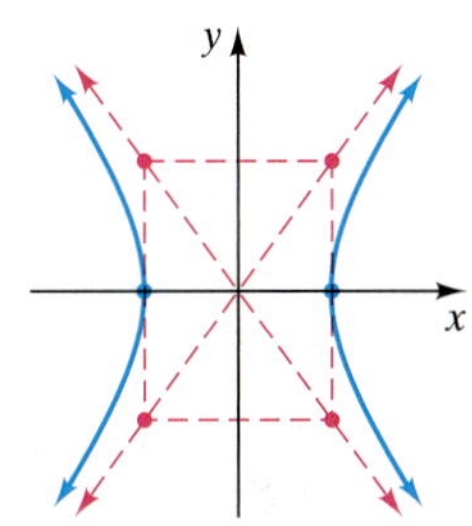 |
| $a < 0$ 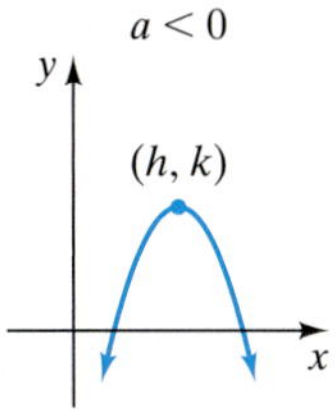 | $(x - h)^2 + (y - k)^2 = r^2$ | $\frac{(x - h)^2}{a^2} + \frac{(y - k)^2}{b^2} = 1$ | $\frac{y^2}{b^2} - \frac{x^2}{a^2} = 1$ |
| $x = a(y - k)^2 + h$ or $x = ay^2 + by + c$ | 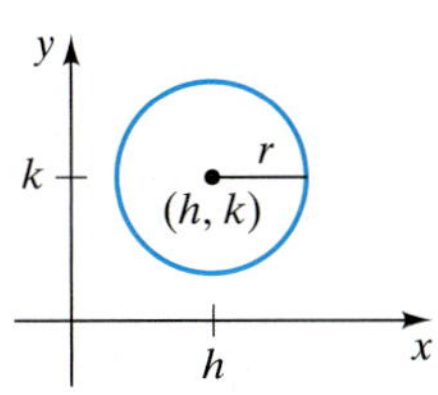 | 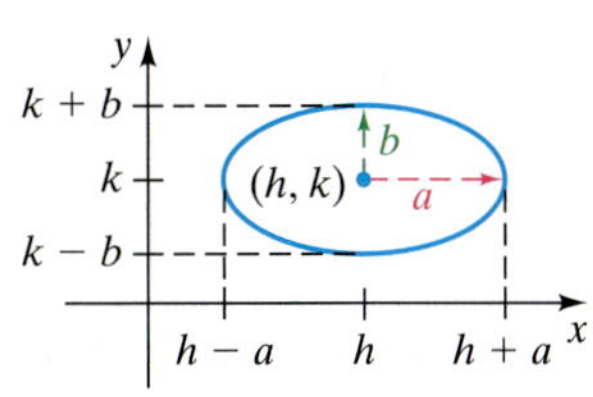 | 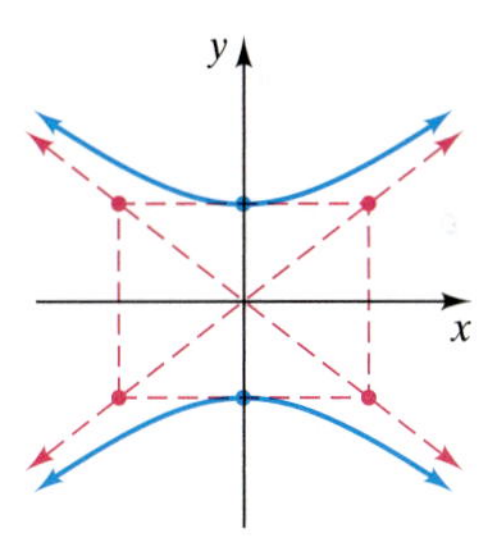 |
| $a > 0$ 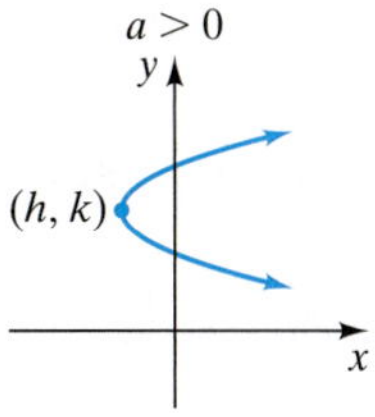 | | | *Asymptotes* $y = \frac{b}{a}x$ and $y = -\frac{b}{a}x$ |
| $a < 0$ 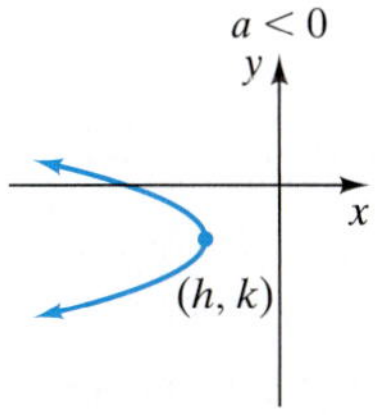 | | | |

**EXAMPLE 3** ▸ Indicate whether each equation represents a parabola, a circle, an ellipse, or a hyperbola.

**a)** $6x^2 = -6y^2 + 48$ **b)** $x - y^2 = 9y + 3$ **c)** $2x^2 = 8y^2 + 72$

*Solution*

**a)** This equation has an $x$-squared term and a $y$-squared term. Let's place all the squared terms on the left side of the equation.

$$6x^2 = -6y^2 + 48$$
$$6x^2 + 6y^2 = 48 \quad \text{Add } 6y^2 \text{ to both sides.}$$

Since the coefficients of both squared terms are the same number, we divide both sides of the equation by this number. Divide both sides by 6.

$$\frac{6x^2 + 6y^2}{6} = \frac{48}{6}$$
$$x^2 + y^2 = 8$$

This equation is of the form $x^2 + y^2 = r^2$ where $r^2 = 8$.

The equation $6x^2 = -6y^2 + 48$ represents a circle.

**b)** This equation has a $y$-squared term, but no $x$-squared term. Let's solve the equation for $x$.

$$x - y^2 = 9y + 3$$
$$x = y^2 + 9y + 3 \quad \text{Add } y^2 \text{ to both sides.}$$

This equation is of the form $x = ay^2 + by + c$ where $a = 1, b = 9$, and $c = 3$.

The equation $x - y^2 = 9y + 3$ represents a parabola that opens to the right.

**c)** This equation has an $x$-squared term and a $y$-squared term. Let's place all the squared terms on the left side of the equation.

$$2x^2 = 8y^2 + 72$$
$$2x^2 - 8y^2 = 72 \quad \text{Subtract } 8y^2 \text{ from both sides.}$$

Since the coefficients of both squared terms are different numbers, we want to divide the equation by the constant on the right side. Divide both sides by 72.

$$\frac{2x^2 - 8y^2}{72} = \frac{72}{72}$$
$$\frac{2x^2}{72} - \frac{8y^2}{72} = 1$$
$$\frac{x^2}{36} - \frac{y^2}{9} = 1$$

This equation is of the form $\frac{x^2}{a^2} - \frac{y^2}{b^2} = 1$ where $a^2 = 36$ (or $a = 6$) and $b^2 = 9$ (or $b = 3$).

The equation $2x^2 = 8y^2 + 72$ represents a hyperbola.

▸ **Now Try Exercise 53**

## EXERCISE SET 11.3

MathXL® MyMathLab

### Concept/Writing Exercises

1. What is the definition of a hyperbola?
2. What are asymptotes? How do you find the equations of the asymptotes of a hyperbola?
3. Discuss the graph of $\frac{x^2}{a^2} - \frac{y^2}{b^2} = 1$ for nonzero real numbers $a$ and $b$. Include the transverse axis, vertices, and asymptotes.

## 2 Solve Nonlinear Systems Using Addition

We can often solve systems of equations more easily using the addition method that was discussed in Section 4.1. As with the substitution method, our objective is to obtain a single equation containing only one variable.

**EXAMPLE 3** ▸ Solve the system of equations using the addition method.

$$x^2 + y^2 = 9$$
$$2x^2 - y^2 = -6$$

*Solution* If we add the two equations, we will obtain one equation containing only one variable.

$$\begin{aligned} x^2 + y^2 &= 9 \\ 2x^2 - y^2 &= -6 \\ \hline 3x^2 &= 3 \\ x^2 &= 1 \\ x &= \pm 1 \end{aligned}$$

Now solve for the variable $y$ by substituting $x = \pm 1$ into *either* of the original equations.

| $x = 1$ | $x = -1$ |
|---|---|
| $x^2 + y^2 = 9$ | $x^2 + y^2 = 9$ |
| $1^2 + y^2 = 9$ | $(-1)^2 + y^2 = 9$ |
| $1 + y^2 = 9$ | $1 + y^2 = 9$ |
| $y^2 = 8$ | $y^2 = 8$ |
| $y = \pm\sqrt{8}$ | $y = \pm\sqrt{8}$ |
| $= \pm 2\sqrt{2}$ | $= \pm 2\sqrt{2}$ |

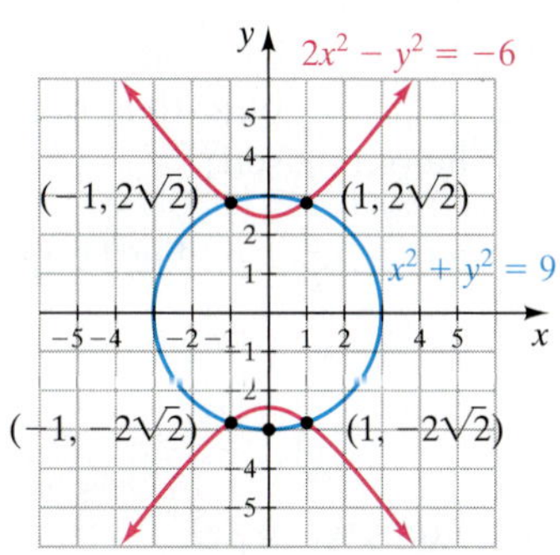


FIGURE 11.37

There are four solutions to this system of equations.

$$(1, 2\sqrt{2}), (1, -2\sqrt{2}), (-1, 2\sqrt{2}), \text{and } (-1, -2\sqrt{2})$$

The graphs of the equations in the system are given in **Figure 11.37**. Notice the four points of intersection of the two graphs.

▸ **Now Try Exercise 25**

It is possible that a system of equations has no real solution (therefore, the graphs do not intersect). Example 4 illustrates such a case.

**EXAMPLE 4** ▸ Solve the system of equations using the addition method.

$$x^2 + 4y^2 = 16 \quad (eq.\ 1)$$
$$x^2 + y^2 = 1 \quad (eq.\ 2)$$

*Solution* Multiply (*eq.* 2) by −1 and add the resulting equation to (*eq.* 1).

$$\begin{aligned} x^2 + 4y^2 &= 16 \\ -x^2 - y^2 &= -1 \quad (eq.\ 2) \text{ multiplied by } -1 \\ \hline 3y^2 &= 15 \\ y^2 &= 5 \\ y &= \pm\sqrt{5} \end{aligned}$$

Now solve for $x$.

$$y = \sqrt{5} \qquad\qquad y = -\sqrt{5}$$

$$x^2 + y^2 = 1 \qquad\qquad x^2 + y^2 = 1$$

$$x^2 + (\sqrt{5})^2 = 1 \qquad\qquad x^2 + (-\sqrt{5})^2 = 1$$

$$x^2 + 5 = 1 \qquad\qquad x^2 + 5 = 1$$

$$x^2 = -4 \qquad\qquad x^2 = -4$$

$$x = \pm\sqrt{-4} \qquad\qquad x = \pm\sqrt{-4}$$

$$x = \pm 2i \qquad\qquad x = \pm 2i$$

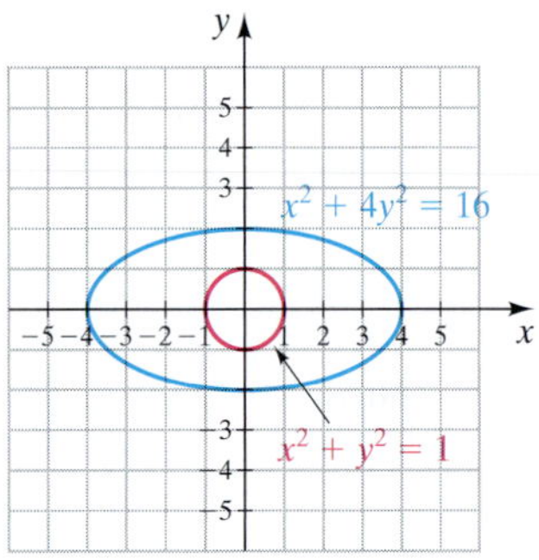


FIGURE 11.38

Since $x$ is an imaginary number for both values of $y$, this system of equations has no real solution. In solving nonlinear systems of equations, we are interested in finding all real number solutions.

The graphs of the equations are shown in **Figure 11.38**. Notice that the two graphs do not intersect; therefore, there is no real solution. This agrees with the answer we obtained algebraically.

▸ Now Try Exercise 37

## 3 Solve Applications

Now we will study some applications of nonlinear systems.

FIGURE 11.39

**EXAMPLE 5 ▸ Flower Garden** Fred and Judy Vespucci want to build a rectangular flower garden behind their house. Fred went to a local nursery and bought enough topsoil to cover 150 square meters of land. Judy went to the local hardware store and purchased 50 meters of fence for the perimeter of the garden. How should they build the garden to use all the topsoil he bought and all the fence she purchased?

*Solution* Understand and Translate We begin by drawing a sketch (see **Fig 11.39**).

Let $x$ = length of garden

$y$ = width of garden.

Since $A = xy$ and Fred bought topsoil to cover 150 square meters, we have

$$xy = 150$$

Since $P = 2x + 2y$ and Judy purchased 50 meters of fence for the perimeter of the garden, we have

$$2x + 2y = 50$$

The system of equations is

$$\begin{aligned} xy &= 150 \\ 2x + 2y &= 50 \end{aligned}$$

Carry Out We will solve the system using substitution. The equation $2x + 2y = 50$ is a linear equation. We will solve this equation for $y$. (We could also solve for $x$.)

$$2x + 2y = 50$$

$$2y = 50 - 2x$$

$$y = \frac{50 - 2x}{2} = \frac{50}{2} - \frac{2x}{2} = 25 - x$$

Now substitute $25 - x$ for $y$ in the equation $xy = 150$.

$$\begin{aligned} xy &= 150 \\ x(25 - x) &= 150 \\ 25x - x^2 &= 150 \\ 0 &= x^2 - 25x + 150 \\ 0 &= (x - 10)(x - 15) \end{aligned}$$

$$x - 10 = 0 \quad \text{or} \quad x - 15 = 0$$
$$x = 10 \qquad\qquad x = 15$$

**Answer** If $x = 10$, then $y = 25 - 10 = 15$. And, if $x = 15$, then $y = 25 - 15 = 10$. Thus, in either case, the dimensions of the flower garden are 10 meters by 15 meters.

▶ Now Try Exercise 43

**EXAMPLE 6 ▶ Bicycles** Hike 'n' Bike Company produces and sells bicycles. Its weekly cost equation is $C = 50x + 400, 0 \leq x \leq 160$, and its weekly revenue equation is $R = 100x - 0.3x^2, 0 \leq x \leq 160$, where $x$ is the number of bicycles produced and sold each week. Find the number of bicycles that must be produced and sold for Hike 'n' Bike to break even.

*Solution* **Understand and Translate** A company breaks even when its cost equals its revenue. When its cost is greater than its revenue, the company has a loss. When its revenue exceeds its cost, the company makes a profit.

The system of equations is

$$\begin{aligned} C &= 50x + 400 \\ R &= 100x - 0.3x^2 \end{aligned}$$

For Hike 'n' Bike to break even, its cost must equal its revenue. Thus, we write

$$\begin{aligned} C &= R \\ 50x + 400 &= 100x - 0.3x^2 \end{aligned}$$

**Carry Out** Writing this quadratic equation in standard form, we obtain

$$0.3x^2 - 50x + 400 = 0, \quad 0 \leq x \leq 160$$

We will solve this equation using the quadratic formula.

$$a = 0.3, \quad b = -50, \quad c = 400$$

$$\begin{aligned} x &= \frac{-b \pm \sqrt{b^2 - 4ac}}{2a} \\ &= \frac{-(-50) \pm \sqrt{(-50)^2 - 4(0.3)(400)}}{2(0.3)} \\ &= \frac{50 \pm \sqrt{2020}}{0.6} \end{aligned}$$

$$x = \frac{50 + \sqrt{2020}}{0.6} \approx 158.2 \quad \text{or} \quad x = \frac{50 - \sqrt{2020}}{0.6} \approx 8.4$$

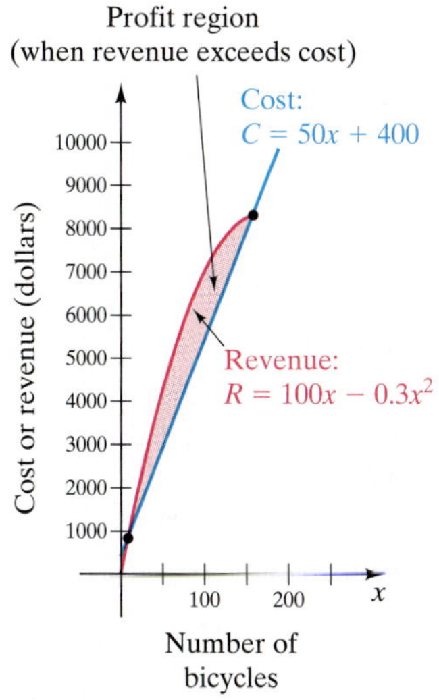


FIGURE 11.40

**Answer** The cost will equal the revenue and the company will break even when approximately 8 bicycles are sold. The cost will also equal the revenue when approximately 158 bicycles are sold. The company will make a profit when between 9 and 158 bicycles are sold. When fewer than 9 or more than 158 bicycles are sold, the company will have a loss (see **Fig. 11.40**).

▶ Now Try Exercise 55

## USING YOUR GRAPHING CALCULATOR

To solve nonlinear systems of equations graphically, graph the equations and find the intersections of the graphs. Consider the system in Example 1, $x^2 + y^2 = 25$ and $3x + 4y = 0$. To graph $x^2 + y^2 = 25$, we use $Y_1 = \sqrt{25 - x^2}$ and $Y_2 = -\sqrt{25 - x^2}$. If we solve $3x + 4y = 0$ for $y$ we obtain $y = -\frac{3}{4}x$. Thus, we use $Y_3 = -\frac{3}{4}x$. Therefore, to solve this system we find the intersection of

$$Y_1 = \sqrt{25 - x^2}$$
$$Y_2 = -\sqrt{25 - x^2}$$
$$Y_3 = -\frac{3}{4}x$$

The system is graphed, using the ZOOM: 5 (ZSquare) feature in **Figure 11.41a**.* In **Figure 11.41b**, we graph the same three equations using the "friendly numbers" shown below the figure. Using the calculator with either the TRACE and ZOOM features, the TABLE feature, or the INTERSECT feature, you will find that the solutions are $(4, -3)$ and $(-4, 3)$.

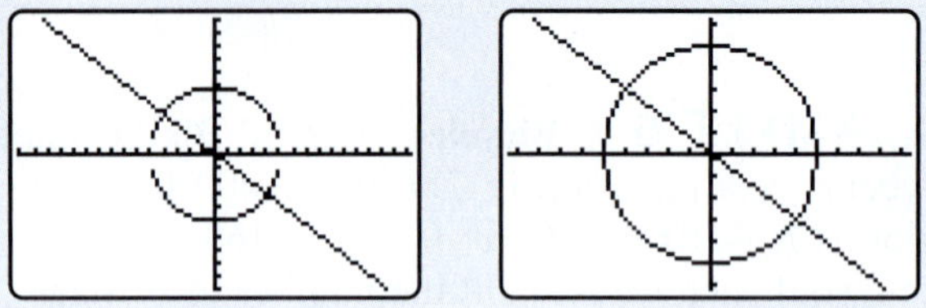

$\approx -15.2, \approx 15.2, 1, -10, 10, 1$ (a) $-9.4, 9.4, 1, -6.2, 6.2, 1$ (b)

FIGURE 11.41

*Start with the standard window, then select ZOOM: 5 to get this graph.

# EXERCISE SET 11.4

MathXL® MyMathLab

## Concept/Writing Exercises

1. What is a nonlinear system of equations?
2. Explain how nonlinear systems of equations may be solved graphically.
3. Can a nonlinear system of equations have exactly one real solution? If so, give an example. Explain.
4. Can a nonlinear system of equations have exactly two real solutions: If so, give an example. Explain.
5. Can a nonlinear system of equations have exactly three real solutions? If so, give an example. Explain.
6. Can a nonlinear system of equations have no real solutions? If so, give an example. Explain.

## Practice the Skills

*Find all real solutions to each system of equations using the substitution method.*

7. $x^2 + y^2 = 18$, $x + y = 0$
8. $x^2 + y^2 = 18$, $x - y = 0$
9. $x^2 + y^2 = 9$, $x + 2y = 3$
10. $x^2 + y^2 = 4$, $x - 2y = 4$
11. $y = x^2 - 5$, $3x + 2y = 10$
12. $x + y = 4$, $x^2 - y^2 = 4$
13. $x^2 + y = 6$, $y = x^2 + 4$
14. $y - x = 2$, $x^2 - y^2 = 4$
15. $2x^2 + y^2 = 16$, $x^2 - y^2 = -4$
16. $x + y^2 = 4$, $x^2 + y^2 = 6$
17. $x^2 + y^2 = 4$, $y = x^2 - 6$
18. $x^2 - 4y^2 = 36$, $x^2 + 2y^2 = 5$
19. $x^2 + y^2 = 9$, $y = x^2 - 3$
20. $x^2 + y^2 = 16$, $y = x^2 - 4$
21. $2x^2 - y^2 = -8$, $x - y = 6$
22. $x^2 + y^2 = 1$, $y - x = 3$

*Find all real solutions to each system of equations using the addition method.*

23. $x^2 - y^2 = 4$, $2x^2 + y^2 = 8$
24. $x^2 + y^2 = 36$, $x^2 - y^2 = 36$
25. $x^2 + y^2 = 16$, $2x^2 - 5y^2 = 25$
26. $x^2 + y^2 = 25$, $x^2 - 2y^2 = 7$

27. $3x^2 - y^2 = 4$
$x^2 + 4y^2 = 10$

28. $3x^2 + 2y^2 = 30$
$x^2 + y^2 = 13$

29. $4x^2 + 9y^2 = 36$
$2x^2 - 9y^2 = 18$

30. $x^2 + 4y^2 = 16$
$-9x^2 + y^2 = 4$

31. $2x^2 - y^2 = 7$
$x^2 + 2y^2 = 6$

32. $5x^2 - 2y^2 = -13$
$3x^2 + 4y^2 = 39$

33. $x^2 + y^2 = 25$
$2x^2 - 3y^2 = -30$

34. $x^2 - 2y^2 = 7$
$x^2 + y^2 = 34$

35. $x^2 + y^2 = 9$
$16x^2 - 4y^2 = 64$

36. $3x^2 + 4y^2 = 35$
$2x^2 + 5y^2 = 42$

37. $x^2 + y^2 = 4$
$16x^2 + 9y^2 = 144$

38. $x^2 + y^2 = 1$
$9x^2 - 4y^2 = 36$

39. $x^2 + 4y^2 = 4$
$10y^2 - 9x^2 = 90$

40. $x^2 + y^2 = 81$
$25x^2 + 4y^2 = 100$

## Problem Solving

41. Make up your own nonlinear system of equations whose solution is the empty set. Explain how you know the system has no solution.

42. If a system of equations consists of an ellipse and a hyperbola, what is the maximum number of points of intersection? Make a sketch to illustrate this.

43. **Dance Floor** Kris Hundley wants to build a dance floor at her gym. The dance floor is to have a perimeter of 84 meters and an area of 440 square meters. Find the dimensions of the dance floor.

44. **Rectangular Region** Ellen Dupree fences in a rectangular area along a riverbank as illustrated. If 20 feet of fencing encloses an area of 48 square feet, find the dimensions of the enclosed area.

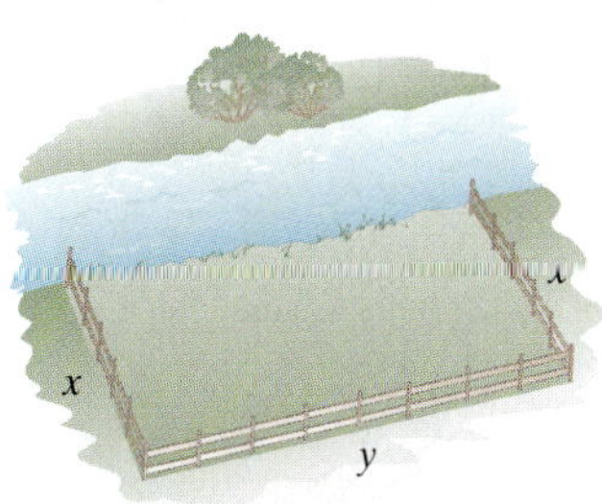


45. **Vegetable Garden** James Cannon is planning to build a rectangular flower garden in his yard. The garden is to have a perimeter of 78 feet and an area of 270 square feet. Find the dimensions of the vegetable garden.

46. **Rectangular Region** A rectangular area is to be fenced along a river as illustrated in Exercise 44. If 20 feet of fencing encloses an area of 50 square feet, find the dimensions of the enclosed area.

47. **Currency** A country's currency includes a bill that has an area of 112 square centimeters with a diagonal of $\sqrt{260}$ centimeters. Find the length and width of the bill.

48. **Ice Rink** A rectangular ice rink has an area of 3000 square feet. If the diagonal across the rink is 85 feet, find the dimensions of the rink.

Rockefeller Plaza, New York City

49. **Piece of Wood** Frank Samuelson, a carpenter, has a rectangular piece of plywood. When he measures the diagonal it measures 34 inches. When he cuts the wood along the diagonal, the perimeter of each triangle formed is 80 inches. Find the dimensions of the original piece of wood.

50. **Sailboat** A sail on a sailboat is shaped like a right triangle with a perimeter of 36 meters and a hypotenuse of 15 meters. Find the length of the legs of the triangle.

**51. Baseball and Football** Paul Martin throws a football upward from the ground. Its height above the ground at any time, $t$, is given by the formula $d = -16t^2 + 64t$. At the same time that the football is thrown, Shannon Ryan throws a baseball upward from the top of an 80-foot-tall building. Its height above the ground at any time, $t$, is given by the formula $d = -16t^2 + 16t + 80$. Find the time at which the two balls will be the same height above the ground. (Neglect air resistance.)

**52. Tennis Ball and Snowball** Robert Snell throws a tennis ball downward from a helicopter flying at a height of 950 feet. The height of the ball above the ground at any time $t$ is found by the formula $d = -16t^2 - 10t + 950$. At the instant the ball is thrown from the helicopter, Ramon Sanchez throws a snowball upward from the top of an 750-foot-tall building. The height above the ground of the snowball at any time, $t$, is found by the formula $d = -16t^2 + 80t + 750$. At what time will the ball and snowball pass each other? (Neglect air resistance.)

**53. Simple Interest** Simple interest is calculated using the simple interest formula, interest = principal · rate · time or $i = prt$. If Seana Hayden invests a certain principal at a specific interest rate for 1 year, the interest she obtains is \$7.50. If she increases the principal by \$25 and the interest rate is decreased by 1%, the interest remains the same. Find the principal and the interest rate.

**54. Simple Interest** If Claire Brooke invests a certain principal at a specific interest rate for 1 year, the interest she obtains is \$72. If she increases the principal by \$120 and the interest rate is decreased by 2%, the interest remains the same. Find the principal and the interest rate. Use $i = prt$.

*For the given cost and revenue equations, find the break-even point(s).*

55. $C = 10x + 300, R = 30x - 0.1x^2$

56. $C = 0.6x^2 + 9, R = 12x - 0.2x^2$

57. $C = 12.6x + 150, R = 42.8x - 0.3x^2$

58. $C = 80x + 900, R = 120x - 0.2x^2$

*Solve the following systems using your graphing calculator. Round your answers to the nearest hundredth.*

59. $3x - 5y = 12$
$x^2 + y^2 = 10$

60. $y = 2x^2 - x + 2$
$4x^2 + y^2 = 36$

## Challenge Problems

**61. Intersecting Roads** The intersection of three roads forms a right triangle, as shown in the figure.

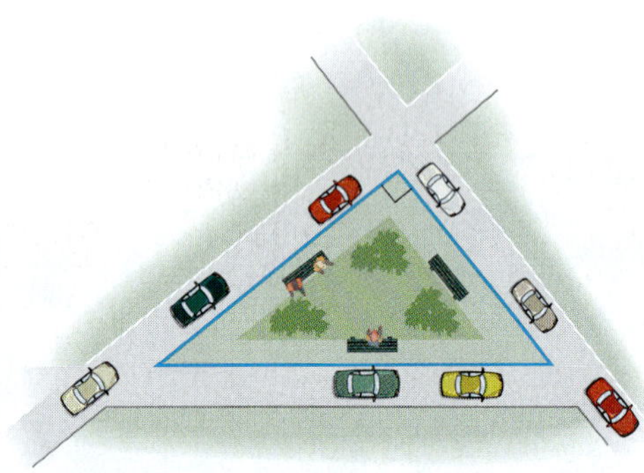

If the hypotenuse is 26 yards and the area is 120 square yards, find the length of the two legs of the triangle.

**62.** In the figure shown, $R$ represents the radius of the larger orange circle and $r$ represents the radius of the smaller orange circles. If $R = 2r$ and if the shaded area is $122.5\pi$, find $r$ and $R$.

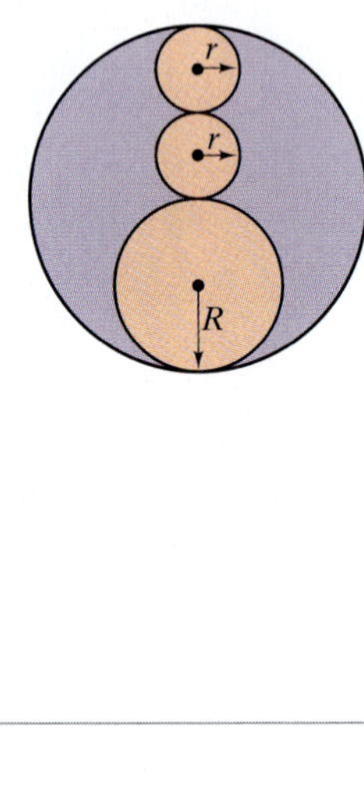


## Cumulative Review Exercises

[1.4] **63.** List the order of operations we follow when evaluating an expression.

[5.6] **64.** Factor $(x + 1)^3 + 1$.

[6.6] **65.** $x$ varies inversely as the square of $P$. If $x = 10$ when $P$ is 6, find $x$ when $P = 20$.

[7.5] **66.** Simplify $\dfrac{5}{\sqrt{x+2} - 3}$.

[9.7] **67.** Solve $A = A_0e^{kt}$ for $k$.

# 11.5 Nonlinear Systems of Inequalities

**1** Graph second-degree inequalities.

**2** Solve a system of inequalities graphically.

## 1 Graph Second-Degree Inequalities

In Sections 11.1 through 11.3 we graphed second-degree equations such as $x^2 + y^2 = 25$. Now we will graph second-degree inequalities, such as $x^2 + y^2 \leq 25$. To graph second-degree inequalities, we use a procedure similar to the one we used to graph linear inequalities. That is, graph the equation using a solid line if the inequality is $\leq$ or $\geq$, and a dashed line if the inequality is $<$ or $>$, and then shade in the region that satisfies the inequality. Examples 1 and 2 illustrate this procedure.

**EXAMPLE 1** ▸ Graph the inequality $\frac{x^2}{4} + \frac{y^2}{9} \geq 1$.

*Solution* First, graph the equation $\frac{x^2}{4} + \frac{y^2}{9} = 1$. Use a solid line when drawing the ellipse since the inequality contains $\geq$ (see **Fig. 11.42a**). Next select a point not on the graph and determine if this point satisfies the inequality. Often the easiest point to use is the point (0, 0). In this example we will select the point (0, 0).

Checkpoint (0, 0)

$$\frac{x^2}{4} + \frac{y^2}{9} \geq 1$$

$$\frac{0^2}{4} + \frac{0^2}{9} \overset{?}{\geq} 1$$

$$0 + 0 \overset{?}{\geq} 1$$

$$0 \geq 1 \quad \text{False}$$

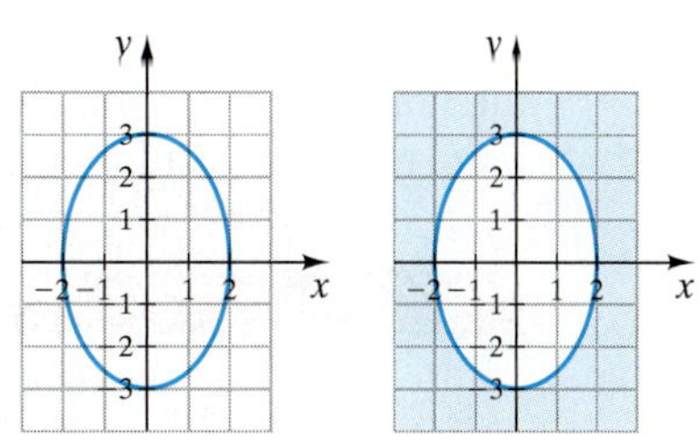


FIGURE 11.42 (a) (b)

The point (0, 0), which is within the ellipse, is not a solution. Thus all the points outside the ellipse will satisfy the inequality. Shade this outer region (see **Fig. 11.42b**). The shaded region, and the ellipse itself, form the solution to the inequality.

▸ **Now Try Exercise 15**

**EXAMPLE 2** ▸ Graph the inequality $\frac{y^2}{25} - \frac{x^2}{9} < 1$.

*Solution* Graph the equation $\frac{y^2}{25} - \frac{x^2}{9} = 1$. Use a dashed line when drawing the hyperbola since the inequality contains $<$ (see **Fig. 11.43a** on the next page). Next select a point not on the graph to use as a checkpoint. We will use (0, 0).

Checkpoint (0, 0)

$$\frac{y^2}{25} - \frac{x^2}{9} < 1$$

$$\frac{0^2}{25} - \frac{0^2}{9} \stackrel{?}{<} 1$$

$$0 - 0 \stackrel{?}{<} 1$$

$$0 < 1 \quad \textit{True}$$

Since (0, 0) is a solution, we shade in the region containing the point (0, 0). The solution is indicated in **Figure 11.43b**.

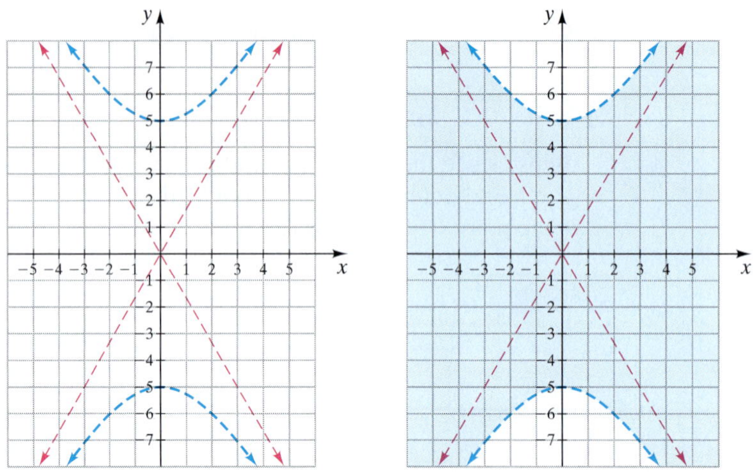

FIGURE 11.43 (a) (b)

▸ Now Try Exercise 21

## 2 Solve a System of Inequalities Graphically

Now we will find the graphical solution to a system of inequalities in which at least one inequality is not linear. The solution to a system of inequalities is the set of ordered pairs that satisfy all inequalities in the system.

**To Solve a System of Two Inequalities Graphically**

1. Graph one inequality.
2. On the same axes, graph the second inequality. Use a different type of shading than was used in the first inequality.
3. The solution is the shaded region from both inequalities.

**EXAMPLE 3** ▸ Solve the system of inequalities graphically.

$$x^2 + y^2 < 25$$
$$2x + y \geq 4$$

*Solution* First, graph the inequality, $x^2 + y^2 < 25$. The inner region of the circle satisfies the inequality (see **Fig. 11.44** on the next page). On the same axes, graph the second inequality, $2x + y \geq 4$. The solution to the system is the region containing both types of shading and that part of the straight line within the boundaries of the circle (see **Fig. 11.45** on the next page).

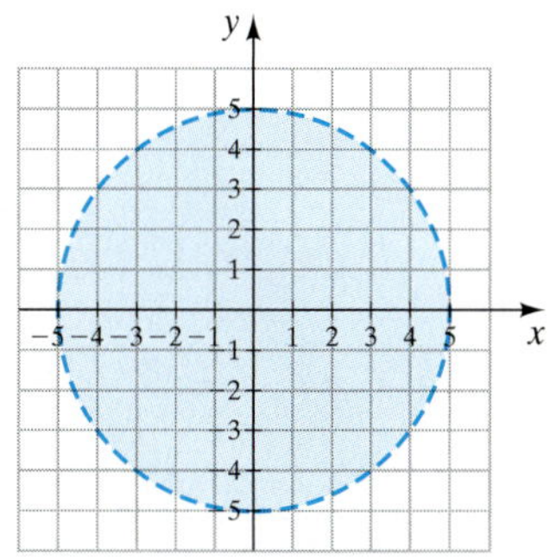

FIGURE 11.44

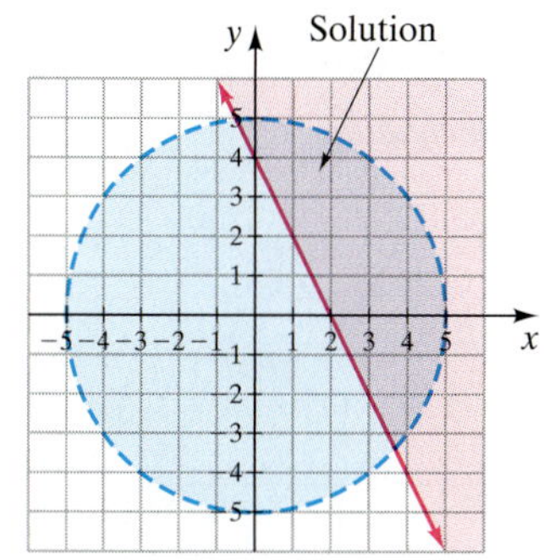

FIGURE 11.45

▸ Now Try Exercise 25

**Helpful Hint** *Study Tip*

In this section we are using a process covered in Section 4.6. If you do not remember this process, review Section 4.6 now.

**EXAMPLE 4** ▸ Solve the system of inequalities graphically.

$$\frac{x^2}{4} - \frac{y^2}{9} \leq 1$$
$$y > (x + 2)^2 - 4$$

*Solution* Graph each inequality on the same axes (see **Fig. 11.46**). The solution is the region containing both types of shading and the solid line within the parabola.

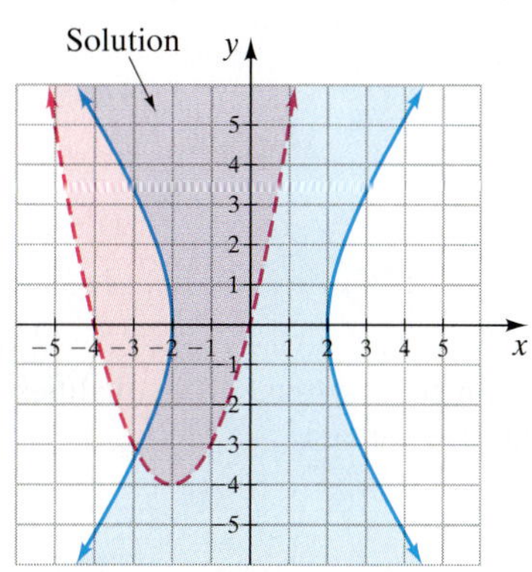

FIGURE 11.46

▸ Now Try Exercise 29

## EXERCISE SET 11.5

MathXL® MyMathLab

### Concept/Writing Exercises

1. When graphing an inequality containing $>$ or $<$, why is the boundary graphed with a dashed line?

2. Can $(0, 0)$ always be used as a check point? Explain.

3. Describe the regions into which the graph of a circle divides the $xy$-plane.

4. Describe the regions into which the graph of a hyperbola divides the $xy$-plane.

## Practice the Skills

*Graph each inequality.*

5. $y < (x - 3)^2 + 2$
6. $y > -(x + 4)^2 + 1$
7. $x \geq (y - 5)^2 - 3$
8. $x \leq 2(y + 3)^2 + 4$
9. $y < -x^2 + 4x - 4$
10. $x^2 + y^2 \leq 16$
11. $x^2 + y^2 > 9$
12. $(x - 1)^2 + (y + 4)^2 < 25$
13. $(x + 2)^2 + (y - 1)^2 \geq 25$
14. $(x - 4)^2 + y^2 \leq 1$
15. $\frac{x^2}{1} + \frac{y^2}{4} \leq 1$
16. $\frac{(x - 2)^2}{16} + \frac{(y + 1)^2}{25} > 1$
17. $\frac{x^2}{64} + \frac{y^2}{49} < 1$
18. $\frac{(x + 5)^2}{25} + \frac{(y + 3)^2}{9} \geq 1$
19. $16x^2 + 4y^2 \geq 64$
20. $\frac{x^2}{9} - \frac{y^2}{4} \geq 1$
21. $\frac{y^2}{9} - \frac{x^2}{25} > 1$
22. $\frac{y^2}{64} - \frac{x^2}{36} < 1$
23. $\frac{x^2}{4} - \frac{y^2}{36} \leq 1$
24. $16y^2 - 25x^2 \leq 400$

*Determine the solution to each system of inequalities.*

25. $x^2 + y^2 \geq 16$
    $x + y < 3$
26. $\frac{x^2}{25} - \frac{y^2}{9} < 1$
    $2x + 3y > 6$
27. $4x^2 + y^2 > 16$
    $y \geq 2x + 2$
28. $x^2 - y^2 > 1$
    $\frac{x^2}{9} + \frac{y^2}{4} \leq 1$
29. $x^2 + y^2 \leq 36$
    $y < (x + 1)^2 - 5$
30. $y \geq x^2 - 2x + 1$
    $x^2 + y^2 > 4$
31. $\frac{y^2}{16} - \frac{x^2}{4} \geq 1$
    $\frac{x^2}{4} - \frac{y^2}{1} < 1$
32. $4x^2 + 9y^2 \leq 36$
    $x > (y - 2)^2 + 1$
33. $xy \leq 6$
    $2x - y \leq 8$
34. $25x^2 - 4y^2 > 100$
    $5x + 3y \leq 15$
35. $(x - 3)^2 + (y + 2)^2 \geq 16$
    $y \leq 4x - 2$
36. $x^2 + (y - 3)^2 \leq 25$
    $y > (x - 2)^2 + 3$

## Problem Solving

37. Is it possible for the solution to a system of inequalities containing 2 linear inequalities to contain
    **a)** no points?
    **b)** exactly 1 point?
    **c)** exactly 2 points?
    **d)** all points on the $xy$-plane?
    Explain your answers.
38. Is it possible for the solution to a system of inequalities containing 1 linear inequality and one second-degree inequality to contain
    **a)** no points?
    **b)** exactly 1 point?
    **c)** exactly 2 points?
    **d)** all points on the $xy$-plane?
    Explain your answers.
39. Is it possible for the solution to a system of inequalities containing 2 second-degree inequalities to contain
    **a)** no points?
    **b)** exactly 1 point?
    **c)** exactly 2 points?
    **d)** all points on the $xy$-plane?
    Explain your answers.
40. Sketch a system of inequalities containing a parabola and line that has
    **a)** no solution.
    **b)** 1 solution.
41. Sketch a system of inequalities containing a circle and an ellipse that has
    **a)** no solution.
    **b)** 1 solution.
    **c)** 2 solutions.
42. Sketch a system of inequalities containing an ellipse and a hyperbola that has
    **a)** no solution.
    **b)** 1 solution.
    **c)** 2 solutions.

## Group Activity

*For Exercises 43 and 44,* **a)** *each group member should graph all of the inequalities on the same coordinate axes,* **b)** *as a group, determine the solution to the system.*

**43.** $y > 4x - 6$
$x^2 + y^2 \geq 36$
$2x + y \leq 8$

**44.** $x^2 + y^2 > 25$
$\dfrac{x^2}{25} + \dfrac{y^2}{9} > 1$
$2x - 3y \leq 12$

## Cumulative Review Exercises

[6.1] **45.** Simplify $\left(\dfrac{4x^{-2}y^3}{2xy^{-4}}\right)^2\left(\dfrac{3xy^{-1}}{6x^4y^{-3}}\right)^{-2}$.

*For Exercises 46 and 47, solve the equation* $x^2 - 2x - 4 = 0$:

[8.1] **46.** By completing the square.

[8.2] **47.** Using the quadratic formula.

[8.6] **48.** Graph the solution to the inequality $(x + 4)(x - 2)(x - 4) \leq 0$ on a number line.

# Chapter 11 Summary

**IMPORTANT FACTS AND CONCEPTS** | **EXAMPLES**

### Section 11.1

The four **conic sections** are the parabola, circle, ellipse, and the hyperbola, which are obtained by slicing a cone.

Parabola Circle Ellipse Hyperbola

The four different forms for equations of parabolas are summarized below.

### Parabola with Vertex at (*h*, *k*)

1. $y = a(x - h)^2 + k, a > 0$ (opens upward)
2. $y = a(x - h)^2 + k, a < 0$ (opens downward)
3. $x = a(y - k)^2 + h, a > 0$ (opens to the right)
4. $x = a(y - k)^2 + h, a < 0$ (opens to the left)

$y = -(x - 2)^2 + 3$

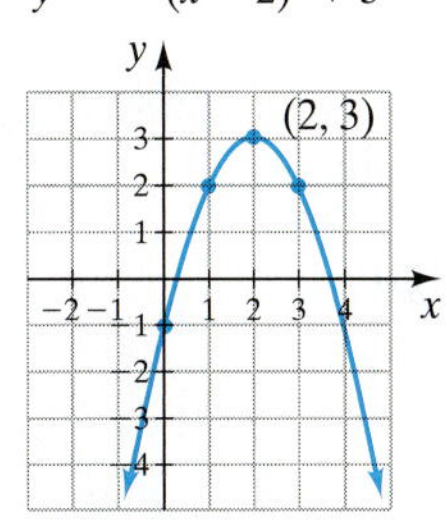


$x = 2(y + 1)^2 - 4$

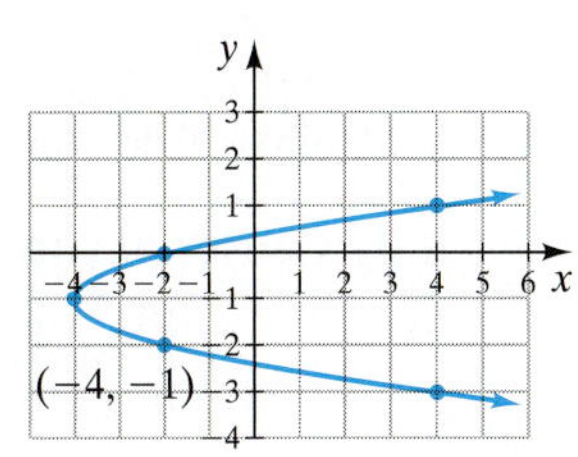


### Distance Formula

The distance, $d$, between any two points $(x_1, y_1)$ and $(x_2, y_2)$ can be found by the distance formula:

$$d = \sqrt{(x_2 - x_1)^2 + (y_2 - y_1)^2}$$

The distance between $(-1, 3)$ and $(4, 15)$ is

$$d = \sqrt{[4 - (-1)]^2 + (15 - 3)^2} = \sqrt{5^2 + 12^2} = \sqrt{169} = 13$$

### Midpoint Formula

Given any two points $(x_1, y_1)$ and $(x_2, y_2)$, the point halfway between the given points can be found by the midpoint formula:

$$\text{midpoint} = \left(\frac{x_1 + x_2}{2}, \frac{y_1 + y_2}{2}\right)$$

The midpoint of the line segment joining $(7, 6)$ and $(-11, 10)$ is

$$\text{midpoint} = \left(\frac{7 + (-11)}{2}, \frac{6 + 10}{2}\right) = \left(\frac{-4}{2}, \frac{16}{2}\right) = (-2, 8)$$

| IMPORTANT FACTS AND CONCEPTS | EXAMPLES |
|---|---|

**Section 11.1 (continued)**

A **circle** is a set of points in a plane that are the same distance from a fixed point called its **center**.

### Circle with Its Center at the Origin and Radius $r$

$$x^2 + y^2 = r^2$$

Sketch the graph of $x^2 + y^2 = 9$.
The graph is a circle with its center at $(0, 0)$ and radius $r = 3$.

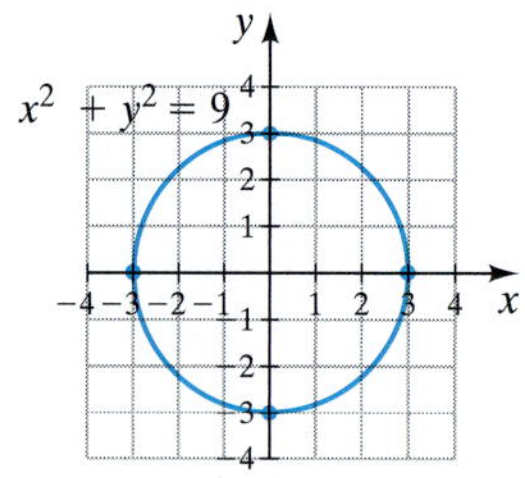


### Circle with Its center at $(h, k)$ and Radius $r$

$$(x - h)^2 + (y - k)^2 = r^2$$

Sketch the graph of $(x - 3)^2 + (y + 5)^2 = 25$.
The graph is a circle with its center at $(3, -5)$ and radius $r = 5$.

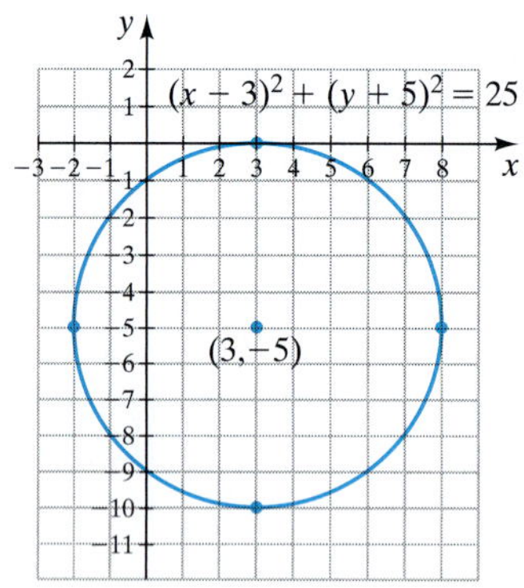


**Section 11.2**

An **ellipse** is a set of points in a plane, the sum of whose distances from two fixed points (called **foci**) is a constant.

### Ellipse with Its Center at the Origin

$$\frac{x^2}{a^2} + \frac{y^2}{b^2} = 1$$

where $(a, 0)$ and $(-a, 0)$ are the $x$-intercepts and $(0, b)$ and $(0, -b)$ are the $y$-intercepts.

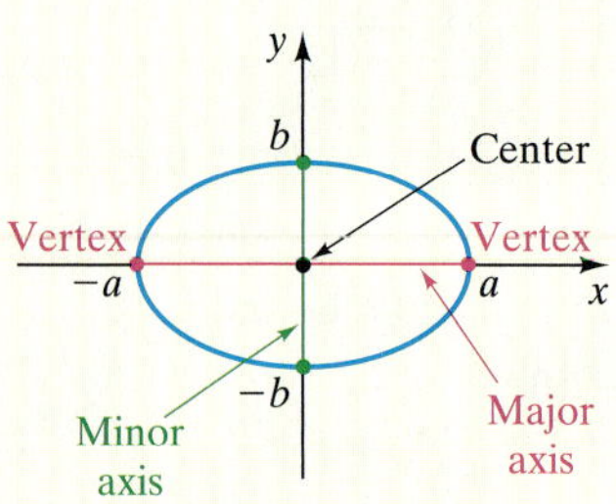


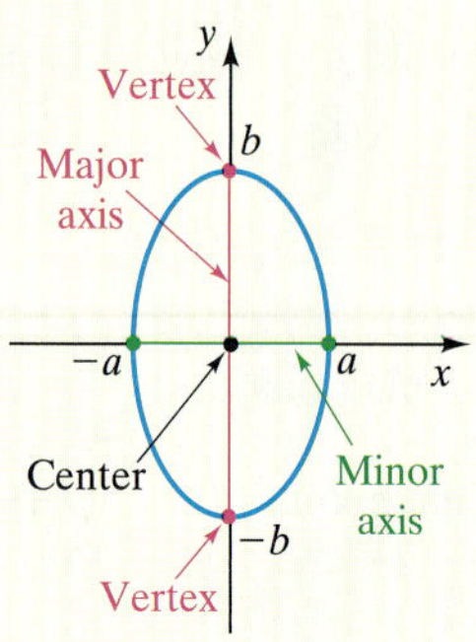


Sketch the graph of $\frac{x^2}{25} + \frac{y^2}{16} = 1$.

The graph is an ellipse. Since $a = 5$, the $x$-intercepts are $(-5, 0)$ and $(5, 0)$. Since $b = 4$, the $y$-intercepts are $(0, -4)$ and $(0, 4)$.

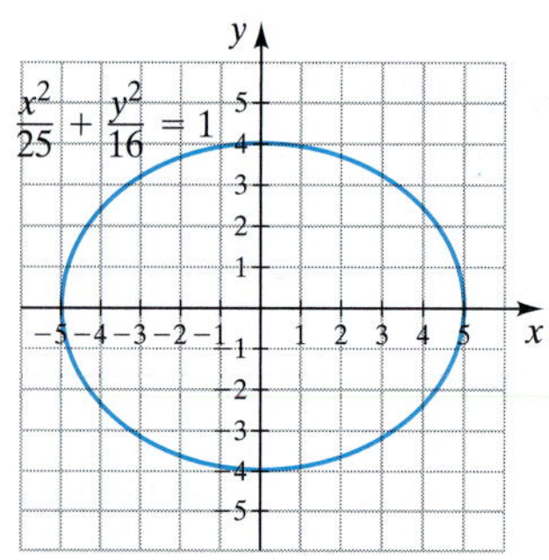

| IMPORTANT FACTS AND CONCEPTS | EXAMPLES |
|---|---|

**Section 11.2 (continued)**

### Ellipse with Its Center at $(h, k)$

$$\frac{(x-h)^2}{a^2}+\frac{(y-k)^2}{b^2}=1$$

Sketch the graph of $\dfrac{(x-2)^2}{9}+\dfrac{(y+1)^2}{16}=1.$

The graph is an ellipse with its center at $(2, -1)$, where $a = 3$ and $b = 4$.

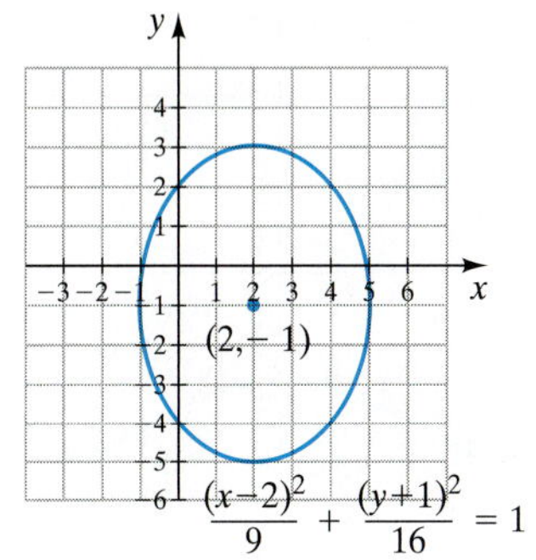


The area, $A$, of an ellipse is $A = \pi ab$.

The area of the ellipse above is

$$A = \pi ab = \pi \cdot 3 \cdot 4 = 12\pi \approx 37.70 \text{ square units.}$$

**Section 11.3**

A **hyperbola** is a set of points in a plane, the difference of whose distances from two fixed points (called **foci**) is a constant.

### Hyperbola with Its Center at the Origin

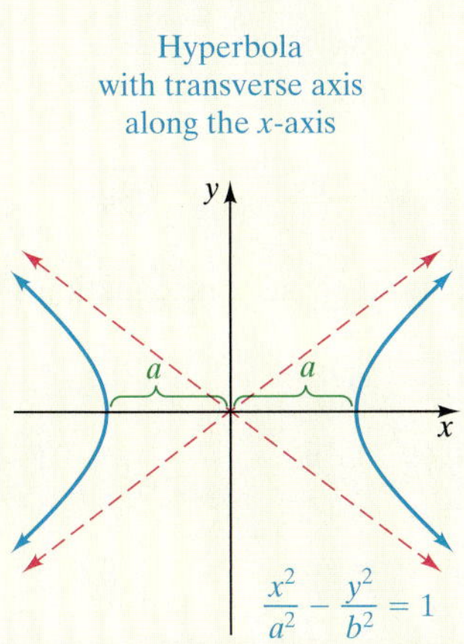


Asymptotes

$$y=\frac{b}{a}x \quad\text{and}\quad y=-\frac{b}{a}x.$$

Determine the equations of the asymptotes and sketch a graph of $\dfrac{x^2}{4}-\dfrac{y^2}{9}=1.$

The graph is a hyperbola with $a = 2$ and $b = 3$.

The equations for the asymptotes are $y = \dfrac{3}{2}x$ and $y = -\dfrac{3}{2}x$.

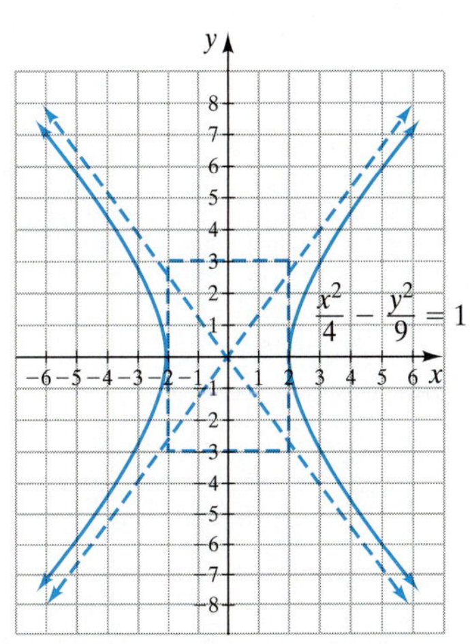

| IMPORTANT FACTS AND CONCEPTS | EXAMPLES |
| --- | --- |

**Section 11.3 (continued)**

### Hyperbola with Its Center at the Origin (*continued*)

Hyperbola with transverse axis along the $y$-axis

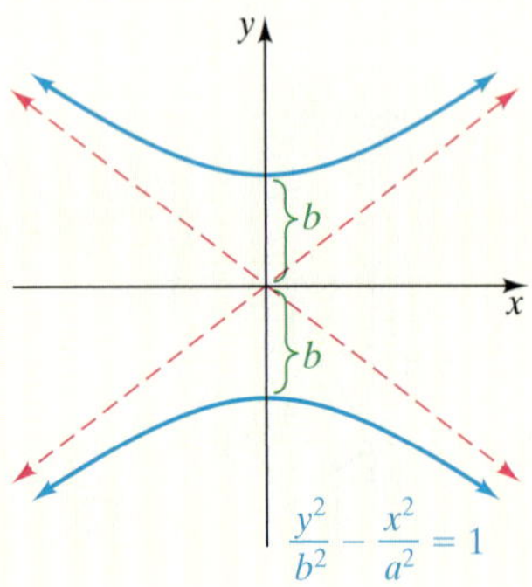


Asymptotes

$$y = \frac{b}{a}x \quad \text{and} \quad y = -\frac{b}{a}x.$$

Determine the equations of the asymptotes and sketch a graph of $\frac{y^2}{25} - \frac{x^2}{16} = 1$.

The graph is a hyperbola with $a = 4$ and $b = 5$. The equations of the asymptotes are $y = \frac{5}{4}x$ and $y = -\frac{5}{4}x$.

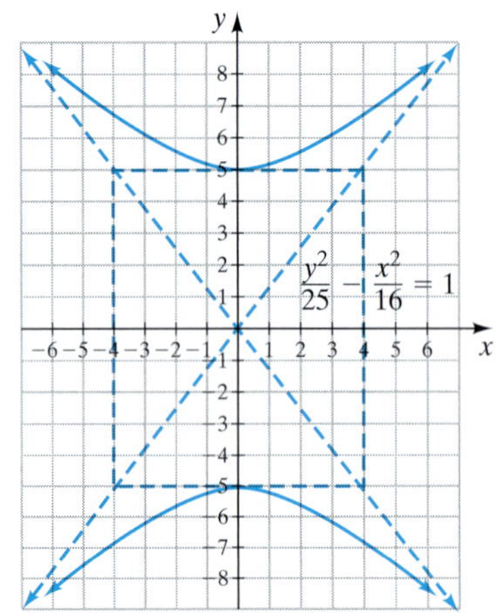


**Section 11.4**

A **nonlinear system of equations** is a system of equations where at least one equation is not linear. The solution to a nonlinear system of equations is the point or points that satisfy all equations in the system.

Solve the system of equations.

$$x^2 + y^2 = 14$$
$$5x^2 - y^2 = -2$$

We will solve this system using the addition method.

$$\begin{aligned} x^2 + y^2 &= 14 \\ \underline{5x^2 - y^2} &\underline{= -2} \\ 6x^2 \quad\quad &= 12 \\ x^2 &= 2 \\ x &= \pm\sqrt{2} \end{aligned}$$

To obtain the value(s) for $y$, use the equation $x^2 + y^2 = 14$.

| $x = \sqrt{2}$ | $x = -\sqrt{2}$ |
| --- | --- |
| $x^2 + y^2 = 14$ | $x^2 + y^2 = 14$ |
| $(\sqrt{2})^2 + y^2 = 14$ | $(-\sqrt{2})^2 + y^2 = 14$ |
| $2 + y^2 = 14$ | $2 + y^2 = 14$ |
| $y^2 = 12$ | $y^2 = 12$ |
| $y = \pm\sqrt{12}$ | $y = \pm\sqrt{12}$ |
| $= \pm 2\sqrt{3}$ | $= \pm 2\sqrt{3}$ |

The system has four solutions:

$$(\sqrt{2}, 2\sqrt{3}), (\sqrt{2}, -2\sqrt{3}), (-\sqrt{2}, 2\sqrt{3}), (-\sqrt{2}, -2\sqrt{3})$$

| IMPORTANT FACTS AND CONCEPTS | EXAMPLES |
|---|---|
| **Section 11.5** | |

**To Solve a System of Two Inequalities Graphically**

1. Graph one inequality.
2. On the same axes, graph the second inequality. Use a different type of shading than was used in the first inequality.
3. The solution is the shaded region from both inequalities.

Solve the system graphically.

$$x^2 + (y - 1)^2 \le 25$$
$$y > x - 2$$

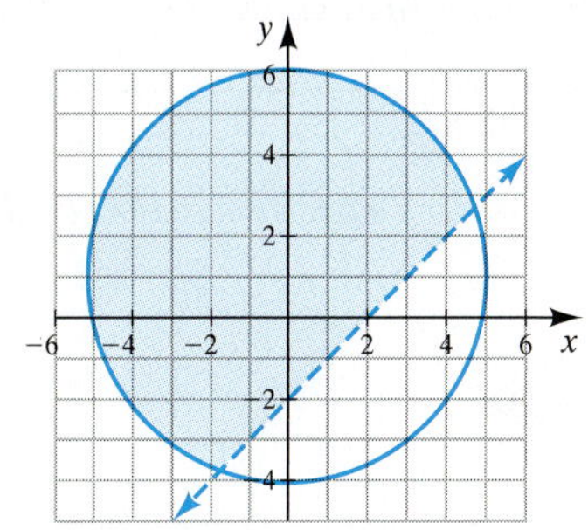

## Chapter 11 Review Exercises

[11.1] *Find the length and the midpoint of the line segment between each pair of points.*

**1.** $(0, 0), (5, -12)$ **2.** $(-4, 1), (-1, 5)$ **3.** $(-9, -5), (-1, 10)$ **4.** $(-4, 3), (-2, 5)$

*Graph each equation.*

**5.** $y = (x - 2)^2 + 1$ **6.** $y = (x + 3)^2 - 4$ **7.** $x = (y - 1)^2 + 4$ **8.** $x = -2(y + 4)^2 - 3$

*In Exercises 9–12,* **a)** *write each equation in the form $y = a(x - h)^2 + k$ or $x = a(y - k)^2 + h$.* **b)** *Graph the equation.*

**9.** $y = x^2 - 8x + 22$ **10.** $x = -y^2 - 2y + 5$ **11.** $x = y^2 + 5y + 4$ **12.** $y = 2x^2 - 8x - 24$

*In Exercises 13–18,* **a)** *write the equation of each circle in standard form.* **b)** *Draw the graph.*

**13.** Center $(0, 0)$, radius 4 **14.** Center $(-3, 4)$, radius 1 **15.** $x^2 + y^2 - 4y = 0$

**16.** $x^2 + y^2 - 2x + 6y + 1 = 0$ **17.** $x^2 - 8x + y^2 - 10y + 40 = 0$ **18.** $x^2 + y^2 - 4x + 10y + 17 = 0$

*Graph each equation.*

**19.** $y = \sqrt{9 - x^2}$ **20.** $y = -\sqrt{36 - x^2}$

*Determine the equation of each circle.*

**21.**

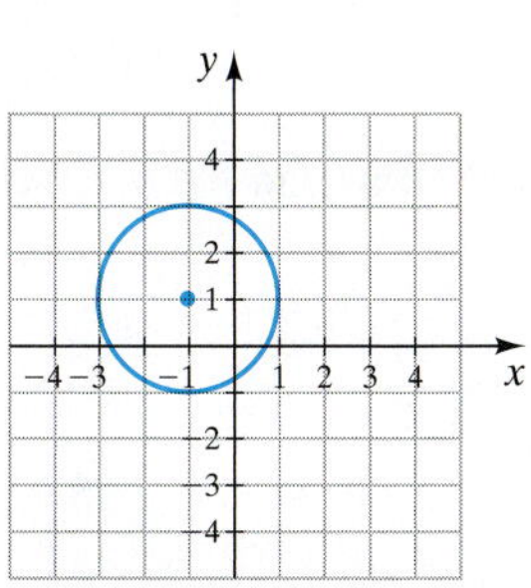

**22.**

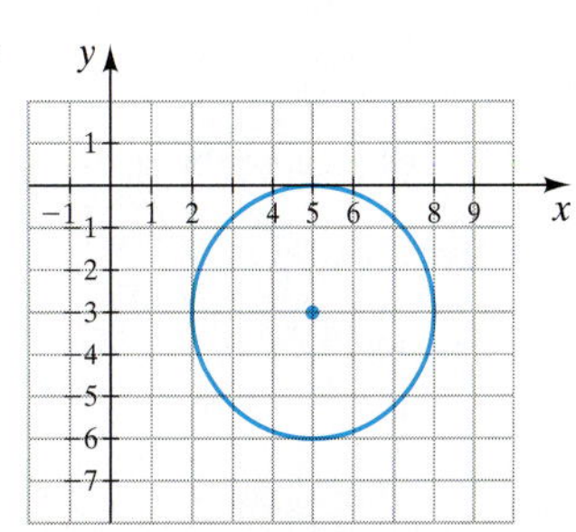

[11.2] *Graph each equation.*

**23.** $\dfrac{x^2}{4} + \dfrac{y^2}{9} = 1$

**24.** $\dfrac{x^2}{36} + \dfrac{y^2}{64} = 1$

**25.** $4x^2 + 9y^2 = 36$

**26.** $9x^2 + 16y^2 = 144$

**27.** $\dfrac{(x-3)^2}{16} + \dfrac{(y+2)^2}{4} = 1$

**28.** $\dfrac{(x+3)^2}{9} + \dfrac{y^2}{25} = 1$

**29.** $25(x-2)^2 + 9(y-1)^2 = 225$

**30.** For the ellipse in Exercise 23, find the area.

[11.3] *In Exercises 31–34,* **a)** *determine the equations of the asymptotes for each equation.* **b)** *Draw the graph.*

**31.** $\dfrac{x^2}{4} - \dfrac{y^2}{16} = 1$

**32.** $\dfrac{x^2}{4} - \dfrac{y^2}{4} = 1$

**33.** $\dfrac{y^2}{4} - \dfrac{x^2}{36} = 1$

**34.** $\dfrac{y^2}{25} - \dfrac{x^2}{16} = 1$

*In Exercises 35–38,* **a)** *write each equation in standard form.* **b)** *Determine the equations of the asymptotes.* **c)** *Draw the graph.*

**35.** $x^2 - 9y^2 = 9$

**36.** $25x^2 - 16y^2 = 400$

**37.** $4y^2 - 25x^2 = 100$

**38.** $49y^2 - 9x^2 = 441$

[11.1–11.3] *Identify the graph of each equation as a circle, ellipse, parabola, or hyperbola.*

**39.** $\dfrac{x^2}{49} - \dfrac{y^2}{16} = 1$

**40.** $4x^2 + 8y^2 = 32$

**41.** $5x^2 + 5y^2 = 125$

**42.** $4x^2 - 25y^2 = 25$

**43.** $\dfrac{x^2}{18} + \dfrac{y^2}{9} = 1$

**44.** $y = (x-2)^2 + 1$

**45.** $12x^2 + 9y^2 = 108$

**46.** $x = -y^2 + 8y - 9$

[11.4] *Find all real solutions to each system of equations using the substitution method.*

**47.** $x^2 + 2y^2 = 25$
$x^2 - 3y^2 = 25$

**48.** $x^2 = y^2 + 4$
$x + y = 4$

**49.** $x^2 + y^2 = 9$
$y = 3x + 9$

**50.** $x^2 + 2y^2 = 9$
$x^2 - 6y^2 = 36$

*Find all real solutions to each system of equations using the addition method.*

**51.** $x^2 + y^2 = 36$
$x^2 - y^2 = 36$

**52.** $x^2 + y^2 = 25$
$x^2 - 2y^2 = -2$

**53.** $-4x^2 + y^2 = -15$
$8x^2 + 3y^2 = -5$

**54.** $3x^2 + 2y^2 = 6$
$4x^2 + 5y^2 = 15$

**55. Pool Table** Jerry and Denise have a pool table in their house. It has an area of 45 square feet and a perimeter of 28 feet. Find the dimensions of the pool table.

**56. Bottles of Glue** The Dip and Dap Company has a cost equation of $C = 20.3x + 120$ and a revenue equation of $R = 50.2x - 0.2x^2$, where $x$ is the number of bottles of glue sold. Find the number of bottles of glue the company must sell to break even.

**57. Savings Account** If Kien Kempter invests a certain principal at a specific interest rate for 1 year, the interest is \$120. If he increases the principal by \$2000 and the interest rate is decreased by 1%, the interest remains the same. Find the principal and interest rate. Use $i = prt$.

**58. Persian Carpet** Tanya Richardson just purchased a rectangular Persian carpet. If the carpet has an area of 300 square feet and the diagonal of the carpet is 25 feet, find the length and width of the carpet.

[11.5] *Graph each inequality.*

**59.** $y > -x^2 + 6x - 8$

**60.** $(x+3)^2 + (y-2)^2 \geq 16$

**61.** $\dfrac{(x-1)^2}{36} + \dfrac{(y-3)^2}{16} \leq 1$

**62.** $\dfrac{y^2}{25} - \dfrac{x^2}{49} \geq 1$

*Graph each system of nonlinear inequalities.*

**63.** $2x + y \geq 6$
$x^2 + y^2 < 9$

**64.** $xy > 5$
$y < 3x + 4$

**65.** $4x^2 + 9y^2 \leq 36$
$x^2 + y^2 > 25$

**66.** $\dfrac{x^2}{4} - \dfrac{y^2}{9} > 1$
$y \geq (x-3)^2 - 4$

## Chapter 11 Practice Test

*To find out how well you understand the chapter material, take this practice test. The answers, and the section where the material was initially discussed, are given in the back of the book. Each problem is also fully worked out on the* **Chapter Test Prep Video CD**. *Review any questions that you answered incorrectly.*

**1.** Why are parabolas, circles, ellipses, and hyperbolas called conic sections?

**2.** Determine the length of the line segment whose endpoints are $(-1, 8)$ and $(6, 7)$.

**3.** Determine the midpoint of the line segment whose endpoints are $(-9, 4)$ and $(7, -1)$.

**4.** Determine the vertex of the graph of $y = -2(x + 3)^2 + 1$, and then graph the equation.

**5.** Graph $x = y^2 - 2y + 4$.

**6.** Write the equation $x = -y^2 - 4y - 5$ in the form $x = a(y - k)^2 + h$, and then draw the graph.

**7.** Write the equation of a circle with center at $(2, 4)$ and radius 3 and then draw the graph of the circle.

**8.** Find the area of the circle whose equation is $(x + 2)^2 + (y - 8)^2 = 9$.

**9.** Write the equation of the circle shown.

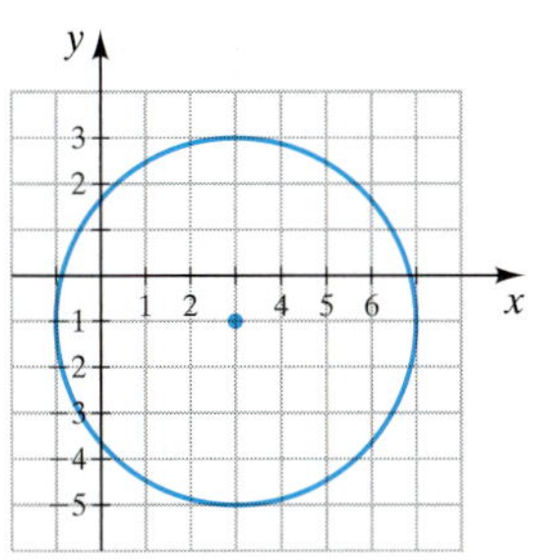


**10.** Graph $y = -\sqrt{16 - x^2}$.

**11.** Write the equation $x^2 + y^2 + 2x - 6y + 1 = 0$ in standard form, and then draw the graph.

**12.** Graph $4x^2 + 25y^2 = 100$.

**13.** Is the following graph the graph of $\frac{(x + 2)^2}{4} + \frac{(y + 1)^2}{16} = 1$? Explain your answer.

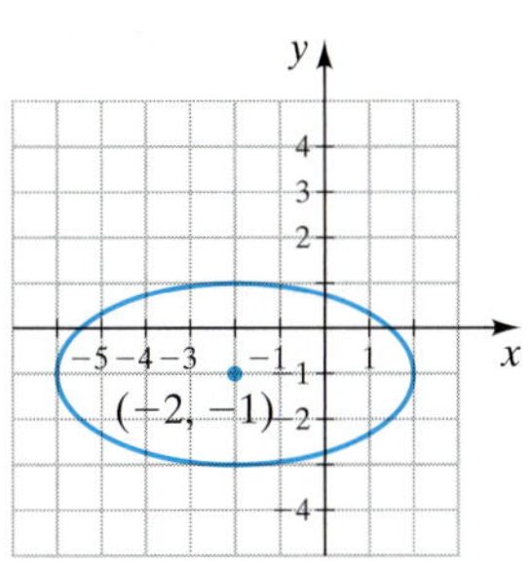


**14.** Graph $4(x - 4)^2 + 36(y + 2)^2 = 36$.

**15.** Find the center of the ellipse given by the equation $3(x - 8)^2 + 6(y + 7)^2 = 18$.

**16.** Explain how to determine whether the transverse axis of a hyperbola lies on the $x$- or $y$-axis.

**17.** What are the equations of the asymptotes of the graph of $\frac{x^2}{16} - \frac{y^2}{49} = 1$?

**18.** Graph $\frac{y^2}{25} - \frac{x^2}{1} = 1$.

**19.** Graph $\frac{x^2}{4} - \frac{y^2}{9} = 1$.

*In Exercises 20 and 21, determine whether the graph of the equation is a parabola, circle, ellipse, or hyperbola.*

**20.** $4x^2 - 15y^2 = 30$

**21.** $25x^2 + 4y^2 = 100$

*Solve each system of equations.*

**22.** $x^2 + y^2 = 7$

$2x^2 - 3y^2 = -1$

**23.** $x + y = 8$

$x^2 + y^2 = 4$

**24. Vegetable Garden** Tom Wilson has a rectangular vegetable garden on his farm that has an area of 1500 square meters. Find the dimensions of the garden if the perimeter is 160 meters.

**25.** Graph the system of nonlinear inequalities.

$$\frac{x^2}{9} - \frac{y^2}{25} < 1$$

$$x^2 + y^2 \le 25$$

## Cumulative Review Test

*Take the following test and check your answers with those given in the back of the book. Rewiew any questions that you answered incorrectly. The section where the material was covered is indicated after the answer.*

**1.** Simplify $(9x^2y^5)(-3xy^4)$.

**2.** Solve $4x - 2(3x - 7) = 2x - 5$.

**3.** Find the solution set: $2(x - 5) + 2x = 4x - 7$.

**4.** Find the solution set: $|3x + 1| > 4$

**5.** Graph $y = -2x + 2$.

**6.** If $f(x) = x^2 + 3x + 9$, find $f(10)$.

**7.** Solve the system of equations.

$$\frac{1}{2}x - \frac{1}{3}y = 2$$

$$\frac{1}{4}x + \frac{2}{3}y = 6$$

**8.** Factor $x^4 - x^2 - 42$.

**9.** A large triangular sign has a height that is 6 feet less than its base. If the area of the sign is 56 square feet, find the length of the base and the height of the sign.

**10.** Multiply $\dfrac{3x^2 - x - 4}{4x^2 + 7x + 3} \cdot \dfrac{2x^2 - 5x - 12}{6x^2 + x - 12}$.

**11.** Subtract $\dfrac{x}{x + 3} - \dfrac{x + 5}{2x^2 - 2x - 24}$.

**12.** Solve $\dfrac{3}{x + 3} + \dfrac{5}{x + 4} = \dfrac{12x + 19}{x^2 + 7x + 12}$.

**13.** Simplify $\left(\dfrac{18x^{1/2}y^3}{2x^{3/2}}\right)^{1/2}$.

**14.** Simplify $\dfrac{6\sqrt{x}}{\sqrt{x} - y}$.

**15.** Solve $3\sqrt[3]{2x + 2} = \sqrt[3]{80x - 24}$.

**16.** Solve $3x^2 - 4x + 5 = 0$ by the quadratic formula.

**17.** Solve $\log(3x - 4) + \log 4 = \log(x + 6)$.

**18.** Find the rational zeros of $f(x) = 4x^4 - 8x^3 - 7x^2 + 17x - 6$.

**19.** Graph $9x^2 + 4y^2 = 36$.

**20.** Graph $\dfrac{y^2}{25} - \dfrac{x^2}{16} = 1$.

# 12 Sequences, Series, and Probability

## GOALS OF THIS CHAPTER

Sequences and series are discussed in this chapter. A sequence is a list of numbers in a specific order and a series is the sum of the numbers in a sequence. In this book, we discuss two types of sequences and series: arithmetic and geometric. Sequences and series can be used to solve many real-life problems as illustrated in this chapter.

In this chapter, we introduce the summation symbol, $\Sigma$, which is often used in statistics and other mathematics courses. We also discuss, in Section 12.5, the binomial theorem for expanding an expression of the form $(a + b)^n$. In Sections 12.6 and 12.7 we discuss topics from probability.


12.1 Sequences and Series

12.2 Arithmetic Sequences and Series

12.3 Geometric Sequences and Series

Mid-Chapter Test: Sections 12.1–12.3

12.4 Mathematical Induction

12.5 The Binomial Theorem

12.6 The Counting Principle, Permutations, and Combinations

12.7 Probability

Chapter 12 Summary

Chapter 12 Review Exercises

Chapter 12 Practice Test

Cumulative Review Test

IF A BALL REBOUNDS 4 feet when dropped from 6 feet, it has rebounded $66\frac{2}{3}\%$ of its original height. Theoretically, every rebound will have a rebound and the ball will never stop bouncing. In Exercise 105 on page 787, you will calculate the total distance traveled by a bouncing ball.

# 12.1 Sequences and Series

1 Find the terms of a sequence.

2 Write a series.

3 Find partial sums.

4 Use summation notation, Σ.

## 1 Find the Terms of a Sequence

Many times we see patterns in numbers. For example, suppose you are given a job offer with a starting salary of \$30,000. You are given two options for your annual salary increases. One option is an annual salary increase of \$2000 per year. The salary you would receive under this option is shown below.

| Year | 1 | 2 | 3 | 4 | ... |
|---|---|---|---|---|---|
| | ↓ | ↓ | ↓ | ↓ | |
| Salary | \$30,000 | \$32,000 | \$34,000 | \$36,000 | ... |

Each year the salary is \$2000 greater than the previous year. The three dots on the right of the lists of numbers indicate that the list continues in the same manner.

The second option is a 5% salary increase each year. The salary you would receive under this option is shown below.

| Year | 1 | 2 | 3 | 4 | ... |
|---|---|---|---|---|---|
| | ↓ | ↓ | ↓ | ↓ | |
| Salary | \$30,000 | \$31,500 | \$33,075 | \$34,728.75 | ... |

With this option, the salary in a given year after year 1 is 5% greater than the previous year's salary.

The two lists of numbers that illustrate the salaries are examples of sequences. A **sequence** of numbers is a list of numbers arranged in a specific order. Consider the list of numbers given below, which is a sequence.

$$5, 10, 15, 20, 25, 30, \ldots$$

The first term is 5. We indicate this by writing $a_1 = 5$. Since the second term is 10, $a_2 = 10$, and so on. The three dots, called an ellipsis, indicate that the sequence continues indefinitely and is an **infinite sequence**.

**Infinite Sequence**

An **infinite sequence** is a function whose domain is the set of natural numbers.

Consider the infinite sequence 5, 10, 15, 20, 25, 30, 35, ...

$$\begin{array}{lllllllllll} \text{Domain:} & \{1, & 2, & 3, & 4, & 5, & 6, & 7, & \ldots, & n, & \ldots\} \\ & \downarrow & \downarrow & \downarrow & \downarrow & \downarrow & \downarrow & \downarrow & & \downarrow & \\ \text{Range:} & \{5, & 10, & 15, & 20, & 25, & 30, & 35, & \ldots, & 5n, & \ldots\} \end{array}$$

Note that the terms of the sequence 5, 10, 15, 20, ... are found by multiplying each natural number by 5. For any natural number, $n$, the corresponding term in the sequence is $5 \cdot n$ or $5n$. The **general term of the sequence**, $a_n$, which defines the sequence, is $a_n = 5n$.

$$a_n = f(n) = 5n$$

To find the twelfth term of the sequence, substitute 12 for $n$ in the general term of the sequence: $a_{12} = 5 \cdot 12 = 60$. Thus, the twelfth term of the sequence is 60. Note that the terms in the sequence are the function values, or the numbers in the range of the function. When writing the sequence, we do not use set braces. The general form of a sequence is

$$a_1, a_2, a_3, a_4, \ldots, a_n, \ldots$$

For the infinite sequence $2, 4, 8, 16, 32, \ldots, 2^n, \ldots$ we can write

$$a_n = f(n) = 2^n$$

Notice that when $n = 1, a_1 = 2^1 = 2$; when $n = 2, a_2 = 2^2 = 4$; when $n = 3, a_3 = 2^3 = 8$; when $n = 4, a_4 = 2^4 = 16$; and so on. What is the seventh term of this sequence? The answer is $a_7 = 2^7 = 128$.

A sequence may also be **finite**.

**Finite sequence**

A **finite sequence** is a function whose domain includes only the first $n$ natural numbers.

A finite sequence has only a finite number of terms.

Examples of Finite Sequences

| | |
|---|---|
| 5, 10, 15, 20 | domain is $\{1, 2, 3, 4\}$ |
| 2, 4, 8, 16, 32 | domain is $\{1, 2, 3, 4, 5\}$ |

**EXAMPLE 1** ▸ Write the finite sequence defined by $a_n = 2n + 3$, for $n = 1, 2, 3, 4$.

*Solution*

$$a_n = 2n + 3$$
$$a_1 = 2(1) + 3 = 5$$
$$a_2 = 2(2) + 3 = 7$$
$$a_3 = 2(3) + 3 = 9$$
$$a_4 = 2(4) + 3 = 11$$

Thus, the sequence is 5, 7, 9, 11.

▸ Now Try Exercise 17

Since each term of the sequence in Example 1 is larger than the preceding term, it is called an **increasing sequence**.

**EXAMPLE 2** ▸ Given $a_n = \dfrac{2n + 3}{n^2}$,

**a)** find the first term in the sequence.
**b)** find the third term in the sequence.
**c)** find the fifth term in the sequence.
**d)** find the tenth term in the sequence.

*Solution*

**a)** When $n = 1$, $a_1 = \dfrac{2(1) + 3}{1^2} = \dfrac{5}{1} = 5$.

**b)** When $n = 3$, $a_3 = \dfrac{2(3) + 3}{3^2} = \dfrac{9}{9} = 1$.

**c)** When $n = 5$, $a_5 = \dfrac{2(5) + 3}{5^2} = \dfrac{13}{25} = 0.52$.

**d)** When $n = 10$, $a_{10} = \dfrac{2(10) + 3}{10^2} = \dfrac{23}{100} = 0.23$.

▸ Now Try Exercise 33

Note in Example 2 that since there is no restriction on $n$, $a_n$ is the general term of an infinite sequence.

In Example 2, the first four terms of the sequence are $5, \frac{7}{4} = 1.75, 1, \frac{11}{16} = 0.6875$. Since each term of the sequence generated by $a_n = \dfrac{2n + 3}{n^2}$ will be smaller than the preceding term, the sequence is called a **decreasing sequence**.

**EXAMPLE 3** ▸ Find the first four terms of the sequence whose general term is $a_n = (-1)^n(n)$.

*Solution*

$$a_n = (-1)^n(n)$$
$$a_1 = (-1)^1(1) = -1$$
$$a_2 = (-1)^2(2) = 2$$
$$a_3 = (-1)^3(3) = -3$$
$$a_4 = (-1)^4(4) = 4$$

If we write the sequence, we get $-1, 2, -3, 4, \ldots, (-1)^n(n)$. Notice that each term alternates in sign. We call this an **alternating sequence**.

▸ Now Try Exercise 25

## 2 Write a Series

A **series** is the expressed sum of the terms of a sequence. A series may be finite or infinite, depending on whether the sequence it is based on is finite or infinite.

Examples

**Finite Sequence**

$$a_1, a_2, a_3, a_4, a_5$$

**Finite Series**

$$a_1 + a_2 + a_3 + a_4 + a_5$$

**Infinite Sequence**

$$a_1, a_2, a_3, a_4, a_5, \ldots, a_n, \ldots$$

**Infinite Series**

$$a_1 + a_2 + a_3 + a_4 + a_5 + \cdots + a_n + \cdots$$

**EXAMPLE 4** ▸ Write the first eight terms of the sequence; then write the series that represents the sum of that sequence if

**a)** $a_n = \left(\frac{1}{2}\right)^n$ **b)** $a_n = (-2)^n$

*Solution*

**a)** We begin with $n = 1$; thus, the first eight terms of the sequence whose general term is $a_n = \left(\frac{1}{2}\right)^n$ are

$$\left(\frac{1}{2}\right)^1, \left(\frac{1}{2}\right)^2, \left(\frac{1}{2}\right)^3, \left(\frac{1}{2}\right)^4, \left(\frac{1}{2}\right)^5, \left(\frac{1}{2}\right)^6, \left(\frac{1}{2}\right)^7, \left(\frac{1}{2}\right)^8$$

or

$$\frac{1}{2}, \frac{1}{4}, \frac{1}{8}, \frac{1}{16}, \frac{1}{32}, \frac{1}{64}, \frac{1}{128}, \frac{1}{256}$$

The series that represents the sum of the sequence is

$$\frac{1}{2} + \frac{1}{4} + \frac{1}{8} + \frac{1}{16} + \frac{1}{32} + \frac{1}{64} + \frac{1}{128} + \frac{1}{256} = \frac{255}{256}$$

**b)** We again begin with $n = 1$; thus, the first eight terms of the sequence whose general term is $a_n = (-2)^n$ are

$$(-2)^1, (-2)^2, (-2)^3, (-2)^4, (-2)^5, (-2)^6, (-2)^7, (-2)^8$$

or

$$-2, 4, -8, 16, -32, 64, -128, 256$$

The series that represents the sum of this sequence is

$$-2 + 4 + (-8) + 16 + (-32) + 64 + (-128) + 256 = 170$$

▶ Now Try Exercise 49

## 3 Find Partial Sums

For an infinite sequence with the terms $a_1, a_2, a_3, \ldots, a_n, \ldots$, a **partial sum** is the sum of a finite number of consecutive terms of the sequence, beginning with the first term.

$$s_1 = a_1 \qquad \text{First partial sum}$$
$$s_2 = a_1 + a_2 \qquad \text{Second partial sum}$$
$$s_3 = a_1 + a_2 + a_3 \qquad \text{Third partial sum}$$
$$\vdots$$
$$s_n = a_1 + a_2 + a_3 + \cdots + a_n \qquad n\text{th partial sum}$$

The sum of all the terms of the infinite sequence is called an **infinite series** and is given by the following:

$$s = a_1 + a_2 + a_3 + \cdots + a_n + \cdots$$

**EXAMPLE 5** ▶ Given the infinite sequence defined by $a_n = \frac{3 + n^2}{n}$, find the indicated partial sums.

**a)** $s_1$ and **b)** $s_4$

*Solution*

**a)** $s_1 = a_1 = \frac{3 + 1^2}{1} = \frac{3 + 1}{1} = 4$

**b)** $s_4 = a_1 + a_2 + a_3 + a_4$

$$= \frac{3 + 1^2}{1} + \frac{3 + 2^2}{2} + \frac{3 + 3^2}{3} + \frac{3 + 4^2}{4}$$

$$= 4 + \frac{7}{2} + \frac{12}{3} + \frac{19}{4}$$

$$= \frac{48}{12} + \frac{42}{12} + \frac{48}{12} + \frac{57}{12}$$

$$= \frac{195}{12} \text{ or } 16\frac{1}{4}$$

▶ Now Try Exercise 39

## 4 Use Summation Notation, Σ

When the general term of a sequence is known, the Greek letter **sigma**, Σ, can be used to write a series. The sum of the first $n$ terms of the sequence whose $n$th term is $a_n$ is represented by

$$\sum_{i=1}^{n} a_i = a_1 + a_2 + a_3 + \cdots + a_n$$

where $i$ is called the **index of summation** or simply the **index**, $n$ is the **upper limit of summation**, and 1 is the **lower limit of summation**. In this illustration, we used $i$ for the index; however, any letter can be used for the index.

Consider the sequence $7, 9, 11, 13, \ldots, 2n + 5, \ldots$. The sum of the first five terms can be represented using **summation notation**.

$$\sum_{i=1}^{5} (2i + 5)$$

This notation is read "the sum as $i$ goes from 1 to 5 of $2i + 5$."

To evaluate the series represented by $\sum_{i=1}^{5}(2i+5)$, we first substitute 1 for $i$ in $2i+5$ and list the value obtained. Then we substitute 2 for $i$ in $2i+5$ and list the value. We follow this procedure for the values 1 through 5. We then sum these values to obtain the series value.

$$\begin{aligned}\sum_{i=1}^{5}(2i+5) &= (2\cdot 1+5)+(2\cdot 2+5)+(2\cdot 3+5)+(2\cdot 4+5)+(2\cdot 5+5)\\ &= 7+9+11+13+15\\ &= 55\end{aligned}$$

**EXAMPLE 6** ▸ Write out the series $\sum_{i=1}^{6}(i^2+1)$ and evaluate it.

*Solution*

$$\begin{aligned}\sum_{i=1}^{6}(i^2+1) &= (1^2+1)+(2^2+1)+(3^2+1)+(4^2+1)+(5^2+1)+(6^2+1)\\ &= 2+5+10+17+26+37\\ &= 97\end{aligned}$$

▸ **Now Try Exercise 61**

---

**EXAMPLE 7** ▸ Consider the general term of a sequence $a_n = 2n^2 - 9$. Represent the third partial sum, $s_3$, in summation notation.

*Solution* The third partial sum will be the sum of the first three terms, $a_1 + a_2 + a_3$. We can represent the third partial sum as $\sum_{i=1}^{3}(2i^2-9)$.

▸ **Now Try Exercise 69**

---

**EXAMPLE 8** ▸ For the following set of values $x_1 = 3$, $x_2 = 4$, $x_3 = 5$, $x_4 = 6$, and $x_5 = 7$, does $\sum_{i=1}^{5}(x_i)^2 = \left(\sum_{i=1}^{5}x_i\right)^2$?

*Solution*

$$\begin{aligned}\sum_{i=1}^{5}(x_i)^2 &= (x_1)^2+(x_2)^2+(x_3)^2+(x_4)^2+(x_5)^2\\ &= 3^2+4^2+5^2+6^2+7^2\\ &= 9+16+25+36+49 = 135\end{aligned}$$

$$\begin{aligned}\left(\sum_{i=1}^{5}x_i\right)^2 &= (x_1+x_2+x_3+x_4+x_5)^2\\ &= (3+4+5+6+7)^2 = (25)^2 = 625\end{aligned}$$

Since $135 \neq 625$, $\sum_{i=1}^{5}(x_i)^2 \neq \left(\sum_{i=1}^{5}x_i\right)^2$.

▸ **Now Try Exercise 75**

---

When a summation symbol is written without any upper and lower limits, it means that all the given data are to be summed.

**EXAMPLE 9** A formula used to find the arithmetic mean, $\bar{x}$ (read $x$ bar), of a set of data is $\bar{x} = \frac{\Sigma x}{n}$, where $n$ is the number of pieces of data.

Joan Sally's five test scores are 70, 95, 83, 74, and 92. Find the arithmetic mean of her scores.

*Solution*

$$\bar{x} = \frac{\Sigma x}{n} = \frac{70 + 95 + 83 + 74 + 92}{5} = \frac{414}{5} = 82.8$$

Now Try Exercise 79

## EXERCISE SET 12.1

MathXL® MyMathLab

### Concept/Writing Exercises

1. What is a sequence?
2. What is an infinite sequence?
3. What is a finite sequence?
4. What is an increasing sequence?
5. What is a decreasing sequence?
6. What is an alternating sequence?
7. What is a series?
8. What is the $n$th partial sum of a series?
9. Write the following notation in words: $\sum_{i=1}^{5}(i + 4)$.
10. Consider the summation $\sum_{k=1}^{5}(k + 3)$.
    a) What is the 1 called?
    b) What is the 5 called?
    c) What is the $k$ called?
11. Let $a_n = 2n - 1$. Is this an increasing sequence or a decreasing sequence? Explain.
12. Let $a_n = -3n + 7$. Is this an increasing sequence or a decreasing sequence? Explain.
13. Let $a_n = 1 + (-2)^n$. Is this an alternating sequence? Explain.
14. Let $a_n = (-1)^{2n}$. Is this an alternating sequence? Explain.

### Practice the Skills

*Write the first five terms of the sequence whose nth term is shown.*

15. $a_n = 6n$
16. $a_n = -5n$
17. $a_n = 4n - 1$
18. $a_n = 2n + 5$
19. $a_n = \frac{7}{n}$
20. $a_n = \frac{8}{n^2}$
21. $a_n = \frac{n + 2}{n + 1}$
22. $a_n = \frac{n - 5}{n + 6}$
23. $a_n = (-1)^n$
24. $a_n = (-1)^{2n}$
25. $a_n = (-2)^{n+1}$
26. $a_n = 3^{n-1}$

*Find the indicated term of the sequence whose nth term is shown.*

27. $a_n = 2n + 7$, twelfth term
28. $a_n = 3n + 2$, sixth term
29. $a_n = \frac{n}{4} + 8$, sixteenth term
30. $a_n = \frac{n}{2} - 13$ fourteenth term
31. $a_n = (-1)^n$, eighth term
32. $a_n = (-2)^n$, fourth term
33. $a_n = n(n + 2)$, ninth term
34. $a_n = (n - 1)(n + 4)$, fifth term
35. $a_n = \frac{n^2}{2n + 7}$, ninth term
36. $a_n = \frac{n(n + 6)}{n^2}$, tenth term

*Find the first and third partial sums, $s_1$ and $s_3$, for each sequence.*

37. $a_n = 3n - 1$
38. $a_n = 2n + 3$
39. $a_n = 2^n + 1$
40. $a_n = 3^n - 8$
41. $a_n = \frac{n - 1}{n + 2}$
42. $a_n = \frac{n}{n + 3}$
43. $a_n = (-1)^n$
44. $a_n = (-3)^n$
45. $a_n = \frac{n^2}{2}$
46. $a_n = \frac{n^2}{n + 4}$

*Write the next three terms of each sequence.*

**47.** $2, 4, 8, 16, 32, \ldots$

**48.** $10, 15, 20, 25, 30, \ldots$

**49.** $7, 9, 11, 13, 15, \ldots$

**50.** $\frac{1}{2}, \frac{1}{3}, \frac{1}{4}, \frac{1}{5}, \ldots$

**51.** $1, \frac{1}{2}, \frac{1}{3}, \frac{1}{4}, \frac{1}{5}, \ldots$

**52.** $\frac{2}{3}, \frac{3}{4}, \frac{4}{5}, \frac{5}{6}, \frac{6}{7}, \ldots$

**53.** $-1, 1, -1, 1, -1, \ldots$

**54.** $-10, -20, -30, -40, \ldots$

**55.** $1, \frac{1}{3}, \frac{1}{9}, \frac{1}{27}, \ldots$

**56.** $\frac{1}{4}, \frac{2}{4}, \frac{3}{4}, \frac{4}{4}, \ldots$

**57.** $1, -\frac{1}{2}, \frac{1}{4}, -\frac{1}{8}, \ldots$

**58.** $\frac{1}{3}, \frac{1}{6}, \frac{1}{12}, \frac{1}{24}, \ldots$

**59.** $37, 32, 27, 22, \ldots$

**60.** $7, -1, -9, -17, \ldots$

*Write out each series, then evaluate it.*

**61.** $\sum_{i=1}^{5} (3i - 1)$

**62.** $\sum_{i=1}^{4} (4i + 5)$

**63.** $\sum_{i=1}^{6} (i^2 + 1)$

**64.** $\sum_{i=1}^{5} (2i^2 - 7)$

**65.** $\sum_{i=1}^{4} \frac{i^2}{2}$

**66.** $\sum_{i=1}^{3} \frac{i^2}{5}$

**67.** $\sum_{i=4}^{9} \frac{i^2 + i}{i + 1}$

**68.** $\sum_{i=2}^{5} \frac{i^3}{i + 1}$

*For the given general term $a_n$, write an expression using $\Sigma$ to represent the indicated partial sum.*

**69.** $a_n = n + 8$, fifth partial sum

**70.** $a_n = n^2 + 3$, fourth partial sum

**71.** $a_n = \frac{n^2}{4}$, third partial sum

**72.** $a_n = \frac{n^2 + 13}{n + 9}$, third partial sum

*For the set of values $x_1 = 2$, $x_2 = 3$, $x_3 = 5$, $x_4 = -1$, and $x_5 = 4$, find each of the following.*

**73.** $\sum_{i=1}^{5} x_i$

**74.** $\sum_{i=1}^{5} (x_i + 5)$

**75.** $\left(\sum_{i=1}^{5} x_i\right)^2$

**76.** $\sum_{i=1}^{5} 2x_i$

**77.** $\sum_{i=1}^{5} (x_i)^2$

**78.** $\sum_{i=1}^{4} (x_i^2 + 3)$

*Find the arithmetic mean, $\bar{x}$, of the following sets of data.*

**79.** 15, 20, 25, 30, 35

**80.** 16, 22, 96, 18, 28

**81.** 72, 83, 4, 60, 18, 20

**82.** 5, 13, 9, 12, 23, 36, 70

## Problem Solving

*In Exercises 83 and 84, consider the following rectangles. For the nth rectangle, the length is 2n and the width is n.*

**83. Perimeter**

**a)** Find the perimeters for the first four rectangles, and then list the perimeters in a sequence.

**b)** Find the general term for the perimeter of the $n$th rectangle in the sequence. Use $p_n$ for perimeter.

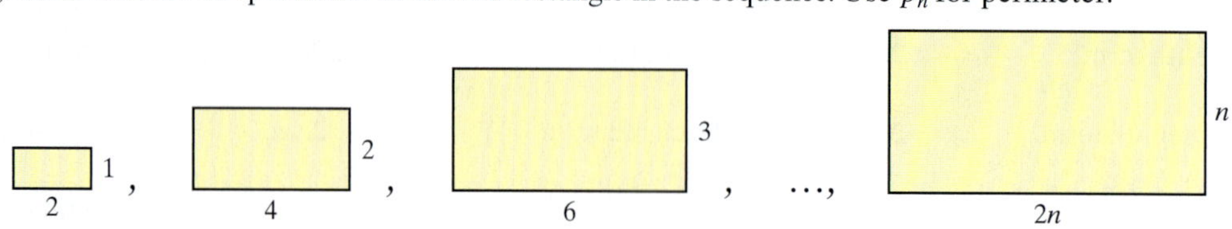


**84. Area**

**a)** Find the areas for the four rectangles, and then list the areas in a sequence.

**b)** Find the general term for the area of the $n$th rectangle in the sequence. Use $a_n$ for area.

**85.** Create your own sequence that is an increasing sequence and list the first five terms.

**86.** Create your own sequence that is a decreasing sequence and list the first five terms.

**87.** Create your own sequence that is an alternating sequence and list the first five terms.

**88.** Write

**a)** $\sum_{i=1}^{n} x_i$ as a sum of terms and

**b)** $\sum_{j=1}^{n} x_j$ as a sum of terms.

**c)** For a given set of values of $x$, from $x_1$ to $x_n$, will $\sum_{i=1}^{n} x_i = \sum_{j=1}^{n} x_j$? Explain.

**89.** Solve $\bar{x} = \frac{\Sigma x}{n}$ for $\Sigma x$.

**90.** Solve $\overline{x} = \dfrac{\Sigma x}{n}$ for $n$.

**91.** Is $\sum_{i=1}^{n} 4x_i = 4\sum_{i=1}^{n} x_i$? Illustrate your answer with an example.

**92.** Is $\sum_{i=1}^{n} \dfrac{x_i}{3} = \dfrac{1}{3}\sum_{i=1}^{n} x_i$? Illustrate your answer with an example.

**93.** Let $x_1 = 3, x_2 = 5, x_3 = 2$, and $y_1 = 4, y_2 = 1, y_3 = 6$. Find the following. Note that $\Sigma x = x_1 + x_2 + x_3$, $\Sigma y = y_1 + y_2 + y_3$, and $\Sigma xy = x_1y_1 + x_2y_2 + x_3y_3$.

**a)** $\Sigma x$,

**b)** $\Sigma y$,

**c)** $\Sigma x \cdot \Sigma y$,

**d)** $\Sigma xy$,

**e)** Is $\Sigma x \cdot \Sigma y = \Sigma xy$?

## Cumulative Review Exercises

[2.6] **94.** Solve $\left|\dfrac{1}{2}x + \dfrac{3}{5}\right| = \left|\dfrac{1}{2}x - 1\right|$.

[5.6] **95.** Factor $8y^3 - 64x^6$.

[7.6] **96.** Solve $\sqrt{x+5} - 1 = \sqrt{x-2}$.

[8.3] **97.** Solve $V = \pi r^2 h$ for $r$.

# 12.2 Arithmetic Sequences and Series

1. Find the common difference in an arithmetic sequence.
2. Find the $n$th term of an arithmetic sequence.
3. Find the $n$th partial sum of an arithmetic sequence.

## 1 Find the Common Difference in an Arithmetic Sequence

In the previous section, we started our discussion by assuming you got a job with a starting salary of \$30,000. One option for salary increases was an increase of \$2000 each year. This would result in the sequence

$$\$30{,}000, \$32{,}000, \$34{,}000, \$36{,}000, \ldots$$

This is an example of an arithmetic sequence.

**Arithmetic Sequence**

An **arithmetic sequence** is a sequence in which each term after the first differs from the preceding term by a constant amount.

The constant amount by which each pair of successive terms differs is called the **common difference**, $d$. The common difference can be found by subtracting any term from the term that directly follows it.

| Arithmetic Sequence | Common Difference |
|---|---|
| $1, 3, 5, 7, 9, \ldots$ | $d = 3 - 1 = 2$ |
| $5, 1, -3, -7, -11, -15, \ldots$ | $d = 1 - 5 = -4$ |
| $\frac{7}{2}, \frac{2}{2}, -\frac{3}{2}, -\frac{8}{2}, -\frac{13}{2}, -\frac{18}{2}, \ldots$ | $d = \frac{2}{2} - \frac{7}{2} = -\frac{5}{2}$ |

Notice that the common difference can be a positive number or a negative number. If the sequence is increasing, then $d$ is a positive number. If the sequence is decreasing, then $d$ is a negative number.

**EXAMPLE 1** ▸ Write the first five terms of the arithmetic sequence with

**a)** first term 6 and common difference 4.

**b)** first term 3 and common difference $-2$.

**c)** first term 1 and common difference $\dfrac{1}{3}$.

*Solution*

**a)** Start with 6 and keep adding 4. The sequence is 6, 10, 14, 18, 22.

**b)** $3, 1, -1, -3, -5$

**c)** $1, \dfrac{4}{3}, \dfrac{5}{3}, 2, \dfrac{7}{3}$

▸ Now Try Exercise 13

## 2 Find the *n*th Term of an Arithmetic Sequence

In general, an arithmetic sequence with first term, $a_1$, and common difference, $d$, has the following terms:

$$a_1 = a_1, \quad a_2 = a_1 + d, \quad a_3 = a_1 + 2d, \quad a_4 = a_1 + 3d, \quad \text{and so on}$$

If we continue this process, we can see that the $n$th term, $a_n$, can be found by the following formula:

**$n$th Term of an Arithmetic Sequence**

$$a_n = a_1 + (n - 1)d$$

**EXAMPLE 2**

**a)** Write an expression for the general (or $n$th) term, $a_n$, of the arithmetic sequence whose first term is $-3$ and whose common difference is 2.

**b)** Find the twelfth term of the sequence.

*Solution*

**a)** The $n$th term of the sequence is $a_n = a_1 + (n - 1)d$. Substituting $a_1 = -3$ and $d = 2$ we obtain

$$\begin{aligned} a_n &= a_1 + (n - 1)d \\ &= -3 + (n - 1)2 \\ &= -3 + 2(n - 1) \\ &= -3 + 2n - 2 \\ &= 2n - 5 \end{aligned}$$

Thus, $a_n = 2n - 5$.

**b)** $a_n = 2n - 5$

$a_{12} = 2(12) - 5 = 24 - 5 = 19$

The twelfth term in the sequence is 19.

Now Try Exercise 11

**EXAMPLE 3** Find the number of terms in the arithmetic sequence $5, 9, 13, 17, \ldots, 41$.

*Solution* The first term, $a_1$, is 5; the $n$th term is 41, and the common difference, $d$, is 4. Substitute the appropriate values into the formula for the $n$th term and solve for $n$.

$$\begin{aligned} a_n &= a_1 + (n - 1)d \\ 41 &= 5 + (n - 1)4 \\ 41 &= 5 + 4n - 4 \\ 41 &= 4n + 1 \\ 40 &= 4n \\ 10 &= n \end{aligned}$$

The sequence has 10 terms.

Now Try Exercise 51

## 3 Find the *n*th Partial Sum of an Arithmetic Sequence

An **arithmetic series** is the sum of the terms of an arithmetic sequence. A finite arithmetic series can be written

$$s_n = a_1 + (a_1 + d) + (a_1 + 2d) + (a_1 + 3d) + \cdots + (a_n - 2d) + (a_n - d) + a_n$$

If we consider the last term as $a_n$, the term before the last term will be $a_n - d$, the second before the last term will be $a_n - 2d$, and so on.

A formula for the $n$th partial sum, $s_n$, can be obtained by adding the reverse of $s_n$ to itself.

$$\begin{array}{rcl} s_n &=& a_1 + (a_1 + d) + (a_1 + 2d) + \cdots + (a_n - 2d) + (a_n - d) + a_n \\ s_n &=& a_n + (a_n - d) + (a_n - 2d) + \cdots + (a_1 + 2d) + (a_1 + d) + a_1 \\ \hline 2s_n &=& (a_1 + a_n) + (a_1 + a_n) + (a_1 + a_n) + \cdots + (a_1 + a_n) + (a_1 + a_n) + (a_1 + a_n) \end{array}$$

Since the right side of the equation contains $n$ terms of $(a_1 + a_n)$, we can write

$$2s_n = n(a_1 + a_n)$$

Now divide both sides of the equation by 2 to obtain the following formula.

**$n$th Partial Sum of an Arithmetic Sequence**

$$s_n = \frac{n(a_1 + a_n)}{2}$$

**EXAMPLE 4** ▸ Find the sum of the first 25 natural numbers.

*Solution* The arithmetic sequence is $1, 2, 3, 4, 5, 6, \ldots, 25$. The first term, $a_1$, is 1; the last term, $a_n$, is 25. There are 25 terms; thus, $n = 25$. Using the formula for the $n$th partial sum, we have

$$s_n = \frac{n(a_1 + a_n)}{2} = \frac{25(1 + 25)}{2} = \frac{25(26)}{2} = 25(13) = 325$$

The sum of the first 25 natural numbers is 325. Thus, $s_{25} = 325$.

▸ **Now Try Exercise 57**

**EXAMPLE 5** ▸ The first term of an arithmetic sequence is 4, and the last term is 31. If $s_n = 175$, find the number of terms in the sequence and the common difference.

*Solution* We substitute the appropriate values, $a_1 = 4$, $a_n = 31$, and $s_n = 175$, into the formula for the $n$th partial sum and solve for $n$.

$$\begin{aligned} s_n &= \frac{n(a_1 + a_n)}{2} \\ 175 &= \frac{n(4 + 31)}{2} \\ 175 &= \frac{35n}{2} \\ 350 &= 35n \\ 10 &= n \end{aligned}$$

There are 10 terms in the sequence. We can now find the common difference by using the formula for the $n$th term of an arithmetic sequence.

$$\begin{aligned} a_n &= a_1 + (n - 1)d \\ 31 &= 4 + (10 - 1)d \\ 31 &= 4 + 9d \\ 27 &= 9d \\ 3 &= d \end{aligned}$$

The common difference is 3. The sequence is 4, 7, 10, 13, 16, 19, 22, 25, 28, 31.

▸ **Now Try Exercise 31**

Examples 6–7 illustrate some applications of arithmetic sequences and series.

**EXAMPLE 6** ▸ **Salary** Mary Tufts is given a starting salary of \$35,000 and is promised a \$1200 raise after each of the next 8 years. Find her salary during her eighth year of work.

*Solution* **Understand** Her salaries during the first few years would be

$$\$35{,}000, \$36{,}200, \$37{,}400, \$38{,}600, \ldots$$

Since we are adding a constant amount each year, this is an arithmetic sequence. The general term of an arithmetic sequence is $a_n = a_1 + (n - 1)d$.

**Translate** In this example, $a_1 = 35{,}000$ and $d = 1200$. Thus, for $n = 8$, Mary's salary would be

$$a_8 = 35{,}000 + (8 - 1)1200$$

**Carry Out**

$$\begin{aligned} &= 35{,}000 + 7(1200) \\ &= 35{,}000 + 8400 \\ &= 43{,}400 \end{aligned}$$

**Answer** During her eighth year of work, Mary's salary would be \$43,400. If we listed all the salaries for the 8-year period, they would be \$35,000, \$36,200, \$37,400, \$38,600, \$39,800, \$41,000, \$42,200, \$43,400.

▸ Now Try Exercise 83

**EXAMPLE 7** ▸ **Pendulum** Each swing of a pendulum (left to right or right to left) is 3 inches shorter than the preceding swing. The first swing is 8 feet.

**a)** Find the length of the twelfth swing.

**b)** Determine the distance traveled by the pendulum during the first 12 swings.

*Solution* **a) Understand** Since each swing is decreasing by a constant amount, this problem can be represented as an arithmetic series. Since the first swing is given in feet and the decrease in swing in inches, we will change 3 inches to 0.25 feet $(3 \div 12 = 0.25)$. The twelfth swing can be considered $a_{12}$. The difference, $d$, is negative since the distance is decreasing with each swing.

**Translate**

$$\begin{aligned} a_n &= a_1 + (n - 1)d \\ a_{12} &= 8 + (12 - 1)(-0.25) \end{aligned}$$

**Carry Out**

$$\begin{aligned} &= 8 + 11(-0.25) \\ &= 8 - 2.75 \\ &= 5.25 \text{ feet} \end{aligned}$$

**Answer** The twelfth swing is 5.25 feet.

**b) Understand and Translate** The distance traveled during the first 12 swings can be found using the formula for the $n$th partial sum. The first swing, $a_1$, is 8 feet and the twelfth swing, $a_{12}$, is 5.25 feet.

$$s_n = \frac{n(a_1 + a_n)}{2}$$

$$s_{12} = \frac{12(a_1 + a_{12})}{2}$$

**Carry Out**

$$= \frac{12(8 + 5.25)}{2} = \frac{12(13.25)}{2} = 6(13.25) = 79.5 \text{ feet}$$

**Answer** The pendulum travels 79.5 feet during its first 12 swings.

▸ Now Try Exercise 75

## EXERCISE SET 12.2

MathXL® MyMathLab

### Concept/Writing Exercises

1. What is an arithmetic sequence?
2. What is an arithmetic series?
3. What do we call the constant amount by which each pair of successive terms in an arithmetic sequence differs?
4. How can the common difference in an arithmetic sequence be found?
5. If an arithmetic sequence is increasing, is the value for $d$ a positive number or a negative number?
6. If an arithmetic sequence is decreasing, is the value for $d$ a positive number or a negative number?
7. Can an arithmetic sequence consist of only negative numbers? Explain.
8. Can an arithmetic sequence consist of only odd numbers? Explain.
9. Can an arithmetic sequence consist of only even numbers? Explain.
10. Can an alternating sequence be an arithmetic sequence? Explain.

### Practice the Skills

*Write the first five terms of the arithmetic sequence with the given first term and common difference. Write the expression for the general (or nth) term, $a_n$, of the arithmetic sequence.*

11. $a_1 = 4, d = 3$
12. $a_1 = -11, d = 4$
13. $a_1 = 7, d = -2$
14. $a_1 = 3, d = -5$
15. $a_1 = \frac{1}{2}, d = \frac{3}{2}$
16. $a_1 = -\frac{5}{3}, d = -\frac{1}{3}$
17. $a_1 = 100, d = -5$
18. $a_1 = \frac{7}{4}, d = -\frac{3}{4}$

*Find the indicated quantity of the arithmetic sequence.*

19. $a_1 = 5, d = 3$; find $a_4$
20. $a_1 = 10, d = -3$; find $a_5$
21. $a_1 = -9, d = 4$; find $a_{10}$
22. $a_1 = -1, d = -2$; find $a_{12}$
23. $a_1 = -8, d = \frac{5}{3}$; find $a_{13}$
24. $a_1 = 5, a_8 = -21$; find $d$
25. $a_1 = 11, a_9 = 27$, find $d$
26. $a_1 = \frac{1}{2}, a_7 = \frac{19}{2}$; find $d$
27. $a_1 = 4, a_n = 28, d = 3$; find $n$
28. $a_1 = -9, a_n = -27, d = -3$; find $n$
29. $a_1 = 82, a_n = 42, d = -8$; find $n$
30. $a_1 = -\frac{4}{3}, a_n = -\frac{14}{3}, d = -\frac{2}{3}$; find $n$

*Find the sum, $s_n$, and common difference, $d$, of each sequence.*

31. $a_1 = 1, a_{10} = 19, n = 10$
32. $a_1 = -8, a_7 = 10, n = 7$
33. $a_1 = \frac{3}{5}, a_8 = 2, n = 8$
34. $a_1 = 12, a_8 = -23, n = 8$
35. $a_1 = -5, a_6 = 13.5, n = 6$
36. $a_1 = \frac{7}{5}, a_5 = \frac{23}{5}, n = 5$
37. $a_1 = 7, a_{11} = 67, n = 11$
38. $a_1 = 14.25, a_{31} = 18.75, n = 31$

*Write the first four terms of each sequence; then find $a_{10}$ and $s_{10}$.*

39. $a_1 = 4, d = 3$
40. $a_1 = 11, d = -6$
41. $a_1 = -6, d = 2$
42. $a_1 = -7, d = -4$
43. $a_1 = -8, d = -5$
44. $a_1 = -15, d = 4$
45. $a_1 = \frac{7}{2}, d = \frac{5}{2}$
46. $a_1 = \frac{9}{5}, d = \frac{3}{5}$
47. $a_1 = 100, d = -7$
48. $a_1 = 35, d = 6$

*Find the number of terms in each sequence and find $s_n$.*

**49.** $1, 4, 7, 10, \ldots, 43$

**50.** $-10, -8, -6, -4, \ldots, 40$

**51.** $-9, -5, -1, 3, \ldots, 31$

**52.** $6, 13, 20, 27, \ldots, 62$

**53.** $\frac{1}{2}, \frac{2}{2}, \frac{3}{2}, \frac{4}{2}, \frac{5}{2}, \ldots, \frac{17}{2}$

**54.** $-\frac{5}{6}, -\frac{7}{6}, -\frac{9}{6}, -\frac{11}{6}, \ldots, -\frac{21}{6}$

**55.** $7, 10, 13, 16, \ldots, 91$

**56.** $-11, -15, -19, \ldots, -51$

## Problem Solving

**57.** Find the sum of the first 50 natural numbers.

**58.** Find the sum of the first 50 even numbers.

**59.** Find the sum of the first 50 odd numbers.

**60.** Find the sum of the first 40 multiples of 5.

**61.** Find the sum of the first 30 multiples of 3.

**62.** Find the sum of the numbers between 50 and 150, inclusive.

**63.** Determine how many numbers between 7 and 1610 are divisible by 6.

**64.** Determine how many numbers between 14 and 1470 are divisible by 8.

*Pyramids occur everywhere. At athletic events, cheerleaders may form a pyramid where the people above stand on the shoulders of the people below. The illustration on the right shows a pyramid with 1 cheerleader on the top row, 2 cheerleaders in the middle row, and 3 cheerleaders in the bottom row. Notice that $a_1 = 1$, $a_2 = 2$, and $a_3 = 3$. Also, observe that $d = 1$, $n = 3$, and $s_3 = 6$.*

*At a bowling alley, the pins at the end of the bowling lane form a pyramid. The first row has 1 pin, the second row has 2 pins, the third row has 3 pins, and the fourth row has 4 pins. Thus, $a_1 = 1$, $d = 1$, $n = 4$, and $s_4 = 10$.*

*Use the idea of a pyramid to solve Exercises 65–70.*

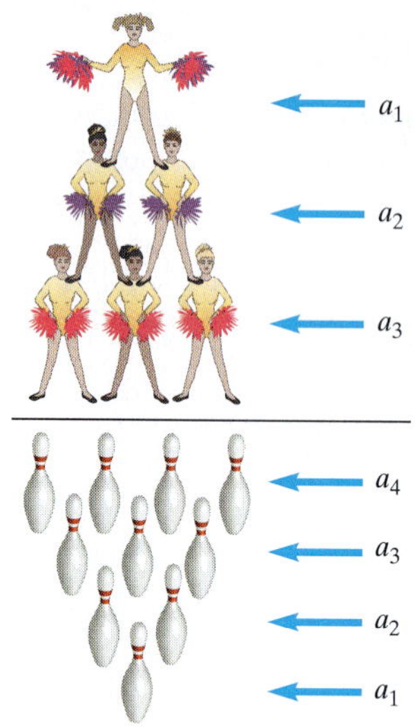


**65. Auditorium** An auditorium has 20 seats in the first row. Each successive row has two more seats than the previous row. How many seats are in the twelfth row? How many seats are in the first 12 rows?

**66. Auditorium** An auditorium has 22 seats in the first row. Each successive row has four more seats than the previous row. How many seats are in the ninth row? How many seats are in the first nine rows?

**67. Logs** Wolfgang Schmidt stacks logs so that there are 26 logs in the bottom layer, and each layer contains one log less than the layer below it. How many logs are in the pile?

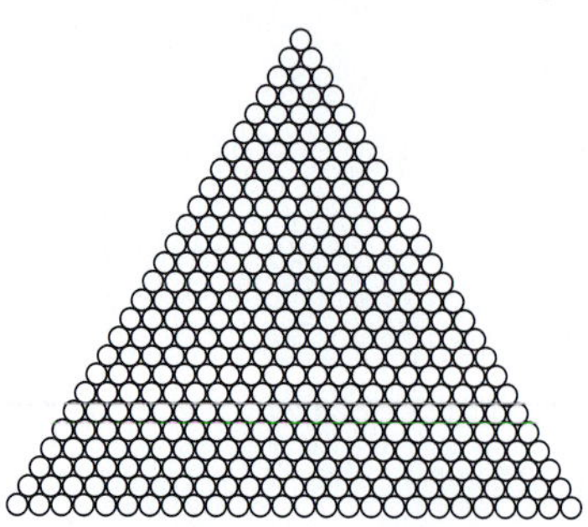

**68. Logs** Suppose Wolfgang, in Exercise 67, stopped stacking the logs after completing the layer containing eight logs. How many logs are in the pile?

**69. Glasses in a Stack** At their fiftieth wedding anniversary, Mr. and Mrs. Carlson are about to pour champagne into the top glass in the photo shown below. The top row has 1 glass, the second row has 3 glasses, the third row has 5 glasses, and so on. Each row has 2 more glasses than the row above it. This pyramid has 14 rows.

**a)** How many glasses are in the fourteenth row (bottom row)?

**b)** How many glasses are there total?

70. **Candies in a Stack** Individually wrapped candies are stacked in rows such that the top row has 1 candy, the second row has 3 candies, the third row has 5 candies, and so on. Each row has 2 more candies than the row above it. There are 7 rows of candies.

a) How many candies are in the seventh row (bottom row)?

b) How many candies are there total?

71. **Sum of Numbers** Karl Friedrich Gauss (1777–1855), a famous mathematician, as a child found the sum of the first 100 natural numbers quickly in his head $(1 + 2 + 3 + \cdots + 100)$. Explain how he might have done this and find the sum of the first 100 natural numbers as you think Gauss might have. (*Hint:* $1 + 100 = 101, 2 + 99 = 101$, etc.)

72. **Sum of Numbers** Use the same process from Exercise 71 to find the sum of the numbers from 101 to 150.

73. **Sum of Numbers** Find a formula for the sum of the first $n$ consecutive odd numbers starting with 1.

$$1 + 3 + 5 + \cdots + (2n - 1)$$

74. **Sum of Even Numbers** Find a formula for the sum of the first $n$ consecutive even numbers starting with 2.

$$2 + 4 + 6 + 8 + \cdots + 2n$$

75. **Swinging on a Vine** A long vine is attached to the branch of a tree. Sally Wynn swings from the vine, and each swing (left to right or right to left) is $\frac{1}{2}$ foot less than the previous swing. If her first swing is 22 feet, find

a) the length of the seventh swing, and

b) the distance traveled during her seven swings.

76. **Pendulum** Each swing of a pendulum is 2 inches shorter than the previous swing (left to right or right to left). The first swing is 6 feet. Find

a) the length of the eighth swing, and

b) the distance traveled by the pendulum during the eight swings.

77. **Bouncing Ball** Frank Holyton drops a ball from a second-story window. Each time the ball bounces, the height reached is 6 inches less than on the previous bounce. If the first bounce reaches a height of 6 feet, find the height attained on the ninth bounce.

78. **Ping-Pong Ball** A Ping-Pong ball falls from the table and bounces to a height of 3 feet. If each successive bounce is 3 inches less than the previous bounce, find the height attained on the tenth bounce.

79. **Packages** On Monday, March 17, Brian Nguyen started a new job at a packing company. On that day, he was able to prepare 105 packages for shipment. His boss expects Brian to be more productive with experience. Each day for six days, Brian is expected to prepare 10 more packages than the previous day's total.

a) How many packages is Brian expected to prepare on March 22?

b) How many packages is Brian expected to prepare during his first six days of employment?

80. **Salary** Marion Nickelson is making an annual salary of \$37,500 at the Thompson Frozen Food Factory. Her boss has promised her an increase of \$1500 in her salary each year over the next 10 years.

a) What will be Marion's salary 10 years from now?

b) What will be her total salary for these 11 years?

81. **Money** If Craig Campanella saves \$1 on day 1, \$2 on day 2, \$3 on day 3, and so on, how much money, in total, will he have saved on day 31?

82. **Money** If Dan Currier saves 50¢ on day 1, \$1.00 on day 2, \$1.50 on day 3, and so on, how much, in total, will he have saved by the end of 1 year (365 days)?

83. **Money** Carrie Dereshi recently retired and met with her financial planner. She arranged to receive \$42,000 the first year. Because of inflation, each year she will get \$400 more than she received the previous year.

a) What income will she receive in her tenth year of retirement?

b) How much money will she have received in total during her first 10 years of retirement?

84. **Salary** Susan Forman is given a starting salary of \$23,000 and is told she will receive a \$1000 raise at the end of each year.

a) Find her salary during year 12.

b) How much will she receive in total during her first 12 years?

85. **Angles** The sum of the interior angles of a triangle, a quadrilateral, a pentagon, and a hexagon are 180°, 360°, 540°, and 720°, respectively. Use the pattern here to find the formula for the sum of the interior angles of a polygon with $n$ sides.

86. Another formula that may be used to find the $n$th partial sum of an arithmetic series is

$$s_n = \frac{n}{2}[2a_1 + (n - 1)d]$$

Derive this formula using the two formulas presented in this section.

## Group Activity

*In calculus, a topic of importance is limits. Consider* $a_n = \frac{1}{n}$. *The first five terms of this sequence are* $\frac{1}{1}, \frac{1}{2}, \frac{1}{3}, \frac{1}{4}, \frac{1}{5}$. *Since the value of* $\frac{1}{n}$ *gets closer and closer to 0 as n gets larger and larger, we say that the limit of* $\frac{1}{n}$ *as n approaches infinity is 0. We write this as* $\lim_{n\to+\infty} \frac{1}{n} = 0$ *or* $\lim_{n\to+\infty} a_n = 0$. *Notice that* $\frac{1}{n}$ *can never equal 0, but its value approaches 0 as n gets larger and larger.*

**a)** *Group member 1: Find* $\lim_{n\to+\infty} a_n$ *for Exercises 87 and 88.*

**b)** *Group member 2: Find* $\lim_{n\to+\infty} a_n$ *for Exercises 89 and 90.*

**c)** *Group member 3: Find* $\lim_{n\to+\infty} a_n$ *for Exercises 91 and 92.*

**d)** *Exchange work and check each other's answers.*

**87.** $a_n = \dfrac{1}{n-2}$

**88.** $a_n = \dfrac{n}{n+1}$

**89.** $a_n = \dfrac{1}{n^2+2}$

**90.** $a_n = \dfrac{2n+1}{n}$

**91.** $a_n = \dfrac{4n-3}{3n+1}$

**92.** $a_n = \dfrac{n^2}{n+1}$

## Cumulative Review Exercises

[2.2] **93.** Solve $A = P + Prt$ for $r$.

[4.1] **94.** Solve the system of equations.

$$y = 2x + 1$$
$$3x - 2y = 1$$

[5.4] **95.** Factor $12n^2 - 6n - 30n + 15$.

[11.1] **96.** Graph $(x+4)^2 + y^2 = 25$.

# 12.3 Geometric Sequences and Series

1. Find the common ratio in a geometric sequence.
2. Find the *n*th term of a geometric sequence.
3. Find the *n*th partial sum of a geometric sequence.
4. Identify infinite geometric series.
5. Find the sum of an infinite geometric series.
6. Study applications of geometric series.

## 1 Find the Common Ratio in a Geometric Sequence

In Section 12.1, we assumed you got a job with a starting salary of \$30,000. We also mentioned that an option for salary increases was a 5% salary increase each year. This would result in the following sequence.

$$\$30{,}000,\quad \$31{,}500,\quad \$33{,}075,\quad \$34{,}728.75, \ldots$$

This is an example of a geometric sequence.

**Geometric Sequence**

A **geometric sequence** is a sequence in which each term after the first is a multiple of the preceding term.

The common multiple is called the **common ratio**.

The common ratio, $r$, in any geometric sequence can be found by dividing any term, except the first, by the preceding term. The common ratio of the previous geometric sequence is $\frac{31{,}500}{30{,}000} = 1.05$ (or 105%).

Consider the geometric sequence

$$1, 3, 9, 27, 81, \ldots, 3^{n-1}, \ldots$$

The common ratio is 3 since $3 \div 1 = 3$ (or $9 \div 3 = 3$, and so on).

| Geometric Sequence | Common Ratio |
|---|---|
| $4, 8, 16, 32, 64, \ldots, 4(2^{n-1}), \ldots$ | $2$ |
| $3, 12, 48, 192, 768, \ldots, 3(4^{n-1}), \ldots$ | $4$ |
| $7, \frac{7}{2}, \frac{7}{4}, \frac{7}{8}, \frac{7}{16}, \ldots, 7\left(\frac{1}{2}\right)^{n-1}, \ldots$ | $\frac{1}{2}$ |
| $5, -\frac{5}{3}, \frac{5}{9}, -\frac{5}{27}, \frac{5}{81}, \ldots, 5\left(-\frac{1}{3}\right)^{n-1}, \ldots$ | $-\frac{1}{3}$ |

**EXAMPLE 1** ▸ Determine the first five terms of the geometric sequence if $a_1 = 6$ and $r = \frac{1}{2}$.

*Solution* $a_1 = 6,\ a_2 = 6 \cdot \frac{1}{2} = 3,\ a_3 = 3 \cdot \frac{1}{2} = \frac{3}{2},\ a_4 = \frac{3}{2} \cdot \frac{1}{2} = \frac{3}{4},\ a_5 = \frac{3}{4} \cdot \frac{1}{2} = \frac{3}{8}$

Thus, the first five terms of the geometric sequence are

$$6, 3, \frac{3}{2}, \frac{3}{4}, \frac{3}{8}$$

▸ **Now Try Exercise 15**

## 2 Find the *n*th Term of a Geometric Sequence

In general, a geometric sequence with first term, $a_1$, and common ratio, $r$, has the following terms:

| $a_1,$ | $a_1r,$ | $a_1r^2,$ | $a_1r^3,$ | $a_1r^4, \ldots,$ | $a_1r^{n-1}, \ldots$ |
|---|---|---|---|---|---|
| ↑ | ↑ | ↑ | ↑ | ↑ | ↑ |
| 1st term, $a_1$ | 2nd term, $a_2$ | 3rd term, $a_3$ | 4th term, $a_4$ | 5th term, $a_5$ | nth term, $a_n$ |

Thus, we can see that the *n*th term of a geometric sequence is given by the following formula:

**nth Term of a Geometric Sequence**

$$a_n = a_1r^{n-1}$$

**EXAMPLE 2** ▸

**a)** Write an expression for the general (or *n*th) term, $a_n$, of the geometric sequence with $a_1 = 3$ and $r = -2$.

**b)** Find the twelfth term of this sequence.

*Solution*

**a)** The *n*th term of the sequence is $a_n = a_1r^{n-1}$. Substituting $a_1 = 3$ and $r = -2$, we obtain

$$a_n = a_1r^{n-1} = 3(-2)^{n-1}$$

Thus, $a_n = 3(-2)^{n-1}$.

**b)**
$$a_n = 3(-2)^{n-1}$$
$$a_{12} = 3(-2)^{12-1} = 3(-2)^{11} = 3(-2048) = -6144$$

The twelfth term of the sequence is $-6144$. The first 12 terms of the sequence are $3, -6, 12, -24, 48, -96, 192, -384, 768, -1536, 3072, -6144$.

▸ **Now Try Exercise 35**

**Helpful Hint** *Study Tip*

In this chapter, you will be working with exponents and using rules for exponents. The rules for exponents were discussed in Section 1.5 and again in Chapter 6. If you do not remember the rules for exponents, now is a good time to review Section 1.5.

**EXAMPLE 3** ▸ Find $r$ and $a_1$ for the geometric sequence with $a_2 = 12$ and $a_5 = 324$.

*Solution* The sequence can be represented with blanks for the missing terms.

$$\_\_, 12, \_\_, \_\_, 324$$
$$\uparrow a_2 \qquad \uparrow a_5$$

If we assume that $a_2$ is the first term of a sequence with the same common ratio, we obtain

$$\underset{\substack{\text{1st}\\\text{term}}}{\underset{\uparrow}{12}}, \_\_, \_\_, \underset{\substack{\text{4th}\\\text{term}}}{\underset{\uparrow}{324}}$$

Now we use the formula for the $n$th term of a geometric sequence to find $r$. We let the first term, $a_1$, be 12 and the number of terms, $n$, be 4.

$$\begin{aligned} a_n &= a_1r^{n-1} \\ 324 &= 12r^{4-1} \\ 324 &= 12r^3 \\ \frac{324}{12} &= r^3 \\ 27 &= r^3 \\ 3 &= r \end{aligned}$$

Thus, the common ratio is 3.

The first term of the original sequence must be $12 \div 3$ or 4. Thus, $a_1 = 4$. The first term, $a_1$, could also be found using the formula with $a_n = 324$, $r = 3$, and $n = 5$. Find $a_1$ by the formula now.

▸ Now Try Exercise 83

## 3 Find the *n*th Partial Sum of a Geometric Sequence

A **geometric series** is the sum of the terms of a geometric sequence. The sum of the first $n$ terms, $s_n$, of a geometric sequence can be expressed as

$$s_n = a_1 + a_1r + a_1r^2 + a_1r^3 + \cdots + a_1r^{n-2} + a_1r^{n-1} \qquad (eq. 1)$$

If we multiply both sides of the equation by $r$, we obtain

$$rs_n = a_1r + a_1r^2 + a_1r^3 + \cdots + a_1r^{n-1} + a_1r^n \qquad (eq. 2)$$

Now we subtract the corresponding sides of $(eq. 2)$ from $(eq. 1)$. The red-colored terms drop out, leaving

$$s_n - rs_n = a_1 - a_1r^n$$

Now we solve the equation for $s_n$.

$$\begin{aligned} s_n(1 - r) &= a_1(1 - r^n) && \text{Factor.} \\ s_n &= \frac{a_1(1 - r^n)}{1 - r} && \text{Divide both sides by } 1 - r. \end{aligned}$$

Thus, we have the following formula for the $n$th partial sum of a geometric sequence.

**nth Partial Sum of a Geometric Sequence**

$$s_n = \frac{a_1(1 - r^n)}{1 - r}, \qquad r \neq 1$$

**EXAMPLE 4** ▸ Find the seventh partial sum of a geometric sequence whose first term is 16 and whose common ratio is $-\frac{1}{2}$.

*Solution* Substitute the appropriate value for $a$, $r$, and $n$.

$$s_n = \frac{a_1(1 - r^n)}{1 - r}$$

$$s_7 = \frac{16\left[1 - \left(-\frac{1}{2}\right)^7\right]}{1 - \left(-\frac{1}{2}\right)} = \frac{16\left(1 + \frac{1}{128}\right)}{\frac{3}{2}} = \frac{16\left(\frac{129}{128}\right)}{\frac{3}{2}} = \frac{\frac{129}{8}}{\frac{3}{2}} = \frac{129}{8} \cdot \frac{2}{3} = \frac{43}{4}$$

Thus, $s_7 = \frac{43}{4}$.

▸ Now Try Exercise 41

**EXAMPLE 5** ▸ Given $s_n = 93$, $a_1 = 3$, and $r = 2$, find $n$.

*Solution*

$$s_n = \frac{a_1(1 - r^n)}{1 - r}$$

$$93 = \frac{3(1 - 2^n)}{1 - 2} \qquad \text{Substitute values for } s_n, a_1, \text{ and } r.$$

$$93 = \frac{3(1 - 2^n)}{-1}$$

$$-93 = 3(1 - 2^n) \qquad \text{Both sides were multiplied by } -1.$$

$$-31 = 1 - 2^n \qquad \text{Both sides were divided by 3.}$$

$$-32 = -2^n \qquad \text{1 was subtracted from both sides.}$$

$$32 = 2^n \qquad \text{Both sides were divided by } -1.$$

$$2^5 = 2^n \qquad \text{Write 32 as } 2^5.$$

Therefore, $n = 5$.

▸ **Now Try Exercise 65**

When working with a geometric series, $r$ can be a positive number as we saw in Example 5 or a negative number as we saw in Example 4.

## 4 Identify Infinite Geometric Series

All the geometric sequences that we have examined thus far have been finite since they have had a last term. The following sequence is an example of an infinite geometric sequence.

$$1, \frac{1}{2}, \frac{1}{4}, \frac{1}{8}, \frac{1}{16}, \ldots, \left(\frac{1}{2}\right)^{n-1}, \ldots$$

Note that the three dots at the end of the sequence indicate that the sequence continues indefinitely. The sum of the terms in an infinite geometric sequence forms an **infinite geometric series**. For example,

$$1 + \frac{1}{2} + \frac{1}{4} + \frac{1}{8} + \frac{1}{16} + \cdots + \left(\frac{1}{2}\right)^{n-1} + \cdots$$

is an infinite geometric series. Let's find some partial sums.

| Partial Sum | Series | Sum |
|---|---|---|
| Second | $1 + \frac{1}{2}$ | 1.5 |
| Third | $1 + \frac{1}{2} + \frac{1}{4}$ | 1.75 |
| Fourth | $1 + \frac{1}{2} + \frac{1}{4} + \frac{1}{8}$ | 1.875 |
| Fifth | $1 + \frac{1}{2} + \frac{1}{4} + \frac{1}{8} + \frac{1}{16}$ | 1.9375 |
| Sixth | $1 + \frac{1}{2} + \frac{1}{4} + \frac{1}{8} + \frac{1}{16} + \frac{1}{32}$ | 1.96875 |

With each successive partial sum, the amount being added is less than with the previous partial sum. Also, the sum seems to be getting closer and closer to 2. In Example 6, we will show that the sum of this infinite geometric series is indeed 2.

## 5 Find the Sum of an Infinite Geometric Series

Consider the formula for the sum of the first $n$ terms of an infinite geometric series:

$$s_n = \frac{a_1(1 - r^n)}{1 - r}, \qquad r \neq 1$$

What happens to $r^n$ if $|r| < 1$ and $n$ gets larger and larger? Suppose that $r = \frac{1}{2}$; then

$$\left(\frac{1}{2}\right)^1 = 0.5, \quad \left(\frac{1}{2}\right)^2 = 0.25, \quad \left(\frac{1}{2}\right)^3 = 0.125, \quad \left(\frac{1}{2}\right)^{20} \approx 0.000001$$

We can see that when $|r| < 1$, the value of $r^n$ gets exceedingly close to 0 as $n$ gets larger and larger. Thus, when considering the sum of an infinite geometric series, symbolized $s_\infty$, the expression $r^n$ approaches 0 when $|r| < 1$. Therefore, replacing $r^n$ with 0 in the formula $s_n = \frac{a_1(1 - r^n)}{1 - r}$ leads to the following formula.

**Sum of an Infinite Geometric Series**

$$s_\infty = \frac{a_1}{1 - r} \quad \text{where} \quad |r| < 1$$

**EXAMPLE 6** ▸ Find the sum of the infinite geometric series

$$1 + \frac{1}{2} + \frac{1}{4} + \frac{1}{8} + \cdots + \left(\frac{1}{2}\right)^{n-1} + \cdots.$$

*Solution* $a_1 = 1$ and $r = \frac{1}{2}$. Note that $\left|\frac{1}{2}\right| < 1$.

$$s_\infty = \frac{a_1}{1 - r} = \frac{1}{1 - \frac{1}{2}} = \frac{1}{\frac{1}{2}} = 2$$

Thus, $1 + \frac{1}{2} + \frac{1}{4} + \frac{1}{8} + \frac{1}{16} + \cdots + \left(\frac{1}{2}\right)^{n-1} + \cdots = 2$.

▸ Now Try Exercise 69

**EXAMPLE 7** ▸ Find the sum of the infinite geometric series

$$3 - \frac{6}{5} + \frac{12}{25} - \frac{24}{125} + \frac{48}{625} + \cdots$$

*Solution* The terms of the corresponding sequence are $3, -\frac{6}{5}, \frac{12}{25}, -\frac{24}{125}, \ldots$ Note that $a_1 = 3$. To find the common ratio, $r$, we can divide the second term, $-\frac{6}{5}$, by the first term, 3.

$$r = -\frac{6}{5} \div 3 = -\frac{6}{5} \cdot \frac{1}{3} = -\frac{2}{5}.$$

Since $\left|-\frac{2}{5}\right| < 1$,

$$s_\infty = \frac{a_1}{1 - r}$$

$$= \frac{3}{1 - \left(-\frac{2}{5}\right)} = \frac{3}{1 + \frac{2}{5}} = \frac{3}{\frac{7}{5}} = \frac{15}{7}$$

▸ Now Try Exercise 71

**EXAMPLE 8** ▸ Write 0.343434... as a ratio of integers.

*Solution* We can write this decimal as

$$0.34 + 0.0034 + 0.000034 + \cdots + (0.34)(0.01)^{n-1} + \cdots$$

This is an infinite geometric series with $r = 0.01$. Since $|r| < 1$,

$$s_\infty = \frac{a_1}{1-r} = \frac{0.34}{1-0.01} = \frac{0.34}{0.99} = \frac{34}{99}$$

If you divide 34 by 99 on a calculator, you will see .34343434 displayed.

▸ **Now Try Exercise 81**

---

What is the sum of a geometric series when $|r| > 1$? Consider the geometric sequence in which $a_1 = 1$ and $r = 2$.

$$1, 2, 4, 8, 16, 32, \ldots, 2^{n-1}, \ldots$$

The sum of its terms is

$$1 + 2 + 4 + 8 + 16 + 32 + \cdots + 2^{n-1} + \cdots$$

What is the sum of this series? As $n$ gets larger and larger, the sum gets larger and larger. We therefore say that the sum "does not exist." For $|r| > 1$, the sum of an infinite geometric series does not exist.

## 6 Study Applications of Geometric Series

Now let's look at some applications of geometric sequences and series.

**EXAMPLE 9** ▸ **Savings Account** Jean Simmons invests \$1000 at 5% interest compounded annually in a savings account. Determine the amount in his account and the amount of interest earned at the end of 10 years.

*Solution* **Understand** Suppose we let $P$ represent any principal invested. At the beginning of the second year, the amount grows to $P + 0.05P$ or $1.05P$. This amount will be the principal invested for year 2. At the beginning of the third year the second year's principal will grow by 5% to $(1.05P)(1.05)$, or $(1.05)^2P$. The amount in Jean's account at the beginning of successive years is

| Year 1 | Year 2 | Year 3 | Year 4 |
|---|---|---|---|
| $P$ | $1.05P$ | $(1.05)^2P$ | $(1.05)^3P$ |

and so on. This is a geometric series with $r = 1.05$. The amount in his account at the end of 10 years will be the same as the amount in his account at the beginning of year 11. We will therefore use the formula

$$a_n = a_1r^{n-1}, \quad \text{with} \quad r = 1.05 \quad \text{and} \quad n = 11$$

**Translate** We have a geometric sequence with $a_1 = 1000$, $r = 1.05$, and $n = 11$. Substituting these values into the formula, we obtain the following.

$$a_n = a_1r^{n-1}$$

**Carry Out**

$$\begin{aligned} a_{11} &= 1000(1.05)^{11-1} \\ &= 1000(1.05)^{10} \\ &\approx 1000(1.62889) \\ &\approx 1628.89 \end{aligned}$$

**Answer** After 10 years, the amount in the account is about \$1628.89. The amount of interest is $\$1628.89 - \$1000 = \$628.89$.

▸ **Now Try Exercise 95**

**EXAMPLE 10 ▸ Money** Suppose someone offered you \$1000 a day for each day of a 30-day month. Or, you could elect to take a penny on day 1, 2¢ on day 2, 4¢ on day 3, 8¢ on day 4, and so on. The amount would continue to double each day for 30 days.

**a)** Without doing any calculations, take a guess at which of the two offerings would provide the greater total return for 30 days.

**b)** Calculate the total amount you would receive by selecting \$1000 a day for 30 days.

**c)** Calculate the amount you would receive on day 30 by selecting 1¢ on day 1 and doubling the amount each day for 30 days.

**d)** Calculate the total amount you would receive for 30 days by selecting 1¢ on day 1 and doubling the amount each day for 30 days.

## Solution

**a)** Each of you will have your own answer to part **a)**.

**b)** If you received \$1000 a day for 30 days, you would receive 30(\$1000) = \$30,000.

**c)** Understand Since the amount is doubled each day, this represents a geometric sequence with $r = 2$. The chart that follows shows the amount you would receive in each of the first 7 days. We also show the amounts written with base 2, the common ratio.

| ***Day*** | 1 | 2 | 3 | 4 | 5 | 6 | 7 |
|---|---|---|---|---|---|---|---|
| ***Amount (cents)*** | 1 | 2 | 4 | 8 | 16 | 32 | 64 |
| ***Amount (cents)*** | $2^0$ | $2^1$ | $2^2$ | $2^3$ | $2^4$ | $2^5$ | $2^6$ |

Notice that for any given day, the exponent on 2 is 1 less than the given day. For example, on day 7, the amount is $2^6$. In general, the amount on day $n$ is $2^{n-1}$.

Translate To find the amount received on day 30, we evaluate $a_n = a_1r^{n-1}$ for $n = 30$.

$$a_n = a_1r^{n-1}$$

$$a_{30} = 1(2)^{30-1}$$

Carry Out

$$a_{30} = 1(2)^{29}$$

$$= 1(536,870,912)$$

$$= 536,870,912$$

Answer On day 30, the amount that you would receive is 536,870,912 cents or \$5,368,709.12

**d)** Understand and Translate To find the total amount received over the 30 days, we find the thirtieth partial sum.

$$s_n = \frac{a_1(1 - r^n)}{1 - r}$$

$$s_{30} = \frac{1(1 - 2^{30})}{1 - 2}$$

Carry Out

$$= \frac{1(1 - 1,073,741,824)}{-1}$$

$$= 1,073,741,823$$

Answer Therefore, over 30 days the total amount you would receive by this method would be 1,073,741,823 cents or \$10,737,418.23. The amount received by this method greatly surpasses the \$30,000 received by selecting \$1000 a day for 30 days.

▸ Now Try Exercise 87

**EXAMPLE 11** ▸ **Pendulum** On each swing (left to right or right to left), a certain pendulum travels 90% as far as on its previous swing. For example, if the swing to the right is 10 feet, the swing back to the left is $0.9 \times 10 = 9$ feet (see **Fig. 12.1**). If the first swing is 10 feet, determine the total distance traveled by the pendulum by the time it comes to rest.

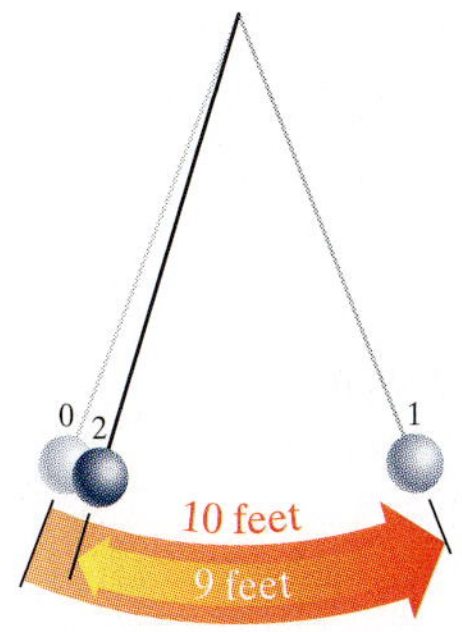

FIGURE 12.1

*Solution* **Understand** This problem may be considered an infinite geometric series with $a_1 = 10$ and $r = 0.9$. We can therefore use the formula $s_\infty = \dfrac{a_1}{1 - r}$ to find the total distance traveled by the pendulum.

**Translate and Carry Out**

$$s_\infty = \frac{a_1}{1 - r} = \frac{10}{1 - 0.9} = \frac{10}{0.1} = 100 \text{ feet}$$

**Answer** By the time the pendulum comes to rest, it has traveled 100 feet.

▸ **Now Try Exercise 99**

## EXERCISE SET 12.3

MathXL® MyMathLab

### Concept/Writing Exercises

1. What is a geometric sequence?
2. What is a geometric series?
3. Explain how to find the common ratio in a geometric sequence.
4. What is an infinite geometric series?
5. In a geometric series, if $|r| < 1$, what does $r^n$ approach as $n$ gets larger and larger?
6. Does the sum of an infinite geometric series when $|r| > 1$ exist?
7. In a geometric sequence, can the value of $r$ be a negative number?
8. In a geometric sequence, can the value of $r$ be a positive number?
9. In a geometric series, if $a_1 = 6$ and $r = 1/4$, does $s_\infty$ exist? If so, what is its value? Explain.
10. In a geometric series, if $a_1 = 6$ and $r = -2$, does $s_\infty$ exist? If so, what is its value? Explain.

### Practice the Skills

*Determine the first five terms of each geometric sequence.*

11. $a_1 = 2, r = 3$
12. $a_1 = 2, r = -3$
13. $a_1 = 6, r = -\dfrac{1}{2}$
14. $a_1 = 6, r = \dfrac{1}{2}$
15. $a_1 = 72, r = \dfrac{1}{3}$
16. $a_1 = \dfrac{1}{8}, r = 4$
17. $a_1 = 90, r = -\dfrac{1}{3}$
18. $a_1 = 32, r = -\dfrac{1}{4}$
19. $a_1 = -1, r = 3$
20. $a_1 = -1, r = -3$
21. $a_1 = 5, r = -2$
22. $a_1 = -13, r = -1$
23. $a_1 = \dfrac{1}{3}, r = \dfrac{1}{2}$
24. $a_1 = \dfrac{1}{2}, r = -\dfrac{1}{3}$
25. $a_1 = 3, r = \dfrac{3}{2}$
26. $a_1 = 60, r = -\dfrac{2}{5}$

*Find the indicated term of each geometric sequence.*

27. $a_1 = 4, r = 2$; find $a_6$
28. $a_1 = 4, r = -2$; find $a_6$
29. $a_1 = -12, r = \dfrac{1}{2}$; find $a_9$
30. $a_1 = 27, r = \dfrac{1}{3}$; find $a_7$
31. $a_1 = \dfrac{1}{4}, r = 2$; find $a_{10}$
32. $a_1 = 3, r = 3$; find $a_6$
33. $a_1 = -3, r = -2$; find $a_{12}$
34. $a_1 = -10, r = -2$; find $a_{10}$
35. $a_1 = 2, r = \dfrac{1}{2}$; find $a_8$

**36.** $a_1 = 5, r = \frac{2}{3}$; find $a_9$

**37.** $a_1 = 50, r = \frac{1}{3}$; find $a_7$

**38.** $a_1 = -7, r = -\frac{3}{4}$; find $a_7$

*Find the indicated sum.*

**39.** $a_1 = 5, r = 2$; find $s_5$

**40.** $a_1 = 7, r = -3$; find $s_5$

**41.** $a_1 = 2, r = 5$; find $s_6$

**42.** $a_1 = 9, r = \frac{1}{2}$; find $s_6$

**43.** $a_1 = 80, r = 2$; find $s_7$

**44.** $a_1 = 2, r = -2$; find $s_{12}$

**45.** $a_1 = -15, r = -\frac{1}{2}$; find $s_9$

**46.** $a_1 = \frac{3}{4}, r = 3$; find $s_7$

**47.** $a_1 = -9, r = \frac{2}{5}$; find $s_5$

**48.** $a_1 = 35, r = \frac{1}{5}$; find $s_{12}$

*For each geometric sequence, find the common ratio, r, and then write an expression for the general (or nth) term, $a_n$.*

**49.** $3, \frac{3}{2}, \frac{3}{4}, \frac{3}{8}, \ldots$

**50.** $3, -\frac{3}{2}, \frac{3}{4}, -\frac{3}{8}, \ldots$

**51.** $9, 18, 36, 72, \ldots$

**52.** $2, 6, 18, 54, \ldots$

**53.** $2, -6, 18, -54, \ldots$

**54.** $-1, -3, -9, -18, \ldots$

**55.** $\frac{3}{4}, \frac{1}{2}, \frac{1}{3}, \frac{2}{9}$

**56.** $\frac{4}{3}, \frac{8}{3}, \frac{16}{3}, \frac{32}{3}, \ldots$

*Find the sum of the terms in each geometric sequence.*

**57.** $1, \frac{1}{2}, \frac{1}{4}, \frac{1}{8}, \frac{1}{16}, \ldots$

**58.** $1, -\frac{1}{2}, \frac{1}{4}, -\frac{1}{8}, \frac{1}{16}, \ldots$

**59.** $1, \frac{1}{5}, \frac{1}{25}, \frac{1}{125}, \frac{1}{625}, \ldots$

**60.** $1, -\frac{1}{5}, \frac{1}{25}, -\frac{1}{125}, \frac{1}{625}, \ldots$

**61.** $6, 3, \frac{3}{2}, \frac{3}{4}, \frac{3}{8}, \ldots$

**62.** $\frac{1}{3}, \frac{1}{9}, \frac{1}{27}, \frac{1}{81}, \ldots$

**63.** $5, 2, \frac{4}{5}, \frac{8}{25}, \ldots$

**64.** $-\frac{4}{3}, -\frac{4}{9}, -\frac{4}{27}, -\frac{4}{81}, \ldots$

*Given $s_n$, $a_1$, and r, find n in each geometric series.*

**65.** $s_n = 93, a_1 = 3$, and $r = 2$

**66.** $s_n = 80, a_1 = 2$, and $r = 3$

**67.** $s_n = \frac{189}{32}, a_1 = 3$, and $r = \frac{1}{2}$

**68.** $s_n = \frac{121}{9}, a_1 = 9$, and $r = \frac{1}{3}$

*Find the sum of each infinite geometric series.*

**69.** $2 + 1 + \frac{1}{2} + \frac{1}{4} + \frac{1}{8} + \cdots$

**70.** $8 + 4 + 2 + 1 + \cdots$

**71.** $8 + \frac{16}{3} + \frac{32}{9} + \frac{64}{27} + \cdots$

**72.** $6 - 2 + \frac{2}{3} - \frac{4}{9} + \cdots$

**73.** $-60 + 20 - \frac{20}{3} + \frac{20}{9} - \cdots$

**74.** $2 + \frac{4}{3} + \frac{8}{9} + \frac{16}{27} + \cdots$

**75.** $-12 - \frac{12}{5} - \frac{12}{25} - \frac{12}{125} - \cdots$

**76.** $5 - 1 + \frac{1}{5} - \frac{1}{25} + \cdots$

*Write each repeating decimal as a ratio of integers.*

**77.** 0.242424...

**78.** 0.454545...

**79.** 0.8888...

**80.** 0.375375...

**81.** 0.515151...

**82.** 0.742742...

## Problem Solving

**83.** In a geometric sequence, $a_2 = 15$ and $a_5 = 405$; find $r$ and $a_1$.

**84.** In a geometric sequence, $a_2 = 27$ and $a_5 = 1$; find $r$ and $a_1$.

**85.** In a geometric sequence, $a_3 = 28$ and $a_5 = 112$, find $r$ and $a_1$.

**86.** In a geometric sequence, $a_2 = 12$ and $a_5 = -324$; find $r$ and $a_1$.

**87. Loaf of Bread** A loaf of bread currently costs $1.40. Determine the cost of a loaf of bread after 8 years (the start of the 9th year) if inflation were to grow at a constant rate of 3% per year. *Hint*: After year 1 (the start of year 2), the cost of a loaf of bread is $1.40(1.03). After year 2 (the start of year 3), the would be $\$1.40(1.03)^2$, and so on.

**88. Bicycle** A specific bicycle currently costs $400. Determine the cost of the bicycle after 12 years if inflation were to grow at a constant rate of 4% per year.

**89. Mass** A substance loses half its mass each day. If there are initially 600 grams of the substance, find

**a)** the number of days after which only 37.5 grams of the substance remain.

**b)** the amount of the substance remaining after 9 days.

**90. Bacteria** The number of a certain type of bacteria doubles every hour. If there are initially 1000 bacteria, after how many hours will the number of bacteria reach 64,000?

**91. Population** On July 1, 2005, the population of the United States was about 296.5 million people. If the population grows at a rate of 1.1% per year, find

**a)** the population after 10 years.

**b)** the number of years for the population to double.

**92. Farm Equipment** A piece of farm equipment that costs \$105,000 decreases in value by 15% per year. Find the value of the equipment after 4 years.

**93. Filtered Light** The amount of light filtering through a lake diminishes by one-half for each meter of depth.

**a)** Write a sequence indicating the amount of light remaining at depths of 1, 2, 3, 4, and 5 meters.

**b)** What is the general term for this sequence?

**c)** What is the remaining light at a depth of 7 meters?

**94. Pendulum** On each swing (left to right or right to left), a pendulum travels 80% as far as on its previous swing. If the first swing is 10 feet, determine the total distance traveled by the pendulum by the time it comes to rest.

**95. Investment** You invest \$10,000 in a savings account paying 6% interest annually. Find the amount in your account at the end of 8 years.

**96. Injected Dye** A tracer dye is injected into Mark Damion for medical reasons. After each hour, two-thirds of the previous hour's dye remains. How much dye remains in Mark's system after 10 hours?

**97. Bungee Jumping** Shawna Kelly goes bungee jumping off a bridge above water. On the initial jump, the bungee cord stretches to 220 feet. Assume the first bounce reaches a height of 60% of the original jump and that each additional bounce reaches a height of 60% of the previous bounce.

**a)** What will be the height of the fourth bounce?

**b)** Theoretically, Shawna would never stop bouncing, but realistically, she will. Use the infinite geometric series to estimate the total distance Shawna travels in a *downward* direction.

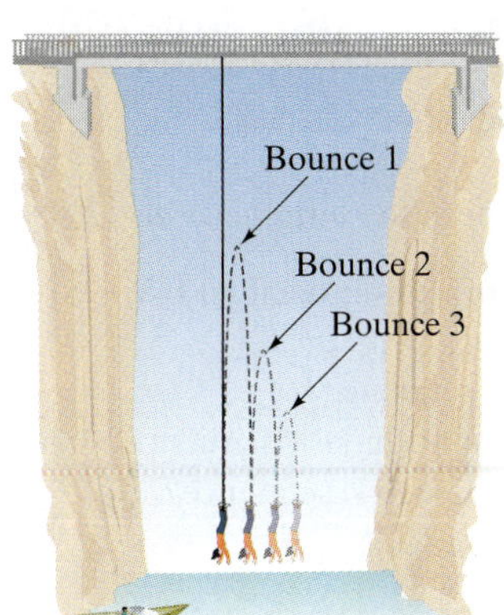


**98. Bungee Jumping** Repeat Exercise 97 **b)**, but this time find the total distance traveled in an *upward* direction.

**99. Ping-Pong Ball** A Ping-Pong ball falls off a table 30 inches high. Assume that the first bounce reaches a height of 70% of the distance the ball fell and each additional bounce reaches a height of 70% of the previous bounce.

**a)** How high will the ball bounce on the third bounce?

**b)** Theoretically, the ball would never stop bouncing, but realistically, it will. Estimate the total distance the ball travels in the *downward* direction.

**100. Ping-Pong Ball** Repeat Exercise 99 **b)**, but this time find the total distance traveled in the *upward* direction.

**101. Stack of Chips** Suppose that you form stacks of blue chips such that there is one blue chip in the first stack and in each successive stack you double the number of chips. Thus, you have stacks of 1, 2, 4, 8, and so on, of blue chips. You also form stacks of red chips, starting with one red chip and then tripling the number in each successive stack. Thus the stacks will contain 1, 3, 9, 27 and so on, red chips. How many more would the sixth stack of red chips have than the sixth stack of blue chips?

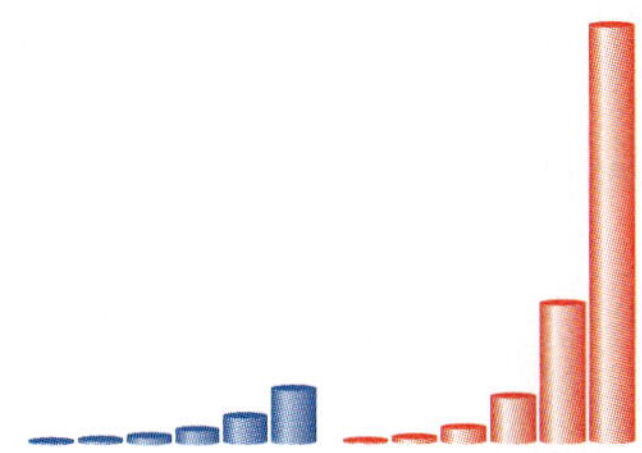

**102. Stack of Money** If you start with \$1 and double your money each day, how many days will it take to surpass \$1,000,000?

**103. Depreciation** One method of depreciating an item on an income tax return is the declining balance method. With this method, a given percent of the cost of the item is depreciated each year. Suppose that an item has a 5-year life and is depreciated using the declining balance method. Then, at the end of its first year, it loses $\frac{1}{5}$ of its value and $\frac{4}{5}$ of its value remains. At the end of the second year it loses $\frac{1}{5}$ of the remaining $\frac{4}{5}$ of its value, and so on. A car has a 5-year life expectancy and costs \$15,000.

**a)** Write a sequence showing the value of the car remaining for each of the first 3 years.

**b)** What is the general term of this sequence?

**c)** Find the value of the car at the end of 5 years.

**104. Scrap Value** On page 635, Exercise Set 9.6, Exercise 77, a formula for scrap value was given. The scrap value, $S$, is found by $S = c(1 - r)^n$ where $c$ is the original cost, $r$ is the annual depreciation rate and $n$ is the number of years the object is depreciated.

**a)** If you have not already done so, do Exercise 103 above to find the value of the car remaining at the end of 5 years.

**b)** Use the formula given to find the scrap value of the car at the end of 5 years and compare this answer with the answer found in part **a)**.

**105. Bouncing Ball** A ball is dropped from a height of 10 feet. The ball bounces to a height of 9 feet. On each successive bounce, the ball rises to 90% of its previous height. Find the *total vertical distance* traveled by the ball when it comes to rest.

**106. Wave Action** A particle follows the path indicated by the wave shown. Find the *total vertical distance* traveled by the particle.

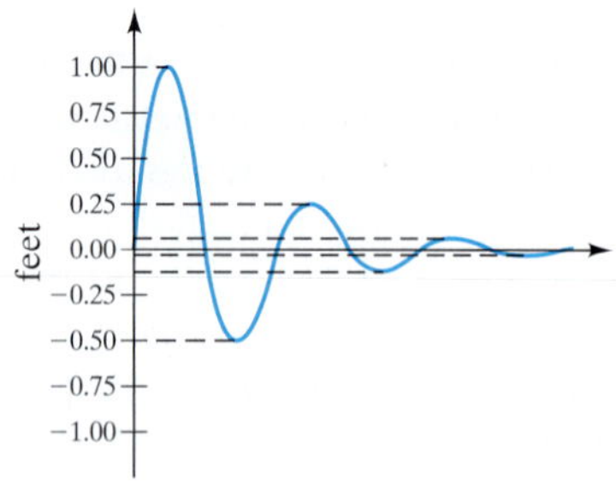


**107.** The formula for the $n$th term of a geometric sequence is $a_n = a_1 r^{n-1}$. If $a_1 = 1$, $a_n = r^{n-1}$.

**a)** How do you think the graphs of $y_1 = 2^{n-1}$ and $y_2 = 3^{n-1}$ will compare?

**b)** Graph both $y_1$ and $y_2$ and determine whether your answer to part **a)** was correct.

**108.** Use your grapher to decide the value of $n$ to the nearest hundredth, where $100 = 3 \cdot 2^{n-1}$.

## Challenge Problem

**109.** Find the sum of the sequence $1, 2, 4, 8, \ldots, 1{,}048{,}576$ and the number of terms in the sequence.

## Cumulative Review Exercises

[3.6] **110.** Let $f(x) = x^2 - 4$ and $g(x) = x - 3$. Find $(f \cdot g)(4)$.

[5.2] **111.** Multiply $(2x - 3y)(3x^2 + 4xy - 2y^2)$.

[6.4] **112.** Solve $S = \dfrac{2a}{1 - r}$ for $r$.

[9.1] **113.** Let $g(x) = x^3 + 9$. Find $g^{-1}(x)$.

[9.6] **114.** Solve $\log x + \log(x - 1) = \log 20$.

[11.4] **115. Sail on a Sailboat** A sail on a sailboat is shaped like a right triangle with a perimeter of 36 meters and a hypotenuse of 15 meters. Find the length of each leg of the triangle.

## Mid-Chapter Test: 12.1–12.3

*To find out how well you understand the chapter material to this point, take this brief test. The answers, and the section where the material was initially discussed, are given in the back of the book. Review any questions you answered incorrectly.*

**1.** Write the first five terms of the sequence whose $n$th term is $a_n = -3n + 5$.

**2.** If $a_n = n(n + 6)$, find the seventh term.

**3.** Find the first and third partial sums, $s_1$ and $s_3$, for the sequence whose $n$th term is $a_n = 2^n - 1$.

**4.** Write the next three terms of the sequence $5, 1, -3, -7, -11, \ldots$.

**5.** Evaluate the series $\sum_{i=1}^{5} (4i - 3)$.

**6.** If the general term of a sequence is $a_n = \frac{1}{3}n + 7$, write an expression using $\Sigma$ to represent the fifth partial sum.

**7.** Write the first four terms of the arithmetic sequence with $a_1 = -6$ and $d = 5$. Find an expression for the general term $a_n$.

**8.** Find $d$ for the arithmetic sequence with $a_1 = \frac{11}{2}$ and $a_7 = -\frac{1}{2}$.

**9.** Find $n$ for the arithmetic sequence with $a_1 = 22$, $a_n = -3$, and $d = -5$.

**10.** Find the common difference, $d$, and the sum, $s_6$, for the arithmetic sequence with $a_1 = -8$ and $a_6 = 7$.

**11.** Find $s_{10}$ for the arithmetic sequence with $a_1 = \frac{5}{2}$ and $d = \frac{1}{2}$.

**12.** Find the number of terms in the arithmetic sequence $-7, 0, 7, 14, \ldots, 63$.

**13.** Logs are stacked in a pile with 16 logs on the bottom row, 15 on the next row, 14 on the next row, and so on to the top, with the top row having one log. Each row has one log less than the row below it. How many logs are on the pile?

**14.** Write the first five terms of the geometric series with $a_1 = 80$ and $r = -\frac{1}{2}$.

**15.** Find $a_7$ for the geometric sequence with $a_1 = 81$ and $r = \frac{1}{3}$.

**16.** Find $s_6$ for the geometric sequence with $a_1 = 5$ and $r = 2$.

**17.** For the geometric sequence $8, -\frac{16}{3}, \frac{32}{9}, -\frac{64}{27}, \ldots$, find $r$.

**18.** Find the sum of the infinite series $12, 4, \frac{4}{3}, \frac{4}{9}, \ldots$.

**19.** Write the repeating decimal $0.878787\ldots$ as a ratio of two integers.

**20.** **a)** What is a sequence?
**b)** What is an arithmetic sequence?
**c)** What is geometric sequence?
**d)** What is a series?

# 12.4 Mathematical Induction

**1** Prove statements using mathematical induction.

## 1 Prove Statements Using Mathematical Induction

There are times when we need formulas or rules to aid in determining sums of large quantities of numbers. Consider the case of adding up the first few natural numbers.

The sum of the first two is

$$1 + 2 = 3$$

The sum of the first three is

$$1 + 2 + 3 = 6$$

The sum of the first five is

$$1 + 2 + 3 + 4 + 5 = 15$$

These sums are easy to compute when only a few natural numbers are being added. To add up the first 100 natural numbers would take some time, and if no mistakes are made, you would find the sum is 5050. You would probably agree that it is somewhat cumbersome to find sums this way.

Fortunately, there is a formula which gives the sum of the first $n$ natural numbers. It is

$$1 + 2 + 3 + 4 + \cdots + n = \frac{n(n + 1)}{2}$$

We can see that it works for some of the examples we just covered. When $n = 2$, the left side is $1 + 2 = 3$ and the right side is $\frac{2(2 + 1)}{2} = \frac{2(3)}{2} = \frac{6}{2} = 3$. Here we see the two sides produce the same value.

When $n = 3$, the left side is $1 + 2 + 3 = 6$ and the right side is $\frac{3(3 + 1)}{2} = \frac{3(4)}{2} = \frac{12}{2} = 6$. Here both values are the same. When $n = 100$, the left side is $1 + 2 + 3 + \cdots + 99 + 100$ which adds up to 5050 (after considerable work). The right side is $\frac{100(101)}{2} = \frac{10{,}100}{2} = 5050$. Again the values are the same.

As $n$ becomes larger, it should be clear that it is easier to substitute a number into the right side to obtain the sum than to add up the numbers on the left side.

If we were to substitute values from 1 to 100 for $n$ in the formula, and if in each case the formula was true, it would suggest that the formula is true for all values of $n$. However, we cannot assume that a formula is true just because it is true for 100 or even 1000 cases. We know that a formula is true for *all values of $n$* only when it has been proven true. Fortunately, we can do this using a process called mathematical induction. (In Example 4 we prove that the formula $1 + 2 + 3 + 4 + \ldots + n = \frac{n(n + 1)}{2}$ is true.)

**Mathematical induction** is a powerful method for proving that a mathematical statement involving natural numbers is true for all natural numbers. To prove a statement using mathematical induction, we must show that the statement is true for $n = 1$ and that if it is true for some natural number $k$, then it is also true for the next natural number, $k + 1$. Recall that the natural numbers are also called the positive integers.

**The Principle of Mathematical Induction**

Let $S_n$ be a mathematical statement concerning the positive integer $n$. If the following two conditions are true, then $S_n$ is true for every positive integer $n$.

**Condition 1.** $S_1$ is true and

**Condition 2.** for any positive integer $k$, $k \leq n$, if $S_k$ is true, then $S_{k+1}$ is true.

We will not prove this theorem, but we will try to give you some insight into why this theorem works. Suppose you have infinitely many dominos lined up in a row. Now suppose we know the following two facts are true:

1. The first domino is pushed over.
2. If one of the dominos falls over, say the $k$th domino, then so will the next one, the $k + 1$ domino.

Is it safe to conclude that *all* the dominos will fall over? The answer is yes, because if the first one falls (Condition 1 in the Principle of Mathematical Induction), then the second one falls (by Condition 2); and if the second one falls, then so does the third (by Condition 2), and so on.

Mathematically, proving that this statement is true for $S_1$ and for $S_{k+1}$ when $S_k$ is true, we know that it is true for $S_2$. Similarly, by knowing $S_2$ is true and that $S_{k+1}$ is true when $S_k$ is true, we know that $S_3$ is true. Continuing this way, we know that $S_n$ is true for every positive integer $n$.

Condition 1 is shown to be true by substituting 1 for $n$ in the expression and then evaluating the expression. If the result is a true statement, then Condition 1 holds. If you are asked to prove a formula is true by induction, Condition 2 may sometimes be accomplished by beginning with the formula using the variable $k$, then adding the $k + 1$ term to both sides of the formula. Then, using algebraic techniques, if you can show that the formula with the $k + 1$ term added to both sides simplifies to the formula obtained by replacing each $k$ with $k + 1$, you have illustrated Condition 2. Example 1 illustrates this process. In this example we use the formula

$$1 + 3 + 5 + \cdots + (2n - 1) = n^2$$

For example, when $n = 1$ (only 1 term), we evaluate $(2n - 1)$ and $n^2$ to get

$$\begin{aligned} 2(1) - 1 &= 1^2 \\ 1 &= 1 \end{aligned}$$

When $n = 2$, (the sum of 2 terms), we get

$$\begin{aligned} 1 + [2(2) - 1] &= 2^2 \\ 1 + 3 &= 2^2 \end{aligned}$$

When $n = 3$, we get

$$\begin{aligned} 1 + 3 + [2(3) - 1] &= 3^2 \\ 1 + 3 + 5 &= 3^2 \text{ and so on.} \end{aligned}$$

In Example 1 we prove that the sum of the first $n$ odd integers is $n^2$.

**EXAMPLE 1** ▸ Prove $1 + 3 + 5 + \cdots + (2n - 1) = n^2$ for all positive integers $n$.

*Solution* **Condition 1**: Show that $1 + 3 + 5 + \cdots + (2n - 1) = n^2$ for $n = 1$. For $n = 1$, we get $1 = 1^2$ (see above). This is a true statement and Condition 1 is satisfied.

**Condition 2**: Assume that $1 + 3 + 5 + \cdots + (2k - 1) = k^2$ is true for some positive integer $k$. We need to show that the statement holds for $k + 1$. That is, we wish to show that

$$1 + 3 + 5 + \cdots + (2k - 1) + [2(k + 1) - 1] = (k + 1)^2.$$

Notice the $k + 1$ term $[2(k + 1) - 1]$, which was added to the left side of the equal sign, is obtained by substituting $k + 1$ for $k$ in the term $(2k - 1)$. On the right side we substitute $k + 1$ for $k$ to obtain $(k + 1)^2$. To continue our proof by induction, we begin with $1 + 3 + 5 + \cdots + (2k - 1) = k^2$, which we assume to be true, and add the $k + 1$ term, $[2(k + 1) - 1]$ to both sides of the formula. Then we simplify the right side of the formula to show that we obtain

$$1 + 3 + 5 + \cdots + (2k - 1) + [2(k + 1) - 1] = (k + 1)^2$$

We will now show the process.

$$1 + 3 + 5 + \cdots + (2k - 1) = k^2$$

$$1 + 3 + 5 + \cdots + (2k - 1) + [2(k + 1) - 1] = k^2 + [2(k + 1) - 1] \quad \text{Add } 2(k + 1) - 1 \text{ to both sides.}$$

$$1 + 3 + 5 + \cdots + (2k - 1) + [2(k + 1) - 1] = k^2 + 2k + 2 - 1 \quad \text{Simplify the expression on the right.}$$

$$1 + 3 + 5 + \cdots + (2k - 1) + [2(k + 1) - 1] = k^2 + 2k + 1 \quad \text{Combine terms.}$$

$$1 + 3 + 5 + \cdots + (2k - 1) + [2(k + 1) - 1] = (k + 1)(k + 1) \quad \text{Factor.}$$

$$1 + 3 + 5 + \cdots + (2k - 1) + [2(k + 1) - 1] = (k + 1)^2 \quad \text{Write factored expression as } (k + 1)^2.$$

We have now shown Condition 2 holds. Since Conditions 1 and 2 hold, we have proven this formula true for all positive integers by mathematical induction.

▸ **Now Try Exercise 7**

As we show in Example 2, mathematical induction can also be used to prove statements about inequalities.

**EXAMPLE 2 ▸** Prove $3^n > n$ for all positive integers $n$.

*Solution* **Condition 1**: Show that $3^n > n$ for $n = 1$.

$$3^n > n$$

$$3^1 \overset{?}{>} 1$$

$$3 > 1$$

Since this is a true statement, Condition 1 holds.

**Condition 2**: Assume $3^k > k$ for some positive integer $k$. We wish to show that the statement holds for $k + 1$; that is, we wish to show that $3^{k+1} > k + 1$. In our proof we use the fact that $3 \cdot 3^k = 3^{k+1}$.

$$3^k > k$$

$$3 \cdot 3^k > 3 \cdot k \quad \text{Multiply both sides of the inequality by 3.}$$

$$3^{k+1} > k + k + k \quad \text{Write } 3k \text{ as } k + k + k.$$

$$3^{k+1} > k + 1 \quad \text{since } k \geq 1, \text{ this inequality must also be true}$$

Since we have shown that Conditions 1 and 2 hold, we have proven, using induction, that the inequality statement is true for all positive integers.

▸ **Now Try Exercise 37**

**EXAMPLE 3 ▸** Prove $6n$ is an even number for all positive integers.

*Solution* If $6n$ is an even number, then it is divisible by 2. If a number, $a$, is divisible by 2 then $a \div 2$ is an integer.

**Condition 1**: To show that $6n$ is an even number for $n = 1$, we need to show that $6(1) \div 2$ is an integer:

$$6(1) \div 2 = \frac{6}{2} = 3$$

Since 3 is an integer, Condition 1 holds.

**Condition 2**: Assume $6k \div 2$ is an integer for some integer $k$. We need to show that this statement holds for $k + 1$, that is, that $6(k + 1) \div 2$ is an integer.

Assume $\frac{6k}{2} = m$ where $m$ is an integer. Now subsitute $k + 1$ for $k$ to obtain the $k + 1$ term.

$$\begin{aligned}\frac{6(k+1)}{2} &= \frac{6k+6}{2} && \text{Subsitute } k + 1 \text{ for } k \text{ and simplify.}\\ &= \frac{6k}{2} + \frac{6}{2} \\ &= m + 3 && \text{since } 6k \div 2 = m\end{aligned}$$

Since $m$ is an integer, $m + 3$ is also an integer, so $6(k + 1) \div 2$ is an integer. Thus, Condition 2 is satisfied. As a result $6n$ is an even number for all positive integers, $n$.

▸ **Now Try Exercise 39**

**EXAMPLE 4** ▸ Prove $1 + 2 + 3 + \cdots + n = \frac{n(n+1)}{2}$ for all positive integers $n$.

*Solution* **Condition 1**: We show the formula is true for $n = 1$ as follows.

$$\begin{aligned}1 &= \frac{1(1+1)}{2}\\ 1 &= 1\end{aligned}$$

**Condition 2**: We assume the formula holds for some $k$, and show that the formula holds for $k + 1$. Thus, we assume that for some $k$

$$1 + 2 + 3 + \cdots + k = \frac{k(k+1)}{2}$$

We need to show the statement is true for $k + 1$. That is, we wish to show

$$1 + 2 + 3 + \cdots + k + (k+1) = \frac{(k+1)[(k+1)+1]}{2}$$

We begin with our original formula in $k$.

$$1 + 2 + 3 + \cdots + k = \frac{k(k+1)}{2}$$

and add the $k + 1$ term to each side.

$$1 + 2 + 3 + \cdots + k + (k+1) = \frac{k(k+1)}{2} + (k+1) \quad \text{Add the } k + 1 \text{ term to both sides.}$$

Now factor out the common factor $k + 1$ on the right to get

$$\begin{aligned}&= (k+1)\left(\frac{k}{2} + 1\right)\\ &= (k+1)\left(\frac{k+2}{2}\right)\\ 1 + 2 + 3 + \cdots + k + (k+1) &= \frac{(k+1)[(k+1)+1]}{2} && \text{Write } k + 2 \text{ as } (k+1) + 1.\end{aligned}$$

This is the statement we wanted to show. Since we have shown that the statement we assumed true for $k$ is also true for $k + 1$, we have shown Condition 2 true. We have therefore proven the given formula by induction.

▸ **Now Try Exercise 9**

**EXAMPLE 5 ▸ Pennies in a Piggy Bank** For a third grade project, Bob Henderson has decided to deposit pennies into a piggy bank by dropping in one penny on the first day, two pennies on the second day, three pennies on the third day, and so on, for 30 days.

**a)** How many pennies are in the piggy bank at the end of 30 days?

**b)** If Bob continues this plan for an additional 30 days, how many pennies are in the piggy bank at the end of 60 days?

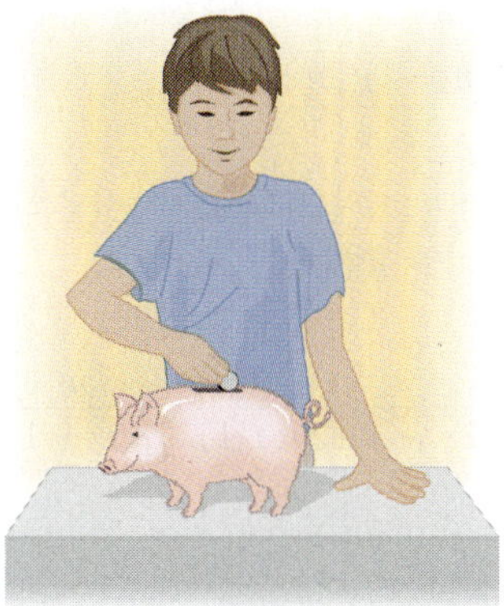

### Solution

**a)** Use the formula from Example 4 with $n = 30$.

$$1 + 2 + 3 + \cdots + n = \frac{n(n+1)}{2} \quad \text{Formula}$$
$$1 + 2 + 3 + \cdots + 29 + 30 = \frac{30(30+1)}{2} \quad \text{Substitute 30 for } n.$$
$$= \frac{30(31)}{2} \quad \text{Simplify.}$$
$$= 465$$

At the end of 30 days, Bob has 465 pennies ($4.65) in the piggy bank.

**b)** For 60 days, use the same formula with $n = 60$.

$$1 + 2 + 3 + \cdots + n = \frac{n(n+1)}{2} \quad \text{Formula}$$
$$1 + 2 + 3 + \cdots + 59 + 60 = \frac{60(60+1)}{2} \quad \text{Substitute 60 for } n.$$
$$= \frac{60(61)}{2} \quad \text{Simplify.}$$
$$= 1830$$

At the end of 60 days, Bob has 1830 pennies ($18.30) in the piggy bank.

▸ Now Try Exercise 47

## EXERCISE SET 12.4

MathXL® MyMathLab

### Concept/Writing Exercises

**1.** Can we prove a statement $S_n$ is true for all $n$ by testing a finite number of cases? Explain.

**2.** What is mathematical induction?

**3.** What two conditions must be proven to be true in order to prove $S_n$ is true for all $n$?

**4.** If $S_1$ is true and, if $S_k$ is true then $S_{k+1}$ is true, how do we know that $S_3$ is true?

### Practice the Skills

*Use mathematical induction to prove the following statements for all positive intergers n.*

**5.** $1 + 5 + 9 + \cdots + (4n - 3) = n(2n - 1)$

**6.** $1 + 4 + 7 + \cdots + (3n - 2) = \frac{n}{2}(3n - 1)$

**7.** $2 + 4 + 6 + \cdots + 2n = n(n + 1)$

**8.** $2 + 5 + 8 + \cdots + (3n - 1) = \frac{n}{2}(3n + 1)$

**9.** $3 + 4 + 5 + \cdots + (n + 2) = \frac{n}{2}(n + 5)$

**10.** $3 + 5 + 7 + \cdots + (2n + 1) = n(n + 2)$

**11.** $3 + 6 + 9 + \cdots + 3n = \dfrac{3n(n+1)}{2}$

**12.** $4 + 8 + 12 + \cdots + 4n = \dfrac{4n(n+1)}{2}$

**13.** $5 + 10 + 15 + \cdots + 5n = \dfrac{5n(n+1)}{2}$

**14.** $7 + 14 + 21 + \cdots + 7n = \dfrac{7n(n+1)}{2}$

**15.** $1 + 2 + 2^2 + \cdots + 2^{n-1} = 2^n - 1$

**16.** $1 + 3 + 3^2 + \cdots + 3^{n-1} = \dfrac{3^n - 1}{2}$

**17.** $1 + 4 + 4^2 + \cdots + 4^{n-1} = \dfrac{4^n - 1}{3}$

**18.** $1 + 6 + 6^2 + \cdots + 6^{n-1} = \dfrac{6^n - 1}{4}$

**19.** $2 + 2^2 + 2^3 + \cdots + 2^n = 2^{n+1} - 2$

**20.** $3 + 3^2 + 3^3 + \cdots + 3^n = 3\left(\dfrac{3^n - 1}{2}\right)$

**21.** $4 + 4^2 + 4^3 + \cdots + 4^n = 4\left(\dfrac{4^n - 1}{3}\right)$

**22.** $1^2 + 2^2 + 3^2 + \cdots + n^2 = \dfrac{n(n+1)(2n+1)}{6}$

**23.** $1^3 + 2^3 + 3^3 + \cdots + n^3 = \left[\dfrac{n(n+1)}{2}\right]^2$

**24.** $1^3 + 3^3 + 5^3 + \cdots + (2n-1)^3 = n^2(2n^2 - 1)$

**25.** $\dfrac{1}{1\cdot 2} + \dfrac{1}{2\cdot 3} + \dfrac{1}{3\cdot 4} + \cdots + \dfrac{1}{n(n+1)} = \dfrac{n}{n+1}$

**26.** $\dfrac{1}{1\cdot 3} + \dfrac{1}{3\cdot 5} + \dfrac{1}{5\cdot 7} + \cdots + \dfrac{1}{(2n-1)(2n+1)} = \dfrac{n}{2n+1}$

**27.** $1 + \dfrac{1}{2} + \dfrac{1}{2^2} + \cdots + \dfrac{1}{2^{n-1}} = 2\left(1 - \dfrac{1}{2^n}\right)$

**28.** $1 + \dfrac{1}{3} + \dfrac{1}{3^2} + \cdots + \dfrac{1}{3^{n-1}} = \dfrac{3}{2}\left(1 - \dfrac{1}{3^n}\right)$

**29.** $1 + \dfrac{1}{4} + \dfrac{1}{4^2} + \cdots + \dfrac{1}{4^{n-1}} = \dfrac{4}{3}\left(1 - \dfrac{1}{4^n}\right)$

**30.** $1 + \dfrac{1}{5} + \dfrac{1}{5^2} + \cdots + \dfrac{1}{5^{n-1}} = \dfrac{5}{4}\left(1 - \dfrac{1}{5^n}\right)$

**31.** $1 + \dfrac{3}{4} + \left(\dfrac{3}{4}\right)^2 + \cdots + \left(\dfrac{3}{4}\right)^{n-1} = 4\left[1 - \left(\dfrac{3}{4}\right)^n\right]$

**32.** $1 + \dfrac{2}{5} + \left(\dfrac{2}{5}\right)^2 + \cdots + \left(\dfrac{2}{5}\right)^{n-1} = \dfrac{5}{3}\left[1 - \left(\dfrac{2}{5}\right)^n\right]$

**33.** $\left(\dfrac{3}{4}\right)^{n+1} < \dfrac{3}{4}$

**34.** $\left(\dfrac{2}{3}\right)^{n+1} < \dfrac{2}{3}$

**35.** $\left(\dfrac{1}{4}\right)^{n+1} < \dfrac{1}{4}$

**36.** $2^{n+4} > 4(n+4)$

**37.** $3^n \geq 2n + 1$

**38.** $3^{n+3} > 9(n+3)$

**39.** $n^2 + n$ is even.

**40.** $n^2 - n + 4$ is even.

**41.** $n^3 + 2n$ is divisible by 3.

**42.** $n(n+1)(n+2)$ is divisible by 6.

## Problem Solving

**43.** Use mathematical induction to prove that when $r \neq 1$, $a + ar + ar^2 + \cdots + ar^{n-1} = \dfrac{a(1 - r^n)}{1 - r}$ for all positive integers $n$.

**44.** Use mathematical induction to prove $a + (a + d) + (a + 2d) + \cdots + [a + (n-1)d] = \dfrac{2na + d(n^2 - n)}{2}$ for all positive integers $n$.

**Pennies** *Suppose pennies are dropped into a piggy bank by dropping one penny on the first day, two pennies on the second day, three pennies on the third day and so on (See Example 5). How many pennies are in the piggy bank at the end of*

**45.** 8 days?

**46.** 16 days?

**47.** 22 days?

**48.** 28 days?

**49.** 32 days?

**50.** 40 days?

**51.** 45 days?

**52.** 52 days?

**53.** 70 days?

**54.** 90 days?

## Group Activity

**55. Chords** A chord is a line segment that connects two points on a circle. The figure below is a circle with two chords. Notice that the two chords divide the circle into 4 regions.

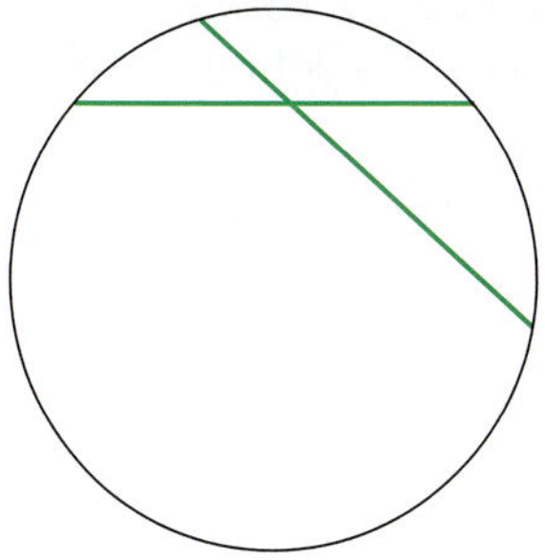

**a)** Draw a circle with no chords. How many regions does the circle have?

**b)** Draw a circle with 1 chord. How many regions does the circle have?

**c)** Draw your own diagram of a circle with 2 chords. How many regions does the circle have? Will all circles with 2 chords have the same number of regions?

**d)** Draw a circle with 3 chords. How many regions does the circle have? Will all circles with 3 chords have the same number of regions? How do you construct a diagram with the maximum number of regions?

**e)** Complete the table.

| Number of Chords on a Circle | Maximum Number of Regions in the Circle |
|---|---|
| 0 | 1 |
| 1 | 2 |
| 2 | 4 |
| 3 | 7 |
| 4 | 11 |

**f)** What is the pattern in the sequence of numbers in the Maximum Number of Regions in the Circle column?

**g)** Write a formula for the maximum number of regions, $r$, in a circle with $n$ chords.

**h)** Use your formula to find the maximum number of regions in a circle with 5 chords. Use a diagram to find the maximum number of regions in a circle with 5 chords.

**i)** Do you think your formula is true for all positive integers?

**j)** Explain how this activity relates to mathematical induction.

**56. Nodes** A node is a designated point in a diagram. The figure below is a circle with 3 nodes connected with line segments. Notice that the line segments connect each node to every other node. The line segments divide the circle into 4 regions.

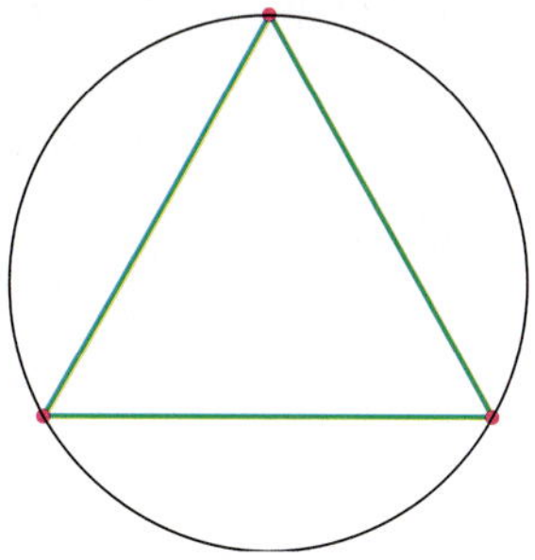

**a)** Draw a circle with one node. How many regions does the circle have?

**b)** Draw a circle with two nodes and connect them with a line segment. How many regions does the circle have?

**c)** Draw your own circle with three nodes and connect each node to every other node. How many regions does the circle have?

**d)** Draw a circle with four nodes and connect each node to every other node. How many regions does the circle have?

**e)** Draw a circle with five nodes and connect each node to every other node. How many regions does the circle have?

**f)** Complete the table.

| Number of Nodes | Number of Regions |
|---|---|
| 1 | 1 |
| ? | ? |
| 3 | 4 |
| 4 | 8 |
| 5 | 16 |

**g)** Consider the formula $R = 2^{n-1}$ where $R$ is the number of regions and $n$ is the number of nodes. Show that this formula works for $n = 1, 2, 3, 4, 5$.

**h)** Use the formula in part **g)** to predict the number of regions in a circle with 6 nodes.

**i)** Draw a circle with six nodes and connect each node to every other node. What is the maximum number of regions in the circle? Is this what you expected? Explain. Hint: Don't let 3 line segments intersect inside the circle.

## Cumulative Review Exercises

[5.2] *Expand.*

**57.** $(2x + 3y)^2$

**58.** $(5x^2 - 2y)^2$

**59.** $[4x + (y - 3)]^2$

**60.** $(2x + 3y)^3$

# 12.5 The Binomial Theorem

1 Evaluate factorials.

2 Use Pascal's triangle.

3 Use the binomial theorem.

## 1 Evaluate Factorials

To understand the binomial theorem, you must have an understanding of what **factorials** are. The notation $n!$ is read "$n$ factorial." Its definition follows.

**n Factorial**

$$n! = n(n-1)(n-2)(n-3)\cdots(1)$$

for any positive integer $n$.

Examples

$$6! = 6 \cdot 5 \cdot 4 \cdot 3 \cdot 2 \cdot 1 = 720$$

$$7! = 7 \cdot 6 \cdot 5 \cdot 4 \cdot 3 \cdot 2 \cdot 1 = 5040$$

Note that by definition **0! is 1.**

Below, we explain how to find factorials on a calculator.

**USING YOUR CALCULATOR**

Scientific Calculator

Factorials can be found on calculators that contain an [n!] or [x!] key. Often, the factorial key is a second function key. In the following examples, the answers appear after [n!].

Evaluate 6! 6 [2nd] [n!] 720

Evaluate 9! 9 [2nd] [n!] 362880

Graphing Calculator

Graphing calculators do not have a factorial key. On some graphing calculators, factorials are found under the [MATH], Probability function menu.

On the TI-84 Plus calculator, to get the probability function, PRB, menu, you press [MATH], and then scroll to the right using the right arrow key, [▶], three times, until you get to PRB. The $n!$ (or !) is the fourth item on the menu.

To find 5! or 6!, the keystrokes are as follows.

| KEYSTROKES | ANSWER |
|---|---|
| [5] [MATH] [▶] [▶] [▶] [4] [ENTER] | 120 |
| [6] [MATH] [▶] [▶] [▶] [4] [ENTER] | 720 |

## 2 Use Pascal's Triangle

Using polynomial multiplication, we can obtain the following expansions of the binomial $a + b$:

$$(a+b)^0 = 1$$

$$(a+b)^1 = a + b$$

$$(a+b)^2 = a^2 + 2ab + b^2$$

$$(a+b)^3 = a^3 + 3a^2b + 3ab^2 + b^3$$

$$(a+b)^4 = a^4 + 4a^3b + 6a^2b^2 + 4ab^3 + b^4$$

$$(a+b)^5 = a^5 + 5a^4b + 10a^3b^2 + 10a^2b^3 + 5ab^4 + b^5$$

$$(a+b)^6 = a^6 + 6a^5b + 15a^4b^2 + 20a^3b^3 + 15a^2b^4 + 6ab^5 + b^6$$

Blaise Pascal

Note that when expanding a binomial of the form $(a + b)^n$,

1. There are $n + 1$ terms in the expansion.
2. The first term is $a^n$ and the last term is $b^n$.
3. Reading from left to right, the exponents on $a$ decrease by 1 from term to term, while the exponents on $b$ increase by 1 from term to term.
4. The sum of the exponents on the variables in each term is $n$.
5. The coefficients of the terms equidistant from the ends are the same.

If we examine just the variables in $(a + b)^5$, we have $a^5$, $a^4b$, $a^3b^2$, $a^2b^3$, $ab^4$, and $b^5$.

The numerical coefficients of each term in the expansion of $(a + b)^n$ can be found by using **Pascal's triangle**, named after Blaise Pascal, a seventeenth-century French mathematician. For example, if $n = 5$, we can determine the numerical coefficients of $(a + b)^5$ as follows.

| Exponent on Binomial | Pascal's Triangle |
|---|---|
| $n = 0$ | 1 |
| $n = 1$ | 1 1 |
| $n = 2$ | 1 2 1 |
| $n = 3$ | 1 3 3 1 |
| $n = 4$ | 1 4 6 4 1 |
| $n = 5$ | 1 5 10 10 5 1 |
| $n = 6$ | 1 6 15 20 15 6 1 |

Examine row 5 ($n = 4$) and row 6 ($n = 5$).

$$\begin{array}{cccccccccc} & 1 & + & 4 & + & 6 & + & 4 & + & 1 \\ 1 & & 5 & & 10 & & 10 & & 5 & & 1 \end{array}$$

Notice that the first and last numbers in each row are 1, and the inner numbers are obtained by adding the two numbers in the row above (to the right and left). The numerical coefficients of $(a + b)^5$ are 1, 5, 10, 10, 5, and 1. Thus, we can write the expansion of $(a + b)^5$ by using the information in 1–5 above for the variables and their exponents, and by using Pascal's triangle for the coefficients.

$$(a + b)^5 = a^5 + 5a^4b + 10a^3b^2 + 10a^2b^3 + 5ab^4 + b^5$$

This method of expanding a binomial is not practical when $n$ is large.

## 3 Use the Binomial Theorem

We will shortly introduce a more practical method, called the binomial theorem, to expand expressions of the form $(a + b)^n$. However, before we introduce this formula, we need to explain how to find *binomial coefficients* of the form $\binom{n}{r}$.

**Binomial Coefficients**

For $n$ and $r$ nonnegative integers, $n \geq r$,

$$\binom{n}{r} = \frac{n!}{r! \cdot (n - r)!}$$

The binomial coefficient $\binom{n}{r}$ is read "the number of *combinations* of $n$ items taken $r$ at a time." Combinations are used in many areas of mathematics, including the study of probability.

**EXAMPLE 1** ▸ Evaluate $\binom{6}{2}$.

*Solution* Using the definition, if we substitute 6 for $n$ and 2 for $r$, we obtain

$$\binom{6}{2} = \frac{6!}{2!\cdot(6-2)!} = \frac{6!}{2!\cdot 4!} = \frac{6\cdot 5\cdot \cancel{4\cdot 3\cdot 2\cdot 1}}{(2\cdot 1)\cdot(\cancel{4\cdot 3\cdot 2\cdot 1})} = 15$$

Thus, $\binom{6}{2}$ equals 15.

▸ **Now Try Exercise 9**

**EXAMPLE 2** ▸ Evaluate.

**a)** $\binom{7}{4}$ **b)** $\binom{8}{8}$ **c)** $\binom{5}{0}$

*Solution*

**a)** $\binom{7}{4} = \dfrac{7!}{4!\cdot(7-4)!} = \dfrac{7!}{4!\cdot 3!} = \dfrac{7\cdot 6\cdot 5\cdot \cancel{4\cdot 3\cdot 2\cdot 1}}{(\cancel{4\cdot 3\cdot 2\cdot 1})(3\cdot 2\cdot 1)} = 35$

**b)** $\binom{8}{8} = \dfrac{8!}{8!\,(8-8)!} = \dfrac{\cancel{8!}}{\cancel{8!}\cdot 0!} = \dfrac{1}{1} = 1$ *Remember that* $0! = 1$.

**c)** $\binom{5}{0} = \dfrac{5!}{0!\cdot(5-0)!} = \dfrac{\cancel{5!}}{0!\cdot \cancel{5!}} = \dfrac{1}{1} = 1$

▸ **Now Try Exercise 17**

By studying Examples 2 **b)** and **c)**, you can reason that, for any positive integer $n$,

$$\binom{n}{n} = 1 \quad \text{and} \quad \binom{n}{0} = 1$$

### USING YOUR GRAPHING CALCULATOR

All graphing calculators can evaluate binomial coefficients. On most graphers, the notation ${}_nC_r$ is used instead of $\binom{n}{r}$. Thus, $\binom{7}{4}$ would be represented as ${}_7C_4$ on a grapher.

On the TI-84 Plus calculator, the notation ${}_nC_r$ can be found under the probability function, PRB, menu. This time it is item 3, ${}_nC_r$. To find ${}_7C_4$ or ${}_8C_2$ use the following keystrokes:

| | Keystrokes | Answer |
|---|---|---|
| ${}_7C_4$ | 7 MATH ▶ ▶ ▶ 3 4 ENTER | 35 |
| ${}_8C_2$ | 8 MATH ▶ ▶ ▶ 3 2 ENTER | 28 |

If you are using a different graphing calculator, consult the manual to learn to evaluate combinations.

Now we introduce the binomial theorem.

**Binomial Theorem**

For any positive integer $n$,

$$(a+b)^n = \binom{n}{0}a^nb^0 + \binom{n}{1}a^{n-1}b^1 + \binom{n}{2}a^{n-2}b^2 + \binom{n}{3}a^{n-3}b^3 + \cdots + \binom{n}{n}a^0b^n$$

Notice in the binomial theorem that the sum of the exponents on the variables in each term is $n$. In the combination, the top number is always $n$ and the bottom number is always the same as the exponent on the second variable in the term.

For example, if we consider the term $\binom{n}{3}a^{n-3}b^3$, the sum of the exponents on the variables is $(n - 3) + 3 = n$. Also, the exponent on the variable $b$ is 3, and the bottom number in the combination is also 3.

If the variables and exponents on one term of the binomial theorem are $a^7b^5$, then $n$ must be $7 + 5$ or 12. Also, the combination preceding $a^7b^5$ must be $\binom{12}{5}$. Thus, the term would be $\binom{12}{5}a^7b^5$.

Now we will expand $(a + b)^5$ using the binomial theorem and see if we get the same expression as we did when we used polynomial multiplication and Pascal's triangle to obtain the expansion.

$$\begin{aligned}(a+b)^5 &= \binom{5}{0}a^5b^0 + \binom{5}{1}a^{5-1}b^1 + \binom{5}{2}a^{5-2}b^2 + \binom{5}{3}a^{5-3}b^3 + \binom{5}{4}a^{5-4}b^4 + \binom{5}{5}a^{5-5}b^5 \\ &= \binom{5}{0}a^5b^0 + \binom{5}{1}a^4b^1 + \binom{5}{2}a^3b^2 + \binom{5}{3}a^2b^3 + \binom{5}{4}a^1b^4 + \binom{5}{5}a^0b^5 \\ &= \frac{5!}{0!\cdot 5!}a^5 + \frac{5!}{1!\cdot 4!}a^4b + \frac{5!}{2!\cdot 3!}a^3b^2 + \frac{5!}{3!\cdot 2!}a^2b^3 + \frac{5!}{4!\cdot 1!}ab^4 + \frac{5!}{5!\cdot 0!}b^5 \\ &= a^5 + 5a^4b + 10a^3b^2 + 10a^2b^3 + 5ab^4 + b^5\end{aligned}$$

This is the same expression as we obtained earlier.

In the binomial theorem, the first and last terms of the expansion contain a factor raised to the zero power. Since any nonzero number raised to the 0th power equals 1, we could have omitted those factors. These factors were included so that you could see the pattern better.

**EXAMPLE 3** ▸ Use the binomial theorem to expand $(2x + 3)^6$.

*Solution* If we use $2x$ for $a$ and 3 for $b$, we obtain

$$\begin{aligned}(2x+3)^6 &= \binom{6}{0}(2x)^6(3)^0 + \binom{6}{1}(2x)^5(3)^1 + \binom{6}{2}(2x)^4(3)^2 + \binom{6}{3}(2x)^3(3)^3 + \binom{6}{4}(2x)^2(3)^4 + \binom{6}{5}(2x)^1(3)^5 + \binom{6}{6}(2x)^0(3)^6 \\ &= 1(2x)^6 + 6(2x)^5(3) + 15(2x)^4(9) + 20(2x)^3(27) + 15(2x)^2(81) + 6(2x)(243) + 1(729) \\ &= 64x^6 + 576x^5 + 2160x^4 + 4320x^3 + 4860x^2 + 2916x + 729\end{aligned}$$

▸ **Now Try Exercise 19**

**EXAMPLE 4** ▸ Use the binomial theorem to expand $(5x - 2y)^4$.

*Solution* Write $(5x - 2y)^4$ as $[5x + (-2y)]^4$. Use $5x$ in place of $a$ and $-2y$ in place of $b$ in the binomial theorem.

$$\begin{aligned}[5x + (-2y)]^4 &= \binom{4}{0}(5x)^4(-2y)^0 + \binom{4}{1}(5x)^3(-2y)^1 + \binom{4}{2}(5x)^2(-2y)^2 + \binom{4}{3}(5x)^1(-2y)^3 + \binom{4}{4}(5x)^0(-2y)^4 \\ &= 1(5x)^4 + 4(5x)^3(-2y) + 6(5x)^2(-2y)^2 + 4(5x)(-2y)^3 + 1(-2y)^4 \\ &= 625x^4 - 1000x^3y + 600x^2y^2 - 160xy^3 + 16y^4\end{aligned}$$

▸ **Now Try Exercise 25**

## EXERCISE SET 12.5

MathXL® MyMathLab

### Concept/Writing Exercises

1. Explain how to construct Pascal's triangle. Construct the first five rows of Pascal's triangle.
2. Explain how to find $n!$ for any whole number $n$.
3. Give the value of 1!
4. Give the value of 0!
5. Can you evaluate $(-3)!$? Explain.
6. Can you evaluate $(-6)!$? Explain.
7. How many terms are there in the expansion of $(a + b)^{13}$? Explain.
8. How many terms are there in the expansion of $(x + y)^{20}$? Explain.

### Practice the Skills

*Evaluate each combination.*

9. $\binom{5}{2}$
10. $\binom{6}{3}$
11. $\binom{5}{5}$
12. $\binom{9}{3}$
13. $\binom{7}{0}$
14. $\binom{10}{7}$
15. $\binom{8}{4}$
16. $\binom{12}{3}$
17. $\binom{8}{2}$
18. $\binom{11}{4}$

*Use the binomial theorem to expand each expression.*

19. $(x + 4)^3$
20. $(x - 4)^3$
21. $(2x - 3)^3$
22. $(2x + 3)^3$
23. $(a - b)^4$
24. $(2r + s^2)^4$
25. $(3a - b)^5$
26. $(x + 2y)^5$
27. $\left(2x + \frac{1}{2}\right)^4$
28. $\left(\frac{2}{3}x + \frac{3}{2}\right)^4$
29. $\left(\frac{x}{2} - 3\right)^4$
30. $(3x^2 + y)^5$

*Write the first four terms of each expansion.*

31. $(x + 10)^{10}$
32. $(2x + 3)^8$
33. $(3x - y)^7$
34. $(3p - 2q)^{11}$
35. $(x^2 - 3y)^8$
36. $\left(2x + \frac{y}{7}\right)^9$

### Problem Solving

37. Is $n!$ equal to $n \cdot (n - 1)!$? Explain and give an example to support your answer.
38. Is $(n + 1)!$ equal to $(n + 1) \cdot n!$? Explain and give an example to support your answer.
39. Is $(n - 3)!$ equal to $(n - 3)(n - 4)(n - 5)!$ for $n \geq 5$? Explain and give an example to support your answer.
40. Is $(n + 2)!$ equal to $(n + 2)(n + 1)(n)(n - 1)!$ for $n \geq 1$? Explain and give an example to support your answer.
41. Under what conditions will $\binom{n}{m}$, where $n$ and $m$ are nonnegative integers, have a value of 1?
42. Can $\binom{n}{m}$ ever have a value of 0? Explain.
43. What are the first, second, next to last, and last terms of the expansion $(x + 3)^8$?
44. What are the first, second, next-to-last, and last terms of the expansion $(2x + 5)^6$?
45. Write the binomial theorem using summation notation.
46. Prove that $\binom{n}{r} = \binom{n}{n-r}$ for any whole numbers $n$ and $r$, and $r \leq n$.

### Cumulative Review Exercises

[3.4] 47. Find the $y$-intercept for the line $2x + y = 10$

[4.1] 48. Solve the system of equations.

$$\frac{1}{5}x + \frac{1}{2}y = 4$$
$$\frac{2}{3}x - y = \frac{8}{3}$$

[5.8] 49. Solve $x(x - 11) = -18$.

[7.4] 50. Simplify $\sqrt{20xy^4}\sqrt{6x^5y^7}$.

[9.1] 51. Find $f^{-1}(x)$ if $f(x) = 3x + 8$.

# 12.6 The Counting Principle, Permutations, and Combinations

1. Use the counting principle to determine the number of possible outcomes of an experiment.
2. Determine the number of permutations.
3. Deterimine the number of combinations.

In the last section we calculated binomial coefficients by finding the number of combinations of $n$ items taken $r$ at a time. In this section we develop this idea further and develop other counting methods.

## 1 Use the Counting Principle to Determine the Number of Possible Outcomes of an Experiment

Before we discuss the counting principle we need to learn the meanings of certain terms. An **experiment** is any process that leads to a set of results. If an experiment consists of more than one part, each part is called a **trial**. For example, an experiment may consist of tossing a coin 3 times. Each individual toss is considered a trial. The possible results of each trial are its **outcomes**. When a coin is tossed, the 2 possible outcomes are a head and a tail. When a die is rolled, its possible outcomes are 1, 2, 3, 4, 5, and 6. The set of all possible outcomes of an experiment is called its **sample space**. The sample space when a coin is tossed is $\{H, T\}$. The sample space when a die is rolled is $\{1, 2, 3, 4, 5, 6\}$. Any subset of a sample space is called an **event**. For example, if we roll a die, obtaining the number 2, it is one event. Rolling an even number may be considered another event. Rolling an even number can be satisfied by rolling either a 2, 4, or 6. When the outcome of one event has no effect on the outcome of any other event, we say the events are **independent events**. For example, if we roll a die three times, the outcome of each roll does not affect the outcomes of the other rolls. Therefore, the events are independent. In the experiment of rolling a die and tossing a coin, the events are independent events. What are the possible outcomes when a die is rolled and a coin is tossed? The possible outcomes are 1 with heads, 1 with tails, 2 with heads, 2 with tails, and so on. We can illustrate the possible outcomes with the *tree diagram* shown in **Figure 12.2**.

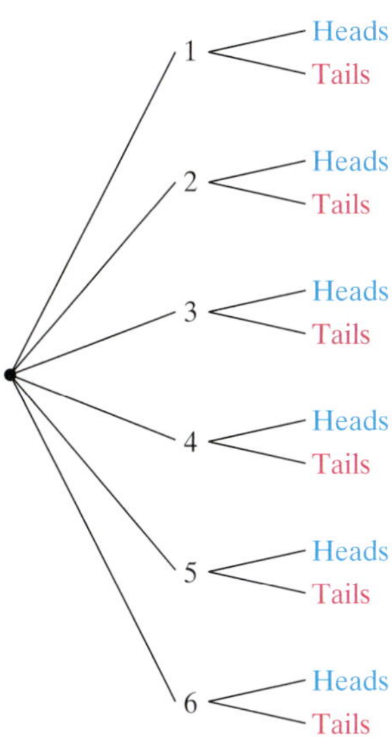

FIGURE 12.2

From the tree diagram, notice that every possible die outcome has 2 possible coin outcomes. Since we have 6 possible die outcomes, our total number of possible outcomes is $6 \cdot 2 = 12$. When we determine the total number of possible outcomes by multiplying the number of possible outcomes of each independent event, we are using the **counting principle**.

**Counting Principle**

If one event can occur in $m$ distinct ways and a second event can occur in $n$ distinct ways, then the two events, in that specific order, can occur in $m \cdot n$ distinct ways.

We will now work some problems using the counting principle.

**EXAMPLE 1 ▸ License Plates** A license plate is to consist of 3 letters followed by 4 digits. Determine the number of distinct license plates possible if the letters and digits can be repeated.

*Solution* Since the letter or digit used in any position has no effect on the letter or number used in any other position, the 7 events are independent events. Each event involving a letter has 26 possible outcomes. Each event involving a digit has 10 possible outcomes. The total number of possible outcomes is

$$\underset{L}{26} \cdot \underset{L}{26} \cdot \underset{L}{26} \cdot \underset{D}{10} \cdot \underset{D}{10} \cdot \underset{D}{10} \cdot \underset{D}{10} = 175{,}760{,}000$$

▸ Now Try Exercise 43

**EXAMPLE 2 ▸ License Plates** A license plate is to consist of 3 letters followed by 4 digits. Determine the number of distinct license plates possible if the letters and digits cannot be repeated.

*Solution* There are 26 possible outcomes for the first letter. Since repetition is not permitted, for the second letter only 25 letters are possible. For the third letter, only 24 letters are possible. For the first digit there are 10 possibilities. Since repetition of the digits is not permitted, for the second digit, 9 digits are possible. There are 8 possibilities for the third digit and 7 possibilities for the fourth digit. The total number of outcomes is therefore

$$\underset{L}{26} \cdot \underset{L}{25} \cdot \underset{L}{24} \cdot \underset{D}{10} \cdot \underset{D}{9} \cdot \underset{D}{8} \cdot \underset{D}{7} = 78{,}624{,}000$$

▸ Now Try Exercise 41

**EXAMPLE 3 ▸ Baseball Cards** At a baseball card show, Sharon Becker wishes to display 5 different Derek Jeter cards.

**a)** In how many different ways can she place the 5 cards in a straight row?

**b)** If she wants to place Derek Jeter's rookie card in the middle, in how many ways can she arrange the cards?

*Solution*

**a)** There are 5 positions to fill, using the 5 cards. In the first position, on the left, she can use any one of the 5 cards. In the second position she can use any of the 4 remaining cards. In the third position she can use any of the 3 remaining cards, and so on. The number of distinct possible arrangements is

$$\underline{5} \cdot \underline{4} \cdot \underline{3} \cdot \underline{2} \cdot \underline{1} \cdot = 120$$

**b)** We begin by satisfying the specified requirement. In this case the rookie card must be placed in the middle. Therefore there is only 1 possibility for the middle position.

$$\underline{\quad}\ \underline{\quad}\ \underline{1}\ \underline{\quad}\ \underline{\quad}$$

For the first position on the left, there are now 4 possibilities. For the second position there will be 3 possibilities. For the fourth position there will be 2 possibilities. Finally, in the last position there is only 1 possibility.

$$4 \cdot 3 \cdot 1 \cdot 2 \cdot 1 = 24$$

Thus there are 24 ways the cards can be arranged.

▸ Now Try Exercise 47

## 2 Determine the Number of Permutations

As we saw in Example 3**a)**, the five baseball cards can be ordered in 120 different ways. Each different arrangement is one permutation of the baseball cards.

**Permutation**

A **permutation** is any ordered arrangement of a given set of objects.

When we find the number of possible arrangements of a set of objects, we are finding the **number of permutations**. In Example 3**a)**, we found that the number of permutations of the 5 baseball cards is $5 \cdot 4 \cdot 3 \cdot 2 \cdot 1$. Recall from Section 12.5 that the product of all positive integers less than or equal to $n$ is $n$ factorial, symbolized $n!$ Thus we write $5 \cdot 4 \cdot 3 \cdot 2 \cdot 1 = 5!$.

We now explain how to determine the number of permutations of $n$ distinct items.

**Number of Permutations**

The number of permutations of $n$ distinct items is $n!$.

**EXAMPLE 4 ▸ Compact Disks** In how many ways can 7 different CDs be stacked in a CD holder with 7 slots?

*Solution* Since there are 7 different CDs, the same as the number of slots available, the number of permutations is 7!.

$$7! = 7 \cdot 6 \cdot 5 \cdot 4 \cdot 3 \cdot 2 \cdot 1 = 5040$$

▸ Now Try Exercise 57

In Example 4 the number of CDs is the same as the number of slots in the CD holder. In Example 5 we find the number of permutations when we have more CDs than our CD holder can hold.

**EXAMPLE 5 ▸ Compact Disks** In how many ways can a CD holder with 7 slots be filled from a collection of 12 different CDs?

*Solution* Since there are 12 CDs, the first slot has 12 options. After 1 CD has been used in the first slot, any of the remaining 11 CDs can be used in the second slot. Then any of the remaining 10 CDs can be used in the third slot, and so on. Following, we use dashes for the 7 slots and show the number of possible outcomes for each slot. The number of permutations is

$$\underline{12} \cdot \underline{11} \cdot \underline{10} \cdot \underline{9} \cdot \underline{8} \cdot \underline{7} \cdot \underline{6} = 3{,}991{,}680$$

▸ Now Try Exercise 49

We use the notation ${}_nP_r$ to indicate the number of permutations of $n$ items taken $r$ at a time. In Example 5 we found the number of permutations of 12 items taken 7 at a time, ${}_{12}P_7$.

We can use the counting principle to evaluate the number of permutations represented by ${}_nP_r$. The $n$ represents the number of distinct possible outcomes for the first event. The $r$ represents the number of events. The number of distinct possible outcomes for each event after the first one will be 1 less than the number of outcomes for the previous event. To evaluate ${}_nP_r$, multiply $n$ by each consecutive integer less than $n$, up to and including $n - r + 1$. For example, when we evaluate ${}_{12}P_7$

$${}_{12}P_7 = 12 \cdot 11 \cdot 10 \cdot 9 \cdot 8 \cdot 7 \cdot \underset{\uparrow\; 12 - 7 + 1 = 6}{6}$$

In general, the number of permutations of $n$ items taken $r$ at a time, ${}_nP_r$, may be found by the formula ${}_nP_r = n(n-1)(n-2)\cdots(n-r+1)$. Now let's look at some other examples.

$${}_7P_4 = 7 \cdot 6 \cdot 5 \cdot \underset{\uparrow\; 7 - 4 + 1 = 4}{4}$$

$${}_9P_2 = 9 \cdot \underset{\uparrow\; 9 - 2 + 1 = 8}{8}$$

$${}_{10}P_3 = 10 \cdot 9 \cdot \underset{\uparrow\; 10 - 3 + 1 = 8}{8}$$

Notice that the last number in the multiplication is always $n - r + 1$. For example, in ${}_{10}P_3$, the last number is $10 - 3 + 1 = 8$. Now let's develop an alternative formula for determining ${}_nP_r$.

$$\begin{aligned}
{}_nP_r &= n(n-1)(n-2)\cdots(n-r+1) \\
&= n(n-1)(n-2)\cdots(n-r+1) \times \frac{(n-r)!}{(n-r)!} \\
&= \frac{n(n-1)(n-2)\cdots(n-r+1)\overbrace{(n-r)(n-r-1)\cdots(3)(2)(1)}^{(n-r)!}}{(n-r)!} \\
&= \frac{n!}{(n-r)!}
\end{aligned}$$

Multiply by $\frac{(n-r)!}{(n-r)!}$ which is equivalent to multiplying by 1.

**Permutation Formula**

The number of permutations of $n$ items taken $r$ at a time is

$${}_nP_r = \frac{n!}{(n-r)!}$$

**EXAMPLE 6** ▶ **Club Officers** A club has 20 members who are eligible to hold office. How many different arrangements of president, vice-president, and secretary are possible?

*Solution* Since there are 20 people and 3 people are to be selected, $n = 20$ and $r = 3$.

$$\begin{aligned}
{}_{20}P_3 &= \frac{20!}{(20-3)!} \\
&= \frac{20!}{17!} \\
&= \frac{20 \cdot 19 \cdot 18 \cdot \cancel{17!}}{\cancel{17!}} \\
&= 20 \cdot 19 \cdot 18 \\
&= 6840
\end{aligned}$$

Therefore, there are 6840 different arrangements, or permutations, of the three officers possible.

▶ **Now Try Exercise 51**

## 3 Determine the Number of Combinations

Sometimes the arrangement of items is not important to the problem. For instance, if you are selecting 7 CDs to be placed in your CD holder, you may be concerned only with the group of CDs selected, not the order in which they are placed in the holder. When we find the number of groups of items that are possible, we are finding the **number of combinations**.

**Combination**

A **combination** is a distinct grouping of objects without regard to their arrangement.

It is important to distinguish a permutation from a combination. The letters $a, b, c$, and $b, c, a$ are 2 different permutations because the ordering of the 3 letters is different, but they represent the same combination of letters because the same letters are used in each case.

**EXAMPLE 7 ▸ Permutation or Combination?** Determine whether the situation represents a permutation or combination problem.

**a)** Two students from a class of 28 are selected to serve on a committee. In how many ways can this be done?

**b)** A group of 8 friends are throwing a surprise party for Shelia Pierce. They draw a name to determine who brings a cake to the surprise party and then draw a name to determine who brings Shelia. How many different assignments are possible?

*Solution*

**a)** Since the order in which the students are selected is not important, the problem is a combination problem.

**b)** Since the two friends have different tasks, one bringing a cake and the other bringing Shelia, the order that names are selected is important. Therefore, this problem is a permutation problem.

▸ Now Try Exercise 61

The notation we use for the number of combinations of $n$ items taken $r$ at a time is ${}_nC_r$. It can also be written as $\binom{n}{r}$ as we did in Section 12.5 when we found binomial coefficients.

Consider the set of elements $\{a, b, c, d, e\}$. The number of permutations of 2 letters from this set of 5 letters, ${}_5P_2$, is 20:

*ab, ac, ad, ae, ba, bc, bd, be, ca, cb, cd, ce, da, db, dc, de, ea, eb, ec, ed*

The number of combinations of 2 letters from this set of 5 letters, ${}_5C_2$, is 10:

*ab, ac, ad, ae, bc, bd, be, cd, ce, de*

Since every group of 2 elements can be written in 2! or 2 ways, the number of permutations is twice the number of combinations. We could calculate the number of combinations by dividing the number of permutations by 2!.

$${}_5C_2 = \frac{{}_5P_2}{2!} = \frac{5 \cdot 4}{2 \cdot 1} = 10$$

From the counting principle we know that a group with $r$ distinct elements can be arranged in $r!$ different ways. Thus a group containing 3 distinct elements can be arranged in 3! or 6 different ways. When determining the number of combinations, we only want to count the same groups of items once. We can calculate the number of combinations by dividing the number of permutations by $r!$, the number of permutations in one group of $r$ distinct elements. Since ${}_nP_r = \dfrac{n!}{(n-r)!}$ we conclude the following.

**Combination Formula**

The number of combinations of $n$ items taken $r$ at a time is

$${}_nC_r = \frac{n!}{(n-r)!\,r!}$$

**EXAMPLE 8 ▸ Nine Drivers Needed** Twelve people have volunteered to drive the swim team to their out-of-town swim meet. Only 9 drivers are needed. How many different groups of drivers can be selected?

*Solution* Since the order in which the drivers are selected is unimportant, this problem is a combination problem. We want to find the number of different groups of 9 that can be formed from 12 people, so $n = 12$ and $r = 9$.

$$\begin{aligned}{}_{12}C_9 &= \frac{12!}{(12-9)!\,9!}\\ &= \frac{12!}{3!\,9!}\\ &= \frac{\overset{4}{\cancel{12}}\cdot 11\cdot \overset{5}{\cancel{10}}\cdot \cancel{9!}}{\cancel{3}\cdot\cancel{2}\cdot 1\cdot\cancel{9!}}\\ &= 4\cdot 11\cdot 5\\ &= 220\end{aligned}$$

Thus there are 220 different combinations of 9 of the 12 drivers.

▸ **Now Try Exercise 65**

**EXAMPLE 9** ▸ **Restaurants** On a 4-day vacation you want to eat dinner at a different local restaurant every evening. The beach town has 10 restaurants. How many different combinations of restaurants could you select?

*Solution* We want to find the number of different groups of 4 that can be formed from 10 restaurants, so $n = 10$ and $r = 4$.

$$\begin{aligned}{}_{10}C_4 &= \frac{10!}{(10-4)!\,4!}\\ &= \frac{10!}{6!\,4!}\\ &= \frac{10\cdot \overset{3}{\cancel{9}}\cdot\cancel{8}\cdot 7\cdot\cancel{6!}}{\cancel{6!}\cdot\cancel{4}\cdot\cancel{3}\cdot\cancel{2}\cdot 1}\\ &= 10\cdot 3\cdot 7\\ &= 210\end{aligned}$$

Thus there are 210 different combinations of 4 of the 10 restaurants.

▸ **Now Try Exercise 67**

Up until now we have used only factorials of positive integers. In Example 10 we use 0!, which is defined to be 1.

**EXAMPLE 10** ▸ **Pizzas** A pizza restaurant offers a cheese pizza with 5 possible toppings: pepperoni, sausage, green peppers, onions, and mushrooms. How many different kinds of pizza can be ordered?

*Solution* A pizza can have 0, 1, 2, 3, 4, or 5 toppings. We must find the sum of the number of combinations for each number of toppings. There is only 1 type of pizza that can be made with no toppings, the plain cheese pizza. We can calculate the number of combinations possible using 0 toppings, ${}_5C_0$, as follows.

$$\begin{aligned}{}_5C_0 &= \frac{5!}{(5-0)!\,0!}\\ &= \frac{5!}{5!\,0!}\\ &= \frac{\cancel{5!}}{\cancel{5!}\cdot 1}\\ &= 1\end{aligned}$$

Thus, there is only 1 pizza possible that uses no toppings.

For 1 topping, we calculate the number of combinations possible using ${}_5C_1$.

$$\begin{aligned}{}_5C_1 &= \frac{5!}{(5-1)!\,1!}\\ &= \frac{5!}{4!\,1!}\\ &= \frac{5\cdot\cancel{4!}}{\cancel{4!}\cdot 1}\\ &= 5\end{aligned}$$

There are 5 types of pizza that can be made using exactly one of the 5 toppings listed.

For 2 toppings, we calculate the number of combinations possible using ${}_5C_2$.

$$\begin{aligned}{}_5C_2 &= \frac{5!}{(5-2)!\,2!}\\ &= \frac{5!}{3!\,2!}\\ &= \frac{5\cdot\overset{2}{\cancel{4}}\cdot\cancel{3!}}{\cancel{3!}\cdot\cancel{2}\cdot 1}\\ &= 5\cdot 2\\ &= 10\end{aligned}$$

There are 10 types of pizza that can be made using exactly 2 of the 5 toppings listed.

For 3 toppings, we calculate the number of combinations possible using ${}_5C_3$.

$$\begin{aligned}{}_5C_3 &= \frac{5!}{(5-3)!\,3!}\\ &= \frac{5!}{2!\,3!}\\ &= \frac{5\cdot\overset{2}{\cancel{4}}\cdot\cancel{3!}}{\cancel{2}\cdot 1\cdot\cancel{3!}}\\ &= 5\cdot 2\\ &= 10\end{aligned}$$

There are 10 types of pizza that can be made using exactly 3 of the 5 toppings listed.

For 4 toppings, we calculate the number of combinations possible using ${}_5C_4$.

$$\begin{aligned}{}_5C_4 &= \frac{5!}{(5-4)!\,4!}\\ &= \frac{5!}{1!\,4!}\\ &= \frac{5\cdot\cancel{4!}}{1\cdot\cancel{4!}}\\ &= 5\end{aligned}$$

There are 5 types of pizza that can be made using exactly 4 of the 5 toppings listed.

For 5 toppings, we calculate the number of combinations possible using ${}_5C_5$.

$$\begin{aligned}{}_5C_5 &= \frac{5!}{(5-5)!\,5!}\\ &= \frac{5!}{0!\,5!}\\ &= \frac{\cancel{5!}}{1\cdot\cancel{5!}}\\ &= 1\end{aligned}$$

There is 1 type of pizza that can be made using all 5 of the 5 toppings listed.

To find the total number of types of pizzas we add the combinations.

$$_5C_0 + {_5C_1} + {_5C_2} + {_5C_3} + {_5C_4} + {_5C_5} = 1 + 5 + 10 + 10 + 5 + 1 = 32$$

Thus, there are 32 types of pizzas that can be ordered.

▶ Now Try Exercise 87

Now we will do an example using combinations and the counting principle.

**EXAMPLE 11** ▶ **Manuscripts** An editor has 8 manuscripts for mathematics books and 5 manuscripts for computer science books. If he is to select 5 mathematics and 3 computer science manuscripts for publication, how many different choices does he have?

*Solution* For the mathematics books, 5 of the 8 manuscripts will be selected. The number of combinations of mathematics manuscripts is $_8C_5 = 56$. For computer science books, 3 of the 5 manuscripts will be selected. The number of combinations of computer science manuscripts is $_5C_3 = 10$. Since each of the combinations of the mathematics manuscripts can be paired with each of the combinations of the computer science manuscripts, we use the counting principle to determine the total number of choices. That is, we multiply the number of combinations of mathematics manuscripts by the number of combinations of computer science manuscripts.

$$_8C_5 \cdot {_5C_3} = 56 \cdot 10 = 560$$

Therefore the editor has 560 different choices.

▶ Now Try Exercise 83

### USING YOUR GRAPHING CALCULATOR

In Section 12.5, we learned how to compute $n!$ and $_nC_r$ on a TI-84 Plus calculator. We can also compute $_nP_r$. To find $_nP_r$ on the TI-84 Plus calculator, use item 2 under the PRB menu. To find $_7P_4$, use the following keystrokes:

| | Keystrokes | Answer |
|---|---|---|
| $_7P_4$ | 7 MATH ▶ ▶ ▶ 2 4 ENTER | 840 |

If you are using a different graphing calculator, consult the manual to learn how to evaluate permutations.

## EXERCISE SET 12.6

MathXL® MyMathLab

### Concept/Writing Exercises

1. What is permutation?
2. What is a combination?
3. What does $_nP_r$ represent?
4. What does $_nC_r$ represent?
5. Explain how to evaluate $n!$.
6. What is 0! equal to?
7. State the counting principle in your own words.
8. Explain the difference between a combination and a permutation. Give an example of a permutation problem and an example of a combination problem.

### Practice the Skills

*Evaluate each expression.*

9. $7!$

10. $5!$

11. $\dfrac{12!}{8!}$

12. $\dfrac{18!}{13!}$

13. $\dfrac{13!}{(13-4)!}$

14. $\dfrac{25!}{(25-5)!}$

15. $\dfrac{9!}{2!\,4!}$

16. $\dfrac{12!}{3!\,4!\,5!}$

17. ${}_5P_0$

18. ${}_5P_1$

19. ${}_5P_2$

20. ${}_5P_3$

21. ${}_5P_4$

22. ${}_5P_5$

23. ${}_6P_4$

24. ${}_8P_5$

25. ${}_9P_4$

26. ${}_8P_7$

27. ${}_{10}P_0$

28. ${}_7P_7$

29. ${}_5C_0$

30. ${}_5C_1$

31. ${}_5C_2$

32. ${}_5C_3$

33. ${}_5C_4$

34. ${}_5C_5$

35. ${}_7C_3$

36. ${}_8C_4$

37. ${}_9C_6$

38. ${}_6C_0$

39. ${}_7C_7$

40. ${}_{10}C_9$

## Problem Solving

*Solve Exercises 41–48 using the counting principle.*

41. **Race Tracks** The daily double at most race tracks consists of selecting the winning horse in both the first and second races. If the first race has 7 entries and the second race has 8 entries, how many daily double tickets must you purchase to guarantee a win?

42. **Teller Machines** To use an automated teller machine, you generally must enter a four-digit code, using the digits 0–9. How many four-digit codes are possible if repetition of digits is permitted?

43. **Security** A secured elevator allows access by pressing the correct sequence of buttons. A display of the 10 buttons by the elevator door is shown below.

The correct sequence of 6 buttons must be pressed to unlock the door. If the same button may be pressed consecutively, how many possible ways can the 6 buttons be pressed?

44. **Elevators** A secured elevator allows access by pressing the correct sequence of letters on a control panel. If the control panel contains the 26 letters of the alphabet and a 4-letter code must be entered (repetition is permitted), how many different codes are possible?

45. **Social Security Numbers** A social security number consists of 9 digits. How many different social security numbers are possible if repetition of digits is permitted?

46. **Identification Numbers** A college identification number consists of a letter followed by 4 digits. How many different identification numbers are possible if repetition is not permitted?

47. **Sound Systems** The operator of the Sounds Great Audio Store is planning a grand opening. He wishes to advertise that the store has many different sound systems available. They stock 10 different tape decks, 12 different receivers, and 9 different speakers. Assuming that a sound system will consist of 1 of each, and that all pieces are compatible, how many different sound systems can they advertise?

48. **Race Tracks** The trifecta at most race tracks consists of selecting the first-, second-, and third-place finishers in a particular race in their proper order. If there are 6 entries in the trifecta race, how many tickets must you purchase to guarantee a win?

*Solve Exercises 49–58 using permutations.*

49. **Door Prizes** Three door prizes—a poster, a can of hair spray, and a bag of jelly beans—are to be awarded (gag gifts) at a get-together of 7 people. If 3 people are to be selected at random and awarded the gifts, how many different ways can the gifts be awarded?

50. **Club Officers** If a club consists of 8 members, how many different arrangements of president and vice-president are possible?

51. **Prizes** A teacher decides to give 6 different prizes to 6 of 30 students in her class. In how many ways can she do so?

52. **Paintings in a Museum** Nine Van Gogh paintings are to be displayed in a museum.

**a)** In how many different ways can they be arranged if they must be next to one another?

**b)** In how many different ways can they be displayed if *Starry Night* is to be in the middle?

53. **Office Security** A night guard at NSB Bank visits 8 different offices every hour. The pattern is varied each night so that the guard will not follow a specific routine. In how many different ways can this pattern be varied?

54. **Permutations of Words** Find the number of permutations of the letters in the word DOUBLE.

55. **Permutations of Words** Find the number of permutations of the letters in the word CURVE.

56. **Permutations of Words** Find the number of permutations of the letters in the word PYRAMID.

57. **History Tests** In one question of a history test the student is asked to match 10 dates with 10 events; each date can only be matched with 1 event. In how many different ways can this question be answered?

58. **Using Flags for Messages** Six different colored flags will be placed on a pole, one beneath another. The arrangement of the colors indicates the message. How many messages are possible if 6 flags are to be selected from 8 different colored flags?

*Solve Exercises 59–70 using combinations.*

59. **Bookcases** A bookcase contains 12 different books. Four books will be selected at random and given away as prizes to 4 individuals. In how many ways can the selection be made?

60. **Banana Splits** John's ice cream parlor has 15 different flavors. Willie orders a banana split and has to select 3 different flavors. How many different selections are possible?

61. **Prizes** Mrs. Suarez, a teacher, decides to give identical prizes to 6 of the 18 students in her class. In how many ways can she do so?

62. **Essay Questions** Amy Drake must select and answer 4 of 5 essay questions on a test. In how many ways can she do so?

63. **Textbook Search Committee** A textbook search committee is considering 6 books for possible adoption. The committee has decided to select 3 of the 6 for further consideration. In how many ways can it do so?

64. **Microwaves** Five of a sample of 18 microwaves will be selected and tested for defects. In how many ways can the quality control engineer do so?

65. **Scholarships** Three of 8 finalists will be selected and awarded a $10,000 scholarship. In how many ways can this selection be made?

66. **Records** Carter Fenton, a disc jockey, has 20 records that he wants to play during the next hour, but he has time to play only 16 of them. How many different ways can he select a group of 16 to play?

67. **Grades in a Class** At the beginning of the semester your instructor informs the class that he marks on a curve and that exactly 6 of the 30 students in the class will receive a final grade of A. In how many ways can this result occur?

68. **Quiniela Bets** A quiniela bet consists of selecting the first- and second-place winners, in any order, in a particular event. For example, suppose you select a 2–5 quiniela. If 2 wins and 5 finishes second, or if 5 wins and 2 finishes second, you win. How many quiniela tickets are necessary to guarantee a win in a race with 8 entries?

69. **Limousines** Zwick's Limousine Service is going to select and purchase 3 different limousines of 6 different limousines on display. How many different selections are possible?

70. **Lotteries** In the California State lottery you must select the 6 winning numbers (in any order) out of 51 possible numbers to win the grand prize. How many combinations of 6 numbers out of the 51 are possible?

*Solve Exercises 71–86 using the counting principle, permutations, combinations, or a combination of these methods.*

71. **ISBN Numbers** Each book registered in the Library of Congress must have an ISBN code number. For an ISBN number of the form D-DD-DDDDDD-D, where D represents a digit from 0 through 9, how many different ISBN numbers are possible if repetition of digits is allowed?

72. **Outfits** Mr. Johnston has 10 pairs of pants, 9 shirts, 12 ties, and 8 sport coats. Assuming he must wear 1 of each, how many different outfits can he wear?

73. **Permutations of Words** Find the number of permutations of the letters in the word NUMBER.

74. **Permutations of Words** Find the number of permutations of the letters in the word EQUATION.

75. **Combination Locks**

**a)** To open a combination lock, you must know the lock's 3-number sequence in its proper order. Repetition of numbers is permitted. Why is this lock more like a permutation lock than a combination lock? Why is it not a true permutation problem?

**b)** Assuming that a combination lock has 40 numbers, determine how many different 3-number arrangements are possible if repetition of numbers is allowed.

**c)** Answer the question in part **b)** if repetition is not allowed

76. **License Plates** A license plate is to consist of 4 digits followed by 2 letters. Determine the number of different license plates possible if

**a)** repetition of numbers and letters is permitted.

**b)** repetition of numbers and letters is not permitted.

**c)** the first and second digits must be odd, and repetition is not permitted.

**d)** the first digit cannot be a zero and repetition is not permitted.

77. **Telephone Numbers** A telephone number consists of 7 digits with the restriction that the first digit cannot be 0 or 1.

**a)** How many distinct telephone numbers are possible?

b) How many distinct telephone numbers are possible with 3-digit area codes preceding the 7-digit number, where the first digit of the area code is not 0 or 1?

c) With the increasing use of cellular phones and paging systems our society is beginning to run out of usable phone numbers. Various phone companies are developing phone numbers that use 11 digits instead of 7. How many distinct phone numbers can be made with 11 digits assuming that the area code remains 3 digits and the first digit of the area code and the phone number cannot be 0 or 1?

**78. Arrangements of Pictures** The 6 pictures shown are to be placed side by side along a wall. In how many ways can they be arranged from left to right if

a) they can be arranged in any order?

b) the bird must be on the far left?

c) the bird must be on the far left and the giraffe must be next to the bird?

d) a four-legged animal must be on the far right?

**79. Wedding Reception** At the reception line of Mr. and Mrs. Doan's wedding, Mrs. Doan, Mr. Doan, the best man, the maid of honor, the 3 ushers, and the 3 bridesmaids must line up to receive the guests.

a) If these individuals can line up in any order, how many arrangements are possible?

b) If Mr. Doan must be the last in line and Mrs. Doan must be next to him, and the others can line up in any order, how many arrangements are possible?

c) If Mr. Doan is to be last in line, Mrs. Doan next to him, and males and females are to alternate, how many arrangements are possible?

**80. Batting Orders** In how many ways can the manager of a National League baseball team arrange his batting order of 9 players if the pitcher must bat last?

**81. History Tests** On a history test, Frank Marsh must write an essay for 3 of the 4 questions in Part 1, and 3 of the 5 questions in Part 2. How many different combinations of questions can he answer?

**82. Food** Miguel is sent to the store to get 4 different boxes of cereal and 3 different containers of ice cream. If there are 8 different types of cereal and 7 different ice creams to choose from, how many different choices does Miguel have?

83. **Committees** How many different committees can be formed from 5 teachers and 30 students if the committee is to consist of 3 teachers and 3 students?

**84. Mathematics Tests** A teacher is constructing a mathematics test consisting of 10 questions. She has a pool of 28 questions, which are classified by level of difficulty as follows: 6 difficult questions, 10 average questions, and 12 easy questions. How many different 10-question tests can she construct from the pool of 28 questions if her test is to have 3 difficult, 4 average, and 3 easy questions?

**85. Cereals** A cereal company is testing 6 oat cereals, 5 wheat cereals, and 4 rice cereals. If it plans to market 3 of the oat cereals, 2 of the wheat cereals, and 2 of the rice cereals, how many different combinations are possible?

**86. Trays of Hors D'oeuvres** Derrick Williams' catering service is making up trays of hors d'oeuvres. The hors d'oeuvres are categorized as inexpensive, average, and expensive. If the client must select 3 of the 7 inexpensive, 5 of the 8 average, and 2 of the 4 expensive hors d'oeuvres, how many different choices are possible?

87. **Pizza** A pizza restaurant offers a cheese pizza with 6 possible toppings. How many different kinds of pizza can be ordered? See Example 10.

**88. Hamburgers** A snack bar offers hamburgers with 3 possible condiments. How many different kinds of hamburgers can be ordered?

## Group Activity

**89. Tapping Glasses**

a) At Drew Archibald's restaurant 4 people at dinner make a toast. If each person is to tap glasses with each other person, how many taps will take place?

b) Repeat part **a)** with 5 people.

c) How many taps will there be if there are $n$ people at the dinner table?

**90. Lotteries** Determine the number of combinations possible in a state lottery where you must select

a) 6 of 46 numbers.

b) 6 of 47 numbers.

c) 6 of 48 numbers.

d) 6 of 49 numbers.

e) Does the number of combinations increase by the same amount going from part **a)** to part **b)** as from part **b)** to part **c)**?

f) Guess the number of combinations possible for 6 of 50 numbers.

g) Determine the number of combinations possible for 6 of 50 numbers. Compare your answer to your answer in part **f)**.

**91. License Plates** Some states in the United States are experiencing tremendous growth and must redesign license plates to accommodate the demand. Each group member will determine the number of distinct license plates possible using digits (0 through 9) or letters from the English alphabet (a through z). Each license plate must fill 6 positions.

**a)** Group Member 1: Determine the number of license plates possible using 6 digits (repetition is permitted).
Group Member 2: Determine the number of license plates possible using a letter in the first position followed by 5 digits (repetition is permitted).
Group Member 3: Determine the number of license plates possible using 3 letters followed by 3 digits (repetition is permitted).

**b)** Share your result with the other members.

**c)** Suppose a state needs 1.5 million license plates. Which of the 3 plans described in part **a)** can be used? Explain.

**d)** Suppose a state needs 25 million license plates. Can any of the plans described in part **a)** be used? If not, as a group, design a license plate containing a combination of 7 letters and digits which will satisfy this need.

## Cumulative Review Exercises

[2.1] **92.** Solve $\frac{3}{4}x - 2 = 4$.

[5.8] **93.** Solve $3x^2 = 18x$.

[6.4] **94.** Solve $\frac{x^2}{x+5} = \frac{25}{x+5}$.

[7.6] **95.** Solve $\sqrt{x-8} + 12 = 19$.

# 12.7 Probability

1 Determine the probability of an event.

2 Determine the probability an event will not occur.

3 Determine the probability of a compound event.

4 Determine the odds against an event.

## 1 Determine the Probability of an Event

In Section 12.6 we discussed outcomes. Recall that the possible results of an experiment are its **outcomes**. For instance the outcomes of rolling a die are 1, 2, 3, 4, 5, and 6.

An **event** is a subset of the possible outcomes of an experiment. For example, when a die is rolled, the event of rolling a number 2 can be satisfied by only one outcome, 2. The event of rolling a number greater than 2 can be satisfied by any of the four outcomes 3, 4, 5, or 6. The event of rolling an even number can be satisfied by any of three outcomes 2, 4, or 6.

When we determine the **probability** of an event, we are finding the likelihood of the event occurring. We use the notation $P(E)$ to represent the probability of an event $E$.

If each outcome of an experiment has the same chance of occurring as any other outcome, they are said to be **equally likely outcomes**. For instance, when rolling a die, the outcomes 1, 2, 3, 4, 5, and 6 are equally likely. An odd number is as likely to be rolled as an even number; therefore, odd and even are another set of equally likely outcomes. If the outcomes of an experiment are equally likely, the probability can be calculated using the probability formula.

**Probability**

$$P(E) = \frac{\text{number of outcomes favorable to } E}{\text{total number of possible outcomes}}$$

**EXAMPLE 1** ▸ **Rolling a Die** A die is rolled. Find the probability of rolling

**a)** a 4.
**b)** an odd number.
**c)** a number less than 3.
**d)** a 7.
**e)** a number less than 7.

*Solution*

**a)** There are 6 possible equally likely outcomes: 1, 2, 3, 4, 5, and 6. The event of rolling a 4 can occur in one way.

$$P(4) = \frac{\text{number of outcomes that result in rolling a 4}}{\text{total number of possible outcomes}}$$
$$= \frac{1}{6}$$

**b)** The event of rolling an odd number can occur in three ways (when a 1, 3, or 5 is rolled).

$$P(\text{odd number}) = \frac{\text{number of outcomes that result an odd number}}{\text{total number of possible outcomes}}$$
$$= \frac{3}{6} = \frac{1}{2}$$

**c)** The event of rolling a number less than 3 can occur in two ways (when a 1 or 2 is rolled).

$$P(\text{number less than } 3) = \frac{\text{number of outcomes that result in a number less than 3}}{\text{total number of possible outcomes}}$$
$$= \frac{2}{6} = \frac{1}{3}$$

**d)** A 7 cannot be rolled on a die. Therefore, the number outcomes that result in a 7 is 0.

$$P(7) = \frac{\text{number of outcomes that result in a 7}}{\text{total number of possible outcomes}}$$
$$= \frac{0}{6} = 0$$

**e)** The event of rolling a number less than 7 can occur in six ways (when a 1, 2, 3, 4, 5, or 6 is rolled).

$$P(\text{number less than } 7) = \frac{\text{number of outcomes that result in a number less than 7}}{\text{total number of possible outcomes}}$$
$$= \frac{6}{6} = 1$$

▸ **Now Try Exercise 11**

Example 1 illustrates two important facts about probabilities. In part **d)** we found that the probability of rolling a 7 on a die is 0. It is impossible to roll a 7 on a die since a die does not contain a 7. The probability of an event that cannot occur is 0. In Part **e)** we found that the probability of rolling a number less than 7 is 1. When a die is rolled, a number less than 7 must be rolled because all of the numbers on a die are less than 7. The probability of an event that must occur is 1.

We now present some important facts about probabilities.

### Important Facts About Probability

1. The probability of an event that cannot occur is 0.
2. The probability of an event that must occur is 1.
3. Every probability will be a number between 0 and 1, inclusive; that is, $0 \le P(E) \le 1$.
4. The sum of the probabilities of all possible outcomes of an experiment is 1.

To illustrate statement 4, consider a die. There are 6 possible outcomes. Each individual outcome has a probability of $\frac{1}{6}$. The sum of the probabilities of all the possible outcomes is 1.

| | Roll a die | | | | | |
|---|---|---|---|---|---|---|
| **Possible outcome** | 1 | 2 | 3 | 4 | 5 | 6 |
| **Probability** | $\frac{1}{6}$ | $\frac{1}{6}$ | $\frac{1}{6}$ | $\frac{1}{6}$ | $\frac{1}{6}$ | $\frac{1}{6}$ |

Sum of Probabilities = 1

## 2 Determine the Probability an Event Will Not Occur

In any experiment an event must either occur or not occur. The sum of the probability that an event will occur and the probability that the event will not occur is 1. Thus, for any event $A$, we conclude

$$P(A) + P(\text{not } A) = 1$$

or

$$P(\text{not } A) = 1 - P(A)$$

To find the probability of an event not happening, subtract the probability of the event happening from 1.

For instance, in Example 1 **c)** we determined the probability of obtaining a number less than 3 when rolling a die. When rolling a die, we will either get a number less than 3 or we will not get a number less than 3. Since we found that the probability of getting a number less than 3 is $\frac{1}{3}$, then the probability that we do not get a number less than 3 must be $1 - \frac{1}{3} = \frac{2}{3}$.

**EXAMPLE 2 ▸ Deck of Cards** A deck of 52 playing cards (see photo) consists of 4 suits: spades, diamonds, hearts, and clubs. Each suit has 13 cards, including numbered cards ace (1) through 10 and 3 face (or picture) cards, the jack, the queen, and the king. Hearts and diamonds are red suits; clubs and spades are black suits. One card is selected at random from the deck of cards. Find the probability that the card selected is

**a)** a 7.

**b)** not a 7.

**c)** a spade.

**d)** a heart *and* a club.

**e)** a card greater than 6 *and* less than 9.

### Solution

**a)** There are four 7s in a deck of 52 cards.

$$P(7) = \frac{4}{52} = \frac{1}{13}$$

**b)** $P(\text{not a } 7) = 1 - P(7) = 1 - \frac{1}{13} = \frac{12}{13}$

**c)** There are 13 spades in the deck.

$$P(\text{spade}) = \frac{13}{52} = \frac{1}{4}$$

**d)** The word *and* means that both events must occur. There is no card that is both a heart and a club.

$$P(\text{heart and club}) = \frac{0}{52} = 0$$

**e)** The cards that are both greater than 6 and less than 9 are the 7s and 8s. There are four 7s and four 8s, or a total of 8 cards.

$$P(\text{greater than 6 and less than 9}) = \frac{8}{52} = \frac{2}{13}$$

▸ **Now Try Exercise 19**

## 3 Determine the Probability of a Compound Event

A **compound event** is one that provides an alternative such as $A$ or $B$, where $A$ and $B$ are events. For example, suppose a die is rolled where $A$ is the event of "rolling a 4" and $B$ is the event of "rolling an odd number." If we are asked to find $P$(rolling a 4 or rolling an odd number), or $P(A \text{ or } B)$, then we are finding the probability of a compound event.

Let's discuss how to find the probability of a compound event using events $A$ and $B$ as given above.

| Possible outcomes for rolling a 4 | Possible outcomes for rolling an odd number |
|---|---|
| $\{4\}$ | $\{1, 3, 5\}$ |

There is one possible outcome for rolling a 4 and three different possible outcomes for rolling an odd number. Thus, there are four outcomes $\{4, 1, 3, 5\}$ that satisfy either one of the events out of six possible outcomes. Thus,

$$P(\text{rolling a 4 or rolling an odd number}) = \frac{4}{6} = \frac{2}{3}$$

Suppose that event $A$ is rolling a 3 rather than rolling a 4. Then $P(A \text{ or } B)$, becomes $P$(rolling a 3 or rolling an odd number). Let's determine the probability.

| Possible outcomes for rolling a 3 | Possible outcomes for rolling an odd number |
|---|---|
| $\{3\}$ | $\{1, 3, 5\}$ |

There is one possible outcome for rolling a 3 and three possible outcomes for rolling an odd number. However, the outcome for rolling a 3 is included in the outcomes for rolling an odd number. Thus, there are only three outcomes $\{1, 3, 5\}$ that satisfy either one of the events out of six possible outcomes.

$$P(\text{rolling a 3 or rolling an odd number}) = \frac{3}{6} = \frac{1}{2}$$

Notice in the last example we had to be careful not to count an outcome that satisfies both events more than once. To find $P(A \text{ or } B)$, we use the addition rule.

**Addition Rule**

For events $A$ and $B$,

$$P(A \text{ or } B) = P(A) + P(B) - P(A \text{ and } B)$$

Sometimes events $A$ and $B$ cannot happen simultaneously. When this happens, we say that events $A$ and $B$ are **mutually exclusive**. The events "rolling a 4" and "rolling an odd number" are mutually exclusive events because it is not possible for both events to occur simultaneously. The events "rolling a 3" and "rolling an odd number" are not mutually exclusive events because it is possible for both events to occur simultaneously. If a 3 were rolled, both events would be satisfied.

If events $A$ and $B$ are mutually exclusive, the $P(A \text{ and } B)$ in the addition formula equals 0. Now let's work an example.

**EXAMPLE 3 ▸ Deck of Cards** One card is selected at random from a deck of cards. Find the probability of selecting

**a)** a queen *or* king.

**b)** a heart *or* a 6.

**c)** a red card *or* a diamond.

*Solution*

**a)** There are 4 queens and 4 kings. Since a card can never be both a queen and a king, the events are mutually exclusive.

$$P(\text{queen or king}) = P(\text{queen}) + P(\text{king}) - P(\text{queen and king})$$
$$= \frac{4}{52} + \frac{4}{52} - \frac{0}{52} = \frac{8}{52} = \frac{2}{13}$$

**b)** There are 13 hearts in the deck and four 6s. One of the 6s is a heart, the 6 of hearts. Since one of the 52 cards is both a heart and a 6, $P(\text{heart and } 6) = \frac{1}{52}$.

$$P(\text{heart or } 6) = P(\text{heart}) + P(6) - P(\text{heart and } 6)$$
$$= \frac{13}{52} + \frac{4}{52} - \frac{1}{52} = \frac{16}{52} = \frac{4}{13}$$

**c)** There are 26 red cards and 13 diamond cards. All of the 13 diamonds are red cards.

$$P(\text{red or diamond}) = P(\text{red}) + P(\text{diamond}) - P(\text{red and diamond})$$
$$= \frac{26}{52} + \frac{13}{52} - \frac{13}{52} = \frac{26}{52} = \frac{1}{2}$$

▸ Now Try Exercise 23

## 4 Determine the Odds Against an Event

We often hear the word "odds." Odds are generally quoted as "odds against an event." The **odds against an event** is a ratio of the probability that the event does not occur to the probability that the event occurs. Thus, to find the odds against an event we first find the probability of the event occurring and the probability of the event not occurring. Recall that the probability that an event does not occur can be calculated from the probability of the event occuring.

**Odds Against an Event**

$$\text{Odds against an event} = \frac{P(\text{Event does not occur})}{P(\text{Event occurs})}$$

**EXAMPLE 4 ▸ Rolling a Die** Find the odds against rolling a 3 on one roll of a die.

*Solution* First determine the probability of rolling a 3 and the probability of not rolling a 3.

$$P(3) = \frac{1}{6} \qquad P(\text{not a } 3) = 1 - P(3)$$
$$= 1 - \frac{1}{6} = \frac{5}{6}$$

Now that we know the probability of rolling a 3 and the probability of not rolling a 3, we can determine the odds against rolling a 3.

$$\text{Odds against rolling a } 3 = \frac{P(\text{not a } 3)}{P(3)}$$
$$= \frac{\frac{5}{6}}{\frac{1}{6}}$$
$$= \frac{5}{\cancel{6}} \cdot \frac{\cancel{6}}{1}$$
$$= \frac{5}{1}$$

When giving the odds, it is common to write the ratio with a colon. The ratio $\frac{5}{1}$ can be written as 5:1 and is read "5 to 1." The odds against rolling 3 are 5:1.

▸ Now Try Exercise 55

**EXAMPLE 5** ▸ **Businesses Going Bankrupt** In a town, an estimated 2 of every 5 businesses go bankrupt within their first year of operation. If you open a new business in this town, what are the odds against it going bankrupt?

*Solution* The probability that your business will go bankrupt is $\frac{2}{5}$. The probability that your business will not go bankrupt is $1 - \frac{2}{5} = \frac{3}{5}$.

$$\text{Odds against business going bankrupt} = \frac{P(\text{business does not go bankrupt})}{P(\text{business goes bankrupt})}$$

$$= \frac{\frac{3}{5}}{\frac{2}{5}}$$

$$= \frac{3}{5} \cdot \frac{5}{2}$$

$$= \frac{3}{2}$$

The odds against the business going bankrupt are 3:2.

▸ Now Try Exercise 61

Although odds are generally given as odds against an event, they may also be given as odds in favor of an event. The **odds in favor of an event** is a ratio of the probability that the event occurs to the probability that the event does not occur.

**Odds in Favor of an Event**

$$\text{Odds in favor of an event} = \frac{P(\text{Event occurs})}{P(\text{Event does not occur})}$$

If the odds against an event are $a:b$, odds in favor of the event are $b:a$.

## EXERCISE SET 12.7

MathXL® MyMathLab

### Concept/Writing Exercises

1. What are equally likely outcomes?
2. How do you determine the probability of an event?
3. What is the sum of $P(A)$ and $P(\text{not } A)$?
4. a) If $P(A) = 1$, what is true about $A$?
   b) If $P(A) = 0$, what is true about $A$?
5. What does it mean for two events to be mutually exclusive?
6. How do you determine the odds against an event?
7. Between what two numbers, inclusive, must all probabilities be?
8. If the odds against an event are 5:3, what are the odds in favor of the event?

## Practice the Skills

*For Exercises 9–16, a marble is taken out of a bag containing 75 marbles—15 red, 20 blue, 10 green, and 30 orange. Find each given probability.*

**9.** $P$(a red marble is chosen)

**10.** $P$(a blue marble is chosen)

**11.** $P$(a green marble is chosen)

**12.** $P$(an orange marble is chosen)

**13.** $P$(a red marble is *not* chosen)

**14.** $P$(a red *and* green marble is chosen)

**15.** $P$(a blue *or* a green marble is chosen)

**16.** $P$(a red *or* an orange marble is chosen)

*For Exercises 17–28, a card is selected at random from a deck of cards. Find each probability.*

**17.** $P$(selecting a 9)

**18.** $P$(selecting a 10)

**19.** $P$(selecting the ace of spades)

**20.** $P$(selecting a black card)

**21.** $P$(selecting a red card)

**22.** $P$(selecting a king)

**23.** $P$(selecting a card that is not a king)

**24.** $P$(selecting a card that is a 9 and a black card)

**25.** $P$(selecting a card that is a 9 or a 10)

**26.** $P$(selecting a card that is a red card and a king)

**27.** $P$(selecting a card that is a red card or a king)

**28.** $P$(selecting a card that is a king or the ace spades)

*For Exercises 29–34, a song is selected at random from a juke box containing 40 rock/pop songs, 27 country songs, and 3 classical pieces. Find each probability.*

**29.** $P$(selecting a rock/pop song)

**30.** $P$(selecting a country song)

**31.** $P$(selecting a classical piece)

**32.** $P$(selecting a song that is *not* a rock/pop song)

**33.** $P$(selecting a country song *or* a classical piece)

**34.** $P$(selecting a song that is a rock/pop song and a classical piece)

*For Exercises 35–40, a spinner is spun. Assuming that the spinner cannot land on a line, find the probability of landing on* **a)** *yellow,* **b)** *purple,* **c)** *green,* **d)** *yellow or purple,* **e)** *a color that is not purple.*

**35.**

**36.**

**37.**

**38.**

39.

40.

*For Exercises 41–46, a traffic light is red for 40 seconds, yellow for 10 seconds, and green for 60 seconds. Find each probability.*

**41.** $P$(the light is red)

**42.** $P$(the light is yellow)

**43.** $P$(the light is green)

**44.** $P$(the light is *not* red)

**45.** $P$(the light is yellow *or* green)

**46.** $P$(the light is *not* green)

**47.** If $P(A) = 0.4$, $P(B) = 0.5$, and $P(A \text{ and } B) = 0.2$, find $P(A \text{ or } B)$.

**48.** If $P(A) = 0.1$, $P(B) = 0.3$, and $P(A \text{ and } B) = 0.1$, find $P(A \text{ or } B)$.

**49.** If $P(A) = 0.8$, $P(B) = 0.2$, and $P(A \text{ and } B) = 0.1$, find $P(A \text{ or } B)$.

**50.** If $P(A) = 0.6$, $P(B) = 0.2$, and $P(A \text{ and } B) = 0.0$, find $P(A \text{ or } B)$.

**51. Blouses** Aimee's closet contains 12 blouses, 4 of which are white. Aimee selects 1 blouse at random.

**a)** Find the probability that it is white.

**b)** Find the probability that it is not white.

**c)** Find the odds against it being white.

**d)** Find the odds in favor of it being white.

**52. Dogs** There are 5 samoyeds, 3 boxers, and 7 poodles in the animal shelter. Paulo selects 1 dog at random.

**a)** Find the probability that it is a boxer.

**b)** Find the probability that it is not a boxer.

**c)** Find the odds in favor of it being a boxer.

**d)** Find the odds against it being a boxer.

*In Exercises 53–56, a die is rolled.*

**53.** Find the odds against rolling a 4.

**54.** Find the odds against rolling an even number.

**55.** Find the odds against rolling a number greater than 5.

**56.** Find the odds against rolling a number less than 4.

*In Exercises 57–60, a card is selected from a deck of cards.*

**57.** Find the odds against selecting a 9.

**58.** Find the odds against selecting a Jack, Queen, or King.

**59.** Find the odds in favor of selecting a red card.

**60.** Find the odds against and the odds in favor of selecting a card greater than 7 (ace is low).

## Problem Solving

**61. Lotteries** One million tickets are sold for a lottery.

**a)** If you purchase a ticket, find your odds against winning.

**b)** If you purchase 10 tickets, find your odds against winning.

**62. Class of Students** One person is selected at random from a class of 16 males and 12 females.

**a)** Find the odds against selecting a female.

**b)** Find the odds against selecting a male.

**63. Odds of an Event** The odds in favor of an event are 7:3.

**a)** Find the probability that the event occurs.

**b)** Find the probability that the event does not occur.

**64. Tennis Tournament** The odds in favor of Erin winning a tennis tournament are 2:7.

**a)** Find the probability that Erin will win the tournament.

**b)** Find the probability that Erin will not win the tournament.

**65. Promotion** The odds against Jamal getting promoted to vice-president are 3:7. Find the probability that Jamal gets promoted to vice-president.

**66. Horse Race** The odds against the horse Speedy winning a race are 5:2.

**a)** Find the probability that Speedy wins.

**b)** Find the probability that Speedy loses.

**67. Buying a House** Suppose that the probability that you buy a house this week is 0.6. Find the odds against buying a house this week.

**68. Probability of Rain** Suppose that the probability that it will rain today is $\frac{3}{8}$. Find the odds against it raining today.

**69. Water Heater** Suppose that the probability that Triangle Pools fixes your pool's water heater correctly the first time is 0.7. Find the odds against your pool's water heater being repaired correctly by Triangle Pools on the first attempt.

**70. Horse Race** Find the probability of each horse winning the race if the odds against the horse winning are as given in the chart below. (Do not be concerned that the sum of the probabilities is not 1.)

| Horse | Odds |
|---|---|
| Old Boy | 8:3 |
| Daylight | 3:1 |
| One For All | 12:1 |
| Sunday | 7:6 |
| Lightning | 1:1 |

## Group Activity

*In this section we have calculated theoretical probabilities using possible outcomes of an experiment. We will now determine empirical (or experimental) probabilities using the results of actual data. Empirical probabilities are determined by dividing the number of times a specific outcome occurs by the total number of trials. For this activity, each group member will need 3 coins.*

**71. Tossing Coins**

**a)** Each member should toss the 3 coins 40 times and record the number of times each toss of the 3 coins results in exactly
3 heads.
2 heads.
1 head.
0 heads.

**b)** Now, each group member should determine the experimental probabilities of obtaining each of the 4 events given in part **a)** by dividing the number of observations of 3 heads, 2 heads, 1 head, and 0 heads by the total number of trials, 40.

**c)** Compare your result with the results of the other members. Are the fractions (probabilities) close? Can you explain why or why not?

**d)** Construct a tree diagram for tossing 3 coins. Use this to determine the theoretical probabilities of getting 3 heads, 2 heads, 1 head, and 0 heads. Write these probabilities with a denominator of 120.

**e)** Combine the results from part **a)** for all 3 group members to find the total number of times each toss of the 3 coins resulted in exactly 3 heads, 2 heads, 1 head, and 0 heads. Divide these numbers by 120 to obtain the experimental probabilities of each of these outcomes.

**f)** Compare the results from parts **d)** and **e)**. Are the experimental probabilities close to the theoretical probabilities? Explain.

## Cumulative Review Exercises

[3.3] **72.** Graph $12x - 16y = 48$ using the $x$- and $y$-intercepts.

[10.4] **73.** Graph $P(x) = x^4 - 2x^3 + 4x^2 - 14x - 21$.

[10.5] **74.** Graph $f(x) = \dfrac{3x^2 - 9x - 12}{2x^2 - 10x + 8}$.

[11.2] **75.** Graph $\dfrac{(x+3)^2}{25} + \dfrac{(y-5)^2}{9} = 1$.

# Chapter 12 Summary

| IMPORTANT FACTS AND CONCEPTS | EXAMPLES |
|---|---|
| **Section 12.1** | |
| A **sequence** of numbers is a list of numbers arranged in a specific order. Each number is called a **term** of the sequence. | $2, 6, 10, 14, 18, 22, \ldots$ is a sequence<br>$7, 14, 21, 28, 35, 42, \ldots$ is a sequence |
| An **infinite sequence** is a function whose domain is the set of natural numbers. | Domain: $\{1, 2, 3, 4, \ldots, n, \ldots\}$<br>$\downarrow \; \downarrow \; \downarrow \; \downarrow \quad \downarrow$<br>Range: $\{7, 14, 21, 28, \ldots 7n, \ldots\}$<br>The infinite sequence is $7, 14, 21, 28, \ldots$. |
| A **finite sequence** is a function whose domain includes only the first $n$ natural numbers. | Domain: $\{1, 2, 3, 4\}$<br>$\downarrow \; \downarrow \; \downarrow \; \downarrow$<br>Range: $\{4, 8, 12, 16\}$<br>The finite sequence is $4, 8, 12, 16$. |
| The **general term of a sequence, $a_n$**, can determine the sequence. | Let $a_n = n^2 - 3$. Write the first three terms of this sequence<br>$a_1 = 1^2 - 3 = -2$<br>$a_2 = 2^2 - 3 = 1$<br>$a_3 = 3^2 - 3 = 6$<br>The first three terms of the sequence are $-2, 1, 6$. |
| An **increasing sequence** is a sequence where each term is larger than the preceding term.<br>A **decreasing sequence** is a sequence where each term is smaller than the preceding term. | $-2, 5, 7, 11$ is an increasing sequence.<br>$50, 48, 46, 44$ is a decreasing sequence. |
| A **series** is the sum of the terms of a sequence. A series may be finite or infinite. | If the sequence is $1, 3, 5, 7, 9$, then the series is $1 + 3 + 5 + 7 + 9 = 25$.<br>If the sequence is $\frac{1}{3}, \frac{1}{9}, \frac{1}{27}, \ldots, \left(\frac{1}{3}\right)^n, \ldots$<br>then the series is $\frac{1}{3} + \frac{1}{9} + \frac{1}{27} + \cdots + \left(\frac{1}{3}\right)^n + \ldots$. |
| A **partial sum**, $s_n$, of an infinite sequence, $a_1, a_2, a_3, \ldots, a_n, \ldots$ is the sum of the first $n$ terms. That is,<br>$s_1 = a_1$<br>$s_2 = a_1 + a_2$<br>$s_3 = a_1 + a_2 + a_3$<br>$\vdots \qquad \vdots$<br>$s_n = a_1 + a_2 + a_3 + \cdots + a_n$ | Let $a_n = \frac{5+n}{n^2}$. Compute $s_1$ and $s_3$.<br>$s_1 = a_1 = \frac{5+1}{1^2} = \frac{6}{1} = 6$<br>$s_3 = a_1 + a_2 + a_3$<br>$= \frac{5+1}{1^2} + \frac{5+2}{2^2} + \frac{5+3}{3^2}$<br>$= \frac{6}{1} + \frac{7}{4} + \frac{8}{9} = 8\frac{23}{36}$ |
| A series can be written using **summation notation**:<br>$\sum_{i=1}^{n} a_i = a_1 + a_2 + a_3 + \cdots + a_n.$<br>$i$ is the **index of summation**, $n$ is the **upper limit of summation**, and 1 is the **lower limit of summation**. | $\sum_{i=1}^{4}(3i - 7) = (3 \cdot 1 - 7) + (3 \cdot 2 - 7) + (3 \cdot 3 - 7) + (3 \cdot 4 - 7)$<br>$= -4 - 1 + 2 + 5 = 2$<br>If $a_n = 6n^2 + 11$, the third partial sum, $s_3$, in summation notation, is written as $\sum_{i=1}^{3}(6i^2 + 11)$. |

| IMPORTANT FACTS AND CONCEPTS | EXAMPLES |
|---|---|
| **Section 12.2** | |
| An **arithmetic sequence** is a sequence in which each term after the first differs from the preceeding term by a **common difference, $d$.** | **Arithmetic Sequence** / **Common Difference, $d$**<br>$3, 8, 13, 18, 23, \ldots$   $d = 8 - 3 = 5$<br>$20, 14, 8, 2, -4, \ldots$   $d = 14 - 20 = -6$ |
| The ***n*th term, $a_n$**, of an arithmetic sequence is<br>$a_n = a_1 + (n-1)d$ | The $n$th term of the arithmetic sequence with $a_1 = 7$ and $d = -5$ is<br>$a_n = 7 + (n-1)(-5) = 7 - 5n + 5 = -5n + 12$<br>For this sequence, the 20th term is<br>$a_{20} = -5(20) + 12 = -100 + 12 = -88$ |
| An **arithmetic series** is the sum of the terms of an arithmetic sequence. The sum of the first $n$ terms, $s_n$ of an arithmetic sequence, also known as the ***n*th partial sum** is<br>$s_n = a_1 + a_2 + a_3 + \cdots + a_n$<br>For an arithmetic series, this sum is determined by the formula<br>$s_n = \dfrac{n(a_1 + a_n)}{2}$ | Find the sum of the first 30 natural numbers. That is, find the sum of<br>$1 + 2 + 3 + \cdots + 30$<br>Since $a_1 = 1$, $a_{30} = 30$, and $n = 30$, the sum is<br>$s_{30} = \dfrac{30(1+30)}{2} = \dfrac{30(31)}{2} = 465$ |
| **Section 12.3** | |
| A **geometric sequence** is a sequence in which each term after the first term is a common multiple of the preceding term. The common multiple is called the **common ratio,** $r$. | **Geometric Sequence** / **Common Ratio, $r$**<br>$2, 6, 18, 54, 162, \ldots$   $r = \dfrac{6}{2} = 3$<br>$8, -2, \dfrac{1}{2}, -\dfrac{1}{8}, \dfrac{1}{32}, \ldots$   $r = \dfrac{-2}{8} = -\dfrac{1}{4}$ |
| The ***n*th term, $a_n$**, of a geometric sequence is<br>$a_n = a_1 r^{n-1}$ | For the geometric sequence with $a_1 = 5$, $r = \dfrac{1}{2}$, and $n = 6$, $a_6$ is found as follows.<br>$a_6 = 5\left(\dfrac{1}{2}\right)^{6-1} = 5\left(\dfrac{1}{2}\right)^5 = \dfrac{5}{32}$ |
| A **geometric series** is the sum of the terms of a geometric sequence. The sum of the first $n$ terms, $s_n$, of a geometric sequence also known as the ***n*th partial sum,** is<br>$s_n = a_1 + a_2 + a_3 + \cdots + a_n.$<br>For a geometric series, this sum is determined by the formula<br>$s_n = \dfrac{a_1(1 - r^n)}{1 - r}, r \neq 1$ | To find the sum of the six terms of a geometric sequence with $a_1 = 12$ and $r = \dfrac{1}{3}$, use the formula with $n = 6$ to obtain<br>$s_6 = \dfrac{12\left[1 - \left(\frac{1}{3}\right)^6\right]}{1 - \frac{1}{3}} = \dfrac{12\left[1 - \frac{1}{729}\right]}{\frac{2}{3}} = \dfrac{12\left(\frac{728}{729}\right)}{\frac{1}{3}}$<br>$= 12\left(\dfrac{728}{729}\right)\left(\dfrac{3}{1}\right) = \dfrac{2912}{81}$ or $35\dfrac{77}{81}$ |
| The sum of an infinite geometric series is<br>$s_\infty = \dfrac{a_1}{1 - r}$ where $\lvert r \rvert < 1$ | To find the sum of the infinite series $4 - 2 + 1 - \dfrac{1}{2} + \dfrac{1}{4} + \cdots$, use the formula with $a_1 = 4$ and $r = -\dfrac{1}{2}$ to obtain<br>$s_\infty = \dfrac{4}{1 - \left(-\frac{1}{2}\right)} = \dfrac{4}{\frac{3}{2}} = 4 \cdot \dfrac{2}{3} = \dfrac{8}{3}$ or $2\dfrac{2}{3}$ |

| IMPORTANT FACTS AND CONCEPTS | EXAMPLES |
|---|---|

## Section 12.4

**The Principle of Mathematical Induction**

Let $S_n$ be a mathematical statement concerning the positive integer $n$. If the following two conditions are true, then $S_n$ is true for every positive integer $n$.

**Condition 1.** $S_1$ is true and

**Condition 2.** for any positive integer $k$, $k \le n$, if $S_k$ is true then $S_{k+1}$ is true.

Prove $5 + 10 + 15 + \ldots + 5n = \frac{5n(n+1)}{2}$ for all positive integers $n$.

Proof:

**Condition 1:**

$$5 \stackrel{?}{=} \frac{5(1)(1+1)}{2}$$

$$5 \stackrel{?}{=} \frac{5(2)}{2}$$

$$5 \stackrel{?}{=} 5 \quad \text{True}$$

**Condition 2:**

Assume $5 + 10 + 15 + \ldots + 5k = \frac{5k(k+1)}{2}$ for some positive integer $k$.

Then

$$5 + 10 + 15 + \ldots + 5k + 5(k+1) = \frac{5k(k+1)}{2} + 5(k+1)$$

$$5 + 10 + 15 + \ldots + 5k + 5(k+1) = \frac{5k^2}{2} + \frac{5k}{2} + 5k + 5$$

$$5 + 10 + 15 + \ldots + 5k + 5(k+1) = \frac{5k^2 + 15k + 10}{2}$$

$$5 + 10 + 15 + \ldots + 5k + 5(k+1) = \frac{5(k^2 + 3k + 2)}{2}$$

$$5 + 10 + 15 + \ldots + 5k + 5(k+1) = \frac{5(k+1)(k+2)}{2}$$

$$5 + 10 + 15 + \ldots + 5k + 5(k+1) = \frac{5(k+1)[(k+1)+1]}{2}$$

## Section 12.5

***n* Factorial**

$$n! = n(n-1)(n-2)(n-3)\cdots(1)$$

for any positive integer $n$.
Note that $0!$ is defined to be 1.

$$5! = 5\cdot4\cdot3\cdot2\cdot1 = 120$$

$$8! = 8\cdot7\cdot6\cdot5\cdot4\cdot3\cdot2\cdot1 = 40{,}320$$

**Binomial Coefficients**

For $n$ and $r$ nonnegative integers, $n \ge r$,

$$\binom{n}{r} = \frac{n!}{r!\cdot(n-r)!}$$

$$\binom{n}{n} = 1 \text{ and } \binom{n}{0} = 1$$

$$\binom{7}{3} = \frac{7!}{3!\cdot(7-3)!} = \frac{7!}{3!\cdot4!} = \frac{7\cdot6\cdot5\cdot\cancel{4\cdot3\cdot2\cdot1}}{3\cdot2\cdot1\cdot\cancel{4\cdot3\cdot2\cdot1}} = 35$$

$$\binom{10}{10} = 1, \quad \binom{10}{0} = 1$$

**Binomial Theorem**

For any positive integer $n$,

$$(a+b)^n = \binom{n}{0}a^nb^0 + \binom{n}{1}a^{n-1}b^1 + \binom{n}{2}a^{n-2}b^2 + \binom{n}{3}a^{n-3}b^3 + \cdots + \binom{n}{n}a^0b^n$$

$$(x+2y)^4 = \binom{4}{0}x^4 + \binom{4}{1}x^3(2y) + \binom{4}{2}x^2(2y)^2 + \binom{4}{3}x(2y)^3 + \binom{4}{4}(2y)^4$$

$$= 1\cdot x^4 + 4\cdot x^3(2y) + 6\cdot x^2(4y^2) + 4\cdot x(8y^3) + 1\cdot16y^4$$

$$= x^4 + 8x^3y + 24x^2y^2 + 32xy^3 + 16y^4$$

| IMPORTANT FACTS AND CONCEPTS | EXAMPLES |
|---|---|
| **Section 12.6** | |
| An **experiment** is any process that leads to a set of results. If an experiment consists of more than one part, each part is called a **trial**. | In rolling a die four times, each roll is considered a trial. |
| The possible results of each trial are called **outcomes**. | In rolling a die, the outcomes are 1, 2, 3, 4, 5, or 6. |
| A **sample space** is the set of all possible outcomes of an experiment. | The sample space for rolling a die is $\{1, 2, 3, 4, 5, 6\}$. |
| An **event** is any subset of a sample space. | Obtaining a 4 on a roll of a die is an event. |
| An **independent event** occurs when the outcome of one event has no effect on the outcome of any other event. | When we roll a die 3 times, each roll is an independent event. When we toss a coin 7 times, each toss is an independent event. |
| **Counting Principle**<br>If one event occurs in $m$ distinct ways and a second event can occur in $n$ distinct ways, then the two events, in that specific order, can occur in $m \cdot n$ distinct ways. | A math class has 20 students. Three students are selected to line up first, second, and third in front of the class. The total number of possible outcomes is<br>$$20 \cdot 19 \cdot 18 = 6840.$$ |
| **Permutations**<br>A **permutation** is any ordered arrangement of a given set of objects. | Each different arrangement of 5 different math books on a shelf is a permutation. |
| The number of permutations of $n$ distinct items is $n!$. | The number of ways to arrange 5 different math books on a shelf is<br>$$5! = 5 \cdot 4 \cdot 3 \cdot 2 \cdot 1 = 120.$$ |
| The number of permutations of $n$ items taken $r$ at a time is $_nP_r = \frac{n!}{(n-r)!}$. | The number of ways to arrange 3 math books on a shelf from a collection of 10 different math books is<br>$$_{10}P_3 = \frac{10!}{(10-3)!} = \frac{10!}{7!} = \frac{10 \cdot 9 \cdot 8 \cdot 7!}{7!} = 10 \cdot 9 \cdot 8 = 720.$$ |
| **Combinations**<br>A **combination** is a distinct grouping of objects without regard to their arrangement. | Selecting 4 students from a class of 22 students to serve on a committee is a combination. |
| The number of combinations of $n$ items taken $r$ at a time is $_nC_r = \frac{n!}{(n-r)!\,r!}$. | The number of ways to select 4 students from a class of 22 students for a committee is<br>$$_{22}C_4 = \frac{22!}{(22-4)!\,4!} = \frac{22!}{18!\,4!} = \frac{22 \cdot 21 \cdot 20 \cdot 19 \cdot 18!}{18! \cdot 4 \cdot 3 \cdot 2 \cdot 1} = 7315.$$ |
| **Section 12.7** | |
| **Probability**<br>The **probability** of an event means the likelihood the event will occur. | A die is rolled. The probability of obtaining a 5 is<br>$$P(5) = \frac{1}{6}.$$ |
| If $E$ represents an event, then the probability that $E$ occurs is<br>$$P(E) = \frac{\text{number of outcomes favorable to } E}{\text{total number of possible outcomes}}$$ | A card is selected from a deck of 52 cards. The probability of obtaining a club is<br>$$P(\text{club}) = \frac{13}{52} = \frac{1}{4}.$$ |

| IMPORTANT FACTS AND CONCEPTS | EXAMPLES |
|---|---|
| **Section 12.7 (continued)** | |
| If the probability that event $A$ occurs is $P(A)$, the probability that event $A$ does not occur is $1 - P(A)$. | A die is rolled. The probability of rolling a 3 is $\frac{1}{6}$. The probability of not rolling a 3 is $1 - P(3) = 1 - \frac{1}{6} = \frac{5}{6}$. |
| **Important Facts About Probability**<br>**1.** The probability of an event that cannot occur is 0.<br>**2.** The probability of an event that must occur is 1.<br>**3.** Every probability will be a number between 0 and 1, inclusive; that is, $0 \le P(E) \le 1$.<br>**4.** The sum of the probabilities of all possible outcomes of an experiment is 1. | A die is rolled. The probabilities are as follows.<br>$P(1) = \frac{1}{6},\quad P(2) = \frac{1}{6},\quad P(3) = \frac{1}{6},$<br>$P(4) = \frac{1}{6},\quad P(5) = \frac{1}{6},\quad P(6) = \frac{1}{6}.$<br>Each probability is between 0 and 1, and the sum of the probabilities is 1. |
| **Mutually exclusive** events are two events that cannot happen simultaneously. If events, $A$ and $B$, are mutually exclusive, then the probability of $A$ or $B$ is<br>$P(A \text{ or } B) = P(A) + P(B).$<br>If the events are not mutually exclusive, then the probability of $A$ or $B$ is<br>$P(A \text{ or } B) = P(A) + P(B) - P(A \text{ and } B).$ | A card is selected from a deck of 52 cards. The probability of obtaining a 4 or a 9 is<br>$P(4 \text{ or } 9) = P(4) + P(9) = \frac{4}{52} + \frac{4}{52} = \frac{8}{52} = \frac{2}{13}.$<br>A card is selected from a deck of 52 cards. The probability of obtaining a 7 or a club is<br>$P(7 \text{ or club}) = P(7) + P(\text{club}) - P(7 \text{ and club})$<br>$= \frac{4}{52} + \frac{13}{52} - \frac{1}{52} = \frac{16}{52} = \frac{4}{13}.$ |
| **Odds**<br>The **odds against an event** is a ratio of the probability that the event does not occur to the probability that the event occurs.<br>$\text{Odds against an event} = \frac{P(\text{Event does not occur})}{P(\text{Event occurs})}$<br>The **odds in favor of an event** is a ratio of the probability that the event occurs to the probability that the event does not occur.<br>$\text{Odds in favor of an event} = \frac{P(\text{Event occurs})}{P(\text{Event does not occur})}$ | The odds against rolling a 6 on one roll of a die is<br>$\text{odds against rolling a 6} = \frac{P(\text{not 6})}{P(6)} = \frac{5/6}{1/6} = \frac{5}{1}$ or 5:1.<br>The odds in favor of selecting an ace from a deck of 52 cards is<br>$\text{odds in favor of an ace} = \frac{P(\text{ace})}{P(\text{not ace})}$<br>$= \frac{4/52}{48/52} = \frac{4}{48} = \frac{1}{12}$ or 1:12. |

## Chapter 12 Review Exercises

[12.1] *Write the first five terms of each sequence.*

**1.** $a_n = n + 5$

**2.** $a_n = n^2 + n - 3$

**3.** $a_n = \frac{6}{n}$

**4.** $a_n = \frac{n^2}{n + 4}$

*Find the indicated term of each sequence.*

**5.** $a_n = 3n - 10$, seventh term

**6.** $a_n = (-1)^n + 5$, seventh term

**7.** $a_n = \frac{n + 17}{n^2}$, ninth term

**8.** $a_n = (n)(n - 3)$, eleventh term

*For each sequence, find the first and third partial sums, $s_1$ and $s_3$.*

**9.** $a_n = 2n + 5$

**10.** $a_n = n^2 + 8$

**11.** $a_n = \frac{n + 3}{n + 2}$

**12.** $a_n = (-1)^n(n + 8)$

*Write the next three terms of each sequence. Then write an expression for the general term, $a_n$.*

**13.** $2, 4, 8, 16, \ldots$

**14.** $-27, 9, -3, 1, \ldots$

**15.** $\frac{1}{7}, \frac{2}{7}, \frac{4}{7}, \frac{8}{7}, \ldots$

**16.** $13, 9, 5, 1, \ldots$

*Write out each series. Then find the sum of the series.*

**17.** $\sum_{i=1}^{3} i^2 + 9$

**18.** $\sum_{i=1}^{4} i(i + 5)$

**19.** $\sum_{i=1}^{5} \frac{i^2}{6}$

**20.** $\sum_{i=1}^{4} \frac{i}{i + 1}$

*For the set of values $x_1 = 3, x_2 = 9, x_3 = 7, x_4 = 10$, evaluate the indicated sum.*

**21.** $\sum_{i=1}^{4} x_i$

**22.** $\sum_{i=1}^{4} (x_i)^2$

**23.** $\sum_{i=2}^{3} (x_i^2 + 1)$

**24.** $\left(\sum_{i=1}^{4} x_i\right)^2$

*In Exercises 25 and 26, consider the following rectangles. For the nth rectangle, the length is $n + 3$ and the width is $n$.*

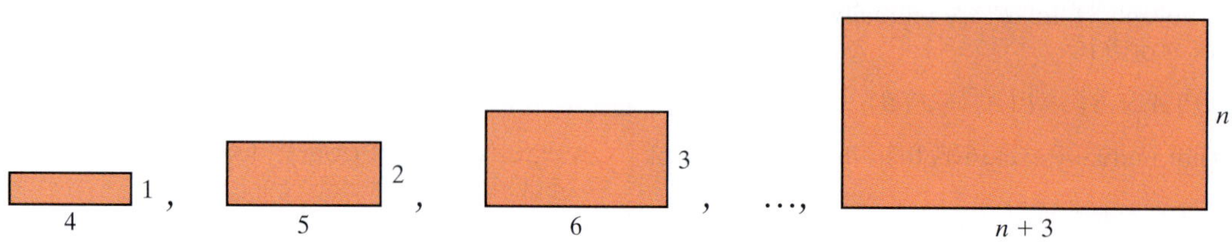


**25. Perimeter**

**a)** Find the perimeters for the first four rectangles, and then list the perimeters in a sequence.

**b)** Find the general term for the perimeter of the $n$th rectangle in the sequence. Use $p_n$ for perimeter.

**26. Area**

**a)** Find the areas for the first four rectangles, and then list the areas in a sequence.

**b)** Find the general term for the area of the $n$th rectangle in the sequence. Use $a_n$ for area.

[12.2] *Write the first five terms of the arithmetic sequence with the indicated first term and common difference.*

**27.** $a_1 = 5, d = 3$

**28.** $a_1 = 5, d = -\frac{1}{3}$

**29.** $a_1 = \frac{1}{2}, d = -2$

**30.** $a_1 = -100, d = \frac{1}{5}$

*For each arithmetic sequence, find the indicated value.*

**31.** $a_1 = 6, d = 3$; find $a_9$

**32.** $a_1 = 10, a_8 = -18$; find $d$

**33.** $a_1 = -3, a_{11} = 2$; find $d$

**34.** $a_1 = 22, a_n = -3, d = -5$; find $n$

*Find $s_n$ and d for each arithmetic sequence.*

**35.** $a_1 = 7, a_8 = 21, n = 8$

**36.** $a_1 = -12, a_7 = -48, n = 7$

**37.** $a_1 = \frac{3}{5}, a_6 = \frac{13}{5}, n = 6$

**38.** $a_1 = -\frac{10}{3}, a_9 = -6, n = 9$

*Write the first four terms of each arithmetic sequence. Then find $a_{10}$ and $s_{10}$.*

**39.** $a_1 = -7, d = 4$

**40.** $a_1 = 4, d = -3$

**41.** $a_1 = \frac{5}{6}, d = \frac{2}{3}$

**42.** $a_1 = -60, d = 5$

*Find the number of terms in each arithmetic sequence. Then find $s_n$.*

**43.** $4, 9, 14, \ldots, 64$

**44.** $-7, -4, -1, \ldots, 11$

**45.** $\frac{6}{10}, \frac{9}{10}, \frac{12}{10}, \ldots, \frac{36}{10}$

**46.** $-9, -3, 3, 9, \ldots, 45$

[12.3] *Determine the first five terms of each geometric sequence.*

**47.** $a_1 = 6, r = 2$

**48.** $a_1 = -12, r = \frac{1}{2}$

**49.** $a_1 = 20, r = -\frac{2}{3}$

**50.** $a_1 = -20, r = \frac{1}{5}$

*Find the indicated term of each geometric sequence.*

**51.** $a_1 = 6, r = \frac{1}{3}$; find $a_5$

**52.** $a_1 = 15, r = 2$; find $a_6$

**53.** $a_1 = -8, r = -3$; find $a_4$

**54.** $a_1 = \frac{1}{12}, r = \frac{2}{3}$; find $a_5$

*Find each sum.*

**55.** $a_1 = 7, r = 2$; find $s_6$

**56.** $a_1 = -84, r = -\frac{1}{4}$; find $s_5$

**57.** $a_1 = 9, r = \frac{3}{2}$; find $s_4$

**58.** $a_1 = 8, r = \frac{1}{2}$; find $s_7$

*For each geometric sequence, find the common ratio, r, and then write an expression for the general term, $a_n$.*

**59.** $6, 12, 24, \ldots$

**60.** $-4, -20, -100, \ldots$

**61.** $10, \frac{10}{3}, \frac{10}{9}, \ldots$

**62.** $\frac{9}{5}, \frac{18}{15}, \frac{36}{45}, \ldots$

*Find the sum of the terms in each infinite geometric sequence.*

**63.** $5, \frac{5}{2}, \frac{5}{4}, \frac{5}{8}, \ldots$

**64.** $\frac{5}{2}, 1, \frac{2}{5}, \frac{4}{25}, \ldots$

**65.** $-8, \frac{8}{3}, -\frac{8}{9}, \frac{8}{27}, \ldots$

**66.** $-6, -4, -\frac{8}{3}, -\frac{16}{9}, \ldots$

*Find the sum of each infinite series.*

**67.** $16 + 8 + 4 + 2 + 1 + \cdots$

**68.** $9 + \frac{9}{3} + \frac{9}{9} + \frac{9}{27} + \cdots$

**69.** $5 - 1 + \frac{1}{5} - \frac{1}{25} + \cdots$

**70.** $-4, -\frac{8}{3} - \frac{16}{9}, -\frac{32}{27}, \ldots$

*Write the repeating decimal as a ratio of integers.*

**71.** $0.363636\ldots$

**72.** $0.621621\ldots$

[12.4] *Use mathematical induction to prove the following statements for all positive integers n.*

**73.** $2 + 5 + 8 + \cdots + (3n - 1) = \frac{n(3n + 1)}{2}$

**74.** $2^2 + 4^2 + 6^2 + \cdots + (2n)^2 = \frac{2n(2n + 1)(n + 1)}{3}$

**75.** $1 \cdot 2 + 2 \cdot 3 + 3 \cdot 4 + \cdots + n(n + 1) = \frac{n(n + 1)(n + 2)}{3}$

**76.** $2 + 6 + 10 + \cdots + (4n - 2) = 2n^2$

[12.5] *Use the binomial theorem to expand the expression.*

**77.** $(3x + y)^4$

**78.** $(2x - 3y^2)^3$

*Write the first four terms of the expansion.*

**79.** $(x - 2y)^9$

**80.** $(2a^2 + 3b)^8$

[12.2]

**81. Sum of Integers** Find the sum of the integers between 101 and 200, inclusive.

**82. Barrels of Oil** Barrels of oil are stacked with 21 barrels in the bottom row, 20 barrels in the second row, 19 barrels in the third row, and so on, to the top row, which has only 1 barrel. How many barrels are there?

**83. Salary** Ahmed Mocanda just started a new job with an annual salary of \$36,000. He has been told that his salary will increase by \$1000 per year for the next 10 years.

**a)** Write a sequence showing his salary for the first 4 years.

**b)** Write a general term of this sequence.

**c)** What will his salary be 6 years from now?

**d)** How much money will he make in the first 11 years?

[12.3]

**84. Money** You begin with \$100, double that to get \$200, double that again to get \$400, and so on. How much will you have after you perform this process 10 times?

**85. Salary** Gertude Dibble started a new job on January 1, 2006 with a monthly salary of \$1600. Her boss has agreed to give her a 4% raise each month for the remainder of the year.

**a)** What is Gertude's salary in July?

**b)** What is Gertude's salary in December?

**c)** How much money does Gertude make in 2006?

**86. Pendulum** On each swing (left to right or right to left), a pendulum travels 92% as far as on its previous swing. If the first swing is 12 feet, find the distance traveled by the pendulum by the time it comes to rest.

[12.6] *Compute each permutation or combination.*

**87. Envelopes** Jeff has paper in 7 different colors and envelopes in 5 different colors. How many different paper and envelope color sets can he create?

**88. Seats in a Row** Eight people are to be seated in a row. In how many ways can they be arranged?

**89. Gifts** At a party 12 unique gifts are randomly distributed to the 12 guests. How many ways can the gifts be distributed?

**90. Dog Sled** Seven of 15 huskies are to be selected to pull a dogsled. They will be arranged in single file. How many arrangements are possible?

**91. Finalists in a Contest** Five finalists remain in a contest. The winner receives \$1000 and the runner up receives \$500. How many different arrangements of prizes are possible?

**92. Employees** In a company with 37 employees, a group of 4 will be selected to attend a training session. How many different groups are possible?

**93. Clubs** A club consists of 6 males and 8 females. Three males and 3 females are to be selected to represent the club at dinner. How many different combinations are possible?

**94. Fitness Books** Ten fitness books are available to be checked out. A maximum of 3 can be checked out at a time. If at least 1 fitness book is checked out, how many different groups of fitness books can be checked out?

[12.7] *In Exercises 95–100 each of the digits 0–9 is written on a piece of paper, and all of the pieces of paper are placed in a hat. One number is selected at random.*

**95.** Find the probability that the number selected is odd.

**96.** Find the probability that the number selected is even and greater than 4.

**97.** Find the probability that the number selected is even or greater than 4.

**98.** Find the probability that the number selected is greater than 6 and less than 4.

**99.** Find the probability that the number selected is a multiple of 3.

**100.** Find the probability that the number selected is less than 10.

**101.** If the probability of rain is 0.75, what are the odds against rain?

**102.** Assume that the odds against winning money at a certain slot machine are 5:4. What is the probability of winning money at the slot machine?

## Chapter 12 Practice Test

*To find out how well you understand the chapter material, take this practice test. The answers, and the section where the material was initially discussed, are given in the back of the book. Each problem is also fully worked-out on the* **Chapter Test Prep Video CD**. *Review any questions that you answered incorrectly.*

**1.** Are the following sequences arithmetic, geometric, or neither? Explain your answer.

**a)** $1, 1, 2, 3, 5, 8, 13, \ldots$

**b)** $4, 1, -2, -5, -8, \ldots$

**c)** $4, -2, 1, -\frac{1}{2}, \frac{1}{4}, -\frac{1}{8}, \ldots$

**2.** Write the first five terms of the sequence if $a_n = \frac{n-2}{3n}$.

**3.** Find the first and third partial sums if $a_n = \frac{2n+1}{n^2}$.

**4.** Write out the following series and find the sum of the series.

$$\sum_{i=1}^{5} (2i^2 + 3)$$

**5.** For $x_1 = 4, x_2 = 2, x_3 = 8$, and $x_4 = 10$ find $\sum_{i=1}^{4} (x_i)^2$.

**6.** Write the general term for the following arithmetic sequence.

$$\frac{1}{3}, \frac{2}{3}, \frac{3}{3}, \frac{4}{3}, \ldots$$

**7.** Write the general term for the following geometric sequence.

$$5, 10, 20, 40, \ldots$$

*In Exercises 8 and 9, write the first four terms of each sequence.*

**8.** $a_1 = 15, d = -6$

**9.** $a_1 = \frac{5}{12}, r = \frac{2}{3}$

**10.** Find $a_{11}$ when $a_1 = 40$ and $d = -8$.

**11.** Find $s_8$ for the arithmetic sequence with $a_1 = 7$ and $a_8 = -12$.

12. Find the number of terms in the arithmetic sequence $-4, -16, -28, \ldots, -136$.

13. Find $a_6$ when $a_1 = 8$ and $r = \frac{2}{3}$.

14. Find $s_7$ when $a_1 = \frac{3}{5}$ and $r = -5$.

15. Find the sum of the following infinite geometric series.

$$4 + \frac{8}{3} + \frac{16}{9} + \frac{32}{27} + \cdots$$

16. Use mathematical induction to prove

$$5 + 10 + 15 + \ldots + 5n = \frac{5n(n+1)}{2}$$

for all positive integers $n$.

17. Use the binomial theorem to expand $(x + 2y)^4$.

18. **Sandwiches** If a deli has 3 choices for bread, 4 choices for meat, and 2 choices for cheese, how many different sandwiches can be prepared if 1 type of bread, 1 type of meat, and 1 type of cheese are to be used?

19. **Letters** If 7 letters arrive in the mail, in how many different ways can they be opened?

20. **Chores** If 8 chores need to be done today and you get to choose the 2 that you will do, how many options do you have?

21. **Photos** If 4 of 11 vacation photos are to be selected to be placed in a display using 4 identical frames, how many ways can this be done?

22. **Rolling a Die** If a die is rolled, what is the probability of rolling a number that is a multiple of 2 *or* a multiple of 3?

23. **Winning a Contest** If the probability of winning a contest is $\frac{1}{500}$, what are the odds against winning the contest?

24. **Saving for Retirement** To save for retirement, Jamie Monroe plans to save \$1000 the first year, \$2000 the second year, \$3000 the third year, and to increase the amount saved by \$1000 in each successive year. How much will she have saved by the end of her twentieth year of savings?

25. **Culture of Bacteria** The number of bacteria in a culture is tripling every hour. If there are initially 500 bacteria in the culture, how many bacteria will be in the culture by the end of the sixth hour?

## Cumulative Review Test

*Take the following test and check your answers with those given in the back of the book. Review any questions that you answered incorrectly. The section where the material was covered is indicated after the answer.*

1. Solve $A = \frac{1}{2}bh$, for $b$.

2. Find an equation of the line through $(4, -2)$ and $(1, 9)$. Write the equation in slope-intercept form.

3. Solve the system of equations.

$$\begin{aligned} x + y + z &= 1 \\ 2x + 2y + 2z &= 2 \\ 3x + 3y + 3z &= 3 \end{aligned}$$

4. Multiply $(5x^3 + 4x^2 - 6x + 2)(x + 5)$.

5. Factor $x^3 + 2x - 6x^2 - 12$.

6. Factor $(a + b)^2 + 8(a + b) + 16$.

7. Subtract $5 - \frac{x - 1}{x^2 + 3x - 10}$.

8. $y$ varies directly as the square of $z$. If $y$ is 80 when $z$ is 20, find $y$ when $z$ is 50.

9. If $f(x) = 2\sqrt[3]{x - 3}$ and $g(x) = \sqrt[3]{5x - 15}$, find all values of $x$ for which $f(x) = g(x)$.

10. Solve $\sqrt{6x - 5} - \sqrt{2x + 6} - 1 = 0$.

11. Solve by completing the square.

$$x^2 + 2x + 15 = 0$$

12. Solve by the quadratic formula.

$$x^2 - \frac{x}{5} - \frac{1}{3} = 0$$

13. **Numbers** Twice the square of a positive number decreased by nine times the number is 5. Find the number.

14. Graph $y = x^2 - 4x$ and label the vertex.

15. Solve $\log_a \frac{1}{64} = 6$ for $a$.

16. Graph $y = 2^x - 1$.

17. Find an equation of a circle with center at $(-6, 2)$ and radius 7.

18. Graph $9x^2 + 16y^2 = 144$.

19. Find the sum of the infinite geometric series.

$$6 + 4 + \frac{8}{3} + \frac{16}{9} + \frac{32}{27} + \cdots$$

20. **Marbles** A bag has 15 red, 20 blue, 10 green, and 30 orange marbles. Find the probability that a blue marble is selected from the bag.

# Appendix

## Geometric Formulas

| Areas and Perimeters | | | |
|---|---|---|---|
| **Figure** | **Sketch** | **Area** | **Perimeter** |
| Square | 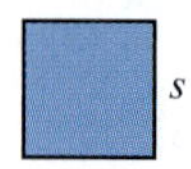 | $A = s^2$ | $P = 4s$ |
| Rectangle | 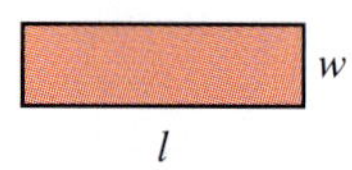 | $A = lw$ | $P = 2l + 2w$ |
| Parallelogram | 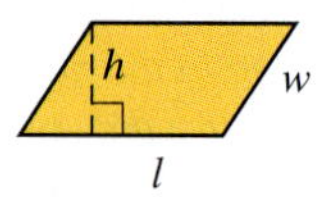 | $A = lh$ | $P = 2l + 2w$ |
| Trapezoid | 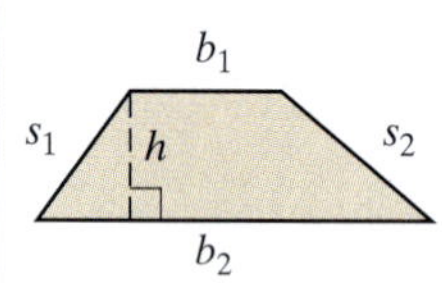 | $A = \frac{1}{2}h(b_1 + b_2)$ | $P = s_1 + s_2 + b_1 + b_2$ |
| Triangle | 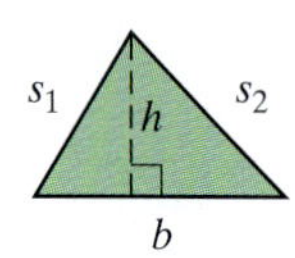 | $A = \frac{1}{2}bh$ | $P = s_1 + s_2 + b$ |

| Area and Circumference of a Circle | | | |
|---|---|---|---|
| Circle | 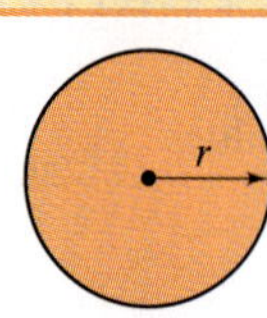 | $A = \pi r^2$ | $C = 2\pi r$ |

## Volumes and Surface Areas of Three-Dimensional Figures

| Figure | Sketch | Volume | Surface Area |
|---|---|---|---|
| Rectangular solid | 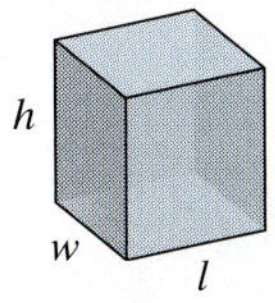 | $V = lwh$ | $s = 2lh + 2wh + 2wl$ |
| Right circular cylinder | 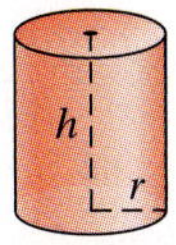 | $V = \pi r^2 h$ | $s = 2\pi rh + 2\pi r^2$ |
| Sphere | 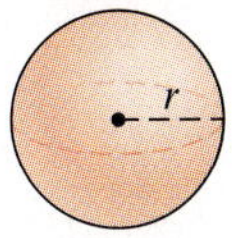 | $V = \frac{4}{3}\pi r^3$ | $s = 4\pi r^2$ |
| Right circular cone | 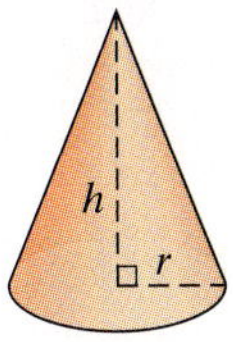 | $V = \frac{1}{3}\pi r^2 h$ | $s = \pi r\sqrt{r^2 + h^2}$ |
| Square or rectangular pyramid | 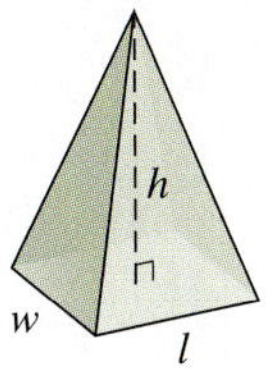 | $V = \frac{1}{3}lwh$ | |

# Answers

## Chapter 1

**Exercise Set 1.1** **1–11.** Answers will vary. **13.** Do all the homework and preview the new material to be covered in class. **15.** See the steps on page 4 of your text. **17.** The more you put into the course, the more you will get out of it. **19.** Answers will vary.

**Exercise Set 1.2** **1.** A variable is a letter used to represent various numbers. **3.** A set is a collection of objects. **5.** A set that contains no elements **7.** $>$, is greater than; $\geq$, is greater than or equal to; $<$, is less than; $\leq$, is less than or equal to; $\neq$, is not equal to **9.** $\{4, 5, 6\}$ **11.** An integer can be written with a denominator of 1. **13.** True **15.** True **17.** False **19.** True **21.** True **23.** $>$ **25.** $>$ **27.** $>$ **29.** $<$ **31.** $>$ **33.** $<$ **35.** $>$ **37.** $>$ **39.** $A = \{0\}$ **41.** $C = \{18, 20\}$ **43.** $E = \{0, 1, 2\}$ **45.** $H = \{0, 7, 14, 21, \ldots\}$ **47.** $J = \{1, 2, 3, 4, \ldots\}$ or $J = N$ **49. a)** 4 **b)** 4, 0 **c)** $-2, 4, 0$ **d)** $-2, 4, \frac{1}{2}, \frac{5}{9}, 0, -1.23, \frac{78}{79}$ **e)** $\sqrt{2}, \sqrt{8}$ **f)** $-2, 4, \frac{1}{2}, \frac{5}{9}, 0, \sqrt{2}, \sqrt{8}, -1.23, \frac{78}{79}$ **51.** $A \cup B = \{1, 2, 3, 4, 5, 6\}$; $A \cap B = \{\ \}$ **53.** $A \cup B = \{-4, -3, -2, -1, 0, 1, 3\}$; $A \cap B = \{-3, -1\}$ **55.** $A \cup B = \{2, 4, 6, 8, 10\}$; $A \cap B = \{\ \}$ **57.** $A \cup B = \{0, 5, 10, 15, 20, 25, 30\}$; $A \cap B = \{\ \}$ **59.** $A \cup B = \{-1, 0, 1, e, i, \pi\}$; $A \cap B = \{-1, 0, 1\}$ **61.** The set of natural numbers **63.** The set of whole number multiples of 3 **65.** The set of odd integers **67. a)** Set $A$ is the set of all $x$ such that $x$ is a natural number less than 7 **b)** $A = \{1, 2, 3, 4, 5, 6,\}$

**69.** 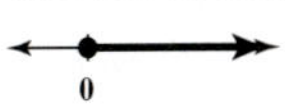

**71.** 

**73.** 

**75.** 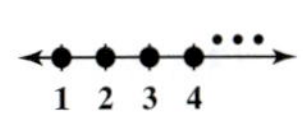

**77.** 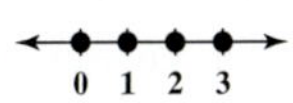

**79.** $\{x|x \geq 1\}$

**81.** $\{x|x < 5 \text{ and } x \in I\}$ or $\{x|x \leq 4 \text{ and } x \in I\}$ **83.** $\{x|-3 < x \leq 5\}$ **85.** $\{x|-2.5 \leq x < 4.2\}$ **87.** $\{x|-3 \leq x \leq 1 \text{ and } x \in I\}$ **89.** Yes **91.** No **93.** Yes **95.** No **97.** One example is $\left\{\frac{3}{2}, \frac{4}{3}, \frac{5}{4}, \frac{6}{5}, \frac{7}{6}\right\}$ **99.** One example is $A = \{2, 4, 5, 8, 9\}$, $B = \{4, 5, 6, 9\}$

**101. a)** {Johnson, Mayfield, Labonte, Gordon, Busch, Earnhardt Jr., Biffle, Stewart, Gaughan} **b)** Union **c)** {Johnson, Gordon, Busch} **d)** Intersection **103. a)** {Albert, Carmen, Frank, Linda, Barbara, Jason, David, Earl, Kate, Ingrid} **b)** Union **c)** {Frank, Linda} **d)** Intersection **105. a)** {China, India, United States, Indonesia, Brazil, Nigeria} **b)** {China, India, United States, Russia, Japan, Indonesia, Nigeria} **c)** {China, India, United States} **d)** {China, India, United States, Indonesia} **e)** {China, India, United States} **107. a)** $A$ = {Alex, James}, $B$ = {Alex, James, George, Connor}, $C$ = {Alex, Stephen}, $D$ = {Alex, George, Connor} **b)** {Alex} **c)** Only Alex **109. a)** $\{1, 3, 4, 5, 6, 7\}$ **b)** $\{2, 3, 4, 6, 8, 9\}$ **c)** $\{1, 2, 3, 4, 5, 6, 7, 8, 9\}$ **d)** $\{3, 4, 6\}$ **111. a)** $\{x|x > 1\}$ includes fractions and decimal numbers which the other set does not contain. **b)** $\{2, 3, 4, 5, \ldots\}$ **c)** No, since it is not possible to list all real numbers greater than 1 in roster form. **113.**

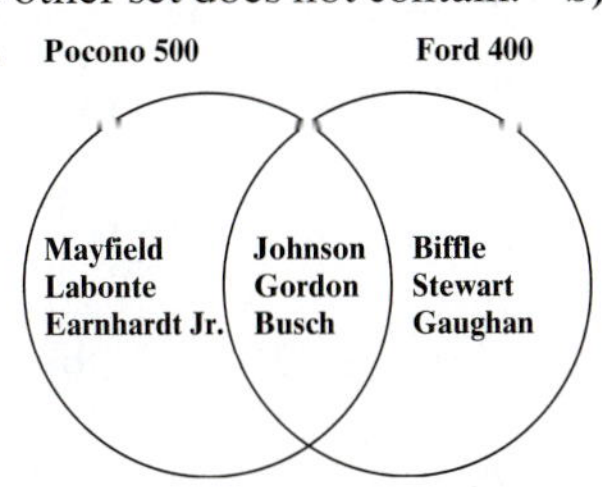


**Exercise Set 1.3** **1.** Two numbers whose sum is zero **3.** No; $|0|$ is not positive **5.** Since $a$ and $-a$ are the same distance from 0 on a number line, $|a| = |-a|$ for all real numbers, $\mathbb{R}$. **7.** Since $|6| = 6$ and $|-6| = 6$, the desired values for $a$ are 6 and $-6$. **9.** $\{\ \}$, the absolute value of any real number must be greater than or equal to 0. **11.** Answers will vary. **13.** Answers will vary. **15.** $-\frac{a}{b}$ or $\frac{-a}{b}$ **17. a)** $a + b = b + a$ **b)** Answers will vary. **19.** Answers will vary. One example is $2 + (3 \cdot 4) \neq (2 + 3) \cdot (2 + 4)$, $14 \neq 30$ **21.** 5 **23.** 7 **25.** $\frac{7}{8}$ **27.** 0 **29.** $-7$ **31.** $-\frac{5}{9}$ **33.** $=$ **35.** $>$ **37.** $>$ **39.** $>$ **41.** $>$ **43.** $<$ **45.** $-|5|, -2, -1, |-3|, 4$ **47.** $-32, -|4|, 4, |-7|, 15$ **49.** $-|-6.5|, -6.1, |-6.3|, |6.4|, 6.8$ **51.** $-2, \frac{1}{3}, \left|-\frac{1}{2}\right|, \left|\frac{3}{5}\right|, \left|-\frac{3}{4}\right|$ **53.** 3 **55.** $-22$ **57.** $-4$ **59.** $-\frac{2}{35}$ **61.** $-0.99$ **63.** 7.92 **65.** $-16.2$ **67.** 2 **69.** $-2$ **71.** $\frac{17}{20}$ **73.** $-40$ **75.** $\frac{5}{4}$ **77.** 12 **79.** 235.9192 **81.** 11 **83.** 1 **85.** $-\frac{3}{64}$ **87.** $\frac{7}{3}$ **89.** $-4$ **91.** 20 **93.** 5 **95.** $-20.6$ **97.** 11 **99.** $-6$ **101.** $\frac{81}{16}$ **103.** $-1$ **105.** $-\frac{17}{45}$ **107.** 77 **109.** $-39$ **111.** 0 **113.** Commutative property of addition **115.** Multiplicative property of zero **117.** Associative property of addition **119.** Identity property of multiplication **121.** Associative property of multiplication **123.** Distributive property **125.** Identity property of addition **127.** Inverse property of addition **129.** Double negative property **131.** $-6, \frac{1}{6}$ **133.** $\frac{22}{7}, -\frac{7}{22}$

**135.** 49°F **137.** 148.2 feet below the starting point, or −148.2 feet **139.** 10.1°F **141.** Gain of $1207 **143.** Answers will vary. **145.** $24,000 **147.** 84 **149.** 1 **150.** True **151.** $\{1, 2, 3, 4, \ldots\}$ **152. a)** 3, 4, −2, 0 **b)** $3, 4, -2, \frac{5}{6}, 0$ **c)** $\sqrt{11}$ **d)** $3, 4, -2, \frac{5}{6}, \sqrt{11}, 0$ **153. a)** $\{1, 4, 7, 9, 12, 15\}$ **b)** $\{4, 7\}$ **154.** 

## Exercise Set 1.4

**1. a)** Base **b)** Exponent **3. a)** Index **b)** Radicand **5.** The positive number whose square equals the radicand **7.** A negative number raised to an odd power is a negative number. **9.** Parentheses, exponents and radicals, multiplication or division from left to right, addition or subtraction from left to right **11. a)** Answers will vary. **b)** 24 **13.** 9 **15.** −9 **17.** 9 **19.** $-\frac{81}{625}$ **21.** 7 **23.** −6 **25.** −3 **27.** 0.1 **29.** 0.015 **31.** 1.897 **33.** 76,183.335 **35.** 2.962 **37.** 3.250 **39.** −0.723 **41. a)** 9 **b)** −9 **43. a)** 100 **b)** −100 **45. a)** 1 **b)** −1 **47. a)** $\frac{1}{9}$ **b)** $-\frac{1}{9}$ **49. a)** 27 **b)** −27 **51. a)** −125 **b)** 125 **53. a)** −8 **b)** 8 **55. a)** $\frac{8}{125}$ **b)** $-\frac{8}{125}$ **57.** −7 **59.** −19 **61.** −22.221 **63.** $-\frac{5}{16}$ **65.** 43 **67.** 25 **69.** 0 **71.** $\frac{1}{2}$ **73.** −10 **75.** 5 **77.** 64 **79.** 16 **81.** $\frac{27}{5}$ **83.** Undefined **85.** −4 **87.** 0 **89.** $-\frac{10}{3}$ **91.** $\frac{242}{5}$ **93.** $\frac{1}{4}$ **95.** 28 **97.** −41 **99.** −9 **101.** −90 **103.** 33 **105.** −5 **107.** $\frac{3}{2}$ **109.** $\frac{7y - 14}{2}$, 14 **111.** $6(3x + 6) - 9$, 81 **113.** $\left(\frac{x + 3}{2y}\right)^2 - 3$, 1 **115. a)** 24.6 miles **b)** 57.4 miles **117. a)** 102 feet **b)** 54 feet **119. a)** $623.05 **b)** $837.97 **121. a)** 9.51 billion trips **b)** 22.51 billion trips **123. a)** $297.83 billion **b)** $405.83 billion **125. a)** 7.62% **b)** 21.78% **127. a)** $1.262 billion **b)** $19.438 billion **129. a)** $A \cap B = \{b, c, f\}$ **b)** $A \cup B = \{a, b, c, d, f, g, h\}$ **130.** All real numbers, ℝ **131.** $a \geq 0$ **132.** 6, −6 **133.** $-|6|, -4, -|-2|, 0, |-5|$ **134.** Associative property of addition

## Mid-Chapter Test*

**1.** Answers will vary. [1.1] **2.** $A \cup B = \{-3, -2, -1, 0, 1, 2, 3, 5\}$, $A \cap B = \{-1, 1\}$ [1.2] **3.** The set of whole number multiples of 5. [1.2] **4.** 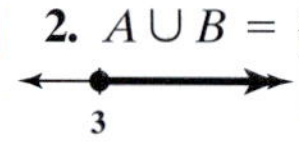 [1.2] **5.** > [1.2] **6.** $\{x \mid -5 \leq x < 2\}$ [1.2] **7.** No [1.2] **8.** $-15, |-6|, 7, |-17|$ [1.2] **9.** 9.2 [1.3] **10.** $\frac{7}{30}$ [1.3] **11.** 256 [1.3] **12.** $-\frac{4}{13}$ [1.3] **13.** −3 [1.3] **14.** Distributive property [1.3] **15.** 0.9 [1.4] **16. a)** 36 **b)** −36 [1.4] **17. a)** 1) Grouping symbols, 2) Exponents and radicals, 3) Multiplications or divisions left to right, 4) Additions or subtractions left to right **b)** −14 [1.4] **18.** 26 [1.4] **19.** 4 [1.4] **20.** $\frac{5}{2}$ [1.4]

## Exercise Set 1.5

**1. a)** $a^m \cdot a^n = a^{m+n}$ **b)** Answers will vary. **3. a)** $a^0 = 1, a \neq 0$ **b)** Answers will vary. **5. a)** $(ab)^m = a^m b^m$ **b)** Answers will vary. **7. a)** $\left(\frac{a}{b}\right)^m = \frac{a^m}{b^m}, b \neq 0$ **b)** Answers will vary. **9.** $x = \frac{1}{5}$, because if $\frac{1}{x} = 5$, then $x = \frac{1}{5}$ **11. a)** The opposite of $x$ is $-x$; the reciprocal of $x$ is $\frac{1}{x}$ **b)** $x^{-1}; \frac{1}{x}$ **c)** $-x$ **13.** 32 **15.** 9 **17.** $\frac{1}{81}$ **19.** 125 **21.** 1 **23.** 64 **25.** 64 **27.** $\frac{16}{49}$ **29. a)** $\frac{1}{9}$ **b)** $\frac{1}{9}$ **c)** $-\frac{1}{9}$ **d)** $-\frac{1}{9}$ **31. a)** 2 **b)** −2 **c)** −2 **d)** 2 **33. a)** 5 **b)** −5 **c)** 1 **d)** −1 **35. a)** $3xy$ **b)** 1 **c)** $3x$ **d)** 3 **37.** $\frac{7}{y^3}$ **39.** $9x^4$ **41.** $2ab^3$ **43.** $\frac{13}{2m^2n^3}$ **45.** $\frac{5z^4}{x^2y^3}$ **47.** $\frac{1}{9xy}$ **49.** $\frac{1}{4}$ **51.** $x^2$ **53.** 64 **55.** $\frac{1}{49}$ **57.** $\frac{1}{m^{11}}$ **59.** $5w^5$ **61.** $\frac{12}{a^8}$ **63.** $3p$ **65.** $-10r^7$ **67.** $8x^7y^2$ **69.** $\frac{3x^2}{y^6}$ **71.** $-\frac{3x^3z^2}{y^5}$ **73. a)** 4 **b)** 8 **c)** 1 **d)** 0 **75. a)** $-\frac{1}{12}$ **b)** $\frac{7}{12}$ **c)** $\frac{11}{10}$ **d)** $\frac{23}{120}$ **77.** 81 **79.** $\frac{1}{81}$ **81.** $b^6$ **83.** $-c^3$ **85.** $\frac{16}{x^6}$ **87.** $\frac{7}{10}$ **89.** $\frac{21}{16}$ **91.** $\frac{9}{16b^2}$ **93.** $\frac{16x^4}{y^4}$ **95.** $\frac{q^{12}}{125p^6}$ **97.** $-\frac{g^{12}}{27h^9}$ **99.** $\frac{9j^2}{16k^4}$ **101.** $8r^6s^{15}$ **103.** $\frac{y^6}{64x^3}$ **105.** $125x^9y^3$ **107.** $\frac{z^3}{8x^3y^3}$ **109.** $\frac{x^{20}}{y^{10}}$ **111.** $\frac{x^4y^8}{4z^{12}}$ **113.** $-\frac{64b^{12}}{a^6c^3}$ **115.** $\frac{27}{8x^{21}y^9}$ **117.** $x^{7a+3}$ **119.** $w^{5a-7}$ **121.** $x^{w+7}$ **123.** $x^{5p+2}$ **125.** $x^{2m+2}$ **127.** $\frac{5m^{2b}}{n^{2a}}$ **129. a)** $x < 0$ or $x > 1$ **b)** $0 < x < 1$ **c)** $x = 0$ or $x = 1$ **d)** Not true for $0 \leq x \leq 1$ **131. a)** The product of an even number of negative factors is positive. **b)** The product of an odd number of negative factors is negative. **133. a)** Yes **b)** Yes, because $x^{-2} = \frac{1}{x^2}$ and $(-x)^{-2} = \frac{1}{(-x)^2} = \frac{1}{x^2}$ **135.** −3, because $(y^{-2}/y^{-3})^2 = y^2$ **137.** −1, 3, because $(x^{-1}/x^4)^{-1} = x^5$, and $(y^5/y^3)^{-1} = 1/y^2$ **139.** $x^{9/8}$ **141.** $\frac{1}{x^{9/2}y^{19/6}}$ **144. a)** $A \cup B = \{1, 2, 3, 4, 5, 6, 9\}$ **b)** $A \cap B = \{\ \}$ **145.**  **146.** −4 **147.** −5

* Numbers in blue brackets after the answer indicates the section where the material was discussed.

## Exercise Set 1.6

**1.** A number greater than or equal to 1 and less than 10 multiplied by a power of 10 **3.** $1 \times 10^{-2}$, because $1 \times 10^{-2} = 0.01$ and $1 \times 10^{-3} = 0.001$. **5.** $3.7 \times 10^3$ **7.** $4.1 \times 10^{-2}$ **9.** $7.6 \times 10^5$ **11.** $1.86 \times 10^{-6}$ **13.** $5.78 \times 10^6$ **15.** $1.06 \times 10^{-4}$ **17.** 31,000 **19.** 0.0000213 **21.** 0.917 **23.** 8,000,000 **25.** 203,000 **27.** 1,000,000 **29.** 240,000,000 **31.** 0.021 **33.** 0.000027 **35.** 11,480 **37.** 0.0003 **39.** 0.0000006734 **41.** $1.5 \times 10^{-5}$ **43.** $5.0 \times 10^3$ **45.** $3.0 \times 10^{-8}$ **47.** $1.645 \times 10^{12}$ **49.** $4.8 \times 10^5$ **51.** $3.0 \times 10^0$ **53.** $9.369 \times 10^{14}$ **55.** $1.056 \times 10^3$ **57.** $5.337 \times 10^2$ **59.** $3.115 \times 10^{-25}$ **61.** $7.604 \times 10^{-27}$ **63.** $3.333 \times 10^{60}$ **65.** $8.5 \times 10^8$ **67.** $2.4 \times 10^6$ **69.** $5.28 \times 10^{10}$ **71.** $9.1 \times 10^{12}$ **73.** $1.0 \times 10^{-5}$ **75.** $1.58 \times 10^{-5}$ **77.** $1.0 \times 10^{-9}$ **79. a)** Subtract 1 from the exponent **b)** Subtract 2 from the exponent **c)** Subtract 6 from the exponent **d)** $6.58 \times 10^{-10}$ **81. a)** $1.0 \times 10^4$ or 10,000 **b)** $4.725 \times 10^5$ or 472,500 **c)** The error in part **b)** because the answer is off by more. **83.** 30,000 hours **85. a)** $\approx 6.1485 \times 10^9$ people **b)** $\approx 4.6\%$ **87. a)** $1.1728 \times 10^{13}$, $2.965 \times 10^8$ **b)** $\approx \$39{,}554.81$ **89.** 132 people/square kilometer **91. a)** $2.1 \times 10^8$ pounds **b)** $3.99 \times 10^9$ pounds **93. a)** 994 million **b)** $\approx 20.01\%$ **c)** $\approx 348.6$ people/square mile **d)** $\approx 81.8$ people/square mile **95. a)** $\$6.9 \times 10^{10}$ **b)** $\$8.28 \times 10^{11}$ **c)** $\$2.139 \times 10^{12}$ **97. a)** $6.03 \times 10^7$ square kilometers **b)** $4.4 \times 10^6$ square kilometers

## Chapter 1 Review Exercises

**1.** $\{4, 5, 6, 7, 8\}$ **2.** $\{0, 3, 6, 9, \ldots\}$ **3.** Yes **4.** Yes **5.** No **6.** Yes **7.** 4, 6 **8.** 4, 6, 0 **9.** $-2, 4, 6, 0$ **10.** $-2, 4, 6, \frac{1}{2}, 0, \frac{15}{27}, -\frac{1}{5}, 1.47$ **11.** $\sqrt{7}, \sqrt{3}$ **12.** $-2, 4, 6, \frac{1}{2}, \sqrt{7}, \sqrt{3}, 0, \frac{15}{27}, -\frac{1}{5}, 1.47$ **13.** False **14.** True **15.** True **16.** True **17.** $A \cup B = \{1, 2, 3, 4, 5, 6, 8, 10\}$; $A \cap B = \{2, 4, 6\}$ **18.** $A \cup B = \{2, 3, 4, 5, 6, 7, 8, 9\}$; $A \cap B = \{\ \}$ **19.** $A \cup B = \{1, 2, 3, 4, \ldots\}$; $A \cap B = \{\ \}$ **20.** $A \cup B = \{3, 4, 5, 6, 9, 10, 11, 12\}$; $A \cap B = \{9, 10\}$ **21.** 

**22.** 

**23.** 

**24.** 

**25.** $<$ **26.** $<$ **27.** $<$ **28.** $=$ **29.** $<$ **30.** $>$ **31.** $>$ **32.** $>$ **33.** $-\pi, -3, 3, \pi$ **34.** $0, \frac{3}{5}, 2.7, |-3|$ **35.** $-2, 3, |-5|, |-10|$ **36.** $-7, -3, |-3|, |-7|$ **37.** $-4, -|-3|, 5, 6$ **38.** $-2, 0, |16|, |-2.3|$ **39.** Distributive property **40.** Commutative property of multiplication **41.** Associative property of addition **42.** Identity property of addition **43.** Associative property of multiplication **44.** Double negative property **45.** Multiplicative property of zero **46.** Inverse property of addition **47.** Inverse property of multiplication **48.** Identity property of multiplication **49.** 14 **50.** 9 **51.** 11 **52.** $-5$ **53.** 1 **54.** 21 **55.** 9 **56.** $-49$ **57.** 15 **58.** 34 **59.** 6 **60.** 64 **61.** Undefined **62.** $\frac{8}{3}$ **63.** 22 **64.** $-67$ **65. a)** \$816.37 million **b)** \$7,223.73 million **66. a)** 944.53 ton-miles **b)** 2135.65 ton-miles **67.** 32 **68.** $x^5$ **69.** $a^8$ **70.** $y^7$ **71.** $b^9$ **72.** $\frac{1}{c^3}$ **73.** $\frac{1}{125}$ **74.** 8 **75.** $81m^6$ **76.** $\frac{7}{4}$ **77.** $\frac{27}{8}$ **78.** $\frac{y^2}{x}$ **79.** $-15x^3y^4$ **80.** $\frac{14}{v^3w^3}$ **81.** $\frac{3y^7}{x^5}$ **82.** $\frac{3}{xy^9}$ **83.** $\frac{g^5}{h^5j^{14}}$ **84.** $\frac{3m}{n^4}$ **85.** $64a^3b^3$ **86.** $\frac{x^{10}}{9y^2}$ **87.** $\frac{p^{14}}{q^{12}}$ **88.** $-\frac{8a^3}{b^9c^6}$ **89.** $\frac{z^4}{25x^2y^6}$ **90.** $\frac{m^9}{27}$ **91.** $\frac{n^6}{4m^4}$ **92.** $\frac{625x^4y^4}{z^{20}}$ **93.** $\frac{9x^{10}}{4y^{14}z^{12}}$ **94.** $-\frac{x^6z^2}{8y^2}$ **95.** $7.42 \times 10^{-5}$ **96.** $4.6 \times 10^5$ **97.** $1.83 \times 10^5$ **98.** $1.0 \times 10^{-6}$ **99.** 30,000 **100.** 0.03 **101.** 200,000,000 **102.** 2000 **103. a)** $\$1.7 \times 10^6$ **b)** $\$4.6 \times 10^6$ **c)** $\approx 1.28$ **104. a)** 14,000,000,000 **b)** 14 billion kilometers **c)** $5.0 \times 10^8$ kilometers or 500,000,000 kilometers **d)** $8.4 \times 10^9$ miles or 8,400,000,000 miles

## Chapter 1 Practice Test

**1.** $A = \{6, 7, 8, 9, \ldots\}$ [1.2] **2.** False [1.2] **3.** True [1.2] **4.** $-\frac{3}{5}, 2, -4, 0, \frac{19}{12}, 2.57, -1.92$ [1.2] **5.** $-\frac{3}{5}, 2, -4, 0, \frac{19}{12}, 2.57, \sqrt{8}, \sqrt{2}, -1.92$ [1.2] **6.** $A \cup B = \{5, 7, 8, 9, 10, 11, 14\}$: $A \cap B = \{8, 10\}$ [1.2] **7.** $A \cup B = \{1, 3, 5, 7, \ldots\}$; $A \cap B = \{3, 5, 7, 9, 11\}$ [1.2] **8.** 

[1.2] **9.** 

[1.2] **10.** $-|4|, -2, |3|, 9$ [1.3] **11.** Associative property of addition [1.3] **12.** Commutative property of addition [1.3] **13.** 2 [1.4] **14.** 33 [1.4] **15.** Undefined [1.4] **16.** $-\frac{37}{22}$ [1.4] **17.** 17 [1.4] **18. a)** 304 feet **b)** 400 feet [1.4] **19.** $\frac{1}{9}$ [1.5] **20.** $\frac{16}{m^6n^4}$ [1.5] **21.** $\frac{4c^2}{5ab^5}$ [1.5] **22.** $-\frac{y^{21}}{27x^{12}}$ [1.5] **23.** $3.89 \times 10^8$ [1.6] **24.** 260,000,000 [1.6] **25. a)** $9.2 \times 10^9$ **b)** 0–14: $1.794 \times 10^9$, 15–64: $5.8052 \times 10^9$, 65 and older: $1.6008 \times 10^9$ [1.6]

# Chapter 2

## Using Your Calculator, 2.1

**1.** No **2.** Yes

## Exercise Set 2.1

**1.** The terms of an expression are the parts added. **3. a)** $\frac{1}{4}$ **b)** $-1$ **c)** $-\frac{3}{5}$ **5. a)** Like terms have the same variables and exponents. **b)** No; the exponent on $x$ is different for each term. **7.** No, 4 does not make the equation true **9.** If $a = b$, then $a + c = b + c$. **11. a)** An infinite number of solutions. **b)** $\mathbb{R}$ **13. a)** Answers will vary. **b)** $-12$ **15.** Symmetric property

**17.** Transitive property **19.** Reflexive property **21.** Addition property of equality **23.** Multiplication property of equality **25.** Multiplication property of equality **27.** Three **29.** Two **31.** Zero **33.** One **35.** Seven **37.** Twelve

**39.** Cannot be simplified **41.** $5x^2 - x - 5$ **43.** $8.7c^2 + 3.6c$ **45.** Cannot be simplified **47.** $-pq + p + q$ **49.** $8d + 2$ **51.** $\frac{8}{3}x + \frac{13}{2}$ **53.** $-17x - 4$ **55.** $11x - 6y$ **57.** $-9b + 93$ **59.** $4r^2 - 2rs + 3r + 4s$ **61.** 3 **63.** $\frac{3}{2}$ **65.** 2 **67.** 16 **69.** 5 **71.** $\frac{3}{5}$ **73.** 1 **75.** 0 **77.** 3 **79.** −1 **81.** 5 **83.** 5 **85.** −1 **87.** $-\frac{1}{2}$ **89.** 6 **91.** 2 **93.** 68 **95.** −64 **97.** −4 **99.** 24 **101.** 10 **103.** −4 **105.** $\frac{15}{16}$ **107.** 5 **109.** 1.00 **111.** 1.18 **113.** 0.43 **115.** 1701.39 **117.** −1.85 **119.** ∅; contradiction **121.** {0}; conditional **123.** ℝ; identity **125.** ℝ; identity **127.** ∅; contradiction **129. a)** ≈85 people per square mile **b)** ≈2026 **131. a)** 58.96% **b)** 2010 **133. a)** ≈2.4 hours **b)** ≈2.08 hours **135.** Answers will vary. One possible answer: $x = \frac{5}{2}, 2x - 4 = 1, 4x = 10$ **137.** Answers will vary. One possible answer: $2x - 4 = 5x - 3(1 + x)$ **139.** Answers will vary. One possible answer: $3p + 3 = \frac{3}{2}p + p + 6$ **141.** −22, substitute −2 for $a$ and solve for $n$. **143.** $\triangle = \frac{\odot + \square}{*}$ **145.** $\odot = \frac{\otimes - \triangle}{\square}$ **147. a)** Answers will vary. **b)** $|a| = \begin{cases} a \text{ if } a \geq 0 \\ -a \text{ if } a < 0 \end{cases}$ **148. a)** −9 **b)** 9 **149.** −5 **150.** $\frac{4}{49}$

## Exercise Set 2.2

**1.** An equation that is a mathematical model of a real-life situation **3.** Understand, translate, carry out, check, answer **5. a)** $l = 5$ **b)** $l = \frac{P - 2w}{2}$ **c)** no **d)** you should obtain the same answer. **7.** 6300 **9.** 300 **11.** 201.06 **13.** 70 **15.** 176 **17.** $\frac{7}{4}$ **19.** 66.67 **21.** 4 **23.** 119.10 **25.** $y = -3x + 5$ **27.** $y = \frac{1}{7}x - \frac{13}{7}$ **29.** $y = 3x - 8$ **31.** $y = \frac{3}{4}x - 5$ **33.** $y = x + 2$ **35.** $y = -\frac{4}{3}x + 11$ **37.** $t = \frac{d}{r}$ **39.** $d = \frac{C}{\pi}$ **41.** $l = \frac{P - 2w}{2}$ **43.** $h = \frac{V}{lw}$ **45.** $r = \frac{A - P}{Pt}$ **47.** $l = \frac{3V}{wh}$ **49.** $m = \frac{y - b}{x}$ **51.** $m = \frac{y - y_1}{x - x_1}$ **53.** $\mu = x - z\sigma$ **55.** $T_2 = \frac{T_1P_2}{P_1}$ **57.** $h = \frac{2A}{b_1 + b_2}$ **59.** $n = \frac{2S}{f + l}$ **61.** $F = \frac{9}{5}C + 32$ **63.** $m_1 = \frac{Fd^2}{km_2}$ **65. a)** $p = 9.11d$ **b)** $d = \frac{p}{9.11}$ **c)** Answers will vary. **67.** \$308 **69.** 6.5 years **71. a)** 3.14 square inches **b)** 78.54 square inches **73. a)** 75 cubic feet **b)** 2.78 cubic yard **c)** \$105 **75.** The cylinder, difference is 0.22 cubic inch **77.** \$11,264.93 **79.** \$4958.41 **81.** ≈4.12% **83. a)** ≈7.08% **b)** ≈6.39% **85. a)** 4 pounds per week **b)** 2500 calories **87. a)** $S = 100 - a$ **b)** 40% **89. a)** $s = \frac{rt^2}{u}$ **b)** $u = \frac{rt^2}{s}$ **90.** −40 **91.** 1 **92.** −125 **93.** $\frac{4}{3}$

## Exercise Set 2.3

**1.** $x - 3$ **3.** $v + 6$ **5.** $d + 2$ **7.** $19.95y$ **9.** $0.096x$ **11.** $x, 12 - x$ **13.** $w, w + 29$ **15.** $p, 165 - p$ **17.** $z, z + 1.3$ **19.** $e, e + 0.22e$ **21.** $A = 72°, B = 18°$ **23.** $A = 36°, B = 144°$ **25.** 40°, 60°, 80°, **27.** \$32 **29.** 25 rides **31.** 225 miles **33.** 13 times **35.** 10 times **37.** \$1600 **39.** Northeast: \$2.145 million; Southeast: \$2.455 million **41.** \$8845.48 **43.** \$3.10 per hour **45.** grasses: 12, weeds: 19, trees: 26 **47.** \$16.25 **49. a)** ≈63.49 months or 5.29 years **b)** First National **51.** ≈28 months or 2.33 years **53.** U.S.: 103, China: 63, Russia: 92, Australia: 49, Germany: 48 **55.** animals: 250,000, plants: 350,000, nonbeetle insects: 540,000, beetles: 360,000 **57.** 9 inches, 12 inches, 15 inches **59.** 10 feet, 24 feet, 26 feet **61.** 13 meters by 13 meters **63.** 3 feet by 6 feet **65.** \$ 60 **67.** 3 **69.** \$ 16 **71. a)** $\frac{88 + 92 + 97 + 96 + x}{5} = 90$ **b)** Answer will vary. **c)** 77 **73. a), b)** Answers will vary. **75.** 220 miles **78.** $\frac{13}{5}$ **79.** −2.7 **80.** $\frac{5}{32}$ **81.** −10 **82.** $\frac{y^{18}}{8x^{12}}$

## Mid-Chapter Test

**1.** 12 [2.1] **2.** $5x^2 - 2x - 11$ [2.1] **3.** $6.4a - 9.6$ [2.1] **4.** −6 [2.1] **5.** 14 [2.1] **6.** $-\frac{11}{3}$ [2.1] **7.** 5 [2.1] **8.** ℝ, identity [2.1] **9.** ∅, contradiction [2.1] **10.** 80 [2.2] **11.** $\frac{100}{3}$ [2.2] **12.** $x = \frac{y - 13}{7}$ [2.2] **13.** $x_3 = nA - 2x_1 - x_2$ [2.2] **14.** \$942.80 [2.2] **15.** $A = 62°, B = 28°$ [2.3] **16.** 10 days [2.3] **17.** 15 feet, 25 feet, 60 feet [2.3] **18.** 4.5% [2.3] **19.** 40 months [2.3] **20.** Multiply both sides by the same number, 12; $-\frac{10}{3}$ [2.3]

## Exercise Set 2.4

**1.** 11.4 miles **3.** 4 hours **5.** 6 hours **7. a)** 6 miles per hour **b)** 12 miles per hour **9. a)** 0.15 hour or 9 minutes **b)** 3.6 miles **11.** 13.8 hours **13.** ≈0.58 hour or 35 minutes **15.** \$12,570 at 3%, \$17,430 at 4.1% **17.** 54 pounds **19. a)** 2200 shares of Johnson & Johnson and 4400 shares of AOL. **b)** \$4480 **21.** 30 ounces **23.** 2.8 teaspoons 30%, 1.2 teaspoons 80% **25.** 35% **27.** 4 pounds leaves, 8 pounds slices **29.** ≈25.77 hours **31.** 500 minutes or $8\frac{1}{3}$ hours **33.** 6 quarts **35. a)** ≈3.71 hours **b)** ≈2971.43 miles **37.** 8 small, 4 large paintings **39.** 9.6 ounces of 80% solution, 118.4 ounces of water

**41.** ≈35.6 ounces of sirloin, ≈28.4 ounces of veal **43.** 3 miles **45.** ≈11.4 ounces **47. a), b), c)** Answers will vary. **49.** ≈149 miles **51.** 6 quarts **52.** $7.0 \times 10^{12}$ **53.** −5.7 **54.** $\frac{21}{4}$ **55.** $y = \frac{x - 42}{30}$ **56.** 140 miles

**Exercise Set 2.5** **1.** It is necessary to reverse the direction of the inequality symbol when multiplying or dividing both sides of the inequality by a negative number **3. a)** When the endpoints are not included **b)** When the endpoints are included **c)** Answers will vary. One example is $x > 4$. **d)** Answers will vary. One example is $x \geq 4$. **5.** $a < x$ and $x < b$

**7. a)** 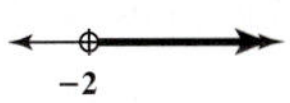 **b)** $(-2, \infty)$ **c)** $\{x|x > -2\}$ **9. a)**  **b)** $(-\infty, \pi]$ **c)** $\{w|w \leq \pi\}$ **11. a)** 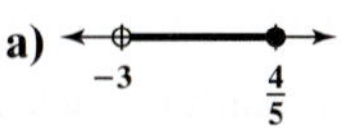 **b)** $\left(-3, \frac{4}{5}\right]$ **c)** $\left\{q \middle| -3 < q \leq \frac{4}{5}\right\}$ **13. a)**  **b)** $(-7, -4]$ **c)** $\{x|-7 < x \leq -4\}$ **15.** 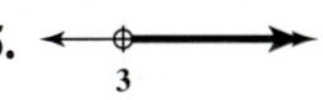

**17.**  **19.**  **21.** 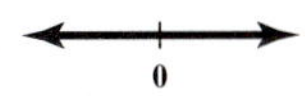 **23.** 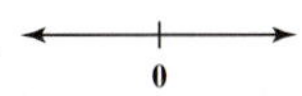 **25.** 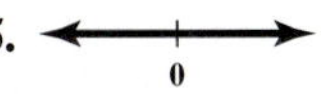 **27.** $\left(-\infty, \frac{3}{2}\right)$ **29.** $[2, \infty)$

**31.** $\left(-\infty, \frac{3}{2}\right]$ **33.** $(-\infty, \infty)$ **35.** $[-5, 1)$ **37.** $[-4, 5]$ **39.** $\left[4, \frac{11}{2}\right)$ **41.** $\left(-\frac{13}{3}, -4\right]$ **43.** $\{x|3 \leq x < 7\}$ **45.** $\{x|0 < x \leq 3\}$

**47.** $\left\{u \middle| 4 \leq u \leq \frac{19}{3}\right\}$ **49.** $\{c|-3 < c \leq 1\}$ **51.** ∅ **53.** $\{x|-5 < x < 2\}$ **55.** $(-\infty, 2) \cup [7, \infty)$ **57.** $[0, 2]$

**59.** $(-\infty, 0) \cup (6, \infty)$ **61.** $[0, \infty)$ **63. a)** $l + g \leq 130$ **b)** $l + 2w + 2d \leq 130$ **c)** 24.5 inches **65.** 11 boxes

**67.** 77 minutes **69.** 1881 books **71.** 41 ounces **73.** For sales over $5,000 per week **75.** 24 **77.** $76 \leq x \leq 100$

**79. a)** $12,885.25 **b)** $79,998.39 **81. a)** $[0, 3]$ **b)** $[3, 10]$ **83. a)** $[0, 5]$ **b)** $[5, 13]$ **85. a)** $[0, 8]$ **b)** None **87.** $6.97 < x < 8.77$

**89. a)** January, February, March, May **b)** March, April, May **c)** April **91.** Answers will vary. **93. a)** $[17.5, 23.5]$ **b)** $[23.5, 31]$ **c)** $[27.2, 36.5]$ **95.** $84 \leq x \leq 100$ **97. a)** Answers will vary. **b)** $(-3, \infty)$ **99. a)** $A \cup B = \{1, 2, 3, 4, 5, 6, 8, 9\}$ **b)** $A \cap B = \{1, 8\}$

**100. a)** 4 **b)** 0, 4 **c)** $-3, 4, \frac{5}{2}, 0, -\frac{13}{29}$ **d)** $-3, 4, \frac{5}{2}, \sqrt{7}, 0, -\frac{13}{29}$ **101.** Associative property of addition

**102.** Commutative property of addition **103.** $V = \frac{R - L + Dr}{r}$

**Exercise Set 2.6** **1.** Set $x = a$ or $x = -a$ **3.** $-a < x < a$ **5.** $x < -a$ or $x > a$ **7.** All real numbers except for 0; the absolute value of every real number except 0 is greater than 0. **9.** Set $x = y$ or $x = -y$ **11. a)** Two **b)** Infinite number **c)** Infinite number **13. a)** D **b)** B **c)** E **d)** C **e)** A **15.** $\{-2, 2\}$ **17.** $\left\{-\frac{1}{2}, \frac{1}{2}\right\}$ **19.** ∅ **21.** $\{-13, 3\}$ **23.** $\{-7\}$ **25.** $\left\{\frac{3}{2}, \frac{11}{6}\right\}$

**27.** $\{-17, 23\}$ **29.** $\{3\}$ **31.** $\{w|-11 < w < 11\}$ **33.** $\{q|-13 \leq q \leq 3\}$ **35.** $\{b|1 < b < 5\}$ **37.** $\{x|-9 \leq x \leq 6\}$

**39.** $\left\{x \middle| \frac{1}{3} < x < \frac{13}{3}\right\}$ **41.** ∅ **43.** $\{j|-22 < j < 6\}$ **45.** $\{x|-1 \leq x \leq 7\}$ **47.** $\{y|y < -2$ or $y > 2\}$ **49.** $\{x|x < -9$ or $x > 1\}$

**51.** $\left\{b \middle| b < \frac{2}{3} \text{ or } b > 4\right\}$ **53.** $\{h|h < 1$ or $h > 4\}$ **55.** $\{x|x < 2$ or $x > 6\}$ **57.** $\{x|x \leq -18$ or $x \geq 2\}$ **59.** ℝ

**61.** $\{x|x < 2$ or $x > 2\}$ **63.** $\{-1, 15\}$ **65.** $\{-3, 1\}$ **67.** $\left\{-23, \frac{13}{7}\right\}$ **69.** $\{10\}$ **71.** $\{-1, 1\}$ **73.** $\{q|q < -8$ or $q > -4\}$

**75.** $\{w|-1 \leq w \leq 8\}$ **77.** $\left\{-\frac{8}{5}, 2\right\}$ **79.** $\left\{x \middle| x < -\frac{5}{2} \text{ or } x > -\frac{5}{2}\right\}$ **81.** $\left\{x \middle| -\frac{13}{3} \leq x \leq \frac{5}{3}\right\}$ **83.** ∅ **85.** $\{w|-16 < w < 8\}$

**87.** ℝ **89.** $\left\{2, \frac{22}{3}\right\}$ **91.** $\left\{-\frac{3}{2}, \frac{9}{7}\right\}$ **93. a)** $[0.085, 0.093]$ **b)** 0.085 inch **c)** 0.093 inch **95. a)** $[132, 188]$ **b)** 132 to 188 feet below sea level, inclusive **97.** $|x| = 5$ **99.** $|x| \geq 5$ **101.** $x = -\frac{b}{a}$; $|ax + b|$ is never less than 0, so set $|ax + b| = 0$ and solve for $x$.

**103. a)** Set $ax + b = -c$ or $ax + b = c$ and solve each equation for $x$. **b)** $x = \frac{-c - b}{a}$ or $x = \frac{c - b}{a}$ **105. a)** Write $ax + b < -c$ or $ax + b > c$ and solve each inequality for $x$. **b)** $x < \frac{-c - b}{a}$ or $x > \frac{c - b}{a}$ **107.** ℝ; Since $3 - x = -(x - 3)$

**109.** $\{x|x \geq 0\}$; by definition of absolute value **111.** $\{2\}$; set $x + 1 = 2x - 1$ or $x + 1 = -(2x - 1)$

**113.** $\{x|x \leq 4\}$; by definition $|x - 4| = -(x - 4)$ if $x \leq 4$ **115.** $\{4\}$ **117.** ∅ **119.** $\frac{29}{72}$ **120.** 25 **121.** ≈1.33 miles **122.** $\{x|x < 4\}$

## Chapter 2 Review Exercises

**1.** Eight **2.** One **3.** Seven **4.** $a^2 - a + 4$ **5.** $7x^2 + 2xy - 13$ **6.** Cannot be simplified **7.** $4x - 3y + 10$ **8.** $-4$ **9.** 20 **10.** $-\frac{13}{3}$ **11.** $-10$ **12.** $-\frac{9}{2}$ **13.** No solution **14.** $\mathbb{R}$ **15.** $-\frac{1}{2}$ **16.** $\frac{1}{4}$ **17.** 69 **18.** $-4$ **19.** $R = \frac{E}{I}$ **20.** $w = \frac{P - 2l}{2}$ **21.** $h = \frac{A}{\pi r^2}$ **22.** $h = \frac{2A}{b}$ **23.** $m = \frac{y - b}{x}$ **24.** $y = \frac{2x - 5}{3}$ **25.** $R_2 = R_T - R_1 - R_3$ **26.** $a = \frac{2S - b}{3}$ **27.** $l = \frac{K - 2d}{2}$ **28.** \$30 **29.** 7 years **30.** \$6800 **31.** 150 miles **32.** \$260 **33.** \$2570 at 3.5%, \$2430 at 4% **34.** 187.5 gallons of 20%, 62.5 gallons of 60% **35.** $6\frac{1}{2}$ hours **36. a)** 3000 miles per hour **b)** 16,500 miles **37.** 15 pounds of \$6.00 coffee; 25 pounds of \$6.80 coffee **38.** \$36 **39. a)** 1 hour **b)** 14.4 miles **40.** 40°; 65°; 75° **41.** 300 gallons per hour; 450 gallons per hour **42.** 40°, 50° **43.** 7.5 ounces **44.** \$4500 at 10%; \$7500 at 6% **45.** More than 5 **46.** 40 miles per hour, 50 miles per hour

**47.** 

**48.** 

**49.** 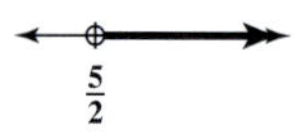

**50.** 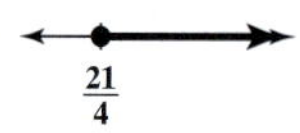

**51.** 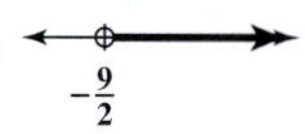

**52.** 

**53.** 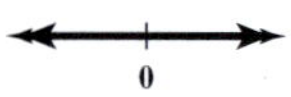

**54.** 

**55.** 6 boxes **56.** 7 minutes **57.** ≈15.67 weeks **58.** $\{x|81 \le x \le 100\}$ **59.** $(5, 11)$

**60.** $(-3, 5]$ **61.** $\left(\frac{7}{2}, 8\right)$ **62.** $\left(\frac{8}{3}, 6\right)$ **63.** $(-3, 1]$ **64.** $(2, 14)$ **65.** $\{h|-3 < h \le 1\}$ **66.** $\mathbb{R}$ **67.** $\{x|x \le -4\}$

**68.** $\{g|g < -6 \text{ or } g \ge 11\}$ **69.** $\{-2, 2\}$ **70.** $\{x|-8 < x < 8\}$ **71.** $\{x|x \le -9 \text{ or } x \ge 9\}$ **72.** $\{-18, 8\}$

**73.** $\{x|x \le -3 \text{ or } x \ge 7\}$ **74.** $\left\{-\frac{1}{2}, \frac{9}{2}\right\}$ **75.** $\{q|1 < q < 8\}$ **76.** $\{-1, 4\}$ **77.** $\{x|-14 < x < 22\}$ **78.** $\left\{-5, -\frac{4}{5}\right\}$ **79.** $\mathbb{R}$

**80.** $\left[-5, -\frac{1}{3}\right]$ **81.** $(4, 8]$ **82.** $\left(-\frac{17}{2}, \frac{27}{2}\right]$ **83.** $[-2, 6)$ **84.** $(-\infty, \infty)$ **85.** $\left(\frac{2}{3}, 10\right]$

## Chapter 2 Practice Test

**1.** Seven [2.1] **2.** $16p - 3q - 4pq$ [2.1] **3.** $10q + 42$ [2.1] **4.** $-26$ [2.1] **5.** $\frac{4}{3}$ [2.1] **6.** $-\frac{35}{11}$ [2.1] **7.** $\varnothing$ [2.1] **8.** $\mathbb{R}$ [2.1] **9.** $\frac{13}{3}$ [2.2] **10.** $b = \frac{a - 2c}{5}$ [2.2] **11.** $b_2 = \frac{2A - hb_1}{h}$ [2.2] **12.** \$625 [2.3–2.4] **13.** 80 visits [2.3–2.4] **14.** 4.2 hours [2.3–2.4] **15.** 6.25 liters [2.3–2.4] **16.** \$7000 at 8%; \$5000 at 7% [2.3–2.4]

**17.** 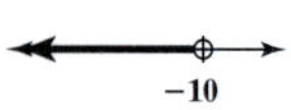

[2.5] **18.** 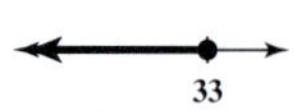

[2.5] **19.** $\left(\frac{9}{2}, 7\right]$ [2.5] **20.** $[13, 16)$ [2.5]

**21.** $\{-7, 2\}$ [2.6] **22.** $\left\{-\frac{14}{3}, \frac{26}{5}\right\}$ [2.6] **23.** $\{-3\}$ [2.6] **24.** $\{x|x < -1 \text{ or } x > 4\}$ [2.6] **25.** $\left\{x\middle|\frac{1}{2} \le x \le \frac{5}{2}\right\}$ [2.6]

## Cumulative Review Test

**1. a)** $\{1, 2, 3, 5, 7, 9, 11, 13, 15\}$ **b)** $\{3, 5, 7, 11, 13\}$ [1.2] **2. a)** Commutative property of addition **b)** Associative property of multiplication **c)** Distributive property [1.3] **3.** $-63$ [1.4] **4.** $-6$ [1.4] **5.** 7 [1.4] **6.** $\frac{1}{25x^8y^6}$ [1.5] **7.** $\frac{16m^{10}}{n^{12}}$ [1.5] **8.** ≈545.8 times [1.6] **9.** 5 [2.1] **10.** 1.15 [2.1] **11.** $\frac{3}{4}$ [2.1] **12.** A conditional linear equation is true for only one value; a linear equation that is an identity is always true; a linear equation that is a contradiction is never true. [2.1] **13.** 3 [2.2] **14.** $x = \frac{y - y_1 + mx_1}{m}$ [2.2] **15. a)** 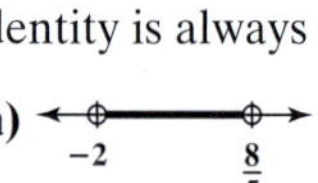

**b)** $\left\{x\middle|-2 < x < \frac{8}{5}\right\}$ **c)** $\left(-2, \frac{8}{5}\right)$ [2.5] **16.** $\left\{-\frac{7}{3}, 3\right\}$ [2.6]
**17.** $\{x|x \le -10 \text{ or } x \ge 14\}$ [2.6] **18.** \$35 [2.3] **19.** 40 miles per hour, 60 miles per hour [2.4]
**20.** Cashews: 15 pounds; peanuts: 25 pounds [2.4]

# Chapter 3

## Exercise Set 3.1

**1. a)** A straight line **b)** Two; Two points uniquely determine a straight line.

**3.** They are in a straight line. **5.** $A(3, 1)$, $B(-6, 0)$, $C(2, -4)$, $D(-2, -4)$, $E(0, 3)$, $F(-8, 1)$, $G\left(\frac{3}{2}, -1\right)$ **7.** 

**9.** I **11.** IV **13.** II **15.** III **17.** No **19.** No **21.** Yes **23.** Yes **25.** No **27.**

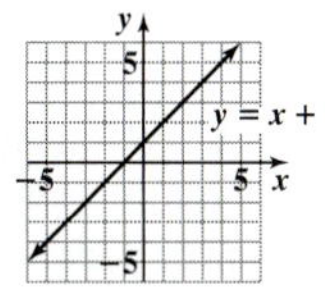


**29.**

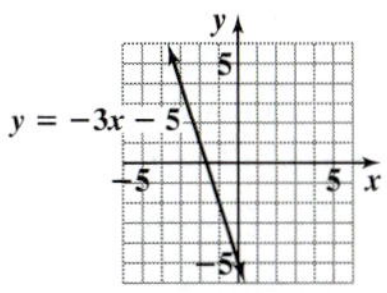


**31.**

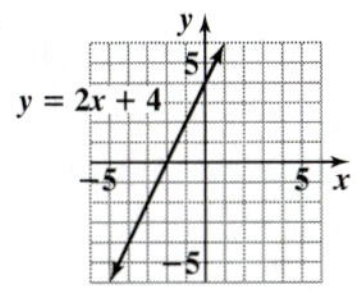


**33.**

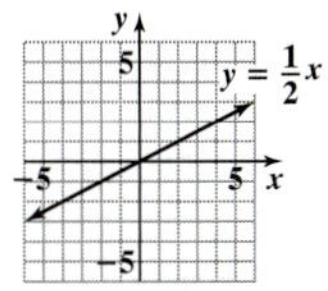


**35.**

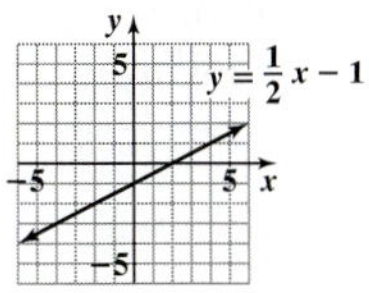


**37.**

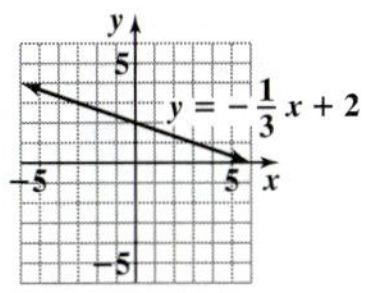


**39.**

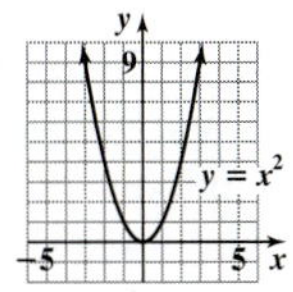


**41.**

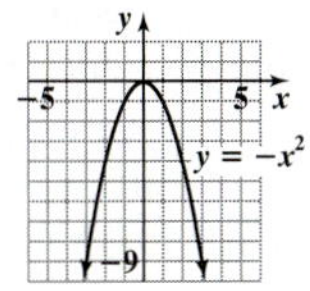


**43.**

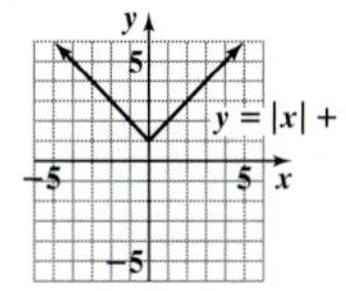


**45.**

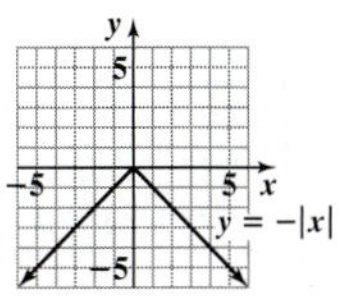


**47.**

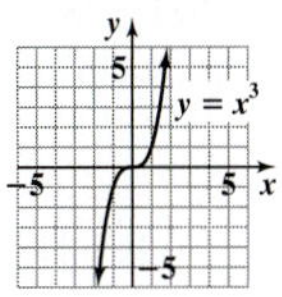


**49.**

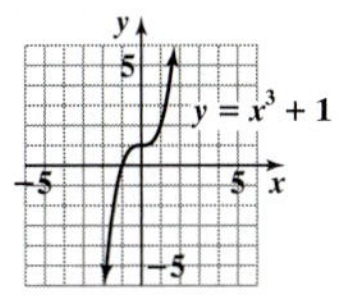


**51.**

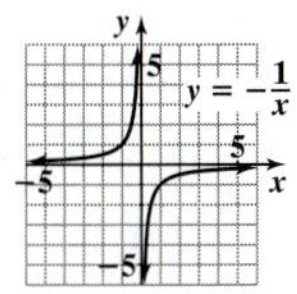


**53.**

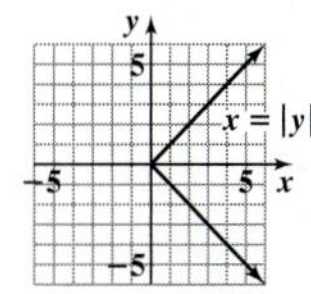


**55.**

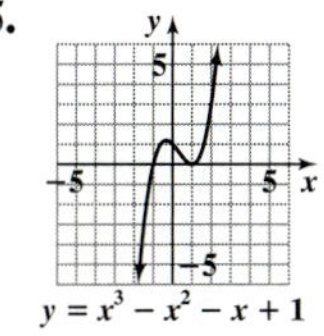


**57.**

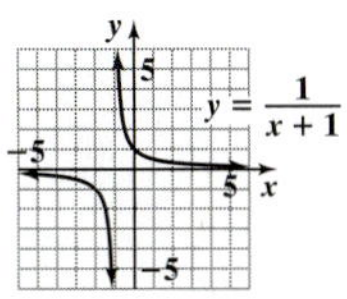


**59.**

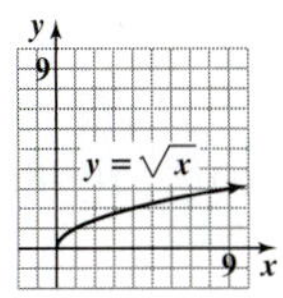


**61.**

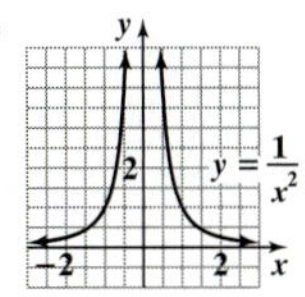


**63.** Yes, the coordinates satisfy the equation

**65. a)**

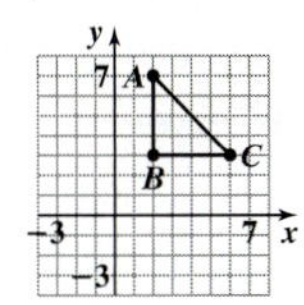

**b)** 8 square units

**67. a)** 6975 yards **b)** 7300 yards **c)** 1990, 2000, 2005 **d)** no

**69.**

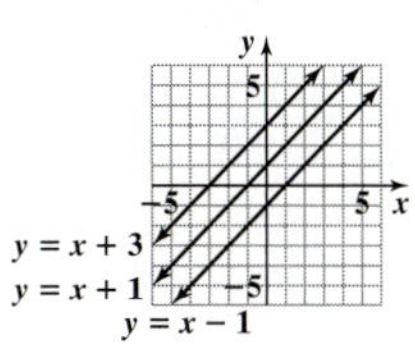


**a)** Each graph crosses the $y$-axis at the point corresponding to the constant term in the graph's equation. **b)** Yes **71.** The rate of change is 2.

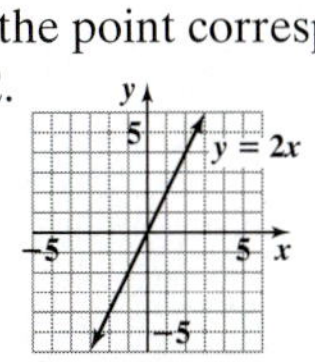


**73.** The rate of change is 3.

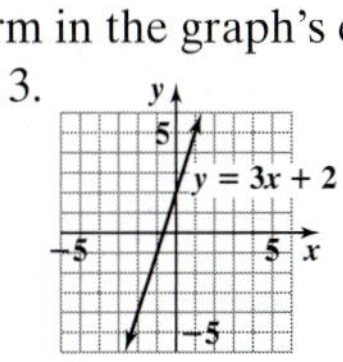


**75.** (4, −3), (5, 1), other answers possible **77.** c **79.** a **81.** d **83.** b **85.** b **87.** d **89.** b **91.** d

**93. a)**

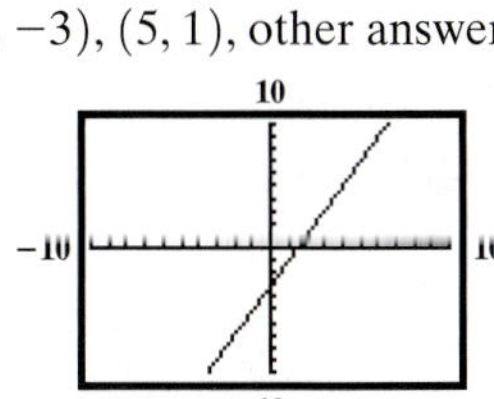

**b)**

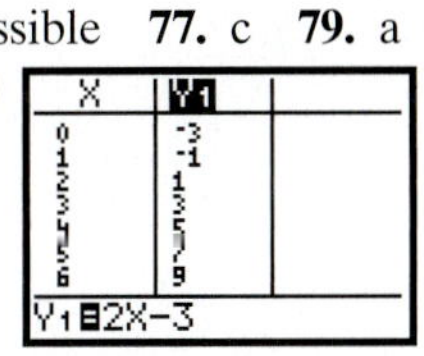


**95. a)**

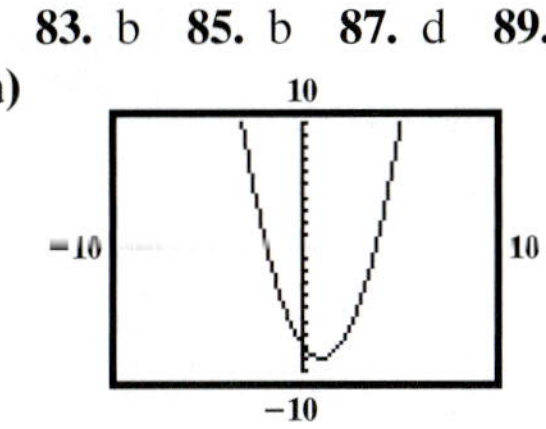

**b)**

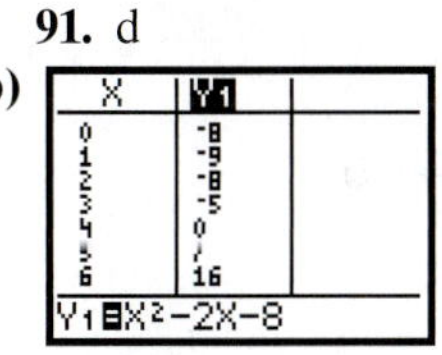


**97. a)**

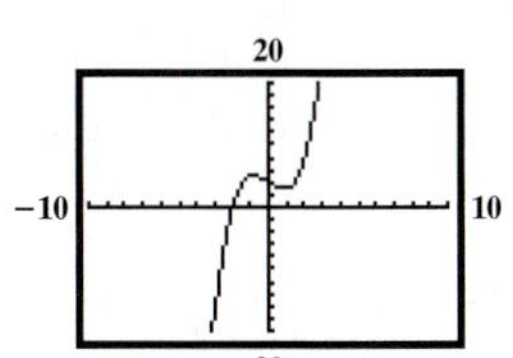

**b)**

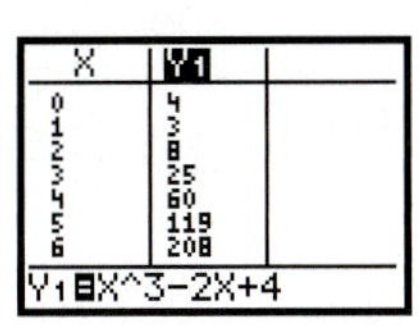


**99.**

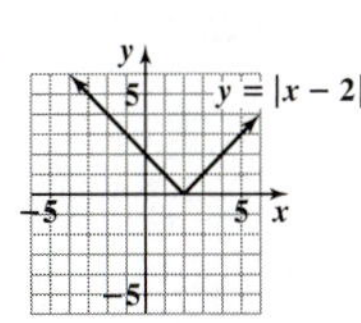


**103.** $\frac{3}{2}$ **104.** ≈71 miles **105.** $\{x \mid -2 < x \le 2\}$ **106.** $\left\{x \,\middle|\, x < -3 \text{ or } x > \frac{5}{3}\right\}$

## Exercise Set 3.2

**1.** A correspondence where each member in the domain corresponds to exactly one member of the range **3.** Yes, a relation is any set of ordered pairs. **5.** If a vertical line drawn through any part of the graph intersects the graph at more than one point, the graph is not a function. **7.** The set of values for the dependent variable **9.** Domain: $\{x \mid x \ne 0\}$, Range: $\{y \mid y \ne 0\}$; $x$ cannot be 0 because you cannot divide by 0. $y$ cannot be 0 because the numerator is 1. **11.** Domain: ℝ, Range: $\{y \mid y \ge 0\}$; $x$ can be any real number, $|x|$ can never be negative **13.** If $y$ depends on $x$, then $x$ is the independent variable. **15. a)** Function **b)** Domain: {3, 5, 11}, Real: {6, 10, 22} **17. a)** Function **b)** Domain: {Cameron, Tyrone, Vishnu}, Range: {3, 6} **19. a)** Not a function **b)** Domain: {1990, 2001, 2002}; Range: {20, 34, 37} **21. a)** Function **b)** Domain: {1, 2, 3, 4, 5}; Range: {1, 2, 3, 4, 5} **23. a)** Function **b)** Domain: {1, 2, 3, 4, 5, 7}; Range: {−1, 0, 2, 4, 9} **25. a)** Not a function **b)** Domain: {1, 2, 3}; Range: {1, 2, 4, 5, 6} **27. a)** Not a function **b)** Domain: {0, 1, 2}; Range: {−7, −1, 2, 3} **29. a)** Function **b)** Domain: ℝ; Range: ℝ **c)** 2 **31. a)** Not a function **b)** Domain: $\{x \mid 0 \le x \le 2\}$, Range: $\{y \mid -3 \le y \le 3\}$ **c)** ≈1.5 **33. a)** Function **b)** Domain: ℝ, Range: $\{y \mid y \ge 0\}$ **c)** −3, −1 **35. a)** Function **b)** Domain: {−1, 0, 1, 2, 3}, Range: {−1, 0, 1, 2, 3} **c)** 2 **37. a)** Not a function **b)** Domain: $\{x \mid x \ge 2\}$, Range: ℝ **c)** 3 **39. a)** Function **b)** Domain: $\{x \mid -2 \le x \le 2\}$, Range: $\{y \mid -1 \le y \le 2\}$ **c)** −2, 2 **41. a)** 3 **b)** 13 **43. a)** −6 **b)** −4 **45. a)** 2 **b)** 2 **47. a)** 7 **b)** 0 **49. a)** 0 **b)** 3 **51. a)** 1 **b)** Undefined **53. a)** 24 square feet **b)** 39 square feet **55. a)** $A(r) = \pi r^2$ **b)** ≈452.4 square yards

**57. a)** $C(F) = \frac{5}{9}(F - 32)$ **b)** $-35°\text{C}$ **59. a)** 18.23°C **b)** 27.68°C **61. a)** 78.32° **b)** 73.04° **63. a)** 91 oranges **b)** 204 oranges

**65.** Answers will vary. One possible interpretation: The person warms up slowly, possibly by walking, for 5 minutes. Then the person begins jogging slowly over a period of 5 minutes. For the next 15 minutes the person jogs. For the next 5 minutes the person walks slowly and his heart rate decreases to his normal heart rate. The rate stays the same for the next 5 minutes. **67.** Answers will vary. One possible interpretation: The man walks on level ground, about 30 feet above sea level, for 5 minutes. For the next 5 minutes he walks uphill to 45 feet above sea level. For 5 minutes he walks on level ground, then walks quickly downhill for 3 minutes to an elevation of 20 feet above sea level. For 7 minutes he walks on level ground. Then he walks quickly uphill for 5 minutes. **69.** Answers will vary. One possible interpretation: Driver is in stop-and-go traffic, then gets on highway for about 15 minutes then stops car for a couple of minutes, then stop-and-go traffic. **71. a)** Yes **b)** Year **c)** \$218,600 **d)** \$865,000 **e)** ≈144.8% **73. a)** Yes **b)** ≈6.0 million **c)** ≈4.4 million **d)** Yes **e)** 2006 to 2007

**75. a)**

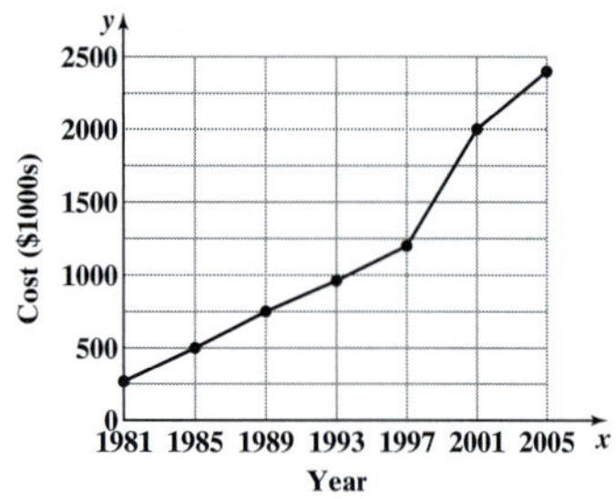


**b)** No. It is not a straight line. **c)** \$2,300,000

**77. a)**

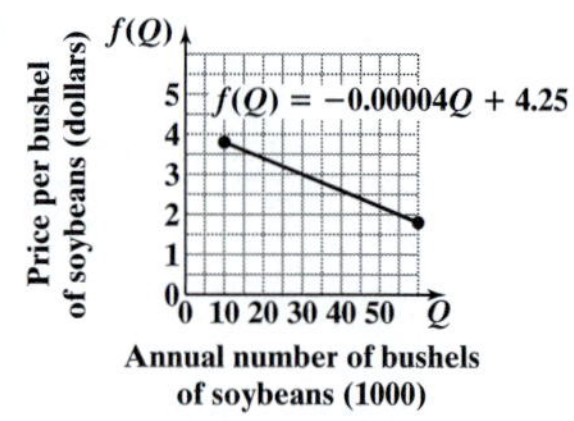


**b)** \$2.65 per bushel **79.**

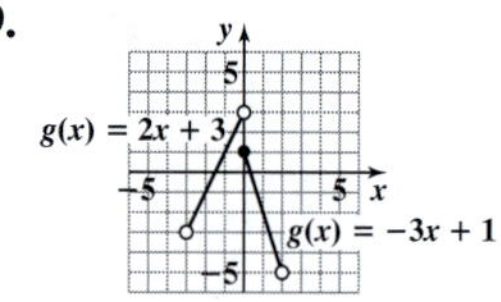


**80.** $\frac{1}{2}$ **81.** $p_2 = \frac{E - a_1p_1 - a_3p_3}{a_2}$ **82. a)**

**b)** $(3, \infty)$ **c)** $\{x|x > 3\}$ **83.** $-2, 10$

## Exercise Set 3.3

**1.** $ax + by = c$ **3.** To find the $x$-intercept, set $y = 0$ and solve for $x$. To find the $y$-intercept, set $x = 0$ and solve for $y$. **5.** Vertical line **7.** Horizontal line **9.** Graph both sides of the equation. The solution is the $x$-coordinate of the point of intersection. **11.** $2x + y = 5$ **13.** $3x - 4y = -14$ **15.**

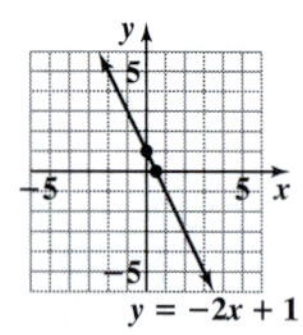


**17.**

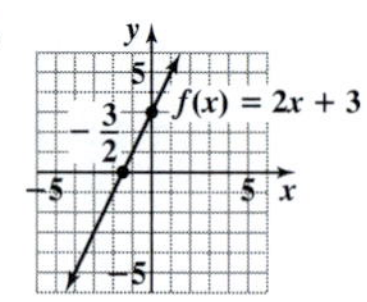


**19.**

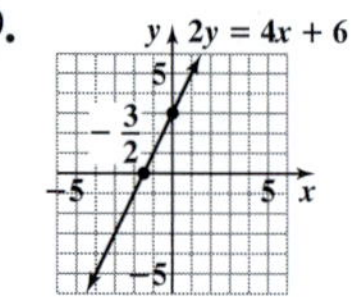


**21.**

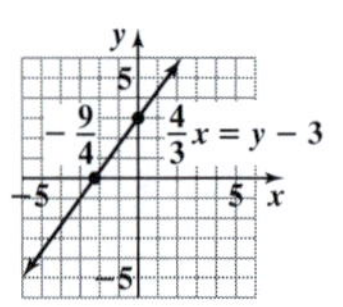


**23.**

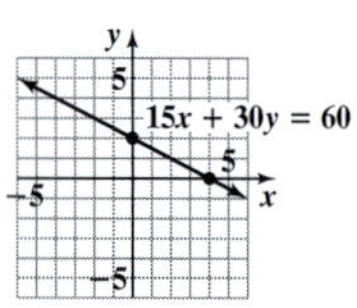


**25.**

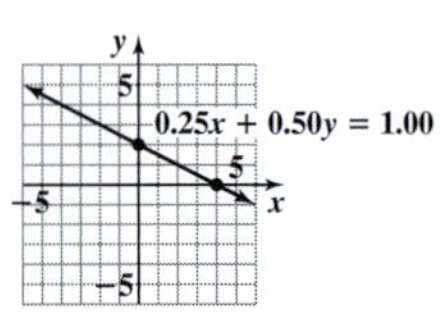


**27.**

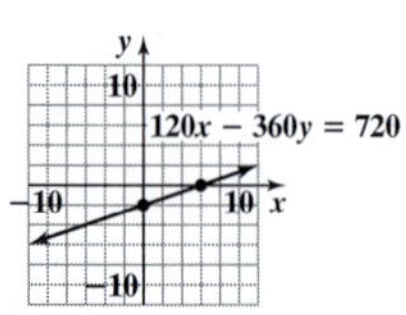


**29.**

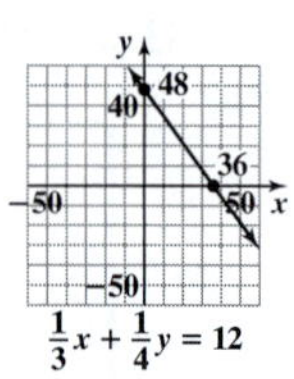


**31.**

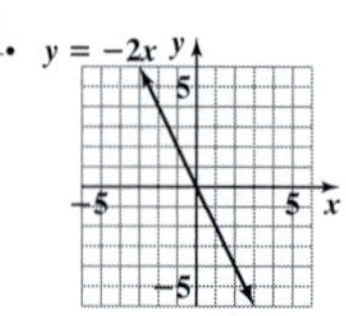


**33.**

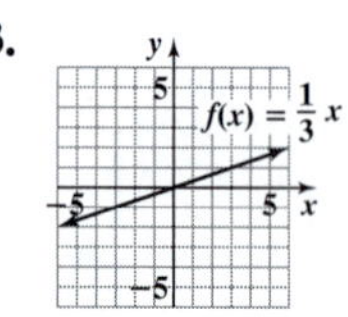


**35.**

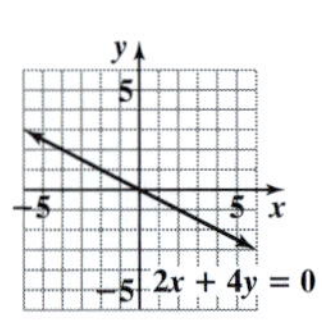


**37.**

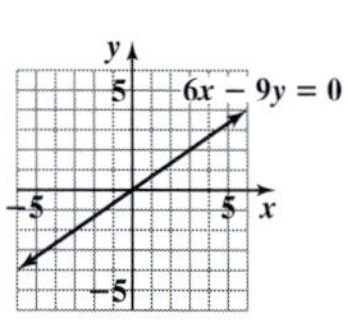


**39.**

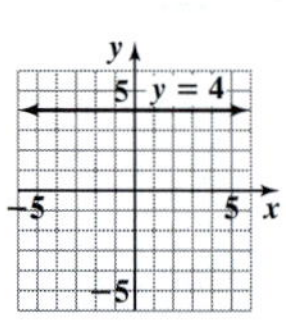


**41.**

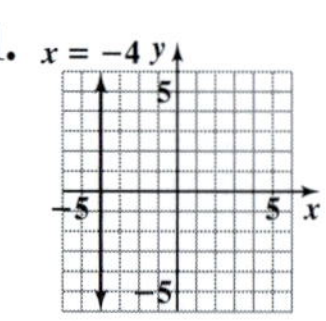


**43.**

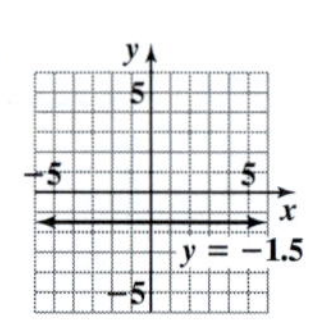


**45.**

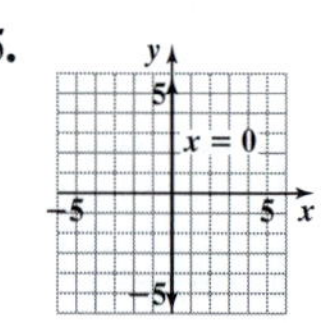


**47.**

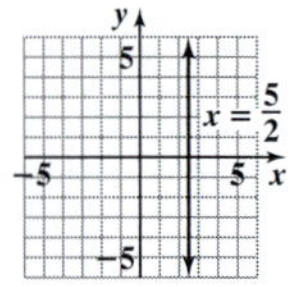


**49.**

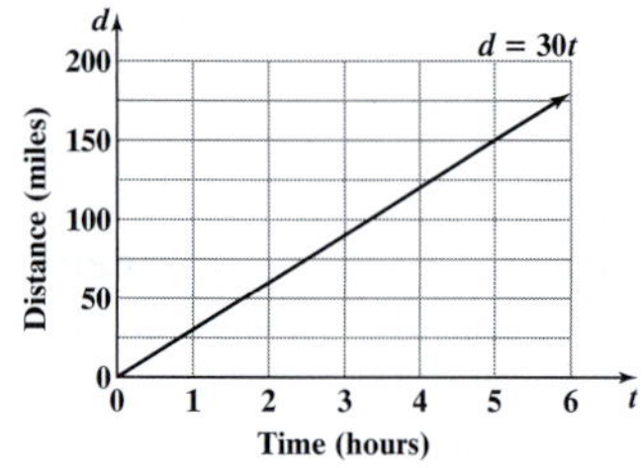


**51. a)**

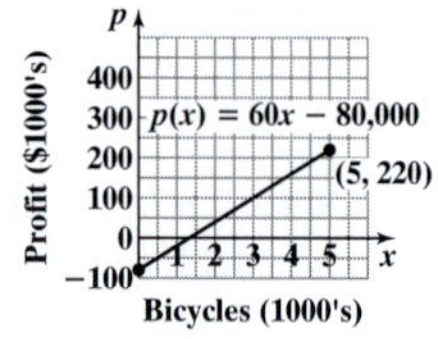


**b)** 1300 bicycles **c)** 3800 bicycles

**53. a)** $s(x) = 500 + 0.15x$ **b)**

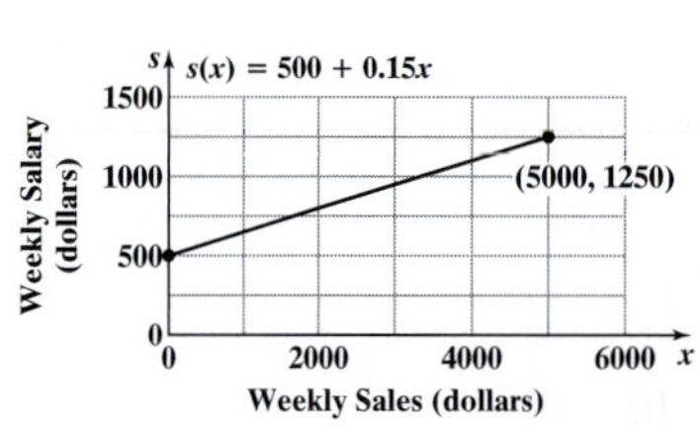


**c)** \$950 **d)** \$4000 **55. a)** There is only one $y$-value for each $x$-value. **b)** Independent: length; Dependent: weight **c)** Yes **d)** 11.5 kilograms **e)** 65 centimeters **f)** 12.0–15.5 kilograms **g)** Increases; yes, as babies get older their weights vary more. **57.** When the graph goes through the origin, because at the origin both $x$ and $y$ are zero.

**59.** Answers will vary. One possible answer is $f(x) = 4$. **61.** Both intercepts will be at 0.

**63. a)** 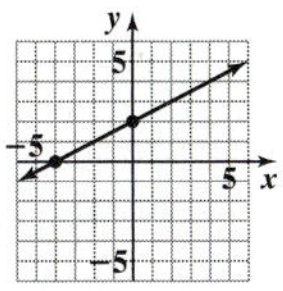

**b)** 2(or $-2$) units **c)** 4 (or $-4$) units **d)** $\frac{1}{2}$; slope

**65.** 1 **67.** $-3$ **69.** $(-3.2, 0), (0, 6.4)$ **71.** $(-2, 0), (0, -2.5)$ **73.** 96 **74.** $-\frac{18}{13}$
**75. a)** Answers will vary. **b)** $x = a + b$ or $x = a - b$ **76. a)** Answers will vary. **b)** $a - b < x < a + b$
**77. a)** Answers will vary. **b)** $x < a - b$ or $x > a + b$ **78.** $\{-2, 2\}$

## Exercise Set 3.4

**1.** Select two points on line; find $\frac{\Delta y}{\Delta x}$. **3.** The line rises going from left to right. **5.** The change in $x$ is 0, and we cannot divide by zero. **7.** Solve for $y$. **9. a)** Moved down 4 units **b)** $(0, -8)$ **11.** The change in $y$ for a unit change in $x$ **13.** $-2$ **15.** $-\frac{1}{2}$ **17.** $-1$ **19.** Undefined **21.** 0 **23.** $-\frac{2}{3}$ **25.** $b = 3$ **27.** $k = -2$ **29.** $x = 6$ **31.** $r = 0$
**33.** $m = -3, y = -3x$ **35.** $m = -\frac{1}{3}, y = -\frac{1}{3}x + 2$ **37.** $m$ is undefined, $x = -2$ **39.** $m = 0, y = 3$
**41.** $m = -\frac{3}{2}, y = -\frac{3}{2}x + 15$ **43.** $y = -x + 2, -1, (0, 2)$ 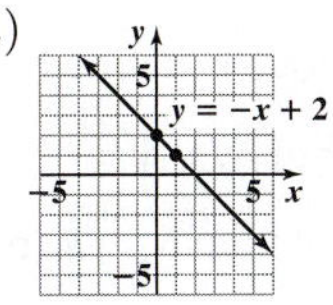


**45.** $y = -\frac{1}{3}x + 2, -\frac{1}{3}, (0, 2)$ 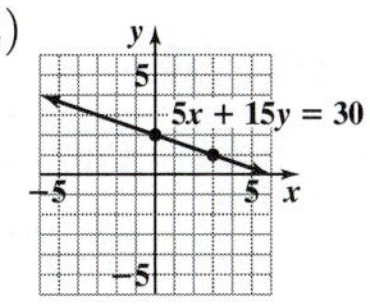


**47.** $y = \frac{5}{2}x + 2, \frac{5}{2}, (0, 2)$ 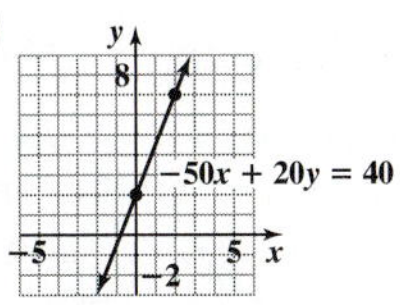


**49.** 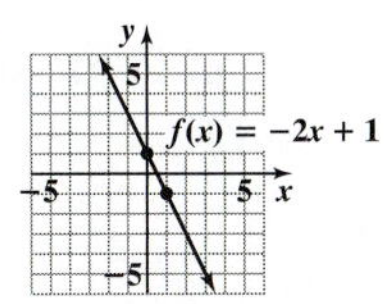


**51.** 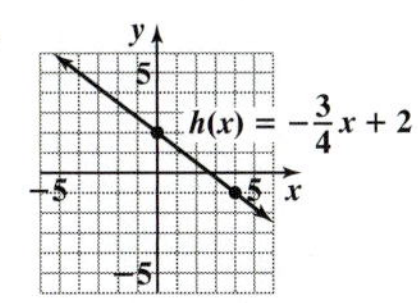


**53. a)** 2 **b)** 4 **c)** 1 **d)** 3 **55.** If the slopes are the same and the $y$-intercepts are different, the lines are parallel. **57.** $(0, -5)$

**59. a)** $y = 3x + 1$ **b)** $y = 3x - 5$ **61. a)** 1 **b)** $(0, 4)$ **c)** $y = x + 4$ **63.** $y = \frac{3}{2}x - 7$ **65.** 0.2 **67. a)** 11.3 **b)** Positive **c)** 7.075 **69. a–b)** 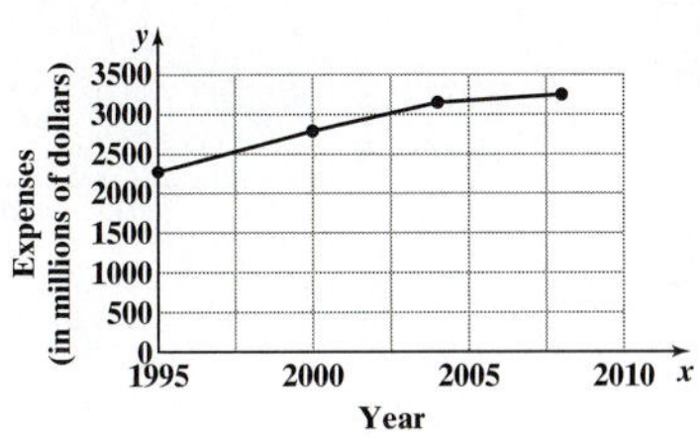


**c)** 123.8, 64.25, 31.75 **d)** 1995–2000, because its line segment has the greatest slope **71. a)** $h(x) = -x + 200$ **b)** 186 beats per minute
**73. a)** $M(t) \approx 19.34t + 159.5$ **b)** \$275.54 billion **c)** \$410.92 billion **d)** 2006
**75. a)** $P(t) = -1.1t + 19.4$ **b)** Negative **c)** 17.2% **d)** 9.5%
**77. a)** $P(t) = 8300t + 110{,}500$ **b)** \$152,000 **c)** \$235,000 **d)** 2005
**79.** The $y$-intercept is wrong. **81.** The slope is wrong. **83.** Height: 14.2 inches, width: 6.4 inches **86.** 19 **87.** 5 **88.** 2.4 **89.** First: 75 miles per hour; second: 60 miles per hour **90. a)** $x < -3$ or $x > 2$ **b)** $-3 < x < 2$

## Mid-Chapter Test

**1.** III [3.1] **2.** 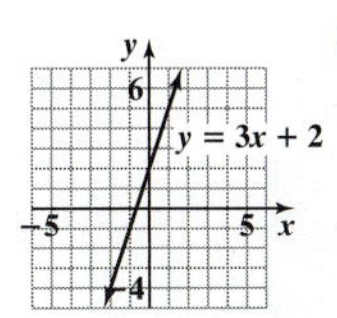

[3.1] **3.** 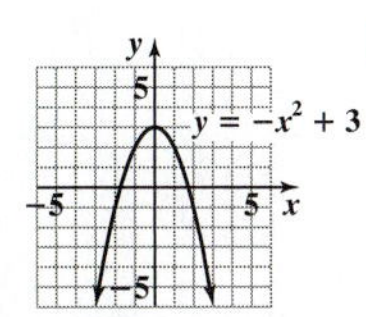

[3.1] **4.** 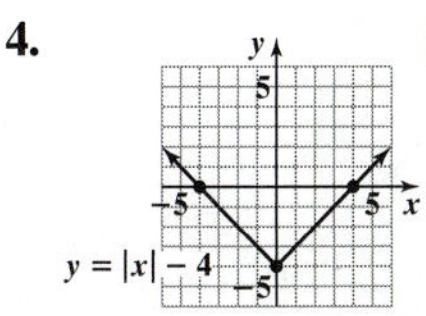

[3.1]

**5.** 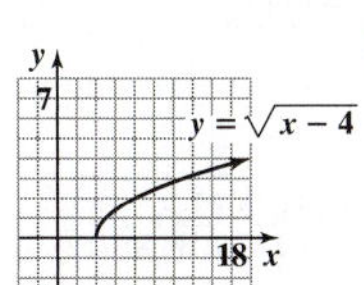

[3.1] **6. a)** A relation is any set of ordered pairs. **b)** A function is a correspondence between a first set of elements, the domain, and a second set of elements, the range, such that each element of the domain corresponds to exactly one element in the range. **c)** No **d)** Yes [3.2] **7.** Function; Domain: $\{1, 2, 7, -5\}$, Range: $\{5, -3, -1, 6\}$ [3.2] **8.** Not a function; Domain: $\{x|-2 \le x \le 2\}$, Range: $\{y|-4 \le y \le 4\}$ [3.2] **9.** Function; Domain: $\{x|-5 \le x \le 3\}$, Range: $\{y|-1 \le y \le 3\}$ [3.2]

**10.** $-21$ [3.2] **11.** 105 feet [3.2] **12.** $7x - y = -6$ [3.3] **13.** 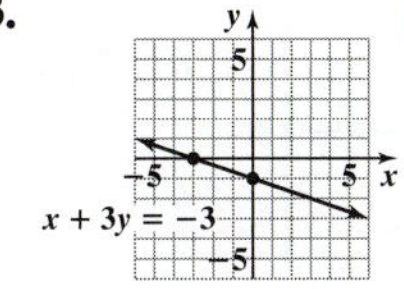

[3.3] **14.** 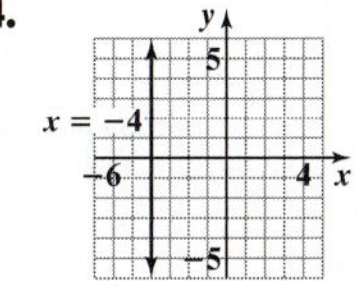

[3.3]

**15.** 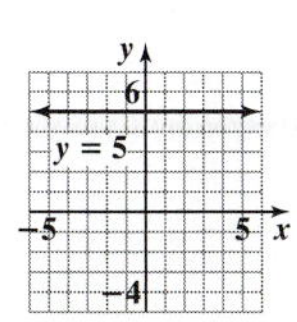

[3.3] **16. a)** 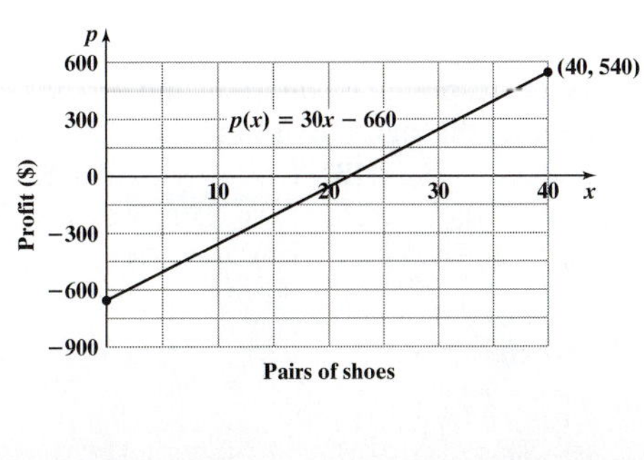


**b)** 22 pairs of shoes **c)** 34 pairs of shoes [3.3]
**17.** $-\frac{5}{8}$ [3.4] **18.** $y = -2x + 2$ [3.4]
**19.** $y = \frac{3}{2}x + 9; \frac{3}{2}; (0, 9)$ [3.4]
**20. a)** 5 **b)** $(0, 1)$ **c)** $y = 5x + 1$ [3.4]

**Exercise Set 3.5** **1.** $y - y_1 = m(x - x_1)$ **3.** Two lines are perpendicular if their slopes are negative reciprocals, or if one line is vertical and the other is horizontal. **5.** $y = 2x - 5$ **7.** $y = -\frac{1}{2}x + 1$ **9.** $y = \frac{1}{2}x - \frac{9}{2}$ **11.** $y = -\frac{3}{2}x$ **13.** $y = \frac{1}{2}x - 5$ **15.** Parallel **17.** Neither **19.** Perpendicular **21.** Perpendicular **23.** Parallel **25.** Neither **27.** Perpendicular **29.** Parallel **31.** Neither **33.** $y = 2x + 1$ **35.** $2x - 5y = 19$ **37.** $y = -\frac{5}{3}x + 5$ **39.** $f(x) = -3x + 13$ **41.** $y = -\frac{2}{3}x + 6$ **43. a)** $C(s) = 45.7s + 95.8$ **b)** 324.3 calories **45. a)** $d(p) = -0.20p + 90$ **b)** 38 DVD players **c)** \$225 **47. a)** $s(p) = 95p - 60$ **b)** 206 kites **c)** \$3.00 **49. a)** $i(t) = 12.5t$ **b)** \$1500 **c)** 176 tickets **51. a)** $r(w) = 0.01w + 10$ **b)** \$46.13 **c)** 5000 pounds **53. a)** $y(a) = -0.865a + 79.25$ **b)** 47.2 years **c)** 62.7 years old **55. a)** $w(a) \approx 0.189a + 10.6$ **b)** 14.758 kilograms **58.** $\left(-\infty, \frac{2}{5}\right)$ **59.** Reverse the direction of the inequality symbol. **60. a)** Any set of ordered pairs **b)** A correspondence where each member of the domain corresponds to a unique member in the range **c)** Answers will vary. **61.** Domain: {3, 4, 5, 6}; Range: {−4, −1, 2, 7}

**Exercise Set 3.6** **1.** Yes, this is how addition of functions is defined. **3.** $g(x) \neq 0$ since division by zero is undefined. **5.** No, subtraction is not commutative. One example is $5 - 3 = 2$ but $3 - 5 = -2$ **7. a)** 2 **b)** −8 **c)** −15 **d)** $-\frac{3}{5}$ **9. a)** $x^2 + 2x + 5$ **b)** $a^2 + 2a + 5$ **c)** 13 **11. a)** $x^3 + x - 4$ **b)** $a^3 + a - 4$ **c)** 6 **13. a)** $4x^3 - x + 4$ **b)** $4a^3 - a + 4$ **c)** 34 **15.** −7 **17.** 29 **19.** −60 **21.** Undefined **23.** 13 **25.** $-\frac{3}{4}$ **27.** $2x^2 - 6$ **29.** 2 **31.** 18 **33.** 0 **35.** $-\frac{3}{7}$ **37.** $-\frac{1}{45}$ **39.** $-2x^2 + 2x - 6$ **41.** 3 **43.** −4 **45.** 1 **47.** Undefined **49.** 0 **51.** 0 **53.** −3 **55.** −2 **57. a)** 2004 **b)** \$800 **c)** \$7900 **d)** \$900 **59. a)** 2003, ≈1.8 million barrels **b)** 1998, 2001 **c)** ≈1.4 million barrels **d)** ≈4.0 million barrels **61. a)** ≈20 **b)** ≈8 **c)** ≈12 **d)** ≈23

**63. a)**

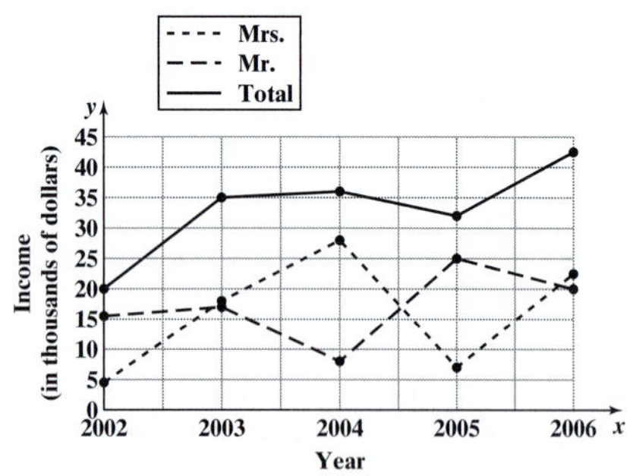


**b)**

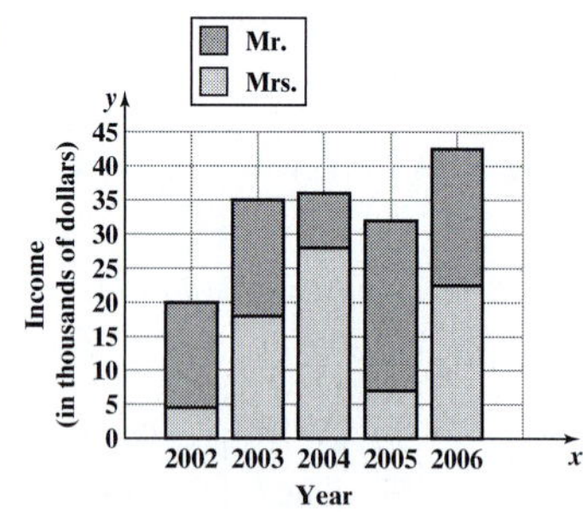


**c)**

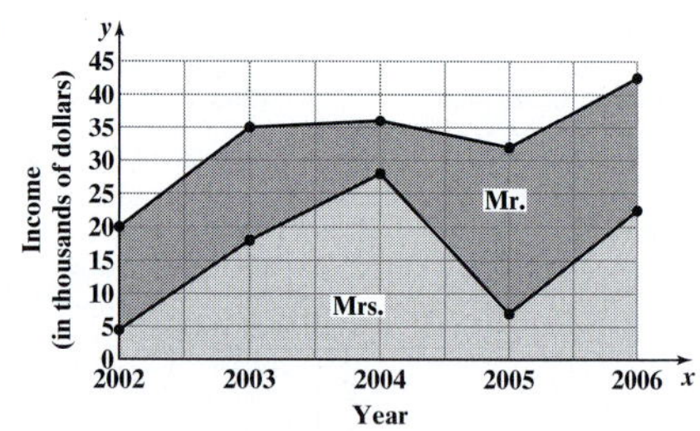


**65. a)**

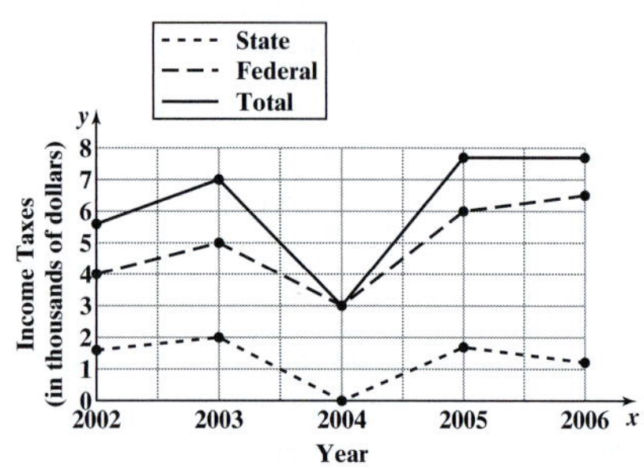


**b)**

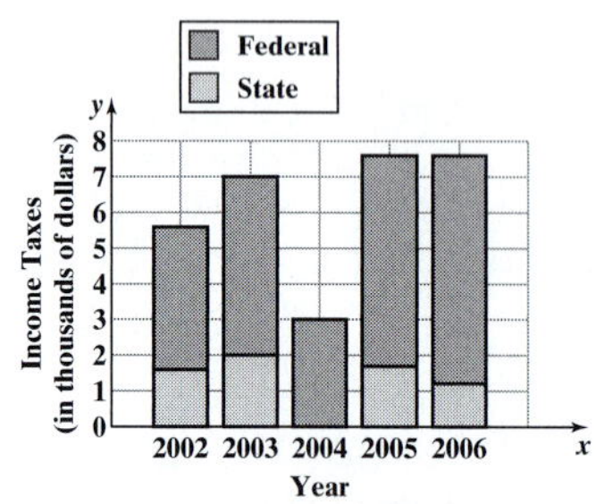


**c)**

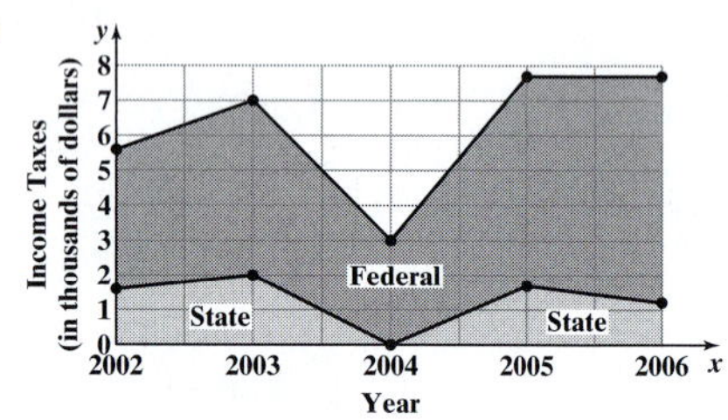


**67.** $f(a)$ and $g(a)$ must either be opposites or both be equal to 0. **69.** $f(a) = g(a)$ **71.** $f(a)$ and $g(a)$ must have opposite signs.

**73.**

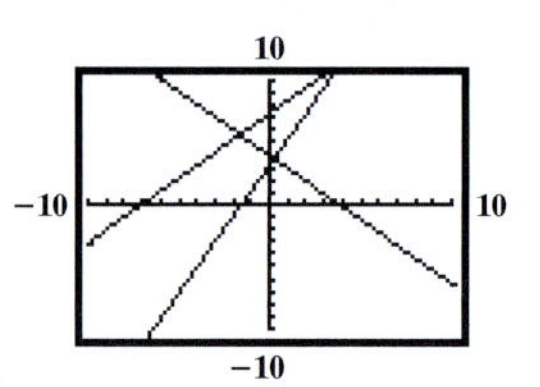

**75.**

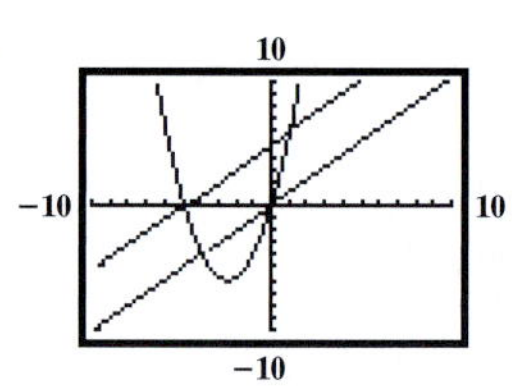

**78.** $-\frac{1}{64}$ **79.** $2.96 \times 10^6$ **80.** $h = \frac{2A}{b}$ **81.** \$450 **82.**

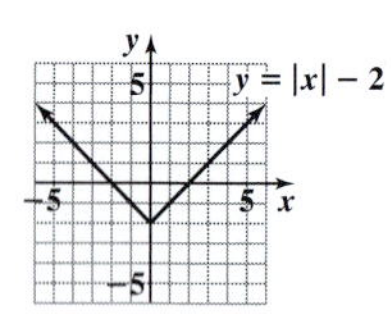


**83.**

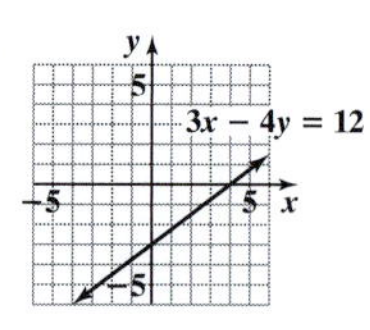


**Exercise Set 3.7** **1.** Points on the line are solutions to the corresponding equation, and are not solutions if the symbol used is $<$ or $>$. **3.** (0, 0) cannot be used as a check point if the line passes through the origin.

**5.**

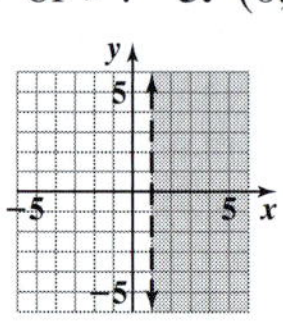

**7.**

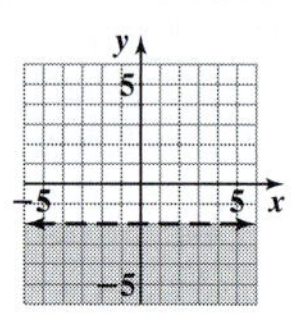

**9.**

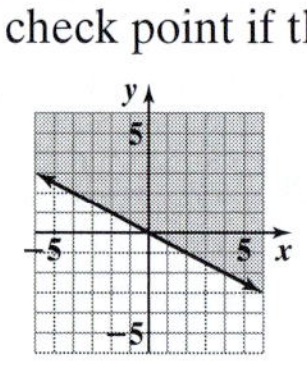

**11.**

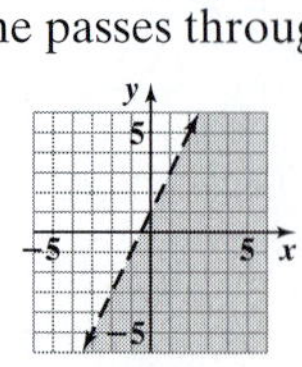

**13.**

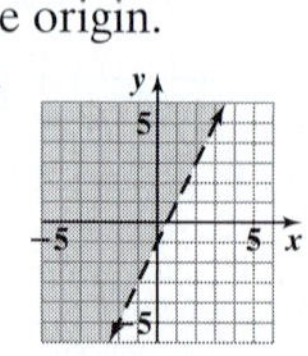

**15.**

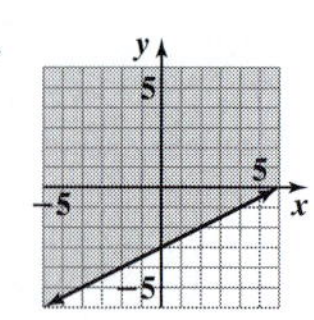

**55.** 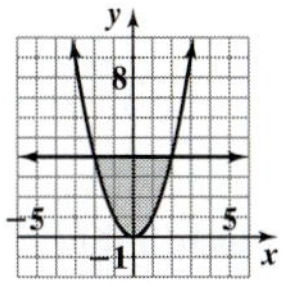 **57.** 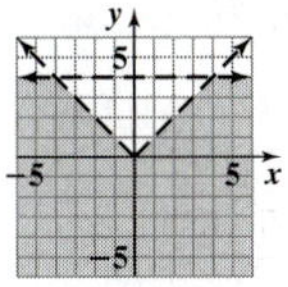 **59.** $f_2 = \dfrac{f_3d_3 - f_1d_1}{d_2}$ **60.** Domain: $\{-1, 0, 4, 5\}$; Range: $\{-5, -2, 2, 3\}$
**61.** Domain: $\mathbb{R}$; Range: $\mathbb{R}$
**62.** Domain: $\mathbb{R}$; Range: $\{y|y \geq -1\}$

## Chapter 4 Review Exercises

**1.** Inconsistent; no solution **2.** Consistent; one solution **3.** Consistent; one solution **4.** Consistent; one solution **5.**

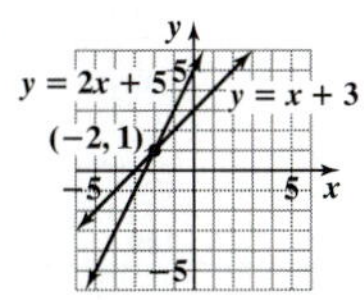


**6.**

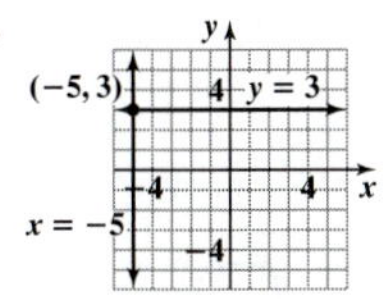


**7.**

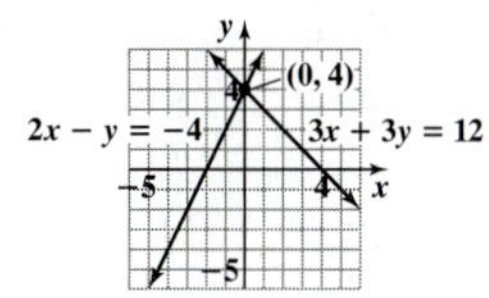


**8.**

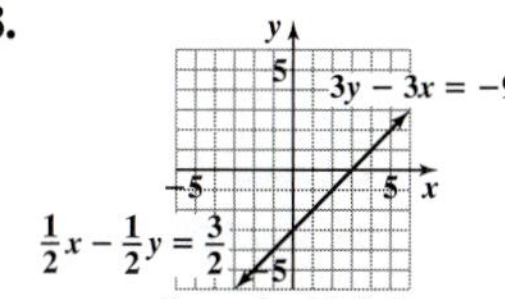


Dependent, infinite number of solutions

**9.** $(2, -6)$ **10.** $(-1, -1)$ **11.** $(2, 5)$ **12.** $(5, 2)$ **13.** $(3, -1)$ **14.** $(-8, 11)$ **15.** $(-1, 3)$ **16.** $(3, -2)$ **17.** $\left(\frac{32}{13}, \frac{8}{13}\right)$ **18.** $\left(-1, \frac{13}{3}\right)$ **19.** $(1, 2)$ **20.** $\left(\frac{7}{5}, \frac{13}{5}\right)$ **21.** $(6, -2)$ **22.** $\left(-\frac{78}{7}, -\frac{48}{7}\right)$ **23.** Infinite number of solutions **24.** No solution **25.** $(1, 2, -4)$ **26.** $(-1, 3, -2)$ **27.** $(-5, 1, 2)$ **28.** $(3, -2, -2)$ **29.** $\left(\frac{8}{3}, \frac{2}{3}, 3\right)$ **30.** $(0, 2, -3)$ **31.** No solution **32.** Infinite number of solutions **33.** Luan: 38, Jennifer: 28 **34.** Airplane: 520 mph, wind: 40 mph **35.** Combine 2 liters of the 20% acid solution with 4 liters of the 50% acid solution. **36.** 410 adult tickets and 240 children tickets were sold. **37.** His ages were 41 years and 77 years. **38.** \$20,000 invested at 7%, \$15,000 invested at 5%, and \$5000 was invested at 3%. **39.** $(11, -2)$ **40.** $(3, 1)$ **41.** Infinite number of solutions **42.** $(2, 1, -2)$ **43.** No solution **44.** Infinite number of solutions **45.** $(2, 3)$ **46.** $(-3, 2)$ **47.** $(-1, 2)$ **48.** $(-2, 3, 4)$ **49.** $(1, 1, 2)$ **50.** No solution **51.**

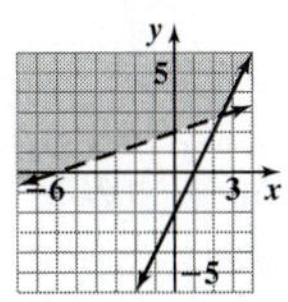

**52.**

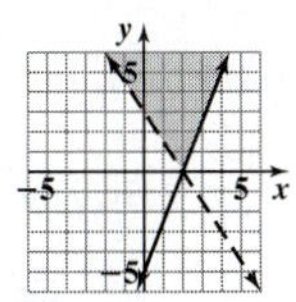

**53.**

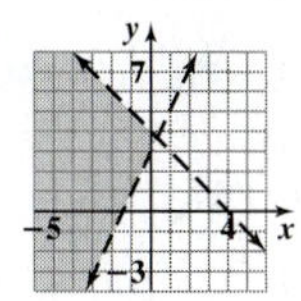

**54.** No solution

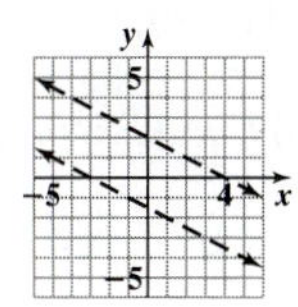

**55.**

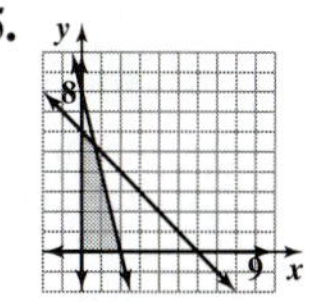

**56.**

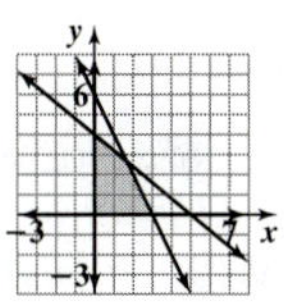

**57.**

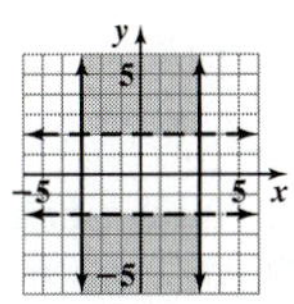

**58.**

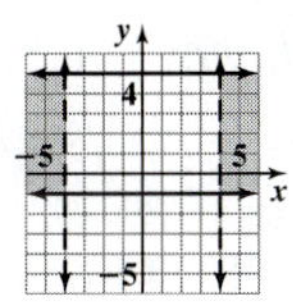

## Chapter 4 Practice Test

**1.** Answers will vary [4.1] **2.** Consistent; one solution [4.1] **3.** Dependent; infinite number of solutions [4.1] **4.** Inconsistent; no solution [4.1] **5.**

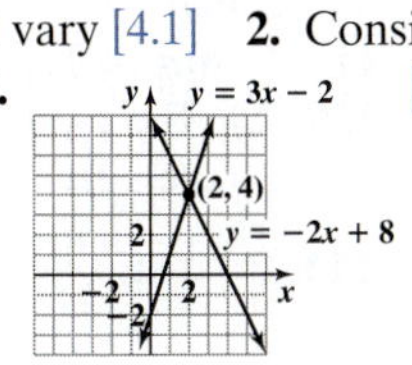


[4.1] **6.**

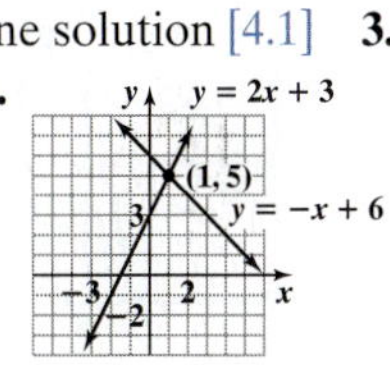


[4.1] **7.** $(1, 1)$ [4.1] **8.** $(-3, 2)$ [4.1]

**9.** $\left(-\frac{1}{2}, 4\right)$ [4.1] **10.** Infinite number of solutions [4.1] **11.** $\left(\frac{44}{19}, \frac{48}{19}\right)$ [4.1] **12.** $(1, -1, 2)$ [4.2] **13.** $\left[\begin{array}{rrr|r} -2 & 3 & 7 & 5 \\ 3 & -2 & 1 & -2 \\ 1 & -6 & 9 & -13 \end{array}\right]$ [4.4]

**14.** $\left[\begin{array}{rrr|r} 6 & -2 & 4 & 4 \\ 0 & 5 & -3 & 12 \\ 2 & -1 & 4 & -3 \end{array}\right]$ [4.4] **15.** $(4, -1)$ [4.4] **16.** $(3, -1, 2)$ [4.4] **17.** $-1$ [4.5] **18.** 165 [4.5] **19.** $(-3, 2)$ [4.5] **20.** $(3, 1, -1)$ [4.5]

**21.** 8 pounds sunflower; 12 pounds bird mix [4.3] **22.** $6\frac{2}{3}$ liters 6% solution; $3\frac{1}{3}$ liters 15% solution [4.3] **23.** 4, 9, and 16 [4.3]

**24.**

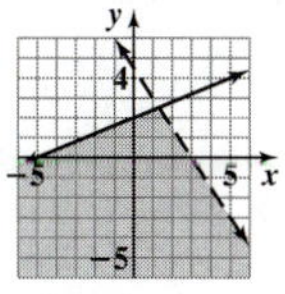

[4.6] **25.**

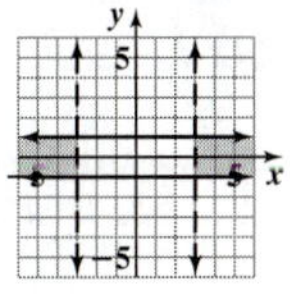

[4.6]

## Cumulative Review Test

**1.** 3 [1.4] **2. a)** 9, 1 **b)** $\frac{1}{2}, -4, 9, 0, -4.63, 1$ **c)** $\frac{1}{2}, -4, 9, 0, \sqrt{3}, -4.63, 1$ [1.2]

**3.** $-|-8|, -1, \frac{5}{8}, \frac{3}{4}, |-4|, |-12|$ [1.3] **4.** 7 [2.1] **5.** $\frac{17}{4}$ [2.1] **6.** 6, $-3$ [2.6] **7.** $x = 2M - a$ [2.2] **8.** $\left\{x \middle| \frac{2}{3} < x \leq \frac{34}{3}\right\}$ [2.5]

**9.** $\frac{y^{10}}{9x^4}$ [1.5] **10.** 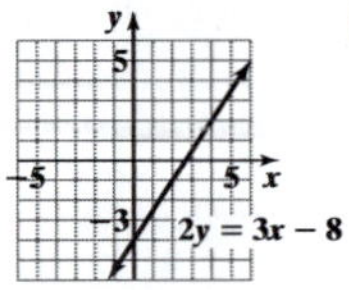

[3.3] **11.** $y = \frac{2}{3}x + \frac{5}{3}$ [3.5] **12.** 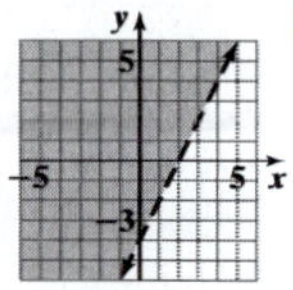

[3.7] **13. a)** function **b)** function **c)** not a function [3.2] **14. a)** $-\frac{1}{7}$ **b)** $\frac{h+3}{h^2-9}$ **c)** undefined [3.2]

**15.** (1, 3) [4.1] **16.** (7, −1) [4.1] **17.** (2, 1, 3) [4.2] **18.** 10°, 80°, 90° [2.3] **19.** 1 hour [2.4] **20.** 600 at \$20, 400 at \$16 [4.3]

# Chapter 5

## Using Your Graphing Calculator, 5.1

**1.** 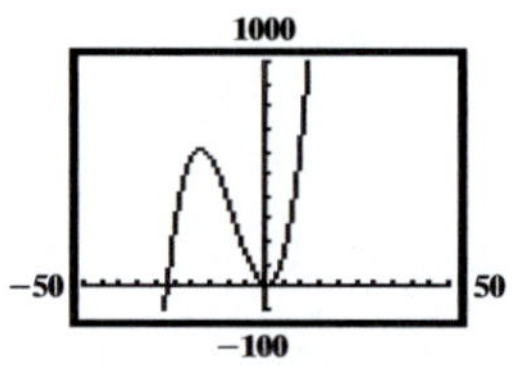

**2.** 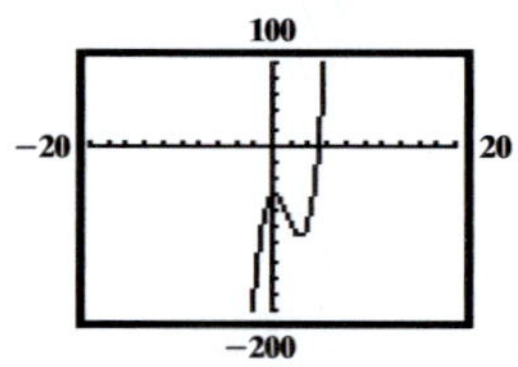


**Exercise Set 5.1** **1.** The terms are the parts that are added. **3.** A polynomial is a finite sum of terms in which all variables have whole number exponents and no variable appears in a denominator. **5.** The leading coefficient is the coefficient of the leading term. **7. a)** It is the same as that of the highest-degree term. **b)** 7 **9. a)** A polynomial is linear if its degree is 0 or 1. **b)** Answers will vary. One example is $x + 4$. **11. a)** A polynomial is cubic if it has degree 3 and is in one variable. **b)** Answers will vary. One example is $x^3 + x - 4$. **13.** Answers will vary. One example is $x^5 + x + 1$. **15.** Monomial **17.** Monomial **19.** Not a polynomial; −3 exponent **21.** Not a polynomial; $\frac{1}{2}$ exponent **23.** $-x^2 + 2x - 5, 2$ **25.** $10x^2 + 3xy + 9y^2, 2$ **27.** In descending order, 4 **29. a)** 6 **b)** 3 **31. a)** 6 **b)** 9 **33. a)** 17 **b)** $-\frac{1}{3}$ **35.** −3 **37.** −7 **39.** −2.0312 **41.** $x^2 + 9x - 6$ **43.** $x^2 - 13x + 2$ **45.** $2y^2 + 9y - 11$ **47.** $-\frac{2}{3}a^2 - \frac{29}{36}a + 5$ **49.** $-3.5x^2 - 2.1x - 19.6$ **51.** $-\frac{4}{3}x^3 - \frac{1}{4}x^2y + 9xy^2$ **53.** $5a - 10b + 13c$ **55.** $8a^2b - 10ab + 11b^2$ **57.** $7r^2 - 4rt - 3t^2$ **59.** $10x^2 - 8x - 9$ **61.** $-3w^2 + 6w$ **63.** $3x + 19$ **65.** $-3x^2 + 2x - 12$ **67.** $-5.4a^2 - 5.7a - 26.4$ **69.** $-\frac{11}{2}x^2y + xy^2 + \frac{2}{45}$ **71.** $5x^{2r} - 10x^r + 3$ **73.** $-x^{2s} - 4x^s + 19$ **75.** $7b^{4n} - 3b^{3n} - 4b^{2n} + 1$ **77.** $4x^2 + 8x + 24$ **79.** $3x^2 + 4x + 19$ **81.** $2x^2 + 12x + 9$ **83.** No, for example $(x^2 + x + 1) + (x^3 - 2x^2 + x) = x^3 - x^2 + 2x + 1$ **85.** No, for example $(x^2 + 3x - 5) + (-x^2 - 4x + 2) = -x - 3$ **87.** 144 square meters **89.** $A \approx 113.10$ square inches **91.** 674 feet **93.** 105 committees **95. a)** \$674 **b)** \$1010 **97. a)** $P(x) = 2x^2 + 360x - 8050$ **b)** \$47,950 **99. c)** The $y$-intercept is (0, −4) and the leading coefficient is positive **101. c)** The $y$-intercept is (0, −6) and the leading coefficient is negative **103. a)** \$120.8 billion **b)** Yes **c)** \$286.4 billion **105.** \$88,210 **107. a)** 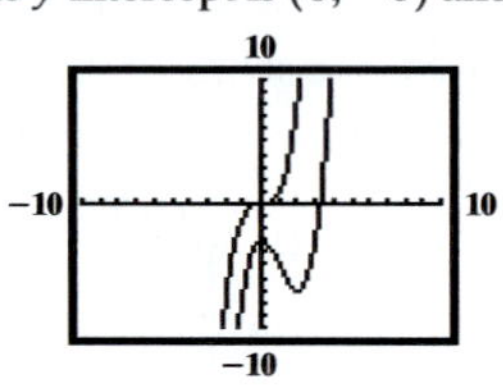

**b)** Increase **c)** Answers will vary. **d)** Decrease **e)** Answers will vary. **109. b)** The $y$-intercept is (0, −5) and the leading coefficient is negative **113.** 3 **114.** $\frac{15}{16}$ **115.** 6 hours **116.** $-\frac{2}{11}$ **117.** (−4, 0, −1)

**Exercise Set 5.2** **1. a)–d)** Answers will vary. **3. a)** Answers will vary. **b)** $x^3 - 2x^2 - 21x + 12$ **5. a)** Answers will vary. **b)** Answers will vary. One possible answer is $(x + 4)(x - 4)$. **c)** Answers will vary. **d)** Answers will vary. One possible answer is $x^2 - 16$. **7.** Yes, for example $(x + 2)(x - 1) = x^2 + x - 2$ **9.** $24x^2y^5$ **11.** $\frac{1}{9}x^7y^8z^2$ **13.** $6x^6y^3 - 15x^3y^4 - 12x^2y$ **15.** $2xyz + \frac{8}{3}y^2z - 8y^3z$ **17.** $0.6x^2 - 1.5x + 3.3y$ **19.** $2.85a^{11}b^5 - 1.38a^9b^7 + 0.36a^6b^9$ **21.** $12x^2 - 38x + 30$ **23.** $-2x^3 + 8x^2 - 3x + 12$ **25.** $x^2 + \frac{23}{6}xy - \frac{2}{3}y^2$ **27.** $0.09a^2 - 0.25b^2$ **29.** $x^3 - x^2 - 11x - 4$ **31.** $2a^3 - 7a^2b + 5ab^2 - 6b^3$ **33.** $x^4 + 2x^2 + 10x + 7$ **35.** $5x^4 + 29x^3 + 14x^2 - 28x + 10$ **37.** $3m^4 - 11m^3 - 5m^2 - 2m - 20$ **39.** $8x^3 - 12x^2 + 6x - 1$ **41.** $10r^4 - 2r^3s - r^2s^2 + rs^3 - 2s^4$ **43.** $x^2 + 4x + 4$ **45.** $4x^2 - 28x + 49$ **47.** $16x^2 - 24xy + 9y^2$ **49.** $25m^4 - 4n^2$ **51.** $y^2 + 8y - 4xy + 16 - 16x + 4x^2$ **53.** $25x^2 + 20xy + 10x + 4y^2 + 4y + 1$ **55.** $a^2 - b^2 - 8b - 16$ **57.** $2x^3y + 2x^2y^2 + 24xy^3$ **59.** $2x^3y^2 + \frac{3}{2}x^2y^3 - \frac{7}{2}xy^6$ **61.** $\frac{3}{5}x^2y^5z^7 + 3x^2y^4z^2 - \frac{1}{15}x^2y^3z^9$ **63.** $21a^2 + 10a - 24$ **65.** $64x^2 - \frac{1}{25}$ **67.** $x^3 - \frac{3}{2}x^2y + \frac{3}{4}xy^2 - \frac{1}{8}y^3$ **69.** $2x^3 + 10x^2 + 9x - 9$ **71.** $6p^3 - p^2q - 16pq^2 + 6q^3$ **73.** $9x^2 + 12x + 4 - y^2$ **75.** $a^4 - 2a^2b^2 + b^4$ **77.** $2x^3 - 4x^2 - 64x + 192$ **79. a)** $x^2 + x - 30$ **b)** −10 **81. a)** $10x^3 + 36x^2 - 2x - 12$ **b)** 1196 **83. a)** $-x^4 + 3x^3 - 2x^2 + 6x$ **b)** −72 **85.** $x^2 + 5x$ **87.** $x^2 + y^2$ **89. a)** and **b)** $x^2 + 7x + 12$ **91.** $36 - x^2$ **93. a)** $11x + 12$ **b)** 117 square inches, 50 square inches. **95.** $(x + 7)(x - 7)$, product of the sum and difference of the same two terms.

**97.** $(x+6)(x+6)$, square of a binomial formula **99.** $a(x-n)(x-n)(x-n)$ **101. a)** Answers will vary. **b)** $a^2+2ab+b^2$ **c)** $a^2+2ab+b^2$ **d)** Same. **103. a)** $A=P(1+r)^t$ **b)** \$1123.60 **105. a)** 110 ways **b)** $P(n)=n^2-n$ **c)** 110 ways **d)** Yes **107.** $a^2+2ab+b^2-3a-3b+5$ **109.** $15x^{3t-1}+18x^{4t}$ **111.** $12x^{3m}-18x^m-10x^{2m}+15$ **113.** $y^{a^2-b^2}$ **115.** $x^4-12x^3y+54x^2y^2-108xy^3+81y^4$ **117. a)** Answers will vary. **b)**

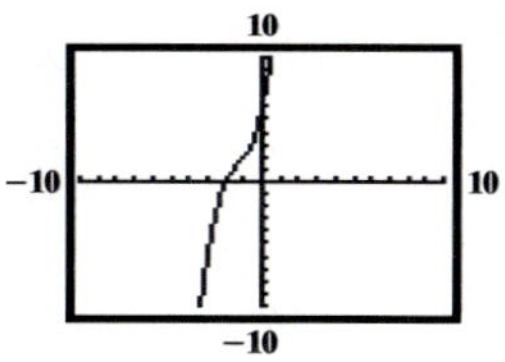


It is correct.

**119.** $y^2-2y-2xy+2x+x^2+1$ **121.** $\frac{43}{60}$ **122.** $8r^6s^{15}$ **123.** $\left(-\frac{7}{3},\frac{4}{3}\right]$ **124.** $\frac{15}{4}$

## Exercise Set 5.3

**1. a)** Answers will vary. **b)** $\frac{5}{3}x^3-2x^2-\frac{4}{3}x-4+\frac{1}{3x}$ **3.** Yes; answers will vary. **5.** Place them in descending order of the variable. **7. a)** Answers will vary. **b)** $x+8+\frac{36}{x-5}$ **9.** No; because the remainder is not equal to 0. **11.** $x^2$ **13.** $a^4$ **15.** $z^8$ **17.** $4r^6s^2$ **19.** $5x^8y^{11}$ **21.** $2x+9$ **23.** $2x+1$ **25.** $\frac{5}{3}y^2+2y-4$ **27.** $x^3-\frac{3}{2}x^2+3x-2$ **29.** $4x^2-5xy-\frac{5}{2y}$ **31.** $\frac{9x}{2y}-6x^2+\frac{15y}{2x}$ **33.** $\frac{z}{2}+z^2-\frac{3}{2}x^2y^4z^7$ **35.** $x+2$ **37.** $2x+4$ **39.** $3x+2$ **41.** $x+5-\frac{2}{x+1}$ **43.** $2b+5+\frac{2}{b-2}$ **45.** $4x+9+\frac{2}{2x-3}$ **47.** $2x+6$ **49.** $x^2+2x+3+\frac{1}{x+1}$ **51.** $2y^2+3y-1-\frac{6}{2y+3}$ **53.** $2a^2+a-2-\frac{2}{2a-1}$ **55.** $3x^3+6x+2$ **57.** $x+4$ **59.** $2c^2-6c+3$ **61.** $x+6$ **63.** $x+3$ **65.** $x-7$ **67.** $x+8+\frac{10}{x-3}$ **69.** $3x+5+\frac{10}{x-4}$ **71.** $4x^2+x+3+\frac{3}{x-1}$ **73.** $3c^2-2c+2+\frac{10}{c+3}$ **75.** $y^3+y^2+y+1$ **77.** $x^3-4x^2+16x-64+\frac{272}{x+4}$ **79.** $x^4-\frac{9}{x+1}$ **81.** $b^4+3b^3-3b^2+3b-3-\frac{11}{b+1}$ **83.** $3x^2+3x-3$ **85.** $2x^3+2x-2+\frac{6}{x-\frac{1}{2}}$ **87.** 12 **89.** 0; factor **91.** $-\frac{19}{4}$ or $-4.75$ **93.** $3x+2$ **95.** 3 times greater, find the areas by multiplying the polynomials, then compare **97.** No, the dividend is a binomial **99.** If the remainder is 0, $x-a$ is a factor. **101.** $x^2-2x-8$ **103.** $x^2+9x+26$ **105.** $2x^2+3xy-y^2$ **107.** $x+\frac{5}{2}+\frac{11}{2(2x-3)}$ **109.** $w=r+1$ **111.** $x^3-6x^2+13x-7$; multiply $(x-3)(x^2-3x+4)$ and then add 5. **113.** $2x+1-\frac{3}{2x}-\frac{1}{2x^2}$ **115.** Not a factor; compute $P(1)$. $P(1)=101$, which is not 0, so $x-1$ is not a factor **117.** Factor; compute $P(-1)$. $P(-1)=0$, so $x+1$ is a factor. **119.** $2.0\times10^{10}$ **120.** 30°, 60°, 90° **121.** $\left\{-1,\frac{11}{5}\right\}$ **122.** $-864$ **123.** $3r+3s-8t$

## Exercise Set 5.4

**1.** Determine if all the terms contain a greatest common factor and, if so, factor it out. **3. a)** Answers will vary. **b)** $2x^2y$ **c)** $2x^2y(3y^4-x+6x^7y^2)$ **5.** $3(x-4)^3$ **7. a)** Answers will vary. **b)** $(3x^2-y^3)(2x+y^2)$ **9.** $7(n+2)$ **11.** $2(x^2-2x+5)$ **13.** $4(3y^2-4y+7)$ **15.** $x^2(9x^2-3x+11)$ **17.** $-3a^2(8a^5-3a^4+1)$ **19.** $3xy(x+2xy+1)$ **21.** $8a^2c(10a^3b^4-2a^2b^2c+3)$ **23.** $3pq^2r(3p^3q^3-pr+4q^3r^2)$ **25.** $-2(11p^2q^2+8pq^3-13r)$ **27.** $-4(2x-1)$ **29.** $-(x^2+4x-22)$ **31.** $-3(r^2+2r-3)$ **33.** $-2rs^3(3r^3-2rs-s^2)$ **35.** $-a^2b(a^2bc-5ac^2-1)$ **37.** $(a+3)(x+1)$ **39.** $(x-4)(9x-8)$ **41.** $-(x-2)(2x-9)$ **43.** $-2(a+2)(a+2)$ or $-2(a+2)^2$ **45.** $(x+4)(x-5)$ **47.** $2(2y-1)(2y-5)$ **49.** $(a+b)(m+n)$ **51.** $(x-3)(x^2+4)$ **53.** $(5m-6n)(2m-5n)$ **55.** $5(a+3)(a^2-2)$ **57.** $c^2(c-1)(c^2+1)$ **59.** $(2x+1)(6x-5)$ **61.** $(x+4)(3x-2)$ **63.** $(3x+2)(9x-5)$ **65. a)** 96 feet **b)** $h(t)=-16t(t-5)$ **c)** 96 feet **67. a)** $\approx$2856.64 square feet **b)** $A=r(\pi r+2l)$ **c)** $\approx$2856.64 square feet **69. a)** \$525 **b)** $A(t)=75(13-t)$ **c)** \$525 **71. a)** $(1-0.06)(x+0.06x)=0.94(1.06x)$ **b)** $0.9964x$; slightly lower than the price of the 2005 model (99.64% of the original cost) **73. a)** $(x+0.15x)-0.20(x+0.15x)=0.80(x+0.15x)$ **b)** $0.80(1.15x)=0.92x$; 92% of the regular price **75.** $(3x+2)^4(15ax+10a+4)$ **77.** $2(x-3)(2x^4-12x^3+15x^2+9x+2)$ **79.** $(x^2+2x-3)(a+b)$ **81.** $x^{4m}(x^{2m}-2)$ **83.** $x^{2m}(3x^{2m}-2x^m+1)$ **85.** $(a^r+c^r)(b^r-d^r)$ **87. a)** Yes **b)** 0; subtracting the same quantity from itself **c)** Answers will vary. **89. a)** They should be the same graph; they represent the same function

**89. b)**

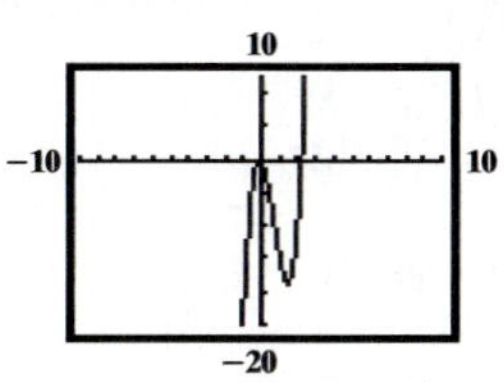


**c)** Answers will vary. **d)** Factoring is not correct. **91.** $-\frac{15}{72}=-\frac{5}{24}$ **92.** 2 **93.**

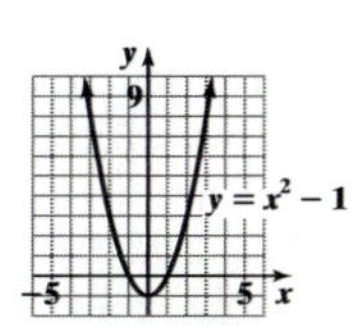


**94.** 0.4 hour **95.** $-14a^3-22a^2+19a-3$

**Mid-Chapter Test** **1.** $5x^4 - 1.5x^3 + 2x - 7$, 4 [5.1] **2.** $\frac{3}{2}$ or $1\frac{1}{2}$ [5.1] **3.** $-n^2 - 7n - 4$ [5.1] **4.** $-16x^2y + 14xy$ [5.1] **5.** $9x^2 - 4x + 13$ [5.1] **6.** $6x^6y^4 + 10x^7 - 14x^8y$ [5.1] **7.** $21x^2 - 4xy - 12y^2$ [5.2] **8.** $6x^4 - x^3 + 14x^2 + 32x + 9$ [5.2] **9.** $64p^2 - \frac{1}{25}$ [5.2] **10.** $12m^3 - m^2n - 30mn^2 + 18n^3$ [5.2] **11.** $x^2 - 14x + 49 = (x - 7)^2$ [5.2] **12.** $2x^2y + 3 - \frac{11}{2xy^2}$ [5.3] **13.** $3x + 5 + \frac{2}{4x + 1}$ [5.3] **14.** $y^2 + y + 5 + \frac{5}{2y - 3}$ [5.3] **15.** $x - 9$ [5.3] **16.** $3a^3 + 4a^2 - 6a - 1$ [5.3] **17.** $8b^2c(4bc^2 + 2 + 3b^3c^3)$ [5.4] **18.** $(2x + 9)(7b - 3c)$ [5.4] **19.** $b^2(b + 2c)(2b - c)$ [5.4] **20.** $(3x - 2)^5(5a - 12x + 8)$ [5.4]

**Using Your Graphing Calculator, 5.5** **1.** Yes **2.** No

**Exercise Set 5.5** **1.** Factor out the greatest common factor if it is present. **3. a)** Answers will vary. **b)** $(2x + 3)(3x - 4)$ **5.** No; $2(x + 3)(x + 1)$; $(2x + 2)$ has a GCF of 2 **7.** No, $3x(x + 4)(x - 2)$; $(3x - 6)$ has a GCF of 3 **9.** Both are + **11.** One is +, one is − **13.** $(x + 3)(x + 4)$ **15.** $(b - 1)(b + 9)$ **17.** $(z + 2)^2$ **19.** $(r + 12)^2$ **21.** $(x + 32)(x - 2)$ **23.** $(x + 2)(x - 15)$ **25.** $-(a - 15)(a - 3)$ **27.** Prime **29.** $-2(m + 2)(m + 5)$ **31.** $4(r + 4)(r - 1)$ **33.** $x(x + 6)(x - 3)$ **35.** $(a - 1)(5a - 3)$ **37.** $3(x - 2)(x + 1)$ **39.** $(3c + 7)(2c - 9)$ **41.** $(2b + 1)(4b - 3)$ **43.** $(3c - 2)(2c + 5)$ **45.** $4(2p - 3q)(2p + q)$ **47.** Prime **49.** $2(3a + 4b)(3a - b)$ **51.** $(4x - 3y)(2x + 9y)$ **53.** $10(5b - 2)(2b - 1)$ **55.** $ab^5(a - 4)(a + 3)$ **57.** $3b^2c(b - 3c)^2$ **59.** $4m^6n^3(m + 2n)(2m - 3n)$ **61.** $(6x - 5)(5x + 4)$ **63.** $8x^2y^5(x + 4)(x - 1)$ **65.** $(x^2 + 3)(x^2 - 2)$ **67.** $(b^2 + 5)(b^2 + 4)$ **69.** $(2a^2 + 5)(3a^2 - 5)$ **71.** $(2x + 5)(2x + 3)$ **73.** $(3a + 1)(2a + 5)$ **75.** $(xy + 7)(xy + 2)$ **77.** $(2xy - 11)(xy + 1)$ **79.** $(2 - y)(y - 1)(2y - 5)$ **81.** $(p - 4)(2p + 3)(p + 2)$ **83.** $(a^3 - 10)(a^3 + 3)$ **85.** $(x + 5)(x + 2)(x + 1)$ **87.** $a^3b^2(5a - 3b)(a - b)$ **89.** $(x + 6)(x + 1)$ **91.** $(x + 6)(x + 3)$ **93.** $2x^2 - 5xy - 12y^2$, multiply $(2x + 3y)(x - 4y)$ **95.** Divide; $x + 7$ **97. a)** Answers will vary. **b)** $(6x - 5)(5x + 8)$; $(7x - 1)(7x - 13)$ **99.** $\pm3, \pm9$ **101.** 6 or −6; $b$ is the sum of the factors of 5 **103. a)** 4 **b)** $(x - 3)(x - 5)$ **105. a)** −8 **b)** Not factorable **107.** Answers will vary. One example is $x^2 + 2x + 1$. **109.** $(2a^n + 3)(2a^n - 5)$ **111.** $(x + y)^2(x - 4y)(x - 3y)$ **113.** $(x^n - 2)(x^n + 5)$ **115. a)** Answers will vary. **b)** Correct **117.** $C = \frac{5}{9}(F - 32)$ **118.**

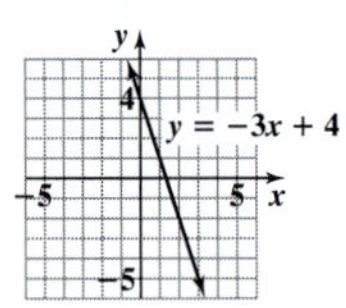


**119.** 4 **120.** $x^2 + 2xy + y^2 + 12x + 12y + 36$ **121.** $(2x^2 - 5)(x + 2)$

**Exercise Set 5.6** **1. a)** Answers will vary. **b)** $(x + 4)(x - 4)$ **3.** Answers will vary. **5.** $a^3 + b^3 = (a + b)(a^2 - ab + b^2)$ **7.** No, $(x + 7)(x - 7) = x^2 - 49$ **9.** No, $(x - 9)^2 = x^2 - 18x + 81$ **11.** $(x + 9)(x - 9)$ **13.** $(a + 10)(a - 10)$ **15.** $(1 + 7b)(1 - 7b)$ **17.** $(5 + 4y^2)(5 - 4y^2)$ **19.** $\left(\frac{1}{10} + y\right)\left(\frac{1}{10} - y\right)$ **21.** $(xy + 11c)(xy - 11c)$ **23.** $(0.2x + 0.3)(0.2x - 0.3)$ **25.** $x(12 - x)$ **27.** $(a + 3b + 2)(a - 3b - 2)$ **29.** $(x + 5)^2$ **31.** $(7 - t)^2$ **33.** $(6pq + 1)^2$ **35.** $(0.9x - 0.2)^2$ **37.** $(y^2 + 2)^2$ **39.** $(a + b + 3)^2$ **41.** $(y + 1)^2$ **43.** $(x + 3 + y)(x + 3 - y)$ **45.** $(x + 7)(3 - x)$ **47.** $(3a - 2b + 3)(3a - 2b - 3)$ **49.** $(y^2 - 3)^2$ **51.** $(a + 5)(a^2 - 5a + 25)$ **53.** $(4 - a)(16 + 4a + a^2)$ **55.** $(p - 3a)(p^2 + 3ap + 9a^2)$ **57.** $(3y - 2x)(9y^2 + 6xy + 4x^2)$ **59.** $2(2a - 3b)(4a^2 + 6ab + 9b^2)$ **61.** $(x^2 + y^3)(x^4 - x^2y^3 + y^6)$ **63.** $(x + 2)(x^2 + x + 1)$ **65.** $(a - b - 3)(a^2 - 2ab + b^2 + 3a - 3b + 9)$ **67.** $-9(b^2 + 3b + 3)$ **69.** $(a^2 + 2b^2)(a^2 - 2b^2)$ **71.** $(7 + 8xy)(7 - 8xy)$ **73.** $(x + y + 4)(x + y - 4)$ **75.** $(x - 4)(x^2 + 4x + 16)$ **77.** $(3xy + 4)^2$ **79.** $(a^2 + b^2)^2$ **81.** $(x - 1 + y)(x - 1 - y)$ **83.** $(x + y + 1)(x^2 + 2xy + y^2 - x - y + 1)$ **85.** $3m(-m + 2n)$ **87.** $(3x - 2)(9x^2 + 6x + 4)$ **89.** $(6a - b)(36a^2 + 6ab + b^2)$ **91.** $(4x + 3a)(16x^2 - 12ax + 9a^2)$ **93. a)** $a^2 - b^2$ **b)** $(a + b)(a - b)$ **95. a)** $6a^3 - 6ab^2$ **b)** $6a(a + b)(a - b)$ **97. a)** $\frac{4}{3}\pi R^3 - \frac{4}{3}\pi r^3$ **b)** $\frac{4}{3}\pi(R - r)(R^2 + Rr + r^2)$ **99.** 12, −12; write $4x^2 + bx + 9$ as $(2x)^2 + bx + (3)^2$; $bx = 2(2x)(3)$ or $bx = -2(2x)(3)$ **101.** $c = 4$; write $25x^2 + 20x + c$ as $(5x)^2 + 20x + (a)^2$ then $20x = 2(5x)(a)$. Since $a = 2$, $c = 4$ **103. a)** Find an expression whose square is $25x^2 - 30x + 9$ **b)** $s(x) = 5x - 3$ **c)** 7 **105.** $(x^2 + 4x + 8)(x^2 - 4x + 8)$ **107.** $h(2a + h)$ **109. a)** 16 **b)** $x^2 + 8x + 16$ **c)** $(x + 4)^2$ **111.** $(8x^{2a} + 3y^{3a})(8x^{2a} - 3y^{3a})$ **113.** $(a^n - 8)^2$ **115.** $(x^n - 2)(x^{2n} + 2x^n + 4)$ **117.** Correct **119. a)** $(x^3 + 1)(x^3 - 1)$ **b)** $(x^2 - 1)(x^4 + x^2 + 1)$ **121.** $4x + 7y - 2$ **122.** −17 **123.** 20°, 30°, 40° **124.** $15y^{10}(3y^2 + 4)$ **125.** $(4x - 3y)(3x + y)$

**Exercise Set 5.7** **1.** Answers will vary. **3.** $3(x + 5)(x - 5)$ **5.** $(5s - 3)(2s + 5)$ **7.** $2x^2y^2(3x + 5y + 7)$ **9.** $0.8(x + 0.3)(x - 0.3)$ **11.** $6x(x^2 + 3)(x^2 - 3)$ **13.** $3x^4(x - 1)(x + 4)$ **15.** $5x^2y^2(x + 4)(x + 3)$ **17.** $x^2(x + y)(x - y)$ **19.** $x^4y^2(x - 1)(x^2 + x + 1)$ **21.** $x(x^2 + 4)(x + 2)(x - 2)$ **23.** $4(x^2 + 2y)(x^4 - 2x^2y + 4y^2)$ **25.** $5(a + b + 2)(a + b - 2)$ **27.** $6(x + 3y)^2$ **29.** $x(x + 4)$ **31.** $3(2x - y)(x + 4y)$ **33.** $(y + 7)^2$ **35.** $(b^2 + 1)^2$ **37.** $\left(x + \frac{1}{4}\right)\left(x^2 - \frac{1}{4}x + \frac{1}{16}\right)$ **39.** $2y(3y + 1)(y + 2)$ **41.** $ab(a + 9b)(a - 9b)$ **43.** $(7 + x + y)(7 - x - y)$ **45.** $2(3x - 2)(4x - 3)$ **47.** $(9x - 3)(2x + 5)$ **49.** $(x^2 + 4)(x + 2)(x - 2)$ **51.** $(b - 2x)(5c - 7y)$ **53.** $(3x^2 - 4)(x^2 + 1)$ **55.** $(z + x - 6)(z - x + 6)$ **57.** $(2y + 5)(y + 8)$ **59.** $(a + 6b + 4c)(a + 6b - 4c)$ **61.** $5x^2y(x + 3)(2x - 1)$ **63.** $(x + y)^2(x - y)^2$ **65. e)** **67. d)** **69. f)** **71. c)** **73.** $2(x + 3)(x + 2)$ **75.** $(x + 6)(x + 2)$ **77.** $(y + 3)(y - 3)$ **79.** $(5x - 3)(25x^2 + 15x + 9)$ **81. a)** $a(a + b) - b(a + b) = a^2 - b^2$ **b)** $(a + b)(a - b)$ **83. a)** $a^2 + 2ab + b^2$ **b)** $(a + b)^2$

**85. a)** $a(a-b)+a(a-b)+b(a-b)+b(a-b)$ or $2a(a-b)+2b(a-b)$ **b)** $2(a+b)(a-b)$ **87. a)** Answers will vary. **b)** Answers will vary. **89. a)** $x^{-5}(x^2-2x-3)$ **b)** $x^{-5}(x-3)(x+1)$ **91. a)** $x^{-3/2}(5x^2+2x-3)$ **b)** $x^{-3/2}(5x-3)(x+1)$ **92.** 1 **93.** $\{z|z<-6 \text{ or } z>0\}$ **94.** $\approx$17.3 pounds at \$5.20; $\approx$12.7 pounds at \$6.30 **95.** $5x^3-x^2+16x+16$ **96.** $(x+3)(2x^2-5)$

## Using Your Graphing Calculator, 5.8

**1.** $y=x^2-6x+5$ **2.** $y=x^2-x-6$ **3.** $y=x^2+4x$

## Exercise Set 5.8

**1.** The degree of a polynomial function is the same as the degree of the leading term. **3.** $ax^2+bx+c=0$ **5. a)** The zero-factor property only holds when one side of the equation is 0. **b)** $-2$ and $-5$ **7. a)** Answers will vary. **b)** $\frac{4}{3},-\frac{5}{4}$ **9. a)** Legs **b)** Hypotenuse **11.** $-8$ and $-2$; at the $x$-intercepts $y=0$ **13.** Yes, if the graph does not cross the $x$-axis **15.** Yes, if the graph crosses the $x$-axis twice **17.** $0,-3$ **19.** $0,1$ **21.** $-1,7$ **23.** $0,-4,9$ **25.** $\frac{2}{3},\frac{1}{7}$ **27.** $0,3$ **29.** $0,-5$ **31.** $0,6$ **33.** $0,9$ **35.** $-1,-5$ **37.** $-4,3$ **39.** $-4$ **41.** $-\frac{1}{2},5$ **43.** $\frac{3}{2},-2$ **45.** $-4,6$ **47.** $0,6,-3$ **49.** $0,-4,3$ **51.** $0,-\frac{1}{2},\frac{4}{3}$ **53.** $5,-5$ **55.** $-\frac{3}{2},\frac{3}{2}$ **57.** $0,-3,3$ **59.** $-11,9$ **61.** $-3,-11$ **63.** $-1,-4$ **65.** $\frac{5}{2},\frac{4}{3}$ **67.** $0,-3,-5$ **69.** $-2,-\frac{1}{3}$ **71.** $\frac{3}{5},\frac{5}{2}$ **73.** $6,-5$ **75.** $(4,0),(6,0)$ **77.** $(-8,0)$ **79.** $(0,0),\left(\frac{4}{3},0\right),\left(\frac{5}{2},0\right)$ **81.** $x=1$ **83.** $x=5$ **85.** $x=9$ **87. d)** **89. b)** **91.** $y=x^2-6x+5$ **93.** $y=x^2-2x-8$ **95.** $y=6x^2-7x-10$ **97.** Width = 2 feet, length = 5 feet **99.** Base = 10 feet, height = 16 feet **101.** 2 feet **103.** 3 feet **105.** 2 seconds **107.** Tim: 5 miles; Bob: 12 miles **109.** 13 feet **111.** 50 bicycles **113.** 13 inches by 13 inches **115. a)** $V=a^3-ab^2$ **b)** $V=a(a+b)(a-b)$ **c)** 3 inches **117. a)** $f(x)=x^2+7x+10$ **b)** $x^2+7x+10=0$ **c)** An infinite number; any function of the form $f(x)=a(x^2+7x+10), a\neq 0$ **d)** An infinite number; any equation of the form $a(x^2+7x+10)=0, a\neq 0$ **119. a)** Answers will vary. Examples are:

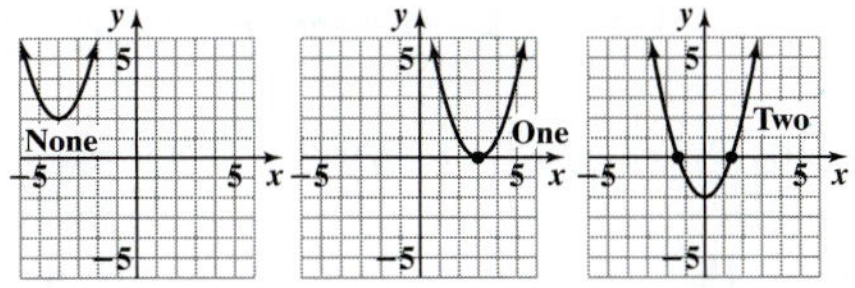


**b)** None (no $x$-intercepts), one (one $x$-intercept), or two (two $x$-intercepts) **121.** $\approx$73.721949 mph **123.** $\pm 2, \pm 3$ **128.** $\frac{x^4}{16y^6}$

**129.** 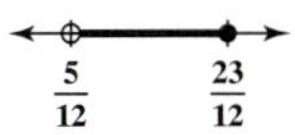


**130.** $(2,-1)$ **131.** $-84$ **132.** $(x+3)(x-2)$

## Chapter 5 Review Exercises

**1. a)** Binomial, **b)** $3x^2+9$, **c)** 2 **2. a)** Trinomial, **b)** $4x^3+5x-7$, **c)** 3 **3.** Not a polynomial **4. a)** Polynomial, **b)** $2x^4-10x^2y+6xy^3-3$, **c)** 4 **5.** $x^2-3x+14$ **6.** $5x^2+11x-4$ **7.** $3a-8b+7$ **8.** $6x^3-9x+13$ **9.** $3x^2y+3xy-9y^2$ **10.** $-3ab+b^2-2a$ **11.** $5x^2+7x+3$ **12.** $-10a^2b+ab$ **13.** 21 **14.** $-76$ **15.** $3x^2+27$ **16.** $2x^2+24x+23$ **17. a)** \$780.46 billion **b)** Yes **18. a)** \$773.13 billion **b)** Yes **19.** $6x^3-14x^2+10x$ **20.** $-3x^4y^2-3x^2y^6+12xy^7$ **21.** $6x^2+17x-45$ **22.** $50a^2-5a-3$ **23.** $x^2+16xy+64y^2$ **24.** $a^2-22ab+121b^2$ **25.** $10x^2y+8xy^2-5x-4y$ **26.** $6p^2q^2+11pqr-7r^2$ **27.** $4a^2+36ab+81b^2$ **28.** $16x^2-24xy+9y^2$ **29.** $49x^2-25y^2$ **30.** $4a^2-25b^4$ **31.** $16x^2y^2-36$ **32.** $81a^4-4b^4$ **33.** $x^2+6xy+9y^2+4x+12y+4$ **34.** $4p^2-4pq+q^2-20p+10q+25$ **35.** $6x^3-x^2-24x+18$ **36.** $4x^4+12x^3+6x^2+16x-6$ **37.** $x^2+8x+10$ **38.** $x^2+xy+4y+xz$ **39. a)** $x^2-2x-3$ **b)** 0 **40. a)** $2x^3-4x^2-6x+12$ **b)** 12 **41. a)** $x^3-x^2-5x+6$ **b)** 9 **42. a)** $x^4-4$ **b)** 77 **43.** $\frac{1}{5}x^6y^2$ **44.** $\frac{1}{4}t^5$ **45.** $9p-5q-3$ **46.** $\frac{7}{4}a^2-4a+8$ **47.** $\frac{x^2}{4y}+x+\frac{3y}{2}$ **48.** $4x-3$ **49.** $x^3-2x^2+3x+7$ **50.** $2a^3+a^2-3a-4$ **51.** $x+4-\frac{10}{x-3}$ **52.** $2x^2+3x-4+\frac{3}{2x+3}$ **53.** $3x^2+7x+21+\frac{73}{x-3}$ **54.** $2y^4-2y^3-8y^2+8y-7+\frac{5}{y+1}$ **55.** $x^4+2x^3+4x^2+8x+16+\frac{14}{x-2}$ **56.** $2x^2+2x+6$ **57.** 10 **58.** $-236$ **59.** $-\frac{53}{9}$ or $-5.\overline{8}$ **60.** 0; factor **61.** $4(x^2+2x+8)$ **62.** $3x^4(5x+2-4xy^3)$ **63.** $2a^2b^3(5a-7b^3)$ **64.** $6xy^2z^2(4y^2z+2xy-5x^2z)$ **65.** $(5x-y)(x+6y)$ **66.** $(3a+2b)(4a+5b)$ **67.** $(2x-5)(x+9)$ **68.** $(3x-7)(16x-21)$ **69.** $(5x+2)(13x-7)$ **70.** $(7x+9)(2x-1)$ **71.** $(17x+3)(9x-7)$ **72.** $(4x+5)(5x-2)$ **73.** $(x+6)(x+3)$ **74.** $(x-2)(x+5)$ **75.** $(x-7)(x+4)$ **76.** $(x-8)(x-2)$ **77.** $-(x-15)(x+3)$ **78.** $-(x-12)(x-1)$ **79.** $x(2x+1)(x+6)$ **80.** $x^2(4x-5)(2x+5)$ **81.** $a^3(4a-5)(a-1)$ **82.** $y^3(12y+1)(y+5)$ **83.** $(x-18y)(x+3y)$ **84.** $(2p-5q)(3p-2q)$ **85.** $(x^2+3)(x^2+7)$ **86.** $(x^2-7)(x^2+9)$ **87.** $(x+9)(x+7)$ **88.** $x(x-9)$ **89.** $(x+10)(x+1)$ **90.** $(x+10)(x+2)$ **91.** $(x+6)(x-6)$ **92.** $(x+11)(x-11)$ **93.** $(x^2+9)(x+3)(x-3)$ **94.** $(x^2+4)(x+2)(x-2)$ **95.** $(2a+1)^2$ **96.** $(4y-3)^2$ **97.** $(x-2)(x+6)$ **98.** $(3y+5)(3y-7)$ **99.** $(p^2+9)^2$ **100.** $(m^2-10)^2$ **101.** $(x+4+y)(x+4-y)$ **102.** $(a+3b+6c)(a+3b-6c)$ **103.** $(4x+y)^2$ **104.** $(6b-5c)^2$ **105.** $(x-3)(x^2+3x+9)$ **106.** $(y+4z)(y^2-4yz+16z^2)$ **107.** $(5x-1)(25x^2+5x+1)$ **108.** $(2a+3b)(4a^2-6ab+9b^2)$ **109.** $(y-4z)(y^2+4yz+16z^2)$ **110.** $(x-5)(x^2-x+7)$ **111.** $(x-1)(x^2+4x+7)$ **112.** $(a+5)(a^2+7a+13)$ **113.** $(x+3)(x-3)$ **114.** $(a+2b)(a-2b)$ **115.** $(2x-y)(4x^2+2xy+y^2)$ **116.** $4a(a+c)(a-c)$ **117.** $y^4(x+3)(x-5)$ **118.** $5x(x-4)(x-2)$ **119.** $3xy^4(x-2)(x+6)$ **120.** $3y(y^2+5)(y^2-5)$ **121.** $4y(x+2)(x^2-2x+4)$ **122.** $5x^2y(x+2)^2$

**123.** $3x(2x+1)(x-4)$ **124.** $(x+5+z)(x+5-z)$ **125.** $5(x+2y)(x^2-2xy+4y^2)$ **126.** $(x+4)(x-1)(x+6)$ **127.** $(4x+1)(4x+5)$ **128.** $(2x^2-1)(2x^2+3)$ **129.** $(x+1)^2(x-2)$ **130.** $(3a-b)(3x+7y)$ **131.** $(2pq-3)(3pq+2)$ **132.** $(3x^2-2)^2$ **133.** $(4y+x+2)(4y-x-2)$ **134.** $2(3a+5)(4a+3)$ **135.** $3x^2y^5(x+3)(2x-3)$ **136.** $\left(x-\frac{2}{3}y^2\right)\left(x^2+\frac{2}{3}xy^2+\frac{4}{9}y^4\right)$ **137.** $(x+9)(x+2)$ **138.** $(y+10)(y+5)$ **139.** $(a+2b)(a-2b)$ **140.** $2b(a+b)$ **141.** $(2a+b)(a+3b)$ **142.** $(a+b)^2$ **143.** $2, -\frac{1}{4}$ **144.** $-\frac{5}{2}, -\frac{10}{3}$ **145.** $0, 2$ **146.** $0, -\frac{4}{3}$ **147.** $-4, -3$ **148.** $5, -6$ **149.** $7, 1$ **150.** $0, 2, 4$ **151.** $4, -4$ **152.** $2, -3$ **153.** $\frac{4}{3}, -\frac{1}{4}$ **154.** $\frac{3}{4}, -\frac{2}{5}$ **155.** $(-3, 0), (6, 0)$ **156.** $\left(\frac{6}{5}, 0\right), \left(\frac{5}{4}, 0\right)$ **157.** $y = x^2 - 2x - 24$ **158.** $y = 12x^2 + 32x + 5$ **159.** Width = 9 feet, length = 12 feet **160.** Height = 4 feet, base = 13 feet **161.** 3 inches, 7 inches **162.** 9 seconds **163.** 9

## Chapter 5 Practice Test

**1. a)** Trinomial **b)** $-6x^4 - 4x^2 + 3x$ **c)** 4 **d)** $-6$ [5.1] **2.** $4x^2y - 14y^2 + 4x + 6y$ [5.1] **3.** $-8x^8y^3 + 24x^6y^4 - 12x^4y^2$ [5.2] **4.** $10a^2 - 13ab - 3b^2$ [5.2] **5.** $4x^3 + 8x^2y - 9xy^2 - 6y^3$ [5.2] **6.** $4x^4 - 5y + \frac{7}{x^2}$ [5.3] **7.** $x - 5 + \frac{24}{2x+3}$ [5.3] **8.** $3x^3 + 3x^2 + 15x + 15 + \frac{76}{x-5}$ [5.3] **9.** $-85$ [5.3] **10.** $2xy(6x^2 + 5xy^3 - 7y^2)$ [5.4] **11.** $x(x-3)(x+1)$ [5.5] **12.** $(a+2b)(2a+3b)$ [5.4] **13.** $(2b^2+9)(b^2-2)$ [5.5] **14.** $4x(x-5)$ [5.5] **15.** $(x+7)(x+3)$ [5.5] **16.** $q^6(3p-2)(9p^2+6p+4)$ [5.6] **17. a)** $3x^2 - 19x + 20$ **b)** $-6$ [5.2] **18.** $4(x+y)(x-y)$ [5.5] **19.** $(x+11)(x+4)$ [5.5] **20.** $\frac{3}{7}, -4$ [5.8] **21.** $0, -5, 2$ [5.8] **22.** $\left(\frac{1}{4}, 0\right), \left(-\frac{3}{2}, 0\right)$ [5.8] **23.** $y = x^2 - 9x + 14$ [5.8] **24.** Height = 4 meters, base = 11 meters [5.8] **25.** 7 seconds [5.8]

## Cumulative Review Test

**1.** $A \cup B = \{2, 3, 4, 5, 6, 8\}$ [1.2] **2.** 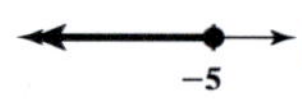

[1.2] **3.** $-\frac{3}{32}$ [1.3] **4.** $-34$ [1.4] **5.** $8r^6s^{15}$ [1.5] **6.** 5 [2.1] **7.** $e = \frac{k - 2d}{2}$ [2.2] **8.** 13 meters by 13 meters [2.3] **9.** 620 pages [2.3] **10.** $34 \le x < 84$ [2.5] **11.** No [3.1] **12.** $6x - 3y = 2$ [3.3] **13.** $-\frac{2}{9}$ [3.4] **14.** $-180$ [3.6] **15.** 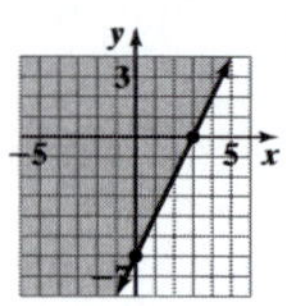

[3.7] **16.** $(10, 4)$ [4.1] **17.** $(4, 1, 2)$ [4.2] **18.** 18 [4.5] **19.** $2x^2 + 12x + 63 + \frac{393}{x-6}$ [5.3] **20.** $(4x - 3y)(16x^2 + 12xy + 9y^2)$ [5.6]

# Chapter 6

## Exercise Set 6.1

**1. a)** A rational expression is an expression of the form $\frac{p}{q}$, $p$ and $q$ polynomials, $q \neq 0$. **b)** Answers will vary. **3. a)** A rational function is a function of the form $f(x) = \frac{p}{q}$, $p$ and $q$ polynomials, $q \neq 0$. **b)** Answers will vary. **5. a)** The domain of a rational function is the set of values that can replace the variable **b)** $\{x|x \neq -5 \text{ and } x \neq 5\}$ **7. a)** Factor $-1$ from numerator or denominator and reduce. **b)** $-1$ **9. a)** Invert second fraction, factor all expressions, simplify, then multiply numerators and multiply denominators. **b)** $\frac{1}{r+6}$ **11.** 4 **13.** $5, \frac{5}{2}$ **15.** None **17.** $9, -9$ **19.** $\{p|p \neq 2\}$ **21.** $\{x|x \neq -3 \text{ and } x \neq 2\}$ **23.** $\left\{a \middle| a \neq \frac{1}{2} \text{ and } a \neq -2\right\}$ **25.** $\{x|x \text{ is a real number}\}$ **27.** $\{a|a \neq -6 \text{ and } a \neq 6\}$ **29.** $1 - y$ **31.** $\frac{x-4y}{3}$ **33.** $x$ **35.** $-1$ **37.** $-(p+4)$ **39.** $\frac{a-5}{a+3}$ **41.** $4x^2 + 10xy + 25y^2$ **43.** $\frac{2x-5}{2}$ **45.** $\frac{a+7}{a+5}$ **47.** $\frac{x-4}{x^2-3x+9}$ **49.** $\frac{xy^2}{15}$ **51.** $12x^3y^2$ **53.** 1 **55.** $\frac{x+5}{4}$ **57.** $\frac{r^3}{r-8}$ **59.** $\frac{7}{x-4}$ **61.** $\frac{(a+1)^2}{9(a+b)^2}$ **63.** $\frac{x-4}{4x+1}$ **65.** $\frac{(x+2)(x-2)}{(x^2+2x+4)(x^2+4)}$ **67.** $\frac{x-y}{x+y}$ **69.** $\frac{x^3}{x+2}$ **71.** $\frac{(a-b)(a+b)}{a^2+ab+b^2}$ **73.** 1 **75.** $\frac{p-q}{p+q}$ **77.** $\frac{r+5s}{2r+5s}$ **79.** One possible answer is $\frac{1}{(x-2)(x+3)}$; denominator 0 at $x = 2$ and $x = -3$ **81.** Numerator is never 0. **83. a)** 4, makes numerator 0 **b)** 6 and $-6$, makes denominator 0 **85.** One possible answer is $f(x) = \frac{x-2}{(x-3)(x+1)}$; numerator 0 at $x = 2$, denominator 0 at $x = 3$ and $x = -1$ **87.** $x + 5$; numerator must be $x + 5$ **89.** $y^2 - 4y - 5$, factors must be $(y-5)(y+1)$ **91.** $x^2 + x - 2$; factors must be $(x-1)(x+2)$ **93.** $2x^2 + x - 6$; factors must be $(x+2)(2x-3)$. **95.** $\frac{3a+b}{2}$ **97.** $2(a+b)$ **99.** $\frac{(x+2)(3x+1)}{(2x-3)(x+1)}$ **101.** $\frac{x-1}{x+3}$ **103.** $\frac{1}{x^4(x-p)^n}$ **105.** $x^y$

**107. a)** $\{x|x \neq 2\}$ **b)**

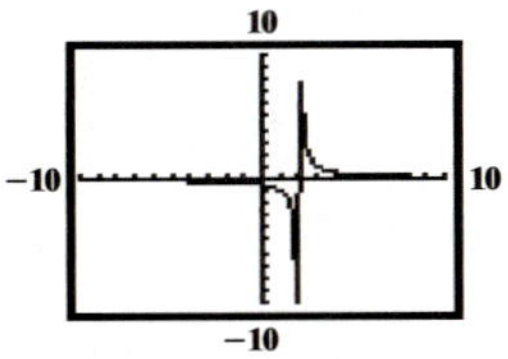


**c)** Decreasing **d)** Increasing

**109. a)** $\{x|x \neq 2\}$ **b)**

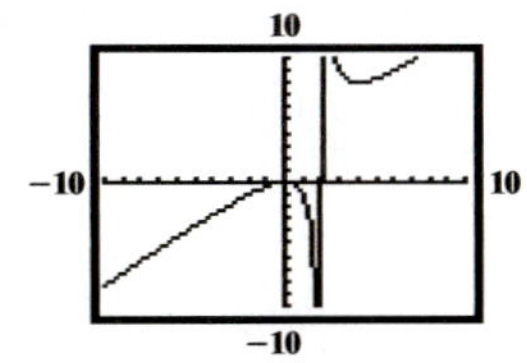


**c)** Decreasing **d)** Increasing

**111. a)** $\{x|x \neq 0\}$ **b)** $-0.1, -1, -2, -10, -100, 100, 10, 2, 1, 0.1$ **c)**

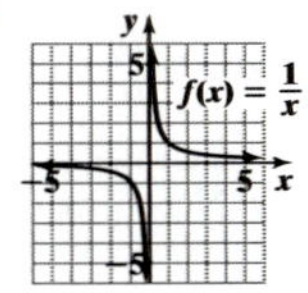


**d)** No; numerator can never be 0 **113.** $y = x + 2$

**114.** $\left(-\infty, \frac{3}{2}\right)$ **115.** $-28, 32$ **116.** 0.1 **117.** $(2, -1)$

**118.** $(3x + y + 2)(3x + y - 2)$

## Exercise Set 6.2

**1. a), b)** Answers will vary. **c)** $(8x + 11)(8x - 11)(x - 2)$ **3. a)** The entire numerator was not subtracted. **b)** $\frac{x^2 - 4x - x^2 - x + 2}{(x + 3)(x - 2)}$ **5.** $\frac{3x + 5}{x + 2}$ **7.** $\frac{7x - 2}{x - 5}$ **9.** $\frac{x + 7}{x + 3}$ **11.** $\frac{7x - 11}{x - 8}$ **13.** $\frac{x - 3}{x - 1}$ **15.** $x - 5$ **17.** $\frac{x + 5}{x + 3}$ **19.** $6a^3$ **21.** $40x^4y^6$ **23.** $6a^4b^5$ **25.** $(x + 3)(x + 9)$ **27.** $z - 6$ **29.** $x^4(x - 2)^3$ **31.** $(a - 8)(a + 3)(a + 8)$ **33.** $(x - 3)(2x - 1)(2x + 3)$ **35.** $\frac{26}{3r}$ **37.** $\frac{5x - 3}{12x^2}$ **39.** $\frac{15y^2 + 8x^2}{40x^4y^3}$ **41.** $\frac{2b^2 - a^2}{b(a - b)}$ **43.** $\frac{2a}{a - b}$ **45.** $\frac{5x^2 + 3x - 12}{(x - 4)(x + 1)}$ **47.** $\frac{6a + 7}{(a + 2)^2}$ **49.** $\frac{2x^2 + 4x + 4}{(x - 1)(x + 4)(x - 2)}$ **51.** $\frac{2x + 3}{(x - 8)(x - 1)}$ **53.** $\frac{4x^2 + 11x - 39}{(x + 5)(x - 2)}$ **55.** $\frac{3a - 1}{4a + 1}$ **57.** $\frac{2x^2 - 4xy + 4y^2}{(x - 2y)^2(x + 2y)}$ **59.** $\frac{16}{r - 4}$ **61.** 0 **63.** $\frac{15x^2 - 70x + 30}{(3x - 2)(x - 4)}$ **65.** $\frac{18r^2 + 11r - 25}{(4r - 5)(2r + 3)}$ **67.** $\frac{x^2 - 18x - 30}{(5x + 6)(x - 2)}$ **69.** $\frac{12m^2 + 7mn}{(2m + 3n)(3m + 2n)(2m + n)}$ **71.** 0 **73.** $\frac{1}{2x + 3y}$ **75.** No **77.** Yes, if you multiply either fraction by $\frac{-1}{-1}$ you get the other fraction. **79. a)** $\{x|x \neq 3\}$ **b)** $\{x|x \neq -4\}$ **c)** $\frac{2x^2 + 3x + 8}{(x - 3)(x + 4)}$ **d)** $\{x|x \neq 3 \text{ and } x \neq -4\}$ **81.** $\frac{2x^2 + 8x - 3}{(x + 1)(x + 2)}$ **83.** $\frac{3x^2 + 19x + 7}{(x + 2)(x + 3)}$ **85.** D: $\{x|x \neq 2\}$, R: $\{y|y \neq 1\}$ **87.** $\frac{x^2 + 5x + 4}{(x + 2)(x - 2)(x + 3)}$ **89.** $\frac{2x}{x^4 + x^3 - 10x^2 - 4x + 24}$ **91.** $\frac{ad}{bd} + \frac{bc}{bd} = \frac{ad + bc}{bd}$ **93. a)** 4 **b)** $\frac{a^2 - b^2}{a^2}$ **95.** $7x^2 - 6x + 6; 5x^2 - (7x^2) = -2x^2, -(-6x) = 6x, -6 - (6) = -12$ **97.** $\frac{3x + 10}{x - 2}$ **99.** $\frac{1}{(a - 5)(a + 3)}$ **101.** $-x^2 + 4x + 5$ **103. a)** $\frac{ax + bn - bx}{n}$ **b)** 79.2 **105.** $\frac{a - b + 1}{(a - b)^2}$ **107.** No **109. a)** $\frac{x + 1}{x}$ **b)** $\frac{x^2 + x + 1}{x^2}$ **c)** $\frac{x^4 + x^3 + x^2 + x + 1}{x^4}$ **d)** $\frac{x^n + x^{n-1} + x^{n-2} + \cdots + 1}{x^n}$ **111.** $\frac{-h}{(a + 1)(a + h + 1)}$ **112. a)** 6 minutes **b)** 960 boxes **113.** $\{x|-2 < x < 8\}$ **114.** $-\frac{2}{3}$ **115.** $-11$ **116.** $3x - 7 + \frac{27}{2x + 3}$ **117.** $\frac{1}{3}, 7$

## Exercise Set 6.3

**1.** One that has a fractional expression in its numerator or denominator or both. **3.** $\frac{75a}{b^5}$ **5.** $\frac{12x^3}{y^6}$ **7.** $\frac{3z^4}{4xy^4}$ **9.** $\frac{y - x}{3xy}$ **11.** $\frac{x(y - 1)}{8 + x}$ **13.** $\frac{xy + 5}{y + x}$ **15.** $\frac{5}{3a^2}$ **17.** $-\frac{a}{b}$ **19.** $\frac{x - y}{y}$ **21.** $-1$ **23.** $\frac{3(x + 2)}{x^5}$ **25.** $\frac{-a + 1}{(a + 1)(2a + 1)}$ **27.** $\frac{x - 1}{x + 1}$ **29.** $\frac{a^2 + 1}{2a}$ **31.** $\frac{x}{5(x - 3)}$ **33.** $\frac{x^2 + 5x - 6}{x(x - 2)}$ **35.** $\frac{a(a + 3)}{(a - 2)(a + 1)}$ **37.** $\frac{2 + a^2b}{a^2}$ **39.** $\frac{ab}{b + a}$ **41.** $\frac{b(1 + a)}{a(1 - b)}$ **43.** $\frac{b^2 - a^3b}{a^3 + ab}$ **45.** $\frac{9a^2 + b}{b(b + 1)}$ **47.** $\frac{(a + b)^2}{ab}$ **49.** $\frac{15y - x}{3xy}$ **51.** $\frac{2y - 8xy + 5y^2}{3y^2 - 4x}$ **53.** $\frac{x^2 + 9x + 14}{x + 1}$ **55.** $\frac{x^2 + 3x - 4}{x + 1}$ **57. a)** $\frac{2}{9}$ **b)** $\frac{1}{5}$ **59.** $R_T = \frac{R_1R_2R_3}{R_2R_3 + R_1R_3 + R_1R_2}$ **61.** $a$ **63.** $\frac{-1}{a(a + h)}$ **65.** $\frac{-1}{(a + 1)(a + h + 1)}$ **67.** $\frac{-2a - h}{a^2(a + h)^2}$ **69.** $\frac{4a^2 + 1}{4a(2a^2 + 1)}$ **71.** $\frac{5}{12}$ **72.** $\frac{13}{48}$ **73.** $\left(-23, -\frac{34}{5}\right)$ **74.** $\left\{3, \frac{5}{3}\right\}$ **75.** Neither

## Exercise Set 6.4

**1.** A number obtained when solving an equation that is not a true solution **3. a)** Multiply both sides by 12 to remove fractions **b)** $-24$ **c)** Write each term with LCD of 12 so you can add and subtract. **d)** $\frac{-x + 24}{12}$

**5.** Similar figures are figures whose corresponding angles are the same and whose corresponding sides are in proportion.
**7.** No, $x = 3$ makes $\frac{7}{x-3}$ undefined **9.** 5 **11.** $\frac{11}{2}$ **13.** 3 **15.** $-5$ **17.** All real numbers. **19.** $\frac{1}{4}$ **21.** $\frac{11}{3}$ **23.** No solution
**25.** $\frac{6}{5}$ **27.** $\approx -1.63$ **29.** 8 **31.** $-\frac{4}{3}, 1$ **33.** 3.76 **35.** $-1, -6$ **37.** $-5$ **39.** $-\frac{5}{2}$ **41.** 5 **43.** No solution **45.** $-5$ **47.** $\frac{17}{4}$
**49.** 12, 2 **51.** 12, 4 **53.** 1, $-2$ **55.** $\frac{25}{2}$ **57.** $\frac{3}{2}$ **59.** $P_2 = \frac{V_1P_1}{V_2}$ **61.** $V_2 = \frac{V_1P_1}{P_2}$ **63.** $y = y_1 + m(x - x_1)$ **65.** $x = zs + \bar{x}$
**67.** $w = \frac{fl - df}{d}$ **69.** $q = \frac{pf}{p - f}$ **71.** $a = \frac{v_2 - v_1}{t_2 - t_1}$ **73.** $d = \frac{a_n - a_1}{n - 1}$ **75.** $G = \frac{Fd^2}{m_1m_2}$ **77.** $T_1 = \frac{T_2P_1V_1}{P_2V_2}$ **79.** $V_0 = \frac{S - S_0 - gt^2}{t}$
**81. a)** $\frac{2x + 9}{(x - 2)(x + 2)}$ **b)** $-\frac{9}{2}$ **83. a)** $\frac{4}{b + 5}$ **b)** No solution **85.** $c \neq 0$; cannot divide by 0 **87.** $f(x)$: graph **b)**; $g(x)$: graph **a)**; $f(x)$ is undefined when $x = 3$ **89. a)** \$ 6250 **b)** $R = \frac{AC}{0.80I}$ **91. a)** 20 feet per minute squared **b)** $t_1 = t_2 + \frac{v_1 - v_2}{a}$
**93. a)** $\approx$22.5% **b)** $D = PR$ **c)** $R = \frac{D}{P}$ **95.** 150 ohms **97.** $\approx$0.101 meter **99. a)** $\approx$9.71% **b)** Tax-Free Money Market Portfolio since 9.71% > 7.68%. **101.** One answer is $\frac{1}{x - 4} + \frac{1}{x + 2} = 0$; 4 and $-2$ make the fraction undefined **103.** One answer is $\frac{1}{x} + \frac{1}{x} = \frac{2}{x}$. **105.** $-1 < x \leq 3$ **106.** $m = -\frac{1}{3}$; y-intercept, $\left(0, \frac{14}{3}\right)$. **107.** $3x^2y - 7xy - 4y^2 - 9x$ **108.** 2 feet

## Mid-Chapter Test

**1.** $\{x|x \neq 0, x \neq -5, \text{and } x \neq 5\}$ [6.1] **2.** $\frac{x + 5}{2x - 3}$ [6.1] **3.** $\frac{55b}{a^2 - ab + b^2}$ [6.1] **4.** $\frac{x - 3}{x + 1}$ [6.1]
**5.** $\frac{(2a + 1)(2a + 3)}{(2a - 1)(a - 9)}$ or $\frac{4a^2 + 8a + 3}{2a^2 - 19a + 9}$ [6.1] **6.** $\frac{4a + 3b}{6}$ [6.1] **7.** $(x + 5)(x - 6)(x + 2)$ [6.2] **8.** 5 [6.2]
**9.** $\frac{20y^2 + ax}{6x^2y^3}$ [6.2] **10.** $\frac{-2x - 7}{(x - 4)(x + 4)(2x - 3)}$ [6.2] **11.** $\frac{9b + a}{3 - c}$ [6.3] **12.** $\frac{5x - 8}{6x^2 - x}$ [6.3] **13.** $y^2$ [6.3]
**14.** An extraneous root is a number obtained when solving an equation that is not a solution to the original equation. Whenever a variable appears in the denominator, you must check the apparent solution. [6.4] **15.** 5. [6.4] **16.** No solution [6.4]
**17.** 4, $-3$ [6.4] **18.** $a = \frac{bc}{b + c}$ [6.4] **19.** $r = \frac{x - 4}{x}$ [6.4] **20.** 14 and 5 [6.4]

## Exercise Set 6.5

**1.** Equal to $\frac{1}{2}$ since it takes exactly the same time for each of them.

**3. a)**

| Worker | Rate of Work | Time Completed | Part of Task Completed |
|---|---|---|---|
| Bill | $\frac{1}{7}$ | $x$ | $\frac{x}{7}$ |
| Bob | $\frac{1}{9}$ | $x$ | $\frac{x}{9}$ |

**b)** $\frac{x}{7} + \frac{x}{9} = 1$ **c)** Less; it should take less time than the fastest person because they are working together.
**5.** 1.5 months **7.** 2 hours **9.** 18.75 minutes **11.** 4 hours **13.** $\approx$2.48 days **15.** 2.4 hours **17.** $\approx$3.08 hours **19.** 100 hours **21.** 7.8 months **23.** 75 minutes **25.** $\approx$15.27 minutes **27.** $\approx$1.62 hours **29.** 12 hours **31.** 5 **33.** 2, 4 **35.** 4, 6

**37.** 20 **39.** $\frac{2}{3}, 1$ **41.** $\approx$0.064 mile per hour **43.** $\approx$1.53 feet per second **45.** 7.5 miles **47.** 36 miles per hour **49.** $\approx$30.59 yards
**51.** Local: $\approx$10.93 miles per hour, express: $\approx$16.13 miles per hour. **53.** Car: 60 miles per hour, train: 30 miles per hour
**55.** 60 miles per hour **57.** 120 kilometers per hour **59.** 2 hours at 6 miles per hour, $\frac{1}{2}$ hour at 10 miles per hour
**61.** 18 feet per minute **63.** 108,000 miles **65.** Answers will vary. **67. a)** 10 minutes **b)** 15 miles **c)** 165 miles per hour
**68.** $\frac{x^8}{72}$ **69.** $9.26 \times 10^9$ **70.** \$ 2500 **71.**

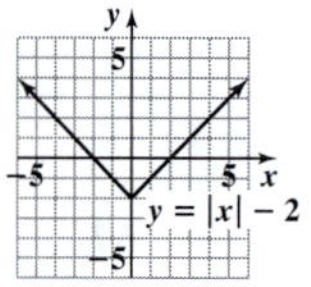


**72.** $a(2a^2 - 5)(a - 1)$

## Exercise Set 6.6

**1. a)** As one quantity increases, the other increases. **b), c)** Answers will vary. **3.** One quantity varies as a product of two or more quantities. **5. a)** Decrease **b)** Inverse variation; by the definition of inverse variation **7.** Direct **9.** Inverse **11.** Direct **13.** Direct **15.** Direct **17.** Inverse **19.** Direct **21.** Inverse **23.** Inverse **25. a)** $x = ky$ **b)** 72
**27. a)** $y = kR$ **b)** 306 **29. a)** $R = \frac{k}{W}$ **b)** $\frac{1}{20}$ **31. a)** $A = \frac{kB}{C}$ **b)** 9 **33. a)** $x = ky$ **b)** 20 **35. a)** $y = kR^2$ **b)** 20

**37. a)** $S = \frac{k}{G}$ **b)** 0.96 **39. a)** $x = \frac{k}{P^2}$ **b)** 25 **41. a)** $F = \frac{kM_1M_2}{d}$ **b)** 40 **43.** Doubled **45.** Halved **47.** Doubled **49.** Unchanged **51.** Doubled **53.** $y = \frac{k}{x}; k = 5$ **55.** \$8814 **57.** 3096 milligrams **59.** 1.05 inches **61.** 6400 cubic centimeters **63.** 3.12 hours **65.** 45 foot-candles **67.** 117.6 feet **69.** 126 cubic meters **71.** 4600 DVDs **73.** ≈133.25 pounds **75.** ≈121,528 calls **77.** $\frac{1}{49}$ of the light of the flash **79. a)** $P = 14.7 + kx$ **b)** 0.43 **c)** ≈337.9 feet **80.** $h = \frac{3V}{4\pi r^2}$ **81.** 132 **82.** $-14x^3 - 22x^2 + 47x - 15$ **83.** $(x + 3)(x - 2)$

## Chapter 6 Review Exercises

**1.** 5 **2.** −1 **3.** None **4.** $\{x|x \neq -3\}$ **5.** $\{x|x \neq 0\}$ **6.** $\{x|x \neq 2 \text{ and } x \neq -6\}$ **7.** $x$ **8.** $x - 6$ **9.** −1 **10.** $\frac{x-1}{x-2}$ **11.** $\frac{x-3}{x+1}$ **12.** $\frac{a^2+2ab+4b^2}{a+2b}$ **13.** $\frac{9x^2-3xy+y^2}{3x-y}$ **14.** $\frac{2x-3}{x^2-2x+4}$ **15.** $x(x+4)$ **16.** $(x+2y)(x-2y)$ **17.** $(x+7)(x-5)(x+2)$ **18.** $(x+2)^2(x-2)(x+3)$ **19.** $12xz^2$ **20.** $-\frac{x}{6}$ **21.** $9x^3z^5$ **22.** $\frac{11x+6}{3x^2}$ **23.** $\frac{(x-y)y^2}{4x^3}$ **24.** $3x+2$ **25.** $\frac{30x+3y^2}{5x^2y}$ **26.** 1 **27.** $\frac{2x+1}{3x+1}$ **28.** $\frac{6a+7}{a+1}$ **29.** $\frac{6b-8}{b-1}$ **30.** $\frac{a^2-b^2}{a^2}$ **31.** $\frac{1}{3(a+3)}$ **32.** $\frac{a^2+c^2}{ac}$ **33.** $\frac{x+1}{2x-1}$ **34.** 1 **35.** $4x(x-5y)$ **36.** $\frac{2a^2+9a+4}{4a(a+2)}$ **37.** $\frac{x^2+5}{(x+5)(x-5)}$ **38.** $-\frac{2(x+1)}{x^2-4}$ **39.** $\frac{x+5}{x+6}$ **40.** $\frac{-x+5}{(x+2)(x-2)(x-3)}$ **41.** $\frac{16(x-2y)}{3(x+2y)}$ **42.** $\frac{3}{a^3}$ **43.** $\frac{22x+5}{(x-5)(x-10)(x+5)}$ **44.** $\frac{2(x-4)}{(x-3)(x-5)}$ **45.** $-\frac{1}{x-3}$ **46.** $\frac{a+3}{a+5}$ **47.** $\frac{x+6}{x-4}$ **48.** $\frac{a+2b^2}{3}$ **49.** $\frac{x^2+6x-24}{(x-1)(x+9)}$ **50.** $\frac{x-4}{x-6}$ **51. a)** $\{x|x \neq -2\}$ **b)** $\{x|x \neq -4\}$ **c)** $\frac{2x^2+7x+4}{(x+2)(x+4)}$ **d)** $\{x|x \neq -2 \text{ and } x \neq -4\}$ **52. a)** $\{x|x \neq 3 \text{ and } x \neq -3\}$ **b)** $\{x|x \neq 3\}$ **c)** $\frac{x^2+8x+12}{(x+3)(x-3)}$ **d)** $\{x|x \neq 3 \text{ and } x \neq -3\}$ **53.** $\frac{3ac^2}{b^3}$ **54.** $\frac{4x+2y}{x^2+xy^3}$ **55.** $\frac{3y-1}{7y^2+1}$ **56.** $\frac{5a+1}{2}$ **57.** $\frac{3x+1}{-x+1}$ **58.** $\frac{3x^2-29x+68}{4x^2-6x-54}$ **59.** $\frac{x^2+3x+2}{x+5}$ **60.** $\frac{x^2+6x+8}{x+3}$ **61.** $\frac{18}{5}$ or $3\frac{3}{5}$ **62.** −2 **63.** 52 **64.** 2.4 **65.** 5 **66.** −9 **67.** −18 **68.** −28 **69.** −6 **70.** −10 **71.** $b = \frac{ac}{a-c}$ **72.** $\bar{x} = x - sz$ **73.** 60 ohms **74.** 2 centimeters **75.** 10, 2 **76.** 21, 3 **77.** ≈17.14 minutes **78.** 14 hours **79.** 3 **80.** $\frac{5}{6}$ **81.** 5 miles per hour **82.** car: 50 miles per hour, plane: 150 miles per hour **83.** 20 **84.** $\frac{25}{2}$ **85.** ≈426.7 **86.** \$8.40 **87.** 1600 feet **88.** 200.96 square units **89.** 2.38 minutes

## Chapter 6 Practice Test

**1.** −7 and 4 [6.1] **2.** $\left\{x \middle| x \neq -4 \text{ and } x \neq \frac{1}{2}\right\}$ [6.1] **3.** $5x^5y + 8 + 11xy^2$ [6.1] **4.** $\frac{x-6y}{x+y}$ [6.1] **5.** $\frac{1}{x^4y^2}$ [6.1] **6.** $\frac{1}{x+2}$ [6.1] **7.** $\frac{7}{a(a+b)}$ [6.1] **8.** $x^2 + y^2$ [6.1] **9.** $\frac{5x^2+2x+2}{x^2(x+1)}$ [6.2] **10.** $\frac{-3x-1}{(x-3)(x+3)(x+1)}$ [6.2] **11.** $\frac{m(6m+n)}{(6m+5n)(2m-n)(2m+3n)}$ [6.2] **12.** $\frac{x(x+10)}{(2x-1)^2(x+3)}$ [6.2] **13.** $x + 3$ [6.1] **14. a)** $\frac{3x^2+2x-9}{(x+5)(2x+3)}$ **b)** $\left\{x \middle| x \neq -5 \text{ and } x \neq -\frac{3}{2}\right\}$ [6.2] **15.** $\frac{x+5}{x+2}$ [6.1] **16.** $\frac{y+2x}{y-3x}$ [6.3] **17.** $\frac{b(a-b)}{a}$ [6.3] **18.** $\frac{7x-6}{4x^2-x}$ [6.3] **19.** 20 [6.4] **20.** 12 [6.4] **21.** $C = \frac{2b+Ad}{A}$ [6.4] **22.** 0.75 watt [6.6] **23.** 6 [6.6] **24.** ≈4.44 hours [6.5] **25.** $6\frac{2}{3}$ miles [6.5]

## Cumulative Review Test

**1.**

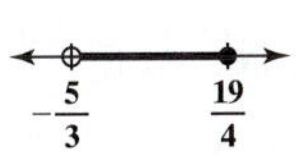


[1.2] **2.** $-27\frac{3}{4}$ [1.4] **3.** −3 [2.1] **4. a)** 28% **b)** ≈44,000 [1.3] **5.** 62 [1.4] **6.** $\frac{x^3}{8y^3}$ [1.5] **7.** $m = \frac{rF}{v^2}$ [2.2] **8.** 6% [2.2] **9.** 11 A.M. [2.4] **10.** $\left\{-\frac{32}{3}, \frac{22}{3}\right\}$ [2.6]

**11.**

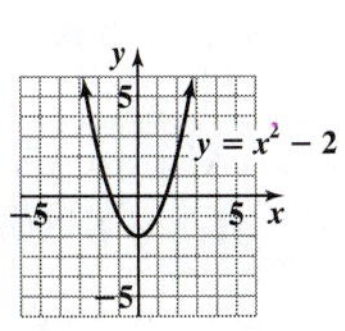


[3.1] **12.** 5 [3.2] **13.** $-\frac{1}{7}$ [3.4] **14.** $2x + 3y = 4$ [3.5] **15.** $\left(\frac{1}{2}, 3\right)$ [4.1] **16.** $9x^4 - 25y^2$ [5.2] **17.** $3(x-5)^2$ [5.6] **18.**

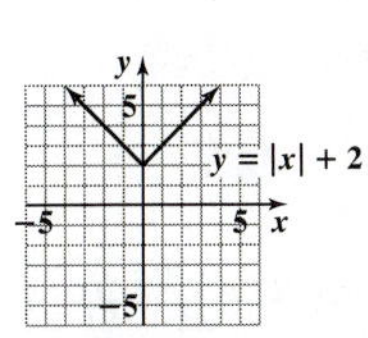


[3.1] **19.** $\frac{3x-4}{(x-1)(x-2)}$ [6.2] **20.** 4 [6.4]

# Chapter 7

**Exercise Set 7.1** **1. a)** Two, positive and negative. **b)** 7, −7 **c)** Principal square root **d)** 7 **3.** There is no real number which, when squared gives −81. **5.** No; if the radicand is negative, the answer is not a real number. **7. a)** 1.3 **b)** 1.3 **9. a)** 3 **b)** −3 **c)** −3 **11.** 6 **13.** −4 **15.** −5 **17.** −1 **19.** 1 **21.** Not a real number **23.** −7 **25.** Not a real number **27.** Not a real number **29.** $\frac{1}{5}$ **31.** $\frac{1}{2}$ **33.** $\frac{2}{7}$ **35.** $-\frac{2}{3}$ **37.** $\approx -2.07$ **39.** 7 **41.** 19 **43.** 119 **45.** 235.23 **47.** 0.06 **49.** $\frac{12}{13}$ **51.** $|x-4|$ **53.** $|x-3|$ **55.** $|3x^2-1|$ **57.** $|6a^3-5b^4|$ **59.** $|a^7|$ **61.** $|z^{16}|$ **63.** $|a-4|$ **65.** $|3a+2b|$ **67.** $7x$ **69.** $4c^3$ **71.** $x+2$ **73.** $2x+y$ **75.** 2 **77.** 8 **79.** 9 **81.** $\approx 9.381$ **83.** $\approx 5.290$ **85.** −3 **87.** 97 **89.** 11 **91.** 45 **93.** Select a value less than $-\frac{1}{2}$. **95.** $x \ge 1$ **97.** $x \ge 3$ **99. a)** All real numbers **b)** $a \ge 0$ **c)** All real numbers **101.** If $n$ is even, you are finding an even root of a positive number. If $n$ is odd, the expression is real. **103.** $x > -5$ **105.** d **107.** a **109.** One answer is $f(x) = \sqrt{x-8}$ **111. a)** No **b)** Yes, when $x = 0$ **c)** Yes **113. a)** $\sqrt{1288} \approx 35.89$ feet per second **b)** $\sqrt{2576} \approx 50.75$ feet per second. **115.**

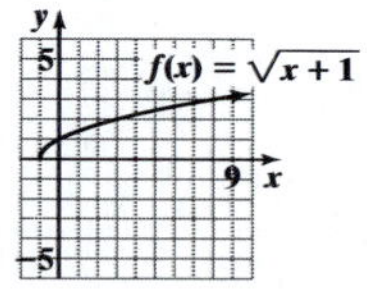


**117.**

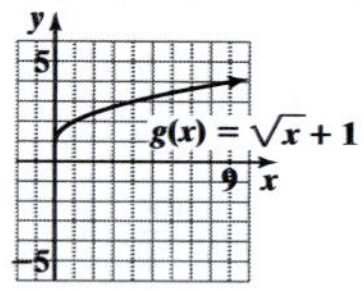


**119.**

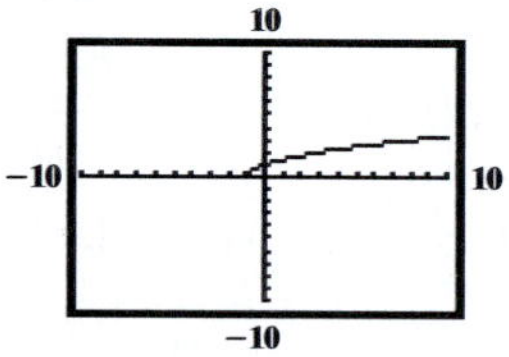


**121.**

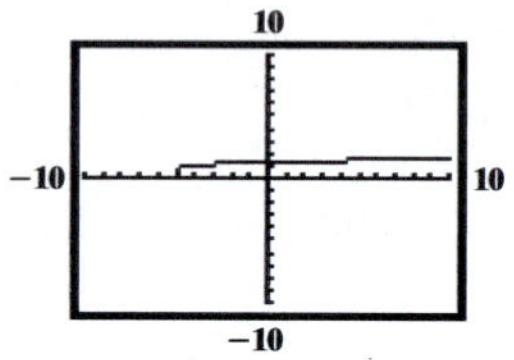


**123.**

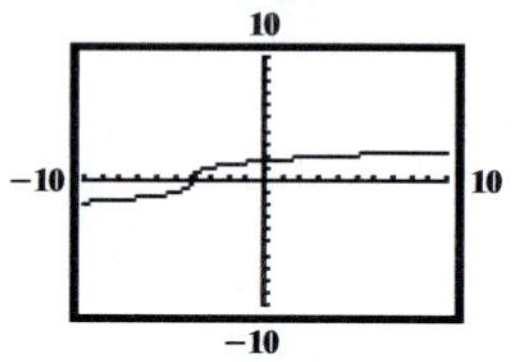


**127.** $(3a-b)(3x+4y)$ **128.** $3x(x-4)(x-2)$ **129.** $(4x^2-1)(2x^2+3)$ **130.** $\left(x-\frac{2}{3}y\right)\left(x^2+\frac{2}{3}xy+\frac{4}{9}y^2\right)$

**Exercise Set 7.2** **1. a)** When $n$ is even and $a \ge 0$, or $n$ is odd **b)** $a^{1/n}$ **3. a)** Always real **b)** $a$ **c)** $a$ **d)** $|a|$ **5. a)** No; $(xy)^{1/2} = x^{1/2}y^{1/2}$ **b)** No; $(xy)^{-1/2} = x^{-1/2}y^{-1/2} = \frac{1}{x^{1/2}y^{1/2}}$ **7.** $a^{3/2}$ **9.** $9^{5/2}$ **11.** $z^{5/3}$ **13.** $7^{10/3}$ **15.** $9^{7/4}$ **17.** $y^{14/3}$ **19.** $(a^3b)^{1/4}$ **21.** $(x^9z^5)^{1/4}$ **23.** $(3a+8b)^{1/6}$ **25.** $\left(\frac{2x^6}{11y^7}\right)^{1/5}$ **27.** $\sqrt{a}$ **29.** $\sqrt{c^5}$ **31.** $\sqrt[3]{18^5}$ **33.** $\sqrt{24x^3}$ **35.** $(\sqrt[5]{11b^2c})^3$ **37.** $\sqrt[5]{6a+5b}$ **39.** $\frac{1}{\sqrt[3]{b^3-d}}$ **41.** $a^3$ **43.** $x^3$ **45.** $\sqrt[3]{y}$ **47.** $\sqrt{y}$ **49.** 19.3 **51.** $x^5y^{10}$ **53.** $\sqrt{xyz}$ **55.** $\sqrt[4]{x}$ **57.** $\sqrt[8]{y}$ **59.** $\sqrt[9]{x^2y}$ **61.** $\sqrt[10]{a^9}$ **63.** 5 **65.** 4 **67.** 16 **69.** Not a real number **71.** $\frac{5}{3}$ **73.** $\frac{1}{2}$ **75.** −9 **77.** −4 **79.** $\frac{1}{4}$ **81.** $\frac{1}{64}$ **83.** $\frac{3}{4}$ **85.** Not a real number **87.** 24 **89.** $\frac{11}{28}$ **91.** $x^{9/2}$ **93.** $x^{1/6}$ **95.** $\frac{1}{x}$ **97.** 1 **99.** $\frac{y^{5/3}}{12}$ **101.** $\frac{12}{x^{11/6}}$ **103.** $\frac{1}{2x^{1/3}}$ **105.** $\frac{121}{x^{1/7}}$ **107.** $\frac{64}{a^{66/5}}$ **109.** $\frac{x}{y^{20}}$ **111.** $8z^{7/2}-4$ **113.** $\frac{5}{x^5}+\frac{20}{x^{3/2}}$ **115.** $12x^{13/6}-18x^2$ **117.** $\approx 13.42$ **119.** $\approx 3.32$ **121.** $\approx 20.53$ **123.** $\approx 0.03$ **125.** $n$ is odd, or $n$ is even and $a \ge 0$. **127.** $(4^{1/2}+9^{1/2})^2 \ne 4+9; 25 \ne 13$ **129.** $(1^{1/3}+1^{1/3})^3 \ne 1+1; 8 \ne 2$ **131.** $x^{1/2}(x+1)$ **133.** $y^{1/3}(1-y)(1+y)$ **135.** $\frac{1+y^2}{y^{2/5}}$ **137. a)** $2^{10} = 1024$ bacteria **b)** $2^{10}\sqrt{2} \approx 1448$ bacteria **139. a)** $2.69\sqrt{7^3} \approx \$49.82$ billion **b)** $2.69\sqrt{16^3} = \$172.16$ billion **141.** 9 **143.** $\{x|x \ge 7\}$ **145. a)** $(x-6)^2$ **b)** $(x-6)^2$ **147.** 2; $z^{\frac{1}{4}\cdot\frac{1}{5}\cdot\frac{1}{a}\cdot\frac{1}{3}} = z^{\frac{1}{60a}}; z^{\frac{1}{60a}} = z^{\frac{1}{120}}, 60a = 120; a = 2$ **149. c)** is a function **150.** $\frac{b^2+a^3b}{a^3-b}$ **151.** 0, 3 **152.** $\approx 441.67$ miles per hour.

**Exercise Set 7.3** **1. a)** Square the natural numbers. **b)** 1, 4, 9, 16, 25, 36 **3. a)** Raise the natural numbers to the fifth power. **b)** 1, 32, 243, 1024, 3125 **5.** If $n$ is even and $a$ or $b$ are negative, the numbers are not real numbers. **7.** If $n$ is even and $a$ or $b$ is negative, the numbers are not real numbers; **9.** $2\sqrt{2}$ **11.** $2\sqrt{6}$ **13.** $4\sqrt{2}$ **15.** $5\sqrt{2}$ **17.** $5\sqrt{3}$ **19.** $2\sqrt{10}$ **21.** $2\sqrt[3]{2}$ **23.** $3\sqrt[3]{2}$ **25.** $2\sqrt[3]{4}$ **27.** $2\sqrt[3]{5}$ **29.** $2\sqrt[4]{3}$ **31.** $-2\sqrt[5]{2}$ **33.** $b^3$ **35.** $x^2$ **37.** $x\sqrt{x}$ **39.** $a^5\sqrt{a}$ **41.** $8z^{10}\sqrt[3]{z^2}$ **43.** $b^5\sqrt[4]{b^3}$ **45.** $x\sqrt[6]{x^3}$ or $x\sqrt{x}$ **47.** $3y^4\sqrt[5]{y^3}$ **49.** $10y^4\sqrt{2y}$ **51.** $xy^2\sqrt[3]{y}$ **53.** $ab^4\sqrt[5]{ab^3}$ **55.** $2x^7y^{10}z^{13}\sqrt{6xz}$ **57.** $3a^2b^2\sqrt[3]{3b^2}$ **59.** $2x^2y^2z^4\sqrt[4]{2yz^3}$ **61.** $3a^2b^2\sqrt[4]{b}$ **63.** $2a^2b^2\sqrt[5]{b^2}$ **65.** 5 **67.** $\frac{9}{10}$ **69.** 3 **71.** $\frac{1}{4}$ **73.** $\frac{1}{2}$ **75.** $\frac{1}{3}$ **77.** $\frac{1}{2}$ **79.** 2 **81.** $\frac{r^2}{2}$ **83.** $\frac{4x^2}{5y^5}$ **85.** $\frac{c^2}{4}$ **87.** $a^2b^6\sqrt[3]{a^2b^2}$ **89.** $2\sqrt{2}$ **91.** $3x^2$ **93.** $2x^2y\sqrt{2y}$ **95.** $\frac{\sqrt[3]{5y}}{2x^4}$ **97.** $\frac{y^2\sqrt[3]{5y}}{x^2}$ **99.** $\frac{x^3\sqrt[4]{10y}}{3}$ **101.** $(a \cdot b)^{1/2} = a^{1/2}b^{1/2} = \sqrt{a}\sqrt{b}$ **103.** No; One example is $\sqrt{18}/\sqrt{2} = 3$.

**105.** **a)** No **b)** When $\sqrt[n]{x}$ is a real number and not equal to 0. **106.** $C = \frac{5}{9}(F - 32)$ **107.** $\{-28, 32\}$ **108.** $3x^6 - x^3 + 4$
**109.** $(x - 1)(x^2 - 8x + 19)$

## Exercise Set 7.4

**1.** Radicals with the same radicand and index **3.** $\approx 5.97$
**5.** No: one example is $\sqrt{9} + \sqrt{16} \neq \sqrt{9 + 16}, 3 + 4 \neq 5, 7 \neq 5.$ **7.** 0 **9.** $4\sqrt{5}$ **11.** $-4\sqrt{3} + 5$ **13.** $-7\sqrt[4]{y}$ **15.** $2\sqrt[3]{x} + 9\sqrt{5}$
**17.** $7\sqrt{x} - 6\sqrt{y}$ **19.** $3\sqrt{5}$ **21.** $-30\sqrt{3} + 25\sqrt{5}$ **23.** $-4\sqrt{10}$ **25.** $18y\sqrt{5x}$ **27.** $-16\sqrt{5x}$ **29.** $-27a\sqrt{2}$ **31.** $5\sqrt[3]{4}$ **33.** $-7$
**35.** $6a\sqrt[3]{ab^2}$ **37.** $3r^3s^2\sqrt{rs}$ **39.** 0 **41.** 9 **43.** $2\sqrt[3]{7}$ **45.** $3m^2n^5\sqrt{3n}$ **47.** $3x^3y^4\sqrt[3]{2x^2y}$ **49.** $x^7y^7z^3\sqrt[5]{x^2y^3z}$ **51.** $x^2y^2\sqrt[3]{4y^2}$
**53.** $5 - \sqrt{15}$ **55.** $2\sqrt[3]{y^2} - y^3$ **57.** $4x^5y^3\sqrt[3]{x} + 4xy^4\sqrt[3]{2x^2y^2}$ **59.** 59 **61.** $6 - x^2$ **63.** $7 - z$ **65.** $23 + 9\sqrt{3}$ **67.** $16 - 10\sqrt{2}$
**69.** $10 - 3\sqrt{6}$ **71.** $29 - 12\sqrt{5}$ **73.** $18x - \sqrt{3xy} - y$ **75.** $8 - 2\sqrt[3]{18} - \sqrt[3]{12}$ **77.** $4x - 8\sqrt{x}$ **79.** $x^2 + x\sqrt[3]{x^2}$
**81.** $x\sqrt[4]{27x^2} - x^2\sqrt[4]{3x}$ **83.** $2\sqrt{6}$ **85.** $3\sqrt{5}$ **87.** $-14 + 11\sqrt{2}$ **89.** $5\sqrt{6} - 2\sqrt{3}$ **91.** $15\sqrt{2}$ **93.** $2x^3\sqrt[3]{10x^2}$ **95.** $2b^2c\sqrt[6]{2ab^5c^3}$
**97.** $4ab\sqrt[4]{b}$ **99.** $x - 2\sqrt[3]{x^2y^2} - \sqrt[3]{xy} + 2y$ **101.** $ab\sqrt[3]{12a^2b^2} - 2a^2b^2\sqrt[3]{3}$ **103.** $2x - 5$ **105.** $2|r - 4|$ **107.** $P = 14\sqrt{5}, A = 60$
**109.** $P = 17\sqrt{5}, A = 52.5$ **111.** No, $-\sqrt{2} + \sqrt{2} = 0$ **113.** **a)** $\approx 45.17$ miles per hour **b)** $\approx 35.33$ miles per hour
**115.** **a)** 37 inches **b)** $\approx 37.97$ inches **117.** **a)**

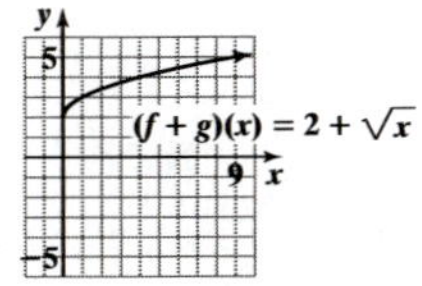


**b)** Raises the graph 2 units

**119.** **a)**

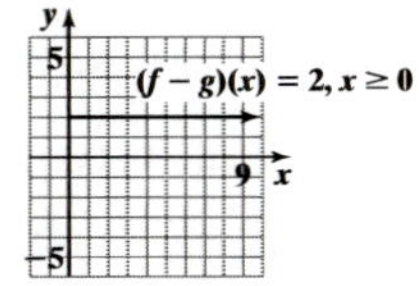


**b)** $\{x|x \geq 0\}$

**121.**

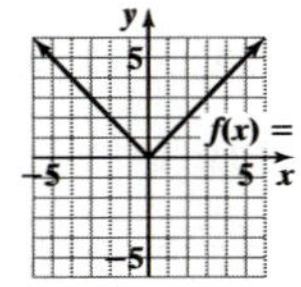


**123.** A quotient of two integers, denominator not 0. **124.** A number that can be represented on a real number line.
**125.** A real number that cannot be expressed as a quotient of two integers. **126.** $|a| = \begin{cases} a, & a \geq 0 \\ -a, & a < 0 \end{cases}$ **127.** $m = \frac{2E}{v^2}$

**128.** **a)**  **b)** $\left(-\frac{1}{2}, 5\right]$ **c)** $\left\{x \middle| -\frac{1}{2} < x \leq 5\right\}$

## Mid-Chapter Test

**1.** 11 [7.1] **2.** $-\frac{3}{4}$ [7.1] **3.** 16.3 [7.1] **4.** $|3a^2 - 4b^3|$ [7.1] **5.** 3 [7.1] **6.** $(7a^4b^3)^{1/5}$ [7.2] **7.** 20 [7.2]
**8.** $a^{10}b^{15}c^5$ [7.3] **9.** $\frac{14}{x}$ [7.3] **10.** $8x + \frac{16}{x^{5/2}}$ [7.3] **11.** $4x^2y^4\sqrt{2y}$ [7.3] **12.** $2a^2b^3c^2\sqrt[6]{ab^5c^3}$ [7.3] **13.** $\frac{1}{3}$ [7.3] **14.** $\frac{y^2\sqrt{y}}{3x^5}$ [7.3]
**15.** $11\sqrt{x} + 12\sqrt{y}$ [7.4] **16.** $27x\sqrt{10y}$ [7.4] **17.** $2x^2 - x\sqrt{5} - 15$ [7.4] **18.** $18a\sqrt{a} - 20a\sqrt{3}$ [7.4] **19.** $7ab\sqrt[4]{ab}$ [7.4]
**20.** Part **a)** will have an absolute value. **a)** $|x - 3|$, **b)** $8x$ [7.1]

## Exercise Set 7.5

**1.** **a)** Same two terms with the sign of the second term changed. **b)** $x + \sqrt{3}$ **3.** **a)** Answers will vary.
**b)** $\frac{4\sqrt{3y}}{3y}$ **5.** (1) No perfect powers are factors of any radicand. (2) No radicand contains fractions. (3) There are no radicals in any denominator. **7.** $\frac{\sqrt{3}}{3}$ **9.** $\frac{4\sqrt{5}}{5}$ **11.** $\sqrt{6}$ **13.** $\frac{\sqrt{z}}{z}$ **15.** $\frac{p\sqrt{2}}{2}$ **17.** $\frac{\sqrt{7y}}{7}$ **19.** $3\sqrt{2}$ **21.** $\frac{\sqrt{xy}}{y}$ **23.** $\frac{\sqrt{10m}}{4}$ **25.** $\frac{\sqrt{2n}}{3}$
**27.** $\frac{3x^2y\sqrt{yz}}{z^2}$ **29.** $\frac{2y^4z\sqrt{15xz}}{3x}$ **31.** $\frac{4x^3y^2\sqrt{yz}}{z^2}$ **33.** $\frac{\sqrt[3]{4}}{2}$ **35.** $\frac{8\sqrt[3]{y^2}}{y}$ **37.** $\frac{\sqrt[4]{27}}{3}$ **39.** $\frac{a\sqrt[4]{2}}{2}$ **41.** $\frac{5\sqrt[4]{z^2}}{z}$ **43.** $\frac{10\sqrt[5]{y^2}}{y}$ **45.** $\frac{2\sqrt[7]{a^3}}{a}$
**47.** $\frac{\sqrt[3]{4x^2}}{2x}$ **49.** $\frac{5m\sqrt[4]{8}}{2}$ **51.** $\frac{\sqrt[4]{135x}}{3x}$ **53.** $\frac{\sqrt[3]{12x^2y}}{2y}$ **55.** $\frac{\sqrt[3]{7xy^2z}}{z}$ **57.** 19 **59.** 62 **61.** $-6$ **63.** $a - b$ **65.** $4x - 9y$
**67.** $\sqrt{3} - 1$ **69.** $2 - \sqrt{3}$ **71.** $\frac{-5\sqrt{2} - 35}{47}$ **73.** $\frac{10 + \sqrt{30}}{14}$ **75.** $\frac{18 - 3\sqrt{x}}{36 - x}$ **77.** $\frac{4x + 4y\sqrt{x}}{x - y^2}$ **79.** $\frac{-13 + 3\sqrt{6}}{23}$ **81.** $a + a^3$
**83.** $\frac{4\sqrt{x + 2} + 12}{x - 7}$ **85.** $\frac{\sqrt{x}}{4}$ **87.** $\frac{\sqrt{2}}{3}$ **89.** 1 **91.** $\frac{2xy^3\sqrt{30xz}}{5z}$ **93.** $\frac{\sqrt{14}}{x}$ **95.** $\frac{\sqrt{a} - 7}{a - 49}$ **97.** $-\frac{\sqrt{2x}}{2}$ **99.** $\frac{\sqrt[4]{24x^3y^2}}{2x}$
**101.** $\frac{2y^4z^3\sqrt[3]{2x^2z}}{x}$ **103.** $\frac{a\sqrt{r} + 2r\sqrt{a}}{a - 4r}$ **105.** $\frac{\sqrt[3]{150y^2}}{5y}$ **107.** $\frac{y^3z\sqrt[4]{54x^2}}{3x}$ **109.** $\sqrt{2}$ **111.** $\frac{3\sqrt{5}}{5}$ **113.** $\frac{8\sqrt{6}}{3}$ **115.** $\frac{19\sqrt{2}}{2}$
**117.** $\frac{21\sqrt{2}}{2}$ **119.** $-\frac{301\sqrt{2}}{20}$ **121.** $\frac{3\sqrt{6}}{4}$ **123.** $\left(-\frac{2}{y} + \frac{3}{x}\right)\sqrt{xy}$ **125.** $2\sqrt{a}$ **127.** $\sqrt[3]{(a + b)^5}$ **129.** $\sqrt[15]{(a + 2b)^2}$ **131.** $\sqrt[6]{rs^5}$
**133.** $\sqrt[15]{x^2y^8}$ **135.** $\approx 3.69$ meters **137.** $\approx 12$ inches **139.** **a)** 6.21 million **b)** $\approx 2.35$ million **141.** $\frac{3}{\sqrt{3}}; \frac{2}{\sqrt{2}} = \sqrt{2}, \frac{3}{\sqrt{3}} = \sqrt{3}$
**143.** $2 + \sqrt{3}$; rationalize the denominator and compare **145.** **a)** 4, 8, 12 **b)** 9, 18, 27 **c)** $x^{(3a+2b)/6}$ **d)** $x^{(3a-2b)/6}$
**147.** $\frac{3\sqrt{2a} - 3b}{2a - 3b}$ **149.** $\frac{10}{15 + 3\sqrt{5}}$ **151.** $\frac{1}{\sqrt{x + h} + \sqrt{x}}$ **154.** $b_2 = \frac{2A}{h} - b_1$ **155.** 40 miles per hour, 50 miles per hour.
**156.** $4x^3 + x^2 - 20x + 4$ **157.** $-8, 1$

## Using Your Graphing Calculator, 7.6

**1.**

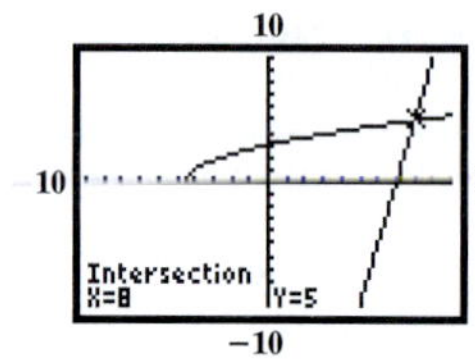


**2.**

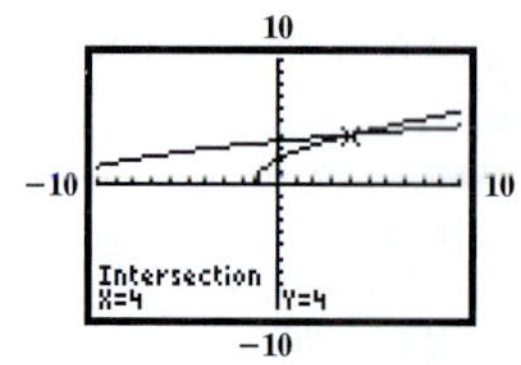


## Exercise Set 7.6

**1. a)** Answers will vary. **b)** 5 **3.** 0 **5.** Answers will vary. **7.** 1; Answers will vary. **9.** 16 **11.** No real solution **13.** −64 **15.** 11 **17.** 9 **19.** −1 **21.** 81 **23.** 71 **25.** No real solution **27.** No real solution **29.** 2, 4 **31.** 8 **33.** 7 **35.** $\frac{2}{3}$ **37.** 16 **39.** 2 **41.** 10 **43.** 6 **45.** 8 **47.** 0 **49.** −3 **51.** $\frac{3}{2}$ **53.** No real solution **55.** 2, 0 **57.** 5, 8 **59.** No real solution **61.** 9 **63.** 3, 7 **65.** −1 **67.** 7 **69.** 4 **71.** 5 **73.** $v = \frac{p^2}{2}$ **75.** $g = \frac{v^2}{2h}$ **77.** $F = \frac{Mv^2}{R}$ **79.** $m = \frac{x^2k}{V_0^2}$ **81.** $A = \pi r^2$ **83.** $\sqrt{87}$ **85.** $2\sqrt{10}$ **87.** 4 **89.** No Solution **91.** 3 **93.** 7 **95.** 1 **97.** 3 **99.** $\sqrt{16{,}200} \approx 127.28$ feet **101.** 13 feet **103. a)** ≈3.14 seconds **b)** $\sqrt{2} \cdot T$; compare $\sqrt{\frac{l}{32}}$ with $\sqrt{\frac{l}{16}}$ **c)** $\sqrt{24} \approx 4.90$ seconds **105.** $R = \frac{8\mu l}{\pi r^4}$ **107.** $0.2(\sqrt{149.4})^3 \approx 365.2$ days **109.** $\sqrt{10{,}000} = 100$ pounds **111.** $\sqrt{320} \approx 17.89$ feet per second **113.** $\sqrt{1649} \approx 40.61$ meters **115.** 2, −2 **117.** 5, −1 **119.** 30 **121.** 5, −5 **123. a)** 3, 7; points of intersection **b)** Yes **c)** 3, 7; yes **125.** At $x = 4$, $g(x)$ or $y = 0$. Therefore the graph must have an $x$-intercept at 4. **127.** $L_1 \approx 0.44$, $L_2 \approx 0.76$ **129.** All real numbers **131.** 1.5 **133.** ≈−3.7; ≈3.7 **135.** No real solution **137.** $n = \frac{z^2\sigma^2}{(\bar{x} - \mu)^2}$ **140.** $P_2 = \frac{P_1P_3}{P_1 - P_3}$ **141.** $x$ **142.** $\frac{3a}{2b(2a + 3b)}$ **143.** $t(t - 5)$ **144.** $\frac{3}{x + 3}$ **145.** 2

## Exercise Set 7.7

**1. a)** $\sqrt{-1}$ **b)** −1 **3.** Yes **5.** Yes **7.** $a - bi$ **9. a)** $\sqrt{2}$ **b)** 1 **c)** $\sqrt{-3}$ or $2i$ **d)** 6 **e)** Every number we have studied is a complex number. **11.** $7 + 0i$ **13.** $5 + 0i$ **15.** $21 - 6i$ **17.** $0 + 2i\sqrt{6}$ **19.** $8 - 2i\sqrt{3}$ **21.** $3 + 7i\sqrt{2}$ **23.** $12 - 5i$ **25.** $0 + (7 - 3\sqrt{5})i$ **27.** $21 + 8i$ **29.** 0 **31.** $-17 - 12i$ **33.** $(4\sqrt{2} + \sqrt{3}) - 2i\sqrt{2}$ **35.** $11 - 4i\sqrt{2}$ **37.** $-3 - 2i\sqrt{5}$ **39.** $6 - 2i$ **41.** $-9 + 4i$ **43.** $-33 + 18i$ **45.** $28 + 4i\sqrt{3}$ **47.** $9 + 9i$ **49.** $1 + 5i$ **51.** 109 **53.** $39 - 9i\sqrt{2}$ **55.** $\frac{25}{72} + \frac{1}{4}i$ **57.** $-\frac{8}{3}i$ **59.** $\frac{3 - 2i}{2}$ **61.** $\frac{12 + 6i}{5}$ **63.** $\frac{3 + 6i}{5}$ **65.** $\frac{9 - 12i}{10}$ **67.** $\frac{3 + i}{5}$ **69.** $\frac{5\sqrt{2} - 2i\sqrt{6}}{37}$ **71.** $\frac{(5\sqrt{10} - 2\sqrt{15}) + (10\sqrt{2} + 5\sqrt{3})i}{45}$ **73.** 5 **75.** $\frac{\sqrt{2}}{3}$ **77.** $12 - 7i$ **79.** $4\sqrt{2} + 2i\sqrt{3}$ **81.** $20.8 - 16.64i$ **83.** $37 - 39i$ **85.** $\frac{4 - 11i}{2}$ **87.** $\frac{6\sqrt{3} + 12i}{7}$ **89.** $7 + \frac{2}{45}i$ **91.** $\frac{1}{4} - \frac{31}{50}i$ **93.** 2 **95.** $-4.33 - 10.91i$ **97.** −1 **99.** 1 **101.** $i$ **103.** $-i$ **105. a)** $-2 - 3i$ **b)** $\frac{2 - 3i}{13}$ **107.** True; $(2i)(2i) = -4$ **109.** False; $(1 + i)(1 + 2i) = -1 + 3i$ **111.** Even values; $i^n$ where $n$ is even will either be −1 or 1 **113.** −4 **115.** $16 - 4i$ **117.** $14 + 8i$ **119.** 0 **121.** 1 **123.** Yes **125.** No **127.** $\approx 0.83 - 3i$ **129.** $\approx 1.5 - 0.33i$ **131.** $-i$ **133.** $1 + i\sqrt{5}, 1 - i\sqrt{5}$ **135.** $6 + 3i\sqrt{3}$ **137.** $-1 + 7i\sqrt{3}$ **139.** 15 pounds at \$5.50, 25 pounds at \$6.30 **140.** $2c - 3 - \frac{8}{4c + 9}$ **141.** $\frac{a^2}{b(a - b)}$ **142.** 4

## Chapter 7 Review Exercises

**1.** 10 **2.** −3 **3.** −5 **4.** 4 **5.** 8 **6.** 38.2 **7.** $|x|$ **8.** $|x - 3|$ **9.** $|x - y|$ **10.** $|x^2 - 4x + 12|$ **11.** 7 **12.** 57 **13.** ≈2.2 **14.** 12 meters **15.** $x^{7/2}$ **16.** $x^{5/3}$ **17.** $y^{13/4}$ **18.** $6^{-2/7}$ **19.** $\sqrt{x}$ **20.** $\sqrt[5]{a^4}$ **21.** $(\sqrt[4]{8m^2n})^7$ **22.** $\frac{1}{(\sqrt[3]{x + y})^5}$ **23.** 16 **24.** $x^6$ **25.** 81 **26.** $\sqrt[4]{a}$ **27.** −6 **28.** Not a real number **29.** $\frac{3}{4}$ **30.** $\frac{3}{8}$ **31.** $x^{4/15}$ **32.** $\frac{4}{y^3}$ **33.** $\frac{1}{a^{16/15}}$ **34.** $\frac{25x^{10}}{y^7}$ **35.** $5a^2 - 3a^{5/2}$ **36.** $\frac{4}{x^{7/6}} + 11$ **37.** $x^{2/5}(1 + x)$ **38.** $\frac{1 + a^2}{a^{1/2}}$ **39.** 5 **40.** ≈2.668 **41.**

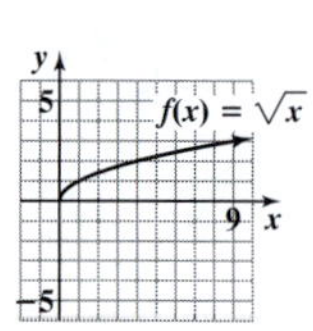


**42.**

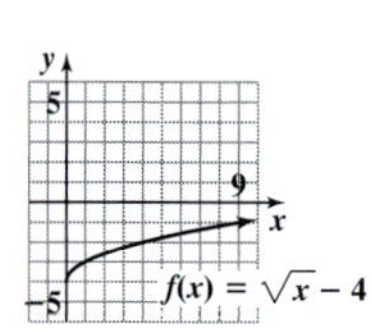


**43.** $4\sqrt{3}$ **44.** $4\sqrt[3]{2}$ **45.** $\frac{7}{3}$ **46.** $\frac{2}{5}$ **47.** $-\frac{9}{7}$ **48.** $-\frac{3}{5}$ **49.** 8 **50.** 4 **51.** $3xyz^2\sqrt{2y}$ **52.** $5xy^3\sqrt{3xy}$ **53.** $3a^2b^3\sqrt[3]{2ab}$ **54.** $5x^2y^3z^5\sqrt[3]{x^2z}$ **55.** $x^{14}y^{21}z^{35}$ **56.** $8a^3b^{12}c^{18}$ **57.** $2x^3\sqrt{10}$ **58.** $2x^3y\sqrt[3]{x^2y^2}$ **59.** $2x^2y^3\sqrt[3]{4x^2}$ **60.** $2x^2y^4\sqrt[4]{x}$ **61.** $6x - 2\sqrt{15x}$ **62.** $2x^2y^2\sqrt[3]{y^2} + x\sqrt[3]{18y}$ **63.** $\sqrt[4]{a^3b^2}$ **64.** $\sqrt[6]{x^5y^2}$ **65.** $\frac{64r^{9/2}}{p^3}$ **66.** $\frac{y^{1/5}}{6xz^{1/3}}$ **67.** $\frac{\sqrt{15}}{5}$ **68.** $\frac{\sqrt[3]{21}}{3}$ **69.** $\frac{\sqrt[4]{20}}{2}$ **70.** $\frac{x\sqrt{10}}{10}$ **71.** $\frac{8\sqrt{x}}{x}$ **72.** $\frac{m\sqrt[3]{5}}{5}$ **73.** $\frac{10\sqrt[3]{y}}{y}$ **74.** $\frac{9\sqrt[4]{z^3}}{z}$ **75.** $\frac{x}{3}$ **76.** $\frac{x}{2}$ **77.** $\frac{4y^2}{x^3}$ **78.** $2x^2y^3$ **79.** $\frac{x^2\sqrt{6y}}{y}$ **80.** $\frac{2\sqrt{21ab}}{7b}$ **81.** $\frac{x^2y^2\sqrt{6yz}}{z}$

**82.** $\dfrac{5xy^2\sqrt{15yz}}{3z}$ **83.** $3xy\sqrt[3]{2y}$ **84.** $\dfrac{\sqrt[3]{75xy^2}}{5y}$ **85.** $\dfrac{y\sqrt[3]{9x^2}}{x}$ **86.** $\dfrac{y^2\sqrt[3]{25x}}{5x}$ **87.** $\dfrac{b^2\sqrt[4]{2ab^2}}{a}$ **88.** $\dfrac{y\sqrt[4]{6x^3y^2}}{2x}$ **89.** 7 **90.** $x - y^2$
**91.** $x^2 - y$ **92.** $7 + 4\sqrt{3}$ **93.** $x + \sqrt{5xy} - \sqrt{3xy} - y\sqrt{15}$ **94.** $\sqrt[3]{6x^2} - \sqrt[3]{4xy} - \sqrt[3]{9xy} + \sqrt[3]{6y^2}$ **95.** $-12 + 6\sqrt{5}$
**96.** $\dfrac{4x - x\sqrt{x}}{16 - x}$ **97.** $\dfrac{4a + a\sqrt{b}}{16 - b}$ **98.** $\dfrac{x\sqrt{y} + 7x}{y - 49}$ **99.** $\dfrac{x - \sqrt{xy}}{x - y}$ **100.** $\dfrac{x - 2\sqrt{xy} - 3y}{x - y}$ **101.** $\dfrac{2\sqrt{a-1} + 4}{a - 5}$
**102.** $\dfrac{5\sqrt{y+2} + 15}{y - 7}$ **103.** $9\sqrt[3]{x}$ **104.** $-4\sqrt{3}$ **105.** $12 - 13\sqrt[3]{2}$ **106.** $\dfrac{45\sqrt{2}}{8}$ **107.** $(9x^2y^3 - 4x^3y^4)\sqrt{x}$
**108.** $(8x^2y^2 - x + 3x^3)\sqrt[3]{xy^2}$ **109.** $3x\sqrt{2} - 3\sqrt{5x}$ **110.** $2x^2 + 2x^2\sqrt[3]{4x}$ **111.** $2x + 7$ **112.** $\sqrt{5}|2a + 5|$ **113.** $\sqrt[6]{x+5}$
**114.** $\sqrt[12]{b^5}$ **115. a)** $12\sqrt{3}$ **b)** 24 **116. a)** $8\sqrt{5} + \sqrt{130}$ **b)** $10\sqrt{13}$ **117. a)**

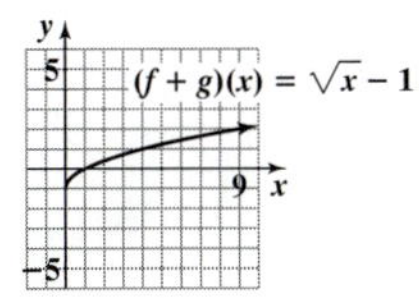


**b)** $x \geq 0$
**118. a)**

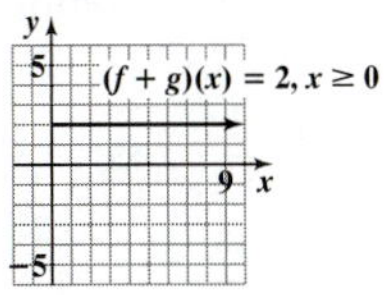


**b)** $x \geq 0$ **119.** 81 **120.** No solution
**121.** 64 **122.** −125 **123.** 9 **124.** 125
**125.** No solution **126.** 4 **127.** −3
**128.** 3 **129.** 0, 9 **130.** 5 **131.** 4 **132.** 6
**133.** $L = \dfrac{V^2w}{2}$ **134.** $A = \pi r^2$ **135.** $2\sqrt{14}$
**136.** $5\sqrt{3}$ **137.** $\sqrt{29} \approx 5.39$ meters **138.** $\sqrt{1280} \approx 35.78$ feet per second **139.** $2\pi\sqrt{2} \approx 2.83\pi \approx 8.89$ seconds
**140.** $\sqrt{\dfrac{90}{0.145}} \approx 24.91$ meters per second **141.** $m \approx 5m_0$. Thus, it is ≈5 times its original mass. **142.** $5 + 0i$ **143.** $-8 + 0i$
**144.** $7 - 16i$ **145.** $9 + 4i$ **146.** $13 + i$ **147.** $6 - 2i$ **148.** $12\sqrt{3} + (\sqrt{5} - \sqrt{7})i$ **149.** $-6 + 6i$ **150.** $17 - 6i$
**151.** $(24 + 3\sqrt{5}) + (4\sqrt{3} - 6\sqrt{15})i$ **152.** $-\dfrac{8i}{3}$ **153.** $\dfrac{(-2 - \sqrt{3})i}{2}$ **154.** $\dfrac{12 - 8i}{13}$ **155.** $\dfrac{5\sqrt{3} + 3i\sqrt{2}}{31}$ **156.** 0 **157.** 7
**158.** $i$ **159.** $-i$ **160.** 1 **161.** −1

## Chapter 7 Practice Test

**1.** $|5x - 3|$ [7.1] **2.** $\dfrac{1}{x^{12/5}}$ [7.2] **3.** $\dfrac{1 + x^2}{x^{2/3}}$ [7.2] **4.**

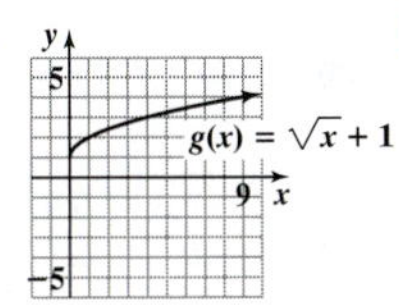


[7.1] **5.** $3x^3y^5\sqrt{6x}$ [7.3]
**6.** $5x^3y^3\sqrt[3]{2x^2y}$ [7.4] **7.** $\dfrac{x^3y\sqrt{14yz}}{4z}$ [7.5] **8.** $\dfrac{9\sqrt[3]{x^2}}{x}$ [7.5] **9.** $\dfrac{3 - \sqrt{3}}{6}$ [7.5]
**10.** $7\sqrt{6}$ [7.3] **11.** $(2xy + 4x^2y^2)\sqrt[3]{y^2}$ [7.4] **12.** $6\sqrt{3} - 2\sqrt{6} - 12 + 4\sqrt{2}$ [7.4]
**13.** $\sqrt[8]{x^5y^3}$ [7.2] **14.** $\sqrt[12]{(7x + 2)^7}$ [7.5] **15.** −5 [7.6] **16.** −3 [7.6] **17.** 9 [7.6]
**18.** 3 [7.6] **19.** $g = \dfrac{8w^2}{h}$ [7.6] **20.** $\sqrt{12{,}880} \approx 113.49$ feet per second [7.6] **21.** 13 feet [7.6] **22.** $2\pi\sqrt{\dfrac{1400}{65{,}000}} \approx 0.92$ second [7.6]
**23.** $20 + 20i$ [7.7] **24.** $\dfrac{33 - 17i}{53}$ [7.7] **25.** [illegible] [7.7]

## Cumulative Review Test

**1.** $\dfrac{57}{9}$ [2.1] **2.** −1 [2.1] **3.** $40 [2.3] **4.** $\{x|-1 < x < 4\}$ [2.6] **5.**

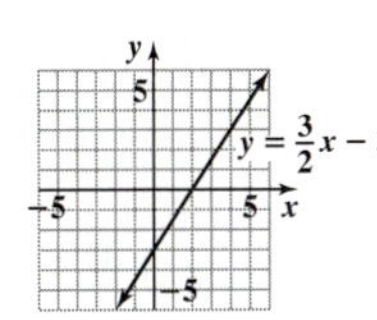


[3.4]
**6.** Parallel [3.5] **7.** $-x^2 + 5x - 13$ [3.6] **8.** $y = -\dfrac{2}{3}x - \dfrac{10}{3}$ [3.5] **9.** $\left(2, 5, \dfrac{34}{5}\right)$ [4.2]
**10.** 40 [4.5] **11.** $w = 2r + 1$ [5.3] **12.** $25x^2y^2 - 9$ [5.2] **13.** 3, −3 [7.6]
**14.** $x(4x - 5)(x - 1)$ [5.5] **15.** $(x - 2)(x^2 + 5x + 13)$ [5.6] **16.** $\dfrac{1}{4}, -\dfrac{3}{2}$ [5.8] **17.** $\dfrac{(x + y)y^2}{3x^3}$ [6.1] **18.** $\dfrac{x + 3}{x + 5}$ [6.2] **19.** 18 [6.4]
**20.** 400 feet [6.6]

# Chapter 8

## Exercise Set 8.1

**1.** ±6 **3.** If $x^2 = a$, then $x = \pm\sqrt{a}$. **5.** $\left(\dfrac{b}{2}\right)^2$ must equal $c$. **7. a)** Yes **b)** No, ±2
**9.** Multiply by $\dfrac{1}{2}$ to make $a = 1$. **11.** $\left(-\dfrac{6}{2}\right)^2 = 9$ **13.** ±5 **15.** $\pm 7i$ **17.** $\pm 2i\sqrt{6}$ **19.** $\pm i\sqrt{61}$ **21.** 8, 0 **23.** $-3 \pm 5i$
**25.** $2 \pm 3i\sqrt{5}$ **27.** $-1, \dfrac{1}{3}$ **29.** $\dfrac{2 \pm 2i}{3}$ **31.** 0.1, −1.7 **33.** $\dfrac{5 \pm 3\sqrt{2}}{2}$ **35.** $-\dfrac{1}{20}, -\dfrac{9}{20}$ **37.** 1, −4 **39.** −3, −5 **41.** −2, −4
**43.** 1, 6 **45.** $-1, \dfrac{1}{2}$ **47.** $-\dfrac{1}{2}, 4$ **49.** 5, 8 **51.** −1, 7 **53.** 4, 5 **55.** 7, −4 **57.** 1, −11 **59.** $2 \pm \sqrt{14}$ **61.** $-4 \pm \sqrt{11}$
**63.** $\dfrac{1 \pm \sqrt{13}}{2}$ **65.** $\dfrac{-3 \pm i\sqrt{15}}{2}$ **67.** 0, 1 **69.** $0, -\dfrac{2}{3}$ **71.** $0, \dfrac{1}{6}$ **73.** 1, −3 **75.** 8, −4 **77.** $\dfrac{-9 \pm \sqrt{73}}{2}$ **79.** $\dfrac{1}{3}, -1$ **81.** $\dfrac{1 \pm i\sqrt{39}}{4}$

**83.** $1 \pm i$ **85. a)** $21 = (x + 2)(x - 2)$ **b)** 5 **87. a)** $18 = (x + 4)(x + 2)$ **b)** $-3 + \sqrt{19}$ **89.** 30 mph **91.** 5, 7
**93.** 5 feet by 12 feet **95.** $\frac{12 + \sqrt{288}}{2} \approx 14.49$ feet by 14.49 feet **97.** $\sqrt{200} \approx 14.14$ inches **99.** $\sqrt{24} \approx 4.90$ feet **101.** 4%
**103.** $\approx 6\%$ **105. a)** $S = 32 + 80\sqrt{\pi} \approx 173.80$ square inches **b)** $r = \frac{4\sqrt{\pi}}{\pi} \approx 2.26$ inches **c)** $r = -5 + \sqrt{\frac{80 + 25\pi}{\pi}} \approx 2.1$ inches
**107.** 2 **108.** \$4200 at 7%, \$5800 at $6\frac{1}{4}\%$ **109.** $\left\{10, \frac{4}{3}\right\}$ **110.** 0 **111.** $4x^3 + x^2 - 21x + 6$

## Exercise Set 8.2

**1.** $x = \frac{-b \pm \sqrt{b^2 - 4ac}}{2a}$ **3.** $a = -3, b = 6, c = 8$ **5.** Yes; if you multiply both sides of one equation by $-1$ you get the other equation **7. a)** $b^2 - 4ac$ **b)** $-84$ **c)** Answers will vary. **9.** Two real solutions **11.** No real solution **13.** Two real solutions **15.** No real solution **17.** One real solution **19.** One real solution **21.** 3, 6 **23.** 2, 4 **25.** 1, −7 **27.** $-2 \pm 2\sqrt{6}$ **29.** $\pm 8$ **31.** $\frac{2 \pm i\sqrt{11}}{3}$ **33.** 0, 5 **35.** $\frac{2 \pm i\sqrt{2}}{2}$ **37.** −1 **39.** $\frac{1}{4}$ **41.** $1 \pm \sqrt{2}$ **43.** $\frac{-3 \pm i\sqrt{15}}{2}$ **45.** $-3, \frac{1}{2}$ **47.** $\frac{1}{2}, -\frac{5}{3}$ **49.** 4, −6 **51.** $\frac{1}{3}, -\frac{2}{3}$ **53.** $\frac{-6 \pm 2\sqrt{6}}{3}$ **55.** $\frac{3 \pm \sqrt{309}}{30}$ **57.** $\frac{3 \pm \sqrt{33}}{2}$ **59.** $\frac{2 \pm i\sqrt{6}}{2}$ **61.** $\frac{-1 \pm i\sqrt{23}}{4}$ **63.** $\frac{-0.6 \pm \sqrt{0.84}}{0.2}$ or $-3 \pm \sqrt{21}$ **65.** 0, 2 **67.** −5, 6 **69.** $\frac{7 \pm \sqrt{17}}{4}$ **71.** No real number **73.** $x^2 - 7x + 10 = 0$ **75.** $x^2 + 8x - 9 = 0$ **77.** $15x^2 - x - 6 = 0$ **79.** $x^2 - 2 = 0$ **81.** $x^2 + 9 = 0$ **83.** $x^2 - 6x + 7 = 0$ **85.** $x^2 - 4x + 13 = 0$ **87. a)** $n(10 - 0.02n) = 450$ **b)** 50 **89. a)** $n(50 - 0.4n) = 660$ **b)** 15 **91.** Answers will vary. **93.** Yes **95.** 3 **97.** $w = 3$ feet, $l = 8$ feet **99.** 2 inches **101.** $\approx 4.39$ seconds **103. a)** $\approx 4.57$ seconds **b)** $\approx 4.79$ seconds. **105.** $2\sqrt{5}, -\sqrt{5}$ **107.** $(-0.12 + \sqrt{14.3952})/1.2 \approx 3.0618$ millimeters **109. a)** $\approx 1.94$ seconds **b)** $\approx 2.74$ seconds **c)** Courtney's **d)** Yes, at 1.5 seconds **110.** $5.0 \times 10^2$ or 500 **111.** 7 **112.** $(2, -1)$ **113.** $\frac{6y - x}{3xy}$ **114.** No real solution

## Exercise Set 8.3

**1.** Answers will vary. **3.** $S = \sqrt{A}$ **5.** $t = \sqrt{\frac{d}{4.9}}$ **7.** $i = \sqrt{\frac{E}{r}}$ **9.** $t = \frac{\sqrt{d}}{4}$ **11.** $c = \sqrt{\frac{E}{m}}$ **13.** $r = \sqrt{\frac{3V}{\pi h}}$ **15.** $W = \sqrt{d^2 - L^2}$ **17.** $b = \sqrt{c^2 - a^2}$ **19.** $H = \sqrt{d^2 - L^2 - W^2}$ **21.** $t = \sqrt{\frac{h - s_0}{-16}}$ or $t = \frac{\sqrt{s_0 - h}}{4}$ **23.** $v = \sqrt{\frac{2E}{m}}$ **25.** $v_1 = \sqrt{v_2^2 - 2ad}$ **27.** $c = \sqrt{(v')^2 + v^2}$ **29. a)** \$10,950 **b)** $\approx 7$ **31. a)** 32°F **b)** 80.8°F **c)** $\approx 2.92$ minutes **33. a)** 0.53 billion **b)** 2007 **35. a)** 11.4 billion tons **b)** 2003 **37. a)** 1.301 million **b)** 2009 **39.** $l = 30$ meters, $w = 20$ meters **41.** 4 feet per hour **43.** Going 6 mph, returning 8 mph **45.** Bonita $\approx 11.52$ hours; Pamela $\approx 12.52$ hours **47.** 130 mph **49.** Chris $\approx 11.76$ hours; John $\approx 12.26$ hours **51.** 75 mph **53.** $l \approx 34.86$ inches, $h \approx 19.61$ inches **55.** Answers will vary. **57.** 6 meters by 3 meters or 2 meters by 9 meters **59.** −16 **60.** $R = \frac{E - Ir}{I}$ **61.** $\frac{8}{r - 4}$ **62.** $\frac{x^2}{y^{32}}$ **63.** No solution.

## Mid-Chapter Test

**1.** $\pm 7\sqrt{2}$ [8.1] **2.** $3 \pm 2i\sqrt{5}$ [8.1] **3.** $-\frac{1}{2}, -\frac{13}{2}$ [8.1] **4.** −6, 2 [8.1] **5.** $2 \pm \sqrt{14}$ [8.1] **6.** $\frac{-1 \pm i\sqrt{143}}{8}$ [8.1] **7.** $(6 + 6\sqrt{2})$ meters [8.1] **8. a)** $b^2 - 4ac$ **b)** Two distinct real solutions: $b^2 - 4ac > 0$; single real solution: $b^2 - 4ac = 0$; no real solution: $b^2 - 4ac < 0$ [8.2] **9.** Two distinct real solutions [8.2] **10.** $-\frac{5}{3}, \frac{3}{2}$ [8.2] **11.** $-2 \pm 2\sqrt{3}$ [8.2] **12.** $\frac{1 \pm i\sqrt{14}}{3}$ [8.2] **13.** $x^2 - 5x - 14 = 0$ [8.2] **14.** $x^2 - 4x - 1 = 0$ [8.2] **15.** 10 lamps [8.2] **16.** $r = \sqrt{x^2 - y}$ [8.3] **17.** $x = \sqrt{\frac{3A}{k}}$ [8.3] **18.** $y = \sqrt{D^2 - x^2}$ [8.3] **19.** 5 feet by 12 feet [8.3] **20.** 5 clocks [8.3]

## Exercise Set 8.4

**1.** Can be written in the form $au^2 + bu + c = 0$. **3.** $u = x^2$; gives equation $3u^2 - 5u + 1 = 0$. **5.** $u = z^{-1}$; gives equation $u^2 - u = 56$. **7.** $\pm 1, \pm 3$ **9.** $\pm i, \pm 4i$ **11.** $\pm 2, \pm 3$ **13.** $\pm 2, \pm\sqrt{3}$ **15.** $\pm\frac{1}{2}, \pm 2$ **17.** $\pm\sqrt{3}, \pm\sqrt{5}$ **19.** $\pm 3, \pm i\sqrt{2}$ **21.** $\pm 1, \pm i\sqrt{5}$ **23.** 4 **25.** 9 **27.** $\frac{1}{9}$ **29.** 1, −9 **31.** $\frac{4}{3}, -\frac{1}{2}$ **33.** $\pm\sqrt{6}, \pm 1$ **35.** $-6, -\frac{5}{2}$ **37.** $\pm\frac{5\sqrt{6}}{6}, \pm\frac{\sqrt{39}}{3}$ **39.** $-\frac{1}{2}$ **41.** 3, 4 **43.** $2, \frac{1}{3}$ **45.** $1, -\frac{1}{10}$ **47.** $-\frac{1}{2}, \frac{1}{6}$ **49.** 1, 27 **51.** 27, 216 **53.** $\frac{1}{4}$ **55.** −32, −1 **57.** (1, 0), (16, 0) **59.** None **61.** $(-4, 0), \left(\frac{1}{5}, 0\right)$ **63.** (−8, 0), (27, 0) **65.** (−1, 0), (4, 0) **67.** $(\pm 2, 0), (\pm 5, 0)$ **69.** Let $u = x^2$ **71.** Let $u = x^{-1}$ **73.** $x^4 - 5x^2 + 4 = 0$; start with $(x - 2)(x + 2)(x - 1)(x + 1) = 0$ **75.** $x^4 - 7x^2 + 10 = 0$; start with $(x + \sqrt{2})(x - \sqrt{2})(x + \sqrt{5})(x - \sqrt{5}) = 0$ **77.** No; imaginary solution always occur in pairs. **79. a)** and **b)** $\frac{1}{5}, -\frac{1}{4}$ **81.** $-\frac{14}{5}, -\frac{8}{3}$ **83.** $2, \frac{1}{4}$ **85.** 2, 1 **87.** −3, 1, 2, −4 **89.** $\pm\sqrt{\frac{3 \pm \sqrt{15}}{2}}$ **91.** $\frac{43}{60}$ **92.** 1 **93.** D: $\mathbb{R}$, R: $\{y|y \geq 0\}$ **94.** $2xy^2\sqrt[3]{2}$ **95.** $9\sqrt{3}$

## Exercise Set 8.5

**1.** The graph of a quadratic equation is called a parabola. **3.** The axis of symmetry of a parabola is the line where, if the graph is folded, the two sides overlap. **5.** $\left(-\frac{b}{2a}, \frac{4ac - b^2}{4a}\right)$ **7. a)** When $a > 0$, $f(x)$ will have a minimum since the graph opens upward. **b)** When $a < 0$, $f(x)$ will have a maximum since the graph opens downward. **9.** Set $x = 0$ and solve for $y$.

**11. a)**  **b)** 

**13.** Minimum value; the graph opens upward

**15. a)** Upward **b)** $(0, 15)$ **c)** $(-4, -1)$ **d)** $(-5, 0), (-3, 0)$ **e)**

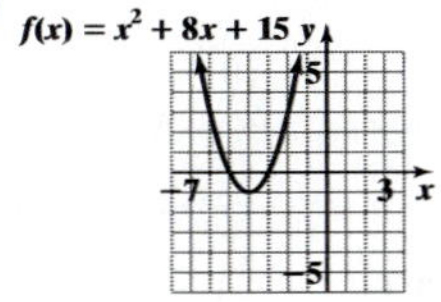


**17. a)** Upward **b)** $(0, 3)$ **c)** $(2, -1)$ **d)** $(1, 0), (3, 0)$ **e)**

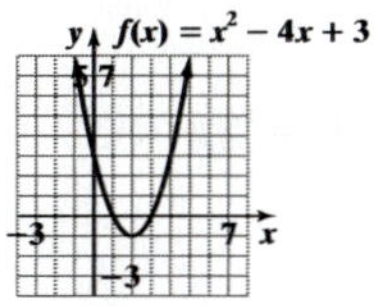


**19. a)** Downward **b)** $(0, 8)$ **c)** $(-1, 9)$ **d)** $(-4, 0), (2, 0)$ **e)**

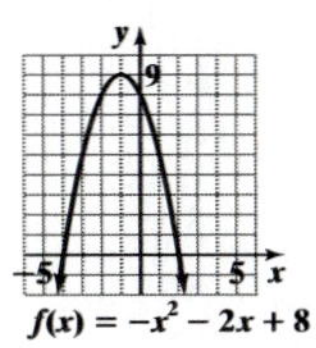


**21. a)** Downward **b)** $(0, 5)$ **c)** $(2, 9)$ **d)** $(-1, 0), (5, 0)$ **e)**

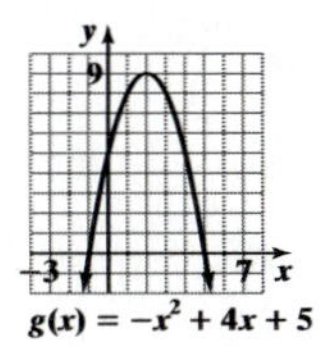


**23. a)** Downward **b)** $(0, -5)$ **c)** $(2, -1)$ **d)** No $x$-intercepts **e)**

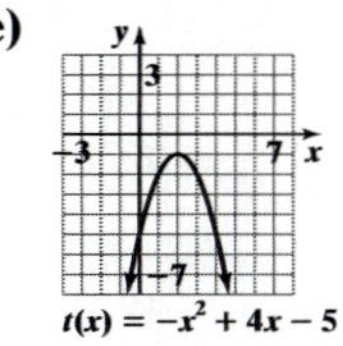


**25. a)** Upward **b)** $(0, 4)$ **c)** $(2, 0)$ **d)** $(2, 0)$ **e)**

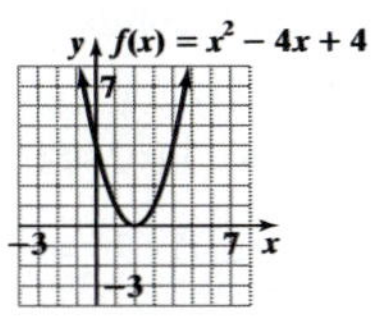


**27. a)** Upward **b)** $(0, 2)$ **c)** $(0, 2)$ **d)** No $x$-intercepts **e)**

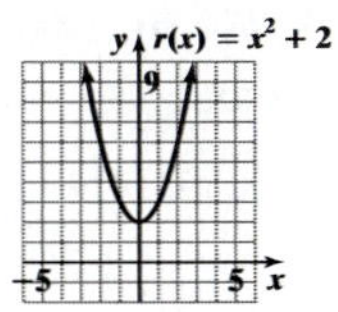


**29. a)** Downward **b)** $(0, 5)$ **c)** $(0, 5)$ **d)** $(-\sqrt{5}, 0), (\sqrt{5}, 0)$ **e)**

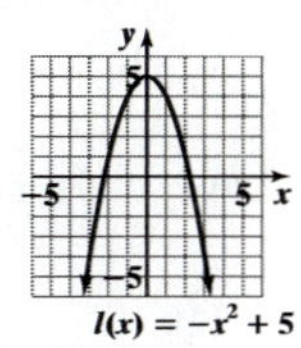


**31. a)** Downward **b)** $(0, -8)$ **c)** $(1, -6)$ **d)** No $x$-intercepts **e)**

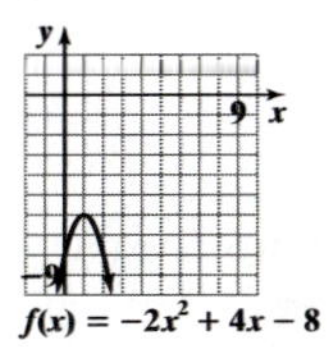


**33. a)** Upward **b)** $(0, 3)$ **c)** $\left(-\frac{2}{3}, \frac{5}{3}\right)$ **d)** No $x$-intercepts **e)**

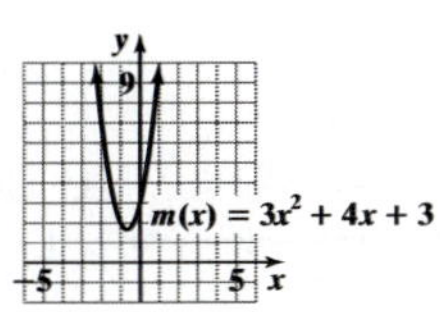


**35. a)** Upward **b)** $(0, -6)$ **c)** $\left(-\frac{2}{3}, -\frac{22}{3}\right)$ **d)** $\left(\frac{-2 + \sqrt{22}}{3}, 0\right), \left(\frac{-2 - \sqrt{22}}{3}, 0\right)$ **e)**

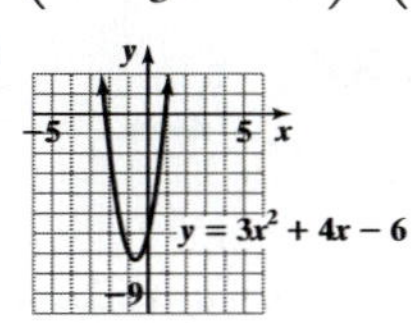


**37. a)** Upward **b)** $(0, -6)$ **c)** $\left(\frac{1}{4}, -\frac{49}{8}\right)$ **d)** $\left(-\frac{3}{2}, 0\right), (2, 0)$ **e)**

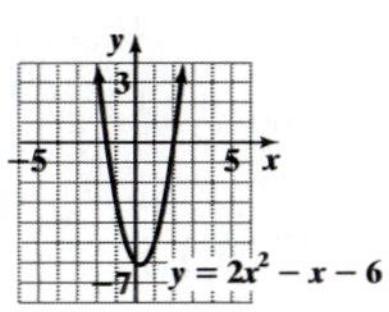


**39. a)** Downward **b)** $(0, -5)$ **c)** $\left(\frac{3}{2}, -\frac{11}{4}\right)$ **d)** No $x$-intercepts **e)**

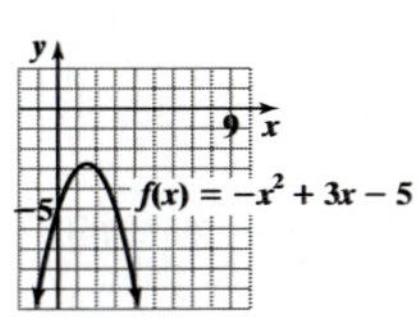


**41.** 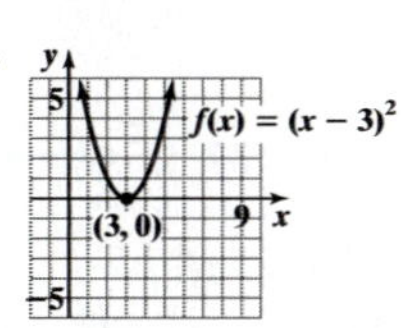


**43.** 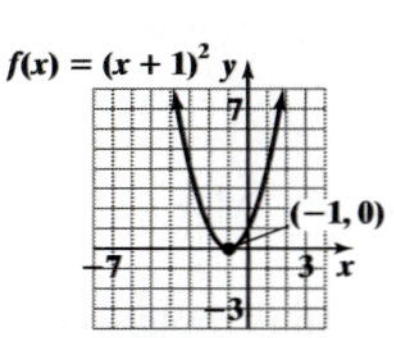


**45.** 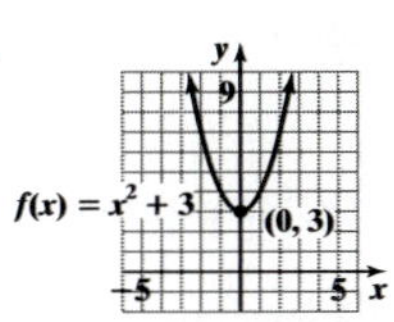


**47.** 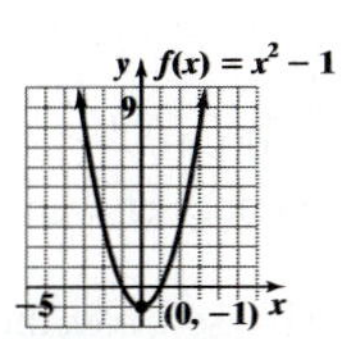


**49.** 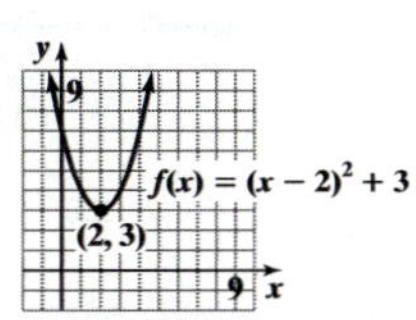


**51.** 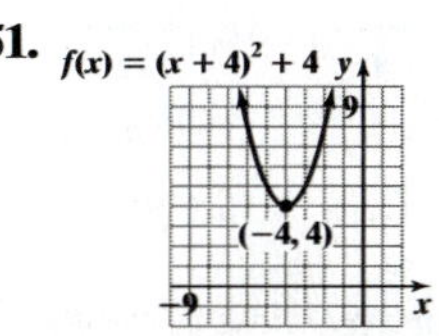

**53.** 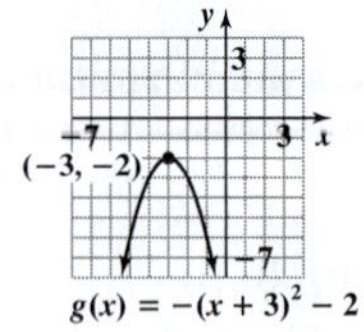


**55.** 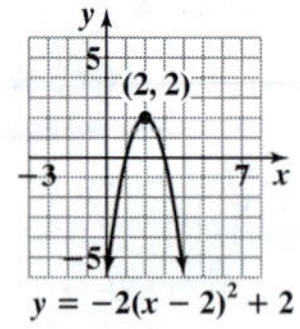


**57.** 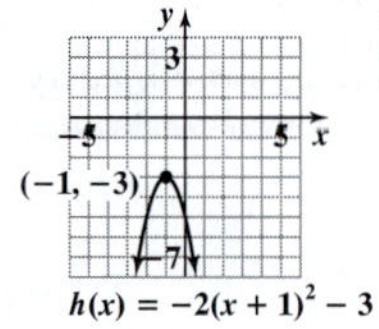


**59. a)** $f(x) = (x - 3)^2 - 1$ **b)** 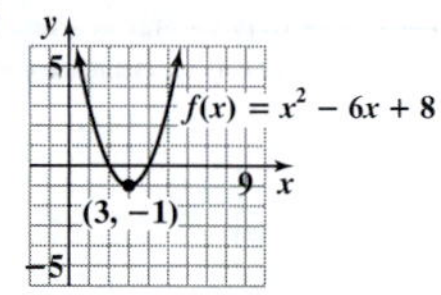


**61. a)** $g(x) = \left(x - \frac{1}{2}\right)^2 - \frac{13}{4}$ **b)** 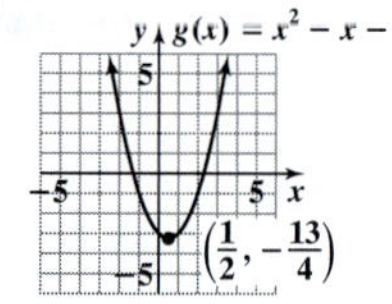


**63. a)** $f(x) = -(x + 2)^2 - 2$ **b)** 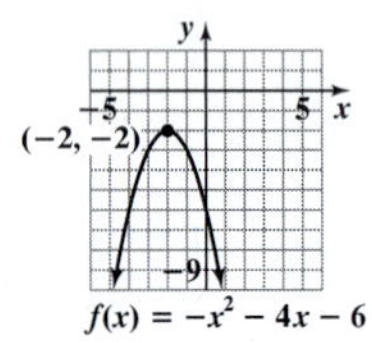


**65. a)** $g(x) = (x - 2)^2 - 5$ **b)** 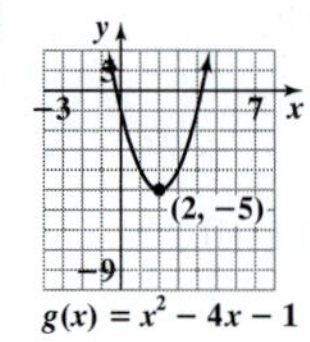


**67. a)** $f(x) = 2\left(x + \frac{5}{4}\right)^2 - \frac{49}{8}$ **b)** 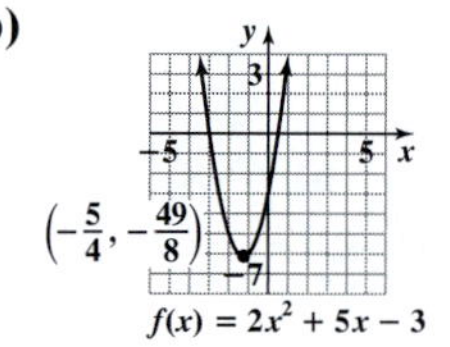


**69.** d) **71.** b) **73. a)** $x = 7$ **b)** $A = 121$ **75. a)** $x = 10.5$ **b)** $A = 240.25$ **77. a)** $n = 200$ **b)** $R = \$800$ **79.** 2010 **81.** 4 units **83.** 3 units **85.** $f(x) = 2(x - 3)^2 - 2$ **87.** $f(x) = -4\left(x + \frac{3}{5}\right)^2 - \sqrt{2}$ **89. a)** The graphs will have the same $x$-intercepts but $f(x) = x^2 - 8x + 12$ will open upward and $g(x) = -x^2 + 8x - 12$ will open downward. **b)** Yes, both at (6, 0) and (2, 0) **c)** No; vertex of $f(x)$ at (4, −4), vertex of $g(x)$ at (4, 4) **d)** 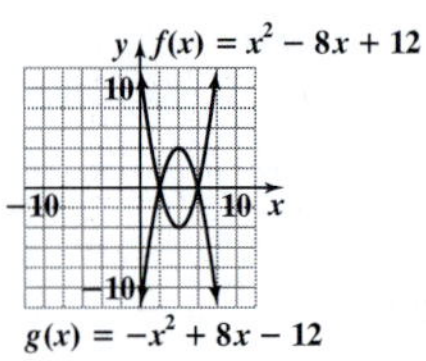


**91. a)** 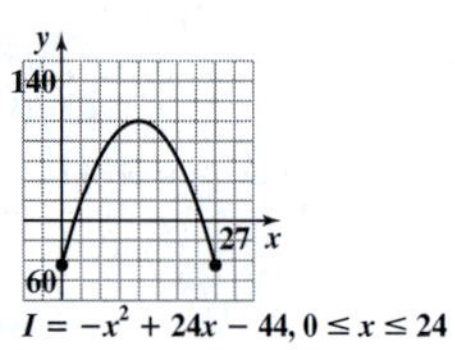


**b)** \$2 **c)** \$22 **d)** \$12 **e)** \$10,000 **93. a)** 100 **b)** \$3800 **95. a)** 40.425 meters **b)** 2.5 seconds **c)** ≈5.37 seconds **97. a)** ≈\$577 **b)** 2000 **99.** 400 square feet **101.** −16, 4 and −4 **103.** 900, 30 and 30 **105. a)** \$142,400 **b)** 380 **107. a)** $f(t) = -16(t - 1.625)^2 + 45.25$ **b)** 45.25 feet, 1.625 seconds **c)** Same

**109.** $200\pi$ square feet **110.** 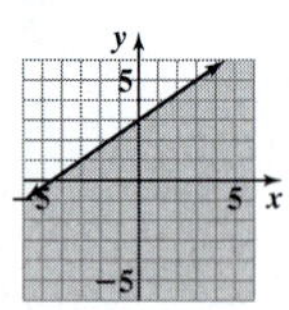 **111.** (−2, 3, 2) **112.** −8 **113.** $\frac{x}{x + 6}$

## Exercise Set 8.6

**1. a)** $x < 2$ or $x > 5$ **b)** $2 < x < 5$ **3.** Yes; ≥ **5.** Yes, −2 and 1 make the fraction 0; No, −1 makes the fraction undefined **7.** 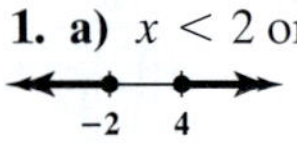

**9.** 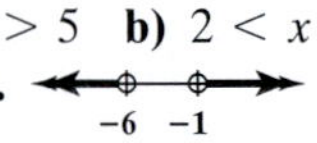

**11.** 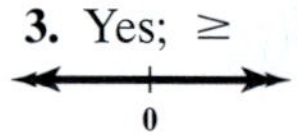

**13.** 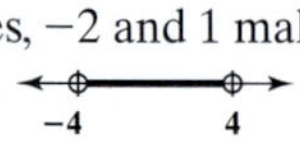

**15.** 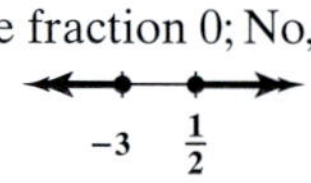

**17.** 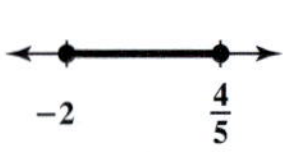

**19.** 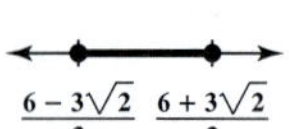

**21.** $[-5, -1] \cup [2, \infty)$ **23.** $(-\infty, -4) \cup (-2, 3)$ **25.** $\left(-6, -\frac{5}{2}\right) \cup (2, \infty)$ **27.** $\left(-1, -\frac{5}{3}\right) \cup (3, \infty)$ **29.** $[-2, -2] \cup \left[\frac{8}{3}, \infty\right)$ **31.** $(-\infty, 0)$ **33.** 

**35.** 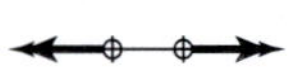

**37.** 

**39.** 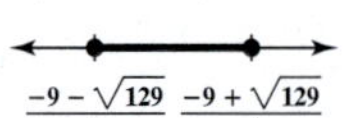

**41.** 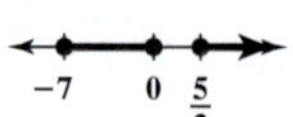

**43.** $\{x|x < -2 \text{ or } x > 4\}$ **45.** $\{x|-5 < x < 1\}$ **47.** $\{x|x \le -3 \text{ or } x > 2\}$ **49.** $\{a|-5 < a < 9\}$ **51.** $\{c|c < 4 \text{ or } c > 10\}$ **53.** $\{y|-4 < y \le -2\}$ **55.** $\left\{a \middle| a \le -2 \text{ or } a > \frac{1}{3}\right\}$ **57.** $\left\{x \middle| -\frac{4}{3} < x < \frac{1}{2}\right\}$ **59.** $\left\{x \middle| -\frac{8}{3} \le x < 2\right\}$ **61.** $(-\infty, -3) \cup (-1, 6)$ **63.** $(-3, 2) \cup (5, \infty)$ **65.** $(-2, 1] \cup [7, \infty)$ **67.** $(-\infty, -8) \cup [0, 3)$ **69.** $(-\infty, -4) \cup (1, 6]$ **71.** $\left[-\frac{5}{2}, 3\right] \cup (4, \infty)$ **73.** 

**75.** 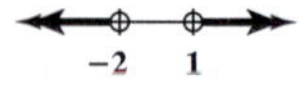

**77.** 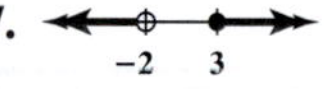

**79.** 

**81.** 

**83.** 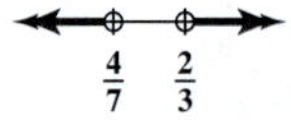

**85. a)** $(4, \infty)$; $y > 0$ in this interval **b)** $(-\infty, 2) \cup (2, 4)$; $y < 0$ in this interval **87.** $x^2 + 2x - 8 > 0$ **89.** $\frac{x + 3}{x - 4} \ge 0$ **91.** All real numbers; for any value of $x$, the expression is ≥0. **93.** All real numbers except −2; for any value of $x$ except −2, the expression is ≥0. **95.** No solution; the graph opens upward and has no $x$-intercepts, so it is always above the $x$-axis. **97.** 

**99.** $x^2 - 3x > 0$; multiply factors containing boundary values. **101.** $x^2 < 0$; $x^2$ is always $\geq 0$.

**103.** $(-\infty, -3) \cup (-1, 1) \cup (3, \infty)$ **105.** $[-2, -1] \cup [2, \infty)$ **109.** 6 quarts **110.** $-\frac{1}{2}$ **111.** $3r + 3s - 9t$ **112.** $\frac{x-3}{x+1}$ **113.** $38 - 9i$

## Chapter 8 Review Exercises

**1.** $5 \pm 2\sqrt{6}$ **2.** $\frac{-1 \pm 2\sqrt{15}}{2}$ **3.** $-\frac{1}{3}, 1$ **4.** $\frac{5}{4}, -\frac{3}{4}$ **5.** $3, 4$ **6.** $4, -8$ **7.** $-1 \pm \sqrt{10}$ **8.** $-3 \pm \sqrt{21}$ **9.** $1 \pm 3i$ **10.** $2 \pm 2i\sqrt{7}$ **11. a)** $32 = (x+1)(x+5)$ **b)** 3 **12. a)** $63 = (x+2)(x+4)$ **b)** 5 **13.** 6, 7 **14.** ≈16.90 ft by ≈16.90 ft **15.** Two real solutions **16.** No real solution **17.** One real solution **18.** No real solution **19.** One real solution **20.** Two real solutions **21.** $0, -\frac{4}{3}$ **22.** 2, 9 **23.** 8, −5 **24.** $0, \frac{9}{7}$ **25.** $\frac{3}{2}, -\frac{5}{3}$ **26.** $\frac{1}{4}, -3$ **27.** $-4 \pm \sqrt{11}$ **28.** $-2 \pm 2\sqrt{3}$ **29.** $\frac{-2 \pm \sqrt{10}}{2}$ **30.** $\frac{3 \pm \sqrt{33}}{3}$ **31.** $\frac{1 \pm i\sqrt{51}}{2}$ **32.** $1 \pm i\sqrt{10}$ **33.** $\frac{5}{2}, -\frac{5}{3}$ **34.** $\frac{1}{4}, -\frac{3}{2}$ **35.** 10, −6 **36.** $\frac{2}{3}, -\frac{3}{2}$ **37.** $\frac{7 \pm \sqrt{89}}{10}$ **38.** $\frac{3 \pm 3\sqrt{3}}{2}$ **39.** $x^2 - 2x - 3 = 0$ **40.** $3x^2 + 4x - 4 = 0$ **41.** $x^2 - 11 = 0$ **42.** $x^2 - 6x + 13 = 0$ **43.** 8 feet by 12 feet **44.** $\sqrt{128} \approx 11.31$ **45.** 4% **46.** 7, 11 **47.** 8 inches by 12 inches **48.** \$540 **49. a)** \$121.8 billion **b)** 2010 **50. a)** 720 feet **b)** 7 seconds **51. a)** 40 milliliters **b)** 150°C **52.** larger ≈23.51 hours; smaller ≈24.51 hours **53.** 50 miles per hour **54.** 1.6 miles per hour **55.** $l = 10$ units, $w = 8$ units **56.** 20 tables **57.** $a = \sqrt{c^2 - b^2}$ **58.** $t = \sqrt{\frac{c-h}{4.9}}$ **59.** $v_y = \sqrt{v^2 - v_x^2}$ **60.** $v_2 = \sqrt{v_1^2 + 2ad}$ **61.** ±2, ±3 **62.** $\pm 4, \pm\sqrt{5}$ **63.** $\pm 2\sqrt{2}, \pm i\sqrt{3}$ **64.** $\frac{3}{2}, -\frac{1}{6}$ **65.** $\frac{1}{9}$ **66.** $\frac{27}{8}, 8$ **67.** $4, \frac{13}{8}$ **68.** $-\frac{1}{5}, -\frac{5}{2}$ **69.** $(\pm 1, 0), (\pm 9, 0)$ **70.** $\left(\frac{4}{25}, 0\right)$ **71.** None **72.** $(3 \pm \sqrt{17}, 0), (3 \pm \sqrt{6}, 0)$

**73. a)** Upward **b)** $(0, 0)$ **c)** $\left(-\frac{5}{2}, -\frac{25}{4}\right)$ **d)** $(0, 0), (-5, 0)$ **e)**

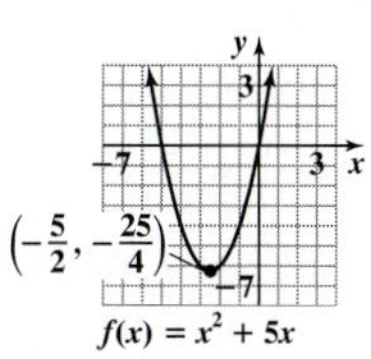


**74. a)** Upward **b)** $(0, -8)$ **c)** $(1, -9)$ **d)** $(-2, 0), (4, 0)$ **e)**

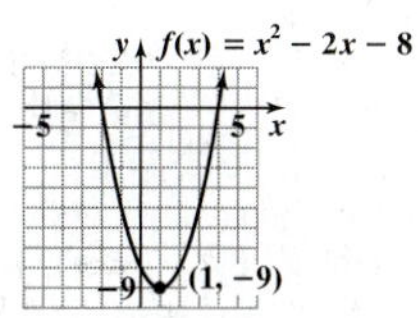


**75. a)** Downward **b)** $(0, -2)$ **c)** $(0, -2)$ **d)** No $x$-intercepts **e)**

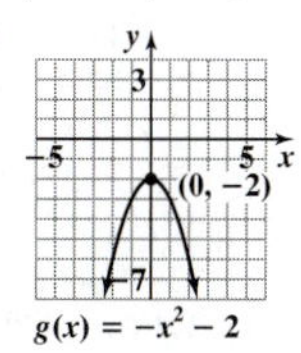


**76. a)** Downward **b)** $(0, 15)$ **c)** $\left(-\frac{1}{4}, \frac{121}{8}\right)$ **d)** $(-3, 0), \left(\frac{5}{2}, 0\right)$ **e)**

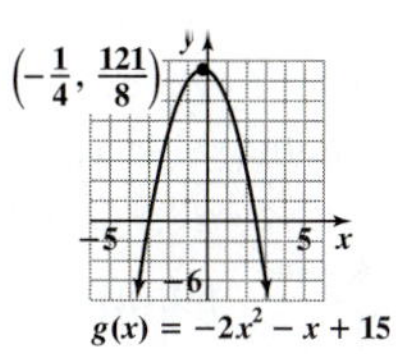


**77. a)** \$11 **b)** \$7600 **78. a)** 2.5 seconds **b)** 175 feet **79.**

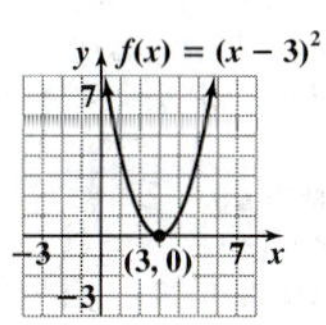


**80.**

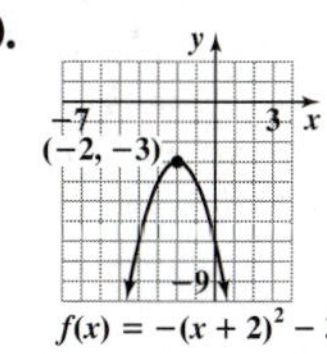


**81.**

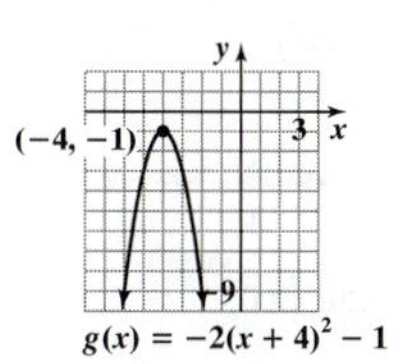


**82.**

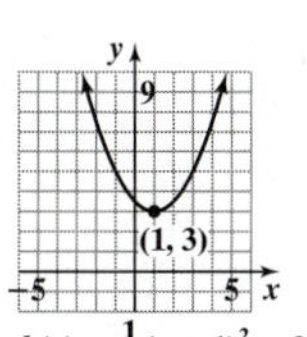


**83.** 

**84.** 

**85.** 

**86.** 

**87.** 

**88.** 

**89.** $\{x | x < -1 \text{ or } x > 5\}$ **90.** $\{x | -2 < x \leq 3\}$ **91.** $\{x | x < -3 \text{ or } x \geq 2\}$ **92.** $\left\{x \middle| -\frac{5}{3} < x < 6\right\}$ **93.** $\{x | -4 < x < -1 \text{ or } x > 2\}$ **94.** $\{x | x \leq 0 \text{ or } 3 \leq x \leq 6\}$ **95.** $\left[-\frac{4}{3}, 1\right] \cup [3, \infty)$ **96.** $(-\infty, -4) \cup (-2, 0)$ **97.** $(-2, 0) \cup (4, \infty)$ **98.** $(-\infty, -3) \cup (2, 8)$ **99.** $(-2, 3] \cup (7, \infty)$ **100.** $(-\infty, -3) \cup [0, 6]$

**101.** 

**102.** 

**103.** 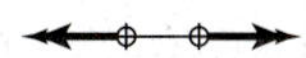


## Chapter 8 Practice Test

**1.** 3, −5 [8.1] **2.** $3 \pm \sqrt{2}$ [8.1] **3.** 8, −2 [8.2] **4.** $2 \pm i\sqrt{7}$ [8.2] **5.** $\frac{2}{3}, -1$ [8.1–8.2] **6.** $\frac{-7 \pm \sqrt{33}}{2}$ [8.1–8.2] **7.** $5x^2 - 18x - 8 = 0$ [8.2] **8.** $v = \sqrt{\frac{2K}{m}}$ [8.3] **9. a)** \$121,200 **b)** ≈2712.57 square feet [8.1–8.3]

**10.** 50 mph [8.1–8.3] **11.** $\pm\frac{\sqrt{10}}{2}, \pm i\sqrt{10}$ [8.4] **12.** $\frac{343}{27}, -216$ [8.4] **13.** $\left(\frac{9}{16}, 0\right)$ [8.4] **14.**

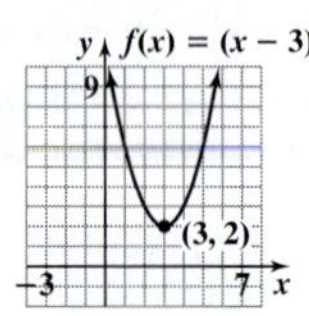


[8.5]

**15.**

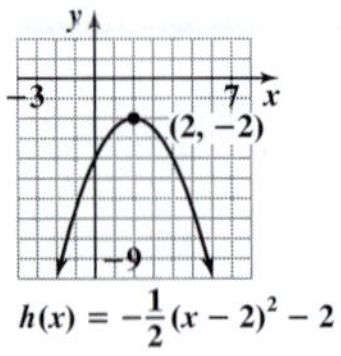


[8.5] **16.** Two real solutions [8.5] **17. a)** Upward **b)** (0, −8) **c)** (−1, −9) **d)** (−4, 0), (2, 0) **e)**

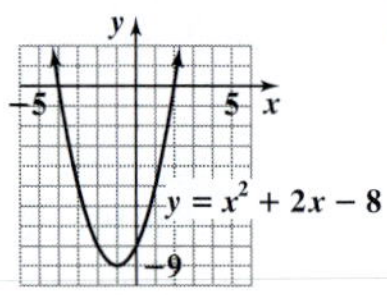


[8.5]

**18.** $2x^2 + 13x - 7 = 0$ [8.5] **19.**

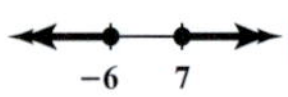


[8.6] **20.**

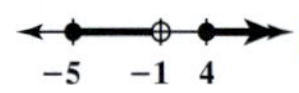


[8.6] **21. a)** $\left[-\frac{5}{2}, -2\right)$ **b)** $\left\{x \middle| -\frac{5}{2} \le x < -2\right\}$ [8.6]
**22.** $w = 5$ feet, $l = 13$ feet [8.5] **23.** 6 seconds [8.5] **24. a)** 20 **b)** $ 490 [8.5] **25.** 30 [8.5]

## Cumulative Review Test

**1.** 13 [1.4] **2.** 18 [1.4] **3.** $2.54 \times 10^6$ [1.6] **4.** $\left\{-\frac{1}{2}, \frac{9}{2}\right\}$ [2.6] **5.** $3x - 7$ [2.1]
**6.** All real numbers, $\mathbb{R}$ [2.1] **7.** (−12, 8) [2.5] **8.** $m = -\frac{9}{7}, \left(0, \frac{15}{7}\right)$ [3.4] **9.** 1500 [3.2] **10.** $y = x - 1$ [3.5]
**11. a)** No, the graph does not pass the vertical line test **b)** Domain: $\{x|x \ge -2\}$, Range: $\mathbb{R}$ [3.2]

**12. a)**

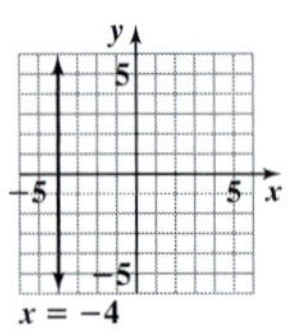


**b)**

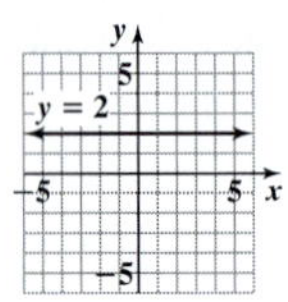


[3.3] **13.** 160 [4.5] **14.** $\left(\frac{5}{2}, 0\right)$ [4.1] **15.** $(x + 9)(x + 7)$ [5.5]
**16. a)** $a^2 + 2ab + b^2$ **b)** $(a + b)^2$ [5.7] **17.** $\frac{2(x - 4)}{(x - 3)(x - 5)}$ [6.2] **18.** $\frac{12}{5}$ [6.4]
**19.** 11.52 watts [6.6] **20.** $\frac{-14 - 23i}{29}$ [7.7]

# Chapter 9

## Exercise Set 9.1

**1.** To find $(f \circ g)(x)$, substitute $g(x)$ for $x$ in $f(x)$. **3. a)** Each $y$ has a unique $x$. **b)** Use the horizontal line test. **5. a)** Yes; each first coordinate is paired with only one second coordinate. **b)** Yes; each second coordinate is paired with only one first coordinate. **c)** $\{(5, 3), (2, 4), (3, -1), (-2, 0)\}$; reverse each ordered pair. **7.** The domain of $f$ is the range of $f^{-1}$ and the range of $f$ is the domain of $f^{-1}$. **9. a)** $x^2 + 4x + 5$ **b)** 37 **c)** $x^2 + 3$ **d)** 19 **11. a)** $x^2 + x - 1$ **b)** 19 **c)** $x^2 + 7x + 8$ **d)** 52 **13. a)** $\frac{1}{2x + 3}$ **b)** $\frac{1}{11}$ **c)** $\frac{2}{x} + 3$ **d)** $3\frac{1}{2}$ **15. a)** $\frac{9}{x} + 1$ **b)** $3\frac{1}{4}$ **c)** $\frac{3}{3x + 1}$ **d)** $\frac{3}{13}$ **17. a)** $x^4 + 10x^2 + 26$ **b)** 442 **c)** $x^4 + 2x^2 + 6$ **d)** 294 **19. a)** $\sqrt{x + 5} - 4$ **b)** −1 **c)** $\sqrt{x + 1}$ **d)** $\sqrt{5}$ **21.** No **23.** Yes **25.** Yes **27.** No **29.** Yes **31.** No **33.** No **35.** Yes **37.** Yes **39.** No **41.** Yes **43.** $f(x)$: Domain: $\{-2, -1, 2, 4, 8\}$; Range: $\{0, 4, 6, 7, 9\}$; $f^{-1}(x)$: Domain: $\{0, 4, 6, 7, 9\}$; Range: $\{-2, -1, 2, 4, 8\}$ **45.** $f(x)$: Domain: $\{-1, 1, 2, 4\}$; Range: $\{-3, -1, 0, 2\}$; $f^{-1}(x)$: Domain: $\{-3, -1, 0, 2\}$; Range: $\{-1, 1, 2, 4\}$ **47.** $f(x)$: Domain: $\{x|x \ge 2\}$; Range: $\{y|y \ge 0\}$; $f^{-1}(x)$; Domain: $\{x|x \ge 0\}$; Range: $\{y|y \ge 2\}$
**49. a)** Yes **b)** $f^{-1}(x) = x + 2$ **51. a)** Yes **b)** $h^{-1}(x) = \frac{x}{4}$ **53. a)** No **55. a)** No **57. a)** Yes **b)** $g^{-1}(x) = \frac{1}{x}$ **59. a)** No
**61. a)** Yes **b)** $g^{-1}(x) = \sqrt[3]{x + 6}$ **63. a)** Yes **b)** $g^{-1}(x) = x^2 - 2, x \ge 0$ **65. a)** Yes **b)** $h^{-1}(x) = \sqrt{x + 4}, x \ge -4$

**67. a)** $f^{-1}(x) = \frac{x - 8}{2}$
**b)**

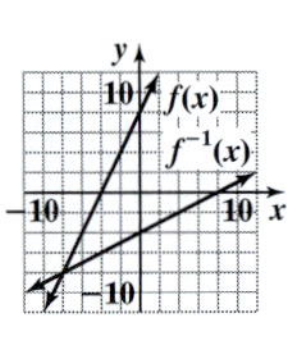


**69. a)** $f^{-1}(x) = x^2, x \ge 0$
**b)**

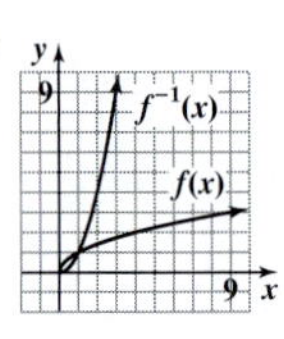


**71. a)** $f^{-1}(x) = x^2 + 1, x \ge 0$
**b)**

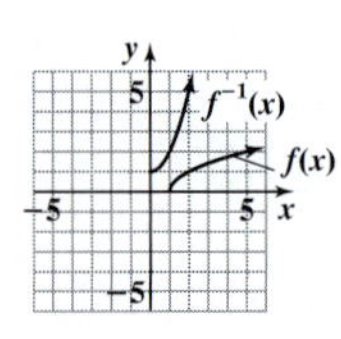


**73. a)** $f^{-1}(x) = x^3$
**b)**

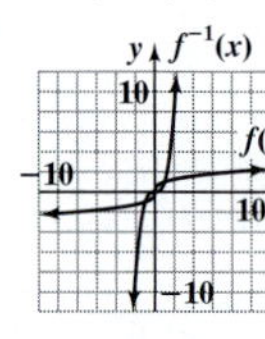


**75. a)** $f^{-1}(x) = \frac{1}{x}, x > 0$
**b)**

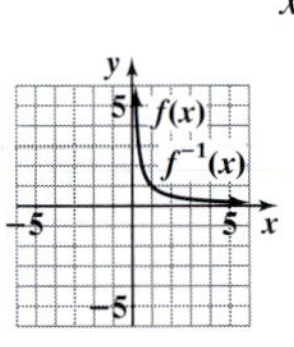


**77.** $(f \circ f^{-1})(x) = x, (f^{-1} \circ f)(x) = x$ **79.** $(f \circ f^{-1})(x) = x, (f^{-1} \circ f)(x) = x$
**81.** $(f \circ f^{-1})(x) = x, (f^{-1} \circ f)(x) = x$ **83.** $(f \circ f^{-1})(x) = x, (f^{-1} \circ f)(x) = x$
**85.** No, composition of functions is not commutative. Let $f(x) = x^2$ and $g(x) = x + 1$. Then $(f \circ g)(x) = x^2 + 2x + 1$, while $(g \circ f)(x) = x^2 + 1$.
**87. a)** $(f \circ g)(x) = x$; $(g \circ f)(x) = x$ **b)** The Domain is $\mathbb{R}$ for all of them.
**89.** The range of $f^{-1}(x)$ is the domain of $f(x)$. **91.** $f^{-1}(x) = \frac{x}{3}$; $x$ is feet and $f^{-1}(x)$ is yards
**93.** $f^{-1}(x) = \frac{9}{5}x + 32$. **95.** $(f \circ g)(x) = 453.6x$, $x$ is pounds, $(f \circ g)(x)$ is grams
**97.** $(f \circ g)(x) = 0.915x$, $x$ is yards, $(f \circ g)(x)$ is meters **99.** Yes **101.** Yes

**103. a)** 6 feet **b)** $36\pi \approx 113.10$ square feet **c)** $A(t) = 4\pi t^2$ **d)** $36\pi \approx 113.10$ square feet **e)** Answers should agree **106.** $\frac{81}{16}$ **107.** $2x + 3y = 10$ **108.** $\frac{18 - 12x}{x^3}$ **109.** $p = \frac{fq}{q - f}$ **110.** $-1 \pm \sqrt{11}$

## Exercise Set 9.2

**1.** Exponential functions are functions of the form $f(x) = a^x, a > 0, a \neq 1$. **3. a)** As $x$ increases, $y$ decreases. **b)** No, $\left(\frac{1}{2}\right)^x$ can never be 0. **c)** No, $\left(\frac{1}{2}\right)^x$ can never be negative. **5. a)** Same; $(0, 1)$. **b)** $y = 3^x$ will be steeper than $y = 2^x$ for $x > 0$.

**7.**
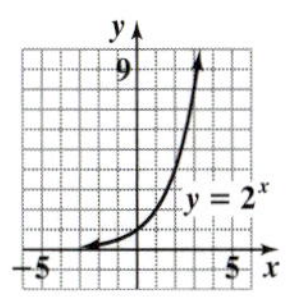


**9.**
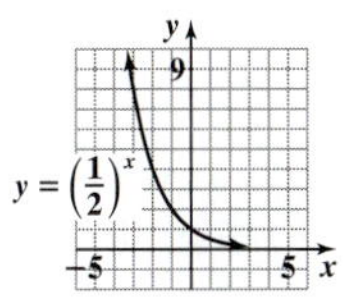


**11.**
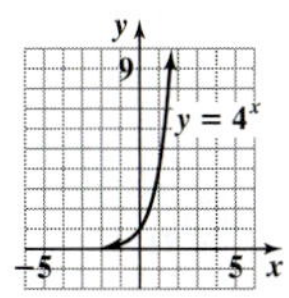


**13.**
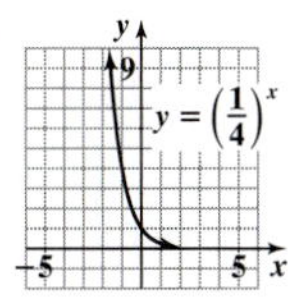


**15.**
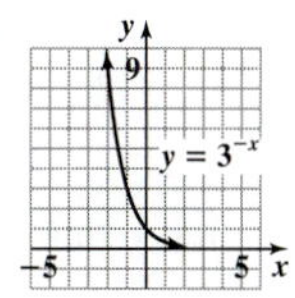


**17.**
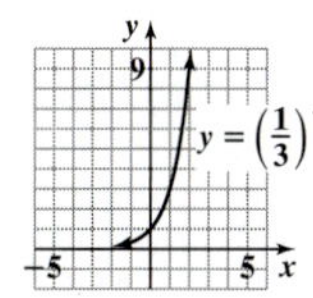


**19.**
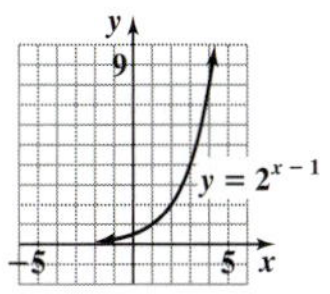


**21.**
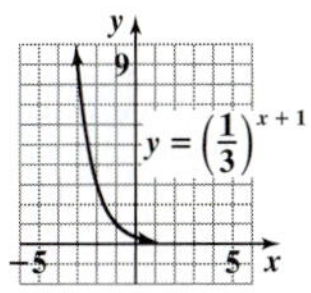


**23.**
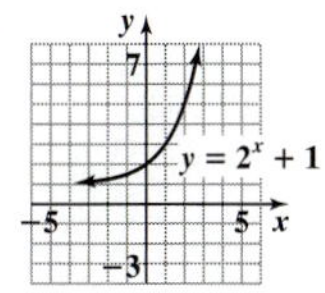


**25.**
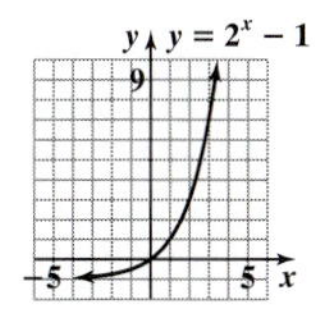


**27. a)** It is a horizontal line through $y = 1$. **b)** Yes **c)** No, not a one-to-one function **29.** $y = a^x - k$ is $y = a^x$ lowered $k$ units. **31.** The graph of $y = a^{x+2}$ is the graph of $y = a^x$ shifted 2 units to the left. **33. a)** $\approx$36.232 million **b)** $\approx$187.846 million **35.** \$512 **37. a)** 14 years **b)** 10 years **c)** \$25 **d)** Increases it **39.** 45 **41.** $\approx$\$6344.93 **43.** $\approx$10.6 grams **45. a)** 5 grams **b)** $\approx 7.28 \times 10^{-11}$ grams **47. a)** 2400 **b)** $\approx$4977 **49.** $\approx$\$10,850.92 **51. a)** Answers will vary. **b)** $\approx$472,414 gallons **53.** $\approx$8.83 kilometers **55. a)** $\approx$\$201.36 **b)** $\approx$\$31.36 **57. a)**

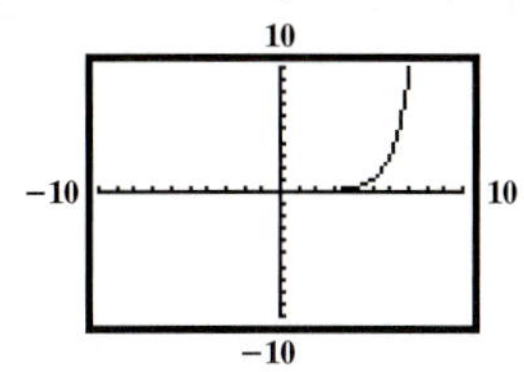


**b)** $\approx$6.26 **59. a)** \$16,384 **b)** \$524,288 **c)** $2^{n-1}$ **d)** $\$2^{29} = \$536{,}870{,}912$ **e)** $2^0 + 2^1 + 2^2 + \cdots + 2^{29}$ **61. a)** $-6.2x^6y^2 + 9.2x^5y^2 + 2.3x^4y$ **b)** 8 **c)** $-6.2$ **62.** $x^3 + 3x^2 - 6x + 20$ **63.** $|a - 4|$ **64.** $\frac{2xy\sqrt[4]{xy^2z^3}}{z}$

## Exercise Set 9.3

**1. a)** $a > 0$ and $a \neq 1$ **b)** $\{x|x > 0\}$ **c)** $\mathbb{R}$ **3.** $\left(\frac{1}{27}, -3\right)\left(\frac{1}{9}, -2\right), \left(\frac{1}{3}, -1\right)$ $(1, 0), (3, 1), (9, 2)$, and $(27, 3)$; the functions $f(x) = a^x$ and $g(x) = \log_a x$ are inverses. **5.** The functions $y = a^x$ and $y = \log_a x$ for $a \neq 1$ are inverses of each other, thus the graphs are symmetric with respect to the line $y = x$. For each ordered pair $(x, y)$ on the graph of $y = a^x$, the ordered pair $(y, x)$ is on the graph of $y = \log_a x$.

**7.**
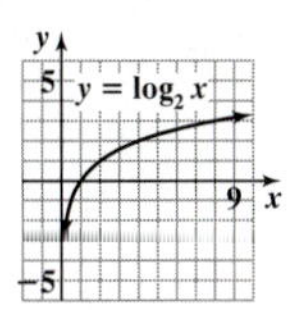


**9.**
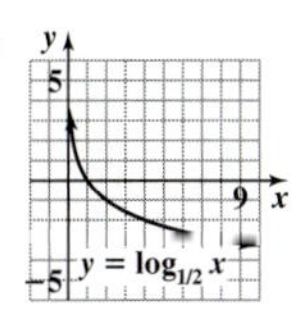


**11.**
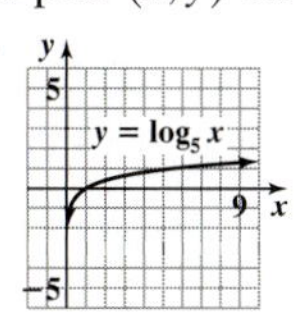


**13.**
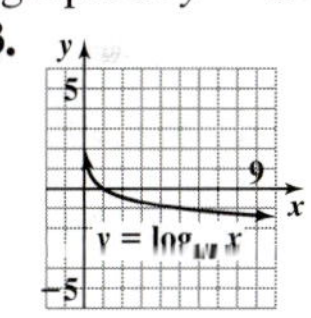

**15.**
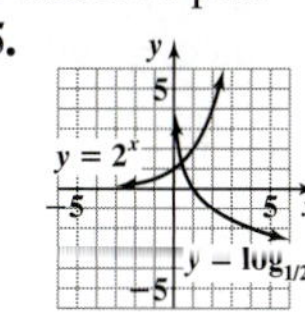


**17.**
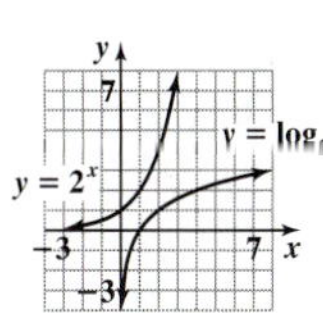


**19.** $\log_2 8 = 3$ **21.** $\log_3 9 = 2$ **23.** $\log_{16} 4 = \frac{1}{2}$ **25.** $\log_8 2 = \frac{1}{3}$ **27.** $\log_{1/2} \frac{1}{32} = 5$ **29.** $\log_2 \frac{1}{8} = -3$ **31.** $\log_4 \frac{1}{64} = -3$ **33.** $\log_{64} 4 = \frac{1}{3}$ **35.** $\log_8 \frac{1}{2} = -\frac{1}{3}$ **37.** $\log_{81} \frac{1}{3} = -\frac{1}{4}$ **39.** $\log_{10} 7 = 0.8451$ **41.** $\log_e 7.3891 = 2$ **43.** $\log_a b = n$ **45.** $2^3 = 8$ **47.** $\left(\frac{1}{3}\right)^3 = \frac{1}{27}$ **49.** $5^{-2} = \frac{1}{25}$ **51.** $49^{1/2} = 7$ **53.** $9^{-2} = \frac{1}{81}$ **55.** $10^{-3} = \frac{1}{1000}$ **57.** $6^3 = 216$ **59.** $10^{-0.2076} = 0.62$ **61.** $e^{1.8749} = 6.52$ **63.** $w^{-p} = s$ **65.** 3 **67.** 5 **69.** 27 **71.** $-4$ **73.** $\frac{1}{64}$ **75.** 3 **77.** 0 **79.** 2 **81.** $-2$ **83.** 4 **85.** 4 **87.** $-4$ **89.** $-2$ **91.** 0 **93.** 1 **95.** 5 **97.** $f^{-1}(x) = \log_5 x$ **99.** 3 and 4, since 62 lies between $3^3 = 27$ and $3^4 = 81$. **101.** 2 and 3, since 425 lies between $10^2 = 100$ and $10^3 = 1000$. **103.** $2^x$; Note that for $x = 10$, $2^x = 1024$ while $\log_{10} x = 1$. **105.** 6 **107.** 8 **109.** 3 **111.** 9 **113.** 10,000,000 **115.** 10,000 **117.**

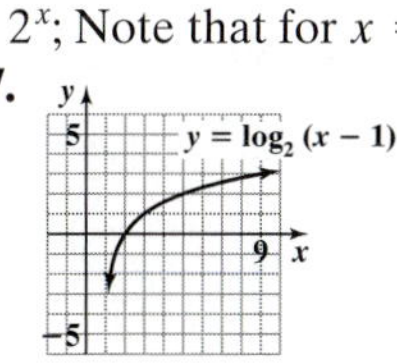


**119.** $2x(x + 3)(x - 6)$ **120.** $(x - 2)(x + 2)(x^2 + 4)$ **121.** $4(2x + 3)(5x - 1)$ **122.** $(3rs - 1)(2rs + 1)$

## Exercise Set 9.4

**1.** Answers will vary. **3.** Answers will vary. **5.** Yes, it is an expansion of property 1. **7.** $\log_4 3 + \log_4 10$ **9.** $\log_8 7 + \log_8(x + 3)$ **11.** $\log_2 27 - \log_2 11$ **13.** $\frac{1}{2}\log_{10} x - \log_{10}(x - 9)$ **15.** $7 \log_6 x$ **17.** $5 \log_4 (r + 7)$ **19.** $\frac{3}{2}\log_4 a - \frac{1}{2}\log_4 (a + 2)$ **21.** $6 \log_3 d - 4 \log_3 (a - 8)$ **23.** $\log_8 (y + 4) - 2 \log_8 y$ **25.** $\log_{10} 9 + \log_{10} m - \log_{10} 8 - \log_{10} n$ **27.** $\log_5 16$ **29.** $\log_2 \frac{9}{5}$ **31.** $\log_4 64$ **33.** $\log_{10} x(x + 3)$ **35.** $\log_9 \frac{z^2}{z - 2}$ **37.** $\log_5 \left(\frac{p}{3}\right)^4$ **39.** $\log_2 \frac{n(n + 4)}{n - 3}$ **41.** $\log_5 \sqrt{\frac{x - 8}{x}}$

**43.** $\log_9 \dfrac{16\sqrt[3]{r} - 6}{\sqrt{r}}$ **45.** $\log_6 \dfrac{81}{(x+3)^2 x^4}$ **47.** 1 **49.** −0.3980 **51.** 1.3980 **53.** 10 **55.** 7 **57.** 3 **59.** 25 **61.** Yes

**63.** $\log_a \dfrac{x}{y} = \log_a\left(x \cdot \dfrac{1}{y}\right) = \log_a x + \log_a \dfrac{1}{y}$ **65.** $\log_a(x-2)$ **67.** Yes, $\log_a(x^2+8x+16) = \log_a(x+4)^2 = 2\log_a(x+4)$

**69.** 0.8640 **71.** 0.1080 **73.** 0.7000 **75.** No, there is no relationship between $\log_{10}(x+y)$ and $\log_{10} xy$ or $\log_{10}\left(\dfrac{x}{y}\right)$

**77.** $\dfrac{1}{4}\log_2 x + \dfrac{1}{4}\log_2 y + \dfrac{1}{3}\log_2 a - \dfrac{1}{5}\log_2(a-b)$ **79.** Answers will vary. **82. a)** $\{x|x > 40\}$ **b)** $(40, \infty)$ **83. a)** $a^2 - 4c^2$ **b)** $(a+2c)(a-2c)$ **84.** 3 **85.** $-26 - 7i$ **86.** 49

## Mid-Chapter Test

**1. a)** In $f(x)$, replace $x$ by $g(x)$. **b)** $6x + 18$ [9.1] **2. a)** $\left(\dfrac{6}{x}\right)^2 + 5$ or $\dfrac{36}{x^2} + 5$ **b)** 9 **c)** $\dfrac{6}{x^2+5}$ **d)** $\dfrac{3}{7}$ [9.1] **3. a)** Answers will vary. **b)** No [9.1] **4. a)** Yes **b)** {(2, −3), (3, 2), (1, 5), (8, 6)} [9.1] **5. a)** Yes **b)** $p^{-1}(x) = 3x + 15$ [9.1] **6. a)** Yes **b)** $k^{-1}(x) = x^2 + 4$, $x \geq 0$ [9.1] **7.** $m^{-1}(x) = -\dfrac{1}{2}x + 2$

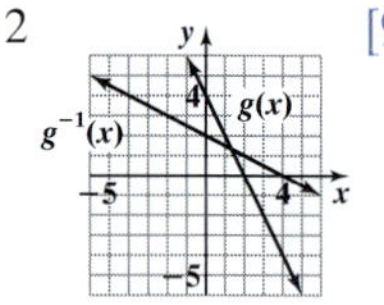


[9.1] **8.**

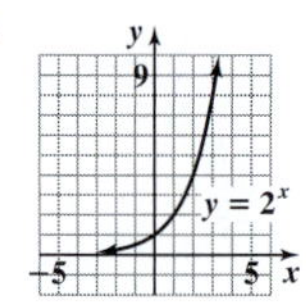


[9.2]

**9.**

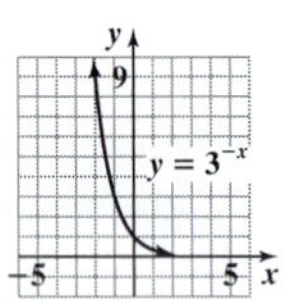


[9.2] **10.**

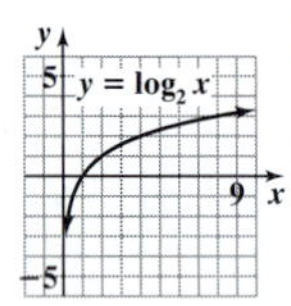


[9.3] **11. a)** 10 **b)** 320 [9.3] **12.** $\log_{27} 9 = \dfrac{2}{3}$ [9.3] **13.** $2^{-6} = \dfrac{1}{64}$ [9.3] **14.** 3 [9.3]

**15.** 2 [9.3] **16.** 4 [9.3] **17.** $2\log_9 x + \log_9(x-5)$ [9.4] **18.** $\log_5 7 + \log_5 m - \dfrac{1}{2}\log_5 n$ [9.4] **19.** $\log_2 \dfrac{x^3(x+7)}{(x+1)^4}$ [9.4]

**20.** $\log_7 \sqrt{\dfrac{x+2}{x}}$ [9.4]

## Exercise Set 9.5

**1.** Common logarithms are logarithms with base 10. **3.** Antilogarithms are numbers obtained by taking 10 to the power of the logarithm. **5.** 1.9345 **7.** 4.2833 **9.** −1.2125 **11.** 2.0000 **13.** 0.5740 **15.** −1.7620 **17.** 1.64 **19.** 42,500 **21.** 0.0196 **23.** 1.00 **25.** 579 **27.** 0.0000726 **29.** 100 **31.** 2410 **33.** 13,800 **35.** 0.0871 **37.** 0.239 **39.** 0.749 **41.** 3.5514 **43.** −1.1385 **45.** 2.3856 **47.** −2.2277 **49.** 679 **51.** 0.303 **53.** 0.0331 **55.** 22.4 **57.** 0 **59.** −1 **61.** −2 **63.** −3 **65.** 7 **67.** 7 **69.** 20.8 **71.** 41.5 **73.** No; $10^2 = 100$ and since $462 > 100$, log 462 must be greater than 2. **75.** No; $10^0 = 1$ and $10^{-1} = 0.1$ and since $1 > 0.163 > 0.1$, log 0.163 must be between 0 and −1. **77.** No; $\log \dfrac{y}{4x} = \log y - \log 4 - \log x$ **79.** 2.0969 **81.** −0.6990 **83.** 2.7958 **85.** 2510 **87.** 501,000 **89. a)** ≈31.62 kilometers **b)** ≈0.50 kilometer **c)** ≈14.68 **91. a)** ≈72% **b)** ≈15% **93.** ≈6310 times more intense **95. a)** $\approx 6.31 \times 10^{20}$ **b)** ≈2.19 **97.** ≈6.2 **99.** $I = \text{antilog } R$ **101.** $t = \text{antilog}\left(\dfrac{26 - R}{41.9}\right) - 1$ **104.** 50 miles per hour, 55 miles per hour **105.** (2, −3) **106.** 0, −4, 3 **107.** $|3x^2 - y|$ **108.** $(-\infty, -4] \cup [2, 5]$

## Exercise Set 9.6

**1.** $c = d$ **3.** Check for extraneous solutions. **5.** log (−2) is not a real number. **7.** 3 **9.** 4 **11.** $\dfrac{1}{2}$ **13.** 2 **15.** $-\dfrac{1}{3}$ **17.** 4 **19.** 3 **21.** 3 **23.** 2.01 **25.** 3.56 **27.** 5.59 **29.** 6.34 **31.** 6 **33.** 5 **35.** $\dfrac{1}{16}$ **37.** 100 **39.** −1 **41.** −6, 4 **43.** 0, −8 **45.** 92 **47.** $\dfrac{4}{5}$ **49.** $\dfrac{3}{2}$ **51.** 4 **53.** 2 **55.** 0.87 **57.** 30 **59.** 5 **61.** 4 **63.** 2 **65.** 3 **67.** 9 **69.** ≈3.47 hours **71.** ≈3.19 years **73.** ≈17.07 years **75. a)** ≈7.95 **b)** ≈5.88 **77.** ≈$7112.09 **79.** ≈19.36 **81. a)** 1,000,000,000,000 times greater **b)** 10,000,000 times greater **83.** 8 **85.** $x = 1$ and $x = 2$ **87.** (3, 1) **89.** (54, 46) **91.** 2.8 **93.** No solution **95.** The box is greater by ≈7.73 cubic feet **96.** −4 **97.**

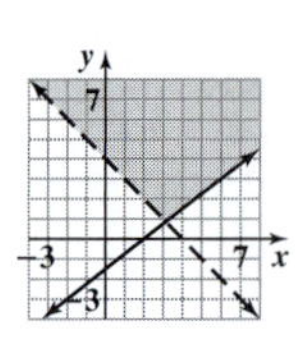


**98.** $\dfrac{x\sqrt{y} - y\sqrt{x}}{x - y}$ **99.** $c = \sqrt{\dfrac{E}{m}}$ **100.** $f(x) = 2(x-3)^2 - 5$

## Exercise Set 9.7

**1. a)** $e$ **b)** ≈2.7183 **3.** $\{x|x > 0\}$ **5.** $\log_a x = \dfrac{\log_b x}{\log_b a}$ **7.** $x$ **9.** $e^x$ **11.** $k < 0$ **13.** 4.1271 **15.** −0.2070 **17.** 4.95 **19.** 0.0578 **21.** 0.972 **23.** 3.6640 **25.** 4.3923 **27.** 1.7297 **29.** 2.7380 **31.** 2.9135 **33.** 3.9318 **35.** −0.4784 **37.** 4 **39.** 1 **41.** 4 **43.** 6 **45.** $P = 4757.5673$ **47.** $P_0 = 224.0845$ **49.** $t = 0.7847$ **51.** $k = 0.2310$ **53.** $k = -0.2888$ **55.** $A = 4719.7672$ **57.** $V_0 = \dfrac{V}{e^{kt}}$ **59.** $t = \dfrac{\ln P - \ln 150}{7}$ **61.** $k = \dfrac{\ln A - \ln A_0}{t}$ **63.** $y = xe^{2.3}$ **65.** $y = (x+6)e^5$ **67.** ≈2.5000; find ln 12.183 **69. a)** ≈$5637.48 **b)** ≈11.55 years **71.** ≈39.98 grams **73. a)** ≈86.47% **b)** ≈34.66 days **75. a)** ≈5.15 feet per second **b)** ≈5.96 feet per second **c)** ≈646,000 **77.** ≈$449,004,412,200,000

**79.** **a)** ≈6.9 billion **b)** ≈53 years **81.** **a)** ≈3.04 million **b)** ≈3.89 million **83.** **a)** ≈\$2788.38 **b)** ≈\$3711.59
**85.** **a)** ≈32.43 inches **b)** ≈35.46 inches **87.** **a)** ≈6626.62 years **b)** ≈5752.26 years **89.** ≈\$6791.91
**91.** **a)** Strontium 90, since it has a higher decay rate **b)** ≈31.66% of original amount **93.** Answers will vary. **95.** 7.286
**97.** $-1.507, 16.659$ **99.** $v_0 = \dfrac{e^{xk} - 1}{kt}$ **101.** $i = Ie^{-t/RC}$ **102.** **a)** 0 **b)** $\dfrac{11}{40}$ or 0.275 **103.** 240 children, 310 adults
**104.** $-9x^2y^3 + 12x^2y^2 - 3xy^2 + 4xy$ **105.** $-20, 20$ **106.** $x + x^2$

## Chapter 9 Review Exercises

**1.** $4x^2 - 26x + 44$ **2.** 2 **3.** $2x^2 - 6x + 3$ **4.** 39 **5.** $6\sqrt{x - 3} + 7, x \geq 3$
**6.** $\sqrt{6x + 4}, x \geq -\dfrac{2}{3}$ **7.** One-to-one **8.** Not one-to-one **9.** One-to-one **10.** Not one-to-one **11.** One-to-one
**12.** Not one-to-one **13.** $f(x)$: Domain: $\{-4, -1, 5, 6\}$; Range: $\{-3, 2, 3, 8\}$; $f^{-1}(x)$: Domain: $\{-3, 2, 3, 8\}$; Range: $\{-4, -1, 5, 6\}$
**14.** $f(x)$: Domain: $\{x|x \geq 0\}$; Range: $\{y|y \geq 4\}$; $f^{-1}(x)$: Domain: $\{x|x \geq 4\}$; Range: $\{y|y \geq 0\}$

**15.** $f^{-1}(x) = \dfrac{x + 2}{4}$,

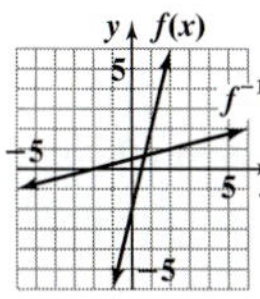

**16.** $f^{-1}(x) = x^3 + 1$;

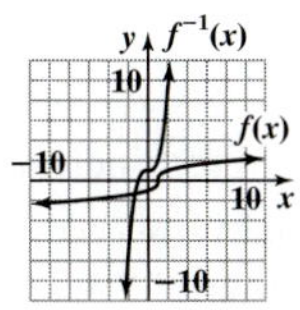

**17.** $f^{-1}(x) = \dfrac{x}{36}$, $x$ is inches, $f^{-1}(x)$ is yards.
**18.** $f^{-1}(x) = \dfrac{x}{4}$, $x$ is quarts, $f^{-1}(x)$ is gallons **19.**

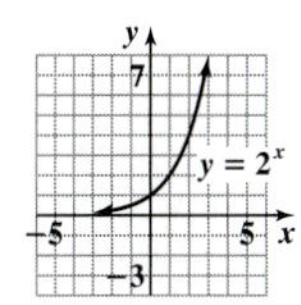


**20.**

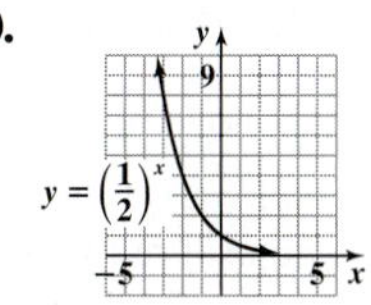


**21.** **a)** 30.23 million **b)** 62.73 million **c)** 187.50 million **22.** $\log_8 64 = 2$ **23.** $\log_{81} 3 = \dfrac{1}{4}$ **24.** $\log_5 \dfrac{1}{125} = -3$ **25.** $2^5 = 32$
**26.** $\left(\dfrac{1}{4}\right)^2 = \dfrac{1}{16}$ **27.** $6^{-2} = \dfrac{1}{36}$ **28.** $4^3 = x; 64$ **29.** $a^4 = 81; 3$ **30.** $\left(\dfrac{1}{5}\right)^{-3} = x; 125$ **31.**

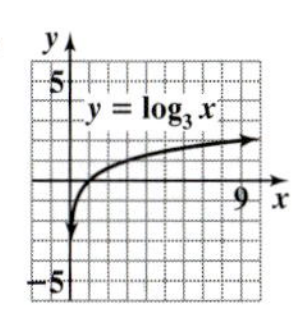


**32.**

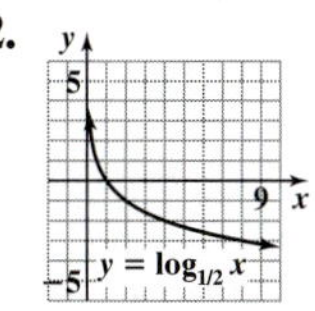


**33.** $8 \log_5 17$ **34.** $\dfrac{1}{2}\log_3 (x - 9)$ **35.** $\log 6 + \log (a + 1) - \log 19$
**36.** $4 \log x - \log 7 - 5 \log (2x + 3)$ **37.** $\log \dfrac{x^5}{(x + 1)^3}$ **38.** $\log \dfrac{(2x)^4}{y}$ **39.** $\ln \dfrac{\sqrt[3]{\dfrac{x}{x + 2}}}{2}$
**40.** $\ln \dfrac{x^3\sqrt{x + 1}}{(x + 4)^6}$ **41.** 10 **42.** 5 **43.** 121 **44.** 3 **45.** 2.9133 **46.** −3.5720 **47.** 1440 **48.** 0.000697 **49.** 11,600 **50.** 0.0594
**51.** 5 **52.** 9 **53.** 22.4 **54.** 9.4 **55.** 4 **56.** $-\dfrac{1}{2}$ **57.** 2 **58.** 5 **59.** 2.582 **60.** 5.968 **61.** 1.353 **62.** 2.240 **63.** 26 **64.** 5
**65.** 1 **66.** 3 **67.** $t \approx 1.155$ **68.** $A_0 \approx 352.542$ **69.** $t = \dfrac{\ln A - \ln A_0}{k}$ **70.** $k = \dfrac{\ln 0.25}{t}$ **71.** $y = xe^6$ **72.** $y = 3x + 23$ **73.** 7.6147
**74.** 3.5046 **75.** ≈\$19,126.18 **76.** ≈17.3 years **77.** **a)** ≈92.88 minutes **b)** ≈118.14 minutes **78.** ≈10.32 pounds per square inch
**79.** **a)** 72 **b)** ≈61.2 **c)** ≈5 months

## Chapter 9 Practice Test

**1.** **a)** Yes **b)** $\{(2, 4), (8, -3), (3, -1), (-7, 6)\}$ [9.1] **2.** **a)** $x^2 + 4x + 1$ **b)** 61 [9.1]
**3.** **a)** $\sqrt{x^2 + 3}$, **b)** $2\sqrt{13}$ [9.1] **4.** **a)** $f^{-1}(x) = -\dfrac{1}{3}(x + 5)$ **5.** **a)** $f^{-1}(x) = x^2 + 1, x \geq 0$

**4. b)** [9.1]

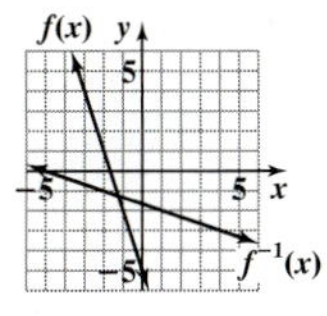

**5. b)** [9.1]

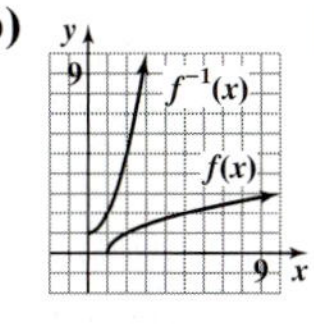

**6.** $\{x|x > 0\}$ [9.3] **7.** −4 [9.4] **8.** [9.2]

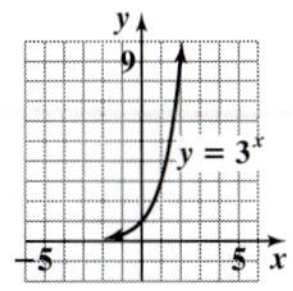


**9.** [9.3]

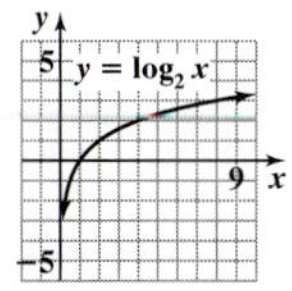


**10.** $\log_2 \dfrac{1}{32} = -5$ [9.4] **11.** $5^3 = 125$ [9.3]
**12.** $2^4 = x + 3, 13$ [9.3] **13.** $64^y = 16, \dfrac{2}{3}$ [9.3]
**14.** $3 \log_2 x + \log_2 (x - 4) - \log_2(x + 2)$ [9.4]
**15.** $\log_6 \dfrac{(x - 4)^7(x + 3)^2}{\sqrt{x}}$ [9.4] **16.** 5 [9.4] **17.** **a)** 3.6646 **b)** −2.6708 [9.5] **18.** ≈2.68 [9.6] **19.** 3 [9.6] **20.** $\dfrac{17}{5}$ [9.6]
**21.** 16.2810 [9.7] **22.** 2.0588 [9.7] **23.** 30.5430 [9.7] **24.** ≈\$5211.02 [9.7] **25.** ≈3364.86 years old [9.7]

**Cumulative Review Test** **1.** $\frac{12xy^5}{z^4}$ [1.5] **2.** 40 [1.4] **3.** 7.5%; [2.3] **4.** $\left\{x \middle| 2 \le x < \frac{15}{2}\right\}, \left[2, \frac{15}{2}\right)$ [2.5] **5.** $y = \frac{2x - 8}{3}$ [2.2] **6.** 0 [3.2] **7.** $m = 2, y = 2x + 3$ [3.4] **8.**

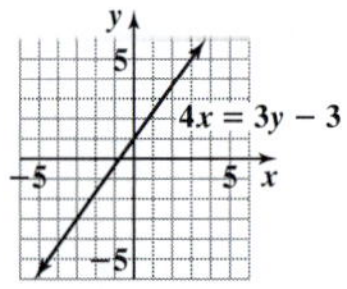


[3.4] **9.**

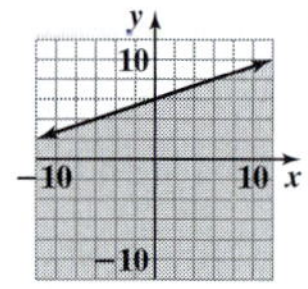

[3.7] **10.** (10, 24) [4.1] **11.** $x^2 + 2x + 3 + \frac{6}{x + 1}$ [5.3] **12.** $(x - y + 8)(x - y - 8)$ [5.6] **13.** 1, −2 [8.1] **14.** −3 [6.4] **15.** $d = \frac{a_n - a_1}{n - 1}$ [6.4] **16.** 6.25 [6.6] **17.** $(12x + 1)\sqrt{5x}$ [7.4] **18.** 8 [7.6] **19.** $0, \pm\sqrt{7}$ [8.4] **20. a)** $g(x) = (x - 2)^2 - 9$ **b)**

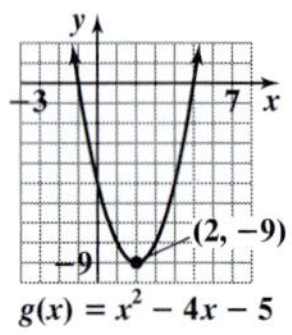


[8.5]

## Chapter 10

**Exercise Set 10.1** **1.** No, a function will only have a local maximum or local minimum if the function has an interval where it changes from increasing to decreasing. If a function never increases or if it never decreases, it would never have an interval where it changes from increasing to decreasing. A linear function is an example of a function that would not have a local maximum or local minimum. **3.** Yes, since $f(x) = x^2 - 4$ is a polynomial function, it is continuous. **5.** No, since the two pieces give different values at $x = -5$, the function is not continuous. **7.** Discontinuous **9.** Continuous **11.** Continuous **13.** Discontinuous **15.** Continuous **17.** Continuous **19.** Continuous **21.** Discontinuous; Point of discontinuity at $x = -5$. **23.** Discontinuous; Points of discontinuity occur at $\ldots, -1, -\frac{1}{2}, 0, \frac{1}{2}, 1, \ldots$ **25.** Continuous **27.** Discontinuous; Point of discontinuity at $x = 3$ **29.** Continuous

**31.**

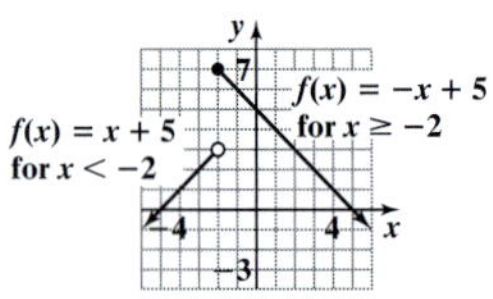


**33.**

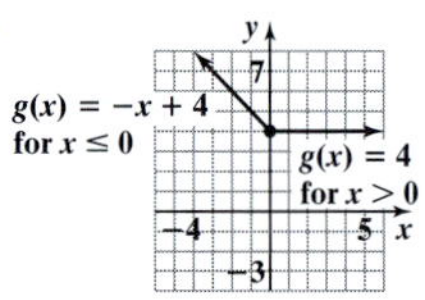


**35.**

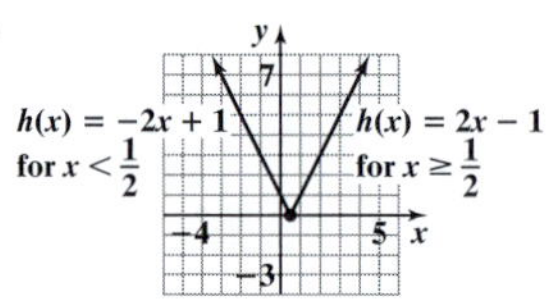


**37.**

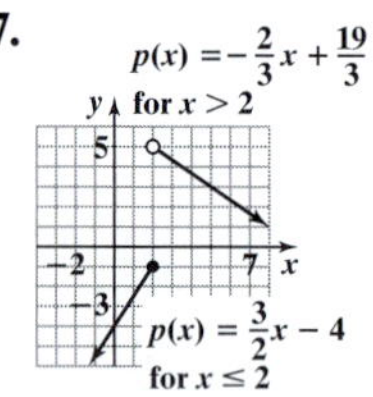


**39.**

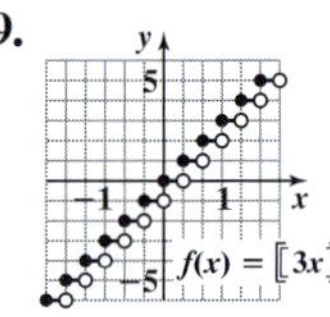


**41.**

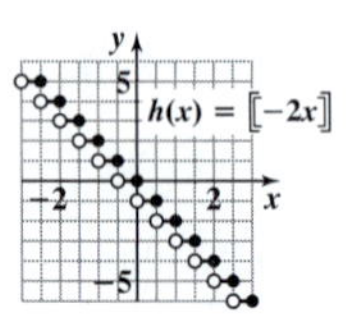


**43.**

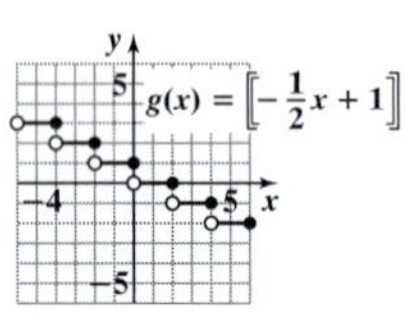


**45.**

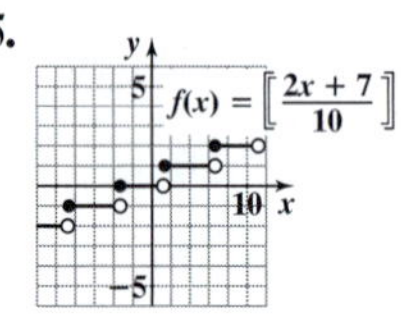


**47.** $f(x) = \left[\!\left[\frac{x}{1.25}\right]\!\right]$ **49.** $f(x) = \left[\!\left[\frac{x}{3.99}\right]\!\right]$ **51.** $f(x) = \begin{cases} 5x & \text{if } x \le 3 \\ 15 + 3(x - 3) & \text{if } x > 3 \end{cases}$ **53.** Since $f(-x) = f(x)$, $f(x) = x^2 - 15$ is an even function. **55.** Since $f(-x) \ne f(x)$ and $f(-x) \ne -f(x)$, $f(x) = 3x^3 + x^2 - 10$ is neither even nor odd. **57.** Since $g(-x) = g(x)$, $g(x) = 3x^6 - 15x^2 + 4$ is an even function. **59.** Since $g(-x) = g(x)$, $g(x) = e^{x^2} + 5$ is an even function. **61.** Since $h(-x) \ne h(x)$ and $h(-x) \ne -h(x)$, $h(x) = x^4 - x^3 + x^2$ is neither even nor odd.

**63.**

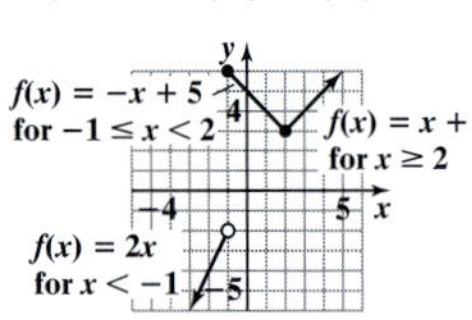


**65.**

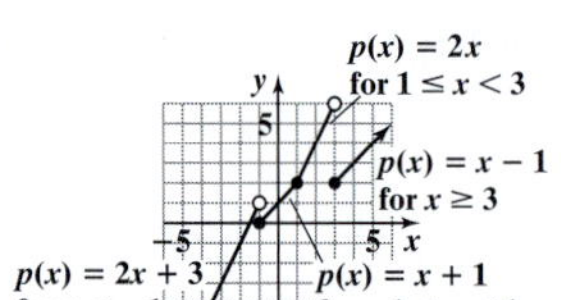


**69.**

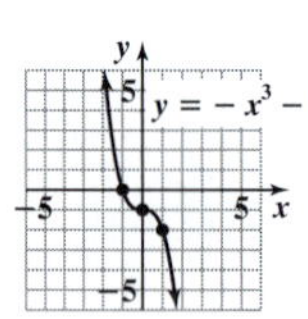


**70.**

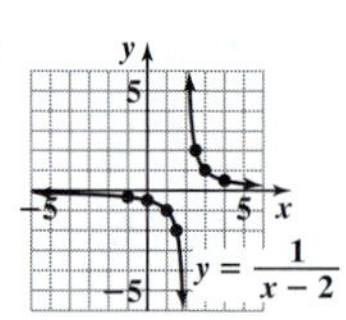


**71. a)** The degree of the polynomial is 5. **b)** The leading coefficient is 4. **72.** $\{x | x \ne 7 \text{ and } x \ne -3\}$

**Exercise Set 10.2** **1. a)** The constant term is negative because the function crosses the $y$-axis below the origin. **b)** The leading coefficient is negative because the value of the function is negative for large positive values of $x$. **c)** The degree of the polynomial function is even because the end behavior for large negative values of $x$ is below the $x$-axis and the end behavior for large positive values of $x$ is also below the $x$-axis. **d)** Yes, the degree of the polynomial function can be 6 because the function has even degree and has fewer than 5 turning points. **e)** No, the degree of the polynomial function cannot be 7 because the degree of the polynomial is even. **f)** The polynomial has 4 real zeros because it crosses the $x$-axis 4 times. **3. a)** $f(x) = x^8$ is an even-degree polynomial function with a positive leading coefficient thus it is above the $x$-axis for large positive values of $x$ and above the $x$-axis for large negative values of $x$. **b)** $f(x) = x^9$ is an odd-degree polynomial function with a positive leading coefficient thus it is above the $x$-axis for large positive values of $x$ and below the $x$-axis for large negative values of $x$. **5.** b; $P(x) = -2x^3 + 5x - 8$ is a third-degree polynomial function with a negative leading coefficient. Thus, large positive values of $x$ will be below the $x$-axis and large negative values of $x$ will be above the $x$-axis. **7.** a; $P(x) = 3x^5 + 5x^3 - 25x^2 + 20$ is a fifth-degree polynomial function with a positive leading coefficient. Thus, large positive values of $x$ will be above the $x$-axis and large negative values of $x$ will be below the $x$-axis.

**9. a)** Above the $x$-axis **b)** Below the $x$-axis **11. a)** Below the $x$-axis **b)** Above the $x$-axis **13. a)** Below the $x$-axis **b)** Below the $x$-axis **15. a)** Below the $x$-axis **b)** Below the $x$-axis **17. a)** Below the $x$-axis **b)** Above the $x$-axis **19.** 5, −3 **21.** 0, −5, −1 **23.** −5, 5 **25.** −7, 2 of multiplicity 3 **27.** $-3, -\frac{1}{2}, \frac{1}{3}$ **29.** −2 of multiplicity 3, 4 of multiplicity 2 **31.** 0, −1 of multiplicity 2, 1 of multiplicity 3, 6 of multiplicity 4 **33.** 4 **35.** 0 **37.** 5 **39.** 2 **41.** 1 **43.** 4

**45.**

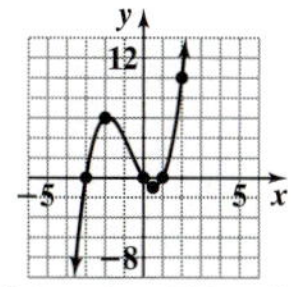

$f(x) = x(x+3)(x-1)$

**47.**

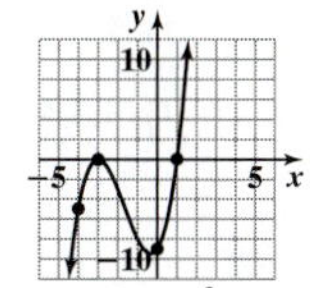

$P(x) = (x+3)^2(x-1)$

**49.**

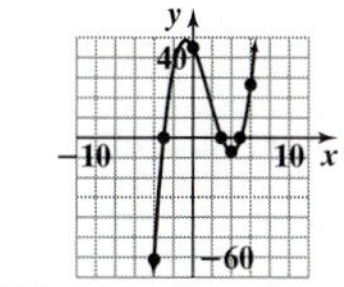

$P(x) = (x+3)(x-3)(x-5)$

**51.**

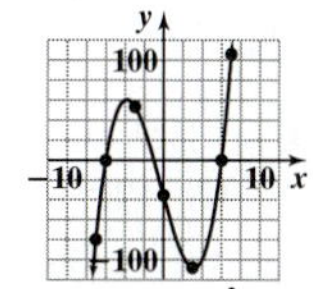

$g(x) = (x+1)(x^2-36)$

**53.** 2 **55.** 3 **57.** 3 **59.** 4 **61.** 4

**63.**

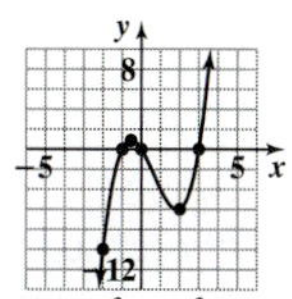

$f(x) = x^3 - 2x^2 - 3x$

**65.**

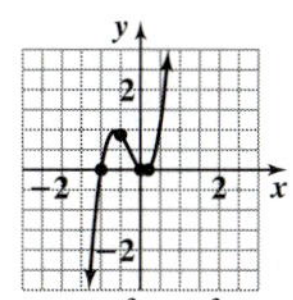

$f(x) = 5x^3 + 4x^2 - x$

**67.**

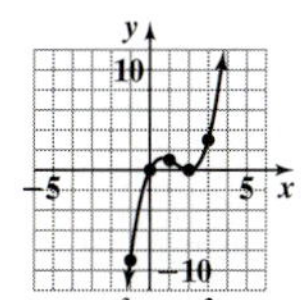

$f(x) = x^3 - 4x^2 + 4x$

**69. a)**

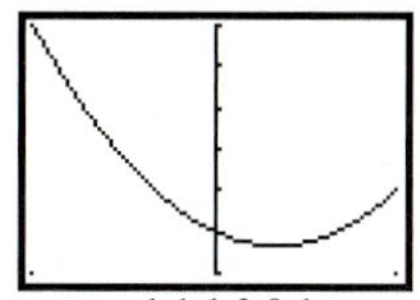
−1, 1, 1, 2, 8, 1

**b)** Local minimum **c)** (0.33, 2.67)

**71. a)**

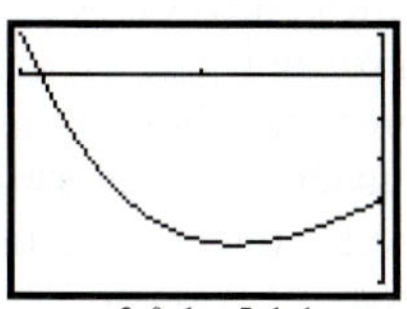
−2, 0, 1, −5, 1, 1

**b)** Local minimum **c)** (−0.82, −4.09) **74.** $4x^3 + 9x^2 + 29x + 86 + \frac{8}{x-3}$ **75.** −479 **76.** $\frac{5x^2+12}{20x}$ **77.** $\frac{5x+4}{y-4}$

**Exercise Set 10.3** **1.** The graph of a fourth-degree polynomial function has a maximum of 4 $x$-intercepts because the maximum number of $x$-intercepts of the graph of a polynomial function is the degree of the polynomial function. **3.** The minimum number of $x$-intercepts of the graph of a sixth-degree polynomial is 0. Because the graph of an even-degree polynomial is either above the $x$-axis on both ends or below the $x$-axis on both ends, it is possible for the graph of the polynomial to be completely above the $x$-axis or completely below the $x$-axis. **5.** If the coefficients of a polynomial are all positive, the coefficients never change sign. By Descartes' Rule of Signs, the graph of the polynomial has no positive $x$-intercepts. **7.** Maximum of 2 $x$-intercepts; Minimum of 0 $x$-intercepts **9.** Maximum of 3 $x$-intercepts; Minimum of 1 $x$-intercept **11.** Maximum of 4 $x$-intercepts; Minimum of 0 $x$-intercepts **13.** Maximum of 5 $x$-intercepts; Minimum of 1 $x$-intercept **15.** Maximum of 6 $x$-intercepts; Minimum of 0 $x$-intercepts **17.** 0 positive zeros; either 3 or 1 negative zeros **19.** Either 2 or 0 positive zeros; 1 negative zero **21.** Either 3 or 1 positive zeros; 1 negative zero **23.** 0 positive zeros; either 4, 2, or 0 negative zeros **25.** Either 3 or 1 positive zeros; either 2 or 0 negative zeros **27.** 0 positive zeros; either 5, 3, or 1 negative zeros **29.** ±1, ±2 **31.** $\pm1, \pm3, \pm\frac{1}{3}$ **33.** ±1, ±2, ±3, ±4, ±6, ±12 **35.** $\pm1, \pm2, \pm4, \pm\frac{1}{3}, \pm\frac{2}{3}, \pm\frac{4}{3}$ **37.** $\pm1, \pm2, \pm4, \pm\frac{1}{2}, \pm\frac{1}{3}, \pm\frac{2}{3}, \pm\frac{4}{3}, \pm\frac{1}{6}$ **39.** $\pm1, \pm3, \pm\frac{1}{2}, \pm\frac{3}{2}, \pm\frac{1}{3}, \pm\frac{1}{6}$ **41.** $\pm1, \pm2, \pm4, \pm\frac{1}{7}, \pm\frac{2}{7}, \pm\frac{4}{7}$ **43.** 2, 4 **45.** $-5, \frac{1}{3}$ **47.** −7, −1, 1 **49.** $-2, -\frac{1}{2}, 2$ **51.** −4, 4 **53.** −2 of multiplicity 2, 2 of multiplicity 2 **55.** −2, 3, −1 of multiplicity 2 **57.** 1 of multiplicity 2, 3 of multiplicity 2 **59. a)** Yes **b)** Yes **61. a)** No **b)** No **63.** 9 **65.** 44 **67.** $i, -i$ **69.** $2, -2, i, -i$ **71.** $1, -1, \frac{-1+i\sqrt{3}}{2}, \frac{-1-i\sqrt{3}}{2}$ **73.** $2, -2, \frac{-3+i\sqrt{3}}{2}, \frac{-3-i\sqrt{3}}{2}$ **75.** 1, $i$ of multiplicity 2, $-i$ of multiplicity 2

**77. a)** $P(x) = x^3 + 4x^2 + x - 6$ **b)** $P(0) = -6$, The company has a loss of \$6000 when no items are produced.; $R(0) = 0^3 + 5(0)^2 + 0 = 0$, The company has no revenue when no items are produced.; $C(0) = 6$, The company has a cost of \$6000 when no items are produced. **c)** The break-even point is $x = 1$ (when 100 items are produced). **d)** $1 < x \le 3$; **e)** $0 \le x < 1$ **f)** The company operates at a profit when $x > 1$ (when more than 100 items are produced). It operates at a loss when $x < 1$ (when less than 100 items are produced). **79. a)** $P(x) = x^3 - 2x^2 - 13x - 10$ **b)** $P(0) = -10$, The company has a loss of \$10,000 when no items are produced.; $R(0) = 0$, The company has no revenue when no items are produced.; $C(0) = 10$, The company has a cost of \$10,000 when no items are produced. **c)** The break-even point is $x = 5$ (when 500 items are produced). **d)** $5 < x \le 8$ **e)** $0 \le x < 5$ **f)** The company operates at a profit when $x > 5$ (when more than 500 items are produced). It operates at a loss when $x < 5$ (when less than 500 items are produced). **81. a)** $P(x) = 2x^3 - x^2 - 5x - 2$ **b)** $P(0) = -2$, The company has a loss of \$2000 when no items are produced.; $R(0) = 0$; The company has no revenue when no items are produced.; $C(0) = 2$; The company has a cost of \$2000 when no items are produced. **c)** The break-even point is $x = 2$ (when 200 items are produced). **d)** $2 < x \le 5$ **e)** $0 \le x < 2$ **f)** The company operates at a profit when $x > 2$ (when more than 200 items are produced). It operates at a loss when $x < 2$ (when less than 200 items are produced). **83.** $y^{8/3}$ **84.** $18x\sqrt{2xy}$ **85.** $-\frac{10+\sqrt{5}}{19}$ **86.** 6

**Mid-Chapter Test** **1.** Continuous [10.1] **2.** Continuos [10.1] **3.** Discontinuous; $x = -3$, $x = 3$ [10.1]

**4.**

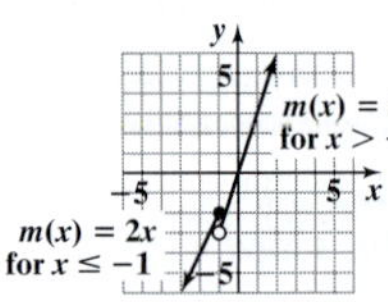


[10.1] **5.**

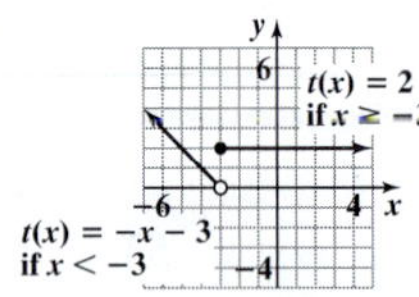


[10.1] **6.**

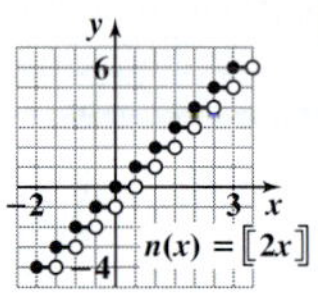


[10.1] **7.**

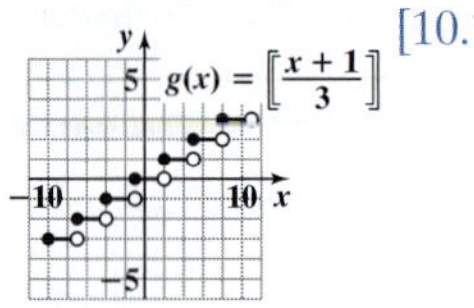


[10.1]

**8. a)** Above $x$-axis. **b)** Above $x$-axis. [10.2] **9. a)** Below $x$-axis. **b)** Above $x$-axis. [10.2] **10.** $-6, 0, 4$ [10.2] **11.** $-5$. $-1$ of multiplicity 2, 3 of multiplicity 4 [10.2] **12.** 2 [10.2] **13.** 5 [10.2]

**14.**

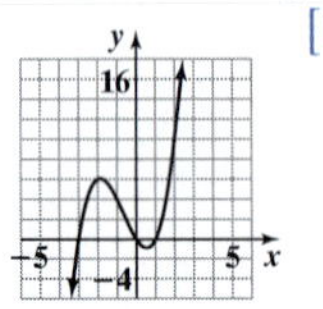

$f(x) = x(x + 3)(x - 1)$

[10.2] **15.**

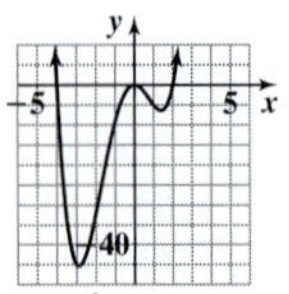

$g(x) = x^2(x + 4)(x - 2)$

[10.2] **16.** Maximum = 3; Minimum = 1 [10.3] **17.** Positive = 2 or 0; Negative = 3 or 1 [10.3] **18.** $\pm1, \pm2, \pm4, \pm\frac{1}{3}, \pm\frac{2}{3}, \pm\frac{4}{3}$ [10.3] **19.** $-4, 5, i, -i$ [10.3] **20.** $-\frac{1}{2}, \frac{2}{5}, 3$ [10.3]

**Exercise Set 10.4** **1. a)** Yes, we know that $x = 5$ is an upper bound for the real zeros of $f(x) = x^3 - 2x^2 - 5x + 2$ because the numbers in the bottom row of the synthetic division are nonnegative. **b)** Yes, we know that $x = 4$ is an upper bound for the real zeros of $f(x) = x^3 - 2x^2 - 5x + 2$ because the numbers in the bottom row of the synthetic division are nonnegative. **c)** No, $x = 3$ is not an upper bound for the real zeros of $f(x) = x^3 - 2x^2 - 5x + 2$. Since the remainder of the synthetic division is $-4$, we know that $f(3) = -4$. Thus the graph of the function is below the $x$-axis at $x = 3$. Since the leading coefficient is positive, we know that the graph of the function is eventually above the $x$-axis as $x$ gets larger. Because the graph is below the $x$-axis at $x = 3$ and it remains above the $x$-axis for all $x$ greater than some $x$-value, there exists a real zero greater than $x = 3$. **3. a)** No. Since the value of the function has the same sign at $x = 1$ that it has at $x = 2$, you cannot tell if it has a zero in the interval $(1, 2)$. **b)** Yes. Since the value of the continuous function is negative at $x = -2$ and positive at $x = -1$, it must have a zero in the interval $(-2, -1)$. **5.** Cannot determine **7.** Yes **9.** Yes **11.** Yes **13.** Yes **15.** Cannot determine **17.** Yes **19.** Cannot determine **21.** Yes **23.** Yes **25.** Cannot determine **27.** Yes **29.** Cannot determine **31.** Yes **33.** Cannot determine

**35.**

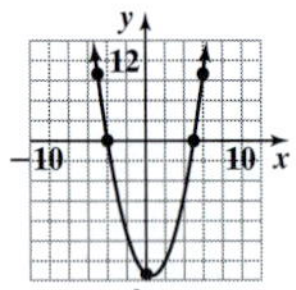

$f(x) = x^2 - x - 20$

**37.**

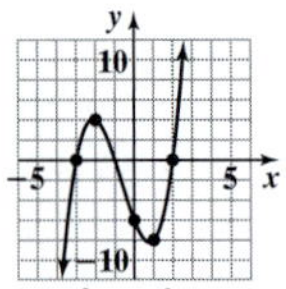

$g(x) = x^3 + 2x^2 - 5x - 6$

**39.**

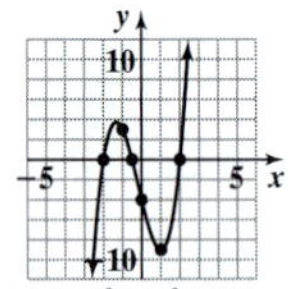

$g(x) = 2x^3 + x^2 - 8x - 4$

**41.**

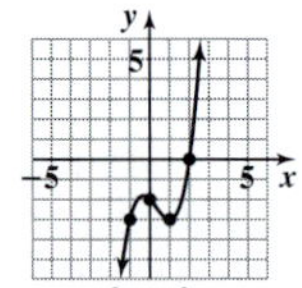

$g(x) = x^3 - x^2 - x - 2$

**43.**

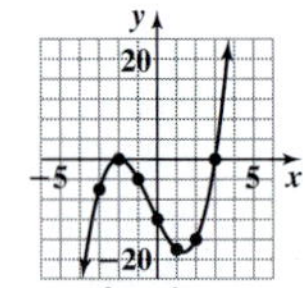

$g(x) = x^3 + x^2 - 8x - 12$

**45.**

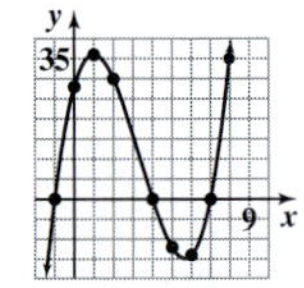

$g(x) = x^3 - 10x^2 + 17x + 28$

**47.**

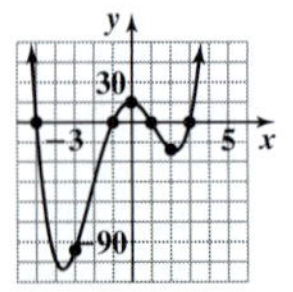

$h(x) = x^4 + 2x^3 - 16x^2 - 2x + 15$

**49.**

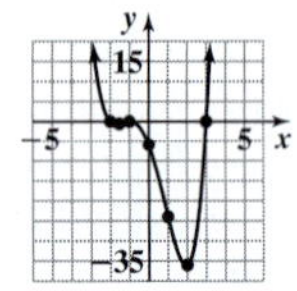

$h(x) = x^4 + x^3 - 7x^2 - 13x - 6$

**51.**

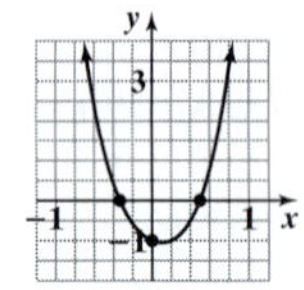

$h(x) = 6x^4 - x^3 + 5x^2 - x - 1$

**53.**

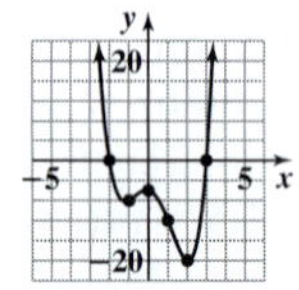

$h(x) = x^4 - x^3 - 5x^2 - x - 6$

**55.** 1.15

**57.** $-1.24, 2$ **59.** $-1, 0.22$ **61.** $-4.05, 0.36, 0.69$ **63.** $\dfrac{1 + 3x - 4x^2y}{2x^2y}$ **64.** $\dfrac{9x + 22}{(x + 2)(x + 3)}$ **65.** $\dfrac{9y - 10x}{12xy}$ **66.** $-7$

**Exercise Set 10.5** **1.** No. If a rational function has no variables in the denominator when it is written in lowest terms, it will not have asymptotes. **3.** No. We have 3 cases to consider. 1) If $f(x)$ is a proper rational function, the values of $f(x)$ approach 0 as $x \to \infty$ and as $x \to -\infty$. 2) If the degree of the numerator of $f(x)$ is the same as the degree of the denominator of $f(x)$, the values of $f(x)$ approach the ratio of the leading coefficients of the numerator and the denominator as $x \to \infty$ and as $x \to -\infty$. 3) If the degree of the numerator is greater than the degree of the denominator, the graph of $f(x)$ does not have a horizontal asymptote. **5.** The point of discontinuity can be found by finding the values of $x$ that make the denominator of the function equal to 0. **7.** A graph of a rational function can have an oblique asymptote when the degree of the numerator is one greater than the degree of the denominator. **9.** $x = 0$: Vertical asymptote **11.** $x = -5$: Vertical asymptote **13.** $x = -5$: Hole **15.** $x = 1$: Hole **17.** $x = 4$: Vertical asymptote; $x = -4$: Vertical asymptote **19.** $x = 6$: Vertical asymptote; $x = 1$: Hole **21.** $x = -3$: Vertical asymptote; $x = -5$: Vertical asymptote **23.** $x = 5$: Vertical asymptote; $x = -5$: Hole **25.** $x = 3$: Vertical asymptote; $x = 1$: Vertical asymptote **27.** No points of discontinuity **29.** Horizontal asymptote; $y = 0$ **31.** Horizontal asymptote; $y = 3$ **33.** Horizontal asymptote; $y = 0$ **35.** Horizontal asymptote; $y = 4$ **37.** Oblique asymptote; $y = x - 10$ **39.** Oblique asymptote; $y = 2x + 12$ **41.** Horizontal asymptote; $y = 0$ **43.** Oblique asymptote; $y = x - 2$

**45.**

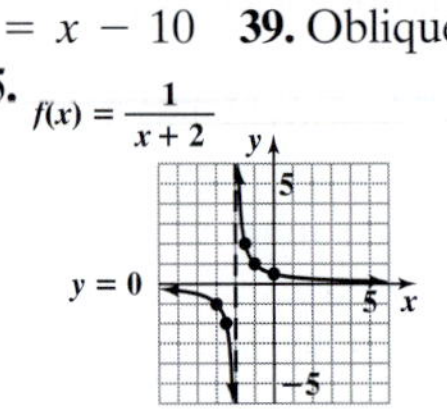


**47.**

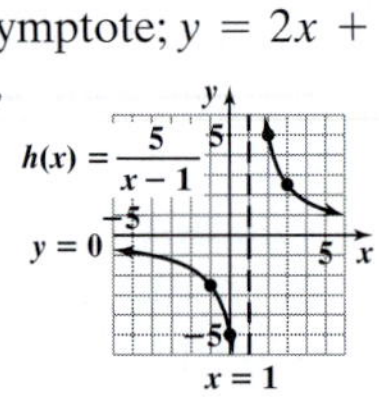


**49.**

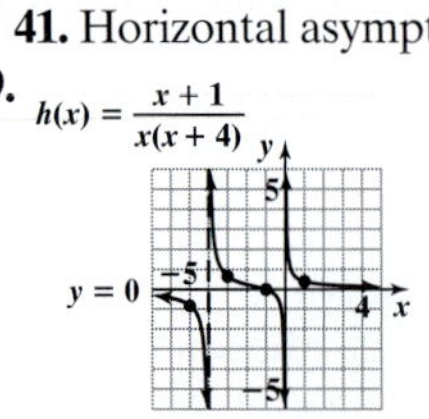


**51.**

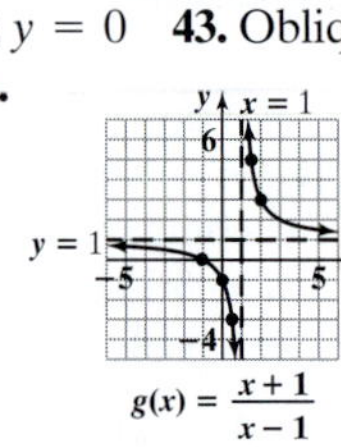


**53.**

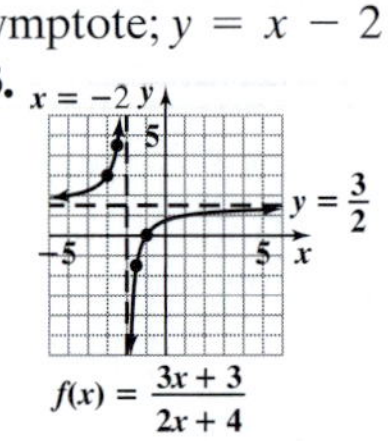

55.
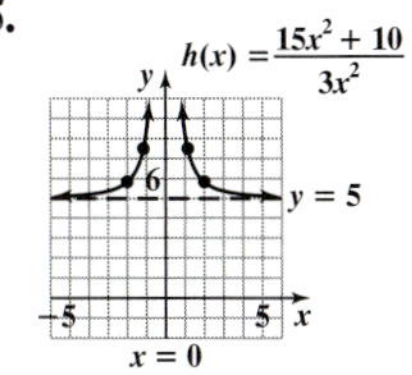


57.
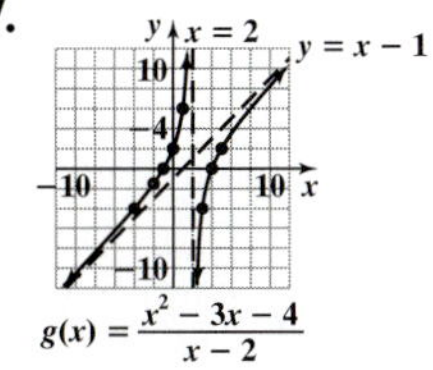


59.
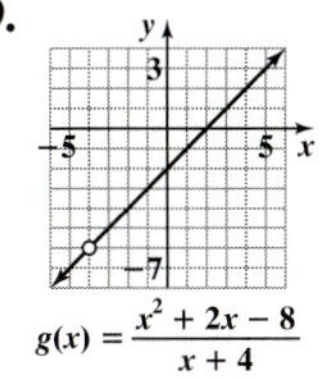


61.
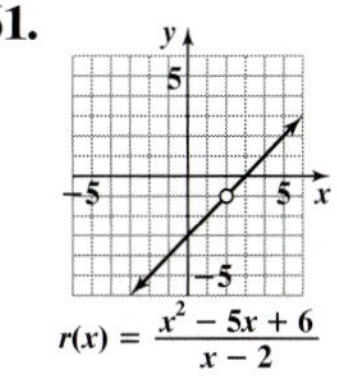


63.
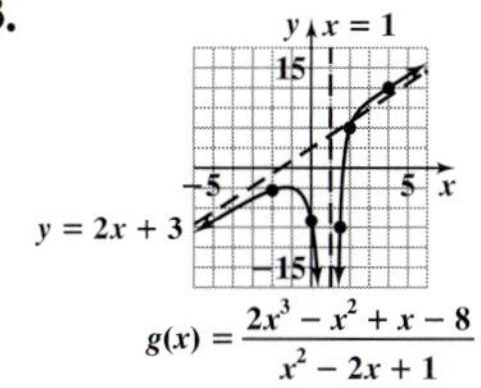


**65.** Answers may vary. Since the function has a vertical asymptote at $x = -4$, the denominator contains the factor $x + 4$. Since the function has a horizontal asymptote at $y = 0$, the function is a proper rational function. A function that meets these requirements is $f(x) = \dfrac{1}{x + 4}$. **67.** Answers may vary. Since the function has a hole at $(4, 3)$, it is undefined at $x = 4$ and has the factor $x - 4$ in its denominator and numerator. A function that meets these requirements is $f(x) = \dfrac{(2x - 5)(x - 4)}{x - 4}$. **69.** Answers may vary. Since the function has a horizontal asymptote at $y = 0$, it is a proper rational function. Since it has no vertical asymptotes, its denominator is never 0. A function that meets these requirements is $f(x) = \dfrac{1}{x^2 + 1}$. **72.** $f^{-1}(x) = 2x - 6$ **73.** $f^{-1}(x) = \dfrac{5}{x}$ **74.** $a = 12$ **75.** $x = \dfrac{1}{32}$

## Chapter 10 Review Exercises

**1.** Continuous **2.** Discontinuous **3.** Continuous **4.** Continuous **5.** Discontinuous; $\ldots, -5, -1, 3, 7, \ldots$ **6.** Continuous

**7.**
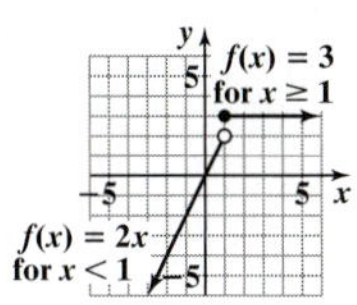


**8.**
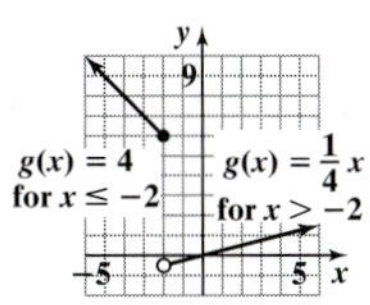


**9.**
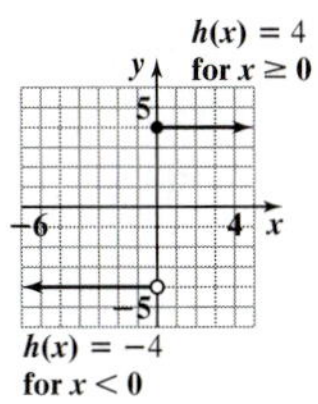


**10.**
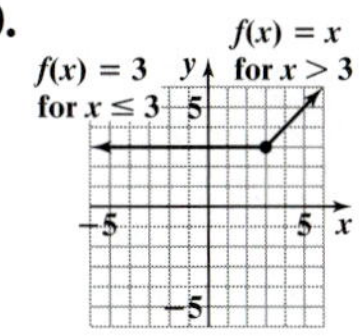


**11.**
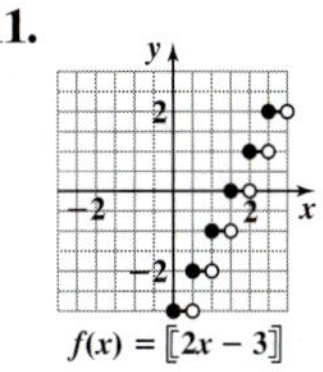


**12.**
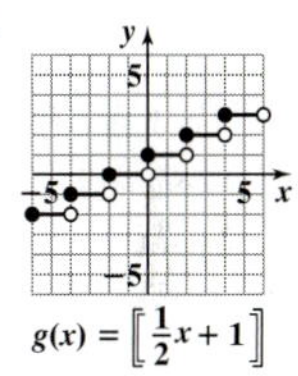


**13.**
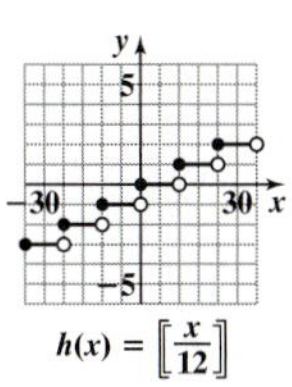


**14.**
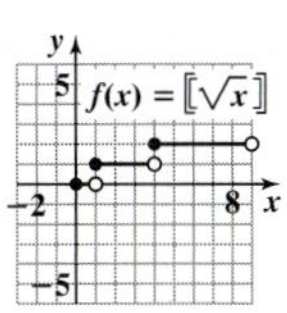


**15.** $f(x) = \left[\!\left[\dfrac{x}{0.37}\right]\!\right]$ **16.** $f(x) = \left[\!\left[\dfrac{x}{2.25}\right]\!\right]$ **17. a)** Below the $x$-axis **b)** Above the $x$-axis **18. a)** Above the $x$-axis **b)** Above the $x$-axis **19. a)** Below the $x$-axis **b)** Below the $x$-axis **20. a)** Above the $x$-axis **b)** Below the $x$-axis **21.** $-8, -3$ **22.** $0, -9, 4$ **23.** $-2$ of multiplicity $3, \dfrac{1}{3}$ **24.** $\sqrt{7}, -\sqrt{7}$ **25.** 1 **26.** 3 **27.** 2 **28.** 5 **29.** 5 **30.** 6

**31.**
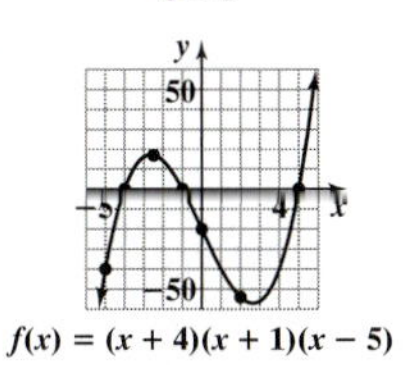


**32.**
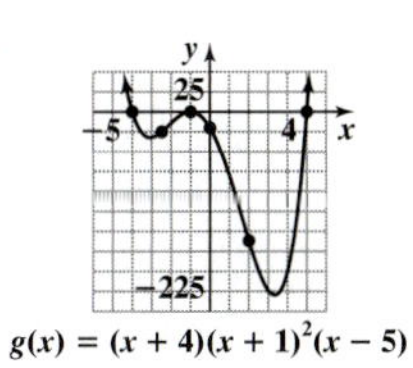


**33.**
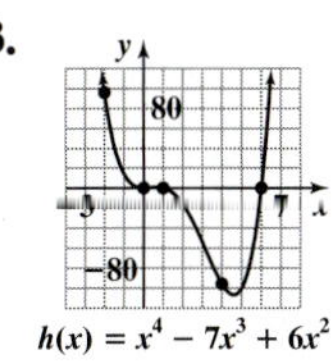


**34.**
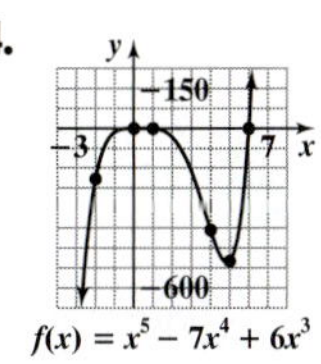


**35.** Maximum: 3; Minimum: 1 **36.** Maximum: 4; Minimum: 2 **37.** Maximum: 5; Minimum: 1 **38.** Maximum: 6; Minimum: 0 **39. a)** 2 or 0 **b)** 0 **40. a)** 1 **b)** 1 **41. a)** 0 **b)** 3 or 1 **42. a)** 2 or 0 **b)** 2 or 0 **43.** $\pm1, \pm2, \pm3, \pm4, \pm6, \pm12$ **44.** $\pm1, \pm5, \pm\dfrac{1}{2}, \pm\dfrac{5}{2}$ **45.** $\pm1, \pm3, \pm9, \pm\dfrac{1}{3}$ **46.** $\pm1, \pm2, \pm4, \pm8, \pm16, \pm32, \pm64$ **47.** $1, -1, -10$ **48.** 2 of multiplicity 2 **49.** $\dfrac{2}{3}, -4$ **50.** $1, 6, -3, -5$ **51.** $5, -1, 4i, -4i$ **52.** $2, -3, 5i, -5i$ **53.** Yes **54.** Cannot be determined **55.** Cannot be determined **56.** Yes **57.** Cannot be determined **58.** Cannot be determined **59.** Yes **60.** Yes

**61.**
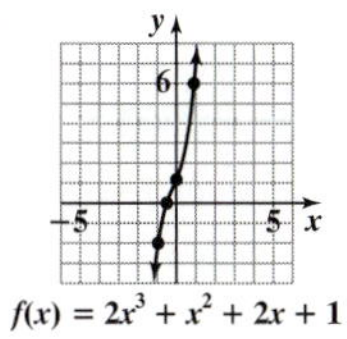


**62.**
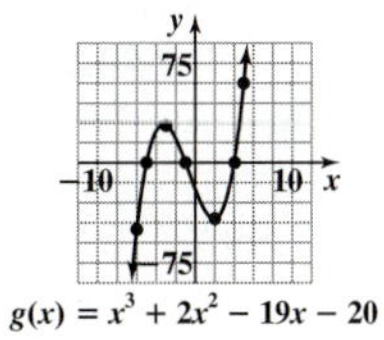


**63.**
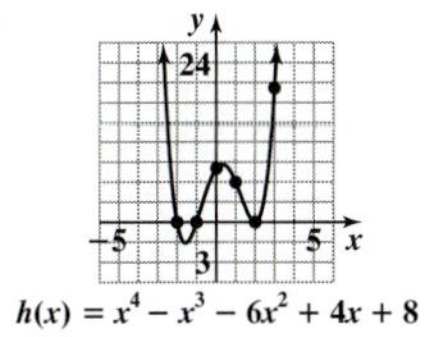


**64.**
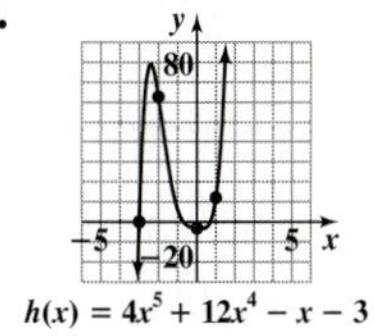


**65.** 5: Vertical asymptote; $-5$: Vertical asymptote **66.** 5: Vertical asymptote; $-5$: Hole **67.** $-4$: Hole **68.** No points of discontinuity **69.** Horizontal asymptote; $y = 0$ **70.** Horizontal asymptote; $y = 3$ **71.** Oblique asymptote; $y = 3x - 7$ **72.** Neither **73.** Oblique asymptote; $y = 2x - 3$ **74.** Horizontal asymptote; $y = \dfrac{3}{2}$ **75.**
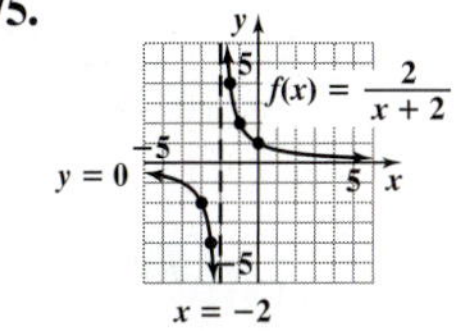


**76.**
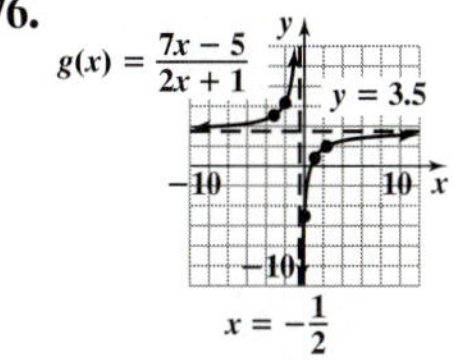

**77.**

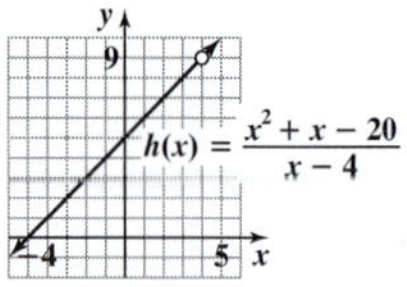


**78.**

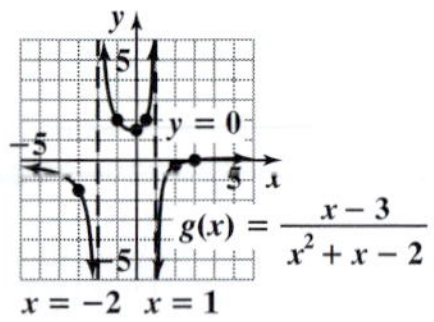


**79.**

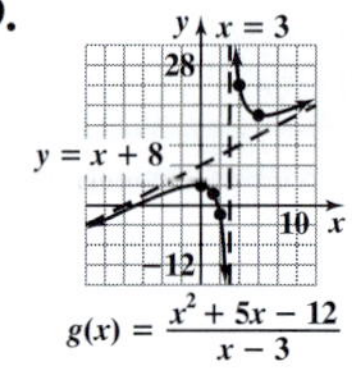


**80.**

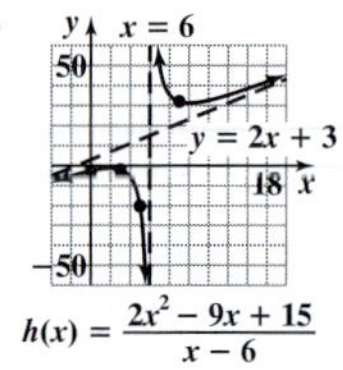


## Chapter 10 Practice Test

**1. a)** Below the $x$-axis **b)** Below the $x$-axis **2.** 0, 4 of multiplicity 3, $-\frac{3}{2}$ of multiplicity 2 **3.** 6 **4.** 4 **5.** Maximum: 5; Minimum: 1 **6. a)** 2 or 0 **b)** 3 or 1 **7.** $\pm 1, \pm 2, \pm 3, \pm 6, \pm \frac{1}{2}, \pm \frac{3}{2}, \pm \frac{1}{4}, \pm \frac{3}{4}$ **8.** $3, -4, 3i, -3i$ **9. a)** Cannot be determined **b)** Yes **10.** Cannot be determined **11.** $-4$: Hole **12.** 5: Vertical asymptote **13.** 0: Vertical asymptote; $-\frac{3}{2}$: Vertical asymptote **14.** Horizontal asymptote; $y = 5$ **15.** Neither **16.** Oblique asymptote; $y = 4x - 3$ **17.** Horizontal asymptote; $y = 0$ **18.**

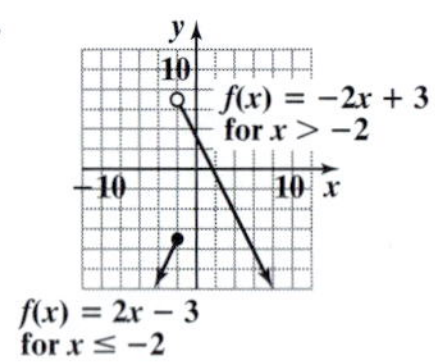


**19.**

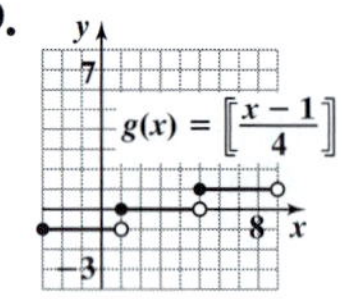


**20.**

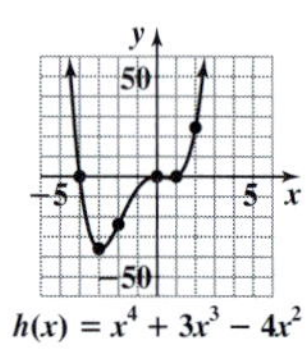


**21.**

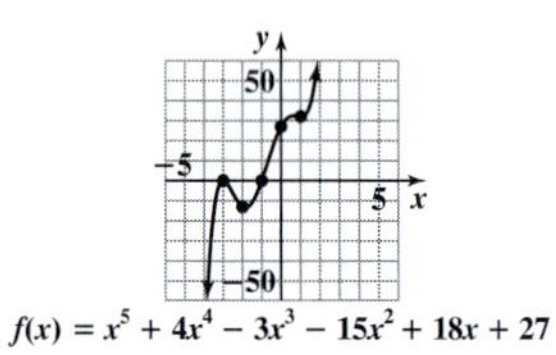


**22.**

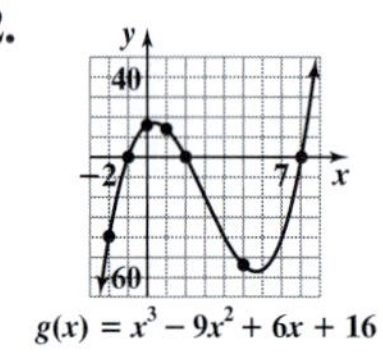


**23.**

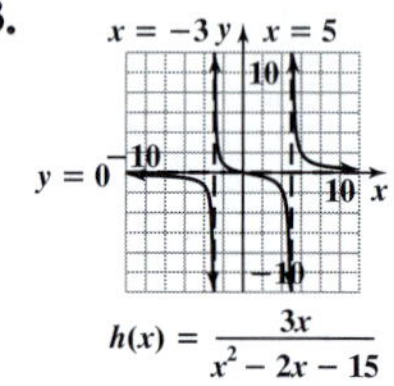


**24.**

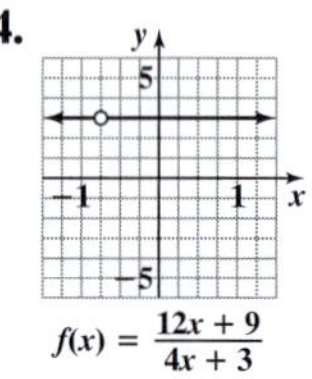


**25.**

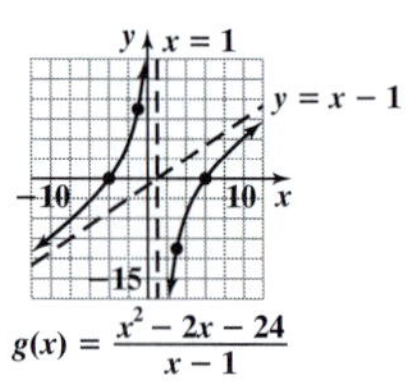


## Cumulative Review Test

**1.** $-56$ [1.4] **2.** 40,000,000,000 [1.6] **3.** $\frac{1}{3}$ [2.1] **4.** $\left\{x \middle| x < -\frac{2}{3} \text{ or } x > 4\right\}$ [2.6] **5.**

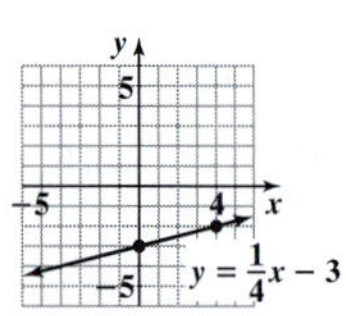


[3.1] **6.** $y = -\frac{3}{2}x - 3$ [3.5] **7.** $(2, 4)$ [4.1] **8.**

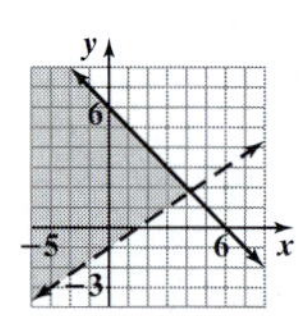

[4.6]

**9.** $4x^2 - 49$ [5.2] **10.** $x^3 + 3x^2 + 9x + 27 + \dfrac{70}{x - 3}$ [5.3] **11.** $(2x + 3y)(3x + 2y)$ [5.4] **12.** $\dfrac{21x + 20}{30x^2}$ [6.2] **13.** 3 [6.4] **14. a)** $\{x|x \geq -3\}$, **b)**

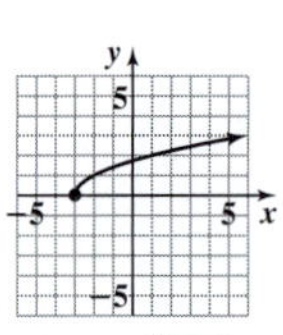


[7.1] **15.** $3 - \sqrt{5}$ [7.5] **16.** $7 \pm 2\sqrt{10}$ [8.2] **17.** 16 [9.4] **18.** 48 [9.6]

**19.**

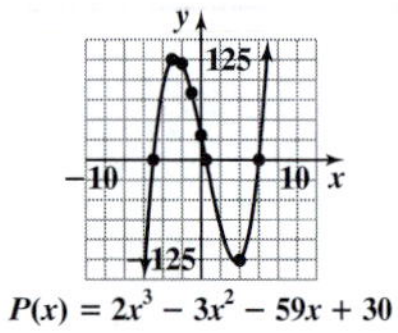


[10.4] **20.**

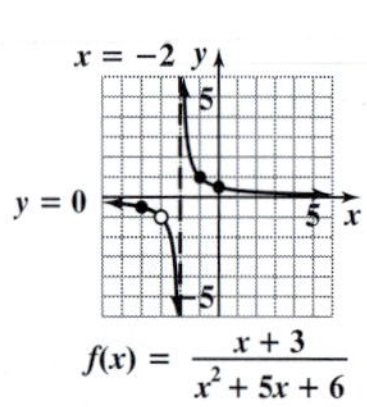


[10.5]

# Chapter 11

**Exercise Set 11.1** **1.** Parabola, circle, ellipse, and hyperbola; see page 658 for picture. **3.** Yes, because each value of $x$ corresponds to only one value for $y$. The domain is $\mathbb{R}$, and the range is $\{y|y \geq k\}$ **5.** The graphs have the same vertex, (3, 4). The first graph opens upward, and the second graph opens downward. **7.** The distance is always a positive number because both differences are squared and we use the principal square root. **9.** A circle is the set of all points in a plane that are the same distance from a fixed point. **11.** No, $x^2 + y^2 = 9$ would be an equation of a circle **13.** No, the coefficients of the $x^2$ and $y^2$ terms would need to be the same **15.** No, if $x^2$ were replaced by $x$ it would be an equation of a parabola

**17.**

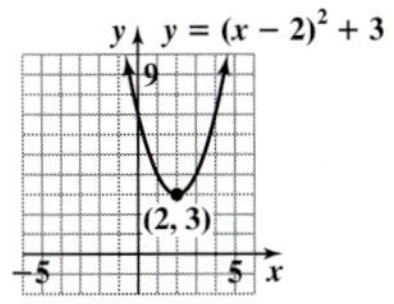


**19.**

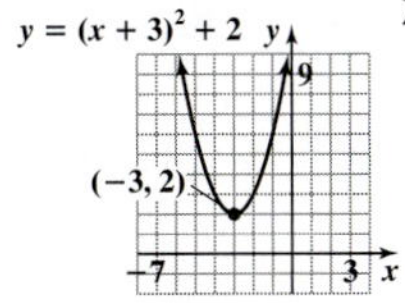


**21.**

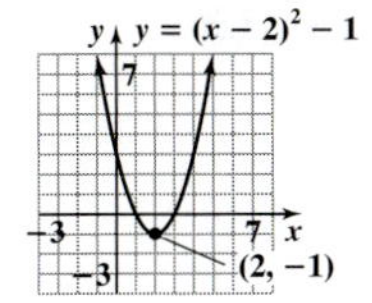


**23.**

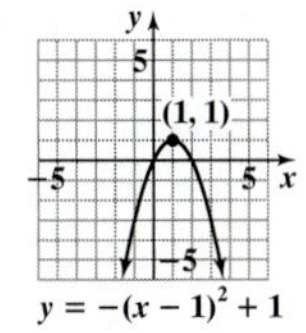


**25.**

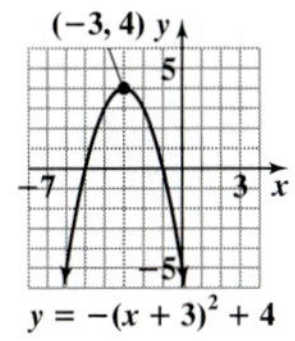


**27.**

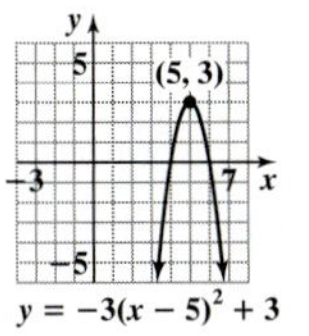


**29.**

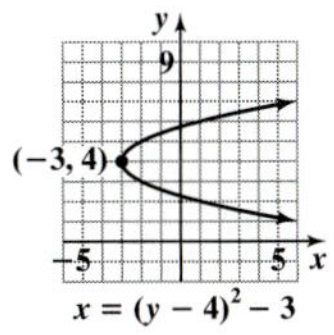


**31.**

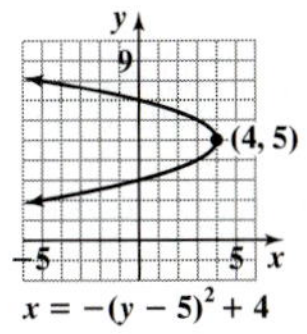


**33.**

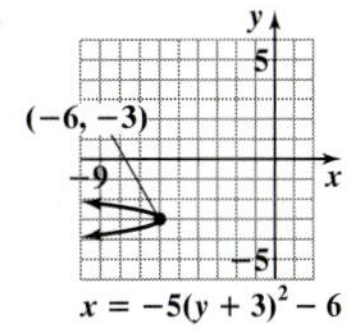


**35.**

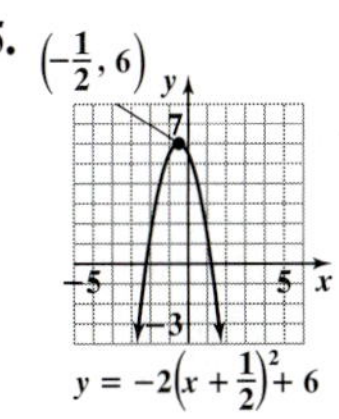


**37. a)** $y = (x + 1)^2 - 1$
**b)**

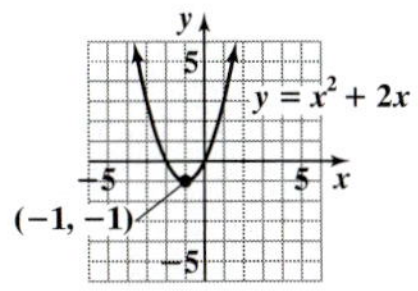


**39. a)** $y = (x + 3)^2 - 9$
**b)**

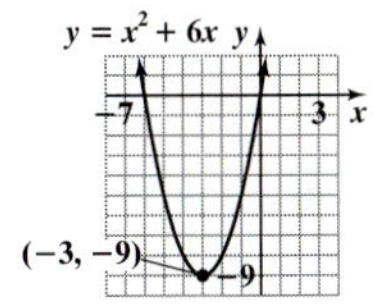


**41. a)** $x = (y + 2)^2 - 4$
**b)**

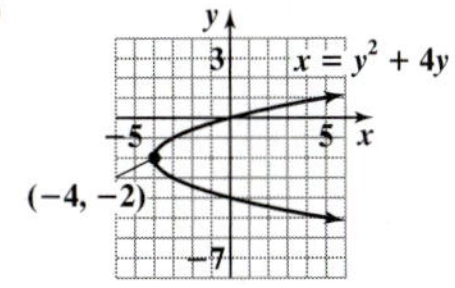


**43. a)** $y = \left(x + \dfrac{7}{2}\right)^2 - \dfrac{9}{4}$
**b)**

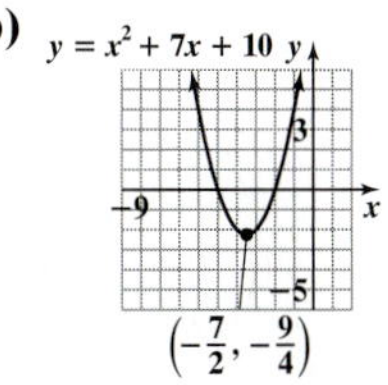


**45. a)** $x = -(y - 3)^2$
**b)**

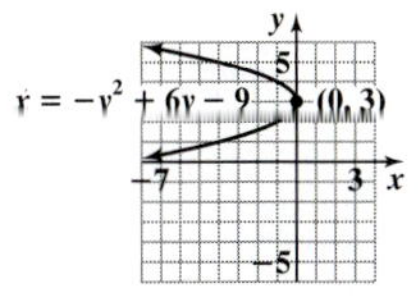


**47. a)** $y = -(x - 2)^2$
**b)**

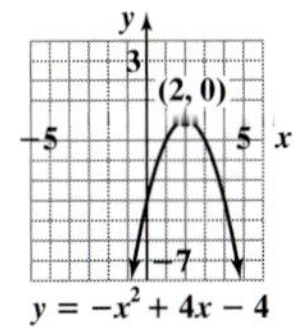


**49. a)** $x = -\left(y - \dfrac{3}{2}\right)^2 - \dfrac{7}{4}$
**b)**

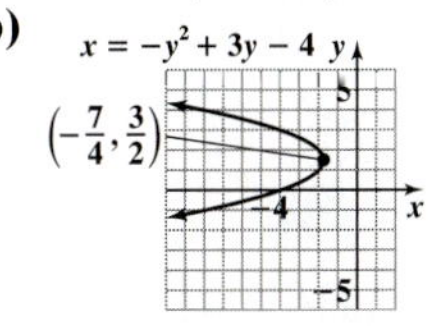


**51.** 5 **53.** 9 **55.** 13 **57.** $\sqrt{90} \approx 9.49$ **59.** $\sqrt{\dfrac{125}{4}} \approx 5.59$ **61.** $\sqrt{34.33} \approx 5.86$

**63.** $\sqrt{10} \approx 3.16$ **65.** (3, 6) **67.** (0, 0) **69.** $\left(\dfrac{3}{2}, 5\right)$ **71.** $\left(\dfrac{5}{2}, -\dfrac{7}{4}\right)$ **73.** $\left(\dfrac{\sqrt{3} + \sqrt{2}}{2}, \dfrac{9}{2}\right)$ **75.** $x^2 + y^2 = 16$

**77.** $(x - 2)^2 + y^2 = 25$ **79.** $x^2 + (y - 5)^2 = 1$ **81.** $(x - 3)^2 + (y - 4)^2 = 64$ **83.** $(x - 7)^2 + (y + 6)^2 = 100$

**85.** $(x - 1)^2 + (y - 2)^2 = 5$ **87.** $x^2 + y^2 = 16$ **89.** $(x - 3)^2 + (y + 2)^2 = 9$

**91.**

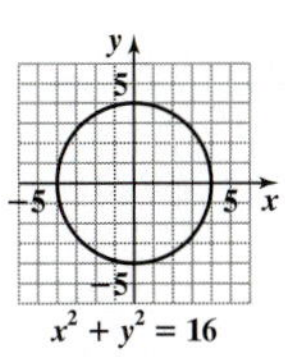


**93.**

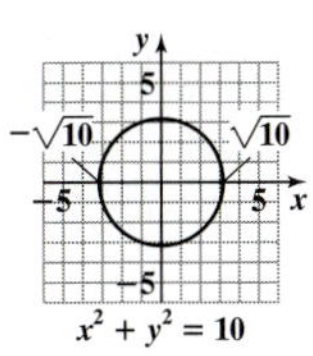


**95.**

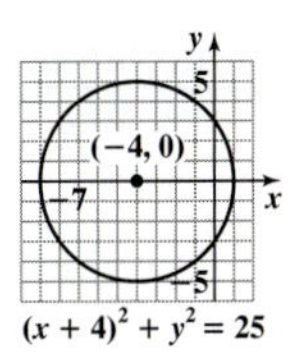


**97.**

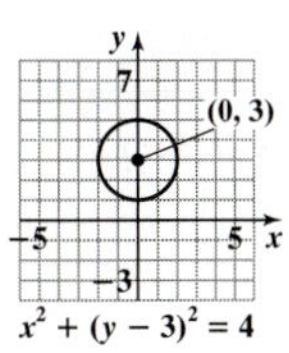


**99.**

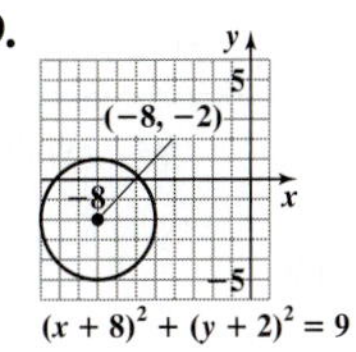


**101.**

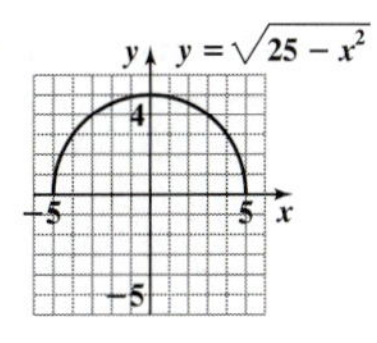


**103.**

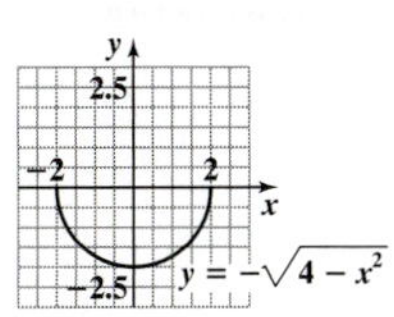


**105. a)** $(x + 4)^2 + y^2 = 1^2$
**b)**

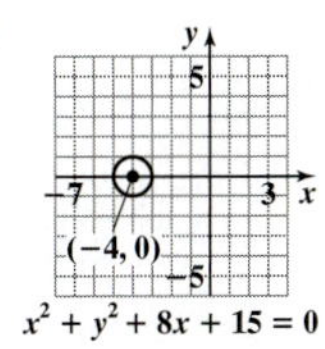


**107. a)** $(x + 3)^2 + (y - 2)^2 = 3^2$
**b)**

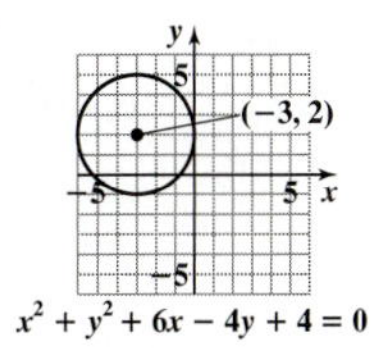


**109. a)** $(x + 3)^2 + (y - 1)^2 = 2^2$
**b)**

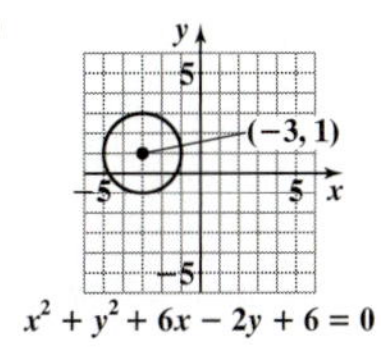

**111. a)** $(x-4)^2+(y+1)^2=2^2$ **b)**

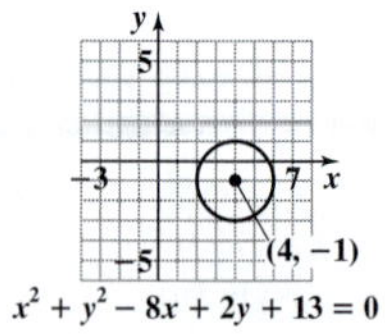


**113.** $25\pi \approx 78.5$ square units **115.** $x$-intercept: $(-7, 0)$; $y$-intercepts: $(0, -1)$, $(0, 7)$ **117.** $x$-intercept: $(24, 0)$; no $y$-intercepts **119.** No, different line segments can have the same midpoint **121.** 10 **123.** $(x+6)^2+(y-2)^2=4$ **125. a)** $2\sqrt{2}$ **b)** $(7, 6)$ **c)** $(x-7)^2+(y-6)^2=8$ **127.** 4, 0, a parabola opening up or down and a parabola opening right or left can be drawn to have a maximum of 4 intersections, or a minimum of 0 intersections.

**129. a)** 13.6 feet **b)** 81.8 feet **c)** $x^2+(y-81.8)^2=4651.24$ **131. a)** $x^2+y^2=16$ **b)** $(x-2)^2+y^2=4$ **c)** $(x+2)^2+y^2=4$ **d)** $8\pi$ square units **133.** $48\pi$ square units **136.** $\dfrac{y}{3x}$ **137.** $(0, 7)$ **138.** 128

**139. a)** 1. $a^2$, 2. $ab$, 3. $ab$, 4. $b^2$ **b)** $(a+b)^2$ **140.**

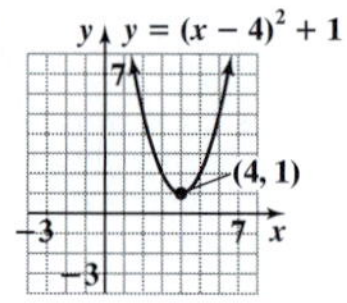


## Exercise Set 11.2

**1.** An ellipse is a set of points in a plane, the sum of whose distances from two fixed points is constant. **3.** $\dfrac{(x-h)^2}{a^2}+\dfrac{(y-k)^2}{b^2}=1$ **5.** If $a=b$, the formula for a circle is obtained. **7.** Divide both sides by 180.

**9.** No, equation for an ellipse is $\dfrac{x^2}{a^2}+\dfrac{y^2}{b^2}=1$. **11.**

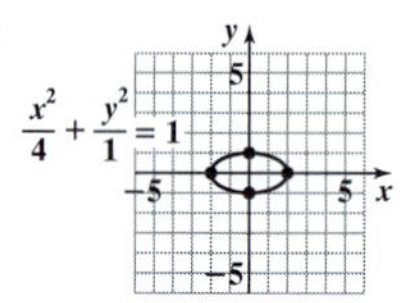


**13.**

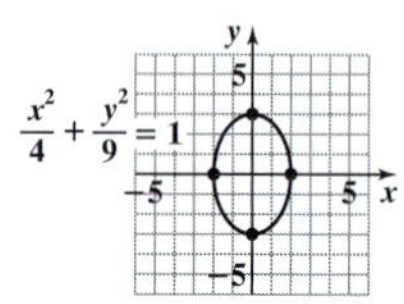


**15.**

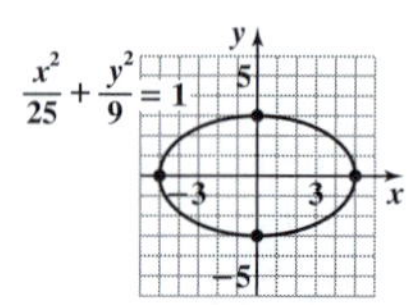


**17.**

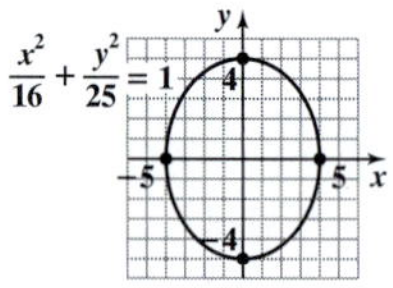


**19.**

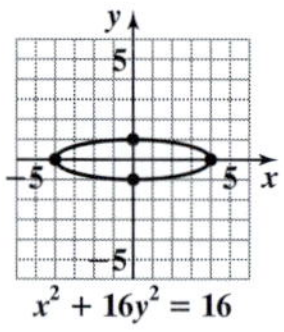


**21.**

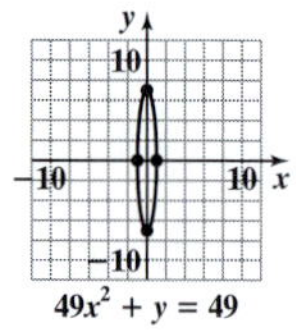


**23.**

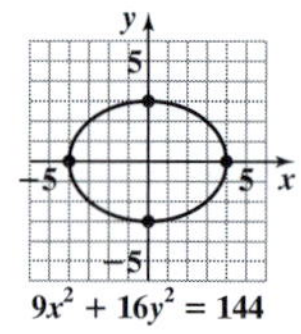


**25.**

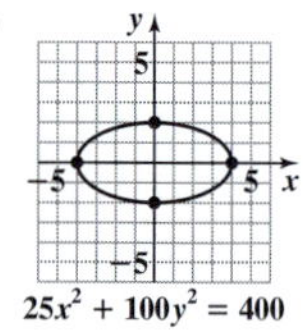


**27.**

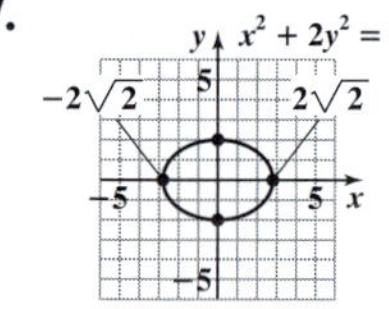


**29.**

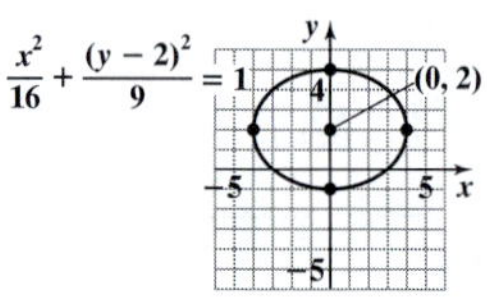


**31.**

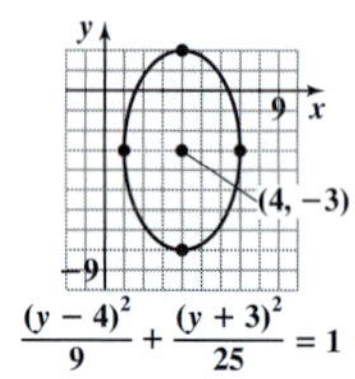


**33.**

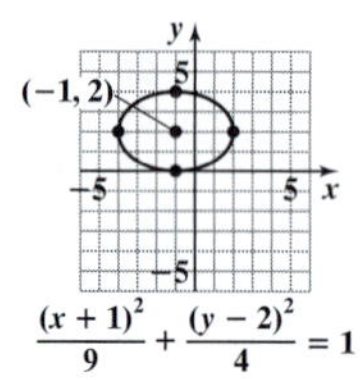


**35.**

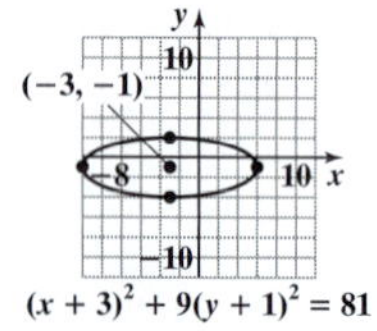


**37.**

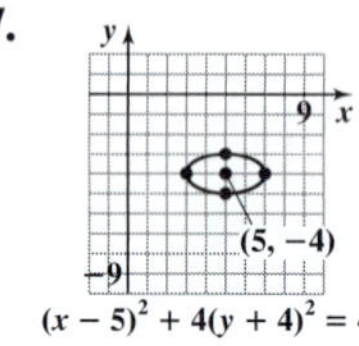


**39.**

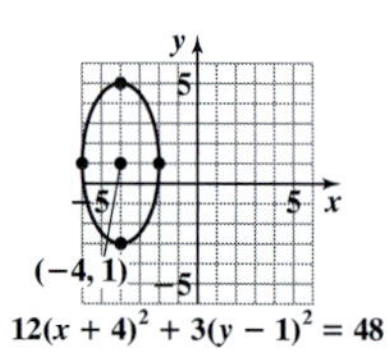


**41.** $2\pi \approx 6.3$ square units **43.** one, at $(0, 0)$, this is the only ordered pair that satisfies the equation **45.** $\dfrac{x^2}{9}+\dfrac{y^2}{16}=1$ **47.** $\dfrac{x^2}{4}+\dfrac{y^2}{9}=1$

**49.** None, the ellipse will be within the circle **51.** $\dfrac{(x+3)^2}{36}+\dfrac{(y+2)^2}{9}=1$; $(-3, -2)$ **53.** 69.5 feet **55. a)** $\dfrac{x^2}{100}+\dfrac{y^2}{576}=1$ **b)** $240\pi \approx 753.98$ square feet **c)** $\approx 376.99$ square feet **57.** $\sqrt{5} \approx 2.24$ feet, in both directions, from the center of the ellipse, along the major axis. **59.** Answers will vary. **61.** Answers will vary. **63.** $\dfrac{(x-4)^2}{49}+\dfrac{(y-2)^2}{9}=1$ **65.** $l=\dfrac{2S-nf}{n}$

**66.** $x+\dfrac{5}{2}+\dfrac{1}{2(2x-3)}$ **67.** 6 **68.** $\left[-\dfrac{5}{3}, 4\right)$ **69.** $\approx 2.7755$

## Mid-Chapter Test

**1.**

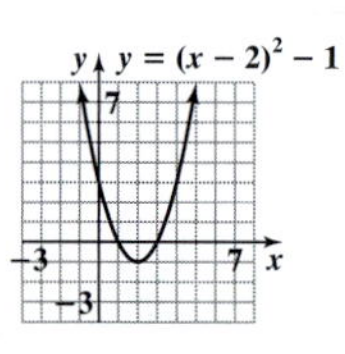


[11.1] **2.**

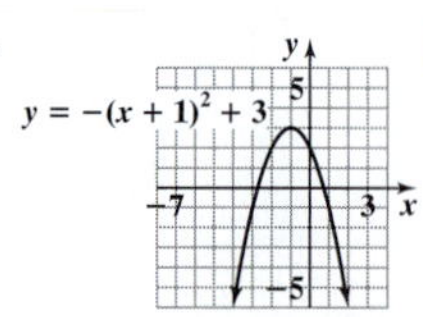


[11.1] **3.**

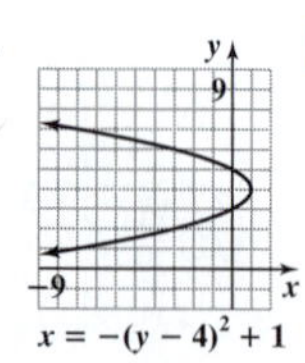


[11.1] **4.**

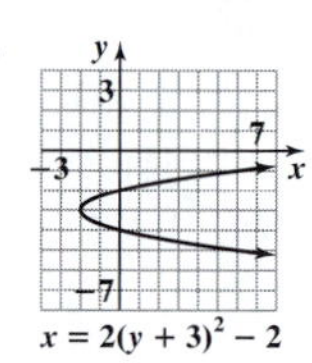


[11.1]

**5.** 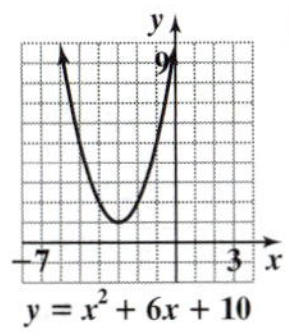


[11.1] **6.** 13 [11.1] **7.** $\sqrt{153} \approx 12.37$ [11.1] **8.** $\left(-1, \frac{5}{2}\right)$ [11.1] **9.** $\left(\frac{11}{4}, \frac{15}{4}\right)$ [11.1] **10.** $(x + 3)^2 + (y - 2)^2 = 25$ [11.1]

**11.** 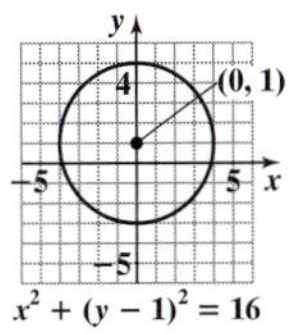


[11.1] **12.** 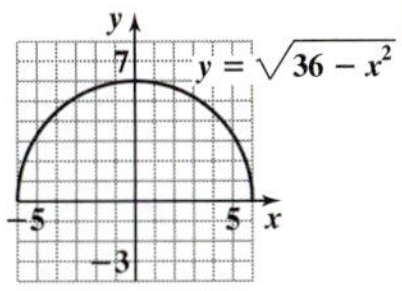


[11.1] **13.** 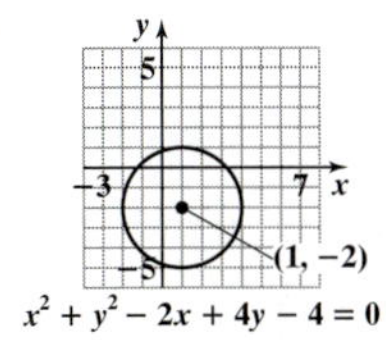


[11.1] **14.** A circle is a set of points in a plane that are the same distance from a fixed point called its center. [11.1]

**15.** 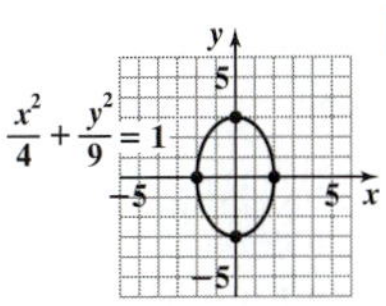


[11.2] **16.** 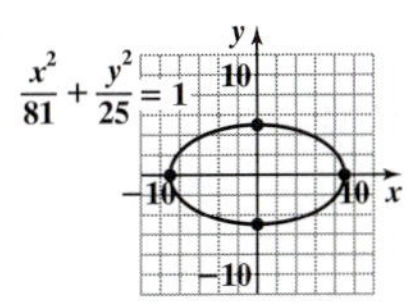


[11.2] **17.** 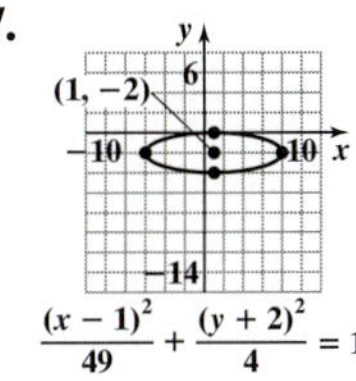


[11.2] **18.** 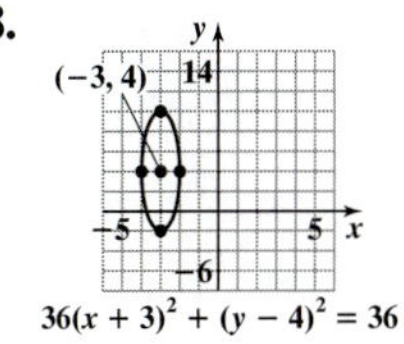


[11.2]

**19.** $6\pi \approx 18.85$ square units [11.2] **20.** $\frac{x^2}{64} + \frac{y^2}{25} = 1$ [11.2]

## Exercise Set 11.3

**1.** A hyperbola is the set of points in a plane the differences of whose distances from two fixed points is a constant. **3.** The graph of $\frac{x^2}{a^2} - \frac{y^2}{b^2} = 1$ is a hyperbola with vertices at $(a, 0)$ and $(-a, 0)$. Its transverse axis lies along the $x$-axis. The asymptotes are $y = \pm\frac{b}{a}x$. **5.** No, the signs of the $x$ and $y$ terms must differ. **7.** Yes, divide both sides of the equation by 100 and you will see the equation is that of a hyperbola. **9.** Divide both sides of the equation by 81.

**11. a)** $y = \pm\frac{2}{3}x$ **b)** 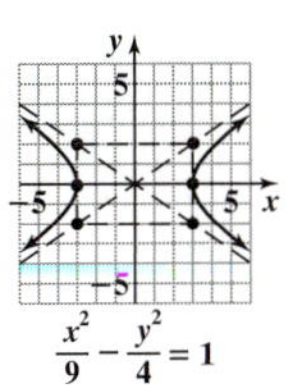


**13. a)** $y = \pm\frac{1}{2}x$ **b)** 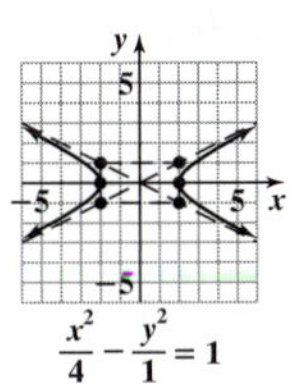


**15. a)** $y = \pm\frac{5}{3}x$ **b)** 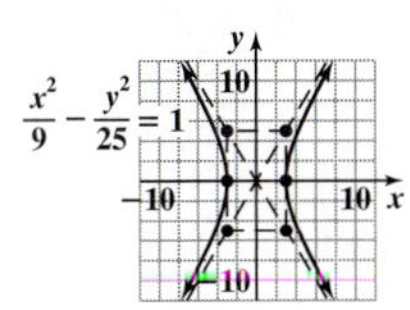


**17. a)** $y = \pm\frac{4}{5}x$ **b)** 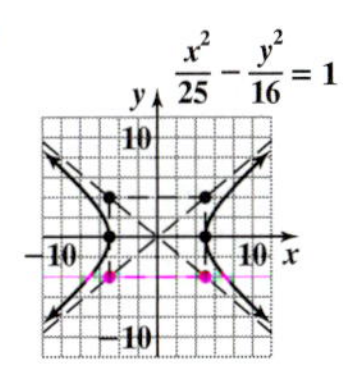


**19. a)** $y = \pm\frac{5}{6}x$ **b)** 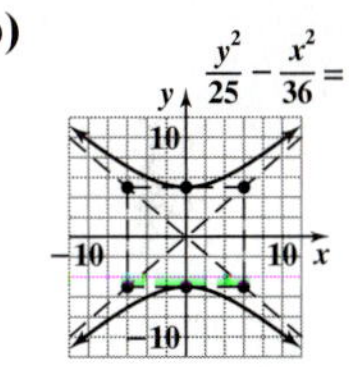


**21. a)** $y = \pm\frac{3}{4}x$ **b)** 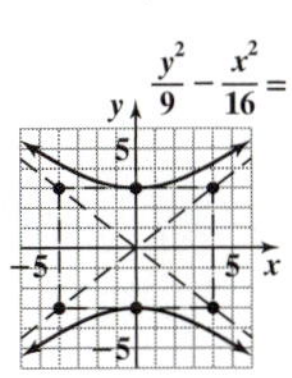


**23. a)** $y = \pm\frac{5}{2}x$ **b)** 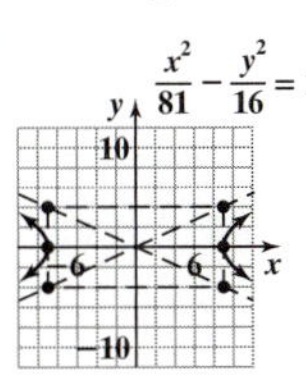


**25. a)** $y = \pm\frac{4}{9}x$ **b)** 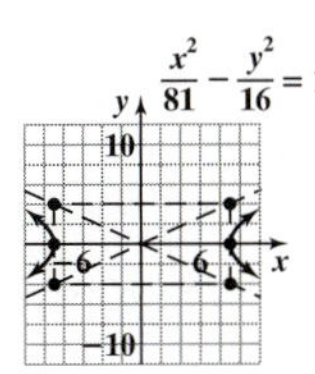


**27. a)** $\frac{x^2}{25} - \frac{y^2}{1} = 1,\ y = \pm\frac{1}{5}x$ **b)** 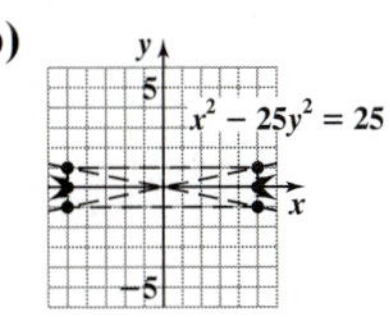


**29. a)** $\frac{y^2}{16} - \frac{x^2}{4} = 1,\ y = \pm 2x$ **b)** 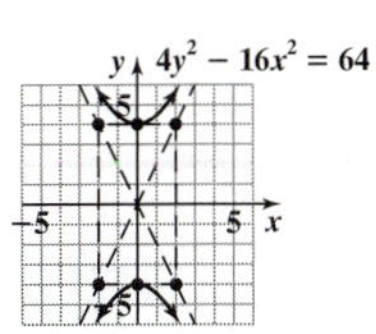


**31. a)** $\frac{y^2}{1} - \frac{x^2}{9} = 1,\ y = \pm\frac{1}{3}x$ **b)** 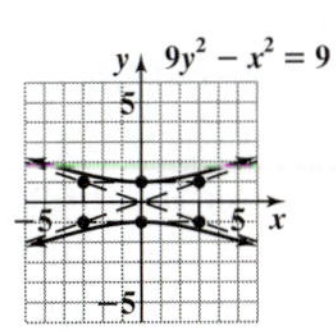


**33. a)** $\frac{x^2}{9} - \frac{y^2}{25} = 1,\ y = \pm\frac{5}{3}x$ **b)** 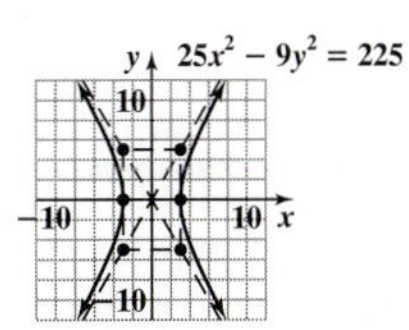


**35. a)** $\frac{y^2}{36} - \frac{x^2}{4} = 1,\ y = \pm 3x$ **b)** 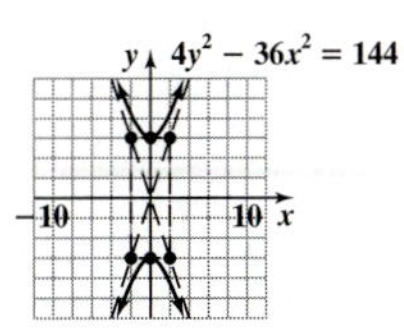


**37.** Circle **39.** Ellipse **41.** Hyperbola **43.** Parabola **45.** Ellipse **47.** Parabola **49.** Circle **51.** Hyperbola **53.** Parabola **55.** Hyperbola **57.** Parabola **59.** Circle **61.** $\frac{y^2}{4} - \frac{x^2}{16} = 1$ **63.** $\frac{x^2}{9} - \frac{y^2}{36} = 1$ **65.** $\frac{x^2}{9} - \frac{y^2}{25} = 1$, no, $\frac{x^2}{18} - \frac{y^2}{50} = 1$ and others will work. The ratio of $\frac{b}{a}$ must be $\frac{5}{3}$. **67.** No, graphs of hyperbolas of this form do not pass the vertical line test **69.** D: $(-\infty, -5] \cup [5, \infty)$; R: $\mathbb{R}$ **71.** The transverse axes of both graphs are along the $x$-axis. The vertices of the second graph will be closer to the origin, and the second graph will open wider. **73.** Answers will vary. **75.** $y = -\frac{1}{2}x + 1$ **76.** $-x^2 - x + 11$ **77.** $\left(-1, \frac{1}{3}\right)$ **78.** $\frac{3x + 2}{2x - 3}$ **79.** $v = \sqrt{\frac{2E}{m}}$ **80.** 1

**Exercise Set 11.4** **1.** A nonlinear system of equations is a system in which at least one equation is nonlinear. **3.** Yes, for example  **5.** Yes, for example  **7.** $(3, -3), (-3, 3)$ **9.** $(3, 0), \left(-\frac{9}{5}, \frac{12}{5}\right)$ **11.** $(-4, 11), \left(\frac{5}{2}, \frac{5}{4}\right)$ **13.** $(-1, 5), (1, 5)$ **15.** $(2, 2\sqrt{2}), (2, -2\sqrt{2}), (-2, 2\sqrt{2}), (-2, -2\sqrt{2})$ **17.** No real solution **19.** $(0, -3), (\sqrt{5}, 2)(-\sqrt{5}, 2)$ **21.** $(2, -4), (-14, -20)$ **23.** $(2, 0), (-2, 0)$ **25.** $(\sqrt{15}, 1)(-\sqrt{15}, 1), (\sqrt{15}, -1), (-\sqrt{15}, -1)$ **27.** $(\sqrt{2}, \sqrt{2}), (\sqrt{2}, -\sqrt{2}), (-\sqrt{2}, \sqrt{2}), (-\sqrt{2}, -\sqrt{2})$ **29.** $(3, 0), (-3, 0)$ **31.** $(2, 1), (2, -1), (-2, 1), (-2, -1)$ **33.** $(3, 4), (3, -4), (-3, 4), (-3, -4)$ **35.** $(\sqrt{5}, 2), (\sqrt{5}, -2), (-\sqrt{5}, 2), (-\sqrt{5}, -2)$ **37.** No real solution **39.** No real solution **41.** Answers will vary. **43.** 20 meters by 22 meters **45.** 9 feet by 30 feet **47.** length: 14 centimeters, width: 8 centimeters **49.** 16 inches by 30 inches **51.** $\approx$1.67 seconds **53.** $r = 6\%$, $p = \$125$ **55.** $\approx$16 and $\approx$184 **57.** $\approx$5 and $\approx$95 **59.** $(-1, -3), (3.12, -0.53)$ **61.** 10 yards, 24 yards **63.** Parentheses, exponents, multiplication or division, addition or subtraction. **64.** $(x + 2)(x^2 + x + 1)$ **65.** 0.9 **66.** $\frac{5\sqrt{x + 2} + 15}{x - 7}$ **67.** $k = \frac{\ln A - \ln A_0}{t}$

**Exercise Set 11.5** **1.** When graphing an inequality containing > or <, the boundary is graphed with a dashed line because it is not part of the solution. **3.** The graph of a circle divides the $xy$-plane into 2 regions: a region inside the circle and a region outside the circle.

**5.** 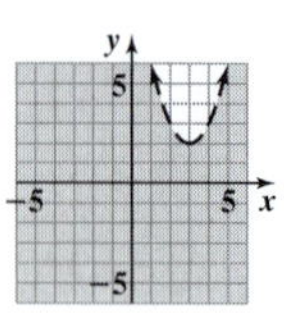
**7.** 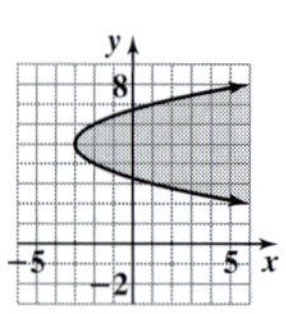
**9.** 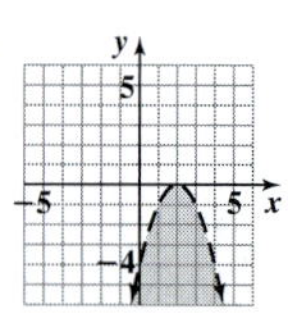
**11.** 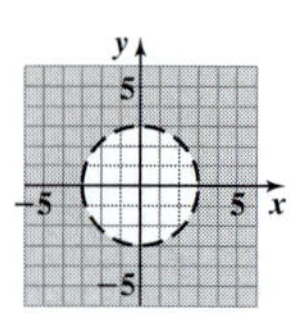
**13.** 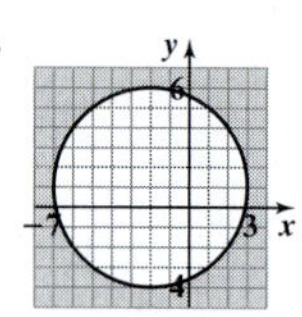
**15.** 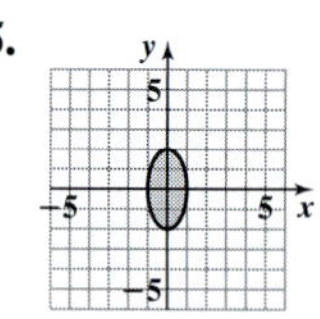

**17.** 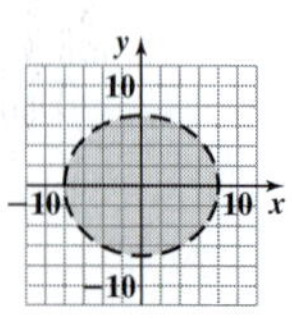
**19.** 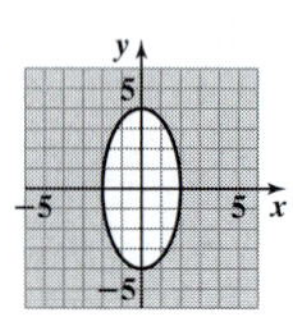
**21.** 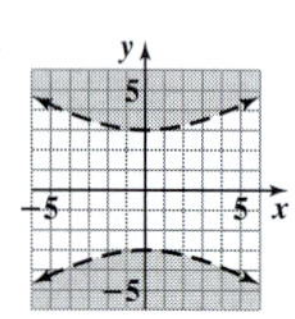
**23.** 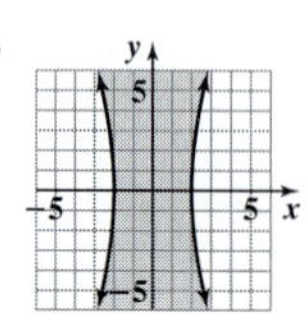
**25.** 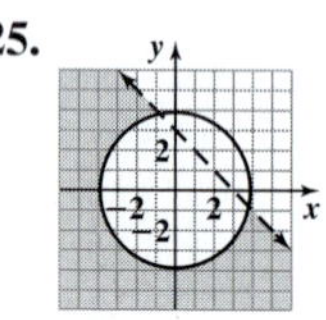
**27.** 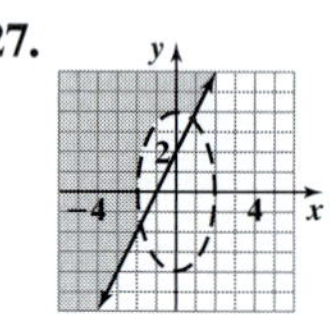

**29.** 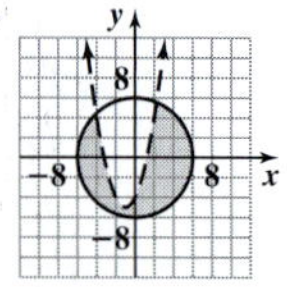
**31.** 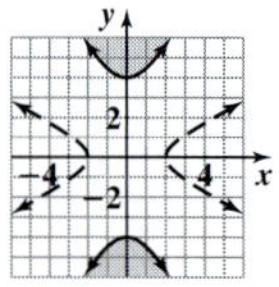
**33.** 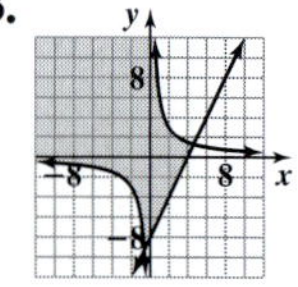
**35.** 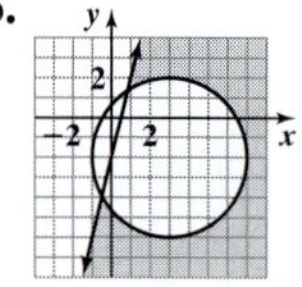

**37. a)** Yes, the solution to a system of linear inequalities containing two linear inequalities can contain no points. For instance, if the lines are parallel and the top line was shaded above and the bottom line was shaded below, the shaded regions would not overlap. Thus, the solution would contain no points. **b)** No, the solution to a system of linear inequalities containing two linear inequalities cannot contain exactly 1 point. If there is a solution, either the lines overlap completely or the shaded regions overlap. In either case, more than 1 solution exists. **c)** No, the solution to a system of linear inequalities containing two linear inequalities cannot contain exactly 2 points. If there is a solution, either the lines overlap completely or the shaded regions overlap. In either case, more than 2 solutions exists. **d)** No, the solution to a system of linear inequalities containing two linear inequalities cannot contain all points on the $xy$-plane. The solution to a system of inequalities is the intersection of the solutions to each inequality. Since the solution to an inequality cannot contain all points on the $xy$-plane, the intersection of two inequalities cannot contain all points on the $xy$-plane.

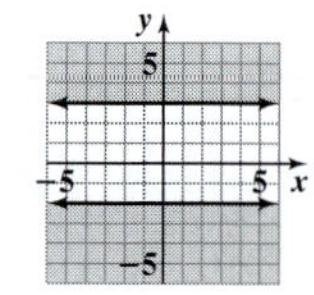

**39. a)** Yes, the solution to a system of inequalities containing two second-degree inequalities can contain no points. For instance, two circles that do not overlap and are shaded within have no overlapping regions. Thus, the solution would contain no points.

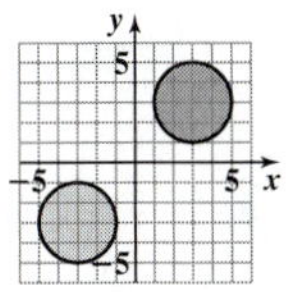

**b)** Yes, the solution to a system of inequalities containing two second-degree inequalities can contain exactly 1 point. For instance two circles which are shaded within and intersect at exactly one point would have exactly one point in common. Thus, the solution would contain exactly 1 point.

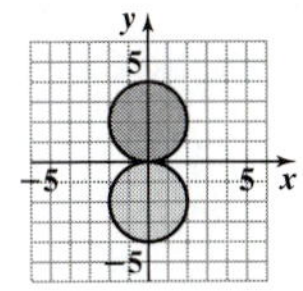

**c)** Yes, the solution to a system of inequalities containing two second-degree inequalities can contain exactly 2 points. For instance an ellipse and a circle can intersect in exactly 2 points. If the ellipse is within the circle and its interior is shaded and the circle's exterior is shaded, they intersect in exactly two points. Thus, the solution would contain exactly 2 points.

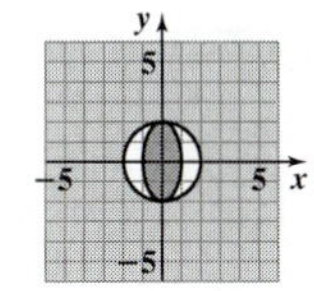

**d)** No, the solution to a system of inequalities containing two second-degree inequalities cannot contain all points on the $xy$-plane. The solution to a system of inequalities is the intersection of the solutions to each inequality. Since the solution to an inequality cannot contain all points on the $xy$-plane, the intersection of two inequalities cannot contain all points on the $xy$-plane.

**41. a)**

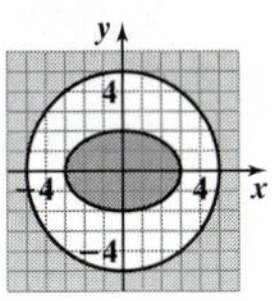

**b)**

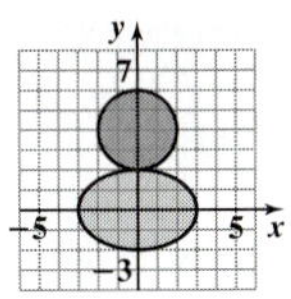

**c)**

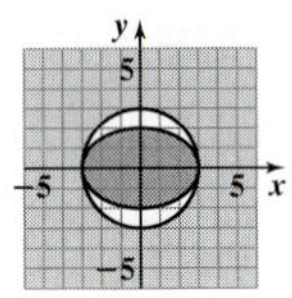

**45.** $16y^{10}$
**46.** $1 \pm \sqrt{5}$
**47.** $1 \pm \sqrt{5}$
**48.**

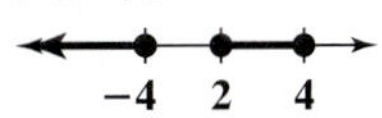


## Chapter 11 Review Exercises

**1.** 13; $\left(\frac{5}{2}, -6\right)$ **2.** 5; $\left(-\frac{5}{2}, 3\right)$ **3.** 17; $\left(-5, \frac{5}{2}\right)$ **4.** $\sqrt{8} \approx 2.83$; $(-3, 4)$

**5.**

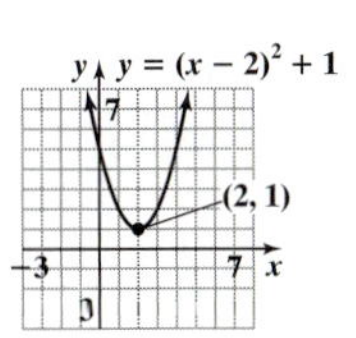


**6.**

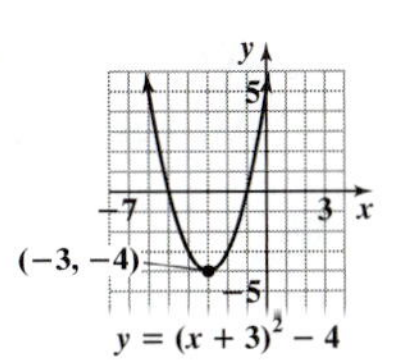


**7.**

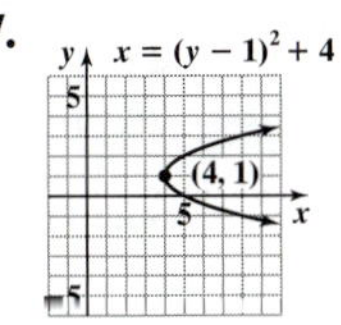


**8.**

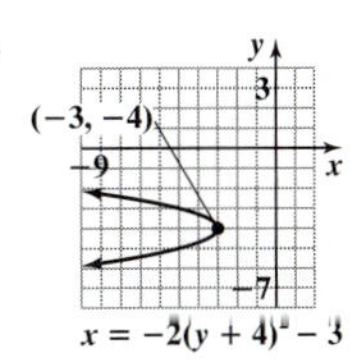


**9. a)** $y = (x - 4)^2 + 6$
**b)**

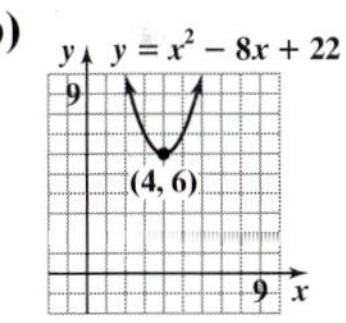


**10. a)** $x = -(y + 1)^2 + 6$
**b)**

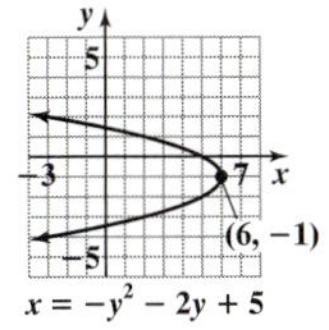


**11. a)** $x = \left(y + \frac{5}{2}\right)^2 - \frac{9}{4}$
**b)**

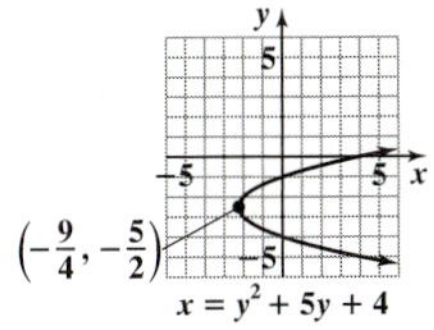


**12. a)** $y = 2(x - 2)^2 - 32$
**b)**

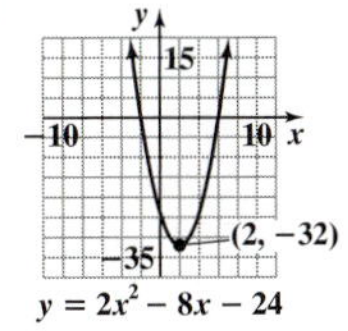


**13. a)** $x^2 + y^2 = 4^2$
**b)**

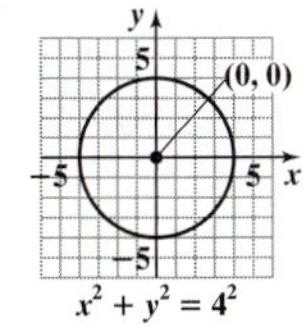


**14. a)** $(x + 3)^2 + (y - 4)^2 = 1^2$
**b)**

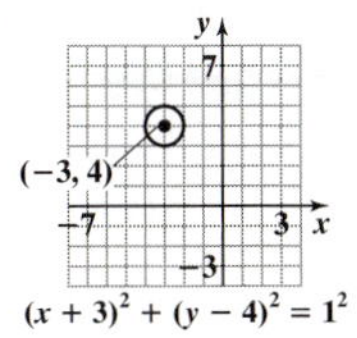


**15. a)** $x^2 + (y - 2)^2 = 2^2$
**b)**

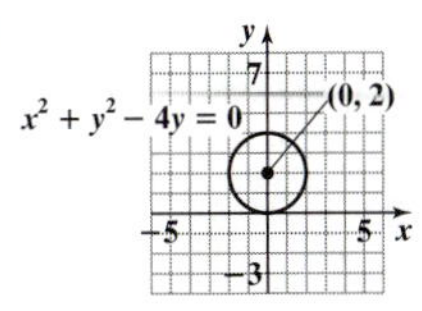


**16. a)** $(x - 1)^2 + (y + 3)^2 = 3^2$
**b)**

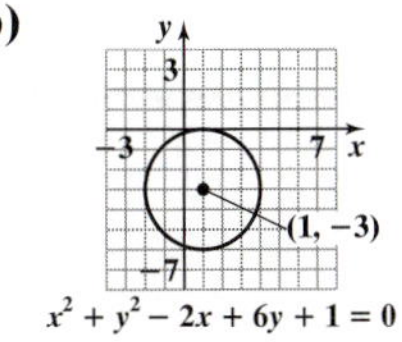


**17. a)** $(x - 4)^2 + (y - 5)^2 = 1^2$
**b)**

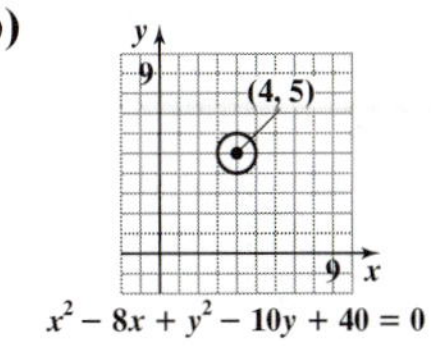

**18. a)** $(x-2)^2 + (y+5)^2 = (\sqrt{12})^2$
**b)**
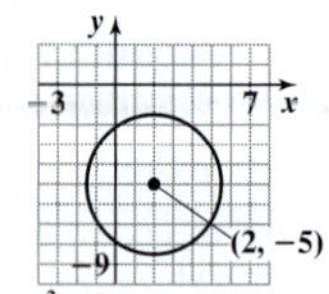

$x^2 + y^2 - 4x + 10y + 17 = 0$

**19.**
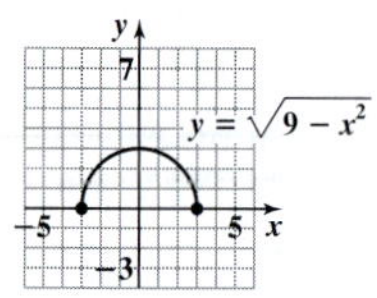


**20.**
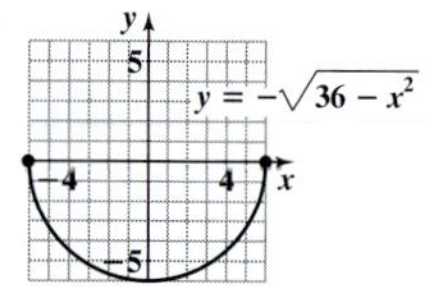


**21.** $(x+1)^2 + (y-1)^2 = 4$ **22.** $(x-5)^2 + (y+3)^2 = 9$

**23.**
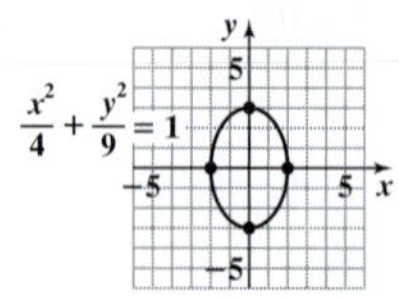


**24.**
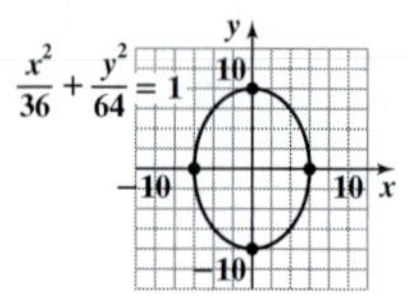


**25.**
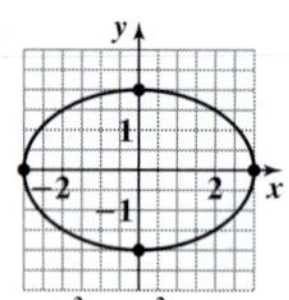
$4x^2 + 9y^2 = 36$

**26.**
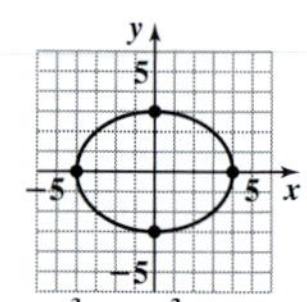
$9x^2 + 16y^2 = 144$

**27.**
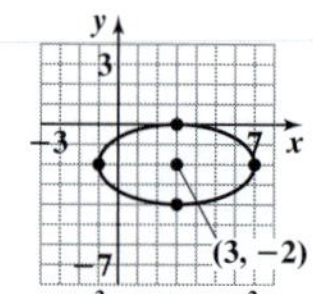

$\frac{(x-3)^2}{16} + \frac{(y+2)^2}{4} = 1$

**28.**
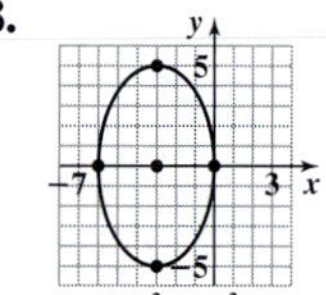
$\frac{(x+3)^2}{9} + \frac{y^2}{25} = 1$

**29.**
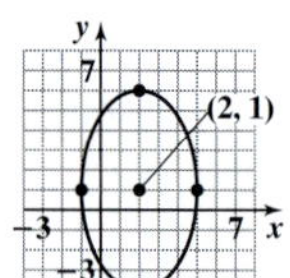

$25(x-2)^2 + 9(y-1)^2 = 225$

**30.** $6\pi \approx 18.85$ square units **31. a)** $y = \pm 2x$
**b)**
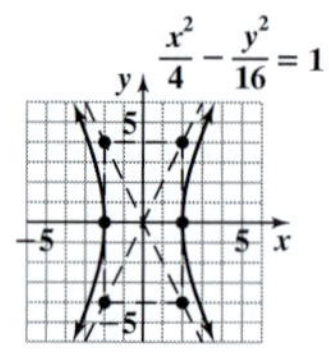


**32. a)** $y = \pm x$
**b)**
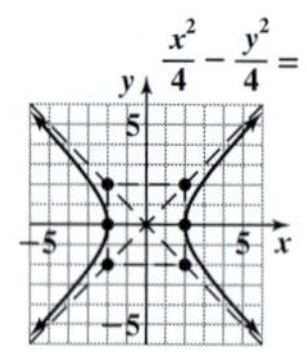


**33. a)** $y = \pm \frac{1}{3}x$
**b)**
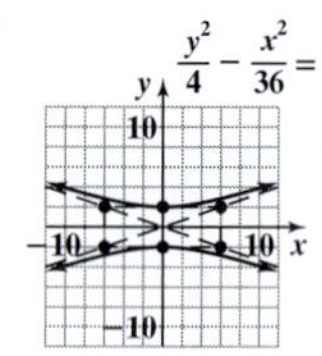


**34. a)** $y = \pm \frac{5}{4}x$
**b)**
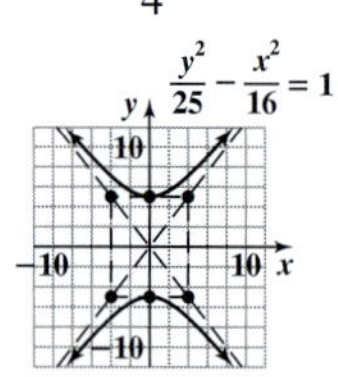


**35. a)** $\frac{x^2}{9} - \frac{y^2}{1} = 1$ **b)** $y = \pm \frac{1}{3}x$
**c)**
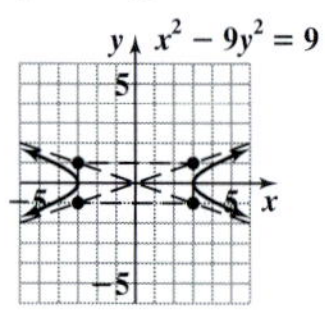


**36. a)** $\frac{x^2}{16} - \frac{y^2}{25} = 1$ **b)** $y = \pm \frac{5}{4}x$
**c)**
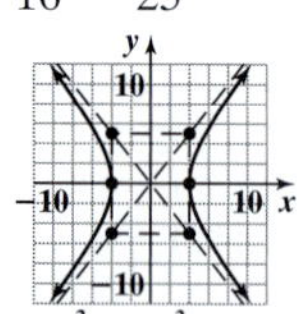
$25x^2 - 16y^2 = 400$

**37. a)** $\frac{y^2}{25} - \frac{x^2}{4} = 1$ **b)** $y = \pm \frac{5}{2}x$
**c)**
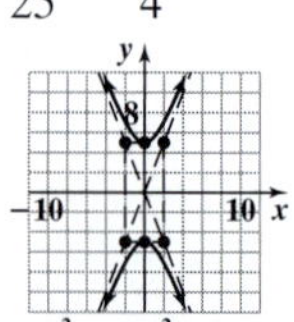
$4y^2 - 25x^2 = 100$

**38. a)** $\frac{y^2}{9} - \frac{x^2}{49} = 1$ **b)** $y = \pm \frac{3}{7}x$
**c)**
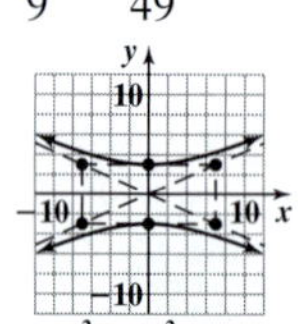
$49y^2 - 9x^2 = 441$

**39.** Hyperbola **40.** Ellipse **41.** Circle
**42.** Hyperbola **43.** Ellipse **44.** Parabola
**45.** Ellipse **46.** Parabola **47.** $(5, 0), (-5, 0)$
**48.** $\left(\frac{5}{2}, \frac{3}{2}\right)$ **49.** $(-3, 0), \left(-\frac{12}{5}, \frac{9}{5}\right)$
**50.** No real solution **51.** $(6, 0), (-6, 0)$
**52.** $(4, 3), (4, -3), (-4, 3), (-4, -3)$

**53.** No real solution **54.** $(0, \sqrt{3}), (0, -\sqrt{3})$ **55.** 5 feet by 9 feet **56.** $\approx 4$ and $\approx 145$ **57.** $r = 3\%$, $p = \$4000$ **58.** 15 feet by 20 feet

**59.**
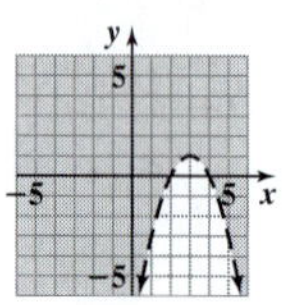

**60.**
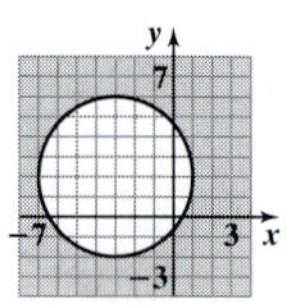

**61.**
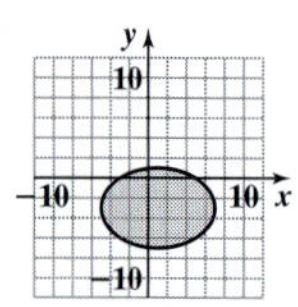

**62.**
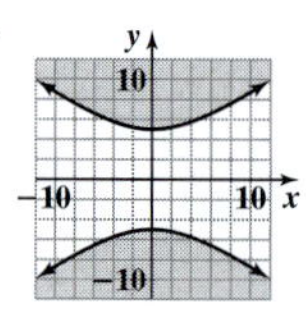

**63.**
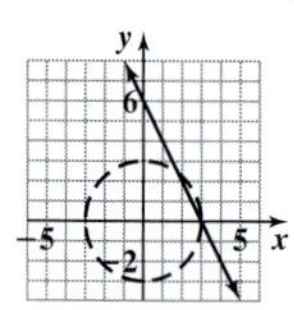

**64.**
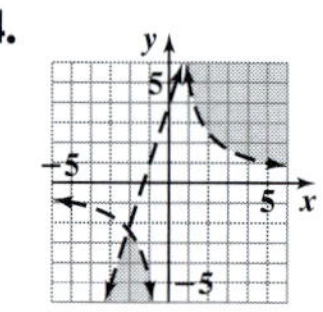

**65.**
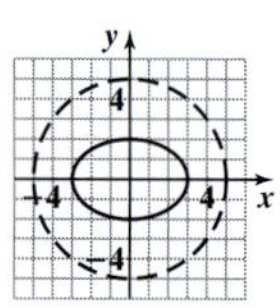

**66.**
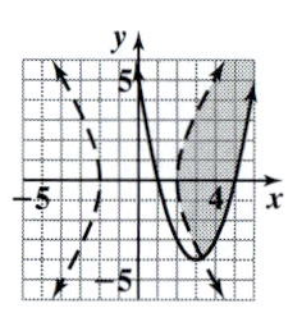

## Chapter 11 Practice Test

**1.** They are formed by cutting a cone or a pair of cones. [11.1] **2.** $\sqrt{50} \approx 7.07$ [11.1] **3.** $\left(-1, \frac{3}{2}\right)$ [11.1]
**4.** $(-3, 1)$, [11.1]
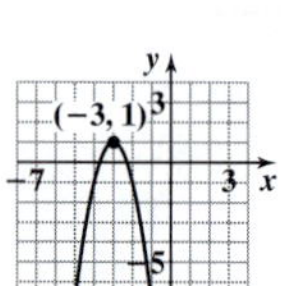

$y = -2(x+3)^2 + 1$

**5.** [11.1]
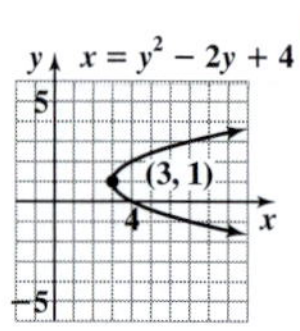


**6.** [11.1]
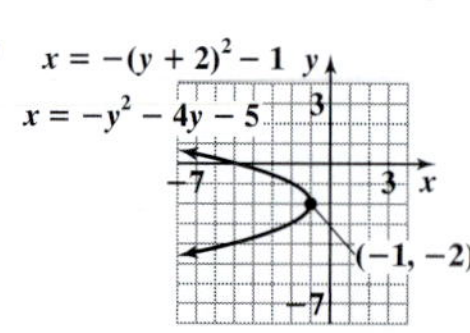


**7.** [11.1]
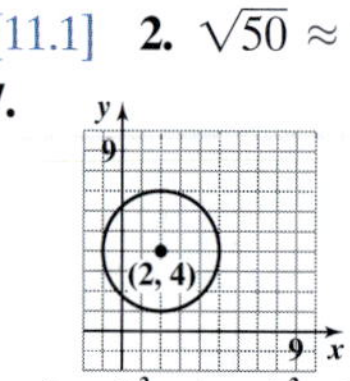

$(x-2)^2 + (y-4)^2 = 9$

**8.** $9\pi \approx 28.27$ square units [11.1] **9.** $(x-3)^2 + (y+1)^2 = 16$ [11.1]

**10.** [11.1]

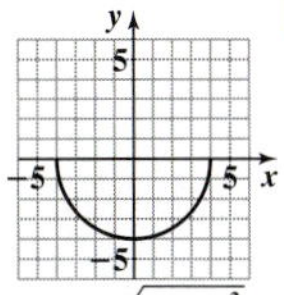


**11.** [11.1]

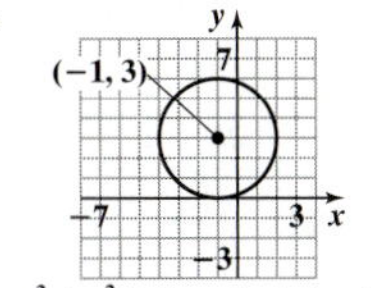


**12.** [11.2]

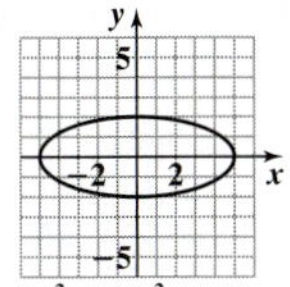


**13.** No, the major axis should be along the $y$-axis. [11.2]

**14.** [11.2]

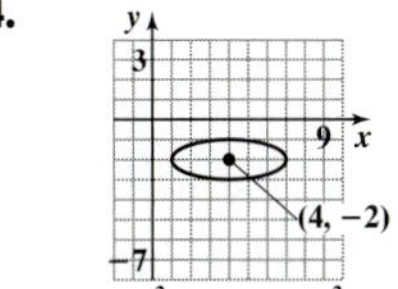


**15.** $(8, -7)$ [11.2] **16.** The transverse axis lies along the axis corresponding to the positive term of the equation in standard form. [11.3]

**17.** $y = \pm\frac{7}{4}x$ [11.3] **18.** [11.3]

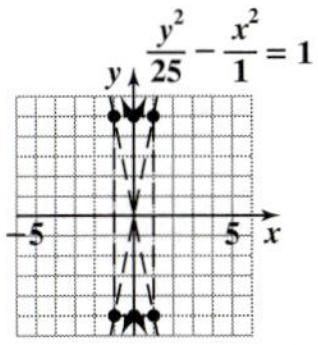


**19.** [11.3]

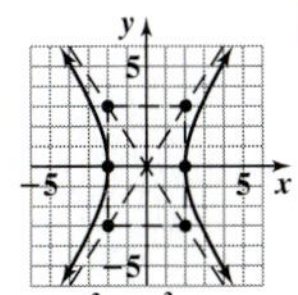


**20.** Hyperbola, divide both sides of the equation by 30. [11.3] **21.** Ellipse, divide both sides of the equation by 100. [11.3] **22.** $(2, \sqrt{3}), (2, -\sqrt{3}), (-2, \sqrt{3}), (-2, -\sqrt{3})$ [11.4] **23.** No real solution [11.4] **24.** 30 meters by 50 meters [11.4] **25.** [11.5]

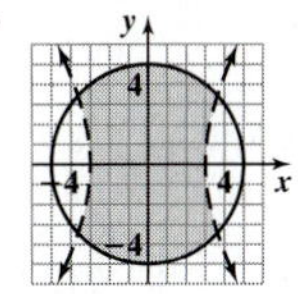

## Cumulative Review Test

**1.** $-27x^3y^9$ [1.5] **2.** $\frac{19}{4}$ [2.1] **3.** $\varnothing$ [2.1] **4.** $\left\{x \middle| x < -\frac{5}{3} \text{ or } x > 1\right\}$ [2.6]

**5.** [3.1]

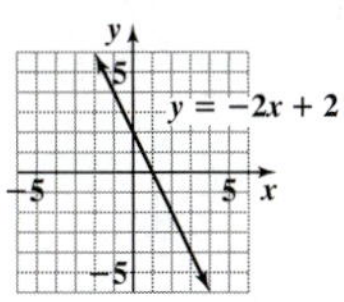


**6.** 139 [3.2] **7.** $(8, 6)$ [4.1] **8.** $(x^2+6)(x^2-7)$ [5.5] **9.** base: 14 feet, height: 8 feet [5.8] **10.** $\frac{x-4}{4x+3}$ [6.1] **11.** $\frac{2x^2-9x-5}{2(x+3)(x-4)}$ [6.2] **12.** 2 [6.4] **13.** $\frac{3y^{3/2}}{x^{1/2}}$ [7.2] **14.** $\frac{6x + 6y\sqrt{x}}{x - y^2}$ [7.5] **15.** 3 [7.6] **16.** $\frac{2 \pm i\sqrt{11}}{3}$ [8.2] **17.** 2 [9.6] **18.** $-\frac{3}{2}, \frac{1}{2}, 1, 2$ [10.4] **19.** [11.2]

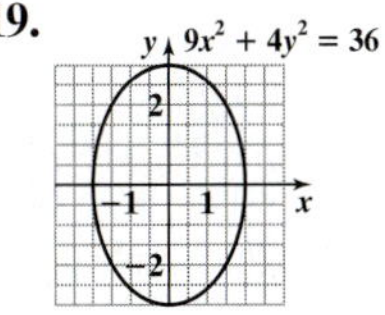


**20.** [11.3]

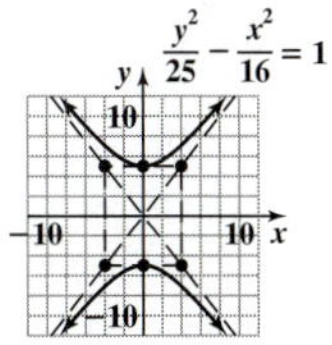


# Chapter 12

## Exercise Set 12.1

**1.** A sequence is a list of numbers arranged in a specific order. **3.** A finite sequence is a function whose domain includes only the first $n$ natural numbers. **5.** In a decreasing sequence, the terms decrease. **7.** A series is the sum of the terms of a sequence. **9.** $\sum_{i=1}^{5}(i+4)$ is the sum as $i$ goes from 1 to 5 of $i + 4$. **11.** It is an increasing sequence. Each number in the sequence is greater than the preceding number **13.** Yes, the signs of the terms alternate. **15.** 6, 12, 18, 24, 30 **17.** 3, 7, 11, 15, 19 **19.** $7, \frac{7}{2}, \frac{7}{3}, \frac{7}{4}, \frac{7}{5}$ **21.** $\frac{3}{2}, \frac{4}{3}, \frac{5}{4}, \frac{6}{5}, \frac{7}{6}$ **23.** $-1, 1, -1, 1, -1$ **25.** $4, -8, 16, -32, 64$ **27.** 31 **29.** 12 **31.** 1 **33.** 99 **35.** $\frac{81}{25}$ **37.** 2, 15 **39.** 3, 17 **41.** $0, \frac{13}{20}$ **43.** $-1, -1$ **45.** $\frac{1}{2}, 7$ **47.** 64, 128, 256 **49.** 17, 19, 21 **51.** $\frac{1}{6}, \frac{1}{7}, \frac{1}{8}$ **53.** $1, -1, 1$ **55.** $\frac{1}{81}, \frac{1}{243}, \frac{1}{729}$ **57.** $\frac{1}{16}, -\frac{1}{32}, \frac{1}{64}$ **59.** 17, 12, 7 **61.** $2 + 5 + 8 + 11 + 14 = 40$ **63.** $2 + 5 + 10 + 17 + 26 + 37 = 97$ **65.** $\frac{1}{2} + 2 + \frac{9}{2} + 8 = 15$ **67.** $4 + 5 + 6 + 7 + 8 + 9 = 39$ **69.** $\sum_{i=1}^{5}(i+8)$ **71.** $\sum_{i=1}^{3}\frac{i^2}{4}$ **73.** 13 **75.** 169 **77.** 55 **79.** 25 **81.** $\approx 42.83$ **83. a)** 6, 12, 18, 24 **b)** $p_n = 6n$ **85.** Answers will vary. **87.** Answers will vary. **89.** $\Sigma x = n\bar{x}$ **91.** Yes, for example if $n = 3$, you obtain $4x_1 + 4x_2 + 4x_3 = 4(x_1 + x_2 + x_3)$ **93. a)** 10 **b)** 11 **c)** 110 **d)** 29 **e)** No **94.** $\frac{2}{5}$ **95.** $8(y - 2x^2)(y^2 + 2x^2y + 4x^4)$ **96.** 11 **97.** $r = \sqrt{\frac{V}{\pi h}}$

## Exercise Set 12.2

**1.** In an arithmetic sequence, each term differs by a constant amount. **3.** It is called the common difference. **5.** Positive number **7.** Yes, for example $-1, -2, -3, \ldots$ **9.** Yes, for example $2, 4, 6, \ldots$ **11.** $4, 7, 10, 13, 16$; $a_n = 3n + 1$ **13.** $7, 5, 3, 1, -1$; $a_n = -2n + 9$ **15.** $\frac{1}{2}, 2, \frac{7}{2}, 5, \frac{13}{2}$; $a_n = \frac{3}{2}n - 1$ **17.** $100, 95, 90, 85, 80$; $a_n = -5n + 105$ **19.** 14 **21.** 27 **23.** 12 **25.** 2 **27.** 9 **29.** 6 **31.** $s_{10} = 100; d = 2$ **33.** $s_8 = \frac{52}{5}; d = \frac{1}{5}$ **35.** $s_6 = 25.5; d = 3.7$ **37.** $s_{11} = 407; d = 6$

**39.** 4, 7, 10, 13; $a_{10} = 31$; $s_{10} = 175$ **41.** $-6, -4, -2, 0$; $a_{10} = 12$; $s_{10} = 30$ **43.** $-8, -13, -18, -23$; $a_{10} = -53$; $s_{10} = -305$ **45.** $\frac{7}{2}, 6, \frac{17}{2}, 11$; $a_{10} = 26$, $s_{10} = 147.5$ **47.** 100, 93, 86, 79; $a_{10} = 37$; $s_{10} = 685$ **49.** $n = 15$, $s_{15} = 330$ **51.** $n = 11$; $s_{11} = 121$ **53.** $n = 17$; $s_{17} = \frac{153}{2}$ **55.** $n = 29$; $s_{29} = 1421$ **57.** 1275 **59.** 2500 **61.** 1395 **63.** 267 **65.** 42,372 **67.** 351 **69. a)** 27 **b)** 196 **71.** $101 \cdot 50 = 5050$ **73.** $s_n = n^2$ **75. a)** 19 feet **b)** 143.5 feet **77.** 2 feet **79. a)** 155 **b)** 780 **81.** \$496 **83. a)** \$45,600 **b)** \$438,000 **85.** $a_n = 180°(n - 2)$ **93.** $r = \frac{A - P}{Pt}$ **94.** $(-3, -5)$ **95.** $3(2n - 5)(2n - 1)$ **96.**

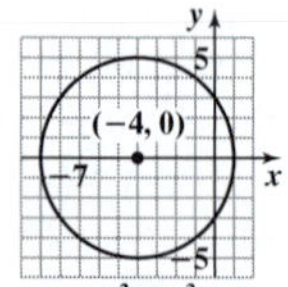


$(x + 4)^2 + y^2 = 25$

## Exercise Set 12.3

**1.** A geometric sequence is a sequence in which each term after the first is the same multiple of the preceding term. **3.** To find the common ratio, take any term except the first and divide by the term that precedes it. **5.** 0 **7.** Yes **9.** Yes, $s_\infty$ exists because $|r| < 1$, $s_\infty = 8$ **11.** 2, 6, 18, 54, 162 **13.** $6, -3, \frac{3}{2}, -\frac{3}{4}, \frac{3}{8}$ **15.** $72, 24, 8, \frac{8}{3}, \frac{8}{9}$ **17.** $90, -30, 10, -\frac{10}{3}, \frac{10}{9}$ **19.** $-1, -3, -9, -27, -81$ **21.** $5, -10, 20, -40, 80$ **23.** $\frac{1}{3}, \frac{1}{6}, \frac{1}{12}, \frac{1}{24}, \frac{1}{48}$ **25.** $3, \frac{9}{2}, \frac{27}{4}, \frac{81}{8}, \frac{243}{16}$ **27.** 128 **29.** $-\frac{3}{64}$ **31.** 128 **33.** 6144 **35.** $\frac{1}{64}$ **37.** $\frac{50}{729}$ **39.** 155 **41.** 7812 **43.** 10,160 **45.** $-\frac{2565}{256}$ **47.** $-\frac{9279}{625}$ **49.** $r = \frac{1}{2}$; $a_n = 3\left(\frac{1}{2}\right)^{n-1}$ **51.** $r = 2$; $a_n = 9(2)^{n-1}$ **53.** $r = -3$; $a_n = 2(-3)^{n-1}$ **55.** $r = \frac{2}{3}$; $a_n = \frac{3}{4}\left(\frac{2}{3}\right)^{n-1}$ **57.** 2 **59.** $\frac{5}{4}$ **61.** 12 **63.** $\frac{25}{3}$ **65.** 5 **67.** 6 **69.** 4 **71.** 24 **73.** $-45$ **75.** $-15$ **77.** $\frac{8}{33}$ **79.** $\frac{8}{9}$ **81.** $\frac{17}{33}$ **83.** $r = 3$; $a_1 = 5$ **85.** $r = 2$ or $r = -2$; $a_1 = 7$ **87.** $\approx\$1.77$ **89. a)** 4 days **b)** $\approx 1.172$ grams **91. a)** $\approx 330.78$ million **b)** $\approx 63.4$ years **93. a)** $\frac{1}{2}, \frac{1}{4}, \frac{1}{8}, \frac{1}{16}, \frac{1}{32}$ **b)** $a_n = \frac{1}{2}\left(\frac{1}{2}\right)^{n-1} = \left(\frac{1}{2}\right)^n$ **c)** $\frac{1}{128} \approx 0.78\%$ **95.** $\approx\$15,938.48$ **97. a)** 28.512 feet **b)** 550 feet **99. a)** 10.29 inches **b)** 100 inches **101.** 211 **103. a)** \$12,000, \$9600, \$7680 **b)** $a_n = 12,000\left(\frac{4}{5}\right)^{n-1}$ **c)** $\approx\$4915.20$ **105.** 190 feet **107. a)** $y_2$ goes up more steeply. **b)**

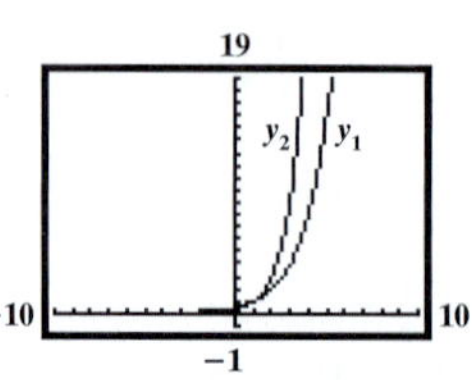


**109.** $n = 21$; $s_n = 2,097,151$ **110.** 12 **111.** $6x^3 - x^2y - 16xy^2 + 6y^3$ **112.** $r = \frac{S - 2a}{S}$ **113.** $g^{-1}(x) = \sqrt[3]{x - 9}$ **114.** 5 **115.** 9 meters, 12 meters

## Mid-Chapter Test

**1.** $2, -1, -4, -7, -10$ [12.1] **2.** 91 [12.1] **3.** 1, 11 [12.1] **4.** $-15, -19, -23$ [12.1] **5.** 45 [12.1] **6.** $\sum_{i=1}^{5}\left(\frac{1}{3}i + 7\right)$ [12.1] **7.** $-6, -1, 4, 9$; $a_n = -11 + 5n$ [12.2] **8.** $-1$ [12.2] **9.** 6 [12.2] **10.** $3, -3$ [12.2] **11.** $47\frac{1}{2}$ [12.2] **12.** 11 [12.2] **13.** 136 [12.2] **14.** $80, -40, 20, -10, 5$ [12.3] **15.** $\frac{1}{9}$ [12.3] **16.** 315 [12.3] **17.** $-\frac{2}{3}$ [12.3] **18.** 18 [12.3] **19.** $\frac{29}{33}$ [12.3] **20. a)** A sequence is a list of numbers arranged in a specific order. **b)** An arithmetic sequence is a sequence where each term differs by a constant amount. **c)** A geometric sequence is a sequence where the terms differ by a common multiple. **d)** A series is the sum of the terms of a sequence. [12.1–12.3]

## Exercise Set 12.4

**1.** No. When we test a case, we are only able to prove that particular case. If there are an infinite number of cases to test, we will never be able to prove all are true by testing a finite number of cases. **3.** $S_n$ is proven to be true for all $n$, if it is proved that $S_1$ is true, and if $S_k$ is true then $S_{k+1}$ is true.

**5.** Prove $1 + 5 + 9 + \ldots + (4n - 3) = n(2n - 1)$ for all positive integers $n$.

Condition 1: For $n = 1$,

$1 \stackrel{?}{=} 1(2(1) - 1)$

$1 \stackrel{?}{=} 1(2 - 1)$

$1 \stackrel{?}{=} 1(1)$

$1 = 1$ True

Condition 2:

Assume $1 + 5 + 9 + \ldots + (4k - 3) = k(2k - 1)$ for some positive integer $k$.

Then

$1 + 5 + 9 + \ldots + (4k - 3) + [4(k + 1) - 3] = k(2k - 1) + [4(k + 1) - 3]$

$1 + 5 + 9 + \ldots + (4k - 3) + [4(k + 1) - 3] = 2k^2 - k + 4k + 4 - 3$

$1 + 5 + 9 + \ldots + (4k - 3) + [4(k + 1) - 3] = 2k^2 + 3k + 1$

$1 + 5 + 9 + \ldots + (4k - 3) + [4(k + 1) - 3] = (k + 1)(2k + 1)$

$1 + 5 + 9 + \ldots + (4k - 3) + [4(k + 1) - 3] = (k + 1)(2k + 2 - 1)$

$1 + 5 + 9 + \ldots + (4k - 3) + [4(k + 1) - 3] = (k + 1)[2(k + 1) - 1]$

**7.** Prove $2 + 4 + 6 + \ldots + 2n = n(n + 1)$ for all positive integers $n$.

Condition 1: For $n = 1$,

$2 \stackrel{?}{=} 1(1 + 1)$

$2 \stackrel{?}{=} 1(2)$

$2 = 2$ True

Condition 2:

Assume $2 + 4 + 6 + \ldots + 2k = k(k + 1)$ for some positive integer $k$.

Then

$$2 + 4 + 6 + \ldots + 2k + 2(k + 1) = k(k + 1) + 2(k + 1)$$
$$2 + 4 + 6 + \ldots + 2k + 2(k + 1) = k^2 + k + 2k + 2$$
$$2 + 4 + 6 + \ldots + 2k + 2(k + 1) = k^2 + 3k + 2$$
$$2 + 4 + 6 + \ldots + 2k + 2(k + 1) = (k + 1)(k + 2)$$
$$2 + 4 + 6 + \ldots + 2k + 2(k + 1) = (k + 1)[(k + 1) + 1]$$

**9.** Prove $3 + 4 + 5 + \ldots + (n + 2) = \frac{n}{2}(n + 5)$ for all positive integers $n$.

Condition 1: For $n = 1$

$3 \stackrel{?}{=} \frac{1}{2}(1 + 5)$

$3 \stackrel{?}{=} \frac{1}{2}(6)$

$3 = 3$ True

Condition 2:

Assume $3 + 4 + 5 + \ldots + (k + 2) = \frac{k}{2}(k + 5)$ for some positive integer $k$.

Then

$$3 + 4 + 5 + \ldots + (k + 2) + [(k + 1) + 2] = \frac{k}{2}(k + 5) + [(k + 1) + 2]$$
$$3 + 4 + 5 + \ldots + (k + 2) + [(k + 1) + 2] = \frac{k^2}{2} + \frac{5k}{2} + k + 3$$
$$3 + 4 + 5 + \ldots + (k + 2) + [(k + 1) + 2] = \frac{k^2 + 7k + 6}{2}$$
$$3 + 4 + 5 + \ldots + (k + 2) + [(k + 1) + 2] = \frac{(k + 1)(k + 6)}{2}$$
$$3 + 4 + 5 + \ldots + (k + 2) + [(k + 1) + 2] = \frac{k + 1}{2}[(k + 1) + 5]$$

**11.** Prove $3 + 6 + 9 + \ldots + 3n = \frac{3n(n + 1)}{2}$ for all positive integers $n$.

Condition: For $n = 1$

$3 \stackrel{?}{=} \frac{3(1)(1 + 1)}{2}$

$3 \stackrel{?}{=} \frac{3(2)}{2}$

$3 = 3$ True

Condition 2:

Assume $3 + 6 + 9 + \ldots + 3k = \frac{3k(k + 1)}{2}$ for some positive integer $k$.

Then

$$3 + 6 + 9 + \ldots + 3k + 3(k + 1) = \frac{3k(k + 1)}{2} + 3(k + 1)$$
$$3 + 6 + 9 + \ldots + 3k + 3(k + 1) = \frac{3k^2}{2} + \frac{3k}{2} + 3k + 3$$
$$3 + 6 + 9 + \ldots + 3k + 3(k + 1) = \frac{3k^2 + 9k + 6}{2}$$
$$3 + 6 + 9 + \ldots + 3k + 3(k + 1) = \frac{3(k^2 + 3k + 2)}{2}$$
$$3 + 6 + 9 + \ldots + 3k + 3(k + 1) = \frac{3(k + 1)(k + 2)}{2}$$
$$3 + 6 + 9 + \ldots + 3k + 3(k + 1) = \frac{3(k + 1)[(k + 1) + 1]}{2}$$

**13.** Prove $5 + 10 + 15 + \ldots + 5n = \frac{5n(n + 1)}{2}$ for all positive integers $n$.

Condition 1:

$5 \stackrel{?}{=} \frac{5(1)(1 + 1)}{2}$

$5 \stackrel{?}{=} \frac{5(2)}{2}$

$5 = 5$ True

Condition 2:

Assume $5 + 10 + 15 + \ldots + 5k = \frac{5k(k + 1)}{2}$ for some positive integer $k$.

Then

$$5 + 10 + 15 + \ldots + 5k + 5(k + 1) = \frac{5k(k + 1)}{2} + 5(k + 1)$$
$$5 + 10 + 15 + \ldots + 5k + 5(k + 1) = \frac{5k^2}{2} + \frac{5k}{2} + 5k + 5$$
$$5 + 10 + 15 + \ldots + 5k + 5(k + 1) = \frac{5k^2 + 15k + 10}{2}$$
$$5 + 10 + 15 + \ldots + 5k + 5(k + 1) = \frac{5(k^2 + 3k + 2)}{2}$$

$$5 + 10 + 15 + \ldots + 5k + 5(k+1) = \frac{5(k+1)(k+2)}{2}$$
$$5 + 10 + 15 + \ldots + 5k + 5(k+1) = \frac{5(k+1)[(k+1)+1]}{2}$$

**15.** Prove $1 + 2 + 2^2 + \ldots + 2^{n-1} = 2^n - 1$ for all positive integers $n$.

Condition 1:

$1 \stackrel{?}{=} 2^1 - 1$
$1 \stackrel{?}{=} 2 - 1$
$1 = 1$ True

Condition 2:

Assume $1 + 2 + 2^2 + \ldots + 2^{k-1} = 2^k - 1$ for some positive integer $k$.

Then

$$1 + 2 + 2^2 + \ldots + 2^{k-1} + 2^{(k+1)-1} = 2^k - 1 + 2^{(k+1)-1}$$
$$1 + 2 + 2^2 + \ldots + 2^{k-1} + 2^{(k+1)-1} = 2^k - 1 + 2^k$$
$$1 + 2 + 2^2 + \ldots + 2^{k-1} + 2^{(k+1)-1} = 2^k + 2^k - 1$$
$$1 + 2 + 2^2 + \ldots + 2^{k-1} + 2^{(k+1)-1} = 2(2^k) - 1$$
$$1 + 2 + 2^2 + \ldots + 2^{k-1} + 2^{(k+1)-1} = 2^{k+1} - 1$$

**17.** Prove $1 + 4 + 4^2 + \ldots + 4^{n-1} = \frac{4^n - 1}{3}$ for all positive integers $n$.

Condition 1:

$1 \stackrel{?}{=} \frac{4^1 - 1}{3}$

$1 \stackrel{?}{=} \frac{4 - 1}{3}$

$1 \stackrel{?}{=} \frac{3}{3}$

$1 = 1$ True

Condition 2:

Assume $1 + 4 + 4^2 + \ldots + 4^{k-1} = \frac{4^k - 1}{3}$ for some positive integer $k$.

Then

$$1 + 4 + 4^2 + \ldots + 4^{k-1} + 4^{(k+1)-1} = \frac{4^k - 1}{3} + 4^{(k+1)-1}$$
$$1 + 4 + 4^2 + \ldots + 4^{k-1} + 4^{(k+1)-1} = \frac{4^k - 1}{3} + \frac{3 \cdot 4^k}{3}$$
$$1 + 4 + 4^2 + \ldots + 4^{k-1} + 4^{(k+1)-1} = \frac{4^k + 3 \cdot 4^k - 1}{3}$$
$$1 + 4 + 4^2 + \ldots + 4^{k-1} + 4^{(k+1)-1} = \frac{4^k(1+3) - 1}{3}$$
$$1 + 4 + 4^2 + \ldots + 4^{k-1} + 4^{(k+1)-1} = \frac{4^k(4) - 1}{3}$$
$$1 + 4 + 4^2 + \ldots + 4^{k-1} + 4^{(k+1)-1} = \frac{4^{k+1} - 1}{3}$$

**19.** Prove $2 + 2^2 + 2^3 + \ldots + 2^n = 2^{n+1} - 2$ for all positive integers $n$.

Condition 1:

$2 \stackrel{?}{=} 2^{1+1} - 2$
$2 \stackrel{?}{=} 2^2 - 2$
$2 \stackrel{?}{=} 4 - 2$
$2 = 2$ True

Condition 2:

Assume $2 + 2^2 + 2^3 + \ldots + 2^k = 2^{k+1} - 2$ for some positive integer $k$.

Then

$$2 + 2^2 + 2^3 + \ldots + 2^k + 2^{k+1} = 2^{k+1} - 2 + 2^{k+1}$$
$$2 + 2^2 + 2^3 + \ldots + 2^k + 2^{k+1} = 2^{k+1} + 2^{k+1} - 2$$
$$2 + 2^2 + 2^3 + \ldots + 2^k + 2^{k+1} = 2(2^{k+1}) - 2$$
$$2 + 2^2 + 2^3 + \ldots + 2^k + 2^{k+1} = 2^{(k+1)+1} - 2$$

**21.** Prove $4 + 4^2 + 4^3 + \ldots + 4^n = 4\left(\frac{4^n - 1}{3}\right)$ for all positive integers $n$.

Condition 1:

$4 \stackrel{?}{=} 4\left(\frac{4^1 - 1}{3}\right)$

$4 \stackrel{?}{=} 4\left(\frac{4 - 1}{3}\right)$

$4 \stackrel{?}{=} 4\left(\frac{3}{3}\right)$

$4 = 4$ True

Condition 2:

Assume $4 + 4^2 + 4^3 + \ldots + 4^k = 4\left(\frac{4^k - 1}{3}\right)$ for some positive integer $k$.

Then

$$4 + 4^2 + 4^3 + \ldots + 4^k + 4^{k+1} = 4\left(\frac{4^k - 1}{3}\right) + 4^{k+1}$$
$$4 + 4^2 + 4^3 + \ldots + 4^k + 4^{k+1} = \frac{4(4^k - 1)}{3} + \frac{3 \cdot 4 \cdot 4^k}{3}$$
$$4 + 4^2 + 4^3 + \ldots + 4^k + 4^{k+1} = \frac{4[(4^k - 1) + 3 \cdot 4^k]}{3}$$
$$4 + 4^2 + 4^3 + \ldots + 4^k + 4^{k+1} = \frac{4(4^k + 3 \cdot 4^k - 1)}{3}$$
$$4 + 4^2 + 4^3 + \ldots + 4^k + 4^{k+1} = \frac{4[4^k(1+3) - 1]}{3}$$
$$4 + 4^2 + 4^3 + \ldots + 4^k + 4^{k+1} = \frac{4(4^k \cdot 4 - 1)}{3}$$
$$4 + 4^2 + 4^3 + \ldots + 4^k + 4^{k+1} = 4\left(\frac{4^{k+1} - 1}{3}\right)$$

**23.** Prove $1^3 + 2^3 + 3^3 + \ldots + n^3 = \left[\dfrac{n(n+1)}{2}\right]^2$ for all positive integers $n$.

Condition 1:

$$1^3 \stackrel{?}{=} \left[\frac{1(1+1)}{2}\right]^2$$

$$1 \stackrel{?}{=} \left[\frac{1(2)}{2}\right]^2$$

$$1 \stackrel{?}{=} (1)^2$$

$1 = 1$ True

Condition 2:

Assume $1^3 + 2^3 + 3^3 + \ldots + k^3 = \left[\dfrac{k(k+1)}{2}\right]^2$ for some positive integer $k$.

Then

$$1^3 + 2^3 + 3^3 + \ldots + k^3 + (k+1)^3 = \left[\frac{k(k+1)}{2}\right]^2 + (k+1)^3$$

$$1^3 + 2^3 + 3^3 + \ldots + k^3 + (k+1)^3 = \frac{k^2(k+1)^2}{4} + \frac{4(k+1)^2(k+1)}{4}$$

$$1^3 + 2^3 + 3^3 + \ldots + k^3 + (k+1)^3 = \frac{(k+1)^2[k^2 + 4(k+1)]}{4}$$

$$1^3 + 2^3 + 3^3 + \ldots + k^3 + (k+1)^3 = \frac{(k+1)^2(k^2 + 4k + 4)}{2^2}$$

$$1^3 + 2^3 + 3^3 + \ldots + k^3 + (k+1)^3 = \frac{(k+1)^2(k+2)^2}{2^2}$$

$$1^3 + 2^3 + 3^3 + \ldots + k^3 + (k+1)^3 = \left[\frac{(k+1)[(k+1)+1]}{2}\right]^2$$

**25.** Prove $\dfrac{1}{1\cdot 2} + \dfrac{1}{2\cdot 3} + \dfrac{1}{3\cdot 4} + \ldots + \dfrac{1}{n(n+1)} = \dfrac{n}{n+1}$ for all positive integers $n$.

Condition 1:

$$\frac{1}{1\cdot 2} \stackrel{?}{=} \frac{1}{1+1}$$

$\dfrac{1}{2} = \dfrac{1}{2}$ True

Condition 2:

Assume $\dfrac{1}{1\cdot 2} + \dfrac{1}{2\cdot 3} + \dfrac{1}{3\cdot 4} + \ldots + \dfrac{1}{k(k+1)} = \dfrac{k}{k+1}$ for some positive integer $k$.

Then

$$\frac{1}{1\cdot 2} + \frac{1}{2\cdot 3} + \frac{1}{3\cdot 4} + \ldots + \frac{1}{k(k+1)} + \frac{1}{(k+1)[(k+1)+1]} = \frac{k}{k+1} + \frac{1}{(k+1)[(k+1)+1]}$$

$$\frac{1}{1\cdot 2} + \frac{1}{2\cdot 3} + \frac{1}{3\cdot 4} + \ldots + \frac{1}{k(k+1)} + \frac{1}{(k+1)[(k+1)+1]} = \frac{k[(k+1)+1] + 1}{(k+1)[(k+1)+1]}$$

$$\frac{1}{1\cdot 2} + \frac{1}{2\cdot 3} + \frac{1}{3\cdot 4} + \ldots + \frac{1}{k(k+1)} + \frac{1}{(k+1)[(k+1)+1]} = \frac{k(k+2) + 1}{(k+1)(k+2)}$$

$$\frac{1}{1\cdot 2} + \frac{1}{2\cdot 3} + \frac{1}{3\cdot 4} + \ldots + \frac{1}{k(k+1)} + \frac{1}{(k+1)[(k+1)+1]} = \frac{k^2 + 2k + 1}{(k+1)(k+2)}$$

$$\frac{1}{1\cdot 2} + \frac{1}{2\cdot 3} + \frac{1}{3\cdot 4} + \ldots + \frac{1}{k(k+1)} + \frac{1}{(k+1)[(k+1)+1]} = \frac{(k+1)^2}{(k+1)(k+2)}$$

$$\frac{1}{1\cdot 2} + \frac{1}{2\cdot 3} + \frac{1}{3\cdot 4} + \ldots + \frac{1}{k(k+1)} + \frac{1}{(k+1)[(k+1)+1]} = \frac{k+1}{k+2}$$

$$\frac{1}{1\cdot 2} + \frac{1}{2\cdot 3} + \frac{1}{3\cdot 4} + \ldots + \frac{1}{k(k+1)} + \frac{1}{(k+1)[(k+1)+1]} = \frac{k+1}{(k+1)+1}$$

**27.** Prove $1 + \dfrac{1}{2} + \dfrac{1}{2^2} + \ldots + \dfrac{1}{2^{n-1}} = 2\left(1 - \dfrac{1}{2^n}\right)$ for all positive integers $n$.

Condition 1:

$$1 \stackrel{?}{=} 2\left(1 - \frac{1}{2^1}\right)$$

$$1 \stackrel{?}{=} 2\left(1 - \frac{1}{2}\right)$$

$$1 \stackrel{?}{=} 2\left(\frac{1}{2}\right)$$

$1 = 1$ True

Condition 2:

Assume $1 + \dfrac{1}{2} + \dfrac{1}{2^2} + \ldots + \dfrac{1}{2^{k-1}} = 2\left(1 - \dfrac{1}{2^k}\right)$ for some positive integer $k$.

Then

$$1 + \frac{1}{2} + \frac{1}{2^2} + \ldots + \frac{1}{2^{k-1}} + \frac{1}{2^{(k+1)-1}} = 2\left(1 - \frac{1}{2^k}\right) + \frac{1}{2^{(k+1)-1}}$$

$$1 + \frac{1}{2} + \frac{1}{2^2} + \ldots + \frac{1}{2^{k-1}} + \frac{1}{2^{(k+1)-1}} = 2 - \frac{2}{2^k} + \frac{1}{2^k}$$

$$1 + \frac{1}{2} + \frac{1}{2^2} + \ldots + \frac{1}{2^{k-1}} + \frac{1}{2^{(k+1)-1}} = 2 - \frac{1}{2^k}$$

$$1 + \frac{1}{2} + \frac{1}{2^2} + \ldots + \frac{1}{2^{k-1}} + \frac{1}{2^{(k+1)-1}} = 2 - \frac{2}{2\cdot 2^k}$$

$$1 + \frac{1}{2} + \frac{1}{2^2} + \ldots + \frac{1}{2^{k-1}} + \frac{1}{2^{(k+1)-1}} = 2 - \frac{2}{2^{k+1}}$$

$$1 + \frac{1}{2} + \frac{1}{2^2} + \ldots + \frac{1}{2^{k-1}} + \frac{1}{2^{(k+1)-1}} = 2\left(1 - \frac{1}{2^{k+1}}\right)$$

**29.** Prove $1 + \frac{1}{4} + \frac{1}{4^2} + \ldots + \frac{1}{4^{n-1}} = \frac{4}{3}\left(1 - \frac{1}{4^n}\right)$ for all positive integers $n$.

Condition 1:

$1 \stackrel{?}{=} \frac{4}{3}\left(1 - \frac{1}{4^1}\right)$

$1 \stackrel{?}{=} \frac{4}{3}\left(1 - \frac{1}{4}\right)$

$1 \stackrel{?}{=} \frac{4}{3}\left(\frac{3}{4}\right)$

$1 = 1$ True

Condition 2:

Assume $1 + \frac{1}{4} + \frac{1}{4^2} + \ldots + \frac{1}{4^{k-1}} = \frac{4}{3}\left(1 - \frac{1}{4^k}\right)$ for some positive integer $k$.

Then

$$1 + \frac{1}{4} + \frac{1}{4^2} + \ldots + \frac{1}{4^{k-1}} + \frac{1}{4^{(k+1)-1}} = \frac{4}{3}\left(1 - \frac{1}{4^k}\right) + \frac{1}{4^{(k+1)-1}}$$

$$1 + \frac{1}{4} + \frac{1}{4^2} + \ldots + \frac{1}{4^{k-1}} + \frac{1}{4^{(k+1)-1}} = \frac{4}{3} - \frac{4}{3 \cdot 4^k} + \frac{1}{4^k}$$

$$1 + \frac{1}{4} + \frac{1}{4^2} + \ldots + \frac{1}{4^{k-1}} + \frac{1}{4^{(k+1)-1}} = \frac{4}{3} - \frac{4}{3 \cdot 4^k} + \frac{3}{3 \cdot 4^k}$$

$$1 + \frac{1}{4} + \frac{1}{4^2} + \ldots + \frac{1}{4^{k-1}} + \frac{1}{4^{(k+1)-1}} = \frac{4}{3} - \frac{1}{3 \cdot 4^k}$$

$$1 + \frac{1}{4} + \frac{1}{4^2} + \ldots + \frac{1}{4^{k-1}} + \frac{1}{4^{(k+1)-1}} = \frac{4}{3} - \frac{4}{3 \cdot 4 \cdot 4^k}$$

$$1 + \frac{1}{4} + \frac{1}{4^2} + \ldots + \frac{1}{4^{k-1}} + \frac{1}{4^{(k+1)-1}} = \frac{4}{3} - \frac{4}{3 \cdot 4^{k+1}}$$

$$1 + \frac{1}{4} + \frac{1}{4^2} + \ldots + \frac{1}{4^{k-1}} + \frac{1}{4^{(k+1)-1}} = \frac{4}{3}\left(1 - \frac{1}{4^{k+1}}\right)$$

**31.** Prove $1 + \frac{3}{4} + \frac{9}{16} + \ldots + \left(\frac{3}{4}\right)^{n-1} = 4\left[1 - \left(\frac{3}{4}\right)^n\right]$ for all positive integers $n$.

Condition:

$1 \stackrel{?}{=} 4\left[1 - \left(\frac{3}{4}\right)^1\right]$

$1 \stackrel{?}{=} 4\left[1 - \frac{3}{4}\right]$

$1 \stackrel{?}{=} 4\left(\frac{1}{4}\right)$

$1 = 1$ True

Condition 2:

Assume $1 + \frac{3}{4} + \frac{9}{16} + \ldots + \left(\frac{3}{4}\right)^{k-1} = 4\left[1 - \left(\frac{3}{4}\right)^k\right]$ for some positive integer $k$.

Then

$$1 + \frac{3}{4} + \frac{9}{16} + \ldots + \left(\frac{3}{4}\right)^{k-1} + \left(\frac{3}{4}\right)^{(k+1)-1} = 4\left[1 - \left(\frac{3}{4}\right)^k\right] + \left(\frac{3}{4}\right)^{(k+1)-1}$$

$$1 + \frac{3}{4} + \frac{9}{16} + \ldots + \left(\frac{3}{4}\right)^{k-1} + \left(\frac{3}{4}\right)^{(k+1)-1} = 4 - \frac{4 \cdot 3^k}{4^k} + \left(\frac{3}{4}\right)^k$$

$$1 + \frac{3}{4} + \frac{9}{16} + \ldots + \left(\frac{3}{4}\right)^{k-1} + \left(\frac{3}{4}\right)^{(k+1)-1} = 4 - \frac{4 \cdot 3^k}{4^k} + \frac{3^k}{4^k}$$

$$1 + \frac{3}{4} + \frac{9}{16} + \ldots + \left(\frac{3}{4}\right)^{k-1} + \left(\frac{3}{4}\right)^{(k+1)-1} = 4 + \frac{-4 \cdot 3^k + 3^k}{4^k}$$

$$1 + \frac{3}{4} + \frac{9}{16} + \ldots + \left(\frac{3}{4}\right)^{k-1} + \left(\frac{3}{4}\right)^{(k+1)-1} = 4 + \frac{3^k(-4 + 1)}{4^k}$$

$$1 + \frac{3}{4} + \frac{9}{16} + \ldots + \left(\frac{3}{4}\right)^{k-1} + \left(\frac{3}{4}\right)^{(k+1)-1} = 4 + \frac{3^k(-3)}{4^k}$$

$$1 + \frac{3}{4} + \frac{9}{16} + \ldots + \left(\frac{3}{4}\right)^{k-1} + \left(\frac{3}{4}\right)^{(k+1)-1} = 4 - \frac{3^{k+1}}{4^k}$$

$$1 + \frac{3}{4} + \frac{9}{16} + \ldots + \left(\frac{3}{4}\right)^{k-1} + \left(\frac{3}{4}\right)^{(k+1)-1} = 4 - \frac{4 \cdot 3^{k+1}}{4 \cdot 4^k}$$

$$1 + \frac{3}{4} + \frac{9}{16} + \ldots + \left(\frac{3}{4}\right)^{k-1} + \left(\frac{3}{4}\right)^{(k+1)-1} = 4\left(1 - \frac{3^{k+1}}{4^{k+1}}\right)$$

$$1 + \frac{3}{4} + \frac{9}{16} + \ldots + \left(\frac{3}{4}\right)^{k-1} + \left(\frac{3}{4}\right)^{(k+1)-1} = 4\left[1 - \left(\frac{3}{4}\right)^{k+1}\right]$$

**33.** Prove $\left(\frac{3}{4}\right)^{n+1} < \frac{3}{4}$ for all positive integers $n$.

Condition 1:

$\left(\frac{3}{4}\right)^{n+1} \overset{?}{<} \frac{3}{4}$

$\left(\frac{3}{4}\right)^{1+1} \overset{?}{<} \frac{3}{4}$

$\left(\frac{3}{4}\right)^{2} \overset{?}{<} \frac{3}{4}$

$\frac{9}{16} < \frac{3}{4}$ True

Condition 2:

Assume $\left(\frac{3}{4}\right)^{k+1} < \frac{3}{4}$ for some positive integer $k$.

Then $\left(\frac{3}{4}\right)\left(\frac{3}{4}\right)^{k+1} < \left(\frac{3}{4}\right)\left(\frac{3}{4}\right)$

$\left(\frac{3}{4}\right)^{(k+1)+1} < \frac{9}{16} < \frac{3}{4}$

**35.** Prove $\left(\frac{1}{4}\right)^{n+1} < \frac{1}{4}$ for all positive integers $n$.

Condition 1:

$\left(\frac{1}{4}\right)^{n+1} \overset{?}{<} \frac{1}{4}$

$\left(\frac{1}{4}\right)^{1+1} \overset{?}{<} \frac{1}{4}$

$\left(\frac{1}{4}\right)^{2} \overset{?}{<} \frac{1}{4}$

$\frac{1}{16} \overset{?}{<} \frac{1}{4}$ True

Condition 2:

Assume $\left(\frac{1}{4}\right)^{k+1} < \frac{1}{4}$ for some positive integer $k$.

Then $\left(\frac{1}{4}\right)\cdot\left(\frac{1}{4}\right)^{k+1} < \frac{1}{4}\cdot\frac{1}{4}$

$\left(\frac{1}{4}\right)^{(k+1)+1} < \frac{1}{16} < \frac{1}{4}$

**37.** Prove $3^n \geq 2n + 1$ for all positive integers $n$.

Condition 1:

$3^n \overset{?}{\geq} 2n + 1$

$3^1 \overset{?}{\geq} 2(1) + 1$

$3 \overset{?}{\geq} 2 + 1$

$3 \geq 3$ True

Condition 2:

Assume $3^k \geq 2k + 1$ for some positive integer $k$.

$3 \cdot 3^k \geq 3(2k + 1)$

$3^{k+1} \geq 6k + 3$

$3^{k+1} \geq 4k + 2k + 2 + 1$

$3^{k+1} \geq 4k + 2(k + 1) + 1 \geq 2(k + 1) + 1$

**39.** Prove $n^2 + n$ is even for all positive integers $n$.

Condition 1:

Is $(n^2 + n) \div 2$ an integer when $n = 1$?

$(1^2 + 1) \div 2$

$(1 + 1) \div 2$

$2 \div 2$

$1$

Condition 2:

Assume $(k^2 + k) \div 2 = m$ for some integers $k$ and $m$.

Then

$[(k + 1)^2 + (k + 1)] \div 2$

$(k^2 + 2k + 1 + k + 1) \div 2$

$(k^2 + 3k + 2) \div 2$

$(k^2 + k + 2k + 2) \div 2$

$(k^2 + k) \div 2 + (2k + 2) \div 2$

$m + 2(k + 1) \div 2$

$m + k + 1$

Since $m$ is an integer and $k$ is an integer, $m + k + 1$ is an integer.

Thus, $[(k + 1)^2 + (k + 1)] \div 2$ is an integer.

**41.** Prove $n^3 + 2n$ is divisible by 3 for all positive integers $n$.

Condition 1:

Is $(n^3 + 2n) \div 3$ an integer when $n = 1$?

$[1^3 + 2(1)] \div 3$

$(1 + 2) \div 3$

$3 \div 3$

$1$

Condition 2:

Assume $(k^3 + 2k) \div 3 = m$ for some integers $k$ and $m$.

Then

$[(k + 1)^3 + 2(k + 1)] \div 3$

$[(k + 1)^2(k + 1) + 2k + 2] \div 3$

$[(k^2 + 2k + 1)(k + 1) + 2k + 2] \div 3$

$(k^3 + k^2 + 2k^2 + 2k + k + 1 + 2k + 2) \div 3$

$(k^3 + 3k^2 + 5k + 3) \div 3$

$(k^3 + 3k^2 + 2k + 3k + 3) \div 3$

$(k^3 + 2k) \div 3 + 3k^2 \div 3 + 3k \div 3 + 3 \div 3$

$m + k^2 + k + 1$

Since $m$ is an integer and $k$ is an integer, $m + k^2 + k + 1$ is an integer. Thus, $[(k + 1)^3 + 2(k + 1)] \div 3$ is an integer.

**43.** Prove $a + ar + ar^2 + \ldots + ar^{n-1} = \dfrac{a(1 - r^n)}{1 - r}$ for all positive integers $n$.

Condition 1:

$$a \stackrel{?}{=} \frac{a(1 - r^1)}{1 - r}$$
$$a \stackrel{?}{=} \frac{a(1 - r)}{1 - r}$$
$$a = a \quad \text{True}$$

Condition 2:

Assume $a + ar + ar^2 + \ldots + ar^{k-1} = \dfrac{a(1 - r^k)}{1 - r}$ for some positive integer $k$.

Then

$$a + ar + ar^2 + \ldots + ar^{k-1} + ar^{(k+1)-1} = \frac{a(1 - r^k)}{1 - r} + ar^{(k+1)-1}$$
$$a + ar + ar^2 + \ldots + ar^{k-1} + ar^{(k+1)-1} = \frac{a - ar^k}{1 - r} + \frac{ar^k(1 - r)}{1 - r}$$
$$a + ar + ar^2 + \ldots + ar^{k-1} + ar^{(k+1)-1} = \frac{a - ar^k + ar^k - ar^{k+1}}{1 - r}$$
$$a + ar + ar^2 + \ldots + ar^{k-1} + ar^{(k+1)-1} = \frac{a - ar^{k+1}}{1 - r}$$
$$a + ar + ar^2 + \ldots + ar^{k-1} + ar^{(k+1)-1} = \frac{a(1 - r^{k+1})}{1 - r}$$

**45.** 36 pennies **47.** 253 pennies **49.** 528 pennies **51.** 1035 pennies **53.** 2485 pennies **57.** $4x^2 + 12xy + 9y^2$ **58.** $25x^4 - 20x^2y + 4y^2$ **59.** $16x^2 + 8xy - 24x + y^2 - 6y + 9$ **60.** $8x^3 + 36x^2y + 54xy^2 + 27y^3$

## Exercise Set 12.5

**1.** Answers will vary. **3.** 1 **5.** No, only factorials of nonnegative numbers can be found. **7.** 14, the number of terms is one more than the exponent **9.** 10 **11.** 1 **13.** 1 **15.** 70 **17.** 28 **19.** $x^3 + 12x^2 + 48x + 64$ **21.** $8x^3 - 36x^2 + 54x - 27$ **23.** $a^4 - 4a^3b + 6a^2b^2 - 4ab^3 + b^4$ **25.** $243a^5 - 405a^4b + 270a^3b^2 - 90a^2b^3 + 15ab^4 - b^5$ **27.** $16x^4 + 16x^3 + 6x^2 + x + \frac{1}{16}$ **29.** $\frac{1}{16}x^4 - \frac{3}{2}x^3 + \frac{27}{2}x^2 - 54x + 81$ **31.** $x^{10} + 100x^9 + 4500x^8 + 120{,}000x^7$ **33.** $2187x^7 - 5103x^6y + 5103x^5y^2 - 2835x^4y^3$ **35.** $x^{16} - 24x^{14}y + 252x^{12}y^2 - 1512x^{10}y^3$ **37.** Yes, $4! = 4 \cdot 3!$ **39.** Yes, $(7 - 3)! = (7 - 3)(7 - 4)(7 - 5)! = 4 \cdot 3 \cdot 2!$ **41.** $m = n$ or $m = 0$ **43.** $x^8, 24x^7, 17{,}496x, 6561$ **45.** $(a + b)^n = \sum_{i=0}^{n} \binom{n}{i} a^{n-i}b^i$ **47.** (0, 10) **48.** (10, 4) **49.** 2, 9 **50.** $2x^3y^5\sqrt{30y}$ **51.** $f^{-1}(x) = \frac{x - 8}{3}$

## Exercise Set 12.6

**1.** A permutation is any ordered arrangement of a given set of objects. **3.** ${}_nP_r$ is a symbol used to represent the number of arrangements of $r$ objects that can be made from a set of $n$ objects. **5.** Answers will vary. Possible answer: Find the product of all natural numbers less than or equal to $n$. **7.** Answers will vary. **9.** 5040 **11.** 11,880 **13.** 17,160 **15.** 7560 **17.** 1 **19.** 20 **21.** 120 **23.** 360 **25.** 3024 **27.** 1 **29.** 1 **31.** 10 **33.** 5 **35.** 35 **37.** 84 **39.** 1 **41.** 56 **43.** 1,000,000 **45.** 1,000,000,000 **47.** 1080 **49.** 210 **51.** 427,518,000 **53.** 40,320 **55.** 120 **57.** 3,628,800 **59.** 495 **61.** 18,564 **63.** 20 **65.** 56 **67.** 593,775 **69.** 20 **71.** 10,000,000,000 **73.** 720 **75. a)** Since the 3-number sequence is ordered this is more like a permutation problem. The fact that repetition is allowed prevents it from being a true permutation problem. **b)** 64,000 **c)** 59,280 **77. a)** 8,000,000 **b)** 6,400,000,000 **c)** 64,000,000,000,000 **79. a)** 3,628,800 **b)** 40,320 **c)** 576 **81.** 40 **83.** 40,600 **85.** 1200 **87.** 64 **92.** 8 **93.** 0, 6 **94.** 5 **95.** 57

## Exercise Set 12.7

**1.** Equally likely outcomes are outcomes that have the same chance of occurring. **3.** Since either $A$ will occur or $A$ will not occur, $P(A) + P(\text{not } A) = 1$. **5.** If two events are mutually exclusive, they cannot occur simultaneously. **7.** All probabilities are between 0 and 1 inclusive. **9.** $\frac{1}{5}$ **11.** $\frac{2}{15}$ **13.** $\frac{4}{5}$ **15.** $\frac{2}{5}$ **17.** $\frac{1}{13}$ **19.** $\frac{1}{52}$ **21.** $\frac{1}{2}$ **23.** $\frac{12}{13}$ **25.** $\frac{2}{13}$ **27.** $\frac{7}{13}$ **29.** $\frac{4}{7}$ **31.** $\frac{3}{70}$ **33.** $\frac{3}{7}$ **35. a)** $\frac{1}{2}$ **b)** $\frac{1}{4}$ **c)** $\frac{1}{4}$ **d)** $\frac{3}{4}$ **e)** $\frac{3}{4}$ **37. a)** $\frac{1}{4}$ **b)** $\frac{1}{2}$ **c)** $\frac{1}{4}$ **d)** $\frac{3}{4}$ **e)** $\frac{1}{2}$ **39. a)** $\frac{1}{3}$ **b)** $\frac{1}{3}$ **c)** $\frac{1}{3}$ **d)** $\frac{2}{3}$ **e)** $\frac{2}{3}$ **41.** $\frac{4}{11}$ **43.** $\frac{6}{11}$ **45.** $\frac{7}{11}$ **47.** 0.7 **49.** 0.9 **51. a)** $\frac{1}{3}$ **b)** $\frac{2}{3}$ **c)** 2:1 **d)** 1:2 **53.** 5:1 **55.** 5:1 **57.** 12:1 **59.** 1:1 **61. a)** 999,999:1 **b)** 99,999:1 **63. a)** $\frac{7}{10}$ **b)** $\frac{3}{10}$ **65.** $\frac{7}{10}$ **67.** 2:3 **69.** 3:7

**72.**

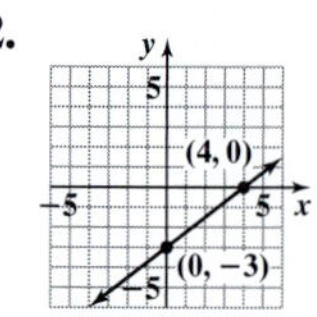


$12x - 16y = 48$

**73.**

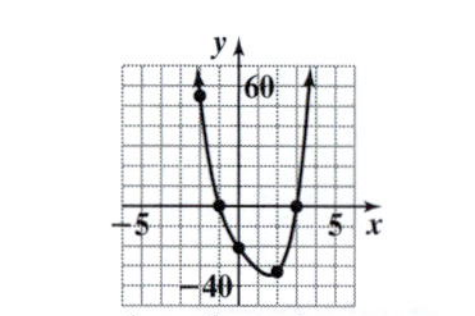


$P(x) = x^4 - 2x^3 + 4x^2 - 14x - 21$

**74.**

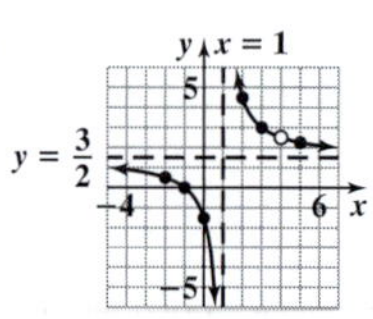


$f(x) = \dfrac{3x^2 - 9x - 12}{2x^2 - 10x + 8}$

**75.**

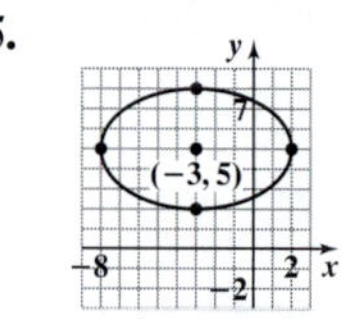


$\dfrac{(x + 3)^2}{25} + \dfrac{(y - 5)^2}{9} = 1$

## Chapter 12 Review Exercises

**1.** 6, 7, 8, 9, 10 **2.** −1, 3, 9, 17, 27 **3.** $6, 3, 2, \frac{3}{2}, \frac{6}{5}$ **4.** $\frac{1}{5}, \frac{2}{3}, \frac{9}{7}, 2, \frac{25}{9}$ **5.** 11 **6.** 4 **7.** $\frac{26}{81}$ **8.** 88 **9.** $s_1 = 7, s_3 = 27$ **10.** $s_1 = 9, s_3 = 38$ **11.** $s_1 = \frac{4}{3}, s_3 = \frac{227}{60}$ **12.** $s_1 = -9, s_3 = -10$ **13.** 32, 64, 128; $a_n = 2^n$

**14.** $-\frac{1}{3}, \frac{1}{9}, -\frac{1}{27}; a_n = (-1)^n(3^{4-n})$ **15.** $\frac{16}{7}, \frac{32}{7}, \frac{64}{7}; a_n = \frac{2^{n-1}}{7}$ **16.** $-3, -7, -11; a_n = 17 - 4n$ **17.** $10 + 13 + 18 = 41$
**18.** $6 + 14 + 24 + 36 = 80$ **19.** $\frac{1}{6} + \frac{4}{6} + \frac{9}{6} + \frac{16}{6} + \frac{25}{6} = \frac{55}{6}$ **20.** $\frac{1}{2} + \frac{2}{3} + \frac{3}{4} + \frac{4}{5} = \frac{163}{60}$ **21.** 29 **22.** 239 **23.** 132 **24.** 841
**25. a)** 10, 14, 18, 22 **b)** $p_n = 4n + 6$ **26. a)** 4, 10, 18, 28 **b)** $a_n = n(n + 3) = n^2 + 3n$ **27.** 5, 8, 11, 14, 17 **28.** $5, \frac{14}{3}, \frac{13}{3}, 4, \frac{11}{3}$
**29.** $\frac{1}{2}, -\frac{3}{2}, -\frac{7}{2}, -\frac{11}{2}, -\frac{15}{2}$ **30.** $-100, -\frac{499}{5}, -\frac{498}{5}, -\frac{497}{5}, -\frac{496}{5}$ **31.** 30 **32.** $-4$ **33.** $\frac{1}{2}$ **34.** 6 **35.** $s_8 = 112; d = 2$
**36.** $s_7 = -210; d = -6$ **37.** $s_7 = \frac{48}{5}; d = \frac{2}{5}$ **38.** $s_9 = -42; d = -\frac{1}{3}$ **39.** $-7, -3, 1, 5; a_{10} = 29, s_{10} = 110$
**40.** $4, 1, -2, -5; a_{10} = -23, s_{10} = -95$ **41.** $\frac{5}{6}, \frac{3}{2}, \frac{13}{6}, \frac{17}{6}; a_{10} = \frac{41}{6}; s_{10} = \frac{115}{3}$ **42.** $-60, -55, -50, -45; a_{10} = -15, s_{10} = -375$
**43.** $n = 13, s_{13} = 442$ **44.** $n = 7, s_7 = 14$ **45.** $n = 11; s_{11} = \frac{231}{10}$ **46.** $n = 10; s_{10} = 180$ **47.** 6, 12, 24, 48, 96
**48.** $-12, -6, -3, -\frac{3}{2}, -\frac{3}{4}$ **49.** $20, -\frac{40}{3}, \frac{80}{9}, -\frac{160}{27}, \frac{320}{81}$ **50.** $-20, -4, -\frac{4}{5}, -\frac{4}{25}, -\frac{4}{125}$ **51.** $\frac{2}{27}$ **52.** 480 **53.** 216 **54.** $\frac{4}{243}$
**55.** 441 **56.** $-\frac{4305}{64}$ **57.** $\frac{585}{8}$ **58.** $\frac{127}{8}$ **59.** $r = 2; a_n = 6(2)^{n-1}$ **60.** $r = 5; a_n = -4(5)^{n-1}$ **61.** $r = \frac{1}{3}; a_n = 10\left(\frac{1}{3}\right)^{n-1}$
**62.** $r = \frac{2}{3}; a_n = \frac{9}{5}\left(\frac{2}{3}\right)^{n-1}$ **63.** 10 **64.** $\frac{25}{6}$ **65.** $-6$ **66.** $-18$ **67.** 32 **68.** $\frac{27}{2}$ **69.** $\frac{25}{6}$ **70.** $-12$ **71.** $\frac{4}{11}$ **72.** $\frac{23}{37}$

**73.** Prove $2 + 5 + 8 + \ldots + (3n - 1) = \frac{n(3n + 1)}{2}$ for all positive integers $n$.

Condition 1: For $n = 1$

$$2 \stackrel{?}{=} \frac{1[3(1) + 1]}{2}$$
$$2 \stackrel{?}{=} \frac{1(3 + 1)}{2}$$
$$2 \stackrel{?}{=} \frac{1(4)}{2}$$
$$2 \stackrel{?}{=} \frac{4}{2}$$
$2 = 2$ True

Condition 2:

Assume $2 + 5 + 8 + \ldots + (3k - 1) = \frac{k(3k + 1)}{2}$ for some positive integer $k$.

Then

$$2 + 5 + 8 + \ldots + (3k - 1) + [3(k + 1) - 1] = \frac{k(3k + 1)}{2} + [3(k + 1) - 1]$$
$$2 + 5 + 8 + \ldots + (3k - 1) + [3(k + 1) - 1] = \frac{3k^2 + k}{2} + 3k + 3 - 1$$
$$2 + 5 + 8 + \ldots + (3k - 1) + [3(k + 1) - 1] = \frac{3k^2 + k + 6k + 6 - 2}{2}$$
$$2 + 5 + 8 + \ldots + (3k - 1) + [3(k + 1) - 1] = \frac{3k^2 + 7k + 4}{2}$$
$$2 + 5 + 8 + \ldots + (3k - 1) + [3(k + 1) - 1] = \frac{(k + 1)(3k + 4)}{2}$$
$$2 + 5 + 8 + \ldots + (3k - 1) + [3(k + 1) - 1] = \frac{(k + 1)[3(k + 1) + 1]}{2}$$

**74.** Prove $2^2 + 4^2 + 6^2 + \ldots + (2n)^2 = \frac{2n(2n + 1)(n + 1)}{3}$ for all positive integers $n$.

Condition 1: For $n = 1$

$$2^2 \stackrel{?}{=} \frac{2(1)[2(1) + 1](1 + 1)}{3}$$
$$4 \stackrel{?}{=} \frac{2(2 + 1)(2)}{3}$$
$$4 \stackrel{?}{=} \frac{2(3)(2)}{3}$$
$4 = 4$ True

Condition 2:

Assume $2^2 + 4^2 + 6^2 + \ldots + (2k)^2 = \frac{2k(2k + 1)(k + 1)}{3}$ for some positive integer $k$.

Then

$$2^2 + 4 + 6^2 + \ldots + (2k)^2 + [2(k + 1)]^2 = \frac{2k(2k + 1)(k + 1)}{3} + [2(k + 1)]^2$$
$$2^2 + 4 + 6^2 + \ldots + (2k)^2 + [2(k + 1)]^2 = \frac{2k(2k^2 + 3k + 1)}{3} + (2k + 2)^2$$
$$2^2 + 4 + 6^2 + \ldots + (2k)^2 + [2(k + 1)]^2 = \frac{4k^3 + 6k^2 + 2k}{3} + \frac{3(4k^2 + 8k + 4)}{3}$$
$$2^2 + 4 + 6^2 + \ldots + (2k)^2 + [2(k + 1)]^2 = \frac{4k^3 + 6k^2 + 2k}{3} + \frac{12k^2 + 24k + 12}{3}$$
$$2^2 + 4 + 6^2 + \ldots + (2k)^2 + [2(k + 1)]^2 = \frac{4k^3 + 18k^2 + 26k + 12}{3}$$
$$2^2 + 4 + 6^2 + \ldots + (2k)^2 + [2(k + 1)]^2 = \frac{(4k^2 + 10k + 6)(k + 2)}{3}$$
$$2^2 + 4 + 6^2 + \ldots + (2k)^2 + [2(k + 1)]^2 = \frac{(2k + 2)(2k + 3)(k + 2)}{3}$$
$$2^2 + 4 + 6^2 + \ldots + (2k)^2 + [2(k + 1)]^2 = \frac{2(k + 1)[2(k + 1) + 1][(k + 1) + 1]}{3}$$

**75.** Prove $1\cdot 2 + 2\cdot 3 + 3\cdot 4 + \ldots + n(n+1) = \dfrac{n(n+1)(n+2)}{3}$ for all positive integers $n$.

Condition 1: For $n = 1$,

$$1\cdot 2 \overset{?}{=} \frac{1(1+1)(1+2)}{3}$$
$$2 \overset{?}{=} \frac{1(2)(3)}{3}$$
$$2 = 2 \quad \text{True}$$

Condition 2:

Assume $1\cdot 2 + 2\cdot 3 + 3\cdot 4 + \ldots + k(k+1) = \dfrac{k(k+1)(k+2)}{3}$ for some positive integer $k$.

Then

$$1\cdot 2 + 2\cdot 3 + 3\cdot 4 + \ldots + k(k+1) + (k+1)[(k+1)+1] = \frac{k(k+1)(k+2)}{3} + (k+1)[(k+1)+1]$$
$$1\cdot 2 + 2\cdot 3 + 3\cdot 4 + \ldots + k(k+1) + (k+1)[(k+1)+1] = \frac{k(k^2+3k+2)}{3} + (k+1)(k+2)$$
$$1\cdot 2 + 2\cdot 3 + 3\cdot 4 + \ldots + k(k+1) + (k+1)[(k+1)+1] = \frac{k^3+3k^2+2k}{3} + \frac{3(k^2+3k+2)}{3}$$
$$1\cdot 2 + 2\cdot 3 + 3\cdot 4 + \ldots + k(k+1) + (k+1)[(k+1)+1] = \frac{k^3+3k^2+2k+3k^2+9k+6}{3}$$
$$1\cdot 2 + 2\cdot 3 + 3\cdot 4 + \ldots + k(k+1) + (k+1)[(k+1)+1] = \frac{k^3+6k^2+11k+6}{3}$$
$$1\cdot 2 + 2\cdot 3 + 3\cdot 4 + \ldots + k(k+1) + (k+1)[(k+1)+1] = \frac{(k^2+3k+2)(k+3)}{3}$$
$$1\cdot 2 + 2\cdot 3 + 3\cdot 4 + \ldots + k(k+1) + (k+1)[(k+1)+1] = \frac{(k+1)(k+2)(k+3)}{3}$$
$$1\cdot 2 + 2\cdot 3 + 3\cdot 4 + \ldots + k(k+1) + (k+1)[(k+1)+1] = \frac{(k+1)[(k+1)+1][(k+1)+2]}{3}$$

**76.** Prove $2 + 6 + 10 + \ldots + (4n-2) = 2n^2$ for all positive integers $n$.

Condition 1: For $n = 1$

$$2 \overset{?}{=} 2\cdot 1^2$$
$$2 = 2 \quad \text{True}$$

Condition 2:

Assume $2 + 6 + 10 + \ldots + (4k-2) = 2k^2$ for some positive integer $k$.

Then

$$2 + 6 + 10 + \ldots + (4k-2) + [4(k+1)-2] = 2k^2 + [4(k+1)-2]$$
$$2 + 6 + 10 + \ldots + (4k-2) + [4(k+1)-2] = 2k^2 + 4k + 4 - 2$$
$$2 + 6 + 10 + \ldots + (4k-2) + [4(k+1)-2] = 2k^2 + 4k + 2$$
$$2 + 6 + 10 + \ldots + (4k-2) + [4(k+1)-2] = 2(k^2 + 2k + 1)$$
$$2 + 6 + 10 + \ldots + (4k-2) + [4(k+1)-2] = 2(k+1)(k+1)$$
$$2 + 6 + 10 + \ldots + (4k-2) + [4(k+1)-2] = 2(k+1)^2$$

**77.** $81x^4 + 108x^3y + 54x^2y^2 + 12xy^3 + y^4$ **78.** $8x^3 - 36x^2y^2 + 54xy^4 - 27y^6$ **79.** $x^9 - 18x^8y + 144x^7y^2 - 672x^6y^3$ **80.** $256a^{16} + 3072a^{14}b + 16{,}128a^{12}b^2 + 48{,}384a^{10}b^3$ **81.** 15,050 **82.** 231 **83. a)** \$36,000, \$37,000, \$38,000, \$39,000 **b)** $a_n = \$35{,}000 + 1000n$ **c)** \$41,000 **d)** \$451,000 **84.** \$102,400 **85. a)** ≈\$2024.51 **b)** ≈\$2463.13 **c)** ≈\$24,041.29 **86.** 150 feet **87.** 35 **88.** 40,320 **89.** 479,001,600 **90.** 32,432,400 **91.** 20 **92.** 66,045 **93.** 1120 **94.** 175 **95.** $\frac{1}{2}$ **96.** $\frac{1}{5}$ **97.** $\frac{4}{5}$ **98.** 0 **99.** $\frac{3}{10}$ **100.** 1 **101.** 1:3 **102.** $\frac{4}{9}$

## Chapter 12 Practice Test

**1. a)** This sequence is neither arithmetic or geometric because the terms do not differ by a constant amount nor by a common multiple. **b)** This sequence is arithmetic because the terms differ by −3. **c)** This sequence is geometric because the terms differ by the multiple $-\frac{1}{2}$. [12.1] **2.** $-\frac{1}{3}, 0, \frac{1}{9}, \frac{1}{6}, \frac{1}{5}$ [12.1] **3.** $s_1 = 3$; $s_3 = \frac{181}{36}$ [12.1] **4.** $5 + 11 + 21 + 35 + 53 = 125$ [12.1] **5.** 184 [12.1] **6.** $a_n = \frac{1}{3} + \frac{1}{3}(n-1) = \frac{1}{3}n$ [12.1] **7.** $a_n = 5(2)^{n-1}$ [12.3] **8.** 15, 9, 3, −3 [12.2] **9.** $\frac{5}{12}, \frac{5}{18}, \frac{5}{27}, \frac{10}{81}$ [12.3] **10.** −40 [12.2] **11.** −20 [12.2] **12.** 12 [12.2] **13.** $\frac{256}{243}$ [12.3] **14.** $\frac{39{,}063}{5}$ [12.3] **15.** 12 [12.3]

**16.** Prove $5 + 10 + 15 + \ldots + 5n = \dfrac{5n(n+1)}{2}$ for all positive integers $n$.

Condition 1: For $n = 1$

$$5 \overset{?}{=} \frac{5(1)(1+1)}{2}$$
$$5 \overset{?}{=} \frac{5(2)}{2}$$
$$5 = 5 \quad \text{True}$$

Condition 2:

Assume $5 + 10 + 15 + \ldots + 5k = \dfrac{5k(k+1)}{2}$ for some positive integer $k$.

Then

$$5 + 10 + 15 + \ldots + 5k + 5(k+1) = \frac{5k(k+1)}{2} + 5(k+1)$$
$$5 + 10 + 15 + \ldots + 5k + 5(k+1) = \frac{5k^2+5k}{2} + 5k + 5$$

$$5 + 10 + 15 + \ldots + 5k + 5(k+1) = \frac{5k^2 + 5k + 10k + 10}{2}$$
$$5 + 10 + 15 + \ldots + 5k + 5(k+1) = \frac{5k^2 + 15k + 10}{2}$$
$$5 + 10 + 15 + \ldots + 5k + 5(k+1) = \frac{5(k^2 + 3k + 2)}{2}$$
$$5 + 10 + 15 + \ldots + 5k + 5(k+1) = \frac{5(k+1)(k+2)}{2}$$
$$5 + 10 + 15 + \ldots + 5k + 5(k+1) = \frac{5(k+1)[(k+1)+1]}{2} \text{ [12.4]}$$

**17.** $x^4 + 8x^3y + 24x^2y^2 + 32xy^3 + 16y^4$ [12.5] **18.** 24 [12.6] **19.** 5040 [12.6] **20.** 28 [12.6] **21.** 330 [12.6] **22.** $\frac{2}{3}$ [12.7] **23.** 499 : 1 [12.7] **24.** $210,000 [12.2] **25.** 364,500 [12.3]

## Cumulative Review Test

**1.** $b = \frac{2A}{h}$ [2.2] **2.** $y = -\frac{11}{3}x + \frac{38}{3}$ [3.5] **3.** Infinite number of solutions [4.2] **4.** $5x^4 + 29x^3 + 14x^2 - 28x + 10$ [5.2] **5.** $(x^2 + 2)(x - 6)$ [5.4] **6.** $(a + b + 4)^2$ [5.5] **7.** $\frac{5x^2 + 14x - 49}{(x+5)(x-2)}$ [6.2] **8.** 500 [6.6] **9.** 3 [7.6] **10.** 5 [7.6] **11.** $-1 \pm i\sqrt{14}$ [8.1] **12.** $\frac{3 \pm \sqrt{309}}{30}$ [8.2] **13.** 5 [5.8]

**14.**

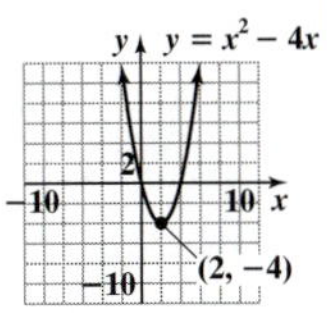


[8.5] **15.** $\frac{1}{2}$ [9.3] **16.**

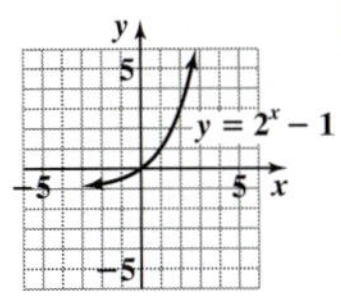


[9.2] **17.** $(x + 6)^2 + (y - 2)^2 = 49$ [11.1] **18.**

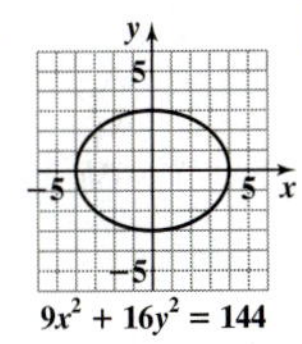


[11.2]

**19.** 18 [12.3] **20.** $\frac{4}{15}$ [12.7]

# Applications Index


**Aeronautical/aerospace**

astronomy, 629
atmospheric pressure, 610, 654
distance to Proxima Centauri, 56
distance to the sun, 55
Earth orbit, 55
flying a plane, 425, 465, 547
gravity on the moon, 538
periods of planets, 420
projectile, 368
return to space, 294
rocket launch, 431
rocket velocity, 379
rockets, 109
Space Shuttle, 139
Voyager, 63

**Agricultural/farming/gardening**

apple orchard, 226
apple production, 546
cornfield, 106
dairy farm, 427
farm divisions, 98
farm equipment, 786
flower garden, 498, 746–747
genetically modified crops, 215–216
grass seed mixture, 107
greatest area, 571
irrigating crops, 428
oil leak, 586
organically grown food, 39
picking apples, 427
picking strawberries, 427
planting a garden, 445
planting flowers, 422, 428
plants and animals, 98
plowing a field, 428
rectangular garden, 367, 526, 538, 586
side of a garden, 497
small orchard, 589
square garden, 367
string beans, picking, 445
sunflower seed, 295
tractors in a field, 427
triangular garden, 98
U.S. farms, 488
vegetable garden, 367, 749
wheat crop, 586
working at a vineyard, 423–424

**Automobile/truck/other vehicle, *See* Transportation**

**Banking**

account comparison, 86
certificate of deposit, 85, 86
checking account, 27
comparing mortgages, 97
compound, 606
compound interest, 183, 316, 524, 609, 647, 654
formula used in banking, 415
home mortgage, 92–93
interest, 227
interest compounded continuously, 642, 645
money market account, 78–79
monthly compounding, 86
mortgage payment, 439
personal loan, 78, 85
refinancing a mortgage, 97
savings accounts, 86, 261, 294, 305, 526, 586, 635, 760
simple interest, 169, 447, 609, 750
teller machines, 809

**Business**

American automakers market share, 76
baby strollers supply, 206
bagel company, 227
battery sales, 537, 570
broom sales, 589
cabinet makers, 428
capital spending, 306
chair sales, 537
changing workforce, 252
clocks, 549
company profits, 539–540
demand:
  for DVD players, 205
  new sandwiches, 205
depreciation, 786
dry cleaning profit, 122
e-commerce, 154
electronics sale, 139
employees, 828
expenditure by oil companies, 586
franchise values, 96
ice cream containers, 85
impact upon businesses, 274
income from department stores, 287
kite supply, 206
lamps, 548
  sales, 537
online advertising, 63
outstanding consumer credit, 609
painting sales, 99
pretzel shop, 436
price reductions, 98
profit, 114, 198, 438, 545, 546, 571, 572, 589
profit graph, 230
promotion, 819
purchasing power of the dollar, 195
rate of discount, 419
sale, 138
small businesses, 27
store sale, 221
supply and demand, 172
SUV sales, 244
table sales, 587
theater ticket sales, 587
toy sales, 643
  comparison, 99
watch sales, 537, 570
working two jobs, 108

**Construction, *See* Housing/construction**

**Education/school**

averaging grades, 122
calculating grades, 119, 124
class of students, 819
classroom grades, 810
college employment, 122
college savings plans, 171
college tuition, 217
distance learning, 447
drug-free schools, 306, 546, 570
earning an A, 99
enrollment, 570
essay questions, 810
exam scores, 140
grade point average, 546
high school play, 206
history tests, 810, 811
learning retention, 629
mathematics tests, 811
minimum grade, 122
passing grade, 122
physics exam average, 99
playground, 546
remembering, 654
SAT scores, 217
scholarships, 810
school enrollment, 545
school theater tickets sales, 570
standardized test, 629
teacher-student committees, 811
test average, 380
test grades, 98

**Environmental**

clean air, 122
disasters, 14–15
Earth days, 497–498
earthquakes, 615–616, 617, 627–628, 629, 636
  shockwaves, 105
expected future population, 610
female population, 229
global population, 216
land area, 57, 91–92, 252–253, 260
most populous countries, 56
natural gas consumption, 231
oil consumption in China, 216
plants and animals, 98
pollen, 97
population density, 55–56, 75
population growth, 138, 786
populous countries, 15
recycling aluminum, 610
recycling plastic, 56
trout in a lake, 645
U.S. and world population, 55
U.S. population, 230, 609
waterfalls, 538
world population, 56, 64, 609, 646

**Farming, *See* Agricultural/farming/gardening**

**Financial/investment**

annuities, 76
balancing a portfolio, 86
bankruptcy, 817
Canadian dollar, 571
comparing debts, 123
comparing investments, 79–80, 86
consumer price index (CPI), 221
currency, 749
depreciation, 635
father and son investing, 86

financing a car, 305
Gross Domestic Product (GDP), 55
inflation, 38, 306
investing a bonus, 138
investing an inheritance, 107
investment comparison, 86
investments, 103–104, 106–107, 108, 261, 264, 420, 786
payment plans, 97
  comparison of, 122
public debt per person, 51–52
retirement:
  account, 215
  funds, 96
  plans, 464
  savings, 829
spending money, 38
stock prices, 27
tax-free investment, 419
two investments, 139, 141

## Gardening, *See* Agricultural/farming/gardening

## Government

CDC spending, 91
federal outlays, 56
Medicaid spending, 195
poverty threshold, 195
social security, 196, 228
  outlays, 375
  receipts, 375

## Health

body mass index (BMI), 86
growth charts, 124
heart disease, 230
heart rate, 170, 194–195
life expectancy, 206, 540–541
sleeping babies, 76
smokers, 221
stress test, 86
typhoid fever, 227
weight loss, 86

## Hobby, *See* Sports/leisure/hobby

## House/home/family

air conditioning, 170
bookcase dimensions, 98
carpeting, 378, 589
cleaning gutters, 428
concrete driveway volume, 85
driveway, 526
fence dimensions, 98
home prices, 171
household expenditures, 172
landscaping, 380, 420
latchkey kids, 39
long distance, 122
long-distance plans, 90
mowing the lawn, 108
painting a wall, 427
patio, 526, 548
picture frame, 367
price:
  of dress, 333
  of mower, 333
pulling weeds, 428
pumping water, 542
rectangular coffee table, 367
rectangular garden, 538
rectangular shed, 367
rent, 571
renting an apartment, 96
room area, 571
room rate, 92
sandbox construction, 85, 98
sauna temperature, 170
shampooing a carpet, 427
silk plant, 368
sump pump rates, 108
telephone bills, 217
water level in a bathtub, 171
water sculpture, 367–368
water use, 610

## Housing/construction

average sale price of homes, 35
buying a house, 820
house sales, 216
installing windows, 429
median home sale price, 196
roofing a house, 429

## Investment, *See* Financial/investment

## Leisure, *See* Sports/leisure/hobby

## Manufacturing

bicycles, 264–265
chairs, 262
demand for steel, 194
furniture, 264
molding machines, 308, 586
soda pop production, 102–103

## Medical/scientific

administering a, 432–433
alternate remedies, 33–34
antibiotic, 438
bacteria, 464, 624, 633–634, 635, 654
bacteria in a Petri dish, 609
blood flowing in an artery, 497
demand for nurses, 646
generic drugs, 646
infant mortality rate, 488, 635
kinetic and potential energy, 514
Medicare premiums, 607–608
medicine mixture, 104–105
optics, 408, 415
physics, 416
prescribing a drug, 646
radioactive decay, 635, 643, 645, 647
radioactive substance, 610
radioisotope, 647
radiometric dating, 648
speed of light, 515
speed of sound, 498
Strontium 90, 645

## Miscellaneous

accidents, 170
acid solutions, 107
advertising, 645
ages, 294
alcohol solution, 108
angle measures, 139
angles:
  of a roof, 274
  of a triangle, 95, 98, 296
antifreeze solution, 108, 109
archeology, 428
area, 332, 333, 377, 378, 379, 446, 548, 549, 587, 602
  of a circle, 169
  of a rectangle, 169, 561–562
  of a triangle, 435
  of circle, 526
  of shaded region, 340–341
army enlistment, 123
arrangements of pictures, 811
art gallery, 734
auctions, 39
auditorium, 776
average acceleration, 419
average velocity, 419
Avon conference, 262
bacteria, 786
  culture of, 829
barrels of oil, 827
billiard table, 734
birdseed, 261, 262
blue dye, 139
board strength, 487
bones and steel, 97
book:
  contract, 27
  profit, 122
bookcases, 810
bouncing ball, 776, 786
bouncing spring, 134
boys' heights, 647
boys' lengths, 207
boys' weights, 207
brass alloy, 262
bridge length/weight, 55
bucket capacity, 85
building a brick wall, 438
building an engine, 547
calendar sale, 138
campaigning, 63
cannonball, 64
  height, 362
car rental comparison, 138
Carbon 14 dating, 607, 609
carbon dating, 464, 647
carbon dioxide, 39
carry-on luggage, 121
castle, 197
cell phones, 534–535
cellular antennas, 363
cellular phones, 39
centenarians, 38, 610
chores, 829
circles, 432
circular steel blade, 369
club officers, 804, 809
clubs, 828
combination locks, 810
committees, 305
complementary angles, 93–94, 95, 139, 260
concentric circles, 728
concert tickets, 264
confidence interval, 499
consecutive numbers, 487
Cub Scouts, 16
cube, 368
currency, 749
currency exchange, 84
current flow, 265
curved mirror, 420
dance floor, 749
diagonal of a box, 498
diagonal of a suitcase, 497, 543
digging a trench, 428
dinner, 655
  bill, 99
dinner seminars, 97
dog sled, 828
dollars and pesos, 438
door prizes, 809
doubling, 609
doubling a penny, 49
drilling a well, 546
drivers needed, 805–806
DVD rentals, 439
economics, 419
electric heater, 547
elevators, 809
elevator limit, 122
energy of an earthquake, 629
envelopes, 828
equation of a parabola, 728
error of estimation, 495
escape velocity, 498
falling ball, 226
falling object, 497, 516, 586
falling rock, 439
Ferris wheel, 728
fertilizer solutions, 138, 261
filling a pool, 428

filling a tank, 429
filling a tub, 423
filling a washtub, 428
filling boxes, 402
filtered light, 786
finalists in a contest, 828
finding the price, 333
firing a cannon, 571
first-class mail, 122
fish tank volume, 487
fitness books, 828
flags, using for messages, 810
flare, 332
focal length, 420
force of attraction, 440
forces on a beam, 265
fractions, 445
free fall, 446
gallons to quarts, 653
Girl Scouts, 96
girls' height, 479, 647
glasses:
in a stack, 776
thickness, 133
glue bottles, 760
goods and services, 16
heating a metal cube, 538
height, 305
height above sea level, 171
helipad area, 85
Heron's formula, 499
history honor society, 95
homicides, 635
Hooke's law, 438
hot dog stand sales, 104
hydrogen peroxide solutions, 107, 231, 264
illumination of a light, 487
immigration, 166
impedance, 504, 507–508
incandescent bulbs, 99
inclined plane, 306
injected dye, 786
inscribed triangle, 526
insurance, 419
policy, 123
rates, 228
integers, 526
intensity:
of illumination, 440
of light, 439, 648
intersecting roads, 750
ISBN numbers, 810
janitorial service, 547
jellyfish tanks, 428
ladder, 516
laundry costs, 95
lavender oil, 261
lawn fertilizer, 261
letters, 829
license plates, 801–802, 810, 812
life insurance, 221
lighting, 435
limousines, 810
line-up, 316
lithotripter machine, 734
logs, 776
lotteries, 810, 811, 819
magnitude of sound, 636
mail volume, 263
making a box, 368
manuscripts, 808
copying, 380
Martha Stewart revenue, 242–243
mass, 786
meeting at a restaurant, 108
meeting for a picnic, 447
melting ice, 434
cubes, 446
microwaves, 810
milk carton machine, 109
milk tank, 428
minimum product, 571
mixing:
coffee, 139
nuts, 107
solutions, 255–256, 294, 295
money, 776, 784, 828
morning shows, 172
moving sidewalk, 429
mowing lawns, 427
museum paintings, 809
new suit, 95
newspaper:
preferences, 17
readers, 39
office security, 809
oil refinery, 428
oranges, 170
ordering lunch, 97
outfits, 810
packages, 777
on a boat, 113–114
packing spaghetti, 106
painting, 428
sales, 108
paper use, 230
parking garage, 122
pencils, 713
pendulum, 776, 785, 786, 828
pennies in a piggie bank, 793
perimeter, 303
and area, 513
of a square, 169
period of a pendulum, 494, 497
Persian rug, 760
pH of a solution, 629
phone calls, 439
phone booth, 140
photos, 538, 829
photocopy costs, 263
piece of wood, 749
Ping-Pong ball, 776, 786
pizzas, 806–808
plywood guarantee, 133
points of intersection, 728
positive numbers, 586
presorted first-class mail, 122
pressure:
and volume, 438
on an object, 440
printing checks, 427
prizes, 809, 810
projectile, 368
pumping water, 429, 542
quadratic mean, 499
records, 810
rectangle, 586
rectangular corral, 562–563
rectangular region, 749
recycling aluminum, 610
restaurants, 806
right angle, 274
rock concert, 170, 296
rolling a die, 829
salt solution, 141
scrap value, 786
seats in a row, 828
security, 809
shaded area, 728
shipments of LCD monitors, 172
silver alloy, 262
similar triangles, 412
six solutions, 538
snail races, 106
social security numbers, 809
sound pressure, 629
spelling bee, 305
springs, 516
square, 379
stack of chips, 786
stack of money, 786
standard deviation, 479
sulfuric acid solutions, 107, 264
sum of numbers, 266, 295
supplementary angles, 94, 95, 260
surface area, 603
and volume, 526
suspension bridge, 431
swimming pool, 139
swinging on a vine, 776
tangent line, 197
tapping glasses, 811
teenagers using illicit drugs, 196
telephone:
numbers, 810–811
pole, 379, 514
textbook search committee, 810
throwing:
balls, 38, 328–329, 535–536, 571, 589
objects, 570–571
rocks, 539
tickets, 648
tossing a ball, 587
tossing coins, 820
totem pole, 427
traffic cones, 544
tree wires, 363
trends, 244
triangle, 264, 361–362
triangle and circle, 586
tunnel, 728, 734
tunnel through a mountain, 734
UPS packages, 121
value of Manhattan, 646
velocity, 38, 123
velocity of an object, 456
videos, 713
vinegar solutions, 107
violins, 207
volume, 316, 332, 376, 377–378, 516
of a cylinder, 170
of a pyramid, 439
wading pool, 526
walkie-talkie range, 106
washing windows, 447
water acidity, 123
water bill, 439
water level, 170
water through a fire hose, 478
wattage rating, 439, 447, 590
wave action, 456
wave motion, 498
weed killer, 261
weight limit, 122
weight of an object, 439
weight of girls, 183
whispering gallery, 734
window washers, 427
winning a contest, 829
working together, 427
writing contests, 15
yards, 653

## Nutrition/food/diet

animal diet, 262
banana splits, 810
bananas, 274
bread loaf, 786
burning calories, 228
candy mixture, 107
candies in a stack, 776
cashews and pecans, 266
cereals, 811
coffee, 261
mixture, 508
fat content, 260
hamburgers, 811
hors d'oeuvres, 811
horseradish sauces, 107
juice, 262
meatloaf, 108
milk, 261
milk mixture, 109
mixed nuts, 142
pizza, 811
quiche lorraine, 262
sandwiches, 829
soft drinks, 645
Splenda® products, 646–647

**Oceanic/boating**

depth of ocean trenches, 20
ocean salinity, 109
sailboats, 399, 431
ships, 547
speed of the Titanic, 85
submarine depth, 133
submarine dive, 27
submarine personnel, 263
wave action, 787

**Salary**

annual salary, 333
bonus, 264
commission salary, 138
income, 216–217
income mark, 216
lecturer salary, 206
minimum wage increase, 97
new job, 242
personal income, 96, 196
retirement income, 229
salary increase, 776
salary plus commission, 183, 261
starting salary, 776, 827, 828
waitress pay, 96
weekly salary plus a commission, 253–254

**Scientific, *See* Medical/scientific**

**School, *See* Education/school**

**Sports/leisure/hobby**

baseball, 379, 560–561, 572, 750
  cards, 802
  diamond, 497
  sale, 142
  hoop, radius of, 497
batting orders, 811
bicycles, 747, 786
  profit, 182
  riding, 38, 105, 368, 431, 547
  shop, 368
  travel, 141
Boston Marathon, 76
bowling alley rates, 114
bungee jumping, 786
canoe:
  speed, 253
  weight limit, 139
canyon hiking, 106
college football bowl games, 263
cross-country skiing, 251
dartboard area, 85
Daytona 500 wins, 260
deck of cards, 814, 815–816
exercise, 262
fitness center, 139, 140
football, 264, 430, 750
  revenue in the NFL, 56–57
golf, 96
  golf club discount, 141
  golf course, 154
Green Monster wall (baseball), 494
health club costs, 141
hiking, 105, 106
hockey tickets, 648
horse racing, 316–317, 819, 820
horse riding, 254–255, 430
ice hockey, 294
ice rink, 749
inclined treadmill, 205
jogging, 139
jump shot, 332
NASCAR Nextel Cup, 14, 16
NASCAR racing, 263
Olympic medals, 97–98
pitching a ball, 439
pool table, 760
quiniela bets, 810
race cars, 109
race tracks, 809
rollerblading, 447
  and jogging, 431
rolling a die, 816–817
rowing speed, 261
runner, 547
running:
  home, 101–102
  races, 15, 438
sailboat, 749
  race, 430
skating rink, 332
Summer Olympics, 263
Super Bowls, 264
  commercials, 172
swimming, 134
tennis ball, 439
  and snowball, 750
tennis tournament, 819
tent wires, 368
theme parks, 260
throwing a horse, 538
top-10 finishes on PGA tour, 263
treadmill, 205
triangular sail, 367
triangular tent, 367
walking, 106, 155–157
  and jogging, 263, 296
  speed, 645
water bike, 425–426

**Tax**

annual tax refund, 647
equivalent taxable rate, 86
estimated taxes, 27
hotel tax rate, 97
income taxes, 122, 217
Internal Revenue Service, 262
sales tax, 96
tax collections, 52–53
tax returns, 287
tax strategy, 107
tax tables, 115

**Technology/electronics**

compact disks (CDs), 803
communications satellite, 734
downloaded songs, 545
electric surcharge, 446
electrical resistance, 439
electrostatic force, 436
Internet sales, 464
photo printers, 260
power gain of an amplifier, 635
purchasing a computer, 333
resistors, 408
smart handheld devices, 653
sound systems, 809
standard TVs, 548
total resistance, 414, 419
widescreen TVs, 548

**Transportation**

air speed, 294
Amtrak:
  auto train, 429
  budget, 97
  expenses, 194
antifreeze, 582
auto registration, 206
automobile jack, 408
balloon flights, 105
bus pass, 95
business jets, 165
car:
  antitheft system, 97
  carrier, 547
  in mud, 368
  travel, 262
driving speed, 106
flying a plane, 424, 445, 465, 547
flying speed, 261
forces on a car, 498
gas mileage of a car, 206
gas price, 98
George Washington Bridge tolls, 96
gondola, 429
helicopter ride, 430–431
inclined railroad, 431
motorboat ride, 541
motorcycle sales, 546
octane ratings, 107
oil leak, 586
price of cars, 333
public transportation, 38
railroad traffic, 63
refueling a jet, 108
road construction, 262
Route 66, 108
ships, 547
ships at sea, 101
skid marks, 478
speed of a car, 171, 226
stopping distance, 170, 369, 439, 526
subway trains, 430
SUVs, 629
  sales, 244
taxi operating costs, 183
toll bridge, 96
trains, 138, 198
  in Alaska, 139
truck rental, 95, 158, 261
two cars, 142, 265
value:
  of a Jeep, 606
  of an SUV, 610

**Travel**

Amtrak fares, 261
boating trip, 430
Buffalo-South Bend trip, 426
bicycle travel, 141
car travel, 430
canoe trip, 587
China travel, 221
Dallas-El Paso travel, 430
distance, 438
distance to Calais, 109
distance traveled, 171
distance versus time, 182
Hawaiian trip, 108
long drive, 546
moving vehicles, 488
Nashville-Baltimore car travel, 547
Red Rock Canyon, 547
San Antonio-Austin car travel, 547
skyline drive, 430
Texas driving, 263
travel time, 586
travelers staying at Marriott, 299–300
traveling by boat, 445
trip in a park, 588
trip to work, 108

**Weather/temperature**

Celsius temperature, 169
extreme temperatures, 21, 27
Fahrenheit temperatures, 545
*Gold* (film), 26
New England snowstorm, 264
probability of rain, 820
temperature change, 26

# Index


**A**

Absolute value inequalities, solving, 284–285
Absolute values, 58
  defined, 18
  equations containing, 125–134
  evaluating, 18–19
  geometric interpretation of, 125
  inequalities containing, 125–134
  of nonzero numbers, 19
  radicals using, 453
  systems of linear equations containing, 283–284, 293
Addition:
  of complex numbers, 502, 511
  of polynomials, 302–303, 370
  of radicals, 472–474, 510
  of rational expressions, 441
    with a common denominator, 391–393, 441
  of real numbers, 19, 58
Addition method, solving systems of linear equations using, 237–240, 289
Addition property of equality, 68, 135
Additive identity element, 23–24
Additive inverses, 17, 58
Algebra:
  applications of, 87–110
  of functions, 208–214
Algebraic equations, translating verbal statements into, 87–89
Algebraic expressions, 6, 57
  translating verbal statements into, 87–89
Alternating sequences, 766
Antilogarithms, 626–628, 651
  graphing calculators, 626
  significant digits, 627
Application problems, study tips, 94
Argument, logarithms, 618
Arithmetic sequence:
  defined, 771, 822
  finding the common difference in, 771, 822
  finding the *n*th term of, 772, 822
Arithmetic series:
  defined, 722, 822
  finding the *n*th partial sum of, 722–724, 822
Associative property, 23–24, 59
Asymptotes, 401
Attending class, 3
Augmented matrix, 266–267, 290
  triangular form, 267, 290
Axis of symmetry, 556, 584

**B**

Base, 28
Base 10 logarithms, 624, 651, *See also* Common logarithms
*between*, use of term, 8
Binomial coefficients, 797–798, 823
Binomial theorem, 796–800, 823
  defined, 798
  using, 797–798
Binomials, 370
  dividing polynomials by, 319–321
  finding the square of, 311–312
  multiplying by polynomials, 309–310
  square of, 311–312, 370
Boundary values, 573
Boundedness Theorem, 690–691, 711
Break-even point, 683

**C**

Calculators, *See also* Graphing calculators; Scientific calculators
  checking solutions by substitution, 70
  learning to use, 5
  reviewing screen for errors, 35
Cartesian coordinate system, defined, 144, 222
CD Lecture Series Videos, 5
Center, circles, 722, 756
Change of base formula, 639–640, 652
Circles:
  center, 722, 756
  with centers at ($h$, $k$), graphing, 724–725, 756
  with centers at the origin and radius $r$, 723, 756
  with centers at the origin, graphing, 722–723, 756
  standard form of the equation of, 722
Class, attending and participating in, 2–3
Coefficient, 66–67, 135
Collinear points, 146
Combinations, determining the number of, 804–808, 824
Combined variation
  defined, 436
  problems, solving, 436
Common difference, 771, 822
Common factor, factoring out, 337
Common logarithms, 624–630, 651
  antilogarithms, 626–628, 651
  defined, 624–625
  and graphing calculators, 625
  of powers of 10, 624–625
Common ratio, 778, 822
Commutative property, 23–24, 59
Complementary angles, 93–94, 136
Completing the square:
  defined, 520
  solving quadratic equations by, 520–525
Complex fractions:
  defined, 403, 442
  main fraction line, 403
  recognizing, 403
  simplifying:
    by multiplying by a common denominator, 404–405, 442
    by simplifying the numerator and denominator, 405–406, 442
Complex numbers, 500–508
  adding, 502, 511
  conjugate of, 511
  defined, 511
  dividing, 503–504, 512
  examples of, 501
  multiplying, 502–503, 511
  recognizing, 500–501
  subtracting, 502, 511
Complex zeros of polynomial functions, 683–687, 710
Composite functions, 592–594, 648
  defined, 592
Compound event:
  defined, 815
  probability of, 815–816
Compound inequality:
  defined, 115, 137
  examples of, 116
  involving *and*, 115–118
  involving *or*, 119–120
  writing the solution to, 118
Compound interest formula, 78–79, 136
Conditional equations, 72, 135
Confidence interval, 499
Conic sections:
  circles, 724
    with centers at ($h$, $k$), graphing, 724–725
    with centers at the origin, graphing, 722–723
  defined, 718, 755
  ellipses, 730–732
    graphing, 729–735
  hyperbolas, 735–739
  parabolas, 557, 560–566, 718–720
Conjugate, 482–483
Conjugate of a complex number, 503, 511
Connected mode, graphing calculators, 383
Consistent system of equations, 233–234, 288
Constant, 6, 57, 67
Constant function, 178
Constant of proportionality, 432
Constraints, 283
Continuous curve, 659, 709
Continuous functions, 658–665
Contradictions, 72, 135
Coordinates, 144
Counting numbers, 6, 10, 58

Counting principle:
- combinations, determining the number of, 804–808, 824
- defined, 801, 824
- permutations, determining the number of, 802–804, 824
- possible outcomes, determining the number of, 801–802

Cramer's rule, 275–277
- for a system of equations in three variables, 292
- for systems of linear equations, 291
- using with systems in three variables, 278–280

Cross multiplication, and proportions, 412
Cube root function, 451, 509
Cube roots, 451, 508
- defined, 60, 451

Cubic equations, 360
Cubic polynomials, 298

## D

Decay formula, 642
Decreasing sequences, 765, 821
Degree of a polynomial equation, 358
Degree of a term, 67, 135, 298, 370
Denominator:
- rational exponents, 458
- rationalizing, 480
  - using the conjugate, 482–483

Dependent system of equations, 288
Dependent systems, recognizing, 248–249
Dependent variables, 151, 222
Descartes, René, 144
Descartes' Rule of Signs, 676–677, 680, 710
Determinants:
- of a $2 \times 2$ matrix, 275, 291
- of a $3 \times 3$ matrix, 277, 291
- defined, 275
- expansion by the minors of the first column, 277, 291
- minor, 277
- solving systems of equations using determinants, 275

Difference of functions, 209
Difference of two cubes, 349, 373
Difference of two squares, 312, 346–347, 373
Direct variation:
- defined, 432, 443
- problems, solving, 432–433

Discontinuity, point of, 659
Discontinuous functions, 658–665, 709
Discriminant, 345, 532
Distance formula, 101, 136, 721–722, 755
Distributive property, 23–24, 59
- extended, 24

Distributive property, expanded form, 309
Division:
- of complex numbers, 503–504, 512
- of polynomials, 317–324, 371
  - by a binomial, 319–321
  - by a monomial, 317–319
  - Remainder Theorem, 323–324, 371
  - using synthetic division, 321–323
- of radicals, 480–488
- of rational expressions, 387–388, 441
- of real numbers, 22–23, 59

Domain, 158–159, 222
- logarithmic functions, 614
- rational exponents, 440
- rational functions, 382–384

Dot mode, graphing calculators, 383
Double negative property, 21, 58

## E

Einstein, Albert, 515
Elements, 6, 57
- of a matrix, 266, 290

Elimination method, solving systems of linear equations using, 237–240, 289
Ellipses, 6
- with centers at $(h, k)$, 731–732, 757
- with centers at the origin, 730, 756
- constructing, 729
- defined, 729
- foci, 756
- graphing, 729–735
- and graphing calculator, 732
- major axis, 730
- minor axis, 730

Empty set, 7, 57, 113
Endpoints, 7–8
Equality:
- addition property of, 68, 135
- multiplication property of, 68–69, 135
- properties of, 66, 135

Equation(s):
- conditional, 72, 135
- consistent system of, 233–234
- containing fractions, 71–72
- containing absolute values, 125–134
- contradictions, 72, 135
- cubic, 360
- defined, 68, 135
- equivalent, 68
- first-degree, 146, 222
- of the form $|x| = \; < a, a > 0$, 126–127
- of the form $|x| = \; > a, a > 0$, 128–129
- of the form $|x| = \; > a$, or $|x| = \; < a, a < 0$, 129–130
- of the form $|x| = \; > a$, or $|x| = |y|$, 130–131
- of the form $|x| = \; > a$, or $|x| = \; > 0$, or $|x| = \; < 0$, 130
- of the form $|x| = a, a > 0$, 126
- of the form $x = a$, and $y = b$, 178
- of a horizontal line, 178
- identities, 72
- inconsistent system of, 233–234, 248–249, 288
- quadratic, 358, 373
- solution set, 68
- solutions, 68
- solving by factoring, 374
- solving for a variable in, 80–82
- third-degree, 360
- translating verbal statements into, 87–89
- understanding the concepts to solve, 73
- of a vertical line, 178

Equations quadratic in form, 549–552, 584
- defined, 549

Equivalent equations, 68
Even index, 452
Even multiplicity:
- zeros of, 669

Even roots, 451
- examples of, 452
- summary of, 453

Even-degree polynomial function, 675–676
Events:
- defined, 812, 824

Examinations:
- studying for, 4
- taking, 4

Expanded form of the distributive property, 309
Expansion by the minors of the first column, 277, 291
Experiment, 801, 824
Exponential equations:
- application, 633–634
- solving, 630–633

Exponential expressions, 28
- changing to radical expressions, 457–464
- converting to a logarithmic expressions, 612
- evaluating, 28–29

Exponential functions:
- comparing graphs of logarithmic functions and, 614–615
- defined, 603–604, 650
- examples of, 604
- graphing, 603–611
- graphs of, 604
- solving applications of, 606–608

Exponential growth (or decay) formula, 642
Exponents, 28, 40–49
- negative exponent rule, 41–42, 60
- power rule, 43–44
- product rule, 60, 308–309
- product rule for, 40–41
- quotient rule, 60
- quotient rule for, 41
- raising a power to a power, 43–44, 46, 60
- raising a product to a power, 44–45, 46, 61
- raising a quotient to a power, 45–46, 61
- rules of, 459–462, 509
- summary of rules of, 46
- writing/copying, 28
- zero exponent rule, 43, 46, 60

Expressions, 6, 57
Extended distributive property, 24
Extraneous roots, 410
Extraneous solutions, 410

## F

Factor by grouping, 330–331, 371

Factor theorem, 678, 710
Factorials, evaluating, 796
Factoring:
checking, 336
compared to multiplying, 327
difference of two squares, 346–347
expressions, with rational exponents, 462–463
general review of, 354–355
perfect square trinomials, 347–348
polynomials, 328–331
using a combination of techniques, 354–355
solving equations by, 374
special factoring formulas, 346–353
sum and difference of two cubes, 348–350
by trial and error, 337–339
trinomials, 335–346
Factors, 28, 60
Finite sequences, 765, 821
Finite series, 766
Finite sets, 6
First-degree equations, 146, 222
FOIL method, 310, 337
Formulas:
defined, 77, 135
distance formula, 101, 136, 721–722, 755
midpoint formula, 721–722, 755
motion formula, 100–101
solving for a variable in, 80–82, 542–543
Fraction bar, as grouping symbol, 32
Fractions:
sign of, 23
solve equations containing, 71–72
in a system of equations, 247
Function notation, 162–164
Functions:
algebra of, 208–214
applications of, 179–180
applications in daily life, 164–167
constant, 178
continuous, 658–665
identifying, 659
decreasing function, 658
defined, 158–160, 222
difference of, 209
discontinuous:
identifying, 659
function defined piecewise (piecewise function), 709
increasing function, 658
local maximum, 658
local minimum, 658
operations on, 209
polynomial, 665–675, 709
product of, 209
quotient of, 209
real zeros of the function, 658
recognizing, 158–160
step, 661, 709
sum of, 209
graphing, 210–214
$f(x) = a(x - h)2 + k$, 566–568

## G

Gap, 659
Gauss, Karl Friedrich, 777
GCF, *See* Greatest common factor (GCF)
General motion problem formula, 136
Geometric sequence:
defined, 778
finding the common ratio in, 778–779
finding the $n$th partial sum of, 780–781
finding the $n$th term of, 779–780
Geometric series, 822
defined, 780
infinite, identifying, 781
studying applications of, 783–785
Graph of a function or relation, defined, 160
Graphing calculators:
antilogarithms, 626
binomial coefficients, 798
checking factoring problems, 340
circles, 724
common logarithms, 625
connected mode, 383
displaying graphs of inequalities, 220
dot mode, 383
ellipses, 732
entering numbers in scientific notation on, 53
estimating intercepts of a graph, 177
evaluating expressions on, 34–35
evaluating roots on, 31–32
exponential equations, 632–633
exponential expressions, 29
exponential functions, 605
factorials, 796
finding roots or expressions with rational exponents on, 462
finding the point of intersection of two functions, 181
functions, 702, 706
graphing a polynomial function on, 301
greatest integer function, 662, 709
hyperbolas, 738–739
INTERSECT feature, 181
intersection of two graphs, 235
inverse functions, 598–599
logarithmic equations, 632–633
and matrices, 272, 279
natural logarithms and natural exponential functions, 644
nonlinear systems of equations, 748
permutations, 808
polynomials, 301
quadratic equations, 363
radical equations, 491
"square the axes" feature, 598
sums, differences, products, and quotients of functions, 214
TABLE feature, 151
TRACE feature, 151, 181
using, 150–151
window, 150
ZOOM feature, 181
Graphs, 144–158
bar graphs, 212–213
Cartesian coordinate system, plotting points in, 144–145
common errors, avoiding, 150
defined, 146, 222
drawing by plotting points, 145–148
of exponential functions, 603–611
graphing calculator, using, 150–151
improving the quality of, 146
interpreting, 152–153
line graph, 211–212
of linear inequalities, 218–221
of logarithmic functions, 613–614
of nonlinear equations, 148–150
of polynomial functions, 300–301
of quadratic functions, 555–572, 558–559
stacked line graph, 213
sum of functions, 210–214
translations of, 184–185
Greatest common factor (GCF), 327, 371
Greatest integer function, 662, 709
Grouping method, factoring trinomials using, 341–342

## H

Higher roots, evaluating, 30–32
Holes, 698–699, 701, 712
Homework, 2–3
Horizontal asymptote, 696–697, 701
defined, 700
locating, 699–700
rational functions, 696–697
at $y = 0$, 699
Horizontal line, equation of, 178
Horizontal line test, 374
Hyperbolas, 735–739
asymptotes, 736, 758
with centers at the origin, 737, 757
defined, 735, 757
foci, 735, 757
and graphing calculator, 738–739
transverse axis, 735, 758
vertices, 735
Hypotenuse, 362

## I

Identities, 72, 136
Identity property, 23–24, 59
Imaginary number, 500, 511
Imaginary unit, 500, 511
Inconsistent systems, 233–234, 288
recognizing, 248–249
Increasing sequences, 765, 821
Independent events, 801, 824
Independent variables, 151, 222
Index, 30
radical expressions, 449
rational exponents, 458
Index of summation, 767, 821

Inequalities:
compound, 115–118
containing absolute values, solving, 125–134
defined, 110
identifying/using, 7–8
interval notation, 9
number line, 111–112
order (sense) of, 110
set builder notation, 8–9
solving, 110–111
properties used for, 110
Inequality symbols, 7, 58, 110
reversing the direction of, 111
Infinite geometric series, *See also* Geometric series
finding the sum of, 782–783
identifying, 781
sum of, 822
Infinite sequences, 764, 766, 821
Infinite series, 766
Infinite sets, 6
Instructor, 94
Integers, 6, 10
Intermediate Value Theorem, 689, 711
Intersection, 9, 58
solution sets, 116, 137
Interval notation, 9, 111–113
Inverse functions, 596–599, 649
Inverse logarithm, *See also* Antilogarithms
defined, 626
Inverse property, 23–24
Inverse variation:
defined, 433, 443
problems, solving, 433–435
Irrational numbers, 58, 449
defined, 10
Irrational zeros, estimating, 694

## J

Joint variation:
defined, 435, 443
problems, solving, 435–436

## L

LCD, *See* Least common denominator (LCD)
LCM, *See* Least common multiple (LCM)
Leading coefficient, 298, 370
Leading term, 370
Least common denominator (LCD), 71–72, 393–394, 416
Least common multiple (LCM), 71–72
Legs, triangles, 362
Like radicals, 472, 510
Like terms, 67, 135
combining, 66–68
Linear equations:
defined, 135, 146, 222
in one variable, 68
solving graphically, 180–181
point-slope form, 199–208, 204
defined, 199, 225
recognizing parallel and perpendicular lines, 201–204
understanding, 199–200
using to construct models from graphs, 200–201
slope-intercept form, 184–198, 204
finding the slope of a line, 185–187
graphing equations using the slope and the *y*-intercept, 189
graphs, translations of, 184–185
slope, recognizing as a rate of change, 187–188
using to construct models from graphs, 190–191
writing equations in, 188–189
solving, 68–71, 135
steps in, 69
solving systems of:
by substitution, 235–237
using the addition method, 237–240, 289
standard form, 204, 223
systems of:
applications and problem solving, 252–266
solving using determinants and Cramer's Rule, 275–281
solving using matrices, 266–274
in three variable, solving systems of, 245–252
in two variable, solving systems of, 233–245
Linear functions, 223
defined, 174
graphing, 173–174
using intercepts, 174–177
Linear graph, 146
Linear inequalities:
graphing, 218–221
solving, 110–124
solving systems of, 282–287
Linear polynomials, 298, 370
Linear programming, 292
defined, 283
problems, solving, 283–284
Linear systems:
sports applications, inflatable boats, 258–259
in three variables, using to solve applications, 257–258
Logarithmic equations, solving, 630–633
Logarithmic functions, 611–618, 650
comparing graphs of exponential functions and, 614–615
domain, 614
graphs of, 613–614
solving applications of, 615–616
Logarithms, 650
argument, 618
avoiding common errors, 621
common, 624–630
defined, 611, 613
power rule, 619–620
product rule, 618, 651
properties of, 618–621, 618–624
quotient rule, 618–619
Lower bound:
for real zeros, 690–691, 711
Lower limit of summation, 767, 821

## M

Main fraction line, 403
Major axis, ellipses, 730
Math anxiety, 2
Mathematical induction:
principle of, 789, 823
proving statements using, 789–795
Mathematical model, 77, 135
Mathematics text, reading, 2
Matrix:
augmented, 266–267, 290
copying numbers from one matrix to another, 269
defined, 266, 289
elements of, 266, 290
square, 266, 290
Midpoint formula, 721–722, 755
Minor axis, ellipses, 730
Minor determinant, 277, 291
Mixture problems, 103–104, 136, 137
Monomials, 370
dividing polynomials by, 317–319
factoring from a polynomial, 371
multiplying by polynomials, 308–309
multiplying by monomials, 308–309
Motion formula, 100–101
Motion problems, 443, 541–542
defined, 101
solving, 100–101, 424–426
Multiplication:
of complex numbers, 502–503, 511
of polynomials, 308–317, 370
of radicals, 474–477, 510
of rational expressions, 385–386, 441
of real numbers, 23
Multiplication property of equality, 68–69, 135
Multiplicative identity element, 23–24
Multiplicative inverse, 23–24
Multiplicative property of zero, 22, 59
Multiplicity, 668–669, 710
even/odd, zeros of, 669
Mutually exclusive events, 815–816

## N

*n* factorials, 796, 823
Natural exponential equations, 640–641
Natural exponential expressions, properties of, 640
Natural exponential function, 652
defined, 637
identifying, 637
Natural logarithm equations, solving, 640–641
Natural logarithmic function, 652
defined, 637
identifying, 637–638
Natural logarithms, 652
change of base formula, 639–640, 652
defined, 637
in exponential form, 637
finding on a calculator, 637–639

properties for, 652
properties of, 640
Natural numbers, 6, 10–11, 58
Negative exponent rule, 41–42, 46, 60
Negative leading coefficient, 301
Negative number, square root of, 500
Negative real zeros, 676–677, 710
Negative reciprocal, 202
Negative slope, 186, 224
Nonlinear equations:
defined, 222
graphs of, 148–150
Nonlinear systems of equations, 742–750
applications, 746–747
defined, 742, 759
and graphing calculator, 748
solving using addition, 745–746
Nonlinear systems of inequalities, 751–755
second-degree inequalities, graphing, 751–752
solving graphically, 752–753, 759
Note-taking, 3
$n$th partial sum, 722–724, 821, 822
$n$th root, defined, 60
Null set, 7, 57, 113
Number line, graphing solutions on, 111–112
Number problems, 443
solving, 424
Numerator, rational exponents, 458
Numerical coefficient, 66–67, 135

## O

Oblique asymptotes, 697, 700, 712
defined, 697, 701
locating, 700–701
Odd index, 452
Odd multiplicity, zeros of, 669
Odd roots, 451
examples of, 452
summary of, 453
Odd-degree polynomial functions, 675
Odds against an event, 816–817
Odds in favor of an event, 817
One-to-one functions, 594–596, 648
finding the inverse function of, 649
Opposites, 17
Order of an inequality, 110
Order of operations, 28–40, 60
evaluate expressions using, 32–33
exponential expressions, evaluating, 28–29
expressions containing variables, evaluating, 33–34
graphing calculator, evaluating expressions on, 34–35
square roots and higher roots, evaluating, 30–32
Ordered pairs, 144, 222
Ordered triple, 233, 245
Origin, 144, 222
Outcomes, 801, 824

## P

Parabola shifts, 563–566
defined, 564
Parabolas, 584
axis of symmetry and vertex of, 565
finding the vertex of, 557
graphing, of the form $x = a(y - k)^2 + h$, 719–720
maximum value, 560
minimum value, 560
translations of, 563–566
with vertex at $(h, k)$, 719, 755
Parallel lines, 185, 201–204
defined, 201, 225
Parenthesis, 112
Partial sum, 767, 821
Pascal, Blaise, 797
Pascal's triangle, 796–797
Perfect cubes, 465, 467, 510
Perfect powers, 465–466
Perfect square trinomials, 347–348, 519–520, 582
Perfect squares, 345, 465, 467, 510
Perpendicular lines, 201–204
defined, 201, 225
Piecewise function, 709
Plotting points, drawing graphs by, 145–146
in the Cartesian coordinate system, 144–145
Point of discontinuity, 659
Point-slope form:
defined, 199, 225
linear equations, 199–208, 204
recognizing parallel and perpendicular lines, 201–204
understanding, 199–200
using to construct models from graphs, 200–201
Polynomial equations, 358–365
degree of, 358
using factoring to find the $x$-intercepts of a quadratic function, 364–365
using factoring to solve applications, 361–363
using factoring to solve equations, 359–361
zero-factor property, 358–359
Polynomial functions, 370
characteristics of graphs of, 665–675
complex zeros of, 683–687
defined, 299, 665
Descartes' Rule of Signs, 676–677, 680, 710
end behavior, 665
evaluating, 299–300
even-degree, 675–676
finding the product of, 313
graphing, 669–671
using its key features, 691–693
graphs of, 300–301
irrational zeros, estimating, 694
maximum number of turning points of the graph of, 669
multiplicity, 668–669
negative leading coefficient, 301
odd-degree, 675
positive leading coefficient, 300
power functions, defined, 665–667, 709
$P(x)$ of degree $n$, 709
rational functions:
asymptotes, 696–697
characteristics of graphs of, 696–708
defined, 696
improper, 696
key features of graphs of, 696–706
proper, 696
rational zeros of, 675–688
determining, 677–678
factor theorem, 678, 710
Remainder Theorem, 678–679
synthetic division, 678, 680–681
Rational Zeros Theorem, 677–678, 710
real zeros of, 688–695
Intermediate Value Theorem, 689, 711
locating/determining, 688–690
upper and lower bounds for, 690–691, 711
turning points, 669
$x$-intercepts and factors of, 667
zero of, 667, 709
Polynomial inequalities, solving, 576–577
Polynomials:
adding, 302–303, 370
common binomial factor, factoring, 329–330
cubic, 298
defined, 298, 370
in descending order, 298
dividing, 317–324, 371
by a binomial, 319–321
by a monomial, 317–319
Remainder Theorem, 323–324, 371
using synthetic division, 321–323
factoring, 373
by grouping, 330–331, 371
monomials from, 328–329
using a combination of techniques, 354–355
finding the product of the sum and difference of the same two terms, 312–313
finding the square of a binomial, 311–312
greatest common factor (GCF), 327, 371
leading coefficient, 298
leading term, 298
linear, 298
multiplying, 308–317, 370
multiplying binomials by, 309–310
multiplying monomials by, 308–309
multiplying polynomials by, 310–311
prime, 372
quadratic, 298
subtracting, 302–303, 370
Positive leading coefficient, 300
polynomial functions, 300

Positive real zeros, 676–677, 710
Positive slope, 186, 224
Positive square root, 30
Power functions, defined, 665–667, 709
Power, raising to a power, 43–44
Power rule, 43–44
  logarithms, 619–620
Powers of *i*, 505
Prime numbers, 393
Prime polynomial, 342, 372
Principal square root, 30, 449, 508
Probability, 812–820, 824–825
  of a compound event, 815–816
  equally likely outcomes, 812
  of an event, 812–813
    failing to occur, 814
  events, defined, 812
  important facts about, 813, 825
  mutually exclusive events, 815–816, 825
  odds against an event, 816–817, 825
  odds in favor of an event, 817, 825
  outcomes, 812
Problem solving, guidelines for, 77, 136
Problem-solving procedure:
  steps in, 90
  using, 77–80, 89–94
Product of functions, 209
Product of the sum and difference of the same two terms, 312–313, 371
Product, raising to a power, 44–45
Product rule, 40–41, 46, 60
  logarithms, 618, 651
  radicals, 465, 510
    simplify radicals using, 466–469
Product rule for exponents, 308–309
Proper rational functions, 699
Proportions, 442
  defined, 412
  solving, 412–413
Pythagorean Theorem, 362–363, 374, 721

## Q

Quadrants, 144, 222
Quadratic equations:
  applications of, 533–534, 539–548
  defined, 358
  determining, given its solutions, 531–532
  problem solving, 539–548
  solutions of, 532, 583
  standard for, 358, 373
  writing, 549–555
Quadratic formula, 523
  avoiding common errors, 529, 530
  deriving, 527–528
  solving a quadratic equation by, 528–531
Quadratic functions:
  graphing, 555–572
    using the axis of symmetry, vertex, and intercepts, 558–559
  solving by completing the square, 518–527, 583
  solving by the quadratic formula, 527–539, 583
Quadratic inequalities, 584
  boundary values, 573
  examples of, 572
  in one variable, 572–582
  sign graph, 573
  solution to, 573, 584
  solving, 573–575
Quadratic mean, 499
Quadratic polynomials, 298
Quotient of functions, 209
Quotient, raising to a power, 45–46
Quotient rule, 41, 46, 60
  logarithms, 618–619
  radicals, 510
Quotient rule for radicals, 469–470
  defined, 469
  examples of, 470

## R

Radical equations:
  containing two radical terms and a nonradical term, 492–493
    avoiding common errors, 493
  containing two radicals, 492
  defined, 489
  examples of, 489
  and graphing calculator, 491
  solving, 489–499
  solving applications using, 494
Radical expressions, 508, *See also* Radicals
  changing to exponential expressions, 457–464
  cube roots, 60, 451
  defined, 449
  dividing with different indices, 485
  even roots, 451
  index, 449
  odd roots, 451
  simplifying, 457–459
Radical sign, 30, 449
Radicals, *See also* Radical expressions
  adding, 472–474
  checking after simplifying, 484
  dividing, 480–488
  like, 472, 510
  multiplying, 472, 474–477
  product rule, 465, 510
    simplify radicals using, 466–469
  quotient rule, 510
  simplifying, 465–472
    using the product rule, 510
    using the quotient rule for radicals, 469–470
  subtracting, 472, 472–474
  unlike, 472, 510
  using absolute value, 453
Radicand, 30, 449
  solving for a variable in, 495
Raising a power to a power, 43–44, 46
Raising a product to a power, 44–45, 46, 61
Raising a quotient to a power, 45–46, 61
Range, 158–159, 222
Rate of change, recognizing slope as, 187–188
Rational equations:
  checking factoring solutions, 409–411
  defined, 381
  motion problems, solving, 424–426
  number problems, solving, 424
  proportions, solving, 412–413
  solving, 409, 442
  work problems, solving, 421–424
Rational exponents, 457–464, 509
  denominator, 458
  factoring expressions with, 462–463
  index, 458
  numerator, 458
  root, 458
  solving equations with, 552–553
Rational expressions, 440
  adding, 441
  applications of, 398–399
  avoiding common errors, 385, 416
  with a common denominator, adding/subtracting, 391–393
  defined, 381, 382
  dividing, 387–388, 441
  examples of, 382
  least common denominator (LCD) of, finding, 393–394
  multiplying, 385–386, 441
  simplifying, 384–385, 441
  solving applications using, 414
  solving for a variable in a formula containing, 415–416
  subtracting, 441
  with unlike denominators, adding/subtracting, 395–398, 441
Rational functions, 440
  asymptotes, 696–697
  characteristics of graphs of, 696–708
  defined, 382, 696
  domains of, 382–384
  examples of, 382
  graphing, 701–708
  holes in graphs of, 698–699, 701, 712
  horizontal asymptote, 696–697, 701
    locating, 699–700
  improper, 696
  key features of graphs of, 696–706
  oblique asymptotes, 697, 700, 712
    locating, 700–701
  proper, 696, 699
  solving problems involving, 413–414
  vertical asymptote, 696–697, 701, 703
    locating, 697–698, 711–712
Rational inequalities, solving, 577–579
Rational numbers, 58, 449
  defined, 10
  examples of, 10

Rational zeros of polynomial functions, 675–688
determining, 677–678
Rational Zeros Theorem, 677–678, 710
Rationalizing the denominator, 480, 510
using in an addition problem, 484
Real numbers, 10–11
absolute values:
defined, 18–19
evaluating, 18–19
adding, 19, 58
additive inverses, 17
defined, 10
dividing, 22–23, 59
important sets of, 10, 58
multiplying, 23
operations with, 17–18
properties of, 17–18, 59
subtracting, 20–21, 59
using the properties of, 23–24
Real zeros of polynomial functions, 688–695
Boundedness Theorem, 690–691, 711
Intermediate Value Theorem, 689, 711
locating/determining, 688–695
upper and lower bounds for, 690–691, 711
Rectangular coordinate system, 144, 222
Reflexive property, examples of, 66
Relations, 158–159, 222
Remainder Theorem, 323–324, 371, 678–679
Repeating decimals, examples of, 10
Reverse FOIL method, 337
Root mean square (RMS), 499
Root, rational exponents, 458
Roster form, 6, 57
Row transformations, 290
defined, 267
procedures for, 267–269
Rules of exponents, 459–462, 509

## S

Sample space, 801, 824
Scientific calculators:
entering numbers in scientific notation on, 53
evaluating roots on, 31
exponential expressions, evaluating on, 29, 30
factorials, 796
finding roots or expressions with rational exponents on, 462
Scientific notation, 50–57
convert a number in, to decimal form, 51, 61
defined, 50
examples of numbers in, 50
and problem solving, 51–53
writing a number in, 50, 61
Sense of an inequality, 110
Sequences, 764–766
alternating, 766
decreasing, 765, 821
defined, 764
find the terms of, 764–766
finite, 765, 766, 821
general term of, 764, 821
increasing, 765, 821
infinite, 764, 821
Series, 766–767
defined, 821
Set builder notation, 7
defined, 8
general form of, 8
using, 8–9
Sets, 6, 57
empty, 7, 57, 113
finite, 6
intersection, 9
null, 7, 57, 113
subsets, 11, 57
union, 9
Sigma (Σ), 767
Sign graph, 573
Sign of a fraction, 23
Significant digits, 627
Similar figures, 412, 442
Simple interest formula, 78, 136
Simplifying an expression, 67, 135
Slope, defined, 186
Slope of a line, defined, 185, 224
Slope-intercept form:
defined, 188, 224
examples of equations in, 188
finding the slope of a line, 185–187
graphing equations using the slope and the *y*-intercept, 189
graphs, translations of, 184–185
linear equations, 204
slope, recognizing as a rate of change, 187–188
using to construct models from graphs, 190–191
writing equations in, 188–189
Smooth curve, 659
Solution sets, 68, 111–113
intersection, 116
union, 119
Solution to a quadratic inequality, 573, 584
Solution to a system of equations, defined, 233, 288
Solutions:
checking by substitution, 70
defined, 135
equations, 68
Special factoring formulas, 346–353
Square matrix, 266, 290
Square of a binomial, 311–312, 350, 371
avoiding common errors, 311
defined, 311
Square root:
avoiding common errors, 471
defined, 60
evaluating, 30–32
finding, 449–450
index, 449
of a negative number, 500
principal square root, 449
simplifying, 467
Square root function, 508
graphing, 450–451
Square root property:
defined, 518, 582
using to solve equations, 518–519
Standard form:
linear equations, 204, 223
quadratic equation, 373
Step functions, 661, 709
Study group, 94
Subscripts, 79
Subsets, 11, 57
Substitution:
factoring by, 342–343, 372
solving systems of linear equations by, 235–237
Subtraction:
of complex numbers, 502, 511
of polynomials, 302–303, 370
of radicals, 472–474, 510
of rational expressions, 441
with a common denominator, 391–393
of real numbers, 20–21, 59
Sum of functions, 209
graphing, 210–214
Sum of two cubes, 373
Summation notation (Σ), 767–769, 821
Supplementary angles, 94, 136
Supplements, using, 4–5
Symmetric property, examples of, 66
Symmetry, 556, 584
Synthetic division, 678, 680–681
defined, 321
dividing polynomials by, 321–323
System of equations in three variables, geometric interpretation of, 248
Systems of linear equations:
defined, 233
solving graphically, 233–234, 282–283
solving using matrices, 267–269
in three variable, solving, 269–271
Systems of linear inequalities, containing absolute value, solving, 284–285, 293

## T

Tangent line, 197
Terminating decimals, examples of, 10
Terms:
defined, 66, 135, 298
degree of, 67, 135
like, 67
unlike, 67
Third-degree equations, 360
Time management, 3
TI-83 Plus calculators, 34
graphing calculator, 150–151
TI-84 Plus calculators, 34
graphing calculator, 150–151
Transitive property, examples of, 66
Translated graphs, 185
Translations of parabolas, 563–566
Transverse axis, 735, 758
Tree diagram, 801
Trial and error, factoring by, 337–339
Trials, 801
Triangular form,, augmented matrix, 267, 290
Trinomials, 370

factoring, 335–346
factor out a common factor, 336
of the form $ax^2 + bx + c, a \neq 1$, using grouping, 341–342, 372
of the form $ax^2 + bx + c, a \neq 1$, using trial and error, 337–339, 372
of the form $x^2 + bx + c$, 335–336, 371
using substitution, 342–343, 372
Turning points, maximum number of, 669
Tutoring, 94
Two numbers with different signs, adding, 19–20
Two numbers with the same sign, adding, 19

## U

Union, 9, 58
solution sets, 119, 137
Unlike denominators, 393–394
adding or subtracting rational expressions with, 395–398, 441
Unlike radicals, 472, 510
Unlike terms, 67, 135
Upper bound, for real zeros, 690–691, 711
Upper limit of summation, 767, 821

## V

Variables, 6, 57
expressions containing, evaluating, 33–34
Variation, 432–440
combined variation problems, solving, 436
defined, 432
direct variation problems, 432–433
inverse variation problems, 433–435
joint variation problems, 435–436
Verbal statement, translating into an algebraic expression/equation, 87–89
Vertex, of a parabola, 557
Vertical asymptote, 696–697, 701 703
locating, 697–698, 711–712
rational functions, 696–697
Vertical line, equation of, 178
Vertical line test, 160–162, 374
defined, 161
Vertices, 735
Video CDs, 94

## W

Whole numbers, 10, 58
Work problems, 443, 541–542
defined, 424
solving, 421–424

## X

$x$-axis, 144, 222
$x$-intercept:
defined, 174, 223
finding, 175
$x$-intercepts of a graph, 364–365

## Y

$y$-axis, 144, 222
$y$-intercept:
defined, 174, 223
finding, 175

## Z

Zero exponent rule, 43, 46, 60
Zero of a polynomial function, 667, 709
Zero-factor property, 374
Zeros of even and odd multiplicity, 669

# Photo Credits

**Chapter 1** **p. 1** Jonathon Ferrey, Getty Images, Inc.; **p. 4** Richard Hutchings, PhotoEdit, Inc.; **p. 14** Jonathon Ferrey, Getty Images, Inc.; **p. 26** Dave Saunders, Getty Images, Inc.–Stone Allstock; **p. 27** Chris Oxley, Corbis/Bettmann; **p. 33** © Image 100/Royalty-Free/Corbis; **p. 38** Allen R. Angel; **p. 39 (left)** Allen R. Angel; **(top right)** Allen R. Angel; **(bottom right)** Ken Chernus, Getty Images, Inc.–Taxi; **p. 50 (left)** National Optical Astronomy Observatories; **(right)** Oliver Meckes/Ottawa, Photo Researchers, Inc.; **p. 54** NASA Jet Propulsion Laboratory; **p. 55 (left)** © Reuters/CORBIS **(right)** Allen R. Angel **p. 56** Agence France Presse/Getty Images; **p. 63** Kim Blaxland, Getty Images, Inc.–Stone Allstock

**Chapter 2** **p. 65** Mark Adams, Getty Images, Inc.; **p. 76 (top left)** Allen R. Angel; **(bottom left)** Allen R. Angel; **(right)** Darren McCollester/CORBIS; **p. 77** AP Wide World Photos; **p. 86** John P. Kelly, Getty Images, Inc.—Image Bank; **p. 91** Centers for Disease Control and Prevention; **p. 92** Allen R. Angel; **p. 95** Allen R. Angel; **p. 96 (top left)** Juan Silva Productions, Getty Images, Inc.–Image Bank; **(middle left)** Allen R. Angel; **(bottom left)** Jeff Greenberg, PhotoEdit, Inc.; **(right)** David Young Wolff, PhotoEdit, Inc.; **p. 97 (left)** Allen R. Angel; **(right)** Allen R. Angel; **p. 98** Itsou Inouye, AP Wide World Photos; **p. 99 (left)** Allen R. Angel; **(right)** Tom Stewart, Corbis/Bettmann; **p. 105** Aimee L. Calhoun; **p. 106 (left)** J.C. Leacock, Image State/ International Stock Photography, Ltd.; **(right)** Allen R. Angel; **p. 107 (left)** Allen R. Angel; **(right)** Allen R. Angel; **p. 108 (left)** Allen R. Angel; **(top right)** George Hall, Corbis/Bettmann; **(bottom right)** Allen R. Angel; **p. 109 (top left)** Allen R. Angel; **(bottom left)** © Royalty-Free/CORBIS; **(right)** NASA Headquarters; **p. 114** Getty Images, Inc.–Photodisc; **p. 119** Getty Images—Photodisc; **p. 121** Allen R. Angel; **p. 122** Allen R. Angel; **p. 123** Allen R. Angel; **p. 138** Allen R. Angel; **p. 139 (left)** Getty Images–Digital Vision **(top right)** Allen R. Angel; **(bottom right)** Allen R. Angel; **p. 141** Allen R. Angel

**Chapter 3** **p. 143** Patrick Molnar, Getty Images, Inc.; **p. 144** Sheila Terry/Science Photo Library/Photo Researchers, Inc.; **p. 165** Paul Bowen, Getty Images, Inc.; **p. 170 (left)** Allen R. Angel; **(right)** Susan Van Etten, PhotoEdit, Inc.; **p. 183** Jeff Greenberg, PhotoEdit, Inc.; **p. 196** Allen R. Angel; **p. 197** Allen R. Angel; **p. 205** Getty Images–Digital Vision; **p. 206 (left)** Allen R. Angel; **(right)** Allen R. Angel; **p. 207** Arne Hodalic, Corbis/Bettmann; **p. 221** Carlos Cortes IV © Reuters/CORBIS

**Chapter 4** **p. 232** Tom Collicott, Masterfile Corporation; **p. 251** David Stoecklein, Corbis/Bettmann; **p. 252** Greg Vaughn/PacificStock.com; **p. 253** Allen R. Angel; **p. 254** Darrell Gulin, Corbis/Bettmann; **p. 260 (top left)** © Yann Arthus-Bertrand/CORBIS; **(bottom left)** Michael T. Sedam, Corbis/Bettmann; **(right)** Dave King © Dorling Kindersley; **p. 261 (left)** Bob Daemmerich, The Image Works; **(right)** Allen R. Angel; **p. 262** Jim Cummins/Taxi/Getty Images; **p. 264** Scott Boehm, Getty Images, Inc.; **p. 295** Allen R. Angel

**Chapter 5** **p. 297** Pete Seaward, Getty Images, Inc.–Stone Allstock; **p. 317** © Jason Szenes/CORBIS, All Rights Reserved; **p. 332** Mug Shots, Corbis/Bettmann; **p. 333** Allen R. Angel; **p. 368 (top left)** Allen R. Angel; **(bottom left)** Allen R. Angel

**Chapter 6** **p. 381** Corbis Royalty Free; **p. 403** Jeff Greenberg, PhotoEdit, Inc.; **p. 422** Allen R. Angel; **p. 423** Allen R. Angel; **p. 425** Allen R. Angel; **p. 427 (left)** Allen R. Angel; **(right)** Allen R. Angel; **p. 428 (top right)** Dennis Marsico, Corbis/Bettmann; **(bottom right)** Allen R. Angel; **p. 429 (left)** Allen R. Angel; **(right)** Allen R. Angel; **p. 430 (left)** Allen R. Angel; **(right)** Kit Houghton, Corbis/Bettmann; **p. 431** Allen R. Angel; **p. 436** Allen R. Angel; **p. 437** Doug Mazell, Index Stock Imagery, Inc.; **p. 438** Bob Daemmrich, PhotoEdit, Inc.; **p. 439 (left)** Bruno Vincent, Getty Images, Inc.; **(right)** Mary Kate Denny, PhotoEdit, Inc.; **p. 440** Dimitri Vervits, Getty Images, Inc.; **p. 445** Allen R. Angel; **p. 447** Allen R. Angel

**Chapter 7** **p. 448** Corbis Royalty Free; **p. 464** Courtesy eBay; **p. 478** Richard Hutching, PhotoEdit, Inc.; **p. 488** Allen R. Angel; **p. 494** Allen R. Angel; **p. 497 (left)** Allen R. Angel; **(right)** NOAA/Phil Degginger, Color-Pic, Inc.; **p. 508** Chris Everard, Getty Images, Inc.–Stone Allstock; **p. 515** Picture Press Bild - und Textagentur GmbH, Munich, Germany; **p. 516** Allen R. Angel

**Chapter 8** **p. 517** Allen R. Angel; **p. 537** Allen R. Angel; **p. 538** NASA, Masterfile Corporation; **p. 547 (left and right)** Allen R. Angel; **p. 548** Allen R. Angel; **p. 570** Tony Freeman, PhotoEdit, Inc.; **p. 586** Melford Inc., Michael, Getty Images, Inc.–Image Bank; **p. 587** Richard Hutchings, PhotoEdit, Inc.; **p. 588** Chris Arend, Alaska Stock

**Chapter 9** **p. 591** © Bob Krist/CORBIS All Rights Reserved; **p. 602** Damir Frkovic, Masterfile Corporation; **p. 610 (top right)** Science Source, Photo Researchers, Inc.; **(bottom right)** Allen R. Angel; **p. 615** Simon Kwong/Reuters New Media Inc., Corbis/Bettmann; **p. 629 (left)** V.C.L., Getty Images, Inc.–Taxi; **(right)** Allen R. Angel; **p. 634** Charles Gupton, Getty Images, Inc.–Stone Allstock; **p. 635** © Roger Ressmeyer/CORBIS, All Rights Reserved; **p. 636** © Bob Krist/CORBIS, All Rights Reserved; **p. 643** Allen R. Angel; **p. 645 (left)** Allen R. Angel; **(right)** Allen R. Angel; **p. 646** Allen R. Angel; **p. 654** Allen R. Angel; **p. 655** Allen R. Angel

**Chapter 10** **p. 657** Ryan McVay, Getty Images, Inc.–Photodisc

**Chapter 11** **p. 717** Santi Visali Inc., Getty Images, Inc., Hulton Archive Photos; **p. 734 (top and bottom left)** Allen R. Angel; **p. 749** Pictor, Image State/International Stock Photography, Ltd.; **p. 760** Allen R. Angel

**Chapter 12** **p. 763** © Richard Megna, Fundamental Photographs, NYC; **p. 774** William James Warren; **p. 776** Allen R. Angel; **p. 777** Corbis/Bettmann; **p. 790** John Matchett/Picture Perfect USA, Inc.; **p. 797** The Granger Collection; **p. 814** Tony Freeman/PhotoEdit

**Allen R. Angel**
**Chapter Test Prep Video t/a Algebra for College Students, 3e**
**0-13-615802-1 / 978-0-13-615802-8**

**Pearson Prentice Hall**
**Pearson Education, Inc.**
**Upper Saddle River, NJ 07458**


YOU SHOULD CAREFULLY READ THE TERMS AND CONDITIONS BEFORE USING THE CD-ROM PACKAGE. USING THIS CD-ROM PACKAGE INDICATES YOUR ACCEPTANCE OF THESE TERMS AND CONDITIONS.

Pearson Education, Inc., provides this program and licenses its use. You assume responsibility for the selection of the program to achieve your intended results, and for the installation, use, and results obtained from the program. This license extends only to use of the program in the United States or countries in which the program is marketed by authorized distributors.

**LICENSE GRANT**

You hereby accept a nonexclusive, nontransferable, permanent license to install and use the program ON A SINGLE COMPUTER at any given time. You may copy the program solely for backup or archival purposes in support of your use of the program on the single computer. You may not modify, translate, disassemble, decompile, or reverse engineer the program, in whole or in part.

**TERM**

The License is effective until terminated. Pearson Education, Inc., reserves the right to terminate this License automatically if any provision of the License is violated. You may terminate the License at any time. To terminate this License, you must return the program, including documentation, along with a written warranty stating that all copies in your possession have been returned or destroyed.

**LIMITED WARRANTY**

THE PROGRAM IS PROVIDED "AS IS" WITHOUT WARRANTY OF ANY KIND, EITHER EXPRESSED OR IMPLIED, INCLUDING, BUT NOT LIMITED TO, THE IMPLIED WARRANTIES OF MERCHANTABILITY AND FITNESS FOR A PARTICULAR PURPOSE. THE ENTIRE RISK AS TO THE QUALITY AND PERFORMANCE OF THE PROGRAM IS WITH YOU. SHOULD THE PROGRAM PROVE DEFECTIVE, YOU (AND NOT PEARSON EDUCATION, INC., OR ANY AUTHORIZED DEALER) ASSUME THE ENTIRE COST OF ALL NECESSARY SERVICING, REPAIR, OR CORRECTION. NO ORAL OR WRITTEN INFORMATION OR ADVICE GIVEN BY PEARSON EDUCATION, INC., ITS DEALERS, DISTRIBUTORS, OR AGENTS SHALL CREATE A WARRANTY OR INCREASE THE SCOPE OF THIS WARRANTY. SOME STATES DO NOT ALLOW THE EXCLUSION OF IMPLIED WARRANTIES, SO THE ABOVE EXCLUSION MAY NOT APPLY TO YOU. THIS WARRANTY GIVES YOU SPECIFIC LEGAL RIGHTS AND YOU MAY ALSO HAVE OTHER LEGAL RIGHTS THAT VARY FROM STATE TO STATE.

Pearson Education, Inc., does not warrant that the functions contained in the program will meet your requirements or that the operation of the program will be uninterrupted or error-free. However, Pearson Education, Inc., warrants the CD-ROM(s) on which the program is furnished to be free from defects in material and workmanship under normal use for a period of ninety (90) days from the date of delivery to you as evidenced by a copy of your receipt. The program should not be relied on as the sole basis to solve a problem whose incorrect solution could result in injury to person or property. If the program is employed in such a manner, it is at the user's own risk and Pearson Education, Inc., explicitly disclaims all liability for such misuse.

**LIMITATION OF REMEDIES**

Pearson Education, Inc.'s entire liability and your exclusive remedy shall be:

1. the replacement of any CD-ROM not meeting Pearson Education, Inc.'s "LIMITED WARRANTY" and that is returned to Pearson Education, or
2. if Pearson Education is unable to deliver a replacement CD-ROM that is free of defects in materials or workmanship, you may terminate this agreement by returning the program.

IN NO EVENT WILL PEARSON EDUCATION, INC., BE LIABLE TO YOU FOR ANY DAMAGES, INCLUDING ANY LOST PROFITS, LOST SAVINGS, OR OTHER INCIDENTAL OR CONSEQUENTIAL DAMAGES ARISING OUT OF THE USE OR INABILITY TO USE SUCH PROGRAM EVEN IF PEARSON EDUCATION, INC., OR AN AUTHORIZED DISTRIBUTOR HAS BEEN ADVISED OF THE POSSIBILITY OF SUCH DAMAGES, OR FOR ANY CLAIM BY ANY OTHER PARTY. SOME STATES DO NOT ALLOW FOR THE LIMITATION OR EXCLUSION OF LIABILITY FOR INCIDENTAL OR CONSEQUENTIAL DAMAGES, SO THE ABOVE LIMITATION OR EXCLUSION MAY NOT APPLY TO YOU.

**GENERAL**

You may not sublicense, assign, or transfer the license of the program. Any attempt to sublicense, assign or transfer any of the rights, duties, or obligations hereunder is void. This Agreement will be governed by the laws of the State of New York. Should you have any questions concerning this Agreement, you may contact Pearson Education, Inc. by writing to:

ESM Media Development
Higher Education Division
Pearson Education, Inc.
1 Lake Street
Upper Saddle River, NJ 07458

Should you have any questions concerning technical support, you may write to:

New Media Production
Higher Education Division
Pearson Education, Inc.
1 Lake Street
Upper Saddle River, NJ 07458

YOU ACKNOWLEDGE THAT YOU HAVE READ THIS AGREEMENT, UNDERSTAND IT, AND AGREE TO BE BOUND BY ITS TERMS AND CONDITIONS. YOU FURTHER AGREE THAT IT IS THE COMPLETE AND EXCLUSIVE STATEMENT OF THE AGREEMENT BETWEEN US THAT SUPERSEDES ANY PROPOSAL OR PRIOR AGREEMENT, ORAL OR WRITTEN, AND ANY OTHER COMMUNICATIONS BETWEEN US RELATING TO THE SUBJECT MATTER OF THIS AGREEMENT.

**System Requirements**

**Windows System Requirements:**

- Pentium II 300-MHz processor-based computer
- Windows 2000 (Service Pack 4) or XP
- 64 MB RAM (128 MB RAM required for Windows XP)
- 7.2 MB of additional hard drive space (if QuickTime installation is needed)
- 800 x 600 resolution
- 8x CD drive
- QuickTime 7.x
- Sound Card
- Internet browser

**Macintosh System Requirements:**

- Power PC G3 233 MHz
- Mac 10.x
- 64 MB RAM
- 19 MB of additional hard drive space if QuickTime installation is needed
- 800 x 600 resolution monitor
- 8x CD drive
- QuickTime 7
- Internet browser

**Support Information**

If you are having problems with this software, call (800) 677-6337 between 8:00 a.m. and 8:00 p.m. EST, Monday through Friday, and 5:00 p.m. through Midnight EST on Sundays. You can also get support by filling out the web form located at: http://247.prenhall.com/mediaform. Our technical staff will need to know certain things about your system in order to help us solve your problems more quickly and efficiently. If possible, please be at your computer when you call for support. You should have the following information ready:

- Textbook ISBN
- CD-ROM ISBN
- corresponding product and title
- computer make and model
- Operating System (Windows or Macintosh) and version
- RAM available
- hard disk space available
- Sound card? Yes or No
- printer make and model
- network connection
- detailed description of the problem, including the exact wording of any error messages.

NOTE: Pearson does not support and/or assist with the following:

- third-party software (i.e. Microsoft including Microsoft Office suite, Apple, Borland, etc.)
- homework assistance
- Textbooks and CD-ROMs purchased used are not supported and are non-replaceable. To purchase a new CD-ROM, contact Pearson Individual Order Copies at 1-800-282-0693.